10	11 IB	12 IIB	13 III IIIA	14 IV IVA	15 V VA	16 VI VIA	VII VIIA	VIII VIIIA
								2 **He** 4.00
			5 **B** 10.81	6 **C** 12.01	7 **N** 14.01	8 **O** 16.00	9 **F** 19.00	10 **Ne** 20.18
			13 **Al** 26.98	14 **Si** 28.09	15 **P** 30.97	16 **S** 32.06	17 **Cl** 35.45	18 **Ar** 39.95
28 **Ni** 58.71	29 **Cu** 63.54	30 **Zn** 65.37	31 **Ga** 69.72	32 **Ge** 72.59	33 **As** 74.92	34 **Se** 78.96	35 **Br** 79.91	36 **Kr** 83.80
46 **Pd** 106.4	47 **Ag** 107.87	48 **Cd** 112.40	49 **In** 114.82	50 **Sn** 118.69	51 **Sb** 121.75	52 **Te** 127.60	53 **I** 126.90	54 **Xe** 131.30
78 **Pt** 195.09	79 **Au** 196.97	80 **Hg** 200.59	81 **Tl** 204.37	82 **Pb** 207.19	83 **Bi** 208.98	84 **Po** 210	85 **At** 210	86 **Rn** 222
110 **Uun**	111 **Uuu**	112 **Uub**	113 **Uut**					

← Metals | Metalloids | Nonmetals →

62 **Sm** 150.35	63 **Eu** 151.96	64 **Gd** 157.25	65 **Tb** 158.92	66 **Dy** 162.50	67 **Ho** 164.93	68 **Er** 167.26	69 **Tm** 168.93	70 **Yb** 173.04	Lanthanides
94 **Pu** 239.05	95 **Am** 241.06	96 **Cm** 247.07	97 **Bk** 249.08	98 **Cf** 251.08	99 **Es** 254.09	100 **Fm** 257.10	101 **Md** 258.10	102 **No** 255	Actinides

CHEMISTRY

Molecules, Matter, and Change

ABOUT THE COVER

Chemistry can be compared to a spider's web. Its interlinked ideas offer an astounding diversity—and a single, fascinating whole. The strength of a spider's web is also an inspiration to chemists, who have determined the structure of its material. Artificial spider's silk can now be made in bulk. It can be spun into thin, tough thread, as on the spools shown here, or wound into cables strong enough to support suspension bridges. It is a new synthetic polymer, a class of materials you will meet in Chapter 11.

Wherever we look, we find chemistry at work. In this textbook, we shall lead you through its intricate but rewarding paths. Before you are through, you will learn to see a spider's web, the color of leaves on a forest floor, and the rest of your world with a chemist's careful but penetrating eye.

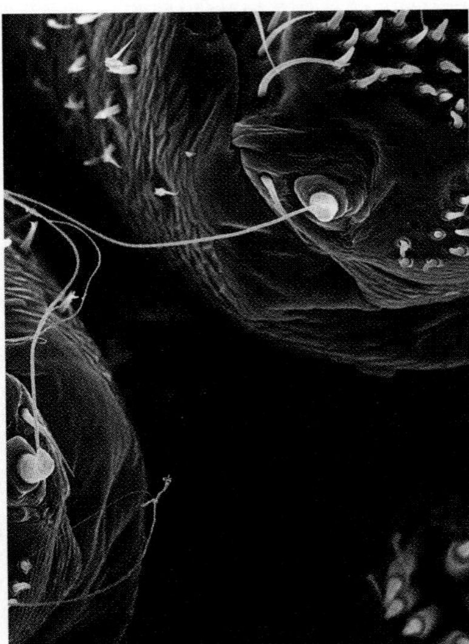

CHEMISTRY

Molecules, Matter, and Change

THIRD EDITION

PETER ATKINS
Oxford University

LORETTA JONES
University of Northern Colorado

W. H. FREEMAN AND COMPANY
NEW YORK

ACQUISITIONS EDITORS: Deborah Allen, Michelle Russel Julet
DEVELOPMENT EDITOR: John Haber
PROJECT EDITOR: Georgia Lee Hadler
DESIGN COORDINATOR: Diana Blume
COVER AND TEXT DESIGNER: Vertigo Design, NYC
ILLUSTRATION COORDINATOR: Bill Page
ILLUSTRATIONS: Peter Atkins with Network Graphics
PHOTO RESEARCHER: Audrey Herbst
PRODUCTION COORDINATOR: Julia De Rosa
COMPOSITION: Black Dot Graphics
MANUFACTURING: RR Donnelley & Sons Company
MARKETING MANAGER: John A. Britch

Library of Congress Cataloging-in-Publication Data
Atkins, P. W. (Peter William), 1940–
Chemistry: molecules, matter, and change. — 3d ed. / Peter W. Atkins, Loretta L. Jones
p. cm.
Includes index.
ISBN 0-7167-2832-X (with CD-ROM)
ISBN 0-7167-3107-X (without CD-ROM)
1. Chemistry. I. Jones, Loretta L. II. Title.
QD31.2.A75 1997
540—dc20 96-21771

PRINTED IN THE UNITED STATES OF AMERICA
Second Printing, 1997

CONTENTS IN BRIEF

1 MATTER 1

2 MEASUREMENTS AND MOLES 39

3 CHEMICAL REACTIONS: MODIFYING MATTER 77

4 REACTION STOICHIOMETRY: CHEMISTRY'S ACCOUNTING 117

5 THE PROPERTIES OF GASES 143

6 THERMOCHEMISTRY: THE FIRE WITHIN 181

7 INSIDE THE ATOM 219

8 INSIDE MATERIALS: CHEMICAL BONDS 267

9 MOLECULES: SHAPE, SIZE, AND BOND STRENGTH 303

10 LIQUID AND SOLID MATERIALS 341

11 CARBON-BASED MATERIALS 389

12 THE PROPERTIES OF SOLUTIONS 437

13 CHEMICAL EQUILIBRIUM 473

14 PROTONS IN TRANSITION: ACIDS AND BASES 509

15 SALTS IN WATER 543

16 ENERGY IN TRANSITION: THERMODYNAMICS 587

17 ELECTRONS IN TRANSITION: ELECTROCHEMISTRY 625

18 KINETICS: THE RATES OF REACTIONS 671

19 THE MAIN-GROUP ELEMENTS: I. THE FIRST FOUR FAMILIES 719

20 THE MAIN-GROUP ELEMENTS: II. THE LAST FOUR FAMILIES 767

21 THE *d*-BLOCK: METALS IN TRANSITION 803

22 NUCLEAR CHEMISTRY 849

TOOLBOXES

2.1	How to use significant figures in calculations	49
2.2	How to convert between mass and moles	55
2.3	How to use molarity	65
2.4	How to calculate the volume of solution to dilute	67
3.1	How to balance chemical equations	80
3.2	How to find oxidation numbers	101
4.1	How to carry out mole-to-mole calculations for a chemical reaction	119
4.2	How to carry out mass-to-mass calculations	120
4.3	How to interpret a titration	123
5.1	How to calculate the effect of changing conditions	152
5.2	How to calculate the volume of gas involved in a reaction	156
6.1	How to devise a reaction sequence to obtain an overall reaction enthalpy	200
6.2	How to use standard enthalpies of formation	209
8.1	How to write the Lewis structure of a polyatomic species	278
9.1	How to use the VSEPR model	311
11.1	How to name hydrocarbons	398
11.2	How to name compounds with functional groups	405
12.1	How to use molality	456
13.1	How to set up and use an equilibrium table	488
14.1	How to calculate the pH of a solution of a weak acid	529
15.1	How to calculate the pH of an electrolyte solution	547
17.1	How to balance complicated redox equations	627
17.2	How to calculate equilibrium constants from electrochemical data	644
22.1	How to identify the products of a nuclear reaction	854

CONTENTS

Preface xxi
To the Student xxxvi

1 MATTER 1

THE ELEMENTS 2
1.1 Atoms 2
1.2 Names of the elements 3
1.3 The nuclear atom 4
Box 1.1 The scientific method 5
1.4 Isotopes 8
Box 1.2 Mass spectrometry 9
1.5 Where did the elements come from? 11
1.6 The periodic table 12
1.7 Metals, nonmetals, and metalloids 14
COMPOUNDS 14
1.8 What are compounds? 15
1.9 Molecular compounds 16
1.10 Ionic compounds and ions 17
MIXTURES 21
1.11 Types of mixtures 22
1.12 Separation techniques 23
Case Study Scientific inquiry on Mars 24
THE NOMENCLATURE OF COMPOUNDS 27
1.13 Names of cations 27
1.14 Names of anions 28
1.15 Names of ionic compounds 29
1.16 Names of molecular compounds 31
Skills you should have mastered 32

2 MEASUREMENTS AND MOLES 39

MEASUREMENTS AND UNITS 40
2.1 The metric system 40
2.2 Prefixes for units 41
2.3 Derived units 41
Case Study Units and trapped atoms 42

2.4 Unit conversions 43
2.5 Temperature 46
2.6 The uncertainty of measurements 47
Toolbox 2.1 How to use significant figures in calculations 49
2.7 Accuracy and precision 50
CHEMICAL AMOUNTS 51
2.8 The mole 51
2.9 Molar mass 53
Toolbox 2.2 How to convert between mass and moles 55
2.10 Measuring out compounds 56
DETERMINATION OF CHEMICAL FORMULAS 58
2.11 Mass percentage composition 59
2.12 Determining empirical formulas 60
2.13 Determining molecular formulas 62
SOLUTIONS IN CHEMISTRY 63
2.14 Molarity 63
Toolbox 2.3 How to use molarity 65
2.15 Dilution 66
Toolbox 2.4 How to calculate the volume of solution to dilute 67
Skills you should have mastered 68

3 CHEMICAL REACTIONS: MODIFYING MATTER 77

CHEMICAL EQUATIONS AND CHEMICAL REACTIONS 78
3.1 Symbolizing chemical reactions 78
3.2 Balancing chemical equations 79
Toolbox 3.1 How to balance chemical equations 80
PRECIPITATION REACTIONS 82
3.3 Aqueous solutions 82
3.4 Reactions between strong electrolyte solutions 85
3.5 Ionic and net ionic equations 85
3.6 Putting precipitation to work 88
THE REACTIONS OF ACIDS AND BASES 90
3.7 Acids and bases in aqueous solution 90
3.8 Strong and weak acids and bases 92
3.9 Acidic and basic character in the periodic table 94
3.10 Neutralization 95
3.11 The formation of gases 97
REDOX REACTIONS 99
3.12 Oxidation and reduction 99
3.13 Keeping track of electrons: Oxidation numbers 100

Toolbox 3.2 How to find oxidation numbers 101
3.14 Oxidizing and reducing agents 102
Box 3.1 Maintaining an artificial atmosphere 103
Case Study From river water to drinking water 106
3.15 Balancing simple redox equations 107
3.16 Classifying reactions 108
Skills you should have mastered 109

4 REACTION STOICHIOMETRY: CHEMISTRY'S ACCOUNTING 117

HOW TO USE REACTION STOICHIOMETRY 118
4.1 Mole-to-mole predictions 118
Toolbox 4.1 How to carry out mole-to-mole calculations for a chemical reaction 119
4.2 Mass-to-mass predictions 120
Toolbox 4.2 How to carry out mass-to-mass calculations 120
4.3 The volume of solution required for reaction 122
Toolbox 4.3 How to interpret a titration 123
THE LIMITS OF REACTION 124
4.4 Reaction yield 124
4.5 Limiting reactants 126
4.6 Combustion analysis 130
Case Study Greenhouse gases 132
Skills you should have mastered 134

5 THE PROPERTIES OF GASES 143

THE NATURE OF GASES 144
5.1 The states of matter 144
5.2 The molecular character of gases 145
5.3 Pressure 146
5.4 Units of pressure 147
THE GAS LAWS 148
5.5 Boyle's law 149
5.6 Charles's law 149
5.7 Avogadro's principle 150
5.8 Using the gas laws 152
Toolbox 5.1 How to calculate the effect of changing conditions 152
5.9 The ideal gas law 153
5.10 Using the ideal gas law to make predictions 154
5.11 Molar volume 155
5.12 The stoichiometry of reacting gases 156
Toolbox 5.2 How to calculate the volume of gas involved in a reaction 156

5.13	Gas density	158
5.14	Mixtures of gases	161
	MOLECULAR MOTION	163
5.15	Diffusion and effusion	163
	Box 5.1 The layers of the atmosphere	164
5.16	The kinetic model of gases	166
5.17	The Maxwell distribution of speeds	167
	Box 5.2 Real gas laws	168
	REAL GASES	169
5.18	The ideal gas law as a limiting law	169
5.19	The Joule-Thomson effect	169
	CASE STUDY Our global metabolism	170
	Skills you should have mastered	173

6	**THERMOCHEMISTRY: THE FIRE WITHIN**	**181**
	ENERGY, HEAT, AND ENTHALPY	182
6.1	Transfer of energy as heat	182
6.2	Exothermic and endothermic processes	184
6.3	Measuring heat transfer	185
6.4	Thermal accounting: Enthalpy	188
6.5	Vaporization	190
6.6	Melting and sublimation	191
6.7	Heating curves	193
	THE ENTHALPY OF CHEMICAL CHANGE	194
6.8	Reaction enthalpies	194
6.9	Standard reaction enthalpies	197
6.10	Combining reaction enthalpies: Hess's law	198
	Box 6.1 The world's enthalpy resources	199
	TOOLBOX 6.1 How to devise a reaction sequence to obtain an overall reaction enthalpy	200
	THE HEAT OUTPUT OF REACTIONS	201
6.11	Enthalpies of combustion	201
6.12	Standard enthalpies of formation	205
	CASE STUDY Food and fitness	206
	TOOLBOX 6.2 How to use standard enthalpies of formation	209
	Skills you should have mastered	209

7	**INSIDE THE ATOM**	**219**
	OBSERVING ATOMS	220
7.1	The characteristics of light	220
7.2	Quanta and photons	223
7.3	Atomic spectra and energy levels	225

CASE STUDY Absorption spectra 226
7.4 The wavelike properties of electrons 229
MODELS OF ATOMS 229
7.5 Atomic orbitals 230
7.6 Quantum numbers and atomic orbitals 232
7.7 Electron spin 236
BOX 7.1 The Stern-Gerlach experiment 236
7.8 The electronic structure of hydrogen 237
THE STRUCTURES OF MANY-ELECTRON ATOMS 237
7.9 Orbital energies 238
7.10 The building-up principle 240
7.11 The electron configurations of atoms 242
BOX 7.2 How the concept of the periodic table was developed 244
7.12 The electron configurations of ions 246
7.13 Electronic structure and the periodic table 247
THE PERIODICITY OF ATOMIC PROPERTIES 248
7.14 Atomic radius 249
7.15 Ionic radius 250
7.16 Ionization energy 251
7.17 Ionization energy and metallic character 254
7.18 The inert-pair effect 254
7.19 Diagonal relationships 255
7.20 Electron affinity 255
CHEMISTRY AND THE PERIODIC TABLE 257
7.21 The *s*-block elements 257
7.22 The *p*-block elements 258
7.23 The *d*-block elements 259
Skills you should have mastered 260

8 INSIDE MATERIALS: CHEMICAL BONDS 267

IONIC BONDS 268
8.1 Lewis symbols for atoms and ions 268
8.2 Lattice enthalpies 270
8.3 The properties of ionic compounds 273
COVALENT BONDS 274
8.4 From atoms to molecules 274
8.5 The octet rule and Lewis structures 275
THE STRUCTURES OF POLYATOMIC SPECIES 276
8.6 Lewis structures 276
TOOLBOX 8.1 How to write the Lewis structure of a polyatomic species 278
8.7 Resonance 280
8.8 Formal charge 282

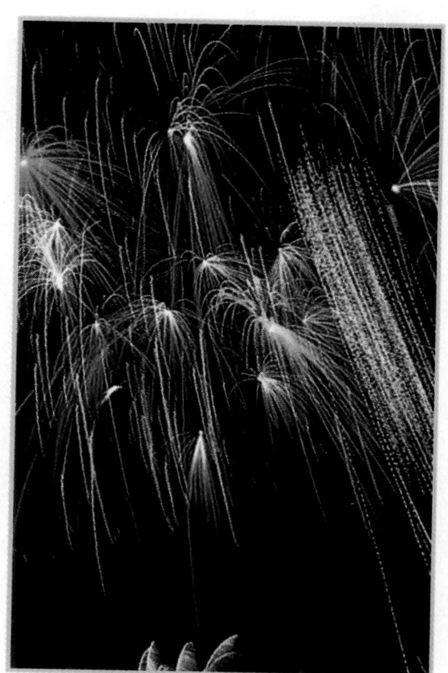

EXCEPTIONS TO THE OCTET RULE 285
8.9 Radicals and biradicals 285
8.10 Expanded valence shells 287
CASE STUDY Smog formers 288
LEWIS ACIDS AND BASES 291
8.11 The unusual structures of Group 13 halides 291
8.12 Lewis acid-base complexes 292
IONIC VERSUS COVALENT BONDS 294
8.13 Correcting the covalent model 294
8.14 Correcting the ionic model 296
Skills you should have mastered 297

9 MOLECULES: SHAPE, SIZE, AND BOND STRENGTH 303

THE SHAPES OF MOLECULES AND IONS 304
9.1 The VSEPR model 304
9.2 Molecules without lone pairs on the central atom 305
9.3 Multiple bonds in the VSEPR model 306
9.4 Molecules with lone pairs on the central atom 308
9.5 The distorting effect of lone pairs 309
TOOLBOX 9.1 How to use the VSEPR model 311
CHARGE DISTRIBUTION IN MOLECULES 312
9.6 Polar bonds 312
9.7 Polar molecules 313
THE STRENGTHS AND LENGTHS OF BONDS 316
9.8 Bond strengths 316
9.9 The variation of bond strength 317
9.10 Bond strengths in polyatomic molecules 318
9.11 Bond lengths 321
ORBITALS AND BONDING 323
9.12 Sigma and pi bonds 323
9.13 Hybridization of orbitals 325
9.14 Hybridization in more complex molecules 327
9.15 Hybrids including *d*-orbitals 329
9.16 Multiple carbon-carbon bonds 330
CASE STUDY Self-assembling materials 332
9.17 Characteristics of double bonds 334
Skills you should have mastered 334

10 LIQUID AND SOLID MATERIALS 341

INTERMOLECULAR FORCES 342
10.1 London forces 342
10.2 Dipole-dipole interactions 344

10.3 Hydrogen bonding 346
LIQUID STRUCTURE 348
10.4 Viscosity 349
10.5 Surface tension 350
SOLID STRUCTURES 352
10.6 Classification of solids 352
Box 10.1 X-ray diffraction 354
10.7 Metallic crystals 355
10.8 Properties of metals 360
Box 10.2 Semiconductors 362
10.9 Alloys 363
Case Study Liquid crystals 366
10.10 Ionic structures 368
10.11 Molecular solids 370
10.12 Network solids 371
PHASE CHANGES 373
10.13 Vapor pressure 373
10.14 Boiling 375
10.15 Freezing and melting 376
10.16 Phase diagrams and cooling curves 377
10.17 Critical properties 380
Skills you should have mastered 381

11 CARBON-BASED MATERIALS 389

HYDROCARBONS 390
11.1 Types of hydrocarbons 390
11.2 Alkanes 391
11.3 Alkenes and alkynes 395
11.4 Aromatic compounds 397
Toolbox 11.1 How to name hydrocarbons 398
FUNCTIONAL GROUPS 399
11.5 Alcohols 399
11.6 Ethers 400
11.7 Phenols 400
11.8 Aldehydes and ketones 401
11.9 Carboxylic acids 402
11.10 Amines and amides 404
Toolbox 11.2 How to name compounds with functional groups 405
ISOMERS 406
11.11 Structural isomers 406
11.12 Geometrical and optical isomers 407
POLYMERS 410
11.13 Addition polymerization 410

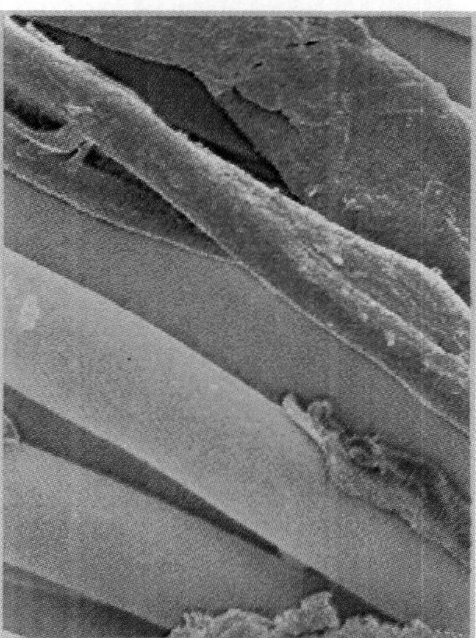

11.14 Condensation polymerization 414
11.15 Copolymers and composites 417
Case Study Conducting polymers 418
11.16 Physical properties 420
EDIBLE POLYMERS 421
11.17 Proteins 421
11.18 Carbohydrates 424
11.19 DNA and RNA 426
Skills you should have mastered 429

12 THE PROPERTIES OF SOLUTIONS 437

SOLUTES AND SOLVENTS 438
12.1 The molecular nature of dissolving 438
12.2 Solubility 439
FACTORS AFFECTING SOLUBILITY 439
12.3 Solubilities of ionic compounds 440
12.4 The like-dissolves-like rule 441
12.5 Pressure and solubility: Henry's law 443
Case Study Colloids in the cafeteria 444
12.6 Temperature and solubility: Thermal pollution 446
WHY DOES ANYTHING DISSOLVE? 447
12.7 The enthalpy of solution 447
12.8 Individual ion hydration enthalpies 449
12.9 Solubility and disorder 450
COLLIGATIVE PROPERTIES 453
12.10 Measures of concentration 453
Toolbox 12.1 How to use molality 456
12.11 Vapor-pressure lowering 457
12.12 Boiling-point elevation and freezing-point depression 459
12.13 Osmosis 462
Skills you should have mastered 465

13 CHEMICAL EQUILIBRIUM 473

EQUILIBRIUM AND COMPOSITION 474
13.1 The reversibility of chemical reactions 474
13.2 The equilibrium constant 475
13.3 Heterogeneous equilibria 480
13.4 Gaseous equilibria 481
USING EQUILIBRIUM CONSTANTS 482
13.5 The extent of reaction 482
13.6 The direction of reaction 485
13.7 Equilibrium tables 486
Toolbox 13.1 How to set up and use an equilibrium table 488

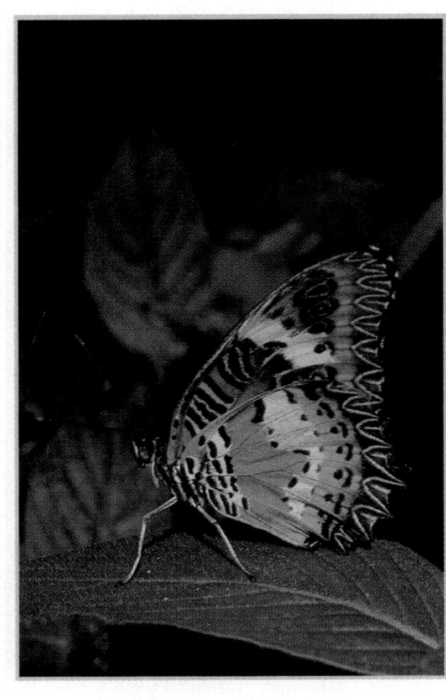

THE RESPONSE OF EQUILIBRIA TO CHANGE 492
13.8 Adding and removing reagents 492
13.9 Compressing a reaction mixture 493
13.10 Temperature and equilibrium 495
CASE STUDY Far from equilibrium 496
13.11 Catalysts and Haber's achievement 499
Skills you should have mastered 500

14 PROTONS IN TRANSITION: ACIDS AND BASES 509
WHAT ARE ACIDS AND BASES? 510
14.1 Brønsted-Lowry acids and bases 510
14.2 Conjugate acids and bases 512
14.3 Proton exchange between water molecules 512
14.4 The pH scale 514
14.5 The pOH of solutions 517
WEAK ACIDS AND BASES 518
14.6 Proton transfer equilibria 518
14.7 The conjugate seesaw 520
14.8 The special role of water 523
14.9 Why are some acids weak and others strong? 524
14.10 The strengths of oxoacids 525
THE pH OF SOLUTIONS OF WEAK ACIDS AND BASES 528
14.11 Solutions of weak acids 528
TOOLBOX 14.1 How to calculate the pH of a solution of a weak acid 529
14.12 Solutions of weak bases 530
14.13 Polyprotic acids and bases 532
CASE STUDY Acid rain 534
Skills you should have mastered 537

15 SALTS IN WATER 543
IONS AS ACIDS AND BASES 544
15.1 Ions as acids 544
15.2 Ions as bases 545
15.3 The pH of salt solutions 546
TOOLBOX 15.1 How to calculate the pH of an electrolyte solution 547
15.4 The pH of mixed solutions 548
TITRATIONS 550
15.5 Strong acid-strong base titrations 550
15.6 Strong acid-weak base and weak acid-strong base titrations 553

15.7	Indicators as weak acids	558
	BUFFER SOLUTIONS	561
15.8	The action of buffers	562
15.9	Selecting a buffer	562
15.10	Buffer capacity	566
	SOLUBILITY EQUILIBRIA	567
	CASE STUDY Burns and blood chemistry	568
15.11	The solubility product	568
15.12	The common-ion effect	572
15.13	Predicting precipitation	573
15.14	Dissolving precipitates	574
15.15	Complex ions and solubilities	575
	Skills you should have mastered	577
16	**ENERGY IN TRANSITION: THERMODYNAMICS**	587
	THE FIRST LAW OF THERMODYNAMICS	588
16.1	Systems and surroundings	588
16.2	Heat and work	589
16.3	Internal energy	590
16.4	Heat transfers at constant volume	592
16.5	Enthalpy	592
	THE DIRECTION OF SPONTANEOUS CHANGE	595
16.6	Spontaneous change	595
16.7	Entropy and disorder	595
16.8	Standard entropies	597
	BOX 16.1 Bridging the macro and micro worlds	598
16.9	The surroundings	600
16.10	The overall change in entropy	602
	FREE ENERGY	603
16.11	Focusing on the system	603
	CASE STUDY Unnatural life	606
16.12	Standard reaction free energies	608
16.13	Using free energies of formation	610
16.14	Free energy and composition	611
16.15	Free energy and equilibrium	613
16.16	The effect of temperature	615
	Skills you should have mastered	616
17	**ELECTRONS IN TRANSITION: ELECTROCHEMISTRY**	625
	TRANSFERRING ELECTRONS	626
17.1	Half-reactions	626

17.2	Balancing redox equations	626
	TOOLBOX 17.1 How to balance complicated redox equations	627
	GALVANIC CELLS	630
17.3	Examples of galvanic cells	630
17.4	The notation for cells	632
17.5	Cell potential	633
17.6	Cell potential and reaction free energy	635
17.7	The significance of standard potentials	639
17.8	The electrochemical series	640
17.9	Standard potentials and equilibrium constants	643
	TOOLBOX 17.2 How to calculate equilibrium constants from electrochemical data	644
17.10	The Nernst equation	645
	BOX 17.1 How pH meters work	646
17.11	Practical cells	647
	CASE STUDY Artificial photosynthesis	650
17.12	Corrosion	652
	ELECTROLYSIS	654
17.13	Electrolytic cells	654
17.14	The potential needed for electrolysis	655
17.15	The products of electrolysis	657
17.16	Applications of electrolysis	659
	Skills you should have mastered	661

18	KINETICS: THE RATES OF REACTIONS	671
	CONCENTRATION AND RATE	672
18.1	The definition of reaction rate	672
18.2	The instantaneous rate of reaction	673
18.3	Rate laws	674
18.4	More complicated rate laws	678
18.5	First-order integrated rate laws	681
18.6	Half-lives for first-order reactions	684
18.7	Second-order reactions	686
	CONTROLLING REACTION RATES	687
18.8	The effect of temperature	688
18.9	Collision theory	688
18.10	Arrhenius behavior	690
18.11	Activated complexes	693
18.12	Catalysis	694
18.13	Living catalysts: Enzymes	696
	REACTION MECHANISMS	697
18.14	Elementary reactions	697

	Case Study Drug therapy and mood regulation	698
18.15	The rate laws of elementary reactions	701
18.16	Chain reactions	704
18.17	Rate and equilibrium	705
	Skills you should have mastered	707

19 THE MAIN-GROUP ELEMENTS: I. THE FIRST FOUR FAMILIES 719

	HYDROGEN	720
19.1	The element	720
19.2	Compounds of hydrogen	723
	GROUP 1: THE ALKALI METALS	725
19.3	The elements	725
19.4	Chemical properties of the alkali metals	727
19.5	Compounds of lithium, sodium, and potassium	729
	GROUP 2: THE ALKALINE EARTH METALS	731
19.6	The elements	732
19.7	Compounds of beryllium, magnesium, and calcium	736
	GROUP 13: THE BORON FAMILY	738
19.8	The elements	738
19.9	Group 13 oxides	742
19.10	Carbides, nitrides, and halides	744
19.11	Boranes and borohydrides	745
	GROUP 14: THE CARBON FAMILY	746
19.12	The elements	747
19.13	The many faces of carbon	748
19.14	Silicon, tin, and lead	751
19.15	Oxides of carbon	752
19.16	Oxides of silicon: The silicates	753
	Case Study Glasses and ceramics	756
19.17	Carbides	759
	Skills you should have mastered	760

20 THE MAIN-GROUP ELEMENTS: II. THE LAST FOUR FAMILIES 767

	GROUP 15: THE NITROGEN FAMILY	768
20.1	The elements	768
20.2	Compounds with hydrogen and the halogens	770
20.3	Nitrogen oxides and oxoacids	772
20.4	Phosphorus oxides and oxoacids	774
	GROUP 16: THE OXYGEN FAMILY	776
20.5	The elements	776

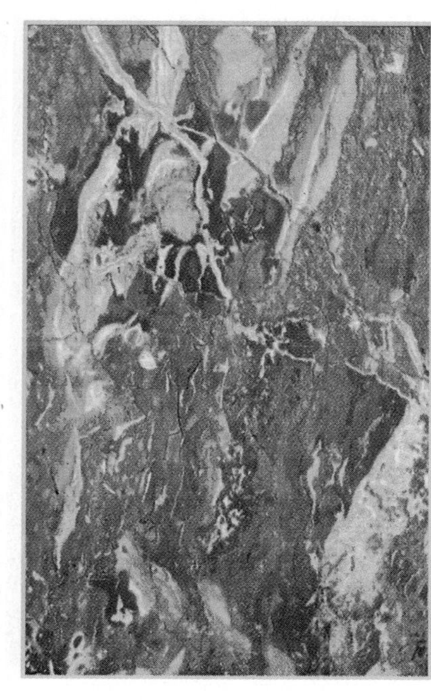

20.6 Compounds with hydrogen 780
20.7 Sulfur oxides and oxoacids 782
20.8 Sulfur halides 784
GROUP 17: THE HALOGENS 785
20.9 The elements 786
20.10 Compounds of the halogens 788
CASE STUDY Rocket fuels 792
GROUP 18: THE NOBLE GASES 794
20.11 The elements 794
20.12 Compounds of the noble gases 795
Skills you should have mastered 796

21 THE *d*-BLOCK: METALS IN TRANSITION 803

THE *d*-BLOCK ELEMENTS AND THEIR COMPOUNDS 804
21.1 Trends in physical properties 804
21.2 Trends in chemical properties 807
21.3 Scandium through nickel 809
21.4 Groups 11 and 12 816
CASE STUDY Photochemical materials 820
COMPLEXES OF THE *d*-BLOCK ELEMENTS 822
21.5 The structures of complexes 823
21.6 Isomers 826
CRYSTAL FIELD THEORY 832
21.7 The effects of ligands on *d*-electrons 832
21.8 The effects of ligands on color 834
21.9 The electronic structures of many-electron complexes 837
21.10 Magnetic properties of complexes 839
Skills you should have mastered 841

22 NUCLEAR CHEMISTRY 849

NUCLEAR STABILITY 850
22.1 Nuclear reactions 850
22.2 Nuclear structure and nuclear radiation 851
22.3 Nuclear decay 852
TOOLBOX 22.1 How to identify the products of a nuclear reaction 854
22.4 The pattern of nuclear stability 855
22.5 Nucleosynthesis 858
RADIOACTIVITY 860
22.6 The effects of radiation 861
CASE STUDY Nuclear medicine: Reducing the risks 862
22.7 Measuring radioactivity 865

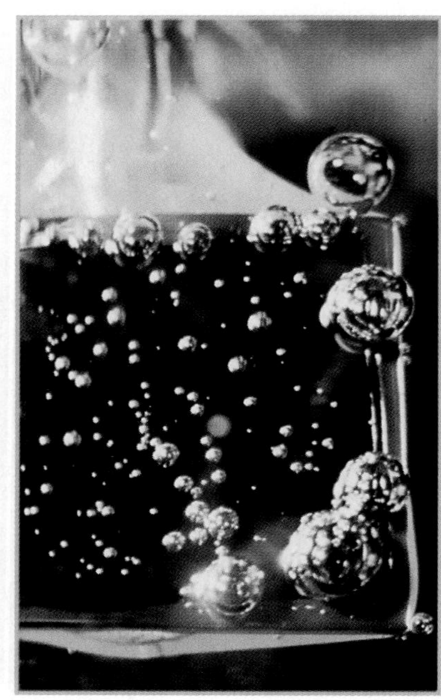

22.8	The law of radioactive decay	867
	Box 22.1 Uses of radioactive isotopes	870
	NUCLEAR ENERGY	872
22.9	Mass-energy conversion	872
22.10	Nuclear fission	874
22.11	Nuclear fusion	878
22.12	The chemistry of nuclear power	879
	Skills you should have mastered	881

APPENDIX 1 MATHEMATICAL INFORMATION A1
- A Algebra rules A1
- B Scientific notation A2
- C Logarithms A3
- D Quadratic and cubic equations A5
- E Graphs A5

APPENDIX 2 EXPERIMENTAL DATA A7
- A Thermodynamic data at 25°C A7
  - Inorganic substances A7
  - Organic compounds A13
- B Standard potentials at 25°C A14
  - Potentials in electrochemical order A14
  - Potentials in alphabetical order A15
- C Ground-state electron configurations A17
- D The elements A18
- E The top 50 chemicals by industrial production in the United States in 1995 A28

APPENDIX 3 NOMENCLATURE A30
- A Polyatomic ions A30
- B Common names of chemicals A31
- C Naming *d*-metal complexes A32

GLOSSARY B1

ANSWERS C1
- Self-Tests B C1
- Odd-Numbered Case Study Questions C8
- Odd-Numbered Exercises C11

ILLUSTRATION CREDITS D1

INDEX E1

PREFACE

Our principal aim is to impart *chemical insight*. From our many years of teaching experience, we know the difficulties students face as they begin to acquire a chemist's view of the world. We strongly believe that a textbook must connect observations—in the everyday world and in the laboratory—to the principles of chemistry. Our approach is always anchored in a student's experience and then goes beyond the perceived, to help students visualize entities they cannot see directly. It is essential to build a bridge between these realms of observation and imagination, so that students can understand phenomena in terms of atoms, molecules, and energy.

Our second major theme is how qualitative concepts lead to quantitative results. Stating and solving problems quantitatively is more than a basic skill in general chemistry. It is also part of the scientific method, and we believe that it too should be guided by chemical insight. These two aims—to convey chemical insight and to build quantitative skills—have remained constant over three editions.

Although this edition builds on the strengths of previous ones, we have introduced important changes to make chemistry more meaningful to students and to help them understand its principles. Most obvious is a new coauthor. Our two backgrounds have given us unique experience with a wide range of students, and we have developed material that works at very different institutions. Together, we have tried to make this edition more helpful to students than ever. We have changed the order of the chapters, rewritten them all, provided a completely new set of drawings and a great many new photographs, relied on important new technologies coming to use in today's classroom, and added many other features designed to make learning chemistry the exciting and rewarding experience we believe it can and should be. We wanted a third edition that is a pleasure to read, yet also is an efficient, innovative resource for teaching and learning.

When we look at changes in the natural world, from the germination of a seed through the unfurling of a leaf to the decay of fall leaves, we may be amazed by the variety of processes that can take place. Yet chemists have discovered that all chemical reactions can be classified into a limited number of types. These shoots of the bloodroot plant (Sanguinaria canadensis)*, a member of the poppy family, make use of the reactions described in this chapter as they convert the debris of the forest floor into new living cells. The sap of the bloodroot was once used as a dye, and it contains medically active compounds.*

A DESIGN FOR LEARNING

We have kept in mind what it is like to be a student grappling with chemistry. We have retained two widely praised features of past editions—the problem-solving *Strategies* and the *Case Studies*. We have also rewritten most of the text to highlight succinctly and clearly the essential principles of the subject, and we have introduced many new features to develop a student's insight into chemistry.

CHAPTER OPENINGS

We typically open chapters by raising observations and applications of current interest. These first paragraphs, NEW to this edition, orient the student and suggest the purpose of the concepts to come. Where appropriate, we return to that opening theme throughout the chapter. We think our readers will see more easily how their accumulating knowledge and developing problem-solving skills help in resolving the important issues that they face every day.

With every breath we take, we inhale carbon dioxide molecules along with the other gases of the air. Each breath connects us with history. Some of the carbon atoms in the molecules were once exhaled by our ancestors; in fact, some of them were exhaled by virtually everyone who has ever lived. One atom

The progress of carbon atoms through the environment is called the **carbon cycle.** For example, some atmospheric carbon dioxide dissolves in seawater, where it is absorbed into the shells of shellfish. After the death of the animal, it is compressed to form limestone (calcium carbonate, $CaCO_3$). It may remain as

CHEMICAL EQUATIONS AND CHEMICAL REACTIONS

3.1 Symbolizing chemical reactions

3.2 Balancing chemical equations

PRECIPITATION REACTIONS

3.3 Aqueous solutions

3.4 Reactions between strong electrolyte solutions

3.5 Ionic and net ionic equations

THE BOTTOM LINE

An aqueous strong electrolyte solution consists of hydrated ions that are free to move through the solvent.

Each section of the text concludes with a one- or two-sentence summary of its principal points. We have found that these summaries, NEW to this edition, help to focus a student's attention on the "bottom line," and they are especially helpful when preparing for a test.

DRAWINGS OF THE MOLECULAR WORLD

Every drawing is NEW to this edition, and they number more than 700. All of them have been drawn—by us—to convey our understanding of the molecular world. By sharing our vision, we can help students develop their own understanding. Many drawings show molecular models, accurately scaled to help students visualize in three dimensions; each of these models is brought further alive on our accompanying CD-ROM.

ANIMATION SEQUENCES

Among other NEW features, certain illustrations look like screens from an animation. Unique to this book, they each bring vividly to life a dynamic sequence. Taking our cue from our CD-ROM, on which many of these illustrations become actual animations, we indicate the time sequence by a scroll bar across the screen.

FIGURE 3.9 This animation sequence shows a series of scenes in a solution of sodium chloride. A sodium ion and a chloride ion move together, linger near each other for a time because of the attraction of their opposite charges, and then move apart and meet other partners elsewhere. The loose, transient association of oppositely charged ions is called an ion pair.

SIDENOTES

Also NEW to this edition, notes in the margin call out further information that might be useful in learning. Sidenotes take readers into our confidence, and they anticipate many of the questions that students often raise. They include the history of terms and ideas, as well as sections to review or anticipate as needed. A thumb button (◀◀ or ▶▶), which also becomes a "hot link" on our accompanying CD-ROM, flags these important cross-references.

▶▶ This definition is due to the Swedish chemist Svante Arrhenius. We shall meet a more general definition in Section 14.1.

CHEMISTRY AT WORK

Among the vast number of photographs NEW to this edition, every chapter highlights *Chemistry at Work*. Many students must take general chemistry courses before they have a clear idea of the relevance of chemistry to their chosen field. We show students that all kinds of people in all kinds of contexts use the techniques that they are struggling to learn. Chemistry, after all, is not just for chemists.

CHEMISTRY AT WORK

A pollution control officer samples water from a river in England. To ensure that the river is safe for drinking, fishing, and recreation, she monitors water quality frequently. She will analyze the acid content of the sample and will add reagents to precipitate heavy metal ions such as lead and mercury. If the pollutant

CASE STUDIES

We have expanded a popular feature of the second edition, the *Case Studies*. A *Case Study* looks in depth at a problem of topical interest, such as water treatment, pollution, the greenhouse effect, and the self-assembling molecules typical of life. We intend *Case Studies* to interest and motivate students, but because they also pull together the concepts of the

FROM RIVER WATER TO DRINKING WATER

EVERY LARGE RIVER is tapped by cities for drinking water. The used water is then treated and returned to the river as sewage, along with industrial waste. As a result, the water in many rivers is growing more and more polluted. Water flowing through underground aquifers (porous rock layers) is relatively pure, but it is still not free of pollution.

Assuring a potable (consumable) drinking water supply is a major concern of today's cities and is the responsibility of water chemists and sanitary engineers. Domestic

and precipitates Mg^{2+}, Fe^{3+}, and heavy metal ions as hydroxides.

Hard water contains relatively high concentrations of Ca^{2+} and Mg^{2+} ions, often in the form of hydrogen carbonates. These cations precipitate in furnace boilers and react with soap to form a scum that is hard to remove from laundry. Magnesium and calcium hydrogen carbonates can ...r by adding

QUESTIONS

1. Identify (a) the physical processes and (b) the chemical processes used for purifying raw water.

2. Classify by type each of the reactions used in the purification of water. If the reaction is a redox reaction, identify the oxidizing agent and the reducing agent.

3. Slaked lime is used to precipitate heavy metal ions.

chapter, each has its own short bank of questions. Many of the *Case Studies* are entirely new, and each now falls within the body of its chapter, rather than within exercise sets, to signal its relevance to important principles.

BOXES

NEW *Boxes* contain useful background, including techniques that chemists use today to determine relevant quantities. These *Boxes* are like islands of relief for tormented and shipwrecked students, for at least here they will find no exercises!

BOX 3.1 MAINTAINING AN ARTIFICIAL ATMOSPHERE

On Earth, our atmosphere is kept at a relatively constant composition by natural processes, such as photosynthesis and the movement of air (see Case Study 5). However, a submarine or space ship must use artificial air purifiers. Otherwise, the crew would suffocate in the carbon dioxide, CO_2, they produce. In such sealed environments, waste gases are removed and oxygen regenerated by a variety of acid-base and redox chemical reactions.

Because the space shuttle serves only for short flights, its oxygen can be stored on-board and need not

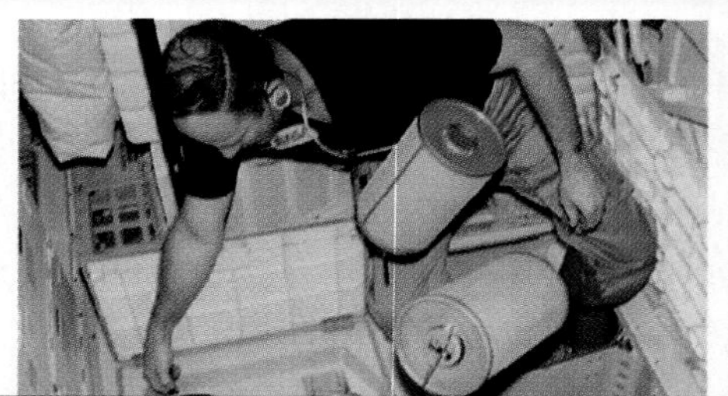

SKILLS YOU SHOULD HAVE MASTERED

The end-of-chapter summaries now take the form of *Skills You Should Have Mastered*. We have also divided these lists into conceptual, problem-solving, and descriptive skills. That division, NEW to this edition, organizes the information in the chapter, making it much less daunting. Acquiring understanding and information involves a constant to-and-fro between learning and review, and these review aids will make the process more effective by allowing students to monitor their progress easily.

SKILLS YOU SHOULD HAVE MASTERED

Conceptual

1. Explain the significance of a balanced chemical equation.
2. Identify solutions as electrolytes or nonelectrolytes on the basis of the formulas of the solutes.
3. Explain the difference between solutions of strong and weak acids and bases.
4. Define oxidation and reduction in terms of oxidation number and electron transfer.
5. Identify the oxidizing and reducing agents in a reaction.

Problem-Solving

1. Write, balance, and label a chemical equation,
3. Determine the oxidation number of an element in an ion or compound.
4. Use the solubility rules to select appropriate solutions which, when mixed, will produce a desired precipitate.

Descriptive

1. Write three balanced chemical equations for reactions that occur in the carbon cycle and describe their roles in the cycle.
2. Describe chemical properties of acids and bases.
3. Recognize common strong and weak acids and bases from their formulas.
4. Describe how acids react with water to form

BUILDING PROBLEM-SOLVING SKILLS

The interplay between the qualitative and the quantitative is central to the scientific method, but it is the hardest part of the course for many students. We provide a great deal of help.

STRATEGIES

Some 250 worked examples teach problem solving. All books have examples, but we use them in a special way. Not only do they illustrate the material, but—most important—they develop a student's ability to reason chemically. We therefore have retained a popular feature of past editions: *Strategies* that precede the solution to every worked example. The *Strategies* encourage students to collect their thoughts before embarking on a problem, they offer reminders of necessary concepts, and they show how to plan the steps toward a solution. Some *Strategies* suggest how to anticipate the magnitude of an answer before developing it in detail; others help in locating equations and data; still others explain when and how to rearrange an equation to find the unknown. In this edition we have rewritten the *Strategies* to make them more general in application, so that students are prepared to solve new problems on their own.

EXAMPLE 3.5 **Devising a means of preparing a gas**

Suggest a method for making the gas phosphine, PH_3. Phosphine is an extremely toxic gas that should be made only in a fume hood.

STRATEGY Because phosphine is a gas, we may be able to devise a reaction in which it is produced by the action of an acid on a suitable salt. To identify the salt to use as starting material, we examine the formula of phosphine,

TOOLBOXES

A NEW feature develops the approach of *Strategies* even further. *Toolboxes* set out how to tackle major calculations. Just as a hammer or a screwdriver serves on many occasions, the procedures described in our *Toolboxes* will come into play time and time again, in many different contexts. A student familiar with them will be well prepared for calculations they will encounter throughout their study of chemistry.

TOOLBOX 3.2 **How to find oxidation numbers**

To assign an oxidation number to an element, we start with two simple rules:

1. The oxidation number of an element uncombined with other elements is zero.
2. The sum of the oxidation numbers of all the atoms in a species is equal to its total charge.

The first of these rules tells us that hydrogen, oxygen, iron, and all other elements have oxidation numbers of 0 in their elemental forms. The second rule implies that the oxidation by using these two rules in conjunction with the following specific values:

The oxidation number of hydrogen is +1 in combination with nonmetals and −1 in combination with metals.

The oxidation number of oxygen is −2 in most of its compounds.

The oxidation number of all the halogens is −1 unless the halogen is in combination with oxygen or another

SELF-TESTS

The *Self-Tests* are also NEW. Like the in-chapter exercises of previous editions, they let students test their ability to do a calculation similar to the one just presented in a worked example. However, now we supply *Self-Tests* in pairs, more than doubling their number. One has its answer in full view, to give the student instant feedback. The second of the pair has its answer hidden discreetly at the back of the book, as many instructors have requested. These

SELF-TEST 3.10A Find the oxidation numbers of sulfur, phosphorus, and nitrogen in (a) H_2S; (b) PO_4^{3-}; (c) NO_3^-, respectively.

[*Answer:* (a) −2; (b) +5; (c) +5]

SELF-TEST 3.10B Find the oxidation numbers of sulfur, nitrogen, and chlorine in (a) SO_3^{2-}; (b) NO_2^-; (c) $HClO_3$.

Self-Tests follow all worked examples and also appear at strategic places in *Toolboxes*. In addition, they occur on their own throughout the text, when it is important to reinforce understanding.

EXERCISES

Instructors will find plenty of material for assignment. There are now a huge number of *Exercises* at the ends of chapters: some 2000, in addition to several hundred *Self-Tests*. As in former editions, we first classify many by topic, leaving others as Supplementary Exercises, for a more balanced survey of each chapter. NEW to this edition are additional Challenging Exercises that require just a little more thought; they are also suitable for collaborative solution in the classroom. All exercises classified by topic are paired. Answers to odd-numbered *Exercises* appear at the back of the book, and all solutions are given in the *Solutions Manual* (the odd-numbered exercises) or the *Instructor's Resource Manual* (the even-numbered exercises). Every answer has been independently solved three times to ensure its accuracy.

CHALLENGING EXERCISES

3.77 Identify all the principal species that exist in aqueous solutions of (a) $HClO_2$; (b) NH_3; (c) CH_3COOH.

3.78 Identify all the principal species that exist in aqueous solutions of (a) CH_3NH_2; (b) HI; (c) $HClO_4$.

3.79 Write a balanced equation for the complete combustion (reaction with oxygen) of octane, C_8H_{18}, the

ORGANIZATION OF THIS EDITION

We have taken care that instructors can cover topics in more than one sequence, as they see fit. Yet we have made several innovations that we consider pedagogically important.

Chapter 1 is not the usual chapter on units. We feel strongly that students ought to be taken to the heart of chemistry on their first encounter with the subject. Therefore, we open the book with an introduction to elements and atoms. Chapter 1 also provides an early introduction to the periodic table, a unifying theme of the book. With this arrangement, students get an immediate introduction to chemistry.

We now discuss units in **Chapter 2**, where we begin the quantitative treatment of data. This order reflects our theme of moving from qualitative understanding to quantitative skills. We include the mole here as simply another unit. Well, it *is* just another unit, and one of exceptional importance to chemists. We think a kid-gloves treatment of moles as something special (and, by implication, subtle and difficult) is misplaced. Given a natural, no-fuss introduction and lots of guidance and practice (which we provide), students should at last find moles easy to understand and deploy.

Another important innovation is **Chapter 11**, an early introduction to organic chemistry. Organic chemistry is so important for our understanding of the world that we consider it regrettable to tuck it away at the end of a text, where it is rarely reached. Although instructors may still defer this chapter if they wish, we believe that it provides superb examples for reviewing the formation and breaking of chemical bonds and an excellent means of reinforcing the ideas of the preceding chapters. It also makes a convenient teaching unit along with the chapter on liquid and solid materials; for that reason, it has an innovative target: synthetic and natural polymers. We introduce very few reactions, just enough to give the flavor of organic chemistry and to show how polymers are made. Students who leave the course after the first semester, and who most likely will never have a formal course in organic chemistry, will have had some exposure to this essential part of a chemical education.

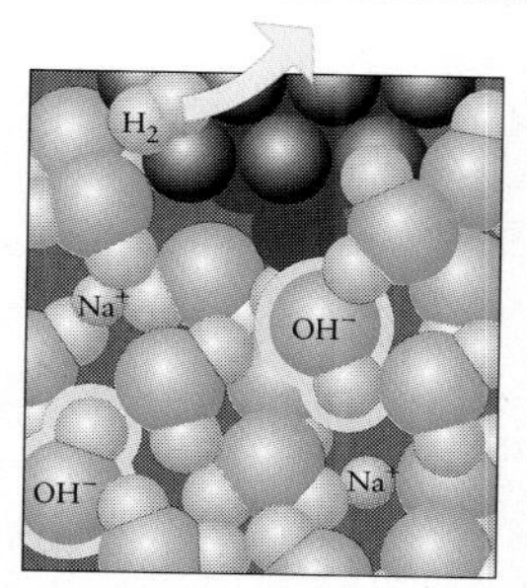

Many of the other chapters will be found in familiar locations, but we have extensively revamped their content with an eye on level, presentation, and readability. We introduce that great organizational principle of chemistry, the periodic table, in **Chapter 1**, for it is so essential to the rationalization—not to mention the remembering—of chemical information, that we believe it should become an intrinsic part of a student's equipment as early as possible. We justify the structure of the periodic table later, providing a focus to **Chapter 7**, the discussion of atomic structure.

Chapter 3 develops an understanding of chemical reaction types early in the course; to maintain this focus, we have deferred balancing redox equations by the half-reaction method to **Chapter 17**. The description of the remaining class of reactions we consider, Lewis acid-base reactions, fits very naturally into the bonding chapter (**Chapter 8**), helping students more easily relate these reactions to the importance of the electron pair in chemistry.

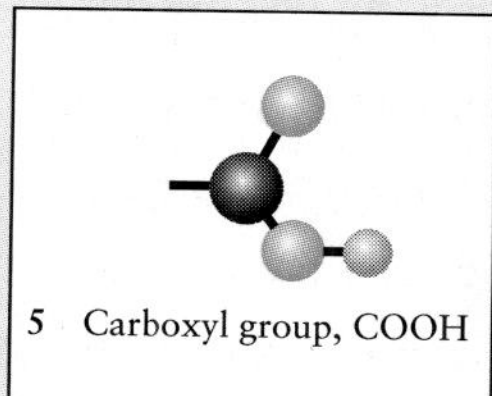

5 Carboxyl group, COOH

Chapter 16 is the second of two bites at thermochemistry and thermodynamics. The first law of thermodynamics is a straightforward account of the production or absorption of heat. The second law accounts for all the equilibria that are encountered in the middle of the course, and we consider that it is correct to place it as a kind of apotheosis of that more quantitative and demanding part of the course. The two laws are presented together in Chapter 16. However, because we know that some instructors prefer to present the first law early in the course we have written the early parts of Chapter 16 so they could quite easily be used in conjunction with the introduction of enthalpy in Chapter 6, where they would certainly illuminate the difference between internal energy and enthalpy. We believe that an understanding of entropy helps to explain many chemical processes. With that in mind, we use the tendency toward disorder in a number of early chapters. In this way, students can build up an intuitive understanding of the concept before having to cope with it quantitatively and before mastering the term *entropy* explicitly.

We have deferred chemical kinetics to **Chapter 18**. Although it too may be taught when the instructor wishes, we chose to set it apart from **Chapters 12–17**, which are essentially the domain of equilibrium and thermodynamics. In short, we think it right to know *where* one is going before worrying about how fast one can get there. Yet, because we concur with the view that a qualitative knowledge of kinetics illuminates dynamic equilibria, we have not hesitated to invoke qualitative rate arguments when discussing equilibria.

As in earlier editions, the book concludes with a sequence of chapters on descriptive chemistry, but **Chapters 19–21** have been extensively revised, rewritten, and rearranged. Here we show how "descriptions" in fact make use of the principles that students are learning. Of course, there is a great deal of descriptive material in earlier chapters, for we always illustrate the principles with real phenomena. These later chapters, as instructors will recognize, provide a different cross section through the subject, where the properties are used as a vehicle for rationalization by principles rather than the other way round. We have had to be very selective in these chapters, as any instructor will appreciate, and our aim is to show that the elements have varied, intriguing personalities. After all, that is what chemistry is all about.

SUPPLEMENTS

Throughout the compilation of this edition, we have been in constant contact with our team of supplements authors, to ensure that the entire package is consistent and mutually enhancing.

CD-ROM

This multimedia learning tool, developed by W. H. Freeman and Company in conjunction with Sumanas, Inc., complements and enriches the textbook. All the features of the CD function within the context of the book's coverage. The entire textbook—with hyperlinked text and the illustrations in an electronic format—serves as the starting point for the CD. All structures in the book's margins are depicted as three-dimensional animations. These and the many other molecular-level simulations and video demonstrations bring the concepts of the book to life. Practice tools, such as interactive quizzes in every chapter, help students review for exams. The CD's customized calculator and dynamic periodic table/data base reduce the tedium of solving problems. Webnotes™ provide direct links to chemistry sites on the World Wide Web. Presentation software for instructors allows them to prepare a series of illustrations, animations, and videos for lecture.

STUDENT'S STUDY GUIDE

This reassuring volume by David Becker, Oakland University and Oakland Community College, reinforces concepts, provides additional practice exercises with answers, and highlights "Pitfalls," common student mistakes, so they can be better avoided.

STUDENT'S SOLUTIONS MANUAL

This manual by Charles Trapp, University of Louisville, contains complete solutions to the odd-numbered end-of-chapter exercises. Additional commentary on problem-solving techniques is included.

CHEMISTRY IN THE LABORATORY

This well-respected manual by Julian Roberts and Leland Hollenberg, University of Redlands, and James M. Postma, California State University at Chico, is now in its fourth edition. It contains 44 lab-tested experiments and a new emphasis on safety and waste disposal. Many new reduced-scale experiments that do not require a full wet lab have been included. All experiments are available as lab separates.

STUDENT COMPANION

The new *Student Companion: New Tools and Techniques for Chemistry*, by Lynn Geiger, Belia Straushein, and Loretta Jones, University of Northern Colorado, is a unique student supplement that offers blueprints for using innovative approaches and emerging technologies to invigorate the teaching and study of general chemistry. Student worksheets and other materials in the *Companion* will aid instructors in incorporating the new techniques of collaborative learning and guided readings in their classrooms. Worksheets for assignments that involve students in the power of the Web and the features of our CD-Rom are also provided.

INSTRUCTOR'S RESOURCE MANUAL

This useful manual by Lowell Parker, Stevens Institute of Technology, contains sample syllabi, lecture outlines, teaching hints and transparency masters. Solutions to the textbook's even-numbered exercises by Charles Trapp, University of Louisville, are included, as is a guide for using the *Student Companion*.

TEST BANK

The test bank by Robert Balahura, University of Guelph, contains two sets of 75 multiple-choice questions and 10 short-answer questions per chapter—double the number of previous editions. The test bank is available in both printed and electronic form, and the test bank software is also new to this edition.

GENERAL CHEMISTRY VIDEODISC

Approximately 100 vivid experiments are again available to the instructor for lecture presentations.

OVERHEAD TRANSPARENCIES

More than 200 transparencies with large-type labels illustrate the key figures and tables.

CHEMISTRY WEB SITE

Students and instructors should visit our site (http://www.whfreeman.com/chemistry3e) for a variety of useful features including study questions, WebNote links to other useful sites, updates, and video lab demonstrations specially formatted for viewing on the Web.

ACKNOWLEDGMENTS

No project as big as this can make progress, let alone come to fruition, without the assistance of large numbers of people. We are both deeply indebted to all those who shared their time with us: we know that we have benefitted greatly from the suggestions that they have made to us. We try to build on the great store of wisdom and experience that the chemical education community collectively represents—and to push forward the frontiers. We would particularly like to thank the following people for their input:

Robert D. Allendoerfer, State University of New York at Buffalo
Lavoir Banks, Elgin Community College
Michael Chetcuti, Notre Dame University
Corinna Czekaj, Oklahoma State University, Stillwater
Dan J. Davis, University of Arkansas, Fayetteville
James E. Davis, Harvard University
Eugene Deardorf, Shippenburg University
T. R. Dickson, Cabrillo College
Philip Dumas, Trenton State University
William J. Evans, University of California, Irvine
Ellen R. Fischer, Colorado State University, Fort Collins
Michele M. Francl, Bryn Mawr College
James F. Garvey, State University of New York at Buffalo
John L. Gland, University of Michigan, Ann Arbor
Michael D. Hampton, University of Central Florida, Orlando
Paul W. W. Hunter, Michigan State University, East Lansing
Ronald C. Johnson, Emory University, Atlanta
Pat Jones, University of the Pacific, Stockton
Richard W. Kopp, East Tennessee State University, Johnson City
John Krenos, Rutgers University
Marlena Maestas, University of Northern Colorado, Greeley
William McMahan, Mississippi State University
D. J. Morrissey, Michigan State University, East Lansing
Lowell Parker, Stevens Institute of Technology, Hoboken
Michael J. Perona, California State University, Stanislaus
Cortlandt G. Pierpont, University of Colorado, Boulder
N. T. Porile, Purdue University, West Lafayette

Martha W. Sellars, Northern Virginia Community College, Annandale

R. Carl Stoufer, University of Florida, Gainesville

Allan Thomas, Temple University, Philadelphia

Holden Thorp, University of North Carolina, Chapel Hill

Garth L. Welch, Weber State University, Ogden

Stan Wittingham, Binghamton University

We wish to acknowledge still others who contributed by their efforts to the first two editions: David L. Adams, Bradford College; John E. Adams, University of Missouri, Columbia; Martin Allen, College of St. Thomas (retired); Norman C. Baenziger, University of Iowa; John Baldwin, Syracuse University; M. C. Banta, Sam Houston State University; John E. Bauman, University of Missouri, Columbia; J. M. Bellama, University of Maryland, College Park; James P. Birk, Arizona State University; Richard Bivens, Allegheny College; P. M. Boorman, University of Calgary; Larry Bray, Miami-Dade Community College; Luther K. Brice, American University; J. Arthur Campbell, Harvey Mudd College; Dewey K. Carpenter, Louisiana State University; Mark Chamberlain, Glassboro State College; Geoffrey Davies, Northeastern University; Walter J. Deal, University of California, Riverside; Walter Dean, Lawrence Technological University; John DeKorte, Northern Arizona University; Fred M. Dewey, Metropolitan State University; James A. Dix, State University of New York, Binghamton; Craig Donahue, University of Michigan, Dearborn; Wendy Elcesser, Indiana University of Pennsylvania; Grover Everett, University of Kansas; Gordon Ewing, New Mexico State University; John H. Forsberg, St. Louis University; Bruce Garatz, Brooklyn Polytechnic University; Marjorie H. Gardner, Lawrence Hall of Science; Gregory D. Gillespie, North Dakota State University; L. Peter Gold, Pennsylvania State University; Michael Golde, University of Pittsburgh; Stanley Grenda, University of Nevada; Thomas J. Greenbowe, Southeastern Massachusetts University; Kevin Grundy, Dalhousie University; Marian Shu Hallada, University of Michigan; Robert N. Hammer, Michigan State University (consultant); Kenneth L. Hardcastle, California State University, Northridge; Joe S. Hayes, Mississippi Valley State University (retired); Henry Heikkinen, University of Northern Colorado; Forest C. Hentz, Jr., North Carolina State University; Larry W. Houk, Memphis State University; Robert Hubbs, De Anza College; Brian Humphrey, Montclair State College; Jeffrey A. Hurlburt, Metropolitan State University; Earl S. Huyser, University of Kansas; Robert Jacobson, Iowa State University; Charles Johnson, University of North Carolina; Delwin Johnson, St. Louis

Community College, Forest Park; Stanley Johnson, Orange Coast College; Murray Johnston, University of Colorado, Boulder; Andrew Jorgenson, University of Toledo; Herbert Kaesz, University of California, Los Angeles; Philip C. Keller, University of Arizona; Manickam Krishnamurthy, Howard University; Robert Loeschtate University, Long Beach; David G. Lovering, Royal Military College of Science, Shrivenham; James G. Malik, San Diego State University; M. Stephen McDowell, South Dakota School of Mines and Technology; Saundra McGuire, Cornell University; Pamela McKinney-Forbes, Northern Virginia Community College; Gloria J. Meadors, University of Tulsa; Patricia Metz, Texas Tech University; Amy E. Stevens Miller, University of Oklahoma; Ronald E. Noftle, Wake Forest University; E. A. Ogryzlo, University of British Columbia; Jane Joseph Ott, Furman University; M. Larry Peck, Texas A&M University; Lee G. Pedersen, University of North Carolina, Chapel Hill; W. D. Perry, Auburn University; Dick Potts, University of Michigan, Dearborn; Everett L. Reed, University of Massachusetts; Don Roach, Miami-Dade Community College; B. Ken Robertson, University of Missouri; Larry Rosenheim, Indiana State University; Patricia Samuel, Boston University; E. A. Secco, St. Francis Xavier University; Henry Shanfield, University of Houston; R. L. Stern, Oakland University; Billy L. Stump, Virginia Commonwealth University; James E. Sturm, Lehigh University; James C. Thompson, University of Toronto; Donald D. Titus, Temple University; Joseph J. Topping, Towson State University; Arlen Viste, Augustana College; Tom Weaver, Northwestern University; Patrick A. Wegner, California State University, Fullerton; Rick White, Sam Houston State University. We remain very grateful for their advice, as we regard those editions as a foundation for this one.

Inevitably, some have contributed more than others. We would particularly like to thank John Krenos, for working through all the Self-Tests so assiduously; Julie Henderleiter, for her careful, thorough checking of exercise solutions; Martha Sellars, for bringing her pedagogical percipience to bear on the entire manuscript; Michael Perona, for reading the entire text with a teacher's eye; and Denis Cullinan, for checking the chemistry throughout. Our supplements authors, particularly Charles Trapp and David Becker, have spread beyond the coasts of their particular responsibilities to give us a great deal of very useful critical advice. We are also grateful to our colleagues, especially David Pringle, Clark Fields, and Robert Baillie, who offered many useful suggestions, and to Patananya Lekhavat, who suggested ideas for conceptual exercises. Our students, of course, have been the unwitting test beds for many of our ideas, and we are grateful to them for the toleration that gradually turned into enjoyment. We especially wish to acknowledge the students in Chemistry 111 and 112 at the University of Northern Colorado who used the book in draft form and contributed many helpful comments.

Our colleagues, both those listed above and others who remain anonymous but present in our files, are but the tip of the enormous iceberg that we have been building over many months. It would be too self-indulgent for the authors to thank each other publicly, but privately we know that each of us has been inspired by the other, and we both acknowledge Jo Beran's lasting contribution to the second edition.

The large number of people who contributed at W. H. Freeman and Company are symbolized but not exhaustively named on the copyright page: that unelaborated listing conceals the deep gratitude that we feel toward both them and their assistants. In fact, we would like to call on to the stage particular people: Deborah Allen, chemistry editor, who initiated this new edition and brought us together; John Haber, who as development editor saw it through from start to finish with a remarkable number of insightful suggestions that we—in retrospect, wisely—interpreted as orders; Jodi Simpson, our awesomely acute coach and trainer (while pretending to be our copy editor), who extracted sense where perhaps we had unwittingly left nonsense; Georgia Lee Hadler, senior project editor, who meticulously wrought the final product from its colossal and (to anyone who could not appreciate the inner fire) seemingly amorphous manuscript; Audrey Herbst, who uses photographs to communicate with images as others communicate with words; and Patrick Shriner, who cunningly orchestrated the varied and novel instruments of the supplements package.

We could go on: our thanks could be never ending. But it is time for our words and images to speak for themselves.

ABOUT THE AUTHORS

PETER ATKINS is professor of chemistry at Oxford University, a fellow of Lincoln College, and the author of more than thirty books for students and a general audience. They have earned him an international reputation for communicating the excitement of chemistry. A frequent lecturer in the United States and throughout the world on chemistry and chemical education, he has held visiting professorships in Israel, France, Japan, China, and New Zealand. He firmly believes that visualization is crucial in developing chemical insight, and he creates the drawings in his books himself.

Professor Atkins worked in industry for two years before studying at Leicester University, where he received his Ph.D., and at the University of California, Los Angeles. His *Physical Chemistry* broke new ground by stressing understanding above the memorization of facts and formulas. Nearly 20 years later, it is still the leading text in its field. His other books include *Inorganic Chemistry* (with D. F. Shriver and C. H. Langford); *Molecules*, *The Second Law*, and *Atoms, Electrons, and Change* (all volumes for the Scientific American Library); and, most recently, *Molecular Quantum Mechanics* (with R. S. Friedman) and *The Periodic Kingdom.* He serves on the International Union of Pure and Applied Chemistry's committees on physical chemistry and on chemical education.

LORETTA JONES is an award-winning chemical educator who is known for her innovative chemistry curricula. After 13 years teaching general chemistry at the University of Illinois at Urbana-Champaign, she became associate professor at the University of Northern Colorado. There she continues to search for ways to involve students actively in learning and to provide them with visual representations of complex phenomena. Her work has contributed to an influential body of multimedia instructional materials, including *Exploring Chemistry* (with S. G. Smith and S. D. Gammon), and she believes that a commitment to how students learn can and must reshape the textbook as well.

Professor Jones earned a bachelor's degree in chemistry at Loyola University, Chicago, and a masters degree at the University of Chicago. For nine years she worked at Argonne National Laboratory and in industry. She then went on to the University of Illinois at Chicago to earn her Ph.D. in chemistry. At the same time she earned a D.A. in chemistry, a preparation for teaching and educational research and development at the college level. She is active in governance in the American Chemical Society, a Fellow of the American Association for the Advancement of Science, and a recipient of the Scholar of the Year Award from the College of Arts and Sciences at the University of Northern Colorado.

TO THE STUDENT

We wrote this book to help you master chemistry. We remember well the jungle chemistry seemed at first. We remember the trouble we had learning new concepts, the frustration of being unable to solve—and sometimes even begin—the homework problems. However, we are very pleased that we stuck to the course, because we now see the world in a very special way. When we look at a leaf in fall or an opening flower, when we feel the texture of textiles, when we hear on the evening news about threats to the environment, we appreciate it much more deeply than we did before, because we can see it through chemists' eyes.

As we explain in the preface, one of our principal aims is to impart insight. Insight means being able to bridge what we *perceive* to what we *imagine* going on. A difficulty with learning chemistry is making that connection—between what we observe going on around us and the explanations that chemists give, in terms of atoms, molecules, and energy. Chemistry is like looking at a city from a thousand miles into space. We know that there is activity down below in the city. However, it is only when we get down to ground level, where we can see all the inhabitants going about their tasks in characteristic ways, that we understand how the city really works. Chemical insight is like knowing those individuals closely. In a sense, it is about knowing the personalities of atoms and molecules so that we can predict how they are going to behave.

As we created the study aids in this book, we kept in mind our own difficulties and tried to help you through them. We know that reading a science or technical textbook requires a special kind of concentration and skill. Like physical exercises, mental exercises are useless unless they stretch you. There is usually a lot of information to process and recall later, and some of the concepts are not easy to understand the first time you encounter them. However, this book is organized so that you can make best use of your study time.

A good way to begin a new section is to skim through it first, read the summary at the end, and then return to read the section in detail. Stop to think about what the sentences mean: science can pack a lot of information into a single sentence. Be especially attentive to unfamiliar terms set in bold type, which are collected together in the *Glossary* at the end of the book.These terms are the basic language of chemistry, so look up words you do not understand. Become familiar with the many study aids built into each chapter, such as the *Self-Tests* and *marginal notes*. We describe them all in the preface, so be sure to take a look at that.

If you have never studied chemistry before, you may need to spend extra time on the course. Even if you have studied chemistry, perhaps some time ago, it would be a good idea to review *Appendix 1*, which reviews the mathematics you will need. If a section has you puzzled, review the sections flagged by a rewind button (◀◀) in the marginal notes. Still other helpful connections are flagged by the fast-forward button (▶▶). All these links become live on the accompanying CD-ROM, which contains the entire book as well as additional hyperlinks, Self-Tests, and other study aids.

Memorizing information about the chemical elements is easier if you look for trends. This textbook emphasizes trends and puts them in the context of the periodic table of the elements. Beginning in Chapter 1 you will recognize the periodic table as one of the great unifying concepts of chemistry.

Problem-solving skills are best developed through thinking about how to approach a problem, including how to assemble the information you need. (That information is usually in the text just before a worked example.) The *Strategy* section in each worked example shows you how to collect your thoughts and assemble the equations you need, and suggests where you can find additional information. The Strategy section also often suggests how you can predict the answer by using the chemical insight you have been developing. It is usually sensible to try to make an estimate of the answer before embarking on a calculation so that you can judge whether the result you obtain is plausible. You will get additional help from the *Toolboxes*, which contain the core calculations and procedures of chemistry—the hammers, screwdrivers, and saws of the calculations and procedures that you use throughout the subject. You may need to come back to these Toolboxes from other chapters, just as in real life you may need a hammer or screwdriver to do many kinds of jobs.

Make good use of the lists of *Skills* at the end of each chapter to monitor your progress. The skills are divided into three groups, which represent three types of learning goals: concepts, problem solving, and descriptive chemistry. You can use these lists to guide you through the chapter and when you return to it for review.

It is very important to keep up with the material in your course. Often topics presented one week are needed to understand topics presented the next week. If you drop behind, you may have trouble catching up again, and the topics will seem unnecessarily difficult. Many general chemistry students have found the following approaches helpful:

Read the relevant sections of the text before attending class.
Review and organize class notes daily.
Form a study group and meet at least twice a week.
Complete all the homework and check your answers.
Read related sections of other books in the library to get another slant on a topic.

Above all, *stay the course*. Even when you feel lost, there are still paths you can take. Go back to read a chapter one more time and work through all the paired Self-Tests. You may check the answers to the second of each pair and to the odd-numbered exercises in the back of the book. Use the learning supplements that your instructor recommends to you. Use the CD-ROM that accompanies and enriches this text. You can also find additional help on our Web page at

http://www.whfreeman.com/chemistry3e

We wish you every success. It may seem that a jungle lies ahead, but you will find the journey one of the great adventures of your mind.

CHAPTER 1

Most of the matter we study in chemistry originated in the stars, where many of the chemical elements were made and then scattered through interstellar space. This photograph was compiled by computer techniques that made use of images collected by the Hubble Space Telescope. It shows a region of the center of the Orion nebula where stars and their planetary systems are in the process of formation. Light emitted by oxygen has been colored blue by the computer treatment of the image, hydrogen emission is shown as green, and nitrogen emission as red. One day, perhaps, these elements will be studied by chemistry students living on the planets around these newly formed stars.

MATTER

Chemistry is the study of matter and the changes it can undergo. **Matter** is anything that has mass and takes up space. Anything you can touch is matter. So are a lot of things you cannot touch, such as the material of flames, rocket exhausts, and stars. Chemistry therefore embraces everything material around you—the stones you stand on, the silicon built into your computers, the food you eat, and even your flesh. Matter appears in a huge variety of forms, such as rocks, people, plants, oceans, cars, computers, clouds, and bacteria, so a knowledge of chemistry is essential if you want to understand how the world works.

Each different pure kind of matter is called a **substance.** By pure, we mean the same throughout, even on a microscopic scale. Thus, iron is one substance; water is another. Notice that the *scientific* meaning of the term substance is a little different from its everyday meaning. A substance in science is a *single, pure* form of matter, not a mixture of several different kinds of matter. According to this definition, water is a substance because it contains only one pure kind of matter, but flesh, soft drinks, and dirt are not single substances in the scientific sense, because they are complex mixtures. You will see that common terms are often given precise meanings when used in science.

THE ELEMENTS

1.1 Atoms
1.2 Names of the elements
1.3 The nuclear atom
1.4 Isotopes
1.5 Where did the elements come from?
1.6 The periodic table
1.7 Metals, nonmetals, and metalloids

COMPOUNDS

1.8 What are compounds?
1.9 Molecular compounds
1.10 Ionic compounds and ions

MIXTURES

1.11 Types of mixtures
1.12 Separation techniques

THE NOMENCLATURE OF COMPOUNDS

1.13 Names of cations
1.14 Names of anions
1.15 Names of ionic compounds
1.16 Names of molecular compounds

FIGURE 1.1 Samples of common elements. Clockwise from the red-brown liquid bromine are the silvery liquid mercury and the solids iodine, cadmium, red phosphorus, and copper.

THE ELEMENTS

Thinkers and philosophers have puzzled over the structure of matter since ancient times. The ancient Greeks developed the concept of *elements* as fundamental substances from which all forms of matter can be built. They identified four elements—earth, air, fire, and water—which they believed could produce all other substances when combined in the right proportions. Their concept of "element" is consistent with what we currently believe, but we now know that there are actually more than 100 **chemical elements** which, in various combinations, make up all the matter on Earth (Fig. 1.1).

1.1 ATOMS

The ancient Greeks first asked what would happen if we went on cutting matter into ever smaller pieces. Is there a point at which we would have to stop because the pieces no longer had the same properties as the whole, or could we go on cutting forever?

We now know that there comes a point at which we have to stop. That is, matter is not continuous but consists of almost unimaginably tiny particles. The smallest particle of an element that can exist is called an **atom**. The first convincing argument for atoms was made in 1807 by the English schoolteacher and chemist John Dalton (Fig. 1.2). He relied on a large number of laborious measurements of the masses of elements that combined together, and assembled arguments that strongly indicated the existence of atoms. Today, modern instrumentation provides much more direct evidence of atoms, and we can now make images of individual atoms (Fig. 1.3). There is no longer any doubt that atoms exist and that they are the units that make up the elements:

The name *atom* comes from the Greek word for "not cuttable."

An **element** is a substance composed of only one kind of atom.

▶▶ The atoms of an element are not all *exactly* the same, because they can differ slightly in mass. We examine this point in Section 1.4.

All the atoms in a block of gold are of the same kind. Similarly, all the atoms in a block of lead are of essentially the same kind (but different from those in a block of gold), and so on for all the elements. When scientists make a new element in one of their giant accelerators, they recognize it as new by checking—in ways to be described—whether its atoms are different from those of all known elements. By 1996, 112 elements had been discovered or created, but in some cases in only very small amounts. For instance, only one or two atoms of the 111th element have been made, and they lasted for only a fraction of a second.

> ***All matter is made up of various combinations of the simple forms of matter called the chemical elements. An element is a substance that consists of only one kind of atom.***

FIGURE 1.2 John Dalton (1766–1844), the English schoolteacher who used experimental measurements to argue that matter consisted of atoms.

1.2 NAMES OF THE ELEMENTS

The names of some elements are very ancient. For example, the name *copper* is derived from Cyprus, where the element was once mined, and the word *gold* is derived from the Old English word meaning "yellow." Some names take note of a characteristic property. Chlorine, for instance, is a yellow-green gas, and its name is derived from the Greek word meaning "yellow-green." Vanadium, which forms attractively colored compounds, is named after Vanadis, the Scandinavian goddess of beauty. More recently, elements have been named by their discoverers. Some elements honor people or places; these include americium, berkelium, californium, einsteinium, and curium. The International Union of Pure and Applied Chemistry (IUPAC) is the international body that currently, among other duties, approves names proposed for elements.

Chemists have a useful system that saves writing out the full names of the elements. Each element is represented by a **chemical symbol** made up of one or two letters. Many of the symbols are the first one or two letters of the element's name:

hydrogen H	carbon C	nitrogen N	oxygen O
helium He	aluminum Al	nickel Ni	silicon Si

Notice that the first letter of a symbol is always uppercase and the second letter, always lowercase (as in He, not HE). Some elements have symbols formed from the first letter of the name and a later letter:

magnesium Mg   chlorine Cl   zinc Zn   plutonium Pu

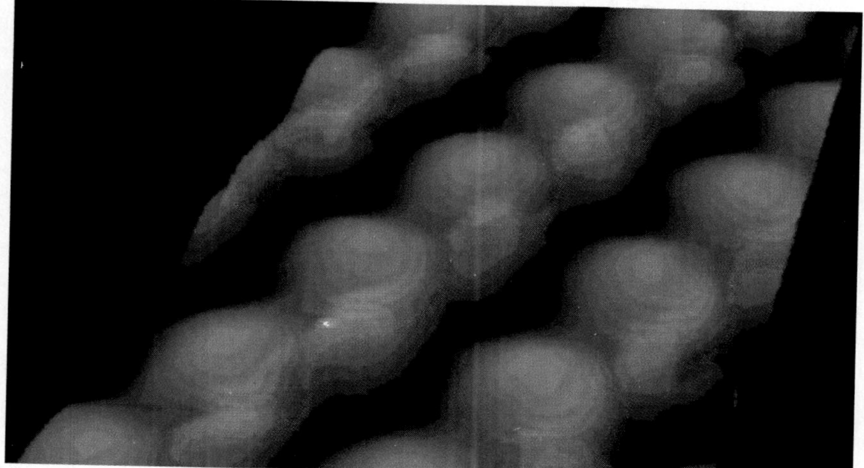

FIGURE 1.3 Individual atoms can be seen as bumps on the surface of a solid by the technique called scanning tunneling microscopy (STM). This is an image of the surface of gallium arsenide. The gallium atoms are shown as blue and the arsenic atoms as red (these are not their actual colors).

▶▶ Appendix 2D lists the names and chemical symbols of all the elements and gives the origins of their names.

Other symbols are taken from the element's name in Latin, German, or Greek. For example, the symbol for iron is Fe, from its Latin name *ferrum.*

Each element has a name and a unique chemical symbol.

1.3 THE NUCLEAR ATOM

Two hundred years ago, Dalton pictured atoms as featureless spheres like billiard balls. Today we know that atoms have an internal structure and are built from even smaller particles. By investigating the internal structure of atoms, we can come to see how one element differs from another.

In this section, we shall see how the process of scientific investigation summarized in Box 1.1 led to our current model of the atom. This model can be summarized as follows:

1. Atoms are made up of **subatomic particles** called electrons, protons, and neutrons.
2. The protons and neutrons form a compact, central body called the **nucleus** of the atom.
3. The electrons are distributed in space like a cloud around the nucleus.

This model of an atom is called the **nuclear atom.** The properties of the three subatomic particles are summarized in Table 1.1. Protons and neutrons have about the same mass, but protons have one unit of positive electric charge and neutrons (as their name suggests) are electrically neutral. An electron is very much less massive than a proton—nearly 2000 times less, in fact—and it has one unit of negative electric charge.

The sign of electric charge tells us whether particles repel or attract one another: like charges (+/+ or −/−) repel one another, and opposite charges (+/−) attract.

The earliest experimental evidence for the nuclear atom was the discovery in 1897 of the first subatomic particle by the British physicist J. J. Thomson (Fig. 1.4). He was investigating "cathode rays," the rays that are emitted when a high potential difference (a high voltage) is applied between two electrodes (metal contacts) in an evacuated glass tube (Fig. 1.5). Thomson showed that cathode rays are streams of negatively charged particles. They came from inside the atoms that made up the electrode called the cathode.

One outcome of Thomson's work was the invention of the cathode-ray tube used in television. Each time you see a television picture, you are watching a (sometimes) more entertaining version of Thomson's experiment.

Thomson found that the charged particles were the same regardless of the metal he used for the cathode. He concluded that they are part of the makeup of *all* atoms. These particles were named electrons and denoted e^-. Later workers, most notably the American Robert Millikan, devised experiments that enabled them to determine the mass of the electron; they found it to be only 9.1×10^{-28} g.

▶▶ If you are unfamiliar with scientific notation for numbers (like 9.1×10^{-28}), see Appendix 1B.

TABLE 1.1 Properties of subatomic particles

Particle	Symbol	Charge*	Mass, g
electron	e^-	−1	9.109×10^{-28}
proton	p	+1	1.673×10^{-24}
neutron	n	0	1.675×10^{-24}

*Charges are given as multiples of the charge on a proton, which in SI units is 1.602×10^{-19} coulomb.

BOX 1.1 THE SCIENTIFIC METHOD

The development of the model of the nuclear atom is an example of the **scientific method,** a systematic process of scientific inquiry. The scientific method can be thought of as a series of steps. The first step is the collection of **data** by making observations and measurements on a small **sample** of matter, that is, a representative piece of the material we want to study. When a pattern is observed in a series of data, it is summarized by formulating a scientific **law,** a succinct summary of observations. For example, the laws of gravity summarize measurements of the effect of gravity on objects of different mass.

In the next step we develop **hypotheses,** possible explanations of the laws in terms of more fundamental concepts. Observation requires careful attention to detail, whereas the development of a hypothesis requires insight, imagination, and creativity. For example, although Dalton could not see individual atoms, he was able to imagine them and formulate his atomic hypothesis. It was a monumental insight that helped others understand the world in a new way.

Once a hypothesis has been developed, we design **experiments,** carefully controlled tests, to verify it. Experiments often require ingenuity and sometimes good luck. If the results of repeated experiments support the hypothesis, then we formulate a **theory,** a formal explanation of a law. A theory is sometimes interpreted through a **model,** a simplified version of the object of study. Theories and models must also be subjected to experiment and revised if they are not supported by experimental results. For example, Rutherford asked his students to perform an experiment he thought would be a simple verification of the atomic model of his time. Instead, they were able to disprove that model and revise it. Our current model of the atom has gone through many formulations and progressive revisions.

Although electrons have a negative charge, atoms overall have zero charge. Therefore, each atom must contain enough positive charge to cancel the negative charge. For years, a lively debate took place over the location of the positive charge. Early in the twentieth century, many scientists thought that atoms were blobs of a positively charged jellylike material, with the electrons suspended in it like raisins in pudding.

FIGURE 1.4 Joseph John Thomson (1856–1940), who discovered the electron, and the apparatus he used.

FIGURE 1.5 A close-up of the glowing path traced by a stream of electrons near the cathode in a simple cathode ray tube, an apparatus like that used by Thomson. Note the deflection of the cathode ray by a magnetic field.

FIGURE 1.6 Ernest Rutherford (1871–1937), who was responsible for many discoveries about the structure of the atom and its nucleus.

In 1908 this model was overthrown by a simple experiment. The New Zealander Ernest Rutherford (Fig. 1.6) was training some of his students in the use of a new piece of apparatus. Rutherford knew that some elements, including radon, emit streams of positively charged particles, which he called **alpha (α) particles**. He asked two students, Hans Geiger and Ernest Marsden, to shoot α particles toward a piece of gold foil only a few atoms thick (Fig. 1.7). If atoms were indeed like blobs of positively charged jelly, then all the α particles would easily pass through the foil, with only an occasional slight deflection in their paths.

α is the first letter in the Greek alphabet.

What Geiger and Marsden observed astonished everyone around them. Although almost all the α particles did pass through and were deflected only very slightly, about 1 in 20,000 was deflected through more than 90°, and a few α particles bounced straight back in the direction from which they had come. "It was almost as incredible," said Rutherford, "as if you had fired a 15-inch shell at a piece of tissue paper and it had come back and hit you."

The explanation had to be that atoms are not blobs of positively charged jelly with electrons suspended in it like raisins. Instead, atoms had to contain massive pointlike centers of positive charge surrounded by a large volume of mostly empty space. Rutherford called the pointlike positively charged region the atomic nucleus. He reasoned that a positively charged α particle that scored a direct hit on one of the minute but heavy gold nuclei was strongly repelled by its positive charge and deflected through a large angle, like a ping-pong ball colliding with a cannonball (Fig. 1.8).

The electrons in an atom are thinly distributed throughout the space around the nucleus. Compared with the size of the nucleus (about 10^{-14} m in diameter), this space is enormous (about 10^{-9} m in diameter; a hundred thousand times greater). If the proton in a hydrogen atom were the size of a fly at the center of a baseball field, then the space occu-

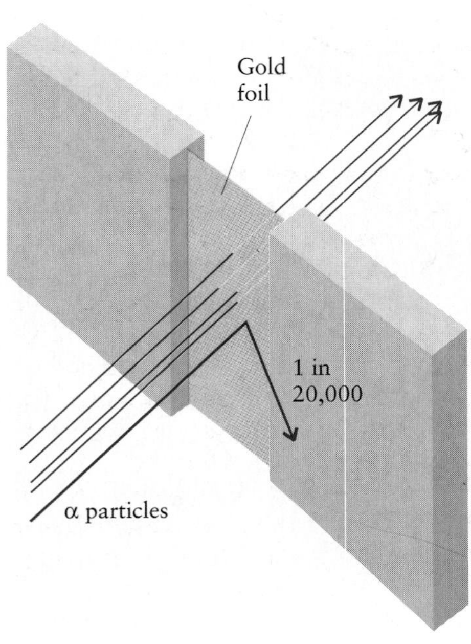

FIGURE 1.7 Part of the experimental arrangement used by Geiger and Marsden. The α particles came from a sample of the radioactive gas radon. Their deflections were measured by observing flashes of light (scintillations) where they struck a zinc sulfide screen. About 1 in 20,000 α particles was deflected through very large angles; most went through the gold foil with very little deflection.

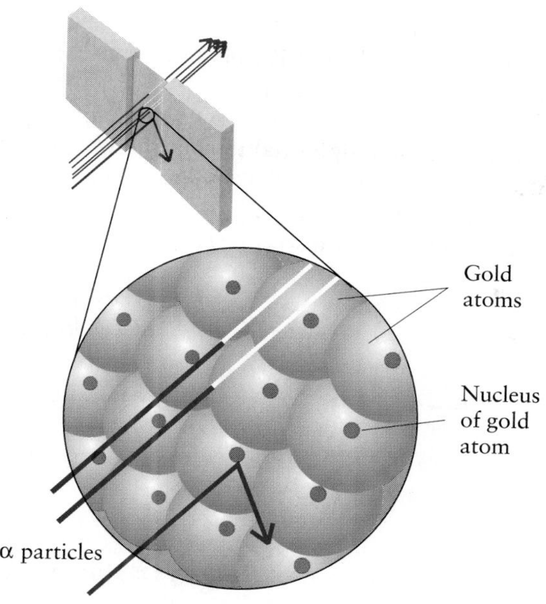

FIGURE 1.8 Rutherford's model of the atom explains why most α particles pass almost straight through, whereas a very few—those scoring a direct hit on the nucleus—undergo very large deflections. The nuclei actually are much smaller relative to the atom than shown here.

pied by the electron would be about the size of the entire baseball stadium (Fig. 1.9).

In an atom the positive charge of the nucleus exactly cancels the negative charge of the surrounding electrons. So, for each electron outside the nucleus, there must be a matching positively charged particle inside the nucleus. The positively charged particles are the protons (denoted p); their properties are given in Table 1.1. Notice that a proton is nearly 2000 times heavier than an electron and that each one has one unit of positive charge.

The number of protons in an atomic nucleus is called the **atomic number,** Z, of the element. Henry Moseley, a young British scientist, was the first to measure atomic numbers accurately, shortly before he was killed in action in World War I. Moseley knew that when elements are bombarded with rapidly moving electrons (the cathode rays of Thomson's experiment), they emit x-rays. These rays are like light rays in some respects, but they are more energetic and can pass through many substances. He found that the energy of the x-rays emitted by an element depends on its atomic number; and by studying the x-rays of many elements, he was able to determine the values of Z for them. Scientists have since determined the atomic numbers of all the known elements (see the table inside the back cover). For example, hydrogen has $Z = 1$, so we know that the nucleus of a hydrogen atom consists of one proton; helium has $Z = 2$, so its nucleus contains two protons. The most recently discovered elements have atomic numbers of over 100, so each of their nuclei contains more than 100 protons.

As noted earlier, an atom is electrically neutral, so the number of electrons in the space around its nucleus must be the same as the number of protons inside the nucleus. Therefore, Moseley's technique for counting the number of protons in the nucleus is an indirect way of counting the number of electrons too. For hydrogen, $Z = 1$, so we know at once

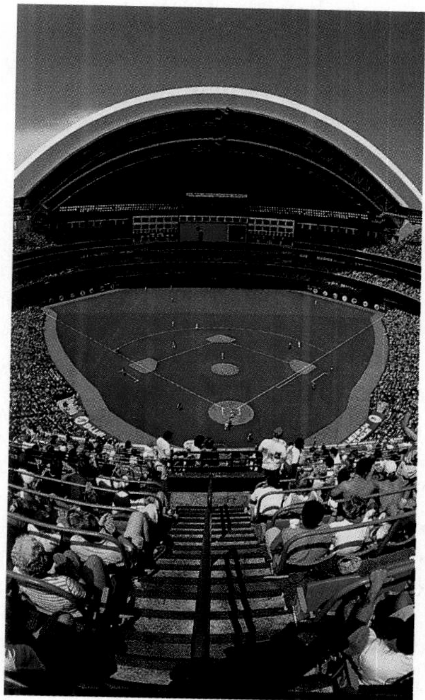

FIGURE 1.9 Think of a fly at the center of this stadium. The fly would be the same size as the nucleus of an atom if the atom were enlarged to the size of the stadium.

that a hydrogen atom must have one electron. A gold atom, with $Z = 79$, has 79 electrons. For uranium, $Z = 92$, so we know that every uranium atom has 92 electrons.

> *In the nuclear atom, all the positive charge and almost all the mass is concentrated in the tiny nucleus, and the negatively charged electrons form a cloud around the nucleus. The atomic number is defined as the number of protons in the nucleus; there is an equal number of electrons outside the nucleus.*

SELF-TEST 1.1A How many electrons are present in an atom of bismuth?
[*Answer:* 83]

SELF-TEST 1.1B How many electrons are present in an atom of sulfur?

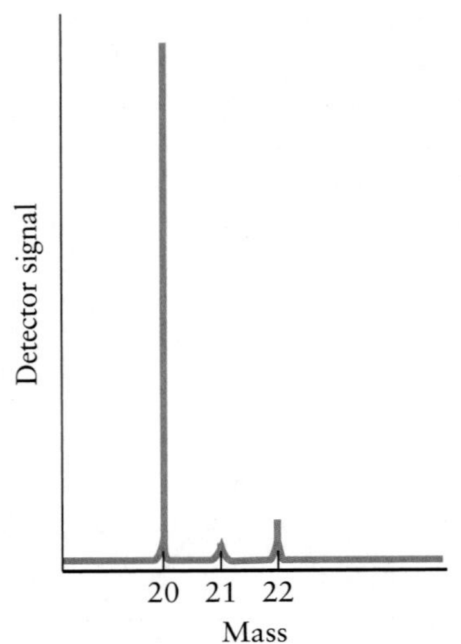

FIGURE 1.10 The mass spectrum of neon. The locations of the peaks tell us the masses of the atoms, and their heights tell us the relative numbers of atoms with each mass.

1.4 ISOTOPES

Technological advances in electronics at the start of the twentieth century led to the invention of the **mass spectrometer,** a device that can be used for determining the mass of a given type of atom (Box 1.2). It was soon applied to all the elements; and we now know, for example, that the mass of a hydrogen atom is 1.67×10^{-24} g and that of a carbon atom is 1.99×10^{-23} g. Even the heaviest atoms have masses of only about 5×10^{-22} g.

As happens so often in science, a new and more precise technique of measurement led to a major discovery. In this case, it was found that not all the atoms of a given element have the same mass. In a sample of perfectly pure neon, for example, most of the atoms have mass 3.32×10^{-23} g, which is about 20 times greater than the mass of a hydrogen atom. Some neon atoms, however, are found to be about 22 times heavier than hydrogen. Others are about 21 times heavier (Fig. 1.10). All three types of atoms have the same atomic number, so they are definitely atoms of neon.

The differences in masses among atoms of a single element suggested to scientists the existence of a third subatomic particle in the nucleus: the neutron (denoted n). Because neutrons have no electric charge, the number of neutrons in the nucleus affects neither the nuclear charge nor the number of electrons in the atom. However, because neutrons have about the same mass as protons, they do add substantially to the mass of the nucleus. Neutrons and protons are very similar particles and are jointly known as **nucleons.**

The total number of protons and neutrons in a nucleus is called the **mass number,** *A,* of the atom. A nucleus of mass number *A* is about *A* times heavier than a hydrogen atom, which has a nucleus that consists of a single proton. Conversely, if we know that a particular atom is a certain

Neon-20 ($^{20}_{10}Ne$)

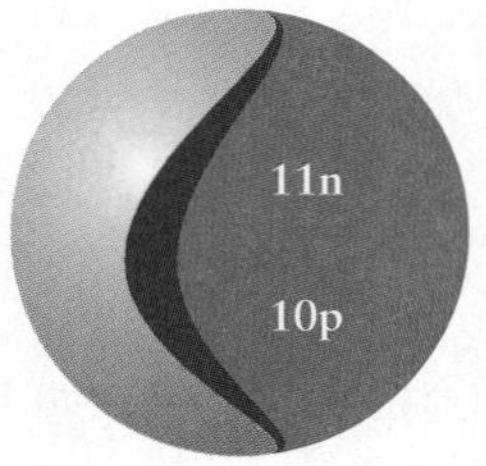

Neon-21 ($^{21}_{10}Ne$)

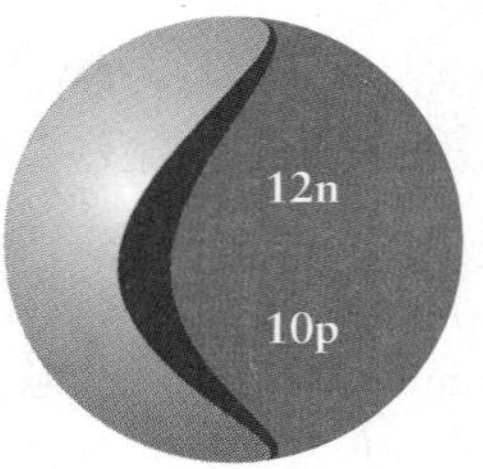

Neon-22 ($^{22}_{10}Ne$)

FIGURE 1.11 The nuclei of isotopes have the same numbers of protons but different numbers of neutrons. These three diagrams show the compositions of the nuclei of the three isotopes of neon. On this scale, the atom itself would be about 1 km in diameter. The arrangement of the protons and neutrons inside the nucleus is not shown.

BOX 1.2 MASS SPECTROMETRY

The masses of atoms and molecules are now measured with a mass spectrometer. All the air is pumped out of the instrument. Then atoms or molecules of the elements—either a gaseous element (neon or oxygen, for instance) or the vapor of a liquid or a solid element (such as mercury or zinc)—are fed into the spectrometer's ionization chamber. There they are exposed to a beam of rapidly moving electrons. When one of the accelerated electrons collides with an atom, it knocks another electron out of it, thereby leaving the atom with a positive charge. In other words, the collision creates a positive ion, an electrically charged atom (or group of atoms). The positive ions are accelerated out of the chamber by a strong electric field applied between two metal grids. The speeds reached by the ions depend on their masses, with lighter ions reaching higher speeds than heavy ones.

As each accelerated ion passes through a magnetic field generated by an electromagnet, its path is bent to an extent that depends on its speed (and hence on its mass). The strength of the magnetic field is slowly changed, and a signal is produced when the magnetic field is just strong enough to bend the beam of ions so that they arrive at the detector. The mass of the type of ion formed is then calculated from the accelerating voltage and the strength of the magnetic field used to produce the signal. The mass spectrum is a plot of the detector signal against the magnetic field. The positions of the peaks are used to calculate the masses of the accelerated ions, and the relative heights of the peaks indicate the proportions of ions of various types.

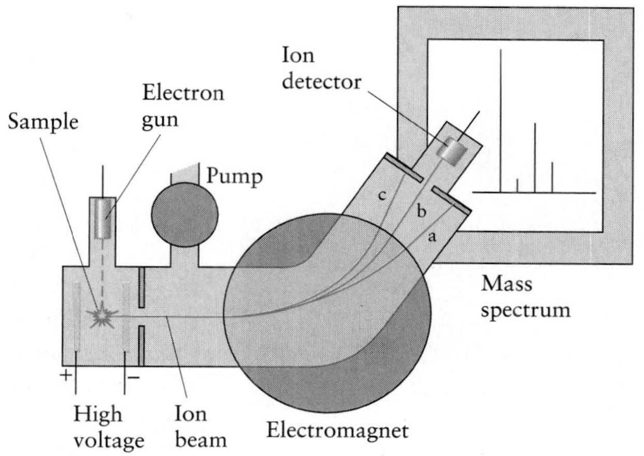

A mass spectrometer is used to measure the mass and abundance of an isotope. As the strength of the magnetic field is changed, the path of the accelerated ions moves from a to c. When the path is at b, the ion detector sends a signal to a recorder. At a fixed magnetic field strength, the three paths represent the trajectories of the ions of isotopes with three different masses, decreasing from a to c.

number of times heavier than a hydrogen atom, then we can infer the mass number. For example, because the three varieties of neon atoms are 20, 21, and 22 times heavier than a hydrogen atom, we know that the mass numbers of the three types of neon atoms are 20, 21, and 22. Because for each of them $Z = 10$, these neon atoms must contain 10, 11, and 12 neutrons, respectively (Fig. 1.11).

Atoms with the same atomic number (belonging to the same element) but with different mass numbers are called **isotopes** of the element. All isotopes of an element have exactly the same atomic number; hence they have the same number of electrons around their nucleus but different numbers of neutrons in the nucleus. An isotope is named by writing its mass number after the name of the element, as in neon-20, neon-21, and neon-22. Its symbol is obtained by writing the mass number as a superscript to the left of the chemical symbol of the element, as in ^{20}Ne, ^{21}Ne, and ^{22}Ne. You will occasionally see the atomic number included as a subscript on the lower left, as in the symbol $^{22}_{10}Ne$ used in Fig. 1.11.

The name *isotope* comes from the Greek words for "equal place."

Because isotopes of the same element have the same number of protons and the same number of electrons, they have essentially the same chemical and physical properties. However, in the case of hydrogen, the mass differences among isotopes are relatively large, leading to noticeable

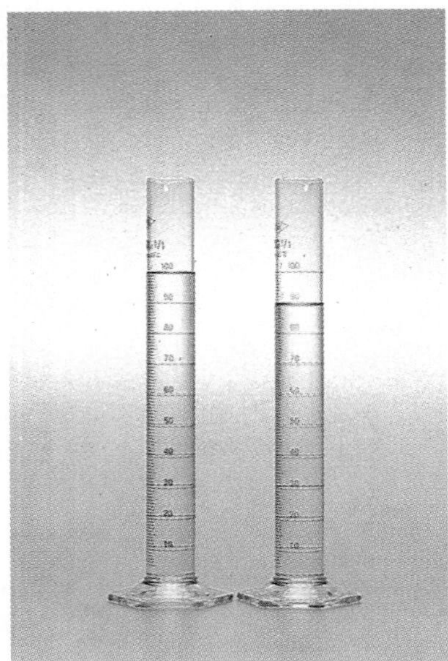

FIGURE 1.12 These two samples, both of which have mass 100 g, illustrate the difference in densities of ordinary water, H_2O, and heavy water, D_2O. The volume occupied by 100 g of heavy water (right) is 11% less than that occupied by the same mass of ordinary water (left).

TABLE 1.2 Some isotopes of common elements

Element	Symbol	Atomic number, *Z*	Mass number, *A*	Abundance, %
hydrogen	^{1}H	1	1	99.985
deuterium	^{2}H or D	1	2	0.015
tritium	^{3}H or T	1	3	—*
carbon-12	^{12}C	6	12	98.90
carbon-13	^{13}C	6	13	1.10
oxygen-16	^{16}O	8	16	99.76

*Radioactive, short-lived.

differences in some physical properties and slight variations in chemical properties. Hydrogen has three isotopes (see Table 1.2). The most common (^{1}H) has no neutrons, so its nucleus is a lone proton. The other two isotopes are less common but nevertheless are so important that they are given special names and symbols. One isotope (^{2}H) is called deuterium (D), and the other (^{3}H) is called tritium (T). A deuterium atom, with a nucleus that consists of one proton and one neutron, has about twice the mass of an ordinary hydrogen atom (more precisely, it is 1.998 times heavier) and, when combined with oxygen, forms "heavy water." A given volume of heavy water is about 11% heavier than the same volume of ordinary water (Fig. 1.12).

Isotopes have the same atomic number but different mass numbers. Their nuclei have the same number of protons but different numbers of neutrons. All the isotopes of a given element have the same number of electrons around the nucleus.

EXAMPLE 1.1 Using atomic numbers and mass numbers

How many protons, neutrons, and electrons are present in an atom of uranium-238?

STRATEGY The number identifying the isotope gives the mass number *A*, the total number of protons and neutrons. The atomic number *Z* (and hence the number of protons) is obtained by reference to the list of elements inside the back cover. The difference $A - Z$ is the number of neutrons. The number of electrons is the same as the number of protons; hence it is equal to *Z*.

SOLUTION The mass number of uranium-238 is 238, so the total number of nucleons is 238. Uranium has $Z = 92$, so its nucleus contains 92 protons. The number of neutrons is then $238 - 92 = 146$. The atom is neutral, so it must contain 92 electrons to cancel the charge of the protons.

SELF-TEST 1.2A How many protons, neutrons, and electrons are present in (a) an atom of nitrogen-14; (b) an atom of iron-56?

[*Answer:* (a) 7, 7, 7; (b) 26, 30, 26]

SELF-TEST 1.2B How many protons, neutrons, and electrons are present in (a) an atom of oxygen-16; (b) an atom of uranium-236?

1.5 WHERE DID THE ELEMENTS COME FROM?

All the elements in the universe except hydrogen and most of the helium were made in the stars. Seconds after the universe came into being with the Big Bang, the only elements present were the two simplest, hydrogen and helium. After millions of years, as the universe cooled, the atoms of hydrogen and helium collected together in large clouds under the influence of gravity. These clouds gradually became hotter and hotter as they contracted, and in due course they burst into incandescence as stars (Fig. 1.13). Within the stars, intense heat causes atoms of hydrogen to smash together, merge, and become atoms of other elements. When two protons and one or two neutrons merge together, the outcome is an atom of helium ($Z = 2$, $A = 3$ or 4). When a third proton and more neutrons join them, we get an atom of lithium ($Z = 3$, $A = 6$ or 7), and so on. This merging releases even more heat, which generates starlight. So starlight—and sunlight—are signs that the heavier elements are still being formed.

Many million years after a star is formed and it begins to cool, its outer layers may collapse, like a falling roof, into its exhausted core. This mighty starquake produces such great shock waves that the star shrugs off its outer layers and sends them into space in a huge explosion called a supernova. Six such explosions have been detected in our galaxy in the past 1000 years, the most recent in 1987. The shock of explosion raises the temperature in the star, making it even brighter than before. The Crab nebula (produced by the supernova of 1054) was visible in broad daylight for three weeks. At such high temperatures, even the heavy atoms collide violently enough to merge and become still heavier atoms. The very heavy elements now found on Earth, including uranium and gold, were made in this way.

The debris of exploding stars gradually collects together under the influence of gravity and gives rise to a new generation of stars. However, not all the debris collects in a single central body; some collects into smaller bodies that go into orbit around the star. These bodies are the planets, and one of them is the Earth. All matter on Earth was formed in

FIGURE 1.13 Stars are born in immense clouds of molecular hydrogen and stardust such as this one in the Eagle Nebula, which is also known as M16. The new stars shining through the dust will emerge as the cloud evaporates.

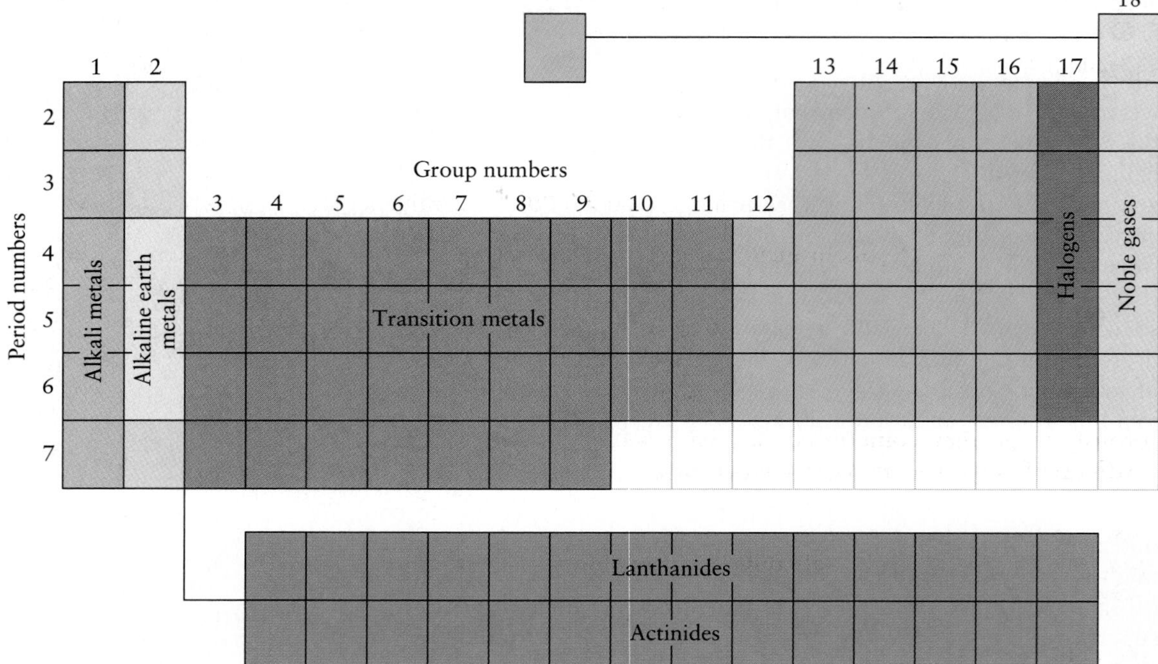

FIGURE 1.14 The structure of the periodic table, showing the names of some regions and groups. The groups are the vertical columns numbered 1 through 18. The periods are the horizontal rows. The main-group elements are those in Groups 1, 2, and 13–18, and hydrogen.

this way in long-dead stars. All the elements—other than hydrogen and most of the helium—from which everything is made were formed inside a star. Even our flesh is stardust.

> ***According to current ideas, hydrogen and helium were formed in the Big Bang. The heavier elements were made inside stars and then scattered throughout space.***

1.6 THE PERIODIC TABLE

It may seem overwhelming to realize that there are over 100 elements. How can we possibly be expected to learn all their properties? How can we be expected to learn all the millions of ways they might combine? Fortunately, chemists have discovered that when the elements are listed in order of their atomic number and arranged in a special way, they form families that show regular trends in properties. As a result, we can obtain a reasonable familiarity with an element by knowing to which family it belongs and which elements are its neighbors.

The arrangement of elements that shows their family relationships is called the **periodic table** (it is printed inside the front cover of this book). Because an element's location in the table is a guide to its properties, we should note the position of every element we meet for the first time.

▶▶ **The periodic table is so called in recognition of the periodic recurrence of family resemblances as *Z* increases. You will learn more of the history and structure of the periodic table in Chapter 7.**

The vertical columns of the periodic table are called **groups** (Fig. 1.14). These groups identify the principal families of elements. The taller columns (Groups 1, 2, and 13 through 18) are called the **main groups** of the table. The horizontal rows are called **periods,** and are numbered from the top down.

The members of each group show a gradual variation in their properties. The properties of sodium (Na) in Group 1, for example, are a good

FIGURE 1.15 The alkali metals react with water, producing gaseous hydrogen and heat. Potassium reacts vigorously, producing so much heat that the hydrogen is ignited.

clue to the properties of the other members of Group 1, namely lithium (Li), potassium (K), rubidium (Rb), cesium (Cs), and francium (Fr). These elements are called the **alkali metals.** All of them are soft, silvery metals that melt at low temperatures. They all produce hydrogen when they come in contact with water (Fig. 1.15)—lithium gently, but cesium with explosive violence.

Next to the alkali metals are the metals of Group 2. These elements resemble the alkali metals in several ways but produce hydrogen less vigorously when they come in contact with water. Calcium (Ca) produces hydrogen from water at room temperature, but magnesium (Mg) does so only if it is heated; and beryllium (Be) does not produce hydrogen from water, even if it is red hot. The elements calcium (Ca), strontium (Sr), and barium (Ba) are called the **alkaline earth metals,** but the name is often extended to all the members of the group.

On the far right of the table, in Group 18, are the elements known as the **noble gases.** They are so called because they combine with very few elements—they are chemically aloof. In fact, until the 1960s they were called the *inert gases* because it was thought that they did not combine with any elements at all. All the Group 18 elements are colorless, odorless gases.

◀◀ In many current versions of the periodic table, you will see a different notation for groups, with the noble gases belonging to Group VIII or Group VIIIA. These alternatives are given in the table printed inside the front cover.

Next to the noble gases are the **halogens** of Group 17. Many of the properties of the halogens show regular variations from fluorine (F) through chlorine (Cl) and bromine (Br) to iodine (I). Fluorine, for instance, is a very pale yellow, almost colorless gas, chlorine a yellow-green gas, bromine a red-brown liquid, and iodine a purple-black solid (Fig. 1.16).

Groups 3 through 11 contain the **transition metals.** They include the important structural elements titanium (Ti) and iron (Fe) and the **coinage metals** copper (Cu), silver (Ag), and gold (Au). The transition metals take their collective name from their role as a bridge between the chemically active metals of Groups 1 and 2 and the much less active metals of Groups 12, 13, and 14.

The long block shown below the main table consists of the **inner transition metals.** It is drawn there simply to save space, for otherwise the periodic table would be inconveniently wide. However, we need to keep in mind where these elements fit into the main part of the table. The upper row of this block, following lanthanum (element 57) in Period 6, consists of the **lanthanides** and the lower row, following actinium (element 89) in Period 7, of the **actinides.**

Right at the head of the periodic table, standing alone, is hydrogen. Some tables place hydrogen in Group 1, others place it in Group 17, and yet others place it in both groups. We shall treat it as a very special element and place it in none of the groups.

FIGURE 1.16 The halogens are colored elements. From left to right, chlorine is a yellow-green gas, bromine is a red-brown liquid (its vapor fills the flask), and iodine is a blue-black solid (note the small crystals) with a violet vapor.

The periodic table is an arrangement of the elements that reflects their family relationships; the members of a group typically show a smooth trend in properties.

FIGURE 1.17 All metals can be deformed by hammering. Gold can be hammered into a sheet so thin that light can pass through it. Here it is possible to see the light of a candle through the sheet of gold.

1.7 METALS, NONMETALS, AND METALLOIDS

The elements in the periodic table can be classified as metals, nonmetals, and metalloids:

> A **metal** conducts electricity, has a metallic luster, and is malleable and ductile.
>
> A **nonmetal** does not conduct electricity and is neither malleable nor ductile.
>
> A **metalloid** has the appearance and some properties of a metal but behaves chemically like a nonmetal.

A **malleable** substance (from the Latin word for "hammer") is one that can be hammered into thin sheets (Fig. 1.17). A **ductile** substance (from the Latin word for "drawing out") is one that can be drawn out into wires.

Copper, for example, is a metal. It conducts electricity, has a luster when polished, and is malleable. It is so ductile that it is readily drawn out to form electrical wires. Sulfur, on the other hand, is a nonmetal. This brittle yellow solid does not conduct electricity, cannot be hammered into thin sheets, and cannot be drawn out into wires. All elements that are gases at room temperature are nonmetals. The distinctions between metals and metalloids and between metalloids and nonmetals are not very precise (and not always made), but the metalloids are often taken to be the six elements shown in Fig. 1.18.

A striking feature of the periodic table becomes clear when we mark the positions of the metals, metalloids, and nonmetals: *all the metallic elements occur on the left side and in the middle of the table and all the nonmetallic elements occur on the right* (see Fig. 1.18). The metalloids lie in a diagonal band between the metals and nonmetals. Thus, with a glance at the table, we can see whether an element is a metal, a metalloid, or a nonmetal. You may never have heard of indium (In), but its position in the periodic table shows you that it is a metal. Because it is close to the right side of the table, far from the alkali metals, you can even guess that it will not be a very reactive metal. This is only one example of the power of the periodic table: you will encounter many more in the following pages.

> ***Metallic elements lie on the left of the periodic table, nonmetallic elements lie on the right, and the two are separated by a diagonal band of metalloids.***

COMPOUNDS

Most of the substances around us are combinations of elements rather than single elements. The ability of the elements to combine with one another is responsible for the extraordinary richness of the world, because, from about 100 elements, countless numbers of combinations can be formed.

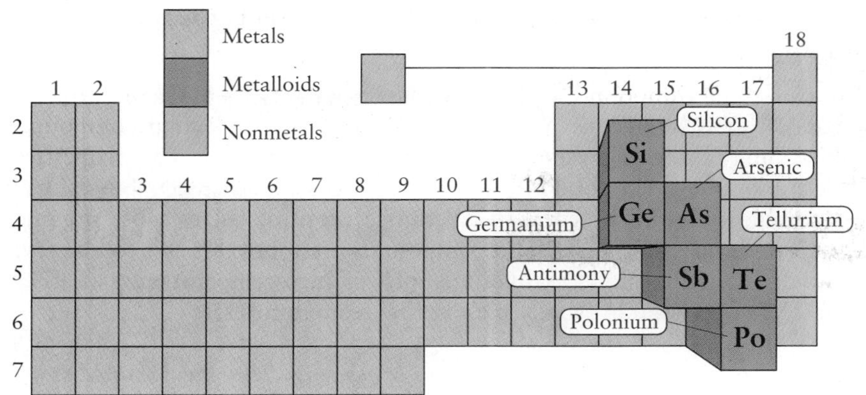

FIGURE 1.18 The location of the six elements commonly regarded as metalloids: these elements have characteristics of both metals and nonmetals. Other elements, notably beryllium, boron, and bismuth, are sometimes included in the classification.

1.8 WHAT ARE COMPOUNDS?

A **compound** is a substance that consists of two or more different elements in a definite ratio. Water, for instance, is a compound of hydrogen and oxygen, with two hydrogen atoms for each oxygen atom. Common table salt, sodium chloride, is another compound. It invariably contains one sodium atom for each chlorine atom. You would find the same ratio in samples of the compound from California, Australia, Siberia, the Antarctic, or Mars. This regularity is called the **law of constant composition.** It was historically important, because it suggested to chemists that compounds consisted of specific combinations of atoms.

Compounds are classified as either organic or inorganic. **Organic compounds** are molecular compounds containing the element carbon, and usually hydrogen too. They include fuels such as methane and propane, and sugars such as glucose; millions of other substances are also organic compounds. These compounds are called organic because it was once believed that they could be formed only by living organisms. That assumption is now known to be incorrect.

Inorganic compounds are all the other compounds; they include water, calcium sulfate, ammonia, silica, hydrochloric acid, and many, many more. In addition, some very simple carbon compounds, particularly carbon dioxide and the carbonates, which include chalk (calcium carbonate), are treated as inorganic compounds.

The elements in a compound are not just mixed together. Their atoms are actually joined, or *bonded,* to one another in a specific way. For example, when a mixture of hydrogen and oxygen is ignited, an explosion takes place. A lot of heat is released; and instead of two colorless gases, we end up with a clear, colorless liquid: water. The atoms of hydrogen and oxygen have formed chemical bonds with one another and each oxygen atom is now joined to two hydrogen atoms. When sulfur is ignited in air, sulfur and oxygen atoms join to form sulfur dioxide. Solid yellow sulfur and odorless oxygen gas have produced a colorless, pungent, and poisonous gas (Fig. 1.19).

The atoms of a compound can be bonded together to form molecules, or they can be present as ions:

A **molecule** is a definite and distinct electrically neutral group of bonded atoms.

FIGURE 1.19 Sulfur burns with a blue flame and produces the dense gas sulfur dioxide.

The name *ion* comes from the Greek word for "go," because charged particles go either toward or away from a charged electrode.

An **ion** is a positively or negatively charged atom or bonded group of atoms.

We classify a compound as **molecular** if it consists of molecules and as **ionic** if it consists of ions. Water is an example of a molecular compound and sodium chloride an example of an ionic compound. The properties of a compound are a good guide to its classification. All compounds that are gases or liquids at room temperature are molecular. Most organic compounds are molecular. Solid compounds are more difficult to classify; a good clue is that ionic compounds melt at high temperatures, whereas molecular compounds usually melt at low temperatures.

▶▶ Bonding is discussed in more detail in Chapters 8–10.

Compounds are specific combinations of elements. They are classified as either molecular or ionic.

1.9 MOLECULAR COMPOUNDS

The **chemical formula** of a compound is a statement of its composition in terms of the chemical symbols of the elements present. There are various types of chemical formulas. For molecular compounds, it is common to give the **molecular formula,** a chemical formula that shows how many atoms of each type of element are present in a molecule. For instance, the molecular formula of water is H_2O: the subscript 2 shows that there are two H atoms present in the molecule, and the single O shows that there is one oxygen atom present. Likewise, the molecular formula for methane, CH_4, another molecular compound, shows that each molecule consists of one C atom and four H atoms. The molecular formula for estrone, a female sex hormone, is $C_{18}H_{22}O_2$, showing that a single molecule of estrone consists of 18 C atoms, 22 H atoms, and 2 O atoms. A molecule of a male sex hormone, testosterone, differs by only a few atoms: its molecular formula is $C_{19}H_{28}O_2$.

Colors are used in molecular graphics to help identify elements; they do not indicate actual colors of atoms. Throughout this text, carbon atoms will be represented by black (or dark gray) spheres, hydrogen by pale gray spheres, and oxygen by red spheres.

A molecule is a specific *arrangement* of bonded atoms. So, to show its structure more fully, we must draw an appropriate diagram. The pictorial representation of molecules that most accurately portrays their shapes comes from the use of computer graphics, like that used to illustrate the *space-filling model* of the ethanol molecule in Fig. 1.20, in which atoms are represented by colored spheres that fit into one another. However, it is hard to distinguish the atoms in this picture without guidance, and not easy to draw a space-filling model freehand. A much sim-

(a)

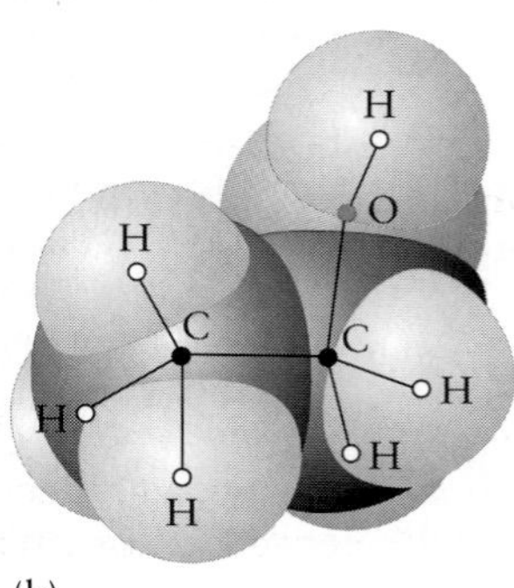

(b)

FIGURE 1.20 (a) A space-filling representation of the ethanol molecule, generated by computer graphics. (b) The superimposed lines and labels identify the atoms and show the pattern of bonds between them. The small circles identify the locations of the centers of the atoms.

pler representation, which can be written down in a few moments, is a **structural formula** like (**1**).

```
 H   O—H
 |   |
H—C — C—H
 |   |
 H   H
```

1 Ethanol, C_2H_6O

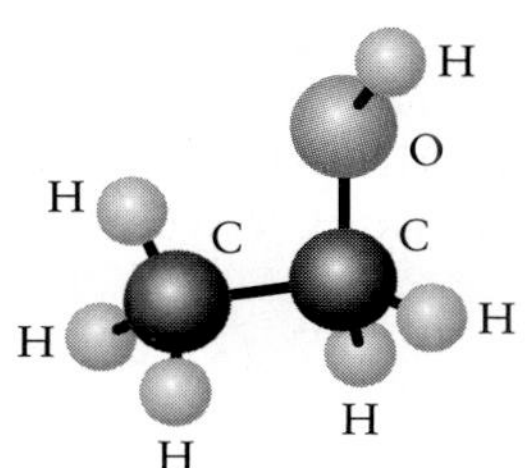

FIGURE 1.21 A ball-and-stick model of the atoms, in which the locations of the atoms are represented by colored balls and the bonds between them by sticks. It is easier to make out the locations of the atoms in this representation than in the space-filling model.

Here the atoms are represented by their chemical symbols, and the lines show which atoms are bonded together. A structural formula gives very little information about the three-dimensional shape of the molecule, but it is compact and very easy to write. Another representation, the *ball-and-stick model* (Fig. 1.21), uses colored balls to depict the atoms and sticks to indicate the bonds. A ball-and-stick model has the advantage of showing us the angles the bonds make with one another, but it is a poor rendering of the shape of the molecule because bonds do not look like sticks. However, we shall usually draw ball-and-stick models, because they convey important aspects of the structure most clearly.

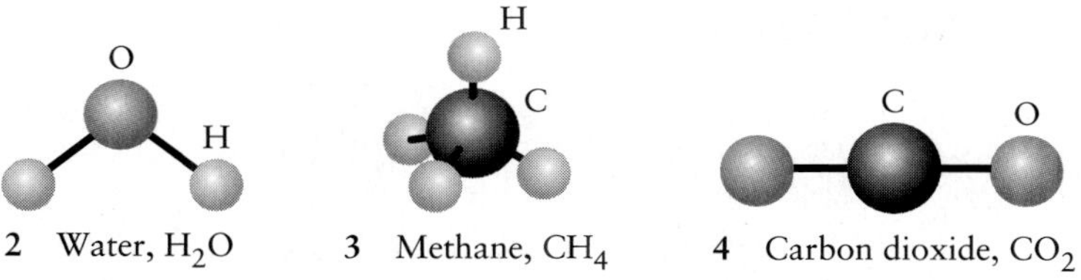

2 Water, H_2O 3 Methane, CH_4 4 Carbon dioxide, CO_2

The molecules of water (**2**), methane (**3**), and carbon dioxide (**4**) are shown above, where the black spheres represent carbon atoms, the pale gray spheres hydrogen atoms, and the red spheres oxygen atoms. As you can see from these diagrams, molecules have definite shapes. All molecules of a given substance have the same arrangement of atoms.

A molecular formula shows the composition of a molecule in terms of the numbers of atoms of each element present. Molecular models indicate the shape of the molecule, the lengths of the bonds, and the angles the bonds make to one another.

1.10 IONIC COMPOUNDS AND IONS

Ionic compounds consist of positive and negative ions held together by the attraction between their opposite charges. An example is sodium chloride, which consists of sodium ions (positively charged sodium atoms, denoted Na^+), and chloride ions (negatively charged chlorine atoms, denoted Cl^-). Each crystal of the compound is an orderly array of a vast number of alternating Na^+ and Cl^- ions (Fig. 1.22). When you take a pinch of salt, you are picking up crystals, each consisting of huge numbers of ions.

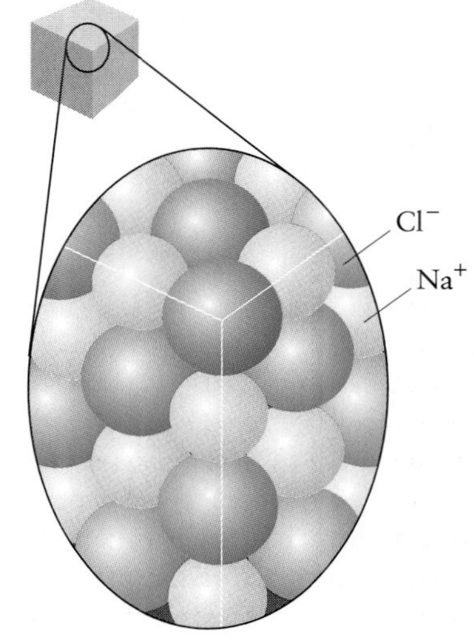

FIGURE 1.22 An ionic solid consists of an array of cations and anions stacked together. This illustration shows the arrangement of sodium cations (Na^+) and chlorine anions (chloride ions, Cl^-) in a crystal of sodium chloride (common table salt). The faces of the crystal are where the stacks of ions come to an end.

The prefixes *cat-* and *an-* come from the Greek words for "down" and "up." Oppositely charged ions travel in opposite directions in an electric field.

Ions are either positively or negatively charged. A positively charged ion is called a **cation**, and a negatively charged ion is called an **anion.** In sodium chloride, the sodium ions are the cations and the chloride ions are the anions.

The nuclear model of the atom readily explains the existence of **monatomic** (single-atom) ions. Because an atom has exactly the same number of electrons and protons, an atom itself has no net charge and is electrically neutral. When an electron is removed, the charge of the remaining electrons no longer cancels the positive charge of the nucleus (Fig. 1.23). Because an electron has one unit of negative charge, removing one electron from a neutral atom leaves behind a cation with one unit of positive charge. For example, a sodium cation, Na^+, has a single positive charge and is a sodium atom that has lost one electron.

Each electron lost from an atom increases the overall positive charge of the atom by one unit. Thus, when a calcium atom loses two electrons, it becomes the doubly positively charged calcium ion, Ca^{2+} (read "calcium two-plus"). When an aluminum atom loses three electrons, it becomes the triply charged aluminum ion, Al^{3+}.

Each electron gained by an atom increases the atom's negative charge by one unit. So, when a chlorine atom gains an electron, it becomes the singly negatively charged chloride ion, Cl^-. When an oxygen atom gains two electrons, it becomes the oxide ion, O^{2-}. When a nitrogen atom gains three electrons, it becomes the triply charged nitride ion, N^{3-}.

The periodic table can help us decide what charge to expect on some cations. For elements in Groups 1 and 2, the charge of the ions the elements can form is equal to the group number. Thus, cesium in Group 1 forms Cs^+ ions; barium in Group 2 forms Ba^{2+} ions. The cations formed by some of the elements are listed in Fig. 1.24. This illustration also shows that atoms of the transition metals and some of the heavier metals of Groups 13 and 14 can form cations with different charges. An iron atom, for instance, can lose two electrons to become Fe^{2+} or three electrons to become Fe^{3+} (Fig. 1.25). Copper can lose either one electron to form Cu^+ or two electrons to become Cu^{2+}.

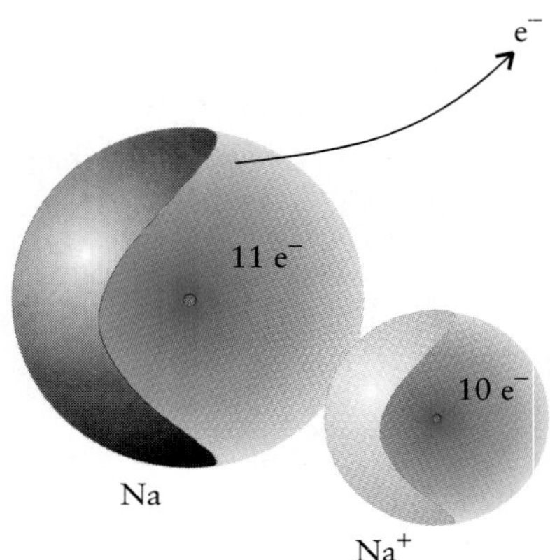

FIGURE 1.23 A neutral sodium atom (left) consists of a nucleus that contains 11 protons and is surrounded by 11 electrons. When 1 electron is lost, the remaining 10 electrons cancel only 10 of the proton charges, and the resulting ion (right) has 1 overall positive charge.

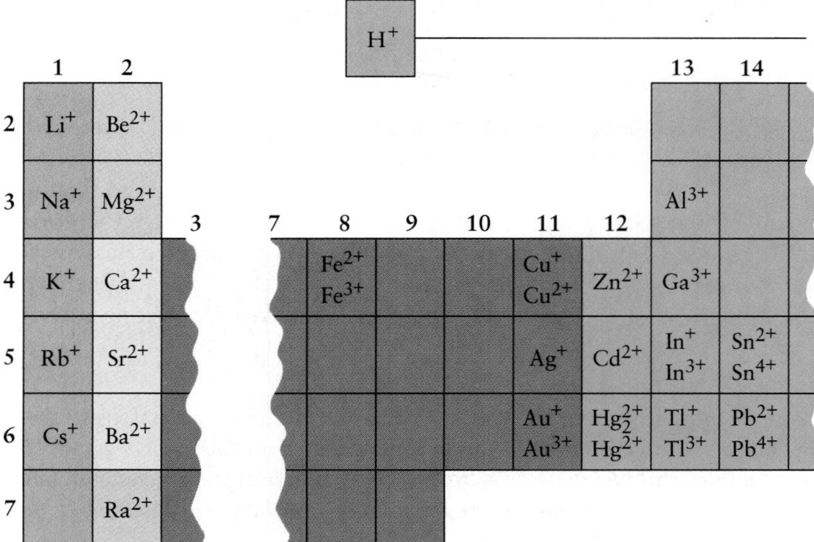

FIGURE 1.24 The typical cations formed by a selection of elements in the periodic table. The transition metals form a wide variety of cations; we have shown only a selection.

Whereas atoms of the metallic elements typically lose electrons and form cations, atoms of nonmetals typically gain electrons and form anions (Fig. 1.26). Some of the anions formed by the elements are listed in Fig. 1.27. Notice another helpful aspect of the periodic table: an element on the right of the table forms an anion with a charge equal to its group number minus 18. Thus, oxygen, Group 16, forms the oxide ion, O^{2-}, with charge $16 - 18 = -2$. Phosphorus, in Group 15, forms the phosphide ion, P^{3-}, with charge $15 - 18 = -3$.

Metallic elements typically form cations, nonmetallic elements typically form anions, and their charges are related to the group to which they belong in the periodic table.

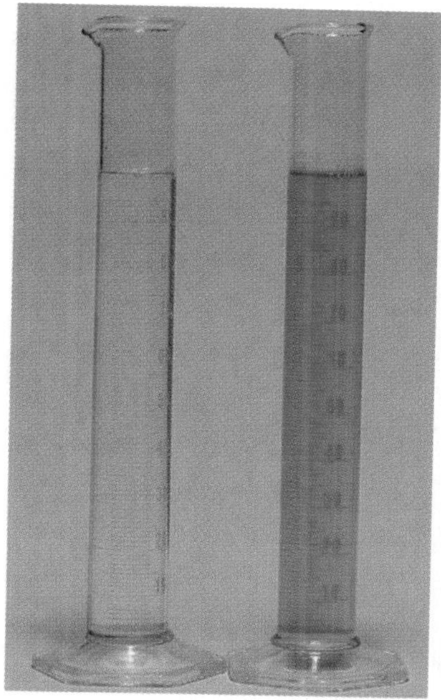

FIGURE 1.25 Iron is an example of an element that can form ions with different charges. Aqueous solutions containing Fe^{2+} are usually pale green and solutions containing Fe^{3+} are usually yellow-brown.

EXAMPLE 1.2 Predicting the ion an element is likely to form

What ions would you expect (a) selenium and (b) calcium to form?

STRATEGY To predict whether the element is likely to form a cation or an anion, decide from its location in the periodic table whether it is a metal or a nonmetal: metals typically form cations, nonmetals typically form anions. To determine the charge of an anion, identify the group to which the element belongs, and then subtract 18. The charge on a cation in Groups 1 and 2 is the same as the group number. Group 13 elements tend to form +3 cations.

SOLUTION (a) Selenium (Se) is a nonmetal, so it can be expected to form an anion. It belongs to Group 16. Because $16 - 18 = -2$, it can be expected to form the anion Se^{2-}. This ion is called the selenide ion. (b) Calcium (Ca) is a metal, so it tends to form cations. It belongs to Group 2 and forms the cation Ca^{2+}, the calcium ion.

SELF-TEST 1.3A What ions are (a) iodine and (b) aluminum likely to form?

[*Answer:* (a) Iodide ion, I^-; (b) aluminum ion, Al^{3+}]

SELF-TEST 1.3B What ions are sulfur and potassium likely to form?

Many ions are **polyatomic**, meaning that they consist of more than one atom bonded together, and have an overall positive or negative charge. An example of a polyatomic cation is the ammonium ion, NH_4^+ (5).

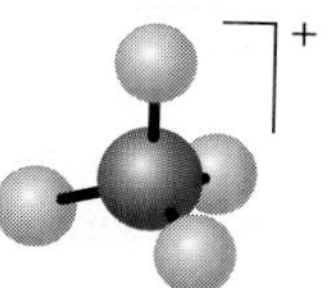

5 Ammonium ion, NH_4^+

As its structure shows, the ammonium ion consists of one N atom bonded to four H atoms. One N atom and four H atoms together would normally have $7 + 4 \times 1 = 11$ electrons. However, the ammonium ion has only 10, so it is a cation with a single positive charge.

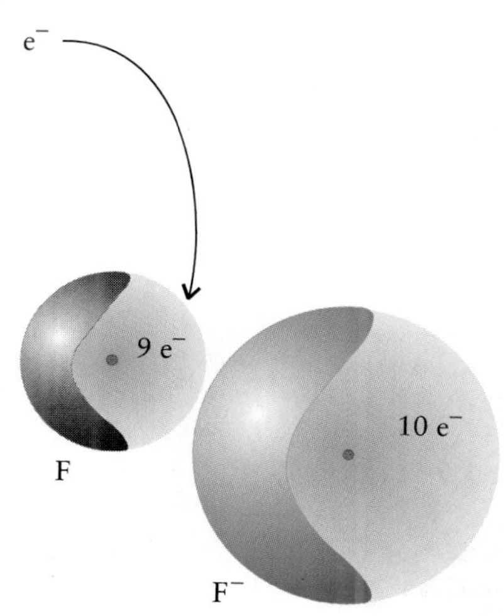

FIGURE 1.26 A neutral fluorine atom (left) consists of a nucleus that contains 9 protons and is surrounded by 9 electrons. When the atom gains 1 electron, the 9 proton charges cancel all but 1 of the 10 electron charges, and the resulting ion (right) has 1 overall negative charge.

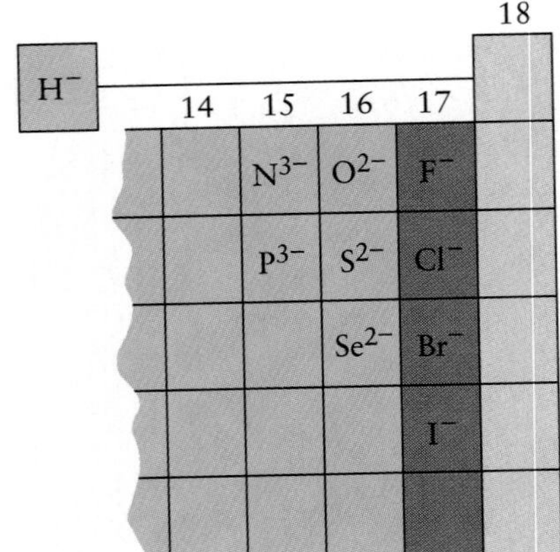

FIGURE 1.27 The typical monatomic anions formed by a selection of elements in the periodic table. Only the nonmetals are shown, for only they form monatomic anions.

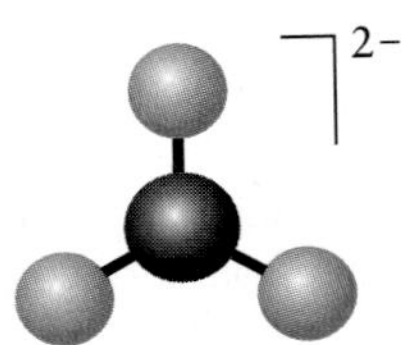

6 Carbonate ion, CO_3^{2-}

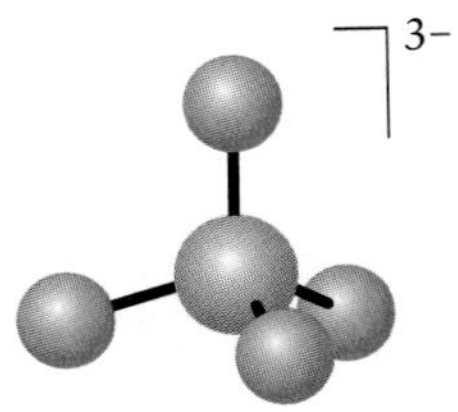

7 Phosphate ion, PO_4^{3-}

The most common polyatomic anions are the **oxoanions,** polyatomic anions that contain oxygen. Among others, they include the carbonate (CO_3^{2-}), nitrate (NO_3^-), phosphate (PO_4^{3-}), and sulfate (SO_4^{2-}) anions. A carbonate ion (**6**), for example, has three O atoms bonded to a carbon atom and an additional two electrons. Similarly, phosphate ion (7) consists of four O atoms bonded to a phosphorus atom, with three additional electrons.

An ionic compound does not consist of individual molecules. Instead, a crystal of an ionic compound (like a crystal of salt) has an arbitrarily large number of ions of each kind (recall Fig. 1.22). So, the formulas of ionic compounds are reported in a different way. Because it is the same whatever the size of the crystal, the *ratio* of the number of cations to the number of anions serves as the formula of the compound. This ratio is given by a chemical formula that shows the relative numbers of each type of ion in terms of the smallest whole numbers. In sodium chloride, for instance, there is one Na^+ cation for each Cl^- ion, so its formula is NaCl regardless of the size of the crystal. In sodium carbonate, there are two Na^+ cations per carbonate ion (CO_3^{2-}), so its formula is Na_2CO_3. When a subscript has to be added to a polyatomic ion, the ion is written within parentheses, as in $(NH_4)_2SO_4$, where $(NH_4)_2$ means that there are two NH_4^+ ions for each SO_4^{2-} ion in ammonium sulfate. As we see from these examples, we omit the charges on the ions when writing a formula of a compound, as in $(NH_4)_2SO_4$. In each case the ions combine in such a way that the positive and negative charges cancel: *all compounds are electrically neutral overall.*

A group of ions with the same number of atoms as that in the formula of an ionic compound is called a **formula unit.** The formula unit of an ionic compound can be thought of as the smallest unit of that compound: it is the equivalent of a "molecule" of the compound. For example, the formula unit of sodium chloride, NaCl, consists of one Na^+ ion and one Cl^- ion; and the formula unit of ammonium sulfate, $(NH_4)_2SO_4$, consists of two NH_4^+ ions and one SO_4^{2-} ion.

The chemical formula of an ionic compound shows the ratio of the numbers of atoms of each element present in the compound in terms of the smallest whole numbers. A formula unit of an ionic compound is a collection of ions with the same numbers of each species as appear in its formula.

EXAMPLE 1.3 Interpreting chemical formulas

How many atoms of each element are represented by the chemical formulas (a) N_2O_4 (a molecular compound) and (b) $(NH_4)_2SO_4$ (an ionic compound)?

STRATEGY Each chemical symbol represents one atom of the element. Find the number of each atom from its subscript. If the atom is part of a polyatomic ion in parentheses, we multiply the subscript of the atom by the subscript outside the parentheses to get the total number of atoms.

SOLUTION (a) The subscripts 2 on the N and 4 on the O tell us that each N_2O_4 molecule has two N atoms and four O atoms. (b) The ionic compound $(NH_4)_2SO_4$ consists of two different polyatomic ions, one with a subscript. Each NH_4^+ ion has one N atom and four H atoms. Because there are two NH_4^+ ions in the formula, there are twice as many N and H atoms—two N atoms and eight

H atoms. There are also one S atom and four O atoms. The N, H, S, and O atoms are therefore present in the ratios 2:8:1:4 in any sample of the compound.

SELF-TEST 1.4A What is the ratio of the numbers of atoms of each element in the mineral azurite, $Cu_3(OH)_2(CO_3)_2$?

[*Answer:* The Cu, O, H, and C atoms are present in azurite in the ratios 3:8:2:2.]

SELF-TEST 1.4B What is the ratio of the numbers of atoms of each element in the mineral mica, $KMg_3Si_3AlO_{10}(OH)_2$?

MIXTURES

We have seen that pure substances can be classified as either elements or compounds. However, most materials are neither single elements nor single compounds, so they are not pure substances; they are **mixtures** of these simpler substances, with one substance mingled with another. Examples of mixtures are air (which consists of nitrogen, oxygen, and small amounts of other gases), seawater (water containing many dissolved substances, particularly sodium chloride), and some kinds of brass (copper and zinc). People and plants are highly complex, highly organized mixtures of mostly organic compounds.

We can see commercial uses of mixtures every day. Gasoline is a mixture of hydrocarbons and additives blended together to achieve efficient combustion. Many alloys, which are mixtures of metals, are formulated for maximum strength and resistance to corrosion. Typically, a medicine is a mixture of various ingredients to achieve an overall biological effect. Much the same can be said of a perfume!

A compound has a fixed composition, whereas a mixture may have any composition. There are invariably two H atoms for each O atom in a sample of the compound water, but sugar and sand, for instance, may be mixed in any proportions. Because the components of a mixture are merely mingled with one another rather than being joined with chemical bonds, they retain their own chemical and physical properties in the mixture. In contrast, a compound may have very different properties from the elements from which it is built. Water is totally unlike the gases hydrogen and oxygen from which it is formed, and the hydrogen and oxygen cannot be easily separated from water. In contrast, a mixture of sugar and sand is both sweet (from the sugar) and gritty (from the sand). When shaken with water, the sugar dissolves but the sand does not, allowing them to be separated easily. The differences between compounds and mixtures are summarized in Table 1.3.

TABLE 1.3 Differences between mixtures and compounds

Mixture	Compound
Components can be separated by using physical techniques	Components cannot be separated by using physical techniques
Composition is variable	Composition is fixed
Properties are related to those of its components	Properties are unlike those of its components

1.10 IONIC COMPOUNDS AND IONS

1.11 TYPES OF MIXTURES

FIGURE 1.28 This piece of granite rock is a heterogeneous mixture of several substances.

We can identify the different components of some mixtures with the unaided eye or a microscope (Fig. 1.28). Such a patchwork of different substances is called a **heterogeneous mixture.** Many of the rocks that form the landscape are heterogeneous mixtures, as are mixtures of sugar and sand, no matter how thoroughly they are mixed. Milk, which looks like a pure substance, is a heterogeneous mixture; through a microscope we can see the individual globules of butterfat.

In some mixtures we cannot make out distinct components even with a very powerful microscope; the molecules or ions of the components are so well mixed that the composition is the same throughout, no matter how large or small the sample. Such a mixture is called a **homogeneous mixture** (Fig. 1.29). We cannot distinguish a homogeneous mixture from a pure substance with a microscope. For example, syrup is a homogeneous mixture of sugar and water. The molecules of sugar are separated and mixed so thoroughly with the water that no separate regions or particles can be seen with a microscope.

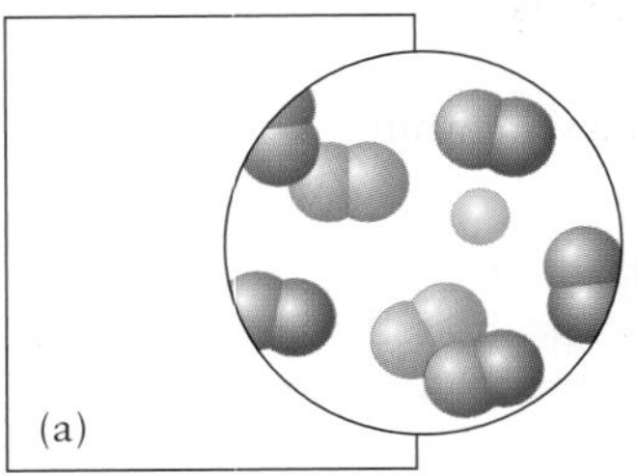

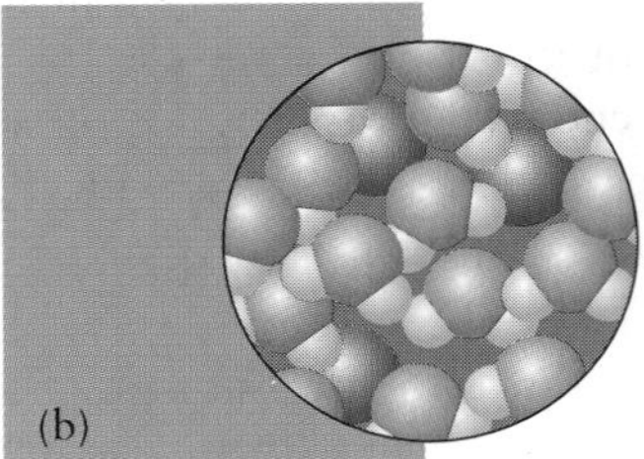

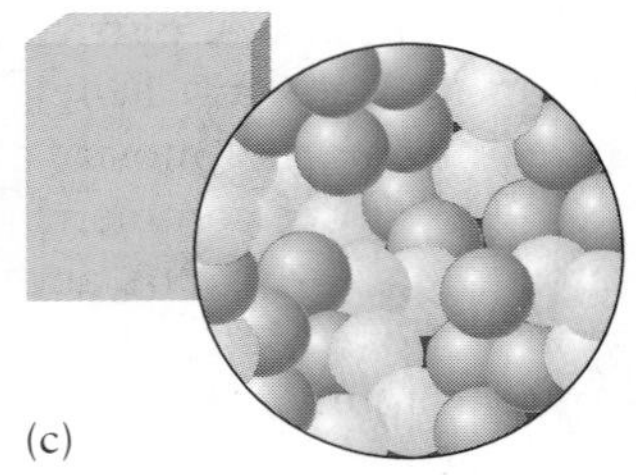

FIGURE 1.29 Three examples of homogeneous mixtures. (a) Air is a homogeneous mixture of many gases, including the nitrogen, oxygen, and argon depicted here. (b) Table salt dissolved in water consists of sodium ions and chloride ions distributed among water molecules. (c) Many alloys are solid mixtures of two or more metals.

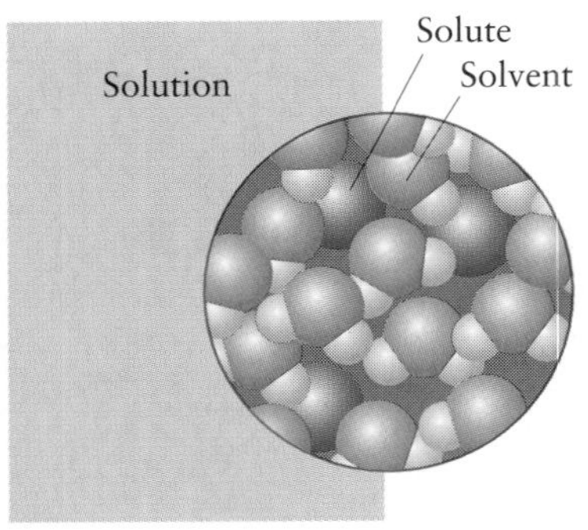

FIGURE 1.30 Solutions are homogeneous mixtures. The solvent (here water) is the substance present in the larger amount. The dissolved substance is called the solute.

Homogeneous mixtures are called **solutions.** Beer is a solution consisting mostly of water, along with alcohol, plant extracts, and various additives. Seawater is a solution of salt (sodium chloride) and many other substances in water. The component of the solution present in the larger amount (water in these examples) is called the **solvent,** and the dissolved substances are the **solutes** (Fig. 1.30). When we use the everyday term "dissolving," we mean the process of producing a solution.

We normally dissolve a solid by shaking it up with the solvent or by stirring, as we do when we dissolve sugar in a cup of coffee. The opposite of dissolving occurs when a component comes out of solution. This process is called **crystallization** if the solute slowly comes out of solution as crystals. For example, salt crystals left when water evaporates line the shores of the Great Salt Lake in Utah. In **precipitation,** a solute comes out of solution as a finely divided powder called a precipitate. Precipitation is often almost instantaneous (Fig. 1.31).

Beverages and seawater are examples of **aqueous solutions,** solutions in which the solvent is water. Aqueous solutions are very common in everyday life and in chemical laboratories; for that reason, most of the solutions in this text will be aqueous. **Nonaqueous solutions** are solutions in which the solvent is not water. Although they are less common than aqueous solutions, they have important uses. In "dry cleaning," the grease and dirt on fabrics are dissolved in tetrachloroethene, C_2Cl_4, a compound of carbon and chlorine. There are also **solid solutions,** in which the solvent is a solid. An example is one form of brass, which is a solution of copper in zinc. The atmosphere could be regarded as a giant gaseous solution of gases in nitrogen (the major component).

FIGURE 1.31 Precipitation occurs when an insoluble substance is formed. Here lead(II) iodide, PbI_2, which is an insoluble yellow solid, precipitates when we mix colorless solutions of lead(II) nitrate, $Pb(NO_3)_2$, and potassium iodide, KI.

Mixtures display the properties of their constituents; they differ from compounds, as summarized in Table 1.3. Mixtures are classified as homogeneous or heterogeneous, and solutions are a special class of homogeneous mixtures.

1.12 SEPARATION TECHNIQUES

To analyze the composition of a sample that we suspect is a mixture, we need to separate its components and identify the individual substances present. The methods all rely on differences in a **physical property** of the components, a characteristic of the substance itself. A physical property must be distinguished from a **chemical property,** which refers to the ability of the substance to change into another substance. For example, a physical property of hydrogen is its occurrence as a colorless gas. A chemical property of hydrogen is that it is converted into water when it burns in oxygen. The elements in a compound can be separated only by chemical change. However, because the components of a mixture retain their individual physical properties, they can be separated by physical means. Thus, another definition of a mixture is *a form of matter that can be separated by making use of the physical properties of the components* (Fig. 1.32).

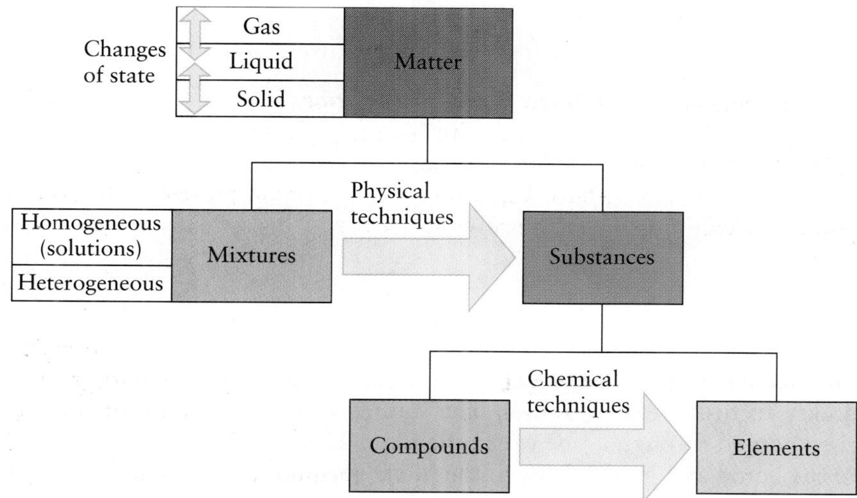

FIGURE 1.32 The hierarchy of materials: matter, mixtures, compounds, and elements. Physical techniques of separation are indicated by the upper horizontal arrow.

CASE STUDY 1 SCIENTIFIC INQUIRY ON MARS

IN 1976 A MAGNIFICENT opportunity was presented to scientists interested in the possibility of extraterrestrial life. The National Aeronautics and Space Administration (NASA) was planning to send the Viking spacecraft to explore Mars and wanted to fit the lander with equipment to test for the presence of life. Imagine the excitement and the tension as the scientists faced their task: find out if life-forms with unknown characteristics exist on a planet with a largely unknown environment.

Direct observations could not be conducted on Mars before hypotheses were made and experiments planned. However, a systematic approach was used. First, the team made some assumptions about life on Mars based on observations of life on Earth. Most living beings depend on sunlight, are affected by temperature changes, and are killed by heating to a very high temperature (sterilization). They also decided that the most likely life-forms to be found would be the simplest ones—microorganisms such as bacteria or algae. So, in addition to direct observation, they would search for indirect evidence of microorganisms in Martian air and soil.

A variety of possible results had to be anticipated in advance—this stage involved making hypotheses about what might be present. It was decided that three criteria for the presence of life on Mars would have to be met for an experiment to be considered positive: the experiment must be reproducible; sterilizing the sample must eliminate any positive response to an experiment; and organic chemicals must be found in the soil. Several experiments were designed, three of which are reported here.

HYPOTHESIS 1 Martians are plentiful and large enough to be visible.

EXPERIMENT 1 Television cameras looked for plants or animals, infrared sensors sought moving, warm objects, a seismometer listened for footsteps, and a mass spectrometer tested the atmosphere for gases associated with life on Earth. The television cameras showed only a barren red landscape. In fact, all the tests but one were negative. The positive result was the finding that the atmosphere is 95% by mass carbon dioxide, which is associated with life on Earth. However, the average temperature of the surface was about −20°C, and the average atmospheric pressure was only about 0.5% that of Earth's. These conditions would be too harsh for life as we know it.

HYPOTHESIS 2 Martian microorganisms undergo photosynthesis in which they absorb carbon dioxide, as many terrestrial microorganisms do.

EXPERIMENT 2 In the "pyrolytic release" experiment, a sample of Martian soil and air was placed into a chamber. Carbon dioxide containing radioactive carbon atoms

View of the Martian surface from the Viking lander. The long, jointed arm is being deployed to collect a soil sample for analysis. The container on the left holds the equipment for the biological experiments that tested for evidence of life.

▶▶ Filtration is a vital step in the treatment of domestic water supplies, where it is used to remove particles of insoluble matter from the water (see Case Study 3).

The technique of **filtration** makes use of differences in solubility (the ability to dissolve in a liquid). The sample is shaken with a liquid and then poured through a fine mesh, the filter. The soluble material dissolves in the liquid and passes through the filter. The insoluble material does not dissolve and is captured by the filter. The technique can be used to separate sugar from sand, because sugar is soluble in water but sand is not.

The technique of **distillation** makes use of differences in boiling points. Distillation separates the components of a mixture by vaporizing one or more of the components (Fig. 1.33). It can be used to remove

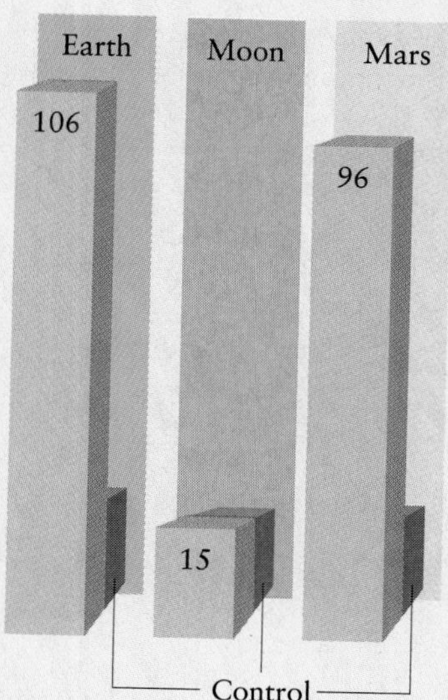

The results of pyrolytic release experiments performed on soil from the Earth, the Moon, and Mars. The control experiments were conducted on sterilized soil. The heights of the bars show the relative number of radioactive carbon nuclei decaying per second in the released gases.

was added and the chamber was heated to 17°C for several days while illuminated by a xenon lamp to simulate sunlight. Then the air was pumped out and the sample was heated strongly to release any gases absorbed by the soil. The released gases were pumped through a radioactivity detector. If radioactive carbon were found, it would mean that something in the soil had absorbed it, possibly lifeforms. In fact, radioactive carbon was absorbed, as shown in the chart. Five more experiments gave similar results.

To distinguish between chemical and biological absorption of the carbon dioxide, a control sample was run, using sterilized soil. If the absorption had been chemical, sterilization should have no effect. If it were biological, sterilization would kill the microorganisms and no activity would be observed. The result supported the biological explanation—there was almost no absorption of carbon dioxide in the control. Sterilized Martian soil responded like soil from the lifeless moon. Raw Martian soil responded like soil from Earth, which is teeming with life. To rule out the possibility that sterilization had destroyed chemicals that might have absorbed the carbon dioxide in the initial experiment, another control was run with raw soil, but without a light source. In the dark, absorption was no better than with sterilized soil. The first two criteria for the presence of life had been met.

HYPOTHESIS 3 Organisms leave traces of organic compounds.

EXPERIMENT 3 A combination gas chromatograph-mass spectrometer (GCMS) was used to test the third criterion: the presence of organic compounds in the soil. A soil sample was heated and its components separated by the gas chromatograph. The mass spectrometer measured the masses of each type of molecule or ion as they emerged. Despite hundreds of trials, no organic compounds were found, so the third criterion was not met.

The official report of the Viking mission stated that no evidence for life on Mars had been found. The discovery in 1996 of the fossilized remains of what may be living organisms in a meteorite believed to come from Mars, however, suddenly and dramatically reopened the question.

QUESTIONS

1. A newspaper headline at the time reported, "No life exists on Mars." Is this statement justified by the Viking evidence? If not, how would you rewrite it?

2. Formulate a hypothesis to explain the results of the pyrolytic release experiment.

3. Some people believe that Mars can be made habitable for humans in a process known as "terraforming," which involves altering the atmosphere and surface temperature. Describe additional information that you would want to collect about Mars before beginning such a project. Suggest a means to gather that information.

water (which boils at 100°C in the pure state) from salt, which boils at a much higher temperature and is therefore left behind when the water evaporates.

One of the most sensitive techniques available for separating and identifying the components of a mixture is **chromatography.** This technique relies on the different abilities of substances to stick to surfaces. There are various versions of the technique. In the simplest, the mixture is washed across a strip of filter paper (Fig. 1.34). Substances that absorb most weakly move further than others and, if they are colored, give rise

The first known distillation apparatus, in the first century AD, is attributed to Mary the Jewess, an alchemist who lived in the Middle East.

FIGURE 1.33 The technique of distillation, which is used to remove a liquid from a solid. The solution is heated, the liquid boils off, condenses in the water-jacketed tube (the condenser), and is collected.

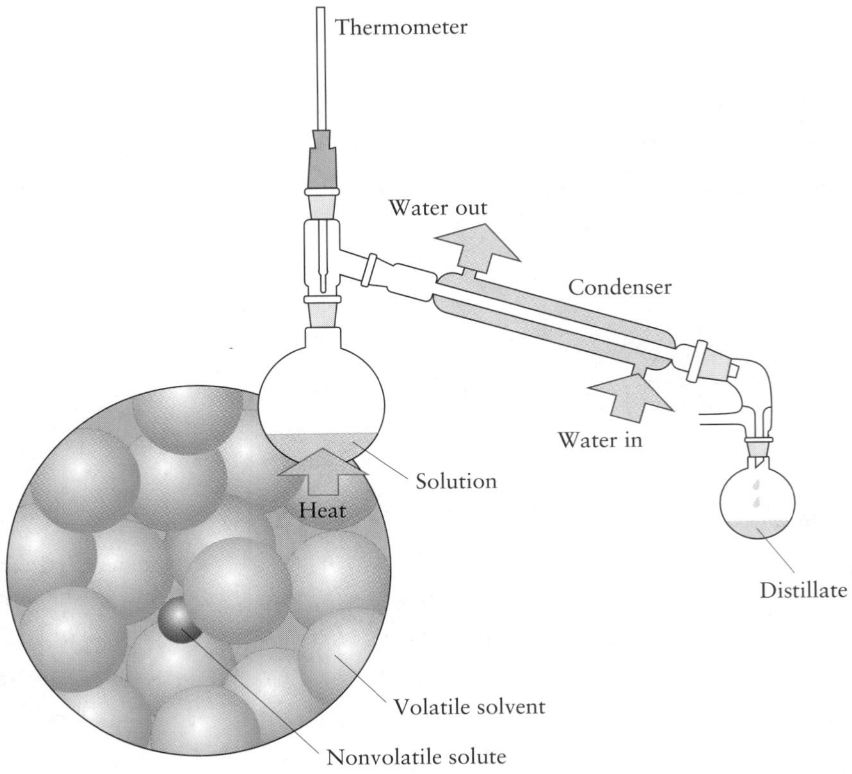

to separate patches of color on the paper. In **gas-liquid chromatography** (GLC), the sample is vaporized and carried in a stream of helium through a long narrow tube. The tube is coated or packed with alumina (aluminum oxide, Al_2O_3) soaked in a liquid that does not vaporize readily. The component of the mixture that sticks to the coated alumina least strongly emerges first from the other end (Fig. 1.35). As time passes, the other components of the original mixture emerge, and each is detected electronically and signaled by a peak on a chart called a **chromatogram.** A chromatogram is a kind of fingerprint of the composition of the mixture (Fig. 1.36). Gas-liquid chromatography is used to detect narcotics and explosives in airline baggage; the equipment "sniffs" the baggage and is highly sensitive to very small amounts of substance (Case Study 1).

The colored patterns that appear gave rise to the name *chromatography,* which comes from the Greek words for "color writing."

Mixtures are separated by making use of the differences in physical properties of the components; techniques include filtration, distillation, and chromatography.

FIGURE 1.34 In paper chromatography, the components of a mixture are separated by washing them along a paper—the support—with a solvent. A primitive form of the technique is shown here. On the left is a dry filter paper to which a drop of food coloring has been applied. The filter paper is then placed in a solvent (middle). The filter paper on the right has been allowed to dry after the solvent spread out to the edges of the paper, carrying two components of the coloring matter to different distances as it spread. The dried support showing the separated components is called a chromatogram.

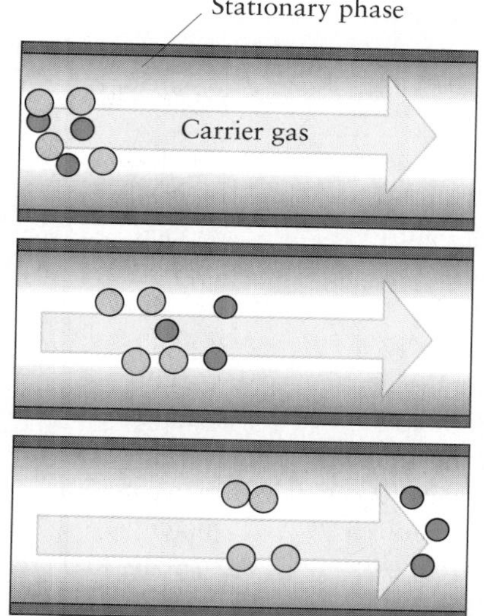

FIGURE 1.35 In gas-liquid chromatography (GLC), the mixture of gases or vapors is separated as it travels through a long coated tube. Some molecules are adsorbed on (stick to) the coating (the stationary phase) more readily than others and therefore emerge later, but eventually all pass through in the stream of carrier gas (typically helium). The less readily adsorbed component (red) emerges first, followed by the more readily adsorbed (yellow).

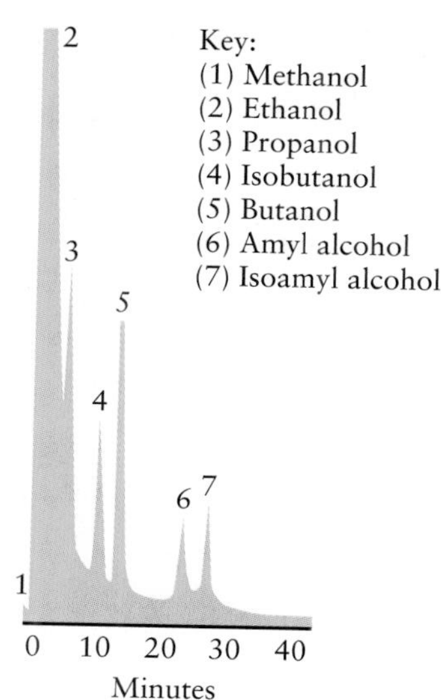

FIGURE 1.36 A gas chromatogram of bourbon whiskey, showing the components that contribute to the flavor. Mixing the chemicals identified here does not, it is said, recreate the flavor: the flavor also depends on hundreds of other compounds present in very small amounts.

THE NOMENCLATURE OF COMPOUNDS

You will meet many compounds in this text and will learn their names as you go along. However, it is useful from the outset to know something about how to form their names. Many compounds were given **common names** before their compositions were known. Common names include water, salt, sugar, ammonia, and quartz. A **systematic name,** on the other hand, reveals which elements are present and, in some cases, how their atoms are arranged. The systematic name of table salt, for instance, is sodium chloride, which indicates at once that it is a compound of sodium and chlorine. The systematic naming of compounds, which is called **chemical nomenclature,** follows a set of rules, so that the name of each compound need not be memorized, only the rules.

1.13 NAMES OF CATIONS

The names of monatomic cations are the same as the name of the element, with the addition of the word *ion,* as in sodium ion for Na^+. When an element can form more than one kind of cation, such as Cu^+ and Cu^{2+} from copper, we use the **Stock number,** a Roman numeral equal to the charge of the cation. Thus, Cu^+ is a copper(I) ion and Cu^{2+} is a copper(II) ion. Similarly, Fe^{2+} is an iron(II) ion and Fe^{3+} is an iron(III) ion. Most transition metals form more than one kind of ion, so it is usually necessary to include a Stock number in the names of their compounds.

The Stock number takes its name from the German chemist Alfred Stock, who devised this numbering system.

An older system of nomenclature is still in use. For example, some cations were once denoted by the endings *-ous* and *-ic* for the ions with lower and higher charges, respectively. In this system, iron(II) ions are called ferrous ions and iron(III) ions are called ferric ions (Table 1.4).

Names of monatomic cations are essentially the same as the names of the elements, except that, for elements that can form more than one cation, the Stock number, a Roman numeral indicating the charge, is included.

TABLE 1.4 Names of some cations with variable charge numbers

Element	Cation	Old style name	Modern name
cobalt	Co^{2+}	cobaltous	cobalt(II)
	Co^{3+}	cobaltic	cobalt(III)
copper	Cu^{+}	cuprous	copper(I)
	Cu^{2+}	cupric	copper(II)
iron	Fe^{2+}	ferrous	iron(II)
	Fe^{3+}	ferric	iron(III)
lead	Pb^{2+}	plumbous	lead(II)
	Pb^{4+}	plumbic	lead(IV)
manganese	Mn^{2+}	manganous	manganese(II)
	Mn^{3+}	manganic	manganese(III)
mercury	Hg_2^{2+}	mercurous	mercury(I)
	Hg^{2+}	mercuric	mercury(II)
tin	Sn^{2+}	stannous	tin(II)
	Sn^{4+}	stannic	tin(IV)

1.14 NAMES OF ANIONS

Monatomic anions are named by adding the suffix *-ide* and the word *ion* to the first part of the name of the element (the "stem" of its name), as shown in Table 1.5. There is no need to give the charge, because most elements that form monatomic anions form only one kind of ion. The ions formed by the halogens are collectively called *halide ions* and include fluoride (F^-), chloride (Cl^-), bromide (Br^-), and iodide ions (I^-).

The names of oxoanions (Table 1.5) are formed by adding the suffix *-ate* to the stem of the name of the element that is not oxygen, as in the carbonate ion, CO_3^{2-}. However, many elements can form a variety of oxoanions, with different numbers of oxygen atoms; nitrogen, for example, forms both NO_2^- and NO_3^-. In such cases, the ion with the larger number of oxygen atoms is given the suffix *-ate,* and that with the smaller number of oxygen atoms is given the suffix *-ite*. Thus, NO_3^- is nitrate and NO_2^- is nitrite.

Think of a m*ite* as being small.

Some elements—particularly the halogens—form more than two oxoanions. The name of the oxoanion with the smallest number of oxygen atoms is formed by adding the prefix *hypo-* to the *-ite* form of the name, as in the *hypo*chlor*ite* ion, ClO^-. The oxoanion with a higher number of oxygen atoms than the *-ate* oxoanion is named with the prefix *per-* added to the *-ate* form of the name. An example is the *per*chlor*ate* ion, ClO_4^-. As can be seen from Table 1.5, chlorine forms oxoanions that span the range.

Hypo comes from the Greek word for "under."

Per is the Latin word for "all over," suggesting that the element's ability to combine with oxygen is finally satisfied.

Some anions include hydrogen, such as HS^- and HCO_3^-. The names of these anions begin with "hydrogen." Thus, HCO_3^- is the hydrogen carbonate ion. In an older system of nomenclature, an anion containing a hydrogen atom was named with the prefix *bi-,* as in bicarbonate ion for HCO_3^-.

The **oxoacids** are molecular compounds that can be regarded as the parents of the oxoanions. The formulas of oxoacids are derived from those of the corresponding oxoanions by adding enough hydrogen ions to balance the charges. This procedure is only a formal way of building the chemical formula, because oxoacids are all *molecular* compounds.

TABLE 1.5 Common anions and their parent acids

Anion	Parent acid	Anion	Parent acid
fluoride ion, F^-	hydrofluoric acid,* HF (hydrogen fluoride)	nitrite ion, NO_2^-	nitrous acid, HNO_2
		nitrate ion, NO_3^-	nitric acid, HNO_3
chloride ion, Cl^-	hydrochloric acid,* HCl (hydrogen chloride)	phosphate ion, PO_4^{3-}	phosphoric acid, H_3PO_4
		hydrogen phosphate ion, HPO_4^{2-}	
bromide ion, Br^-	hydrobromic acid,* HBr (hydrogen bromide)	dihydrogen phosphate ion, $H_2PO_4^-$	
iodide ion, I^-	hydroiodic acid,* HI (hydrogen iodide)	sulfite ion, SO_3^{2-}	sulfurous acid, H_2SO_3
oxide ion, O^{2-}	water, H_2O	sulfate ion, SO_4^{2-}	sulfuric acid, H_2SO_4
hydroxide ion, OH^-		hydrogen sulfate ion, HSO_4^-	
sulfide ion, S^{2-}	hydrosulfuric acid,* H_2S (hydrogen sulfide)	hypochlorite ion, ClO^-	hypochlorous acid, HClO
cyanide ion, CN^-	hydrocyanic acid,* HCN (hydrogen cyanide)	chlorite ion, ClO_2^-	chlorous acid, $HClO_2$
acetate ion, $CH_3CO_2^-$	acetic acid, CH_3COOH	chlorate ion, ClO_3^-	chloric acid, $HClO_3$
carbonate ion, CO_3^{2-}	carbonic acid, H_2CO_3	perchlorate ion, ClO_4^-	perchloric acid, $HClO_4$
hydrogen carbonate (bicarbonate) ion, HCO_3^-			

*The name of the aqueous solution of the compound. The name of the compound itself is in parentheses.

For example, the sulfate ion, SO_4^{2-}, needs two H^+ ions to cancel its negative charge, so sulfuric acid is the molecular compound H_2SO_4. Similarly, the phosphate ion, PO_4^{3-}, needs three H^+ ions, so its parent acid is the molecular compound H_3PO_4, phosphoric acid. As these examples illustrate, the names of the parent oxoacids are derived from those of the corresponding oxoanions by replacing the *-ate* suffix with *-ic acid.* The more important oxoacids are included in Table 1.5. In general, *-ic* oxoacids are the parents of *-ate* oxoanions and *-ous* oxoacids are the parents of *-ite* oxoanions.

▶▶ **Rules for naming polyatomic ions are summarized in Appendix 3A. More halogen oxoacids are listed in Table 20.8.**

Names of monatomic anions end in -ide; oxoanions are anions that contain oxygen. Oxoacids are acids that contain oxygen. The suffixes -ate and -ic acid indicate a greater number of oxygen atoms than the suffixes -ite and -ous acid.

SELF-TEST 1.5A Give (a) the name of HIO and (b) the formula of sulfuric acid.

[*Answer:* (a) Hypoiodous acid; (b) H_2SO_4]

SELF-TEST 1.5B Give (a) the name of H_3PO_4 and (b) the formula of chloric acid.

1.15 NAMES OF IONIC COMPOUNDS

An ionic compound is named with the cation name first, followed by the name of the anion; the word *ion* is omitted in each case. Typical names include potassium chloride (KCl), a compound containing K^+ and Cl^- ions, and ammonium nitrate (NH_4NO_3), which contains NH_4^+ and NO_3^-

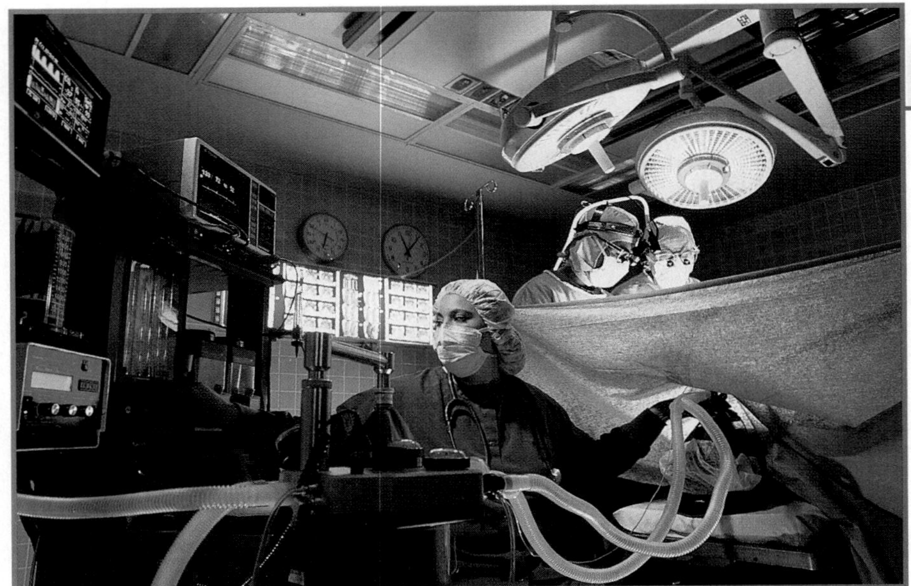

CHEMISTRY AT WORK

An anesthesiologist monitors the condition of an anesthetized patient undergoing microsurgery. The names of some chemicals used as medications can be quite similar, but their properties can be very different. Medical personnel must have a firm grasp of chemical nomenclature to make sure they administer the correct medications to their patients.

ions. The copper chloride that contains Cu^+ ions (CuCl) is called copper(I) chloride, and the chloride that contains Cu^{2+} ions ($CuCl_2$) is called copper(II) chloride.

Some ionic compounds form crystals that incorporate a definite proportion of molecules of water as well as the ions of the compound itself. These compounds are called **hydrates.** For example, copper(II) sulfate normally occurs as blue crystals of composition $CuSO_4 \cdot 5H_2O$. The raised dot in this formula is used to separate the water of hydration from the rest of the formula. This formula indicates that there are five H_2O molecules for each $CuSO_4$ formula unit. Hydrates are named by first giving the name of the compound, then adding the word hydrate with a Greek prefix indicating how many molecules of water are found in each formula unit (Table 1.6). For example, the name of $CuSO_4 \cdot 5H_2O$ is copper(II) sulfate pentahydrate.

TABLE 1.6 Prefixes used for naming compounds

Prefix	Meaning
mono-	1
di-	2
tri-	3
tetra-	4
penta-	5
hexa-	6
hepta-	7
octa-	8
nona-	9
deca-	10
undeca-	11
dodeca-	12

Ionic compounds are named by giving the name of the cation (with the charge indicated by a Roman numeral if more than one charge is possible), followed by the name of the anion; hydrates are named by adding the word "hydrate," preceded by a Greek prefix indicating the number of water molecules.

EXAMPLE 1.4 Naming ionic compounds

Give the systematic names of (a) $CrCl_3 \cdot 6H_2O$ and (b) $Ba(ClO_4)_2$.

STRATEGY In each case, identify the cation and the anion, by referring to Tables 1.4, 1.5, and 1.6 if necessary. If the cation is a transition metal or from Groups 12 through 15, give its Stock number. To identify the Stock number, note the charge of the anion and the number of anions and cations in the formula; then decide what cation charge is required to cancel the total negative charge. If the cation comes from an element that exists in only one charge state (as set out in Fig. 1.24), omit the Stock number. If there is more than one cation present, divide the total positive charge by the number of cations. If water molecules are included in the formula, then the compound is a hydrate, so add the word *hydrate* with a prefix corresponding to the number in front of H_2O in the formula.

SOLUTION (a) The cation is a chromium ion and the anion is a chloride ion, Cl^-. The cation must be Cr^{3+} for electrical neutrality; so chromium (a transition metal) is present as chromium(III). The $6H_2O$ indicates that the compound is a hexahydrate; so the compound is chromium(III) chloride hexahydrate. (b) The cation is a barium ion, Ba^{2+} (the only cation that barium forms). The anion is a perchlorate ion, ClO_4^-. Hence, the compound is barium perchlorate.

SELF-TEST 1.6A Name the compounds (a) $FeCl_2 \cdot 2H_2O$; (b) $AlBr_3$; (c) $Cr(ClO_3)_2$.

[*Answer:* (a) Iron(II) chloride dihydrate; (b) aluminum bromide; (c) chromium(II) chlorate]

SELF-TEST 1.6B Name the compounds (a) $AuCl_3$; (b) CaS; (c) Mn_2O_3.

TABLE 1.7 Common names for some simple molecular compounds

Formula*	Common name
NH_3	ammonia
N_2H_4	hydrazine
NH_2OH	hydroxylamine
PH_3	phosphine
NO	nitric oxide
N_2O	nitrous oxide
C_2H_4	ethylene
C_2H_2	acetylene

*For historical reasons, the molecular formulas of binary hydrogen compounds of Group 15 elements are written with the Group 15 elements first.

1.16 NAMES OF MOLECULAR COMPOUNDS

Many simple molecular compounds are named by using the Greek prefixes in Table 1.6 to indicate the number of each type of atom present. Usually, no prefix is used if only one atom of an element is present; an important exception to this rule is carbon monoxide, CO. Most of the common **binary** molecular compounds—molecular compounds built from two elements—have at least one element from Group 16 or 17. These elements are named second, with their endings changed to *-ide*:

phosphorus trichloride PCl_3 dinitrogen oxide N_2O

sulfur hexafluoride SF_6 dinitrogen pentoxide N_2O_5

Certain binary molecular compounds have common names that are widely used (Table 1.7). The phosphorus oxides are distinguished by a Roman numeral, as though phosphorus were a metal. Thus, P_4O_6 is phosphorus (III) oxide and P_4O_{10} is phosphorus (V) oxide.

Except for organic compounds and a few other cases, both the name and the molecular formula of compounds of hydrogen and nonmetals are written with the H atom first:

hydrogen chloride HCl   hydrogen cyanide HCN

All the oxoacids are written in this way, as we have seen in Table 1.5.

Binary molecular compounds are named by using Greek prefixes to indicate the number of atoms of each element present; the element named second has its ending changed to -ide.

EXAMPLE 1.5 Naming simple molecular compounds

Give the systematic names of the molecular compounds (a) N_2O_5 and (b) ClO_2.

STRATEGY The subscripts tell how many atoms of each element are present in the molecule. For the first element, use the Greek prefix that matches the subscript, followed by the name of the element. For the second element, do the same, but change the ending to *-ide*. Normally, no prefix is used for single atoms.

SOLUTION (a) The molecule N_2O_5 has two nitrogen atoms and five oxygen atoms, so its name is dinitrogen pentoxide. (b) A ClO_2 molecule has one chlorine atom and two oxygen atoms, so its name is chlorine dioxide.

SELF-TEST 1.7A Name the compounds (a) As_2O_5; (b) BCl_3; (c) IF_5.
[*Answer:* (a) Diarsenic pentoxide; (b) boron trichloride; (c) iodine pentafluoride]

SELF-TEST 1.7B Name the compounds (a) PCl_3; (b) SO_3; (c) N_2O_4.

EXAMPLE 1.6 Writing the formula of a binary compound from its name

Write the formulas of (a) magnesium nitride and cobalt(II) chloride hexahydrate, and (b) diboron trisulfide and silicon tetrachloride.

STRATEGY First check to see whether the compounds are ionic or molecular. Many compounds that contain a metal are ionic. The symbol of the metal is written first, followed by the symbol of the nonmetal. The charge on the cation (the metal ion) is determined by its position in the periodic table, if it is a main group metal, and by the Roman numeral in parentheses, if it can have more than one charge. The charge on the anion is determined by its position in the periodic table. Subscripts are chosen to balance charges. Formulas for compounds of two nonmetals are written by listing the symbols of the elements as they are in the name and giving them subscripts corresponding to the Greek prefix used.

SOLUTION (a) Magnesium is a metal in Group 2, so its cation will be Mg^{2+}; nitrogen is a nonmetal in Group 15, so the nitride ion has a charge of $15 - 18 = -3$. We need the -6 charge of two N^{3-} ions to balance the $+6$ charge of three magnesium ions; so the formula must be Mg_3N_2 because $3 \times (+2) + 2 \times (-3) = 0$. The II in cobalt(II) signifies that the metal cobalt has a charge of $+2$ in the compound. Chlorine is in Group 17, so the chloride ion has a charge of $17 - 18 = -1$. We can balance the charge of the $+2$ cation with two -1 anions, so the formula is $CoCl_2$. This compound is a hydrate, so we then add the six water molecules, as indicated by the Greek prefix *hexa-;* the complete formula is therefore $CoCl_2{\cdot}6H_2O$. (b) The subscripts for each atom are taken from the Greek prefixes, so we write B_2S_3 and $SiCl_4$. Notice that the prefix *mono-* is omitted in the case of silicon.

SELF-TEST 1.8A Write the formulas for (a) vanadium(V) oxide and beryllium carbide, and (b) sulfur tetrafluoride and dinitrogen trioxide.
[*Answer:* (a) V_2O_5 and Be_2C; (b) SF_4 and N_2O_3]

SELF-TEST 1.8B Write the formulas for (a) cesium sulfide tetrahydrate and manganese(VII) oxide, and (b) hydrogen sulfide (the gas responsible for the smell of rotten eggs) and disulfur dichloride.

SKILLS YOU SHOULD HAVE MASTERED

Conceptual

1. Describe the structure of the nuclear atom.
2. Predict the characteristics of elements in different regions of the periodic table.
3. Distinguish by their properties metals, metalloids, and nonmetals and locate them in the periodic table.

4. Distinguish molecules, ions, and atoms.
5. Identify physical and chemical changes.
6. Use various means of representing molecules and write formulas from molecular structures.
7. Distinguish mixtures, compounds, and elements.
8. Distinguish heterogeneous and homogeneous mixtures.

Problem-Solving

1. Write the symbols for the elements, given their names, and vice versa.
2. State the numbers of neutrons, protons, and electrons in an isotope, given its atomic number and mass number.
3. Predict the anion or cation a main-group element is likely to form.
4. Interpret chemical formulas in terms of the number of each type of atom present.
5. Name binary compounds, compounds with common polyatomic ions, and hydrates, and write their formulas.

Descriptive

1. Describe the historical development of the model of the nuclear atom.
2. Describe the function and operation of a mass spectrometer.
3. Describe how mixtures are separated by each of the techniques described in Section 1.12.

EXERCISES

The elements are listed in the periodic table inside the front cover and in alphabetical order inside the back cover.

The Nuclear Atom and Isotopes

1.1 (a) What is the charge and mass of the particle in cathode rays? (b) Who first determined the charge of the electron?

1.2 (a) Who proposed the existence of a nucleus in an atom? (b) What experimental evidence motivated the proposal?

1.3 Explain the difference between a law and a theory.

1.4 Rank the following according to increasing credibility among scientists and explain your reasoning: (a) theory; (b) premonition; (c) hypothesis.

1.5 Give the chemical symbol and atomic number of (a) arsenic; (b) sulfur; (c) palladium; (d) gold.

1.6 Give the chemical symbol and atomic number of (a) bromine; (b) barium; (c) bismuth; (d) boron.

1.7 State the number of protons, neutrons, and electrons in an atom of (a) carbon-13; (b) ^{37}Cl; (c) chlorine-35; (d) ^{235}U.

1.8 State the number of protons, neutrons, and electrons in an atom of (a) tritium, ^{3}H; (b) ^{60}Co; (c) oxygen-16; (d) ^{204}Pb.

1.9 Identify the isotope that has atoms with (a) 63 neutrons, 48 protons, and 48 electrons; (b) 46 neutrons, 36 protons, and 36 electrons; (c) 6 neutrons, 5 protons, and 5 electrons.

1.10 Identify the isotope that has atoms with (a) 78 neutrons, 52 protons, and 52 electrons; (b) 108 neutrons, 73 protons, and 73 electrons; (c) 32 neutrons, 28 protons, and 28 electrons.

1.11 (a) What characteristics do carbon-12, carbon-13, and carbon-14 atoms have in common? (b) In what ways are they different?

1.12 (a) What characteristics do argon-40, potassium-40, and calcium-40 atoms have in common? (b) In what ways are they different?

The Periodic Table

1.13 Name the elements (a) Li; (b) Ga; (c) Xe; (d) K. Give their group numbers in the periodic table. Identify each one as a metal, a nonmetal, or a metalloid.

1.14 Name the elements (a) P; (b) Sb; (c) Fe; (d) Ag. Give their group numbers in the periodic table, and identify each one as a metal, a nonmetal, or a metalloid.

1.15 Write the chemical symbol of (a) chlorine; (b) cobalt; (c) arsenic. Classify each one as a metal, a nonmetal, or a metalloid.

1.16 Write the chemical symbol of (a) neon; (b) bismuth; (c) tungsten. Classify each one as a metal, a nonmetal, or a metalloid.

1.17 List the metallic elements of Group 14 by name and chemical symbol.

1.18 List the metalloids of Group 14 by name and chemical symbol.

1.19 Write the chemical symbol of (a) iodine; (b) chromium; (c) mercury; (d) aluminum; and classify each one as a metal, a nonmetal, or a metalloid.

1.20 Write the chemical symbol of (a) carbon; (b) zinc; (c) barium; (d) germanium; and classify each one as a metal, a nonmetal, or a metalloid.

1.21 List the names, chemical symbols, and atomic numbers of the alkali metals. Describe their reactions with water.

1.22 List the names, chemical symbols, and atomic numbers of the halogens. Identify the normal physical state of each.

Compounds

1.23 Explain the meaning of the terms (a) *atom* and (b) *molecule*.

1.24 What is meant by (a) an ionic compound and (b) a molecular compound; what are the typical properties of the two classes of compound?

1.25 State whether or not the following elements are more likely to form a cation or an anion and write the formula for that ion: (a) sulfur; (b) potassium; (c) strontium; (d) chlorine.

1.26 State whether or not the following elements more likely form a cation or an anion and write the formula for that ion: (a) zinc; (b) magnesium; (c) nitrogen; (d) oxygen.

1.27 How many protons, neutrons, and electrons are present in (a) $^{2}H^{+}$; (b) $^{9}Be^{2+}$; (c) $^{80}Br^{-}$; (d) $^{32}S^{2-}$?

1.28 How many protons, neutrons, and electrons are present in (a) $^{40}Ca^{2+}$; (b) $^{115}In^{3+}$; (c) $^{127}Te^{2-}$; (d) $^{14}N^{3-}$?

1.29 Write the symbol of the isotopic ion that has (a) 9 protons, 10 neutrons, and 10 electrons; (b) 12 protons, 12 neutrons, and 10 electrons; (c) 52 protons, 76 neutrons, and 54 electrons; (d) 37 protons, 49 neutrons, and 36 electrons.

1.30 Write the symbol of the isotopic ion that has (a) 11 protons, 13 neutrons, and 10 electrons; (b) 13 protons, 14 neutrons, and 10 electrons; (c) 34 protons, 45 neutrons, and 36 electrons; (d) 24 protons, 28 neutrons, and 22 electrons.

Substances and Mixtures

1.31 Identify the following substances as elements or compounds: (a) gold; (b) chlorine gas; (c) table salt.

1.32 Identify the following substances as mixtures or pure substances: (a) sodium hydrogen carbonate (baking soda); (b) brass; (c) diamond, a crystalline form of carbon.

1.33 The containers pictured at the top of the page each hold either a mixture, a single compound, or a single element. In each case, identify the type of contents:

(a)

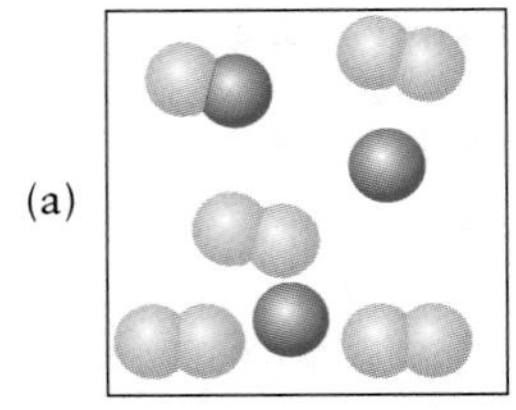

(b)

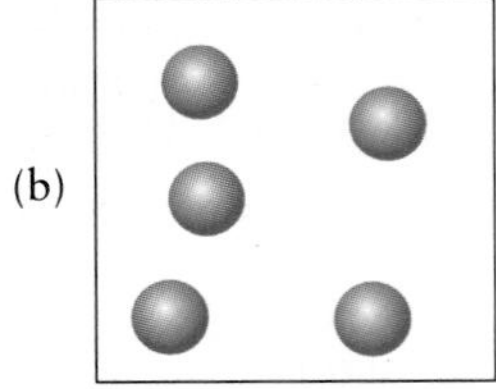

1.34 The containers pictured below each hold either a mixture, a single compound, or a single element. In each case, identify the type of contents:

(a)

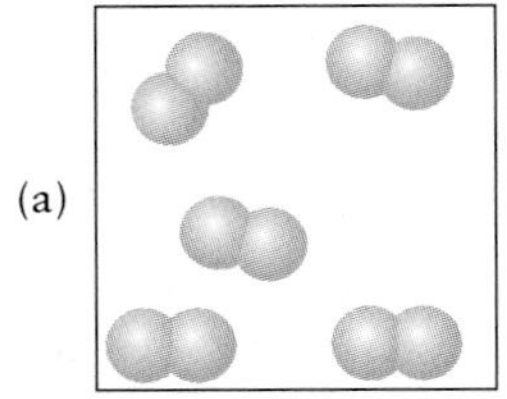

(b)

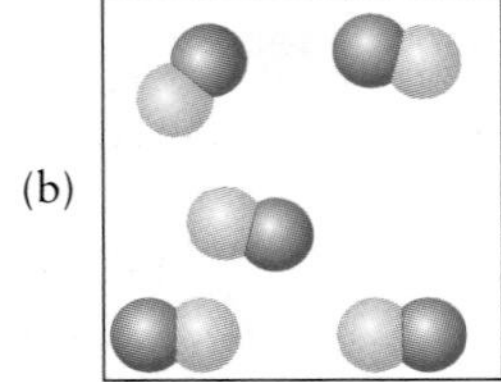

1.35 Explain the distinction between compounds and mixtures.

1.36 Explain the distinction between physical and chemical properties.

1.37 Classify the following as chemical or physical properties: (a) the color of copper(II) sulfate pentahydrate is blue; (b) the melting point of sodium metal is 98°C; (c) iron rusts in areas with high humidity.

1.38 A chemist investigates the boiling point, melting point, and flammability of a compound called acetone, a component of fingernail polish remover. Which of these properties are physical properties and which are chemical properties?

1.39 Classify the following changes as chemical or physical: (a) the freezing of water; (b) the vaporization of alcohol; (c) the corrosion of aluminum when exposed to air.

1.40 A general chemistry student was asked to decide which of the following changes were physical and which were chemical: melting, rusting, and bending of iron. What is the correct answer?

1.41 Determine which of the following properties are chemical and which are physical: (a) the hardness of a metal; (b) the ductility of a metal; (c) the corrosiveness of an acid; (d) the freezing point of a liquid.

1.42 Identify the following as chemical or physical changes: (a) the formation of frost; (b) the tarnishing of silver metal; (c) the production of hydrogen when potassium is in contact with water; (d) the liquefaction of air.

1.43 List all the physical properties and changes in the statement, "The temperature of the land is an important factor for the ripening of oranges, because it affects the evaporation of water and the humidity of the surrounding air."

1.44 List all the chemical properties and changes in the statement, "Copper is a red-brown element obtained from copper sulfide ores by heating in air, which forms copper oxide. Heating the copper oxide with carbon produces impure copper."

1.45 What physical properties are used for the separation of the components of a mixture by (a) filtration; (b) chromatography; (c) distillation?

1.46 Explain how you would distinguish sugar from salt. Identify the physical and chemical properties used in your explanation.

1.47 Identify the following as homogeneous or heterogeneous mixtures and suggest a technique for separating their components: (a) alcohol and water; (b) chalk and table salt; (c) salt water.

1.48 Identify the following as homogeneous or heterogeneous mixtures and suggest a technique for separating their components: (a) gasoline and motor oil; (b) carbonated water; (c) charcoal and sugar.

Chemical Nomenclature

1.49 Name the ions (a) Cl^-; (b) O^{2-}; (c) C^{4-}; (d) P^{3-}.

1.50 Name the ions (a) Se^{2-}; (b) Br^-; (c) F^-; (d) Mg^{2+}.

1.51 Name the ions (a) PO_4^{3-}; (b) SO_4^{2-}; (c) N^{3-}; (d) SO_3^{2-}; (e) PO_3^{3-}; (f) I^-.

1.52 Name the ions (a) ClO_4^-; (b) NO_2^-; (c) S^{2-}; (d) NO_3^-; (e) ClO_2^-; (f) BrO^-.

1.53 Write the formulas of (a) chlorate ion; (b) nitrate ion; (c) carbonate ion; (d) hypochlorite ion; (e) hydrogen sulfate ion.

1.54 Write the formulas of (a) phosphite ion; (b) bromite ion; (c) dihydrogen phosphate ion; (d) periodate ion; (e) hydrogen sulfide ion.

1.55 Write the old and modern names of (a) Pb^{2+}; (b) Fe^{2+}; (c) Co^{3+}; (d) Cu^+.

1.56 Write the names of the following ions, giving both the old and modern names where appropriate: (a) HSO_4^-; (b) Hg_2^{2+}; (c) CN^-; (d) HCO_3^-.

1.57 Write the formulas of (a) copper(II) ion; (b) chlorite ion; (c) phosphide ion; (d) hydride ion.

1.58 Write the formulas of (a) ferric ion; (b) manganese(III) ion; (c) hydrogen ion; (d) sulfate ion.

1.59 Write the formulas of (a) magnesium oxide (magnesia); (b) calcium phosphate (the major inorganic component of bones); (c) aluminum sulfate; (d) calcium nitride.

1.60 Write the formulas of (a) potassium hydroxide; (b) barium sulfate; (c) silver(I) chloride; (d) cupric chloride.

1.61 Name the following ionic compounds. Write both the old and modern names wherever appropriate. (a) K_3PO_4; (b) FeI_2; (c) Nb_2O_5; (d) $CuSO_4$.

1.62 The following ionic compounds are commonly found in laboratories. Write their modern names. (a) $NaHCO_3$ (baking soda); (b) Hg_2Cl_2 (calomel); (c) NaOH (lye); (d) ZnO (calamine).

1.63 Write the names of (a) $Cu(NO_3)_2{\cdot}6H_2O$; (b) $NdCl_3{\cdot}6H_2O$; (c) $NiF_2{\cdot}4H_2O$.

1.64 Write the names of (a) $Na_2CO_3{\cdot}10H_2O$; (b) $Ce(IO_3)_3{\cdot}2H_2O$; (c) $Co(CN)_2{\cdot}3H_2O$.

1.65 Write the formulas of (a) sodium carbonate monohydrate; (b) indium(III) nitrate pentahydrate; (c) copper(II) perchlorate hexahydrate.

1.66 Write the formulas of (a) lithium nitrite monohydrate; (b) vanadium(III) iodide hexahydrate; (c) chromium(II) sulfate heptahydrate.

1.67 Write the formulas of (a) selenium trioxide; (b) carbon tetrachloride; (c) carbon disulfide; (d) sulfur hexafluoride; (e) diarsenic trisulfide; (f) phosphorus pentachloride; (g) dinitrogen oxide; (h) chlorine trifluoride.

1.68 Write the formulas of (a) dinitrogen tetroxide; (b) hydrogen sulfide; (c) dichlorine heptoxide; (d) disulfur dichloride; (e) nitrogen triiodide; (f) sulfur dioxide; (g) hydrogen fluoride; (h) diiodine hexachloride.

1.69 Name the molecular compounds (a) SF_4; (b) N_2O_5; (c) NI_3; (d) XeF_4; (e) $AsBr_3$; (f) ClO_2; (g) P_2O_5.

1.70 The following molecular compounds are often found in chemical laboratories. Name each compound. (a) SiO_2 (silica); (b) SiC (carborundum); (c) N_2O (a general anesthetic); (d) P_4O_{10} (a drying agent for organic solvents); (e) CS_2 (a solvent); (f) SO_2 (a bleaching agent); (g) NH_3 (a common reagent).

1.71 The following aqueous solutions are common laboratory acids. What are their names? (a) HCl(aq); (b) H_2SO_4(aq); (c) HNO_3(aq); (d) CH_3COOH(aq); (e) H_2SO_3(aq); (f) H_3PO_4(aq).

1.72 The following acids are also used in chemical laboratories, although they are less common than those in the preceding exercise. Write the formulas of (a) perchloric acid; (b) hypochlorous acid; (c) hypoiodous acid; (d) hydrofluoric acid; (e) phosphorous acid; (f) periodic acid.

1.73 Write formulas for the ionic compounds formed from (a) sodium and oxide ions; (b) potassium and sulfate ions;

(c) silver and fluoride ions; (d) zinc and nitrate ions; (e) aluminum and sulfide ions.

1.74 Write formulas for the ionic compounds formed from (a) calcium and chloride ions; (b) iron(III) and sulfate ions; (c) ammonium and iodide ions; (d) lithium and sulfide ions; (e) calcium and phosphide ions.

SUPPLEMENTARY EXERCISES

1.75 Distinguish between *atomic number* and *mass number* by referring to the isotopes of an element.

1.76 List three properties of metals. Contrast these properties with the corresponding properties of nonmetals.

1.77 Classify the following as an element, compound, homogeneous mixture, or heterogeneous mixture: (a) iron; (b) sugar water; (c) table salt; (d) helium; (e) filtered seawater; (f) carbon dioxide; (g) gold; (h) dirt.

1.78 Classify the following as chemical or physical changes: (a) decay of wood; (b) burning of wood; (c) sawing of wood; (d) turning of the color of leaves in the fall.

1.79 The pictures below show either a physical or chemical change. In each case, identify the type of change.

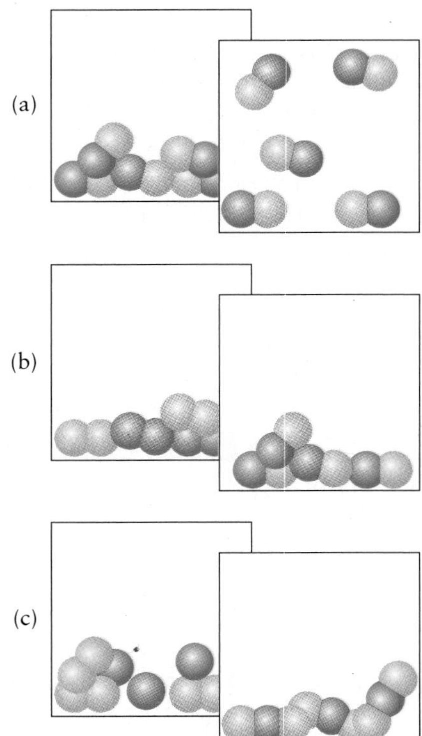

1.80 Sketch the periodic table and mark on it the locations of (a) the alkali metals; (b) the alkaline earth metals; (c) the halogens; (d) the noble gases; (e) the transition metals; (f) the metalloids; (g) the lanthanides; (h) the actinides.

1.81 Which group in the periodic table contains the alkaline earth metals? Write the symbols and atomic numbers of each member of the group.

1.82 Name the compounds (a) Ag_2S; (b) $ZnCl_2$; (c) ClF_5; (d) $Mg(OH)_2$; (e) $NiSO_4 \cdot 6H_2O$; (f) PCl_5; (g) $Cr(H_2PO_4)_3$; (h) As_2O_3; (i) $MnCl_2$.

1.83 Name the compounds (a) $FeCl_3 \cdot 6H_2O$; (b) $Co(NO_3)_2 \cdot 6H_2O$; (c) CuCl; (d) BrCl; (e) MnO_2; (f) $Hg(NO_3)_2$; (g) $Ni(NO_3)_2$; (h) N_2O_4; (i) V_2O_5.

1.84 Write formulas for (a) ammonium phosphite; (b) ferric oxide; (c) copper(II) bromate; (d) phosphine; (e) calcium bicarbonate; (f) hydrogen cyanide; (g) lithium bisulfate; (h) selenium tetrafluoride; (i) iron(II) sulfate heptahydrate.

1.85 Write formulas for (a) aluminum phosphate; (b) barium nitrate dihydrate; (c) silicon disulfide; (d) sodium phosphide; (e) perchloric acid; (f) copper(II) oxide; (g) hydroiodic acid; (h) silver(I) sulfate.

1.86 Water purification, in which impure water is made drinkable, can be achieved by several methods. On the basis of the information in this chapter, suggest a physical process for converting seawater to pure water.

CHALLENGING EXERCISES

1.87 (a) Determine the total number of protons, neutrons, and electrons in one water molecule, H_2O. (b) What are the total masses of protons, neutrons, and electrons in one water molecule? (c) What fraction of your own mass is due to the neutrons in your body, assuming you consist primarily of water?

1.88 Determine the fraction of the total mass of an ^{56}Fe atom that is due to (a) neutrons; (b) protons; (c) electrons. (d) What is the mass of protons in a 1000-kg automobile? Assume the total mass of the vehicle is due to ^{56}Fe.

1.89 Name the following compounds, using analogous compounds with phosphorus and sulfur as a guide: (a) H_2TeO_4; (b) Na_3AsO_4; (c) $CaSeO_3$; (d) $Ba_3(SbO_4)_2$; (e) H_3AsO_4; (f) $Co_2(TeO_4)_3$.

1.90 The British chemist Michael Faraday gave Christmas lectures to children in the mid-nineteenth century. One of his lectures involved investigating what he called the "chemical history of a candle." He would light a candle, then examine the colors and the smoke given off. He weighed the candle before and after burning to see if it lost mass, and it did. He would hold a mirror over it to catch condensation of water and soot. Then he would place a glass tube into the flame and show that thick, white fumes came out. He suggested that the fumes might be candle wax that had been vaporized by the heat. To check for the validity of that suggestion, he lit the fumes and they did indeed burn. Categorize each of Faraday's suggestions and actions as observation, measurement, hypothesis, experiment, or theory.

1.91 Suppose you are a student living in a shared room and that you have been experiencing headaches every evening after dinner when you sit down at your desk to begin your homework. Consider how you might determine the cause of your headaches by using the principles of the scientific method. Discuss what data you could collect, and suggest at least three hypotheses and experiments to test them.

CHAPTER 2

This delicate animal is a tunicate (so called because it has a tunic of cellulose) found in the ocean off Papua, New Guinea. Organisms of all kinds produce a multitude of chemicals, some of which have great potential—as pharmaceuticals—to overcome diseases such as cancer and AIDS. However, to investigate them and, if they are effective, to make them in usable quantities, chemists need to determine their composition. This kind of investigation, and the communication of the results to other scientists, requires a thorough knowledge of the measurements and units we meet in the chapter.

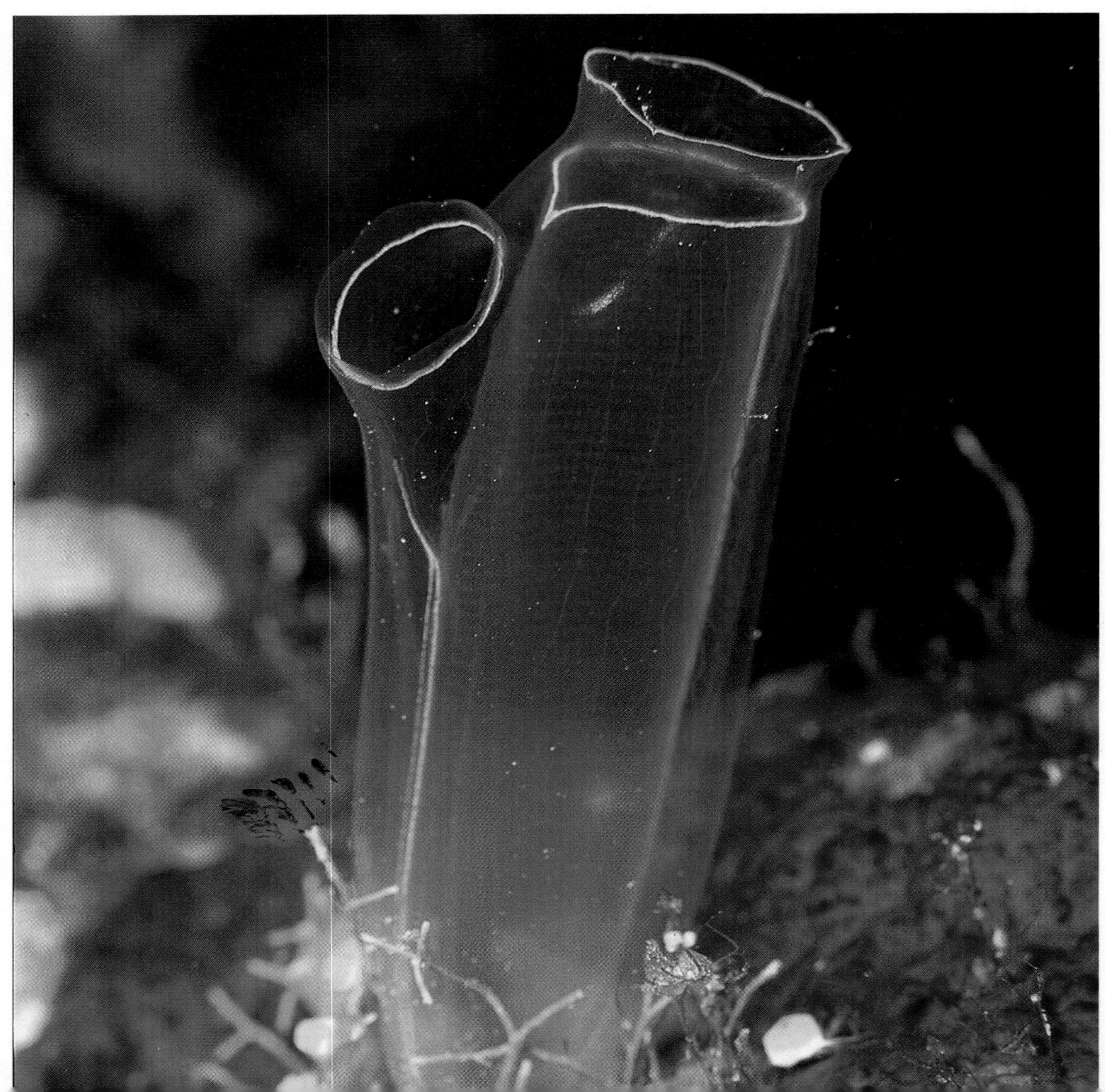

MEASUREMENTS AND MOLES

Some of the luckiest chemists in the world have jobs that require them to dive to the floor of tropical seas. There they collect marine **natural products,** materials obtained from the environment. The oceans are a rich source of natural products, and some have important medical applications. Recognizing the compounds and discovering that they have certain properties constitutes the **qualitative** part of their investigation, the part that concerns the description of properties like color, smell, toxicity, and antiviral activity. For example, extracts of marine animals such as tunicates (sea squirts) are highly active against tumors and viruses.

Only a small amount of the active compound is present in marine animals and it is difficult to collect. **Synthesizing** a powerful antitumor and antiviral agent—making it from simpler components—would make it more widely available. However, before they can start thinking about synthesizing such an agent, chemists need to know its composition, and in particular its chemical formula. The diving chemists are not farming the ocean floor: they are collecting tunicates as a first step in the analysis of active components. The determination of chemical composition is the **quantitative** part of the investigation, the part that involves measuring and using numerical data.

We shall illustrate the use of numerical data by considering one particular natural product, vitamin C. Before vitamin C could be synthesized, sailors on long voyages had to keep a large supply of citrus fruit on hand to avoid diseases such as scurvy. In 1928 Albert Szent-Györgi, a Hungarian-American chemist, isolated the compound from plants and the adrenal glands of animals and found a way to produce it in bulk.

MEASUREMENTS AND UNITS

2.1 The metric system
2.2 Prefixes for units
2.3 Derived units
2.4 Unit conversions
2.5 Temperature
2.6 The uncertainty of measurements
2.7 Accuracy and precision

CHEMICAL AMOUNTS

2.8 The mole
2.9 Molar mass
2.10 Measuring out compounds

DETERMINATION OF CHEMICAL FORMULAS

2.11 Mass percentage composition
2.12 Determining empirical formulas
2.13 Determining molecular formulas

SOLUTIONS IN CHEMISTRY

2.14 Molarity
2.15 Dilution

MEASUREMENTS AND UNITS

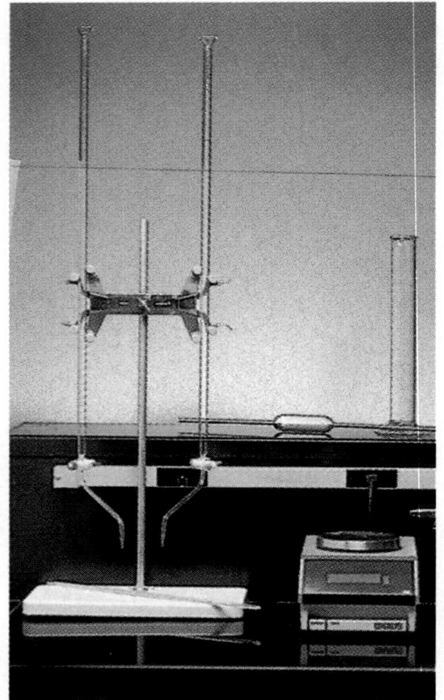

FIGURE 2.1 Some of the laboratory equipment chemists use to make measurements. Clockwise from the upper left: two burets and a pipet for transferring specific volumes of liquids; a graduated cylinder for measuring volume; a balance for determining mass; and a thermometer for measuring temperature.

We are already familiar with the way that measurements are reported in everyday life. For instance, we report the distance along a highway in miles or kilometers, the volume of gasoline in gallons or liters, and the passage of time in seconds, minutes, and hours. Scientists have made these procedures more systematic and more precise through the use of international standards (Fig. 2.1).

2.1 THE METRIC SYSTEM

The metric system was devised in the 1790s when, with revolutionary fervor, the French swept away their old units as well as their politicians and sought to replace both by more rational systems. The metric system has been developed over the past 200 years and is the basis of the **Système International d'Unités** (SI), which is used for almost all modern scientific work.

When we examine an everyday report of a measurement (such as 5 miles or 10 gallons), we see that it has the form

$$\text{Result of a measurement} = \text{number} \times \text{unit}$$

The same is true in the metric system. For example, the mass of a package might be reported as 5.0 kg—or 5.0 times the unit of mass, which is 1 kilogram (1 kg). If we were to measure the length of a pole, then we might report it as 10 m, which means 10 times the unit of length, 1 meter (1 m). In the metric system, the fundamental units for mass, length, and time are the kilogram, the meter, and the second.

The unit of mass (the quantity of matter) is 1 kilogram (1 kg).

The unit of length (the distance between two points) is 1 meter (1 m).

The unit of time (the duration of an event) is 1 second (1 s).

The name *second* came about because the hour is first divided into *minute* parts called "minutes" which are then divided a *second* time into "seconds."

These three units (1 kg, 1 m, and 1 s) are three of the SI **base units,** or fundamental units. All other units are expressed in terms of these and several other base units. The meter and the second are now defined in terms of the properties of light; the kilogram is defined as the mass of an actual physical object, a platinum-iridium cylinder kept in France (see Case Study 2). Note that in science we distinguish between *mass* and *weight.* The mass of an object is a measure of the *quantity of matter* it contains. The weight of an object is a measure of the *gravitational pull* it experiences. Mass and weight are proportional to each other, but they are not identical. For example, an astronaut would have the same mass (contain the same quantity of matter) on Earth and on Mars. However, the astronaut's weight would be less on Mars than on Earth because Mars has a lower gravity. An astronaut might be weightless, but never massless.

A measurement is reported as the numerical multiple of a standard unit.

2.2 PREFIXES FOR UNITS

The Système International also uses combinations of base units and various prefixes. These prefixes denote *multiples* of powers of 10 of the SI units themselves. The centimeter (cm), for instance, is 1/100 of a meter. In common with all the prefixes, the prefix *c* can be used with any unit and always means 1/100, or a multiple of 10^{-2}, of the unit:

$$1\ \text{cm} = 10^{-2} \times (1\ \text{m}), \quad \text{or, more simply, } 1\ \text{cm} = 10^{-2}\ \text{m}$$

In this book we shall use the nine prefixes given in Table 2.1. The prefixes can be used with any of the units to give expressions like 1 pm = 10^{-12} m for picometer (pm) and 1 ms = 10^{-3} s for millisecond (ms). The kilogram is a little out of line: it is a base unit but it already has a prefix (k). When we use prefixes with the unit of mass, we apply them to the gram instead, and note that 1 kg = 10^3 g.

When doing calculations, it is best to replace prefixes by the powers of 10 they represent, and then proceed numerically. For example, if we replace 1 μm by 10^{-6} m in calculations, we can then cancel units and simplify calculations. A second point to remember is that an expression like cm^3 (centimeter cubed) means $(\text{cm})^3$:

$$\begin{aligned} 1\ \text{cm}^3 &= (1\ \text{cm}) \times (1\ \text{cm}) \times (1\ \text{cm}) \\ &= (10^{-2}\ \text{m}) \times (10^{-2}\ \text{m}) \times (10^{-2}\ \text{m}) \\ &= 10^{-6}\ \text{m}^3 \end{aligned}$$

Multiples of units that are powers of 10 are represented by prefixes attached to the symbol for the unit.

TABLE 2.1 Common SI prefixes

Prefix	Name	Meaning
G	giga	10^9
M	mega	10^6
k	kilo	10^3
d	deci	10^{-1}
c	centi	10^{-2}
m	milli	10^{-3}
μ	micro	10^{-6}
n	nano	10^{-9}
p	pico	10^{-12}

2.3 DERIVED UNITS

We multiply and divide the symbols for units just as though they were ordinary numbers. For example, the result of dividing 6 g by 3 g is a pure number with no units, because the units (grams) cancel:

$$\frac{6\not{g}}{3\not{g}} = \frac{6}{3} = 2$$

An example of the multiplication of units is the calculation we do when we want to find the volume of a box of sides 1.0 m, 2.0 m, and 3.0 m:

$$\text{Volume} = (1.0\ \text{m}) \times (2.0\ \text{m}) \times (3.0\ \text{m}) = 6.0\ \text{m}^3$$

The result is shorthand for $6.0 \times (1\ \text{m}^3)$, where 1 m^3 is the SI unit of volume. Units built up in this way are called **derived units.** It is common in chemistry to report volumes of liquids and gases in liters (L), where 1 L is exactly 10^3 cm^3. Note that 1 mL (1 milliliter) is exactly the same as 1 cm^3.

In some instances, a derived unit includes a base unit raised to a negative power. An example is the unit for **density,** the mass of a sample divided by its volume:

$$\text{Density} = \frac{\text{mass of sample}}{\text{volume of sample}}$$

The SI unit of density is formed by dividing the SI unit of mass by the SI unit of volume:

CASE STUDY 2 UNITS AND TRAPPED ATOMS

THE STANDARD FOR the unit of mass, 1 kg, is currently a platinum-iridium cylinder kept at the *Bureau International des Poids et Mesures* in Sèvres, France. Even though the alloy of which it is made does not react easily, a very thin film of oxide can form on its surface and increase the mass. Therefore, the standard cylinder must be carefully polished each month. However, if it is polished too much, atoms will be rubbed off the surface, so decreasing its mass. Clearly, a more stable standard would be preferred, one that will not change with time.

All the other SI base units are defined in terms of properties that do not change and are available everywhere. For example, the meter is defined as the distance light travels in exactly 1/299 792 458 s. This fraction was chosen so that the distance would be very close to the meter already in use. The second is defined as the time required for $9.192\,631\,770 \times 10^9$ oscillations of the wave corresponding to a certain color of light emitted by cesium atoms. Neither the speed of light nor the properties of the light emitted by atoms is subject to change over time. Consequently, the standards will survive for millennia. They can also be made available worldwide; we could even inform extraterrestrials about them should the need ever arise.

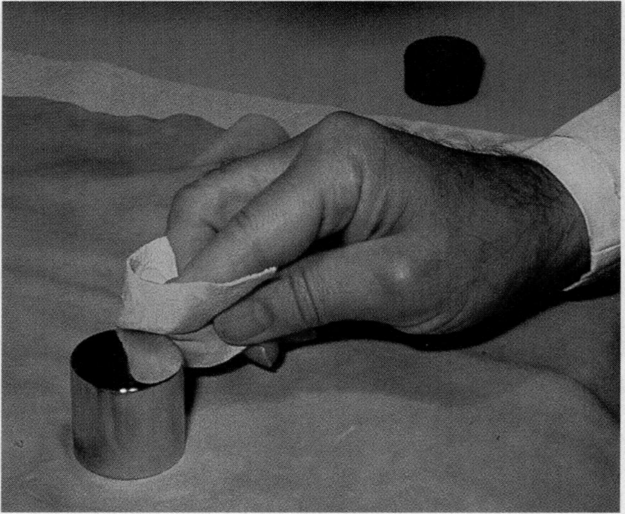

The 1-kg platinum-iridium cylinder mass standard being polished by a skilled caretaker.

If we could measure the mass of a single atom, then we would have a reliable and universal mass standard, because the mass of each atom of a particular isotope is identical and does not change over time. The relative masses of atoms—not the actual masses—were first determined for several elements by John Dalton, who measured the mass of each element that would combine with 1 g of hydrogen. Advances in scientific instrumentation made it possible to measure actual atomic masses, for instance, by

$$\text{Unit of density} = \frac{1\ \text{kg}}{1\ \text{m}^3} = 1\ \text{kg}\cdot\text{m}^{-3}$$

We often see $\text{kg}\cdot\text{m}^{-3}$ written as kg/m^3; similar constructions are used for negative powers of other units, such as m/s (meters per second) instead of $\text{m}\cdot\text{s}^{-1}$.

Note that $\text{m}^{-3} = 1/\text{m}^3$, just as for numbers. The unit is read "one kilogram per meter cubed." For example, if a sample of packing foam has a mass of 12 kg and its volume is 4.0 m^3, then its density would be reported as

$$\text{Density} = \frac{\text{mass}}{\text{volume}} = \frac{12\ \text{kg}}{4.0\ \text{m}^3}$$
$$= \frac{12}{4.0}\frac{\text{kg}}{\text{m}^3} = 3.0\ \text{kg}\cdot\text{m}^{-3}$$

The density of a substance is independent of the size of the sample: doubling the volume also doubles the mass, so the ratio of mass to volume remains the same. Density is therefore an example of an **intensive property,** a property that is independent of the size of the sample (a representative part of the whole). Properties that do depend on the size of the sample, like mass and volume, are called **extensive.**

The units kilogram/meter^3 are often not very convenient, but we can use gram/centimeter^3 instead, with $1\ \text{g}\cdot\text{cm}^{-3} = 10^3\ \text{kg}\cdot\text{m}^{-3}$. The density of water is very close to $1\ \text{g}\cdot\text{cm}^{-3}$. This value implies that the mass of

using a mass spectrometer (see Box 1.2), and to do so with high accuracy and precision.

In 1994 it became possible to measure atomic mass with a precision of 11 significant figures. The instrument that made this possible is the *Penning trap,* a device that can hold a single monatomic ion in an arrangement of electric and magnetic fields. The fields cause the ion to oscillate around a central position at a rate that is proportional to its mass. Thus, the masses of individual atoms of different elements can be determined by measuring their rate of oscillation. In the Penning trap an atom of ^{1}H has been found to have a mass of 1.007 825 031 6 u, an atom of ^{2}H a mass of 2.014 101 777 9 u, and an atom of ^{28}Si a mass of 27.976 926 532 4 u. Every atom of a particular isotope has the same mass. The mass of an atom of a particular isotope, such as ^{28}Si, which can be manufactured in ultrapure crystals, could some day be our new standard of mass.

Because it is convenient to use simple numerical values in calculations, masses of single atoms are often reported in terms of the *atomic mass unit* (currently denoted u, formerly amu). One atomic mass unit, 1 u, is defined as exactly $\frac{1}{12}$ the mass of an atom of carbon-12. Because we know experimentally that the mass of an atom of carbon-12 is 1.9926×10^{-23} g, it follows that 1 u = 1.6605×10^{-24} g. The mass of any other type of atom in grams can be expressed in atomic mass units (and vice versa) by using this relation as a conversion factor.

The mass of an atom of carbon-12 is exactly 12 u. Because the mass of an atom depends on the number of nucleons it contains, the mass of any atom in atomic mass units is *approximately* equal to its mass number. Thus, an atom of ^{24}Mg has a mass of about 24 u and an atom of ^{2}H a mass of about 2 u.

QUESTIONS

Molar mass and the Avogadro constant are discussed in Section 2.9.

1. The natural abundance of ^{1}H is 99.985% and that of ^{2}H is 0.015%. (a) Assuming that no ^{3}H is present, and that the percentages are exact (actually, they are not), use the values given above to calculate the average molar mass of hydrogen atoms to seven significant figures. (b) Assuming that the percentages are *not* exact, how should you report the average molar mass of hydrogen?

2. Use the atomic mass of ^{2}H determined in the Penning trap to calculate the mass of 1.0000 mol ^{2}H in atomic mass units. Convert this mass to grams of ^{2}H.

3. Suggest a reason why it would be important for the mass standard to consist of an isotope (such as silicon-28) that can be made in ultrapure crystals.

4. What is the relation between the atomic unit of mass and the Avogadro constant?

1 cm^3 of water is very close to 1 g. Equivalently, it implies that the mass of 1 L of water at room temperature is very close to 1 kg.

Units are multiplied and divided like numbers; derived units are built up from base units.

2.4 UNIT CONVERSIONS

The units in common use in the United States are called English units. Until the metric system is adopted everywhere, it will be helpful to know how the magnitudes of the fundamental units compare with those of familiar units. Thus, it is helpful to know that 1 m is approximately 3 inches longer than 1 yard, that 1 cm is slightly shorter than half an inch, and that 1 kg is approximately 2.2 lb.

In Table 2.2 we see that

$$1 \text{ inch} = 2.54 \text{ cm}$$

Because 1 inch (1 in.) and 2.54 cm are different ways of expressing the same length, it follows that any object that is $x \times$ (1 inch) long is also $x \times$ (2.54 cm) long. Therefore, to convert inches to centimeters, we simply

TABLE 2.2 Relations between units*

Common unit	SI unit
1 pound (lb)	453.6 grams (g)
1 inch (in.)	**2.54** centimeters (cm)
1 foot (ft)	**30.48** cm
1 quart (qt)[†]	0.946 liter (L)
1 minute (min)	**60** s

*A longer list is given inside the back cover of the text. The numbers in bold type are exact equivalents.

[†]The European and Canadian quart is 1.201 times larger than the American quart used here.

multiply by 2.54. However, to make sure the units are given correctly, we have to cancel the inches and replace them by centimeters. This replacement is taken care of automatically if instead of simply multiplying by 2.54, we multiply by the following factor:

$$\frac{2.54\ \text{cm}}{1\ \text{in.}}$$

Then the in. in the denominator cancels any inches this factor multiplies and the cm in the numerator multiplies the resulting numerical factor. For instance, if the length of a piece of glass tubing is 2.00 inches then its length in centimeters is

$$(2.00\ \cancel{\text{in.}}) \times \frac{2.54\ \text{cm}}{1\ \cancel{\text{in.}}} = 2.00 \times 2.54\ \text{cm} = 5.08\ \text{cm}$$

The general form of this relation is

$$\text{Information required} = \text{information given} \times \text{conversion factor}$$

and the **conversion factor** we use has the form

$$\text{Conversion factor} = \frac{\text{units required}}{\text{units given}}$$

One way to remember the form of this factor is to note that the term in the denominator will cancel the units in the data, and the units in the numerator will replace them with the units required (Fig. 2.2).

SELF-TEST 2.1A A supplier packages hydrochloric acid in liters, but we need 1.85 quarts (qt). What volume in liters corresponds to 1.85 qt of the acid?

[*Answer:* 1.75 L]

SELF-TEST 2.1B Express the mass in ounces (oz) of a 250-g package of breakfast cereal.

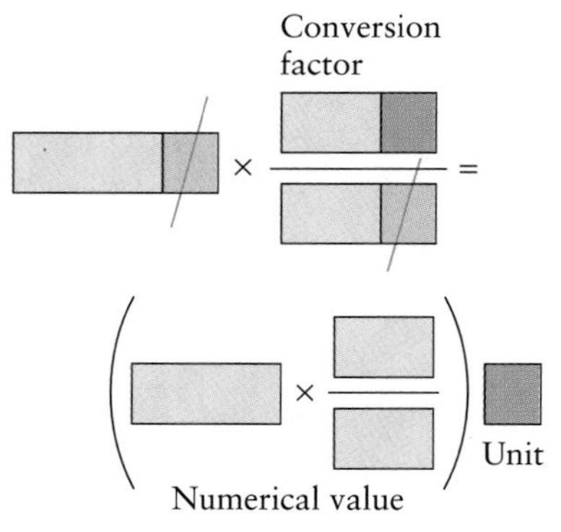

FIGURE 2.2 The effect of a conversion factor is to cancel the old units and replace them with the new units and to make the appropriate numerical conversion.

It is often necessary to convert a unit that is raised to a power. For example, we might wish to convert a surface area expressed as 256 cm^2 into square meters (m^2). The same procedure is employed each time the unit appears, so in this case the conversion factor appears as its square. The explicit calculation follows. First, we note that 1 m = 10^2 cm and express this relation as the conversion factor

$$\frac{\text{Units required}}{\text{Units given}} = \frac{1\ \text{m}}{10^2\ \text{cm}}$$

Then we use the conversion factor squared:

$$\text{Area (m}^2) = (256\ \text{cm}^2) \times \left(\frac{1\ \text{m}}{10^2\ \text{cm}}\right)^2 = \frac{(256\ \cancel{\text{cm}^2}) \times \text{m}^2}{10^4\ \cancel{\text{cm}^2}} = 2.56 \times 10^{-2}\ \text{m}^2$$

Note how the cm^2 of the information given cancels the cm^2 that comes from squaring the conversion factor. The original units always cancel like this when the conversion factor has been set up correctly, so you know immediately if you are carrying out the conversion properly.

SELF-TEST 2.2A A meter monitoring the supply of oxygen gas to an incubator recorded the volume of gas delivered as 25.4 $in.^3$. Express that volume in centimeters cubed.

[*Answer:* 416 cm^3]

SELF-TEST 2.2B The surface area of a sample of copper was reported as 1.22 $in.^2$. What is this area in centimeters squared?

In some cases we have to convert units that appear with a negative power, such as time in a measurement of speed (meters per second, $m \cdot s^{-1}$) or volume in a density measurement (grams per centimeter cubed, $g \cdot cm^{-3}$). The negative power simply means that a unit appears in the denominator of an expression. The conversion is carried out exactly as we have described already, but with the conversion factor raised to the same negative power as the unit. For example, suppose we wanted to convert a speed of 1.5 $km \cdot s^{-1}$ (1.5 kilometers per second) into kilometers per hour ($km \cdot h^{-1}$). We would use the relation 1 h = 3600 s to write the conversion factor

$$\frac{\text{Units required}}{\text{Units given}} = \frac{1\ \text{h}}{3600\ \text{s}}$$

and then raise this factor to the power -1 (the same power to which the seconds are raised in the information given):

$$\text{Speed}\ (\text{km}\cdot\text{h}^{-1}) = (1.5\ \text{km}\cdot\text{s}^{-1}) \times \left(\frac{1\ \text{h}}{3600\ \text{s}}\right)^{-1}$$

It is usually easier to see what is going on if we get rid of the negative powers of the units and write this expression as

$$\begin{aligned}\text{Speed}\ (\text{km}\cdot\text{h}^{-1}) &= \frac{1.5\ \text{km}}{\text{s}} \times \frac{3600\ \text{s}}{1\ \text{h}} \\ &= 5.4 \times 10^3\ \text{km}\cdot\text{h}^{-1}\end{aligned}$$

A more complicated illustration is the conversion of the density of mercury, 13.6 $g \cdot cm^{-3}$, into kilograms per meter cubed ($kg \cdot m^{-3}$). Two conversion factors are involved. We need to convert from grams (units given) to kilograms (units required) by using 1 kg = 10^3 g, so we use the form of the factor that has kilograms in the numerator and grams in the denominator:

$$\frac{\text{Units required}}{\text{Units given}} = \frac{1\ \text{kg}}{10^3\ \text{g}}$$

We also need to convert from centimeters (units given) to meters (units required) by using 10^2 cm = 1 m:

$$\frac{\text{Units required}}{\text{Units given}} = \frac{1\ \text{m}}{10^2\ \text{cm}}$$

Each conversion factor is raised to the same power as the unit it is converting, so the second conversion factor must be raised to the power -3:

$$\begin{aligned}\text{Density}\ (\text{kg}\cdot\text{m}^{-3}) &= (13.6\ \text{g}\cdot\text{cm}^{-3}) \times \left(\frac{1\ \text{kg}}{10^3\ \text{g}}\right) \times \left(\frac{1\ \text{m}}{10^2\ \text{cm}}\right)^{-3} \\ &= \frac{13.6\ \cancel{\text{g}}}{\cancel{\text{cm}^3}} \times \frac{1\ \text{kg}}{10^3\ \cancel{\text{g}}} \times \frac{10^6\ \cancel{\text{cm}^3}}{1\ \text{m}^3} \\ &= 1.36 \times 10^4\ \text{kg}\cdot\text{m}^{-3}\end{aligned}$$

We could do each unit conversion separately in a series of steps, but it is normally more efficient to carry them out all at the same time. To reduce error, you should check that all the original units cancel, leaving the units required.

> ***To convert between units, use a conversion factor of the form "units required/units given."***

SELF-TEST 2.3A Express a density reported as 6.5 $g{\cdot}mm^{-3}$ in micrograms per nanometer cubed ($\mu g{\cdot}nm^{-3}$).

[*Answer:* $6.5 \times 10^{-12}\ \mu g{\cdot}nm^{-3}$]

SELF-TEST 2.3B Express a density of $1.100 \times 10^3\ kg{\cdot}m^{-3}$ in grams per centimeter cubed.

2.5 TEMPERATURE

Daniel Fahrenheit, the German scientist who invented the scale, set body temperature at 100°F (he must have had a slight fever that day), and the lowest temperature he could reach with a mixture of salt and water as 0°F.

There are three common units for reporting temperature. The United States is the only country that still uses the **Fahrenheit scale,** in which water freezes at 32°F and boils at 212°F. This everyday but awkward scale is very rarely used in science. Much more common is the **Celsius scale** devised by Anders Celsius, an eighteenth-century Swedish astronomer. On the Celsius scale, water freezes at 0°C (zero degrees Celsius) and boils at 100°C (Fig. 2.3). The length of a mercury column (in a thermometer) between the freezing and boiling points of water is the same regardless of the scale written beside the column; but on the Celsius scale, this length is divided into 100 degrees, and on the Fahrenheit scale it is divided into 180 degrees and begins at 32. It follows that a temperature on the Celsius scale is converted to a temperature on the Fahrenheit scale, and vice versa, by using the relation

$$\frac{\text{Fahrenheit temperature}}{°\text{F}} = 1.8 \times \frac{\text{Celsius temperature}}{°\text{C}} + 32 \qquad (1)$$

Note that we are treating the temperature units like numbers and will cancel them when it is appropriate to do so (see below).

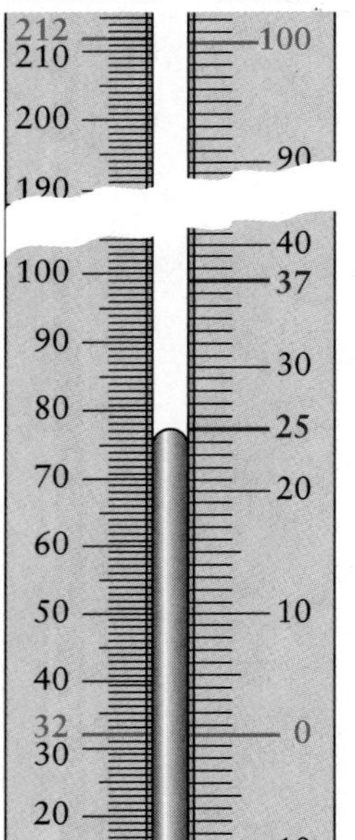

FIGURE 2.3 The Fahrenheit and Celsius temperature scales. The boiling and freezing points of water are marked in red. Two other common temperatures are marked in blue: many data are reported at 25°C, and body temperature is 37°C.

EXAMPLE 2.1 Converting between Celsius and Fahrenheit temperatures

Express body temperature, about 99°F, on the Celsius scale.

STRATEGY First, estimate the required temperature by referring to Fig. 2.3. For the precise value, we rearrange Eq. 1 to give the temperature in degrees Celsius in terms of the temperature in degrees Fahrenheit:

$$\frac{\text{Celsius temperature}}{°\text{C}} = \frac{(\text{Fahrenheit temperature}/°\text{F}) - 32}{1.8}$$

and insert the Fahrenheit temperature.

SOLUTION In Fig. 2.3 we see that a reading of 99 on the Fahrenheit scale should be about the same temperature as 37 on the Celsius scale, so body temperature is about 37°C. From the equation,

$$\frac{\text{Celsius temperature}}{^\circ\text{C}} = \frac{99 - 32}{1.8} = 37$$

When we multiply both sides of this expression by °C, we obtain

$$\text{Celsius temperature} = 37^\circ\text{C}$$

That is, 99°F corresponds to 37°C, which is consistent with our estimate.

SELF-TEST 2.4A Convert (a) −23°C to the Fahrenheit scale and (b) −10°F to the Celsius scale.

[*Answer:* (a) −9°F; (b) −23°C]

SELF-TEST 2.4B At what temperature do the two scales give the same numerical value for the temperature?

The Celsius scale (like the Fahrenheit scale) has an arbitrary zero point. However, in 1848 the British scientist Lord Kelvin showed that there is an **absolute zero** of temperature, a temperature below which it is impossible to cool anything. It is obviously more natural to set the zero of the temperature scale to this lowest possible temperature. The absolute zero of temperature lies at −273.15°C. Scientists have cooled objects *almost* to that temperature, and in 1995 a group of scientists managed to cool a sample to within 0.000 000 002°C of absolute zero, possibly making the sample colder than anywhere else in the universe.

The **Kelvin scale** has the value 0 at absolute zero, and the size of the degrees, which are called *kelvins* and abbreviated as K, are the same size as degrees Celsius. Thus, on the Kelvin scale, water freezes at 273.15 K (that corresponds to 273.15°C above absolute zero) and boils 100 K higher, at 373.15 K (Fig. 2.4). Notice that neither the word degree nor the degree symbol are used for the kelvin. To express a Celsius temperature on the Kelvin scale, we simply add 273.15:

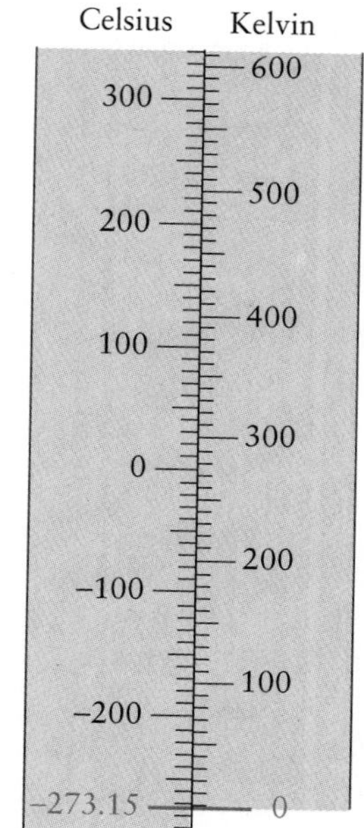

FIGURE 2.4 The Celsius and Kelvin temperature scales. Note that the Kelvin scale does not extend below 0, corresponding to −273.15°C. The Celsius scale is now defined in terms of the Kelvin scale.

$$\frac{\text{Kelvin temperature}}{\text{K}} = \frac{\text{Celsius temperature}}{^\circ\text{C}} + 273.15$$

For example, the temperature on a warm day, 25.00°C, becomes

$$\frac{\text{Kelvin temperature}}{\text{K}} = 25.00 + 273.15 = 298.15$$

That is, the temperature corresponds to 298.15 K. Body temperature, about 37°C, corresponds to about 310 K. The Kelvin scale is awkward for everyday use, but because it is more natural than the Fahrenheit and Celsius scales, its use greatly simplifies many calculations.

In science, temperatures are measured on the Celsius temperature scale; calculations involving temperature use the Kelvin scale, which starts at 0 for the absolute zero of temperature.

2.6 THE UNCERTAINTY OF MEASUREMENTS

One reason for the high credibility of scientists is the honesty with which they must report measurements. Every measurement is limited by the reliability of the measuring instrument and the skill of the operator, and the

Because science is a public, open pursuit, scientists who misrepresent their data are usually soon found out.

uncertainty must be reported correctly. We shall use the convention that the last digit—the least significant figure, the digit on the right—in the data is imprecise to the extent of ±0.5 of that figure. A measurement reported as 1.2 cm^3 actually means that the volume lies between 1.15 cm^3 and 1.25 cm^3, and a mass reported as 1.78 g means that it lies between 1.775 g and 1.785 g (Fig. 2.5).

The digits in a reported measurement are called the **significant figures.** There are two significant figures (written 2 sf) in 1.2 cm^3 and 3 sf in 1.78 g. To find the number of significant figures in a measurement, express it in scientific notation (see Appendix 1B) with one nonzero digit in front of the decimal point, and then count the total number of digits. Thus, 0.0025 kg is written as 2.5×10^{-3} kg, a value with 2 sf.

The significance of zeros in a measurement, such as that in 22.0 mL, can be a source of difficulty, because some zeros are legitimately measured digits, but some zeros serve only to mark the place of the decimal point (Table 2.3). Trailing zeros (the last ones after a decimal point), as in 22.0 mL, are significant, because they were measured. Thus, 22.0 mL has 3 sf; this number signifies that the volume of the sample lies between 21.95 mL and 22.05 mL. The "captive" zero in 80.1 kg is also a measured digit, so 80.1 kg has 3 sf. However, the leading digits in 0.0025 g are not significant; they are only placeholders used to indicate powers of 10, not measured numbers. We can see that they are only placeholders by reporting the mass as 2.5×10^{-3} g, which has 2 sf. What about a length reported as 400 m: does it have 3 sf (4.00×10^2), 2 sf (4.0×10^2), or only 1 sf (4×10^2)? In such cases, the use of scientific notation removes any ambiguity. Throughout this text, all trailing zeros are significant (so 400 g has three significant figures), unless indicated by the context.

Another convention you will occasionally encounter is a number such as 400. (that is, a decimal point but no following digits). This notation indicates that all the zeros are significant.

EXAMPLE 2.2 Counting the number of significant figures

Report the number of significant figures in (a) 50.00 g; (b) 0.00501 m; (c) 0.0100 mm.

STRATEGY In each case, write the information in scientific notation with one nonzero digit in front of the decimal point and all trailing zeros preserved, then count the total number of digits.

SOLUTION (a) A mass of 50.00 g is the same as 5.000×10 g, a quantity with four significant figures (4 sf). (b) A length of 0.00501 m is the same as 5.01×10^{-3} m, a quantity with 3 sf. (c) A length of 0.0100 mm is the same as 1.00×10^{-2} mm, a quantity with 3 sf.

SELF-TEST 2.5A Determine the number of significant figures in (a) 2.1010 kg; (b) 100.000°C; (c) 0.000 000 1 mm.

[*Answer:* (a) 5 sf; (b) 6 sf; (c) 1 sf]

SELF-TEST 2.5B Determine the number of significant figures in (a) 5.110 cm; (b) 0.00500 g; (c) 5.000505 m.

When determining the number of significant figures to report, we distinguish between the results of *measurements,* which are always uncer-

TABLE 2.3 Examples of significant figures

Decimal notation	Scientific notation	Number of sf
0.751	7.51×10^{-1}	3
0.00751	7.51×10^{-3}	3
0.07051	7.051×10^{-2}	4
0.750100	7.50100×10^{-1}	6
7.5010	7.5010	5
7501	7.501×10^{2}	4
7500		ambiguous*

*In this text, treat trailing zeros as significant unless instructed otherwise.

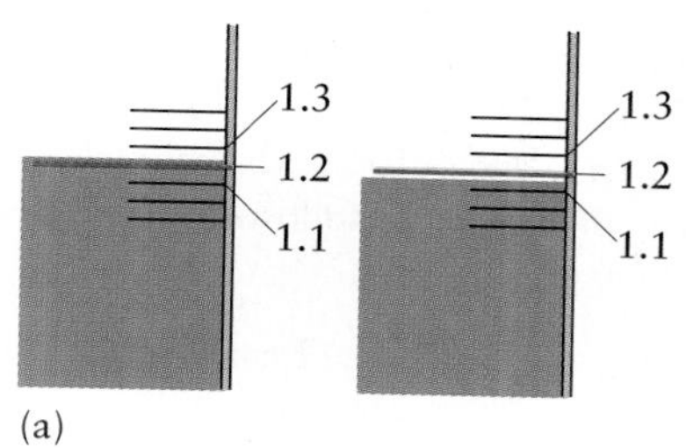

(a)

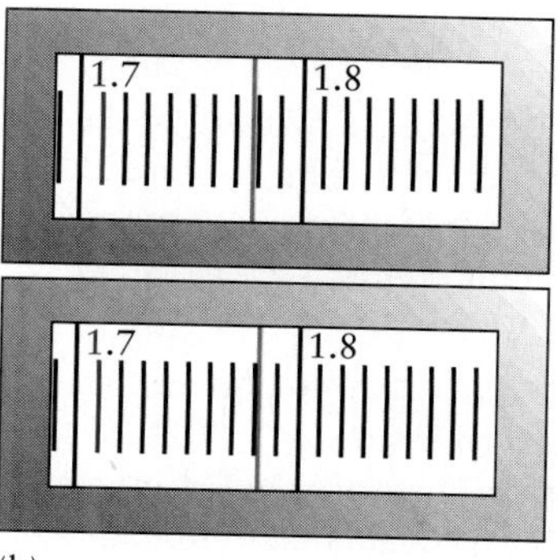

(b)

FIGURE 2.5 Two sets of measurements. (a) The volume of the liquid is reported as 1.23 ± 0.005 mL and 1.17 ± 0.005 mL. (b) The mass (the reading on this balance scale) is reported as 1.779 ± 0.005 g and 1.781 ± 0.005 g.

tain, and the results of *counting,* which are exact. For example, the report "12 eggs" means that there are *exactly* 12 eggs present, not a number somewhere between 11.5 and 12.5.

In science, we do a lot of calculations on data from measurements. So we need to make sure that the number of significant figures that we use to report the results of calculations is correct, given the number of significant figures in the data. For instance, if the mass of a sample of plastic is reported as 1.78 g and its volume as 1.2 cm^3, it is wrong to report its density as

$$\text{Density} = \frac{1.78\ \text{g}}{1.2\ \text{cm}^3} = 1.483333\ \text{g·cm}^{-3}$$

simply by using the display of our calculator. The correct procedure is described in Toolbox 2.1.

The uncertainty of the data determines the uncertainty of the results of calculations based on the data.

TOOLBOX 2.1 How to use significant figures in calculations

Different rules are needed for addition (and its reverse, subtraction) and multiplication (and its reverse, division). Both procedures require us to round off the answers to the correct number of significant figures.

Rounding off In calculations, we round up if the last digit is above 5 and round down if it is below 5. For numbers ending in 5, always round to the nearest even number. For example, 2.35 rounds to 2.4 and 2.65 rounds to 2.6. This strategy avoids compounding round-off errors in the final answer. The quantity 14.348 cm^3 rounds to 14.3 cm^3 if the answer should have 3 sf, but it would have rounded to 14.35 cm^3 if the data had justified 4 sf. Rounding must be carried out in a single step: 14.348 cm^3 should not first be rounded to 14.35 cm^3 and then to 14.4 cm^3. The correct procedure is to round off only at the final stage of the calculation and to carry all digits in the memory of the calculator until that stage.*

Addition and subtraction When we add or subtract,

> The number of decimal places in the result should be the same as the *smallest number of decimal places* in the data.

*In this text we often need to display intermediate rounded results, so the final answer might differ slightly from the answer obtained by rounding only at the end of the calculation.

For instance, suppose we need the total volume of three blocks of copper with measured volumes 11.12 cm^3, 1.2 cm^3, and 3.107 cm^3; then according to this rule, only one digit should follow the decimal point in the total volume:

$$
\begin{array}{r}
11.12\ \text{cm}^3 \\
1.2\ \text{cm}^3 \\
\underline{3.107\ \text{cm}^3}
\end{array}
$$

$$\text{Total: } 15.427\ \text{cm}^3; \quad \text{round to } 15.4\ \text{cm}^3$$

SELF-TEST 2.6A What is (a) the total volume of a sample of water prepared by adding 25.6 mL to 50 mL; (b) the temperature in kelvins corresponding to the boiling point of sulfur, 444.67°C?

[*Answer:* (a) 76 mL; (b) 717.82 K]

SELF-TEST 2.6B Report (a) the total mass of a sample prepared by mixing 1.001 g of sugar, 2.05 g of salt, and 5.0 g of flour; (b) the Celsius temperature corresponding to the melting point of iron, 1813 K.

Multiplication and division When we multiply or divide numbers obtained from measurements,

> The number of significant figures in the result should be the same as *the smallest number of significant figures* in the data.

For example, in the calculation of the density of the sample of plastic mentioned in Section 2.6, we note that the smallest number of figures in the data is 2 (in the measurement of volume, 1.2 cm^3), so we round 1.483333 $g{\cdot}cm^{-3}$ to 1.5 $g{\cdot}cm^{-3}$.

SELF-TEST 2.7A (a) Calculate the mass of 250 mL (3 sf) of carbon dioxide that has a density of 2.095 $g{\cdot}L^{-1}$. (b) What would the density of the gas be (in grams per liter) if the same sample were in a container of volume 155 mL?

[*Answer:* (a) 0.524 g; (b) 3.38 $g{\cdot}L^{-1}$]

SELF-TEST 2.7B (a) Calculate the volume occupied by 1.04 mg of oxygen gas, given that its density is 1.5 $g{\cdot}L^{-1}$. (b) What would its density be (in grams per liter) if the same sample occupied 0.755 mL?

Integers and exact numbers When multiplying or dividing by an integer or an exact number, the uncertainty of the result is determined by the measured value. Some unit conversion factors are defined exactly, even though they are not whole numbers. For example, 1 in. is defined as *exactly* 2.54 cm and the 273.15 in the conversion between Celsius and Kelvin temperatures is also exact, so 100.000°C converts to 373.150 K. All the bold numbers in Table 2.2 are exact; and when they are used in a calculation, the number of significant figures in the answer is the same as the number of significant figures in the data.

2.7 ACCURACY AND PRECISION

The terms *precision* and *accuracy* refer to two types of errors present in a measurement. To make sure of their data, scientists usually repeat their measurements a number of times. The **precision** of a measurement refers to how close to each other these repeated measurements are. The **accuracy** of a series of measurements is the closeness of their average value to the true value. Even precise measurements can give inaccurate values. For instance, if there is a speck of dust on the pan of a chemical balance that we were using to measure the mass of a sample of vitamin C, then even though we might be justified in reporting our measurements to 5 sf, the reported mass of the sample will be inaccurate.

A **systematic error** is an error present in every one of a series of repeated measurements, like a speck of dust on a pan, which distorts the mass of each sample in the same direction (the speck makes each sample appear heavier than it is). A **random error** is an error that varies at random and can average to zero over a series of observations, like drafts of air from an open window moving a balance pan either up or down a bit, decreasing or increasing the mass measurements randomly. Scientists attempt to improve the accuracy of their measurements by making many observations and taking the average of the results. This procedure minimizes random errors; systematic errors are much harder to identify and eliminate.

> *The precision of measurements indicates how close to one another the measurements are; the accuracy of a measurement is its closeness to the true value.*

CHEMICAL AMOUNTS

We can now start to apply these ideas to the determination of the chemical formulas of compounds. One step in the determination is to find out how many atoms of each element are present in a sample. However, the numbers involved are astronomical, often as many as 10^{23} or even more. We need an efficient way of determining these numbers and a compact way of reporting them.

2.8 THE MOLE

The mole (abbreviated mol) is the unit chemists use to keep track of large numbers of atoms, ions, and molecules. The unit was invented to provide a simple way of reporting the huge numbers—the "massive heaps"—of atoms and molecules in visible samples. It would be inconvenient to refer to large numbers like 2.0×10^{25} atoms, just as wholesalers would find it inconvenient to count individual items instead of dozens (12) or gross (144). Chemists have defined a unit that gives a count of the number of atoms in a massive heap of atoms of a particular size:

The name *mole* comes from the Latin word for "massive heap." The animal of the same name makes massive heaps of soil on lawns.

> 1 **mole** (1 mol) is the number of atoms in exactly 12 g of carbon-12.

See Fig. 2.6 for an illustration of this definition. Although the mole is defined in terms of carbon atoms, the unit applies to any chemical species, just as 1 dozen means 12 of anything.

We saw in Box 1.2 that mass spectrometry can be used to determine the masses of atoms of individual isotopes. The mass of a carbon-12 atom has been found to be 1.9926×10^{-23} g. It follows that, to 4 sf, the number of atoms in exactly 12 g of carbon-12 is

$$\text{Number of carbon-12 atoms} = \frac{12\text{ g}}{1.9926 \times 10^{-23}\text{ g}} = 6.022 \times 10^{23}$$

Because the mole gives the number of atoms in a sample, it follows that 1.000 mol of atoms (of any element) is 6.022×10^{23} atoms of the element. The same is true of 1 mol of any objects—atoms, ions, or molecules:

> 1.000 mol of objects *always* means 6.022×10^{23} of those objects.

The number of objects per mole, $6.022 \times 10^{23}\ \text{mol}^{-1}$, is called the **Avogadro constant,** N_A, in honor of the nineteenth-century Italian scientist Amedeo Avogadro (Fig. 2.7), who helped to establish the existence of atoms. Some awe-inspiring illustrations can help you visualize its size. For instance, 1 mol of chemistry textbooks would cover the surface of the Earth to a height of about 300 km. If you won 1 mol of dollars in a lottery the day you were born, and spent a billion dollars a second, you

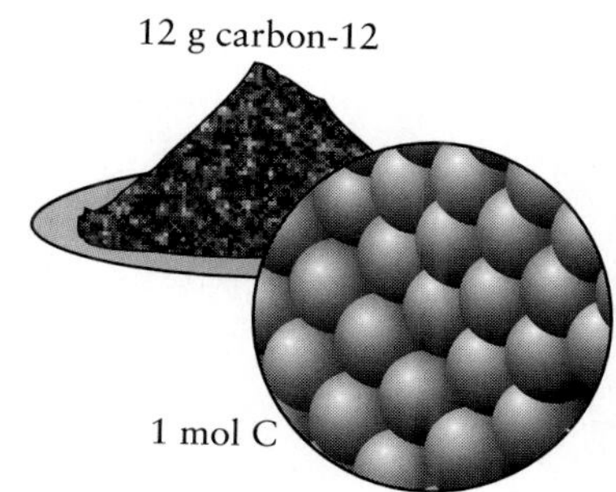

FIGURE 2.6 The definition of the mole. If we measure out exactly 12 g of carbon-12, then we have exactly 1 mol of carbon-12 atoms. There will be exactly an Avogadro number of atoms in the pile.

would still have more than 99.999% of the prize money left when you died at 90.

The Avogadro constant is used to convert between the number of moles and the number of atoms, ions, or molecules:

$$\begin{aligned}\text{Number of objects} &= \text{number of moles} \times \text{number of objects per mole}\\ &= \text{number of moles} \times N_A\end{aligned}$$

Because 1 mol is a unit, it can be used with a prefix. For example, 1 mmol is 10^{-3} mol. As we study chemistry, we shall only rarely find a reference to the actual numbers of atoms, ions, or molecules in a sample. Almost always, we shall find it much easier to refer to the numbers of moles of particles and to do all our calculations using the mole as a unit, just as grocers typically work with dozens and gross rather than with individual eggs and cans.

FIGURE 2.7 Lorenzo Romano Amedeo Carlo Avogadro, count of Quaregna and Cerreto (1776–1856).

The amounts of atoms, ions, or molecules in a sample are reported in moles, and the Avogadro constant, N_A, is used to convert between numbers of these particles and the numbers of moles.

EXAMPLE 2.3 Converting number of atoms to moles

Suppose a sample of vitamin C is known to contain 1.29×10^{24} hydrogen atoms (as well as other kinds of atoms). Express that number as the number of moles of hydrogen atoms.

STRATEGY A good strategy in chemistry is to estimate an approximate answer before going through the calculation, because we can then detect major errors in our work quickly. In this case, because the number of atoms in the sample is greater than 6×10^{23}, we anticipate that *more* than 1 mol of atoms is present. For the calculation, we rearrange the relation given above to give us the quantity we want (the number of moles) in terms of the information given (the number of hydrogen atoms):

$$\text{Number of moles of H atoms} = \frac{\text{number of H atoms}}{N_A}$$

and then substitute the data.

SOLUTION Substituting the data (the number of H atoms) and the value of the Avogadro constant gives

$$\begin{aligned}\text{Number of moles of H atoms} &= \frac{1.29 \times 10^{24}}{6.022 \times 10^{23}\ \text{mol}^{-1}}\\ &= 2.14\ \text{mol}\end{aligned}$$

Notice how much simpler it is to report that 2.14 mol of hydrogen atoms is present rather than the actual number, 1.29×10^{24} atoms.

SELF-TEST 2.8A A sample of vitamin C is known to contain 2.58×10^{24} oxygen atoms. How many moles of O atoms are present in the sample?

[*Answer:* 4.28 mol O atoms]

SELF-TEST 2.8B A sample of water contains 3.14 mol of water molecules. How many H_2O molecules are present?

When reporting the number of moles of atoms, molecules, or ions, we always state explicitly which *species* (that is, which atoms, molecules, or ions) we mean. For example, hydrogen occurs naturally as a gas with each molecule built from two atoms, which is why it is denoted H_2. We write 1 mol H if we mean hydrogen atoms or 1 mol H_2 if we mean hydrogen molecules. We do not refer simply to 1 mol of hydrogen, because it would not be clear whether we meant hydrogen atoms or hydrogen molecules. Alternatively, we could write that the number of H atoms (or H_2 molecules) is 1 mol. The essential point, therefore, is to be unambiguous. Other elements that occur as **diatomic molecules**, or molecules composed of two atoms, are nitrogen (N_2), oxygen (O_2), and the halogens (Cl_2, for example). Phosphorus exists as P_4 molecules and sulfur as S_8.

Always state explicitly (in some unambiguous way) the identity of the particles to which the term moles refers.

2.9 MOLAR MASS

We can readily determine the *mass* of a sample, but how do we convert that mass into the number of moles of atoms, molecules, or ions in the sample? The key concept that acts as a bridge between mass and moles is the **molar mass,** the mass per mole of particles:

The **molar mass of an element** is the mass of the element per mole of its atoms.

The **molar mass of a molecular compound** is the mass of the compound per mole of its molecules.

The **molar mass of an ionic compound** is the mass of the compound per mole of its formula units.

You will still see these properties called atomic weight, molecular weight, and formula weight. However, the properties to which these terms refer are masses per mole, not weights.

For example, the molar mass of the element carbon is the mass per mole of carbon atoms. The molar mass of vitamin C, a molecular compound, is the mass per mole of vitamin C molecules. The molar mass of the ionic compound sodium chloride is the mass per mole of NaCl formula units. The units of molar mass in each case are grams per mole ($g \cdot mol^{-1}$).

The masses of the individual atoms of an isotope can be calculated from mass spectrometry data, so in principle it is easy to determine the molar mass of an element: we simply multiply the mass of one of its atoms by the number of atoms per mole (that is, by the Avogadro constant):

$$\text{Mass per mole of atoms} = \text{mass of one atom} \times \text{number of atoms per mole}$$

For example, the mass of a fluorine atom is 3.155×10^{-23} g, so the molar mass of fluorine is

$$\begin{aligned}\text{Mass per mole of F atoms} &= (3.155 \times 10^{-23}\ \text{g}) \times (6.022 \times 10^{23}\ \text{mol}^{-1}) \\ &= 19.00\ \text{g} \cdot \text{mol}^{-1}\end{aligned}$$

Once we know the molar mass of an element, it is easy to convert the mass of a sample to a number of moles, because we know that

$$\begin{aligned}\text{Mass of sample} &= \text{number of moles} \times \text{mass per mole of atoms} \\ &= \text{number of moles} \times \text{molar mass of element}\end{aligned}$$

For example, we can find the number of moles of F atoms in 22.5 g of fluorine by rearranging this expression into

$$\text{Number of moles of F atoms} = \frac{\text{mass of sample}}{\text{molar mass of fluorine}}$$

and substituting the data:

$$\text{Number of moles of F atoms} = \frac{22.5\text{ g}}{19.00\text{ g}\cdot\text{mol}^{-1}} = 1.18\text{ mol}$$

So, from a straightforward measurement of mass, we know the number of moles of atoms present in the sample (and, if desired, the actual number of atoms by multiplying the number of moles by the Avogadro constant). The procedure is summarized in Toolbox 2.2.

Unfortunately, there is a catch. For most elements, we need to take into account the fact that a naturally occurring sample of an element is a mixture of different isotopes. To cope with this difficulty, we need to calculate the *average* mass of an atom in a sample. Then we multiply this average mass by the Avogadro constant to obtain the *average* molar mass of a natural sample. *All molar masses quoted in this text refer to these average values.* Their values are given in the table inside the back cover and in Appendix 2D. They are also included in the periodic table inside the front cover.

◀◀ Recall from Section 1.4 that isotopes are atoms of the same element, but have different masses.

EXAMPLE 2.4 Evaluating an average molar mass

There are two naturally occurring isotopes of chlorine, chlorine-35 and chlorine-37. The mass of an atom of chlorine-35 is 5.807×10^{-23} g and that of an atom of chlorine-37 is 6.139×10^{-23} g. In a typical natural sample of chlorine, 75.77% of the sample is chlorine-35 and 24.23% is chlorine-37. What is the average molar mass of chlorine?

STRATEGY First, calculate the average mass of a chlorine atom by adding together the individual masses, each multiplied by the fraction that represents its abundance. Then obtain the average molar mass, the mass per mole of atoms, by multiplying the average atomic mass by the Avogadro constant.

SOLUTION The average mass of an atom of chlorine in a natural sample is

$$\begin{aligned}&\text{Average mass of a Cl atom}\\&\quad= \frac{75.77}{100} \times (5.807 \times 10^{-23}\text{ g}) + \frac{24.23}{100} \times (6.139 \times 10^{-23}\text{ g})\\&\quad= 4.400 \times 10^{-23}\text{ g} + 1.487 \times 10^{-23}\text{ g} = 5.887 \times 10^{-23}\text{ g}\end{aligned}$$

It follows that the molar mass of a typical sample of chlorine atoms is

$$\begin{aligned}&\text{Molar mass of chlorine}\\&\quad= \text{average mass of a Cl atom} \times \text{number of Cl atoms per mole}\\&\quad= (5.887 \times 10^{-23}\text{ g}) \times (6.022 \times 10^{23}\text{ mol}^{-1}) = 35.45\text{ g}\cdot\text{mol}^{-1}\end{aligned}$$

SELF-TEST 2.9A In a typical sample of magnesium, 78.99% is magnesium-24 (3.983×10^{-23} g), 10.00% magnesium-25 (4.149×10^{-23} g), and 11.01% magnesium-26 (4.315×10^{-23} g). Calculate the average molar mass of a sample of magnesium, given the atomic masses (in parentheses).

[*Answer:* 24.31 g·mol^{-1}]

SELF-TEST 2.9B Calculate the average molar mass of copper given that a natural sample typically consists of 69.17% copper-63, which has a molar mass of 62.94 $g \cdot mol^{-1}$, and 30.83% copper-65, which has a molar mass of 64.93 $g \cdot mol^{-1}$.

TOOLBOX 2.2 How to convert between mass and moles

The key relation for this toolbox is

$$\text{Mass of sample (grams)} = \text{number of moles} \times \text{molar mass of element (in grams per mole)}$$

Converting from moles to mass The expression above is used when we want to convert a given number of moles of atoms into a mass in grams. For example, suppose we want to know what mass of copper corresponds to 10.0 mol Cu (that is, 10.0 mol of Cu atoms), then we use the molar mass of copper given in the periodic table (63.54 $g \cdot mol^{-1}$) to write

$$\text{Mass of copper} = (10.0\ \text{mol}) \times (63.54\ \text{g}\cdot\text{mol}^{-1}) = 635\ \text{g}$$

This result means that if we wanted a sample that contained 10.0 mol Cu atoms, then we would need to measure out 635 g of copper.

SELF-TEST 2.10A What mass of iron contains 1.23 mol Fe atoms?

[*Answer:* 68.7 g]

SELF-TEST 2.10B We want a sample of uranium that contains 0.26 mol U atoms. What mass of uranium do we need?

Converting from mass to moles To find the number of moles of atoms in a sample of given mass, we rearrange the starting expression so that the given information appears on the right:

$$\text{Number of moles} = \frac{\text{mass of sample (grams)}}{\text{molar mass (grams per mole)}}$$

For example, if we wanted to calculate the number of moles of Cl atoms in 15.0 g of chlorine, then we would write

$$\text{Number of moles Cl atoms} = \frac{15.0\ \text{g}}{35.45\ \text{g}\cdot\text{mol}^{-1}} = 0.423\ \text{mol}$$

SELF-TEST 2.11A In an experimental solar-powered decomposition of water, 2.53 g of hydrogen was obtained for use as a fuel. How many moles of H atoms were produced? (The fact that the atoms were produced as H_2 molecules does not affect the result.)

[*Answer:* 2.51 mol H atoms]

SELF-TEST 2.11B The same experiment produced 20.04 g of oxygen gas as a by-product. How many moles of O atoms were produced in the experiment? (The O atoms are present as O_2 molecules, but that does not affect the result.)

It is now very simple to measure out the "massive heaps" of elements that contain 1 mol of atoms: *to measure out 1 mol of atoms, we measure out the numerical value of the molar mass of the element in grams.* So, to obtain 1.000 mol Cu atoms, of molar mass 63.54 $g \cdot mol^{-1}$, we measure out 63.54 g of copper, for 1.0000 mol Hg atoms, of molar mass 200.59 $g \cdot mol^{-1}$, we measure out 200.59 g of mercury, and so on. This procedure has been used to measure out the "massive heaps" of elements shown in Fig. 2.8. Each of these heaps contains about the same number of atoms of the element (approximately 6.022×10^{23} in each case), but the masses vary because the masses of the atoms are different (Fig. 2.9).

The molar mass of an element, the mass per mole of its atoms, is used to convert between the mass of a sample and the moles of atoms it contains.

FIGURE 2.8 Each sample consists of 1 mol of atoms of the element. Clockwise from the upper right are 32 g of sulfur, 201 g of mercury, 207 g of lead, 64 g of copper, and 12 g of carbon.

2.10 MEASURING OUT COMPOUNDS

Now let's see how to measure out a given number of moles of molecules of a molecular compound or formula units of an ionic compound. To do this, we need to know the molar mass of the compound. In each case, *the molar mass of a compound is the sum of the molar masses of the elements that make up the molecule or the formula unit.* For example, the molar mass of water, which is composed of H_2O molecules, is

$$\begin{aligned}\text{Molar mass of } H_2O &= 2 \times (\text{molar mass of H}) + (\text{molar mass of O}) \\ &= 2 \times (1.008\ \text{g}\cdot\text{mol}^{-1}) + (16.00\ \text{g}\cdot\text{mol}^{-1}) \\ &= 18.02\ \text{g}\cdot\text{mol}^{-1}\end{aligned}$$

Similarly, the molar mass of sodium chloride is the mass per mole of NaCl formula units, which is given by the following sum:

$$\begin{aligned}\text{Molar mass of NaCl} &= (\text{molar mass of Na}) + (\text{molar mass of Cl}) \\ &= (22.99\ \text{g}\cdot\text{mol}^{-1}) + (35.45\ \text{g}\cdot\text{mol}^{-1}) \\ &= 58.44\ \text{g}\cdot\text{mol}^{-1}\end{aligned}$$

Once the molar mass of a compound has been calculated in this way, we can use the technique outlined for elements in Toolbox 2.2 to find out how many moles of molecules or formula units are in a sample of given mass, or what mass of sample we have to measure out to get a given number of moles.

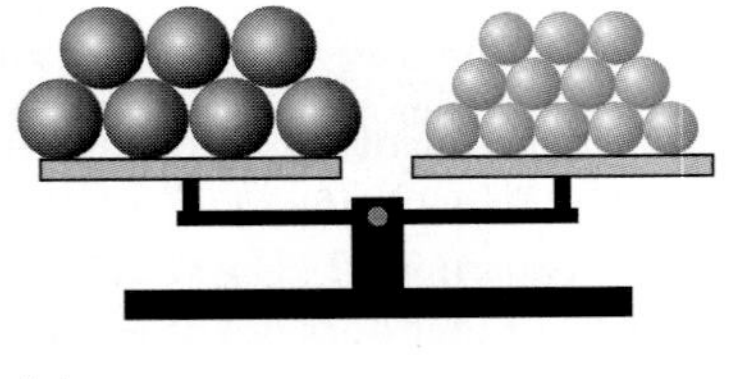

(a)

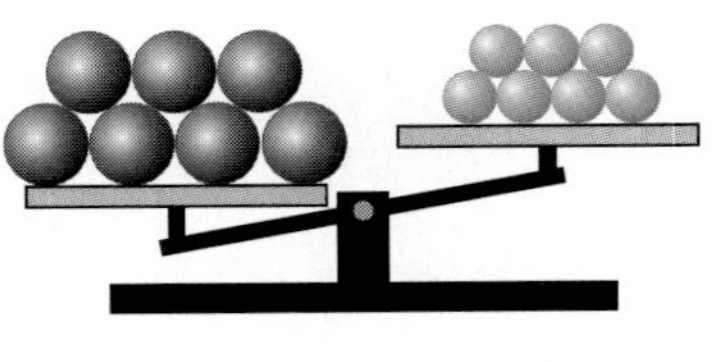

(b)

FIGURE 2.9 (a) The two samples have the same *mass,* but because the atoms on the right are lighter, the sample on the right consists of a greater number of atoms. (b) The two samples contain the same number of moles of atoms, but because the atoms on the right are lighter, the mass of that sample is smaller.

EXAMPLE 2.5 Calculating the molar mass of a compound

Calculate the molar mass of (a) glucose, $C_6H_{12}O_6$; (b) potassium sulfate, K_2SO_4.

STRATEGY To obtain the molar mass of a molecular or ionic compound, add together the molar masses of the elements, with each molar mass multiplied by the number of times an atom of the element appears in the formula. The molar masses of the elements are given in the periodic table inside the front cover and the alphabetical list inside the back cover.

SOLUTION (a) Because the molecular formula of glucose is $C_6H_{12}O_6$, its molar mass is

$$\begin{aligned}&\text{Molar mass of } C_6H_{12}O_6\\&\quad = 6 \times (12.01\ \text{g·mol}^{-1}) + 12 \times (1.008\ \text{g·mol}^{-1}) + 6 \times (16.00\ \text{g·mol}^{-1})\\&\quad = 180.16\ \text{g·mol}^{-1}\end{aligned}$$

(b) For the ionic compound,

$$\begin{aligned}&\text{Molar mass of } K_2SO_4\\&\quad = 2 \times (39.10\ \text{g·mol}^{-1}) + (32.06\ \text{g·mol}^{-1}) + 4 \times (16.00\ \text{g·mol}^{-1})\\&\quad = 174.26\ \text{g·mol}^{-1}\end{aligned}$$

SELF-TEST 2.12A Calculate the molar mass of (a) sulfuric acid, a molecular compound of formula H_2SO_4; (b) aluminum oxide, which is an ionic compound of formula Al_2O_3.

[*Answer:* (a) 98.08 g·mol^{-1}; (b) 101.96 g·mol^{-1}]

SELF-TEST 2.12B Calculate the molar mass of (a) nitric acid, a molecular compound of formula HNO_3; (b) aluminum sulfate, an ionic compound of formula $Al_2(SO_4)_3$.

Just as for elements, we now know that if we measure out 58.44 g of sodium chloride, then we have a sample that contains 1.000 mol NaCl formula units. Similarly, if we want 1.000 mol $C_6H_{12}O_6$ molecules, then we measure out 180.2 g of glucose. In general, to measure out 1 mol of molecules or formula units, we measure out the numerical value of the molar mass of the compound in grams. Some examples of the "massive heaps" that correspond to 1 mol of molecules or formula units of a variety of compounds are illustrated in Figs. 2.10 and 2.11.

The molar mass of a compound is used to relate the mass of a sample to the number of moles of molecules or formula units. The procedure is the same as in Toolbox 2.2, except that we use

$$\text{Number of moles of molecules or formula units} = \frac{\text{mass of compound (grams)}}{\text{molar mass of compound (grams per mole)}}$$

FIGURE 2.10 Each sample contains 1 mol of molecules of a molecular compound. From left to right are 18 g of water (H_2O), 46 g of ethanol (C_2H_6O), 180 g of glucose ($C_6H_{12}O_6$), and 342 g of sucrose ($C_{12}H_{22}O_{11}$).

FIGURE 2.11 Each sample contains 1 mol of formula units of an ionic compound. From left to right are 58 g of sodium chloride (NaCl), 100 g of calcium carbonate ($CaCO_3$), 278 g of iron(II) sulfate heptahydrate ($FeSO_4 \cdot 7H_2O$), and 78 g of sodium peroxide (Na_2O_2).

The molar mass of a compound is determined by adding the molar masses of its constituent elements, with each molar mass multiplied by the subscript of the element in the formula.

SELF-TEST 2.13A Calculate (a) the number of moles of $OC(NH_2)_2$ molecules in 2.3×10^5 kg of urea, which is used in facial creams and, on a somewhat bigger scale, as an agricultural fertilizer; (b) the mass of aluminum oxide corresponding to 6.3 mol Al_2O_3.

[*Answer:* (a) 3.8×10^6 mol; (b) 0.64 kg]

SELF-TEST 2.13B Calculate (a) the number of moles of $Ca(OH)_2$ formula units in 1.00 kg of slaked lime (calcium hydroxide), which is used to adjust the acidity of soils; (b) the mass of sucrose (cane sugar) corresponding to 1.5 mmol $C_{12}H_{22}O_{11}$.

DETERMINATION OF CHEMICAL FORMULAS

Now we come to the final stage of our quest for the chemical formulas of compounds, with vitamin C as our target. There are two steps: the first is to determine the empirical formula, and then the second (if the compound is molecular or contains a polyatomic ion) is to determine the molecular formula. The **empirical formula** of a compound is a chemical formula that shows the *relative* numbers of atoms of each element. For example, the empirical formula of glucose, which is CH_2O, tells us that carbon, hydrogen, and oxygen are present in the ratio 1:2:1. The elements are present in these proportions regardless of the size of the sample. The **molecular formula** tells us the *actual* numbers of atoms of each element in a molecule. The molecular formula for glucose, which is $C_6H_{12}O_6$, tells us that each glucose molecule consists of six carbon atoms, twelve hydrogen atoms, and six oxygen atoms.

2.11 MASS PERCENTAGE COMPOSITION

To determine the empirical formula of a compound, we begin by measuring the *mass* of each element present in a sample. This composition is usually reported as the **mass percentage composition,** that is, the mass of each element expressed as a percentage of the total mass:

$$\text{Mass percentage of element} = \frac{\text{mass of element in the sample}}{\text{total mass of sample}} \times 100\%$$

For instance, a sample of vitamin C of total mass 8.00 g was analyzed to find which elements it contains and how much of each. The following data were collected:

Carbon	3.27 g
Hydrogen	0.366 g
Oxygen	4.36 g

So, the mass percentage of carbon in vitamin C is

$$\begin{aligned}\text{Mass percentage of C} &= \frac{\text{mass of C in sample}}{\text{total mass of sample}} \times 100\% \\ &= \frac{3.27\ \text{g}}{8.00\ \text{g}} \times 100\% = 40.9\%\end{aligned}$$

The same procedure gives the following values for the mass percentages of hydrogen and oxygen:

Carbon	3.27 g	40.9%
Hydrogen	0.366 g	4.58%
Oxygen	4.36 g	54.5%

We now know that *any* sample of vitamin C has mass percentage composition 40.9% C, 4.58% H, and 54.5% O.

Mass percentage composition is found by calculating the fraction of the total mass contributed by each element present in a compound and expressing the fraction as a percentage.

EXAMPLE 2.6 Determining mass percentage composition

For centuries, the Australian aborigines have used the leaves of the eucalyptus tree to alleviate sore throats and other pains. The primary active ingredient has been identified and named eucalyptol. The analysis of a sample of eucalyptol of total mass 3.16 g gave its composition as 2.46 g carbon, 0.373 g hydrogen, and 0.329 g oxygen. Determine the mass percentage of carbon in eucalyptol.

STRATEGY The mass percentage composition is defined as

$$\text{Mass percentage of element} = \frac{\text{mass of element in sample}}{\text{total mass of sample}} \times 100\%$$

To use this definition, we simply substitute the data for each element present in the compound.

SOLUTION The mass percentage of carbon in eucalyptol is

$$\text{Mass percentage of C} = \frac{\text{mass of C in sample}}{\text{total mass of sample}} \times 100\%$$
$$= \frac{2.46\text{ g}}{3.16\text{ g}} \times 100\% = 77.8\%$$

SELF-TEST 2.14A What are the mass percentages of hydrogen and oxygen in eucalyptol?

[*Answer:* 11.8% H and 10.4% O]

SELF-TEST 2.14B The compound α-pinene, a natural antiseptic found in the resin of the piñon tree, has been used since ancient times by Zuni healers. A 7.50-g sample of α-pinene contains 6.61 g carbon and 0.89 g hydrogen. What are the mass percentages of carbon and hydrogen in α-pinene?

2.12 DETERMINING EMPIRICAL FORMULAS

To determine the empirical formula of a compound, we need the relative number of atoms of each element in the sample or, what is the same thing, the relative number of moles of each type of atom. We can get this information by calculating the numbers of moles of atoms present from the mass percentage composition. The easiest procedure is to imagine that we have a sample of mass 100 g exactly. That way, the mass percentage composition tells us the mass in grams of each element. Then we can use the molar masses of the elements to convert these masses into moles.

In vitamin C, the mass of carbon in a sample of mass 100 g is 40.9 g. We find the amount of C atoms in 40.9 g of carbon from the molar mass of carbon, which the periodic table gives as 12.01 $g{\cdot}mol^{-1}$:

$$\text{Number of moles C atoms} = \frac{40.9\text{ g}}{12.01\text{ g}{\cdot}\text{mol}^{-1}}$$
$$= 3.41\text{ mol}$$

In the same way, we find the numbers of moles of H atoms and O atoms in the sample:

Carbon	40.9%	3.41 mol
Hydrogen	4.58%	4.54 mol
Oxygen	54.5%	3.41 mol

Because number of moles is proportional to number of atoms, we now know that atoms of each element are present in the ratio (3.41 C atoms):(4.54 H atoms):(3.41 O atoms), or 3.41:4.54:3.41.

At this point, we might be tempted to write the chemical formula of vitamin C as $C_{3.41}H_{4.54}O_{3.41}$. However, a compound cannot contain fractions of atoms; so to get the empirical formula, we must express the ratios of numbers of atoms as the simplest whole numbers. First, we divide each number by the smallest value (3.41), which gives a ratio of 1.00:1.33:1.00. One number is still not a whole number, hence we must multiply each number by a factor until all numbers are whole numbers or can be rounded off to whole numbers. Because 1.33 is $\frac{4}{3}$ (within experimental error), we multiply through by 3 to obtain 3.00:3.99:3.00, or

CHEMISTRY AT WORK

An ethnobotanist and a pharmacognosist are instructed by a Samoan healer and her apprentice in the properties of a traditional medicinal herb. The two scientists will extract the active ingredients from the herb, conduct biological tests on the separated ingredients, and determine the formulas and structures of substances with promising characteristics. Herbs being sought most avidly are those with antitumor and antiviral properties.

approximately 3:4:3. Now we know that the empirical formula of vitamin C is $C_3H_4O_3$.

The empirical formula of a compound is determined from the mass percentage composition and the molar masses of the elements present.

SELF-TEST 2.15A Use the mass percentage composition of eucalyptol calculated in Example 2.6 and Self-Test 2.14A to determine its empirical formula.

[*Answer:* $C_{10}H_{18}O$]

SELF-TEST 2.15B The mass percentage composition of the compound thionyl difluoride is 18.59% O, 37.25% S, and 44.16% F. Calculate its empirical formula.

EXAMPLE 2.7 Determining the empirical formula of an inorganic compound

A sample of magnesium of mass 0.450 g burns in nitrogen to form 0.623 g of magnesium nitride. Determine the empirical formula of magnesium nitride.

STRATEGY We are given the mass of one element and we know the total mass of the compound; the difference between the two numbers gives us the mass of the second element present. The number of moles of atoms of each element in the compound is determined from the molar mass of the element and the mass present. We express these numbers of moles as an empirical formula.

SOLUTION We know that magnesium belongs to Group 2 and nitrogen belongs to Group 15; so we expect these elements to form Mg^{2+} and N^{3-} ions. Therefore, we can expect the compound to have the formula Mg_3N_2. However, our task is to determine its formula experimentally from the data we are given; some compounds do not have their expected formulas. The molar masses we need are 24.31 $g \cdot mol^{-1}$ for magnesium and 14.01 $g \cdot mol^{-1}$ for nitrogen. The number of moles of magnesium atoms present in the compound is

$$\text{Number of moles of Mg} = \frac{0.450\text{ g}}{24.31\text{ g}\cdot\text{mol}^{-1}} = 0.0185\text{ mol}$$

The mass of nitrogen in the compound is

$$\text{Mass of nitrogen} = 0.623\text{ g} - 0.450\text{ g} = 0.173\text{ g}$$

Hence,

$$\text{Number of moles of N} = \frac{0.173\text{ g}}{14.01\text{ g}\cdot\text{mol}^{-1}} = 0.0123\text{ mol}$$

The elements are therefore present in the ratio 0.0185:0.0123. Division by the smaller number gives a ratio of 1.50:1.00. Multiplying through by 2 gives 3:2. This result suggests that the empirical formula of magnesium nitride is indeed Mg_3N_2.

SELF-TEST 2.16A It is found that a sample of bromine of mass 1.546 g reacts with fluorine to form 2.649 g of a bromine fluoride. Determine the empirical formula of the compound.

[*Answer:* BrF_3]

SELF-TEST 2.16B The first compounds of the noble gases were prepared in the 1960s. In one experiment, it was found that 2.56 g of xenon reacted with fluorine to produce 4.04 g of a xenon fluoride. What is the empirical formula of this compound?

2.13 DETERMINING MOLECULAR FORMULAS

So far, we know that the empirical formula of vitamin C is $C_3H_4O_3$. However, all this formula tells us is that the C, H, and O atoms are present in the sample in the ratio 3:4:3. We do not yet know the numbers of atoms in individual molecules. The same empirical formula would be obtained for $C_3H_4O_3$, $C_6H_8O_6$, $C_9H_{12}O_9$, or any other whole-number multiple of the empirical formula.

To find the molecular formula of a compound, one more piece of information is needed—its molar mass. Once we know the molar mass, we can calculate how many formula units are needed to account for that mass. The molar mass of vitamin C is found to be 176.14 g·mol^{-1} by mass spectrometry. The molar mass of a $C_3H_4O_3$ formula unit is

$$\begin{aligned}&\text{Molar mass of } C_3H_4O_3\\ &\quad = 3 \times (12.01\text{ g}\cdot\text{mol}^{-1}) + 4 \times (1.008\text{ g}\cdot\text{mol}^{-1}) + 3 \times (16.00\text{ g}\cdot\text{mol}^{-1})\\ &\quad = 88.06\text{ g}\cdot\text{mol}^{-1}\end{aligned}$$

We need *two* $C_3H_4O_3$ formula units to account for the observed molar mass of vitamin C:

$$\frac{176.14\text{ g}\cdot\text{mol}^{-1}}{88.06\text{ g}\cdot\text{mol}^{-1}} = 2.000$$

so the molecular formula of vitamin C is $2 \times (C_3H_4O_3)$, or $C_6H_8O_6$.

The molecular formula of a molecular compound is found by determining how many empirical formula units are needed to account for the measured molar mass of the compound.

EXAMPLE 2.8 Determining a molecular formula from an empirical formula

The molar mass of ethyl butanoate, a compound that contributes to the flavor of pineapple, is 116 $g{\cdot}mol^{-1}$. Its empirical formula, determined from its mass percentage composition, is C_3H_6O. What is its molecular formula?

STRATEGY We have to decide how many empirical formula units are required to account for the measured molar mass. Therefore, we calculate the molar mass of the empirical formula unit and then compare it with the measured molar mass.

SOLUTION The molar mass of the formula unit C_3H_6O is

$$\text{Molar mass of } C_3H_6O = 3 \times (12.01\ g{\cdot}mol^{-1}) + 6 \times (1.008\ g{\cdot}mol^{-1}) + (16.00\ g{\cdot}mol^{-1}) = 58.08\ g{\cdot}mol^{-1}$$

Because the measured molar mass is 116 $g{\cdot}mol^{-1}$, it follows that

$$\text{Number of empirical formula units per molecule} = \frac{116\ g{\cdot}mol^{-1}}{58.08\ g{\cdot}mol^{-1}} = 2.00$$

We conclude that the molecular formula of ethyl butanoate is $C_6H_{12}O_2$.

SELF-TEST 2.17A The molar mass of styrene, which is used in the manufacture of the plastic polystyrene, is 104 $g{\cdot}mol^{-1}$ and its empirical formula is CH. Deduce its molecular formula.

[*Answer:* C_8H_8]

SELF-TEST 2.17B The molar mass of oxalic acid, a toxic substance found in rhubarb leaves, is 90.0 $g{\cdot}mol^{-1}$ and its empirical formula is CHO_2. What is its molecular formula?

FIGURE 2.12 A pipet is an accurate means of transferring a fixed volume of solution. Here, a solution containing a reactant is being added to a reaction vessel.

SOLUTIONS IN CHEMISTRY

Vitamin C is usually taken in tablet form, but many medications are administered in solution. The same is true in chemistry. Although we often use a reagent in solid form, in many cases we use it as a solution. The question we have to tackle is how to report the amount of solute in a given volume of solution. This kind of knowledge has great practical importance. It is obviously important in medicine and agriculture, because accidental injury or death can result from incorrect dosages or from overexposure to pesticides. It is also very important in chemistry, because we often need to know how much of a substance we are using when we pour one solution into another (Fig. 2.12).

2.14 MOLARITY

The **molarity,** *M,* of a solution is the number of moles of solute molecules or formula units divided by the volume of the solution (in liters).

The formal term for molarity is the *molar concentration* of the solute in the solution.

$$\text{Molarity} = \frac{\text{number of moles of solute}}{\text{volume of solution (liters)}}$$

The units of molarity are moles per liter ($mol \cdot L^{-1}$). They are often denoted by the symbol M:

$$1 \text{ M} = 1 \text{ mol} \cdot L^{-1}$$

The symbol M is often read "molar." For example, suppose we dissolved 10.0 g of cane sugar in enough water to make 200 mL of water, which we might do if we were making a sweet cup of coffee. Cane sugar is sucrose ($C_{12}H_{22}O_{11}$); and from its molar mass (342 $g \cdot mol^{-1}$), we can calculate that we are using 0.0292 mol of sucrose molecules. The molarity of the solute in the solution is therefore

$$\text{Molarity} = \frac{0.0292 \text{ mol}}{0.200 \text{ L}} = 0.146 \text{ mol} \cdot L^{-1}$$

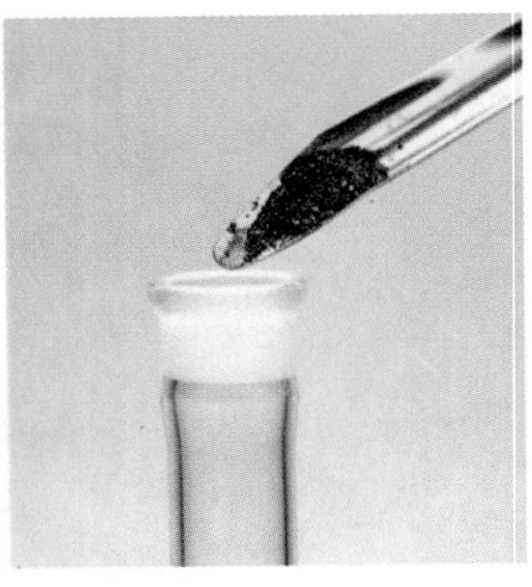

(a)

We report this molarity as 0.146 M $C_{12}H_{22}O_{11}(aq)$. The (aq) indicates an aqueous solution. If we were to dissolve 20.0 g of cane sugar instead of 10.0 g, the sugar in the coffee would be twice as concentrated: its molarity would be 0.292 M $C_{12}H_{22}O_{11}(aq)$.

Molarity is defined in terms of the volume of *solution,* not the volume of solvent used to prepare the solution. This distinction makes sense, because we need to know what volume of solution to measure out to obtain a given amount of solute. The usual way to prepare an aqueous solution of a solid substance is to transfer a known mass of the solid into a volumetric flask, dissolve it in a little water, fill the flask up to the mark with water, and then mix it thoroughly by tipping the flask end over end (Fig. 2.13).

(b)

The molarity of a solution is the number of moles of solute divided by the volume of the solution in liters.

(c)

FIGURE 2.13 The steps involved in making up a solution of known concentration (here, a solution of potassium permanganate, $KMnO_4$). (a) A known mass of the compound is dispensed into a volumetric flask. (b) Some water is added to dissolve it. (c) Finally, water is added up to the mark. The bottom of the solution's meniscus, the curved top surface, should be level with the mark.

EXAMPLE 2.9 Calculating the molarity

A student prepared a solution by dissolving 1.345 g of potassium nitrate, KNO_3, in enough water to prepare 25.00 mL of solution. What is the molarity of the solution?

STRATEGY Because the molarity is the number of moles of formula units divided by the volume of solution, we first convert the mass of solute to moles of formula units by using the molar mass of the solute. Then we calculate the molarity by dividing the moles of solute by the volume of the solution (in liters).

SOLUTION The molar mass of potassium nitrate is 101.11 $g \cdot mol^{-1}$. It follows that the number of moles of KNO_3 formula units in 1.345 g of potassium nitrate is

$$\text{Number of moles } KNO_3 = \frac{1.345 \text{ g}}{101.11 \text{ g} \cdot mol^{-1}} = 1.330 \times 10^{-2} \text{ mol}$$

Because the volume of the solution is 25.00 mL (0.02500 L), the molarity is

$$\text{Molarity} = \frac{1.330 \times 10^{-2} \text{ mol}}{0.02500 \text{ L}} = 0.5320 \text{ mol} \cdot L^{-1}$$

We report this molarity as 0.5320 M $KNO_3(aq)$.

SELF-TEST 2.18A Calculate the molarity of a solution made by dissolving 2.357 g of sodium chloride in enough water to make 75.00 mL of solution.

[*Answer:* 0.5378 M NaCl(aq)]

SELF-TEST 2.18B Calculate the molarity of a solution made by dissolving 1.368 g of glucose ($C_6H_{12}O_6$) in enough water to make 50.00 mL of solution.

Once we know the molarity of a solution, we can calculate the number of moles of solute in a sample of any volume. We can also find out how much solution to use if we want to obtain a given number of moles of solute. These calculations are explained in Toolbox 2.3.

The use of solutions allows us to transfer small amounts of solute readily and precisely; the volume to be transferred is calculated from the molarity.

TOOLBOX 2.3 How to use molarity

The two uses of molarity, *M*, described in this toolbox are based on the definition

$$M\ (\text{mol}\cdot\text{L}^{-1}) = \frac{n\ (\text{moles})}{V\ (\text{liters})}$$

where n is the number of moles of solute and V is the volume of the solution. In each application, we rearrange this formula to give the quantity required on the left and symbols for the information given on the right.

The moles of solute in a given volume of solution Because the volume is given and the moles of solute, n, is the information required, we arrange the expression into

$$n = MV$$

For example, suppose we want to know the number of moles of sucrose molecules in 15 mL (0.015 L) of 0.10 M $C_{12}H_{22}O_{11}(aq)$ solution. Substitution of the data into the right side of this expression gives

$$\begin{aligned}\text{Number of moles } C_{12}H_{22}O_{11} &= (0.10\ \text{mol}\cdot\text{L}^{-1}) \times (0.015\ \text{L}) \\ &= 1.5 \times 10^{-3}\ \text{mol}\end{aligned}$$

SELF-TEST 2.19A How many moles of NaCl formula units are present in 25.00 mL of 1.85 M NaCl(aq)?

[*Answer:* 4.62×10^{-2} mol NaCl]

SELF-TEST 2.19B How many moles of urea molecules are present in 100.0 mL of 1.45×10^{-2} M $(NH_2)_2CO(aq)$?

The volume of solution that contains a given amount of solute When the number of moles of solute is given, we can calculate the volume of solution, *V*, that contains that number of moles by rearranging the expression into

$$V = \frac{n}{M}$$

Suppose, for example, we want to measure out 0.760 mmol $KMnO_4$, where $KMnO_4$ is potassium permanganate, a substance that is used in bleaching straw and purifying water, and we have available a solution of molarity 0.0380 M $KMnO_4(aq)$. To find the correct volume of solution to use, we substitute the data:

$$\begin{aligned}\text{Volume of solution} &= \frac{0.760 \times 10^{-3}\ \text{mol}}{0.0380\ \text{mol}\cdot\text{L}^{-1}} \\ &= \frac{0.760 \times 10^{-3}}{0.0380}\ \text{L} \\ &= 2.00 \times 10^{-2}\ \text{L}\end{aligned}$$

We should therefore transfer 20.0 mL of the permanganate solution by using a buret or pipet. The flask will then contain 0.760 mmol $KMnO_4$.

SELF-TEST 2.20A What volume of a 1.25×10^{-3} M $C_6H_{12}O_6(aq)$ solution should you transfer to obtain a solution that contains 1.44×10^{-6} mol of glucose molecules?

[*Answer:* 1.15 mL]

SELF-TEST 2.20B What volume of 0.358 M HCl(aq) solution should you transfer to have a sample that contains 2.55 mmol HCl?

EXAMPLE 2.10 Preparing a solution of specified molarity

Calculate the mass of potassium permanganate needed to prepare 250 mL of a 0.0380 M $KMnO_4(aq)$ solution.

STRATEGY We need to know the number of moles of solute formula units in the stated volume of solution. For this step, we use the procedure in Toolbox 2.3. Then we convert this number of moles to mass of solute. For this step, we use the procedure in Toolbox 2.2.

SOLUTION The number of moles of $KMnO_4$ in 250 mL (0.250 L) of solution is

$$\begin{aligned}\text{Number of moles } KMnO_4 &= (0.0380\ \text{mol}\cdot\text{L}^{-1}) \times (0.250\ \text{L}) \\ &= 0.00950\ \text{mol}\end{aligned}$$

Because the molar mass of potassium permanganate is 158.04 $g\cdot mol^{-1}$, this number of moles corresponds to the following mass:

$$\begin{aligned}\text{Mass of solute} &= (0.00950\ \text{mol}) \times (158.04\ \text{g}\cdot\text{mol}^{-1}) \\ &= 1.50\ \text{g}\end{aligned}$$

That is, 1.50 g of potassium permanganate should be dissolved in water in a 250-mL volumetric flask, and then water added up to the mark.

SELF-TEST 2.21A Calculate the mass of glucose needed to prepare 150 mL of a 0.442 M $C_6H_{12}O_6(aq)$ solution.

[*Answer:* 11.9 g]

SELF-TEST 2.21B Calculate the mass of oxalic acid needed to prepare 50.00 mL of a 0.125 M $C_2H_2O_4(aq)$ solution.

2.15 DILUTION

Let's go back to the natural products collected from the tunicates mentioned at the start of the chapter. There is a good chance that the research chemists will be able to isolate only a very tiny amount of the active substance. This is where solutions play another very useful role, because they can be used to transfer, not just small, but *very* small amounts of a substance from one container to another. This procedure is important when only a little sample is available, perhaps because of its rarity or its cost. It can also be important when a sample being examined contains very little of the substance of interest, as in the analysis of an ore for a rare metal or of a contaminated food for traces of poison.

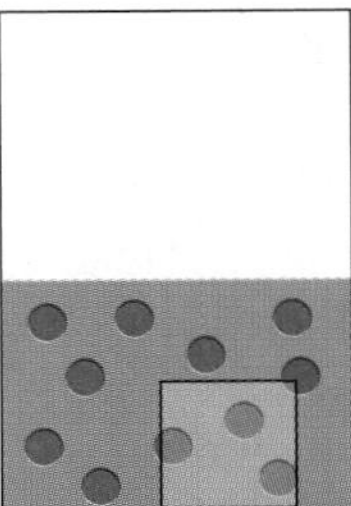

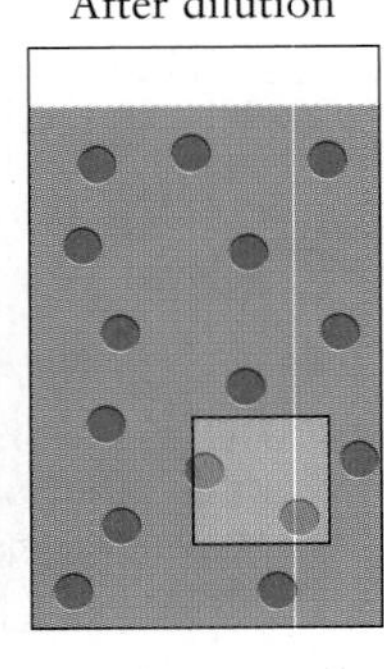

FIGURE 2.14 When a solution is diluted, the same number of solute molecules occupy a larger volume, so there is a smaller number of molecules in a given volume (indicated by the small square).

Suppose we need to add 0.010 mmol $KMnO_4$ to a sample that we think contains the active compound. However, on hand we have available only the 0.0380 M $KMnO_4(aq)$ solution prepared as described in Example 2.10. First, we need to know the volume of solution that contains 0.010 mmol $KMnO_4$:

$$\text{Volume of solution} = \frac{1.0 \times 10^{-5}\ \text{mol}}{0.0380\ \text{mol}\cdot\text{L}^{-1}} = 2.6 \times 10^{-4}\ \text{L}$$

We could measure out 0.010 mmol $KMnO_4$ by measuring out 0.26 mL of the solution. An instrument called a micropipet can be used to measure out very tiny volumes of liquid accurately, perhaps as little as 10^{-3} mL (1 microliter, 1 μL)—about the volume of a pinhead. A much simpler approach, however, does not require such expensive equipment. We simply **dilute,** or reduce the concentration, of the solute by adding more solvent. After a solution has been diluted, the same amount of solute is present but the volume is greater. Less solute is now present in a given volume (Fig. 2.14). An ordinary pipet or buret can then be used for accurate transfer of the solution. If the original solution of molarity 0.0380 M is diluted a hundredfold, then 0.010 mmol $KMnO_4$ would be contained in 26 mL of solution (instead of 0.26 mL).

To dilute a solution in a controlled manner, we need to reduce the molarity from an initial value to some final target value. First, we use a pipet to transfer the appropriate volume of solution to a volumetric flask. Then we add enough solvent to increase the volume of the solution to its final value. Toolbox 2.4 shows how to calculate the correct initial volume of solution to use.

Pipets and burets are narrow tubes used to transfer liquids accurately (see Figs. 2.1 and 2.12).

TOOLBOX 2.4 How to calculate the volume of solution to dilute

This procedure is based on a simple idea: although we may add more solvent to a given volume of solution, we do not change the number of moles of solute. After dilution, the solute simply occupies a larger volume of solution. The procedure has two steps:

Step 1. Calculate the number of moles of formula units in the final solution. The result of this calculation is the amount to transfer into the volumetric flask. As explained in Toolbox 2.3, we use the expression

$$n = M_{\text{final}}V_{\text{final}}$$

where M_{final} is the molarity of the final solution and V_{final} is its volume.

Step 2. Calculate the volume of the *initial* undiluted solution that contains this amount of solute. This step also makes use of the procedure described in Toolbox 2.3. This time, though, we need to use the expression

$$V_{\text{initial}} = \frac{n}{M_{\text{initial}}}$$

Because the moles of solute, n, are the same in these two expressions, they can be combined into

$$V_{\text{initial}} = \frac{M_{\text{final}}V_{\text{final}}}{M_{\text{initial}}}$$

This expression is easy to remember when it is rearranged into the form

$$M_{\text{initial}}V_{\text{initial}} = M_{\text{final}}V_{\text{final}} \qquad (2)$$

That is, the amount of solute in the final solution (the product on the right) is the same as the amount of solute in the initial volume of solution (the product on the left).

As an example, suppose we wanted to know the volume of 0.0380 M $KMnO_4$(aq) solution that we should use to prepare 250 mL of a 1.50×10^{-3} M $KMnO_4$(aq) solution. First, we rearrange Eq. 2 to give the quantities we know on the right and the quantity we want to know on the left:

$$V_{\text{initial}} = \frac{M_{\text{final}}V_{\text{final}}}{M_{\text{initial}}}$$

Then we simply substitute the data:

$$V_{\text{initial}} = \frac{(1.50 \times 10^{-3}\ \text{mol}\cdot\text{L}^{-1}) \times (0.250\ \text{L})}{0.0380\ \text{mol}\cdot\text{L}^{-1}} = 9.87 \times 10^{-3}\ \text{L}$$

We can conclude that 9.87 mL of the original solution should be measured into a 250-mL volumetric flask (using a buret) and water added up to the mark (Fig. 2.15).

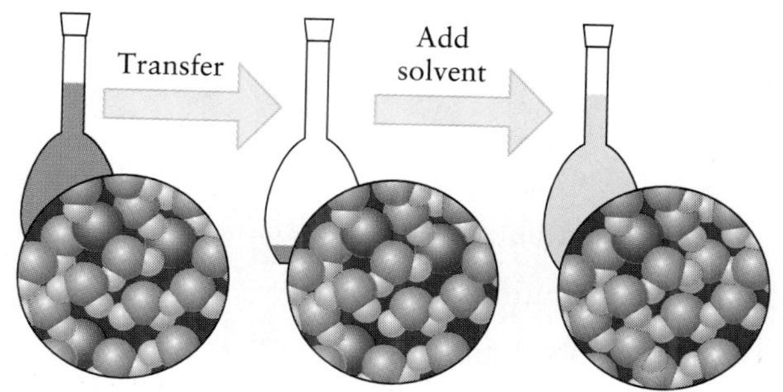

FIGURE 2.15 The steps involved in dilution. A small sample of the original solution is transferred to a volumetric flask, and then water is added up to the mark.

SELF-TEST 2.22A Calculate the volume of 0.0155 M HCl(aq) that we should use to prepare 100 mL of a 5.23×10^{-4} M HCl(aq) solution.

[*Answer:* 3.37 mL]

SELF-TEST 2.22B Calculate the volume of 0.152 M $C_6H_{12}O_6$(aq) that should be used to prepare 25.00 mL of a 1.59×10^{-5} M $C_6H_{12}O_6$(aq) solution.

The use of solutions and of techniques like dilution give chemists very precise control over the amounts of the substances they are handling, even in very small quantities. For example, pipetting 25.0 mL of the 1.50×10^{-3} M $KMnO_4(aq)$ solution prepared in Toolbox 2.4 corresponds to transferring as little as 37.5 μmol $KMnO_4$, or only 5.93 mg of the compound. A mass this small would be difficult to weigh accurately. Furthermore, solutions can be stored in a concentrated form to save space and then diluted to whatever concentration is appropriate for their intended use.

When a small volume of a solution is diluted to a larger volume, the total number of moles of solute in the solution does not change, but the concentration of solute is reduced.

SKILLS YOU SHOULD HAVE MASTERED

Conceptual

1. Explain the significance of units for reporting measurements.
2. Distinguish accuracy and precision.
3. Give the definition of a mole.
4. Explain how diluting a solution affects the concentration of the solute.

Problem-Solving

1. Convert a measurement from one unit to another.
2. Convert between Kelvin, Celsius, and Fahrenheit temperatures.
3. Calculate the density of a substance.
4. Use the correct number of significant figures when reporting measurements and the results of calculations.
5. Use the Avogadro constant to convert between number of moles and the number of atoms, molecules, or ions in a sample.
6. Calculate the average molar mass of an element, given its isotopic composition.
7. Convert between mass and number of moles by using the molar mass.
8. Calculate the empirical formula of a compound from its mass percentage composition.
9. Determine the molecular formula of a compound from its empirical formula and its molar mass.
10. Calculate the molarity of a solute in a solution, volume of solution, and mass of solute, given the other two quantities.
11. Determine the volume of solution needed to prepare a dilute solution of a given molarity.

Descriptive

1. Describe the main steps involved in finding the molecular formula of a compound.
2. Describe the uses of volumetric flasks and pipets.

EXERCISES

International System (SI) of Units

2.1 Express the following quantities in scientific notation: (a) 203 370 000 000 000 dollars, 1994 U.S. budget deficit; (b) 1 169 811, number of AIDS cases reported worldwide, 1980–1995; (c) 0.000006 g, recommended daily allowance of vitamin B_{12} for adults; (d) 0.000 000 1 m, diameter of the smallest living cells.

2.2 Express the following quantities in scientific notation: (a) 56 000 000 km, closest approach of Mars to Earth; (b) 93 900 000 Hz, frequency of a popular FM radio station; (c) 0.000 000 000 000 000 000 000 020 g, mass of one carbon atom; (d) 0.000 000 000 096 m, approximate O—H bond length in water.

2.3 Convert the following quantities into standard scientific notation: (a) 0.0043×10^{2}; (b) 1492×10^{-2}; (c) $0.000\,051 \times 10^{-4}$; (d) 237×10^{12}.

2.4 Convert the following quantities into standard scientific notation: (a) $0.000\,673 \times 10^{9}$; (b) 986×10^{-4}; (c) 0.0018×10^{-3}; (d) 5305×10^{7}.

2.5 Complete the following equalities by using the information in Table 2.1:
(a) 250 g (3 sf) = ______ kg
(b) 25.4 mm = ______ cm (1 in.)
(c) 250 μs (3 sf) = ______ ms
(d) 1.49 cm = ______ dm
(e) 2.48 cg = ______ g
(f) 28.35 g = ______ kg (1 ounce)

2.6 Complete the following equalities by using the information in Table 2.1:
(a) 200 μg (3 sf) = ______ g
(b) 88 nm = ______ pm
(c) 0.0789 mg = ______ μg
(d) 454 g = ______ kg (1 lb)
(e) 25.0 mL = ______ kL
(f) 2000 nm (3sf) = ______ μm

2.7 Complete the following equalities, using scientific notation:
(a) 1 μm = ______ m
(b) 550 nm (3 sf) = ______ mm
(c) 0.10 g = ______ mg
(d) 105 pm = ______ μm

2.8 Express the following measurements in scientific notation:
(a) 186,000 miles·s^{-1} (3 sf), the speed of light
(b) 0.000 000 002 K, the lowest-ever recorded temperature
(c) 1/100 000 000 m
(d) 0.000 000 535 m, the approximate wavelength of green light

2.9 When a piece of metal of mass 3.60 g is dropped into a graduated cylinder containing 8.3 mL of water, the water level rises to 9.8 mL. What is the density of the metal in grams per centimeter cubed (g·cm^{-3})?

2.10 When a piece of metal of mass 5.21 g is dropped into a graduated cylinder containing 16.7 mL of water, the water level rises to 18.2 mL. What is the density of the metal in grams per centimeter cubed (g·cm^{-3})?

2.11 The density of balsa wood is 0.16 g·cm^{-3}. What is the mass of 1.00 ft^3 of balsa wood?

2.12 A supersonic transport (SST) airplane consumes about 18 000 L of kerosene per hour of flight. Kerosene has a density of 0.965 g·mL^{-1}. What mass of kerosene is consumed on a flight of duration 3.0 h?

2.13 The density of diamond is 3.51 g·cm^{-3}. The international (but non-SI) unit for reporting the masses of diamonds is the "carat", with 1 carat = 200 mg. What is the volume of a diamond of mass 0.300 carat?

2.14 What volume (in cm^3) of lead (of density 11.3 g·cm^{-3}) has the same mass as 100 cm^3 of a piece of redwood (of density 0.38 g·cm^{-3})?

Conversion Factors

2.15 Use the conversion factors in Table 2.2 and inside the back cover to express the following measurements in the designated units: (a) 25 L to m^3; (b) 25 g·L^{-1} to mg·dL^{-1}; (c) 1.54 mm·s^{-1} to pm·μs^{-1}; (d) 2.66 g·cm^{-3} to μg·μm^{-3}; (e) 4.2 L·h^{-2} to mL·s^{-2}.

2.16 Use the conversion factors in Table 2.2 and inside the back cover to express the following measurements in the designated units: (a) 4.82 nm to pm; (b) 1.83 mL·min^{-1} to mm^3·s^{-1}; (c) 1.88 ng to kg; (d) 7.01 cm·s^{-1} to km·h^{-1}; (e) 0.044 g·L^{-1} to mg·cm^{-3}; (f) 4.2 °C·s^{-1} to °C·min^{-1}.

2.17 Make the following conversions from one system of units to another: (a) $\frac{4}{5}$ quart bottle of beverage to milliliters; (b) the density of compacted metal from an automobile salvage yard, 450 lb·ft^{-3}, to kg·m^{-3}; (c) \$1.20/gallon to peso/liter (assume 1 dollar ≈ 780 peso); (d) the density of water, 1.0 g·mL^{-1}, to lb·ft^{-3}.

2.18 (a) The distance between the centers of adjacent carbon atoms in a diamond is 154 pm. Express this length in inches. (b) The approximate concentration of sodium in seawater is 10.5 g·L^{-1}. Express this concentration in ounce/gallon. (c) Over 40 billion kilograms (2 sf) of sulfuric acid are produced in the United States annually. Express this production in units of lb/day. (d) The mass of an electron, one of the subatomic particles, is 9.11×10^{-28} g. What is this mass in atomic mass units?

2.19 Convert the following temperatures as indicated: (a) normal body temperature, 98.6°F to °C; (b) −40°C to °F; (c) absolute zero, 0 K to °F; (d) the boiling point of helium, −269°C to K.

2.20 Certain temperatures occur frequently in chemistry, and it is helpful to know their values on various scales. For future convenience, express the following temperatures on the scale indicated: (a) conventional temperature for reporting standard properties, 298.15 K to °C; (b) the boiling point of nitrogen, used as a low-temperature coolant, 77 K to °C; (c) "hot" water that is still comfortable to the skin, 65°C to K and °F; (d) the temperature of a dry ice-acetone mixture, used as a "cold bath" in the laboratory, −78°C to K.

2.21 Convert the following quantities as indicated:
(a) 1 cm^3 = ______ m^3
(b) 30 m·s^{-1} = ______ cm·μs^{-1}

(c) $22\ m^2$ = ______ cm^2
(d) $25\ cm^3$ = ______ mL

2.22 Convert the following quantities to the units designated within the brackets: (a) The density of water at 3.98°C [°F] is 1.0 $g{\cdot}mL^{-1}$ [$mg{\cdot}L^{-1}$]. (b) The density of oxygen gas is 1.43 $g{\cdot}L^{-1}$ [$mg{\cdot}mL^{-1}$] at 0°C [°F]. (c) The volume of a laboratory test tube is 3.0 mL [dL]. (d) A penny has a mass of about 3 g, a diameter of about 1 cm, and a thickness of about 1 mm. Given that the volume of a cylinder of radius r and height h is $\pi r^2 h$, determine the density of a penny in units of $mg{\cdot}mm^{-3}$.

2.23 Rewrite the following statement, using the units in brackets: "A sample of tin of area 1.0 cm^2 [mm^2] was set on a small block of lead of volume 10.0 cm^3 [m^3]. The two metals were placed in a 100-mL [L] flask and 25.0 mL [cm^3] of acid was added."

2.24 Make the appropriate conversions in the following statement: "A chemist recovered a minute sample of the metal iridium of volume 0.5 mm^3 [μm^3] from a land area of 1.5 km^2 [m^2]. The chemist analyzed a 25-mL [dL] soil sample to find the amount of iridium in 1.0 cm^3 [m^3] of soil as a part of a research project to determine whether there had been a major comet impact on Earth at the time of the extinction of the dinosaurs."

Uncertainty of Measurements and Calculations

2.25 What determines the number of significant figures (a) in a measurement; (b) in a calculation involving addition or subtraction; (c) in a calculation involving multiplication or division?

2.26 Explain the difference between *accuracy* and *precision.*

2.27 State the number of significant figures in the following quantities: (a) 2.00 g of silver; (b) 0.0200 s; (c) 2.00×10^2 mL of water; (d) six thermometers; (e) 0.0023°C; (f) 12 $in.{\cdot}ft^{-1}$.

2.28 State the number of significant figures in the following quantities: (a) 3.00100 g of sugar; (b) 12.011 $g{\cdot}mol^{-1}$; (c) $2.998 \times 10^8\ m{\cdot}s^{-1}$ (speed of light); (d) 22 beakers; (e) 0.0001 K; (f) $10^3\ m{\cdot}km^{-1}$.

2.29 Shown below are two pieces of laboratory glassware graduated in milliliters. Report the volumes, using the correct number of significant figures.

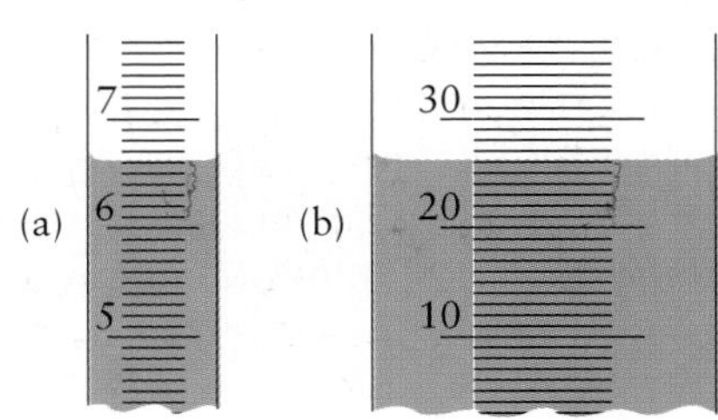

2.30 Record the volumes of solution in milliliters in the two graduated cylinders below. Report the volumes, using the correct number of significant figures.

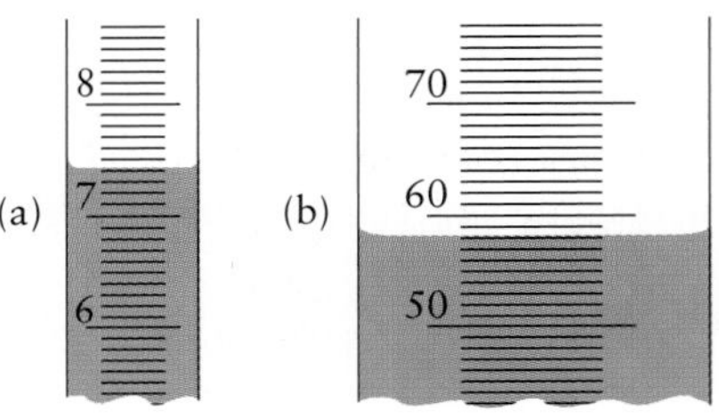

2.31 The masses of copper, zinc, and manganese in a sample of an alloy were measured as 2.011 g, 1.02 g, and 1.4 g, respectively. What is the total mass of the alloy?

2.32 A forensic chemist collected three samples from the scene of a crime. Their masses were 0.220 g, 0.03476 g, and 0.0001 g. What is the total mass of the collected samples?

2.33 A pharmacist made up a capsule consisting of 0.21 g of one drug, 0.124 g of another, and 1.311 g of a "filler." What is the total mass of the contents of the capsule?

2.34 A diamond of mass 2.001 μg was removed from a sample of soil of mass 1.78×10^{-4} g. What is the mass of the remaining sample of soil?

2.35 To how many significant figures should the result of the following calculation be reported?

$$\frac{0.08206 \times (273.15 + 1.2)}{1.23 \times 7.004}$$

2.36 To how many significant figures should the result of the following calculation be reported?

$$\frac{534.71 \times 321.83 \times 0.00186}{7.529 \times 10^{-3}}$$

Fun with Atoms and Moles

2.37 The visible universe is estimated to contain 10^{22} stars. How many moles of stars are there?

2.38 There are approximately 10^{11} neurons in your head. How many heads are needed to have one mole of neurons?

2.39 (a) The approximate population of Earth is 5.7 billion people. How many moles of people inhabit Earth? (b) If all people were pea pickers and pea counters, then how long would it take for the Earth's population to count out 1 mol of peas at the rate of one pea per second, working 24 hours per day, 365 days per year?

2.40 (a) About 1000 metric tons of sand contain about a trillion (10^{12}) grains of sand. How many metric tons of sand are needed to provide 1 mol of sand? (b) Assuming the volume of a grain of sand is 1 mm^3 and the land area of the continental United States is 3,600,000 mi^2, how deep would be the sand pile over the United States if it were evenly covered with 1 mol of grains of sand?

Moles and Molar Masses of Elements

2.41 Calculate the average molar mass of carbon in a natural sample, which consists of 98.89% ^{12}C and 1.11% ^{13}C. The mass of an atom of ^{12}C is 1.9926×10^{-23} g and that of ^{13}C is 2.1593×10^{-23} g.

2.42 The nuclear power industry extracts ^6Li but not ^7Li from natural samples of lithium. As a result, the average molar mass of commercial samples of lithium is increasing. The current abundances of the two isotopes are 7.42% and 92.58%, respectively, and the masses of their atoms are 9.988×10^{-24} g and 1.165×10^{-23} g. (a) What is the current molar mass of a natural sample of lithium? (b) What will be the molar mass when the abundance of ^6Li is reduced to 5.67%?

2.43 Calculate the average molar mass of bromine in a natural sample, which consists of 50.54% ^{79}Br (molar mass 78.918 $g \cdot mol^{-1}$) and 49.46% ^{81}Br (molar mass 80.916 $g \cdot mol^{-1}$).

2.44 Calculate the molar mass of sulfur in a natural sample, which consists of 95.0% ^{32}S (molar mass 31.97 $g \cdot mol^{-1}$), 0.8% ^{33}S (molar mass 32.97 $g \cdot mol^{-1}$), and 4.2% ^{34}S (molar mass 33.97 $g \cdot mol^{-1}$).

2.45 Calculate the amount in moles of (a) 4.82×10^{22} atoms of ^{35}Cl; (b) 2.22 g of copper atoms; (c) 1.11×10^{24} atoms of helium; (d) 8.96 μg of iron atoms.

2.46 Which sample in each of the following pairs contains the greater number of moles of atoms? (a) 25 g of carbon or 35 g of silicon; (b) 1.0 g of Au or 1.0 g of Hg; (c) 2.49×10^{22} atoms of Au or 2.49×10^{22} atoms of Hg.

2.47 Determine the number of atoms in (a) 3.97 mol Xe; (b) 18.3 μg Sc; (c) 12.8 pg Li; (d) 3.78×10^{-4} mol Ar.

2.48 Calculate the mass, in micrograms, of (a) 3.77×10^{18} atoms of Na; (b) 8.22 μmol U; (c) 0.000 006 020 mol K; (d) 6.02×10^{23} boron atoms.

2.49 What mass of nickel contains as many atoms as there are (a) carbon atoms in 12 g of carbon; (b) chromium atoms in 12 g of chromium?

2.50 Determine the mass of aluminum that has the same number of atoms as there are in (a) 6.29 mg of silver; (b) 6.29 mg of gold.

Molar Masses of Compounds

2.51 Determine the molar mass of (a) $CaBr_2$; (b) C_8H_{18}; (c) $NiSO_4 \cdot 6H_2O$; (d) CO_2; (e) CH_4 (methane, the major component of natural gas).

2.52 Determine the molar mass of (a) sulfur tetrafluoride, SF_4; (b) hydrazine, N_2H_4; (c) sodium cyanide, NaCN; (d) sucrose, $C_{12}H_{22}O_{11}$; (e) $CuCl_2 \cdot 4H_2O$.

2.53 Calculate the amount (in moles) and the number of molecules (or atoms, if indicated) in (a) 10.0 g of carbon tetrachloride, CCl_4; (b) 1.65 mg of hydrogen iodide, HI; (c) 3.77 μg of hydrazine, N_2H_4; (d) 500 g of sucrose, $C_{12}H_{22}O_{11}$; (e) 2.33 g of oxygen as O atoms and as O_2 molecules.

2.54 Convert the following masses to amounts (in moles) and to number of molecules (or atoms, if indicated). (a) 1.00 kg of H_2O; (b) 1.00 kg of C_2H_5OH (ethanol); (c) 10.0 g of sulfur, as S atoms and as S_8 molecules; (d) 3.0 g of CO_2; (e) 3.0 g of NO_2.

2.55 Calculate the amount (in moles) of (a) Ag^+ ions in 2.00 g of AgCl; (b) UO_3 in 600 g of UO_3 (3 sf); (c) Cl^- ions in 4.19 mg of $FeCl_3$; (d) H_2O in 1.00 g of $AuCl_3 \cdot 2H_2O$.

2.56 Calculate the amount (in moles) of (a) CN^- in 1.00 g of KCN; (b) H atoms in 200 mg of H_2O; (c) $CaCO_3$ in 500 g of $CaCO_3$ (3 sf); (d) H_2O in 5.00 g of $La_2(SO_4)_3 \cdot 9H_2O$.

2.57 (a) Determine the number of formula units in 0.670 mol $AgNO_3$. (b) What is the mass (in μg) of 2.39×10^{20} formula units of Rb_2SO_4? (c) Estimate the number of formula units in 6.66 kg of $NaHCO_2$, sodium formate, which is used in dyeing and printing fabrics.

2.58 (a) How many NaH formula units are present in 3.61 g of NaH? (b) Determine the mass of 5.78×10^{24} formula units of $NaBF_4$, sodium tetrafluoroborate. (c) Calculate the amount (in moles) of 8.52×10^{20} formula units of CeI_3, cerium(III) iodide, a bright yellow, water-soluble solid.

2.59 (a) Calculate the amount (in moles) of molecules in 1.0 μg of testosterone, $C_{19}H_{28}O_2$, a male sex hormone. (b) What is the mass percentage composition of testosterone?

2.60 (a) Aspartame, $C_{14}H_{18}N_2O_5$, is an artificial sweetener sold as NutraSweet. How many molecules are present in 1.0 mg of aspartame? (b) What is the mass percentage composition of aspartame?

2.61 (a) Calculate the mass, in grams, of a water molecule. (b) Determine the number of H_2O molecules in 1.00 g of H_2O.

2.62 Octane, C_8H_{18}, is typical of the molecules found in gasoline. (a) Calculate the mass of one octane molecule. (b) Determine the number of C_8H_{18} molecules in 1.00 mL of C_8H_{18}, the mass of which is 0.82 g.

2.63 A chemist measured out 5.50 g of copper(II) bromide tetrahydrate, $CuBr_2 \cdot 4H_2O$. (a) How many moles of $CuBr_2 \cdot 4H_2O$ were measured out? (b) How many moles of Br^- are present in the sample? (c) How many water molecules are present in the sample? (d) What fraction of the total mass of the sample was due to copper?

2.64 A chemist wants to extract the gold from 15.0 g of gold(III) chloride dihydrate, $AuCl_3 \cdot 2H_2O$, by electrolysis of an aqueous solution (this technique is described in Chapter 17). What mass of gold could be obtained from the sample?

Determining Chemical Formulas

2.65 Determine the empirical formulas from the following analyses. (a) The mass composition of cryolite, a compound used in the production of aluminum, is 32.79% Na, 13.02% Al, and 54.19% F. (b) A compound used to generate O_2 gas in the laboratory has mass composition 31.91% K and 28.93% Cl, the remainder being oxygen. (c) A fertilizer is found to have the following mass composition: 12.2% N, 5.26% H, 26.9% P, and 55.6% O.

2.66 Determine the empirical formula of each compound from the following data. (a) Talc (used in talcum powder) has mass composition 19.2% Mg, 29.6% Si, 42.2% O, and 9.0% H. (b) Saccharin, a sweetening agent, has mass composition 45.89% C, 2.75% H, 7.65% N, 26.20% O, and 17.50% S. (c) Salicylic acid, used in the synthesis of aspirin, has mass composition 60.87% C, 4.38% H, and 34.75% O.

2.67 In an experiment, 4.14 g of the element phosphorus combined with chlorine to produce 27.8 g of a white solid compound. What is the empirical formula of the compound?

2.68 A chemist found that 4.69 g of sulfur combined with fluorine to produce 15.81 g of a gas. What is the empirical formula of the gas?

2.69 Lindane, used as an insecticide, has mass composition 24.78% C, 2.08% H, and 73.14% Cl and molar mass 290.85 $g \cdot mol^{-1}$. What is the molecular formula of lindane?

2.70 Nicotine has mass composition 74.03% C, 8.70% H, and 17.27% N and molar mass 162.23 $g \cdot mol^{-1}$. Determine the molecular formula of nicotine.

2.71 Caffeine, a primary stimulant in coffee and tea, has molar mass 194.19 $g \cdot mol^{-1}$ and mass composition 49.48% C, 5.19% H, 28.85% N, and 16.48% O. What is the molecular formula of caffeine?

2.72 Cacodyl, which has an intolerable garlicky odor and is used in the manufacture of cacodylic acid, a cotton herbicide, has mass composition 22.88% C, 5.76% H, and 71.36% As, and molar mass 209.96 $g \cdot mol^{-1}$. What is the molecular formula of cacodyl?

Solutions

2.73 (a) An aqueous solution was prepared by dissolving 1.567 mol of $AgNO_3$ in enough water to make 250.0 mL of solution. What is the molarity of silver nitrate in the solution? (b) A 2.11-g sample of NaCl is placed into a 1500-mL volumetric flask. The sample is dissolved and diluted to the mark on the flask with water. What is the molarity of NaCl in the solution?

2.74 (a) A chemist prepared a solution by dissolving 1.230 g KCl in enough water to make 150.0 mL of solution. What molar concentration of potassium chloride should appear on the label? (b) If the chemist had mistakenly used a 500-mL volumetric flask instead of the 150.0-mL flask in (a), what molar concentration of potassium chloride has the chemist actually prepared?

2.75 A chemist studying the properties of photographic emulsions needed to prepare 25.00 mL of 0.155 M $AgNO_3(aq)$. What mass of silver nitrate must be placed into a 25.00-mL volumetric flask and dissolved and diluted to the mark with water?

2.76 What mass of $Na_2CO_3 \cdot 10H_2O$, a compound used in detergents, must be dissolved and diluted to the mark in a 500-mL volumetric flask to prepare a 0.10 M $Na_2CO_3(aq)$ solution?

2.77 A student prepared a solution of barium hydroxide by adding 2.577 g of the solid to a 250.0-mL volumetric flask and adding water to the mark. Some of this solution was transferred to a buret. What volume of solution should the student run into a flask to transfer (a) 1.0 mmol $Ba(OH)_2$; (b) 3.5 mmol OH^-; (c) 50.0 mg $Ba(OH)_2$?

2.78 A student investigating the properties of solutions containing carbonate ions prepared a solution containing 7.112 g of Na_2CO_3 in a 250.0-mL volumetric flask. Some of the solution was transferred to a buret. What volume of solution should be dispensed from the buret to provide (a) 5.112×10^{-3} mol Na_2CO_3; (b) 3.451×10^{-3} mol CO_3^{2-}?

2.79 Explain how you would prepare an aqueous solution of 0.010 M $KMnO_4(aq)$ starting with (a) solid $KMnO_4$; (b) 0.050 M $KMnO_4(aq)$.

2.80 (a) A 12.56-mL sample of 1.345 M $K_2SO_4(aq)$ is diluted to 250.0 mL. What is the molar concentration of K_2SO_4 in the diluted solution? (b) A 25.00-mL sample of 0.366 M HCl(aq) is drawn from a reagent bottle with a pipet. The sample is transferred to a 125.00-mL volumetric flask and diluted to the mark with water. What is the molar concentration of the dilute hydrochloric acid solution?

2.81 (a) What volume of a 0.778 M $Na_2CO_3(aq)$ solution should be diluted to 150.0 mL with water to reduce its concentration to 0.0234 M $Na_2CO_3(aq)$? (b) An experiment requires the use of 60.0 mL of 0.50 M NaOH(aq). The stockroom assistant can only find a reagent bottle of 2.5 M NaOH(aq). How is the 0.50 M NaOH(aq) solution to be prepared?

2.82 0.094 g of $CuSO_4 \cdot 5H_2O$ is dissolved and diluted to the mark in a 500.0-mL volumetric flask. A 2.000-mL sample of this solution is transferred to a second 500.0-mL volumetric flask and diluted. (a) What is the molarity of $CuSO_4$ in the final solution? (b) To prepare the solution directly, what mass of $CuSO_4 \cdot 5H_2O$ would need to be weighed out?

SUPPLEMENTARY EXERCISES

2.83 Select a convenient SI unit from Table 2.1 for recording (a) your mass; (b) the diameter of an atom; (c) the mass of a coin; (d) the tolerance (of the order of 0.00001 m) on a finely machined tool.

2.84 The angstrom unit (1 Å $= 10^{-10}$ m) is still widely used to report measurements of the sizes of atoms and molecules. Express the following data in angstroms: (a) the radius of a sodium atom is 180 pm (2 sf); (b) the wavelength of yellow light is 550 nm (2 sf). (c) Write a (single) conversion factor between angstroms and nanometers.

2.85 Express the volume in milliliters of a "1.00-cup" sample of milk given that 2 cups = 1 pint, 2 pints = 1 quart.

2.86 The distance for a marathon is 26 miles and 385 yards. Convert this distance to kilometers given that 1 mile = 1760 yd.

2.87 The diameter of a typical aerosol smog particle is 0.1 μm. Express this diameter in inches.

2.88 Convert (a) the density of bismuth, 8.90 $g \cdot cm^{-3}$, into $mg \cdot mm^{-3}$; (b) the density of gold, 19.3 $g \cdot cm^{-3}$, into $kg \cdot m^{-3}$.

2.89 An international committee of distinguished scientists met to establish the standard snail's pace, P_{sn}. The average snail covered 1 inch in 2 minutes. Express the average P_{sn} in $cm \cdot s^{-1}$.

2.90 Rewrite the following statements, using the temperature scales indicated in brackets: (a) The melting point of gold is 1064°C [K]. (b) Sulfur boils at 445°C [°F and K]. (c) Neon boils at −411°F [K] (3 sf). (d) The temperature of outer space is 2.7 K [°C].

2.91 The temperature on the day side of the lunar surface is 127°C and the temperature on the night side of the lunar surface is −183°C. Convert each temperature to K and °F.

2.92 The density of gold is 19.3 $g \cdot cm^{-3}$. What volume of water will a gold nugget of mass 16.7 mg displace when placed into a graduated cylinder? The density of water at room temperature is 1.00 $g \cdot cm^{-3}$.

2.93 A chemist determined in a set of four experiments that the density of magnesium metal was 1.68 $g \cdot cm^{-3}$, 1.67 $g \cdot cm^{-3}$, 1.69 $g \cdot cm^{-3}$, 1.69 $g \cdot cm^{-3}$. The accepted value for its density is 1.74 $g \cdot cm^{-3}$. What can you conclude about the precision and accuracy of the chemist's data?

2.94 Determine the average mass, in grams, of (a) one gold atom, and the number of atoms in 1.0 g of gold; (b) one krypton atom, and the number of atoms in 5.0 L of krypton of mass 18.7 g; (c) one deuterium atom, ^2H (molar mass 2.01 $g \cdot mol^{-1}$), and the number of atoms in 1.0 mg of deuterium.

2.95 A chemical reaction requires at least 0.683 mol of sulfur atoms to react with 0.683 mol of copper atoms. (a) How many S atoms are required? (b) How many sulfur molecules, S_8, are necessary? (c) What mass of sulfur is needed for the reaction?

2.96 Copper metal can be extracted from a copper(II) sulfate solution by electrolysis (as described in Chapter 17). If 10.0 g of copper(II) sulfate pentahydrate, $CuSO_4 \cdot 5H_2O$, is dissolved in 100 mL of water and all the copper is electroplated out, what mass of copper would be obtained?

2.97 Epsom salts consists of magnesium sulfate heptahydrate. Write its formula. (a) How many atoms of magnesium are in 2.00 g of Epsom salts? (b) How many formula units of the compound are present in 2.00 g? (c) How many moles of water molecules are in 2.00 g of Epsom salts?

2.98 L-Dopa, a drug used for the treatment of Parkinson's disease, is 54.82% carbon, 5.62% hydrogen, 7.10% nitrogen, and 32.46% oxygen. What is the empirical formula of the compound?

2.99 A chemical analysis of a complex carbohydrate is 40.0% C, 6.72% H, and 53.5% O. Its molar mass is approximately 860 $g \cdot mol^{-1}$. (a) What is the empirical formula of the carbohydrate? (b) What is the molecular formula of the carbohydrate?

2.100 A certain chemical that causes a severe skin disease is composed of 44.3% C, 43.5% Cl, 9.82% O, and 2.47% H. It has a molar mass of 326 $g \cdot mol^{-1}$. (a) What is its empirical formula? (b) Determine its molecular formula.

2.101 The concentration of gold in seawater is 0.011 $\mu g \cdot L^{-1}$. What mass of gold (in kg) is present in the Atlantic Ocean, of volume 3.23×10^{11} km^3?

2.102 What mass of $NiSO_4 \cdot 6H_2O$ must be dissolved and diluted to the mark in a 500-mL volumetric flask to prepare a 0.15 M $NiSO_4$(aq) solution?

2.103 (a) Determine the mass of anhydrous copper(II) sulfate that must be used to prepare 250 mL of a 0.20 M $CuSO_4$(aq) solution. (b) Determine the mass of $CuSO_4 \cdot 5H_2O$ that must be used to prepare 250 mL of a 0.20 M $CuSO_4$(aq) solution.

2.104 The sulfuric acid solution that is purchased for a stockroom has a molarity of 17.8 M; all sulfuric acid solutions for experiments are prepared by dilution of this stock solution. (a) Determine the volume of 17.8 M H_2SO_4 that must be diluted to 250 mL to prepare a 2.0 M H_2SO_4(aq) solution. (b) An experiment requires a 0.50 M H_2SO_4(aq) solution. The stockroom manager estimates that 6.0 L of the acid is needed. What volume of 17.8 M H_2SO_4(aq) must be used for the preparation?

CHALLENGING EXERCISES

2.105 Devise an SI version of the measurement of time in which the basic unit is the second and all other units are expressed in seconds with the appropriate prefix. (a) Express 1 hour in seconds in this system, using scientific notation and the appropriate prefix. (b) Express 1.00 day in seconds.

2.106 The reference points on a newly proposed temperature scale expressed in °X are the freezing and boiling points of water, set equal to 50°X and 250°X. (a) Derive a formula for converting temperatures on the Celsius scale to the new scale. (b) Comfortable room temperature is 22°C. What is that temperature in °X?

2.107 Derive the SI units of pressure by finding the units of each of the following steps: (a) velocity = distance/time; (b) acceleration = velocity/time; (c) force = mass × acceleration; (d) pressure = force/area.

2.108 Calculate the average density of a single carbon atom by assuming that it is a uniform sphere of radius 77 pm. The volume of a sphere is $\frac{4}{3}\pi r^3$, where r is its radius. Express the density of a carbon atom in grams per cubic centimeter. The density of diamond, a crystalline form of carbon, is 3.5 $g \cdot cm^{-3}$. What does your answer suggest about the way the atoms are packed together in diamond?

2.109 Assume the entire mass of an atom is concentrated in its nucleus, a sphere of radius 1.5×10^{-5} pm. (a) What is the density of a carbon nucleus? The volume of a sphere is $\frac{4}{3}\pi r^3$, where r is its radius. (b) What would be the radius of the Earth if its matter were compressed to the same density as a carbon nucleus? (Its average radius is 6.4×10^3 km and its average density is 5.5 $g \cdot cm^{-3}$.)

2.110 A 1.0-cm^3 cube of uranium (of density 18.95 $g \cdot cm^{-3}$) is placed next to a 2.0-cm^3 cube of niobium (of density 8.57 $g \cdot cm^{-3}$). Which cube contains the greater number of atoms?

2.111 The "acre" (1 acre = 4840 yd^2) was originally the area of land a team of oxen could plow in 1 working day. What is this rate in $m^2 \cdot h^{-1}$? (Assume that oxen can work for 8.0 h in 1.0 day.)

2.112 The molar mass of boron atoms in a natural sample is 10.81 $g \cdot mol^{-1}$. The sample is known to consist of ^{10}B (molar mass 10.013 $g \cdot mol^{-1}$) and ^{11}B (molar mass 11.093 $g \cdot mol^{-1}$). What are the percentage abundances of the two isotopes?

2.113 Calculate the molar mass of the noble gas krypton in a natural sample, which consists of 0.3% ^{78}Kr (molar mass 77.92 $g \cdot mol^{-1}$), 2.3% ^{80}Kr (molar mass 79.91 $g \cdot mol^{-1}$), 11.6% ^{82}Kr (molar mass 81.91 $g \cdot mol^{-1}$), 11.5% ^{83}Kr (molar mass 82.92 $g \cdot mol^{-1}$), 56.9% ^{84}Kr (molar mass 83.91 $g \cdot mol^{-1}$), and 17.4% ^{86}Kr (molar mass 85.91 $g \cdot mol^{-1}$).

2.114 In 1978, scientists extracted a compound with antitumor and antiviral properties from marine animals in the Caribbean Sea. A 2.52-mg sample of the compound, didemnin-C, was analyzed and found to have the following composition: 1.55 mg C, 0.204 mg H, 0.209 mg N, and 0.557 mg O. The molar mass of didemnin-C was found to be 1014 $g \cdot mol^{-1}$. What is its molecular formula?

2.115 The alcohol content in most wines is 12% by volume, corresponding to 9.7% by mass. The formula of ethanol, the alcohol from fermented grains, is C_2H_5OH. How many molecules of ethanol are in 100 mL of wine? The density of wine is approximately 0.98 $g \cdot mL^{-1}$.

2.116 A chemist prepared an aqueous solution by mixing 2.50 g of ammonium phosphate trihydrate, $(NH_4)_3PO_4 \cdot 3H_2O$ and 1.50 g of potassium phosphate, K_3PO_4, with 500 g water. (a) Determine the number of moles of formula units of each compound that was measured. (b) How many moles of PO_4^{3-} are present in solution? (c) Calculate the mass of phosphate ions present in the solution. (d) What is the total mass of the water present in the solution?

2.117 To prepare a very dilute solution, it is advisable to perform successive dilutions of a single prepared reagent solution, rather than weighing a very small mass or measuring a very small volume of stock chemical. A solution was prepared by transferring 0.661 g of $K_2Cr_2O_7$ to a 250.0-mL volumetric flask and adding water to the mark. A 1.000-mL sample of this solution was transferred to a 500-mL volumetric flask and diluted to the mark with water. Then 10 mL of the diluted solution was transferred to a 250-mL flask and diluted to the mark with water. (a) What is the final concentration of $K_2Cr_2O_7$ in solution? (b) What mass of $K_2Cr_2O_7$ is in this final solution? (The answer to the last question gives the amount that would have had to have been weighed out if the solution had been prepared directly.)

2.118 Certain chemicals used in research are extremely rare and expensive. Suppose you wrote an order for 2.0 g of a chemical at $500 per gram. When your sample arrives, you weigh it and observe that the actual mass is only 1.96 g. How much did you overpay for your compound? Considering how you wrote your order, was the chemical company justified in sending the amount they did?

CHAPTER 3

When we look at changes in the natural world, from the germination of a seed through the unfurling of a leaf to the decay of fall leaves, we may be amazed by the variety of processes that can take place. Yet chemists have discovered that all chemical reactions can be classified into a limited number of types. These shoots of the bloodroot plant (Sanguinaria canadensis), *a member of the poppy family, make use of the reactions described in this chapter as they convert the debris of the forest floor into new living cells. The sap of the bloodroot was once used as a dye, and it contains medically active compounds.*

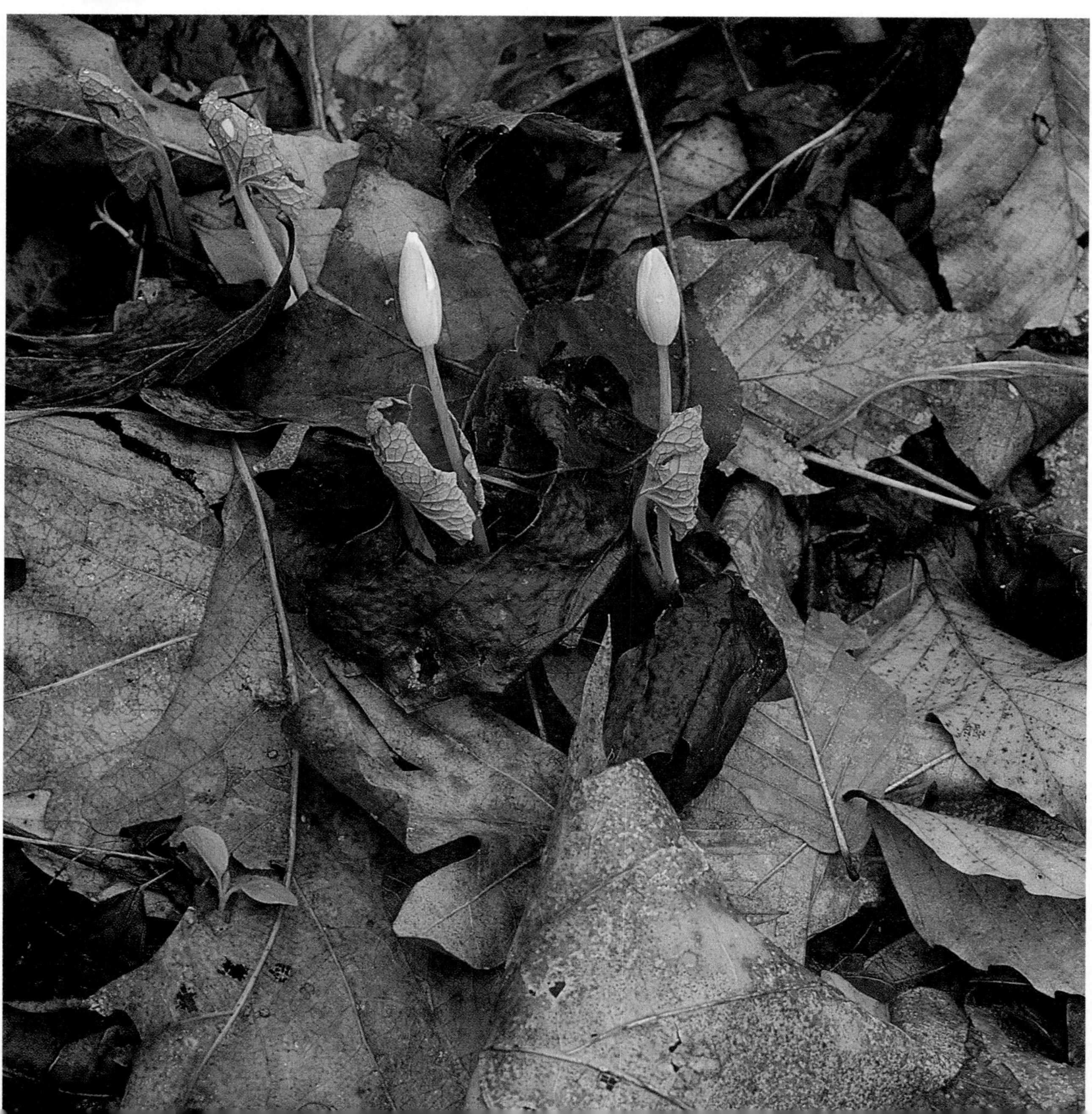

CHEMICAL REACTIONS: MODIFYING MATTER

With every breath we take, we inhale carbon dioxide molecules along with the other gases of the air. Each breath connects us with history. Some of the carbon atoms in the molecules were once exhaled by our ancestors; in fact, some of them were exhaled by virtually everyone who has ever lived. One atom might have come from Julius Caesar, another from Mohammed, and another from Jesus Christ. Think of *anyone:* we are inhaling a part of them. We are also contributing to forests, to prairies, and to future generations. The air that we exhale contributes to the reservoir of carbon dioxide needed by plants. In the presence of sunlight and water, they use it to produce the compounds called carbohydrates, which include sugars, starches, and cellulose. It is startling to realize that all the vegetation of the world has, in effect, been plucked from the skies.

The progress of carbon atoms through the environment is called the **carbon cycle.** For example, some atmospheric carbon dioxide dissolves in seawater, where it is absorbed into the shells of shellfish. After the death of the animal, it is compressed to form limestone (calcium carbonate, $CaCO_3$). It may remain as limestone or marble (a denser form of limestone) for centuries until, one day, someone converts it into mortar for buildings, uses it to reduce the acidity of soils and feed nations, or quarries it and carves a statue.

The life story of carbon atoms includes both physical and chemical changes. In this chapter we shall introduce several kinds of chemical change, in which one substance changes into another. The process that brings about a chemical change is called a **chemical reaction.** All the processes of life involve chemical reactions in one way or another;

CHEMICAL EQUATIONS AND CHEMICAL REACTIONS

3.1 Symbolizing chemical reactions
3.2 Balancing chemical equations

PRECIPITATION REACTIONS

3.3 Aqueous solutions
3.4 Reactions between strong electrolyte solutions
3.5 Ionic and net ionic equations
3.6 Putting precipitation to work

THE REACTIONS OF ACIDS AND BASES

3.7 Acids and bases in aqueous solution
3.8 Strong and weak acids and bases
3.9 Acidic and basic character in the periodic table
3.10 Neutralization
3.11 The formation of gases

REDOX REACTIONS

3.12 Oxidation and reduction
3.13 Keeping track of electrons: Oxidation numbers
3.14 Oxidizing and reducing agents
3.15 Balancing simple redox equations
3.16 Classifying reactions

and to understand how our bodies grow, repair themselves, and reproduce, we need to know about reactions. The chemical industry depends on chemical reactions to make the products essential to modern life. Pollution is a sign of uncontrolled chemical reactions; clearing up pollution depends on the wiser use of other chemical reactions.

There are millions of different chemical reactions, but they all fall into a limited number of types. Once we have learned the different types of reactions that are likely to occur when substances are mixed or heated, we can begin to make predictions. In this chapter, we shall meet three of the most common types of reactions.

CHEMICAL EQUATIONS AND CHEMICAL REACTIONS

A chemical reaction is the *process* of chemical change. The starting materials are called the **reactants.** The substances formed are called the **products.** Bottles of chemicals kept on hand in a laboratory are called **reagents.** Only when a reagent is being used in a particular reaction is it called a reactant.

3.1 SYMBOLIZING CHEMICAL REACTIONS

You can read the arrow as "go to," "form," or "react to form."

A chemical reaction is symbolized by an arrow:

$$\text{Reactants} \longrightarrow \text{products}$$

For example, sodium is a soft, shiny metal that reacts vigorously with water. When we drop a small lump of sodium metal into a container of water, hydrogen gas forms rapidly and sodium hydroxide is left in solution (Fig. 3.1). To summarize this reaction, we could write

$$Na + H_2O \longrightarrow NaOH + H_2$$

FIGURE 3.1 When a small piece of sodium is dropped into water, a vigorous reaction takes place. Hydrogen gas and sodium hydroxide are formed, and the heat released melts the sodium, which then assumes a rounded shape. A dye that changes color in the presence of sodium hydroxide causes the pink color.

This expression is called a **skeletal equation** because it shows the bare bones of the reaction (the identities of the reactants and products) in terms of chemical formulas. A skeletal equation is a *qualitative* summary of a chemical reaction.

Chemists have found a way to report reactions in a way that lets us understand them *quantitatively.* To do so, we recognize that atoms are neither created nor destroyed in a chemical reaction: they simply change their partners. The principal evidence for this conclusion is that if a reaction is carried out in a sealed container, then there is no change of mass. The preservation of mass during a chemical reaction is called the **law of conservation of mass.** To express the fact that atoms are neither created nor destroyed, chemists report reactions so that the same numbers of atoms of each element are shown on each side of the arrow. For example, there are two H atoms on the left of the skeletal equation above, but three H atoms on the right. So, we change the expression to

$$2\,Na + 2\,H_2O \longrightarrow 2\,NaOH + H_2$$

Now there are four H atoms, two Na atoms, and two O atoms on each side.

The numbers multiplying *entire* chemical formulas in chemical equations (for example, the 2 multiplying H_2O) are called the **stoichiometric coefficients** of the substances. A coefficient of 1 (for H_2) is not written explicitly. When the same numbers of atoms of each element appear on each side of the arrow, the expression is said to be **balanced,** and it is called a **chemical equation.** We distinguish between the chemical reaction (the actual process) and the chemical equation, which summarizes the reaction in terms of chemical formulas.

The word *stoichiometric* comes from the Greek words for "element" and "measure."

Let's look at a simple chemical equation and interpret what it is telling us. We see from the chemical equation written above that for every two Na atoms and two H_2O molecules that react, two NaOH formula units and one H_2 molecule are formed. Therefore, for every two *moles* of Na atoms and two *moles* of H_2O molecules that react, two *moles* of NaOH formula units and one *mole* of H_2 molecules are formed. In other words, *the stoichiometric coefficients multiplying the chemical formulas in any balanced chemical equation tell us the relative number of moles of each substance that reacts or is produced in the reaction.*

We can add additional useful information to a chemical equation by indicating the physical state of each reactant and product with a label called a **state symbol:**

(s): solid (l): liquid (g): gas (aq): aqueous solution

For the reaction between solid sodium and water, the complete, balanced chemical equation is therefore

$$2\,Na(s) + 2\,H_2O(l) \longrightarrow 2\,NaOH(aq) + H_2(g)$$

A pictorial representation of this reaction is shown in Fig. 3.2.

When we want to show that a reaction requires high temperatures, we include the Greek letter Δ (delta) over the arrow. For example, the conversion of limestone to quicklime takes place at about 800° C, and we write

$$CaCO_3(s) \xrightarrow{\Delta} CaO(s) + CO_2(g)$$

The carbon dioxide is released into the atmosphere. Quicklime (calcium oxide, CaO) is widely used in industry, particularly in steelmaking, glassmaking, and the manufacture of cement.

A chemical equation uses chemical formulas to symbolize both the qualitative change that occurs in a chemical reaction and also the quantitative information that no atoms are created or destroyed. The stoichiometric coefficients tell us the relative numbers of moles of reactants and products taking part in the reaction.

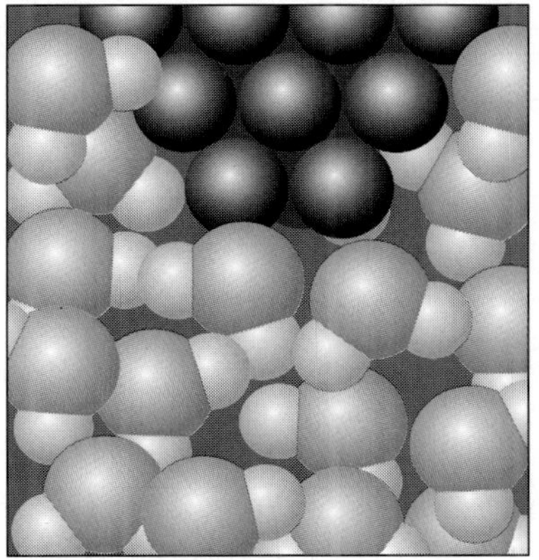

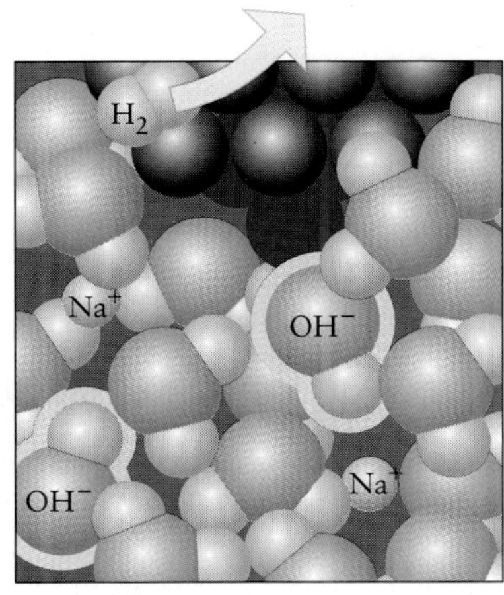

FIGURE 3.2 A representation of the reaction between sodium (the dark gray atoms) and water. Note that two sodium atoms give rise to two sodium ions (light purple), and that two water molecules give rise to one hydrogen molecule (which escapes as a gas) and two hydroxide ions (outlined in yellow). There is a rearrangement of partners, not a creation or annihilation of atoms.

3.2 BALANCING CHEMICAL EQUATIONS

Simple reactions can usually be balanced almost at a glance; others need more work. Consider the reaction in which hydrogen and oxygen gases combine to form water. To write the balanced equation, we always start

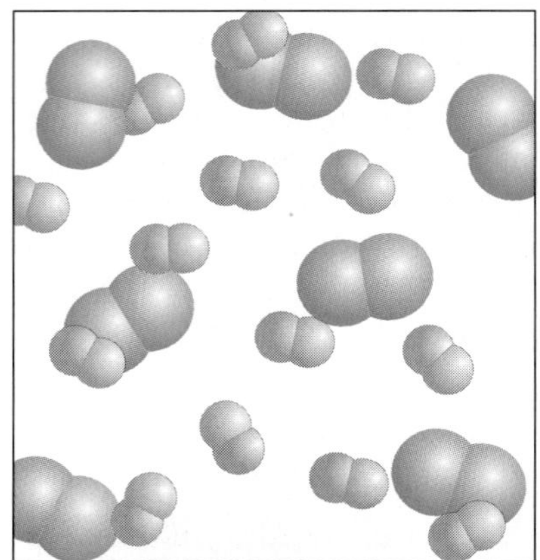

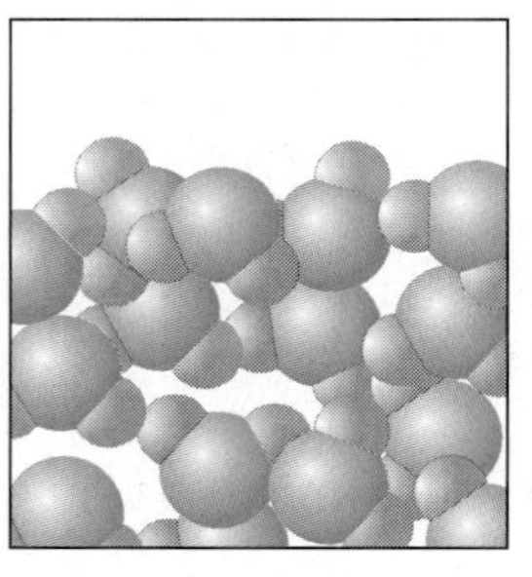

FIGURE 3.3 A representation of the reaction between hydrogen and oxygen that leads to the production of water. No atoms are created or destroyed: they simply change partners. For every two hydrogen molecules that react, one oxygen molecule is consumed and two water molecules are formed.

by summarizing the qualitative information as a skeletal equation:

$$H_2 + O_2 \longrightarrow H_2O \quad ⚠$$

We shall use the international *Hazard!* road sign (⚠) to warn that a skeletal equation is not balanced. Then we find the stoichiometric coefficients that balance all the elements:

$$2\,H_2 + O_2 \longrightarrow 2\,H_2O$$

Once the equation has been balanced, we add the state symbols:

$$2\,H_2(g) + O_2(g) \longrightarrow 2\,H_2O(l)$$

A pictorial representation of this reaction at the molecular level is shown in Fig. 3.3.

An equation must never be balanced by changing the subscripts in the chemical formulas. That change would suggest that different substances were taking part in the reaction. For example, changing H_2O to H_2O_2 in the skeletal equation and writing

$$H_2 + O_2 \longrightarrow H_2O_2$$

certainly results in a balanced equation. However, it is a summary of a *different* reaction—the formation of hydrogen peroxide (H_2O_2) from its elements. Nor should we write

$$2\,H + O \longrightarrow H_2O$$

Although this equation is balanced, it summarizes the reaction between hydrogen and oxygen *atoms,* not the molecules that are the actual starting materials.

Some chemical equations are more difficult to balance. Toolbox 3.1 describes a procedure to use in these cases.

A chemical equation expresses a chemical reaction in terms of chemical formulas; the stoichiometric coefficients are chosen to show that atoms are neither created nor destroyed in the reaction.

TOOLBOX 3.1 How to balance chemical equations

To balance a chemical equation, consider one element at a time. In many cases, when we multiply a chemical formula by a coefficient to balance a particular element, the balance of the other elements in the formula is upset. It is therefore wise to reduce the amount of work by proceeding systematically. One procedure can reduce the number of steps:

Balance first the element that occurs in the fewest formulas.

Balance last the element that is found in the greatest number of formulas.

For example, methane, CH_4, is the main ingredient of natural gas (Fig. 3.4). When methane burns in air, it forms carbon dioxide and water, so the skeletal equation for the reaction is

$$CH_4 + O_2 \longrightarrow CO_2 + H_2O \quad ⚠$$

Because C and H occur in two formulas and O occurs in three, we begin with C and H. The C atoms are already balanced. We balance the H atoms by using a stoichiometric coefficient of 2 for H_2O to give four H atoms on each side:

$$CH_4 + O_2 \longrightarrow CO_2 + 2\,H_2O \quad ⚠$$

Now only O remains to be balanced. Because there are four O atoms on the right but only two on the left, the O_2 needs a stoichiometric coefficient of 2. The result is

$$CH_4 + 2\,O_2 \longrightarrow CO_2 + 2\,H_2O$$

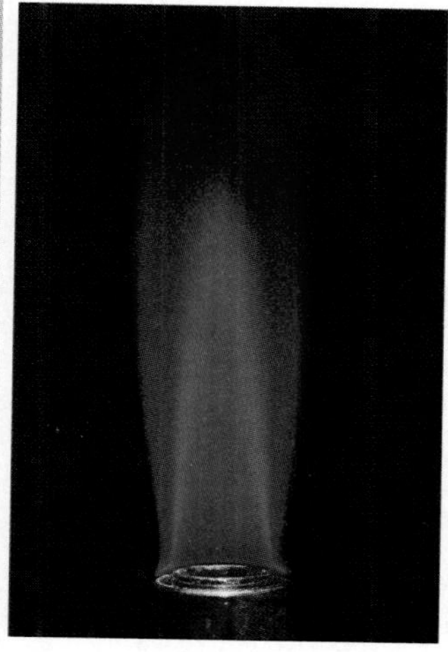

FIGURE 3.4 When methane burns, it forms carbon dioxide and water. The blue color is due to the presence of C_2 molecules in the flame. If the oxygen supply is inadequate, these molecules can stick together and form soot, producing a smokey flame.

We verify that the equation is balanced by counting the numbers of atoms of each element on each side of the arrow. At this stage, we specify the states. If water is produced as a vapor, we write

$$CH_4(g) + 2\,O_2(g) \longrightarrow CO_2(g) + 2\,H_2O(g)$$

A pictorial representation of this reaction is shown in Fig. 3.5.

In some cases, this balancing procedure leads to fractional stoichiometric coefficients, as in the equation for the combustion of butane, C_4H_{10}:

$$C_4H_{10}(g) + \tfrac{13}{2}\,O_2(g) \longrightarrow 4\,CO_2(g) + 5\,H_2O(g)$$

There is nothing wrong with fractional coefficients, but it is common to clear the fractions by multiplying the equation by a numerical factor. An *entire* chemical equation can always be multiplied by a factor without affecting its validity, because multiplication of *both* sides does not affect the balance of each type of atom. In this case, we can clear the fraction $\tfrac{13}{2}$ by multiplying by 2:

$$2\,C_4H_{10}(g) + 13\,O_2(g) \longrightarrow 8\,CO_2(g) + 10\,H_2O(g)$$

SELF-TEST 3.1A When aluminum is melted and heated with solid barium oxide, BaO, a vigorous reaction takes place, and elemental molten barium and solid aluminum oxide, Al_2O_3, are formed. Write the chemical equation for the reaction.

[*Answer:* $2\,Al(l) + 3\,BaO(s) \xrightarrow{\Delta} Al_2O_3(s) + 3\,Ba(l)$]

SELF-TEST 3.1B Write the balanced chemical equation for the combustion of propane gas, C_3H_8, to carbon dioxide and liquid water.

Polyatomic ions (such as NH_4^+ and PO_4^{3-}) often remain intact during a reaction. When that is so, they can be balanced as single entities. Remember that a subscript outside parentheses tells us how many of the polyatomic ions within the parentheses are present. For example, suppose we have to balance the following skeletal equation, which describes the reaction between calcium hydroxide and phosphoric acid to form water and calcium phosphate:

$$Ca(OH)_2(aq) + H_3PO_4(aq) \longrightarrow H_2O(l) + Ca_3(PO_4)_2(s) \quad \triangle$$

The PO_4 group is present on both sides of the arrow (on the right as the PO_4^{3-} ion), so we treat the whole group as a single entity and use a coefficient 2 for H_3PO_4 to balance it. Then to balance calcium, we use a coefficient 3 for $Ca(OH)_2$. We are left with 12 hydrogen atoms on the left and two on the right, so we use a coefficient 6 for H_2O. Oxygen is now balanced too.

$$3\,Ca(OH)_2(aq) + 2\,H_3PO_4(aq) \longrightarrow 6\,H_2O(l) + Ca_3(PO_4)_2(s)$$

SELF-TEST 3.2A Balance the following skeletal equation:

$$Mg(NO_3)_2(aq) + (NH_4)_2CO_3(aq) \longrightarrow MgCO_3(s) + NH_4NO_3(aq) \quad \triangle$$

[*Answer:* $Mg(NO_3)_2(aq) + (NH_4)_2CO_3(aq) \rightarrow MgCO_3(s) + 2\,NH_4NO_3(aq)$]

SELF-TEST 3.2B Balance the following skeletal equation:

$$HF(g) + SiO_2(s) \longrightarrow SiF_4(g) + H_2O(l) \quad \triangle$$

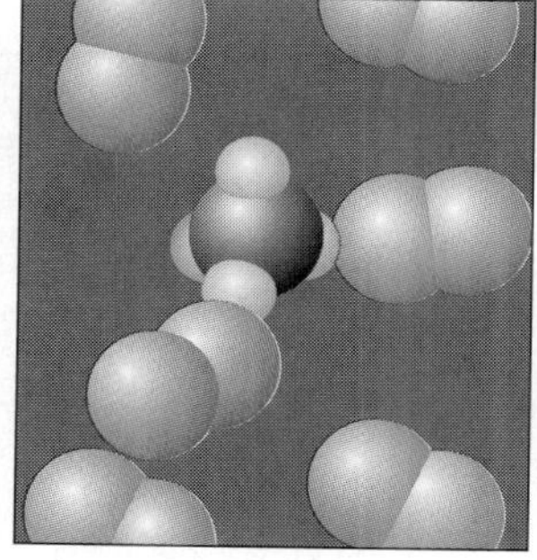
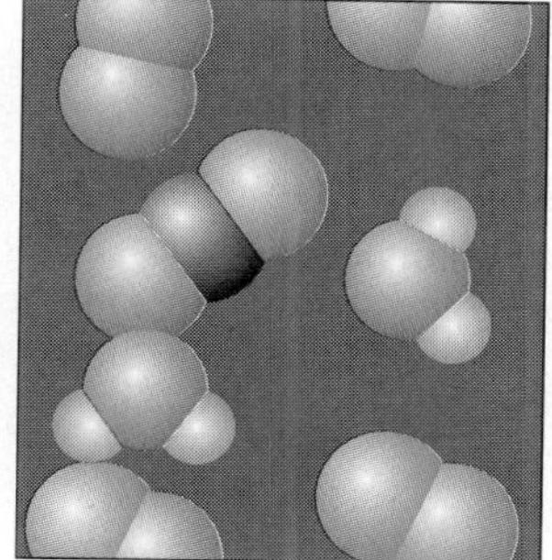

FIGURE 3.5 A representation of the combustion of methane. Note that one carbon dioxide molecule and two water molecules are produced for each methane molecule that is consumed. The hydrogen atoms in a water molecule do not necessarily both come from the same methane molecule: the illustration depicts the overall outcome, not the specific outcome of the reaction of one molecule.

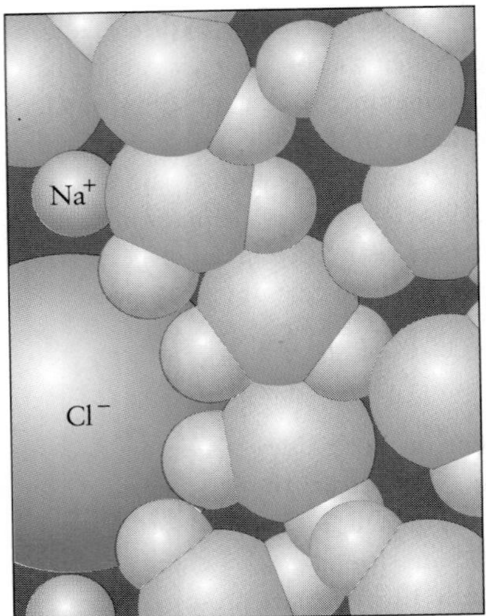

FIGURE 3.6 Sodium chloride consists of sodium ions and chloride ions. When it is in contact with water (top), the ions are able to separate, and they spread through the solvent (bottom). The solution consists of water molecules, sodium ions, and chloride ions. There are no NaCl molecules present at any stage.

PRECIPITATION REACTIONS

At the mouth of every river, where it feeds into the sea, is a large area of muddy mineral deposits called a delta. The processes responsible for the formation of river deltas include precipitation reactions. To understand these chemical reactions, we need to know something about the nature of solutions.

3.3 AQUEOUS SOLUTIONS

A **soluble substance** is one that dissolves in a specified solvent. The solvent is often water; and whenever we refer to solubility in this chapter, we mean "soluble in water." An **insoluble substance** is one that does not dissolve significantly in a specified solvent: substances are often regarded as "insoluble" if they do not dissolve to more than about $0.1\ mol \cdot L^{-1}$. Throughout this chapter, we shall use the term *insoluble* to mean "insoluble in water."

It is important to know whether the solute is present as ions or as molecules (see Sections 1.9 and 1.10). The distinction is made experimentally by determining whether the solution conducts an electric current. Because a current is a flow of electric charge, only solutions that contain ions conduct electricity. An **electrolyte solution** is a solution that does conduct electricity, so we can conclude that ions are present. Electrolyte solutions are often aqueous solutions of ionic compounds, such as sodium chloride and potassium nitrate. They also might be aqueous solutions of acids, such as hydrochloric acid and sulfuric acid.

Ionic compounds produce electrolyte solutions. When they dissolve, the ions become separated from one another by water molecules. The ions are not *formed* when an ionic solid dissolves but become free to

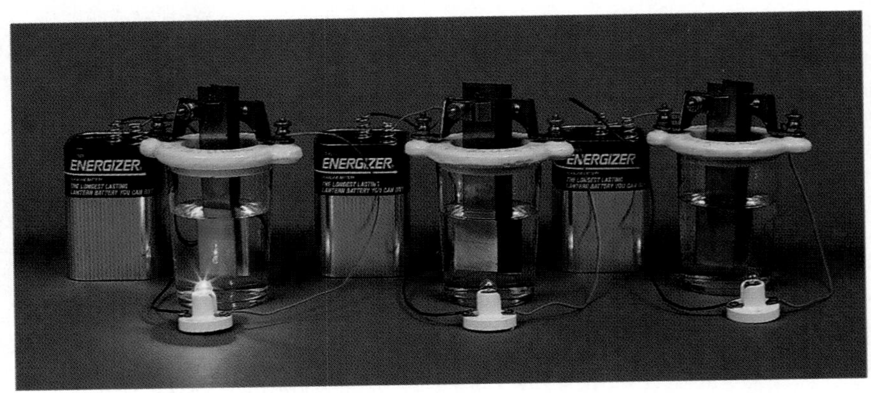

FIGURE 3.7 Pure water is a poor conductor of electricity, as shown by the very dim glow of the bulb in the circuit (right). However, when ions are present, as in an electrolyte solution, the solution does conduct. The ability of the solution to conduct is low if it is a weak electrolyte (center), but significant if it is a strong electrolyte (left) even if the solute concentration is the same in each case.

move apart in the presence of water (Fig. 3.6). Some molecular compounds form electrolyte solutions. In such cases, the ions are not present until the substance dissolves. One example is hydrogen chloride, which exists as molecules of HCl when pure, but in solution is present as hydrogen ions and chloride ions.

▶▶ We shall treat these species in more detail in Sections 3.7 and 3.8.

In a **strong electrolyte solution,** the solute is present entirely as ions. In a **weak electrolyte solution,** the solute is only incompletely ionized in solution; in other words, some molecules survive. Aqueous acetic acid is a weak electrolyte: in aqueous solution, only a small fraction of CH_3COOH molecules ionizes to hydrogen ions and acetate ions, $CH_3CO_2^-$. Because ions have electrical charges, they can carry a current by moving through a solution. One consequence of the difference between solutions of strong and weak electrolyte solutions is illustrated in Fig. 3.7.

A **nonelectrolyte solution** is a solution that does not conduct electricity because ions are not present. Nonelectrolyte solutions include aqueous solutions of acetone (**1**) and glucose (**2**). Sweetened water consists of intact sugar molecules moving among the water molecules. Because there are no ions present, the solution does not conduct electricity. If we could see the individual molecules in a nonelectrolyte solution, we would see the solute molecules dispersed among the solvent molecules (Fig. 3.8).

There are a few ions present because H_2O forms some H_3O^+ and OH^- ions. However, these ions are present in *very* low concentrations.

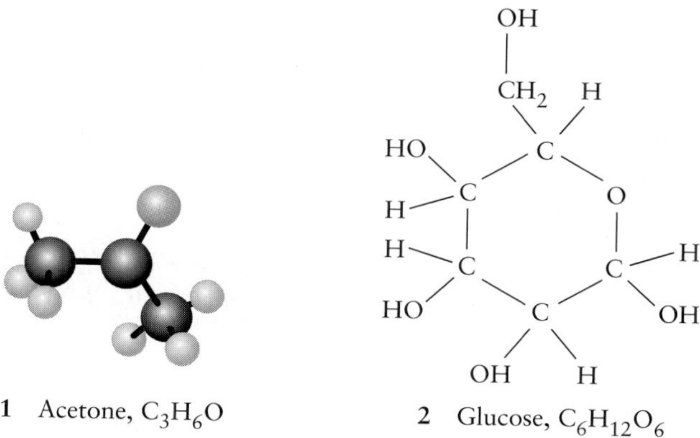

1 Acetone, C_3H_6O

2 Glucose, $C_6H_{12}O_6$

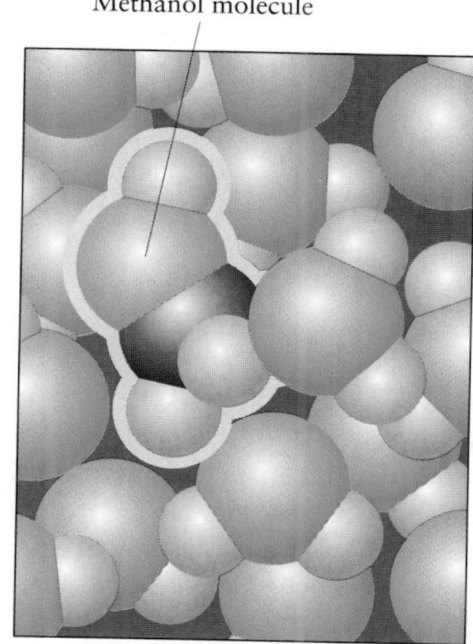

FIGURE 3.8 In a nonelectrolyte solution, the solute remains as molecules and does not break up into ions. Methanol, CH_3OH, is a nonelectrolyte and is present as molecules when it is dissolved in water.

For the rest of this section we shall consider only strong electrolyte solutions. If we could see the individual ions in a strong electrolyte solution, we would see each ion jostling through the solvent, meeting another ion only occasionally, but sometimes lingering near an ion of opposite charge, then moving off again (Fig. 3.9). Each ion has a number of water molecules closely associated with it (Fig. 3.10). We say that each ion is **hydrated.** A sodium cation, for example, may have about half a dozen water molecules stuck to it, and it carries this coating of water molecules through the solution as it moves. A chloride ion is also hydrated in water. The coating of water molecules on each ion helps to prevent the ions from coming out of solution and forming a solid again.

An aqueous strong electrolyte solution consists of hydrated ions that are free to move through the solvent.

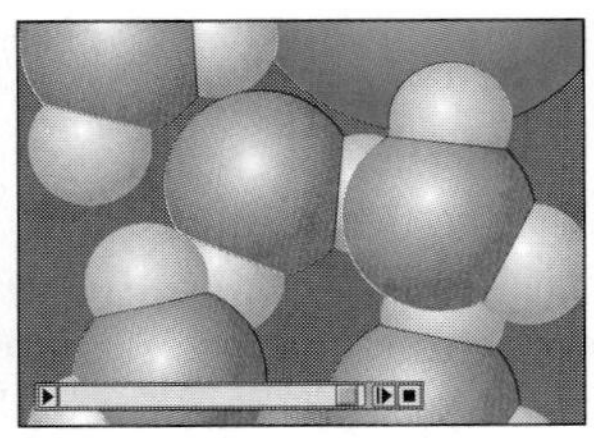

FIGURE 3.9 This animation sequence shows a series of scenes in a solution of sodium chloride. A sodium ion and a chloride ion move together, linger near each other for a time because of the attraction of their opposite charges, and then move apart and meet other partners elsewhere. The loose, transient association of oppositely charged ions is called an ion pair.

EXAMPLE 3.1 Predicting conductivity

Which of the following compounds gives aqueous solutions that conduct electricity? (a) Na_2SO_4; (b) O_2; (c) $C_{12}H_{22}O_{11}$ (sucrose).

STRATEGY Compounds that are present as ions in aqueous solution will conduct electricity. So, we need to identify the species that are present as ions in water. Organic compounds usually form nonelectrolyte solutions, except for the carboxylic acids, which form weak electrolyte solutions.

SOLUTION (a) Sodium sulfate, Na_2SO_4, is an ionic compound, so its solution in water conducts electricity. (b) Oxygen, O_2, is a gas that dissolves sparingly, as molecules, in water; its solution does not conduct electricity. (c) Sucrose, $C_{12}H_{22}O_{11}$, is an organic compound: its molecules remain intact in solution, so its solution does not conduct electricity.

SELF-TEST 3.3A Identify the following solutions as electrolytes or nonelectrolytes and predict which will conduct electricity: (a) NaOH; (b) Br_2.

[*Answer:* (a) Strong electrolyte, conducts electricity; (b) nonelectrolyte, does not conduct electricity]

SELF-TEST 3.3B Identify the following solutions as electrolytes or nonelectrolytes and predict which will conduct electricity: (a) ethanol, $CH_3CH_2OH(aq)$; (b) $Pb(NO_3)_2(aq)$.

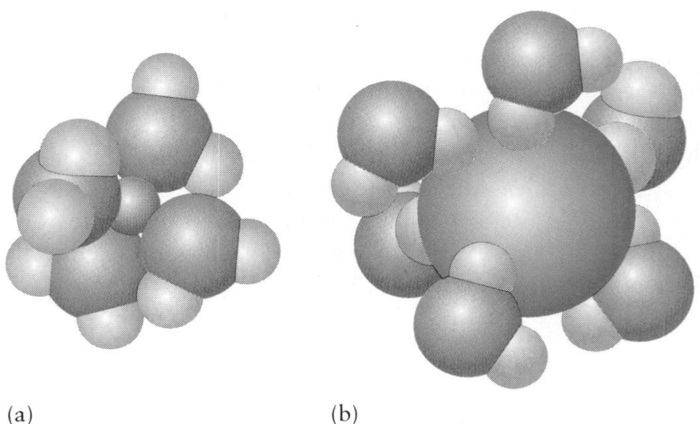

FIGURE 3.10 In water, ions are hydrated; that is, they are surrounded by a cluster of water molecules bound loosely to the ion. Note that a cation (a) is surrounded by water molecules, with the O atom in the water molecule closest to the ion, whereas an anion (b) has water molecules attached through their hydrogen atoms. The number of hydrating molecules depends on the size of the ion, but for most ions it is approximately six.

3.4 REACTIONS BETWEEN STRONG ELECTROLYTE SOLUTIONS

Many ionic solids dissolve in water to give strong electrolyte solutions. Sodium chloride, NaCl, dissolves to give a colorless solution consisting of Na^+ cations and Cl^- anions. Similarly, silver nitrate, $AgNO_3$, dissolves to give a colorless solution consisting of Ag^+ cations and NO_3^- anions. However, if we mix these two aqueous solutions, we immediately get a white **precipitate,** a cloudy deposit. Analysis shows that the precipitate is silver chloride, AgCl, an insoluble white solid (Fig. 3.11). When an insoluble substance is formed in water, it immediately precipitates. The colorless solution remaining above the precipitate in our example contains Na^+ cations and NO_3^- anions. These ions remain in solution because sodium nitrate ($NaNO_3$) is soluble in water.

FIGURE 3.11 The formation of a silver chloride precipitate occurs immediately when silver nitrate solution is added to a solution of sodium chloride.

A **precipitation reaction** is a reaction in which an insoluble solid product is formed when two electrolyte solutions are mixed. We write the chemical equation for the precipitation reaction between sodium chloride and silver nitrate as follows, with (aq) indicating substances that are dissolved in water and (s) a solid that has precipitated:

$$AgNO_3(aq) + NaCl(aq) \longrightarrow AgCl(s) + NaNO_3(aq)$$

A more colorful precipitation reaction is the one that occurs when solutions of lead(II) nitrate, $Pb(NO_3)_2$, and potassium chromate, K_2CrO_4, are mixed. We get an immediate precipitate of insoluble yellow lead(II) chromate, $PbCrO_4$ (Fig. 3.12):

$$Pb(NO_3)_2(aq) + K_2CrO_4(aq) \longrightarrow PbCrO_4(s) + 2\,KNO_3(aq)$$

Lead(II) chromate is mixed with lead(II) sulfate, $PbSO_4$, and lead(II) molybdate, $PbMoO_4$, and blended into paint to add a bright, golden yellow color to traffic markings on highways.

A precipitation reaction occurs when solutions of two strong electrolytes are mixed and result in the formation of an insoluble precipitate.

3.5 IONIC AND NET IONIC EQUATIONS

A **complete ionic equation** for a precipitation reaction shows all the species of ions in solution explicitly. For example, the complete ionic equation for the silver chloride precipitation shown in Fig. 3.11 is

$$Ag^+(aq) + NO_3^-(aq) + Na^+(aq) + Cl^-(aq) \longrightarrow AgCl(s) + Na^+(aq) + NO_3^-(aq)$$

In the complete ionic equation it is easy to see which compounds are dissolved, because they are shown as separate ions.

FIGURE 3.12 In this precipitation reaction, yellow lead(II) chromate is formed when lead(II) nitrate solution is added to a solution of potassium chromate.

Because the Na^+ and NO_3^- ions appear as both reactants and products, they play no direct role in the reaction. They are **spectator ions,** ions that are present while the reaction takes place but remain unchanged. Because spectator ions remain unchanged, we can cancel them on each side of the arrow in the ionic equation:

$$Ag^+(aq) + \cancel{NO_3^-}(aq) + \cancel{Na^+}(aq) + Cl^-(aq) \longrightarrow AgCl(s) + \cancel{Na^+}(aq) + \cancel{NO_3^-}(aq)$$

This cancellation leaves the **net ionic equation** for the reaction, the chemical equation that displays the net change that occurs in the reaction:

$$Ag^+(aq) + Cl^-(aq) \longrightarrow AgCl(s)$$

In other words, a net ionic equation shows only the chemical change that occurs (Fig. 3.13). It shows that the Ag^+ ions supplied by one solution combine with the Cl^- ions supplied by the other solution, and that these ions precipitate as solid silver chloride, AgCl. A net ionic equation focuses our attention on the actual process taking place and emphasizes the change.

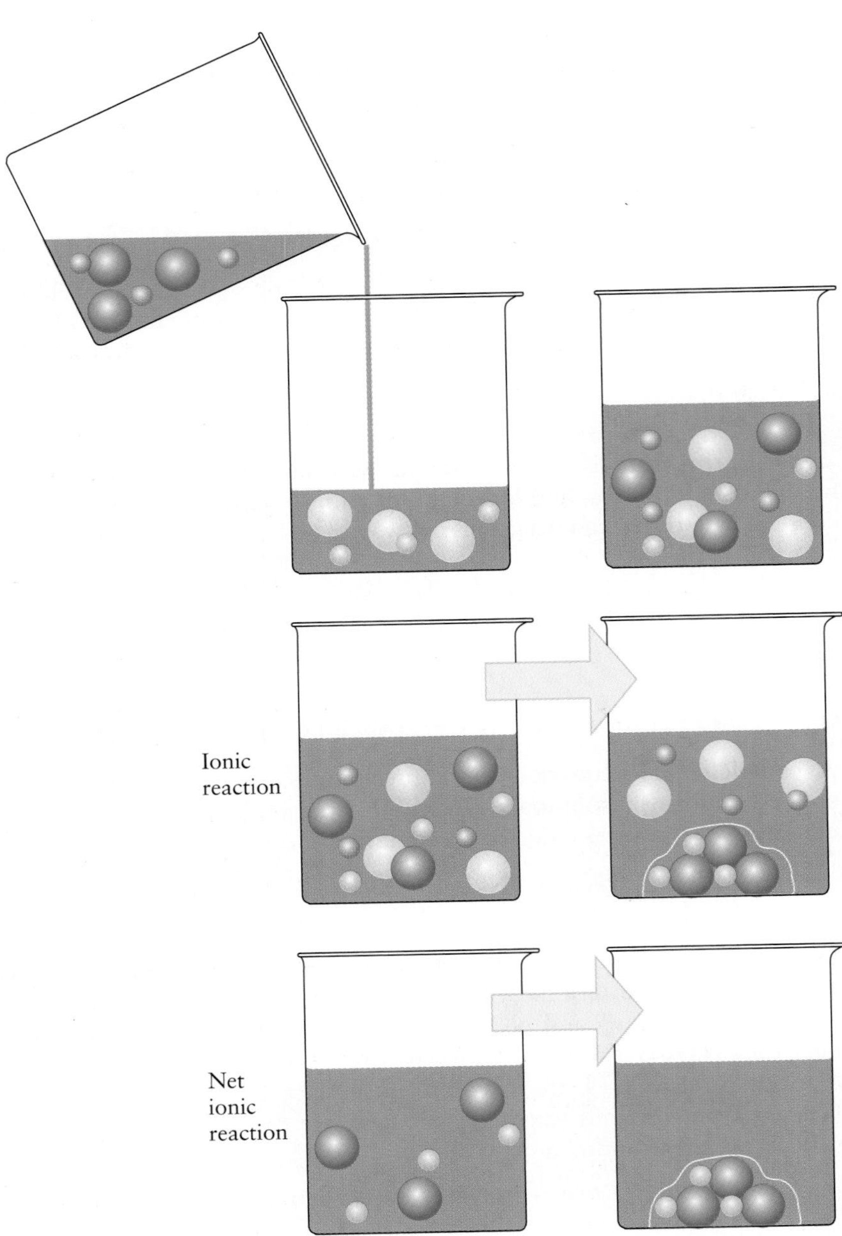

FIGURE 3.13 Two depictions of a precipitation reaction that results when the ions in two electrolyte solutions are mixed (top beakers). By imagining the ionic reaction without the spectator ions (bottom beakers), we can focus on the essential process described by the net ionic equation.

We have imagined what a solution would look like if we could see the individual molecules and ions. What would we observe if we could see a precipitation reaction in progress? Suppose we mix solutions of $AgNO_3$ and NaCl. In the two separate solutions, the hydrated cations and anions each jostle through the solvent. In the mixed solution, when Ag^+ ions encounter Cl^- ions, they stick together and form clumps of ions. The resulting compound, silver chloride, is insoluble and comes out of solution. These encounters occur throughout the solution as soon as the two solutions are mixed, and there is no time for large crystals of silver chloride to grow. Solid particles are formed everywhere in the solution where Ag^+ ions and Cl^- ions meet, so the product is formed as a cloudy, finely divided solid.

A complete ionic equation expresses a reaction in terms of the ions that are present in solution; a net ionic equation is the chemical equation that remains after the cancellation of the spectator ions.

EXAMPLE 3.2 Writing a net ionic equation

The chemical equation for the precipitation reaction that occurs when aqueous solutions of barium nitrate, $Ba(NO_3)_2$, and ammonium iodate, NH_4IO_3, are mixed to form barium iodate, $Ba(IO_3)_2$, is

$$Ba(NO_3)_2(aq) + 2\,NH_4IO_3(aq) \longrightarrow Ba(IO_3)_2(s) + 2\,NH_4NO_3(aq)$$

Write the net ionic equation for the reaction.

STRATEGY First, we write and balance the complete ionic equation, showing all the dissolved ions as they actually exist in solution—as separate, charged ions. Insoluble solids are shown as complete compounds. Next, we cancel the spectator ions—the ions that remain in solution on both sides of the arrow.

SOLUTION The complete ionic equation, with all the dissolved ions written as they exist in the separate solutions, is

$$Ba^{2+}(aq) + \cancel{2\,NO_3^-}(aq) + \cancel{2\,NH_4^+}(aq) + 2\,IO_3^-(aq) \longrightarrow Ba(IO_3)_2(s) + \cancel{2\,NH_4^+}(aq) + \cancel{2\,NO_3^-}(aq)$$

The spectator ions are NH_4^+ and NO_3^-. When they are canceled, we obtain the net ionic equation:

$$Ba^{2+}(aq) + 2\,IO_3^-(aq) \longrightarrow Ba(IO_3)_2(s)$$

Notice how much easier it is to see which species actually react by looking at the net ionic equation rather than either the complete ionic equation or the chemical equation itself.

SELF-TEST 3.4A Write the net ionic equation for the reaction in which aqueous solutions of colorless silver nitrate and yellow sodium chromate react to give a precipitate of red silver chromate.

[*Answer:* $2\,Ag^+(aq) + CrO_4^{2-}(aq) \rightarrow Ag_2CrO_4(s)$]

SELF-TEST 3.4B The mercury(I) ion, Hg_2^{2+}, consists of two Hg^+ ions joined together. Write the net ionic equation for the reaction in which colorless aqueous solutions of mercury(I) nitrate, $Hg_2(NO_3)_2$, and potassium phosphate, K_3PO_4, react to give a white precipitate of mercury(I) phosphate. Mercury(I) phosphate was used to treat syphilis before its deadly side effects were known.

3.6 PUTTING PRECIPITATION TO WORK

We can make insoluble compounds by choosing starting solutions that will result in a precipitate of the desired insoluble compound when they are mixed. Then we can separate it from the reaction mixture by filtration. This strategy is commonly used in environmental monitoring, where it may be necessary to find out how much lead or mercury is in a sample of water. Precipitating the lead or mercury ions as insoluble compounds allows us to separate them and measure their amounts. The **solubility rules** given in Table 3.1 summarize the observed solubility patterns of common ionic compounds in water. Notice that all nitrates and all compounds of the common Group 1 metals are soluble, so they make useful starting solutions.

The mercury(I) ion is Hg_2^{2+}, with two Hg^+ ions bonded together.

Any spectator ions can be used, provided they remain in solution and do not otherwise react. For example, the table shows that mercury(I) iodide, Hg_2I_2, is insoluble. It is formed as a precipitate whenever solutions containing Hg_2^{2+} ions and I^- ions are mixed:

$$Hg_2^{2+}(aq) + 2\,I^-(aq) \longrightarrow Hg_2I_2(s)$$

Mercury(I) nitrate, $Hg_2(NO_3)_2$, is soluble (like all nitrates) and potassium iodide, KI, is soluble (like all common compounds of Group 1 metals), so we can predict that mixing solutions of mercury(I) nitrate and potassium iodide will result in the precipitation of mercury(I) iodide (Fig. 3.14). The chemical equation for this reaction is

$$Hg_2(NO_3)_2(aq) + 2\,KI(aq) \longrightarrow Hg_2I_2(s) + 2\,KNO_3(aq)$$

and the complete ionic equation is

$$Hg_2^{2+}(aq) + 2\,NO_3^-(aq) + 2\,K^+(aq) + 2\,I^-(aq) \longrightarrow Hg_2I_2(s) + 2\,NO_3^-(aq) + 2\,K^+(aq)$$

FIGURE 3.14 Another example of a precipitation reaction. This time, a solution of mercury(I) nitrate is being added to a solution of potassium iodide, and the insoluble product, mercury(I) iodide, is precipitated.

The K^+ and NO_3^- ions are spectators and remain in solution. The same precipitate results when we mix *any* soluble mercury(I) compound with *any* soluble iodide, because in every case the mercury(I) and iodide ions will form insoluble mercury(I) iodide. For example, we could also form mercury(I) iodide from aqueous solutions of mercury(I) acetate and sodium iodide, both of which are soluble:

$$Hg_2(CH_3CO_2)_2(aq) + 2\,NaI(aq) \longrightarrow Hg_2I_2(s) + 2\,NaCH_3CO_2(aq)$$

This chemical equation has the same net ionic equation as the equation for the reaction of $Hg_2(NO_3)_2$ and KI:

$$Hg_2^{2+}(aq) + 2\,I^-(aq) \longrightarrow Hg_2I_2(s)$$

EXAMPLE 3.3 Predicting the outcome of a precipitation reaction

Predict the likely product when aqueous solutions of sodium phosphate, Na_3PO_4, and lead(II) nitrate, $Pb(NO_3)_2$, are mixed. Write the net ionic equation for the reaction.

STRATEGY Decide which ions are present in the mixed solutions and consider all possible combinations. Use the solubility rules to decide which combination corresponds to an insoluble compound, and write the net ionic equation to match.

TABLE 3.1 Solubility rules for inorganic compounds

Soluble compounds	Insoluble compounds
compounds of Group 1 elements	carbonates (CO_3^{2-}), chromates (CrO_4^{2-}), oxalates ($C_2O_4^{2-}$), and phosphates (PO_4^{3-}), **except** those of the Group 1 elements and NH_4^+
ammonium (NH_4^+) compounds	sulfides (S^{2-}), **except** those of the Group 1 and 2 elements and NH_4^+
chlorides (Cl^-), bromides (Br^-), and iodides (I^-), **except** those of Ag^+, Hg_2^{2+}, and Pb^{2+}*	hydroxides (OH^-) and oxides (O^{2-}), **except** those of the Group 1 and 2 elements†
nitrates (NO_3^-), acetates ($CH_3CO_2^-$), chlorates (ClO_3^-), and perchlorates (ClO_4^-)	
sulfates (SO_4^{2-}), **except** those of Ca^{2+}, Sr^{2+}, Ba^{2+}, Pb^{2+}, Hg_2^{2+}, and Ag^+‡	

*$PbCl_2$ is slightly soluble.

†$Ca(OH)_2$ and $Sr(OH)_2$ are sparingly (slightly) soluble; $Mg(OH)_2$ is only very slightly soluble.

‡Ag_2SO_4 is slightly soluble.

SOLUTION The mixed solution will contain Na^+, PO_4^{3-}, Pb^{2+}, and NO_3^- ions. All nitrates and all compounds of Group 1 metals are soluble, but phosphates of other elements are generally insoluble. Hence, we can predict that Pb^{2+} and PO_4^{3-} ions will form an insoluble compound and that lead(II) phosphate, $Pb_3(PO_4)_2$, will precipitate:

$$3\,Pb^{2+}(aq) + 2\,PO_4^{3-}(aq) \longrightarrow Pb_3(PO_4)_2(s)$$

SELF-TEST 3.5A Predict the identity of any precipitate that forms when aqueous solutions of ammonium sulfide, $(NH_4)_2S$, and copper(II) sulfate, $CuSO_4$, are mixed, and write the net ionic equation for the reaction.

[*Answer:* Copper(II) sulfide; $Cu^{2+}(aq) + S^{2-}(aq) \rightarrow CuS(s)$]

SELF-TEST 3.5B Suggest two solutions that can be mixed to prepare strontium sulfate, and write the net ionic equation for the reaction.

FIGURE 3.15 The shell of this shellfish has grown as a result of a precipitation reaction in which the organism has secreted calcium ions, which have reacted with carbonate ions in the surrounding water. The colors of the shell are due to iron impurities that have been captured in the solid precipitate as it formed.

Nature also makes use of precipitation reactions. For instance, shellfish form their shells from calcium carbonate. The organism secretes calcium ions from cells that are in contact with seawater, which contains dissolved carbon dioxide, some of which is present as carbonate ions, CO_3^{2-}. The ions combine to give a precipitate of calcium carbonate:

$$Ca^{2+}(aq) + CO_3^{2-}(aq) \longrightarrow CaCO_3(s)$$

The organism constructs its shell by secreting the calcium ions at the right spots (Fig. 3.15).

The solubility rules in Table 3.1 are used to predict and rationalize precipitation reactions.

3.6 PUTTING PRECIPITATION TO WORK

THE REACTIONS OF ACIDS AND BASES

Every major city in the United States has an environmental monitoring site. Every second throughout the day, acid concentrations in air are measured at these sites and the result is sent electronically to a central station. There the measurements are compared with the maximum safe concentrations. In industry, acidity not only must be monitored but also must be controlled. Even minor items are monitored: each can of tomato soup, for instance, must have exactly the same acidity level, or the taste will not be right. Our bodies also function as efficient chemical factories that control the acidity of our blood automatically. If this automatic control were to fail, we would die. Clearly, acids and bases are compounds of great importance to humanity.

3 Acetic acid, CH_3COOH

Early chemists applied the term *acid* to substances that have a sharp, or sour, taste. Vinegar, for instance, contains acetic acid, CH_3COOH (**3**). Aqueous solutions of substances they called *bases*, which are also known as **alkalis,** were recognized by their soapy feel. However, tasting and feeling are very dangerous ways to determine classes of compounds: you must *never* taste or touch solutions of chemicals in a laboratory. Fortunately, there are less hazardous ways of recognizing acids and bases. For instance, acids and bases change the color of certain dyes known as indicators (Fig. 3.16). One of the most famous indicators is litmus, a vegetable dye obtained from lichen. Aqueous solutions of acids turn litmus red; aqueous solutions of bases (alkalis) turn it blue.

3.7 ACIDS AND BASES IN AQUEOUS SOLUTION

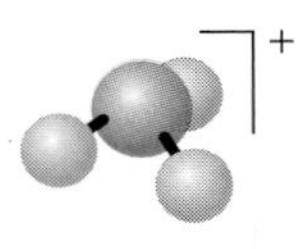

4 Hydronium ion, H_3O^+

When a molecule of an acid dissolves in water, it loses a hydrogen ion, H^+, to one of the water molecules, and forms a hydronium ion, H_3O^+ (**4**). For example, when hydrogen chloride, HCl, dissolves in water, it

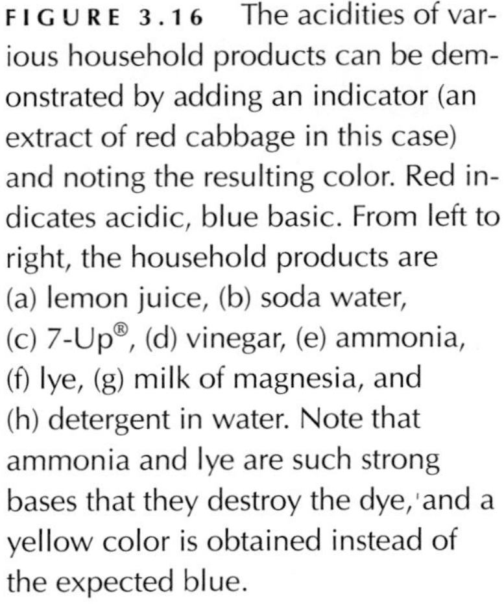

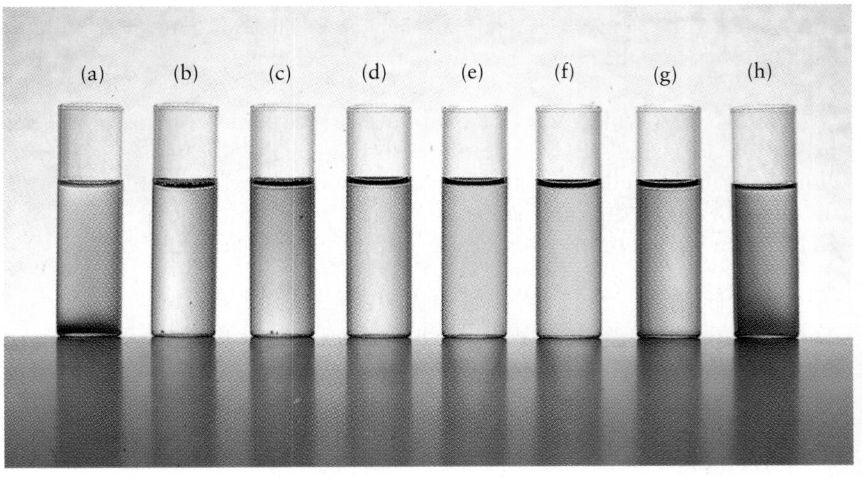

FIGURE 3.16 The acidities of various household products can be demonstrated by adding an indicator (an extract of red cabbage in this case) and noting the resulting color. Red indicates acidic, blue basic. From left to right, the household products are (a) lemon juice, (b) soda water, (c) 7-Up®, (d) vinegar, (e) ammonia, (f) lye, (g) milk of magnesia, and (h) detergent in water. Note that ammonia and lye are such strong bases that they destroy the dye, and a yellow color is obtained instead of the expected blue.

releases a hydrogen ion, and the resulting solution consists of hydronium ions and chloride ions:

$$HCl(aq) + H_2O(l) \longrightarrow H_3O^+(aq) + Cl^-(aq)$$

All acids produce hydronium ions in solution, and we now classify a species as an acid if it can act in this way:

An **acid** is a molecule or polyatomic ion that contains hydrogen and reacts with water to produce hydronium ions.

▶▶ This definition is due to the Swedish chemist Svante Arrhenius. We shall meet a more general definition in Section 14.1.

Although many acids are molecular compounds before they dissolve, in water they give electrolyte solutions.

Hydrogen chloride, HCl, and nitric acid, HNO_3, are acids. Both contain hydrogen and form hydronium ions in water; both turn litmus red. An aqueous solution of hydrogen chloride, HCl(aq), is called hydrochloric acid; the other hydrogen halides have analogous names, for example, HBr is hydrobromic acid. Methane, CH_4, is not an acid. Although it contains hydrogen, it does not release hydrogen ions in water and has no effect on litmus. Acetic acid, CH_3COOH, releases *one* hydrogen ion in water (from the hydrogen atom attached to the oxygen atom), so it is an acid.

The **acidic hydrogen atom** in a compound is the hydrogen atom that can be released in water as a hydrogen ion. It is often written as the first element in a molecular formula, as in HCl, HNO_3, and $HC_2H_3O_2$. However, for organic acids, such as acetic acid, it is more informative to write the formulas to show the *carboxyl group,* COOH (5), explicitly to make it easier to remember that the H atom in this group of atoms is the acidic one. Organic compounds that contain a carboxyl group are called *carboxylic acids.*

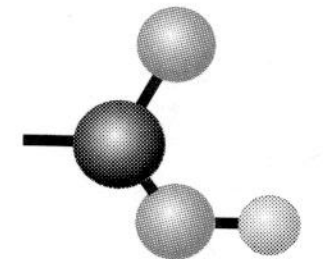

5 Carboxyl group, COOH

We can usually recognize which compounds are acids by noting whether their molecular formulas begin with H or—if they are organic compounds—whether they contain a carboxyl group. Thus, we can immediately recognize that HCl, H_2CO_3 (carbonic acid), H_2SO_4 (sulfuric acid), and C_6H_5COOH (benzoic acid) are acids, but that CH_4, NH_3 (ammonia), and $CH_3CO_2^-$ (the acetate ion) are not. The common oxoacids, acids containing oxygen, introduced in Section 1.14, are listed in Table 1.5.

Now let's consider bases. All bases produce hydroxide ions in solution, and we now classify a species as a base if it can act this way:

A **base** is a molecule or ion that produces hydroxide ions in water.

▶▶ Arrhenius suggested this definition too. Once again, a more general definition will be given in Section 14.1.

When the ionic compound sodium hydroxide, NaOH, dissolves in water, its ions separate and the concentration of hydroxide ions is increased; so sodium hydroxide is a base. The molecular compound ammonia, NH_3, does not contain hydroxide ions. However, when it dissolves, it attracts a hydrogen ion from a neighboring water molecule and forms an ammonium ion, NH_4^+:

$$NH_3(aq) + H_2O(l) \longrightarrow NH_4^+(aq) + OH^-(aq)$$

As a result of this transfer of a hydrogen ion from water to ammonia, the concentration of hydroxide ions in the solution is increased, so we classify ammonia as a base.

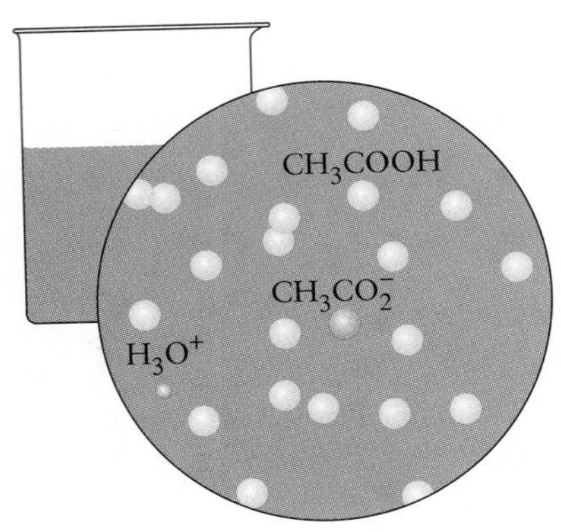

FIGURE 3.17 Acetic acid, like all carboxylic acids, is a weak acid in water. That classification indicates that most of it remains as acetic acid molecules, CH_3COOH, but a small proportion of these molecules donate a hydrogen ion to a water molecule to give hydronium ions, H_3O^+, and acetate ions, $CH_3CO_2^-$.

Acids are molecules or ions that contain hydrogen and produce hydronium ions in water. Bases are molecules or ions that produce hydroxide ions in solution.

SELF-TEST 3.6A Which of the following are acids or bases? (a) HNO_3; (b) C_6H_6; (c) KOH; (d) C_3H_5COOH.

[*Answer:* (a) and (d) are acids; (b) is neither an acid nor a base; (c) is a base.]

SELF-TEST 3.6B Which of the following are acids or bases? (a) KCl; (b) HClO; (c) HF; (d) $Ca(OH)_2$.

3.8 STRONG AND WEAK ACIDS AND BASES

Aqueous solutions of acids and bases are electrolytes and are classified in similar terms:

A **strong acid** is almost completely ionized in solution.

A **weak acid** is incompletely ionized in solution.

In this context "ionized" means that a hydrogen ion has been transferred from the acid molecule to a water molecule to form a hydronium ion. By "completely ionized," we mean that each acid molecule or ion has lost its acidic hydrogen atom by transferring it as a hydrogen ion to a water molecule. By "incompletely ionized," we mean that only a fraction (and usually a very small fraction) of the acid molecules or ions has lost their acidic hydrogen atoms.

Hydrochloric acid is a strong acid. A solution of hydrogen chloride and water contains hydronium ions, chloride ions, and virtually no HCl molecules:

$$HCl(aq) + H_2O(l) \longrightarrow H_3O^+(aq) + Cl^-(aq)$$

Acetic acid, on the other hand, is a weak acid. Because it is an acid, it undergoes ionization in water according to the equation

$$CH_3COOH(aq) + H_2O(l) \longrightarrow H_3O^+(aq) + CH_3CO_2^-(aq)$$

However, because it is a weak acid, the solution consists principally of CH_3COOH molecules, with only a small proportion of hydronium ions and acetate ions, $CH_3CO_2^-$ (Fig. 3.17).

There are very few strong acids, and all the common ones are listed in Table 3.2. As you can see, they include three acids that are often found as reagents in laboratories, namely, hydrochloric acid, nitric acid, and sulfuric acid. Most acids are weak. All carboxylic acids are weak. For example, a 0.1 M solution of acetic acid in water contains only about one acetate ion out of a hundred acetic acid molecules that were added.

Strong acids (the acids listed in Table 3.2) are completely ionized in water; weak acids (all other acids) are not.

Bases are also classified as strong or weak:

A **strong base** is completely ionized in water.

A **weak base** is not completely ionized in water.

In this context, *ionized* means the attachment of a hydrogen ion, as in the formation of NH_4^+ from NH_3.

TABLE 3.2 The strong acids and bases in water

Strong acids	Strong bases
hydrobromic acid, HBr(aq)	Group 1 hydroxides
hydrochloric acid, HCl(aq)	alkaline earth metal hydroxides*
hydroiodic acid, HI(aq)	
nitric acid, HNO_3	
perchloric acid, $HClO_4$	
chloric acid, $HClO_3$	
sulfuric acid, H_2SO_4 (to HSO_4^-)	

*$Ca(OH)_2$, $Sr(OH)_2$, $Ba(OH)_2$.

The strong bases are easy to remember: they are the alkali metal and alkaline earth metal oxides and hydroxides, such as sodium hydroxide, calcium oxide, and barium hydroxide. Metal oxides contain oxide ions, O^{2-}. However, in solution, oxide ions react completely with water to form hydroxide ions:

Recall that the true alkaline earth metals are calcium, strontium, and barium.

$$O^{2-}(aq) + H_2O(l) \longrightarrow 2\,OH^-(aq)$$

Calcium oxide is called quicklime because it reacts so aggressively with water:

$$CaO(s) + H_2O(l) \longrightarrow Ca^{2+}(aq) + 2\,OH^-(aq)$$

The reaction is so vigorous and gives out so much heat that the wooden boats once used to transport quicklime sometimes caught fire when water seeped into their holds. Solid calcium hydroxide, $Ca(OH)_2$, is called slaked lime—calcium hydroxide is the product when the "thirst" of quicklime (CaO) for water has been quenched (slaked), that is, when quicklime has reacted with all the water it can. Slaked lime is widely used in agriculture to adjust the acidity of the soil and to break up clay. Metal hydroxides are bases because they release hydroxide ions when they dissolve in water.

Ammonia is a weak base. When it dissolves in water, it gives a solution in which it exists almost entirely as NH_3 molecules, with just a small proportion—usually less than one in a hundred molecules—of NH_4^+ cations and OH^- anions. Other common weak bases are the **amines,** smelly compounds that are derived from ammonia by replacement of one or more of its hydrogen atoms by an organic group. For example, the replacement of one hydrogen atom in NH_3 by a methyl group, CH_3 (**6**), gives methylamine, CH_3NH_2 (**7**). The replacement of all three hydrogen atoms gives trimethylamine, $(CH_3)_3N$ (**8**), a substance found in decomposing fish and on unwashed dogs.

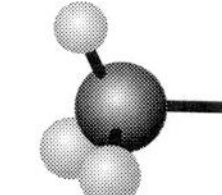

6 Methyl group, CH_3

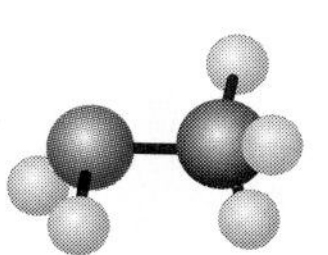

7 Methylamine, CH_3NH_2

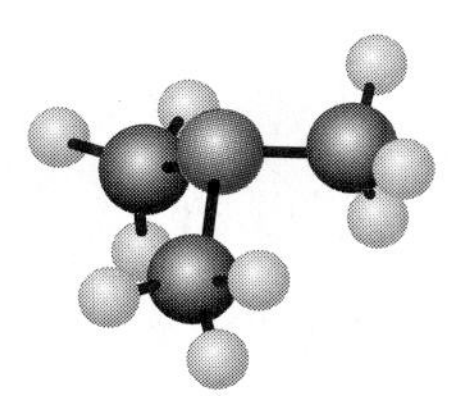

8 Trimethylamine, $(CH_3)_3N$

Whenever we imagine a solution of a weak base, such as ammonia, we should picture it as consisting almost entirely of the original molecules. Only a very tiny proportion of base is present as cations (such as NH_4^+) and a matching number of OH^- ions. These ions result from the transfer of a hydrogen ion from the acid or water molecule to the ammonia molecule:

$$NH_3(aq) + HCl(aq) \longrightarrow NH_4^+(aq) + Cl^-(aq)$$

Amines react with water and acids in the same way as ammonia.

> ***Strong bases (metal oxides and hydroxides) are completely ionized in solution; oxide ions are converted to hydroxide ions. Weak bases (ammonia and its organic derivatives, the amines) are only partially ionized in solution.***

SELF-TEST 3.7A What, if anything, is wrong with the following statement: "6 M HCl is a stronger acid than 1 M HCl"?

[*Answer:* 6 M HCl is more concentrated than 1 M HCl, but both are solutions of a strong acid. Hydrogen chloride is completely ionized in solution.]

SELF-TEST 3.7B What, if anything, is wrong with the following statement: "A 0.1 M solution of methylamine (CH_3NH_2) contains hydroxide ions at a concentration of 0.1 M"?

3.9 ACIDIC AND BASIC CHARACTER IN THE PERIODIC TABLE

We have seen that most soluble metal oxides are strong bases in water. Conversely, many nonmetal oxides react with water to give acids:

$$CO_2(g) + H_2O(l) \longrightarrow H_2CO_3(aq)$$

$$SO_2(g) + H_2O(l) \longrightarrow H_2SO_3(aq)$$

$$P_4O_{10}(s) + 6\,H_2O(l) \longrightarrow 4\,H_3PO_4(aq)$$

Sulfur dioxide, sulfur trioxide, and various oxides of nitrogen in the atmosphere are responsible for the enhanced acidity of rainwater downwind of industrial centers. These oxides dissolve in water to produce a mixture of acids, sulfuric acid and nitric acid among them, that increases the normal acidity of rain (which is due to dissolved carbon dioxide). In other words, they produce "acid rain," which eats away at the limestone of buildings, damages plants, and threatens delicate ecosystems.

We can summarize the properties of binary oxides:

> Most soluble metal oxides form basic solutions in water, whereas many nonmetal oxides form acidic solutions.

Selenium, for instance, is a nonmetal, so we expect its oxides (SeO_2 and SeO_3) to be acidic. Barium is a metal, so we expect its oxide (BaO) to be basic. Acidic oxides are also recognized by the fact that they react with bases; similarly, basic oxides react with acids. For instance, magnesium oxide, a basic oxide, reacts with hydrochloric acid:

$$MgO(s) + 2\,HCl(aq) \longrightarrow MgCl_2(aq) + H_2O(l)$$

and carbon dioxide, an acidic oxide, reacts with sodium hydroxide:

$$2\,NaOH(aq) + CO_2(g) \longrightarrow Na_2CO_3(aq) + 2\,H_2O(l)$$

The white crust often seen on pellets of sodium hydroxide is a mixture of sodium hydrogen carbonate and sodium carbonate formed in this way.

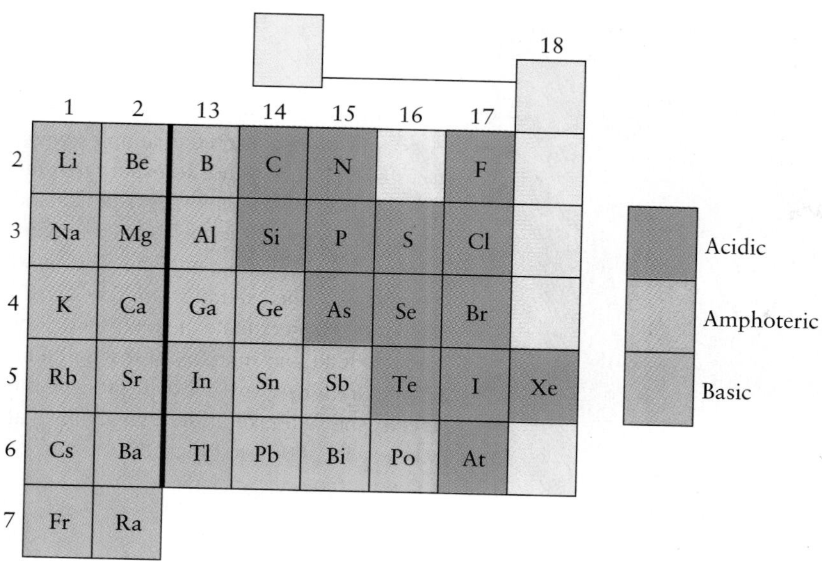

FIGURE 3.18 The location of acidic, amphoteric, and basic oxides of elements in the main groups of the periodic table. Metals form basic oxides, nonmetals form acidic oxides. The diagonal band of amphoteric oxides closely matches the diagonal band of metalloids (recall Fig. 1.18).

Recall from Fig. 1.18 that there is a diagonal frontier in the periodic table lying between the metals and nonmetals. On this frontier, metallic character blends into nonmetallic character. The oxides of the elements on this frontier therefore have both acidic and basic character and are classified as **amphoteric,** from the Greek word for "both." For example, aluminum oxide, Al_2O_3, reacts with acids, a reaction showing that it is basic:

$$Al_2O_3(s) + 6\,HCl(aq) \longrightarrow 2\,AlCl_3(aq) + 3\,H_2O(l)$$

However, it also reacts with bases, a reaction showing that it is acidic as well:

$$2\,NaOH(aq) + Al_2O_3(s) + 3\,H_2O(l) \longrightarrow 2\,Na[Al(OH)_4](aq)$$

The product in this case is sodium aluminate, a compound that contains the aluminate ion, $[Al(OH)_4]^-$. Because aluminum oxide reacts with both acids and bases, it is classified as amphoteric. Some other main-group elements that form amphoteric oxides are shown in Fig. 3.18. As you can see, they lie in a diagonal band across the table from beryllium to polonium.

Most metallic elements form basic oxides and most nonmetallic elements form acidic oxides. The elements lying in a diagonal line in the periodic table from beryllium to polonium form amphoteric oxides.

3.10 NEUTRALIZATION

The reaction between an acid and a base is called a **neutralization reaction.** The ionic compound produced in the reaction is called a **salt.** The general form of a neutralization reaction is

$$\text{Acid} + \text{base} \longrightarrow \text{salt} + \text{water}$$

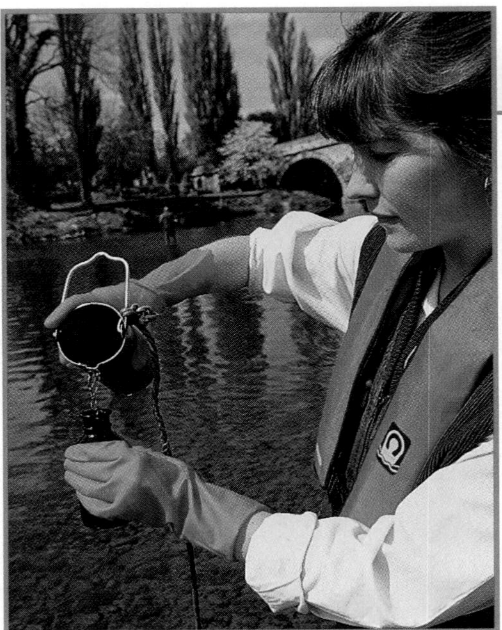

CHEMISTRY AT WORK

A pollution control officer samples water from a river in England. To ensure that the river is safe for drinking, fishing, and recreation, she monitors water quality frequently. She will analyze the acid content of the sample and will add reagents to precipitate heavy metal ions such as lead and mercury. If the pollutant concentrations are above acceptable levels, she will take additional samples at various locations to track the source of the pollutants.

The next time you have a saltshaker handy, pour out a few crystals and examine them closely. You will see that each crystal is a tiny cube, a shape that reflects the orderly arrangement of ions in the crystal.

The name *salt* is taken from ordinary table salt, sodium chloride, the ionic product of the reaction between hydrochloric acid and sodium hydroxide:

$$\mathrm{HCl(aq) + NaOH(aq) \longrightarrow NaCl(aq) + H_2O(l)}$$

In chemistry the term *salt* means "any ionic compound that can be formed by the neutralization of an acid with a base." The cation of the salt is provided by the base and the anion is provided by the acid. An example is the reaction between nitric acid and magnesium hydroxide, which produces a salt, magnesium nitrate, composed of the metal cation (magnesium) provided by the base and the anion provided by the acid (the nitrate ion):

$$\mathrm{2\,HNO_3(aq) + Mg(OH)_2(s) \longrightarrow Mg(NO_3)_2(aq) + 2\,H_2O(l)}$$

The magnesium nitrate remains in solution as Mg^{2+} and NO_3^- ions.

Notice that each hydroxide ion reacts with one hydrogen ion. The net ionic equation of this reaction is

At first sight, we might write $2\,H^+(aq) + 2\,OH^-(aq) \rightarrow 2\,H_2O(l)$; but remember that H^+ forms H_3O^+ in water.

$$\mathrm{2\,H_3O^+(aq) + 2\,OH^-(aq) \longrightarrow 4\,H_2O(l)}$$

which simplifies to

$$\mathrm{H_3O^+(aq) + OH^-(aq) \longrightarrow 2\,H_2O(l)}$$

The net outcome of the neutralization is the formation of water from hydronium ions and hydroxide ions. This is the net ionic equation for all neutralization reactions between strong acids and strong bases that form soluble salts.

In a neutralization reaction, an acid reacts with a base to produce a salt and water.

EXAMPLE 3.4 Predicting the outcome of a neutralization reaction

What salt is produced by the reaction between aqueous solutions of barium hydroxide and sulfuric acid?

STRATEGY We need to identify the base (the metal oxide or hydroxide) that provides the cation. The acid provides the anion. If the salt formed is insoluble in water, it will precipitate, so we must check the solubility of the salt in the solubility rules in Table 3.1 before we write the chemical equation.

SOLUTION The base is barium hydroxide, so we expect a barium salt to form. Because the acid is sulfuric acid, we obtain barium sulfate, $BaSO_4$. This salt is insoluble (Table 3.1), so we expect it to precipitate. The chemical equation for the reaction is

$$H_2SO_4(aq) + Ba(OH)_2(aq) \longrightarrow BaSO_4(s) + 2\,H_2O(l)$$

SELF-TEST 3.8A What acid and base solutions could you use to prepare rubidium nitrate? Write the chemical equation for the neutralization.

[*Answer:* $HNO_3(aq) + RbOH(aq) \rightarrow RbNO_3(aq) + H_2O(l)$]

SELF-TEST 3.8B Write the chemical equation for a neutralization reaction in which calcium phosphate is produced.

FIGURE 3.19 Carbon dioxide is formed when a dilute acid is dropped onto the surface of limestone. This reaction is a convenient test for limestone.

3.11 THE FORMATION OF GASES

Some reactions of acids result in the formation of gases. An example is the action of dilute sulfuric acid on calcium carbonate, a reaction used to prepare carbon dioxide in a chemistry laboratory (Fig. 3.19):

$$CaCO_3(s) + H_2SO_4(aq) \longrightarrow CaSO_4(s) + H_2O(l) + CO_2(g)$$

The same reaction occurs when acid rain falls on limestone and marble.

Similar reactions are used to prepare gaseous compounds of hydrogen. One example is the preparation of hydrogen sulfide, H_2S, by the action of a dilute acid on iron(II) sulfide:

$$FeS(s) + 2\,HCl(aq) \longrightarrow FeCl_2(aq) + H_2S(g)$$

Hydrogen sulfide is an intensely poisonous gas and smells like rotten eggs. In fact, as eggs decompose, they give off hydrogen sulfide. Another example is the production of hydrogen halides by the action of acid on metal halides:

$$2\,NaCl(s) + H_2SO_4(l) \longrightarrow Na_2SO_4(s) + 2\,HCl(g)$$

This reaction is used to prepare hydrogen chloride from table salt.

The common feature of these reactions is the transfer of hydrogen ions from a strong acid to the anion of the salt. Each sulfide ion, S^{2-}, picks up two hydrogen ions from the acid and escapes as the gas H_2S. In the second reaction, each Cl^- ion in the solid picks up one hydrogen ion and escapes as the gas HCl.

Even water can act as an acid strong enough to produce a gas. An example is the action of water on calcium carbide, CaC_2, which contains

Yes, a molecule of water can act as an acid. It can release one of its hydrogen atoms as a hydrogen ion.

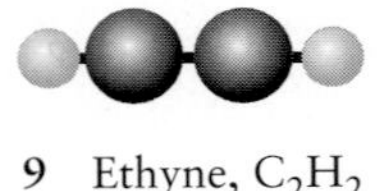

9 Ethyne, C_2H_2

the carbide ion, C_2^{2-}. The reaction produces ethyne, C_2H_2 (**9**), which is commonly called *acetylene*. It is a highly flammable gas that is used in welding because of the very high temperature (about 3000°C) of its flame. The chemical equation for the formation of acetylene is

$$CaC_2(s) + 2\,H_2O(l) \longrightarrow Ca(OH)_2(aq) + C_2H_2(g)$$

This reaction was once used in automobile and mining lamps (Fig. 3.20).

In some gas-formation reactions, the hydrogen compound is not itself a gas but decomposes into one as soon as it is formed. One example is the action of acids on carbonates described at the beginning of this section. The action of the acid is first to convert the carbonate ions into carbonic acid, H_2CO_3, which then decomposes into carbon dioxide and water:

$$CaCO_3(s) + H_2SO_4(aq) \longrightarrow CaSO_4(s) + H_2CO_3(aq)$$

$$H_2CO_3(aq) \longrightarrow H_2O(l) + CO_2(g)$$

The attack of acids on salts may produce a gas or a compound that decomposes into a gas.

EXAMPLE 3.5 Devising a means of preparing a gas

Suggest a method for making the gas phosphine, PH_3. Phosphine is an extremely toxic gas that should be made only in a fume hood.

STRATEGY Because phosphine is a gas, we may be able to devise a reaction in which it is produced by the action of an acid on a suitable salt. To identify the salt to use as starting material, we examine the formula of phosphine,

(a)

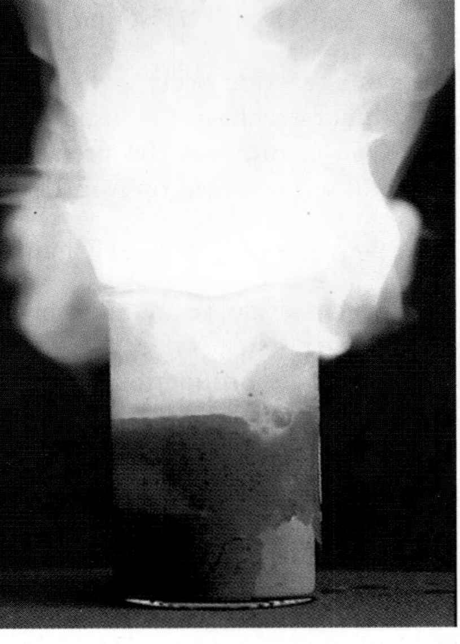

(b)

FIGURE 3.20 (a) An acetylene lamp. Impure calcium carbide is a dirty white solid formed from limestone and carbon. (b) When water is poured on it, ethyne (acetylene), which burns with a brilliant white, hot flame, is produced.

remove hydrogen ions (these will be supplied by the acid), and identify a compound that contains the resulting anion.

SOLUTION Removal of three hydrogen ions from phosphine leaves the phosphide ion, P^{3-}. Because any phosphide contains this ion, we can expect to produce phosphine by the action of acid on sodium phosphide:

$$Na_3P(s) + 3\,HCl(aq) \longrightarrow 3\,NaCl(aq) + PH_3(g)$$

SELF-TEST 3.9A Devise a method for producing hydrogen iodide that uses phosphoric acid.

[*Answer:* $3\,NaI(s) + H_3PO_4(aq) \rightarrow Na_3PO_4(aq) + 3\,HI(g)$]

SELF-TEST 3.9B Write the chemical equation for the reaction of hydrochloric acid with sodium hydrogen carbonate in aqueous solution.

REDOX REACTIONS

One of the most important chemical reactions in the world is photosynthesis, in which carbon dioxide is removed from the atmosphere and converted into carbohydrates:

$$6\,CO_2(g) + 6\,H_2O(l) \longrightarrow C_6H_{12}O_6(s, \text{glucose}) + 6\,O_2(g)$$

This equation summarizes the overall outcome of many individual reactions taking place in a complicated sequence inside green leaves.

Another very important reaction in technological societies is the combustion of methane, the principal component of natural gas:

$$CH_4(g) + 2\,O_2(g) \longrightarrow CO_2(g) + 2\,H_2O(l)$$

At first sight, there seems to be little in common between these two reactions and even less between them and the reaction between magnesium and oxygen,

$$2\,Mg(s) + O_2(g) \longrightarrow 2\,MgO(s)$$

the reaction of magnesium with chlorine,

$$Mg(s) + Cl_2(g) \longrightarrow MgCl_2(s)$$

and the reaction that occurs when dilute acid is poured on zinc:

$$Zn(s) + 2\,HCl(aq) \longrightarrow ZnCl_2(aq) + H_2(g)$$

Nevertheless, all these reactions *do* belong to the same family. They are all examples of *oxidation-reduction reactions,* which are commonly called *redox reactions.*

FIGURE 3.21 An example of an oxidation reaction: magnesium burning brightly in air. Magnesium also burns brightly in water and carbon dioxide; consequently magnesium fires are very difficult to extinguish.

3.12 OXIDATION AND REDUCTION

To find the common feature, consider the reaction between magnesium and oxygen to produce magnesium oxide (Fig. 3.21). For a long time, chemists have called reaction with oxygen "oxidation" and have referred to this reaction as the oxidation of magnesium. We shall now see that the term *oxidation* can be given a much broader meaning (Fig. 3.22). First, note that in the course of the reaction, the Mg atoms in the solid magnesium lose electrons to form Mg^{2+} ions and the O atoms in the molecular

FIGURE 3.22 When bromine is poured on red phosphorus, a vigorous reaction takes place. Although it is not immediately obvious, this reaction is the same type as the reaction shown in Fig. 3.21.

oxygen gain electrons to form O^{2-} ions:

$$2\,Mg(s) + O_2(g) \longrightarrow 2\,Mg^{2+}(s) + 2\,O^{2-}(s) \qquad (\text{as } 2\,MgO(s))$$

Two electrons have been transferred from each magnesium atom to an oxygen atom. *The essential step in the reaction is the transfer of electrons from one species to another.* In the reaction between magnesium and chlorine, the overall outcome is also a transfer of electrons, in this case from Mg atoms to Cl atoms:

$$Mg(s) + Cl_2(g) \longrightarrow Mg^{2+}(s) + 2\,Cl^-(s) \qquad (\text{as } MgCl_2(s))$$

In the reaction of magnesium with oxygen, the magnesium atoms lose electrons. The same is true of the reaction of magnesium with chlorine. Chemists now define **oxidation** as the loss of electrons, even if no oxygen is involved.

The increase in positive charge of an iron ion from Fe^{2+} to Fe^{3+} shows that a negative charge—an electron—has been removed. We see at once that the Fe^{2+} ion has been "oxidized." The increase in charge of a species always indicates that it has undergone oxidation, that it has been oxidized. This rule also applies to the conversion of an anion to a neutral species, as in the conversion of chloride ions (charge -1) to chlorine gas (charge 0).

Now let's consider reduction. The name *reduction* originally referred to extraction of a metal from its oxide, often by reaction with hydrogen, carbon, or carbon monoxide. One example is

$$Fe_2O_3(s) + 3\,H_2(g) \xrightarrow{\Delta} 2\,Fe(l) + 3\,H_2O(g)$$

In the manufacture of steel, the reduction is brought about by using carbon monoxide in place of hydrogen:

$$Fe_2O_3(s) + 3\,CO(g) \xrightarrow{\Delta} 2\,Fe(l) + 3\,CO_2(g)$$

What is the feature common to both reactions?

In both reactions of iron(III) oxide, Fe^{3+} ions present in Fe_2O_3 are converted into Fe atoms. For this conversion to take place, the Fe^{3+} ions must gain electrons to neutralize their positive charges. The common process in both reactions is electron gain by the Fe^{3+} ions: **reduction** is electron gain.

In the conversion of Cu^{2+} to Cu, the charge decreases from $+2$ to 0. There is a gain of electrons (to cancel the positive charge), so the copper ions have been reduced. The same rule applies if the charge is negative. For example, when bromine is converted to bromide ions, the charge decreases from 0 (in Br_2) to -1 (in Br^-). The conversion of Br_2 to Br^- is therefore a reduction.

Oxidation is electron loss; reduction is electron gain.

3.13 KEEPING TRACK OF ELECTRONS: OXIDATION NUMBERS

The change in charge, and the corresponding loss or gain of electrons, is easy to identify when we are dealing with monatomic ions. To extend the procedure to molecules and polyatomic ions, chemists have found a way of keeping track of electrons by assigning an **oxidation number** to each element. The oxidation number is defined so that:

Oxidation corresponds to an increase in oxidation number.

Reduction corresponds to a decrease in oxidation number.

The oxidation number of an element in a monatomic ion is the same as its charge. So, the oxidation number of magnesium is +2 when it is present as Mg^{2+} ions, and the oxidation number of chlorine is −1 when it is present as Cl^- ions. The oxidation number of the elemental form of an element is 0, so magnesium metal has oxidation number 0 and chlorine in the form of Cl_2 molecules also has 0. With these definitions in mind, we can see that the conversion of Mg to Mg^{2+} ions is an oxidation (the oxidation number increases from 0 to +2), whereas the conversion of Cl_2 to Cl^- ions is a reduction (the oxidation number of chlorine decreases from 0 to −1). We shall use the term **oxidation state** to refer to the actual condition of a species with a specified oxidation number. So, Mg^{2+} is an oxidation state of magnesium with oxidation number +2, and Cl^- is an oxidation state of chlorine with oxidation number −1.

▶▶ See Section 8.8 for a deeper understanding of oxidation numbers.

When an element is part of a compound, we assign its oxidation number by using the procedure in Toolbox 3.2. These rules mirror the fact that an increase in the number of oxygen atoms in an electrically neutral binary compound (for instance, from SO_2 to SO_3) results in an increase in oxidation number of the remaining element and therefore corresponds to an oxidation. However, we have to be careful if the overall charge changes, as in the conversion of NO_2 to NO_3^-, because a change in charge indicates a change in the number of electrons present. As illustrated in Toolbox 3.2, the rules also apply to ions.

Oxidation increases the oxidation number of an element; reduction decreases the oxidation number.

TOOLBOX 3.2 How to find oxidation numbers

To assign an oxidation number to an element, we start with two simple rules:

1. The oxidation number of an element uncombined with other elements is zero.
2. The sum of the oxidation numbers of all the atoms in a species is equal to its total charge.

The first of these rules tells us that hydrogen, oxygen, iron, and all other elements have oxidation numbers of 0 in their elemental forms. The second rule implies that the oxidation number of H^+ is +1, that of Al^{3+} is +3, and that of Br^- is −1.

The oxidation numbers of elements in the compounds we are likely to meet at this stage in the text are assigned by using these two rules in conjunction with the following specific values:

The oxidation number of hydrogen is +1 in combination with nonmetals and −1 in combination with metals.

The oxidation number of oxygen is −2 in most of its compounds.

The oxidation number of all the halogens is −1 unless the halogen is in combination with oxygen or another halogen higher in the group. The oxidation number of fluorine is −1 in all its compounds.

We shall see how to use the rules by determining the oxidation numbers of sulfur in (a) SO_2 and (b) SO_4^{2-}. Let's represent the oxidation number of sulfur by x; the oxidation number of oxygen is -2 and there are two oxygen atoms.

(a) By rule 2, the sum of oxidation numbers of the atoms in the neutral compound must be 0:

$$\text{Oxidation number of S} + [2 \times (\text{oxidation number of O})] = 0$$

In terms of x:

$$x + [2 \times (-2)] = 0$$

Therefore, the oxidation number of sulfur is $x = +4$.

(b) By rule 2, the sum of oxidation numbers of the atoms in the ion is -2, so

$$x + [4 \times (-2)] = -2$$

Therefore, $x = +6$.

These two values tell us that sulfur is more highly oxidized in the sulfate ion than in sulfur dioxide.

SELF-TEST 3.10A Find the oxidation numbers of sulfur, phosphorus, and nitrogen in (a) H_2S; (b) PO_4^{3-}; (c) NO_3^-, respectively.

[*Answer:* (a) -2; (b) $+5$; (c) $+5$]

SELF-TEST 3.10B Find the oxidation numbers of sulfur, nitrogen, and chlorine in (a) SO_3^{2-}; (b) NO_2^-; (c) $HClO_3$.

3.14 OXIDIZING AND REDUCING AGENTS

We have seen that the oxidation or reduction of an element corresponds to an increase or decrease in its oxidation number. Specifically, a **redox reaction** is a reaction in which oxidation numbers change. We can verify that the combustion of methane—one of the reactions that we mentioned at the start of this section—is a redox reaction because the oxidation number of oxygen changes from 0 in O_2 to -2 in both CO_2 and H_2O. In fact, all combustion reactions are redox reactions. Now we can see that photosynthesis is also a redox reaction:

$$6\,CO_2(g) + 6\,H_2O(l) \longrightarrow C_6H_{12}O_6(s, \text{glucose}) + 6\,O_2(g)$$

The oxidation number of oxygen in glucose, water, and carbon dioxide is -2, but it is 0 in the O_2 produced as a by-product. Both acid-base and redox reactions are used to purify air in closed environments (Box 3.1).

The species that *causes* oxidation in a redox reaction is called the **oxidizing agent.** Oxygen and chlorine are two we have met so far. Oxidizing agents can be elements, ions, or compounds. We can look at the changes occurring in the oxidizing agents in redox reactions in three different ways:

Electron transfer: Electrons go to the oxidizing agent from the species being oxidized.

Oxidation numbers: The oxidizing agent contains an element that undergoes a decrease in oxidation number.

Reaction: The oxidizing agent is the species that is reduced.

For instance, chlorine removes electrons from magnesium. Because it accepts those electrons, its oxidation number decreases from 0 to -1 (a reduction). Chlorine is therefore the oxidizing agent in this reaction.

The species that brings about reduction is called the **reducing agent.** Again the process can be described in three different ways:

Electron transfer: Electrons go from the reducing agent to the species being reduced.

Oxidation numbers: The reducing agent contains an element that undergoes an increase in oxidation number.

Reaction: The reducing agent is the species that is oxidized.

Magnesium metal supplies electrons to the chlorine, causing the reduction of chlorine. In doing so, the oxidation number of magnesium increases from 0 to +2 (an oxidation). It is the reducing agent in the reaction of magnesium and chlorine.

To identify the oxidizing and reducing agents in a redox reaction, we examine the oxidation numbers of the elements in the reactants. For

BOX 3.1 MAINTAINING AN ARTIFICIAL ATMOSPHERE

On Earth, our atmosphere is kept at a relatively constant composition by natural processes, such as photosynthesis and the movement of air (see Case Study 5). However, a submarine or space ship must use artificial air purifiers. Otherwise, the crew would suffocate in the carbon dioxide, CO_2, they produce. In such sealed environments, waste gases are removed and oxygen regenerated by a variety of acid-base and redox chemical reactions.

Because the space shuttle serves only for short flights, its oxygen can be stored on-board and need not be replaced. To remove carbon dioxide from the air, chemists use the fact that it is an acidic oxide and reacts with bases (see Section 3.9). Canisters on the space shuttle are filled with solid lithium hydroxide, which reacts with the carbon dioxide:

$$CO_2(g) + 2\,LiOH(s) \longrightarrow Li_2CO_3(s) + H_2O(l)$$

Lithium hydroxide is preferred to the hydroxides of the other alkali metals because of its low molar mass. A by-product of this reaction is water, which can be a valuable commodity in space.

During extended missions in a shuttle or on board a space station, the crew must regenerate oxygen, O_2, directly from CO_2. Because the oxidation number of oxygen is −2 in CO_2 but 0 in O_2, an oxidizing agent must be used. On the Russian Salyut space station, potassium superoxide, KO_2, reacts with the carbon dioxide:

$$4\,KO_2(s) + 2\,CO_2(g) \longrightarrow 2\,K_2CO_3(s) + 3\,O_2(g)$$

In this redox reaction the oxidation numbers of oxygen have changed from $-\frac{1}{2}$ and −2 in O_2^- and CO_2 to −2 and 0 in CO_3^{2-} and O_2.

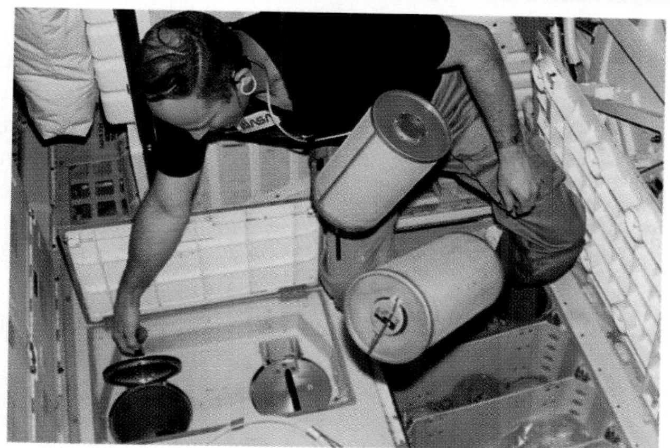

Astronauts on the space shuttle must change the canisters of lithium hydroxide daily to prevent the accumulation of carbon dioxide in the air. Two canisters are used and one is changed every 12 hours so that the capacity to remove carbon dioxide is never diminished.

Of course, eventually all the KO_2 is used up. Very long space flights therefore need processes that recycle all the oxygen. For example, carbon dioxide can react with hydrogen to form carbon and water in a series of redox reactions for which the net equation is:

$$CO_2(g) + 2\,H_2(g) \longrightarrow C(s) + 2\,H_2O(l)$$

Each element can be recovered and reused. The carbon can be used to filter cabin odors and hydrogen and oxygen gases can be regenerated in another redox reaction by passing an electrical current through the water:

$$2\,H_2O(l) \xrightarrow{\text{electrical current}} 2\,H_2(g) + O_2(g)$$

Such processes require energy, but the energy can usually be supplied with solar panels, so it is renewable as well.

example, when a piece of zinc is placed in a copper(II) solution (Fig. 3.23), the reaction is

$$Zn(s) + Cu^{2+}(aq) \longrightarrow Zn^{2+}(aq) + Cu(s)$$

We see that the oxidation number of zinc changes from 0 to +2 (an oxidation) whereas that of copper decreases from +2 to 0 (a reduction). Therefore, because zinc is oxidized, zinc metal is the reducing agent in this reaction. Conversely, because copper is reduced, we can regard the copper(II) ions as the oxidizing agent.

The hydrogen ions provided by acids can also act as oxidizing agents. That is their role when hydrochloric acid is poured onto zinc:

$$Zn(s) + 2\,H^{+}(aq) \longrightarrow Zn^{2+}(aq) + H_2(g)$$

We see that the oxidation number of hydrogen decreases from +1 to 0 (a reduction), which tells us that it is the oxidizing agent in this reaction. Note that although hydrogen ions are present as hydronium ions, H_3O^+, it is convenient to write them as $H^+(aq)$ in redox reactions.

An *acidic solution* is a solution to which an acid has been added.

Some oxoanions (NO_3^- and MnO_4^-, for instance) are powerful oxidizing agents, particularly in acidic solution. For example, when copper metal is added to concentrated nitric acid, the product includes nitrogen dioxide (Fig. 3.24):

$$Cu(s) + 4\,H^{+}(aq) + 2\,NO_3^{-}(aq) \longrightarrow Cu^{2+}(aq) + 2\,NO_2(g) + 2\,H_2O(l)$$

Because the oxidation number of copper has increased from 0 to +2, copper has been oxidized; it is therefore the *reducing* agent. The oxidation number of nitrogen has been reduced from +5 in NO_3^- to +4 in NO_2, so nitric acid has been reduced: it is therefore the *oxidizing* agent. Sulfuric acid is also an oxidizing agent. The hot, concentrated acid oxidizes copper and some of the sulfate ions are reduced to sulfur dioxide:

$$Cu(s) + 2\,H_2SO_4(aq, conc.) \xrightarrow{\Delta} Cu^{2+}(aq) + SO_4^{2-}(aq) + SO_2(g) + 2\,H_2O(l)$$

(a)

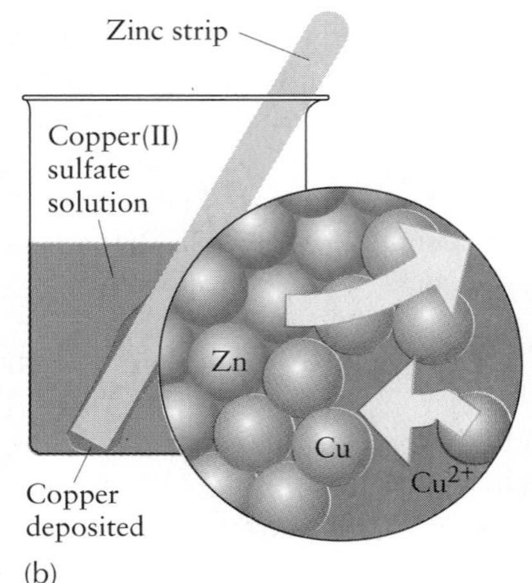

(b)

FIGURE 3.23 When a strip of zinc is placed in a solution that contains Cu^{2+} ions, the blue solution slowly becomes colorless and copper metal is deposited on the zinc. In this redox reaction, the zinc metal is reducing the Cu^{2+} ions to copper and the Cu^{2+} ions are oxidizing the zinc metal to Zn^{2+} ions. (a) The reaction. (b) A visualization of the process.

(a)

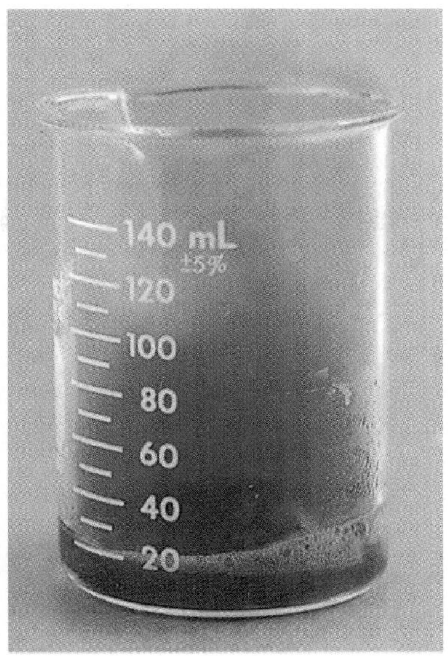

(b)

FIGURE 3.24 (a) Copper reacts with dilute nitric acid to give blue Cu^{2+} ions and the colorless gas nitric oxide, NO. (b) When copper reacts with *concentrated* nitric acid, nitrogen dioxide, NO_2, is produced instead of NO. The blue solution is turned green by this brown gas.

Oxidation is brought about by an oxidizing agent, a species that contains an element that undergoes reduction; reduction is brought about by a reducing agent, a species that contains an element that undergoes oxidation. Certain oxoanions can act as oxidizing agents, particularly in acidic solution.

EXAMPLE 3.6 Identifying oxidizing agents and reducing agents

When an acidic solution of potassium dichromate, $K_2Cr_2O_7$, is mixed with an iron(II) chloride solution, iron(III) ions and chromium(III) ions form. Which species is oxidized and which is reduced? Identify the oxidizing agent and the reducing agent.

STRATEGY First, write the formulas for the species present and determine the oxidation numbers of the elements they contain. Oxidation is signaled by an increase in oxidation number; reduction by a decrease in oxidation number. The oxidizing agent is the species that, in the course of carrying out the oxidation, is itself reduced.

SOLUTION Consider the fate of each reactant: $K_2Cr_2O_7$ produces Cr^{3+} ions; $FeCl_2$ produces Fe^{3+} ions. No changes to the potassium or chloride ions are mentioned, so we assume they are spectator ions. The oxidation number, x, of chromium in the $Cr_2O_7^{2-}$ ion is calculated from

$$2x + [7 \times (-2)] = -2$$

Therefore, $x = +6$. In the reaction, the oxidation number of chromium decreases from $+6$ to $+3$. The dichromate ion therefore undergoes reduction, so it is the oxidizing agent. At the same time, the oxidation number of iron increases from $+2$ to $+3$, so it undergoes oxidation. Fe^{2+} is the reducing agent.

CASE STUDY 3 FROM RIVER WATER TO DRINKING WATER

EVERY LARGE RIVER is tapped by cities for drinking water. The used water is then treated and returned to the river as sewage, along with industrial waste. As a result, the water in many rivers is growing more and more polluted. Water flowing through underground aquifers (porous rock layers) is relatively pure, but it is still not free of pollution.

Assuring a potable (consumable) drinking water supply is a major concern of today's cities and is the responsibility of water chemists and sanitary engineers. Domestic water must be free of color, odor, suspended solids, toxic compounds, and bacteria. The extent of municipal water treatment depends on the purity of the raw (untreated) water. Water treatment makes use of some of the physical separation techniques discussed in Section 1.12 as well as chemical treatments.

The illustration below shows some of the steps used in water treatment. Raw water is aerated by bubbling air through it to remove foul-smelling dissolved gases such as H_2S, to oxidize some organic compounds to CO_2, and to add oxygen and nitrogen. Nonaerated water, such as well water, tastes flat. Aeration also oxidizes any Fe^{2+} ions to Fe^{3+} ions. Adding slaked lime, $Ca(OH)_2$, reduces acidity and precipitates Mg^{2+}, Fe^{3+}, and heavy metal ions as hydroxides.

Hard water contains relatively high concentrations of Ca^{2+} and Mg^{2+} ions, often in the form of hydrogen carbonates. These cations precipitate in furnace boilers and react with soap to form a scum that is hard to remove from laundry. Magnesium and calcium hydrogen carbonates can be precipitated and removed from hard water by adding calcium ions in the form of slaked lime:

$$Mg^{2+}(aq) + 2\,HCO_3^-(aq) + Ca(OH)_2(aq) \longrightarrow Mg(OH)_2(s) + Ca^{2+}(aq) + 2\,HCO_3^-(aq)$$

$$Ca^{2+}(aq) + 2\,HCO_3^-(aq) + Ca(OH)_2(aq) \longrightarrow 2\,CaCO_3(s) + 2\,H_2O(l)$$

It seems ironic that calcium ions can be removed by adding more calcium ions!

After the lime is added, the water is pumped into a primary settling basin. The precipitates tend to form as a very fine powder that remains suspended in the water, so either $Fe_2(SO_4)_3$ or an alum, $Al_2(SO_4)_3{\cdot}18H_2O$, is added to *coagulate* and *flocculate* the precipitate so that it can be filtered. Coagulation involves pulling the ions together to form larger particles. Flocculation involves joining particles

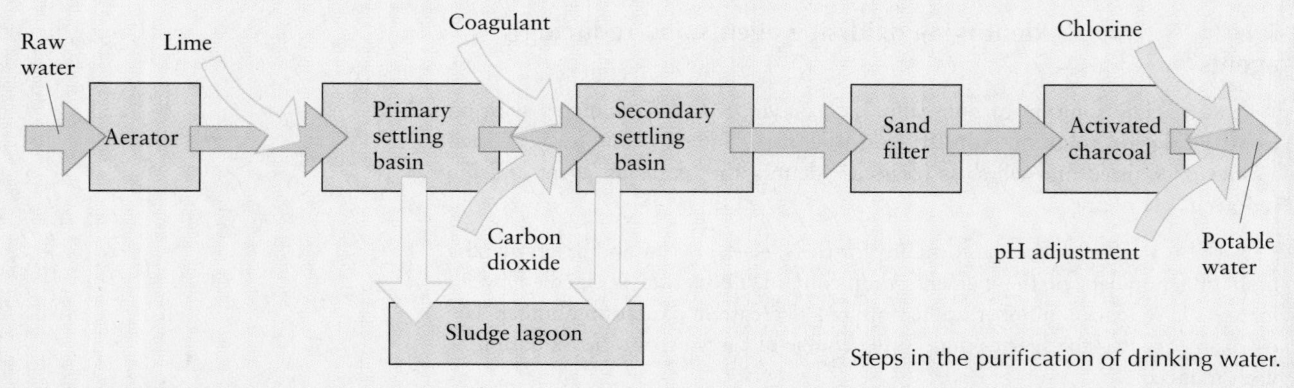

Steps in the purification of drinking water.

SELF-TEST 3.11A In the Claus process for the recovery of sulfur from natural gas and petroleum, hydrogen sulfide, H_2S, reacts with sulfur dioxide, SO_2, to form elemental sulfur and water: $2\,H_2S(g) + SO_2(g) \rightarrow 3\,S(s) + 2\,H_2O(l)$. Identify the oxidizing agent and the reducing agent.

[*Answer:* H_2S is the reducing agent; SO_2 is the oxidizing agent]

SELF-TEST 3.11B When sulfuric acid reacts with sodium iodide, sodium iodate and sulfur dioxide are produced. Identify the oxidizing and reducing agents in this reaction.

together in a fluffy gel that can be filtered. Aluminum is amphoteric and forms $[Al(OH)_4]^-$ ions in basic solution. Carbon dioxide is often added to raise the acidity of the water, which promotes precipitation of the aluminum as $Al(OH)_3$.

As the precipitate settles slowly in a secondary basin, it *adsorbs* (attracts to its surface) any remaining suspended $CaCO_3$, bacteria, and other suspended particles, such as dirt and algae. Precipitates from the primary and secondary basins are combined in a sludge lagoon for disposal.

Wastewater contains a high level of organic matter that must be oxidized or removed before the water is released. Thus, it requires additional purification steps. This sewage treatment plant uses biological digestion as well as aeration to reduce the level of organic matter in wastewater. The methane produced as a by-product of the digestion is stored in large tanks and used to power some of the plant operations.

The clear water is then passed through a sand filter to remove any remaining suspended particles.

Often river water contains organic compounds, which come both from natural sources such as soil and from certain industrial wastes. An activated charcoal filter is used to remove the organic compounds. Activated charcoal is made of finely divided carbon with a highly porous surface that adsorbs organic compounds as the water passes through it.

The acidity level of the water is checked again, and the water is made slightly basic to reduce acid corrosion of the pipes. At this point, a disinfectant, usually chlorine, is added. In the United States, by law, the chlorine level is required to be greater than 1 g of Cl_2 per 10^3 kg of water at the point of consumption. In water, chlorine forms hypochlorous acid, which is highly toxic to bacteria:

$$Cl_2(g) + 2\,H_2O(l) \longrightarrow H_3O^+(aq) + Cl^-(aq) + HClO(aq)$$

Depending on the source and condition of the water, additional water purification steps, such as ion exchange and reverse osmosis (Section 12.13) may be required.

QUESTIONS

1. Identify (a) the physical processes and (b) the chemical processes used for purifying raw water.

2. Classify by type each of the reactions used in the purification of water. If the reaction is a redox reaction, identify the oxidizing agent and the reducing agent.

3. Slaked lime is used to precipitate heavy metal ions. Write the complete chemical equations and the net ionic equations for the precipitation of Pb^{2+} and Fe^{3+} ions by slaked lime.

4. Write the balanced chemical equation for the oxidation of ethanol, C_2H_5OH, to CO_2 and H_2O by O_2 during aeration.

3.15 BALANCING SIMPLE REDOX EQUATIONS

Electrons cannot be lost or created, so when one species is oxidized, another species taking part in the reaction must be reduced. Indeed, *all* the electrons lost in an oxidation step must be gained by a reduction step. Because electrons are charged, we can make sure that the chemical equation reflects this requirement by ensuring that the total charge of the reactants is equal to the total charge of the products.

For example, consider the equation for the oxidation of copper to copper(II) ions by silver ions:

$$Cu(s) + Ag^+(aq) \longrightarrow Cu^{2+}(aq) + Ag(s)$$ ⚠

At first glance, the equation appears to be balanced, because we see the same number of each kind of atom on each side. However, each copper atom has lost two electrons, whereas each silver atom has gained only one. To balance the electrons, we have to balance the charge. In this example, balancing requires a stoichiometric coefficient of 2 for the Ag^+ ion and therefore also for the Ag atom:

$$Cu(s) + 2\,Ag^+(aq) \longrightarrow Cu^{2+}(aq) + 2\,Ag(s)$$

Now all the electrons lost by copper have been gained by silver. Everything is in balance.

SELF-TEST 3.12A When tin is placed in contact with a solution of Fe^{3+} ions, it reduces the iron to iron(II) and is itself oxidized to tin(II) ions. Write the chemical equation for the reaction.

[*Answer:* $Sn(s) + 2\,Fe^{3+}(aq) \rightarrow Sn^{2+}(aq) + 2\,Fe^{2+}(aq)$]

SELF-TEST 3.12B Cerium(IV) ions oxidize iodide ions to iodine and are themselves reduced to cerium(III) ions. Write the chemical equation for the reaction.

Some redox reactions, particularly those involving oxoanions, have complex equations that require special balancing procedures. We shall meet some examples and see how to balance them in Chapter 17.

When balancing the chemical equation for a redox reaction involving ions, the total charge on each side must be balanced.

3.16 CLASSIFYING REACTIONS

It is important to know what type of reaction a given chemical reaction is if we want to predict its products. We can identify a reaction as belonging to one of the three types described in this chapter by looking for the characteristics of each reaction type:

A *precipitation reaction* involves the formation of an insoluble solid from electrolyte solutions.

A *neutralization reaction* involves the transfer of a hydrogen ion.

A *redox reaction* involves a change in oxidation number.

All three types of reactions are encountered in industrial processes and in procedures used to purify air and water (Case Study 3).

SELF-TEST 3.13A Write and balance the equation for the following reaction and identify the reaction type: aqueous copper sulfate and aqueous silver nitrate solution are mixed and solid silver sulfate is produced.

[*Answer:* $CuSO_4(aq) + 2\,AgNO_3(aq) \rightarrow Ag_2SO_4(s) + Cu(NO_3)_2(aq)$; precipitation]

SELF-TEST 3.13B Write and balance the equation for the following reaction and identify the reaction type: ammonia gas reacts with solid copper(II) oxide to form nitrogen gas, copper metal, and liquid water.

We started this chapter by sketching the complex life story of carbon dioxide. We can now look back at that story and see that it illustrated the three simple types of reactions described in this chapter. The combustion reactions that convert organic materials into carbon dioxide are redox reactions. So is the important reaction of photosynthesis, in which carbon dioxide is converted into carbohydrate. Shellfish use precipitation reactions to create their shells; and the reactions in which carbon dioxide forms carbonic acid are neutralization reactions between acids and bases in which hydrogen ions are transferred. The attack of acid rain on limestone and marble, another result of the transfer of hydrogen ions, results in the formation of carbon dioxide gas. The rich web of carbon's activities in the world are largely the interplay of these three types of reactions—precipitation, hydrogen ion transfer, and redox. The same themes will recur throughout the remainder of the text, because a high proportion of the chemical activities of the elements and their compounds can be discussed in terms of the same three fundamental types of reactions.

Reactions can be classified by identifying the characteristic features of a type of reaction.

SKILLS YOU SHOULD HAVE MASTERED

Conceptual

1. Explain the significance of a balanced chemical equation.
2. Identify solutions as electrolytes or nonelectrolytes on the basis of the formulas of the solutes.
3. Explain the difference between solutions of strong and weak acids and bases.
4. Define oxidation and reduction in terms of oxidation number and electron transfer.
5. Identify the oxidizing and reducing agents in a reaction.

Problem-Solving

1. Write, balance, and label a chemical equation, given the information in a sentence.
2. Construct the balanced complete ionic and net ionic equations for reactions involving ions.
3. Determine the oxidation number of an element in an ion or compound.
4. Use the solubility rules to select appropriate solutions which, when mixed, will produce a desired precipitate.

Descriptive

1. Write three balanced chemical equations for reactions that occur in the carbon cycle and describe their roles in the cycle.
2. Describe chemical properties of acids and bases.
3. Recognize common strong and weak acids and bases from their formulas.
4. Describe how acids react with water to form acidic solutions.
5. Describe how ammonia, amines, and oxides react with water to form basic solutions.

6. Identify an oxide as acidic or basic from the position of the element in the periodic table.

7. Identify a reaction as precipitation, neutralization, or redox.

8. Describe how a gas is formed by reaction of an acid with an ionic compound and devise a means of preparing a gas in this way.

EXERCISES

Balancing Equations

3.1 Balance the following skeletal chemical equations:
(a) $P_2O_5(s) + H_2O(l) \longrightarrow H_3PO_4(l)$
(b) $Cd(NO_3)_2(aq) + Na_2S(aq) \longrightarrow CdS(s) + NaNO_3(aq)$
(c) $KClO_3(s) \xrightarrow{\Delta} KClO_4(s) + KCl(s)$
(d) $HCl(aq) + Ca(OH)_2(aq) \longrightarrow CaCl_2(aq) + H_2O(l)$

3.2 Balance the following skeletal chemical equations:
(a) $Al(s) + H_2SO_4(aq) \longrightarrow Al_2(SO_4)_3(s) + H_2(g)$
(b) $Pb(NO_3)_2(aq) + Na_3PO_4(aq) \longrightarrow Pb_3(PO_4)_2(s) + NaNO_3(aq)$
(c) $KClO_3(s) \xrightarrow{\Delta} KCl(s) + O_2(g)$
(d) $H_3PO_4(aq) + Na_2CO_3(aq) \longrightarrow Na_3PO_4(aq) + CO_2(g) + H_2O(l)$

3.3 Write balanced chemical equations for the following reactions:
(a) Sodium metal reacts with water to produce hydrogen gas and sodium hydroxide.
(b) The reaction of sodium oxide, Na_2O, and water produces sodium hydroxide.
(c) Hot lithium metal reacts in a nitrogen atmosphere to produce lithium nitride, Li_3N.
(d) The reaction of calcium metal with water leads to the evolution of hydrogen gas and the formation of calcium hydroxide, $Ca(OH)_2$.

3.4 Write balanced chemical equations for the following reactions:
(a) One process by which nickel is recovered from nickel(II) sulfide ores is to heat the ore in air. During this "roasting" process, oxygen reacts with the nickel(II) sulfide to produce solid nickel(II) oxide and sulfur dioxide gas.
(b) The diamondlike abrasive silicon carbide, SiC, is made by reacting silicon dioxide with elemental carbon at 2000°C to produce silicon carbide and carbon monoxide.
(c) The reaction of elemental hydrogen and nitrogen is used for the commercial production of ammonia in the Haber process.
(d) Magnesium metal reacts with boron oxide, B_2O_3, to form elemental boron and magnesium oxide.

3.5 In one stage in the commercial production of iron metal in a blast furnace, the iron(III) oxide, Fe_2O_3, reacts with carbon monoxide to form Fe_3O_4 and carbon dioxide. In a second stage, the Fe_3O_4 reacts further with carbon monoxide to produce elemental iron and carbon dioxide. Write the balanced equation for each stage in the process.

3.6 One of the roles of stratospheric ozone, O_3, is to remove damaging ultraviolet radiation from sunlight. One result is the eventual dissociation of ozone into molecular oxygen. Write the balanced equation for the reaction.

3.7 When nitrogen and oxygen react in the cylinder of an automobile engine, nitric oxide, NO, is formed. After it escapes into the atmosphere with the other exhaust gases, the nitric oxide reacts with oxygen to produce nitrogen dioxide, one of the precursors of acid rain. Write the two balanced equations for the reactions leading to the formation of nitrogen dioxide.

3.8 The reaction of boron trifluoride, $BF_3(g)$, with sodium borohydride, $NaBH_4(s)$, leads to the formation of sodium tetrafluoroborate, $NaBF_4(s)$, and diborane gas, $B_2H_6(g)$. The diborane reacts with the oxygen in air, forming boron oxide, $B_2O_3(s)$, and water. Write the two balanced equations leading to the formation of boron oxide.

3.9 Hydrofluoric acid is used to etch grooves in glass because it reacts with the silica, $SiO_2(s)$, in glass. The products of the reaction are silicon tetrafluoride and water. Write the balanced equation for the reaction.

3.10 The compound, $Sb_4O_5Cl_2(s)$, which has been investigated because of its interesting electrical properties, can be prepared by warming a mixture of antimony(III) oxide and antimony(III) chloride, both of which are solids. Write the balanced equation for the reaction.

Precipitation Reactions

3.11 Explain how a net ionic equation is constructed.

3.12 Explain what is meant by a spectator ion. What are the spectator ions in (a) a precipitation reaction; (b) an acid-base neutralization reaction?

3.13 Classify aqueous solutions of the following compounds as strong electrolytes, weak electrolytes, or nonelectrolytes: (a) CH_3OH; (b) $CaBr_2$; (c) KI.

3.14 Classify aqueous solutions of the following

compounds as strong electrolytes, weak electrolytes, or nonelectrolytes: (a) HCl; (b) KOH; (c) CH_3COOH.

3.15 Use the information in Table 3.1 to classify the following salts as soluble or insoluble in water: (a) lead(II) nitrate, $Pb(NO_3)_2$; (b) lead(II) chloride, $PbCl_2$; (c) silver nitrate, $AgNO_3$; (d) sodium sulfate, Na_2SO_4.

3.16 Use the information in Table 3.1 to classify the following salts as soluble or insoluble in water: (a) zinc acetate, $Zn(CH_3CO_2)_2$; (b) iron(III) chloride, $FeCl_3$; (c) silver chloride, AgCl; (d) copper(II) hydroxide, $Cu(OH)_2$.

3.17 What are the principal ions that form when the following substances dissolve in water: (a) KCl; (b) $CuCl_2$; (c) $CsHSO_4$?

3.18 What are the principal ions that form when the following substances dissolve in water: (a) $KMnO_4$; (b) Na_2S; (c) $CaHPO_4$?

3.19 What are the principal species present in aqueous solutions of (a) NaI; (b) Ag_2CO_3; (c) $(NH_4)_3PO_4$; (d) $NaHCO_3$; (e) $FeSO_4$?

3.20 What are the principal species present in aqueous solutions of (a) $PbSO_4$; (b) K_2CO_3; (c) K_2CrO_4; (d) $MgHPO_4$; (e) Hg_2Cl_2?

3.21 (a) When aqueous solutions of iron(III) sulfate, $Fe_2(SO_4)_3$, and sodium hydroxide were mixed, a precipitate formed. Write the formula of the precipitate. (b) Does a precipitate form when aqueous solutions of silver nitrate, $AgNO_3$, and potassium carbonate are mixed? If so, write the formula of the precipitate. (c) Aqueous solutions of lead(II) nitrate, $Pb(NO_3)_2$, and sodium acetate are mixed. Does a precipitate form? If so, write the formula of the precipitate.

3.22 (a) Solid ammonium nitrate, NH_4NO_3, and solid calcium chloride were placed in water and the mixture stirred. Is the formation of a precipitate expected? If so, write the formula of the precipitate. (b) Solid magnesium carbonate, $MgCO_3$, and solid sodium nitrate were mixed and placed in water and stirred. What is observed? If a solid is present, write its formula. (c) Aqueous solutions of sodium sulfate, Na_2SO_4, and barium chloride are mixed and a precipitate forms. What is the formula of the precipitate?

3.23 When the solution in Beaker 1 is mixed with the solution in Beaker 2, a precipitate forms. Write three equations describing the formation of the precipitate in terms of the overall reaction, the full ionic form of the reaction, and the net ionic reaction, and then identify the spectator ions.

	Beaker 1	Beaker 2
(a)	$FeCl_3(aq)$	$NaOH(aq)$
(b)	$AgNO_3(aq)$	$KI(aq)$
(c)	$Pb(NO_3)_2(aq)$	$K_2SO_4(aq)$
(d)	$Na_2CrO_4(aq)$	$Pb(NO_3)_2(aq)$
(e)	$Hg(NO_3)_2(aq)$	$K_2CrO_4(aq)$

3.24 The contents of Beaker 1 are mixed with those of Beaker 2. If a reaction occurs, write the net ionic equation and indicate the spectator ions.

	Beaker 1	Beaker 2
(a)	$H_2SO_4(aq)$	$KOH(aq)$
(b)	$H_3PO_4(aq)$	$CuCl_2(aq)$
(c)	$K_2S(aq)$	$AgNO_3(aq)$
(d)	$NiSO_4(aq)$	$(NH_4)_2CO_3(aq)$
(e)	$HNO_3(aq)$	$Ba(OH)_2(aq)$

3.25 The five procedures below each result in the formation of a precipitate. For each reaction, write the chemical equations describing the formation of the precipitate: the overall equation, the overall ionic equation, and the net ionic equation. Identify the spectator ions.
(a) $(NH_4)_2CrO_4(aq)$ is mixed with $BaCl_2(aq)$
(b) $CuSO_4(aq)$ is mixed with $Na_2S(aq)$
(c) $FeCl_2(aq)$ is mixed with $(NH_4)_3PO_4(aq)$
(d) Potassium oxalate, $K_2C_2O_4(aq)$, is mixed with $Ca(NO_3)_2(aq)$
(e) $NiSO_4(aq)$ is mixed with $Ba(NO_3)_2(aq)$

3.26 The five procedures below each result in the formation of a precipitate. For each reaction, write the chemical equations describing the formation of the precipitate: the overall equation, the overall ionic equation, and the net ionic equation. Identify the spectator ions.
(a) $AgNO_3(aq)$ is mixed with $Na_2CO_3(aq)$
(b) $Pb(NO_3)_2(aq)$ is mixed with $KI(aq)$
(c) $Ba(OH)_2(aq)$ is mixed with $H_2SO_4(aq)$
(d) $(NH_4)_2S(aq)$ is mixed with $Cd(NO_3)_2(aq)$
(e) $KOH(aq)$ is mixed with $CuCl_2(aq)$

3.27 Write the full ionic form of the equation and the net ionic equation corresponding to each of the following reactions:

(a) $Pb(ClO_4)_2(aq) + 2\,NaBr(aq) \longrightarrow PbBr_2(s) + 2\,NaClO_4(aq)$

(b) $AgNO_3(aq) + NH_4Cl(aq) \longrightarrow AgCl(s) + NH_4NO_3(aq)$

(c) $2\,NaOH(aq) + Cu(NO_3)_2(aq) \longrightarrow Cu(OH)_2(s) + 2\,NaNO_3(aq)$

3.28 Write the full ionic form of the equation and the net ionic equation corresponding to each of the following reactions:

(a) $3\,BaCl_2(aq) + 2\,K_3PO_4(aq) \longrightarrow Ba_3(PO_4)_2(s) + 6\,KCl(aq)$

(b) $Hg_2(NO_3)_2(aq) + 2\,KCl(aq) \longrightarrow Hg_2Cl_2(s) + 2\,KNO_3(aq)$

(c) $Mg(C_2H_3O_2)_2(aq) + Ba(OH)_2(aq) \longrightarrow Mg(OH)_2(s) + Ba(C_2H_3O_2)_2(aq)$

3.29 Suggest two soluble salts that, when mixed together in water, result in the following net ionic equations:

(a) $2\,Ag^+(aq) + CrO_4^{2-}(aq) \longrightarrow Ag_2CrO_4(s)$

(b) $Ca^{2+}(aq) + CO_3^{2-}(aq) \longrightarrow CaCO_3(s)$, the reaction responsible for the deposition of chalk hills and sea urchin spines

(c) $Cd^{2+}(aq) + S^{2-}(aq) \longrightarrow CdS(s)$, one of the substances used to color glass yellow

3.30 Suggest two soluble salts that, when mixed together in water, result in the following net ionic equations:

(a) $2\,Ag^+(aq) + SO_4^{2-}(aq) \longrightarrow Ag_2SO_4(s)$

(b) $Mg^{2+}(aq) + 2\,OH^-(aq) \longrightarrow Mg(OH)_2(s)$, the suspension present in milk of magnesia

(c) $3\,Ca^{2+}(aq) + 2\,PO_4^{3-}(aq) \longrightarrow Ca_3(PO_4)_2(s)$, the major component of phosphate rock

3.31 Write the net ionic equation for the formation of each salt in aqueous solution: (a) lead(II) sulfate, $PbSO_4$, a precipitate formed in a lead-acid battery; (b) copper(II) sulfide, CuS; (c) cobalt(II) carbonate, $CoCO_3$. (d) Select two soluble salts that, when mixed in solution, form each of the above insoluble salts. Identify the spectator ions.

3.32 Write the net ionic equation for the formation of each of the following compounds in aqueous solution: (a) lead(II) carbonate, $PbCO_3$, the white pigment in putty; (b) aluminum hydroxide, $Al(OH)_3$; (c) zinc chromate, $ZnCrO_4$. (d) Select two soluble salts that, when mixed in solution, form each of the above insoluble salts. Identify the spectator ions.

Acids and Bases

3.33 Give simple definitions for an acid and a base in aqueous solution.

3.34 Explain how to choose an acid and a base to prepare a specified salt.

3.35 Identify the following as either an acid or a base: (a) $NH_3(aq)$; (b) HCl(aq); (c) NaOH(aq); (d) $H_2SO_4(aq)$; (e) $Ba(OH)_2(aq)$.

3.36 Classify each of the following as either an acid or a base: (a) $HNO_3(aq)$; (b) $CH_3NH_2(aq)$, a derivative of ammonia; (c) $CH_3COOH(aq)$; (d) KOH(aq); (e) $HClO_4(aq)$

3.37 Complete and write the overall equation, the overall ionic equation, and the net ionic equation for the acid-base reactions given below. If the substance is a weak acid or base, leave it in its molecular form in writing the equations.

(a) $HCl(aq) + NaOH(aq) \longrightarrow$

(b) $NH_3(aq) + HNO_3(aq) \longrightarrow$

(c) $CH_3NH_2(aq) + HI(aq) \longrightarrow$

3.38 Complete and write the overall equation, the overall ionic equation, and the net ionic equation for the acid-base reactions given below. If the substance is a weak acid or base, leave it in its molecular form in writing the equations.

(a) $H_2SO_4(aq) + KOH(aq) \longrightarrow$

(b) $Ba(OH)_2(aq) + HCN(aq) \longrightarrow$

(c) $NH_3(aq) + HClO_4(aq) \longrightarrow$

3.39 Select an acid and a base for a neutralization reaction that results in the formation of (a) potassium bromide, KBr; (b) zinc nitrite, $Zn(NO_2)_2$; (c) calcium cyanide, $Ca(CN)_2$; (d) potassium phosphate, K_3PO_4. Write the overall and net ionic equations for each reaction.

3.40 Determine the salt that is produced from the acid-base neutralization reactions between (a) potassium hydroxide and acetic acid, CH_3COOH; (b) ammonia, NH_3, and hydroiodic acid; (c) barium hydroxide and sulfuric acid, H_2SO_4 (both H atoms react); (d) sodium hydroxide and hydrocyanic acid, HCN. Write the full ionic equation for each reaction.

3.41 Identify the acid and the base in the following reactions:

(a) $CH_3NH_2(aq) + H_3O^+(aq) \longrightarrow CH_3NH_3^+(aq) + H_2O(l)$

(b) $C_2H_5NH_2(aq) + HCl(aq) \longrightarrow C_2H_5NH_3^+(aq) + Cl^-(aq)$

(c) $CaO(s) + 2\,HI(aq) \longrightarrow CaI_2(aq) + H_2O(l)$

3.42 Identify the acid and the base in the following reactions:

(a) $CH_3COOH(aq) + NH_3(aq) \longrightarrow NH_4^+(aq) + CH_3CO_2^-(aq)$

(b) $(CH_3)_3N(aq) + HCl(aq) \longrightarrow (CH_3)_3NH^+(aq) + Cl^-(aq)$

(c) $O^{2-}(aq) + H_2O(l) \longrightarrow 2\,OH^-(aq)$

3.43 Use the periodic table to determine which oxides form acidic solutions in water and which form basic solutions: (a) CaO; (b) SO_3; (c) N_2O_3; (d) Tl_2O.

3.44 Use the periodic table to determine which oxides form acidic solutions in water and which form basic solutions: (a) P_2O_5; (b) Na_2O; (c) CO_2; (d) MgO.

Redox Reactions

3.45 Define the terms oxidation and reduction in terms of electron transfer.

3.46 How can we tell from a chemical equation if a reaction is a redox reaction?

3.47 Write the balanced equations for the following skeletal redox reactions:
(a) $P(s) + Br_2(l) \longrightarrow PBr_3(s)$
(b) $Fe^{2+}(aq) + Sn^{4+}(aq) \longrightarrow Fe^{3+}(aq) + Sn^{2+}(aq)$
(c) $H_2(g) + S_8(s) \longrightarrow H_2S(g)$
(d) $NO(g) + O_2(g) \longrightarrow NO_2(g)$

3.48 Write the balanced equations for the following skeletal redox equations:
(a) $Fe(s) + H_2O(l) + O_2(aq) \longrightarrow Fe(OH)_2(s)$
(b) $KNO_3(s) \longrightarrow KNO_2(s) + O_2(g)$
(c) $Al(s) + Cu(NO_3)_2(aq) \longrightarrow Cu(s) + Al(NO_3)_3(aq)$
(d) $Na(s) + H_2O(l) \longrightarrow NaOH(aq) + H_2(g)$

3.49 Write the balanced equations for the following redox reactions:

(a) Displacement of copper(II) ion from solution by magnesium metal:

$$Mg(s) + Cu^{2+}(aq) \longrightarrow Mg^{2+}(aq) + Cu(s)$$

(b) Formation of iron(III) ion in the following reaction:

$$Fe^{2+}(aq) + Ce^{4+}(aq) \longrightarrow Fe^{3+}(aq) + Ce^{3+}(aq)$$

(c) Synthesis of hydrogen chloride from its elements:

$$H_2(g) + Cl_2(g) \longrightarrow HCl(g)$$

(d) Formation of rust (a simplified equation):

$$Fe(s) + O_2(g) \longrightarrow Fe_2O_3(s)$$

3.50 Write the balanced equations for the following redox equations:
(a) Displacement of silver ion from solution by copper metal:

$$Ag^+(aq) + Cu(s) \longrightarrow Cu^{2+}(aq) + Ag(s)$$

(b) Production of titanium metal by magnesium metal:

$$TiCl_4(g) + Mg(l) \longrightarrow MgCl_2(s) + Ti(s)$$

(c) Production of copper metal by the smelting of copper(II) sulfide:

$$CuS(s) + O_2(g) \longrightarrow Cu(l) + SO_2(g)$$

(d) Industrial production of elemental bromine from brine:

$$Cl_2(g) + Br^-(aq) \longrightarrow Br_2(l) + Cl^-(aq)$$

Oxidation Numbers

3.51 What is meant by the *oxidation number* of an element?

3.52 Define *oxidation* and *reduction* in terms of oxidation numbers.

3.53 Determine the oxidation number of the italicized element in the following compounds: (a) *Xe*F_4; (b) H*Cl*O; (c) *N*O; (d) H*N*O_3; (e) *S*O_2; (f) H_2*S*.

3.54 Determine the oxidation number of the italicized element in the following compounds: (a) H_2*S*O_3; (b) *B*$_2O_3$; (c) *N*H_3; (d) *N*$_2O_3$; (e) *S*O_3; (f) H_3*P*O_3.

3.55 Identify the oxidation number of the italicized element in each ion: (a) *Mn*O_4^-; (b) *S*$_2O_3^{2-}$; (c) *S*O_4^{2-}; (d) *Mn*O_4^{2-}; (e) *Cr*$_2O_7^{2-}$.

3.56 Identify the oxidation number of the italicized element in each ion: (a) *I*O_3^-; (b) *Cr*O_4^{2-}; (c) *V*O^{2+}; (d) *Br*O_4^-; (e) *I*O_2^-.

3.57 For the following reactions that are redox reactions, identify the substance oxidized and the substance reduced by the change in oxidation numbers.
(a) $CH_3Br(aq) + OH^-(aq) \longrightarrow CH_3OH(aq) + Br^-(aq)$
(b) $BrO_3^-(aq) + 5\,Br^-(aq) + 6\,H^+(aq) \longrightarrow 3\,Br_2(l) + 3\,H_2O(l)$
(c) $2\,F_2(g) + 2\,H_2O(l) \longrightarrow 4\,HF(aq) + O_2(g)$

3.58 In each of the following reactions, use oxidation numbers to identify the substance oxidized and the substance reduced.
(a) Production of iodine from seawater:

$$Cl_2(g) + 2\,I^-(aq) \longrightarrow I_2(aq) + 2\,Cl^-(aq)$$

(b) Reaction to prepare bleach:

$$Cl_2(g) + 2\,NaOH(aq) \longrightarrow NaCl(aq) + NaOCl(aq) + H_2O(l)$$

(c) Reaction that destroys ozone in the stratosphere:

$$NO(g) + O_3(g) \longrightarrow NO_2(g) + O_2(g)$$

Oxidizing and Reducing Agents

3.59 Explain why the definitions of *oxidizing agent* and *reducing agent* in terms of oxidation numbers are consistent with the definitions in terms of electron transfer.

3.60 Explain how you can identify oxidizing and reducing agents from a chemical equation. Does the equation need to be balanced?

3.61 Which do you expect to be the stronger oxidizing agent? Explain your reasoning. (a) Cl_2 or Cl^-; (b) N_2O_5 or N_2O.

3.62 Which do you expect to be the stronger oxidizing agent? Explain your reasoning. (a) KBrO or $KBrO_3$; (b) MnO_4^- or Mn^{2+}.

3.63 Identify the oxidizing agent and the reducing agent in each of the following reactions:
(a) $Zn(s) + 2\,HCl(aq) \longrightarrow ZnCl_2(aq) + H_2(g)$, a simple means of preparing H_2 gas in the laboratory.
(b) $2\,H_2S(g) + SO_2(g) \longrightarrow 3\,S(s) + 2\,H_2O(l)$, a reaction used to produce sulfur from hydrogen sulfide, the "sour gas" in natural gas

(c) $B_2O_3(s) + 3\,Mg(s) \longrightarrow 2\,B(s) + 3\,MgO(s)$, a preparation of elemental boron

3.64 Identify the oxidizing agent and the reducing agent in each of the following reactions:
(a) $2\,Al(s) + Cr_2O_3(s) \longrightarrow Al_2O_3(s) + 2\,Cr(s)$, an example of a thermite reaction used to obtain some metals from their ores
(b) $6\,Li(s) + N_2(g) \longrightarrow 2\,Li_3N(s)$, a reaction that shows the similarity of lithium and magnesium
(c) $2\,Ca_3(PO_4)_2(s) + 6\,SiO_2(s) + 10\,C(s) \longrightarrow P_4(g) + 6\,CaSiO_3(s) + 10\,CO(g)$, a reaction for the preparation of elemental phosphorus

3.65 Decide whether to choose an oxidizing agent or a reducing agent to bring about each of the following changes:
(a) $Br^-(aq) \longrightarrow BrO_3^-(aq)$
(b) $S_2O_6^{2-}(aq) \longrightarrow SO_4^{2-}(aq)$
(c) $NO_3^-(aq) \longrightarrow NO(g)$
(d) HCHO (formaldehyde) $\longrightarrow$ CH_3OH (methanol)

3.66 Would you choose an oxidizing agent or a reducing agent to make the following conversions?
(a) $ClO_3^-(aq) \longrightarrow ClO_2(g)$
(b) $SO_4^{2-}(aq) \longrightarrow S^{2-}(aq)$
(c) $Mn^{2+}(aq) \longrightarrow MnO_2(s)$
(d) HCHO (formaldehyde) $\longrightarrow$ HCOOH (formic acid)

3.67 The industrial production of sodium metal and chlorine gas makes use of the Downs process, in which molten sodium chloride is electrolyzed (Chapter 17). Write the balanced equation for the production of the two elements from molten sodium chloride. Which element is produced by oxidation and which by reduction?

3.68 When water is electrolyzed (Chapter 17), hydrogen is produced at the electrode called the cathode and oxygen is produced at the electrode called the anode. At which electrode does reduction occur and at which electrode does oxidation occur?

SUPPLEMENTARY EXERCISES

3.69 Identify the following as a strong acid, a weak acid, a base, a soluble ionic compound, or an insoluble ionic compound in water: (a) HNO_3; (b) KOH; (c) NH_3; (d) $CuSO_4$; (e) HCOOH; (f) $Ca_3(PO_4)_2$; (g) ZnS; (h) H_2SO_4.

3.70 Select two soluble ionic compounds that, when mixed in solution, produce the precipitates (a) $MgCO_3$; (b) Ag_2SO_4; (c) $Zn(OH)_2$; (d) $PbCrO_4$.

3.71 Select an acid and a base that, when mixed in solution, produce (a) $Ba(NO_3)_2(aq)$; (b) $Na_2SO_4(aq)$; (c) $KClO_4(aq)$; (d) $NiCl_2(aq)$.

3.72 Aqueous solutions of potassium chloride conduct electricity. Which picture below best represents KCl in aqueous solution?

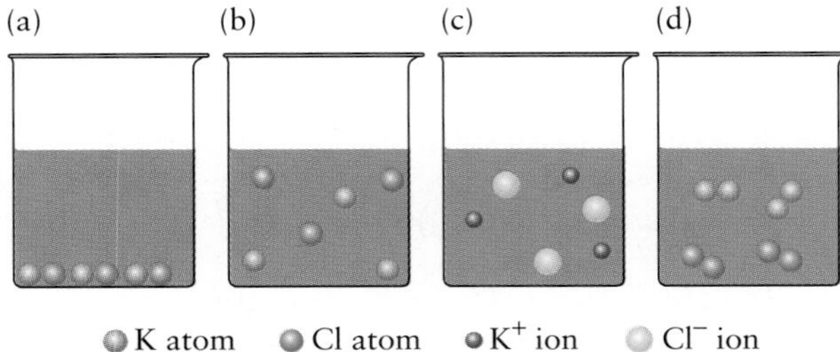

K atom Cl atom K^+ ion Cl^- ion

3.73 Classify the following reactions as precipitation, acid-base neutralization, and/or redox. If a precipitation reaction, write a net ionic equation; if a neutralization reaction, identify the acid and the base; if a redox reaction, identify the oxidizing agent and the reducing agent.
(a) Formation of magnesium chloride by the action of hydrochloric acid on magnesium hydroxide solution:

$$Mg(OH)_2(aq) + 2\,HCl(aq) \longrightarrow MgCl_2(aq) + 2\,H_2O(l)$$

(b) Formation of barium sulfate by the action of sulfuric acid on barium hydroxide:

$$Ba(OH)_2(aq) + H_2SO_4(aq) \longrightarrow BaSO_4(s) + 2\,H_2O(l)$$

(c) Production of sulfur trioxide from sulfur dioxide in the manufacture of sulfuric acid:

$$2\,SO_2(g) + O_2(g) \longrightarrow 2\,SO_3(g)$$

3.74 Classify the following reactions as precipitation, acid-base neutralization, or redox. If a precipitation reaction, write a net ionic equation; if a neutralization reaction, identify the acid and the base; if a redox reaction, identify the oxidizing agent and the reducing agent.
(a) Formation of nitrogen and water vapor when ammonia burns in air:

$$4\,NH_3(g) + 3\,O_2(g) \longrightarrow 2\,N_2(g) + 6\,H_2O(g)$$

(b) Reaction used to produce elemental phosphorus from its oxide:

$$P_4O_{10}(s) + 10\,C(s) \longrightarrow P_4(s) + 10\,CO(g)$$

(c) Formation of silver chromate, Ag_2CrO_4, from sodium chromate, Na_2CrO_4, and silver nitrate solutions:

$$2\,AgNO_3(aq) + Na_2CrO_4(aq) \longrightarrow Ag_2CrO_4(s) + 2\,NaNO_3(aq)$$

3.75 Classify the following reactions as precipitation, acid-base neutralization, or redox. If a precipitation reaction, write a net ionic equation; if a neutralization reaction, identify the acid and the base; if a redox reaction, identify the oxidizing agent and the reducing agent.

(a) Reaction used to measure the concentration of carbon monoxide in a gas stream:

$$5\,CO(g) + I_2O_5(s) \longrightarrow I_2(s) + 5\,CO_2(g)$$

(b) Reaction often used to monitor the amount of iodine in a sample:

$$I_2(aq) + 2\,S_2O_3^{2-}(aq) \longrightarrow 2\,I^-(aq) + S_4O_6^{2-}(aq)$$

(c) Test for bromide ions in solution:

$$AgNO_3(aq) + Br^-(aq) \longrightarrow AgBr(s) + NO_3^-(aq)$$

(d) Heating of uranium tetrafluoride with magnesium, one stage in the purification of uranium metal:

$$UF_4(g) + 2\,Mg(s) \longrightarrow U(s) + 2\,MgF_2(s)$$

3.76 Identify the oxidizing agent and the reducing agent for each of the following reactions:
(a) Production of tungsten metal from its oxide by the reaction

$$WO_3(s) + 3\,H_2(g) \longrightarrow W(s) + 3\,H_2O(l)$$

(b) Generation of hydrogen gas in the laboratory:

$$Zn(s) + 2\,HCl(aq) \longrightarrow H_2(g) + ZnCl_2(aq)$$

(c) Production of metallic tin from tin(IV) oxide, the mineral cassiterite:

$$SnO_2(s) + 2\,C(s) \xrightarrow{\Delta} Sn(l) + 2\,CO(g)$$

(d) Hydrazine has a low molar mass and releases a significant amount of energy in its reaction with dinitrogen tetroxide. The combination is used as a rocket propellant:

$$2\,N_2H_4(g) + N_2O_4(g) \longrightarrow 3\,N_2(g) + 4\,H_2O(g)$$

CHALLENGING EXERCISES

3.77 Identify all the principal species that exist in aqueous solutions of (a) $HClO_2$; (b) NH_3; (c) CH_3COOH.

3.78 Identify all the principal species that exist in aqueous solutions of (a) CH_3NH_2; (b) HI; (c) $HClO_4$.

3.79 Write a balanced equation for the complete combustion (reaction with oxygen) of octane, C_8H_{18}, the major component of gasoline to carbon dioxide and water vapor.

3.80 Energy is obtained in the body by a series of reactions which, overall, is equivalent to the combustion of sucrose (cane sugar), $C_{12}H_{22}O_{11}$, to carbon dioxide and liquid water. Write a balanced equation for its combustion.

3.81 The psychoactive drug sold as methamphetamine ("speed"), $C_{10}H_{15}N$, undergoes a series of reactions in the body; the net result of these reactions is the oxidation of solid methamphetamine by oxygen to produce carbon dioxide, liquid water, and nitrogen. Write the balanced equation for this net reaction.

3.82 Write the balanced equation for the combustion of the analgesic (pain killer) sold as Tylenol, $C_8H_9O_2N$, to carbon dioxide, liquid water, and nitrogen.

3.83 Sodium thiosulfate, which as the pentahydrate, $Na_2S_2O_3{\cdot}5H_2O$, forms the large white crystals used as "photographer's hypo," can be prepared by bubbling oxygen through a solution of sodium polysulfide, Na_2S_5, in alcohol and adding water. Sulfur dioxide is a by-product. Sodium polysulfide is made by the action of hydrogen sulfide on a solution of sodium sulfide, Na_2S, in alcohol, which, in turn, is made by the neutralization of hydrogen sulfide, H_2S, with sodium hydroxide. Write the three chemical equations that show how hypo is prepared from hydrogen sulfide and sodium hydroxide.

3.84 The first stage in the production of nitric acid by the Ostwald process is the reaction of ammonia with oxygen, producing nitric oxide, NO, and water. The nitric oxide further reacts with oxygen to produce nitrogen dioxide, which, when dissolved in water, produces nitric acid and nitric oxide. Write the three balanced equations that lead to the production of nitric acid.

3.85 Some compounds of hydrogen and oxygen are exceptions to the common observation that H has an oxidation number of +1 and O has an oxidation number of −2. Assuming that each metal has the oxidation number of its most common ion, find the oxidation numbers of H and O in the following compounds: (a) KO_2; (b) $LiAlH_4$; (c) Na_2O_2; (d) NaH; (e) KO_3.

CHAPTER 4

Life on the space shuttle may seem all upside-down, but considerations of how much fuel to load, for reaching orbit and for survival in the hostile environment of space, are matters of life and death that depend on calculations like those in this chapter. We shall see how to predict how much oxygen needs to be carried to react with a given amount of hydrogen. We shall also see how much water is produced when hydrogen and oxygen are used on board to generate electricity. Back on Earth, much of industry relies on similar calculations.

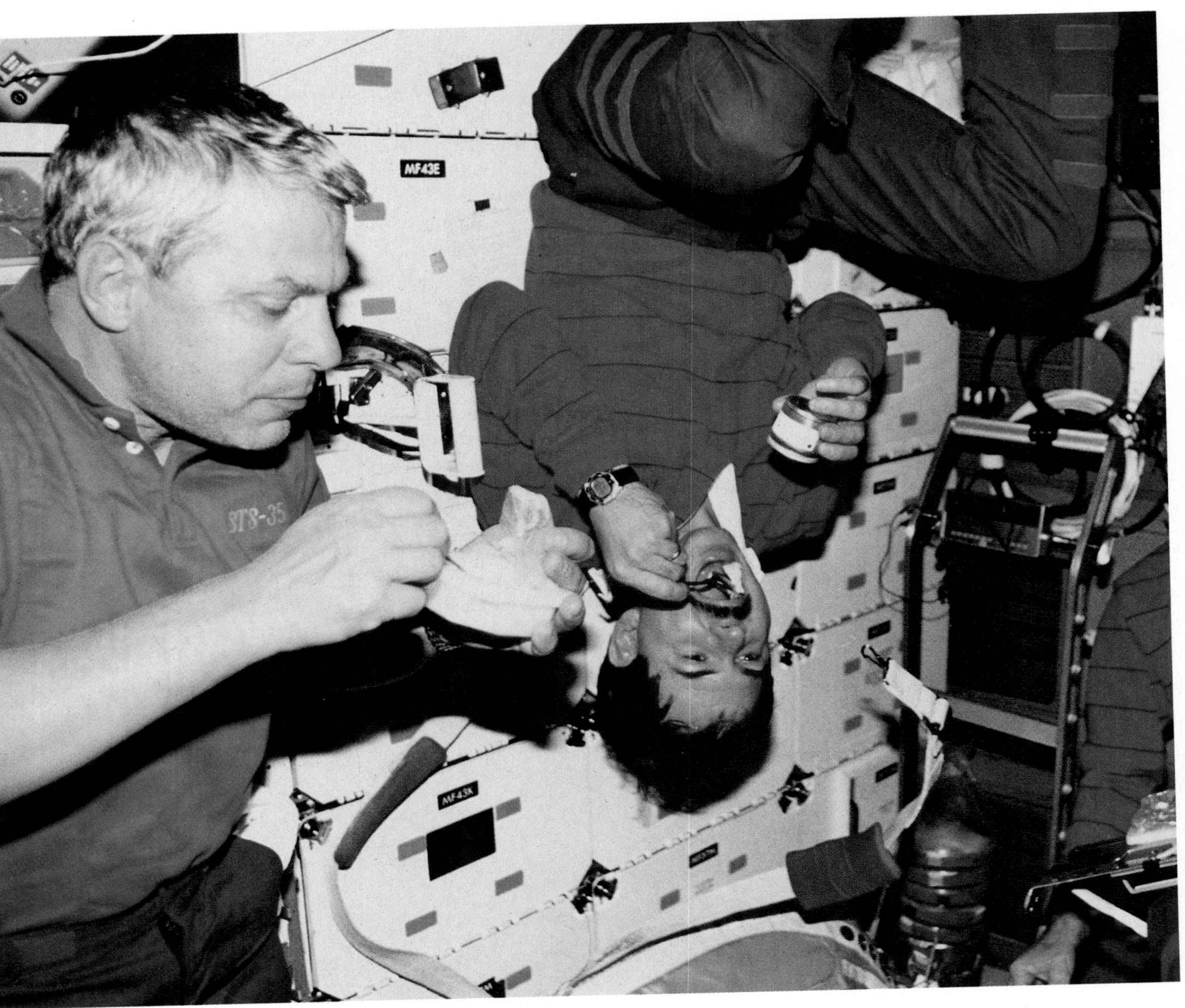

REACTION STOICHIOMETRY: CHEMISTRY'S ACCOUNTING

One day you might wake up in a space shuttle, make coffee, conduct experiments, read a book, and call home. All the power that you require for your work and to stay alive will come from the reaction between two elements, hydrogen and oxygen. Liquid hydrogen and oxygen power the main engine of the space shuttle, and the gases react in its fuel cells to generate electricity. You cannot afford to run out of either element, or you could be stranded in space. On the other hand, you cannot afford to carry more fuel than you need, because so much energy is needed to get into orbit that every kilogram must be counted and justified. We already know that hydrogen is oxidized by oxygen: that is part of our *qualitative* knowledge about the reaction. Now we have to find out how much of one reactant combines with the other: this is a part of the *quantitative* information about the reaction.

In this chapter we shall see how to combine two important concepts we met in the two preceding chapters: the number of moles and balanced equations. The ability to interpret chemical reactions in terms of moles will enable us to make all kinds of predictions. For instance, we shall be able to calculate the mass of carbon dioxide a liter of gasoline contributes to the atmosphere. We shall be able to calculate the mass of ammonia that we can expect to make from a given supply of nitrogen and hydrogen (Fig. 4.1) and the mass of urea that we can then go on to make from that ammonia. We shall see how to calculate how much oxygen we need to react with a given mass of hydrogen in a fuel cell, and how to calculate the mass of water we can expect to obtain from the reaction.

HOW TO USE REACTION STOICHIOMETRY

4.1 Mole-to-mole predictions
4.2 Mass-to-mass predictions
4.3 The volume of solution required for reaction

THE LIMITS OF REACTION

4.4 Reaction yield
4.5 Limiting reactants
4.6 Combustion analysis

FIGURE 4.1 A modern plant for producing ammonia by the Haber process. The quantities of nitrogen and hydrogen that must be supplied to produce a given amount of ammonia can be calculated by using the techniques described in this chapter.

HOW TO USE REACTION STOICHIOMETRY

We saw in Section 3.1 that the stoichiometric coefficients multiplying the chemical formulas in a balanced chemical equation reflect the fact that no atoms are created or destroyed in the reaction. For instance, from the chemical equation

$$2\,H_2(g) + O_2(g) \longrightarrow 2\,H_2O(l)$$

we know that 2 mol H_2O are formed if 1 mol O_2 reacts. This interpretation of the coefficients is the basis of the calculations described in this chapter.

4.1 MOLE-TO-MOLE PREDICTIONS

The numbers of moles of reactants actually used in a reaction are usually different from the stoichiometric coefficients in the equation. For example, we might want to know how much water is formed when 0.25 mol O_2 reacts with hydrogen. How do we take different starting amounts into account?

The calculation is based on a conversion factor just like the ones we used to convert units in Chapter 2. First, we summarize the information in the chemical equation—that 1 mol O_2 reacts to form 2 mol H_2O—by writing

$$1 \text{ mol } O_2 \simeq 2 \text{ mol } H_2O$$

The sign $\simeq$ is read "is chemically equivalent to," and the expression is called a **stoichiometric relation.** In calculations, the chemical equivalence sign is treated just like an equal sign. However, the stoichiometric relation written above applies only to the chemical equation for the formation of water and not necessarily to other reactions in which O_2 and H_2O appear.

Next, the stoichiometric relation is used to set up a conversion factor relating the substance for which the amount has been given (oxygen, in our example) to the substance for which the amount is required (water):

$$\frac{\text{Substance required}}{\text{Substance given}} = \frac{2 \text{ mol } H_2O}{1 \text{ mol } O_2}$$

This factor, which is commonly called the **mole ratio** for the reaction, allows us to relate the moles of oxygen molecules consumed to the moles of water molecules produced. We use it in the same way we use a conversion factor when we are converting units:

$$\begin{aligned}\text{Moles of } H_2O &= (0.25 \cancel{\text{ mol } O_2}) \times \frac{2 \text{ mol } H_2O}{1 \cancel{\text{ mol } O_2}} \\ &= 0.50 \text{ mol } H_2O\end{aligned}$$

Note that the unit mol *and* the species (in this case, O_2 molecules) cancel. This procedure is summarized in Toolbox 4.1.

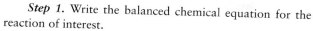

TOOLBOX 4.1 How to carry out mole-to-mole calculations for a chemical reaction

The strategy for this type of calculation is much the same as that for the conversion of units; it is summarized in (**1**). The calculation is done in three steps:

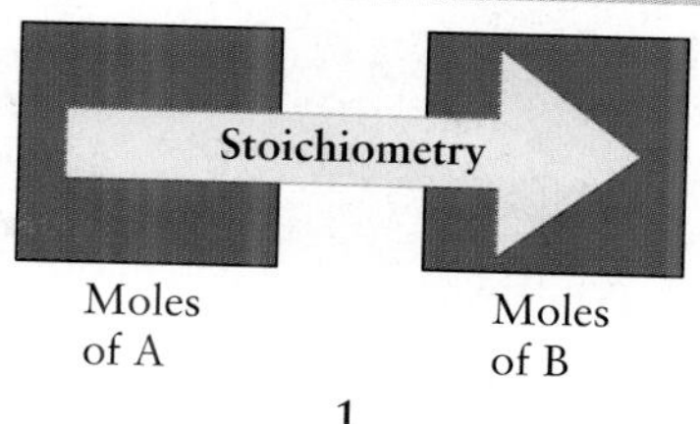

1

Step 1. Write the balanced chemical equation for the reaction of interest.

Step 2. Find the stoichiometric relation between the two substances of interest and use it to write the mole ratio in the form

$$\frac{\text{Substance required}}{\text{Substance given}}$$

The numbers in the mole ratios are exact, so they have no effect on the number of significant figures in the result.

Step 3. Apply this conversion factor to the information given and obtain the information required.

For example, suppose we want to find the moles of N_2 needed to produce 5.0 mol NH_3 by reaction with H_2. We work through the steps as follows:

Step 1. The balanced chemical equation is

$$N_2(g) + 3\,H_2(g) \longrightarrow 2\,NH_3(g)$$

Step 2. It follows that 1 mol $N_2 \simeq$ 2 mol NH_3 and therefore that the mole ratio is

$$\frac{\text{Substance required}}{\text{Substance given}} = \frac{1 \text{ mol } N_2}{2 \text{ mol } NH_3}$$

Step 3. We now apply this conversion factor to the information given and obtain the information required:

$$\text{Chemical amount of } N_2 = (5.0\ \cancel{\text{mol NH}_3}) \times \frac{1 \text{ mol } N_2}{2\ \cancel{\text{mol NH}_3}}$$
$$= 2.5 \text{ mol } N_2$$

Stoichiometric relations are used to convert from moles of reactant to moles of a product or another reactant or from moles of a product to moles of a reactant or another product. For the synthesis of ammonia, for instance, we can write the stoichiometric relation between the two reactants as

$$1 \text{ mol } N_2 \simeq 3 \text{ mol } H_2$$

and use it in exactly the way we used it in the preceding calculations to find the number of moles of hydrogen molecules needed to react with a given number of moles of nitrogen molecules.

SELF-TEST 4.1A How many moles of NH_3 can be produced from 2.0 mol H_2 if all the hydrogen reacts?

[*Answer:* 1.3 mol NH_3]

SELF-TEST 4.1B How many moles of Fe atoms can be extracted from 25 mol Fe_2O_3?

Although it is used in the same way, a stoichiometric relation is a little different from the relation between pairs of units. A relation between units, such as 1 in. = 2.54 cm, is *universally* true. A stoichiometric relation, such as 1 mol $N_2 \simeq$ 3 mol H_2, applies *only to the reaction we are considering* (the formation of ammonia). A different reaction may have a different stoichiometric relation. For example, consider the reaction

$$2\,H_2(g) + N_2(g) \longrightarrow N_2H_4(l)$$

where N_2H_4 is hydrazine, a rocket fuel. For this reaction

$$1 \text{ mol } N_2 \simeq 2 \text{ mol } H_2$$

The balanced chemical equation for a reaction is used to set up the conversion factor from one substance to another; and that conversion factor, the mole ratio for the reaction, is applied to the moles given to calculate the moles required.

4.2 MASS-TO-MASS PREDICTIONS

We have seen how to relate the moles of reactants to the moles of products. We already know how to use molar masses to convert between moles and masses. So, by combining the two calculations, we can determine the *masses* of the substances that can be produced in a reaction from the masses of the reactants.

Suppose we want to know the mass of iron that can be obtained from 10.0 kg of iron(III) oxide, Fe_2O_3, present in iron ore by the reaction

$$Fe_2O_3(s) + 3\,CO(g) \longrightarrow 2\,Fe(s) + 3\,CO_2(g)$$

This redox reaction takes place inside a blast furnace, with carbon monoxide as the reducing agent. We already know how to convert moles of Fe_2O_3 to moles of Fe: we use the stoichiometric relation

$$1\text{ mol } Fe_2O_3 \simeq 2\text{ mol Fe}$$

in the form of the mole ratio

$$\frac{2\text{ mol Fe}}{1\text{ mol } Fe_2O_3}$$

However, before we can use this ratio as a conversion factor, we have to convert the given mass of iron(III) oxide from kilograms to grams, then to moles of Fe_2O_3. The molar mass of iron(III) oxide is 159.69 $g \cdot mol^{-1}$, so

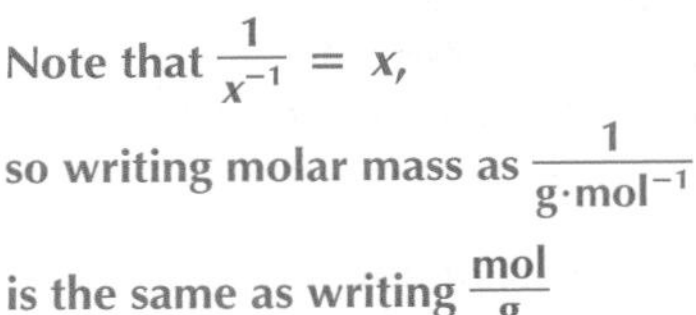

Note that $\frac{1}{x^{-1}} = x$, so writing molar mass as $\frac{1}{g \cdot mol^{-1}}$ is the same as writing $\frac{mol}{g}$

$$\text{Moles of } Fe_2O_3 = \frac{10.0\text{ kg}}{159.69\text{ g}\cdot\text{mol}^{-1}} \times \frac{10^3\text{ g}}{1\text{ kg}} = 62.6\text{ mol}$$

At this stage, we apply the conversion factor to calculate the moles of Fe atoms produced in the reaction:

$$\text{Moles of Fe} = (62.6\text{ mol } Fe_2O_3) \times \frac{2\text{ mol Fe}}{1\text{ mol } Fe_2O_3} = 125\text{ mol Fe}$$

The formula of a substance appears as part of the unit in the result of a calculation when the formula appears in the calculation and does not cancel.

Finally, we convert moles of Fe atoms into mass of iron by using the molar mass of iron, which is 55.85 $g \cdot mol^{-1}$:

$$\text{Mass of iron} = (125\text{ mol}) \times (55.85\text{ g}\cdot\text{mol}^{-1}) = 6.98 \times 10^3\text{ g}$$

The mass of iron we can expect to extract from 10.0 kg of the ore is therefore 6.98 kg. This type of calculation is summarized in Toolbox 4.2.

In a mass-to-mass calculation, convert the given mass to moles, apply the mole-to-mole conversion factor to obtain the moles required, and finally convert moles required to mass.

TOOLBOX 4.2 How to carry out mass-to-mass calculations

The general procedure for mass-to-mass calculations is:

Step 1. Convert the given mass of one substance (in grams) to number of moles by using its molar mass.

Step 2. Write the balanced chemical equation and use the stoichiometric relation to convert from the given number of moles of one substance to the number of moles of the other substance.

Step 3. Convert from number of moles of the second substance to mass (in grams) by using the molar mass of the substance.

The first step has the form

$$\text{Number of moles of substance A} = \frac{\text{mass of A (grams)}}{\text{molar mass of A (grams per mole)}}$$

The second step makes use of the mole-ratio conversion factor

$$\frac{\text{Substance required}}{\text{Substance given}}$$

The third step makes use of the relation

$$\begin{aligned}\text{Mass of B required (grams)} &= \text{number of moles of B} \\ &\quad \times \text{molar mass of B (grams per mole)}\end{aligned}$$

If the masses are in units other than grams, we first convert to grams. This procedure is summarized by diagram (2).

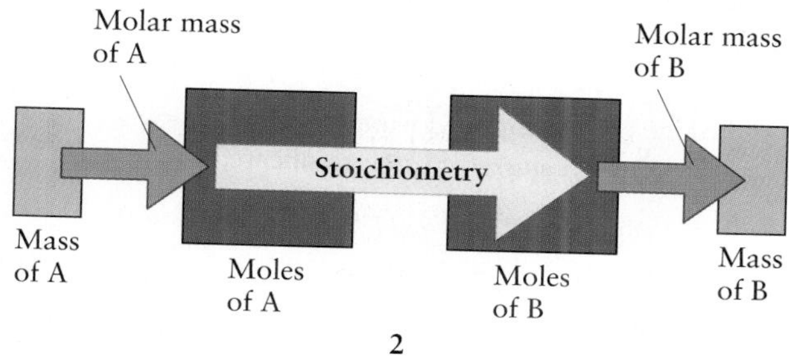

2

Suppose we are asked to calculate the mass of carbon dioxide produced when 1.00×10^2 g of propane (C_3H_8) is burned in a camp stove.

Step 1. We convert from mass of propane to moles of C_3H_8 molecules by using the molar mass of propane, which is 44.09 $g{\cdot}mol^{-1}$:

$$\text{Number of moles of } C_3H_8 = \frac{100\text{ g}}{44.09\text{ g}{\cdot}\text{mol}^{-1}} = 2.27\text{ mol}$$

Step 2. To find the stoichiometric relation, we write the combustion equation:

$$C_3H_8(g) + 5\,O_2(g) \longrightarrow 3\,CO_2(g) + 4\,H_2O(g)$$

and find that

$$1\text{ mol } C_3H_8 \simeq 3\text{ mol } CO_2$$

This relation lets us write

$$\text{Number of moles of } CO_2 = (2.27\text{ mol } C_3H_8) \times \frac{3\text{ mol } CO_2}{1\text{ mol } C_3H_8} = 6.81\text{ mol } CO_2$$

Step 3. Now we convert this amount of CO_2 into mass by using the molar mass of carbon dioxide, which is 44.01 $g{\cdot}mol^{-1}$:

$$\text{Mass of carbon dioxide} = (6.81\text{ mol}) \times (44.01\text{ g}{\cdot}\text{mol}^{-1}) = 3.00 \times 10^2\text{ g}$$

We conclude that the combustion of 100 g of propane produces 300 g of carbon dioxide.

Another way to proceed is to write down the string of conversions in a single line:

$$\begin{aligned}\text{Mass of product (grams)} &= \frac{\text{mass of reactant (grams)}}{\text{molar mass of reactant (grams per mole)}} \\ &\quad \times \text{conversion from reactants to products} \\ &\quad \times \text{molar mass of product (grams per mole)}\end{aligned}$$

This approach is used in Example 4.1.

SELF-TEST 4.2A Calculate the mass of ammonia that can be produced from 50.0 g of hydrogen in the reaction

$$N_2(g) + 3\,H_2(g) \longrightarrow 2\,NH_3(g)$$

[*Answer:* 2.8×10^2 g]

SELF-TEST 4.2B Calculate the mass of sulfuric acid, H_2SO_4, that can be produced from 25 kg of sulfur in the reaction

$$4\,S(s) + 3\,O_2(g) + 2\,H_2O(l) \longrightarrow 2\,H_2SO_4(aq)$$

EXAMPLE 4.1 Calculating the mass of reactant required to react with another reactant

What mass of aluminum is needed to react with 10.0 kg of chromium(III) oxide to produce chromium metal? The chemical equation for the reaction is

$$2\,Al(l) + Cr_2O_3(s) \xrightarrow{\Delta} Al_2O_3(s) + 2\,Cr(l)$$

STRATEGY The procedure is the same as that used in Toolbox 4.2, except that we use the stoichiometric relation between two reactants. We shall set up the solution to show the numerical calculation in a single step.

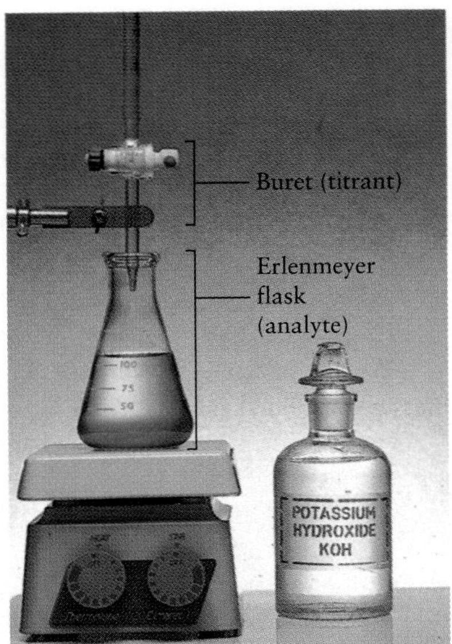

FIGURE 4.2 The apparatus typically used for a titration: magnetic stirrer; flask containing the analyte; clamp and buret containing the titrant.

SOLUTION First, we convert 10.0 kg to grams. Then we obtain the number of moles of Cr_2O_3 in 10.0 kg (1.00×10^4 g) of chromium(III) oxide from its molar mass, which is 152.00 g·mol^{-1}. The stoichiometric relation between the two reactants is

$$1 \text{ mol } Cr_2O_3 \simeq 2 \text{ mol Al}$$

Once we have converted to the number of moles of Al atoms, we express the answer in terms of the mass of aluminum by using the molar mass of aluminum, which is 26.98 g·mol^{-1}. Therefore,

$$\begin{aligned}\text{Mass of aluminum} \\ &= \left(\frac{1.00 \times 10^4 \text{ g } Cr_2O_3}{152.00 \text{ g·mol}^{-1}}\right) \times \left(\frac{2 \text{ mol Al}}{1 \text{ mol } Cr_2O_3}\right) \times (26.98 \text{ g·mol}^{-1}) \\ &= 3.55 \times 10^3 \text{ g Al}\end{aligned}$$

We see that we need to use 3.55 kg of aluminum.

SELF-TEST 4.3A Calculate the mass of potassium needed to react with 0.450 g of hydrogen gas to produce potassium hydride in the reaction $2\,K(s) + H_2(g) \rightarrow 2\,KH(s)$.

[*Answer:* 17.5 g]

SELF-TEST 4.3B Carbon dioxide can be removed from power plant exhaust gases by combining it with an aqueous slurry of calcium silicate:

$$2\,CO_2(g) + H_2O(l) + CaSiO_3(s) \longrightarrow SiO_2(s) + Ca(HCO_3)_2(aq)$$

What mass of $CaSiO_3$ (of molar mass 116.17 g·mol^{-1}) is needed to react completely with 3.00×10^2 g CO_2?

4.3 THE VOLUME OF SOLUTION REQUIRED FOR REACTION

A very similar procedure can be used to calculate the volume of solution required for reaction. One of the most common laboratory techniques for determining the concentration of a solute is called **titration.** Titrations are used daily to monitor water purity and blood composition, and for quality control in the food industry. The solution being titrated is called the **analyte.** A known volume is transferred into a flask. Then a solution containing a reactant is measured into the flask from a buret until all the analyte has reacted (Fig. 4.2). The solution in the buret is called the **titrant.**

The name *titration* comes from an old French word meaning "assay"—a test of quality.

In a typical acid-base titration, the analyte is a base and the titrant is an acid, or vice versa. We have seen already (in Chapter 3) that acids and bases can be detected by noting the colors of indicators, such as the dyes litmus (red in acidic solution, blue in basic solution) or phenolphthalein (colorless in acidic solution, pink in basic solution). An indicator therefore lets us detect the **stoichiometric point,** the stage at which the volume of titrant added is exactly that required by the stoichiometric relation between titrant and analyte. For example, if we titrate hydrochloric acid containing a few drops of the indicator phenolphthalein, the solution is initially colorless. After the stoichiometric point, when excess base is present, the analyte solution is basic and the indicator is pink. The indicator

The stoichiometric point is often referred to as the equivalence point.

color change is sudden, so it is easy to detect the stoichiometric point. We then note the exact volume of titrant needed to cause the indicator to change color (Fig. 4.3).

To interpret a titration, we need to know the stoichiometric relation from the chemical equation for the reaction. This relation is used to write the mole ratio in the usual way. The only new step is to use the molarities of the solutions to convert between the moles of reactants and the volumes of the analyte and titrant. Toolbox 4.3 shows how to combine this information; the procedure is summarized in diagram (**3**).

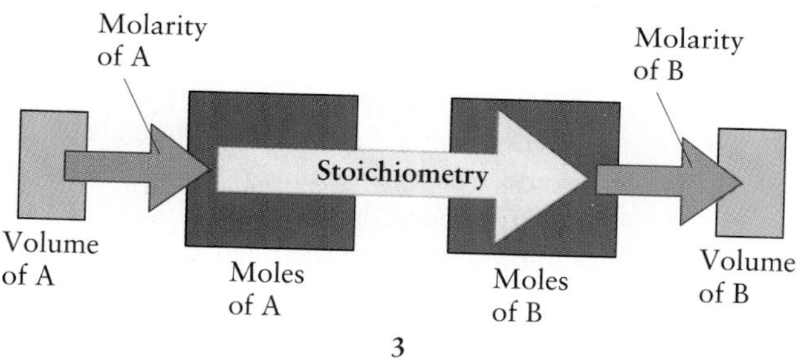

3

The stoichiometric relation between analyte and titrant species, together with the molarity of the titrant, is used in titrations to determine the molarity of the analyte.

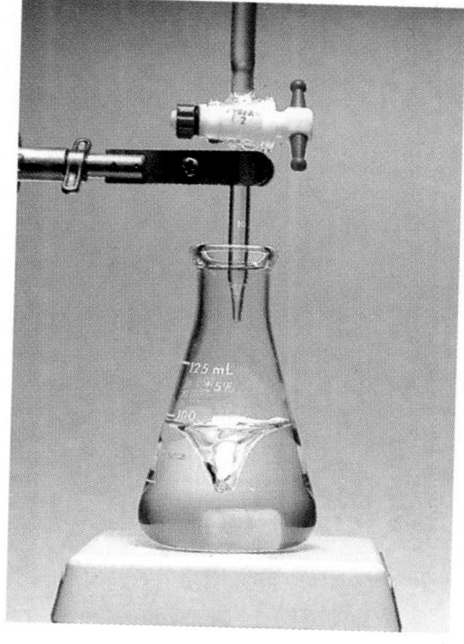

FIGURE 4.3 An acid-base titration in progress. The indicator is phenolphthalein.

TOOLBOX 4.3 How to interpret a titration

The data in a typical titration are the volume of analyte solution, the molarity of the titrant, and the volume of titrant solution needed to reach the stoichiometric point. Our goal is to determine the molarity of the analyte. The strategy for reaching this goal consists of three steps:

Step 1. Use the volume of titrant and its molarity to calculate the number of moles of titrant species added from the buret (see Toolbox 2.3):

Number of moles of titrant
= volume of titrant used (liters)
× molarity of titrant (moles per liter)

Step 2. Write the stoichiometric relation between the titrant and analyte species from the chemical equation for the reaction and use it to convert the moles of titrant species to moles of analyte species. This step gives us a mole ratio of the form

$$\frac{\text{Substance required}}{\text{Substance given}}, \quad \text{which in this case is} \quad \frac{\text{Analyte}}{\text{Titrant}}$$

We then multiply the moles of titrant by this ratio:

Number of moles of analyte present
= number of moles of titrant used
× conversion factor

Step 3. Calculate the molarity of the analyte by dividing the number of moles of solute by the initial volume of the solution:

$$\text{Molarity of analyte} = \frac{\text{number of moles}}{\text{volume of solution (liters)}}$$

For example, suppose that 25.00 mL of a solution of oxalic acid, $H_2C_2O_4$ (**4**), was titrated with 0.500 M NaOH(aq) and that the stoichiometric point was reached when 38.0 mL of the solution of base had been added. What is the molarity of the oxalic acid solution?

Step 1. The moles of NaOH added is

Number of moles of NaOH used
$= (38.0 \times 10^{-3}\ \text{L}) \times (0.500\ \text{mol·L}^{-1})$
$= 1.90 \times 10^{-2}\ \text{mol}$

Step 2. The neutralization reaction is

$$H_2C_2O_4(aq) + 2\,NaOH(aq) \longrightarrow Na_2C_2O_4(aq) + 2\,H_2O(l)$$

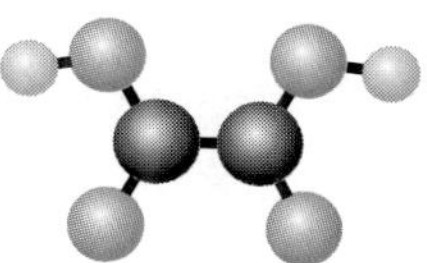

4 Oxalic acid, $(COOH)_2$

It follows that the stoichiometric relation we require is

$$2 \text{ mol NaOH} \simeq 1 \text{ mol } H_2C_2O_4$$

Therefore, the moles of $H_2C_2O_4$ in the original analyte solution is

$$\text{Number of moles of } H_2C_2O_4 = (1.90 \times 10^{-2} \text{ mol NaOH}) \times \frac{1 \text{ mol } H_2C_2O_4}{2 \text{ mol NaOH}} = 9.50 \times 10^{-3} \text{ mol } H_2C_2O_4$$

Step 3. The molarity of the acid is therefore

$$\text{Molarity of } H_2C_2O_4(aq) = \frac{9.50 \times 10^{-3} \text{ mol}}{25.00 \times 10^{-3} \text{ L}} = 0.380 \text{ mol}\cdot\text{L}^{-1}$$

That is, the solution is 0.380 M $H_2C_2O_4(aq)$.

SELF-TEST 4.4A A student prepared a sample of hydrochloric acid that contained 0.72 g of hydrogen chloride in 500 mL of solution. This solution was used to titrate 25.0 mL of a solution of calcium hydroxide, and the stoichiometric point was reached when 15.1 mL of acid had been added. What was the molarity of the calcium hydroxide solution?

[*Answer:* 0.012 M $Ca(OH)_2(aq)$]

SELF-TEST 4.4B Many abandoned mines have exposed nearby communities to the problem of acid mine drainage. Certain minerals, such as pyrite (FeS_2), decompose when exposed to air, forming solutions of sulfuric acid. The acidic mine water then drains into lakes and creeks, killing fish and other animals. At a mine in Colorado, a sample of mine water of volume 16.45 mL was neutralized with 25.00 mL of 0.255 M KOH. What is the concentration of H_2SO_4 in the water?

THE LIMITS OF REACTION

We can now predict how much product a given reaction can yield, such as how much iron we can extract from a given mass of iron(III) oxide. However, each calculation has assumed that all the starting material reacts in the way we have described. In practice, that might not be so.

4.4 REACTION YIELD

Some of the starting materials may be consumed in a competing reaction, a reaction occurring at the same time as the one we are interested in and using some of the same reactants. Even if there is only one reaction going on, some of the starting material might not have reacted at the time we make our measurements.

We can see what is involved by considering one of the major contributors to the supply of CO_2 in the atmosphere, the combustion of fossil fuels—coal, petroleum, and natural gas. An important source of carbon dioxide is the burning of gasoline in automobiles. So, one question we ought to tackle is how much CO_2 is likely to be produced from a liter of gasoline.

Octane, C_8H_{18}, is a representative compound in gasoline. The chemical equation for its combustion is

$$2\,C_8H_{18}(l) + 25\,O_2(g) \longrightarrow 16\,CO_2(g) + 18\,H_2O(l)$$

The density of octane is 0.702 $g\cdot cm^{-3}$ at room temperature. It follows that 1.00 L (1.00×10^3 cm^3) of octane has the following mass:

$$(1.00 \times 10^3 \text{ cm}^3) \times (0.702 \text{ g}\cdot\text{cm}^{-3}) = 702 \text{ g}$$

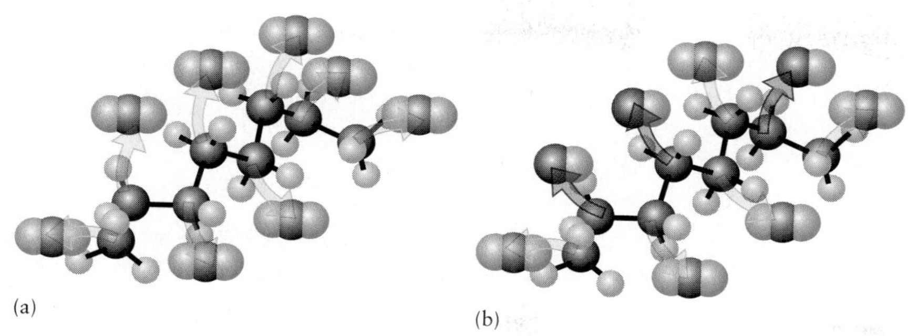

FIGURE 4.4 (a) When an octane molecule undergoes complete combustion, it forms carbon dioxide and water: one CO_2 molecule is formed for each carbon atom present. (b) However, in a limited supply of oxygen, some of the carbon atoms end up as carbon monoxide molecules, CO, so the yield of carbon dioxide is reduced.

We can calculate (see Toolbox 4.2) that when 702 g of octane burns in a plentiful supply of oxygen, 2.16 kg of carbon dioxide should be produced:

$$\text{Mass of } CO_2 = \frac{702 \text{ g } C_8H_{18}}{114.2 \text{ g·mol}^{-1}} \times \frac{16 \text{ mol } CO_2}{2 \text{ mol } C_8H_{18}} \times 44.01 \text{ g·mol}^{-1}$$
$$= 2.16 \times 10^3 \text{ g } CO_2$$

This mass is about 8 kg of CO_2 per gallon of fuel. The average car thus emits about 1 lb of CO_2 per mile.

This 2.16 kg is the theoretical yield of carbon dioxide that can be produced in the reaction. In general, the **theoretical yield** is the maximum mass of product that can be obtained from a given mass of reactant.

Carbon monoxide is formed as well as carbon dioxide when octane burns in a limited supply of oxygen. So, in addition to the reaction written above, the following reaction also takes place:

$$2\, C_8H_{18}(l) + 17\, O_2(g) \longrightarrow 16\, CO(g) + 18\, H_2O(l)$$

Because this reaction consumes some of the octane, the amount of carbon dioxide formed is less than theoretically possible (Fig. 4.4). To express the extent to which this reaction cuts down the production of carbon dioxide, we speak of the **percentage yield** of the product, the fraction of its theoretical yield actually produced, expressed as a percentage (Fig. 4.5):

$$\text{Percentage yield} = \frac{\text{actual yield}}{\text{theoretical yield}} \times 100\%$$

Suppose, for instance, we find that in a test of an automobile engine to monitor the combustion of 1.00 L of octane under various conditions, only 1.14 kg of carbon dioxide is produced, not the 2.16 kg of the theoretical yield, then

$$\text{Percentage yield of } CO_2 = \frac{1.14 \text{ kg}}{2.16 \text{ kg}} \times 100\% = 52.8\%$$

The theoretical yield of a product is the maximum mass that can be expected on the basis of the stoichiometry of a single chemical equation. The percentage yield is the percentage of the theoretical yield actually achieved.

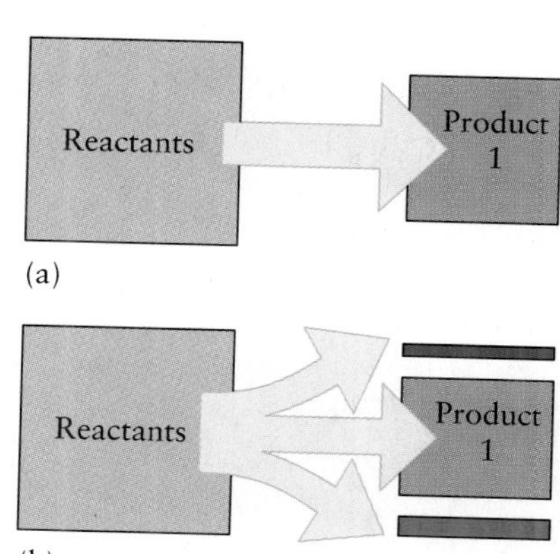

FIGURE 4.5 (a) The yield of a product would be 100% if no competing reactions were taking place. (b) However, if a reactant can take part in more than one reaction at the same time, then the yield of a particular product will be less than 100% because other products will also form.

EXAMPLE 4.2 Calculating the percentage yield of a product

When a large volume of water was poured on 100 g of calcium carbide, 28.3 g of ethyne (C_2H_2, acetylene) was produced. Calculate the percentage yield of ethyne for the reaction

$$CaC_2(s) + 2\,H_2O(l) \longrightarrow Ca(OH)_2(aq) + C_2H_2(g)$$

STRATEGY First, we need to know the theoretical yield (in grams) of the product. It is calculated by the technique explained in Toolbox 4.2. We perform the three steps together in the following solution. Then we express the actual yield as a percentage of the theoretical yield.

SOLUTION The molar mass of calcium carbide is 64.10 g·mol^{-1} and that of ethyne is 26.04 g·mol^{-1}. The theoretical yield of ethyne is

$$\text{Mass of ethyne} = \frac{100\text{ g }CaC_2}{64.10\text{ g·mol}^{-1}} \times \frac{1\text{ mol }C_2H_2}{1\text{ mol }CaC_2} \times 26.04\text{ g·mol}^{-1} = 40.6\text{ g }C_2H_2$$

Because the actual yield of ethyne is 28.3 g, its percentage yield is

$$\text{Percentage yield of ethyne} = \frac{28.3\text{ g}}{40.6\text{ g}} \times 100\% = 69.7\%$$

SELF-TEST 4.5A When 24.0 g of potassium nitrate was heated with lead, 13.8 g of potassium nitrite was formed in the reaction

$$Pb(s) + KNO_3(s) \xrightarrow{\Delta} PbO(s) + KNO_2(s)$$

Calculate the percentage yield of potassium nitrite.

[*Answer:* 68.3%]

SELF-TEST 4.5B Reduction of 15 kg of iron(III) oxide in a blast furnace resulted in the production of 8.8 kg of iron. What is the percentage yield of iron?

4.5 LIMITING REACTANTS

We saw in Example 4.1 that 3.55 kg of aluminum is needed to react with 10.0 kg of chromium (III) oxide. Suppose, though, that we did not have that much aluminum available. Then less of the oxide can be reduced, so less of the product will form. The supply of aluminum has limited the quantity of product formed. The aluminum is an example of a **limiting reactant,** a reactant that governs the maximum yield of product in a given reaction. A limiting reactant is like a part in an automobile factory: if there are 1000 headlights and 600 car bodies, then the maximum number of automobiles is limited by the number of headlights. Because each body requires two headlights, there are enough headlights for only 500 cars, so the headlights play the role of the limiting reactant. When all the headlights have been used, 100 bodies remain unused.

The limiting reactant is sometimes referred to as the limiting reagent.

EXAMPLE 4.3 Calculating the mass of reactant remaining

How much chromium(III) oxide of the original 10.0-kg sample in Example 4.1 would remain if only 2.54 kg of aluminum had been used for its reduction?

STRATEGY The mass of reactant remaining is the difference between the mass supplied and the mass that reacts. We can calculate the mass that reacts from the masses of reducing agent supplied and the stoichiometric relation between the two reactants. All the data we need are in Example 4.1; the chemical equation is given there too.

SOLUTION From the information in Example 4.1, we calculate the mass of chromium(III) oxide reduced by 2.54 kg of aluminum:

$$\begin{aligned}\text{Mass of chromium(III) oxide reduced}\\ &= \frac{2.54 \times 10^3 \text{ g Al}}{26.98 \text{ g·mol}^{-1}} \times \frac{1 \text{ mol } Cr_2O_3}{2 \text{ mol Al}} \times 152.00 \text{ g·mol}^{-1}\\ &= 7.15 \times 10^3 \text{ g } Cr_2O_3\end{aligned}$$

That is, 7.15 kg of chromium(III) oxide can be reduced. The mass of Cr_2O_3 remaining after the reaction is therefore

$$\text{Mass remaining} = 10.0 \text{ kg} - 7.15 \text{ kg} = 2.8 \text{ kg}$$

SELF-TEST 4.6A Suppose in Self-Test 4.3A that 25.0 g of potassium was supplied to react with the same amount (0.450 g) of hydrogen gas. What mass of potassium would remain at the end of the reaction?

[*Answer:* 7.5 g]

SELF-TEST 4.6B In a particular experiment, 15.0 g of chlorine was allowed to react with 43.6 g of phosphorus to produce PCl_3:

$$2\,P(s) + 3\,Cl_2(g) \longrightarrow 2\,PCl_3(l)$$

What mass of phosphorus will remain unreacted if all the chlorine reacts in this way?

In some cases, we must determine which is the limiting reactant. For example, in the reaction between nitrogen and hydrogen used to make ammonia,

$$N_2(g) + 3\,H_2(g) \xrightarrow{\text{Fe}} 2\,NH_3(g)$$

we have

$$1 \text{ mol } N_2 \simeq 3 \text{ mol } H_2$$

Symbols written above the arrow in a chemical equation indicate the conditions for the reaction. In this case, iron facilitates the reaction.

CHEMISTRY AT WORK

A chemical engineer surveys his realm, a large-scale aluminum manufacturing process. The engineer will decide which conditions produce the highest percentage yield. Maintaining a high yield reduces costs and waste while conserving valuable resources.

Either nitrogen or hydrogen will be the limiting reactant unless we happen to supply exactly 3 mol H_2 for each mole of N_2 present. To decide which reactant is the limiting one, we compare the moles of reactants supplied with the stoichiometric coefficients. For example, suppose we had only 2 mol H_2 available for every 1 mol N_2. Because this amount of hydrogen is less than is required by the stoichiometric relation, the hydrogen is the limiting reactant. Once we have identified the limiting reactant, we can calculate the amount of product that can be formed. We can also calculate the amount of excess reactant that remains at the end of the reaction.

The limiting reactant in a reaction is the species supplied in an amount smaller than that required by the stoichiometric relation between the reactants.

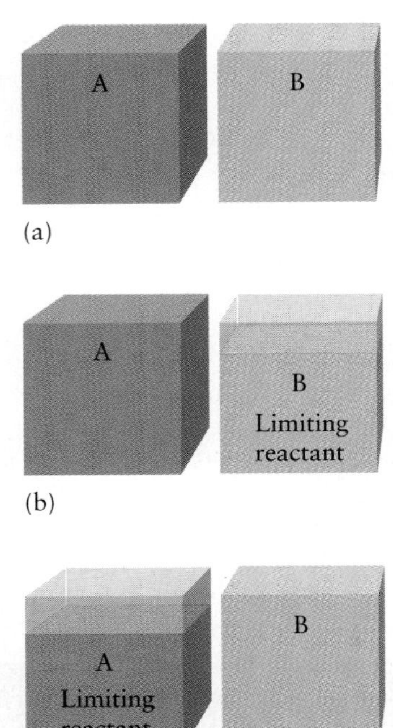

FIGURE 4.6 How to decide which is the limiting reactant. (a) The orange and green cubes depict the relative amounts of each reactant that are required by the stoichiometric relation. (b) If the amount of reactant B is less than that required for all A to react, then B is the limiting reactant. (c) If the amount of A is less than that required for all B to react, then A is the limiting reactant.

EXAMPLE 4.4 Identifying the limiting reactant

As we saw in Example 4.2, calcium carbide reacts with water to form calcium hydroxide and the flammable gas ethyne (acetylene). Which is the limiting reactant when 1.00×10^2 g of water reacts with 1.00×10^2 g of calcium carbide? The chemical equation is

$$CaC_2(s) + 2\,H_2O(l) \longrightarrow Ca(OH)_2(aq) + C_2H_2(g)$$

STRATEGY We are given the masses of the reactants, not the amounts in moles. Therefore, we convert masses to moles by using the molar masses of the substances. Then we choose one of the substances and calculate (Toolbox 4.1) how much of the second reactant is needed for complete reaction. If the actual amount of the second reactant is greater than the amount needed, then it is present in excess and not all of it will react; in this case the *first* substance is the limiting reactant. If the actual amount of the second reactant is less than that calculated, then all of it will react, and it is the limiting reactant (Fig. 4.6).

SOLUTION The molar mass of calcium carbide is 64.10 $g{\cdot}mol^{-1}$ and that of water is 18.02 $g{\cdot}mol^{-1}$. Therefore, the numbers of moles of the reactants are

$$\text{Number of moles of } CaC_2 = \frac{100\ \cancel{g}}{64.10\ \cancel{g}{\cdot}mol^{-1}} = 1.56\ mol$$

$$\text{Number of moles of } H_2O = \frac{100\ \cancel{g}}{18.02\ \cancel{g}{\cdot}mol^{-1}} = 5.55\ mol$$

Next, we work out from the stoichiometric relation

$$1\ mol\ CaC_2 \simeq 2\ mol\ H_2O$$

the amount of H_2O that is needed to react with 1.56 mol CaC_2:

$$\begin{aligned}\text{Number of moles of } H_2O &= (1.56\ \cancel{mol\ CaC_2}) \times \frac{2\ mol\ H_2O}{1\ \cancel{mol\ CaC_2}}\\ &= 1.56 \times 2\ mol\ H_2O = 3.12\ mol\ H_2O\end{aligned}$$

Because 3.12 mol H_2O is required and 5.55 mol H_2O is supplied, all the calcium carbide can react; so the calcium carbide is the limiting reactant and water is present in excess.

SELF-TEST 4.7A Which is the limiting reactant when 100 kg of hydrogen reacts with 800 kg of nitrogen in the synthesis of ammonia?

[*Answer:* Hydrogen]

SELF-TEST 4.7B Suppose that 28 g NO_2 and 18 g of water are supplied in the reaction used to produce nitric acid from nitrogen dioxide:

$$3\,NO_2(g) + H_2O(l) \longrightarrow 2\,HNO_3(l) + NO(g)$$

Which substance is the limiting reactant?

EXAMPLE 4.5 Using the limiting reactant

Freons, a type of chlorofluorocarbon, at one time were used extensively in spray cans and as coolants in refrigerators and air conditioners. Unfortunately, they contribute to the greenhouse effect (Case Study 4) and attack the ozone layer in the stratosphere. The ozone layer protects us from harmful solar radiation. One of the most promising substitutes is $C_2H_2F_4$, which is called HFC-134a in industry. The following reaction,

$$C_2HF_3(l) + HF(g) \longrightarrow C_2H_2F_4(l)$$

can be used to make HFC-134a. In a certain reaction, 100.0 g of C_2HF_3 is mixed with 30.12 g of HF. What mass of HFC-134a can be produced and what mass of excess reactant remains at the end of the reaction?

STRATEGY As in Example 4.4, the limiting reactant must be identified. This limiting reactant determines the theoretical yield of the reaction. So, we use it to calculate the mass of product (Toolbox 4.2). The mass of excess reactant is the difference between the mass supplied and the mass that reacts.

SOLUTION The additional information required for the calculation are the molar masses of C_2HF_3 (82.03 $g \cdot mol^{-1}$), hydrogen fluoride (20.01 $g \cdot mol^{-1}$), and HFC-134a (102.04 $g \cdot mol^{-1}$). The amount of C_2HF_3 provided is

$$\text{Number of moles of } C_2HF_3 = \frac{100.0\ g}{82.03\ g \cdot mol^{-1}} = 1.219\ mol$$

The amount of HF needed to react with 1.219 mol C_2HF_3 is calculated from the relation

$$1\ mol\ C_2HF_3 \simeq 1\ mol\ HF$$

as follows:

$$\begin{aligned}\text{Number of moles of HF} &= (1.219\ mol\ C_2HF_3) \times \frac{1\ mol\ HF}{1\ mol\ C_2HF_3} \\ &= 1.219\ mol\ HF\end{aligned}$$

The amount of HF actually supplied is

$$\text{Number of moles of HF} = \frac{30.12\ g}{20.01\ g \cdot mol^{-1}} = 1.505\ mol$$

Because 1.219 mol HF is needed to react with all the C_2HF_3 and 1.505 mol HF is provided, the C_2HF_3 is the limiting reactant. The reaction stoichiometry implies that

$$1\ mol\ C_2HF_3 \simeq 1\ mol\ C_2H_2F_4$$

It follows that the amount of HFC-134a that can be formed from the C_2HF_3 available is

$$\text{Number of moles of HFC-134a} = (1.219 \text{ mol } C_2HF_3) \times \frac{1 \text{ mol } C_2H_2F_4}{1 \text{ mol } C_2HF_3}$$
$$= 1.219 \text{ mol } C_2H_2F_4$$

Hence, the mass of HFC-134a that can be produced is

$$\text{Mass of HFC-134a} = (1.219 \text{ mol}) \times (102.04 \text{ g·mol}^{-1}) = 124.4 \text{ g}$$

The mass of excess reactant (HF) is the difference between the mass supplied and the mass that reacts. We have calculated that 1.219 mol HF reacts, so the mass that reacts is

$$\text{Mass of HF} = (1.219 \text{ mol}) \times (20.01 \text{ g·mol}^{-1}) = 24.39 \text{ g}$$

Hence, the mass of hydrogen fluoride remaining after the reaction is

$$\text{Mass remaining} = 30.12 \text{ g} - 24.39 \text{ g} = 5.73 \text{ g}$$

SELF-TEST 4.8A Identify the limiting reactant in the reaction

$$6\,Na(l) + Al_2O_3(s) \longrightarrow 2\,Al(l) + 3\,Na_2O(s)$$

when 5.52 g of sodium is mixed with 5.10 g of Al_2O_3. What mass of aluminum can be produced, and what mass of excess reactant remains at the end of the reaction?

[*Answer:* Sodium; 2.16 g Al; 1.02 g Al_2O_3]

SELF-TEST 4.8B What is the limiting reactant for the preparation of urea from ammonia in the reaction

$$2\,NH_3(g) + CO_2(g) \longrightarrow OC(NH_2)_2(s) + H_2O(l)$$

when 14.5 kg of ammonia is available to react with 22.1 kg of carbon dioxide? What mass of urea can be produced, and what mass of excess reactant remains at the end of the reaction?

4.6 COMBUSTION ANALYSIS

Combustion analysis is a technique based on the concept of limiting reactants. It is widely used for the determination of the empirical formulas of organic compounds containing carbon, hydrogen, and oxygen (Fig. 4.7) and involves determining the masses of water and carbon dioxide that

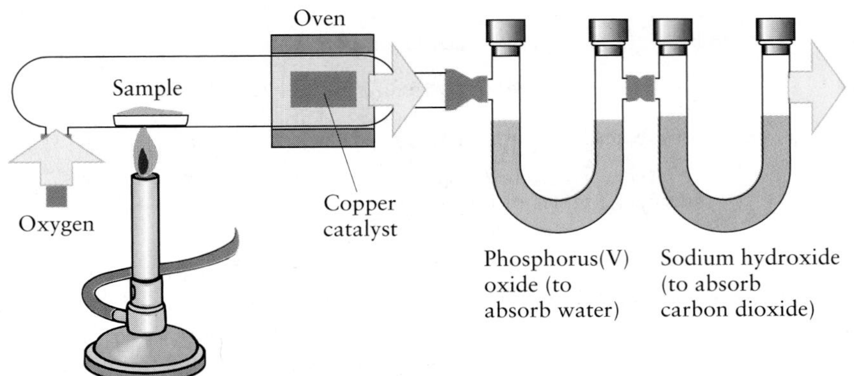

FIGURE 4.7 The apparatus used for a combustion analysis. The masses of carbon dioxide and water produced are obtained from the differences of the masses of the collecting tubes before and after the experiment. The catalyst ensures that any CO produced is oxidized to CO_2.

are formed when an organic compound—the limiting reactant—is burned in a plentiful supply of oxygen.

The sample is ignited in a tube through which there is a flow of oxygen. All the hydrogen in the compound is converted to water and all the carbon is converted to carbon dioxide. The product gases flow through two more tubes that trap them. The water produced is absorbed in the first tube by phosphorus(V) oxide, P_4O_{10}. The increase in mass of this tube is equal to the mass of water absorbed in the reaction with phosphorus(V) oxide to form phosphoric acid:

This compound is still widely called by its common name, phosphorus pentoxide, a name based on its empirical formula, P_2O_5.

$$P_4O_{10}(s) + 6\,H_2O(l) \longrightarrow 4\,H_3PO_4(l)$$

The carbon dioxide is trapped (as sodium hydrogen carbonate) in the second tube, which contains sodium hydroxide:

$$NaOH(s) + CO_2(g) \longrightarrow NaHCO_3(s)$$

The increase in mass of this tube is equal to the mass of carbon dioxide produced in the combustion.

In the presence of excess oxygen, each carbon atom in the compound ends up in one molecule of carbon dioxide (Fig. 4.8). For example, in the combustion of vitamin C,

$$C_6H_8O_6(s) + 5\,O_2(g) \longrightarrow 6\,CO_2(g) + 4\,H_2O(g)$$

As predicted, each C atom present in the compound becomes a CO_2 molecule. Therefore, we can write

$$1 \text{ mol C} \simeq 1 \text{ mol } CO_2$$

for any organic compound, even if we do not yet know its formula. Hence, by measuring the mass of carbon dioxide produced and converting that mass to moles, we can find the number of moles of C atoms in the CO_2 produced and thus in the original sample.

Now we turn to the mass of water produced in the combustion. In the presence of excess oxygen, each hydrogen atom in a compound contributes to a water molecule when the compound burns. We can infer from the chemical equation for the combustion of vitamin C that

$$2 \text{ mol H} \simeq 1 \text{ mol } H_2O$$

This relation is true for all combustions of compounds containing hydrogen. Therefore, if we measure the mass of water produced when the compound burns in plenty of oxygen, we can find the number of moles of H atoms in the sample.

Many organic compounds also contain oxygen. We can calculate the mass of oxygen originally present by subtracting the masses of carbon and hydrogen in the sample from the original mass of the sample.

In a combustion analysis, the numbers of moles of C, H, and O atoms in a sample are determined from the masses of carbon dioxide and water produced when the compound burns in excess oxygen.

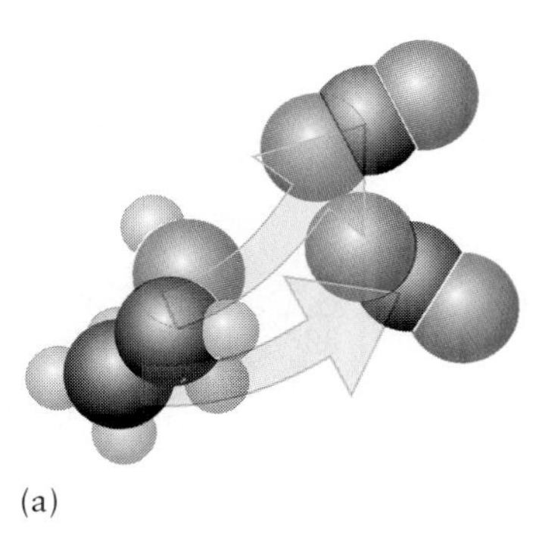

(a)

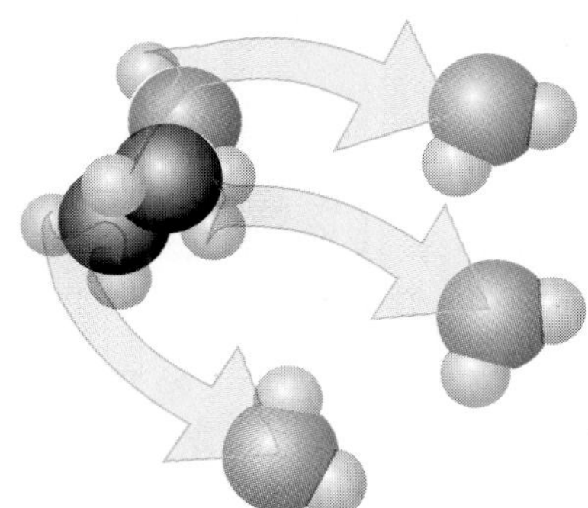

(b)

FIGURE 4.8 When an organic compound burns in a plentiful supply of oxygen, (a) each C atom produces one CO_2 molecule and (b) each pair of H atoms ends up as one H_2O molecule (not necessarily in the same molecule).

CASE STUDY 4 GREENHOUSE GASES

EVERY YEAR OUR planet receives from the Sun more than enough radiant energy to supply all our energy needs. About 55% of solar radiation is reflected away or used in natural processes. The remaining 45% is converted to heat, most of which escapes as infrared radiation (heat radiation). This energy balance has resulted in a stable surface temperature on Earth. However, we are now faced with the possibility that the balance has been upset.

The so-called *greenhouse gases* in the atmosphere act like glass in a greenhouse. They are transparent to visible light but absorb infrared radiation, reflecting it back to Earth rather than allowing it to escape. This energy retention warms the Earth, as if the entire planet were enclosed in a huge greenhouse. The greenhouse gases include water vapor, carbon dioxide, methane, dinitrogen oxide, and certain chlorofluorocarbons. Because the concentration of CO_2 in the atmosphere is relatively high, it makes a considerable contribution to the *greenhouse effect*; and for an initial look at the problem, we can concentrate on it.

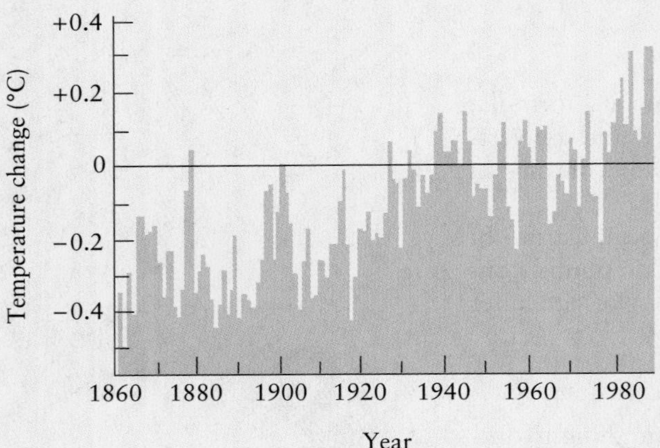

The average surface temperature of the Earth from 1860 to 1990.

Studies of air pockets in ancient ice deposits show that changes in the concentration of atmospheric CO_2 over the past 150,000 years correlate with changes in the global surface temperature. The fact that the concentration of CO_2 rose about 25% between 1850 and 1990 is thus of some concern. The Earth's average surface temperature over a period of 100 years has been reconstructed and found to have risen by about 0.5°C (see the graph below, left).

Where is the additional CO_2 coming from? Some is generated when limestone, $CaCO_3$, is heated and decomposed. Large amounts of CO_2 are also released to the atmosphere by deforestation, which involves burning large areas of brush and trees. However, most of it comes from the burning of fossil fuels, which began on a large scale after 1850. The rate of burning of fossil fuels, including coal, petroleum products, and natural gas, increased by about 5.3% per year from 1945 until 1975.

Computer projections of the CO_2 concentration for the next 200 years are shown in the graph on the next page. The concentration of CO_2 depends on its annual rate of growth. Limiting growth through energy-efficient lifestyles and capture of CO_2 released in industrial processes can reduce the contribution of CO_2 to global warming. However, these reductions must be severe to halt the increase completely. All the projected curves represent growth rates (rates of change) that are slower than the 5.3% growth rate that existed in 1975, yet only a "negative growth" scenario would not involve an increase in CO_2 concentration. Negative growth would involve actually decreasing our energy usage by about 2% annually.

What might happen if atmospheric CO_2 concentration continues to rise? If it rose to about twice its present concentration, the average temperature of the surface of the Earth might rise by as much as 5°C. The prediction of specific changes in climate is difficult, because the Earth is such a complicated system. However, there is general

EXAMPLE 4.6 Determining an empirical formula by combustion analysis

A combustion analysis was carried out on 1.621 g of a newly synthesized compound, which was known to contain only C, H, and O. The masses of water and carbon dioxide produced were 1.902 g and 3.095 g, respectively. What is the empirical formula of the compound?

STRATEGY We need to use the stoichiometric relations given on the previous page and the procedures set out in Toolbox 4.1 to find the number of moles of carbon and hydrogen atoms in the sample, and then convert those moles to

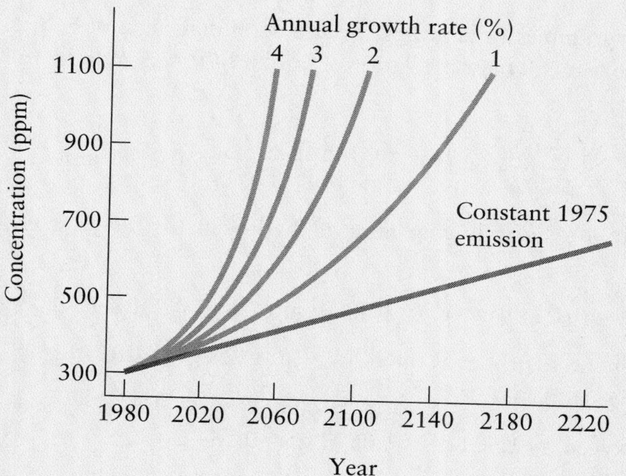

The concentration of atmospheric carbon dioxide projected from various scenarios.

agreement that there would be a rise in sea level of between 0.5 m and 1.5 m over the next 50 to 100 years. This rise would wipe out many coastal communities and farms and lead to further changes in climate that might result in the spread of deserts into agricultural areas.

Fortunately, there are measures we can take now both to reduce the production of CO_2 and to capture industrial CO_2 emissions. Alternatives to fossil fuels, such as hydrogen, are explored in Section 6.11. Coal, which is mostly carbon, can also be converted into fuels with a lower proportion of carbon. Its conversion to methane, CH_4, for instance, would reduce CO_2 emissions for a given release of energy. Carbon dioxide can also be removed chemically from power plant exhaust gases. (See Self-Test 4.3B.)

QUESTIONS

1. Methanol, CH_3OH, is a clean-burning liquid fuel being considered as a replacement for gasoline. Calculate the theoretical yield in kilograms of CO_2 produced by the combustion of 1.00 L of methanol (of density 0.791 $g \cdot cm^{-3}$) and compare it with the 2.16 kg of CO_2 generated by the combustion of 1.00 L of octane. Which fuel contributes more CO_2 per liter to the atmosphere when burned? What other factors would you take into consideration when deciding which of the two fuels to use?

2. What mass of CO_2 is released during the conversion of 1.00 metric ton (1.00×10^3 kg) of limestone to quicklime?

3. In 1992 the concentration of methane was estimated to be 1.74 ppm (parts per million, mg (solute)/kg (solution)). Assuming a total atmospheric mass of 5.1×10^{21} g, what was the mass of methane in the atmosphere in 1992?

4. What is the mass percentage of carbon in methane?

Although these smokestacks are emitting many tons of carbon dioxide into the air, most of the growth in atmospheric carbon dioxide concentration is caused by automobiles. Nevertheless, industrial CO_2 emissions may also need to be controlled.

masses. The mass of oxygen in the sample is obtained by subtraction of the total mass of carbon and hydrogen from the mass of the original sample. It is then converted to the number of moles of O atoms. Finally, the relative numbers of atoms are expressed as an empirical formula.

SOLUTION To convert the mass of carbon dioxide to moles of C atoms, we use the molar mass of carbon dioxide (44.01 $g \cdot mol^{-1}$) and the stoichiometric relation 1 mol C $\simeq$ 1 mol CO_2:

$$\text{Number of moles of C} = \left(\frac{3.095}{44.01}\text{ mol } CO_2\right) \times \frac{1 \text{ mol C}}{1 \text{ mol } CO_2} = 0.07032 \text{ mol C}$$

The mass of carbon in the sample is therefore

$$\text{Mass of carbon} = (0.07032 \text{ mol}) \times (12.01 \text{ g·mol}^{-1}) = 0.8445 \text{ g}$$

To convert mass of water to moles of H atoms, we use the molar mass of water (18.01 g·mol^{-1}) and the stoichiometric relation 1 mol H_2O ≏ 2 mol H to obtain

$$\text{Number of moles of H} = \left(\frac{1.902}{18.02} \text{ mol } H_2O\right) \times \frac{2 \text{ mol H}}{1 \text{ mol } H_2O} = 0.2111 \text{ mol H}$$

This amount corresponds to the following mass of hydrogen (of molar mass 1.008 g·mol^{-1}):

$$\text{Mass of hydrogen} = (0.2111 \text{ mol}) \times (1.008 \text{ g·mol}^{-1}) = 0.2128 \text{ g}$$

The total mass of carbon and hydrogen is 0.8445 g + 0.2128 g = 1.0573 g, so the mass of oxygen in the sample is

$$\text{Mass of oxygen} = 1.621 \text{ g} - 1.0573 \text{ g} = 0.564 \text{ g}$$

Therefore,

$$\text{Number of moles of O} = \frac{0.564 \text{ g}}{16.00 \text{ g·mol}^{-1}} = 0.0352 \text{ mol}$$

At this stage we know that the relative numbers of moles, and thus the relative number of atoms of each element, are

$$\text{C:H:O} = 0.07032:0.2111:0.0352$$

Division by the smallest number, 0.0352, gives

$$\text{C:H:O} = 2.00:6.00:1.00$$

We conclude that the empirical formula of the new compound is C_2H_6O.

SELF-TEST 4.9A When 0.528 g of sucrose (a compound of carbon, hydrogen, and oxygen) is burned, 0.306 g of water and 0.815 g of carbon dioxide are formed. Deduce the empirical formula of sucrose.

[*Answer:* $C_{12}H_{22}O_{11}$]

SELF-TEST 4.9B When 0.236 g of aspirin is burned in oxygen, 0.519 g of carbon dioxide and 0.0945 g of water are formed. Deduce the empirical formula of aspirin.

SKILLS YOU SHOULD HAVE MASTERED

Conceptual

1. Explain the significance of the stoichiometric coefficients in a chemical equation.
2. Explain what is meant by a mole ratio and state how it is used.
3. Explain the significance of the stoichiometric (equivalence) point of a titration.

Problem-Solving

1. Carry out mole-to-mole, mass-to-mole, and mass-to-mass calculations for any two species involved in a chemical reaction.
2. Calculate the volume of solution required for a reaction, given the corresponding balanced chemical equation and the reactant concentration.

3. Calculate the molar concentration (molarity) of a solute from titration data.
4. Calculate the theoretical and percentage yields of the products of a reaction, given the mass of starting material.
5. Identify the limiting reactant of a reaction and calculate the amount of excess reactant present, given the initial mass of each reactant.
6. Use the limiting reactant to calculate the yield of a product.
7. Determine the empirical formula of an organic compound by combustion analysis.

Descriptive

1. Describe the purpose and procedure of a titration.
2. Describe how a combustion analysis is conducted and interpreted.

EXERCISES

Mole Calculations

4.1 (a) How many moles of H_2 are needed to convert 5.0 mol O_2 to water? (b) Determine the amount (in moles) of H_2 needed to convert 5.0 mol O_2 to hydrogen peroxide, H_2O_2.

4.2 (a) Calculate the number of moles of product gases produced from the explosion of 2.0 mol TNT molecules by the reaction

$$4\,C_7H_5O_6N_3(s) + 21\,O_2(g) \longrightarrow 28\,CO_2(g) + 6\,N_2(g) + 10\,H_2O(g)$$

(b) Calculate the number of moles of product gases produced from the detonation of 3.2 mol of nitroglycerin molecules by the reaction

$$4\,C_3H_5O_9N_3(l) \longrightarrow 12\,CO_2(g) + 6\,N_2(g) + O_2(g) + 10\,H_2O(g)$$

4.3 (a) How many moles of NaOH are needed to neutralize 3.7 mol HCl in the reaction

$$HCl(aq) + NaOH(aq) \longrightarrow NaCl(aq) + H_2O(l)$$

(b) In the commercial manufacture of nitric acid, how many moles of NO_2 produce 7.33 mol HNO_3 in the reaction

$$3\,NO_2(g) + H_2O(l) \longrightarrow 2\,HNO_3(aq) + NO(g)$$

(c) The concentration of iodide ions in a solution can be determined by its oxidation to iodine with permanganate ion according to the reaction

$$2\,MnO_4^-(aq) + 16\,H^+(aq) + 10\,I^-(aq) \longrightarrow 2\,Mn^{2+}(aq) + 5\,I_2(aq) + 8\,H_2O(l)$$

How many moles of permanganate ion react with 0.042 mol I^- ions?

4.4 (a) The amount of red-orange silver chromate precipitate produced is used to measure the amount of Ag^+ ions in a sample by the reaction

$$2\,Ag^+(aq) + CrO_4^{2-}(aq) \longrightarrow Ag_2CrO_4(s)$$

Determine the moles of Ag_2CrO_4 that form from 0.40 mol Ag^+ ions.
(b) The removal of hydrogen sulfide gas from "sour" natural gas occurs by the reaction

$$2\,H_2S(g) + SO_2(g) \longrightarrow 3\,S(s) + 2\,H_2O(l)$$

How many moles of S atoms form from the reaction of 5.0 mol H_2S?
(c) The neutralization of phosphoric acid with potassium hydroxide occurs by the reaction

$$H_3PO_4(aq) + 3\,KOH(aq) \longrightarrow K_3PO_4(aq) + 3\,H_2O(l)$$

Calculate the amount (in moles) of KOH that reacts with 0.22 mol H_3PO_4.

4.5 Calculate the number of moles of CO_2 produced when 1.5 mol of hexane molecules, C_6H_{14}, burn in air by the reaction

$$2\,C_6H_{14}(l) + 19\,O_2(g) \longrightarrow 12\,CO_2(g) + 14\,H_2O(g)$$

4.6 A method used by the Environmental Protection Agency (EPA) for determining the concentration of ozone in air is to pass the air sample through a "bubbler" containing iodide ions, which remove the ozone according to the equation

$$O_3(g) + 2\,I^-(aq) + H_2O(l) \longrightarrow O_2(g) + I_2(aq) + 2\,OH^-(aq)$$

How many moles of iodide ions are needed to remove 3.5×10^{-5} mol O_3?

Mass-Mole Relationships

4.7 Ammonia burns in oxygen according to the equation

$$4\,NH_3(g) + 3\,O_2(g) \longrightarrow 2\,N_2(g) + 6\,H_2O(g)$$

(a) Calculate the number of moles of H_2O produced in the combustion of 1.0 g of NH_3. (b) How many grams of O_2 are required for the complete reaction of 13.7 mol NH_3?

4.8 Sodium thiosulfate $Na_2S_2O_3$ (photographer's hypo), reacts with unexposed silver bromide in a film emulsion to form sodium bromide and a soluble compound of formula $Na_3[Ag(S_2O_3)_2]$:

$$2\,Na_2S_2O_3(aq) + AgBr(s) \longrightarrow NaBr(aq) + Na_3[Ag(S_2O_3)_2](aq)$$

(a) How many moles of $Na_2S_2O_3$ are needed to dissolve 1.0 mg of AgBr? (b) Calculate the mass of silver bromide that will produce 0.033 mol $Na_3[Ag(S_2O_3)_2]$.

4.9 Small bottles of propane gas are sold in hardware stores as convenient, portable heat sources (for soldering, for example). The combustion reaction of propane is

$$C_3H_8(g) + 5\,O_2(g) \longrightarrow 3\,CO_2(g) + 4\,H_2O(l)$$

(a) What mass of CO_2 is produced from the combustion of 1.55 mol C_3H_8? (b) How many moles of water molecules accompany the production of 4.40 g of CO_2?

4.10 Impure phosphoric acid for use in the manufacture of fertilizers is produced by the reaction of sulfuric acid on phosphate rock, of which a principal component is $Ca_3(PO_4)_2$. The reaction is

$$Ca_3(PO_4)_2(s) + 3\,H_2SO_4(aq) \longrightarrow 3\,CaSO_4(s) + 2\,H_3PO_4(aq)$$

(a) How many moles of H_3PO_4 are produced from the reaction of 200 kg of H_2SO_4? (b) Determine the mass of calcium sulfate that is produced as a by-product of the reaction of 200 mol $Ca_3(PO_4)_2$.

Mass-Mass Relationships

4.11 The surface atoms of aluminum metal corrode in air to form an impervious aluminum oxide coating that prevents further corrosion of the lower layers of atoms. The oxidation reaction is

$$4\,Al(s) + 3\,O_2(g) \longrightarrow 2\,Al_2O_3(s)$$

(a) Calculate the mass of aluminum oxide formed from the corrosion of 10.0 g of aluminum. (b) In the reaction of 10.0 g of aluminum, what mass of oxygen is needed?

4.12 A typical problem in the iron industry is to determine the mass of iron that can be obtained from a mixture of iron(III) oxide, the principal component of the ore hematite, and carbon obtained from coal. The iron reduction reaction is

$$Fe_2O_3(s) + 3\,C(s) \xrightarrow{\Delta} 2\,Fe(l) + 3\,CO(g)$$

(a) Calculate the mass of iron that can be produced from 1.0 metric ton (1.0 t) of Fe_2O_3 (where 1 t = 10^3 kg). (b) What mass of carbon is needed for the reduction of Fe_2O_3 to produce 500 kg of iron?

4.13 The solid fuel in the booster stage of the space shuttle is a mixture of ammonium perchlorate and aluminum powder. Upon ignition, one reaction that occurs is

$$6\,NH_4ClO_4(s) + 10\,Al(s) \longrightarrow 5\,Al_2O_3(s) + 3\,N_2(g) + 6\,HCl(g) + 9\,H_2O(g)$$

(a) What mass of aluminum should be mixed with 1.5×10^4 kg of NH_4ClO_4 for this reaction? (b) Determine the mass of Al_2O_3 (alumina, a finely divided white powder that is produced as billows of white smoke) formed in the reaction of 5000 kg of aluminum.

4.14 The compound diborane, B_2H_6, was at one time considered for use as a rocket fuel. Its combustion reaction is

$$B_2H_6(g) + 3\,O_2(l) \longrightarrow 2\,HBO_2(g) + 2\,H_2O(l)$$

The fact that HBO_2, a reactive compound, was produced rather than the relatively inert B_2O_3 was a factor in the discontinuation of the investigation of diborane as a fuel. (a) What mass of liquid oxygen (LOX) would be needed to burn 50.0 g of B_2H_6? (b) Determine the mass of HBO_2 produced from the combustion of 30.0 g of B_2H_6.

4.15 The camel stores the fat tristearin, $C_{57}H_{110}O_6$, in its hump. As well as being a source of energy, the fat is also a source of water, because, when it is used, the reaction

$$2\,C_{57}H_{110}O_6(s) + 163\,O_2(g) \longrightarrow 114\,CO_2(g) + 110\,H_2O(l)$$

takes place. (a) What mass of water is available from the oxidation of 2.5 kg of this fat? (b) What mass of oxygen is needed to oxidize 2.5 g of tristearin?

4.16 Potassium superoxide, KO_2, is utilized in closed-system breathing apparatus to remove carbon dioxide and water from exhaled air. The removal of water generates oxygen for breathing by the reaction

$$4\,KO_2(s) + 2\,H_2O(l) \longrightarrow 3\,O_2(g) + 4\,KOH(s)$$

The potassium hydroxide removes carbon dioxide from the apparatus by the reaction

$$KOH(s) + CO_2(g) \longrightarrow KHCO_3(s)$$

(a) What mass of potassium superoxide generates 20.0 g of O_2? (b) What mass of CO_2 can be removed from the apparatus by 100 g of KO_2?

4.17 When a hydrocarbon burns, water is produced as well as carbon dioxide. (For this reason, clouds of condensed water droplets are often seen coming from automobile exhausts, especially on a cold day.) The density of gasoline is 0.79 $g \cdot mL^{-1}$. Assume gasoline to be only octane, C_8H_{18}, for which the combustion reaction is

$$2\,C_8H_{18}(l) + 25\,O_2(g) \longrightarrow 16\,CO_2(g) + 18\,H_2O(l)$$

Calculate the mass of water produced from the combustion of 1.0 L of gasoline.

4.18 The density of oak wood is 0.72 $g \cdot cm^{-3}$. Assuming oak wood to have the formula $C_6H_{12}O_6$, calculate the mass of water produced when a log of dimensions 12 cm × 14 cm × 25 cm is burned.

Titrations

4.19 What is meant by the *stoichiometric point* of a reaction? Give an example.

4.20 Outline the procedure used for an acid-base titration, and describe how the data are interpreted.

4.21 A 15.00-mL sample of sodium hydroxide was titrated to the stoichiometric point with 17.40 mL of 0.234 M HCl(aq). (a) What is the molarity of the NaOH solution? (b) Calculate the mass of NaOH in the solution.

4.22 A 15.00-mL sample of oxalic acid, $H_2C_2O_4$ (with two acidic protons), was titrated to the stoichiometric point with 17.02 mL of 0.288 M NaOH(aq). (a) What is the molarity of the oxalic acid solution? (b) Determine the mass of oxalic acid in the sample.

4.23 A 9.670-g sample of barium hydroxide is dissolved and diluted to the mark in a 250.0-mL volumetric flask. It was found that 11.56 mL of this solution was needed to reach the stoichiometric point in a titration of 25.00 mL of a nitric acid solution. (a) Calculate the molarity of the HNO_3 solution. (b) What mass of HNO_3 is in solution?

4.24 A 10.0-mL sample of 3.0 M KOH(aq) is transferred to a 250.0-mL volumetric flask and diluted to the mark. It was found that 38.5 mL of this diluted solution was needed to reach the stoichiometric point in a titration of 10.0 mL of a phosphoric acid solution, according to the reaction

$$3\,KOH(aq) + H_3PO_4(aq) \longrightarrow K_3PO_4(aq) + 3\,H_2O(l)$$

(a) Calculate the molarity of the H_3PO_4 solution. (b) What mass of H_3PO_4 is in solution?

4.25 In a titration, a 3.25-g sample of an acid, HX, requires 68.8 mL of a 0.750 M NaOH(aq) solution for complete reaction. What is the molar mass of the acid?

4.26 In a titration, 16.02 mL of 0.100 M NaOH(aq) was required to titrate 0.2011 g of an unknown acid, HX. What is the molar mass of the acid?

Reaction Yield

4.27 The theoretical yield of sodium perxenate, Na_4XeO_4, from a certain reaction is 1.25 mg. In a certain laboratory preparation, only 1.07 mg was obtained. What was the percentage yield?

4.28 The theoretical production yield of ammonia in an industrial synthesis is 95 metric tons per day. The production manager recorded a production of only 62.9 metric tons on a given day. What was the percentage yield for that day?

4.29 When limestone rock, which is principally $CaCO_3$, is heated, carbon dioxide and quicklime, CaO, are produced by the reaction

$$CaCO_3(s) \longrightarrow CaO(s) + CO_2(g)$$

If 11.7 g of CO_2 was produced from the thermal decomposition of 30.7 g of $CaCO_3$, what is the percentage yield of the reaction?

4.30 Phosphorus trichloride, PCl_3, is produced from the reaction of white phosphorus, P_4, and chlorine:

$$P_4(s) + 6\,Cl_2(g) \longrightarrow 4\,PCl_3(g)$$

A 16.4-g sample of PCl_3 was collected from the reaction of 5.00 g of P_4 with excess chlorine. What is the percentage yield of the reaction?

Limiting Reactants

4.31 Explain what is meant by a *limiting reactant* and describe the importance of the concept of limiting reactant in combustion analysis and for the determination of reaction yield.

4.32 Describe a strategy for determining the limiting reactant in a reaction.

4.33 The souring of wine occurs when ethanol, C_2H_5OH, is converted by oxidation into acetic acid:

$$C_2H_5OH(aq) + O_2(g) \longrightarrow CH_3COOH(aq) + H_2O(l)$$

If 2.00 g of ethanol and 1.00 g of oxygen were sealed in a wine bottle, which would be the limiting reactant for the oxidation?

4.34 Slaked lime, $Ca(OH)_2$, is formed from quicklime, CaO, by the addition of water:

$$CaO(s) + H_2O(l) \longrightarrow Ca(OH)_2(s)$$

What mass of slaked lime can be produced from a mixture of 30.0 g of CaO and 10.0 g of H_2O?

4.35 A reaction vessel contains 5.77 g of white phosphorus and 5.77 g of oxygen. The first reaction to occur is the formation of phosphorus(III) oxide, P_4O_6:

$$P_4(s) + 3\,O_2(g) \longrightarrow P_4O_6(s)$$

If enough oxygen is present, it can further react with this oxide to produce phosphorus(V) oxide, P_4O_{10}:

$$P_4O_6(s) + 2\,O_2(g) \longrightarrow P_4O_{10}(s)$$

(a) What is the limiting reactant for the formation of P_4O_{10}? (b) How much P_4O_{10} is produced? (c) How many grams of the excess reactant remain in the reaction vessel?

4.36 A mixture of 7.45 g of iron(II) oxide and 0.111 mol Al as aluminum metal is placed in a crucible and heated in a high-temperature oven, where a reduction of the oxide occurs:

$$3\,FeO(s) + 2\,Al(l) \xrightarrow{\Delta} 3\,Fe(l) + Al_2O_3(s)$$

(a) Which is the limiting reactant? (b) Determine the maximum amount of iron (in moles of Fe) that can be produced. (c) Calculate the mass of excess reactant remaining in the crucible.

4.37 Phosphorus trichloride, PCl_3, reacts with water to form phosphorous acid, H_3PO_3, and hydrochloric acid:

$$PCl_3(l) + 3\,H_2O(l) \longrightarrow H_3PO_3(aq) + 3\,HCl(aq)$$

(a) Which is the limiting reactant when 12.4 g of PCl_3 is mixed with 10.0 g of H_2O? (b) What masses of phosphorous acid and hydrochloric acid are formed?

4.38 A solution containing 3.44 g of $AgNO_3$ is mixed with a solution containing 4.22 g of K_3PO_4. A precipitate of Ag_3PO_4 forms. (a) Write the balanced equation for the reaction. (b) What mass of Ag_3PO_4 can form in the mixture?

4.39 4.0 L of SO_2 is mixed with 4.0 L of H_2S. Upon heating, a reaction occurs, forming elemental sulfur and water:

$$SO_2(g) + 2\,H_2S(g) \longrightarrow 3\,S(s) + 2\,H_2O(g)$$

Under the conditions of the experiment, the density of SO_2 is $2.86\ g{\cdot}L^{-1}$ and that of H_2S is $1.52\ g{\cdot}L^{-1}$. (a) What is the limiting reactant in the formation of sulfur? (b) Calculate the mass of excess reactant in the system. (c) What masses of sulfur and water can form in this experiment? (d) Compare the sum of the masses of SO_2 and H_2S before reaction with the sum of the masses of the excess reactant (sulfur) and water after reaction.

4.40 Lithium metal is the only member of Group 1 that reacts directly with nitrogen to produce a nitride, Li_3N:

$$6\,Li(s) + N_2(g) \longrightarrow 2\,Li_3N(s)$$

(a) What mass of lithium nitride can form from 48.0 g of lithium and 1.0×10^{24} molecules of N_2? (b) Determine the mass of excess reactant in the system. (c) Confirm that the sum of the masses of lithium and nitrogen before reaction is the same as the sum of the masses of the excess reactant and lithium nitride after reaction.

4.41 The compound aluminum chloride is widely used to control chemical reactions in industry. It is made by the following reaction:

$$Al_2O_3(s) + 3\,C(s) + 3\,Cl_2(g) \xrightarrow{\Delta} 2\,AlCl_3(s) + 3\,CO(g)$$

(a) Which is the limiting reactant if 160 g Cl_2 is added to a mixture of 100 g Al_2O_3 and 40 g C? (b) How many grams of $AlCl_3$ can be produced from these reactants?

4.42 The highly toxic compound hydrogen cyanide is used in the preparation of cyanimid fertilizers and in gas chambers for executions. It is made by the following reaction:

$$2\,CH_4(g) + 2\,NH_3(g) + 3\,O_2(g) \xrightarrow{100°C,\ Pt} 2\,HCN(g) + 6\,H_2O(g)$$

(a) Which is the limiting reactant if 600 g O_2 is added to a mixture of 400 g NH_3 and 300 g CH_4? (b) How many grams of HCN can be produced from these reactants?

Combustion Analysis

4.43 A chemical analysis of a 4.39-g sample of benzoic acid, a common carboxylic acid used as a food preservative, is 3.03 g C, 1.14 g O, and the remainder hydrogen. What is the empirical formula of benzoic acid?

4.44 Oxalic acid occurs in rhubarb leaves. Analysis of a sample of oxalic acid of mass 10.0 g showed that it contained 0.22 g H, 2.7 g C, and the remainder oxygen. What is the empirical formula of oxalic acid?

4.45 The analysis of 0.922 g of aniline, a common organic base used in some varnishes, is 0.714 g C, 0.138 g N, and the remainder hydrogen. What is the empirical formula of aniline?

4.46 Urea is used as a commercial fertilizer because of its nitrogen content. An analysis of 25.0 mg of urea showed that it contained 5.0 mg C, 11.68 mg N, 6.65 mg O, and the remainder hydrogen. What is the empirical formula of urea?

4.47 In a combustion analysis of a 0.152-g sample of the artificial sweetener aspartame, it was found that 0.318 g of carbon dioxide, 0.084 g of water, and 0.0145 g of nitrogen were produced. What is the empirical formula of aspartame? The molar mass of aspartame is $294\ g{\cdot}mol^{-1}$. What is its molecular formula?

4.48 The bitter-tasting compound quinine is a component of tonic water and is used as a protection against malaria. When a sample of mass 0.487 g was burned, 1.321 g of carbon dioxide, 0.325 g of water, and 0.0421 g of nitrogen were produced. The molar mass of quinine is 324 $g \cdot mol^{-1}$. Determine the empirical and molecular formulas of quinine.

4.49 The stimulant in coffee and tea is caffeine, a substance of molar mass 194 $g \cdot mol^{-1}$. When 0.376 g of caffeine was burned, 0.682 g of carbon dioxide, 0.174 g of water, and 0.110 g of nitrogen were formed. Determine the empirical and molecular formulas of caffeine, and write the equation for its combustion.

4.50 Nicotine, the stimulant in tobacco, causes a very complex set of physiological effects in the body. It is known to have a molar mass of 162 $g \cdot mol^{-1}$. When a sample of mass 0.385 g was burned, 1.072 g of carbon dioxide, 0.307 g of water, and 0.068 g of nitrogen were produced. What are the empirical and molecular formulas of nicotine? Write the equation for its combustion.

SUPPLEMENTARY EXERCISES

4.51 Define the terms *theoretical yield* and *percentage yield,* and explain why the latter may be less than 100%.

4.52 Sodium thiosulfate, $Na_2S_2O_3$, is often used in the laboratory to determine the molarity of solutions of iodine (as I_3^-):

$$2\,S_2O_3^{2-}(aq) + I_3^-(aq) \longrightarrow 3\,I^-(aq) + S_4O_6^{2-}(aq)$$

If 1.134 g of $Na_2S_2O_3$ was needed to react with all the iodine in a sample, how many moles of I_3^- were present?

4.53 Ultrapure silicon used in solid-state electronics is produced by the zone refining of impure silicon from the high-temperature reduction of silicon dioxide (SiO_2, the principal component of sand and quartz) with carbon:

$$SiO_2(s) + 2\,C(s) \longrightarrow Si(l) + 2\,CO(g)$$

(a) How many Si atoms are produced in conjunction with 5.00 μg of CO? (b) How many C atoms react with 7.33×10^{22} SiO_2 formula units?

4.54 When aqueous solutions of calcium nitrate and phosphoric acid are mixed, a white solid precipitates. (a) What is the formula of the solid? (b) How many grams of the solid can be formed from 206 g calcium nitrate and 150 g phosphoric acid?

4.55 A mixture of hydrogen peroxide and hydrazine can be used as a rocket propellant:

$$7\,H_2O_2(g) + N_2H_4(l) \longrightarrow 2\,HNO_3(aq) + 8\,H_2O(l)$$

(a) How many moles of H_2O_2 react with 0.477 mol N_2H_4? (b) How many moles of HNO_3 can be produced in a reaction of 6.77 g of H_2O_2 with excess N_2H_4? (c) What mass of H_2O can be produced in a reaction of 49.6 mg H_2O_2 with excess N_2H_4?

4.56 Nitrogen dioxide contributes to the formation of acid rain as a result of its reaction with water in the air, according to the reaction:

$$3\,NO_2(g) + H_2O(g) \longrightarrow 2\,HNO_3(aq) + NO(g)$$

(a) What mass (in milligrams) of HNO_3 can be produced from the reaction of 2.93 mg of NO_2, assuming an excess of H_2O? (b) If only 1.91 mg of HNO_3 was produced in the reaction, what is the percentage yield?

4.57 Manganese metal can be prepared by the thermite reaction

$$4\,Al(s) + 3\,MnO_2(s) \longrightarrow 3\,Mn(l) + 2\,Al_2O_3(s)$$

(a) What mass of manganese metal (in milligrams) can be produced from the reaction of 2.935 mg of aluminum with an excess of MnO_2? (b) If, in the reaction, only 2.386 mg of manganese was produced, what is the percentage yield?

4.58 Small amounts of chlorine gas can be generated in the laboratory from the reaction of manganese(IV) oxide with hydrochloric acid:

$$4\,HCl(aq) + MnO_2(s) \longrightarrow 2\,H_2O(l) + MnCl_2(s) + Cl_2(g)$$

(a) What mass of Cl_2 can be produced from 42.7 g of MnO_2 with an excess of HCl(aq)? (b) What volume of chlorine gas (of density 3.17 $g \cdot L^{-1}$) will be produced from the reaction of 300 mL of 0.100 M HCl(aq) with an excess of MnO_2? (c) Suppose only 150 mL of chlorine was produced in the reaction in (b). What is the percentage yield of the reaction?

4.59 The first stage in the production of nitric acid is the oxidation of ammonia to nitric oxide:

$$4\,NH_3(g) + 5\,O_2(g) \xrightarrow{Pt} 4\,NO(g) + 6\,H_2O(g)$$

Calculate the mass (in grams) of NO that is produced from the reaction of 600 L of $O_2(g)$ with an excess of ammonia, assuming 90% yield. The density of O_2 under the conditions of the reactions is 1.43 $g \cdot L^{-1}$.

4.60 Consider the reaction

$$10\,N_2O(g) + C_3H_8(g) \longrightarrow 10\,N_2(g) + 3\,CO_2(g) + 4\,H_2O(g)$$

What volume of carbon dioxide gas can be produced from the reaction of 8.40 mg of N_2O? Assume an excess of C_3H_8 and take the density of carbon dioxide to be 1.96 $g \cdot L^{-1}$ under the conditions of the experiment.

4.61 Octane, the principal component of gasoline, burns in excess air by the reaction

$$2\,C_8H_{18}(l) + 25\,O_2(g) \longrightarrow 16\,CO_2(g) + 18\,H_2O(l)$$

(a) Calculate the volume of oxygen gas needed to react with 2.27 mg of C_8H_{18}, given that the density of oxygen is 1.43 $g{\cdot}L^{-1}$ under the conditions of the experiment.
(b) What volume of air is required in (a), given that air is 21% O_2 by volume?

4.62 The equation describing the reaction used for the analysis of ozone in the atmosphere is

$$O_3(g) + 2\,KI(aq) + H_2O(l) \longrightarrow O_2(g) + I_2(aq) + 2\,KOH(aq)$$

(a) Determine the amount (in moles) of KI needed for the reaction of 0.20 mol of O_3. (b) Calculate the mass of I_2 produced from the reaction of 49.7 μg of O_3. (c) Suppose in (b) 0.118 mg I_2 was produced. What is the percentage yield in the reaction?

4.63 Nitrogen and hydrogen combine to form ammonia according to this reaction:

$$N_2(g) + 3\,H_2(g) \longrightarrow 2\,NH_3(g)$$

Assume that you start with the numbers of molecules shown in the accompanying illustration of the reaction container and that the reaction is brought to completion. Draw the flask as it would look at the end of the reaction. (a) Show each molecule of NH_3 that forms and draw all reactant molecules that remain in excess. (b) Which would be the limiting reactant if you have three N_2 molecules and three H_2 molecules?

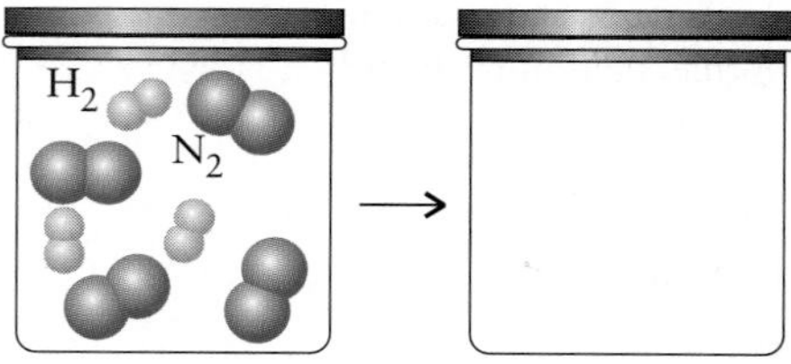

4.64 What is the limiting reactant in a reaction mixture of 12.0 kg of SO_2 and 8.0 kg H_2S for the reaction

$$SO_2(g) + 2\,H_2S(g) \longrightarrow 3\,S(s) + 2\,H_2O(l)$$

4.65 Sodium metal can reduce aluminum oxide according to the equation

$$6\,Na(s) + Al_2O_3(s) \longrightarrow 2\,Al(s) + 3\,Na_2O(s)$$

A mixture of 10.0 g of sodium and 10.0 g of Al_2O_3 is heated in an inert atmosphere (so that oxygen does not react with the sodium metal). (a) Which reactant limits the amount of aluminum that can be produced in the reaction? (b) What is the mass of aluminum that can be produced? (c) What is the mass of excess reactant remaining in the system? (d) If 1.77 g of aluminum were produced in the reaction, what is the percentage yield?

4.66 Butane, C_4H_{10}, is used as a relatively cheap, portable heat source (for example, in cigarette lighters). From the combustion of a mixture of 4.66 g of butane and 11.1 L of O_2 (of density 1.43 $g{\cdot}L^{-1}$),

$$2\,C_4H_{10}(g) + 13\,O_2(g) \longrightarrow 8\,CO_2(g) + 10\,H_2O(g)$$

12.7 g of CO_2 was collected. (a) Which is the limiting reactant? (b) What is the percentage yield? (c) Calculate the mass of excess reactant.

4.67 A mixture of 4.94 g of 85.0% pure phosphine, PH_3, and 0.110 kg of $CuSO_4{\cdot}5H_2O$ (of molar mass 250 $g{\cdot}mol^{-1}$) is placed in a reaction vessel. Calculate the mass (in grams) of Cu_3P_2 (of molar mass 252 $g{\cdot}mol^{-1}$) if a 6.31% yield were produced in the reaction

$$3\,CuSO_4{\cdot}5H_2O(s) + 2\,PH_3(g) \longrightarrow Cu_3P_2(s) + 3\,H_2SO_4(aq) + 15\,H_2O(l)$$

4.68 What mass of the excess reactant remains in a reaction mixture consisting of 1.54 g of $Cr(NO_3)_3$ dissolved in 120 mL of 0.10 M $H_2S(aq)$? The reaction is

$$2\,Cr(NO_3)_3(aq) + 3\,H_2S(aq) \longrightarrow Cr_2S_3(s) + 6\,HNO_3(aq)$$

4.69 Teflon is the DuPont trade name for a class of synthetic fluorocarbon plastics. (Fluorocarbons are compounds of fluorine and carbon.) A chemist from a rival laboratory analyzed a sample and found it to have a mass percentage composition of 24% C and 76% F. What is the empirical formula of the plastic?

4.70 Fructose is a type of sugar that occurs in fruit and honey. Analysis of a 2.0-g sample showed it to contain 0.80 g carbon, 1.06 g oxygen, and the remainder hydrogen. (a) What is the empirical formula of fructose? (b) Given that the molar mass of fructose is 180 $g{\cdot}mol^{-1}$, determine the molecular formula of the compound.

4.71 The mass percentage of sulfur in a fuel, such as coal, can be determined by burning the fuel in air and passing the resulting SO_2 and SO_3 gases into dilute H_2O_2, which converts the oxides to sulfuric acid. The amount of the sulfuric acid can be determined by titration with a base. In one experiment, 6.43 g of coal was burned in excess air and the resulting H_2SO_4 was titrated to the stoichiometric point with 17.4 mL of 0.100 M NaOH. (a) Determine the amount (in moles) of H_2SO_4 that was produced. (b) What is the mass percentage of sulfur in the coal sample?

4.72 A chemist was analyzing the concentration of sulfuric acid in water draining from an abandoned mine. A solution of sodium hydroxide was used to titrate the sample. In the process, the chemist made several mistakes. How would each mistake affect the results of the analysis if it were the only mistake made? In each case, state whether the reported result will show a concentration that is too high or too low, or whether there will be no effect. (a) The flask used to contain the acid was cleaned but not dried, and a small pool of water remained in the flask when the sample was added. (b) The buret used to contain the NaOH was cleaned but not dried. The inside of the buret was wet when the base was added, and the buret was not rinsed with base first. (c) During the titration, the base was added too rapidly, so that two extra drops of base were added.

CHALLENGING EXERCISES

4.73 A solution of hydrochloric acid was prepared by measuring 10.00 mL of the concentrated acid into a 1.000-L volumetric flask and adding water up to the mark. Another solution was prepared by adding 0.530 g of anhydrous sodium carbonate to a 100.0-mL volumetric flask and adding water up to the mark. Then, 25.00 mL of the latter solution was pipetted into a flask and titrated with the diluted acid. The stoichiometric point was reached after 26.50 mL of acid had been added. (a) Write the balanced equation for the reaction of HCl(aq) with Na_2CO_3(aq). (b) What is the molarity of the original hydrochloric acid?

4.74 The reduction of iron(III) oxide to iron metal in a blast furnace takes place in these two steps:

$$2\,C(s) + O_2(g) \longrightarrow 2\,CO(g)$$

$$Fe_2O_3(s) + 3\,CO(g) \longrightarrow 2\,Fe(l) + 3\,CO_2(g)$$

Assume that all the CO generated in the first step reacts in the second. (a) How many C atoms are needed to react with 600 Fe_2O_3 formula units? (b) What volume of carbon dioxide (taken to have a density of 1.25 $g{\cdot}L^{-1}$) is generated in the production of 1.0 metric ton of iron? (c) Assuming a 67.9% yield, what volume of carbon dioxide is generated in the production of 1.0 metric ton of iron? (d) How many kilograms of O_2 are required for the production of 5.00 kg of Fe?

4.75 A vitamin C tablet was analyzed to determine whether it did in fact contain, as the manufacturer claimed, 1.0 g of the vitamin. A tablet was dissolved in water to form a 100.00-mL solution, and a 10.0-mL sample was titrated with iodine (as potassium triiodide). It required 10.1 mL of 0.0521 M I_3^- to reach the stoichiometric point in the titration. Given that 1 mol $I_3^- \simeq$ 1 mol vitamin C in the reaction, is the manufacturer's claim correct? The molar mass of vitamin C is 176 $g{\cdot}mol^{-1}$.

4.76 A forensic chemist needed to determine the concentration of HCN in the blood of a suspected homicide victim and decided to titrate a diluted sample of the blood with iodine, using the reaction

$$HCN(aq) + I_3^-(aq) \longrightarrow ICN(aq) + 2\,I^-(aq) + H^+(aq)$$

A diluted blood sample of volume 15.00 mL was titrated to the stoichiometric point with 5.21 mL of an I_3^- solution. The molarity of I_3^- in the solution was determined by titrating it against arsenic(III) oxide, which in solution forms arsenious acid, H_3AsO_3. 10.42 mL of the triiodide solution was needed to reach the stoichiometric point on a 1.222-g sample of As_4O_6 in the reaction

$$H_3AsO_3(aq) + I_3^-(aq) + H_2O(l) \longrightarrow H_3AsO_4(aq) + 3\,I^-(aq) + 2\,H^+(aq)$$

(a) What is the molarity of the triiodide solution? (b) What is the molar concentration of HCN in the blood sample?

CHAPTER 5

We are kept alive by the gases of the atmosphere, which provide oxygen for us to breathe, nitrogen for plants to convert into compounds, and ozone to protect us from the damaging effects of solar radiation. But the atmosphere is a fragile system, and we need to monitor its composition and properties by sampling it with equipment carried to high altitudes by balloons like these being prepared by a joint French, German, and American team at a launching site in Sweden. The data obtained are interpreted by using information like that described in this chapter.

THE PROPERTIES OF GASES

The atmosphere is a precious layer of gas held by gravity to the surface of the Earth. Although it is approximately 300 km thick, relative to the size of the Earth it is very thin. When you are so far out in space that the Earth looks the size of a basketball, the atmosphere looks only 1 mm thick (Fig. 5.1). Yet this thin, delicate layer is vital to life: it shields us from harmful radiation and supplies chemicals needed for life, such as oxygen, nitrogen, carbon dioxide, and water. In the past the atmosphere was considered vast and unchangeable, so smoke, noxious gases, and exhaust fumes have been released to it indiscriminately since humans first learned to use fire. Now we know that human activities do have an effect on the atmosphere. The changes are small, but they accumulate over decades; some are so important that they could even threaten our survival.

The gases in the atmosphere are not the only ones we are likely to encounter. Eleven elements are gases under normal conditions (Fig. 5.2), as are many compounds. All gases are molecular except the six noble gases, which are monatomic (they consist of single atoms). Many low-molar-mass organic

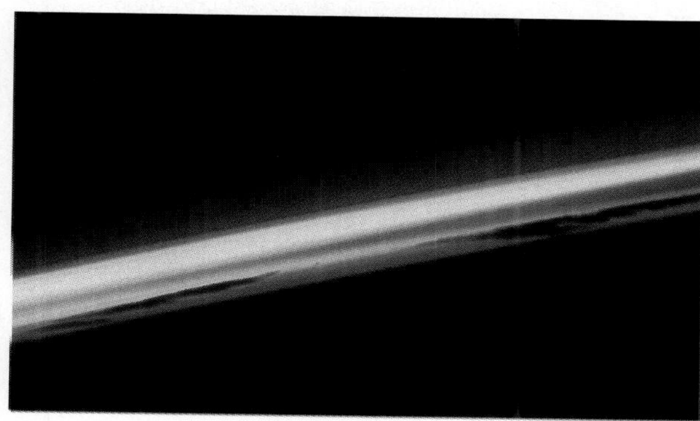

FIGURE 5.1 The delicate film of the Earth's atmosphere, as seen from space.

THE NATURE OF GASES

5.1 The states of matter
5.2 The molecular character of gases
5.3 Pressure
5.4 Units of pressure

THE GAS LAWS

5.5 Boyle's law
5.6 Charles's law
5.7 Avogadro's principle
5.8 Using the gas laws
5.9 The ideal gas law
5.10 Using the ideal gas law to make predictions
5.11 Molar volume
5.12 The stoichiometry of reacting gases
5.13 Gas density
5.14 Mixtures of gases

MOLECULAR MOTION

5.15 Diffusion and effusion
5.16 The kinetic model of gases
5.17 The Maxwell distribution of speeds

REAL GASES

5.18 The ideal gas law as a limiting law
5.19 The Joule-Thomson effect

compounds are gases at room temperature. They include the methane, CH_4, of natural gas and the propane, C_3H_8, and butane, C_4H_{10}, of camping fuel. Nearly all gases have low molar masses, because only small molecules can escape from one another easily.

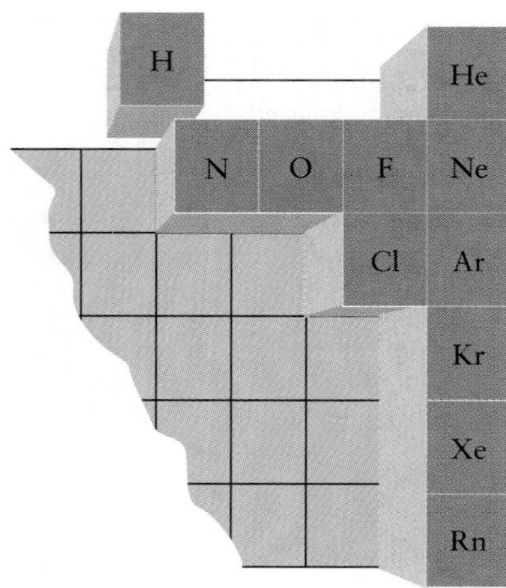

FIGURE 5.2 The 11 elements that are gases under normal conditions. Note how they lie toward the upper right of the periodic table.

THE NATURE OF GASES

To understand the atmosphere we need to study smaller samples of gases. The atmosphere itself is difficult to study, largely because it is not uniform in composition, temperature, and density (Table 5.1). One reason for this nonuniformity is the effect of solar radiation, which causes different chemical reactions at different altitudes. The density of air varies with altitude partly because the temperature varies, but also as a result of the *compressibility* of air, its ability to be squashed into a smaller volume. The weight of air above compresses the air at sea level, causing it to be denser than the air at higher altitudes. Half the mass of the atmosphere lies below 5.5 km, and an airplane at a typical cruising altitude of 10 km is flying above 70% of the mass of the atmosphere.

We begin our study of gases by seeing how they differ from the other two **states of matter**, solids and liquids. This comparison will suggest a molecular model of gases which we shall draw on later to understand why gases behave in their characteristic ways.

5.1 THE STATES OF MATTER

Suppose we take an ice cube from the freezer and place it in a closed container. Initially, the ice maintains its shape (Fig. 5.3a). This behavior is characteristic of a **solid,** a rigid form of matter that maintains its own shape whatever the shape of its container. If the ice is left standing in a room, its temperature begins to rise. Eventually, its temperature reaches the melting point of ice. As the ice melts, water runs down its sides into the bottom of the container (Fig. 5.3b). It is now a **liquid,** a fluid form of matter that takes the shape of the part of the container it occupies. Some of the water evaporates, becoming water vapor. The water vapor spreads rapidly throughout the container. If the container is fitted with a piston, we could **compress** the gas—confine it to a smaller volume—with little

A fluid is a form of matter that can flow.

A vapor is a gas, but it usually means the gaseous state of a substance that is ordinarily a liquid or solid.

TABLE 5.1 Composition of dry air at sea level

Constituent	Molar mass,* $g \cdot mol^{-1}$	Composition, %	
		Volume	Mass
nitrogen, N_2	28.02	78.09	75.52
oxygen, O_2	32.00	20.95	23.14
argon, Ar	39.95	0.93	1.29
carbon dioxide, CO_2	44.01	0.03	0.05

*The average molar mass of molecules in the air, allowing for their different abundances, is 28.97 $g \cdot mol^{-1}$. The percentage of water vapor in ordinary air varies with the humidity.

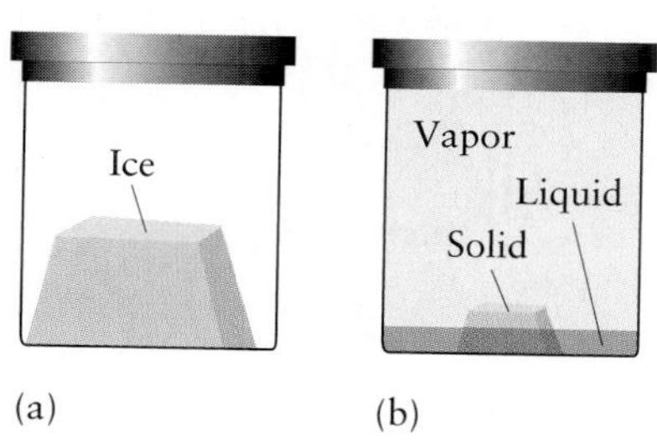

FIGURE 5.3 (a) A solid (in this illustration, an ice cube) retains its original shape in a container. (b) When the ice melts, the liquid formed occupies the lower part of the container and adapts to that shape. Any water vapor present fills the container.

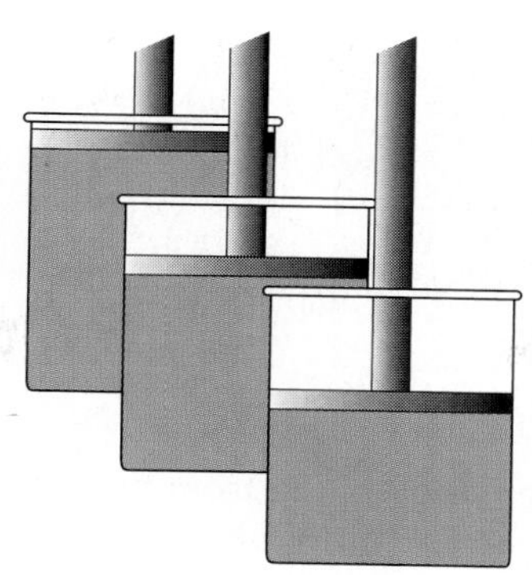

FIGURE 5.4 A gas can easily be compressed into a smaller volume by pushing in a piston. This property suggests that there is a lot of space between the molecules. Liquids and solids are almost incompressible, which suggests that their molecules are in contact.

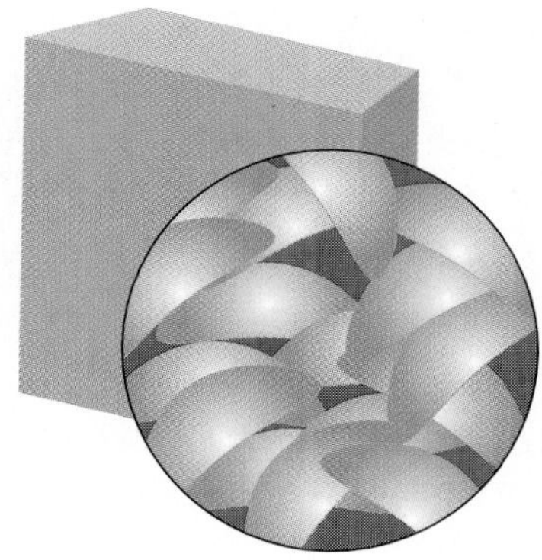

FIGURE 5.5 In a solid, molecules (here represented by the banana-shaped objects) are packed together closely and are unable to move past one another.

effort (Fig. 5.4). These features are characteristic of a **gas,** a fluid form of matter that fills the container it occupies and can easily be compressed.

> ***A gas is a fluid state of matter that fills the container it occupies and can easily be compressed.***

5.2 THE MOLECULAR CHARACTER OF GASES

We can use the distinctive properties of a gas to construct a model of a gas in terms of its molecules. Because the gas expands to fill its container, we know that its molecules do not interact with one another very strongly. If they did interact strongly, they would stick together and form a liquid or a solid (Fig. 5.5). The fact that gases are highly compressible suggests that there is a lot of empty space between the molecules. In liquids and solids, the molecules are already in contact, so it is very difficult to compress these states of matter any further.

Because a gas immediately fills its container, its molecules must be moving very rapidly. We can estimate just how fast the molecules travel on average by noting that a sound wave moving through a gas consists of alternating regions where molecules are closer together and further apart on average (Fig. 5.6). As the wave moves through the gas, molecules have to move closer to one another or further apart. Because the speed of sound in air is about 300 $m \cdot s^{-1}$ (about 700 mph), we can infer that the molecules must travel at about that speed.

The word *kinetic* is from the Greek word for movement.

The molecular model of gases that takes these features into account is called the **kinetic model.** It pictures a gas as widely separated molecules that are in ceaseless random motion. In the kinetic model, molecules are

The name *gas* comes from the same Greek word as our word *chaos.*

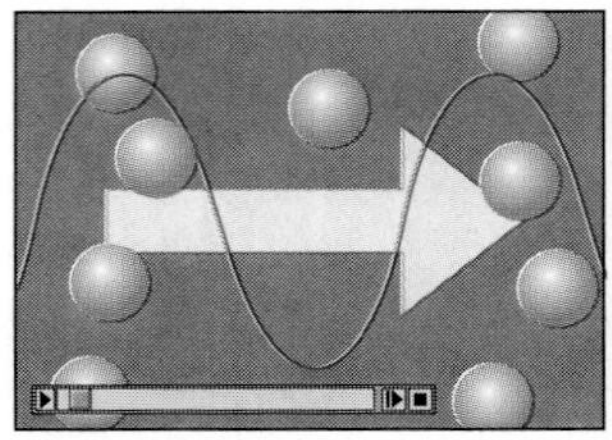

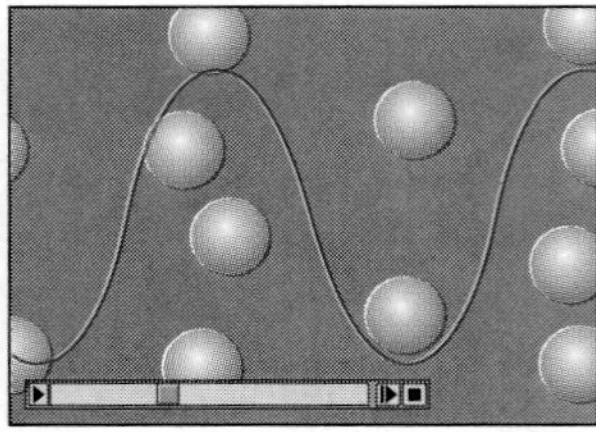

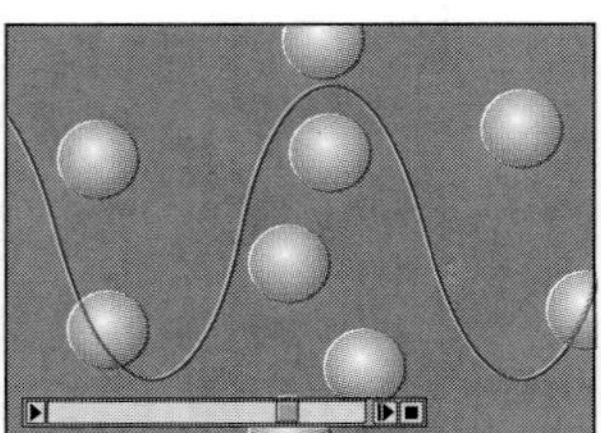

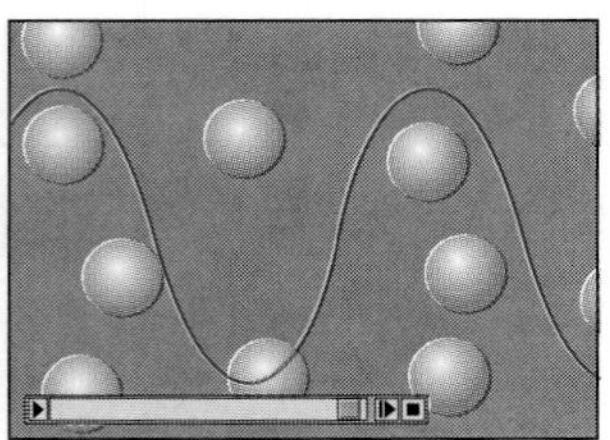

FIGURE 5.6 These four scenes from a computer animation show a pressure wave—a sound wave—traveling through a gas. The red curve depicts the variation in pressure. For the wave to move through the gas, the molecules must adjust their positions so that the regions rich and poor in molecules move across the field of view. As a result, the speed at which the wave moves—the speed of sound—is close to the speed at which the molecules move.

pictured as zooming from place to place, always in straight lines, changing direction only when they collide with a wall of the container or another molecule. The collisions change the speed and direction of the molecules, just like balls in a three-dimensional cosmic game of pool. At one instant, a molecule may be traveling at the speed of sound. At the next, it might be brought to a standstill in a collision. Then, at the next instant, another molecule may collide with it and send it hurtling off again in a different direction.

Although the molecules are pictured as having a wide range of speeds, they have a characteristic *average speed*. Experimentally, it is known that the speed of sound in gases increases as the temperature is raised. This observation strongly suggests that as the temperature of the gas is raised the average speed of its molecules increases too. In fact, *the temperature of a gas is a measure of the average speed of its molecules.* In a cold gas, the molecules move slowly on average; in a hot gas, they move on average much more quickly. If we could cool a gas to absolute zero, translational molecular motion would cease entirely.

The molecules of a gas are widely separated and in ceaseless chaotic motion. The temperature of a gas is a measure of the average speed of its molecules: the higher the temperature, the faster the molecules travel on average.

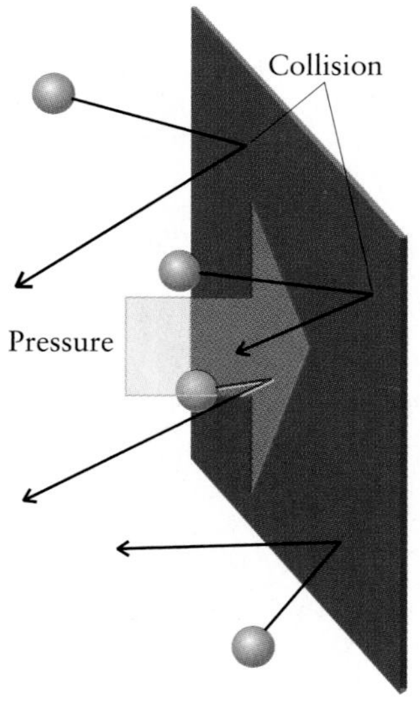

FIGURE 5.7 The pressure of a gas arises from the collisions that its molecules make with the walls of the container. The storm of collisions exerts an almost steady force on the walls.

5.3 PRESSURE

Everyone who has pumped up a bicycle tire or squeezed an inflated balloon has experienced an opposing force arising from the confined air. The **pressure,** P, is the force exerted by the gas divided by the area on which the force is exerted:

$$\text{Pressure} = \frac{\text{force}}{\text{area}}$$

The pressure a gas exerts on the walls of its container results from the collisions of its molecules with the container's surface (Fig. 5.7). A vigorous, chaotic storm of molecules on a surface results in a strong force and hence a high pressure. A gentle storm of molecules on the same area results in a low pressure. When we inflate a tire, we pack more and more molecules into the tire with each stroke of the pump. The rising number of molecules inside the tire exerts a greater force on the inner surface, and hence a higher pressure. The pressure of the atmosphere arises in the same way. We stand in an invisible storm of molecules that beat on us incessantly and exert a force all over our bodies.

The pressure exerted by the atmosphere is measured with a **barometer.** This instrument was invented in the seventeenth century by Evangelista Torricelli, a student of Galileo. Torricelli (whose name coincidentally means "little tower") formed a little tower of mercury, a dense liquid. He sealed a long glass tube at one end, filled it with mercury, and inverted it into a beaker (Fig. 5.8). The column of mercury fell until the pressure exerted by the liquid mercury matched the pressure exerted by the atmosphere. The final height of the column is proportional to the atmospheric pressure.

The precise relation is $P = dgh$, where P is the pressure, d the density of the liquid, g the acceleration of free fall ($9.81\ m{\cdot}s^{-2}$), and h the height of the column.

When you measure the pressure of the air in a tire by using a pressure gauge, you are actually measuring the "gauge pressure," the *difference* between the pressure inside the tire and the atmospheric pressure. A flat tire registers a gauge pressure of zero, because the pressure inside the tire is the same as the pressure of the atmosphere.

The pressure of a gas, the force it exerts divided by the area subjected to the force, arises from the impact of its molecules.

5.4 UNITS OF PRESSURE

The height of the mercury column in a barometer on a typical day at sea level is about 760 mm (corresponding to the 30 inches referred to in weather forecasts). Hence a pressure of 760 millimeters of mercury (written 760 mmHg) corresponds to normal atmospheric pressure at sea level.

The unit 1 mmHg has been largely replaced by the unit 1 Torr (named for Torricelli), with 1 Torr = 1 mmHg. Atmospheric pressure varies with altitude and weather. The pressure of the atmosphere at the cruising height of a commercial jetliner (10 km) is only about 200 Torr, so airplane cabins must be pressurized. A gas at low pressure, such as a gas sample being used to study reactions high in the atmosphere, might have a pressure of only a few Torr (a few millimeters of mercury).

Several units of pressure are in common use (Table 5.2). Because we shall often deal with substances close to atmospheric pressure, it is convenient to use the **atmosphere** (atm), where 1 atm = 760 Torr. A very low-pressure atmospheric region (an "area of low pressure" on a weather chart, Fig. 5.9) typically has a pressure of about 0.98 atm at sea level, and the pressure in the eye of a hurricane might be as low as 0.90 atm. A typical region of high pressure is about 1.03 atm. The SI unit of pressure is the **pascal** (Pa), with

$$1\ \text{Pa} = 1\ \text{kg}{\cdot}\text{m}^{-1}{\cdot}\text{s}^{-2}$$

The pascal is a very small unit, and atmospheric pressure is close to 1×10^5 Pa.

The units for reporting pressure are Torr and atmospheres; the SI unit is pascal. Pressure units can be interconverted by using Table 5.2.

SELF-TEST 5.1A Express a pressure of 500 Torr in atmospheres.

[*Answer:* 0.658 atm]

SELF-TEST 5.1B The atmospheric pressure in Denver, Colorado, on a certain day was 630 Torr (2 sf). Express this pressure in pascals.

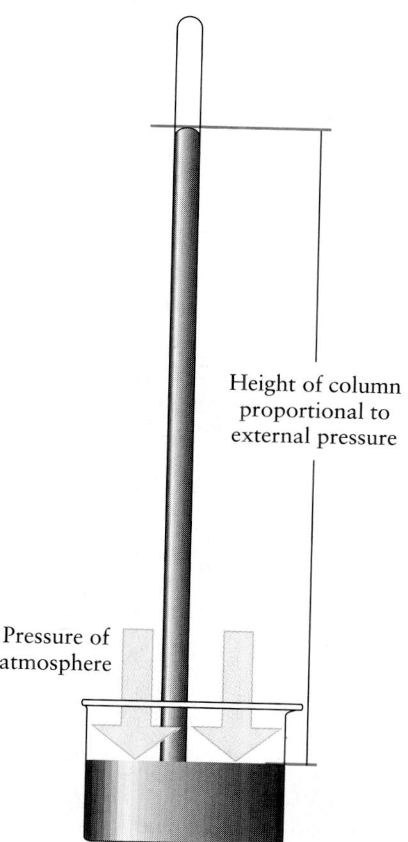

FIGURE 5.8 A barometer is used to measure the pressure of the atmosphere. The pressure of the atmosphere is balanced by the pressure exerted by the column of mercury. The height of the column is proportional to the pressure.

TABLE 5.2 Pressure units*

SI UNIT: PASCAL (Pa)
1 Pa = 1 $kg \cdot m^{-1} \cdot s^{-2}$ = 1 $N \cdot m^{-2}$
CONVENTIONAL UNITS
1 bar = 10^5 Pa = **100** kPa
1 atm = **1.01325** × **10^5** Pa
= **101.325** kPa
1 atm = **760** Torr
1 atm = 14.7 $lb \cdot inch^{-2}$

*Figures in bold are exact. See inside the back cover for more relations. N, newtons.

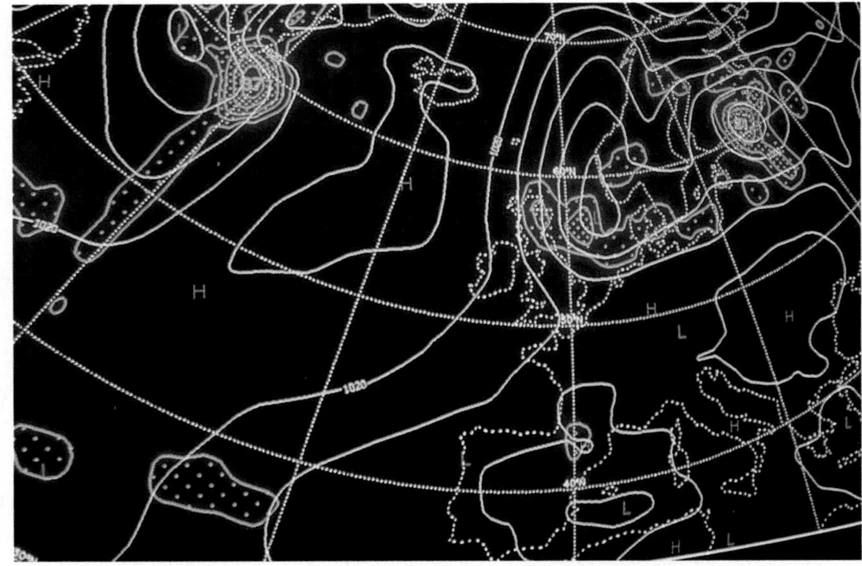

FIGURE 5.9 A typical weather map showing Europe and the North Atlantic. The closed curves are called isobars and are contours of constant atmospheric pressure. A red H indicates a region of high pressure; a blue L a region of low pressure.

THE GAS LAWS

Gay-Lussac used his knowledge to establish a world hot-air balloon altitude record of about 23,000 feet in 1804. Charles built the first hydrogen balloon, which became known as the *charlière.*

The first reliable measurements of the properties of gases were made by the Anglo-Irish scientist Robert Boyle in the seventeenth century. Nearly two centuries later, a new pastime, hot-air ballooning, motivated two French scientists, Jacques Charles and Joseph-Louis Gay-Lussac, to discover additional gas laws. Charles and Gay-Lussac measured how the temperature of a gas affected its pressure, volume, and density.

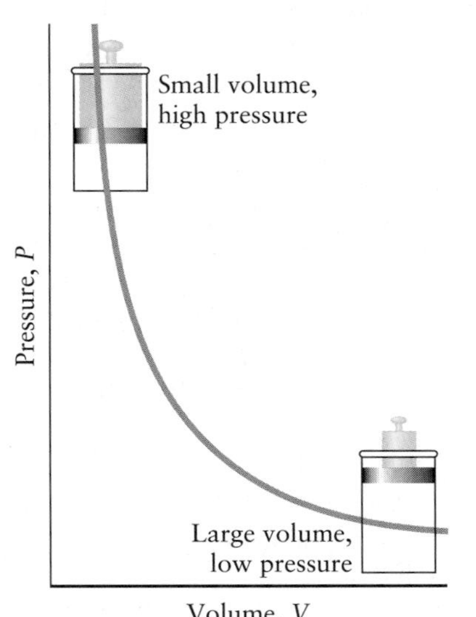

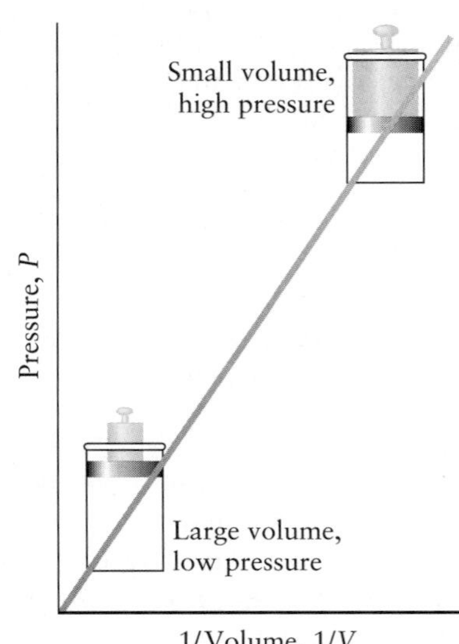

FIGURE 5.10 (left) Boyle's law summarizes the effect of pressure on the volume of a fixed amount of gas at constant temperature.

FIGURE 5.11 (right) When the pressure is plotted against 1/volume, a straight line is obtained. Boyle's law breaks down at high pressures, and a straight line is not obtained in these regions.

5.5 BOYLE'S LAW

Boyle observed that when he compressed a fixed amount of gas at constant temperature, the pressure of the gas increased. A graph of the dependence is shown in Fig. 5.10. The curve shown there is an **isotherm,** a plot of a property at constant temperature. Scientists often look for ways of plotting data in a manner that gives straight lines, because such graphs are easier to analyze and interpret. The data used to construct Fig. 5.10 give a straight line if the pressure is plotted against 1/volume (Fig. 5.11). The implication of this result is that for a fixed amount of gas at constant temperature, *pressure is inversely proportional to volume:*

▶▶ The interpretation of graphs is reviewed in Appendix 1E.

$$P \propto \frac{1}{V} \qquad \text{or} \qquad P = \frac{\text{constant}}{V}$$

This relation is known as **Boyle's law.** It implies that if we compress a gas into half its initial volume, from 1 L to 0.5 L, or from 10 mL to 5 mL, then the pressure will double. The graph in Fig. 5.11 illustrates how the pressure rises as a gas is compressed: it describes what happens to the pressure of the air inside a pump chamber as we press in the piston and decrease the volume. Boyle's law also tells us that the product of the pressure and volume of a sample of gas at constant temperature is a constant:

$$PV = \text{constant}$$

We can feel Boyle's law in action with every step we take if we wear athletic shoes with sealed air pockets (Fig. 5.12). When we take a step, we compress the air pockets in the sole and increase the pressure of the air inside them. The pockets of air are not flattened, but they have become smaller; and the increased pressure inside them makes them resist further compression. The "give" of the air pockets as they are gradually compressed and decompressed cushions our feet.

The kinetic model of gases is consistent with Boyle's law. As a gas is compressed, its molecules are confined to a smaller volume. The interior of the container becomes more crowded (Fig. 5.13) and more of the molecules collide with the walls in a given time. As a result, they exert a higher total force on the walls, thus giving rise to a higher pressure.

The pressure of a fixed amount of gas at constant temperature is inversely proportional to the volume.

(a)

(b)

FIGURE 5.12 A section through the sole of an athletic shoe. (a) The resilience of the shoe is partly due to the compressibility of the air trapped in the holes. (b) The trapped air is compressed when weight is placed on the sole.

5.6 CHARLES'S LAW

Charles and Gay-Lussac found that the volume of a gas in a container fitted with a movable wall (to keep the pressure constant) increases as its temperature is raised. When the volume is plotted against the temperature, we get a straight-line graph (Fig. 5.14). The implication of this result is that, for a fixed amount of gas under constant pressure, *the volume varies linearly with the temperature.* This relation is called **Charles's law.**

Experiments like those done by Charles and Gay-Lussac have revealed another remarkable feature. When graphs like those in Fig. 5.14 are plotted for measurements made on different gases and at different pressures, the straight lines extend to the same point (Fig. 5.15). The extension of a graph to a region outside the region where data have been

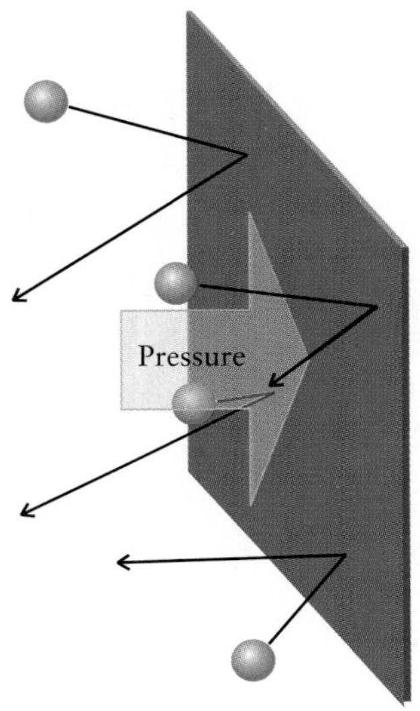

(a)

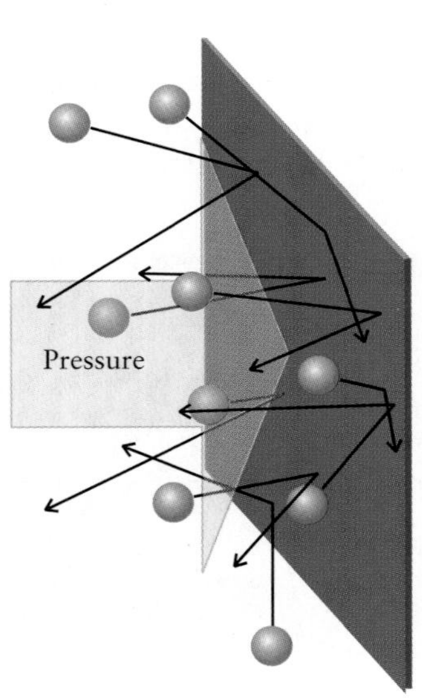

(b)

FIGURE 5.13 (a) As we have seen, the pressure of a gas arises from the impact of its molecules on the walls of the container. (b) When the volume of the sample is decreased, there are more molecules in a given volume, so there are more collisions with the walls. Because the impact on the walls is now greater, so is the pressure. In other words, decreasing the volume occupied by a gas increases its pressure.

obtained is called **extrapolation.** Therefore, we say that the straight lines extrapolate to the same point. This unique point corresponds to zero volume and −273°C. Because the volume cannot be negative, this temperature must be the lowest possible temperature. Indeed, it is the value set at zero on the Kelvin scale, as we saw in Section 2.5. It follows that if we use temperatures on the Kelvin scale, then we can write Charles's law as

$$V \propto T \qquad \text{or} \qquad V = \text{constant} \times T$$

This expression tells us that if the temperature of a fixed amount of gas at constant pressure is doubled, then the volume doubles. For example, if a gas is heated from 300 K to 600 K in a container fitted with a piston, then its volume doubles. Charles's law also tells us that V/T is constant for a gas at constant pressure.

A similar expression summarizes the results of measuring the variation in pressure of a sample of gas in a container of fixed volume. Once again, it is found experimentally that the pressure varies linearly with temperature and that the data extrapolate to zero pressure at −273°C (Fig. 5.16). Therefore, we can write

$$P \propto T \qquad \text{or} \qquad P = \text{constant} \times T$$

It follows that doubling the temperature *on the Kelvin scale* doubles the pressure of a gas, provided the volume is constant.

We can explain the effect of temperature on the pressure by remembering the effect of temperature on the average speed of the molecules. As the temperature of a gas is raised, the average molecule moves more rapidly. As a result, each molecule strikes the walls more often and with greater force. Therefore, the gas exerts a greater pressure.

The volume of a fixed amount of gas in a container at constant pressure is directly proportional to the absolute temperature. The pressure of a fixed amount of gas in a container of constant volume is proportional to the absolute temperature.

5.7 AVOGADRO'S PRINCIPLE

A third contribution to our understanding of gases was made by the Italian scientist Amedeo Avogadro. The **molar volume,** V_m, of a substance—*any* substance, not only a gas—is the volume it occupies per mole of molecules, n:

$$V_m = \frac{\text{volume occupied}}{\text{amount of substance}} = \frac{V}{n}$$

Some results of measurements of molar volume at the same pressure and temperature are illustrated in Fig. 5.17. We see that regardless of the gas, all the molar volumes are very similar. Results like these suggest that under the same conditions of temperature and pressure, a given number of molecules occupy the same volume regardless of their chemical identity. This observation is now known as **Avogadro's principle.**

It follows from the definition of the molar volume that

$$V \propto n \qquad \text{or} \qquad V = \text{constant} \times n$$

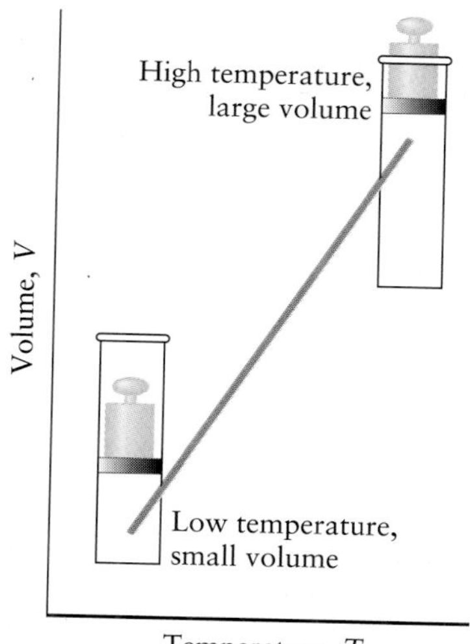

FIGURE 5.14 When the temperature of a gas is increased and it is free to change its volume at constant pressure (as depicted by the constant weight acting on the piston), the volume increases. A graph of volume against temperature is a straight line.

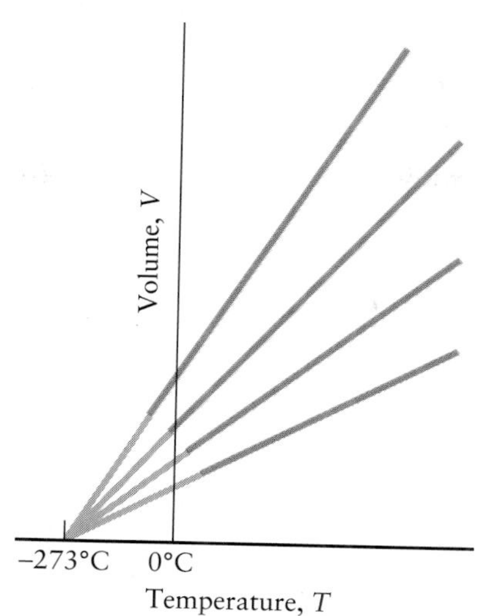

FIGURE 5.15 The extrapolation of data like that in Fig. 5.14 for a number of gases suggests that the volume of all gases should become zero at $T = 0$ ($-273°C$). All gases condense to liquids well before that temperature is reached.

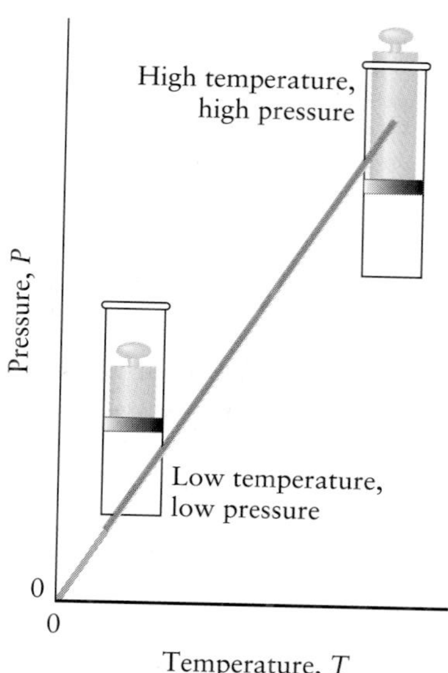

FIGURE 5.16 The pressure of a fixed amount of gas in a vessel of constant volume is proportional to the absolute temperature. The pressure extrapolates to 0 at $T = 0$ on the Kelvin scale.

According to this expression, doubling the number of moles of molecules at constant temperature and pressure doubles the volume occupied by a gas. This behavior is consistent with the kinetic model of gases. In this model, gas molecules are in constant motion and beat against the walls of their container. To keep the pressure constant as more molecules are added, the size of the container must increase.

> ***The volume occupied by a sample of gas at constant pressure and temperature is proportional to the number of moles of molecules present.***

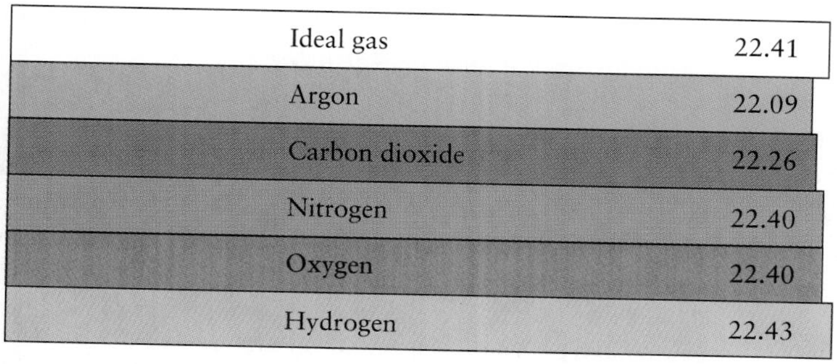

Gas	Molar volume
Ideal gas	22.41
Argon	22.09
Carbon dioxide	22.26
Nitrogen	22.40
Oxygen	22.40
Hydrogen	22.43

FIGURE 5.17 The molar volumes (in liters per mole) of various gases at 0°C and 1 atm: all are very similar. An ideal gas is introduced in Section 5.9.

5.8 USING THE GAS LAWS

The gas laws are used to calculate changes in the pressure, volume, or temperature of a fixed amount of gas. The procedure is described and illustrated in Toolbox 5.1.

TOOLBOX 5.1 How to calculate the effect of changing conditions

Compression at constant temperature To calculate the effect on the pressure of changing the volume from V_1 to V_2, we use Boyle's law to write $P_2V_2 = P_1V_1$, and hence

$$P_2 = P_1 \times \frac{V_1}{V_2}$$

To see that the factor V_1/V_2 is correct, note that if the gas is compressed, then V_2 is less than V_1. Therefore, the factor V_1/V_2 is greater than 1, so the pressure P_2 will be greater than P_1, as we expect.

For example, suppose a certain amount of nitrogen occupies 1.00×10^2 L at 1.00 atm, and we are asked to calculate the pressure it exerts when it is compressed into 10.0 L at the same temperature. First, we expect an increase in pressure (because the gas is compressed). The ratio of volumes to use will therefore be larger than 1; so we use the factor 1.00×10^2 L/10.0 L and write

$$P_2 = (1.00 \text{ atm}) \times \frac{100 \not{\text{L}}}{10.0 \not{\text{L}}} = 10.0 \text{ atm}$$

As we could have anticipated from Boyle's law, a tenfold increase in the pressure results from a tenfold reduction in the volume. We can use a similar approach to find the volume when the pressure acting on the gas is changed at constant temperature.

SELF-TEST 5.2A What pressure, in Torr, would a sample of argon exert if it were compressed from 500 mL to 300 mL at constant temperature, given that initially its pressure was 750 Torr? (All data have 3 sf.)

[*Answer:* 1.25 kTorr]

SELF-TEST 5.2B A small cylinder of helium gas has a volume of 2.0 L and a pressure of 10.0 atm at 26°C. If all the gas remaining in the cylinder is used to fill a single large balloon at 2.0 atm and the same temperature, to what volume does the balloon inflate?

Heating at constant volume If the temperature is changed from T_1 to T_2, then from Charles's law we know that the pressure will change by a factor of T_2/T_1. Therefore, we can write

$$P_2 = P_1 \times \frac{T_2}{T_1}$$

To see that the factor T_2/T_1 is correct, note that if the gas is heated, then T_2 is greater than T_1. The factor T_2/T_1 is then greater than 1, so the pressure P_2 will be greater than P_1, as we expect. A similar expression is used to calculate the effect of temperature on the volume of a gas at constant pressure.

For example, suppose we were interested in the pressure of helium gas in a sealed rigid container being carried in the unheated cargo hold of a weather-monitoring airplane as it rises from sea level. At sea level, the temperature is 20°C and the pressure of the sample is 745 Torr. At a height of 10 km, the temperature is −40°C. The decrease in temperature (from 293 K to 233 K) results in a reduction in pressure by the factor 233 K/293 K. Therefore, to find the pressure of the sample, we write

$$P_2 = (745 \text{ Torr}) \times \frac{233 \not{\text{K}}}{293 \not{\text{K}}} = 592 \text{ Torr}$$

SELF-TEST 5.3A The pressure of 100 L of compressed nitrogen gas in an industrial process is 250 kPa at 18°C, but it is then heated to 322°C in a strong, constant-volume vessel. What is its final pressure? (All data have 3 sf.)

[*Answer:* 511 kPa]

SELF-TEST 5.3B A nearly empty aerosol can has a pressure of 1.01 atm at a temperature of 25°C. What would the pressure in the can be if it were tossed into an incinerator for disposal, which would raise its temperature to 1300°C (2 sf)? Why is the incineration of aerosol cans not recommended?

The effect of simultaneous heating and expansion To take into account the effect of changes in temperature and volume on a fixed amount of gas, we apply two factors. The Charles's law factor, $P_2 = P_1 \times T_2/T_1$, takes into account the change in temperature. The Boyle's law factor, $P_2 = P_1 \times V_1/V_2$, takes into account the effect of volume. Consequently, overall,

$$P_2 = P_1 \times \frac{V_1}{V_2} \times \frac{T_2}{T_1}$$

For example, suppose a sample of methane gas is stored in a flask of volume 227 mL at 600 Torr (3 sf) and 10°C (2 sf) and then is allowed to expand into a second flask so that its total volume is 1.78 L at 27°C. What is the pressure of methane after the expansion? The expansion in volume (from 0.227 L to 1.78 L) results in a decrease in pressure, so we need the factor 0.227 L/1.78 L. The increase in temperature (from 283 K to 300 K) results in an increase in pressure, so we need the factor 300 K/283 K. Therefore, we write

$$P_2 = (600 \text{ Torr}) \times \frac{0.227 \text{ L}}{1.78 \text{ L}} \times \frac{300 \text{ K}}{283 \text{ K}}$$
$$= 81.1 \text{ Torr}$$

Although the temperature increase raises the pressure, the effect of the increase in volume outweighs the effect of temperature, and the net effect is a lowering of pressure.

SELF-TEST 5.4A In an industrial process, 500 L (3 sf) of hydrogen initially at 101 kPa and 22°C is compressed into a vessel of volume 15.0 L and heated to 420°C (3 sf). What is its final pressure?

[*Answer:* 7.91 MPa]

SELF-TEST 5.4B What would be the effect on the pressure occupied by 1 mol of gas molecules if the volume were halved and the absolute temperature were doubled?

5.9 THE IDEAL GAS LAW

The observations made by Boyle, Charles, Gay-Lussac, and Avogadro have been combined into one simple relation between the pressure (P), volume (V), temperature (T), and number of moles (n) of a gas:

$$PV = nRT \qquad (1)$$

where R is a constant, the **gas constant.** A gas that obeys this law under all conditions is called an **ideal gas,** and the expression itself is called the **ideal gas law.** All gases are found to obey this expression with increasing accuracy as the pressure is reduced toward zero. It is reasonably reliable at normal pressures.

Let's verify that Eq. 1 does, in fact, summarize the individual gas laws. If we take the amount of gas and its temperature to be constant, then Eq. 1 has the form

$$PV = \text{constant} \quad \text{or} \quad P = \text{constant} \times \frac{1}{V}$$

CHEMISTRY AT WORK

Four kilometers above the Pacific Ocean on Mauna Loa, Hawaii, an atmospheric chemist collects a sample of air. Once the temperature and pressure of the sample have been measured, the chemist can calculate the amounts of gases such as carbon dioxide, chlorofluorocarbons, and nitrogen oxides present. The data are used to monitor global warming and the destruction of the ozone layer. In 1996 the first reports emerged that the concentration of CFCs is at last starting to decline.

That is, the pressure is inversely proportional to the volume of the gas, just as Boyle observed.

To see how the ideal gas law includes Charles's law, we keep the pressure (P) and amount (n) constant. Then Eq. 1 has the form

$$\text{Constant} \times V = \text{constant} \times T$$

which shows that the volume is proportional to the absolute temperature, in accord with Charles's law.

When the pressure and temperature are constant, Eq. 1 reduces to

$$V = \text{constant} \times n$$

and the volume is proportional to the number of moles of molecules present, in accord with Avogadro's principle.

The gas constant, R, is found to have the same value for all gases. We find its value by measuring P, V, n, and T and substituting their values into

$$R = \frac{PV}{nT}$$

For pressure in atmospheres and volume in liters,

$$R = 8.206 \times 10^{-2}\ \text{L·atm·K}^{-1}\text{·mol}^{-1}$$

TABLE 5.3 The gas constant, R

$8.205\,78 \times 10^{-2}$ L·atm·K^{-1}·mol^{-1}
$8.314\,51 \times 10^{-2}$ L·bar·K^{-1}·mol^{-1}
$8.314\,51$ J·K^{-1}·mol^{-1}
62.364 L·Torr·K^{-1}·mol^{-1}

Unless you are informed otherwise, there is no need to try to remember this value, because it is usually supplied in tests and examinations.

The values of R for different units of pressure are given in Table 5.3. When doing calculations that involve R, select from the table the value that has units matching the ones you need.

The ideal gas law contains all the relations describing the response of ideal gases to changes in pressure, volume, temperature, and moles of molecules.

5.10 USING THE IDEAL GAS LAW TO MAKE PREDICTIONS

The ideal gas law has many uses in chemistry. In the following example, we see how to predict the properties of a gas when we are given some information about the sample. For example, we can use it to predict the pressure of a gas when we are given its amount (or mass), temperature, and volume.

EXAMPLE 5.1 Calculating the pressure of a given sample

Estimate the pressure (in atmospheres) inside a television picture tube, given that its volume is 5.0 L, its temperature is 23°C, and it contains 0.010 mg of nitrogen gas.

STRATEGY To calculate the pressure, we rearrange Eq. 1 into

$$P = \frac{nRT}{V}$$

Then we substitute the values of n, V, and T (remembering to convert temperatures to the Kelvin scale). Select the value of R from Table 5.3 that uses the same units of pressure. For the number of moles of gas molecules, use the mass of gas and its molar mass.

SOLUTION The molar mass of N_2 is 28.02 g·mol^{-1}, and the temperature, 23°C, corresponds to T = (273.15 + 23) K = 296 K. We want the pressure in atmospheres, so we take from Table 5.3 the value of R that is expressed in those units. Therefore, the pressure is

$$P = \left(\frac{1.0 \times 10^{-5}\,\text{g}}{28.02\,\text{g·mol}^{-1}}\right) \times \left(\frac{(8.206 \times 10^{-2}\,\text{L·atm·K}^{-1}\text{·mol}^{-1}) \times (296\,\text{K})}{5.0\,\text{L}}\right)$$
$$= 1.7 \times 10^{-6}\,\text{atm}$$

A very low pressure is necessary to minimize the collisions between the electrons in the beam and the gas molecules. Collisions and the resulting deflections of the electrons would give a blurred, dim picture.

SELF-TEST 5.5A Calculate the pressure (in kilopascals: 1 kPa = 10^3 Pa) exerted by 1.0 g of carbon dioxide in a flask of volume 1.0 L at 300°C.

[*Answer:* 1.1×10^2 kPa]

SELF-TEST 5.5B An idling, badly tuned automobile engine can release as much as 1.00 mol CO per minute into the atmosphere. At 27°C and 1.00 atm, what volume of CO, adjusted to 1.00 atm, is emitted per minute?

5.11 MOLAR VOLUME

The ideal gas law can also be used to predict the molar volume of an ideal gas under any conditions of temperature and pressure. To do so, we write

$$V_m = \frac{V}{n} = \frac{nRT/P}{n} = \frac{RT}{P} \qquad (2)$$

For a pressure of 1.000 atm and the temperature used commonly for chemistry calculations, 25.00°C (298.15 K), the molar volume is

$$V_m = \frac{(8.2058 \times 10^{-2}\,\text{L·atm·K}^{-1}\text{·mol}^{-1}) \times (298.15\,\text{K})}{1.000\,\text{atm}} = 24.47\,\text{L·mol}^{-1}$$

It should be easy to remember this value as about 24 L·mol^{-1}, which is about the volume of a cube 1 ft on a side (Fig. 5.18). The expression **standard temperature and pressure** (STP) means 0°C and 1 atm. At STP, the molar volume of an ideal gas is 22.41 L·mol^{-1}.

To obtain the volume of a known amount of gas at a specified temperature and pressure, we simply multiply the molar volume at that temperature and pressure by the number of moles. When we want to know the volume occupied by a given *mass* of gas, we first convert the mass to moles by using the molar mass.

The molar volume of an ideal gas at 1 atm and 25°C is about 24 L·mol^{-1}.

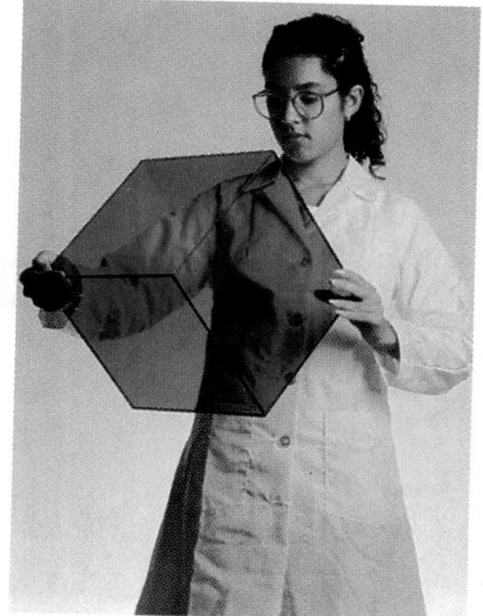

FIGURE 5.18 The blue cube is the volume occupied by 1 mol of ideal gas molecules at 25°C and 1 atm (24 L).

EXAMPLE 5.2 Calculating the volume of a given mass of gas

Calculate the volume occupied by 10 g (2 sf) of carbon dioxide at 25°C and 1.0 atm by using the ideal gas molar volume.

STRATEGY We need to know the number of moles of molecules in the sample; to calculate that, we use the mass of the sample and the molar mass of the compound. Then we multiply the molar volume, 24.47 L·mol^{-1}, by the number of moles to convert to the volume of the actual sample.

SOLUTION The molar mass of carbon dioxide is 44.01 $g \cdot mol^{-1}$. Therefore, the total volume of the sample is

$$V = n \times V_m = \left(\frac{10\ g}{44.01\ g\cdot mol^{-1}}\right) \times 24.47\ L\cdot mol^{-1}$$
$$= \frac{10 \times 24.47}{44.01} L = 5.6\ L$$

SELF-TEST 5.6A Calculate the volume occupied by 1.0 kg of hydrogen at 25°C and 1.0 atm.

[*Answer:* 1.2×10^4 L]

SELF-TEST 5.6B Calculate the volume occupied by 2.0 g of helium at 25°C and 1.0 atm.

5.12 THE STOICHIOMETRY OF REACTING GASES

We might want to know how much gas is consumed or produced in a reaction. To answer this kind of question, we have to combine the material we learned in Chapter 4 with the material presented in this chapter. One part of the calculation is a mole-to-mole calculation of the type described in Section 4.1. Then we convert moles of gas to volume by using the molar volume for the given temperature and pressure. There are no new principles involved in this type of calculation, and only one additional step: the conversion from moles of gas molecules to liters of gas. We can attach a new arrow to the stoichiometry diagram (1) to represent this step and use the technique in Toolbox 5.2.

The molar volume (at the specified temperature and pressure) is used to convert the amount of a reactant or product in a chemical reaction into a volume of gas.

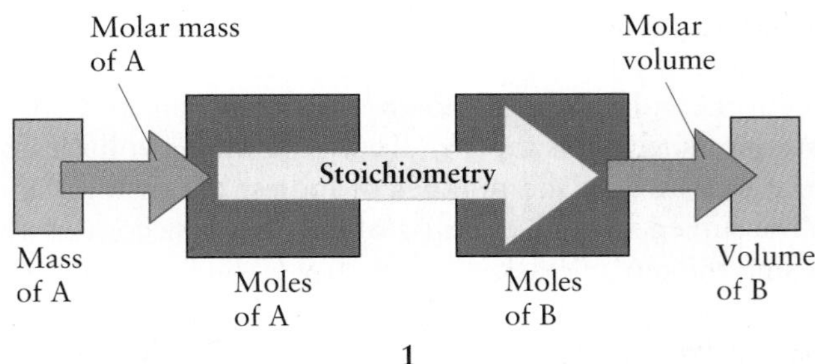

1

TOOLBOX 5.2 How to calculate the volume of gas involved in a reaction

To calculate the volume of gas produced from a given mass of reactant, we adopt the following procedure:

Step 1. Use the molar mass of the reactant to convert the mass of reactant (in grams) to moles of reactant molecules. See Toolbox 4.2.

Step 2. Identify the stoichiometric relation between the reactant and the gaseous product from the chemical equation and use it to convert from moles of reactant to moles of gaseous product. See Toolbox 4.1.

Step 3. Multiply the number of moles of gaseous product by the molar volume of an ideal gas (at the specified temperature and pressure) to obtain the volume of gas produced.

It is good practice to write these conversions in a string, because that reduces rounding errors. When the temperature is 25°C and the pressure 1.000 atm, the molar volume of an ideal gas is 24.47 $L\cdot mol^{-1}$.

As an illustration, let's calculate the volume of sulfur dioxide produced at 25°C and 1.00 atm by the combustion of 10 g of sulfur, according to the reaction

$$S_8(s) + 8\,O_2(g) \longrightarrow 8\,SO_2(g)$$

The first step is to convert grams of sulfur to moles of S_8. For this step, we use the molar mass of S_8, which is 256.48 $g\cdot mol^{-1}$. For the mole-to-mole step, we use the stoichiometric relation

$$1\text{ mol } S_8 \triangleq 8\text{ mol } SO_2$$

In the third stage of the calculation, we multiply the number of moles of SO_2 by the molar volume, 24.47 $L\cdot mol^{-1}$.

It follows that

$$\text{Volume of gas (L)} = \left(\frac{10\text{ g } S_8}{256.48\text{ g}\cdot\text{mol}^{-1}}\right) \times \frac{8\text{ mol } SO_2}{1\text{ mol } S_8} \times (24.47\text{ L}\cdot\text{mol}^{-1}) = 7.6\text{ L } SO_2$$

SELF-TEST 5.7A Calculate the volume of ethyne (acetylene), C_2H_2, produced at 25°C and 1.00 atm when 10 g (2 sf) of calcium carbide reacts completely with water in the reaction

$$CaC_2(s) + 2\,H_2O(l) \longrightarrow Ca(OH)_2(s) + C_2H_2(g)$$

[*Answer:* 3.8L]

SELF-TEST 5.7B Methanol, CH_3OH, is added to some automobile fuels to improve burning and reduce pollution. What volume of O_2 is consumed in the combustion of 1.00 L of methanol at 25°C and 1.00 atm? The density of methanol is 0.791 $g\cdot mL^{-1}$.

$$2\,CH_3OH(l) + 3\,O_2(g) \longrightarrow 2\,CO_2(g) + 4\,H_2O(l)$$

EXAMPLE 5.3 Calculating the mass of reagent needed to react with a specified volume of gas

The carbon dioxide generated by the personnel in the artificial atmosphere of submarines and spacecraft must be removed from the air and the oxygen recovered. Submarine design teams are investigating the use of potassium superoxide, KO_2, as an air purifier, because this compound reacts with carbon dioxide and releases oxygen (Fig. 5.19):

$$4\,KO_2(s) + 2\,CO_2(g) \longrightarrow 2\,K_2CO_3(s) + 3\,O_2(g)$$

Calculate the mass of KO_2 needed to react with 50 L (2 sf) of carbon dioxide at 25°C and 1.0 atm.

STRATEGY The strategy for the calculation is the same as before, but carried out in reverse. We convert from volume of gas to moles of molecules (by using the molar volume), then to moles of reactant molecules or formula units (by using a mole ratio), and then to the mass of reactant (by using its molar mass).

SOLUTION The molar volume under the stated conditions is 24.47 $L\cdot mol^{-1}$. The stoichiometric relation between CO_2 and KO_2 from the balanced chemical equation is

$$2\text{ mol } CO_2 \triangleq 4\text{ mol } KO_2$$

and the molar mass of KO_2 is 71.1 $g\cdot mol^{-1}$. Therefore, we can write

$$\text{Mass of reactant (g)} = \left(\frac{50\text{ L } CO_2}{24.47\text{ L}\cdot\text{mol}^{-1}}\right) \times \frac{4\text{ mol } KO_2}{2\text{ mol } CO_2} \times (71.1\text{ g}\cdot\text{mol}^{-1}) = 2.9 \times 10^2\text{ g } KO_2$$

SELF-TEST 5.8A Calculate the volume of carbon dioxide, adjusted to 25°C and

FIGURE 5.19 When carbon dioxide is passed over potassium superoxide (the yellow solid), it reacts to form colorless potassium carbonate and oxygen gas. The reaction is used to remove carbon dioxide from air in a closed-system breathing environment.

1.0 atm, that plants need to make 1.00 g of glucose, $C_6H_{12}O_6$, by photosynthesis in the reaction

$$6\,CO_2(g) + 6\,H_2O(l) \longrightarrow C_6H_{12}O_6(s) + 6\,O_2(g)$$

[*Answer:* 0.81 L]

SELF-TEST 5.8B The reaction of H_2 and O_2 gases to produce liquid H_2O is used in fuel cells on the space shuttles to provide electricity. What mass of water is produced in the reaction of 100.0 L of oxygen stored at 25°C and 1.00 atm?

There may be a thousandfold increase in volume when liquids or solids react to form a gas. The molar volumes of gases are close to 25 $L\cdot mol^{-1}$ under normal conditions, whereas liquids and solids occupy only about a few tens of milliliters per mole. The molar volume of liquid water, for instance, is only 18 $mL\cdot mol^{-1}$. In other words, 1 mol of gas molecules occupies about a thousand times the volume of 1 mol of molecules in a liquid or solid.

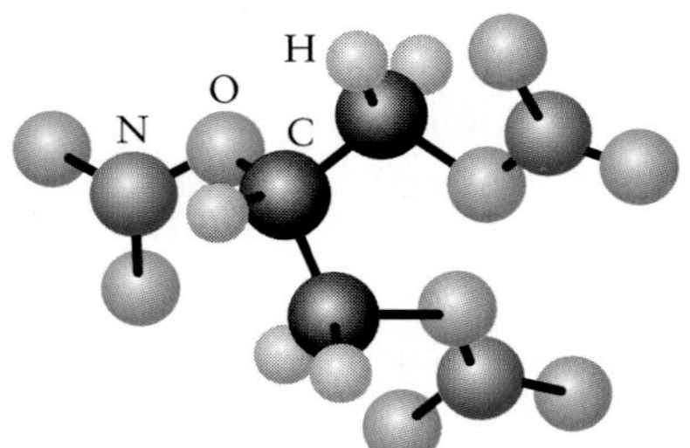

2 Nitroglycerin, $C_3H_5O_9N_3$

The increase in volume as gaseous products are formed is even larger if several gas molecules are produced from each reactant molecule (Fig. 5.20). The explosive action of liquid nitroglycerin, $C_3H_5(NO_3)_3$ (**2**), illustrates the point. When subjected to a shock wave from a detonator, nitroglycerin decomposes into many small gaseous molecules:

$$4\,C_3H_5(NO_3)_3(l) \longrightarrow 6\,N_2(g) + O_2(g) + 12\,CO_2(g) + 10\,H_2O(g)$$

In this reaction, 4 mol $C_3H_5(NO_3)_3$, which corresponds to a little over 570 mL of the liquid, produces 29 mol of gas molecules of various kinds, giving a total of about 710 L of gas under normal conditions. The pressure wave from the sudden 1300-fold expansion is the destructive shock of the explosion. The detonator, which is typically lead azide, $Pb(N_3)_2$, works on a similar principle; it releases a large volume of nitrogen gas when it is struck:

$$Pb(N_3)_2(s) \longrightarrow Pb(s) + 3\,N_2(g)$$

FIGURE 5.20 An explosion caused by the ignition of coal dust. A shock wave is created by the tremendous expansion of volume as large numbers of gas molecules form.

The sudden shock of expansion stimulates the explosive it is being used to detonate (dynamite, for instance) to react. An explosion of the same kind, but using sodium azide, NaN_3, is used in air bags in automobiles (Fig. 5.21). The explosive release of nitrogen is detonated electrically when the vehicle decelerates abruptly in a collision.

The molar volumes of gases are about 1000 times greater than those of liquids and solids.

5.13 GAS DENSITY

When we compress a gas, its density increases because the same number of molecules are being confined into a smaller volume. Similarly, heating a gas that is free to expand reduces the density of a gas: fewer molecules will be found in a given volume. The lower density of a warmer gas is one reason air rises over warm land; glider pilots ride these "thermals" to reach high altitudes.

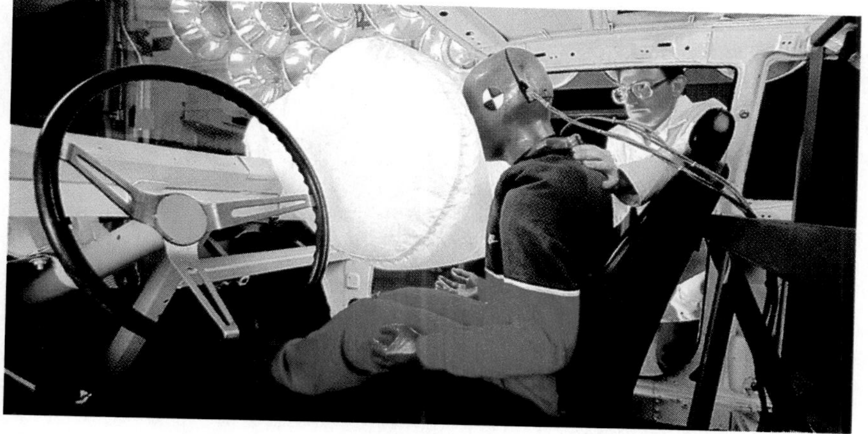

FIGURE 5.21 The rapid decomposition of sodium azide, NaN_3, results in the formation of a large volume of nitrogen gas. The reaction is triggered electrically in this air bag.

We can predict the density, d, of a gas by using the ideal gas law. Density is the mass of the sample of gas divided by its volume:

$$d = \frac{\text{mass}}{\text{volume}}$$

We can write the mass in terms of the moles, n, of molecules present in the sample and the molar mass of the compound: mass = $n \times$ molar mass. We have already seen that the volume is related to the molar volume by $V = n \times$ molar volume. When these two expressions are substituted into the equation for the density, the number of moles cancel, and we are left with

$$d = \frac{\text{mass}}{\text{volume}} = \frac{n \times \text{molar mass}}{n \times \text{molar volume}} = \frac{\text{molar mass}}{\text{molar volume}} \qquad (3)$$

Because the molar volume decreases as the pressure is raised (Boyle's law), this expression tells us that the density increases as the pressure is raised. According to Charles's law, the molar volume increases as the temperature is raised, so the density of the gas decreases as the temperature is raised (Fig. 5.22). The expression also tells us that for a given

(a)

(b)

FIGURE 5.22 (a) Hot air balloonists continue a version of the sport that led Charles and Gay-Lussac to discover some important properties of gases. The balloons rise because the air inside them is hotter than the surrounding air, and thus less dense. (b) To loft a balloon, first a fan is used to blow air into the envelope to inflate it; then a propane heater is used to heat the enclosed air.

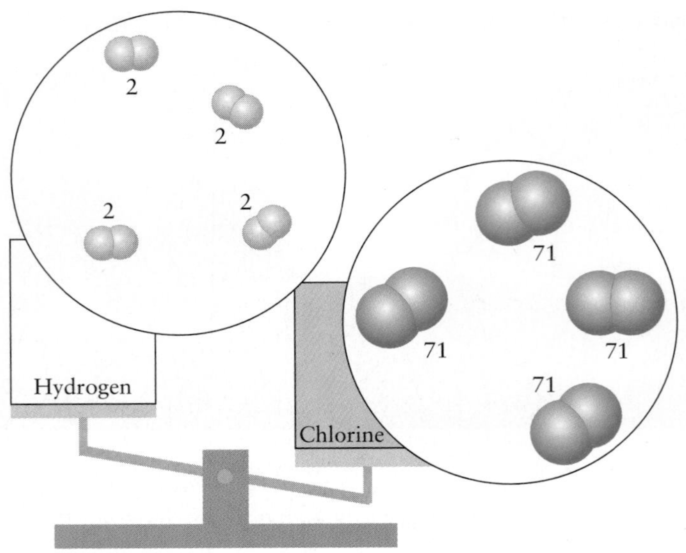

FIGURE 5.23 At the same temperature and pressure, a molecule occupies the same volume in any gas. Hence, the greater the mass of each molecule, the greater the density of the gas. The illustration shows samples of hydrogen and chlorine at the same pressure, volume, and temperature. The numbers are the molar masses in grams per mole.

pressure and temperature (which fixes the molar volume at the same value for all gases), a gas with a high molar mass has a higher density than one with a low molar mass. Chlorine (molar mass 70.9 $g{\cdot}mol^{-1}$), for example, is much denser than hydrogen (molar mass 2.02 $g{\cdot}mol^{-1}$) under the same set of conditions (Fig. 5.23). At 298.15 K and 1.00 atm,

$$d_{Cl_2} = \frac{70.9\ g{\cdot}mol^{-1}}{24.47\ L{\cdot}mol^{-1}} = 2.90\ g{\cdot}L^{-1}$$

$$d_{H_2} = \frac{2.02\ g{\cdot}mol^{-1}}{24.47\ L{\cdot}mol^{-1}} = 0.0826\ g{\cdot}L^{-1}$$

The densities of gases increase with increasing pressure and decreasing temperature, and are proportional to the molar mass.

EXAMPLE 5.4 Determining molar mass from gas density

The volatile organic compound geraniol, a component of oil of roses, is used in perfumery. The density of the vapor at 260°C is 0.480 $g{\cdot}L^{-1}$ when the pressure is 103 Torr. What is the molar mass of geraniol?

STRATEGY We can rearrange Eq. 3 into an expression for the molar mass in terms of the density, which we are given:

$$\text{Molar mass} = d \times \text{molar volume}$$

To use this relation, we need to calculate the molar volume under the conditions of the experiment. For that step, we use Eq. 2, with R in the units that match the data (Table 5.3). Because the pressure is given in Torr, we use $R = 62.364$ $L{\cdot}Torr{\cdot}K^{-1}{\cdot}mol^{-1}$.

SOLUTION The data are as follows:

$d = 0.480\ g{\cdot}L^{-1}$ $\quad P = 103$ Torr $\quad T = (273.15 + 260)\ K = 533\ K$

The molar volume at the stated pressure and temperature is

$$\text{Molar volume} = \frac{RT}{P} = \frac{(62.364\ \text{L·Torr·K}^{-1}\text{·mol}^{-1}) \times (533\ \text{K})}{103\ \text{Torr}}$$
$$= 323\ \text{L·mol}^{-1}$$

Therefore,

$$\text{Molar mass} = (0.480\ \text{g·L}^{-1}) \times (323\ \text{L·mol}^{-1}) = 155\ \text{g·mol}^{-1}$$

SELF-TEST 5.9A The oil produced from eucalyptus leaves contains the volatile organic compound eucalyptol. At 190°C and 60.0 Torr, a sample of eucalyptol vapor had a density of 0.320 g·L^{-1}. Calculate the molar mass of eucalyptol.

[*Answer:* 154 g·mol^{-1}]

SELF-TEST 5.9B The *Codex Ebers,* an Egyptian medical papyrus, describes the use of garlic to treat many ailments, including the cleansing of wounds. Chemists today continue to study the healing properties of garlic. They have discovered that the oxide of diallyl disulfide (the volatile compound responsible for garlic odor) is a powerful antibacterial agent more potent against typhoid than penicillin. At 177°C and 200 Torr, a sample of diallyl disulfide vapor had a density of 1.04 g·L^{-1}. What is the molar mass of diallyl disulfide?

5.14 MIXTURES OF GASES

All gases respond in the same way to changes in pressure, volume, and temperature. Therefore, for calculations of the type we are doing in this chapter, it is unimportant whether all the molecules in a sample are the same. A mixture of gases that do not react with one another behaves like a single pure gas. For instance, we can treat air as a single gas when we want to use the ideal gas law to predict its properties.

John Dalton, whose contribution to atomic theory is described in Chapter 1, showed how to calculate the pressure of a mixture of gases. His reasoning was something like this. Imagine that we introduce a certain amount of oxygen into a container, resulting in a pressure of 0.60 atm. Then we evacuate the container so that it is empty of all gas. Now introduce into it enough nitrogen gas to give a pressure of 0.40 atm at the same temperature. Dalton wondered what the total pressure would be if these same amounts of the two gases were present in the container simultaneously. From some fairly crude measurements, he concluded that the total pressure resulting from the pressure of both gases in the same container would be 1.00 atm, the sum of the individual pressures.

Dalton summarized his observations in terms of what he called the **partial pressure** of each gas, the pressure the gas would exert if it occupied the container alone. In our example, the partial pressures of oxygen and nitrogen in the mixture are 0.60 atm and 0.40 atm, respectively, because those are the pressures the gases exert when each one is in the container alone. Dalton then described the behavior of gaseous mixtures by his **law of partial pressures:**

> The total pressure of a mixture of gases is the sum of the partial pressures of its components.

If we write the partial pressures of the gases A, B, . . . as P_A, P_B, . . . and

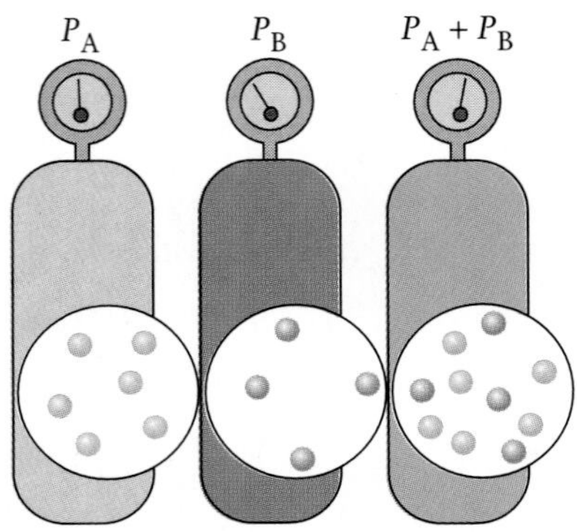

FIGURE 5.24 According to Dalton's law, the total pressure P of a mixture of gases is the sum of the partial pressures P_A and P_B of the components. These partial pressures are the pressures the gases would exert if they were alone in the container (at the same temperature).

the total pressure of the mixture as P, then Dalton's law can be written

$$P = P_A + P_B + \cdots \qquad (4)$$

The law is illustrated in Fig. 5.24. It is *exactly* true only for gases that behave ideally, but it is a good approximation for nearly all gases under normal conditions.

We can use partial pressures to help describe the composition of a humid gas, such as the air in our lungs:

$$P = P_{\text{dry air}} + P_{\text{water vapor}}$$

In a closed container, water vaporizes until its partial pressure has reached a certain value, called its **vapor pressure.** When there is another gas, such as air, in the container, water vaporizes in the same way, until its partial pressure equals its vapor pressure. At this point, the gas holds as much water vapor as it can and is said to be **saturated** with water vapor. The vapor pressure of water varies with temperature, and some values are given in Table 5.4. At blood temperature, the temperature in our lungs, the vapor pressure of water is 47 Torr. Therefore, provided the air in our lungs is saturated with water vapor, the partial pressure of the air itself in our lungs is

$$P_{\text{dry air}} = P - 47\ \text{Torr}$$

On a typical day, the total pressure at sea level is 760 Torr, so the pressure due to all the molecules other than H_2O in our lungs is 713 Torr.

The partial pressure of a gas is the pressure it would exert if it were alone in the container; the total pressure of a mixture of gases is the sum of the partial pressures of the components.

TABLE 5.4 Vapor pressure of water

Temperature, °C	Vapor pressure, Torr
0	4.58
10	9.21
20	17.54
21	18.65
22	19.83
23	21.07
24	22.38
25	23.76
30	31.83
37*	47.08
40	55.34
60	149.44
80	355.26
100	760.00

*Body temperature.

EXAMPLE 5.5 Calculating partial pressures

A certain sample of dry air of total mass 1.00 g consists of approximately 0.76 g of nitrogen and 0.24 g of oxygen. Calculate the partial pressures of these gases and the total pressure (in atmospheres) when this sample occupies a vessel of volume 1.00 L at 20°C.

STRATEGY The partial pressures of the gases are given by the ideal gas law (Eq. 1) applied to each gas in turn, so we need the number of moles of each gas. To convert mass to number of moles, we use the molar masses of the gases. The choice of R from inside the back cover will determine the units of partial pressure: we need to use the value for pressure expressed in atmospheres ($R = 8.206 \times 10^{-2}\ \text{L·atm·K}^{-1}\text{·mol}^{-1}$). Once the two partial pressures have been calculated, the total pressure is obtained by adding them together.

SOLUTION The molar masses of N_2 and O_2 are 28.02 g·mol^{-1} and 32.00 g·mol^{-1}, respectively. It then follows from Eq. 1 that

$$P_{N_2} = \left(\frac{0.76\ \text{g}}{28.02\ \text{g·mol}^{-1}}\right) \times \frac{(8.206 \times 10^{-2}\ \text{L·atm·K}^{-1}\text{·mol}^{-1}) \times (293\ \text{K})}{1.00\ \text{L}}$$
$$= 0.65\ \text{atm}$$

$$P_{O_2} = \left(\frac{0.24\ \text{g}}{32.00\ \text{g mol}^{-1}}\right) \times \frac{(8.206 \times 10^{-2}\ \text{L·atm·K}^{-1}\text{·mol}^{-1}) \times (293\ \text{K})}{1.00\ \text{L}}$$
$$= 0.18\ \text{atm}$$

The total pressure is the sum of these two partial pressures:

$$P = P_{N_2} + P_{O_2} = (0.65 + 0.18)\ \text{atm} = 0.83\ \text{atm}$$

SELF-TEST 5.10A The composition of gases in a "neon" advertising sign of volume 0.75 L is 0.10 g of neon and 0.20 g of xenon. Calculate their partial pressures and the total pressure (in atmospheres) when the tube is operating at 40°C.

[*Answer:* $P_{Ne} = 0.17$ atm, $P_{Xe} = 0.052$ atm, $P = 0.22$ atm]

SELF-TEST 5.10B A sample of oxygen gas of volume 1.00 L was collected over water at 25°C and a total pressure of 1.00 atm. The vapor pressure of water at 25°C is 23.8 Torr. What is the mass of the oxygen collected?

MOLECULAR MOTION

We can obtain additional information about gases by examining the motion of their molecules in more detail.

5.15 DIFFUSION AND EFFUSION

The gradual dispersal of one substance through another substance is called **diffusion.** The escape of one substance (particularly a gas) through a small hole into a vacuum is called **effusion** (Fig. 5.25).

The diffusion of one gas into another explains the spread of pheromones (chemical signals between animals) and perfumes through air. Diffusion helps to keep the composition of the atmosphere approximately constant, because abnormally high concentrations of one gas diffuse away and disperse. At low altitudes, the large-scale motion of air that we call **convection** and that we experience as wind is a bigger factor, but diffusion is more important at high altitudes. One reason why the

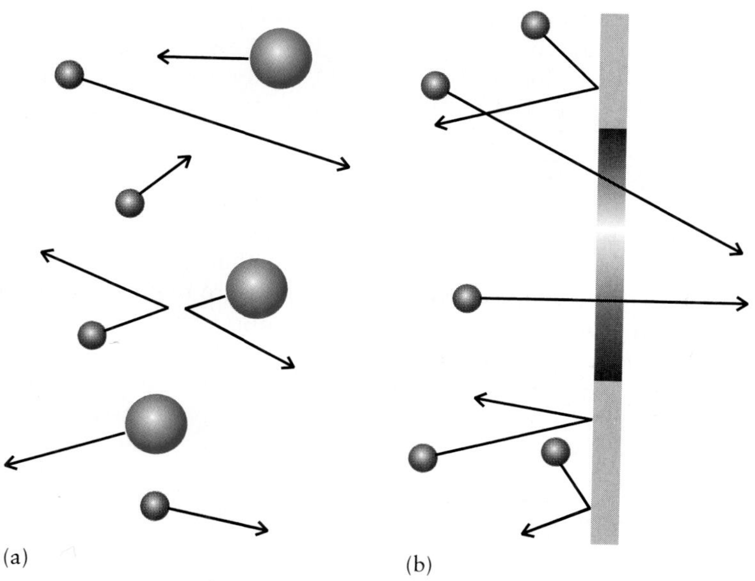

FIGURE 5.25 (a) In diffusion, the molecules of one substance spread into the region occupied by molecules of another in a series of random steps, undergoing collisions as they move. (b) In effusion, the molecules of one substance escape through a small hole in a barrier. In both cases the rate increases with increasing temperature and decreases with increasing molar mass.

BOX 5.1 THE LAYERS OF THE ATMOSPHERE

The atmosphere is divided into several regions that reflect the variation in temperature caused by the chemical reactions taking place in it. These chemical reactions are stimulated by solar radiation, which they convert into heat.

The lowest region of the atmosphere is called the *troposphere*. In this region the temperature decreases with increasing altitude. Between about 11 and 16 km lies the *tropopause*, a region where the temperature remains constant at about −55°C. Above about 16 km, the *stratosphere* begins. Here the temperature rises, until it reaches about 0°C in the *stratopause*, a region lying at about 45 km above sea level. Above that altitude, in the *mesosphere*, the temperature falls again, until the *mesopause* is reached.

After the mesopause, in the *thermosphere*, the temperature rises, and at very high altitudes, it can be over 1000°C. However, astronauts are not burned to a cinder during a space walk because there are so few molecules at that altitude.

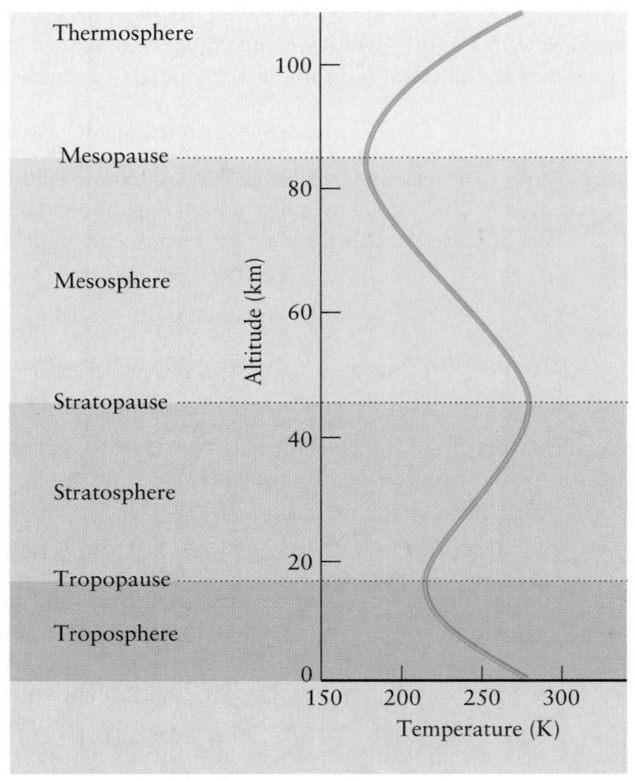

The variation of temperature with altitude in the atmosphere and the various zones into which the atmosphere is divided. The temperature profile represents the consequences of the differing interactions of molecules with solar radiation.

presence of chorofluorocarbons in the ozone layer is viewed as a serious problem is that the stratosphere is warmer than the layers below it, so there is little mixing (Box 5.1). Chemicals that diffuse into the stratosphere can stay there for years. Even at low altitudes, diffusion helps to disperse the substances responsible for the aromas of food, flowers, and perfumes.

A tiny puncture in a tire results in a kind of effusion, in this case from a high-pressure region to a low-pressure region.

Effusion occurs whenever a gas is separated from a vacuum by a porous barrier (a barrier that contains microscopic holes) or a single pinhole. A gas escapes through a pinhole because there are more "collisions" with the hole on the high-pressure side than on the low-pressure side, so more molecules pass from the high-pressure region into the low-pressure region than pass in the opposite direction.

Thomas Graham, a nineteenth-century Scottish chemist, did a series of experiments on the rate of effusion of gases and found that the rate of effusion is inversely proportional to the square root of its molar mass. This observation is now known as **Graham's law:**

$$\text{Rate of effusion} \propto \sqrt{\frac{1}{\text{molar mass}}}$$

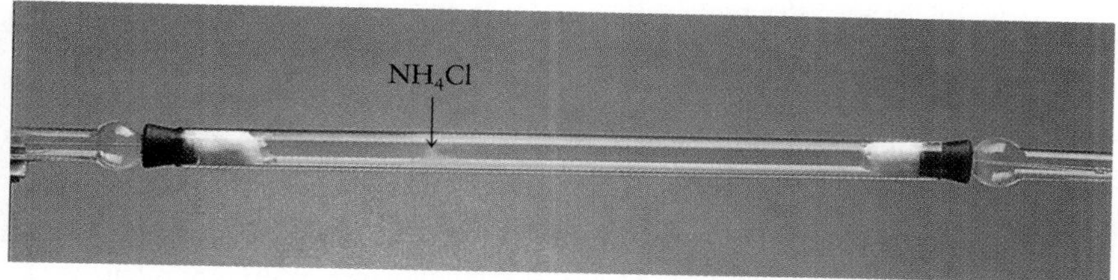

FIGURE 5.26 The plug on the left is soaked in hydrochloric acid and that on the right in aqueous ammonia. Formation of ammonium chloride occurs where gaseous hydrogen chloride and ammonia meet. The reaction occurs closer to the HCl plug because HCl has the greater molar mass and thus its molecules diffuse more slowly.

The diffusion of a gas has also been found to follow an approximate inverse square-root dependence of the same kind, with heavy molecules diffusing more slowly than light molecules (Fig. 5.26). Graham's law strongly suggests that molecules with small molar mass travel most rapidly on average and that their average speeds are inversely proportional to the square root of the molar mass.

Because the rate of effusion is *inversely* proportional to the time a given amount of gas takes to escape, Graham's law implies that the time, t_{effuse}, required for the effusion of a given number of moles of gas molecules is *directly* proportional to the square root of the molar mass:

$$t_{\text{effuse}} \propto \sqrt{\text{molar mass}}$$

On average, heavy molecules move more slowly than light molecules do, so they take longer to escape through a hole. The ratio of the times it takes the same amounts of two gases, A and B, to effuse is therefore

$$\frac{t_{\text{effuse}}(\text{A})}{t_{\text{effuse}}(\text{B})} = \sqrt{\frac{\text{molar mass of A}}{\text{molar mass of B}}} \qquad (5)$$

The simple expression above explains why isotope separation plants used to produce enriched uranium for nuclear reactors are so large. Nuclear power generation depends on our ability to separate uranium-235 from the much more abundant uranium-238. One process uses a series of reactions to convert uranium ore into the volatile solid uranium hexafluoride. The UF_6 vapor is then allowed to effuse through a series of porous barriers. The UF_6 molecules containing uranium-235, which are lighter than those containing uranium-238, effuse more quickly and thus can be separated from the rest. However, the ratio of the times that the same amounts of $^{235}UF_6$ and $^{238}UF_6$ take to effuse is only 1.004, so very little separation occurs at each stage. To improve the separation, the vapor is passed through many effusion stages; consequently, the plants must be very large. The original plant in Oak Ridge, Tennessee (called, somewhat loosely, the gaseous diffusion plant), uses 4000 stages and covers an area of 43 acres (Fig. 5.27). You can see why such plants, which are vital to nuclear defense and nuclear power generation, are difficult to hide from foreign powers who want to monitor one another's nuclear capabilities.

FIGURE 5.27 The individual diffusion stages in the original uranium-235 diffusion plant at Oak Ridge, Tennessee. There are thousands of such stages in the entire plant. Note the size of the components relative to that of the technician.

The time it takes the molecules of a gas to effuse through an opening or diffuse through one another is directly proportional to the square root of their molar mass. Differences in molar mass can be used to separate gases by diffusion.

SELF-TEST 5.11A If it takes a certain amount of helium atoms 10 s to effuse through a porous barrier, how long does it take the same amount of methane molecules, CH_4, under the same conditions?

[*Answer:* 20 s]

SELF-TEST 5.11B Which effuse more slowly, NO_2 molecules or O_3 molecules?

5.16 THE KINETIC MODEL OF GASES

The kinetic model of an ideal gas introduced in Section 5.2 provides an explanation of these observations. The model is based on four assumptions:

1. A gas consists of a collection of molecules in continuous random motion.
2. Gas molecules are infinitely small.
3. These pointlike particles move in straight lines until they collide.
4. The molecules do not influence one another except during collisions.

The fourth assumption means that there are no attractive or repulsive forces between ideal gas molecules except during collisions (Fig. 5.28).

The average distance that a molecule travels between collisions is called the **mean free path.** For a gas like air at 1 atm, the mean free path is about 70 nm, or about 200 molecular diameters. For a molecule magnified to the size of a tennis ball, the mean free path would be about 15 m (about two-thirds as long as a tennis court).

The kinetic model leads to the conclusion that the average speed of molecules in a gas is related to the temperature and the molar mass by

$$\text{Average speed} \propto \sqrt{\frac{\text{temperature}}{\text{molar mass}}} \quad (6)$$

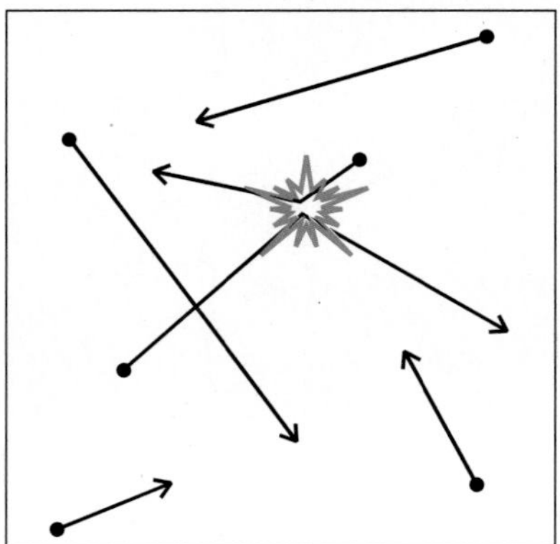

FIGURE 5.28 In the kinetic model of gases, the molecules are regarded as infinitesimal points that travel in straight lines until they undergo instantaneous collisions.

Although we know that the average speed of gas molecules increases as the temperature is raised, the kinetic model expresses this dependence more quantitatively. It predicts that *the average speed, v, is proportional to the square root of the absolute temperature:*

$$v \propto \sqrt{T}$$

It follows that doubling the temperature of *any* gas (from 200 K to 400 K, for example) increases the average speed of its molecules by a factor of $\sqrt{2} = 1.4$.

Equation 6 also shows that at a given temperature the average speed of molecules in a gas is inversely proportional to the square root of the molar mass. This conclusion is exactly what we need to explain Graham's law, which is therefore strong experimental support for the kinetic model.

We can use the kinetic model to refine our picture of the atmosphere. At any given temperature, CO_2 molecules have an average speed that is only 64% that of the lighter H_2O molecules. A sample of humid air at 25°C can therefore be pictured as a storm of molecules, with the CO_2 molecules, the heaviest molecules present, lumbering along at 410 $m \cdot s^{-1}$ and H_2O molecules, the lightest molecules present, zipping along at about 640 $m \cdot s^{-1}$ (Fig. 5.29). The temperature of the atmosphere varies with altitude, so the average molecular speeds also vary (see Box 5.1).

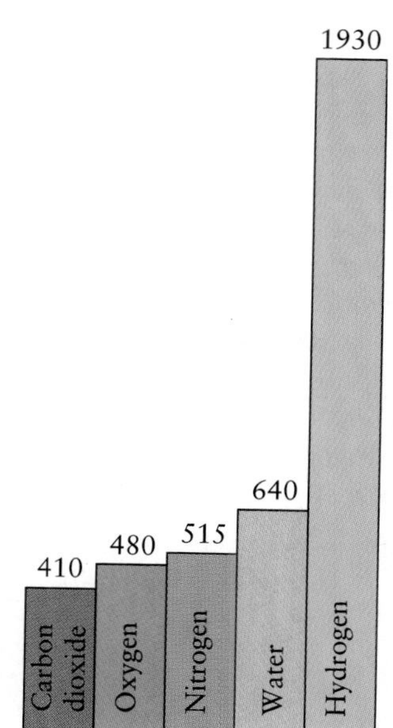

FIGURE 5.29 The average speeds of gas molecules at 25°C in meters per second. The gases are some of the components of air; hydrogen is included to show that light molecules travel much more rapidly on average than heavy molecules.

The average speed of the molecules in a gas sample is directly proportional to the square root of the temperature and inversely proportional to the square root of the molar mass.

SELF-TEST 5.12A At 25°C the average speed of the molecules of an unknown gas is twice that of the average speed of helium atoms. What is the molar mass of the unknown gas?

[*Answer:* 16.0 $g \cdot mol^{-1}$]

SELF-TEST 5.12B In which gas do the molecules have the higher average speed, CH_4 or Cl_2, if they have the same temperature?

5.17 THE MAXWELL DISTRIBUTION OF SPEEDS

Equation 6 gives only the *average* speeds of gas molecules. Like cars in traffic, individual molecules have speeds that vary over a wide range. Moreover, as in a collision between two cars, a molecule might be brought almost to a standstill when it collides with another. Then, in the next instant (but now unlike a collision between cars), it might be struck to one side by another and move off at the speed of sound. An individual molecule undergoes several billion changes of speed and direction each second.

The fraction of gas molecules moving at each speed at any instant is called the **distribution of molecular speeds.** The general formula was first calculated by the Scottish scientist James Maxwell. His conclusions are summarized in Fig. 5.30. The graphs show that heavy molecules (CO_2) travel with speeds close to their average values. Light molecules (H_2) not only have a higher average speed, but the speeds of many of them are very different from their average speed. This wide range of speeds implies that light molecules are more likely than heavy molecules to have such high speeds that they escape from the gravitational pull of small planets and go off into space. Consequently, hydrogen molecules and helium atoms, which are both very light, are very rare in the Earth's atmosphere, although they are abundant on massive planets like Jupiter.

Hydrogen is also reactive, so it is trapped on Earth in compounds.

The curves in the illustration also show that the spread of speeds widens as the temperature increases. At low temperatures, most molecules have speeds close to their average speed. At high temperatures, a

FIGURE 5.30 (a) The range of molecular speeds for several gases, as given by the Maxwell distribution. All the curves correspond to the same temperature. The greater the molar mass, the narrower the spread of speeds. (b) The Maxwell distribution again, but now the curves correspond to the speeds of a single substance at different temperatures. The higher the temperature, the broader the spread of speeds.

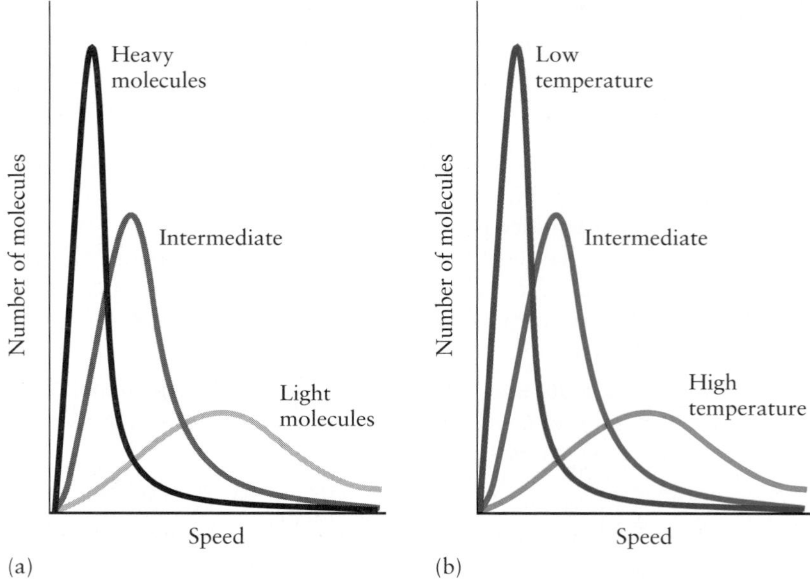

high proportion have speeds widely different from their average speed. Hot gases have a high proportion of very fast molecules.

The molecules of all gases have a wide range of speeds. As the temperature increases, the average speed and the range of speeds increase.

BOX 5.2 REAL GAS LAWS

Many scientists have suggested ways of changing the ideal gas law so that it takes into account the effect of intermolecular forces. One of the earliest and most useful of these improved equations was proposed by Johannes van der Waals, a nineteenth-century Dutch scientist:

$$\left(P + \frac{an^2}{V^2}\right)(V - nb) = nRT$$

This expression is called the **van der Waals equation.** The constant a represents the effect of attractions between molecules, and the constant b represents the effect of repulsions. We can interpret b as a measure of the volume taken up by the gas molecules themselves. Both constants are 0 in an ideal gas, in which the pointlike molecules have zero volume and do not attract one another. The two constants are determined for each gas individually by adjusting their values until the van der Waals equation fits the observed dependence of the pressure on the volume, temperature, and amount of gas being studied. The values of a and b differ from gas to gas because the sizes of molecules differ (which affects b), and their ability to attract each other differs too (which affects a).

At low pressures, a gas occupies a very large volume. The term $V - nb$ is then approximately equal to V itself. Moreover, the term an^2/V^2 is very small and can be neglected relative to P, because low pressure corresponds to a situation in which the molecules are generally so far apart that they do not interact with one another. Consequently, the attractions and repulsions are unimportant. Therefore, an approximate form of the van der Waals equation is

$$PV \approx nRT$$

which is the ideal gas law. Thus, we see that the equation for a real gas becomes the ideal gas law when the pressure is very low.

REAL GASES

Boyle's, Charles's, and Gay-Lussac's experiments were all carried out more than two centuries ago. Since then, chemists have found that their laws are only *approximate* descriptions of actual gases.

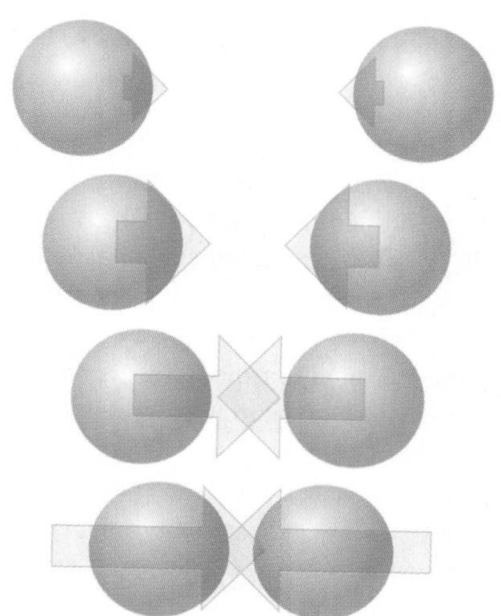

FIGURE 5.31 Molecules attract one another over several molecular diameters. The strength of the attraction (as depicted by the lengths of the arrows) increases as they approach. This attraction is responsible for the condensation of a gas to a liquid.

5.18 THE IDEAL GAS LAW AS A LIMITING LAW

Real gases, the gases that make up the atmosphere and that we use in laboratories, behave like ideal gases, provided their pressures are low (below about 2 atm). The ideal gas law is found to be increasingly accurate as the pressure of the gas is reduced. In the limit of zero pressure, every real gas behaves like an ideal gas and the law describes a gas exactly. We say that the ideal gas law is a **limiting law:** it is valid at a certain limit, the limit of zero pressure.

The fact that the ideal gas law is only a limiting law is also consistent with the kinetic model. In that model we assume that the molecules are infinitely small and do not interact with one another (except when they collide). Deviations from ideal behavior arise because molecules have definite volumes and attract and repel one another. Their **intermolecular forces,** the attractions and repulsions between the molecules (Fig. 5.31), become important at high pressures, when the molecules are closer together on average. At very low pressures, the molecules are so far apart on average that these intermolecular forces are unimportant. Under these conditions, a real gas acts like an ideal gas (Box 5.2).

The ideal gas law is a limiting law in the sense that it becomes increasingly accurate as the pressure of a gas approaches zero.

5.19 THE JOULE-THOMSON EFFECT

One of the most important consequences of intermolecular attractions is that real gases can be liquefied. At low temperatures, gas molecules move so slowly that their intermolecular attractions may cause one molecule to be captured by others and stick to them instead of moving freely. When the temperature is reduced to below the boiling point of the substance, the gas condenses to a liquid: the molecules move too slowly to escape from one another, and the entire sample condenses to a jostling crowd of molecules held together by the attractions between them.

The simplest method of liquefying a gas is to immerse a sample in a bath kept at a lower temperature than the boiling point of the substance (Fig. 5.32). Temperatures as low as about 196 K can be reached by adding chips of solid carbon dioxide to a low-freezing-point liquid, such as acetone.

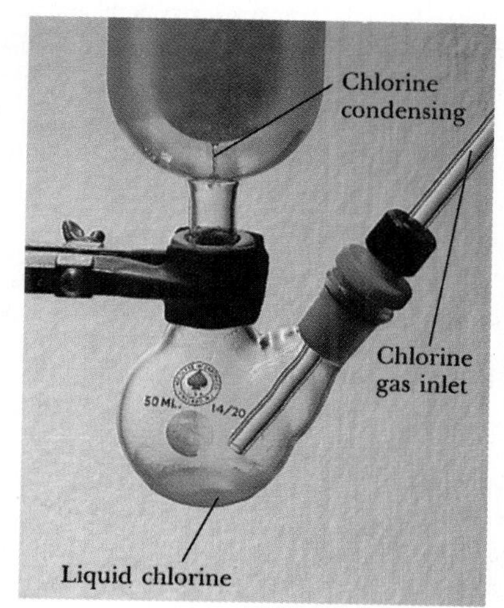

FIGURE 5.32 Chlorine can be condensed to a liquid at atmospheric pressure by cooling it to −35°C or lower. The upper tube contains a "cold finger," a smaller tube filled with a very cold mixture of dry ice and acetone.

CASE STUDY 5 OUR GLOBAL METABOLISM

THE ABUNDANCE OF life on Earth is unique in our solar system. Life as we know it requires sunlight, a temperate climate, a plentiful supply of water, and wide availability of certain elements, particularly carbon, hydrogen, oxygen, nitrogen, phosphorus, and sulfur. These elements must be recycled through a complex web of natural cycles centered on the atmosphere. Each plant and animal has a metabolism and circulatory system that regulates its internal chemical balance. The metabolic systems of all living things together with the natural circulation of the atmosphere can be thought of as our "global metabolism." Oxygen and nitrogen are good examples of how this circulation functions.

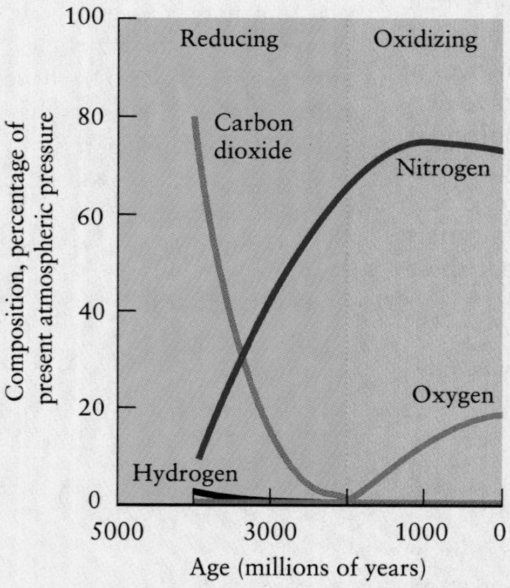

This speculative history of the atmosphere shows a change from a predominantly reducing to a predominantly oxidizing atmosphere about 2 billion years ago when photosynthetic plants started to appear. Hydrogen has been lost to compound formation and to outer space. Earth is the only planet in the solar system with an oxidizing atmosphere.

Billions of years ago, most atmospheric oxygen was in the form of carbon dioxide and water. Some molecular oxygen, O_2, came from the outgassing of rocks and volcanoes and the effect of solar radiation on water; but, as green plants appeared, they produced much more by photosynthesis. Today, oxygen makes up 21% of the volume of dry air and the oxygen cycle involves both plants and animals.

During electrical storms, some O_2 forms ozone, O_3, a reactive form of oxygen. In the troposphere (see Box 5.1), ozone is a pollutant that reacts with organic components of engine exhausts to produce the eye irritants in smog. However, the presence of ozone in the stratosphere is vital to our existence.

Ozone is formed in the stratosphere in two steps. First, molecules of O_2 or another oxygen-containing compound are broken apart into atoms by sunlight, then the O atoms react with other O_2 molecules, to form ozone:

$$O_2(g) \xrightarrow{\text{sunlight}} 2\,O(g)$$

$$O(g) + O_2(g) \longrightarrow O_3(g)$$

The second of these two reactions releases heat, which raises the temperature of the stratosphere. Some of the ozone is decomposed by ultraviolet light:

$$O_3(g) \xrightarrow{\text{uv}} O(g) + O_2(g)$$

Because this reaction absorbs ultraviolet radiation, it helps to shield the Earth from radiation damage. The atmospheric history chart shows that land plants did not appear until there was sufficient molecular oxygen in the atmosphere to produce the protective ozone shield.

Nitrogen gas accounts for 78% of the volume of dry air, but molecular nitrogen is very unreactive. For it to play its part in the global metabolism, nitrogen gas must be converted into compounds that plants and animals can use. This process is called the "fixation" of nitrogen. Atmospheric nitrogen is fixed by lightning, which oxidizes it to oxides such as NO and NO_2. It is also fixed in the soil by bacteria, which oxidize it to the nitrate ion (Section 20.1). Fertilizer factories that use the Haber process (Section 13.11) fix nitrogen industrially by reducing it to ammonia with hydrogen gas.

Nitrogen monoxide, NO (nitric oxide), is also formed in automobile engines and then oxidized by the oxygen in air to nitrogen dioxide, NO_2. The NO_2 dissolves in water to form nitric acid, which delivers nitrogen to the soil in the form of acid rain (see Case Study 14):

$$3\,NO_2(g) + H_2O(g) \longrightarrow 2\,HNO_3(aq) + NO(g)$$

Nitrogen oxides that diffuse up to the stratosphere can threaten the ozone layer, because they are decomposed by solar radiation to species that react with ozone.

The normal concentration of gases in the stratosphere

An experimental pod being prepared for its ascent to the stratosphere in the research balloon pictured on the first page of this chapter. The equipment can measure the concentration of 15 pollutants suspected of reducing atmospheric ozone levels.

is such that ozone formation is balanced by ozone destruction. In the heart of the ozone layer, the concentration of O_3 molecules is about 0.1 $\mu mol \cdot L^{-1}$. However, this balance is in danger of being upset by the activities of the civilized world. Large amounts of nitrogen oxides are generated by jet engines, by automobiles, and by the degradation of nitrogen-based fertilizers. There is also evidence that chlorofluorocarbons (CFCs) may represent the most serious threat to stratospheric ozone. Use of CFCs is now restricted, but because it takes years for CFC molecules to diffuse into the stratosphere, much of the CFCs used during the last decade may still be on its way up. Currently, we are losing about 2% of the ozone in the stratosphere every 10 years.

QUESTIONS

1. Molecular speed is an important factor in the atmospheric diffusion of gases. Rank O_2, O_3, NO_2, and NO in order of increasing average molecular speed.

2. How many moles of $HNO_3(aq)$ can be produced from 1.000×10^3 L of air at 25.0°C containing NO_2 at a partial pressure of 12.5 Pa?

3. Which gas is most dense at 298 K and 1 atm: O_2, N_2, NH_3, or NO?

4. What is the partial pressure of O_2 in the air at a ski resort 3000 m above sea level, where the atmospheric pressure is 520 Torr and 21 out of 100 molecules are O_2?

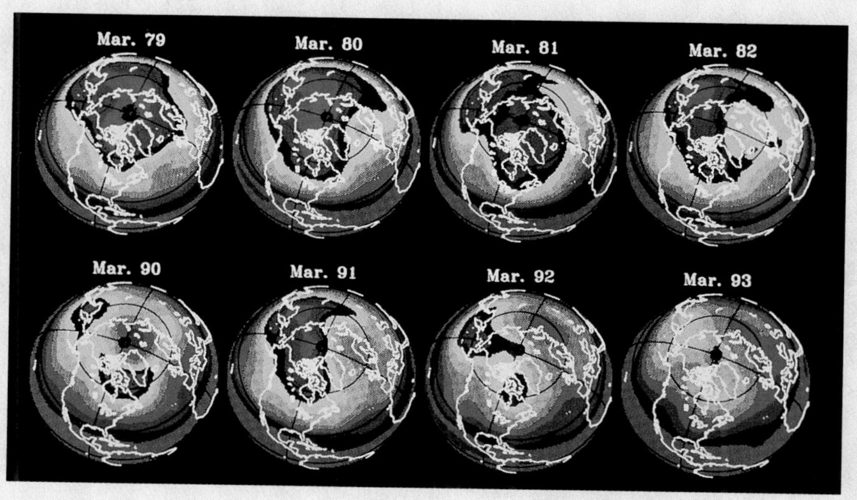

These maps of stratospheric ozone concentration over the North Pole show how the ozone has been depleted since 1979. Red areas represent ozone concentrations greater than 500 Dobson units (DU); the concentrations decrease through green, yellow, and blue, to purple at less than 270 DU. Normal ozone concentration at temperate latitudes is about 350 DU.

FIGURE 5.33 Cooling by the Joule-Thomson effect can be visualized as a slowing of the molecules as they climb away from each other against the force of attraction between them.

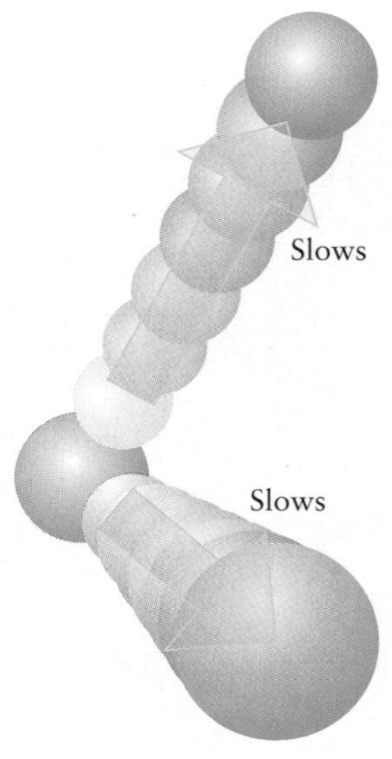

Gases can also be liquefied by making use of the relation between temperature and molecular speed. Because lower average speed corresponds to lower temperature, slowing the molecules is equivalent to cooling the gas. Molecules can be slowed by making use of the attractions between them: the molecules have to climb away from one another when a gas expands; therefore, increasing the volume of a gas and hence the average separation of molecules lowers their average speed (Fig. 5.33). In other words, a real gas cools as it expands. This observation is called the **Joule-Thomson effect** in honor of the scientists who first studied it, James Joule and William Thomson (later to become Lord Kelvin of the temperature scale).

Some real gases get hot as they expand unless they are already quite cool.

The Joule-Thomson effect is used in some commercial refrigerators to liquefy gases. The gas to be liquefied is compressed and then allowed to expand through a small hole, called the throttle. The gas cools as it expands, and the cooled gas circulates past the incoming compressed gas (Fig. 5.34). This contact cools the incoming gas before it expands and cools still further. As the gas is continually recompressed and recirculated, its temperature progressively falls until finally it condenses to liquid.

If the gas is a mixture, like air, then the liquid it forms can be distilled to separate its components. This is the technique that is used for harvesting nitrogen, oxygen, neon, argon, krypton, and xenon from the atmosphere.

Real gases consist of atoms or molecules with intermolecular attractions and repulsions. Gases can be liquefied by making use of the Joule-Thomson effect.

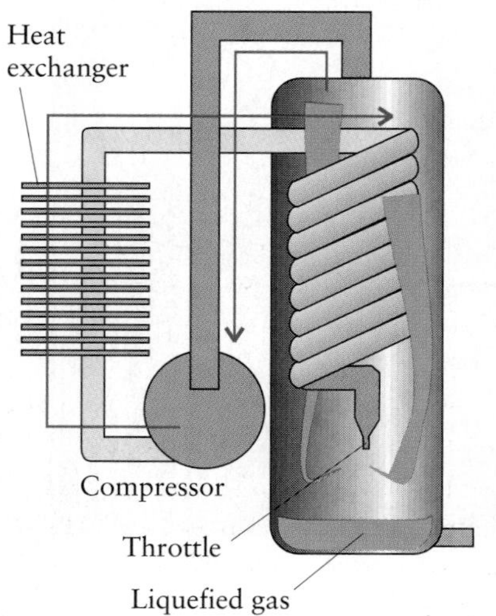

FIGURE 5.34 A Linde refrigerator for liquefying gases. The compressed gas gives up heat to the surroundings in the heat exchanger and is cooled further as it passes through the coil past which cooled gas circulates. The gas is cooled still further by the Joule-Thomson effect as it emerges through the throttle. This gas cools the incoming gas and is recirculated through the system. Eventually, the temperature of the incoming gas is so low that it condenses to a liquid.

SKILLS YOU SHOULD HAVE MASTERED

Conceptual

1. Explain the origin of pressure in molecular terms.
2. Describe the relation of molecular speed to temperature.
3. Explain the ideal gas law in terms of the kinetic model of gases.
4. List and explain the assumptions of the kinetic theory of gases.
5. Describe the effect of molar mass and temperature on the distribution of molecular speeds.
6. Explain how real gases differ from ideal gases.

Problem-Solving

1. Use the ideal gas law to calculate P, V, T, and n for given conditions or after a change in conditions.
2. Calculate the mass of reactant needed to react with a specified volume of gas.
3. Calculate the molar volume of a gas from its mass.
4. Calculate the volume of a gas involved in a reaction.
5. Determine molar mass from gas density and vice versa.
6. Calculate partial pressures of gases and the total pressure of a gas mixture.
7. Use Graham's law to account for relative rates of effusion.

Descriptive

1. Describe and give names to changes of state.
2. Describe the Joule-Thomson effect.
3. Describe the structure of the atmosphere.

EXERCISES

The Nature of a Gas

5.1 Describe the kinetic model of a gas.

5.2 Indicate how observations of gas behavior verify the kinetic model of a gas.

5.3 Define the temperature of a gas in terms of molecular motion.

5.4 How is the speed of sound in a gas affected by an increase in temperature? Explain your answer in terms of molecular motion.

Pressure

5.5 Explain how a barometer measures atmospheric pressure.

5.6 Is the space above the mercury in a barometer truly a vacuum? Would the height of the mercury column change if air were introduced into that space?

5.7 Gas pressures are expressed in different units, depending on the application. Make the following conversions:
(a) 1.00 bar to atmosphere
(b) 1.00 Torr to atmosphere
(c) 1.00 mmHg to pascal
(d) 1.00 kPa to atmosphere

5.8 Several different units of pressure will be used in these exercises. Complete the following conversions:
(a) 725 Torr to atmosphere
(b) 150 kPa to atmosphere
(c) 10.7 lb·inch^{-2} to Torr
(d) 600 mmHg to atmosphere

5.9 The pressure needed to make synthetic diamonds from graphite is 80,000 atm (1 sf). Express this pressure in (a) kilobars; (b) pascals.

5.10 An argon gas cylinder measures a pressure of 22.5 lb·inch^{-2}. Convert this pressure to (a) kilopascals; (b) Torr.

5.11 Suppose you were marooned on a tropical island and had to make a primitive barometer, using seawater (density 1.10 g·cm^{-3}). What height would the water reach in your barometer when a mercury barometer would reach 77.5 cm? The density of mercury is 13.6 g·cm^{-3}.

5.12 Assume that the width of your body (across your shoulders) is 20 inches and the depth of your body (chest to back) is 10 inches. If atmospheric pressure is 14.7 lb·inch^{-2}, what mass of air does your body support when you are in an upright position?

The Gas Laws

5.13 Calculate the final volume for the following constant-temperature processes: (a) the pressure on 1.00 L H_2 changes from 2.20 kPa to 3.00 atm; (b) the pressure on 25.0 mL of CO_2 changes from 200 Torr to 0.500 atm.

5.14 (a) A helium balloon has a volume of 22 L at sea level, where the atmospheric pressure is 0.951 atm. When the balloon has risen to a height at which the atmospheric pressure is 550 Torr, the balloon bursts. What is the volume of the balloon just before it breaks? (b) A 1.0-mm^3 bubble at the bottom of a lake where the pressure is 3.5 atm rises to the surface. What is the volume of the bubble when it hits the top where the atmospheric pressure is 780 Torr? Assume a constant temperature in each case.

5.15 Determine the final pressure when (a) 1.0 mL of krypton at 105 kPa is transferred to a 1.0-L vessel; (b) 30.0 cm^3 of O_2 at 600 Torr is compressed to 5.0 cm^3. Assume constant temperature.

5.16 (a) 2.4 L of methane at a pressure of 1220 Torr is transferred to a 4.1-L vessel. What is the final pressure of methane if the change occurs at constant temperature? (b) A fluorinated organic gas in a cylinder is compressed from an initial volume of 500 mL at 300 Pa to 175 mL at the same temperature. What is the final pressure?

5.17 (a) An outdoor storage vessel for hydrogen gas with a volume of 300 m^3 is at 1.5 atm and 10°C at 2:00 AM. By 2:00 PM, the temperature has risen to 30°C. What is the new pressure of the hydrogen gas in the vessel? (b) A chemist had prepared a sample of hydrogen bromide and found that it occupied 255 mL at 85°C and 600 Torr. What volume would it occupy at 0°C at the same pressure?

5.18 (a) A 250-mL aerosol can at 25°C and 1.10 atm was thrown into an incinerator. When the temperature in the can reached 600°C, it exploded. What was the pressure in the can before it exploded? (b) A helium balloon has a volume of 22 L at sea level, where the pressure is 0.951 atm and the temperature is 18°C. The balloon is cooled at a constant pressure until the temperature is −15°C. What is the volume of the balloon at this stage?

5.19 The pressure in an automobile tire is 30 $lb\cdot inch^{-2}$ at 20°C and the atmospheric pressure is 14.7 $lb\cdot inch^{-2}$. The automobile is driven for several hundred kilometers and the pressure in the tire then reads 34 $lb\cdot inch^{-2}$. Assuming a constant volume and no leaks in the tire, calculate the new temperature of the air in the tire.

5.20 A relief valve on an industrial storage tank operates if the pressure of the enclosed carbon dioxide gas exceeds 115 atm. In December, the tank was filled with carbon dioxide at 100 atm when the ambient temperature was −10°C. On a hot summer day, the temperature rose to 35°C. Should the relief valve have operated?

5.21 A chemist prepares a sample of helium gas at a certain pressure, temperature, and volume, and then removes half the gas molecules. How must the temperature be changed to keep the pressure and volume the same?

5.22 A chemist prepares 0.100 mol of neon gas at a certain pressure and temperature in an expandable container. Another 0.010 mol of neon atoms is then added to the same container. How must the volume be changed to keep the pressure and temperature the same?

5.23 A sample of carbon monoxide has a volume of 150 mL at 10°C and 750 Torr. What pressure will be exerted by the gas if the temperature is increased to 522°C and the volume changed to 500 mL?

5.24 In a certain chemical process, 2500 L of sulfur dioxide gas is fed from a storage tank at 2000 kPa and 20°C into a 6000-L chemical reactor at 150°C. What is the pressure of the SO_2 in the reactor prior to reaction?

5.25 A lung-full of air (350 cm^3) is exhaled into a machine that measures lung capacity. If the air is exhaled from the lungs at a pressure of 1.08 atm at 37°C but the machine is at ambient conditions of 0.958 atm and 23°C, what is the volume of air measured by the machine?

5.26 A 20-mL sample of xenon exerts a pressure of 0.48 atm at −15°C. (a) What volume does the sample occupy at 1.00 atm and 298 K? (b) What pressure would it exert if it were transferred to a 12-mL flask at 20°C? (c) Calculate the temperature needed for the xenon to exert a pressure of 500 Torr in the 12-mL flask.

The Ideal Gas Law

5.27 Show how the ideal gas law summarizes Boyle's law, Charles's law, and Avogadro's principle.

5.28 Which of the following plots of functions of an ideal gas will not be linear? (a) P against T; (b) V against T; (c) P against V; (d) V against n. Assume all other variables are constant.

5.29 (a) A 100-mL flask contains argon at 1.3 atm and 77°C. What amount of Ar is present (in moles)? (b) A 120-mL flask contains 2.7 μg of O_2 at 17°C. What is the pressure (in Torr)? (c) A 20-L flask at 200 K and 20 Torr contains nitrogen. What mass of nitrogen is present (in grams)? (d) 16.7 g of krypton exert a pressure of 100 mTorr at 44°C. What is the volume of the container (in liters)? (e) A 2.6-μL ampule of xenon has a pressure of 2.00 Torr at 15°C. How many Xe atoms are present?

5.30 (a) A 350.0-mL flask contains 0.1500 mol Ar at 25°C. What is the pressure of the gas (in kilopascals)?

(b) 23.9 mg of bromine trifluoride exerts a pressure of 10 Torr at 100°C. What is the volume of the container (in milliliters)? (c) A 100-mL flask contains sulfur dioxide at 0.77 atm and 30°C. What mass of gas is present? (d) A 6000-m^3 storage tank contains methane at 129 kPa and 15°C. What amount of CH_4 is present (in moles)? (e) A 1.0-μL ampule of helium has a pressure of 2.00 kPa at −115°C. How many He atoms are present?

5.31 What mass of ammonia will exert the same pressure as 12 mg of hydrogen sulfide, H_2S, in the same container under the same conditions?

5.32 2.00 mg of argon is confined to a 50-mL vial at 20°C; 2.00 mg of krypton is also confined to another 50-mL vial. What must the temperature of the krypton be if it is to have the same pressure as the argon?

5.33 (a) Calculate the mass of air (which has an average molar mass of 28.97 $g \cdot mol^{-1}$ in the troposphere) that is needed to fill a 250-mL flask to a pressure of 1.22 atm at 25°C. (b) Calculate the mass of nitrogen needed to fill the same 250-mL flask to a pressure of 1.22 atm at 25°C.

5.34 (a) The mass of ammonia in a 500-mL flask is 14.6 mg at 28°C. What is the pressure of the ammonia? (b) What mass of ammonia would need to be added to change the pressure to 50 Torr?

5.35 A sample of helium gas occupies a spherical tank measuring 12.0 inches in diameter and exerts a pressure of 100.0 $lb \cdot inch^{-2}$ at 15.0°C. What mass of helium is in the tank? (The volume of a sphere of radius r is $V = \frac{4}{3}\pi r^3$.)

5.36 A bottle of oxygen gas measures 3.0 cm in diameter and 30.0 cm in length. If the oxygen pressure is 200.0 kPa at 27.0°C, calculate the mass of oxygen in the bottle. (The volume of a cylinder of radius r and length l is $V = \pi r^2 l$.)

Molar Volume

5.37 Assume ideal gas behavior and calculate the volume at 1.00 atm and 298 K occupied by (a) 1.00 mol of H_2; (b) 27.0 g of Cl_2; (c) 3.00 mol of O_2; (d) 14.8 mg of SO_2.

5.38 Assume ideal gas behavior and calculate the volume at 1.00 atm and 298 K occupied by (a) 33.9 μg of SiF_4; (b) 0.572 mg of Xe; (c) 0.822 mol of BrF_3; (d) 3.55 g of air (of average molar mass 28.97 $g \cdot mol^{-1}$).

5.39 Calculate the mass of gas in (a) 2.45 L of O_2; (b) 1.94 mL of SO_3; (c) 6000 L of CH_4; (d) 1.44 mL of CO_2 (all measured at 1.00 atm and 298 K).

5.40 Calculate the number of molecules in (a) 6.99 nL of nitrogen; (b) a spherical flask of ammonia with a diameter of 10 cm ($V = \frac{4}{3}\pi r^3$); (c) an ampule of krypton with a diameter of 1.4 mm and a length of 16 mm ($V = \pi r^2 l$). (All measurements are at 1.00 atm and 298 K.)

Stoichiometry of Reacting Gases

5.41 Oxygen gas is generated in the laboratory by the thermal decomposition of potassium chlorate. What volume of O_2 (at 1.00 atm and 298 K) is generated from 1.00 g of potassium chlorate in the reaction

$$2\,KClO_3(s) \longrightarrow 2\,KCl(s) + 3\,O_2(g)$$

5.42 Calculate the mass of ammonium nitrate that should be heated to obtain 100 mL of dinitrogen oxide, N_2O, at 1.00 atm and 298 K in the reaction

$$NH_4NO_3(s) \longrightarrow N_2O(g) + 2\,H_2O(g)$$

5.43 What volume of carbon dioxide at 1.00 atm and 298 K is produced when an excess of hydrochloric acid is added to 10.0 g of calcium carbonate and the reaction

$$CaCO_3(s) + 2\,HCl(aq) \longrightarrow CaCl_2(aq) + CO_2(g) + H_2O(l)$$

takes place?

5.44 One industrial process for the removal of hydrogen sulfide from natural gas is its reaction with sulfur dioxide:

$$2\,H_2S(g) + SO_2(g) \longrightarrow 3\,S(s) + 2\,H_2O(g)$$

(a) What volume of sulfur dioxide at 1.00 atm and 298 K is needed to prepare 1.00 kg of sulfur? (b) What volume of sulfur dioxide is needed in (a) if it is supplied at 5.0 atm and 250°C?

5.45 The Haber process for the synthesis of ammonia is one of the most significant industrial processes for the well-being of humanity. (a) What volume of hydrogen at 1.00 atm and 298 K must be supplied to produce 1.0 metric ton (1 t = 10^3 kg) of NH_3? (b) What volume of hydrogen is needed in (a) if it is supplied at 200 atm and 400°C?

5.46 Nitroglycerin is a shock-sensitive liquid that detonates by the reaction

$$4\,C_3H_5(NO_3)_3(l) \longrightarrow 6\,N_2(g) + 10\,H_2O(g) + 12\,CO_2(g) + O_2(g)$$

Calculate the total volume of product gases at 150 kPa and 100°C from the detonation of 1.0 g of nitroglycerin.

5.47 Urea, $CO(NH_2)_2$, is used as a fertilizer and is made by the reaction of carbon dioxide and ammonia:

$$CO_2(g) + 2\,NH_3(g) \longrightarrow CO(NH_2)_2(s) + H_2O(g)$$

What volumes of CO_2 and NH_3 at 200 atm and 450°C are needed to produce 2.50 kg of urea?

5.48 Xenon and fluorine react at 350°C to produce a mixture of XeF_2, XeF_4, and XeF_6. What volumes of xenon and fluorine at 200 kPa and 350°C are needed to produce 1.00 μg of XeF_4, assuming a 100% yield in the reaction

$$Xe(g) + 2\,F_2(g) \longrightarrow XeF_4(g)$$

5.49 Which starting condition would produce the larger volume of carbon dioxide by combustion of $CH_4(g)$ with an excess of oxygen gas to produce carbon dioxide and water? Justify your answer. The system is maintained at a temperature of 75°C and 1.00 atm. (a) 2.00 L of $CH_4(g)$; (b) 2.00 g of $CH_4(g)$.

5.50 Which starting condition would produce the larger volume of carbon dioxide by combustion of $C_2H_4(g)$ with an excess of oxygen gas to produce carbon dioxide and water? Justify your answer. The system is maintained at a temperature of 45°C and 2.00 atm. (a) 1.00 L of $C_2H_4(g)$; (b) 1.20 g of $C_2H_4(g)$.

Density

5.51 Which gas is the most dense at 1.00 atm and 298 K? (a) N_2; (b) NH_3; (c) NO_2.

5.52 Which gas is the most dense at 1.00 atm and 298 K? (a) CO; (b) CO_2; (c) H_2O.

5.53 What is the density (in $g \cdot L^{-1}$) of chloroform ($CHCl_3$) vapor at (a) 1.00 atm and 298 K; (b) 100.0°C and 1.00 atm?

5.54 Calculate the density (in $g \cdot L^{-1}$) of hydrogen sulfide (H_2S) at (a) 1.00 atm and 298 K; (b) 25°C and 0.962 atm.

5.55 A sample of a gas with a mass of 21.3 g is confined to a vessel of volume 7.73 L at 0.880 atm and 30°C. (a) What is the density of the gas at 1.00 atm and 298 K. (b) What is the molar mass of the gas?

5.56 A sample of halogen gas has a mass of 0.239 g and exerts a pressure of 600 Torr at 14°C in a 100-mL flask. (a) Calculate the density of the gas at 1.00 atm and 298 K. (b) What is the molar mass of the halogen? (c) Identify the halogen.

5.57 The density of a gaseous compound of phosphorus is 3.60 $g \cdot L^{-1}$ at 420 K when its pressure is 727 Torr. (a) What is the molar mass of the compound? (b) What is the density of the gas at 1.00 atm and 298 K?

5.58 A gaseous fluorinated methane compound has a density of 8.0 $g \cdot L^{-1}$ at 2.81 atm and 300 K. (a) What is the molar mass of the compound? (b) What is the density of the gas at 1.00 atm and 298 K?

5.59 The analysis of a hydrocarbon revealed that it was 85.7% C and 14.3% H by mass. When 1.77 g of the gas was stored in a 1.500-L flask at 17°C, it exerted a pressure of 508 Torr. What is the molecular formula of the hydrocarbon?

5.60 A compound used in the manufacture of Saran is 24.7% C, 2.1% H, and 73.2% Cl by mass. The storage of 3.557 g of the compound in a 750-mL vessel at 0°C results in a pressure of 1.10 atm. What is the molecular formula of the compound?

Gaseous Mixtures

5.61 Calculate the partial pressure of each gas and the total pressure of the following mixtures, each of which occupies a 500-mL vessel at 0°C: (a) 0.020 mol N_2 and 2.33 g of O_2; (b) 0.015 mol H_2, 4.22 mg of He, and 0.030 mol NH_3.

5.62 The following gases were placed in a 3500-mL flask at 25°C: (a) 0.0195 g of CH_4 and 0.0195 g of CO_2; (b) 1.00 mg of Ar, 2.00 mg of Kr, and 3.00 mg of Xe. Calculate the partial and the total pressure of each gas.

5.63 A sample of damp air in a 1.00-L container exerts a pressure of 762.0 Torr at 20°C; but when it is cooled to −10.0°C, the pressure falls to 607.1 Torr as the water condenses. What mass of water was present? Assume the vapor pressure of the ice at −10°C to be negligible.

5.64 A 2.00-L flask contains carbon dioxide at 20°C and 606.0 Torr. After 1.0 g of nitrogen is added to the flask and heated to 200°C, what is the expected pressure in the flask?

5.65 An apparatus consists of a 4.0-L flask containing nitrogen gas at 25°C and 803 kPa, joined by a tube to a 10.0-L flask containing argon gas at 25°C and 47.2 kPa. The stopcock on the connecting tube is opened and the gases mix. (a) What is the partial pressure of each gas after mixing? (b) What is the total pressure of the gas mixture?

5.66 The total pressure of a mixture of sulfur dioxide and nitrogen gases at 25°C in a 500-mL vessel is 1.09 atm. The mixture is passed over warm calcium oxide powder, which removes the sulfur dioxide by the reaction $CaO(s) + SO_2(g) \rightarrow CaSO_3(s)$ and is then transferred to a 150-mL vessel, where the pressure is 1.09 atm at 50°C. (a) What was the partial pressure of the SO_2 in the initial mixture? (b) What is the mass of SO_2 in the initial mixture?

5.67 In the course of a redox reaction, 220 mL of hydrogen was collected over water at 20°C when the external pressure was 756.7 Torr. (See Table 5.4.) (a) What is the partial pressure of hydrogen? (b) What mass of oxygen was also produced in the reaction?

5.68 When a potassium chlorate sample was heated in the presence of MnO_2 (a catalyst for this reaction), 25.7 mL of $O_2(g)$ was collected over water at 14°C and 97.6 kPa. What was the mass of $KClO_3$ in the sample? The decomposition reaction is

$$2\,KClO_3(s) \xrightarrow{MnO_2} 2\,KCl(s) + 3\,O_2(g)$$

5.69 A 37.6-mL sample of dry hydrogen gas was collected at 24°C over a nonvolatile liquid when the atmospheric pressure was 107 kPa. What volume would the hydrogen gas have occupied had it been collected over water? (See Table 5.4.)

5.70 Dinitrogen oxide, N_2O, gas was generated from

the thermal decomposition of ammonium nitrate and collected over water. The wet gas occupied 126 mL at 21°C when the atmospheric pressure was 755 Torr. What volume would the same amount of dry dinitrogen oxide have occupied if collected at 755 Torr and 21°C? (See Table 5.4.)

Molecular Motion

5.71 Describe the assumptions of the kinetic model of gases.

5.72 Describe Boyle's law in terms of the kinetic model of gases.

5.73 Use the kinetic model to explain why the pressure of a gas is proportional to the temperature.

5.74 Account for Dalton's law of partial pressures in terms of the kinetic model of gases.

5.75 Which gas effuses more rapidly: (a) argon or sulfur dioxide? (b) D_2 (of molar mass 4.018 $g \cdot mol^{-1}$) or H_2?

5.76 Which gas effuses more rapidly: (a) HF or H_2O? (b) CO_2 or CO?

5.77 What is the molar mass of a compound that takes 2.7 times longer to effuse through a porous plug than it did for the same amount of XeF_2 at the same temperature and pressure?

5.78 What is the molecular formula of a compound of empirical formula CH that diffuses 3.22 times more slowly than krypton at the same temperature and pressure?

Real Gases

5.79 Identify the conditions of temperature and pressure under which real gases are most likely to behave ideally. Explain your conclusions.

5.80 Under what conditions would you expect a real gas to be (a) more compressible than an ideal gas? (b) less compressible than an ideal gas?

5.81 Describe how intermolecular forces vary with distance.

5.82 The pressure of a sample of hydrogen fluoride is lower than expected and rises more quickly as the temperature is increased than the ideal gas law predicts. Suggest an explanation.

5.83 Imagine that a 1.0-L vessel contains 1.0 mol of an ideal gas at 273 K and that another 1.0-L vessel contains 1.0 mol of the real gas C_2H_4 at 273 K. (a) Which vessel has the greater pressure? (b) Which vessel has more free space between molecules?

5.84 Imagine that a 1.0-L vessel contains 1.0 mol of an ideal gas at 273 K and that another 1.0-L vessel contains 1.0 mol of the real gas SO_2 at 273 K. (a) In which flask will P reach zero first if the flasks are cooled at the same rate? (b) Which gas will have the greatest volume at the absolute zero of temperature?

SUPPLEMENTARY EXERCISES

5.85 Explain how you would convert the volume of a gas sample at 0°C and 1 atm to the volume of the same sample at another temperature and pressure.

5.86 Does an ideal gas always have a higher density at high pressure than at low pressure?

5.87 Explain what is meant by the Joule-Thomson effect. Describe how it can be used to liquify air.

5.88 Which of the following statements are true? Explain your answer in each case. (a) Real gases act more like ideal gases as the temperature is raised. (b) If n and T are held constant, an increase in P will result in an increase in V. (c) At 1.00 atm and 298 K, every molecule in a sample of a gas has exactly the same speed. (d) At the same temperature and pressure, N_2 gas is denser than NH_3 gas. (e) At constant P and T, a decrease in n will result in a decrease in V.

5.89 A 1.000-L flask of He and a 1.000-L flask of Cl_2 at 1.00 atm and 298 K differ in which *three* properties? (a) number of atoms; (b) pressure; (c) average molecular speed; (d) temperature; (e) density; (f) volume.

5.90 (a) The weather report states that the current atmospheric pressure is 28.92 inches of mercury. Convert this pressure to Torr and atmospheres. (b) The pressure of nitrogen gas in a steel cylinder is 59.5 kPa. What is this pressure in Torr?

5.91 Low-pressure gauges in research laboratories are occasionally calibrated in inches of water ("inH_2O"). Considering that the density of mercury at 15°C is 13.6 $g \cdot cm^{-3}$ and the density of water at 5°C is 1.0 $g \cdot cm^{-3}$, what is the pressure (in Torr) inside a gas cylinder that reads 1.5 inH_2O at 15°C?

5.92 Originally at −20°C and 759 Torr, a 1.00-L air sample is heated to 100°C, the pressure is decreased to 300 Torr, then it is heated to 1000°C, and finally the pressure is decreased to 120 Torr. What is the final volume of the air?

5.93 An air-fuel mixture in an internal combustion engine occupies 500 cm^3 at 764 Torr. If the volume is reduced to 58.8 cm^3 at the same temperature, what is the resulting pressure of the mixture prior to ignition?

5.94 At the cruising altitude of an airplane (10 km), the atmospheric pressure is about 25 kPa. Disregarding the effect of temperature, calculate the volume of a sample of air at that altitude, which at the surface of the Earth would occupy 1.0 L.

5.95 How much *additional* pressure would you need to exert on a sample of helium at 765 Torr to compress it from 555 mL to 125 mL at constant temperature?

5.96 A natural gas well with an approximate volume of 2.0×10^9 L has a reservoir pressure of 1.20 atm at 40°C. If all the gas is to be transferred to a steel tank where the pressure is not to exceed 6.00 atm on a very cold day (−25°C), what must the volume of the tank be?

5.97 (a) Calculate the final volume of a 100.0-L sample of air that, originally at 0°C and 2.00 atm, is heated to 100°C, compressed to 1.55 atm, and then heated again to 175°C at constant pressure. (b) What is the mass of air in the sample? (The average molar mass of the air is 28.97 $g{\cdot}mol^{-1}$.)

5.98 At the top of the troposphere, the temperature is about −50°C and the pressure is about 0.25 atm. If a balloon that is filled with helium to a volume of 10.0 L at 2.00 atm pressure and 27.0°C is released at ground level, what will the volume of the balloon be at the top of the troposphere?

5.99 A 2.10-mL bubble of methane gas formed at the bottom of a deep lake where the temperature is 8.1°C and the pressure is 6.4 atm, and rises to the top of the swamp, where the temperature is 25°C and the pressure is 1.00 atm. What is the volume of the methane bubble before it bursts?

5.100 A 0.297-g sample of an unidentified gas occupies 250 mL at 300 K and 670 Torr. (a) What is the density of the gas at 1.00 atm and 298 K? (b) What is the density of the gas if it is transferred to a vessel where its pressure is 170 kPa and 70.0°C? (c) What is the molar mass of the gas?

5.101 Carbon monoxide gas is purchased in a 425-mL bottle at a pressure of 5.00 atm and 23°C. (a) What mass (in grams) of the gas has been purchased? (b) What is the density of the gas in the bottle? (c) If the temperature of the bottle changed to 35.0°C (on a hot day), would the density of the gas change? Explain your answer.

5.102 0.466 g of an unidentified gas occupies 500 mL at 16°C and 740 Torr. (a) What is the density of the gas at 1.00 atm and 298 K? (b) What is the molar mass of the gas?

5.103 An air pollution chemist collects gaseous wastes in an initially empty 22.1-L steel tank until the pressure reaches 0.84 atm at 26°C. Assuming the gases to have an average molar mass of 33.8 $g{\cdot}mol^{-1}$, what mass of gas has been collected?

5.104 Calculate the volumes of hydrogen and oxygen at 1.00 atm and 298 K that are produced in the decomposition of 8.81 g of water into the elements.

5.105 When titanium tetrachloride is mixed with water, voluminous clouds of a white smoke of titanium dioxide particles are produced by the reaction

$$TiCl_4(g) + 2\,H_2O(l) \longrightarrow TiO_2(s) + 4\,HCl(g)$$

(a) What mass of TiO_2 is produced from the reaction of 200 L of $TiCl_4$, measured at 500 kPa and 30°C?
(b) Determine the volume of hydrogen chloride gas produced at 1.00 atm and 298 K.

5.106 The first stage in the production of nitric acid by the Ostwald process is the oxidation of ammonia:

$$4\,NH_3(g) + 5\,O_2(g) \longrightarrow 4\,NO(g) + 6\,H_2O(g)$$

(a) Calculate the mass of nitric oxide that can be produced from the reaction of 150 L of ammonia at 15.0 atm and 200°C with an excess of oxygen. (b) If the water produced in (a) is condensed to a liquid with a density of 1.0 $g{\cdot}mL^{-1}$, what volume will it occupy?

5.107 An initial volume of 250 L of butane at 150°C and 2.33 atm is burned with an excess of oxygen:

$$2\,C_4H_{10}(g) + 13\,O_2(g) \longrightarrow 8\,CO_2(g) + 10\,H_2O(g)$$

(a) Calculate the mass of carbon dioxide released to the atmosphere. (b) What pressure would the carbon dioxide exert if it were collected and stored in a 4000-L vessel at 16°C?

5.108 200 mL of hydrogen chloride at 690 Torr and 20°C is dissolved in 100 mL of water. The solution was titrated to the stoichiometric point with 15.7 mL of a sodium hydroxide solution. What is the molar concentration of the NaOH in solution?

5.109 Through a series of enzymatic steps, carbon dioxide and water undergo photosynthesis to produce glucose and oxygen according to the equation

$$6\,CO_2(g) + 6\,H_2O(l) \longrightarrow C_6H_{12}O_6(s) + 6\,O_2(g)$$

Given that the partial pressure of carbon dioxide in the troposphere is 0.26 Torr and that the temperature is 25°C, calculate the volume of air that is needed to support the production of 10.0 g of glucose.

5.110 Small quantities of hydrogen gas can be generated in the laboratory by the action of dilute hydrochloric acid on zinc metal. When 0.40 g of impure zinc was reacted with an excess of hydrochloric acid, 127 mL of hydrogen was collected over water at 17°C. The external pressure was 737.7 Torr. (a) What volume would the dry hydrogen occupy at 1.00 atm and 298 K? (b) What amount (in moles) of H_2 was collected? (c) What is the percentage purity of the zinc?

5.111 0.473 g of a gas that occupies 200 mL at 1.81 atm and 25°C was analyzed and found to contain 0.414 g nitrogen and 0.0591 g hydrogen. What is the molecular formula of the compound?

5.112 A compound used to make polyvinyl chloride (PVC) has composition 38.4% C, 4.82% H, and 56.8% Cl by mass. It took 7.73 min for a given mass of the compound to effuse through a porous plug, but it took only 6.18 min for the same amount of Ar to diffuse at the same temperature and pressure. What is the molecular formula of the compound?

5.113 A sample of argon gas effuses through a porous plug in 147 s. Calculate the time required for the same number of moles of (a) CO_2; (b) C_2H_4; (c) H_2; (d) SO_2 to effuse under the same conditions of pressure and temperature.

5.114 Two identical flasks are each filled with a gas at 0°C. One flask contains 1 mol CO_2 and the other 1 mol Ne. In which flask do the particles have more collisions per second with the walls of the container?

5.115 A hydrocarbon of empirical formula C_2H_3 takes 349 s to effuse through a porous plug; under the same conditions of temperature and pressure, it took 210 s for the same amount of argon to effuse. What is the molar mass and molecular formula of the hydrocarbon?

5.116 When 2.36 g of phosphorus was burned in chlorine, 10.5 g of a phosphorus chloride was produced. Its vapor took 1.8 times longer to diffuse than the same amount of CO_2 under the same conditions of temperature and pressure. What is the molar mass and molecular formula of the phosphorus chloride?

5.117 An artificial atmosphere for a commercial fruit storage facility is being designed. If 4.0% of the molecules are CO_2 and the total atmospheric pressure is to be 640 Torr, what is the partial pressure of CO_2 in the storage facility?

CHALLENGING EXERCISES

5.118 Suggest reasons why there is so little hydrogen and helium gas in the Earth's atmosphere. Venus and Earth are similar planets, but the atmosphere of Venus is mainly carbon dioxide. Where has the CO_2 of the Earth's atmosphere gone?

5.119 Yellow sulfur consists of S_8 molecules. When a sample of sulfur was vaporized, it gave rise to a pressure that was about three times larger than that expected; and as heating continued, the pressure rose more quickly than the ideal gas law predicts. Suggest an explanation.

5.120 At 180.0°C, 27.20 mg of the vapor of the anti-malarial drug quinine in a 250.0-mL container gave rise to a pressure of 9.48 Torr. In a combustion experiment, 17.2 mg of quinine produced 47.0 mg of carbon dioxide, 11.5 mg of water, and 1.51 mg of nitrogen. What is the molecular formula of quinine?

5.121 A 1.509-g sample of an osmium oxide (which melts at 40°C and boils at 130°C) is placed into a cylinder with a movable piston that can expand against the atmospheric pressure of 745 Torr. When the sample is heated to 200°C, it is completely vaporized and the volume of the cylinder expands by 235 mL. What is the molar mass of the oxide? Assuming that the oxide is OsO_x, what is the value of x?

5.122 Iron pyrite, FeS_2, is the form in which much of the sulfur occurs in coal. In the combustion of the coal, oxygen reacts with iron pyrite to produce iron(III) oxide and sulfur dioxide:

$$4\,FeS_2(s) + 11\,O_2(g) \longrightarrow 2\,Fe_2O_3(s) + 8\,SO_2(g)$$

(a) Calculate the mass of Fe_2O_3 that is produced from the reaction of 75.0 L of oxygen at 2.33 atm and 150°C with an excess of iron pyrite. (b) If the sulfur dioxide that is generated in (a) is dissolved to form 5.00 L of aqueous solution, what is the molar concentration of the resulting sulfurous acid, H_2SO_3, solution?

5.123 15 mL of ammonia at 100 Torr and 30°C is mixed with 25 mL of hydrogen chloride at 150 Torr and 25°C, and the reaction

$$NH_3(g) + HCl(g) \longrightarrow NH_4Cl(s)$$

occurs. (a) Calculate the mass of NH_4Cl that forms. (b) Identify the gas in excess and determine the partial pressure (in the combined volume of the original two flasks) of the excess gas at 27°C after the reaction is complete.

5.124 A 2.00-L sample of ethene gas, C_2H_4, at 1.00 atm and 298 K is burned in 2.00 L oxygen gas at the same pressure and temperature to form carbon dioxide gas and liquid water. Ignoring the volume of water, what is the final volume of the reaction mixture at 1.00 atm and 298 K if reaction goes to completion?

5.125 A group of chemistry students have injected a 32.5-g sample of a gas into an evacuated, constant-volume container at 22°C and atmospheric pressure. Now the students will heat the gas at constant pressure by allowing some of the gas to escape during the heating. What mass of gas must be released if the temperature is raised to 212°C?

5.126 At 280.0°C and in a 500.0-mL container, 115 mg of eugenol, the compound responsible for the odor of cloves, caused a pressure of 48.3 Torr. In a combustion experiment, 18.8 mg of eugenol burned to 50.0 mg of carbon dioxide and 12.4 mg of water. What is the molecular formula of eugenol?

CHAPTER 6

The ultimate source of much of the energy we use on Earth is the Sun. Solar radiation causes reactions high in the atmosphere, which give out heat. It also causes chemical reactions on the surface of the Earth, particularly the reactions responsible for photosynthesis in green plants. These reactions effectively store the energy for future use, sometimes thousands of years later as fossil fuels but sometimes for almost immediate consumption as food. In this chapter we see how to keep track of the energy changes that accompany physical and chemical changes.

THERMOCHEMISTRY: THE FIRE WITHIN

People lived simply in the Stone Age. Their energy needs were supplied by the Sun and by the **biomass,** plants and trees that could be burned. Because the biomass is renewed every year, and populations were small, Stone Age energy sources were plentiful and renewable. However, this simple lifestyle has almost completely vanished, and modern civilization cannot survive without abundant fuel. If all transport came to an end, our communications would still need electricity generated somewhere by the combustion of fuel. Even if we could do away with communication, too, our bodies and brains would still need the fuel we call food.

Most of the fuels used by modern civilization are **fossil fuels:** natural gas, coal, and products derived from petroleum, such as gasoline, diesel fuel, and kerosene. Fossil fuels are not renewable. They are extracted from organic matter that decayed and was deposited in the Earth millions of years ago. One day, they will run out. To manage our energy resources and develop new fuels, we need to study energy and understand how it is released or used in chemical reactions.

ENERGY, HEAT, AND ENTHALPY

6.1 Transfer of energy as heat
6.2 Exothermic and endothermic processes
6.3 Measuring heat transfer
6.4 Thermal accounting: Enthalpy
6.5 Vaporization
6.6 Melting and sublimation
6.7 Heating curves

THE ENTHALPY OF CHEMICAL CHANGE

6.8 Reaction enthalpies
6.9 Standard reaction enthalpies
6.10 Combining reaction enthalpies: Hess's law

THE HEAT OUTPUT OF REACTIONS

6.11 Enthalpies of combustion
6.12 Standard enthalpies of formation

ENERGY, HEAT, AND ENTHALPY

The reactions by which we obtain energy from fuels are usually combustions in which a hydrocarbon fuel burns in oxygen to form carbon dioxide and water. A typical reaction is the combustion of natural gas, which is largely methane, CH_4:

$$CH_4(g) + 2\,O_2(g) \longrightarrow CO_2(g) + 2\,H_2O(l)$$

Living organisms make use of a very sophisticated form of combustion to obtain the energy they require. Glucose, a typical component of energy-giving carbohydrates, is the fuel that powers all our thoughts and actions:

$$C_6H_{12}O_6(aq) + 6\,O_2(g) \longrightarrow 6\,CO_2(g) + 6\,H_2O(l)$$

This reaction does not proceed like a familiar combustion: the glucose does not burn with a flame inside us. Instead, reactions governed by a complex interplay of enzymes ensure that it takes place to the extent required effectively and safely. Nevertheless, the overall outcome of the reaction is very much like a combustion, and the process can be discussed in a similar way.

In this chapter we shall see how to predict the heating or cooling effect of any reaction, not just combustions. The calculations are a straightforward extension of the stoichiometry calculations we have been doing.

6.1 TRANSFER OF ENERGY AS HEAT

We have been using a term that we often encounter in daily life: *energy.* Because we use the term in conversation, it is familiar. But what does it really mean?

The scientific definition of **energy** is "the capacity to do work or supply heat." A wound-up spring releases energy as it unwinds, and this energy can be used to raise a weight against the pull of gravity. As it burns, fuel releases energy that heats the surroundings or does the work of propelling a car along a highway. Doing work includes raising weights, propelling vehicles along roads, generating electric currents, and even building complicated molecules like proteins inside living cells. We shall deal with work in Chapter 16. Here we concentrate on energy transferred as heat. The branch of chemistry that deals with the heat released or absorbed in chemical reactions is called **thermochemistry.**

The **law of the conservation of energy** tells us that energy cannot be created or destroyed. However, it can be transformed from one form into another and transferred from place to place. To keep track of these changes, it is convenient to distinguish between a system and its surroundings. By **system,** we mean the substance or reaction mixture in which we have an interest (Fig. 6.1). A chemical system may consist of the reaction mixture inside a flask, or it may be a single pure system such as a block of copper. The **surroundings** consist of the container and everything else outside the system. We normally consider only the imme-

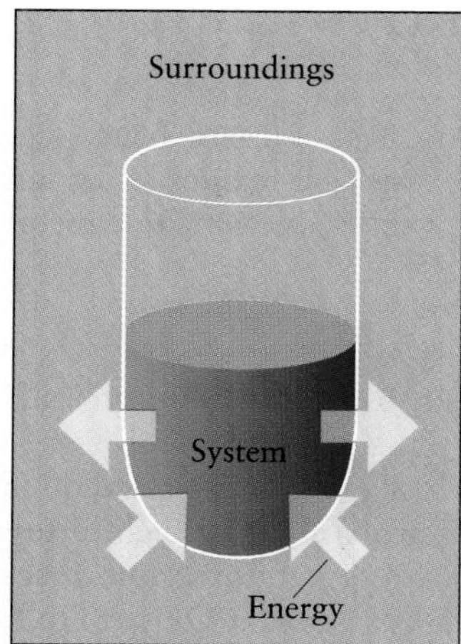

FIGURE 6.1 The system is the sample or reaction mixture that we are interested in. Outside the system are the surroundings. The system plus its surroundings is sometimes called the universe. In many cases, the surroundings consist of a stirred water bath.

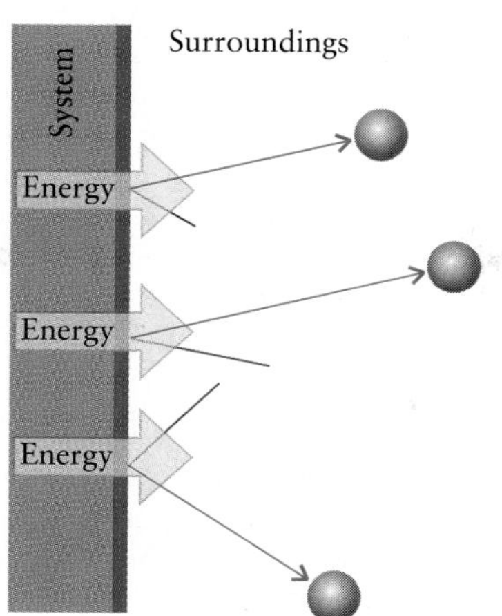

FIGURE 6.2 When energy leaves a system as a result of a temperature difference between the system and the surroundings, we say that the system has lost energy as heat. This transfer of energy stimulates random thermal motion of molecules in the surroundings.

diate surroundings of the system, such as the water bath in which a flask is immersed.

We say that energy has left a system as **heat** if the energy flows out as a result of a difference in temperature between the system and the surroundings. When we touch a cool object, energy flows out of us as heat. Energy flows into us as heat when we relax in a hot bath.

On a molecular scale, the outward flow of energy takes place as heat stirs up chaotic motion of the molecules next to the system. These rapidly moving molecules collide with others and pass on their extra energy during the collisions, spreading the energy throughout the surroundings (Fig. 6.2). Random molecular motion is called **thermal motion,** so we can say that heat stirs up the thermal motion of molecules in the surroundings. When we say that heat flows from a high temperature region to a low temperature region we mean that energy flows from a region where the thermal motion is vigorous to a region where there is less thermal motion.

Temperature is a measure of thermal motion.

The energy of a sample depends on the size of the sample. For example, a block of iron of mass 2 kg contains twice as much energy as a block of iron of mass 1 kg at the same temperature. Therefore, energy is an extensive property. The temperature of a sample of water taken from a uniformly heated water bath is the same, however, no matter what size sample is taken. Temperature, therefore, is an intensive property.

◀◀ Intensive and extensive properties were introduced in Section 2.3.

Energy is the capacity to do work or release heat. Heat is the transfer of energy from regions of high temperature to regions of low temperature. Heat stimulates thermal motion.

6.2 EXOTHERMIC AND ENDOTHERMIC PROCESSES

An **exothermic reaction** is a chemical reaction that releases heat (Fig. 6.3). All combustions are exothermic. One striking example of an exothermic reaction is the "thermite reaction," the reduction of a metal oxide by aluminum:

$$2\,Al(s) + Fe_2O_3(s) \longrightarrow Al_2O_3(s) + 2\,Fe(s)$$

The reaction is so exothermic that it has been used to weld iron rails together (Fig. 6.4).

An **endothermic reaction** absorbs heat. Most endothermic reactions absorb only a small amount of heat, so there is probably not much future for chemical refrigerators. However, one moderately endothermic reaction takes place when barium hydroxide octahydrate, $Ba(OH)_2{\cdot}8H_2O$, and ammonium thiocyanate, NH_4SCN, are ground together:

$$Ba(OH)_2{\cdot}8H_2O(s) + 2\,NH_4SCN(s) \longrightarrow Ba(SCN)_2(aq) + 2\,NH_3(g) + 10\,H_2O(l)$$

The reaction mixture becomes so cold that moisture from the air forms a layer of frost on the outside of the beaker (Fig. 6.5).

Dissolving may be either exothermic or endothermic and may involve only a little or a lot of heat, depending on the solute and the solvent. For example, instant cold packs contain one packet of water and another of ammonium nitrate. When the barrier between the salt and the water is broken, the ammonium nitrate dissolves, absorbing a lot of heat.

FIGURE 6.3 The decomposition of ammonium dichromate, $(NH_4)_2Cr_2O_7$, to chromium(III) oxide, Cr_2O_3, is so exothermic that, once ignited, the heat generated produces a miniature volcano of sparks and solid product.

FIGURE 6.4 The thermite reaction is another highly exothermic reaction—one that can melt the metal it produces. In this reaction, aluminum metal is reacting with iron(III) oxide, Fe_2O_3, causing a shower of molten iron sparks.

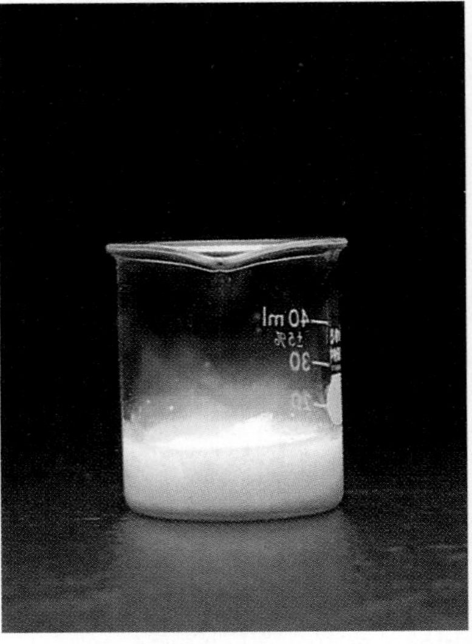

FIGURE 6.5 The reaction between ammonium thiocyanate, NH_4SCN, and barium hydroxide octahydrate, $Ba(OH)_2{\cdot}8H_2O$, absorbs a lot of heat and can cause water vapor in the air to freeze on the outside of the beaker.

The dissolution of ammonium nitrate in a cold pack is an example of an endothermic process. However, when sulfuric acid is mixed with water, it generates a lot of heat, enough to boil the solution, so the dissolution of sulfuric acid in water is classified as an exothermic process.

Always add acid to water when diluting the acid. Then, if the mixture does boil and splash out of the container, the spatters will be mainly water.

Changes of state can also be classified as exothermic or endothermic. All vaporizations are endothermic: a liquid absorbs heat from its surroundings as it changes to vapor. For example, heat is needed to boil the water in a kettle. Condensation and freezing are exothermic: a freezer must remove heat from the water in an ice tray for it to freeze.

Exothermic processes release heat; endothermic processes absorb heat.

6.3 MEASURING HEAT TRANSFER

The unit of energy and heat is the **joule,** J. Each beat of the human heart uses about 1 J of energy. To lift this book from the floor to a table, you need to use about 15 J of energy. Most energy transfers in chemistry are expressed in kilojoules:

The unit is named for James Joule, a nineteenth-century English scientist who made many contributions to the study of heat. Formally, 1 J = 1 kg·m^2·s^{-2}.

$$1 \text{ kJ} = 10^3 \text{ J}$$

A unit of energy still widely used in biochemistry and related fields is the **calorie** (cal): 1 cal is the energy needed to raise the temperature of 1 g of water by 1°C. The exact relation between the two units is

$$1 \text{ cal} = 4.184 \text{ J}$$

It follows that 1.0 J = 0.24 cal; so 1.0 kJ of heat can raise the temperature of 240 g (2 sf) of water (about the amount in a cup of coffee) by 1.0°C.

We measure the heat released or absorbed in a reaction by monitoring changes in temperature (in degrees Celsius). Then we convert the change in temperature to joules by using the **heat capacity,** the heat required to raise the temperature of an object by 1°C:

$$\text{Heat capacity} = \frac{\text{heat supplied}}{\text{temperature rise}}$$

Suppose, for example, that we measure a temperature rise of 2.0°C when we supply 98 kJ of heat to a sample of ethanol. Then we would report that

$$\text{Heat capacity} = \frac{98 \text{ kJ}}{2.0°\text{C}} = 49 \text{ kJ}\cdot(°\text{C})^{-1}$$

As we see from this example, the units of heat capacity are joules per degree Celsius (J·(°C)$^{-1}$). More heat is needed to raise the temperature of a large sample of a substance by 1°C than that of a small sample: the larger the sample, the greater its heat capacity (Fig. 6.6). It is therefore common to report the **specific heat capacity** (often called just specific heat), which is the heat capacity divided by the mass of the sample in grams. The specific heat capacities of some common substances are given in Table 6.1. We can calculate the heat capacity of a substance from its mass and its specific heat capacity by using the expression

$$\text{Heat capacity} = \text{mass} \times \text{specific heat capacity}$$

Heat capacity is an extensive property; specific heat capacity is intensive.

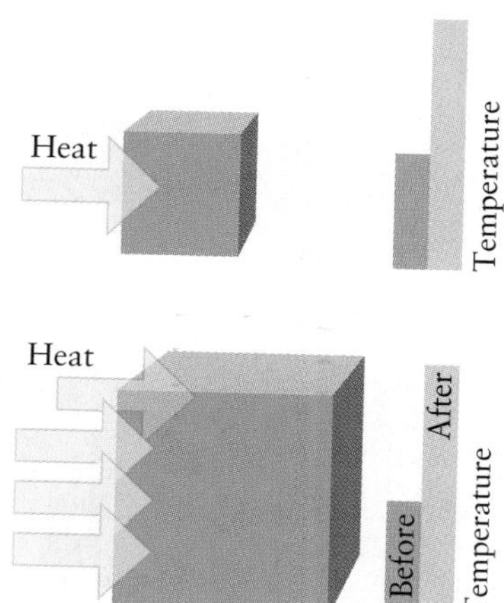

FIGURE 6.6 The heat capacity of an object determines the change in temperature brought about by a given quantity of heat: an object with a large heat capacity requires a lot of heat to bring about a given rise in temperature. Heat capacity is an extensive property, so a large object (bottom) has a larger heat capacity than a small object (top) made of the same material.

TABLE 6.1 Specific heat capacities of common materials

Material	Specific heat capacity, $J\cdot(°C)^{-1}\cdot g^{-1}$
air	1.01
benzene	1.05
brass	0.37
copper	0.38
ethanol	2.42
glass (Pyrex)	0.78
granite	0.80
marble	0.84
polyethylene	2.3
stainless steel	0.51
water: solid	2.03
liquid	4.18
vapor	2.01

Heat transfers are measured with a **calorimeter,** an insulated container fitted with a thermometer. A calorimeter can be as simple as a Styrofoam cup (Fig. 6.7) or as sophisticated as a bomb calorimeter (Fig. 6.8). The latter consists of a heavy-walled steel container. The sample is placed inside the container, which is then sealed and placed in a water bath. An electric current is passed through the sample to ignite it, and the heat given off by the reaction raises the temperature of the entire apparatus. We measure the increase in temperature and then use the heat capacity of the calorimeter (the entire assembly) to convert the change in temperature into heat released by the reaction.

EXAMPLE 6.1 Measuring specific heat capacity with a simple calorimeter

Suppose we have 50 g of water at 20.0°C in a calorimeter like that in Fig. 6.7. Then we drop 21 g of iron at 90.2°C into the calorimeter and measure a final temperature of 23.2°C. What is the specific heat capacity of the iron?

STRATEGY All the heat lost by the iron is transferred to the water and the rest of the calorimeter. The heat lost or gained by each substance equals its heat capacity multiplied by its temperature rise, and the heat capacity is its specific heat capacity multiplied by its mass. The heat capacity of the Styrofoam is so small that it can be neglected. Therefore, we can write

$$\begin{aligned}&(\text{Specific heat})_{\text{metal}} \times (\text{mass})_{\text{metal}} \times (\text{fall in temperature})_{\text{metal}}\\ &\quad = (\text{specific heat})_{\text{water}} \times (\text{mass})_{\text{water}} \times (\text{rise in temperature})_{\text{water}}\end{aligned}$$

There are enough data to find the one unknown, the specific heat capacity of the metal.

SOLUTION We substitute the data and solve for the specific heat capacity of the metal. Note that the temperature of the iron falls from 90.2°C to 23.2°C, or 67.0°C, and the temperature of the water rises from 20.0°C to 23.2°C, or 3.2°C. Therefore,

$$\begin{aligned}&(\text{Specific heat})_{\text{metal}} \times (21\text{ g}) \times (67.0°\text{C})\\ &\quad = (4.184\text{ J}\cdot(°\text{C})^{-1}\cdot\text{g}^{-1}) \times (50\text{ g}) \times (3.2°\text{C})\end{aligned}$$

That is,

$$\begin{aligned}(\text{Specific heat})_{\text{metal}} &= \frac{(4.184\text{ J}\cdot(°\text{C})^{-1}\cdot\text{g}^{-1}) \times (50\text{ g}) \times (3.2°\text{C})}{(21\text{ g}) \times (67.0°\text{C})}\\ &= 0.48\text{ J}\cdot(°\text{C})^{-1}\cdot\text{g}^{-1}\end{aligned}$$

SELF-TEST 6.1A A piece of copper of mass 19.0 g was heated to 87.4°C and then placed in a calorimeter that contained 55.5 g of water at 18.3°C. The temperature of the water rose to 20.4°C. What is the specific heat capacity of copper?

[*Answer:* $0.38\text{ J}\cdot(°\text{C})^{-1}\cdot\text{g}^{-1}$]

SELF-TEST 6.1B An alloy of mass 25.0 g was heated to 88.6°C and then placed in a calorimeter that contained 61.2 g of water at 19.6°C. The temperature of the water rose to 21.3°C. What is the specific heat capacity of the alloy?

In practice, we first **calibrate** the calorimeter: we measure the temperature rise brought about by a reaction with a *known* heat output. Then we use that value to convert the measured temperature rise into a heat

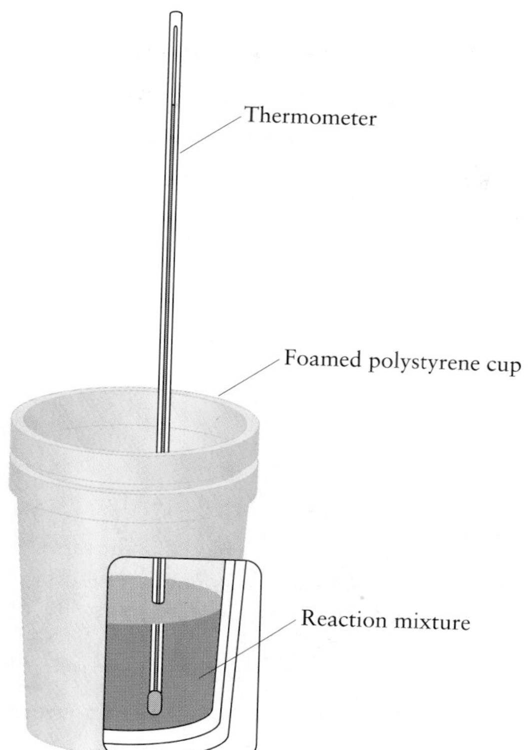

FIGURE 6.7 The quantity of heat released or absorbed by a reaction can be measured in this primitive version of a calorimeter. The outer polystyrene cup acts as an extra layer of insulation to ensure that no heat enters or leaves the inner cup. The quantity of heat released or absorbed is proportional to the change in temperature of the calorimeter.

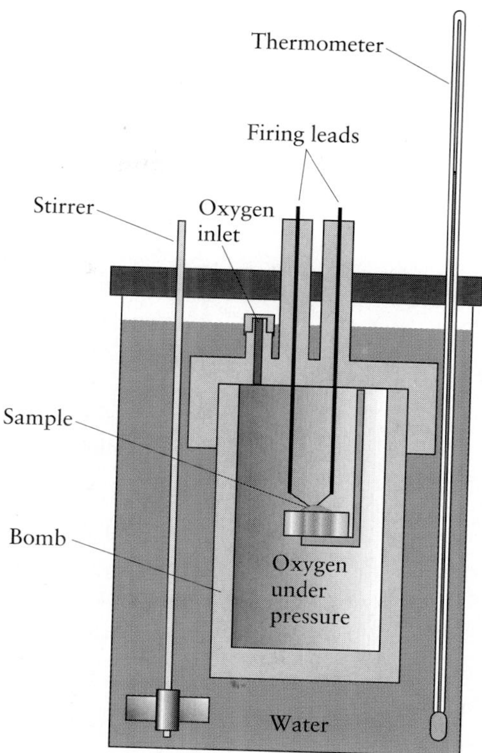

FIGURE 6.8 A bomb calorimeter. The combustion is started with an electrically ignited fuse. Once the reaction has begun, energy is released as heat that spreads through the walls of the bomb into the water. The heat released is proportional to the temperature change of the entire assembly.

output. The calorimeter should contain the same volume of solution in each case, as it is important to ensure that the heat capacity is the same in the two measurements.

Heat transfers are measured by using a calibrated calorimeter and are reported in joules or kilojoules. The heat capacity of an object is the ratio of the heat supplied to the temperature rise produced.

EXAMPLE 6.2 Determining the heat output of a reaction

A reaction known to release 1.78 kJ of heat takes place in a calorimeter containing 0.100 L of solution, like that shown in Fig. 6.7. The temperature rose by 3.65°C. Next, 50 mL of hydrochloric acid and 50 mL of aqueous sodium hydroxide were mixed in the same calorimeter and the temperature rose by 1.26°C. What is the heat output of the neutralization reaction?

STRATEGY First, we calibrate the calorimeter. To do so, we calculate the heat capacity from the information on the first reaction. For this step, we use the expression

$$\text{Heat capacity} = \frac{\text{heat supplied}}{\text{temperature rise}}$$

Then we use the value of that heat capacity to convert the temperature rise caused by the neutralization reaction into a heat output. For this step, we use the same equation rearranged to

$$\text{Heat supplied} = \text{heat capacity} \times \text{temperature rise}$$

Note that the calorimeter contains the same volume of liquid in both cases, so its heat capacity is the same.

SOLUTION The heat capacity of the calorimeter is

$$\text{Heat capacity} = \frac{1.78\text{ kJ}}{3.65°\text{C}} = 0.488\text{ kJ}\cdot(°\text{C})^{-1}$$

It follows that the heat supplied by the neutralization is

$$\text{Heat supplied} = 1.26°\text{C} \times 0.488\text{ kJ}\cdot(°\text{C})^{-1} = 0.614\text{ kJ}$$

SELF-TEST 6.2A A small piece of calcium carbonate was placed in the same calorimeter, and 0.100 L of dilute hydrochloric acid was poured over it. The temperature of the calorimeter rose by 3.57°C. What is the heat released by the reaction?

[*Answer:* 1.74 kJ]

SELF-TEST 6.2B A calorimeter was calibrated by mixing two 0.100 L aqueous solutions together. The reaction that occurred was known to release 4.16 kJ of heat, and the temperature of the calorimeter rose by 3.24°C. Calculate the heat capacity of this calorimeter when it contains 0.200 L of water.

6.4 THERMAL ACCOUNTING: ENTHALPY

The name *enthalpy* comes from the Greek words for "heat inside."

To keep track of energy changes during chemical reactions, we use a quantity called the **enthalpy,** *H: a change in the enthalpy of a system is equal to the heat released or absorbed at constant pressure.* We can picture the enthalpy as a measure of the energy of a system that can be released as heat: When we transfer heat to a system at constant pressure, its enthalpy goes up. When energy leaves the system as heat, its enthalpy goes down. Whenever a reaction takes place in a container open to the atmosphere, its pressure is constant and equal to the atmospheric pressure. Therefore, under these conditions, we can equate the heat transferred with the change in enthalpy of the system.

For example, the formation of zinc iodide from its elements is an exothermic reaction that releases 208 kJ of heat to the surroundings for each mole of ZnI_2 formed:

$$Zn(s) + I_2(s) \longrightarrow ZnI_2(s)$$

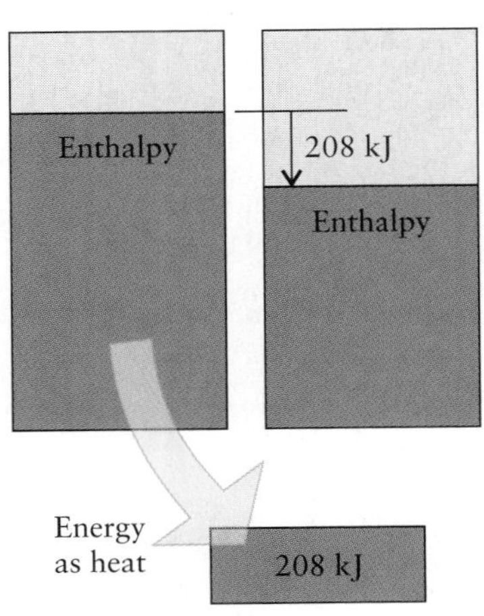

FIGURE 6.9 The enthalpy of a system is like a measure of the height of water in a reservoir. When a reaction releases 208 kJ of heat, the reservoir falls by 208 kJ.

The enthalpy of the reaction mixture decreases by 208 kJ in this reaction (Fig. 6.9).

An endothermic process absorbs heat, so when barium hydroxide octahydrate reacts with ammonium thiocyanate in the reaction described in Section 6.2, the enthalpy of the reaction mixture increases (Fig. 6.10).

Enthalpy is a **state property.** That means it has a value that depends on the current state of the system (its temperature, for instance) and is independent of how the system was prepared in that state. For example, 100 g of water at 25°C always has exactly the same enthalpy. It doesn't matter whether the water has previously been heated to 100°C and cooled, or frozen and then melted again. Pressure, volume, temperature, density, and heat capacity are other state properties. Their values depend only on the state of the system, not the changes it went through to get to that state.

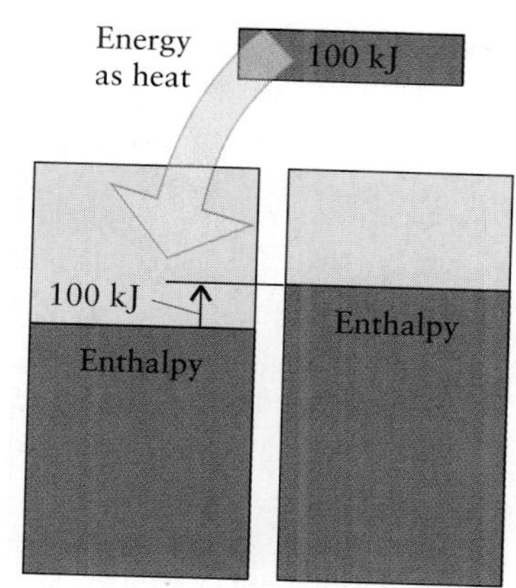

FIGURE 6.10 If an endothermic reaction absorbs 100 kJ of heat, the height of the enthalpy "reservoir" rises by 100 kJ.

A state property is like altitude. Several roads might lead to a mountain cabin. One is a direct ascent; another rises to an overlook before it descends to the cabin (Fig. 6.11). No matter which route we take, the altitude of the cabin is the same. So is the net change in altitude between our starting and final positions. For any state function, X, that has a value X_{initial} initially and X_{final} in the final state of the system, we can write

$$\Delta X = X_{\text{final}} - X_{\text{initial}}$$

where Δ is the Greek uppercase letter delta; this symbol is often used to indicate a change. When X is a state function, the value of ΔX depends only on the initial and final states, and is independent of how the system is taken between them.

The difference in enthalpy between two states of any system is

$$\text{Change in enthalpy:} \quad \Delta H = H_{\text{final}} - H_{\text{initial}}$$

When heat is *absorbed* by a system at constant pressure, the final state of the system has a higher enthalpy than its initial state, so the enthalpy change is positive (Fig. 6.12). Heat is absorbed in an endothermic reaction, so *all endothermic reactions have positive enthalpy changes:*

$$\text{For an endothermic change:} \quad \Delta H > 0$$

For example, if dissolving a certain amount of ammonium nitrate absorbs 2 kJ of heat, then we write $\Delta H = +2$ kJ. In this case, heat is

Always include the plus sign when reporting a positive change (do not write simply $\Delta H = 2$ kJ).

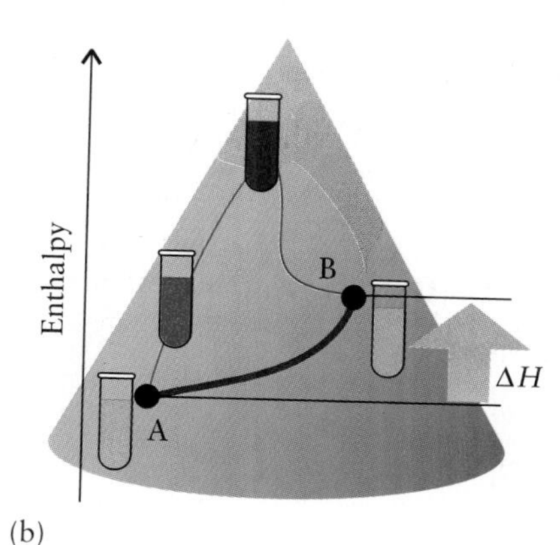

FIGURE 6.11 (a) The altitude of a location on a mountain is like a state property: it does not matter what route you take between two points, the net change in altitude is the same. (b) Enthalpy is a state property: if a system changes from state A to state B (as depicted highly diagrammatically here), the net change in enthalpy is the same whatever the route—the sequence of chemical or physical changes—between the two states.

Endothermic process, $\Delta H > 0$

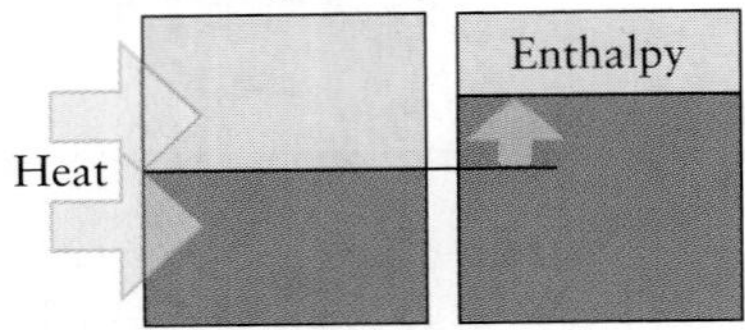

Exothermic process, $\Delta H < 0$

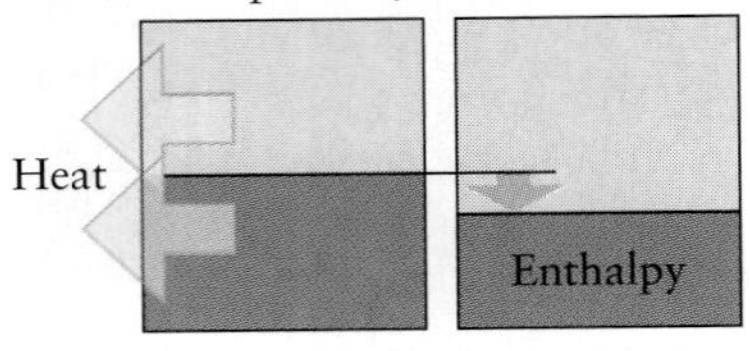

FIGURE 6.12 The enthalpy of a system increases ($\Delta H > 0$) if heat is absorbed (in an endothermic process); it decreases ($\Delta H < 0$) if heat is released (in an exothermic process).

drawn into the system as the solid dissolves, so the level of the energy reservoir in the system rises (hence the positive sign).

When heat is *released* in a process, the enthalpy change is negative because the level of the energy reservoir has fallen as a result of the escape of energy as heat. All exothermic reactions release heat, so *all exothermic reactions have negative enthalpy changes:*

$$\text{For an exothermic change:} \quad \Delta H < 0$$

For example, if the combustion of a certain mass of a fuel results in the release of 100 kJ into the surroundings, then we write $\Delta H = -100$ kJ.

The enthalpy of a system, a state property, is a measure of the energy of a system that is available as heat at constant pressure. For an endothermic process, $\Delta H > 0$; for an exothermic process, $\Delta H < 0$.

6.5 VAPORIZATION

Vaporization is an endothermic process, because energy has to be supplied to overcome the forces that hold molecules together as a liquid. The absorption of heat from the skin when perspiration evaporates is one of the body's strategies for keeping us cool. The cold we feel when we step out of a shower is another sign that vaporization is endothermic. Cosmetic "cold creams" feel cool because they contain water, which evaporates after the mixture is applied to our skin. Because vaporization is endothermic, water vapor at 100°C has a higher enthalpy—stores more energy—than the same mass of liquid water at 100°C. We experience this stored energy when we are scalded by steam: as the steam condenses, it releases energy as heat as it changes back to liquid water.

The difference in enthalpy *per mole of molecules* between the vapor and liquid states of a substance is called the **enthalpy of vaporization,** ΔH_{vap}. For water at 100°C,

$$\Delta H_{vap} = H_{vapor} - H_{liquid} = +40.7 \text{ kJ}\cdot\text{mol}^{-1}$$

This value means that to vaporize 1.00 mol $H_2O(l)$ (18.0 g of water) at

TABLE 6.2 Standard enthalpies of physical change*

Substance	Formula	Freezing point, K	ΔH°_{fus}, kJ·mol^{-1}	Boiling point, K	ΔH°_{vap}, kJ·mol^{-1}
acetone	CH_3COCH_3	177.8	5.72	329.4	29.1
ammonia	NH_3	195.4	5.65	239.7	23.4
argon	Ar	83.8	1.2	87.3	6.5
benzene	C_6H_6	278.6	10.59	353.2	30.8
ethanol	C_2H_5OH	158.7	4.60	351.5	43.5
helium	He	3.5	0.021	4.22	0.084
mercury	Hg	234.3	2.292	629.7	59.3
methane	CH_4	90.7	0.94	111.7	8.2
methanol	CH_3OH	175.2	3.16	337.2	35.3
water	H_2O	273.2	6.01	373.2	40.7 (44.0 at 25°C)

*Values correspond to the temperature of the phase change. The superscript ° signifies that the change takes place at 1 atm and that the substance is pure.

100°C, we have to supply 40.7 kJ of energy as heat. To vaporize twice that amount of water, we need to supply twice the energy. Values for some other substances are given in Table 6.2.

Vaporization is endothermic, as molecules need energy to escape from one another.

EXAMPLE 6.3 Measuring the enthalpy of vaporization

Suppose we used an electric heater to heat a sample of water to its boiling point in a constant pressure calorimeter (like the simple Styrofoam cup). We brought the water to its boiling point and then continued heating until 35.0 g of water had vaporized. We calculated from the power rating of the heater and the time taken that the vaporization alone required 79 kJ of heat. What is the enthalpy of vaporization of water at 100°C?

STRATEGY Because vaporization occurs at constant atmospheric pressure, the heat absorbed by the system (the water) is equal to the change in its enthalpy. We convert this enthalpy change to an enthalpy change per mole of molecules by dividing the observed enthalpy change by the number of moles of molecules vaporized.

SOLUTION We are told that the heat supplied is 79 kJ, so the change in enthalpy of 35.0 g of water is +79 kJ. The number of moles of H_2O (of molar mass 18.02 $g \cdot mol^{-1}$) in 35.0 g of water is

$$\text{Moles of } H_2O = \frac{35.0 \text{ g}}{18.02 \text{ g} \cdot \text{mol}^{-1}} = 1.94 \text{ mol}$$

The enthalpy change per mole of H_2O is therefore

$$\Delta H_{vap} = \frac{79 \text{ kJ}}{1.94 \text{ mol}} = +41 \text{ kJ} \cdot \text{mol}^{-1}$$

SELF-TEST 6.3A The same heater was used to heat a sample of benzene, C_6H_6, to 80°C, its boiling point. The heating was continued until 28 kJ had been supplied; as a result, 71 g of boiling benzene was vaporized. What is the enthalpy of vaporization of benzene at its boiling point?

[*Answer:* +31 $kJ \cdot mol^{-1}$]

SELF-TEST 6.3B The same heater was used to heat a sample of ethanol, C_2H_5OH, of mass 23 g to its boiling point. It was found that 22 kJ was required to vaporize all the ethanol. What is the enthalpy of vaporization of ethanol at its boiling point?

6.6 MELTING AND SUBLIMATION

A solid melts when heat is supplied at its melting point. The added energy causes the molecules to oscillate more and more vigorously until they can move past one another and flow as a liquid (Fig. 6.13). The enthalpy change that accompanies melting per mole of molecules is called the **enthalpy of fusion, ΔH_{fus},** of the substance:

$$\Delta H_{fus} = H_{liquid} - H_{solid}$$

Melting is endothermic, so all enthalpies of fusion are positive. The enthalpy of fusion of water at 0°C is +6.0 $kJ \cdot mol^{-1}$: to melt 1.0 mol

FIGURE 6.13 Melting (fusion) is an endothermic process. As molecules acquire energy, they begin to struggle past their neighbors. Finally the sample changes from a solid with ordered molecules (top) to a liquid with disordered molecules (bottom).

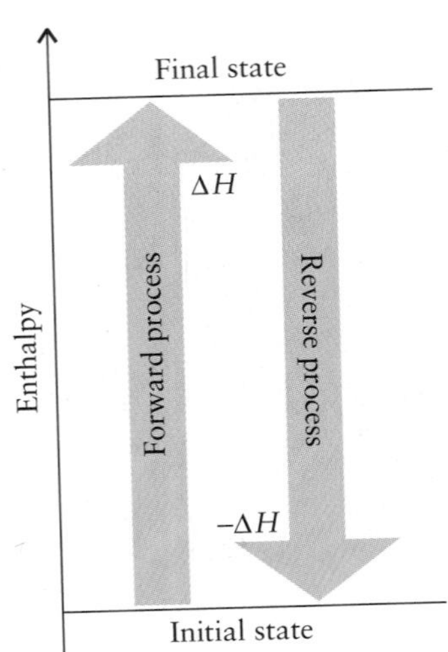

FIGURE 6.14 The enthalpy change for a reverse process is the negative of the enthalpy change for the forward process at the same temperature.

H_2O(s) (18 g of ice) at 0°C, we have to supply 6.0 kJ of heat. Vaporizing the same amount of water takes much more energy (over 40 kJ, in fact), because when water is vaporized to a gas, its molecules must be separated completely. The enthalpies of fusion of various other substances are included in Table 6.2.

The **enthalpy of freezing** is the change in enthalpy per mole when a liquid turns into a solid. For water at 0°C, the enthalpy of freezing is -6.0 kJ·mol^{-1}, because 6.0 kJ of heat is released when 1 mol H_2O(s) freezes. This is the heat we remove when making ice in a freezer.

The enthalpy of freezing of a substance is the negative of its enthalpy of fusion. This relation follows from the fact that enthalpy is a state property. The enthalpy of water must be the same after being frozen and then melted as it was before it was frozen. Therefore, the same quantity of heat is released in freezing as is absorbed in melting. To obtain the enthalpy change for the reverse of *any* process, we just take the negative of the enthalpy change for the original process:

$$\Delta H(\text{reverse process}) = -\Delta H(\text{forward process})$$

This relation is illustrated in Fig. 6.14. If we find, for example, that the enthalpy of vaporization of mercury is $+59$ kJ·mol^{-1} at its boiling point, then we immediately know that the enthalpy change occurring when mercury vapor condenses at that temperature is -59 kJ·mol^{-1}. This value tells us that 59 kJ of heat is released when 1 mol Hg(g) condenses to a liquid.

Enthalpies of phase changes can be put to practical use in solar-powered homes (Fig. 6.15). Hot air from solar collectors on the roof is circulated through a container of a solid salt with a melting point just above room temperature, such as calcium chloride hexahydrate, $CaCl_2 \cdot 6H_2O$.

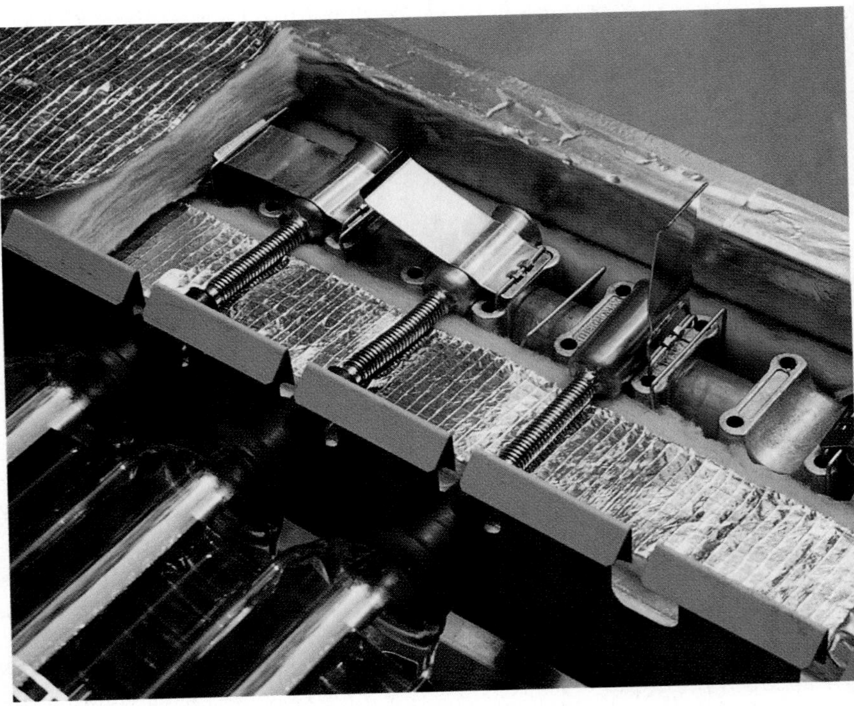

FIGURE 6.15 The heat exchangers in a phase-change heating installation. In this case, a liquid flowing through tubes exposed to sunlight is vaporized. The vapor condenses inside the house and releases the stored energy as heat. The glass tubes in the foreground contain the liquid; its vapor condenses in the heat exchangers above.

The hot air melts some of the salt (its melting point is only 29°C), forming a liquid aqueous solution of calcium chloride: in effect, the salt dissolves in its own water of hydration. At night, the salt solidifies again, releasing energy to warm the air of the house.

Sublimation is the direct conversion of a solid into its vapor. Frost disappears on a cold, dry morning as the ice sublimes directly into water vapor. Solid carbon dioxide also sublimes, which is why it is called dry ice. Each winter on Mars, solid carbon dioxide is deposited as polar frost, which sublimes when the feeble summer arrives (Fig. 6.16). The **enthalpy of sublimation,** ΔH_{sub}, is the enthalpy change per mole when a solid sublimes:

$$\Delta H_{sub} = H_{vapor} - H_{solid}$$

We can use the fact that enthalpy is a state property to calculate the change in enthalpy for the direct conversion of a solid to vapor. We simply add the enthalpy change for melting and vaporization at the same temperature (Fig. 6.17):

$$\Delta H_{sub} = \Delta H_{fus} + \Delta H_{vap}$$

For example, at 25°C, the enthalpy of fusion of sodium metal is $+2.6\ \text{kJ}\cdot\text{mol}^{-1}$ and the enthalpy of vaporization of liquid sodium is $+98\ \text{kJ}\cdot\text{mol}^{-1}$. Therefore, the enthalpy of sublimation of solid sodium at 25°C is

$$\begin{aligned}\Delta H_{sub} &= \Delta H_{fus} + \Delta H_{vap}\\ &= +2.6\ \text{kJ}\cdot\text{mol}^{-1} + 98\ \text{kJ}\cdot\text{mol}^{-1} = +101\ \text{kJ}\cdot\text{mol}^{-1}\end{aligned}$$

We can add enthalpy changes like this only if they correspond to the same temperature.

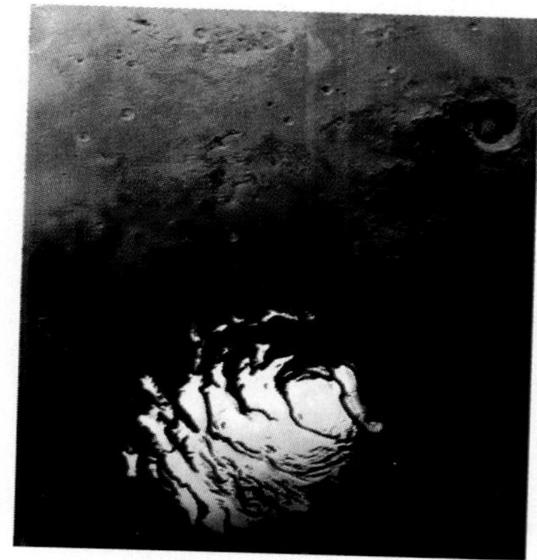

FIGURE 6.16 The polar ice caps on Mars extend and recede with the seasons. They are solid carbon dioxide and form by direct conversion of the gas to a solid. They disappear by sublimation. Although some water ice is also present in the polar caps, the temperature on Mars never becomes high enough to melt or sublime it. On Mars, ice is just another rock.

Enthalpies of phase transitions are reported in kilojoules per mole of molecules. The enthalpy change for a reverse reaction is the negative of the enthalpy change for the forward reaction. Enthalpy changes can be added to obtain the value for an overall process.

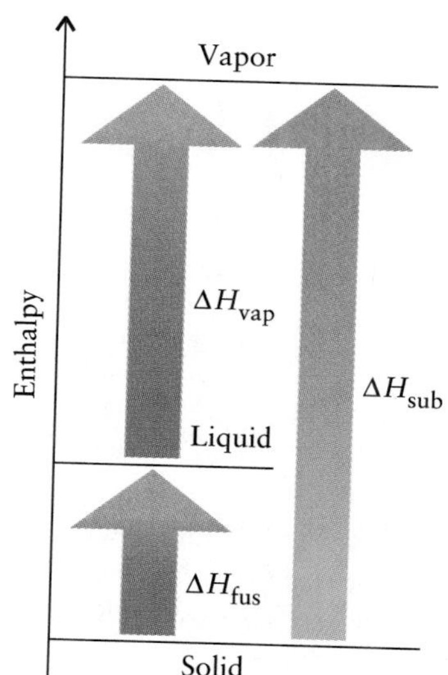

FIGURE 6.17 Because enthalpy is a state property, the enthalpy of sublimation can be expressed as the sum of the enthalpies of fusion and vaporization.

6.7 HEATING CURVES

The enthalpies of fusion and vaporization affect the appearance of the **heating curve** of a substance, the graph showing the variation of the temperature of a sample as it is heated at a constant rate.

Consider what happens when we heat a sample of very cold ice. As we see in Fig. 6.18, its temperature rises steadily. Although the molecules are still locked in a solid mass, they are oscillating faster and faster. At the melting point, the molecules have enough energy to move past one another. At this temperature, all the added energy is used to overcome the attractive forces between molecules. Thus, the temperature remains constant at the melting point while the heating continues, until all the ice has melted. Only then does the temperature rise again, and the rise continues right up to the boiling point. At the boiling point, the temperature rise again comes to a halt. Now the water molecules have enough energy to escape into the vapor state, and all the heat supplied is used to form the

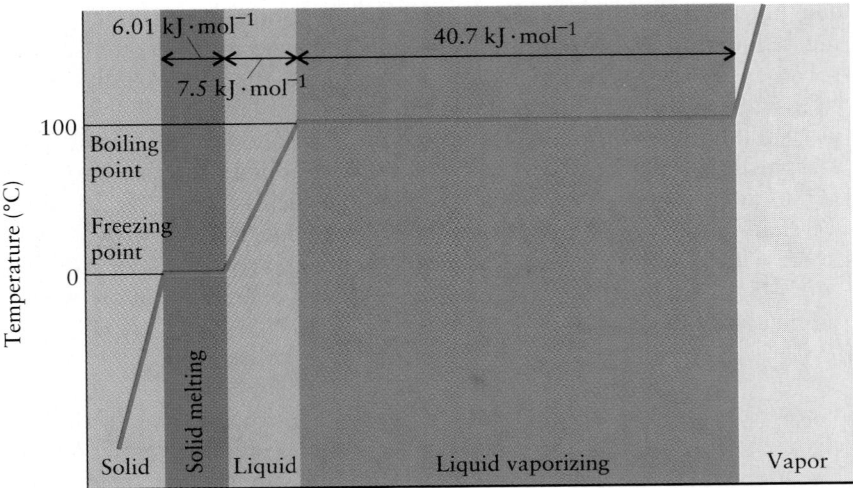

FIGURE 6.18 The temperature of a solid rises as heat is supplied. At the melting point, the temperature remains constant and the heat is used to melt the sample. When enough heat has been supplied to melt all the solid, the temperature begins to rise again. A similar pause occurs at the boiling point. The data are for water.

vapor. After the sample has evaporated and heating continues, the temperature of the vapor rises again.

> ***The temperature of a sample is constant at its melting and boiling points, even though heat is being supplied.***

THE ENTHALPY OF CHEMICAL CHANGE

We can apply the same principles used to study the energy changes of physical processes to the study of chemical reactions.

6.8 REACTION ENTHALPIES

This arrangement models a person, where the body maintains a constant temperature when fuel—food—is digested.

For simplicity, we shall first consider a combustion reaction that takes place in a vessel that is immersed in a constant-temperature water bath. The system is kept at constant temperature so that any change in enthalpy is due to the chemical reaction, not to heating or cooling. Consider the combustion of methane:

$$CH_4(g) + 2\,O_2(g) \longrightarrow CO_2(g) + 2\,H_2O(l)$$

Calorimetry shows that burning 1 mol $CH_4(g)$ produces 890 kJ of heat at 298 K and 1 atm. That is, when 1 mol of methane and 2 mol of oxygen at 298 K react completely and the products have given off the excess heat and cooled back down to 298 K, the enthalpy of the system has

CHEMISTRY AT WORK

An agricultural researcher studying photosynthesis assesses the growth rate of a seedling. He will use thermochemical methods to try to find conditions that increase the efficiency of photosynthesis. In most plants, photosynthesis is only about 3% efficient.

decreased by 890 kJ for each mole of CH_4 consumed. To report this value, we write

$$CH_4(g) + 2\,O_2(g) \longrightarrow CO_2(g) + 2\,H_2O(l) \qquad \Delta H = -890 \text{ kJ}$$

This entire expression is a **thermochemical equation,** a chemical equation together with the enthalpy change. The stoichiometric coefficients in a thermochemical equation are interpreted as the number of moles that react to give the reported change in enthalpy. In this case, the stoichiometric coefficient of CH_4 is 1 and that of O_2 is 2, so the enthalpy change is that resulting from the complete reaction of 1 mol CH_4 and 2 mol O_2.

The **reaction enthalpy,** the enthalpy change in a reaction, refers to the reaction exactly as written. For example, if the same reaction is written with the coefficients all multiplied by 2, then the reaction enthalpy will be twice as great:

$$2\,CH_4(g) + 4\,O_2(g) \longrightarrow 2\,CO_2(g) + 4\,H_2O(l) \qquad \Delta H = -1780 \text{ kJ}$$

This makes sense, because the equation now represents the burning of twice as much methane.

Just as we saw in Section 6.6 for physical changes, the enthalpy change for the reverse of a chemical reaction is the negative of the enthalpy change of the forward reaction:

$$CO_2(g) + 2\,H_2O(l) \longrightarrow CH_4(g) + 2\,O_2(g) \qquad \Delta H = +890 \text{ kJ}$$

This also makes sense. If methane and oxygen give off heat when they burn, they must contain more enthalpy than the products. To go back from products to reactants, we would have to restore that additional enthalpy.

A thermochemical equation combines a chemical equation and the reaction enthalpy. The reaction enthalpy is the change in enthalpy for the stoichiometric numbers of moles of reactants in the chemical equation.

EXAMPLE 6.4 Calculating the reaction enthalpy

When 0.125 g of cyclohexane, C_6H_{12}, burns in excess oxygen in a calibrated constant-pressure calorimeter with a heat capacity of 551 $J\cdot(°C)^{-1}$, the temperature of the calorimeter rises by 10.6°C. Calculate the reaction enthalpy for

$$C_6H_{12}(l) + 9\,O_2(g) \longrightarrow 6\,CO_2(g) + 6\,H_2O(l)$$

and write the corresponding thermochemical equation.

STRATEGY First, we note whether the temperature rises or falls. If it rises, heat is given off, the reaction is exothermic, and ΔH is negative. If the temperature falls, heat is absorbed, the reaction is endothermic, and ΔH is positive. The enthalpy change is calculated from the temperature change times the heat capacity. Once we know the enthalpy change for the masses of reactants used, we convert to the reaction enthalpy for moles of reactants equal to the stoichiometric coefficients. We convert the given mass of reactant to moles by using the molar mass. and express the heat output as an enthalpy change per mole of reactant molecules.

SOLUTION The temperature rises, so we know that the reaction is exothermic. The calorimeter has been calibrated and is known to have a heat capacity of 551 $J\cdot(°C)^{-1}$, so

$$\begin{aligned}\text{Heat supplied} &= (10.6°C) \times (551\ J\cdot(°C)^{-1}) \\ &= 5.84 \times 10^3\ J\end{aligned}$$

Because the molar mass of cyclohexane is 84.16 $g\cdot mol^{-1}$, the number of moles of C_6H_{12} that reacts is

$$\text{Moles of } C_6H_{12} = \frac{0.125\ g}{84.16\ g\cdot mol^{-1}} = 1.49 \times 10^{-3}\ mol$$

The enthalpy change per mole of C_6H_{12} that reacts is therefore

$$\begin{aligned}\Delta H &= -\frac{5.84 \times 10^3\ J}{1.49 \times 10^{-3}\ mol} \\ &= -3.92 \times 10^6\ J\cdot mol^{-1} = -3.92 \times 10^3\ kJ\cdot mol^{-1}\end{aligned}$$

The negative sign has been included because the reaction is exothermic. Because the equation represents the enthalpy change for the reaction of 1 mol C_6H_{12}, the expression $\Delta H = -3.92 \times 10^3$ kJ becomes part of the thermochemical equation:

$$C_6H_{12}(l) + 9\,O_2(g) \longrightarrow 6\,CO_2(g) + 6\,H_2O(l) \qquad \Delta H = -3.92 \times 10^3\ kJ$$

SELF-TEST 6.4A When 0.231 g of phosphorus reacts with chlorine to form phosphorus trichloride, PCl_3, in a calorimeter of heat capacity 216 $J\cdot(°C)^{-1}$, the temperature of the calorimeter rises by 11.06°C. Write the thermochemical equation for the reaction.

[*Answer:* $2\,P(s) + 3\,Cl_2(g) \rightarrow 2\,PCl_3(l)$, $\Delta H = -641$ kJ]

SELF-TEST 6.4B When a solution of 3.382 g of silver nitrate is mixed with an excess of sodium chloride to form silver chloride in the same calorimeter used in Self-Test 6.4A, the temperature rises 6.06°C. Write the thermochemical equation for the reaction.

6.9 STANDARD REACTION ENTHALPIES

The heat released or absorbed by a reaction depends on the physical states of the reactants and products. Let's consider once again the combustion of methane. Two different thermochemical equations for slightly different reactions are

$$CH_4(g) + 2\,O_2(g) \longrightarrow CO_2(g) + 2\,H_2O(g) \qquad \Delta H = -802\ \text{kJ}$$

$$CH_4(g) + 2\,O_2(g) \longrightarrow CO_2(g) + 2\,H_2O(l) \qquad \Delta H = -890\ \text{kJ}$$

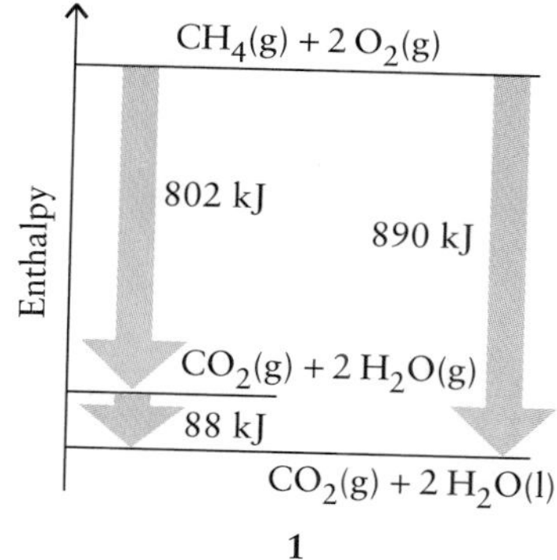

1

In the first reaction, all the water is produced as a vapor; in the second, it is produced as a liquid. The heat produced is different in each case. The enthalpy of water vapor is 44 kJ·mol^{-1} higher than that of liquid water at 25°C (see Table 6.2). As a result, an additional 88 kJ (for 2 mol H_2O) remains stored in the system if water vapor is formed rather than the liquid (**1**). If the 2 mol $H_2O(g)$ condense, then 88 kJ is given off as heat.

Enthalpy changes also depend on pressure. All the tables in this book give data for reactions in which the reactants and products are in their **standard state,** that is, their pure form at 1 atm pressure. The standard state of liquid water is pure water at 1 atm. The standard state of ice is pure ice at 1 atm. A solute in a liquid solution is in its standard state when its concentration is 1 mol·L^{-1}.

The most recent definition of standard state replaces 1 atm by 1 bar, where 1 bar = 10^5 Pa. The values of thermochemical quantities are changed only very slightly by this new definition.

A reaction enthalpy based on standard states is called a **standard reaction enthalpy,** $\Delta H°$, the reaction enthalpy when reactants in their standard states change to products in their standard states. For example, we can write

A superscript circle is commonly used to denote a standard value.

$$CH_4(g) + 2\,O_2(g) \longrightarrow CO_2(g) + 2\,H_2O(l) \qquad \Delta H° = -890\ \text{kJ}$$

Here pure methane gas at 1 atm is allowed to react with pure oxygen gas at 1 atm, giving pure carbon dioxide gas and pure liquid water, both at 1 atm (Fig. 6.19). Reaction enthalpies do not change very much with pressure, so the standard value gives a good indication of the change in enthalpy for pressures near 1 atm.

Most thermochemical data are reported for 25°C (more precisely, for 298.15 K). Temperature is not part of the definition of standard states: we can have a standard state at any temperature; 298.15 K is simply the most common temperature used in tables of data. All reaction enthalpies used in this text are for 298.15 K unless another temperature is indicated.

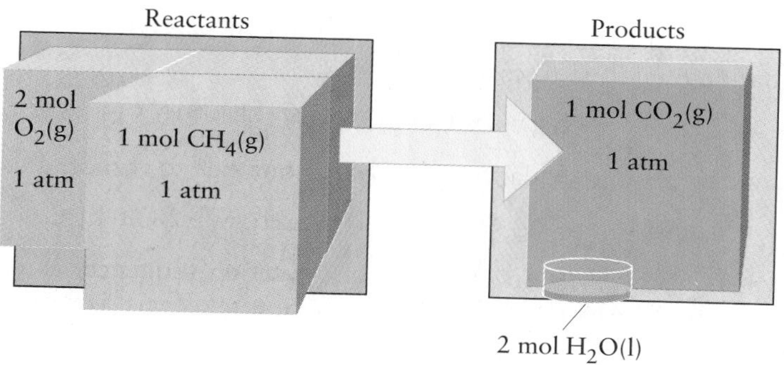

FIGURE 6.19 The standard reaction enthalpy is the difference in enthalpy between the pure products, each at 1 atm, and the pure reactants at the same pressure and the specified temperature (which is commonly but not necessarily 25°C). The scheme here is for the combustion of methane.

Standard reaction enthalpies refer to reactions in which the reactants and products are in their standard states; they are usually reported for a temperature of 298.15 K.

6.10 COMBINING REACTION ENTHALPIES: HESS'S LAW

We know from comparing enthalpies of fusion and freezing (Section 6.6) that the enthalpy change of any reverse process is the negative of the change for the forward process. The standard reaction enthalpy of the reverse reaction can therefore be obtained from that of the forward reaction by reversing the sign:

$$P_4(s) + 6\,Cl_2(g) \longrightarrow 4\,PCl_3(l) \qquad \Delta H° = -1279\text{ kJ}$$

$$4\,PCl_3(l) \longrightarrow P_4(s) + 6\,Cl_2(g) \qquad \Delta H° = +1279\text{ kJ}$$

it follows that a reaction that is exothermic in one direction is endothermic in the reverse direction. The combustion of glucose is exothermic:

$$C_6H_{12}O_6(aq) + 6\,O_2(g) \longrightarrow 6\,CO_2(g) + 6\,H_2O(l) \qquad \Delta H° = -2808\text{ kJ}$$

We know then that the reverse of this reaction, the formation of glucose from carbon dioxide and water, must be endothermic:

$$6\,CO_2(g) + 6\,H_2O(l) \longrightarrow C_6H_{12}O_6(aq) + 6\,O_2(g) \qquad \Delta H° = +2808\text{ kJ}$$

This is the process that takes place in photosynthesis, and its energy is supplied by the Sun (Box 6.1). It would be very difficult to measure that energy directly in living plants, but it is easy to measure the enthalpy change for the combustion of glucose and then to reverse the sign. The reaction enthalpies show how glucose acts as a store of energy: forming glucose by photosynthesis is like filling a reservoir with energy, and breaking down glucose releases the energy it has stored.

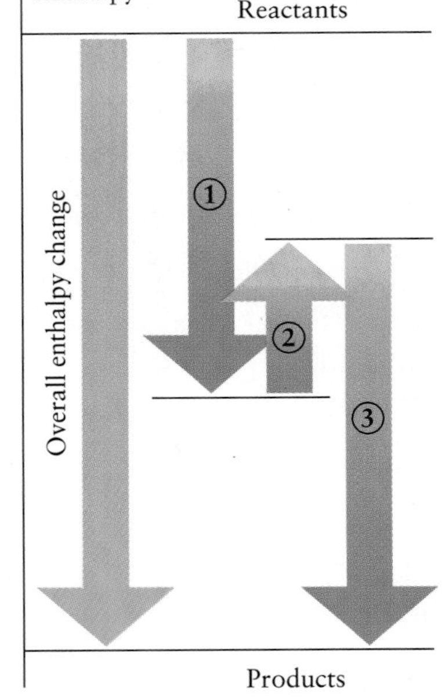

FIGURE 6.20 If the overall reaction can be broken down into a series of three steps, then the corresponding overall reaction enthalpy is the sum of the reaction enthalpies of those steps. None of the steps need be a reaction that can actually be carried out in the laboratory.

We saw in Section 6.6 that the enthalpy of an overall physical process can be expressed as the sum of the enthalpies for individual steps. The same rule applies to chemical reactions and, in this context, is known as **Hess's law:** *the overall reaction enthalpy is the sum of the reaction enthalpies of the steps into which the reaction can be divided, even if the division is only theoretical.* The intermediate reactions need not be ones that can actually be carried out. So long as they balance and add up to the equation for the reaction of interest, a reaction enthalpy can be calculated from any convenient sequence of reactions (Fig. 6.20).

Take, as an example, the oxidation of carbon to carbon dioxide:

$$C(s) + O_2(g) \longrightarrow CO_2(g)$$

This reaction can be thought of as the outcome of two steps. One step is the oxidation of carbon to carbon monoxide:

$$C(s) + \tfrac{1}{2}\,O_2(g) \longrightarrow CO(g) \qquad \Delta H° = -110.5\text{ kJ}$$

The second step is the oxidation of carbon monoxide to carbon dioxide:

$$CO(g) + \tfrac{1}{2}\,O_2(g) \longrightarrow CO_2(g) \qquad \Delta H° = -283.0\text{ kJ}$$

This two-step process is an example of a **reaction sequence,** a series of reactions in which the products of one reaction take part as reactants in

BOX 6.1 THE WORLD'S ENTHALPY RESOURCES

Because fossil fuel reserves are limited, they must be extracted wherever they are found. This platform is used to pump petroleum from beneath the ocean; however, the natural gas accompanying it cannot be easily transported and so is burned off.

The energy of all the solar radiation absorbed by the vegetation on Earth is enough to build about 6×10^{14} kg of glucose a year. Most of this glucose is converted to starch and cellulose. As long as its carbohydrates do not decay completely back to carbon dioxide and water, dead vegetation remains a store of energy. If losses through forest fires are ignored, this store, called the biomass, increases by about 10^{19} kJ each year, which is about 30 times the current annual global industrial demand for energy. Therefore, the quantity of energy seems more than adequate to meet our needs. However, the biomass is too widely dispersed to harvest the energy economically. We need high concentrations of energy-rich substances such as the deposits of fossil fuels. Favorable geological formations have kept these deposits out of contact with the atmosphere, so they have not been oxidized back to carbon dioxide and water. Coal, for example, originated as matter that collected at the bottom of swamps, marshes, and bogs.

Of the 10^{19} kJ of energy stored annually by photosynthesis, less than 0.01% survives without being oxidized. Moreover, only a tiny fraction of that (estimated as 0.07%) forms deposits large enough to be economical sources of fuels. The rate at which energy is stored in a useful form is therefore only about 10^{11} kJ per year. We are using it at the rate of 3×10^{17} kJ per year, which is 30 million times more rapidly than it is being stored.

Conservation of energy can help ease the dilemma, as can alternative energy sources such as nuclear power (see Chapter 22) and renewable energy resources. Renewable energy resources are those produced and replenished by the sun; for example, biomass, hydroelectric plants, and wind turbines. Solar radiation itself can be harnessed to generate electricity by using *photovoltaic cells*, in which silicon crystals convert solar radiation into an electrical current. Because an energy resource is practical only if it generates more energy than required to produce it, much energy research goes into increasing the efficiency of both producing fuels and burning them.

We can reduce our use of fossil fuels if we can find ways to use solar radiation directly. However, photovoltaic solar panels, such as the ones that power this remote telephone booth in Australia, still require the use of fossil fuels for their manufacture.

another reaction. The equation for the **overall reaction,** the net outcome of the sequence, is the sum of the equations for the intermediate steps:

$$C(s) + \tfrac{1}{2}O_2(g) \longrightarrow \cancel{CO(g)} \qquad \Delta H^\circ = -110.5\ \text{kJ} \qquad \text{(a)}$$

$$\cancel{CO(g)} + \tfrac{1}{2}O_2(g) \longrightarrow CO_2(g) \qquad \Delta H^\circ = -283.0\ \text{kJ} \qquad \text{(b)}$$

$$C(s) + O_2(g) \longrightarrow CO_2(g) \qquad \Delta H^\circ = -393.5\ \text{kJ} \qquad \text{(a + b)}$$

The same procedure is used to predict the enthalpies of reactions that we cannot measure directly in the laboratory. The procedure is described in Toolbox 6.1.

Thermochemical equations for the individual steps of a reaction sequence may be combined to give the thermochemical equation for the overall reaction.

TOOLBOX 6.1 How to devise a reaction sequence to obtain an overall reaction enthalpy

The aim is to find a sequence of reactions, the sum of which is the reaction of interest. In many cases, we can see at a glance what reactions to use. When the answer is not obvious, the following procedure is helpful:

Step 1. Select one of the reactants in the overall reaction and write down a chemical equation in which it also appears as a reactant.

Step 2. Select one of the products in the overall reaction and write down a chemical equation in which it also appears as a product. Add this equation to the equation written in step 1.

Step 3. Cancel unwanted species in the sum obtained in step 2 by adding an equation that has the same substance or substances on the opposite side of the arrow.

Step 4. Once the sequence is complete, combine the standard reaction enthalpies.

In each step we may need to reverse the equation or multiply it by a factor. Recall from Section 6.8 that if we want to reverse a chemical equation, then we have to change the sign of the reaction enthalpy. If we multiply the stoichiometric coefficients by a factor, then we must multiply the reaction enthalpy by the same factor.

Because enthalpies of combustion are readily available, combustion reactions are often useful for obtaining the enthalpies of reactions involving organic compounds. As an illustration, consider the synthesis of propane, C_3H_8, a gas used as camping fuel:

$$3\,C(s) + 4\,H_2(g) \longrightarrow C_3H_8(g)$$

We have the following experimental information:

$$C_3H_8(g) + 5\,O_2(g) \longrightarrow 3\,CO_2(g) + 4\,H_2O(l) \qquad \Delta H^\circ = -2220\ \text{kJ} \qquad \text{(a)}$$

$$C(s) + O_2(g) \longrightarrow CO_2(g) \qquad \Delta H^\circ = -394\ \text{kJ} \qquad \text{(b)}$$

$$H_2(g) + \tfrac{1}{2}O_2(g) \longrightarrow H_2O(l) \qquad \Delta H^\circ = -286\ \text{kJ} \qquad \text{(c)}$$

Step 1. Only (b) and (c) have at least one of the reactants. In both cases, they are on the correct side of the arrow in the overall chemical equation. We select (b) and multiply it through by 3 to give carbon the coefficient it will have in the final equation:

$$3\,C(s) + 3\,O_2(g) \longrightarrow 3\,CO_2(g) \qquad \Delta H^\circ = 3 \times (-394\ \text{kJ}) = -1182\ \text{kJ}$$

Step 2. To obtain C_3H_8 on the right, we reverse equation (a), changing the sign of its reaction enthalpy, and add it to the equation we have just derived:

$$3\,C(s) + 3\,O_2(g) \longrightarrow 3\,CO_2(g) \qquad \Delta H^\circ = -1182\ \text{kJ}$$

$$3\,CO_2(g) + 4\,H_2O(l) \longrightarrow C_3H_8(g) + 5\,O_2(g) \qquad \Delta H^\circ = +2220\ \text{kJ}$$

The sum of these two equations is

$$3\,C(s) + 3\,O_2(g) + 3\,CO_2(g) + 4\,H_2O(l) \longrightarrow C_3H_8(g) + 5\,O_2(g) + 3\,CO_2(g) \qquad \Delta H^\circ = +1038\ \text{kJ}$$

This equation simplifies to

$$3\,C(s) + 4\,H_2O(l) \longrightarrow C_3H_8(g) + 2\,O_2(g) \qquad \Delta H^\circ = +1038\ \text{kJ}$$

Step 3. To cancel the unwanted reactant H_2O and product O_2, we add equation (c) after multiplying it by 4:

$$3\,C(s) + 4\,H_2O(l) \longrightarrow C_3H_8(g) + 2\,O_2(g) \qquad \Delta H^\circ = +1038\ \text{kJ}$$

$$4\,H_2(g) + 2\,O_2(g) \longrightarrow 4\,H_2O(l) \qquad \Delta H^\circ = 4 \times (-286\ \text{kJ}) = -1144\ \text{kJ}$$

The sum of these two reactions is

$$3\,C(s) + 4\,H_2(g) + \cancel{4\,H_2O(l)} + \cancel{2\,O_2(g)} \longrightarrow$$
$$C_3H_8(g) + \cancel{2\,O_2(g)} + \cancel{4\,H_2O(l)} \qquad \Delta H^\circ = -106\text{ kJ}$$

which simplifies to

$$3\,C(s) + 4\,H_2(g) \longrightarrow C_3H_8(g) \qquad \Delta H^\circ = -106\text{ kJ}$$

giving us the quantity we require.

SELF-TEST 6.5A Gasoline, which contains octane as one component, may burn to carbon monoxide if the air supply is restricted. Derive the standard reaction enthalpy for the incomplete combustion of octane:

$$2\,C_8H_{18}(l) + 17\,O_2(g) \longrightarrow 16\,CO(g) + 18\,H_2O(l)$$

from the standard reaction enthalpies for the combustions of octane and carbon monoxide:

$$2\,C_8H_{18}(l) + 25\,O_2(g) \longrightarrow 16\,CO_2(g) + 18\,H_2O(l)$$
$$\Delta H^\circ = -10{,}942\text{ kJ}$$

$$2\,CO(g) + O_2(g) \longrightarrow 2\,CO_2(g) \qquad \Delta H^\circ = -566.0\text{ kJ}$$

[*Answer:* −6414 kJ]

SELF-TEST 6.5B Methanol is a clean-burning liquid fuel being considered as a replacement for gasoline. It can be made from the methane in natural gas:

$$2\,CH_4(g) + O_2(g) \longrightarrow 2\,CH_3OH(l)$$

Find the standard reaction enthalpy for the formation of methanol from methane, given the following three equations:

$$CH_4(g) + H_2O(g) \longrightarrow CO(g) + 3\,H_2(g)$$
$$\Delta H^\circ = +206.10\text{ kJ}$$

$$2\,H_2(g) + CO(g) \longrightarrow CH_3OH(l) \qquad \Delta H^\circ = -128.33\text{ kJ}$$

$$2\,H_2(g) + O_2(g) \longrightarrow 2\,H_2O(g) \qquad \Delta H^\circ = -483.64\text{ kJ}$$

THE HEAT OUTPUT OF REACTIONS

Once we know the reaction enthalpy, we can calculate the enthalpy change for any amount, mass, or volume of reactant consumed or product formed. To do this kind of calculation, we carry out a stoichiometry calculation like those described in Chapter 4. We proceed as if heat were a reactant or product (**2**).

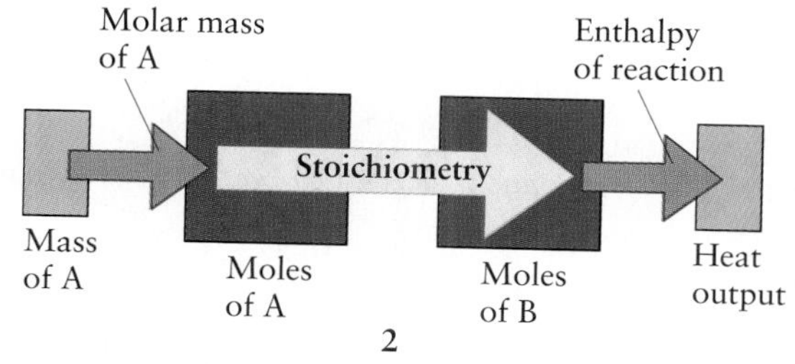

2

6.11 ENTHALPIES OF COMBUSTION

The heat released in an exothermic reaction is an important measure of the value of a fuel. For example, suppose we want to know the heat output from the combustion of 150 g of methane. The thermochemical equation

$$CH_4(g) + 2\,O_2(g) \longrightarrow CO_2(g) + 2\,H_2O(l) \qquad \Delta H = -890\text{ kJ}$$

allows us to write the following relation

$$1\text{ mol } CH_4 \simeq 890\text{ kJ}$$

and to use it as a conversion factor in the usual way. We already know how to use the molar mass of methane ($16.04\ g\cdot mol^{-1}$) to work out the number of moles of CH_4 corresponding to 150 g. From these relations, we can write

$$\begin{aligned}\text{Heat output} &= 150\ \text{g} \times \frac{1\ \text{mol CH}_4}{16.04\ \text{g}} \times \frac{890\ \text{kJ}}{1\ \text{mol CH}_4} \\ &= 8.32 \times 10^3\ \text{kJ}\end{aligned}$$

The second factor converts from grams to moles, and the third factor converts from moles to kilojoules.

The heat absorbed or given off by a reaction can be treated like a reactant or product in a stoichiometric relation.

EXAMPLE 6.5 Calculating the heat output of a fuel

How much propane should a backpacker carry: do we *really* need to carry a kilogram of gas? Calculate the mass of propane that you would need to burn to obtain 350 kJ of heat, which is just enough energy to heat 1 L of water from room temperature (20°C) to boiling at sea level (if we ignore all heat losses). The thermochemical equation is

$$\text{C}_3\text{H}_8(\text{g}) + 5\ \text{O}_2(\text{g}) \longrightarrow 3\ \text{CO}_2(\text{g}) + 4\ \text{H}_2\text{O}(\text{l}) \qquad \Delta H° = -2220\ \text{kJ}$$

STRATEGY Because we are given the heat output, the first step is to convert heat output required to moles of fuel molecules by using the thermochemical equation. Then we convert from moles of fuel molecules to grams by using the molar mass of the fuel.

SOLUTION The thermochemical equation tells us that

$$2220\ \text{kJ} \simeq 1\ \text{mol C}_3\text{H}_8$$

The molar mass of propane is $44.09\ g\cdot mol^{-1}$. It follows that

$$\begin{aligned}\text{Mass of C}_3\text{H}_8\text{ required} &= 350\ \text{kJ} \times \frac{1\ \text{mol C}_3\text{H}_8}{2220\ \text{kJ}} \times (44.09\ \text{g}\cdot\text{mol}^{-1}) \\ &= 6.95\ \text{g C}_3\text{H}_8\end{aligned}$$

That is, just under 7 g of propane is needed to boil the water (and more if we allow for heat losses).

SELF-TEST 6.6A The thermochemical equation for the combustion of butane is

$$2\ \text{C}_4\text{H}_{10}(\text{g}) + 13\ \text{O}_2(\text{g}) \longrightarrow 8\ \text{CO}_2(\text{g}) + 10\ \text{H}_2\text{O}(\text{l}) \qquad \Delta H° = -5756\ \text{kJ}$$

What mass of butane must be burned to supply 350 kJ of heat? Would it be easier to pack butane rather than propane?

[*Answer:* 7.07 g. Propane would be very slightly lighter to carry.]

SELF-TEST 6.6B Ethanol trapped in a gel is another common camping fuel. What mass of ethanol must be burned to supply 350 kJ of heat? The thermochemical equation for the combustion of ethanol is

$$\text{C}_2\text{H}_5\text{OH}(\text{l}) + 3\ \text{O}_2(\text{g}) \longrightarrow 2\ \text{CO}_2(\text{g}) + 3\ \text{H}_2\text{O}(\text{l}) \qquad \Delta H° = -1368\ \text{kJ}$$

Because combustion reactions are so important, their standard enthalpies are given a special symbol, $\Delta H_c°$. This symbol stands for the **standard enthalpy of combustion,** the change in enthalpy per mole of a

TABLE 6.3 Standard enthalpies of combustion at 25°C*

Substance	Formula	ΔH_c°, kJ·mol^{-1}
benzene	$C_6H_6(l)$	−3268
carbon	C(s, graphite)	−394
ethanol	$C_2H_5OH(l)$	−1368
ethyne (acetylene)	$C_2H_2(g)$	−1300
glucose	$C_6H_{12}O_6(s)$	−2808
hydrogen	$H_2(g)$	−286
methane	$CH_4(g)$	−890
octane	$C_8H_{18}(l)$	−5471
propane	$C_3H_8(g)$	−2220
urea	$CO(NH_2)_2(s)$	−632

*In a combustion, carbon is converted to carbon dioxide, hydrogen to liquid water, and nitrogen to nitrogen gas. More values are given in Appendix 2A.

substance when the substance reacts completely with oxygen under standard conditions. In a combustion of an organic compound, carbon forms carbon dioxide and hydrogen forms water; any nitrogen present is released as N_2, unless we specify that nitrogen oxides are formed. For example, from Self-Test 6.6A, we know that the reaction enthalpy for the combustion of 2 mol C_4H_{10} is −5756 kJ. Therefore, the enthalpy of combustion of butane is

$$\Delta H_c^\circ = \frac{-5756 \text{ kJ}}{2 \text{ mol}} = -2878 \text{ kJ·mol}^{-1}$$

Standard enthalpies of combustion are listed in Table 6.3 and Appendix 2A. We have seen how to use enthalpies of combustion to obtain the standard enthalpies of reactions (see Toolbox 6.1). Here we shall consider another practical application—the choice of a fuel.

One practical measure of a fuel's value is its **specific enthalpy,** the enthalpy of combustion per gram. Fuels with a high specific enthalpy release a lot of heat per gram when they burn. Hence, specific enthalpy is an important criterion when mass is important, as it is in airplanes. When the *volume* occupied by a fuel is important, as it is in automobiles, we assess its value in terms of its **enthalpy density,** the enthalpy of combustion per liter. A fuel with a low enthalpy density releases very little heat per liter when it burns (Fig. 6.21).

When reporting specific enthalpies and enthalpy densities, it is conventional to omit the negative sign.

Three readily available fuels are hydrogen, methane, and octane. Methane is the major component of natural gas, and octane is a typical component of gasoline. The data in Table 6.4 illustrate one advantage of using gasoline: its high enthalpy density of 38 MJ·L^{-1} (where 1 MJ = 10^6 J). That high value means that a fuel tank need not be large to store a lot of energy. Where mass is important, as it is in a rocket, hydrogen may be an attractive fuel because of its high specific enthalpy.

Like octane, most of the compounds in gasoline are hydrocarbons with about eight carbon atoms.

A great advantage of hydrogen over hydrocarbon fuels is that it produces no carbon dioxide when it burns. Carbon dioxide is potentially harmful, because it contributes to the greenhouse effect (see Case Study 4). However, although hydrogen does not contribute to the greenhouse effect, its use is limited by a low enthalpy density, because it is a gas. Chemists are working to synthesize solid compounds that concentrate

TABLE 6.4 Thermochemical properties of four fuels

Fuel	Combustion equation	ΔH_c°, $kJ \cdot mol^{-1}$	Specific enthalpy, $kJ \cdot g^{-1}$	Enthalpy density,* $kJ \cdot L^{-1}$
hydrogen	$2\,H_2(g) + O_2(g) \longrightarrow 2\,H_2O(l)$	−286	142	13
methane	$CH_4(g) + 2\,O_2(g) \longrightarrow CO_2(g) + 2\,H_2O(l)$	−890	55	40
octane	$2\,C_8H_{18}(l) + 25\,O_2(g) \longrightarrow 16\,CO_2(g) + 18\,H_2O(l)$	−5471	48	3.8×10^4
methanol	$2\,CH_3OH(l) + 3\,O_2(g) \longrightarrow 2\,CO_2(g) + 4\,H_2O(l)$	−726	23	1.8×10^4

*At atmospheric pressure and room temperature.

hydrogen and release it as needed (Fig. 6.22). Candidates include the hydrides formed when titanium, copper, and other metals are heated in hydrogen. These compounds occupy a much smaller volume than the equivalent amount of hydrogen gas and release hydrogen when heated or treated with acid. One example is an iron titanium hydride of approximate formula $FeTiH_2$. Its enthalpy density is high, but its iron and titanium content make the compound dense, so its specific enthalpy is low. Food is just a special kind of fuel. Case Study 6 looks at the fuel value of foods such as hamburgers, potatoes, and bread.

The specific enthalpy tells how much heat can be obtained per gram of fuel. The enthalpy density indicates how much heat can be obtained per liter of fuel.

FIGURE 6.21 During World War II, fuel was in short supply and all manner of ingenious solutions were sought. However, as we can see from this photograph of a vehicle powered by coal gas (a mixture of carbon monoxide and hydrogen) in London, the low enthalpy density of gases creates storage problems.

FIGURE 6.22 The range and speed of this electric-powered car depend on the type of battery it uses. For example, metal-hydride devices have a longer range than lead-acid storage batteries. However, regardless of the type of battery used, recharging is generally a slow process. The cable attached to this car may look like a gasoline hose, but it is actually delivering electricity while its owner waits.

SELF-TEST 6.7A From the data in Table 6.4, determine whether methane or methanol would make a better rocket fuel and explain why.

[*Answer:* Methane. Its specific enthalpy is higher than methanol's, so less mass would be required for the same heat output.]

SELF-TEST 6.7B From the data in Table 6.4, determine whether methanol or hydrogen gas would be more useful in a small automobile.

6.12 STANDARD ENTHALPIES OF FORMATION

Chemists have devised an ingenious way of reporting enthalpies of reactions. They report the **standard enthalpy of formation** of each individual substance, ΔH_f°. This quantity is the standard reaction enthalpy for the formation of the substance from its elements in their most stable form; it is expressed in kilojoules per mole of the substance ($kJ \cdot mol^{-1}$). Then, to obtain the standard enthalpy of any reaction, they just subtract the standard enthalpies of formation of the reactants from those of the products.

We obtain ΔH_f° for ethanol, for instance, from the thermochemical equation for its formation starting from graphite (the most stable form of carbon) and gaseous hydrogen and oxygen:

$$4\,C(s) + 6\,H_2(g) + O_2(g) \longrightarrow 2\,C_2H_5OH(l) \qquad \Delta H^\circ = -555.38\ kJ$$

Because the thermochemical equation refers to the formation of 2 mol $C_2H_5OH(l)$, the standard reaction enthalpy per mole of ethanol molecules is

$$\Delta H_f^\circ(C_2H_5OH,\ l) = \frac{-555.38\ kJ}{2\ mol} = -277.69\ kJ \cdot mol^{-1}$$

Note how the substance and its state are used to label the enthalpy change, so we know which species we are talking about.

FOOD AND FITNESS

ON A LONG journey, we often stop to refuel both the automobile and ourselves. Although the two types of fuel will both be burned to generate energy, there are some differences. For example, we usually give the automobile the same kind of fuel each time, but we eat a wide variety of foods; we cannot overfill the gas tank, but we sometimes overeat; and the food in the diner is usually (but not always) tastier than gasoline.

Automobiles are powered by hydrocarbons with formulas similar to that of octane, C_8H_{18}, but we get our energy primarily from carbohydrates, proteins, and fats (see Chapter 11). Over 50% by mass of the food in a typical diet is taken in the form of carbohydrates. But not all carbohydrates are digestible. Cellulose, for instance, which is the structural material of wood and plant stalks, is a primary "fiber" in our diet. As dietary fiber, it helps move material through the intestines. The digestible carbohydrates are the starches and sugars. Our bodies break them down into glucose, which is soluble in the bloodstream and can be transported into cells. In animal cells, glucose is used as fuel:

$$C_6H_{12}O_6(aq) + 6\,O_2(g) \longrightarrow 6\,CO_2(g) + 6\,H_2O(l)$$

The standard enthalpy of combustion of glucose is -2.8 MJ·mol^{-1}, which corresponds to a specific enthalpy of 16 kJ·g^{-1}. Therefore, the oxidation of 1.0 g of glucose to carbon dioxide and water produces 16 kJ. That is enough energy to heat 1 L of water by about 4°C. We burn 1 g of glucose in about 1 min when bicycling, or in about 2 min when studying chemistry.

The *average* specific enthalpy of digestible carbohydrates, including starches, is actually a little higher, about 17 kJ·g^{-1}. We often see enthalpies reported in calories. The average specific enthalpy of carbohydrates is therefore about 4 kcal·g^{-1}. In nutrition, food Calories are denoted Cal, with 1 Cal = 1 kcal. Be very careful with this distinction, or you might overeat a thousandfold! The average specific enthalpy of carbohydrates in food Calories is about 4 Cal·g^{-1}.

Regular exercise is not only good for the metabolism, it can be fun, too, when we make it a part of daily life.

The second major type of nutritional fuel is proteins. These complex compounds of carbon, hydrogen, oxygen, and nitrogen are made up of long chains of small units called amino acids. Proteins are large molecules, with molar masses that sometimes reach 10^6 g·mol^{-1}. They carry out many of the functions of a living cell and build muscle, and are too important to be used only as fuel. Much of the protein we eat is taken apart and reassembled. However, some of the proteins are oxidized to form urea, $CO(NH_2)_2$. This oxidation corresponds to a specific enthalpy of about 17 kJ·g^{-1}, which is about the same as that of carbohydrates.

Fats are the third major energy source. These compounds have long $-CH_2-CH_2-CH_2-$ chains, a characteristic they share with the hydrocarbons in gasoline. As a result, they behave in some ways like hydrocarbon fuels. One example is tristearin, $C_{57}H_{110}O_6$, a component of beef fat. Tristearin consists of three long hydrocarbon chains

Glucose

$(H_2N)_2C{=}O$

Urea

$H_2C-O-C(=O)-CH_2(CH_2)_{15}CH_3$
$HC-O-C(=O)-CH_2(CH_2)_{15}CH_3$
$H_2C-O-C(=O)-CH_2(CH_2)_{15}CH_3$

Tristearin

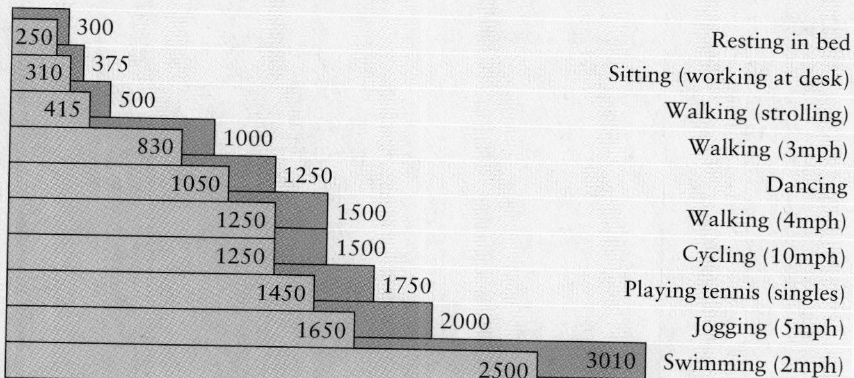

Energy consumed (in kilojoules per hour) in typical activities: blue for a 70-kg male and pink for a 58-kg female.

attached to a partially oxidized group. The specific enthalpy of combustion of tristearin is about 38 $kJ \cdot g^{-1}$, or 9 $kcal \cdot g^{-1}$. This value is nearly twice the specific enthalpy for carbohydrates and, in fact, is closer to the value for gasoline (48 $kJ \cdot g^{-1}$). Because fats can be regarded as only very slightly oxidized hydrocarbons, they release more heat during burning than do proteins and carbohydrates, which already contain many oxygen atoms.

When we take in more fuel than our bodies can use, the extra fuel is stored in the form of fat for future use. Animals, automobiles, and airplanes all need to maximize their efficiency by storing energy in the smallest mass possible. Body fat is the human equivalent of petroleum—energy stored compactly until needed.

The enthalpy values of some typical foods are given in the table. The recommended daily consumption for 18- to 20-year-old males is 12 MJ, or about 2800 kcal; for females of the same age, it is 9 MJ, or about 2100 kcal. However, the amount of fuel required by an individual depends on fitness and activity levels. The more active we are, the more fuel we require. In fact, a very active person may need up to twice the enthalpy input that a sedentary person needs. The illustration above shows the average energy requirements for some common activities.

Beyond simply expending energy, physical activity tends to build muscle and decrease the amount of body fat. Because muscle tissue requires more energy for maintenance than fat tissue, a physically fit and active person burns more fuel just breathing than a sedentary person does, and is less likely to gain unwanted weight.

QUESTIONS

1. A person of average mass burns about 30 kJ per minute playing tennis. Use the table below to find the time you would have to spend playing tennis to burn up the energy in a 2-ounce serving of cheese.

2. A premium ice cream contains 16 g fat and 20 g sugar per serving (4 oz). A brand of frozen yogurt contains 4 g fat and 32 g sugar per serving. Calculate the fuel value of a serving of each item. Both contain 5 g protein.

3. Use the structures to explain why the specific enthalpies of fats are greater than those of carbohydrates.

Thermochemical properties of some foods

Food	Percentage composition				Specific enthalpy, $kJ \cdot g^{-1}$
	Water	Protein	Fat	Carbohydrate	
apples	84.3	0.3	0	11.9	2.5
beef	54.3	23.6	21.1	0	13.1
bread	39.0	7.8	1.7	49.7	12.6
cheese	37.0	26.0	33.5	0	17.0
cod	76.6	21.4	1.2	0	3.1
hamburger	40.9	15.8	14.2	29.1	17.3
milk	87.6	3.3	3.8	4.7	2.6
potatoes	80.5	1.4	0.1	19.7	3.5

TABLE 6.5 Standard enthalpies of formation at 25°C*

Substance	Formula	ΔH_f°, kJ·mol^{-1}	Substance	Formula	ΔH_f°, kJ·mol^{-1}
INORGANIC COMPOUNDS			ORGANIC COMPOUNDS		
ammonia	$NH_3(g)$	−46.11	benzene	$C_6H_6(l)$	+49.0
carbon dioxide	$CO_2(g)$	−393.51	ethanol	$C_2H_5OH(l)$	−277.69
carbon monoxide	$CO(g)$	−110.53	ethyne (acetylene)	$C_2H_2(g)$	+226.73
dinitrogen tetroxide	$N_2O_4(g)$	+9.16	glucose	$C_6H_{12}O_6(s)$	−1268
hydrogen chloride	$HCl(g)$	−92.31	methane	$CH_4(g)$	−74.81
hydrogen fluoride	$HF(g)$	−271.1			
nitrogen dioxide	$NO_2(g)$	+33.18			
nitric oxide	$NO(g)$	+90.25			
sodium chloride	$NaCl(s)$	−411.15			
water	$H_2O(l)$	−285.83			
	$H_2O(g)$	−241.82			

*A much longer list is given in Appendix 2A.

The standard enthalpy of formation of an element in its most stable form is zero. For example, in the reaction

$$C(s, \text{graphite}) \longrightarrow C(s, \text{graphite})$$

nothing happens, and there is clearly zero change of enthalpy. However, the enthalpy of formation of an element in a form other than its most stable one is nonzero. For example, the conversion of carbon from graphite to diamond is endothermic:

$$C(s, \text{graphite}) \longrightarrow C(s, \text{diamond}) \qquad \Delta H^\circ = +1.9 \text{ kJ}$$

It follows that the standard enthalpy of formation of diamond is $+1.9$ kJ·mol^{-1}.

When a compound cannot be synthesized from its elements directly (or the reaction is too difficult to study), its enthalpy of formation can still be found from its enthalpy of combustion, as explained in Toolbox 6.1. Some values are listed in Table 6.5 and in Appendix 2A.

Now let's see how standard enthalpies of formation are combined. First, we calculate the total standard enthalpy of formation of all the products of a reaction. Then we calculate the total standard enthalpy of formation of all the reactants. The difference between these two totals is the enthalpy of the reaction (Fig. 6.23):

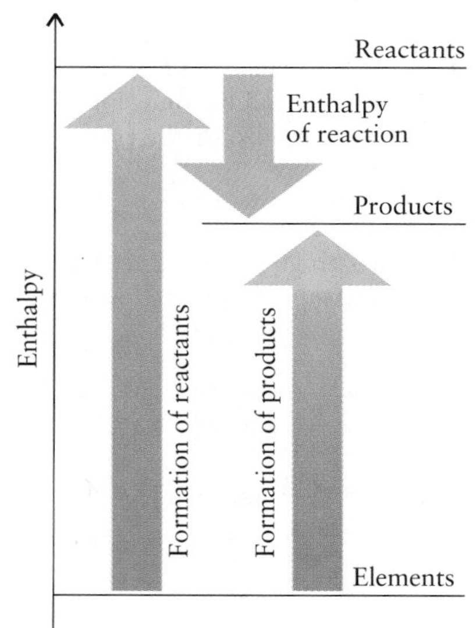

FIGURE 6.23 The reaction enthalpy can be constructed from enthalpies of formation by imagining the formations of both the reactants and the products from their respective elements. The reaction enthalpy is the difference between the two.

$$\begin{aligned}\Delta H^\circ &= H^\circ_{\text{final}} - H^\circ_{\text{initial}}\\ &= \sum n\Delta H_f^\circ(\text{products}) - \sum n\Delta H_f^\circ(\text{reactants})\end{aligned}$$

Here $\sum$ (sigma) means a sum, and n stands for the stoichiometric coefficient of each substance. The first sum is the total enthalpy of formation of the products multiplied by their stoichiometric coefficients. The second sum is the similar total for the reactants. Toolbox 6.2 shows how to use this expression.

Standard enthalpies of formation can be combined to obtain the standard enthalpy of any reaction.

TOOLBOX 6.2 How to use standard enthalpies of formation

This toolbox is based on the expression given above for the standard reaction enthalpy. To use that expression,

Step 1. Write down the chemical equation for the reaction of interest.

Step 2. Add up the individual enthalpies of formation of the products. Each value is multiplied by the numbers of moles, as given by the stoichiometric coefficients. Remember that the standard enthalpy of formation of an element in its most stable form is 0.

Step 3. Calculate the total enthalpy of formation of the reactants in the same way.

Step 4. Subtract the second sum from the first.

For example, let's use the information in Table 6.5 to predict the enthalpy of combustion of benzene, one of the components of gasoline.

Step 1. The combustion reaction to consider is

$$2\,C_6H_6(l) + 15\,O_2(g) \longrightarrow 12\,CO_2(g) + 6\,H_2O(l)$$

Step 2. The total enthalpy of formation of the products is

$$\begin{aligned}\sum n\Delta H_f^\circ(\text{products}) &= (12\text{ mol}) \times \Delta H_f^\circ(CO_2, g) + (6\text{ mol}) \times \Delta H_f^\circ(H_2O,l)\\ &= (12\text{ mol}) \times (-393.51\text{ kJ}\cdot\text{mol}^{-1}) + (6\text{ mol}) \times (-285.83\text{ kJ}\cdot\text{mol}^{-1})\\ &= -4722.12\text{ kJ} - 1714.98\text{ kJ} = -6437.10\text{ kJ}\end{aligned}$$

The first term on the right is the enthalpy of formation of 12 mol $CO_2(g)$. The second term is the total enthalpy of formation of 6 mol $H_2O(l)$.

Step 3. The total enthalpy of formation of the reactants is

$$\begin{aligned}\sum n\Delta H_f^\circ(\text{reactants}) &= (2\text{ mol}) \times \Delta H_f^\circ(C_6H_6,l) + (15\text{ mol}) \times \Delta H_f^\circ(O_2,g)\\ &= (2\text{ mol}) \times (+49.0\text{ kJ}\cdot\text{mol}^{-1}) + (15\text{ mol}) \times (0)\\ &= +98.0\text{ kJ}\end{aligned}$$

Step 4. The difference between the two totals is

$$\Delta H^\circ = (-6437.10\text{ kJ}) - (+98.0\text{ kJ}) = -6535.1\text{ kJ}$$

Suppose we have been asked to find the standard enthalpy of combustion of benzene. We have already described the combustion of 2 mol C_6H_6; so we must divide by 2 mol to find the enthalpy of combustion per mole of benzene molecules:

$$\Delta H_c^\circ = \frac{-6535.1\text{ kJ}}{2\text{ mol}} = -3267.6\text{ kJ}\cdot\text{mol}^{-1}$$

SELF-TEST 6.8A Calculate the standard enthalpy of combustion of glucose from the information in Table 6.5 and Appendix 2A. This result is discussed further in Case Study 6.

[*Answer:* -2808 kJ·mol^{-1}]

SELF-TEST 6.8B You have an inspiration: maybe diamonds would make a great fuel! Calculate the standard enthalpy of combustion of diamonds from the information in Appendix 2A.

SKILLS YOU SHOULD HAVE MASTERED

Conceptual

1. Distinguish between a system and its surroundings.
2. Distinguish exothermic and endothermic reactions by the direction of heat flow and by the sign of ΔH.
3. Explain the meanings of heat capacity and specific heat capacity.
4. Define enthalpy and explain how Hess's law depends on the fact that enthalpy is a state property.
5. Describe the various parts of a heating curve and explain its features.

Problem-Solving

1. Determine the heat output of a reaction, given the temperature change of a calibrated calorimeter.
2. Calculate enthalpy changes from calorimetry data.
3. Calculate an overall reaction enthalpy from the enthalpies of the reactions in a reaction sequence.
4. Use standard enthalpies of formation to calculate the standard enthalpy of a reaction.

Descriptive

1. Describe the function of calorimeters and how they are calibrated.
2. Describe the energy changes that take place during changes of state and the direction of energy flow during these changes.
3. Define the standard state of a substance and recognize when a substance is in its standard state.
4. Define and use the specific enthalpy and the enthalpy density of fuels.

EXERCISES

Energy and Heat

6.1 Define the terms *system* and *surroundings.*

6.2 Distinguish between an *endothermic* process and an *exothermic* process.

6.3 One tablespoon of a low-fat salad dressing is reported to have an "energy content" of 16 kcal. What is this value in kilojoules?

6.4 It requires about 100 kJ of energy to heat a cup of water from room temperature to 100°C. Express this energy requirement in kilocalories.

6.5 How does an increase in temperature affect the motion of the molecules in a system?

6.6 If energy is absorbed by a system as heat, what happens to the molecular motion of the *surroundings?*

6.7 What is meant by the property called "enthalpy"? What does the statement "enthalpy is a state property" mean?

6.8 What does the term *heat capacity* mean?

6.9 (a) Near room temperature, the specific heat capacity of benzene is 1.05 $J\cdot(°C)^{-1}\cdot g^{-1}$. Calculate the heat needed to raise the temperature of 50.0 g of benzene from 25.3°C to 37.2°C. (b) A 1.0-kg block of aluminum is supplied with 490 kJ of heat. What is the temperature change of the aluminum? The specific heat capacity of aluminum is 0.90 $J\cdot(°C)^{-1}\cdot g^{-1}$.

6.10 (a) Near room temperature, the specific heat capacity of ethanol is 2.42 $J\cdot(°C)^{-1}\cdot g^{-1}$. Calculate the heat needed to reduce the temperature of 150.0 g of C_2H_5OH from 50.0°C to 16.6°C. (b) What mass of copper can be heated from 10°C to 200°C when 400 kJ of energy is available?

6.11 The specific heat capacity of water is 4.18 $J\cdot(°C)^{-1}\cdot g^{-1}$ and that of stainless steel is 0.51 $J\cdot(°C)^{-1}\cdot g^{-1}$. Calculate the heat that must be supplied to a 500.0-g stainless steel vessel containing 450.0 g of water to raise its temperature from 25°C to the boiling point of water, 100°C. What percentage of the heat is used to raise the temperature of the water?

6.12 The specific heat capacity of water is 4.18 $J\cdot(°C)^{-1}\cdot g^{-1}$ and that of copper is 0.38 $J\cdot(°C)^{-1}\cdot g^{-1}$. Calculate the heat that must be supplied to a 500.0-g copper kettle containing 450.0 g of water to raise its temperature from 25°C to the boiling point of water, 100°C. What percentage of the heat is used to raise the temperature of the water? Compare these answers with those of the preceding exercise.

6.13 (a) Calculate the energy needed to raise the temperature of 10.0 g of iron (of specific heat capacity 0.45 $J\cdot(°C)^{-1}\cdot g^{-1}$) from 25°C to 500°C?, (b) What mass of gold (of specific heat capacity 0.13 $J\cdot(°C)^{-1}\cdot g^{-1}$) can be heated through the same temperature difference when supplied with the same amount of energy as in (a)?

6.14 (a) How much energy must be removed to lower the temperature of a 25.0-g block of ice (of specific heat capacity 2.03 $J\cdot(°C)^{-1}\cdot g^{-1}$) from −12°C to −28°C? (b) If the same energy were removed from 25.0 g of steam (of specific heat capacity 2.01 $J\cdot(°C)^{-1}\cdot g^{-1}$), what would be the temperature change?

6.15 A piece of metal of mass 20.0 g at 100°C is placed in a calorimeter containing 50.7 g of water at 22.0°C. The final temperature of the mixture is 25.7°C. What is the specific heat capacity of the metal? Assume that all the energy lost by the metal is gained by the water.

6.16 A piece of copper of mass 20.0 g at 100°C is placed in a vessel of negligible heat capacity but containing 50.7 g of water at 22.0°C. Calculate the final temperature of the water. Assume that all the energy lost by the copper is gained by the water. The specific heat capacity of copper is 0.38 $J\cdot(°C)^{-1}\cdot g^{-1}$.

Enthalpy Change

6.17 Classify the following processes as exothermic or endothermic: (a) the formation of ethyne, which is used for oxyacetylene welding:

$$2\,C(s) + H_2(g) \longrightarrow C_2H_2(g) \qquad \Delta H° = +227\text{ kJ}$$

(b) the freezing of water:

$$H_2O(l) \longrightarrow H_2O(s) \qquad \Delta H° = -6.0\text{ kJ}$$

6.18 State whether the temperature will rise or fall when the following reactions are carried out in an insulated calorimeter: (a) the reaction used in industry to produce carbon disulfide from natural gas:

$$CH_4(g) + 4\,S(s) \longrightarrow CS_2(l) + 2\,H_2S(g) \quad \Delta H° = -106 \text{ kJ}$$

(b) the dissolution of table salt in water:

$$NaCl(s) \longrightarrow NaCl(aq) \qquad \Delta H° = +3.9 \text{ kJ}$$

6.19 What is the enthalpy change when 20.0 g of water is cooled from 20.0°C to 4.00°C? Is the process endothermic or exothermic?

6.20 What is the enthalpy change when 50.3 g of copper is heated from 10°C to 150°C? Is the process endothermic or exothermic?

6.21 100.0 g of water at 62.5°C was poured into a calorimeter containing 100.0 g water at 19.8°C. The final temperature was 40.1°C. How much heat leaked to the surroundings during this process?

6.22 50.0 g of water at 60.5°C was poured into a calorimeter containing 50.0 g water at 18.7°C. The final temperature was 35°C. How much heat leaked to the surroundings during this process?

6.23 The enthalpy of combustion of benzoic acid, C_6H_5COOH, which is often used to calibrate calorimeters, is $-3227\ \text{kJ·mol}^{-1}$. When 1.236 g of benzoic acid was burned in a calorimeter, the temperature increased by 2.345°C. What is the heat capacity of the calorimeter?

6.24 A calorimeter was calibrated with an electric heater. It was used to supply 22.5 kJ of energy to the calorimeter, which increased the temperature of the calorimeter and its water bath from 22.45°C to 23.97°C. What is the heat capacity of the calorimeter?

6.25 The heat capacity of a calorimeter was measured as $5.24\ \text{kJ·(°C)}^{-1}$. With the combustion of a piece of asparagus, the temperature rose from 22.45°C to 23.17°C. What is the enthalpy change for the combustion?

6.26 A calorimeter has a measured heat capacity of $6.27\ \text{kJ·(°C)}^{-1}$. The combustion of 1.84 g of magnesium led to a temperature change from 21.30°C to 28.56°C. Calculate the enthalpy change of the reaction

$$2\,Mg(s) + O_2(g) \longrightarrow 2\,MgO(s)$$

6.27 50.0 mL of 0.500 M NaOH(aq) and 50.0 mL of 0.500 M HNO_3(aq), both initially at 18.6°C, were mixed and stirred in a calorimeter having a heat capacity equal to $525.0\ \text{J·(°C)}^{-1}$. The temperature of the mixture rose to 21.3°C. (a) What is the change in enthalpy for the neutralization reaction? (b) What is the change in enthalpy for the neutralization in kilojoules per mole of HNO_3?

6.28 The heat capacity of a certain calorimeter is $488.1\ \text{J·(°C)}^{-1}$. When 25.0 mL of 0.700 M NaOH(aq) was mixed in that calorimeter with 25.0 mL of 0.700 M HCl(aq), both initially at 20.0°C, the temperature increased to 22.1°C. Calculate the enthalpy of neutralization in kilojoules per mole of HCl.

Enthalpy of Physical Change

6.29 How does the enthalpy change indicate whether a process is endothermic or exothermic?

6.30 Describe the appearance of a heating curve when a solid is heated until it melts and then becomes a gas.

6.31 (a) The vaporization of 0.235 mol of liquid CH_4 requires 1.93 kJ of heat. What is the enthalpy of vaporization of methane? (b) An electric heater was immersed in a flask of boiling ethanol, C_2H_5OH, and 22.45 g of ethanol was vaporized when 21.2 kJ of energy was supplied. What is the enthalpy of vaporization of ethanol?

6.32 (a) When 25.23 g of methanol, CH_3OH, froze, 4.01 kJ of heat was released. What is the enthalpy of fusion of methanol? (b) A sample of benzene was vaporized at a reduced pressure at 25°C. When 37.5 kJ of heat was supplied, 95 g of the liquid benzene vaporized. What is the enthalpy of vaporization of benzene at 25°C?

6.33 Use the information in Table 6.2 to calculate the enthalpy change for (a) the vaporization of 100.0 g of water at 373.2 K; (b) the melting of 600 g of solid ammonia at its freezing point (195.3 K).

6.34 Use the information in Table 6.2 to calculate the enthalpy change that occurs when (a) 200 g of methanol condenses at its boiling point (337.2 K); (b) 17.7 g of acetone, CH_3COCH_3, freezes at its freezing point (177.8 K).

6.35 How much heat is needed to melt 50.0 g of ice at 0°C and then heat the water to 25°C? (See Tables 6.1 and 6.2.)

6.36 If we start with 150 g of water at 30°C, how much heat must we add to convert all the liquid to steam at 100°C? (See Tables 6.1 and 6.2.)

Enthalpy of Chemical Change

6.37 What is meant by the *standard state* of a substance?

6.38 Define the term *standard reaction enthalpy.* Identify two methods by which it can be determined without actually carrying out the reaction.

6.39 Carbon disulfide can be prepared from coke (an impure form of carbon) and elemental sulfur:

$$4\,C(s) + S_8(s) \longrightarrow 4\,CS_2(l) \qquad \Delta H° = +358.8 \text{ kJ}$$

(a) How much heat is absorbed in the reaction of 0.20 mol S_8? (b) Calculate the heat absorbed in the reaction of 20.0 g of carbon with an excess of sulfur. (c) If the heat absorbed in the reaction was 217 kJ, how much CS_2 was produced?

6.40 The oxidation of nitrogen in the hot exhaust of jet engines and automobiles occurs by the reaction

$$N_2(g) + O_2(g) \longrightarrow 2\,NO(g) \qquad \Delta H^\circ = +180.6\text{ kJ}$$

(a) How much heat is absorbed in the formation of 0.70 mol NO? (b) How much heat is absorbed in the oxidation of 17.4 L of nitrogen measured at 1.00 atm and 273 K? (c) When the oxidation of N_2 to NO was completed in a bomb calorimeter, the heat absorbed was measured as 790 J. What mass of nitrogen gas was oxidized?

6.41 The combustion of octane is expressed by the thermochemical equation

$$2\,C_8H_{18}(l) + 25\,O_2(g) \longrightarrow 16\,CO_2(g) + 18\,H_2O(l) \qquad \Delta H^\circ = -10{,}942\text{ kJ}$$

(a) Calculate the mass of octane that must be burned to produce 12 MJ of heat. (b) How much heat will be evolved from the combustion of 1.0 gal of gasoline (assumed to be exclusively octane)? The density of octane is 0.70 $g{\cdot}mL^{-1}$.

6.42 Suppose that coal, of density 1.5 $g{\cdot}cm^{-3}$, is carbon (it is in fact much more complicated, but this is a reasonable first approximation). The combustion of carbon is described by the equation

$$C(s) + O_2(g) \longrightarrow CO_2(g) \qquad \Delta H^\circ = -394\text{ kJ}$$

(a) Calculate the heat produced when a lump of coal of size 7.0 cm × 6.0 cm × 5.0 cm is burned. (b) Estimate the mass of water that can be heated from 15°C to 100°C with this piece of coal.

6.43 How much heat can be produced from a reaction mixture of 50.0 g of iron(III) oxide and 25.0 g of aluminum in the thermite reaction:

$$Fe_2O_3(s) + 2\,Al(s) \longrightarrow Al_2O_3(s) + 2\,Fe(s) \qquad \Delta H^\circ = -851.5\text{ kJ}$$

6.44 Calculate the heat evolved from a reaction mixture of 13.4 L of sulfur dioxide at 1.00 atm and 273 K and 15.0 g oxygen in the reaction

$$2\,SO_2(g) + O_2(g) \longrightarrow 2\,SO_3(g) \qquad \Delta H^\circ = -198\text{ kJ}$$

Hess's Law

6.45 How does the validity of Hess's law depend on the fact that enthalpy is a state function?

6.46 Distinguish between the enthalpy of combustion and the enthalpy of formation for a compound. Explain why the enthalpy of combustion is not the negative of the enthalpy of formation of a compound.

6.47 The standard enthalpies of combustion of graphite and diamond are −393.51 and −395.41 $kJ{\cdot}mol^{-1}$, respectively. Calculate the change in enthalpy for the graphite → diamond transition.

6.48 Elemental sulfur occurs in several forms, with rhombic sulfur the most stable under normal conditions and monoclinic sulfur slightly less stable. The standard enthalpies of combustion of the two forms to sulfur dioxide are −296.83 and −297.16 $kJ{\cdot}mol^{-1}$, respectively. Calculate the change in enthalpy for the rhombic → monoclinic transition.

6.49 Two successive stages in the industrial manufacture of sulfuric acid are the combustion of sulfur and the oxidation of sulfur dioxide to sulfur trioxide. From the standard reaction enthalpies

$$S(s) + O_2(g) \longrightarrow SO_2(g) \qquad \Delta H^\circ = -296.83\text{ kJ}$$

$$2\,S(s) + 3\,O_2(g) \longrightarrow 2\,SO_3(g) \qquad \Delta H^\circ = -791.44\text{ kJ}$$

calculate the reaction enthalpy for the oxidation of sulfur dioxide to sulfur trioxide in the reaction

$$2\,SO_2(g) + O_2(g) \longrightarrow 2\,SO_3(g)$$

6.50 In the manufacture of nitric acid by the oxidation of ammonia, the first product is nitric oxide, which is then oxidized to nitrogen dioxide. From the standard reaction enthalpies

$$N_2(g) + O_2(g) \longrightarrow 2\,NO(g) \qquad \Delta H^\circ = +180.5\text{ kJ}$$

$$N_2(g) + 2\,O_2(g) \longrightarrow 2\,NO_2(g) \qquad \Delta H^\circ = +66.4\text{ kJ}$$

calculate the standard reaction enthalpy for the oxidation of nitric oxide to nitrogen dioxide:

$$2\,NO(g) + O_2(g) \longrightarrow 2\,NO_2(g)$$

6.51 Calculate the enthalpy of the reaction

$$P_4(s) + 10\,Cl_2(g) \longrightarrow 4\,PCl_5(s)$$

from the reactions

$$P_4(s) + 6\,Cl_2(g) \longrightarrow 4\,PCl_3(l) \qquad \Delta H^\circ = -1278.8\text{ kJ}$$

$$PCl_3(l) + Cl_2(g) \longrightarrow PCl_5(s) \qquad \Delta H^\circ = -124\text{ kJ}$$

6.52 Calculate the reaction enthalpy for the reduction of hydrazine to ammonia

$$N_2H_4(l) + H_2(g) \longrightarrow 2\,NH_3(g)$$

from the following data:

$$N_2(g) + 2\,H_2(g) \longrightarrow N_2H_4(l) \qquad \Delta H^\circ = +50.63\text{ kJ}$$

$$N_2(g) + 3\,H_2(g) \longrightarrow 2\,NH_3(g) \qquad \Delta H^\circ = -92.22\text{ kJ}$$

6.53 Determine the reaction enthalpy for the hydrogenation of ethyne to ethane,

$$C_2H_2(g) + 2\,H_2(g) \longrightarrow C_2H_6(g)$$

from the following data:

$$2\,C_2H_2(g) + 5\,O_2(g) \longrightarrow 4\,CO_2(g) + 2\,H_2O(l) \qquad \Delta H^\circ = -2600 \text{ kJ}$$

$$2\,C_2H_6(g) + 7\,O_2(g) \longrightarrow 4\,CO_2(g) + 6\,H_2O(l) \qquad \Delta H^\circ = -3120 \text{ kJ}$$

$$H_2(g) + \tfrac{1}{2}O_2(g) \longrightarrow H_2O(l) \qquad \Delta H^\circ = -286 \text{ kJ}$$

6.54 Determine the reaction enthalpy for the partial combustion of methane to carbon monoxide:

$$2\,CH_4(g) + 3\,O_2(g) \longrightarrow 2\,CO(g) + 4\,H_2O(l)$$

Use the following data:

$$CH_4(g) + 2\,O_2(g) \longrightarrow CO_2(g) + 2\,H_2O(l) \qquad \Delta H^\circ = -890 \text{ kJ}$$

$$2\,CO(g) + O_2(g) \longrightarrow 2\,CO_2(g) \qquad \Delta H^\circ = -566.0 \text{ kJ}$$

6.55 Calculate the reaction enthalpy for the synthesis of hydrogen chloride gas

$$H_2(g) + Cl_2(g) \longrightarrow 2\,HCl(g)$$

from the following data:

$$NH_3(g) + HCl(g) \longrightarrow NH_4Cl(s) \qquad \Delta H^\circ = -176.0 \text{ kJ}$$

$$N_2(g) + 3\,H_2(g) \longrightarrow 2\,NH_3(g) \qquad \Delta H^\circ = -92.22 \text{ kJ}$$

$$N_2(g) + 4\,H_2(g) + Cl_2(g) \longrightarrow 2\,NH_4Cl(s) \qquad \Delta H^\circ = -628.86 \text{ kJ}$$

6.56 Calculate the reaction enthalpy for the formation of anhydrous aluminum chloride

$$2\,Al(s) + 3\,Cl_2(g) \longrightarrow 2\,AlCl_3(s)$$

from the following data:

$$2\,Al(s) + 6\,HCl(aq) \longrightarrow 2\,AlCl_3(aq) + 3\,H_2(g) \qquad \Delta H^\circ = -1049 \text{ kJ}$$

$$HCl(g) \longrightarrow HCl(aq) \qquad \Delta H^\circ = -74.8 \text{ kJ}$$

$$H_2(g) + Cl_2(g) \longrightarrow 2\,HCl(g) \qquad \Delta H^\circ = -185 \text{ kJ}$$

$$AlCl_3(s) \longrightarrow AlCl_3(aq) \qquad \Delta H^\circ = -323 \text{ kJ}$$

Enthalpy of Formation

6.57 Write the thermochemical equations that give the values of the standard enthalpies of formation for (a) $KClO_3(s)$, potassium chlorate; (b) $H_2NCH_2COOH(s)$, glycine; (c) $Al_2O_3(s)$, alumina.

6.58 Write the thermochemical equations that give the values of the standard enthalpies of formation for (a) $CH_3COOH(l)$; (b) $SO_3(g)$; (c) $CO_2(g)$.

6.59 Calculate the standard enthalpy of formation of dinitrogen pentoxide from the data

$$2\,NO(g) + O_2(g) \longrightarrow 2\,NO_2(g) \qquad \Delta H^\circ = -114.1 \text{ kJ}$$

$$4\,NO_2(g) + O_2(g) \longrightarrow 2\,N_2O_5(g) \qquad \Delta H^\circ = -110.2 \text{ kJ}$$

and the standard enthalpy of formation of nitric oxide.

6.60 An important reaction that occurs in the atmosphere is $NO_2(g) \rightarrow NO(g) + O(g)$, which is brought about by sunlight. How much energy must be supplied by the Sun to cause it? Calculate the standard enthalpy of the reaction from the following information:

$$O_2(g) \longrightarrow 2\,O(g) \qquad \Delta H^\circ = +498.4 \text{ kJ}$$

$$NO(g) + O_3(g) \longrightarrow NO_2(g) + O_2(g) \qquad \Delta H^\circ = -200 \text{ kJ}$$

and additional information from Appendix 2A.

6.61 Calculate the standard enthalpy of formation of $PCl_5(s)$ from the enthalpy of formation of $PCl_3(l)$ (Appendix 2A) and

$$PCl_3(l) + Cl_2(g) \longrightarrow PCl_5(s) \qquad \Delta H^\circ = -124 \text{ kJ}$$

6.62 When 1.92 g of magnesium reacts with nitrogen to form magnesium nitride, the heat evolved is 12.2 kJ. Calculate the standard enthalpy of formation of Mg_3N_2.

6.63 Use the information in Appendix 2A to calculate the enthalpy of reaction of (a) the oxidation of 10.0 g of sulfur dioxide:

$$2\,SO_2(g) + O_2(g) \longrightarrow 2\,SO_3(g)$$

(b) the reduction of 1.00 mol CuO(s) with hydrogen:

$$CuO(s) + H_2(g) \longrightarrow Cu(s) + H_2O(l)$$

6.64 Use the enthalpies of formation from Appendix 2A to determine the enthalpy of reaction of (a) the hydrogenation of 50.0 g of benzene to cyclohexane:

$$C_6H_6(l) + 3\,H_2(g) \longrightarrow C_6H_{12}(l)$$

(b) the hydrogenation of 50.0 g of ethene to ethane:

$$C_2H_4(g) + H_2(g) \longrightarrow C_2H_6(g)$$

6.65 Use the enthalpies of formation in Appendix 2A to calculate the standard enthalpy of the following reactions: (a) the replacement of deuterium by ordinary hydrogen in heavy water:

$$H_2(g) + D_2O(l) \longrightarrow H_2O(l) + D_2(g)$$

(b) the removal of hydrogen sulfide from natural gas:

$$2\,H_2S(g) + SO_2(g) \longrightarrow 3\,S(s) + 2\,H_2O(l)$$

(c) the oxidation of ammonia:

$$4\,NH_3(g) + 5\,O_2(g) \longrightarrow 4\,NO(g) + 6\,H_2O(g)$$

6.66 Using standard enthalpies of formation from Appendix 2A, calculate the standard reaction enthalpy for each reaction: (a) the final stage in the production of nitric acid, when nitrogen dioxide dissolves in and reacts with water:

$$3\,NO_2(g) + H_2O(l) \longrightarrow 2\,HNO_3(aq) + NO(g)$$

(b) the formation of boron trifluoride, which is widely used in the chemical industry:

$$B_2O_3(s) + 3\,CaF_2(s) \longrightarrow 2\,BF_3(g) + 3\,CaO(s)$$

(c) the formation of a sulfide by the action of hydrogen sulfide on an aqueous solution of a base:

$$H_2S(aq) + 2\,KOH(aq) \longrightarrow K_2S(aq) + 2\,H_2O(l)$$

Fuels

6.67 A minor component of gasoline is heptane (C_7H_{16}), which has a standard enthalpy of combustion of $-4854\ kJ \cdot mol^{-1}$ and a density of $0.68\ g \cdot mL^{-1}$. Calculate the specific enthalpy of heptane and its enthalpy density.

6.68 Another minor component of gasoline is toluene (C_7H_8), with a standard enthalpy of combustion of $-3910\ kJ \cdot mol^{-1}$ and a density of $0.867\ g \cdot mL^{-1}$. Calculate the specific enthalpy of toluene and its enthalpy density.

6.69 Calculate the specific enthalpy of magnesium from its enthalpy of combustion to magnesium oxide. Use the standard enthalpy of formation in Appendix 2A. Would aluminum, which burns to aluminum oxide, be a better fuel if mass were the only consideration?

6.70 Calculate the specific enthalpy of phosphorus from its enthalpy of combustion in oxygen to P_4O_{10}. Use the standard enthalpy of formation in Appendix 2A. Would sulfur burned to either SO_2 or SO_3 be a more efficient fuel than phosphorus if mass were the only consideration?

6.71 One problem with fuels containing carbon is that they produce carbon dioxide when they burn, so one consideration governing the selection of a fuel could be the heat per mole of CO_2 produced. (a) Calculate this quantity for methane and octane. (b) Calculate the heat produced per mole of CO_2 from the combustion of glucose.
(c) Which process produces more carbon dioxide in the environment for each kilojoule generated, eating (resulting in the combustion of glucose) or burning octane?

6.72 The booster rockets of the space shuttle use a mixture of powdered aluminum and ammonium perchlorate in the exothermic redox reaction

$$2\,Al(s) + 2\,NH_4ClO_4(s) \longrightarrow Al_2O_3(s) + 2\,HCl(g) + 2\,NO(g) + 3\,H_2O(g)$$

(a) Calculate the specific enthalpy of a stoichiometric mixture of aluminum and ammonium perchlorate.
(b) Would it be better to use magnesium in place of aluminum? Explain your conclusion.

SUPPLEMENTARY EXERCISES

6.73 Interpret the meaning of "the standard enthalpy of formation of a compound."

6.74 Classify the following properties as intensive or extensive: (a) heat capacity; (b) specific heat capacity; (c) enthalpy; (d) standard enthalpy of formation.

6.75 The heat capacity of a certain calorimeter is $8.92\ kJ \cdot (°C)^{-1}$. The combustion of 30 g of cheese produces about 460 kJ. What mass of cheese will produce a temperature change in the calorimeter of 2.37°C?

6.76 A 100-g serving of shrimp provides 91 kcal. If a 20.0-g sample is burned in a calorimeter with a heat capacity of $4.66\ kJ \cdot (°C)^{-1}$, what is the expected temperature change?

6.77 A slice of bread has a mass of about 30 g. If there are 24 slices of bread in one loaf and each slice supplies 85 kcal, how much energy (in kilojoules) can be obtained from a loaf of bread?

6.78 When 25.0 g of a metal at a temperature of 90.0°C is added to 50.0 g of water at 25.0°C, the water temperature rises to 29.8°C. The specific heat of water is $4.184\ J \cdot (°C)^{-1} \cdot g^{-1}$. What is the specific heat of the metal?

6.79 Strong sunshine bombards the Earth with about $1\ kJ \cdot m^{-2}$ in 1 s. Calculate the maximum mass of pure ethanol that can be vaporized in 10 min from a beaker left in strong sunshine, assuming the surface area of the ethanol to be $50\ cm^2$. Assume all the heat is used for vaporization, not to increase the temperature.

6.80 Explain the difference between a chemical equation and a thermochemical equation. Illustrate your answer by referring to the equation $2\,C_8H_{18}(l) + 25\,O_2(g) \rightarrow 16\,CO_2(g) + 18\,H_2O(l)$, for which $\Delta H° = -11$ MJ.

6.81 Calculate the heat required (a) to melt 10 g of solid ethanol at its freezing point at 158.7 K; (b) to vaporize it at its boiling point of 351.5 K. Use Table 6.2. Sketch the heating curve.

6.82 A 50.0-g ice cube at 0°C is added to a glass containing 400 g of water at 45°C. What is the final temperature of the system? Assume that no heat is lost to the surroundings.

6.83 The following reaction can be used for the production of manganese:

$$3\,MnO_2(s) + 4\,Al(s) \longrightarrow 2\,Al_2O_3(s) + 3\,Mn(s)$$

(a) Use the information in Appendix 2A and the enthalpy of formation of $MnO_2(s)$, which is $-521\ kJ\cdot mol^{-1}$, to calculate the standard enthalpy of reaction. (b) What is the enthalpy change in the production of 10.0 g of manganese?

6.84 The enthalpy of combustion of methanol, CH_3OH, is the enthalpy change of the following reaction:

$$CH_3OH(l) + \tfrac{3}{2}O_2(g) \longrightarrow CO_2(g) + 2\,H_2O(l) \qquad \Delta H^\circ = -726\ kJ$$

What mass of methanol must be burned to heat 200 g of water in a 50-g Pyrex beaker from 20°C to 100°C? (For additional information, see Table 6.1.)

6.85 Calculate the standard enthalpy of formation of methanol, $CH_3OH(l)$, from the data in Appendix 2A and the equation for its standard enthalpy of combustion given in Exercise 6.84.

6.86 A 0.922-g sample of naphthalene, $C_{10}H_8$, a major component of moth balls, is burned in a calorimeter that has a heat capacity of $9.44\ kJ\cdot(°C)^{-1}$. The temperature of the calorimeter rose from 15.73°C to 19.66°C. Calculate the enthalpy of combustion for naphthalene.

6.87 Would the designer of an industrial plant need to supply heat or remove it in the final step in the industrial preparation of urea, a commercial fertilizer? The reaction is

$$CO_2(g) + 2\,NH_3(g) \longrightarrow H_2O(l) + CO(NH_2)_2(s)$$

6.88 When 3.245 g of lead(IV) oxide is formed from lead metal and oxygen, 3.76 kJ of heat is released. What is the enthalpy of formation of $PbO_2(s)$?

6.89 The standard enthalpy of combustion of sulfur to sulfur dioxide is $-2374.4\ kJ\cdot(mol\ S_8)^{-1}$. What is the standard enthalpy of formation of $SO_2(g)$?

6.90 Calculate the standard reaction enthalpy of the reduction of iron(II) oxide by carbon monoxide, a step in the production of iron,

$$FeO(s) + CO(g) \longrightarrow Fe(s) + CO_2(g)$$

given the following thermochemical equations:

$$3\,Fe_2O_3(s) + CO(g) \longrightarrow 2\,Fe_3O_4(s) + CO_2(g) \qquad \Delta H^\circ = -47.2\ kJ$$

$$Fe_2O_3(s) + 3\,CO(g) \longrightarrow 2\,Fe(s) + 3\,CO_2(g) \qquad \Delta H^\circ = -24.7\ kJ$$

$$Fe_3O_4(s) + CO(g) \longrightarrow 3\,FeO(s) + CO_2(g) \qquad \Delta H^\circ = +35.9\ kJ$$

6.91 Calculate the enthalpy of formation of ethyne,

$$2\,C(s) + H_2(g) \longrightarrow C_2H_2(g)$$

from the following information:

$$2\,C_2H_2(g) + 5\,O_2(g) \longrightarrow 4\,CO_2(g) + 2\,H_2O(l) \qquad \Delta H^\circ = -2600\ kJ$$

$$C(s) + O_2(g) \longrightarrow CO_2(g) \qquad \Delta H^\circ = -394\ kJ$$

$$2\,H_2(g) + O_2(g) \longrightarrow 2\,H_2O(g) \qquad \Delta H^\circ = -483.6\ kJ$$

$$H_2O(l) \longrightarrow H_2O(g) \qquad \Delta H^\circ = +44\ kJ$$

6.92 Calculate the standard reaction enthalpy for the hydrogenation of ethyne to ethene:

$$C_2H_2(g) + H_2(g) \longrightarrow C_2H_4(g)$$

This reaction cannot be performed easily in the laboratory because of the formation of many by-products in the reaction. Use the following data:

$$2\,C_2H_2(g) + 5\,O_2(g) \longrightarrow 4\,CO_2(g) + 2\,H_2O(l) \qquad \Delta H^\circ = -2600\ kJ$$

$$2\,C_2H_4(g) + 6\,O_2(g) \longrightarrow 4\,CO_2(g) + 4\,H_2O(l) \qquad \Delta H^\circ = -2822\ kJ$$

$$2\,H_2(g) + O_2(g) \longrightarrow 2\,H_2O(l) \qquad \Delta H^\circ = -572\ kJ$$

6.93 (a) Is the production of water gas, a cheap, low-grade industrial fuel, exothermic or endothermic? The reaction is

$$C(s) + H_2O(g) \longrightarrow CO(g) + H_2(g)$$

(b) Calculate the enthalpy change in the production of 200 L (at 500 Torr and 65°C) of hydrogen by this reaction.

6.94 (a) Acetic acid can be produced by the reaction of carbon monoxide with methanol in the presence of a catalyst:

$$CO(g) + CH_3OH(l) \longrightarrow CH_3COOH(l)$$

Use the information in Appendix 2A to determine whether this reaction is exothermic or endothermic by calculating its standard reaction enthalpy. (b) Acetic acid can also be formed by the oxidation of ethanol

$$C_2H_5OH(l) + O_2(g) \longrightarrow CH_3COOH(l) + H_2O(l)$$

This reaction occurs when wine goes sour. Decide whether this reaction is exothermic or endothermic by calculating its standard reaction enthalpy.

6.95 Why is the heat of formation of gaseous water less negative than that of liquid water?

6.96 Use the information in Appendix 2A to determine the standard enthalpy for the reaction of pure nitric acid and hydrazine:

$$4\,HNO_3(l) + 5\,N_2H_4(l) \longrightarrow 7\,N_2(g) + 12\,H_2O(l)$$

6.97 Use the information in Appendix 2A to determine how much heat is evolved in the dissolution of 20.0 g of sodium hydroxide. The process is

$$NaOH(s) \longrightarrow Na^+(aq) + OH^-(aq)$$

6.98 Use the information in Appendix 2A to calculate the standard reaction enthalpy for the reaction of calcium carbonate in the form of calcite with hydrochloric acid:

$$CaCO_3(s) + 2\,HCl(aq) \longrightarrow CaCl_2(aq) + H_2O(l) + CO_2(g)$$

6.99 Consider the reaction of iron(II) sulfide with hydrochloric acid,

$$FeS(s) + 2\,H^+(aq) \longrightarrow Fe^{2+}(aq) + H_2S(g)$$

Use the information in Appendix 2A to calculate the enthalpy change for the production of 30.0 L of hydrogen sulfide at 1.00 atm and 298 K.

6.100 A "silver tree" can be made in the laboratory by cutting a piece of copper metal in the shape of a tree and placing it into a silver nitrate solution, when the reaction

$$2\,Ag^+(aq) + Cu(s) \longrightarrow 2\,Ag(s) + Cu^{2+}(aq)$$

deposits sparkling silver crystals on the tree. Use the information in Appendix 2A to find the change in enthalpy for the formation of 1.88 g of silver. Is the reaction an endothermic or exothermic process?

6.101 What considerations are appropriate to the selection of a fuel?

6.102 If you had to choose between carrying propane or butane on a camping expedition, which would you choose on the basis of its specific enthalpy?

6.103 Suppose that, of the heat produced by the combustion of 100 mg of octane, C_8H_{18}, only 70% is useful in heating. What will be the resulting temperature change if the fuel is used to heat 250.0 g of ethanol?

CHALLENGING EXERCISES

6.104 During a heavy rainstorm, 2.13×10^9 L of rain fell. If the density of the rainwater is $1.00\ g\cdot cm^{-3}$, how much heat is released when this quantity of water condenses from water vapor to rain?

6.105 A 20.0-g ice cube at −14°C was heated until it became steam at 110°C. Using values from Table 6.2 and the specific heat capacities of ice ($2.03\ J\cdot(°C)^{-1}\cdot g^{-1}$) and water vapor ($2.01\ J\cdot(°C)^{-1}\cdot g^{-1}$), determine the total heat that must have been supplied.

6.106 A solar heat storage system consists of 60 sealed pipes. Each pipe is loaded with 45.6 kg $CaCl_2\cdot 6H_2O$. Heat from the Sun is absorbed by the salt and stored when the salt melts at 30°C to form an aqueous solution of $CaCl_2$. At night, the heat is released when the salt recrystallizes. How many liters of water can be heated to 25°C by the recrystallization in this heat storage system if the water temperature is initially at 15°C? Assume the density of water to be $1.00\ g\cdot mL^{-1}$ and $\Delta H_{fus}(CaCl_2\cdot 6H_2O)$ to be $27\ kJ\cdot mol^{-1}$.

6.107 The enthalpy of formation of trinitrotoluene (TNT) is $-67\ kJ\cdot mol^{-1}$ and the density of TNT is $1.65\ g\cdot cm^{-3}$. In principle, it could be used as a rocket fuel, with the gases resulting from its decomposition streaming out of the rocket to give the required thrust. In practice, of course, it would be extremely dangerous as a fuel because it is sensitive to shock. Explore its potential as a rocket fuel by calculating its enthalpy density for the reaction

$$4\,C_7H_5N_3O_6(s) + 21\,O_2(g) \longrightarrow 28\,CO_2(g) + 10\,H_2O(g) + 6\,N_2(g)$$

6.108 A natural gas mixture is burned in a furnace at a power-generating station at a rate of 13.0 mol per minute. If the fuel consists of 9.3 mol CH_4, 3.1 mol C_2H_6, 0.40 mol C_3H_8, and 0.20 mol C_4H_{10}, what mass of CO_2 is produced per minute? How much heat is released per minute?

6.109 Calculate the enthalpy of vaporization of solid sodium chloride to a gas of ions,

$$NaCl(s) \longrightarrow Na^+(g) + Cl^-(g)$$

from the following information and the data in Appendix 2A.

Atomization of sodium:

$$Na(s) \longrightarrow Na(g) \qquad \Delta H° = +108.4\ kJ$$

Ionization of sodium:

$$Na(g) \longrightarrow Na^+(g) + e^-(g) \qquad \Delta H° = +495.8\ kJ$$

Dissociation of chlorine:

$$Cl_2(g) \longrightarrow 2\,Cl(g) \qquad \Delta H° = +242\ kJ$$

Electron attachment to chlorine:

$$Cl(g) + e^-(g) \longrightarrow Cl^-(g) \qquad \Delta H° = -348.6\ kJ$$

6.110 Calculate the enthalpy of vaporization of solid potassium bromide to a gas of ions, the process

$$KBr(s) \longrightarrow K^+(g) + Br^-(g)$$

from the following information, and the enthalpy of formation of KBr(s), which is $-394\ kJ\cdot mol^{-1}$.

Atomization of potassium:

$$K(s) \longrightarrow K(g) \qquad \Delta H° = +89.2 \text{ kJ}$$

Ionization of potassium:

$$K(g) \longrightarrow K^+(g) + e^-(g) \qquad \Delta H° = +425.0 \text{ kJ}$$

Vaporization of bromine:

$$Br_2(l) \longrightarrow Br_2(g) \qquad \Delta H° = +30.9 \text{ kJ}$$

Dissociation of bromine:

$$Br_2(g) \longrightarrow 2\,Br(g) \qquad \Delta H° = +192.9 \text{ kJ}$$

Electron attachment to bromine:

$$Br(g) + e^-(g) \longrightarrow Br^-(g) \qquad \Delta H° = -331.0 \text{ kJ}$$

6.111 Use reactions (a), (b), and (c) to determine the enthalpy change of this reaction:

$$CH_4(g) + \tfrac{3}{2}\,O_2(g) \longrightarrow CO(g) + 2\,H_2O(g)$$

(a) $CH_4(g) + 2\,O_2(g) \longrightarrow CO_2(g) + 2\,H_2O(g)$ $\quad \Delta H° = -802 \text{ kJ}$

(b) $CH_4(g) + CO_2(g) \longrightarrow 2\,CO(g) + 2\,H_2(g)$ $\quad \Delta H° = +206 \text{ kJ}$

(c) $CH_4(g) + H_2O(g) \longrightarrow CO(g) + 3\,H_2(g)$ $\quad \Delta H° = +247 \text{ kJ}$

CHAPTER 7

The sky over New York City is not normally so colorful, but in this composite view, taken from across the river in New Jersey, we see what can be achieved with chemicals. The red, white, and blue colors of the fireworks are light given out by energetically excited atoms and molecules. Similar effects give rise to the everyday colors of street lighting, where an electric current is used to excite atoms of sodium and mercury. Chemists use these effects in laboratories: they analyze colors emitted by substances to identify them and to discover how electrons are arranged inside atoms.

INSIDE THE ATOM

OBSERVING ATOMS
7.1 The characteristics of light
7.2 Quanta and photons
7.3 Atomic spectra and energy levels
7.4 The wavelike properties of electrons

MODELS OF ATOMS
7.5 Atomic orbitals
7.6 Quantum numbers and atomic orbitals
7.7 Electron spin
7.8 The electronic structure of hydrogen

THE STRUCTURES OF MANY-ELECTRON ATOMS
7.9 Orbital energies
7.10 The building-up principle
7.11 The electron configurations of atoms
7.12 The electron configurations of ions
7.13 Electronic structure and the periodic table

THE PERIODICITY OF ATOMIC PROPERTIES
7.14 Atomic radius
7.15 Ionic radius
7.16 Ionization energy
7.17 Ionization energy and metallic character
7.18 The inert-pair effect
7.19 Diagonal relationships
7.20 Electron affinity

CHEMISTRY AND THE PERIODIC TABLE
7.21 The *s*-block elements
7.22 The *p*-block elements
7.23 The *d*-block elements

Colors often fill the night sky. On Earth, their source may be a fireworks display. Far above, the colors are those of stars, luminous interstellar gas, or a rare supernova. All these colors have the same origin: they come from energetically excited atoms. Atoms are excited to high energies in burning fireworks and the even hotter surfaces of stars; they then throw off that excess energy as light of different colors. The precise colors emitted by an atom depend on how its electrons are arranged. So, by investigating the colors atoms emit, we can determine their internal structure. A knowledge of this structure is the key to understanding the periodic table, the properties of elements, what compounds they can form, the reactions they undergo, molecular shapes—and even the colors of fireworks and stars.

In Section 1.3 we learned about the nuclear atom. We saw that an atom of atomic number Z consists of a central, tiny, dense, positively charged nucleus surrounded by Z electrons. By the **electronic structure** of an atom, we mean how its electrons are arranged around its nucleus. When Rutherford proposed the nuclear atom, he expected to be able to use **classical mechanics**, the laws of motion proposed by Newton in the seventeenth century, to describe its electronic structure. However, it soon became clear that classical mechanics fails when it is applied to electrons in atoms. New laws, which came to be known as **quantum mechanics**, were developed in the early part of the twentieth century. Their introduction caused an intellectual earthquake that shook science to its foundations. We shall see a little of that earthquake in this chapter.

OBSERVING ATOMS

The laboratory version of watching fireworks is called **spectroscopy,** the analysis of the light emitted or absorbed by substances.

7.1 THE CHARACTERISTICS OF LIGHT

The name *spectroscopy* has Latin and Greek origins. It means "a viewer of appearances."

The instrument used for spectroscopy is called a **spectrometer.** Spectroscopy makes use of the fact that atoms emit light of many colors when their compounds are burned or exposed to an electric discharge (Fig. 7.1). The light emitted by a sample in a spectrometer is first reduced to a narrow beam by passing it through a slit (Fig. 7.2). That beam is then separated into its different colors by using a device such as a prism. Finally, the separated colors are recorded photographically. The photographic image shows a picture of the slit the rays of light passed through for each color of light emitted by the atoms; in other words, the individual colors are recorded as **spectral lines.** This set of lines, the **spectrum,** is unique for the atoms of each element and is like a fingerprint of the element. Astronomers use this fingerprint to identify the elements present in distant stars.

We can think of a field as a region of influence, like the gravitational field near the Earth.

To interpret the results of spectroscopy, we need to know about the properties of light. Light is **electromagnetic radiation,** a wave of electric and magnetic fields. All electromagnetic radiation travels through empty space at $3.00 \times 10^8\ \text{m}\cdot\text{s}^{-1}$, or at just over 670 million miles per hour. This speed is denoted *c*.

An electric field pushes on charged particles such as electrons. If we were to monitor the electric field of light as it passed by us, we would find that it pushed in one direction, then the opposite direction, over and over again (Fig. 7.3). That is, the field oscillates in direction and strength. The number of cycles—complete reversals of direction—per second is called the **frequency,** ν (the Greek letter nu), of the radiation. The unit of frequency is the **hertz** (Hz), which is defined as 1 cycle per second:

The unit honors Heinrich Hertz, one of the pioneers of the study of electromagnetic radiation.

$$1\ \text{Hz} = 1\ \text{s}^{-1}$$

Electromagnetic radiation of frequency 1 Hz would push in one direction, then the opposite direction, and return to the original direction in 1 s. The frequency of electromagnetic radiation we see as light is close to

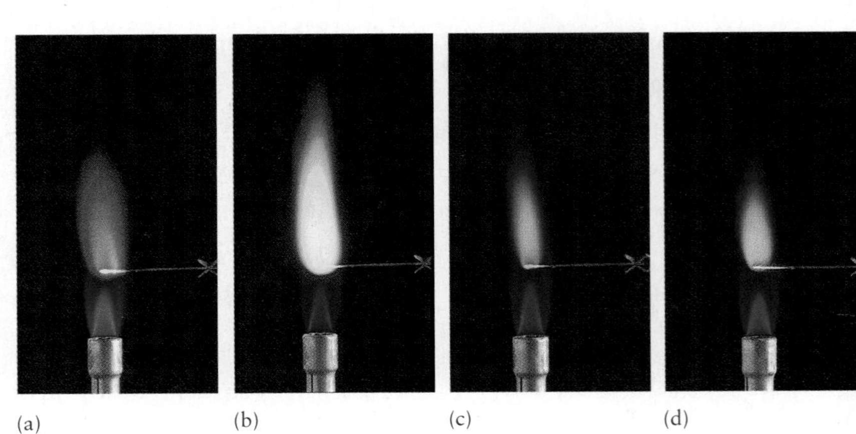

(a) (b) (c) (d)

FIGURE 7.1 Flame tests are used to identify the elements in a compound. In particular, they provide an easy way of distinguishing the alkali metals: (a) lithium; (b) sodium; (c) potassium; (d) rubidium. In each case except lithium, the colors come from energetically excited atoms. In lithium's case, LiOH molecules are responsible for the color of the lithium emissions.

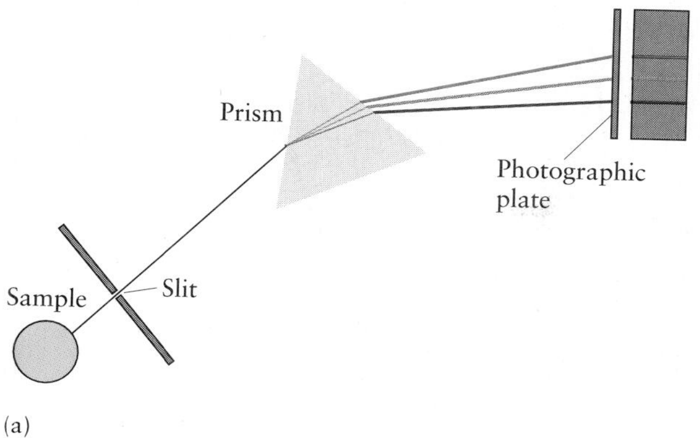

FIGURE 7.2 (a) In a spectrometer, the light emitted by an energetically excited sample of an element is passed through a slit, to give a narrow ray, and then through a prism. The prism separates the ray into different colors, which are recorded photographically. The spectral lines on the photograph are the separate images of the slit. (b) A rainbow is formed when white light from the Sun is split into its component colors by raindrops that act as tiny prisms. The light enters the front of the raindrop, reflects from the back, and emerges from the front. Secondary rainbows are formed when the light reflects a second time inside the drop.

5×10^{15} Hz, so its electric field changes direction more than a thousand trillion (10^{15}) times a second as it travels past us.

The frequency determines the color of light (see Table 7.1). Our eyes detect different colors because they respond in different ways to light of different frequencies. When the electric field oscillates at about 6.4×10^{14} Hz, for example, we see blue light. The light from a traffic signal changes frequency from about 5.7×10^{14} Hz to 5.2×10^{14} Hz and then to 4.3×10^{14} Hz as it changes from green to yellow and then to red.

An instantaneous snapshot of a wave of electromagnetic radiation spread through space would look like Fig. 7.4. This wave is characterized by its **wavelength,** λ (the Greek letter lambda), the peak-to-peak distance. The wavelengths of visible light are close to 500 nm, where 1 nm = 10^{-9} m. Although 500 nm is only half of one-thousandth of a millimeter, it is much longer than the diameters of atoms, which are typically close to 0.2 nm.

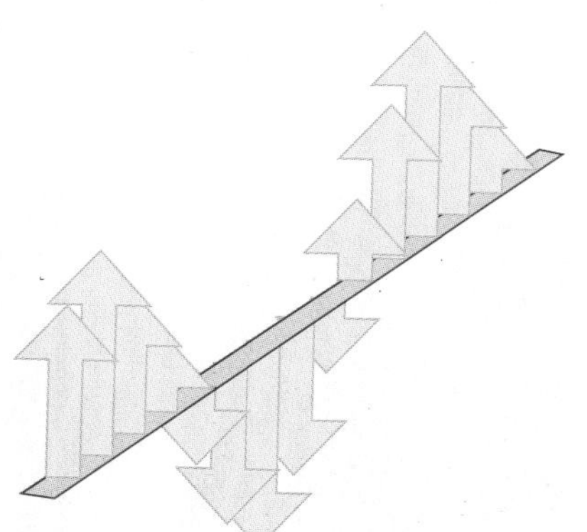

FIGURE 7.3 The electric field of electromagnetic radiation oscillates in space and time. The length of an arrow at any point at a given instant represents the strength of the force that the field exerts on a charged particle at that point. Note the wavelike distribution of the field.

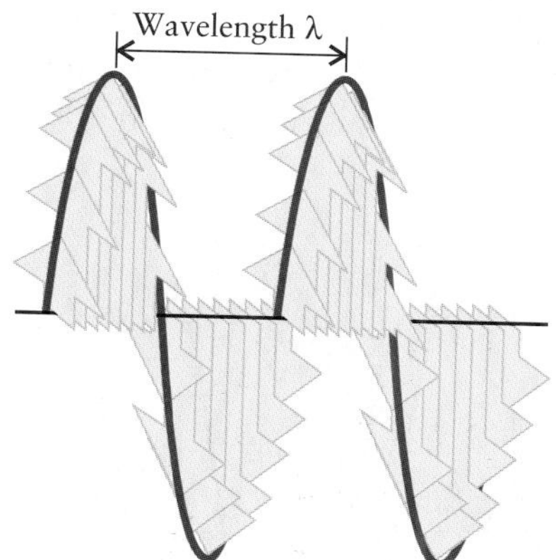

FIGURE 7.4 This diagram represents an instantaneous snapshot of an electromagnetic wave at a given instant. The distance between the peaks is the wavelength of the radiation. The height of the peaks depends on the intensity of the radiation.

FIGURE 7.5 Wavelength and frequency are closely related. The two parts of this illustration show the electric field at three instants that you might experience as a wave flashed by a single point at the speed of light from left to right. (a) Short-wavelength radiation: note how the electric field changes markedly at the three successive instants. (b) For the same three instants, the electric field of the long-wavelength radiation changes much less. Short-wavelength radiation has a high frequency, whereas long-wavelength radiation has a low frequency.

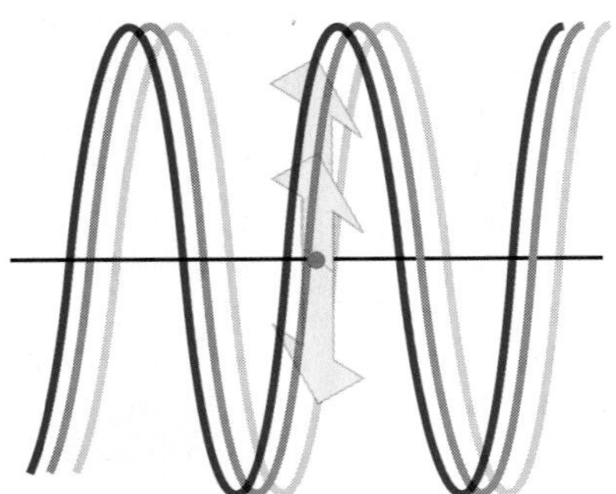

(a) Short wavelength, high frequency (b) Long wavelength, low frequency

The light wave in Fig. 7.4 is zooming along at a speed c. At any given point, this motion results in an oscillation of the field (Fig. 7.5). If the wavelength of the light is very short, very many oscillations pass this point in a second. If the wavelength is long, the light is still traveling at the same speed c, but fewer oscillations would pass the point in a second. That is, *short wavelength corresponds to high frequency; long wavelength corresponds to low frequency.* The precise relation is

$$\lambda \times \nu = c \qquad (1)$$

For example, the wavelength of blue light is

$$\lambda = \frac{c}{\nu} = \frac{3.00 \times 10^{8}\ \text{m·s}^{-1}}{6.4 \times 10^{14}\ \text{s}^{-1}} = 4.7 \times 10^{-7}\ \text{m}$$

or about 470 nm. Blue light, which has a relatively high frequency, has a shorter wavelength than low-frequency red light (700 nm). The light from the three lamps of a traffic signal changes from about 530 nm to 580 nm and then to 700 nm as they change from green through yellow to red (Fig. 7.6).

Our eyes detect electromagnetic radiation with wavelengths in the range 700 nm (red light) to 400 nm (violet light). Radiation in this range

FIGURE 7.6 The color of electromagnetic radiation is determined by its wavelength (and frequency). (a) The three rays shown here are drawn to scale and show that the wavelengths of green, yellow, and red light increase in that order. The perception of color arises from the effect of the radiation on our eyes and the response of our brain. (b) Each lamp in a traffic signal generates white light, a mixture of all colors, but the tinted glass screens allow only certain wavelengths to pass through.

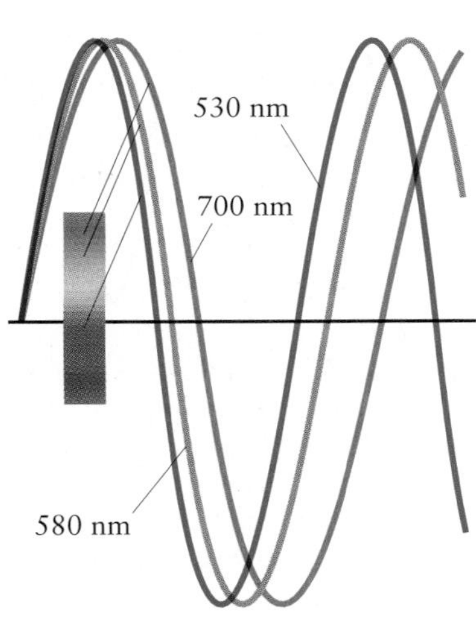

(a)

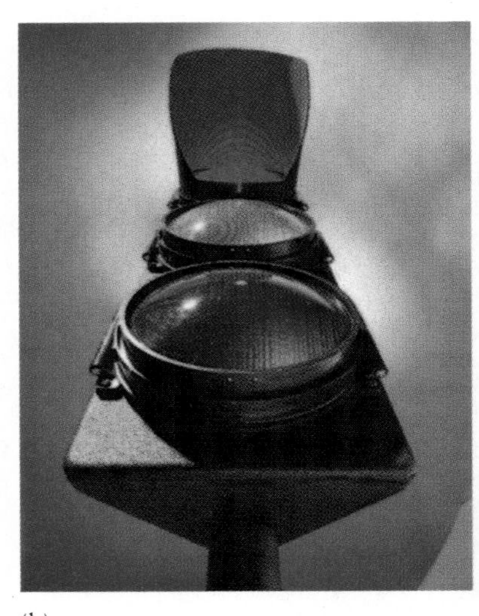

(b)

is called **visible light.** White light, which includes sunlight, is a mixture of all frequencies of visible light. Electromagnetic radiation can also include wavelengths ranging from less than a picometer to more than several kilometers (Fig. 7.7). **Ultraviolet radiation** is radiation at higher frequency than violet light; its wavelength is less than about 400 nm. This damaging radiation is responsible for sunburn and tanning. It is largely prevented from reaching the surface of the Earth by the ozone layer. **Infrared radiation,** the radiation we experience as heat, has a frequency lower than that of red light; its wavelength is greater than about 800 nm. **Microwaves,** which are used in radar and microwave ovens, have wavelengths in the millimeter to centimeter range.

There is no lower or upper limit to the wavelength of electromagnetic radiation.

The color of light depends on its frequency or wavelength; long-wavelength radiation has a lower frequency than short-wavelength radiation.

SELF-TEST 7.1A What is the wavelength of orange light of frequency 4.8×10^{14} Hz?

[*Answer:* 620 nm (2 sf)]

SELF-TEST 7.1B Which color has the higher frequency, orange or yellow?

7.2 QUANTA AND PHOTONS

According to classical mechanics, an object can have any total energy—high, low, and anything in between. For example, a pendulum seems to lose energy continuously by friction, and we seem to be able to give it any energy simply by pushing it harder. However, very precise experiments have shown that classical mechanics is wrong: objects can accept energy only in discrete amounts, or **quanta.** We say that the energy of an object is **quantized,** or restricted to a series of discrete values. Quantization is like pouring water into a bucket. Water *seems* to be a continuous fluid, and it seems that any amount can be transferred. However, the smallest amount of water that we can transfer is one H_2O molecule. Classical mechanics gives excellent agreement between theory and observation for large objects, such as baseballs and buildings, jet planes and planets. However, only quantum mechanics can explain the behavior of objects as small as electrons.

The word *quantum* comes from the Latin for amount—literally, "How much?".

A light ray is a stream of energy emitted by the source. According to quantum mechanics, we should regard the ray as being made up of discrete packets of electromagnetic energy. These packets are called **photons.** The more intense the light, the greater the number of photons passing a given point. A dim source of light emits relatively few photons; a bright source of light emits a dense stream of photons. We can feel the energy of the photons in the infrared radiation emitted by the Sun as we stand in warm sunlight.

Experiments have shown that the energy of a photon is proportional to the frequency of the radiation. So, photons of ultraviolet radiation are more energetic, and hence more damaging, than photons of visible light.

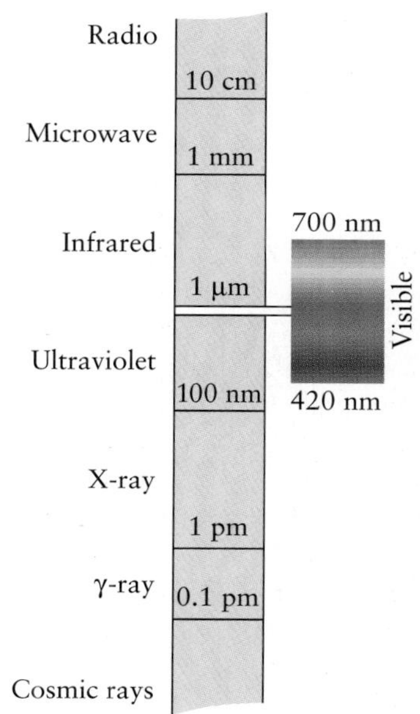

FIGURE 7.7 The electromagnetic spectrum and the names of its regions. Note that the region we call "visible light" occupies a very narrow range of wavelengths. The regions are not shown to scale.

We write

$$E = h\nu \tag{2}$$

The constant is named for Max Planck, the German scientist who introduced the idea that energy is transferred in packets.

where h is the **Planck constant:**

$$h = 6.63 \times 10^{-34}\ \text{J·s}$$

For example, in blue light of frequency 6.4×10^{14} Hz, each photon has an energy

$$E = (6.63 \times 10^{-34}\ \text{J·s}) \times (6.4 \times 10^{14}\ \text{Hz}) = 4.2 \times 10^{-19}\ \text{J}$$

To derive this value, we have used 1 Hz = 1 s^{-1}, so

$$(\text{J·s}) \times \text{Hz} = (\text{J·s}) \times \text{s}^{-1} = \text{J}$$

A ray of red light also consists of a stream of photons, but because the frequency of the light is lower, each of its photons has less energy (2.8×10^{-19} J). We see different colors because photons of different energies cause different effects in our eyes. The photon energies of light of various colors are included in Table 7.1. Ultraviolet light is damaging because its photons have enough energy to break chemical bonds when they strike matter. X-ray photons are even more energetic, which is why we have to take special precautions when using them.

A typical lamp emits huge numbers of photons. For example, suppose a lamp emits 25 J of yellow (580 nm) light in 1 s. The energy of one photon of yellow light is 3.4×10^{-19} J, so the number of photons that accounts for 25 J of energy is

$$\begin{aligned}\text{Number of photons} &= \frac{\text{total energy}}{\text{energy of one photon}}\\ &= \frac{25\ \text{J}}{3.4 \times 10^{-19}\ \text{J}} = 7.4 \times 10^{19}\end{aligned}$$

In other words, when you turn on an electric lamp, it generates about 10^{20} photons of yellow light each second.

TABLE 7.1 Color, frequency, and wavelength of electromagnetic radiation

Radiation type	Frequency, 10^{14} Hz	Wavelength, nm (2 sf)	Energy per photon, 10^{-19} J
x-rays and γ-rays	$\geq 10^3$	≤ 3	$\geq 10^3$
ultraviolet	8.6	350	5.7
visible light			
violet	7.1	420	4.7
blue	6.4	470	4.2
green	5.7	530	3.8
yellow	5.2	580	3.4
orange	4.8	620	3.2
red	4.3	700	2.8
infrared	3.0	1000	2.0
microwaves and radio waves	$\leq 10^{-3}$	$\geq 3 \times 10^6$	$\leq 10^{-3}$

Light can be regarded as consisting of discrete packets of energy called photons. The energy of a photon is directly proportional to the frequency of the radiation.

SELF-TEST 7.2A A certain lamp produces 5.0 J of energy per second in the blue region of the spectrum. How many photons of blue (470 nm) light would it generate if it were left on for 8.5 s?

[*Answer:* 1.0×10^{20}]

SELF-TEST 7.2B A certain lamp produces 25 J of energy per second in a certain region of the spectrum. In 1.0 s it emits 5.5×10^{19} photons of light in that region. What is the energy of one photon?

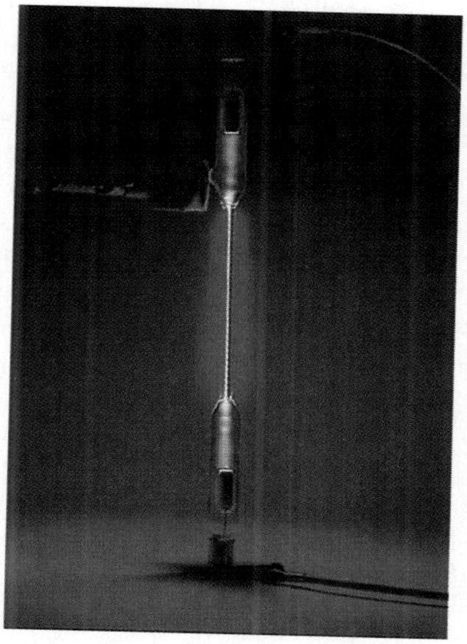

7.3 ATOMIC SPECTRA AND ENERGY LEVELS

Now let's see how these ideas help us to interpret atomic spectra. Many frequencies of light are emitted when an electric current is passed through a low-pressure sample of hydrogen gas (Fig. 7.8). The current—which is like a storm of electrons—breaks up the H_2 molecules and excites the resulting hydrogen atoms to higher energies. These atoms discard their excess energy by giving off electromagnetic radiation; then they combine to form H_2 molecules again. The visible region of the spectrum of atomic hydrogen consists of four lines. The brightest line (at 656 nm) is red, and the excited atoms in the gas glow with this red light. Energy-rich hydrogen atoms also emit ultraviolet and infrared radiation, which can be detected electronically and photographically.

FIGURE 7.8 The red glow from this hydrogen discharge lamp comes from excited hydrogen atoms that are returning to a lower energy state and emitting the excess energy as visible radiation.

The spectrum of atomic hydrogen is astonishing (Fig. 7.9): it consists of a series of discrete lines. How can an atom emit only particular frequencies of radiation and not all possible frequencies? The answer must be that it can lose energy only in certain, discrete amounts. This, in turn, suggests that the electron in the atom can exist only in a series of discrete states, called **energy levels.** When an electron makes a **transition,** it changes from one of these energy levels to another, and the difference in energy, ΔE, is carried away as a photon. Because the energy of a photon is $h\nu$, where h is Planck's constant and ν is the frequency of the radiation to which the photon contributes, it follows that

$$\Delta E = h\nu \qquad (3)$$

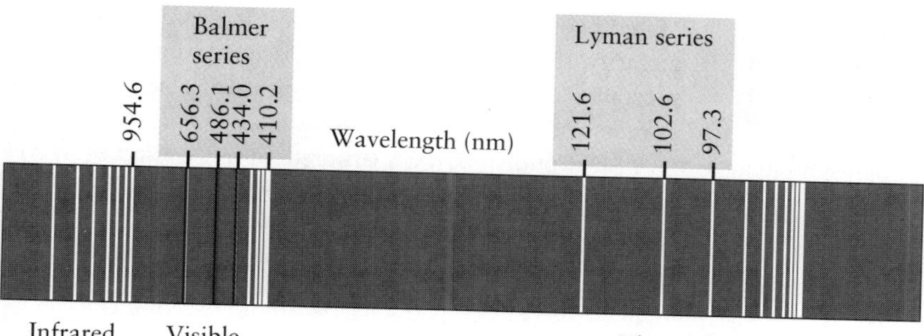

FIGURE 7.9 The spectrum of atomic hydrogen. The spectral lines have been assigned to various groups called series, two of which are shown with their names.

CASE STUDY 7 ABSORPTION SPECTRA

IN 1835 THE philosopher Auguste Comte remarked in reference to the Sun, stars, and planets, "we understand the possibility of determining their shapes, their distances, their sizes and motions, whereas never by any means will we be able to study their chemical composition . . .". Well, science has moved on, and we now know a great deal about their chemical composition. Apart from the visits by space probes to the planets, the major technique used to investigate the composition of heavenly bodies is spectroscopy. Spectroscopy is also used to determine the concentrations of solutes in laboratory samples of solutions.

The amount of light absorbed by a sample of a solution is called the *absorbance, A,* of the solution. The absorbance is the logarithm of the ratio of the intensity of incident light (the light that shines on the sample), I_0, to that of transmitted light (the light that was not absorbed), I:

$$A = \log\left(\frac{I_0}{I}\right)$$

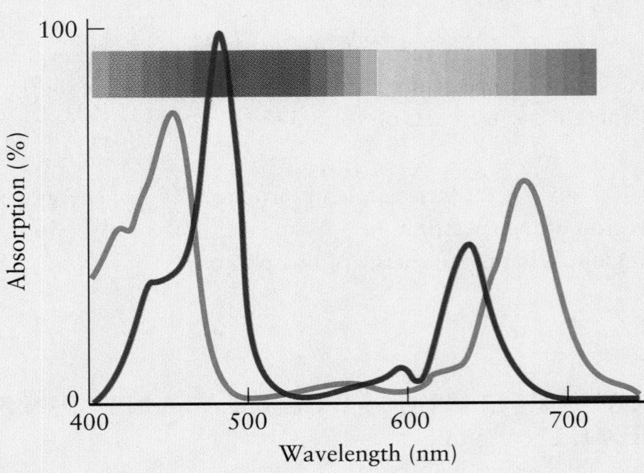

The optical absorption spectrum of chlorophyll as a plot of percentage absorption against wavelength. Chlorophyll *a* is shown in red, chlorophyll *b* in blue.

The side of this spectrophotometer has been opened so that we can see the spectrum generated by the diffraction grating inside. The grating is rotated so that the light of the desired wavelength falls on the sample.

The absorbance is proportional to the molar concentration of the absorbing species and the thickness of the sample it passes through. This is *Beer's law:*

$$A = \text{constant} \times (\text{molar concentration}) \times (\text{sample thickness})$$

The constant depends on the identity of the absorbing substance and is called the *molar absorption coefficient.* Beer's law allows us to determine the concentration of a substance in solution in a sample container of known width by measuring how much of certain wavelengths of light it absorbs.

A *spectrophotometer* is an instrument used to monitor the absorption of light. It consists of a light source, a diffraction grating, a sample holder, and a detector. The grating is used to select the wavelength of the incoming light. The absorbance is proportional to the molarity of the solution. Therefore, after the spectrophotometer has been calibrated by measuring the absorbance of a sample of known molarity, the molarity of an unknown solution of the same substance can be determined by measuring its absorbance of radiation of the same wavelength.

Absorption can be measured over a range of wavelengths by gradually rotating the grating to select a different wavelength of incident light. The graph of absorbance against the wavelength of the light is called the *absorption spectrum* of the substance. The illustration shows the absorption spectrum of two forms of chlorophyll. Note how the compounds absorb red and blue light strongly, but not green light. That is why vegetation looks green to us, because only green light is reflected from it.

When the ratio of transmitted to incident light is expressed as a percentage, it is called the *percentage transmittance, T,* of the sample:

$$T = \frac{I}{I_0} \times 100\%$$

Percentage transmittance can also be measured over a range of wavelengths, to produce spectra.

QUESTIONS

1. Suppose a sample of a solution of a substance of molarity 0.10 M gives a 65.0% transmittance of light at a certain wavelength. What would be the concentration of the same species when the percentage transmittance was measured as 40.0% at the same wavelength and in the same sample holder?

2. (a) Plot the following data obtained from a spectrophotometer set at 532 nm and equipped with a cell 1.0 cm thick. Use the graph to determine the molar absorption coefficient of the absorbing species:

A	Molarity, $mol \cdot L^{-1}$
0.12	3.2×10^{-3}
0.27	7.2×10^{-3}
0.43	1.1×10^{-2}
0.66	1.8×10^{-2}
0.94	2.5×10^{-2}

(b) A solution containing an unknown concentration of the same absorbing species has an absorbance of 0.78. What is the molar concentration of the absorbing species?

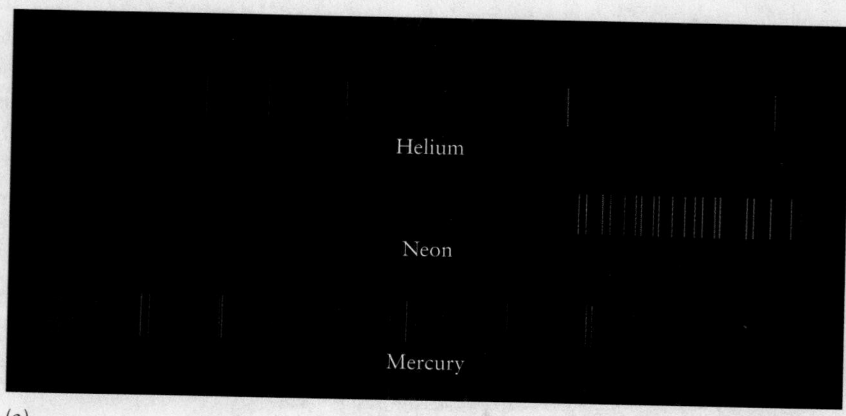

(a)

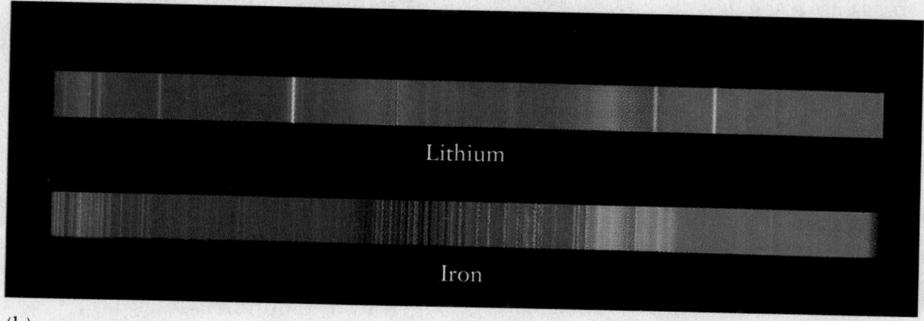

(b)

(a) The emission spectra of the elements display the wavelengths of light emitted by their excited atoms. (b) The absorption spectra of the elements are the reverse of their emission spectra: the wavelengths of light absorbed by the atoms appear as dark lines.

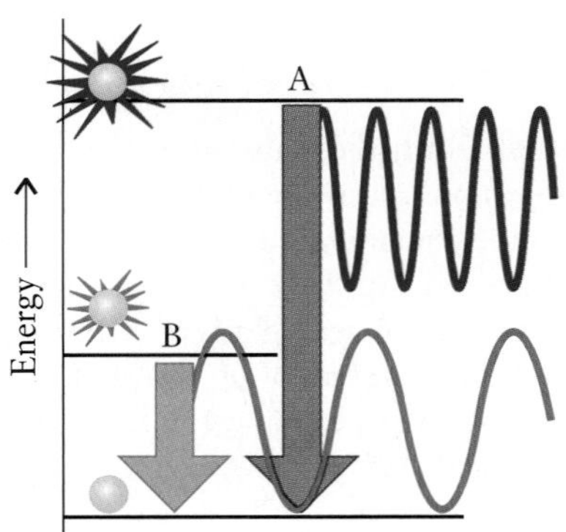

FIGURE 7.10 When an atom undergoes a transition from a state of higher energy to one of lower energy, it loses energy that is carried away as a photon. The greater the energy loss, the higher the frequency (and the shorter the wavelength) of the radiation emitted. Thus, transition A generates light with a higher frequency than transition B.

Niels Bohr was a Danish physicist. He devised the first model of atomic structure that could explain spectral lines.

This relation is called the **Bohr frequency condition.** Each spectral line arises from a specific transition (Fig. 7.10). Because only certain energy differences can occur, only certain frequencies are present in the spectrum. These frequencies are given by

$$\nu = \frac{\Delta E}{h} \tag{4}$$

By analyzing the appearance of the spectrum, we can build up a picture of the ladder of energy levels for the atom (Fig. 7.11). The ladder is rather peculiar, because the rungs get closer together the higher we climb!

The observation of discrete spectral lines suggests that an electron in an atom can have only certain energies. Transitions between these energy levels generate photons in accord with the Bohr frequency condition.

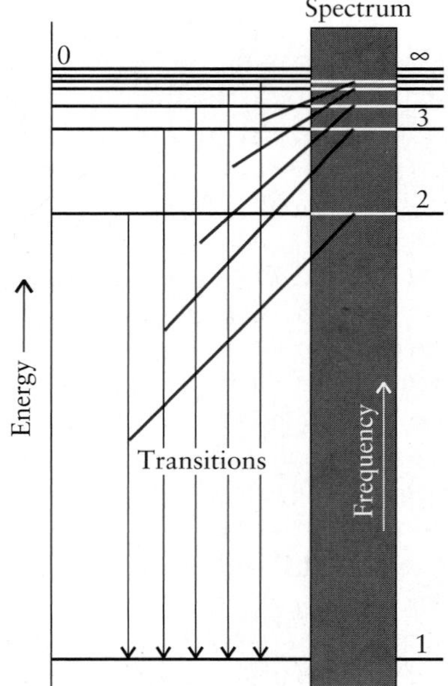

FIGURE 7.11 The spectrum of atomic hydrogen (reproduced on the right) tells us the arrangement of the energy levels of the atom because the frequency of the radiation emitted in a transition is proportional to the energy difference between the two energy levels involved. The 0 on the energy scale corresponds to the completely separated proton and electron. The numbers on the right label the energy levels: they are examples of quantum numbers (see Section 7.6).

7.4 THE WAVELIKE PROPERTIES OF ELECTRONS

Our next step is to come up with an explanation of the ladder of energy levels revealed by spectroscopy. It appears that electrons can have only certain, discrete energies in atoms. We should be reminded of a bell, which emits only specific frequencies of sound when it is struck, or a violin string. Like a string of fixed length bound in a violin, a bound electron is confined within an atom. Can its confinement somehow lead to its giving off only certain energies, like the notes of a violin?

According to quantum mechanics, an electron behaves like a wave as well as like a particle, and it can have only certain wavelengths inside an atom, just as a taut violin string can support only certain wavelengths. In other words, we have a paradox: a particle, an electron, can behave like a wave. We have seen that a light ray, which classically is treated as a wave, should also be thought of as a stream of photons. The same **wave-particle duality,** or combined wavelike and particlelike character of light, applies to electrons, too.

In 1924 the French scientist Louis de Broglie suggested that all matter—not just electrons—has wavelike properties. The **de Broglie relation** states that the wavelength, λ, of a particle is related to its mass and velocity by

$$\lambda = \frac{h}{\text{mass} \times \text{velocity}} \qquad (5)$$

According to this formula, a heavy particle traveling rapidly has a short wavelength and a slow-moving particle of small mass has a relatively long wavelength. An electron traveling at 1 $m{\cdot}s^{-1}$ has a wavelength of 700 m; but when it is traveling at high speed in an atom, its wavelength becomes comparable to the diameter of the atom, about 10^{-12} m.

One of the first experiments to confirm the wavelike character of electrons was carried out by the American scientists Clinton Davisson and Lester Germer in 1927. They knew that when light waves pass through a grid with a spacing comparable to the wavelength, characteristic patterns of light and dark intensity, called **diffraction patterns,** are obtained. Davisson and Germer showed that electrons reflected from a crystal also give a diffraction pattern on a photographic plate. In this case, the layers of atoms act as the grid. They also found that the pattern corresponds exactly to that expected for electrons with a wavelength given by the de Broglie relation (Fig. 7.12).

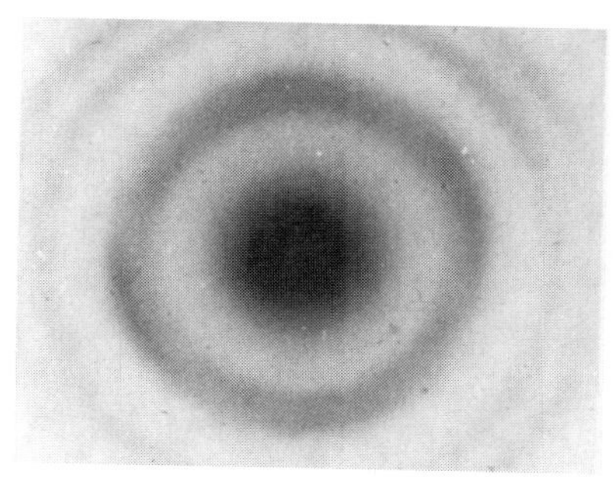

FIGURE 7.12 Davisson and Germer showed that electrons gave a diffraction pattern when reflected from a crystal. G. P. Thomson working in Aberdeen, Scotland, showed that they also give a diffraction pattern when they pass through a very thin gold foil. The latter is shown here. G. P. Thomson was the son of J. J. Thomson, who identified the electron (Section 1.3). Both received Nobel prizes—J. J. for showing that the electron is a particle and G. P. for showing that it is a wave.

Electrons have both wavelike and particlelike properties; their wavelike properties must be taken into account when describing the structure of atoms.

MODELS OF ATOMS

Our task is to combine the wavelike properties of electrons with the nuclear model of the atom and hope to explain the strange ladder of energy levels in a hydrogen atom observed experimentally. The current model of a hydrogen atom combines all these features. It was proposed in

1926 by the Austrian scientist Erwin Schrödinger, who devised an equation that describes electrons in terms of quantum mechanics.

7.5 ATOMIC ORBITALS

In classical mechanics we speak of the path of a particle, such as the orbit of a planet around the Sun. However, because of the wavelike properties of electrons, we cannot say that an electron *will* be found at a certain point in the atom. Instead, Schrödinger's equation lets us calculate the *probability* that an electron is at a particular point in space.

The region of space in which an electron is most likely to be found is called an **atomic orbital.** To visualize an atomic orbital, we could think of a cloud surrounding the nucleus, with the density of the cloud representing the probability of finding an electron there. Denser regions of the cloud represent locations where the electron is more likely to be found.

Each different atomic orbital corresponds to an energy level of the electron. The higher the energy, the larger the orbital, so atoms expand somewhat when their electrons are excited to higher levels. When a hydrogen atom is excited, the shape and extent of its cloudlike atomic orbital change to a higher energy form. When the electron falls back to a lower energy orbital, it gives off energy in the form of electromagnetic radiation, which appears as a line in the hydrogen spectrum. We can begin to see that the colors of fireworks and stars are due to the collapse of electron clouds from shapes corresponding to high energies to shapes corresponding to lower energies (Fig. 7.13). The variety of different frequencies emitted shows that transitions take place between a number of different energy levels.

Spectroscopic lines were once classified as *s*harp, *p*rincipal, *d*iffuse, and *f*undamental.

The various shapes of atomic orbitals can be classified into four main types, which are labeled *s*, *p*, *d*, and *f*. There are many orbitals of each

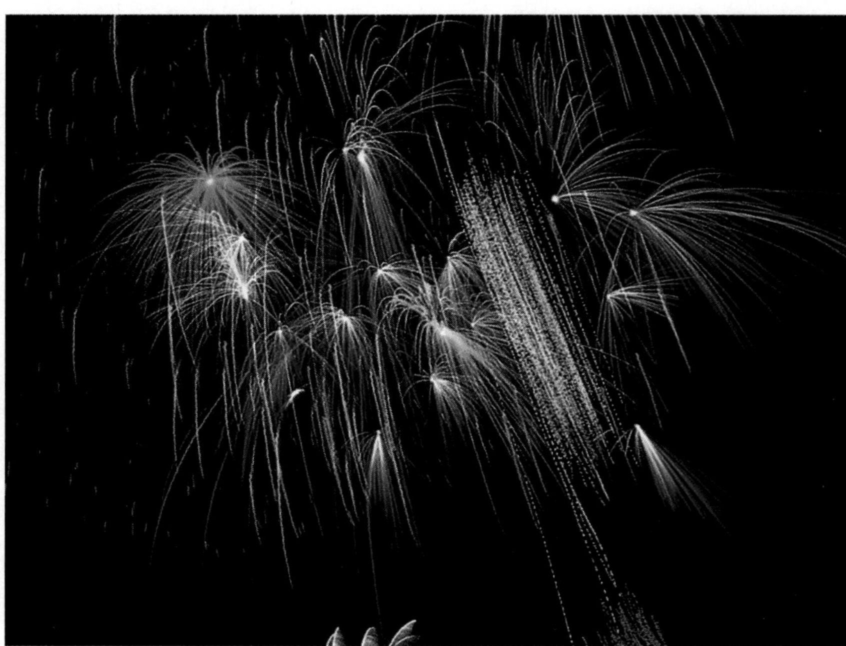

FIGURE 7.13 The oxidation of magnesium powder is responsible for much of the brilliance of a fireworks display. Magnesium gives rise to a bright white light; yellows arise from sodium compounds, reds from strontium compounds, and blue from copper chloride molecules.

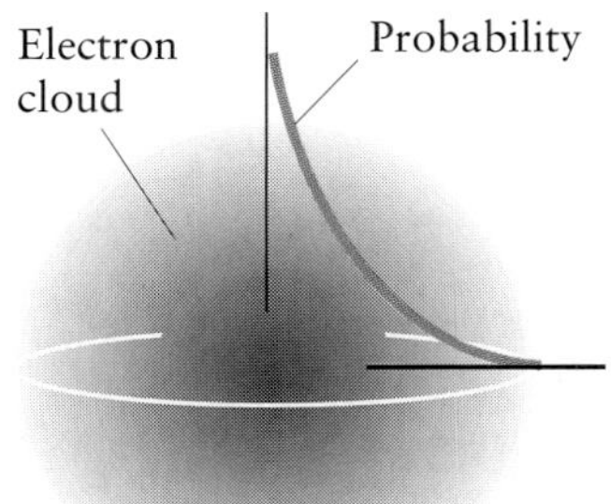

FIGURE 7.14 The three-dimensional electron cloud corresponding to an electron in a 1*s*-orbital of hydrogen. The density of shading represents the probability of finding the electron at any point. The superimposed graph shows how the probability varies with the distance of the point from the nucleus along any radius.

FIGURE 7.15 The simplest way of drawing an atomic orbital is as a boundary surface, a surface within which there is a high probability (typically 90%) of finding the electron. The sphere here represents the boundary surface of an *s*-orbital. We shall use blue to denote *s*-orbitals, but that color is only an aid to their identification.

type: they differ principally in the size of the clouds.

An ***s*-orbital** is a spherical cloud that becomes less dense as the distance from the nucleus increases (Fig. 7.14). In principle, the cloud never thins to exactly zero, so you could think of an atom as being bigger than the Earth! However, there is virtually no chance of finding an electron farther from the nucleus than about 100 pm, so atoms are, in fact, very small. As you can see from the high density of the cloud at the nucleus, an electron in an *s*-orbital has a nonzero probability of being found right at the nucleus itself.

Usually, instead of drawing the *s*-orbital as a cloud, we draw its **boundary surface,** a surface that encloses the densest regions of the cloud. The electron is likely to be found only inside the boundary surface of the orbital. An *s*-orbital has a spherical boundary surface (Fig. 7.15), because the electron cloud is spherical. *s*-Orbitals with higher energies have spherical boundary surfaces of bigger diameter.

A ***p*-orbital** is a cloud with two lobes on opposite sides of the nucleus (Fig. 7.16). The two lobes are separated by a planar region called a **nodal**

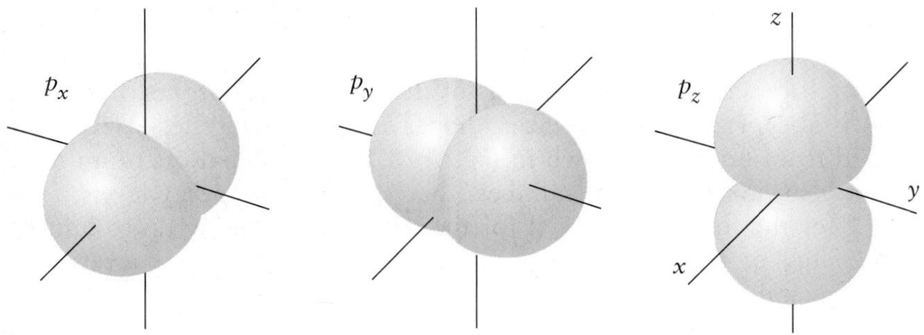

FIGURE 7.16 The boundary surface of a *p*-orbital has two lobes; the nucleus lies on the plane that divides the two lobes, and an electron will, in fact, never be found at the nucleus itself if it is in a *p*-orbital. There are three *p*-orbitals of a given energy, and they lie along three perpendicular axes. We shall use yellow to indicate *p*-orbitals.

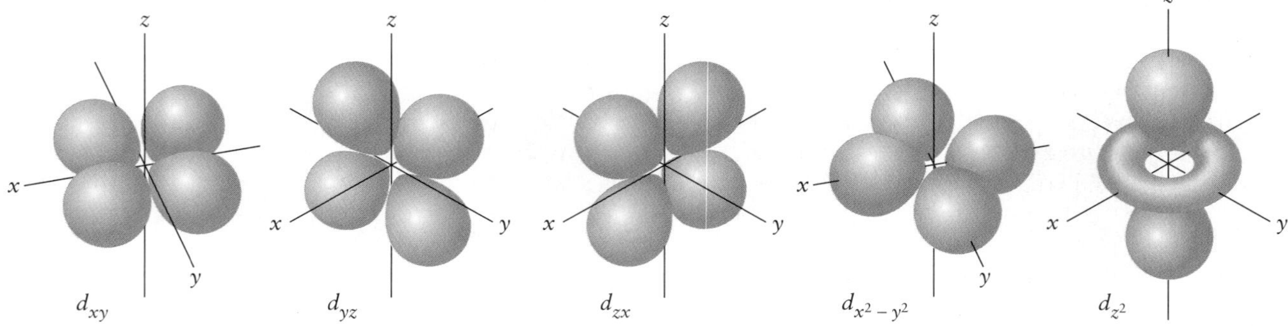

FIGURE 7.17 The boundary surface of a *d*-orbital is more complicated than that of an *s*- or a *p*-orbital. There are five *d*-orbitals of a given energy; four of them have four lobes, one is slightly different and is orientated along an axis. In each case, an electron that occupies a *d*-orbital will not be found at the nucleus. We shall use orange to indicate *d*-orbitals.

plane, which cuts through the nucleus. A *p*-electron will never be found on this plane, so it is never found at the nucleus. This difference from *s*-orbitals is of major importance for understanding the structure of the periodic table (Sections 7.9 through 7.13).

The ***d*-orbitals** and ***f*-orbitals** have more complicated shapes. The five possible orientations of the *d*-orbitals are shown in Fig. 7.17. The shapes of the *f*-orbitals are rarely needed to explain chemical properties, so they are not included here.

> ***The location of an electron in an atom is best described as a cloud of probable locations. In chemistry, the most important shapes of the clouds are the spherical s-orbitals, the two-lobed p-orbitals, and the more complicated d-orbitals.***

7.6 QUANTUM NUMBERS AND ATOMIC ORBITALS

Schrödinger found that each atomic orbital is identified by three numbers called **quantum numbers.** One quantum number is called the principal quantum number, n; the other two are the azimuthal quantum number, l, and the magnetic quantum number, m_l. These quantum numbers have another job: as well as labeling the orbital, they tell us about the properties of the electron that occupies a given orbital.

The **principal quantum number** specifies the energy of an electron in a hydrogen atom. Thus, Schrödinger found that the allowed energies of the electron in a hydrogen atom are

$$E = -\frac{h\mathcal{R}}{n^2} \qquad n = 1, 2, \ldots \tag{6}$$

where h is the Planck constant and $\mathcal{R}$ is a constant called the **Rydberg constant;** its value is 3.29×10^{15} Hz. The value of the Rydberg constant was first obtained from an analysis of the spectrum of atomic hydrogen, but it is also predicted by the Schrödinger equation. The energy levels calculated from Eq. 6 are shown in Fig. 7.18. The principal quantum number, n, is an integer (whole number) that labels these energy levels, from $n = 1$ for the first level, $n = 2$ for the second, continuing up to infinity. The energy of the bound electron climbs up the ladder of levels as n increases and reaches the top of the ladder, corresponding to $E = 0$ and freedom, only when n has reached infinity. The minus sign in Eq. 6

means that the energy of an electron bound in a hydrogen atom is *lower* than when it is free.

The energy levels calculated from Eq. 6 are exactly those required to account for the spectrum of atomic hydrogen (recall Fig. 7.11). The lowest energy of all is obtained when $n = 1$. This lowest energy state is called the **ground state** of the atom. A hydrogen atom is normally found in its ground state, with its electron in the level with $n = 1$.

EXAMPLE 7.1 Identifying a line in the hydrogen spectrum

Calculate the wavelength of a photon emitted by a hydrogen atom when an electron makes a transition between the third ($n = 3$) and second ($n = 2$) principal quantum levels. Identify in Fig. 7.9 the spectral line produced by this transition.

STRATEGY Calculate the energy difference between two principal quantum levels from Eq. 6, and then calculate the wavelength corresponding to that energy difference. The energy difference between a level with quantum number n_u and energy $-h\mathcal{R}/n_u^2$ and another level with quantum number n_l and energy $-h\mathcal{R}/n_l^2$ is

$$\Delta E = -\frac{h\mathcal{R}}{n_u^2} - \left(\frac{h\mathcal{R}}{n_l^2}\right) = h\mathcal{R} \times \left(\frac{1}{n_l^2} - \frac{1}{n_u^2}\right)$$

To use this expression, substitute the values of the principal quantum numbers and the value for $\mathcal{R}$. For n_l use the lower value of n; for n_u use the higher value. It is best to leave the expression as a multiple of h, because that constant will cancel in the next step, in which this energy is expressed as a frequency by using Eq. 3. Finally, convert that frequency to a wavelength by using Eq. 1. In the last step, use $1\ \text{Hz} = 1\ \text{s}^{-1}$.

SOLUTION The energy difference between the two states is

$$\Delta E = h \times (3.29 \times 10^{15}\ \text{Hz}) \times \left(\frac{1}{2^2} - \frac{1}{3^2}\right)$$
$$= h \times (4.57 \times 10^{14}\ \text{Hz})$$

Then, from Eq. 3, the frequency of the emitted light is

$$\Delta E = h\nu = h \times (4.57 \times 10^{14}\ \text{Hz})$$

The h cancels on each side, to give $\nu = 4.57 \times 10^{14}$ Hz. Therefore, the wavelength of the radiation is

$$\lambda = \frac{c}{\nu} = \frac{3.00 \times 10^{8}\ \text{m}\cdot\text{s}^{-1}}{4.57 \times 10^{14}\ \text{Hz}} = 6.56 \times 10^{-7}\ \text{m, or 656 nm}$$

The transition gives the red line in the spectrum.

SELF-TEST 7.3A Repeat the calculation for the transition from the state with $n = 4$ to $n = 2$ and identify the spectral line in Fig. 7.9

[*Answer:* 486 nm; blue line]

SELF-TEST 7.3B Repeat the calculation for the transition from the state with $n = 5$ to $n = 2$ and identify the spectral line in Fig. 7.9

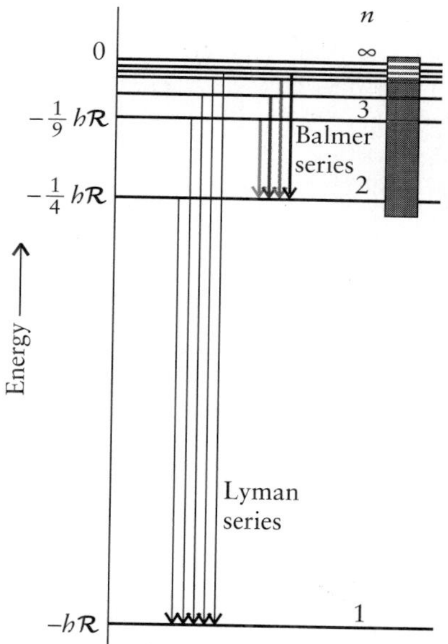

FIGURE 7.18 The permitted energy levels of a hydrogen atom as calculated from Eq. 6. The 0 on the energy scale corresponds to the completely separated proton and electron; the lowest energy state lies at $h\mathcal{R}$ below the zero of energy. The levels are labeled with the principal quantum number *n*, which ranges from 1 (for the lowest state) to infinity (for the separated proton and electron). Compare this diagram with the experimentally determined array of levels shown in Fig. 7.11.

Because the clouds representing the orbitals get bigger as n increases, the average distance of an electron from the nucleus also increases as n

Azimuthal quantum number, l	Subshell label
0	s
1	p
2	d
3	f
4	g
5	h

increases. On average, an electron is closest to the nucleus in the ground state ($n = 1$). The orbitals form a series of thick **shells,** something like fuzzy layers of an imaginary onion. Shells of higher n surround the inner shells of lower n.

The second quantum number, the **azimuthal quantum number,** l, governs the shape of the orbital. It can have the following values:

$$l = 0, 1, 2, \ldots, n - 1$$

Each value of l corresponds to one of the orbital shapes. Thus, $l = 0$ corresponds to s; $l = 1$ to p; $l = 2$ to d; and $l = 3$ to f; as shown in the table in the margin. The energy expression, Eq. 6, does not involve l, so all the orbitals of a given shell, regardless of the value of l, have exactly the same energy in a hydrogen atom.

Notice in Table 7.2 that l stops at $n - 1$. If $n = 1$, then the only value of l allowed is 0, corresponding to an s-orbital. It follows that the lowest energy level, corresponding to $n = 1$, must be an s-orbital. When $n = 2$, corresponding to the next higher energy level, l can have two values, 0 or 1. Therefore, when a hydrogen atom has been excited to an energy level of the second shell, its electron may be found in either the s- or p-orbitals of that shell. Only when n has reached 3 can there be orbitals with $l = 2$ (d-orbitals) as well as s- and p-orbitals. When n has reached 4, f-orbitals ($l = 3$) occur.

All the orbitals with a given value of the azimuthal quantum number, l, are said to belong to the same **subshell** of a given shell. For example, in the shell with $n = 3$, the s-orbital forms one subshell, the three p-orbitals form another subshell, and the five d-orbitals form a third subshell. Because l can take the values $0, 1, 2, \ldots, n - 1$, corresponding to n different values in all, there are n subshells in a given shell. There is only one subshell of the shell with $n = 1$ (the subshell with $l = 0$), two subshells of the shell with $n = 2$ (those with $l = 0$ and 1), and so on.

The third quantum number, the **magnetic quantum number,** m_l, labels the different orbitals of a given subshell. The allowed values of m_l are

$$m_l = l, l - 1, l - 2, \ldots, -l$$

For example, for the subshell with $l = 1$, which consists of p-orbitals, m_l can have the values $m_l = +1$, 0, and -1, so there are three p-orbitals in the subshell. These three orbitals are normally denoted p_x, p_y, and p_z.

TABLE 7.2 Quantum numbers for electrons in atoms

Name	Symbol	Values	Meaning	Indicates
principal	n	$1, 2, \ldots$	labels shell, specifies energy	size
azimuthal*	l	$0, 1, \ldots, n - 1$	labels subshell: $l = 0, 1, 2, 3, 4, \ldots$ $s, p, d, f, g, \ldots$	shape
magnetic	m_l	$l, l - 1, \ldots, -l$	labels orbitals of subshell	direction
spin magnetic	m_s	$+\frac{1}{2}, -\frac{1}{2}$	labels spin state	spin direction

*Also called the *orbital angular momentum quantum number.*

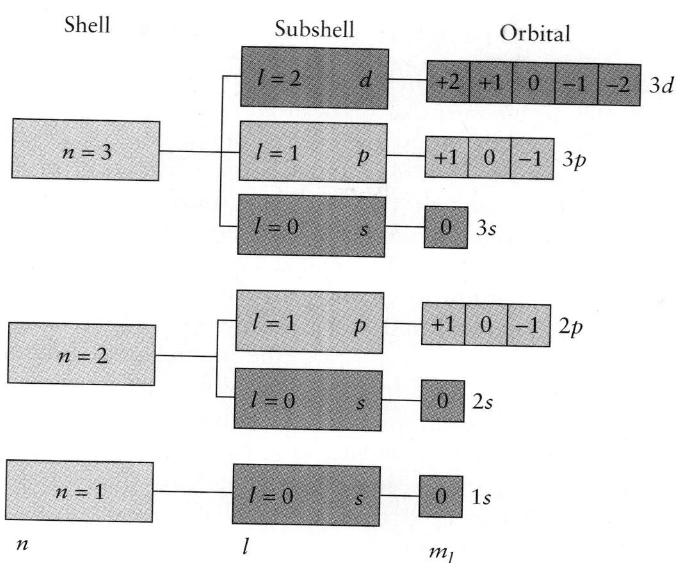

FIGURE 7.19 A summary of the arrangement of shells, subshells, and orbitals in an atom and the corresponding quantum numbers. Note that the quantum number m_l is an alternative label for the individual orbitals: in chemistry it is more common to use x, y, and z as labels, as shown in Figs. 7.16 and 7.17. There is no direct correspondence between axis designation and m_l.

Each label corresponds to a possible orientation of the lobes of the orbitals (recall Fig. 7.16).

In general, a subshell with quantum number l consists of $2l + 1$ individual orbitals of that type (Fig. 7.19). There is one s-orbital ($l = 0$) in a subshell, three p-orbitals ($l = 1$), five d-orbitals ($l = 2$), and seven f-orbitals ($l = 3$).

An orbital is specified by three quantum numbers; orbitals are organized into shells and subshells. The relations between shells, subshells, and orbitals are summarized in Fig. 7.19.

EXAMPLE 7.2 Identifying the number of orbitals in a shell

How many orbitals are there in the shell with $n = 4$?

STRATEGY Decide which subshells are present in the shell, write the number of orbitals in each one, and then add these numbers together. From the discussion above, l has whole-number values from 0 up to $n - 1$ and the number of orbitals in a subshell for a given value of l is $2l + 1$. Combine the two pieces of information.

SOLUTION For $n = 4$, there are four subshells with $l = 0, 1, 2, 3$, consisting of one s-orbital, three p-orbitals, five d-orbitals, and seven f-orbitals, respectively. There are therefore $1 + 3 + 5 + 7 = 16$ orbitals in the shell with $n = 4$ (Fig. 7.20).

SELF-TEST 7.4A Calculate the total number of orbitals in a shell with $n = 6$.

[*Answer:* 36]

SELF-TEST 7.4B What is the general formula for the number of orbitals in a principal quantum level?

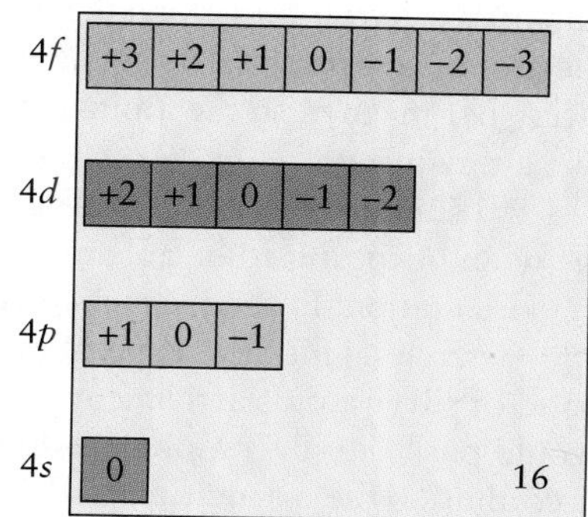

FIGURE 7.20 The solution to Example 7.2. The boxes show the individual orbitals in each subshell and the numbers of each. There are 16 orbitals in a shell with $n = 4$.

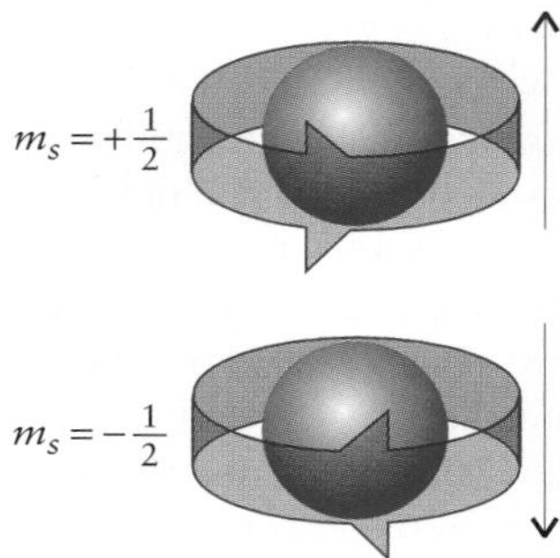

FIGURE 7.21 The two spin states of an electron can be represented as clockwise or counterclockwise rotation around an axis passing through the electron. The two states are labeled by the quantum number m_s and depicted by the arrows shown on the right.

7.7 ELECTRON SPIN

Schrödinger's calculation of the energies of the hydrogen orbitals was a milestone in the development of modern atomic theory. Yet the spectral lines did not have exactly the frequencies he predicted. Two Dutch-American physicists, Samuel Goudsmit and George Uhlenbeck, proposed an explanation for these tiny deviations. They suggested that an electron behaves in some respects like a spinning sphere, something like a planet rotating on its axis. This property is called **spin.**

According to quantum mechanics, an electron has two spin states, represented by the arrows ↑ and ↓. We can think of an electron as being able to spin clockwise at a certain rate (the ↑ state) or counterclockwise at exactly the same rate (the ↓ state). These two spin states are distinguished by a fourth quantum number, the **spin magnetic quantum number, m_s.** This quantum number can have only two values: $+\frac{1}{2}$ indicates an ↑ electron and $-\frac{1}{2}$ indicates a ↓ electron (Fig. 7.21). Box 7.1 describes how the discovery of spin states helped to explain the mysterious results of an important experiment.

An electron has the property of spin; the spin is described by the quantum number m_s, which may have one of two values.

BOX 7.1 THE STERN-GERLACH EXPERIMENT

Electron spin is a very strange property, particularly the idea that it can take one of only two orientations. The experimental confirmation of this fact was obtained by two German scientists in 1920.

Otto Stern and Walter Gerlach shot a stream of silver atoms through a peculiarly shaped magnetic field. A silver atom has one unpaired electron (all the rest of its 47 electrons are paired), so it behaves like a single unpaired electron riding on a heavy platform, the rest of the atom. As a result of its spin, an electron behaves like a tiny bar magnet, so the atom as a whole also behaves like a tiny magnet.

If a spinning electron behaved like a spinning ball, we would expect it to be able to spin with its axis in any orientation. In the magnetic field they used, Stern and Gerlach might have expected to observe a broad band of silver atoms arriving at their detector. The field would push the silver atoms by different amounts according to the orientation of the spin. Indeed, that is what they observed when they carried out the experiment. However, it is a difficult experiment to do properly, because the atoms collide with one another in the beam, and their paths are affected randomly. When Stern and Gerlach repeated their experiment, they used a much less dense beam of atoms, thereby reducing the number of collisions between the atoms. In this experiment they observed, not a single wide blurred band of atoms, but *two* narrow bands (see the illustration). One band corresponds to all the atoms flying through the magnetic field with one orientation of their spin; the other band corresponds to the atoms with opposite spin. The fact that only two bands are observed confirms that an electron spin can have only two orientations.

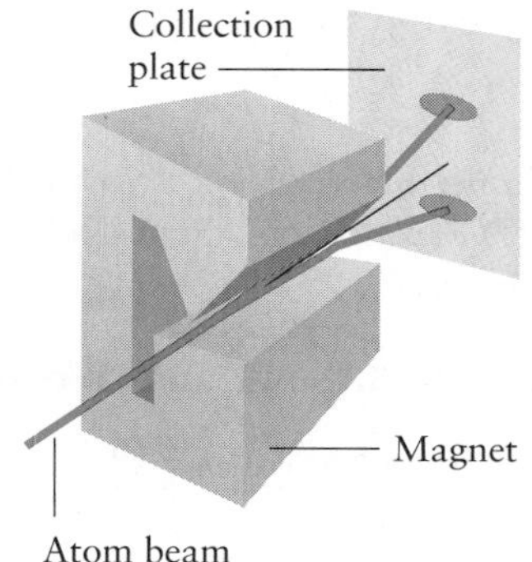

The quantization of electron spin is confirmed by the Stern-Gerlach experiment, in which a stream of atoms splits into two as it passes between the poles of a magnet. The atoms in one stream have an odd ↑ electron, and those in the other an odd ↓ electron.

7.8 THE ELECTRONIC STRUCTURE OF HYDROGEN

Let's review what we now know about the hydrogen atom by imagining what happens to its electron as the atom acquires energy. Initially, the electron is in the lowest energy level, the ground state of the atom, with $n = 1$. The only orbital with this energy is an *s*-orbital, and it is called the 1*s*-orbital. The electron in the ground state of a hydrogen atom is described by the following values of the four quantum numbers:

$$n = 1 \qquad l = 0 \qquad m_l = 0 \qquad m_s = +\tfrac{1}{2} \text{ or } -\tfrac{1}{2}$$

Either spin state is allowed.

When the atom acquires enough energy for its electron to reach the larger $n = 2$ shell, it can occupy any of the four orbitals in that shell. There is one 2*s*- and three 2*p*-orbitals in this shell, and they all have the same energy.

When the atom acquires even more energy, the electron moves into the even larger $n = 3$ shell. There it can occupy any of nine orbitals (one 3*s*, three 3*p*, and five 3*d* orbitals). Still more energy raises it further from the nucleus to the $n = 4$ shell, where 16 orbitals are available (one 4*s*, three 4*p*, five 4*d*, and seven 4*f* orbitals).

The electron is ejected from the atom when it absorbs enough energy to overcome the attraction of the nucleus. It then escapes from the atom, and we say that the atom has been **ionized.** To ionize a hydrogen atom from its ground state, we have to supply enough energy to move the electron from the $n = 1$ shell (which lies at $-h\mathcal{R}$) up to the energy of the separated proton and electron (which is defined as the zero of energy). Therefore, we have to supply $h\mathcal{R}$, or 2.18×10^{-18} J. To ionize 1 mol H atoms therefore requires 6.02×10^{23} times this energy, or 1.31×10^{3} kJ.

The state of a hydrogen atom depends on the energy it possesses; the energy $h\mathcal{R}$ must be provided to ionize a ground-state atom.

SELF-TEST 7.5A The three quantum numbers for an electron in a hydrogen atom in a certain state are $n = 4$, $l = 2$, and $m_l = +1$. In what type of orbital is the electron located?

[*Answer:* 4*d*]

SELF-TEST 7.5B The three quantum numbers for an electron in a hydrogen atom in a certain state are $n = 3$, $l = 1$, and $m_l = -1$. In what type of orbital is the electron located?

THE STRUCTURES OF MANY-ELECTRON ATOMS

All neutral atoms other than hydrogen have more than one electron. The helium atom ($Z = 2$) has two, the lithium atom ($Z = 3$) has three, and a neutral atom of an element with atomic number Z has Z electrons. All these atoms are examples of **many-electron atoms,** or atoms with more than one electron. We can use the concept of atomic orbitals to describe their electronic structures, building on what we have learned about the hydrogen atom. These electronic structures are the key to the form of the periodic table, periodic properties of the elements, and the abilities of atoms to form chemical bonds.

7.9 ORBITAL ENERGIES

The electrons in a many-electron atom occupy orbitals like those of hydrogen. However, the energies of these orbitals are not the same as those of hydrogen. The nucleus of a many-electron atom is more highly charged than the hydrogen nucleus, so it attracts electrons more strongly and hence lowers their energy. The electrons also repel one another, which raises their energy. The combination of these two factors affects the electronic structures and properties of the elements. Both contributions to the energy depend on the relative positions of the nuclei and electrons, so they are contributions to the **potential energy.** The distance dependence of energy is given by **Coulomb's law,** which states that the potential energy of two charges (such as two electrons or an electron and a nucleus) is proportional to their charges and inversely proportional to their separation, *r:*

$$\text{Coulomb potential energy} \propto \frac{\text{charge}_1 \times \text{charge}_2}{r}$$

In general, a potential energy is a contribution to the energy that depends on position. The contribution to the energy that depends on speed is called **kinetic energy,** and for a particle of mass m is given by

$$\text{Kinetic energy} = \tfrac{1}{2} \times m \times (\text{speed})^2$$

The total energy of an atom is the sum of the kinetic energies and potential energies of all its electrons.

In the hydrogen atom, in which there are no electron-electron repulsions, all the orbitals of a given shell have the same energy. For instance, the 2*s*-orbital and all three 2*p*-orbitals have the same energy. The same is true of the 3*s*-orbital, the three 3*p*-orbitals, and the five 3*d*-orbitals of the shell with $n = 3$, which all have the same energy in a hydrogen atom. In many-electron atoms, however, electron-electron repulsions cause the energy of a 2*p*-orbital to be higher than that of a 2*s*-orbital. Similarly, in the $n = 3$ shell, the three 3*p*-orbitals lie higher than the 3*s*-orbital, and the five 3*d*-orbitals lie higher still (Fig. 7.22). The orbitals of a given subshell, though, continue to have the same energy. For example, all three 2*p*-orbitals have the same energy.

The differences in energy of orbitals in different subshells of the same shell is due to the combined effects of the attraction between the electrons and the nucleus and the repulsions between the electrons. As well as being attracted by the nucleus, each electron is repelled by all the other electrons in the atom. As a result, it is less tightly bound to the nucleus than it would be if those other electrons were absent. We say that each

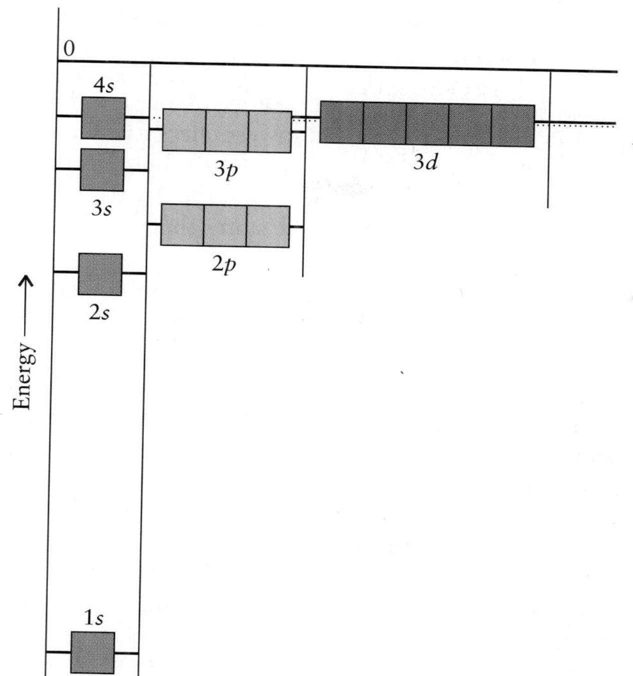

FIGURE 7.22 The relative energies of the shells, subshells, and orbitals in a many-electron atom. Each of the boxes can hold up to two electrons. The energies of the $3d$-orbitals lie below that of the $4s$-orbital after $Z = 20$.

electron is **shielded** from the full attraction of the nucleus by the other electrons in the atom. The shielding effectively reduces the pull of the nucleus on an electron. The **effective nuclear charge** experienced by the electron is always less than the actual charge because the electron-electron repulsions work against the pull of the nucleus.

An *s*-electron of any shell can be found very close to the nucleus, so we say that it can **penetrate** through the inner shells. A *p*-electron penetrates much less. We have seen that its orbital has a node passing through the nucleus, where there is zero probability of finding that electron. This node means that *p*-electrons are more effectively shielded from the nucleus than are *s*-electrons. Because an *s*-electron is less strongly shielded from the nucleus by other electrons, it feels a stronger effective nuclear charge than a *p*-electron. It is bound more tightly: an *s*-electron has a slightly lower (more negative) energy than a *p*-electron of the same shell and is harder to remove from the atom. *d*-Electrons are bound less tightly than *p*-electrons of the same shell because they are even less able to penetrate through to the nucleus.

The effects of penetration and shielding can be large. A $4s$-electron may be much lower in energy than a $4p$- or $4d$-electron; it may even be lower in energy than a $3d$-electron of the same atom (see Fig. 7.22). It all depends on the number of electrons in the atom. We can assess these energies experimentally by using spectroscopy, because the energies of the electrons show up in the frequencies of the radiation emitted by excited many-electron atoms. Indeed, the colors of fireworks are a sign of shielding and penetration at work.

In a many-electron atom, because of the effects of penetration and shielding, s-electrons lie at a lower energy than p-electrons of the same shell; the order is s < p < d < f.

7.10 THE BUILDING-UP PRINCIPLE

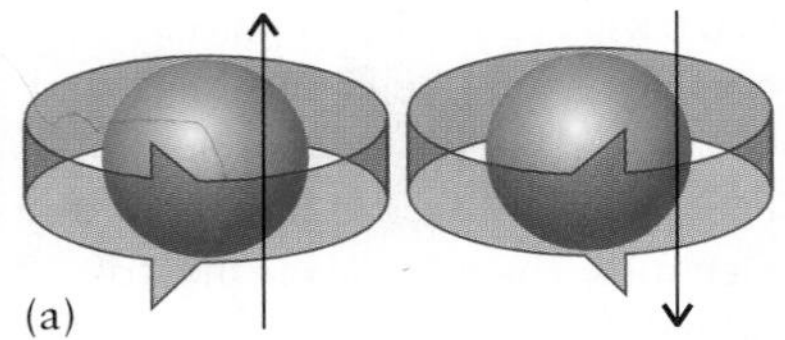

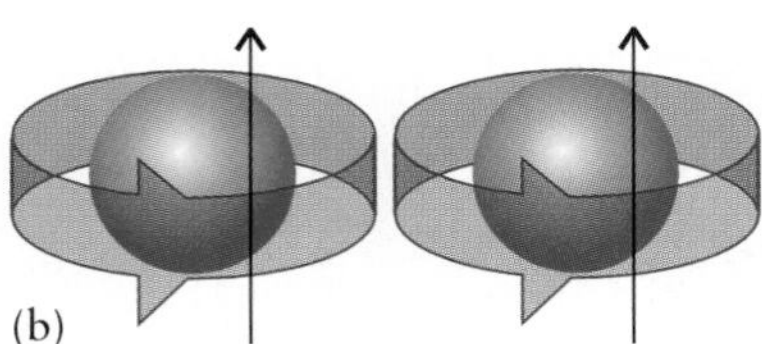

FIGURE 7.23 (a) Two electrons are said to be paired if they have opposite spins (one clockwise, the other counterclockwise). (b) Two electrons are classified as parallel if their spins are in the same direction; in this case, both ↑.

In the ground state of a many-electron atom, the electrons occupy atomic orbitals in such a way that the total energy of the atom is a minimum. We might expect an atom to have its lowest energy when all its electrons are in the lowest energy orbital (the 1*s*-orbital), but except for hydrogen and helium, that can never happen. In 1925, the Austrian Wolfgang Pauli discovered why—his **exclusion principle** states that

> No more than two electrons may occupy any given orbital. When two electrons do occupy one orbital, their spins must be paired.

The spins of two electrons are said to be **paired** if one is ↑ and the other is ↓ (Fig. 7.23). Paired spins are denoted ↑↓ and have spin magnetic quantum numbers of opposite sign: $+\frac{1}{2}$ and $-\frac{1}{2}$. Because an atomic orbital is designated by three quantum numbers (n, l, and m_l) and the two spin states are specified by a fourth quantum number, m_s, another way of expressing the exclusion principle for atoms is

> No two electrons in an atom can have the same set of four quantum numbers.

The exclusion principle implies that each orbital in the energy level diagram in Fig. 7.22 can hold no more than two electrons. For atoms with $Z > 2$, the electrons cannot all enter the 1*s*-orbital, so some must occupy higher energy orbitals.

We report the electronic structure of an atom by writing its **electron configuration,** a list of all its occupied orbitals, with the numbers of electrons that each one contains. The hydrogen atom in its ground state has one electron in the 1*s*-orbital. To show this structure, we use a single arrow in the 1*s*-orbital (see diagram (**1**), which is a fragment of Fig. 7.22) and report its configuration as $1s^1$ ("one s one"). The box in this diagram represents one orbital and can hold up to two electrons.

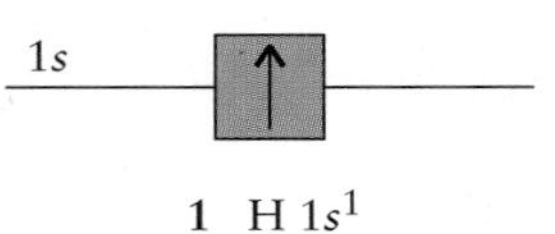

1 H $1s^1$

In the ground state of a helium atom ($Z = 2$), both electrons are in a 1*s*-orbital, which is reported as $1s^2$ ("one s two"). As we see in (**2**), the two electrons are paired. At this point, the 1*s*-orbital holds its maximum number of electrons. The shell with $n = 1$ is now complete, and no more electrons can occupy it. We say that the helium atom contains a **closed shell,** a shell containing the maximum number of electrons allowed by the exclusion principle.

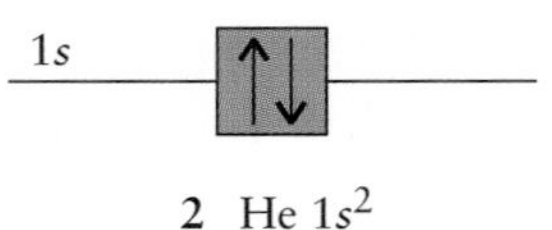

2 He $1s^2$

Lithium ($Z = 3$) has three electrons. Two electrons can occupy the 1*s*-orbital and complete the $n = 1$ shell. The third electron must occupy the next available orbital up the ladder of energy levels, which according to Fig. 7.22 is the 2*s*-orbital. The ground state of a lithium atom is therefore $1s^22s^1$ (**3**). We can think of this atom as consisting of a **core** made up of the inner 1*s* closed shell surrounded by an outer shell containing a higher energy electron. Electrons in the outermost shell are called **valence electrons.** In general, only valence electrons can be lost in chemical reactions, because core electrons are too tightly bound. We are therefore beginning to see that lithium will lose only one electron when it forms compounds; hence it will be present as Li^+ ions, rather than as Li^{2+} or Li^{3+} ions.

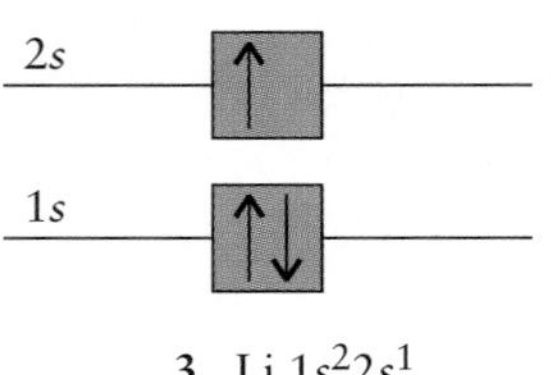

3 Li $1s^22s^1$

The element with $Z = 4$ is beryllium, Be, with four electrons. The first three electrons form $1s^22s^1$, like Li. The fourth electron pairs with the 2*s*-electron, giving $1s^22s^2$. A Be atom therefore has a heliumlike core surrounded by a valence shell of two paired electrons. Like lithium—and for the same reason—a Be atom will lose its two valence electrons to form Be^{2+} ions.

EXAMPLE 7.3 Writing the electron configurations of atoms

Predict the ground-state electron configuration of a boron atom.

STRATEGY First, decide how many electrons are present (from the atomic number). Then add arrows representing electrons to the boxes in Fig. 7.22. Start at the 1*s*-orbital, and add a pair of electrons to each orbital in a subshell before moving to an orbital in a subshell of higher energy.

SOLUTION The atomic number of boron is 5, so a boron atom has five electrons. Two enter the 1*s*-orbital and complete the $n = 1$ shell. The 2*s*-orbital can accommodate two of the remaining electrons, so the fifth electron must occupy an orbital of the next available subshell, which Fig. 7.22 shows is a 2*p*-subshell. This arrangement of electrons is reported as the configuration $1s^22s^22p^1$ (4).

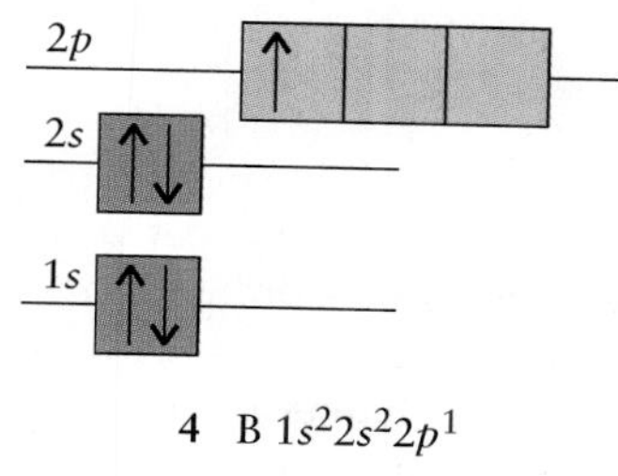

4 B $1s^22s^22p^1$

SELF-TEST 7.6A Predict the ground-state configuration of a sodium atom.

[*Answer:* $1s^22s^22p^63s^1$]

SELF-TEST 7.6B Predict the ground-state configuration of a magnesium atom.

We need to make another decision at carbon: does the sixth electron join the one already in the 2*p*-orbital or does it enter a different 2*p*-orbital? (Remember, there are three *p*-orbitals in the subshell.) To answer this question, we note that electrons that occupy different *p*-orbitals are farther from one another than when they occupy the same orbital. Therefore, they repel each other less than if they occupied the same orbital. So the sixth electron goes into an empty *p*-orbital and the ground state of carbon is $1s^22s^22p_x^12p_y^1$ (5). We write out the individual subshells like this only when we need to emphasize that electrons occupy different orbitals in a subshell. In most cases, we can write the shorter form, such as $1s^22s^22p^2$.

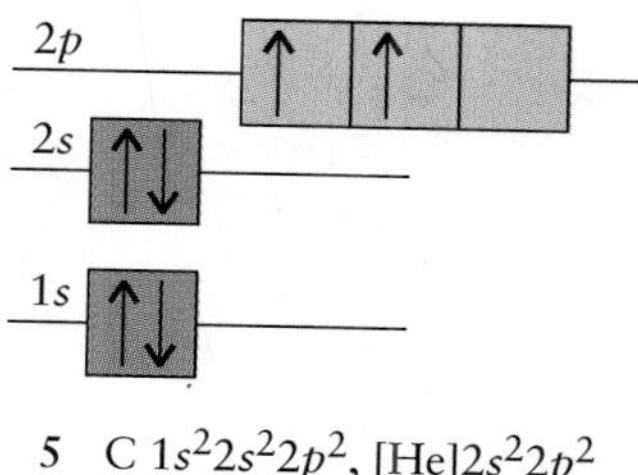

5 C $1s^22s^22p^2$, [He]$2s^22p^2$

Note that in the diagram we have drawn the two 2*p*-electrons with **parallel spins,** (↑↑), indicating that they have the same spin magnetic quantum numbers. This arrangement turns out to have slightly lower energy than a paired arrangement. However, it is allowed only when the electrons occupy different orbitals.

The procedure we have been using is called the **building-up principle.** To assign a configuration to an element with atomic number *Z*,

Some people call the building-up principle by its equivalent German name, the *Aufbau* principle.

Step 1. Add *Z* electrons, one after the other, to the orbitals in the order shown in Fig. 7.24, but with no more than two electrons in any one orbital.

Step 2. If more than one orbital in a subshell is available, add electrons with parallel spins to different orbitals of that subshell rather than pairing two electrons in one of them.

The second step is called **Hund's rule,** for the German spectroscopist Fritz Hund, who first proposed it. This procedure gives the configuration

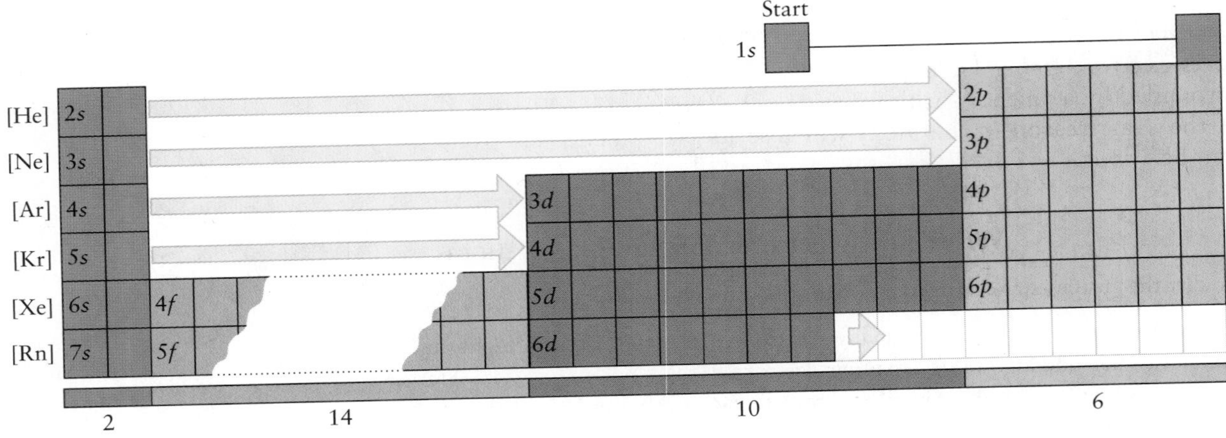

FIGURE 7.24 The order in which atomic orbitals are occupied according to the building-up principle. Each time an electron is added, we move one place to the right. At the end of a period, we move to the start of the next period.

of the atom that corresponds to its lowest *total* energy, taking into account the attraction of the electrons to the nucleus and their repulsion from one another.

> ***The ground-state electron configuration of an atom of an element with atomic number Z is predicted by adding electrons to available orbitals so as to obtain the lowest total energy.***

7.11 THE ELECTRON CONFIGURATIONS OF ATOMS

We have already seen that the configuration of lithium is $1s^22s^1$. Because it consists of a single $2s$-electron outside a heliumlike $1s^2$ core, its configuration is more simply written $[He]2s^1$. We can think of an atom as having a noble-gas core surrounded by electrons in the **valence shell,** the outermost occupied shell. The valence shell is the occupied shell with the largest value of n.

The underlying organization of the periodic table begins to unfold once we know the electron configurations of the atoms (Box 7.2). All the atoms in a given period (a horizontal row of the periodic table) have their valence electrons in the same shell. The principal quantum number, n, of an element's valence shell is equal to its period number. For example, the valence shell of elements in Period 2 (lithium to neon) is the shell with $n = 2$. All the atoms in one period have the same type of core. Thus the atoms of Period 2 elements all have a heliumlike $1s^2$ core; and those of Period 3 elements have a neonlike $1s^22s^22p^6$ core, written [Ne]. In Periods 1 to 3, atoms of elements within each period differ only in their number of valence electrons. Elements belonging to the same group have analogous valence electron configurations.

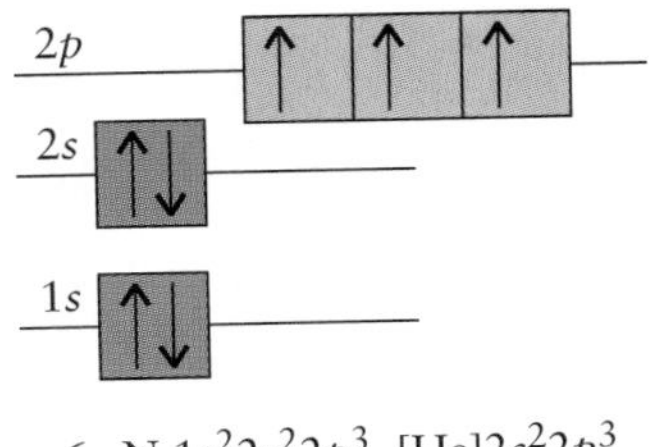

6 N $1s^22s^22p^3$, $[He]2s^22p^3$

We have constructed electron configurations for Li, Be, B, and C. Let's continue building up the electron configurations across Period 2. Nitrogen has $Z = 7$ and one more electron than carbon, giving $[He]2s^22p^3$. Each p-electron occupies a different orbital, and the three have parallel spins (6). Oxygen has $Z = 8$ and one more electron than nitrogen; therefore its configuration is $[He]2s^22p^4$ (7). Similarly, fluorine,

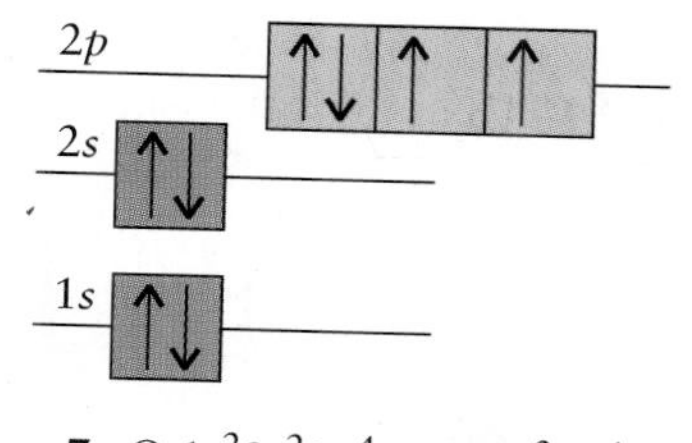

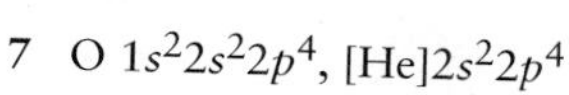
7 O $1s^22s^22p^4$, [He]$2s^22p^4$

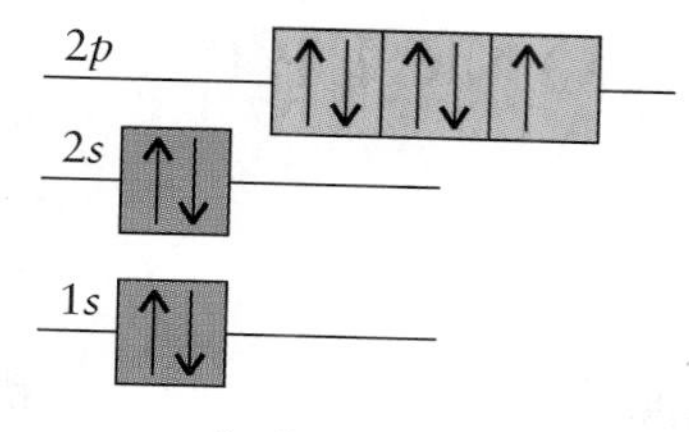

8 F $1s^22s^22p^5$, [He]$2s^22p^5$

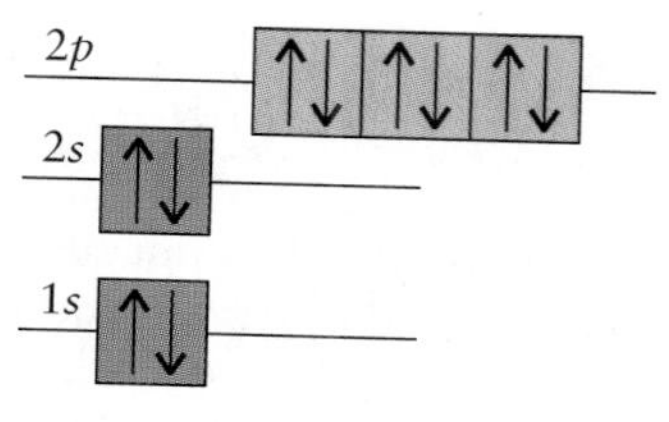

9 Ne $1s^22s^22p^6$, [He]$2s^22p^6$

with $Z = 9$ and one more electron than oxygen, has the configuration [He]$2s^22p^5$ (**8**). The fluorine atom can be pictured as having a heliumlike core surrounded by a valence shell that is complete except for one *p*-electron. Neon, with $Z = 10$, has one more electron than fluorine. This electron completes the 2*p*-subshell, giving [He]$2s^22p^6$ (**9**). According to Fig. 7.24, the next electron enters the 3*s*-orbital, the lowest energy orbital of the next shell. The configuration of sodium is therefore [He]$2s^22p^63s^1$, or more briefly [Ne]$3s^1$, where [Ne] is the neonlike core.

The *s*- and *p*-orbitals of the shell with $n = 3$ are full by the time we get to argon, [Ne]$3s^23p^6$, which is a colorless, odorless, unreactive gas resembling neon. Argon completes the third period. The fourth period begins by filling the 4*s*-orbitals. Hence, the next two electron configurations are [Ar]$4s^1$ for potassium and [Ar]$4s^2$ for calcium. At this point, the 3*d*-orbitals are the next to be filled and there is a change in the rhythm of the periodic table.

According to Fig. 7.24, the next 10 electrons (for scandium, with $Z = 21$, through zinc, with $Z = 30$) enter the 3*d*-orbitals. The ground-state electron configuration of scandium, for example, is [Ar]$3d^14s^2$, and that of its neighbor titanium is [Ar]$3d^24s^2$. Note that, beginning at scandium, we write the 4*s*-electrons after the 3*d*-electrons: the 3*d*-orbitals now lie lower in energy than the 4*s*-orbital.

For transition metals, the electrons are added to the *d*-orbitals as Z increases: [Ar]$3d^14s^2$ for scandium, [Ar]$3d^24s^2$ for titanium, and so on. However, there are two exceptions: the half-complete subshell configuration d^5 and the complete subshell configuration d^{10} turn out experimentally to have a lower energy than simple theory suggests. In some cases, the neutral atom has a lower total energy if the 3*d*-subshell is half-full (d^5) or full (d^{10}) as a result of transferring a 4*s*-electron into it. For example, the experimental electron configuration of chromium is [Ar]$3d^54s^1$ and that of copper is [Ar]$3d^{10}4s^1$. Other exceptions to the building-up principle can be found in Appendix 2C.

Because the transition metals in the same period differ mainly in the number of *d*-electrons, their properties are very similar. Most of them form ions in more than one oxidation state, because the *d*-electrons have similar energies and a variable number can be lost.

Electrons occupy 4*p*-orbitals once the 3*d*-orbitals are full. The configuration of germanium, [Ar]$3d^{10}4s^24p^2$, for example, is obtained by adding two electrons to the 4*p*-orbitals outside the completed 3*d*-subshell. The fourth period of the table contains 18 elements, because the 4*s*- and 4*p*-orbitals can accommodate a total of 8 electrons and the 3*d*-orbitals can accommodate 10. It is the first **long period** of the periodic table.

BOX 7.2 HOW THE CONCEPT OF THE PERIODIC TABLE WAS DEVELOPED

The periodic table is one of the most important concepts in chemistry. Its development is an example of how scientific discoveries can come from using insight to organize data collected by a large number of scientists over a period of many years. There are often simultaneous discoveries in science, because discoveries tend to be made when enough of the right kind of data has been collected.

Until the middle of the nineteenth century, the separation and identification of elements and the determination of their relative masses was a primary occupation of chemists. As the number of known elements grew, scientists observed curious similarities in certain groups of elements. The German chemist Johann Döbereiner described triads of elements with similar properties; moreover, the mass of the central member of the triad was approximately the mean of the masses of the two other members. The English chemist John Newlands responded with his law of octaves, in which he organized elements in groups of eight by their properties. Observations like these suggested an intrinsic organization in matter, but no one could find the basis for it. The 63 known elements were simply displayed in long lists.

In 1860, the Congress of Karlsruhe brought together many prominent chemists in an attempt to come to some agreement on issues such as the existence of atoms, the correct atomic masses, and how the elements are related to one another. No agreement was reached, but many major ideas were presented in an atmosphere of heated debate. One concept discussed

Dmitri Ivanovich Mendeleev (1834–1907).

was Avogadro's principle—that the number of molecules of different gases in samples of equal volume, pressure, and temperature, is the same. Two scientists attending the congress were the German Lothar Meyer and the Russian Dmitri Mendeleev, both of whom left with copies of Avogadro's paper. In 1869, Meyer and Mendeleev discovered independently that a regular repeating pattern of properties could be observed when the elements were arranged in order of increasing atomic mass. Mendeleev called this observation the Periodic Law.

Mendeleev, who taught chemistry and wrote general chemistry textbooks, was looking for a way to organize his discussion of the elements. He made up cards with the names and properties of the elements and arranged them in different ways, looking for

Next in line for occupation at the beginning of Period 5 is the 5*s*-orbital, followed by the 4*d*-orbitals. As in Period 4, the energy of the 4*d*-orbitals falls below that of the 5*s*-orbital after two electrons have been accommodated in the 5*s*-orbital. A similar effect is seen in Period 6, but now another set of inner orbitals, the 4*f*-orbitals, begins to be occupied. Cerium, for example, has the configuration $[Xe]4f^1 5d^1 6s^2$. Electrons then continue to occupy the seven 4*f*-orbitals, which are complete after 14 electrons have been added, at ytterbium, $[Xe]4f^{14} 6s^2$. Next, the 5*d*-orbitals are occupied. The 6*p*-orbitals are occupied only after the 6*s*- and 5*d*-orbitals are filled at mercury; thallium, for example, has the configuration $[Xe]4f^{14} 5d^{10} 6s^2 6p^1$.

We account for the ground-state electron configuration of an atom by using the building-up principle in conjunction with Fig. 7.24, the Pauli exclusion principle, and Hund's rule.

relationships among elements. According to legend, not finding success, he fell asleep discouraged. He then awoke with a new plan, to arrange the elements in rows by increasing atomic mass, beginning a new row when the cycle of properties repeated itself. This arrangement put elements with similar properties into columns, forming a table of repeating, or periodic, properties. Meyer found a similar organization, but it was Mendeleev who realized that the arrangement was the key to the fundamental organization of matter.

Mendeleev's chemical insight led him to leave gaps for elements that would be needed to complete the pattern but were unknown at the time. When they were discovered later, he turned out to be strikingly correct. For example, his pattern required an element he named "eka-silicon" below silicon and between gallium and arsenic. He predicted that the element would have a relative atomic mass of 72 (taking the mass of hydrogen as 1) and properties similar to those of silicon. This prediction spurred the German chemist Clemens Winkler in 1886 to discover eka-silicon, which he named germanium. It has a relative atomic mass of 72.6 and properties similar to those of silicon, as shown in the accompanying table.

One problem with Mendeleev's table was that some elements seemed to be out of place. For example, when argon was isolated, it didn't seem to have the correct mass for its location. Its relative atomic mass of 40 is the same as calcium's, but argon is an inert gas and calcium a reactive metal. Such anomalies led scientists to question the use of relative atomic mass as the basis for organizing the elements. When Henry Moseley examined x-ray spectra of the elements in the early twentieth century (as described in Section 1.3), he realized that all atoms of the same element had identical nuclear charge and, therefore, the same number of protons, which gives the element's atomic number. It was soon discovered that elements fall into the uniformly repeating pattern of the periodic table if they are organized according to atomic *number,* rather than atomic mass.

Mendeleev's predictions for eka-silicon (germanium)

Property	Eka-silicon (E)	Germanium
molar mass, $g \cdot mol^{-1}$	72	72.59
density, $g \cdot cm^{-3}$	5.5	5.32
melting point, °C	high	937
appearance	dark gray	gray-white
oxide	EO_2; white solid; amphoteric; density 4.7 $g \cdot cm^{-3}$	GeO_2; white solid; amphoteric; density 4.23 $g \cdot cm^{-3}$
chloride	ECl_4; boils below 100°C; density 1.9 $g \cdot cm^{-3}$	$GeCl_4$; boils at 84°C; density 1.84 $g \cdot cm^{-3}$

EXAMPLE 7.4 Predicting a ground-state configuration of a heavy atom

Predict the ground-state configuration of a lead atom.

STRATEGY Because lead is in the same group as carbon (Group 14), we can anticipate that it will have an analogous valence configuration of the form ns^2np^2. To proceed systematically, use the two rules of the building-up principle. However, there is a quicker method that is particularly useful for elements with large numbers of electrons. First, note the group number, which tells us the number of valence electrons in the ground state of the atom, and the period number, which tells us the value of the principal quantum number of the valence shell. The core consists of the preceding noble-gas configuration together with any completed *d*- and *f*-subshells.

SOLUTION Lead belongs to Group 14 and Period 6. It therefore has four electrons in its valence shell, two in a 6*s*-orbital and two in different 6*p*-orbitals.

The period has complete $5d$- and $4f$-subshells, and the preceding noble gas is xenon. The electron configuration of lead is therefore $[Xe]4f^{14}5d^{10}6s^26p^2$.

SELF-TEST 7.7A Write the ground-state configuration of a bismuth atom.

[*Answer:* $[Xe]4f^{14}5d^{10}6s^26p^3$]

SELF-TEST 7.7B Write the ground-state configuration of an arsenic atom.

7.12 THE ELECTRON CONFIGURATIONS OF IONS

A neutral atom becomes a cation when it loses one or more electrons. If the principal quantum number of the valence shell is n, then any np-electrons are lost first, because that subshell has the highest energy. Then ns-electrons are lost, and finally the d-electrons of the previous shell, if any, until the appropriate number of electrons has been lost. Thus, for the Fe^{3+} ion, we work out the configuration of the Fe atom, which is $[Ar]3d^64s^2$, and remove three electrons from it. There are no $4p$-electrons, so the first two electrons removed are $4s$-electrons. The third electron comes from the $3d$-subshell, giving $[Ar]3d^5$.

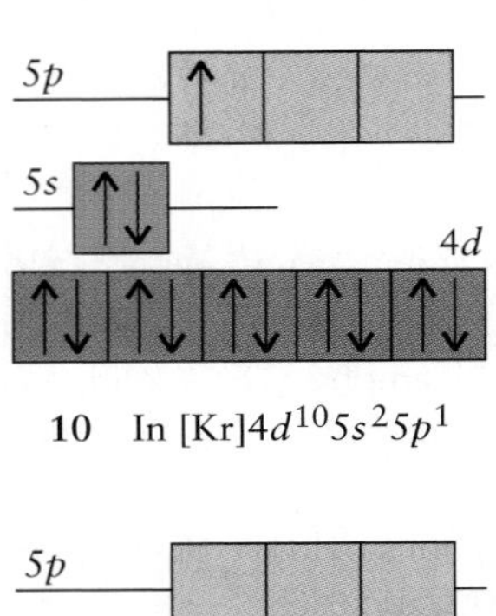

10 In $[Kr]4d^{10}5s^25p^1$

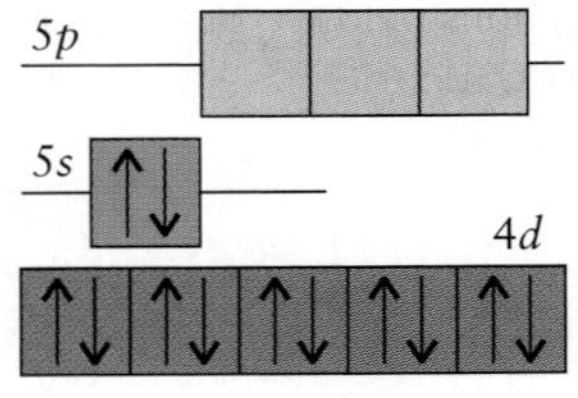

11 In^+ $[Kr]4d^{10}5s^2$

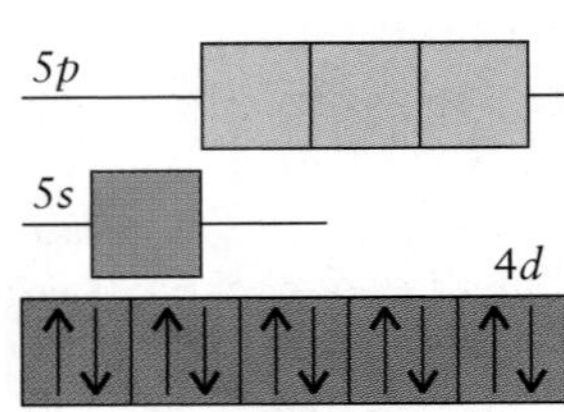

12 In^{3+} $[Kr]4d^{10}$

EXAMPLE 7.5 Writing the electron configurations of cations

Write the electron configurations of the In^+ and In^{3+} ions.

STRATEGY Determine the configuration of the neutral atom from the group and period of the element. Remove one electron to obtain a singly charged ion, two for a doubly charged ion, and so on. Remove electrons from the valence shell p-orbitals first, then from the s-orbitals, and finally, if necessary, from the d-orbitals in the next lower shell.

SOLUTION Indium, in Group 13, has three valence electrons. It is in Period 5, so its valence shell has $n = 5$. It is preceded by the Period 4 noble gas krypton (so its noble-gas core is [Kr]). Its ground-state configuration is therefore $[Kr]4d^{10}5s^25p^1$ (**10**). One electron can be lost from the $5p$-orbital, giving In^+ as $[Kr]4d^{10}5s^2$ (**11**). The next two electrons are lost from the $5s$-orbital, giving In^{3+} as $[Kr]4d^{10}$ (**12**).

SELF-TEST 7.8A Write the electron configurations of the copper(I) and copper(II) ions.

[*Answer:* $[Ar]3d^{10}$, $[Ar]3d^9$]

SELF-TEST 7.8B Write the electron configurations of a manganese(II) ion and manganese in oxidation state +5.

To form monatomic anions, we add enough electrons to complete the valence shell. For example, nitrogen has five valence electrons, so three more electrons are needed to reach a noble-gas configuration, that of neon. Therefore the ion will be N^{3-}. The configuration of the nitrogen atom is $[He]2s^22p^3$, with room for three more electrons in the $2p$-subshell. Thus, the N^{3-} ion has the configuration $[He]2s^22p^6$ (**13**), the same as that of neon.

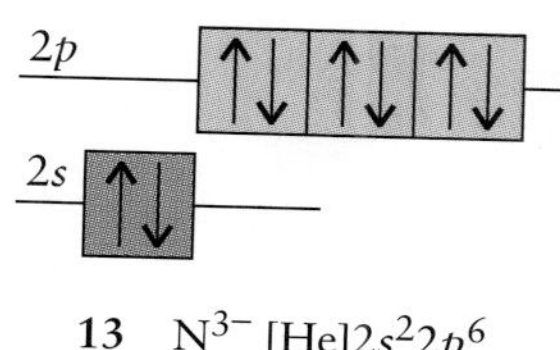

13 N^{3-} $[He]2s^22p^6$

To predict the electron configuration of a cation, remove outermost electrons in the order np, ns, and (n − 1)d; for an anion, add electrons until the next noble-gas configuration has been reached.

EXAMPLE 7.6 Writing the electron configurations of anions

Write the formula and electron configuration of the oxide ion.

STRATEGY First, establish the ground-state configuration of the neutral oxygen atom by using the building-up principle. Then add enough electrons to complete the vacant orbitals and achieve the closed-shell configuration of the next noble-gas atom (Ne).

SOLUTION We have already seen that the electron configuration of an O atom is $[He]2s^22p^4$. The 2*p*-subshell can accommodate two more electrons, so we expect the oxide ion to be O^{2-}, with configuration $[He]2s^22p^6$ (**14**), the same configuration neon has.

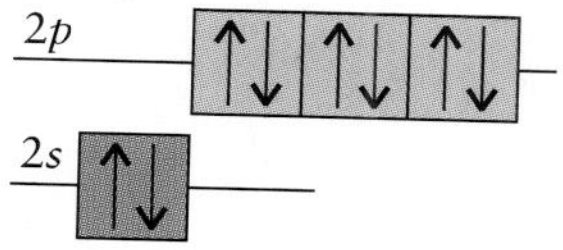

14 O^{2-} $[He]2s^22p^6$

SELF-TEST 7.9A Predict the chemical formula and electron configuration of the phosphide ion.

[*Answer:* P^{3-}, $[Ne]3s^23p^6$]

SELF-TEST 7.9B Predict the chemical formula and electron configuration of the iodide ion.

7.13 ELECTRONIC STRUCTURE AND THE PERIODIC TABLE

We can understand the organization of the periodic table now that we know about electron configurations. The table is divided into *s*-, *p*-, *d*-, and *f*-blocks, named for the last subshell that is occupied according to the building-up principle (Fig. 7.25). Two elements are exceptions. Strictly, helium belongs in the *s*-block, but it is shown in the *p*-block. It is a gas with properties matching those of the noble gases in Group 18, rather than the reactive metals in Group 2. Its place in Group 18 is justified by the fact that, like all the other Group 18 elements, it has a filled valence shell. Hydrogen occupies a unique position in the periodic table. It

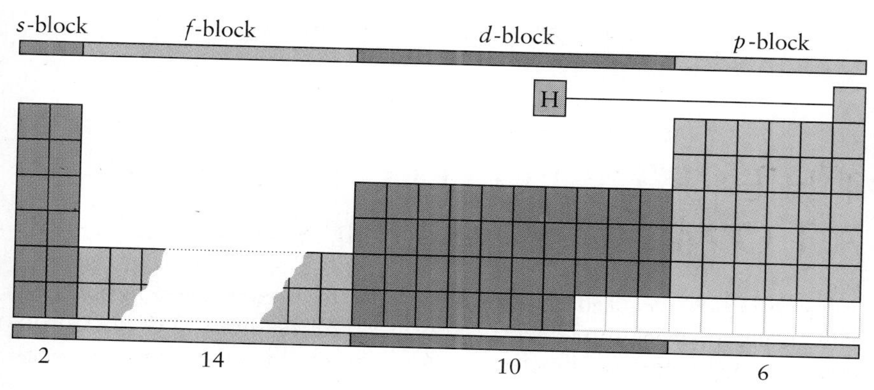

FIGURE 7.25 The names of the blocks of the periodic table are based on the last subshell occupied in an atom of an element according to the building-up principle. The number of electrons that each type of orbital can accommodate is written below the bottom of the block. The colors of the blocks match the colors used in the illustrations of the corresponding orbitals.

Some periodic tables place hydrogen in Group 1, some place it in Group 17, and some place it in both groups.

has one *s*-electron, so it belongs in Group 1. But it is also one electron short of a noble-gas configuration, so it can also act like a member of Group 17.

The *s*- and *p*-blocks comprise the **main groups.** The atoms of elements in the same main group have analogous valence-shell electron configurations, which differ only in the value of n. The similar electron configurations for the elements in a main group are the reason for their similar properties. The group number tells us how many valence-shell electrons are present. In the *s*-block, the group number (1 or 2) is the same as the number of valence electrons. In the *p*-block, we have to subtract 10 from the group number to find the number of valence electrons. For example, fluorine in Group 17 has seven valence electrons.

Each new period corresponds to the occupation of a shell with a new principal quantum number. This feature explains the different lengths of the periods. Period 1 consists of only two elements, H and He, in which the single 1*s*-orbital of the $n = 1$ shell is being filled with its two electrons. Period 2 consists of the eight elements Li through Ne, in which the one 2*s*- and three 2*p*-orbitals are being filled with eight more electrons. In Period 3 (Na through Ar) the 3*s*- and 3*p*-orbitals are being occupied by eight additional electrons. In Period 4, not only are the 8 electrons of the 4*s*- and 4*p*-orbitals being added, but so are the 10 electrons of the 3*d*-orbitals. Hence there are 18 elements in Period 4.

Period 5 elements add another 18 electrons as the 5*s*-, 4*d*-, and 5*p*-orbitals are filled. In Period 6, a total of 32 electrons are added, because 14 electrons are also being added to the 3*f*-orbitals. The *f*-block elements in Period 6 are called the lanthanides, or rare earths. They have very similar properties, because their electron configurations differ only in the population of inner *f*-orbitals, and these electrons do not participate much in bond formation. The same is true of the *f*-block elements in Period 7; the actinides, which are all radioactive metals, have very similar properties.

The blocks of the periodic table are named for the last orbital to be occupied according to the building-up principle. The periods are numbered according to the principal quantum number of the valence shell.

SELF-TEST 7.10A Give the valence-electron configuration of the atoms of the group that contains tin.

[*Answer:* ns^2np^2]

SELF-TEST 7.10B Give the valence-electron configuration of the atoms of the group that contains arsenic.

THE PERIODICITY OF ATOMIC PROPERTIES

In Section 1.6 we used the periodic table to predict the physical and chemical properties of elements. Now we shall see how to predict properties we cannot see directly, such as the radii of atoms and ions.

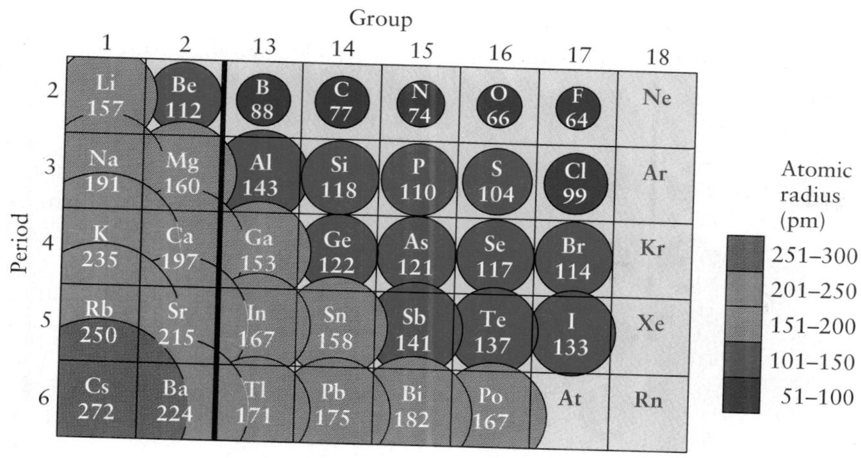

FIGURE 7.26 The atomic radii (in picometers, 1 pm = 10^{-12} m) of the main-group elements. The radii decrease from left to right within a period and increase down a group. The colors used here and in subsequent charts represent the general magnitude of the property, as indicated by the key on the right.

These properties will turn out to be crucial for understanding bonding (Chapter 8).

7.14 ATOMIC RADIUS

Electron clouds do not have sharp edges. However, when atoms pack together in solids and molecules, their centers are found at definite distances from one another. The **atomic radius** of an element is defined as half the distance between the centers of neighboring atoms (**15**). If the element is a metal, then the distance is between the centers of neighboring atoms in a solid sample. Because the distance between neighboring nuclei in solid copper is 256 pm, the atomic radius of copper is 128 pm (remember that 1 pm = 10^{-12} m). If the element is a nonmetal, then the appropriate distance is that between the centers of atoms joined by a chemical bond; this radius is also called the **covalent radius** of the element. The distance between the nuclei in a Cl_2 molecule is 198 pm, so the atomic (covalent) radius of chlorine is 99 pm.

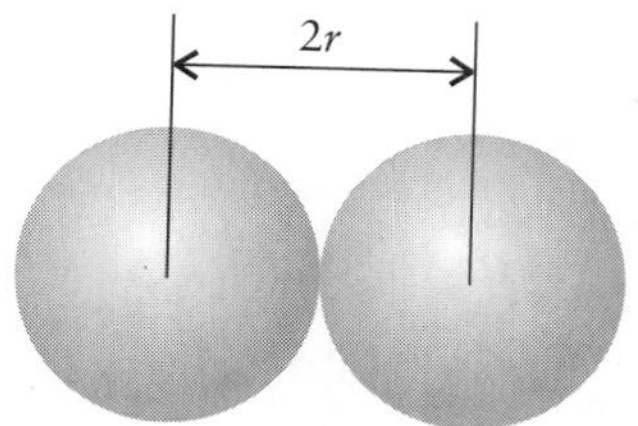

15 Atomic radius

Figure 7.26 shows some atomic radii, and Fig. 7.27 shows the variation in atomic radius with atomic number. Note the periodic, sawtooth

▶▶ To see how the values are measured, see Box 10.1.

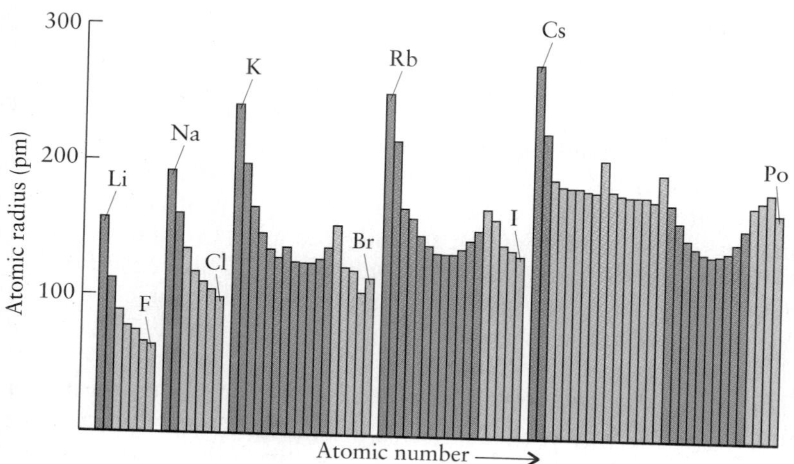

FIGURE 7.27 The periodic variation of the atomic radii of the elements. This variation can be explained in terms of the effect of increasing atomic number and the occupation of shells with increasing principal quantum number. Numerical values are given in Appendix 2D.

FIGURE 7.28 The ionic radii (in picometers) of the ions of the main-group elements. Note that cations (on the left) are typically smaller than anions (on the right), and in some cases very much smaller.

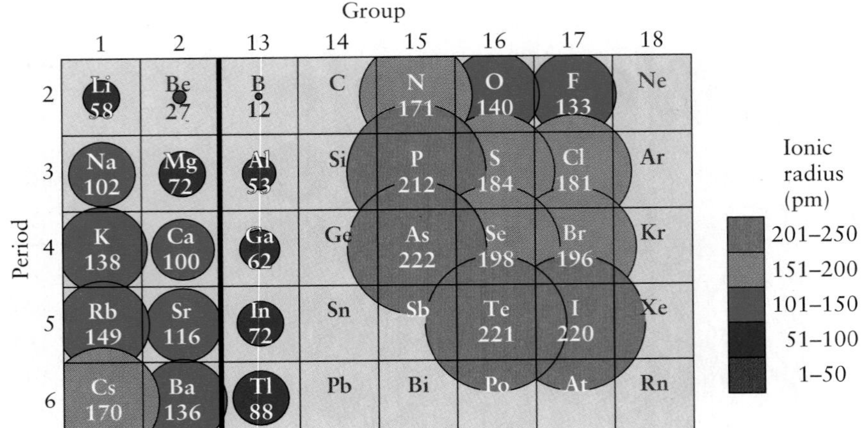

pattern. Atomic radius generally decreases from left to right across a period and increases down a group.

The increase down a group, like that from Li to Cs, makes sense: with each new period, the outermost electrons occupy shells that lie farther from the nucleus. But how can we account for the decrease across a period, like that from Li to Ne, despite the fact that the number of electrons is increasing? The explanation is that the new electrons are in the same shell of the atom, and about as close to the nucleus as other electrons in the same shell. The increasing nuclear charge draws them in; and, as a result, the atom is more compact.

Atomic radii generally increase down a group and decrease across a period.

7.15 IONIC RADIUS

$r_{anion} + r_{cation}$

−

+

16 Ionic radius

The **ionic radius** of an element is its share of the distance between neighboring ions in an ionic solid (**16**). The distance between the nuclei of a neighboring cation and anion is the sum of the two ionic radii. In practice, we use the radius of the oxide ion, 140 pm, to calculate the radii of other ions. For example, because the distance between Mg and O nuclei in magnesium oxide is 205 pm, the radius of the Mg^{2+} ion is calculated from 205 pm − 140 pm = 65 pm.

Figure 7.28 illustrates some ionic radii, and Fig. 7.29 shows the relative sizes of some ions and their parent atoms. All cations are smaller than their parent atoms, because the atom loses one or more electrons to form the cation and exposes its core, which is generally much smaller than the parent atom. For example, the atomic radius of Li, with the configuration $1s^2 2s^1$, is 157 pm, but the ionic radius of Li^+, the bare helium-like $1s^2$ core of the parent atom, is only 58 pm. This size difference is comparable to that between a cherry and its pit. Like atomic radii, cation radii increase down each group because electrons are occupying shells with higher principal quantum numbers.

Figure 7.29 shows that anions are *larger* than their parent atoms. The reason can be traced to the increased number of electrons in the valence shell of the anion and the repulsive effects the electrons exert on one another. The variation in the radii of anions is the same as that for atoms and cations, with the smallest at the upper right of the periodic table, close to fluorine.

Atoms and ions with the same number of electrons are called **isoelectronic.** For example, Na^+, F^-, and Mg^{2+} are isolectronic. All three ions have the same electron configuration, $[He]2s^2p^6$, but their radii differ because they have different nuclear charges. The Mg^{2+} ion has the largest positive nuclear charge, so it has the strongest attraction for the electrons and thus the smallest radius. The F^- ion has more electrons than protons. The net repulsion of the electrons outweighs the nuclear charge, which allows the ion to expand. As a result, it has the largest radius.

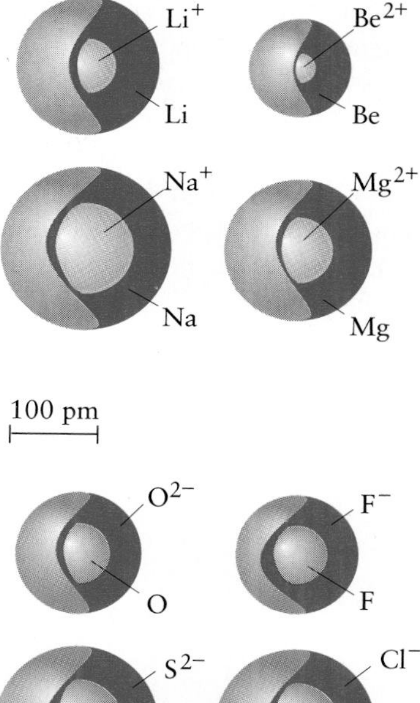

FIGURE 7.29 The relative sizes of cations, anions, and their parent atoms for selected elements. Note that cations are smaller than their parent atoms, whereas anions are larger.

Ionic radii generally increase down a group and decrease across a period. Cations are smaller than their parent atoms and anions are larger.

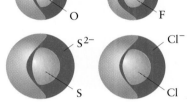

EXAMPLE 7.7 Predicting the relative sizes of ions

Arrange each pair of ions, (a) Mg^{2+} and Ca^{2+}, and (b) O^{2-} and F^-, in order of increasing ionic radius.

STRATEGY Because ions and atoms show the same pattern of radii, the smaller member of a pair with the same number of electrons will be an ion of an element that lies further to the right in a period. If the two ions are in the same group, the smaller ion will be the one that lies higher in the group.

SOLUTION (a) Because Mg lies above Ca in Group 2, it will have the smaller ionic radius; the actual values are 72 pm and 100 pm, respectively. (b) Because F lies to the right of O in Period 2, it will have the smaller ionic radius. The actual values are 133 pm and 140 pm, respectively.

SELF-TEST 7.11A Arrange each pair of ions, (a) Mg^{2+} and Al^{3+}, and (b) O^{2-} and S^{2-}, in order of increasing ionic radius.

[*Answer:* $r(Al^{3+}) < r(Mg^{2+})$; $r(O^{2-}) < r(S^{2-})$]

SELF-TEST 7.11B Arrange each pair of ions, (a) Ca^{2+} and K^+, and (b) S^{2-} and Cl^-, in order of increasing ionic radius.

7.16 IONIZATION ENERGY

Copper wire is the most common material used to conduct electricity. But why does it conduct at all? An electric current is a flow of electrons, so the reason must lie in the ease with which copper atoms lose electrons. The same must be true of the new high-temperature superconductors that are currently being developed (Fig. 7.30), although no one is sure how these new materials conduct electricity without any resistance. Many chemical properties—particularly the types and numbers of bonds that atoms form—also depend on the ease with which atoms give up some of their electrons.

The **ionization energy** is the energy needed to remove an electron from an atom in the gas phase. For the *first* ionization energy, I_1, we start with the neutral atom. For example, for copper,

$$Cu(g) \longrightarrow Cu^+(g) + e^-(g) \qquad \text{energy required} = I_1 \qquad (745.1\ kJ\cdot mol^{-1})$$

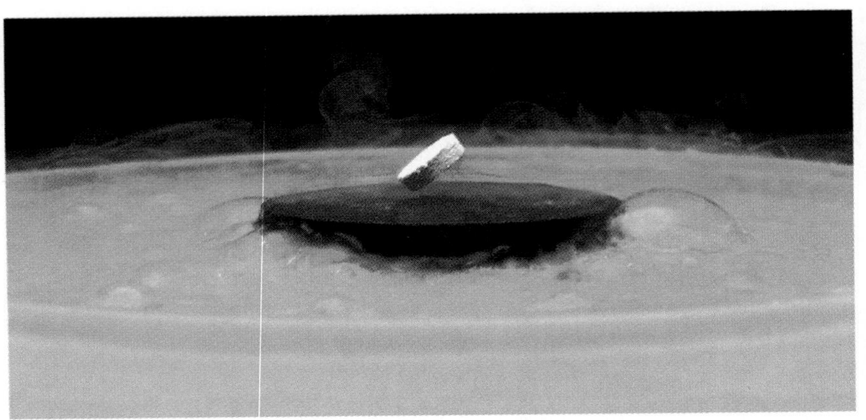

FIGURE 7.30 A sample of a high-temperature superconductor, a type of material first produced in 1987. This sample is repelled strongly by a magnetic field when cooled by liquid nitrogen. When the material becomes warm, superconductivity is lost and the sample no longer hovers over the magnet.

The *second* ionization energy, I_2, of an element is the energy needed to remove an electron from a singly charged gas-phase cation. For copper,

$$Cu^{+}(g) \longrightarrow Cu^{2+}(g) + e^{-}(g) \qquad \text{energy required} = I_2 \qquad (1955\ \text{kJ·mol}^{-1})$$

Figure 7.31 shows that ionization energies generally decrease down a group. The decrease means that it takes less energy to remove an electron from a cesium atom, for instance, than from a sodium atom. Figure 7.32 shows the variation of first ionization energy with atomic number, and we see a periodic sawtooth pattern like that shown by atomic radii. With few exceptions, first ionization energy rises from left to right across a period. Then it falls back to a lower value at the start of the following period. The lowest values occur at the bottom left of the periodic table (near cesium) and the highest at the upper right (near helium). In other words, less energy is needed to remove an electron from atoms of elements near cesium and more energy is needed to remove an electron from atoms of elements near helium. Elements with low ionization energies can be expected to form cations readily and to conduct electricity in their solid forms. Elements with high ionization energies are unlikely to form cations and are unlikely to conduct electricity.

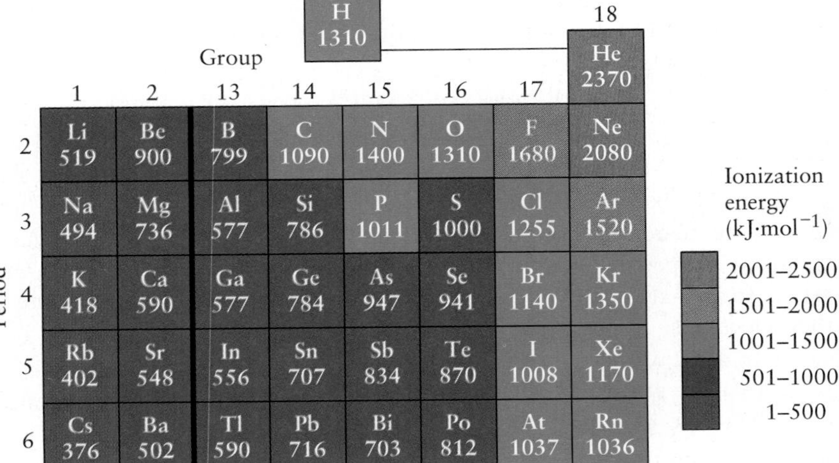

H 1310
Group
1 2 13 14 15 16 17 18
He 2370
Period
2 Li 519 Be 900 B 799 C 1090 N 1400 O 1310 F 1680 Ne 2080
3 Na 494 Mg 736 Al 577 Si 786 P 1011 S 1000 Cl 1255 Ar 1520
4 K 418 Ca 590 Ga 577 Ge 784 As 947 Se 941 Br 1140 Kr 1350
5 Rb 402 Sr 548 In 556 Sn 707 Sb 834 Te 870 I 1008 Xe 1170
6 Cs 376 Ba 502 Tl 590 Pb 716 Bi 703 Po 812 At 1037 Rn 1036
Ionization energy (kJ·mol⁻¹)
2001–2500
1501–2000
1001–1500
501–1000
1–500

FIGURE 7.31 The first ionization energies of the main-group elements, in kilojoules per mole (kJ·mol^{-1}). In general, low values are found toward the lower left and high values are found to the upper right. More values are given in Appendix 2D.

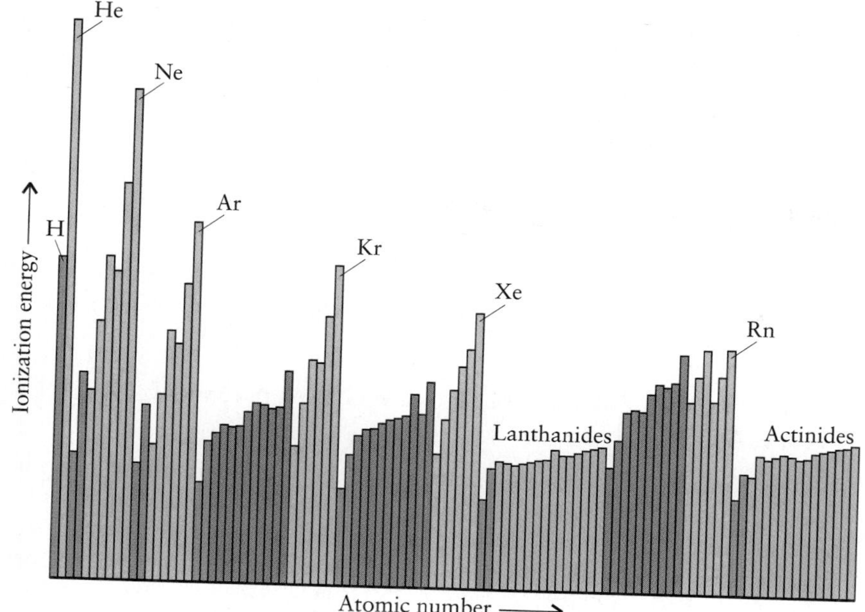

FIGURE 7.32 The periodic variation of the first ionization energies of the elements.

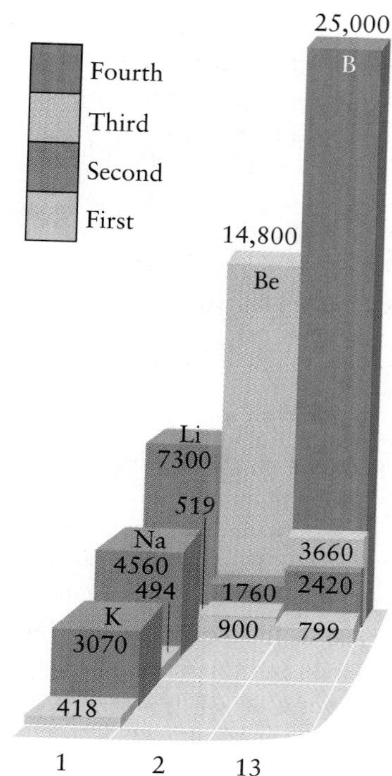

FIGURE 7.33 The successive ionization energies of a selection of main-group elements. Note the great increase in energy required to remove an electron from an inner shell.

Ionization energy decreases down a group because the outermost electron occupies a shell that is farther from the nucleus and is therefore less tightly bound. The effective nuclear charge increases as we go from left to right across a given period. As a result, the outermost electron is gripped more tightly and the ionization energies generally increase. The small departures from these trends can usually be traced to repulsions between electrons, particularly electrons occupying the same orbital.

Figure 7.33 shows that the second ionization energy of an element is always higher than its first ionization energy. It takes more energy to remove an electron from a positively charged ion than from a neutral atom (Fig. 7.34). For the Group 1 elements, the second ionization energy is considerably larger than the first, but in Group 2 the two ionization energies are close in value. This difference makes sense, because the Group 1 elements have an ns^1 valence-shell electron configuration. Although the removal of the first electron requires only a small energy, a second electron must come from the noble-gas core. The core electrons have lower principal quantum numbers and are much closer to the nucleus. They are strongly attracted to it and a lot of energy is needed to remove them.

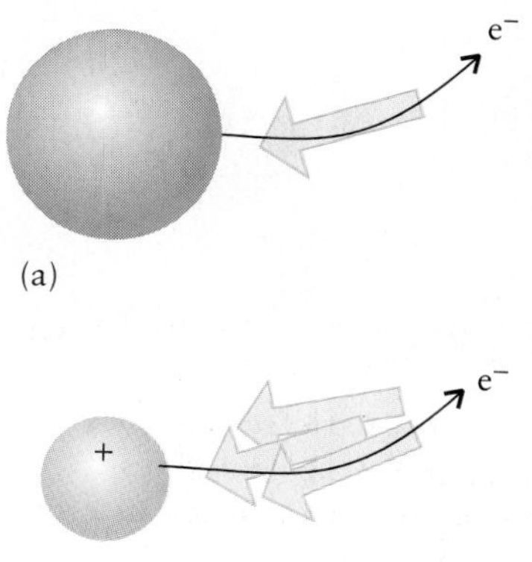

FIGURE 7.34 (a) To remove an electron from a neutral atom, energy has to be supplied to drag the electron away from the attraction (as represented by the arrow) of the partially shielded nucleus. (b) Much more energy has to be supplied to drag a second electron away from a cation, because now the nucleus is less shielded and exerts a greater attraction (represented by more arrows).

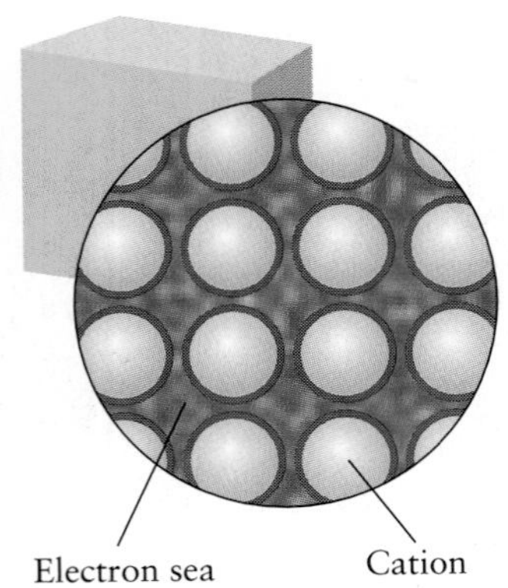

FIGURE 7.35 A block of metal consists of an array of cations (the spheres) surrounded by a sea of electrons. The charge of the electron sea cancels the charges of the cations. The electrons of the sea are mobile and can move past the cations quite easily and hence conduct an electric current.

The first ionization energy is highest for elements close to helium and is lowest for elements close to cesium. Second ionization energies are higher than first ionization energies (of the same element), and very much higher if the electron is to be expelled from a closed shell.

SELF-TEST 7.12A Account for the decrease in ionization energy between beryllium and boron.

[*Answer:* Boron loses an electron from a higher energy subshell than beryllium.]

SELF-TEST 7.12B Account for the fact that the two Group 13 elements aluminum and gallium have approximately the same ionization energies.

7.17 IONIZATION ENERGY AND METALLIC CHARACTER

The low ionization energies of elements at the lower left of the periodic table account for their metallic character. A block of metal consists of a collection of cations of the element surrounded by a sea of valence electrons that the atoms have lost (Fig. 7.35). For example, a piece of copper consists of a stack of Cu^+ ions held together by a sea of electrons, each of which comes from one of the atoms in the sample. Only elements with low ionization energies—the members of the *s*-block, the *d*-block, the *f*-block, and the lower left of the *p*-block—can form metallic solids, because only they can lose electrons easily.

The elements at the upper right of the periodic table have high ionization energies, so they do not readily lose electrons and, hence, are not metals. Our knowledge of electronic structure has helped us to understand a major feature of the periodic table, in this case, why the metals are found at the lower left and the nonmetals are found toward the upper right.

Metals are found toward the lower left of the periodic table because these elements have low ionization energies and can readily lose their electrons.

FIGURE 7.36 When tin(II) oxide is heated in air, it becomes incandescent as it reacts to form tin(IV) oxide. Even without being heated, it smolders and can ignite.

7.18 THE INERT-PAIR EFFECT

Aluminum and indium are in Group 13. Aluminum forms Al^{3+} ions, but indium forms both In^{3+} and In^+ ions. The tendency to form ions two units lower in charge than expected from the group number is called the **inert-pair effect.** Another example of the inert-pair effect is found in Group 14; tin forms tin(IV) oxide when heated in air but the heavier lead atom loses only its two *p*-electrons and forms lead(II) oxide. Tin(II) oxide can be prepared, but it is readily oxidized to tin(IV) oxide (Fig. 7.36).

The inert-pair effect is due in part to the different energies of the valence *p*- and *s*-electrons. In the later periods of the periodic table, valence *s*-electrons are very low in energy because of penetration and shielding. They may therefore remain attached to the atom. The inert-pair effect is most pronounced among the heaviest members of a group, where the difference in energy between *s*- and *p*-electrons is greatest (Fig. 7.37). Even so, the pair of *s*-electrons can be removed from the atom under sufficiently vigorous conditions. An inert pair would be better called a "lazy pair" of electrons.

The inert-pair effect is the tendency to form ions two units lower in charge than expected from the group number; it is most pronounced for heavy elements in the p-block.

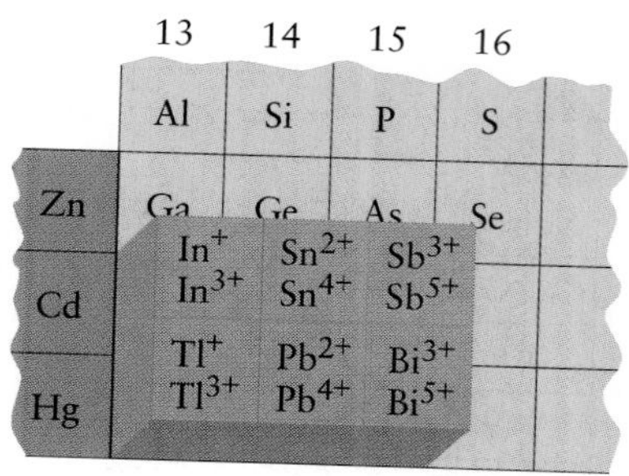

FIGURE 7.37 The typical ions formed by the heavy elements in Groups 13–15 show the influence of the inert pair. These elements have the tendency to form compounds in which the oxidation numbers differ by 2.

7.19 DIAGONAL RELATIONSHIPS

A **diagonal relationship** is a similarity in properties between diagonal neighbors in the main groups of the periodic table. A part of the reason for this similarity can be seen in Figs. 7.26 and 7.31 by concentrating on the colors that show the general trends in atomic radius and ionization energy. The colored bands of similar values lie in diagonal stripes across the table. Because these properties lie in a diagonal pattern, it is not surprising to find that the elements within a diagonal band show similar chemical properties. Diagonal relationships are helpful for making predictions about the properties of elements and their compounds.

The diagonal band of metalloids dividing the metals from the nonmetals is one example of a diagonal relationship (Fig. 7.38). So is the chemical similarity of lithium and magnesium and of beryllium and aluminum (Fig. 7.39). For example, both lithium and magnesium react directly with nitrogen to form nitrides. Like aluminum, beryllium reacts with both acids and bases. We shall see many examples of this diagonal similarity when we look at the elements in detail (in Chapter 19).

Diagonally related pairs of elements often show similar chemical properties.

7.20 ELECTRON AFFINITY

The **electron affinity,** E_{ea}, of an element is the energy released when an electron is added to a gas-phase atom. A high electron affinity means that a lot of energy is released when an electron attaches to an atom. A negative electron affinity means that energy must be supplied to push an electron on to an atom.

Figure 7.40 shows the variation in electron affinity through the periodic table. It is much less periodic than the variation in radius and ionization energy. However, one broad trend is clearly visible: electron affinities are highest toward the upper right of the periodic table close to the triangle of oxygen, fluorine, and chlorine. In these atoms the incoming electron occupies a *p*-orbital close to a highly charged nucleus and can experience its attraction quite strongly.

Once an electron has entered the single vacancy in the valence shell of a Group 17 atom, the shell is complete and any additional electron would have to begin a new shell. In that shell it would not only be farther from the nucleus but would also feel the repulsion of the negative charge already present. As a result, the second electron affinity of fluorine is strongly negative, meaning that a lot of energy has to be expended to form F^{2-} from F^-. As a consequence, the ionic compounds of the halogens are built from singly charged ions such as F^- and never from doubly charged ions such as F^{2-}.

A Group 16 atom, such as O or S, has two vacancies in its valence shell *p*-orbitals and can accommodate two additional electrons. The first electron affinity is positive. However, attachment of the second electron requires energy because of the repulsion by the negative charge already

FIGURE 7.38 Boron (top) and silicon (bottom) have a diagonal relationship. They are both brittle solids with high melting points. They also show a number of chemical similarities.

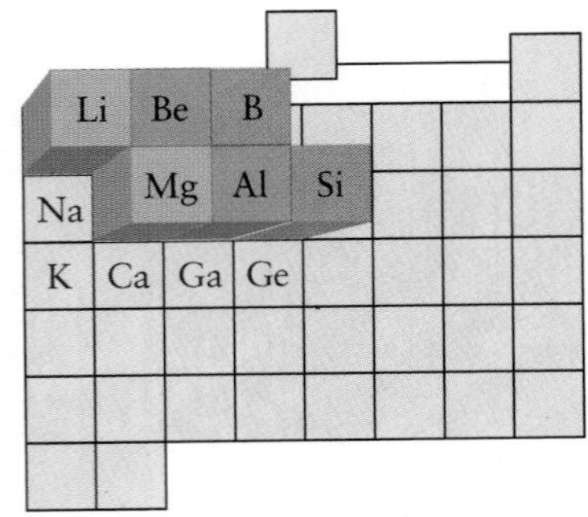

FIGURE 7.39 The pairs of elements represented by boxes of the same color show a strong diagonal relationship to each other.

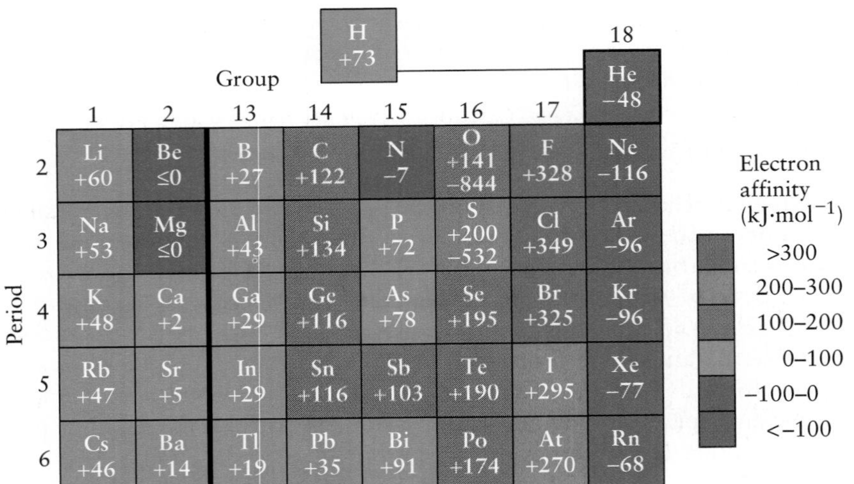

Group
Period
H +73
18
He −48
1
2
13
14
15
16
17
2
Li +60
Be ≤0
B +27
C +122
N −7
O +141 −844
F +328
Ne −116
3
Na +53
Mg ≤0
Al +43
Si +134
P +72
S +200 −532
Cl +349
Ar −96
4
K +48
Ca +2
Ga +29
Ge +116
As +78
Se +195
Br +325
Kr −96
5
Rb +47
Sr +5
In +29
Sn +116
Sb +103
Te +190
I +295
Xe −77
6
Cs +46
Ba +14
Tl +19
Pb +35
Bi +91
Po +174
At +270
Rn −68
Electron affinity (kJ·mol−1)
>300
200–300
100–200
0–100
−100–0
<−100

FIGURE 7.40 The variation of electron affinity in kilojoules per mole ($kJ \cdot mol^{-1}$) of the main-group elements. Where two values are given, the first refers to the formation of a singly charged anion and the second refers to the additional energy needed to produce a doubly charged anion. The variation is less systematic than for atomic radius and ionization energy, but high values tend to be found close to fluorine (except for the noble gases).

present in O^- or S^-. The difference from the halogen case, however, is that there is room to accommodate the additional electron in the valence shell, which still has a single vacancy in the singly charged ion. Therefore, we expect that less energy is needed to make O^{2-} from O^- than is required to make F^{2-} from F^-, where no such vacancy exists (compare diagrams 7 and 8). In fact, 141 $kJ \cdot mol^{-1}$ is released when the first electron adds to the neutral atom to form O^-, but 844 $kJ \cdot mol^{-1}$ must be *supplied* to add a second electron to form O^{2-}, so the total energy required to make O^{2-} from O is 703 $kJ \cdot mol^{-1}$. As we shall see in Chap-

CHEMISTRY AT WORK

Two materials scientists test a superconductor they have just prepared with a new formulation. The periodic table is the keyboard of the materials scientist, who must select elements based on their properties. The organization of the periodic table makes trends in properties evident and greatly facilitates the work of designing new materials.

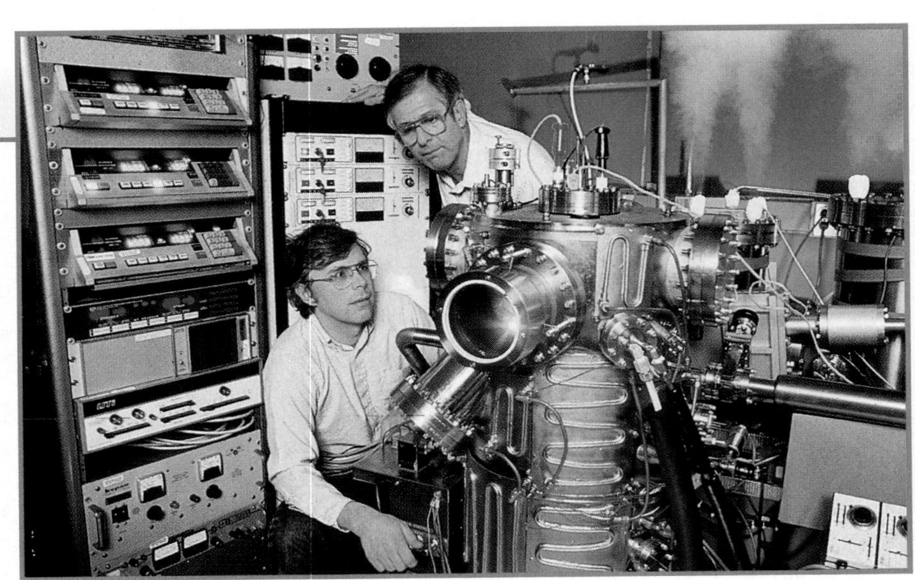

ter 8, this energy can be achieved in chemical reactions, and O^{2-} ions are typical of metal oxides.

Elements with the highest electron affinities are those close to oxygen, fluorine, and chlorine. Group 17 atoms can acquire one electron with a release of energy and Group 16 atoms can accept two electrons with chemically attainable energies; so the halogens typically form X^- ions and Group 16 elements typically form X^{2-} ions.

CHEMISTRY AND THE PERIODIC TABLE

Because the periodic table is a summary of electron configurations, we can use it to predict the properties of elements, once we locate them in the table.

7.21 THE *s*-BLOCK ELEMENTS

An *s*-block element has a low ionization energy, which means its outermost electrons can be lost easily (Fig. 7.41). A Group 1 element is likely to form +1 ions, such as Li^+, Na^+, and K^+. Group 2 elements similarly form +2 ions, such as Mg^{2+}, Ca^{2+}, and Ba^{2+}. An *s*-block element is likely to be a reactive metal with all the features that the name *metal* implies (Table 7.3). Because ionization energies are lowest at the bottom of each group, and the elements there lose their valence electrons most easily, the heavy elements cesium and barium react most vigorously of all Group 2 elements.

Beryllium, at the head of Group 2, has the highest ionization energy of the block. It therefore loses its valence electrons less readily than other Group 2 elements and has the least pronounced metallic character. The compounds of all the *s*-block elements (with the exception of beryllium) are ionic.

The elements in the s-block are all reactive metals that form basic oxides.

FIGURE 7.41 All the alkali metals are soft, silvery metals. Sodium is so reactive that it is kept under paraffin oil to protect it from air, and a freshly cut surface soon becomes covered with oxides.

TABLE 7.3 Characteristics of metals and nonmetals

Metals	Nonmetals
PHYSICAL PROPERTIES	
good conductors of electricity	poor conductors of electricity
ductile	not ductile
malleable, lustrous	not malleable
typically: solid	solid, liquid, or gas
high melting point	low melting point
good conductors of heat	poor conductors of heat
CHEMICAL PROPERTIES	
react with acids	do not react with acids
form basic oxides (which react with acids)	form acidic oxides (which react with bases)
form cations	form anions
form ionic halides	form covalent halides

7.22 THE *p*-BLOCK ELEMENTS

Elements on the left of the *p*-block have ionization energies that are low enough to confer on these elements some of the metallic properties of the *s*-block elements, especially the heavier elements. The metalloids, the elements that lie along the frontier between *p*-block metals and nonmetals, run diagonally across the table (Fig. 7.42). However, the ionization energies of the *p*-block metals are quite high, and they are less reactive than those in the *s*-block.

In Group 14, lead and tin are metals; but, although they are malleable and conduct electricity, they are not nearly as reactive as the *s*-block elements and many *d*-block elements (Fig. 7.43). That is why steel cans are tin-plated and why tin has been used for centuries in alloys such as bronze and pewter: to protect them against corrosion. Lead, too, was used for about 2000 years to make pipes for domestic water supplies, for it is very resistant to corrosion. It is no longer used, because lead compounds are now known to be toxic.

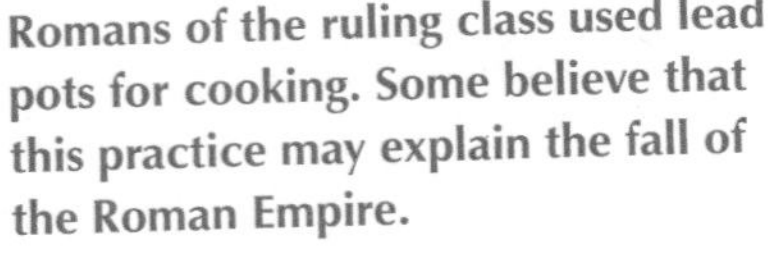
Romans of the ruling class used lead pots for cooking. Some believe that this practice may explain the fall of the Roman Empire.

Elements at the right of the *p*-block have characteristically high electron affinities: they gain electrons to complete closed shells. Except for the metalloids tellurium and polonium, the members of Groups 16 and 17 are nonmetals (Fig. 7.44). They typically form molecular compounds with each other and react with metals to form the anions in ionic compounds. Fluorine forms ionic or molecular compounds with every element except helium, neon, and argon.

The p-block elements tend to gain electrons to complete closed shells; they range from metals through metalloids to nonmetals.

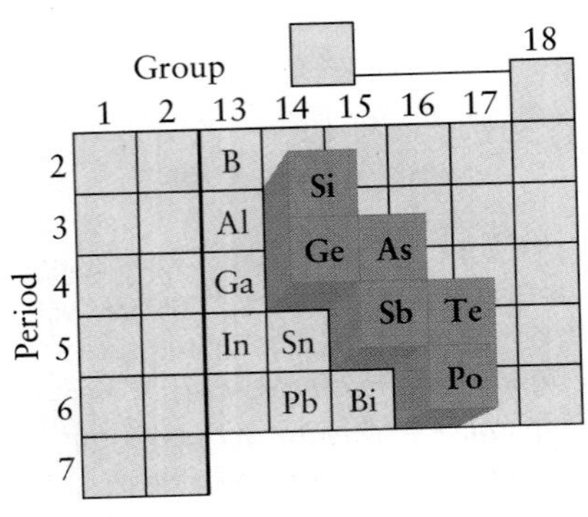

FIGURE 7.42 The metalloids fall in a diagonal band across the periodic table, between the metals and the nonmetals. The location of the metalloids reflects the diagonal relationship between elements.

FIGURE 7.43 The Group 14 elements. From left to right: carbon (as graphite), silicon, germanium, tin, and lead.

FIGURE 7.44 The Group 16 elements. From left to right: oxygen, sulfur, selenium, and tellurium. Note the trend from nonmetal to metal.

7.23 THE *d*-BLOCK ELEMENTS

All *d*-block elements are metals (Fig. 7.45). Their properties are transitional between the *s*- and *p*-block elements, which accounts for their alternative name, the transition metals.

Zinc, cadmium, and mercury, in Group 12, are not usually considered to be transition metals.

One characteristic feature of the *d*-block elements is that many of them form compounds with a variety of oxidation states. Iron, we have seen, forms iron(II) and iron(III) ions, Fe^{2+} and Fe^{3+}. Copper forms copper(I) and copper(II), Cu^{+} and Cu^{2+}. Potassium, an *s*-block metal, has a $4s^1$ valence configuration like copper, but forms only one type of ion, K^{+}. The reason for the difference between copper and potassium can be seen by comparing their second ionization energies, which are 1960 $kJ \cdot mol^{-1}$ and 3070 $kJ \cdot mol^{-1}$, respectively. To form Cu^{2+}, an electron is removed from the *d*-subshell of $[Ar]3d^{10}$; but to form K^{2+}, the electron would have to be removed from potassium's argonlike core. Because such huge amounts of energy are not readily available in chemical reactions, potassium can lose only its 4*s*-electron.

All d-block elements are metals with properties between those of s-block and p-block metals. Many d-block elements form cations with different oxidation numbers.

FIGURE 7.45 The elements in the first row of the *d*-block. Top row (left to right): scandium, titanium, vanadium, chromium, and manganese. Bottom row: iron, cobalt, nickel, copper, and zinc.

SKILLS YOU SHOULD HAVE MASTERED

Conceptual

1. Describe the nature of light as electromagnetic radiation and as a stream of photons.
2. Use the Bohr frequency condition to explain the origin of the lines in the spectrum of an element.
3. Distinguish and sketch the boundary surfaces of *s*-, *p*-, and *d*-orbitals.
4. Describe the interpretation of atomic orbitals in terms of probability.
5. Name and explain the relationship of each of the four quantum numbers to the properties of electrons in orbitals.
6. List the allowed energy levels of a bound electron in terms of the quantum numbers n, l, and m_l.
7. State how many orbitals of each type are contained in the shell corresponding to a given principal quantum number and how many electrons can be accommodated in those orbitals.
8. Describe the factors affecting the energy of an electron in a many-electron atom.

Problem-Solving

1. Calculate the wavelength or frequency of light from the relation $\lambda \times \nu = c$.
2. Use the relation $E = h\nu$ to calculate the energy, frequency, or number of photons emitted from a light source.
3. Use the expression for the energy levels of a hydrogen atom to correlate the lines in a spectrum with specific energy transitions.
4. Write the ground-state electron configuration for an element or ion.
5. Use the periodic table to predict and explain trends in atomic radius, ionic radius, ionization energy, and electron affinity.

Descriptive

1. Describe the general characteristics of elements in the *s*-, *p*-, and *d*-blocks.
2. Describe and explain diagonal relationships and the inert-pair effect.

EXERCISES

Frequency and Wavelength

7.1 (a) The approximate wavelength range of the visible spectrum is 400 nm to 700 nm. Express this range in hertz (Hz). (b) The wavelength of an AM radio band is about 250 m. What is the frequency of this band?

7.2 (a) Infrared radiation has wavelengths greater than about 800 nm. What is the frequency of 800-nm infrared radiation? (b) Microwaves, such as those used for radar and to heat food in a microwave oven, have wavelengths greater than 3 mm. What is the frequency of 3.0-mm radiation?

7.3 (a) Violet light has a frequency of 7.1×10^{14} Hz. What is the wavelength (in nanometers) of violet light? (b) When an electron beam strikes a block of copper, x-rays with a frequency of 2.0×10^{18} Hz are emitted. What is the wavelength (in picometers) of these x-rays?

7.4 (a) Radio waves for the FM station "Rock 99" at "99.3" on the FM dial are generated at 99.3 MHz. What is the wavelength of this station? (b) Radioastronomers use 1420-MHz waves to look at interstellar clouds of hydrogen atoms. What is the wavelength of this radiation?

Quanta and Photons

7.5 Sodium vapor lamps, used for public lighting, emit 589-nm yellow light. How much energy is emitted by (a) an excited sodium atom when it generates a photon? (b) 1.00 mol of excited sodium atoms at this wavelength?

7.6 When an electron beam strikes a block of copper, x-rays with a frequency of 2.0×10^{18} Hz are emitted. How much energy is emitted by (a) an excited copper atom when it generates an x-ray photon? (b) 1.0 mol of excited copper atoms at this wavelength?

7.7 (a) Gamma radiation emitted by the nucleus of an iron-57 atom has a wavelength of 86 pm. Calculate the energy of one photon of gamma radiation. (b) A mixture of argon and mercury vapor used in blue advertising signs emits 470-nm light. Calculate the energy change resulting from the emission of 1.0 mol of photons at this wavelength.

7.8 (a) The frequency for the FM radio station "Z-95" is 95.5 MHz. Calculate the energy produced in the transmission of 1.00 mol of photons at this frequency. (b) Ultraviolet radiation has wavelengths less than about

350 nm. What is the energy accompanying the emission of 1.00 mol of photons at this wavelength?

Atomic Spectra

7.9 What is the energy change of a sodium atom that emits a yellow photon with a wavelength of 589 nm? This is commonly called the sodium "D line."

7.10 A transition in a mercury atom emits a blue photon with a wavelength of 435.8 nm. What is the energy change of the atom?

7.11 (a) Use the Rydberg formula for atomic hydrogen to calculate the wavelength for the transition from $n = 6$ to $n = 2$. (b) Use Table 7.1 to determine the color in the visible spectrum to which this transition corresponds.

7.12 (a) What is the energy of a photon emitted from a hydrogen atom when an electron makes a transition from $n = 7$ to $n = 2$? (b) What is the wavelength of this photon?

7.13 (a) What is the highest frequency photon that can be emitted from the hydrogen atom? (b) What is the wavelength of this photon? In what part of the electromagnetic spectrum does this photon appear?

7.14 The violet line in the hydrogen spectrum is at 434.0 nm. What is the principal quantum number of the orbital corresponding to the upper energy state of the electron that produces a photon of this wavelength? (*Hint:* For the visible spectrum of hydrogen, the principal quantum number of the lower state is $n = 2$.)

Particles and Waves

7.15 Calculate the wavelengths of (a) an electron and (b) a neutron, both with a velocity of $1.5 \times 10^7\ \text{m·s}^{-1}$ (one-twentieth the speed of light).

7.16 (a) A baseball must weigh between 5.0 and 5.25 ounces (1 oz = 28.3 g). What is the wavelength of a 5.0-oz baseball thrown at 92 mph? (b) The velocity of an electron that is emitted from a metallic surface by a photon is $2.2 \times 10^3\ \text{km·s}^{-1}$. What is the wavelength of the electron?

Atomic Orbitals and Quantum Numbers

7.17 (a) Explain what is meant by an atomic orbital. (b) Sketch the shape of an *s*-orbital, a *p*-orbital, and a *d*-orbital.

7.18 In the early nineteenth century, John Dalton described atoms as hard spheres. Explain how and why his description should be revised.

7.19 How many subshells are there for n equal to (a) 2? (b) 3? (c) What are the permitted values of l when $n = 3$?

7.20 (a) How many subshells are there for the principal quantum number $n = 5$? (b) Identify the subshells in the form 5*s*, ... (c) How many orbitals are there in the shell with $n = 5$?

7.21 How many orbitals are in subshells with l equal to (a) 0? (b) 2? (c) 1? (d) 3?

7.22 How many electrons can occupy a subshell with l equal to (a) 0? (b) 2? (c) 1? (d) 3?

7.23 State the number of (a) subshells with $n = 2$; (b) orbitals with $n = 4$ and $l = 2$; (c) orbitals with $n = 2$; (d) orbitals in a 3*d* subshell.

7.24 State the number of (a) orbitals in a 4*s*-subshell; (b) subshells with $n = 1$; (c) orbitals in a 5*p*-subshell; (d) orbitals in a 6*d*-subshell.

7.25 Write the subshell notation (in the form 3*d*, for instance) and the number of orbitals having these quantum numbers in an atom: (a) $n = 3, l = 2$; (b) $n = 1, l = 0$; (c) $n = 6, l = 3$; (d) $n = 2, l = 1$.

7.26 Identify the values of the principal quantum number, n, and the azimuthal quantum number, l, for the following subshells: (a) 2*p*; (b) 5*f*; (c) 3*s*; (d) 4*d*.

7.27 How many electrons can have the following quantum numbers in an atom: (a) $n = 2, l = 1$; (b) $n = 4, l = 2, m_l = -2$; (c) $n = 2$; (d) $n = 3, l = 2, m_l = +1$?

7.28 How many electrons can occupy the (a) 4*p*; (b) 3*d*; (c) 1*s*; (d) 4*f* subshells?

7.29 Which of the following subshells cannot exist in an atom: (a) 2*d*; (b) 4*d*; (c) 4*g*; (d) 6*f*?

7.30 Which of the following subshells cannot exist in an atom: (a) 1*p*; (b) 3*f*; (c) 7*p*; (d) 5*d*?

7.31 Identify the sets of four quantum numbers $\{n, l, m_l, m_s\}$ that cannot exist for an electron in an atom and explain why not: (a) $\{4, 2, -1, +\frac{1}{2}\}$; (b) $\{5, 0, -1, +\frac{1}{2}\}$; (c) $\{4, 4, -1, +\frac{1}{2}\}$.

7.32 Identify the sets of four quantum numbers $\{n, l, m_l, m_s\}$ that cannot exist for an electron in an atom and explain why not: (a) $\{2, 2, -1, +\frac{1}{2}\}$; (b) $\{6, 0, 0, +\frac{1}{2}\}$; (c) $\{5, 4, +5, +\frac{1}{2}\}$.

7.33 Identify the possible sets of four quantum numbers $\{n, l, m_l, m_s\}$ for the starred electron in the diagrams below. Select the values of m_l by numbering from $-l$ to $+l$ from left to right.

(a) 2*p* ⇅ ⇅* ↑

(b) 5*d* ⇅ ⇅ ⇅ ↑* ↑

7.34 Identify the possible sets of four quantum numbers $\{n, l, m_l, m_s\}$ for the starred electron in the diagrams below. Select the values of m_l by numbering from $-l$ to $+l$ from left to right.

(a) 6*p* ↑*↓ ⇅ ↑

(b) 5*f* ⇅ ⇅ ⇅* ⇅ ⇅ ⇅ ⇅

Orbital Energies

7.35 Explain why, in a many-electron atom, a 2*s*-electron is bound more strongly than a 3*s*-electron.

7.36 Explain why, in a many-electron atom, a 3*s*-electron is bound more strongly than a 3*p*-electron.

7.37 (a) List the shells with n ranging from 1 to 4 in order of increasing energy. (b) For the shell with $n = 4$, list the subshells with l ranging from 0 to $n - 1$ in order of increasing energy.

7.38 For electrons in the *s*-, *p*-, *d*-, and *f*-subshells, list them in order of (a) the probability of being found close to the nucleus; (b) the effectiveness with which they can shield electrons in higher shells.

Electron Configurations and the Building-Up Principle

7.39 Refer to the periodic table and list the subshells 3*s*, 5*d*, 1*s*, 2*p*, 3*d* in order of increasing energy when empty.

7.40 Refer to the periodic table and list the subshells 2*s*, 4*d*, 2*p*, 1*s* in order of increasing energy when empty.

7.41 Use the periodic table and the Pauli exclusion principle to determine the set of quantum numbers $\{n, l, m_l, m_s\}$ that, according to the building-up principle, correspond to the (a) first; (b) ninth; (c) twentieth electron added to an atom.

7.42 Use the periodic table and the Pauli exclusion principle to determine the set of quantum numbers $\{n, l, m_l, m_s\}$ that, according to the building-up principle, correspond to the (a) sixth; (b) thirty-third; (c) fifty-fourth electron added to an atom.

7.43 Predict the ground-state electron configuration of an atom of (a) calcium; (b) nitrogen; (c) bromine; (d) uranium.

7.44 Predict the ground-state electron configuration of an atom of (a) iron; (b) gallium; (c) zirconium; (d) sodium.

7.45 Starting with the previous noble gas, write the ground-state electron configuration of an atom of (a) nickel; (b) cadmium; (c) lead; (d) silver.

7.46 Starting with the previous noble gas, write the ground-state electron configuration of an atom of (a) vanadium; (b) osmium; (c) tellurium; (d) mercury.

7.47 The valence electron configuration for Group 13 elements is ns^2np^1. Write the valence electron configuration for the elements of (a) Group 2; (b) Group 18.

7.48 The valence electron configuration for Group 13 elements is ns^2np^1. Write the valence electron configuration for the elements of (a) Group 1; (b) the titanium group.

7.49 Starting with the previous noble gas, write the electron configurations of the following ions: (a) Fe^{2+}; (b) Cl^-; (c) Tl^+.

7.50 Starting with the previous noble gas, write the electron configurations of the following ions: (a) Ni^{2+}; (b) Pb^{2+}; (c) N^{3-}.

7.51 Identify the element having the following description and then write its electron configuration, starting with the previous noble gas: (a) Group 14, Period 3; (b) Group 18, Period 2; (c) Group 1, Period 6; (d) Group 16, Period 3.

7.52 Identify the element having the following description and then write its electron configuration, starting with the previous noble gas: (a) Group 2, Period 5; (b) Group 18, Period 4; (c) Group 13, Period 3; (d) Group 17, Period 3.

7.53 Use the periodic table to determine the number of electrons in the *s*-, *p*-, and *d*-orbitals outside the previous noble-gas core for atoms of the following elements: (a) silicon; (b) chlorine; (c) manganese; (d) cobalt.

7.54 Use the periodic table to determine the number of electrons in the *s*-, *p*-, and *d*-orbitals outside the previous noble-gas core for atoms of the following elements: (a) tin; (b) iron; (c) sulfur; (d) bismuth.

Atomic and Ionic Radius

7.55 Explain why atomic radii increase down a group in the periodic table.

7.56 Use the concepts of penetration and shielding to explain why atomic radii decrease across a period.

7.57 (a) From Fig. 7.27, decide which group of elements has the largest atomic radii. (b) From Figs. 7.26 and 7.28, what general statement can be made regarding the ionic radius of a cation relative to its neutral atom? (c) What general statement can be made regarding the ionic radii of cations having the same number of electrons, but a different atomic number (for example, Al^{3+} versus Na^+)?

7.58 (a) From Fig. 7.27, decide which group of elements shown has the smallest atomic radii. (b) How may the relative sizes of the ionic radii of an anion and its parent neutral atom be summarized? (c) Summarize the relative ionic radii of anions having the same number of electrons, but a different atomic number (for example, N^{3-} versus F^-).

7.59 Identify the species with the larger radius in each of the following pairs: (a) Cl or S; (b) Cl^- or S^{2-}; (c) Na or Mg; (d) Mg^{2+} or Al^{3+}.

7.60 Which species has the smaller radius: (a) Li^+ or Na^+; (b) Cl or Cl^-; (c) Al or Al^{3+}; (d) N^{3-} or O^{2-}?

Ionization Energy

7.61 Explain why ionization energies decrease down a group and increase across a period.

7.62 (a) Which group of elements has the highest ionization energies? (b) Which group of elements has the lowest ionization energies?

7.63 Predict which atom of each pair has the greater first ionization energy (be alert for any exceptions to the general trend): (a) Na or Mg; (b) C or N; (c) P or S.

7.64 Predict which atom of each pair has the greater first ionization energy (be alert for any exceptions to the general trend): (a) Ba or Ca; (b) Be or B; (c) Ar or Xe.

7.65 In terms of electron configurations obtained from the building-up principle, explain why the ionization energies of Group 16 elements are smaller than those of Group 15 elements.

7.66 In terms of electron configurations obtained from the building-up principle, explain why the ionization energies of Group 13 elements are smaller than those of Group 2 elements.

7.67 Explain why the second ionization energy of sodium ($4560\ kJ\cdot mol^{-1}$) is significantly higher than the second ionization energy of magnesium ($1450\ kJ\cdot mol^{-1}$), even though the first ionization energy of sodium is less than that of magnesium (see Fig. 7.31).

7.68 Explain why the second ionization energy of sodium ($4560\ kJ\cdot mol^{-1}$) is larger than the third ionization energy of aluminum ($2740\ kJ\cdot mol^{-1}$), even though the first ionization energy of sodium is lower than that of aluminum (see Fig. 7.31).

Electron Affinity

7.69 (a) Which group of elements tends to have high electron affinities? (b) What is the general trend in electron affinities across a period?

7.70 (a) Explain why the electron affinity of chlorine is greater than that of bromine. (b) Explain why the electron affinity of the S atom is greater than that of the S^- ion.

7.71 Select the atom that has the greater electron affinity in the following pairs (be alert for any exceptions to the general trend): (a) S or Cl; (b) C or O; (c) Cl or Br.

7.72 Select the atom that has the greater electron affinity in the following pairs (be alert for any exceptions to the general trend): (a) S or Se; (b) Si or P; (c) N or O.

7.73 Describe the correlation in the periodic trends of atomic radii and ionization energies, both across a period and down a group.

7.74 Describe the correlation in the periodic trends of electron affinities and ionization energies, both across a period and down a group.

Trends in Chemical Properties

7.75 What is the diagonal relationship? Give two examples to illustrate the concept.

7.76 Give an example of the manner in which the inert-pair effect shows up in Group 13.

7.77 State three chemical properties that are typical of (a) metals; (b) nonmetals.

7.78 Describe the general trend in chemical reactivity of metals (a) across a period; (b) down a group.

7.79 Why are the *s*-block metals more reactive than the *p*-block metals?

7.80 Explain the trend in chemical properties down Group 14.

7.81 Identify the following elements as metals, nonmetals, or metalloids: (a) aluminum; (b) carbon; (c) germanium; (d) arsenic.

7.82 Identify the following elements as metals, nonmetals, or metalloids: (a) lead; (b) sulfur; (c) zinc; (d) silicon.

7.83 Which of the following pairs of elements do *not* exhibit the diagonal relationship? (a) B and Si; (b) Be and Al; (c) Al and Ge; (d) Na and Ca.

7.84 Which of the following pairs of elements do *not* exhibit the diagonal relationship? (a) O and Cl; (b) Li and Mg; (c) F and Ar; (d) H and Be.

SUPPLEMENTARY EXERCISES

7.85 Name the regions of the electromagnetic spectrum adjacent to the visible spectrum at (a) high energy; (b) low energy.

7.86 (a) Identify the product of the speed of light and the inverse of the wavelength. (b) Name the SI unit that is convenient for reporting the wavelength of visible light.

7.87 The wavelength of radar is about 3 cm. What is the frequency of radar waves?

7.88 X-rays of wavelength 149 pm are generated from a particle accelerator called a synchrotron. What is the frequency of these x-rays?

7.89 The average speed of a hydrogen molecule at 20°C is $1930\ m\cdot s^{-1}$. What is the wavelength of the H_2 molecule at this temperature? (Remember to calculate the mass of one H_2 molecule in kilograms, and that $1\ J = 1\ kg\cdot m^2\cdot s^{-2}$.)

7.90 How much energy in kilojoules per mole is released when an electron makes a transition from $n = 5$ to $n = 2$

in a hydrogen atom? Is this energy sufficient to break the H—H bond? (436 $kJ\cdot mol^{-1}$ is needed to break this bond.)

7.91 The energy required to break C—C bonds in a molecule is 348 $kJ\cdot mol^{-1}$. Will violet light of wavelength 420 nm be able to break the bond?

7.92 Many fireworks mixtures depend on the highly exothermic combustion of magnesium to magnesium oxide, in which the heat causes the oxide to become incandescent and to give out a bright white light. The color of the light can be changed by including nitrates and chlorides of elements that have spectra in the visible region (Fig. 7.1). Barium nitrate is often added to produce a yellow-green color. The excited barium ions generate 487-nm, 524-nm, 543-nm, 553-nm, and 578-nm light. Calculate the change in energy of the atom in each case, and the change in energy per mole of atoms.

7.93 (a) How many values of the quantum number l are possible when $n = 6$? (b) How many values of m_l are allowed for an electron in a 4*d*-subshell? (c) How many values of m_l are allowed for a 3*p*-subshell? (d) How many subshells are there in the shell with $n = 4$?

7.94 (a) How many orbitals are there in an *h*-subshell? (b) How many 5*f*-orbitals are there? (c) What is the maximum number of electrons that can have $n = 5$ and $l = 2$? (d) How many 5*p*-electrons are there in an antimony atom?

7.95 (a) How many unpaired electrons are there in the ground state of an iron atom? (b) How many *p*-electrons (allowing for all shells) are there in the ground state of a phosphorus atom? (c) What is the maximum number of electrons that can be accommodated in a shell with $n = 3$? (d) Which group of elements has a [noble gas]ns^2 electron configuration?

7.96 State the following:
(a) The number of orbitals with the quantum numbers $n = 3$, $l = 2$, and $m_l = 0$.
(b) The number of valence electrons in the outermost *p*-subshell of an S atom.
(c) The number of unpaired electrons in a Mn^{2+} ion.
(d) The subshell with the quantum numbers $n = 4$, $l = 2$.
(e) The m_l values allowed for a *d*-orbital.
(f) The allowed values of l for the shell with $n = 4$.
(g) The number of unpaired electrons in the cobalt atom.
(h) The number of orbitals in a shell with $n = 3$.
(i) The number of orbitals with $n = 3$ and $l = 1$.
(j) The maximum number of electrons with quantum numbers $n = 3$ and $l = 2$.
(k) The allowed values of l when $n = 2$.
(l) The possible values of m_l when $l = 2$.
(m) The number of electrons with $n = 4$ and $l = 1$.
(n) The quantum number that characterizes the angular shape of an atomic orbital.
(o) The designation of the subshell with $n = 3$ and $l = 1$.
(p) The lowest value of n for which a *d*-subshell can occur.

7.97 Which sets of quantum numbers are unacceptable?
(a) $n = 3, l = -2, m_l = 0, m_s = +\frac{1}{2}$
(b) $n = 2, l = 2, m_l = -1, m_s = -\frac{1}{2}$
(c) $n = 6, l = 2, m_l = -2, m_s = +\frac{1}{2}$
(d) $n = 4, l = 0, m_l = 0, m_s = -\frac{1}{2}$

7.98 Identify the group in the periodic table the members of which have the following ground-state valence electron configurations: (a) ns^2np^3; (b) $ns^2(n-1)d^{10}$; (c) ns^1; (d) ns^2np^6; (e) $ns^2(n-1)d^6$; (f) ns^2.

7.99 Write the ground-state electron configurations (starting with the previous noble gas) for (a) Zr; (b) Se; (c) Rb; (d) Cl; (e) Sb; (f) Pu; (g) Si; (h) Ar.

7.100 Write the ground-state electron configuration (starting with the previous noble gas) for (a) Zn^{2+}; (b) Se^{2-}; (c) I^-; (d) Y; (e) P; (f) In; (g) As; (h) Ir.

7.101 Write the values for the quantum numbers for the starred electron, as in Exercise 7.33:
(a) 3*p* ⥮ ⥮ ↑*
(b) 3*s* ⥮*

7.102 Starting with the previous noble gas, write the electron configurations of (a) V; (b) Cl^-; (c) Sn^{2+}; (d) Ni; (e) N^{3-}.

7.103 Select the best of the three choices:

(a) highest first ionization energy	Se	S	Te
(b) smallest radius	Cl^-	Br^-	F^-
(c) lowest electron affinity	K	Rb	Cs
(d) largest first ionization energy	O	S	F
(e) lowest second ionization energy	Ar	K	Ca
(f) greatest number of unpaired electrons	Fe	Co	Ni
(g) largest ionic radius	Ca^{2+}	Mg^{2+}	Ba^{2+}
(h) largest radius	S^{2-}	Cl^-	Cl
(i) highest first ionization energy	C	N	O
(j) highest electron affinity	P	S	Cl
(k) smallest atomic radius	Sn	I	Bi
(l) lowest first ionization energy	K	Na	Ca
(m) impossible subshell designation	4*g*	5*d*	4*p*
(n) number of orbitals with $n = 2$	2	4	8
(o) number of 5*f*-orbitals	14	7	9

7.104 Explain why the elements zinc, cadmium, and mercury have higher ionization energies than the coinage metals (copper, silver, and gold).

7.105 Identify the blocks to which the following elements belong: (a) Zr; (b) Sr; (c) Kr; (d) Au; (e) Pb; (f) Fe.

7.106 In July of 1994, when the Shoemaker-Levy comet struck Jupiter, astronomers focused telescopes fitted with diffraction gratings on Jupiter to collect spectra for future analysis. What would astronomers look for in the spectra, and what did they hope to learn about Jupiter by this means?

7.107 (a) Atomic sodium emits light very weakly at 389 nm when an excited electron moves from a 4*s*-orbital to a 3*s*-orbital, and at 300 nm when an electron moves from a 4*p*-orbital to the same 3*s*-orbital. What is the energy separation (in joules and kilojoules per mole) between the 4*s*- and 4*p*-orbitals? (b) Atomic potassium emits light very weakly at 365 nm when an excited electron moves from a 4*d*-orbital to a 4*s*-orbital, and at 689 nm when an electron moves from a 4*d*-orbital to a 4*p*-orbital. What is the energy separation (in joules and kilojoules per mole) between the 4*s*- and 4*p*-orbitals? (c) Why is the energy separation between the potassium orbitals larger than for the two corresponding orbitals in sodium?

7.108 In some forms of the periodic table, zinc, cadmium, and mercury are classified as belonging to "Group IIB", with the alkaline earth metals in "Group IIA." (a) What is the justification of this classification in terms of electronic structure? (b) How are any differences between "Group IIA" and "Group IIB" to be explained?

7.109 The radius of the ammonium ion is 137 pm. (a) If NH_4^+ were actually the cation of the "element" NH_4, where would NH_4 be located in the periodic table, considering its total number of protons? (b) As a result of its placement, describe the properties of the "element" ammonium. (c) Compare its ionic radius to that of the element actually found in the predicted location and explain any difference.

7.110 The German physicist Lothar Meyer observed a periodicity in the physical properties of the elements at about the same time as Mendeleev was working on their chemical properties. Some of Meyer's observations can be reproduced by plotting the molar volume, the volume occupied per mole of atoms, for the solid forms of the elements against atomic number. Do this for Periods 2 and 3, given the following densities of the elements in their solid forms:

Element	Density, $g{\cdot}cm^{-3}$	Element	Density, $g{\cdot}cm^{-3}$
Li	0.53	Na	0.97
Be	1.85	Mg	1.74
B	2.34	Al	2.70
C	2.26	Si	2.33
N	0.88	P	1.82
O	1.14	S	2.07
F	1.11	Cl	2.03
Ne	1.21	Ar	1.66

Suggest a reason for the sharp change in molar volume between the *s*-block and *p*-block elements.

7.111 In the spectroscopic technique known as "photoelectron spectroscopy" (PES), ultraviolet light is directed at an atom or molecule. Electrons are ejected from the valence shell, and their kinetic energies are measured. Because the energy of the incoming ultraviolet photon is known and the kinetic energy of the outgoing electron is measured, the ionization energy, *I*, can be deduced from the fact that the total energy is conserved. (a) Show that the speed v of the ejected electron and the frequency (ν) of the incoming radiation are related by

$$h\nu = I + \tfrac{1}{2}m_e v^2$$

(b) Use this relation to calculate the ionization energy of a rubidium atom, given that light of wavelength 58.4 nm produces electrons with a speed of $2450\ km{\cdot}s^{-1}$; recall that $1\ J = 1\ kg{\cdot}m^2{\cdot}s^{-2}$.

7.112 Suppose that in some other universe a rule corresponding to our Pauli exclusion principle reads "as many as two electrons in the same atom may have the same set of four quantum numbers." Suppose further that all other factors affecting electron configurations are unchanged. (a) Give the electron configuration of the element in the other universe that has five protons. (b) What is the most likely charge on the ion of this element? (c) Give the value of *Z* for the second inert gas in the other universe. Explain your reasoning.

CHAPTER 8

In the ancient legend, Icarus built wings from the finest materials then available; but he flew too close to the Sun and fell into the ocean. Modern chemistry has provided the world with materials that help us to achieve Icarus's ambition, and others like it, with a greater margin of safety. All these materials are based on the properties of bonds between atoms of the elements, and in this chapter we see what is involved in bond formation.

INSIDE MATERIALS: CHEMICAL BONDS

A quiet revolution has swept across the face of the Earth in the past half-century. Materials that were fantasies just a few decades ago have become reality. Synthetic fibers and sturdy yet flexible CDs have transformed our leisure time. Polyethylene, polystyrene, and polyvinylchloride (PVC) have made possible cellular phones and electronic goods. Composite materials have made possible fast, lightweight airplanes and automobiles. Chemists are providing new materials to meet the needs of technicians, designers, engineers, physicians, farmers, and anyone else who contributes to the quality of our lives.

The production of new materials has become possible because scientists now understand how atoms stick together in certain definite ways. They also understand the properties of compounds in terms of the types of bonds that hold atoms together. Why, for example, is calcium phosphate so rigid that nature has adopted it for the formation of bones? Can we make better bones? Why is it so difficult to make compounds from the nitrogen in air? Can we find an easy way? How can we explain the ability of hemoglobin to form a loose compound with oxygen, transport it to another part of the body, and then release it in response to a metabolic need? Can we make artificial blood? Even thinking about the answers to these questions requires a knowledge of chemical bonds.

A chemical bond forms if the resulting arrangement of atoms has a lower energy than the separate atoms. Sodium and chlorine atoms bond together as ions because solid sodium chloride has a lower energy than a gas of widely separated sodium and chlorine atoms. Hydrogen and nitrogen atoms bond together to form ammonia, NH_3, because a gas consisting of NH_3 molecules

IONIC BONDS

8.1 Lewis symbols for atoms and ions
8.2 Lattice enthalpies
8.3 The properties of ionic compounds

COVALENT BONDS

8.4 From atoms to molecules
8.5 The octet rule and Lewis structures

THE STRUCTURES OF POLYATOMIC SPECIES

8.6 Lewis structures
8.7 Resonance
8.8 Formal charge

EXCEPTIONS TO THE OCTET RULE

8.9 Radicals and biradicals
8.10 Expanded valence shells

LEWIS ACIDS AND BASES

8.11 The unusual structures of Group 13 halides
8.12 Lewis acid-base complexes

IONIC VERSUS COVALENT BONDS

8.13 Correcting the covalent model
8.14 Correcting the ionic model

has a lower energy than a gas consisting of the same number of separate nitrogen and hydrogen atoms. Moreover, if the lowest energy can be achieved by ion formation, then ions result. If the lowest energy can be achieved by sharing electrons, then molecules are formed.

All the changes in energy that occur when bonds form result from changes in the locations of the valence electrons of atoms, the electrons in the outermost shells of atoms. We can therefore expect to explain bond formation in terms of the electronic structures of atoms we studied in Chapter 7. Because the electronic structures of atoms are related to their locations in the periodic table, we can predict the ability of an atom to form bonds from its group and period.

The name *valence,* comes from the Latin word for "strength." "Vale!" ("Be strong!") was what the Romans said on parting. The valence electrons determine the number and strengths of bonds that an atom can form.

The American chemist, G. N. Lewis (Fig. 8.1), one of the greatest of all chemists, laid the foundations of our current understanding of bond formation. Most of the content of this chapter was built on his insights. If there were a single individual to whom it should be dedicated, it would be Lewis.

IONIC BONDS

An **ionic bond** is the attraction between the opposite charges of cations and anions. For example, the attraction between Na^+ ions and Cl^- ions results in the ionic compound sodium chloride. No bond is purely ionic, but the **ionic model,** the description of bonding in terms of ions, works well for many compounds. It is particularly appropriate for describing compounds of metallic elements, especially those of the *s*-block metals, with nonmetals. Metals have a low ionization energy, so they can give up their valence electrons relatively easily to form cations.

8.1 LEWIS SYMBOLS FOR ATOMS AND IONS

Lewis devised a very simple way to keep track of valence electrons when atoms form ionic bonds. The procedure is also useful for predicting the chemical formula of an ionic compound.

A **Lewis symbol** consists of the chemical symbol of an element and a dot for each of its valence electrons. Typical examples are

H· He: :N· ·O· :Cl· K· Mg:

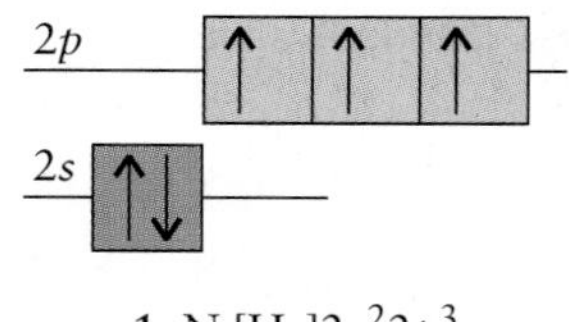

1 N $[He]2s^2 2p^3$

A single dot represents an electron alone in an orbital. A pair of dots represents two paired electrons in the same orbital. The Lewis symbol for nitrogen, for example, represents the valence electron configuration $2s^2 2p_x^1 2p_y^1 2p_z^1$ (**1**), with two electrons paired in an *s*-orbital and three unpaired electrons in separate *p*-orbitals. The Lewis symbol is a visual summary of the valence-shell electron configuration of an atom and allows us to see what happens to the electrons when an ion forms.

To work out the formula of an ionic compound, we first remove the dots from the Lewis symbol for the metal atom. Then we transfer the dots to the Lewis symbol for the nonmetal atom and complete its valence shell. Then we adjust the numbers of atoms of each kind so that all the dots removed from the metal atom symbols are accommodated by the nonmetal atom symbols. Finally, we add the charge of the ion. A simple example is the formation of the ionic solid potassium chloride:

$$K\cdot + :\ddot{\underset{\cdot\cdot}{Cl}}\cdot \longrightarrow K^+[:\ddot{\underset{\cdot\cdot}{Cl}}:]^-$$

The collection of symbols $K^+[:\ddot{\underset{\cdot\cdot}{Cl}}:]^-$ is the **Lewis formula** for potassium chloride. The electron supplied by one metal atom is accommodated in the single vacancy of one Cl atom, so the chemical formula of potassium chloride is KCl. We need to combine ions in such a way as to achieve a neutral formula. For calcium chloride, we need two Cl atoms to accommodate the two electrons supplied by one Ca atom:

$$Ca: + :\ddot{\underset{\cdot\cdot}{Cl}}\cdot + :\ddot{\underset{\cdot\cdot}{Cl}}\cdot \longrightarrow Ca^{2+}[:\ddot{\underset{\cdot\cdot}{Cl}}:]^-[:\ddot{\underset{\cdot\cdot}{Cl}}:]^-$$

It follows that the chemical formula for calcium chloride is $CaCl_2$.

In each case, the transfer of electrons results in the formation of an **octet** of electrons, an s^2p^6 electron configuration, on each of the atoms. Hydrogen, lithium, and beryllium are exceptions to this rule, because the $n = 1$ shell is complete with only two electrons. Hydrogen either loses an electron to form a bare proton or gains one to form a heliumlike $1s^2$ configuration, a **duplet.** Similarly, lithium ($[He]2s^1$) and beryllium ($[He]2s^2$) lose electrons to form a heliumlike duplet when they become Li^+ and Be^{2+}.

When an *s*-block metal atom forms a cation, it loses all its valence electrons and is left with only its core. That core has the octet configuration of the *preceding* noble gas atom (Fig. 8.2). The calcium cation, for instance, has an argonlike configuration. The metals on the left of the *p*-block may also lose their *s*- and *p*-electrons. However, when they do so, the atoms in Period 4 and below leave a noble-gas core surrounded by an additional, complete subshell of *d*-electrons. For instance, gallium forms the ion Ga^{3+} with the configuration $[Ar]3d^{10}$. The *d*-electrons of the *p*-block atoms are gripped tightly by the nucleus and, in most cases, cannot be lost.

For anions, the s^2p^6 configuration (or $1s^2$ in the case of the hydride ion, H^-) is that of the completed valence shell. It corresponds to the configuration of an atom of the *following* noble gas (Fig. 8.3). The oxide ion, for instance, has a neonlike configuration, and the sulfide ion (S^{2-}) has an argonlike configuration.

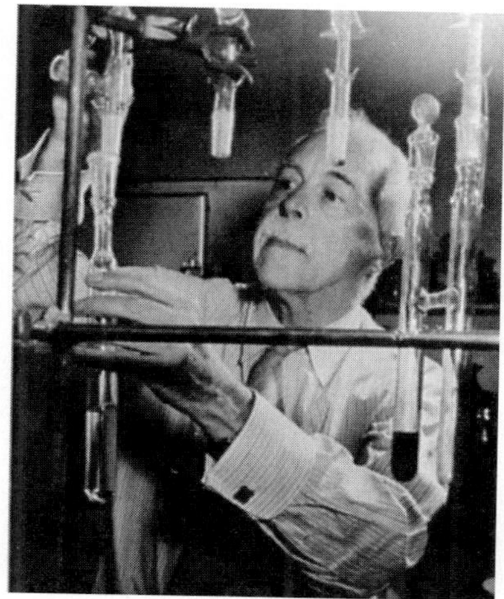

FIGURE 8.1 Gilbert Newton Lewis (1875–1946).

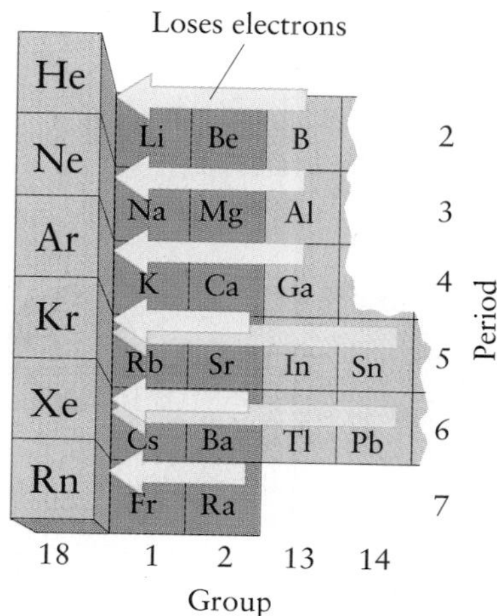

FIGURE 8.2 When a main-group metal atom forms a cation, it loses its *s* and *p* valence electrons and acquires the electron configuration of the preceding noble-gas atom. The heavier atoms in Groups 13 and 14 behave similarly, but the resulting core consists of the noble-gas configuration and an additional complete subshell of *d*-electrons.

EXAMPLE 8.1 Writing the chemical formula of a binary ionic compound

Write the chemical formula of aluminum oxide with Lewis symbols.

STRATEGY We write the Lewis symbols of Al and O atoms by noting that, as main-group elements in Groups 13 and 16, respectively, they have a number of valence electrons (represented as dots) equal to their group numbers minus 10. We then follow the procedure described earlier and ensure that all the dots are accommodated.

SOLUTION Aluminum belongs to Group 13 and has three valence electrons ($:Al\cdot$). Oxygen belongs to Group 16 and has six valence electrons ($\cdot\ddot{\underset{\cdot\cdot}{O}}\cdot$). The $2 \times 3 = 6$ dots released by the two Al atoms can be accommodated by three O atoms, which have $3 \times 2 = 6$ gaps in their valence shells.

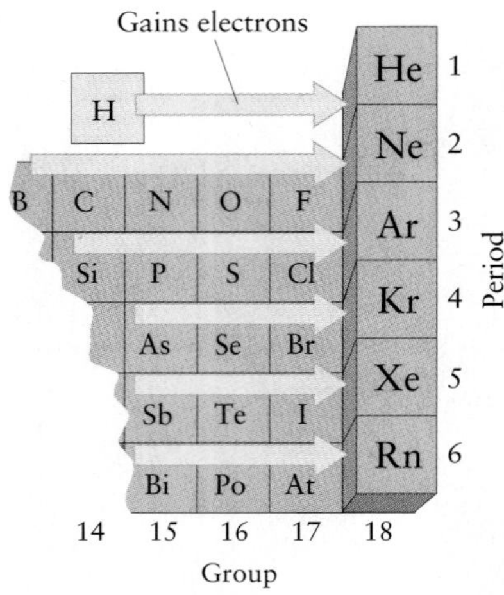

FIGURE 8.3 When *p*-block elements acquire electrons and form anions, they do so until they have reached the electron configuration of the following noble gas.

$$\cdot\ddot{\underset{\cdot\cdot}{O}}\cdot + :\text{Al}\cdot + \cdot\ddot{\underset{\cdot\cdot}{O}}\cdot + :\text{Al}\cdot + \cdot\ddot{\underset{\cdot\cdot}{O}}\cdot \longrightarrow [:\ddot{\underset{\cdot\cdot}{O}}:]^{2-}\ \text{Al}^{3+}\ [:\ddot{\underset{\cdot\cdot}{O}}:]^{2-}\ \text{Al}^{3+}\ [:\ddot{\underset{\cdot\cdot}{O}}:]^{2-}$$

Hence, the chemical formula of aluminum oxide is Al_2O_3.

SELF-TEST 8.1A Derive the chemical formulas of (a) calcium nitride and (b) sodium telluride by using Lewis symbols.

[*Answer:* (a) Ca_3N_2; (b) Na_2Te]

SELF-TEST 8.1B Derive the chemical formulas of (a) aluminum sulfide and (b) strontium phosphide by using Lewis symbols.

We saw in Section 7.18 that the inert-pair effect implies that the elements listed in Fig. 7.37 can lose either their valence *p*-electrons or all their valence *p*- and *s*-electrons. Similarly, we saw in Section 7.23 that many members of the *d*-block can lose a variable number of *d*-electrons. In each case, different ionic compounds can be obtained, such as SnO and SnO_2 for tin. We call this property—the ability to form a variety of compounds—**variable valence.**

The formation of ionic bonds is represented in terms of Lewis symbols by the loss or gain of electrons (dots) until both species have reached an octet of electrons.

SELF-TEST 8.2A Use Lewis symbols to write the formulas of the two chlorides that indium can form.

[*Answer:* $In^{3+}\ [:\ddot{\underset{\cdot\cdot}{Cl}}:]^-\ [:\ddot{\underset{\cdot\cdot}{Cl}}:]^-\ [:\ddot{\underset{\cdot\cdot}{Cl}}:]^-$ and $[:In]^+\ [:\ddot{\underset{\cdot\cdot}{Cl}}:]^-$]

SELF-TEST 8.2B Use Lewis symbols to write the formulas of lead(II) oxide and lead(IV) oxide.

8.2 LATTICE ENTHALPIES

The completion of an octet is not an end in itself: atoms do not "want" to acquire a noble-gas configuration. Atoms of metallic elements lose electrons down to their core—and a considerable investment of ionization energy is needed for them to do so. For example, to form K^+ ions from K atoms requires an input of 418 kJ·mol^{-1}. The K^+ ions cannot lose more electrons in a chemical reaction because the ionization energies of core electrons are too high. Nonmetals acquire electrons up to a completed octet. They cannot gain more, because that would involve accepting electrons into a higher energy shell. Even the completion of a noble-gas configuration may require considerable energy. For instance, to form an O^{2-} ion from an O atom requires 703 kJ·mol^{-1} (see Fig. 7.40).

The driving force for the formation of ionic bonds is the considerable lowering of energy that takes place when ions pack together closely as a solid and their opposite charges attract each other strongly. Ion formation is an energy-intensive investment; the strong attraction between

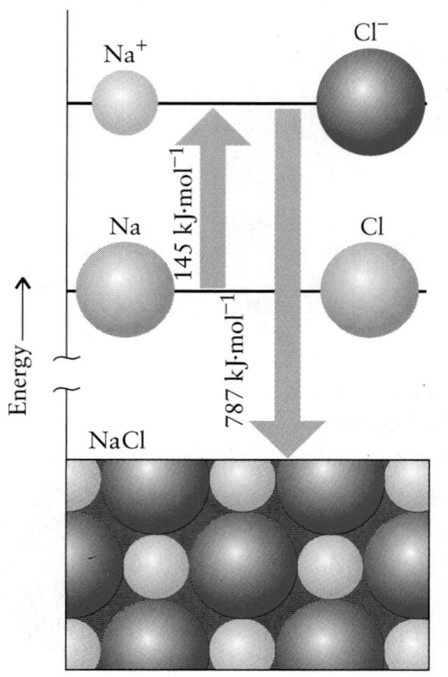

FIGURE 8.4 Considerable energy is needed to produce cations and anions from neutral atoms: the ionization energy of the metal atoms must be supplied, and it is only partly recovered from the electron affinity of the nonmetal atoms. The overall lowering of energy that drags the ionic solid into existence is due to the strong attraction between cations and anions that occurs in the solid.

opposite charges is the payoff (Fig. 8.4). The heat required for the reverse process, the vaporization of the solid to a gas of ions, is called the **lattice enthalpy** of the compound. Heat equal to the lattice enthalpy is released when the solid lattice forms from gaseous ions.

We cannot measure the lattice enthalpy of a compound directly, but we can obtain it by combining other measurements. The procedure uses a **Born-Haber cycle,** a series of steps starting at the elements and ending at the elements again, and including the formation of a solid lattice from the gaseous ions. The sum of the enthalpy changes for a complete Born-Haber cycle is 0, because the enthalpy of the system must be the same at the start and finish. Therefore, a knowledge of all the other enthalpy changes in the cycle can be used to find the unknown lattice enthalpy. The lattice enthalpies of other compounds obtained in this way are listed in Table 8.1.

The overall enthalpy change around a Born-Haber cycle must be 0 because enthalpy is a state function (Section 6.4).

EXAMPLE 8.2 Using a Born-Haber cycle to calculate a lattice enthalpy

Use a Born-Haber cycle to calculate the lattice enthalpy of potassium chloride.

STRATEGY Start with the elements in the appropriate amounts to form the compound (Fig. 8.5). Atomize the metal and the nonmetal (that is, vaporize each element from its solid form to its separated atoms), and write the corresponding enthalpies of atomization next to the upward-pointing arrows in the diagram. Form the cation of the metal. This step requires the ionization energy of the element and possibly the sum of the first and higher ionization energies (Fig. 7.31). The corresponding arrow points upward. Form the anion of the nonmetal. This step *releases* an energy equal to the electron affinity of the element. If the electron affinity is positive, the corresponding arrow points downward because energy is released (so ΔH is negative). If it is negative, then the arrow points upward because energy must be supplied (so ΔH is positive). Let the gas of ions form the solid compound. This step is the reverse of the formation of the ions from the solid, so its enthalpy change is the negative of the lattice enthalpy, $-\Delta H_L$. Denote it by an arrow pointing downward. Complete the cycle with an arrow pointing from the compound to the elements: the enthalpy

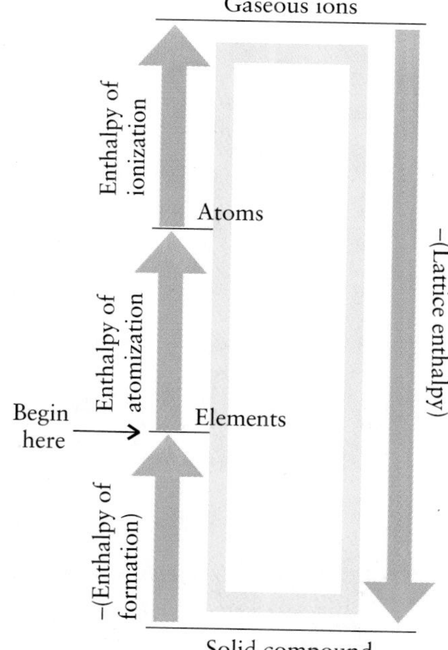

FIGURE 8.5 In a Born-Haber cycle, we select a sequence of steps that starts and ends at the same point (the elements, for instance). One of the steps is the formation of the solid from a gas of ions. The sum of enthalpy changes for the complete cycle is 0 because enthalpy is a state property.

TABLE 8.1 Lattice enthalpies at 25°C in kilojoules per mole

HALIDES							
LiF	1046	LiCl	861	LiBr	818	LiI	759
NaF	929	NaCl	787	NaBr	751	NaI	700
KF	826	KCl	717	KBr	689	KI	645
AgF	971	AgCl	916	AgBr	903	AgI	887
$BeCl_2$	3017	$MgCl_2$	2524	$CaCl_2$	2260	$SrCl_2$	2153
		MgF_2	2961	$CaBr_2$	1984		
OXIDES							
MgO	3850	CaO	3461	SrO	3283	BaO	3114
SULFIDES							
MgS	3406	CaS	3119	SrS	2974	BaS	2832

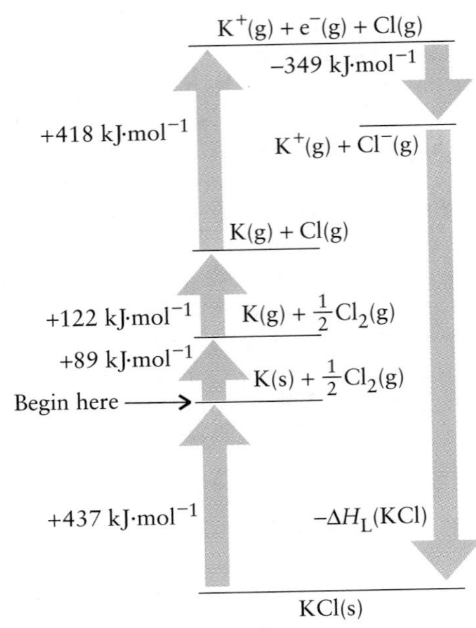

FIGURE 8.6 The Born-Haber cycle used to determine the lattice enthalpy of potassium chloride.

change for this step is the negative of the enthalpy of formation of the compound (Appendix 2A). Finally, calculate ΔH_L from the fact that the sum of all the enthalpy changes for the complete cycle is 0.

SOLUTION The Born-Haber cycle for KCl is shown in Fig. 8.6. The sum of the enthalpy changes for the complete cycle is 0, so we can write

$$(89 + 122 + 418 - 349 + 437)\ kJ{\cdot}mol^{-1} - \Delta H_L = 0$$

and hence

$$\Delta H_L = (89 + 122 + 418 - 349 + 437)\ kJ{\cdot}mol^{-1} = +717\ kJ{\cdot}mol^{-1}$$

Therefore, the lattice enthalpy of potassium chloride is 717 $kJ{\cdot}mol^{-1}$. *All lattice enthalpies are positive,* so there is no need to write the sign.

SELF-TEST 8.3A Calculate the lattice enthalpy of calcium chloride, $CaCl_2$, by using the data in Appendix 2A and Appendix 2D. The second ionization energy of calcium is 1150 $kJ{\cdot}mol^{-1}$.

[*Answer:* 2259 $kJ{\cdot}mol^{-1}$]

SELF-TEST 8.3B Calculate the lattice enthalpy of magnesium bromide, $MgBr_2$. The second ionization energy of magnesium is 1450 $kJ{\cdot}mol^{-1}$.

Lattice enthalpies are large when the ions are small and highly charged: compare, for example, the lattice enthalpies of MgO and KCl in Table 8.1. We expect this to be the case from Coulomb's law (Section 7.9). Highly charged ions attract one another strongly. Moreover, the potential energy is low if they can get close together, which they can if they are small.

The interaction between ions accounts for the formation of ionic compounds; its strength is measured by the lattice enthalpy. The lattice enthalpy is large for compounds formed from small, highly charged ions.

8.3 THE PROPERTIES OF IONIC COMPOUNDS

Ionic solids are assemblies of cations and anions stacked together in a regular array. In the case of sodium chloride, positive sodium ions alternate with negative chloride ions, and large numbers of oppositely charged ions are lined up in all three dimensions (Fig. 8.7). Ionic solids are examples of **crystalline solids,** or solids that consist of atoms, molecules, or ions stacked together in a regular array.

The strong attraction between oppositely charged ions in ionic solids accounts for their typical properties, such as high melting points, high boiling points, and brittleness. It requires a high temperature to get the ions to move past one another to form a liquid, and an even higher temperature to drive them apart into a gas of ions. Ionic solids are brittle because of those strong attractions and repulsions. We cannot just push a block of ions past another block; instead, when we strike an ionic solid, it tends to shatter into fragments (Fig. 8.8).

One way to separate the ions from an ionic solid without heating it is to dissolve it in water. Many ionic solids are soluble in water. As we saw in Section 3.3, when an ionic solid dissolves in water, the ions are separated by the water molecules and give an electrolyte solution. The mobile ions conduct electricity. Some ionic solids, such as MgO and $AgCl$, are insoluble in water: the attraction between their cations and anions is too strong to allow them to be separated from one another.

We can now see why nature has adopted an ionic solid, calcium phosphate, for our skeletons: the doubly charged small calcium ions and the triply charged phosphate ions attract one another very strongly and clamp together to form a rigid, insoluble solid. It is fortunate for us that calcium phosphate does not dissolve in water: if it were soluble, our skeletons would dissolve in our own body fluids! Nevertheless, calcium

FIGURE 8.7 An ionic solid—here we see a fragment of sodium chloride, with the sodium ions represented by red spheres and the chloride ions by green spheres—consists of an almost infinite array of cations and anions stacked together to give the lowest energy arrangement. The pattern shown here is repeated throughout the crystal.

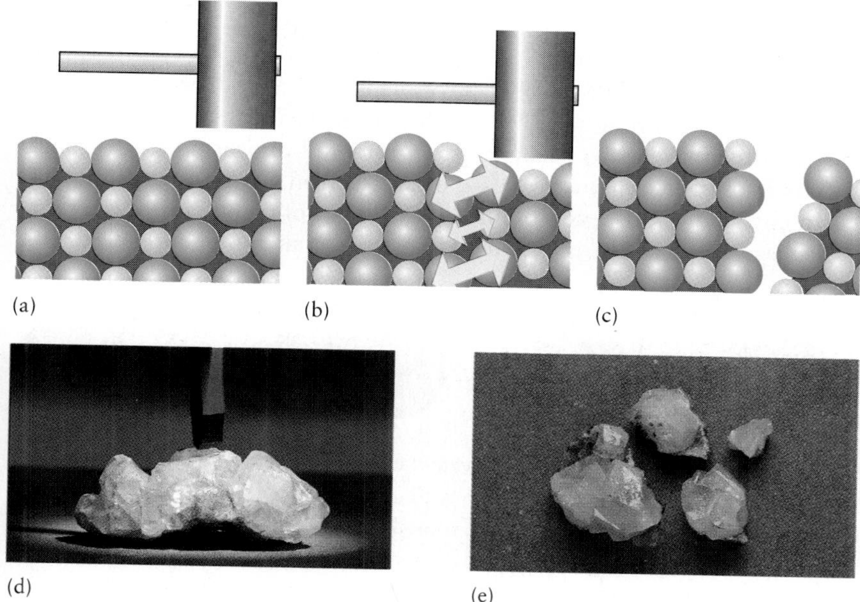

FIGURE 8.8 This sequence of images illustrates why ionic solids are brittle. (a) The original solid consists of an orderly array of cations and anions. (b) A hammer blow can push the ions into positions where cations are next to cations and anions are next to anions; there are now strong repulsive forces acting (as depicted by the double-headed arrows). (c) As a result of these repulsive forces, the solid springs apart in fragments. (d) This chunk of calcite consists of several large crystals joined together. (e) The blow of a hammer has shattered the crystal, leaving flat, regular surfaces consisting of planes of ions.

phosphate is not completely insoluble, and each day a small amount of the calcium in our skeleton dissolves and must be replaced or osteoporosis (the disintegration of bone) results. The replacement of bone calcium is one of the reasons why we need calcium in our daily diet.

> ***Ions stack together in regular crystalline structures. Ionic solids typically have high melting and boiling points, are brittle, and form electrolyte solutions if they dissolve in water.***

COVALENT BONDS

Nonmetallic elements cannot form monatomic cations, because their ionization energies are too high. However, we know that nonmetals do combine with one another: the existence of millions of different compounds of carbon, hydrogen, and oxygen alone is evidence enough! The nature of the bond between nonmetals puzzled scientists until 1916, when Lewis found an explanation. With brilliant insight, and before anyone knew about quantum mechanics or electron orbitals, Lewis proposed that a **covalent bond** is a pair of electrons shared between two atoms.

8.4 FROM ATOMS TO MOLECULES

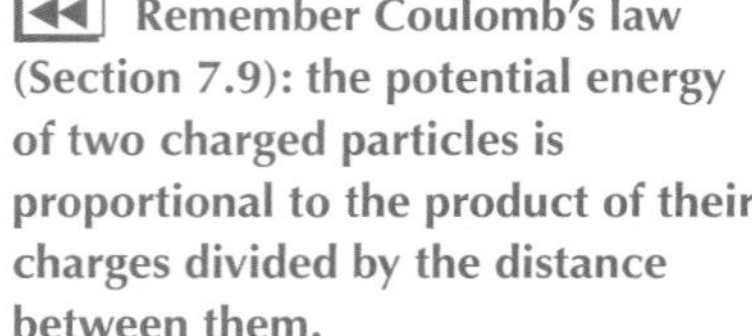

Remember Coulomb's law (Section 7.9): the potential energy of two charged particles is proportional to the product of their charges divided by the distance between them.

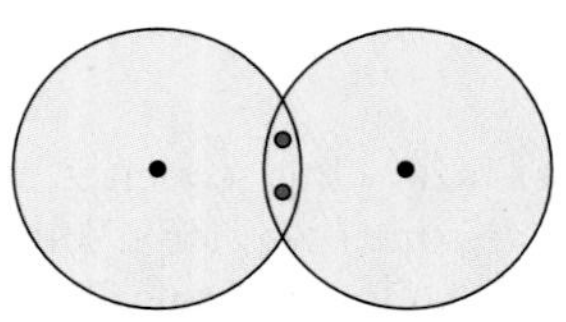

2 Shared electron pair

Lewis knew that the two electrons of a shared pair can attract the nuclei they lie between. Attractions between the electron pair and the nuclei pull the two atoms together (**2**). He had no way of knowing why it had to be a *pair* of electrons and not some other number. The explanation came only with the development of quantum mechanics in the 1920s and the discovery of the Pauli exclusion principle.

The simplest example is the H_2 molecule, illustrated in Fig. 8.9. Initially, the two hydrogen atoms each have one valence electron in a 1*s*-orbital. As the atoms approach, their orbitals merge and the two electrons that occupy the merged orbitals pair with each other. Once the orbitals have merged, the two 1*s*-electrons are equally attracted to and shared by the two nuclei. The electron density *between* the two nuclei is greater than it is on either side. Neither atom has to release an electron totally, so neither has to be supplied with its full ionization energy. Sharing corresponds to a *partial* release of an electron, so less energy is needed. In covalent bonds, each bonded atom may contribute one electron to the shared electron pair, or one atom may contribute both electrons.

(a) 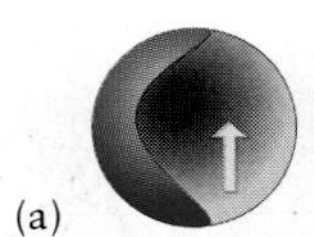  (b) 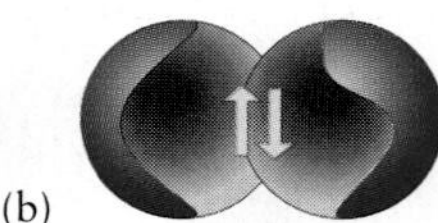 (c) 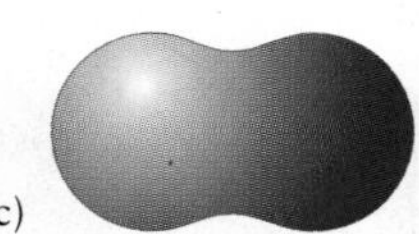

FIGURE 8.9 The formation of a covalent bond between two hydrogen atoms. (a) Two separate hydrogen atoms, each with one electron. (b) The electron cloud that forms when the spins pair and the orbitals merge. Note the dense region lying between the two nuclei. (c) The boundary surface that we shall use to depict a covalent bond.

Either way, the shared pair of electrons lies between the two neighboring atoms and binds them together.

> ***Nonmetals form covalent bonds to one another by sharing pairs of electrons.***

8.5 THE OCTET RULE AND LEWIS STRUCTURES

How many covalent bonds can an atom form? When an ionic bond forms, one atom loses electrons and the other gains them until both atoms have reached a noble-gas configuration—a duplet for elements close to helium and an octet for all other elements. In covalent bonds, atoms *share* electrons until they too reach a noble-gas configuration. Lewis called this the **octet rule:**

> In covalent bond formation, atoms go as far as possible toward completing their octets by sharing electron pairs.

Nitrogen (:Ṇ̇·) has five valence electrons and needs three more to complete its octet. Chlorine (:C̤̈l·) has seven valence electrons and needs ~~a are in~~ one more to complete its octet. Argon (:Ä̤r:) already has a complete octet and has no tendency to share any more electrons. Hydrogen (H·) needs one more electron to reach its heliumlike duplet.

The **valence** of an element is the number of covalent bonds an atom of the element forms. This number is most easily found by using the same Lewis symbols that we introduced for the discussion of ionic bonds. Consider molecular hydrogen, H_2. Each atom completes its heliumlike duplet by sharing its electron with the other:

$$\mathrm{H\cdot + \cdot H \longrightarrow H{:}H \quad or \quad H{-}H}$$

The symbol H—H, in which the line represents the shared electron pair (the covalent bond between the two atoms), is the simplest example of a **Lewis structure,** a diagram showing how electron pairs are shared between atoms in a molecule. Because hydrogen completes its duplet by sharing one pair of electrons, it has a valence of 1 in all its compounds.

In a Lewis structure, a line between two atoms, as in X—Y, represents a shared pair of electrons.

A fluorine atom has seven valence electrons and needs one more to complete its octet. It can achieve an octet by accepting a share in an electron supplied by another fluorine atom:

$$:\ddot{\underset{\cdot\cdot}{\mathrm{F}}}\cdot + \cdot\ddot{\underset{\cdot\cdot}{\mathrm{F}}}: \longrightarrow (:\ddot{\underset{\cdot\cdot}{\mathrm{F}}})(\ddot{\underset{\cdot\cdot}{\mathrm{F}}}:) \quad \text{or} \quad :\ddot{\underset{\cdot\cdot}{\mathrm{F}}}{-}\ddot{\underset{\cdot\cdot}{\mathrm{F}}}:$$

The circles have been drawn around each F atom to show how each one gets an octet by sharing one electron pair. The valence of fluorine is therefore 1, the same as that of hydrogen.

From its Lewis structure, we see that F_2 possesses **lone pairs** of electrons, that is, pairs of valence electrons that are not involved in bonding. The lone pairs on neighboring F atoms repel each other, and this repulsion is almost enough to overcome the favorable attractions of the bonding pair that holds the F_2 molecule together. This repulsion is one of the reasons why fluorine gas is so reactive: the atoms are bound together as F_2 molecules only very weakly. Many other molecules (not only diatomic molecules) have lone pairs of electrons. Among the common diatomic molecules, only H_2 does not have any lone pairs at all.

Nonmetal atoms share electrons with one another until they have completed their octet (or duplet); a Lewis structure shows the arrangement of electrons as shared pairs (or lines) and lone pairs (pairs of dots).

SELF-TEST 8.4A Write the Lewis structure for the "interhalogen" compound chlorine monofluoride, ClF, and state how many lone pairs each atom possesses in the compound.

[*Answer:* :C̤̈l—F̤̈:; three on each atom]

SELF-TEST 8.4B Write the Lewis structure for the compound HBr and state how many lone pairs each atom in the compound possesses.

THE STRUCTURES OF POLYATOMIC SPECIES

We have seen how *pairs* of atoms form bonds. The atoms in polyatomic molecules and ions are also bound together by covalent bonds. Each atom completes its octet (or duplet for hydrogen) by sharing pairs of electrons with its immediate neighbors. Each shared pair counts as one covalent bond and is represented by a line between the two atoms.

8.6 LEWIS STRUCTURES

The simplest organic molecule is methane, CH_4. To write its Lewis structure, we count the valence electrons available from all the atoms in the molecule. For methane, the Lewis symbols are

:Ċ̣   H·   H·   H·   H·

so there are $4 + (4 \times 1) = 8$ valence electrons. The next step is to arrange the dots representing the electrons so that the C atom has an octet and each H atom has a duplet. We draw the arrangement shown on the left in (**3**). The Lewis structure of methane is then redrawn as shown on the right in (**3**). Because the carbon atom is linked by four bonds to other atoms, we say that it is *tetravalent:* it has a valence of 4. Almost all compounds of carbon are tetravalent.

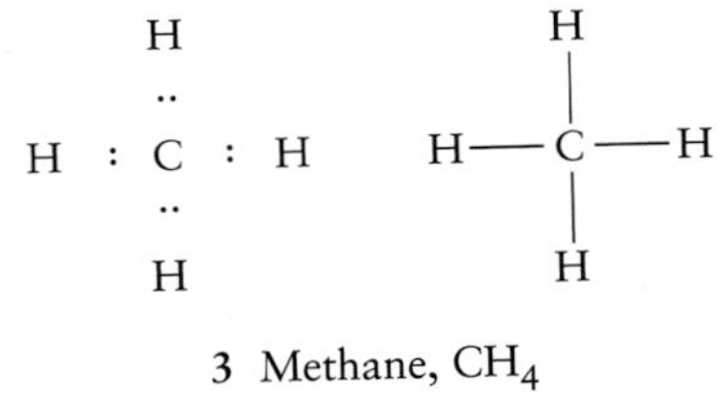

3 Methane, CH_4

The only common exception to the tetravalence of carbon is CO.

A single shared pair of electrons is called a **single bond.** However, atoms can share two or three electron pairs. Two shared electron pairs form a **double bond,** and three shared electron pairs form a **triple bond.** For instance, a double bond between a carbon atom and an oxygen atom, C::O appears as C=O in a Lewis structure. Similarly, a triple bond, such as C:::C, appears as C≡C. Double and triple bonds are collectively called **multiple bonds.** As before, each line represents a pair of electrons. Thus, a double bond involves a total of four electrons; a triple bond involves six electrons.

Sometimes we may be uncertain about the arrangement of atoms in a polyatomic ion. Just which is the central atom? A good rule of thumb for this situation is to *choose the atom with the lowest ionization energy for the central atom.* This arrangement often leads to the lowest energy because the designated central atom requires the smallest amount of ener-

CHEMISTRY AT WORK

An analytical chemist uses an auger electron spectrometer to determine the composition of a sample of copper iridium selenide. Once the atomic arrangement has been determined, he will study the response of the material to light of different wavelengths to assess its potential for generating electricity from solar radiation.

gy to give up its electrons and share them with its neighbors. The atoms with higher ionization energies can hold on to their electrons as lone pairs. However, hydrogen is never central, because it can form only one bond.

Another rule of thumb is to *arrange the atoms symmetrically around the central atom.* For instance, SO_2 is OSO, not SOO. In chemical formulas, especially simple ones, the central atom is often written first, followed by the atoms attached to it. For example, in the compound with the chemical formula OF_2 (**4**), the arrangement of the atoms is actually FOF not OFF; and in PO_4^{3-} the P atom lies at the center of four O atoms. If the compound is an oxoacid, then the acidic hydrogen atoms are attached to oxygen atoms, which in turn are attached to the central atom. Two examples are sulfuric acid, H_2SO_4, which has the structure $(HO)_2SO_2$, and chlorous acid, $HClO_2$, which has the structure (HO)ClO.

Recognizing other characteristic patterns helps to reduce the work we have to do. For example, a "terminal" halogen atom, that is, a halogen atom linked to just one other atom, always has a single bond and three lone pairs:

$$-\ddot{\underset{..}{F}}: \quad -\ddot{\underset{..}{Cl}}: \quad -\ddot{\underset{..}{Br}}: \quad -\ddot{\underset{..}{I}}:$$

Except for CO, terminal oxygen and sulfur atoms form either a single bond with three lone pairs or a double bond with two lone pairs:

$$-\ddot{\underset{..}{O}}: \text{ or } =\ddot{\underset{..}{O}} \qquad -\ddot{\underset{..}{S}}: \text{ or } =\ddot{\underset{..}{S}}$$

The marginal structures contain several examples of these patterns.

The same general procedure is also used for polyatomic ions, such as the ammonium ion, NH_4^+, or the sulfate ion, SO_4^{2-}. We simply adjust the total number of dots to represent the overall charge. For a cation, subtract one dot for each positive charge on the overall structure, because each positive charge corresponds to the loss of an electron. For an anion, add one dot for each negative charge. For example, a sulfate ion, SO_4^{2-},

▶▶ After we have met the concept of electronegativity (Section 8.13), we shall be able to express this rule in a different manner: choose the atom with the lowest electronegativity to be the central atom.

$$:\ddot{\underset{..}{F}}-\ddot{\underset{..}{O}}-\ddot{\underset{..}{F}}:$$

4 Oxygen difluoride, OF_2

TOOLBOX 8.1 How to write the Lewis structure of a polyatomic species

We begin to write Lewis structures for polyatomic species just as we did for methane:

Step 1. Count the total number of valence electrons on each atom and determine the number of electron pairs in the molecule.

Each hydrogen atom supplies one electron. Each element in Groups 1 and 2 supplies a number of electrons equal to its group number. Each element in Groups 13 through 18 supplies a number of electrons equal to its group number minus 10. For example, carbon in Group 14 supplies four electrons. Once we know the total number of electrons, we divide by 2 to obtain the number of electron pairs. For example, the HCN molecule has 1 + 4 + 5 = 10 valence electrons, so it has five electron pairs.

Step 2. Write the chemical symbols of the atoms to show their layout in the molecule.

When there are three or more atoms in a molecule, we need to decide how to arrange them. We can predict the most likely arrangements of atoms by using common patterns and the clues given earlier. For example, carbon has a lower ionization energy than nitrogen, so the HCN molecule is written

$$\mathrm{H\ C\ N} \qquad :::::$$

The dots representing the electron pairs are parked on the right.

Step 3. Place one electron pair between each pair of bonded atoms.

Each bonded pair of atoms must have at least a single bond between them. For example, HCN has five pairs of electrons. We use two pairs to form bonds between the atoms:

$$\mathrm{H{:}C{:}N} \qquad :::$$

At this point, three of the five electron pairs remain unused.

Step 4. Complete the octet (or duplet, in the case of H) of each atom by placing any remaining electron pairs around the atoms. If there are not enough electron pairs, form multiple bonds.

For HCN, we could try to put all three remaining pairs on the N atom:

$$\mathrm{H{:}C{:}\underset{\cdot\cdot}{\ddot{N}}{:}}$$

However, this arrangement does not complete carbon's octet. If we use the electrons to complete the octet on the C atom, then we do not complete nitrogen's octet:

$$\mathrm{H{:}\underset{\cdot\cdot}{\ddot{C}}{:}N}$$

Therefore, we rearrange the electron pairs to form a triple bond between carbon and nitrogen:

$$\mathrm{H{:}C{:}{:}{:}N{:}} \quad \text{or} \quad \mathrm{H{-}C{\equiv}N{:}}$$

Each line represents a bonded electron pair. To check on the validity of a Lewis structure, verify that each atom has an octet or duplet. With a triple bond, both the nitrogen atom and the carbon atom achieve octets by sharing electron pairs.

SELF-TEST 8.5A Write the Lewis structure for OF_2.
[*Answer:* See (4).]

SELF-TEST 8.5B Write the Lewis structure for CH_3Cl.

has 6 (from the S atom) + 4 × 6 (from the 4 O atoms) + 2 (for the −2 charge) = 32 electrons; so it has 16 electron pairs. The cation and the anion must be treated separately: they are individual ions and are not linked by shared pairs. The Lewis structure of ammonium sulfate, $(NH_4)_2SO_4$, for instance, is written as three bracketed ions (5):

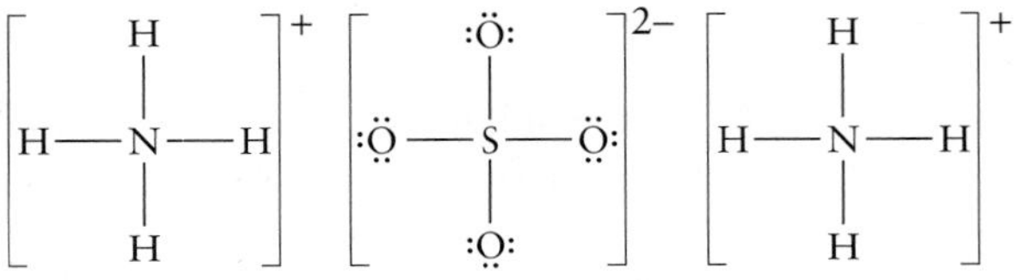

5 Ammonium sulfate, $(NH_4)_2SO_4$

The sign at the top right of each bracket shows the charge of each ion: it belongs to the whole ion, not to any particular atom.

The Lewis structure of a polyatomic species is obtained by using all the valence electrons to complete the octets (or duplets) of the atoms present by forming single or multiple bonds and leaving some electrons as lone pairs.

EXAMPLE 8.3 Writing Lewis structures for polyatomic molecules

Write the Lewis structure for acetic acid, CH_3COOH. In the $-COOH$ group, both O atoms are attached to the same C atom and one of them is bonded to the final H atom. The two C atoms are bonded to each other.

STRATEGY Acetic acid seems to consist of two groups with central C atoms joined together: a CH_3- group and a $-COOH$ group. We anticipate that the CH_3- group, by analogy with methane, will consist of a C atom joined to three H atoms by single bonds. It is often very helpful to look for fragments of molecules with Lewis structures that you already know and to build up the complete structure by combining them. The full Lewis structure is obtained by working through the four-step procedure outlined earlier.

SOLUTION The total number of valence electrons is

C	2×4	$= 8$
H	4×1	$= 4$
O	2×6	$= 12$
	Total	$= 24$

so the molecule has 12 valence electron pairs. The atomic arrangement in the molecule, which is suggested by the way the molecular formula is written, is shown in (**6a**); the linked atoms are indicated by the pale gray rectangles. We use seven electron pairs to link neighboring atoms, as shown in (**6b**). Five pairs remain. To complete octets, we arrange electron pairs so that each atom has an octet of electrons: this can be achieved by adding two lone pairs to each oxygen atom and allowing the terminal oxygen atom to form a double bond to the carbon atom, as depicted in (**6c**). The final Lewis structure is shown in (**6d**).

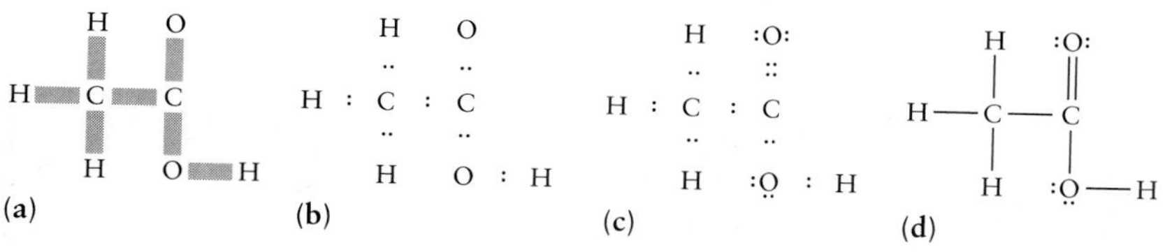

(a) (b) (c) (d)

6 Acetic acid, CH_3COOH

SELF-TEST 8.6A Write a Lewis structure for the urea molecule, $(NH_2)_2CO$.

[*Answer:* See (7).]

SELF-TEST 8.6B Suggest a reason why, based on Lewis structures, the C atom in CO can bind to hemoglobin, causing carbon monoxide poisoning, but the C atom in CO_2 cannot.

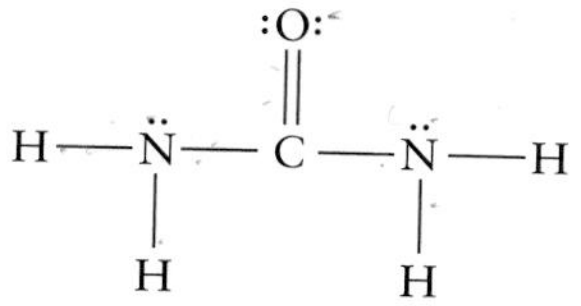

7 Urea, $(NH_2)_2CO$

8.7 RESONANCE

In some Lewis structures, the multiple bonds can be written in several equivalent locations. Consider the nitrate ion, NO_3^-. The three Lewis structures shown in (8) differ only in the position of the double bond. All are valid structures, and all have exactly the same energy. There is no reason to choose one over the other, so which one is correct? The answer is that none *alone* is correct. If one of the pictured structures were correct, we would expect two long single bonds and one short double bond. However, the experimental evidence is that the bond lengths in a nitrate ion are all the same (at 124 pm). They are longer than a typical N=O double bond (120 pm), but shorter than a typical N—O single bond (140 pm). The bonds in the nitrate ion have a character intermediate between a single bond and a full double bond.

The sign on the right angle at the top right of the structure of an ion shows the ion's charge.

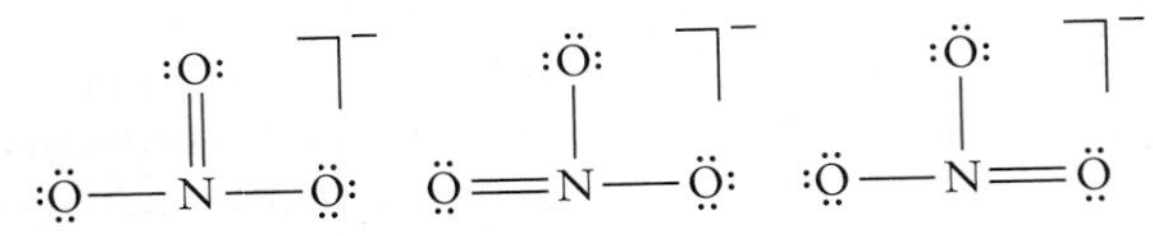

8

Because all three bonds are identical, a better model of the nitrate ion is a *blend* of all three Lewis structures. This blending of structures, which is called **resonance,** is depicted in (**9**) by double-headed arrows. The blended structure is a **resonance hybrid** of the contributing Lewis structures. Resonance should be thought of as a *blend* of the individual Lewis structures rather than as the flickering of a molecule between different structures, just as a mule is a blend of a horse and a donkey, not a creature that flickers between the two.

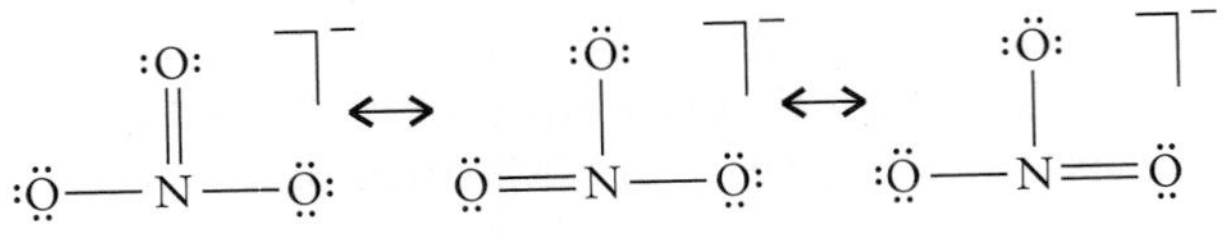

9 Nitrate ion, NO_3^-

Electrons involved in resonance structures are said to be **delocalized.** Delocalization means that the additional electron density due to the second pair of electrons in a double bond is not shared by two particular atoms but is spread out over several atoms.

EXAMPLE 8.4 Writing a resonance structure

Suggest two Lewis structures that contribute to the resonance structure for the O_3 molecule. Experimental data show that the two bond lengths are the same.

STRATEGY Write a Lewis structure for the molecule, using the method outlined in Toolbox 8.1. Decide whether there is another equivalent structure that results from the interchange of a single bond and a double bond. Write the actual structure as a resonance hybrid of these Lewis structures.

SOLUTION Oxygen is a member of Group 16, so each atom has six valence electrons. The total number of valence electrons in the molecule is $3 \times 6 = 18$. One structure is therefore $\ddot{O}{=}\ddot{O}{-}\ddot{O}:$. Interchanging the bonds gives $:\ddot{O}{-}\ddot{O}{=}\ddot{O}$. The overall structure is the following resonance hybrid:

$$\ddot{O}{=}\ddot{O}{-}\ddot{O}: \longleftrightarrow :\ddot{O}{-}\ddot{O}{=}\ddot{O}$$

SELF-TEST 8.7A Write Lewis structures contributing to the resonance hybrid for the acetate ion, $CH_3CO_2^-$. Recall that the structure of CH_3COOH is described in Example 8.3.

[*Answer:* See (**10**).]

SELF-TEST 8.7B Write Lewis structures contributing to the resonance hybrid for the nitrite ion, NO_2^-.

10 Acetate ion, $CH_3CO_2^-$

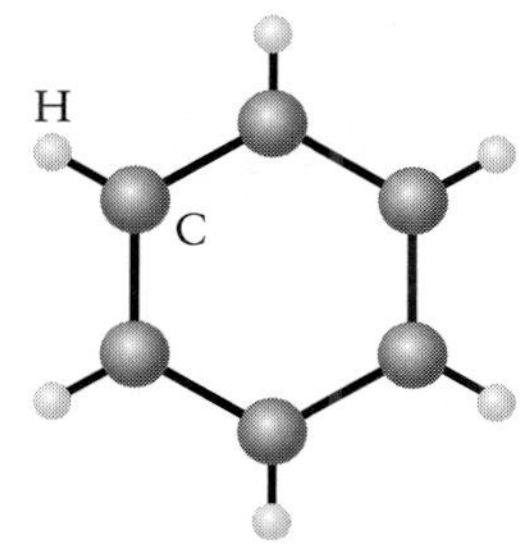

11 Benzene, C_6H_6

Benzene, C_6H_6, is another molecule best described as a resonance hybrid. It consists of a hexagonal ring of six carbon atoms, with a hydrogen atom attached to each one (**11**). A Lewis structure that contributes to the resonance hybrid is shown in (**12**). It is called a **Kekulé structure** for the German chemist Friedrich Kekulé, who first proposed (in 1865) that benzene had a cyclic structure with alternating single and double bonds. The structure is normally abbreviated by a simple hexagon (**13**).

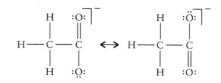

12 Kekulé structure

13 Kekulé structure, shortened form

The H atoms in structure 13 are not shown, but there is one attached to each corner (that is, each C atom).

However, a Kekulé structure does not fit all the evidence that chemists have since collected. For one thing, benzene does not undergo reactions typical of compounds with double bonds. Also, a Kekulé structure suggests that benzene should have two different bond lengths: three longer single bonds (154 pm) and three shorter double bonds (134 pm). Instead, all the bonds are found experimentally to have the same length (139 pm). Moreover, all six locations on the ring are the same; if the Kekulé structures were correct, there would be two distinct dichlorobenzenes in which the chlorine atoms are neighbors, as shown in (**14**). However, only one dichlorobenzene is known.

14 Dichlorobenzene

These characteristics of the benzene molecule can be explained if we note that there are in fact *two* Kekulé structures: they differ only in the positions of the double bonds (**15**). These two structures have exactly the same energy and are blended together. As a result of this resonance, the electron density of the C—C double bonds is spread over the whole

15 Benzene resonance structure

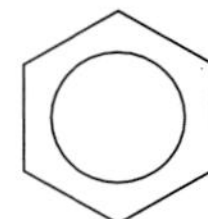

16 Benzene, C_6H_6

Cl

Cl

17 1,2-Dichlorobenzene, $C_6H_4Cl_2$

molecule, thus giving each bond a length and electron density intermediate between that of a single and a double bond. All six C—C bonds are identical; this equivalence is implied by the circle inside the hexagon in (**16**). We can see from (**17**) why there can be only one dichlorobenzene with Cl atoms on adjacent C atoms.

Resonance stabilizes a molecule by lowering its total energy. This stabilization, which can be explained only by quantum mechanics, makes benzene less reactive than expected for a molecule with three carbon-carbon double bonds. Resonance results in the greatest lowering of energy when the contributing structures have equal energies. However, a molecule is a blend of *all reasonable* Lewis structures, including those with different energies. As we shall see, even structures with more than eight electrons in a valence shell can contribute to resonance. The lowest energy structures contribute most strongly to the overall structure in the sense that if we think of a resonance hybrid as a blend of Lewis structures, then the structures with lowest energy contribute most to the mixture.

Resonance occurs only between structures with the same arrangement of atoms but with different arrangements of electrons pairs. For example, although we might be able to write two hypothetical structures for the dinitrogen oxide molecule, NNO and NON, there is no resonance between them, because the atoms lie in different locations.

Resonance is a blending of structures with the same arrangement of atoms but different arrangements of electrons. It spreads multiple bond character over a molecule and also lowers its energy.

(a)

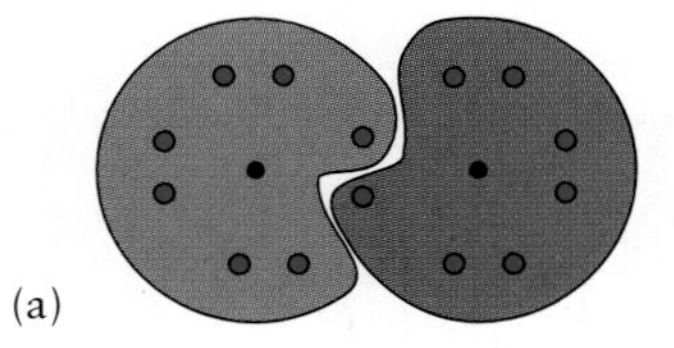

(b)

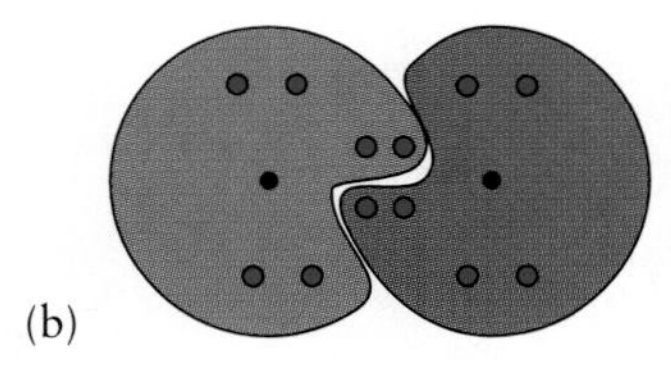

FIGURE 8.10 The formal charge on an atom is calculated by dividing the electrons up in a special way, as shown by the boundaries. Each atom owns all the electrons of its own lone pairs and owns one electron of any bonding pair it shares. The boundaries show the ownership of electrons for (a) a single bond and (b) a double bond in diatomic molecules.

8.8 FORMAL CHARGE

The overall electric charge on a polyatomic ion belongs to the ion as a whole. It is possible, however, to divide up the overall charge artificially and to use the resulting charges to help decide which contributions to resonance structure are most important. The same division of charge can be made for molecules with zero charge: in their case, the sum of any charges on individual atoms is 0.

To assign a **formal charge** to an atom, we decide how many electrons each atom "owns." We suppose that each atom owns one electron of each bonding pair attached to it. We also imagine that it owns its lone pairs completely (Fig. 8.10). We count the number of electrons assigned to an atom this way and then compare this number with the number of electrons on the free atom. If the atom has more electrons in the molecule than when it is a free, neutral atom, then the atom has a negative formal charge, like a monatomic anion. If the assignment of electrons leaves the atom with fewer electrons than when it is free, then the atom has a positive formal charge, as if it were a monatomic cation. Mathematically, we write

$$\text{FC} = V - (L + \tfrac{1}{2}S)$$

where FC is the formal charge, V is the number of valence electrons in the free atom, L is the number of electrons present as lone pairs, and S is the number of shared electrons.

Formal charge can be regarded as an exaggeration of the purely covalent character of bonds and the perfect sharing of electrons that that model implies. Conversely, oxidation number (Section 3.13) is an exaggeration of the *ionic* character of bonds. It is based on a representation in which the atoms are pictured as ions. Formal charges depend on the particular Lewis structure we write; oxidation numbers do not.

EXAMPLE 8.5 Calculating the formal charges on a structure

Calculate the formal charges on the atoms in the three Lewis structures of a sulfate ion shown in (**18**).

18 Sulfate ion, SO_4^{2-}

STRATEGY To find the formal charge of each individual atom, draw the Lewis structures showing each electron pair as dots (not as a line). Then draw a closed curve that captures all the atom's lone pairs of electrons, plus one electron from each of its bonding pairs. Decide on the number of valence electrons (V) possessed by each free atom by noting the number of its group in the periodic table. Then subtract the number of electrons assigned to the atom in the Lewis structure ($L + \frac{1}{2}S$) from the number of valence electrons in the free atom. That is, calculate FC $= V - (L + \frac{1}{2}S)$. We need do only one calculation for equivalent atoms, such as the oxygen atoms in the first diagram, because they all have the same formal charge. A good check on the values calculated is that their sum is equal to the overall charge of the molecule or ion.

SOLUTION Both O and S belong to Group 16, so the neutral atoms each have six valence electrons and the ion has two additional electrons. The closed curves that capture the electrons are shown in (**19**).

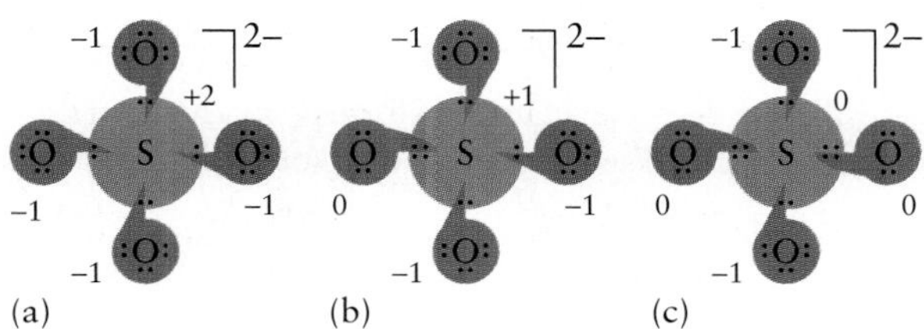

19 Sulfate ion, SO_4^{2-}

To calculate the formal charges that are shown in these diagrams, we draw up the following table:

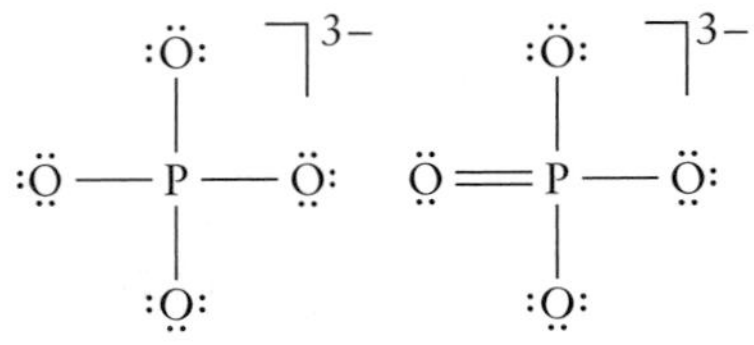

Lewis structure	Number of electrons: On free atom (V)	Number of electrons: Assigned to bonded atom ($L + \frac{1}{2}S$)	Formal charge (FC)
S in (**19a**)	6	4	+2
S in (**19b**)	6	5	+1
S in (**19c**)	6	6	0
—Ö:	6	7	−1
=Ö	6	6	0

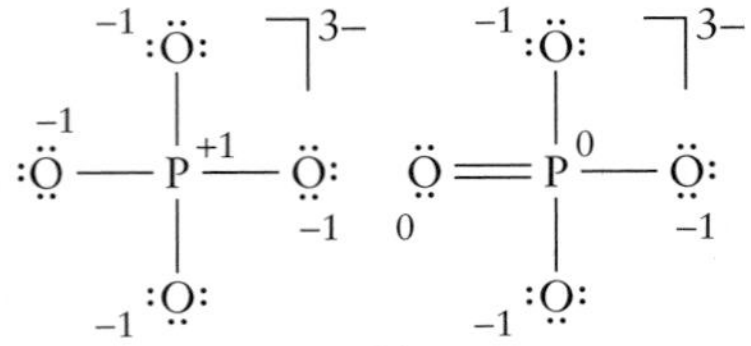

20

21

SELF-TEST 8.8A Calculate the formal charges for the two Lewis structures of the phosphate ion shown in (**20**).

[*Answer:* The formal charges are shown in (**21**).]

SELF-TEST 8.8B Calculate the formal charges for the three oxygen atoms in one of the Lewis structures of the ozone resonance structure (Example 8.4).

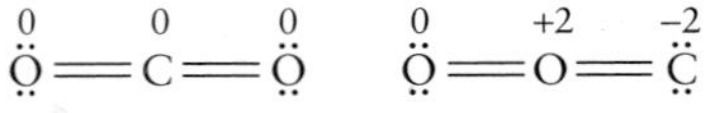

Lewis structures typically have the lowest energy when the formal charges of the individual nonmetal atoms are closest to 0. The atoms in the preferred structures have undergone the least redistribution of electrons relative to the free atoms. For example, the formal charge rule suggests that the structure OCO is more likely for carbon dioxide than COO, as shown in (**22**). Similarly, it also suggests that the structure NNO is more likely for dinitrogen oxide than NON, as shown in (**23**). Likewise, when considering resonance, the Lewis structures with the lowest formal charges usually make the greatest contribution to the overall resonance hybrid.

22

23

The formal charge gives an indication of the extent to which atoms have gained or lost electrons in the process of covalent bond formation; structures with lowest formal charges are likely to have the lowest energy.

EXAMPLE 8.6 Judging the plausibility of a structure

Write three Lewis structures with different atomic arrangements for the cyanate ion, NCO^-, and suggest which one is likely to be the most plausible structure.

STRATEGY We need to calculate the formal charges on the three possible arrangements of atoms and select the one with formal charges closest to 0.

SOLUTION The formal charges of the Lewis structures corresponding to the three possible arrangements are shown in (**24**). The individual formal charges are closest to 0 in the first. That structure is therefore the likely one.

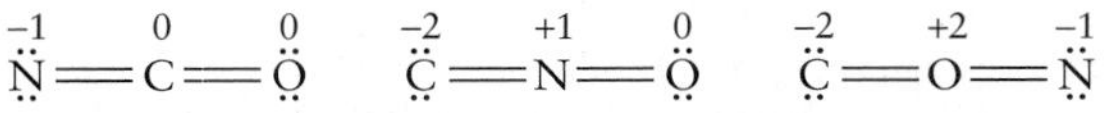

24

SELF-TEST 8.9A Suggest a plausible structure for the poisonous gas phosgene, $COCl_2$. Write its Lewis structure and formal charges.

[*Answer:* See (25).]

SELF-TEST 8.9B Suggest a plausible structure for the AsO_4^{3-} ion. Write its Lewis structure and formal charges.

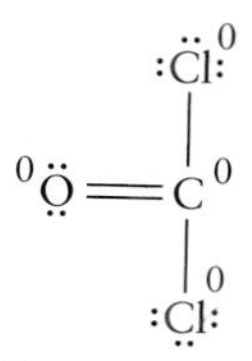

25 Phosgene, $COCl_2$

EXCEPTIONS TO THE OCTET RULE

The octet rule predicts the valence of the elements and the structures of many compounds. Boron, carbon, nitrogen, oxygen, and fluorine obey the octet rule rigorously—when it is arithmetically possible for them to do so; that is, there must be enough electrons to go round. Phosphorus, sulfur, and chlorine atoms, however, can accommodate more than eight electrons. In this section we shall learn to recognize exceptions to the octet rule.

8.9 RADICALS AND BIRADICALS

Some species have an odd number of valence electrons, so at least one of their atoms cannot have an octet. Odd-electron species are called **radicals.** They are generally highly reactive. One example is the seven valence-electron methyl radical, $\cdot CH_3$, which is so reactive that it cannot be stored. It occurs in the flames of burning hydrocarbon fuels. Under enough stress, a carbon-carbon bond may break, as happens when ethane burns:

The older name *free radicals* is still widely used.

Most radicals are very reactive because they can use their unpaired electron to form a new bond.

$$H_3C{-}CH_3 \longrightarrow H_3C\cdot + \cdot CH_3$$

The electron pair that formed the carbon-carbon bond has split into two separate electrons, as indicated by the dot on the C atom in each $\cdot CH_3$.

A second example is the hydroxyl radical, $\cdot OH$. This radical forms briefly when a mixture of hydrogen and oxygen is ignited by a spark. It is also present in the upper atmosphere as a result of the action of the Sun's radiation on water molecules and in smoggy areas of the troposphere (Case Study 8).

Like most radicals, $\cdot CH_3$ and $\cdot OH$ are highly reactive and survive only for very short times under normal conditions. The methyl radical is partly responsible for the explosion that occurs when a mixture of ethane and air is ignited. At very low pressures, however, a radical can survive for long periods. Hydroxyl radicals have been detected in interstellar gas clouds, where they can survive for millions of years. In that environment they collide very rarely with other molecules, so they do not have an opportunity to react.

Radicals control the chemistry of the upper atmosphere, where they contribute to the formation and decomposition of ozone (see Case Study 5). Radicals can also cause problems. They are responsible for the rancidity of foods and the degradation of plastics in sunlight. Damage from radicals can be delayed by an additive called an **antioxidant,** which reacts rapidly with radicals to form species in which all electrons are paired before the radicals have a chance to do their damage. It is believed that human aging is partly due to the action of radicals, and antioxidants such as vitamins C and E may delay the process.

Radicals are used in the manufacture of many polymers. An initial process generates a radical. This radical reacts with a small organic molecule, creating a larger radical, which goes on to attack still another molecule. The result is a radical with an even longer chain of atoms (Fig. 8.11). The polymerization process continues until long chains of repeating organic units have formed.

▶▶ There is more information about radical polymerization in Section 11.13.

Nitrogen monoxide (nitric oxide, $:\dot{N}=\ddot{O}$), which has 5 + 6 = 11 valence electrons, is a radical. It is formed by the direct reaction of nitrogen and oxygen in the hot gases of automobile exhausts and jet engines:

$$N_2(g) + O_2(g) \longrightarrow 2\,NO(g)$$

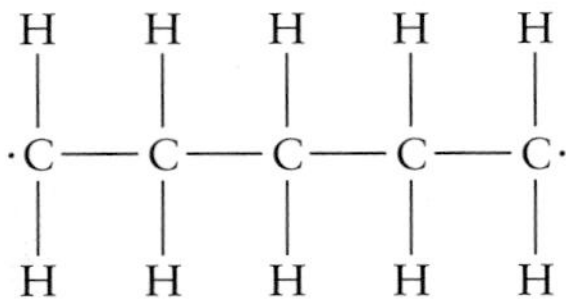

26 A biradical

Nitric oxide also occurs naturally in our neurons: it is one of the *neurotransmitters,* small molecules that transmit nerve signals across a synapse. It plays a role in blood flow and sexual arousal.

A **biradical** is a molecule with *two* unpaired electrons. The unpaired electrons are usually on different atoms, as depicted in (**26**). One unpaired electron is on one carbon atom of the chain, and the second is on another carbon atom several bonds away. In some cases, though, both electrons are on the same atom. One of the most important examples is the oxygen atom itself. Its electron configuration is $[He]2s^22p_x^22p_y^12p_z^1$ and its Lewis symbol is $\cdot\ddot{O}\cdot$. The O atom has two unpaired electrons, so it can be regarded as a special case of a biradical.

Although it is not obvious from its Lewis structure, molecular oxygen, O_2, is a biradical! In fact, experiments have shown that the most plausible Lewis structure, $\ddot{O}=\ddot{O}$, gives a false impression of the arrangement of electrons. In molecular oxygen, two of the electrons that the Lewis structure implies are responsible for the bonds do not in fact pair with one another. The molecule is really a biradical with an unpaired electron on each O atom. For this reason, its Lewis structure is often written as shown in (**27**).

$\ddot{O}\cdot\!-\!\cdot\ddot{O}$

27 Dioxygen, O_2

A radical is a species with an unpaired electron; a biradical has two unpaired electrons on either the same or different atoms.

EXAMPLE 8.7 Writing the Lewis structure of an odd-electron molecule

Write the Lewis structure of the radical nitrogen dioxide, NO_2. When gaseous nitrogen dioxide is cooled, it forms dinitrogen tetroxide, N_2O_4. Suggest a reason why this happens.

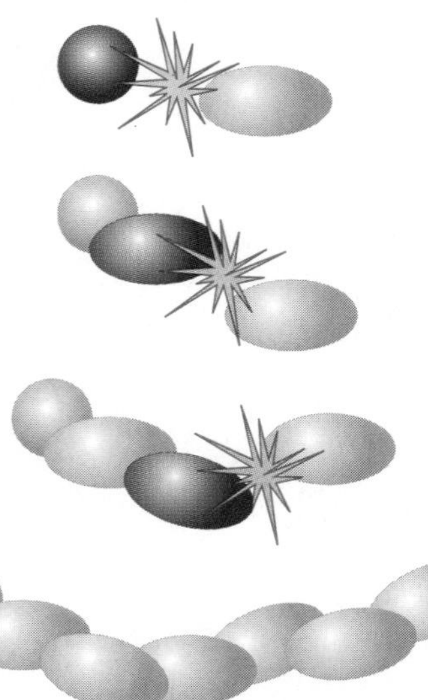

FIGURE 8.11 One method of forming a polymer is to make use of the attack of a radical on a monomer molecule. The initiating molecule (purple) attacks a monomer molecule, attaches to it, and converts it into a radical. That radical then attacks another monomer, and the chain of events continues until all the monomer has been converted into polymer.

STRATEGY We expect a covalent bond. Therefore, the first step is to write a Lewis structure for the radical. A radical has an odd number of electrons, so one of the atoms must have an incomplete octet.

SOLUTION The number of valence electrons present in NO_2 is $5 + (2 \times 6) = 17$, corresponding to eight pairs and one remaining electron. Nitrogen has the lower ionization energy, so it is found in the middle, ONO. The next step is to form single bonds:

O:N:O   ·::::::

There are still six pairs and one electron to be accommodated. One Lewis structure that accounts for the six pairs is

Ö::N:Ö: ·

The nitrogen atom has an incomplete octet. It can accommodate the remaining one electron to give

Ö::Ṅ:Ö: or Ö=Ṅ—Ö:

The alternative structure :Ö—Ṅ=Ö has the same energy, because it differs only in the position of the double bond. Consequently, we conclude that nitrogen dioxide has the resonance structure

Ö=Ṅ—Ö: ⟷ :Ö—Ṅ=Ö

We could also draw structures with the unpaired electron on an O atom, but the formal charges would be higher than those of the structures we have selected. An N_2O_4 molecule forms when the two unpaired electrons on two $\cdot NO_2$ radicals pair to form a single covalent bond, as shown (for one of several contributions to the resonance structure) in (**28**).

SELF-TEST 8.10A Write a Lewis structure for the hydrogenperoxyl radical, HOO, which plays an important role in atmospheric chemistry and which in the body has been implicated in the degeneration of neurons. (Think of the radical as the molecule remaining after the loss of an H atom from hydrogen peroxide, H_2O_2.)

[*Answer:* See (**29**).]

SELF-TEST 8.10B Write a Lewis structure for ClO_2.

28 Dinitrogen tetroxide, N_2O_4

29 Hydrogenperoxyl, $HO_2\cdot$

8.10 EXPANDED VALENCE SHELLS

The octet rule tells us that eight electrons fill a shell to give a noble-gas s^2p^6 configuration. However, when the central atom in a molecule has empty *d*-orbitals, it may be able to accommodate 10, 12, or even more electrons. The electrons in such an **expanded octet** may be present as lone pairs. They may also allow the central atom to form additional bonds.

Because the additional electrons must be accommodated in orbitals, only nonmetal atoms in Period 3 or higher can have expanded octets. For these elements, there are empty *d*-orbitals in the valence shell of the atom. Another factor—possibly the main factor—in determining whether more atoms than are allowed by the octet rule can bond to a central atom is the size of that atom. A P atom is big enough for up to six Cl

CASE STUDY SMOG FORMERS

FOSSIL FUELS SUPPLY the energy for our industry and technology-based society. Yet their combustion, especially in automobiles, releases incompletely burned chemicals and oxidized species known as *primary pollutants* into the atmosphere. The dangers they pose range from eye and throat irritation to global warming. Many of the primary pollutants undergo further reaction under the influence of sunlight. The products of these photochemical reactions are called *secondary pollutants*. Primary and secondary pollutants—along with *aerosols*, which are suspended fine particles such as water droplets, dust, and soot—contribute to the brown haze we call smog.

One critical primary pollutant is nitric oxide, NO, which is produced in the high-temperature combustion cylinders of automobiles:

$$N_2(g) + O_2(g) \xrightarrow{\Delta} 2\,NO(g)$$

In the atmosphere, nitric oxide is oxidized to nitrogen dioxide, a major constituent of smog:

$$2\,NO(g) + O_2(g) \longrightarrow 2\,NO_2(g)$$

Smog appears brown partly because it absorbs sunlight at wavelengths less than 400 nm. Light of these wavelengths dissociates nitrogen dioxide into NO molecules and O atoms:

$$NO_2(g) \xrightarrow{h\nu,\ \lambda < 400\text{ nm}} NO(g) + \cdot O \cdot (g)$$

Light with a wavelength of 400 nm is at the violet end of the visible spectrum; so when it is absorbed, the remaining transmitted light appears yellow-orange (see Section 21.8

The gas NO (left) is colorless. When it is exposed to air (right), it is rapidly oxidized to brown NO_2.

The equipment in the illustration monitors air quality from a rooftop in Los Angeles. The concentration of NO_2 in the air due to automobile traffic increases during the day, contrbuting to the typical brown color of the afternoon sky.

atoms to fit comfortably around it, and PCl_5 is a common laboratory chemical. An N atom, though, is too small, and NCl_5 is unknown (Fig. 8.12).

Elements that can expand their octets show **variable covalence,** the ability to form different numbers of covalent bonds. Variable covalence means that an element can form one number of bonds in some compounds and a different number in others. Phosphorus is one example. It reacts directly with a limited supply of chlorine to form the toxic, colorless liquid phosphorus trichloride:

$$P_4(g) + 6\,Cl_2(g) \longrightarrow 4\,PCl_3(l)$$

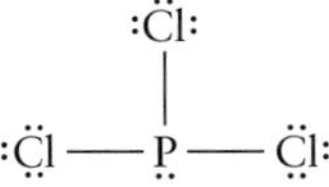

30 Phosphorus trichloride, PCl_3

The Lewis structure of the PCl_3 molecule is shown in (**30**), and we see that it obeys the octet rule. However, when excess chlorine is present or

and Case Study 7). Light scattered by dust particles in the atmosphere also contributes to the brown color of smog.

Lachrymators, the eye irritants that cause tears, are oxidation products of primary pollutants. The oxygen atom, ·O·, produced by the dissociation of NO_2 is a highly reactive biradical. It is a strong oxidizing agent that can oxidize many of the primary pollutants and even reacts with O_2. The most evident products of its action are ozone and the peroxyacetylnitrates (PAN). These compounds are all highly toxic substances, even at concentrations of less than 0.1 ppm; and exposure to them at concentrations of 0.5 ppm for a few minutes can cause eye irritation.

Many reaction paths lead to the formation of a PAN. In one common path, an ·O· atom first abstracts an ·H atom from a molecule of unburned hydrocarbon fuel:

$$CH_4 + \cdot O\cdot \longrightarrow H_3C\cdot + \cdot OH$$

The hydroxyl radical, ·OH, is a common oxidizing agent in smog. The methyl radical, $H_3C\cdot$, is also highly reactive and goes on to cause further havoc with the atmosphere.

Hydroxyl radicals react with many substances, including fuel molecules that were only partially oxidized in an engine. For example, acetaldehyde, CH_3CHO, reacts with the highly reactive hydroxyl radical to produce yet another radical:

$$CH_3CHO + \cdot OH \longrightarrow CH_3CO\cdot + H_2O$$

The $CH_3CO\cdot$ radical is also highly reactive and combines with molecular oxygen to form a peroxyl radical:

$$CH_3CO\cdot + O_2 \longrightarrow CH_3{-}\overset{\overset{\displaystyle O}{\|}}{C}{-}O{-}O\cdot$$

The peroxyl radical combines with NO_2 (also a radical, as described in Section 8.9) to form peroxyacetylnitrate:

In almost every country of the world, traditional modes of transport have given way to motorized vehicles. All of them, even this tiny cart in Bangkok, contribute to air pollution.

$$CH_3{-}\overset{\overset{\displaystyle O}{\|}}{C}{-}O{-}O\cdot + NO_2 \longrightarrow CH_3{-}\overset{\overset{\displaystyle O}{\|}}{C}{-}O{-}O{-}NO_2$$

Catalytic converters in automobiles help to control smog by converting the nitrogen oxides back to nitrogen and oxygen. They also complete the oxidation of unburned and partially burned hydrocarbons. Even so, the automobile remains the primary contributor to smog.

QUESTIONS

1. Write the Lewis structures for the following components of smog: (a) NO; (b) OH (hydroxyl radical); (c) CH_3CO_2 (acetyl radical); (d) NO_2.

2. Draw the Lewis structure for acetaldehyde, CH_3CHO. Is resonance likely to be important for this molecule?

3. By analogy with the reaction between an oxygen atom and methane, suggest a chemical equation for the reaction of the hydroxyl radical with ethane, CH_3CH_3.

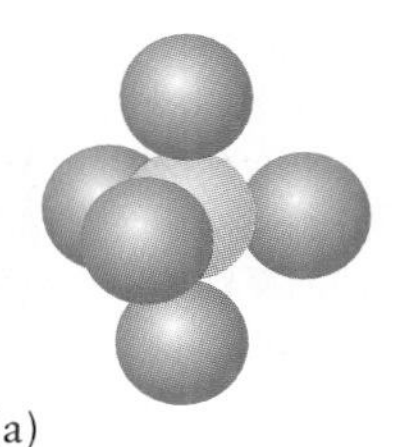

(a)

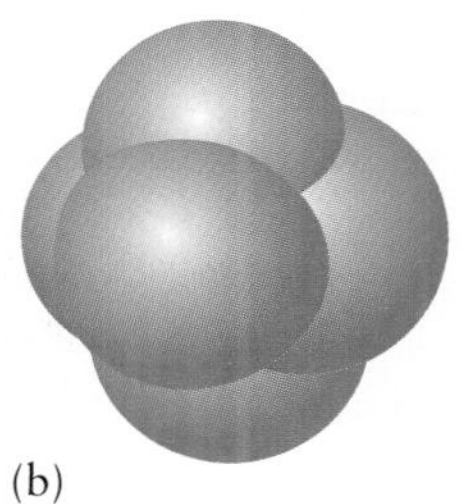

(b)

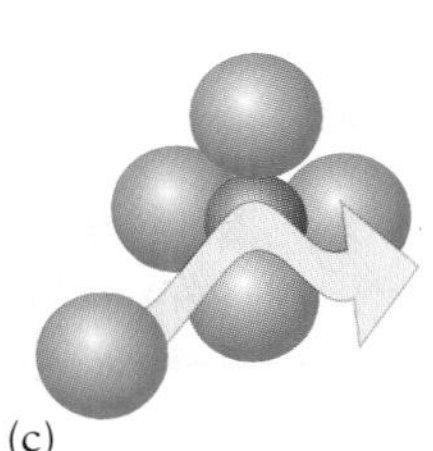

(c)

FIGURE 8.12 (a) A model using small spheres to represent atoms and (b) a space-filling model of PCl_5, showing how closely the chlorine atoms must pack around the central phosphorus atom. (c) A nitrogen atom is quite a lot smaller than a phosphorus atom, and five chlorine atoms cannot pack around it.

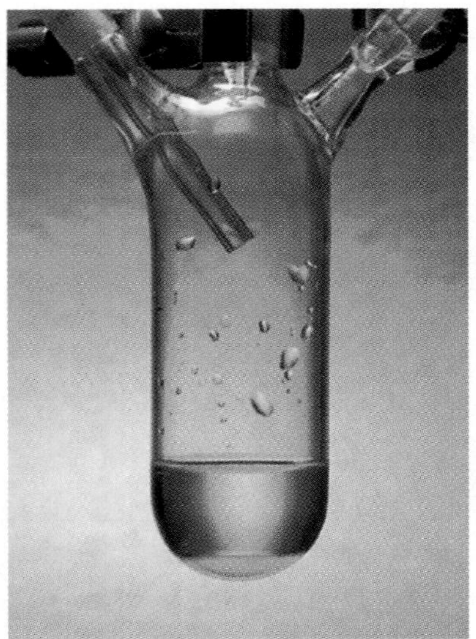

FIGURE 8.13 Phosphorus trichloride is a colorless liquid. When it reacts with chlorine (the pale yellow-green gas in the flask), it forms the very pale yellow solid phosphorus pentachloride (at the bottom of the flask).

when phosphorus trichloride reacts with more chlorine (Fig. 8.13), phosphorus pentachloride, a pale yellow crystalline solid, is produced:

$$P_4(s) + 10\,Cl_2(g) \longrightarrow 4\,PCl_5(s)$$

$$PCl_3(l) + Cl_2(g) \longrightarrow PCl_5(s)$$

Phosphorus pentachloride is an ionic solid consisting of PCl_4^+ cations and PCl_6^- anions, and sublimes at 160°C to a gas of PCl_5 molecules. This peculiar behavior underlines the subtlety of the energy balances that determine the details of chemical bonding. The Lewis structures of the polyatomic ions and the molecule are shown in (**31**). Although the cation is a polyatomic ion in which the P atom does not need to expand its octet, in the anion the P atom has expanded its octet to 12, with the extra four electrons in two of its 3*d*-orbitals. In PCl_5, the P atom has expanded its octet to 10 by using one 3*d*-orbital.

(a) PCl_4^+ (b) PCl_6^- (c) PCl_5

31

EXAMPLE 8.8 Writing a structure with an expanded valence shell

A fluoride of composition SF_4 is formed when fluorine diluted with nitrogen is passed over a film of sulfur at −75°C in the absence of oxygen and moisture. Write the Lewis structure of sulfur tetrafluoride.

STRATEGY A fluorine atom forms only single bonds, so we anticipate that the Lewis structure consists of a shared pair between the central S atom and each of the four surrounding F atoms. However, because each F atom has three lone pairs and supplies one bonding electron, and the S atom already has six electrons in its valence shell, there are two extra electrons. Because sulfur is in Period 3 and has empty 3*d*-orbitals available, we should have octet expansion in mind.

SOLUTION Sulfur (·S̤̈·) supplies six valence electrons, and each fluorine atom(:F̤̈·) supplies seven. Hence there are $6 + (4 \times 7) = 34$ electrons, or 17 electron pairs, to accommodate. We write each F atom with three lone pairs and a bonding pair shared with the central S atom, then place the two extra electrons on the S atom, as shown in (**32**). All 17 electron pairs are now accommodated. Because the S atom has 10 electrons, which require at least five orbitals, it needs to use one 3*d*-orbital in addition to the four 3*s*- and 3*p*-orbitals.

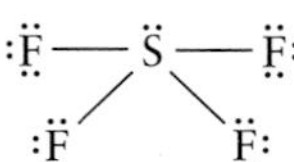

32 Sulfur tetrafluoride, SF_4

33 Xenon tetrafluoride, XeF_4

SELF-TEST 8.11A Write the Lewis structure for xenon tetrafluoride, XeF_4, and give the number of electrons in the expanded octet.

[*Answer:* See (33); 12 electrons.]

SELF-TEST 8.11B Write the Lewis structure for the I_3^- ion.

The sulfate ion, SO_4^{2-}, is a resonance hybrid of octet and expanded-octet Lewis structures. The structures shown as (**34**) are just two of the many Lewis structures. The first structure follows the octet rule, but the additional multiple bonds in the second expand the octet to 12. These and the other contributions to the resonance structure have the same arrangement of atoms, so we can expect the true structure to be a resonance hybrid. Computer calculations show that the expanded-octet structures have a lower energy than the octet structure, so they may make the greatest contribution to the hybrid.

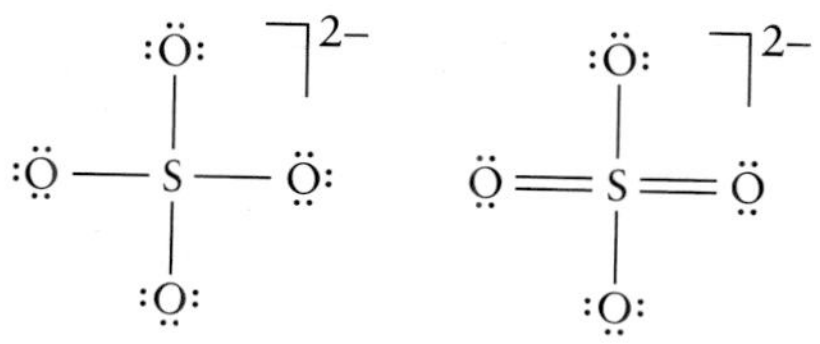

34 Sulfate ion, SO_4^{2-}

Sulfur dioxide, SO_2, is another molecule in which the expanded-octet Lewis structures are important (**35**).

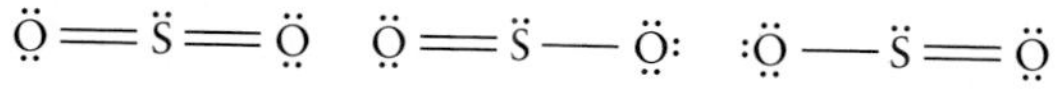

35 Sulfur dioxide, SO_2

The first Lewis structure makes a bigger contribution to the resonance hybrid than the other two, in which there is no octet expansion.

Octet expansion can occur in elements of Period 3 and later and enables them to form variable numbers of covalent bonds.

LEWIS ACIDS AND BASES

Boron and aluminum atoms need five electrons to complete their octets. However, a lot of energy is needed to gain shares in that many electrons, and the compounds they form have special chemical characteristics. These compounds are of special interest because they introduce a whole new class of reactions called Lewis acid-base reactions, that we have not previously described.

8.11 THE UNUSUAL STRUCTURES OF GROUP 13 HALIDES

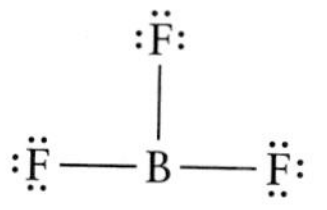

To introduce this new class of reactions, let's investigate the molecular structure of the colorless gas boron trifluoride, BF_3. An important feature of the Lewis structure (**36**) is that the boron atom has an incomplete octet: its valence shell consists of only six electrons. The molecule could complete its octet by sharing more electrons with fluorine, as depicted in (**37**), but the fluorine has such a high ionization energy that this arrangement is unlikely. Experimental evidence, such as the relatively short B—F bond lengths, has led some chemists to suggest that the true structure of BF_3 may be a resonance hybrid of both types of Lewis structures, with the singly bonded structure making the major contribution.

36 Boron trifluoride, BF_3

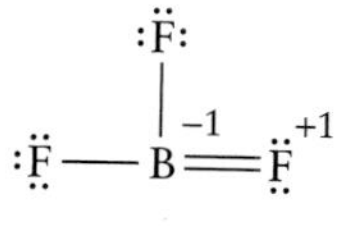

37

The boron octet can be completed if *another* atom or ion forms a bond. For example, tetrafluoroborate anion, BF_4^- (**38**), forms when boron trifluoride is passed over a metal fluoride. Now, all the fluorine atoms

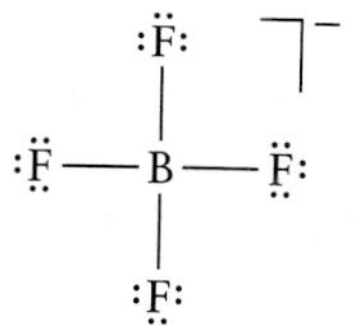

38 Tetrafluoroborate, BF_4^-

have their normal valence of 1 and B has an octet. Another example is the compound formed when boron trifluoride reacts with ammonia:

$$BF_3(g) + NH_3(g) \longrightarrow NH_3BF_3(s)$$

A molecular model of the product, a white molecular solid, is shown in (**39**). The lone pair on the nitrogen atom of ammonia can be regarded as completing boron's octet by forming a covalent bond to give the Lewis structure shown in (**40**). A bond in which both electrons come from one of the atoms is called a **coordinate covalent bond.**

▶▶ Coordinate covalent bonds are described in more detail in Section 21.5.

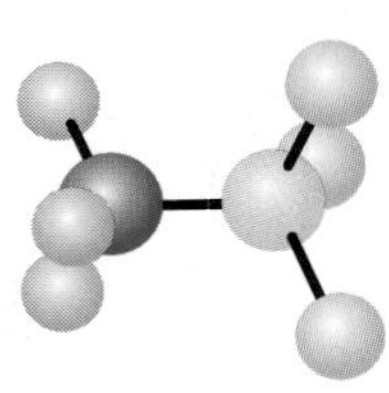

39 NH_3BF_3

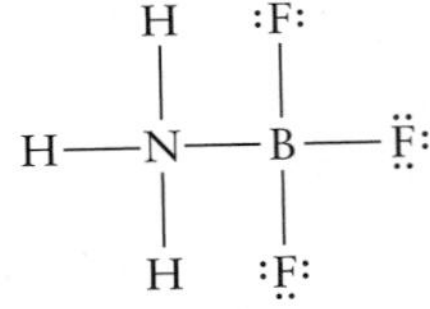

40 NH_3BF_3

Boron trichloride, a colorless, reactive gas of BCl_3 molecules, behaves chemically like BF_3. However, the trichloride of aluminum, which is in the same group as boron, is a volatile white solid that sublimes at 180°C to a gas of Al_2Cl_6 molecules. These molecules survive in the gas up to about 200°C, and only then fall apart into $AlCl_3$ molecules. The Al_2Cl_6 molecule exists because a Cl atom of one $AlCl_3$ molecule uses one of its lone pairs to form a bond to the Al atom of a neighboring $AlCl_3$ molecule. The resulting molecule has the structure shown in (**41**).

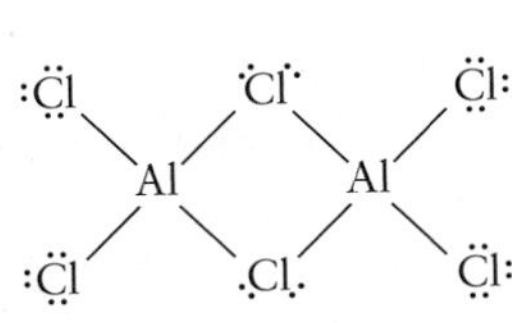

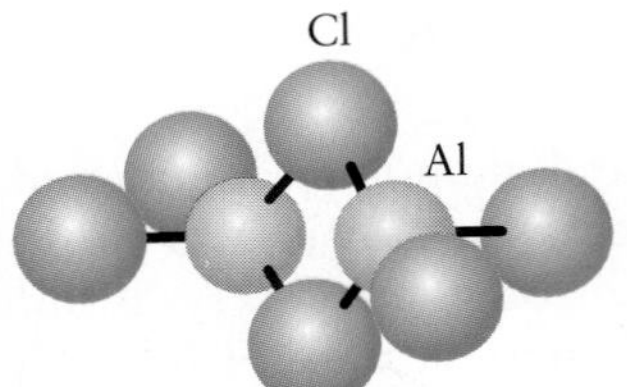

41 Aluminum chloride, Al_2Cl_6

Compounds of boron and aluminum may have unusual Lewis structures in which boron and aluminum have incomplete octets or halogen atoms act as bridges.

8.12 LEWIS ACID-BASE COMPLEXES

The pattern of coordinate covalent bond formation, in which one species provides a lone pair and the other species accepts it, is common. The species that provides the lone pair is called a **Lewis base,** and the species that accepts it is called a **Lewis acid.** In other words, a Lewis acid is an electron pair acceptor, and a Lewis base is an electron pair donor. The product of the reaction between a Lewis acid and a Lewis base, BF_4^- in our first example, is called a **complex.** The form of the reaction is therefore

$$\text{Acid} + \text{:base} \longrightarrow \text{complex}$$

The BF_3 is the Lewis acid, the F^- ion is the Lewis base, and the BF_4^- ion is the Lewis acid-base complex. All bonds formed in a Lewis acid-base reaction are coordinate covalent bonds.

Why do we use the terms *acid* and *base* in this context? Recall from Section 3.7 that acids and bases are normally discussed in terms of the behavior of a hydrogen ion, H^+ (a proton), and that in water an acid such as HCl donates its proton to a neighboring water molecule. We could imagine this reaction as taking place in two steps: first, an HCl molecule releases a hydrogen ion as soon as it dissolves in water, and then that hydrogen ion immediately bonds to a neighboring water molecule. The second step is

$$H^+ + :\underset{\cdot\cdot}{\overset{\displaystyle H \atop |}{O}}\!-\!H \longrightarrow \left[H\!-\!\underset{\cdot\cdot}{\overset{\displaystyle H \atop |}{O}}\!-\!H \right]^+$$

This step in the reaction has exactly the same form as the Lewis acid-base reaction, with H^+ playing the role of the Lewis acid, H_2O the role of the Lewis base, and H_3O^+ the resulting complex. It is this analogy to the behavior of H^+ (the species typical of acids) that led to the use of the term *acid* to describe an electron pair acceptor like BF_3.

Similarly, just as the hydroxide ion, OH^-, is a typical base in the conventional sense, it acts as an electron pair donor in the reaction

$$H^+ + [:\ddot{\underset{\cdot\cdot}{O}}\!-\!H]^- \longrightarrow :\underset{\cdot\cdot}{\overset{\displaystyle H \atop |}{O}}\!-\!H$$

In the Lewis system, the term *base* is used for any species that can act as an electron pair donor. An ammonia molecule, $:NH_3$ is another example of a typical conventional base that has a lone pair to donate (for example, to BF_3, as described earlier), so it can also be regarded as a Lewis base.

A Lewis acid is an electron pair acceptor; a Lewis base is an electron pair donor; they react to form a Lewis acid-base complex.

EXAMPLE 8.9 Identifying a Lewis acid and a Lewis base

When water is added to calcium oxide (quicklime), a vigorous reaction takes place and calcium hydroxide (slaked lime) is formed. Describe the reaction in terms of the formation of a Lewis acid-base complex.

STRATEGY Look for the species that has a lone pair of electrons it can donate to form a covalent bond: such a species is a Lewis base. The species to which the lone pair is donated is the Lewis acid. Bear in mind (as explained earlier) that the water molecule can act as a source of the Lewis acid H^+.

SOLUTION The oxide ion, $[:\ddot{\underset{\cdot\cdot}{O}}:]^{2-}$, is a Lewis base. A water molecule, H_2O, can be regarded as a Lewis acid-base complex in which H^+ is the Lewis acid and OH^- is the Lewis base. When water is poured on calcium oxide, the Lewis base O^{2-} pulls the Lewis acid H^+ out of the molecule H_2O:

$$[:\ddot{\underset{\cdot\cdot}{O}}:]^{2-} + :\underset{\cdot\cdot}{\overset{\displaystyle H \atop |}{O}}\!-\!H \longrightarrow [:\ddot{\underset{\cdot\cdot}{O}}\!-\!H]^- + [:\ddot{\underset{\cdot\cdot}{O}}\!-\!H]^-$$

An oxide ion is such a strong Lewis base—a strong electron pair donor—that it never exists as such in water but always reacts to form hydroxide ions.

SELF-TEST 8.12A Account for the formation of ammonium and hydroxide ions when ammonia dissolves in water.

[*Answer:* $H_3N\colon + H{-}\ddot{\underset{\cdot\cdot}{O}}(H)\colon \longrightarrow NH_4^+ + [\colon\ddot{\underset{\cdot\cdot}{O}}{-}H]^-$]

SELF-TEST 8.12B Account for the formation of Al_2Cl_6 from $AlCl_3$ molecules in terms of Lewis acids and bases.

Lewis acid-base reactions occur widely in chemistry. In fact, they complete our collection of fundamental reaction types that we began to assemble in Chapter 3. The four types of reactions—precipitation, acid-base, redox, and Lewis acid-base reactions—are the procedures that chemists use to make the new materials of the modern world.

IONIC VERSUS COVALENT BONDS

Ionic and covalent bonding are two extreme descriptions. Actual bonds lie somewhere between purely ionic and purely covalent. When describing bonds between nonmetals, covalent bonding is a good model. When a metal is present, ionic bonding is a good model for most simple compounds. But just how good are these initial descriptions, and how can they be improved?

8.13 CORRECTING THE COVALENT MODEL

An electron pair is subject to a tug-of-war between the two atoms that share it. The covalent bond it forms acquires some ionic character if one atom has a greater pulling power than the other atom, because then the electron pair is more likely to be found closer to one atom than to the other. The electron-pulling power of an atom when it is part of a bond is called **electronegativity.** Electronegativities are denoted as χ, the Greek letter chi (pronounced "kye"). When a chemical bond forms between two atoms, the atom of the element with the higher electronegativity has a stronger pulling power on electrons and tends to pull them away from the atom of the element with lower electronegativity (Fig. 8.14).

The simplest procedure is to think of χ as the average of the ionization energy and electron affinity of the element. If the ionization energy is high, then electrons are given up reluctantly. If the electron affinity is

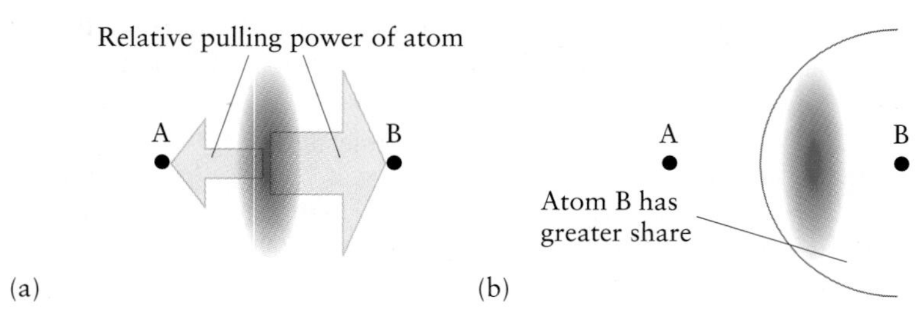

FIGURE 8.14 The electronegativity of an element is its electron-pulling power when it is part of a compound. (a) An atom (B) with a high electronegativity has a strong pulling power on electrons (as represented by the large arrow), particularly for the electron pair it shares with its neighbor. (b) The outcome of the tug-of-war is that the more electronegative atom has a greater share in the electron pair of the covalent bond.

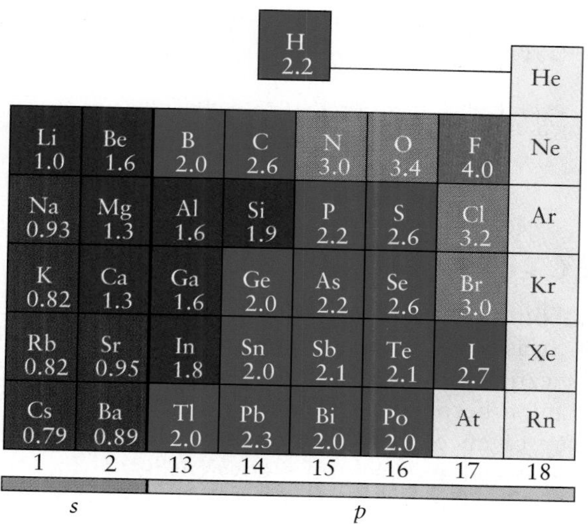

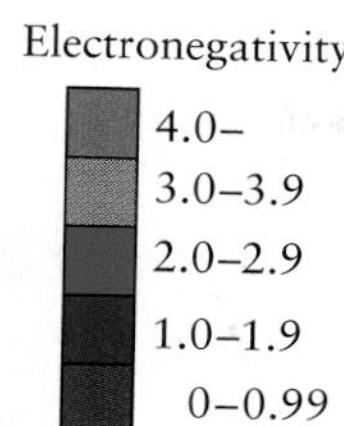

FIGURE 8.15 The variation of the electronegativity of the main-group elements (with the exception of the noble gases). Electronegativity tends to be high toward the upper right corner of the periodic table and low on the lower left. Elements with low electronegativities, such as the *s*-block metals, are often called electropositive.

high, then it is energetically favorable to attach electrons to an atom. Elements with both these properties are reluctant to lose their electrons and tend to gain them; hence, they are classified as highly electronegative. Conversely, if the ionization energy and the electron affinity are both low, then it takes very little energy for the element to give up its electrons and it has little tendency to gain more; hence the electronegativity is low.

Electronegativity was first defined by the American chemist Linus Pauling. The approach described here was developed by another American chemist, Robert Mulliken.

Figure 8.15 shows the variation of electronegativity for the main-group elements of the periodic table. Because ionization energies and electron affinities are highest at the top right of the periodic table (close to fluorine), it is not surprising to find that nitrogen, oxygen, bromine, chlorine, and fluorine are the elements with the highest electronegativities. Whenever these elements are present in compounds, we can expect their atoms to pull strongly on electrons shared with their neighbors. Electrons will still be shared with the less electronegative atom, but the sharing is unequal, and the electron cloud will be denser on the atom of the more electronegative element.

Electronegativities are most useful for nonmetals and are hardly ever used for *d*-block metals.

If the electronegativities of the two elements in a bond are the same, the atoms have equal pulling power on the electron pair they share, and neither wins the tug-of-war. The covalent model is then a good description of the bonding. If the electronegativities are very different, then one atom can acquire the lion's share of the electron pair. Because it has largely robbed the other atom of its share of the electrons, the highly electronegative element resembles an anion and the other atom resembles a cation. We say such a bond has a lot of ionic character (Fig. 8.16).

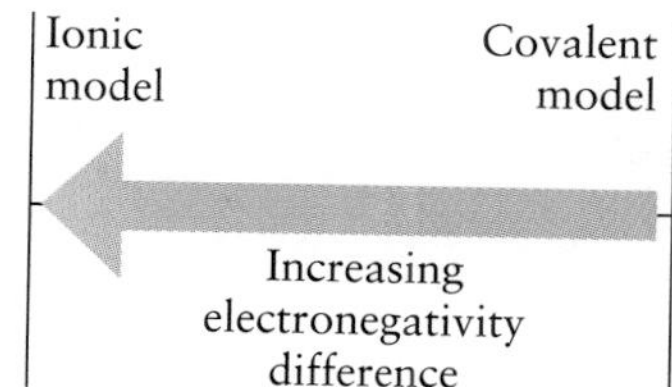

FIGURE 8.16 The electronegativity difference between two elements is an indication of the validity of the covalent model of a chemical bond. When the difference is small (less than about 1.5), the covalent model is good. When the difference is large (greater than about 2), it is most accurate to express the bonding in terms of the ionic model. Note that the numbers we have quoted are only a guide.

There is no hard-and-fast dividing line between ionic and covalent bonding. However, it is a good rule of thumb that an electronegativity difference of about 2 means that so much ionic character is present in a bond that it is best regarded as ionic. For electronegativity differences smaller than about 1.5, a covalent description of the bond is probably reasonably reliable.

Electronegativity is a measure of the pulling power of an atom on an electron pair in a molecule. Compounds composed of elements with a large difference in electronegativity tend to have significant ionic character in their bonding.

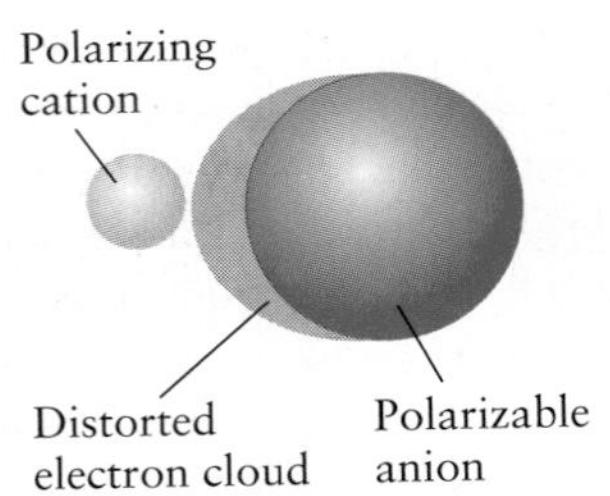

FIGURE 8.17 When a small, highly charged cation is close to a large anion, the electron cloud of the latter is distorted in the process we call polarization. Small, highly charged cations are highly polarizing. Large, electron-rich anions are highly polarizable.

SELF-TEST 8.13A In which of the following compounds do the bonds have greater ionic character: (a) P_2O_5 or (b) PCl_3?

[*Answer:* (a)]

SELF-TEST 8.13B In which of the following compounds do the bonds have greater ionic character: (a) CO_2 or (b) NO_2?

8.14 CORRECTING THE IONIC MODEL

In practice, all ionic bonds have some covalent character. Consider a monatomic anion (such as a chloride ion) next to a cation (such as a sodium cation). As the cation pulls on the anion's electrons, the spherical electron cloud of the anion becomes distorted. We can think of this distortion as the tendency of an electron pair to move into the bonding region between the two nuclei and to form a covalent bond (Fig. 8.17). We can expect ionic bonds to have more covalent character as the distortion of the electron clouds increases.

Atoms and ions that readily undergo a large distortion are said to be highly **polarizable.** Ions that can *cause* large distortions are said to have a high **polarizing power.** An anion can be expected to be highly polarizable if it is large, like an iodide ion, I^-. In such a large, highly polarizable ion, the nucleus exerts only relatively weak control over its outermost electrons, because they are so far away. As a result, the electron cloud of the large anion is easily distorted. Conversely, a cation can be expected to have a strong polarizing power if it is small and highly charged, like an Al^{3+} cation. A small radius means that the center of charge of a highly charged cation can get very close to the anion, where it can exert a strong pull on the anion's electrons. Compounds composed of a small, highly charged cation and a large, polarizable anion tend to have bonds with significant covalent character (Fig. 8.18).

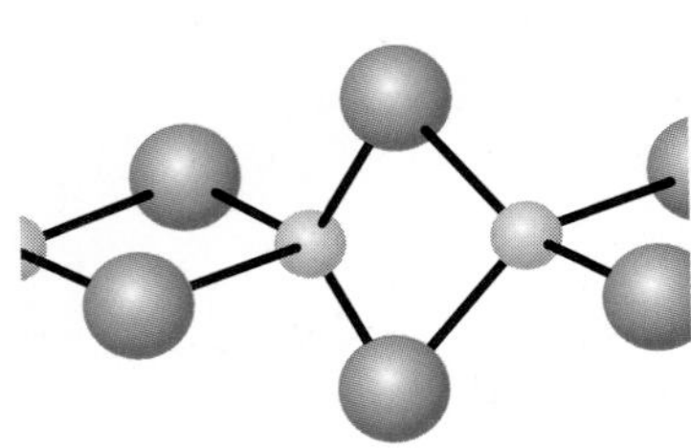

42 Beryllium chloride

For example, the very small radius of a beryllium atom gives the element strikingly different properties from the other Group 2 elements. A Be—Cl bond is significantly covalent; and $BeCl_2$ consists of long chains of covalently bonded $BeCl_2$ units in the solid, as shown in (**42**), which form individual $BeCl_2$ molecules in the vapor. Another observation we can now explain is the decreasing solubility in water of the silver halides as we go from AgCl to AgI (Fig. 8.19). The covalent character of the silver halides increases from AgCl to AgI as the anion becomes larger and more polarizable. Silver fluoride, which contains the small and almost unpolarizable fluoride ion, is freely soluble in water because it is predominantly ionic.

Compounds composed of highly polarizing cations and highly polarizable anions have a significant covalent character in their bonding.

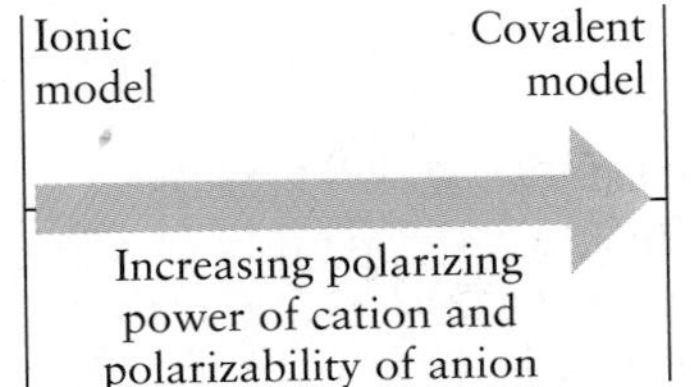

FIGURE 8.18 The polarizability of an anion and the polarizing power of a cation can be used as a guide to judge whether an ionic model of a bond is likely to be valid. As the polarizing power and polarizability increase, the distortion of the anion becomes so great that a covalent description of the bond becomes more reasonable.

FIGURE 8.19 The silver halides, which were formed here by precipitation from silver nitrate and silver halide solutions, become increasingly insoluble down the group from AgCl to AgI. The figure shows, from left to right, AgCl, AgBr, and AgI.

SELF-TEST 8.14A In which of the following compounds do the bonds have greater covalent character: (a) NaBr or (b) $MgBr_2$?

[*Answer:* (b)]

SELF-TEST 8.14B In which of the following compounds do the bonds have greater covalent character: (a) CaS or (b) CaO?

SKILLS YOU SHOULD HAVE MASTERED

Conceptual

1. Distinguish between ionic and covalent bonds.
2. Define resonance hybrid and explain its relationship to the individual Lewis structures that contribute.
3. Show how formal charges can be used to evaluate alternative Lewis structures.
4. Explain the characteristics of Lewis acids and bases and how they form bonds.
5. Predict and explain periodic trends in the polarizability of anions and the polarizing power of cations.
6. Explain the significance of electronegativity and what is meant by the ionic or covalent character of a bond.

Problem-Solving

1. Predict the chemical formula of a binary ionic compound and write its formula with Lewis structures.
2. Draw the Lewis structure of a molecule or ion.
3. Write the resonance structure for a molecule.
4. Calculate the formal charges on the atoms in a Lewis structure.
5. Use a table of electronegativities to predict which of two bonds has greater ionic or covalent character.

Descriptive

1. Describe the general characteristics of radical species and how some radicals affect our lives.
2. Predict which atoms are likely to form molecules in which they have an expanded octet.
3. Predict whether a compound can act as a Lewis acid or a Lewis base.

EXERCISES

Ionic Bonds

8.1 Write the most likely charge for the ions formed by each element: (a) Li; (b) S; (c) Ca; (d) Al.

8.2 Write the most likely charge for the ions formed by each element: (a) F; (b) Ba; (c) Se; (d) O.

8.3 Explain why sodium occurs as Na^+ and not Na^{2+} in ionic compounds.

8.4 Explain why chlorine occurs as Cl^- and not Cl^{2-} in ionic compounds.

8.5 Write the Lewis symbols of (a) calcium; (b) sulfur; (c) oxide ion; (d) nitride ion.

8.6 Write the Lewis symbols of (a) chlorine; (b) arsenic; (c) carbon; (d) chloride ion.

8.7 Write the Lewis formulas (for the ionic model of the compounds) of (a) potassium fluoride; (b) aluminum sulfide; (c) calcium nitride.

8.8 Write the Lewis formulas (for the ionic model of the compounds) of (a) cesium carbide; (b) sodium oxide; (c) calcium phosphide.

Lattice Enthalpies

8.9 Explain why the lattice enthalpy of magnesium oxide ($3850\ kJ\cdot mol^{-1}$) is greater than that of barium oxide ($3114\ kJ\cdot mol^{-1}$). See Appendix 2D.

8.10 Explain why the lattice enthalpy of magnesium oxide ($3850\ kJ\cdot mol^{-1}$) is greater than that of magnesium sulfide ($3406\ kJ\cdot mol^{-1}$). See Appendix 2D.

8.11 Calculate the lattice enthalpy of silver fluoride from the data in Fig. 7.40 and the following information:

Enthalpy of formation of Ag(g):	$+284\ kJ\cdot mol^{-1}$
Ionization energy of Ag(g):	$+731\ kJ\cdot mol^{-1}$
Enthalpy of formation of F(g):	$+79\ kJ\cdot mol^{-1}$
Enthalpy of formation of AgF(s):	$-205\ kJ\cdot mol^{-1}$

8.12 Calculate the lattice enthalpy of calcium sulfide from the data in Appendix 2D and the following information:

Enthalpy of formation of Ca(g):	$+178\ kJ\cdot mol^{-1}$
Enthalpy of formation of S(g):	$+279\ kJ\cdot mol^{-1}$
Enthalpy of formation of CaS(s):	$-482\ kJ\cdot mol^{-1}$
Second ionization energy of Ca(g):	$+1150\ kJ\cdot mol^{-1}$

8.13 Identify each of the following as endothermic or exothermic processes: (a) ionization of copper; (b) electron attachment to oxygen; (c) sublimation of copper; (d) formation of copper(II) oxide from its ions; (e) formation of copper(II) oxide from the elements.

8.14 Identify each of the following as endothermic or exothermic processes: (a) ionization of silver; (b) electron attachment to chlorine; (c) dissociation of Cl_2 molecules; (d) formation of silver chloride from its ions; (e) formation of silver chloride from its elements.

Lewis Structures

8.15 Write the Lewis structures of (a) hydrogen fluoride; (b) ammonia; (c) methane.

8.16 Write the Lewis structures of (a) hydrogen sulfide; (b) nitrogen trichloride; (c) iodine chloride.

8.17 Write the Lewis structures of (a) ammonium ion, NH_4^+; (b) hypochlorite ion, ClO^-; (c) tetrafluoroborate ion, BF_4^-.

8.18 Write the Lewis structures of (a) nitronium ion, ONO^+; (b) chlorite ion, ClO_2^-; (c) peroxide ion, O_2^{2-}.

8.19 Write the complete Lewis structures for each of the following compounds: (a) ammonium chloride; (b) potassium phosphate; (c) sodium hypochlorite.

8.20 Write the complete Lewis structures for each of the following compounds: (a) zinc cyanide; (b) potassium tetrafluoroborate; (c) barium peroxide (the peroxide ion is O_2^{2-}).

8.21 Write the Lewis structures of the following organic compounds: (a) formaldehyde, H_2CO, which, as its aqueous solution called "formalin," is used to preserve biological specimens; (b) methanol, CH_3OH, the toxic compound also called wood alcohol; (c) glycine, $CH_2(NH_2)COOH$, the simplest of the amino acids, the building blocks of proteins.

8.22 Write the Lewis structures of the following organic compounds: (a) ethanol, CH_3CH_2OH, which is also called grain alcohol; (b) methylamine, CH_3NH_2, a putrid-smelling substance formed when flesh decays; (c) formic acid, HCOOH, a component of the venom injected by ants.

Resonance

8.23 Use Lewis structures to describe what is meant by resonance and name two consequences of resonance.

8.24 Do $H{-}C{\equiv}N$ and $H{-}N{\equiv}C$ form a pair of resonance structures? Explain your answer.

8.25 Write the Lewis structures that contribute to the resonance hybrid for (a) the nitrite ion, NO_2^-; (b) nitryl chloride, $ClNO_2$.

8.26 Write the Lewis structures that contribute to the resonance hybrid of (a) ozone, O_3, (b) the formate ion, HCO_2^-.

8.27 Write Lewis structures, including typical contributions to the resonance structure (where appropriate, allow for the possibility of octet expansion) for (a) dihydrogen phosphate ion; (b) sulfite ion; (c) chlorate ion; (d) nitrate ion.

8.28 Write Lewis structures, including typical contributions to the resonance (where appropriate, allow for the possibility of octet expansion as well as double bonds in different locations) for (a) phosphite ion; (b) hydrogen sulfite ion; (c) perchlorate ion; (d) nitrite ion.

Formal Charge

8.29 Justify the definition of formal charge and summarize how the concept is used.

8.30 Why are formal charge and oxidation number artificial measures of the distributions of electrons in molecules and ions?

8.31 Determine the formal charges of each atom in the following Lewis structures:

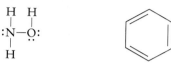

$\ddot{O}=C=\ddot{O}$

(a) Hydrazine (b) Carbon dioxide

8.32 Determine the formal charges of each atom in the following Lewis structures:

H H
:N—Ö:
H

(a) Hydroxylamine (b) Kekulé structure of benzene

8.33 Two contributions to the resonance structure are shown for each species. Determine the formal charge on each atom and then, if possible, identify the Lewis structure of lowest energy for each species.

(a) :Ö—S=Ö Ö=S=Ö

(b) (:O: double-bonded to S) Ö=S=Ö (:Ö: single-bonded to S) Ö=S=Ö

8.34 Two contributions to the resonance structure are shown for each species. Determine the formal charge on each atom and then, if possible, identify the Lewis structure of lowest energy for each species.

(a) [:Ö:—S(=Ö)—Ö:—H]$^-$ [:Ö:—S(—:Ö:)—Ö:—H]$^-$

(b) [Ö=S(:Ö:)(:Ö:)—Ö:—H]$^-$ [Ö=S(=:O:)(=:O:)—Ö:—H]$^-$

8.35 Use formal charge arguments to identify the lower energy contribution to the resonance structure of sulfurous acid.

(a) H:Ö: H / :O—S—O: (b) H :O: H / :O—S—O: (S=O)

8.36 Use formal charge arguments to identify the lower energy contribution to the resonance structure of sulfuric acid.

(a) H :Ö: H / :O—S—O: / :O: (b) H :O: H / :O—S—O: / :O: (S=O double bonds)

8.37 Select from each of the following pairs of Lewis structures the one that is likely to make the dominant contribution to a resonance hybrid.

(a) :F̈—Xë—F̈: :F̈—Xë=F̈

(b) Ö=C=Ö :Ö—C≡O:

8.38 Select from each of the following pairs of Lewis structures the one that is likely to make the dominant contribution to a resonance hybrid.

(a) N̈=N=Ö :N≡N—Ö:

(b) [Ö=P(—:Ö:)(—:Ö:)—Ö:]$^{3-}$ [Ö=P(—:Ö:)(—:Ö:)=Ö]$^{3-}$

Ionic and Covalent Bonding

8.39 Decide which of the following compounds are likely to be ionic and justify your choice: (a) magnesium oxide; (b) nitrogen triiodide; (c) iron(II) oxide.

8.40 Decide which of the following compounds are likely to be ionic and justify your choice: (a) sulfur tetrachloride; (b) cobalt(III) oxide; (c) oxygen difluoride.

8.41 Use the information in Fig. 8.15 to label the following bonds as ionic, covalent, or a significant mix of covalent and ionic character: (a) Ba—Cl; (b) Bi—I; (c) Si—H.

8.42 Use the information in Fig. 8.15 to label the following bonds as ionic, covalent, or a significant mix of covalent and ionic character: (a) N—H; (b) C—S; (c) P—F.

8.43 Arrange the cations Rb^+, Be^{2+}, Sr^{2+} in order of increasing polarizing power. Give an explanation of your arrangement.

8.44 Arrange the cations K^+, Mg^{2+}, Al^{3+}, Cs^+ in order of increasing polarizing power. Give an explanation of your arrangement.

8.45 Arrange the anions Cl^-, Br^-, N^{3-}, O^{2-} in order of increasing polarizability, giving reasons for your decisions.

8.46 Arrange the anions N^{3-}, P^{3-}, I^-, At^- in order of increasing polarizability, giving reasons for your decisions.

8.47 In each pair, determine which compound has bonds with greater ionic character: (a) HCl or HI; (b) CH_4 or CF_4; (c) CO_2 or CS_2.

8.48 In each pair, determine which compound has bonds with greater ionic character: (a) PH_3 or NH_3; (b) SO_2 or NO_2; (c) SF_6 or IF_5.

8.49 Classify the bonding in the following compounds as mainly ionic or significantly covalent: (a) AgF; (b) AgI; (c) $AlCl_3$; (d) AlF_3. See Appendix 2D.

8.50 Classify the bonding in the following compounds as mainly ionic or significantly covalent: (a) $BeCl_2$; (b) $CaCl_2$; (c) TlCl; (d) $InCl_3$.

Exceptions to the Octet Rule

8.51 What is meant by the term *radical?* Give three examples of radicals, showing their Lewis structures.

8.52 Why are radicals generally highly reactive?

8.53 Write the Lewis structures and note the number of lone pairs on the central atom for (a) sulfur tetrachloride; (b) iodine trichloride; (c) IF_4^-.

8.54 Write the Lewis structures and note the number of lone pairs on the central atom for (a) PF_4^-; (b) ICl_4^+; (c) XeF_4.

8.55 Determine the numbers of electron pairs (both bonding and lone pairs) on the iodine atom in (a) ICl_2^+; (b) ICl_4^-; (c) ICl_3; (d) ICl_5.

8.56 Determine the numbers of electron pairs (both bonding and lone pairs) on the phosphorus atom in (a) PCl_3; (b) PCl_5; (c) PCl_4^+; (d) PCl_6^-.

8.57 Write the Lewis structures of the following reactive species found to contribute to the destruction of the ozone layer and indicate which are radicals: (a) chlorine monoxide, ClO; (b) dichloroperoxide, Cl—O—O—Cl; (c) chlorine nitrate, $ClONO_2$ (the central O atom is attached to the Cl atom and to the N atom of the $—NO_2$ group); (d) chlorine peroxide, Cl—O—O.

8.58 Write the Lewis structures of the following species and indicate which are radicals: (a) the superoxide ion, O_2^-; (b) the methoxy group, CH_3O; (c) XeO_4; (d) $HXeO_4^-$.

8.59 Write Lewis structures and state the number of lone pairs on xenon, the central atom of the following compounds: (a) $XeOF_2$; (b) XeF_2; (c) $HXeO_4^-$.

8.60 Write Lewis structures and state the number of lone pairs on the central atom of the following compounds: (a) ClF_3; (b) AsF_5; (c) SF_6.

Lewis Acids and Bases

8.61 Explain the terms *Lewis acid* and *Lewis base* and give three examples of each type of species.

8.62 Express the formation of magnesium carbonate from carbon dioxide and magnesium oxide as a reaction between a Lewis acid and a Lewis base.

8.63 Explain how an acid-base neutralization reaction in water can be expressed in terms of Lewis acids and bases.

8.64 Explain why (a) precipitation reactions and (b) redox reactions are not Lewis acid-base reactions.

8.65 Identify the following species as Lewis acids or Lewis bases: (a) NH_3; (b) BF_3; (c) Ag^+; (d) F^-.

8.66 Identify the following species as Lewis acids or Lewis bases: (a) H^+; (b) Al^{3+}; (c) CN^-; (d) NO_2^-.

8.67 Write the Lewis structure of the methoxide ion, CH_3O^-. Would you expect it to be a Lewis acid or a Lewis base? Explain your answer.

8.68 Write the Lewis structure of the hydride ion. Would you expect it to be a Lewis acid or a Lewis base? Explain your answer.

8.69 Write the Lewis structures of each reactant, identify the Lewis acid and the Lewis base, and then write the product (a complex) for the following Lewis acid-base reactions:

(a) $PF_5 + F^- \longrightarrow$

(b) $SO_2 + Cl^- \longrightarrow$

(c) $Cu^{2+} + 4\ NH_3 \longrightarrow$

8.70 Write the Lewis structures of each reactant, identify the Lewis acid and the Lewis base, and then write the product (a complex) for the following Lewis acid-base reactions:

(a) $GaCl_3 + Cl^- \longrightarrow$

(b) $SF_4 + F^- \longrightarrow$

(c) $Ag^+ + 2\ CN^- \longrightarrow$

SUPPLEMENTARY EXERCISES

8.71 Suggest a reason why it is reasonable to write Lewis structures for boron compounds in which boron does not have a complete octet of electrons.

8.72 What is the relation between the electronegativities of two atoms and the type of bond (covalent, polar, or ionic) they are likely to form?

8.73 List the following bonds in order of increasing degree of ionic character: C—Cl, Na—Cl, Al—Cl, Br—Cl.

8.74 Write the Lewis symbols of (a) N^{3-}; (b) O^{2-}; (c) Sn^{2+}.

8.75 Write the Lewis formulas for the following compounds (assuming an ionic model of their bonding): (a) lithium hydride; (b) copper(II) chloride; (c) barium nitride; (d) gallium oxide.

8.76 Complete the following table (all values are in kilojoules per mole):

Compound (MX)	ΔH_f° (M,g)	Ionization energy (M)	ΔH_f° (X,g)	Electron affinity (X)	ΔH_L° (MX)	ΔH_f° (MX,s)
(a) NaCl	108	494	122	−349	787	?
(b) KBr	89	418	97	−325	?	−394
(c) RbF	?	402	79	−328	774	−558

8.77 Use Appendix 2A, Appendix 2D, and the following data to calculate the lattice enthalpy of (a) Na_2O; (b) $AlCl_3$.

First, second, and third ionization energies of Al are 557, 1820, and 2740 $kJ \cdot mol^{-1}$, respectively.

$$\Delta H_f^\circ(Na_2O) = -409\ kJ \cdot mol^{-1}$$
$$\Delta H_f^\circ(O, g) = +249\ kJ \cdot mol^{-1}$$
$$\Delta H_f^\circ(Al, g) = +326\ kJ \cdot mol^{-1}$$

8.78 Draw the (noted) number of Lewis structures that contribute to the resonance structures for the following compounds:
(a) CN_2^{2-} (two)
(b) N_2O_3 (three, for the arrangement $ONNO_2$)
(c) OCN^- (three)
(d) C_3O_2 (two, for the arrangement OCCCO)

8.79 Identify each of the following as a Lewis acid or a Lewis base: (a) BCl_3; (b) Fe^{3+}; (c) GaI_3.

8.80 Write the Lewis structure for each reactant, identify the Lewis acid and Lewis base, and then write the formula of the product (a complex) for the following reactions:
(a) $AlCl_3 + Cl^- \longrightarrow$
(b) $I_2 + I^- \longrightarrow$
(c) $Cr^{3+} + 6\,NH_3 \longrightarrow$
(d) $SnCl_4 + 2\,Cl^- \longrightarrow$

8.81 In the reaction of sulfur dioxide with water, SO_2 functions as a Lewis acid. Write the balanced chemical equation for the reaction, using Lewis structures.

8.82 Which of the following pairs of species is the stronger Lewis acid: (a) BF_3 or NF_3; (b) Al^{3+} or K^+? Explain your reasoning.

8.83 Which of the following pairs of species is the stronger Lewis base: (a) CH_3^- or CH_4; (b) H_2O or H_2S? Explain your reasoning.

8.84 Ionic compounds tend to have higher boiling points and lower vapor pressures than covalent compounds. Predict which compound in the following pairs has the lower vapor pressure: (a) Cl_2O or Na_2O; (b) $InCl_3$ or $SbCl_3$; (c) LiH or HCl; (d) $MgCl_2$ or PCl_3.

8.85 Determine the formal charge of each atom in the following species. Where two or more Lewis structures are given, identify the structure of lowest energy.

(a) :Ö=Cl(=O:)(H)–Ö: :Ö–Cl(–Ö:)(H)–Ö:

(b) Ö=C=S̈

(c) H–C≡N:

(d) $[$N̈=C=N̈$]^{2-}$ $[$:N≡C–N̈:$]^{2-}$

(e) $[$:Ö–As(–Ö:)(–Ö:)–Ö:$]^{3-}$ $[$:Ö–As(=O:)(–Ö:)–Ö:$]^{3-}$

8.86 Determine the maximum oxidation number that each of the following elements can have in compounds: (a) lead; (b) vanadium; (c) sulfur; (d) chlorine.

CHALLENGING EXERCISES

8.87 Could you support the view that there is only one type of chemical bond? Explain how covalent bonds, ionic bonds, complex formation, and the octet rule would fit into this view.

8.88 Explain the following observations: (a) The difference in electronegativity between magnesium and iodine is 1.4, yet magnesium iodide exhibits the properties of ionic bonding. (b) The difference in electronegativity between silicon and fluorine is 2.1, yet silicon tetrafluoride exhibits the properties of covalent bonding.

8.89 By assuming that the lattice enthalpy of $NaCl_2$ would be the same as that of $MgCl_2$, use enthalpy arguments based on data in Appendix 2A, Appendix 2D, and Self-Test 8.3B to explain why $NaCl_2$ is an unlikely compound.

8.90 By assuming that the lattice enthalpy of MgCl would be the same as that of KCl, use enthalpy arguments based on data in Appendix 2A, Appendix 2D, and Self-Test 8.3B to explain why MgCl is an unlikely compound.

8.91 Phosgene is a highly toxic gas that has been used in chemical warfare. A chemical analysis of phosgene is 12.1% carbon, 16.2% oxygen, and 71.7% chlorine by mass, and its molar mass is 98.9 $g \cdot mol^{-1}$. Write the Lewis structure of phosgene.

8.92 Limestone (calcium carbonate) is used to remove impurities from iron ore in a blast furnace. When limestone is heated, it decomposes into calcium oxide and carbon dioxide, and the calcium oxide combines with impurities such as silica, SiO_2, to form calcium silicate, $CaSiO_3$. The latter is commonly called "slag." Write the chemical equation for the Lewis acid-base decomposition of carbonate ions (the complex) and a Lewis acid-base reaction for the formation of calcium silicate. Identify the Lewis acid and Lewis base in each reaction.

8.93 Show how resonance can occur in the following organic ions:
(a) The acetate ion, $CH_3CO_2^-$
(b) An enolate ion, $CH_2COCH_3^-$, which has one resonance structure with a C=C double bond and an $-O^-$ group on the central carbon atom
(c) An allyl cation, $CH_2CHCH_2^+$
(d) An amidate ion, CH_3CONH^-

CHAPTER 9

A section through the retina of a human eye, showing the rods and cones that act as light collectors. If we could magnify the molecular events taking place inside the rods and cones, we would see molecules changing their shapes when light strikes them. That change of shape is the first step in the process we call vision. But what controls the shapes of molecules? We explore that question in this chapter.

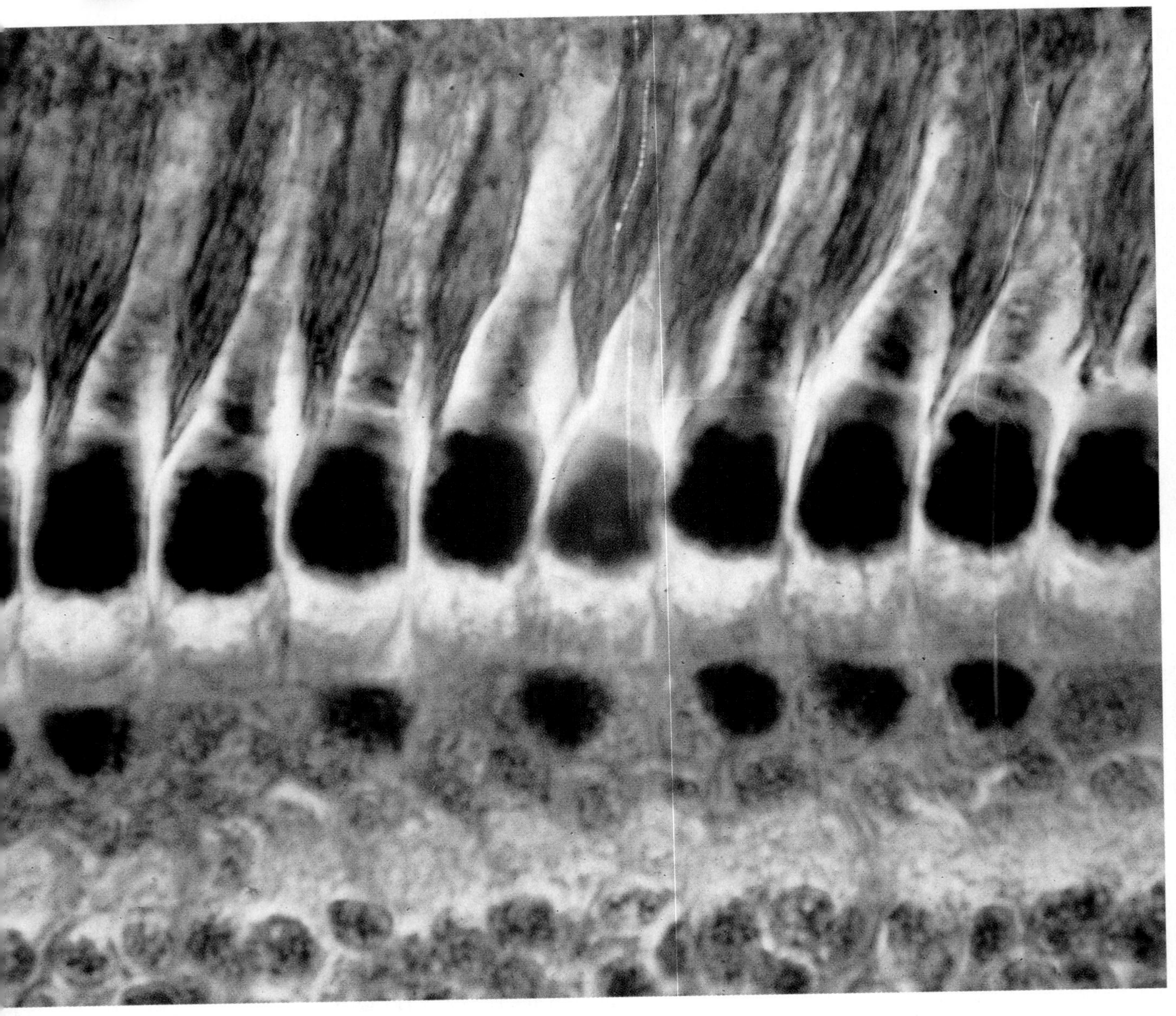

MOLECULES: SHAPE, SIZE, AND BOND STRENGTH

How do we see these words? Scientists are starting to understand that perception, thinking, and learning depend on the shapes of molecules and how they change. Seeing depends on changes in the shapes of molecules in our eyes. So does the signal processing that turns those images—in a most amazing and complicated and almost completely unknown way—into thoughts, memories, actions, and deeds. If we are ever to know what it means to be conscious, then we have to understand why molecules have certain shapes.

The shapes of molecules also determine their odors, their tastes, and their actions as drugs. They govern the reactions that occur throughout our bodies and contribute to the process of being alive. Molecular shapes also govern the properties of the materials around us, including their color and their solubility. They help to determine whether a substance is a solid, a liquid, or a gas.

THE SHAPES OF MOLECULES AND IONS

9.1 The VSEPR model
9.2 Molecules without lone pairs on the central atom
9.3 Multiple bonds in the VSEPR model
9.4 Molecules with lone pairs on the central atom
9.5 The distorting effect of lone pairs

CHARGE DISTRIBUTION IN MOLECULES

9.6 Polar bonds
9.7 Polar molecules

THE STRENGTHS AND LENGTHS OF BONDS

9.8 Bond strengths
9.9 The variation of bond strength
9.10 Bond strengths in polyatomic molecules
9.11 Bond lengths

ORBITALS AND BONDING

9.12 Sigma and pi bonds
9.13 Hybridization of orbitals
9.14 Hybridization in more complex molecules
9.15 Hybrids including *d*-orbitals
9.16 Multiple carbon-carbon bonds
9.17 Characteristics of double bonds

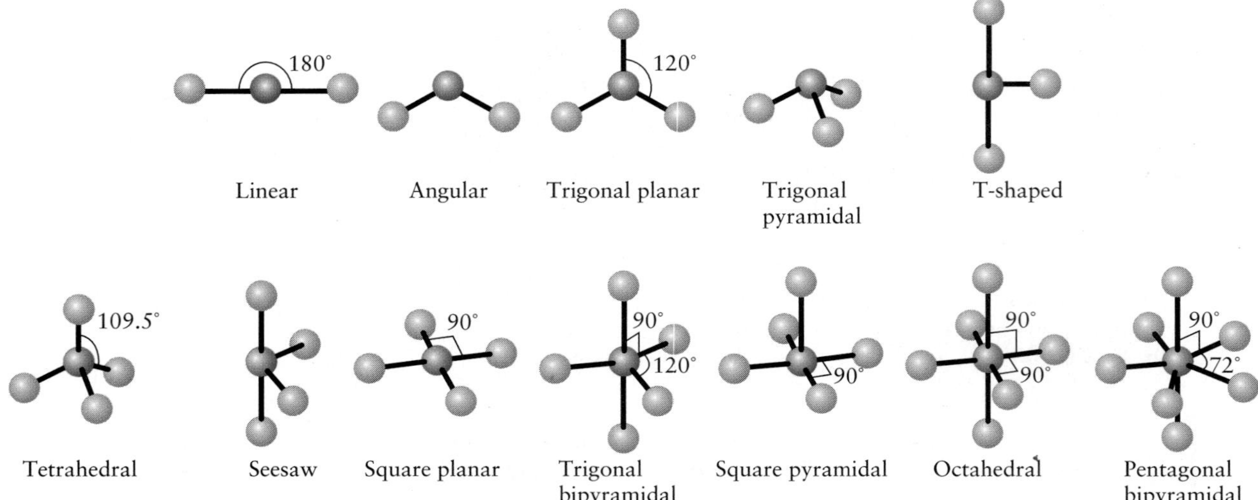

FIGURE 9.1 The names of the shapes of simple molecules and the bond angles certain shapes imply. To use this table, compare the locations of the atoms alone, *not* lone pairs, with the locations shown here.

THE SHAPES OF MOLECULES AND IONS

A Lewis structure shows the approximate locations of bonding electrons and lone pairs in a molecule. Because it is only a two-dimensional diagram of the links between atoms, it does not show the arrangement of atoms in space. In this section we see how to extend Lewis's ideas to account for the shapes of molecules.

9.1 THE VSEPR MODEL

When we report the shape of a molecule, we note the locations of its atoms and ignore any lone pairs it may possess. Many molecules have the shapes of the geometrical figures shown in Fig. 9.1. For instance, CH_4 (**1**) is tetrahedral, SF_6 (**2**) is octahedral, and H_2O (**3**) is angular.

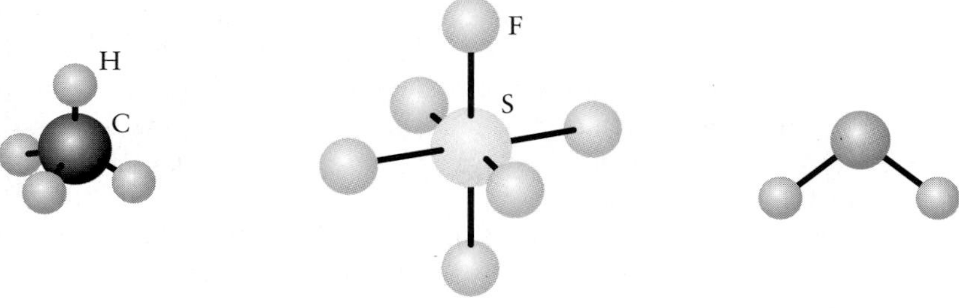

1 Methane, CH_4 **2** Sulfur hexafluoride, SF_6 **3** Water, H_2O

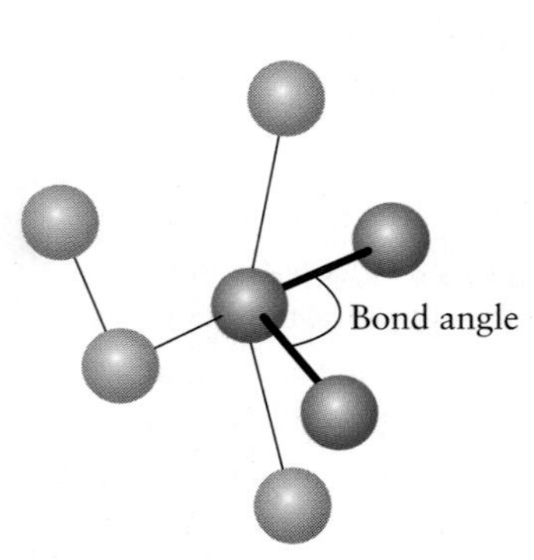

FIGURE 9.2 The bond angle in a molecule is the angle between the two straight lines joining the centers of the atoms concerned.

To report shapes more precisely, we give the **bond angles,** visualizing bonds as straight lines that join the atom centers (Fig. 9.2). The bond angle in H_2O, for instance, is the angle (104.5°) between the two O—H bonds.

To account for these molecular shapes, we need just one addition to Lewis's model: *regions of high electron concentration repel one another.*

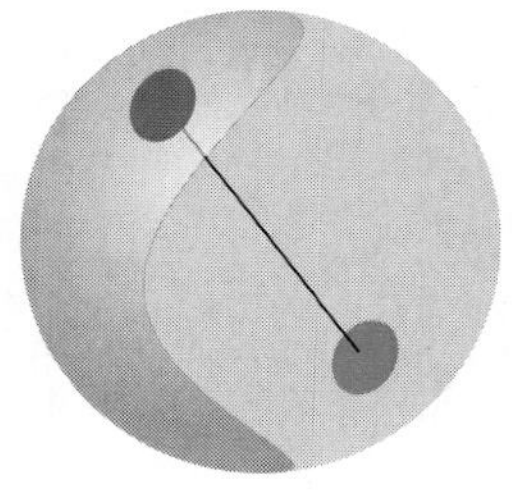
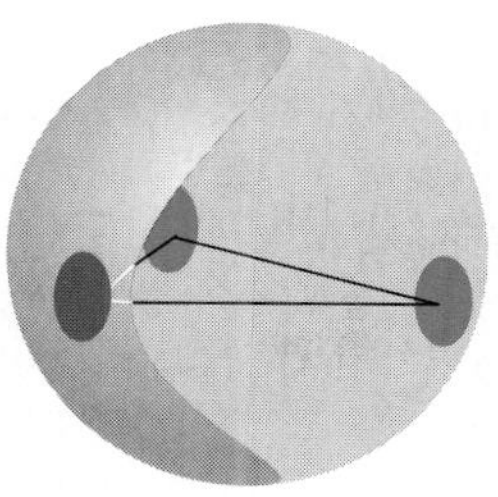
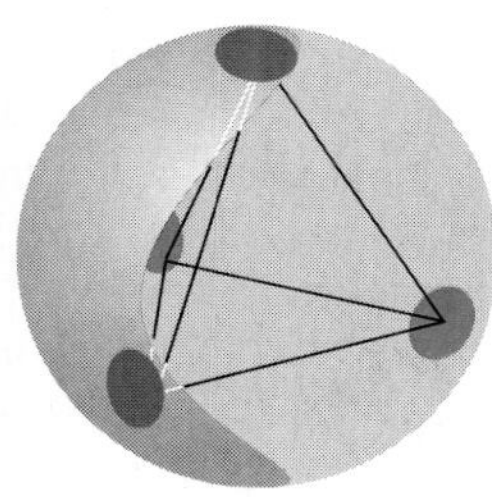

FIGURE 9.3 The positions taken up by regions of high electron concentration (shown in green) around a central atom, for three simple cases. These positions minimize the repulsions between the regions and result in the lowest energy. The regions of high electron concentration represent either atoms or lone pairs.

Bonding electrons and lone pairs take up positions as far from one another as possible, for then they repel each other the least.

This idea was first explored by the British chemists Nevil Sidgwick and Herbert Powell, and has been developed by the Canadian chemist Ronald Gillespie.

In the **valence-shell electron-pair repulsion model,** or VSEPR model, we focus attention on the central atom of a molecule, such as the B atom in BF_3 or the C atom in CO_2. We then imagine that all the electrons involved in bonds to atoms and the electrons of lone pairs belonging to the central atom lie on the surface of an invisible sphere that surrounds the central atom (Fig. 9.3). These bonding electrons and lone pairs are regions of high electron concentration, and they repel one another. To minimize their repulsions, these regions move as far apart as possible on the surface of the sphere. Once we have identified the "most distant" locations of the electron pairs (Fig. 9.4), we identify the shape of the molecule by noting where the *atoms* lie and look up the name of the shape in Fig. 9.1.

9.2 MOLECULES WITHOUT LONE PAIRS ON THE CENTRAL ATOM

First we shall look at molecules in which there are from two to six atoms and no lone pairs attached to the central atom. We deal with lone pairs in Section 9.4.

A molecule with only two atoms attached to the central atom is $BeCl_2$. The Lewis structure is :C̤̈l—Be—C̤̈l:, and there are no lone pairs on the central atom. To be as far apart as possible, the bonding pairs, and therefore the Cl atoms, lie on opposite sides of the Be atom. The $BeCl_2$ molecule is therefore expected to be linear, with a bond angle of 180° (**4**). That shape is confirmed by experiment.

A boron trifluoride molecule, BF_3, has the Lewis structure shown in (**5**). There are three bonding pairs attached to the central atom and no

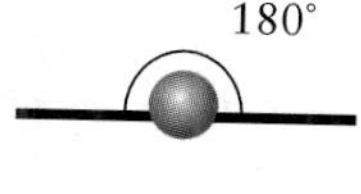

Linear

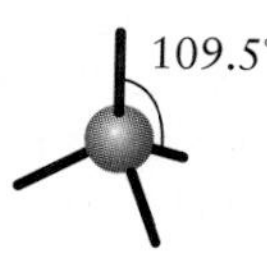

Trigonal planar

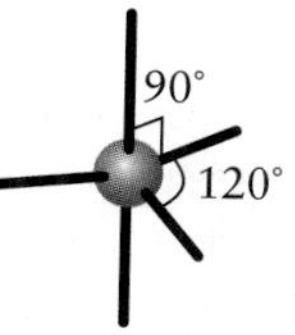

Tetrahedral

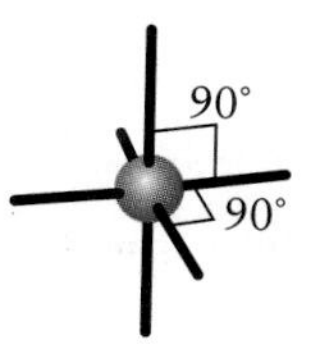

Trigonal bipyramidal

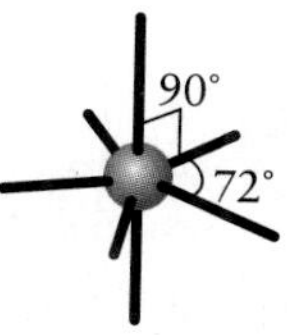

Octahedral

Pentagonal bipyramidal

FIGURE 9.4 A summary of the positions taken up by regions of high electron concentration (atoms and lone pairs) around a central atom. The regions are denoted by the straight lines sticking out of the central atom. Use this chart to identify the arrangement of a given number of atoms and lone pairs, and then use Fig. 9.1 to identify the shape of the molecule from the location of its atoms.

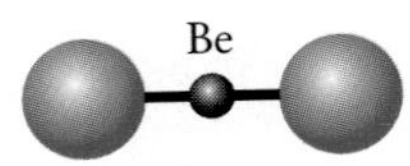

4 Beryllium chloride, $BeCl_2$

:F̤̈:
|
:F̤—B—F̤:

5 Boron trifluoride, BF_3

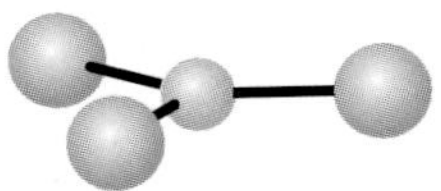

6 Boron trifluoride, BF_3

:C̤̈l:
|
:C̤̈l—P—C̤̈l:
/ \
:C̤̈l C̤̈l:

7 Phosphorus pentachloride, PCl_5

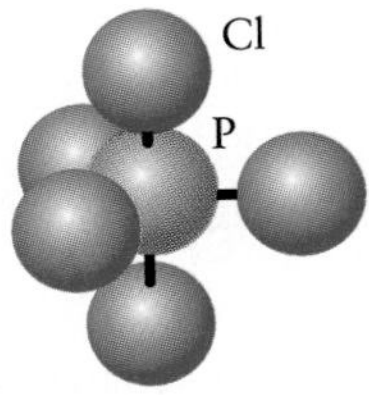

8 Phosphorus pentachloride, PCl_5

lone pairs. According to the VSEPR model, as illustrated in Fig. 9.4, the three bonding pairs, and the fluorine atoms they link, lie at the corners of an equilateral triangle. Looking up the structure in Fig. 9.1, we see that the molecule is trigonal planar and all three F—B—F angles are 120° (**6**).

Methane, CH_4, has four bonding pairs on the central atom, each connecting a hydrogen atom to the carbon atom. According to Fig. 9.4, we can expect the four H atoms to take up positions at the corners of a tetrahedron with the C atom at the center. We therefore expect the molecule to be tetrahedral (see **1**), with a bond angle of 109.5°. That is the shape found experimentally.

There are five bonding pairs on the central atom in a phosphorus pentachloride molecule, PCl_5, and no lone pairs (**7**). According to the VSEPR model, the five pairs and the atoms they carry are farthest apart in a trigonal bipyramidal arrangement (see Fig. 9.4). In this arrangement, three atoms lie at the corners of an equilateral triangle and the other two atoms lie above and below the plane of the triangle (**8**). This structure has two different bond angles. Around the "equator" formed by the triangle, bond angles are 120°. The angle between the axis and the equatorial atoms is 90°. This structure is also confirmed experimentally.

A sulfur hexafluoride molecule, SF_6, has six atoms attached to the central S atom and no lone pairs on that atom. According to the VSEPR model, the fluorine atoms are farthest apart when four lie in a square around the equator and the remaining two above and below the plane of the square (see Fig. 9.4). By referring to Fig. 9.1, we see that the molecule should be classified as octahedral. All its bond angles are either 90° or 180°, and each position is equivalent.

According to the VSEPR model, bonding pairs and lone pairs take up positions around an atom that maximize their separations; the shape of the molecule is reported by noting the locations of the atoms attached to the central atom.

SELF-TEST 9.1A Predict the shape of the PCl_6^- ion present in solid PCl_5.

[*Answer:* Octahedral]

SELF-TEST 9.1B Predict the shape of a $SiCl_4$ molecule.

9.3 MULTIPLE BONDS IN THE VSEPR MODEL

When using the VSEPR model to predict molecular shapes, we do not need to distinguish between single and multiple bonds. A multiple bond is treated simply as another region of high electron concentration. The two electron pairs in a double bond stay together and repel other bonds or lone pairs as a unit. The three electron pairs in a triple bond also stay together and act like a single region of high electron concentration. For instance, a carbon dioxide molecule has a structure similar to that of $BeCl_2$, except for the presence of double bonds, as in Ö̤=C=Ö̤. The two pairs of electrons that make up one double bond are treated as a unit: the two double bonds lie on opposite sides of the C atom and bind the O atoms to it. Therefore, the molecule is linear (**9**). One of the Lewis structures of a carbonate ion, CO_3^{2-}, is shown in (**10**). The two pairs of electrons in the double bond are treated as a unit and the resulting shape (**11**) is the same as that for BF_3 (see **6**).

9 Carbon dioxide, CO_2

:O:]2−
||
:Ö—C—Ö:

10 Carbonate ion, CO_3^{2-}

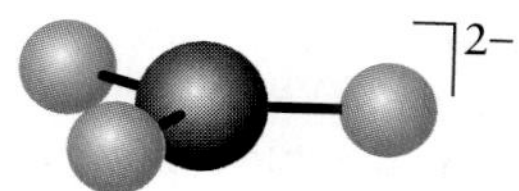

11 Carbonate ion, CO_3^{2-}

Now let's predict the shape of an ethene (ethylene) molecule, $CH_2{=}CH_2$. There are two centers in ethene to consider. The first step is to write the Lewis structure (**12**). Each carbon atom has three regions of electron concentration on it: two single bonds and one double bond. There are no lone pairs. The arrangement around each carbon atom is therefore trigonal planar. We predict that the HCH and HCC angles will both be 120° (**13**), and our prediction is confirmed experimentally.

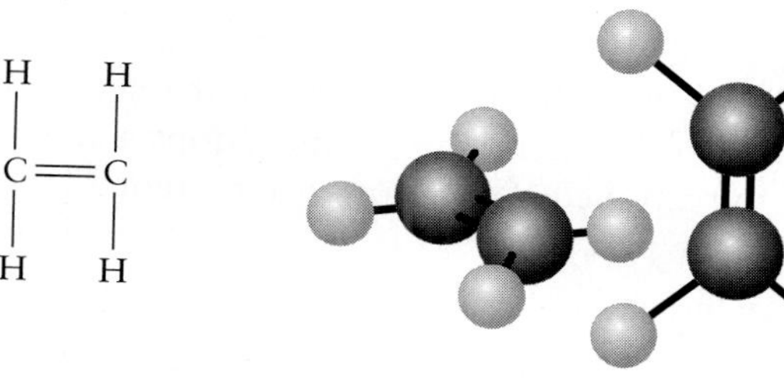

12 Ethene, C_2H_4

13 Ethene, C_2H_4

▶▶ **The VSEPR model cannot explain why both CH_2 groups lie in the same plane. We shall see in Section 9.17 that double bonds between carbon atoms always result in a planar arrangement of atoms.**

EXAMPLE 9.1 Predicting the shape of a molecule with multiple bonds

Suggest a shape for the ethyne (acetylene) molecule, $HC{\equiv}CH$.

STRATEGY Write down the Lewis structure and decide how the electron pairs can be arranged around each "central" atom (two C atoms, in this case) to minimize repulsions. Treat each multiple bond as a single unit. Then identify the overall shape of the molecule (refer to Fig. 9.1 if necessary).

SOLUTION The Lewis structure of the molecule is $H{-}C{\equiv}C{-}H$. Each C atom is attached to two other atoms (one H atom and one C atom), and there are no lone pairs. The arrangement of atoms around each C atom is therefore predicted to be linear (**14**).

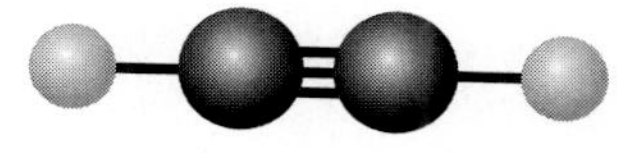

14 Ethyne, C_2H_2

SELF-TEST 9.2A Predict the shape of a carbon suboxide molecule, C_3O_2, in which the atoms lie in the order OCCCO.

[*Answer:* Linear]

SELF-TEST 9.2B Predict the shape of a formaldehyde molecule, CH_2O.

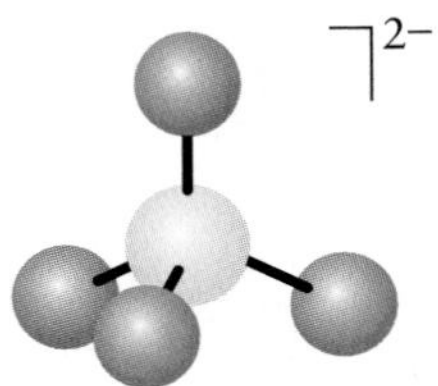

15 Sulfate ion, SO_4^{2-}

Because we treat single bonds and multiple bonds as equivalent in the VSEPR model, it does not matter which of the Lewis structures contributing to the resonance structure we consider. For example, although we can write several different Lewis structures for a sulfate ion, all of them have four regions of electron concentration around the central S atom, and in each case we expect a tetrahedral structure (**15**).

According to the VSEPR model, electron pairs in multiple bonds are treated as a single unit equivalent to one electron pair.

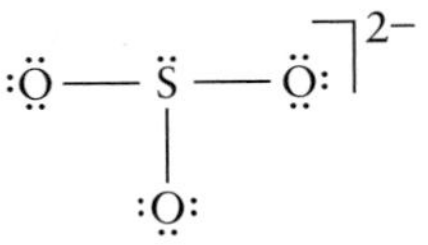

16 Sulfite ion, SO_3^{2-}

We shall use this notation several times: remember that A is a central atom, X an attached atom, and E a lone pair.

9.4 MOLECULES WITH LONE PAIRS ON THE CENTRAL ATOM

The lone pairs on the central atom of a molecule help to *govern* its shape, but we ignore them when we come to *name* the shape. To see what this rule means, let's consider a sulfite ion, SO_3^{2-}. Its Lewis structure is shown in (**16**). If we let A represent a central atom, X an attached atom, and E a lone pair, then the sulfite ion is an example of an AX_3E species. The S atom has three O atoms and one lone pair attached to it.

When deciding on the arrangement of electrons about the central atom, we do not distinguish between bonding pairs and lone pairs: single bonds, multiple bonds, and lone pairs are all regions where electrons are concentrated. Figure 9.4 tells us that these four regions of high electron concentration will be farthest apart if they adopt a tetrahedral arrangement. Three of the tetrahedral locations are occupied by atoms, and one is occupied by a lone pair. By referring to Fig. 9.1, we find that the shape of a SO_3^{2-} ion is trigonal pyramidal (**17**).

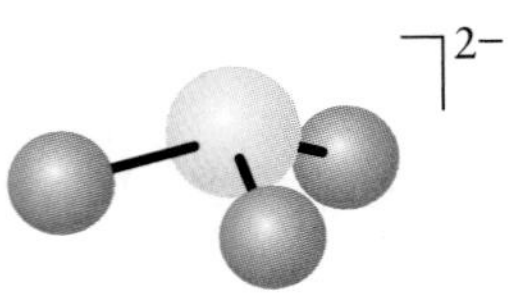

17 Sulfite ion, SO_3^{2-}

It is important to distinguish between the arrangement of electrons and the molecular shape. All electrons, whether bonded or not, are included in a description of the arrangement of electrons. However, only the positions of atoms are considered when describing the shape of a molecule; at that stage, lone pairs are ignored. For example, SO_3^{2-} has a tetrahedral arrangement of electron pairs, but a trigonal pyramidal shape. Radicals such as NO_2 have a single nonbonding electron. Such an electron is treated like a lone pair when determining molecular shape.

If a molecule has lone pairs on the central atom, they contribute to the shape of the molecule but are ignored when we name the shape.

EXAMPLE 9.2 Predicting the shape of a simple molecule with lone pairs on the central atom

Predict the electronic arrangement and the shape of a nitrogen trifluoride molecule, NF_3.

STRATEGY Draw the Lewis structure and then decide how the bonding pairs and lone pairs are arranged around the central (nitrogen) atom (consult Fig. 9.4 if necessary). Ignore lone pairs and identify the molecule's shape in Fig. 9.1.

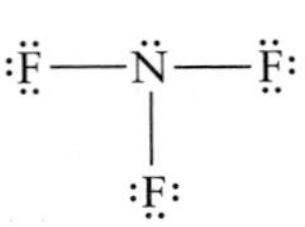

18 Nitrogen trifluoride, NF_3

SOLUTION The Lewis structure of nitrogen trifluoride is shown in (**18**): we see that the central N atom has four electron pairs. According to the VSEPR model, these four electron-rich regions adopt a tetrahedral arrangement. Because one of the pairs is a lone pair, the molecule is expected to be trigonal pyramidal. That is, in fact, the case.

SELF-TEST 9.3A Predict (a) the electronic arrangement and (b) the shape of an IF_5 molecule.

[*Answer:* (a) Octahedral; (b) square pyramidal]

SELF-TEST 9.3B Predict (a) the electronic arrangement and (b) the shape of an SO_2 molecule.

9.5 THE DISTORTING EFFECT OF LONE PAIRS

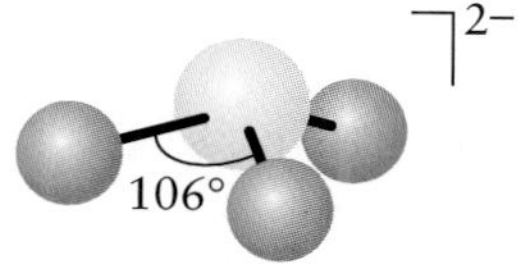

19 Sulfite ion, SO_3^{2-}

We have predicted that an SO_3^{2-} ion is trigonal pyramidal. Because the four regions of high electron concentration (the bonds and the lone pair) are tetrahedral around the S atom, we would also predict OSO angles of 109.5°. Experiments have shown that the sulfite ion is indeed trigonal pyramidal, but its bond angle is only 106° (**19**). Experimental information like this tells us that the VSEPR model as we have described it so far is incomplete.

We have treated lone pairs on the same footing as bonds to atoms. However, according to the VSEPR model, lone pairs should be treated as having a more strongly repelling effect than bonding pairs. No one is really sure why this is so. One possible explanation is that the electron cloud of a bonding pair cannot spread over such a large volume, because a bonding pair is pinned down by two atoms, not one (Fig. 9.5).

Whatever the true reason, the VSEPR model needs another rule: the strengths of repulsions are in the order

Lone pair-lone pair > lone pair-bonding pair > bonding pair-bonding pair

It is best for lone pairs to be as far from each other as possible. It is also best for atoms to be far from lone pairs, even though that might bring them close to other atoms.

Our revised model helps to account for the bond angle of the AX_3E sulfite ion (see **19**). The shared electron pairs adopt a tetrahedral arrangement around the S atom. However, the lone pair exerts a strong repulsion on the bonding electrons, forcing them to move together slightly. As a result of this adjustment in positions, the OSO angle is reduced from the 109.5° of the regular tetrahedron to the 106° observed experimentally.

Note that, although the VSEPR model can predict the *direction* of the distortion, it cannot predict its extent. We can predict that in any AX_3E species the angle will be less than 109.5°, but we cannot predict its actual value.

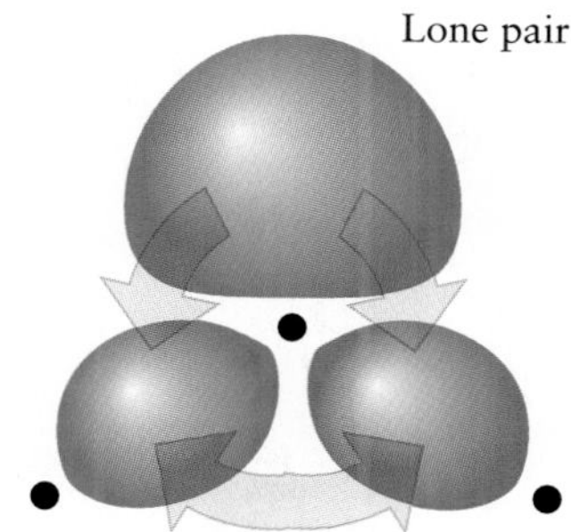

FIGURE 9.5 One possible explanation of why lone pairs have a greater repelling effect than atoms. A lone pair is more extensive than the bonding pairs that hold the atoms in place, so the peripheral atoms move away from the lone pair in an attempt to lower the repulsion they experience.

EXAMPLE 9.3 Using the VSEPR model to predict shapes

Account for the shape of a water molecule (and of AX_2E_2 molecules in general).

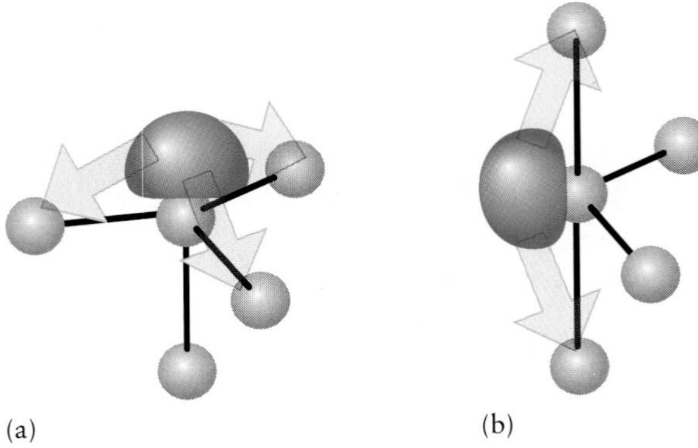

FIGURE 9.6 (a) A lone pair in an axial position is close to three equatorial atoms, whereas (b) in an equatorial position it is close to only two atoms. The latter arrangement is therefore more favorable.

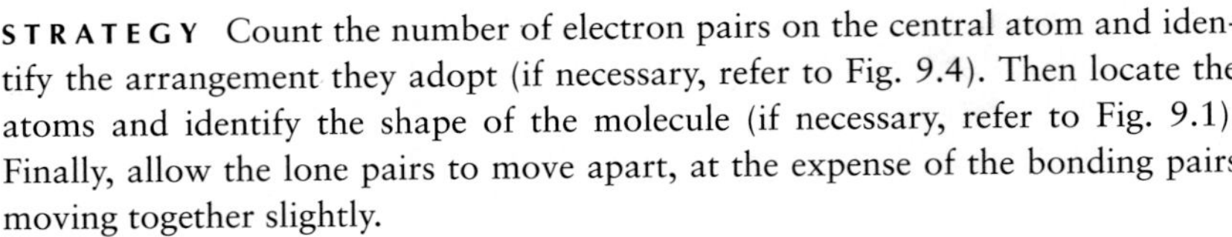

STRATEGY Count the number of electron pairs on the central atom and identify the arrangement they adopt (if necessary, refer to Fig. 9.4). Then locate the atoms and identify the shape of the molecule (if necessary, refer to Fig. 9.1). Finally, allow the lone pairs to move apart, at the expense of the bonding pairs moving together slightly.

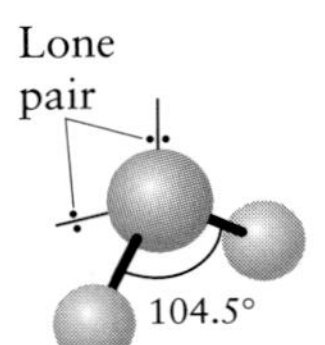

20 Water, H_2O

SOLUTION The central atom (O) has four electron pairs: two single bonds and two lone pairs. These four pairs adopt a tetrahedral arrangement, as illustrated in Fig. 9.4. Two of these tetrahedral locations are occupied by atoms, so according to Fig. 9.1 the molecule is classified as angular or "bent." The lone pairs push apart from each other, forcing the bonding pairs together, so we predict an HOH bond angle of less than 109.5°. This prediction is in agreement with the experimental value of 104.5° (**20**).

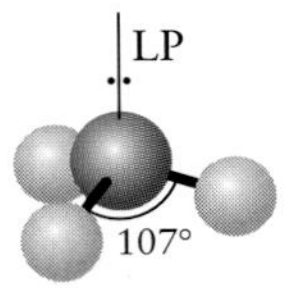

21 Ammonia, NH_3

SELF-TEST 9.4A Predict the shape of an NH_3 molecule (and of AX_3E molecules in general).

[*Answer:* Trigonal pyramidal (**21**); HNH angle less than 109.5°]

SELF-TEST 9.4B Predict the shape of an H_3O^+ ion.

Larger numbers of lone pairs have a greater effect on the bond angle. For example, NH_3 has one lone pair and a bond angle of 107°, whereas H_2O has two lone pairs and a bond angle of 104.5°.

In AX_4E molecules, which have one lone pair, there are two different locations a lone pair could take. An **axial lone pair** lies on the axis of the molecule. An **equatorial lone pair** lies on the molecule's "equator." An axial lone pair repels three electron pairs strongly but an equatorial lone pair repels only two strongly (Fig. 9.6). Therefore, it is best for a lone pair to be equatorial, producing a seesaw-shaped molecule.

Now consider an AX_4E_2 molecule, which has two lone pairs. It is best for the two lone pairs to lie opposite each other. Then the atoms attached to the remaining pairs will all lie in a plane pointing to the corners of a square. This shape is called square planar (Fig. 9.7).

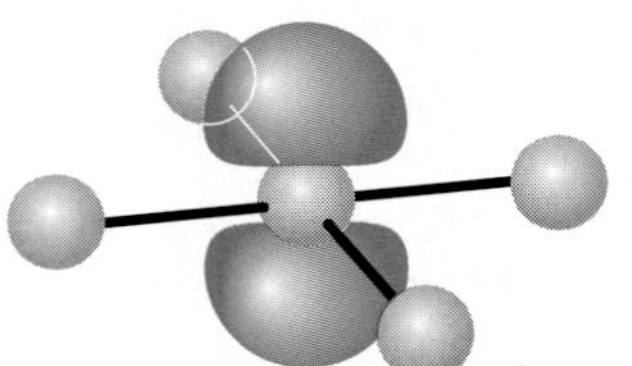

FIGURE 9.7 The square-planar arrangement of atoms taken up in AX_4E_2 molecules: the two lone pairs are farthest apart when they are on opposite sides of the central atom.

Lone pairs distort the shape of a molecule to reduce lone pair–bonding pair repulsions.

TOOLBOX 9.1 How to use the VSEPR model

The general procedure for predicting the shape of a molecule is as follows:

Step 1. Decide how many electron pairs are present on the central atom by writing a Lewis structure of the molecule. Treat a multiple bond as the equivalent of a single electron pair.

Step 2. Identify the arrangement of electron pairs (as in Fig. 9.4).

Step 3. Locate the atoms and classify the molecule (as in Fig. 9.1).

Step 4. Allow the molecule to distort so that lone pairs are as far from one another and from bonding pairs as possible. The repulsions are in the order

Lone pair-lone pair > lone pair-bonding pair > bonding pair-bonding pair

To predict the shape of a sulfur tetrafluoride molecule, SF_4, for example, we first write its Lewis structure (**22**) (recall Example 8.8). The molecule is an example of an AX_4E species, with four atoms on the central S atom and one lone pair. According to Fig. 9.4, the five electron pairs adopt a trigonal bipyramidal arrangement. As shown in (**23**), the lone pair can be either (a) axial or (b) equatorial, but the latter arrangement is likely to have the lower energy. The energy is lowered still further if the atoms move away from the lone pair to give the structure shown in (c). This shape (which resembles a slightly bent seesaw when viewed from the side) is the one found experimentally.

SELF-TEST 9.5A Predict the shape of a chlorine trifluoride molecule, ClF_3.

[*Answer:* T-shaped]

SELF-TEST 9.5B Predict the shape of a xenon tetrafluoride molecule, XeF_4.

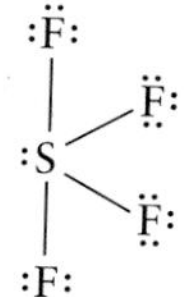

22 Sulfur tetrafluoride, SF_4

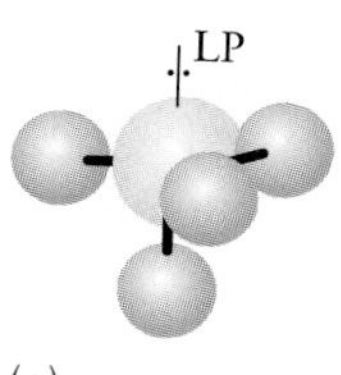

(a)

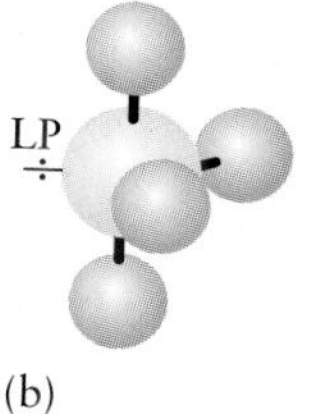

(b)

LP

(c)

23 Sulfur tetrafluoride, SF_4

It is important to be alert to the presence of lone pairs in molecules, because they help to determine the molecular shape. Do not conclude that every molecule with three atoms attached to a central atom is trigonal planar. Similarly, do not conclude that every molecule with four atoms attached to a central atom is tetrahedral. In each case, one or more lone pairs may affect the shape of the molecule. All AX_3 molecules (with no lone pairs) can be expected to have the same shape, as can all AX_3E molecules (with one lone pair) and all AX_3E_2 molecules (with two lone pairs).

Molecules with the same general formula all have the same general shape, although their bond angles generally differ slightly. For example, O_3 is an AX_2E species; it has a trigonal planar arrangement of electron pairs and bonds and an angular molecular shape (**24**). The nitrite ion, NO_2^-, has the same general formula and the same shape (**25**); so too does sulfur dioxide, SO_2 (**26**).

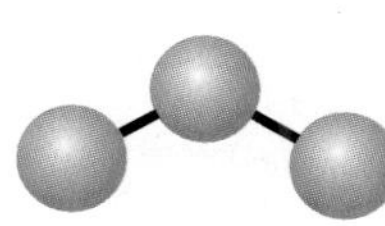

24 Ozone, O_3

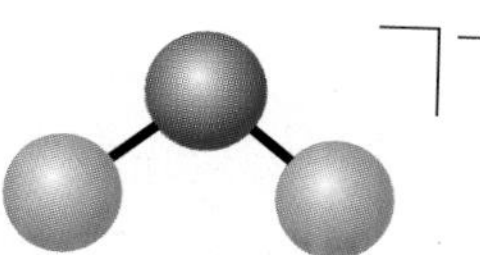

25 Nitrite ion, NO_2^-

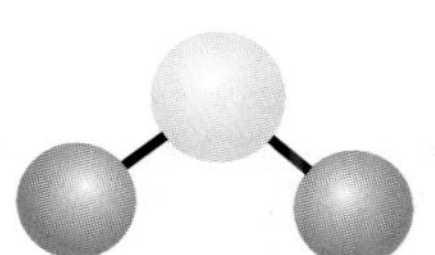

26 Sulfur dioxide, SO_2

To determine the shape of a molecule, write the Lewis structure, then identify the arrangement of electron pairs and bonds in which the lone pairs are farthest from each other and from bonds; name the molecular shape by considering only the locations of the atoms.

9.5 THE DISTORTING EFFECT OF LONE PAIRS

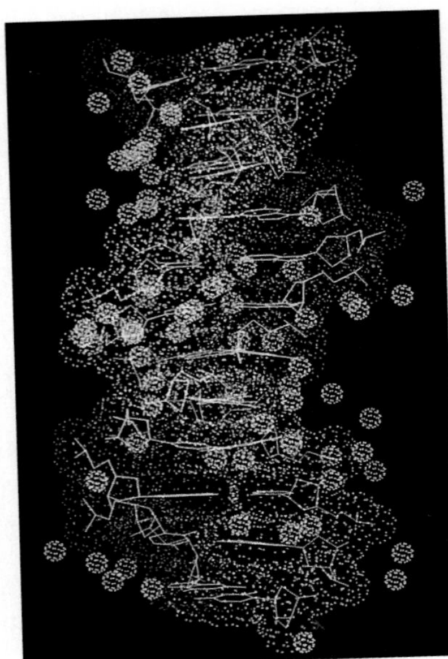

FIGURE 9.8 DNA molecules consist of two very long chains of four units that repeat in a specific order that determines our genetic makeup. The shapes of the units and their charge distributions cause the chains to wind around each other to form a double helix. The dots indicate electron concentration.

The Greek letter δ, delta, is often used to denote a small quantity.

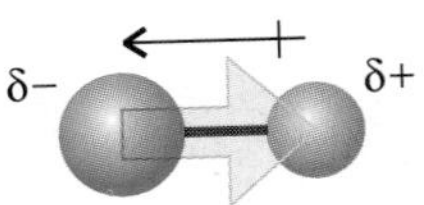

27 Dipole moment

In an older notation for dipoles, an arrow of the form ↔ is used: the + marks the positive end of the dipole.

The unit is named for the Dutch scientist Peter Debye, who made extensive investigations of dipole moments in the early twentieth century. The SI unit is the coulomb-meter (1 C·m).

CHARGE DISTRIBUTION IN MOLECULES

We need to know where electrons are most likely to be found in molecules if we want to understand a molecule's properties. Even life itself depends on the locations of electrons, because they control the shape of the DNA helix and the way it unwinds in the course of reproduction (Fig. 9.8). Electron distribution also controls the shapes of our individual proteins and enzymes, and shape is crucial to their function. In fact, when proteins lose their shape—for instance, when we suffer burns—they cease to function and we suffer tissue damage and possibly death. The same kind of information is essential for understanding other properties that we shall meet later in the text, such as the ability of water to dissolve ionic compounds.

9.6 POLAR BONDS

So far, we have assumed that an electron pair is shared equally in a covalent bond. However, the pair may be shared unequally. A covalent bond in which the electron pair is shared unequally is called a **polar covalent bond.** The extent to which an atom has a greater or lesser share of electrons in a bond is determined by its electronegativity (Section 8.13).

An O—H bond is polar because oxygen is more electronegative than hydrogen and gains a greater share in the bonding electron pair. Its greater share of electrons means that oxygen has a **partial negative charge,** which we denote δ−. Because the electron pair has been pulled away from the hydrogen atom, that atom has a **partial positive charge,** δ+. We show the partial charges on the atoms by writing $^{\delta+}H—O^{\delta-}$.

The more electronegative element in a polar covalent bond almost always has the negative partial charge. The only important exception is CO, which is $^{\delta-}CO^{\delta+}$. When the two atoms have only a small electronegativity difference, the partial charges are very small. As the difference in electronegativities increases, so do the partial charges. The electronegativities of carbon and hydrogen are 2.6 and 2.2 (see Fig. 8.15), and C—H bonds are best regarded as nonpolar.

The two atoms in a polar covalent bond are said to possess an **electric dipole,** a positive charge next to an equal but opposite negative charge. A dipole is represented by an arrow that points toward the positive partial charge (**27**). The size of an electric dipole—which is essentially a measure of the magnitude of the partial charges—is reported as the **electric dipole moment,** μ (the Greek letter mu), in units called **debye** (D). The debye is defined so that a single negative charge (an electron) separated by 100 pm from a single positive charge (a proton) has a dipole moment of 4.80 D. The dipole moment associated with an O—H bond is about 1.2 D. We can think of this dipole as arising from a partial charge of about 25% of an electron's charge on the O atom and an equivalent positive charge on the H atom.

A polar covalent bond is a bond between two atoms with partial electric charges arising from their difference in electronegativity. Partial charges give rise to an electric dipole moment.

TABLE 9.1 Dipole moments of selected molecules

Molecule	Dipole moment, D	Molecule	Dipole moment, D
HF	1.91	PH_3	0.58
HCl	1.08	AsH_3	0.20
HBr	0.80	SbH_3	0.12
HI	0.42	O_3	0.53
CO	0.12	CO_2	0
ClF	0.88	BF_3	0
NaCl*	9.00	CH_4	0
CsCl*	10.42	*cis*-CHCl=CHCl	1.90
H_2O	1.85	*trans*-CHCl=CHCl	0
NH_3	1.47		

*These two species consist of pairs of ions in the gas phase.

9.7 POLAR MOLECULES

A **polar molecule** is a molecule with a nonzero dipole moment. An HCl molecule, with its polar covalent bond ($^{\delta+}H—Cl^{\delta-}$), is a polar molecule, and its dipole moment is 1.1 D. That is a typical value for a polar diatomic molecule (Table 9.1). All diatomic molecules composed of atoms of different elements are at least slightly polar. A **nonpolar molecule** is a molecule that has zero electric dipole moment. All diatomic molecules built from the same atoms are nonpolar: a chlorine molecule is nonpolar because there are no partial charges on its atoms.

For polyatomic molecules, it becomes very important to distinguish between a polar *molecule* and a polar *bond*. Although each bond in a polyatomic molecule may be polar, the molecule *as a whole* will be nonpolar if the dipoles of the individual bonds cancel one another. For example, the two $^{\delta+}C—O^{\delta-}$ dipoles in carbon dioxide point in opposite directions, so they cancel each other (28). As a result, CO_2 is a nonpolar molecule even though its bonds are polar. In contrast, the two $^{\delta-}O—H^{\delta+}$ dipoles in H_2O lie at an angle and do not cancel each other, so H_2O is a polar molecule.

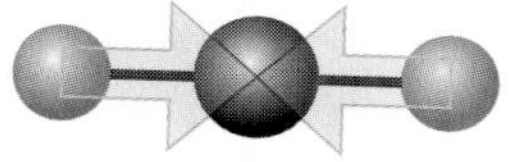

28 Carbon dioxide, CO_2

The shape of a molecule governs whether or not it is polar. The atoms and bonds are all the same in *cis*-dichloroethene (29) and *trans*-dichloroethene (30), but in the latter the C—Cl bonds point in opposite directions and the dipoles (represented by the arrows) cancel. Thus, whereas *cis*-dichloroethene is polar, *trans*-dichloroethene is nonpolar:

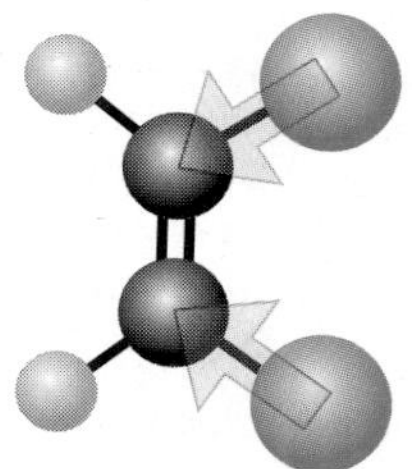

29 *cis*-Dichloroethene, $C_2H_2Cl_2$

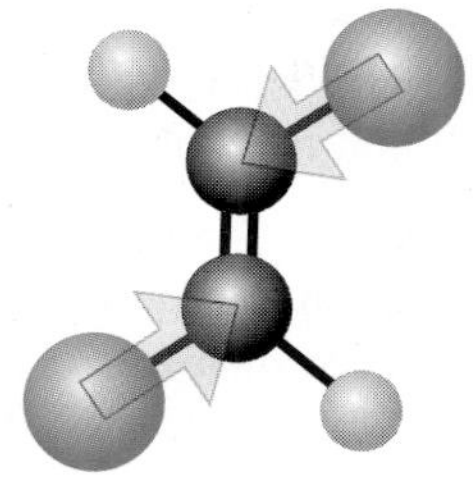

30 *trans*-Dichloroethene, $C_2H_2Cl_2$

FIGURE 9.9 Whether or not a liquid is composed of polar molecules can be shown by running a stream of liquid past a charged rod. (a) A nonpolar fluid (such as *trans*-dichloroethene) is undeflected, but (b) a polar fluid (*cis*-dichloroethene) is deflected.

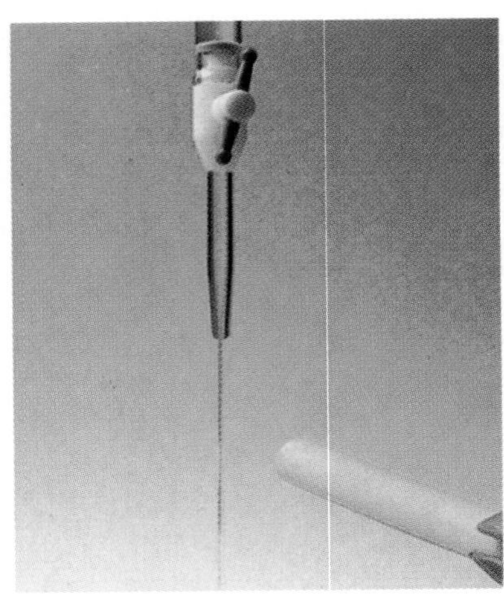

(a)

(b)

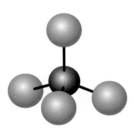

31 Tetrachloromethane, CCl_4

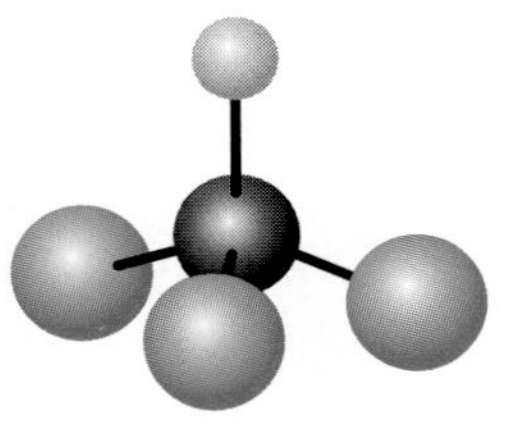

32 Trichloromethane, $CHCl_3$

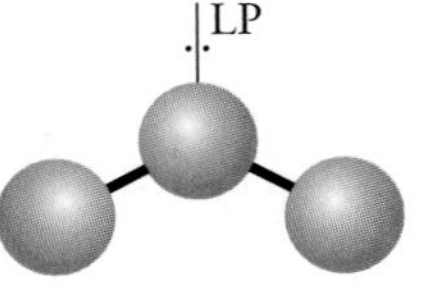

33 Ozone, O_3

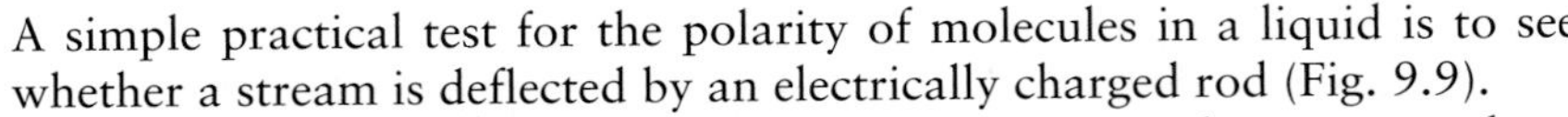

A simple practical test for the polarity of molecules in a liquid is to see whether a stream is deflected by an electrically charged rod (Fig. 9.9).

Tetrahedral molecules with the same atom at each corner, such as tetrachloromethane (carbon tetrachloride), CCl_4 (**31**), are nonpolar because the dipoles of the four bonds cancel in three dimensions. However, if one or two atoms are replaced by different atoms, as in trichloromethane (chloroform), $CHCl_3$, or by lone pairs, then the bond dipoles do not cancel and the molecule is polar (**32**). Examples of polar molecules of this kind are NH_3, PH_3, SO_2, and—in the notation used earlier in the chapter—any other AX_3E or AX_2E molecule. Highly symmetrical arrangements of polar bonds result in molecules that are nonpolar, because in these molecules the bond dipoles cancel (Fig. 9.10).

A diatomic molecule is polar if its bond is polar. A polyatomic molecule is polar if it has polar bonds arranged in space in such a way that their dipoles do not cancel.

EXAMPLE 9.4 Predicting the polarity of a molecule

Predict whether (a) a boron trifluoride molecule, BF_3, and (b) an ozone molecule, O_3, are polar.

STRATEGY In each case we must decide on the shape of the molecule by using the VSEPR model and then refer to Fig. 9.10. If the shape is not listed in the figure, decide whether the symmetry of the molecule allows the bond dipoles to cancel. If not, the molecule is polar.

SOLUTION (a) According to the VSEPR model, BF_3 is a trigonal planar AX_3 molecule. This shape is listed in Fig. 9.10 and classified there as nonpolar (see also **6**). (b) According to the VSEPR model, the O_3 molecule is angular, with one lone pair on the central O atom (**33**). Two positions around that central O atom are linked to O atoms, but the third holds a lone pair. Because one position is occupied by a lone pair, making it an AX_2E molecule, the dipoles do not

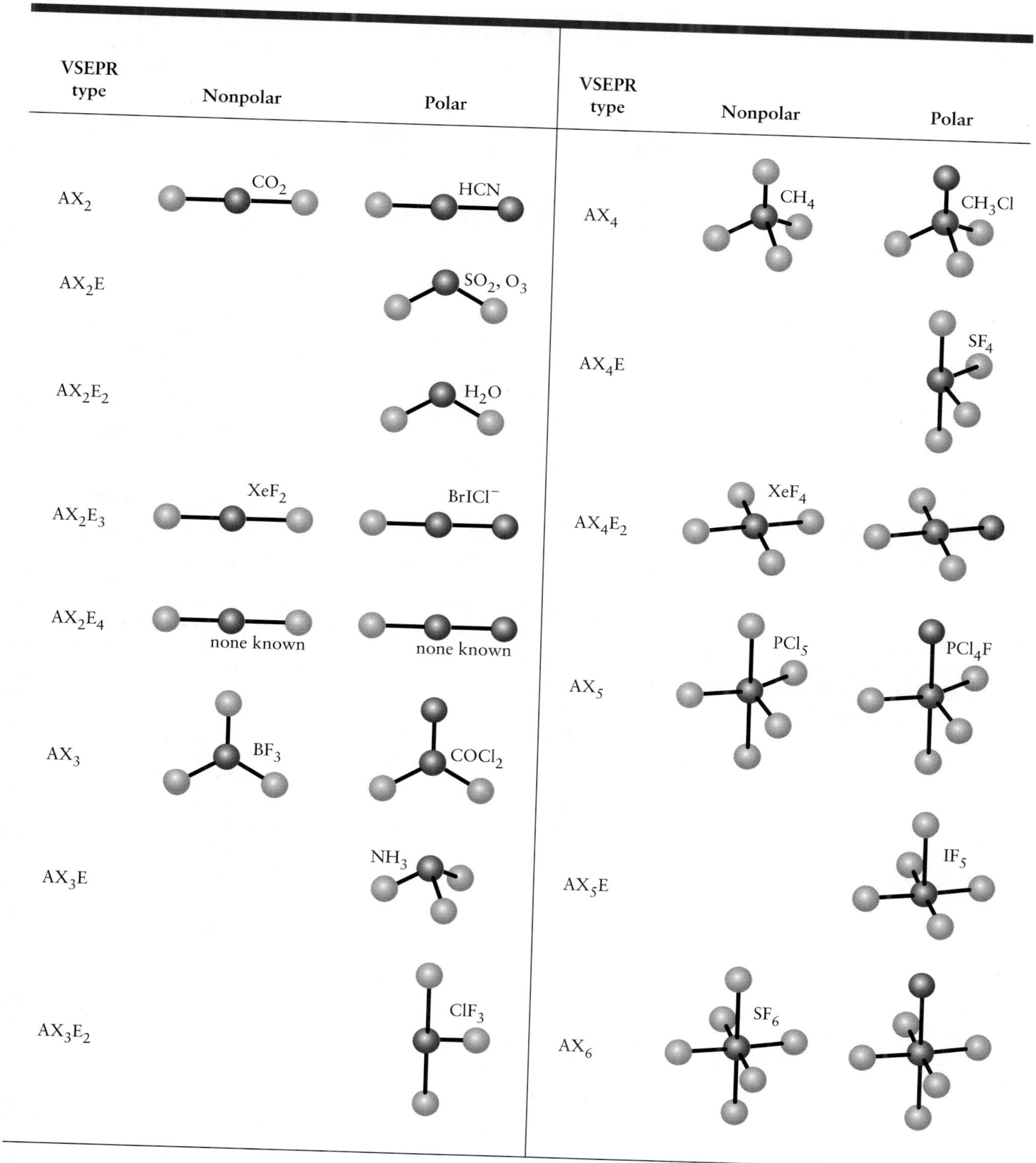

VSEPR type	Nonpolar	Polar
AX_2	CO_2	HCN
AX_2E		SO_2, O_3
AX_2E_2		H_2O
AX_2E_3	XeF_2	$BrICl^-$
AX_2E_4	none known	none known
AX_3	BF_3	$COCl_2$
AX_3E		NH_3
AX_3E_2		ClF_3

VSEPR type	Nonpolar	Polar
AX_4	CH_4	CH_3Cl
AX_4E		SF_4
AX_4E_2	XeF_4	
AX_5	PCl_5	PCl_4F
AX_5E		IF_5
AX_6	SF_6	

FIGURE 9.10 The arrangements of atoms that give rise to polar and nonpolar molecules. In the notation, A stands for a central atom, X for an attached atom, and E for a lone pair. Atoms of the same color are identical; X atoms colored differently represent different elements.

cancel; as a result, the molecule is polar. This example shows that a molecule can be polar even if all the atoms belong to the same element.

SELF-TEST 9.6A Predict whether (a) SF_6 and (b) SF_4 are polar or nonpolar.

[*Answer:* (a) Nonpolar; (b) polar]

SELF-TEST 9.6B Predict whether (a) PCl_5 and (b) IF_5 are polar or nonpolar.

THE STRENGTHS AND LENGTHS OF BONDS

Given any two atoms, we can say quite a lot about the bond between them. For instance, the length of the bond and its strength are always approximately the same, regardless of the actual molecule we are considering. For example, polytetrafluoroethylene (Teflon) is resistant to chemical attack. To understand why, we can study the character of C—F bonds in a much simpler compound, such as tetrafluoromethane, CF_4. We can be confident that the C—F bonds in the polymer are similar.

9.8 BOND STRENGTHS

The term *homolytic dissociation* is sometimes used to emphasize that each atom carries off one electron from the bonding pair.

The strength of a chemical bond is measured by the **bond enthalpy, ΔH_B,** the change in enthalpy that occurs when the bond in a gaseous molecule is broken. In all the cases we shall consider, we shall suppose that when a covalent bond is broken, each atom carries away one of the electrons in the bonding pair: bond breaking of this kind is called **dissociation** of the bond. Bond breaking always requires energy, so all bond enthalpies are positive. For example, the bond enthalpy of H_2 is obtained from the change in enthalpy for the dissociation

$$H_2(g) \longrightarrow 2\,H(g) \qquad \Delta H^\circ = +436\text{ kJ}$$

We write $\Delta H_B(\text{H—H}) = +436\text{ kJ·mol}^{-1}$ to report this value. A large, positive bond enthalpy means that the enthalpy of the separated atoms is much higher than that of the original molecule—in other words, a lot of energy has to be supplied to break the bond.

The greater the bond strength, the harder it is to break the bond. Typically, high bond strength indicates that the molecule is unlikely to be very reactive.

Bonds between different atoms have different strengths. Compare the dissociation of the C—O and O—H bonds of methanol. For the dissociation of the C—O bond,

$$CH_3\text{—O—H}(g) \longrightarrow CH_3(g) + \text{O—H}(g) \qquad \Delta H^\circ = +377\text{ kJ}$$

For the C—O bond, $\Delta H_B(CH_3\text{—OH}) = +377\text{ kJ·mol}^{-1}$. For the dissociation of the O—H bond,

$$CH_3\text{—O—H}(g) \longrightarrow CH_3\text{—O}(g) + H(g) \qquad \Delta H^\circ = +437\text{ kJ}$$

This equation shows that $\Delta H_B(CH_3O\text{—H}) = +437\text{ kJ·mol}^{-1}$. We see that it takes more energy to break an O—H bond in methanol than it does to break a C—O bond, so an O—H bond is stronger than a C—O single bond.

Single bonds between carbon and fluorine are much stronger than those between carbon and chlorine: $\Delta H_B(\text{F—CF}_3) = +452\text{ kJ·mol}^{-1}$ and $\Delta H_B(\text{Cl—CCl}_3) = +293\text{ kJ·mol}^{-1}$. The high bond enthalpy of a carbon-fluorine bond explains why polymers of fluorocarbons are very resistant

FIGURE 9.11 A Teflon coating is being applied to a gasket that will be used in an industrial chemical reactor. Because of its strong bonds, Teflon is very resistant to chemical attack.

TABLE 9.2 Bond enthalpies of diatomic molecules, kJ·mol^{-1}

Molecule	ΔH_B
H_2	436
N_2	944
O_2	496
CO	1074
F_2	158
Cl_2	242
Br_2	193
I_2	151
HF	565
HCl	431
HBr	366
HI	299

to chemical attack. They are used to construct valves for corrosive gases and to line the interiors of chemical reactors (Fig. 9.11). Polymers made from chlorocarbons are more reactive and can be destroyed by concentrated acids and bases, whereas Teflon cannot. Nevertheless, polyvinylchloride is sufficiently resistant to chemical attack that it is used to make drain pipes and coatings for electrical wires.

The strength of a bond between two atoms is measured by the bond enthalpy; all bond enthalpies are positive.

9.9 THE VARIATION OF BOND STRENGTH

Bond enthalpy values for typical diatomic molecules range from +151 kJ·mol^{-1} (for I_2) up to +1074 kJ·mol^{-1} (for CO, the strongest bond known) (Table 9.2).

Some of the trends shown in Table 9.2 are explained by the Lewis structures for the molecules. Consider, for example, the diatomic molecules of nitrogen, oxygen, and fluorine (Fig. 9.12). Note the decline in bond enthalpy as the number of bonds between the atoms decreases—from three in N_2 to one in F_2. A multiple bond is almost always stronger than a single bond because more electrons bind the multiply bonded atoms together.

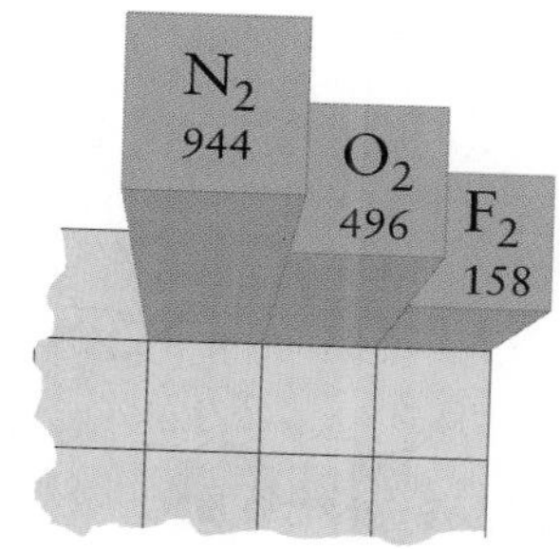

FIGURE 9.12 The bond enthalpies, in kilojoules per mole (kJ·mol^{-1}), of nitrogen, oxygen, and fluorine molecules. Note how the bonds weaken as they change from a triple bond in N_2 to a single bond in F_2.

In chemistry there is rarely a *single* reason for a trend in properties. The Lewis structure of a nitrogen molecule is :N≡N:, with one lone pair on each atom. The Lewis structure of fluorine, :F̤̈—F̤̈:, shows that it has three lone pairs on each atom. Lone pairs repel one another, even across the bond. Therefore, the energy needed to separate the F atoms in F_2 is less than that needed to separate the N atoms in N_2. This repulsion between lone pairs also explains why the bond in F_2 is weaker than the bond in H_2, for the latter molecule has no lone pairs.

Variations in atomic radii also contribute to the trend in bond strengths. If the nuclei of the bonded atoms cannot get very close to the electron pair lying between them, then the two atoms will be only weakly bonded together. For example, the bond enthalpies of the hydrogen halides decrease from HF to HI, as shown in Fig. 9.13. All the structures are singly bonded, and there are no lone pairs on hydrogen atoms. Because atomic radii increase from fluorine to iodine (see Fig. 7.27), the hydrogen atom is progressively farther away from the nucleus of the halogen atom, and the bond becomes weaker.

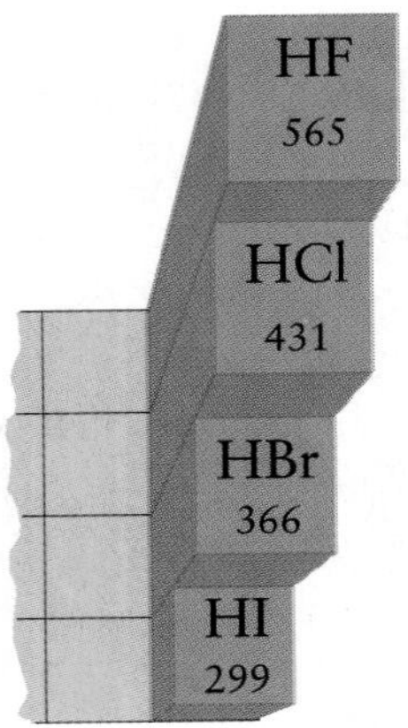

FIGURE 9.13 The bond enthalpies of the hydrogen halide molecules in kilojoules per mole ($kJ \cdot mol^{-1}$). Note how the bonds weaken as the halogen atom becomes larger.

The bond enthalpy increases as the number of bonds increases, decreases as the number of lone pairs on neighboring atoms increases, and decreases as the atomic radii increase.

9.10 BOND STRENGTHS IN POLYATOMIC MOLECULES

The first step in predicting the strength of a bond in a polyatomic molecule is to look at the identities of the atoms. However, all the other atoms in the molecule exert a pull—through their electronegativities—on all the electrons in the molecule (Fig. 9.14). As a result, the bond strength between a given pair of atoms varies slightly from one compound to another. Because these variations in strength are not very great, the **average bond enthalpy,** which we also denote ΔH_B, is a good guide to the strength of a bond in any molecule. Some average bond enthalpies are given in Table 9.3.

Assume that ΔH_B is the average bond enthalpy unless it is specified as the enthalpy of a specific bond.

As we can see from these average values, a triple bond between two atoms is always stronger than a double bond between the same two atoms, and a double bond is always stronger than a single bond between the same two atoms. However, a double bond between two carbon atoms is not twice as strong as a single bond, and a triple bond is a lot less than three times as strong:

TABLE 9.3 Average bond enthalpies, $kJ \cdot mol^{-1}$

Bond	Average bond enthalpy	Bond	Average bond enthalpy
C—H	412	C—I	238
C—C	348	N—H	388
C=C	612	N—N	163
C···C*	518	N=N	409
C≡C	837	N—O	210
C—O	360	N=O	630
C=O	743	N—F	195
C—N	305	N—Cl	381
C—F	484	O—H	463
C—Cl	338	O—O	157
C—Br	276		

*In benzene.

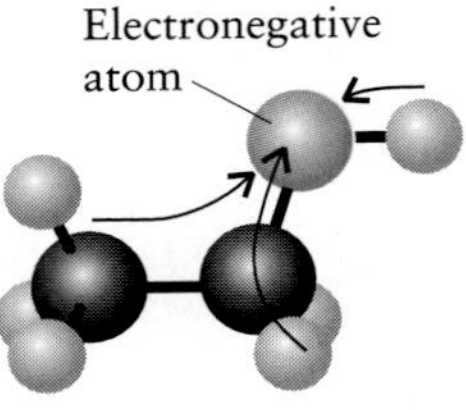

FIGURE 9.14 An electronegative atom can pull electrons toward itself from more distant parts of a molecule. Hence it can influence the strengths of bonds even between atoms to which it is not directly attached.

$$\Delta H_B(\text{C}{=}\text{C}) = +612\ \text{kJ·mol}^{-1} \qquad \Delta H_B(\text{C}{\equiv}\text{C}) = +837\ \text{kJ·mol}^{-1}$$

$$2 \times \Delta H_B(\text{C}{-}\text{C}) = +696\ \text{kJ·mol}^{-1} \qquad 3 \times \Delta H_B(\text{C}{-}\text{C}) = +1044\ \text{kJ·mol}^{-1}$$

Stronger bonds result in lower energy, more stable molecules, because it takes more energy to break strong bonds than weak bonds. Because it takes more energy to break two C—C bonds than one C=C bond, it can be energetically advantageous to replace multiple bonds by more single bonds. The electron pairs in a double bond repel each other slightly, so each pair is not quite as effective at bonding as a pair of electrons in a single bond (Fig. 9.15).

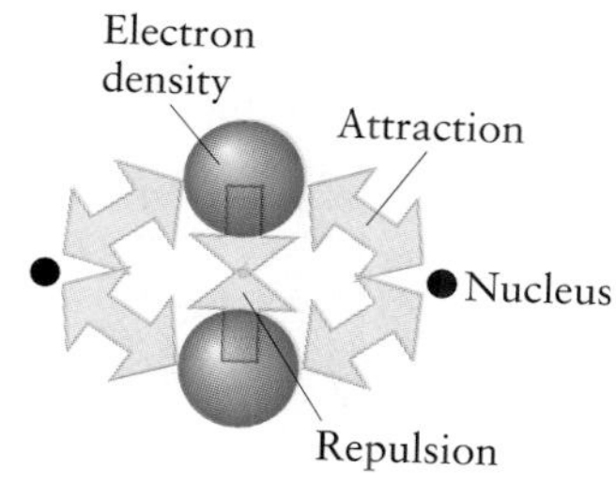

FIGURE 9.15 The two pairs of electrons in a multiple bond repel each other and can weaken the bond. As a result, a double bond between carbon atoms is not twice as strong as two single bonds would be.

For example, consider the reaction

$$\text{CH}_2{=}\text{CH}_2\text{(g)} + \text{HCl(g)} \longrightarrow \text{CH}_3{-}\text{CH}_2\text{Cl(g)} \qquad \Delta H = -72\ \text{kJ}$$

The negative value of ΔH tells us that the reaction is exothermic. The decrease in enthalpy means that the energy of the products is lower than that of the reactants. The reaction releases energy even though the H—Cl bond and the C=C bonds have to be broken before the C—C, C—H, and C—Cl bonds can form.

The values in Table 9.3 also show how resonance lowers the energy of a molecule. For example, the strength of the carbon-carbon bonds in benzene is intermediate between that of a single and a double bond. Benzene has a resonance structure, and because the strength of the resonance hybrid bond is closer to that of a double bond, a molecule with six hybrid bonds has a lower energy than one with three single and three double bonds.

Average bond enthalpies also help explain the relative stabilities of families of compounds. For example, the enthalpy of the bond between hydrogen and a Group 4 element decreases down the group, from carbon (+412 kJ·mol^{-1}) to lead (+205 kJ·mol^{-1}), as shown in Fig. 9.16. This

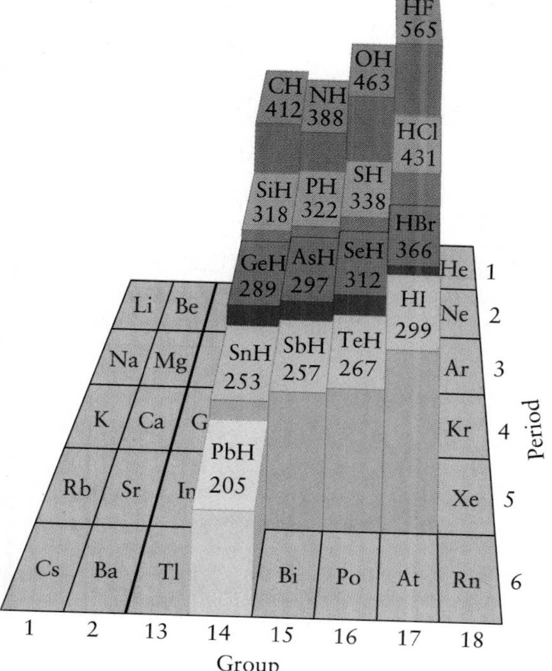

FIGURE 9.16 The bond enthalpies for bonds between hydrogen and the *p*-block elements. The bond strengths decrease from top to bottom of each group as the atoms increase in size.

weakening of the bond parallels a decrease in the stability of the hydrides down the group. Methane (CH_4) can be kept indefinitely in air at room temperature. Silane (SiH_4) bursts into flame on contact with air. Stannane (SnH_4) decomposes into tin and hydrogen. Plumbane (PbH_4) has never been prepared, except perhaps in trace amounts.

When accurate bond enthalpies are not available, reaction enthalpies can be estimated by using average bond enthalpies compiled from measurements on a series of related compounds.

EXAMPLE 9.5 Using average bond enthalpies for substitution reactions

Decide whether the reaction

$$CH_3CH_2I(g) + H_2O(g) \longrightarrow CH_3CH_2OH(g) + HI(g)$$

is exothermic or endothermic.

STRATEGY Decide which bonds are broken and which bonds are formed. Use the average bond enthalpies in Table 9.3 to calculate the change in enthalpy when the bonds are broken in the reactants and when the new bonds are formed to produce the products. For diatomic molecules, use the information in Table 9.2 for the specific molecule. Calculate the bond formation enthalpies from the bond dissociation enthalpies by reversing the sign. Finally, add the enthalpy change required to break the reactant bonds (a positive value) to the enthalpy change that occurs when the product bonds form (a negative value).

SOLUTION We need to break a CH_3CH_2—I bond (238 $kJ\cdot mol^{-1}$) and an HO—H bond (463 $kJ\cdot mol^{-1}$), so the total increase in enthalpy for the dissociation steps per mole of CH_3CH_2I that reacts is

$$\Delta H° = 238 + 463 \text{ kJ} = +701 \text{ kJ}$$

To break one CH_3CH_2—OH bond (360 $kJ\cdot mol^{-1}$) and one H—I bond (299 $kJ\cdot mol^{-1}$) gives a change in enthalpy of

$$\Delta H° = 360 + 299 \text{ kJ} = +659 \text{ kJ}$$

The enthalpy change when the product bonds are formed is therefore −659 kJ.

TABLE 9.4 Average and actual bond lengths

Bond	Average bond length, pm	Molecule	Bond length, pm
C—H	109	H_2	74
C—C	154	N_2	110
C=C	134	O_2	121
C⋯C*	139	F_2	142
C≡C	120	Cl_2	199
C—O	143	Br_2	228
C=O	112	I_2	268
O—H	96		
N—H	101		

*In benzene.

The overall enthalpy change is the sum of these two changes:

$$\Delta H° = 701 + (-659) \text{ kJ} = +42 \text{ kJ}$$

Therefore, the reaction is endothermic, primarily because a relatively large amount of energy is needed to break an O—H bond in water.

SELF-TEST 9.7A Which of the following reactions are exothermic?
(a) $CCl_3CHCl_2(g) + HF(g) \rightarrow CCl_3CHClF(g) + HCl(g)$
(b) $CCl_3CHCl_2(g) + HF(g) \rightarrow CCl_3CCl_2F(g) + H_2(g)$

[*Answer:* (a)]

SELF-TEST 9.7B Estimate the standard enthalpy of the reaction in which CH_4 reacts with gaseous F_2 to form gaseous CH_2F_2 and HF.

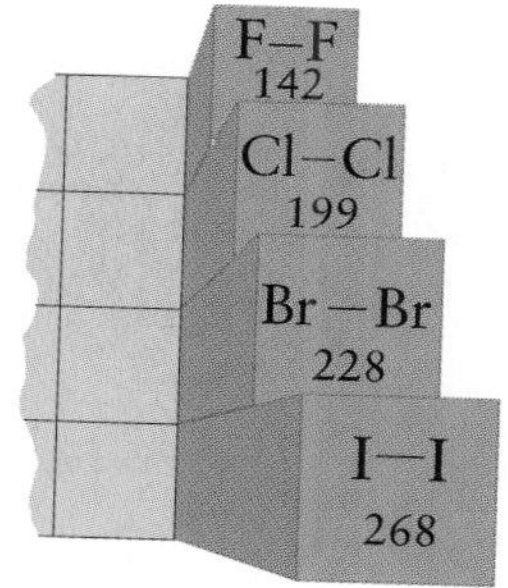

FIGURE 9.17 Bond lengths (in picometers) of the diatomic halogen molecules. Notice how the bond lengths increase from top to bottom of the group as the atomic radii become larger.

9.11 BOND LENGTHS

A **bond length** is the distance between the centers of two atoms joined by a chemical bond. Bond lengths help determine the overall shape of a molecule. The transmission of hereditary information in DNA, for instance, depends on bond lengths because the two strands of the double helix must fit together like pieces of a jigsaw puzzle (see Fig. 9.8). Bond lengths are also crucial to the action of enzymes, because only if a molecule is the right size—which includes having the right bond lengths—will it fit into the active site of the enzyme molecule (see Section 18.13). As we see from Table 9.4, the lengths of bonds between the elements of Period 2 typically lie in the range 100 ppm to 150 pm. Bond lengths are determined experimentally either by using spectroscopy or x-ray diffraction (see Box 10.1).

Bonds between heavy atoms tend to be longer than those between light atoms because heavier atoms have larger radii than lighter ones (Fig. 9.17). Multiple bonds are shorter than single bonds between the same pair of atoms because the additional bonding electrons pull the atoms closer together: compare the lengths of the various carbon-carbon bonds in Table 9.4. We can also see the averaging effect of resonance. The length of the bond in benzene is intermediate between the lengths of the single and double bonds of a Kekulé structure (but closer to that of a double bond). For bonds between the same pair of atoms, *the stronger*

CHEMISTRY AT WORK

A genetic researcher examines the sequence of individual units in a DNA molecule. She will alter that sequence so that the DNA will produce a chemical that destroys cancer cells (tumor necrosis factor). The DNA of white blood cells can then be altered in this way and used to treat cancer patients. Because the shapes of biological molecules are critical to their function, she must design enzyme molecules with specific shapes to control the splicing of the DNA.

the bond, the shorter it is. Thus, a C≡C triple bond is both stronger and shorter than a C=C double bond. Similarly, a C=O double bond is both stronger and shorter than a C—O single bond.

Each atom makes a characteristic contribution, called its **covalent radius,** to the length of a bond (Fig. 9.18). A bond length is approximately the sum of the covalent radii of the two atoms (**34**). The O—H bond length in ethanol, for example, is the sum of the covalent radii of H and O, 37 pm + 74 pm = 111 pm. We see from Fig. 9.18 that the covalent radius of an atom taking part in a multiple bond is smaller than that for a single bond of the same atom.

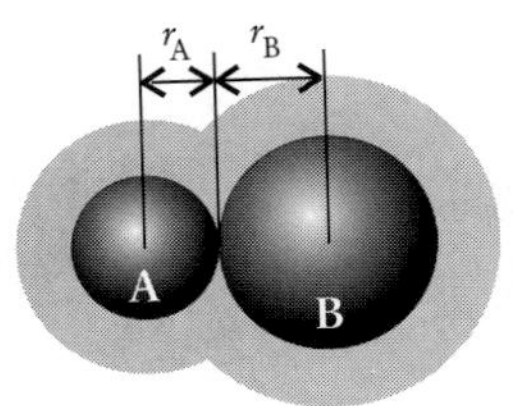

34 Covalent radii

Covalent single bond radii typically decrease from left to right across a period. The same trend is observed for atomic radii (Section 7.14), and the reason is the same: the increasing nuclear charge draws in the electrons and makes the atom more compact. Because the atom is smaller, it can get closer to any atom to which it bonds. Again like atomic radii, covalent radii increase down a group because, in successive periods, the valence electrons occupy shells that are more distant from the nucleus and better shielded by a larger inner core of electrons. Such fatter atoms cannot approach their neighbors very closely; hence they form long, weak bonds.

The covalent radius of an atom is the contribution it makes to the length of a covalent bond; covalent radii are added together to estimate the lengths of bonds in molecules.

SELF-TEST 9.8A Compare the bond enthalpies of the N—O and N—H bonds. Which bond would you expect to be shorter? Why?

[*Answer:* N—H. It is a stronger bond and H has a smaller atomic radius than O.]

SELF-TEST 9.8B Which bond would you expect to be shorter, N—O or N=O? Why?

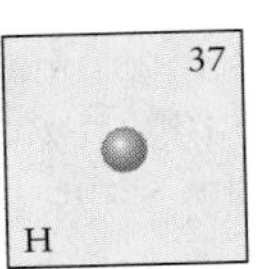

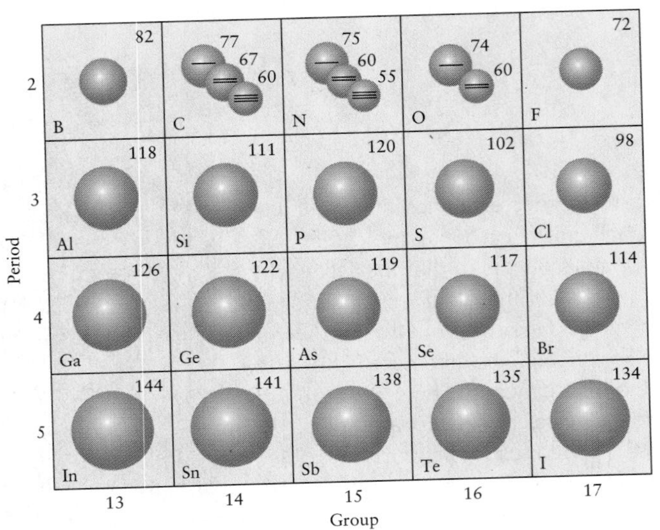

FIGURE 9.18 Covalent radii of hydrogen and the *p*-block elements (in picometers). Where more than one value is given, they refer to single, double, and triple bonds. Covalent radii tend to become smaller toward fluorine. A bond length is approximately the sum of the covalent radii of the two atoms involved.

FIGURE 9.19 When electrons (depicted as ↑ and ↓) in two hydrogen 1*s*-orbitals pair and the *s*-orbitals overlap, they form a σ-bond, which is depicted here by the boundary surface of the electron cloud. The cloud has cylindrical symmetry around the internuclear axis and spreads over both nuclei.

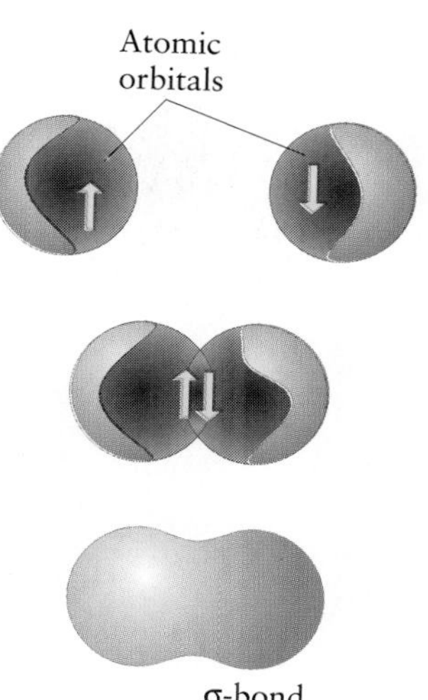

ORBITALS AND BONDING

The Lewis model of the chemical bond assumes that each bonding electron pair is located between the two bonded atoms. However, we know from Chapter 7 that the location of an electron in an atom cannot be described in terms of a precise position, but only in terms of the probability of finding it somewhere in a region of space called an orbital. The description of covalent bonding in terms of atomic orbitals is called **valence-bond theory.**

9.12 SIGMA AND PI BONDS

Let's begin with the simplest molecule of all, H_2. A single hydrogen atom consists of an electron in a spherical 1*s*-orbital that surrounds the nucleus. When two H atoms come together to form H_2, their 1*s*-electrons pair (denoted ⇅ in the discussion of atomic structure), and the atomic orbitals they occupy merge together (Fig. 9.19). The resulting sausage-shaped distribution of electrons is called a **σ-bond**. A hydrogen molecule is held together by one such σ-bond. The merging of the 1*s*-orbitals is called the **overlap** of atomic orbitals.

Much the same kind of σ-bond formation ("σ-bonding") occurs in the hydrogen halides. For example, the electron configurations of hydrogen and fluorine are shown in (**35**): the notation is the same as that used in Section 7.11.

As explained in Section 7.7, the symbols ↑ and ↓ represent the spin states of electrons.

The Greek letter σ, sigma, is the equivalent of our letter *s*. It reminds us that, looking along the internuclear axis, the electron distribution resembles that of an *s*-orbital. All σ-bonds will be colored blue.

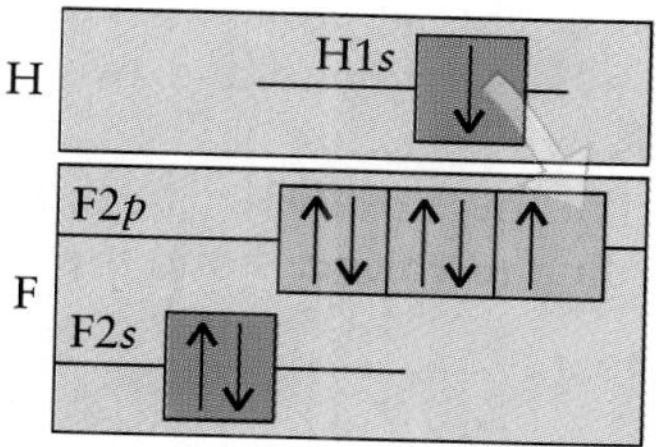

35 Hydrogen fluoride, HF

The unpaired electron on the fluorine atom occupies a $2p_z$-orbital, and the unpaired electron on the hydrogen atom occupies a 1*s*-orbital. These two electrons are the ones that pair to form a bond. They pair as the orbitals they occupy overlap and merge into a cloud that spreads over both atoms (Fig. 9.20). Although the resulting bond has a more complicated shape than the σ-bond in H_2 when viewed from the side, it looks much the same when viewed along the internuclear (z) axis; hence it is also designated a σ-bond.

We identify the bonding orbital on the F atom as a p_z-orbital only by convention. We usually define the *z*-axis as the one lying along the internuclear axis.

All *single* covalent bonds consist of a σ-bond in which two paired electrons lie between the two bonded atoms. A σ-bond can be formed by the pairing of electrons in two *s*-orbitals (as in H_2), an *s*-orbital and a

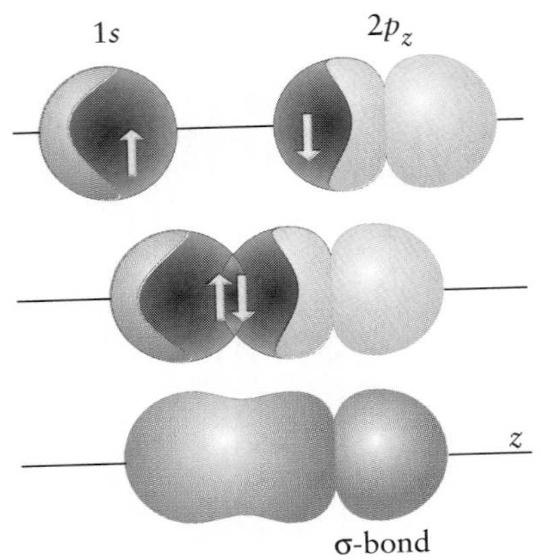

FIGURE 9.20 A σ-bond can also be formed when electrons in 1*s*- and $2p_z$-orbitals pair (*z* is the direction along the internuclear axis). The two electrons in the bond are concentrated within the entire region of space surrounded by the boundary surface.

The Greek letter pi, π, is the equivalent of our letter *p*. When we imagine looking along the internuclear axis, a π-bond resembles a pair of electrons in a *p*-orbital. All π-bonds will be colored yellow.

p-orbital (as in the hydrogen halides), or two *p*-orbitals (as in a diatomic halogen molecule).

We meet a different type of bond when we consider the structure of a nitrogen molecule, N_2. Suppose we follow the same procedure as before. First, we write the atomic configurations, as shown in (**36**).

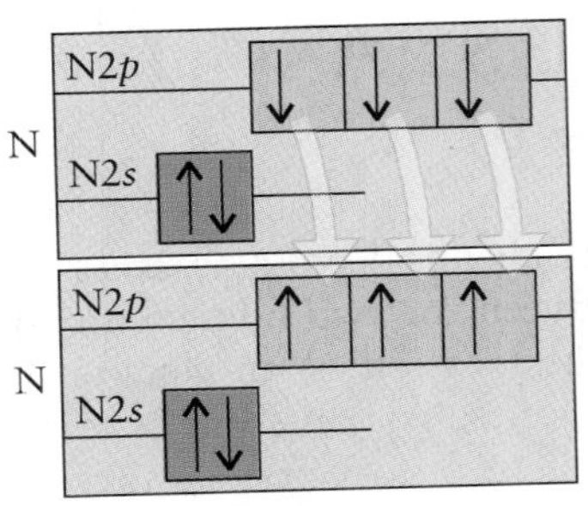

36 Nitrogen, N_2

There is an electron in each of the three 2*p*-orbitals on each atom. However, when we try to pair them and form three bonds, only one of the three orbitals can overlap end-to-end to form a σ-bond (Fig. 9.21). Two of the 2*p*-orbitals on each atom ($2p_x$ and $2p_y$) are perpendicular to the internuclear axis, and each one contains an unpaired electron (Fig. 9.22, top). When the remaining 2*p*-electrons pair, their orbitals can overlap only in a side-by-side arrangement and form a **π-bond**, a bond in which the two electrons lie in two lobes, one on each side of the internuclear axis (Fig. 9.22, bottom). Although a π-bond has electron density on each side of the internuclear axis, it is only *one* bond with the electron cloud in the form of *two* lobes, just as a *p*-orbital is one orbital with two lobes.

Now we can describe bonding in the nitrogen molecule. A σ-bond (of two electrons) is formed from the end-to-end overlap of two $2p_z$-orbitals directed along the internuclear axis, and two π-bonds (of two electrons each) are formed from the side-by-side overlap of the remaining 2*p*-orbitals on the two atoms. This approach leads to the conclusion that there are three bonds in all; one σ-bond and two π-bonds (Fig. 9.23), in agreement with the Lewis structure :N≡N:.

We can generalize from these examples to the description of a multiply bonded species according to valence-bond theory:

A **single bond** is a σ-bond.

A **double bond** is a σ-bond plus one π-bond.

A **triple bond** is a σ-bond plus two π-bonds.

When using valence-bond theory, first identify the valence-shell atomic orbitals that contain unpaired electrons, then allow these electrons to pair and the atomic orbitals they occupy to overlap end-to-end to form σ-bonds or side-by-side to form π-bonds.

SELF-TEST 9.9A How many σ-bonds and how many π-bonds are there in (a) CO_2? (b) CO?

[*Answer:* (a) Two σ, two π; (b) one σ, two π]

SELF-TEST 9.9B How many σ-bonds and how many π-bonds are there in (a) NH_3? (b) HCN?

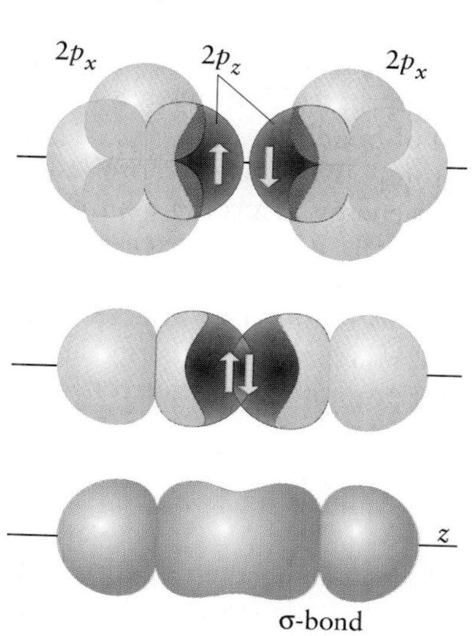

FIGURE 9.21 A σ-bond is formed by the pairing of electron spins in two $2p_z$-orbitals on neighboring atoms. At this stage we are ignoring the effect of the $2p_x$- (and $2p_y$-) orbitals that may also contain unpaired electrons. The electron pair is concentrated within the boundary surface shown in the bottom diagram. Although the shape is quite complex, the bond has cylindrical symmetry.

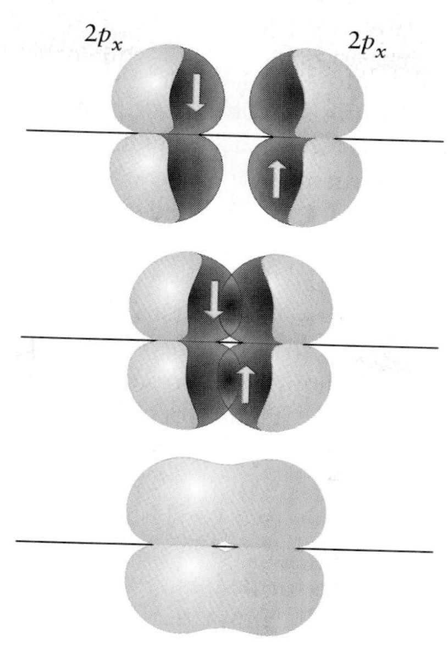

FIGURE 9.22 A π-bond is formed when electrons in two 2*p*-orbitals pair and overlap side-by-side. The middle diagram shows the region in which the resulting electron cloud is concentrated, and the bottom diagram shows the corresponding boundary surface. Even though the bond has a complicated shape, with two lobes, it is occupied by one pair of electrons and counts as one bond.

9.13 HYBRIDIZATION OF ORBITALS

The valence-bond model as described so far cannot account for bonding in polyatomic molecules like methane, CH_4. For example, if we tried to apply valence-bond theory to methane, we might write down the electron

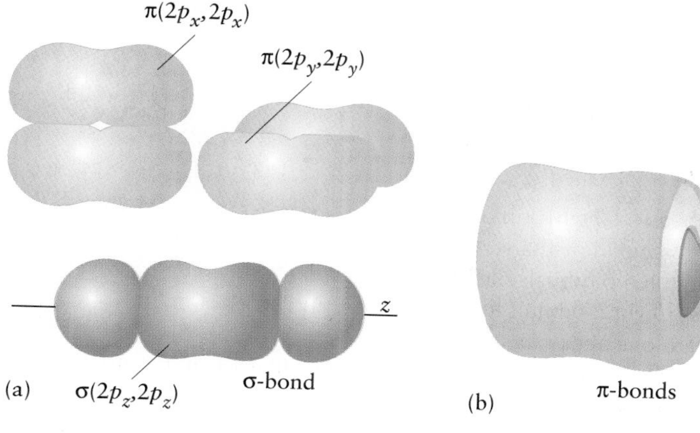

FIGURE 9.23 The bonding pattern in a nitrogen molecule, N_2. (a) The two atoms are bonded together by one σ-bond (blue) and two perpendicular π-bonds (yellow). (b) When the three orbitals are put together, the two π-bonds merge to form a long donut-shaped cloud surrounding the σ-bond cloud, so the overall structure resembles a cylindrical hot dog.

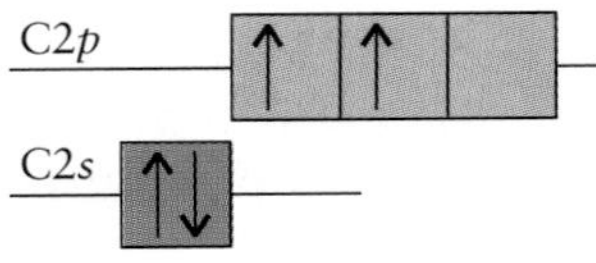

37 Carbon, [He]$2s^2 2p_x^1 2p_y^1$

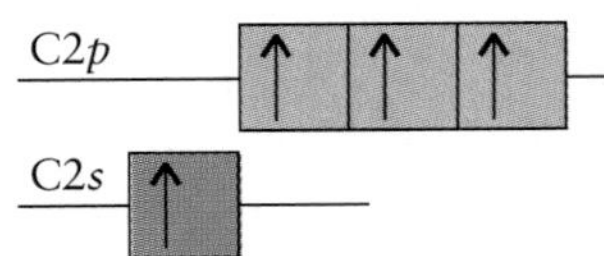

38 Carbon, [He]$2s^1 2p_x^1 2p_y^1 2p_z^1$

configuration of the carbon atom and look for half-filled orbitals that we can use for bond formation (**37**). However, there are only two half-filled 2*p*-orbitals on the carbon atom. The other valence electrons are already paired. It looks as though carbon can form only two bonds and should have a valence of 2. However, we know that carbon almost always has a valence of 4, so we need to revise our model of bonding.

Notice that carbon has an empty *p*-orbital. We can get more half-filled orbitals in carbon if we invest enough energy to **promote** an electron, that is, excite it to a higher energy orbital. When we excite a 2*s*-electron into an empty 2*p*-orbital, we get the configuration shown in (**38**). Without promotion, a carbon atom can form only two bonds; after promotion, it can form four bonds. Each bond releases energy as it forms. Therefore, despite the energy cost of promoting the electron, the overall energy of the CH_4 molecule is lower than it would be if carbon formed only two C—H bonds. Promotion of an electron is possible if the overall change, taking account of the greater number of bonds that can thereby be formed, is toward lower energy.

We still cannot explain methane's tetrahedral shape and its four identical bonds. It looks as though we should obtain two different types of bonds in CH_4: one from the overlap of a hydrogen 1*s*-orbital and a carbon 2*s*-orbital, and three more bonds from the overlap of hydrogen 1*s*-orbitals with each of the three carbon 2*p*-orbitals. The overlap with the 2*p*-orbitals should result in three σ-bonds at 90° to one another.

A hybrid orbital is a blend of atomic orbitals on a single atom in a molecule. Do not confuse this use of the term *hybrid* with resonance hybrid, which refers to a blend of Lewis structures of entire molecules.

To modify this description, we have to visualize *s*- and *p*-orbitals as waves of electron density centered on the nucleus of an atom. Like waves in water, the four orbitals produce new patterns where they intersect; these new patterns are called **hybrid orbitals,** or mixed orbitals. The four hybrid orbitals are identical and point toward the corners of a tetrahedron (Fig. 9.24). Each orbital has a node close to the nucleus and a small "tail" on the other side where the *s*- and *p*-orbitals do not completely cancel.

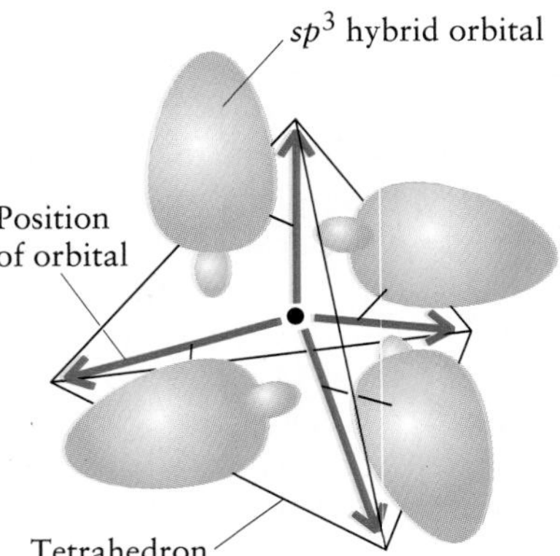

FIGURE 9.24 The hybrid orbitals of a carbon atom in methane. One *s*-orbital and three *p*-orbitals blend into four sp^3 hybrid orbitals that each point toward one apex of a tetrahedron. For clarity, the orbitals are "exploded" away from the center.

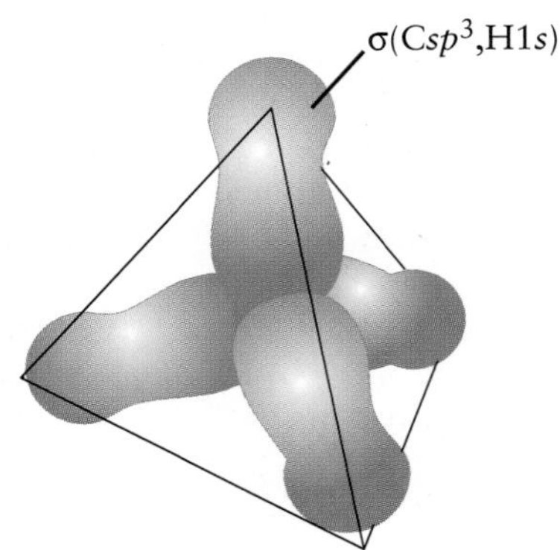

FIGURE 9.25 Each C—H bond in methane is formed by the pairing of an electron in a hydrogen 1*s*-orbital and an electron in one of the four sp^3 hybrid orbitals of carbon. As a result, there are four equivalent σ-bonds in a tetrahedral arrangement.

The four hybrid orbitals in methane are called **sp^3 hybrids** because they are formed from one *s*-orbital and three *p*-orbitals. In an orbital-energy diagram, we represent the hybridization as the formation of four orbitals of equal energy intermediate between the energies of the *s*- and *p*-orbitals from which they are constructed (**39**). The hybrids are colored green to remind us that they are a blend of (blue) *s*-orbitals and (yellow) *p*-orbitals.

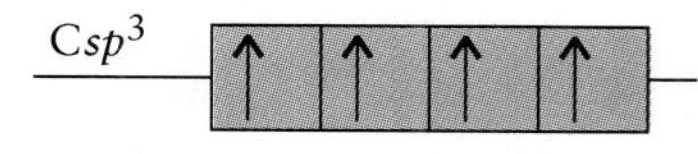

39 sp^3 hybridized carbon

With our revised bonding model, we can now account for the bonding in a methane molecule. An unpaired electron occupies each of carbon's sp^3 hybrid orbitals. Each of these four electrons can pair with an electron in a hydrogen 1*s*-orbital. Their overlapping orbitals form σ-bonds (Fig. 9.25). Because the four hybrid orbitals point toward the corners of a tetrahedron, so do the σ-bonds. All four bonds are identical because they are formed from the same blend of atomic orbitals. The valence-bond description is now consistent with experiment.

It is important to realize that methane is not tetrahedral *because* carbon has sp^3 hybrid orbitals. Hybridization is only a theoretical way of describing the bonds that are needed for a given molecular structure. Hybridization is an interpretation of molecular shape; shape is not a consequence of hybridization.

Hybrid orbitals are constructed on an atom to reproduce the electron arrangement characteristic of the experimentally determined shape of a molecule.

9.14 HYBRIDIZATION IN MORE COMPLEX MOLECULES

We can also use hybridization to describe bonds in molecules in which there is no single central atom. All we have to do is concentrate on the atoms one at a time. Let's use this technique to discuss ethane.

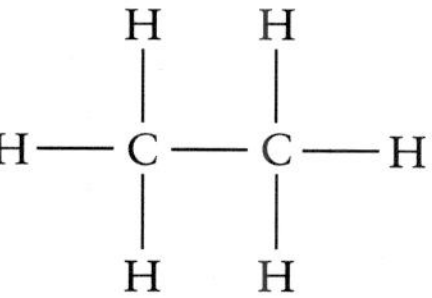

40 Ethane, CH_3CH_3

The Lewis structure of ethane is shown in (**40**). According to the VSEPR model, the four electron pairs around each atom take up a tetrahedral arrangement. This arrangement suggests sp^3 hybridization of the carbon atoms. Each C atom has one unpaired electron in each of its four sp^3 hybrid orbitals and, as a result, can form four σ-bonds pointing toward the corners of a regular tetrahedron. We denote the C—C bond $\sigma(Csp^3, Csp^3)$ to show its composition: Csp^3 denotes an sp^3 hybrid orbital on a carbon atom, and the parentheses show which orbitals overlap to give the σ-bond. The remaining electrons in the three hybrid orbitals on each C atom can be used to form σ-bonds by pairing with the electrons in the 1*s*-orbitals of the six H atoms (denoted H1*s*). The pattern of bonds is illustrated in Fig. 9.26.

The bonding pattern in ammonia illustrates how to describe a molecule with a lone pair. The four electron pairs in NH_3 take up a tetrahedral arrangement (according to the VSEPR model), so we describe the nitrogen atom in terms of four sp^3 hybrid orbitals. However, because nitrogen has five valence electrons, one of these hybrid orbitals is already doubly occupied (**41**). The 1*s*-electrons of the three hydrogen atoms pair with the three unpaired electrons in the remaining sp^3 hybrid orbitals. This pairing and overlap results in the formation of three σ-bonds. The lone pair on the nitrogen is said to occupy a **nonbonding orbital.**

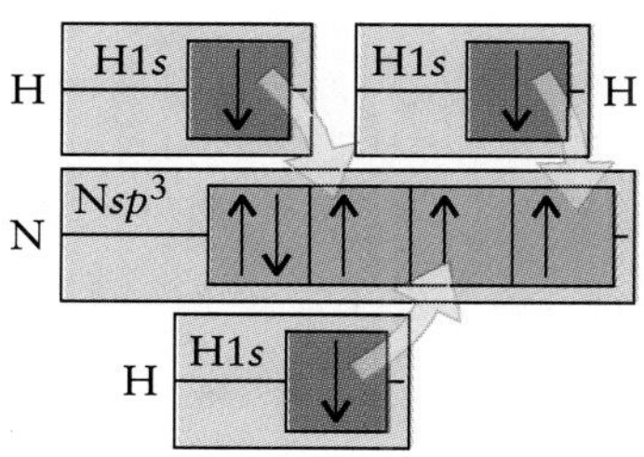

41 Ammonia, NH_3

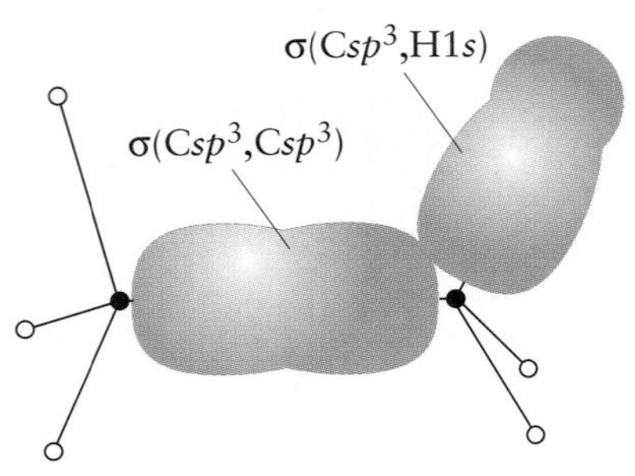

FIGURE 9.26 The valence-bond description of the bonding in an ethane molecule, C_2H_6. Only two of the bonds are shown in terms of their boundary surfaces. Each pair of neighboring atoms is linked by a σ-bond formed by the pairing of electrons in either H1*s*-orbitals or C*sp*^{3} hybrid orbitals. All the bond angles are close to 109.5° (the tetrahedral angle).

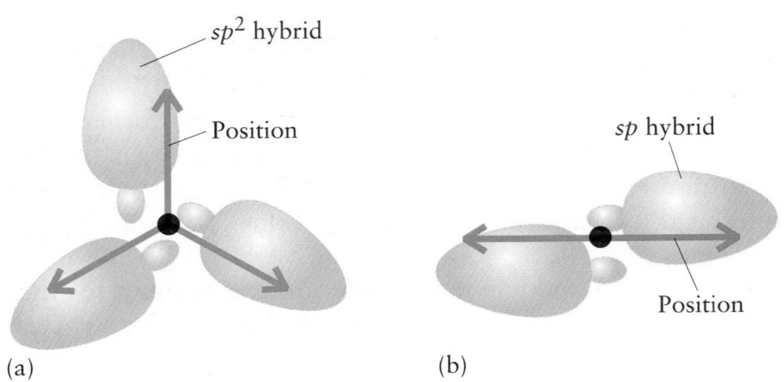

FIGURE 9.27 Two common hybridization schemes. (a) An *s*- and two *p*-orbitals can blend together to give three *sp*^{2} hybrid orbitals that point toward the corners of an equilateral triangle. (b) An *s*-orbital and a *p*-orbital hybridize into two *sp* hybrid orbitals that point in opposite directions. In each case, the hybrid orbital is shown alongside the arrow indicating its location and direction.

We use different hybridization schemes to describe molecules with shapes other than tetrahedral. For example, to explain a trigonal planar structure, we say that mixing one *s*-orbital with two *p*-orbitals results in three ***sp*^{2} hybrid orbitals.** These hybrid orbitals point toward the corners of an equilateral triangle (Fig. 9.27a). We describe a linear structure as consisting of two ***sp* hybrid orbitals**, the result of mixing one *s*-orbital and one *p*-orbital (Fig. 9.27b). The three most important types of hybridization in chemistry are

Tetrahedral hybridization: four *sp*^{3} hybridized orbitals, angle 109.5°

Trigonal-planar hybridization: three *sp*^{2} hybridized orbitals, angle 120°

Linear hybridization: two *sp* hybridized orbitals, angle 180°

The notation *sp*^{3} tells us that each of four hybrid orbitals is built from one *s*-orbital and three *p*-orbitals. The notation *sp*^{2} tells us that each of three hybrid orbitals is built from one *s*-orbital and three *p*-orbitals, leaving one unhybridized *p*-orbital. An atom with *sp* hybridization has two hybrid orbitals, each built from one *s*- and one *p*-orbital, and two unhybridized *p*-orbitals. No matter how many atomic orbitals we mix together, the number of hybrid orbitals is always the same as the number of atomic orbitals we started with. Other characteristic hybridization schemes are listed in Table 9.5.

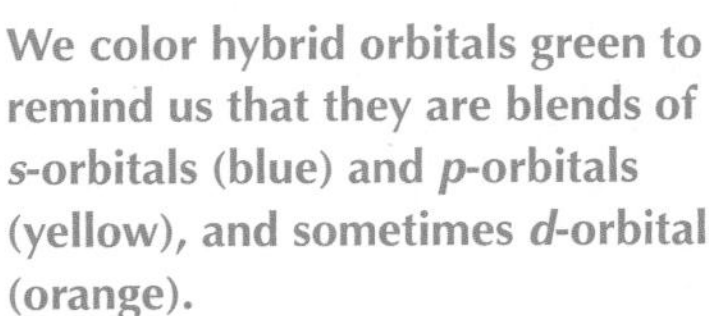
We color hybrid orbitals green to remind us that they are blends of *s*-orbitals (blue) and *p*-orbitals (yellow), and sometimes *d*-orbitals (orange).

TABLE 9.5 Hybridization and molecular shape

Number of atomic orbitals blended	Pattern of hybrid orbitals	Hybridization type	Number of hybrid orbitals around the central atom
2	linear	*sp*	2
3	trigonal planar	*sp*^{2}	3
4	tetrahedral	*sp*^{3}	4
5	trigonal bipyramidal	*dsp*^{3}	5
6	octahedral	*d*^{2}*sp*^{3}	6

In valence-bond theory, bonds are built by pairing electrons in orbitals on neighboring atoms; a hybridization scheme is adopted to match the shape of the molecule and the arrangement of its electron pairs.

SELF-TEST 9.10A Suggest a structure in terms of hybrid orbitals for BF_3.
[*Answer:* Three σ-bonds formed from Bsp^2 hybrids and F$2p_z$-orbitals in a trigonal-planar arrangement.]

SELF-TEST 9.10B Suggest a structure in terms of hybrid orbitals for dichloromethane, CH_2Cl_2.

9.15 HYBRIDS INCLUDING *d*-ORBITALS

An atom of an element in Period 3 or later can accommodate more than four electron pairs by using its *d*-orbitals. When five pairs of valence electrons are present on a central atom, five valence-shell orbitals are needed. In this case, we visualize an atom that uses one *d*-orbital in addition to the four *s*- and *p*-orbitals of the valence shell. If we need to account for a trigonal-bipyramidal arrangement of electron pairs, then Table 9.5 tells us that we should hybridize these five orbitals. The resulting orbitals are called dsp^3 hybrid orbitals (Fig. 9.28). We use this hybridization when we want to describe a trigonal-bipyramidal molecule, such as PCl_5.

To accommodate six electron pairs around an atom, we need six orbitals, so we use two *d*-orbitals in addition to the *s*- and *p*-orbitals. Table 9.5 shows that we should form six d^2sp^3 hybrid orbitals from these six atomic orbitals, because these hybrids point toward the corners of a regular octahedron (Fig. 9.29). We use this hybridization when we want

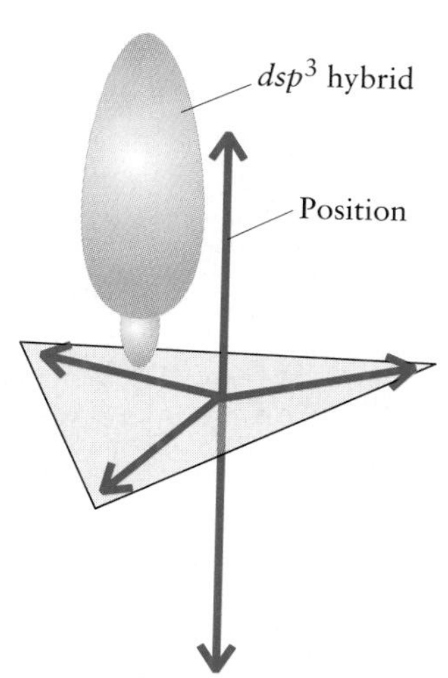

FIGURE 9.28 One of the five dsp^3 hybrid orbitals, and their five directions, that may be formed when *d*-orbitals are also available and we need to reproduce a trigonal-bipyramidal arrangement of electron pairs. Each arrow represents the location of a bond or electron pair.

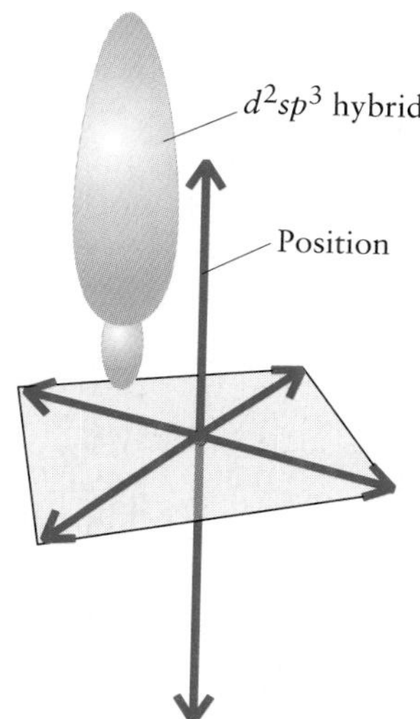

FIGURE 9.29 One of the six d^2sp^3 hybrid orbitals, and their six directions, that may be formed when *d*-orbitals are also available and we need to reproduce an octahedral arrangement of electron pairs. Each arrow represents the location of a bond or electron pair.

to describe a molecule based on an octahedral arrangement of electron pairs, such as SF_6 or XeF_4.

Octet expansion corresponds to the involvement of d-orbitals, which mix with s- and p-orbitals to give hybridization schemes that are consistent with the predictions of the VSEPR model and experiment.

SELF-TEST 9.11A Describe the molecular shape and the hybridization of the central atom in phosphorus pentachloride, PCl_5.

[*Answer:* Trigonal bipyramidal; five dsp^3 hybrid orbitals]

SELF-TEST 9.11B Describe the molecular shape and the hybridization of the central atom in sulfur hexafluoride, SF_6.

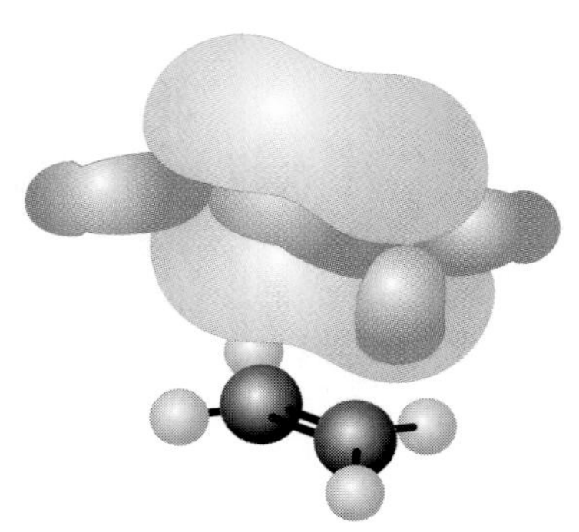

FIGURE 9.30 A view of the bonding pattern in ethene (ethylene), showing the framework of σ-bonds and the single π-bond formed by side-to-side overlap of unhybridized C2*p*-orbitals. The double bond is resistant to twisting because twisting would reduce the overlap between the two C2*p*-orbitals and weaken the π-bond.

9.16 MULTIPLE CARBON-CARBON BONDS

To describe carbon-carbon double bonds, we use the pattern provided by ethene, $CH_2{=}CH_2$. We know from experimental data that all six atoms in ethene lie in the same plane, with H—C—H and C—C—H bond angles of 120°. We need to use sp^2 hybridization to recreate the electron distribution this angle suggests. There is one electron in each of the three hybrid orbitals. The fourth valence electron of each atom must therefore occupy an unhybridized 2*p*-orbital to give the arrangement in (**42**). The unhybridized 2*p*-orbital lies perpendicular to the plane formed by the hybrids (Fig. 9.30). The two carbon atoms form a σ-bond by overlap of an sp^2 hybrid orbital on each atom. The H atoms form σ-bonds with the remaining lobes of the sp^2 hybrids. This arrangement of orbitals leaves the electrons in the two unhybridized 2*p*-orbitals free to pair and form a π-bond by side-to-side overlap.

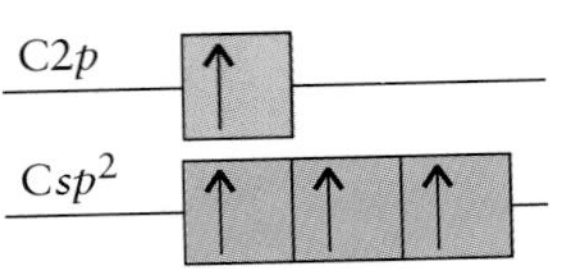

42 sp^2 hybridized carbon

To describe the bonding in the Kekulé structures of benzene, we need hybrid orbitals that match the 120° bond angles of the hexagonal ring. Therefore, we take each carbon atom to be sp^2 hybridized as in ethene (Fig. 9.31). There is one electron in each hybrid orbital and one electron in an unhybridized 2*p*-orbital perpendicular to the plane of the hybrids. The ring formed by overlap of the Csp^2 hybrid orbitals brings the unhybridized carbon 2*p*-orbitals close to one another. If one of the Kekulé structures were correct, each 2*p*-orbital would overlap side-by-side with *one* of its neighbors to form a π-bond to that atom with only a σ-bond to the other neighboring carbon atom (Fig. 9.32). The actual structure of the benzene molecule is a resonance hybrid of the two alternative bonding patterns. The six carbon-carbon bonds making up the ring are identical, and the electrons in the π-bonds are spread around the entire ring (Fig. 9.33).

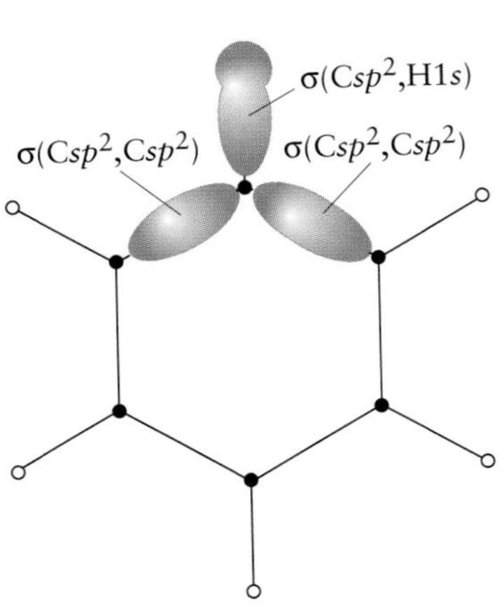

FIGURE 9.31 The framework of σ-bonds in benzene: each carbon atom is sp^2 hybridized, and the array of hybrid orbitals neatly matches the bond angles (of 120°) in the hexagonal molecule. Only bonding around one carbon atom is shown explicitly; all the others are the same.

The Lewis structure of the linear molecule ethyne (acetylene) is H—C≡C—H. To describe the bonding in a linear molecule, we need a hybridization scheme that produces two equivalent orbitals at 180° from each other. This is *sp* hybridization. Each C atom has one electron in each of its two *sp* hybrid orbitals and one electron in each of its two perpendicular unhybridized 2*p*-orbitals. The electrons in an *sp* hybrid orbital on each C atom pair and form a carbon-carbon σ-bond. The electrons in the remaining *sp* hybrid orbitals pair with hydrogen 1*s*-electrons to form

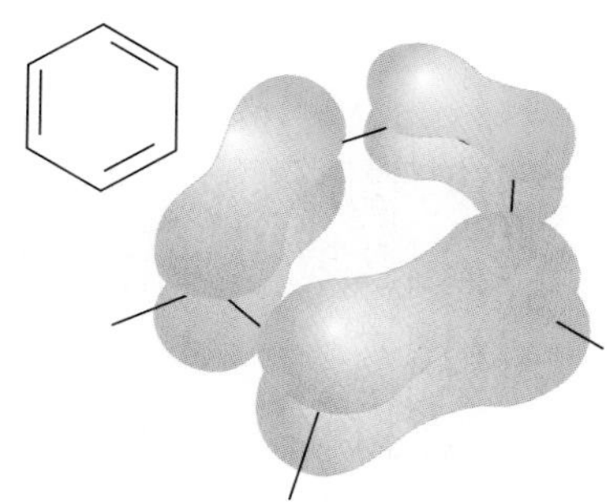

FIGURE 9.32 The remaining carbon atomic orbitals are unhybridized and can form a π-bond with either of their immediate neighbors. Two arrangements are possible, each one corresponding to one Kekulé structure. One Kekulé structure and the corresponding π-bonds are shown here.

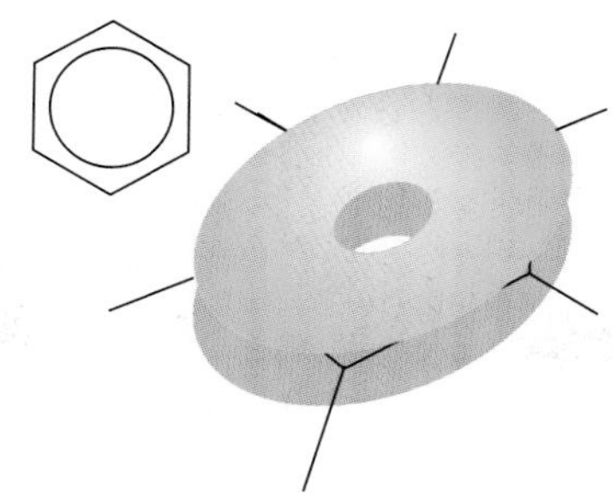

FIGURE 9.33 As a result of resonance between two structures like the one shown in the preceding illustration (corresponding to resonance of the two Kekulé structures), the π-electrons form double donut-shaped clouds, one above and one below the plane of the ring.

two carbon-hydrogen σ-bonds. The electrons in the perpendicular 2*p*-orbitals pair with a side-by-side overlap, forming two π-bonds at 90° to each other. The resulting bonding pattern is shown in Fig. 9.34.

Multiple bonds are formed when an atom forms a σ-bond by using a hybrid orbital and one or more π-bonds by using unhybridized p-orbitals.

EXAMPLE 9.6 Accounting for the structure of a multiply bonded molecule

Describe the structure of a formic acid (methanoic acid) molecule, HCOOH, in terms of hybrid orbitals, bond angles, and σ- and π-bonds.

STRATEGY First, draw the Lewis structure of the molecule and use the VSEPR model to determine the arrangement of the electron pairs around each

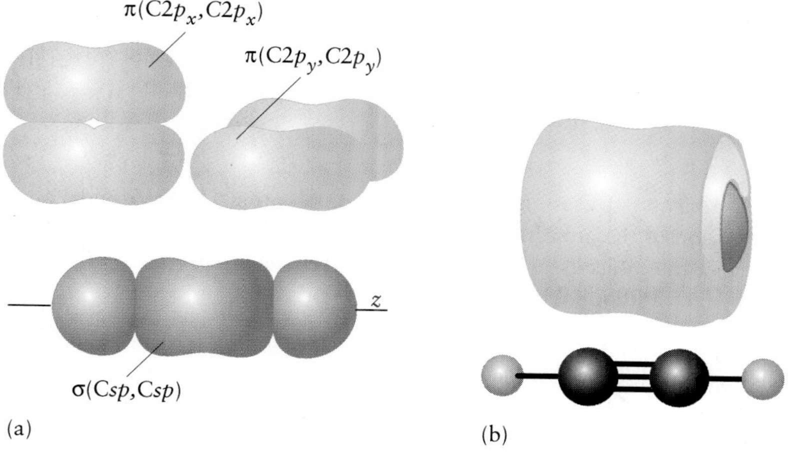

FIGURE 9.34 The pattern of bonding in ethyne (acetylene). (a) The carbon atoms are *sp* hybridized, and the two remaining *p*-orbitals on each ring form two π-bonds. (b) The resulting pattern is very similar to that for nitrogen (see Fig. 9.23), but two C—H groups replace the N atoms.

CASE STUDY 9 SELF-ASSEMBLING MATERIALS

IN 1991 A GROUP of scientists sealed two graphite rods inside a container of helium gas. Then they sent an electric discharge from one rod to the other. Much of one rod evaporated, but out of the inferno some amazing structures emerged (see accompanying illustrations). As well as the tiny 60-atom carbon spheres known as buckminsterfullerene—which had been known since 1985—long, hollow, perfectly straight tubes were also detected. Finding these orderly structures arising from a hot, turbulent electron discharge is almost as astonishing to chemists as the sight of a skyscraper erecting itself from a bonfire of bricks and concrete would be to construction engineers.

Nature's fabrication skills are the envy of human scientists and engineers. If we can learn to understand and control those forces, we might be able to produce materials that assemble themselves into desired shapes.

In *coded self-assembly,* instructions stored in the structure of a molecule such as DNA are used to construct other molecules or to duplicate the parent molecule. In *thermodynamic self-assembly,* atoms and molecules take positions that result in the lowest energy. For example, ions form regular three-dimensional arrays that maximize attractions between oppositely charged ions and reduce repulsions between like-charged ions. Carbon atoms form the six-membered rings of graphite because that is a lower energy arrangement than the tetrahedral structure of diamond.

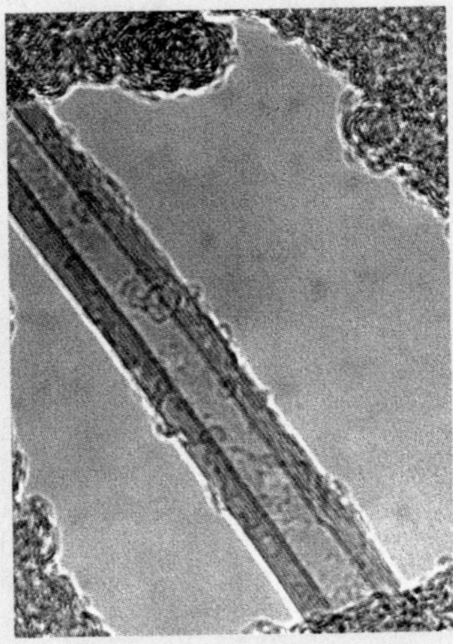

This cross section view of a carbon nanotube shows that it consists of several concentric tubes. Such layered structures are very strong. The surrounding material in the photograph consists of fullerenes and other carbon structures.

Buckminsterfullerene, C_{60}, is so called because of its resemblance to the geodesic domes designed by the American architect R. Buckminster Fuller (see figure). The family of related molecules are known as the fullerenes. The carbon atoms in buckminsterfullerene form both six- and five-membered rings, whereas only six-membered rings are found in graphite. The five-membered rings pucker the otherwise planar structure into a sphere. The rounded shape of the molecule may make it an excellent lubricant, perhaps after some of the double bonds have been replaced by bonds to fluorine atoms. Graphite is composed of flat layers of six-membered benzenelike rings, linked like chicken wire. The π-bonds of these sheets are connected, so electrons are delocalized and can travel through them and conduct electricity.

The tiny tubes of carbon are called nanotubes, because they are approximately 1 nm in diameter. Nanotubes have the same bonding structure as graphite, but under the conditions of the arc discharge, large, flat planes of graphite

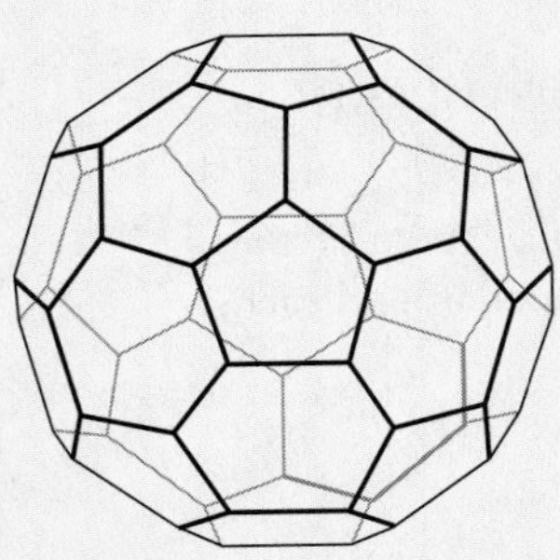

Buckminsterfullerene, C_{60}, consists of five- and six-membered rings. The structure shown here is a stick structure, in which only the bonds are shown.

atom that is bonded to more than one other atom. Next, decide on the hybridization of the atom that reproduces that arrangement. Then describe the bonding in the molecule with a σ-bond between each pair of atoms and a π-bond for each additional bond in a multiple bond.

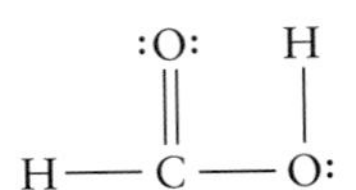

43 Formic acid, HCOOH

SOLUTION The Lewis structure of formic acid is shown in (**43**); according to the VSEPR model, the three electron pairs in σ-bonds around the carbon atom take up a trigonal-planar arrangement, which suggests sp^2 hybridization. The C atom has one unpaired electron in each of its three sp^2 hybrid orbitals and, as a result, can form three σ-bonds pointing toward the corners of an equilateral tri-

cannot be formed and small fragments of graphite have many "dangling bonds," locations on carbon atoms where the valence is not fulfilled. Coiling into a tube allows more bonds to form and lowers the energy of the structure.

Carbon nanotubes conduct electricity because of the extended network of π-bonds that runs from one end of the tube to the other. They are very strong and, in fact, along the length of the tube their strength is the greatest of any known material. Because they have a very low density, their strength-to-mass ratio is 40 times that of steel! Their great strength and conductivity suggest that carbon nanotubes could be used as submicroscopic electronic components. Their rigidity may also allow them to be used as molds for other elements. For example, they could be filled with molten lead to create lead wires one atom in diameter. It has even been suggested that hydrogen molecules could be adsorbed into the tubes, creating a high-specific-enthalpy storage medium for hydrogen.

The molecules of some covalent compounds have also been found to assemble themselves into three-dimensional structures. For example, narrow molecules with "sticky" (highly polar) groups at one end have been found to form a monolayer (a layer one molecule thick) on a polar surface. The polar ends of the molecules in the monolayer point down toward the surface and the nonpolar tails point up. Polar molecules are not attracted to the nonpolar tails, so the top surface appears slippery rather than sticky. Thus, it may be possible to create extremely thin sheets of oriented molecules that are sticky on one side and slippery on the other. Such materials could be used to coat the inner walls of artificial blood vessels to inhibit the buildup of deposits.

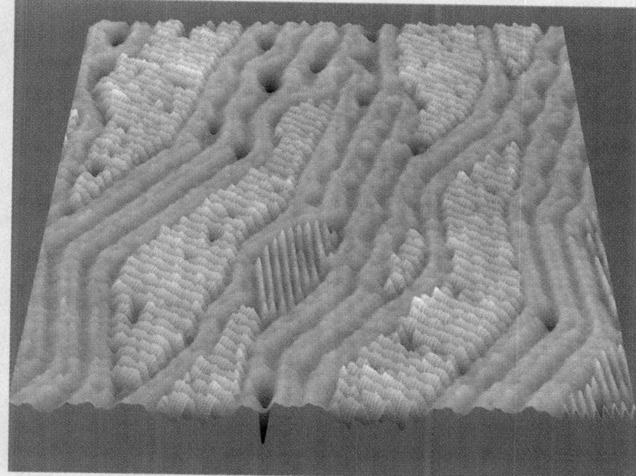

A partial monolayer of molecules on a surface of gold, as seen with a scanning tunneling microscope. Long, narrow molecules with polar groups on one end are lined up in rows, with the polar end attached to the gold. The rows of molecules appear as striped regions. The zig-zag features are the atoms of the bare gold surface.

QUESTIONS

1. What is the hybridization of the carbon atoms in (a) buckminsterfullerene? (b) carbon nanotubes?

2. How many different bond lengths do you expect to find in (a) buckminsterfullerene? (b) carbon nanotubes? Explain your reasoning and give an estimate of each bond length.

3. Which of the following substances found in blood would be attracted to the "sticky" polar ends of molecules used to create self-assembling layers: (a) nitric oxide (nitrogen monoxide), NO; (b) carbon dioxide, CO_2; (c) ethanol, CH_3CH_2OH; (d) glycine, $H_2NCH_2{-}COOH$; (e) dioxygen, O_2?

4. Carbon nanotubes have been found to have an electrical conductivity somewhat less than that of graphite. Explain the difference in terms of orbital overlap.

angle. The bond angles about the C atom are therefore about 120°. Carbon's fourth valence electron is in a *p*-orbital perpendicular to the triangle.

The O atom forming a single bond to the C atom and a single bond to an H atom has two bonds and two lone pairs of electrons, a total of four electron pairs. These pairs lie in a tetrahedral arrangement. We therefore expect a bond angle of about 109.5° and sp^3 hybridization. The remaining O atom has two lone pairs and a double bond to the C atom, formed by the side-by-side overlap of *p*-orbitals on the two atoms.

On the basis of this analysis, we can describe the structure of formic acid as follows: C is sp^2 hybridized, and the oxygen atom in the OH group is sp^3

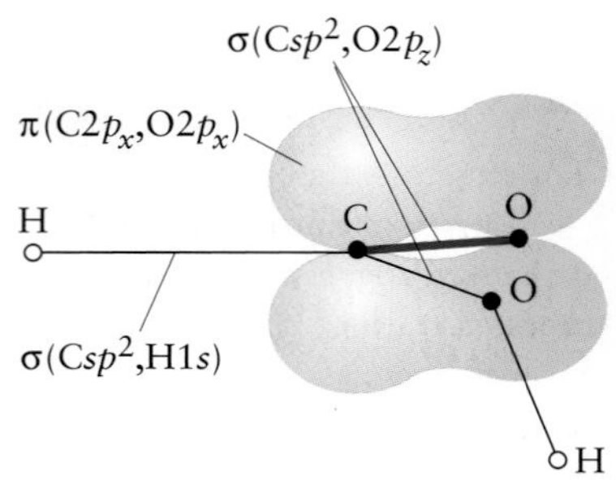

FIGURE 9.35 The pattern of bonds in formic acid (methanoic acid).

hybridized. There are four σ-bonds and one π-bond in the molecule. All the atoms except for the H of the OH group are in the same plane. The pattern of bonds is illustrated in Fig. 9.35.

SELF-TEST 9.12A Describe the structure of the carbon dioxide molecule, CO_2, in terms of hybrid orbitals, bond angles, and σ- and π-bonds.

[*Answer:* Linear; bond angle 180°; C is *sp* hybridized; it forms one σ-bond and one π-bond to each O atom. The π-bonds are at 90° to each other.]

SELF-TEST 9.12B Describe the structure of the propene molecule, $CH_3—CH=CH_2$, in terms of hybrid orbitals, bond angles, and σ- and π-bonds.

9.17 CHARACTERISTICS OF DOUBLE BONDS

We have seen that a carbon-carbon double bond is stronger than a carbon-carbon single bond, but weaker than the sum of two single bonds. A carbon-carbon triple bond is also weaker than the sum of three single bonds. A part of the reason for the relative weakness of carbon-carbon multiple bonds is that, whereas a single bond is a σ-bond, the additional bonds in a multiple bond are π-bonds. Carbon-carbon π-bonds are about 84 $kJ\cdot mol^{-1}$ weaker than carbon-carbon σ-bonds because the side-by-side overlap puts their electrons in less favorable locations for holding the atoms together.

A double bond prevents one part of a molecule from rotating relative to another part. The double bond of ethene, for example, holds the entire molecule flat, as shown in (**13**). Figure 9.30 shows that the two 2*p*-orbitals overlap best if the two CH_2 groups lie in the same plane. The π-bond could not form if the CH_2 groups were twisted about the C—C bond.

Double bonds help you to read these words. Vision depends on the shape of the molecule *retinal* in the retina of the eye. Retinal is held rigid by its double bonds (**44**). When light enters the eye, it excites an electron out of the π-bond marked by the arrow. The double bond is now weaker, and the molecule is free to rotate around the remaining σ-bond. When the excited electron falls back, the molecule is trapped in its new shape (**45**). This change of shape triggers a signal along the optic nerve and is interpreted by the brain as the sensation of vision.

44 *cis*-Retinal

45 *trans*-Retinal

▶▶ Structures 44 and 45 are "stick structures." Each unlabeled angle represents a C atom. See Section 11.1.

The side-by-side overlap that forms a π-bond makes a molecule resistant to twisting.

SKILLS YOU SHOULD HAVE MASTERED

Conceptual

1. Explain why lone pairs are more likely to be found in certain locations around a central atom and how and why they affect the bond angles in a molecule.
2. Define the electric dipole moment of a bond and explain how it depends on the electronegativities of the two atoms in a bond.
3. Explain how bond enthalpy is related to bond multiplicity, atomic radius, and the presence of unpaired electrons.

4. List the factors affecting bond length and explain the effect of each.

5. Distinguish σ- and π-bonds by their shapes, properties, and component orbitals.

6. Explain how hybridization arises from atomic orbitals.

Problem-Solving

1. Predict the shape of a molecule or polyatomic ion from its formula, giving each bond angle approximately.

2. Predict the polarity of a molecule.

3. Use average bond enthalpies to estimate a reaction enthalpy.

4. Account for the structure of a molecule in terms of hybrid orbitals and σ- and π-bonds.

Descriptive

1. Describe π-bonding in benzene, according to valence-bond theory.

2. Describe the structure and bonding in molecules of ethane, ethene, and ethyne.

EXERCISES

The Shapes of Molecules and Ions

9.1 What is the expected shape and bond angle or angles in a molecule of VSEPR type (a) AX_5; (b) AX_2; (c) AX_3E; (d) AX_2E_2; (e) AX_4?

9.2 What is the expected shape and bond angle or angles in a molecule of VSEPR type (a) AX_4E; (b) AX_3; (c) AX_6; (d) AX_3E_2; (e) AX_2E?

9.3 Using Lewis structures and VSEPR theory, predict the shape of each of the following molecules: (a) HClO (for which the structure is HOCl); (b) PF_3; (c) N_2O; (d) O_3.

9.4 Using Lewis structures and VSEPR theory, predict the shape of each of the following molecules: (a) H_2S; (b) CS_2; (c) NO_2; (d) Cl_2O.

9.5 Using Lewis structures and VSEPR theory, predict the shape of each of the following ions: (a) H_3O^+; (b) SO_4^{2-}; (c) IF_4^+; (d) NO_3^-.

9.6 Using Lewis structures and VSEPR theory, predict the shape of each of the following ions: (a) $CH_3NH_3^+$; (b) ClO_3^-; (c) ClO_4^-; (d) PF_4^+.

9.7 Using Lewis structures and VSEPR theory, predict the shape of each of the following species: (a) sulfur tetrachloride; (b) iodine trichloride; (c) IF_4^-; (d) xenon trioxide.

9.8 Using Lewis structures and VSEPR theory, predict the shape of each of the following species: (a) PF_4^-; (b) ICl_4^+; (c) phosphorus pentachloride; (d) xenon tetrafluoride.

9.9 Predict the molecular shape and bond angles in each of the following: (a) I_3^-; (b) IF_3; (c) IO_4^-; (d) TeF_6.

9.10 Predict the molecular shape and bond angles in each of the following: (a) I_3^+; (b) PCl_3; (c) SeO_3^{2-}; (d) GeH_4.

9.11 Write the Lewis structures and the VSEPR formula, list the shape, and predict the approximate bond angles in (a) CF_3Cl; (b) GaI_3; (c) CH_3^-; (d) SF_6; (e) $XeOF_4$.

9.12 Write the Lewis structures and the VSEPR formula, list the shape, and predict the approximate bond angles in (a) PCl_3F_2; (b) SnF_4; (c) SnF_6^{2-}; (d) IF_5; (e) XeO_4.

9.13 Estimate the bond angles marked with arcs and lowercase letters in acrolein, an eye irritant in smoke with a pungent smell:

9.14 Estimate the bond angles marked with arcs and lowercase letters in peroxyacetylnitrate, an eye irritant in smog:

9.15 When BF_3 reacts with NH_3, a white solid of molecular formula $F_3B{-}NH_3$ is formed. Draw Lewis structures of the reactants and the product, and estimate the bond angles.

9.16 Hydrogen fluoride reacts with antimony pentafluoride to give a salt in which the cation is H_2F^+ and the anion is SbF_6^-. Draw Lewis structures of the reactants and the product, and estimate the bond angles.

Charge Distribution in Molecules

9.17 What do the terms *electric dipole* and *electric dipole moment* mean?

9.18 Explain why a molecule may be nonpolar even though it has polar bonds.

9.19 Indicate the direction of the electric dipole moment in the following bonds: (a) O—H; (b) O—F; (c) F—Cl; (d) O—S.

9.20 Indicate the direction of the electric dipole moment in the following bonds: (a) N—O; (b) C—O; (c) C—N; (d) N—H.

9.21 Identify in the following molecules the bonds that are polar and nonpolar: (a) Br_2; (b) H_2NNH_2; (c) CH_4; (d) O_3.

9.22 Identify in the following molecules the bonds that are polar and nonpolar: (a) I_2; (b) S_8; (c) CH_3CH_3; (d) H_2O_2.

9.23 Classify the following molecules as polar or nonpolar. Use the VSEPR model to determine their shapes, and assume that all Xs are atoms of the same element. (a) AX_3; (b) AX_4E_2; (c) AX_3E; (d) AX_2E_2.

9.24 Classify the following molecules as polar or nonpolar. Use the VSEPR model to determine their shapes, and assume that all Xs are atoms of the same element. (a) AX_3E_2; (b) AX_4E; (c) AX_5; (d) AX_4.

9.25 Write the Lewis structures and predict whether each of the following molecules is polar or nonpolar: (a) CCl_4; (b) CS_2; (c) PCl_5; (d) XeF_4.

9.26 Write the Lewis structures and predict whether each of the following molecules is polar or nonpolar: (a) CH_2Cl_2; (b) H_2S; (c) PCl_3; (d) SF_4.

9.27 Organic molecules are generally nonpolar or only slightly polar. Predict whether the following molecules are likely to be either polar or nonpolar: (a) C_6H_6 (benzene); (b) CH_3OH; (c) H_2CO (formaldehyde, used in aqueous solution to preserve biological specimens).

9.28 Predict whether the following molecules are likely to be either polar or nonpolar: (a) C_6H_5Cl (chlorobenzene); (b) CH_3—CHOH—CH_3 (2-propanol, rubbing alcohol); (c) CCl_2F_2 (Freon-12, a refrigerant).

9.29 There are three different dichlorobenzenes, $C_6H_4Cl_2$, which differ in the relative positions of the chlorine atoms in the benzene ring. (a) Which of the three forms are polar and which are nonpolar? (b) Which has the largest dipole moment?

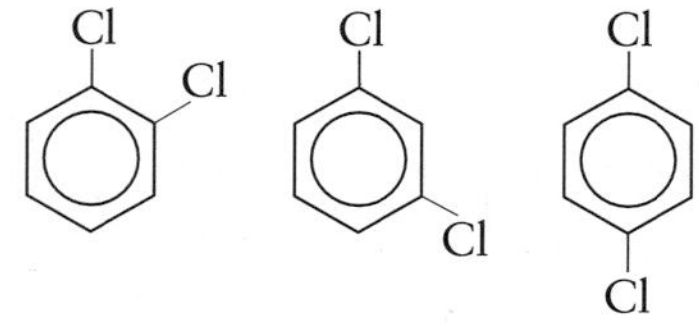

9.30 There are three different dichloroethenes, $C_2H_2Cl_2$, which differ in the locations of the chlorine atoms. (a) Which of the forms are polar and which are nonpolar? (b) Which has the largest dipole moment?

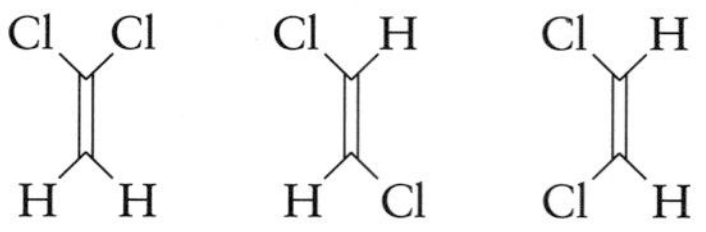

Bond Strengths and Bond Lengths

Refer to Tables 9.2, 9.3, and 9.4 for the following exercises.

9.31 Estimate the enthalpy change that occurs when each molecule is dissociated into its atoms: (a) H_2O; (b) CO_2; (c) CH_3COOH; (d) CH_3NH_2 (methylamine).

9.32 Estimate the enthalpy change that occurs when each molecule is dissociated into its atoms: (a) NH_3; (b) C_2H_4 (ethene); (c) C_2H_5OH (ethanol); (d) C_6H_6 (benzene).

9.33 Estimate the standard enthalpy of formation, ΔH_f°, of the following molecules in the gas phase: (a) HCl; (b) H_2O_2; (c) CCl_4; (d) NH_3.

9.34 Estimate the standard enthalpy of formation, ΔH_f°, of the following molecules in the gas phase: (a) HF; (b) CH_2Cl_2; (c) N_2O (NNO); (d) NH_2OH.

9.35 Use the information in Tables 6.2, 9.2, and 9.3 to estimate the enthalpy of formation of the following compounds in the *liquid* state, given that the standard enthalpy of sublimation of carbon is $+717\ kJ{\cdot}mol^{-1}$: (a) H_2O; (b) CH_3OH (methanol); (c) benzene, C_6H_6 (assume *no* resonance); (d) benzene, C_6H_6 (assume resonance).

9.36 Use the information in Tables 6.2, 9.2, and 9.3 to estimate the enthalpy of formation of the following compounds in the *liquid* state, given that the standard enthalpy of sublimation of carbon is $+717\ kJ{\cdot}mol^{-1}$: (a) NH_3; (b) C_2H_5OH (ethanol); (c) CH_3COCH_3 (acetone).

9.37 Use the bond enthalpies in Tables 9.2 and 9.3 to estimate the reaction enthalpy for

(a) $HCl(g) + F_2(g) \longrightarrow HF(g) + ClF(g)$, given that $\Delta H_B(Cl{-}F) = 256\ kJ{\cdot}mol^{-1}$

(b) $C_2H_4(g) + HCl(g) \longrightarrow CH_3CH_2Cl(g)$

(c) $C_2H_2(g) + 2H_2(g) \longrightarrow CH_3CH_3(g)$

9.38 Use the bond enthalpies in Tables 9.2 and 9.3 to estimate the enthalpy of reaction for

(a) $N_2(g) + 3\,F_2(g) \longrightarrow 2\,NF_3(g)$

(b) $CH_3CH{=}CH_2(g) + H_2O(g) \longrightarrow CH_3CH(OH)CH_3(g)$

(c) $CH_4(g) + Cl_2(g) \xrightarrow{h\nu} CH_3Cl(g) + HCl(g)$

9.39 List the carbon-oxygen bonds in the following compounds in order of increasing length: (a) CH_3CH_2OH; (b) H_2CO; (c) CO.

9.40 List the nitrogen-nitrogen bonds in the following

compounds in order of increasing length: (a) H_2NNH_2 (hydrazine); (b) N_2; (c) HNNH (nitrogen hydride).

9.41 Use the information in Fig. 9.18 to estimate bond lengths in (a) H_2NNH_2 (hydrazine); (b) CO_2; (c) $OC(NH_2)_2$ (urea); (d) HNNH (nitrogen hydride).

9.42 Use the information in Fig. 9.18 to estimate bond lengths in (a) H_2CO (formaldehyde); (b) CH_3OCH_3 (dimethyl ether); (c) CH_3OH; (d) CH_3SH.

Orbitals and Bonds

9.43 State whether σ-bonds, π-bonds, or neither are formed by the overlap of the given orbitals on neighboring atoms, where the internuclear distance lies along the z-axis: (a) ($1s$, $1s$); (b) ($2p_x$, $2p_x$); (c) ($2s$, $2p_y$); (d) ($2p_z$, $2p_z$).

9.44 State whether σ-bonds, π-bonds, or neither are formed by the overlap of the given orbitals on neighboring atoms, where the internuclear distance lies along the z-axis: (a) ($1s$, $2p_z$); (b) ($2p_y$, $2p_y$); (c) ($3s$, $4p_z$); (d) ($2p_x$, $2p_z$).

9.45 What atomic orbitals overlap to form a bond when the Cl_2 molecule forms? Classify the bond as σ or π.

9.46 What atomic orbitals overlap to form a bond when the HBr molecule forms? Classify the bond as σ or π.

9.47 State the relative orientations of the following hybrid orbitals: (a) sp^3; (b) sp; (c) d^2sp^3; (d) sp^2.

9.48 The relative orientation of bonds on a central atom of a molecule that has no lone pairs can be any of those listed below. What is the hybridization of the orbitals used by each central atom for its bonding pairs? (a) Tetrahedral; (b) trigonal bipyramidal; (c) octahedral; (d) linear.

9.49 State the hybridization of the atom in bold type in the following molecules: (a) **S**F_4; (b) **B**Cl_3; (c) **N**H_3; (d) $(CH_3)_2$**Be**. (*Hint:* Lone-pair electrons occupy hybrid orbitals that are very similar to those occupied by bonding electrons.)

9.50 State the hybridization of the atom in bold type in the following molecules: (a) **S**F_6; (b) O_3Cl-**O**$-ClO_3$; (c) H_2N-**C**H_2-COOH (glycine); (d) O**C**$(NH_2)_2$ (urea). (See the hint in Exercise 9.49.)

9.51 Use an orbital box diagram to identify the hybrid orbitals used by the central atoms for bonding in the following species: (a) CH_3^+; (b) $AlCl_4^-$; (c) ClO_2; (d) BI_3.

9.52 Use an orbital box diagram to identify the hybrid orbitals used by the central atoms for bonding in the following species: (a) CH_3^-; (b) BiI_4^-; (c) XeF_5^+; (d) NH_2^-.

9.53 Identify hybrid orbitals used by C, O, and P in the following molecules: (a) C_2H_6; (b) C_2H_2; (c) PCl_5; (d) HOCl.

9.54 Identify the hybrid orbitals used by C, N, O, S, and I in the following molecules and ion: (a) N_2H_4; (b) CH_3OH; (c) SF_4; (d) IF_4^+.

SUPPLEMENTARY EXERCISES

9.55 Write the Lewis structure of each of the following compounds, predict the shape about each central atom, and state whether it is polar or nonpolar: (a) O_2SCl_2 (S is the central atom); (b) F_3SSF; (c) $AsCl_5$; (d) SiF_4; (e) SF_4; (f) H_2O; (g) HCN; (h) ICl_3.

9.56 Predict the types of hybrid orbitals used in bonding by the central atom, the shape, and the bond angles in (a) $In(CH_3)_3$; (b) PCl_3; (c) ICl_2^-; (d) SiF_6^{2-}; (e) CH_3^+; (f) H_3O^+; (g) SO_3; (h) SiH_4.

9.57 Arrange the carbon-nitrogen bonds in the following compounds in order of increasing length: (a) $CH_3CH_2NH_2$; (b) CH_3CN; (c) CH_3CHNH.

9.58 For each molecule or ion, write the Lewis structure, list the number of lone pairs on the central atom, identify the shape, and estimate the bond angle: (a) XeF_5^+; (b) $XeOF_2$; (c) SF_5^+; (d) TeH_2; (e) BrO_3^-.

9.59 Predict the shapes and estimate the bond angles of (a) the thiosulfate ion, $S_2O_3^{2-}$; (b) $(CH_3)_2Be$; (c) BH_2^-; (d) $SnCl_2$.

9.60 Write the Lewis structures and predict the shapes of (a) TeF_4; (b) NH_2^-; (c) CS_2; (d) NH_4^+; (e) GeH_4; (f) OCS.

9.61 Write the Lewis structures and give the approximate bond angles of (a) C_2H_4; (b) ClCN; (c) $OPCl_3$; (d) N_2H_4.

9.62 Write the Lewis structures and predict the shapes of (a) $OCCl_2$; (b) $OSCl_2$; (c) $OSbCl_3$. In each case the formula expresses the atomic arrangement.

9.63 The production of Freon-12, CCl_2F_2, makes use of the reaction between carbon tetrachloride and hydrogen fluoride:

$$CCl_4(g) + 2\,HF(g) \longrightarrow CCl_2F_2(g) + 2\,HCl(g)$$

Estimate the standard enthalpy of this reaction from the bond enthalpies in Tables 9.2 and 9.3.

9.64 Investigate whether the replacement of a carbon-carbon double bond by single bonds is energetically favored by using Tables 9.2 and 9.3 to calculate the reaction enthalpy for the conversion of ethene, C_2H_4, to ethane, C_2H_6. The reaction is

$$H_2C{=}CH_2 + H_2 \longrightarrow CH_3{-}CH_3$$

9.65 Use covalent radii to calculate the bond lengths in (a) CF_4; (b) SiF_4; (c) SnF_4. Account for the trend in the values you calculate.

9.66 Polar molecules attract other polar molecules through dipole-dipole intermolecular forces. Polar solutes tend to have higher solubilities than nonpolar solutes in polar solvents. Which of the following compounds will have the higher solubility in water? (a) SiF_4 or PF_3; (b) SF_6 or SF_4; (c) IF_5 or AsF_5.

9.67 The halogens form compounds among themselves. These compounds, called the interhalogens, have the formulas XX′, XX'_3, and XX'_5, where X is the heavier halogen. Predict their structures and bond angles. Which of them are polar?

9.68 Nonpolar solutes typically have higher solubilities than polar solutes in nonpolar solvents. Which solute is likely to have the greater solubility in benzene: (a) CCl_4 or CH_3OH; (b) H_2CO (formaldehyde) or C_8H_{18} (octane); (c) the three dichlorobenzenes shown in Exercise 9.29?

9.69 Oxygen difluoride is almost the only compound in which oxygen is assigned a positive oxidation number. Write its Lewis structure and then determine the composition of the hybrid orbitals it uses to form bonds to the fluorine atoms. Predict the bond angle.

9.70 Write three equal-energy resonance structures of the carbonate ion, CO_3^{2-}. Determine the formal charge on each atom of the structure. What are the hybrid orbitals that the carbon atom uses to form bonds to the oxygen atoms?

9.71 Estimate the bond angles and determine the hybrid orbitals used by the carbon atoms in the following compounds: (a) acetaldehyde (ethanal), $CH_3{-}CHO$; (b) acetic acid, CH_3COOH.

9.72 Acrylonitrile, CH_2CHCN, is used in the synthesis of acrylic fibers (polyacrylonitriles), such as Orlon. Write the Lewis structure of acrylonitrile and describe the hybrid orbitals on each carbon atom. What is the approximate value of the CCC bond angle?

9.73 Write the Lewis structure of ClF_3 and identify the orbitals used for bonding by the Cl and F atoms. What is the shape of the molecule? Estimate the value of the FClF bond angle.

9.74 Xenon forms XeO_3, XeO_4^{2-}, and XeO_6^{4-}, all of which are powerful oxidizing agents. Give their Lewis structures, their bond angles, and the hybridization of the xenon atom.

CHALLENGING EXERCISES

9.75 An organic compound distilled from wood was found to have a molar mass of 32.04 $g{\cdot}mol^{-1}$ and the following composition by mass: 37.5% C, 12.6% H, and 49.9% O. (a) Write the Lewis structure of the compound and determine the bond angles about the carbon and oxygen atoms. (b) Give the hybridization of the carbon and oxygen atoms. (c) Predict whether the molecule is polar or not.

9.76 Confirm, using trigonometry, that the dipoles of the three bonds in a trigonal-planar AB_3 molecule cancel and that the molecule is nonpolar. Go on to show that the dipoles of the four bonds in a tetrahedral AB_4 molecule cancel and that the molecule is nonpolar.

9.77 A two-step reaction in the stratosphere that may be responsible for the depletion of the ozone layer is

Step 1: $Cl(g) + O_3(g) \longrightarrow ClO(g) + O_2(g)$

Step 2: $ClO(g) + O(g) \longrightarrow Cl(g) + O_2(g)$

Given that the enthalpy of the Cl—O bond is 270 $kJ{\cdot}mol^{-1}$, (a) determine the enthalpy of reaction for each step; (b) determine the enthalpy of the overall reaction and write its thermochemical equation; (c) because O_3 is stabilized by resonance, comment on your calculated value for the overall reaction compared with the experimental value, which is −392 kJ.

9.78 The bond enthalpy in NO is 632 $kJ{\cdot}mol^{-1}$ and that of each N—O bond in NO_2 is 469 $kJ{\cdot}mol^{-1}$. Using Lewis structures and the average bond enthalpies in Table 9.3, explain (a) the difference in bond enthalpies between the two molecules; (b) the fact that the bond enthalpies of the two bonds in NO_2 are the same.

9.79 Benzene is more stable and less reactive than would be predicted from its Kekulé structures. Use the average bond enthalpies in Table 9.3 to calculate the lowering in energy per mole that occurs when resonance is allowed between the Kekulé structures of benzene.

9.80 The following molecules are bases that are part of the nucleic acids involved in the genetic code. Identify the hybridizations of the atoms, the composition of the bonds, and the orbitals occupied by lone pairs. Each unlabeled angle represents a C atom.

1 2

9.81 Nitrogen, phosphorus, oxygen, and sulfur exist as N_2, tetrahedral P_4, O_2, and ringlike S_8 molecules. Rationalize this fact in terms of the abilities of the atoms to form different types of bonds with each other. In the case of P_4 and S_8, describe the bonding in terms of the hybridization of the atoms.

9.82 Borazine, $B_3N_3H_6$, a compound that has been called "inorganic benzene" on account of its similar hexagonal structure (but with alternating B and N atoms in place of C atoms), is the basis of a large class of boron-nitrogen compounds. Write its Lewis structure and predict the composition of the hybrid orbitals used by each B and N atom.

9.83 Knowing that carbon has a valence of four in nearly all its compounds and can form chains and rings of C atoms, (a) draw two possible structures for C_3H_4; (b) determine all bond angles in each structure; (c) determine the hybridization of each carbon atom in the two structures; (d) ascertain whether the two structures are resonance structures or not and explain your reasoning.

9.84 Iodine heptafluoride, IF_7, has a pentagonal-bipyramidal structure:

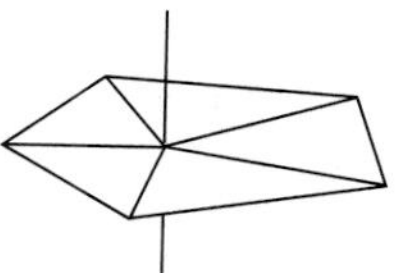

(a) How many lone pairs, if any, are on the I atom in this molecule? (b) What are the bond angles about the I atom? (c) What are the hybrid orbitals that I uses in this compound? (d) Is it likely that ClF_7 exists? Why or why not?

CHAPTER 10

This lava explosion at Kilauea, Hawaii, shows that transformations between liquid and solid forms of matter can have dramatic consequences and that what we commonly regard as solid rocks can occur as liquids. But what grips molecules and ions together as a liquid or a solid, why do solids melt, and what determines the properties of the resulting liquid? In this chapter, we see how the properties of individual molecules and ions control the properties of bulk materials.

LIQUID AND SOLID MATERIALS

Much of our oceans, lakes, rivers, and clouds arrived from outer space in the form of comets when the Earth was young. Comets are like huge dirty snowballs, and each one left its precious load of water when it smashed into the planet. Another source of water was the rocks from which the young Earth formed. Water was locked up as hydrates and was released when the rocks melted. Even today, when volcanoes erupt, vast quantities of water burst out. Released from its rocky prison deep underground, this water adds to our surface supplies.

Water has a number of unique properties that are critical for the survival of life. For instance, ice is less dense than liquid water. When lakes and rivers freeze, ice floats on the surface rather than sinking to the deeps. The surface layer helps to insulate the water below, so fish and other aquatic life can survive harsh winters. Water also has an unusually high boiling point and heat capacity. These properties allow it to moderate the Earth's climate. For example, on hot days, some of the heat goes into warming and evaporating water. Water's high enthalpy of vaporization also helps to keep us cool by evaporating in the process of perspiration.

Water dissolves a wide variety of compounds, a property allowing it to serve as the solvent in blood and cells and to transport nutrients and waste. Water is sometimes called the "universal solvent." That is an exaggeration, but it does dissolve a huge number of molecular and ionic compounds.

But why is water a liquid? Indeed, why do any liquids form, and why do liquids freeze to solids? The atoms in their molecules can form no additional chemical bonds, so how do molecules stick together?

INTERMOLECULAR FORCES

10.1 London forces
10.2 Dipole-dipole interactions
10.3 Hydrogen bonding

LIQUID STRUCTURE

10.4 Viscosity
10.5 Surface tension

SOLID STRUCTURES

10.6 Classification of solids
10.7 Metallic crystals
10.8 Properties of metals
10.9 Alloys
10.10 Ionic structures
10.11 Molecular solids
10.12 Network solids

PHASE CHANGES

10.13 Vapor pressure
10.14 Boiling
10.15 Freezing and melting
10.16 Phase diagrams and cooling curves
10.17 Critical properties

TABLE 10.1 Interionic and intermolecular forces*

Type of interaction	Typical energy, $kJ \cdot mol^{-1}$	Interacting species
ion-ion	250	ions only
dipole-dipole	2	stationary polar moleules
	0.3	rotating polar molecules
London (dispersion)	2	all types of molecules
hydrogen bonding	20	N, O, F; the link is a shared H atom

*The total force experienced by a species is the sum of all the forces in which it can participate.

INTERMOLECULAR FORCES

Molecules are drawn together by **intermolecular forces.** These forces are sometimes called "van der Waals forces," after Johannes van der Waals, who studied the effects of intermolecular forces on the behavior of real gases. We shall see that there are three main types of intermolecular forces; their names and strengths are summarized in Table 10.1. The total intermolecular force acting between two molecules is the sum of all the types of the forces they exert on each other.

Fritz London was the German-American physicist who first explained the force. The London force is also called the "dispersion force."

Only a movie filmed at the impossible rate of about 10^{16} frames a second would show the swirling motion.

10.1 LONDON FORCES

All molecules interact by the kind of intermolecular force called the **London force.** For nonpolar molecules, it is the only force that acts between them. To find the origin of this force, we have to realize that the electron clouds of atoms and molecules are like a swirling fog (Fig. 10.1). If we could take a snapshot of an atom at a given instant, the atom might look deformed. As the electrons move about, they produce a fleeting **instantaneous dipole moment.** Electrons may pile up at one end of a molecule, leaving the nucleus at the other end partially exposed. One end of the molecule will have a fleeting partial negative charge and the other end a fleeting partial positive charge. Even a noble gas atom can have a small instantaneous dipole moment. The instantaneous partial charges on different molecules attract one another and the molecules stick together.

London forces act between *all* types of molecules, polar as well as nonpolar. London forces hold benzene molecules together as a liquid at normal temperatures. They pull carbon dioxide molecules together into a

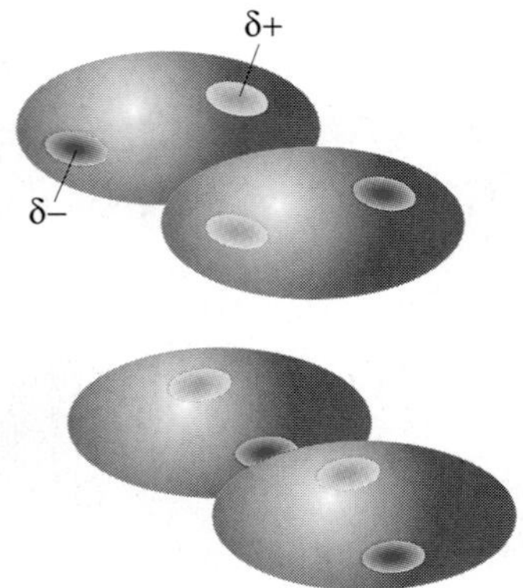

FIGURE 10.1 London forces arise from the attraction between two instantaneous dipoles. The dipoles are due to fluctuations in electron locations in the molecules. Although the instantaneous dipoles on two molecules are continuously changing direction, they remain in step long enough to attract each other.

solid at low temperature. They pull atoms of noble gases together when they condense to liquids.

The strength of the London force increases with molar mass (Fig. 10.2). Heavier molecules have more electrons, and there are bigger fluctuations in partial charges as the electrons flicker between different positions. An H_2 molecule has only two electrons; as a result, the molecules interact so weakly that hydrogen gas condenses to a liquid only if the temperature is lowered to 20 K (−253°C) at 1 atm. The variation of strength of London forces with molar mass explains why F_2 and Cl_2 are gases, Br_2 a liquid, and I_2 a solid at room temperature.

The strength of London forces is also determined in part by the shapes of molecules. Both pentane (**1**) and 2,2-dimethylpropane (**2**), for instance, have the molecular formula C_5H_{12}, so they each have the same number of electrons. However, they have different boiling points (36°C and 10°C, respectively).

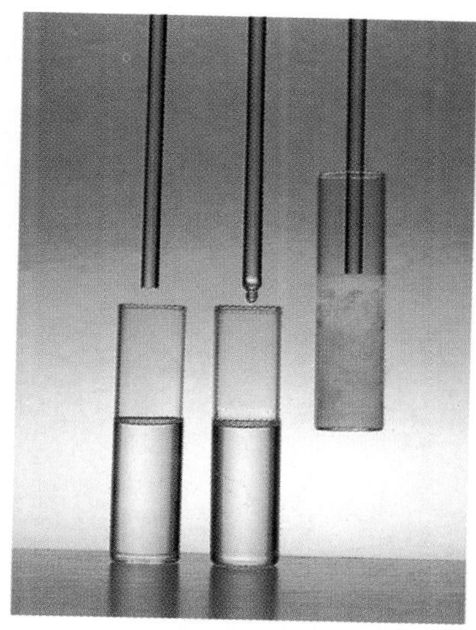

FIGURE 10.2 Hydrocarbons show how London forces increase in strength with molar mass. Pentane, C_5H_{12}, is a mobile fluid (left); pentadecane, $C_{15}H_{32}$, a viscous liquid (middle); and octadecane, $C_{18}H_{38}$, a waxy solid (right). The effect of increasing intermolecular forces is enhanced by the ability of long-chain molecules to tangle with one another.

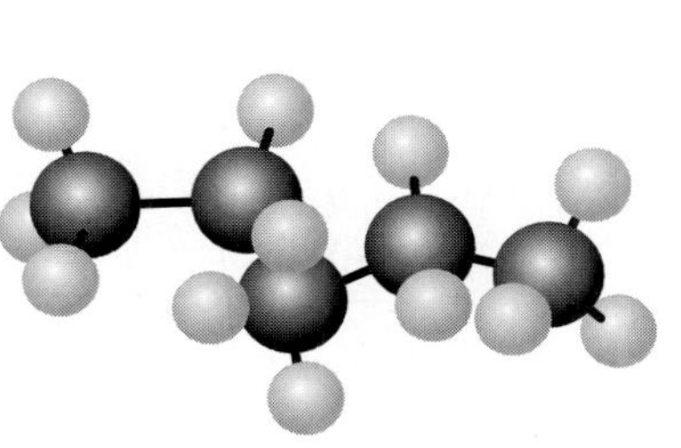

1 Pentane, C_5H_{12}

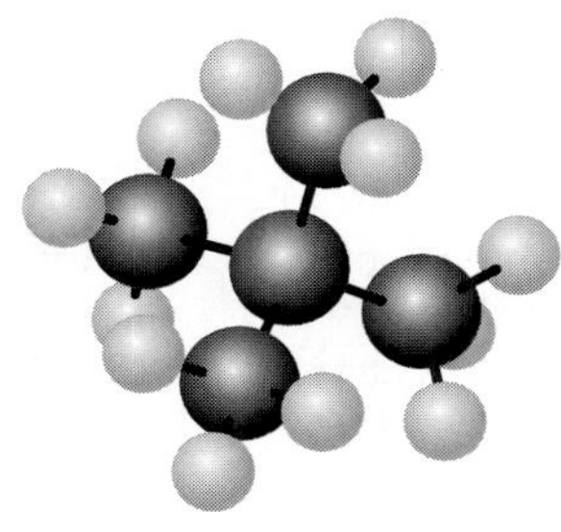

2 2,2-Dimethylpropane

Molecules of pentane are relatively long and rod-shaped. The instantaneous partial charges on adjacent rod-shaped molecules can interact strongly. In contrast, the instantaneous partial charges on more spherical molecules like 2,2-dimethylpropane cannot get so close to one another because only a small region of each molecule can be in contact (Fig. 10.3). As a result, the London forces between rod-shaped molecules are stronger than those between spherical molecules of the same mass.

The increase in strength of the London interaction is striking when heavier atoms are substituted for hydrogen atoms. Methane boils at −162°C, but tetrachloromethane (carbon tetrachloride, CCl_4) is a liquid at room temperature and boils at 77°C. Tetrabromomethane, CBr_4, has even more electrons: it is a solid at room temperature. Most covalent fluorides boil at lower temperatures than the corresponding covalent chlorides do, because fluorides have fewer electrons and hence smaller electron clouds. Compare, for example, the values for CF_4 and CCl_4 in Table 10.2.

We can now begin to see why polytetrafluoroethylene (Teflon) is used for nonstick surfaces. The relatively high effective nuclear charge of the fluorine nucleus exerts firm control over the surrounding electrons. The electrons in the atom do not undergo large fluctuations, so fluorocarbons and hydrocarbons have only weak intermolecular forces. The

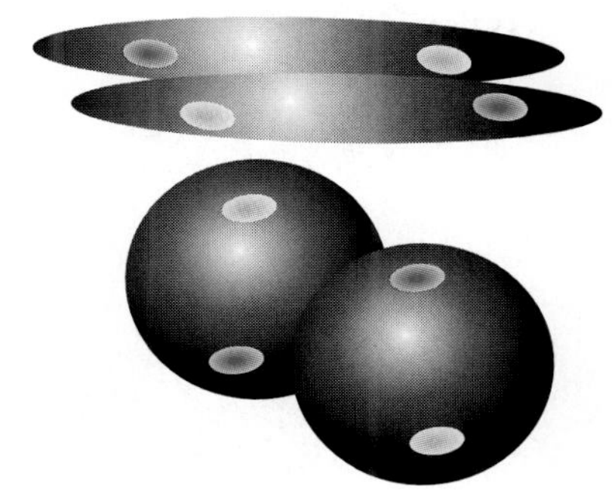

FIGURE 10.3 The instantaneous dipole moments in two neighboring rod-shaped molecules tend to be close together and to interact strongly. Those on neighboring spherical molecules tend to be far apart and to interact weakly.

TABLE 10.2 Normal melting and boiling points of substances

Substance	Melting point, °C	Boiling point, °C	Substance	Melting point, °C	Boiling point, °C
Noble gases			**Small inorganic species**		
He	−270 (3.5 K)*	−269 (4.2 K)	H_2	−259	−253
Ne	−249	−246	N_2	−210	−196
Ar	−189	−186	O_2	−218	−183
Kr	−157	−153	H_2O	0	100
Xe	−112	−108	H_2S	−86	−60
			NH_3	−78	−33
Halogens			CO_2	—	−78s†
F_2	−220	−188	SO_2	−76	−10
Cl_2	−101	−34			
Br_2	−7	59	**Organic compounds**		
I_2	114	184	CH_4	−182	−162
			CF_4	−150	−129
Hydrogen halides			CCl_4	−23	77
HF	−93	20	C_6H_6	6	80
HCl	−114	−85	CH_3OH	−94	65
HBr	−89	−67	glucose	142	d†
HI	−51	−35	sucrose	184d†	—

*Under pressure.

†s, solid sublimes; d, solid decomposes.

nonstick character of fluorocarbon surfaces is partly due to the feeble London forces between these compounds and the mixture of organic compounds present in fat and grease. The coating sticks firmly to pot surfaces, but only because it is applied while molten. The molten solid penetrates crevices in the metal and sets to a solid (Fig. 10.4).

The London force arises from the attraction between instantaneous electric dipoles on neighboring molecules and acts between all types of molecules; its strength increases with molar mass.

10.2 DIPOLE-DIPOLE INTERACTIONS

Polar molecules have *permanent* partial charges in addition to the instantaneous partial charges that arise from fluctuations in their electron clouds. The partial charges of one polar molecule can interact with the partial charges of a neighboring molecule and give rise to a **dipole-dipole interaction** (Fig. 10.5). This interaction is in addition to the London force. The latter is often more important, particularly when the molecules are rotating, for that weakens the dipole-dipole force.

To judge the strengths of the dipole-dipole interaction, let's compare the enthalpy of vaporization of a solid composed of polar molecules with

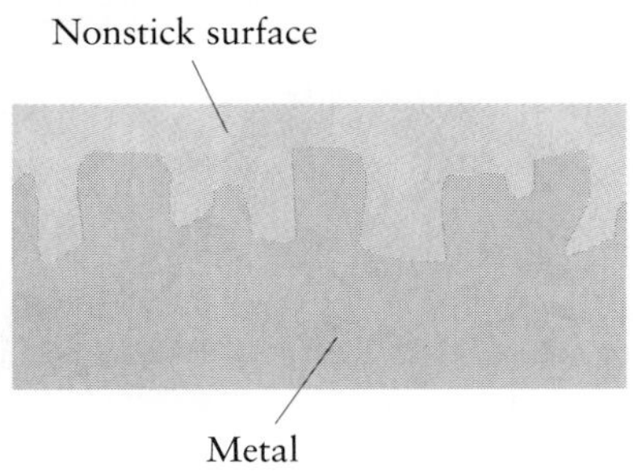

FIGURE 10.4 The nonstick surfaces on kitchenware are fluorocarbon polymers. They adhere to the metal surface by penetrating into cracks and gaps in the metal. Some adhesives act in the same way to bind two surfaces.

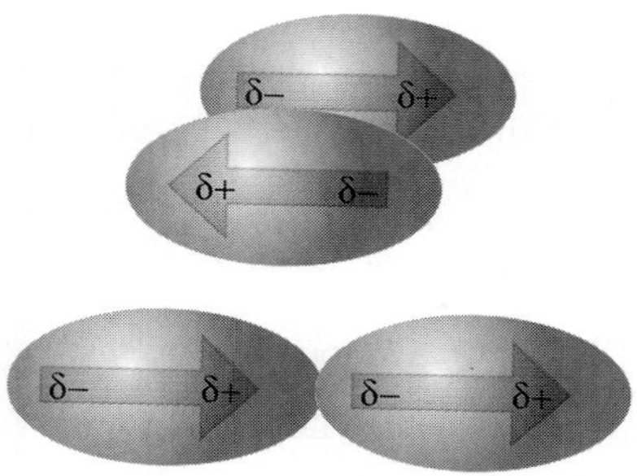

FIGURE 10.5 Polar molecules attract each other by the interaction between the partial charges of their electric dipoles (represented by the arrows). Both the relative orientations shown (end-to-end or side-by-side) are energetically favorable.

the lattice enthalpy of an ionic solid. In both cases the enthalpy change is a measure of the energy needed to separate the molecules or ions. The enthalpy of vaporization of solid hydrogen chloride is 18 kJ·mol^{-1}; the lattice enthalpy of sodium chloride is 787 kJ·mol^{-1}, nearly 50 times bigger. The dipole-dipole interaction is weaker than the ion-ion interactions in ionic solids because only the partial charges of the dipoles are involved in dipole-dipole interactions.

◀◀ **Lattice enthalpies were introduced in Section 8.2; they measure the strength of the coulombic (electrostatic) interactions between ions.**

The strength of dipole-dipole interactions depends on the magnitudes of the bond dipoles and the shape of the molecule. In diatomic molecules, the dipole-dipole interactions depend on the difference in electronegativity between the two atoms. In a polyatomic molecule, the shape of the molecule must also be taken into consideration. If bond dipoles cancel within a molecule (see Section 9.7), then the molecule itself has no dipole moment and experiences only London forces.

EXAMPLE 10.1 Predicting the variation in boiling points on the basis of dipole-dipole interactions

Which would you expect to have the higher boiling point, *p*-dichlorobenzene (**3**) or *o*-dichlorobenzene (**4**)?

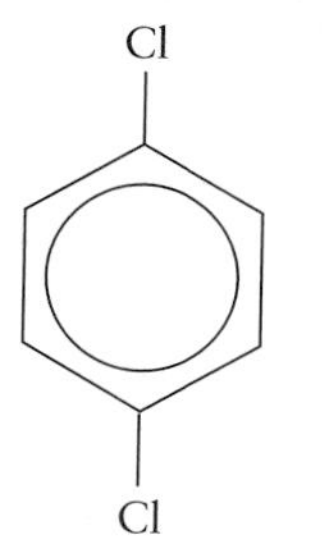

3 *p*-Dichlorobenzene

Cl
Cl

4 *o*-Dichlorobenzene

STRATEGY When two molecules have different dipole moments but are otherwise very similar, we expect the molecules with the bigger electric dipole moment to interact more strongly. Therefore, assign the higher boiling point to the more strongly polar compound. To decide whether a molecule is polar, decide whether or not the dipole moments of the bonds cancel one another, as explained in Section 9.7.

SOLUTION The two C—Cl bonds in *p*-dichlorobenzene lie directly across the ring and their dipole moments cancel. An *o*-dichlorobenzene molecule is

H Cl
C=C
H Cl

5 *cis*-Dichloroethene

H Cl
C=C
Cl H

6 *trans*-Dichloroethene

Cl Cl
C=C
H H

7 1,1-Dichloroethene

polar, because the two C—Cl dipole moments do not cancel. We therefore predict that *o*-dichlorobenzene should have a higher boiling point than *p*-dichlorobenzene. The experimental values are 180°C for *o*-dichlorobenzene and 174°C for *p*-dichlorobenzene.

SELF-TEST 10.1A Which will have the higher boiling point, *cis*-dichloroethene (**5**) or *trans*-dichloroethene (**6**)?

[*Answer: cis*-Dichloroethene]

SELF-TEST 10.1B Which will have the higher boiling point, 1,1-dichloroethene (**7**) or *trans*-dichloroethene?

We can use trends in physical properties to decide whether a trend in the strengths of intermolecular forces within a series of similar compounds can be explained in terms of the strengths of London forces or their dipole-dipole interactions.

Polar molecules form liquids and solids partly as a result of dipole-dipole interactions, the attraction between the partial charges of their molecules.

EXAMPLE 10.2 Accounting for a trend in boiling points

Explain the trend in the boiling points of the hydrogen halides: HCl, −85°C; HBr, −67°C; HI, −35°C. The electronegativity difference between hydrogen and the halogen *decreases* from HCl to HI.

STRATEGY Stronger intermolecular forces result in higher boiling points. The dipole moments, and thus the strength of the dipole-dipole interactions, increase with the difference in electronegativity between the hydrogen and halogen atoms. The strength of London forces increases with the number of electrons. Use the data to decide which is the dominant effect.

SOLUTION Because electronegativity differences decrease from HCl to HI, the dipole moments decrease in that order. Therefore, dipole-dipole forces decrease too, a trend suggesting that the boiling point should decrease from HCl to HI. This prediction conflicts with the data, so we examine the London forces. The number of electrons in a molecule increases from HCl to HI, so the London interaction increases too. Hence, the boiling points should increase from HCl to HI, in accord with the data. This analysis suggests that London forces dominate dipole-dipole interactions for these molecules.

SELF-TEST 10.2A Account for the trend in boiling points of the noble gases, which increase from helium to radon.

[*Answer:* London interactions increase as the number of electrons increases.]

SELF-TEST 10.2B Suggest a reason why trifluoromethane, CHF_3, has a higher boiling point than tetrafluoromethane, CF_4.

10.3 HYDROGEN BONDING

The third intermolecular interaction, called **hydrogen bonding,** gives water its unusual properties. We see the effect of this force when we plot the boiling points of the hydrides of the elements in Groups 14 to 17 (Fig. 10.6). The trend in Group 14 is what we expect for similar compounds that differ in their number of electrons—the boiling points increase with molar mass because their London forces increase. However, the hydrides of nitrogen, oxygen, and fluorine show anomalous behavior. Water boils at a much higher temperature (100°C) than hydrogen sulfide (−60°C). In fact, hydrogen sulfide is a gas at room temperature, even though an H_2S molecule has many more electrons—and hence stronger London forces—than an H_2O molecule. Ammonia and hydrogen fluoride have higher boiling points than phosphine, PH_3, and hydrogen chloride, HCl, respectively. The exceptionally high boiling points of water, ammonia, and hydrogen fluoride suggest that there are unusually strong attractive forces between their molecules.

A hydrogen bond forms when a hydrogen atom lies between two small, strongly electronegative atoms with lone pairs of electrons, specifically N, O, or F. A hydrogen bond is denoted by dots to distinguish it from a true covalent bond, as in O—H···N, for example. To understand the formation of the bond, let's imagine what happens when one water molecule comes close to another. Each O—H bond is polar. The electronegative O atom exerts a strong pull on the electrons in the bond, and the proton is almost completely unshielded. Because it is so small, the hydrogen atom, with its partial positive charge, can get very close to one of the lone pairs of electrons on the O atom of another water molecule. The lone pair and the partial charge attract each other strongly and form a bond. Hydrogen bonding is strongest when the hydrogen atom is on a straight line between the two oxygen atoms (**8**). Note that the graphic shows only the atoms directly involved in the hydrogen bond.

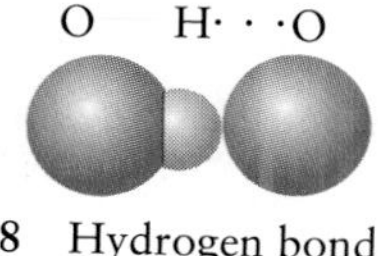

8 Hydrogen bond

SELF-TEST 10.3A Which of the following intermolecular links can be made by hydrogen bonds: (a) CH_3NH_2 to CH_3NH_2; (b) CH_3OCH_3 to CH_3OCH_3; (c) HBr to HBr?

[*Answer:* Only (a) has H bonded directly to N, O, or F.]

SELF-TEST 10.3B Which of the following molecules can take part in hydrogen bonding in the pure state: (a) CH_3OH; (b) PH_3; (c) HClO (which has the structure Cl—O—H)?

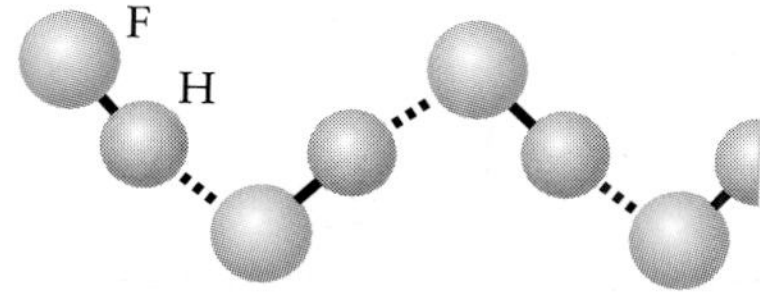

9 Hydrogen fluoride, $(HF)_n$

When it can occur, a hydrogen bond is strong enough to dominate all the other types of intermolecular interactions. In fact, hydrogen bonding is so strong that in some cases it survives even in a vapor. Liquid hydrogen fluoride, for instance, contains zigzag chains of HF molecules (**9**),

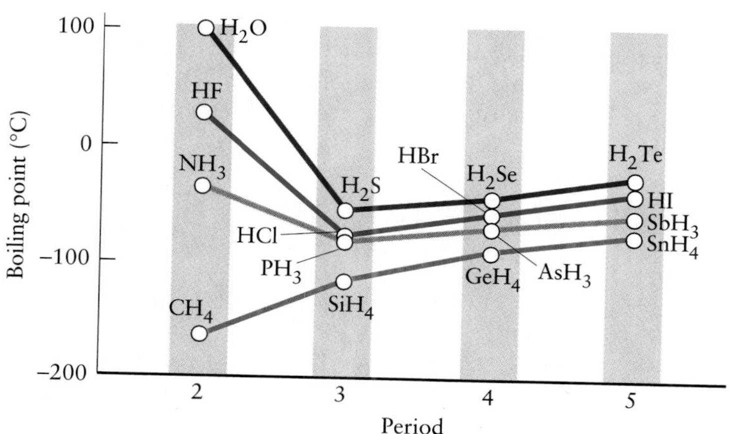

FIGURE 10.6 The boiling points of the hydrogen compounds of most of the *p*-block elements show a smooth variation and increase with molar mass in each group. However, three compounds—ammonia, water, and hydrogen fluoride—are strikingly out of line.

FIGURE 10.7 Many adhesives act by forming hydrogen bonds to the OH groups in the surfaces they are being used to bond. Other adhesives set to a solid polymer after they have spread into the cracks and gaps in the surfaces and hence lock the two surfaces together.

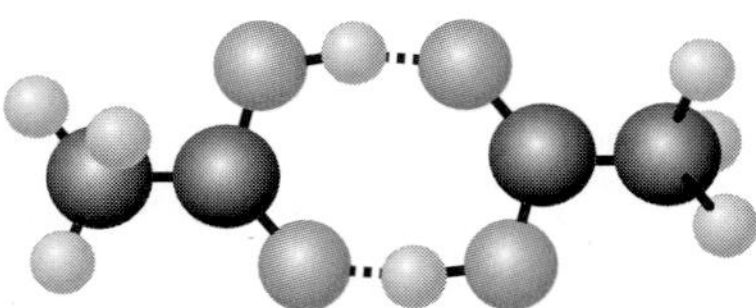

10 Acetic acid dimer

▶▶ **The role of hydrogen-bonding in protein molecules is described in Section 11.17.**

and the vapor contains short fragments of the chains and $(HF)_6$ rings. Acetic acid vapor contains **dimers,** or pairs of molecules, linked by two hydrogen bonds (**10**).

The shape of a protein molecule is governed partly by hydrogen bonds; and once the bonds are broken, the delicately organized protein molecule loses its function. When we cook an egg, the clear albumen becomes milky white, because the heat breaks the hydrogen bonds in its molecules, which collapse into a random jumble. Trees stand upright on account of hydrogen bonds. Cellulose molecules (which have many OH groups) can form many hydrogen bonds with one another, and the strength of wood is due in large part to the strength of the hydrogen bonds between neighboring ribbonlike cellulose molecules. Many wood and paper glues are substances that form hydrogen bonds to the two surfaces they join (Fig. 10.7). Hydrogen bonding also binds the two chains of a DNA molecule together, so hydrogen bonding is a key to understanding reproduction (see Section 11.19).

Hydrogen bonding, which occurs between oxygen, nitrogen, and fluorine atoms, is the strongest type of intermolecular force.

LIQUID STRUCTURE

The molecules in a liquid are in contact with their neighbors, but they can move past and tumble over one another (Fig. 10.8). When we imagine a liquid, we can think of a jostling crowd of molecules, each one con-

FIGURE 10.8 The structure of a liquid. Although the molecules (represented by the spheres in this series of diagrams) remain in contact with their neighbors, they have enough energy to push through to a new neighborhood, and the entire substance is fluid.

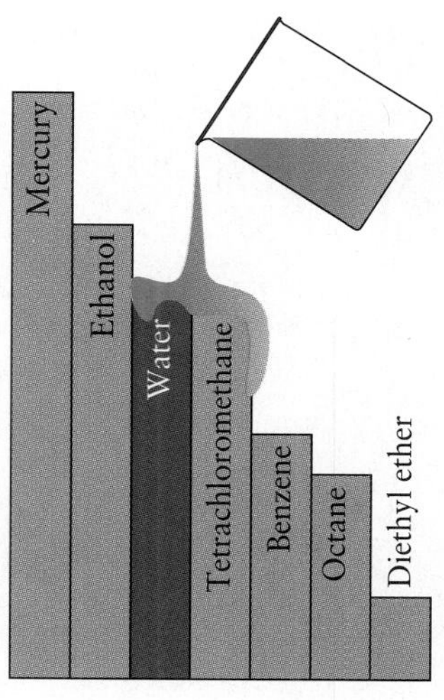

FIGURE 10.9 The relative viscosities of several liquids, compared with water. Liquids composed of molecules that cannot form hydrogen bonds are generally less viscous than those that can form hydrogen bonds. Mercury is an exception: its atoms stick together by a kind of metallic bonding, and its viscosity is relatively high.

stantly changing places with its neighbors. A liquid at rest is like a crowd of people milling about in a stadium; a flowing liquid is like a crowd of people leaving a stadium.

In the next two sections we shall look at two of the most characteristic properties of liquids: their ability to flow and their possession of a sharply defined surface. In Section 10.13 we shall also meet their ability to vaporize and thus exert a vapor pressure. All three properties are related to the strengths of the intermolecular forces within the liquid.

10.4 VISCOSITY

The **viscosity** of a liquid is its resistance to flow: the higher the viscosity, the more sluggish the flow. A liquid with a high viscosity (like molasses at room temperature, or molten glass) is said to be viscous. Figure 10.9 shows the relative viscosities of several liquids.

Viscosity arises from the forces between molecules: strong intermolecular forces hold molecules together and do not let them move past one another easily. Therefore, we should be able to explain the trends shown in Fig. 10.9 in terms of intermolecular forces. Because hydrogen bonding is the strongest kind of intermolecular bonding, water has a greater viscosity than benzene.

Phosphoric acid (H_3PO_4) and glycerol ($C_3H_8O_3$) are very viscous at room temperature because of the numerous hydrogen bonds their molecules can form. London forces can also be strong enough to cause high viscosity. The long chains in hydrocarbon oils with high molar masses are tangled like cooked spaghetti (Fig. 10.10), and because of London forces, they slide past one another only with difficulty.

FIGURE 10.10 Heavy hydrocarbon oils, which consist of long hydrocarbon chains, tend to get tangled together like a plate of cooked spaghetti. As a result, the molecules do not move past one another very readily, and the liquid is very viscous.

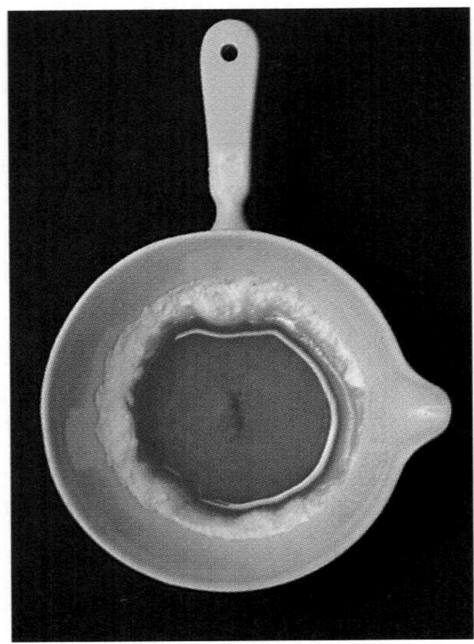

(a)

(b)

(c)

FIGURE 10.11 (a) When sulfur is heated, it melts to form a mobile, straw-colored liquid of S_8 rings. (b) As the temperature is increased, the rings break open and form chains that become tangled. As a result, the viscosity increases. (c) At higher temperatures, the S_8 chains break up into S_2 and S_3 molecules and the viscosity falls again.

Viscosity usually decreases as the temperature rises. Molecules have more kinetic energy at high temperatures and can wriggle past their neighbors more readily. The viscosity of water at 100°C, for instance, is only one-sixth of its value at 0°C, so the same amount of water flows through a tube six times as fast at the higher temperature. In some cases, though, a change in molecular structure takes place on heating, and the viscosity increases. For example, just above 113°C, rhombic sulfur is a mobile, straw-colored liquid (Fig. 10.11). Its S_8 rings can move past one another readily. The viscosity increases when the liquid is heated further because the S_8 rings break open into chains that become tangled. The viscosity falls again at still higher temperatures, and the color changes to deep red-brown: now the S_8 chains are breaking up into smaller, more mobile, highly colored, S_2 and S_3 molecules.

The greater the viscosity of a liquid, the more slowly it flows; hydrogen-bonded liquids typically have high viscosities. Viscosity usually decreases with increasing temperature.

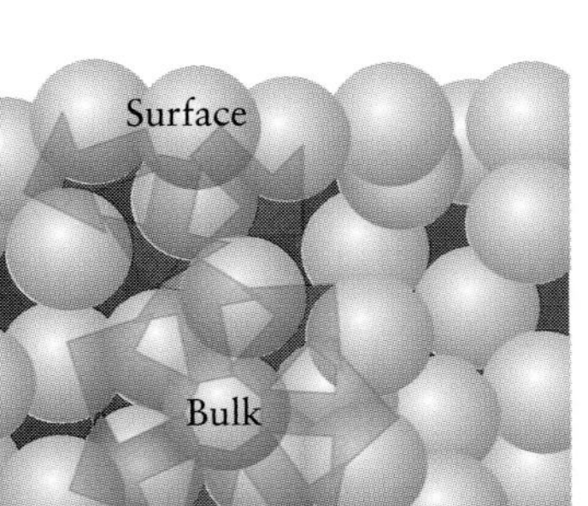

FIGURE 10.12 Surface tension arises from the attractive forces acting on the molecules at the surface. A molecule within the liquid experiences forces from all directions, but a molecule at the surface experiences a net inward force.

10.5 SURFACE TENSION

The surface of a liquid is smooth because intermolecular forces tend to pull the water molecules together and inward (Fig. 10.12). The **surface tension** of a liquid is the net inward pull. It packs the molecules together and forms a smooth surface. The surface tension of water is about three

FIGURE 10.13 The nearly spherical shape of these beads of water on the waxy surface of a leaf arises from the effect of surface tension. The beads are flattened slightly by the effect of the Earth's gravity.

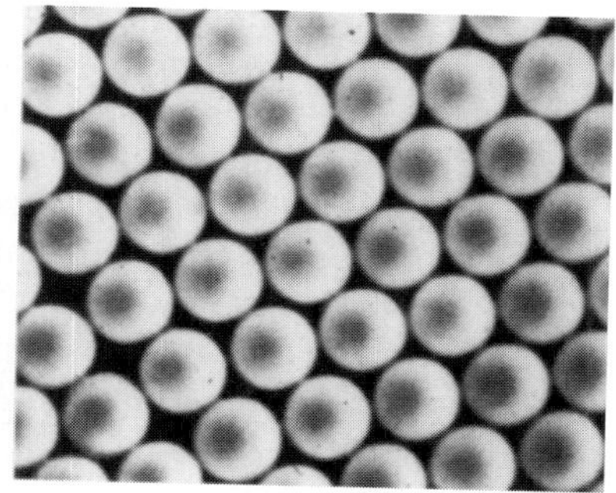

FIGURE 10.14 Latex spheres produced on one of the space shuttle missions. Each one is a perfect sphere of diameter 0.01 mm, formed by surface tension in the absence of gravity.

times higher than that of most other common liquids, because of its strong hydrogen bonds. The surface tension of mercury is higher still—more than six times that of water. This difference suggests that there are strong bonds between mercury atoms in the liquid.

A droplet of liquid suspended in air is spherical because the surface tension pulls the molecules into the most compact shape, a sphere. The droplet is approximately spherical even when it rests on a waxed surface (Fig. 10.13). The attractive forces between water molecules are greater than those between water and wax, which is largely hydrocarbon. On Earth, the droplets are slightly flattened by gravity, but in the gravity-free environment of an orbiting space shuttle, the shape of droplets is governed by surface tension alone, and tiny (0.01-mm diameter), perfect spheres of polymers that have been formed in space are now commercially available (Fig. 10.14). The spheres are used to calibrate particle sizes for powdered pharmaceuticals.

Water has strong interactions with paper, wood, or cloth, because the molecules in their surfaces form hydrogen bonds. As a result, water spreads over them; in other words, water wets them.

A duck's feathers are coated with oils that repel water, so its feathers never actually get wet, even when it is swimming in water.

The attraction between water and materials such as glass accounts for **capillary action,** the rise of liquids up narrow tubes. The liquid rises because there are favorable attractions between its molecules and the tube's inner surface. These are forces of **adhesion,** forces that bind a substance to a surface, as distinct from the forces of **cohesion,** the forces that bind the molecules of a substance together to form a bulk material. Narrow tubes and high adhesive forces result in tall columns of liquid; wide

FIGURE 10.15 When the adhesive forces between a liquid and glass are stronger than the cohesive forces within the liquid, the liquid forms the meniscus shown here for water in glass (left). When the cohesive forces are stronger than the adhesive forces (as they are for mercury in glass), the surface is curved downward (right).

tubes and low adhesive forces result in short columns of liquid. Capillary action is partly responsible for the ability of a paper towel to mop up a water spill, because the narrow channels between the fibers of the paper act like capillaries.

The **meniscus** of a liquid is the curved surface it forms in a narrow tube (Fig. 10.15). The meniscus of water in a glass capillary is curved upward at the edges (forming a ⌣ shape) because the adhesive forces between water molecules and the oxygen atoms and OH groups of the glass surface are stronger than the cohesive forces between water molecules. The water therefore tends to spread over the greatest possible area of glass. The meniscus of mercury curves downward in glass (forming a ⌢ shape). This shape is a sign that cohesive forces between mercury atoms are stronger than their adhesion to glass, for in this case the liquid tends to reduce its contact with the glass.

Surface tension arises from the imbalance of intermolecular forces at the surface of a liquid. It is responsible for the tendency of liquids to form droplets and for capillary action.

SOLID STRUCTURES

Almost all substances form solids when the temperature is low enough. However, the nature of the solid depends on the types of forces that hold the atoms, ions, or molecules together in a tightly packed array.

10.6 CLASSIFICATION OF SOLIDS

The word *amorphous* comes from the Greek words for "without form."

Solids are classified as crystalline or amorphous. A **crystalline solid** is a solid in which the atoms, ions, or molecules lie in an orderly array (Fig. 10.16). An **amorphous solid** is one in which the atoms, ions, or molecules lie in a random jumble, as in butter, rubber, and glass (Fig. 10.17). Crystalline solids typically have flat, well-defined surfaces called **faces,** that

FIGURE 10.16 Crystalline solids have well-defined faces and an orderly internal structure. Each face is the edge of a stack of atoms, molecules, or ions. The crystal in the photograph is galena, PbS.

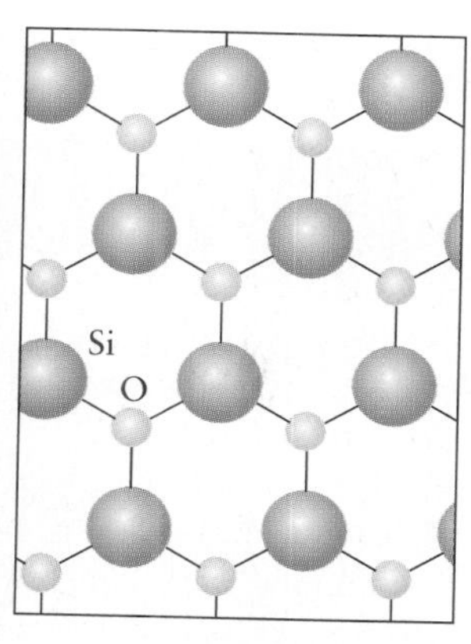

(a)

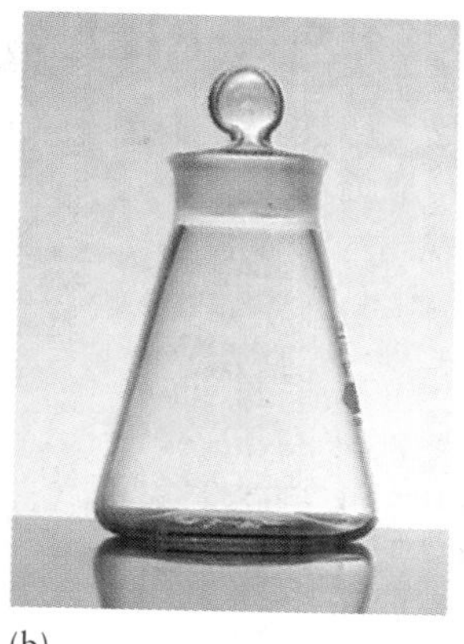

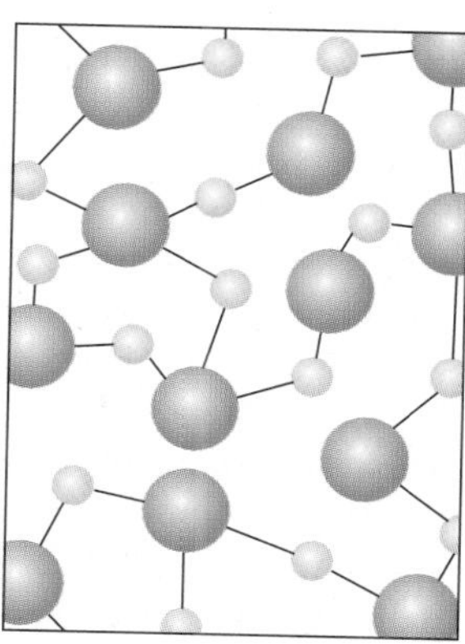

(b)

FIGURE 10.17 (a) Quartz is a crystalline form of silica, SiO_2; and the atoms are arranged in an orderly network, represented here in two dimensions. (b) When molten silica solidifies rapidly it becomes glass. Now the atoms form a disorderly network.

have definite angles at their edges. These faces are formed by orderly stacks of atoms. Amorphous solids do not have well-defined faces unless they have been molded or cut. When some solids melt, they become **liquid crystals,** a phase of matter with some properties of both liquids and crystalline solids. The properties of liquid crystals are discussed in Case Study 10.

Many substances form crystalline solids. These materials include metallic elements, such as copper and iron, and alloys such as brass, in which the atoms of the elements are stacked in regular arrays (Fig. 10.18). The solid forms of the nonmetallic elements, such as sulfur, phosphorus, iodine, and solid argon are also crystalline. Ionic compounds, such as sodium chloride and potassium nitrate, crystallize into structures that allow cations and anions to stack together in electrostatically favorable arrangements, sometimes with hydrated water molecules (as in $CuSO_4 \cdot 5H_2O$). In many substances, the atoms, ions, or molecules can adopt more than one arrangement, depending on the conditions under which they formed. These different arrangements are the different **solid phases** of the substance. Thus, diamond and graphite are two different solid phases of carbon: the carbon atoms are arranged differently in each phase. Different solid phases have different physical properties, such as melting point and density. The specific arrangement of atoms, ions, and molecules within a crystal can be determined by x-ray diffraction (Box 10.1).

▶▶ Alloys are discussed in Section 10.9.

We classify crystalline solids according to the bonds that hold their atoms, ions, or molecules in place (Table 10.3):

FIGURE 10.18 The stacks of apples, oranges, and other produce in a grocery store form faces with different slopes. They illustrate how atoms stack together in metals to give single crystals with flat faces.

BOX 10.1 X-RAY DIFFRACTION

X-ray diffraction is one of the most powerful techniques for determining the structures of solids and the molecules they contain. We can use it not only to determine the separations of layers of atoms but also to determine the locations of the atoms in enzymes containing thousands of atoms. The technique is primarily responsible for the enormous advances in molecular biology over the past few decades. With it, we can understand the processes of life more deeply than we have ever done before.

To understand the principles involved, we need to know that waves may **interfere** with one another. Imagine two waves of electromagnetic radiation in the same region of space. Where the peaks of one wave coincide with the peaks of the other wave, they add together to give a stronger wave, as shown in the illustration below. This strengthening is called **constructive interference.** When the combined wave is detected photographically or electronically, the spot obtained is brighter than would be obtained by either ray alone. However, the waves partially cancel each other when the peaks of one coincide with the troughs of the other; the result is a weaker wave. This cancellation is called **destructive interference.** When the combined wave is detected, we see a dimmer spot than would be obtained from either ray alone. No spot at all is detected when the peaks and troughs match exactly and cancellation is complete.

Diffraction is the interference between waves caused by an object in their path. The resulting pattern of bright spots against a dark background is called a **diffraction pattern.** A crystal can cause diffraction in a beam of x-rays, and a bright spot of constructive interference is obtained when the crystal is held at a certain angle to the beam. The angle θ (the Greek letter theta) at which constructive interference occurs is related to the wavelength λ and the distance between the atoms (see the illustration below) by

$$2d \sin \theta = \lambda$$

This formula, which is called the **Bragg equation,** lets us measure the spacing between layers of atoms. For example, if we find a spot at 17.5° when we use x-rays of wavelength 154 pm, we can conclude that there are layers of atoms separated by a distance

$$d = \frac{\lambda}{2 \sin \theta} = \frac{154 \text{ pm}}{2 \sin 17.5^\circ} = 256 \text{ pm}$$

Because $\sin \theta$ cannot be greater than 1, the smallest distance d that can be measured in this way is $\frac{1}{2}\lambda$. That is why x-rays need to be used: only their wavelengths are short enough to measure distances comparable to the separations of atoms in crystals.

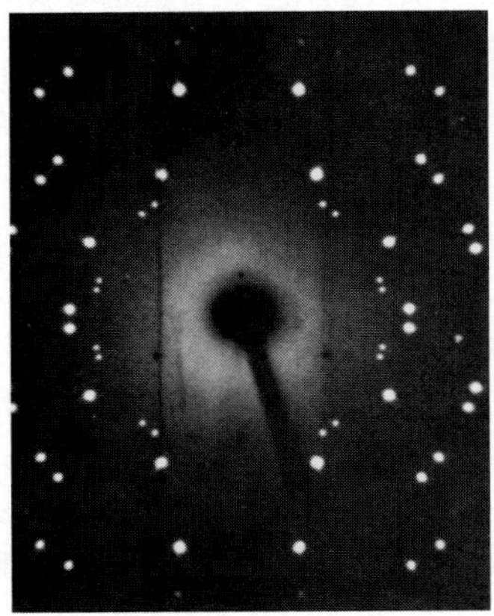

The diffraction pattern formed when x-rays pass through a single crystal of sodium chloride. The bright spots are the photographic record of the pattern formed where x-rays interfere constructively.

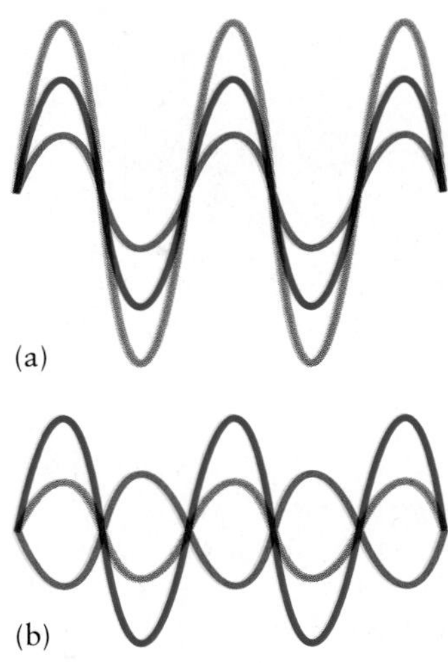

Interference between waves (the red and green lines) may be either (a) constructive, to give a wave of greater amplitude (the orange line), or (b) destructive, to give a wave of smaller amplitude.

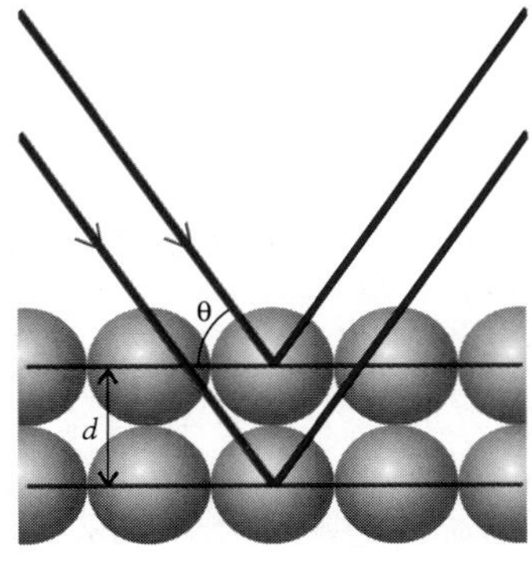

The definition of the distance d between layers of atoms and the angle θ that occur in the Bragg equation.

TABLE 10.3 Typical characteristics of solids

Class	Examples	Characteristics
metallic	*s*- and *d*-block elements	malleable, ductile, lustrous, electrically and thermally conducting
ionic	NaCl, KNO_3, $CuSO_4 \cdot 5H_2O$	hard, rigid, brittle; high melting and boiling points; those soluble in water give conducting solutions
network	B, C, black P, BN, SiO_2	hard, rigid, brittle; very high melting points; insoluble in water
molecular	$BeCl_2$, S_8, P_4, I_2, ice, glucose, naphthalene	relatively low melting and boiling points; brittle if pure

Metals, also called metallic solids, consist of cations held together by a sea of electrons.

Ionic solids are built from the mutual attractions of cations and anions.

Network solids consist of atoms bonded to their neighbors covalently throughout the extent of the solid.

Molecular solids are collections of distinct molecules held in place only by intermolecular forces.

Crystalline solids have a regular internal arrangement of atoms; solids are classified as metallic, ionic, network, or molecular.

10.7 METALLIC CRYSTALS

All the atoms in a crystal of a metallic element have identical electronic structures. To describe them, we can use identical spheres to represent the atoms and stack them together like oranges in a grocery display.

In a **close-packed structure,** the atoms stack together with the least waste of space. To see how to stack identical spheres together to give a close-packed structure, look at Fig. 10.19. The spheres in the first layer (A) touch their six neighbors. The spheres of the second (upper) layer (B) lie in the dips of the first layer. The third layer of spheres (not shown in Fig. 10.19) lies in the dips of the second layer, and so on.

The third layer of spheres may lie in the dips that are directly above the atoms of the first layer (Fig. 10.20). The third layer then duplicates layer A, the next layer duplicates B, and so on. This results in an ABABAB . . . pattern of layers called the **hexagonal close-packed** (hcp) **structure.** It is possible to make out a hexagonal pattern in the arrangement of atoms (Fig. 10.21). Magnesium and zinc are examples of metals that crystallize in this way.

As can be seen from the illustration, each sphere has three nearest neighbors in the plane below, six in its own plane, and three in the one above, giving 12 in all. This arrangement is reported by saying that the **coordination number,** the number of nearest neighbors of each atom, in

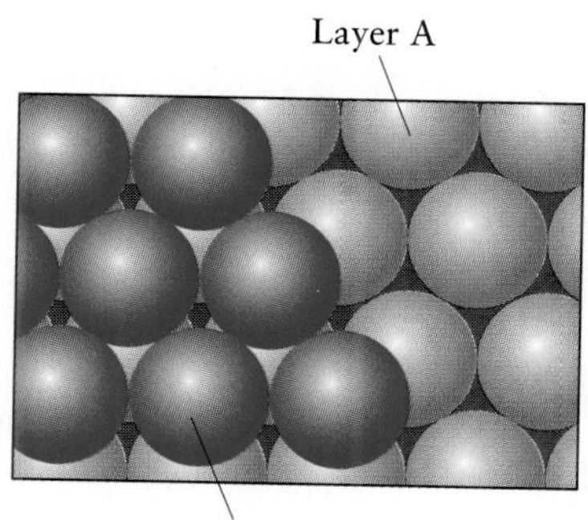

FIGURE 10.19 A close-packed structure can be built up in stages. The first layer (A) is laid down with minimum waste of space, and the second layer (B) lies in the dips—the depressions—of the first.

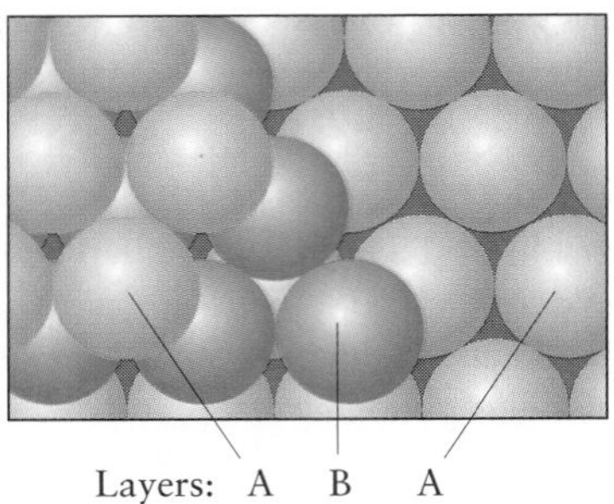

FIGURE 10.20 The third layer of spheres can lie above the atoms of the first layer to give an ABABAB . . . structure. Notice that you can see all the way through the holes in the layers.

FIGURE 10.21 A fragment of the structure formed as described in Fig. 10.20 shows the hexagonal symmetry of the arrangement, and the origin of its name "hexagonal close-packed."

the solid is 12. It is impossible to pack identical spheres together with a coordination number greater than 12.

In an alternative arrangement, the spheres of the third layer lie in the dips of the second layer that do not lie directly over the atoms of the first layer (Fig. 10.22). If we call this third layer C, then the resulting structure has an ABCABC . . . pattern of layers to give a **cubic close-packed** (ccp) **structure.** This structure has a cubic pattern (Fig. 10.23). The coordination number is also 12: each sphere has three nearest neighbors in the layer below, six in its own layer, and three in the layer above. Aluminum, copper, silver, and gold are examples of metals that crystallize in this way.

The holes between groups of neighboring atoms are known as *interstices.*

Which close-packed structure—if either—a metal adopts depends on which gives the lower energy, and that depends on details of its electronic

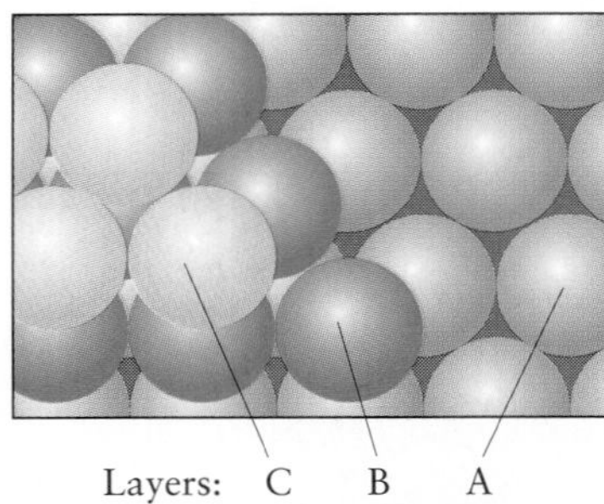

FIGURE 10.22 As an alternative to the scheme shown in Fig. 10.20, the spheres of the third layer can lie above the dips in the first layer to give an ABCABC . . . arrangement of layers. In this case you cannot see through the layers because each interstice has an atom either on top of it or below it.

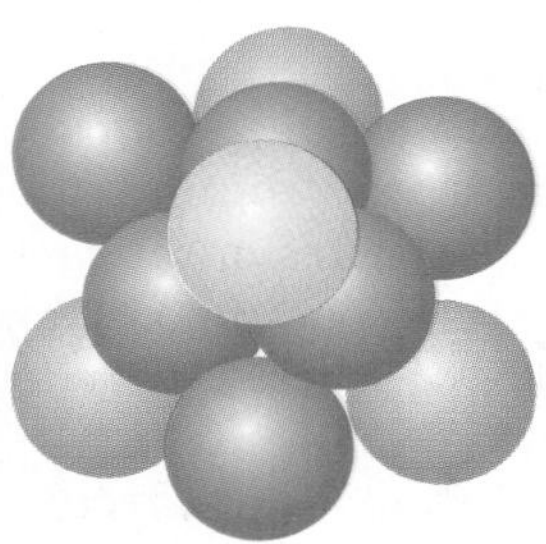

FIGURE 10.23 A fragment of the structure formed, as described in Fig. 10.22, shows the origin of the names "cubic close-packed" or "face-centered cubic" for this arrangement. The layers A, B, C can be seen along the diagonals.

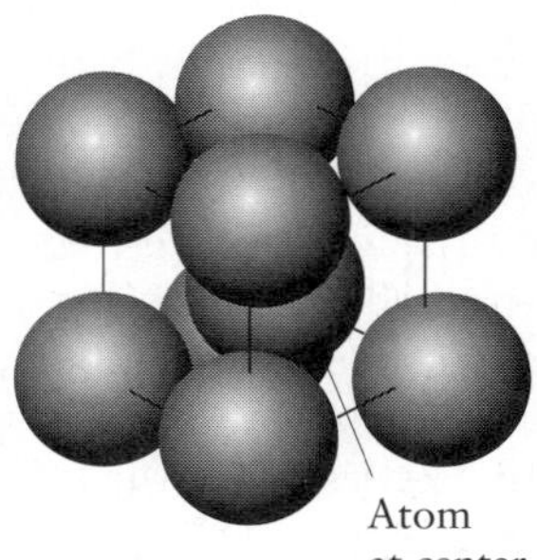

FIGURE 10.24 The body-centered cubic (bcc) structure. This structure is not packed as closely as the others we have illustrated. It is less common among metals than the close-packed structures, but some ionic structures are based on it.

structure. In fact, some atoms achieve a lower energy by adopting a different arrangement entirely. In a **body-centered cubic** (bcc) **structure,** for instance, a single atom lies at the center of a cube formed by eight other atoms (Fig. 10.24). This structure is not close packed. Iron, sodium, and potassium are examples of metals that crystallize in this way.

Many metals have close-packed structures, with the atoms stacked in either a hexagonal or a cubic arrangement; close-packed atoms have a coordination number of 12.

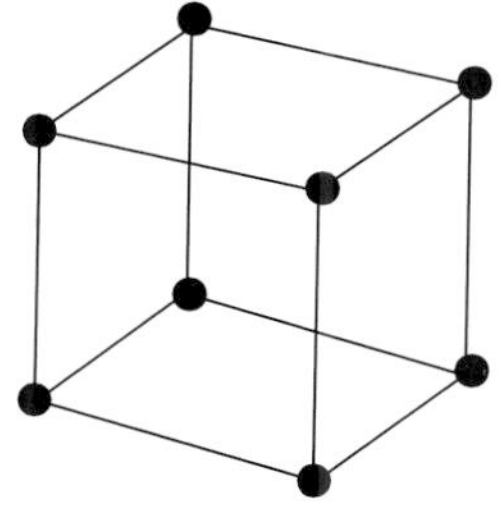

11 Primitive cubic cell

SELF-TEST 10.4A What is the coordination number of an atom in a body-centered cubic (bcc) structure?

[*Answer:* 8]

SELF-TEST 10.4B What is the coordination number of an atom in a primitive cubic structure, in which there is one atom at each corner of a cube (**11**)? (*Hint:* Take into account the atoms in the neighboring cubes (Fig. 10.25).)

Each of the small units shown in Figs. 10.21 and 10.23–10.25 is a **unit cell,** the smallest hypothetical unit that, when stacked together repeatedly without any gaps, can reproduce the entire crystal. A unit cell is to a crystal in three dimensions as a tile is to a tiled floor in two dimensions (Fig. 10.26). Unit cells are sometimes drawn by representing each atom by a dot that marks the location of its center (Fig. 10.27). A cubic close-packed unit cell has a dot at the center of each face; for this reason, it is also called a **face-centered cubic** (fcc) **structure.**

The number of atoms in a unit cell is counted by noting how they are shared between neighboring cells. For example, an atom at the center of a cell belongs entirely to that cell, but one on a face is shared between two cells and counts as one-half an atom. For an fcc structure, each of the eight corner atoms is shared by eight cells, so overall they contribute

FIGURE 10.25 A primitive cubic unit cell. To count the number of nearest neighbors of an atom, we have to imagine the cell of interest with its neighbors stacked up around it.

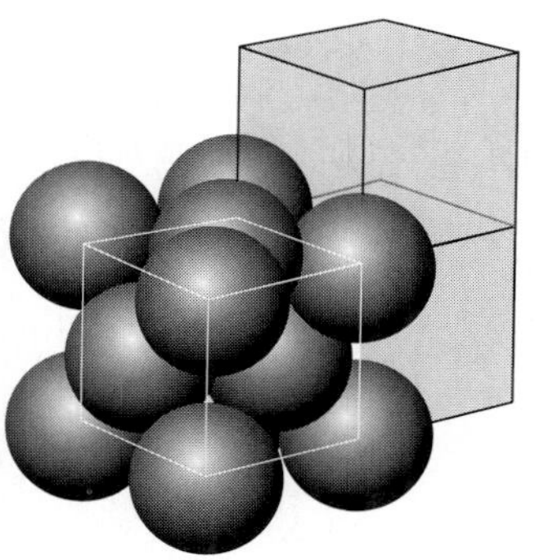

FIGURE 10.26 The entire crystal structure is constructed from a unit cell by stacking the cells together without any gaps in between. Each corner atom is shared by eight cubes that touch at the corner.

FIGURE 10.27 The unit cells of the (a) ccp (or fcc) and (b) bcc structures in which the locations of the centers of the spheres are marked by dots.

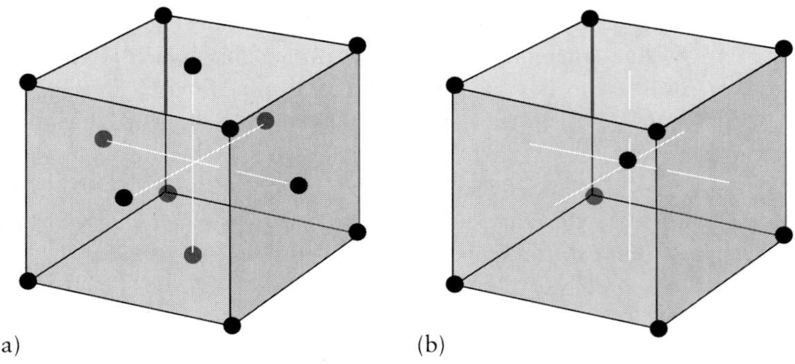

$8 \times \frac{1}{8} = 1$ atom to the cell. Each atom at the center of each of the six faces contributes one-half an atom, so jointly they contribute $6 \times \frac{1}{2} = 3$ atoms (Fig. 10.28). The total number of atoms in an fcc unit cell is therefore $1 + 3 = 4$, and the mass of the unit cell is four times the mass of one atom.

EXAMPLE 10.3 Counting the number of atoms in a unit cell

How many atoms are there in the bcc cell shown in Fig. 10.27b?

STRATEGY As explained above, count 1 for each atom that is not shared by other cells, and $\frac{1}{2}$ for each atom on a face. When an atom is at an edge, it counts $\frac{1}{4}$ (because it is shared by four cells). Each atom at a corner (where it is shared by eight cells) counts as $\frac{1}{8}$.

SOLUTION Inspection of Fig. 10.27b shows that there is one central atom (count 1) and eight corner atoms (count $8 \times \frac{1}{8} = 1$). Therefore, each cell has two atoms.

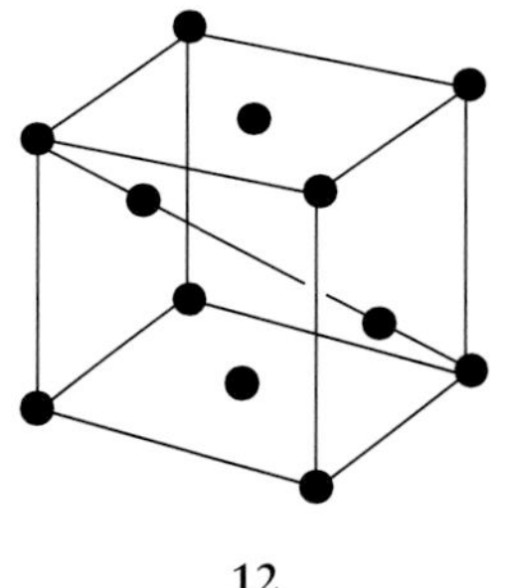

12

SELF-TEST 10.5A How many atoms are there in a primitive cubic cell (see Fig. 10.25)?

[*Answer:* 1]

SELF-TEST 10.5B How many atoms are there in the structure made up of unit cells like the one shown in (**12**), which has an atom at each corner, two on opposite faces, and two inside the cell on a diagonal?

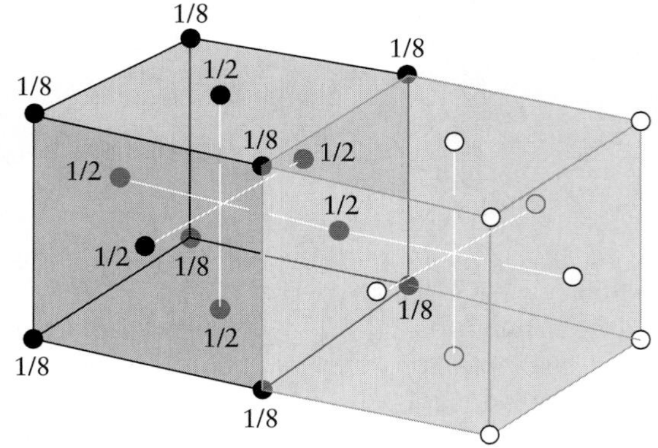

FIGURE 10.28 The contribution ($\frac{1}{8}$ or $\frac{1}{2}$) of the individual atoms in a face-centered cell to the total number of atoms in a unit cell.

Because many metals are close packed, they are generally denser than ionic solids. Metals that are not close packed, such as those that have a body-centered cubic structure, can often be forced under pressure into a close-packed form.

The atoms in a unit cell are counted by determining what fraction of each atom resides within the cell.

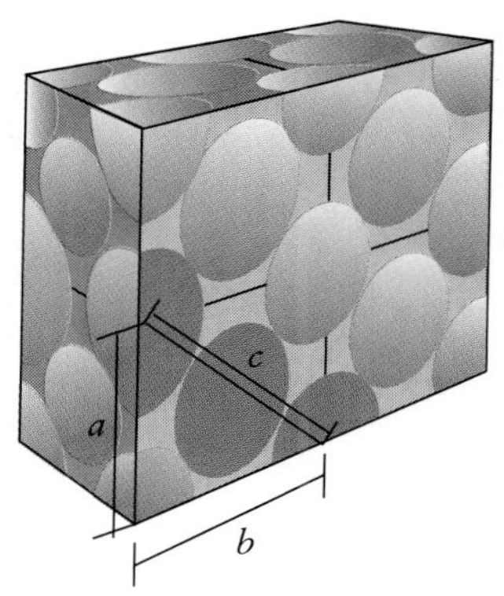

FIGURE 10.29 A part of the structure of a crystal of copper, showing the arrangement of atoms (four fcc unit cells are shown) and the relation of the dimensions of each cell to the radius of an atom, r. The length b, the length of the side of a cell, is $\sqrt{8}r$.

EXAMPLE 10.4 Deducing the structure of a metal from its density

The atomic radius of copper is 128 pm, and its density is 8.92 $g{\cdot}cm^{-3}$. Is it close packed?

STRATEGY Calculate the density of a unit cell of copper with a close-packed structure and compare the calculated value with the actual density. A cubic close-packed structure has the same density as a hexagonal close-packed structure because they are both close packed, and the former is easier to treat. The mass of a unit cell is the sum of the masses of the atoms it contains. The mass of one atom is its molar mass divided by the Avogadro constant. The volume of a cubic unit cell is the cube of the length of one of its sides. That length can be obtained from the radius of the metal atom, the Pythagorean theorem ($a^2 + b^2 = c^2$), and the geometry of the cell. A cubic close-packed structure has a face-centered cubic unit cell with four atoms, in which the length of the diagonal of one face of the cube is four times the atomic radius, r (Fig. 10.29). Because the density is given in grams per cubic centimeter, convert lengths to centimeters.

SOLUTION From Fig. 10.29 and the Pythagorean theorem,

$$\text{Side}^2 + \text{side}^2 = (4 \times r)^2 = 16r^2$$

Therefore,

$$\text{Side}^2 = \tfrac{1}{2} \times (16r^2) = 8r^2$$

so

$$\text{Side} = \sqrt{8}r$$

with

$$r = 128 \text{ pm} = 1.28 \times 10^{-10} \text{ m} = 1.28 \times 10^{-8} \text{ cm}$$

The volume of the unit cell is therefore

$$V = (\sqrt{8}r)^3 = (\sqrt{8} \times 1.28 \times 10^{-8} \text{ cm})^3 = 4.75 \times 10^{-23} \text{ cm}^3$$

The mass, m, of one atom of copper is

$$m = \frac{63.54 \text{ g}{\cdot}\text{mol}^{-1}}{6.022 \times 10^{23} \text{ mol}^{-1}} = 1.055 \times 10^{-22} \text{ g}$$

Each unit cell contains $8 \times \frac{1}{8} + 6 \times \frac{1}{2} = 4$ copper atoms. The total mass of a close-packed unit cell is four times the mass of one atom. Therefore,

$$\text{Density} = \frac{\text{mass}}{\text{volume}} = \frac{4 \times (1.055 \times 10^{-22} \text{ g})}{4.75 \times 10^{-23} \text{ cm}^3} = 8.89 \text{ g}{\cdot}\text{cm}^{-3}$$

The value is close to the experimental value for a close-packed structure.

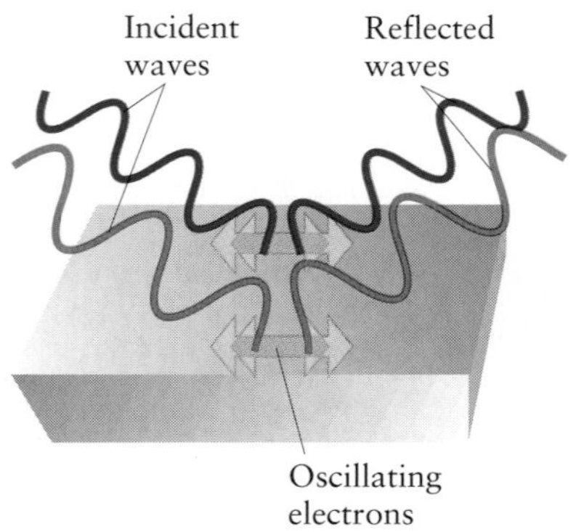

(a)

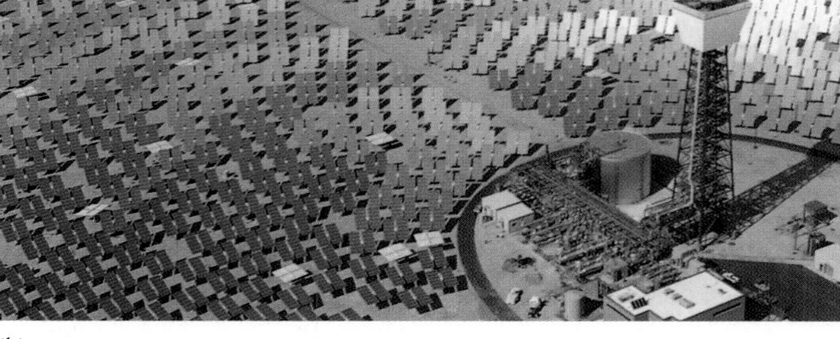

(b)

FIGURE 10.30 (a) When light of a particular color shines on the surface of a metal, the electrons at the surface oscillate in step. This oscillating motion gives rise to an electromagnetic wave that we perceive as the reflection of the source. (b) Each of these solar mirrors in California is positioned at the best angle to reflect sunlight into a collector that uses it to generate electricity.

SELF-TEST 10.6A The atomic radius of silver is 144 pm and its density is 10.5 $g \cdot cm^{-3}$. Is the structure close packed?

[*Answer:* Yes]

SELF-TEST 10.6B The atomic radius of iron is 124 pm and its density is 7.87 $g \cdot cm^{-3}$. Is this density consistent with a body-centered cubic structure?

10.8 PROPERTIES OF METALS

In a metallic solid, cations lie in a regular array and are surrounded by a sea of electrons. This structure gives unique properties to metals. Their characteristic luster is due to the mobility of their electrons. An incident light wave is an oscillating electric field (Section 7.1). When it strikes the surface, the electric field pushes the mobile electrons backward and forward. These oscillating electrons radiate light, and we see it as a luster—essentially a reflection of the incident light (Fig. 10.30). The electrons oscillate in step with the incident light, so they give out light of the same frequency as the incident light. In other words, red light reflected from a metallic surface is red and blue light is blue. That is why an image in a mirror—a thin metallic coating on glass—is a faithful portrayal of the reflected object. When we look at ourselves in a mirror, what we see are the oscillations of the mobile electrons in the metal film, with different parts of the film oscillating at different frequencies according to the color being reflected there.

The malleability and ductility of metals are also due to the mobility of their electrons. Because the cations are surrounded by an electron sea, there is very little directional character in the bonding. As a result, a cation can be pushed past its neighbors without too much effort. A blow from a hammer can drive large numbers of cations past their neighbors. The electron sea immediately adjusts to ensure that the atoms are not simply broken off but remain attached in their new positions (Fig. 10.31).

One of the most striking properties of a metal is its ability to conduct an electric current, the flow of electric charge. In **electronic conduction,** the type of electrical conduction that occurs in metals (so it is also called *metallic conduction*), the charge is carried by electrons. In **ionic conduction,** the charge is carried by ions. This is the mechanism of electrical

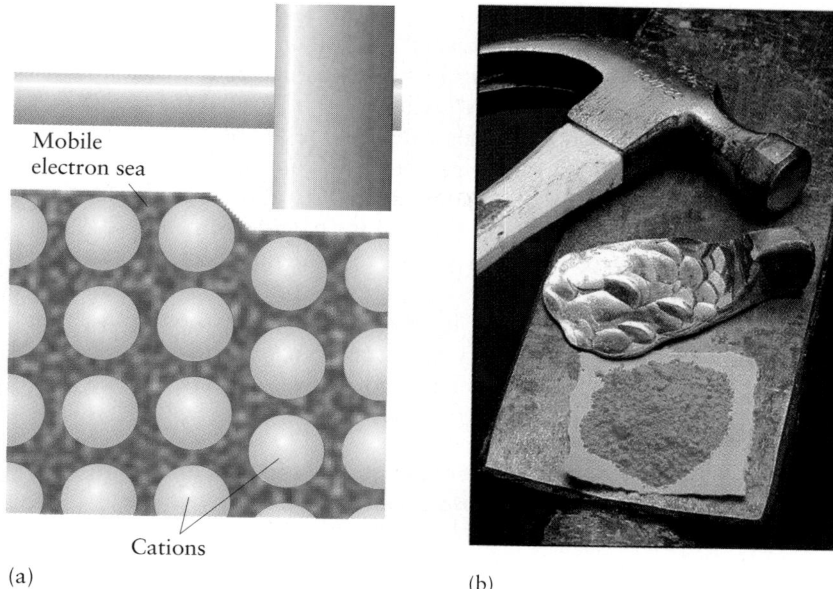

FIGURE 10.31 (a) A metal is malleable because, when cations are displaced by a blow from a hammer, the mobile electrons can immediately respond and follow the cations to their new positions. (b) This piece of lead has been flattened by hammering whereas crystals of ionic compounds, such as the lead(II) oxide shown here, shatter when struck.

conduction in a molten salt or an electrolyte solution. Because ions are too bulky to travel easily through most solids, the flow of charge through solids is almost always a result of electronic conduction.

The ability of a substance to conduct electricity is measured by its resistance: the lower the resistance, the better it conducts. We can classify substances according to the resistance they give the passage of a current and how the resistance varies with temperature (Fig. 10.32):

An **insulator** is a substance that does not conduct electricity.

A **metallic conductor** is an electronic conductor with a resistance that increases as the temperature is raised.

A **semiconductor** is an electronic conductor with a resistance that decreases as the temperature is raised.

A **superconductor** is an electronic conductor that conducts electricity with zero resistance.

Insulators include gases, most ionic solids, most network solids, almost all organic compounds, and all molecular and covalent liquids and solids. Metallic conductors include all metals and certain other solids, such as graphite. An example of a semiconductor is a crystal of pure silicon containing a tiny amount of arsenic or indium.

Until 1987, most superconductors were metals (such as lead) or compounds cooled to close to absolute zero. However, in 1987 the first *high-temperature superconductors* were reported. They could be used at about

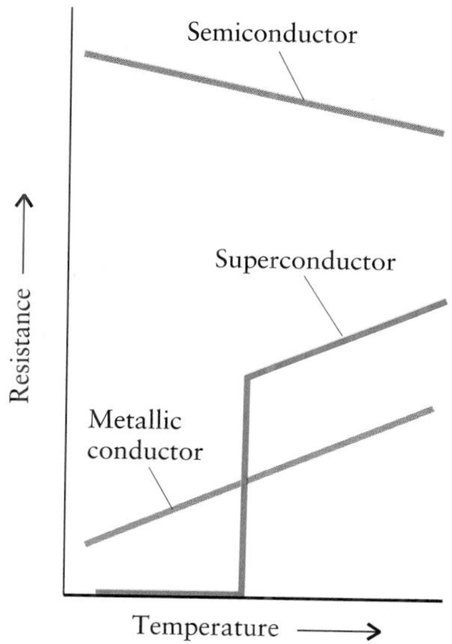

FIGURE 10.32 A metallic conductor is a substance with a resistance that increases with temperature. A semiconductor is a substance with a resistance that decreases with increasing temperature. A superconductor is a substance that has zero resistance below a certain temperature. An insulator behaves like a semiconductor with a very high resistance.

100 K, which means that liquid nitrogen (which boils at 77 K) can be used to cool them rather than the very much more expensive liquid helium (see Fig. 7.30). High-temperature superconductors are ceramics with complicated ionic oxide structures, such as $YbLa_2Cu_3O_7$.

Electrical conduction in metals can be thought of as resulting from electrons in the mobile sea moving past the cations. The resistance to their motion increases as the temperature is raised because the ions vibrate more violently and hinder the passage of the electrons.

Semiconductors are solids in which the valence electrons, although not free, can be removed from their parent atoms by a modest rise in temperature. As the temperature is raised, more electrons are liberated, so the ability of the solid to carry an electric current increases. In other words, its resistance decreases. The properties of semiconductors are described in more detail in Box 10.2.

The mobility of the valence electrons in a metal accounts for its luster, malleability, ductility, and electrical conductivity. The resistance of conductors increases with temperature, but the resistance of semiconductors decreases with temperature.

BOX 10.2 SEMICONDUCTORS

The atoms in a solid are so close together that the orbitals of large numbers of them merge together to form a continuum of energy levels that spread throughout the solid. In a metal, only about half the available levels are filled, so an electron needs very little extra energy to escape from an occupied level and move to an empty level just above. Once it is in the empty part of the band, an electron is free to move through the solid. A band of energy levels that is not completely full of electrons is called a **conduction band.** As a metal is heated, its atoms vibrate more vigorously and help to block the movement of the conducting electrons through the solid. As a result, the resistance of a metal increases with temperature.

In an insulating solid, the adjacent band of energy levels is completely full, and the next higher empty band is so high in energy that the electrons cannot be excited into it. As a result, they cannot move through the solid and it does not conduct electricity. A full band is called a **valence band.** So, in an insulator, we can say that the (empty) conduction band lies well above the (full) valence band.

In a semiconductor, an empty conduction band lies quite close in energy to a full valence band. As a result, as the solid is warmed, electrons can be excited from the valence band into the conduction band, where they can travel through the solid. So the resistance of a semiconductor falls as it is heated. The ability of a semiconductor to carry an electric current can also be enhanced by adding electrons to the conduction band or removing some from the valence band. This modification is carried out chemically by **doping**

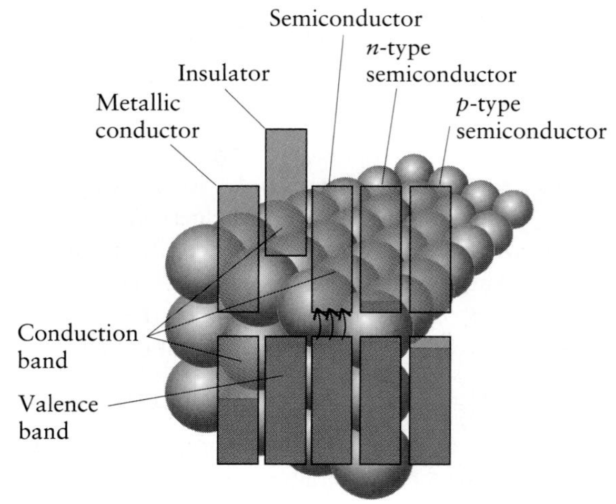

The band structure typical of a variety of solids. The occupied regions of the bands are shaded. Note that the distinction between a solid insulator and a semiconductor is the width of the gap between the valence and conduction bands.

10.9 ALLOYS

Alloys are mixtures of two or more metals. In homogeneous alloys, atoms of the different elements are distributed uniformly. Examples are brass, bronze, and the coinage alloys. Heterogeneous alloys, like tin-lead solder and the mercury amalgam used to fill teeth, consist of a mixture of crystalline phases with different compositions. Examples of common alloys are listed in Table 10.4.

The structures of alloys are more complicated than those of pure metals, because the two or more types of metal atoms have different radii. The packing problem is now akin to that of a storekeeper trying to stack oranges and grapefruit in the same pile.

Because the metallic radii of the *d*-block elements are all quite similar, for them the stacking problem is easily solved. They form an extensive range of alloys with each other because one type of atom can take the place of another with little distortion of the crystal structure. One example is the copper-zinc alloy used for some "copper" coins. Copper atoms crystallize in a cubic close-packed structure. Because zinc atoms are nearly the same size as copper atoms and have similar electronic

---

the solid, or spreading small amounts of impurities through it.

In an ***n*-type semiconductor,** a minute amount of a Group 15 element such as arsenic is added to very pure silicon. The arsenic increases the number of electrons in the solid: each Si atom (Group 14) has four valence electrons, whereas each As atom (Group 15) has five. The additional electrons enter the upper, normally empty conduction band of silicon and allow the solid to conduct. This type of material is called an *n*-type semiconductor because it contains excess *n*egatively charged electrons.

When silicon (Group 14) is doped with indium (Group 13) instead of arsenic, the solid has fewer valence electrons than pure silicon, so the valence band is no longer completely full. We say that the band now contains "holes." Because the valence band is no longer full, it has been turned into a conduction band; so now an electric current can flow. This type of semiconductor is called a ***p*-type semiconductor,** because the absence of negatively charged electrons is equivalent to the presence of positively charged holes. The solid is overall electrically neutral, because the nuclei of the doping atoms have a lower charge and hence compensate for the absence of electrons.

Solid-state electronic devices such as diodes, transistors, and integrated circuits contain *p-n* junctions in which a *p*-type semiconductor is in contact with an

This solar-powered vehicle makes use of silicon-based semiconductors to convert sunlight into electrical energy.

n-type semiconductor. An electric current can flow in only one direction across a *p-n* junction. The excess electrons in the *n*-type material can move across into the holes of the *p*-type material, but electrons cannot move in the opposite direction because the *n*-type material is already electron-rich and cannot accommodate them. Moreover, because solar radiation can excite electrons into a conduction band, semiconductors can be used to generate an electrical current by solar radiation. There is a hope that one day they will be used to power pollution-free transportation, such as the automobile in the illustration above.

TABLE 10.4 Compositions of typical alloys

Alloy	Mass percentage composition
brass	up to 40% zinc in copper
bronze	a metal other than zinc or nickel in copper (casting bronze: 10% Sn and 5% Pb)
cupronickel	nickel in copper (coinage cupronickel: 25% Ni)
pewter	6% antimony and 1.5% copper in tin
solder	tin and lead
stainless steel*	over 12% chromium in iron

*For more detailed information on steels, see Tables 21.2 and 21.3.

properties, they can take the place of some of the copper atoms in the crystal lattice. An alloy in which atoms of one metal are substituted for atoms of another metal is called a **substitutional alloy** (Fig. 10.33). Elements that can form substitutional alloys have atoms with atomic radii that differ by no more than about 15%. Because there are slight differences in size and electronic structure, the solute atoms in a substitutional alloy distort the shape of the lattice and hinder the flow of electrons. Because the lattice is distorted, it is harder for one plane of atoms to slip past another. Thus, although a substitutional alloy has lower electrical and thermal conductivity than the pure element, it is harder and stronger.

Although carbon is not a metal, it is conventional to regard steel as an alloy because the iron is modified by its presence.

Steel is an alloy of about 2% or less carbon in iron. Carbon atoms are much smaller than iron atoms and so cannot substitute for them in the crystal lattice. They are so small that they can fit into the gaps, or interstices, in the iron lattice. The resulting material is called an **interstitial alloy** (Fig. 10.34). For two elements to form an interstitial alloy, the atomic radius of the solute element must be less than 60% of the atomic radius of the host metal. The interstitial atoms interfere with electrical conductivity and with the movement of the atoms forming the lattice.

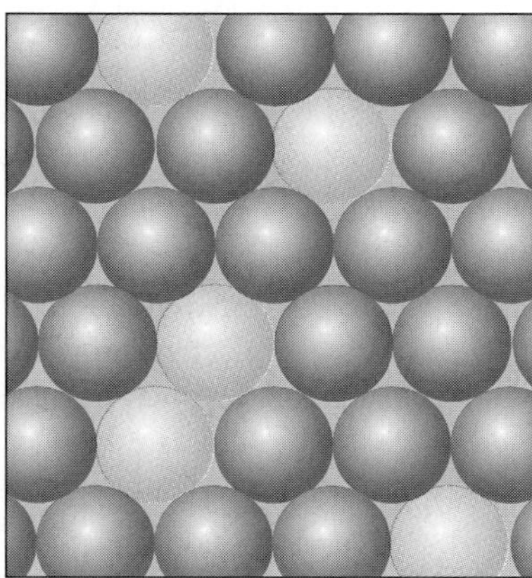

FIGURE 10.33 In a substitutional alloy, the positions of some of the atoms of one metal are taken by atoms of another metal. The two elements must have similar atomic radii.

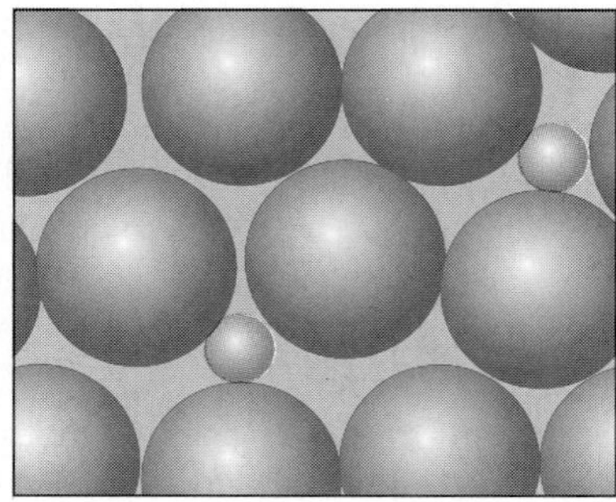

FIGURE 10.34 In an interstitial alloy, the atoms of one metal lie in the gaps between the atoms of another metal. The two elements must have markedly different atomic radii.

CHEMISTRY AT WORK

A computer scientist adjusts a computer-aided design and manufacturing (CADCAM) system that is used to produce computer chips. To produce a chip with the desired characteristics he must specify materials with the optimal electronic and crystalline properties.

This restricted motion makes the alloy harder and stronger than the pure host metal would be. For example, 10/20 steel, which has only 0.2% carbon, is more ductile than 10/90 steel, which has 0.9% carbon. Only low-carbon steel is soft enough to be stamped or formed. High-carbon steel is very hard and brittle unless it is subjected to heat treatment. Stainless steel is an alloy of iron with metals such as chromium and nickel, which help it resist corrosion. Adding other metals to stainless steel produces a range of desired properties. For example, steel that contains manganese and molybdenum has greater resistance to mechanical shock and is used in racing cycles (Fig. 10.35).

Some alloys are softer than the component metals. The presence of big bismuth atoms helps to soften a metal and lower its melting point, much as melons destabilize a stack of oranges because they just do not fit together well. A low-melting-point alloy of lead, tin, and bismuth is employed to control water sprinklers used in certain fire-extinguishing systems. The heat of the fire melts the alloy, which activates the sprinklers before the fire can spread.

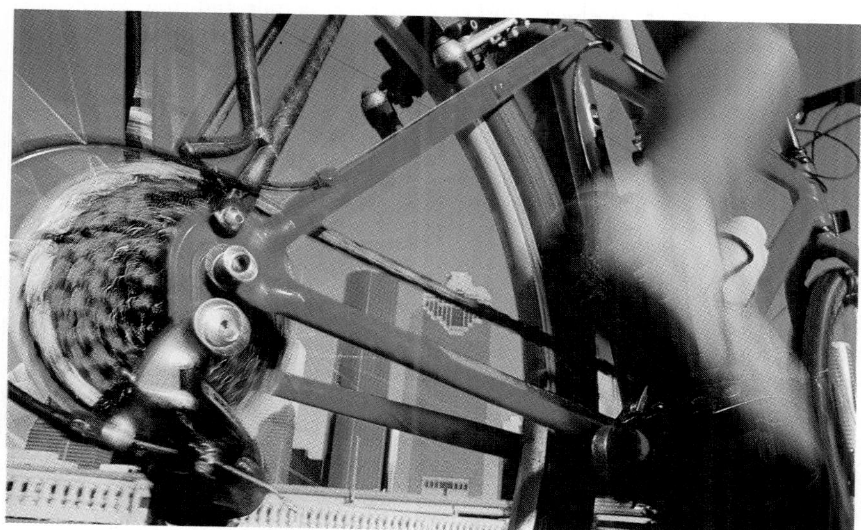

FIGURE 10.35 Racing bicycle frames like this one are built from high-strength, low-density steels formed by alloying iron with metals such as manganese, molybdenum, and titanium.

CASE STUDY 10 LIQUID CRYSTALS

Nematic phase

Smectic phase

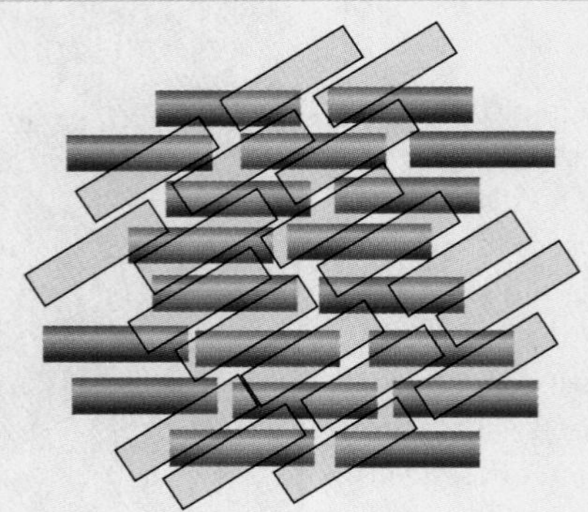

Cholesteric phase

ULTRATHIN, FLEXIBLE COMPUTER and television displays and fast-reacting thermometers have been made possible because of a special kind of material that seems to be neither solid nor liquid. Liquid crystals are substances that flow like viscous liquids, but their molecules lie in a moderately orderly array, like those in a crystal. They are substances in an intermediate state of matter with the fluidity of a liquid and some of the molecular order of a solid.

A typical liquid crystal molecule is long and rodlike; *p*-azoxyanisole is an example:

$$H_3C{-}O{-}C_6H_4{-}\underset{\underset{O}{|}}{N}{=}N{-}C_6H_4{-}O{-}CH_3$$

The rodlike shape causes the molecules to stack together like raw spaghetti: they lie parallel but are free to slide past one another along their long axes. Liquid crystals are anisotropic because of this ordering. *Anisotropic* materials have properties that depend on the direction of measurement. The viscosity of liquid crystals is least in the direction parallel to the molecules. It takes less energy for the long rod-shaped molecules to slip past one another along their axes than to roll sideways. *Isotropic* materials are materials with properties that do not depend on the direction of measurement. For example, ordinary liquids are isotropic: their viscosities are the same in every direction. Liquid crystals become isotropic liquids when they are heated above a characteristic temperature because then the molecules have enough energy to overcome the attractions that restrict their movement.

There are three classes of liquid crystals, which differ in the arrangement of their molecules. In the *nematic phase* the molecules lie together as shown in the accompanying figure, all in the same direction, but staggered, like cars on a busy multilane highway. In the *smectic phase* the molecules line up like soldiers on parade and form layers. Cell membranes are composed mainly of smectic liquid crystals. In the *cholesteric phase,* the molecules form nematiclike layers, but neighboring layers have molecules at different angles, so the liquid crystal has a helical arrangement of molecules.*

We can also classify liquid crystals by their mode of preparation. *Thermotropic* liquid crystals are made by melting the solid phase. The liquid crystal phase exists over a short temperature range between the solid and liquid

An STM image of liquid crystal molecules aligned in a smectic phase.

**Nematic* comes from the Greek word for "thread"; *smectic* comes from the Greek words for "soapy"; *cholesteric* is related to the word *cholesterol,* which comes from the Greek for "bile solid."

states. *p*-Azoxyanisole is a thermotropic liquid crystal. Thermotropic liquid crystals are highly viscous and can be either translucent or opaque. They are used in applications such as watches, computer screens, and thermometers. *Lyotropic* liquid crystals are layered structures that result from the action of a solvent on a solid. Examples are aqueous solutions of detergents, lipids (fats), and cell membranes. These molecules, like the detergent sodium lauryl sulfate, have long, nonpolar hydrocarbon chains attached to polar heads:

$$\text{(zigzag chain)}\text{OSO}_3^- \quad \text{Na}^+$$

Dilute solutions of detergents—more formally, *surfactant molecules* (surface-active molecules)—tend to form layers on the surface of water. For example, detergent molecules line up next to one another on the water surface, with their heads in the water and their nonpolar tails in the air. In high concentrations, detergents from *micelles,* small clusters in which the polar heads point out toward the water and the hydrocarbon tails point inward. The hydrocarbon tails of the detergent molecules are attracted to greasy dirt and form micellelike structures around it. The polar heads surrounding the micelle keep it suspended in water, and it can be washed away. The lipids that form cell membranes have structures similar to those of micelles. When mixed with water, they spontaneously form sheets in which the molecules are aligned in rows, forming a double layer, with their polar heads facing outward on each side of the sheet. These sheets form the protective membranes of our cells.

Electronic displays make use of the fact that the orientation of the molecules in liquid crystals changes in the presence of an electric field. This reorientation causes a change in their optical properties, making them opaque or transparent, and hence forming a pattern on a screen. Cholesteric liquid crystals are also of interest because the helical structure unwinds slightly as the temperature is changed. Because the twist of the helical structure affects the optical properties of the liquid crystal, these properties change with temperature. The effect is utilized in liquid crystal thermometers.

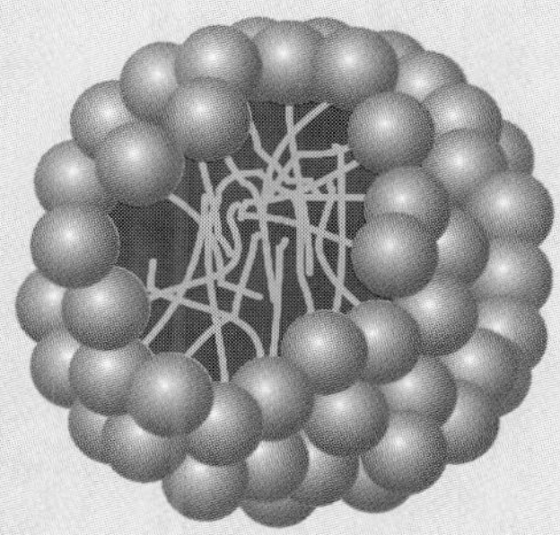

The polar head groups of this spherical micelle are represented by spheres and the hydrophobic parts are represented by the long tails.

A cross section through one of the two layers of a cell membrane. The long, narrow molecules are aligned with their polar ends toward the surfaces of the membrane.

QUESTIONS

1. Identify the following substances as isotropic or anisotropic: (a) a crystal of NaCl; (b) an aqueous solution of NaCl; (c) a smectic liquid crystal; (d) a snowflake; (e) a cell membrane.

2. Explain how a nematic liquid crystal differs from (a) a solid glass material; (b) an isotropic liquid.

3. Nonpolar molecules with shapes similar to that of *p*-azoxyanisole do not form liquid crystals. Explain how polar groups contribute to the properties of a liquid crystal.

4. Pentanol, an organic compound with the formula $CH_3CH_2CH_2CH_2CH_2OH$, has molecules with a nonpolar hydrocarbon chain of medium length attached to a polar —OH group. It is insoluble in water. However, if detergent is added, a lyotropic liquid crystal forms that suspends the pentanol in the water. Propose a structure for the lyotropic liquid crystal, describing the arrangement of the water, pentanol, and detergent molecules.

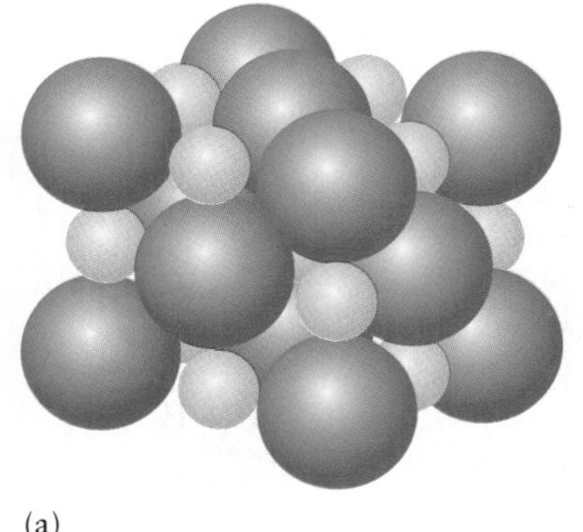

(a)

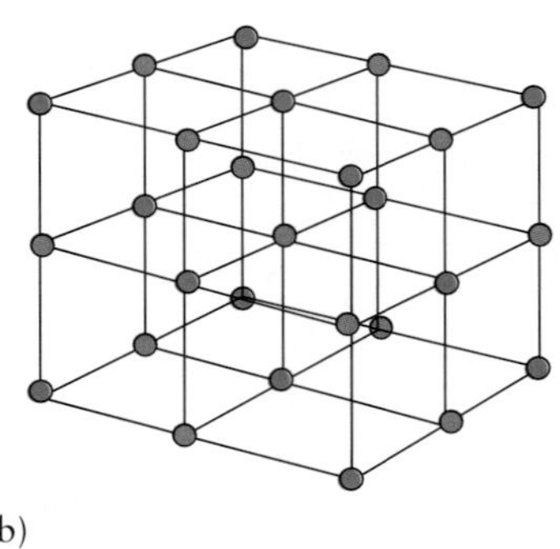

(b)

FIGURE 10.36 The arrangement of ions in the rock-salt structure. (a) The unit cell, showing the packing of the individual ions. (b) A representation of the structure in terms of dots that identify the centers of the ions. The pink spheres are cations and the green spheres are anions.

Alloys of metals tend to be harder and have lower electrical conductivity than pure metals. In substitutional alloys, atoms of the solute metal take the place of some atoms of a metal of similar atomic radius. In interstitial alloys, atoms of the solute element fit into the gaps in the lattice formed by atoms of a metal with a larger atomic radius.

10.10 IONIC STRUCTURES

Ionic solids are formed from oppositely charged ions with different radii, and both kinds must pack together. Overall, the crystal is electrically neutral, and each unit cell must reflect the stoichiometry of the compound. No wonder ionic solids can have quite complicated structures! However, when the ions are both singly charged (as in NaCl) or both doubly charged (as in MgO), one of two simple structures is commonly found.

The **rock-salt structure** takes its name from the mineral form of sodium chloride. In it, the chloride anions lie at the corners and in the centers of the faces of a cube, forming a face-centered cubic unit cell. The smaller sodium cations fit snugly between the anions. If we look carefully at Fig. 10.36, we can see that each cation is surrounded by six anions and each

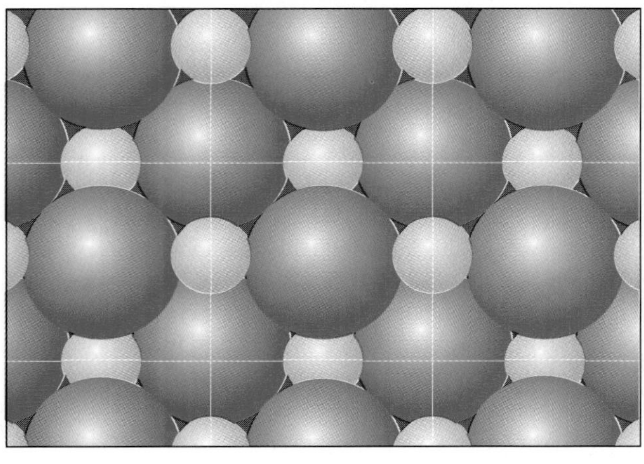

(a)

(b)

FIGURE 10.37 (a) Billions of unit cells are stacked together to recreate the structure of an actual crystal of sodium chloride. Here we see some of the stacked unit cells, viewed from one side of the crystal. (b) The orderly arrangement of ions creates the smooth faces of a crystal, as shown in this micrograph of sodium chloride.

anion is surrounded by six cations. The pattern repeats over and over, and each ion is surrounded by six other ions of the opposite charge (Fig. 10.37). The three-dimensional array of a vast number of these little cubes gives rise to a crystal of sodium chloride.

In an ionic solid, the coordination number means the number of ions of opposite charge immediately surrounding it. In rock salt the coordination numbers of the cations and the anions are both 6, and the structure overall is described as having (6,6)-coordination. In this notation, the first number is the cation coordination number and the second is that of the anion. The rock-salt structure is found for a number of other minerals with ions of the same charge type, including KBr, RbI, MgO, CaO, and AgCl. It is common whenever the cations and anions have very different radii. Specifically, a rock-salt structure is expected when the **radius ratio,** the radius of the cation divided by that of the anion, is in the range 0.4 to 0.7. For sodium chloride,

$$\text{Radius ratio} = \frac{102\ \text{pm}}{181\ \text{pm}} = 0.564$$

When the radii of the cations and anions are similar and the radius ratio is greater than 0.7, the cations cannot fit into the spaces between the anions in the rock-salt structure. Instead, the ions adopt the **cesium-chloride structure,** which has a body-centered cubic unit cell. As illustrated in Fig. 10.38, the cesium cation is at the center of the cell, with eight chloride ions at its corners. Equivalently, each chloride ion can be regarded as being at the center of a body-centered unit cell, with eight cesium cations at its corners. The coordination number of each type of ion is 8, and overall the structure has (8,8)-coordination. This pattern repeats throughout the crystal (Fig. 10.39). The cesium-chloride structure is much less common than the rock-salt structure, but it is found for CsI as well as CsCl.

In these and many other ionic structures, we treat ions as spheres of the appropriate radii and stack them together in the arrangement that has

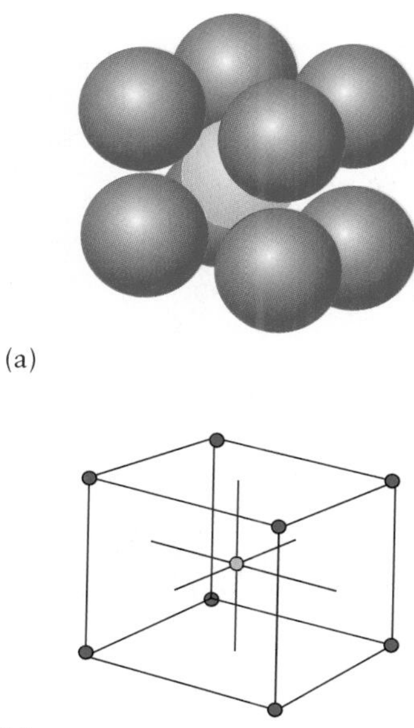

FIGURE 10.38 The cesium-chloride structure: (a) the unit cell and (b) the location of the centers of the ions. The pink spheres are cations and the green spheres are anions.

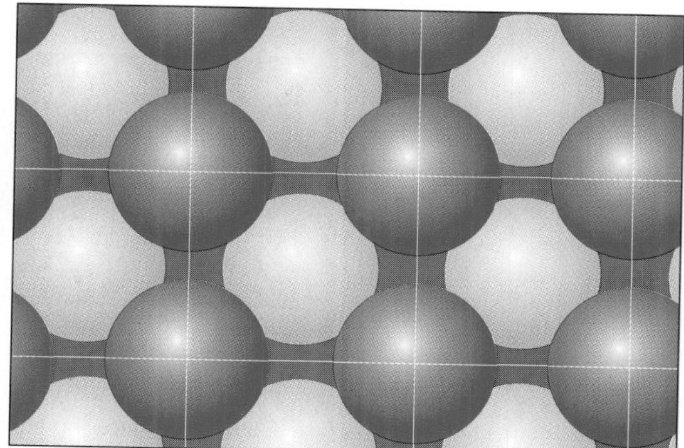

FIGURE 10.39 The repetition of the cesium-chloride unit cell recreates the entire crystal. This view is from one side of the crystal and shows six unit cells.

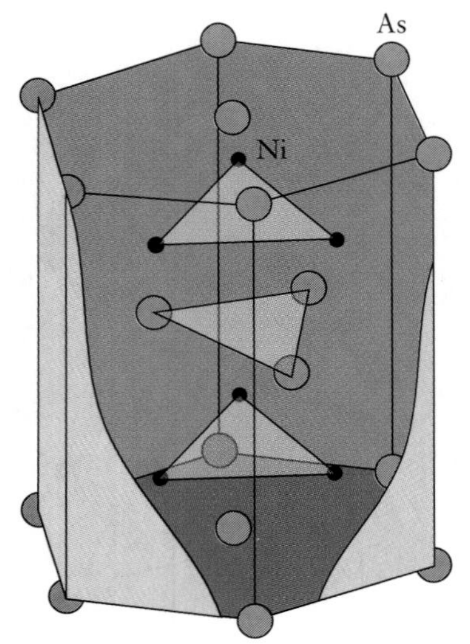

FIGURE 10.40 The structure of nickel arsenide. This structure is found when covalent bonding is starting to become important and the ions have to take up specific positions relative to one another to maximize their bonding.

the lowest total energy. However, if the bonding is not purely ionic, then the simple sphere-packing arrangement may break down. Because the bonding has partial covalent character, the ions may have to lie in specific positions around one another. An example is nickel arsenide, Ni_3As_2. In this solid, the small Ni^{2+} cations polarize the big As^{3-} ions, and the bonds have some covalent character. The ions pack together in an arrangement that is quite different from a purely ionic, sphere-packing model (Fig. 10.40).

Ions stack together in the lowest energy regular crystalline structure; two common structures are the rock-salt structure and the cesium-chloride structure. Covalent character in an ionic bond imposes a directional character on the bonding.

SELF-TEST 10.7A How many Na^+ ions are present in the NaCl unit cell? Account for each one.

[*Answer:* 4; 1 center atom and $\frac{1}{4} \times 1$ on each of 12 edges]

SELF-TEST 10.7B How many Cl^- ions are there in one unit cell of CsCl?

10.11 MOLECULAR SOLIDS

Molecular solids—solids composed of individual molecules—often have low melting points: only a small amount of thermal motion is needed to overcome the relatively weak intermolecular forces that hold the molecules in place. Titanium tetrachloride ($TiCl_4$), for instance, is a liquid that boils at 136°C and freezes at −25°C to a molecular solid (Fig. 10.41). The solid is composed of tetrahedral $TiCl_4$ units, held together largely by London forces between the electron-rich chlorine atoms.

The physical properties of molecular solids depend on the strengths of their intermolecular forces. Some molecular solids are as soft as paraffin wax, which is a mixture of long-chain hydrocarbons. These molecules lie together in a disorderly way, and the forces between them are so weak that they can be pushed past one another very easily. Other molecular

FIGURE 10.41 Titanium tetrachloride, $TiCl_4$, is a liquid that reacts with water to form the white solid titanium dioxide, TiO_2. When squirted out of appropriate containers, it can be used for producing smoke screens or, as here, for skywriting.

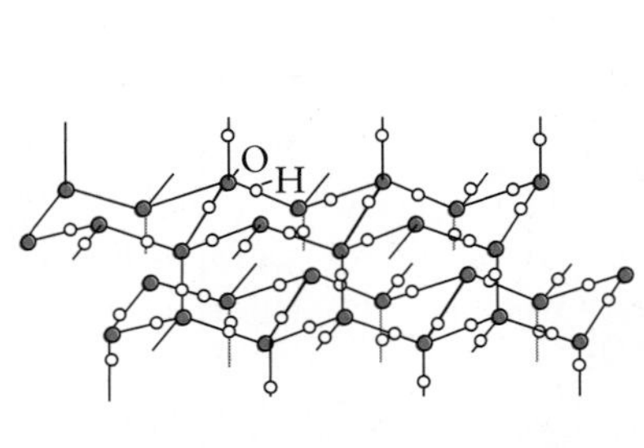

(a)

(b)

FIGURE 10.42 (a) Ice is made up of water molecules that are held together by hydrogen bonds in a relatively open structure. Each O atom (red) is surrounded tetrahedrally by four hydrogen atoms, two of which are σ-bonded to it and two of which are hydrogen-bonded to it. (b) As the crystalline structure of ice forms, intricate regular designs can result.

solids are brittle and hard. For example, sucrose molecules, $C_{12}H_{22}O_{11}$, are held together by hydrogen bonds between their numerous OH groups. Hydrogen bonding between sucrose molecules is so strong that by the time the melting point has been reached (at 184°C), the molecules themselves have started to decompose. The partly decomposed mixture of products, called caramel, is used to add flavor and color to food.

Because molecules have such widely varying shapes, they stack together in a wide variety of different ways. In ice, each O atom is surrounded by four H atoms. Two of these H atoms are linked to the O atom through σ-bonds. The other two belong to neighboring H_2O molecules and are linked to the O atom by hydrogen bonds. As a result, the structure of ice is an open network of H_2O molecules held in place by hydrogen bonds (Fig. 10.42). Some of the hydrogen bonds break when ice melts; and as the orderly arrangement collapses, the molecules pack less uniformly but more densely (Fig. 10.43). The openness of the network in ice compared with that in the liquid explains why it has a lower density than liquid water (0.92 $g \cdot cm^{-3}$ and 1.00 $g \cdot cm^{-3}$, respectively, at 0°C). Solid benzene and solid carbon dioxide, in contrast, have higher densities than their liquids. Their molecules are held in place by London forces, which are much less directional than hydrogen bonds, so they can pack together more closely in the solid than in the liquid (Fig. 10.44).

Molecular solids are typically less hard than ionic solids and melt at lower temperatures.

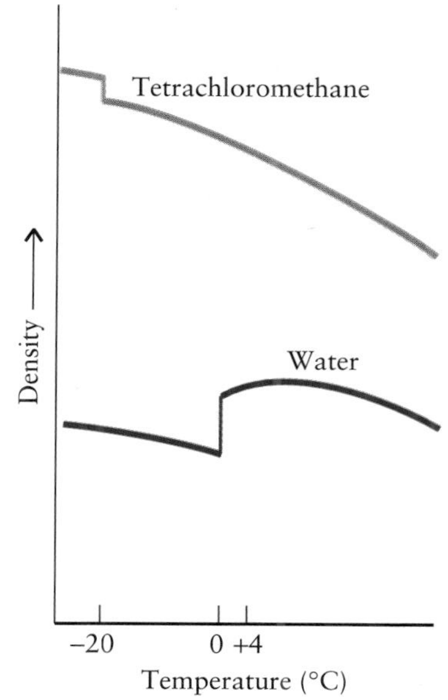

FIGURE 10.43 The variation of the densities of water and tetrachloromethane with temperature. Note that ice is less dense than liquid water at its freezing point and that water itself has its maximum density at 4°C.

10.12 NETWORK SOLIDS

The atoms in network solids are joined to their neighbors by covalent bonds. These bonds form a network that extends throughout the crystal. Network solids exhibit the strength of the covalent bonds that bind them by being very hard, rigid materials with high melting and boiling points.

FIGURE 10.44 As a result of its open structure, ice is less dense than water and floats in it (left). Solid benzene is denser than liquid benzene, and "benzenebergs" sink in liquid benzene (right).

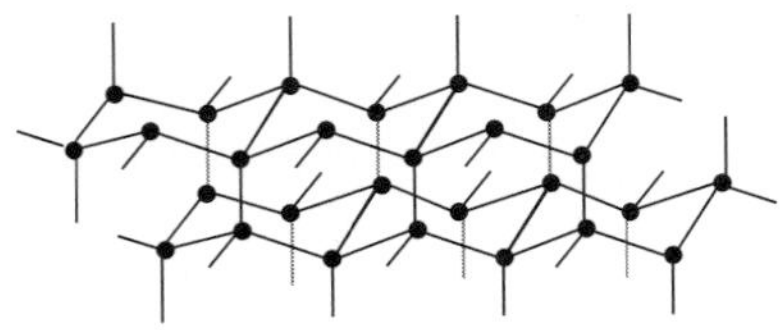

FIGURE 10.45 Part of the structure of a diamond. Each dot represents the location of the center of a carbon atom. Each atom forms an sp^3-hybrid covalent bond to each of its four neighbors.

Diamond and graphite are network solids. These two forms of carbon are **allotropes,** meaning forms of an element that differ in the way the atoms are linked. Each C atom in diamond is covalently bonded to four neighbors through sp^3 hybrid σ-bonds (Fig. 10.45). The tetrahedral framework extends throughout the solid like the steel framework of a large building. This structure accounts for the great hardness of the solid. Diamond is so hard it is used to protect drill bits and as a long-lasting abrasive. Diamond is also one of the best conductors of heat and is used as a base for some integrated circuits so that they do not overheat. Its high thermal conductivity results from the rigid network structure of the crystal, because the vigorous vibration of an atom in a hot part of the crystal is rapidly transmitted to distant, cooler parts through the covalent bonds, rather like the effect of slamming a door in a steel-framed building.

Graphite, the "lead" of pencils, is a black, lustrous, electrically conducting, slippery solid that sublimes at 3700°C. It consists of flat sheets of sp^2 hybridized carbon atoms bonded covalently into hexagons like chicken wire (Fig. 10.46). There are also weak bonds between the sheets. In the commercially available forms of graphite, there are many impurity atoms trapped between the sheets; these atoms weaken the already weak intersheet bonds and let the sheets of atoms slide over one another. So, in contrast to the hardness of diamonds, graphite is soft and slippery. When we write with a pencil, the mark left on the paper consists of rubbed-off layers of graphite. Electrons can move within the sheets but much less readily from one sheet to another. Hence, graphite conducts electricity better parallel to the sheets than perpendicular to them.

In nature, diamond is found embedded in a soft rock called kimberlite. This rock rises in columns from deep in the Earth, where the diamonds are formed under intense pressure. To make synthetic industrial diamonds, we need to recreate the geological conditions that produce natural diamonds by compressing graphite at pressures over 80,000 atm and temperatures above 1500°C (Fig. 10.47). Small amounts of metals such as chromium and iron are included. It is thought that the molten metals dissolve the graphite and then, as they cool, deposit crystals of diamond, which are less soluble than graphite in the molten metal. Graphite is produced naturally as a result of changes in ancient organic

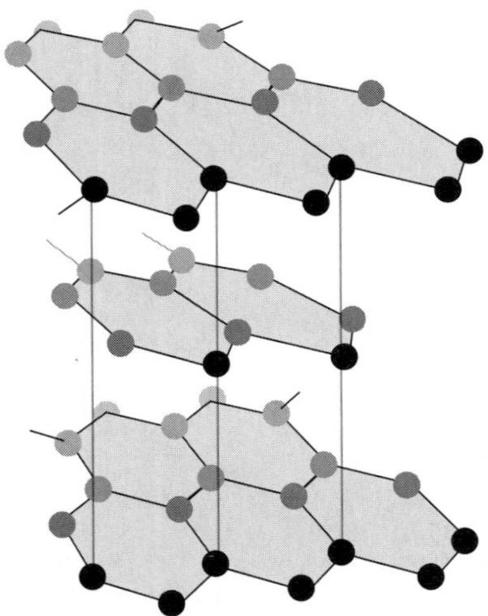

FIGURE 10.46 Graphite consists of layers of hexagonal rings of sp^2-hybridized carbon atoms. The slipperiness of graphite is due to the ease with which the layers can slide over one another when there are impurity atoms lying between the planes.

FIGURE 10.47 Tiny synthetic diamonds about 350 μm in diameter.

remains. Most commercial graphite is produced by heating carbon rods to a high temperature in an electric furnace for several days.

Network solids are typically hard and rigid, and have high melting and boiling points.

PHASE CHANGES

A **phase** is a physical state of a substance. It may be solid, liquid, or gas, or one of several different solid forms. Thus, carbon and graphite are two solid phases of carbon. Ice, liquid water, and water vapor are three phases of water. The conversion of one phase to another, such as the melting of ice, the vaporization of water, or the conversion of graphite into diamond, is called a **phase transition.**

10.13 VAPOR PRESSURE

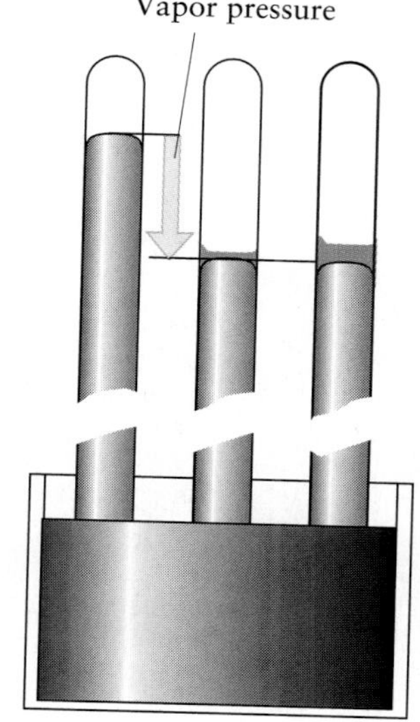

FIGURE 10.48 The vapor pressure exerted by the liquid (in Torr) is equal to the distance by which the column of mercury is lowered. The vapor pressure is the same, however much liquid water is present.

Suppose we have a long glass tube full of mercury (like a barometer) and upend it in a dish of mercury. The mercury inside the tube falls to a height corresponding to the external atmospheric pressure, leaving it about 76 cm high at sea level. The space above the mercury is almost a vacuum.

Now suppose we introduce just a trace of water into that space. The water immediately evaporates and fills the space with vapor. This vapor exerts a pressure that pushes the surface of the mercury down a few centimeters. The pressure exerted by the vapor—as measured by the fall in the height of mercury—depends on the amount of water added. However, suppose we add so much water that some liquid remains on the surface of the mercury in the tube. The pressure of the vapor soon stops increasing, no matter how much more liquid water is added (Fig. 10.48). At a fixed temperature, the vapor exerts the same, characteristic pressure. For example, at 40°C the mercury falls 55 mm, so the pressure exerted by the vapor is 55 Torr; the outcome is the same whether there is 0.1 mL or 1 mL of liquid water present. This characteristic pressure is the **vapor pressure** of the liquid (Table 10.5). Liquids with high vapor pressures are said to be **volatile.** Solids also exert a vapor pressure, but it is usually much lower than the vapor pressures of liquids because the molecules in a solid are gripped more tightly than they are in a liquid.

A lot of liquid water will press the mercury down under its weight. We are disregarding that effect here.

A vapor is formed as molecules leave the surface of the liquid. In a closed container, as the number of molecules in the vapor increases, more of them strike the surface of the liquid. Eventually the number of molecules returning to the liquid each second exactly matches the number escaping. The vapor is now condensing as fast as the liquid is vaporizing. We say that the liquid and vapor are in **dynamic equilibrium,** a condition in which a forward process and its reverse take place at equal rates (Fig. 10.49). We write

$$\text{Rate of evaporation} = \text{rate of condensation}$$

The word *dynamic* is important: it implies continuous activity, unlike the static equilibrium of a ball resting at the foot of a hill. A liquid in dynamic

TABLE 10.5 Vapor pressures at 25°C

Substance	Vapor pressure, Torr
benzene	94.6
ethanol	58.9
mercury	0.0017
methanol	122.7
water*	23.8

*For values at other temperatures, see Table 5.4.

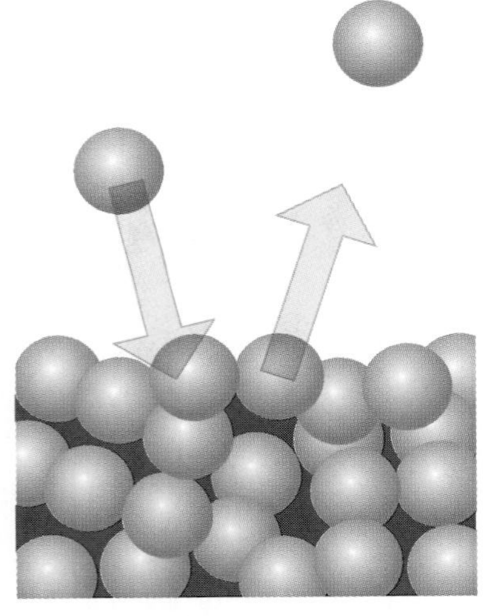

FIGURE 10.49 When a liquid and its vapor are in dynamic equilibrium inside a closed container, the rate at which the molecules leave the liquid is equal to the rate at which they return.

equilibrium with its vapor is like a busy store that has customers continually arriving and leaving so that the number of customers inside the store is constant.

The dynamic equilibrium between liquid water and its vapor is denoted

$$H_2O(l) \rightleftharpoons H_2O(g)$$

Whenever we see the symbol $\rightleftharpoons$, it means that the species on both sides are in dynamic equilibrium. Although products are being formed from reactants, the products are changing back into reactants at a matching rate. The vapor pressure of a liquid (or a solid) is the pressure exerted by its vapor when the vapor and the liquid (or the solid) are in dynamic equilibrium. Most evaporation takes place from the surface of the liquid or solid phase because the molecules there are least strongly bound and can escape to the vapor more easily than those in the bulk.

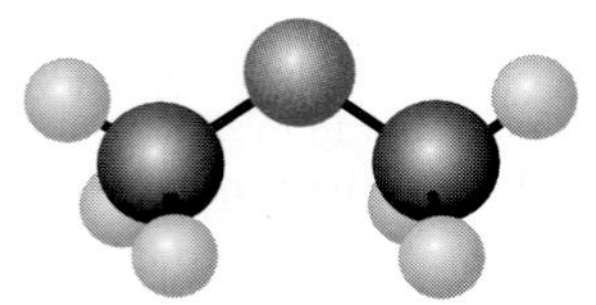

13 Dimethylether, CH_3OCH_3

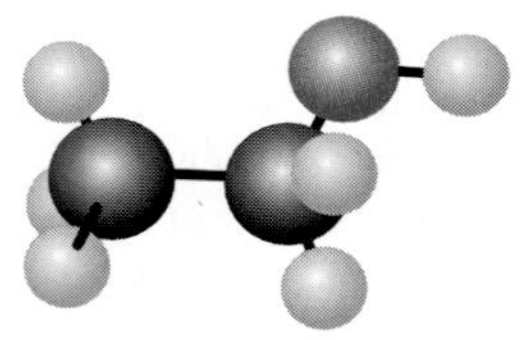

14 Ethanol, CH_3CH_2OH

Vapor pressure increases with temperature because molecules in the heated liquid move more energetically and can escape more readily from their neighbors (Fig. 10.50). The vapor pressure of a liquid also depends on the strength of the intermolecular forces that bind it together. We therefore expect liquids composed of molecules capable of forming hydrogen bonds to be less volatile than others (see Fig. 10.50). We can see the effect of hydrogen bonding clearly when we compare dimethylether (**13**) and ethanol (**14**), both of which have the molecular formula C_2H_6O and so might be expected to have similar vapor pressures. However, ethanol molecules each have an —OH group and can form hydrogen bonds to one another. The ether molecules cannot form hydrogen bonds to one another because their hydrogen atoms are all attached to carbon atoms. As a result, ethanol is a liquid at room temper-

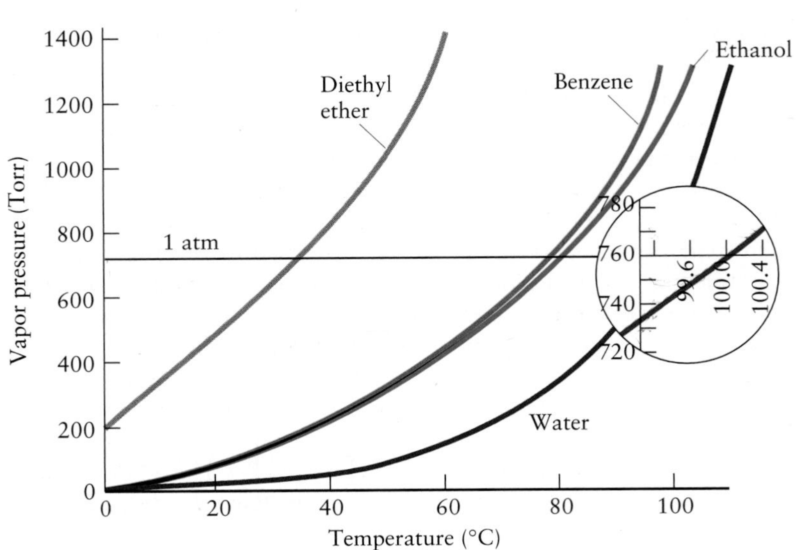

FIGURE 10.50 The variation of the vapor pressure of liquids with temperature, for diethyl ether (orange), ethanol (green), benzene (red), and water (blue). The normal boiling point is the temperature at which the vapor pressure is 1 atm (760 Torr).

ature, whereas dimethyl ether is a gas. Even diethyl ether, $(C_2H_5)_2O$, which has a much higher molar mass than ethanol, has a higher vapor pressure than ethanol (see Fig. 10.50). Benzene has a slightly higher vapor pressure than ethanol: although it has heavier molecules and its London forces are relatively strong, it cannot form hydrogen bonds.

EXAMPLE 10.5 Predicting the relative vapor pressures of substances

Which liquid do you expect to have the higher vapor pressure at room temperature, *cis*-dibromoethene (**15**) or *trans*-dibromoethene (**16**)?

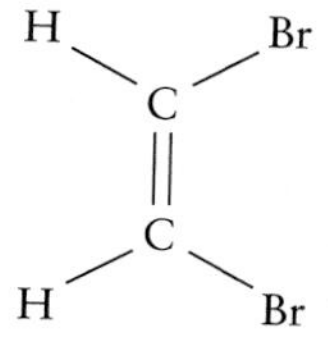

15 *cis*-Dibromoethene

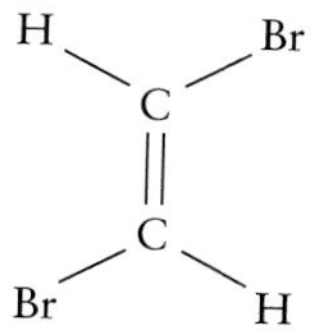

16 *trans*-Dibromoethene

STRATEGY The two molecules have the same molar masses and differ only in the arrangement of their atoms. Decide which intermolecular interactions they can have and then decide which has the stronger interactions. That substance is likely to have the lower vapor pressure.

SOLUTION Both substances can interact by London forces, and the strengths will be about the same in each case because they are composed of the same atoms. However, only the cis compound is polar, so only it can have dipole-dipole interactions. Therefore, we can expect *cis*-dibromoethene to have stronger intermolecular forces than *trans*-dibromoethene. It follows that we should expect *trans*-dibromoethene to be the more volatile.

SELF-TEST 10.8A Which do you expect to have the higher vapor pressure at room temperature, tetrabromomethane, CBr_4, or tetrachloromethane, CCl_4? Give reasons.

[*Answer:* CCl_4; weaker London forces]

SELF-TEST 10.8B Which do you expect to have the higher vapor pressure at −50°C, CH_3CHO or $CH_3CH_2CH_3$?

10.14 BOILING

Now let's consider what happens when a liquid is left in an open container. The vapor formed spreads away from the liquid. Little, if any, is recaptured at the surface, so the rate of condensation never increases to the point at which it matches the rate of vaporization. As a result, dynamic equilibrium is never reached and, given enough time, the liquid vaporizes completely.

Vaporization can occur *throughout* the liquid when its temperature is raised to the point at which its vapor pressure matches the pressure exerted by the atmosphere. At this temperature, any vapor formed can drive back the atmosphere and make room for itself. Thus, bubbles of vapor form in the liquid and rise to the surface. We call this rapid vaporization "boiling." The **normal boiling point,** T_b, of a liquid is the temperature at which a liquid boils when the atmospheric pressure is 1 atm.

Boiling occurs at a temperature higher than the normal boiling point when the pressure is greater than 1 atm, as it is in a pressure cooker. Boiling occurs at a temperature lower than T_b when the pressure is lower than 1 atm. At the summit of Mt. Everest—where the pressure is about 240 Torr—water would boil at only 71°C, which is nearly cool enough to hold your hand in.

The maximum temperature we can tolerate is about 60°C.

A high normal boiling point is a sign of strong intermolecular forces, because high temperatures are needed to break the forces to raise the vapor pressure to 1 atm. For example, water has a much higher boiling point than hydrogen sulfide, which cannot form hydrogen bonds.

Boiling occurs when the vapor pressure of a liquid is equal to the atmospheric pressure. Strong intermolecular forces usually lead to high normal boiling points.

EXAMPLE 10.6 **Predicting the boiling temperature at a given pressure**

Use Fig. 10.50 to predict the temperature of boiling water at a high altitude where the atmospheric pressure is 400 Torr.

STRATEGY Because the atmospheric pressure (400 Torr) is lower than 760 Torr (the equivalent of 1.00 atm), we expect boiling to occur below the normal boiling point. Use Fig. 10.50 to find the temperature at which the vapor pressure of water reaches 400 Torr.

SOLUTION From Fig. 10.50, the vapor pressure of water is 400 Torr at 84°C. Hence its boiling point at that pressure is 84°C.

SELF-TEST 10.9A Predict the temperature of boiling water on a day when the pressure is 770 Torr.

[*Answer:* 100.3°C]

SELF-TEST 10.9B Predict the temperature of boiling water on a day when the pressure is 740 Torr.

10.15 FREEZING AND MELTING

The freezing point of a liquid is the same as the melting point of the solid.

A liquid solidifies when its molecules have such low energies that they are unable to wriggle past their neighbors. In the solid, the molecules vibrate about their average positions but rarely move from place to place. The freezing temperature varies slightly as the pressure is changed, and the **normal freezing point,** T_f, of a liquid is the temperature at which it freezes at 1 atm. In practice, a liquid sometimes does not freeze until the temperature is below its freezing point. A liquid below its freezing point is said to be *supercooled.*

Most liquids freeze at higher temperatures when subjected to pressure because the solid phase is denser with molecules packed more closely together than in the liquid phase, and the pressure helps to hold the molecules together. However, except at very high pressures, the effect of pressure is usually quite small. Iron, for example, melts at 1800 K under 1 atm, and only a few degrees higher when the pressure is a thousand times as great. At the center of the Earth, the pressure is high enough for iron to be solid despite the high temperatures there, so the Earth's inner core is believed to be solid.

At 1000 atm, water freezes at −5°C.

Water is an exception to the general rule: it freezes at a *lower* temperature under pressure. Ice melts under pressure because water has a higher density than ice: many of the hydrogen bonds in ice are broken when ice melts, thereby allowing ice to shrink into a smaller volume. The melting of ice under pressure is thought to contribute to the advance of glaciers. The weight of ice pressing on the edges of rocks deep under the glacier results in very high local pressures. The liquid forms despite the low temperature, and the glacier slides slowly downhill on a film of water.

Most liquids freeze at a higher temperature under pressure. Water's hydrogen bonds make it anomalous: it freezes at a lower temperature under pressure.

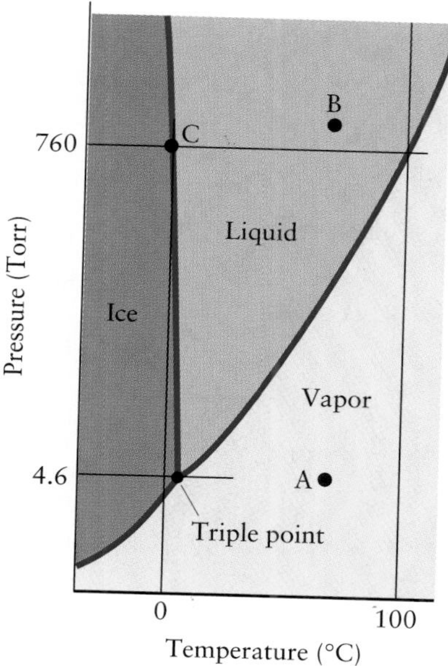

FIGURE 10.51 Phase diagram for water (not to scale). The solid lines define the boundaries of the regions of pressure and temperature at which each phase is the most stable. Note that the freezing point decreases with increasing pressure. The letters A and B are referred to in Example 10.7.

10.16 PHASE DIAGRAMS AND COOLING CURVES

A **phase diagram** is a summary of the temperatures and pressures at which the various solid, liquid, and gaseous phases of a pure substance exist. Figure 10.51 shows the phase diagram for water. In the regions marked "Liquid," the liquid phase is the most stable: we should expect to find liquid water at the pressures and temperatures corresponding to these points on the graph. If we raise the temperature at constant pressure, we cross into the region marked "Vapor," and now the vapor phase is more stable. For example, if the pressure is 760 Torr, we cross the frontier at 100°C, the boiling point of water. The vapor phase is the most stable at all values of pressure and temperature corresponding to the points in the "vapor" region. Any sample of water under these conditions will exist as a vapor. A phase diagram is a map divided into regions that tell us which phase is the most stable under the corresponding conditions. For example, to find the stable phase of water at room temperature and pressure, we see in which region the corresponding point lies.

SELF-TEST 10.10A A sample of water is kept at 20°C and 760 Torr. Which phase is the most stable?

[*Answer:* Liquid]

SELF-TEST 10.10B A sample of water is kept at 150°C and 750 Torr. Which phase is the most stable?

The lines separating the areas in the diagram, the frontiers between the regions, are called **phase boundaries.** The points *on* a phase boundary show the conditions under which two phases coexist in dynamic equilibrium. The solid and liquid are in equilibrium along the line showing the dependence of melting point on pressure. For example, ice and liquid water are in equilibrium at 0°C and 760 Torr (point C in Fig. 10.51). The phase boundary between liquid and vapor is the line showing the pressures at which vapor and liquid are in equilibrium at a given temperature. It is therefore a plot of the vapor pressure of the liquid against temperature. The solid-vapor phase boundary is the plot of the vapor pressure of the solid, because it shows the pressures at which the solid is in equilibrium with its vapor.

The slope of a line helps us predict how the phase change it represents responds to compression. If we compress a sample of water vapor at 50°C, it will begin to condense to a liquid when the pressure reaches the vapor pressure of water at 50°C. Only when the vapor has been completely condensed does the pressure of the sample rise again.

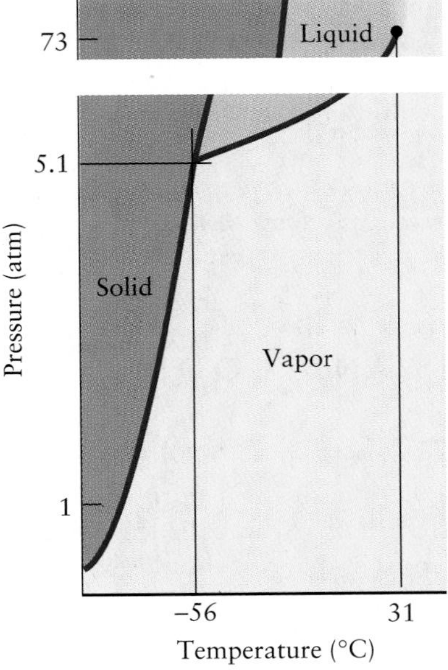

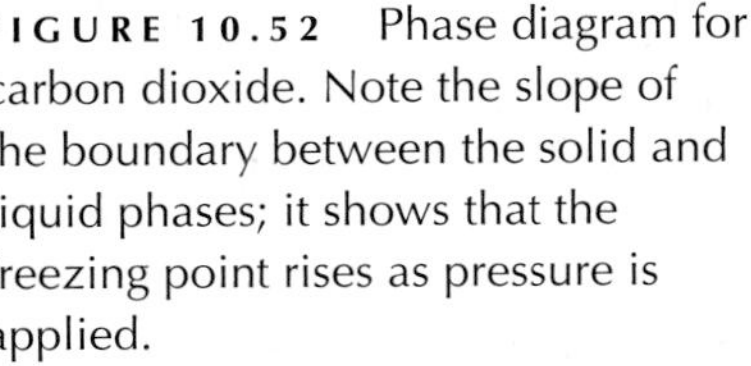

FIGURE 10.52 Phase diagram for carbon dioxide. Note the slope of the boundary between the solid and liquid phases; it shows that the freezing point rises as pressure is applied.

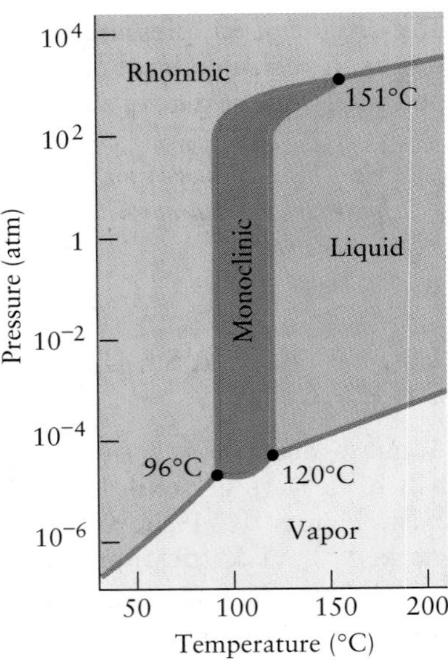

FIGURE 10.53 Phase diagram for sulfur. Notice that there are three triple points and that the pressure scale covers a very wide range of values.

Water's solid-liquid phase boundary leans over slightly to the left. This slope signifies that pressure favors the denser liquid. Therefore, if we increase the pressure on a sample of ice just below 0°C it will eventually melt, just as we predicted from the relative densities of ice and water in Section 10.15.

SELF-TEST 10.11A From the phase diagram for carbon dioxide (Fig. 10.52), predict which is more dense, the solid or liquid phase. Why?

[*Answer:* The solid, because it is more stable at higher pressures.]

SELF-TEST 10.11B From the phase diagram for sulfur (Fig. 10.53), predict which solid phase is more dense. Why?

We can use a phase diagram to predict how the temperature of a liquid changes as it cools. For example, suppose we place 100 mL of water at 1 atm in a freezer and monitor its temperature. The initial state of the sample might be point A in Fig. 10.54. The cooling occurs at constant pressure, so the state of the sample is represented by the points on the horizontal line in the phase diagram starting at A. The temperature of the water falls steadily as it loses energy to its surroundings, until point B is reached on the solid-liquid phase boundary. Then, instead of the temperature falling as energy is lost, the liquid begins to freeze. As this point, the temperature remains constant, despite the fact that the sample is giving up energy to its surroundings. The energy released by the molecules as they take their places in the solid keeps the temperature of the sample constant. The solid and liquid are in equilibrium with each other until all the water has frozen to ice. So the sample stays at that one point on the solid-liquid boundary until no more liquid is left. Only then does the temperature start to fall again.

If we were to follow the temperature of the sample of water as a function of time as it cooled, we could draw a plot like the cooling curve superimposed on Fig. 10.54. As energy is lost from the liquid sample, its temperature falls. As soon as it begins to freeze, solid and liquid are in equilibrium and the temperature of the sample stops falling as long as both liquid and solid are present. The horizontal portion of the cooling curve shows that the temperature is constant during freezing. As soon as all the liquid has frozen, the temperature begins to fall again. Cooling curves like this can be used to construct phase diagrams.

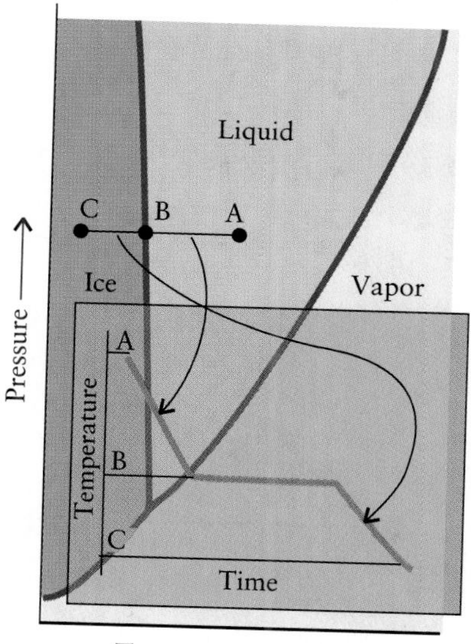

FIGURE 10.54 Phase diagram for water (like that in Fig. 10.51) and the cooling curve for a sample initially at point A. The sample cools at constant pressure through B to ice at C. The pause in the decline of the cooling curve at B is due to the release of heat when the liquid freezes.

A phase diagram summarizes the regions of pressure and temperature at which a phase is most stable. The phase boundaries show the conditions under which two phases can coexist in equilibrium with each other.

EXAMPLE 10.7 Interpreting a phase diagram

Use the phase diagram in Fig. 10.51 to describe the physical states and phase changes of water as the pressure on it is increased from 5 Torr to 800 Torr at 70°C.

STRATEGY First, locate the points corresponding to the initial and final conditions on the phase diagram. The region in which each point lies shows the stable phase of the sample under those conditions. If a point lies on one of the curves, then both phases are present in mutual equilibrium. If the path crosses a curve, then both phases are at equilibrium during the phase transition.

SOLUTION Although the phase diagram is not to scale, we can find the approximate locations of the points we need. Point A is at 5 Torr and 70°C, so it lies in the vapor region. Increasing the pressure takes the system to the liquid-vapor phase boundary, at which point liquid begins to form. At this pressure, liquid and vapor are in equilibrium. The pressure is increased further to 800 Torr, which takes it to point B, in the liquid region. The sample is a liquid in this region because the applied pressure is greater than the vapor pressure and all the molecules have been pushed back into the condensed phase.

SELF-TEST 10.12A The phase diagram for carbon dioxide is shown in Fig. 10.52. Describe the physical states and phase changes of carbon dioxide as it is heated at 2 atm from −155°C to 25°C.

[*Answer:* Solid CO_2 is heated until it begins to sublime at the solid-vapor boundary. The temperature remains constant until all the CO_2 has vaporized. The vapor is then heated to 25°C.]

SELF-TEST 10.12B Describe what happens when liquid carbon dioxide in a container at 60 atm and 25°C is released into a room at 1 atm pressure and the same temperature.

A **triple point** is a point on a phase diagram where three phase boundaries meet. For water, it occurs at 4.6 Torr and 0.01°C. Under these conditions (and only these) all three phases (ice, liquid, and vapor) coexist in dynamic equilibrium. Under these conditions, water molecules leave ice to become liquid and return to form ice at the same rate; liquid vaporizes and vapor condenses at the same rate; and ice sublimes and vapor condenses directly to ice again at the same rate.

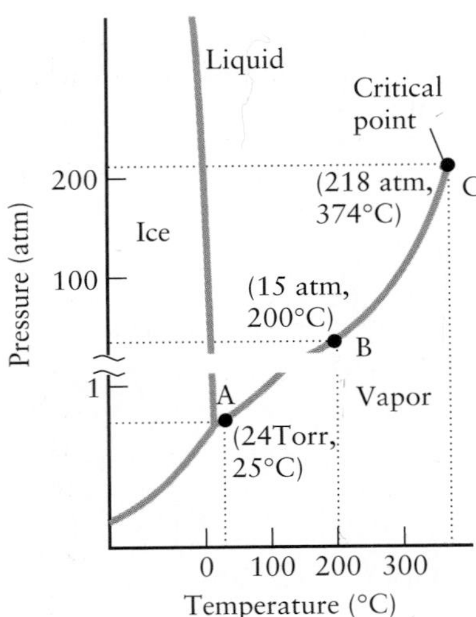

FIGURE 10.55 The phase diagram for water shown with more detail. C is the critical point.

Unlike the freezing or boiling points of a substance, which depend on the applied pressure, the triple point of water is a fixed property. It is in fact used to define the size of the kelvin: by definition, there are exactly 273.16 kelvins between absolute zero and the triple point of water. The freezing point of water is found to lie 0.01 K below the triple point.

Three phases coexist at equilibrium at the triple point.

10.17 CRITICAL PROPERTIES

The liquid-vapor boundary terminates at a point (the point labeled C in Fig. 10.55). To see what happens at that point, suppose that the tube shown in Fig. 10.56 contains liquid water and water vapor at 25°C and 24 Torr (its vapor pressure at 25°C). This system lies on the liquid-vapor curve (point A in Fig. 10.55). As the water and vapor are heated, the vapor pressure increases, with the system continuing to lie higher and higher on the liquid-vapor curve. However, the water never boils, even at 100°C, because the volume is constant and the vapor cannot escape. Instead, the pressure continues to rise as the water vaporizes. At 200°C, the vapor pressure has risen to 11,700 Torr (15.4 atm, point B); liquid and vapor are still in dynamic equilibrium, but now the vapor is very dense because it is at such a high pressure.

At the end of the boundary line (at point C), the system has reached 374°C and a vapor pressure of 218 atm—the container must be very strong! The density of the vapor is now so great that it is equal to that of the remaining liquid. Suddenly the surface separating the liquid from vapor vanishes. A liquid phase can no longer be identified because there is no longer a dividing surface. Instead, a single uniform substance fills the container (as shown on the right in Fig. 10.56). Because a substance that fills any container it occupies is by definition a gas, we conclude that we have reached a temperature above which the liquid state does not exist. We have reached the **critical temperature,** T_c, of water, the temperature above which it cannot be condensed to a liquid. The corresponding pressure is called the **critical pressure,** P_c, of the substance. The critical pressure of water is 218 atm; that of carbon dioxide is 73 atm.

A substance can be liquefied by applying pressure only if it is below its critical temperature. For example, the critical temperature of carbon dioxide is 31°C, and we can liquefy a sample only if its temperature is lower than 31°C. The critical temperatures and pressures of several substances are listed in Table 10.6. Their values are important in practice because it is useless to try to liquefy a gas by applying pressure if it is above its critical temperature. For example, the critical temperature of oxygen is −118°C, so we know that it cannot be compressed to a liquid at room temperature. Oxygen tanks contain only gaseous oxygen. Because the critical temperature of helium is 5.2 K, the gas must be cooled almost to absolute zero before it can be liquefied by pressure.

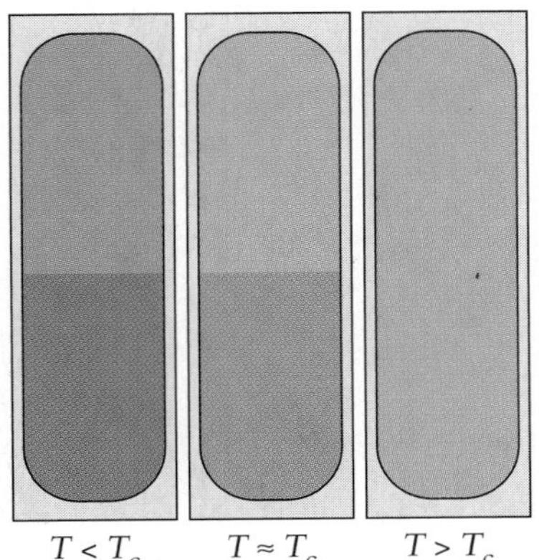

FIGURE 10.56 When the temperature of a liquid in a sealed, constant-volume container is raised, the density of the liquid decreases and the density of the vapor increases as the liquid evaporates. At the critical temperature, T_c, the density of the vapor becomes the same as the density of the liquid; above that temperature, a single uniform phase fills the container.

A substance just above its critical pressure is called a **supercritical fluid.** It may be so dense that, although it is a gas, it can act as a solvent for liquids and solids. Supercritical carbon dioxide can dissolve organic compounds. It is used to remove caffeine from coffee beans and to extract perfumes from flowers without contaminating the extracts with potentially harmful solvents. The carbon dioxide is easily separated by allowing it to evaporate at a lower pressure. Supercritical hydrocarbons are used to dissolve coal and separate it from ash, and they have been proposed for extracting oil from oil-rich tar sands.

A substance can be converted to a liquid by the application of pressure only if it is below its critical temperature.

SELF-TEST 10.13A Look for trends in the data in Table 10.6 that help you determine the effect of London forces on the critical temperature.

[*Answer:* Noble gases have higher critical temperatures as molar mass increases, so critical temperature increases with the strength of London forces.]

SELF-TEST 10.13B Look for trends in the data in Table 10.6 that help you determine the effect of hydrogen bonding on the critical temperature.

TABLE 10.6 Critical temperatures and pressures of substances

Substance	Critical temperature, °C	Critical pressure, atm
He	−268 (5.2 K)	2.3
Ne	−229	27
Ar	−123	48
Kr	−64	54
Xe	17	58
H_2	−240	13
O_2	−118	50
H_2O	374	218
N_2	−147	34
NH_3	132	111
CO_2	31	73
CH_4	−83	46
C_6H_6	289	49

SKILLS YOU SHOULD HAVE MASTERED

Conceptual

1. Explain how the enthalpy of vaporization and the boiling point of a compound are related to the strength of its intermolecular forces.
2. Explain how London forces arise and how they vary with the polarizability of an atom and the size and shape of a molecule.
3. Describe hydrogen bonds and explain why they are stronger than other kinds of intermolecular forces.
4. Describe how viscosity and surface tension vary with temperature and the strength of intermolecular forces.
5. Distinguish metals, ionic solids, network solids, and molecular solids by their structures and by their properties.
6. Distinguish cubic and hexagonal close-packing.
7. Distinguish face-centered cubic and body-centered cubic, and the rock-salt and cesium-chloride structures.
8. Define dynamic equilibrium and explain the relationship between the rate of evaporation and vapor pressure of a substance.
9. Interpret the major features of a phase diagram.
10. Describe why liquid water is denser than ice.

Problem-Solving

1. Predict the relative boiling points, vapor pressures, and enthalpies of vaporization of two substances from the strengths of their intermolecular forces.
2. Identify molecules that can experience hydrogen bonding.
3. Give the coordination number of an atom or ion in a given crystal lattice.
4. Find the number of atoms or ions in a given unit cell.
5. Identify the stable phase or phases of a substance at a given temperature and pressure from its phase diagram.
6. Predict the relative densities of the liquid and solid phases of a substance from its phase diagram.
7. Interpret the cooling curve of a substance.

Descriptive

1. List properties of water that are anomalous as a result of hydrogen bonding.

2. Describe the structures of the carbon allotropes graphite and diamond, and explain how their properties are affected by their structures.

EXERCISES

Intermolecular Forces

10.1 Identify the kinds of intermolecular forces that might arise between molecules of the following substances: (a) Cl_2; (b) HCl; (c) C_6H_6; (d) C_6H_5Cl.

10.2 Identify the kinds of intermolecular forces that might arise between molecules of the following substances: (a) N_2; (b) H_2O; (c) $CH_2{=}CH_2$; (d) CHCl=CHCl (both isomers).

10.3 Write the Lewis structure and indicate the dominant type of intermolecular forces for (a) CO; (b) H_2O; (c) NF_3; (d) CH_4.

10.4 Write the Lewis structure and indicate the dominant type of intermolecular forces for (a) CH_3Cl; (b) OF_2; (c) Xe; (d) SO_2Cl_2 (O_2SCl_2).

10.5 Explain the structure of a hydrogen bond.

10.6 List four consequences of hydrogen bonding and explain why they occur.

10.7 Which of the following molecules are likely to form hydrogen bonds: (a) HF; (b) CH_4; (c) NH_3; (d) CH_3OH?

10.8 Which of the following molecules are likely to form hydrogen bonds: (a) D_2O; (b) CH_3COOH; (c) CH_3CH_2OH; (d) H_3PO_4?

10.9 Suggest, giving reasons, which substance in each pair is likely to have the higher normal melting point (Lewis structures may help your arguments): (a) HCl or NaCl; (b) CH_4 or SiH_4; (c) HF or HCl; (d) H_2O (to compare, write as HOH) or CH_3OH.

10.10 Suggest, giving reasons, which substance in each pair is likely to have the higher normal melting point (Lewis structures may help your arguments): (a) H_2S or Na_2S; (b) NH_3 or PH_3; (c) KBr or CH_3Br; (d) $C_2H_5OC_2H_5$ (diethyl ether) or C_4H_9OH (butanol), both molecules have the same molar mass.

10.11 Account for the following observations in terms of the type and strength of intermolecular forces. (a) The melting point of xenon is −112°C and that of argon is −189°C. (b) The critical temperature of HI is 151°C and that of HCl is 52°C. (c) The vapor pressure of diethyl ether ($C_2H_5OC_2H_5$) is greater than that of water.

10.12 Account for the following observations in terms of the type and strength of intermolecular forces. (a) The boiling point of NH_3 is −33°C and that of PH_3 is −87°C. (b) The critical temperature of H_2O is 374°C and that of HF is 188°C. (c) The vapor pressure of CH_3OH is greater than that of CH_3SH.

10.13 Write the Lewis structures and give the shapes of each of the following molecules; predict which substance of each pair has the higher boiling point: (a) SF_4 or SF_6; (b) BF_3 or ClF_3; (c) SF_4 or CF_4; (d) *cis*-CHCl=CHCl or *trans*-CHCl=CHCl.

10.14 Write the Lewis structures and give the shapes of each of the following molecules; predict which substance of each pair has the higher boiling point: (a) PF_3 or PCl_3; (b) SO_2 or CO_2; (c) BF_3 or BCl_3; (d) $AsCl_3$ or $AsCl_5$.

Liquid Structure

10.15 Water "beads" on a waxed surface. What physical property accounts for this observation?

10.16 Explain why water forms a concave meniscus in a glass graduated cylinder but a convex meniscus in a plastic graduated cylinder.

10.17 Predict which substance has the greater viscosity in its liquid form at 0°C: (a) ethanol, CH_3CH_2OH, or dimethyl ether, $CH_3{-}O{-}CH_3$; (b) butane, C_4H_{10}, or propanone, CH_3COCH_3.

10.18 Predict which liquid has the greater surface tension: (a) *cis*-dichloroethene or *trans*-dichloroethene; (b) benzene at 20°C or benzene at 60°C.

Classification of Solids

10.19 Use Table 10.3 to classify each of the following solids according to its bonding type: (a) sodium chloride; (b) solid nitrogen; (c) sugar (sucrose); (d) copper.

10.20 Use Table 10.3 to classify each of the following solids according to its bonding type: (a) quartz; (b) sulfur; (c) iron(II) sulfate; (d) brass.

10.21 Three unknown substances were tested in order to classify them. The following table shows the results of the tests. Use Table 10.3 to classify substances A, B, and C as metallic solids, ionic solids, network solids, or molecular solids.

Substance	Appearance	Melting point, °C	Electrical conductivity	Solubility in water
A	hard, white	800	only if melted or dissolved in water	soluble
B	lustrous, malleable	1500	high	insoluble
C	soft, yellow	113	none	insoluble

10.22 Three unknown substances were tested in order to classify them. The following table shows the results of the tests. Use Table 10.3 to classify substances X, Y, and Z as metallic, ionic, network, or molecular solids.

Substance	Appearance	Melting point, °C	Electrical conductivity	Solubility in water
X	hard, white	146	none	soluble
Y	very hard, colorless	1600	none	insoluble
Z	hard, orange	398	only if melted or dissolved in water	soluble

Close-Packed Structures

10.23 Determine (a) the number of atoms per unit cell and (b) the coordination number of an atom in a body-centered cubic (bcc) structure.

10.24 Determine (a) the number of atoms per unit cell and (b) the coordination number of an atom in a face-centered cubic (fcc) structure.

10.25 Aluminum crystallizes in a cubic close-packed structure. Its metallic radius is 125 pm. (a) What is the length of the side of the unit cell? (b) How many unit cells are there in 1.00 cm^3 of aluminum?

10.26 Calcium crystallizes in a cubic close-packed structure. Its metallic radius is 180 pm. (a) What is the length of the side of the unit cell? (b) How many unit cells are there in 1.00 cm^3 of calcium?

10.27 The metal polonium (which was named by Marie Curie after her homeland, Poland) crystallizes in a primitive cubic structure, with an atom at each corner of a cubic unit cell. The atomic radius of polonium is 190 pm. Sketch the unit cell and determine (a) the number of atoms per unit cell; (b) the coordination number; (c) the length of the side of the unit cell.

10.28 Potassium crystallizes in a bcc structure. The atomic radius of potassium is 220 pm. Sketch the unit cell and determine (a) the number of atoms per unit cell; (b) the coordination number of the lattice; (c) the length of the side of the unit cell. (*Note:* For a bcc structure, the length of a side is $4r/\sqrt{3}$.)

10.29 Determine the density of each of the following metals from the data: (a) nickel (fcc structure), atomic radius 125 pm; (b) rubidium (bcc structure), atomic radius 250 pm. (*Note:* For a bcc structure, side $= 4r/\sqrt{3}$.)

10.30 Calculate the density of each of the following metals from the data: (a) platinum (fcc structure), atomic radius 138 pm; (b) cesium (bcc structure), atomic radius 266 pm. (*Note:* For a bcc structure, side $= 4r/\sqrt{3}$.)

10.31 Calculate the atomic radius of the following elements from the data: (a) gold (fcc structure), density 19.3 $g{\cdot}cm^{-3}$; (b) vanadium (bcc structure), density 6.11 $g{\cdot}cm^{-3}$. (*Note:* For a bcc structure, side $= 4r/\sqrt{3}$.)

10.32 Calculate the atomic radius of each of the following elements from the data: (a) iridium (fcc structure), density 22.6 $g{\cdot}cm^{-3}$, the densest of all known elements; (b) molybdenum (bcc structure), density 10.2 $g{\cdot}cm^{-3}$. (*Note:* For a bcc structure, side $= 4r/\sqrt{3}$.)

Metals and Alloys

10.33 Explain the difference between electronic and ionic conduction.

10.34 Explain the difference between a superconductor and a semiconductor.

10.35 Silicon can be "doped" with small amounts of phosphorus to create a semiconductor used in transistors. (a) Is the alloy interstitial or substitutional? Justify your answer. (b) How do you expect the doping process to change the properties of the silicon?

10.36 When iron surfaces are exposed to ammonia at high temperatures, "nitriding," the incorporation of nitrogen into the iron lattice, occurs. The atomic radius of iron is 124 pm. (a) Is the alloy interstitial or substitutional? Justify your answer. (b) How do you expect nitriding to change the properties of the iron?

10.37 Zinc oxide is a semiconductor that loses oxygen atoms in a vacuum but gains additional oxygen atoms when heated in oxygen. Its conductivity increases when it is heated in a vacuum but decreases when it is heated in oxygen. Account for these observations.

10.38 The electrical conductivity of graphite parallel to its planes is different from that perpendicular to them. Parallel to the planes, the conductivity decreases as the temperature is raised; but perpendicular to them, it rises. In what sense is graphite a metallic conductor or a semiconductor?

Ionic Solids

10.39 Calculate the number of cations, anions, and formula units per unit cell in the following solids: (a) the rock-salt unit cell shown in Fig. 10.36; (b) the fluorite, CaF_2, unit cell shown below. What are the coordination numbers of the ions in fluorite?

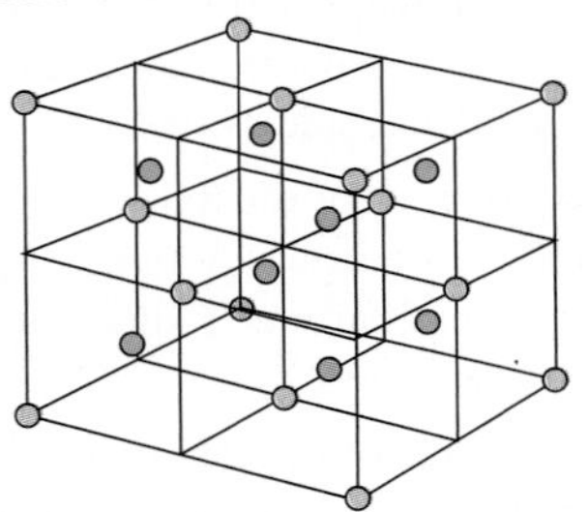

10.40 Calculate the number of cations, anions, and formula units per unit cell in the following solids: (a) the cesium-chloride unit cell shown in Fig. 10.38; (b) the rutile, TiO_2, unit cell shown below. What are the coordination numbers of the ions in rutile?

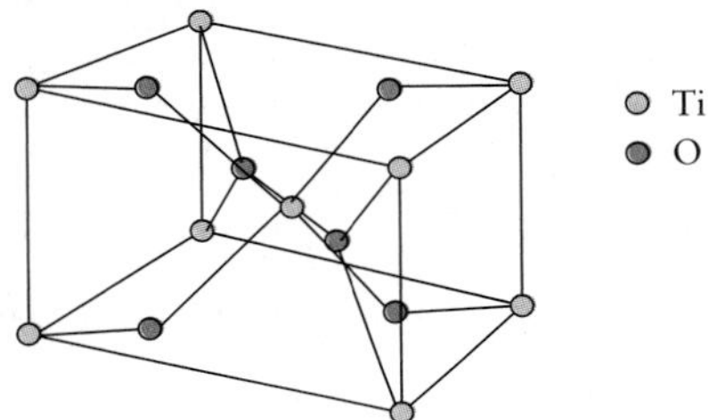

10.41 A unit cell of the mineral perovskite is drawn below. What is its formula?

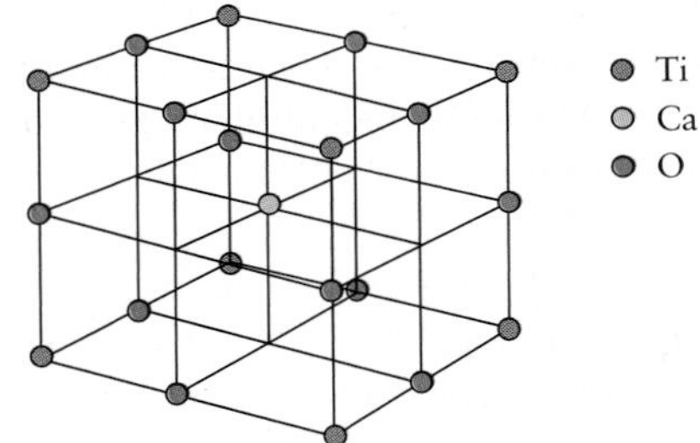

10.42 A unit cell of one of the new high-temperature superconductors is shown below. What is its formula?

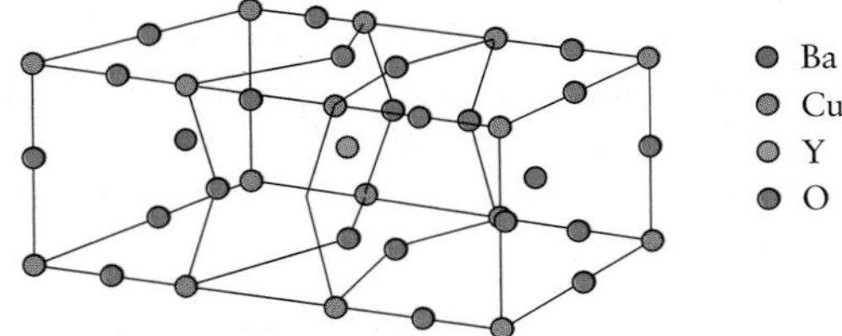

10.43 How many unit cells of the kind shown in Fig. 10.36 are there in a 1.00-mm^3 grain of table salt? (The density of sodium chloride is 2.16 $g{\cdot}cm^{-3}$.)

10.44 How many unit cells of the kind shown in Fig. 10.36 are there in a 1.00-mm^3 grain of potassium bromide? (The density of potassium bromide is 2.75 $g{\cdot}cm^{-3}$.)

10.45 Use radius ratios to predict the coordination number of the cation in (a) KBr; (b) LiBr; (c) BaO. (See Fig. 7.28.)

10.46 Use radius ratios to predict the coordination number of the cation in (a) RbF; (b) MgO; (c) NaBr. (See Fig. 7.28.)

10.47 Calculate the density of the following solids from the data in Fig. 7.28: (a) sodium iodide (rock-salt structure, Fig. 10.36); (b) cesium iodide (cesium-chloride structure, Fig. 10.38).

10.48 Calculate the density of the following solids from the data in Fig. 7.28: (a) calcium oxide (rock-salt structure, Fig. 10.36); (b) cesium bromide (cesium-chloride structure, Fig. 10.38).

Network Solids

10.49 Name four elements that, in at least one allotropic form, are network solids.

10.50 Name two compounds that exist as network solids at room temperature and pressure.

10.51 The enthalpy of sublimation of diamond is 713 $kJ{\cdot}mol^{-1}$. What is the C—C bond enthalpy in the solid?

10.52 The enthalpy of formation of BN(s) is -254 $kJ{\cdot}mol^{-1}$, and that of BN(g) is $+647$ $kJ{\cdot}mol^{-1}$. Given that the enthalpies of formation of gaseous B and N atoms are $+563$ $kJ{\cdot}mol^{-1}$ and $+473$ $kJ{\cdot}mol^{-1}$, respectively, calculate the B—N bond enthalpy in the solid.

Molecular Solids

10.53 Identify the types of intermolecular forces that are responsible for the existence of each of the following molecular solids: (a) solid argon; (b) ice; (c) solid HCl.

10.54 Identify the types of intermolecular forces that are responsible for the existence of each of the following molecular solids: (a) iodine; (b) oxalic acid, $H_2C_2O_4$; (c) solid benzene.

10.55 The figure below shows a unit cell of a two-dimensional molecular compound of elements A, B, and C. What is its molecular formula?

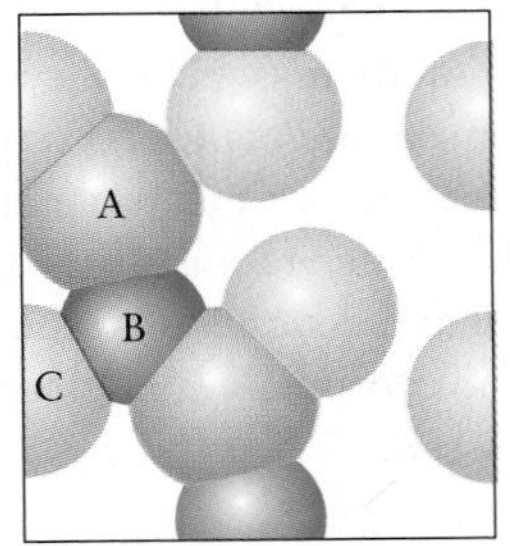

10.56 State the molecular formula of the element with the two-dimensional unit cell shown below.

Vapor Pressure

10.57 What is meant by "dynamic equilibrium"? How does it differ from static equilibrium?

10.58 How, using isotopes, could you show that an equilibrium was dynamic and not static?

10.59 Suppose you were to collect 1.0 L of air by slowly passing it through water at 20°C and into a container. Estimate the mass of water vapor in the collected air, assuming that the air is saturated with it. At 20°C, the vapor pressure of water is 17.5 Torr.

10.60 Suppose you were to collect 500 mL of nitrogen gas by slowly passing it through ethanol at 25°C, at which temperature ethanol's vapor pressure is 58.9 Torr. Estimate the mass of ethanol vapor in the collected gas, assuming that the nitrogen is saturated with it.

10.61 What mass of water vapor can you expect to find in the air of a bathroom of dimensions 4.0 m by 3.0 m by 3.0 m when water has been left in the bathtub at 40°C? The vapor pressure of water at that temperature is 7.4 kPa. Assume that the air is saturated with water vapor.

10.62 A bottle of mercury at 25°C was left unstoppered in a chemical supply room measuring 3.0 m by 2.0 m by 2.0 m. What mass of mercury vapor would be present if the air became saturated with it? The vapor pressure of mercury at 25°C is 0.227 kPa.

10.63 Use the vapor pressure curve in Fig. 10.50 to estimate the boiling point of water when the atmospheric pressure is (a) 735 Torr; (b) 750 Torr.

10.64 Use the vapor pressure curve in Fig. 10.50 to estimate the boiling point of ethanol when the atmospheric pressure is (a) 600 Torr; (b) 400 Torr.

Phase Diagrams

10.65 Use Fig. 10.55 (the phase diagram for water) to predict the state of a sample of water under the following sets of conditions: (a) 1 atm, 200°C; (b) 100 atm, 200°C; (c) 3 Torr, 25°C; (d) 218 atm, 375°C.

10.66 Using Fig. 10.52 (the phase diagram for carbon dioxide) to predict the state of a sample of CO_2 under the following sets of conditions. (a) 6 atm, −80°C; (b) 1 atm, −56°C; (c) 80 atm, 25°C; (d) 5.1 atm, −56°C.

10.67 The phase diagram for helium is shown below. Use it to answer the following questions: (a) What is the maximum temperature at which superfluid helium-II can exist? (b) What is the minimum pressure at which solid helium can exist? (c) What is the normal boiling point of helium-I? (d) Can solid helium sublime?

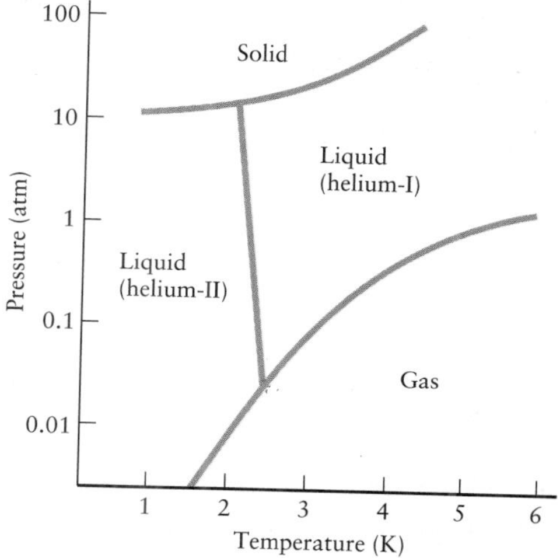

10.68 The phase diagram for carbon is shown below: it indicates the extreme conditions that are needed to form diamonds from graphite. Use the diagram to answer the following questions: (a) At 2000 K, what is the minimum pressure needed before graphite changes to diamond? (b) What is the minimum temperature at which liquid carbon can exist at pressures below 10,000 atm? (c) At what pressure does graphite melt at 3000 K? (d) Are diamonds stable under normal conditions? If not, why is it that people can wear them without having to be compressed and heated?

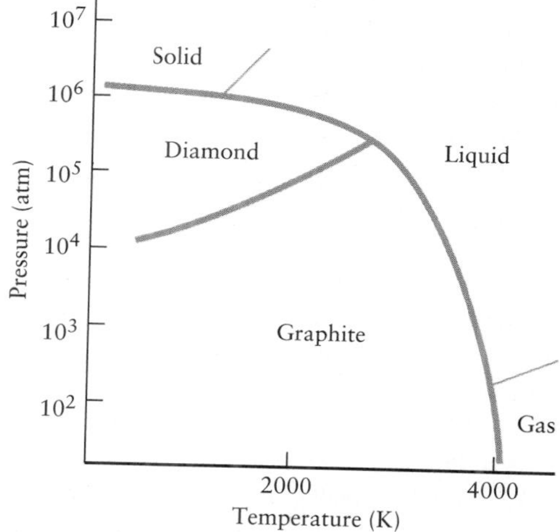

10.69 Use the phase diagram for helium (in Exercise 10.67) (a) to describe the phases in equilibrium at each of helium's two triple points. (b) to state which liquid phase is more dense, helium-I or helium-II.

10.70 Use the phase diagram for carbon in Exercise 10.68 (a) to describe the phase transitions that carbon would experience if compressed at a constant temperature of 2000 K from 100 atm to 1×10^6 atm; (b) to rank the diamond, graphite, and liquid phases of carbon in order of increasing density.

10.71 Naphthalene sublimes at room temperature and pressure. What can be said about the triple point of naphthalene?

10.72 Under what conditions can liquid carbon dioxide exist at room temperature?

SUPPLEMENTARY EXERCISES

10.73 Account for vapor pressure in terms of the rates at which molecules can leave and return to a liquid.

10.74 Explain in terms of intermolecular forces why the skin feels cooler when wet with rubbing alcohol (isopropanol) than when wet with water.

10.75 Why do the solid-liquid phase boundaries in the water and carbon dioxide phase diagrams slope in opposite directions? What are the practical consequences of the slopes?

10.76 Complete the following statements about the effect of intermolecular forces on the physical properties of a substance: (a) The higher the boiling point of a liquid, the (stronger, weaker) its intermolecular forces. (b) Substances with strong intermolecular forces have (high, low) vapor pressures. (c) Substances with strong intermolecular forces typically have (high, low) surface tensions. (d) The higher the vapor pressure of a liquid, the (stronger, weaker) its intermolecular forces. (e) Because nitrogen, N_2, has (strong, weak) intermolecular forces, it has a (high, low) critical temperature. (f) Substances with high vapor pressures have correspondingly (high, low) boiling points. (g) Because water has a relatively high boiling point, it must have (strong, weak) intermolecular forces and a correspondingly (high, low) enthalpy of vaporization.

10.77 Select the one substance that has the corresponding property. Justify your answer in each case: (a) The strongest hydrogen bonding: H_2O, H_2S, CH_3OH (as liquids). (b) The greatest surface tension: CH_3OH, C_2H_5OH, C_3H_7OH (as liquids). (c) The highest vapor pressure: CO_2, SO_2, SiO_2 (as solids). (d) The lowest viscosity: HCl, HBr, HI (as liquids). (e) The lowest enthalpy of vaporization: H_2O, H_2S, H_2Te (as liquids). (f) The lowest critical temperature: O_2, N_2, H_2 (as gases). (g) The highest boiling point: AsH_3, PH_3, NH_3 (as liquids). (h) The highest enthalpy of fusion: Na_2O, H_2O, Cl_2O (as solids). (i) The strongest dipole-dipole forces: H_2S, SCl_2, SF_2. (j) The strongest London forces: CH_4, SiF_4, GeF_4.

10.78 Hydrogen peroxide, H_2O_2, is a syrupy liquid with a vapor pressure lower than that of water and a boiling point of 152°C. Account for the differences between these properties and those of water.

10.79 Explain the effect an increase in temperature has on each of the following properties: (a) viscosity; (b) surface tension; (c) vapor pressure; (d) evaporation rate.

10.80 Explain how the vapor pressure of a liquid is affected by each of the following changes in conditions: (a) an increase in temperature; (b) an increase in surface area of the liquid; (c) an increase in volume above the liquid; (d) the addition of air to the volume above the liquid.

10.81 The normal boiling temperature of water is 100°C. Suppose a cyclonic region (a region of low pressure) moves into the area. State and explain what happens to the boiling point of the water.

10.82 A vacuum pump is attached to a flask of water at 0°C and 2 atm and the pressure on the liquid is decreased to 5 Torr at constant temperature. (a) Explain what would be observed, based on the phase diagram of water, Fig. 10.55. (b) Explain what would be observed if the water were at 50°C instead of 0°C.

10.83 Use the phase diagram for carbon dioxide (Fig. 10.52) to predict what would happen to a sample of carbon dioxide gas at −50°C and 1 atm if its pressure were suddenly increased to 73 atm. What would be the final physical state of the carbon dioxide?

10.84 An air mass is saturated with water vapor at 10°C. Given that at 10°C the vapor pressure of water is 9.21 Torr, explain what happens to the water in the air mass if it meets (a) a warm weather front; (b) a cold front.

10.85 Relative humidity at a particular temperature is defined as

$$\text{Relative humidity} = \frac{\text{partial pressure of water}}{\text{vapor pressure of water}} \times 100\%$$

The vapor pressure of water at various temperatures is given in Table 5.4. (a) What is the relative humidity at 30°C when the partial pressure of water is 25.0 Torr? (b) Explain what would be observed if the temperature of the air fell to 25°C.

10.86 The 5:00 PM weather report states that the ambient temperature is 94°F (34°C) and the relative humidity (see the preceding exercise) is 64%; an overnight low of 77°F

(25°C) is expected. Can fog or rain showers be predicted? Explain your reasoning.

10.87 Calculate the relative number of atoms of each element contained in each alloy: (a) coinage cupronickel, which is 25% Ni by mass in copper; (b) pewter, which is about 7% Sb and 3% Cu by mass in tin; (c) Wood's metal, which is a low-melting-point alloy used to trigger automatic sprinkler systems and is 12.5% Sn, 12.5% Cd, and 25% Pb by mass in bismuth.

10.88 Classify the following solids as ionic, network, or molecular. (a) quartz, SiO_2; (b) limestone, $CaCO_3$; (c) dry ice, CO_2; (d) sucrose, $C_{12}H_{22}O_{11}$; (e) polyethylene, a polymer of repeating $—CH_2CH_2—$ units.

10.89 Sodium metal has a density of 0.97 $g{\cdot}cm^{-3}$ and forms a body-centered lattice. (a) Calculate the length (in pm) of the edge of the unit cell. (b) What is the atomic radius of sodium? (c) What is the mass of one unit cell of sodium? (See Exercise 10.28 for more information.)

10.90 Lead metal has a density of 11.34 $g{\cdot}cm^{-3}$ and crystallizes in a face-centered lattice. (a) Use the Avogadro constant to determine the length of the edge of the unit cell. (b) Determine the volume of one unit cell. (c) What is the mass of one unit cell of lead? (d) Calculate the atomic radius of the lead atom.

10.91 Krypton crystallizes with a face-centered cubic unit cell of edge 559 pm. (a) What is the density of solid krypton? (b) What is the atomic radius of krypton? (c) What is the volume of one krypton atom? (The volume of a sphere of radius r is $\frac{4}{3}\pi r^3$.) (d) What percentage of the unit cell is empty space if each atom is treated as a hand sphere?

CHALLENGING EXERCISES

10.92 Paper towels are effective in cleaning up water spills. Synthetic towels made from a synthetic fiber (such as a polyester) may be more economical to produce. Explain why paper towels are effective. Could a synthetic towel be more effective? If so, what properties should the synthetic fiber exhibit?

10.93 "Graphite bisulfates" are formed by heating graphite with a mixture of sulfuric and nitric acids. In the reaction the graphite planes are partially oxidized (so that there is one positive charge shared between about 24 carbon atoms) and the HSO_4^- anions are distributed between the planes. What effects should this have on the electrical conductivity?

10.94 Aluminum metal has a density of 2.70 $g{\cdot}cm^{-3}$ and crystallizes in a lattice with an edge of 404 pm. (a) What type of cubic unit cell is formed by aluminum? (b) What is the coordination number of aluminum?

10.95 All the noble gases except helium crystallize with cubic close-packed structures. Find an equation relating the atomic radius to the density of a cubic close-packed solid of given molar mass and apply it to deduce the atomic radii of the noble gases, given the following densities (in $g{\cdot}cm^{-3}$): Ne, 1.20; Ar, 1.40; Kr, 2.16; Xe, 2.83; Rn, 4.4

10.96 All the alkali metals crystallize into bcc structures. (a) Find an equation relating the metallic radius to the density of a bcc solid of an element in terms of its molar mass and use it to deduce the atomic radii of the elements, given the following densities (in $g{\cdot}cm^{-3}$): Li, 0.53; Na, 0.97; K, 0.86; Rb, 1.53; Cs, 1.87. Develop this question further by combining it with the preceding exercise: (b) Find a factor for converting a bcc density to a cubic close-packed density of the same element. (c) Calculate what the densities of the alkali metals would be if they were cubic close-packed. (d) Which, if any, would float on water?

10.97 Metals with bcc structures, like tungsten, are not close packed. Therefore, their densities would be greater if they were to change to a cubic close-packed structure (under pressure, for instance). What would the density of tungsten be if it had a cubic close-packed structure rather than bcc? Its actual density is 19.3 $g{\cdot}cm^{-3}$.

10.98 A new substance developed in a laboratory has the following properties: normal melting point, 83.7°C; normal boiling point, 177°C; triple point, 200 Torr and 38.6°C. (a) Sketch the approximate phase diagram and label the solid, liquid, and gaseous phases and the solid-liquid, liquid-gas, and solid-gas phase boundaries. (b) Sketch an approximate cooling curve for a sample at constant pressure, beginning at 500 Torr and 25°C and ending at 200°C.

10.99 The vapor pressure of ethanol at 25°C is 58.9 Torr. A sample of ethanol vapor at 25°C and 58.9 Torr partial pressure is in equilibrium with a very small amount of liquid ethanol in a 10-L container also containing dry air, at a total pressure of 750 Torr. The volume of the container is then reduced to 5.0 L. (a) What is the partial pressure of ethanol in the smaller volume? Explain your reasoning. (b) What is the total pressure of the mixture?

CHAPTER 11

In this chapter we see how chemists use the information described in the previous three chapters to form the covalent bonds that produce the materials we need. No element is more versatile at forming bonds than carbon, and here we see the awesome position this element adopts in nature and in industry. Its compounds range from very simple methane to amazingly complex DNA, a part of which is shown in this scanning tunneling micrograph. The long strands of the DNA molecule contain repeating units that interact with each other to produce a double helix; this helix unwinds to reproduce itself.

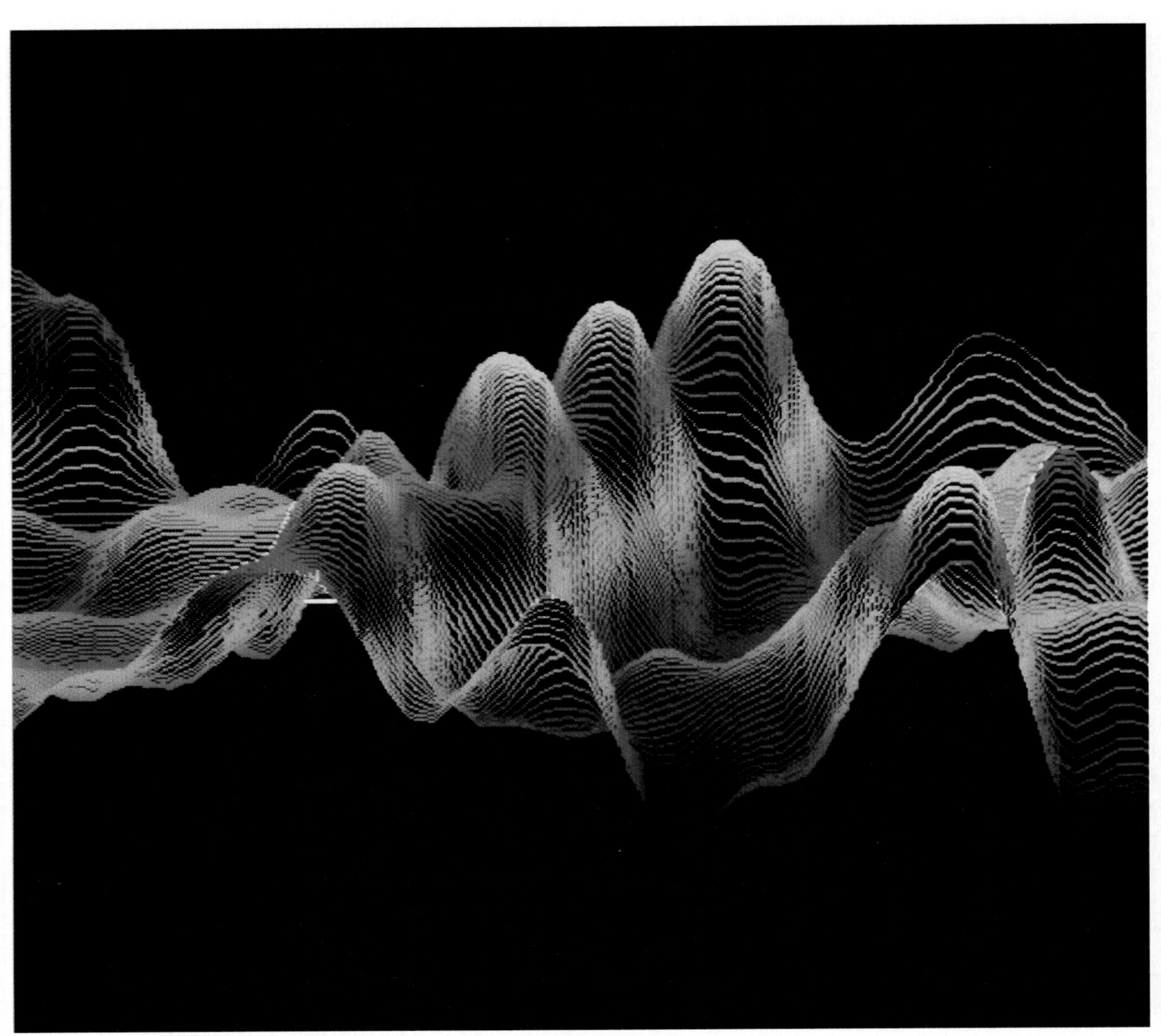

CARBON-BASED MATERIALS

Carbon is a most extraordinary element. All life on Earth is based on it. So is the fuel we burn, our food, and the clothes we wear. Materials based on carbon have properties ranging from the softness of artificial skin and cartilage to the tough composite materials used in experimental aircraft. Polymer chemists are learning how to create custom materials for medicine, industry, and research. Many of these materials are based on carbon.

The unique feature that makes carbon so adaptable is that carbon atoms can string together—by forming covalent bonds—to form chains and rings of endless variety. This versatility allows carbon to form not only synthetic materials but also the complicated molecules that are needed to carry out the intricate processes we associate with being alive.

Despite the variety of molecules that carbon forms and the reactions they undergo, the properties of organic compounds can be understood in terms of small groups of atoms. The same is true of the enzymes that make life possible and of the synthetic materials that make it more agreeable. Organic chemistry—the chemistry of the compounds of carbon—is the chemistry of families of compounds and of patterns shown by the groups of atoms these compounds contain.

HYDROCARBONS

11.1 Types of hydrocarbons
11.2 Alkanes
11.3 Alkenes and alkynes
11.4 Aromatic compounds

FUNCTIONAL GROUPS

11.5 Alcohols
11.6 Ethers
11.7 Phenols
11.8 Aldehydes and ketones
11.9 Carboxylic acids
11.10 Amines and amides

ISOMERS

11.11 Structural isomers
11.12 Geometrical and optical isomers

POLYMERS

11.13 Addition polymerization
11.14 Condensation polymerization
11.15 Copolymers and composites
11.16 Physical properties

EDIBLE POLYMERS

11.17 Proteins
11.18 Carbohydrates
11.19 DNA and RNA

HYDROCARBONS

Toolbox 11.1 in Section 11.4 explains how to name hydrocarbons.

The simplest family of carbon compounds are the **hydrocarbons,** which contain only carbon and hydrogen. Even though hydrocarbons consist of only two elements, there are enormous numbers of them because carbon can form an amazing variety of chains, rings, and networks of atoms.

11.1 TYPES OF HYDROCARBONS

Many organic molecules are quite complicated, so we use a simple way to represent their structures. **Stick structures** represent a chain of carbon atoms as a zigzag line, where the end of each short line in the zigzag represents a carbon atom. Because carbon nearly always has a valence of four in organic compounds, we do not need to show any C—H bonds. We just fill in the correct number of hydrogen atoms mentally, as we see for methylbutane (**1**), isoprene (**2**), and propyne (**3**).

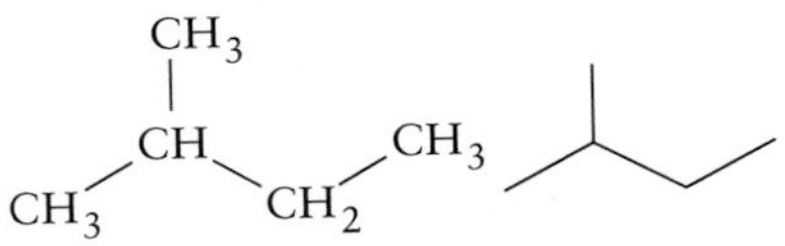

1 Methylbutane, $(CH_3)_2CH_2CH_3$

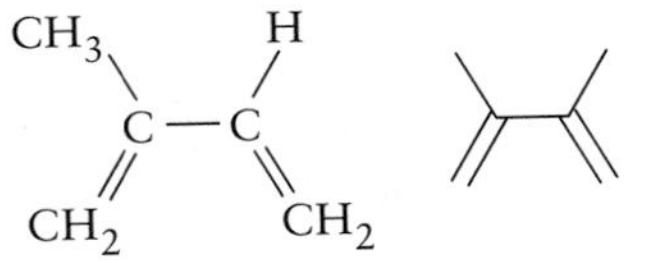

2 Isoprene, $CH_2C(CH_3)CHCH_2$

3 Propyne, CH_3CCH

SELF-TEST 11.1A Draw the stick structure of aspirin (**4a**).

[*Answer:* (**4b**)]

SELF-TEST 11.1B Write the molecular structure of carvone (**5**).

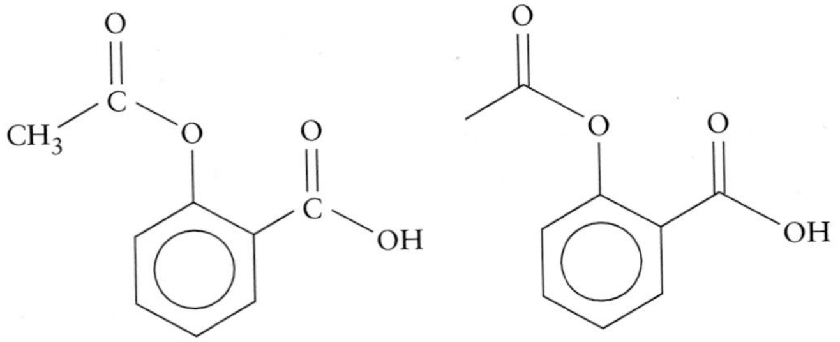

(a)

(b)

4 Acetylsalicylic acid (aspirin)

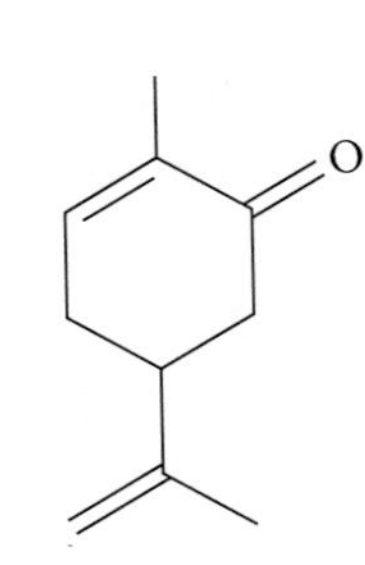

5 Carvone

Aromatic hydrocarbons have a benzene ring as a part of their molecular structure; **aliphatic hydrocarbons** do not. The compound shown as (**6**) is aromatic; the compound in (**7**) is aliphatic. In complex molecules, we speak of aliphatic parts and aromatic parts, rather than trying to classify the molecule as a whole. A **saturated hydrocarbon** is an aliphatic hydrocarbon with no multiple bonds; an **unsaturated hydrocarbon** has

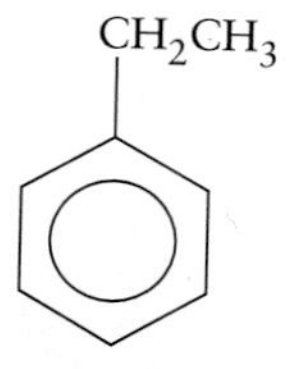

6 Ethylbenzene, $C_6H_5CH_2CH_3$

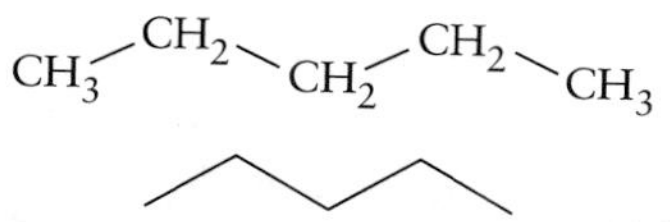

7 Pentane, C_5H_{12}

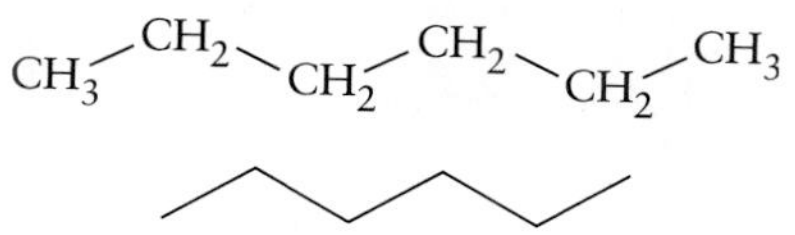

8 Hexane, C_6H_{14}

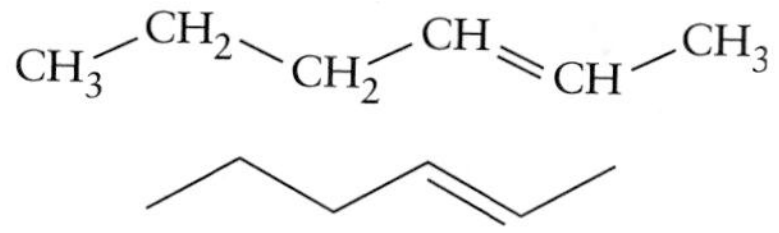

9 2-Hexene, C_6H_{12}

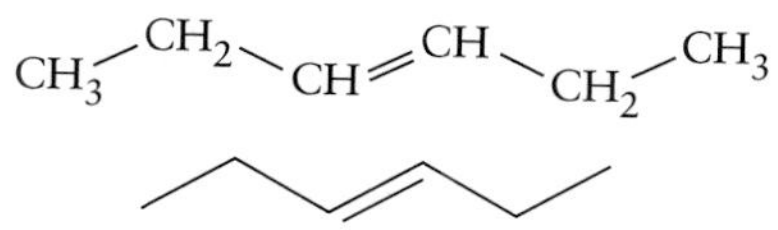

10 3-Hexene, C_6H_{12}

one or more double or triple bonds. More hydrogen can be added to compounds in which there are multiple bonds, but compounds with only single bonds are "saturated" with hydrogen. Compound (**8**) is saturated; compounds (**9**) and (**10**) are unsaturated.

Aromatic hydrocarbons contain benzene rings; aliphatic hydrocarbons do not. Saturated hydrocarbons have only single bonds.

11.2 ALKANES

Saturated hydrocarbons are called **alkanes.** The simplest alkane is methane, CH_4 (**11**). The other alkanes form a family in which the formula of each succeeding member is derived from CH_4 by inserting CH_2 groups between pairs of atoms. Alkanes are named by combining the prefix for the number of carbon atoms with the suffix *-ane* (Table 11.1). Alkanes that have ringlike structures are called **cycloalkanes.** Cyclopropane, C_3H_6 (**12**), and cyclohexane, C_6H_{12} (**13**), are cycloalkanes.

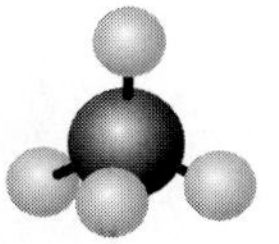

11 Methane, CH_4

TABLE 11.1 Alkane nomenclature

Number of carbon atoms	Formula	Name of alkane	Name of alkyl group
1	CH_4	methane	methyl
2	CH_3CH_3	ethane	ethyl
3	$CH_3CH_2CH_3$	propane	propyl
4	$CH_3(CH_2)_2CH_3$	butane	butyl
5	$CH_3(CH_2)_3CH_3$	pentane	pentyl
6	$CH_3(CH_2)_4CH_3$	hexane	hexyl
7	$CH_3(CH_2)_5CH_3$	heptane	heptyl
8	$CH_3(CH_2)_6CH_3$	octane	octyl
9	$CH_3(CH_2)_7CH_3$	nonane	nonyl
10	$CH_3(CH_2)_8CH_3$	decane	decyl
11	$CH_3(CH_2)_9CH_3$	undecane	undecyl
12	$CH_3(CH_2)_{10}CH_3$	dodecane	dodecyl

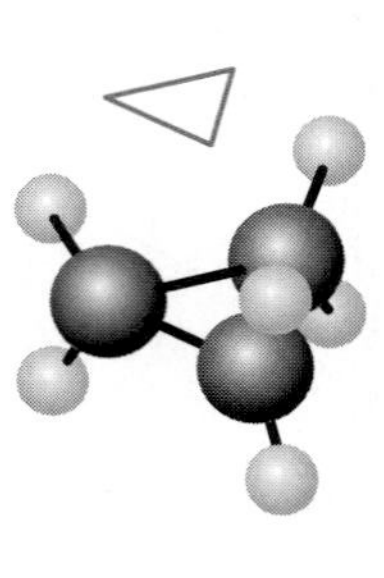

12 Cyclopropane, C_3H_6

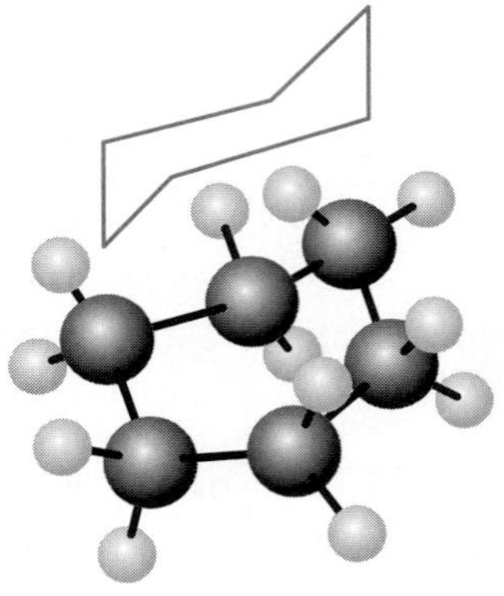

13 Cyclohexane, C_6H_{12}

As noted in Section 9.13, the bonds to each carbon atom in alkanes lie in a tetrahedral arrangement, with sp^3 hybridization. Because all the carbon-carbon bonds are single, one part of a molecule can easily rotate relative to the rest of the molecule around any C—C bond. As a result, in liquids and gases the chains are normally in constant motion, freely twisting about, rolling up into a ball (**14**), or stretching out into a zigzag arrangement (**15**).

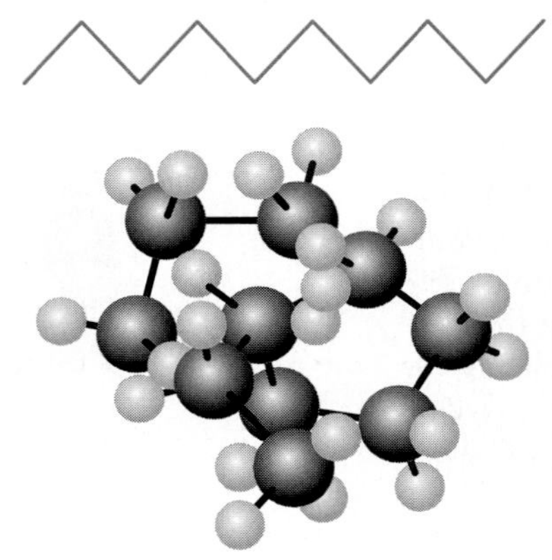

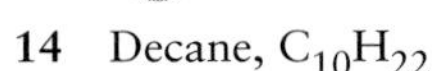

14 Decane, $C_{10}H_{22}$

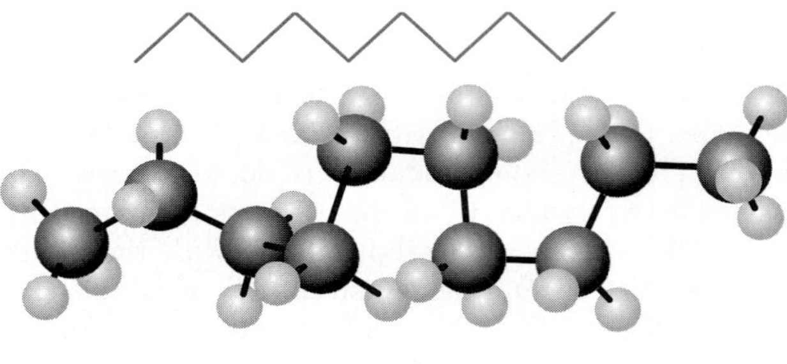

15 Decane, $C_{10}H_{22}$

The dominant force between alkane molecules is the London force. Therefore, the alkanes become less volatile with increasing molar mass (Fig. 11.1). The lightest members, methane through butane, are gases at

TABLE 11.2 Hydrocarbon constituents of petroleum

Hydrocarbons	Boiling range, °C	Name
C_1 to C_4	−160 to 0	gas
C_5 to C_{10}	30 to 200	gasoline
C_{11} to C_{16}	180 to 400	kerosene, fuel oil
C_{17} to C_{22}	350 and above	lubricants
C_{23} to C_{34}	low-melting solids	paraffin wax
C_{35} upward	soft solids	asphalt

room temperature. Pentane is a volatile liquid, and hexane through undecane ($C_{11}H_{24}$) are moderately volatile liquids that are present in gasoline (Table 11.2). Kerosene, a fuel used in jet and diesel engines, contains a number of alkanes in the range C_{10} to C_{16}. Lubricating oils are mixtures in the range C_{17} to C_{22}. The heavier members of the series include the paraffin waxes and asphalt. All alkanes are insoluble in water.

As soon as there are four carbon atoms in a chain, at C_4H_{10}, we start to see another reason for the variety of compounds that carbon can form. Four carbon atoms can link together in a chain to form butane (**16**) or in a Y-shape to form the alkane called methylpropane (**17**).

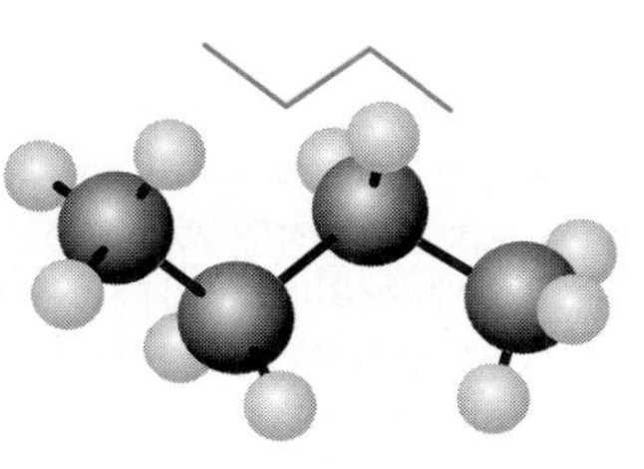

16 Butane, C_4H_{10}

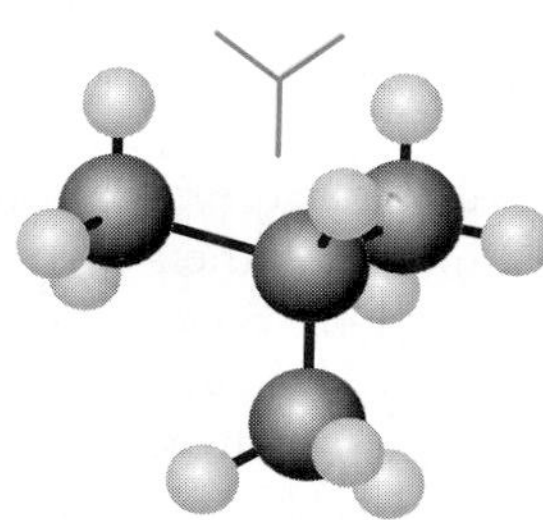

17 Methylpropane, C_4H_{10}

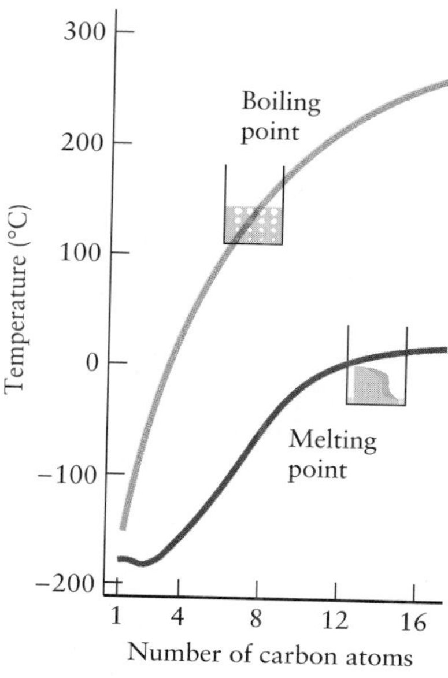

FIGURE 11.1 The melting and boiling points of the unbranched alkanes from CH_4 to $C_{16}H_{34}$.

The name *isomer* comes from the Greek words for "equal parts." We can think of isomeric molecules as being built from the same kit of parts.

Compounds with the same molecular formula, but with their atoms in different arrangements are called **isomers.** Although they have the same molecular formula, isomers have different physical properties. Alkanes with long, unbranched chains tend to have higher melting points, boiling points, and enthalpies of vaporization than their branched isomers, which may have one or more **side chains,** shorter chains attached to the longest chain. This difference can be traced to the fact that the atoms of neighboring branched molecules cannot get as close together as their unbranched isomers can, so the attractive forces between them are weaker (Fig. 11.2).

You can tell if a group in parentheses is a side chain or part of the main chain by recalling that carbon has a valence of four. Thus, $(CH_3)_2$ must be side chains, but $(CH_2)_2$ must be part of the main chain.

There are several ways of writing structural formulas. However, it is often sufficient to show only how the atoms are grouped together. For instance, we write $CH_3CH_2CH_2CH_3$ for butane and $CH_3CH(CH_3)CH_3$

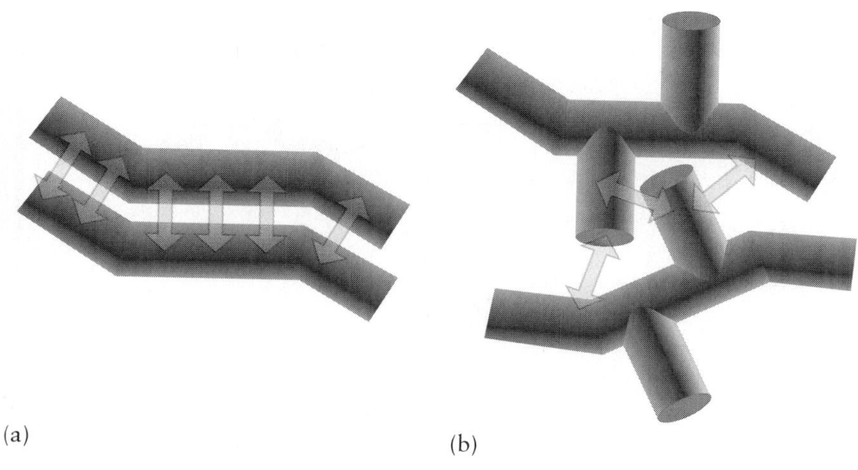

FIGURE 11.2 (a) The atoms in neighboring straight-chain alkanes, represented by the tubelike structures, can lie close together. (b) Fewer of the atoms of neighboring branched alkane molecules can get so close together, so the London forces (represented by double-headed arrows) are weaker and branched alkanes are more volatile.

for methylpropane. The parentheses around the CH_3 group show that it is attached to the carbon atom on its left. In general, when writing a formula we look for a short, unambiguous way to describe the compound. For instance, when several CH_2 groups are repeated, we collect them together; thus, we can write butane as $CH_3(CH_2)_2CH_3$. This notation is an example of a *condensed structural formula.*

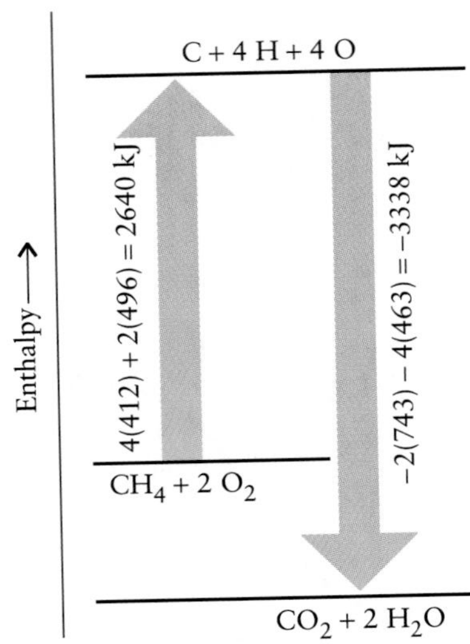

FIGURE 11.3 The enthalpy changes accompanying the combustion of methane. Although the bonds in the reactants are strong, they are even stronger in the products; and the overall process is exothermic.

SELF-TEST 11.2A Which compound has the higher boiling point: (a) $CH_3CH_2CH_2CH_2CH_3$ or (b) $CH_3C(CH_3)CH_2CH_3$? Why?

[*Answer:* (a), because it is not branched.]

SELF-TEST 11.2B Which compound has the higher boiling point: (a) $CH_3CH_2CH_2CH_2CH_3$ or (b) $CH_3CH_2CH_2CH_3$? Why?

The alkanes were once called the *paraffins,* from the Latin words for "little affinity." As their name suggests, they are not very reactive. They are unaffected by concentrated sulfuric acid, by boiling nitric acid, by strong oxidizing agents like potassium permanganate, and by boiling aqueous sodium hydroxide. One reason for their resistance to this chemical onslaught is thermodynamic: the C—C and C—H bonds are strong (Table 9.3) and there is little energy advantage in replacing them with other bonds, except most notably by C=O, C—OH, and C—F bonds.

The most familiar reaction of alkanes is their combustion to carbon dioxide and water:

$$CH_4(g) + 2\,O_2(g) \longrightarrow CO_2(g) + 2\,H_2O(g)$$

In this reaction, the strong carbon-hydrogen bonds (average bond enthalpy 412 $kJ\cdot mol^{-1}$) are replaced by the even stronger hydrogen-oxygen bonds (463 $kJ\cdot mol^{-1}$), and the oxygen-oxygen bond (496 $kJ\cdot mol^{-1}$) is replaced by two very strong carbon-oxygen bonds (743 $kJ\cdot mol^{-1}$). The energy released is given off as heat (Fig. 11.3). Similar exothermic reactions make the alkanes valuable fuels (Section 6.11).

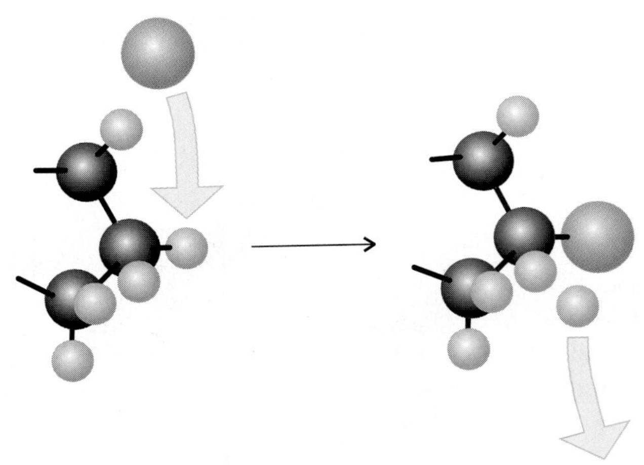

FIGURE 11.4 In an alkane substitution reaction, an incoming atom or group of atoms (represented by the orange sphere) replaces a hydrogen atom in the alkane molecule.

In a **substitution reaction,** an atom or group of atoms replaces an atom in the original molecule. Substitution reactions are typical of alkanes, and the displaced atom is a hydrogen atom (Fig. 11.4). An example is the reaction between methane and chlorine. A mixture of these two gases survives unchanged in the dark; but when exposed to ultraviolet radiation or heated, they react:

$$CH_4(g) + Cl_2(g) \longrightarrow CH_3Cl(g) + HCl(g)$$

Chloromethane, CH_3Cl, is only one of the products of this reaction, because it also forms dichloromethane (CH_2Cl_2), trichloromethane ($CHCl_3$), and tetrachloromethane (CCl_4). Trichloromethane is better known as *chloroform* and was one of the early anesthetics. Tetrachloromethane, which is commonly called *carbon tetrachloride,* is nonflammable; it was used in fire extinguishers and as a solvent until its toxicity became known.

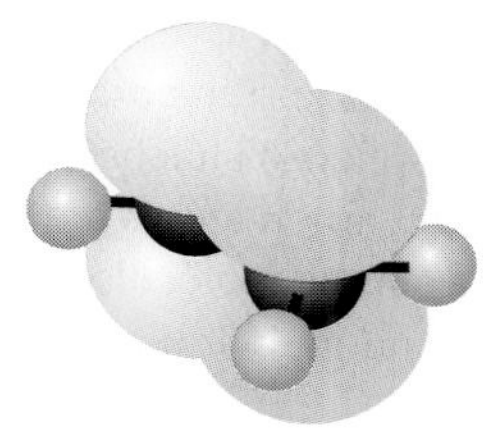

FIGURE 11.5 The π-bond (yellow electron clouds) in an alkene molecule makes the molecule resistant to twisting around a double bond, so all six atoms lie in the same plane.

The strength of the London forces between alkane molecules increases as the molar mass of the molecules increases. Alkanes are not very reactive but do undergo oxidation and substitution.

11.3 ALKENES AND ALKYNES

The simplest unsaturated hydrocarbon is ethene, C_2H_4, commonly called ethylene (**18**). Ethene is the parent of the **alkenes,** a series of molecules with formulas derived from ethene by adding $-CH_2-$ groups. For example, the next member of the family is propene, $HC_3CH{=}CH_2$. The names of the alkenes are the same as the names of the corresponding alkanes, except that they end in *-ene*. The **alkynes** are hydrocarbons that have at least one carbon-carbon triple bond. The simplest is ethyne, $HC{\equiv}CH$, which is commonly called acetylene (**19**). Alkynes are named with the suffix *-yne*.

▶▶ **See Toolbox 11.1 for more information on the nomenclature of alkenes and alkynes.**

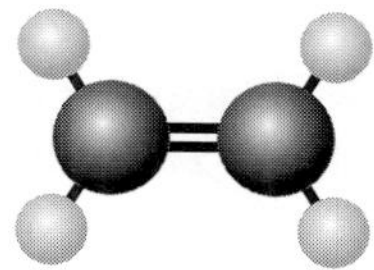

18 Ethene, C_2H_4

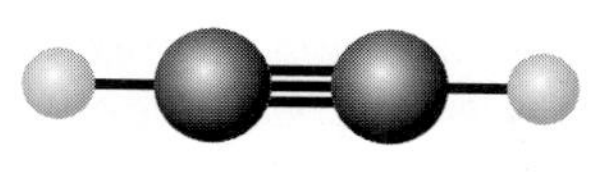

19 Ethyne, C_2H_2

As we saw in Section 9.17, the C=C double bond of alkenes consists of a σ-bond and a π-bond. The double bond makes alkenes more rigid than alkanes and explains their chemical differences. All four atoms attached to the C=C group lie in the same plane and are locked into that arrangement by the π-bond (Fig. 11.5). Because alkene molecules cannot roll up into a ball so compactly they cannot pack together as closely as alkanes can; so they have lower melting points (Fig. 11.6).

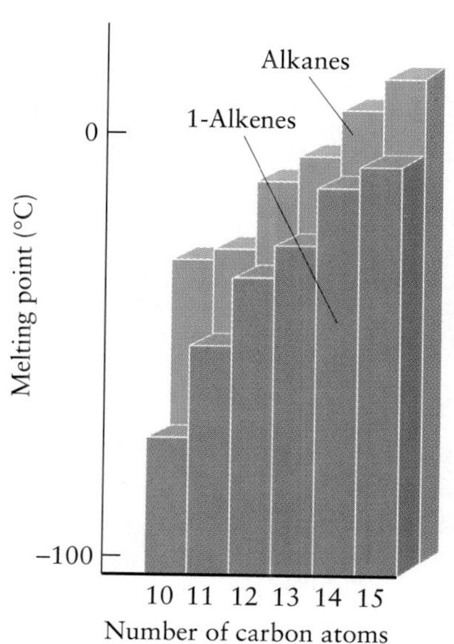

FIGURE 11.6 The melting point of an alkene is usually lower than that of the alkane with the same number of carbon atoms. The values shown are for unbranched alkanes and 1-alkenes (that is, alkenes in which the double bond is at the end of the carbon chain).

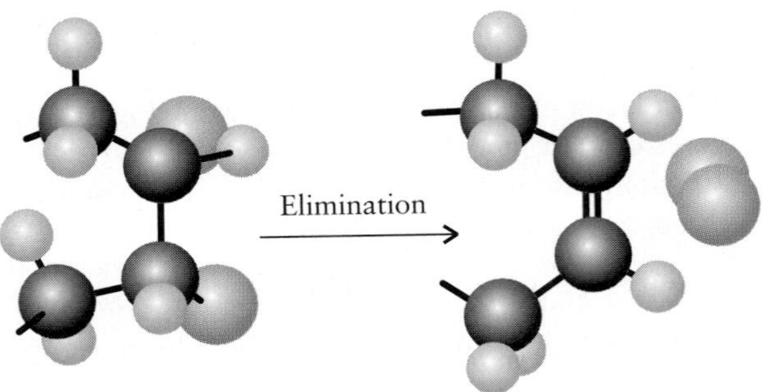

FIGURE 11.7 In an elimination reaction, two atoms (the orange and purple spheres) attached to neighboring carbon atoms are removed from the molecule, leaving a double bond in their place.

Alkenes are important feedstocks for the production of polymers, so one of the first steps in the petrochemical industry is to convert some of the abundant alkanes into alkenes. This conversion is achieved by removing hydrogen atoms from neighboring carbon atoms:

$$CH_3CH_3(g) \xrightarrow{Cr_2O_3,\ 500°C} CH_2{=}CH_2(g) + H_2(g)$$

This reaction is an example of an **elimination reaction,** a reaction in which two groups or two atoms on neighboring carbon atoms are removed, or eliminated, from a molecule, leaving a multiple bond between the carbon atoms (Fig. 11.7). The formula for chromium(III) oxide written above the arrow indicates that it is being used to make the reaction go faster. Chromium(III) oxide is an example of a catalyst, a substance that facilitates a reaction but remains unchanged itself.

The most characteristic chemical reaction of an alkene is an **addition reaction,** in which atoms supplied by the reactant are attached to the two atoms joined by the double bond (Fig. 11.8). An example is the addition of a halogen (Fig. 11.9). The two atoms of the halogen molecule are added to the structure of the molecule, as in the formation of 1,2-dichloroethane:

The states of the reactants and products are rarely given for organic reactions.

$$CH_2{=}CH_2 + Cl_2 \longrightarrow CH_2Cl{-}CH_2Cl$$

We can see why addition reactions are typical of alkenes by considering the energy changes in this reaction. As shown in Fig. 11.10, for this reaction to occur, we need to invest 264 $kJ{\cdot}mol^{-1}$ to convert the carbon-carbon double bond into a single bond and another 242 $kJ{\cdot}mol^{-1}$ to break the chlorine-chlorine bond (see Table 9.3). However, as the prod-

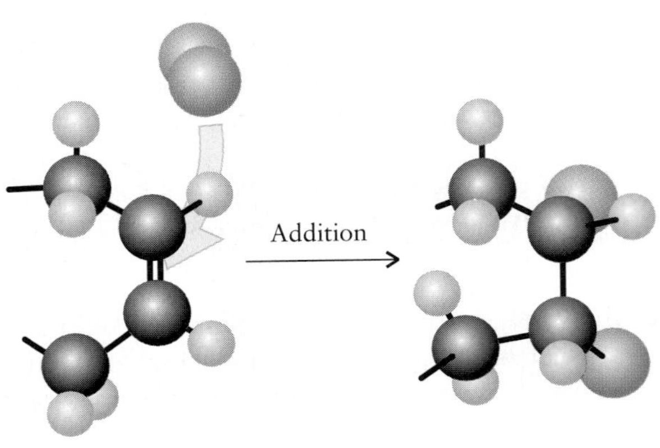

FIGURE 11.8 In an addition reaction, the atoms provided by an incoming molecule are attached to the carbon atoms originally joined by a multiple bond. Addition is the reverse of elimination.

FIGURE 11.9 When bromine dissolved in a solvent (the brown liquid) is mixed with an alkene (the colorless liquid), the bromine atoms add to the molecule at the double bond, giving a colorless product.

ucts are formed, *two* carbon-chlorine bonds are formed, which releases $2 \times (352\ kJ \cdot mol^{-1})$. Overall, a net amount of $198\ kJ \cdot mol^{-1}$ is released in the reaction. Almost all addition reactions are exothermic.

Hydrocarbons containing a double bond are called alkenes, those with a triple bond are called alkynes. Alkenes and alkynes undergo addition reactions at the multiple bond.

SELF-TEST 11.3A Write the structural formula of the product of the addition of hydrogen to butene:

$$CH_3CH{=}CHCH_3 + H_2 \xrightarrow{\text{Ni, 500°C}} \text{product}$$

[*Answer:* $CH_3CH_2CH_2CH_3$]

SELF-TEST 11.3B Write the structural formula of the compound formed by the addition of hydrogen bromide to ethene.

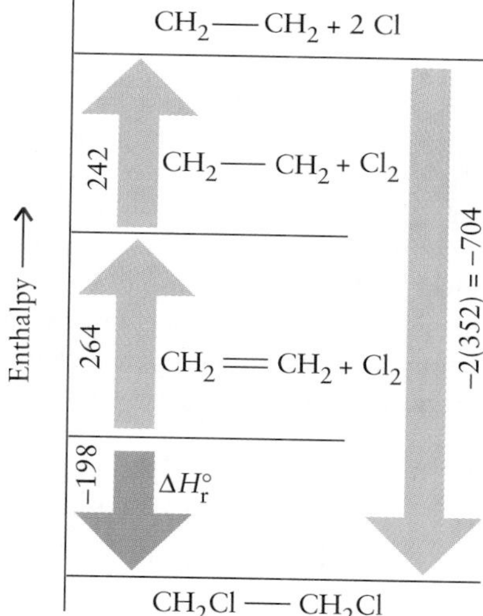

FIGURE 11.10 The enthalpy changes accompanying the addition of chlorine to a double bond. The upward-pointing arrows on the left show the enthalpy changes for the conversion of a double bond to a single bond and the dissociation of chlorine. The arrow on the right shows the return on this investment. Overall, the reaction is exothermic. Note that the actual C—Cl bond enthalpy differs somewhat from the average C—Cl bond enthalpies in Tables 9.2 and 9.3.

11.4 AROMATIC COMPOUNDS

Aromatic hydrocarbons as a group are called **arenes.** The parent compound of aromatic compounds is benzene, C_6H_6 (**20**). Aromatic compounds also include the fused-ring analogues of benzene, such as naphthalene, $C_{10}H_8$ (**21**), and anthracene, $C_{14}H_{10}$ (**22**). Both compounds can be obtained by the distillation of coal. Indeed, coal itself, which is a very complex mixture of complicated molecules, consists of very large networks containing regions in which aromatic rings can be identified (Fig. 11.11). When coal is destructively distilled—heated in the absence of oxygen so that it decomposes and vaporizes—the sheetlike molecules break up, and the fragments include the aromatic hydrocarbons and their derivatives. The complicated liquid mixture that results is called *coal tar.*

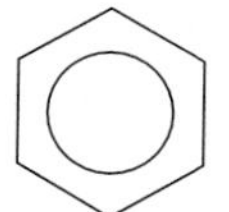

20 Benzene, C_6H_6

21 Naphthalene, $C_{10}H_8$

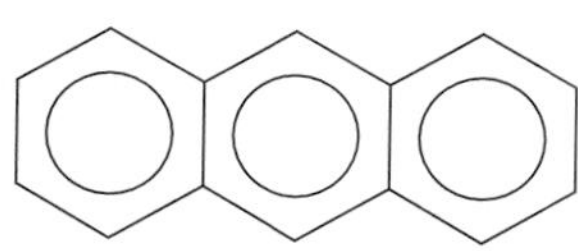

22 Anthracene, $C_{14}H_{10}$

Arenes predominantly undergo substitution reactions; the π-bonds of the ring are usually left intact. As an example, bromine immediately adds to a double bond of an alkene; however, it reacts with benzene only in the presence of a catalyst—typically iron(III) bromide—and it does not affect the bonding in the ring. Instead, one of the bromine atoms substitutes for a hydrogen atom:

$$C_6H_5\text{–}H + Br_2 \xrightarrow{FeBr_3} C_6H_5\text{–}Br + HBr$$

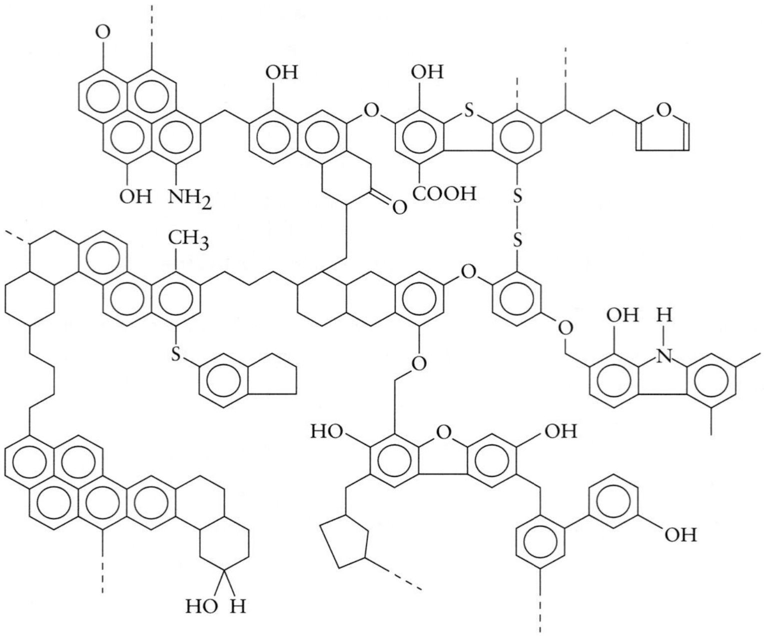

FIGURE 11.11 A highly schematic representation of a portion of the structure of coal. When coal is heated in the absence of oxygen, the structure breaks up and a complex mixture of products—many of them aromatic—is obtained as coal tar.

Aromatic rings are much less reactive than their double bond character would suggest; they commonly undergo substitution rather than addition.

TOOLBOX 11.1 How to name hydrocarbons

Alkanes The names of straight-chain alkanes are given in Table 11.1: they all end in *-ane*. To name a branched alkane, we treat the side chains as substituents. First, identify the longest unbranched chain of carbon atoms and give it the name of the corresponding alkane. Thus

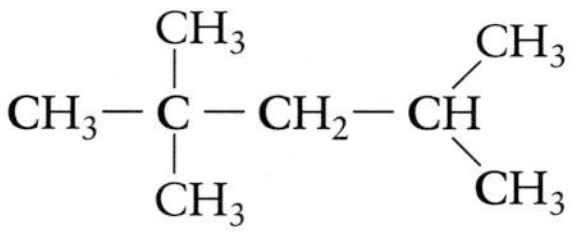

$$\begin{array}{ccccccc} & & CH_3 & & & & CH_3 \\ & & | & & & \diagup & \\ CH_3 & - & C & - & CH_2 & - & CH \\ & & | & & & \diagdown & \\ & & CH_3 & & & & CH_3 \end{array}$$

is a substituted pentane because its longest carbon chain (in bold type) has five C atoms. Next, identify the substituent groups by referring to Table 11.1 again. In this example there are three methyl groups (CH_3—). The names of these alkyl groups are obtained by changing the *-ane* suffix to *-yl*. The molecule is therefore a trimethylpentane. When different groups are present, list them in alphabetical order. Finally, denote the locations of the substituents by numbering the backbone C atoms from whichever end of the molecule results in the lowest numbered locations for the substituents:

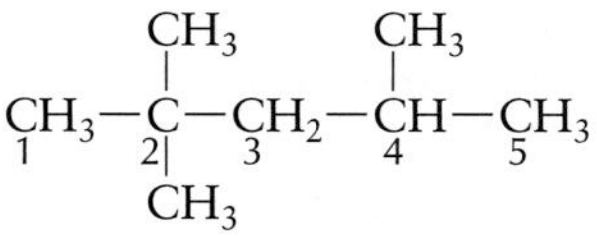

$$\begin{array}{ccccccccc} & & CH_3 & & & & CH_3 & & \\ & & | & & & & | & & \\ \underset{1}{CH_3} & - & \underset{2}{C} & - & \underset{3}{CH_2} & - & \underset{4}{CH} & - & \underset{5}{CH_3} \\ & & | & & & & & & \\ & & CH_3 & & & & & & \end{array}$$

The locations are then listed before the substituents. The molecule is 2,2,4-trimethylpentane.

SELF-TEST 11.4A Name the compound

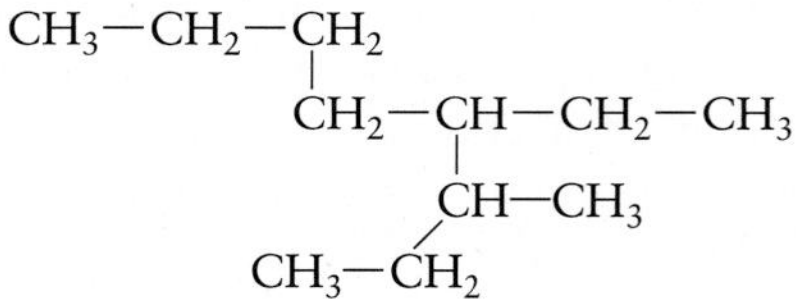

$$\begin{array}{l} CH_3-CH_2-CH_2 \\ \qquad\qquad\quad | \\ \qquad\qquad CH_2-CH-CH_2-CH_3 \\ \qquad\qquad\qquad\quad\;\; | \\ \qquad\qquad\qquad\;\; CH-CH_3 \\ \qquad\qquad\quad\;\; \diagup \\ \qquad\quad CH_3-CH_2 \end{array}$$

and write the formula of 5-ethyl-2,2-dimethyloctane.

[*Answer:* 4-Ethyl-3-methyloctane;

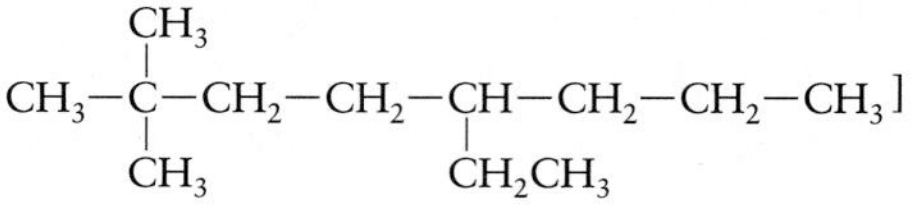

$$\begin{array}{ccccccccccccccc} & & CH_3 & & & & & & & & & & & & \\ & & | & & & & & & & & & & & & \\ CH_3 & - & C & - & CH_2 & - & CH_2 & - & CH & - & CH_2 & - & CH_2 & - & CH_3 \\ & & | & & & & & & | & & & & & & \\ & & CH_3 & & & & & & CH_2CH_3 & & & & & & \end{array}$$]

SELF-TEST 11.4B Name the compound $(CH_3)_2CHCH_2CH(CH_2CH_3)_2$ and give the formula of 3,3,5-triethylheptane.

Alkenes and alkynes Change the suffix *-ane* to either *-ene* (for alkenes) or *-yne* (for alkynes). In all cases except ethene (ethylene) and propene (propylene) and the corresponding alkynes, it is necessary to specify the location of the multiple bond. Do so by numbering the C atoms in the backbone and then reporting the lower of the two numbers of the two atoms joined by the multiple bond. The numbering starts at the end of the backbone that results in the lower number and has priority over the numbering of substituents. For example, $CH_3CH_2CH{=}CH_2$ is 1-butene.

SELF-TEST 11.5A Name the compound $(CH_3)_2CHCH{=}CH_2$.

[*Answer:* 3-Methyl-1-butene]

SELF-TEST 11.5B Name the compound $(CH_3CH_2)_2CHCH{=}CHCH_3$.

Arenes When the benzene ring, as C_6H_5—, is treated as a substituent, it is called a phenyl group. For derivatives of benzene, the locations of substituents on the ring are denoted by numbers that run clockwise or counterclockwise. Select the direction that corresponds to the lowest numbers for the substituents. For example, the compound

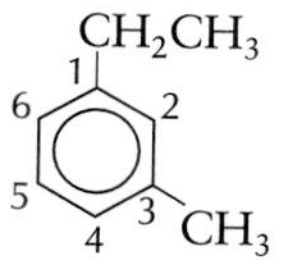

CH_2CH_3 (1), CH_3 (3); ring positions 1–6

is 1-ethyl-3-methylbenzene. In an older but still widely used system of nomenclature, substituents in location 1 and locations 2, 3, or 4 are denoted *ortho-* (abbreviated *o*-), *meta-* (*m*-), and *para-* (*p*-), respectively. The three dimethylbenzenes, which are also known as xylenes, would therefore be named

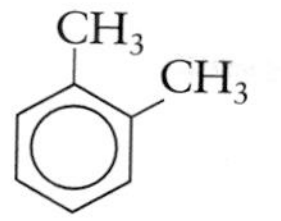

1,2-Dimethylbenzene *o*-Xylene — 1,3-Dimethylbenzene *m*-Xylene — 1,4-Dimethylbenzene *p*-Xylene

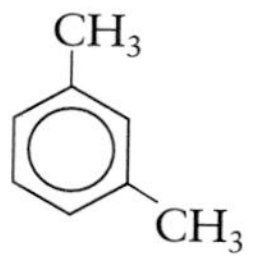

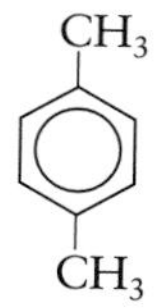

SELF-TEST 11.6A Name the compound

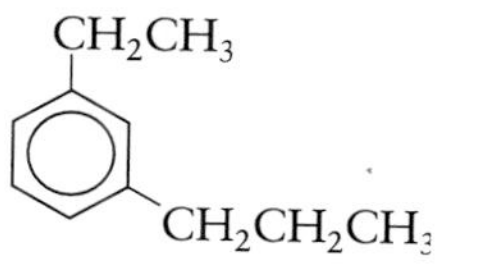

CH_2CH_3, $CH_2CH_2CH_3$ (benzene ring)

[*Answer:* 1-Ethyl-3-propylbenzene]

SELF-TEST 11.6B Name the compound

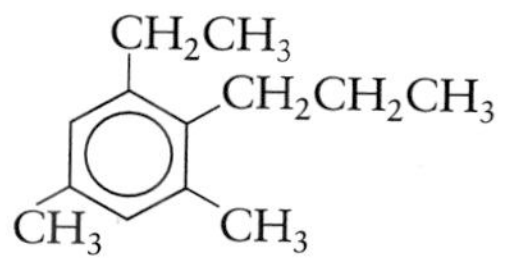

CH_2CH_3, $CH_2CH_2CH_3$, CH_3, CH_3 (benzene ring)

FUNCTIONAL GROUPS

Certain groups of atoms dominate the properties and reactions (the "functions") of many organic compounds. These **functional groups** are attached to the carbon chains and rings. Examples are the chlorine in chloroethane and the ⟩CO group in acetone (propanone), CH_3COCH_3.

Table 11.3 lists the more common functional groups. These are the groups that chemists use to build new pharmaceuticals, plastics, and biologically active materials. Living cells use them to form proteins and participate in the biochemical processes of life. Functional groups are like a kit of parts from which the whole range of organic molecules, and ultimately living beings, are constructed. They are also the key to the synthesis of many polymers.

▶▶ Toolbox 11.2 in Section 11.10 shows how to name compounds containing these groups.

TABLE 11.3 Common functional groups

Group	Class of compound
—OH	alcohol, phenol
—N⟨	amine
—CHO	aldehyde
—CO—	ketone
—COOH	carboxylic acid
—COOR	ester
—CO—N⟨	amide
—O—	ether
—X	halide (X = F, Cl, Br, or I)

11.5 ALCOHOLS

The **hydroxyl group,** —OH, is an —O—H group covalently bonded to a carbon atom. The hydroxyl group must be distinguished from the hydroxide ion (OH^-) of inorganic hydroxides, which is a diatomic ion. An **alcohol** is an organic compound that contains a hydroxyl group not connected directly to a benzene ring or to a ⟩C=O group. One of the best-known organic compounds is ethanol, CH_3CH_2OH, which is also

The name *alcohol* comes from the Arabic for "fine powder." The term gradually came to mean the "essence" of a thing and, in particular, the liquid obtained by distilling wine.

called *ethyl alcohol* and *grain alcohol.* Alcohols are divided into three classes on the basis of the number of other organic groups, denoted R, attached to the carbon atom with the —OH group. A **primary alcohol** has the form $RCH_2—OH$; a **secondary alcohol,** the form $R_2CH—OH$; and a **tertiary alcohol,** the form $R_3C—OH$. The R groups need not all be the same.

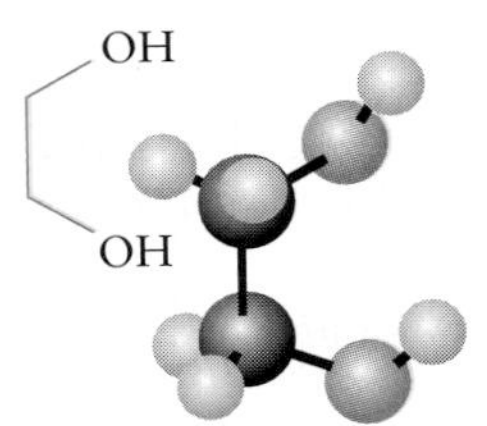

23 1,2-Ethanediol, $HOCH_2CH_2OH$

A further aspect of carbon's variety is that its molecules can contain more than one functional group. Ethylene glycol, or 1,2-ethanediol, $HOCH_2CH_2OH$ (**23**), is an example of a **diol,** a compound with two hydroxyl groups. It is used as a component of antifreeze and in the manufacture of some synthetic fibers. Its action as an antifreeze stems from its high solubility—on account of its hydroxyl groups, which can form hydrogen bonds with water—coupled with the disruptive effect its short hydrocarbon backbone has on the structure of ice. The presence of ethylene glycol makes it difficult for the water molecules to form the hydrogen bonds that are needed to form the ice structure. It is also not very volatile, so it does not boil away readily.

Alcohols with low molar masses are liquids. This property is a sign of the importance of hydrogen bonding, for alcohols have much lower vapor pressures than hydrocarbons with approximately the same molar mass. For example, ethanol is a liquid at room temperature, but propane, which has a higher molar mass than ethanol, is a gas.

Alcohols are derived from water by replacing one of the hydrogen atoms with an organic group. Like water, they form intermolecular hydrogen bonds.

11.6 ETHERS

Just as we can think of an alcohol as being derived from HOH by the replacement of one H atom with an alkyl group, so we can think of an **ether** as an HOH molecule in which both H atoms have been replaced by alkyl groups (denoted R) to give a compound of the form R—O—R:

Diethyl ether is the "ether" once used as an anesthetic.

H—O—H	$CH_3CH_2—O—H$	$CH_3CH_2—O—CH_2CH_3$
Water	Ethanol	Diethyl ether

Ethers are more volatile than the alcohols of the same molar mass because they do not form hydrogen bonds (Fig. 11.12).

Because ethers are not very reactive, they are useful solvents for other organic compounds. However, ethers are flammable; diethyl ether is easily ignited and must be used with great care.

Ethers are more volatile and much less reactive than the corresponding alcohols.

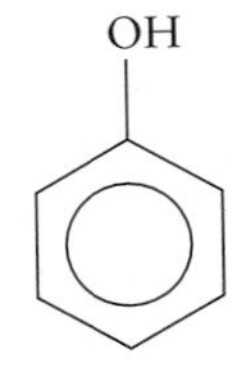

24 Phenol, C_6H_5OH

11.7 PHENOLS

In a **phenol,** a hydroxyl group is attached directly to an aromatic ring. The parent compound, phenol itself, C_6H_5OH (**24**), is a white, crystalline, molecular solid. It was once obtained from the distillation of coal tar, but now it is mainly produced synthetically from benzene. Many phe-

FIGURE 11.12 The boiling points of ethers (given on each column, in degrees Celsius) are lower than those of isomeric alcohols, because hydrogen bonding occurs in alcohols but not in ethers. All the molecules referred to here are unbranched.

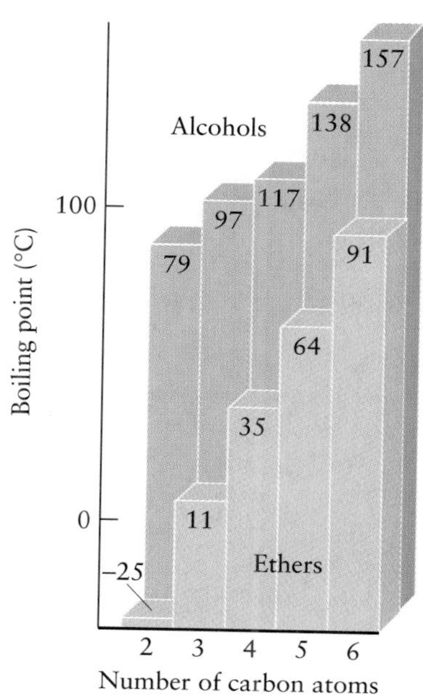

nols occur naturally, and some are responsible for the fragrances of plants. They are often components of **essential oils,** the oils that can be distilled from flowers and leaves. Thymol (**25**), for instance, is the active ingredient of oil of thyme, and eugenol (**26**) provides most of the scent and flavor of oil of cloves.

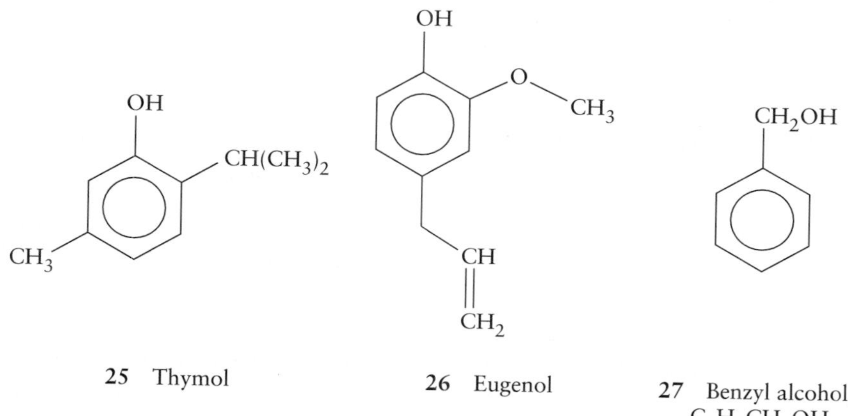

25 Thymol

26 Eugenol

27 Benzyl alcohol, $C_6H_5CH_2OH$

Phenols differ from alcohols in that they are weak acids. Benzyl alcohol (**27**) is so weak a proton donor that it is best regarded as a neutral compound; phenol, though, is definitely acidic and is in fact also known as *carbolic acid.*

Phenols are weak acids.

11.8 ALDEHYDES AND KETONES

The **carbonyl group,** $>C{=}O$, occurs in two closely related families of compounds:

Aldehydes are compounds of the form $R{-}\overset{\displaystyle O}{\overset{\|}{C}}{-}H$.

Ketones are compounds of the form $R{-}\overset{\displaystyle O}{\overset{\|}{C}}{-}R$.

Notice that in an aldehyde the carbonyl group is found at the end of the carbon chain, whereas in a ketone it is at an intermediate position. The R groups may be either aliphatic or aromatic. The group characteristic of the aldehydes is normally written —CHO, as in HCHO (formaldehyde), the simplest aldehyde. The liquid *formalin,* which is used to preserve biological specimens, is an aqueous solution of formaldehyde. Wood smoke contains formaldehyde, and formaldehyde's destructive effect on bacteria is one of the reasons why smoking food helps to preserve it.

Aldehydes occur naturally in essential oils and are often responsible for the flavors of fruits and the odors of plants. Benzaldehyde, C_6H_5CHO (**28**), contributes to the characteristic aroma of cherries and almonds. Cinnamaldehyde (**29**) occurs in oil of cinnamon, and vanillin (**30**) in oil of vanilla. Ketones can be fragrant; carvone (**5**), for example, the essential oil of spearmint, is used to flavor chewing gum.

H, C, O, CHO, OH, OCH_3, CHO

28 Benzaldehyde, C_6H_5CHO

29 Cinnamaldehyde

30 Vanillin

Formaldehyde is prepared industrially (for the manufacture of phenol-formaldehyde resins) by the catalytic oxidation of methanol:

$$2\,CH_3OH(g) + O_2(g) \xrightarrow{600°C,\ Ag} 2\,HCHO(g) + 2\,H_2O(g)$$

When oxidizing a primary alcohol to an aldehyde, we can avoid further oxidation of the product (to a carboxylic acid) by using a mild oxidizing agent. There is less risk of further oxidation for ketones than for aldehydes, because a C—C bond would have to be broken for a carboxylic acid to form. Dichromate oxidation of secondary alcohols produces the ketone in good yield, with little additional oxidation. Schematically,

Chemical equations for organic reactions are commonly left unbalanced. The compounds of interest are shown, but all products are not always shown.

$$CH_3CH_2CH(OH)CH_3 \xrightarrow{Na_2Cr_2O_7(aq),\ H_2SO_4(aq)} CH_3CH_2COCH_3$$

One indication of the difference between the ease of oxidation of aldehydes and that of ketones is the ability of aldehydes but not ketones to produce a silver mirror—a coating of silver on the test tube—with *Tollens reagent,* a solution of Ag^+ ions in aqueous ammonia (Fig. 11.13):

Aldehydes: $CH_3CH_2CHO + Ag^+(\text{from Tollens reagent}) \longrightarrow CH_3CH_2COOH + Ag(s)$

Ketones: $CH_3COCH_3 + Ag^+(\text{from Tollens reagent}) \longrightarrow$ no reaction

Aldehydes and ketones can be prepared by the oxidation of alcohols. Aldehydes are reducing agents; ketones are not.

FIGURE 11.13 An aldehyde (left) produces a silver mirror with Tollens reagent, but a ketone (right) does not.

11.9 CARBOXYLIC ACIDS

The carboxyl group is $-\overset{O}{\overset{\|}{C}}-OH$. It is the functional group found in **carboxylic acids,** which are weak acids of the form R—COOH. Although the carboxyl group is normally abbreviated to —COOH, it must be remembered that the group does not contain an O—O bond. The simplest carboxylic acid is formic acid, HCOOH, the acid in ant venom; one of the most common carboxylic acids is acetic acid, CH_3COOH. Acetic acid, the acid of vinegar, is formed when the ethanol in wine is oxidized by air:

$$CH_3{-}CH_2{-}OH(aq) + O_2(g) \longrightarrow CH_3{-}\overset{\overset{\displaystyle O}{\|}}{C}{-}OH(aq) + H_2O(l)$$

Carboxylic acids can be prepared by oxidizing primary alcohols and aldehydes with acidified aqueous potassium permanganate solution. In some cases, alkyl groups can be oxidized directly to carboxyl groups. This process is very important industrially and, among other applications, is used for the oxidation of *p*-xylene:

$$CH_3{-}C_6H_4{-}CH_3 + 3\,O_2(g) \xrightarrow{\text{Co(III) catalyst}} HOOC{-}C_6H_4{-}COOH + 2\,H_2O$$

The product, terephthalic acid, is used for the production of artificial fibers.

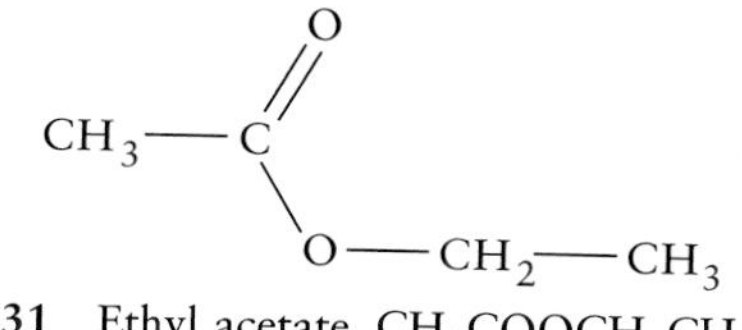

31 Ethyl acetate, $CH_3COOCH_2CH_3$

The product of the reaction between a carboxylic acid and an alcohol is called an **ester** (**31**). Acetic acid and ethanol, for example, react when heated to about 100°C in the presence of a strong acid. The products of this **esterification** are ethyl acetate, a fragrant liquid, and water:

$$CH_3{-}C({=}O){-}OH + HO{-}CH_2CH_3 \xrightarrow{H^+,\,\Delta} CH_3{-}C({=}O){-}OCH_2CH_3 + H_2O$$

Many esters have fragrant odors and contribute to the flavors of fruits. Other naturally occurring esters include fats and oils. For example, the animal fat tristearin (**32**), which is a component of beef fat, is an ester formed from glycerol and stearic acid.

$CH_2O{-}C({=}O){-}(CH_2)_{16}CH_3$
$CHO{-}C({=}O){-}(CH_2)_{16}CH_3$
$CH_2O{-}C({=}O){-}(CH_2)_{16}CH_3$

32 Tristearin, $C_{57}H_{110}O_6$

Ester formation is an example of a **condensation reaction.** In this type of reaction, two molecules combine to form a larger one and a small molecule is eliminated (Fig. 11.14), thus "condensing" the two molecules into one. In an esterification, the eliminated molecule is H_2O. Notice that, although the alcohol and carboxylic acid can each form hydrogen bonds to themselves, the ester cannot because it does not have a hydrogen atom bonded to an oxygen atom.

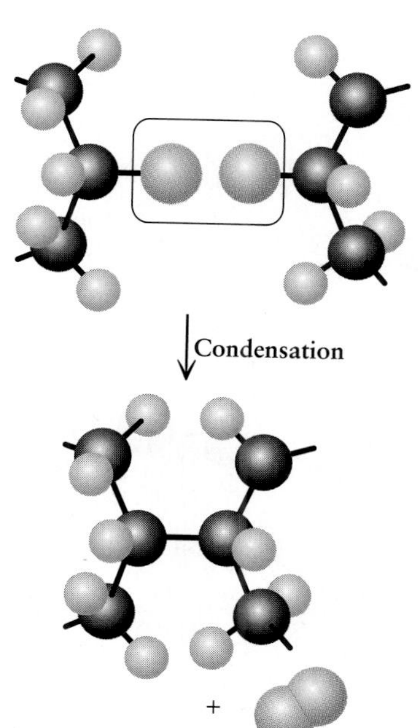

FIGURE 11.14 In a condensation reaction, two molecules are linked as a result of removing two atoms or groups of atoms (the orange and purple spheres) as a small molecule (typically water).

Carboxylic acids have an —OH group attached to a carbonyl group. Alcohols condense with carboxylic acids to form esters. Alcohols and carboxylic acids have —OH groups that take part in hydrogen bonding; esters cannot form hydrogen bonds to themselves.

SELF-TEST 11.7A (a) Write the structural formula of the ester formed from the reaction between propanoic acid, CH_3CH_2COOH, and methanol, CH_3OH. (b) Write the structural formulas of the acid and alcohol that react to form 1-pentyl ethanoate, $CH_3COOC_5H_{11}$, a contributor to the flavor of bananas.

[*Answer:* (a) $CH_3CH_2COOCH_3$; (b) CH_3COOH and $CH_3(CH_2)_3CH_2OH$]

SELF-TEST 11.7B (a) Write the structural formula of the ester formed from the reaction between formic acid, HCOOH, and ethanol, CH_3CH_2OH. (b) Write the structural formulas of the acid and alcohol that react to form methyl butanoate, $CH_3(CH_2)_2COOCH_3$, a contributor to the flavor of apples.

11.10 AMINES AND AMIDES

An **amine** is a compound derived from ammonia by replacing various numbers of H atoms by organic groups:

H—N(—H)—H	CH_3—N(—H)—H	CH_3—N(—H)—CH_3	CH_3—N(—CH_3)—CH_3
Ammonia	Methylamine	Dimethylamine	Trimethylamine

In each case, the N atom is sp^3 hybridized, with one lone pair of electrons and three σ-bonds. Like ammonia, amines are also weak bases (Section 3.8). An exception is the **quaternary ammonium ion,** a tetrahedral ion of the form R_4N^+, such as the tetramethylammonium ion, $(CH_3)_4N^+$. Quaternary ammonium ions have little effect on pH.

Amines are widespread naturally. Many of them have a pungent, often unpleasant odor. Because proteins are organic polymers containing nitrogen, amines are present in the decomposing remains of living matter and, together with sulfur compounds, are responsible for the stench of decaying flesh. The common names of two diamines—putrescine, $NH_2(CH_2)_4NH_2$, and cadaverine, $NH_2(CH_2)_5NH_2$—speak for themselves.

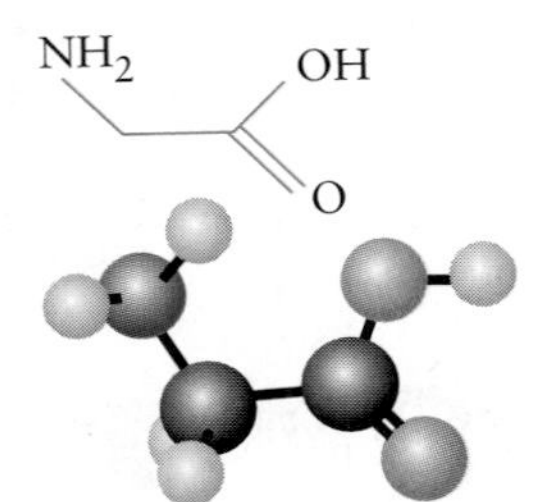

33 Glycine, NH_2CH_2COOH

An **amino acid** is a carboxylic acid that contains an amino group ($—NH_2$) as well as a carboxyl group. The simplest example is glycine, NH_2CH_2COOH (**33**). Notice that an amino acid has both a basic group ($—NH_2$) and an acidic group (—COOH) in the same molecule. We can predict that the tying together of a basic group and an acidic group in the same molecule—separated by a single carbon atom—leads to some remarkable properties. This is, in fact, the case, as we shall see in Section 11.17.

Like alcohols, amines also condense with carboxylic acids:

$$CH_3—C(=O)—OH + H—N(—H)—CH_3 \longrightarrow CH_3—C(=O)—NHCH_3 + H_2O$$

The product is an **amide.** One example of an amide is the pain-relieving drug sold as Tylenol (**34**). Amides have N—H bonds that can take part in

34 Tylenol®

hydrogen bonding, so the intermolecular forces between their molecules are relatively strong.

Amines are derived from ammonia by replacing hydrogen atoms with organic groups. Amides result from the condensation of amines with carboxylic acids. Amines and amides take part in hydrogen bonding.

SELF-TEST 11.8A Predict whether an amino acid or an amine of the same molar mass would have the higher boiling point and explain why.

[*Answer:* The amino acid; it can form hydrogen bonds with both the $-NH_2$ group and the $-COOH$ group. The amine has only one functional group that can form hydrogen bonds.]

SELF-TEST 11.8B Predict whether an ester or an amide of the same molar mass would have the higher boiling point and explain why.

TOOLBOX 11.2 How to name compounds with functional groups

In general, the names of compounds containing functional groups follow the same conventions and numbering system as do the names of hydrocarbons (Toolbox 11.1). However, in general, the ending of the name is changed to indicate the functional group.

Alcohols To form the systematic name, we identify the parent alkane and replace the ending by *-ol*. The location of the hydroxyl group is denoted by numbering the C atoms of the backbone, as described in Toolbox 11.1, starting at the end of the chain that results in a low number for the $-OH$ group location. Thus, the systematic name of isopropanol, $CH_3CHOHCH_3$, is 2-propanol. When $-OH$ is named as a substituent, it is called *hydroxy*.

SELF-TEST 11.9A Name the alcohol with the formula $CH_3CH(CH_2CH_2OH)CH_3$.

[*Answer:* 3-Methyl-1-butanol]

SELF-TEST 11.9B Name the alcohol with the formula $CH_3CH_2CHOHCH_2CH_3$.

Aldehydes and ketones For aldehydes, identify the parent alkane: include the C of $-CHO$ in the count of carbon atoms. Then change the final *-e* of the alkane name to *-al*. Thus, CH_3CH_2CHO is propanal. The $-CHO$ group can occur only at the end of a carbon chain and is given the number 1 if other substituents need to be located. For ketones, change the *-e* of the parent alkane to *-one*. The location of the $\rangle CO$ group is denoted by selecting a numbering order that gives it the lowest number. Thus, $CH_3CH_2CH_2COCH_3$ is 2-pentanone.

SELF-TEST 11.10A Name the compound with the formula $CH_3CH(CHO)CH_2CH_3$.

[*Answer:* 2-Methylbutanal]

SELF-TEST 11.10B Name the compound with the formula $CH_3CH_2COCH_2CH_3$.

Carboxylic acids Change the *-e* of the parent alkane to *-oic acid*. To identify the parent acid, include the C atom of the $-COOH$ group when counting carbon atoms. Thus, $CH_3CH_2CH_2COOH$ is butanoic acid.

Esters Change the *-ol* of the alcohol to *-yl* and the *-oic acid* of the parent acid to *-oate*. Thus, $CH_3CH_2COOCH_3$ is methyl propanoate.

Amines Amines are named systematically by specifying the groups attached to the nitrogen atom in alphabetical order, followed by the suffix *-amine*. Thus, diethylpropylamine is $(CH_3CH_2)_2NCH_2CH_2CH_3$. Amines with two amino groups are called diamines. The $-NH_2$ group is called *amino-* when it is a substituent.

SELF-TEST 11.11A Name the compound with the formula $(C_6H_5)_3N$.

[*Answer:* Triphenylamine]

SELF-TEST 11.11B Name the compound with the formula $(CH_3CH_2)CH_3NH$.

Halides Name the halogen atom as a substituent by changing the *-ine* part of its name to *-o*. Thus, CH_3Br is bromomethane.

SELF-TEST 11.12A Name the compound with the formula $CH_3CH_2CHFCH_3$.

[*Answer:* 2-Fluorobutane]

SELF-TEST 11.12B Name the compound with the formula $CH_2ClCH(CH_3)CH_2Cl$.

Same atoms, different partners

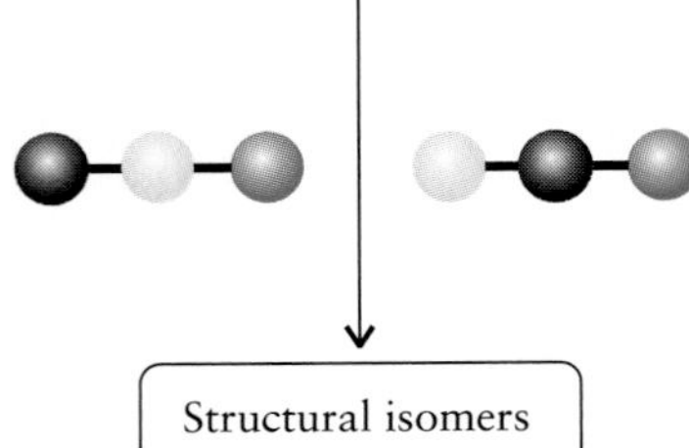

Same atoms, same partners, different arrangement in space

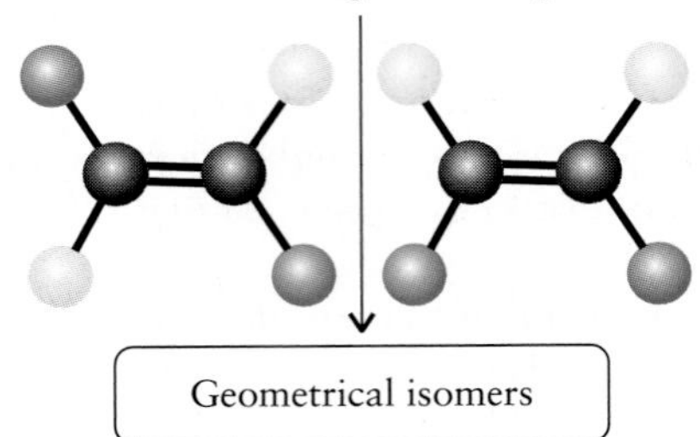

Same atoms, same partners, same neighbors, different mirror image

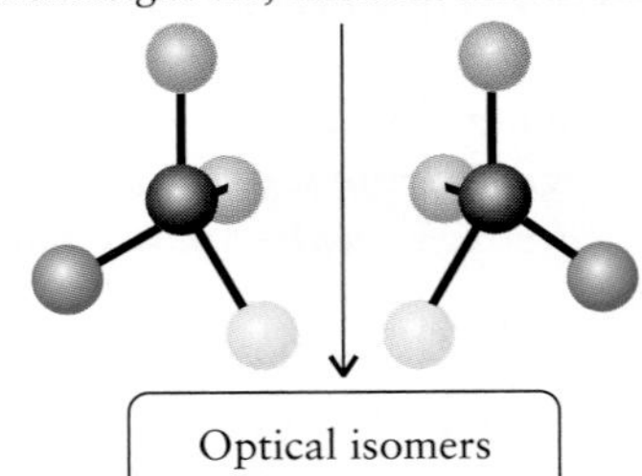

FIGURE 11.15 A summary of the various types of isomerism that are encountered in molecular compounds.

ISOMERS

Carbon's ability to form an enormously wide variety of compounds does not end with its various functional groups. We have seen that isomers are different compounds built from the same kit of atoms. The various types of isomerism form a kind of family tree (Fig. 11.15).

11.11 STRUCTURAL ISOMERS

Molecules built from the same atoms but connected differently are called **structural isomers.** For example, we can insert the C atom and two H atoms of a $-CH_2-$ group into the C_3H_8 molecule in two different ways to give two different compounds with the formula C_4H_{10}:

The bonds have been drawn with different lengths only for convenience; in the actual molecules all carbon-carbon bonds are the same length.

```
  H H H H              H
  | | | |              |
H-C-C-C-C-H          H-C-H
  | | | |            H |  H
  H H H H            | |  |
                   H-C-C--C-H
                     | |  |
                     H H  H
```

$CH_3CH_2CH_2CH_3$
Butane

$CH(CH_3)_3$
Methylpropane

$CH_3-CH_2-CH_2-CH_3$

35 Butane, C_4H_{10}

$CH(CH_3)_3$

36 Methylpropane, C_4H_{10}

Although the $-CH_2-$ group could be inserted in other places, the resulting molecules can all be twisted into one or the other of these two isomers. Both compounds are gases, but butane (**35**) condenses at $-1°C$, whereas methylpropane (**36**) condenses at $-12°C$. These singly bonded structures are continuously writhing and twisting, and the $-CH_3$ groups in methylpropane rotate like tiny propellers.

Structural isomers have identical molecular formulas, but their atoms are linked to different neighbors.

EXAMPLE 11.1 Writing the formulas of isomeric molecules

Draw two-dimensional molecular structures for all the alkanes of formula C_5H_{12} and write the short condensed structural formulas for each.

STRATEGY Decide which distinct molecules can be constructed by inserting one $-CH_2-$ group into the formulas of the two C_4H_8 molecules given earlier. Distinct molecules cannot be changed into one another simply by rotating either the entire formula or parts of the formula on the page. One approach is to insert $-CH_2-$ groups into different parts of the molecule, and then to discard formulas that repeat those already obtained.

SOLUTION From butane, we can form

```
                                      H
                                      |
                                    H-C-H
     H  H  H  H  H               H  H |  H
     |  |  |  |  |               |  | |  |
   H-C--C--C--C--C-H           H-C--C-C--C-H
     |  |  |  |  |               |  | |  |
     H  H  H  H  H               H  H H  H
```

(a) $CH_3CH_2CH_2CH_2CH_3$ (b) $CH_3CH_2CH(CH_3)_2$

From methylpropane, we can form

```
          H                 H                 H
          |                 |                 |
        H-C-H             H-C-H             H-C-H  H
     H  H |  H          H   |   H         H   |    |
     |  | |  |          |   |   |         |   |  H-C-H
   H-C--C-C--C-H      H-C---C---C-H     H-C---C    |
     |  | |  |          |   |   |         |   |----C-H
     H  H H  H          H   |   H         H   H    |
                          H-C-H                    H
                            |
                            H
```

(c) $CH_3CH_2CH(CH_3)_2$ (d) $C(CH_3)_4$ (e) $(CH_3)_2CHCH_2CH_3$

Molecules (b) and (c) are the same. The atoms of molecule (e) are joined together in the same arrangement as (b) and (c) even though they look different as drawn on the page; so (b), (c), and (e) are the same. There are therefore only three distinct isomers with formula C_5H_{12}: (a), (b), and (d).

SELF-TEST 11.13A Write the structural formulas for the five isomeric alkanes of formula C_6H_{14}.

[*Answer:* $CH_3(CH_2)_4CH_3$; $CH_3(CH_2)_2CH(CH_3)CH_3$; $CH_3CH_2CH(CH_3)CH_2CH_3$; $CH_3CH_2C(CH_3)_2CH_3$; $CH_3CH(CH_3)CH(CH_3)CH_3$]

SELF-TEST 11.13B Write the structural formulas for the four isomers with the molecular formula C_4H_9Br.

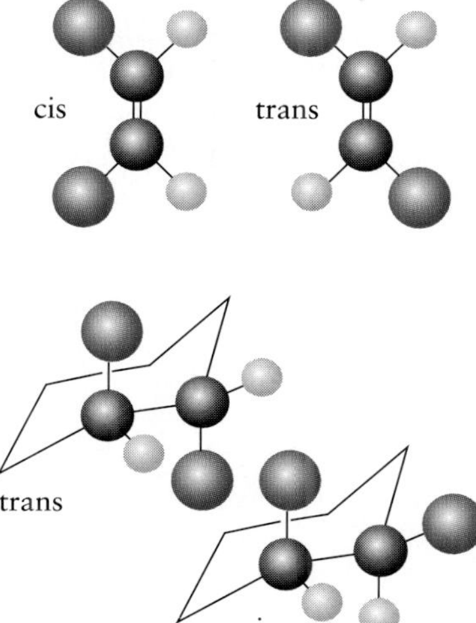

FIGURE 11.16 An example of two pairs of geometrical isomers. Note that the bonded neighbors of each atom are the same in both cases, but nevertheless the arrangements of the atoms in space are different. The prefix *trans* means "across" and *cis* means "on this side."

11.12 GEOMETRICAL AND OPTICAL ISOMERS

There are two different compounds with the name 2-butene. They have the same molecular formula and the same structural formula, yet they have different properties. They are called **geometrical isomers** because their atoms are bonded to the same neighbors but have different arrangements in space on either side of a double bond or above and below the ring of a cycloalkane (Fig. 11.16). Geometrical isomers of organic molecules are distinguished by the prefixes *cis* and *trans.*

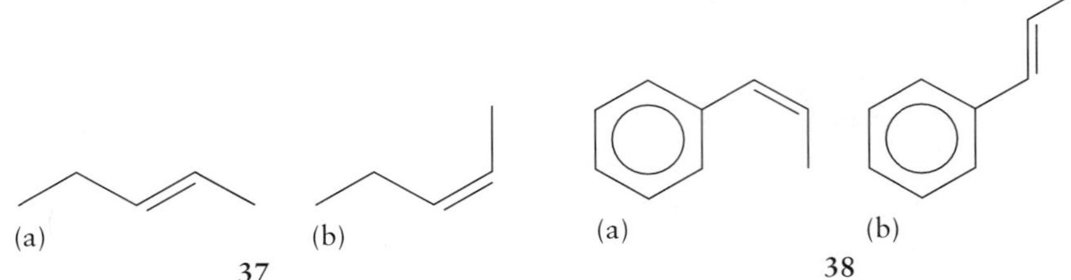

SELF-TEST 11.14A Identify (**37a**) and (**37b**) as cis or trans:
[***Answer:*** 37a is *trans*-2-pentene; 37b is *cis*-2-pentene.]

SELF-TEST 11.14B Identify (**38a**) and (**38b**) as cis or trans.

Optical isomers are each other's mirror image. In organic compounds, optical isomers occur whenever four different groups are attached to a carbon atom. To appreciate this statement, consider the amino acid alanine, $NH_2CH(CH_3)COOH$ and its mirror image. If we look at Fig. 11.17 carefully (or, better, build models), then we see that no matter how we twist and turn the molecules, we cannot superimpose the mirror image molecule on the original molecule. It is like trying to superimpose your right hand on your left hand. The two molecules are, in fact, two distinct compounds.

The name *chiral* comes from the Greek word for "hand." A human hand is chiral, and the mirror image of a left hand is a right hand.

The word *enantiomer* is the Greek word for "both." Your two hands are an enantiomeric pair of objects.

A **chiral molecule** is one that is not identical to its mirror image. We say that alanine is chiral, and that a chiral molecule and its mirror image form a pair of **enantiomers.** Glycine, NH_2CH_2COOH (**33**), does not have four different groups attached to the central carbon atom. Its mirror image can therefore be rotated and superimposed on the original molecule. We say that glycine is an **achiral** molecule.

Enantiomers have identical chemical properties, except when they react with other chiral compounds. One consequence of this difference in reactivity is that enantiomers may have different odors and pharmacological activities. The molecule has to fit into a cavity, or slot, of a certain shape, either in an odor receptor in the nose or in an enzyme. Only one member of the enantiomeric pair may be able to fit.

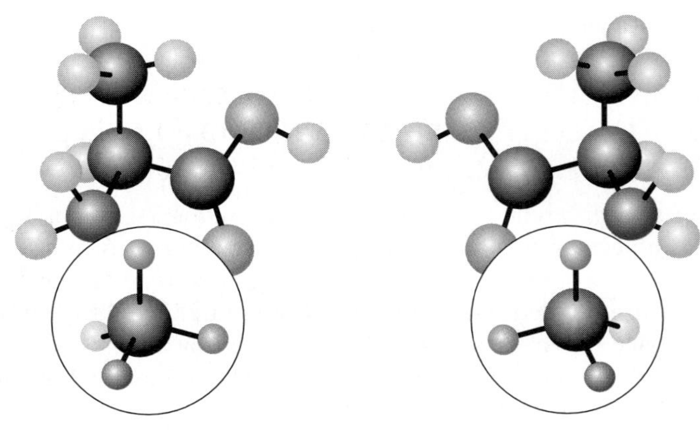

FIGURE 11.17 The molecule on the right is the mirror image of the molecule on the left, as can be seen more clearly by inspecting the simplified representations in the circles. Because the two molecules cannot be superimposed, they are distinct optical isomers.

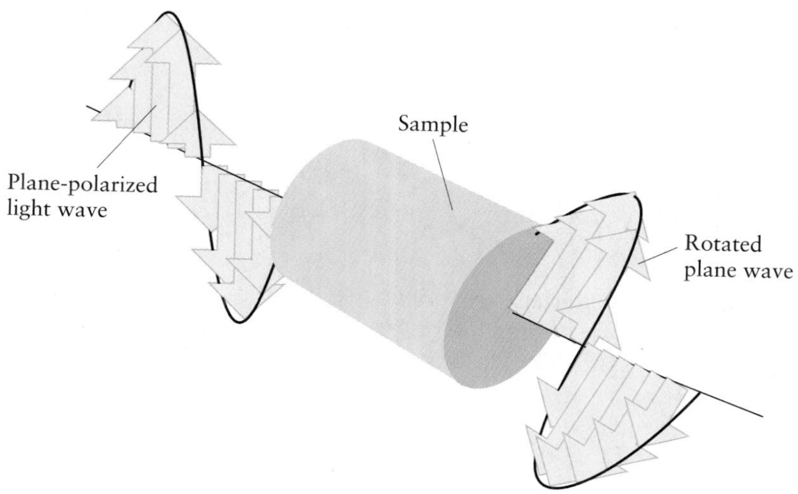

FIGURE 11.18 Plane-polarized light consists of radiation in which all the wave motion lies in one plane (as represented by the orange arrows on the left). When such light passes through a solution of an optically active substance, the plane of the polarization is rotated through a characteristic angle that depends on the concentration of the solution and the length of the path through it.

Enantiomers differ in one physical property: chiral molecules display **optical activity,** the ability to rotate the plane of polarization of light. In ordinary light, the wave motion lies at random directions around the direction of travel. In plane-polarized light, the wave lies in a single plane (Fig. 11.18). Plane-polarized light can be prepared by passing ordinary light through a special filter, like the material used to make polarized sunglasses. One type of chiral molecule rotates the plane of polarization clockwise; its mirror image partner rotates it by the same amount the other way.

Amino acids synthesized in the laboratory are **racemic mixtures,** or mixtures of enantiomers in equal proportions. Because enantiomers rotate the plane of polarization of light in opposite directions, a racemic sample is not optically active. In contrast, reactions in living cells lead to only one enantiomer. It is a remarkable feature of nature (which is not yet fully understood) that all naturally occurring amino acids in animals have the same handedness.

Geometrical isomers have the same molecular and structural formulas but different arrangements in space. Molecules with four different groups attached to a single carbon atom are chiral and show optical isomerism; chiral molecules are optically active.

EXAMPLE 11.2 Predicting whether a compound is chiral

Predict whether the isomeric amino acids (a) $CH_3CH_2CH(NH_2)COOH$ and (b) $(CH_3)_2C(NH_2)COOH$ are chiral.

STRATEGY To decide whether a compound is chiral, we should check for a carbon atom with four different groups attached.

SOLUTION The two amino acids are (**39**) and (**40**). We see that (**39**) contains a carbon atom that has four different groups attached and hence is chiral. However, (**40**) does not have such a carbon atom, it is identical to its mirror image, and therefore it is not chiral.

NH_2 (top), CH_3CH_2—C—COOH, H (bottom)

39 $CH_3CH_2CH(NH_2)COOH$

NH_2 (top), CH_3—C—COOH, CH_3 (bottom)

40 $(CH_3)_2C(NH_2)COOH$

SELF-TEST 11.15A Which of the following chlorofluorocarbons is chiral: (a) CH_3CF_2Cl; (b) CH_3CHFCl; (c) CH_2FCl?

[*Answer:* (b)]

SELF-TEST 11.15B Which of the following alcohols is chiral: (a) CH_3CH_2OH; (b) $CH_3CH(OH)CH_3$; (c) $CH_3CH(OH)CH_2CH_3$?

POLYMERS

Poly is from the Greek word for "many" and _mer_ stands for "unit."

Many synthetic materials are **polymers,** giant molecules formed from chains or networks of small repeating units. Some, like polypropylene and Teflon, are synthetic. Plastics are simply polymers that can be molded into shape.

Although their enormous size gives them special properties, polymers are built of the same types of atoms and bonds found in smaller molecules. Big as these molecules may be, they are built by using reactions like the ones we have been discussing so far. Furthermore, whereas a small organic molecule might have only one or two functional groups, a polymer molecule might have thousands. Nevertheless, the properties we have already met also apply to these complicated polymer molecules. This is one of the pleasures of organic chemistry: once we have mastered the properties of a functional group, we can expect the same pattern of behavior in any molecule containing that functional group.

11.13 ADDITION POLYMERIZATION

Alkenes react with themselves in a process called **addition polymerization.** For example, an ethene molecule may form a bond to another ethene molecule, another ethene molecule may add to that, and so on, until a long hydrocarbon chain has grown. The original alkene, such as ethene, is called the **monomer.** The product, the chain of covalently linked monomers, is a polymer. The simplest addition polymer is polyethylene, $\text{+}CH_2CH_2\text{+}_n$, which consists of long chains of CH_2CH_2 units. Many polymer molecules also have a number of branches, which are new chains that sprout from intermediate points in the original chain.

TABLE 11.4 Addition polymers

Monomer name	Formula	Polymer formula	Common name
ethene*	$CH_2{=}CH_2$	$\text{+}CH_2{-}CH_2\text{+}_n$	polyethylene
vinyl chloride	$CHCl{=}CH_2$	$\text{+}CHCl{-}CH_2\text{+}_n$	polyvinyl chloride
styrene	$CH(C_6H_5){=}CH_2$	$\text{+}CH(C_6H_5)CH_2\text{+}_n$	polystyrene
acrylonitrile	$CH(CN){=}CH_2$	$\text{+}CH(CN){-}CH_2\text{+}_n$	Orlon, Acrilan
propene*	$CH(CH_3){=}CH_2$	$\text{+}CH(CH_3){-}CH_2\text{+}_n$	polypropylene
methyl methacrylate	$CH_3OC(=O)C(CH_3){=}CH_2$	$\left(\text{-}C(CH_3)(COOCH_3){-}CH_2\text{-}\right)_n$	Plexiglas, Lucite
tetrafluoroethene*	$CF_2{=}CF_2$	$\text{+}CF_2{-}CF_2\text{+}_n$	Teflon, PTFE†

*The suffix *-ene* is replaced by *-ylene* in the common names of these compounds, hence the names of the corresponding polymers.
†PTFE, polytetrafluoroethylene.

TABLE 11.5 Recycling codes

Recycling code		Polymer
1	PET (E)	polyethylene terephthalate
2	HDPE	high-density polyethylene
3	PVC or VC	polyvinyl chloride
4	LDPE	low-density polyethylene
5	PP	polypropylene
6	PS	polystyrene

The plastics industry has developed polymers from a number of monomers with the formula $CHX{=}CH_2$, where X is a single atom (such as the Cl in vinyl chloride, $CHCl{=}CH_2$) or a group of atoms (such as the CH_3 in propene). These substituted alkenes give polymers of formula $\text{-(}CHXCH_2\text{)-}_n$ and include polyvinyl chloride (PVC), $\text{-(}CHClCH_2\text{)-}_n$, and polypropylene, $\text{-(}CH(CH_3)CH_2\text{)-}_n$. Some of these polymers are listed in Table 11.4. They differ in appearance, rigidity, transparency, and resistance to weathering. Many plastic materials made of addition polymers can be recycled by melting and reprocessing. The recycling code on the bottom of a plastic container indicates the polymer used to make the container (Table 11.5).

A widely used polymerization procedure is **radical polymerization,** in which the reaction depends on the presence of radicals (Section 8.9). In a typical procedure, a monomer (such as ethene) is compressed to about 1000 atm and heated to 100°C in the presence of a small amount of an organic peroxide (a compound of formula R—O—O—R, where R is an organic group). The reaction is initiated by dissociation of the O—O bond, giving two radicals:

$$R{-}O{-}O{-}R \longrightarrow R{-}O\cdot + \cdot O{-}R$$

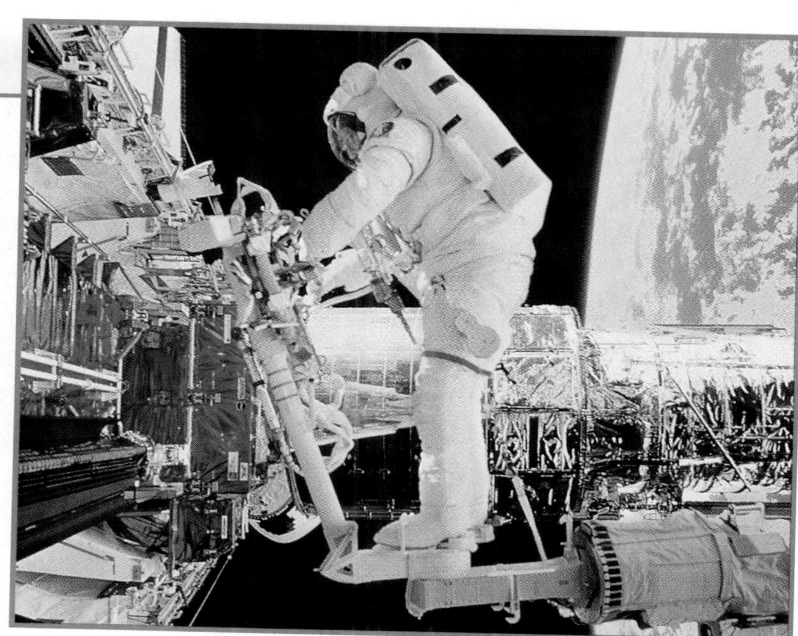

CHEMISTRY AT WORK

An aerospace engineer completes the adjustment of equipment in the payload of the space shuttle in Earth orbit. For success in his work and for his own survival he must be aware of the physical and electrical properties of the different types of material on the space shuttle. Many of these materials, including the fabric of his spacesuit, are polymers. If it were not for the development of strong, tough polymers, spacesuits would be bulky constructions of jointed metal.

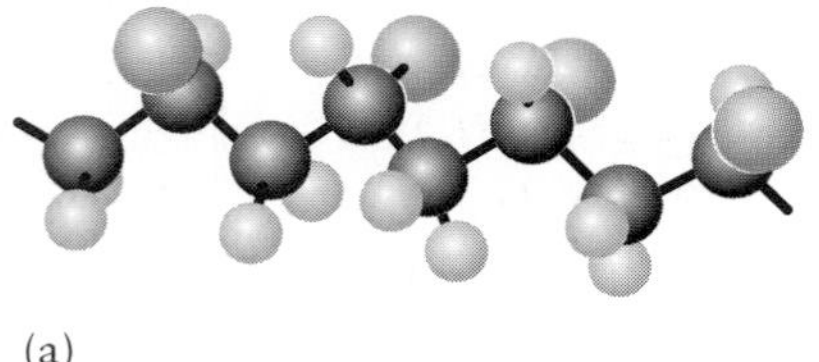

(a)

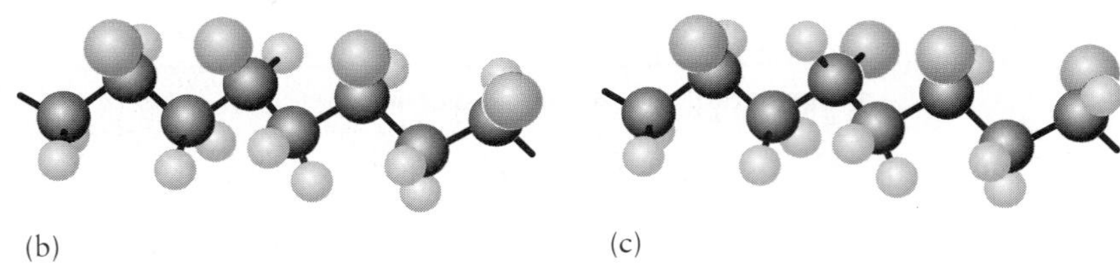

(b) (c)

FIGURE 11.19 (a) In an atactic polymer, the substituents lie on random sides of the chain. The stereoregular polymers produced by using Ziegler-Natta catalysts may be (b) isotactic (all on one side) or (c) syndiotactic (alternating).

These radicals attack monomer molecules $CHX{=}CH_2$ (with X = H for ethene itself) and form a new, highly reactive radical:

$$\mathrm{R{-}O\cdot + CH_2{=}\underset{\displaystyle X}{\overset{\displaystyle H}{C}} \longrightarrow R{-}O{-}CH_2{-}\underset{\displaystyle X}{\overset{\displaystyle H}{C}}\cdot}$$

This radical attacks other monomer molecules and the chain grows longer:

$$\mathrm{R{-}O{-}CH_2{-}\underset{\displaystyle X}{\overset{\displaystyle H}{C}}\cdot + CH_2{=}\underset{\displaystyle X}{\overset{\displaystyle H}{C}} \longrightarrow R{-}O{-}CH_2{-}\underset{\displaystyle X}{\overset{\displaystyle H}{C}}{-}CH_2{-}\underset{\displaystyle X}{\overset{\displaystyle H}{C}}\cdot}$$

The reaction continues until all the monomer has been used up, or until pairs of chains have linked together into single nonradical species. The product consists of long chains of formula $\text{-}\!\!(CH_2CH_2)\!\!\text{-}_n$ in which n can reach many thousands.

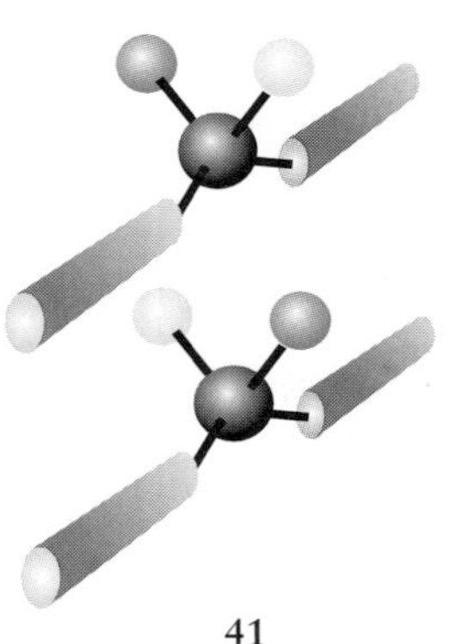

41

The backbone carbon atoms of some polymers are chiral, with structures like those shown in (**41**). This chirality was an early problem with polypropylene, because the relative orientations of the H and CH_3 groups on each chiral C atom were random and the resulting material was amorphous, sticky, and nearly useless. Now, however, the chirality of the

FIGURE 11.20 Isotactic polypropylene is used in sportswear, such as these cross-country ski outfits.

FIGURE 11.21 Collecting latex from a rubber tree in Malaysia, one of its principal producers.

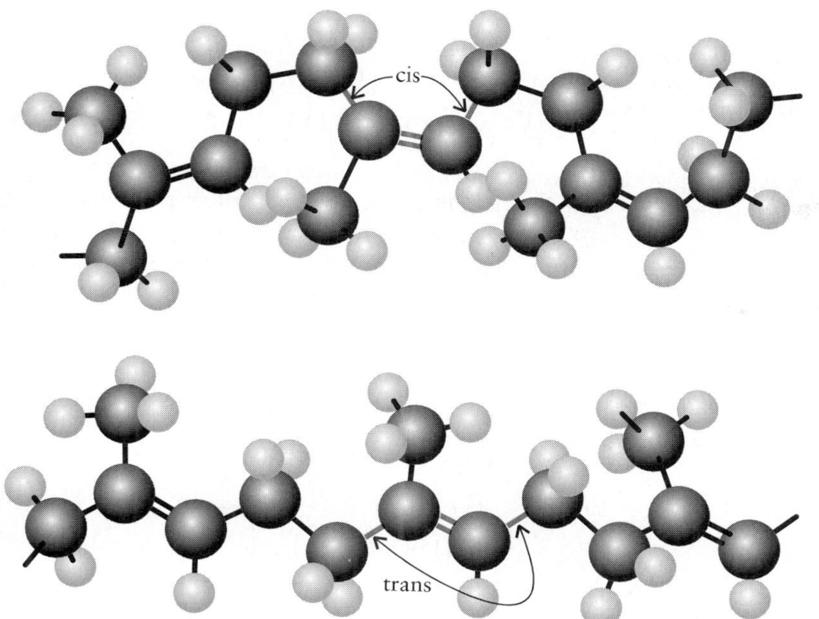

FIGURE 11.22 In natural rubber, the isoprene units are polymerized to be all cis. The harder material, gutta-percha, is the all-trans polymer.

atoms can be controlled by using a Ziegler-Natta catalyst in the polymerization. In **isotactic** polypropylene, the carbon atoms forming the chain all have the same handedness, so all the methyl groups are on the same side of the chain. A polymer in which the units all take the same handedness or direction is called *stereoregular.* In **syndiotactic** polypropylene, the carbon atoms in the chain alternate in handedness, so the methyl groups alternate in direction. In **atactic** polypropylene the methyl groups are attached randomly, on either side of the chain (Fig. 11.19).

This catalyst is named for the German chemist Karl Ziegler and the Italian chemist Giulio Natta, who developed it. It is typically $TiCl_4$ and $Al(CH_2CH_3)_3$.

The high density of polymers produced by Ziegler-Natta catalysts results from the regular arrangement of groups along the polymer chain. As a result of this regular arrangement in space, the chains can pack together well and form a highly crystalline, dense material (Fig. 11.20). Catalysts are also used to control the structure of conducting polymers (see Case Study 11), to enhance their electrical characteristics.

Rubber is a natural polymer with isoprene monomers (2). Natural rubber is obtained from the bark of the rubber tree as a milky white liquid, called latex (Fig. 11.21), that consists of a suspension of rubber particles in water. The rubber itself is a soft white solid that becomes even softer when warm. It is used for pencil erasers and has been used as crepe for the soles of shoes.

Latex, from the Latin word for "liquid," is also the white fluid inside dandelion and milkweed stalks.

Chemists were unable to synthesize rubber for a long time, even though they knew it was a polymer of isoprene. The enzymes in the rubber tree produce a polymer in which all the links between monomers are in a cis arrangement (Fig. 11.22); straightforward radical polymerization, however, produces a random mixture of cis and trans links and a sticky, useless product. The stereoregular polymer was achieved with a Ziegler-Natta catalyst, and almost pure, rubbery *cis*-polyisoprene can now be produced. *trans*-Polyisoprene, in which all the links are trans, is also known and produced naturally: it is the hard material *gutta-percha,* once used inside golf balls and still used to fill root canals in teeth.

Alkenes undergo addition polymerization; if a Ziegler-Natta catalyst is used, the polymer is stereoregular and has a relatively high density.

FIGURE 11.23 Synthetic fibers are made by extruding liquid polymer from small holes in an industrial version of the spider's spinneret (see the illustrations on page ii).

11.14 CONDENSATION POLYMERIZATION

Polymers formed by linking together monomers that have carboxylic acid groups with those that have alcohol groups are called **polyesters.** Polymers of this type are widely used for making artificial fibers and are examples of **condensation polymers,** which are polymers formed by a series of condensation reactions. A typical polyester is Dacron, or Terylene, a polymer produced from the esterification of terephthalic acid with ethylene glycol: its technical name is polyethylene terephthalate. The first condensation is

$$\text{HOOC}-\text{C}_6\text{H}_4-\text{COOH} + \text{HO}-\text{CH}_2-\text{CH}_2-\text{OH} \longrightarrow \text{HOOC}-\text{C}_6\text{H}_4-\overset{\text{O}}{\overset{\|}{\text{C}}}-\text{O}-\text{CH}_2-\text{CH}_2-\text{OH}$$

A new ethylene glycol molecule can condense with the carboxyl group on the left of the product, and another terephthalic acid can condense with the hydroxyl group on the right. As a result, the polymer grows at each end and in due course becomes

$$\text{HOOC}-\text{C}_6\text{H}_4-\overset{\text{O}}{\overset{\|}{\text{C}}}\left[\text{O}-\text{CH}_2-\text{CH}_2-\text{O}-\overset{\text{O}}{\overset{\|}{\text{C}}}-\text{C}_6\text{H}_4-\overset{\text{O}}{\overset{\|}{\text{C}}}\right]_n\text{O}-\text{CH}_2-\text{CH}_2-\text{OH}$$

In the radical addition polymerization of alkenes, side chains can start to grow out from the main chain. However, in condensation polymerization, growth can occur only at the functional groups, so chain-branching is much less likely. Thus, polyester molecules make good fibers, because they can be made to lie side by side by stretching the heated product and forcing it through a small hole (Fig. 11.23). The fibers produced can then be spun into yarn (Fig. 11.24). On the other hand, because many monomers can initiate the reaction simultaneously, the chains of condensation polymers are typically shorter than those of addition polymers.

Polyesters have many uses besides fibers. For example, Dacron can be molded and used in surgical implants, such as artificial hearts, or made into thin films for cassette tapes.

Condensation polymerization of amines with carboxylic acids leads to the **polyamides,** substances more commonly known as *nylons.* A typical polyamide is *nylon-66,* which is a polymer of 1,6-diaminohexane, $H_2N(CH_2)_6NH_2$, and adipic acid, $HOOC(CH_2)_4COOH$. The 66 in the name indicates the numbers of C atoms in the two monomers.

For condensation polymerization, it is necessary to have *two* functional groups on each monomer and to mix stoichiometric amounts of the reactants. In the case of polyamide formation, the starting materials form "nylon salt" by proton transfer, as in

$$HOOC(CH_2)_4COOH + H_2N(CH_2)_6NH_2 \longrightarrow {}^-O_2C(CH_2)_4CO_2^- + {}^+H_3N(CH_2)_6NH_3^+$$

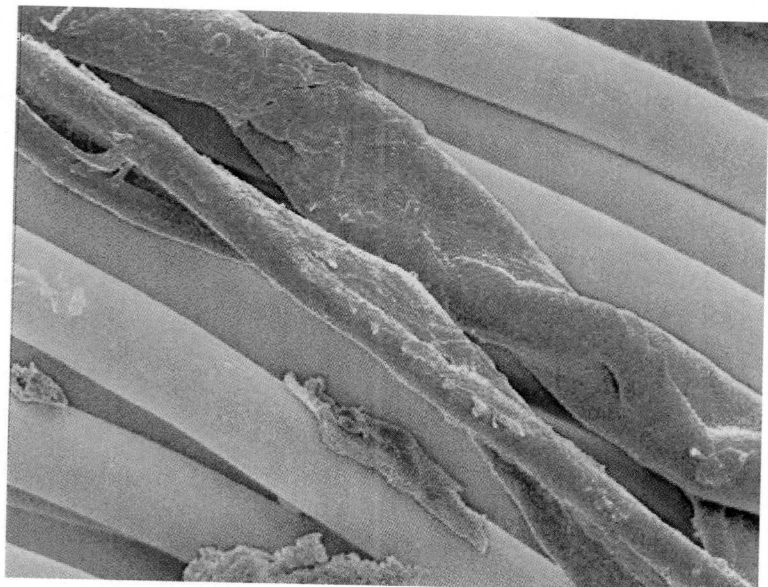

FIGURE 11.24 A scanning electron micrograph of Dacron polyester and cotton fibers in a blended shirt fabric. The cotton fibers have been colored green. Compare the smooth cylinders of the polyester with the irregular surface of cotton. The smooth polyester fibers resist wrinkles, and the irregular cotton fibers produce a more comfortable and absorbent texture.

At this point, the excess acid or amine can be removed. Then, when the nylon salt—an ammonium salt of a carboxylic acid—is heated, the condensation begins. The first step is

$$^{-}O(O{=})C-(CH_2)_4-C(=O)O^{-} + {}^{+}H_3N-(CH_2)_6-NH_3^{+} \longrightarrow$$

$$^{-}O(O{=})C-(CH_2)_4-C(=O)-NH-(CH_2)_6-NH_3^{+} + H_2O$$

The amide grows at both ends by further condensations (Fig. 11.25), and the final product is

$$^{-}O(O{=})C-(CH_2)_4-\overset{O}{\overset{\|}{C}}\!\left[NH-(CH_2)_6-NH-\overset{O}{\overset{\|}{C}}-(CH_2)_4-\overset{O}{\overset{\|}{C}}\right]_n\!NH-(CH_2)_6-NH_3^{+}$$

The long polyamide (nylon) chains can be spun into fibers (like polyesters) or molded. Some of the strength of nylon fibers arises from the $\rangle N-H\cdots O{=}C\langle$ hydrogen bonding that can occur between neighboring chains (Fig. 11.26). However, this ability to form hydrogen bonds also accounts for nylon's tendency to absorb moisture, because H_2O molecules can hydrogen bond to the chains and worm their way in among them. The ability of an —NH— group to become charged, forming $-NH_2^{+}-$, for example, accounts for the buildup of electrostatic charge on nylon fabrics and carpets that are exposed to friction, such as when we walk across them.

FIGURE 11.25 A rather crude nylon fiber can be made by dissolving the salt of the amine in water and dissolving the acid in a layer of hexane, which floats on the water. The polymer forms at the interface of the two layers, and a long string can be slowly pulled out.

Condensation polymers are formed by condensing a carboxylic acid with an alcohol to form a polyester or with an amine to form a polyamide. They tend to have shorter chains than addition polymers and properties that depend on the structures of their monomers.

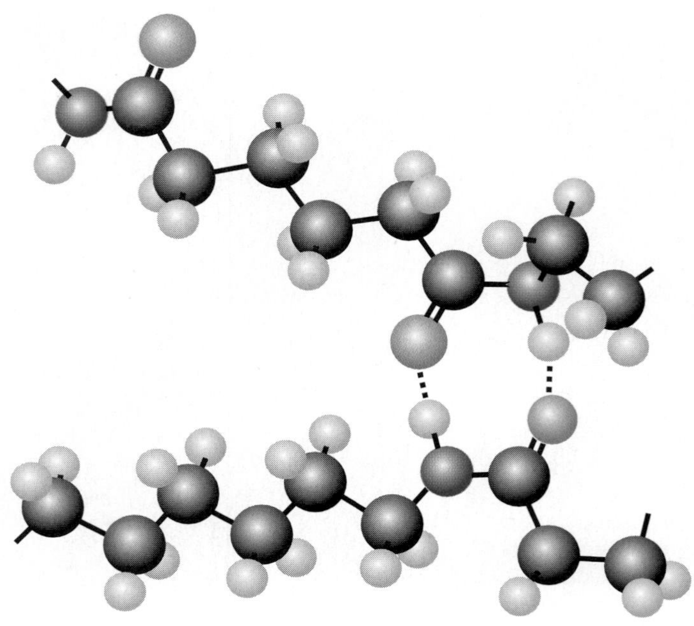

FIGURE 11.26 The strength of nylon fibers is yet another sign of the presence of hydrogen bonds, this time between neighboring polyamide chains.

EXAMPLE 11.3 Determining the formulas of polymers and monomers

Write the formulas of (a) the monomers of Kevlar, a strong fiber used to make bullet proof vests:

$$\left(-\overset{\overset{\displaystyle O}{\|}}{C}-C_6H_4-\overset{\overset{\displaystyle O}{\|}}{C}-NH-C_6H_4-NH- \right)_n$$

(b) two repeating units of the polymer that is formed when peroxides are added to $CH_3CH_2CH{=}CH_2$.

STRATEGY (a) Look at the "backbone" of the polymer, the long chain to which the other groups are attached. If the atoms are all carbon atoms, it is an addition polymer. If ester groups are present in the backbone, the polymer is a polyester and the monomers will be an acid and an alcohol. If the backbone contains amide groups, the polymer is a polyamide and the monomers will be an acid and an amine. If all the polyester or polyamide groups face in the same direction, both groups are on the same molecule and there is only one monomer. Monomers of condensation polymers have two functional groups in each molecule. (b) If the monomer is an alkene or alkyne, the monomers will add to one another; a π-bond will be replaced by new σ-bonds between the monomers. If the monomers are an acid and an ester or amine, a condensation polymer forms. Draw the ester or amide that would result from the loss of a molecule of water.

SOLUTION (a) Amide groups are present in the backbone, so the polymer is a polyamide. The amide groups face in opposite directions, so there are two different monomers, one with two acid groups and one with two amine groups. We split each amide group apart and add a molecule of water:

$$\cdots-\overset{\overset{O}{\|}}{C}-C_6H_4-\overset{\overset{O}{\|}}{C}-NH-C_6H_4-NH-\cdots \quad (H_2O,\ H_2O) \longrightarrow$$

$$HO-\overset{\overset{O}{\|}}{C}-C_6H_4-\overset{\overset{O}{\|}}{C}-OH + NH_2-C_6H_4-NH_2$$

(b) The monomer is an alkene, so it forms an addition polymer. Replace the π-bond by two additional σ-bonds, one to each adjacent monomer:

$$\underset{\underset{CH_2}{|}}{CH_3}\!\!-CH{=}CH_2 \quad \underset{\underset{CH_2}{|}}{CH_3}\!\!-CH{=}CH_2 \longrightarrow -\overset{\overset{\overset{CH_3}{|}}{CH_2}}{\overset{|}{CH}}-CH_2-\overset{\overset{\overset{CH_3}{|}}{CH_2}}{\overset{|}{CH}}-CH_2-$$

SELF-TEST 11.16A (a) Write the formula for the monomer of Teflon, $-\!\!(CF_2-CF_2)\!\!-_n$; (b) the polymer of lactic acid (**42**) is used in surgical sutures that dissolve in the body; write the formula for two repeating units of this polymer.

[*Answer:* (a) $CF_2{=}CF_2$; (b) $-\!\!(O-CH(CH_3)-CO)\!\!-_n$]

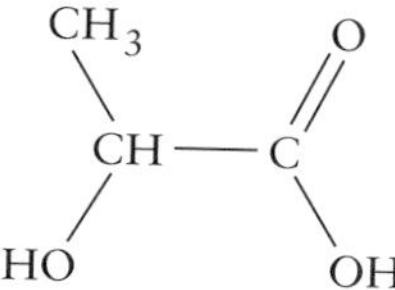

42 Lactic acid, $CH_3CH(OH)COOH$

SELF-TEST 11.16B Write the formula for (a) the monomer of polymethylmethacrylate,

$$\left(-CH_2-\underset{\underset{COOCH_3}{|}}{\overset{\overset{CH_3}{|}}{C}}-\right)_n$$

used in contact lenses; (b) two repeating units of polyalanine, the polymer of the amino acid alanine, $CH_3CH(NH_2)COOH$.

11.15 COPOLYMERS AND COMPOSITES

Some polymers are **copolymers,** that is, polymers made up of more than one type of repeating unit (Fig. 11.27). For example, as noted earlier, in nylon-66 the repeating units are formed from 1,6-diaminohexane, $H_2N(CH_2)_6NH_2$, and adipic acid, $HOOC(CH_2)_4COOH$. They form an **alternating copolymer,** in which acid and amine monomers alternate. Another type of copolymer is a **block copolymer,** which consists of long segments in which the repeating unit is formed from one monomer, followed by segments formed from the other. One example is high-impact polystyrene. Pure polystyrene is a transparent, brittle material that is easily broken. Polybutadiene is a synthetic rubber that is very resilient, but soft. A block copolymer of the two monomers produces a material that is a durable, strong, yet transparent plastic. A different formulation of the two polymers produces styrene-butadiene rubber (SBR), which is used mainly for automobile tires; it is also used in chewing gum.

In **random copolymers,** different monomers are mixed in no particular order. **Graft copolymers** consist of long chains of one monomer, with the other monomer attached as side groups. For example, the polymer used to make hard contact lenses is a nonpolar hydrocarbon that repels water. On the other hand, the polymer used to make soft contact lenses is a graft copolymer with similar chains of nonpolar monomers, but it also has side groups of a different monomer that absorbs water. So much water is absorbed that 50% of the volume of the contact lens is water, which makes the lens very pliable, soft, and more comfortable.

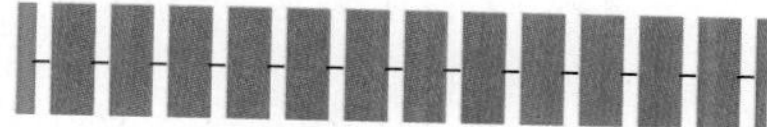

(a) Simple polymer

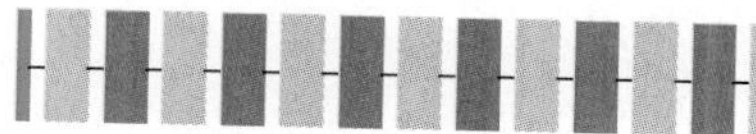

(b) Alternating copolymer

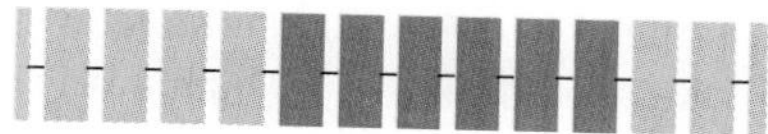

(c) Block copolymer

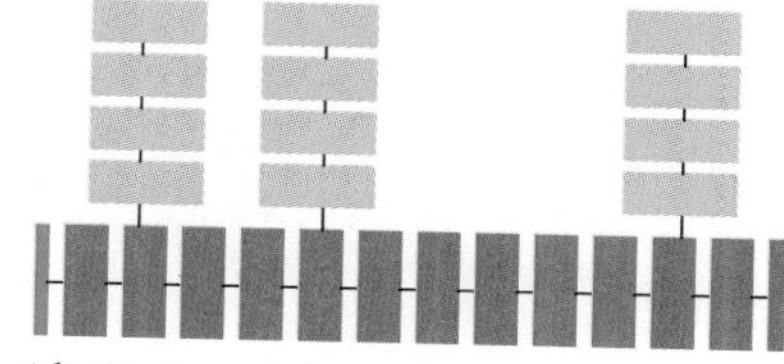

(d) Graft copolymer

FIGURE 11.27 The classification of copolymers. (a) A polymer made from a single monomer, represented by the red rectangles. (b) An alternating copolymer of two monomers, represented by the red and yellow rectangles. (c) A block copolymer. (d) A graft copolymer.

CASE STUDY 11 CONDUCTING POLYMERS

SOME MORNING YOU might stop at a newsstand to purchase a plastic sheet rolled into a narrow tube. Unrolling the plastic, you will activate a tiny microprocessor and the daily news will scroll past on the single compact sheet. The remarkable material of this "newssheet" already exists: an early form of it was discovered by accident in the early 1970s when a chemist who was polymerizing ethyne (acetylene) added a thousand times too much catalyst. Instead of a synthetic rubber, he had made a thin, flexible film. It looked like metal tinted pink (see the photograph), and—very much like a metal—it conducted electricity.

An electric current is the flow of electrons. Metals conduct electricity because their valence electrons move easily from atom to atom. Most covalently bonded solids do not conduct electricity because their valence electrons are locked into individual bonds and are not free to move. The exceptions, like graphite, have connected systems of π-bonds through which electrons can move freely. For example, graphite can be thought of as vast planes of benzene-like rings linked to one another. Because the π-electron clouds of neighboring rings merge with one another, the electrons can travel through the solid. However, a disadvantage is that commercial graphite is fragile and brittle.

This flexible polyacetylene sheet was peeled from the walls of the reaction flask in which it was made from acetylene gas.

Conducting polymers provide a new and exciting alternative. They do not rust and they have low densities. They can be molded or drawn into rust-free shells, fibers, or thin plastic sheets, but can act like metallic conductors. They can be made to glow with almost any color and to change conductivity with conditions. Imagine polymer tags on cases of food that change conductivity when the cases are left too long unrefrigerated. Because conducting polymers can also be designed to change shape with the level of electrical current, they could serve as artificial muscles or give flexibility of movement to robots.

All conducting polymers have a common feature: a long chain of sp^2 hybridized carbon atoms, often with nitrogen or sulfur atoms included in the chains. Polyacetylene, the first conducting polymer, is also the simplest: it consists of thousands of $\left(CH{=}CH\right)_n$ units:

The double bonds alternate, meaning that each C atom has an unhybridized *p*-orbital that can overlap with the *p*-orbital on either side, thus allowing electrons to be delocalized along the entire chain like a one-dimensional version of graphite.

Butyl rubber is a copolymer of a little isoprene with isobutylene, $(CH_3)_2C{=}CH_2$. Its advantage over natural rubber stems from the fact that isobutylene has only one double bond; because this polymer has fewer double bonds in the chain than natural rubber has, it is less likely to be attacked and degraded.

SELF-TEST 11.17A Use Fig. 11.27 to identify the type of copolymer formed by monomers A and B: —A—A—A—A—B—B—B—B—B—.

[*Answer:* Block copolymer]

SELF-TEST 11.17B Identify the type of copolymer formed by monomers A and B: —A—B—A—B—A—B—A—B—.

One conducting polymer, polypyrrole,

has been used in "smart" windows, which darken from a transparent yellow-green to a nearly opaque blue-black in bright sunlight. Fibers of polypyrrole are also woven into radar camouflage cloth. Because it does not reflect microwaves back to their source, on radar the cloth appears to be a patch of empty space.

Polyaniline is being used in flexible coaxial cable; rechargeable, flat, buttonlike batteries; and laminated, rolled films that could be used as flexible computer or television displays.

It could one day serve both as a nonmetallic solder for the printed circuit boards inside a computer and as electrical shielding for the case. All-plastic transistors are also being studied, raising the possibility of an all-plastic computer that could survive highly corrosive conditions, such as those at marine research sites.

Thin films of poly-*p*-phenylenevinylene, PPV, give off light when exposed to an electric field. By varying the composition of the polymer, scientists have coaxed it to glow in a wide range of colors. Such multicolor light-emitting diode (LED) displays could be as bright as fluorescent ones. Current PPV displays last only about 10% as long as fluorescent displays, but as their longevity grows, they could begin to replace the computer and television screens we use today.

QUESTIONS

1. (a) Write the Lewis structure of ethyne and sketch its valence bond structure, showing σ- and π-bonds. What is the hybridization of the carbon atoms in ethyne? (b) Draw two resonance structures for three units of polyacetylene, $-(CHCH)_3-$. (c) Sketch the valence bond picture of one of the structures, showing all σ- and π-bonds, and indicating the hybridization of each carbon atom. Are polyacetylene molecules likely to be fairly stiff, or do the molecules rotate and coil freely about their carbon-carbon bonds? Explain your answer.

2. (a) What functional groups are present in polypyrrole? (b) What is the hybridization of the N atom in polypyrrole? (c) Do the N atoms in polypyrrole or in polyaniline help to carry the current? Explain your reasoning.

3. Draw three of the repeating units of the polyacetylene formed from $CH_3C{\equiv}CH$.

4. The monomer of polyaniline is the compound aniline (aminobenzene). Draw the structural formula of aniline.

Copolymers have molecular structures built from two or more monomers, but **composite materials** contain two or more separate materials that have been solidified together. Composite materials like fiberglass have inorganic solids mixed into a polymer matrix. The result is a material that has great strength yet remains flexible. Some lightweight composites, like the graphite composite used for tennis rackets and the body of the space shuttle and high performance aircraft, can have three times the strength-to-density ratio of steel (Fig. 11.28).

Composite materials and copolymers can give materials the advantage of more than one component material. Copolymers contain more than one type of monomer.

FIGURE 11.28 This high-performance race car is made of a composite material that is stronger than steel.

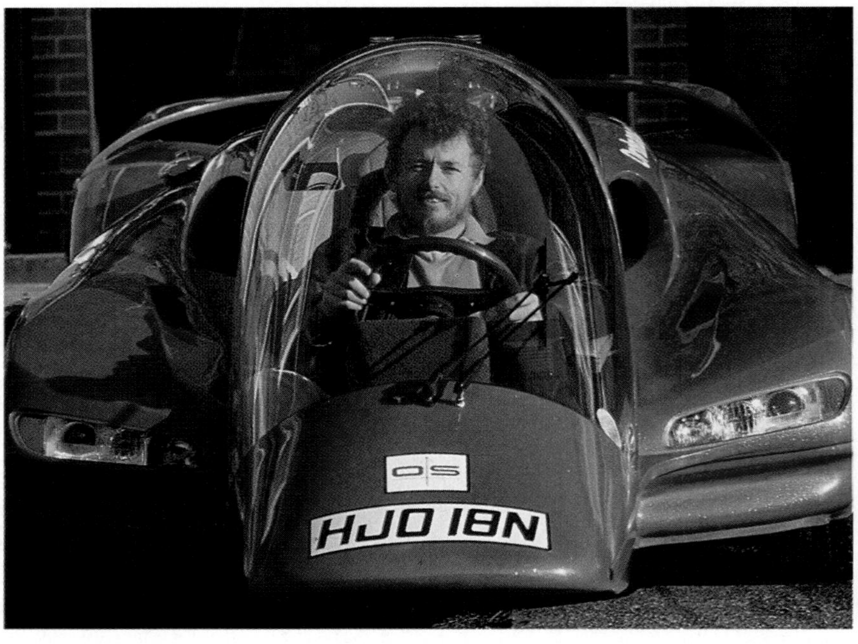

11.16 PHYSICAL PROPERTIES

Synthetic polymers do not have definite molar masses because they consist of molecules with different lengths. So, we speak of the *average* molar mass and the *average* chain length of a polymer. The presence of a mixture of molecules means that polymers do not have definite melting points but soften gradually as the temperature is raised. The viscosity of a polymer, its ability to flow when molten (Section 10.4), depends on its chain length. The longer the chains, the more tangled together they may be, and hence the slower their flow.

If we think of polymer molecules as cooked spaghetti, it is easy to understand how chain length affects strength. Spaghetti that has been chopped into small pieces represents short chains. It can be easily torn apart. However, long strands of cooked spaghetti, which represent long-chain polymers, are very difficult to separate (and even to ladle out of a bowl), because the chains become tangled with one another. Tearing a tangled long-chain polymer apart means breaking backbone covalent bonds, which are much stronger than intermolecular forces, the only forces holding very short chains together. Longer chains thus produce higher mechanical strength.

FIGURE 11.29 The two samples of polyethylene in the test tube were produced by different processes. The floating, low-density polymer was produced by high-pressure polymerization. The high-density polymer at the bottom was produced with a Ziegler-Natta catalyst. It has a higher density because its chains pack together better.

The polarity of the functional groups in a polymer also affects the strength of intermolecular forces, which in turn affects mechanical strength. For example, nylon is a polyamide, which can take part in hydrogen bonding. Polyethylene has only London forces holding its molecules together. For chains of the same average length, stronger intermolecular forces produce greater mechanical strength.

The way the chains of a polymer pack together also affects polymer strength. Packing arrangements that maximize intermolecular contact also maximize intermolecular forces and increase the strength of the material. Long, unbranched chains can line up next to one another like raw spaghetti and form crystalline regions that allow maximum interaction between chains. The closer the contact, the denser the material and the stronger the forces. Branched polymer chains cannot fit together as closely (Fig. 11.29).

The **elasticity** of a polymer refers to its ability to return to its original shape when stretched. Natural rubber has low elasticity and is easily softened by heating. These properties limited the use of rubber until the early nineteenth century, when Charles Goodyear was searching for a means to reduce the heat-induced softening of rubber and improve its elasticity. In 1839 he discovered, partly by accident, that vulcanizing rubber produces the desired properties. In vulcanization, rubber is heated with sulfur. The sulfur atoms form cross-links between the polyisoprene chains and produce a three-dimensional network of atoms (Fig. 11.30). Because the chains are linked together, vulcanized rubber does not soften as much as natural rubber as the temperature is raised. It is also much more resistant to deformation when stretched, because the cross-links pull it back. Materials that return to their original shapes after stretching are called **elastomers.** However, extensive cross-linking can produce a rigid network that resists stretching. For example, high concentrations of sulfur result in very extensive cross-linking and the hard material called *ebonite.*

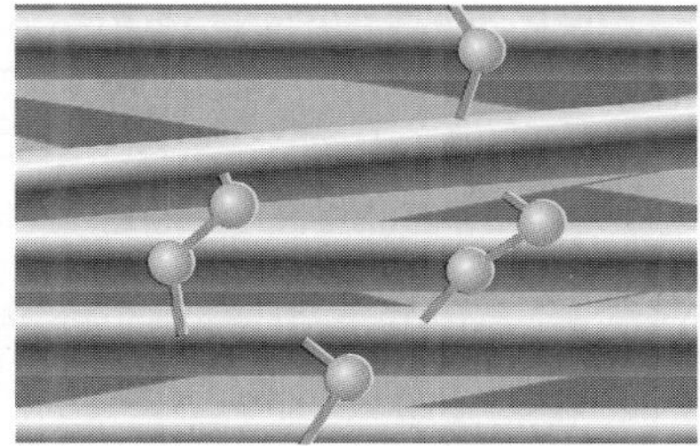

(a)

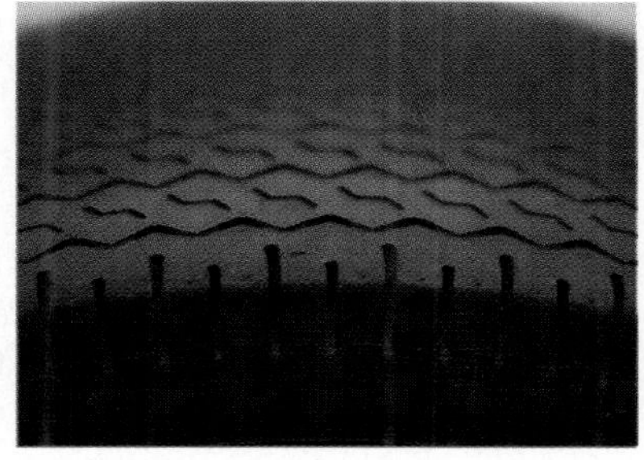

(b)

FIGURE 11.30 (a) The gray cylinders represent polyisoprene molecules and the beaded yellow strings represent disulfide (—S—S—) links that are introduced when the rubber is vulcanized, or heated with sulfur. These cross-links increase the resilience of the treated rubber and make it more useful than natural rubber. (b) Automobile tires are made of vulcanized rubber and a number of additives, including carbon.

Polymers melt over a range of temperatures, and polymers consisting of long chains tend to have high viscosities. Polymer strength increases with increasing chain length, polarity of functional groups in the polymer, and the extent of crystallization. Elastomers contain cross-links that increase their strength and elasticity.

EDIBLE POLYMERS

We would be surrounded by polymers even if there were no polymer industry. Many natural materials are polymers, including the cellulose of wood, natural fibers like cotton and silk, the proteins and carbohydrates in our food, and the nucleic acids of our genes.

11.17 PROTEINS

Amino acids are the building blocks of proteins. In a sense, proteins are a very elaborate form of nylon, because they result from condensation reactions that produce amide links. However, whereas nylon is a monotonous alternation of the same two monomers, the polypeptide chains of proteins are condensation copolymers of up to 20 different naturally occurring amino acids distinguished only by their side chains (Table 11.6), so there are billions of possible combinations. Each one of the cells in our body contains more than 5000 different kinds of proteins, each kind with a specific function to perform. Our bodies can synthesize 11 amino acids in sufficient amounts for our needs. It is essential to ingest the other nine, which are known as the **essential amino acids.** The uniqueness of every individual stems from the small differences in body proteins that result from the way these 20 amino acid building blocks are arranged.

TABLE 11.6 The naturally occurring amino acids, X—CH(NH$_2$)COOH

—X	Name	Abbreviation
$-H$	glycine	Gly
$-CH_3$	alanine	Ala
$-CH_2-C_6H_5$ (benzene ring)	phenylalanine*	Phe
$-CH(CH_3)_2$	valine*	Val
$-CH_2CH(CH_3)_2$	leucine*	Leu
$-CH(CH_3)CH_2CH_3$	isoleucine*	Ile
$-CH_2OH$	serine	Ser
$-CH(OH)CH_3$	threonine*	Thr
$-CH_2-C_6H_4-OH$ (benzene ring)	tyrosine	Tyr
$-CH_2COOH$	aspartic acid	Asp
$-CH_2CH_2COOH$	glutamic acid	Glu
$-CH_2SH$	cysteine	Cys
$-CH_2CH_2SCH_3$	methionine*	Met
$-CH_2(CH_2)_3NH_2$	lysine*	Lys
$-CH_2(CH_2)_2NH-C(=NH)-NH_2$	arginine	Arg
$-CH_2-$ (imidazole ring: N, NH)	histidine*	His
$-CH_2-$ (indole ring: N—H)	tryptophan*	Trp
$-CH_2CONH_2$	asparagine	Asn
$-CH_2CH_2CONH_2$	glutamine	Gln
(pyrrolidine ring: NH, COOH)	proline†	Pro

*Essential amino acids for humans.

†The entire amino acid is shown.

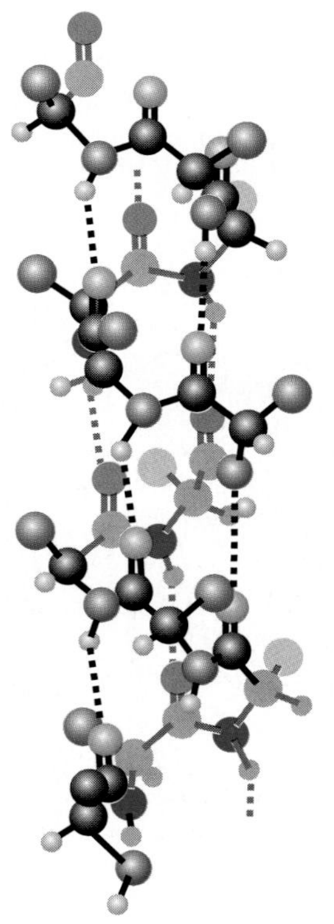

FIGURE 11.31 A representation of part of an α helix, one of the secondary structures adopted by polypeptide chains. The purple spheres represent substituents in the amino acids.

A molecule formed from two or more amino acids is called a **peptide.** An example is the combination of glycine and alanine, denoted Gly-Ala:

$$H_2N-CH_2-\overset{\overset{\displaystyle O}{\|}}{C}-NH-CH(CH_3)COOH$$

The —CO—NH— link shown in the gray box is called a **peptide bond,** and each amino acid in a peptide is called a **residue,** because H_2O has been lost in the formation of the peptide link. A typical protein is a polypeptide chain of more than a hundred residues joined through peptide bonds and arranged in a strict order. When only a few amino acid residues are present, we call the molecule an **oligopeptide.** The artificial sweetening agent aspartame is a type of oligopeptide called a **dipeptide** because it has two residues.

The sequence of residues in the peptide chain is the protein's **primary structure.** Aspartame consists of phenylalanine (Phe) and aspartic acid (Asp), so its primary structure is Phe-Asp. A fragment of the primary structure of human hemoglobin is

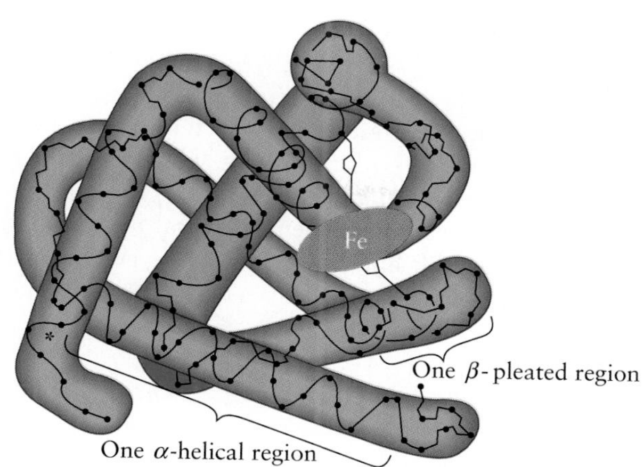

FIGURE 11.32 One of the four polypeptide chains that make up the human hemoglobin molecule. Each dot represents an amino acid monomer. The chain consists of alternating regions of α helix and β-pleated sheet. The significance of the asterisk (toward the lower left) is explained in Fig. 11.33. The oxygen molecules we inhale attach to the iron atom, and are carried through the bloodstream to be released where they are needed.

Leu-Ser-Pro-Ala-Asp-Lys-Thr-Asn-Val-Lys-···
···Val-Lys-Gly-Trp-Ala-Ala-···
··· Ser-Thr-Val-Leu-Thr-Ser-Lys-Ser-Lys-Tyr-Arg

The determination of the primary structures of proteins is a very demanding analytical task but, thanks to automated procedures, many of these structures are now known.

The way polypeptide chains coil together or line up next to one another is called the **secondary structure** of the molecule. Because almost all naturally occurring amino acids have the same handedness, the structures are stereoregular. The most common secondary structure in animal proteins is the **α helix**, a specific conformation of a polypeptide chain held in place by hydrogen bonds between residues (Fig. 11.31). An alternative secondary structure is the **β-pleated sheet** form of the protein we know as *silk*. In this structure, the protein molecules lie side by side to form nearly flat sheets. The molecules of many proteins consist of alternating regions of α helix and β-pleated sheet (Fig. 11.32).

Protein molecules also adopt a specific **tertiary structure,** the shape into which the α helix, β-pleated sheet, and other regions are folded as a result of interactions between residues lying in different parts of the primary structure. The globular form of each chain in hemoglobin is an example. One important type of link responsible for tertiary structure is the **disulfide link,** —S—S—, between amino acids containing sulfur.

Proteins may also have a **quaternary structure,** in which neighboring polypeptide units stack together in a specific arrangement. The hemoglobin molecule, for example, has a quaternary structure of four polypeptide units like the one shown in Fig. 11.32.

Any modification of the primary structure of a protein—the replacement of one amino acid residue by another—may lead to the malfunction we call congenital disease. Even one wrong amino acid in the chain can disrupt the normal function of the molecule (Fig. 11.33).

The loss of structure of a protein is called **denaturation.** This structural change may be a loss of quaternary, tertiary, or secondary structure, or even the degradation of the primary structure by cleavage of the peptide bonds. Even mild heating can cause irreversible denaturation, as happens when we cook an egg and the albumen denatures into a white mass.

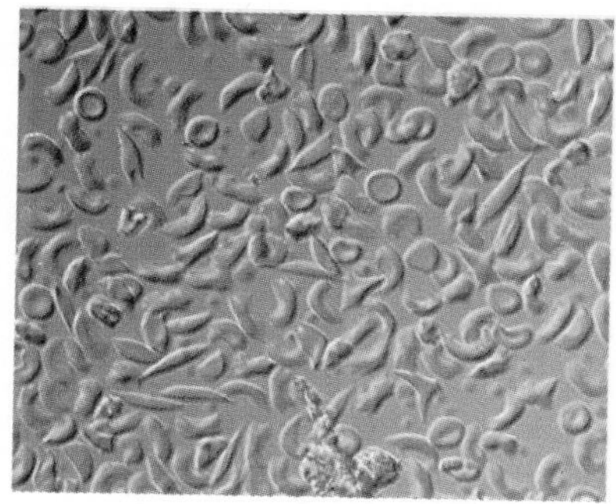

FIGURE 11.33 The sickle-shaped red blood cells that form when the glutamic acid marked with an asterisk in Fig. 11.32 is replaced by valine.

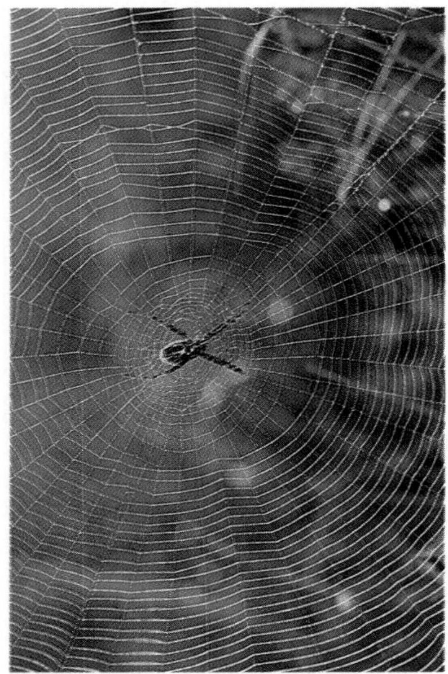

FIGURE 11.34 The protein made by spiders to produce a web is a form of silk that can be exceptionally strong (see the illustrations on page ii).

The permanent waving of hair—which consists primarily of long α helices of the protein keratin—is a result of partial denaturation. Mild reducing agents are applied to the hair to sever the disulfide links between the protein strands that make up hair. The links are then reformed by applying a mild oxidizing agent while the hair is stretched and twisted into the desired arrangement.

We wear proteins as well as eat them. Natural protein-based materials include fibers such as silk and wool and some adhesives. Some animals, such as spiders, also use proteins as a structural material (Fig. 11.34).

The primary structure of a polypeptide is the sequence of amino acid residues; the secondary structure is their geometrical arrangement in space; the tertiary structure is the folding of helices and sheets; the quaternary structure is the packing of individual polypeptide chains together.

11.18 CARBOHYDRATES

The **carbohydrates** are so called because they often have the empirical formula CH_2O, which suggests a hydrate of carbon. They include starches, cellulose, and sugars like glucose, which is an aldehyde, $C_6H_{12}O_6$ (**43**), and fructose (fruit sugar), a structural isomer of glucose that is a ketone (**44**). Carbohydrates have many —OH groups and form numerous hydrogen bonds with one another and with water.

$$\begin{array}{c} \mathrm{H{-}C{=}O} \\ | \\ \mathrm{H{-}C{-}OH} \\ | \\ \mathrm{HO{-}C{-}H} \\ | \\ \mathrm{H{-}C{-}OH} \\ | \\ \mathrm{H{-}C{-}OH} \\ | \\ \mathrm{CH_2OH} \end{array} \qquad\qquad \begin{array}{c} \mathrm{CH_2OH} \\ | \\ \mathrm{C{=}O} \\ | \\ \mathrm{HO{-}C{-}H} \\ | \\ \mathrm{H{-}C{-}OH} \\ | \\ \mathrm{H{-}C{-}OH} \\ | \\ \mathrm{CH_2OH} \end{array}$$

43 Glucose, $C_6H_{12}O_6$ **44** Fructose, $C_6H_{12}O_6$

The **polysaccharides** are polymers of glucose. They include starch, which we can digest, and cellulose, which we cannot. Starch is made up of two components, amylose and amylopectin. Amylose, which makes up about 20 to 25% of most starches, is composed of chains made up of several thousand glucose units linked together (Fig. 11.35). Amylopectin is also made up of glucose chains (Fig. 11.36), but its chains are linked

FIGURE 11.35 The amylose molecule, one component of starch, is a polysaccharide, a polymer of glucose. It consists of glucose units linked together to give a structure like this but with moderate degree of branching.

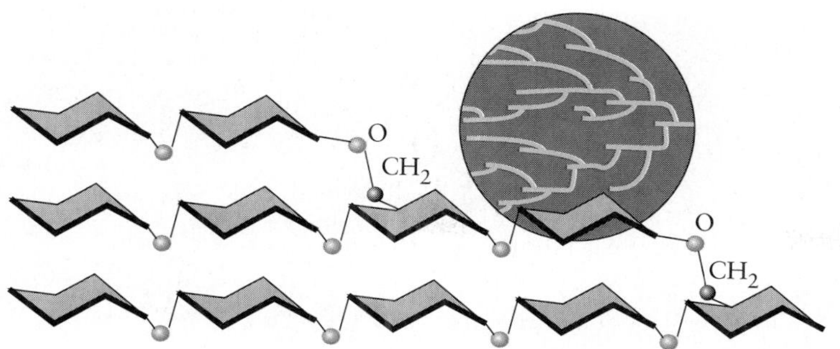

FIGURE 11.36 The amylopectin molecule is another component of starch. It has a more highly branched structure than amylose.

into a branched structure and its molecules are much larger. Each molecule consists of about a million glucose units.

Cellulose is the structural material of plants. Like starch, it is a polymer of glucose, but the units link differently, forming flat, ribbonlike strands (Fig. 11.37). These strands can lock together through hydrogen bonds into a rigid structure that for us (but not for termites) is indigestible. The difference between cellulose and starch shows nature at its most economical and elegant, for only a small modification of the linking between glucose units results, on the one hand, in an important foodstuff and, on the other, in a versatile construction material. Cellulose is the most abundant organic chemical in the world, and billions of tons of it are produced annually by photosynthesis.

Glucose is an alcohol and an aldehyde that polymerizes to form starch and cellulose.

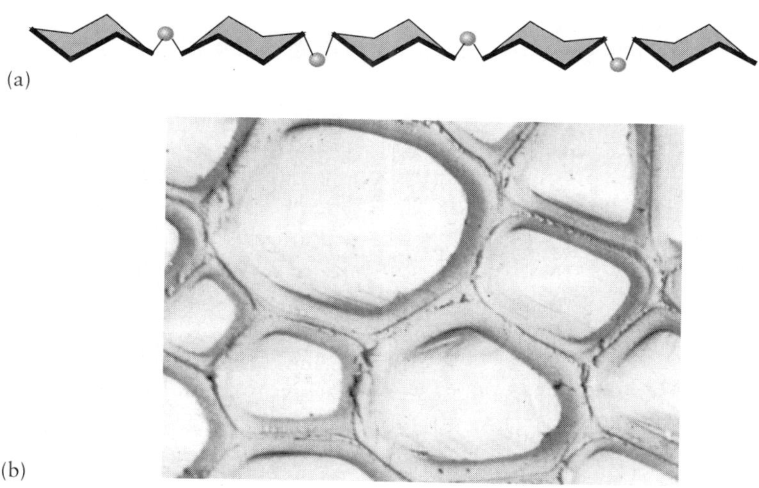

FIGURE 11.37 (a) Cellulose is yet another polysaccharide constructed from glucose units. However, the linking between the units in cellulose results in long, flat ribbons that can produce a fibrous material through hydrogen bonding. (b) These long tubes of cellulose formed the structural material of an aspen tree.

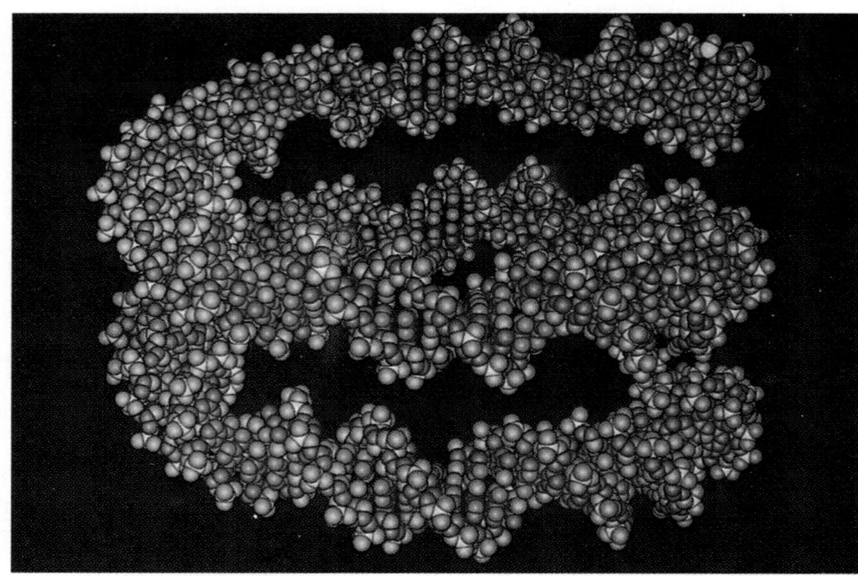

FIGURE 11.38 A computer graphics image of a short section of a DNA molecule, which consists of two entwined helices. In this illustration, the double helix is also coiled around itself in a shape called a superhelix.

11.19 DNA AND RNA

People have used materials to store information since they made the first cave paintings. Today, we can speculate about the possibility of storing information in single molecules. Arrays of such molecules could serve as data storage devices of enormous capacity. Nature, however, has used this technique for millions of years. It uses the molecule called deoxyribonucleic acid (DNA) to store the genetic information that enables life forms to reproduce (Fig. 11.38).

Every living cell contains at least one DNA molecule to control the production of proteins and carry genetic information from one generation of cells to the next. Human DNA molecules are immense: if one could be extracted without damage from a cell nucleus and drawn out straight from its highly coiled natural shape, it would be about 2 m long (Fig. 11.39). The ribonucleic acid (RNA) molecule is closely related to DNA. One of its functions is to carry information stored by DNA to a region of the cell where that information can be used in protein synthesis.

The best way to understand the structure of DNA is to see how it gets its name. It is a polymer built up of repeating units derived from the sugar ribose (**45**). For DNA, the ribose molecule has been modified by removing the oxygen atom at carbon atom 2, the second carbon atom from the ether oxygen atom in the five-membered ring. Therefore, the repeating unit—the monomer—is called deoxyribose (**46**).

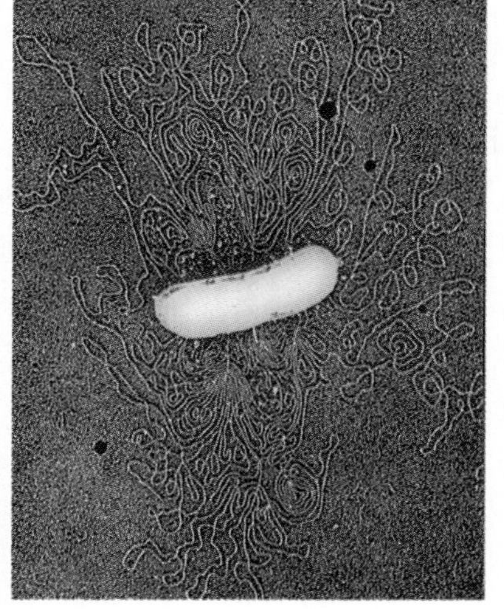

FIGURE 11.39 A DNA molecule is very large, even in bacteria. In this micrograph, a DNA molecule has spilled out through the damaged cell wall of a bacterium.

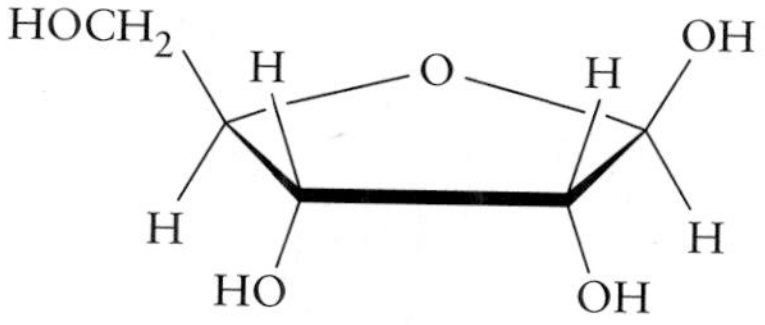

45 Ribose, $C_5H_{10}O_5$

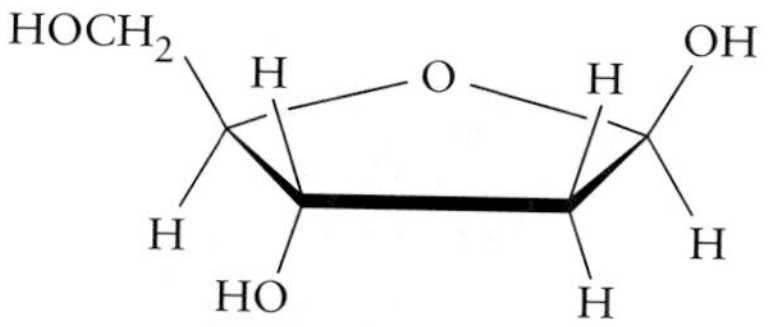

46 Deoxyribose, $C_5H_{10}O_4$

Attached by a covalent bond to carbon atom 1 of the deoxyribose ring is an amine (and therefore a base), which may be adenine, A (**47**), guanine, G (**48**), cytosine, C (**49**), or thymine, T (**50**). In RNA, uracil, U (**51**), occurs in place of thymine. The base bonds to carbon atom 1 of deoxyribose through the nitrogen of the >NH group (printed in red); and the compound so formed is called a *nucleoside.* All nucleosides have a similar structure, which we can summarize as the shape shown in (**52**).

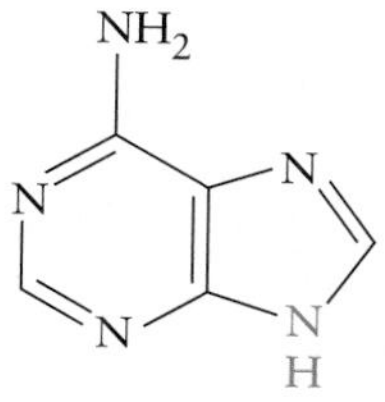

47 Adenine

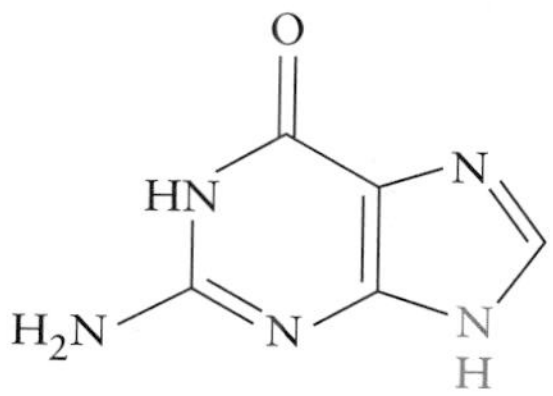

48 Guanine

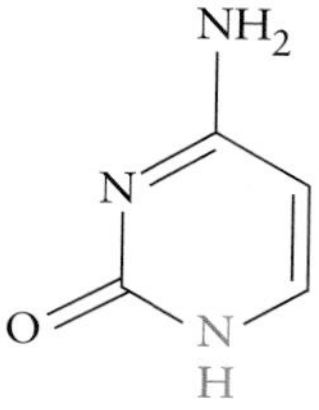

49 Cytosine

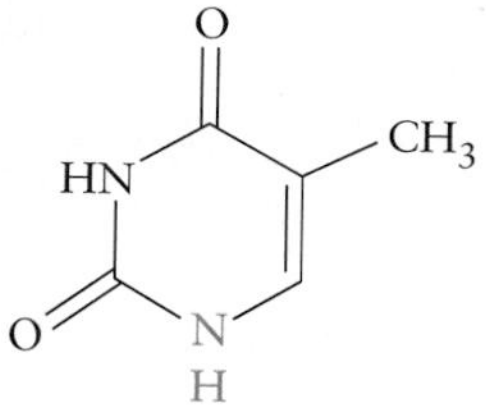

50 Thymine

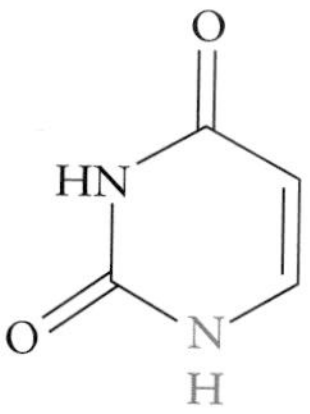

51 Uracil

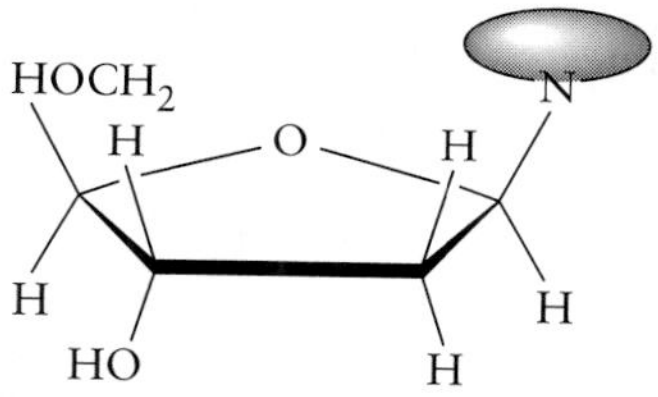

52 A nucleoside

At this stage, the DNA monomers are each completed by a phosphate group, $—O—PO_3^{2-}$, covalently bonded to carbon atom 5 to give a compound called a *nucleotide* (**53**). Because there are four possible nucleoside monomers (one for each base) there are four possible nucleotides in each type of nucleic acid.

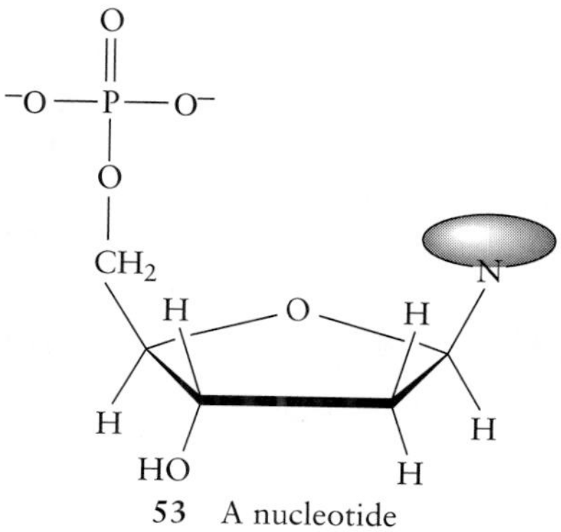

53 A nucleotide

FIGURE 11.40 The condensation of nucleotides that leads to the formation of a nucleic acid—a polynucleotide.

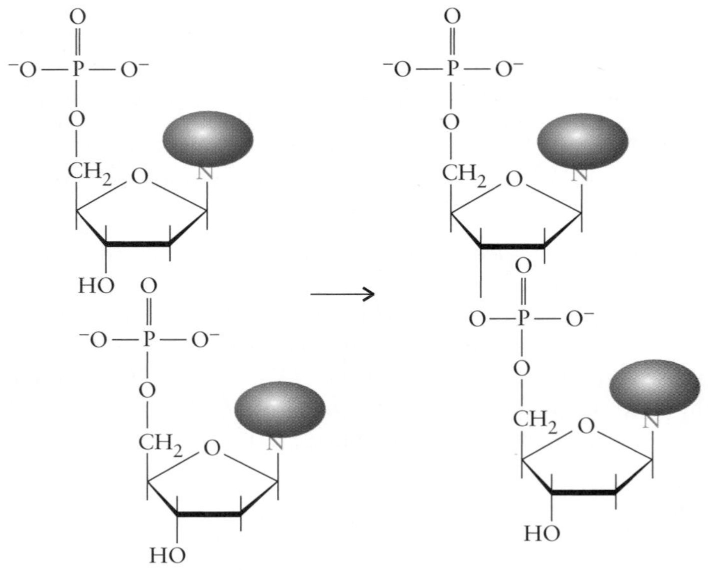

Molecules of DNA and RNA are *polynucleotides,* polymeric species built from nucleotide units. Polymerization occurs when the phosphate group of one nucleotide condenses with the —OH group on carbon atom 3 of another nucleotide, thereby releasing a molecule of water. As this condensation continues, it results in a structure like that shown in Fig. 11.40, a compound known as a **nucleic acid.** The DNA molecule itself is a double helix in which two long nucleic acid strands are wound around each other.

The ability of DNA to replicate lies in its double helical structure. There is a precise correspondence between the bases in the two strands.

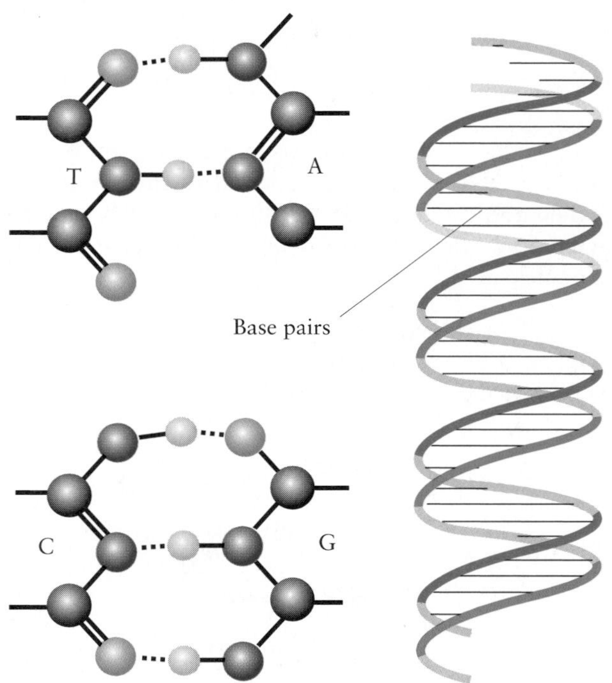

FIGURE 11.41 The bases in the DNA double helix fit together by virtue of the hydrogen bonds that they can form as shown on the left. Once formed, the AT and GC pairs are almost identical in size and shape. As a result, the turns of the helix shown on the right are regular and consistent.

Adenine in one strand always forms hydrogen bonds to thymine in the other, and guanine always forms hydrogen bonds to cytosine, so the base pairs are always AT and GC (Fig. 11.41).

As well as replication—the production of copies of itself for reproduction and cell division—DNA governs the production of proteins by generating molecules of RNA. These molecules, with U in place of T, carry information about segments of the genetic message out of the animal or plant cell nucleus to the locations where protein synthesis takes place.

Nucleic acids are copolymers of four nucleotides linked with phosphate groups. The nucleotide sequence stores all genetic information.

SKILLS YOU SHOULD HAVE MASTERED

Conceptual

1. Distinguish alkanes, alkenes, alkynes, and arenes by differences in bonding, structure, and reactivity.
2. Recognize which functional groups allow hydrogen bonding between molecules and explain how the presence of hydrogen bonds affects the physical properties of compounds.
3. Distinguish between addition and condensation polymers.
4. Explain the role of chain length, crystallinity, network formation, cross-linking, and intermolecular forces in determining the physical properties of polymers.

Problem-Solving

1. Predict the products of given elimination, addition, and substitution reactions.
2. Predict the oxidation products of aldehydes and ketones.
3. Write the structural formula of an ester or an amide formed from the condensation reaction of a given carboxylic acid with a given alcohol or amine.
4. Identify two molecules, given their structural formulas, as structural, geometrical, or optical isomers.
5. Judge whether a compound is optically active.
6. Predict the type of polymer a given monomer can form and identify monomers, given the repeating unit of a polymer.

Descriptive

1. Describe general trends in the physical properties of alkanes.
2. Recognize a simple haloalkane, alcohol, ether, phenol, aldehyde, ketone, carboxylic acid, amine, amide, or ester, given a molecular structure.
3. Name simple hydrocarbons, alcohols, ethers, aldehydes, ketones, carboxylic acids, amines, and esters.
4. Distinguish the various types of copolymers.
5. Describe the composition of proteins and distinguish their primary, secondary, tertiary, and quaternary structures.
6. Describe the structures and functions of DNA and RNA.

EXERCISES

Structures and Reactions of Aliphatic Compounds

11.1 Why do branched-chain alkanes have lower melting points and boiling points than unbranched alkanes with the same number of carbon atoms?

11.2 Why do alkenes have lower melting points and boiling points than alkanes with the same number of carbon atoms?

11.3 Identify the type and number of bonds on carbon atom 2 in (a) pentane; (b) 2-pentene; (c) 2-pentyne.

11.4 Predict the geometry and hybridization of the orbitals used in bonding on carbon atom 2 in (a) pentane; (b) 2-pentene; (c) 2-pentyne.

11.5 Classify each of the following reactions as addition or substitution and write its chemical equation: (a) chlorine reacts with methane when exposed to light; (b) bromine reacts with ethene in the absence of light.

11.6 Classify each of the following reactions as addition or substitution and write its chemical equation: (a) hydrogen reacts with 2-pentene in the presence of a nickel catalyst; (b) hydrogen chloride reacts with propene.

11.7 Classify as addition or substitution and write the chemical equations for the reactions that occur (if any) when chlorine gas is mixed with (a) ethane and exposed to sunlight; (b) ethene; (c) ethyne.

11.8 Classify as addition or substitution and write the chemical equations for the reactions that occur (if any) when hydrogen is mixed in the presence of a platinum catalyst with (a) 2-methylpropene; (b) propyne; (c) propane.

Nomenclature of Aliphatic Compounds

11.9 Name each of the following as an unbranched alkane: (a) C_2H_6; (b) C_6H_{14}; (c) C_8H_{18}; (d) CH_4.

11.10 Name each of the following as an unbranched alkane: (a) $C_{10}H_{22}$; (b) C_4H_{10}; (c) C_9H_{20}; (d) C_5H_{12}.

11.11 Name the following substituents: (a) $CH_3—$; (b) $CH_3(CH_2)_3CH_2—$; (c) $CH_3CH_2CH_2—$.

11.12 Name the following substituents: (a) $Cl—$; (b) $CH_3CH_2—$; (c) $CH_3CH_2CH_2CH_2—$.

11.13 Give the systematic name of
(a) $CH_3CH_2CH_3$
(b) CH_3CH_3
(c) $CH_3(CH_2)_3CH_3$
(d) $(CH_3)_2CHCH(CH_3)_2$

11.14 Give the systematic name of
(a) $CH_3CH_2CH_2CH_2CH_2CH_3$
(b) $CH_3CH(CH_2CH_3)C(CH_3)_2CH_3$
(c) $(CH_3)_3CCH_2CH(CH_3)CH_3$
(d) $CH_3C(CH_3)_2CH_2CH_3$

11.15 Name the compounds
(a) $CH_3CH{=}CHCH(CH_3)_2$
(b) $CH_3CH_2CH(CH_3)C(C_6H_5)(CH_3)_2$

11.16 Name the compounds
(a) $CH_2{=}CHCH(C_6H_5)(CH_2)_4CH_3$
(b) $(CH_3)_2CHCH(CH_3)CHClC{\equiv}CH$

11.17 Give the systematic name of the following compounds; if geometrical isomers are possible, write the names of each one:
(a) $CH_2{=}CHCH_3$
(b) $CH_3CH{=}CH(CH_2)_2CH_3$
(c) $CH{\equiv}CCH_2CH_3$
(d) $CH_3C{\equiv}CCH_3$

11.18 Give the systematic name of the following compounds; if geometrical isomers are possible, write the names of each one:
(a) $CH_3CH_2CH{=}CHCH_3$
(b) $CH_3CH_2CH{=}CH_2$
(c) $(CH_3)_2C{=}CHCH_2CH_3$
(d) $CH_3C{\equiv}CCH(CH_3)CH_2CH_3$

11.19 Write the shortened (condensed) structural formula of (a) 3-methyl-1-pentene; (b) 4-ethyl-3,3-dimethylheptane; (c) 5,5-dimethyl-1-hexyne; (d) 3-ethyl-2,4-dimethylpentane.

11.20 Write the shortened (condensed) structural formula of (a) 4-ethyl-2-methylhexane; (b) 3,3-dimethyl-1-pentene; (c) *cis*-4-methyl-2-hexene; (d) *trans*-2-butene.

11.21 Write the structural formula of (a) 4,4-dimethylnonane; (b) 4-propyl-5,5-diethyl-1-decyne; (c) 2,2,4-trimethylpentane; (d) *trans*-3-hexene.

11.22 Write the structural formula of (a) 2,7,8-trimethyldecane; (b) 2,3,5-trimethyl-4-propylheptane; (c) 1,3-dimethylcyclohexane; (d) 5-ethyl-3-heptene.

Aromatic Compounds

11.23 Name the following compounds:

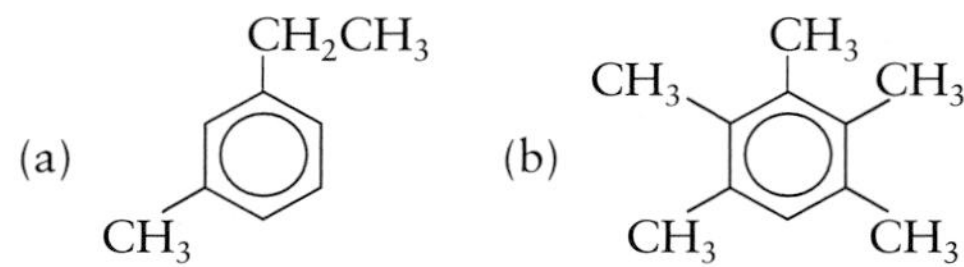

11.24 Name the following compounds:

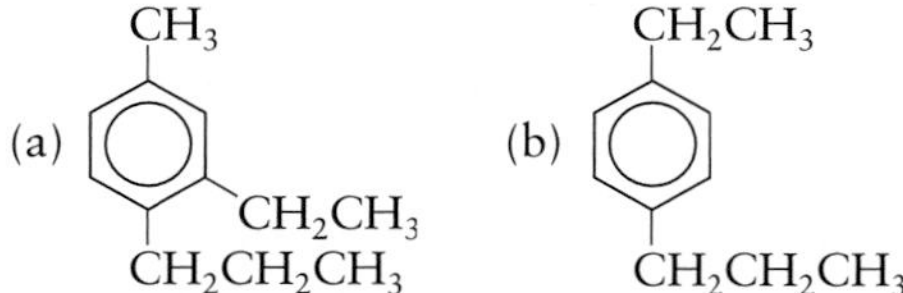

11.25 The common name of methylbenzene is toluene. Write the structural formulas of (a) toluene; (b) *p*-chlorotoluene; (c) 1,3-dimethylbenzene; (d) 4-chloromethylbenzene.

11.26 Write the structural formulas of (a) *p*-xylene; (b) 1,2-dichlorobenzene; (c) 3-phenylpropene; (d) 2-ethyl-1,4-dimethylbenzene.

11.27 (a) Draw all the isomeric dichlorobenzenes. (b) Name each one and indicate which are polar and which are nonpolar.

11.28 (a) Draw all the isomeric trichlorobenzenes. (b) Name each one and indicate which are polar and which are nonpolar.

Functional Groups

11.29 Write the general formula of each of the following types of compounds, using R to denote an organic group: (a) amine; (b) alcohol; (c) carboxylic acid; (d) aldehyde.

11.30 Write the general formula of each of the following types of compounds, using R to denote an organic group: (a) ether; (b) ketone; (c) ester; (d) amide.

11.31 Name each type of compound: (a) R—O—R; (b) R—CO—R; (c) $R-NH_2$; (d) R—COOR.

11.32 Name each type of compound: (a) R—CHO; (b) R—COOH; (c) R—CONHR; (d) R—OH.

11.33 Write the formulas of the following compounds and state whether each one is a primary, secondary, or tertiary alcohol or a phenol: (a) 1-butanol; (b) 2-butanol; (c) *p*-hydroxymethylbenzene.

11.34 Write the formulas of the following compounds and state whether each one is a primary, secondary, or tertiary alcohol or a phenol: (a) 2-methyl-2-propanol; (b) 1-propanol; (c) *o*-hydroxytoluene (see Exercise 11.25).

11.35 (a) Name the compound $CH_3CH_2OCH_2CH_3$. (b) Write the formula of methylethyl ether.

11.36 (a) Name the compound $CH_3CH_2CH_2OCH_2CH_3$. (b) Write the formula of dimethyl ether.

11.37 Identify each compound as an aldehyde or ketone and give its systematic name: (a) CH_3CHO; (b) CH_3COCH_3; (c) $(CH_3CH_2)_2CO$.

11.38 Identify each compound as an aldehyde or ketone and give its systematic name:

(a) CH_3CH_2CHO; (b) CH_3—(benzene ring)—CHO

(c) $(CH_3CH_2)_2CHCH_2COCH_3$

11.39 Write the structural formulas of (a) methanal; (b) propanone; (c) 2-heptanone.

11.40 Write the structural formulas of (a) 2-ethyl-2-methylpentanal; (b) 3,5-dihydroxy-4-octanone; (c) octanal.

11.41 Suggest an alcohol that could be used for the preparation of each of the following compounds and indicate how the reaction would be carried out: (a) ethanal; (b) 2-octanone; (c) 5-methyloctanal.

11.42 Suggest an alcohol that could be used for the preparation of each of the following compounds and indicate how the reaction would be carried out: (a) methanal; (b) propanone; (c) 6-methyl-5-decanone.

11.43 Give the systematic names of (a) CH_3COOH; (b) $CH_3CH_2CH_2COOH$; (c) $CH_2(NH_2)COOH$.

11.44 Give the systematic names of (a) CH_3CH_2COOH; (b) $CHCl_2COOH$; (c) $CH_3(CH_2)_7COOH$.

11.45 Write the structural formulas of (a) benzoic acid (C_6H_5COOH); (b) 2-methylbutanoic acid; (c) propanoic acid.

11.46 Write the structural formulas of (a) 2-methyl propanoic acid; (b) 2,2-dichlorobutanoic acid; (c) 2,2,2-trifluoroethanoic acid.

11.47 Give the systematic names of the following amines: (a) CH_3NH_2; (b) $(CH_3CH_2)_2NH$; (c) *o*-$CH_3C_6H_4NH_2$.

11.48 Give the systematic names of the following amines: (a) $CH_3CH_2CH_2NH_2$; (b) $(CH_3CH_2)_4N^+$; (c) *p*-$ClC_6H_4NH_2$.

11.49 Write the structural formulas of the following amines: (a) *o*-methylphenylamine; (b) triethylamine; (c) tetramethylammonium ion.

11.50 Write the structural formulas of the following amines: (a) methylpropylamine; (b) dimethylamine; (c) *m*-methylphenylamine.

11.51 Name the following compounds: (a) $CH_3CH(OH)CH_3$; (b) CH_3OCH_3; (c) HCHO; (d) $(CH_3CH_2)_2CO$; (e) CH_3NH_2.

11.52 Name the following compounds: (a) CH_3COOCH_3; (b) CH_3COCH_3; (c) HCOOH; (d) $CH_3CH_2CH_2CH_2OH$; (e) $(CH_3CH_2)_3N$.

11.53 Identify all the functional groups in the following compounds: (a) vanillin, the compound responsible for vanilla flavor

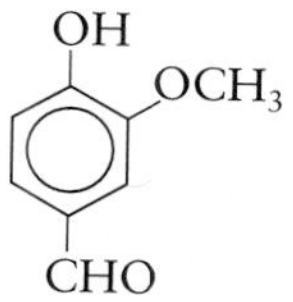

OH, OCH_3, CHO

(b) carvone, the compound responsible for spearmint flavor

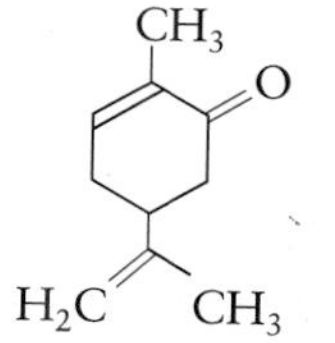

CH_3, O, H_2C, CH_3

(c) caffeine, the stimulant in coffee, tea, and cola drinks

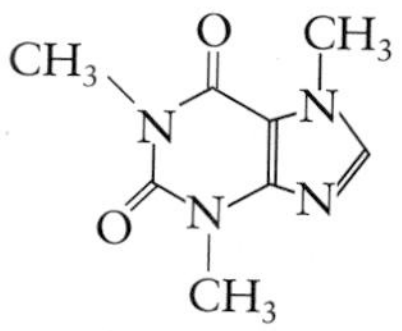

CH_3, O, CH_3, N, N, N, N, O, CH_3

11.54 Identify all the functional groups in the following compounds: (a) zingerone, the pungent, hot component of ginger

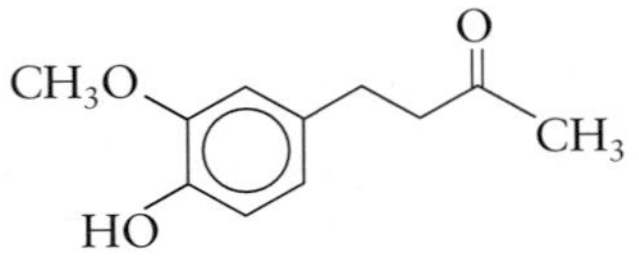

CH_3O HO O CH_3

(b) capsaicin, the biting, hot component of chili peppers

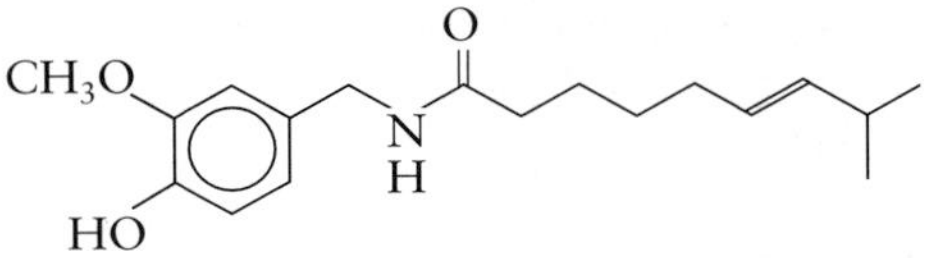

CH_3O HO O N H

(c) procaine, a local anesthetic

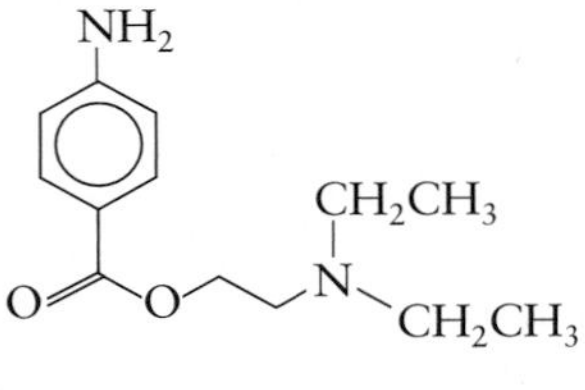

NH_2 O O N CH_2CH_3 CH_2CH_3

11.55 Write the structural formulas of the principal products formed from the following condensation reactions: (a) butanoic acid with 2-propanol; (b) ethanoic acid with 1-pentanol; (c) hexanoic acid with methylethylamine; (d) ethanoic acid with propylamine.

11.56 Write the structural formulas of the principal products formed from the following condensation reactions: (a) octanoic acid with methanol; (b) propanoic acid with ethanol; (c) propanoic acid with methylamine; (d) methanoic acid with diethylamine.

11.57 You are given samples of propanal, 2-propanone, and ethanoic acid. Describe how you would use chemical tests, such as acid-base indicators and oxidizing agents, to distinguish between the three compounds.

11.58 You are given samples of 1-propanol, pentane, and ethanoic acid. Describe how you would use chemical tests, such as aqueous solubility and acid-base indicators, to distinguish between the three compounds.

Isomerism

11.59 Write the structural formulas and name three alkene isomers having the formula C_4H_8.

11.60 Write the structural formulas and name two cycloalkane structural isomers having the formula C_4H_8.

11.61 Write the structural formulas for and name all the isomers of the alkanes (a) C_4H_{10}; (b) C_5H_{12}.

11.62 Write the structural formulas for and name all the isomers (including geometrical isomers) of the alkenes (a) C_4H_8; (b) C_5H_{10}.

11.63 Identify each of the following pairs as structural isomers, geometrical isomers, or not isomers: (a) butane and cyclobutane; (b) pentane and 2,2-dimethylpropane; (c) cyclopentane and pentene;

(d) 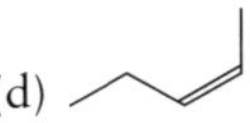 and 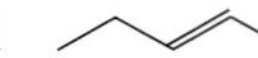 ;

(e) 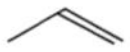 and  .

11.64 Identify each of the following pairs as structural isomers, geometrical isomers, or not isomers: (a) ethanol and dimethyl ether; (b) 1-hydroxy-2-pentene and 3-methylbutanal; (c) 1-chlorohexane and chlorocyclohexane;

(d) 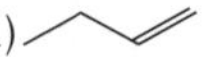 and  ;

(e) 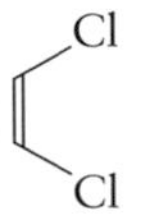 Cl Cl and Cl Cl .

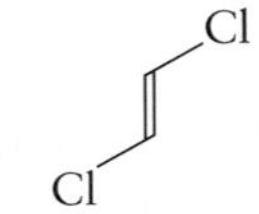

11.65 The branched hydrocarbon C_4H_{10} reacts with chlorine in the presence of light to give two branched structural isomers with the formula C_4H_9Cl. Write the structural formulas of (a) the hydrocarbon; (b) the isomeric products.

11.66 The hydrocarbon C_6H_{14} reacts with chlorine in the presence of light to give only two structural isomers with the formula $C_6H_{13}Cl$. Write the structural formulas of (a) the hydrocarbon; (b) the two isomeric products.

11.67 Indicate which molecules display optical activity and identify the chiral carbons in those that do:
(a) $CH_3CHBrCH_2CH_3$; (b) $CH_3CH_2CHCl_2$;
(c) 1-bromo-2-chloropropane;
(d) 1,2-dichloropentane.

11.68 Indicate which molecules display optical activity and identify the chiral carbons in those that do:
(a) $CH_3CHOHCOOH$; (b) $CH_3CH_2CHClCH_3$;
(c) 2-bromo-2-chloropropane;
(d) 1,2-dibromobutane.

11.69 Identify each chiral carbon in the following compounds: (a) camphor, used in cooling salves

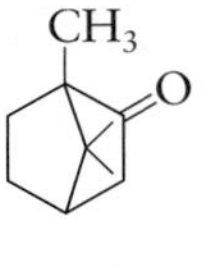

CH_3 O

(b) testosterone, a male sex hormone

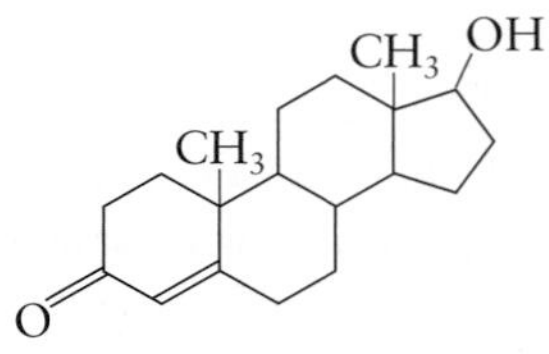

CH_3 OH CH_3 O

11.70 Identify each chiral carbon in the following compounds: (a)menthol, the flavor of peppermint

(b) estradiol, a female sex hormone

Polymers

11.71 Sketch three repeating units of the polymer formed from (a) $CH_2{=}C(CH_3)_2$; (b) $CH_2{=}CHC{\equiv}N$ (only the double bond polymerizes); (c) isoprene

11.72 Sketch three repeating units of each of the polymers formed from (a) tetrafluoroethene; (b) phenylethene; (c) $CH_3CH{=}CHCH_3$.

11.73 Write the structural formulas of the monomers of the following polymers, for which one repeating unit is shown: (a) polyvinylchloride (PVC), $\text{-}\!\left(CHClCH_2\right)\!\text{-}_n$; (b) Kel-F, $\text{-}\!\left(CFClCF_2\right)\!\text{-}_n$.

11.74 Write the structural formulas of the monomers of the following polymers, for which one repeating unit is shown: (a) a polymer used to make carpets, $\text{-}\!\left(OC(CH_3)_2{-}CO\right)\!\text{-}_n$; (b) $\text{-}\!\left(CH(CH_3)CH_2\right)\!\text{-}_n$; (c) the polypeptide $\text{-}\!\left(NHCH_2CO\right)\!\text{-}_n$

11.75 Write the structural formula of two units of the polymer formed from (a) the reaction of oxalic acid (ethanedioic acid, HOOCCOOH) with 1,4-diaminobutane, $H_2NCH_2CH_2CH_2CH_2NH_2$; (b) the polymerization of the amino acid alanine (2-aminopropanoic acid).

11.76 Write the structural formula of two units of the polymer formed from (a) the reaction of terephthalic acid (see below) with 1,2-diaminoethane, $H_2NCH_2CH_2NH_2$; (b) the polymerization of 4-hydroxybenzoic acid (see below).

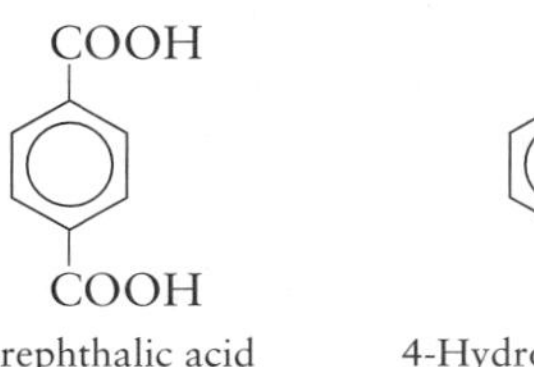
Terephthalic acid

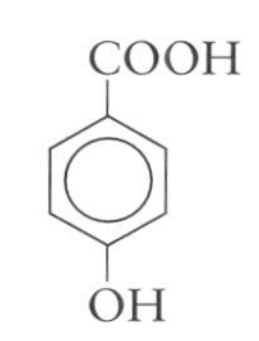
4-Hydroxybenzoic acid

11.77 Distinguish between an isotactic polymer, a syndiotactic polymer, and an atactic polymer.

11.78 Explain why a syndiotactic polymer is stronger than an atactic polymer.

11.79 Identify the type of copolymer formed by monomers A and B: —BBBBAA—.

11.80 Identify the type of copolymer formed by monomers A and B: —AABABBAA—.

11.81 How does average molar mass affect the following polymer characteristics: (a) softening point; (b) viscosity; (c) strength?

11.82 How does the polarity of the functional groups in a polymer affect the following characteristics of the polymer: (a) softening point; (b) viscosity; (c) strength?

11.83 Describe how the linearity of the polymer chain affects polymer strength.

11.84 Describe how cross-linking affects the elasticity and rigidity of a polymer.

Edible Polymers

11.85 (a) Draw the structure of the peptide bond that links the amino acids in proteins. (b) Identify the functional group formed. (c) Identify the type of polymer formed (addition or condensation).

11.86 (a) Draw the structure of the link between glucose units that creates amylose. (b) Identify the functional group formed. (c) Identify the type of polymer formed (addition or condensation).

11.87 Name the amino acids in Table 11.6 that contain side groups capable of hydrogen bonding. This interaction contributes to the tertiary structures of proteins.

11.88 Name the amino acids in Table 11.6 that contain nonpolar side groups. These groups can contribute to the tertiary structure of a protein by avoiding contact with water.

11.89 Write the structural formula of the peptide formed from the reaction of the acid group of tyrosine with the amine group of glycine.

11.90 Write the structural formula of the peptide formed from the reaction of the acid group of glycine with the amine group of tyrosine.

11.91 Identify the functional groups in the glucose molecule shown in (**43**).

11.92 Identify the chiral carbon atoms in the glucose molecule shown in (**43**).

11.93 Write the complementary nucleic acid sequence that would pair with the following DNA sequences: (a) CATGAGTTA; (b) TGAATTGCA.

11.94 Write the complementary nucleic acid sequence that would pair with the following DNA sequences:
(a) GGATCTCAG; (b) CTAGCCTGT.

SUPPLEMENTARY EXERCISES

11.95 Explain why there is a large number and variety of organic compounds.

11.96 Explain the difference between saturated and unsaturated hydrocarbons.

11.97 Explain the difference in chemical bonding in aromatic and aliphatic hydrocarbons.

11.98 Spiropentane has the structural formula

$$\begin{array}{c} CH_2-CH_2 \\ \diagdown\ \diagup \\ C \\ \diagup\ \diagdown \\ CH_2-CH_2 \end{array}$$

(a) Draw the three-dimensional structure of the molecule.
(b) Identify the type of hybridization at each carbon atom.
(c) Estimate the bond angles.

11.99 In a combustion experiment, a 2.14-g sample of a hydrocarbon formed 3.32 g of water and 6.48 g of carbon dioxide. Deduce its empirical formula and state whether it is likely to be an alkane, an alkene, or an alkyne. Explain your reasoning.

11.100 In a combustion experiment, 3.69 g of a hydrocarbon formed 4.50 g of water and 11.7 g of carbon dioxide. Deduce its empirical formula and state whether it is likely to be an alkane, an alkene, or an alkyne. Explain your reasoning.

11.101 Write the structural formulas of
(a) methylcyclopropane; (b) 2,4,6-trimethyltoluene (toluene is the common name of methylbenzene); (c) 2-phenyl-2-methylpentane; (d) cyclohexene; (e) 2-methyl-2-butene; (f) *m*-bromomethylbenzene

11.102 Write the systematic names of
(a) $CH_3(CH_2)_2CH(CH_2CH_2CH_3)CH_3$
(b) $HC{\equiv}CCl$
(c) $CH_3(CH_2)_2C(CH_2CH_2CH_3){=}CH_2$
(d) $(CH_3)_2C{=}CHCH_3$

11.103 Write a balanced equation for the production of 2,3-dibromobutane from 2-butyne.

11.104 The reaction of glycerol (1,2,3-trihydroxypropane) with stearic acid, $CH_3(CH_2)_{16}COOH$, produces a saturated fat. Write the structural formula of the fat.

11.105 (a) Write the structural formulas of diethyl ether and 1-butanol (note that they are isomers). (b) Whereas the solubility of both compounds in water is about 8 g per 100 mL, the boiling point of 1-butanol is 117°C, but that of diethyl ether is 35°C. Account for these observations.

11.106 Pheromones are commonly called sex attractants, although they have more complex signaling functions. The structure of a pheromone in the queen bee is *trans*-$CH_3CO(CH_2)_5CH{=}CHCOOH$. (a) Write the structural formula of the pheromone. (b) Identify and name the functional groups in the molecule.

11.107 Write the structural formula of the product of the oxidation of 4-hydroxybenzyl alcohol by sodium dichromate in an acidic organic solvent.

11.108 Write the formulas of the principal products of the reaction that occurs when (a) ethylene glycol (1,2-ethanediol) is heated with stearic acid, $CH_3(CH_2)_{16}COOH$ (b) ethanol is heated with oxalic acid HOOCCOOH;
(c) 1-butanol is heated with propanoic acid.

11.109 Two structural isomers can result when hydrogen bromide reacts with 2-pentene. Write their structural formulas.

11.110 Why do polymers not have definite molar masses? How does the fact that polymers have average molar masses affect their melting points?

11.111 Rank these polymers according to increasing value as fibers: polyesters, polyamides, polyalkenes. Explain your reasoning.

11.112 A certain polyester has the structural formula $\text{-}\!(OCH_2{-}C_6H_4{-}CH_2O{-}CO{-}C_6H_4{-}CO)\!\text{-}_n$. Identify the monomers of the polyester.

11.113 (a) Explain the differences among the primary, secondary, tertiary, and quaternary structures of a protein. (b) Identify the forces holding each structure together as covalent bonds or primarily intermolecular forces.

11.114 (a) Describe the structure of one of the four monomers of the DNA molecule. (b) What functional group links the monomers of a single DNA strand together? (c) In the monomer structure you have drawn, identify the functional groups that are involved in holding the two strands of the double helix together.

CHALLENGING EXERCISES

11.115 Concentrated sulfuric acid is a strong dehydrating agent, in the sense that it can remove the elements of water, HOH, from a compound. What principal product results from the reaction of $CH_3(CH_2)_2CH(OH)CH_3$ with concentrated sulfuric acid at 120°C?

11.116 A hydrocarbon contains 90% by mass of carbon and 10% by mass of hydrogen and has molar mass

40 g·mol^{-1}. It decolorizes bromine water, and 0.73 g of the hydrocarbon reacts with 800 mL of hydrogen at 1.0 atm and 273 K in the presence of a nickel catalyst. Write the molecular formula of the hydrocarbon and the structural formulas of three possible isomers.

11.117 Dopamine is a neurotransmitter in the central nervous system, and dopamine deficiency is related to Parkinson's disease. Two systematic names for dopamine are 4-(2-aminoethyl)-1,2-benzenediol and 3,4-dihydroxyphenylethylamine. Write the structural formula of dopamine.

11.118 Suggest an experimental method, using radioisotopes, of showing that in ester formation, the oxygen atom in the eliminated water molecule comes from the carboxyl group rather than from the alcohol. Does the process of amide formation shed any light on the details of the esterification reaction?

11.119 Explain the process of condensation polymerization. How might the polymer obtained from benzene-1,2-dicarboxylic acid and ethylene glycol differ from Dacron?

11.120 The protein beef insulin consists of two chains in which the groups —$(Leu)_2Ile(Cys)_4Arg$— and —$Pro(Phe)_3(Cys)_2Arg$— occur. (a) What could be the nature of the interaction between the chains? (b) The chains can be separated by adding a reducing agent. Suggest a reason.

CHAPTER 12

A great deal of chemistry takes place in aqueous solution, for ions and molecules can move through the solvent with relative freedom, meet and react. The presence of solutes can alter the physical properties of the solvent. One very important solution is the plasma that carries the red blood cells (erythrocytes) through our bodies, some of which are shown in this false-color scanning electron micrograph. If you look carefully, you can see that one of the cells in the picture has lost its shape—and hence its function—perhaps because the concentrations of solutes in the plasma drifted from their proper values.

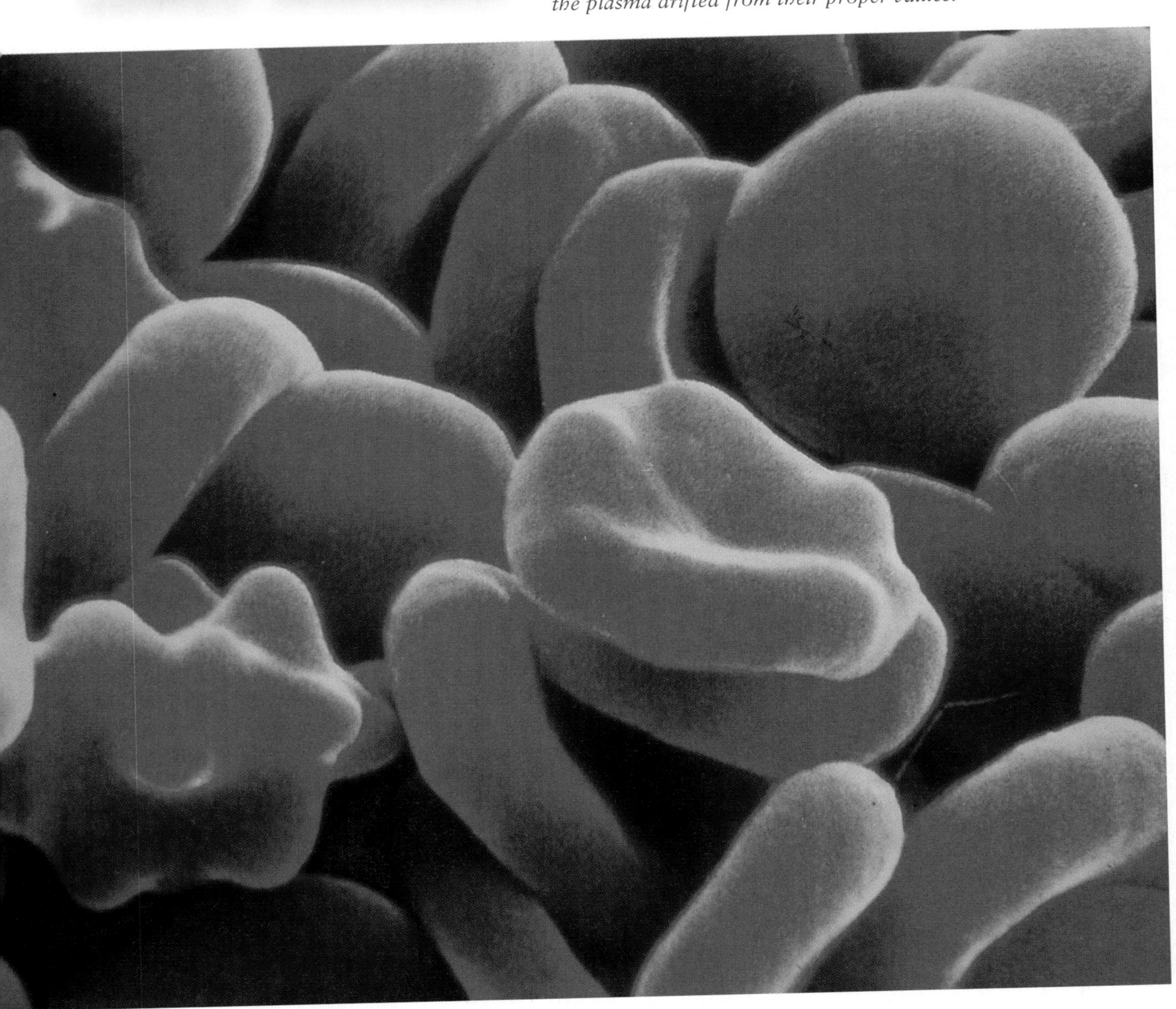

THE PROPERTIES OF SOLUTIONS

Every time we take a breath, oxygen is taken up by our red blood cells and carried through the bloodstream to the cells that need it. Red blood cells and white blood cells are washed through our arteries, capillaries, and veins in plasma—a solution of salts, sugars, and other nutrients and waste products in water.

Other solutions occur on a much vaster scale. The largest liquid solutions we are likely to encounter on Earth are the oceans, which account for 1.4×10^{12} kg of the Earth's surface water. That amounts to nearly 300,000 tons of water for each inhabitant, more than enough to supply all the drinking water needed for a thirsty world. However, seawater cannot be used directly, because of the high concentrations of dissolved salts. The main solutes in seawater are Na^+ and Cl^- ions (Table 12.1), but it also contains huge amounts of other substances and at least a trace of every naturally occurring element. Transforming seawater into potable water is a major challenge to water engineers. A number of techniques have been proposed. One of the most promising is reverse osmosis, which uses one of the properties of solutions that we meet in this chapter.

TABLE 12.1 Principal ions found in seawater

Element	Principal form	Concentration, $g \cdot L^{-1}$
Cl	Cl^-	19.0
Na	Na^+	10.5
Mg	Mg^{2+}	1.35
S	SO_4^{2-}	0.89
Ca	Ca^{2+}	0.40
K	K^+	0.38
Br	Br^-	0.065
C	HCO_3^-, H_2CO_3, CO_2	0.028

SOLUTES AND SOLVENTS

12.1 The molecular nature of dissolving
12.2 Solubility

FACTORS AFFECTING SOLUBILITY

12.3 Solubilities of ionic compounds
12.4 The like-dissolves-like rule
12.5 Pressure and solubility: Henry's law
12.6 Temperature and solubility: Thermal pollution

WHY DOES ANYTHING DISSOLVE?

12.7 The enthalpy of solution
12.8 Individual ion hydration enthalpies
12.9 Solubility and disorder

COLLIGATIVE PROPERTIES

12.10 Measures of concentration
12.11 Vapor-pressure lowering
12.12 Boiling-point elevation and freezing-point depression
12.13 Osmosis

SOLUTES AND SOLVENTS

Although we shall concentrate mainly on liquid solutions, solutions can be solids, liquids, or gases. Solid solutions of metalloids and nonmetals, such as silicon with a tiny amount of phosphorus as solute, are the primary materials of the electronics industry. Gaseous solutions provide special environments for deep-sea divers and food storage facilities.

Chemists carry out many reactions in liquid solutions, partly because molecules and ions of solid substances are more mobile in solution than they would be in the solid state and so can react with one another more quickly. Chemists use both aqueous solutions—solutions in water—and nonaqueous solutions, such as solutions in hydrocarbons, alcohols, and ethers. We shall concentrate largely on aqueous solutions because they are so important, but much of this material also applies to nonaqueous solutions.

FIGURE 12.1 The events that take place at the interface of a solid solute and a solvent (water). The ions at the surface of the solid become hydrated; and as they do, they move off into the solution.

12.1 THE MOLECULAR NATURE OF DISSOLVING

In Section 1.11 a solution was described as a homogeneous mixture, with its components mixed at the molecular level. Let's imagine what happens to the molecules of two substances as they form a solution. Suppose we add a crystal of glucose to some water. At the surface of the crystal, the glucose molecules are in contact with water molecules, and hydrogen bonds begin to form between the two. The surface glucose molecules are pulled into solution by water molecules but are simultaneously held back by other glucose molecules. When the water molecules are successful, the surface glucose molecules break away from the crystal and drift off into the solvent, surrounded by water molecules.

A similar process occurs when an ionic solid dissolves. The oxygen atoms in H_2O molecules have negative partial charges. These partially charged atoms surround the cations and pry them away from the crystal lattice. At the same time, water molecules form hydrogen bonds to

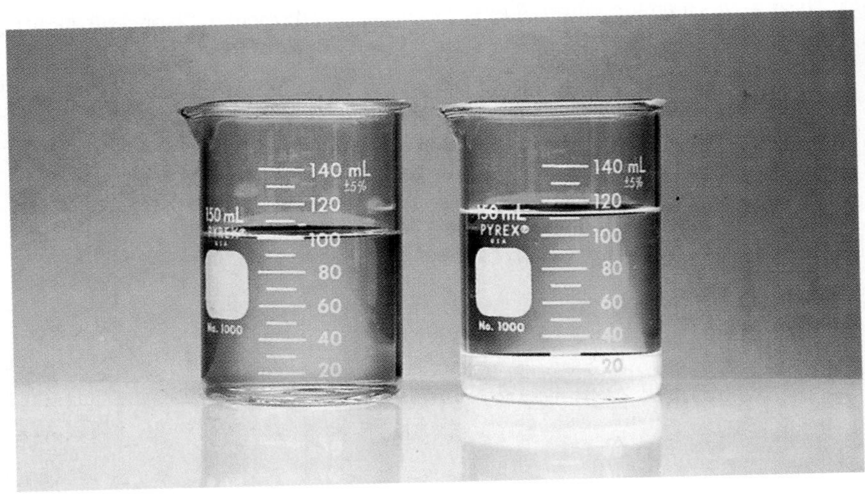

FIGURE 12.2 When a little glucose is stirred into 100 mL of water, it all dissolves (left). However, when a large amount is added, some undissolved glucose remains (right).

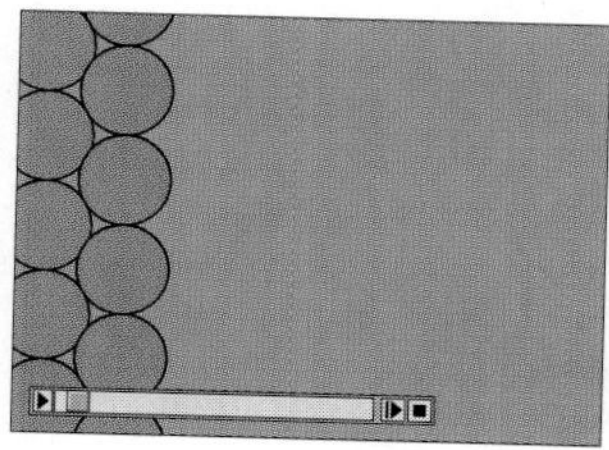

FIGURE 12.3 The solute in a saturated solution is in dynamic equilibrium with the undissolved solute. If we could follow the history of a solute particle (orange spheres), we should sometimes find it in solution and at other times forming the solid. Red, green, and blue lines represent the paths of individual solute particles. The solvent molecules are not shown.

anions at the crystal surface and begin to attract the anions away from their cation partners (Fig. 12.1). Stirring or shaking speeds the process because it brings more free water molecules to the surface of the solid and sweeps the hydrated ions away.

Solids dissolve in water when individual molecules or ions are attracted to water molecules and break away from the solid.

12.2 SOLUBILITY

A given quantity of solvent can dissolve only so much solute. If we add 20 g of glucose to 100 mL of water at room temperature, then all of it dissolves. However, if we add 300 g, some remains undissolved (Fig. 12.2). A solution is said to be **saturated** when the solvent has dissolved all the solute it can and some undissolved solute remains.

The concentration of solid solute in a saturated solution has reached its greatest value and no more can dissolve. In other words, a saturated solution represents the limit of a solute's ability to dissolve in a given quantity of solvent. The **molar solubility** of a substance is its molar concentration in a saturated solution. However, although there is no further rise in concentration, any solid solute present still continues to dissolve, but the rate at which it dissolves exactly matches the rate at which the solute returns to the solid (Fig. 12.3). In other words, *in a saturated solution the dissolved and undissolved solute are in dynamic equilibrium.* For example, in a saturated solution of calcium hydroxide, the dynamic equilibrium is

$$Ca(OH)_2(s) \rightleftharpoons Ca^{2+}(aq) + 2\,OH^-(aq)$$

with Ca^{2+} and OH^- ions leaving and returning to the surface of the solid at matching rates.

The molar solubility of a substance is its molar concentration in a saturated solution; a saturated solution is one in which the dissolved and undissolved solute are in dynamic equilibrium with one another.

FACTORS AFFECTING SOLUBILITY

The solubility of a substance in a given solvent depends on a number of factors. For example, the solubility of any substance depends on the identity of the solvent, the solubility of a gas depends on its partial pressure, and the solubilities of all substances, including gases, vary with temperature.

FIGURE 12.4 This saltpeter (impure sodium nitrate) has survived in the arid region of Chile because there is too little groundwater and rainwater to dissolve it and wash it away.

12.3 SOLUBILITIES OF IONIC COMPOUNDS

We can predict whether a given ionic solid is likely to be soluble in water by using the solubility rules in Table 3.1. In Chapter 3 we used these rules to choose reagents for precipitation reactions. However, the rules also help us understand the behavior of some everyday substances and the properties of minerals. For example, nitrates are rarely found in mineral deposits because they are soluble in groundwater, the water that trickles through the ground and washes away soluble substances. There are exceptions, such as the large deposit of impure sodium nitrate in the arid coastal region of Chile, where groundwater is absent (Fig. 12.4). A part of the reason for the solubility of nitrates is the fact that the NO_3^- ion is quite bulky but has only a single negative charge. As a result, it is easy for water to detach the ions from solid nitrates.

Phosphate ions are even more bulky than nitrate ions, but they are triply charged. As a result, they are more strongly attracted to cations than the nitrate ions are, and it is much harder for water to break them out of solids. The low solubility of most phosphates is an advantage for animals with skeletons, because bone consists largely of calcium phosphate (much of the rest is the protein collagen). However, this insolubility is inconvenient for agriculture, because it means that phosphorus is slow to circulate through the ecosystem. One of chemistry's achievements has been the development of manufacturing processes to turn phosphates into soluble fertilizers. Phosphate fertilizers are obtained from phosphate rocks, principally the apatites—hydroxyapatite, $Ca_5(PO_4)_3OH$, and fluorapatite, $Ca_5(PO_4)_3F$—by treating them with concentrated sulfuric acid:

$$Ca_5(PO_4)_3OH(s) + 5\,H_2SO_4(aq) \longrightarrow 3\,H_3PO_4(aq) + 5\,CaSO_4(s) + H_2O(l)$$

The phosphate rocks themselves were once alive, for they are the crushed and compressed remains of the skeletons of prehistoric animals. Calcium hydrogen phosphate, $CaHPO_4$, is more soluble than calcium phosphate because the HPO_4^{2-} ion has only a double charge. It is used in commercial phosphate fertilizers.

Just as hydrogen phosphates are more soluble than phosphates, so hydrogen carbonates (bicarbonates, HCO_3^-) are more soluble than carbonates. This solubility difference is responsible for the behavior of "hard water," which is water that contains dissolved calcium and magnesium salts. Hard water originates as rainwater, which dissolves carbon dioxide from the air and forms carbonic acid:

$$CO_2(g) + H_2O(l) \longrightarrow H_2CO_3(aq)$$

As the water runs over and through the ground, the carbonic acid reacts with the calcium carbonate of limestone or chalk and forms the more soluble hydrogen carbonate:

$$CaCO_3(s) + H_2CO_3(aq) \longrightarrow Ca^{2+}(aq) + 2\,HCO_3^-(aq)$$

These two reactions are reversed when water containing $Ca(HCO_3)_2$ is heated in a kettle or furnace:

$$Ca^{2+}(aq) + 2\,HCO_3^-(aq) \xrightarrow{\Delta} CaCO_3(s) + H_2O(l) + CO_2(g)$$

The carbon dioxide is driven off and the almost insoluble calcium carbonate is deposited as scale.

Ionic compounds formed from large anions with low charges are often soluble.

FIGURE 12.5 The molecular solid sulfur does not dissolve in water (left), but it does dissolve in carbon disulfide (right), with which the S_8 molecules have favorable London interactions.

12.4 THE LIKE-DISSOLVES-LIKE RULE

The solubility of a substance depends on the choice of solvent as well as on the substance itself. For instance, sodium chloride is very soluble in water but insoluble in benzene. Conversely, grease (a mixture of long-chain hydrocarbonlike molecules) dissolves in benzene but not in water.

Suppose we need to remove some spilled wax. How can we know what solvent to use? A good guide is the rule that *like dissolves like.* A polar liquid, such as water, is generally the best solvent for ionic and polar compounds. Conversely, nonpolar liquids, including benzene and the tetrachloroethene ($Cl_2C{=}CCl_2$) used for dry cleaning, are often better solvents for nonpolar compounds like wax, which are held together by London forces.

The like-dissolves-like rule reflects the fact that attractions between solute molecules and some attractions between solvent molecules must be replaced by solute-solvent attractions when a solution forms. If the new attractions are similar to those replaced, very little energy is required for the solution to form. For example, when the main cohesive forces in a solute are hydrogen bonds, the solute is more likely to dissolve in a hydrogen-bonded solvent than in other solvents. The molecules can slip into solution only if they can replace solute-solute hydrogen bonds with solute-solvent hydrogen bonds. Glucose, for example, has hydrogen bond-forming $-OH$ groups and dissolves readily in water but not in benzene.

If the principal cohesive forces between solute molecules are London forces, then the best solvent is likely to be one held together by the same kind of forces. For example, a good solvent for nonpolar substances is nonpolar carbon disulfide. It is a far better solvent than water for sulfur because solid sulfur is a molecular solid of S_8 molecules held together by London forces (Fig. 12.5). The sulfur molecules slip easily between the nonpolar carbon disulfide molecules. They cannot penetrate into the strongly hydrogen-bonded structure of water because they cannot replace these bonds with interactions of similar strength.

Soaps and detergents are a practical application of the like-dissolves-like rule. Soaps are the sodium salts of long-chain carboxylic acids, including sodium stearate (**1**):

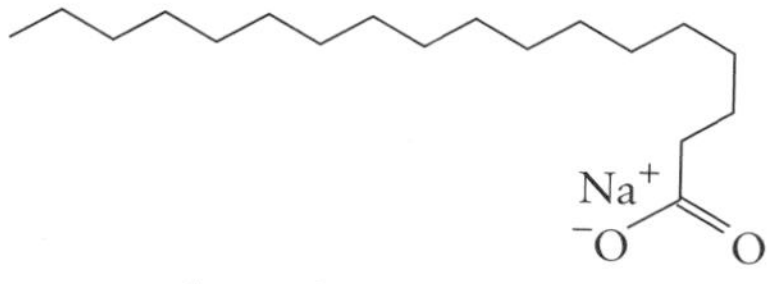

1 Sodium stearate, $NaCH_3(CH_2)_{16}CO_2$

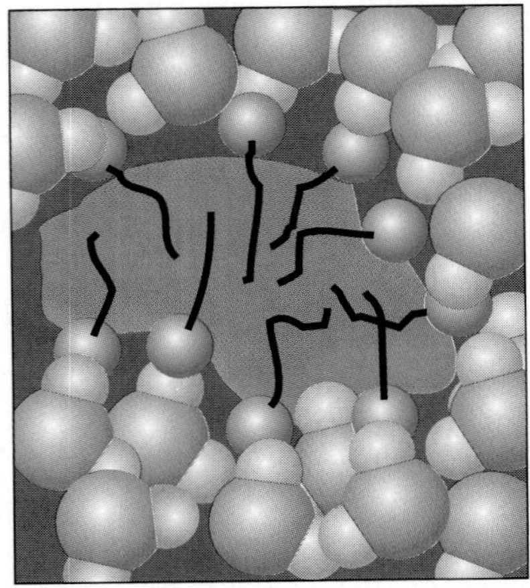

FIGURE 12.6 The hydrocarbon tails of soap or surfactant molecules dissolve in grease, leaving the water-attracting head groups on the surface where they can interact favorably with water. As a result, the whole blob can dissolve in water and be washed away.

The anions of these acids have a polar carboxylate ($—CO_2^-$) group, called the head group, at one end of a nonpolar hydrocarbon chain. The head group is **hydrophilic,** or water-attracting, whereas the nonpolar hydrocarbon tails are **hydrophobic,** or water-repelling. The like-dissolves-like rule suggests that the hydrophilic head group of the anion ought to dissolve in water and the hydrophobic hydrocarbon tails ought to dissolve in grease. This dual ability makes soap very effective at removing grease. The hydrocarbon tails sink into the blob of grease up to the head groups. These hydrophilic groups remain on the surface of the blob (Fig. 12.6) and enable the entire unit to dissolve in water. The grease is carried away in a casing of soap molecules called a **micelle.** The hydrophobic tail of each soap molecule sticks into the grease blob at the center and its hydrophilic head points out, toward the surrounding water molecules.

Early soaps were made by heating sodium hydroxide with beef fat, which contains an ester formed between glycerol and stearic acid. The sodium hydroxide attacks the ester and releases the stearic acid as sodium stearate. A problem with soaps, however, is that they form a scum in hard water. The scum is a precipitate of calcium stearate, which forms because calcium salts are less soluble than sodium salts:

$$Ca^{2+}(aq) + 2\,C_{17}H_{35}CO_2^-(aq) \longrightarrow Ca(C_{17}H_{35}CO_2)_2(s)$$

Case Study 3 describes how hard water is softened in water treatment plants.

One way to avoid the formation of scum is to precipitate the Ca^{2+} ions from the solution before adding the soap. For example, sodium carbonate (washing soda) can be added to water to precipitate Ca^{2+} ions as calcium carbonate:

$$Ca(HCO_3)_2(aq) + Na_2CO_3(aq) \longrightarrow CaCO_3(s) + 2\,NaHCO_3(aq)$$

Surfactants are discussed further in Case Study 10.

Modern commercial detergents are mixtures. Their most important component is a **surfactant,** or surface active agent, which takes the place of the soap. Surfactant molecules are organic compounds that resemble the one shown as (**2**):

SO_3^-

2 A typical surfactant ion

Like the stearate ion, they have a hydrophilic head group and a hydrophobic tail and act similarly. Most commercial detergents also contain additives such as fillers (to generate bulk), foam-reducing agents, and bleach. Often, additives that fluoresce (absorb ultraviolet radiation and then give out visible light) are added to detergents to give the impression of greater cleanliness.

Colloids are suspensions of large particles (20–100 μm in diameter) in a solvent. They differ from solutions in that the suspended particles are large enough to scatter light (see Case Study 12).

A general guide to the suitability of a solvent is the rule that like dissolves like. Soaps and detergents contain surfactant molecules that have both hydrophobic and hydrophilic regions.

12.5 PRESSURE AND SOLUBILITY: HENRY'S LAW

Knowing about gas solubility could save your life. The certification examination for scuba diving, for instance, tests knowledge of the concepts in this section that are essential for deep-sea survival.

We have already seen (in Section 5.16) that the pressure of a gas arises from the impact of its molecules. When the gas is in contact with a liquid, its molecules can burrow into a liquid like meteorites plunging into the ocean. Because the number of impacts increases as its pressure increases, we should expect the solubility of the gas to increase as its pressure increases. This is in fact the case, as shown by Fig. 12.7.

If the gas above the liquid is a mixture (like air), then the solubility of each component depends on its partial pressure. Deep-sea divers breathe compressed air, which contains nitrogen that is normally insoluble in blood. However, at great depths, the divers' bodies are exposed to very high pressures, and the nitrogen becomes more soluble in their blood. The dissolved nitrogen comes out of solution rapidly when the divers return to the surface, and numerous small bubbles form in the bloodstream. These bubbles can burst the capillaries—the narrow vessels that distribute the blood—or block them and starve the tissues of oxygen. This condition is called "the bends" (Fig. 12.8). In serious cases the bends can be fatal. The risk is reduced if helium is used instead of nitrogen to dilute the diver's oxygen supply, because helium is less soluble in plasma than nitrogen. Moreover, because helium atoms are so small, they

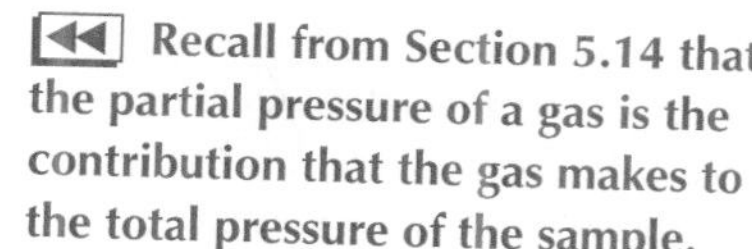

Recall from Section 5.14 that the partial pressure of a gas is the contribution that the gas makes to the total pressure of the sample.

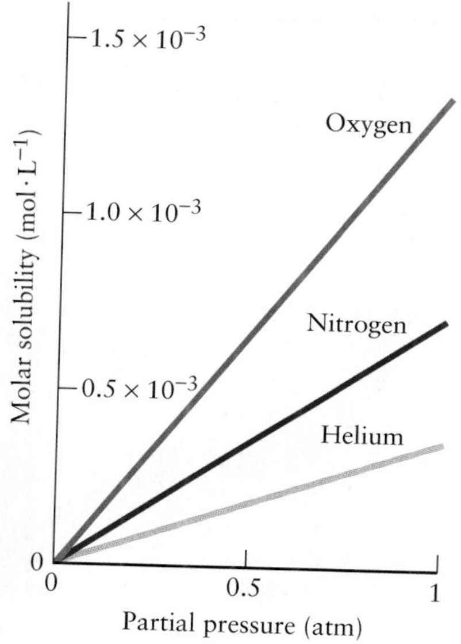

FIGURE 12.7 The variation of the molar solubilities of oxygen, nitrogen, and helium with partial pressure. Note that the solubility of each gas is doubled when its partial pressure is doubled.

CASE STUDY 12 COLLOIDS IN THE CAFETERIA

THE NEXT TIME you are in a cafeteria, look closely at the colorful gelatin dessert. It appears to be a transparent, wobbly solid, yet it consists primarily of water. The chocolate pudding nearby is also mainly water. Is it a solid or a liquid? When you fill your glass with milk, can you determine whether it is a solution or a mixture? Like most foods, these are *colloids,* suspensions of particles ranging from 20 μm to 100 μm in diameter, in a solvent. Colloidal particles are much larger than most molecules but are too small to be seen with a microscope. Colloids thus are often classified between homogeneous solutions and heterogeneous mixtures. The small particles give the colloid a homogeneous appearance but are large enough to scatter light. The light scattering explains why milk is white, not transparent.

A colloid that is a suspension of solids in a liquid is called a *sol,* and a suspension of one liquid in another is called an *emulsion.* For example, skim milk is a suspension of solids, mainly proteins, in water, so it is a sol; mayonnaise has small droplets of water suspended in oil, so it is an emulsion. When we whip cream, milk rich in butterfat, or beat egg whites to form meringues, we make *foams,* suspensions of a gas in a liquid or solid. When we separate the fat from milk and churn it into butter, we create a *solid emulsion,* a suspension of a liquid, in this case, milk, in a solid, butterfat. Gelatin desserts are a type of solid emulsion called a *gel,* which is soft, but holds its shape.

Aqueous colloids can be classified as hydrophilic or hydrophobic. Suspensions of fat in water, such as milk and mayonnaise, are hydrophobic colloids, because fat molecules have little attraction for water molecules. Gels and puddings are examples of hydrophilic colloids. The macromolecules of the proteins in gelatin and the starch in pudding have many hydrophilic groups that attract water. The giant protein molecules in gelatin uncoil in hot water. Their abundant amide groups (see Section 11.10) form hydrogen bonds with water. When the mixture cools, the protein chains link together again, but now they enclose many water molecules within themselves, as well as molecules of sugar, dye, and flavoring agents. The result is an open network of protein chains that hold the water in a flexible solid structure.

The action of starch in the preparation of pudding is similar. Starch contains granules of amylose molecules (see Section 11.18), which have long chains covered with —OH groups. In hot water, the granules absorb water and break open. Water wriggles in and forms hydrogen bonds with the amylose. On cooling, the amylose molecules attempt to refold into their original shapes, but this time they take the water, sugar, and cocoa with them.

Hydrophilic colloids are held together by strong intermolecular attractions to the solvent, but care is needed to

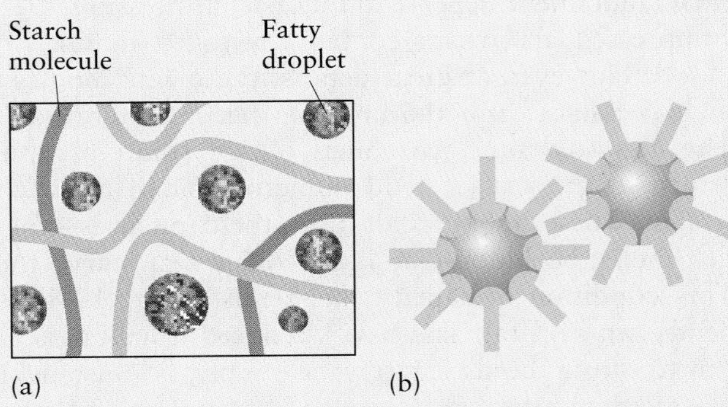

(a) An emulsion can be stabilized by long starch molecules or protein chains. These chains wind among the droplets and prevent them from merging together. (b) An emulsion can also be stabilized by large molecules (yellow) that stick to the droplets (red) and prevent their close approach.

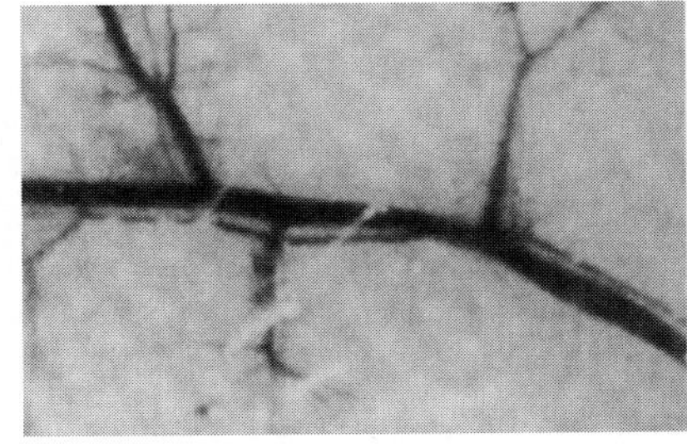
(a)

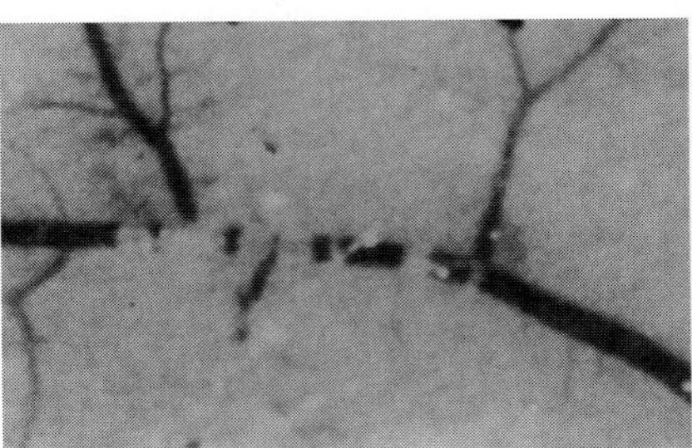
(b)

FIGURE 12.8 Small bubbles of air are responsible for the bends. (a) Normal blood vessels; (b) catastrophic collapse as bubbles escape from solution in the blood plasma.

keep hydrophobic colloids from separating. Until recent years, butterfat rose to the surface of whole milk and had to be skimmed off or mixed in before drinking. Today we can drink homogenized milk, which is prepared by forcing the milk through a very fine sieve. This procedure breaks up the fat globules into very tiny particles that are more easily suspended. The tiny particles are kept from settling out by *Brownian motion,* the motion of small particles resulting from constant bombardment by solvent molecules. The emulsion is further stabilized by the adsorption of ions on the surfaces of the particles. The ions attract a layer of water molecules that prevents the particles from adhering to each other.

Mayonnaise is an emulsion of water in oil that is held together with the aid of the cholesterol and lecithin in egg yolk. These large molecules have very polar groups at one end. They form micelle-like clusters (see Case Study 10) around the oil globules, with their polar ends toward the water and their nonpolar groups in the oil. Other proteins from egg yolk separate the clusters. This separation helps to keep the oil suspended for long periods and over a wide temperature range.

The importance of shape to the function and taste of proteins is evident in the kitchen. Meringues are possible because the beating of egg whites partially denatures the proteins, allowing the strands to unwind and trap air while tangling with other strands. Heating proteins also denatures them. When we cook egg white, its partially denatured proteins form a firm gel.

Most cafeterias offer various frozen emulsions, some of which can legally be called ice cream. In the United States, ice cream is required to contain at least 10% butterfat and is limited to no more than 0.5% stabilizers. These stabilizers are usually natural carbohydrates like guar gum and carrageenan, which have large molecules that slow the crystallization of ice. The volume of ice cream can be expanded up to twice its original volume by beating in air. This expansion partly explains price differences among ice cream brands. Expensive ice creams have the least amount of added air, so a pint of a premium brand may have a mass of up to twice that of ordinary brands.

Ice cream is a solid emulsion of fat in an aqueous solution of sugar, proteins, and flavor. This premium ice cream is 16% butterfat.

QUESTIONS

1. Classify these foods by type of colloid: (a) yogurt; (b) marshmallows; (c) orange juice.

2. Which type of emulsion do you expect will take more energy to form, a hydrophobic or a hydrophilic emulsion? Explain your answer.

3. What kinds of intermolecular forces are primarily responsible for the formation of the colloid in (a) a gelatin dessert, (b) a starch-based pudding, (c) meringue?

4. Explain how the addition of cream of tartar, which is an acid, might help to stabilize meringues.

can pass through cell walls without damaging them more readily than nitrogen molecules can.

The production of soft drinks and champagne is a familiar illustration of the pressure dependence of solubility. In each case, carbon dioxide is dissolved in a liquid under pressure. When a bottle is opened, the partial pressure of carbon dioxide above the solution is suddenly reduced and the gas **effervesces,** or bubbles out of solution. For champagne, the gas blows out the cork with a festive pop.

In champagne, the CO_2 results from fermentation in the sealed bottle. In soft drinks, the liquid is carbonated by dissolving CO_2 in it under pressure.

The straight lines in Fig. 12.7 show that the solubility of a gas is directly proportional to its partial pressure, *P.* This observation was first

TABLE 12.2 Henry's constants for gases in water at 20°C

Gas	k_H, mol·L^{-1}·atm^{-1}
air	7.9×10^{-4}
argon	1.5×10^{-3}
carbon dioxide	2.3×10^{-2}
helium	3.7×10^{-4}
hydrogen	8.5×10^{-4}
neon	5.0×10^{-4}
nitrogen	7.0×10^{-4}
oxygen	1.3×10^{-3}

made in 1801 by the English chemist William Henry and is now known as **Henry's law.** Molar solubility is denoted by S, and Henry's law is normally written

$$S = k_H \times P$$

where k_H, which is called **Henry's constant,** depends on the gas, the solvent, and the temperature (Table 12.2). The law implies that, at constant temperature, doubling the partial pressure of a gas doubles its solubility.

The solubility of a gas is proportional to its partial pressure, because an increase in pressure corresponds to an increase in the rate at which gas molecules strike the surface of the solvent.

EXAMPLE 12.1 Using Henry's law to predict solubilities

The lowest concentration of oxygen that can support aquatic life is about 1.3×10^{-4} mol·L^{-1}. The partial pressure of oxygen is 0.21 atm at sea level. Is the solubility of oxygen adequate to maintain aquatic life?

STRATEGY Because the solubility of a gas is proportional to its partial pressure, use Henry's law and the information from Table 12.2.

SOLUTION The molar solubility of oxygen is

$$S = (1.3 \times 10^{-3}\ \text{mol·L}^{-1}\text{·atm}^{-1}) \times (0.21\ \text{atm}) = 2.7 \times 10^{-4}\ \text{mol·L}^{-1}$$

This molar concentration corresponds to about 8.6 mg of oxygen per liter of water (8.6 mg·L^{-1}), which is more than adequate to sustain life.

SELF-TEST 12.1A At the elevation of Bear Lake in Rocky Mountain National Park, 2900 m, the partial pressure of oxygen is 0.14 atm. What is the solubility of oxygen in Bear Lake at 20°C?

[*Answer:* 1.8×10^{-4} mol·L^{-1}]

SELF-TEST 12.1B Use the information in Table 12.2 to calculate the number of moles of CO_2 that will dissolve in 900 g of water at 20°C if the partial pressure of CO_2 is 1.00 atm. Assume a total of 900 mL.

12.6 TEMPERATURE AND SOLUBILITY: THERMAL POLLUTION

Most substances dissolve more quickly at higher temperatures. However, that does not necessarily mean that they are more soluble at higher temperatures. In a number of cases the solubility turns out to be lower at higher temperatures. The effect of temperature on the rate of a process must always be distinguished from its effect on the final outcome.

Most gases become less soluble as the temperature is raised. The lower solubility of gases in warm water is responsible for the tiny bubbles that appear when cool water from the faucet is left to stand in a warm room. The bubbles consist of air that dissolved when the water was cooler; it comes out of solution as the temperature rises.

Thermal pollution is the damage caused to the environment by the waste heat of an industrial process. Oxygen is less soluble in the warm water discharged from power stations. Moreover, the lower density of warm water causes it to rise to the surface of rivers, where it acts like a suffocating blanket that prevents absorption of oxygen and its penetration to the cooler water below. As a result, marine life (principally, fish) may die.

Most ionic and molecular solids are more soluble in hot water than in cold (Fig. 12.9). We make use of this characteristic in the laboratory to dissolve a substance and to grow crystals by letting a saturated solution cool slowly. However, a few solids, such as lithium sulfate, are less soluble at high temperatures than at low. A smaller number of compounds show a mixed behavior. For example, the solubility of sodium sulfate increases up to 32°C but then decreases as the temperature is raised further.

The variation of solubility with temperature is difficult to predict, except in the case of gases, where raising the temperature nearly always lowers the solubility.

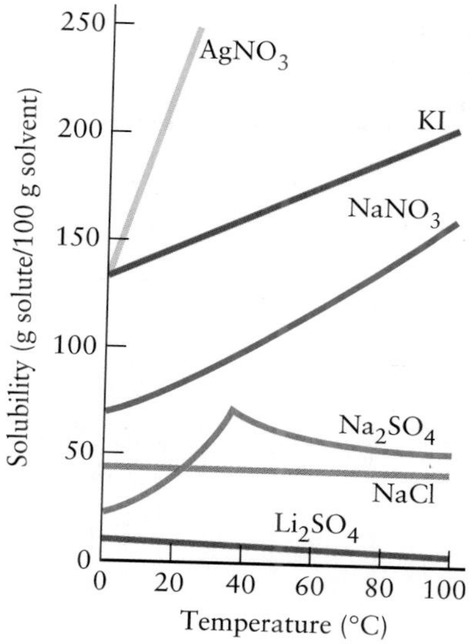

FIGURE 12.9 The variation with temperature of the solubilities of six substances in water.

WHY DOES ANYTHING DISSOLVE?

We have seen that like dissolves like because good solvents mimic the surroundings of the solute molecules: partial charges on polar solvent molecules play the role of ionic charges, hydrogen bonds replace hydrogen bonds, and London forces exerted by a solvent mimic those of the solute molecules themselves. We can discover the relative strengths of these forces by measuring how much heat is released or absorbed when a solute dissolves.

12.7 THE ENTHALPY OF SOLUTION

The heat released or absorbed per mole when a substance dissolves at constant pressure to form a very dilute solution is called the **enthalpy of solution.** It can be measured by calorimetry (Section 6.3). The experimental values in Table 12.3 show that some solids dissolve exothermically

TABLE 12.3 Enthalpies of solution, ΔH_{sol}, at 25°C for very dilute aqueous solutions, in kilojoules per mole*

	Anion							
Cation	**fluoride**	**chloride**	**bromide**	**iodide**	**hydroxide**	**carbonate**	**nitrate**	**sulfate**
lithium	+4.9	−37.0	−48.8	−63.3	−23.6	−18.2	−2.7	−29.8
sodium	+1.9	+3.9	−0.6	−7.5	−44.5	−26.7	+20.4	−2.4
potassium	−17.7	+17.2	+19.9	+20.3	−57.1	−30.9	+34.9	+23.8
ammonium	−1.2	+14.8	+16.0	+13.7	—	—	+25.7	+6.6
silver	−22.5	+65.5	+84.4	+112.2	—	+41.8	+22.6	+17.8
magnesium	−12.6	−160.0	−185.6	−213.2	+2.3	−25.3	−90.9	−91.2
calcium	+11.5	−81.3	−103.1	−119.7	−16.7	−13.1	−19.2	−18.0
aluminum	−27	−329	−368	−385	—	—	—	−350

*The value for silver iodide, for example, is the entry found where the row labeled "silver" intersects the column labeled "iodide." A positive value of ΔH_{sol} indicates an endothermic process.

and others dissolve endothermically. Lithium chloride has a negative enthalpy of solution, indicating an exothermic process: when it is added to water, the solution becomes noticeably warm (Fig. 12.10). In contrast, when ammonium nitrate is added to water, the temperature falls, indicating that ammonium nitrate dissolves endothermically.

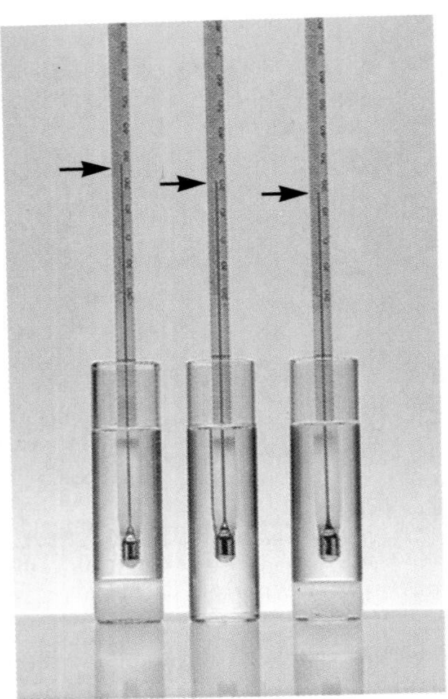

FIGURE 12.10 The exothermic dissolution of lithium chloride (left) is shown by the rise in temperature above that of the original water (center); in contrast, ammonium nitrate (right) dissolves endothermically.

To understand the energy changes involved, we can think of dissolving as a two-step process. In the first step we imagine the solid breaking up to form a gas of molecules or ions. In the second step we imagine the separated molecules or ions entering the solvent (Fig. 12.11).

Suppose we have an ionic solid and some water. In the first step, we imagine the solute ions separating to form a gas of ions. As noted in Section 8.2, the change of enthalpy for ionic solids in this endothermic step is the lattice enthalpy, ΔH_L, of the solid (Table 8.1). For sodium chloride, for instance, it is the molar enthalpy change for the process

$$NaCl(s) \longrightarrow Na^+(g) + Cl^-(g)$$

Compounds formed from small, highly charged ions (such as Mg^{2+} and O^{2-}) have high lattice enthalpies, a property indicating that a lot of energy is needed to break up the solid. Because it is hard for solvent molecules to pull apart such strongly attached ions, these compounds are usually insoluble.

Heat is released as the ions become solvated by the solvent molecules: when the solvent is water, this step is called **hydration.** The molar enthalpy of the step

$$Na^+(g) + Cl^-(g) \longrightarrow Na^+(aq) + Cl^-(aq)$$

is called the **enthalpy of hydration,** ΔH_{hyd}, of the compound, and values are given in Table 12.4. The water molecules form hydrogen bonds to anions, and the negative partial charge on the oxygen atoms is attracted to cations. The latter interaction is called an **ion-dipole interaction.** Hydration is always exothermic for ionic compounds. It is also exothermic for molecules that can form hydrogen bonds with water; examples include sucrose, glucose, acetone, and ethanol.

The enthalpy of solution is the sum of the lattice enthalpy and the enthalpy of hydration of the compound. For sodium chloride, we can write

$$NaCl(s) \longrightarrow Na^+(g) + Cl^-(g) \qquad \Delta H = +787 \text{ kJ}$$

$$Na^+(g) + Cl^-(g) \longrightarrow Na^+(aq) + Cl^-(aq) \qquad \Delta H = -784 \text{ kJ}$$

The sum of these two thermochemical equations is

$$NaCl(s) \longrightarrow Na^+(aq) + Cl^-(aq) \qquad \Delta H = +3 \text{ kJ}$$

Because the enthalpy of solution is positive, there is a net inflow of energy as heat. Sodium chloride therefore dissolves endothermically, but only to the extent of $+3 \text{ kJ·mol}^{-1}$. As this example shows, the overall change in enthalpy depends on a very delicate balance between the lattice enthalpy and the enthalpy of hydration. If sodium chloride had only a 0.5% smaller lattice enthalpy ($+783$ instead of $+787 \text{ kJ·mol}^{-1}$), it would dissolve exothermically instead of endothermically.

Because of experimental uncertainties, the calculated value of $+3 \text{ kJ·mol}^{-1}$ varies slightly from the measured value, $+3.9 \text{ kJ·mol}^{-1}$, given in Table 12.3.

Predictions of enthalpies of solution based on tabulated values of lattice enthalpies and enthalpies of hydration are not very reliable. A small

TABLE 12.4 Enthalpies of hydration, ΔH_{hyd}, at 25°C, of some halides, in kilojoules per mole*

Cation	Anion			
	F^-	Cl^-	Br^-	I^-
H^+	−1613	−1470	−1439	−1426
Li^+	−1041	−898	−867	−854
Na^+	−927	−784	−753	−740
K^+	−844	−701	−670	−657
Ag^+	−993	−850	−819	−806
Ca^{2+}	—	−2337	—	—

*The entry where the row labeled Na^+ intersects the column labeled Cl^-, for instance, is the enthalpy change, −784 kJ·mol^{-1}, for the process $Na^+(g) + Cl^-(g) \rightarrow Na^+(aq) + Cl^-(aq)$; the values here apply only when the resulting solution is very dilute.

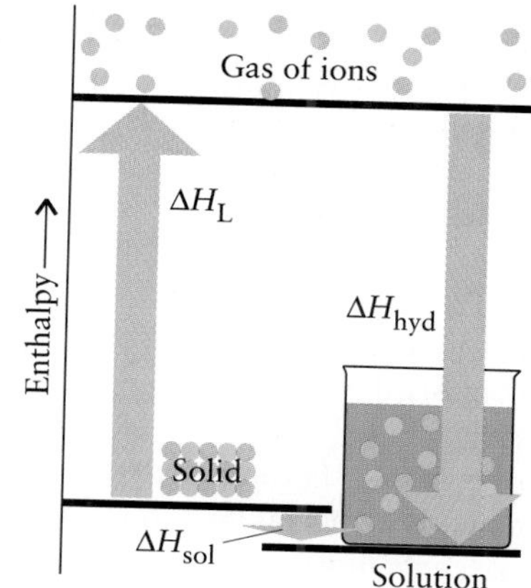

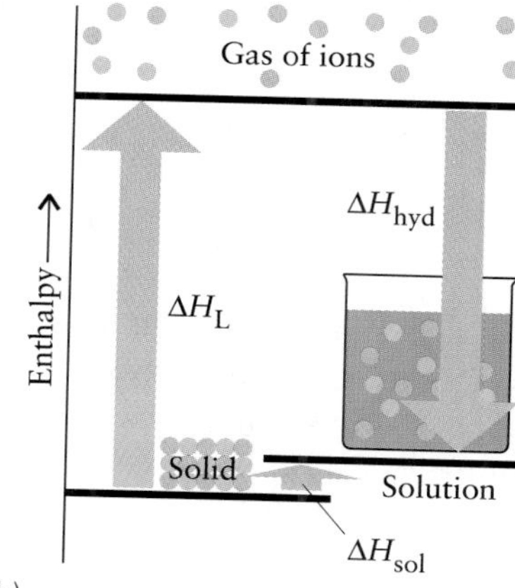

FIGURE 12.11 The enthalpy of solution, ΔH_{sol}, is the sum of the enthalpy changes required to separate the molecules or ions of the solute (the lattice enthalpy, ΔH_L) and the enthalpy change accompanying their hydration, ΔH_{hyd}. The outcome is finely balanced: (a) in some cases it is exothermic, (b) in others it is endothermic. For gaseous solutes, the lattice enthalpy is 0, because the molecules are already widely separated.

inaccuracy in either quantity can lead to an answer with the wrong sign. It is like trying to calculate the mass of the captain of a ship by measuring the mass of the ship before and after the captain comes aboard. However, it is reasonably safe to expect a positive enthalpy of solution if the lattice enthalpy is very high. Dissolving is then endothermic. Similarly, we can expect a negative enthalpy of solution if the lattice enthalpy is low and the enthalpy of hydration is large. Dissolving is then exothermic.

Enthalpies of solution in dilute solution can be expressed as the sum of the lattice enthalpy and the enthalpy of hydration of the compound.

12.8 INDIVIDUAL ION HYDRATION ENTHALPIES

We can break down the process of dissolving still further. The enthalpies of hydration in Table 12.4 are the enthalpy changes that occur when widely separated cations and anions in a gas form a very dilute solution. It would be useful to know the values for the cations and anions separately, for then we could predict the enthalpy of hydration of NaCl(g), for instance, by adding together the values for Na^+(g) and Cl^-(g). Unfortunately, we cannot measure enthalpies of hydration of cations without anions being present too. However, the value for the hydrogen ion has been measured by a special technique and it has the following value:

$$H^+(g) \longrightarrow H^+(aq) \qquad \Delta H = -1130 \text{ kJ}$$

That is, when 1 mol H^+(g) ions dissolves in a very large sample of water, 1130 kJ of energy is released as heat. This large value is about the same as the energy obtained by burning about 50 mL of gasoline. We can combine this known value with the experimentally determined values in Table 12.4 to obtain the individual hydration enthalpies for other ions. For example, the enthalpy of hydration of HCl (as H^+ and Cl^- ions) is −1470 kJ·mol^{-1}. The ion hydration enthalpy of Cl^- is therefore −1470 − (−1130) = −340 kJ·mol^{-1}. Values for a selection of ions are given in Table 12.5: all the values are negative.

CHEMISTRY AT WORK

A pharmacologist has just added a solution of a new drug to a container of living tissue that she has cultured. Now she is adjusting the water jacket that controls the temperature of the tissue. In order to keep the tissue alive, she must carefully control temperature and the concentrations of nutrients.

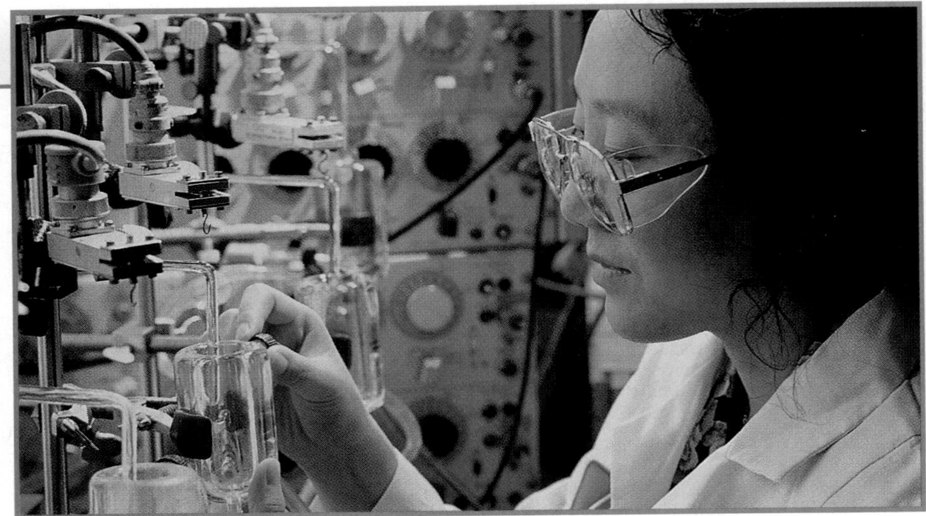

The ion hydration enthalpies in Table 12.5 show patterns:

1. The transfer of an ion from gas to water is more exothermic if the ion is highly charged.

Compare, for instance, Li^+ (-558 kJ·mol^{-1}), Be^{2+} (-1435 kJ·mol^{-1}), and Al^{3+} (-2537 kJ·mol^{-1}). The increasingly large negative values reflect the much greater strength of the interaction between a highly charged ion and the polar water molecules.

2. The transfer of an ion from gas to water is more exothermic if its radius is small.

Let's consider Li^+ (-558 kJ·mol^{-1}), Na^+ (-444 kJ·mol^{-1}), and K^+ (-361 kJ·mol^{-1}). The reason for this pattern is that a water molecule can approach the center of charge of a small ion more closely and hence interact with it more strongly. There are exceptions: Ag^+ is bigger than Na^+, but its hydration is more exothermic. The explanation of this anomaly may be that the Ag^+ ion can form covalent bonds with the hydrating water molecules.

Small, highly charged ions have high enthalpies of hydration.

TABLE 12.5 Individual ion hydration enthalpies

Ion	Hydration enthalpy, kJ·mol^{-1}
CATIONS	
H^+	-1130
Li^+	-558
Na^+	-444
K^+	-361
Ag^+	-510
Be^{2+}	-1435
Mg^{2+}	-2003
Ca^{2+}	-1657
Sr^{2+}	-1524
Al^{3+}	-2537
ANIONS	
F^-	-483
Cl^-	-340
Br^-	-309
I^-	-296

SELF-TEST 12.2A Why does Cl^- have a more negative hydration enthalpy than I^-?

[*Answer:* Cl^- has the smaller radius.]

SELF-TEST 12.2B Would you expect Rb^+ to have a more negative or less negative hydration enthalpy than Sr^{2+}? Explain your answer.

12.9 SOLUBILITY AND DISORDER

A negative enthalpy of solution is a sign that the energy released during hydration is more than enough to compensate for the energy required to break apart the ions or molecules of the solute. We can expect such sub-

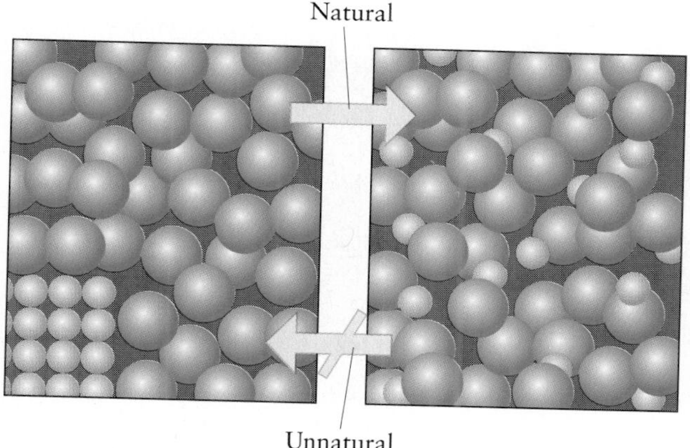

FIGURE 12.12 Matter has a natural tendency to disperse in a disorderly way. The reverse process is unnatural and has no tendency to occur. The dispersal of matter accounts for the expansion of gases and the spreading of matter through the environment, and contributes to the tendency of solids to dissolve. We shall see later that it also contributes to the tendency of chemical reactions to occur.

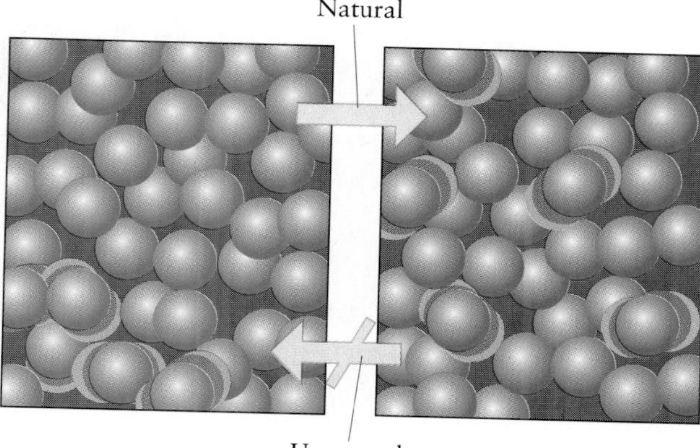

FIGURE 12.13 Energy has a natural tendency to disperse in a disorderly way. The reverse process is unnatural, and has no tendency to occur. The molecules with the greater energy are shown in the drawing by suggesting a shaking motion; think of the vigorous shaking motion as being passed on from molecule to molecule as they bump into one another.

stances to be soluble because the ions or molecules have a lower energy in the solution than in the solid. However, not all substances with a negative enthalpy of solution are soluble, so there must be another factor at work. Moreover, some very soluble substances like ammonium nitrate have positive enthalpies of solution in water, so the ions have a *higher* energy in solution than in the solid. How can we explain endothermic dissolving? The answer lies in a very simple idea:

Energy and matter tend to disperse.

By disperse, we mean spread out in a disorderly way (Figs. 12.12 and 12.13). The tendency of energy and matter to disperse is a fundamental scientific principle. As we shall see again and again, it explains why *any* change occurs.

Let's apply this idea to dissolving. The disorder of the solution normally increases when a substance dissolves, because the ions or molecules become more disordered in the solution. Moreover, if the dissolving is exothermic, energy spreads into the surroundings, stirring it up into disorder (Fig. 12.14). Therefore, exothermic dissolving is natural, because disorder increases in both the system and the surroundings when it takes place.

In a few cases the presence of a solute causes the solvent molecules to take up a more organized arrangement than they had in the pure liquid (Fig. 12.15). Now the disorder of the solution is *lower* than that in the pure liquid. Even though energy is released into the surroundings, the increase in the disorder of the surroundings might not be enough to overcome the decrease in disorder of the solution (Fig. 12.16). As a result, the solution does not form. This situation is illustrated by the inability of some hydrocarbons to dissolve in water, even though they have weakly negative enthalpies of solution.

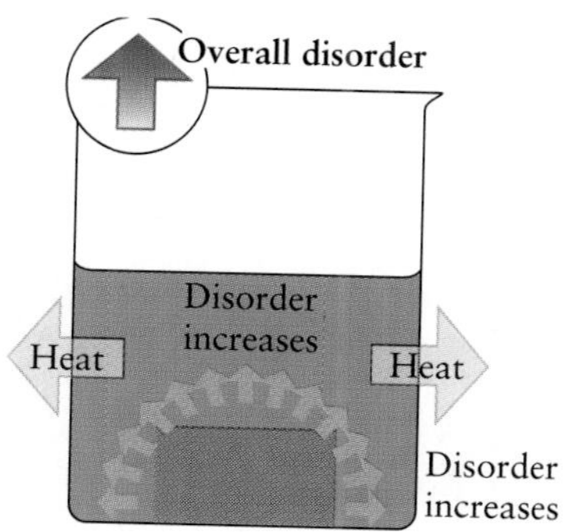

FIGURE 12.14 Many substances dissolve to give a system with more disorder than was present initially. If their dissolving is also exothermic, we can be confident that the dissolving is a natural process.

FIGURE 12.15 When a nonpolar compound (the yellow sphere) dissolves in water, the water molecules may become organized around it. As a result, the disorder of the solvent is reduced when the solution forms.

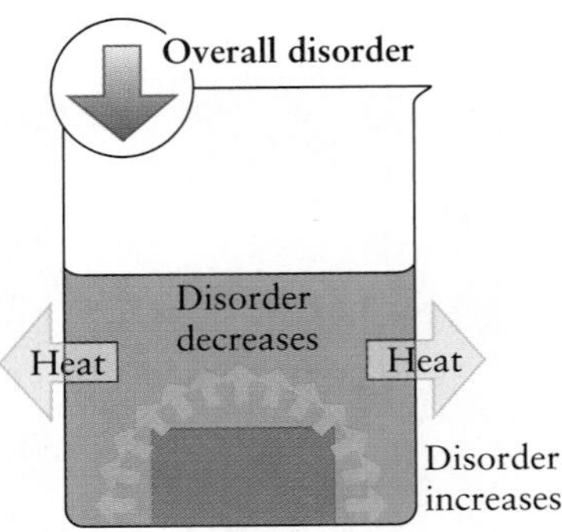

FIGURE 12.16 Even though dissolving may be weakly exothermic, and hence increase the disorder of the surroundings, the reduced disorder of the solvent illustrated in Fig. 12.15 might cancel the increase. As a result, dissolution does not take place.

The two signposts to increased disorder may also point in opposite directions when a substance dissolves endothermically. The solution typically becomes more disordered as species dissolve. However, as energy flows into the solution from the surroundings, their disorder decreases (Fig. 12.17). Whether or not dissolving can take place now depends on the balance of these two effects. If the dissolution is only slightly endothermic (as for sodium chloride or even ammonium nitrate), the disorder of the solution increases more than the disorder of the surroundings decreases. In contrast, suppose we start with magnesium oxide, which has a very high lattice enthalpy. Now the dissolution is highly endothermic, and a great deal of energy must be absorbed from the surroundings. This removal of energy results in a large decrease in the disorder of the surroundings. In fact, the change in disorder is so great that the small increase in disorder of the solution cannot compensate for it. Hence, a substance with a strongly endothermic enthalpy of solution is likely to be insoluble, because dissolving would correspond to an overall decrease in disorder.

▶▶ Chapter 16 identifies disorder with the property called *entropy* and treats these changes in more depth.

Dissolving depends on the balance between the dispersal of matter in the solution and the dispersal of energy in the surroundings.

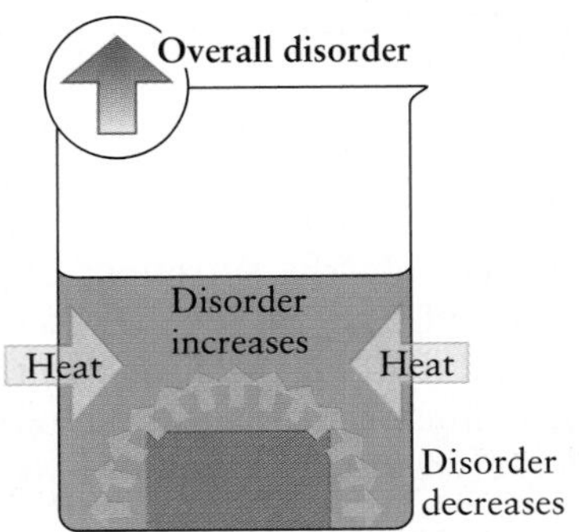

FIGURE 12.17 If dissolution is endothermic, it will occur only if the disorder caused by the solute dispersing is great enough to produce an overall increase in disorder.

COLLIGATIVE PROPERTIES

Now let's see how the presence of a solute affects the physical properties of the solvent. It is found experimentally that four physical properties are affected in the same way by solutes, regardless of the identity of the solute. Properties that depend on the number of solute molecules and not on their chemical identity are called **colligative properties.** The four colligative properties we consider are the lowering of the vapor pressure of the solvent, the raising of its boiling point, the lowering of its freezing point, and the tendency of a solvent to flow through a membrane into a solution. The last property is of enormous importance, because it contributes to the flow of nutrients through biological cell walls and the movement of solutions through plants.

Only the concentration of the solute, not its identity, is important for predicting the magnitude of these colligative properties. For example, the water in an aqueous glucose solution in which 1 out of every 100 molecules is a glucose molecule has the same vapor pressure, boiling point, and freezing point as the water in an aqueous sucrose solution in which 1 out of every 100 molecules is a sucrose molecule.

12.10 MEASURES OF CONCENTRATION

To predict the magnitudes of colligative properties, we need a measure of the relative numbers of solute and solvent molecules. For convenience, Table 12.6 summarizes the measures of concentration that are commonly encountered. There are two measures useful for our purposes: the mole fraction and the molality.

Be careful to distinguish molality from molarity.

The **mole fraction,** x, is the ratio of the number of moles of a species to the total number of moles of all the species present in a mixture. For solute molecules in a nonelectrolyte solution,

$$x_{\text{solute}} = \frac{\text{moles of solute molecules}}{\text{total moles of solute and solvent molecules}}$$

TABLE 12.6 Concentration units

Measure	Units	Note
molarity (molar concentration)	moles per liter, $\text{mol}\cdot\text{L}^{-1}$, M	moles of solute per liter of solution
molality	moles per kilogram, $\text{mol}\cdot\text{kg}^{-1}$, m	moles of solute per kilogram of solvent
volume percentage	%	volume of component expressed as a percentage of the total volume, both measured prior to mixing
mass percentage	%	mass of component expressed as a percentage of the total mass
mole fraction (x)	—	number of moles of component expressed as a fraction of the total number of moles
parts per million by volume	ppm (by volume)	volume in milliliters per kiloliter of sample
parts per million by mass	ppm (by mass)	mass in milligrams per kilogram of sample

$x_{RED} = 0.1$

$x_{RED} = 0.5$

$x_{RED} = 0.9$

FIGURE 12.18 A mole fraction, *x*, tells us the fraction of molecules of a particular kind in a mixture of two or more kinds of molecules.

For a two-component mixture,

$$x_{solute} = \frac{n_{solute}}{n_{solvent} + n_{solute}} \quad (1)$$

where n_X is the number of moles of species X. A mole fraction lies in the range $x_{solute} = 0$, corresponding to no solute, to $x_{solute} = 1$, corresponding to no solvent (Fig. 12.18).

The mole fractions of a solvent and all the solute species always sum to 1:

$$x_{solute} + x_{solvent} = 1$$

This relation follows directly from the definition of x. For a nonelectrolyte solution,

$$\begin{aligned} x_{solute} + x_{solvent} &= \frac{n_{solute}}{n_{solvent} + n_{solute}} + \frac{n_{solvent}}{n_{solvent} + n_{solute}} \\ &= \frac{n_{solute} + n_{solvent}}{n_{solvent} + n_{solute}} = 1 \end{aligned}$$

Because a colligative property depends only on the relative number of solute particles present, cations and anions in an electrolyte solution contribute separately and equally to the total. Therefore, if 1 mol NaCl is added to a solvent and the solution is so dilute that the NaCl is fully dissociated, the solution contains 2 mol of ions (1 mol Na^+ ions and 1 mol Cl^- ions). If 1 mol $CaCl_2$ is added, the solution contains 3 mol of ions (1 mol Ca^{2+} ions and 2 mol Cl^- ions). That is, for a dilute electrolyte solution, we treat the cations and anions as individual species:

$$x_{cations} = \frac{n_{cations}}{n_{cations} + n_{anions} + n_{solvent}}$$

$$x_{anions} = \frac{n_{anions}}{n_{cations} + n_{anions} + n_{solvent}}$$

Again, the mole fractions add up to 1:

$$x_{cations} + x_{anions} + x_{solvent} = 1$$

EXAMPLE 12.2 Calculating the mole fractions of species in a solution

Calculate mole fractions of the nonelectrolyte sucrose, $C_{12}H_{22}O_{11}$, and water in a solution prepared by dissolving 5.00 g of sucrose in 100.0 g of water.

STRATEGY To find the numbers of moles of solute and solvent molecules in the solution, convert the masses of solute and solvent to moles by using molar masses. The total number of moles is the sum of the moles of solute and solvent. Use Eq. 1 to calculate the mole fractions as ratios of the moles of the species divided by the total number of moles of species present.

SOLUTION The molar masses of sucrose and water are 342.3 $g{\cdot}mol^{-1}$ and 18.02 $g{\cdot}mol^{-1}$, respectively. Therefore, the numbers of moles of each species are

$$\text{Moles of } C_{12}H_{22}O_{11} = \frac{5.00\ g}{342.3\ g{\cdot}mol^{-1}} = 0.0146\ mol$$

$$\text{Moles of } H_2O = \frac{100.0 \text{ g}}{18.02 \text{ g·mol}^{-1}} = 5.549 \text{ mol}$$

The total number of moles of molecules in the solution is the sum, 5.564 mol. The two mole fractions are therefore

$$x_{\text{sucrose}} = \frac{0.0146 \text{ mol}}{5.564 \text{ mol}} = 0.00262$$

$$x_{\text{water}} = \frac{5.549 \text{ mol}}{5.564 \text{ mol}} = 0.9973$$

Note that $0.00262 + 0.9973 = 0.9999$, or virtually 1.

SELF-TEST 12.3A Calculate the mole fractions of H_2O and CH_3CH_2OH in a mixture of equal masses of water and ethanol.

[*Answer:* $x_{\text{water}} = 0.719$; $x_{\text{ethanol}} = 0.281$]

SELF-TEST 12.3B Calculate the mole fractions of the species present in a solution prepared by adding 5.00 g of sodium chloride to 100 g of water.

A second useful measure of composition is the **molality,** the number of moles of solute in a solution divided by the mass of the solvent in kilograms:

$$\text{Molality} = \frac{\text{moles of solute (mol)}}{\text{mass of solvent (kg)}}$$

The unit of molality is 1 mole per kilogram of solvent (1 mol·kg^{-1}); this unit is often denoted *m*—for example, a 1 *m* $NiSO_4$(aq) solution—and read "molal." Note the emphasis on *solvent* in the definition. Therefore, to prepare a 1 *m* $NiSO_4$(aq) solution, we dissolve 1 mol $NiSO_4$ in 1 kg of water (Fig. 12.19). Toolbox 12.1 shows how to use molality.

To emphasize the relative numbers of solute and solvent particles in a solution, express the composition in terms of either mole fractions or molality.

(a)

(b)

FIGURE 12.19 (a) A solution of given molarity is prepared by measuring the required mass of solute, and then adding it to a flask of known volume. Solvent is added up to the mark. (b) The steps taken to prepare a solution of given molality. First (left), the required masses of solute and solvent are measured out. Then (right), the solute is dissolved in the solvent.

OOLBOX 12.1 How to use molality

This toolbox shows how to calculate the moles of solute present in a given mass of solvent from the molality, convert between molality and mole fraction, and convert between molarity and molality.

Calculating the mass of solute First, calculate the moles of solute molecules present in a given mass of solvent by rearranging the equation defining molality into

$$\text{Moles of solute} = \text{molality} \times \text{mass of solvent (kg)}$$

Then, to find the mass of solute present, multiply by the molar mass.

For example, suppose we wanted to know the mass of potassium nitrate to add to 250 g of water to prepare a 0.200 *m* $KNO_3(aq)$ solution. We first calculate the moles of KNO_3 to use, and then convert that number of moles into a mass by using the molar mass of the compound ($101.1\ g{\cdot}mol^{-1}$):

$$\begin{aligned}\text{Mass of } KNO_3 &= (0.200\ mol{\cdot}kg^{-1}) \times (0.250\ kg) \\ &\quad \times (101.1\ g{\cdot}mol^{-1}) \\ &= 5.06\ g\end{aligned}$$

SELF-TEST 12.4A What mass of potassium permanganate is needed to prepare a 0.150 *m* $KMnO_4(aq)$ solution with 500 g of water?

[*Answer:* 11.9 g]

SELF-TEST 12.4B What mass of glucose, $C_6H_{12}O_6$, is needed to prepare a 0.255 *m* $C_6H_{12}O_6(aq)$ solution with 250 g of water?

Calculating the molality, given a mole fraction The mole fraction tells us that in a total of 1 mol of molecules in a sample, there is x_{solute} mol of solute molecules and $1 - x_{solute}$ mol of solvent molecules. To convert this number of moles of solvent molecules to mass of solvent, we multiply it by the molar mass of the solvent. At this stage, we know both the moles of solute molecules and the mass of solvent in a total of 1 mol of molecules, so we can work out the molality from the definition.

For example, suppose we have a solution of benzene, C_6H_6, in toluene, $C_6H_5CH_3$, in which $x_{benzene} = 0.150$. From the definition of mole fraction, 1.000 mol of solution molecules contains 0.150 mol C_6H_6 and 0.850 mol $CH_3C_6H_5$. The mass of this amount of toluene (of molar mass $92.13\ g{\cdot}mol^{-1}$) is

$$\begin{aligned}\text{Mass of solvent} &= (0.850\ mol) \times (92.13\ g{\cdot}mol^{-1}) \\ &= 78.31\ g\end{aligned}$$

or 7.831×10^{-2} kg. The molality of benzene in the solution is therefore

$$\text{Molality} = \frac{0.150\ mol}{7.831 \times 10^{-2}\ kg} = 1.92\ mol{\cdot}kg^{-1}$$

To convert from molality to mole fraction, we convert 1 kg of solvent to moles of solvent molecules by using the molar mass of the solvent. We already know the number of moles of solute molecules in the solution (from the molality). Once we know the numbers of moles of both species, we can use Eq. 1 to find their mole fractions.

SELF-TEST 12.5A Calculate the molality of a solution of toluene in benzene, given that the mole fraction of toluene is 0.150.

[*Answer:* $2.26\ mol{\cdot}kg^{-1}$]

SELF-TEST 12.5B Calculate the molality of a solution of methanol in water, given that the mole fraction of methanol is 0.250.

Calculating the molality given the molarity Molality is defined as moles of solute divided by the mass of solvent, whereas molarity is defined as moles of solute divided by the volume of the solution. Thus, we need to find the mass of solvent in the solution of given molarity. To find this mass, we use the following procedure:

Step 1. Calculate the total mass of exactly 1 L of solution by using the density of the solution. (The density of a solution cannot be calculated from the molarity; it must be measured.)

Step 2. From the molarity, calculate the mass of solute in 1 L of solution.

Step 3. Subtract the mass of solute from the total mass to find the mass of solvent in 1 L of solution.

Step 4. Divide the moles of solute by the mass of the solvent in kilograms to obtain the molality.

For example, let's consider a 1.06 M $C_{12}H_{22}O_{11}(aq)$ solution, which is known to have density $1.14\ g{\cdot}cm^{-3}$, or $1.14\ g{\cdot}mL^{-1}$.

Step 1. The mass of exactly 1 L of solution is

$$\begin{aligned}\text{Mass of solution} &= \text{density} \times \text{volume} \\ &= (1.14\ g{\cdot}mL^{-1}) \times (1 \times 10^3\ mL) \\ &= 1.14\ kg\end{aligned}$$

Step 2. The moles of solute in 1.00 L of the 1.06 M solution is 1.06 mol, and the molar mass of sucrose is $342.3\ g{\cdot}mol^{-1}$. Thus, the mass of sucrose in 1.00 L of solution is

$$\text{Mass of sucrose} = (1.06 \text{ mol}) \times (342.3 \text{ g·mol}^{-1}) = 1.06 \times 342.3 \text{ g} = 363 \text{ g}$$

or 0.363 kg.

Step 3. The mass of water present in 1.00 L of solution is therefore

$$\text{Mass of water} = 1.14 \text{ kg} - 0.363 \text{ kg} = 0.78 \text{ kg}$$

Step 4. Because 1.06 mol $C_{12}H_{22}O_{11}$ is dissolved in 0.78 kg of water, we conclude that the molality of the solution is

$$\text{Molality of } C_{12}H_{22}O_{11} = \frac{1.06 \text{ mol}}{0.78 \text{ kg}} = 1.4 \text{ mol·kg}^{-1}$$

SELF-TEST 12.6A Battery acid is 4.27 M H_2SO_4(aq) and has a density of 1.25 g·cm^{-3}. What is the molality of H_2SO_4 in the solution?

[*Answer:* 5.14 mol·kg^{-1} H_2SO_4(aq)]

SELF-TEST 12.6B The density of 1.83 M NaCl(aq) is 1.07 g·cm^{-3}. What is the molality of NaCl in the solution?

EXAMPLE 12.3 Converting between molality and mass percentage composition

What is the mass percentage of sodium chloride in a solution reported as 0.500 m NaCl(aq)?

STRATEGY The molality is the number of moles of solute per kilogram of solvent. Therefore, consider a solution that contains exactly 1 kg of solvent, and work out the mass of solute present from the number of moles and the molar mass. Then calculate the mass percentage of solute by dividing the mass of solute by the *total* mass of the solution and multiply by 100%.

SOLUTION The amount of NaCl in a sample of the solution that contains 1.00 kg of solvent is 0.500 mol. The molar mass of NaCl is 58.44 g·mol^{-1}; therefore the mass of NaCl present is

$$\text{Mass of NaCl} = (0.500 \text{ mol}) \times (58.44 \text{ g·mol}^{-1}) = 29.2 \text{ g}$$

The total mass of the solution is 1000.0 g + 29.2 g = 1029.2 g, so the mass percentage of NaCl present is

$$\text{Mass percentage of NaCl} = \frac{29.2 \text{ g}}{1029.2 \text{ g}} \times 100\% = 2.84\%$$

SELF-TEST 12.7A Calculate the mass percentage of *solvent* in 1.20 m HCl(aq).

[*Answer:* 95.9%]

SELF-TEST 12.7B Calculate the molality of an aqueous solution that is 4.27% by mass Na_2CO_3.

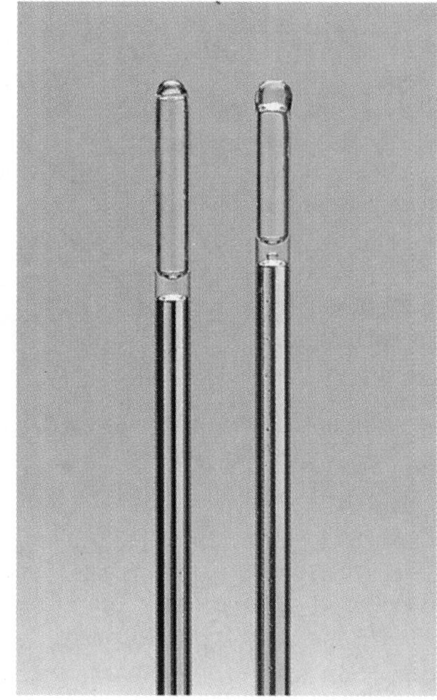

FIGURE 12.20 The vapor pressure of a solvent is lowered by a nonvolatile solute. The barometer tube on the left has a small volume of pure water floating on the mercury. That on the right has a small volume of 10 mol·kg^{-1} NaCl(aq) solution, and a lower vapor pressure. Note that the column on the right is depressed less by the vapor in the space above the mercury than the one on the left, showing that the vapor pressure is lower when the solute is present.

12.11 VAPOR-PRESSURE LOWERING

It has long been known that salt lakes evaporate more slowly than freshwater lakes. Small salt lakes scattered through the desert regions of the southwestern United States remain liquid when all other water has dried up, tempting unwary animals to drink their poisonous waters. The lower rate of evaporation in salt lakes implies that the presence of the salt has reduced the vapor pressure. We see the same phenomenon in the laboratory. The vapor pressure of a solvent is found to be lower when a nonvolatile solute is present (Fig. 12.20). For example, the vapor pressure of

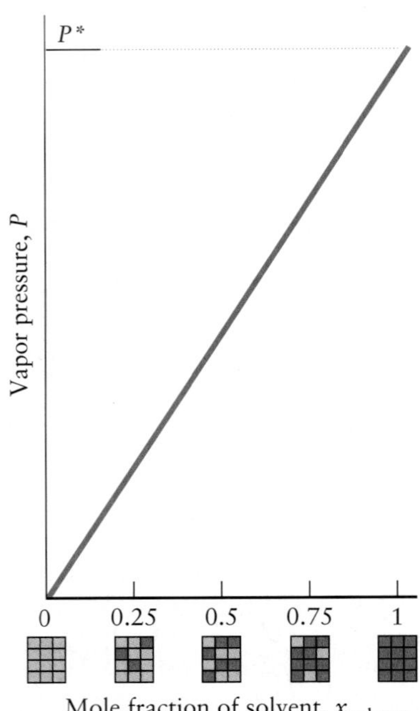

FIGURE 12.21 Raoult's law predicts that the vapor pressure of a solvent in a solution should be proportional to the mole fraction of the solvent molecules. The vapor pressure of the pure solvent is P^*.

pure water at 40°C is 55 Torr but that of a 0.1 *m* NaCl(aq) solution is only 44 Torr at the same temperature.

The French scientist François-Marie Raoult spent much of his life measuring vapor pressures. He found that the effect of a solute could be summarized as follows:

> **Raoult's law:** The vapor pressure of a solvent in the presence of a nonvolatile solute is proportional to the mole fraction of the solvent.

Raoult's law is normally written

$$P = x_{\text{solvent}} P_{\text{pure}}$$

where P_{pure} is the vapor pressure of the pure solvent, x_{solvent} is the mole fraction of the solvent, and P is the vapor pressure of the solvent in the solution (Fig. 12.21). Notice that at any given temperature, the vapor pressure of the solvent is directly proportional to its mole fraction in the solution.

EXAMPLE 12.4 Using Raoult's law

Calculate the vapor pressure of water at 100°C of a solution prepared by dissolving 5.00 g of sucrose in 100 g of water.

STRATEGY Once we know the mole fraction of the solvent (water) in the solution, the calculation is a straightforward application of Raoult's law. That mole fraction was calculated in Example 12.2. To use Raoult's law, we need the vapor pressure of the pure solvent. We could get the value from Table 5.4, but because the normal boiling point of water is 100°C, we know that its vapor pressure at that temperature is 760 Torr. Expect a lower vapor pressure when the solute is present.

SOLUTION From Example 12.2, we know that $x_{\text{water}} = 0.9973$. Therefore, because $P_{\text{pure}} = 760$ Torr, the vapor pressure of the water in the solution is

$$P = 0.9973 \times (760 \text{ Torr}) = 758 \text{ Torr}$$

SELF-TEST 12.8A Calculate the vapor pressure of water at 90°C for a solution prepared by dissolving 5.00 g of glucose ($C_6H_{12}O_6$) in 100 g of water. The vapor pressure of pure water at 90°C is 524 Torr.

[*Answer:* 521 Torr]

SELF-TEST 12.8B Calculate the vapor pressure of ethanol in kilopascals (kPa) at 19°C for a solution prepared by dissolving 2.00 g of cinnamaldehyde, C_9H_8O, in 50.0 g of ethanol, C_2H_5OH. The vapor pressure of pure ethanol at that temperature is 5.3 kPa.

To understand why a nonvolatile solute lowers the vapor pressure of the solvent, think about the molecules at the surface of the solution. The solute molecules block part of the surface and hence reduce the rate at which the solvent molecules escape from the solution (Fig. 12.22). However, the solute molecules have no effect on the rate at which the solvent molecules return, because the returning molecules can stick to any part of the surface, even if a solute molecule is there. At equilibrium, the rate of evaporation equals the rate of condensation, so the lower rate of escape is matched by a lower rate of condensation. The pressure of the vapor is therefore lower than in the pure solvent, because a lower pressure corresponds to a lower rate of impact on the surface.

A hypothetical solution that obeys Raoult's law exactly at all concentrations is called an **ideal solution.** In an ideal solution, the forces between solute and solvent molecules are the same as the forces between solvent molecules, so the solute molecules mingle freely and almost invisibly with the solvent molecules. Solutes that form nearly ideal solutions are often similar in composition and structure to the solvent molecules. For instance, methylbenzene (toluene) forms nearly ideal solutions with benzene.

Real solutions do not obey Raoult's law at all concentrations, but they resemble ideal solutions more closely the lower the solute concentration. A solution that does not obey Raoult's law at a particular solute concentration is called a **nonideal solution.** Real solutions are approximately ideal at solute concentrations below about 10^{-1} mol·kg^{-1} for nonelectrolyte solutions and 10^{-2} mol·kg^{-1} for electrolyte solutions. The problem with electrolyte solutions is that interactions between ions occur over a long distance and hence have a pronounced effect. We can safely assume that all the solutions we meet are ideal, because we shall treat only dilute solutions.

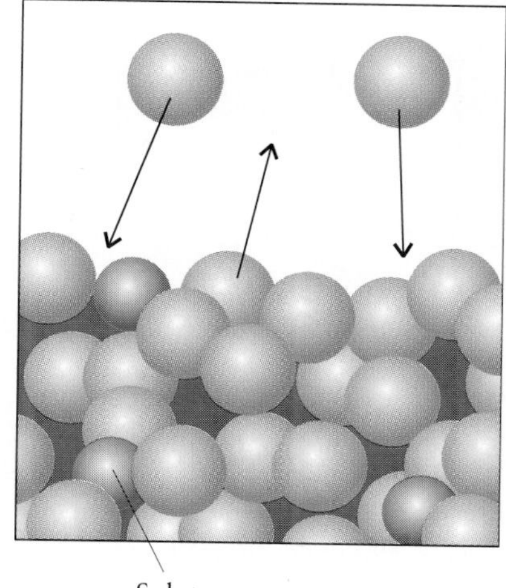

FIGURE 12.22 A nonvolatile solute particle can block the escape of solvent molecules but has no effect on the rate of return of the solvent particles from the vapor to the solution.

The vapor pressure of a solvent is reduced by the presence of a nonvolatile solute: in an ideal solution, the lowering is proportional to the mole fraction of solvent.

12.12 BOILING-POINT ELEVATION AND FREEZING-POINT DEPRESSION

The lowering of vapor pressure explains another colligative property: the raising of the boiling point by a nonvolatile solute. This increase is called the **boiling-point elevation.** Because the vapor pressure of a solvent in a solution is lower than that of the pure solvent at the same temperature, a higher temperature is needed to raise the vapor pressure up to atmospheric pressure, when boiling begins (Fig. 12.23). However, the increase is usually quite small and is of little practical importance. A 0.1 mol·kg^{-1} aqueous sucrose solution, for instance, boils at 100.05°C.

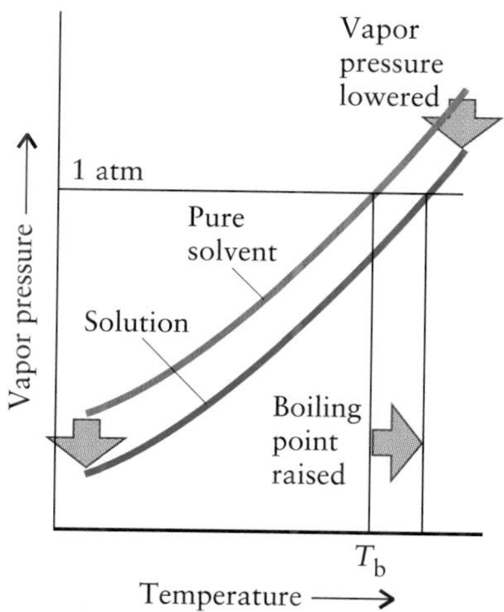

FIGURE 12.23 The lowering of vapor pressure of the solvent in a solution results in an increase in its boiling point because a higher temperature is needed to restore the vapor pressure to 1 atm.

More important is the **freezing-point depression,** the lowering of the freezing point of the solvent. For example, seawater, an aqueous solution rich in Na^+ and Cl^- ions, freezes about 1°C lower than fresh water. In regions with cold winters, salt is spread on highways to lower the freezing point of water and delay ice formation. Chemists make use of the depression of the freezing point to judge the purity of a solid compound: its melting point is lower than the accepted value if impurities are present.

At the freezing point of a pure solvent, the rates at which its molecules stick together to form the solid and leave it to return to the liquid are equal. When a solute is present, fewer solvent molecules in the solution are in contact with the surface of the solid, because solute particles get in the way (Fig. 12.24). As a result, the solvent molecules attach to the surface more slowly. However, the rate at which molecules leave the solid, which is pure solvent, is unchanged. Therefore, even at the melting point of the pure solvent, there is a net flow of molecules away from the

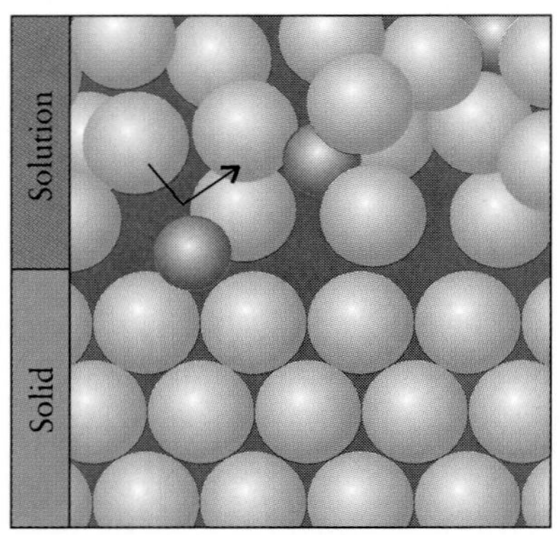

FIGURE 12.24 The rate at which solvent molecules leave the pure solid solvent is unaffected by the presence of solute particles nearby in the solution, but their rate of return is reduced.

solid, and the solid melts. Only if the temperature is lowered will that flow be stopped and the equilibrium restored.

The freezing-point depression in an ideal solution is proportional to the molality of the solute. For a nonelectrolyte solution,

$$\text{Freezing-point depression} = k_f \times \text{molality}$$

The constant k_f is called the **freezing-point constant** of the solvent, and it has the units $K{\cdot}kg{\cdot}mol^{-1}$. It is different for each solvent and must be determined experimentally (Table 12.7). A 0.1 $mol{\cdot}kg^{-1}$ aqueous sucrose solution, for instance, has

$$\begin{aligned}\text{Freezing-point depression} &= (1.86\ K{\cdot}kg{\cdot}mol^{-1}) \times (0.1\ mol{\cdot}kg^{-1}) \\ &= 0.2\ K\end{aligned}$$

Hence it freezes at −0.2°C.

In an electrolyte solution, each formula unit contributes two or more ions. Sodium chloride, for instance, dissolves to give Na^+ and Cl^- ions, and both kinds of ions contribute to freezing-point depression. The cations and anions contribute nearly independently in very dilute solutions, so the total solute molality is twice the molality in terms of NaCl formula units. In more concentrated solutions, the ions do not move independently. For instance, some stick together to form **ion pairs** and other aggregates of small numbers of ions. The effect of the solute on the freezing point in these solutions is very difficult to predict. We can write

$$\text{Freezing-point depression} = ik_f \times \text{molality}$$

where i is known as the **van't Hoff i factor;** it is determined experimentally. In a *very* dilute solution, when all ions are independent, $i = 2$ for MX salts and $i = 3$ for MX_2 salts such as $CaCl_2$, and so on. For dilute nonelectrolyte solutions, $i = 1$. The i factor is so unreliable that it is best to confine quantitative treatments of freezing-point depression to nonelectrolyte solutions. Even these solutions must be dilute enough to be approximately ideal.

The *i* factor is named for Jacobus van't Hoff, a Dutch chemist who in 1901 was awarded the first Nobel prize for chemistry (but not for introducing this simple factor).

The i factor can be used to help determine the extent to which a substance is ionized in solution. For example, in dilute solution HCl has an i factor of 1 in toluene and 2 in water. These values suggest that HCl retains its molecular form in toluene, but is ionized in water. A weak acid that is 5% ionized in water would have an i factor of

Note that percentage ionization is a mole percentage, not a mass percentage.

TABLE 12.7 Boiling-point and freezing-point constants

Solvent	Freezing point, °C	k_f, $K{\cdot}kg{\cdot}mol^{-1}$	Boiling point, °C	k_b, $K{\cdot}kg{\cdot}mol^{-1}$
acetone	−95.35	2.40	56.2	1.71
benzene	5.5	5.12	80.1	2.53
camphor	179.8	39.7	204	5.61
carbon tetrachloride	−23	29.8	76.5	4.95
cyclohexane	6.5	20.1	80.7	2.79
naphthalene	80.5	6.94	217.7	5.80
phenol	43	7.27	182	3.04
water	0	1.86	100.0	0.51

$(0.05 \times 2) + 0.95 = 1.05$, because the ionized molecules produce two ions each.

SELF-TEST 12.9A How many moles of ions are present in a solution containing 0.10 mol Na_2SO_4, assuming complete dissociation? Estimate the i factor.

[*Answer:* 0.30 mol (0.20 mol Na^+ ions and 0.10 mol SO_4^{2-} ions); $i = 3$]

SELF-TEST 12.9B How many moles of ions are present in a solution containing 0.25 mol $CoCl_3$, assuming complete dissociation? Estimate the i factor.

Freezing-point depression can be used to measure the molar mass of the solute. Camphor is often used as the solvent for organic compounds because it has a large freezing-point constant, so solutes depress its freezing point significantly.

EXAMPLE 12.5 Determining molar mass from a freezing-point depression

The addition of 0.24 g of sulfur to 100 g of carbon tetrachloride lowers the latter's freezing point by 0.28°C. What is the molar mass and molecular formula of sulfur?

STRATEGY First, convert the observed freezing-point depression into solute molality by writing

$$\text{Molality of solute} = \frac{\text{freezing-point depression}}{k_f}$$

Take the freezing-point constant from Table 12.7. Use this molality to calculate the moles of solute in the sample by multiplying it by the mass of solvent in kilograms. At this stage, determine the molar mass of the solute by dividing the given mass of solute by the number of moles present. For the molecular formula, decide how many atoms of sulfur are needed in each molecule to account for the molar mass.

SOLUTION From the data,

$$\text{Molality of solute} = \frac{0.28\ \text{K}}{29.8\ \text{K·kg·mol}^{-1}} = 0.0094\ \text{mol·kg}^{-1}$$

Because the mass of carbon tetrachloride is 100 g (0.100 kg), we find

$$\text{Moles of } S_x = (0.100\ \text{kg}) \times (0.0094\ \text{mol·kg}^{-1}) = 9.4 \times 10^{-4}\ \text{mol}$$

It then follows that

$$\text{Molar mass of } S_x = \frac{0.24\ \text{g}}{9.4 \times 10^{-4}\ \text{mol}} = 2.6 \times 10^2\ \text{g·mol}^{-1}$$

Because the molar mass of atomic sulfur is 32.1 g·mol^{-1}, we can conclude that the value of x in the molecular formula S_x is

$$x = \frac{2.6 \times 10^2\ \text{g·mol}^{-1}}{32.1\ \text{g·mol}^{-1}} = 8.1$$

Elemental sulfur is therefore composed of S_8 molecules.

SELF-TEST 12.10A When 250 mg of eugenol, the compound responsible for the odor of oil of cloves, was added to 100 g of camphor, it lowered the freezing point of camphor by 0.62°C. Calculate the molar mass of eugenol.

[*Answer:* 1.6×10^2 g·mol^{-1} (actual: 164.2 g·mol^{-1})]

SELF-TEST 12.10B When 200 mg of linalool, a fragrant compound extracted from Ceylon cinnamon oil, was added to 100 g of camphor, it lowered the freezing point of camphor by 0.51°C. What is the molar mass of linalool?

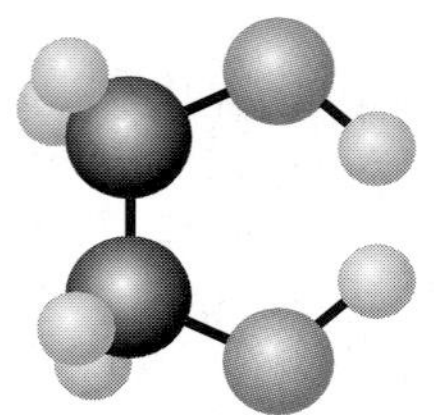

3 Ethylene glycol, $CH_2(OH)CH_2OH$

An aqueous solution of antifreeze, typically the organic compound ethylene glycol (**3**), has a much lower freezing point than that of pure water (Fig. 12.25). However, antifreeze is used in such high concentrations that the solutions cannot be considered ideal and the freezing-point depression is not really an example of a colligative property. A better explanation is that ethylene glycol and water molecules pack together so badly that it is difficult for them to form a rigid structure.

The presence of a solute lowers the freezing point of a solvent; if the solute is nonvolatile, the boiling point is also raised. The freezing point depression can be used to calculate the molar mass of the solute. If the solute is an electrolyte, the extent of its dissociation and ionization must also be taken into account.

12.13 OSMOSIS

The experiment shown in Fig. 12.26 seems to defy nature. A solution in the tube is separated from the pure solvent in the flask by a thin sheet of cellulose acetate. Initially the heights of the solution and the pure solvent are the same. However, the level of the solution inside the tube begins to rise as pure solvent passes through the membrane into the solution. At equilibrium, the pressure exerted by the rising column of solution forces solvent molecules back through the membrane and the rate at which molecules enter the column matches the rate at which they leave.

Cellulose acetate is widely used as a transparent wrapper on candy boxes, so this experiment is easy to repeat.

The flow of solvent through a membrane into a more concentrated solution is called **osmosis.** The membrane is **semipermeable,** which means that only certain types of molecules or ions can pass through it. Cellulose acetate allows water molecules to pass through it, but not solute molecules or ions with their bulky coating of hydrating water molecules. The pressure needed to stop the flow of solvent is called the **osmotic pressure,** Π (the Greek uppercase letter pi). The greater the osmotic pressure, the greater the height of the solution needed to stop the net flow. The pressure exerted by a column of aqueous solution or water is called the **hydrostatic pressure.**

The name *osmosis* comes from the Greek word for "push."

Osmosis plays an important role in maintaining life. Biological cell walls act as semipermeable membranes that allow water, small molecules, and hydrated ions to pass (Fig. 12.27). However, they block the passage of the enzymes and proteins that have been synthesized within the cell. The difference in concentrations of solute inside and outside a cell gives rise to an osmotic pressure, and water passes into the more concentrated solution in the interior of the cell, carrying small nutrient molecules with

FIGURE 12.25 A mixture of water and a commercial antifreeze at −13.3°C. The mixture is still a liquid; pure water would be frozen solid at this temperature.

FIGURE 12.26 An experiment to illustrate osmosis. Initially, the tube contained a sucrose solution and the beaker contained pure water: the initial heights of the two liquids were the same. At the stage shown here, water has passed into the solution through the membrane by osmosis and its level has risen above that of the pure water.

it. This influx of water also keeps the cell turgid (swollen). When the supply is cut off, the turgidity is lost and the cell becomes dehydrated. In a plant, this dehydration results in wilting. Salted meat is preserved from bacterial attack by osmosis. In this case the concentrated salt solution dehydrates—and kills—the bacteria by causing water to flow out of them.

To understand the origin of osmosis, think about the motion of molecules. Solvent molecules can pass readily through pores in the membrane into the solution. However, the solute molecules block the return of some solvent molecules into the pure solvent (Fig. 12.28). As a result, the flow from the pure solvent into the solution is initially faster than the return flow. The return flow increases as the growing pressure forces more solvent molecules back through the membrane more quickly.

The same van't Hoff responsible for the i factor showed that the osmotic pressure of a nonelectrolyte solution is related to the molarity of the solute:

$$\Pi = RT \times \text{molarity}$$

where R is the gas constant and T is the temperature. This expression is now known as the **van't Hoff equation.** It follows that the osmotic pressure of a solution of any 0.010 M nonelectrolyte solution at 25°C (298 K) is

$$\Pi = (0.0821\ \text{L·atm·K}^{-1}\text{·mol}^{-1}) \times (298\ \text{K}) \times (0.010\ \text{mol·L}^{-1}) = 0.24\ \text{atm}$$

This pressure is enough to push a column of water to a height of over 2 m (the height needed to exert a hydrostatic pressure of 0.24 atm). The van't Hoff equation is also used to interpret osmotic pressure measurements in terms of the molar mass of the solute. This technique, which is called **osmometry,** is commonly used to determine the molar masses of polymer molecules.

Osmosis is the flow of solvent through a semipermeable membrane into a solution; the osmotic pressure is proportional to the molar concentration of the solute in the solution.

(a)

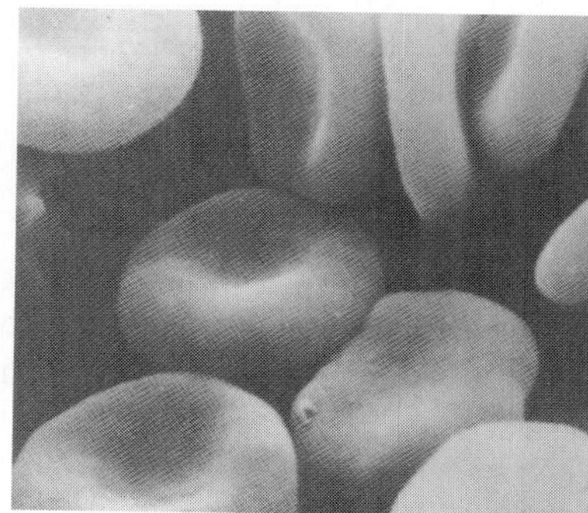

(b)

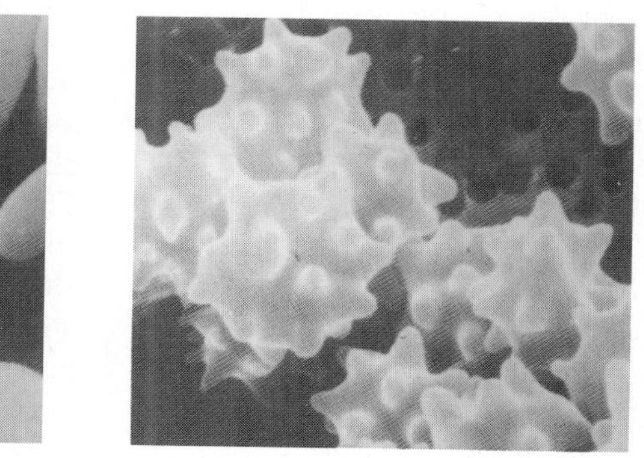

(c)

FIGURE 12.27 (a) Red blood cells need to be in a solution of the correct solute concentration if they are to function properly. (b) When the solution is too dilute, water passes into them and they burst. (c) When it is too concentrated, water flows out of them and they shrivel up.

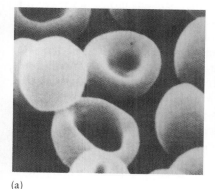

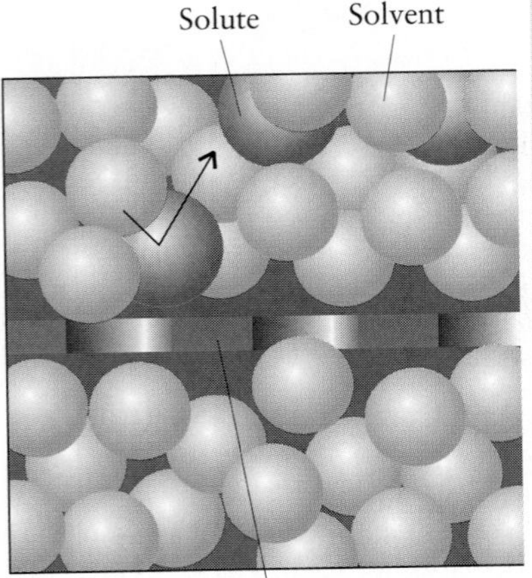

FIGURE 12.28 During osmosis, the presence of solute particles on one side of the membrane hinders solvent molecules from passing through the membrane into the side containing pure solvent.

EXAMPLE 12.6 Using osmometry to determine molar mass

The osmotic pressure due to 2.20 g of polyethylene (PE) dissolved in enough benzene to produce 100 mL of solution was 1.10×10^{-2} atm at 25°C. Calculate the average molar mass of the polymer.

STRATEGY Use the relation between the osmotic pressure and the molarity to determine the latter from the data. To do so, rearrange the equation to

$$\text{Molarity of solution} = \frac{\text{osmotic pressure}}{RT}$$

Then use this molarity as a conversion factor to calculate the number of moles of solute in the stated volume of solution. The mass of solute is given and the number of moles of solute present is now known; therefore, to find the molar mass of the solute, divide the mass by the amount.

SOLUTION From the data,

$$\begin{aligned}\text{Molarity} &= \frac{1.10 \times 10^{-2}\ \text{atm}}{(0.0821\ \text{L·atm·K}^{-1}\text{·mol}^{-1}) \times (298\ \text{K})} \\ &= 4.50 \times 10^{-4}\ \text{mol·L}^{-1}\end{aligned}$$

It follows that in 0.100 L of solution

$$\begin{aligned}\text{Moles of PE} &= (0.100\ \text{L}) \times (4.50 \times 10^{-4}\ \text{mol·L}^{-1}) \\ &= 4.50 \times 10^{-5}\ \text{mol}\end{aligned}$$

Because this amount corresponds to 2.20 g, we conclude that

$$\text{Average molar mass of PE} = \frac{2.20\ \text{g}}{4.50 \times 10^{-5}\ \text{mol}} = 4.9 \times 10^{4}\ \text{g·mol}^{-1}$$

SELF-TEST 12.11A The osmotic pressure of 3.0 g of polystyrene dissolved in enough benzene to produce 150 mL of solution was 1.21 kPa at 25°C. Calculate the average molar mass of the sample of polystyrene.
[*Answer:* 41 kg·mol^{-1}]

SELF-TEST 12.11B The osmotic pressure of 1.50 g of polymethyl methacrylate dissolved in enough methylbenzene to produce 175 mL of solution was 2.11 kPa at 20°C. Calculate the average molar mass of the sample of polymethyl methacrylate.

In **reverse osmosis,** a pressure greater than the osmotic pressure is applied to the solution side of the semipermeable membrane. This application of pressure increases the rate at which solvent molecules leave the solution and thus reverses the flow of solvent, forcing it to flow from the solution to pure solvent. Reverse osmosis is used to remove salts from seawater to produce fresh water for drinking and irrigation. The water is almost literally squeezed out of the salt solution through the membrane. The technological challenge is to fabricate new membranes that are strong enough to withstand high pressures and that do not easily become clogged. Commercial plants use cellulose acetate at pressures of up to 70 atm. Each cubic meter of membrane can produce about 250,000 L of pure water a day.

Osmometry is used to determine the molar masses of polymers and natural macromolecules; osmosis helps to transport nutrients in plants; reverse osmosis is used in water purification.

SKILLS YOU SHOULD HAVE MASTERED

Conceptual

1. Explain the basis for the like-dissolves-like rule.
2. Explain how solutions result from the tendency of energy and matter to disperse.
3. Explain how a nonvolatile solute lowers the vapor pressure, raises the boiling point, and lowers the freezing point of a solvent.
4. Interpret a measured value of the van't Hoff i factor.

Problem-Solving

1. Calculate the solubility of a gas at a given pressure, given its Henry's constant.
2. Predict the relative hydration enthalpies of two ions.
3. Calculate the mole fraction and molality of a solute, given the masses or mass percentages of solute and solvent.
4. Calculate the amount of solute present in a given mass of solvent from the molality.
5. Convert between molality and mole fraction or molarity.
6. Calculate the vapor pressure of a solvent in a solution by using Raoult's law.
7. Find the molar mass of a substance from a freezing-point depression.
8. Find the molar mass of a substance from its osmotic pressure in a solution.

Descriptive

1. Describe how surfactants allow grease to be suspended in water.

EXERCISES

Solubility

12.1 Explain why CH_3OH is completely soluble in water but not completely soluble in toluene, $C_6H_5CH_3$.

12.2 Explain why calcium hydrogen phosphate is soluble in water, whereas calcium phosphate is not.

12.3 Which would be the better solvent, water or benzene, for each of the following: (a) KCl; (b) CCl_4; (c) CH_3COOH?

12.4 Which would be the better solvent, H_2O or CCl_4, for each of the following: (a) NH_3; (b) HCl; (c) I_2?

12.5 The following groups are found in some organic molecules. Which are hydrophilic and which are hydrophobic: (a)—NH_2; (b)—CH_3; (c)—C_6H_5; (d)—COOH?

12.6 The following groups are found in some organic molecules. Which are hydrophilic and which are hydrophobic: (a)—OH; (b)—CH_2CH_3; (c)—$CONH_2$; (d)—Cl?

12.7 (a) Why is grease soluble in gasoline but not in water? (b) How can grease be suspended in water by the addition of a soap? Draw a molecular picture to show how the detergent suspends the grease; identify the hydrophilic and hydrophobic regions of the detergent molecules.

12.8 Water is insoluble in oil, but when detergent is added to the oil, water becomes suspended in the oil. (a) Explain the role of the detergent in this process. (b) The micelles that form in the suspension of water and detergent in oil are called *inverted micelles*. Draw a picture of one of these inverted micelles, showing the locations of the detergent, water, and oil molecules, and identifying the hydrophilic and hydrophobic regions of the detergent molecules. (Micelles are described in Section 12.4.)

Gas Solubility

12.9 State the molar solubility (in moles per liter) in water of (a) O_2 at 50 kPa; (b) CO_2 at 500 Torr; (c) CO_2 at 0.10 atm. The temperature in each case is 20°C and the pressures are partial pressures of the gas. Use the information in Table 12.2.

12.10 Calculate the aqueous solubility (in milligrams per liter) of (a) air at 1.0 atm; (b) He at 1.0 atm; (c) He at 25 kPa. The temperature is 20°C in each case and the pressures are partial pressures of the gas. Use the information in Table 12.2.

12.11 A soft drink is made by dissolving CO_2 at 3.00 atm in a flavored solution and sealing the solution in an aluminum can at 20°C. What volume of CO_2 is released when a 355-mL can is opened to 1.00 atm at 20°C and all the CO_2 is allowed to escape?

12.12 The volume of blood in the body of a certain deep-sea diver is about 6.00 L. Blood cells make up about 55% of the blood volume, and the remaining 45% is the aqueous solution called plasma. What is the maximum volume of nitrogen measured at 1.00 atm and 37°C that could dissolve in the diver's blood plasma at a depth of 93 m, where the pressure is 10.0 atm? (This is the volume that could come out of solution suddenly, causing the bends if the diver ascended too quickly.) Assume that the Henry's constant for nitrogen at 37°C (body temperature) is 5.8×10^{-4} mol· L^{-1}· atm^{-1}.

12.13 The carbon dioxide gas dissolved in a sample of water in a partially filled, sealed container has reached equilibrium with its partial pressure in the air above the solution. Explain what happens to the solubility of the CO_2 if (a) the partial pressure of the CO_2 gas is doubled by the addition of more CO_2; (b) the total pressure of the gas above the liquid is doubled by the addition of nitrogen.

12.14 Explain what happens to the solubility of the CO_2 in Exercise 12.13 if (a) the partial pressure of the CO_2 gas is increased by compressing the gas to a third of its original volume; (b) the temperature is raised.

12.15 The minimum mass concentration of oxygen required for fish life is 4 mg·L^{-1}. (a) Assuming the density of the solution to be 1.00 g·mL^{-1}, express this concentration in parts per million (mg·kg^{-1}; see Table 12.6).
(b) What is the minimum partial pressure of O_2 that would supply this concentration in water at 20°C?
(c) What is the minimum atmospheric pressure that would give this partial pressure, assuming oxygen exerts about 21% of the atmospheric pressure?

12.16 The solubility of oxygen in rivers and streams depends on various factors, such as temperature, salt concentration, depth, and biological activity. Its solubility equilibrium is $O_2(g) \rightleftharpoons O_2(aq)$. Compare the dissolved oxygen concentration at 1.0 atm total pressure (0.21 atm partial pressure) with its concentration for pure oxygen at 1.0 atm.

Enthalpy of Solution

12.17 Describe and explain the trend in the ion hydration enthalpies for the halide ions. See Table 12.5.

12.18 Describe and explain the trend in ion hydration enthalpies for the isoelectronic cations Na^+, Mg^{2+}, and Al^{3+}. See Table 12.5.

12.19 Lithium sulfate dissolves exothermically in water. (a) Is the enthalpy of solution for Li_2SO_4 positive or negative? (b) Write the chemical equation for the dissolving process. (c) Which is larger for lithium sulfate, the lattice enthalpy or the enthalpy of hydration?

12.20 The enthalpy of solution of ammonium nitrate in water is positive. (a) Does NH_4NO_3 dissolve endothermically or exothermically? (b) Write the chemical equation for the dissolving process. (c) Which is larger for NH_4NO_3, the lattice enthalpy or the enthalpy of hydration?

12.21 Calculate the heat evolved or absorbed when 10.0 g (a) NaCl; (b) NaBr; (c) $AlCl_3$; (d) NH_4NO_3 is dissolved in 100.0 g of water. Assume that the enthalpies of solution in Table 12.3 are applicable.

12.22 Determine the temperature change when 10.0 g of (a) KCl; (b) $MgBr_2$; (c) KNO_3; (d) NaOH is dissolved in 100.0 g of water. Assume that the specific heat capacity of the solution is 4.18 J·K^{-1}·g^{-1} and that the enthalpies of solution in Table 12.3 are applicable.

12.23 Estimate the enthalpy of solution of $SrCl_2$ from the data for the ion enthalpies of hydration and the lattice enthalpy in Tables 8.1 and 12.5.

12.24 (a) Calculate the enthalpy of hydration of Br^- by using the appropriate data from Table 12.4 and the enthalpy of hydration of H^+ (-1130 kJ·mol^{-1}). (b) Use the value obtained in (a) to determine the enthalpy of hydration of Rb^+, given that the enthalpy of solution of RbBr is $+22$ kJ·mol^{-1} and its lattice enthalpy is 651 kJ·mol^{-1}.

12.25 Consider the enthalpy of solution of (a) LiCl; (b) NaCl; (c) AgCl in Table 12.3. Identify and explain any trends you find. Compare the trend in enthalpies of solution to the trend in solubilities of these salts and explain any relationship between solubility and ΔH_{sol}.

12.26 Consider the enthalpy of solution of (a) NaF; (b) NaCl; (c) NaBr; (d) NaI in Table 12.3. Identify and explain the trend in the enthalpy of solution on proceeding down the group of halides.

12.27 Describe the changes that take place in the energy and disorder of (a) the system and (b) the surroundings when a solute dissolves exothermically.

12.28 Describe the changes that take place in the energy and disorder of (a) the system and (b) the surroundings when a solute dissolves endothermically.

Measures of Concentration

See Table 12.6 for definitions of concentration units.

12.29 Calculate the molarity of (a) 50.0 g of $KNO_3(aq)$ dissolved in 1.50 L of solution and (b) 10.0 g of glucose, $C_6H_{12}O_6(aq)$, dissolved in 250 mL of solution.

12.30 A stockroom attendant was asked to prepare the following solutions: (a) 1.63 g anhydrous sodium carbonate, Na_2CO_3, dissolved in water and diluted to 200 mL and (b) 6.29 g anhydrous calcium chloride, $CaCl_2$, dissolved in water and diluted to 750 mL. Calculate the molarities of the two solutions.

12.31 What mass (in grams) of anhydrous solute is needed to prepare each of the following solutions: (a) 1.0 L of 0.10 M NaCl(aq); (b) 250 mL of 0.10 M $CaCl_2$(aq); (c) 500 mL of 0.63 M $C_6H_{12}O_6$(aq).

12.32 Describe the preparation of each solution, starting with the anhydrous solute and water and using the corresponding volumetric flask: (a) 25.0 mL of 6.0 M NaOH(aq); (b) 1.0 L of 0.10 M $BaCl_2$(aq); (c) 500 mL of 0.0010 M $AgNO_3$(aq).

12.33 Determine the mass percentage of each solute in the following solutions: (a) 4.0 g of NaCl dissolved in enough water to form a total of 100 g of aqueous solution; (b) 4.0 g of NaCl dissolved in 100 g of water; (c) 1.66 g of $C_{12}H_{22}O_{11}$ dissolved in 200 g of water.

12.34 What mass (in grams) of each solute should be added to make the following solutions: (a) to 200 g water to form a 3.0% by mass KBr solution? (b) to 25.0 g water to form a 6.0% by mass $AgNO_3$ solution? (c) to 500 g water to form a 10.5% by mass C_2H_5OH solution?

12.35 Calculate the mole fraction of each component in the following solutions: (a) 25.0 g of water and 50.0 g of ethanol, C_2H_5OH; (b) 25.0 g of water and 50.0 g of methanol, CH_3OH; (c) a glucose solution that is 0.10 *m* $C_6H_{12}O_6$(aq).

12.36 Calculate the mole fraction of each component in the following solutions: (a) 25.0 g of benzene, C_6H_6, and 50.0 g of toluene, $C_6H_5CH_3$; (b) 25.0 g of benzene, 10.0 g of carbon tetrachloride, CCl_4, and 50.0 g of naphthalene, $C_{10}H_8$; (c) a sucrose solution that is 0.020 *m* $C_{12}H_{22}O_{11}$(aq).

12.37 Calculate the molality of the solute in each of the following solutions: (a) 10.0 g of NaCl dissolved in 250 g water; (b) 0.48 mol of KOH dissolved in 50.0 g water; (c) 1.94 g of urea, $CO(NH_2)_2$, dissolved in 200 g water.

12.38 (a) What mass (in grams) of NaOH must be mixed with 250 g of water to prepare a 0.22 *m* NaOH solution? (b) Calculate the amount (in moles) of ethylene glycol, HOC_2H_4OH, that should be added to 2.0 kg of water to prepare a 0.44 *m* HOC_2H_4OH(aq) solution. (c) Determine the amount (in moles) of HCl that must be dissolved in 500 g water to prepare a 0.0010 *m* HCl(aq) solution.

12.39 Calculate the mass of solute required: (a) $ZnCl_2$ to dissolve in 150 g of water to prepare a 0.200 *m* $ZnCl_2$(aq) solution; (b) $KClO_3$ to dissolve in 20.0 g of water to prepare a 3.0% by mass $KClO_3$(aq) solution; (c) $KClO_3$ to dissolve in 20.0 g of water to prepare a 3.0 *m* $KClO_3$(aq) solution.

12.40 Calculate the mass of solute required: (a) $CuSO_4$ to dissolve in 2.0 kg of water to prepare a 0.066 *m* $CuSO_4$(aq) solution; (b) $Na_2Cr_2O_7$ to dissolve in 200 g of water to prepare a 1.5% by mass $Na_2Cr_2O_7$(aq) solution; (c) $Na_2Cr_2O_7$ to dissolve in 200 g water to prepare a 1.5 *m* $Na_2Cr_2O_7$(aq) solution.

12.41 In a laboratory exercise, a student mixes 25.0 g of ethanol, C_2H_5OH, with 150 g of water. (a) What is the mole fraction of ethanol in the solution? (b) What is the molality of ethanol in the solution?

12.42 In the laboratories of an oil company, a research chemist dissolved 1.0 g of octane, C_8H_{18}, in 50.0 g of benzene, C_6H_6. (a) What is the mole fraction of octane in the solution? (b) What is the molality of octane in the solution?

12.43 Calculate the mole fractions of the cations, anions, and water in (a) 0.10 *m* NaCl(aq); (b) 0.20 *m* Na_2CO_3(aq); (c) 10.0% by mass KNO_3(aq).

12.44 What are the mole fractions of the cations and anions in (a) 0.10 *m* $MgSO_4$; (b) 0.55% $MgBr_2$; (c) 0.72 *m* $Al_2(SO_4)_3$; (d) 0.72% by mass $Al_2(SO_4)_3$?

12.45 The density of a 0.35 M $(NH_4)_2SO_4$(aq) solution is 1.027 $g \cdot mL^{-1}$. Determine (a) the molality; (b) the mole fraction of ammonium sulfate in the solution.

12.46 A 5.00% K_3PO_4 aqueous solution has a density of 1.043 $g \cdot mL^{-1}$. Determine (a) the molality; (b) the molarity of potassium sulfate in solution.

Vapor-Pressure Lowering

12.47 Calculate the vapor pressure of the solvent in each solution listed below. Use the data in Table 5.4 for the vapor pressure of water at various temperatures. (a) An aqueous solution at 100°C in which the mole fraction of sucrose is 0.100. (b) An aqueous solution at 100°C in which the molality of sucrose is 0.100 $mol \cdot kg^{-1}$.

12.48 What is the vapor pressure of the solvent in each of the solutions listed below? Use the data in Table 5.4 for the vapor pressure of water at various temperatures. (a) The mole fraction of glucose is 0.050 in an aqueous solution at 80°C. (b) An aqueous solution at 25°C is 0.10 *m* urea, $CO(NH_2)_2$, a nonelectrolyte.

12.49 (a) Calculate the vapor pressure of a 1.0% by mass aqueous ethylene glycol, $C_2H_4(OH)_2$, solution at 0°C. (b) What is the vapor pressure of 0.10 *m* NaOH(aq) at 80°C? (c) Determine the *change* in the vapor pressure of water when 6.6 g of $CO(NH_2)_2$ is dissolved in 100 g water at 10°C. For data, see Tables 5.4 and 12.6.

12.50 (a) Calculate the *change* in vapor pressure from that of pure water for an aqueous solution at 40°C in which the mole fraction of fructose is 0.22. (b) Determine the vapor pressure of a saturated magnesium fluoride solution at 20°C. The solubility of MgF_2 at 20°C is 8 mg/100 g water. (c) What is the vapor pressure of a 0.010 *m* $Fe(NO_3)_3$ solution at 0°C? For data, see Tables 5.4 and 12.6.

12.51 When 8.05 g of an unknown compound X was dissolved in 100 g of benzene, C_6H_6, the vapor pressure of the benzene decreased from 100 Torr to 94.8 Torr at 26°C. What is (a) the mole fraction and (b) the molar mass of X?

12.52 The normal boiling point of ethanol, C_2H_5OH, is 78.4°C. When 9.15 g of a soluble nonelectrolyte was dissolved in 100 g ethanol, the vapor pressure of the solution at that temperature was 740 Torr. (a) What are the mole fractions of ethanol and solute? (b) What is the molar mass of the solute?

Boiling-Point Elevation

12.53 Estimate the boiling-point elevation and the normal boiling points of (a) 0.10 *m* $C_{12}H_{22}O_{11}$(aq); (b) 0.22 *m* NaCl(aq); (c) a saturated solution of LiF, of solubility 230 mg/100 g water at 100°C. Assume complete dissociation.

12.54 Estimate the boiling-point elevation and the normal boiling points of (a) 0.22 *m* $CaCl_2$(aq); (b) a saturated solution of Li_2CO_3, of solubility 0.72 mg/100 g water at 100°C; (c) 1.7% by mass $CO(NH_2)_2$(aq). Assume complete dissociation.

12.55 (a) What is the normal boiling point of an aqueous solution that has a vapor pressure of 751 Torr at 100°C? (b) Determine the normal boiling point of a benzene solution that has a vapor pressure of 740 Torr at 80.1°C, the normal boiling point of pure benzene.

12.56 (a) What is the normal boiling point of an aqueous solution that has a freezing point of −1.04°C? (b) The freezing point of a benzene solution is 2.0°C. The normal freezing point of benzene is 5.5°C. What is the expected normal boiling point of the solution? The normal boiling point of benzene is 80.1°C.

12.57 A 1.05-g sample of a molecular compound is dissolved in 100 g of carbon tetrachloride, CCl_4. The normal boiling point of the solution is 61.51°C; the normal boiling point of CCl_4 is 61.20°C. What is the molar mass of the compound?

12.58 When 2.25 g of an unknown compound is dissolved in 150 g of cyclohexane, the boiling point increased by 0.481°C. Determine the molar mass of the compound.

Freezing-Point Depression

12.59 Estimate the freezing-point depression and the freezing point of each of the following aqueous solutions: (a) 0.10 *m* $C_{12}H_{22}O_{11}$(aq); (b) 0.22 *m* NaCl(aq); (c) a saturated solution of LiF, of solubility 120 mg/100 g water at 0°C. Assume complete dissociation.

12.60 Estimate the freezing point of the following aqueous solutions: (a) 0.10 *m* $C_6H_{12}O_6$(aq); (b) 0.22 *m* $CaCl_2$(aq); (c) a saturated solution of Li_2CO_3, of solubility 1.54 mg/100 g water at 0°C. Assume complete dissociation.

12.61 A 1.14-g sample of a molecular substance dissolved in 100 g camphor (freezing point 179.8°C) freezes at 177.3°C. What is the molar mass of the substance?

12.62 When 2.11 g of a nonpolar solute was dissolved in 50.0 g phenol, the latter's freezing point was lowered by 1.753°C. Calculate the molar mass of the solute.

12.63 (a) What would be the freezing point of a benzene solution that boils at 82.0°C, rather than 80.1°C, the normal boiling point of pure benzene? The freezing point of benzene is 5.5°C. (b) An aqueous solution freezes at −3.04°C. What is the molality of the solution? (c) The freezing point of an aqueous solution containing a nonelectrolyte dissolved in 200 g water is −1.94°C. How many moles of the solute are present?

12.64 (a) A benzene solution has a vapor pressure of 740 Torr at 80.1°C, the normal boiling point of pure benzene. What is the expected freezing point of the solution? The freezing point of benzene is 5.5°C. (b) How many moles of $CO(NH_2)_2$ are present in 1200 g water, given that the freezing point of the solution is −4.02°C? (c) A 1.0-g sample of a protein (of molar mass $1.0 \times 10^5\ g \cdot mol^{-1}$) is dissolved in 1.0 kg of water. What is the expected freezing point of the solution?

12.65 A 1.00% NaCl(aq) by mass solution has a freezing point of −0.593°C. (a) Estimate the van't Hoff *i* factor from the data. (b) Determine the total molality of all solute species. (c) Calculate the percentage dissociation of NaCl in this solution. (*Hint:* The molality calculated from the freezing-point depression is the sum of the molalities of the undissociated ion pairs, the Na^+ ions, and the Cl^- ions.)

12.66 A 1.00% by mass $MgSO_4$(aq) solution has a freezing point of −0.192°C. (a) Estimate the van't Hoff *i* factor for the solution. (b) Determine the total molality of all the solute species. (c) Calculate the percentage dissociation of the $MgSO_4$ in this solution. (*Hint:* The molality calculated from the freezing-point depression is the sum of the molalities of the undissociated ion pairs, the Mg^{2+} ions, and the SO_4^{2-} ions.)

12.67 Determine the freezing point of a 0.10 $mol \cdot kg^{-1}$ aqueous solution of a weak electrolyte that is 7.5% dissociated into two ions.

12.68 A 0.124 *m* $CCl_3COOH(aq)$ solution has a freezing point of $-0.423°C$. What is the percentage ionization of the acid?

Osmosis and Osmometry

12.69 What is the osmotic pressure of (a) 0.010 M $C_{12}H_{22}O_{11}(aq)$? (b) 1.0 M HCl(aq)? (c) 0.010 M $CaCl_2(aq)$ at 20°C?

12.70 Which of the following solutions has the highest osmotic pressure at 50°C: (a) 0.10 M KCl(aq); (b) 0.60 M $CO(NH_2)_2(aq)$; (c) 0.30 *m* $K_2SO_4(aq)$? Justify your answer by calculating the osmotic pressure of each solution.

12.71 A 0.40-g sample of a polypeptide dissolved in 1.0-L of an aqueous solution at 27°C has an osmotic pressure of 3.74 Torr. What is the molar mass of the polypeptide?

12.72 When 0.10 g of insulin is dissolved in 200 mL of water, the osmotic pressure is 2.30 Torr at 20°C. What is the molar mass of insulin?

12.73 A 0.10-g sample of a polymer, dissolved in 100 mL of toluene has an osmotic pressure of 5.4 Torr at 20°C. What is the molar mass of the polymer?

12.74 A solution prepared by adding 0.50 g of a polymer to 200 mL of toluene (an organic solvent) showed an osmotic pressure of 0.582 Torr at 20°C. What is the molar mass of the polymer?

12.75 Calculate the osmotic pressure at 20°C of each of the solutions listed below; assume complete dissociation for any ionic solutes. (a) 0.050 M $C_{12}H_{22}O_{11}(aq)$; (b) 0.0010 M NaCl(aq); (c) a saturated solution of AgCN of solubility 2.3×10^{-5} g/100 g water.

12.76 Calculate the osmotic pressure at 20°C of the solutions specified below. Assume complete dissociation of ionic compounds. (a) 3.0×10^{-3} M $C_6H_{12}O_6(aq)$; (b) 2.0×10^{-3} M $CaCl_2(aq)$; (c) 0.010 M $K_2SO_4(aq)$.

SUPPLEMENTARY EXERCISES

12.77 A saturated solution of copper(II) sulfate was prepared by adding an excess of small crystals of $CuSO_4$ to a flask of water. The flask, which contained the saturated solution and many small crystals of $CuSO_4$, was sealed and placed in a cabinet for several months. When it was removed, most of the small crystals were gone and several large crystals had appeared. However, the concentration of $CuSO_4$ in the solution was the same. Explain, describing the events that occur at the molecular level, how it is possible for the crystals to change without a change in concentration of the solution.

12.78 (a) Calculate the mass of $CaCl_2 \cdot 6H_2O$ needed to prepare a 0.10 *m* $CaCl_2(aq)$ solution, using 250 g water. (b) What mass of $NiSO_4 \cdot 6H_2O$ must be dissolved in 500 g water to produce a 0.22 *m* $NiSO_4(aq)$ solution?

12.79 A student helper was asked by the stockroom director to label the concentration of each aqueous solution as a mass percentage. What labels should the student have prepared for (a) 0.33 *m* $CO(NH_2)_2(aq)$; (b) $x_{C_2H_5OH} = 0.28$; (c) 1.04 *m* $K_2SO_4(aq)$; (d) $x_{HCl} = 0.11$? (See Table 12.6.)

12.80 A saturated magnesium chloride solution is 34.6% $MgCl_2(aq)$ and has a density of 1.27 $g \cdot cm^{-3}$. Calculate (a) the molality of $MgCl_2$; (b) the mole fraction of water; (c) the molarity of $MgCl_2$ in the solution. (See Table 12.6.)

12.81 A 10.0% $H_2SO_4(aq)$ solution has a density of 1.07 $g \cdot cm^{-3}$. (a) How many milliliters of solution contain 6.32 g of H_2SO_4? (b) What is the molality of H_2SO_4 in solution? (c) What mass (in grams) of H_2SO_4 is in 300 mL of solution?

12.82 The density of a 16.0% by mass $C_{12}H_{22}O_{11}(aq)$ solution is 1.0635 $g \cdot cm^{-3}$ at 20°C. (a) What is the molarity of $C_{12}H_{22}O_{11}(aq)$? (b) What is the vapor pressure of the solution at 20°C? (c) Predict the boiling point of the solution. (See Table 12.6.)

12.83 Nitric acid is purchased from chemical suppliers as a solution that is 70% HNO_3 by mass. What mass (in grams) of a 70% $HNO_3(aq)$ solution is necessary to prepare 250 mL of a 2.0 *m* $HNO_3(aq)$ solution? The density of 70% $HNO_3(aq)$ is 1.42 $g \cdot cm^{-3}$. (See Table 12.6.)

12.84 Explain in terms of events at a molecular level why a solute raises the boiling point and lowers the freezing point of a solution.

12.85 Water temperatures in the Gulf of Mexico can be as high as 30°C along the coast during the summer. Use Table 5.4 to estimate the vapor pressure of seawater at that temperature and at 100°C and 0°C, assuming that a 0.50 *m* NaCl(aq) solution simulates seawater.

12.86 Organic chemists sometimes use freezing-point and boiling-point measurements to determine the molar mass of compounds they have synthesized. When 0.30 g of a nonvolatile solute is dissolved in 30.0 g CCl_4, the boiling point increases from 76.54°C (the normal boiling point of CCl_4) to 77.19°C. What is the molar mass of the compound?

12.87 A 10.0-g sample of *p*-dichlorobenzene, a component of mothballs, is dissolved in 80.0 g benzene, C_6H_6. The freezing point of the solution is 1.20°C. The freezing point of pure benzene is 5.48°C. (a) What is an approximate molar mass of *p*-dichlorobenzene? (b) An elemental analysis of *p*-dichlorobenzene indicated that the empirical formula is C_3H_2Cl. What is the molecular formula for *p*-dichlorobenzene? (c) Using the atomic molar masses from the periodic table, calculate a more accurate molar mass of *p*-dichlorobenzene.

12.88 When determining a molar mass from freezing-point depression, it is possible to make each of the following errors (among others). In each case predict whether the error would cause the reported molar mass to be greater or less than the actual molar mass: (a) There was dust on the balance, causing the mass of solute to appear greater than it actually was. (b) The water was measured by volume, assuming a density of 1.00 $g \cdot cm^{-3}$, but the water was warmer and less dense than assumed. (c) The thermometer was not calibrated accurately, so the temperature of the freezing point was actually 0.5°C higher than recorded. (d) The solution was not stirred sufficiently, so that not all of the solute dissolved.

12.89 An elemental analysis of adrenaline is 59.0% carbon, 26.2% oxygen, 7.10% hydrogen, and 7.60% nitrogen by mass. When 0.64 g of adrenaline was dissolved in 36.0 g benzene, the freezing point decreased by 0.50°C. (a) Determine the empirical formula of adrenaline. (b) What is the molar mass of adrenaline? (c) Deduce the molecular formula of adrenaline.

12.90 Intravenous medications are often administered in 5.0% glucose, $C_6H_{12}O_6$, aqueous solutions by mass. What is the osmotic pressure of such solutions at 37°C (body temperature)?

12.91 Catalase, a liver enzyme, dissolves in water. A 10-mL solution containing 0.166 g of catalase exhibits an osmotic pressure of 1.2 Torr at 20°C. What is the molar mass of catalase?

12.92 Suggest a reason why saltwater fish die when they are suddenly transferred to a freshwater aquarium.

12.93 Solutions that are injected into the bloodstream (for blood transfusions and intravenous feeding) must be "isotonic" with the blood (that is, have the same osmotic pressure). If the injected solution is too dilute, it is termed "hypotonic" and if too concentrated, "hypertonic." Explain the effects of hypotonic and hypertonic solutions on the cells in the bloodstream.

12.94 Interpret the following verse from Coleridge's *The Rime of the Ancient Mariner:*

> Water, water, every where,
> And all the boards did shrink;
> Water, water every where,
> Nor any drop to drink.

12.95 Explain how osmometry is used to measure molar mass. What are the advantages of osmometry over measurements of boiling points and freezing points of solutions?

12.96 Describe the procedure of "reverse osmosis" for the purification of salt water for drinking water.

CHALLENGING EXERCISES

12.97 The reaction $Ca(HCO_3)_2(aq) \rightarrow CaCO_3(s) + CO_2(g) + H_2O(l)$ takes place during the formation of stalagmites and stalactites in caves. The soluble calcium hydrogen carbonate seeps through the water in the soil until an air space is encountered. Explain, using Henry's law and principles of solubility, the formation of the stalagmites and stalactites, the major component of which is calcium carbonate.

12.98 (a) Determine the mass of $Na_2CO_3 \cdot 10H_2O$ needed to prepare a 1.0 *m* $Na_2CO_3(aq)$ solution, using 250 g water. (b) Find a general expression for the mass of solute needed to prepare a solution of the solute B of molality *m* when the hydrated solute has the formula $B \cdot xH_2O$.

12.99 The freezing point of a 5.00% by mass $CH_3COOH(aq)$ solution is −1.576°C. Determine the experimental van't Hoff *i* factor for this solution. Account for its value on the basis of your understanding of intermolecular forces.

12.100 What is the osmotic pressure of each of the following aqueous solutions at the given temperature? Assume the density of each solution is 1.00 $g \cdot cm^{-3}$. (a) A saturated solution of Li_2CO_3, which has a solubility of 1.54 g/100 g H_2O at 0°C. (b) A solution that has a vapor pressure of 751 Torr at 100°C. (c) A solution that has a boiling point of 101°C.

12.101 The height of a column of liquid that can be supported by a given pressure is inversely proportional to its density. An aqueous solution of 0.010 g of a protein in 10 mL of water at 20°C shows a 5.22-cm rise in the apparatus shown in Fig. 12.26. Assume the density of the solution to be 0.998 $g \cdot cm^{-3}$ and the density of mercury to be 13.6 $g \cdot cm^{-3}$. (a) What is the molar mass of the protein?

(b) What is the freezing point of the solution? (c) Which colligative property is best for measuring the molar mass of these large molecules? Give reasons for your answer.

12.102 A 0.020 M $C_6H_{12}O_6$(aq) solution is separated from a 0.050 M $CO(NH_2)_2$(aq) solution by a semipermeable membrane at 25°C. (a) Which solution has the higher osmotic pressure? (b) Which solution becomes more dilute with the passage of H_2O molecules through the membrane? (c) To which solution should an external pressure be applied in order to maintain an equilibrium flow of H_2O molecules across the membrane? (d) What external pressure (in atm) should be applied in (c)?

12.103 (a) A solution of a nonvolatile solute in benzene has a vapor pressure of 740 Torr at 80.1°C, the normal boiling point of benzene. What is the osmotic pressure of this solution at 20°C? Assume that the density of the solution is the same as that of benzene, 0.88 $g\cdot cm^{-3}$. (b) The freezing point of a different benzene solution was 5.4°C (the normal freezing point of benzene is 5.5°C). Determine the osmotic pressure of the solution at 10°C. (c) Determine the height to which osmosis will force the solution in (b) to rise in an arrangement like that shown in Fig. 12.26. The density of mercury is 13.6 $g\cdot cm^{-3}$. See Exercise 12.101 for more information.

CHAPTER 13

Did you ever think that chemical equilibrium could be as beautiful as a butterfly's wing? The markings on the wing of this butterfly (incidentally, a Malay Lacewing, Cethosia hypsea*) are the outcome of many different chemical reactions. The result of these reactions is a pattern of concentrations of chemicals that affect the properties of the wing. In this chapter we examine much simpler reactions, but the techniques we meet can be used to discuss the patterns and shapes we see in nature as well as the way chemists improve the efficiency of the chemical industry.*

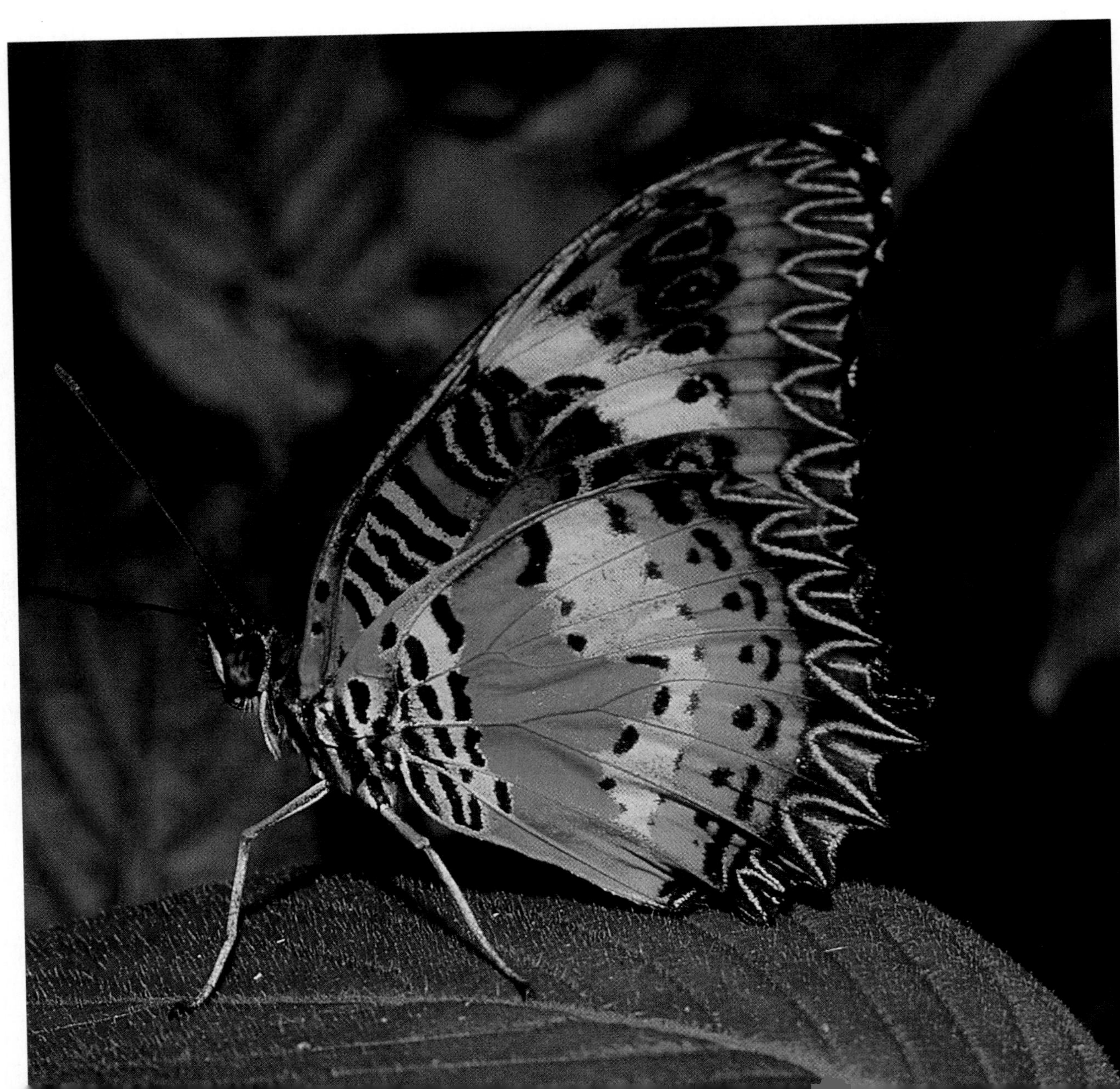

CHEMICAL EQUILIBRIUM

The world was in desperate need of nitrogen-based fertilizers at the opening of the twentieth century. Almost all the nitrates used for fertilizers and explosives were quarried from deposits in Chile, and the limited supply could not keep up with the demand. All attempts to convert the abundant atmospheric nitrogen into compounds that bacteria can use were too costly to employ on a large scale. Finally, though, determination, application, and a modicum of luck enabled the German chemist Fritz Haber to find an economical way to capture the nitrogen of the air for agriculture, a feat that won him a Nobel prize. In part, his success was due to his application of the principles that are introduced in this chapter.

Here we are concerned with **chemical equilibrium,** the stage of a reaction when there is no further tendency for the composition to change. All chemical equilibria are *dynamic* equilibria, with the forward and reverse reactions occurring at the same rate. We have already considered several physical processes, including vaporizing and dissolving, that reach dynamic equilibrium. This chapter shows how to apply the same ideas to chemical changes and how to treat equilibria quantitatively. The key idea is that dynamic equilibria are responsive to changes in the conditions. In other words, a dynamic equilibrium is a responsive, living equilibrium.

EQUILIBRIUM AND COMPOSITION

13.1 The reversibility of chemical reactions

13.2 The equilibrium constant

13.3 Heterogeneous equilibria

13.4 Gaseous equilibria

USING EQUILIBRIUM CONSTANTS

13.5 The extent of reaction

13.6 The direction of reaction

13.7 Equilibrium tables

THE RESPONSE OF EQUILIBRIA TO CHANGE

13.8 Adding and removing reagents

13.9 Compressing a reaction mixture

13.10 Temperature and equilibrium

13.11 Catalysts and Haber's achievement

EQUILIBRIUM AND COMPOSITION

Recall that the Lewis structure of nitrogen is :N≡N: .

Haber was looking for a way to "fix" nitrogen—that is, to turn atmospheric nitrogen into compounds. However, nitrogen has a strong triple bond, and a correspondingly high bond enthalpy (944 kJ·mol^{-1}). As a result, it is very unreactive. Chemists considered that a sensible way to proceed was to convert nitrogen into ammonia:

$$N_2(g) + 3\,H_2(g) \longrightarrow 2\,NH_3(g)$$

The average N—H bond enthalpy in ammonia is only 388 kJ·mol^{-1}, and ammonia is much more reactive than nitrogen. Microorganisms in the roots of certain plants fix nitrogen, but they use complicated enzymes that chemists are still trying to replicate. Haber was looking for a simple, efficient, and economical way to produce ammonia, using the technological resources of the early twentieth century. To do that, he had to understand what happens when a reaction reaches equilibrium.

13.1 THE REVERSIBILITY OF CHEMICAL REACTIONS

◀◀ Haber did this analysis by dissolving the gaseous mixture in water and titrating it with hydrochloric acid, using procedures like those described in Chapter 4.

In one series of experiments, Haber started with known amounts of nitrogen and hydrogen maintained at high temperature and pressure and, at regular intervals, determined the amount of ammonia present. He worked out from the reaction stoichiometry how much of the nitrogen and hydrogen remained (Fig. 13.1). As the graph shows, after a certain time the composition of the mixture remains the same, even though some of the reactants are still present. The fact that the composition has stopped changing indicates that the reaction has reached equilibrium.

At a molecular level, the chemical reactions never actually stop even though the concentrations remain constant. Let's imagine carrying out two ammonia syntheses with exactly the same starting conditions, but using D_2 (deuterium) in place of H_2 in one of them (Fig. 13.2). The two reaction mixtures reach equilibrium with the same composition, except that D_2 and ND_3 are present in one of the systems instead of H_2 and NH_3. Suppose we now mix the two mixtures together and leave them for a while. When later we analyze the mixture, we find that the concentration of ammonia is just the same as before. However, when we analyze the sample with a mass spectrometer, we find that all isotopic forms of ammonia (NH_3, NH_2D, NHD_2, and ND_3) and all isotopic forms of hydrogen (H_2, HD, and D_2) are present. This scrambling of H and D atoms in the molecules must result from a continuation of the forward and reverse reactions in the mixture. If reactions had simply stopped when they reached equilibrium, then there would have been no mixing of isotopes in this way.

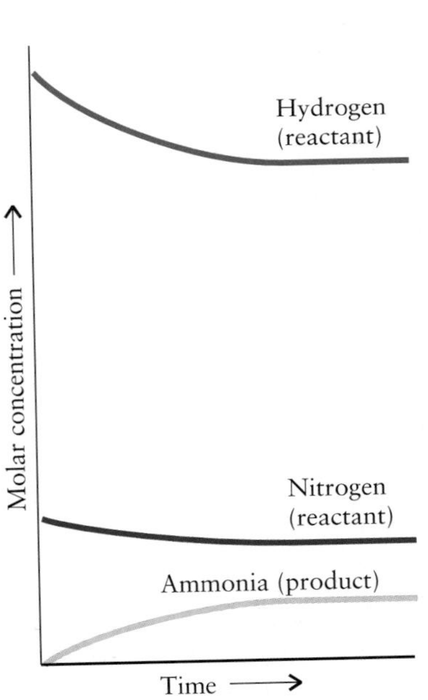

FIGURE 13.1 In the synthesis of ammonia, the molar concentrations of N_2, H_2, and NH_3 change with time until they finally settle into values corresponding to a mixture in which all three are present and there is no further net change.

To imagine what is going on, we have to consider both the forward reaction

$$N_2(g) + 3\,H_2(g) \longrightarrow 2\,NH_3(g)$$

and the reverse reaction:

$$N_2(g) + 3\,H_2(g) \longleftarrow 2\,NH_3(g)$$

An NH_3 molecule forms by a complicated series of steps in which N_2 and H_2 molecules collide and exchange atoms. The NH_3 molecules that form also collide with one another and with the N_2 and H_2 molecules still present. As a result of these collisions, the NH_3 molecules decompose back into nitrogen and hydrogen. As the concentration of the product grows, collisions involving product molecules take place more frequently and the reverse reaction goes faster. At the same time, reactants are being depleted and the forward reaction slows down. The system is at equilibrium when the two reactions take place at the same rate.

The forward and reverse reactions continue even after equilibrium has been reached. However, they continue to occur at matching rates. To symbolize this condition, we write

$$N_2(g) + 3\,H_2(g) \rightleftharpoons 2\,NH_3(g) \qquad (1)$$

The symbol $\rightleftharpoons$ always signifies a condition of chemical equilibrium in which the forward and reverse reactions still continue, but at matching rates.

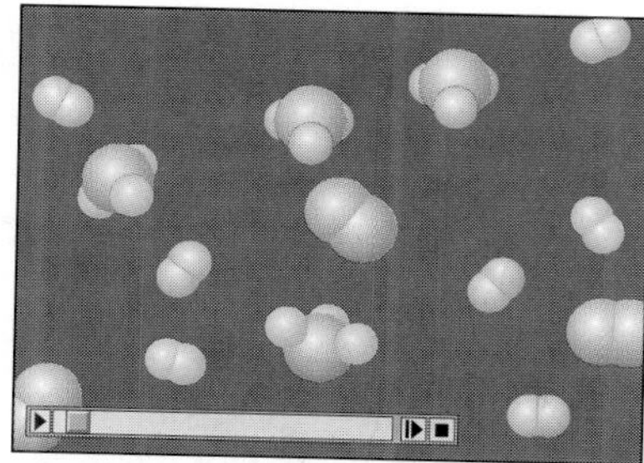

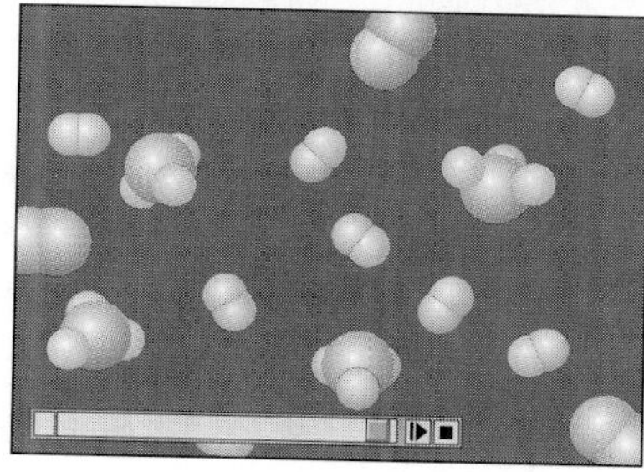

FIGURE 13.2 In an experiment showing that equilibrium is dynamic, a reaction mixture in which N_2 (pairs of blue spheres), D_2 (pairs of yellow spheres), and ND_3 have reached equilibrium is mixed with one with the same concentrations of N_2, H_2, and NH_3. After some time, the concentrations of nitrogen, hydrogen, and ammonia are found to be the same, but the D atoms are distributed among the hydrogen and ammonia molecules.

Chemical reactions reach a state of dynamic equilibrium in which the rates of forward and reverse reactions are equal and there is no net change in composition.

13.2 THE EQUILIBRIUM CONSTANT

The relation needed to describe chemical equilibria quantitatively was identified by the Norwegians Cato Guldberg (a mathematician) and Peter Waage (a chemist) in 1864, long before Haber began his work. We can appreciate their discovery by referring to the data in Table 13.1 for the reaction between SO_2 and O_2:

$$2\,SO_2(g) + O_2(g) \rightleftharpoons 2\,SO_3(g)$$

This reaction is used in another industrial process of great importance, the production of sulfuric acid.

In the experiments reported in Table 13.1, several different mixtures with

TABLE 13.1 Equilibrium data and the equilibrium constant for the reaction $2\,SO_2(g) + O_2(g) \rightleftharpoons 2\,SO_3(g)$ at 1000 K

$[SO_2]$, mol·L^{-1}	$[O_2]$, mol·L^{-1}	$[SO_3]$, mol·L^{-1}	K_c
0.660	0.390	0.0840	0.0415
0.0380	0.220	0.00360	0.0409
0.110	0.110	0.00750	0.0423
0.950	0.880	0.180	0.0408
1.44	1.98	0.410	0.0409
		Average:	0.0413

different initial compositions were prepared at 1000 K and allowed to reach equilibrium. At first sight, there seems to be no pattern in the data. However, suppose we calculate the value of the following quantity:

$$K_c = \frac{[SO_3]^2}{[SO_2]^2[O_2]}$$

The subscript c on K_c tells us that K is defined in terms of molar concentrations, and [X] is the molar concentration of X, with the units struck out. For example, if the molar concentration of SO_2 happens to be 2.3×10^{-3} mol·L^{-1}, then we write $[SO_2] = 2.3 \times 10^{-3}$. The remarkable result is that, within experimental error, the same value of K_c is obtained in each case. This result shows that K_c is characteristic of the composition of the reaction at equilibrium. It is known as the **equilibrium constant** for the reaction.

The equilibrium composition of any reaction can be expressed in a similar way. The relationship is called the **law of mass action:**

> The composition of the equilibrium mixture of the reaction
>
> $$a\text{A} + b\text{B} \rightleftharpoons c\text{C} + d\text{D}$$
>
> is characterized by an equilibrium constant calculated from the expression
>
> $$K_c = \frac{[\text{C}]^c[\text{D}]^d}{[\text{A}]^a[\text{B}]^b}$$
>
> with [X] representing the molar composition (with units omitted) of X at equilibrium.

The products (C and D) occur in the numerator, and the reactants (A and B) in the denominator. Each [X] is raised to a power equal to the stoichiometric coefficient in the balanced chemical equation for the reaction. For example, the equilibrium constant for the reaction

$$4\,NH_3(g) + 5\,O_2(g) \rightleftharpoons 4\,NO(g) + 6\,H_2O(g)$$

is

$$K_c = \frac{[NO]^4[H_2O]^6}{[NH_3]^4[O_2]^5}$$

with the molar concentrations measured at equilibrium.

SELF-TEST 13.1A Write the equilibrium constant for the reaction $3\,ClO_2^-(aq) \rightleftharpoons 2\,ClO_3^-(aq) + Cl^-(aq)$.

[*Answer:* $K_c = [ClO_3^-]^2[Cl^-]/[ClO_2^-]^3$]

SELF-TEST 13.1B Write the equilibrium constant for the reaction $2\,O_3(g) \rightleftharpoons 3\,O_2(g)$.

Each reaction has its own characteristic equilibrium constant, with a value that can be changed only by varying the temperature (Table 13.2). Whatever the initial composition of the reaction mixture, at a given temperature its equilibrium composition will always correspond (in practice, within about 5%) to the value of K_c for that reaction. Therefore, to find the numerical value of an equilibrium constant, we can take any conve-

TABLE 13.2 Equilibrium constants, K_c, for various reactions

Reaction	Temperature, K	K_c
$H_2(g) + Cl_2(g) \rightleftharpoons 2\,HCl(g)$	300	4.0×10^{31}
	500	4.0×10^{18}
	1000	5.1×10^{8}
$H_2(g) + Br_2(g) \rightleftharpoons 2\,HBr(g)$	300	1.9×10^{17}
	500	1.3×10^{10}
	1000	3.8×10^{4}
$H_2(g) + I_2(g) \rightleftharpoons 2\,HI(g)$	298	794
	500	160
	700	54
$2\,BrCl(g) \rightleftharpoons Br_2(g) + Cl_2(g)$	300	377
	500	32
	1000	5
$2\,HD(g) \rightleftharpoons H_2(g) + D_2(g)$	100	0.52
	500	0.28
	1000	0.26
$F_2(g) \rightleftharpoons 2\,F(g)$	500	7.3×10^{-13}
	1000	1.2×10^{-4}
	1200	2.7×10^{-3}
$Cl_2(g) \rightleftharpoons 2\,Cl(g)$	1000	1.2×10^{-7}
	1200	1.7×10^{-5}
$Br_2(g) \rightleftharpoons 2\,Br(g)$	1000	4.1×10^{-7}
	1200	1.7×10^{-5}
$I_2(g) \rightleftharpoons 2\,I(g)$	800	3.1×10^{-5}
	1000	3.1×10^{-3}
	1200	6.8×10^{-2}

nient initial mixture of reagents for the reaction, allow the reaction to reach equilibrium at the temperature of interest, measure the concentrations of the reactants and products, and substitute them into the expression for K_c.

EXAMPLE 13.1 Determining the equilibrium constant, given equilibrium concentrations

Haber mixed some nitrogen and hydrogen and allowed it to react at 500 K until the mixture reached equilibrium with the product, ammonia. When he analyzed the equilibrium mixture, he found it to consist of 0.796 mol·L^{-1} NH_3, 0.305 mol·L^{-1} N_2, and 0.324 mol·L^{-1} H_2. What is the equilibrium constant for the reaction?

STRATEGY Always begin equilibrium constant calculations by writing the chemical equation and the expression for the equilibrium constant. Then substitute the equilibrium concentrations of each substance into the expression for K_c. Raise each concentration to a power equal to the stoichiometric coefficient of that species in the chemical equation.

SOLUTION The chemical equation and the expression for the equilibrium constant are as follows:

$$N_2(g) + 3\,H_2(g) \rightleftharpoons 2\,NH_3(g) \qquad K_c = \frac{[NH_3]^2}{[N_2][H_2]^3}$$

Now substitute the equilibrium molar concentrations (without their units) into the expression for K_c:

$$K_c = \frac{[NH_3]^2}{[N_2][H_2]^3} = \frac{(0.796)^2}{0.305 \times (0.324)^3} = 61.0$$

SELF-TEST 13.2A A reaction important in the gasification of coal is

$$2\,CO(g) + 2\,H_2(g) \rightleftharpoons CH_4(g) + CO_2(g)$$

Determine the equilibrium constant for this reaction at 298 K, given the following equilibrium concentrations: CO, 4.30×10^{-6} mol·L^{-1}; H_2, 1.15×10^{-5} mol·L^{-1}; CH_4, 5.14×10^{4} mol·L^{-1}; and CO_2, 4.12×10^{4} mol·L^{-1}.

[*Answer:* 8.66×10^{29}]

SELF-TEST 13.2B Determine the equilibrium constant for $2\,BrCl(g) \rightleftharpoons Br_2(g) + Cl_2(g)$ at 500 K, given the following equilibrium concentrations: BrCl, 0.131 mol·L^{-1}; Br_2, 3.51 mol·L^{-1}; and Cl_2, 0.156 mol·L^{-1}.

Equilibrium constants are normally given for chemical equations written with the smallest whole numbers for the stoichiometric coefficients. However, if we change the stoichiometric coefficients in a chemical equation (for instance, by multiplying through by a factor), then we must make sure that the equilibrium constant reflects that change. The law of mass action is still valid, but the form of the expression and the calculated value of K_c reflects the way we write the chemical equation. For example, at 700 K,

$$H_2(g) + I_2(g) \rightleftharpoons 2\,HI(g) \qquad K_{c1} = \frac{[HI]^2}{[I_2][H_2]} = 54$$

If we write the equation as

$$2\,H_2(g) + 2\,I_2(g) \rightleftharpoons 4\,HI(g)$$

then the equilibrium constant becomes

$$K_{c2} = \frac{[HI]^4}{[I_2]^2[H_2]^2} = \left(\frac{[HI]^2}{[I_2][H_2]}\right)^2 = (K_{c1})^2 = (54)^2 = 2.9 \times 10^3$$

In general, if we multiply a chemical equation by a factor n, then we raise K_c to the nth power.

Now suppose we reverse the original equation for the reaction:

$$2\,HI(g) \rightleftharpoons H_2(g) + I_2(g)$$

This equation still describes the same equilibrium, but how is the equilibrium constant related to the one we wrote earlier? To find out, we simply write the equilibrium constant from this equation and obtain

$$K_{c3} = \frac{[H_2][I_2]}{[HI]^2}$$

This expression is the inverse ($1/K_c$, or K_c^{-1}) of the expression we obtained previously, so

$$K_{c3} = (54)^{-1} = 1.8 \times 10^{-2}$$

We summarize this conclusion by saying that the equilibrium constant for an equilibrium written in one direction is the inverse of the equilibrium constant for the equilibrium written in the opposite direction.

If a chemical equation can be expressed as the sum of two or more chemical equations, then the equilibrium constant for the overall reaction is the *product* of the equilibrium constants for the component reactions. For example, consider the three gas-phase reactions

$$2\,P(g) + 3\,Cl_2(g) \rightleftharpoons 2\,PCl_3(g) \qquad K'_c = \frac{[PCl_3]^2}{[P]^2[Cl_2]^3}$$

$$PCl_3(g) + Cl_2(g) \rightleftharpoons PCl_5(g) \qquad K''_c = \frac{[PCl_5]}{[PCl_3][Cl_2]}$$

$$2\,P(g) + 5\,Cl_2(g) \rightleftharpoons 2\,PCl_5(g) \qquad K_c = \frac{[PCl_5]^2}{[P]^2[Cl_2]^5}$$

The third reaction is the following sum:

$$\begin{array}{l} 2\,P(g) + 3\,Cl_2(g) \rightleftharpoons 2\,PCl_3(g) \\ PCl_3(g) + Cl_2(g) \rightleftharpoons PCl_5(g) \\ PCl_3(g) + Cl_2(g) \rightleftharpoons PCl_5(g) \\ \hline 2\,P(g) + 5\,Cl_2(g) \rightleftharpoons 2\,PCl_5(g) \end{array}$$

and its equilibrium constant, K_c, can be written

$$K_c = \frac{[PCl_5]^2}{[P]^2[Cl_2]^5} = \frac{[PCl_3]^2}{[P]^2[Cl_2]^3} \times \frac{[PCl_5]}{[PCl_3][Cl_2]} \times \frac{[PCl_5]}{[PCl_3][Cl_2]}$$
$$= K'_c \times K''_c \times K''_c$$

Notice that because we used twice the second reaction in the sum, its equilibrium constant appears twice in the product.

Equilibrium constants are written by dividing the product concentrations (raised to powers equal to their stoichiometric coefficients) by the reactant concentrations (raised to powers equal to their coefficients). Table 13.3 summarizes the relations between equilibrium constants.

TABLE 13.3 Relations between equilibrium constants*

Chemical equation	Equilibrium constant
$aA + bB \rightleftharpoons cC + dD$	K_1
$cC + dD \rightleftharpoons aA + bB$	$K_2 = \frac{1}{K_1} = K_1^{-1}$
$naA + nbB \rightleftharpoons ncC + ndD$	$K_3 = K_1^n$

*In this table, K is either K_c or K_p, with $K_c = [C]^c[D]^d/[A]^a[B]^b$ or $K_p = P_C^c P_D^d / P_A^a P_B^b$. ($K_p$ is introduced in Section 13.4.)

SELF-TEST 13.3A At 500 K, K_c for $H_2(g) + D_2(g) \rightleftharpoons 2\,HD(g)$ is 3.6. What is K_c for $2\,HD(g) \rightleftharpoons H_2(g) + D_2(g)$?

[*Answer:* 0.28]

SELF-TEST 13.3B At 500 K, K_c for $F_2(g) \rightleftharpoons 2\,F(g)$ is 7.3×10^{-13}. What is K_c for $\frac{1}{2}\,F_2(g) \rightleftharpoons F(g)$?

13.3 HETEROGENEOUS EQUILIBRIA

Chemical equilibria in which all reactants and products are in the same phase are called **homogeneous equilibria.** All the equilibria described so far in this chapter are homogeneous. Equilibria in systems having more than one phase are called **heterogeneous equilibria.** For example, the equilibrium between water vapor and liquid water in a closed system is heterogeneous:

$$H_2O(l) \rightleftharpoons H_2O(g)$$

In this reaction there is a gas phase and a liquid phase. Likewise, the equilibrium between a solid and its saturated solution,

$$Ca(OH)_2(s) \rightleftharpoons Ca^{2+}(aq) + 2\,OH^-(aq)$$

is a heterogeneous equilibrium. Here the phases are solid and liquid.

Like the equilibria in these examples, heterogeneous equilibria often involve a pure solid or liquid. We can simplify the equilibrium expression for heterogeneous equilibria involving a pure liquid or a pure solid because *the molar concentration of a pure solid or liquid is a constant, independent of the amount present.* To see why, we need to realize that the molar concentration of a substance is measured by dividing the number of moles of the substance by the volume it occupies. However, moles divided by volume is proportional to mass divided by volume, or the density of the substance. Because density is an intensive property, it follows that the molar concentration must be intensive too. Therefore, the molar concentration of a pure solid or liquid always has the same value, no matter how much or how little of the substance is present.

We do not divide by the *total* volume of the system: we divide by the volume the phase actually occupies.

◀◀ Recall from Section 2.3 that an intensive property is a property with a value that is independent of the size of the sample.

Because the molar concentration of a pure solid or a pure liquid does not change as a reaction approaches equilibrium, we can ignore it in all equilibrium constant calculations. Therefore, when we write an equilibrium constant for a reaction involving pure solids or liquids, we ignore substances in separate solid or liquid phases. For example, we write

$$Ca(OH)_2(s) \rightleftharpoons Ca^{2+}(aq) + 2\,OH^-(aq) \qquad K_c = [Ca^{2+}][OH^-]^2$$

Similarly, in the equilibrium between nickel, carbon monoxide, and nickel carbonyl that is used in the purification of nickel,

$$Ni(s) + 4\,CO(g) \rightleftharpoons Ni(CO)_4(g) \qquad K_c = \frac{[Ni(CO)_4]}{[CO]^4}$$

The pure substances must be present for the equilibrium to exist, but they do not appear in the expression for the equilibrium constant. The concentrations of the gases appear, because their concentrations change during the reaction.

Some reactions in solution involve the solvent as a reactant or product. When the solution is very dilute, the change in solvent concentration due to the reaction is insignificant. In such cases, the solvent is treated as a pure substance and ignored when writing K_c.

Pure liquids and solids are ignored when writing expressions for equilibrium constants.

SELF-TEST 13.4A Write the equilibrium constant for $Ag_2O(s) + 2\,HNO_3(aq) \rightleftharpoons 2\,AgNO_3(aq) + H_2O(l)$.

[*Answer:* $K_c = [AgNO_3]^2/[HNO_3]^2$]

SELF-TEST 13.4B Write the equilibrium constant for $P_4(s) + 5\,O_2(g) \rightleftharpoons P_4O_{10}(s)$.

13.4 GASEOUS EQUILIBRIA

An example of an equilibrium that involves gases is

$$CaCO_3(s) \rightleftharpoons CaO(s) + CO_2(g)$$

In such cases we can replace the molar concentration of each gas in the equilibrium constant by its partial pressure, because the two are proportional to each other. The equilibrium constant is then denoted K_p (Table 13.4). To keep K_p a pure number, we write all pressures in atmospheres, and then strike out the units. For the calcium carbonate decomposition, we write

◀◀ Partial pressures were introduced in Section 5.14.

$$K_p = P_{CO_2}$$

In this case, there are two pure solids ($CaCO_3$ and CaO) in the chemical equation, and neither appears in the expression for K_p. It follows from this expression that for the decomposition of calcium carbonate, the equilibrium partial pressure of the CO_2 in the system is equal to K_p. This relation gives us a very easy way to measure K_p: we simply measure the pressure of carbon dioxide in equilibrium with calcium carbonate and calcium oxide at the temperature of interest. At 800°C, for instance, that pressure is 0.22 atm, so $K_p = 0.22$ at that temperature.

TABLE 13.4 Equilibrium constants, K_p, for various reactions

Reaction	Temperature, K	K_p
$N_2(g) + 3\,H_2(g) \rightleftharpoons 2\,NH_3(g)$	298	6.8×10^5
	400	41
	500	3.6×10^{-2}
$H_2(g) + I_2(g) \rightleftharpoons 2\,HI(g)$	298	794
	500	160
	700	54
$2\,SO_2(g) + O_2(g) \rightleftharpoons 2\,SO_3(g)$	298	4.0×10^{24}
	500	2.5×10^{10}
	700	3.0×10^4
$N_2O_4(g) \rightleftharpoons 2\,NO_2(g)$	298	0.98
	400	47.9
	500	1700

SELF-TEST 13.5A Write the expression for K_p for the reaction $N_2(g) + 3\,H_2(g) \rightleftharpoons 2\,NH_3(g)$.

[*Answer:* $K_p = P_{NH_3}^2/P_{N_2}P_{H_2}^3$]

SELF-TEST 13.5B Write the expression for K_p for the reaction $S(s) + \frac{3}{2}\,O_2(g) \rightleftharpoons SO_3(g)$.

The numerical values of K_p and K_c may be different, so it is important to specify which one is being used. The two quantities are related at a temperature T as follows:

The factor T/K means the numerical value of the temperature in kelvins; it denotes the temperature divided by the unit kelvin.

$$K_p = (RT)^{\Delta n}K_c = \{0.08206(T/K)\}^{\Delta n}K_c$$

where Δn is the difference in the number of gas-phase molecules (products − reactants) and R is the gas constant. This relation follows from the fact that the molar concentration is n/V. It assumes that the gases are all ideal gases; it then follows from the ideal gas law that for each gas A, $P_A = RT(n_A/V) = RT[A]$.

Equilibrium constants for gaseous reactions can be written by using either molar concentrations or partial pressures.

SELF-TEST 13.6A Find the relation between K_p and K_c for the reaction $2\,NO_2(g) \rightleftharpoons N_2O_4(g)$.

[*Answer:* $K_p = (RT)^{-1}K_c$]

SELF-TEST 13.6B Find the relation between K_p and K_c for the reaction $H_2(g) + I_2(g) \rightleftharpoons 2\,HI(g)$.

USING EQUILIBRIUM CONSTANTS

An equilibrium constant tells us—virtually at a glance—whether we can expect a reaction mixture to contain a high concentration of product at equilibrium or whether there will be only a small concentration. It also allows us to predict the *direction* in which the reaction will proceed. This information helps us to predict, for instance, whether an industrial process will produce enough product to be worthwhile. It is important to realize that an equilibrium constant tells us nothing about the *rate* at which equilibrium is reached. Many of the conclusions apply equally to K_c and K_p, so we can write simply K when either equilibrium constant is meant.

13.5 THE EXTENT OF REACTION

The product concentrations appear in the numerator of K, and the reactant concentrations appear in the denominator. If the products are relatively abundant at equilibrium, the numerator will be large and the denominator small. Therefore, K is large when the equilibrium mixture consists mostly of products. In contrast, when the numerator is small and the denominator large, K is small and equilibrium is reached after very little reaction has occurred. Specifically (Fig. 13.3),

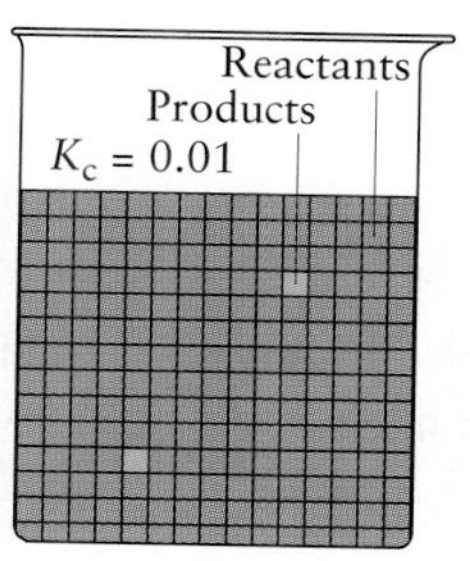

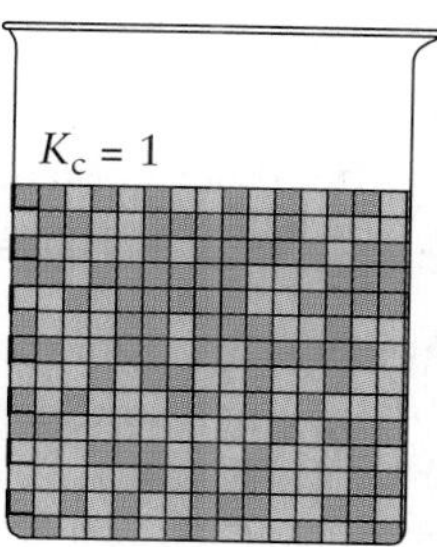

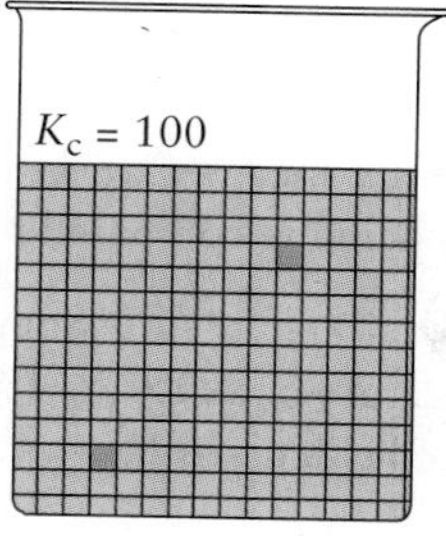

FIGURE 13.3 The size of the equilibrium constant indicates whether the reactants or the products are favored. In this diagram, the reactants are represented by blue squares and the products by yellow squares. Note that reactants are favored when K_c is small (left), products are favored when K_c is large (right), and reactants and products are in almost equal abundance when K_c is close to 1 (middle).

Large values of K (larger than about 10^3): the equilibrium favors the products strongly.

Intermediate values of K (approximately in the range 10^{-3} to 10^3): neither reactants nor products are strongly favored at equilibrium.

Small values of K (smaller than about 10^{-3}): the equilibrium favors the reactants strongly.

For instance, consider the reaction

$$H_2(g) + Cl_2(g) \rightleftharpoons 2\,HCl(g) \qquad K_c = \frac{[HCl]^2}{[H_2][Cl_2]} \tag{2}$$

Experiment shows that $K_c = 4.0 \times 10^{31}$ at 300 K. Such a large value for K_c tells us that the system does not reach equilibrium until most of the reactants have been converted to HCl. We say that the equilibrium favors HCl. In fact, this reaction is fast as well as favorable, because it goes explosively as soon as the reactants are mixed and exposed to sunlight.

Now consider the equilibrium

$$N_2(g) + O_2(g) \rightleftharpoons 2\,NO(g) \qquad K_c = \frac{[NO]^2}{[N_2][O_2]} \tag{3}$$

It is found that $K_c = 4.8 \times 10^{-31}$ at 298 K. The very small value of K_c implies that the reactants N_2 and O_2 will be the dominant species in the system at equilibrium.

EXAMPLE 13.2 Calculating the equilibrium composition

Suppose that in the equilibrium between HCl, Cl_2, and H_2, the concentration of H_2 is 1.0×10^{-17} mol·L^{-1} and that of Cl_2 is 2.0×10^{-16} mol·L^{-1}. What is the equilibrium molar concentration of HCl, given $K_c = 4.0 \times 10^{31}$ for the reaction in Eq. 2?

STRATEGY At equilibrium, the molar concentrations of the reactants and products satisfy the expression for K_c. Therefore, rearrange the expression to give the one unknown concentration, and substitute the data.

SOLUTION The expression for K_c is given in Eq. 2; to find [HCl] we rearrange it into

$$[HCl] = \sqrt{K_c[H_2][Cl_2]}$$

and substitute the data:

$$[HCl] = \sqrt{(4.0 \times 10^{31})(1.0 \times 10^{-17})(2.0 \times 10^{-16})} = 0.28$$

That is, the molar concentration of HCl at equilibrium is 0.28 mol·L^{-1}. This result means that, at equilibrium, the amount of product in the system is overwhelming compared to the amounts of reactants.

SELF-TEST 13.7A Suppose that the equilibrium molar concentrations of H_2 and Cl_2 are both 1.0×10^{-16} mol·L^{-1}. What is the equilibrium molar concentration of HCl, given $K_c = 4.0 \times 10^{31}$?

[*Answer:* [HCl] = 0.63, about 10^{16} times greater than the concentrations of H_2 and Cl_2.]

SELF-TEST 13.7B Suppose that the equilibrium molar concentrations of N_2 and O_2 at 298 K are 0.0010 mol·L^{-1}. What is the equilibrium molar concentration of NO in the reaction in Eq. 3?

An example of a reaction with an equilibrium constant of intermediate value is

$$H_2(g) + I_2(g) \rightleftharpoons 2\,HI(g) \qquad K_c = \frac{[HI]^2}{[H_2][I_2]} \tag{4}$$

for which $K_c = 46$ at 783 K. The equilibrium mixture consists of hydrogen and hydrogen iodide gases and iodine vapor. In this system, if the equilibrium concentrations of two of the species are known, then the same type of calculation used in Example 13.2 can be used to calculate the third. For example, if the concentrations of H_2 and I_2 are each 1.0×10^{-3} mol·L^{-1}, then the equilibrium concentration of HI is 6.8×10^{-3} mol·L^{-1}. In this case, the equilibrium concentrations of the reactants and products are all approximately the same.

If K is large, then products are favored at equilibrium; if K is small, then reactants are favored.

EXAMPLE 13.3 Using K_p to determine a partial pressure

The equilibrium constant for the gas-phase reaction $PCl_5(g) \rightleftharpoons PCl_3(g) + Cl_2(g)$ is $K_p = 25$ at 298 K. The partial pressures of PCl_5 and Cl_2 at equilibrium are 0.0021 atm and 0.48 atm, respectively. What is the equilibrium partial pressure of PCl_3?

STRATEGY The equilibrium constant has an intermediate value, so we should suspect that the equilibrium partial pressures of all the components will be comparable to one another. To calculate the value of the one unknown, write the chemical equation for the reaction and the expression for the equilibrium constant in terms of the partial pressures. Rearrange the expression to give the one unknown on the left, and then substitute the data.

SOLUTION The equilibrium is

$$PCl_5(g) \rightleftharpoons PCl_3(g) + Cl_2(g) \qquad K_p = \frac{P_{PCl_3}P_{Cl_2}}{P_{PCl_5}}$$

The unknown quantity is the partial pressure of PCl_3, so we rearrange the expression for K_p into

$$P_{PCl_3} = \frac{K_p P_{PCl_5}}{P_{Cl_2}}$$

CHEMISTRY AT WORK

A biophysicist is producing ethanol for fuel from waste biomass in an experimental bioreactor. Although living systems do not reach equilibrium, they do experience homeostasis, an oscillating regulation of bodily processes. By using genetic modification, she is developing a strain of bacteria that convert biomass into ethanol without oxidizing it further.

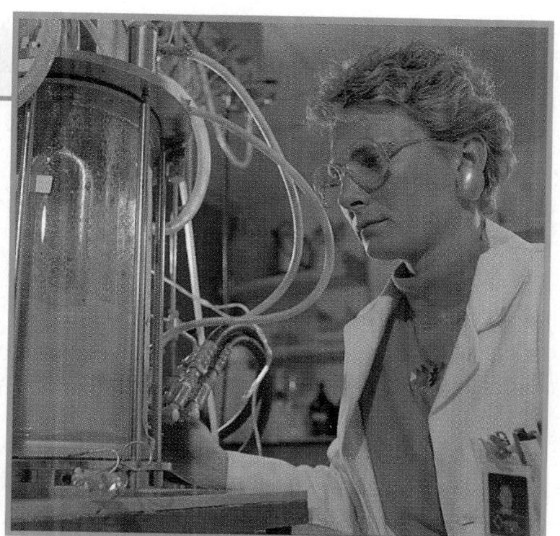

Substitution of the data gives

$$P_{PCl_3} = \frac{25 \times 0.0021}{0.48} = 0.11$$

That is, the partial pressure of PCl_3 at equilibrium in the mixture is 0.11 atm.

SELF-TEST 13.8A Nitrosyl chloride, NOCl, decomposes into NO and Cl_2 according to the equation $2\,NOCl(g) \rightleftharpoons 2\,NO(g) + Cl_2(g)$, with $K_p = 1.8 \times 10^{-2}$ at 500 K. An analysis of a reaction mixture at equilibrium indicates that the partial pressures of NO and Cl_2 are 0.11 atm and 0.84 atm, respectively. What is the equilibrium partial pressure of NOCl?

[*Answer:* 0.75 atm]

SELF-TEST 13.8B Carbon monoxide reacts with oxygen to produce carbon dioxide: $2\,CO(g) + O_2(g) \rightleftharpoons 2\,CO_2(g)$. At 1000 K, $K_p = 2.8 \times 10^{20}$ for this reaction. An analysis of a reaction mixture at equilibrium indicates that the partial pressures of O_2 and CO_2 are 1.4×10^{-9} atm and 75 atm, respectively. What is the equilibrium partial pressure of CO?

13.6 THE DIRECTION OF REACTION

The equilibrium constant is also used to decide whether an arbitrary reaction mixture will tend to form more products or decompose into reactants. We first calculate the **reaction quotient,** Q: this quantity is defined in exactly the same way as the equilibrium constant, but with the molar concentrations (to give Q_c) or partial pressures (to give Q_p) at any stage of the reaction. For example, if at a certain stage of the hydrogen iodide reaction (Eq. 4) we know the molar concentrations of H_2, I_2, and HI are 0.1 mol·L^{-1}, 0.2 mol·L^{-1}, and 0.4 mol·L^{-1}, respectively, then the reaction quotient at that stage of the reaction is

$$Q_c = \frac{[HI]^2}{[H_2][I_2]} = \frac{(0.4)^2}{0.1 \times 0.2} = 8$$

Note that, like K, a reaction quotient is a pure (unitless) number.

To predict whether a particular mixture of reactants and products will tend to produce more products or more reactants, we compare Q

with K. If $Q > K$, then the concentrations of the products are too high (or the concentrations of the reactants too low) for equilibrium. Hence, the reaction has a tendency to proceed in the reverse direction, toward reactants. If $Q < K$, then the reaction tends to go forward and form products. If $Q = K$, then the mixture has its equilibrium composition and has no tendency to change in either direction. This pattern is summarized in Fig. 13.4.

Notice that we say there is a *tendency* toward either reactants or products. The reaction may be so slow that it never reaches equilibrium in the time we are prepared to wait. A mixture of hydrogen and oxygen has a tendency to form water; nevertheless, at room temperature and in the absence of an initiating spark, the reaction is almost infinitely slow. A bottle of benzene has a tendency to decompose into diamonds and hydrogen gas, but that change will never happen, even if we waited a million years.

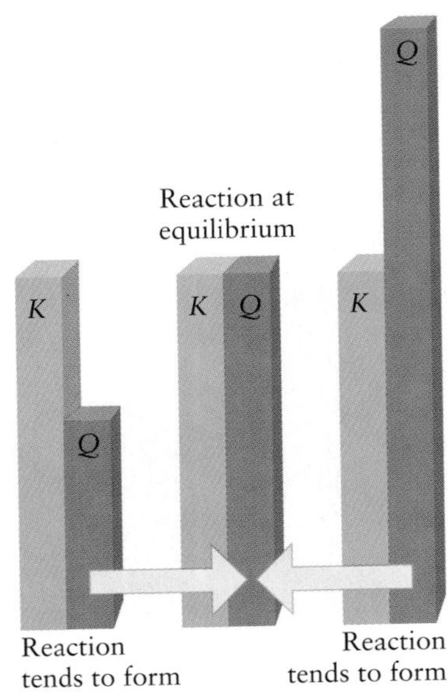

FIGURE 13.4 The relative sizes of the reaction quotient Q and the equilibrium constant K indicate the direction in which a reaction mixture tends to change. The arrows point from reactants to products (left) or from products to reactants (right). There is no tendency to change once the reaction quotient has become equal to the equilibrium constant.

A reaction has a tendency to form products if $Q < K$ and to form reactants if $Q > K$.

EXAMPLE 13.4 Predicting the direction of reaction

A mixture of hydrogen, iodine, and hydrogen iodide, each at 0.0020 mol·L^{-1}, was introduced into a container heated to 783 K. At this temperature, $K_c = 46$ for the reaction in Eq. 4. Predict whether or not more HI has a tendency to form.

STRATEGY Because the equilibrium constant is given in terms of molar concentrations, calculate Q_c and compare it with K_c. If $Q_c > K_c$, the products need to decompose until their concentrations match K_c. The opposite is true if $Q_c < K_c$: in that case, more products need to form.

SOLUTION The reaction quotient is

$$Q_c = \frac{[HI]^2}{[H_2][I_2]} = \frac{(0.0020)^2}{0.0020 \times 0.0020} = 1.0$$

Because $Q_c < K_c$, we conclude that the reaction will tend to form more product and consume reactants.

SELF-TEST 13.9A A mixture of H_2, N_2, and NH_3 with molar concentrations 3.0×10^{-3} mol·L^{-1}, 1.0×10^{-3} mol·L^{-1}, and 2.0×10^{-3} mol·L^{-1}, respectively, was prepared and heated to 500 K, at which temperature $K_c = 61$ for the reaction in Eq. 1. Decide whether ammonia tends to form or decompose.

[*Answer:* $Q_c = 1.5 \times 10^5$; tends to decompose]

SELF-TEST 13.9B For the reaction $N_2O_4(g) \rightleftharpoons 2\,NO_2(g)$ at 298 K, $K_p = 0.98$. A mixture of N_2O_4 and NO_2 with initial partial pressures of 2.4 atm and 1.2 atm, respectively, was prepared at 298 K. Which compound will tend to increase its partial pressure?

13.7 EQUILIBRIUM TABLES

Equilibrium constants can also be used to calculate the equilibrium composition of a reaction mixture for any initial composition. The key idea is that the changes in concentration or partial pressure of each component are linked by the reaction stoichiometry. The easiest way to proceed is to draw up an **equilibrium table** that shows the initial composition, the changes needed to reach equilibrium, and the final equilibrium composition.

EXAMPLE 13.5 Calculating the equilibrium constant for a reaction

Haber started one experiment with a mixture consisting of 0.500 $mol{\cdot}L^{-1}$ N_2 and 0.800 $mol{\cdot}L^{-1}$ H_2 and allowed it to reach equilibrium with the product, ammonia. He found that, at equilibrium at a certain temperature, the molar concentration of NH_3 is 0.150 $mol{\cdot}L^{-1}$. Calculate the equilibrium constant for the reaction at that temperature.

STRATEGY Always begin equilibrium calculations by writing the chemical equation and the expression for the equilibrium constant. Identify the change in concentration of one substance, and use the reaction stoichiometry to calculate the implied changes in the other substances. To keep track of changes, it is convenient to draw up a table with columns headed by each substance and rows giving the initial concentration, the change in concentration, and the final, equilibrium concentration. Finally, substitute the equilibrium concentrations into the expression for K_c.

SOLUTION The reaction is

$$N_2(g) + 3\,H_2(g) \rightleftharpoons 2\,NH_3(g) \qquad K_c = \frac{[NH_3]^2}{[N_2][H_2]^3}$$

The reaction implies that 1 mol $N_2 \simeq$ 2 mol NH_3 and 3 mol $H_2 \simeq$ 2 mol NH_3. Therefore, because the molar concentration of NH_3 increases by 0.150 $mol{\cdot}L^{-1}$ to reach equilibrium, the concentration of N_2 decreases by half that much, 0.075 $mol{\cdot}L^{-1}$, and the concentration of H_2 by 1.5 times, 0.225 $mol{\cdot}L^{-1}$. The equilibrium table, with all molar concentrations in moles per liter, is therefore

	Species		
	N_2	H_2	NH_3
1. Initial molar concentration	0.500	0.800	0
2. Change in molar concentration	−0.075	−0.225	+0.150
3. Equilibrium molar concentration	0.425	0.575	0.150

It follows that

$$K_c = \frac{[NH_3]^2}{[N_2][H_2]^3} = \frac{(0.150)^2}{0.425 \times (0.575)^3} = 0.278$$

SELF-TEST 13.10A A sample of gaseous N_2O_4 with an initial partial pressure of 3.0 atm is prepared at a certain temperature and allowed to react as follows:

$$N_2O_4(g) \rightleftharpoons 2\,NO_2(g)$$

Once equilibrium has been established, it is found that the partial pressure of N_2O_4 has dropped to 1.0 atm. Use an equilibrium table to find the equilibrium partial pressure of NO_2, then calculate K_p.

[*Answer:* P_{NO_2} = 4.0 atm; K_p = 16]

SELF-TEST 13.10B Nitrogen and hydrogen are allowed to react in a container of volume 1.0 L, which initially contained no ammonia. When equilibrium is reached at a certain temperature, the equilibrium concentrations of NH_3, N_2, and H_2 are 0.40 $mol\cdot L^{-1}$, 0.20 $mol\cdot L^{-1}$, and 0.30 $mol\cdot L^{-1}$, respectively. What were the initial concentrations of N_2 and H_2?

Much more commonly, we are likely to be given the equilibrium constant and asked to calculate the composition. The procedure is exactly the same, but we write one of the unknown changes as x and use the reaction stoichiometry to find the other changes. The procedure is summarized in Toolbox 13.1 and illustrated in the examples that follow.

To calculate the equilibrium composition of a reaction mixture, set up an equilibrium table in terms of a change x in one of the species, express the equilibrium constant in terms of x, and solve the resulting equation for x.

TOOLBOX 13.1 How to set up and use an equilibrium table

Begin by writing the balanced chemical equation for the equilibrium and the corresponding expression for the equilibrium constant. Then draw up an equilibrium table by working through the following steps:

Step 1. Set up a table with columns labeled by the species taking part in the reaction. In the first row, show the initial molar concentrations or partial pressures of each species.

This step shows how the reaction system is prepared. As always, strike out the units from molar concentrations (in moles per liter) and partial pressures (in atmospheres). Because the molar concentrations of pure solids and liquids are unchanged by reaction, there is no need to include them in the table.

Step 2. Write the changes in the molar concentrations or partial pressures that are needed for the reaction to reach equilibrium.

It is often the case that we do not know the changes, so write one of them as x and then use the reaction stoichiometry to express the other changes in terms of that x.

Step 3. Write the equilibrium molar concentrations or partial pressures by adding the change in molar concentration or partial pressure (from step 2) to the initial value for each substance (from step 1).

Although a change in concentration may be positive (an increase) or negative (a decrease), the value of the concentration itself must always be positive.

Step 4. Use the equilibrium constant to determine the value of x, the unknown molar concentration or partial pressure at equilibrium.

In some cases the equation for x is a quadratic equation of the form

$$ax^2 + bx + c = 0$$

The two possible solutions of this equation are

$$x = \frac{-b \pm \sqrt{b^2 - 4ac}}{2a}$$

We have to decide which of the two solutions given by this expression is valid (the one with the + sign or the one with the − sign in front of the square root) by seeing which solution is chemically possible. See Example 13.6 for an illustration of this procedure.

In other cases the equation for x in terms of K may be quite complicated. There are then two possible ways forward. One is to use a computer.* The alternative is to look for an approximate solution. One approximation technique can greatly simplify calculations when the change in molar concentration (x) is less than about 5% of the initial concentrations. To use it, assume that x is negligible when added to or subtracted from a number. Thus, we can replace all expressions like $A - x$ or $A - 2x$, for example, by A. When x occurs on its own (not added to or subtracted from another number), it is left unchanged. So, an expression like $(0.1 - x)x$ simplifies to $0.1x$. At the end of the calculation, it is important to verify that the calculated value of x is indeed smaller than about 5% of the initial values. If it is, then the approximation is valid. If not, then we must solve the equation without making an approximation. The approximation procedure is illustrated in Example 13.7.

*You can use the plotting software or the calculator on the CD-ROM that accompanies this text.

EXAMPLE 13.6 Calculating the equilibrium composition

Suppose that 1.50 mol PCl_5 is placed in a reaction vessel of volume 500 mL and allowed to reach equilibrium with its decomposition products phosphorus trichloride and chlorine at 250°C, when $K_c = 1.80$. What is the composition of the equilibrium mixture? All three substances are gases at 250°C.

STRATEGY The general procedure is like that set out in Toolbox 13.1: write the chemical equation and the expression for its equilibrium constant, then set up an equilibrium table with x used to denote the change in molar concentration of the decomposing substance. Use the reaction stoichiometry to express the molar concentrations of the decomposition products in terms of x. Calculate the molar concentrations by dividing the number of moles of each substance by the volume of the container.

SOLUTION The chemical equation is

$$PCl_5(g) \rightleftharpoons PCl_3(g) + Cl_2(g) \qquad K_c = \frac{[PCl_3][Cl_2]}{[PCl_5]}$$

The initial molar concentration of PCl_5 is

$$\text{Molar concentration of } PCl_5 = \frac{1.50}{0.500} = 3.00$$

and we suppose that the change in its molar concentration is $-x$ mol·L^{-1}. Using the chemical equation above, we draw up the following table, with all concentrations in moles per liter:

	Species		
	PCl_5	PCl_3	Cl_2
Step 1. Initial molar concentration	3.00	0	0
Step 2. Change in molar concentration	$-x$	$+x$	$+x$

The stoichiometry of the reaction implies that if the molar concentration of PCl_5 decreases by x, then the molar concentrations of PCl_3 and Cl_2 both increase by x.

Step 3. Equilibrium molar concentration	$3.00 - x$	x	x

These values are the sums of the initial concentrations, step 1, and the changes in concentration brought about by reaction, step 2.

Step 4. Substitution of these equilibrium values into the expression for the equilibrium constant gives

$$K_c = \frac{[PCl_3][Cl_2]}{[PCl_5]} = \frac{x \times x}{3.00 - x}$$

and because we are told that $K_c = 1.80$, the equation we have to solve is

$$1.80 = \frac{x^2}{3.00 - x}$$

This expression rearranges to the quadratic equation

$$x^2 + 1.80x - 5.40 = 0$$

The solutions of this equation (obtained by using the quadratic formula in Toolbox 13.1) are

$$x = \frac{-1.80 \pm \sqrt{(1.80)^2 - 4(1)(-5.40)}}{2} = 1.59 \text{ and } -3.39$$

Because the concentrations must be positive and because (from step 3) x is the molar concentration of each of the products, we select 1.59 as the solution. It follows that at equilibrium

$$[PCl_5] = 3.00 - x = 3.00 - 1.59 = 1.41$$

$$[PCl_3] = x = 1.59$$

$$[Cl_2] = x = 1.59$$

that is, the equilibrium concentrations of PCl_5, PCl_3, and Cl_2 are 1.41 $mol{\cdot}L^{-1}$, 1.59 $mol{\cdot}L^{-1}$, and 1.59 $mol{\cdot}L^{-1}$, respectively.

SELF-TEST 13.11A Bromine monochloride, BrCl, decomposes into bromine and chlorine and reaches the equilibrium $2\,BrCl(g) \rightleftharpoons Br_2(g) + Cl_2(g)$, for which $K_c = 32$ at 500 K. If initially pure BrCl is present at a concentration of 3.30×10^{-3} $mol{\cdot}L^{-1}$, what is its molar concentration in the mixture at equilibrium?

[*Answer:* 3×10^{-4} $mol{\cdot}L^{-1}$]

SELF-TEST 13.11B Chlorine and fluorine react at 2500 K to produce ClF: $Cl_2(g) + F_2(g) \rightleftharpoons 2\,ClF(g)$, $K_c = 20$. If 0.200 mol Cl_2 and 0.100 mol F_2 are added to a 1.00-L container and allowed to come to equilibrium at 2500 K, what is the molar concentration of ClF in the equilibrium mixture?

EXAMPLE 13.7 Calculating the equilibrium composition by approximation

Under certain conditions, nitrogen and oxygen react to form dinitrogen oxide, N_2O. Suppose that a mixture of 0.482 mol N_2 and 0.933 mol O_2 is placed in a reaction vessel of volume 10.0 L and allowed to form N_2O at a temperature for which $K_c = 2.0 \times 10^{-37}$. What will the composition of the equilibrium mixture be?

STRATEGY Proceed exactly as set out in Toolbox 13.1. However, because K_c is so small, we can expect that very little N_2O is produced, suggesting that terms like $A - x$ can be approximated by A itself.

SOLUTION The equilibrium is

$$2\,N_2(g) + O_2(g) \rightleftharpoons 2\,N_2O(g) \qquad K_c = \frac{[N_2O]^2}{[N_2]^2[O_2]}$$

The initial molar concentrations are

$$\text{Molar concentration of } N_2 = \frac{0.482 \text{ mol}}{10.0 \text{ L}} = 0.0482 \text{ mol}{\cdot}L^{-1}$$

$$\text{Molar concentration of } O_2 = \frac{0.933 \text{ mol}}{10.0 \text{ L}} = 0.0933 \text{ mol}{\cdot}L^{-1}$$

Because there is no N_2O present initially, the equilibrium table, with all molar concentrations in moles per liter, is

	Species		
	N_2	O_2	N_2O
Step 1. Initial molar concentration	0.0482	0.0933	0
Step 2. Change in molar concentration	$-2x$	$-x$	$+2x$
Step 3. Equilibrium molar concentration	$0.0482 - 2x$	$0.0933 - x$	$2x$

Step 4. Substitution of the values in the last line of the table into the expression for K_c gives

$$K_c = \frac{[N_2O]^2}{[N_2]^2[O_2]} = \frac{(2x)^2}{(0.0482 - 2x)^2(0.0933 - x)}$$

When rearranged, this equation is a cubic equation (an equation in x^3). It is difficult to solve cubic equations exactly. However, because K_c is very small, we suppose that x will turn out to be so small that we can replace $0.0482 - 2x$ by 0.0482 and $0.0933 - x$ by 0.0933. These approximations simplify the equation above to

$$K_c \approx \frac{4x^2}{(0.0482)^2 \times 0.0933}$$

This equation is easily solved by rearranging it to

$$x \approx \sqrt{\frac{(0.0482)^2 \times 0.0933 \times K_c}{4}}$$

Because $K_c = 2.0 \times 10^{-37}$, we obtain

$$x \approx \sqrt{\frac{(0.0482)^2 \times 0.0933 \times (2.0 \times 10^{-37})}{4}} = 3.3 \times 10^{-21}$$

The value of x is very small compared with 0.0482 (far smaller than 5% of it), so the approximation is valid. We conclude that at equilibrium

$$[N_2] = 0.0482 - 2x \approx 0.0482$$

$$[O_2] = 0.0933 - x \approx 0.0933$$

$$[N_2O] = 2x \approx 6.6 \times 10^{-21}$$

Therefore, the equilibrium concentrations of N_2, O_2, and N_2O are reported as $0.0482\ mol{\cdot}L^{-1}$, $0.0933\ mol{\cdot}L^{-1}$, and $6.6 \times 10^{-21}\ mol{\cdot}L^{-1}$, respectively.

SELF-TEST 13.12A The initial concentrations of nitrogen and hydrogen are $0.010\ mol{\cdot}L^{-1}$ and $0.020\ mol{\cdot}L^{-1}$, respectively. The mixture is heated to a temperature at which $K_c = 0.11$ for the reaction $N_2(g) + 3\,H_2(g) \rightleftharpoons 2\,NH_3(g)$. What is the equilibrium composition of the mixture?

[*Answer:* $0.010\ mol{\cdot}L^{-1}\ N_2$; $0.020\ mol{\cdot}L^{-1}\ H_2$; $9.4 \times 10^{-4}\ mol{\cdot}L^{-1}\ NH_3$]

SELF-TEST 13.12B Hydrogen chloride gas is added to a reaction vessel containing solid iodine until its concentration reaches $0.012\ mol{\cdot}L^{-1}$. At the temperature of the experiment, $K_c = 3.5 \times 10^{-32}$ for the reaction $2\,HCl(g) + I_2(s) \rightleftharpoons 2\,HI(g) + Cl_2(g)$. What is the equilibrium composition of the mixture?

FIGURE 13.5 Henri Le Chatelier (1850–1936).

THE RESPONSE OF EQUILIBRIA TO CHANGE

Chemical equilibria are dynamic and therefore respond to changes in the conditions. When we add or remove a reactant, the equilibrium composition shifts to compensate. The composition also might change in response to a change in pressure. In the following sections we shall see how Haber used this responsiveness to manipulate the equilibrium between ammonia, nitrogen, and hydrogen to increase the production of ammonia.

The French chemist Henri Le Chatelier (Fig. 13.5) found a general principle that lets us predict how the composition of a reaction mixture at equilibrium tends to change when the conditions are changed:

> **Le Chatelier's principle:** When a stress is applied to a system in dynamic equilibrium, the equilibrium tends to adjust to minimize the effect of the stress.

Le Chatelier's principle only suggests an outcome; it does not provide an explanation or lead to a quantitative prediction.

13.8 ADDING AND REMOVING REAGENTS

Let's suppose that the following reaction has reached equilibrium:

$$N_2(g) + 3\,H_2(g) \rightleftharpoons 2\,NH_3(g)$$

Now suppose we pump in more hydrogen gas. According to Le Chatelier's principle, the reaction will tend to minimize the increase in the number of hydrogen molecules. Hydrogen will tend to react with nitrogen; and, as a result, additional ammonia will be formed. If, instead of hydrogen, we were to add some ammonia, then the reaction would tend instead to form reactants at the expense of the added ammonia (Fig. 13.6).

The explanation of this application of Le Chatelier's principle is found in the discussion of the relative sizes of Q and K as summarized in Fig. 13.4. When reactants are added, the reaction quotient Q falls below K momentarily, because the reactant concentrations appear in the denominator of Q. As we have seen, when $Q < K$, the reaction mixture responds by forming products. Likewise, when products are added, Q rises above K momentarily, because products appear in the numerator. Then, because $Q > K$, the reaction mixture responds by forming reactants at the expense of the products until $Q = K$ again.

EXAMPLE 13.8 Predicting the effect on a chemical equilibrium of adding or removing reactants and products

Consider the equilibrium

$$4\,NH_3(g) + 3\,O_2(g) \rightleftharpoons 2\,N_2(g) + 6\,H_2O(g)$$

Predict the effect on each equilibrium concentration of (a) the addition of N_2; (b) the removal of NH_3; (c) the removal of H_2O.

STRATEGY Le Chatelier's principle implies that an equilibrium mixture tends to minimize the effect of a change. Therefore, when a substance is added, the reaction tends to reduce the concentration of that substance. When a substance is removed, the reaction tends to replace it.

SOLUTION (a) The addition of N_2 to the equilibrium mixture causes the reaction to shift in the direction that minimizes the increase in N_2. Therefore, the reaction shifts toward reactant formation. Forming more reactants increases the amounts of NH_3 and O_2 while decreasing the amount of H_2O. The concentration of N_2 remains slightly higher than its original equilibrium value, but lower than its concentration immediately after the additional N_2 was supplied. (b) When NH_3 is removed from the system at equilibrium, the reaction shifts to minimize its loss. The reaction tends to increase the amount of O_2 and decrease the amounts of N_2 and H_2O. The concentration of NH_3 will be somewhat lower than its original equilibrium value, but not as low as it was immediately after the removal of NH_3. (c) The removal of H_2O causes the equilibrium to shift in favor of products to restore (partially) the amount removed. This shift increases the amount of N_2 while decreasing the amounts of NH_3 and O_2. The concentration of H_2O is somewhat lower than its original value.

SELF-TEST 13.13A Consider the equilibrium $SO_3(g) + NO(g) \rightleftharpoons SO_2(g) + NO_2(g)$. Predict the effect on the equilibrium of (a) the addition of NO; (b) the removal of SO_2; (c) the addition of NO_2.

[*Answer:* The equilibrium tends to shift toward (a) products; (b) products; (c) reactants.]

SELF-TEST 13.13B Consider the equilibrium $CO(g) + 2\,H_2(g) \rightleftharpoons CH_3OH(g)$. Predict the effect on the equilibrium of (a) the addition of H_2; (b) the removal of CH_3OH; (c) the removal of CO.

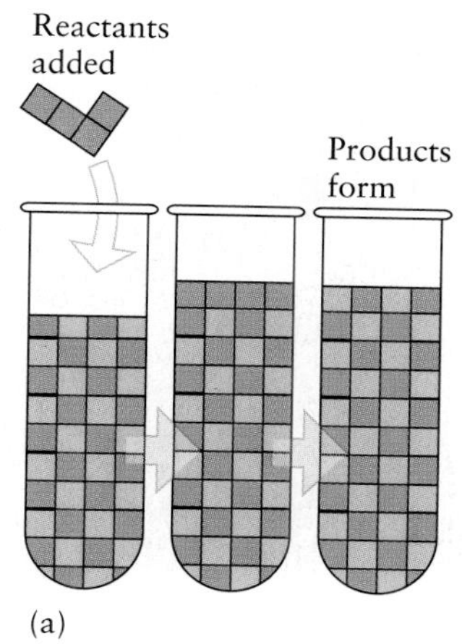

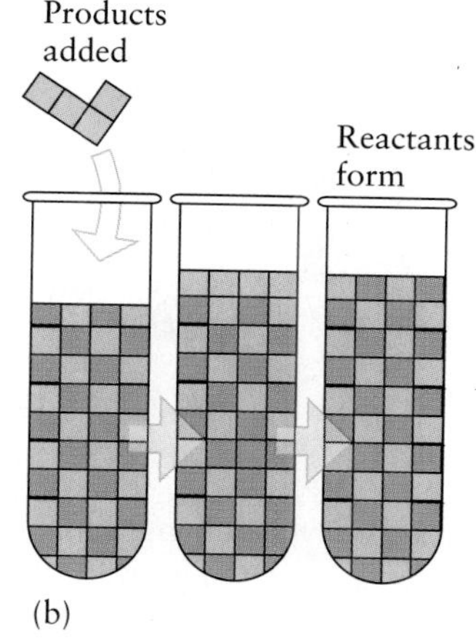

FIGURE 13.6 (a) When a reactant (blue) is added to a reaction mixture at equilibrium, the products have a tendency to form. (b) When a product (yellow) is added instead, the reactants tend to be formed. For this reaction, we have used $K_c = 1$.

A good way of ensuring that a reaction goes on generating a substance is to remove products as they are formed. Industrial processes rarely reach equilibrium. Instead, the product is removed as soon as it is formed: then, in its continuing hunt for equilibrium, the reaction generates more product. Removal of the product is an important feature of the Haber process. Haber's ammonia synthesis never reaches equilibrium; instead, ammonia is continually removed to encourage its further production. Reactions in living systems also tend to be far from equilibrium, for the same reason: there is a ceaseless exchange of products and reactants with the surroundings (Case Study 13).

When a reactant is added to a reaction mixture at equilibrium, the reaction tends to form products; when a reactant is removed, more reactant tends to form; when a product is added, the reaction tends to form reactants; when a product is removed, more product is formed.

13.9 COMPRESSING A REACTION MIXTURE

A gas-phase equilibrium responds to compression—a reduction in volume—of the reaction vessel. According to Le Chatelier's principle, the composition will tend to change in a way that minimizes the increase in pressure resulting from the compression. For instance, the formation of NH_3 from N_2 and H_2 decreases the number of gas-phase molecules in the

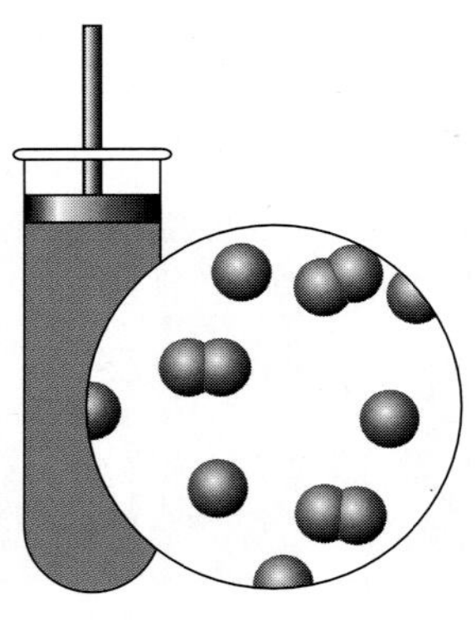

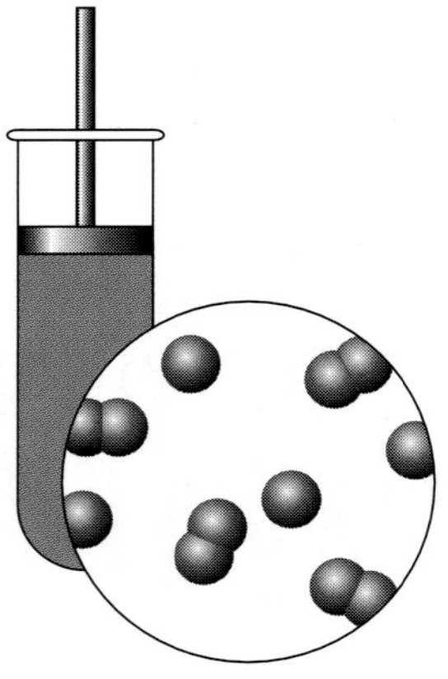

FIGURE 13.7 The same comments apply to any reaction in which there is a change in the number of gas-phase molecules. Le Chatelier's principle predicts that when a reaction at equilibrium is compressed, the number of molecules in the gas phase will tend to decrease. For example, in the reaction pictured, the dissociation of a diatomic molecule, note the increase in the number of diatomic molecules in this system as it is compressed.

container because 4 mol of reactant molecules produce 2 mol of product molecules. The forward reaction therefore decreases the pressure the mixture exerts. When the mixture is compressed, the equilibrium composition will tend to shift in favor of product, for that minimizes the increase in pressure (Fig. 13.7). The opposite response, a tendency for products to decompose, occurs in an expansion. Haber realized that in order to increase the yield of ammonia, he needed to carry out the synthesis with highly compressed gases. The actual industrial process uses pressures of 250 atm and more (Fig. 13.8).

EXAMPLE 13.9 Predicting the effect of compression on an equilibrium

Predict the effect of compression on the equilibrium composition of the reaction mixture in which the equilibria (a) $N_2O_4(g) \rightleftharpoons 2\,NO_2(g)$ and (b) $H_2(g) + I_2(g) \rightleftharpoons 2\,HI(g)$ have been established.

STRATEGY A glance at the chemical equation shows which direction corresponds to a decrease in the number of gas-phase molecules. The composition of the equilibrium mixture will tend to shift in that direction when the reaction mixture is compressed.

SOLUTION (a) In the reverse reaction, two NO_2 molecules combine to form one N_2O_4 molecule. Hence, compression favors the formation of N_2O_4. (b) Because neither direction corresponds to a reduction of gas-phase molecules, compressing the mixture should have little effect on the composition of the equilibrium mixture.

SELF-TEST 13.14A Predict the effect on the equilibrium compositions of compressing $CH_4(g) + H_2O(g) \rightleftharpoons CO(g) + 3\,H_2(g)$.

[*Answer:* Reactants favored]

SELF-TEST 13.14B Predict the effect on the equilibrium compositions of compressing a mixture in which diamond and graphite are in equilibrium. Consider the relative densities of the two solids, which are 3.5 $g\cdot cm^{-3}$ for diamond and 2.0 $g\cdot cm^{-3}$ for graphite.

We can find the explanation of this application of Le Chatelier's principle in the fact that K is a constant at constant temperature. Compressing a system will momentarily change Q, and the reaction tends to adjust in the direction that restores Q to K.

For example, suppose we want to discover the effect of compression on the equilibrium

$$N_2O_4(g) \rightleftharpoons 2\,NO_2(g) \qquad K_c = \frac{[NO_2]^2}{[N_2O_4]}$$

First, we express K_c in terms of the volume of the system. The molar concentration of each gas is the amount, n, of that gas divided by the volume, V, of the reaction vessel:

$$\text{Molar concentration of } NO_2 = \frac{n_{NO_2}}{V}$$

$$\text{Molar concentration of } N_2O_4 = \frac{n_{N_2O_4}}{V}$$

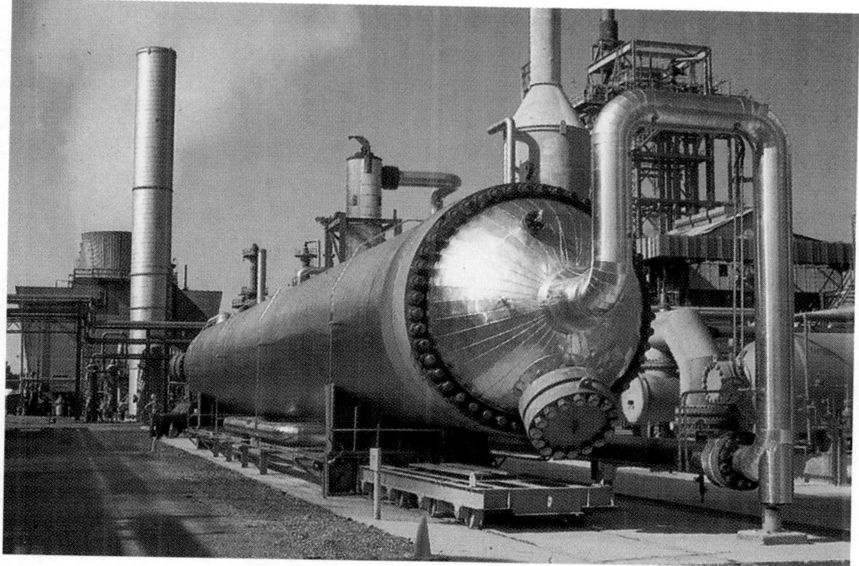

FIGURE 13.8 One of the high-pressure vessels inside which the catalytic synthesis of ammonia takes place.

Hence, the expression for the equilibrium constant has the form

$$K_c = \frac{(n_{NO_2}/V)^2}{(n_{N_2O_4}/V)} = \frac{n_{NO_2}^2}{n_{N_2O_4}} \times \frac{1}{V}$$

For this expression to remain constant when the volume of the system is reduced, the ratio $n_{NO_2}^2/n_{N_2O_4}$ must decrease. That is, the amount of NO_2 must decrease and the amount of N_2O_4 must increase. Therefore, as we have seen before, as the volume of the system is decreased, the equilibrium shifts to a smaller total number of gas molecules.

Suppose that, instead of decreasing the volume, we were to increase the total pressure inside a reaction vessel by pumping in argon or some other inert gas. In this case the equilibrium composition is unaffected. The reacting gases continue to occupy the same volume, so their individual molar concentrations and partial pressures remain unchanged despite the presence of an inert gas.

Compression of a reaction mixture at equilibrium tends to drive the reaction in the direction that reduces the number of gas-phase molecules; increasing the pressure by introducing an inert gas has no effect on the equilibrium composition.

13.10 TEMPERATURE AND EQUILIBRIUM

We can see from Tables 13.2 and 13.4 that the equilibrium constant depends on the temperature. It is found experimentally that for an exothermic reaction, the formation of products is favored by lowering the temperature. Conversely, for an endothermic reaction, the products are favored by an increase in temperature. Le Chatelier's principle is consistent with these observations because we can imagine the heat generated in an exothermic reaction as helping to offset the effect of lowering the temperature. Similarly, we can imagine the heat absorbed in an endothermic reaction as helping to offset the increase in temperature.

CASE STUDY 13 FAR FROM EQUILIBRIUM

ALTHOUGH ALL REACTIONS proceed in the direction of equilibrium, some do not reach it smoothly and uniformly. The beautiful patterns of butterfly wings, the stripes on the pelts of zebras, the myriad colors of tropical fish, the orderly spiral of rose petals, and even the assembly of an organism—a plant, animal, or human being—from a single cell all result from spontaneous chemical reactions that move toward equilibrium in complex paths and as a result generate complex patterns.

Three stages in the progress of the Belousov-Zhabotinskii reaction.

If a reaction mixture is kept far from equilibrium, it may have time for regions of different concentration to develop and disappear. One of the ways to keep a reaction mixture far from equilibrium is to supply it with new reactants and remove products at a balancing rate. Organisms maintain this kind of steady-state condition by eating and excreting. In fact, an organism can be thought of as an open reaction vessel in which reactants are added as fast as they are used up. Our bodies sink into equilibrium only after death, when metabolic processes cease.

The protective spots on this Amur leopard cub result from reactions that were not able to reach equilibrium during the formation of its pelt; thus, the composition of the reaction mixture varied in different regions.

Chemical reactions like those that result in patterns on pelts can be studied in a much simpler form in the laboratory. In 1951 the Russian scientist R. P. Belousov discovered a complex reaction mixture in which the concentrations of reactants and products oscillated in time. The significance of Belousov's findings was not recognized until 1964, when another Russian scientist, A. H. Zhabotinskii, was able to reproduce the reaction. Today it is known as the Belousov-Zhabotinskii, or BZ, reaction. If the reactants are not stirred, the BZ reaction varies spatially, producing beautiful spirals and circles of color, an apparent generation of order out of disorder. In an animal pelt, an oscillating color-generating reaction occurs in the skin of the animal embryo, resulting in characteristic patterns.

Oscillating reactions are the subject of intensive study, because they explain many processes that support life. For example, our heartbeat is maintained by an oscillating

The explanation of the effect of temperature is that high temperatures result in atoms being more likely to be found in high-energy arrangements. In an exothermic reaction, the reactants have more energy than the products. So, as we raise the temperature of a reaction at equilibrium, the atoms are more likely to be found as reactant molecules if the reaction is exothermic. They are more likely to be found as product molecules if the reaction is endothermic (Fig. 13.9). As we lower the temperature, it is more likely that the atoms will be found in low-energy arrangements, so the exothermic direction of reaction, the reverse of the endothermic process, is favored.

An example is the decomposition of carbonates. A reaction such as

$$CaCO_3(s) \rightleftharpoons CaO(s) + CO_2(g)$$

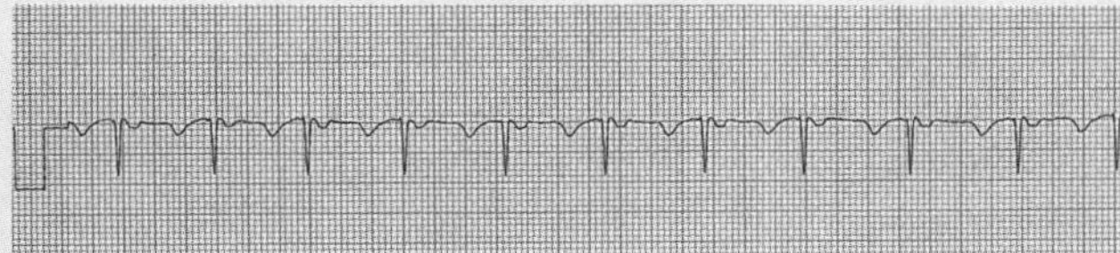

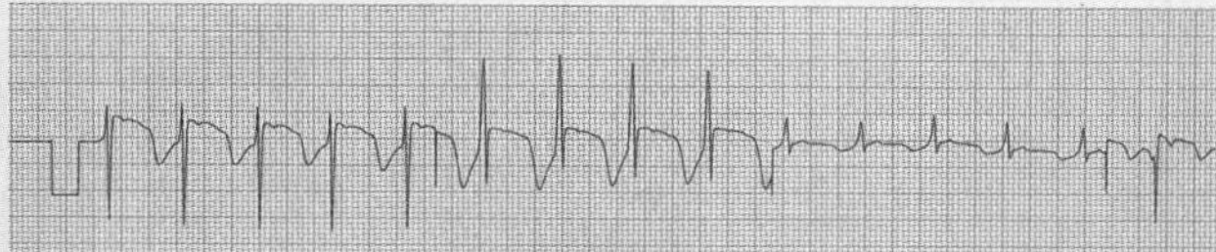

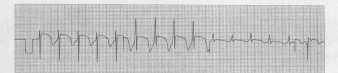

Portions of two electrocardiograms. The upper one shows the regular beating of a normal heart. The lower one shows fibrillation.

process that results from an intricate balance of concentrations of various chemicals. Normally our body maintains this balance automatically and adjusts quickly to changes in rhythm and bodily needs for oxygen. However, if the oscillations become too irregular, then the heart begins to fibrillate, or beat rapidly and irregularly. Then blood is not pumped properly and a heart attack may result.

In an oscillating reaction, several linked reactions occur simultaneously, with the products of one reaction acting as reactants in another reaction. One reaction might increase the concentration of a substance and then another reaction might use that substance up, so the concentrations would vary in time and space. The oscillations continue as long as the reactants continue to be supplied, just as our body continues to work and play and our heart to beat as long as we breathe, drink, eat, and excrete. It is the tendency toward rather than the achievement of equilibrium that keeps oscillating reactions going, generates patterns, and maintains life. In biology, equilibrium is death.

QUESTIONS

1. Many processes contribute to the BZ reaction: three are

$$2\,H^+(aq) + BrO_2^-(aq) + BrO_3^-(aq) \longrightarrow 2\,BrO_2(aq) + H_2O(l)$$

$$2\,Ce^{3+}(aq) + 2\,BrO_2(aq) \longrightarrow 2\,Ce^{4+}(aq) + 2\,BrO_2^-(aq)$$

$$2\,BrO_2^-(aq) \longrightarrow BrO_3^-(aq) + BrO^-(aq)$$

(a) When these reactions are isolated from the other processes in the BZ reaction, they can go to equilibrium. Write the expression for K_c for each step as written above. (b) Write the overall chemical equation for the three steps, and write K_c for the overall reaction. (c) What is the relation between the three K_cs of the individual reaction steps and K_c for the overall reaction?

2. If the first reaction in Question 1 were allowed to reach equilibrium in the absence of the other reactions, in what direction would the reaction shift if the following changes were made to the equilibrium system: (a) some BrO_2 is removed; (b) half the reaction mixture is poured out; (c) some sodium hydroxide is added to neutralize the solution?

3. Like many reactions in living cells, all three reactions specified in Question 1 are redox reactions. Identify the species being oxidized and the species being reduced in each reaction.

is strongly endothermic in the forward direction, and an appreciable partial pressure of carbon dioxide is present at equilibrium only if the temperature is high. For instance, at 800°C the partial pressure is 0.22 atm. If the heating takes place in an open container, this partial pressure is never reached because equilibrium is never reached. The gas drifts away, and the calcium carbonate decomposes completely, leaving a solid residue

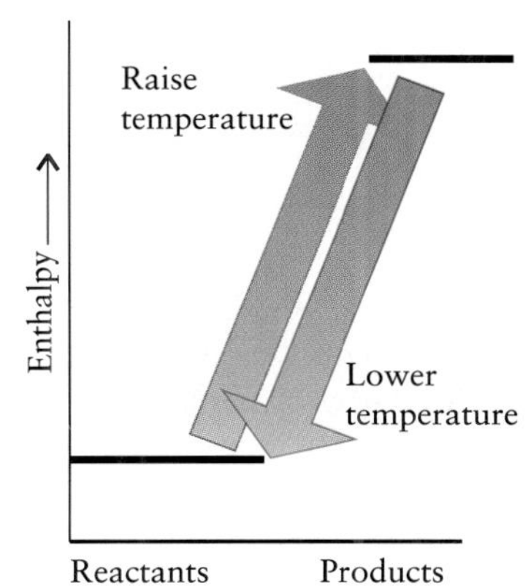

FIGURE 13.9 Le Chatelier's principle predicts that when the temperature is raised, a reaction at equilibrium will tend to shift in the endothermic direction as the atoms move into higher energy arrangements. When the temperature is lowered, a reaction at equilibrium will tend to shift in the exothermic direction as the atoms move into lower energy arrangements.

of CaO—like water evaporating in an open vessel. However, if the surroundings are already so rich in carbon dioxide that its partial pressure exceeds 0.22 atm, then virtually no decomposition occurs: for every CO_2 molecule that is formed, one is converted back to carbonate. This is probably what happens on the hot surface of Venus (Fig. 13.10), where the partial pressure of carbon dioxide is about 87 atm. This high value of the pressure has led to speculation that the planet's surface is rich in carbonates, in spite of its high temperature (about 500°C).

Raising the temperature of an exothermic reaction favors the formation of reactants; raising the temperature of an endothermic reaction favors the formation of products.

EXAMPLE 13.10 Predicting the effect of temperature on an equilibrium

One stage in the manufacture of sulfuric acid is the formation of sulfur trioxide by the reaction of SO_2 with O_2 in the presence of a vanadium (V) oxide catalyst. Predict how the equilibrium composition for the sulfur trioxide synthesis will tend to change when the temperature is raised.

STRATEGY Heating an equilibrium mixture will tend to shift its composition in the endothermic direction of the reaction. A positive reaction enthalpy indicates that the reaction is endothermic in the forward direction. A negative reaction enthalpy indicates that the reaction is endothermic in the *reverse* direction. To find the standard reaction enthalpy, use the standard enthalpies of formation of the species given in Appendix 2A.

SOLUTION The chemical equation is

$$2\,SO_2(g) + O_2(g) \rightleftharpoons 2\,SO_3(g)$$

The standard reaction enthalpy is therefore

$$\begin{aligned}\Delta H^\circ &= (2\text{ mol}) \times \Delta H_f^\circ(SO_3,g) - (2\text{ mol}) \times \Delta H_f^\circ(SO_2,g)\\ &= 2 \times (-395.72\text{ kJ}) - 2 \times (-296.83\text{ kJ})\\ &= -197.78\text{ kJ}\end{aligned}$$

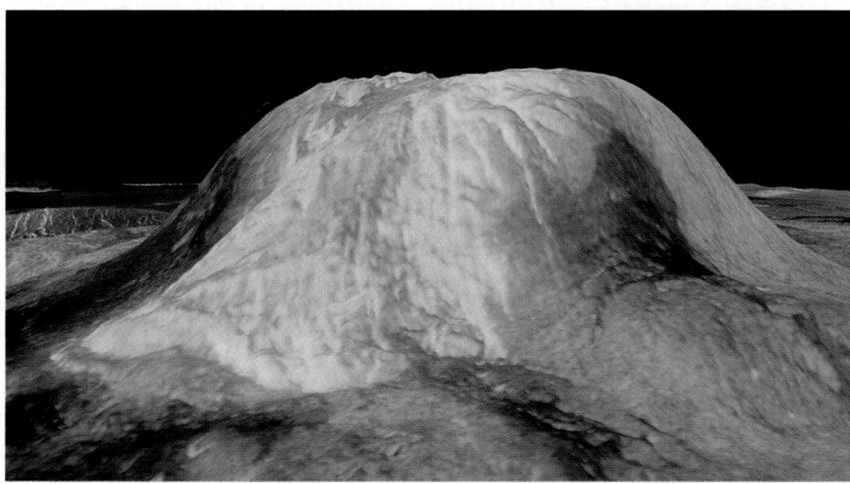

FIGURE 13.10 A radar image of the surface of Venus. Although the rocks are very hot, the partial pressure of carbon dioxide in the atmosphere is so great that carbonates may be abundant.

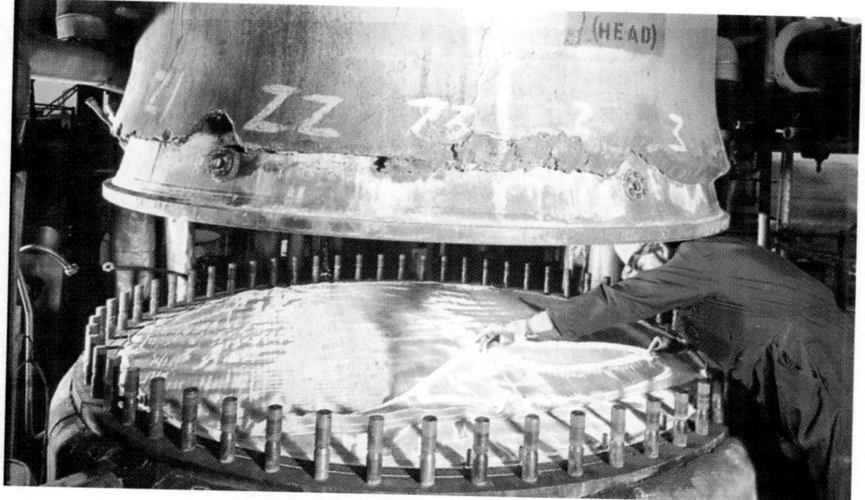

FIGURE 13.11 Installing a rhodium gauze catalyst for the production of nitric acid by the oxidation of ammonia. The catalyst will accelerate the rate of the reaction without affecting the equilibrium composition.

Because the formation of SO_3 is exothermic, the reverse reaction is endothermic. Hence, raising the temperature of the equilibrium mixture favors the decomposition of SO_3.

SELF-TEST 13.15A Predict the effect of raising the temperature on the equilibrium composition of the reaction $N_2O_4(g) \rightleftharpoons 2\ NO_2(g)$. See Appendix 2A for data.

[*Answer:* NO_2 favored]

SELF-TEST 13.15B Predict the effect of lowering the temperature on the equilibrium composition of the reaction $2\ CO(g) + O_2(g) \rightleftharpoons 2\ CO_2(g)$. See Appendix 2A for data.

13.11 CATALYSTS AND HABER'S ACHIEVEMENT

We have already met the term *catalyst,* a substance that increases the rate of a chemical reaction without being consumed itself (Fig. 13.11). We shall see a lot more of catalysts later, when we discuss reaction rates. However, it is important to be aware at this stage that *a catalyst has no effect on the equilibrium composition of a reaction mixture.* A catalyst can speed up the rate at which a reaction reaches equilibrium, but it does not affect the composition at equilibrium. A catalyst acts by providing a faster route to the same destination.

▶▶ The action of catalysts is described in Section 18.12.

At a molecular level, we can think of a catalyst as speeding up both the forward and reverse reactions by the same amount. Therefore, the dynamic equilibrium is unaffected.

We can use the material presented in this chapter to understand one aspect of Haber's achievement. He realized that he had to compress the gases and remove the ammonia as it was formed. As we have seen, compression shifts the equilibrium composition in favor of ammonia, which increases the yield of product. Removing the ammonia also encourages more to be formed. In addition, Haber wanted to work at low temperatures: the synthesis reaction is exothermic, and low temperature favors the formation of product. However, nitrogen and hydrogen combine too

FIGURE 13.12 Fritz Haber (1868–1934) (left) and Carl Bosch (1874–1940).

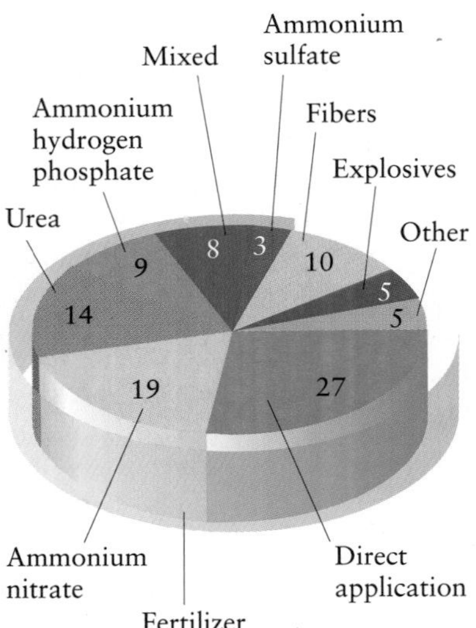

slowly at low temperatures, so Haber had to find a way to increase the rate of the reaction. We shall see how that part of his problem was solved—by selecting the appropriate catalyst—in Chapter 18. The process Haber devised in collaboration with the chemical engineer Carl Bosch (Fig. 13.12) is still in use throughout the world. In the United States alone it accounts for almost the entire annual production of over 1.6×10^{10} kg of ammonia (Fig. 13.13).

A catalyst does not affect the equilibrium composition of a reaction mixture.

FIGURE 13.13 The Haber process is still used to produce almost all the ammonia manufactured in the world. This pie chart shows how the ammonia is used. The figures are percentages. Note that 80 percent—as shown by the green band—is used as fertilizer, either directly or after conversion to another compound.

SKILLS YOU SHOULD HAVE MASTERED

Conceptual

1. Interpret chemical equilibrium as a dynamic process involving change at the molecular level.
2. Describe how the dynamic nature of equilibrium can be verified experimentally.
3. Explain how the value of the reaction quotient, Q, allows the direction of reaction to be predicted.
4. Use Le Chatelier's principle to predict how the equilibrium composition of a reaction mixture is affected by adding or removing reagents, compressing or expanding the mixture, or changing the temperature.

Problem-Solving

1. Write the equilibrium constant for a chemical reaction, given its balanced equation.
2. Determine the equilibrium constant given equilibrium concentrations.

3. Calculate the effect on K of reversing a reaction or multiplying the chemical equation by a factor.

4. Determine an equilibrium concentration or partial pressure, given K and all other equilibrium concentrations.

5. Determine the direction of a reaction, given K and the concentrations of all reactants and products.

6. Use an equilibrium table to calculate K, equilibrium composition, and the initial concentrations of reactants and products.

7. Determine the direction in which a temperature change will shift an equilibrium from the standard enthalpies of formation of the reactants and products.

Descriptive

1. Distinguish homogeneous and heterogeneous equilibria and write equilibrium constants for both types.

EXERCISES

Equilibrium and Equilibrium Constants

13.1 Explain what is wrong with the following statements: (a) Once a reaction has reached equilibrium, all reaction stops. (b) If more reactant is used, the equilibrium constant will have a larger value.

13.2 Explain what is wrong with the following statements: (a) If we can make a reaction go faster, we can increase the amount of product at equilibrium. (b) The reverse reaction does not begin until all the reactants have formed products.

13.3 Figure 13.1 shows how the concentrations of reactants and product change as a function of time for the synthesis of ammonia. Why do the concentrations of the reactants and product stop changing?

13.4 Copy Fig. 13.1 onto another piece of paper and on the same plot draw a second set of curves for the same reaction at the same temperature, but starting with pure NH_3. In what ways do your plots show the reversibility of reactions?

13.5 A 0.10-mol sample of pure ozone, O_3, is placed in a sealed 1.0-L container and the reaction $2\,O_3(g) \rightleftharpoons 3\,O_2(g)$ is allowed to reach equilibrium. A 0.50-mol sample of pure ozone is placed in a second 1.0-L container at the same temperature and allowed to reach equilibrium. Without doing any calculations, predict which of the following will be different in the two containers at equilibrium. Which will be the same? Explain each of your answers: (a) amount of O_2; (b) concentration of O_2; (c) the ratio $[O_2]/[O_3]$; (d) the ratio $[O_2]^3/[O_3]^2$; (e) the ratio $[O_3]^2/[O_2]^3$; (f) the time it takes for equilibrium to be established.

13.6 A 0.10-mol sample of H_2 gas and a 0.10-mol sample of Br_2 gas are placed into a 2.0-L container. The reaction $H_2(g) + Br_2(g) \rightleftharpoons 2\,HBr(g)$ is then allowed to come to equilibrium. A 0.20-mol sample of HBr is placed into a second 2.0-L sealed container at the same temperature and allowed to reach equilibrium with H_2 and Br_2. Which of the following will be different in the two containers at equilibrium? Which will be the same? Explain each of your answers: (a) amount of Br_2; (b) concentration of H_2; (c) the ratio $[HBr]/[H_2][Br_2]$; (d) the ratio $[HBr]/[Br_2]$; (e) the ratio $[HBr]^2/[H_2][Br_2]$; (f) the total pressure in the container.

13.7 Write the equilibrium expressions (in terms of molar concentrations) for the following reactions:
(a) $CO(g) + Cl_2(g) \rightleftharpoons COCl(g) + Cl(g)$
(b) $H_2(g) + Br_2(g) \rightleftharpoons 2\,HBr(g)$
(c) $2\,H_2S(g) + 3\,O_2(g) \rightleftharpoons 2\,SO_2(g) + 2\,H_2O(g)$

13.8 Write the equilibrium expressions (in terms of partial pressures) for the following reactions:
(a) $2\,NO(g) + O_2(g) \rightleftharpoons 2\,NO_2(g)$
(b) $SbCl_5(g) \rightleftharpoons SbCl_3(g) + Cl_2(g)$
(c) $N_2(g) + 2\,H_2(g) \rightleftharpoons N_2H_4(g)$

13.9 For the reaction $N_2(g) + 3\,H_2(g) \rightleftharpoons 2\,NH_3(g)$ at 400 K, $K_p = 41$. Find the value of K_p for each of the following reactions at the same temperature:
(a) $2\,NH_3(g) \rightleftharpoons N_2(g) + 3\,H_2(g)$
(b) $\frac{1}{2}\,N_2(g) + \frac{3}{2}\,H_2(g) \rightleftharpoons NH_3(g)$
(c) $2\,N_2(g) + 6\,H_2(g) \rightleftharpoons 4\,NH_3(g)$

13.10 The equilibrium constant for the reaction $2\,SO_2(g) + O_2(g) \rightleftharpoons 2\,SO_3(g)$ has the value $K_p = 2.5 \times 10^{10}$ at 500 K. Find the value of K_p for each of the following reactions at the same temperature:
(a) $SO_2(g) + \frac{1}{2}\,O_2(g) \rightleftharpoons SO_3(g)$
(b) $SO_3(g) \rightleftharpoons SO_2(g) + \frac{1}{2}\,O_2(g)$
(c) $3\,SO_2(g) + \frac{3}{2}\,O_2(g) \rightleftharpoons 3\,SO_3(g)$

Equilibrium Constants from Equilibrium Amounts

13.11 Use the following data, which were collected at 460°C, and are equilibrium molar concentrations, to determine K_c for the reaction $H_2(g) + I_2(g) \rightleftharpoons 2\,HI(g)$:

$[H_2]$, mol·L^{-1}	$[I_2]$, mol·L^{-1}	$[HI]$, mol·L^{-1}
6.47×10^{-3}	0.594×10^{-3}	0.0137
3.84×10^{-3}	1.52×10^{-3}	0.0169
1.43×10^{-3}	1.43×10^{-3}	0.0100

13.12 Determine K_p from the following equilibrium data collected at 24°C for the reaction $NH_4HS(s) \rightleftharpoons NH_3(g) + H_2S(g)$:

P_{NH_3}, atm	P_{H_2S}, atm
0.309	0.309
0.364	0.258
0.539	0.174

13.13 Determine K_c for each of the following equilibria from the value of K_p:
(a) $2\,NOCl(g) \rightleftharpoons 2\,NO(g) + Cl_2(g)$, $K_p = 1.8 \times 10^{-2}$ at 500 K.
(b) $CaCO_3(s) \rightleftharpoons CaO(s) + CO_2(g)$, $K_p = 167$ at 1073 K

13.14 Determine K_c for each of the following equilibria from the value of K_p:
(a) $2\,SO_2(g) + O_2(g) \rightleftharpoons 2\,SO_3(g)$, $K_p = 3.4$ at 1000 K
(b) $NH_4HS(s) \rightleftharpoons NH_3(g) + H_2S(g)$, $K_p = 9.4 \times 10^{-2}$ at 24°C

Heterogeneous Equilibria

13.15 Explain why the following equilibria are heterogeneous and write the reaction quotient Q_p for each one.
(a) $NH_4HS(s) \rightleftharpoons NH_3(g) + H_2S(g)$
(b) $NH_4(NH_2CO_2)(s) \rightleftharpoons 2\,NH_3(g) + CO_2(g)$
(c) $2\,KNO_3(s) \rightleftharpoons 2\,KNO_2(s) + O_2(g)$

13.16 Explain why the following equilibria are heterogeneous and write the reaction quotient Q_p for each one.
(a) $NH_4Cl(s) \rightleftharpoons NH_3(g) + HCl(g)$
(b) $Na_2CO_3 \cdot 10H_2O(s) \rightleftharpoons Na_2CO_3(s) + 10\,H_2O(g)$
(c) $2\,KClO_3(s) \rightleftharpoons 2\,KCl(s) + 3\,O_2(g)$

13.17 Write the reaction quotients Q_c for
(a) $Cu(s) + Cl_2(g) \rightleftharpoons CuCl_2(s)$
(b) $NH_4NO_3(s) \rightleftharpoons N_2O(g) + 2\,H_2O(g)$
(c) $MgCO_3(s) \rightleftharpoons MgO(s) + CO_2(g)$

13.18 Write the reaction quotients Q_c for
(a) $NH_4HS(s) \rightleftharpoons NH_3(g) + H_2S(g)$
(b) $CuSO_4(s) + 5\,H_2O(l) \rightleftharpoons CuSO_4 \cdot 5H_2O(s)$
(c) $ZnO(s) + CO(g) \rightleftharpoons Zn(s) + CO_2(g)$

The Extent and Direction of Reactions

13.19 In a gas-phase equilibrium mixture of H_2, I_2, and HI at 500 K, $[HI] = 2.21 \times 10^{-3}$ mol·L^{-1} and $[I_2] = 1.46 \times 10^{-3}$ mol·L^{-1}. Given the value of the equilibrium constant in Table 13.2, calculate the concentration of H_2.

13.20 In a gas-phase equilibrium mixture of H_2, Cl_2, and HCl at 1000 K, $[HCl] = 1.45 \times 10^{-3}$ mol·L^{-1} and $[Cl_2] = 2.45 \times 10^{-3}$ mol·L^{-1}. Use the information in Table 13.2 to calculate the concentration of H_2.

13.21 In a gas-phase equilibrium mixture of PCl_5, PCl_3, and Cl_2 at 500 K, $P_{PCl_5} = 0.15$ atm and $P_{Cl_2} = 0.20$ atm. What is the partial pressure of PCl_3 given that $K_p = 25$ for the reaction $PCl_5(g) \rightleftharpoons PCl_3(g) + Cl_2(g)$?

13.22 In a gas-phase equilibrium mixture of $SbCl_5$, $SbCl_3$, and Cl_2 at 500 K, $P_{SbCl_5} = 0.15$ atm and $P_{SbCl_3} = 0.20$ atm. Calculate the equilibrium partial pressure of Cl_2 given that $K_p = 3.5 \times 10^{-4}$ for the reaction $SbCl_5(g) \rightleftharpoons SbCl_3(g) + Cl_2(g)$.

13.23 For the reaction $H_2(g) + I_2(g) \rightleftharpoons 2\,HI(g)$, $K_c = 160$ at 500 K. An analysis of a reaction mixture at 500 K showed that it had the composition 4.8×10^{-3} mol·L^{-1} H_2, 2.4×10^{-3} mol·L^{-1} I_2, and 2.4×10^{-3} mol·L^{-1} HI. (a) Calculate the reaction quotient. (b) Is the reaction mixture at equilibrium? (c) If not, is there a tendency to form more reactants or more products?

13.24 Analysis of a reaction mixture showed that it had the composition 0.417 mol·L^{-1} N_2, 0.524 mol·L^{-1} H_2, and 0.122 mol·L^{-1} NH_3 at 800 K, at which temperature $K_c = 0.278$ for $N_2(g) + 3\,H_2(g) \rightleftharpoons 2\,NH_3(g)$. (a) Calculate the reaction quotient. (b) Is the reaction mixture at equilibrium? (c) If not, is there a tendency to form more reactants or more products?

13.25 A 500-mL reaction vessel at 700 K contains 1.20×10^{-3} mol $SO_2(g)$, 5.0×10^{-4} mol $O_2(g)$, and 1.0×10^{-4} mol $SO_3(g)$. At 700 K, $K_c = 1.7 \times 10^{6}$ for the equilibrium $2\,SO_2(g) + O_2(g) \rightleftharpoons 2\,SO_3(g)$. (a) Calculate the reaction quotient Q_c. (b) Will more $SO_3(g)$ tend to form?

13.26 Given that $K_c = 61$ for the reaction $N_2(g) + 3\,H_2(g) \rightleftharpoons 2\,NH_3(g)$ at 500 K, calculate whether more ammonia will tend to form when a mixture of composition 2.23×10^{-3} mol·L^{-1} N_2, 1.24×10^{-3} mol·L^{-1} H_2, and 1.12×10^{-4} mol·L^{-1} NH_3 is present in a container at 500 K.

Equilibrium Constants from Initial Amounts

13.27 When 0.0172 mol of HI is heated to 500 K in a 2.00-L sealed container, the resulting equilibrium mixture contains 1.90 g of HI. Calculate K_c for the decomposition reaction $2\ HI(g) \rightleftharpoons H_2(g) + I_2(g)$.

13.28 When 1.00 g of gaseous I_2 is heated to 1000 K in a 1.00-L sealed container, the resulting equilibrium mixture contains 0.830 g of I_2. Calculate K_c for the dissociation equilibrium $I_2(g) \rightleftharpoons 2\ I(g)$.

13.29 A 25.0-g sample of ammonium carbamate, $NH_4(NH_2CO_2)$, was placed in an evacuated 250-mL flask and kept at 25°C. At equilibrium, 17.4 mg of CO_2 was present. What is the value of K_c for the decomposition of ammonium carbamate into ammonia and carbon dioxide? The reaction is $NH_4(NH_2CO_2)(s) \rightleftharpoons 2\ NH_3(g) + CO_2(g)$.

13.30 Carbon monoxide and water vapor, each at 200 Torr, were introduced into a 250-mL container. When the mixture had reached equilibrium at 700°C, the partial pressure of CO_2 was 88 Torr. Calculate the value of K_p for the equilibrium $CO(g) + H_2O(g) \rightleftharpoons CO_2(g) + H_2(g)$.

13.31 Data for the equilibrium $CH_3COOH + C_2H_5OH \rightleftharpoons CH_3COOC_2H_5 + H_2O$ at 100°C were obtained for reaction in an organic solvent. The initial concentrations of the reactants are reported in columns 1 and 2 in the following table and the equilibrium concentration of $CH_3COOC_2H_5$ is reported in column 3. Calculate $[H_2O]$ and determine the value of K_c for each experiment.

Initial concentration		Equilibrium concentration
$[CH_3COOH]$, $mol \cdot L^{-1}$	$[C_2H_5OH]$, $mol \cdot L^{-1}$	$[CH_3COOC_2H_5]$, $mol \cdot L^{-1}$
1.00	0.180	0.171
1.00	1.00	0.667
1.00	8.00	0.966

13.32 A 1.0-L reaction vessel was filled to the partial pressures of N_2O_4 given in column 1 in the following table, and the gas was allowed to decompose into NO_2. At 25°C, the total pressure ($P = P_{N_2O_4} + P_{NO_2}$) of the system at equilibrium is listed in column 2. Write the chemical equation and calculate K_p for each set of data.

$P_{N_2O_4}$, atm	P, atm
0.154	0.212
0.333	0.425

Equilibrium Table Calculations

13.33 (a) A sample of 2.0 mmol Cl_2 was sealed into a 2.0-L reaction vessel and heated to 1000 K to study its dissociation into Cl atoms. Use the information in Table 13.2 to calculate the equilibrium composition of the mixture. What is the percentage decomposition of the Cl_2? (b) If 2.0 mmol F_2 were placed into the reaction vessel instead of the chlorine, what would be its equilibrium composition at 1000 K? (c) Use your results from (a) and (b) to determine which is more stable at 1000 K, Cl_2 or F_2.

13.34 (a) A sample of 5.0 mmol Cl_2 was sealed into a 2.0-L reaction vessel and heated to 1200 K, and the dissociation equilibrium was established. What is the equilibrium composition of the mixture? What is the percentage decomposition of the Cl_2? Use the information in Table 13.2. (b) If 5.0 mmol of Br_2 were placed into the reaction vessel instead of the chlorine, what would be the equilibrium composition at 1200 K? (c) Use your results from (a) and (b) to determine which is more stable at 1200 K, Cl_2 or Br_2.

13.35 The initial concentration of HBr in a reaction vessel is $1.2 \times 10^{-3}\ mol \cdot L^{-1}$. If the vessel is heated to 500 K, what is the percentage decomposition of the HBr and the equilibrium composition of the mixture? For $2\ HBr(g) \rightleftharpoons H_2(g) + Br_2(g)$, $K_c = 7.7 \times 10^{-11}$.

13.36 The initial concentration of BrCl in a reaction vessel is $1.4 \times 10^{-3}\ mol \cdot L^{-1}$. If the vessel is heated to 500 K, what is the percentage decomposition of the BrCl and what is the equilibrium composition of the mixture? See Table 13.2 for data on the reaction.

13.37 The equilibrium constant $K_c = 1.1 \times 10^{-2}$ for the reaction $PCl_5(g) \rightleftharpoons PCl_3(g) + Cl_2(g)$ at 400 K. (a) Given that 1.0 g of PCl_5 was initially placed in a 250-mL reaction vessel, determine the molar concentrations in the mixture at equilibrium. (b) What percentage of PCl_5 is decomposed at 400 K?

13.38 For the reaction $PCl_5(g) \rightleftharpoons PCl_3(g) + Cl_2(g)$, $K_c = 0.61$ at 500 K. (a) Calculate the equilibrium molar concentrations of the components in the mixture when 2.0 g of PCl_5 is placed in a 300-mL reaction vessel and allowed to come to equilibrium. (b) What percentage of PCl_5 is decomposed at 500 K?

13.39 When solid NH_4HS and 0.400 mol of gaseous NH_3 were placed into a 2.0-L vessel at 24°C, the equilibrium $NH_4HS(s) \rightleftharpoons NH_3(g) + H_2S(g)$, for which $K_c = 1.6 \times 10^{-4}$, was reached. What are the equilibrium concentrations of NH_3 and H_2S?

13.40 When solid NH_4HS and 0.200 mol of gaseous NH_3 were placed into a 2.0-L vessel at 24°C, the equilibrium $NH_4HS(s) \rightleftharpoons NH_3(g) + H_2S(g)$, for which $K_c = 1.6 \times 10^{-4}$, was reached. What are the equilibrium concentrations of NH_3 and H_2S?

13.41 At 760°C, $K_c = 33.3$ for the reaction $PCl_5(g) \rightleftharpoons PCl_3(g) + Cl_2(g)$. If a mixture that consists of

0.200 mol PCl_5 and 0.600 mol PCl_3 is placed in a 4.00-L reaction vessel and heated to 760°C, what is the equilibrium composition of the system?

13.42 At 760°C, $K_c = 33.3$ for the reaction $PCl_5(g) \rightleftharpoons PCl_3(g) + Cl_2(g)$. If a mixture that consists of 0.200 mol PCl_3 and 0.600 mol Cl_2 is placed in a 8.00-L reaction vessel and heated to 760°C, what is the equilibrium composition of the system?

13.43 The equilibrium constant K_c for the reaction $N_2(g) + O_2(g) \rightleftharpoons 2\,NO(g)$ at 1200°C is 1.00×10^{-5}. Calculate the equilibrium molar concentrations of NO, N_2, and O_2 in a 1.00-L reaction vessel that initially held 0.114 mol N_2 and 0.114 mol O_2.

13.44 The equilibrium constant K_c for the reaction $N_2(g) + O_2(g) \rightleftharpoons 2\,NO(g)$ at 1200°C is 1.00×10^{-5}. Calculate the equilibrium concentrations of NO, N_2, and O_2 in a 10.0-L reaction vessel that initially held 0.014 mol N_2 and 0.214 mol O_2.

13.45 A reaction mixture that consisted of 0.400 mol H_2 and 1.60 mol I_2 was prepared in a 3.00-L flask and heated. At equilibrium, 60.0% of the hydrogen gas had reacted. What is the equilibrium constant for the reaction $H_2(g) + I_2(g) \rightleftharpoons 2\,HI(g)$ at this temperature?

13.46 A reaction mixture that consisted of 0.20 mol N_2 and 0.20 mol H_2 was placed into a 25.0-L reactor and heated. At equilibrium, 5.0% of the nitrogen gas had reacted. What is the value of the equilibrium constant K_c for the reaction $N_2(g) + 3\,H_2(g) \rightleftharpoons 2\,NH_3(g)$ at this temperature?

13.47 The equilibrium constant K_c of the reaction $2\,CO(g) + O_2(g) \rightleftharpoons 2\,CO_2(g)$ is 0.66 at 2000°C. If 0.28 g of CO and 0.032 g of O_2 are placed in a 2.0-L reaction vessel and heated to 2000°C, what will the equilibrium composition of the system be? (You will need a graphing program or calculator to solve the cubic equation in x.)

13.48 In the Haber process for ammonia synthesis, $K_P = 0.036$ for $N_2(g) + 3\,H_2(g) \rightleftharpoons 2\,NH_3(g)$ at 500 K. If a 2.0-L reactor is charged with 0.010 atm of N_2 and 0.010 atm of H_2, what will the equilibrium partial pressures in the mixture be?

13.49 An ester is formed in the reaction of an organic acid with an alcohol. For example, in the reaction of acetic acid, CH_3COOH, with ethanol, C_2H_5OH, in an organic solvent, the ester ethyl acetate, $CH_3COOC_2H_5$, and water form, the reaction being

$$CH_3COOH + C_2H_5OH \rightleftharpoons CH_3COOC_2H_5 + H_2O \qquad K_c = 4.0 \text{ at } 100°C$$

If the initial concentrations of CH_3COOH and C_2H_5OH are 0.32 $mol{\cdot}L^{-1}$ and 6.3 $mol{\cdot}L^{-1}$, respectively, and no products are present initially, what will the equilibrium concentration of the ester be?

13.50 In the esterification reaction specified in Exercise 13.49, if the initial concentrations of CH_3COOH and C_2H_5OH are 0.024 $mol{\cdot}L^{-1}$ and 0.059 $mol{\cdot}L^{-1}$, respectively, but in addition the concentration of the water in the initial mixture is 0.015 $mol{\cdot}L^{-1}$, what will the equilibrium concentration of the ester be?

Effect of Added Reagents

13.51 Consider the equilibrium $CO(g) + H_2O(g) \rightleftharpoons CO_2(g) + H_2(g)$. (a) If the partial pressure of CO_2 is increased, what happens to the partial pressure of H_2? (b) If the partial pressure of CO is decreased, what happens to the partial pressure of CO_2? (c) If the concentration of CO is increased, what happens to the concentration of H_2? (d) If the concentration of H_2O is decreased, what happens to the equilibrium constant for the reaction?

13.52 Consider the equilibrium $CH_4(g) + 2\,O_2(g) \rightleftharpoons CO_2(g) + 2\,H_2O(g)$. (a) If the partial pressure of CO_2 is increased, what happens to the partial pressure of CH_4? (b) If the partial pressure of CH_4 is decreased, what happens to the partial pressure of CO_2? (c) If the concentration of CH_4 is increased, what happens to the equilibrium constant for the reaction? (d) If the concentration of H_2O is decreased, what happens to the concentration of CO_2?

13.53 The four gases NH_3, O_2, NO, and H_2O are mixed in a reaction vessel and allowed to reach equilibrium in the reaction $4\,NH_3(g) + 5\,O_2(g) \rightleftharpoons 4\,NO(g) + 6\,H_2O(g)$. Certain changes (see the following table) are then made to this mixture. Considering each change separately, state the effect (increase, i; decrease, d; or no change, nc) that the change has on the original equilibrium value of the quantity in the second column (or K_c, if that is specified). The temperature and volume are constant unless otherwise noted.

Change	Quantity	Effect		
add NO	amount of H_2O	i	d	nc
add NO	amount of O_2	i	d	nc
remove H_2O	amount of NO	i	d	nc
remove O_2	amount of NH_3	i	d	nc
add NH_3	K_c	i	d	nc
remove NO	amount of NH_3	i	d	nc
add NH_3	amount of O_2	i	d	nc

13.54 The four substances HCl, I_2, HI, and Cl_2 are mixed in a reaction vessel and allowed to reach equilibrium in the reaction $2\,HCl(g) + I_2(s) \rightleftharpoons 2\,HI(g) + Cl_2(g)$. Certain changes (which are specified in the first column in the following table) are then made to this mixture. Considering each change separately, state the effect (increase, i; decrease, d; or no change, nc) that the change has on the original equilibrium value of the quantity in the second column (or K_c, if that is specified). The temperature and volume are constant unless otherwise noted.

Change	Quantity	Effect		
add HCl	amount of HI	i	d	nc
add I_2	amount of Cl_2	i	d	nc
remove HI	amount of Cl_2	i	d	nc
remove Cl_2	amount of HCl	i	d	nc
add HCl	K_c	i	d	nc
remove HCl	amount of I_2	i	d	nc
add I_2	K_c	i	d	nc

Response to Pressure

13.55 State whether reactants or products will be favored by an increase in the total pressure on each of the following equilibria. If no change occurs, explain why that is so.
(a) $2\,O_3(g) \rightleftharpoons 3\,O_2(g)$
(b) $H_2O(g) + C(s) \rightleftharpoons H_2(g) + CO(g)$
(c) $4\,NH_3(g) + 5\,O_2(g) \rightleftharpoons 4\,NO(g) + 6\,H_2O(g)$
(d) $2\,HD(g) \rightleftharpoons H_2(g) + D_2(g)$
(e) $Cl_2(g) \rightleftharpoons 2\,Cl(g)$

13.56 State what happens to the concentration of the indicated substance when the total pressure on each of the following equilibria is increased:
(a) $NO_2(g)$ in $2\,Pb(NO_3)_2(s) \rightleftharpoons 2\,PbO(s) + 4\,NO_2(g) + O_2(g)$
(b) $NO(g)$ in $3\,NO_2(g) + H_2O(l) \rightleftharpoons 2\,HNO_3(aq) + NO(g)$
(c) $HI(g)$ in $2\,HCl(g) + I_2(s) \rightleftharpoons 2\,HI(g) + Cl_2(g)$
(d) $SO_2(g)$ in $2\,SO_2(g) + O_2(g) \rightleftharpoons 2\,SO_3(g)$
(e) $NO_2(g)$ in $2\,NO(g) + O_2(g) \rightleftharpoons 2\,NO_2(g)$

13.57 Consider the equilibrium $4\,NH_3(g) + 5\,O_2(g) \rightleftharpoons 4\,NO(g) + 6\,H_2O(g)$. (a) What happens to the partial pressure of NH_3 when the partial pressure of NO is increased? (b) Does the partial pressure of O_2 increase or decrease when the partial pressure of NH_3 decreases?

13.58 Consider the equilibrium $2\,SO_2(g) + O_2(g) \rightleftharpoons 2\,SO_3(g)$. (a) What happens to the partial pressure of SO_3 when the partial pressure of SO_2 is decreased? (b) If the partial pressure of SO_2 increases, what happens to the partial pressure of O_2?

13.59 The density of quartz (SiO_2) is greater than that of the glassy form of silica (also SiO_2). Would glass or quartz be favored as the pressure is increased?

13.60 The density of red phosphorus is 2.34 $g{\cdot}cm^{-3}$ and that of its white allotrope is 1.82 $g{\cdot}cm^{-3}$. Would you expect the red or the white allotrope to be favored as the pressure is increased?

Response to Temperature

13.61 Predict whether each of the following equilibria will shift toward products or reactants with a temperature increase:
(a) $N_2O_4(g) \rightleftharpoons 2\,NO_2(g)$ $\quad \Delta H° = +57$ kJ
(b) $X_2(g) \rightleftharpoons 2\,X(g)$, where X is a halogen
(c) $Ni(s) + 4\,CO(g) \rightleftharpoons Ni(CO)_4(g)$ $\quad \Delta H° = -161$ kJ
(d) $CO_2(g) + 2\,NH_3(g) \rightleftharpoons CO(NH_2)_2(s) + H_2O(g)$ $\quad \Delta H° = -90$ kJ

13.62 Predict whether each of the following equilibria will shift toward products or reactants with a temperature increase:
(a) $CH_4(g) + H_2O(g) \rightleftharpoons CO(g) + 3\,H_2(g)$ $\quad \Delta H° = +206$ kJ
(b) $CO(g) + H_2O(g) \rightleftharpoons CO_2(g) + H_2(g)$ $\quad \Delta H° = -41$ kJ
(c) $2\,SO_2(g) + O_2(g) \rightleftharpoons 2\,SO_3(g)$ $\quad \Delta H° = -198$ kJ

13.63 A mixture consisting of 2.23×10^{-3} mol N_2 and 6.69×10^{-3} mol H_2 in a 500-mL container was heated to 600 K and allowed to reach equilibrium. Will more ammonia be formed if that equilibrium mixture is then heated to 700 K? For $N_2(g) + 3\,H_2(g) \rightleftharpoons 2\,NH_3(g)$, $K_p = 1.7 \times 10^{-3}$ at 600 K and 7.8×10^{-5} at 700 K.

13.64 A mixture consisting of 1.1 mmol SO_2 and 2.2 mmol O_2 in a 250-mL container was heated to 500 K and allowed to reach equilibrium. Will more sulfur trioxide be formed if that equilibrium mixture is cooled to 25°C? For the reaction $2\,SO_2(g) + O_2(g) \rightleftharpoons 2\,SO_3(g)$, $K_p = 2.5 \times 10^{10}$ at 500 K and 4.0×10^{24} at 25°C.

SUPPLEMENTARY EXERCISES

13.65 Write the reaction quotients Q_c and Q_p for the following reactions:
(a) $S(s) + O_2(g) \rightleftharpoons SO_2(g)$
(b) $SO_3(g) + H_2(g) \rightleftharpoons SO_2(g) + H_2O(g)$
(c) $W(s) + 6\,HCl(g) \rightleftharpoons WCl_6(g) + 3\,H_2(g)$

13.66 The value of the equilibrium constant K_c for the reaction $F_2(g) \rightleftharpoons 2\,F(g)$ is 2.7×10^{-3} at 1200 K. Determine the value of K_p for the reactions (a) $F_2(g) \rightleftharpoons 2\,F(g)$ and (b) $2\,F(g) \rightleftharpoons F_2(g)$.

13.67 The value of the equilibrium constant K_p for the

reaction $2\,SO_2(g) + O_2(g) \rightleftharpoons 2\,SO_3(g)$ is 3.0×10^4 at 700 K. Determine the value of K_c for the reactions
(a) $2\,SO_2(g) + O_2(g) \rightleftharpoons 2\,SO_3(g)$
(b) $SO_3(g) \rightleftharpoons SO_2(g) + \frac{1}{2}\,O_2(g)$

13.68 At 500°C, $K_c = 0.061$ for $N_2(g) + 3\,H_2(g) \rightleftharpoons 2\,NH_3(g)$. Calculate the value of K_c at 500°C for the reactions
(a) $\frac{1}{6}\,N_2(g) + \frac{1}{2}\,H_2(g) \rightleftharpoons \frac{1}{3}\,NH_3(g)$
(b) $NH_3(g) \rightleftharpoons \frac{1}{2}\,N_2(g) + \frac{3}{2}\,H_2(g)$
(c) $4\,NH_3(g) \rightleftharpoons 2\,N_2(g) + 6\,H_2(g)$

13.69 At 500 K, $K_c = 0.061$ for $N_2(g) + 3\,H_2(g) \rightleftharpoons 2\,NH_3(g)$. If analysis shows that the composition is 3.00 $mol{\cdot}L^{-1}$ N_2, 2.00 $mol{\cdot}L^{-1}$ H_2, and 0.500 $mol{\cdot}L^{-1}$ NH_3, is the reaction at equilibrium? If not, in which direction does the reaction tend to proceed to reach equilibrium?

13.70 At 2500 K, the equilibrium constant is $K_c = 20$ for the reaction $Cl_2(g) + F_2(g) \rightleftharpoons 2\,ClF(g)$. An analysis of a reaction vessel at 2500 K revealed the presence of 0.18 $mol{\cdot}L^{-1}$ Cl_2, 0.31 $mol{\cdot}L^{-1}$ F_2, and 0.92 $mol{\cdot}L^{-1}$ ClF. Will ClF tend to form or to decompose as the reaction proceeds toward equilibrium?

13.71 At 500 K, the equilibrium constant is $K_c = 0.031$ for the reaction $Cl_2(g) + Br_2(g) \rightleftharpoons 2\,BrCl(g)$. If the equilibrium composition is 0.22 $mol{\cdot}L^{-1}$ Cl_2 and 0.097 $mol{\cdot}L^{-1}$ BrCl, what is the equilibrium concentration of Br_2?

13.72 The equilibrium constant is $K_p = 3.5 \times 10^4$ for reaction $PCl_3(g) + Cl_2(g) \rightleftharpoons PCl_5(g)$ at 760°C. At equilibrium, the partial pressure of PCl_5 was 2.2×10^{-4} atm and that of PCl_3 was 1.33 atm. What was the equilibrium partial pressure of Cl_2?

13.73 A reaction mixture consisting of 2.00 mol CO and 3.00 mol H_2 is placed into a 10.0-L reaction vessel and heated to 1200 K. At equilibrium, 0.478 mol CH_4 was present in the system. Determine the value of K_c for the reaction $CO(g) + 3\,H_2(g) \rightleftharpoons CH_4(g) + H_2O(g)$.

13.74 A mixture consisting of 1.0 mol $H_2O(g)$ and 1.0 mol CO(g) is placed in a 10.0-L reaction vessel at 800 K. At equilibrium, 0.665 mol $CO_2(g)$ is present as a result of the reaction $CO(g) + H_2O(g) \rightleftharpoons CO_2(g) + H_2(g)$. What are (a) the equilibrium concentrations for all substances and (b) the value of K_c?

13.75 A reaction mixture was prepared by mixing 0.100 mol SO_2, 0.200 mol NO_2, 0.100 mol NO, and 0.150 mol SO_3 in a 5.00-L reaction vessel. The reaction $SO_2(g) + NO_2(g) \rightleftharpoons NO(g) + SO_3(g)$ is allowed to reach equilibrium at 460°C, when $K_c = 85.0$. What is the equilibrium concentration of each substance?

13.76 A 0.100-mol sample of H_2S is placed in a 10.0-L reaction vessel and heated to 1132°C. At equilibrium, 0.0285 mol H_2 is present. Calculate the value of K_c for the reaction $2\,H_2S(g) \rightleftharpoons 2\,H_2(g) + S_2(g)$.

13.77 A mixture of 0.0560 mol O_2 and 0.0200 mol N_2O is placed in a 1.00-L reaction vessel at 25°C. When the reaction $2\,N_2O(g) + 3\,O_2(g) \rightleftharpoons 4\,NO_2(g)$ is at equilibrium, 0.0200 mol NO_2 is present. (a) What are the equilibrium concentrations of O_2 and N_2O? (b) What is the value of K_c?

13.78 At 500 K, 1.0 mol NOCl is 9.0% dissociated in a 1.0-L vessel. Calculate the value of K_c for the reaction $2\,NOCl(g) \rightleftharpoons 2\,NO(g) + Cl_2(g)$.

13.79 The equilibrium constant $K_c = 0.56$ for the reaction $PCl_3(g) + Cl_2(g) \rightleftharpoons PCl_5(g)$ at 250°C. Upon analysis, it was found that 1.5 mol PCl_5, 3.0 mol PCl_3, and 0.50 mol Cl_2 were present in a 500-mL reaction vessel at 250°C. (a) Is the reaction at equilibrium? (b) If not, in which direction does it tend to proceed and (c) what is the equilibrium composition of the reaction system?

13.80 A 1.50-mol sample of PCl_5 is placed into a 500-mL reaction vessel. What is the concentration of each substance present in the system when the reaction $PCl_5(g) \rightleftharpoons PCl_3(g) + Cl_2(g)$ has reached equilibrium at 250°C (when $K_c = 1.80$)?

13.81 At 25°C, $K_p = 3.2 \times 10^{-34}$ for the reaction $2\,HCl(g) \rightleftharpoons H_2(g) + Cl_2(g)$. If a 1.0-L reaction vessel is charged with HCl at a pressure of 0.22 atm, what are the equilibrium partial pressures of HCl, H_2, and Cl_2?

13.82 If 4.00 L HCl(g) at 1.00 atm and 273 K and 26.0 g $I_2(s)$ are transferred to a 12.00-L reaction vessel and heated to 25°C, what will the equilibrium concentrations of HCl, HI, and Cl_2 be? $K_c = 1.6 \times 10^{-34}$ at 25°C for $2\,HCl(g) + I_2(s) \rightleftharpoons 2\,HI(g) + Cl_2(g)$.

13.83 A 30.1-g sample of NOCl is placed into a 200-mL reaction vessel and heated to 500 K. The value of K_p for the decomposition of NOCl at 500 K in the reaction $2\,NOCl(g) \rightleftharpoons 2\,NO(g) + Cl_2(g)$ is 1.13×10^{-3}. (a) What are the equilibrium partial pressures NOCl, NO, and Cl_2? (b) What is the percentage decomposition of NOCl at this temperature?

13.84 At 25°C, $K_c = 4.01 \times 10^{-2}$ for $N_2O_4(g) \rightleftharpoons 2\,NO_2(g)$. If 2.50 g of N_2O_4 and 0.33 g of NO_2 are placed in a 2.0-L reaction vessel, what are the equilibrium concentrations of N_2O_4 and NO_2?

13.85 The photosynthesis reaction is $6\,CO_2(g) + 6\,H_2O(l) \longrightarrow C_6H_{12}O_6(s) + 6\,O_2(g)$, and $\Delta H° = +2802$ kJ. Suppose that the reaction is at equilibrium. State the effect that each of the following changes will have on the equilibrium composition (tend to shift toward the formation of reactants, tend to shift toward the formation of products, or have no effect). (a) The partial pressure of O_2 is increased. (b) The system is compressed. (c) The amount of CO_2 is increased. (d) The temperature is increased. (e) Some of the $C_6H_{12}O_6$ is removed. (f) Water is added. (g) The partial pressure of CO_2 is decreased.

13.86 Use Le Chatelier's principle to predict the effect (increase i; decrease d; no change, nc) that the change given in the first column of the table below has on the quantity in the second column for the following equilibrium system:

$$5\,CO(g) + I_2O_5(s) \rightleftharpoons I_2(g) + 5\,CO_2(g) \qquad \Delta H° = -1175\ kJ$$

Each change is applied separately to the system.

Change	Quantity	Effect
decrease volume	K_c	i d nc
increase volume	amount of CO	i d nc
raise temperature	K_c	i d nc
add I_2	amount of CO_2	i d nc
add I_2O_5	amount of I_2	i d nc
remove CO_2	amount of I_2	i d nc
compress	amount of CO	i d nc
reduce temperature	amount of CO_2	i d nc
add CO_2	amount of I_2O_5	i d nc
add CO_2	amount of CO_2	i d nc

13.87 A 3.00-L reaction vessel is filled with 0.150 mol CO, 0.0900 mol H_2, and 0.180 mol CH_3OH. Equilibrium is reached in the presence of a zinc oxide-chromium(III) oxide catalyst, and at 300°C, $K_c = 1.1 \times 10^{-2}$ for the reaction, $CO(g) + 2\,H_2(g) \rightleftharpoons CH_3OH(g)$. (a) As the reaction approaches equilibrium, will the molar concentration of CH_3OH increase, decrease, or remain unchanged? Explain your answer. (b) What is the equilibrium composition of the mixture?

13.88 $K_c = 0.395$ at 350°C for the reaction $2\,NH_3(g) \rightleftharpoons N_2(g) + 3\,H_2(g)$. A 15.0-g sample of NH_3 is placed in a 5.00-L reaction vessel and heated to 350°C. What are the equilibrium concentrations of NH_3, N_2, and H_2?

13.89 At 25°C, $K_c = 4.66 \times 10^{-3}$ for the reaction $N_2O_4(g) \rightleftharpoons 2\,NO_2(g)$. If a 2.50-g N_2O_4 sample is placed in a 2.00-L reaction flask, what will be the composition of the equilibrium mixture?

CHALLENGING EXERCISES

13.90 Let the equilibrium constants for the reactions $2\,H_2O(g) \rightleftharpoons 2\,H_2(g) + O_2(g)$ and $2\,CO_2(g) \rightleftharpoons 2\,CO(g) + O_2(g)$ be K_{p1} and K_{p2}, respectively. Show that the equilibrium constant for the reaction $CO_2(g) + H_2(g) \rightleftharpoons H_2O(g) + CO(g)$ is $K_{p3} = (K_{p2}/K_{p1})^{1/2}$, and evaluate it at 1565 K, at which temperature $K_{p1} = 1.6 \times 10^{-11}$ and $K_{p2} = 1.3 \times 10^{-10}$.

13.91 Suppose that in an esterification reaction in which an organic acid reacts with an alcohol in an organic solvent to produce an ester (see Exercise 13.49), an amount A mol of acid and B mol of alcohol are mixed and heated to 100°C. Find an expression for the amount (in moles) of ester that is present at equilibrium, in terms of A, B, and K_c. Evaluate the expression for $A = 1.0$, $B = 0.50$, and $K_c = 3.5$.

13.92 Let α be the fraction of PCl_5 molecules that have decomposed to PCl_3 and Cl_2 in the reaction $PCl_5(g) \rightleftharpoons PCl_3(g) + Cl_2(g)$ in a constant-volume container, so that the amount of PCl_5 at equilibrium is $n(1 - \alpha)$, where n is the amount present initially. Derive an equation for K_p in terms of α and the total pressure P, and solve it for α in terms of P. Calculate the fraction decomposed at 556 K, at which temperature $K_p = 4.96$, and the total pressure is (a) 0.50 atm; (b) 1.0 atm.

13.93 The reaction $N_2O_4(g) \rightleftharpoons 2\,NO_2(g)$ is allowed to reach equilibrium in chloroform solution at 25°C. The equilibrium concentrations are 0.405 $mol{\cdot}L^{-1}$ N_2O_4 and 2.13 $mol{\cdot}L^{-1}$ NO_2. (a) Calculate K_c for the reaction. (b) An additional 1.00 mol NO_2 is added to 1.00 L of the solution and the system is allowed to reach equilibrium again at the same temperature. Use Le Chatelier's principle to predict the direction of change (increase, decrease, or no change) for N_2O_4, NO_2, and K_c after the addition of NO_2. (c) Calculate the final equilibrium concentrations after the addition of NO_2 and confirm that your predictions in (b) were valid. If they do not agree, check your procedure and repeat it if necessary.

13.94 The van't Hoff equation relates the equilibrium constant K_p' at a temperature T' to its value K_p at T:

$$\ln\left(\frac{K_p'}{K_p}\right) = \frac{\Delta H°}{R}\left(\frac{1}{T} - \frac{1}{T'}\right)$$

where $\Delta H°$ is the enthalpy of the forward process. The temperature dependence of the equilibrium constant of the reaction $N_2(g) + O_2(g) \rightleftharpoons 2\,NO(g)$, which makes an important contribution to atmospheric nitrogen oxides, can be expressed as $\ln K_p = 2.5 - 21700/(T/K)$. What is the standard enthalpy of the forward reaction?

13.95 The dissociation vapor pressure of ammonium hydrogen sulfide (NH_4HS) is 501 Torr at 298.3 K and 919 Torr at 308.8 K. Using the information in Exercise 13.94, estimate its enthalpy of dissociation and the temperature at which it would "boil" when the external pressure is 1 atm.

CHAPTER 14

Thunderstorms—like this one in Australia—stimulate chemical reactions in the atmosphere. The products of these reactions dissolve in water, which falls from the sky as acid rain. In this chapter we see why some substances are acids, why some acids are stronger than others, and why acids—like their partners the bases—are such important chemicals.

PROTONS IN TRANSITION: ACIDS AND BASES

Acid rain has been blamed for outbreaks of severe respiratory ailments, destruction of forests, pollution of lakes, and the erosion of marble and limestone (see Case Study 14). Yet rain is naturally slightly acidic: it contains dissolved carbon dioxide, which reacts with water to give carbonic acid, H_2CO_3. Why, then, the concern? A part of the answer is that automobile engines and coal-fired power plants generate oxides of sulfur and nitrogen in great abundance. These oxides react with the water in raindrops to form acids that are much stronger than carbonic acid. But what does it mean to say that some acids are "stronger" than others? What difference does it make if rain contains strong or weak acids? Indeed, why do the oxides of many nonmetals form acidic solutions in water?

Acid rain is less of a problem in the western and southwestern United States. There, the acids in rain are often neutralized before they fall. Mineral dust from the dry soils of the region contains metal oxides and hydroxides and is blown up into the clouds. These compounds—which are bases—react with the acids. But how do metal oxides and hydroxides react with acids? What do we really mean by a base?

In this chapter we build on the information we have gathered since we first met acids and bases in Chapter 3. In particular, we shall see how to use equilibrium calculations to discuss their properties quantitatively. We shall also see how the strengths of acids are related to their molecular structures.

WHAT ARE ACIDS AND BASES?

14.1 Brønsted-Lowry acids and bases
14.2 Conjugate acids and bases
14.3 Proton exchange between water molecules
14.4 The pH scale
14.5 The pOH of solutions

WEAK ACIDS AND BASES

14.6 Proton transfer equilibria
14.7 The conjugate seesaw
14.8 The special role of water
14.9 Why are some acids weak and others strong?
14.10 The strengths of oxoacids

THE pH OF SOLUTIONS OF WEAK ACIDS AND BASES

14.11 Solutions of weak acids
14.12 Solutions of weak bases
14.13 Polyprotic acids and bases

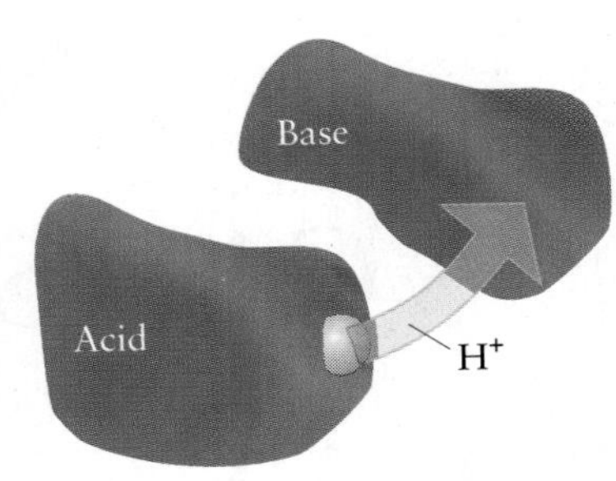

FIGURE 14.1 A Brønsted acid is a proton donor, and a Brønsted base is a proton acceptor. In this context, a proton is an H^+ ion.

WHAT ARE ACIDS AND BASES?

In Chapter 3 we saw that an acid is a molecular compound, such as HCl or CH_3COOH, that contains hydrogen and produces hydronium ions (H_3O^+) in water. We also saw that a base is a compound, such as NaOH, CaO, or NH_3, that produces hydroxide ions (OH^-) in water. However, these definitions were improved early in the twentieth century when chemists realized that they could discuss acids and bases by focusing on the ability of ions and molecules to donate or accept protons (hydrogen ions, H^+). The new definitions are much broader in scope and more helpful for understanding acids and bases.

14.1 BRØNSTED-LOWRY ACIDS AND BASES

In 1923 the Danish chemist Johannes Brønsted proposed that an **acid** is a proton donor and a **base** is a proton acceptor. The same definitions were proposed independently by the English chemist Thomas Lowry, and the theory based on them is widely called the **Brønsted-Lowry theory** of acids and bases (Fig. 14.1). Whenever we refer to acids and bases in this chapter, we use the Brønsted-Lowry definitions.

Brønsted acids include all the substances that we commonly regard as acids. For example, when a hydrogen chloride molecule dissolves in water, it donates a proton to a neighboring H_2O molecule (Fig. 14.2):

$$HCl(aq) + H_2O(l) \longrightarrow H_3O^+(aq) + Cl^-(aq)$$

Because HCl donates a proton, it is an acid according to the Brønsted definition. As we saw in Section 3.8, HCl is a strong acid in the sense that almost every HCl molecule loses its proton in water. The Brønsted definition also includes the possibility that an ion is an acid. For instance, a hydrogen carbonate ion, HCO_3^-, one of the species present in natural waters, can act as a proton donor (Fig. 14.3):

$$HCO_3^-(aq) + H_2O(l) \longrightarrow H_3O^+(aq) + CO_3^{2-}(aq)$$

Very few HCO_3^- ions give up their protons, so the hydrogen carbonate ion is an example of a weak acid.

Brønsted bases include all the substances that we commonly regard as bases. For instance, when a metal oxide (such as CaO) dissolves in water, the strong negative charge of the small O^{2-} ion pulls a proton out of a neighboring H_2O molecule (Fig. 14.4). The oxide ion forms a cova-

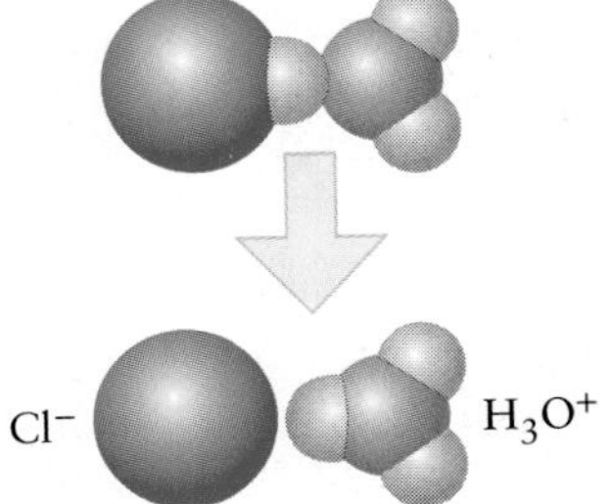

FIGURE 14.2 When an HCl molecule dissolves in water, a hydrogen bond forms between the H atom of HCl and the O atom of a neighboring H_2O molecule, and the nucleus of the hydrogen atom (H^+) is pulled out of the HCl molecule.

FIGURE 14.3 Stalactites hang from the roof of a cave and stalagmites grow from the floor. They are caused by the formation of carbonates from the more soluble hydrogen carbonates carried into the cave by groundwater.

lent bond to the proton by providing both the electrons in the bond and becomes a hydroxide ion:

$$O^{2-}(aq) + H_2O(l) \longrightarrow 2\,OH^-(aq)$$

Because the oxide ion accepts a proton, it is classified as a base. Every oxide ion accepts a proton in water, so O^{2-} is an example of a strong base. Many molecular compounds that contain nitrogen are also bases because the lone pair of electrons on the nitrogen atom can attract a proton. For example, when ammonia, NH_3, dissolves in water, some of the molecules undergo the following reaction:

Recall from Section 3.8 that a strong base is completely protonated in water.

$$NH_3(aq) + H_2O(l) \longrightarrow NH_4^+(aq) + OH^-(aq)$$

In this reaction an NH_3 molecule behaves like an O^{2-} ion. However, whereas all the O^{2-} ions are converted to OH^- ions in water, the lone pair of electrons on the N atom in NH_3 has much less proton-pulling power than the full negative charge of an ion, and only a very small proportion of the NH_3 molecules are converted into NH_4^+ ions (Fig. 14.5). Therefore, ammonia is an example of a weak base. All amines (Section 11.10) are weak bases in water.

Recall from Section 8.12 that a Lewis acid is defined as an electron pair acceptor and a Lewis base as an electron pair donor. A proton (H^+) is an electron pair acceptor, and therefore a Lewis acid, because it can attach to a lone pair of electrons on, for instance, H_2O. We now see that a Brønsted acid is a supplier of one particular Lewis acid, a proton. A Brønsted base is a special kind of Lewis base, one that can use a lone pair

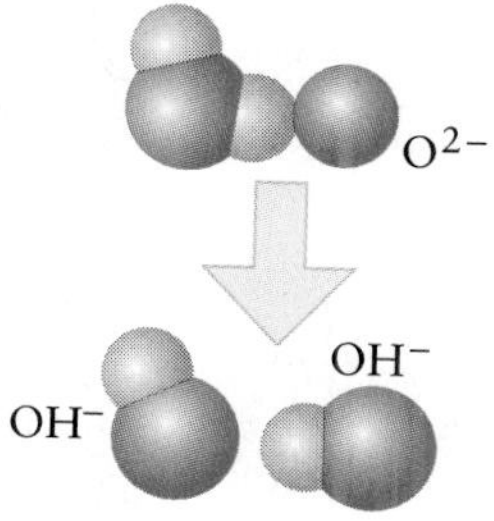

FIGURE 14.4 When an oxide ion is present in water, it exerts such a strong attraction on the nucleus of a hydrogen atom in a neighboring water molecule that the nucleus is pulled out of the molecule. As a result, the oxide ion forms two hydroxide ions.

to bond to a proton. In this chapter the Brønsted-Lowry definitions are all we need, but we shall find the broader Lewis definitions helpful later (Section 15.1).

An acid is a proton donor and a base is a proton acceptor.

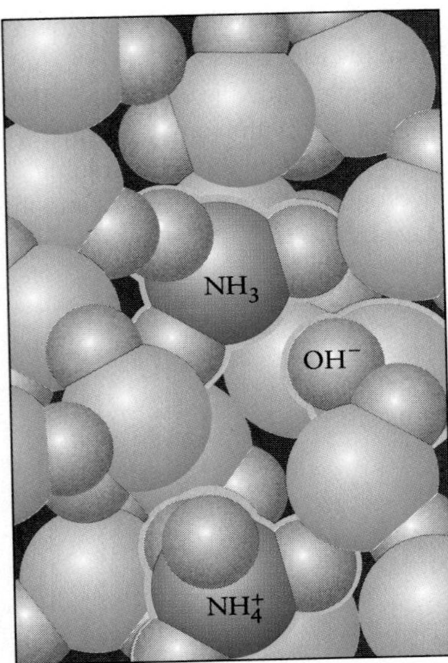

FIGURE 14.5 In this portrayal of the structure of a solution of ammonia in water, we see that there are NH_3 molecules still present, because not all of them have been protonated by transfer of hydrogen ions from water. In practice, only about 1 in 100 NH_3 molecules is protonated in a typical solution.

14.2 CONJUGATE ACIDS AND BASES

When CH_3COOH dissolves in water, it forms hydronium ions and acetate ions:

$$CH_3COOH(aq) + H_2O(l) \longrightarrow H_3O^+(aq) + CH_3CO_2^-(aq) \quad (1)$$

However, an acetate ion can accept a proton from a proton donor and be converted into an acetic acid molecule:

$$H_3O^+(l) + CH_3CO_2^-(aq) \longrightarrow CH_3COOH(aq) + H_2O(l) \quad (2)$$

Therefore, an acetate ion is a base. Because the acetate ion, a base, is derived from acetic acid by proton loss, it is called the **conjugate base** of acetic acid. In general,

$$\text{acid} \xrightarrow{\text{donates } H^+} \text{conjugate base}$$

According to this definition, Cl^- is the conjugate base of HCl and O^{2-} is the conjugate base of OH^-.

Because CH_3COOH is an acid that we can regard as formed by attaching a proton to an acetate ion, it is the **conjugate acid** of the base $CH_3CO_2^-$. In general,

$$\text{base} \xrightarrow{\text{accepts } H^+} \text{conjugate acid}$$

Likewise, HCl is the conjugate acid of Cl^- and OH^- is the conjugate acid of O^{2-}.

The conjugate base of an acid is the base formed when the acid has donated a proton. The conjugate acid of a base is the acid that forms when the base has accepted a proton.

SELF-TEST 14.1A Write chemical formulas for (a) the conjugate acids of CH_3NH_2 and CN^-; (b) the conjugate bases of HSO_4^- and HI.
[*Answer:* (a) $CH_3NH_3^+$ and HCN; (b) SO_4^{2-} and I^-]

SELF-TEST 14.1B Write chemical formulas for (a) the conjugate acids of NH_3 and CO_3^{2-}; (b) the conjugate bases of NH_3 and HNO_3.

14.3 PROTON EXCHANGE BETWEEN WATER MOLECULES

Is water an acid, a base, both, or neither? We have seen that a water molecule accepts a proton from an acid molecule to form an H_3O^+ ion. So, water is a base. However, a water molecule can donate a proton to a base and become an OH^- ion. So, water is also an acid. We say that water is **amphiprotic,** meaning that it can act as both a proton donor and a proton acceptor.

The prefix *amphi-* comes from the Greek word for "both."

Protons migrate between water molecules, even in the absence of another acid or base:

$$2\,H_2O(l) \longrightarrow H_3O^+(aq) + OH^-(aq)$$

We already know that H_3O^+ is an acid and OH^- is a base, so the reverse reaction

$$2\,H_2O(l) \longleftarrow H_3O^+(aq) + OH^-(aq)$$

can also occur. The transfer of protons is very rapid, and the equilibrium

$$2\,H_2O(l) \rightleftharpoons H_3O^+(aq) + OH^-(aq)$$

is always present in water and aqueous solutions. This type of reaction, in which one molecule transfers a proton to another molecule of the same kind, is called **autoprotolysis** (Fig. 14.6).

The equilibrium constant for the autoprotolysis of water is

$$K_c = \frac{[H_3O^+][OH^-]}{[H_2O]^2}$$

In water itself and in dilute aqueous solutions (the only ones we consider), the solvent, water, is very nearly pure, so the molar concentration of water can be treated as a constant and combined with K_c. The resulting expression is called the **autoprotolysis constant** of water and is written K_w:

Another name for this quantity is the *autoionization constant.*

$$K_w = [H_3O^+][OH^-]$$

The molarities of H_3O^+ and OH^- in pure water at 25°C are known by experiment to be $1.0 \times 10^{-7}\ mol{\cdot}L^{-1}$, so

$$K_w = (1.0 \times 10^{-7}) \times (1.0 \times 10^{-7}) = 1.0 \times 10^{-14}$$

The concentrations of H_3O^+ and OH^- are very low in pure water, which explains why pure water is such a poor conductor of electricity. To imagine the very tiny extent of autoprotolysis, think of each letter in this book as a water molecule. We would need to search through more than 50 books to find one ionized water molecule.

Because K_w is an equilibrium constant, the product of the molarities of H_3O^+ and OH^- ions is *always* equal to K_w. We cannot increase the molarities of *both* hydronium ions and hydroxide ions. We can increase the concentration of $[H_3O^+]$ ions by adding acid, in which case the concentration of OH^- ions must decrease to preserve the value of K_w. Alternatively, we can increase the concentration of OH^- ions by adding base, but then the concentration of H_3O^+ ions must decrease. The autoprotolysis equilibrium links the molarities of H_3O^+ and OH^- ions rather like a seesaw: when one goes up, the other goes down (Fig. 14.7).

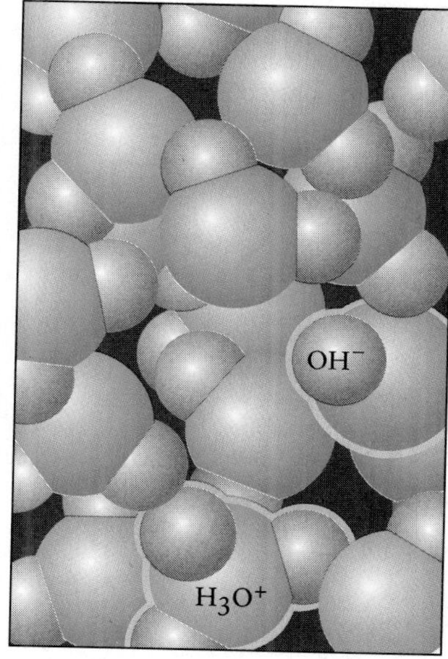

FIGURE 14.6 As a result of autoprotolysis, pure water consists of hydronium ions and hydroxide ions as well as water molecules. The concentration of ions that results from autoprotolysis is only about $10^{-7}\ mol{\cdot}L^{-1}$, so only about 1 molecule in 200 million is ionized.

In aqueous solutions, the molarities of H_3O^+ and OH^- ions are related by the autoprotolysis equilibrium; if one concentration is increased, then the other must decrease to maintain the value of K_w.

EXAMPLE 14.1 Calculating the molarities of ions in a solution of a strong base

What are the molarities of H_3O^+ and OH^- in a 0.0030 M $Ba(OH)_2(aq)$ solution at 25°C?

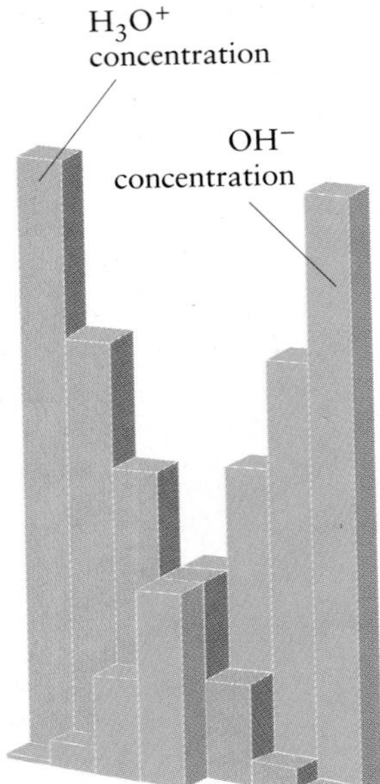

FIGURE 14.7 The product of the concentrations of hydronium and hydroxide ions in water (pure water and aqueous solutions) is a constant. If the concentration of one type of ions increases, the other must decrease to keep the product of concentrations constant.

STRATEGY Strong acids are almost completely deprotonated in water. Similarly, strong bases are virtually entirely present as OH^- in water. First, decide whether the base is strong; if so, decide from the chemical formula how many OH^- ions are provided by each formula unit. Then calculate the molarity of these ions in the solution. To find the molarity of H_3O^+ ions, use the water autoprotolysis constant $K_w = [H_3O^+][OH^-]$.

SOLUTION Barium hydroxide is a hydroxide of an alkaline earth metal, so it is a strong base (see Section 3.8). The equation

$$Ba(OH)_2(s) \longrightarrow Ba^{2+}(aq) + 2\,OH^-(aq)$$

tells us that $Ba(OH)_2 \simeq 2\,OH^-$. Because the molarity of $Ba(OH)_2(aq)$ is $0.0030\ mol{\cdot}L^{-1}$, it follows that the molarity of OH^- is twice that value, or $0.0060\ mol{\cdot}L^{-1}$. Then, for the molarity of H_3O^+ ions, we write

$$[H_3O^+] = \frac{K_w}{[OH^-]} = \frac{1.0 \times 10^{-14}}{0.0060} = 1.7 \times 10^{-12}$$

That is, the molarity of H_3O^+ ions in the solution is only $1.7 \times 10^{-12}\ mol{\cdot}L^{-1}$.

SELF-TEST 14.2A Estimate the molarities of H_3O^+ and OH^- in 6.0×10^{-5} M HI(aq).

[*Answer:* $6.0 \times 10^{-5}\ mol{\cdot}L^{-1}$; $1.7 \times 10^{-10}\ mol{\cdot}L^{-1}$]

SELF-TEST 14.2B Estimate the molarities of H_3O^+ and OH^- in 2.2×10^{-3} M NaOH(aq).

14.4 THE pH SCALE

An "order of magnitude" means a power of 10.

The pH scale was introduced by the Danish chemist Søren Sørensen in 1909 for his quality control work at a brewery.

▶▶ Logarithms are reviewed in Appendix 1C.

In industry, research institutions, and hospitals it is often necessary to report the acidity of a solution. In an emergency room or in a manufacturing plant it may be necessary to report concentrations and make comparisons quickly and easily, in a manner that reduces error. However, the molarities of H_3O^+ and OH^- vary over many orders of magnitude; in some solutions they may be as high as $1\ mol{\cdot}L^{-1}$ and in others as low as $10^{-14}\ mol{\cdot}L^{-1}$. The awkwardness of dealing with such a wide range of values is avoided by using logarithms, which condense these values into a much smaller and more convenient range (Fig. 14.8). Chemists therefore usually report hydronium ion molarities in terms of the **pH** of the solution:

$$pH = -\log[H_3O^+]$$

The logarithm in this definition is a common logarithm, to the base 10. As usual, $[H_3O^+]$ is the molarity of H_3O^+ ions, with the units of moles per liter struck out. For example, the pH of pure water, in which the molarity of H_3O^+ ions is $1.0 \times 10^{-7}\ mol{\cdot}L^{-1}$ at 25°C, is

$$pH = -\log(1.0 \times 10^{-7}) = 7.00$$

The common logarithm of 1 is 0 and that of 10^{-14} is -14.

Most solutions we meet have a pH in the range 0 to 14, but values outside this range are possible. The pH scale is used throughout chemistry, biochemistry, environmental chemistry, biology, geology, medicine, agriculture, and many other technical fields.

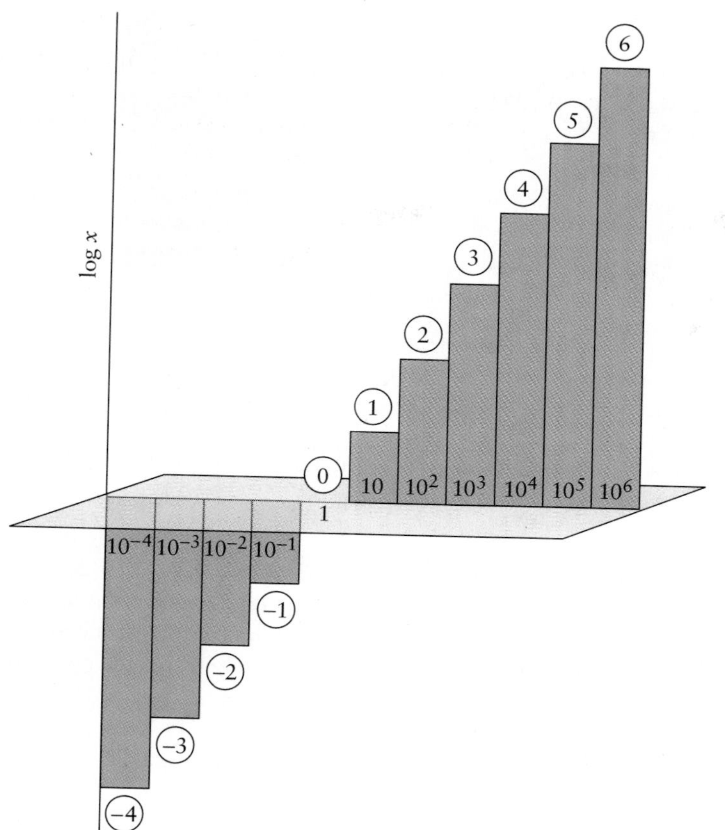

FIGURE 14.8 Very large ranges of numbers are difficult to represent graphically. However, their logarithms span a much smaller range and can be represented easily. Note how the numbers shown here range over 10 orders of magnitude (from 10^{-4} to 10^6), but their logarithms range over 10 units (from −4 to 6). Negative values of logarithms correspond to numbers between 0 and 1; positive values correspond to numbers greater than 1.

The negative sign in the definition of pH means that *the higher the H_3O^+ molarity, the lower the pH.* Therefore (Fig. 14.9),

The pH of pure water is 7.

The pH of an acidic solution is less than 7.

The pH of a basic solution is greater than 7.

Because pH is the negative of the common logarithm of the concentration, a change of one pH unit means the molarity of H_3O^+ ions has changed by a factor of 10. For example, when the pH decreases from 5 to 4, the H_3O^+ molarity increases by a factor of 10, from 10^{-5} mol·L^{-1} to 10^{-4} mol·L^{-1}.

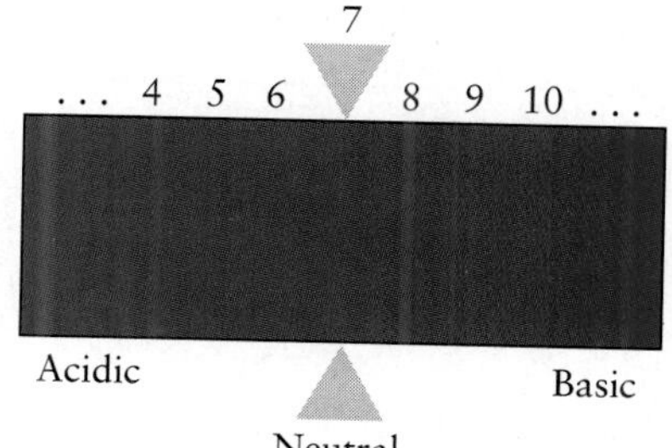

FIGURE 14.9 The numbers along the top of the rectangle are values of pH. Note that neutrality corresponds to a pH of 7; smaller values correspond to acidic solutions, larger values to basic solutions. Most values of pH lie in the range 1 to 14, but in principle pH values can lie outside this range and even be negative.

EXAMPLE 14.2 Calculating a pH

What is the pH of (a) human blood, in which the molarity of H_3O^+ ions is 4.0×10^{-8} mol·L^{-1}; (b) 0.020 M HCl(aq); (c) 0.040 M KOH(aq)?

STRATEGY The definition of pH is given above. To use it, take the H_3O^+ molarity, strike out the units, take the logarithm (make sure it is to the base 10), and change the sign. For strong acids, the molarity of H_3O^+ is equal to the molarity of the acid; for acids, expect pH < 7. For strong bases, carry out a calculation like that illustrated in Example 14.1: first find the molarity of OH^-,

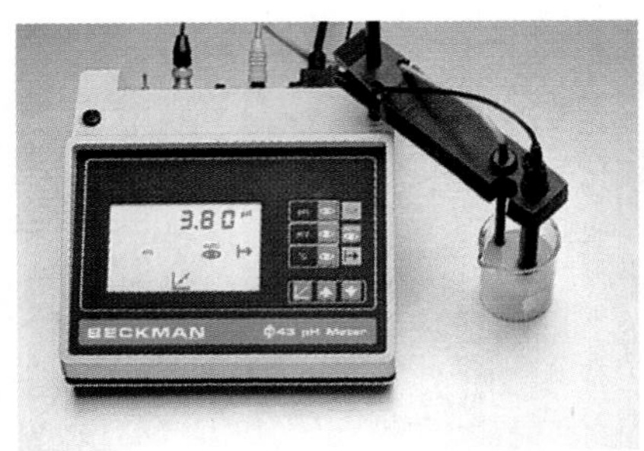

(a)

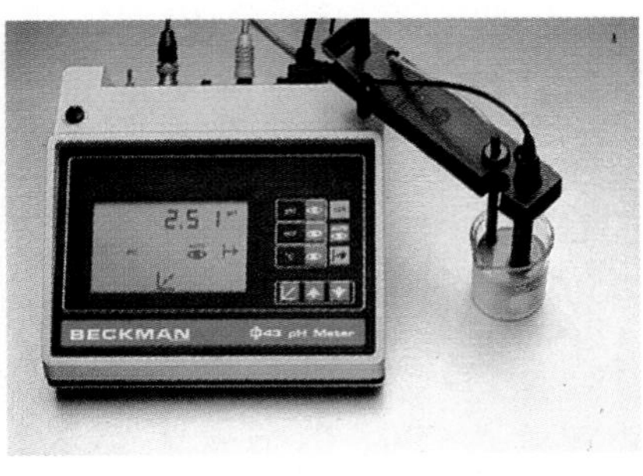

(b)

FIGURE 14.10 A pH meter is a voltmeter that measures the potential difference (the "voltage") between the two electrodes that are immersed in the solution. The display is calibrated to give the pH directly. The two samples are (a) orange juice, (b) lemon juice. Notice that the latter has a lower pH and hence a higher concentration of hydronium ions.

pH
1
2 Corrosive
3 Lemon juice Vinegar
4 Soda Wine
5 Beer Acid rain
6 Tomato juice
7 Milk Tap water Urine
8 Blood Saliva
9
10 Detergents
11 Household ammonia
12
13
14 Corrosive

FIGURE 14.11 Typical pH values of common aqueous solutions. The orange regions indicate the pH for liquids regarded as corrosive.

then convert that molarity to $[H_3O^+]$ by using $[H_3O^+][OH^-] = K_w$; for bases, expect pH > 7. The number of digits following the decimal point in a pH value is equal to the number of significant figures in the corresponding molarity, because the digits preceding the decimal point simply report the power of 10 in the data (as in $\log 10^5 = 5$).

SOLUTION (a) For a solution in which the molarity of H_3O^+ ions is 4.0×10^{-8} mol·L^{-1}, we write $[H_3O^+] = 4.0 \times 10^{-8}$ and obtain

$$\text{pH} = -\log(4.0 \times 10^{-8}) = 7.40$$

(b) Because HCl is a strong acid, the molarity of H_3O^+ is 0.020 mol·L^{-1}. Hence,

$$\text{pH} = -\log 0.020 = 1.70$$

(c) Each formula unit of KOH (a strong base) provides one OH^- ion; therefore the molarity of OH^- is 0.040 mol·L^{-1} and

$$[H_3O^+] = \frac{K_w}{[OH^-]} = \frac{1.0 \times 10^{-14}}{0.040} = 2.5 \times 10^{-13}$$

Hence

$$\text{pH} = -\log(2.5 \times 10^{-13}) = 12.60$$

SELF-TEST 14.3A Calculate the pH of (a) household ammonia, for which the OH^- molarity is about 3×10^{-3} mol·L^{-1}; (b) 6.0×10^{-5} M $HClO_4$(aq).

[*Answer:* (a) 11.5; (b) 4.22]

SELF-TEST 14.3B Calculate the pH of 0.077 M NaOH(aq).

The pH of an aqueous solution can be estimated with a strip of universal indicator paper, which turns different colors at different pH values. More precise measurements are made with a pH meter (Fig. 14.10). This instrument consists of a voltmeter connected to two electrodes that dip into the solution. The difference in electrical potential across the electrodes is proportional to the pH (as explained in Box 17.1); so once the scale on the meter has been calibrated, the pH can be read directly. In the United States, the Environmental Protection Agency (EPA) defines waste as corrosive if its pH is either lower than 3.0 (highly acidic) or higher than 12.5 (highly basic). The results of measuring the pH of a selection of liquids and beverages to assess their relative acidities are shown in Fig. 14.11. Fresh lemon juice has a pH of 2.5, corresponding to an H_3O^+ molarity of 3×10^{-3} mol·L^{-1}, more than five times more acidic than orange juice. Lemon juice also tastes sharper because H_3O^+ ions stimulate the "sour" taste receptors on the surface of the tongue. Natural rain, with an acidity due largely to dissolved carbon dioxide, has a pH of about 5.7. This pH corresponds to an H_3O^+ ion molarity of 2×10^{-6} mol·L^{-1}; the pH of some of the rainfall experienced by Scandinavian countries has been as low as 4, which corresponds to an H_3O^+ molarity of 1×10^{-4} mol·L^{-1}, or nearly 50 times more concentrated.

The pH scale is used to report H_3O^+ molarity: high pH denotes a basic solution, low pH an acidic solution; a neutral solution has pH = 7.

SELF-TEST 14.4A The pH of stomach fluids is about 1.7. What is the H_3O^+ molarity in the stomach?

[*Answer:* 2×10^{-2} mol·L^{-1}]

SELF-TEST 14.4B The pH of pancreatic fluids, which help to digest food once it has left the stomach, is about 8.2. What is the approximate H_3O^+ molarity of pancreatic fluids?

14.5 THE pOH OF SOLUTIONS

Many expressions involving acids and bases are greatly simplified by expressing quantities in terms of their logarithms, and the values are much easier to remember. The quantity "pX" is a generalization of pH:

The p in pH, and pX in general, stands for the negative *power* of 10.

$$\mathrm{p}X = -\log X$$

For example, **pOH** is defined as

$$\mathrm{pOH} = -\log[\mathrm{OH}^-]$$

The pOH is a convenient scale for reporting the molarities of OH^- ions in solution. For example, in pure water, where the molarity of OH^- ions is 1.0×10^{-7} mol·L^{-1}, the pOH is 7.0. Similarly, by pK_w, we mean

$$\mathrm{p}K_w = -\log K_w = -\log(1.0 \times 10^{-14}) = 14.00$$

Suppose we take logarithms of both sides of the expression for the autoprotolysis constant of water, $K_w = [H_3O^+][OH^-]$. Because $\log ab = \log a + \log b$,

$$\log[\mathrm{H_3O^+}] + \log[\mathrm{OH^-}] = \log K_w$$

Multiplication of both sides of the equation by -1 gives

$$-\log[\mathrm{H_3O^+}] - \log[\mathrm{OH^-}] = -\log K_w$$

which is the same as

$$\mathrm{pH} + \mathrm{pOH} = \mathrm{p}K_w$$

Because pK_w = 14.00 at 25°C, at that temperature

$$\mathrm{pH} + \mathrm{pOH} = 14.00$$

In fact, it is simpler to use this relation to calculate the pH of a solution of a strong base than the procedure used in Example 14.1. Because $[OH^-] = 0.0060\ mol\cdot L^{-1}$ for the solution treated there, $pOH = -\log(0.0060) = 2.22$; hence $pH = 14.00 - 2.22 = 11.78$.

The pH and pOH of a solution are related by $pH + pOH = pK_w$.

FIGURE 14.12 Equal masses of magnesium metal have been added to solutions of HCl, a strong acid (top) and CH_3COOH, a weak acid (bottom). Although the acids have the same concentrations, the rate of hydrogen evolution, which depends on the concentration of hydronium ions, is much greater in the strong acid.

WEAK ACIDS AND BASES

It is found experimentally that 0.10 M $CH_3COOH(aq)$ has a higher pH, and hence a lower H_3O^+ molarity, than 0.10 M HCl(aq). Similarly, it is found that 0.10 M $NH_3(aq)$ has a lower pH than 0.10 M NaOH(aq). The explanation must be that CH_3COOH is not fully ionized and NH_3 is not fully protonated in water. That is, they are examples of the weak acids and bases that we met in Section 3.8. Incomplete ionization is a part of the reason why the carbonic acid in rain is not as harmful as strong acids like nitric acid. It also explains why solutions of HCl and CH_3COOH react at different rates even though they have the same molarities (Fig. 14.12).

14.6 PROTON TRANSFER EQUILIBRIA

Proton transfer is so fast that we can always be confident that conjugate acids and bases, such as CH_3COOH and $CH_3CO_2^-$ or NH_4^+ and NH_3, are in equilibrium with each other in water. Therefore, forward and reverse reactions like

$$CH_3COOH(aq) + H_2O(l) \longrightarrow H_3O^+(aq) + CH_3CO_2^-(aq) \qquad (3)$$

and

$$CH_3COOH(aq) + H_2O(l) \longleftarrow H_3O^+(aq) + CH_3CO_2^-(aq) \qquad (4)$$

can be combined into

$$CH_3COOH(aq) + H_2O(l) \rightleftharpoons H_3O^+(aq) + CH_3CO_2^-(aq)$$

This equilibrium is a **proton transfer equilibrium.** It is common to all acids and bases in water, and we met it earlier in the autoprotolysis equilibrium of water. Another example is

$$H_2O(l) + NH_3(aq) \rightleftharpoons NH_4^+(aq) + OH^-(aq)$$

In each proton transfer equilibrium, a proton is transferred from an acid to a base, producing the conjugate base and conjugate acid.

SELF-TEST 14.5A Identify (a) the Brønsted acid and base in the following reaction, and (b) the conjugate base and acid formed:

$$HNO_3(aq) + HPO_4^{2-}(aq) \rightleftharpoons NO_3^-(aq) + H_2PO_4^-(aq)$$

[*Answer:* (a) Acid, HNO_3; base, HPO_4^{2-}; (b) conjugate base, NO_3^-; conjugate acid, $H_2PO_4^-$]

SELF-TEST 14.5B Identify (a) the Brønsted acid and base in the following reaction, and (b) the conjugate base and acid formed:

$$HCO_3^-(aq) + NH_4^+(aq) \rightleftharpoons H_2CO_3(aq) + NH_3(aq)$$

Proton transfer equilibria are described by equilibrium constants. For example, for acetic acid in water,

$$CH_3COOH(aq) + H_2O(l) \rightleftharpoons H_3O^+(aq) + CH_3CO_2^-(aq)$$

$$K_c = \frac{[H_3O^+][CH_3CO_2^-]}{[CH_3COOH][H_2O]}$$

Because the solutions we consider are dilute and the solvent is almost pure water, the value of $[H_2O]$ can be treated as a constant and combined with the equilibrium constant K_c. The resulting expression is called an **acidity constant,** denoted K_a:

$$K_a = \frac{[H_3O^+][CH_3CO_2^-]}{[CH_3COOH]}$$

The acidity constant is also widely called the *ionization constant* and sometimes the *dissociation constant* of the acid. The constants K_a and K_c are related by $K_a = K_c[H_2O]$.

The experimental value of K_a for acetic acid at 25°C is 1.8×10^{-5}. This small value tells us that only a small proportion of CH_3COOH molecules lose their protons when dissolved in water. Later in the chapter, we shall see how to estimate the fraction of molecules ionized and will see that, depending on the concentration, about 99 out of 100 CH_3COOH molecules remain intact. This value is typical of weak acids in water (Fig. 14.13).

An equilibrium constant can also be written for the proton transfer equilibrium of a base in water. For aqueous ammonia, for instance,

$$NH_3(aq) + H_2O(l) \rightleftharpoons NH_4^+(aq) + OH^-(aq) \qquad K_c = \frac{[NH_4^+][OH^-]}{[H_2O][NH_3]}$$

In dilute solutions, the water is almost pure, and its nearly constant molar concentration can be combined with K_c to form the **basicity constant,** K_b:

$$K_b = \frac{[NH_4^+][OH^-]}{[NH_3]}$$

The experimental value of K_b for ammonia in water at 25°C is 1.8×10^{-5}. This small value tells us that only a small proportion of NH_3 molecules are present as NH_4^+ under normal conditions. Calculations of the type described in Section 14.12 suggest that only about 1 in 100 molecules is protonated in a typical solution (Fig. 14.14).

The fact that K_b for ammonia has the same numerical value as K_a for acetic acid is a coincidence.

Acidity and basicity constants are commonly reported as their negative logarithms, by defining

$$pK_a = -\log K_a \qquad pK_b = -\log K_b$$

The higher the value of K_a, the lower the value of pK_a. Hence, we can conclude that the lower the value of pK_a, the stronger the acid. Thus, the pK_a of acetic acid is $-\log(1.8 \times 10^{-5}) = 4.75$ and that of trichloroacetic acid, a much stronger acid, is 0.5.

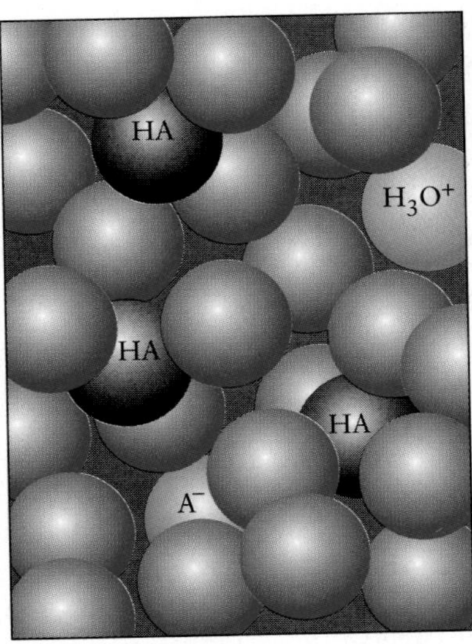

FIGURE 14.13 In a solution of a weak acid, only some of the ionizable hydrogen atoms are present as hydronium ions (the red spheres), and the solution contains a high proportion of the original acid molecules (HA, gray spheres).

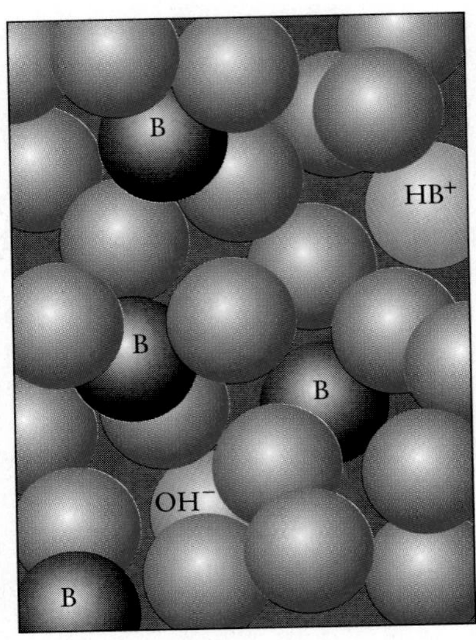

FIGURE 14.14 In a solution of a weak base, only a small proportion of the base molecules (B, represented here by the gray spheres) have accepted protons from water (the blue spheres) to form HB^+ ions (the red spheres) and OH^- ions.

The proton-donating strength of an acid is measured by its acidity constant; the proton-accepting strength of a base is measured by its basicity constant. The larger the constants, the greater the respective strengths.

14.7 THE CONJUGATE SEESAW

Trichloroacetic acid, CCl_3COOH, is a stronger acid than CH_3COOH, so its conjugate base, $CCl_3CO_2^-$, must be weaker than $CH_3CO_2^-$. That is, $CCl_3CO_2^-$ is less able than $CH_3CO_2^-$ to remove a proton from H_3O^+. We can conclude that *the stronger the acid, the weaker its conjugate base.* Hydrochloric acid is strong because its conjugate base, Cl^-, is unable to remove protons from H_3O^+. As a result, HCl is fully ionized in water. Conversely, acetic acid is a weak acid because its conjugate base, the acetate ion, $CH_3CO_2^-$, is a relatively good proton acceptor and forms CH_3COOH molecules in water.

Similarly, *the stronger the base, the weaker its conjugate acid.* For instance, methylamine, CH_3NH_2, is a stronger base than ammonia (Table 14.1). The conjugate acid of methylamine—the methylammonium ion, $CH_3NH_3^+$—must therefore be less able to donate a proton to H_2O than NH_4^+, the conjugate acid of ammonia.

We can express these ideas more quantitatively. Because the strengths of conjugate acids and bases have a seesaw relation, we should expect the

K_b of a base (such as NH_3) to be related to the K_a of its conjugate acid (here NH_4^+). To find the relation, consider the proton transfer equilibrium of the base NH_3 in water,

$$NH_3(aq) + H_2O(l) \rightleftharpoons NH_4^+(aq) + OH^-(aq) \qquad K_b = \frac{[NH_4^+][OH^-]}{[NH_3]}$$

and the proton transfer equilibrium of ammonia's conjugate acid, NH_4^+, in water,

$$NH_4^+(aq) + H_2O(l) \rightleftharpoons H_3O^+(aq) + NH_3(aq) \qquad K_a = \frac{[H_3O^+][NH_3]}{[NH_4^+]}$$

When we multiply these two equilibrium constants together, we obtain

$$K_a \times K_b = \frac{[H_3O^+]\cancel{[NH_3]}}{\cancel{[NH_4^+]}} \times \frac{\cancel{[NH_4^+]}[OH^-]}{\cancel{[NH_3]}} = [H_3O^+][OH^-]$$

That is,

$$K_a \times K_b = K_w$$

This important relation applies to all conjugate acid-base pairs.

TABLE 14.1 Conjugate acid-base pairs arranged by strength

Acid name	Formula	Base formula	Name
STRONG ACID			**VERY WEAK BASE**
hydroiodic acid	HI	I^-	iodide ion
perchloric acid	$HClO_4$	ClO_4^-	perchlorate ion
hydrobromic acid	HBr	Br^-	bromide ion
hydrochloric acid	HCl	Cl^-	chloride ion
sulfuric acid	H_2SO_4	HSO_4^-	hydrogen sulfate ion
chloric acid	$HClO_3$	ClO_3^-	chlorate ion
nitric acid	HNO_3	NO_3^-	nitrate ion
hydronium ion	H_3O^+	H_2O	*water*
hydrogen sulfate ion	HSO_4^-	SO_4^{2-}	sulfate ion
hydrofluoric acid	HF	F^-	fluoride ion
nitrous acid	HNO_2	NO_2^-	nitrite ion
acetic acid	CH_3COOH	$CH_3CO_2^-$	acetate ion
carbonic acid	H_2CO_3	HCO_3^-	hydrogen carbonate ion
hydrosulfuric acid	H_2S	HS^-	hydrogen sulfide ion
ammonium ion	NH_4^+	NH_3	ammonia
hydrocyanic acid	HCN	CN^-	cyanide ion
hydrogen carbonate ion	HCO_3^-	CO_3^{2-}	carbonate ion
methylammonium ion	$CH_3NH_3^+$	CH_3NH_2	methylamine
water	H_2O	OH^-	*hydroxide ion*
ammonia	NH_3	NH_2^-	amide ion
hydrogen	H_2	H^-	hydride ion
methane	CH_4	CH_3^-	methide ion
hydroxide ion	OH^-	O^{2-}	oxide ion
VERY WEAK ACID			**STRONG BASE**

TABLE 14.2 Acid and base ionization constants at 25°C*

Acid	K_a	pK_a	Base	K_b	pK_b
trichloroacetic acid, CCl_3COOH	3.0×10^{-1}	0.52	urea, $CO(NH_2)_2$	1.3×10^{-14}	13.90
benzene sulfonic acid, $C_6H_5SO_3H$	2.0×10^{-1}	0.70	aniline, $C_6H_5NH_2$	4.3×10^{-10}	9.37
iodic acid, HIO_3	1.7×10^{-1}	0.77	pyridine, C_5H_5N	1.8×10^{-9}	8.75
sulfurous acid, H_2SO_3	1.5×10^{-2}	1.81	hydroxylamine, NH_2OH	1.1×10^{-8}	7.97
chlorous acid, $HClO_2$	1.0×10^{-2}	2.00	nicotine, $C_{10}H_{14}N_2$	1.0×10^{-6}	5.98
phosphoric acid, H_3PO_4	7.6×10^{-3}	2.12	morphine, $C_{17}H_{19}O_3N$	1.6×10^{-6}	5.79
chloroacetic acid, $CH_2ClCOOH$	1.4×10^{-3}	2.85	hydrazine, NH_2NH_2	1.7×10^{-6}	5.77
lactic acid, $CH_3CH(OH)COOH$	8.4×10^{-4}	3.08	ammonia, NH_3	1.8×10^{-5}	4.75
nitrous acid, HNO_2	4.3×10^{-4}	3.37	trimethylamine, $(CH_3)_3N$	6.5×10^{-5}	4.19
hydrofluoric acid, HF	3.5×10^{-4}	3.45	methylamine, CH_3NH_2	3.6×10^{-4}	3.44
formic acid, HCOOH	1.8×10^{-4}	3.75	dimethylamine, $(CH_3)_2NH$	5.4×10^{-4}	3.27
benzoic acid, C_6H_5COOH	6.5×10^{-5}	4.19	ethylamine, $C_2H_5NH_2$	6.5×10^{-4}	3.19
acetic acid, CH_3COOH	1.8×10^{-5}	4.75	triethylamine, $(C_2H_5)_3N$	1.0×10^{-3}	2.99
carbonic acid, H_2CO_3	4.3×10^{-7}	6.37			
hypochlorous acid, HClO	3.0×10^{-8}	7.53			
hypobromous acid, HBrO	2.0×10^{-9}	8.69			
boric acid, $B(OH)_3$†	7.2×10^{-10}	9.14			
hydrocyanic acid, HCN	4.9×10^{-10}	9.31			
phenol, C_6H_5OH	1.3×10^{-10}	9.89			
hypoiodous acid, HIO	2.3×10^{-11}	10.64			

*The K_a and K_b listed here have been calculated from pK_a and pK_b values with more significant figures than shown so as to minimize rounding errors. Values for polyprotic acids—those capable of donating more than one proton—refer to the first ionization.
†The proton transfer equilibrium is $B(OH)_3(aq) + 2H_2O(l) \rightleftharpoons H_3O^+(aq) + B(OH)_4^-(aq)$.

The equation we have just derived is the quantitative version of the seesaw relation between conjugate acid and base strengths. If K_b of a base is large (that is, the base is relatively strong), then K_a of its conjugate acid must be small (that is, the conjugate acid is relatively weak) to keep the product $K_a \times K_b$ equal to K_w.

If we take logarithms of both sides of the relation $K_a \times K_b = K_w$, we obtain

$$\log K_a + \log K_b = \log K_w$$

When we multiply through by -1, this expression becomes

$$pK_a + pK_b = pK_w$$

For example, because the pK_b of NH_3 is 4.75, the pK_a of NH_4^+ at 25°C is

$$pK_a = pK_w - pK_b = 14.00 - 4.75 = 9.25$$

This value shows that NH_4^+ is a weaker proton donor than acetic acid ($pK_a = 4.75$) but stronger than hypoiodous acid (HIO, $pK_a = 10.64$). The conjugate acid of urea, $NH_2CONH_3^+$, has $pK_a = 0.10$, showing that it is a stronger proton donor than any of the acids listed in Table 14.2 (Fig. 14.15).

The lower the value of K_a (the higher the value of pK_a), the weaker the proton-donating power of an acid; the lower the value of K_b (the higher the value of pK_b), the weaker the proton-accepting power of a base.

EXAMPLE 14.3 Predicting relative strengths of acids and bases

Use the information in Table 14.2 to decide which member of each of the following pairs of species is the stronger acid or base in water: (a) HF or HIO_3; (b) NO_2^- or CN^-.

STRATEGY To determine the relative strengths of acids and bases, bear in mind that the higher the K_a of a weak acid, the stronger the acid and the weaker its conjugate base. The higher the K_b of a weak base, the stronger the base and the weaker its conjugate acid.

SOLUTION (a) Because $K_a(HIO_3) > K_a(HF)$, it follows that HIO_3 is the stronger acid. (b) Because $K_a(HNO_2) > K_a(HCN)$, and because the stronger acid has the weaker conjugate base, it follows that NO_2^- is a weaker base than CN^-. Hence, CN^- is the stronger base.

SELF-TEST 14.6A Use Table 14.2 to decide which species of each pair is the stronger acid or base: (a) HF or HIO; (b) $C_6H_5CO_2^-$ or $CH_2ClCO_2^-$; (c) $C_6H_5NH_2$ or $(CH_3)_3N$; (d) $C_6H_5NH_3^+$ or $(CH_3)_3NH^+$.

[*Answer:* Stronger acids: (a) HF; (d) $C_6H_5NH_3^+$. Stronger bases: (b) $C_6H_5CO_2^-$; (c) $(CH_3)_3N$]

SELF-TEST 14.6B Use Table 14.2 to decide which species of each pair is the stronger acid or base: (a) C_5H_5N or NH_2NH_2; (b) $C_5H_5NH^+$ or $NH_2NH_3^+$; (c) HIO_3 or $HClO_2$; (d) ClO_2^- or HSO_3^-.

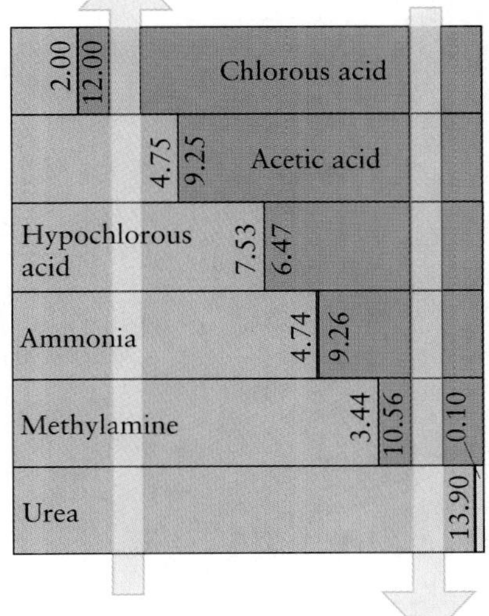

FIGURE 14.15 The sum of the pK_a of an acid (pink) and the pK_b of its conjugate base (blue) is constant, and equal to pK_w, which is 14.00 at 25°C. The values of the constants let us arrange all acids and bases in order of strength on a single chart.

14.8 THE SPECIAL ROLE OF WATER

Strong acids lose their acidic protons almost completely when they dissolve in water. We can picture the process as a tug-of-war between the proton accepting powers of H_2O and A^-, the conjugate base of the strong acid (Cl^- or NO_3^-, for instance). The conjugate base of a strong acid is a weaker proton acceptor than water. Thus, the battle is resolved in favor of H_2O, and the solution contains only H_3O^+ ions and A^- ions. In other words, the only proton donor that survives in an aqueous solution of a strong acid is the H_3O^+ ion. Because all strong acids behave as though they were solutions of the acid H_3O^+, we say that strong acids are **leveled** to the strength of the acid H_3O^+ in water.

We use A^- as a symbol to represent the conjugate base of an acid HA and B as a symbol for any base. However, note that a specific base can be either electrically neutral or negatively charged. Either A^- or B may represent NH_3, NH_2^-, or PO_4^{3-}.

Suppose that the base A^- is a stronger proton acceptor than H_2O. Now the competition for the proton favors the protonated base—its conjugate acid HA. Such an acid is weak, for it is only slightly ionized in aqueous solution. That is, *the strength of H_2O as a proton acceptor marks the frontier between strong and weak acids in water.* Any acid with a conjugate base that lies above H_2O in the listing of bases in Table 14.1 (H_2SO_4 is an example) is a strong acid in aqueous solution because water is a better proton acceptor than the conjugate base of the acid. These acids are the ones listed in Table 3.2. Any acid with a conjugate base that lies below water in Table 14.1, such as CH_3COOH, is a weak acid because the conjugate base is a better proton acceptor than water.

Similar considerations apply to bases. Now the tug-of-war is between the proton-*donating* power of H_2O and the conjugate acid, BH^+, of the base. If H_2O is a stronger proton donor than BH^+, then no B species will survive in solution and we classify B as a strong base. Thus, O^{2-} is a strong base because its conjugate acid, OH^-, is a much weaker proton donor than H_2O. Conversely, if H_2O is a weaker donor than BH^+, then BH^+ will be able to donate its proton, and a high proportion of B will survive in solution. That is, B is a weak base. Any base that has a conjugate acid lying below H_2O in the listing of acids in Table 14.1 is strong.

Any species that is a stronger proton donor than H_3O^+ is a strong acid in water. Any species that lies below H_3O^+ in proton-donating strength is a weak acid in water. Any substance that is a weaker proton acceptor than OH^- is a weak base in water. Any species that is a stronger proton acceptor than OH^- is a strong base in water.

SELF-TEST 14.7A Label each of the following species as a strong or a weak acid: (a) $HClO_4$; (b) NH_4^+, (c) HNO_2.

[*Answer:* (a) Strong acid; (b) weak acid; (c) weak acid]

SELF-TEST 14.7B Label each of the following species as a strong or a weak base: (a) ClO_4^-; (b) NH_2^-; (c) $CH_3CO_2^-$.

14.9 WHY ARE SOME ACIDS WEAK AND OTHERS STRONG?

The actual values of K_a and K_b are difficult to predict because they depend on a number of factors. These factors include the strength of the H—A bond, the strength of the O—H bond in H_3O^+, and the extent to which the conjugate base, A^-, of the acid is hydrated in water. However, we can spot trends among series of compounds with similar structures.

The ease with which an acid molecule, HA, donates a proton to a water molecule depends in part on the strength of the hydrogen bond it forms with the O atom of the H_2O molecule:

$$H_2O\cdots H{-}A \longrightarrow H_2O{-}H^+ + A^-$$

The stronger the hydrogen bond, the easier it is for the water molecule to pull the proton from the acid and form an H_3O^+ ion. We know that the more polar the H—A bond, the greater the partial positive charge on H and the stronger the O···H—A hydrogen bond. Therefore, we expect an acid HA with a very polar H—A bond to be stronger than one with a less polar H—A bond. Because the polarity of the H—A bond increases with the electronegativity of A, we can also predict that the greater the electronegativity of A, the stronger the acid HA.

For example, the electronegativity difference of the atoms is 0.8 in the N—H bond and 1.8 in F—H. Therefore, the H—F bond is markedly more polar than the N—H bond. Experimentally, HF is acidic and NH_3 is basic in aqueous solutions. For these two compounds at least, the bond polarity dominates the trend of acid strengths. In general, bond polarity dominates the trend of acid strengths for binary acids of elements of the same period.

Another factor affecting acid strength is the strength of the H—A bond. The weaker the H—A bond, the easier it is for the proton to leave and the stronger the acid HA. An example is the anomalous position of HF among the hydrohalic acids. Even though the H—F bond is the most polar of the group, HF is a weak acid in water, whereas all other hydrohalic acids are strong. A part of the reason is the great strength of the H—F bond, which means that the proton is lost from HF only with difficulty.

The relative strengths of the hydrohalic acids can be measured in a solvent that is a poorer proton acceptor than water (such as pure acetic acid), because they are not leveled in this solvent. The strengths so found lie in the order HF $<$ HCl $<$ HBr $<$ HI. This order is consistent with the weakening of the H—A bond down the group. The same trend is also found for the Group 16 acids in aqueous solution; the acid strengths lie in the order $H_2O < H_2S < H_2Se < H_2Te$. The bond strengths and bond polarities both decrease down the group. Therefore, it appears that bond strength dominates the trend in acid strength in these two sets of binary acids. In general, bond polarity dominates the trend in acid strengths for binary acids of elements in the same group.

The bond strengths of binary hydrides decrease down a group of the periodic table.

The more polar or the weaker the H—A bond, the stronger the acid.

14.10 THE STRENGTHS OF OXOACIDS

The high polarity of the O—H bond is one reason why the proton of an —OH group in an oxoacid molecule is acidic. For example, phosphorous acid, H_3PO_3, has the structure $(HO)_2PHO$ (**1**): it can donate the protons from its two —OH groups but not the proton attached directly to the phosphorus atom.

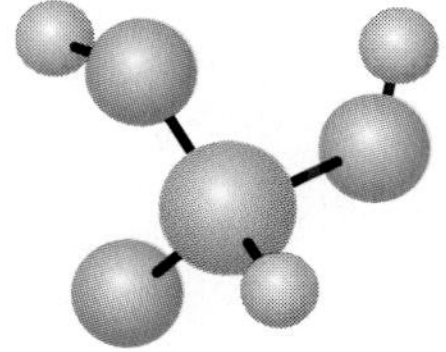

1 Phosphorous acid, H_3PO_3

Let's consider a family of oxoacids in which the number of oxygen atoms varies, as in the chlorine oxoacids HClO, $HClO_2$, $HClO_3$, and $HClO_4$, or the sulfur oxoacids H_2SO_3 and H_2SO_4. It is found experimentally that the greater the number of oxygen atoms attached to the central atom, the stronger the acid. Because the oxidation number of the central atom increases as the number of O atoms increases, we can also conclude that the greater the oxidation number of the central atom, the stronger the acid (Table 14.3).

Now let's consider a family of oxoacids in which the number of O atoms is constant, as in the hypohalous acids HClO, HBrO, and HIO. It is found that the greater the electronegativity of the central atom (the halogen atoms in these examples), the stronger the oxoacid (Table 14.4). A partial explanation of this trend is that electrons are withdrawn slightly from the O—H bond as the electronegativity of the central atom increases. As these bonding electrons move toward the O atom, the O—H bond becomes more polar. As the bond becomes more polar, the —OH proton can be donated to H_2O more readily, so the molecule becomes a stronger acid. A central atom with high electronegativity also makes the conjugate base weaker, because electronegative atoms do not readily share their electrons.

TABLE 14.3 Correlation of acid strength and oxidation number

Acid	Structure*	Oxidation number of chlorine atom	pK_a
hypochlorous acid, $HClO$	:C̈l̤–Ö̤–H	+1	7.53
chlorous acid, $HClO_2$	O=Cl–O–H	+3	2.00
chloric acid, $HClO_3$	(O=)₂Cl–O–H	+5	strong
perchloric acid, $HClO_4$	(O=)₃Cl–O–H	+7	strong

*The red arrows indicate the direction of the shift of electron density away from the O—H bond.

EXAMPLE 14.4 Predicting the relative strengths of acids

In the following pairs, predict which acid is stronger and explain why: (a) H_2S and H_2Se; (b) H_2SO_4 and H_2SO_3; (c) H_2SO_4 and H_3PO_4.

STRATEGY Recall from Sections 14.9 and 14.10 that

For binary acids:

1. The more polar the H—A bond, the stronger the acid. This effect is dominant for acids of the same period.
2. The weaker the H—A bond, the stronger the acid. This effect is dominant for acids of the same group.

For oxoacids:

3. The greater the number of O atoms attached to the central atom (or the greater the oxidation number of the central atom), the stronger the acid.
4. For the same number of O atoms attached to the central atom, then the greater the electronegativity of the central atom, the stronger the acid.

SOLUTION (a) Sulfur and selenium are in the same group, and we expect the H—Se bond to be weaker than the H—S bond. Thus, H_2Se can be expected to be the stronger acid. (b) H_2SO_4 has the greater number of O atoms bonded to the S atom and the oxidation number of sulfur is +6, whereas in H_2SO_3 the sulfur has an oxidation number of only +4. Thus, H_2SO_4 is expected to be the stronger acid. (c) Both acids have four O atoms bonded to the central atom; because the electronegativity of sulfur is greater than that of phosphorus, H_2SO_4 is expected to be the stronger acid.

SELF-TEST 14.8A In the following pairs, predict which acid is stronger and explain why: (a) H_2S and HCl; (b) HNO_2 and HNO_3; (c) H_2SO_3 and $HClO_3$.
[*Answer:* (a) HCl, rule 1; (b) HNO_3, rule 3; (c) $HClO_3$, rule 4]

SELF-TEST 14.8B In the following pairs, predict which acid is stronger and explain why: (a) $HClO$ and $HClO_2$; (b) HBr and HI.

TABLE 14.4 Correlation of acid strength and electronegativity

Acid, HXO	Structure*	Electronegativity of atom X	pK_a
hypochlorous acid, HClO	:Cl—O—H	3.2	7.53
hypobromous acid, HBrO	:Br—O—H	3.0	8.69
hypoiodous acid, HIO	:I—O—H	2.7	10.64

*The red arrows indicate the direction and magnitude of the shift of electron density away from the O—H bond.

The effect of the number of O atoms on the strengths of organic acids can be seen by comparing alcohols and carboxylic acids. Alcohols are organic compounds in which an —OH group is attached to a carbon atom, as in ethanol (**2**). Carboxylic acids have another O atom bonded to the carbon atom to which the —OH group is attached, as in acetic acid (**3**). Although carboxylic acids are weak acids, they are much stronger acids than alcohols on account, in part, of the electron-withdrawing power of the second O atom.

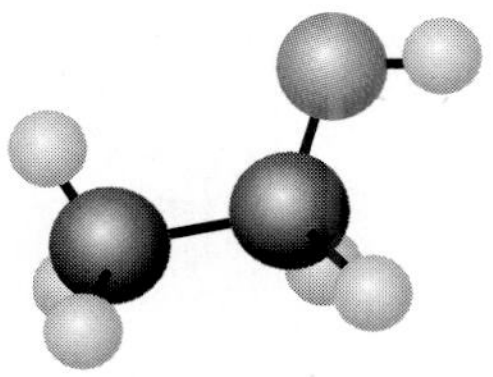

2 Ethanol, CH_3CH_2OH

The strength of a carboxylic acid is also increased relative to that of an alcohol by electron delocalization in the conjugate base. The second O atom of the carboxyl group provides an additional electronegative atom over which the negative charge of the conjugate base can spread: this electron delocalization stabilizes the anion by resonance (**4**). Moreover, because the charge is spread over several atoms, it is less effective at attracting a proton. A carboxylate ion is therefore a weaker base than the conjugate base of an alcohol (for example, the ethoxide ion, $CH_3CH_2O^-$). Hence, carboxylic acids are stronger acids than alcohols.

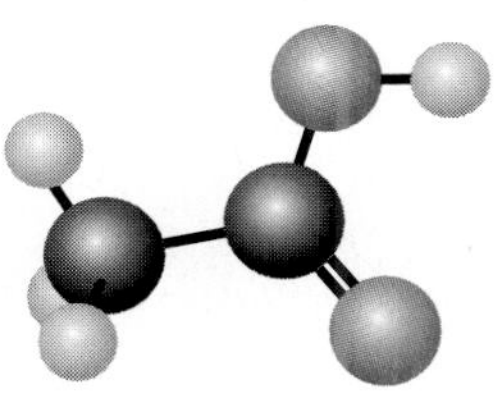

3 Acetic acid, CH_3COOH

The strengths of carboxylic acids also vary with total electron-withdrawing power of the atoms bonded to the carboxyl group. Because hydrogen is less electronegative than chlorine, the $—CH_3$ group bonded to —COOH in acetic acid is less electron withdrawing than the $—CCl_3$ group in trichloroacetic acid. Therefore, we expect CCl_3COOH to be a stronger acid than CH_3COOH. In agreement with this prediction, the pK_a of acetic acid is 4.75, whereas that of trichloroacetic acid is 0.52.

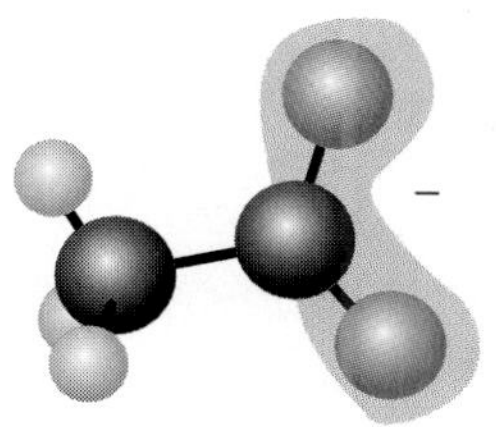

4 Acetate ion, $CH_3CO_2^-$

Finally, the acid strengths of organic compounds also vary with the electronegativity of the atoms attached to an —OH group. For example, let's compare the structure of ethanol, $CH_3CH_2—OH$, with that of HClO, hypochlorous acid, which we can write as Cl—OH. Because the Cl atom has a higher electronegativity than the atoms in the $CH_3CH_2—$ group, it follows that HClO should be the stronger acid. Indeed, the pK_a of CH_3CH_2OH is 16, whereas that of HClO is 7.53. Alcohols have such weak proton-donating power that they are not usually regarded as oxoacids.

The greater the number of oxygen atoms and the more electronegative the atoms present in a molecule, the stronger the acid.

EXAMPLE 14.5 Predicting the trend in strengths of carboxylic acids

List the following carboxylic acids in order of increasing strength: CH_3COOH, CH_2FCOOH, CHF_2COOH, and CF_3COOH.

STRATEGY The greater the electron-withdrawing power of the group attached to the carboxyl group, the stronger the acid. Therefore, judge the electron-withdrawing power of that group by comparing the electronegativities of the atoms.

SOLUTION Because fluorine is more electronegative than hydrogen, the groups attached to —COOH increase in electron-attracting power in the order $CH_3 < CH_2F < CHF_2 < CF_3$. Therefore the acid strengths increase accordingly, and we predict that they lie in the order $CH_3COOH < CH_2FCOOH < CHF_2COOH < CF_3COOH$.

SELF-TEST 14.9A List the following carboxylic acids in order of increasing strength: $CH_2ClCOOH$, $CH_2BrCOOH$, and CH_2FCOOH.

[*Answer:* $CH_2BrCOOH < CH_2ClCOOH < CH_2FCOOH$]

SELF-TEST 14.9B List the following carboxylic acids in order of increasing strength: $CHCl_2COOH$, CH_3COOH, and $CH_2ClCOOH$.

THE pH OF SOLUTIONS OF WEAK ACIDS AND BASES

A solution of a weak acid in water consists of H_3O^+ ions, the conjugate base of the acid, and acid molecules that have not donated their protons to water molecules. To find the H_3O^+ molarity, and hence the pH of the solution, we need to take into account the equilibrium constant K_a. Similarly, to calculate the pH of a solution of a weak base, we have to use K_b.

14.11 SOLUTIONS OF WEAK ACIDS

Weak acids produce a lower concentration of H_3O^+ ions in aqueous solution than do strong acids of the same **nominal concentration**, the concentration of the acid as prepared, as if none had ionized. For example, a 0.01 M HCl(aq) solution has a pH of 2; a 0.01 M CH_3COOH(aq) solution has a much lower concentration of H_3O^+ ions and it is found that its pH is 3.

Remember that an increase in pH by 1 corresponds to a tenfold decrease in concentration of hydronium ions.

Once we know the H_3O^+ concentration, we can calculate the percentage of HA molecules that are ionized:

$$\text{Percentage ionized} = \frac{\text{molarity of ionized HA}}{\text{nominal molarity of HA}} \times 100\%$$

A small percentage of ionized molecules indicates that the solute consists primarily of the acid HA. The concentrations of H_3O^+ ions and conjugate base ions must then be very low.

To find the H_3O^+ molarity in a solution of a weak acid, we have to take into account the equilibrium between the acid HA and its conjugate base A^-. Because it is an equilibrium system, we can use the same methods used in Chapter 13 to calculate the equilibrium composition of the solution. The procedure is described in Toolbox 14.1.

TOOLBOX 14.1 How to calculate the pH of a solution of a weak acid

We set up an equilibrium table as described in Toolbox 13.1 and follow the steps outlined there.

Step 1. Set up a table with columns labeled by the acid HA, its conjugate base A^-, and H_3O^+. In the first row show the initial molarities of each species.

For the initial values, assume that no acid molecules have donated a proton.

Step 2. Write the changes in the molarities that are needed for the reaction to reach equilibrium.

We do not know the number of acid molecules that ionize, so we assume that the molarity of the acid decreases by x mol·L^{-1} as a result of ionization. Use the reaction stoichiometry to express the other changes in terms of x.

Step 3. Write the equilibrium molarities by adding the change in molarities (step 2) to the initial values for each substance (step 1).

Although a change in concentration may be positive (an increase) or negative (a decrease), the value of the concentration must always be positive.

Step 4. Use the value of K_a to calculate the value of x.

The calculation of x can often be simplified, as shown in Toolbox 13.1, by neglecting changes of less than 5% of the initial molarity of the acid. However, at the end of the calculation, we must check the value of x to ensure that it is consistent with the approximation, by calculating the percentage of acid ionized. If this percentage is greater than 5%, then the exact expression for K_a must be solved for x; an exact calculation often involves solving a quadratic equation, as explained in Toolbox 13.1.

As an example, let's calculate the pH and percentage ionization of 0.10 M CH_3COOH(aq), given that K_a for acetic acid is 1.8×10^{-5}. Acetic acid is a weak acid; consequently, we expect the molarity of H_3O^+ ions to be less than 0.10 mol·L^{-1} and, therefore, its pH to be greater than 1.0 (because $-\log 0.10 = 1.0$). To find the actual value, we set up an equilibrium table with the initial molarity of acid equal to 0.10 mol·L^{-1} and allow the molarity of acid to decrease by x mol·L^{-1} to reach equilibrium. If x is less than about 5% of the initial molarity of acid, we simplify the expression for the equilibrium constant by ignoring x relative to the initial molarity of the acid.

The proton transfer equilibrium we consider is

$$CH_3COOH(aq) + H_2O(l) \rightleftharpoons H_3O^+(aq) + CH_3CO_2^-(aq)$$

$$K_a = \frac{[H_3O^+][CH_3CO_2^-]}{[CH_3COOH]}$$

and the equilibrium table, with the concentrations in moles per liter, is

	Species		
	CH_3COOH	H_3O^+	$CH_3CO_2^-$
Step 1. Initial molarity	0.10	0	0
Step 2. Change in molarity	$-x$	$+x$	$+x$
Step 3. Equilibrium molarity	$0.10 - x$	x	x

Step 4. Substitute these equilibrium molarities into the expression for the acidity constant:

$$K_a = 1.8 \times 10^{-5} = \frac{x \times x}{0.10 - x}$$

If we anticipate that $x \ll 0.1$, we can approximate this expression to

$$1.8 \times 10^{-5} \approx \frac{x^2}{0.10}$$

Solving for x gives

$$x \approx \sqrt{(0.10) \times (1.8 \times 10^{-5})} = 1.3 \times 10^{-3}$$

From step 3, $x = [H_3O^+] = 1.3 \times 10^{-3}$, so

$$pH = -\log(1.3 \times 10^{-3}) \approx 2.89$$

We have assumed that x is less than about 5% of 0.10. The percentage ionized is

$$\text{Percentage ionized} = \frac{[H_3O^+]}{[CH_3COOH]_{initial}} \times 100\% = \frac{1.3 \times 10^{-3}}{0.10} \times 100\% = 1.3\%$$

and the approximation is valid.

SELF-TEST 14.10A Calculate the pH of 0.20 M lactic acid. See Table 14.2 for K_a. Be sure to check any approximation to see whether it is valid.

[*Answer:* 1.90 (must use exact solution)]

SELF-TEST 14.10B Calculate the pH of 0.22 M chloroacetic acid.

CHEMISTRY AT WORK

U.S. Forest Service scientists test the acidity of a remote wilderness lake near Aspen, Colorado. They will measure the pH of the water at various depths to monitor the acidity of the lake. Even remote areas such as this can be affected by acid rain.

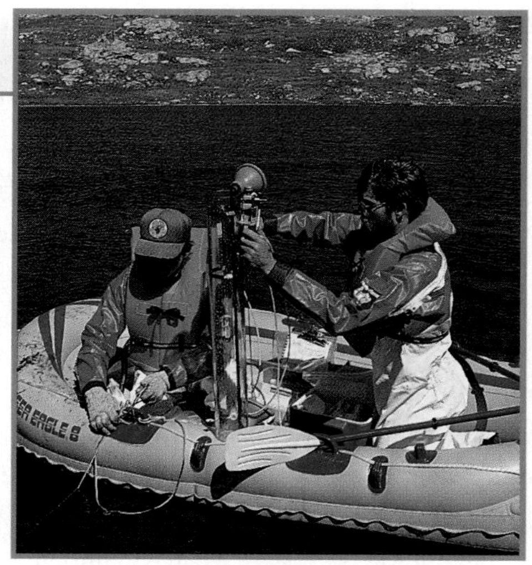

The procedure in Toolbox 14.1 ignores the H_3O^+ ions that result from the autoprotolysis of water. Suppose the acid is so dilute or weak that the calculation predicts an H_3O^+ molarity of less than 10^{-7} mol·L^{-1}. In this case we must not report its pH as more than 7 because the autoprotolysis already provides H_3O^+ ions at a molarity of 10^{-7} mol·L^{-1}. For example, the pH of a 1.0×10^{-8} M HCl(aq) solution is not 8, because the autoprotolysis of water leads to $[H_3O^+] = 1.0 \times 10^{-7}$, which is 10 times greater than provided by the acid, so the pH of the solution is approximately 7. We can ignore the contribution of the autoprotolysis of water only when the calculated H_3O^+ molarity is substantially (about three times) higher than 10^{-7} mol·L^{-1}.

To calculate the pH of a solution of a weak acid, set up an equilibrium table and determine the H_3O^+ molarity by using the acidity constant.

14.12 SOLUTIONS OF WEAK BASES

A weak base has a lower pH than a strong base of the same nominal molarity: a weak base establishes a dynamic equilibrium with its conjugate acid and therefore produces fewer OH^- ions than a comparable strong base. A 0.01 M NaOH(aq) solution has a pH of 12, but the pH of a 0.01 M NH_3(aq) solution is 11. We calculate the pH of solutions of weak bases in the same way that we calculate the pH of solutions of weak acids—by using the equilibrium table technique. To calculate the pH of the solution, we first calculate the molarity of OH^- ions at equilibrium, express that as pOH, and then calculate pH from the relation pH + pOH = 14.00 at 25°C.

In some applications it is important to know the fraction of base molecules that have been protonated. We report this fraction by calculating the **percentage protonated:**

$$\text{Percentage protonated} = \frac{\text{molarity of protonated base}}{\text{nominal molarity of base}} \times 100\%$$

The percentage protonated varies with concentration and depends on the strength of the base, as illustrated in the following example.

To calculate the pH of a solution of a weak base, use the equilibrium table to calculate pOH from the value of K_b and convert that pOH to pH by subtracting it from 14.

EXAMPLE 14.6 Calculating the pH of a solution of a weak base

Calculate the pH and percentage protonation of a 0.20 M aqueous solution of methylamine, CH_3NH_2. The K_b for CH_3NH_2 is 3.6×10^{-4}.

STRATEGY Expect pH > 7, because the solution is basic. Calculate the molarity of OH^- ions by using the equilibrium table as explained in Toolbox 14.1, but use K_b instead of K_a. Calculate $[OH^-]$, convert it to pOH, and then convert that pOH to pH by using the relation pH + pOH = 14.00.

SOLUTION The proton transfer equilibrium is

$$H_2O(l) + CH_3NH_2(aq) \rightleftharpoons CH_3NH_3^+(aq) + OH^-(aq)$$

$$K_b = \frac{[CH_3NH_3^+][OH^-]}{[CH_3NH_2]}$$

The equilibrium table, with all concentrations in moles per liter, is

	Species		
	CH_3NH_2	$CH_3NH_3^+$	OH^-
Step 1. Initial molarity	0.20	0	0
Step 2. Change in molarity	$-x$	$+x$	$+x$
Step 3. Equilibrium molarity	$0.20 - x$	x	x

Step 4. Substituting the equilibrium molarities into the expression for the basicity constant and using $K_b = 3.6 \times 10^{-4}$ yields

$$3.6 \times 10^{-4} = \frac{x \times x}{0.20 - x}$$

We now anticipate that x is less than 5% of 0.20 and approximate this expression by

$$3.6 \times 10^{-4} \approx \frac{x^2}{0.20}$$

Therefore,

$$x \approx \sqrt{(0.20) \times (3.6 \times 10^{-4})} = 8.5 \times 10^{-3}$$

According to step 3, $[OH^-] = x = 8.5 \times 10^{-3}$, so

$$\text{pOH} = -\log(8.5 \times 10^{-3}) \approx 2.1$$

and hence

$$\text{pH} \approx 14.00 - 2.1 = 11.9$$

The percentage of base molecules protonated is

$$\text{Percentage protonated} = \frac{\text{molarity of protonated } CH_3NH_2}{\text{nominal molarity of } CH_3NH_2} \times 100\%$$

$$= \frac{8.5 \times 10^{-3}}{0.20} \times 100\% = 4.2\%$$

That is, 4.2% of the methylamine is present as the protonated form, $CH_3NH_3^+$, and the approximation is valid.

SELF-TEST 14.11A Estimate the pH and percentage of protonated base in 0.15 M $NH_2OH(aq)$ (aqueous hydroxylamine).

[*Answer:* 9.61; 0.027%]

SELF-TEST 14.11B Estimate the pH and percentage of protonated base in 0.012 M $C_{10}H_{14}N_2(aq)$ (nicotine).

14.13 POLYPROTIC ACIDS AND BASES

Brønsted acids that can donate more than one proton are called **polyprotic acids.** Common examples of polyprotic acids are sulfuric acid, H_2SO_4, and carbonic acid, H_2CO_3, each of which can donate two protons, and phosphoric acid, H_3PO_4, which can donate three protons. A **polyprotic base** is a species that can accept more than one proton. Examples include the CO_3^{2-} anion and the oxalate anion, $C_2O_4^{2-}$, both of which can accept two protons, and the PO_4^{3-} anion, which can accept three protons.

Carbonic acid has the following ionization equilibria:

$$H_2CO_3(aq) + H_2O(l) \rightleftharpoons H_3O^+(aq) + HCO_3^-(aq) \qquad K_{a1} = \frac{[H_3O^+][HCO_3^-]}{[H_2CO_3]}$$

$$HCO_3^-(aq) + H_2O(l) \rightleftharpoons H_3O^+(aq) + CO_3^{2-}(aq) \qquad K_{a2} = \frac{[H_3O^+][CO_3^{2-}]}{[HCO_3^-]}$$

The conjugate base of H_2CO_3 in the first equilibrium, HCO_3^-, acts as an acid in the second equilibrium, producing in turn its own conjugate base, CO_3^{2-}. The equation for the overall ionization,

$$H_2CO_3(aq) + 2\,H_2O(l) \rightleftharpoons 2\,H_3O^+(aq) + CO_3^{2-}(aq)$$

is the sum of the two individual equations. It follows from the discussion in Section 13.2 that the overall acidity constant is the product of the two individual acidity constants:

$$K_a = K_{a1} \times K_{a2}$$

TABLE 14.5 Acidity constants of polyprotic acids

Acid	K_{a1}	pK_{a1}	K_{a2}	pK_{a2}	K_{a3}	pK_{a3}
sulfuric acid, H_2SO_4	strong		1.2×10^{-2}	1.92		
oxalic acid, $(COOH)_2$	5.9×10^{-2}	1.23	6.5×10^{-5}	4.19		
sulfurous acid, H_2SO_3	1.5×10^{-2}	1.81	1.2×10^{-7}	6.91		
phosphorous acid, H_3PO_3	1.0×10^{-2}	2.00	2.6×10^{-7}	6.59		
phosphoric acid, H_3PO_4	7.6×10^{-3}	2.12	6.2×10^{-8}	7.21	2.1×10^{-13}	12.68
tartaric acid, $C_2H_4O_2(COOH)_2$	6.0×10^{-4}	3.22	1.5×10^{-5}	4.82		
carbonic acid, H_2CO_3	4.3×10^{-7}	6.37	5.6×10^{-11}	10.25		
hydrosulfuric acid, H_2S	1.3×10^{-7}	6.89	7.1×10^{-15}	14.15		

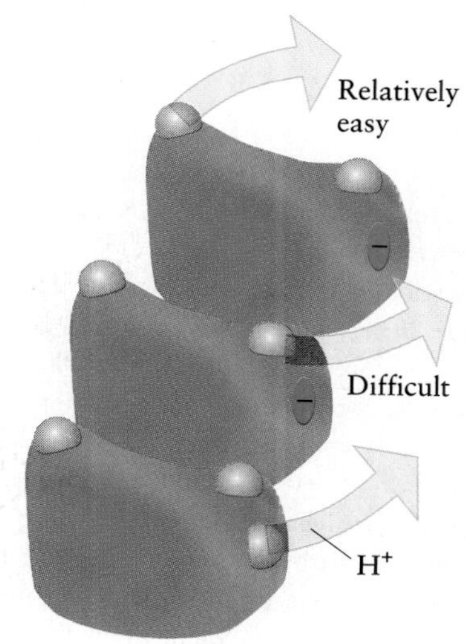

FIGURE 14.16 In this illustration, the three blobs represent enzyme molecules. The loss of a second proton from a location close to the site of one that has been lost already is difficult because of the attraction between opposite charges. However, the loss of the first proton has relatively little effect on the ease with which a more distant proton can be lost.

Table 14.5 gives the acidity constants of some polyprotic acids. The data show that the strengths of polyprotic acids decrease as protons are lost:

$$K_{a1} > K_{a2} > \ldots$$

The decrease is reasonable: it is harder to lose a positively charged proton from a negatively charged ion (such as HCO_3^-) than from the original uncharged molecule (H_2CO_3). Sulfuric acid, for example, is a strong acid, and H_2SO_4 loses its first proton to give its conjugate base, the hydrogen sulfate ion, HSO_4^-; this ion, however, is a weak acid. Some proteins can donate dozens of protons in reactions with little decrease in acid strength. Because the protons are donated from widely separated sites, the loss of one proton has little influence on the next (Fig. 14.16).

EXAMPLE 14.7 Calculating the pH of a solution of sulfuric acid

Calculate the pH of 0.010 M H_2SO_4(aq) at 25°C. Use the appropriate information from Table 14.5.

STRATEGY The pH depends on the total H_3O^+ molarity, so both ionization steps must be taken into account. Sulfuric acid is the only common polyprotic acid for which the first ionization is complete. The second ionization adds to the H_3O^+ molarity slightly, so the overall pH will be slightly less than that due to the first ionization alone. To find its contribution, set up the equilibrium table in the usual way but use as the initial H_3O^+ and HSO_4^- molarities the values obtained by assuming that the first proton is lost completely. Because K_{a2} is relatively large (0.012), there are no shortcuts: it will be necessary to solve a quadratic equation.

A graphing calculator or plotting program can be used to solve the quadratic equation more quickly.

SOLUTION The first ionization,

$$H_2SO_4(aq) + H_2O(l) \longrightarrow H_3O^+(aq) + HSO_4^-(aq)$$

results in an H_3O^+ molarity equal to the original molarity of the acid before ionization, 0.010 mol·L^{-1}. This value corresponds to pH = 2.0. The second proton transfer equilibrium is

$$HSO_4^-(aq) + H_2O(l) \rightleftharpoons H_3O^+(aq) + SO_4^{2-}(aq) \qquad K_{a2} = \frac{[H_3O^+][SO_4^{2-}]}{[HSO_4^-]}$$

and the equilibrium table, with all concentrations in moles per liter, is

	Species		
	HSO_4^-	H_3O^+	SO_4^{2-}
Step 1. Initial molarity	0.010	0.010	0
Step 2. Change in molarity	$-x$	$+x$	$+x$
Step 3. Equilibrium molarity	$0.010 - x$	$0.010 + x$	x

CASE STUDY 14 ACID RAIN

IN 1989 SCIENTISTS in the Netherlands noticed that the great tit, a forest songbird, was producing eggs with thin, porous shells. The insecticide DDT had caused a similar problem in the 1960s and 1970s, but no evidence of toxic pesticides could be found. Scientists therefore investigated the birds' supply of calcium, which is needed for strong shells. Great tits normally get their calcium from the snails that make up most of their diet. However, the snails had virtually vanished from the forests. Dry forest soil normally contains 5–10 g calcium per kilogram; the calcium content of this soil had fallen to 0.3 $g \cdot kg^{-1}$, too low for snails to survive. Without snails to eat, the birds had begun raiding chicken houses and picnic grounds, seeking discarded eggshells to supplement their diets.

The fall in the calcium content of soil in Europe and the United States has been traced to acid rain, in particular to rain containing sulfuric acid. As the first illustration (below) shows, acid rain is a regional phenomenon. The lines on the map are contours showing the pH of rain, which decreases downwind of heavily populated areas. The low pH in heavily industrialized and populated areas is thought to be caused by the acidic oxides sulfur dioxide, SO_2, and the nitrogen oxides NO and NO_2. The second illustration shows that there is a strong correlation between the levels of these oxides and pH.

Rain unaffected by human activity contains mostly weak acids and has a pH of 5.7. The primary acid present is carbonic acid, H_2CO_3, which results from the dissolving of atmospheric carbon dioxide, an acidic oxide, in water.

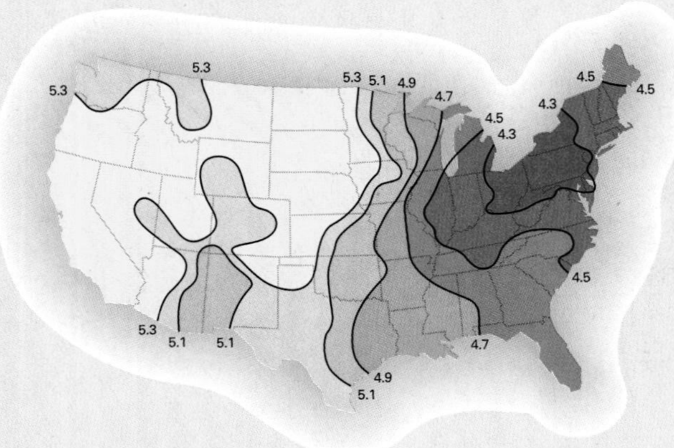

The curves on this map represent pH isopleths, regions in which the precipitation has the same pH. Notice that the average pH of precipitation decreases from west to east.

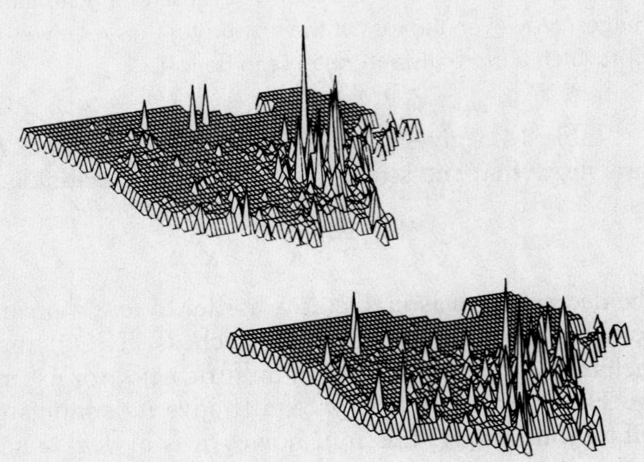

These maps of North America show the distribution of sulfur oxides, SO_x (top), and nitrogen oxides, NO_x (bottom), in the atmosphere. The peaks represent the relative concentrations of SO_x and NO_x in each region. Note the increase in concentration from west to east.

The serious pollutants in acid rain are strong acids. Atmospheric nitrogen and oxygen can react to form NO at the high temperatures of automobile internal combustion engines and electrical power stations:

$$N_2(g) + O_2(g) \longrightarrow 2\,NO(g)$$

Nitric oxide, NO, is not very soluble in water, but it can be oxidized further in air to form nitrogen dioxide:

$$2\,NO(g) + O_2(g) \longrightarrow 2\,NO_2(g)$$

The NO_2 reacts with water, forming nitric acid and nitric oxide:

$$3\,NO_2(g) + 3\,H_2O(l) \longrightarrow 2\,H_3O^+(aq) + 2\,NO_3(aq) + NO(g)$$

Catalytic converters in automobiles reduce the nitrogen in NO to N_2 and are required in the United States for all new cars and trucks.

Sulfur dioxide is produced as a by-product of the burning of fossil fuels. It may combine with water directly, to form sulfurous acid, a weak acid:

$$SO_2(g) + 2\,H_2O(l) \longrightarrow H_3O^+(aq) + HSO_3^-(aq)$$

Alternatively, in the presence of particulate matter and aerosols, sulfur dioxide may react with atmospheric oxy-

One consequence of acid rain is vividly illustrated by these two photographs of a forest site in Germany. The one on the left was taken in 1970, the one on the right in 1983.

gen to form sulfur trioxide, which forms sulfuric acid in water:

$$2\,SO_2(g) + O_2(g) \longrightarrow 2\,SO_3(g)$$

$$SO_3(g) + 2\,H_2O(l) \longrightarrow H_3O^+(aq) + HSO_4^-(aq)$$

Sulfuric acid is a strong acid that is especially damaging to soil because it causes the leaching of calcium ions. Most soil contains clay particles, which are surrounded by layers of ions, including Ca^{2+}. However, calcium ions on the clay particles can be replaced by hydrogen ions from sulfuric acid. Because calcium sulfate is insoluble in water, it can no longer circulate through the soil or be taken up by plants. If the calcium leached from soil is not replaced, plants suffer and entire forests can be affected.

Research on the impact of air pollution on forests is difficult. Forests cover such vast areas and the interplay of regional air pollutants is so subtle that it may take many years to clarify the role of environmental stresses. However, control of acidic oxide emissions can be improved to maintain our quality of life without losing our precious natural heritage.

QUESTIONS

1. The partial pressure of CO_2 in air saturated with water vapor at 25°C and 1.00 atm is 3.04×10^{-4} atm. Henry's constant for CO_2 in water is 2.3×10^{-2} mol·L^{-1}·atm^{-1}, and for carbonic acid, $pK_{a1} = 6.37$. Verify by calculation that the pH of "normal" rainwater is about 5.7.

2. One metric ton (1.0×10^3 kg) of coal that contains 2.5% sulfur is burned in a coal-fired plant. (a) What mass of SO_2 is produced? (b) What is the pH of rainwater when the SO_2 dissolves in a volume of water equivalent to 2.0 cm of rainfall over 2.6 km^2? (The pK_{a1} of sulfurous acid is 1.81. Consider the water to be initially pure and at a pH of 7.) (c) If the SO_2 is first oxidized to SO_3 before the rainfall occurs, what would the pH of the same rainwater be?

3. If the rainfall described in Question 2(c) fell on soil and half the sulfuric acid molecules in the rain leached calcium ions from clay particles in the soil by replacing them with hydrogen ions, what mass of calcium could be leached?

4. One process used to clean SO_2 from the emissions of coal-fired plants is to pass the stack gases through a wet calcium carbonate slurry, where the following reaction occurs:

$$CaCO_3(s) + SO_2(g) + \tfrac{1}{2}O_2(g) \longrightarrow CaSO_4(s) + CO_2(g)$$

What mass of limestone ($CaCO_3$) is needed to remove the sulfur dioxide produced in Question 2(a) if the removal process is 90% efficient?

FIGURE 14.17 The effervescence that occurs when an antacid tablet containing hydrogen carbonate ions is dissolved in acid is the result of a chain of effects that links several equilibria.

Step 4. Now substitute the molarities in step 3 into the expression for the second acidity constant, and use $K_{a2} = 0.012$, which gives

$$0.012 = \frac{(0.010 + x) \times x}{0.010 - x}$$

To find x, rearrange this expression into a quadratic equation:

$$x^2 + 0.022x - (1.2 \times 10^{-4}) = 0$$

Then solve this equation by using the quadratic formula in Toolbox 13.1:

$$x = \frac{-0.022 \pm \sqrt{(0.022)^2 - 4(-1.2 \times 10^{-4})}}{2}$$
$$= 4.5 \times 10^{-3} \text{ and } -2.7 \times 10^{-2}$$

Because $x = [SO_4^{2-}]$ (step 3), the value of x cannot be negative, so we select the solution $x = 4.5 \times 10^{-3}$. The total H_3O^+ molarity is

$$[H_3O^+] = 0.010 + x = 0.010 + (4.5 \times 10^{-3}) = 1.4 \times 10^{-2}$$

so

$$pH = -\log(1.4 \times 10^{-2}) \approx 1.9$$

The pH of the solution is less than 2.0, as was predicted.

SELF-TEST 14.12A Estimate the pH of a 0.050 M sulfuric acid solution.

[*Answer:* 1.2]

SELF-TEST 14.12B Estimate the pH of a 0.10 M sulfuric acid solution.

Almost all polyprotic acids other than sulfuric are weak for all their ionization stages. To calculate their pH, we take only the first ionization into account: subsequent ionizations are much less important than the first and can be ignored. For carbonic acid, H_2CO_3, for instance, $K_{a2} = 5.6 \times 10^{-11}$, which is much smaller than $K_{a1} = 4.3 \times 10^{-7}$. We just use K_{a1} and do the calculation illustrated in Toolbox 14.1.

If we wanted to find the concentration of CO_3^{2-} in a solution of carbonic acid, we would have to use both K_as.

We estimate the pH of a polyprotic acid for which all ionizations are weak by using only the first deprotonation equilibrium and assuming that further ionization is insignificant. An exception is sulfuric acid, the only polyprotic acid that is a strong acid in its first ionization.

SELF-TEST 14.13A Estimate the pH of a 0.10 M phosphorous acid solution. Refer to Table 14.5 for K_a values. (*Hint:* Treat the first ionization as a weak acid and ignore the second ionization.)

[*Answer:* 1.6]

SELF-TEST 14.13B Estimate the pH of a 0.10 M sulfurous acid solution, using the data in Table 14.5.

Proton transfer explains the effervescence of carbon dioxide in antacid tablets (Fig. 14.17). When an acid is added to a carbonate or a hydrogen carbonate, the equilibrium shifts in accord with Le Chatelier's principle. At first, the added H_3O^+ ions shift the equilibrium composition to the left, favoring the formation of H_2CO_3:

$$H_2CO_3(aq) + H_2O(l) \rightleftharpoons H_3O^+(aq) + HCO_3^-(aq)$$

However, carbonic acid in solution is also in equilibrium with dissolved CO_2 molecules:

$$H_2CO_3(aq) \rightleftharpoons H_2O(l) + CO_2(aq)$$

Thus, when the H_2CO_3 concentration increases as a result of the addition of acid, the CO_2 concentration increases in response. The increase in concentration of CO_2 is so great that the CO_2 bubbles out of solution with the familiar fizz. As in this everyday example, chemical effects are often transmitted along chains of equilibria, the disturbance of one equilibrium affecting another. This effect helps to explain the impact of acid rain. The acid enables other reactions to occur in lakes, rivers, and soil (see Case Study 14).

SKILLS YOU SHOULD HAVE MASTERED

Conceptual

1. Explain how the pH of a solution is related to its hydronium ion and hydroxide ion concentrations.
2. Explain why solutions of weak acids have higher pH values than solutions of strong acids at the same concentration.
3. Show how the acidity constant of an acid is related to the basicity constant of its conjugate base.
4. Use K_a values to predict the relative strengths of two acids or two bases.

Problem-Solving

1. Identify species as Brønsted acids or bases.
2. Write the formulas for conjugate acids and bases.
3. Calculate the hydroxide ion concentration from a hydronium ion concentration, and vice versa.
4. Calculate the pH and pOH of a solution of a strong or weak acid or base.
5. Calculate the percentage ionization of a weak acid or base.
6. Calculate the pH of a solution of a polyprotic acid.

Descriptive

1. Identify the common strong acids and bases.
2. Describe and explain trends in acid strength in series of binary acids, oxoacids, and carboxylic acids.

EXERCISES

Unless stated otherwise, assume that all solutions are aqueous and that the temperature is 25°C.

Brønsted Acids and Bases

14.1 Classify each of the following species as an acid, a base, or amphiprotic in aqueous solution: (a) H_2O; (b) CH_3NH_2; (c) PO_4^{3-}; (d) $C_6H_5NH_3^+$.

14.2 Classify each of the following species as an acid, a base, or amphiprotic in aqueous solution: (a) HSO_4^-; (b) HI; (c) IO^-; (d) OH^-.

14.3 Write the formulas for the conjugate acids of (a) CH_3NH_2 (methylamine); (b) NH_2NH_2; (c) HCO_3^- and the conjugate bases of (d) HCO_3^-; (e) C_6H_5OH (phenol); (f) CH_3COOH.

14.4 Write the formulas for the conjugate acids of (a) H_2O; (b) OH^-; (c) $C_6H_5NH_2$ (aniline) and the conjugate bases of (d) H_2S; (e) HPO_4^{2-} (hydrogen phosphate ion); (f) $HClO_4$ (perchloric acid).

14.5 Write the proton transfer equilibria for the following acids in aqueous solution and identify the conjugate acid-base pairs in each one: (a) H_2SO_4; (b) $C_6H_5NH_3^+$ (anilinium ion); (c) $H_2PO_4^-$ (dihydrogen phosphate ion); (d) HCOOH (formic acid); (e) $NH_2NH_3^+$ (hydrazinium ion).

14.6 Write the proton transfer equilibria for the following bases in aqueous solution and identify the conjugate acid-base pairs in each one: (a) CN^-; (b) NH_2NH_2 (hydrazine); (c) CO_3^{2-}; (d) HPO_4^{2-}; (e) NH_2CONH_2 (urea).

14.7 Write the two proton transfer equilibria that demonstrate the amphiprotic character of (a) HCO_3^-; (b) HPO_4^{2-}, and identify the conjugate acid-base pairs in each equilibrium.

14.8 Write the two proton transfer equilibria that show the amphiprotic character of (a) $H_2PO_4^-$; (b) $HC_2O_4^-$ (hydrogen oxalate ion), and identify the conjugate acid-base pairs in each equilibrium.

14.9 Identify (a) the Brønsted acid and base in the following reaction, and (b) the conjugate base and acid formed: $HNO_3(aq) + HPO_4^{2-}(aq) \rightleftharpoons NO_3^-(aq) + H_2PO_4^-(aq)$.

14.10 Identify (a) the Brønsted acid and base in the following reaction, and (b) the conjugate base and acid formed: $HSO_3^-(aq) + NH_4^+(aq) \rightleftharpoons NH_3(aq) + H_2SO_3(aq)$.

Autoprotolysis of Water

14.11 Calculate the molarity of OH^- in solutions with the following H_3O^+ concentrations: (a) 3.1×10^{-2} M H_3O^+; (b) 1.0×10^{-4} M H_3O^+; (c) 0.20 M H_3O^+.

14.12 Estimate the molarity of H_3O^+ in solutions with the following OH^- concentrations: (a) 5.6×10^{-3} M OH^- (b) 8.5×10^{-5} M OH^-; (c) 0.12 M OH^-.

14.13 The value of K_w for water at body temperature (37°C) is 2.5×10^{-14}. (a) What is the molarity of H_3O^+ ions and the pH of neutral water at 37°C? (b) What is the molarity of OH^- in neutral water at 37°C?

14.14 The molarity of H_3O^+ ions at the freezing point of water is 3.9×10^{-8} mol·L^{-1}. (a) Calculate K_w and pK_w at 0°C. (b) What is the pH of neutral water at 0°C?

pH and pOH of Strong Acids and Bases

14.15 Calculate the nominal molarity of HCl and the molarities of H_3O^+, Cl^-, and OH^- in an aqueous solution that contains 0.48 mol of HCl in 500 mL of solution.

14.16 Calculate the nominal molarity of HNO_3 and the molarities of H_3O^+, NO_3^-, and OH^- in an aqueous solution that contains 0.062 mol of HNO_3 in 250 mL of solution.

14.17 Calculate the nominal molarity of $Ba(OH)_2$ and the molarities of Ba^{2+}, OH^-, and H_3O^+ in an aqueous solution that contains 0.50 g of $Ba(OH)_2$ in 100 mL of solution.

14.18 Calculate the nominal molarity of KNH_2 and the molarities of K^+, NH_2^-, OH^-, and H_3O^+ in an aqueous solution that contains 1.0 g of KNH_2 in 250 mL of solution.

14.19 The pH of several solutions was measured in the research laboratories of a food company; convert each of the following pH values to the molarity of H_3O^+ ions: (a) 3.3 (the pH of sour orange juice); (b) 6.7 (the pH of a saliva sample); (c) 4.4 (the pH of beer); (d) 5.3 (the pH of a coffee sample). (e) List the samples in order of increasing acidity.

14.20 The pH of several solutions was measured in a hospital laboratory; convert each of the following pH values to the molarity of H_3O^+ ions: (a) 5.0 (the pH of a urine sample); (b) 2.3 (the pH of a sample of lemon juice); (c) 7.4 (the pH of blood); (d) 10.5 (the pH of milk of magnesia). (e) List the samples in order of increasing acidity.

14.21 The molarity of H_3O^+ ions in the following solutions was measured at 25°C. Calculate the pH and pOH of the solutions: (a) 2.0×10^{-5} mol·L^{-1} (sample of rainwater); (b) 1.0 mol·L^{-1}; (c) 5.0×10^{-14} mol·L^{-1}; (d) 5.02×10^{-5} mol·L^{-1}.

14.22 The molarity of OH^- ions in the following solutions was measured at 25°C. Calculate the pH and pOH of the solutions: (a) 1.5×10^{-7} mol·L^{-1} (a sample of milk); (b) 2.2×10^{-3} mol·L^{-1}; (c) 1.00×10^{-6} mol·L^{-1} (a sample of tap water); (d) 7.09×10^{-4} mol·L^{-1}.

14.23 Calculate the pH and pOH of each of the following aqueous solutions of strong acid or base: (a) 0.010 M HNO_3; (b) 0.22 M HCl; (c) 1.0×10^{-3} M $Ba(OH)_2$; (d) 10.0 mL of 0.022 M KOH after dilution to 250 mL; (e) 14.0 mg of NaOH dissolved in 250 mL of solution; (f) 50.0 mL of 0.000 43 M HBr after dilution to 250 mL.

14.24 Calculate the pH and pOH of each of the following aqueous solutions of strong acid or base: (a) 0.0149 M HI; (b) 0.0602 M HCl; (c) 1.73×10^{-3} M $Ba(OH)_2$; (d) 4.4 mg of KOH dissolved in 10.0 mL of solution; (e) 10.0 mL of 0.0022 M NaOH after dilution to 250 mL; (f) 5.0 mL of 0.000 43 M $HClO_4$ after dilution to 25.0 mL.

Acidity and Basicity Constants

14.25 Refer to Table 14.2. Name the following acids, write their K_a and pK_a values, and list the acids in order of increasing strength: (a) HCOOH; (b) CH_3COOH; (c) CCl_3COOH; (d) C_6H_5COOH.

14.26 Refer to Table 14.2. Name the following acids, write their K_a and pK_a values, and list the acids in order of increasing strength: (a) HCN; (b) HIO_3; (c) HNO_2; (d) HF.

14.27 Give the K_a values and list the following acids in order of increasing strength: (a) phosphoric acid, H_3PO_4, pK_{a1} = 2.12; (b) phosphorous acid, H_3PO_3, pK_{a1} = 2.00; (c) selenous acid, H_2SeO_3, pK_{a1} = 2.46; (d) selenic acid, H_2SeO_4, pK_{a1} = 1.92.

14.28 Give the pK_b values and list the following bases in order of increasing strength: (a) ammonia, NH_3, K_b = 1.8×10^{-5}; (b) deuterated ammonia, ND_3, K_b = 1.1×10^{-5}; (c) hydrazine, NH_2NH_2, K_b = 1.7×10^{-6}; (d) hydroxylamine, NH_2OH, K_b = 1.1×10^{-8}.

14.29 (a) Write the names and formulas for the strongest and weakest conjugate bases of the acids listed in Exercise 14.27. (b) What are the K_b values for the two bases? (c) Which base, dissolved in water to a given concentration, would produce a solution with the highest pH?

14.30 (a) Write the names and formulas for the strongest and weakest conjugate acids of the bases listed in Exercise 14.28. (b) What are the K_a values for the two acids? (c) Which acid, dissolved in water to a given concentration, would produce a solution with the highest pH?

Structures and Strengths of Acids

14.31 The pK_a for HIO (hypoiodous acid) is 10.64 and that for HIO_3 (iodic acid) is 0.77. Account for the differences in acid strength.

14.32 The pK_a for HClO (hypochlorous acid) is 7.53 and that for HBrO (hypobromous acid) is 8.69. Account for the differences in acid strength.

14.33 Determine which acid is stronger and explain why: (a) HF or HCl; (b) HClO or $HClO_2$; (c) $HBrO_2$ or $HClO_2$; (d) $HClO_4$ or H_3PO_4; (e) HNO_3 or HNO_2; (f) H_2CO_3 or H_2GeO_3.

14.34 Determine which acid is stronger and explain why: (a) H_3AsO_4 or H_3PO_4; (b) $HBrO_3$ or HBrO; (c) H_3PO_4 or H_3PO_3; (d) H_2Te or H_2Se; (e) H_2S or HCl; (f) HClO or HIO.

14.35 Suggest an explanation for the difference between the strengths of (a) acetic acid and trichloroacetic acid and (b) acetic acid and formic acid.

14.36 Suggest an explanation of the difference in base strengths between (a) ammonia and methylamine and (b) hydrazine and hydroxylamine.

14.37 The values of K_a for phenol and 2,4,6-trichlorophenol (see following structures) are 1.3×10^{-10} and 1.0×10^{-6}, respectively. Which is the stronger acid? Account for the differences in acid strength.

OH OH Cl Cl Cl

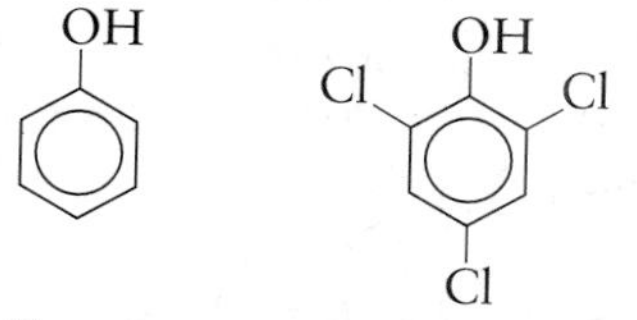

Phenol 2,4,6-Trichlorophenol

14.38 The values of pK_b for aniline is 9.37 and that for 4-chloroaniline is 9.85 (see following structures). Which is the stronger base? Account for the differences in base strength.

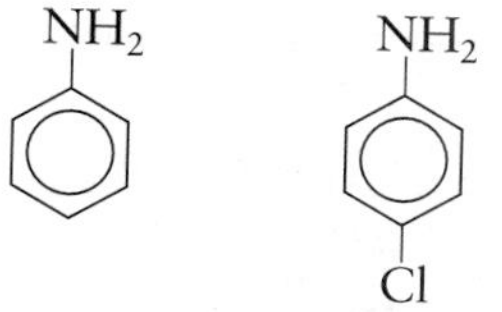

NH_2 NH_2 Cl

Aniline 4-Chloroaniline

14.39 Arrange the following bases in order of increasing strength on the basis of the pK_a values of their conjugate acids, which are given in parentheses: (a) ammonia (9.26); (b) methylamine (10.56); (c) ethylamine (10.81); (d) aniline (4.63). Is there a simple pattern of strengths?

14.40 Arrange the following bases in order of increasing strength on the basis of the pK_a values of their conjugate acids, which are given in parentheses: (a) aniline (4.63); (b) 2-hydroxyaniline (4.72); (c) 3-hydroxyaniline (4.17); (d) 4-hydroxyaniline (5.47). Is there a simple pattern of strengths?

NH_2 OH NH_2 OH NH_2 OH

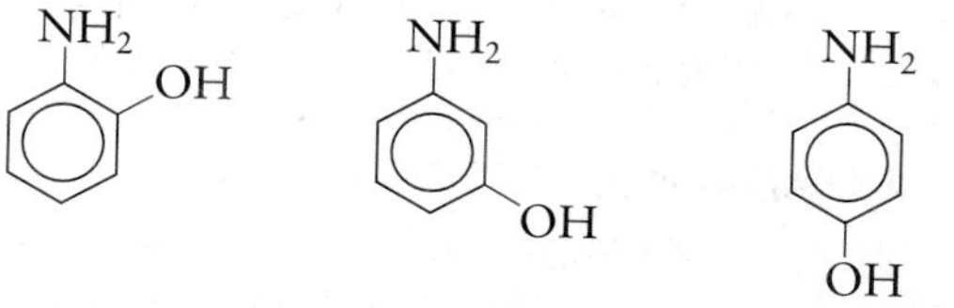

2-Hydroxyaniline 3-Hydroxyaniline 4-Hydroxyaniline

Weak Acid and Weak Base Calculations

Refer to Table 14.2 for the appropriate K_a and K_b values for the following exercises.

14.41 Determine the concentration of H_3O^+ and OH^- in (a) 0.20 M C_6H_5COOH; (b) 0.20 M NH_2NH_2 (hydrazine); (c) 0.20 M $(CH_3)_3N$ (trimethylamine).

14.42 Determine the concentration of H_3O^+ and OH^+ in (a) 0.15 M HCOOH; (b) 0.10 M $C_6H_5SO_3H$ (benzenesulfonic acid); (c) 0.10 M NH_3.

14.43 Calculate the pH of the following solutions: (a) 0.20 M HCOOH(aq); (b) 0.12 M NH_2NH_2(aq) (hydrazine); (c) 0.15 M C_6H_5COOH(aq) (benzoic acid); (d) 0.0034 M $C_{10}H_{14}N_2$(aq) (nicotine, a base).

14.44 Calculate the pH of the following solutions: (a) 0.0477 M HCN(aq) (hydrocyanic acid); (b) 1.5×10^{-5} M CH_3COOH(aq); (c) 0.023 M HBrO(aq).

14.45 Calculate the pH and pOH of the following aqueous solutions: (a) 0.15 M CH_3COOH(aq); (b) 0.15 M CCl_3COOH(aq); (c) 0.15 M HCOOH(aq).

14.46 Calculate the pH and pOH of the following aqueous solutions: (a) 0.20 M $CH_3CH(OH)COOH$(aq) (lactic acid); (b) 1.0×10^{-5} M $CH_3CH(OH)COOH$(aq); (c) 0.10 M $C_6H_5SO_3H$(aq) (benzenesulfonic acid).

14.47 Calculate the pH, pOH, and percentage protonation of solute in the following aqueous solutions: (a) 0.10 M NH_3(aq); (b) 0.017 M NH_2OH(aq); (c) 0.20 M $(CH_3)_3N$(aq); (d) 0.020 M codeine, given that the pK_a of its conjugate acid is 8.21. Codeine, a cough suppressant, is extracted from opium.

14.48 Calculate the pOH, pH, and percentage protonation of solute in the following aqueous solutions: (a) 0.11 M C_5H_5N (pyridine); (b) 0.0058 M $C_{10}H_{14}N_2$ (nicotine); (c) 0.020 M quinine, given that the pK_a of its conjugate acid is 8.52; (d) 0.011 M strychnine, given that the K_a of its conjugate acid is 5.49×10^{-9}.

14.49 (a) When the pH of a 0.10 M $HClO_2$ aqueous solution was measured, it was found to be 1.2. What are the values of K_a and pK_a of chlorous acid? (b) The pH of a 0.10 M propylamine, $C_3H_7NH_2$, aqueous solution was measured as 11.86. What are the values of K_b and pK_b of propylamine?

14.50 (a) The pH of a 0.015 M HNO_2 aqueous solution was measured as 2.63. What are the values of K_a and pK_a of nitrous acid? (b) The pH of a 0.10 M butylamine, $C_4H_9NH_2$, aqueous solution was measured as 12.04. What are the percentage protonation and the values of K_b and pK_b of butylamine?

14.51 Use the information in Table 14.2 to find the nominal concentration of the weak acid or base in each of the following aqueous solutions: (a) a solution of HClO with pH = 4.6; (b) a solution of hydrazine, NH_2NH_2, with pH = 10.2.

14.52 Use the information in Table 14.2 to find the nominal concentration of the weak acid or base in each of the following aqueous solutions: (a) a solution of HCN with pH = 5.3; (b) a solution of pyridine, C_5H_5N, with pH = 8.8.

14.53 The percentage ionization of benzoic acid in a 0.110 M solution is 2.4%. What is the pH of the solution and the K_a of benzoic acid?

14.54 The percentage ionization of veronal (diethylbarbituric acid) in a 0.020 M aqueous solution is 0.14%. What is the pH of the solution and the K_a of veronal?

14.55 The percentage ionization of octylamine (an organic base) in a 0.10 M aqueous solution is 6.7%. What is the pH of the solution and the K_b of octylamine?

14.56 Cacodylic acid is used as a cotton defoliant. A 0.011 M cacodylic acid solution is 0.77% ionized in water. What is the pH of the solution and the K_a of cacodylic acid?

Polyprotic Acids

Refer to Table 14.5 for the K_a values needed for the following exercises.

14.57 Write the stepwise proton transfer equilibria for the ionization of (a) sulfuric acid, H_2SO_4; (b) arsenic acid, H_3AsO_4; (c) phthalic acid, $C_6H_4(COOH)_2$.

14.58 Write the stepwise proton transfer equilibria for the ionization of (a) phosphoric acid, H_3PO_4; (b) adipic acid; $(CH_2)_4(COOH)_2$; (c) succinic acid; $(CH_2)_2(COOH)_2$.

14.59 Calculate the pH of 0.15 M H_2SO_4(aq) at 25°C.

14.60 Calculate the pH of 0.010 M H_2SeO_4(aq) given that K_{a1} is very large and $K_{a2} = 1.2 \times 10^{-2}$.

14.61 Calculate the pH of the following diprotic acid solutions at 25°C, ignoring second ionizations only when the approximation is justified: (a) 0.0010 M H_2CO_3; (b) 0.10 M $(COOH)_2$; (c) 0.20 M H_2S.

14.62 Calculate the pH of the following diprotic acid solutions at 25°C; ignore second ionizations only when that approximation is justified: (a) 0.10 M H_2S; (b) 0.15 M $H_2C_4H_4O_6$ (tartaric acid); (c) 1.1×10^{-3} M H_2TeO_4 (telluric acid, for which $K_{a1} = 2.1 \times 10^{-8}$ and $K_{a2} = 6.5 \times 10^{-12}$).

SUPPLEMENTARY EXERCISES

14.63 What condition is required for an aqueous solution to be considered neutral?

14.64 Write the formula for the conjugate acid of (a) the carbonate ion, CO_3^{2-}; (b) the acetate ion, $CH_3CO_2^-$, and the conjugate base of (c) the dihydrogen phosphate ion, $H_2PO_4^-$; (d) the hydroxide ion, OH^-.

14.65 Identify the conjugate acid-base pairs and write the proton transfer equilibria for (a) propionic acid, C_2H_5COOH; (b) chloric acid, $HClO_3$; (c) acetylsalicylic acid (aspirin), $C_8H_7O_2COOH$; (d) caffeine (a base), $C_8H_{10}N_4O_2$; (e) pyridine (a base), C_5H_5N.

14.66 The value of K_w at 40°C is 3.8×10^{-14}. What is the pH of pure water at 40°C?

14.67 Under what conditions can the pH of solutions be (a) negative; (b) greater than 14?

14.68 Calculate the pH and pOH of (a) 0.026 M HNO_3(aq); (b) 0.012 M $Ba(OH)_2$(aq); (c) 1.47×10^{-4} M HBr(aq); (d) 3.19×10^{-3} M KOH(aq).

14.69 Calculate the molarity of H_3O^+ ions in a solution having a pH of (a) 9.33; (b) 7.95; (c) 0.01; (d) 4.33; (e) 1.99; (f) 11.95.

14.70 1.00 g of NaOH was dissolved in 100 mL of solution and then diluted to 500 mL. What is the pH of the resulting solution?

14.71 Decide on the basis of the information in Table 14.2 which of the acids CH_3COOH or HNO_2 will have the greater percentage of solute ionized in water.

14.72 Use the information in Table 14.1 to decide whether carbonic acid is a strong or weak acid in liquid ammonia solvent. Explain your answer.

14.73 The K_a of phenol (C_6H_5OH, which is also called carbolic acid) is 1.3×10^{-10} and that for the ammonium ion is 5.6×10^{-10}. (a) Write the proton transfer equilibria for each in aqueous solution. (b) Calculate the pK_a for each. (c) Which acid is the stronger acid?

14.74 Determine the value of K_b for the following anions (all of which are the conjugate bases of acids) and list them in order of increasing base strength (refer to Tables 14.2 and 14.5 for information): F^-, $CH_2ClCO_2^-$, CO_3^{2-}, IO_3^-, Cl^-.

14.75 Identify the stronger acid in each of the following pairs, and give reasons for your choice: (a) $HBrO_4$ or HIO_4; (b) HF or HI; (c) HIO_2 or HIO_3; (d) H_3AsO_4 or H_2SeO_4.

14.76 Calculate the pOH and pH of each of the following aqueous solutions: (a) 0.029 M $C_6H_5NH_2$; (b) 0.10 M $C_6H_{11}NH_2$ (cyclohexamine, for which $pK_b = 3.36$); (c) 0.0194 M $C_{17}H_{19}O_3N$ (morphine, a base); (d) 0.015 M $NH_2CH_2CH_2NH_2$ (ethylenediamine), given that the K_a of its conjugate acid is 1.9×10^{-11}.

14.77 When 150 mg of an organic base of molar mass 31.06 $g \cdot mol^{-1}$ is dissolved in 50.0 mL of water, the pH is found to be 10.05. What is the percentage protonation of the base? Calculate the pK_b of the base and the pK_a of its conjugate acid.

14.78 Write the stepwise proton transfer equilibria for the ionization of (a) hydrosulfuric acid, H_2S; (b) tartaric acid, $C_2H_4O_2(COOH)_2$; (c) malonic acid, $CH_2(COOH)_2$.

CHALLENGING EXERCISES

14.79 Heavy water, D_2O, is used in some nuclear reactors (see Chapter 22). The K_w for heavy water at 25°C is 1.35×10^{-15}. (a) Write the chemical equation for the autoprotolysis of D_2O. (b) Evaluate pK_w for D_2O at 25°C. (c) Calculate the molarities of D_3O^+ and OD^- in neutral heavy water at 25°C. (d) Evaluate the pD and pOD of neutral heavy water at 25°C. (e) Find the relation between pD, pOD, and pK_w.

14.80 Although many chemical reactions take place in water, it is often necessary to use other solvents instead, and liquid ammonia (b.p. −33°C) has been used extensively. Many of the reactions that occur in water have analogous reactions in liquid ammonia. (a) Write the chemical equation for the autoprotolysis of NH_3. (b) What are the formulas of the acid and base species that result from the autoprotolysis of liquid ammonia? (c) The autoionization constant, K_{am}, of liquid ammonia has the value 1×10^{-33} at −35°C. What is the value of pK_{am} at that temperature? (d) What is the molarity of NH_4^+ ions in neutral liquid ammonia? (e) Evaluate pNH_4 and pNH_2 in neutral liquid ammonia at −35°C. (f) Determine the relation between pNH_4, pNH_2, and pK_{am}.

14.81 (a) The pH of a 0.025 M aqueous solution of a base was 11.6. What is the pK_b of the base and the pK_a of its conjugate acid? (b) The percentage protonation of thiazole (an organic base) in a 0.0010 M solution is 5.2×10^{-3}%. What is the pH of the solution and K_b of thiazole? (In this case you must take into consideration the autoprotolysis of water.)

14.82 Calculate the pH of a 0.020 M succinic acid solution at 25°C, taking into account both ionizations ($K_{a1} = 6.9 \times 10^{-5}$, $K_{a2} = 2.5 \times 10^{-6}$).

14.83 Calculate the pH of the following acid solutions at 25°C; ignore second ionizations only when that approximation is justified. (a) 1.0×10^{-4} M $H_3BO_3(aq)$ (boric acid acts as a monoprotic acid); (b) 0.015 M $H_3PO_4(aq)$; (c) 0.10 M $H_2SO_3(aq)$.

14.84 (a) Calculate the molarities of $(COOH)_2$, $HOOCCO_2^-$, $(CO_2)_2^{2-}$, H_3O^+, and OH^- in a 0.10 M $(COOH)_2$ solution. (b) Calculate the molarities of H_2S, HS^-, S^{2-}, H_3O^+, and OH^- in a 0.050 M H_2S solution.

14.85 Calculate the molarities of H_2SO_3, HSO_3^-, SO_3^{2-}, H_3O^+, and OH^- in a 0.10 M H_2SO_3 solution.

14.86 Estimate the enthalpy of ionization of formic acid at 25°C given that K_a is 1.765×10^{-4} at 20°C and 1.768×10^{-4} at 30°C. (*Hint:* Use the van't Hoff equation, Exercise 13.94.)

14.87 Convert the van't Hoff equation (stated in Exercise 13.94) for the temperature dependence of an equilibrium constant to an expression for the temperature dependence of pK_w. Estimate the value of pK_w at the normal boiling point of water from the enthalpy of the water autoprotolysis reaction, which is +57 kJ for the reaction $2\,H_2O(l) \rightarrow H_3O^+(aq) + OH^-(aq)$. What is the pH of pure water at that temperature?

14.88 Use the van't Hoff equation (see Exercise 13.94) to estimate the enthalpy of the ionization of heavy water, D_2O, given the following data on its autoprotolysis:

Temperature, °C	pK_w
10	15.44
20	15.05
25	14.87
30	14.70
40	14.39
50	14.10

What is the pD ($pD = -\log[D_3O^+]$) of pure heavy water at 40°C? What would be the values of pOD, pK_{HW}, and K_{HW} (HW denotes heavy water) at 40°C?

CHAPTER 15

Aqueous solutions of salts are often acidic or basic. The pH can have profound consequences for marine life. This environmental chemist is measuring the pH of cleaned industrial effluent on behalf of the fish in this tank. The pH of aqueous solutions can be affected by many of the ionic compounds released into our waterways by industry and agriculture. But why are such solutions acidic or basic? We explore that and related questions in this chapter and come to a deeper understanding of salts in water. As we shall see, they can be a matter of life and death.

SALTS IN WATER

You are likely to die if the pH of your blood plasma falls by more than 0.4 from its normal value of 7.4. The pH can fall as a result of disease or shock, both of which would generate acidic conditions in your body. You are also likely to die if the pH of your blood plasma rises to 7.8, as could happen during the early stages of recovery from severe burns. To survive, your body must constantly control its own pH; and if your control systems fail, then medical professionals must intervene rapidly. Intravenous administration of electrolyte solutions is often the first treatment in emergency rooms.

The pH of aqueous solutions—not only blood plasma, but seawater, detergents, sap, and reaction mixtures—is controlled by the transfer of protons between ions and water molecules. In this chapter we see how different ions affect pH and how they can be used to control it.

IONS AS ACIDS AND BASES

15.1 Ions as acids
15.2 Ions as bases
15.3 The pH of salt solutions
15.4 The pH of mixed solutions

TITRATIONS

15.5 Strong acid-strong base titrations
15.6 Strong acid-weak base and weak acid-strong base titrations
15.7 Indicators as weak acids

BUFFER SOLUTIONS

15.8 The action of buffers
15.9 Selecting a buffer
15.10 Buffer capacity

SOLUBILITY EQUILIBRIA

15.11 The solubility product
15.12 The common-ion effect
15.13 Predicting precipitation
15.14 Dissolving precipitates
15.15 Complex ions and solubilities

FIGURE 15.1 Solutions of salts in water give rise to acidic, basic, and neutral solutions, as shown here by the color of the indicator bromothymol blue (see Table 15.3). From left to right are solutions of ammonium chloride, sodium chloride, and sodium acetate.

Throughout this chapter we use the Brønsted definitions of acids and bases.

IONS AS ACIDS AND BASES

What happens when we dissolve a salt in water? Figure 15.1 shows a simple experiment to find out. The pH rises above 7 if the salt provides ions that are bases. It falls below 7 if the salt provides ions that are acids. For example, sodium acetate, $NaCH_3CO_2$, provides the acetate ion, a base, so we correctly expect a pH greater than 7. On the other hand, NH_4Cl provides the NH_4^+ ion, an acid, so we expect a pH of less than 7. A salt such as sodium chloride, NaCl, provides ions that are such weak acids and bases that we can ignore their acidic or basic character, and the pH of the solution is close to 7.

15.1 IONS AS ACIDS

All cations that are the conjugate acids of weak bases function as acids and lower the pH of the solution. For example, the ammonium ion, NH_4^+, the conjugate acid of the weak base NH_3, is an acid:

$$NH_4^+(aq) + H_2O(l) \rightleftharpoons H_3O^+(aq) + NH_3(aq)$$

Although K_a for NH_4^+ is small (5.6×10^{-10}), enough of the NH_4^+ ions provided by the salt donate protons for there to be an appreciable lowering of pH. Table 15.1 lists some cations that are acidic in water.

Small, highly charged metal cations that can act as Lewis acids in water, such as Al^{3+} and Fe^{3+}, also produce acidic solutions, even though the cations themselves have no hydrogen ions to donate (Fig. 15.2). The protons come from the water molecules that hydrate the ions in solution

TABLE 15.1 Acidic character and K_a values of common cations in water*

Character	Examples	K_a	pK_a
ACIDIC			
conjugate acids of weak bases	anilinium ion, $C_6H_5NH_3^+$	2.3×10^{-5}	4.64
	pyridinium ion, $C_5H_5NH^+$	5.6×10^{-6}	5.24
	ammonium ion, NH_4^+	5.6×10^{-10}	9.25
	methylammonium ion, $CH_3NH_3^+$	2.8×10^{-11}	10.56
small, highly charged metal cations	$Fe^{3+}(aq)$	3.5×10^{-3}	2.46
	$Cr^{3+}(aq)$	1.3×10^{-4}	3.89
	$Al^{3+}(aq)$	1.4×10^{-5}	4.85
	$Fe^{2+}(aq)$	1.3×10^{-6}	5.89
	$Cu^{2+}(aq)$	3.2×10^{-8}	7.49
	$Ni^{2+}(aq)$	9.3×10^{-10}	9.03
NEUTRAL			
Group 1 and 2 cations; metal cations with charge +1	Li^+, Na^+, K^+		
	Mg^{2+}, Ca^{2+}, Ag^+		
BASIC	none		

*As in Table 14.2, the experimental pK_a values have more significant figures than shown here, and the K_a values have been calculated from these better data.

TABLE 15.2 Acidic and basic character of common anions in water

Character	Examples
ACIDIC	
very few	HSO_4^-, $H_2PO_4^-$
NEUTRAL	
conjugate bases of strong acids	Cl^-, Br^-, I^-, NO_3^-, ClO_4^-
BASIC	
conjugate bases of weak acids	F^-, O^{2-}, OH^-, S^{2-}, HS^-, CN^-, CO_3^{2-}, PO_4^{3-}, NO_2^-, $CH_3CO_2^-$, other carboxylate ions

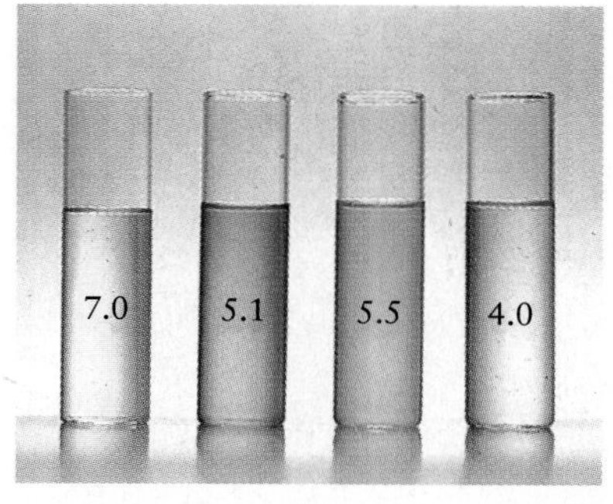

FIGURE 15.2 These four solutions show that hydrated cations can be significantly acidic. The flasks contain, from left to right, pure water, 0.1 M $Al_2(SO_4)_3$(aq), 0.1 M $Ti_2(SO_4)_3$(aq), and 0.1 M CH_3COOH(aq). All four tubes contain a few drops of Fisher universal indicator; the numbers give the pH of each solution.

(Fig. 15.3). These molecules act as Lewis bases and share electrons with the metal cation. This partial loss of electrons weakens the O—H bonds and allows one or more hydrogen ions to be lost. Small, highly charged ions exert the greatest pull on the electrons. Some cations that act as acids in this way are included in Table 15.1 (Fig. 15.4).

The cations of Group 1 and 2 metals, and those of charge +1 from other groups, are such weak Lewis acids that they do not act as acids when hydrated. These metal cations are too large or have too low a charge to have an appreciable polarizing effect on the hydrating water molecules that surround them, so the water molecules do not readily release their protons.

Only a few anions are acids (Table 15.2); it is difficult for a positively charged proton to leave a negatively charged anion. Acid anions include HSO_4^- and $H_2PO_4^-$, which take part in equilibria like

$$HSO_4^-(aq) + H_2O(l) \rightleftharpoons H_3O^+(aq) + SO_4^{2-}(aq)$$

Salts that contain the conjugate acids of weak bases produce acidic aqueous solutions; so do salts that contain small, highly charged metal cations.

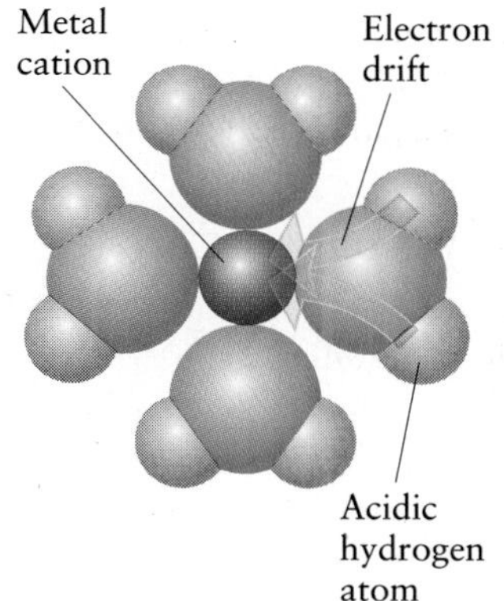

FIGURE 15.3 In water, Al^{3+} cations exist as hydrated ions that can act as Brønsted acids. Although for clarity only four water molecules are shown here, metal cations typically have six H_2O molecules attached to them.

15.2 IONS AS BASES

Cations cannot readily accept protons: their positive charge repels the incoming proton. However, all anions that are the conjugate bases of weak acids do act as proton acceptors, so we can expect them to give basic solutions. For example, formic acid, HCOOH, the acid in ant venom, is a weak acid, so the formate ion acts as a base in water:

$$H_2O(l) + HCO_2^-(aq) \rightleftharpoons HCOOH(aq) + OH^-(aq)$$

Acetate ions and the other ions listed in Table 15.2 act as bases in water.

Anions derived by complete loss of protons from polyprotic acids also accept protons and give rise to a basic solution. Examples are the PO_4^{3-} ion and the CO_3^{2-} ion. Both act as bases in aqueous solution:

$$H_2O(l) + PO_4^{3-}(aq) \rightleftharpoons HPO_4^{2-}(aq) + OH^-(aq)$$

$$H_2O(l) + CO_3^{2-}(aq) \rightleftharpoons HCO_3^-(aq) + OH^-(aq)$$

FIGURE 15.4 A solution of titanium(III) sulfate is so acidic that it can release H_2S from some sulfides.

We saw in Chapter 14 that the stronger the acid, the weaker its conjugate base (Fig. 15.5). The anions of strong acids—which include Cl^-, Br^-, I^-, NO_3^-, and ClO_4^-—are *very* weak bases. They have no significant effect on the pH of a solution.

Salts that contain the conjugate bases of weak acids produce basic aqueous solutions.

SELF-TEST 15.1A Use Tables 15.1 and 15.2 to decide whether aqueous solutions of the salts (a) $Ba(NO_2)_2$; (b) $CrCl_3$; and (c) NH_4NO_3 are acidic, neutral, or basic.

[*Answer:* (a) Basic; (b) acidic; (c) acidic]

SELF-TEST 15.1B Decide whether aqueous solutions of (a) Na_2CO_3; (b) $AlCl_3$; (c) KNO_3 are acidic, neutral, or basic.

15.3 THE pH OF SALT SOLUTIONS

It is sometimes necessary to know the pH of a salt solution. For example, in hospitals salt solutions are administered intravenously to regulate the pH of bodily fluids. (Case Study 15 contains additional information on how salt solutions are used in emergency medicine.) The pH of swimming pools and aquaria is also regulated by adding salts.

To calculate the pH of a salt solution there is nothing new to learn: we can use exactly the same procedures described in Chapter 14—an acid cation is treated as a weak acid and a basic anion as a weak base. So, as we did in Chapter 14, we set up an equilibrium table. Initially, the instant the solution is prepared, we suppose that it consists of only water molecules and the cations and anions of the salt. Proton transfer between the cations and anions and water immediately adjusts the composition to equilibrium. The procedure is reviewed in Toolbox 15.1.

Aqueous solutions of salts with acidic cations have a pH lower than 7; salts with basic anions produce a pH higher than 7 in aqueous solution.

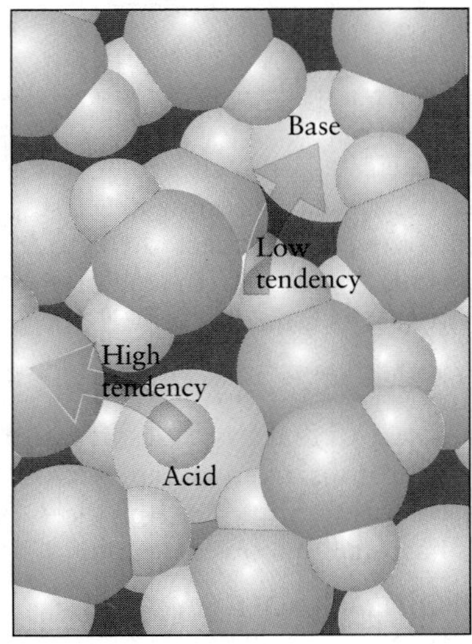

(a)

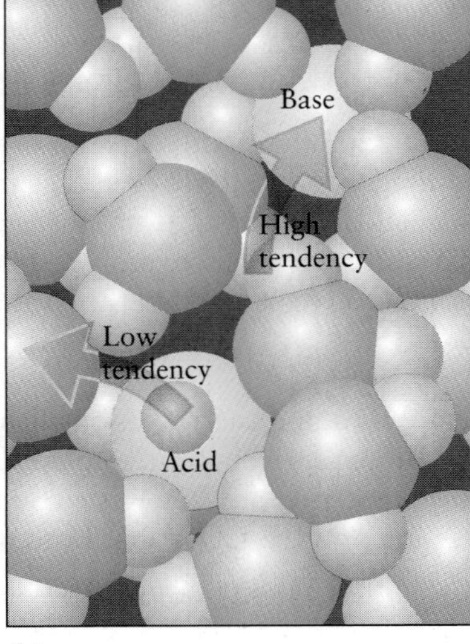

(b)

FIGURE 15.5 The relative strengths of conjugate acids and bases have a reciprocal relation. (a) When a species has a high tendency to donate a proton, the resulting conjugate base has a low tendency to accept one. (b) When a species has a low tendency to donate a proton, the resulting conjugate base has a strong tendency to accept one.

TOOLBOX 15.1 How to calculate the pH of an electrolyte solution*

The procedure is very similar to that set out in Toolbox 14.1, the only difference being that the acid or base is an ion added as a salt.

Although we shall calculate pH and pK values to the number of significant figures appropriate to the data, the answers are often considerably less reliable than that. For instance, we might calculate the pH of a solution as 8.82, but in practice the answer is unlikely to be reliable to more than one decimal place (pH = 8.8). One reason is that we are ignoring interactions between the ions in solution.

A salt with an acidic cation First, write the chemical equation for proton transfer to water and the expression for K_a or K_b. Then set up an equilibrium table:

Step 1. The initial (nominal) molarity of the acidic cations is equal to the molarity of the cations that the salt would produce if completely ionized. The initial molarities of its conjugate base and H_3O^+ are zero.

Step 2. Write the decrease in molarity of the acidic cation as $-x$ mol·L^{-1} and use the reaction stoichiometry to write the corresponding changes for the conjugate base and H_3O^+. Ignore H_3O^+ from the autoprotolysis of water; this approximation is valid if $[H_3O^+]$ is substantially (about three times) higher than 1×10^{-7}.

Step 3. Write the equilibrium molarities of the species in terms of x.

Step 4. Express the acidity constant for the acidic cation in terms of x and solve the equation for x. If K_a is not available, obtain it from the value of K_b for the conjugate base by using $K_a = K_w/K_b$ (Section 14.7). Because x mol·L^{-1} is the H_3O^+ molarity, the pH of the solution is $-\log x$.

As an example, let's estimate the pH of 0.15 M NH_4Cl(aq). Because NH_4^+ cations are acids and Cl^- anions are neutral, we expect pH < 7. The equilibrium to consider is

$$NH_4^+(aq) + H_2O(l) \rightleftharpoons H_3O^+(aq) + NH_3(aq)$$

$$K_a = \frac{[H_3O^+][NH_3]}{[NH_4^+]}$$

We construct the following equilibrium table, with all concentrations in moles per liter:

	Species		
	NH_4^+	H_3O^+	NH_3
Step 1. Initial molarity	0.15	0	0
Step 2. Change in molarity	$-x$	$+x$	$+x$
Step 3. Equilibrium molarity	$0.15 - x$	x	x

Step 4. The acidity constant K_a for NH_4^+ is obtained from the value of K_b for NH_3:

$$K_a = \frac{K_w}{K_b} = \frac{1.0 \times 10^{-14}}{1.8 \times 10^{-5}} = 5.6 \times 10^{-10}$$

Substitution of this value and the information from step 3 into

$$K_a = \frac{[H_3O^+][NH_3]}{[NH_4^+]}$$

gives

$$5.6 \times 10^{-10} = \frac{x \times x}{0.15 - x}$$

We now suppose that x is less than 5% of 0.15 and simplify this expression to

$$\frac{x^2}{0.15} \approx 5.6 \times 10^{-10}$$

The solution of this equation is

$$x \approx \sqrt{0.15 \times (5.6 \times 10^{-10})} = 9.2 \times 10^{-6}$$

The approximation that x is less than 5% of 0.15 is valid. Moreover, the H_3O^+ molarity (9.2×10^{-6} mol·L^{-1}) is also much larger than that generated by the autoprotolysis of water (1.0×10^{-7} mol·L^{-1}), so the neglect of the latter contribution is also valid. Because the H_3O^+ molarity is 9.2×10^{-6} mol·L^{-1}, the pH of the solution is

$$\text{pH} = -\log(9.2 \times 10^{-6}) = 5.04$$

or about 5.0.

SELF-TEST 15.2A Estimate the pH of 0.10 M CH_3NH_3Cl(aq), where CH_3NH_3Cl is methylammonium chloride; the cation is $CH_3NH_3^+$.

[*Answer:* 5.8]

SELF-TEST 15.2B Estimate the pH of 0.10 M NH_4NO_3(aq).

A salt with a basic anion In this case we expect pH > 7. The main difference in the procedure is that the proton transfer to the anion results in the formation of OH^- ions, so the equilibrium table leads to a value for pOH. Convert pOH to pH by using pH + pOH = 14.00.

*Recall from Section 3.3 that an electrolyte solution is one that conducts electricity and contains ions. All salt solutions are electrolyte solutions.

As an example, let's estimate the pH of 0.075 M $Ca(CH_3CO_2)_2(aq)$. The proton transfer equilibrium is

$$H_2O(l) + CH_3CO_2^-(aq) \rightleftharpoons CH_3COOH(aq) + OH^-(aq)$$

$$K_b = \frac{[CH_3COOH][OH^-]}{[CH_3CO_2^-]}$$

The initial molarity of $CH_3CO_2^-$ is $2 \times 0.075\ mol{\cdot}L^{-1} = 0.15\ mol{\cdot}L^{-1}$, because each formula unit of salt provides two $CH_3CO_2^-$ ions.

	Species		
	$CH_3CO_2^-$	CH_3COOH	OH^-
Step 1. Initial molarity	0.15	0	0
Step 2. Change in molarity	$-x$	$+x$	$+x$
Step 3. Equilibrium molarity	$0.15 - x$	x	x

Step 4. Because Table 14.2 gives the K_a of CH_3COOH as 1.8×10^{-5}, the K_b of its conjugate base, the $CH_3CO_2^-$ ion, is

$$K_b = \frac{K_w}{K_a} = \frac{1.0 \times 10^{-14}}{1.8 \times 10^{-5}} = 5.6 \times 10^{-10}$$

Next, we insert the information from step 3 and the value of K_b into

$$K_b = \frac{[CH_3COOH][OH^-]}{[CH_3CO_2^-]}$$

which gives

$$5.6 \times 10^{-10} = \frac{x \times x}{0.15 - x}$$

Because K_b is so small, we can anticipate that x is less than 5% of 0.15; hence we simplify this expression to

$$\frac{x^2}{0.15} \approx 5.6 \times 10^{-10}$$

It follows that

$$x \approx \sqrt{0.15 \times (5.6 \times 10^{-10})} = 9.2 \times 10^{-6}$$

which is far less than 5% of 0.15. The OH^- molarity ($9.2 \times 10^{-6}\ mol{\cdot}L^{-1}$, from step 3) arising from the proton transfer equilibrium is much larger than that arising from the autoprotolysis of water ($1.0 \times 10^{-7}\ mol{\cdot}L^{-1}$), so the neglect of autoprotolysis is also valid. Because the molarity of OH^- ions is $9.2 \times 10^{-6}\ mol{\cdot}L^{-1}$, it follows that

$$pOH = -\log(9.2 \times 10^{-6}) = 5.04$$

and hence that

$$pH = 14.00 - 5.04 = 8.96$$

or about 9.0. The solution is basic, as expected.

SELF-TEST 15.3A Estimate the pH of 0.10 M $KC_6H_5CO_2(aq)$, potassium benzoate. (The conjugate acid of the benzoate ion is benzoic acid, C_6H_5COOH, $K_a = 6.5 \times 10^{-5}$.)

[*Answer:* 8.6]

SELF-TEST 15.3B Estimate the pH of 0.020 M KF(aq); see Table 14.2.

15.4 THE pH OF MIXED SOLUTIONS

We have seen in Section 14.11 how to estimate the pH of a solution of a weak acid when it is the only solute. But suppose some salt of the acid were also added: how would that affect the pH? For example, we might want to find the pH of a solution that contains equal amounts of sodium acetate and acetic acid. We can predict what happens: because acetate ions are bases, increasing their concentration increases the pH of the solution. Similarly, suppose we have a solution of a base and add a salt of the base to it (for instance, NH_4Cl added to aqueous ammonia). Because the salt provides the conjugate acid of the base, we can expect that the pH will fall.

◀◀ The effect of increasing the concentration of a product on equilibrium is discussed in Section 13.8.

To estimate the pH of mixed solutions, we set up an equilibrium table and use the acidity or basicity constant to calculate the unknown concentration. The only difference is that the initial concentrations of the conjugate acid and base are as specified in the problem.

The pH of a solution of a weak acid increases when a salt containing its conjugate base is added. The pH of a solution of a weak base decreases when a salt containing its conjugate acid is added.

EXAMPLE 15.1 Calculating the pH of a solution of a weak acid and its salt

Calculate the pH of a solution that is 0.500 M $HNO_2(aq)$ and 0.100 M $KNO_2(aq)$. From Table 14.2, $K_a = 4.3 \times 10^{-4}$ for HNO_2.

STRATEGY The solution contains NO_2^-, a base, so we expect the pH to be higher than that of nitrous acid alone. The K^+ ion supplied by the salt has no effect on the pH of the solution. To calculate the pH of the mixed solution, set up an equilibrium table and consider the initial molarity of HNO_2 to be 0.500 $mol{\cdot}L^{-1}$. Because nitrite ions have also been added to the solution, set their initial molarity equal to the molarity of salt (each KNO_2 formula unit supplies one NO_2^- anion). Then proceed as described in Toolbox 15.1.

SOLUTION The proton transfer equilibrium to consider is

$$HNO_2(aq) + H_2O(l) \rightleftharpoons H_3O^+(aq) + NO_2^-(aq) \qquad K_a = \frac{[H_3O^+][NO_2^-]}{[HNO_2]}$$

The equilibrium table, with all concentrations in moles per liter, is

	Species		
	HNO_2	H_3O^+	NO_2^-
Step 1. Initial molarity	0.500	0	0.100
Step 2. Change in molarity	$-x$	$+x$	$+x$
Step 3. Equilibrium molarity	$0.500 - x$	x	$0.100 + x$

Step 4. Substitution of the information from step 3 and $K_a = 4.3 \times 10^{-4}$ into the expression for K_a gives

$$4.3 \times 10^{-4} = \frac{x \times (0.100 + x)}{0.500 - x}$$

Assuming that x is less than 5% of 0.100 (and therefore also less than 5% of 0.500), we write

$$4.3 \times 10^{-4} \approx \frac{x \times 0.100}{0.500}$$

The solution of this equation is $x \approx 2.2 \times 10^{-3}$. To check the validity of the approximation, we evaluate

$$\frac{2.2 \times 10^{-3}}{0.100} \times 100\% = 2.2\%$$

Because the value of x is 2.2% of 0.100, it is only 0.44% of 0.500, so the approximations are valid. It follows that the equilibrium molarity of H_3O^+ ions is 2.2×10^{-3} $mol{\cdot}L^{-1}$. This is much larger than the molarity arising from the autoprotolysis of water, so the neglect of autoprotolysis is also valid. It follows that the pH of the solution is

$$pH = -\log(2.2 \times 10^{-3}) = 2.66$$

or about 2.7. The pH of 0.500 M HNO_2(aq) is 1.8; so, as anticipated, the pH of the mixed solution is higher.

SELF-TEST 15.4A Calculate the pH of a solution that is 0.300 M CH_3NH_2(aq) and 0.146 M CH_3NH_3Cl(aq). From Table 14.2, the K_b of CH_3NH_2 is 3.6×10^{-4}.

[*Answer:* 10.87]

SELF-TEST 15.4B Calculate the pH of a solution that is 0.010 M HClO(aq) and 2.0×10^{-4} M NaClO(aq).

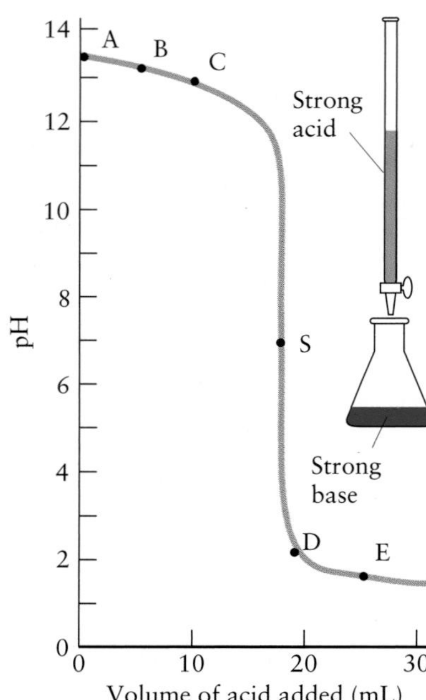

FIGURE 15.6 The variation of pH during the titration of a strong base: 25.00 mL of 0.250 M NaOH(aq) with a strong acid, 0.340 M HCl(aq). The stoichiometric point occurs at pH = 7 (point S). The other points are discussed in the text.

TITRATIONS

Sometimes we need to know not only the pH, but how much acid or base is present in a sample. For example, an environmental chemist studying a lake in which fish are dying must know exactly how much acid is present in a sample of the water. We saw in Section 4.3 that titration involves adding a solution called the titrant from a buret to a flask containing the sample, called the analyte. The success of an acid-base titration depends on how accurately we can detect the stoichiometric point. At that point, the number of moles of H_3O^+ (or OH^-) added as titrant is equal to the number of moles of OH^- (or H_3O^+) initially present in the analyte. At the stoichiometric point the solution consists of a salt in water. The solution is acidic if acidic ions are present, basic if basic ions are present.

15.5 STRONG ACID-STRONG BASE TITRATIONS

A plot of the pH of the analyte solution as titrant is added during a titration is called a **pH curve** (Fig. 15.6). The shape of the pH curve in the illustration is typical of titrations in which a strong acid is added to a strong base, and the points on the curve can be calculated as illustrated in the example below. Initially the pH falls slowly. When the stoichiometric point is passed, there is a sudden fall through pH = 7 as the H_3O^+ molarity increases sharply to a value typical of a dilute solution of the added acid. At this point, an indicator changes color or an automatic titrator detects the stoichiometric point by responding electronically to this sudden fall in pH. The pH then falls slowly toward the value of the acid itself as the dilution caused by the original analyte solution becomes less and less important.

The pH curve for the titration of a strong acid (the analyte) with a strong base (the titrant) mirrors that of a strong base with a strong acid. As shown in Fig. 15.7, the initial pH is low, it increases only slightly until just before the stoichiometric point, and then it increases sharply. The pH continues to increase after the stoichiometric point, but then levels off as a result of the presence of excess strong base in the solution.

EXAMPLE 15.2 Calculating points on the pH curve for a strong acid-strong base titration

Suppose we were carrying out a titration in which 0.340 M HCl(aq) was the titrant and the analyte initially consisted of 25.0 mL of 0.250 M NaOH(aq).

What is the pH of the analyte solution after the addition of 5.00 mL of titrant?

STRATEGY After the addition of 5.00 mL of the acid titrant, we can expect the pH to have decreased from its initial value. First, calculate the moles of H_3O^+ ions (if the analyte is a strong acid) or OH^- ions (if the analyte is a strong base) in the original analyte solution from its molarity and its volume. Then calculate the moles of OH^- ions (if the titrant is a strong base) or H_3O^+ ions (if the titrant is a strong acid) in the volume of titrant added, and use the reaction stoichiometry to find the moles of H_3O^+ ions (or OH^- ions) that remain in the analyte solution. Each mole of H_3O^+ ions reacts with 1 mole of OH^- ions; therefore, subtract the number of moles of H_3O^+ or OH^- ions added from the initial number of H_3O^+ or OH^- ions. Divide the remaining number of moles by the total volume of the combined solutions, which gives the molarity of the H_3O^+ (or OH^-) ions in the solution, and take the negative logarithm of the relevant molarity to find the pH or the pOH. Convert pOH to pH by using the relation $\text{pH} + \text{pOH} = 14.00$.

SOLUTION Initially, the pOH of the analyte is $\text{pOH} = -\log 0.250 = 0.602$, so the pH of the solution is $\text{pH} = 14.00 - 0.602 = 13.40$. This is point A in Fig. 15.6. The amount of OH^- ions initially present is

$$\text{Moles of OH}^- \text{(from the base)} = (25.00 \times 10^{-3}\ \text{L}) \times (0.250\ \text{mol·L}^{-1}) = 6.25 \times 10^{-3}\ \text{mol}$$

The amount of H_3O^+ ions supplied by the titrant is

$$\text{Moles of H}_3\text{O}^+ \text{(from the acid)} = (5.00 \times 10^{-3}\ \text{L}) \times (0.340\ \text{mol·L}^{-1}) = 1.70 \times 10^{-3}\ \text{mol}$$

After reaction of all the H_3O^+ ions added, the amount of OH^- remaining is

$$(6.25 \times 10^{-3} - 1.70 \times 10^{-3})\ \text{mol} = 4.55 \times 10^{-3}\ \text{mol}$$

Because the total volume of the solution is now 30.0 mL, or 0.0300 L, the molarity of OH^- is

$$\text{Molarity of OH}^- = \frac{4.55 \times 10^{-3}\ \text{mol}}{0.0300\ \text{L}} = 0.152\ \text{mol·L}^{-1}$$

Hence, $\text{pOH} = -\log 0.152 = 0.82$, so $\text{pH} = 13.18$, point B. Note that the pH has fallen, as expected, but only by a very small amount. This small change is consistent with the shallow slope of the pH curve at the start of the titration.

SELF-TEST 15.5A What is the pH of the solution that results from the addition of a further 5.00 mL of the HCl titrant to the analyte?

[*Answer:* 12.91, point C]

SELF-TEST 15.5B What is the pH of the solution that results from the addition of another 2.00 mL of titrant to the analyte?

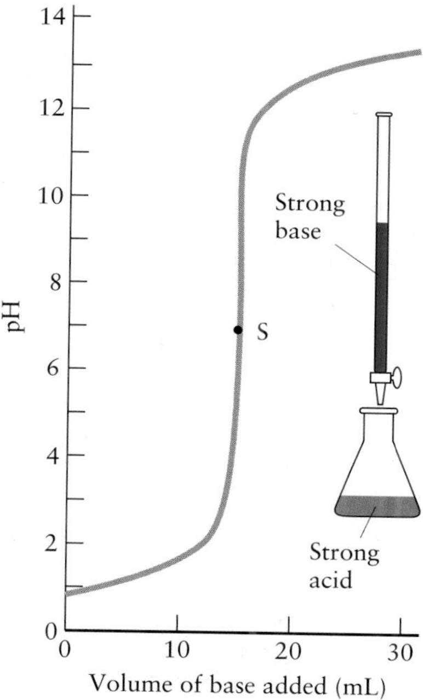

FIGURE 15.7 The variation of pH during a typical titration of a strong acid (the analyte) with a strong base (the titrant). The stoichiometric point occurs at pH = 7.

The technique described in Example 15.2 can be applied at any stage of the calculation. For example, to demonstrate the sharp change in pH close to the stoichiometric point, we could calculate the pH of the solution at the stoichiometric point itself (that is easy: the solution is neutral, so pH = 7) and at a nearby point. The volume of titrant needed to reach the stoichiometric point is calculated from the stoichiometry of the reaction by using the techniques described in Toolbox 4.3:

$$\text{Volume of acid} = \frac{\text{moles of HCl}}{\text{molarity of HCl}} = \frac{6.25 \times 10^{-3}\ \text{mol}}{0.340\ \text{mol·L}^{-1}}$$
$$= 18.4 \times 10^{-3}\ \text{L}$$

The stoichiometric point occurs when 18.4 mL of titrant has been added. In the following example, we calculate the pH for a solution to which 19.4 mL has been added, which takes it 1.00 mL beyond the stoichiometric point.

In the titration of a strong acid with a strong base, or a strong base with a strong acid, the pH changes slowly initially, changes rapidly through pH = 7 at the stoichiometric point, and then changes slowly again.

EXAMPLE 15.3 Calculating the pH after the stoichiometric point of a strong base-strong acid titration

Calculate the pH of the solution in Example 15.2 after 19.4 mL of the acid has been added.

STRATEGY After the stoichiometric point, all the base in the analyte has reacted with acid. Because the amount of acid added now exceeds the amount of base in the analyte, we expect a pH of less than 7. We can use the reaction stoichiometry to determine how much of the acid added remains after neutralization. Then we use the total volume of solution to find the molar concentration of H_3O^+ and convert it to pH.

SOLUTION The moles of H_3O^+ ions supplied by the solution is

$$\text{Moles of } H_3O^+ \text{ supplied} = (19.4 \times 10^{-3}\ \text{L}) \times (0.340\ \text{mol·L}^{-1})$$
$$= 6.60 \times 10^{-3}\ \text{mol}$$

The number of moles of OH^- ions initially present (from Example 15.2) is 6.25×10^{-3} mol. More moles of acid have been supplied than there were moles of base originally, so we know that we are past the stoichiometric point. After all the OH^- ions have reacted:

$$\text{Moles of } H_3O^+ \text{ remaining} = (6.60 \times 10^{-3} - 6.25 \times 10^{-3})\ \text{mol}$$
$$= 3.5 \times 10^{-4}\ \text{mol}$$

Because the total volume of the solution is 25.0 mL + 19.4 mL, or 44.4 mL, which is 0.0444 L, the molar concentration of H_3O^+ is

$$\text{Molarity of } H_3O^+ = \frac{3.5 \times 10^{-4}\ \text{mol}}{0.0444\ \text{L}} = 7.9 \times 10^{-3}\ \text{mol·L}^{-1}$$

Hence,

$$\text{pH} = -\log(7.9 \times 10^{-3}) = 2.1$$

A pH of 2.1 (point D) is well below the pH at the stoichiometric point, although only one more milliliter of acid has been added.

SELF-TEST 15.6A Calculate the pH of the solution after the addition of 20.4 mL of titrant.

[*Answer:* 1.8, point E]

SELF-TEST 15.6B Calculate the pH of the solution after the addition of 25.0 mL of titrant.

15.6 STRONG ACID–WEAK BASE AND WEAK ACID–STRONG BASE TITRATIONS

At the stoichiometric point of the titration of formic acid, HCOOH (**1**), with sodium hydroxide, the solution consists of sodium formate, $NaHCO_2$, and water. Formate ion, HCO_2^- (**2**), is a base and the Na^+ ions have virtually no effect on pH; so overall the solution is basic, even though the acid has been completely neutralized. Because the solute is a salt with a basic anion, we can therefore expect the pH to be greater than 7 at the stoichiometric point. A good rule of thumb is that a strong base dominates a weak acid, so the solution is basic.

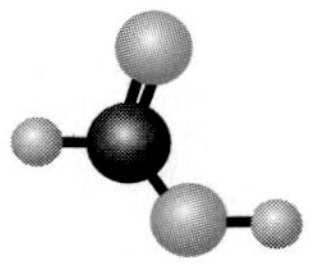

1 Formic acid, HCOOH

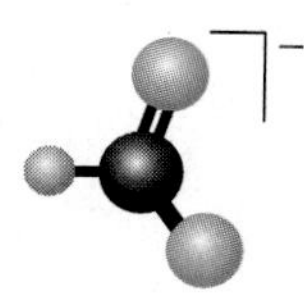

2 Formate ion, HCO_2^-

At the stoichiometric point of the titration of aqueous ammonia with hydrochloric acid, the solute is ammonium chloride. Because NH_4^+ is an acid, we expect the solution to be acidic with a pH of less than 7. The same is true for the titration of any weak base and strong acid. The rule of thumb is that a strong acid dominates a weak base, so the solution is acidic at the stoichiometric point.

EXAMPLE 15.4 Estimating the pH at the stoichiometric point of the titration of a weak acid with a strong base

Estimate the pH at the stoichiometric point of the titration of 25.00 mL of 0.100 M HCOOH(aq) with 0.150 M NaOH(aq).

STRATEGY The number of moles of HCO_2^- ions in the solution at the stoichiometric point is equal to the initial number of moles of HCOOH in the analyte, which can be calculated from the initial volume of analyte solution and its molarity. The molarity of HCO_2^- ions is this number of moles of ions divided by the *total* volume of the solution. To find this total volume, calculate the volume of titrant solution needed to reach the stoichiometric point and add it to the initial volume of analyte. The pH of the salt solution is then calculated as described in Toolbox 15.1. The K_b of HCO_2^- is related to the K_a of its conjugate acid HCOOH by $K_a \times K_b = K_w$; K_a is listed in Table 14.2.

SOLUTION From Table 14.2, $K_a = 1.8 \times 10^{-4}$ for formic acid; therefore, $K_b = K_w/K_a = 5.6 \times 10^{-11}$. The number of moles of HCOOH in the initial analyte solution of volume 25.00 mL (2.500×10^{-2} L) is

$$\begin{aligned}\text{Moles of HCOOH} &= (2.500 \times 10^{-2}\ \text{L}) \times (0.100\ \text{mol·L}^{-1}) \\ &= 2.50 \times 10^{-3}\ \text{mol}\end{aligned}$$

The number of moles of HCO_2^- at the stoichiometric point is therefore also 2.50×10^{-3} mol. The stoichiometry of the reaction requires 1 mol OH^- for 1 mol HCOOH. So, we need the volume of 0.150 M NaOH(aq) that contains that number of moles of OH^- ions:

$$\text{Volume} = \frac{2.50 \times 10^{-3}\ \text{mol}}{0.150\ \text{mol·L}^{-1}} = 16.7 \times 10^{-3}\ \text{L}$$

or 16.7 mL. Therefore, the total volume of the combined solutions at the stoichiometric point is (25.00 + 16.7) mL = 41.7 mL. It follows that at the stoichiometric point

$$\text{Molarity of } HCO_2^- = \frac{2.50 \times 10^{-3}\ \text{mol}}{41.7 \times 10^{-3}\ \text{L}} = 0.0600\ \text{mol·L}^{-1}$$

The equilibrium to consider is

$$HCO_2^-(aq) + H_2O(l) \rightleftharpoons HCOOH(aq) + OH^-(aq)$$

Now set up the equilibrium table, with all concentrations in moles per liter:

	Species		
	HCO_2^-	HCOOH	OH^-
Step 1. Initial molarity	0.0600	0	0
Step 2. Change in molarity	$-x$	$+x$	$+x$
Step 3. Equilibrium molarity	$0.0600 - x$	x	x

Step 4. Substitution of the values from step 3 into the expression for K_b gives

$$5.6 \times 10^{-11} = \frac{[HCO_2H][OH^-]}{[HCO_2^-]} = \frac{x \times x}{0.0600 - x}$$

If we suppose that x is less than 5% of 0.0600, we can write

$$5.6 \times 10^{-11} \approx \frac{x^2}{0.0600}$$

and obtain

$$x \approx \sqrt{0.0600 \times K_b} = \sqrt{0.0600 \times (5.6 \times 10^{-11})} = 1.8 \times 10^{-6}$$

It follows from step 3 that the molarity of OH^- is 1.8×10^{-6} mol·L^{-1}, which is about 18 times greater than the molarity of OH^- ions that come from the autoprotolysis of water (1.0×10^{-7} mol·L^{-1}), so the neglect of the latter is reasonable. At this stage we can write

$$pOH = -\log(1.8 \times 10^{-6}) = 5.74$$

and therefore

$$pH = 14.00 - 5.74 = 8.26$$

or about 8.3.

Notice that the pH at the stoichiometric point is greater than 7, because the solution at that point consists of the salt that has been formed. The salt is basic, because it contains a basic anion (formate ion) and a neutral cation. Conversely, in the titration of a weak base with a strong acid, we would expect a pH less than 7 at the stoichiometric point, because the salt formed will have an acidic cation (the conjugate acid of the weak base) and a neutral anion.

SELF-TEST 15.7A Calculate the pH at the stoichiometric point of the titration of 25.00 mL of 0.020 M NH_3(aq) with 0.015 M HCl(aq). (For NH_4^+, $K_a = 5.6 \times 10^{-10}$.)

[*Answer:* 5.7]

SELF-TEST 15.7B Calculate the pH at the stoichiometric point of the titration of 25.00 mL of 0.010 M HClO(aq) with 0.020 M KOH(aq). See Table 14.2 for K_a.

The complete pH curves for two typical types of titration are shown in Figs. 15.8 and 15.9. As for strong acid-strong base titrations, there is a sharp change in pH close to the stoichiometric point. The rapid change

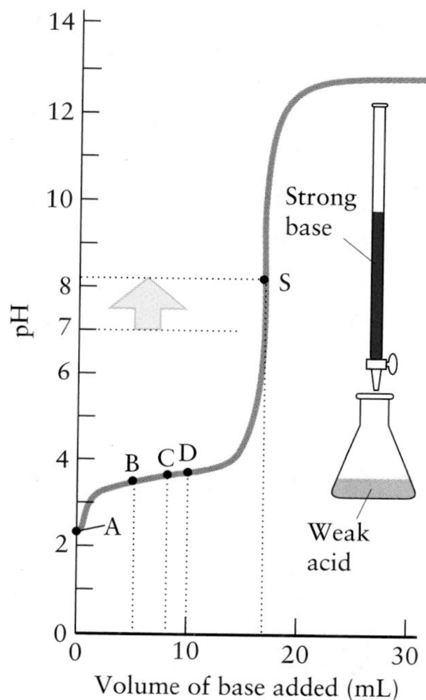

FIGURE 15.8 The pH curve for the titration of a weak acid with a strong base: 25.00 mL of 0.1 M HCOOH(aq) with 0.150 M NaOH(aq). The stoichiometric point (S) occurs on the basic side of pH = 7 because the anion HCO_2^- is a base.

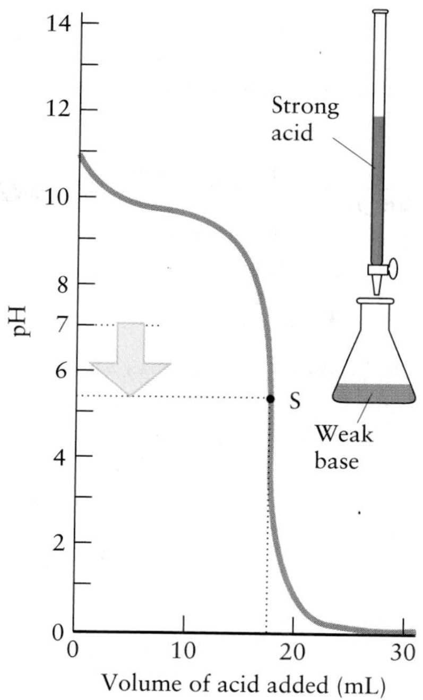

FIGURE 15.9 A typical pH curve for the titration of a weak base with a strong acid. The stoichiometric point (S) occurs on the acidic side of pH = 7 because the salt formed by the neutralization reaction has an acid cation.

acts like a switch and gives a clear signal that the titration has passed through the stoichiometric point. The slow change in pH about halfway to the stoichiometric point is another characteristic feature. This feature is also of great practical importance, because it accounts for the stabilization of the pH of solutions, such as that of oceans, lakes, cell fluids, and blood plasma.

We have already seen how to estimate the pH of the initial analyte when only weak acid or weak base is present (point A in Fig. 15.8, for instance), and the pH at the stoichiometric point (point S). Other points correspond to a mixed solution, of some weak acid (or base) and some salt. We can therefore use the techniques described in Toolbox 15.1 and Example 15.1 to predict the shape of the entire curve.

EXAMPLE 15.5 Calculating the pH before the stoichiometric point in a weak acid-strong base titration

Calculate the pH of the solution resulting when 5.00 mL of 0.150 M NaOH(aq) is added to 25.00 mL of 0.100 M HCOOH(aq). Use $K_a = 1.8 \times 10^{-4}$ for HCOOH.

STRATEGY First, write the chemical equation for the reaction and use the reaction stoichiometry to calculate the number of moles of HCO_2^- ions formed by the reaction of the acid with added base and the moles of HCOOH remaining. Then find the molarities of these species in the solution by dividing the number of moles by the total volume of the solution. Finally, use an equilibrium table to find the pH.

SOLUTION The initial amount of HCOOH in the analyte is $(25.00 \times 10^{-3}\ \text{L}) \times (0.100\ \text{mol·L}^{-1}) = 2.50 \times 10^{-3}$ mol HCOOH. The amount of OH^- in 5.00 mL of the titrant is $(5.00 \times 10^{-3}\ \text{L}) \times (0.150\ \text{mol·L}^{-1}) = 7.50 \times 10^{-4}$ mol OH^-. The chemical equation

$$HCOOH(aq) + OH^-(aq) \longrightarrow HCO_2^-(aq) + H_2O(l)$$

shows that 1 mol $OH^- \simeq$ 1 mol HCOOH and 1 mol $OH^- \simeq$ 1 mol HCO_2^-. So, 7.50×10^{-4} mol OH^- ions produces 7.50×10^{-4} mol HCO_2^- ions and leaves $2.50 \times 10^{-3} - 7.50 \times 10^{-4} = 1.75 \times 10^{-3}$ mol HCOOH. The total volume of the solution at this stage is $(25.00 + 5.00)$ mL $= 30.00$ mL, so the molarities of acid and conjugate base are

$$\text{Molarity of HCOOH} = \frac{1.75 \times 10^{-3}\ \text{mol}}{30.00 \times 10^{-3}\ \text{L}} = 0.0583\ \text{mol·L}^{-1}$$

$$\text{Molarity of } HCO_2^- = \frac{7.50 \times 10^{-4}\ \text{mol}}{30.00 \times 10^{-3}\ \text{L}} = 0.0250\ \text{mol·L}^{-1}$$

The proton transfer equilibrium for HCOOH in water is

$$HCOOH(aq) + H_2O(l) \rightleftharpoons H_3O^+(aq) + HCO_2^-(aq) \qquad K_a = \frac{[H_3O^+][HCO_2^-]}{[HCOOH]}$$

The equilibrium table, with concentrations in moles per liter, is

	Species		
	HCOOH	**H_3O^+**	**HCO_2^-**
Step 1. Initial molarity	0.0583	0	0.0250
Step 2. Change in molarity	$-x$	$+x$	$+x$
Step 3. Equilibrium molarity	$0.0583 - x$	x	$0.0250 + x$

Step 4. Substitution of the values from step 3 and $K_a = 1.8 \times 10^{-4}$ into the expression for K_a gives

$$1.8 \times 10^{-4} = \frac{x \times (0.0250 + x)}{0.0583 - x}$$

Assuming that x is less than 5% of 0.0250 (and therefore also less than 5% of 0.0583), we write

$$1.8 \times 10^{-4} \approx \frac{x \times 0.0250}{0.0583}$$

The solution is $x \approx 4.2 \times 10^{-4}$. This value is less than 5% of 0.0250, so the approximations are valid. It follows that the equilibrium H_3O^+ molarity is 4.2×10^{-4} mol·L^{-1}. Therefore, the pH of the solution is

$$\text{pH} = -\log(4.2 \times 10^{-4}) = 3.38$$

or about 3.4. This is point B in Fig. 15.8. As anticipated, the pH of the mixed solution is higher than that of the original acid.

SELF-TEST 15.8A Calculate the pH of the solution after the addition of another 5.00 mL of 0.150 M NaOH(aq).

[*Answer:* 3.92, point D]

SELF-TEST 15.8B Calculate the pH of the solution after the addition of yet another 5.00 mL of 0.150 M NaOH(aq).

In the region of the pH curve between points B and D in Fig. 15.8, the pH changes very slowly as base is added. Both HCOOH molecules and HCO_2^- ions are present in similar concentrations in solutions corresponding to the slowly changing region of the pH curve. When base is added from the buret, the OH^- ions take protons from HCOOH molecules. The presence of the acid molecules acts as a reservoir of protons to satisfy the demands of the OH^- ions being added, and the pH of the solution barely changes.

When $[HCO_2^-] = [HCOOH]$, the expression for the acidity constant becomes

$$K_a = \frac{[H_3O^+][HCO_2^-]}{[HCOOH]} = [H_3O^+]$$

Therefore, by taking the negative logarithm of both sides, we see that at the halfway point of a weak acid-strong base titration,

$$pH = pK_a$$

For the formic acid titration, $pK_a = 3.75$, so halfway to the stoichiometric point, pH = 3.75 (point C in Fig. 15.8). The same expression applies to the titration of a weak base with a strong acid, but now K_a refers to the conjugate acid of the weak base. For example, if we were to titrate NH_3 with HCl, halfway to the stoichiometric point, pH = pK_a of NH_4^+.

It is now simple to determine pK_a: all we need to do is plot the pH curve during a titration and then identify the pH halfway to the stoichiometric point (Fig. 15.10). To obtain the pK_b of a weak base, we find pK_a

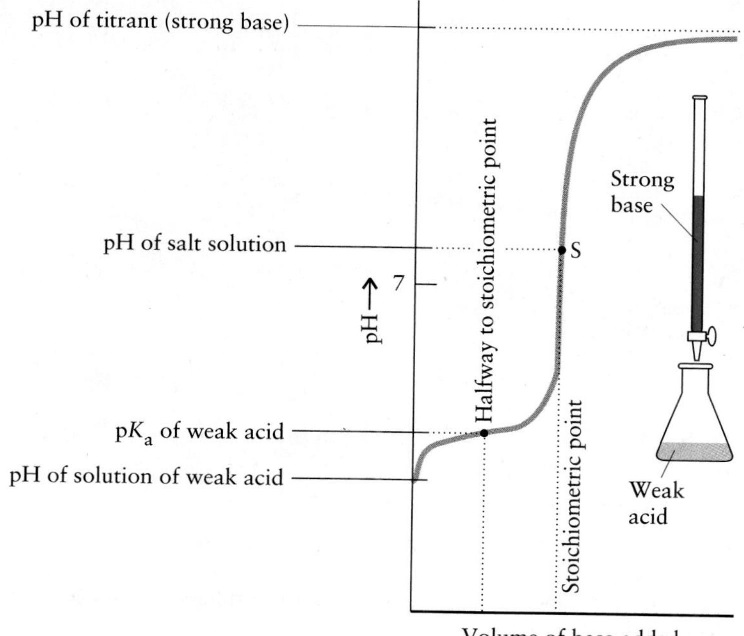

FIGURE 15.10 The pK_a of an acid can be determined by carrying out a titration of the weak acid with a strong base (or vice versa) and locating the pH of the solution after the addition of half the volume of titrant needed to reach the stoichiometric point.

the same way but go on to use $pK_a + pK_b = pK_w$. The values recorded in Table 14.2 were obtained by this method.

Well after the stoichiometric point in the titration of a weak acid with a strong base, the pH depends only on the concentration of excess strong base. For example, suppose we went on to add several liters of strong base from an enormous buret. The presence of salt—the product of the neutralization reaction—would be negligible relative to the concentration of excess base. The pH would be that of the nearly pure titrant (the original base solution).

The changes in pH of a solution during a titration of a weak acid are summarized in Fig. 15.10. Halfway to the stoichiometric point, the pH is equal to the pK_a of the acid present in the solution. The pH is greater than 7 at the stoichiometric point of the titration of a weak acid and strong base. The pH is less than 7 at the stoichiometric point of the titration of a weak base and strong acid, as shown in Fig. 15.9.

15.7 INDICATORS AS WEAK ACIDS

An automatic titrator monitors the pH of the analyte solution continuously and detects the stoichiometric point by the characteristic rapid change in pH (Fig. 15.11). When automatic equipment is not available, the stoichiometric point can be detected with a pH meter or by observing the change in color of an indicator. We first met indicators just before Section 3.7, and we can now explain their color changes in the context of Brønsted-Lowry theory and proton transfer equilibria.

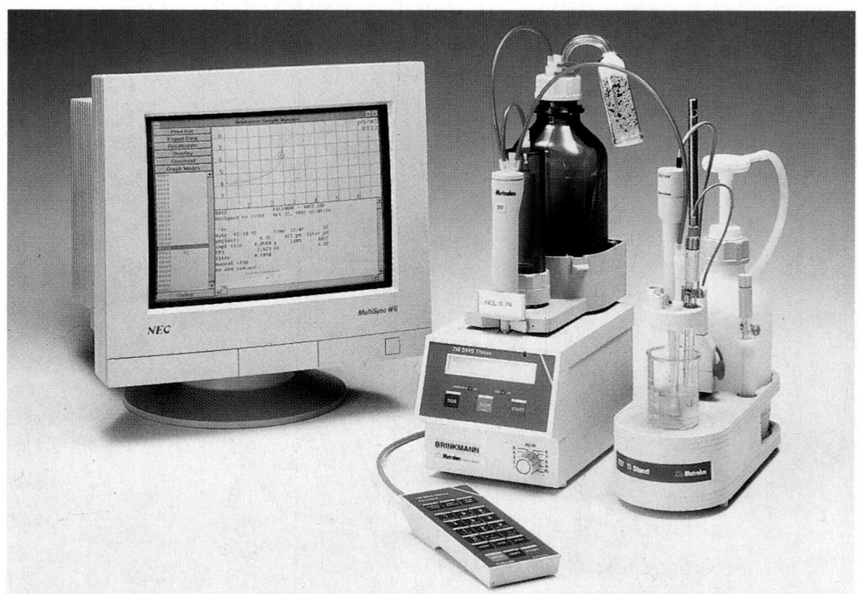

FIGURE 15.11 A commercially available automatic titrator. The stoichiometric point of the titration is detected by a sudden change in pH; the pH is monitored electrically, using a technique described in Box 17.1. The pH can be plotted as the reaction proceeds, as shown on the left.

FIGURE 15.12 The stoichiometric point of an acid-base titration may be detected by the color change of an indicator. Here we see the colors of solutions containing a few drops of phenolphthalein at (from left to right) pH of 7.0, 8.5, 9.4 (its end point), 9.8, and 12.0. At the end point, equal amounts of the conjugate acid and base forms of the indicator are present.

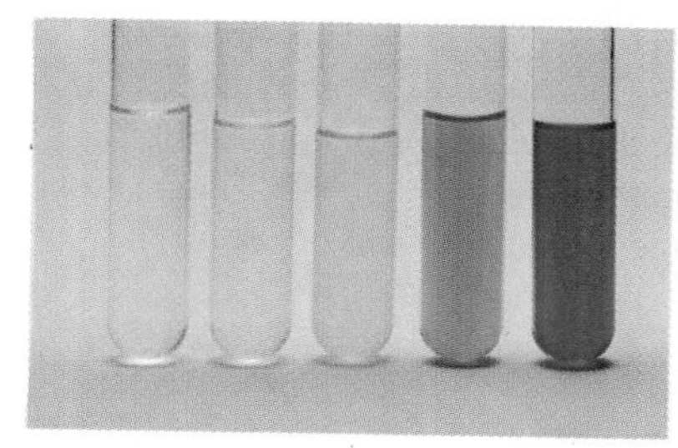

Only one or two drops of indicator are added to the solution, so as not to upset the accuracy of the titration.

An indicator is a weak acid that has one color in its acid form (HIn, where In stands for indicator) and another color in its conjugate base form (In^-). The color change results from the effect of the acidic proton in the acid form HIn: it changes the structure of the molecule in such a way that the light absorption characteristics of HIn are different from those of In^-. When the concentration of HIn is much greater than that of In^-, the solution has the color of the acid form of the indicator. When the concentration of HIn is much less than that of In^-, the solution has the color of the base form of the indicator.

Because it is a weak acid, in solution an indicator takes part in a proton transfer equilibrium:

$$\mathrm{HIn(aq)} + \mathrm{H_2O(l)} \rightleftharpoons \mathrm{H_3O^+(aq)} + \mathrm{In^-(aq)} \qquad K_{\mathrm{In}} = \frac{[\mathrm{H_3O^+}][\mathrm{In^-}]}{[\mathrm{HIn}]}$$

The **end point** of a titration is defined as the point at which the concentrations of the acid and base forms of the indicator are equal: $[\mathrm{HIn}] = [\mathrm{In^-}]$. If we substitute this equality into the expression above, we see that at the end point $[\mathrm{H_3O^+}] = K_{\mathrm{In}}$. That is, the color change occurs when

$$\mathrm{pH} = \mathrm{p}K_{\mathrm{In}}$$

Be careful to distinguish the end point of a titration from the stoichiometric point. The former refers to the color change of the indicator; the latter refers to the stage at which reaction is complete.

One common indicator is phenolphthalein. The acid form of this large organic acid molecule (**3**) is colorless; its conjugate base form (**4**) is pink (Fig. 15.12). The $\mathrm{p}K_{\mathrm{In}}$ of phenolphthalein is 9.4, so the end point occurs in slightly basic solution. The change from colorless to pink starts at pH = 8.2 and is complete by pH = 10. Litmus, another well-known indicator, has $\mathrm{p}K_{\mathrm{In}} = 6.5$; it is red for pH < 5 and blue for pH > 8.

3 Phenolphthalein (Acid form, colorless)

4 Phenolphthalein (Base form, pink)

There are many naturally occurring indicators. A single compound is responsible for the colors of red poppies and blue cornflowers (Fig. 15.13): the pH of the sap is different in the two plants. The color of

FIGURE 15.13 The same dye is responsible for the red of poppies (a) and the blue of cornflowers (b). The color difference is a consequence of the more acidic sap of poppies.

(a)

(b)

hydrangeas also depends on the acidity of their sap and can be controlled by modifying the acidity of the soil. Hydrangeas grown in volcanic areas, for instance, where the soil is acidic, are blue. Hydrangeas grown in soil with a high lime content are pink.

It is important to select an indicator with an end point close to the stoichiometric point of a titration (Fig. 15.14). In practice, the pK_{In} of the indicator should be within about 1 pH unit of the stoichiometric point of the titration:

$$pK_{In} \approx \text{pH(stoichiometric point)} \pm 1$$

Phenolphthalein can be used for titrations with a stoichiometric point near pH = 9, such as a titration of a weak acid with a strong base (Fig. 15.15). Methyl orange changes color between pH = 3.2 and pH = 4.4 and can be used in the titration of a weak base with a strong acid (Fig. 15.16). Ideally, indicators for strong acid-strong base titrations should have end points close to pH = 7; however, in strong acid-strong base titrations, the pH changes rapidly over several pH units, and even phen-

TABLE 15.3 Indicator color changes

Indicator	Color of acid form	pH range of color change	pK_{In}	Color of base form
thymol blue	red	1.2 to 2.8	1.7	yellow
	yellow	8.0 to 9.6		blue
methyl orange	red	3.2 to 4.4	3.4	yellow
bromophenol blue	yellow	3.0 to 4.6	3.9	blue
bromocresol green	yellow	3.8 to 5.4	4.7	blue
methyl red	red	4.8 to 6.0	5.0	yellow
bromothymol blue	yellow	6.0 to 7.6	7.1	blue
litmus	red	5.0 to 8.0	6.5	blue
phenol red	yellow	6.6 to 8.0	7.9	red
thymol blue	yellow	8.0 to 9.6	8.9	blue
phenolphthalein	colorless	8.2 to 10.0	9.4	pink
alizarin yellow R	yellow	10.1 to 12.0	11.2	red
alizarin	red	11.0 to 12.4	11.7	purple

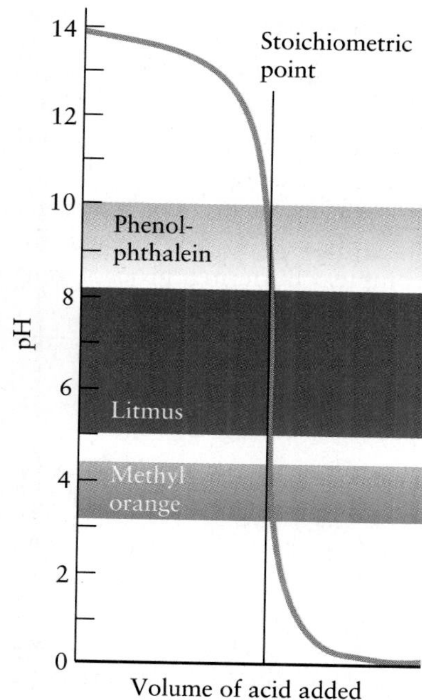

FIGURE 15.14 Ideally, an indicator should have a sharp color change close to the stoichiometric point of the titration, which is at pH = 7 for a strong acid-strong base titration. However, the change in pH is so abrupt that phenolphthalein can also be used. The color change of methyl orange, however, would give a less accurate result.

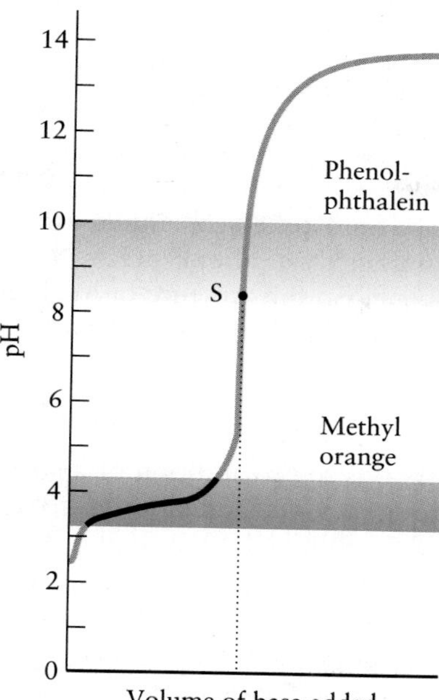

FIGURE 15.15 Phenolphthalein can be used to detect the stoichiometric point of a weak acid-strong base titration, but methyl orange would give a very inaccurate indication of the stoichiometric point. The pH curves are superimposed on approximations to the colors of the indicators in the neighborhoods of their end points.

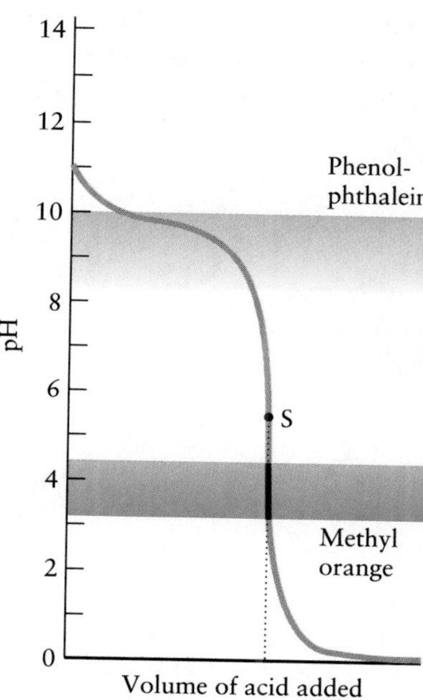

FIGURE 15.16 Methyl orange can be used for a weak base-strong acid titration. Phenolphthalein would be inappropriate because its color change occurs well away from the stoichiometric point. The pH curves are superimposed on approximations to the colors of the indicators in the neighborhoods of their end points.

olphthalein can be used. The properties of several indicators are listed in Table 15.3.

Acid-base indicators are weak acids that change color close to pH = pK_{In}; an indicator should be chosen so that its end point is close to the stoichiometric point of the titration.

BUFFER SOLUTIONS

Figures 15.9 and 15.10 show how slowly the pH changes halfway to the stoichiometric point, close to pH = pK_a. When a small amount of acid or base is added to a solution corresponding to this region, the pH changes much less than when the same amount is added to pure water. Solutions that resist changes in pH when small amounts of acid or base are added are called **buffers.**

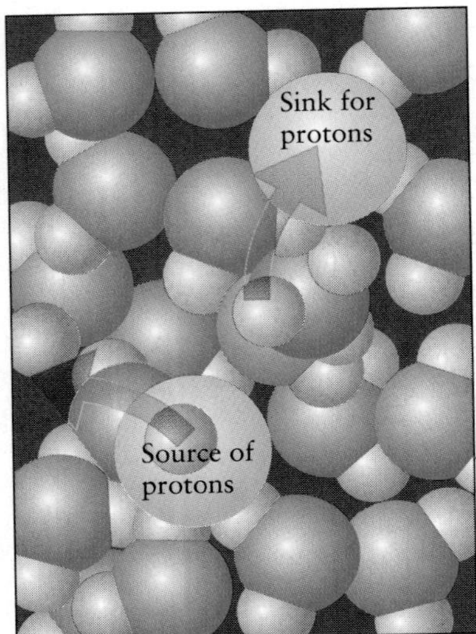

FIGURE 15.17 A buffer solution contains a sink for protons supplied when a strong acid is added and a source of protons to supply to a strong base that is added. The joint action of the source and sink keeps the pH constant when strong acid or strong base is added.

15.8 THE ACTION OF BUFFERS

A buffer solution consists of a weak acid (to supply protons to a strong base) and its conjugate base (to accept protons from a strong acid). Alternatively, it may consist of a weak base (to accept protons from a strong acid) and its conjugate acid (to transfer protons to a strong base). Buffer solutions are particularly important in biological systems. For example, blood and other cell fluids are buffered at pH = 7.4. The oceans are maintained at about pH = 8.4 by a complex buffering process that depends on the presence of hydrogen carbonates and silicates.

The action of buffers can be understood by considering how the acetic acid equilibrium

$$CH_3COOH(aq) + H_2O(l) \rightleftharpoons H_3O^+(aq) + CH_3CO_2^-(aq)$$

responds to the addition of a strong acid or base (Fig. 15.17). Suppose a strong acid is added to a solution that contains $CH_3CO_2^-$ ions and CH_3COOH molecules in about equal concentrations. As we see from Fig. 15.18, the newly arrived H_3O^+ ions transfer protons to the $CH_3CO_2^-$ ions to form CH_3COOH molecules. Only a small fraction of these newly formed CH_3COOH molecules ionize, so the pH is left nearly unchanged. If a small amount of strong base is added instead, the OH^- ions remove protons from the CH_3COOH molecules. As a result, the concentration of $CH_3CO_2^-$ increases, but the OH^- concentration remains nearly unchanged. Consequently, the H_3O^+ concentration (and the pH) is also left nearly unchanged. In other words, the weak acid acts as a source of hydrogen ions to react with OH^- (from added strong base), whereas its conjugate base (the anion of the weak acid) accepts protons from H_3O^+ (from added strong acid).

The flattest part of the pH curve of a solution of a weak acid and its conjugate base occurs where their molarities are equal. The pH is on the acid side of neutral and equal to the pK_a of the weak acid. Such a mixture is an **acid buffer** that stabilizes the solution at pH $\approx pK_a$. Similarly, a mixture of a weak base and its conjugate acid stabilizes the pH on the basic side of neutrality, at pH $\approx pK_a$, where K_a is the acidity constant of the conjugate acid of the base. The latter mixtures are called **base buffers.** Mixtures in which the salt and acid (or base) have unequal molarities are also buffers but are usually less effective than those in which the molarities are nearly equal. Some typical buffer systems are listed in Table 15.4.

A buffer is a mixture of weak conjugate acids and bases that stabilizes the pH of a solution by providing a source or sink for protons.

TABLE 15.4 Typical buffer systems

Composition	pK_a
ACID BUFFERS	
$CH_3COOH/CH_3CO_2^-$	4.74
HNO_2/NO_2^-	3.37
$HClO_2/ClO_2^-$	2.00
BASE BUFFERS	
NH_4^+/NH_3	9.25
$(CH_3)_3NH^+/(CH_3)_3N$	9.81
$H_2PO_4^-/HPO_4^{2-}$	7.21

15.9 SELECTING A BUFFER

Buffer solutions are used to calibrate pH meters, to culture bacteria, and to control the pH of solutions in which chemical reactions are taking place. They are also administered intravenously to critically ill patients. Once we know how to calculate the pH of a buffer solution, we can design a buffer with the desired pH.

Because buffer solutions are mixed solutions in the sense that they contain acid and salt, we can calculate their pH, as illustrated in Example

15.1. For a solution containing a weak acid HA and a salt that provides the conjugate base anion A^-, the proton transfer equilibrium is

$$HA(aq) + H_2O(l) \rightleftharpoons H_3O^+(aq) + A^-(aq) \qquad K_a = \frac{[H_3O^+][A^-]}{[HA]}$$

We now rearrange the expression for K_a into

$$[H_3O^+] = K_a \times \frac{[HA]}{[A^-]}$$

and take the negative logarithm of both sides:

$$-\log[H_3O^+] = -\log K_a - \log\left(\frac{[HA]}{[A^-]}\right)$$

That is,

$$pH = pK_a - \log\left(\frac{[HA]}{[A^-]}\right) = pK_a + \log\left(\frac{[A^-]}{[HA]}\right)$$

We are using the actual equilibrium molarities of acid and base in the solution when we write [HA] and $[A^-]$. However, a weak acid HA typically loses only a tiny fraction of its protons and only a tiny fraction of the anions of the weak base A^- accept protons. Thus, we can approximate their molarities by the nominal molarities of the acid and the salt as added initially. Then

$$pH \approx pK_a + \log\left(\frac{[\text{base}]_{\text{nominal}}}{[\text{acid}]_{\text{nominal}}}\right)$$

This relation is called the **Henderson-Hasselbalch equation.** Because so many chemical reactions in our bodies occur in buffered environments, biochemists commonly use the equation for quick estimates of pH. In practice, the equation is used to estimate the pH of a buffer, and then the pH is adjusted to the value required by adding more acid or base and monitoring the effect with a pH meter.

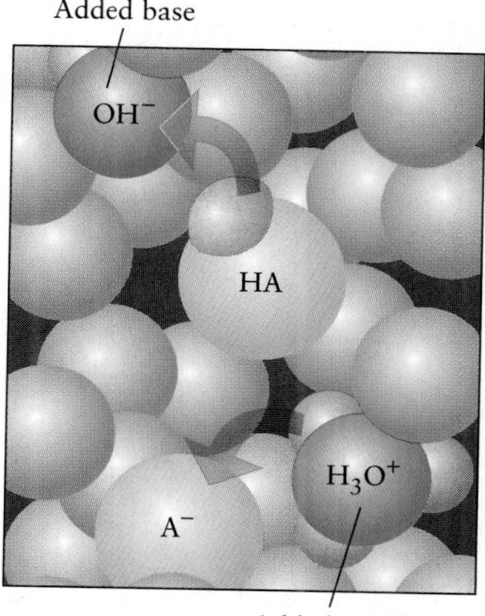

FIGURE 15.18 Buffer action depends on the donation of protons by the weak acid molecules, HA, when a strong base is added and the acceptance of protons by the conjugate base ions, A^-, when a strong acid is added.

EXAMPLE 15.6 Calculating the pH of a buffer solution

Calculate the pH of a buffer solution that is 0.040 M $NaCH_3CO_2(aq)$ and 0.080 M $CH_3COOH(aq)$ at 25°C.

STRATEGY Begin by identifying the weak acid and its conjugate base (the acid has one more H atom than the base). Then write the proton transfer equilibrium between them, rearrange the expression for K_a to give $[H_3O^+]$, and find the pH, as in the derivation of the Henderson-Hasselbalch equation. Alternatively, use that equation directly.

SOLUTION The acid is CH_3COOH and its conjugate base is $CH_3CO_2^-$. The equilibrium to consider is

$$CH_3COOH(aq) + H_2O(l) \rightleftharpoons H_3O^+(aq) + CH_3CO_2^-(aq)$$

$$K_a = \frac{[H_3O^+][CH_3CO_2^-]}{[CH_3COOH]}$$

From Table 14.2, $pK_a = 4.75$. Rearranging the expression for K_a, we obtain first

$$[H_3O^+] = K_a \times \frac{[CH_3COOH]}{[CH_3CO_2^-]}$$

and then

$$pH = pK_a + \log\left(\frac{[CH_3CO_2^-]}{[CH_3COOH]}\right)$$

The actual molarities of acid and base are now approximated by their initial values:

$$pH \approx 4.75 + \log\left(\frac{0.040}{0.080}\right) = 4.45$$

That is, the solution acts as a buffer close to pH = 4.4.

SELF-TEST 15.9A Calculate the pH of a buffer solution that is 0.040 M $NH_4Cl(aq)$ and 0.030 M $NH_3(aq)$.

[*Answer:* 9.1]

SELF-TEST 15.9B Calculate the pH of a buffer solution that is 0.15 M $HNO_2(aq)$ and 0.20 M $NaNO_2(aq)$.

EXAMPLE 15.7 Calculating the pH change when acid or base is added to a buffer solution

Suppose that 1.2 g of sodium hydroxide (0.030 mol NaOH) is dissolved in 500 mL of the buffer solution described in Example 15.6. Calculate the pH of the resulting solution and the change in pH. Assume that the volume is constant.

STRATEGY Once we know the molar concentrations of the acid and its conjugate base, we can rearrange the expression for K_a into the form of the Henderson-Hasselbalch equation to obtain the pH of the solution. The

CHEMISTRY AT WORK

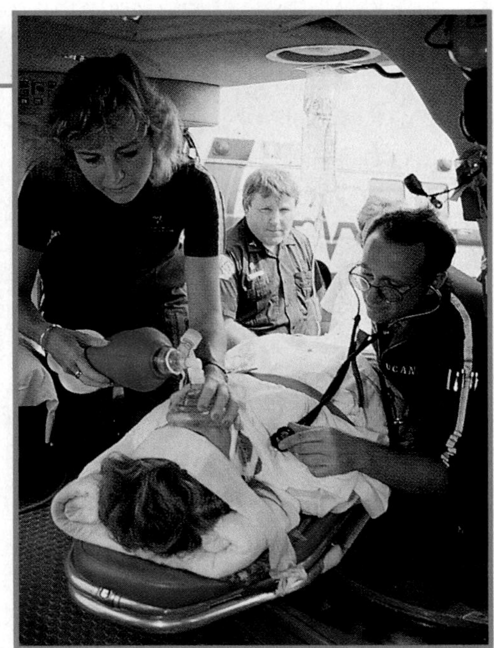

A medical helicopter team provides emergency treatment to an accident victim. As soon as the patient is secured, intravenous administration of electrolyte solutions will begin to combat the symptoms of shock. These solutions will help to maintain the pH of the blood and its buffer capacity.

0.030 mol OH^- that is added to the buffer solution reacts with the acid of the buffer system, CH_3COOH, thereby decreasing its amount by 0.030 mol and increasing the amount of its conjugate base, $CH_3CO_2^-$, by 0.030 mol. There is no significant change in the volume of solution.

SOLUTION The number of moles of the acid CH_3COOH in the solution initially (from the data in Example 15.6) is

$$\text{Moles of } CH_3COOH = (0.500\ L) \times (0.080\ mol{\cdot}L^{-1}) = 0.040\ mol$$

The added sodium hydroxide reacts with the acetic acid:

$$CH_3COOH(aq) + OH^-(aq) \rightleftharpoons CH_3CO_2^-(aq) + H_2O(aq)$$

The amount of CH_3COOH that reacts is:

$$0.030\ mol\ OH^- \times \frac{1\ mol\ CH_3COOH}{1\ mol\ OH^-} = 0.030\ mol\ CH_3COOH$$

Therefore, the amount of CH_3COOH remaining is 0.040 mol − 0.030 mol = 0.010 mol. The resulting molarity of CH_3COOH is now:

$$\text{Molarity of } CH_3COOH = \frac{0.010\ mol}{0.500\ L} = 0.020\ mol{\cdot}L^{-1}$$

Similarly, the initial number of moles of the base $CH_3CO_2^-$ in the 500 mL of solution (using data from Example 15.6) is

$$\text{Moles of } CH_3CO_2^- = (0.500\ L) \times (0.040\ mol{\cdot}L^{-1}) = 0.020\ mol$$

The addition of 0.030 mol OH^- ions increases this amount to (0.020 + 0.030) mol = 0.050 mol through the reaction with CH_3COOH. Hence the molarity of the base $CH_3CO_2^-$ is increased to

$$\text{Molarity of } CH_3CO_2^- = \frac{0.050\ mol}{0.500\ L} = 0.10\ mol{\cdot}L^{-1}$$

The proton transfer equilibrium is

$$CH_3COOH(aq) + H_2O(l) \rightleftharpoons H_3O^+(aq) + CH_3O_2^-(aq)$$

$$K_a = \frac{[H_3O^+][CH_3O_2^-]}{[CH_3COOH]}$$

We can now set up the equilibrium table with all concentrations in moles per liter:

	Species		
	CH_3COOH	H_3O^+	$CH_3CO_2^-$
Step 1. Initial molarity	0.020	0	0.10
Step 2. Change in molarity	$-x$	$+x$	$+x$
Step 3. Equilibrium molarity	$0.020 - x$	x	$0.10 + x$

We assume that x is very small compared to 0.10 and 0.020 and obtain

$$pH = pK_a + \log\left(\frac{[CH_3COO_2^-]}{[CH_3COOH]}\right) = 4.75 + \log\left(\frac{0.10}{0.020}\right) = 5.45$$

or about 5.4. That is, the pH of the solution changes from about 4.4 to about 5.4.

SELF-TEST 15.10A Suppose that 0.0100 mol HCl(g) is dissolved in 500 mL of the buffer solution of Example 15.6. Calculate the pH of the resulting solution and the change in pH.

[*Answer:* 4.1, a decrease of 0.3]

SELF-TEST 15.10B Suppose that 0.0200 mol NaOH(s) is dissolved in 300 mL of the buffer solution of Example 15.6. Calculate the pH of the resulting solution and the change in pH.

Commercially available buffer solutions can be purchased for virtually any desired pH. For example, pH meters are often calibrated by using a mixed solution of 0.025 M $Na_2HPO_4(aq)$ and 0.025 M $KH_2PO_4(aq)$, which has pH = 6.87 at 25°C. The method demonstrated in Example 15.6 predicts pH = 7.2 for this solution; however, as noted at the start of the chapter, these calculations ignore ion-ion interactions, which modify the pH slightly.

The pH of a buffer solution is close to the pK_a of the weak acid component when the acid and base have similar concentrations.

15.10 BUFFER CAPACITY

The **buffer capacity** is an indication of the amount of acid or base that can be added before the buffer loses its ability to resist the change in pH. A buffer with a high capacity can maintain its buffering action after the addition of more strong acid or base than can one with a small capacity. The buffer becomes exhausted when most of the weak base has been converted to acid or when most of the weak acid has been converted to base.

A buffer has a high capacity when the amount of base present is about 10% or more of the amount of acid, for otherwise the base gets used up quickly. Similarly, the amount of acid present should be about 10% or more of the amount of base, for otherwise the acid gets used up quickly. The Henderson-Hasselbalch equation then shows us that the corresponding pH range of the buffer is from

$$\text{pH} = \text{p}K_\text{a} + \log\left(\frac{[\text{A}^-]}{10[\text{A}^-]}\right) = \text{p}K_\text{a} + \log\left(\frac{1}{10}\right) = \text{p}K_\text{a} - 1$$

when the acid is 10 times more abundant than the base to

$$\text{pH} = \text{p}K_\text{a} + \log\left(\frac{10[\text{HA}]}{[\text{HA}]}\right) = \text{p}K_\text{a} + \log\left(\frac{10}{1}\right) = \text{p}K_\text{a} + 1$$

when the base is 10 times more abundant than the acid. The buffer can act effectively within this range (Fig. 15.19). Buffer capacity also depends on the absolute amount of conjugate acid and base available. If a buffer solution is diluted, it will have a lower buffer capacity than the same volume of the more concentrated solution.

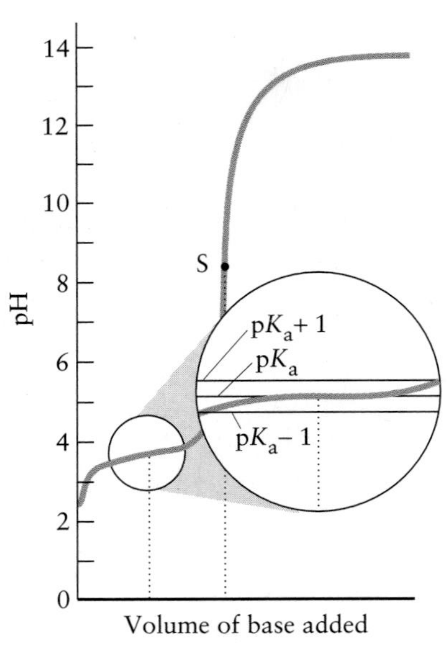

FIGURE 15.19 When conjugate acid and base are present at similar concentrations, the pH changes very little as more strong base (or strong acid) is added. As the inset shows, the pH lies between $pK_a \pm 1$ for a wide range of concentrations.

SELF-TEST 15.11A Which of the buffer systems listed in Table 15.4 would be a good choice to prepare a buffer with pH = 5 ± 1?

[*Answer:* $CH_3COOH/CH_3CO_2^-$]

SELF-TEST 15.11B Which of the buffer systems listed in Table 15.4 would be a good choice to prepare a buffer with pH = 10 ± 1?

EXAMPLE 15.8 Determining the ratio of molarities for a buffer solution with a given pH

Calculate the ratio of the molarities of CO_3^{2-} and HCO_3^- ions required to achieve buffering at pH = 9.50. The pK_{a2} of H_2CO_3 is 10.25.

STRATEGY The pH of an HA/A$^-$ buffer is calculated from the Henderson-Hasselbalch equation:

$$\text{pH} = \text{p}K_a + \log\left(\frac{[A^-]}{[HA]}\right)$$

Identify the acid and base, and rearrange this expression to give the desired concentration ratio.

SOLUTION The base is CO_3^{2-} and the acid is HCO_3^-; hence,

$$\log\left(\frac{[CO_3^{2-}]}{[HCO_3^-]}\right) = \text{pH} - \text{p}K_a = 9.50 - 10.25 = -0.75$$

The antilogarithm of -0.75 is 0.18, so the ratio of molarities is

$$\frac{[CO_3^{2-}]}{[HCO_3^-]} = 0.18$$

Therefore, the solution will act as a buffer with a pH close to 9.50 if it is prepared by mixing the solutes in the ratio 0.18 mol CO_3^{2-} to 1.0 mol HCO_3^-.

SELF-TEST 15.12A Calculate the ratio of the molarities of acetate ions and acetic acid needed to buffer a solution at pH = 5.25. The pK_a of CH_3COOH is 4.75.

[*Answer:* 3.2:1]

SELF-TEST 15.12B Calculate the ratio of the molarities of benzoate ions and benzoic acid (C_6H_5COOH) needed to buffer a solution at pH = 3.50. The pK_a of C_6H_5COOH is 4.19.

The composition of blood plasma, with the concentration of HCO_3^- ions about 20 times that of H_2CO_3, seems to be outside the range for optimum buffering. However, the principal waste products of living cells are carboxylic acids, such as lactic acid. Plasma, with its relatively high concentration of HCO_3^- ions, can absorb a significant surge of hydrogen ions from the carboxylic acids. The high proportion of HCO_3^- also helps us to withstand disturbances that lead to excess acid, such as disease and shock due to burns (Case Study 15).

The pH of a buffer solution is close to the pK_a of the weak acid component when the acid and base have similar concentrations and is effective in the range pK_a ± 1.

SOLUBILITY EQUILIBRIA

Many salts are sparingly soluble in water. By learning to predict just how soluble, we can control when precipitates form and use precipitation in

CASE STUDY 15 BURNS AND BLOOD CHEMISTRY*

WHEN MEDICS ARRIVE at the side of a severely burned patient, one of the first lifesaving procedures they use is the intravenous administration of fluids. This treatment is necessary because the most immediate threat to the survival of a badly burned person is a change in blood pH.

The pH of human blood is normally maintained within a narrow range (7.35–7.45) by several buffer systems, primarily the carbonic acid-hydrogen carbonate (bicarbonate) ion system. The normal ratio of HCO_3^- to H_2CO_3 in the blood is 20:1. When the concentration of HCO_3^- increases relative to that of H_2CO_3, blood pH rises. If this increase is sufficient to cause blood pH to exceed the normal range, the condition is called *alkalosis*. Conversely, blood pH decreases when the ratio decreases; and when blood pH falls below the normal range, the condition is called *acidosis*.

Blood carbonic acid and bicarbonate concentrations are controlled by different mechanisms. Carbonic acid concentration is controlled by respiration: as we exhale, we deplete our system of CO_2, and thus H_2CO_3. Breathing faster and more deeply increases the amount of CO_2 exhaled and lowers the carbonic acid concentration in the blood. Hydrogen carbonate ion concentration is controlled by its rate of excretion in urine.

When a person is burned, blood plasma leaks from the circulatory system into the injured area, thereby producing swelling (edema) and reducing the blood volume. If the burned area is large, this loss of blood volume may be sufficient to reduce blood flow and oxygen supply to all the body's tissues. Lack of oxygen, in turn, causes the tissues to produce an excessive amount of lactic acid, which enters the bloodstream, reacts with hydrogen carbonate ion to produce H_2CO_3, and shifts the ratio of HCO_3^- to H_2CO_3 to a lower value. To minimize the consequent decrease in pH (called *metabolic acidosis*), the injured individual breathes harder, increasing the respiratory excretion of CO_2. However, if blood volume drops below levels for which the body can compensate, blood pressure falls, CO_2 excretion diminishes, and acidosis becomes more severe. Individuals in this state are said to be in *shock* and will die if not treated promptly.

Shock is avoided or treated by intravenous infusion of large volumes of a salt-containing solution, usually one known as *lactated Ringer's solution*. The added liquid increases blood volume and blood flow, which improves

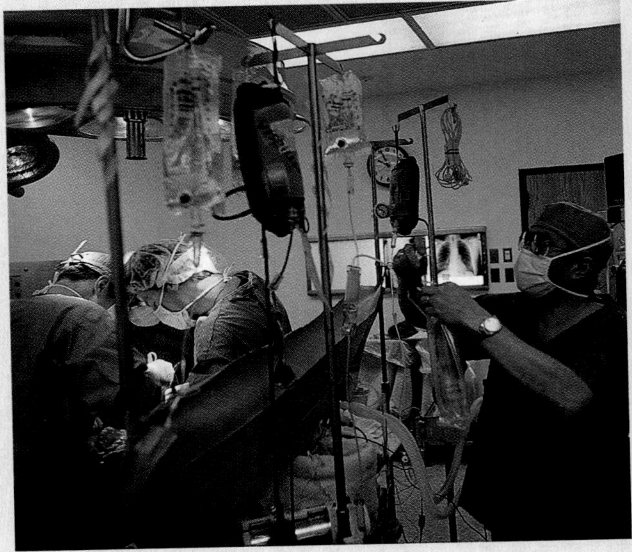

Patients undergoing surgery often need several kinds of intravenous solutions. These doctors are operating on a patient's liver, a procedure that requires the administration of three different intravenous solutions and a blood transfusion.

*This case study is an adaptation of a contribution by B. A. Pruitt, M.D., and A. D. Mason, M.D., U.S. Army Institute of Surgical Research.

the laboratory to separate mixtures. The calculations involved are applications of equilibria: no matter how insoluble a salt, it is always in equilibrium with its dissolved ions in a saturated solution.

15.11 THE SOLUBILITY PRODUCT

The equilibrium constant for a solubility equilibrium is called the **solubility product,** K_{sp}, of the solute. For example, the solubility product for bismuth sulfide, Bi_2S_3, is defined

oxygen delivery; the $[HCO_3^-]/[H_2CO_3]$ ratio then increases toward normal, thereby allowing the severely injured person to survive.

The danger increases when smoke is inhaled at the time of burn injury. Chemicals contained in the smoke cause lung tissues to swell, a change that may interfere with the normal excretion of CO_2. The consequent accumulation of H_2CO_3 in the blood decreases the $[HCO_3^-]/[H_2CO_3]$ ratio and blood pH and leads to *respiratory acidosis*. The condition is treated by using a mechanical ventilator to assist the victim's breathing.

Other blood components involved in the body's response to burn injury include sodium and potassium ions. The plasma lost from the bloodstream into the burn wound contains sodium ions, a loss that must be replaced by the same intravenous infusion of salt-containing solution used to combat shock. Even though this infusion increases the total amount of sodium ions in the body, the

Three solutions commonly given intravenously: lactated Ringer's solution, 0.9% sodium chloride, and 5% dextrose (glucose). The first two help to control electrolyte levels, the third maintains the blood sugar level, and all three help to maintain blood volume.

sodium concentration in the added liquid is still lower than that of normal blood plasma, and mild *hyponatremia,* reduced plasma sodium ion concentration, can result. In severe cases, hyponatremia can lead to disorientation or even coma. Burned patients may also develop *hypernatremia,* or increased blood sodium concentration, because the protective layer of skin is damaged and salt-free water evaporates through the burn wounds.

To guard against damage, medical personnel monitor plasma sodium concentrations frequently in burned patients and measure body weight and urinary sodium daily. This information is used to calculate appropriate doses of salt and water to keep plasma sodium concentration within an optimal range.

Most of the body's potassium resides within cells. After burn injury, heat-damaged cells release potassium ions into the bloodstream. The resulting mild increase in potassium concentration in the blood plasma is called *hyperkalemia.* Later, as healing begins, treated patients may experience a reduced plasma potassium concentration, or *hypokalemia.* If necessary, this deficit can be corrected with intravenous solutions of potassium salts.

QUESTIONS

1. Very rapid, deep breathing called hyperventilation causes dizziness and disturbs the $[HCO_3^-]/[H_2CO_3]$ equilibrium. Predict in which direction the equilibrium is driven by hyperventilation and explain your answer.

2. Blood pH is controlled by a complex mixture of acids and bases. What would be the pH of a buffer with a 20:1 ratio of HCO_3^- to H_2CO_3 if no other acids and bases were present? Take $pK_a = 6.1$ for H_2CO_3 in blood.

3. What is the effective pH range of the $[HCO_3^-]/[H_2CO_3]$ buffer in water?

$$Bi_2S_3(s) \rightleftharpoons 2\,Bi^{3+}(aq) + 3\,S^{2-}(aq) \qquad K_{sp} = [Bi^{3+}]^2[S^{2-}]^3$$

Solid Bi_2S_3 does not appear in the expression for K_{sp} because it is a pure solid (Section 13.3).

A solubility product is no different from any other equilibrium constant, and it is used in the same way. However, because ion-ion interactions in concentrated electrolyte solutions can complicate its interpretation, a solubility product is generally used only for sparingly soluble salts. One of the simplest ways to determine K_{sp} is to measure the molar

TABLE 15.5 Solubility products at 25°C

Compound	Formula	K_{sp}
aluminum hydroxide	$Al(OH)_3$	1.0×10^{-33}
antimony sulfide	Sb_2S_3	1.7×10^{-93}
barium carbonate	$BaCO_3$	8.1×10^{-9}
fluoride	BaF_2	1.7×10^{-6}
sulfate	$BaSO_4$	1.1×10^{-10}
bismuth sulfide	Bi_2S_3	1.0×10^{-97}
calcium carbonate	$CaCO_3$	8.7×10^{-9}
fluoride	CaF_2	4.0×10^{-11}
hydroxide	$Ca(OH)_2$	5.5×10^{-6}
sulfate	$CaSO_4$	2.4×10^{-5}
copper(I) bromide	$CuBr$	4.2×10^{-8}
chloride	$CuCl$	1.0×10^{-6}
iodide	CuI	5.1×10^{-12}
sulfide	Cu_2S	2.0×10^{-47}
copper(II) iodate	$Cu(IO_3)_2$	1.4×10^{-7}
oxalate	$Cu(C_2O_4)$	2.9×10^{-8}
sulfide	CuS	1.3×10^{-36}
iron(II) hydroxide	$Fe(OH)_2$	1.6×10^{-14}
sulfide	FeS	6.3×10^{-18}
iron(III) hydroxide	$Fe(OH)_3$	2.0×10^{-39}
lead(II) bromide	$PbBr_2$	7.9×10^{-5}
chloride	$PbCl_2$	1.6×10^{-5}
fluoride	PbF_2	3.7×10^{-8}
iodate	$Pb(IO_3)_2$	2.6×10^{-13}
iodide	PbI_2	1.4×10^{-8}
sulfate	$PbSO_4$	1.6×10^{-8}
sulfide	PbS	8.8×10^{-29}
magnesium		
ammonium phosphate	$MgNH_4PO_4$	2.5×10^{-13}
carbonate	$MgCO_3$	1.0×10^{-5}
fluoride	MgF_2	6.4×10^{-9}
hydroxide	$Mg(OH)_2$	1.1×10^{-11}
mercury(I) chloride	Hg_2Cl_2	1.3×10^{-18}
iodide	Hg_2I_2	1.2×10^{-28}
mercury(II) sulfide, black	HgS	1.6×10^{-52}
sulfide, red	HgS	1.4×10^{-53}
nickel(II) hydroxide	$Ni(OH)_2$	6.5×10^{-18}
silver bromide	$AgBr$	7.7×10^{-13}
carbonate	Ag_2CO_3	6.2×10^{-12}
chloride	$AgCl$	1.6×10^{-10}
hydroxide	$AgOH$	1.5×10^{-8}
iodide	AgI	1.5×10^{-16}
sulfide	Ag_2S	6.3×10^{-51}
zinc hydroxide	$Zn(OH)_2$	2.0×10^{-17}
sulfide	ZnS	1.6×10^{-24}

◄◄ Molar solubility was introduced in Section 12.5.

solubility, S, of the compound, but more advanced and accurate methods are also available. Table 15.5 gives some experimental values.

The solubility product is the equilibrium constant for the equilibrium between an undissolved salt and its ions in a saturated solution.

EXAMPLE 15.9 Determining the solubility product

The molar solubility, S, of silver chromate, Ag_2CrO_4, is 6.5×10^{-5} mol·L^{-1}. Determine the value of K_{sp}.

STRATEGY Write the expression for the equilibrium constant. The molarity of the pure solid does not appear in the expression for K_{sp}. To evaluate K_{sp}, we need to know the molarities of each type of ion formed by the salt. Determine them from the relation between the molarity of each type of ion and the molar solubility, S, of the salt. For this step, use the chemical equation for the equilibrium and the stoichiometric relations between the species.

SOLUTION The chemical equation and the expression for the solubility product are

$$Ag_2CrO_4(s) \rightleftharpoons 2\,Ag^+(aq) + CrO_4^{2-}(aq) \qquad K_{sp} = [Ag^+]^2[CrO_4^{2-}]$$

Because 2 mol Ag^+ ≏ 1 mol Ag_2CrO_4 and 1 mol CrO_4^{2-} ≏ 1 mol Ag_2CrO_4,

$$[Ag^+] = 2S = 2 \times (6.5 \times 10^{-5}) = 1.30 \times 10^{-4}$$

$$[CrO_4^{2-}] = S = 6.5 \times 10^{-5}$$

Therefore

$$K_{sp} = [Ag^+]^2[CrO_4^{2-}] = (2S)^2(S)$$
$$= (1.30 \times 10^{-4})^2 \times (6.5 \times 10^{-5}) = 1.1 \times 10^{-12}$$

SELF-TEST 15.13A The molar solubility of lead(II) iodate, $Pb(IO_3)_2$, at 26°C is 4.0×10^{-5} mol·L^{-1}. What is the value of K_{sp} for lead(II) iodate?

[*Answer:* 2.6×10^{-13}]

SELF-TEST 15.13B The molar solubility of calcium iodate, $Ca(IO_3)_2$, at 10°C is 3.8×10^{-3} mol·L^{-1}. What is the value of K_{sp} for calcium iodate?

EXAMPLE 15.10 Calculating the molar solubility from the solubility product

According to Table 15.5, $K_{sp} = 1.4 \times 10^{-8}$ for lead(II) iodide in water. Estimate its molar solubility.

STRATEGY Write the chemical equation for the solubility equilibrium and the expression for K_{sp}. The molar solubility, S, is the molarity of formula units in the saturated solution. Because each formula unit produces a known number of cations and anions in solution, express the molarities of the cations and anions in terms of S. Then express K_{sp} in terms of S and solve for S.

SOLUTION The solubility equilibrium is

$$PbI_2(s) \rightleftharpoons Pb^{2+}(aq) + 2\,I^-(aq) \qquad K_{sp} = [Pb^{2+}][I^-]^2$$

Because 1 mol PbI_2 ≏ 1 mol Pb^{2+} and 1 mol PbI_2 ≏ 2 mol I^-,

$$[Pb^{2+}] = S \qquad [I^-] = 2S$$

It follows that

$$K_{sp} = [Pb^{2+}][I^-]^2 = S \times (2S)^2 = 4S^3$$

This expression is easily solved for S:

$$S = \left(\tfrac{1}{4}K_{sp}\right)^{1/3}$$

The expression $x^{1/3}$ means the cube root of x: it is easily found by using the y^x command on a calculator.

Now we can substitute the numerical value of K_{sp}:

$$S = \left(\tfrac{1}{4} \times (1.4 \times 10^{-8})\right)^{1/3} = 1.5 \times 10^{-3}$$

The molar solubility of PbI_2 is therefore 1.5×10^{-3} mol·L^{-1}.

SELF-TEST 15.14A The solubility product of silver sulfate, Ag_2SO_4, is 1.4×10^{-5}. Calculate the molar solubility of the salt.

[*Answer:* 1.5×10^{-2} mol·L^{-1}]

SELF-TEST 15.14B The solubility product of magnesium ammonium phosphate, $MgNH_4PO_4$, is 2.5×10^{-13}. Calculate the molar solubility of the salt, which forms Mg^{2+}, NH_4^+, and PO_4^{3-} ions in water.

FIGURE 15.20 If the concentration of one of the ions in solution is increased, then the concentration of the other is decreased to maintain a constant value of K_{sp}. (a) The cations (pink) and anions (green) in solution. (b) When more anions are added (together with their accompanying spectator ions, which are not shown), the concentration of cations decreases. In other words, the solubility of the original compound is reduced by the presence of a common ion.

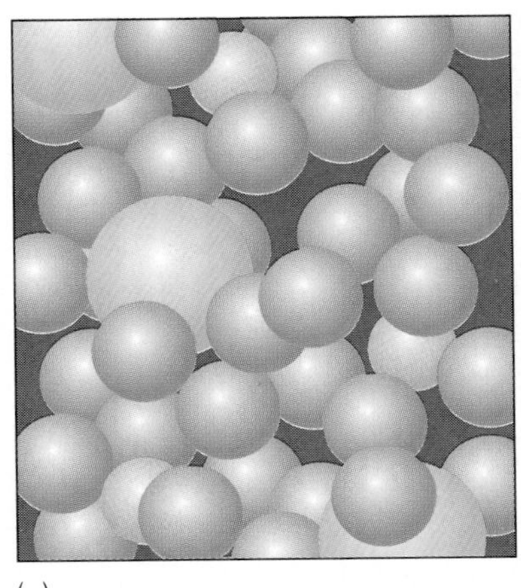

(a)

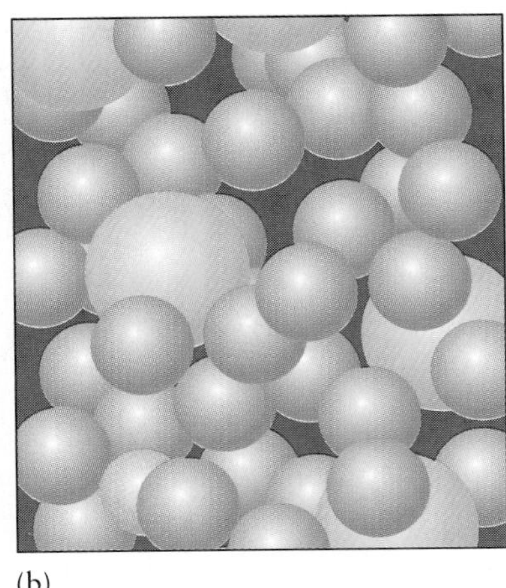

(b)

15.12 THE COMMON-ION EFFECT

Sometimes we must find a way to precipitate a sparingly soluble salt. Because the ions are in dynamic equilibrium with the solid salt, we can use Le Chatelier's principle: if a second salt that contains one of the same ions is added to a saturated solution, then the solubility of the other ion is decreased and the salt precipitates (Fig. 15.20). The decrease in solubility is called the **common-ion effect.**

Suppose we have a saturated solution of silver chloride in water:

$$AgCl(s) \rightleftharpoons Ag^+(aq) + Cl^-(aq) \qquad K_{sp} = [Ag^+][Cl^-]$$

Experimentally, $K_{sp} = 1.6 \times 10^{-10}$ at 25°C, and the molar solubility of AgCl in water is 1.3×10^{-5} mol·L^{-1}. If we add sodium chloride to the solution, the concentration of Cl^- ions increases. The system responds by

FIGURE 15.21 (a) A saturated solution of zinc acetate in water. (b) When acetate ions are added (as solid sodium acetate in the spatula shown in (a)), the solubility of the zinc acetate is significantly reduced, and additional zinc acetate precipitates.

(a)

(b)

tending to minimize the effect of adding more Cl^- ions, so some of the Cl^- ions form solid AgCl, taking some Ag^+ ions out of solution. Because there is now less Ag^+ in solution, the solubility of AgCl is lower in a solution of NaCl than it is in pure water. A similar effect occurs whenever two salts have an ion in common (Fig. 15.21).

Predicting the common-ion effect quantitatively is more difficult. Because ions interact with one another strongly, simple equilibrium calculations are rarely valid. However, we can still get an idea of the size of the common-ion effect by rearranging the expression for K_{sp} into

$$[Ag^+] = \frac{K_{sp}}{[Cl^-]}$$

Suppose we try to dissolve silver chloride in 0.10 M NaCl(aq), in which $[Cl^-] = 0.10$. The silver chloride dissolves until the concentration of Ag^+ ions is given by

$$[Ag^+] = \frac{1.6 \times 10^{-10}}{0.10} = 1.6 \times 10^{-9}$$

The molarity of Ag^+ ions, and hence the solubility of AgCl formula units, rises only to 1.6×10^{-11} mol·L^{-1}. That is a million-fold decrease from its solubility in pure water.

The common-ion effect is the reduction in the solubility of a sparingly soluble salt by the addition of a soluble salt that has an ion in common with it.

SELF-TEST 15.15A What is the molar solubility of calcium carbonate in 0.20 M $CaCl_2$(aq)?

[*Answer:* 4.4×10^{-8} mol·L^{-1}]

SELF-TEST 15.15B What is the molar solubility of silver bromide in 0.10 M $CaBr_2$(aq)?

15.13 PREDICTING PRECIPITATION

In Section 13.6 we saw how to predict the direction in which a reaction would occur by using Q, the reaction quotient. Exactly the same techniques can be used to decide whether a precipitate is likely to form from the ions in two electrolyte solutions that are mixed. In this case, the equilibrium constant is the solubility product, K_{sp}, and the quotient is denoted Q_{sp}. Precipitation occurs when Q_{sp} is equal to or greater than K_{sp} (Fig. 15.22).

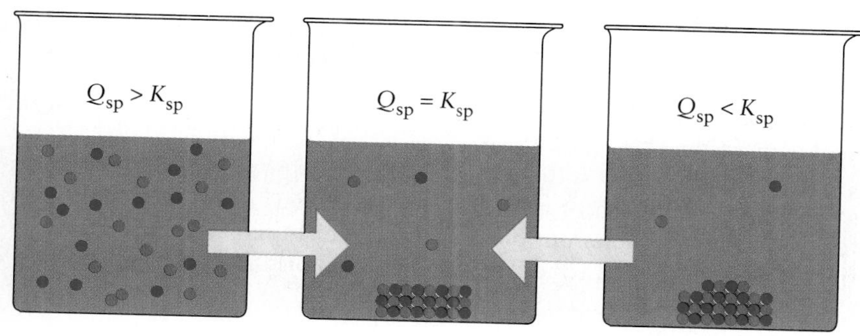

FIGURE 15.22 The relative magnitudes of the solubility quotient, Q_{sp}, and the solubility constant, K_{sp}, are used to decide whether a salt will precipitate (left) or dissolve (right).

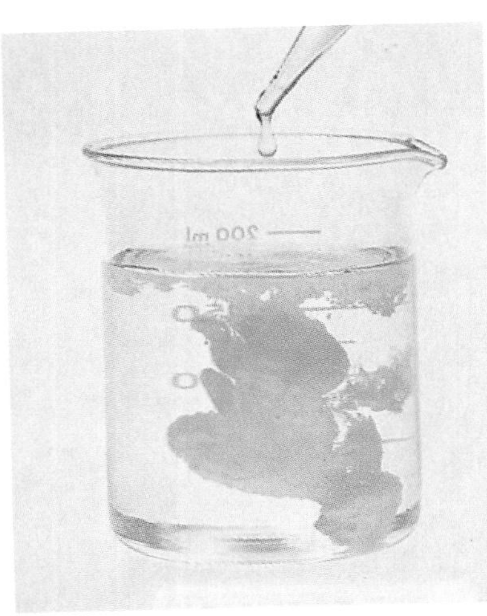

FIGURE 15.23 When a few drops of lead(II) nitrate solution are added to a solution of potassium iodide, yellow lead(II) iodide immediately precipitates.

For example, we might want to know whether a precipitate of PbI_2 will form when equal volumes of 0.2 M solutions of lead(II) nitrate and potassium iodide are mixed. The overall equation for the reaction is

$$Pb(NO_3)_2(aq) + 2\,KI(aq) \longrightarrow 2\,KNO_3(aq) + PbI_2(s)$$

and the net ionic equation is

$$Pb^{2+}(aq) + 2\,I^-(aq) \rightleftharpoons PbI_2(s)$$

The reverse of this equation is the equation for the dissolution of PbI_2:

$$PbI_2(s) \rightleftharpoons Pb^{2+}(aq) + 2\,I^-(aq) \qquad K_{sp} = [Pb^{2+}][I^-]^2$$

We know from Table 15.5 that $K_{sp} = 1.4 \times 10^{-8}$ at 25°C. The instant the solutions are mixed, the molarities of the Pb^{2+} and I^- ions are very high. Because equal volumes were mixed together, the new volume is twice as large, so the new molarities are half (0.1 $mol{\cdot}L^{-1}$ $Pb^{2+}(aq)$ and 0.1 $mol{\cdot}L^{-1}$ $I^-(aq)$) their original values. Therefore,

$$Q_{sp} = [Pb^{2+}][I^-]^2 = 0.1 \times (0.1)^2 = 1 \times 10^{-3}$$

This value is considerably higher than K_{sp}, so a precipitate will form. Precipitations occur immediately: we never have to wait more than a millisecond or so, because ions in electrolyte solutions are very mobile and clump together rapidly (Fig. 15.23).

A salt precipitates if Q_{sp} is greater than or equal to K_{sp}.

SELF-TEST 15.16A Does a precipitate of silver chloride form when 200 mL of 1.0×10^{-4} M $AgNO_3(aq)$ and 900 mL of 1.0×10^{-6} M KCl(aq) are mixed?

[*Answer:* No ($Q_{sp} = 1.5 \times 10^{-11} < K_{sp}$)]

SELF-TEST 15.16B Does a precipitate of barium fluoride form when 100 mL of 1.0×10^{-3} M $Ba(NO_3)_2(aq)$ is mixed with 200 mL of 1.0×10^{-3} M KF(aq)? Ignore possible protonation of F^-.

15.14 DISSOLVING PRECIPITATES

It may be necessary to dissolve a precipitate if we want to identify it or to use it in another reaction. For example, once a silver salt precipitates, it may be necessary to dissolve it again so that it can be electroplated as metallic silver. One strategy is to remove one of the ions from the solubility equilibrium, so that the precipitate continues to dissolve.

Suppose, for example, a solid hydroxide is in equilibrium with its ions in solution:

$$Fe(OH)_3(s) \rightleftharpoons Fe^{3+}(aq) + 3\,OH^-(aq)$$

To dissolve more of the solid, we can add acid. The H_3O^+ ions from the acid convert the OH^- into water.

Many carbonate, sulfite, and sulfide precipitates can also be dissolved by the addition of acid, because the anions react with the acid to form a gas that bubbles out of solution. For example, in a saturated solution of zinc carbonate, solid $ZnCO_3$ is in equilibrium with its ions:

$$ZnCO_3(s) \rightleftharpoons Zn^{2+}(aq) + CO_3^{2-}(aq)$$

The CO_3^{2-} ions react with the acid to form CO_2:

$$CO_3^{2-}(aq) + 2\,HCl(aq) \longrightarrow CO_2(g) + H_2O(l) + 2\,Cl^-(aq)$$

An alternative procedure to remove an ion from solution is to change its oxidation number. The metal ions in very insoluble heavy metal sulfide precipitates can be dissolved by oxidizing the sulfide ion to elemental sulfur. For example, copper(II) sulfide, CuS, takes part in the equilibrium

$$CuS(s) \rightleftharpoons Cu^{2+}(aq) + S^{2-}(aq)$$

However, when nitric acid is added, the sulfide ions are oxidized to elemental sulfur:

$$3\,S^{2-}(aq) + 8\,HNO_3(aq) \longrightarrow 3\,S(s) + 2\,NO(g) + 4\,H_2O(l) + 6\,NO_3^-(aq)$$

This complicated oxidation removes the sulfide ions from the equilibrium, and Cu^{2+} ions dissolve as $Cu(NO_3)_2$.

The solubility of a solid can be increased by removing an ion from solution; acid can be used to dissolve a hydroxide or carbonate precipitate, and nitric acid can be used to oxidize metal sulfides to sulfur and a soluble salt.

15.15 COMPLEX IONS AND SOLUBILITIES

The formation of a complex can also remove an ion and disturb the solubility equilibrium until more solid dissolves. We first met complexes in Section 8.12, where we saw that they were species formed by the reaction of a Lewis acid and a Lewis base. In this section, we consider complexes in which the Lewis acid is a metal cation, such as Ag^+. An example is the formation of $[Ag(NH_3)_2]^+$ when aqueous ammonia is added to a solution of silver chloride:

$$Ag^+(aq) + 2\,NH_3(aq) \longrightarrow Ag(NH_3)_2^+(aq)$$

If enough ammonia is present, all the precipitate dissolves. A similar procedure is used to remove the silver halide emulsion from an exposed photographic film after it has been developed. In this case, the reagent used to form the complex ion is the thiosulfate ion, $S_2O_3^{2-}$:

$$Ag^+(aq) + 2\,S_2O_3^{2-}(aq) \longrightarrow Ag(S_2O_3)_2^{3-}(aq)$$

Complex formation removes some of the Ag^+ ions from solution. As a result, to preserve the value of K_{sp}, more silver chloride will dissolve. We can therefore conclude that the qualitative effect of complex formation is to *increase* the solubility of a sparingly soluble compound.

To treat complex formation quantitatively, we note that complex formation and dissolving are at equilibrium; so we can write

$$AgCl(s) \rightleftharpoons Ag^+(aq) + Cl^-(aq) \qquad K_{sp} = [Ag^+][Cl^-] \qquad (1)$$

$$Ag^+(aq) + 2\,NH_3(aq) \rightleftharpoons Ag(NH_3)_2^+(aq) \qquad K_f = \frac{[Ag(NH_3)_2^+]}{[Ag^+][NH_3]^2} \qquad (2)$$

The equilibrium constant for the formation of the complex ion is called the **formation constant**, K_f. At 25°C, $K_f = 1.6 \times 10^7$ for this reaction.

TABLE 15.6 Formation constants in water at 25°C

Equilibrium	K_f
$Ag^+(aq) + 2\,CN^-(aq) \rightleftharpoons Ag(CN)_2^-(aq)$	5.6×10^8
$Ag^+(aq) + 2\,NH_3(aq) \rightleftharpoons Ag(NH_3)_2^+(aq)$	1.6×10^7
$Au^+(aq) + 2\,CN^-(aq) \rightleftharpoons Au(CN)_2^-(aq)$	2.0×10^{38}
$Cu^{2+}(aq) + 4\,NH_3(aq) \rightleftharpoons Cu(NH_3)_4^{2+}(aq)$	1.2×10^{13}
$Hg^{2+}(aq) + 4\,Cl^-(aq) \rightleftharpoons HgCl_4^{2-}(aq)$	1.2×10^5
$Fe^{2+}(aq) + 6\,CN^-(aq) \rightleftharpoons Fe(CN)_6^{4-}(aq)$	7.7×10^{36}
$Ni^{2+}(aq) + 6\,NH_3(aq) \rightleftharpoons Ni(NH_3)_6^{2+}(aq)$	5.6×10^8

Values for other equilibria are given in Table 15.6 and the example below shows how to use them.

Salts are more soluble if a complex ion can be formed.

EXAMPLE 15.11 Calculating molar solubility in the presence of complex formation

Calculate the molar solubility of silver chloride in 0.10 M $NH_3(aq)$, given that $K_{sp} = 1.6 \times 10^{-10}$ for silver chloride and $K_f = 1.6 \times 10^7$ for the ammonia complex of Ag^+ ions, $Ag(NH_3)_2^+$.

STRATEGY Write the chemical equation for the equilibrium between the solid solute and the complex in solution and express this equilibrium as the sum of the solubility and complex formation equilibria. The equilibrium constant for the overall equilibrium is thus the product of the equilibrium constants for the two contributing processes. Then set up an equilibrium table, express the equilibrium constant in terms of the unknown concentrations, and solve for the equilibrium concentrations of ions in solution.

SOLUTION The overall equilibrium is

$$AgCl(s) + 2\,NH_3(aq) \rightleftharpoons Ag(NH_3)_2^+(aq) + Cl^-(aq)$$

This equation is the sum of Eqs. 1 and 2. The equilibrium constant for this equilibrium,

$$K = \frac{[Ag(NH_3)_2^+][Cl^-]}{[NH_3]^2}$$

is therefore the product of the equilibrium constants for the two reactions:

$$K = K_{sp} \times K_f$$

Because 1 mol AgCl $\simeq$ 1 mol $Ag(NH_3)_2^+$, the molar solubility of AgCl, which is given by $[Cl^-]$, is equal to the molarity of $Ag(NH_3)_2^+$ in the saturated solution. The equilibrium table, with all concentrations in moles per liter, is

	Species		
	NH_3	$Ag(NH_3)_2^+$	Cl^-
Step 1. Initial molarity	0.10	0	0
Step 2. Change in molarity	$-2x$	$+x$	$+x$
Step 3. Equilibrium molarity	$0.10 - 2x$	x	x

Step 4. We find K and substitute the information from step 3 for the equilibrium concentrations:

$$K = K_{sp} \times K_f = (1.6 \times 10^{-10}) \times (1.6 \times 10^{7}) = 2.6 \times 10^{-3}$$

$$K = \frac{[Ag(NH_3)_2^+][Cl^-]}{[NH_3]^2} = \frac{x \times x}{(0.10 - 2x)^2} = 2.6 \times 10^{-3}$$

Take the square root of each side and obtain:

$$\frac{x}{0.10 - 2x} = 5.1 \times 10^{-2}$$

Rearranging and solving for x gives $x = 4.6 \times 10^{-3}$. Therefore, from step 3, $[Ag(NH_3)_2^+] = 4.6 \times 10^{-3}$ mol·L^{-1}. Hence, the molar solubility of silver chloride is 4.6×10^{-3} mol·L^{-1}. The molar solubility of silver chloride in pure water is 1.3×10^{-5} mol·L^{-1}, different by a factor of over 100 from the molar solubility of silver chloride in an aqueous solution of ammonia.

SELF-TEST 15.17A Calculate the molar solubility of silver bromide in 1.0 M $NH_3(aq)$.

[*Answer:* 3.5×10^{-3} mol·L^{-1}]

SELF-TEST 15.17B Calculate the molar solubility of copper(II) sulfide in 1.2 M $NH_3(aq)$.

SKILLS YOU SHOULD HAVE MASTERED

Conceptual

1. Explain why salts of weak bases produce acidic solutions and salts of weak acids produce basic solutions.
2. Interpret the features of the pH curve for the titration of a strong or weak acid with a strong base, or a strong or weak base with a strong acid.
3. Select an appropriate indicator for a given titration.
4. Explain how buffer solutions resist changes in pH.
5. Select an appropriate buffer for a given pH.
6. Explain what is meant by the buffer capacity of a solution.

Problem-Solving

1. Calculate the pH of a salt solution.
2. Calculate the pH at any point in a strong base-strong acid, strong base-weak acid, or weak base-strong acid titration.
3. Calculate the pH of a buffer solution.
4. Calculate the pH change when acid or base is added to a buffer solution.
5. Determine the relative concentrations of conjugate acid and base needed to prepare a buffer solution with a given pH.
6. Determine a solubility constant from molar solubility and vice versa.

7. Predict whether a salt will precipitate, given the concentrations of its ions in water.

8. Calculate molar solubility in the presence of complex ion formation.

Descriptive

1. Show how small, highly charged hydrated metal cations produce acidic solutions.

2. Describe how indicators change color with pH.

3. Describe complex ions and explain how they can affect the solubility of a sparingly soluble salt.

4. Describe techniques used to dissolve precipitates.

EXERCISES

The values for the acidity and basicity constants are listed in Tables 14.2, 14.5, and 15.1. Unless stated otherwise, take the solutions to be aqueous and at 25°C.

Ions as Acids and Bases

15.1 Determine whether aqueous solutions of the following salts have a pH equal to, greater than, or less than 7. If pH $>$ 7 or pH $<$ 7, write a chemical equation to justify your answer. (a) NH_4Br; (b) Na_2CO_3; (c) KF; (d) KBr; (e) $AlCl_3$; (f) $Cu(NO_3)_2$.

15.2 Determine whether aqueous solutions of the following salts have a pH equal to, greater than, or less than 7. If pH $>$ 7 or pH $<$ 7, write a chemical equation to justify your answer. (a) $K_2C_2O_4$ (potassium oxalate); (b) $Ca(NO_3)_2$; (c) CH_3NH_3Cl; (d) K_3PO_4; (e) $FeCl_3$; (f) C_5H_5NHCl (pyridinium chloride).

15.3 Determine the value of the acidity or basicity constants of the following ions: (a) NH_4^+; (b) CO_3^{2-}; (c) F^-; (d) ClO^-; (e) HCO_3^-; (f) $(CH_3)_3NH^+$ (trimethylammonium ion).

15.4 Determine the value of the acidity or basicity constants of the following ions: (a) $N_2H_5^+$; (b) PO_4^{3-}; (c) NO_2^-; (d) NO_3^-; (e) HPO_4^{2-}; (f) $C_5H_5NH^+$.

15.5 Calculate the pH of the following solutions: (a) 0.20 M $NaCH_3CO_2(aq)$; (b) 0.10 M $NH_4Cl(aq)$; (c) 0.10 M $AlCl_3(aq)$; (d) 0.15 M KCN(aq).

15.6 Calculate the pH of the following solutions: (a) 0.15 M $CH_3NH_3Cl(aq)$; (b) 0.20 M $Na_2SO_3(aq)$; (c) 0.30 M $FeCl_3(aq)$; (d) 0.10 M $KClO_2(aq)$.

15.7 (a) A 10.0-g sample of potassium acetate, KCH_3CO_2, is dissolved in 250 mL of solution. What is the pH of the solution? (b) What is the pH of a solution resulting from the dissolution of 5.75 g of ammonium bromide, NH_4Br, in 100 mL of solution?

15.8 (a) A 1.00-g sample of sodium hydrogen sulfite, $NaHSO_3$, is dissolved in 50.0 mL of solution. What is the pH of the solution? (b) What is the pH of a solution resulting from the dissolution of 100 mg of silver nitrate, $AgNO_3$, in 10.0 mL of solution?

15.9 (a) A 200-mL sample of 0.200 M $NaCH_3CO_2(aq)$ is diluted to 500 mL. What is the concentration of acetic acid at equilibrium? (b) What is the pH of a solution resulting from the dissolution of 5.75 g of ammonium bromide, NH_4Br, in 400 mL of solution?

15.10 (a) A 50.0-mL sample of 0.630 M KCN(aq) is diluted to 125 mL. What is the concentration of hydrocyanic acid present at equilibrium? (b) A 1.00-g sample of sodium hydrogen carbonate, $NaHCO_3$, is dissolved in 150.0 mL of solution. What is the pH of the solution?

Mixed Solutions

15.11 Explain what happens to (a) the concentration of H_3O^+ ions in an acetic acid solution when solid sodium acetate is added; (b) the percentage deprotonation of benzoic acid in a benzoic acid solution when hydrochloric acid is added; (c) the pH of the solution when solid ammonium chloride is added to an ammonia solution.

15.12 Explain what happens to (a) the pH of a phosphoric acid solution after the addition of solid sodium dihydrogen phosphate; (b) the percentage deprotonation of HCN in a hydrocyanic acid solution after the addition of hydrobromic acid; (c) the concentration of H_3O^+ ions when pyridinium chloride is added to a pyridine solution.

15.13 A solution of equal concentrations of lactic acid and sodium lactate was found to have pH $=$ 3.08. (a) What are the values of pK_a and K_a of lactic acid? (b) What would the pH be if the acid had twice the concentration of the salt?

15.14 A solution containing equal concentrations of saccharin and its sodium salt was found to have pH $=$ 11.68. (a) What are the values of pK_a and K_a of saccharin? (b) What would the pH be if the salt had twice the concentration of the acid?

15.15 Calculate the concentration of hydronium ions in

(a) a solution that is 0.20 M $HBrO(aq)$ and 0.10 M $KBrO(aq)$; (b) a solution that is 0.010 M $(CH_3)_2NH(aq)$ and 0.150 M $(CH_3)_2NH_2Cl(aq)$; (c) a solution that is 0.10 M $HBrO(aq)$ and 0.20 M $KBrO(aq)$; (d) a solution that is 0.020 M $(CH_3)_2NH(aq)$ and 0.030 M $(CH_3)_2NH_2Cl(aq)$.

15.16 What is the concentration of hydronium ions in (a) a solution that is 0.050 M $HCN(aq)$ and 0.030 M $NaCN(aq)$; (b) a solution that is 0.10 M $NH_2NH_2(aq)$ and 0.50 M $NaCl(aq)$; (c) a solution that is 0.030 M $HCN(aq)$ and 0.050 M $NaCN(aq)$; (d) a solution that is 0.15 M $NH_2NH_2(aq)$ and 0.15 M $NH_2NH_3Br(aq)$?

15.17 Determine the pH and pOH of (a) a solution that is 0.40 M $NaHSO_4(aq)$ and 0.80 M $Na_2SO_4(aq)$; (b) a solution that is 0.40 M $NaHSO_4(aq)$ and 0.20 M $Na_2SO_4(aq)$; (c) a solution that is 0.40 M $NaHSO_4(aq)$ and 0.40 M $Na_2SO_4(aq)$.

15.18 Calculate the pH and pOH of (a) a solution that is 0.17 M $Na_2HPO_4(aq)$ and 0.25 M $Na_3PO_4(aq)$; (b) a solution that is 0.66 M $Na_2HPO_4(aq)$ and 0.42 M $Na_3PO_4(aq)$; (c) a solution that is 0.12 M $Na_2HPO_4(aq)$ and 0.12 M $Na_3PO_4(aq)$.

15.19 The pH of 0.40 M $HF(aq)$ is 1.93. Calculate the change in pH when 1.0 g of sodium fluoride is added to 25.0 mL of the solution. Ignore any volume change.

15.20 The pH of 0.50 M $HBrO(aq)$ is 4.50. Calculate the change in pH when 10.0 g of sodium hypobromite is added to 250 mL of the solution. Ignore any volume change.

15.21 Calculate the pH of the solution that results from mixing (a) 20.0 mL of 0.050 M $HCN(aq)$ with 80.0 mL of 0.030 M $NaCN(aq)$; (b) 80.0 mL of 0.030 M $HCN(aq)$ with 20.0 mL of 0.050 M $NaCN(aq)$; (c) 25.0 mL of 0.105 M $HCN(aq)$ with 25.0 mL of 0.105 M $NaCN(aq)$.

15.22 Calculate the pH of the solution that results from mixing (a) 100 mL of 0.020 M $(CH_3)_2NH(aq)$ with 300 mL of 0.030 M $(CH_3)_2NH_2Cl(aq)$; (b) 65.0 mL of 0.010 M $(CH_3)_2NH(aq)$ with 10.0 mL of 0.150 M $(CH_3)_2NH_2Cl(aq)$; (c) 50.0 mL of 0.015 M $(CH_3)_2NH(aq)$ with 125 mL of 0.015 M $(CH_3)_2NH_2Cl(aq)$.

Strong Acid–Strong Base Titrations

15.23 Calculate the pH of the following solutions: (a) 25.0 mL of 0.30 M HCl was added to 25.0 mL of 0.20 M NaOH; (b) 25.0 mL of 0.15 M HCl was added to 50.0 mL of 0.15 M KOH; (c) 21.7 mL of 0.27 M HNO_3 was added to 10.0 mL of 0.30 M NaOH.

15.24 Calculate the pH of the following solutions: (a) 17.3 mL of 0.25 M HCl was added to 15.0 mL of 0.33 M NaOH; (b) 21.8 mL of 0.15 M NaOH was added to 50.0 mL of 0.073 M HCl; (c) 15.94 mL of 0.101 M NaOH was added to 25.0 mL of 0.094 M HNO_3.

15.25 A 14.0-g sample of NaOH was dissolved in 250 mL of solution, and then 25.0 mL was pipetted into 50.0 mL of 0.20 M HBr. What is the pH of the resulting solution?

15.26 A 0.150-g sample of $Ba(OH)_2$ was dissolved in 50.0 mL of solution, and then 25.0 mL was pipetted into 100 mL of 0.0010 M HCl. What is the pH of the resulting solution?

15.27 What volume of 0.0631 M HCl is required to neutralize 25.0 mL of an 0.0497 M KOH solution?

15.28 It was found that 24.7 mL of 0.184 M HI solution was required to neutralize 20.0 mL of a $Ba(OH)_2$ solution. What is the molarity of the $Ba(OH)_2$ solution?

15.29 Sketch reasonably accurately the pH curve for the titration of 20.0 mL of 0.10 M HCl with 0.20 M KOH. Mark on the curve (a) the initial pH; (b) the pH at the stoichiometric point.

15.30 Sketch reasonably accurately the pH curve for the titration of 20.0 mL of 0.10 M $Ba(OH)_2$ with 0.20 M HCl. Mark on the curve (a) the initial pH; (b) the pH at the stoichiometric point.

15.31 Calculate the volume of 0.150 M HCl required to neutralize (a) one-half and (b) all the hydroxide ions in 25.0 mL of 0.110 M NaOH. (c) What is the molarity of Na^+ ions at the stoichiometric point? (d) Calculate the pH of the solution after the addition of 20.0 mL of 0.150 M HCl to 25.0 mL of 0.110 M NaOH.

15.32 Calculate the volume of 0.116 M HCl required to neutralize (a) one-half and (b) all the hydroxide ions in 25.0 mL of 0.215 M KOH. (c) What is the molarity of Cl^- ions at the stoichiometric point? (d) Calculate the pH of the solution after the addition of 40.0 mL of 0.116 M HCl to 25.0 mL of 0.215 M KOH.

15.33 (a) 2.54 g of NaOH was dissolved in 25.0 mL of solution and titrated with HCl(aq). What volume (in milliliters) of 0.150 M HCl is required to reach the stoichiometric point? (b) What is the molarity of Cl^- ions at the stoichiometric point?

15.34 (a) 2.88 g of KOH was dissolved in 25.0 mL of solution and titrated with HCl(aq). What volume of 0.200 M HNO_3 is required to reach the stoichiometric point? (b) What is the molarity of NO_3^- ions at the stoichiometric point?

15.35 A 0.968-g sample of impure sodium hydroxide was dissolved in 200 mL of aqueous solution. A 20.0-mL portion of this solution was titrated to the stoichiometric point with 15.8 mL of 0.107 M HCl. What is the percentage purity of the original sample?

15.36 A 1.331-g sample of impure barium hydroxide was dissolved in 250 mL of aqueous solution. A 35.0-mL portion of this solution was titrated to the stoichiometric point with 17.6 mL of 0.0935 M HCl. What is the percentage purity of the original sample?

15.37 Calculate the pH at each stage in the titration for the addition of 0.150 M HCl to 25.0 mL of 0.110 M NaOH (a) initially; (b) after the addition of 5.0 mL of acid; (c) after the addition of a further 5.0 mL; (d) at the stoichiometric point; (e) after the addition of 5.0 mL of acid beyond the stoichiometric point; (f) after the addition of 10 mL of acid beyond the stoichiometric point.

15.38 Calculate the pH at each stage in the titration in which 0.116 M HCl is added to 25.0 mL of 0.215 M KOH (a) initially; (b) after the addition of 5.0 mL of acid; (c) after the addition of a further 5.0 mL; (d) at the stoichiometric point; (e) after the addition of 5.0 mL of acid beyond the stoichiometric point; (f) after the addition of 10 mL of acid beyond the stoichiometric point.

Weak Acid-Strong Base and Weak Base-Strong Acid Titrations

15.39 A 25.0-mL sample of 0.10 M $CH_3COOH(aq)$ is titrated with 0.10 M NaOH(aq). K_a for CH_3COOH is 1.8×10^{-5}. (a) What is the initial pH of the 0.10 M $CH_3COOH(aq)$ solution? (b) What is the pH after the addition of 10.0 mL of 0.10 M NaOH(aq)? (c) What volume of 0.10 M NaOH(aq) is required to reach halfway to the stoichiometric point? (d) Calculate the pH at that halfway point. (e) What volume of 0.10 M NaOH(aq) is required to reach the stoichiometric point? (f) Calculate the pH at the stoichiometric point. (g) Suggest a suitable indicator.

15.40 A 30.0-mL sample of 0.20 M $C_6H_5COOH(aq)$ solution is titrated with 0.30 M KOH(aq). K_a for C_6H_5COOH is 6.5×10^{-5}. (a) What is the initial pH of the 0.20 M $C_6H_5COOH(aq)$ solution? (b) What is the pH after the addition of 15.0 mL of 0.30 M KOH(aq)? (c) What volume of 0.20 M NaOH(aq) is required to reach halfway to the stoichiometric point? (d) Calculate the pH at the halfway point. (e) What volume of 0.20 M KOH(aq) is required to reach the stoichiometric point? (f) Calculate the pH at the stoichiometric point. (g) Suggest a suitable indicator.

15.41 A 15.0-mL sample of 0.15 M $NH_3(aq)$ solution is titrated with 0.10 M HCl(aq). K_b for NH_3 is 1.8×10^{-5}. (a) What is the initial pH of the 0.15 M $NH_3(aq)$ solution? (b) What is the pH after the addition of 15.0 mL of 0.10 M HCl(aq)? (c) What volume of 0.10 M HCl(aq) is required to reach halfway to the stoichiometric point? (d) Calculate the pH at the halfway point. (e) What volume of 0.10 M HCl(aq) is required to reach the stoichiometric point? (f) Calculate the pH at the stoichiometric point. (g) Suggest a suitable indicator.

15.42 A 50.0-mL sample of 0.25 M $CH_3NH_2(aq)$ solution is titrated with 0.35 M HCl(aq). K_b for CH_3NH_2 is 3.6×10^{-4}. (a) What is the initial pH of the 0.25 M $CH_3NH_2(aq)$ solution? (b) What is the pH after the addition of 15.0 mL of 0.35 M HCl(aq)? (c) What volume of 0.35 M HCl(aq) is required to reach halfway to the stoichiometric point? (d) Calculate the pH at the halfway point. (e) What volume of 0.35 M HCl(aq) is required to reach the stoichiometric point? (f) Calculate the pH at the stoichiometric point. (g) Suggest a suitable indicator.

15.43 Calculate the pH of 25.0 mL of 0.110 M aqueous lactic acid being titrated with 0.150 M NaOH(aq) (a) initially; (b) after the addition of 5.0 mL of base; (c) after the addition of a further 5.0 mL of base; (d) at the stoichiometric point; (e) after the addition of 5.0 mL of base beyond the stoichiometric point; (f) after the addition of 10 mL of base beyond the stoichiometric point. (g) Suggest a suitable indicator.

15.44 Calculate the pH of 25.0 mL of 0.215 M chloroacetic acid being titrated with 0.116 M NaOH(aq) (a) initially; (b) after the addition of 5.0 mL of acid; (c) after the addition of a further 5.0 mL of acid; (d) at the stoichiometric point; (e) after the addition of 5.0 mL of acid beyond the stoichiometric point; (f) after the addition of 10 mL of acid beyond the stoichiometric point. (g) Suggest a suitable indicator.

Indicators

For the pH ranges over which common indicators change color, see Table 15.3.

15.45 Over what pH range can each of the following indicators be used for detecting the stoichiometric point in a titration: (a) methyl orange; (b) litmus; (c) methyl red; (d) phenolphthalein?

15.46 Over what pH range can each of the following indicators be used for detecting the stoichiometric point in a titration: (a) thymol blue; (b) phenol red; (c) bromophenol blue; (d) alizarin.

15.47 Which indicators could you use for a titration of 0.20 M acetic acid with 0.20 M NaOH: (a) methyl orange; (b) litmus; (c) thymol blue; (d) phenolphthalein? Explain your selections.

15.48 Which indicators could you use for a titration of 0.20 M ammonia with 0.20 M HCl: (a) bromocresol green; (b) methyl red; (c) phenol red; (d) thymol blue? Explain your selections.

Buffers

15.49 Identify which of the following mixed systems can function as a buffer solution and write an equilibrium equation for each buffer system: (a) equal volumes of 0.10 M HCl(aq) and 0.10 M NaCl(aq); (b) a solution that is 0.10 M HClO(aq) and 0.10 M NaClO(aq); (c) a solution that is 0.10 M $(CH_3)_3N$(aq) and 0.10 M $(CH_3)_3NHCl$(aq); (d) equal volumes of 0.20 M CH_3COOH(aq) and 0.10 M NaOH(aq); (e) equal volumes of 0.20 M HNO_3(aq) and 0.20 M NaOH(aq).

15.50 Identify which of the following mixed systems can function as a buffer solution. Write an equilibrium equation for each buffer system: (a) equal volumes 0.10 M C_6H_5COOH(aq) and 0.10 M $NaC_6H_5CO_2$(aq); (b) a solution that is 0.10 M HNO_3(aq) and 0.10 M $NaNO_3$(aq); (c) a solution that is 0.10 M C_5H_5N(aq) and 0.10 M C_5H_5NHCl(aq); (d) equal volumes of 0.10 M NH_3(aq) and 0.10 M HCl(aq); (e) a solution that is 0.10 M HNO_2(aq) and 0.10 M $NaNO_2$(aq).

15.51 Sodium hypochlorite, NaClO, is the active ingredient in many bleaches. Calculate the ratio of the concentrations of ClO^- and HClO in a bleach solution having a pH adjusted to 6.50 by using strong acid or strong base.

15.52 Aspirin (shown below) is a derivative of salicylic acid, which has a $K_a = 1.1 \times 10^{-3}$. Calculate the ratio of the concentrations of the salicylate ion (its conjugate base) to salicylic acid in a solution that has a pH adjusted to 2.50 by using strong acid or strong base.

COOH O—C(=O)CH$_3$

15.53 Predict the pH region in which each of the following buffers will be effective, assuming equal molarities of the acid and its conjugate base: (a) sodium lactate and lactic acid; (b) sodium benzoate and benzoic acid; (c) potassium hydrogen phosphate and potassium phosphate; (d) potassium hydrogen phosphate and potassium dihydrogen phosphate; (e) hydroxylamine and hydroxylammonium chloride.

15.54 Predict the pH region in which each of the following buffers will be effective, assuming equal molarities of the acid and its conjugate base: (a) sodium nitrite and nitrous acid; (b) sodium formate and formic acid; (c) sodium carbonate and sodium hydrogen carbonate; (d) ammonia and ammonium chloride; (e) pyridine and pyridinium chloride.

15.55 Use Tables 14.2 and 14.5 to suggest a conjugate acid-base system that would be an effective buffer at a pH close to (a) 2; (b) 7; (c) 3; (d) 12.

15.56 Use Tables 14.2 and 14.5 to suggest a conjugate acid-base system that would be an effective buffer at a pH close to (a) 4; (b) 9; (c) 5; (b) 11.

15.57 (a) What must be the ratio of the concentrations of CO_3^{2-} and HCO_3^- ions in a buffer solution having a pH of 11.0? (b) What mass of K_2CO_3 must be added to 1.00 L of 0.100 M $KHCO_3$(aq) to prepare a buffer solution with a pH of 11.0? (c) What mass of $KHCO_3$ must be added to 1.00 L of 0.100 M K_2CO_3(aq) to prepare a buffer solution with a pH of 11.0? (d) What volume of 0.200 M K_2CO_3(aq) must be added to 100 mL of 0.100 M $KHCO_3$(aq) to prepare a buffer solution with a pH of 11.0?

15.58 (a) What must be the ratio of the molarities of PO_4^{3-} and HPO_4^{2-} ions in a buffer solution having a pH of 12.0? (b) What mass of K_3PO_4 must be added to 1.00 L of 0.100 M K_2HPO_4(aq) to prepare a buffer solution with a pH of 12.0? (c) What mass of K_2HPO_4 must be added to 1.00 L of 0.100 M K_3PO_4(aq) to prepare a buffer solution with a pH of 12.0? (d) What volume of 0.150 M K_3PO_4(aq) must be added to 50.0 mL of 0.100 M K_2HPO_4(aq) to prepare a buffer solution with a pH of 12.0?

15.59 A 100-mL buffer solution is 0.10 M CH_3COOH(aq) and 0.10 M $NaCH_3CO_2$(aq). (a) What is the pH of the buffer solution? (b) What are the pH and the pH change resulting from the addition of 3.0 mmol NaOH to the buffer solution? (c) What are the pH and the pH change resulting from the addition of 6.0 mmol of HNO_3 to the initial buffer solution?

15.60 A 100-mL buffer solution is 0.15 M Na_2HPO_4(aq) and 0.10 M KH_2PO_4(aq). From Table 14.5, the K_{a2} for phosphoric acid is 6.2×10^{-8}. (a) What is the pH of the buffer solution? (b) What are the pH and the pH change resulting from the addition of 8.0 mmol NaOH to the buffer solution? (c) What are the pH and the pH change resulting from the addition of 10.0 mmol of HNO_3 to the initial buffer solution?

15.61 A 100-mL buffer solution is 0.10 M CH_3COOH(aq) and 0.10 M $NaCH_3CO_2$(aq). (a) What are the pH and the pH change resulting from the addition of 10.0 mL of 0.950 M NaOH(aq) to the buffer solution? (b) What are the pH and the pH change resulting from the addition of 20.0 mL of 0.10 M HNO_3(aq) to the initial buffer solution? (*Hint:* Consider the dilution stemming from the addition of strong base or acid.)

15.62 A 100-mL buffer solution is 0.15 M Na_2HPO_4(aq) and 0.10 M KH_2PO_4(aq). (a) What are the pH and the pH change resulting from the addition of 80.0 mL of 0.010 M NaOH(aq) to the buffer solution? (b) What are the pH and the pH change resulting from the addition of 10.0 mL of 1.0 M HNO_3(aq) to the initial buffer solution? (*Hint:* Consider the dilution stemming from the addition of strong base or acid.)

Solubility Products

The values for the solubility products of some sparingly soluble salts are listed in Table 15.5.

15.63 Write the expression for the solubility product of the following substances: (a) AgBr; (b) Ag_2S; (c) $Ca(OH)_2$; (d) Ag_2CrO_4.

15.64 Write the expressions for the solubility products of the following substances: (a) AgI; (b) AgSCN; (c) Sb_2S_3; (d) $Mg_3(PO_4)_2$.

15.65 Determine the K_{sp} for the following sparingly soluble substances, given their molar solubilities: (a) AgBr, 8.8×10^{-7} mol·L^{-1}; (b) $PbCrO_4$, 1.3×10^{-7} mol·L^{-1}; (c) $Ba(OH)_2$, 0.11 mol·L^{-1}; (d) MgF_2, 1.2×10^{-3} mol·L^{-1}.

15.66 Determine the K_{sp} for the following sparingly soluble compounds, given their molar solubilities: (a) AgI, 9.1×10^{-9} mol·L^{-1}; (b) $Ca(OH)_2$, 0.011 mol·L^{-1}; (c) Ag_3PO_4, 2.7×10^{-6} mol·L^{-1}; (d) Hg_2Cl_2, 5.2×10^{-7} mol·L^{-1}.

15.67 Use the data in Table 15.5 to calculate the molar solubility of (a) Ag_2S; (b) CuS; (c) $CaCO_3$.

15.68 Use the data in Table 15.5 to determine the molar solubility of (a) $PbSO_4$; (b) Ag_2CO_3; (c) $Fe(OH)_2$.

15.69 The molarity of CrO_4^{2-} in a saturated Tl_2CrO_4 solution is 6.3×10^{-5} mol·L^{-1}. What is the K_{sp} of Tl_2CrO_4?

15.70 The molar solubility of cerium(III) hydroxide, $Ce(OH)_3$, is 5.2×10^{-6} mol·L^{-1}. What is the K_{sp} of cerium(III) hydroxide?

Common-Ion Effect

15.71 Use the data in Table 15.5 to calculate the molar solubility of each sparingly soluble substance in its respective solution: (a) silver chloride in a 0.20 mol·L^{-1} NaCl solution; (b) mercury(I) chloride in a 0.10 M NaCl(aq) solution; (c) lead(II) chloride in a 0.10 M $CaCl_2$(aq) solution; (d) iron(II) hydroxide in a 1.0×10^{-4} M $FeCl_2$(aq) solution.

15.72 Use the data in Table 15.5 to calculate the solubility of each sparingly soluble substance in its respective solution: (a) silver bromide in a 1.0×10^{-3} M NaBr(aq) solution; (b) magnesium carbonate in a 4.2×10^{-5} M Na_2CO_3(aq) solution; (c) lead(II) sulfate in a 0.10 M Na_2SO_4(aq) solution; (d) nickel hydroxide in a 3.7×10^{-5} M $NiSO_4$(aq) solution.

15.73 (a) What molarity of Ag^+ ions is required for the formation of a precipitate when added to a 1.0×10^{-5} M NaCl(aq) solution? (b) What mass (in micrograms) of $AgNO_3$ needs to be added for the onset of precipitation in 100 mL of the solution in (a)?

15.74 It is necessary to add iodide ions to precipitate lead(II) ion from a 0.0020 M $Pb(NO_3)_2$(aq) solution. (a) What (minimum) iodide ion concentration is required for the onset of PbI_2 precipitation? (b) What mass (in grams) of KI must be added for PbI_2 to form?

15.75 Determine the pH required for the onset of precipitation of $Ni(OH)_2$ from a 0.010 M $NiSO_4$(aq) solution.

15.76 At what minimum molarity of Fe^{3+} ions will the ion precipitate as $Fe(OH)_3$ from a solution with pH = 12.0?

15.77 The concentration of Mg^{2+} ions in seawater is about 1.3 μg·L^{-1}. In a commercial (Dow Chemical) recovery process, the magnesium is precipitated as the hydroxide. At what pH does magnesium hydroxide precipitate?

15.78 Limestone is composed primarily of calcium carbonate. A 1.0-mm^3 chip of limestone was accidentally dropped into a water-filled swimming pool, measuring 10 m × 7 m × 2 m. Assuming that the carbonate ion does not function as a Brønsted base and that the pH of the water is 7, will the pebble dissolve entirely? The density of calcium carbonate is 2.71 g·cm^{-3}.

Predicting Precipitation Reactions

15.79 Decide whether a precipitate will form when the following solutions are mixed: (a) 27.0 mL of 0.0010 M NaCl(aq) and 73.0 mL of 0.0040 M $AgNO_3$(aq); (b) 1.0 mL of 1.0 M K_2SO_4(aq), 10.0 mL of 0.0030 M $CaCl_2$(aq), and 100 mL of water.

15.80 Decide whether a precipitate will form when the following solutions are mixed: (a) 5.0 mL of 0.10 M K_2CO_3(aq) and 1.00 L of 0.010 M $AgNO_3$(aq); (b) 3.3 mL of 1.0 M HCl(aq), 4.9 mL of 0.0030 M $AgNO_3$(aq), and enough water to dilute the solution to 50.0 mL.

15.81 Suppose that there are typically 20 average-sized drops in 1 mL of an aqueous solution. Will a precipitate form when 1 drop of 0.010 M NaCl(aq) is added to 10.0 mL of (a) 0.0040 M $AgNO_3$(aq) solution? (b) 0.0040 M $Pb(NO_3)_2$(aq) solution?

15.82 Assuming 20 drops per milliliter, will a precipitate form if (a) 7 drops of 0.0029 M K_2CO_3(aq) are added to

25.0 mL of a 0.0018 M $CaCl_2(aq)$ solution; (b) 10 drops of 0.010 M Na_2CO_3 are added to 10.0 mL of 0.0040 M $AgNO_3(aq)$ solution?

15.83 The concentrations of magnesium, calcium, and nickel(II) ions in an aqueous solution are 0.0010 $mol \cdot L^{-1}$. (a) In what order do they precipitate when a KOH solution is added? (b) Determine the pH at which each salt precipitates.

15.84 Suppose that two hydroxides MOH and $M'(OH)_2$ both have $K_{sp} = 1.0 \times 10^{-12}$ and that initially both cations are present in a solution at concentrations of 0.0010 $mol \cdot L^{-1}$. Which hydroxide precipitates first, and at what pH, when a NaOH solution is added?

Dissolving Precipitates

15.85 Calculate the solubility of silver bromide in a 0.10 M KCN(aq) solution. Refer to Tables 15.5 and 15.6.

15.86 Precipitated silver chloride dissolves in ammonia solutions as a result of the formation of the $Ag(NH_3)_2^+$ ion. What is the solubility of silver chloride in 1.0 M $NH_3(aq)$?

15.87 Use the data in Table 15.5 to calculate the solubility of each sparingly soluble substance in its respective solution: aluminum hydroxide at (a) pH = 7.0; (b) pH = 4.5; zinc hydroxide at (c) pH = 7.0; (d) pH = 6.0.

15.88 Use the data in Table 15.5 to calculate the solubility of each sparingly soluble compound in its respective solution: iron(III) hydroxide at (a) pH = 11.0; (b) pH = 3.0; iron(II) hydroxide at (c) pH = 8.0; (d) pH = 6.0.

15.89 Consider the two equilibria

$$CaF_2(s) \rightleftharpoons Ca^{2+}(aq) + 2\,F^-(aq) \qquad K_{sp} = 4.0 \times 10^{-11}$$

$$F^-(aq) + H_2O(l) \rightleftharpoons HF(aq) + OH^-(aq) \qquad K_b(F^-) = 2.9 \times 10^{-11}$$

(a) Write the chemical equation for the overall equilibrium and determine the corresponding equilibrium constant. (b) Determine the solubility of CaF_2 at pH = 7.0. (c) Determine the solubility of CaF_2 at pH = 5.0.

15.90 Consider the two equilibria

$$BaF_2(s) \rightleftharpoons Ba^{2+}(aq) + 2\,F^-(aq) \qquad K_{sp} = 1.7 \times 10^{-6}$$

$$F^-(aq) + H_2O(l) \rightleftharpoons HF(aq) + OH^-(aq) \qquad K_b(F^-) = 2.9 \times 10^{-11}$$

(a) Write the chemical equation for the overall equilibrium and determine the corresponding equilibrium constant. (b) Determine the solubility of BaF_2 at pH = 7.0. (c) Determine the solubility of BaF_2 at pH = 4.0.

SUPPLEMENTARY EXERCISES

15.91 Predict whether aqueous solutions of the following salts will be acidic, basic, or neutral (and justify your prediction): (a) KI; (b) CsF; (c) CrI_3; (d) $C_6H_5NH_3Cl$; (e) Na_2CO_3; (f) $Cu(NO_3)_2$.

15.92 Write an equilibrium that shows that (a) a $CrCl_3$ solution is acidic; (b) a $(CH_3)_3NCl$ (trimethylammonium chloride) solution is acidic; (c) a $NaC_2H_5CO_2$ (sodium propionate) solution is basic; (d) a Na_3PO_4 solution is basic.

15.93 When 10.0 mg of sodium barbituate are dissolved in 250 mL of solution, the resulting pH is 7.71. The molar mass of sodium barbituate is 150 $g \cdot mol^{-1}$. Determine (a) the percentage protonation of barbituate ions; (b) the K_a of barbituric acid.

15.94 Determine (a) the pH of a 0.0240 M hydroxylammonium chloride (more commonly called hydroxylamine hydrochloride) solution; (b) the pH of 0.010 M $Na_2CO_3(aq)$.

15.95 A solution is prepared by mixing 200 mL of 0.27 M $Na_3PO_4(aq)$ and 150 mL of 0.62 M KCl(aq). What is the pH of the mixed solution?

15.96 (a) What is the pH of 0.037 M $NaCH_3CO_2(aq)$? (b) 0.020 M $CH_3COOH(aq)$? (c) 200 mL of 0.020 M $CH_3COOH(aq)$ is added to 150 mL of the solution in part (a). What is the pH of the mixed solution?

15.97 A 60.0-mL sample of 0.10 M $NaHCO_2(aq)$ is mixed with 4.0 mL of 0.070 M HCl(aq). Calculate the pH and the molarity of HCOOH in the mixed solution.

15.98 Describe the principal features that distinguish the pH curve of a strong acid-strong base titration from that of a weak acid-strong base titration.

15.99 Distinguish between the end point and the stoichiometric point of an acid-base titration.

15.100 The pH at the stoichiometric point in the titration of a weak base with a strong acid occurs near pH = 4.0. Which indicators in Table 15.3 would have a suitable end point?

15.101 In the determination of the heat evolved in the neutralization of a strong acid by a strong base, 21.0 mL of a 3.0 M HNO_3 solution was mixed with 25.2 mL of a 2.50 M NaOH solution. What is the pH of the resulting solution?

15.102 A 25.0-mL sample of 6.0 M HCl is diluted to 1.0 L, and 15.7 mL of this diluted solution was required to reach the stoichiometric point in the titration of 25.0 mL of a KOH solution. What is the molarity of the KOH solution?

15.103 A 20-mL sample of 0.020 M HCl solution was titrated with 0.035 M KOH. Calculate the pH at the following points in the titration and sketch the pH curve: (a) no KOH added; (b) 5.00 mL KOH added; (c) an additional 5.00 mL KOH (for a total of 10.0 mL) added; (d) another 5.0 mL KOH added; (e) another 5.00 mL KOH added. (f) Determine the volume of KOH required to reach the stoichiometric point.

15.104 An old bottle labeled "Standardized 6.0 M NaOH" was found on the back of a shelf in the stockroom. Over time, some of the NaOH had reacted with the glass and the solution was no longer 6.0 M. To determine its purity, 5.0 mL of the solution was diluted to 100 mL and titrated to the stoichiometric point with 11.8 mL of 2.05 M HCl. What is the molarity of the sodium hydroxide solution?

15.105 (a) What volume of 0.0400 M NaOH(aq) is required to reach the stoichiometric point in the titration of 10.00 mL of 0.0633 M HBrO(aq)? (b) What is the pH at the stoichiometric point? (c) Use Table 15.3 to suggest a suitable indicator.

15.106 The narcotic cocaine is a weak base with $pK_b = 5.59$. Calculate the ratio of the concentration of cocaine and its conjugate acid in a solution of pH = 8.00.

15.107 Novocaine, which is used by dentists as a local anesthetic, is a weak base with $pK_b = 5.05$. Blood has a pH of 7.4. What is the ratio of concentrations of novocaine to its conjugate acid in the bloodstream?

15.108 A buffer solution is prepared by mixing 50.0 mL of 0.022 M C_6H_5COOH(aq) and 20.0 mL of 0.032 M $NaC_6H_5CO_2$(aq). (a) What is the pH of the buffer solution? (b) What are the pH and the change in pH after the addition of 0.054 mmol HCl to the buffer solution? (c) What would be the pH change if the 0.054 mmol HCl had been added to pure water instead of the buffer solution? (d) What are the pH and the change in pH after the addition of 10.0 mL of 0.054 M HCl to the original buffer solution?

15.109 For 100 mL of a buffer solution that is 0.150 M CH_3COOH(aq) and 0.50 M $NaCH_3CO_2$(aq), what is the pH before and after adding (a) 10.0 mL of 1.2 M HCl(aq)? (b) 50.0 mL of 0.094 M NaOH(aq)?

15.110 To simulate blood conditions, a phosphate buffer system with a pH = 7.40 is desired. (a) What must be the ratio of the concentrations of HPO_4^{2-} to $H_2PO_4^-$ ions? (b) What mass of Na_2HPO_4 must be added to 500 mL of 0.10 M NaH_2PO_4(aq) in the preparation of the buffer?

15.111 Describe, with accompanying calculations, the procedure for preparing a buffer solution for pH = 10.0, starting with solid Na_2CO_3 and solid $NaHCO_3$.

15.112 What is the ideal pH range for a buffer solution that uses HBrO and NaBrO as the acid-base pair?

15.113 (a) What is the molar solubility of Ag_2S, the black tarnish on silverware? (b) What is its molar solubility in 2.0×10^{-4} M $AgNO_3$(aq)? (c) What mass of Ag_2S will dissolve in 10.0 L of the solution in (b)?

15.114 What is the molar solubility of $Al(OH)_3$ in a solution with pH = 6.0?

15.115 Fluoridation of city water supplies produces a fluoride ion concentration close to 5×10^{-5} mol·L^{-1}. Will CaF_2 precipitate in hard water in which the Ca^{2+} ion concentration is 2×10^{-4} mol·L^{-1}?

15.116 The fluoride ions in drinking water convert the hydroxyapatite, $Ca_5(PO_4)_3OH$, of teeth into fluorapatite, $Ca_5(PO_4)_3F$. The K_{sp} of the two compounds are 1.0×10^{-36} and 1.0×10^{-60}, respectively. What are the molar solubilities of each substance? The solubility equilibria to consider are

$$Ca_5(PO_4)_3OH(s) \rightleftharpoons 5\,Ca^{2+}(aq) + 3\,PO_4^{3-}(aq) + OH^-(aq)$$

$$Ca_5(PO_4)_3F(s) \rightleftharpoons 5\,Ca^{2+}(aq) + 3\,PO_4^{3-}(aq) + F^-(aq)$$

15.117 Milk of magnesia, taken internally for acid indigestion, is a saturated solution of magnesium hydroxide. What is the pH of milk of magnesia?

15.118 Limewater is a saturated aqueous calcium hydroxide solution. (a) What is the pH of limewater? (b) What volume of 0.010 M HCl(aq) is required to titrate 25.0 mL of limewater to the phenolphthalein end point?

15.119 Will Ag_2CO_3 precipitate from a solution formed from a mixture of 100 mL of 1.0×10^{-4} M $AgNO_3$(aq) and 100 mL of 1.0×10^{-4} M Na_2CO_3(aq)?

15.120 To 500 mL of an aqueous solution adjusted to a pH of 8.00, a student adds 1.36 mg of $ZnCl_2$. Does a precipitate of $Zn(OH)_2$ form?

CHALLENGING EXERCISES

15.121 A 25.0-mL sample of tartaric acid solution is titrated with 0.100 M KOH(aq), and a pH curve was constructed from the data. A sharp rise in the pH occurred after the addition of 17.0 mL of titrant, and a second sharp rise occurred after the addition of 34.0 mL. (a) Explain why there were two rapid increases in pH. (b) What is the molarity of the tartaric acid solution? What is the pH of the solution after the addition of (c) 17.0 mL; (d) 34.0 mL of 0.100 M NaOH(aq)? (e) What is the tartrate ion concentration after the addition of 34.0 mL of titrant? (f) What is the pH of the solution after the addition of 8.5 mL of titrant?

15.122 A 25.0-mL sample of 0.20 M $(COOH)_2(aq)$, oxalic acid, is titrated with 0.20 M NaOH(aq). For oxalic acid, $K_{a1} = 5.9 \times 10^{-2}$ and $K_{a2} = 6.5 \times 10^{-5}$ (see Table 14.5). (a) What volume of 0.20 M NaOH(aq) is required to reach the first stoichiometric point? (b) What is the salt present at that point? (c) Calculate the pH at the first stoichiometric point. (d) What (total) volume of 0.20 M NaOH(aq) is required to reach the second stoichiometric point? What is the salt present at that point? (e) Calculate the pH at the second stoichiometric point. (f) Suggest a suitable indicator to detect the first stoichiometric point and a second indicator for the second stoichiometric point.

15.123 What is the pH at each stoichiometric point in the titration of 0.20 M $H_2SO_4(aq)$ with 0.20 M NaOH(aq)?

15.124 What volume (in liters) of a saturated mercury(II) sulfide, HgS, solution contains an average of one mercury(II) ion, Hg^{2+}?

15.125 A chemist decides to prepare a buffer solution from acetic acid, CH_3COOH, and hydrazine, N_2H_4. The equilibrium representing this buffer could be

$$CH_3COOH(aq) + N_2H_4(aq) \rightleftharpoons N_2H_5^+(aq) + CH_3CO_2^-(aq)$$

(a) Write the expression for the equilibrium constant and use the information in Table 14.2 to determine its value. (b) Derive a Henderson-Hasselbalch equation for this buffer system.

15.126 It is often useful to know whether two ions can be separated by selective precipitation from a solution. Generally, a 99% separation is considered "separated." A solution is 0.010 M Pb^{2+} and 0.010 M Ag^+. Chloride ions are added from a sodium chloride solution. (a) Determine the chloride ion concentration required for the precipitation of each cation. (b) Which cation precipitates first? (c) What is the molarity of the first cation that precipitates when the second cation begins to precipitate? (d) Determine the percentage of the first cation that remains in solution when the second cation begins to precipitate.

15.127 Consider the two equilibria

$$ZnS(s) \rightleftharpoons Zn^{2+}(aq) + S^{2-}(aq) \qquad K_{sp} = 1.6 \times 10^{-24}$$

$$S^{2-}(aq) + 2\,H_2O(l) \rightleftharpoons H_2S(aq) + 2\,OH^-(aq) \qquad K_{a1}K_{a2}(H_2S) = 9.3 \times 10^{-22}$$

(a) Write the chemical equation for the overall equilibrium and determine the corresponding equilibrium constant.
(b) Determine the solubility of ZnS in a saturated H_2S (0.1 M $H_2S(aq)$) solution adjusted to pH = 7.0.
(c) Determine the solubility of ZnS in a saturated H_2S (0.1 M $H_2S(aq)$) solution adjusted to pH = 10.0.

CHAPTER 16

A raging forest fire is a particularly fearsome example of a natural chemical process. But what does it mean for a process to be "natural"? Why does a combustion reaction like the one shown here take place at all? There must be a driving force for chemical reactions, a reason why reactions take place in one direction rather than another and why they tend to approach equilibrium. We explore these questions in this chapter, which brings together and explains many of the concepts we have met in earlier chapters.

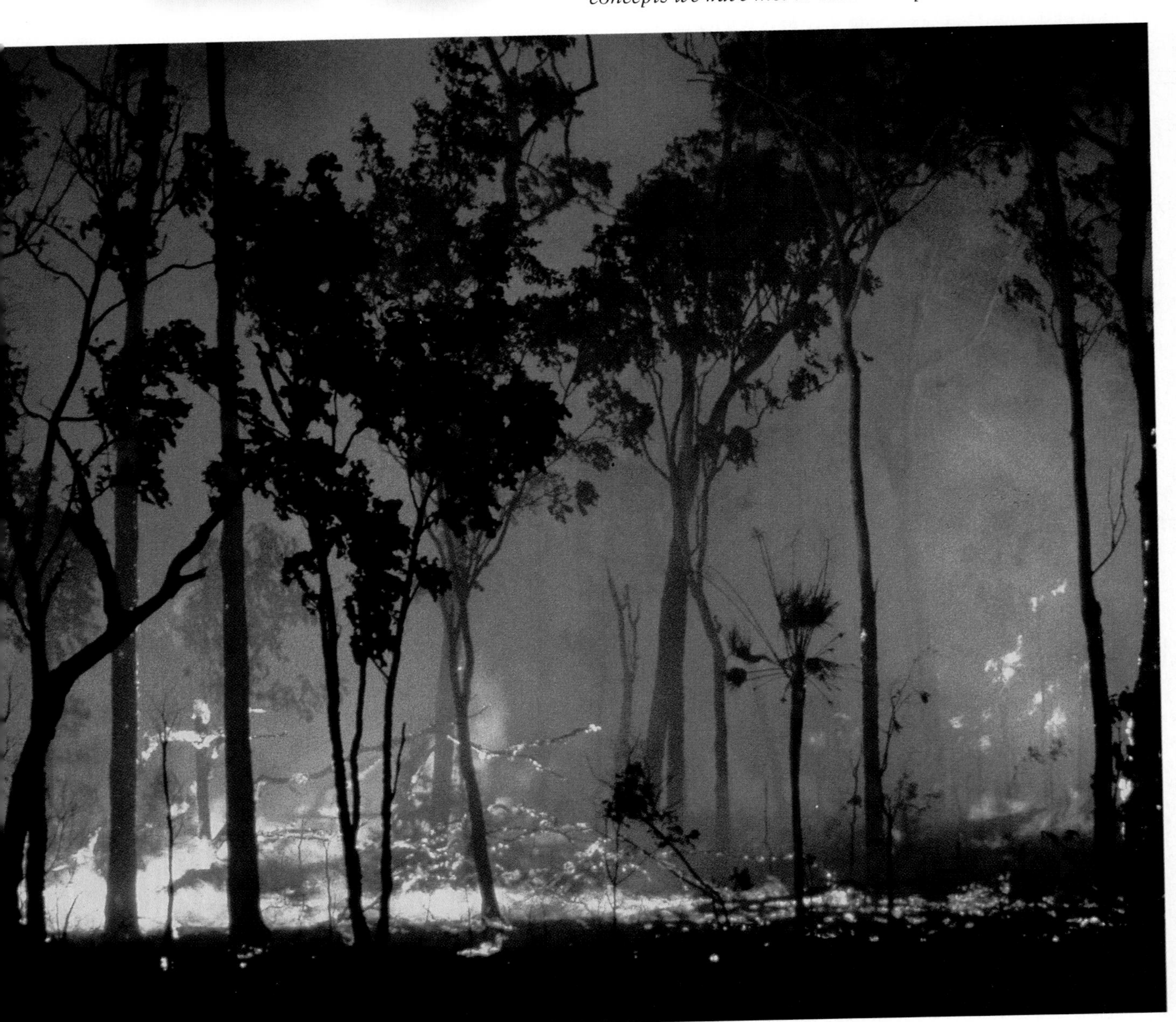

ENERGY IN TRANSITION: THERMODYNAMICS

Have you ever wondered why *anything* happens? Why water evaporates, why hot objects cool, why hydrogen combines with oxygen, why leaves turn red in the fall, why we stay alive? A part of the answer must be related to energy. We need energy to think, to move, and to live. Landscapes emerge and change as a result of the energy associated with volcanos, hurricanes, and earthquakes. Agriculture depends on energy, not only to run machinery but also because plants and animals use energy to grow. Every chemical reaction in a test tube and every metabolic process makes use of energy to rearrange the bonds between atoms. We have seen how to measure changes in energy, but we have not yet tackled the major problem of why certain changes occur. Why does life go on? In what sense do we live off the energy of the Sun? How is energy involved in making things happen?

The branch of chemistry that brings us close to an answer is called **thermodynamics**, the science of the transformations of energy. In Chapter 6 we met one branch of thermodynamics—thermochemistry, the study of the heat released and absorbed by reactions. Equilibrium, the subject of the past three chapters, is also a branch of thermodynamics. So is electrochemistry, the subject of the next chapter. Thermodynamics helps us to understand all processes, whether in a laboratory or in our everyday world: it provides us with a basis for understanding the mechanism of life itself.

THE FIRST LAW OF THERMODYNAMICS

16.1 Systems and surroundings
16.2 Heat and work
16.3 Internal energy
16.4 Heat transfers at constant volume
16.5 Enthalpy

THE DIRECTION OF SPONTANEOUS CHANGE

16.6 Spontaneous change
16.7 Entropy and disorder
16.8 Standard entropies
16.9 The surroundings
16.10 The overall change in entropy

FREE ENERGY

16.11 Focusing on the system
16.12 Standard reaction free energies
16.13 Using free energies of formation
16.14 Free energy and composition
16.15 Free energy and equilibrium
16.16 The effect of temperature

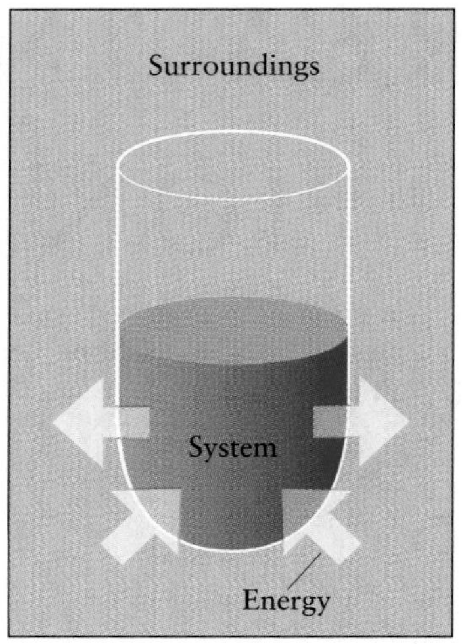

FIGURE 16.1 In thermodynamics, the world is divided into a system (the object of interest) and the surroundings (everything else). In practice, the surroundings may be a constant-temperature water bath.

THE FIRST LAW OF THERMODYNAMICS

The principles of thermodynamics are at work in any event that involves the transfer of energy. We can use these principles to study and understand simple, everyday events like the freezing of water in an ice tray, or more complicated events like using a battery to operate a radio or finding the minimum temperature at which iron can be reduced from its ore by carbon in a blast furnace.

16.1 SYSTEMS AND SURROUNDINGS

We met the terms *system* and *surroundings* in Section 6.1. A **system** is the part of the world we want to study; in chemistry, the system is usually a reaction mixture. The **surroundings** consist of everything else outside the system. In principle, the surroundings are the laboratory, the building, the country, and the whole planet. However, it is not practical to measure the changes in such a vast region, so in practice the surroundings may consist only of a water bath used to keep the reaction flask at constant temperature (Fig. 16.1).

A system can be open, closed, or isolated. An **open system** can exchange both matter and energy with the surroundings. Examples are an automobile engine and the human body. A **closed system** has a fixed amount of matter, but it can exchange energy with the surroundings. Examples are electric batteries and the cold packs used for athletic injuries. An **isolated system** has no contact with its surroundings. We can think of an isolated system as sealed inside rigid, thermally insulated walls (Fig. 16.2).

In thermodynamics, the world is divided into a system and its surroundings. An open system can exchange both matter and energy with the surroundings, a closed system can exchange only energy, and an isolated system can exchange nothing.

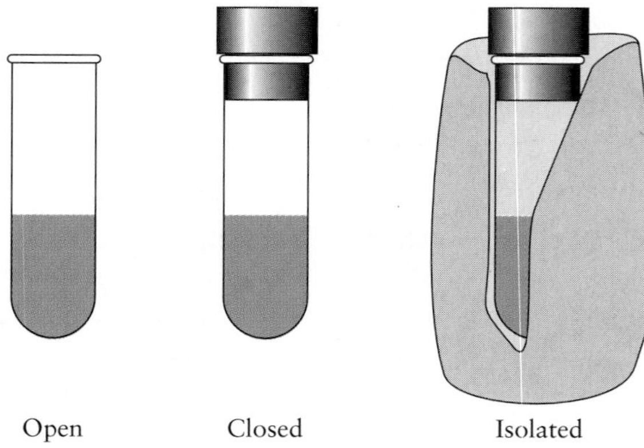

FIGURE 16.2 We classify systems into one of three kinds. An open system can exchange matter and energy with its surroundings. A closed system can exchange energy but not matter. An isolated system can exchange neither matter nor energy.

SELF-TEST 16.1A Identify the following as open, closed, or isolated systems: (a) a space shuttle booster rocket; (b) hot coffee in a perfectly insulated pot.

[*Answer:* (a) Open; (b) isolated]

SELF-TEST 16.1B Identify the following as open, closed, or isolated systems: (a) propane gas in a sealed bottle; (b) a burning candle.

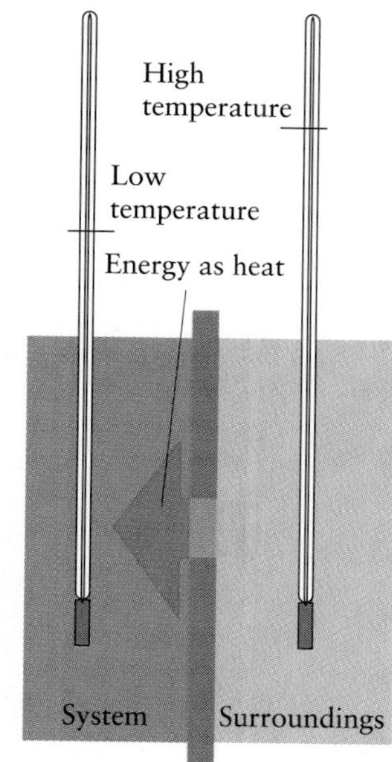

FIGURE 16.3 When we heat a system, we make use of a difference in temperature between it and the surroundings to induce energy to flow through the walls of the system. Heat flows from high temperature to low.

16.2 HEAT AND WORK

In Section 6.1 energy was defined as the capacity of a system to do work. If a system possesses a lot of energy, like a tightly coiled spring, a fully charged battery, or the very hot steam of a turbine, then it can do a lot of work. If it contains only a little energy, like an uncoiled spring, an exhausted battery, or cold water, then it can do little work.

There are three ways of changing the energy of a system—of changing how much work it can do. If we have an open system, we can add more material to it or take material away. To change the energy of an automobile, for example, we put gasoline in its tank. In most of the systems we consider, the energy cannot be changed in this way because they are closed systems. A closed system is prepared by adding reagents, and then it is sealed. All the atoms remain inside; they can only exchange partners as a result of the chemical reactions inside the system.

A second way to change the energy of a system is by heating or cooling it: hot steam can be used to do more work (driving a turbine, for example) than cold water. **Heating** is a transfer of energy between a system and its surroundings that occurs because of a temperature difference between them (Fig. 16.3). A hot flame in the surroundings, for instance, has a higher temperature than the water in a kettle, so energy flows as heat from the flame to the water. Heating can raise the temperature of a system, or it can cause a change of state without affecting the temperature. In each case the energy of the system has been increased. For example, if the water in the kettle is boiling, then continued heating simply evaporates the water; the temperature remains the same until all the water is gone.

The third way of changing the energy is by doing work on the system or letting the system do work on the surroundings. **Work** is done when an object is moved against an opposing force. A gas inside a container fitted with a piston does work when it pushes the piston out against the pressure of the atmosphere (Fig. 16.4). For example, the gases produced by burning gasoline push out the pistons that transfer energy to the wheels of our cars. In power plants, steam produced by boiling water is used to generate electricity in turbines. We do work when we raise a

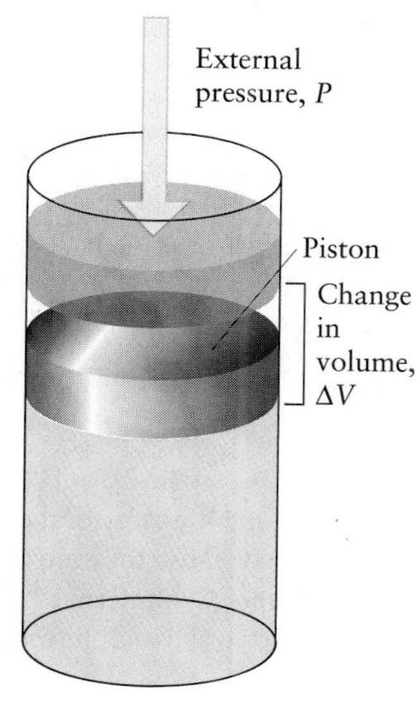

FIGURE 16.4 A system does work when it expands against an external pressure. Here we see a gas that pushes a piston out against a pressure P. We shall see shortly that the work done is proportional to P and the change in volume, ΔV, that the system undergoes.

weight. If a spring is part of a system, then we increase the energy of the system by winding—doing work on—the spring.

> *There are three ways of increasing the energy of a system: by adding material to it, by heating it, or by doing work on it.*

16.3 INTERNAL ENERGY

◀◀ Remember from Section 7.9 that kinetic energy is the energy due to motion and potential energy is the energy due to position.

◀◀ The value of a state property depends only on the current state of the system and is independent of the changes it underwent to get into its present condition (Section 6.4).

The total energy of a system is called the **internal energy,** U. It is the sum of the kinetic energies of all the particles plus the potential energy arising from their interactions with one another. If the molecules are moving rapidly, as they do in a hot gas, then their kinetic energy is high and the internal energy of the sample is high too. If the particles are close together, as they are in a solid—for example, the atoms in a metal spring—then squashing them even closer together can raise their potential energy and hence the internal energy of the solid (Fig. 16.5).

The internal energy is a state property in the same sense that the enthalpy is. Changes in the internal energy are denoted ΔU, with

$$\Delta U = U_{\text{final}} - U_{\text{initial}}$$

A positive value of ΔU means that the internal energy of the final state of the system is higher than that of the initial state (Fig. 16.6). This is the case when water is heated or a spring is wound. A negative value of ΔU means that the final state of the system has a lower energy than the initial state. This is the case when water cools or a spring unwinds.

We can increase the internal energy of a closed system in two ways: by heating it or by doing work on it. Like energy, heat and work are measured in joules (J; see Section 6.3). Work is denoted w and heat is denoted q. When we do work on the system—for instance, by winding a spring inside it—its internal energy rises and we write $\Delta U = w$. Suppose we do 100 kJ of work; then the system increases its store of energy by 100 kJ and we write $\Delta U = +100$ kJ. When we supply heat to the system, its internal energy increases by the corresponding amount and we write $\Delta U = q$. Suppose we transfer 40 kJ to a beaker of water by standing it on a hot surface; then the internal energy of the water increases by 40 kJ and we write $\Delta U = +40$ kJ.

When we supply energy to the system both by doing work and by heating, the total change in internal energy is

Change in internal energy
= energy transferred by heating + energy transferred by doing work

In symbols,

$$\Delta U = q + w \qquad (1)$$

For example, if we supply 40 kJ of heat and do 25 kJ of work on the system, then the total change in internal energy is

$$\Delta U = 40\text{ kJ} + 25\text{ kJ} = +65\text{ kJ}$$

If, in another process, 40 kJ of heat leaks out of the system while we are doing 25 kJ of work on it, we would write $q = -40$ kJ, the minus sign indicating that heat has left the system, and $w = +25$ kJ. In this case

FIGURE 16.5 When we wind a spring, the potential energy of the atoms changes because they are squashed together. The internal energy of the spring rises as a result of this increase in potential energy.

$$\Delta U = -40 \text{ kJ} + 25 \text{ kJ} = -15 \text{ kJ}$$

Now there is a net decrease of 15 kJ in the internal energy of the system; in other words, the capacity of the system to do work has been reduced by 15 kJ.

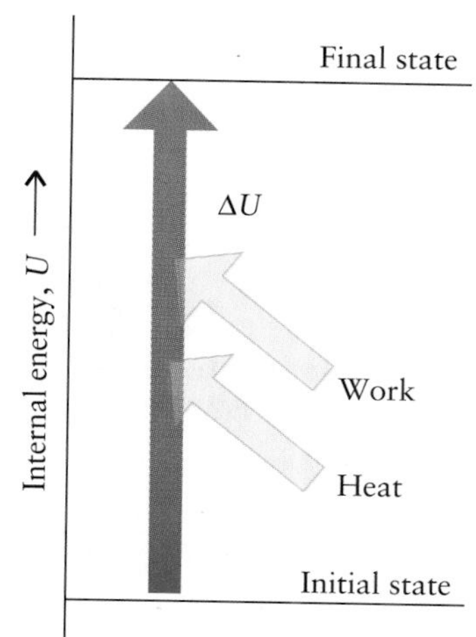

FIGURE 16.6 The internal energy of a system can be changed either by doing work or by heating. The change in internal energy is positive (U increases) when energy is supplied in either way. When energy leaves the system as heat or work, the internal energy falls and ΔU is negative.

EXAMPLE 16.1 Calculating the change in internal energy

Suppose that a battery drives an electric motor in an aquarium pump, and we consider the battery plus the motor as the system. During a certain period, the system does 555 kJ of work on the pump and releases 124 kJ of heat into the surroundings. What is the change in internal energy of the system?

STRATEGY All losses of energy as heat or work reduce the internal energy of the system, so they occur in Eq. 1 with negative signs. All gains of energy increase the internal energy, so they occur in Eq. 1 with positive signs.

SOLUTION Both energy transfers are losses, so we write $q = -124$ kJ and $w = -555$ kJ. Therefore, from Eq. 1,

$$\Delta U = -124 - 555 \text{ kJ} = -679 \text{ kJ}$$

That is, the internal energy of the system decreases by 679 kJ while the motor is running.

SELF-TEST 16.2A A certain system gains 250 kJ of energy as heat while it is doing 500 kJ of work. What is the change in internal energy?

[*Answer:* -250 kJ]

SELF-TEST 16.2B An electric motor and its battery together do 500 kJ of work; the battery also releases 250 kJ of energy as heat and the motor releases 50 kJ as heat due to friction. What is the change in internal energy of the system, with the system regarded as (a) the battery alone; (b) the motor alone; (c) the battery and the motor together? For (a) and (b), assume that the battery does 500 kJ of work on the motor, which then does the same amount of work on the surroundings.

A change in internal energy can be brought about *either* by heat *or* by work. We can increase the internal energy by 100 kJ by doing 100 kJ of work on it or by supplying 100 kJ of energy as heat. Any combination of the two will do, provided the net transfer of energy is 100 kJ. *Heat and work are equivalent ways of changing the energy of a system.* The internal energy is like the reserves of a bank: the bank accepts deposits and withdrawals in either of two currencies (heat or work) but stores them as a common fund, the internal energy (Fig. 16.7).

Now suppose we consider an *isolated* system. No matter can be added to or removed from it. No work can be done on or by it, because we cannot compress the system or get any mechanical instruments through the walls. We cannot heat the system, because the walls insulate the interior from the exterior. Because $q = 0$ and $w = 0$, it follows that

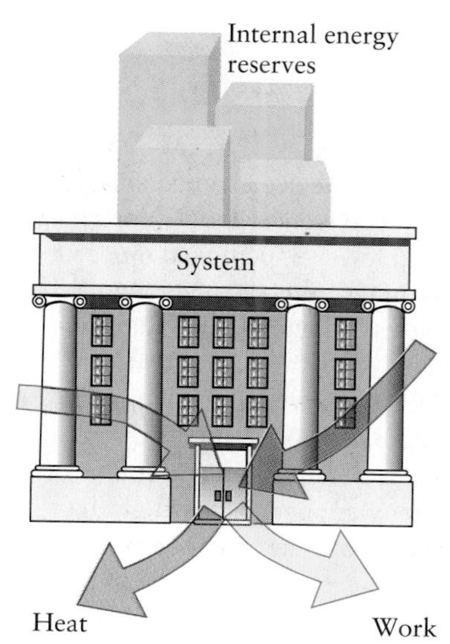

FIGURE 16.7 The internal energy is like the reserves of a bank that makes transactions in two types of currency, work and heat. Once inside the bank, the currency is stored as internal energy and can be withdrawn in either currency.

$\Delta U = 0$. If we could completely isolate a system from its surroundings, then it would have exactly the same energy forever. This *conservation of energy* is stated as the first law of thermodynamics:

> **First law of thermodynamics:** The internal energy of an isolated system is constant.

If we consider the whole universe to be an isolated system, then we could state the first law as "the internal energy of the universe is constant."

In other words, *we cannot create or destroy internal energy.* We can change the internal energy of a system only by transferring energy between the system and its surroundings.

A lot of people have tried to become rich by designing systems intended to disprove the first law of thermodynamics. If their devices could create internal energy out of nothing, then they would have an inexhaustible supply of energy, with no fuel required. Many of these "perpetual motion machines" have been announced, but every case has proved to be either a deliberate deceit or a technologically faulty device (Fig. 16.8).

Heat and work are two equivalent ways of changing the internal energy of a system; the internal energy of an isolated system cannot change.

FIGURE 16.8 The inventor of this elaborate device claimed that it produced work without consuming fuel. However, perpetual motion machines have all been found to be hoaxes. Falling water (A) turns the wheel (C), which does work. Some of the work is used to pump the water back up (E). The inventor claimed that no external source of water was needed; however, the amount of water pumped up is less than the amount that falls, so the system runs down quickly. The first law is always obeyed.

16.4 HEAT TRANSFERS AT CONSTANT VOLUME

One common way of letting a system do work is by letting it expand and push back a piston or the atmosphere (Fig. 16.9). That is what happens in an internal combustion engine. However, if a reaction takes place inside a sturdy container, like a bomb calorimeter (Section 6.3), it cannot expand and therefore cannot do any work (Fig. 16.10). Because $w = 0$, it follows that any change in internal energy in a constant volume system must arise from any energy transferred as heat, so we write

$$\text{At constant volume:} \quad \Delta U = q \qquad (2)$$

This equation shows us how to measure ΔU: we measure the heat released or absorbed when a reaction takes place at constant volume. For example, if a reaction takes place inside a bomb calorimeter and releases 100 kJ through its walls, then we know that $\Delta U = -100$ kJ. Conversely, if 100 kJ of heat flows into the calorimeter, then we know that $\Delta U = +100$ kJ.

When a reaction takes place in a constant-volume system, ΔU is equal to the energy transferred as heat.

16.5 ENTHALPY

We first met the enthalpy, H, of a system in Section 6.4 and saw that a change in enthalpy, ΔH, is equal to the heat transferred at constant pressure:

$$\text{At constant pressure:} \quad \Delta H = q \qquad (3)$$

For example, in a combustion reaction taking place in an open vessel, 100 kJ of energy might escape into the surroundings as heat. We conclude that the enthalpy of the system has decreased by 100 kJ and write $\Delta H = -100$ kJ.

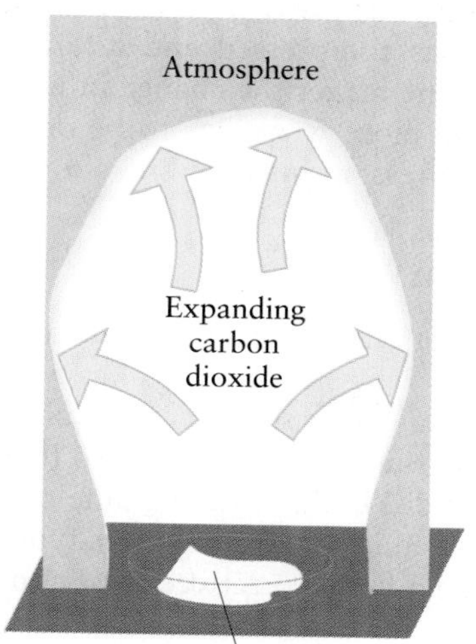

FIGURE 16.9 A reaction does work when it generates a gas, as depicted in this example, in which carbon dioxide is formed from the thermal decomposition of calcium carbonate. As the gas is formed, it drives back the surrounding atmosphere.

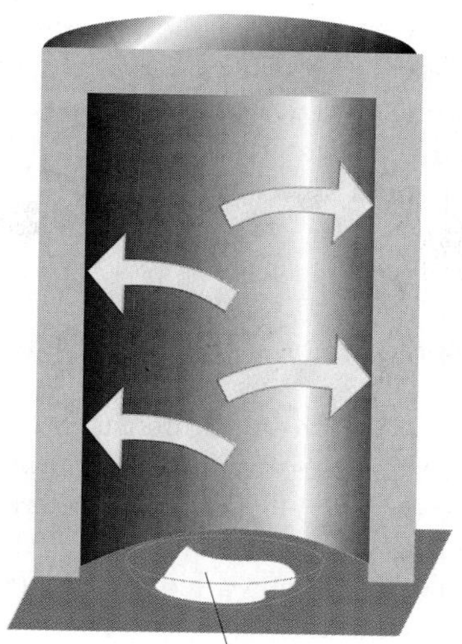

FIGURE 16.10 When a reaction (such as the thermal decomposition of calcium carbonate) takes place in a closed, constant-volume container, the gas fills the container but cannot expand against the surrounding atmosphere. As a result, it does no work on the surroundings.

FIGURE 16.11 When a system is free to expand against an external pressure, some of the energy supplied to it as heat escapes back into the surroundings as work. As a result, the change in internal energy is less than the energy supplied as heat.

When we use the enthalpy, we effectively ignore the work that a system must do to make room for the products of a reaction. A bank inspector who wanted to account for all transactions of energy would keep track of the bank's reserves in terms of the internal energy of the bank vault. On the other hand, a bank inspector who disregards the cost of expanding or contracting the vault, will examine the enthalpy books.

To see this interpretation, think about what happens when we supply 100 kJ of heat to a container full of gas. When the volume cannot change, all the heat supplied to it is used to raise the internal energy of the system, so $\Delta U = +100$ kJ. However, when the gas is free to change its volume and stay at the same pressure, some of the 100 kJ is used to drive back the atmosphere as the system expands (Fig. 16.11). Therefore, at constant pressure, only *some* of the energy supplied as heat is used to increase the internal energy of the system—the rest leaks away back into the surroundings as work. Although $\Delta H = +100$ kJ, the change in internal energy would be only 90 kJ if 10 kJ is used up in expanding the gas, so $\Delta U = +90$ kJ. In general,

$$\Delta U = \Delta H - \text{energy used as work of expansion}$$

If the only work done by a system during a process is expansion work, with a change in volume ΔV, then

$$\text{Energy used as work of expansion} = P \times \Delta V \qquad (4)$$

The formal expression is $w = -P\Delta V$.

where P is the external pressure. If ΔV is positive, the system has done the work of expansion and its internal energy falls. If ΔV is negative, work is being done on the system as it is compressed and its internal energy rises. This expression tells us that the bigger the change in volume and the greater the external pressure, the more work has to be done to expand a system. When we insert this expression into the one above, we get

$$\Delta U = \Delta H - P\Delta V \tag{5}$$

for the relation between ΔH and ΔU for a reaction taking place at constant pressure.

Pressure-volume work is converted to energy units by the exact relation 1 L·atm = 101.325 J.

If the volume increases by 30 L in a particular reaction, and the external pressure is 1.00 atm, $P\Delta V$ works out as about 3.0 kJ, so there is an appreciable difference between ΔH and ΔU when the volume change is so large. However, if no gases are involved in a reaction, ΔV is typically very small (only a few milliliters), and the difference between ΔH and ΔU is negligible.

SELF-TEST 16.3A In a certain exothermic reaction at constant pressure, 50 kJ of heat left the system and 20 kJ of energy left the system as expansion work to make room for the products. What are its values of (a) ΔH, (b) ΔU?

[*Answer:* (a) −50 kJ, (b) −70 kJ]

SELF-TEST 16.3B In a certain endothermic reaction at constant pressure, 50 kJ of heat entered the system. The products took up less volume than the reactants, and 20 kJ of energy entered the system as the outside atmosphere pressed down on it. What are its values of (a) ΔH, (b) ΔU?

We saw in Section 6.12 how to calculate the standard reaction enthalpy, $\Delta H°$, for a chemical reaction from standard enthalpies of formation, $\Delta H_f°$, and how to write it as part of a thermochemical equation, with units of kilojoules. Here we shall use the symbol $\Delta H_r°$ to denote the quantity

$$\Delta H_r° = \sum n\Delta H_f°(\text{products}) - \sum n\Delta H_f°(\text{reactants}) \tag{6}$$

In this case, n represents the stoichiometric coefficients as pure numbers, rather than as numbers of moles. Because the enthalpies of formation are in kilojoules per mole, $\Delta H_r°$ works out in kilojoules per mole too, whereas $\Delta H°$ is in kilojoules. To convert $\Delta H_r°$ to $\Delta H°$ for a particular thermochemical equation, simply delete the mol^{-1} in its units.

A change in enthalpy is equal to the heat supplied at constant pressure. The change ignores the work that a system must do to make room for its products.

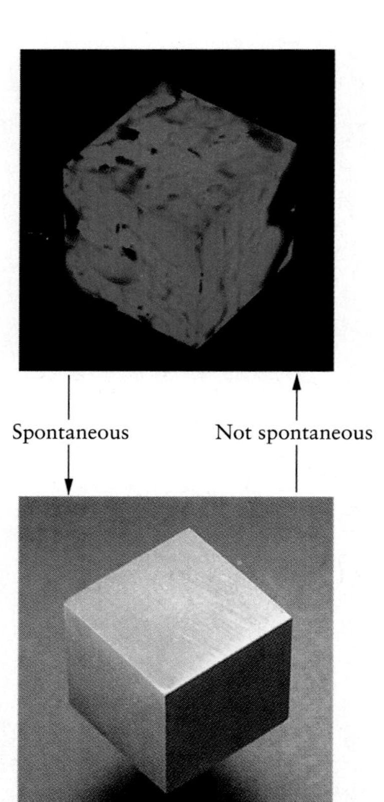

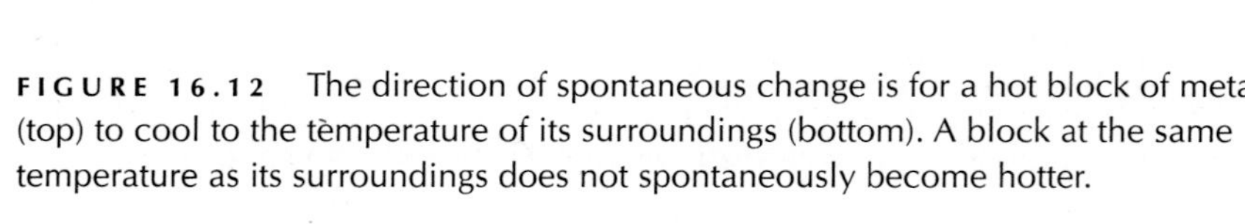
FIGURE 16.12 The direction of spontaneous change is for a hot block of metal (top) to cool to the temperature of its surroundings (bottom). A block at the same temperature as its surroundings does not spontaneously become hotter.

THE DIRECTION OF SPONTANEOUS CHANGE

What controls the natural direction of chemical reactions? Why does methane form carbon dioxide and water, but carbon dioxide and water not form methane? Why do reactions tend to run in the direction that leads to equilibrium? Why does *anything* happen at all? To answer this question, we need to take a further step into thermodynamics and learn about another property of energy beyond the fact that it is conserved.

16.6 SPONTANEOUS CHANGE

The technical term for a natural change is a **spontaneous change;** a spontaneous change tends to occur without needing to be driven by an external influence. One simple example is the cooling of a block of hot metal to the temperature of its surroundings (Fig. 16.12). The reverse change, a block of metal growing hotter than its surroundings, has never been observed. Such nonspontaneous changes can be made to happen—for example, by forcing an electric current through the metal to heat it—but they will not occur without being driven by an external source of energy. The expansion of a gas into a vacuum is also spontaneous (Fig. 16.13). A gas has no tendency to contract spontaneously into one part of a container. However, we can drive a gas into a smaller volume and thus bring about this nonspontaneous change by pushing in a piston.

A spontaneous change need not be fast. Molasses has a spontaneous tendency to flow out of an overturned can, but at low temperatures that flow may be very slow. The reaction in which water is formed from a mixture of hydrogen and oxygen gases is spontaneous, but the mixture can be kept safely for centuries, provided we do not ignite it with a spark. A spontaneous process has a natural *tendency* to occur; it does not necessarily take place at a significant rate.

A spontaneous change has a tendency to occur without being driven by an external influence; spontaneous changes need not be fast.

16.7 ENTROPY AND DISORDER

The key idea that accounts for spontaneous change is that *energy and matter tend to become more disordered.* A hot block of metal tends to cool because the energy of its atoms tends to spread into the surroundings. The reverse change is very unlikely to be observed, because it is very unlikely that energy will collect inside a small block of metal. Similarly,

FIGURE 16.13 The direction of spontaneous change for a gas is toward filling its container. A gas that already fills its container does not collect spontaneously in a small region of the container. A glass cylinder containing a brown gas (upper piece of glassware in the top illustration) is attached to an empty flask. When the stopcock between them is opened, the brown gas fills both upper and lower vessels (bottom illustration). The brown gas is nitrogen dioxide.

although it is natural for randomly moving gas molecules to spread out all over their container, it is very unlikely that their random motion will bring them all simultaneously back into one corner. How, though, do we make this idea precise and use it to account for more complicated changes, like chemical reactions?

In thermodynamics, a measure of the disorder of a system is its **entropy,** *S*. The internal energy is a measure of the *quantity* of energy; the entropy is a measure of how that energy is stored. Low entropy means little disorder; high entropy means great disorder. Therefore, we can summarize the direction of natural change as follows:

The second law of thermodynamics: Entropy tends to increase.

In short, things get worse.

At this stage, this law is just expressing the fact that matter and energy tend to become more disordered. However, once we have defined entropy quantitatively, we shall see that the law can be used to make quantitative predictions.

To develop insight into entropy, we need a sense of when to expect high entropy—high disorder, with energy and atoms or molecules widely dispersed—and when to expect low entropy—an orderly arrangement, with energy and atoms or molecules highly localized.

The entropy of a substance can be increased two ways:

1. Entropy is increased by heating, which increases the motion and thus the relative disorder of the molecules.
2. Entropy is increased by providing more locations into which the molecules can spread.

The relative entropies of different physical states of the same substance are easy to predict. When a solid melts, its molecules have more freedom to move; the liquid is more disordered than the solid, so its entropy is higher (Fig. 16.14). Conversely, the entropy of a liquid falls when it freezes to a solid and the molecules settle into orderly arrays. An even bigger increase in entropy is expected when a substance vaporizes, because now the molecules are almost completely free of one another and occupy a much greater volume than in the condensed state. We can expect gases to have much higher entropies than solids and liquids.

TABLE 16.1 Standard molar entropies at 25°C*

Substance	S°_m, $J \cdot K^{-1} \cdot mol^{-1}$
GASES	
ammonia, NH_3	192.4
carbon dioxide, CO_2	213.7
hydrogen, H_2	130.7
nitrogen, N_2	191.6
oxygen, O_2	205.1
LIQUIDS	
benzene, C_6H_6	173.3
ethanol, C_2H_5OH	160.7
water, H_2O	69.9
SOLIDS	
calcium oxide, CaO	39.8
calcium carbonate, $CaCO_3$[†]	92.9
diamond, C	2.4
graphite, C	5.7
lead, Pb	64.8

*Additional values are given in Appendix 2A.

†Calcite

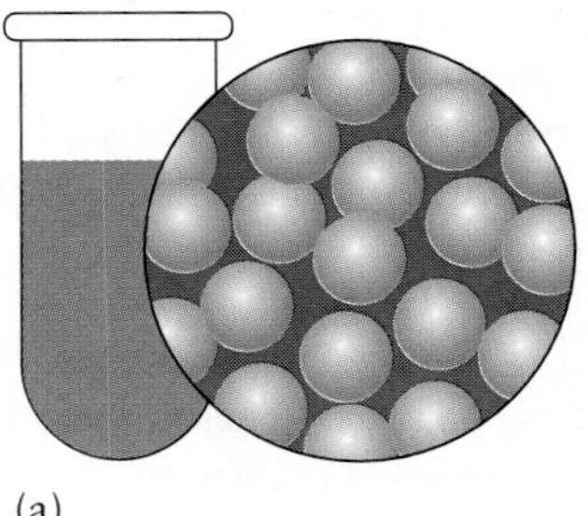

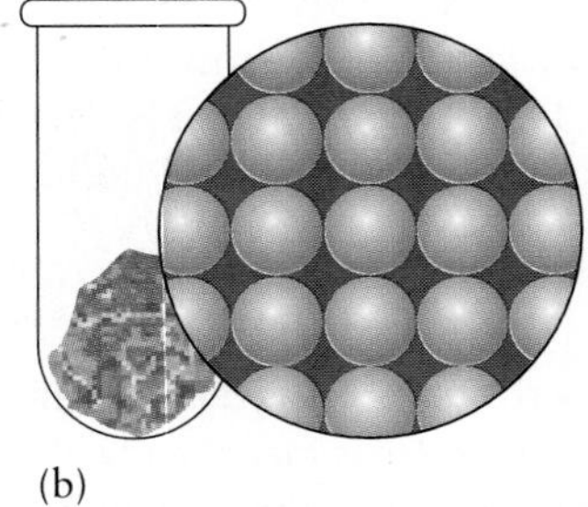

FIGURE 16.14 A representation of the arrangement of molecules in (a) a liquid and (b) a solid. When the solid melts, there is an increase in the disorder of the system and hence a rise in entropy.

We can interpret all entropy changes in terms of changes at the molecular level. At $T = 0$ in a perfect crystal, as depicted in Fig. 16.14b, there is perfect order and the entropy is 0. As the substance is heated, more orientations are available to the molecules and the entropy of a substance increases. Thus we can expect the entropy of any substance at room temperature to be greater than 0.

The entropy of a substance can be increased by heating it or by allowing its particles greater freedom of movement.

16.8 STANDARD ENTROPIES

The entropies of substances can be found experimentally at any temperature. A few of these **standard molar entropies,** S_m°, the molar entropy of the pure substance at 1 atm, are given in Table 16.1, and a longer list can be found in Appendix 2A. Note that all values are positive: all substances are more disordered at 298 K than at $T = 0$. Box 16.1 explains how the entropy of a substance is related to the arrangement of its molecules.

To measure entropies, we need to measure heat capacities down to very low temperatures, and then do a calculation, using these experimentally determined values.

We can understand some of the differences in the standard molar entropies in terms of differences in structure. For example, let's compare the molar entropy of diamond, 2.4 $J \cdot K^{-1} \cdot mol^{-1}$, with the much higher value for lead, 64.8 $J \cdot K^{-1} \cdot mol^{-1}$. The low entropy of diamond is what we should expect for a solid that has rigid bonds: at room temperature, its atoms are not able to jiggle around as much as the loosely bonded atoms of lead can. We also see that substances built from heavy atoms have higher entropies than lighter ones (compare H_2, N_2, and O_2), and that large, complex molecules have higher entropies than smaller, simpler ones (compare $CaCO_3$ with CaO or NH_3 with H_2O).

Entropies increase as the temperature is raised, as we see from the data for water in Table 16.2. As the temperature is raised, the molecules have more thermal motion and the sample becomes more disordered. The large increase in entropy at the boiling point of water, from 87 $J \cdot K^{-1} \cdot mol^{-1}$ for the liquid to 197 $J \cdot K^{-1} \cdot mol^{-1}$ for the vapor, shows the increase in disorder that occurs when a liquid changes to a much more chaotic gas. A smaller increase occurs when solids melt, because a liquid is only slightly more disordered than a solid (Fig. 16.15).

The standard molar entropies of gases are higher than those of comparable solids and liquids at the same temperature. The entropy of a substance increases when it melts, when it vaporizes, and as its temperature is raised.

SELF-TEST 16.4A Which substance in each pair has the higher molar entropy: (a) CO_2 at 25°C and 1 atm or CO_2 at 25°C and 3 atm? (b) He at 25°C and 1 atm or He at 100°C and 1 atm? (c) $Br_2(l)$ or $Br_2(g)$?

[*Answer:* (a) CO_2 at 1 atm; (b) He at 100°C; (c) $Br_2(g)$]

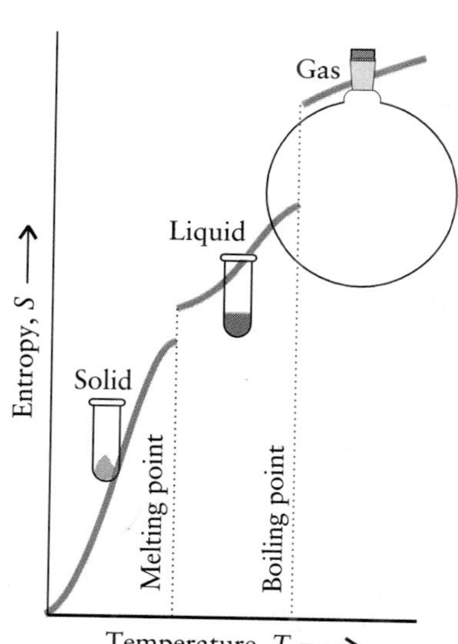

FIGURE 16.15 The entropy of a solid increases as its temperature is raised. The entropy increases sharply when the solid melts to form the more disordered liquid and then gradually increases again up to the boiling point. A second, larger jump in entropy occurs when the liquid turns into a vapor.

BOX 16.1 BRIDGING THE MACRO AND MICRO WORLDS

In the mid-nineteenth century there was an explosion of knowledge about how matter and energy interact. This knowledge provided the foundations of thermodynamics. However, although thermodynamic properties could be predicted, they could not be explained. The Austrian physicist Ludwig Boltzmann finally established the link between measurable thermodynamic properties and the behavior of the vast collection of atoms, molecules, and ions that make up matter. Unfortunately, the existence of atoms was not yet universally accepted, and his work was actively discredited. He eventually took his own life in despair.

Ludwig Boltzmann (1844–1906). His formula for entropy (using an earlier notation for natural logarithms) became his epitaph.

Boltzmann proposed that the reason systems move spontaneously toward increased entropy is simply that disordered states are more numerous and thus more probable. He proposed that entropy is a measure of the probability that a system is in a certain state and found a formula for relating the numerical value of the entropy of a substance to the number of arrangements of atoms that correspond to the same state:

$$S = k \ln W$$

The fundamental constant k, which is called the **Boltzmann constant,** is 1.3807×10^{-23} J·K^{-1}. The term W is the number of ways that the atoms or molecules in the sample can be arranged and yet have the same total energy.*

Let's use the Boltzmann formula to find W for a very simple system, a tiny solid made up of 20 diatomic molecules of a binary compound such as carbon monoxide, CO. Suppose that the 20 molecules have formed a perfectly ordered crystal and that, because $T = 0$, all motion has ceased (see left panel of figure).

*The term $\ln W$ is the natural logarithm of W (see Appendix 1C).

TABLE 16.2 Standard molar entropy of water at various temperatures

Phase	Temperature, °C	S_m°, J·K^{-1}·mol^{-1}
solid	−273 (0 K)	3.4
	0	43.2
liquid	0	65.2
	20	69.6
	50	75.3
	100	86.8
vapor	100	196.9
	200	204.1

SELF-TEST 16.4B Use the information in Table 16.1 or Appendix 2A to calculate the changes in molar entropy for the following transitions between allotropes; in each case indicate which is the more ordered form: (a) white tin (Fig. 16.16) changes to gray tin at 25°C; (b) diamonds change to graphite at 25°C.

We can use the data in Table 16.1 to calculate the entropy change of a chemical reaction. The **standard reaction entropy, ΔS_r°,** is the difference in standard molar entropies of the products and reactants, taking into account their stoichiometric coefficients:

$$\Delta S_r^\circ = \sum nS_m^\circ\,(\text{products}) - \sum nS_m^\circ\,(\text{reactants}) \qquad (7)$$

The first term on the right is the total standard molar entropy of the products and the second term is that of the reactants; n denotes the various stoichiometric coefficients in the chemical equation.

Because the molar entropy of a gas is so much bigger than that of solids and liquids, a change in the number of moles of gas normally dom-

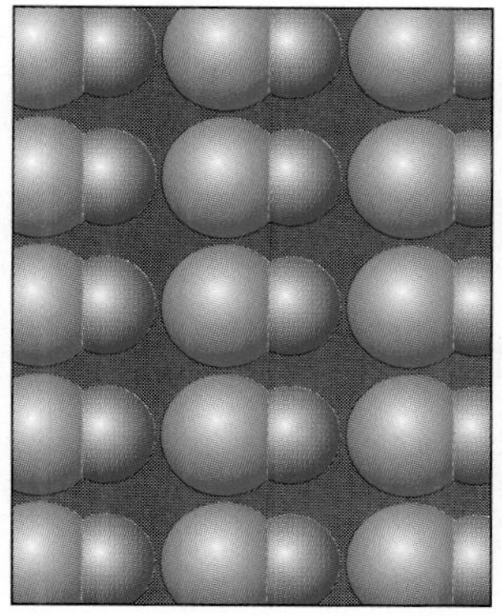

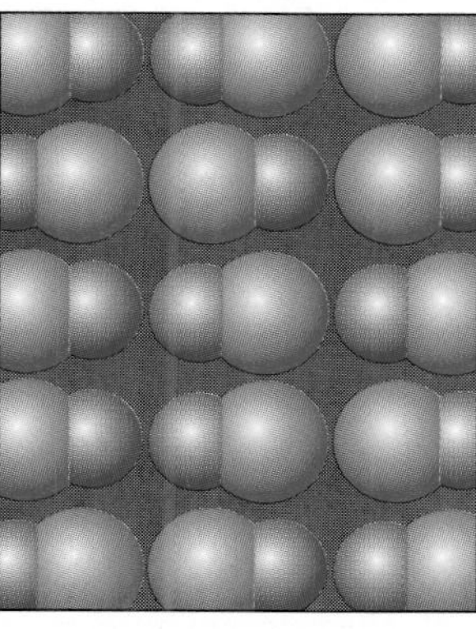

Some of the 20 heteronuclear diatomic molecules in a perfectly ordered arrangement at $T = 0$ (left). The sample has zero spatial and thermal disorder, and hence zero entropy ($S = 0$). This sample represents a perfect crystal at absolute zero. Each of the molecules in the sample on the right (15 of which are shown here) can take up either of two orientations without affecting the energy. There are $2 \times 2 \times \ldots = 2^{20}$, or 1,048,576, different possible arrangements, and this illustration shows just one of them.

We expect the sample to have zero entropy, because there is no disorder in either location or energy. This expectation is confirmed by the Boltzmann formula: because there is only one way of arranging the molecules in the perfect crystal, $W = 1$ and (because $\ln 1 = 0$)

$$S = k \ln 1 = 0$$

Now suppose that the compound is frozen in a disordered state, such that each molecule can point in either of two directions in the solid, yet still have the same energy (see right panel of figure). Because each of the 20 molecules can be in one of two orientations, the total number of ways of arranging the molecules is

$$W = (2 \times 2 \times 2 \ldots)_{\text{twenty factors}} = 2^{20}$$

or over 1 million different arrangements. The entropy of this tiny disorderly solid is therefore

$$\begin{aligned} S &= k \ln 2^{20} = (1.38 \times 10^{-23}\ \mathrm{J{\cdot}K^{-1}}) \times (20 \ln 2) \\ &= 1.9 \times 10^{-22}\ \mathrm{J{\cdot}K^{-1}} \end{aligned}$$

The entropy of the disordered solid is higher than that of the perfectly ordered solid. For a solid that contained 1.00 mol CO molecules that could point in either of two directions, the entropy would be

$$\begin{aligned} S &= (1.38 \times 10^{-23}\ \mathrm{J{\cdot}K^{-1}}) \times (6.02 \times 10^{23} \ln 2) \\ &= 5.76\ \mathrm{J{\cdot}K^{-1}} \end{aligned}$$

Chemists now use calculations like this one to determine the entropies of more complicated substances and get very good agreement with experimental values. In some cases experimental values of S are not available, and the Boltzmann formula has to be used to obtain their values.

inates any other entropy change in a reaction. A net increase in the amount of gas usually results in a positive reaction entropy. Conversely, a net consumption of gas usually results in a negative reaction entropy. Entropy changes are much more finely balanced for reactions in which there is no net change in the amount of gas, and for these reactions we have to use numerical data to predict the sign of a reaction entropy. We also need tables of numerical data to predict the numerical value of a reaction entropy, whether or not gases are involved.

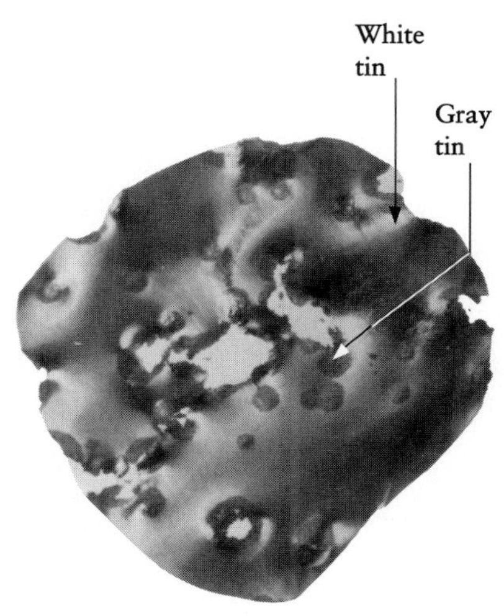

FIGURE 16.16 Gray tin and white tin are two solid forms of tin. The denser white metallic form is the more stable phase above 13°C, and the powdery gray form is more stable below that temperature. The formation of gray tin is said to have contributed to Napoleon's defeat at Moscow, when his soldier's tin buttons fell off their clothes at the low temperatures they encountered there.

The standard reaction entropy is positive (an increase in entropy) if there is a net production of gas in a reaction; it is negative (a decrease) if there is a net consumption of gas.

EXAMPLE 16.2 Calculating the standard reaction entropy

Calculate the standard entropy for the reaction

$$N_2(g) + 3\,H_2(g) \longrightarrow 2\,NH_3(g)$$

STRATEGY We expect a decrease in entropy because there is a net reduction in the amount of gas. To find the numerical value, we use the chemical equation to write an expression for ΔS_r°, and then substitute values from Table 16.1.

SOLUTION From the chemical equation we can write

$$\begin{aligned}\Delta S_r^\circ &= 2S_m^\circ(NH_3, g) - \{S_m^\circ(N_2, g) + 3S_m^\circ(H_2, g)\}\\ &= 2 \times 192.4 - \{191.6 + (3 \times 130.7)\}\ J\cdot K^{-1}\cdot mol^{-1}\\ &= -198.9\ J\cdot K^{-1}\cdot mol^{-1}\end{aligned}$$

Because ΔS_r° is negative, the product is less disordered than the reactants, as we expected.

SELF-TEST 16.5A Use data from Appendix 2A to calculate the standard entropy of the reaction $N_2O_4(g) \rightarrow 2\,NO_2(g)$ at 25°C.

[*Answer:* $+175.83\ J\cdot K^{-1}\cdot mol^{-1}$]

SELF-TEST 16.5B Use data from Appendix 2A to calculate the standard entropy of the reaction $C_2H_4(g) + H_2(g) \rightarrow C_2H_6(g)$ at 25°C.

16.9 THE SURROUNDINGS

We can see from Table 16.2 that at 0°C the entropy of liquid water is 22.0 $J\cdot K^{-1}\cdot mol^{-1}$ higher than the entropy of ice. This difference is expected, because liquid water is more disordered than ice. Freezing, the reverse of melting, therefore corresponds to water becoming less disordered as it forms ice, and the change in entropy is negative. This conclusion, though, may seem puzzling at first sight, because water freezes spontaneously below 0°C, and an entropy *increase* is a signal of spontaneous change.

What have we overlooked? Whenever we use arguments based on entropy, we must always take into account *all* the changes in entropy:

$$\Delta S_{tot} = \Delta S + \Delta S_{surr}$$

Here ΔS is the entropy change of the system and ΔS_{surr} that of the surroundings. The entropy of the substance that is freezing (the system) decreases. However, because freezing is an exothermic process, heat passes into the surroundings. This heat stirs up the thermal motion of the atoms in the surroundings, which increases their disorder and hence their entropy (Fig. 16.17). There is an overall increase in the total disorder when the increase in disorder of the surroundings is greater than the decrease in the disorder of the system. The total entropy increases and the change is spontaneous.

Whenever a symbol appears without a subscript, such as ΔS, it refers to a property of the system.

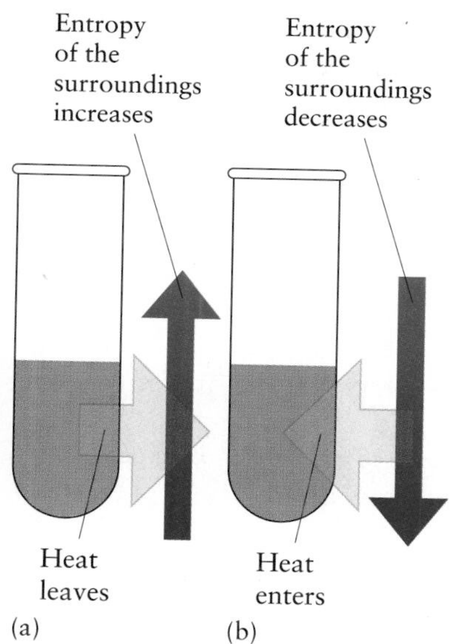

FIGURE 16.17 (a) In an exothermic process, heat escapes into the surroundings and increases their entropy. (b) In an endothermic process, the entropy of the surroundings decreases. The blue-green arrows indicate the direction of entropy change in the surroundings.

To see how to calculate the change in entropy of the surroundings, suppose 1 mol H_2O freezes in the system. The change in enthalpy is -6 kJ; so 6 kJ of heat flows into the surroundings and stirs up thermal motion there. In general, if the enthalpy change of the system is ΔH (in this case, -6 kJ), then the heat that flows into the surroundings is $-\Delta H$ (that is, $+6$ kJ). The sign changes because heat that *leaves* the system *enters* the surroundings. We assume that the disorder stirred up in the surroundings is proportional to the heat transferred to them. At constant pressure and temperature,

$$\Delta S_{\text{surr}} \propto -\Delta H$$

where ΔH is the change in enthalpy of the *system.*

The change in entropy of the surroundings caused by a given transfer of heat also depends on the temperature. When the surroundings are hot, they are already very chaotic and a small inflow of heat from the system has very little impact (Fig. 16.18). On the other hand, when the surroundings are cool, they are relatively ordered and the disorder caused by the same amount of heat is relatively large. Think of the effect of sneezing in a crowded street, which may pass unnoticed, and sneezing in a quiet library, which will not. This argument suggests that the change in entropy is inversely proportional to the temperature at which the transfer takes place and that we should write

$$\Delta S_{\text{surr}} = -\frac{\Delta H}{T} \tag{8}$$

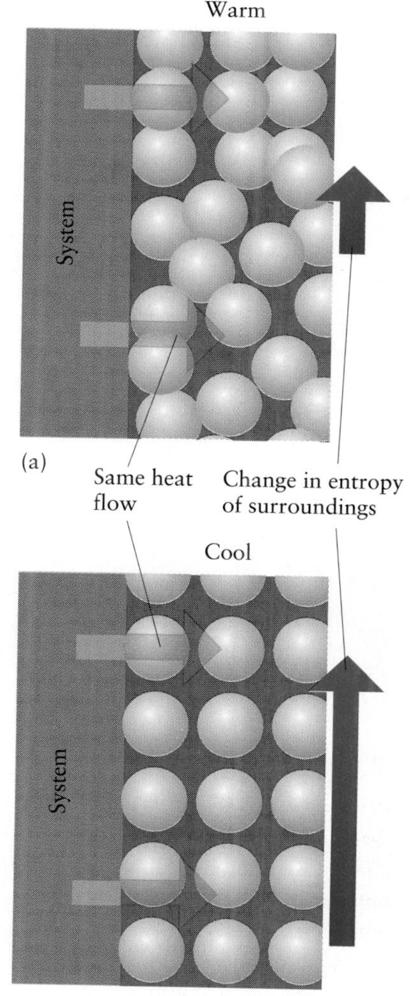

FIGURE 16.18 (a) When a given quantity of heat flows into hot surroundings, it produces very little additional chaos and the increase in entropy is quite small. (b) When the surroundings are cool, however, the same quantity of heat can make a considerable contribution to the disorder of the relatively ordered surroundings, and the change in entropy of the surroundings is correspondingly large.

This formula, which can be derived from the laws of thermodynamics, applies when the change takes place at constant pressure.

The same calculation applies to chemical reactions. For example, for the synthesis of NH_3:

$$N_2(g) + 3\,H_2(g) \longrightarrow 2\,NH_3(g) \qquad \Delta H^\circ = -92.22\text{ kJ}$$

This reaction is exothermic, so we expect the entropy of the surroundings to increase as heat spreads into them. In fact, if the reaction were to go to completion,

$$\Delta S_{surr} = -\frac{(-92.22 \times 10^3\text{ J})}{298\text{ K}} = +309\text{ J}\cdot\text{K}^{-1}$$

The entropy change in the surroundings due to a process taking place at constant pressure and temperature is equal to $-\Delta H/T$.

SELF-TEST 16.6A An exothermic reaction releases 35.7 kJ of heat to the surroundings at 300 K. What is the change in entropy of the surroundings for this reaction?

[*Answer:* $+119\text{ J}\cdot\text{K}^{-1}$]

SELF-TEST 16.6B An endothermic reaction absorbs 71.5 kJ of heat from the surroundings at 150 K. What is the change in entropy of the surroundings for this reaction?

16.10 THE OVERALL CHANGE IN ENTROPY

As we have seen, when we want to express the direction of spontaneous change in terms of the entropy, we must consider changes in the entropy of the system plus the entropy change in the surroundings. Then, if the total entropy change is positive (an increase), the process being discussed is spontaneous. If the total entropy change is negative (a decrease), then the *reverse* process is spontaneous.

We can restate the second law in a manner that allows us to use it to make quantitative predictions of spontaneity:

Second law of thermodynamics: A spontaneous change is accompanied by an increase in the total entropy of the system and its surroundings.

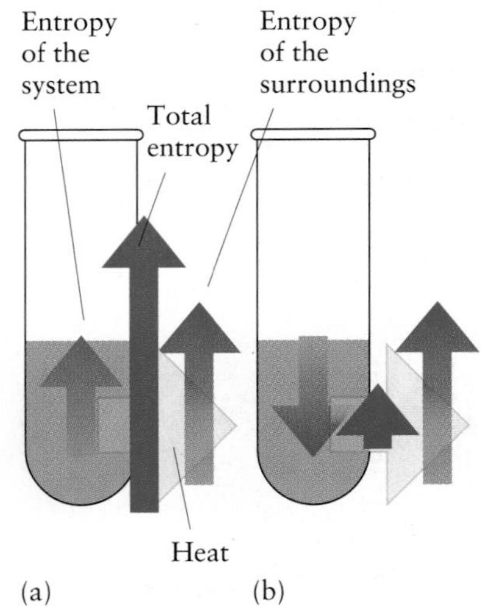

FIGURE 16.19 In an exothermic reaction, (a) the overall entropy change is certainly positive when the entropy increases. (b) The overall entropy change may also be positive when the entropy of the system decreases. The reaction is spontaneous in both cases.

The second law explains why spontaneous exothermic reactions are so common. In an exothermic reaction, such as the synthesis of ammonia or a combustion reaction, the heat released by the reaction increases the disorder of the surroundings. In some cases the entropy of the system may decrease, as in the combustion of magnesium, where a gas is converted to a solid:

$$2\,Mg(s) + O_2(g) \longrightarrow 2\,MgO(s)$$

$$\Delta S_r^\circ = -217\text{ J}\cdot\text{K}^{-1}\cdot\text{mol}^{-1} \qquad \Delta H_r^\circ = -1202\text{ kJ}\cdot\text{mol}^{-1}$$

However, provided the reaction is strongly exothermic, the increased disorder of the surroundings is so great that overall the total entropy increases. For this reaction,

$$\Delta S_{surr} = -\frac{(-1202 \times 10^3\text{ J}\cdot\text{mol}^{-1})}{298\text{ K}} = +4.03 \times 10^3\text{ J}\cdot\text{K}^{-1}\cdot\text{mol}^{-1}$$

a massive increase, so the total change is

$$\Delta S_{tot} = -217 + 4.03 \times 10^3\ \text{J·K}^{-1}\text{·mol}^{-1} = +3.81 \times 10^3\ \text{J·K}^{-1}\text{·mol}^{-1}$$

and the reaction is spontaneous. In many spontaneous exothermic reactions, the disorder of the system may also increase, as happens in the formation of hydrogen fluoride:

$$H_2(g) + F_2(g) \longrightarrow 2\,HF(g)$$

$$\Delta H_r^\circ = -546.4\ \text{kJ·mol}^{-1} \qquad \Delta S_r^\circ = +14.1\ \text{J·K}^{-1}\text{·mol}^{-1}$$

This reaction increases the overall disorder through both the change in the system and the change in its surroundings. As long as ΔH_r° is negative, with a reasonably large numerical value, a reaction with ΔS_r° either positive or negative may be spontaneous (Fig. 16.19). In fact, many exothermic reactions are spontaneous because entropy changes of the system are usually quite small compared with the large increase in the entropy of the surroundings.

Endothermic reactions were a puzzle for nineteenth-century chemists, who believed that reactions ran only in the direction of decreasing energy of the system. An endothermic reaction is one in which the products have a higher enthalpy than the reactants, so it seemed to these chemists that in endothermic reactions reactants were spontaneously rising to higher energies, like a weight suddenly leaping up from the floor to a table. However, *whenever we are considering spontaneity, we have to consider the entropy, not the energy.* Because heat flows from the surroundings into the system, the entropy of the surroundings decreases in the course of an endothermic reaction. However, there can still be an overall increase in entropy, provided the disorder of the system increases enough. Every endothermic reaction must be accompanied by increased disorder *within* the system if it is to be spontaneous (Fig. 16.20), but this constraint does not apply to exothermic reactions.

A chemical reaction is spontaneous if it is accompanied by an increase in the total entropy of the system and the surroundings. Spontaneous exothermic reactions are common because they release heat that increases the entropy of the surroundings. Endothermic reactions are spontaneous only if the reaction mixture undergoes a large increase in entropy.

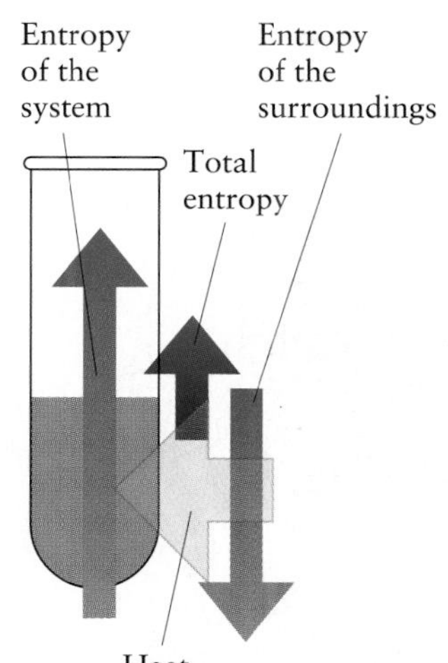

FIGURE 16.20 An endothermic reaction is spontaneous only if the entropy of the system increases enough to overcome the decrease in entropy of the surroundings, as it does here.

FREE ENERGY

We can use the second law to study chemical equilibria and predict the equilibrium constant of a reaction. It would be much easier to predict equilibrium constants, though, if we could use a property of the system that combines the entropy calculations for the system and surroundings, for then we could use a single table of data.

16.11 FOCUSING ON THE SYSTEM

As we have seen, the total entropy change, ΔS_{tot}, is the sum of the changes in the system, ΔS, and its surroundings, ΔS_{surr}:

$$\Delta S_{tot} = \Delta S + \Delta S_{surr}$$

For a process at constant pressure, the change in the entropy of the surroundings is given by Eq. 8, $\Delta S_{surr} = -\Delta H/T$. Therefore,

$$\Delta S_{tot} = \Delta S - \frac{\Delta H}{T}$$

This equation shows how to calculate the total entropy change from information about the system alone—its temperature and the entropy and enthalpy changes it undergoes.

It turns out to be even more convenient to rearrange the last equation into

$$-T\Delta S_{tot} = \Delta H - T\Delta S$$

The Gibbs free energy itself (as distinct from its change) is defined as $G = H - TS$.

We can now express the change in the total entropy as the change in a property called the **Gibbs free energy,** G. We write

$$\Delta G = -T\Delta S_{tot} \qquad (9)$$

so

$$\Delta G = \Delta H - T\Delta S \qquad (10)$$

This rearrangement has important consequences: instead of the change in the *total* entropy, we can now make predictions about the spontaneity of a process by focusing on the change in the free energy *of the system.*

The minus sign in Eq. 9 means that an increase in total entropy corresponds to a *decrease* in free energy. Therefore, *at constant pressure and temperature, the direction of spontaneous change is the direction of decreasing free energy.* The state property G was named for Josiah Willard Gibbs (Fig. 16.21), a nineteenth-century Yale professor who was responsible for turning thermodynamics from an abstract theory into a subject of great usefulness.

The expression for ΔG shows very clearly the factors that govern the direction of spontaneous change at constant pressure. We use Eq. 10 and look for values of ΔH, ΔS, and T that result in a negative value of ΔG. These factors are summarized in Table 16.3. One condition is a large negative ΔH, such as that for a strongly exothermic process like combustion. A large negative ΔH corresponds to a large increase in the entropy of the surroundings. Unless $T\Delta S$ is also very large and negative, very exothermic processes such as combustion are spontaneous because of the thermal disorder they create in the surroundings.

A negative ΔG can occur even if ΔH is positive (an endothermic reaction) provided that $T\Delta S$ is large and positive. The size and sign of $T\Delta S$ depends on two factors. One is the size and sign of ΔS itself, the change

FIGURE 16.21 Josiah Willard Gibbs (1839–1903).

TABLE 16.3 Factors that favor spontaneity

Enthalpy change	Entropy change	Spontaneous?
exothermic ($\Delta H < 0$)	increase ($\Delta S > 0$)	yes, $\Delta G < 0$
exothermic ($\Delta H < 0$)	decrease ($\Delta S < 0$)	yes, if $\lvert T\Delta S\rvert < \lvert\Delta H\rvert$*, $\Delta G < 0$
endothermic ($\Delta H > 0$)	increase ($\Delta S > 0$)	yes, if $T\Delta S > \Delta H$, $\Delta G < 0$
endothermic ($\Delta H > 0$)	decrease ($\Delta S < 0$)	no, $\Delta G > 0$

*The symbol $\lvert x\rvert$ means the absolute value of x, its value disregarding its sign.

of entropy of the system. The other factor is the temperature, T. If disorder is created in the system (as in vaporization or the decomposition of calcium carbonate to calcium oxide and gaseous carbon dioxide), then ΔS is positive and the $-T\Delta S$ term in Eq. 10 contributes a negative term to ΔG. However, the size of its contribution depends on the temperature, and even a large increase in disorder within the system can have a negligible influence if the temperature is low. The role of the entropy of the system becomes more important the higher the temperature. In other words, for reactions in which the entropy change is large (in either direction), the temperature plays an important role in determining whether or not the reaction is spontaneous.

Because ΔH and ΔS change only very slightly with temperature, we can consider them approximately constant over small temperature ranges.

A nonspontaneous reaction can be made to occur by driving it with energy from an external source. For example, water naturally flows downhill, but we can carry it uphill. We can also make nonspontaneous reactions occur. Sending an electrical current through water generates hydrogen and oxygen gases, although ΔG for the reaction

$$2\,H_2O(l) \longrightarrow 2\,H_2(g) + O_2(g)$$

is +494 kJ. Similarly, many of the reactions that are the basis of life are nonspontaneous and must be driven by an external energy source (Case Study 16).

The free energy change is a measure of the change in the total entropy of a system and its surroundings at constant pressure; spontaneous processes are accompanied by a decrease in free energy.

SELF-TEST 16.7A Does an exothermic process with a negative ΔS become more or less spontaneous as the temperature increases?

[*Answer:* Less spontaneous]

SELF-TEST 16.7B Does an endothermic process with a positive ΔS become more or less spontaneous as the temperature increases?

Now let's see how equilibrium is described in terms of the free energy. When a system is at equilibrium, there is no tendency for spontaneous change in either direction, so

$$\Delta G = -T\Delta S_{tot} = 0$$

The condition $\Delta G = 0$ applies to any phase change and any chemical reaction at equilibrium. For example, $\Delta G = 0$ for the transfer of molecules from a liquid into the vapor when the two phases are in equilibrium. The same condition applies to a solid in equilibrium with its liquid, a solute in equilibrium with its saturated solution, and a chemical reaction at equilibrium, provided the pressure and temperature are constant.

◀◀ Recall from Section 10.16 that the lines on a phase diagram show the conditions under which two phases are in equilibrium.

We can verify that $\Delta G = 0$ for the freezing of water at 0°C (273.15 K) and 1 atm. For water,

$$\Delta H_{freeze} = -\Delta H_{fus} = -6.00\ kJ\cdot mol^{-1}$$

$$\Delta S_{freeze} = -\Delta S_{fus} = -21.97\ J\cdot K^{-1}\cdot mol^{-1} = -21.97 \times 10^{-3}\ kJ\cdot K^{-1}\cdot mol^{-1}$$

When we substitute these values and the temperature into

$$\Delta G_{freeze} = \Delta H_{freeze} - T\Delta S_{freeze}$$

we obtain

CASE STUDY 16 UNNATURAL LIFE

HOW CAN WE explain life? Why do the molecules in our body form a highly organized, complex structure, rather than slime, ooze, or gas? Every cell of a living being is organized to an extraordinary extent. Thousands of different compounds, each one having a specific function to perform, move in the intricately choreographed dance we call life. We are examples of systems with *very* low entropy. In fact, our existence seems at first thought a contradiction of the second law of thermodynamics.

The ultimate example of low entropy is in our genes. Every one of the billions of atoms in the long double spiral of a DNA molecule has an appointed place, and that location contributes to the blueprint for our bodies. These molecules carry all the information needed to create a human being. The protein molecules that act as enzymes or structural molecules (as in our hair and nails) are also highly organized. If a protein loses its highly organized shape, then it ceases to function. Heightened entropy—a loss of order—means disease and perhaps even death.

One of the most important processes that supports life is *photosynthesis,* the series of reactions that uses sunlight to convert carbon dioxide and water into carbohydrates and oxygen. Without photosynthesis, the Earth would be a large, wet rock with no green plants or animals. The reaction is itself nonspontaneous:

$$6\,CO_2(g) + 6\,H_2O(l) \longrightarrow C_6H_{12}O_6(s) + 6\,O_2(g)$$

It is accompanied by a decrease in entropy of the system because many small molecules must be assembled into the larger glucose molecule. It is also accompanied by a decrease in entropy of the surroundings because it is endothermic. So how can it occur at all, let alone on a megaton scale worldwide? The answer lies in sunlight. Solar radiation floods a leaf with energy, and the energy is captured by chlorophyll. This captured energy is used by subsequent chemical reactions to generate even more entropy than is lost in the glucose-construction reaction itself.

The Sun's energy also accounts for nonspontaneous reactions in our bodies; we get our energy not directly from the Sun but from chemicals—food—that have stored the energy of sunlight. In each case, one biochemical process, which generates a lot of entropy as it runs in its spontaneous direction, drives other reactions in a nonspontaneous direction, perhaps to build a protein or contribute to the construction of a DNA molecule. In other words, biochemical processes are coupled: one may be driven uphill in free energy by another reaction that rolls downhill. Staying alive is very much like the effect of a heavy weight tied to another weight by a string that passes over a pulley. The lighter weight could never fly up into the air on its own.

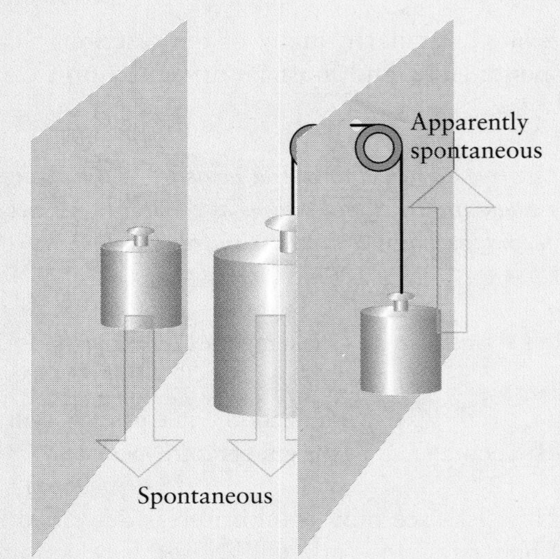

A weight with a small mass can be lifted into the air by another weight of the same or greater mass. What would appear unnatural if we saw it by itself (a weight rising) is actually part of a spontaneous event overall. The "natural" fall of the heavier weight causes the "unnatural" rise of the smaller weight.

$$\begin{aligned}\Delta G_{\text{freeze}} &= (-6.00\ \text{kJ·mol}^{-1}) - \{(273.15\ \text{K}) \times (-21.97 \times 10^{-3}\ \text{kJ·K}^{-1}\text{·mol}^{-1})\} \\ &= 0.00\ \text{kJ·mol}^{-1}\end{aligned}$$

The fact that $\Delta G = 0$ at 0°C signifies that water and ice are in equilibrium at that temperature. Therefore, if liquid is converted to solid, the decrease in disorder of the system is counterbalanced by the increase in disorder of the surroundings as energy is released into them.

For a reaction that is at equilibrium at constant pressure and temperature, $\Delta G = 0$.

The bubbles on the leaves of this underwater plant are oxygen produced by photosynthesis. Molecules such as the chlorophyll that colors the leaves green capture sunlight to begin the transformation of carbon dioxide and water to glucose and oxygen.

However, when it is connected to a heavier weight falling downward on the other side of a pulley, it can soar upward.

When we eat food containing glucose, we consume a fuel. Like all fuels, it has a spontaneous tendency to form products. If we were simply to burn glucose, it would give off a great deal of heat and light. However, in our body the combustion is slower and represents a highly controlled and sophisticated version of burning. Burning glucose in an open container does no work other than pushing back the atmosphere, so it gives off a lot of heat. Oxidizing glucose in our cells, under the careful control of enzymes and all the intricate biochemical mechanisms of the body, does much more work and, in fact, makes us very efficient at utilizing the energy in our food.

To harness as much free energy as possible, a process must proceed slowly and carefully. In such a controlled reaction, the work the process can do approaches ΔG_r. Human cells are thus like small, efficient power plants. We copy nature when we design practical fuel cells. Hydrogen and oxygen normally form an explosive mixture. When combined in a controlled fashion in a fuel cell, however, they react slowly, generating an electrical current that can be used to do work.

When we die, we no longer ingest the second-hand sunlight stored in molecules of carbohydrate, protein, and fat. Now the natural direction of change becomes dominant, and our intricate molecules start to decompose to the ooze and slime that we managed to avoid becoming during our lives. Life is a constant battle to generate enough entropy in our surroundings to go on building and maintaining our intricate interiors. As soon as we stop the battle, we stop generating that external entropy, and our bodies decay.

QUESTIONS

1. Calculate the standard reaction enthalpy, entropy, and free energy for the burning of glucose (the reverse of the photosynthetic reaction). Does the energy available to do work (ΔG_r°) increase or decrease as the temperature is raised? Explain.

2. Calculate the standard reaction free energy of a fuel cell in which methane is oxidized by oxygen to carbon dioxide and water.

3. The key metabolic reaction by which free energy is generated in living systems is the hydrolysis of adenosine triphosphate (ATP) to adenosine diphosphate (ADP). If ΔG_r for the hydrolysis of ATP to ADP is $-30.5\ \text{kJ}\cdot\text{mol}^{-1}$, how many moles of ATP would have to be hydrolyzed in the body to provide the free energy needed for the production of 1 mol of glucose molecules by photosynthesis under standard conditions?

EXAMPLE 16.3 Predicting the boiling point of a substance

Liquid metals, such as mixtures of sodium and potassium, are used as coolants in some nuclear reactors. Predict the normal boiling point of liquid sodium, given that the entropy of vaporization of liquid sodium is $84.8\ \text{J}\cdot\text{K}^{-1}\cdot\text{mol}^{-1}$ and that its enthalpy of vaporization is $98.0\ \text{kJ}\cdot\text{mol}^{-1}$.

STRATEGY At the normal boiling point, a liquid and its vapor are in equilibrium at 1 atm. In other words, there is no change in free energy when one phase changes into the other. We are given the entropy and enthalpy changes of the system. Use these values to find the temperature at which $\Delta G = \Delta H - T\Delta S$ is 0.

SOLUTION We want the temperature at which $\Delta G = 0$:

$$\Delta H_{vap} - T_b\Delta S_{vap} = 0$$

This equation can be solved for T:

$$T_b = \frac{\Delta H_{vap}}{\Delta S_{vap}}$$

from which we find that

$$T_b = \frac{98.0 \times 10^3\ J\cdot mol^{-1}}{84.8\ J\cdot K^{-1}\cdot mol^{-1}} = 1.16 \times 10^3\ K$$

or about 890°C. The experimental value is 883°C.

SELF-TEST 16.8A Predict the melting point of chlorine, given that its enthalpy of fusion is 6.41 $kJ\cdot mol^{-1}$ and its entropy of fusion is 37.3 $J\cdot K^{-1}\cdot mol^{-1}$.

[*Answer:* 172 K (experimental: 172 K)]

SELF-TEST 16.8B Predict the boiling point of methanol, given that its enthalpy of vaporization is 35.3 $kJ\cdot mol^{-1}$ and its entropy of vaporization is 104.7 $J\cdot K^{-1}\cdot mol^{-1}$.

16.12 STANDARD REACTION FREE ENERGIES

We can calculate ΔH_r° and ΔS_r° for a reaction, and from them find the **standard free energy of reaction,** ΔG_r°, where:

$$\Delta G_r^\circ = \Delta H_r^\circ - T\Delta S_r^\circ$$

However, there is a simpler way to find ΔG_r for many reactions. In Section 6.12 we assigned each compound a standard enthalpy of formation, ΔH_f°. We saw that standard enthalpies of formation are handy means of tabulating thermochemical data and that they can be combined to obtain the standard reaction enthalpy of any reaction. The same approach can be used to tabulate free energies of substances.

◀◀ Recall from Section 6.12 that ΔH_f° is the change in enthalpy per mole for the formation of a substance from its elements in their most stable form and under standard conditions.

The **standard free energy of formation,** ΔG_f°, of a compound is the standard reaction free energy per mole for the formation of a compound from its elements in their most stable form and under standard conditions. The most stable form of an element is the state with the lowest free energy (Table 16.4). For example, the standard free energy of formation of hydrogen iodide gas at 25°C is $\Delta G_f^\circ = +1.70\ kJ\cdot mol^{-1}$. It is the standard reaction free energy for

$$\tfrac{1}{2} H_2(g) + \tfrac{1}{2} I_2(s) \longrightarrow HI(g)$$

The standard free energies of formation of elements in their most stable forms are 0.

Standard free energies of formation can be determined in a variety of ways. One straightforward way is to combine the enthalpy and entropy data from Tables 6.5 and 16.1. A list of the resulting values is given in Table 16.5, and a more extensive one appears in Appendix 2A.

TABLE 16.4 Examples of the most stable forms of elements

Element	Most stable form at 25°C and 1 atm
H_2, O_2, Cl_2, Xe	gas
Br_2, Hg	liquid
C	graphite
Na, Fe, I_2	solid

EXAMPLE 16.4 Calculating a standard free energy of formation

Calculate the standard free energy of formation of HI(g) at 25°C from its standard entropy and its standard enthalpy of formation.

STRATEGY First, we write the chemical equation, with the compound of interest having a stoichiometric coefficient of 1. Then we calculate the standard reaction free energy from $\Delta G_r^\circ = \Delta H_r^\circ - T\Delta S_r^\circ$. The standard reaction enthalpy is found from the standard enthalpies of formation by using data from Appendix 2A. The standard reaction entropy is found as in Example 16.2, by using the data from Table 16.1 or Appendix 2A.

SOLUTION The chemical equation is

$$\tfrac{1}{2}\,H_2(g) + \tfrac{1}{2}\,I_2(s) \longrightarrow HI(g)$$

From data in Appendix 2A, and $\Delta H_f^\circ = 0$ for each of the elements,

$$\begin{aligned}\Delta H_r^\circ &= \Delta H_f^\circ(\text{HI, g}) - \{\tfrac{1}{2}\Delta H_f^\circ(\text{H}_2\text{, g}) + \tfrac{1}{2}\Delta H_f^\circ(\text{I}_2\text{, s})\}\\ &= +26.48\ \text{kJ·mol}^{-1}\end{aligned}$$

From data in Appendix 2A, noting that the standard entropies of elements are not 0,

$$\begin{aligned}\Delta S_r^\circ &= S_m^\circ(\text{HI, g}) - \{\tfrac{1}{2}S_m^\circ(\text{H}_2\text{, g}) + \tfrac{1}{2}S_m^\circ(\text{I}_2\text{, s})\}\\ &= 206.6 - \{(\tfrac{1}{2}\times 130.7) + (\tfrac{1}{2}\times 116.1)\}\ \text{J·K}^{-1}\text{·mol}^{-1}\\ &= +83.2\ \text{J·K}^{-1}\text{·mol}^{-1}\end{aligned}$$

Therefore, because $T = 298$ K (corresponding to 25°C),

$$\begin{aligned}\Delta G_r^\circ &= \Delta H_r^\circ - T\Delta S_r^\circ\\ &= 26.48\ \text{kJ·mol}^{-1} - (298\ \text{K}) \times (83.2\times 10^{-3}\ \text{kJ·K}^{-1}\text{·mol}^{-1})\\ &= +1.69\ \text{kJ·mol}^{-1}\end{aligned}$$

By definition, this value is the standard free energy of formation of HI(g), which we write ΔG_f°(HI, g).

SELF-TEST 16.9A Calculate the standard free energy of formation of NH_3(g) at 25°C.

[*Answer:* -16.5 kJ·mol^{-1}]

SELF-TEST 16.9B Calculate the standard free energy of formation of C_3H_6(g), cyclopropane, at 25°C.

The standard free energy of formation of a compound is a measure of its stability relative to its elements. If $\Delta G_f^\circ < 0$ at a certain temperature, then the elements have a spontaneous tendency to form the compound at that temperature. Under standard conditions, the compound is more stable than its elements at that temperature (Fig. 16.22). If ΔG_f° is positive, the reverse reaction, the decomposition of the compound, is

TABLE 16.5 Standard free energies of formation at 25°C*

Substance	ΔG_f°, kJ·mol^{-1}
GASES	
ammonia, NH_3	−16.45
carbon dioxide, CO_2	−394.4
nitrogen dioxide, NO_2	+51.3
water, H_2O	−228.6
LIQUIDS	
benzene, C_6H_6	+124.3
ethanol, C_2H_5OH	−174.8
water, H_2O	−237.1
SOLIDS	
calcium carbonate, $CaCO_3$[†]	−1128.8
silver chloride, AgCl	−109.8

*Additional values are given in Appendix 2A.
[†]Calcite.

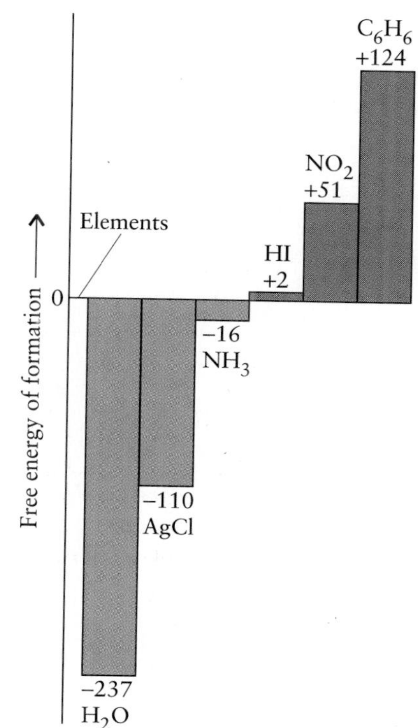

FIGURE 16.22 The standard free energies of formation of compounds are defined as the standard reaction free energy for their formation from the elements. They represent a "thermodynamic altitude" with respect to the elements at "sea level." The numerical values are in kilojoules per mole.

spontaneous. Under standard conditions, the compound has a tendency to decompose into its elements. In the latter case, there is no point in trying to synthesize the compound from its elements at that temperature under standard conditions.

For example, the standard free energy of formation of benzene is $+124\ \text{kJ}\cdot\text{mol}^{-1}$ at 25°C. Therefore, the reaction

$$6\,C(s) + 3\,H_2(g) \longrightarrow C_6H_6(l)$$

(with all the species in their standard states) is not spontaneous. There is no point in trying to make benzene by exposing carbon to hydrogen gas at 25°C and 1 atm, because pure benzene has a higher free energy than graphite and hydrogen at 1 atm. However, the reverse reaction

$$C_6H_6(l) \longrightarrow 6\,C(s) + 3\,H_2(g)$$

has $\Delta G_f^\circ = -124\ \text{kJ}\cdot\text{mol}^{-1}$ and is spontaneous at 25°C. The system and its surroundings become more disordered if pure benzene decomposes into graphite and pure hydrogen gas at 1 atm. We say, therefore, that benzene is thermodynamically unstable with respect to its elements under standard conditions.

A **thermodynamically unstable compound** is a compound with a positive standard free energy of formation. Such a compound has a thermodynamic tendency to decompose into its elements. However, that tendency may not be realized in practice because the decomposition may be very slow. Benzene can, in fact, be kept indefinitely without decomposing at all.

The standard free energy of formation of a substance is the standard free energy of reaction per mole of compound when it is formed from its elements in their most stable forms. The sign of ΔG_f° tells us whether a compound is stable or unstable with respect to its elements.

SELF-TEST 16.10A Is glucose stable relative to its elements at 25°C and under standard conditions? Explain your answer.

[*Answer:* Yes; for glucose, $\Delta G_f^\circ = -910\ \text{kJ}\cdot\text{mol}^{-1}$]

SELF-TEST 16.10B Is methylamine, CH_3NH_2, stable relative to its elements at 25°C and under standard conditions? Explain your answer.

16.13 USING FREE ENERGIES OF FORMATION

◀◀ Compare this expression with that used to find the standard reaction enthalpy (Eq. 6 in Section 16.5): it has the same form.

Standard free energies of formation can be combined to obtain standard reaction free energies by writing

$$\Delta G_r^\circ = \sum n\Delta G_f^\circ(\text{products}) - \sum n\Delta G_f^\circ(\text{reactants}) \qquad (11)$$

where n denotes the unitless stoichiometric coefficients in the chemical equation. For example, for the oxidation of ammonia

$$4\,NH_3(g) + 5\,O_2(g) \longrightarrow 4\,NO(g) + 6\,H_2O(g)$$

we have (from Appendix 2A)

$$\begin{aligned}\Delta G_r^\circ &= \{4\Delta G_f^\circ(\text{NO, g}) + 6\Delta G_f^\circ(\text{H}_2\text{O, g})\} \\ &\quad - \{4\Delta G_f^\circ(\text{NH}_3\text{, g}) + 5\Delta G_f^\circ(\text{O}_2\text{, g})\} \\ &= \{4 \times 86.55 + 6 \times (-228.57)\} - \{4 \times (-16.45) + 0\}\ \text{kJ·mol}^{-1} \\ &= -959.42\ \text{kJ·mol}^{-1}\end{aligned}$$

Standard free energies of formation can be combined to calculate the standard free energy of a reaction.

EXAMPLE 16.5 Calculating the standard reaction free energy from free energies of formation

Calculate the standard free energy of the following reaction from standard free energies of formation:

$$2\,\text{CO(g)} + \text{O}_2\text{(g)} \longrightarrow 2\,\text{CO}_2\text{(g)}$$

STRATEGY Find the standard free energy of reaction from the free energies of formation in Appendix 2A by using Eq. 11, with the stoichiometric coefficients given in the chemical equation.

SOLUTION We substitute the free energies of formation from Appendix 2A for each of the reactants and products:

$$\begin{aligned}\Delta G_r^\circ &= 2\Delta G_f^\circ(\text{CO}_2\text{, g}) - \{2\Delta G_f^\circ(\text{CO, g}) + \Delta G_f^\circ(\text{O}_2\text{, g})\} \\ &= \{2 \times (-394.36)\} - \{2 \times (-137.17) + 0\}\ \text{kJ·mol}^{-1} \\ &= -514.38\ \text{kJ·mol}^{-1}\end{aligned}$$

SELF-TEST 16.11A Calculate the standard free energy for the following reaction from free energies of formation: $2\,\text{SO}_2\text{(g)} + \text{O}_2\text{(g)} \rightarrow 2\,\text{SO}_3\text{(g)}$.

[*Answer:* -141.74 kJ·mol^{-1}]

SELF-TEST 16.11B Calculate the standard free energy for the following reaction for glucose formation from free energies of formation:

$$6\,\text{CO}_2\text{(g)} + 6\,\text{H}_2\text{O(l)} \longrightarrow \text{C}_6\text{H}_{12}\text{O}_6\text{(s)} + 6\,\text{O}_2\text{(g)}$$

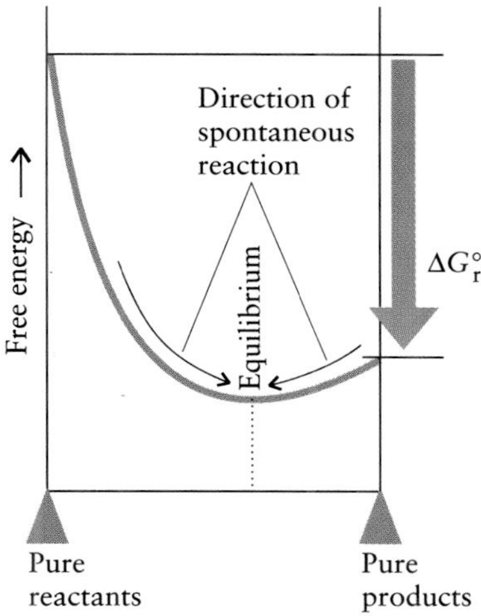

FIGURE 16.23 At constant temperature and pressure, the direction of spontaneous change is toward lower free energy. The equilibrium composition of a reaction mixture corresponds to the lowest point on the curve. In this example, substantial quantities of both reactants and products are present at equilibrium, and K is close to 1.

16.14 FREE ENERGY AND COMPOSITION

When we plot the free energy of a reaction mixture against its changing composition, we get a ⌣-shaped curve (Fig. 16.23). The reaction tends to proceed toward the composition at the lowest point of the curve, because that is the direction of decreasing free energy. *The composition at the lowest point of the curve—the point of minimum free energy—corresponds to equilibrium.* For a system at equilibrium, any change—either the forward or the reverse reaction—would lead to an increase in free energy, so neither change is spontaneous. When the free-energy minimum lies very close to the pure products, the equilibrium composition strongly favors products and "goes to completion" (Fig. 16.24). When the free-energy minimum lies very close to the pure reactants, the equilibrium composition strongly favors the reactants and the reaction "does not go"

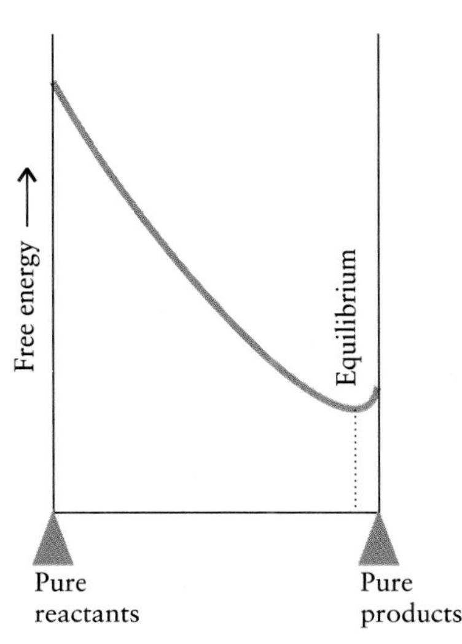

FIGURE 16.24 In this reaction, the free energy is a minimum when products are much more abundant than reactants. The equilibrium lies in favor of the products, and $K \gg 1$. This reaction effectively goes almost to completion.

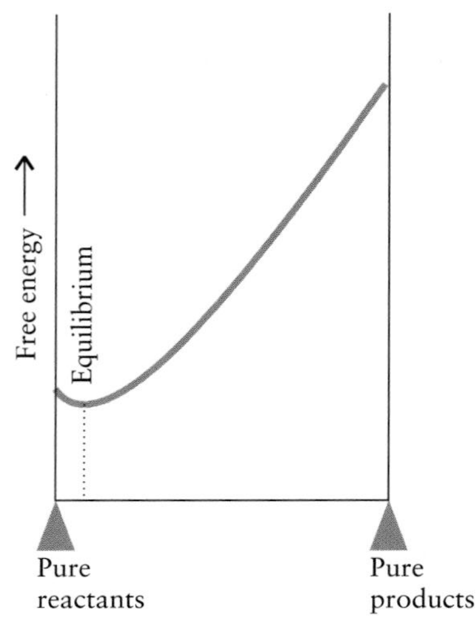

FIGURE 16.25 In this reaction, the free energy is a minimum when the reactants are much more abundant than the products. The equilibrium lies in favor of the reactants, and $K \ll 1$. This reaction "does not go."

(Fig. 16.25). When the minimum lies close to reactants, $K \ll 1$; when it lies close to products, $K \gg 1$. When the minimum lies approximately halfway between pure products and pure reactants, so that all the species are present in similar abundances at equilibrium, K is close to 1 (Fig. 16.23).

To make these ideas quantitative, we need to distinguish between the standard reaction free energy, ΔG_r°, and the reaction free energy, ΔG_r:

> ΔG_r° is the difference in standard free energies of formation of the pure products and the pure reactants.

> ΔG_r is the reaction free energy at a definite, fixed composition of the reaction mixture.

The value of ΔG_r° is the difference in free energy between the pure products and the pure reactants in their standard states. Pressures of gases are all taken to be 1 atm and concentrations of substances in solution are all taken to be 1 $mol{\cdot}L^{-1}$. The value of ΔG_r refers to *any* chosen composition of the reaction mixture and represents the difference in free energy between the products and reactants at the concentrations present at that point in the reaction.

To fix the meaning of ΔG_r, and to see how a change in a quantity can refer to a system at fixed composition, think of a very large reaction mixture—a swimming pool full, for instance—and consider the change in free energy when 1 mol of reactant molecules changes into products. The amount of reaction is so small compared with the huge amount of material present that the change has a negligible effect on the composition of the reaction mixture. For example, suppose nitrogen and hydrogen react to form ammonia in a large container, with 10,000 mol of each substance present. The free energy of reaction is the change in free energy when 1 mol N_2 and 3 mol H_2 are consumed in such an environment.

Recall that ΔG_r changes as the composition of the reaction mixture changes. If the composition of the swimming pool of reaction mixture contains relatively more reactants than it would have at equilibrium, then ΔG_r at that composition will be negative, and the reaction will have a tendency to form more products. If the reaction mixture happens to contain a higher proportion of products than at equilibrium, then ΔG_r will be positive, and the reverse reaction, the formation of reactants, will be spontaneous. However, if the reaction mixture happens to have its equilibrium composition, then ΔG_r will be 0 and there is no tendency for the reaction to run in either direction.

The reaction free energy is related to the composition of the reaction mixture and standard reaction free energy by the formula

The logarithm, ln, is the *natural* logarithm; it is usually labeled "ln" on calculators, but sometimes "$\log_e$."

$$\Delta G_r = \Delta G_r^\circ + RT \ln Q \qquad (12)$$

where Q is the reaction quotient and R is the gas constant, with the value 8.31451 $J{\cdot}K^{-1}{\cdot}mol^{-1}$. If the species are gases, the reaction quotient in Eq. 12 is in terms of partial pressures (Q_p, Section 13.4). If the species are in solution, the reaction quotient is in terms of molar concentrations (Q_c, Section 13.2). Because most standard free energies are tabulated for 25°C, most calculations are carried out for that temperature. We can then use

The fact that the gas constant appears does not mean that we can use this expression only for gases: *R* is a fundamental constant.

$$RT = (8.31451\ J{\cdot}K^{-1}{\cdot}mol^{-1}) \times (298.15\ K) = 2.4790\ kJ{\cdot}mol^{-1}$$

The reaction free energy, ΔG_r, is the difference in free energy between the products and reactants in a reaction mixture with a specific composition; it is calculated from Eq. 12.

EXAMPLE 16.6 Calculating the reaction free energy from the reaction quotient

The standard reaction free energy for

$$N_2(g) + 3\,H_2(g) \rightleftharpoons 2\,NH_3(g)$$

is $\Delta G_r^\circ = -32.90\ \text{kJ·mol}^{-1}$ at 25°C. What is the reaction free energy when the partial pressure of each gas is 100 atm? What is the spontaneous direction of the reaction under these conditions?

STRATEGY Calculate the reaction quotient and substitute it and the data into Eq. 12. If ΔG_r turns out to be negative, then the formation of ammonia is spontaneous at the given composition; if ΔG_r is positive, then ammonia has a spontaneous tendency to decompose into its elements at the given composition.

SOLUTION The reaction quotient is

$$Q_p = \frac{P_{NH_3}^2}{P_{N_2}P_{H_2}^3} = \frac{(100)^2}{100 \times (100)^3} = 1.00 \times 10^{-4}$$

Therefore, the reaction free energy at this composition is

$$\begin{aligned}\Delta G_r &= \Delta G_r^\circ + RT\ln Q\\ &= (-32.90\ \text{kJ·mol}^{-1}) + (2.479\ \text{kJ·mol}^{-1})\ln(1.00 \times 10^{-4})\\ &= -55.7\ \text{kJ·mol}^{-1}\end{aligned}$$

Because the reaction free energy is negative, the formation of products is spontaneous at this composition and temperature.

SELF-TEST 16.12A The standard reaction free energy for $H_2(g) + I_2(g) \rightleftharpoons 2\,HI(g)$ is $\Delta G_r^\circ = -21.1\ \text{kJ·mol}^{-1}$ at 500 K (when $RT = 4.16\ \text{kJ·mol}^{-1}$). What is the value of ΔG_r when the partial pressures of the gases are $P_{H_2} = 1.5$ atm, $P_{I_2} = 0.88$ atm, and $P_{HI} = 0.065$ atm? What is the spontaneous direction of the reaction?

[*Answer:* $-45\ \text{kJ·mol}^{-1}$; toward products]

SELF-TEST 16.12B The standard reaction free energy for $N_2O_4(g) \rightleftharpoons 2\,NO_2(g)$ is $\Delta G_r^\circ = +4.73\ \text{kJ·mol}^{-1}$ at 298 K. What is the value of ΔG_r when the partial pressures of the gases are $P_{N_2O_4} = 0.80$ atm and $P_{NO_2} = 2.10$ atm? What is the spontaneous direction of the reaction?

16.15 FREE ENERGY AND EQUILIBRIUM

Let's see what Eq. 12, $\Delta G_r = \Delta G_r^\circ + RT\ln Q$, becomes at equilibrium. The reaction free energy, ΔG_r, is 0 because the reaction mixture has no tendency to change in either direction. Therefore, at equilibrium, the left-hand side of Eq. 12 is equal to 0. Moreover, at equilibrium, the reaction quotient is the equilibrium constant, K, for the reaction. Therefore, Q can be replaced by K. The equation then becomes

$$0 = \Delta G_r^\circ + RT\ln K$$

and then

$$\Delta G_r^\circ = -RT \ln K \tag{13}$$

Equation 13 is probably one of the most important equations in the whole of chemical thermodynamics. It is the link between the equilibrium constant for any change—physical or chemical—and the standard reaction free energy. We can use Eq. 13 to calculate K whenever free energy data are available.

Notice that ΔG_r° is negative if $K > 1$.

EXAMPLE 16.7 Calculating an equilibrium constant from free energies

Calculate K_p at 25°C for the equilibrium

$$N_2O_4(g) \rightleftharpoons 2\,NO_2(g)$$

STRATEGY Because gases are present, the required value can be found by solving Eq. 13 for K_p. First, however, the standard reaction free energy must be calculated from the standard free energies of formation in Appendix 2A.

SOLUTION From the data in Appendix 2A:

$$\begin{aligned}\Delta G_r^\circ &= 2\Delta G_f^\circ(NO_2, g) - \Delta G_f^\circ(N_2O_4, g)\\ &= 2 \times 51.31 - 97.89\ \text{kJ·mol}^{-1} = +4.73\ \text{kJ·mol}^{-1}\end{aligned}$$

After rearranging Eq. 13 and substituting the value of ΔG_r° (at 25°C) and $RT = 2.4790\ \text{kJ·mol}^{-1}$, we have

$$\ln K_p = -\frac{\Delta G_r^\circ}{RT} = -\frac{4.73\ \text{kJ·mol}^{-1}}{2.4790\ \text{kJ·mol}^{-1}} = -1.91$$

The natural antilogarithm of -1.91 is 0.15, so for this reaction $K_p = 0.15$.

Press the e^x button on a calculator to obtain the natural antilogarithm of x. This key is sometimes labeled "inv ln."

SELF-TEST 16.13A Calculate K_p for the reaction $N_2(g) + 3\,H_2(g) \rightleftharpoons 2\,NH_3(g)$ at 25°C.

[*Answer:* 5.8×10^5]

SELF-TEST 16.13B Calculate K_p for the reaction $2\,NO(g) + O_2(g) \rightleftharpoons 2\,NO_2(g)$ at 25°C.

Because Eq. 13 gives the connection between K and ΔG_r°, we can begin to understand why some reactions have large equilibrium constants and others have small ones. If ΔG_r° is positive, then $\ln K$ is negative, which means that K is less than 1. For such a reaction, the reactants are favored. They are strongly favored if, as well as being positive, ΔG_r° is large. Because $\Delta G_r^\circ = \Delta H_r^\circ - T\Delta S_r^\circ$ we see that ΔG_r° depends on the value of ΔH_r°, and hence is likely to be large and positive for strongly endothermic reactions. These reactions are therefore likely to have K much smaller than 1 and are unlikely to form much product. Only if $T\Delta S_r^\circ$ is large and positive is the equilibrium constant K likely to exceed 1 for endothermic reactions.

Conversely, if a reaction is strongly exothermic, then ΔH_r° is large and negative, and ΔG_r° is likely to be large and negative too. Now $\ln K$ will be large and positive, which means in turn that K will be much larger than 1. Therefore, we can expect a strongly exothermic reaction to have a large K, and hence go to completion. Only if the enthalpy change is small and the entropy of the system decreases strongly is K likely to fall below

CHEMISTRY AT WORK

In her quest for renewable fuels, a biotechnologist tests a sample of oils produced by microalgae. She will determine the standard free energies of formation for the various oils to decide which can be converted into fuels that burn in diesel engines with the minimum of pollution.

1 for exothermic reactions. These features have already been summarized in Table 16.3, but now we see how they relate to K.

The equilibrium constant of a reaction is related to the standard free energy of reaction by $\Delta G_r^\circ = -RT \ln K$.

16.16 THE EFFECT OF TEMPERATURE

A reaction that is not spontaneous at one temperature can become spontaneous at a different temperature. For an exothermic reaction with a negative reaction entropy (so ΔH_r° and ΔS_r° are both negative), a temperature can be found above which the reaction is no longer spontaneous, in the sense that ΔG_r° becomes positive and K falls below 1. For such a reaction, ΔG_r° is negative at low temperatures, but it may become positive at higher temperatures when $T\Delta S_r^\circ$ is more negative than ΔH_r°. The reverse is true for an endothermic reaction with a positive reaction entropy. In this case, ΔG_r° is positive at low temperatures but may become negative when the temperature is raised to the point that $T\Delta S_r^\circ$ is larger than ΔH_r°. Such a reaction may become spontaneous, in the sense that ΔG_r° is negative and K is greater than 1, when the temperature is raised sufficiently. This point is illustrated in the following example. For an endothermic reaction with a negative reaction entropy, there is no temperature at which the reaction has K greater than 1 (provided the reaction enthalpy or entropy do not change sign at some temperature).

The free energy increases with temperature for reactions with a negative ΔS_r° and decreases with temperature for reactions with a positive ΔS_r°.

EXAMPLE 16.8 Estimating the minimum temperature at which a reaction for which ΔH and ΔS are both positive can occur spontaneously.

Estimate the temperature at which it is thermodynamically possible for carbon to reduce iron(III) oxide to iron by the reaction

$$2\,Fe_2O_3(s) + 3\,C(s) \longrightarrow 4\,Fe(s) + 3\,CO_2(g)$$

STRATEGY The reaction becomes feasible in the sense that $K > 1$ when the reaction free energy, $\Delta G_r^\circ = \Delta H_r^\circ - T\Delta S_r^\circ$, becomes negative. An endothermic reaction therefore becomes spontaneous at temperatures above

$$T = \frac{\Delta H_r^\circ}{\Delta S_r^\circ}$$

Calculate the standard reaction enthalpy and entropy from the data in Appendix 2A, assuming that neither changes appreciably with temperature.

SOLUTION From data in Appendix 2A, and recalling that the standard enthalpies of formation of the elements are 0,

$$\begin{aligned}\Delta H_r^\circ &= 3\Delta H_f^\circ(CO_2, g) - 2\Delta H_f^\circ(Fe_2O_3, s)\\ &= 3 \times (-393.5) - 2 \times (-824.2)\ kJ\cdot mol^{-1} = +467.9\ kJ\cdot mol^{-1}\end{aligned}$$

Similarly, but noting that the standard molar entropies of elements are not 0,

$$\begin{aligned}\Delta S_r^\circ &= \{4S_m^\circ(Fe, s) + 3S_m^\circ(CO_2, g)\} - \{2S_m^\circ(Fe_2O_3, s) + 3S_m^\circ(C, s)\}\\ &= (4 \times 27.3 + 3 \times 213.7) - (2 \times 87.4 + 3 \times 5.7)\ J\cdot K^{-1}\cdot mol^{-1}\\ &= +558.4\ J\cdot K^{-1}\cdot mol^{-1}\end{aligned}$$

The temperature at which ΔG_r° changes from positive to negative is

$$T = \frac{\Delta H_r^\circ}{\Delta S_r^\circ} = \frac{467.9 \times 10^3\ J\cdot mol^{-1}}{558.4\ J\cdot K^{-1}\cdot mol^{-1}} = 838\ K$$

The minimum temperature at which reduction occurs is therefore about 565°C.

SELF-TEST 16.14A What is the minimum temperature at which magnetite (Fe_3O_4) can be reduced to iron by using carbon (to produce CO_2)?

[*Answer:* 943 K]

SELF-TEST 16.14B Estimate the temperature at which magnesium carbonate can be expected to decompose to magnesium oxide and carbon dioxide.

SKILLS YOU SHOULD HAVE MASTERED

Conceptual

1. Distinguish the three types of thermodynamic systems.
2. Distinguish heat and work.
3. State and explain the implications of the first and second laws of thermodynamics.
4. Distinguish ΔU and ΔH and show how they are related.
5. Explain how temperature, volume, and state of matter affect the entropy of a substance.
6. Show how ΔS_{surr} is related to ΔH for a change at constant temperature and pressure and justify the relationship.
7. Show how the free energy change accompanying a process is related to the direction of spontaneous reaction.

Problem-Solving

1. Calculate the change in internal energy due to heat and work.
2. Predict which of two systems has the greater entropy, given their compositions and conditions.
3. Calculate the standard reaction entropy from standard molar entropies.
4. Predict the boiling point and melting point of a substance from the changes in entropy and enthalpy of the substance.

5. Predict the temperature at which a process with a known ΔH and ΔS becomes spontaneous.

6. Calculate a standard reaction free energy from the standard enthalpy of reaction and standard molar entropies.

7. Calculate the standard reaction free energy from free energies of formation.

8. Calculate the reaction free energy from ΔG_r° and the composition of the reaction mixture.

9. Calculate an equilibrium constant from ΔG_r°.

Descriptive

1. Describe the criteria for spontaneity of a reaction.

2. Identify thermodynamically unstable compounds from their free energies of formation.

EXERCISES

The First Law of Thermodynamics

16.1 Identify the following systems as open, closed, or isolated: (a) coffee in a perfectly insulated thermos; (b) coolant in a refrigerator coil; (c) the combustion of benzene in a bomb calorimeter.

16.2 Identify the following systems as open, closed, or isolated: (a) gasoline burning in an automobile engine; (b) mercury in a thermometer; (c) a living plant.

16.3 A gas sample is heated in a cylinder, using 550 kJ of heat. A piston compresses the gas, using 700 kJ of work. What is the change in internal energy of the gas during this process?

16.4 Calculate the work for a system that absorbs 150 kJ of heat in a process for which the increase in internal energy is 120 kJ. Is work done on or by the system during this process?

16.5 A 100-W electric heater ($1\ W = 1\ J{\cdot}s^{-1}$) operates for 20 min to heat the gas in a cylinder. The gas expands from 2.0 L to 2.5 L against an atmospheric pressure of 1.0 atm. What is the change in internal energy of the gas? (1 L·atm = 101.325 J.)

16.6 A gas in a cylinder was placed in a heater and gained 7000 kJ of heat. If the cylinder increased in volume from 700 mL to 1450 mL against an atmospheric pressure of 750 Torr during this process, what is the change in internal energy of the gas in the cylinder? (1L·atm = 101.325 J.)

16.7 The change in internal energy for the combustion of 1 mol CH_4 in a cylinder according to the reaction $CH_4(g) + 2\,O_2(g) \rightarrow CO_2(g) + 2\,H_2O(g)$ is −892.4 kJ. If a piston connected to the cylinder performs 492 kJ of expansion work due to the combustion, how much heat is lost from the system (the reaction mixture)?

16.8 In a combustion cylinder, the total internal energy change produced from the burning of a fuel is −1740 kJ. The cooling system that surrounds the cylinder absorbs 470 kJ as heat. How much work can be done by the fuel in the cylinder?

16.9 For a certain reaction at constant pressure, $\Delta U = -65$ kJ and 28 kJ of expansion work is done by the system. What is ΔH for this process?

16.10 For a certain reaction at constant pressure, $\Delta H = -15$ kJ and 22 kJ of expansion work is done *on* the system. What is ΔU for this process?

Entropy

16.11 Which substance in each pair has the higher molar entropy at 298 K: (a) HBr(g), HF(g); (b) NH_3(g), Ne(g); (c) I_2(s), I_2(l); (d) Ar(g) at 1.00 atm or Ar(g) at 2.00 atm?

16.12 Which substance in each pair has the higher molar entropy? (Assume the temperature is 298 K unless otherwise specified.) (a) CH_4(g), C_3H_8(g); (b) KCl(aq), KCl(s); (c) He(g), Kr(g); (d) O_2(g) at 273 K and 1.00 atm, O_2(g) at 450 K and 1.00 atm.

16.13 List the following substances in order of increasing molar entropy at 298 K: H_2O(l), H_2O(g), H_2O(s), C(s). Explain your reasoning.

16.14 List the following substances in order of increasing molar entropy at 298 K: CO_2(g), Ar(g), H_2O(l), Ne(g). Explain your reasoning.

16.15 Without performing any calculations, state whether the entropy of the system increases or decreases during each of the following processes: (a) the oxidation of nitrogen $N_2(g) + 2\,O_2(g) \rightarrow 2\,NO_2(g)$; (b) the sublimation of dry ice: $CO_2(s) \rightarrow CO_2(g)$; (c) the cooling of water from 50°C to 4°C. Explain your reasoning.

16.16 Without performing any calculations, state whether the entropy of the system increases or decreases during each of the following processes: (a) the dissolution of table salt, NaCl(s) → NaCl(aq); (b) the photosynthesis of glucose: $6\,CO_2(g) + 6\,H_2O(l) \rightarrow C_6H_{12}O_6(s) + 6\,O_2(g)$; (c) the evaporation of water from damp clothes. Explain your reasoning.

16.17 Use data from Table 6.2 or Appendix 2A to

calculate the entropy change for (a) the freezing of 1 mol H_2O at 0°C; (b) the vaporization of 50.0 g ethanol, C_2H_5OH, at 351.5 K.

16.18 Use data from Table 6.2 or Appendix 2A to calculate the entropy change for (a) the vaporization of 1 mol H_2O at 100°C and 1 atm; (b) the freezing of 3.33 g NH_3 at 195.3 K.

16.19 Use data from Table 16.1 or Appendix 2A to calculate the standard reaction entropy for each of the following reactions at 25°C. For each reaction, interpret the sign and magnitude of the reaction entropy.
(a) The combustion of hydrogen:
$2\,H_2(g) + O_2(g) \longrightarrow 2\,H_2O(g)$
(b) The oxidation of carbon monoxide:
$2\,CO(g) + O_2(g) \longrightarrow 2\,CO_2(g)$
(c) The decomposition of calcite:
$CaCO_3(s) \longrightarrow CaO(s) + CO_2(g)$
(d) The following reaction of potassium chlorate:
$4\,KClO_3(s) \longrightarrow 3\,KClO_4(s) + KCl(s)$

16.20 Use data from Table 16.1 or Appendix 2A to calculate the standard reaction entropy for each of the following reactions at 25°C. For each reaction, interpret the sign and magnitude of the reaction entropy.
(a) The synthesis of carbon disulfide from natural gas (methane): $2\,CH_4(g) + S_8(s) \longrightarrow 2\,CS_2(l) + 4\,H_2S(g)$
(b) The production of acetylene from calcium carbide:
$CaC_2(s) + 2\,H_2O(l) \longrightarrow Ca(OH)_2(s) + C_2H_2(g)$
(c) The oxidation of ammonia, the first step in the production of nitric acid:
$4\,NH_3(g) + 5\,O_2(g) \longrightarrow 4\,NO(g) + 6\,H_2O(l)$
(d) The industrial synthesis of urea, a common fertilizer:
$CO_2(g) + 2\,NH_3(g) \longrightarrow CO(NH_2)_2(s) + H_2O(l)$

Entropy Change in the Surroundings

16.21 Account for the spontaneity of the following processes in terms of the entropy changes in the system and the surroundings: (a) hot coffee cools; (b) an organic compound dissolves in water with no change in temperature; (c) gasoline burns; (d) helium gas diffuses through argon gas.

16.22 Account for the spontaneity of the following processes in terms of the entropy changes in the system and the surroundings: (a) sodium reacts with chlorine to produce sodium chloride, giving off heat and light; (b) water boils at 100.1°C at 1 atm; (c) water freezes at −0.01°C; (d) milk spills out of an overturned bottle.

16.23 Calculate the change in the entropy of the surroundings when (a) 120 kJ is released to the surroundings at 25°C; (b) 120 kJ is released into the surroundings at 100°C; (c) 100 J is absorbed from the surroundings at 50°C.

16.24 Calculate the change in the entropy of the surroundings when (a) 1 mJ is released to the surroundings at 2×10^{-7} K; (b) 1 J, the energy of a single heartbeat, is released to the surroundings at 37°C (normal body temperature); (c) 20 J, the energy released when 1 mol He freezes at 3.5 K, is absorbed by the surroundings.

16.25 A human body generates heat at the rate of about 100 W (1 W = 1 $J{\cdot}s^{-1}$). (a) At what rate does your body heat generate entropy in your surroundings, taken to be at 20°C? (b) How much entropy do you generate each day? (c) Would the entropy generated be greater or less if you were in a room kept at 30°C? Explain your answer.

16.26 An electric heater is rated at 2 kW (1 W = 1 $J{\cdot}s^{-1}$). (a) At what rate does it generate entropy in a room maintained at 28°C? (b) How much entropy does it generate in the course of a day? (c) Would the entropy generated be greater or less if the room were maintained at 25°C? Explain your answer.

16.27 (a) Calculate the change in entropy of a block of copper at 25°C that absorbs 5 J of energy from a heater. (b) If the block of copper is at 100°C and it absorbs 5 J of energy from the heater, what is its entropy change? (c) Explain any difference in entropy change.

16.28 (a) Calculate the change in entropy of 1.0 L of liquid water at 0°C when it absorbs 500 J of energy from a heater. (b) If the 1.0 L of water is at 99°C, what is its entropy change? (c) Explain any difference in entropy change.

16.29 Consider the reaction for the production of formaldehyde:

$$H_2(g) + CO(g) \longrightarrow HCHO(g)$$

$$\Delta H° = +1.96\ kJ,\ \Delta S° = -109.58\ J{\cdot}K^{-1}$$

Calculate the change in entropy of the surroundings and predict the spontaneity of the reaction at 25°C.

16.30 Consider the reaction in which deuterium exchanges with ordinary hydrogen in water:

$$D_2(g) + H_2O(l) \longrightarrow H_2(g) + D_2O(l)$$

$$\Delta H° = -7.38\ kJ,\ \Delta S° = +114.15\ J{\cdot}K^{-1}$$

Calculate the change in entropy of the surroundings and predict the spontaneity of the reaction at 25°C.

16.31 Why are many exothermic reactions spontaneous?

16.32 Explain how an endothermic reaction can be spontaneous.

Free Energy

16.33 Predict the sign of ΔG for a reaction that is (a) exothermic and accompanied by an increase in entropy; (b) endothermic and accompanied by an increase in entropy. (c) Can a temperature change affect the sign of ΔG in (a) or (b)? If so, how?

16.34 Predict the sign of ΔG for a reaction that is (a) exothermic and accompanied by a decrease in entropy; (b) endothermic and accompanied by a decrease in entropy. (c) Can a temperature change affect the sign of ΔG in (a) or (b)? If so, how?

16.35 Use the information in Table 6.2 to calculate the change in entropy of the surroundings and the system for (a) the vaporization of 1.0 mol CH_4 at its normal boiling point; (b) the melting of 1.0 mol C_2H_5OH at its normal melting point; (c) the freezing of 1.0 mol C_2H_5OH at its normal freezing point.

16.36 Use the information in Table 6.2 to calculate the change in entropy of the surroundings and the system for (a) the melting of 1.0 mol NH_3 at its normal melting point; (b) the freezing of 1.0 mol CH_3OH at its normal freezing point; (c) the vaporization of 1.0 mol H_2O at its normal boiling point.

16.37 Use the information in Table 6.2 to calculate the entropy change of the surroundings and the system when 1.0 mol of each of the following substances evaporates at its normal boiling point: (a) water, H_2O; (b) ammonia, NH_3; (c) methanol, CH_3OH.

16.38 Use the information in Table 6.2 to calculate the entropy change of the surroundings and the system when 1.0 mol of each of the following substances evaporates at its normal boiling point: (a) benzene, C_6H_6; (b) ethanol, C_2H_5OH; (c) argon, Ar.

16.39 The entropy of vaporization of liquid chlorine is $85.4\ J\cdot K^{-1}\cdot mol^{-1}$ and its enthalpy of vaporization is $20.4\ kJ\cdot mol^{-1}$. What is its boiling point?

16.40 Estimate the boiling point of fluorine by assuming that its enthalpy and entropy of vaporization are similar to the values for argon (see Table 6.2).

16.41 The entropy of vaporization of benzene is approximately $85\ J\cdot K^{-1}\cdot mol^{-1}$. (a) Estimate the enthalpy of vaporization of benzene at its normal boiling point of 80°C. (b) What is the entropy change of the surroundings when 10 g of benzene, C_6H_6, vaporizes at its normal boiling point?

16.42 The entropy of vaporization of acetone is approximately $85\ J\cdot K^{-1}\cdot mol^{-1}$. (a) Estimate the enthalpy of vaporization of acetone at its normal boiling point of 56.2°C. (b) What is the entropy change of the surroundings when 10 g of acetone, CH_3COCH_3, condenses at its normal boiling point?

Standard Reaction Free Energies

16.43 Calculate the standard free energy for each reaction, using the equation $\Delta G_r^\circ = \Delta H_r^\circ - T\Delta S_r^\circ$ and data from Appendix 2A. (Note that ΔS_r° for each reaction was calculated in Exercise 16.19.) For each reaction, interpret the sign and magnitude of the free energy.
(a) The combustion of hydrogen:
$2\,H_2(g) + O_2(g) \longrightarrow 2\,H_2O(g)$
(b) The oxidation of carbon monoxide:
$2\,CO(g) + O_2(g) \longrightarrow 2\,CO_2(g)$
(c) The decomposition of calcite:
$CaCO_3(s) \longrightarrow CaO(s) + CO_2(g)$
(d) The following reaction of potassium chlorate:
$4\,KClO_3(s) \longrightarrow 3\,KClO_4(s) + KCl(s)$

16.44 Calculate the standard free energy for each reaction, using the equation $\Delta G_r^\circ = \Delta H_r^\circ - T\Delta S_r^\circ$ and data from Appendix 2A. (Note that ΔS_r° for each reaction was calculated in Exercise 16.20.) For each reaction, interpret the sign and magnitude of the free energy.
(a) The synthesis of carbon disulfide from natural gas (methane): $2\,CH_4(g) + S_8(s) \longrightarrow 2\,CS_2(l) + 4\,H_2S(g)$
(b) The production of acetylene from calcium carbide:
$CaC_2(s) + 2\,H_2O(l) \longrightarrow Ca(OH)_2(s) + C_2H_2(g)$
(c) The oxidation of ammonia, the first step in the production of nitric acid:
$4\,NH_3(g) + 5\,O_2(g) \longrightarrow 4\,NO(g) + 6\,H_2O(l)$
(d) The industrial synthesis of urea, a common fertilizer:
$CO_2(g) + 2\,NH_3(g) \longrightarrow CO(NH_2)_2(s) + H_2O(l)$

16.45 Write a chemical equation for the formation reaction and then calculate the standard free energy of formation of each of the following compounds from the enthalpies of formation and the standard molar entropies, using $\Delta G_r^\circ = \Delta H_r^\circ - T\Delta S_r^\circ$: (a) $NH_3(g)$; (b) $H_2O(g)$; (c) $CO(g)$; (d) $NO_2(g)$.

16.46 Write a chemical equation for the formation reaction and then calculate the standard free energy of formation of each of the following compounds from the enthalpies of formation and the standard molar entropies, using $\Delta G_r^\circ = \Delta H_r^\circ - T\Delta S_r^\circ$: (a) $SO_3(g)$; (b) $C_6H_6(l)$; (c) $C_2H_5OH(l)$; (d) $CaCO_3(s)$.

16.47 Use the standard free energies of formation in Appendix 2A to calculate the standard free energy of each of the following reactions at 25°C. Comment on the spontaneity of each reaction at 25°C.
(a) $2\,SO_2(g) + O_2(g) \longrightarrow 2\,SO_3(g)$
(b) $CaCO_3(s, \text{calcite}) \longrightarrow CaO(s) + CO_2(g)$
(c) $SbCl_5(g) \longrightarrow SbCl_3(g) + Cl_2(g)$
(d) $2\,C_8H_{18}(l) + 25\,O_2(g) \longrightarrow 16\,CO_2(g) + 18\,H_2O(l)$

16.48 Use the standard free energies of formation in Appendix 2A to calculate the standard free energy of each of the following reactions at 25°C. Comment on the spontaneity of each reaction at 25°C.
(a) $NH_4Cl(s) \longrightarrow NH_3(g) + HCl(g)$

(b) $H_2(g) + D_2O(l) \longrightarrow D_2(g) + H_2O(l)$
(c) $2\,NO_2(g) \longrightarrow N_2O_4(g)$
(d) $2\,CH_3OH(g) + 3\,O_2(g) \longrightarrow 2\,CO_2(g) + 4\,H_2O(l)$

Thermodynamic Stability

16.49 Determine which of the following compounds are stable with respect to decomposition into their elements at 25°C (see Appendix 2A): (a) $PCl_5(g)$; (b) $HCN(g)$; (c) $NO(g)$; (d) $SO_2(g)$.

16.50 Determine which of the following compounds are stable with respect to decomposition into their elements at 25°C (see Appendix 2A): (a) $CuO(s)$; (b) $C_6H_{12}(l)$, cyclohexane; (c) $SbCl_3(g)$; (d) $N_2H_4(l)$.

16.51 On the basis of the equation $\Delta G_r^\circ = \Delta H_r^\circ - T\Delta S_r^\circ$, which of the following compounds becomes more unstable with respect to its elements as the temperature is raised: (a) $PCl_5(g)$; (b) $HCN(g)$; (c) $NO(g)$; (d) $SO_2(g)$?

16.52 On the basis of the equation $\Delta G_r^\circ = \Delta H_r^\circ - T\Delta S_r^\circ$, which of the following compounds becomes more unstable with respect to its elements as the temperature is raised: (a) $CuO(s)$; (b) $C_6H_{12}(l)$, cyclohexane; (c) $PCl_3(g)$; (d) $N_2H_4(l)$?

16.53 Does potassium chlorate have a thermodynamic tendency to form potassium perchlorate and potassium chloride at 25°C? (See your answer to Exercise 16.43d.) Would the tendency be stronger at a higher or lower temperature?

16.54 Does methanol have a thermodynamic tendency to decompose into carbon monoxide and hydrogen at 25°C and 1 atm? Would the tendency be stronger at a higher or lower temperature?

Free Energy and Equilibrium

16.55 How does the free energy of a reaction change as it approaches equilibrium?

16.56 Use entropy and free energy arguments to account for the observation that ice melts if the temperature is raised above 0°C and water freezes if the temperature is cooled below 0°C.

16.57 Calculate the equilibrium constant at 25°C for each of the following reactions. (Note that the standard free energy of each reaction was determined in Exercise 16.43.)
(a) The combustion of hydrogen:
$2\,H_2(g) + O_2(g) \longrightarrow 2\,H_2O(g)$
(b) The oxidation of carbon monoxide:
$2\,CO(g) + O_2(g) \longrightarrow 2\,CO_2(g)$
(c) The decomposition of limestone:
$CaCO_3(s) \longrightarrow CaO(s) + CO_2(g)$

16.58 Calculate the equilibrium constant at 25°C for each reaction. (Note that the standard free energy of each reaction was determined in Exercise 16.44.)
(a) The synthesis of carbon disulfide from natural gas (methane): $2\,CH_4(g) + S_8(s) \longrightarrow 2\,CS_2(l) + 4\,H_2S(g)$
(b) The production of acetylene from calcium carbide:
$CaC_2(s) + 2\,H_2O(l) \longrightarrow Ca(OH)_2(s) + C_2H_2(g)$
(c) The oxidation of ammonia, the first step in the production of nitric acid:
$4\,NH_3(g) + 5\,O_2(g) \longrightarrow 4\,NO(g) + 6\,H_2O(l)$
(d) The industrial synthesis of urea, a common fertilizer:
$CO_2(g) + 2\,NH_3(g) \longrightarrow CO(NH_2)_2(s) + H_2O(l)$

16.59 Calculate the standard free energy of each of the following reactions:
(a) $N_2(g) + 3\,H_2(g) \rightleftharpoons 2\,NH_3(g)$ $\quad K_P = 41$ at 400 K
(b) $2\,SO_2(g) + O_2(g) \rightleftharpoons 2\,SO_3(g)$
$K_P = 3.0 \times 10^4$ at 700 K

16.60 Calculate the standard free energy for each of the following reactions:
(a) $H_2(g) + I_2(g) \rightleftharpoons 2\,HI(g)$ $\quad K_P = 160$ at 500 K
(b) $N_2O_4(g) \rightleftharpoons 2\,NO_2(g)$ $\quad K_P = 47.9$ at 400 K

16.61 If $Q_P = 1.0$ for the reaction $N_2(g) + O_2(g) \rightleftharpoons 2\,NO(g)$ at 25°C, will the reaction have a tendency to form products or reactants, or will it be at equilibrium?

16.62 If $Q_P = 1.0 \times 10^{50}$ for the reaction $C(s) + O_2(g) \rightleftharpoons CO_2(g)$ at 25°C, will the reaction have a tendency to form products or reactants, or will it be at equilibrium?

16.63 Calculate the reaction free energy of $I_2(g) \rightarrow 2\,I(g)$ at 1200 K ($K_c = 6.8 \times 10^{-2}$) when the concentrations of I_2 and I are 0.026 $mol \cdot L^{-1}$ and 0.0084 $mol \cdot L^{-1}$, respectively. What is the spontaneous direction of the reaction?

16.64 Calculate the reaction free energy of $PCl_3(g) + Cl_2(g) \rightarrow PCl_5(g)$ at 230°C when the concentrations of PCl_3, Cl_2, and PCl_5 are 0.22 $mol \cdot L^{-1}$, 0.41 $mol \cdot L^{-1}$, and 1.33 $mol \cdot L^{-1}$, respectively. What is the spontaneous direction of reaction, given that $K_c = 49$ at 230°C?

16.65 (a) Calculate the reaction free energy of $N_2(g) + 3\,H_2(g) \rightleftharpoons 2\,NH_3(g)$ when the partial pressures of N_2, H_2, and NH_3 are 1.0 atm, 4.2 atm, and 63 atm, respectively, and the temperature is 400 K. For this reaction, $K_P = 41$ at 400 K. (b) Indicate whether this reaction mixture is likely to form reactants, is likely to form products, or is at equilibrium.

16.66 (a) Calculate the reaction free energy of $H_2(g) + I_2(g) \rightleftharpoons 2\,HI(g)$ when the concentrations are 0.026 $mol \cdot L^{-1}$ (H_2), 0.33 $mol \cdot L^{-1}$ (I_2), and 1.84 $mol \cdot L^{-1}$ (HI) and the temperature is 700 K. For this reaction $K_c = 54$ at 700 K. (b) Indicate whether this reaction mixture is likely to form reactants, is likely to form products, or is at equilibrium.

16.67 Use data from Appendix 2A to calculate K_{sp} for each of the following compounds: (a) AgI; (b) $CaCO_3$ (aragonite); (c) FeS, (d) $Ca(OH)_2$.

16.68 Use data from Appendix 2A to calculate K_{sp} for each of the following compounds: (a) AgBr; (b) AgCl; (c) $PbBr_2$; (d) $MgCO_3$.

16.69 Calculate the standard reaction free energy for the autoprotolysis of water, $2\,H_2O(l) \rightarrow H_3O^+(aq) + OH^-(aq)$, from K_w at 25°C.

16.70 Calculate the standard reaction free energy for the ionization of formic acid, $HCOOH(aq) + H_2O(l) \rightarrow H_3O^+(aq) + HCO_2^-(aq)$, from its K_a at 25°C.

16.71 For the reaction, $N_2O_4(g) \rightleftharpoons 2\,NO_2(g)$, $\Delta H_r^\circ = +57.2\ kJ{\cdot}mol^{-1}$ and $\Delta S_r^\circ = +175.83\ J{\cdot}K^{-1}{\cdot}mol^{-1}$ at 25°C. (a) What is the standard reaction free energy at 25°C? (b) What is the standard reaction free energy at 75°C? (Assume that ΔH_r° and ΔS_r° are unaffected by temperature changes.) (c) Determine the temperature at which $\Delta G_r^\circ = 0$ and the value of K_P at that temperature.

16.72 For the reaction, $2\,NO(g) + O_2(g) \rightleftharpoons 2\,NO_2(g)$, $\Delta H_r^\circ = -114.1\ kJ{\cdot}mol^{-1}$ and $\Delta S_r^\circ = -146.54\ J{\cdot}K^{-1}{\cdot}mol^{-1}$.
(a) What is the standard reaction free energy at 25°C?
(b) What is the reaction free energy at 700°C? (Assume that ΔH_r° and ΔS_r° are unaffected by temperature changes.)
(c) Temperatures in an internal combustion cylinder approach 2500°C. In the presence of an ample supply of oxygen, which nitrogen oxide is favored at that temperature? (d) Which nitrogen oxide is favored at the temperature of the exhaust gases, which is approximately 50°C?

16.73 (a) Do $Cu^+(aq)$ ions have a thermodynamic tendency to form $Cu^{2+}(aq)$ ions and copper metal in water at room temperature? (b) Is the tendency greater at higher temperatures or lower temperatures?

16.74 (a) At what temperature does $NO_2(g)$ have a thermodynamic tendency to form $N_2O_4(g)$? (b) Is the tendency greater at higher temperatures or lower temperatures?

16.75 Determine whether iron(III) oxide can be reduced by carbon at 1000 K in each of the following reactions:
(a) $Fe_2O_3(s) + 3\,C(s) \longrightarrow 2\,Fe(s) + 3\,CO(g)$
(b) $2\,Fe_2O_3(s) + 3\,C(s) \longrightarrow 4\,Fe(s) + 3\,CO_2(g)$
given that at 1000 K $\Delta G_f^\circ(CO, g) = -200\ kJ{\cdot}mol^{-1}$, $\Delta G_f^\circ(CO_2, g) = -396\ kJ{\cdot}mol^{-1}$, and $\Delta G_f^\circ(Fe_2O_3, s) = -562\ kJ{\cdot}mol^{-1}$.

16.76 Determine whether manganese(IV) oxide can be reduced by carbon at 1000 K in each of the following reactions:
(a) $MnO_2(s) + 2\,C(s) \longrightarrow Mn(s) + 2\,CO(g)$
(b) $MnO_2(s) + C(s) \longrightarrow Mn(s) + CO_2(g)$
given that at 1000 K $\Delta G_f^\circ(CO, g) = -200\ kJ{\cdot}mol^{-1}$, $\Delta G_f^\circ(CO_2, g) = -396\ kJ{\cdot}mol^{-1}$, and $\Delta G_f^\circ(MnO_2, s) = -405\ kJ{\cdot}mol^{-1}$.

SUPPLEMENTARY EXERCISES

16.77 (a) Describe three ways in which you could increase the internal energy of an open system. (b) Which of these methods could you use to increase the internal energy of a closed system? (c) Which, if any, of these methods could you use to increase the internal energy of an isolated system?

16.78 The internal energy of a system increased by 400 J when it absorbed 600 J of heat. (a) Was work done by or on the system? (b) How much work was done?

16.79 (a) Distinguish between ΔU and ΔH for a reaction. (b) Under what circumstances are ΔH and ΔU equal?

16.80 How does entropy differ from enthalpy? Why are the two thermodynamic properties both needed?

16.81 How would the entropy of water vapor change if it were compressed at 150°C?

16.82 The standard molar entropy of Cl(g) is 165.2 $J{\cdot}K^{-1}{\cdot}mol^{-1}$ and that of $Cl_2(g)$ is 223.07 $J{\cdot}K^{-1}{\cdot}mol^{-1}$. Suggest a reason why the standard molar entropy of the diatomic molecules is greater than that of the atoms at 25°C.

16.83 Without performing any calculations, state whether an increase or decrease in entropy occurs for each of the following processes:
(a) $Cl_2(g) + H_2O(l) \longrightarrow HCl(aq) + HClO(aq)$
(b) $Cu_3(PO_4)_2(s) \longrightarrow 3\,Cu^{2+}(aq) + 2\,PO_4^{3-}(aq)$
(c) $SO_2(g) + Br_2(g) + 2\,H_2O(g) \longrightarrow H_2SO_4(aq) + 2\,HBr(g)$
(d) $2\,Fe(s) + 3\,Cl_2(g) \longrightarrow 2\,FeCl_3(s)$

16.84 Use data from Table 16.1 and Appendix 2A to calculate the standard reaction entropy for each of the following reactions and interpret the sign and magnitude of each answer.
(a) The decomposition of hydrogen peroxide: $2\,H_2O_2(l) \longrightarrow 2\,H_2O(l) + O_2(g)$
(b) The preparation of hydrofluoric acid from fluorine and water: $2\,F_2(g) + 2\,H_2O(l) \longrightarrow 4\,HF(aq) + O_2(g)$
(c) The production of "synthesis gas," a low-grade industrial fuel: $CH_4(g) + H_2O(g) \longrightarrow CO(g) + 3\,H_2(g)$
(d) The thermal decomposition of ammonium nitrate: $NH_4NO_3(s) \longrightarrow N_2O(g) + 2\,H_2O(g)$

16.85 Determine whether titanium dioxide can be reduced by carbon at 1000 K in the reactions
(a) $TiO_2(s) + 2\,C(s) \longrightarrow Ti(s) + 2\,CO(g)$
(b) $TiO_2(s) + C(s) \longrightarrow Ti(s) + CO_2(g)$

given that at 1000 K, $\Delta G_f^\circ(CO, g) = -200\ kJ\cdot mol^{-1}$, $\Delta G_f^\circ(CO_2, g) = -396\ kJ\cdot mol^{-1}$, and $\Delta G_f^\circ(TiO_2, s) = -762\ kJ\cdot mol^{-1}$.

16.86 Is the reaction $4\ Fe_3O_4(s) + O_2(g) \rightarrow 6\ Fe_2O_3(s)$ spontaneous at 25°C?

16.87 For the reaction $PCl_3(g) + Cl_2(g) \rightarrow PCl_5(g)$ at 25°C, $\Delta G_r^\circ = -37.2\ kJ\cdot mol^{-1}$ and $\Delta H_r^\circ = -87.9\ kJ\cdot mol^{-1}$. (a) Calculate ΔS_r° for the reaction at 25°C. (b) Does the sign of the entropy change agree with your expectations?

16.88 The reaction for the production of the synthetic fuel "water gas" from coal is $C(s, graphite) + H_2O(g) \rightleftharpoons CO(g) + H_2(g)$. (a) Calculate the standard reaction free energy at 25°C. (b) Assuming that ΔH_r° and ΔS_r° are unaffected by temperature changes, calculate the temperature at which $\Delta G_r^\circ = 0$ and the value of K_P at that temperature.

16.89 (a) Calculate ΔH_r° and ΔS_r° at 25°C for the reaction $C_2H_2(g) + 2\ H_2(g) \rightarrow C_2H_6(g)$. (b) Calculate the standard reaction free energy from the equation $\Delta G_r^\circ = \Delta H_r^\circ - T\Delta S_r^\circ$. (c) Interpret the calculated values for ΔH_r° and ΔS_r°. (d) Determine the temperature at which $\Delta G_r^\circ = 0$. What is the significance of that value? (Assume that ΔH_r° and ΔS_r° are unaffected by temperature changes.)

16.90 (a) Using values from Appendix 2A, calculate ΔH_r° and ΔG_r° at 25°C for the combustion of methane: $CH_4(g) + 2\ O_2(g) \rightarrow CO_2(g) + 2\ H_2O(g)$. (b) Calculate the standard reaction entropy from the equation $\Delta G_r^\circ = \Delta H_r^\circ - T\Delta S_r^\circ$. (c) Interpret the calculated values of ΔH_r° and ΔS_r°.

16.91 For the phase transition $M(s_1) \rightarrow M(s_2)$ of a metal M from solid(1) to solid(2), $\Delta H^\circ = -14.07$ kJ and $\Delta S^\circ = -0.0480\ kJ\cdot K^{-1}$ at −84°C. (a) What is the value of ΔG° for the transition? (b) At what temperature does $\Delta G^\circ = 0$ and what is the significance of that value? (c) Which phase is favored at 0°C?

16.92 For the reaction $CO(g) + Cl_2(g) \rightleftharpoons COCl(g) + Cl(g)$, $K_P = 9.1 \times 10^{-30}$ at 25°C. Calculate the standard free energy of the reaction.

16.93 Assuming that ΔH° and ΔS° are unaffected by temperature changes, use data from Appendix 2A to estimate (a) the normal boiling point of $PCl_3(l)$; (b) the transition temperature of the phase change S(s, rhombic) → S(s, monoclinic); (c) the sublimation temperature of iodine; (d) the normal boiling point of heavy water, D_2O.

16.94 Calculate the equilibrium constant at 25°C for each of the following reactions. (Note that the standard free energy of each reaction was determined in Exercise 16.48.)
(a) $NH_4Cl(s) \longrightarrow NH_3(g) + HCl(g)$
(b) $H_2(g) + D_2O(l) \longrightarrow D_2(g) + H_2O(l)$
(c) $2\ NO_2(g) \longrightarrow N_2O_4(g)$
(d) $2\ CH_3OH(g) + 3\ O_2(g) \longrightarrow 2\ CO_2(g) + 4\ H_2O(l)$

16.95 The depletion of the ozone in the stratosphere can be summarized by the net equation, $2\ O_3(g) \longrightarrow 3\ O_2(g)$. (a) From values in Appendix 2A, determine the standard reaction free energy and the standard reaction entropy for the reaction. (b) What is the equilibrium constant of the reaction? (c) What is the significance of your answers with regard to ozone depletion?

16.96 Determine the equilibrium constant of the reaction $SbCl_5(g) \rightleftharpoons SbCl_3(g) + Cl_2(g)$ at 25°C.

16.97 Calculate the entropy change for (a) the vaporization of 1.0 g of water at its normal boiling point and (b) the melting of 1.0 g of ice at its normal melting point. (c) Why is the entropy change for the vaporization so much larger?

16.98 The enthalpy of vaporization of heavy water is $41.6\ kJ\cdot mol^{-1}$ at its normal boiling point of 101.4°C. (a) Calculate the entropy change of the surroundings when 10.0 g of heavy water vaporizes at its normal boiling point. (b) Use data in Appendix 2A to determine the vapor pressure of heavy water at 25°C. (*Hint:* For a vaporization equilibrium, K_p is the same as the vapor pressure in atmospheres.)

CHALLENGING EXERCISES

16.99 A system undergoes a two-step process. In step 1 it absorbs 50 J of heat at constant volume. In step 2 it gives off 5 J of heat at a constant pressure of 1 atm as it is returned to its original state. Using the relation 1 L·atm = 101.325 J, find the change in volume of the system during the second step and identify it as an expansion or compression.

16.100 (a) Using the relation 1 L·atm = 101.325 J, calculate the work that must be done at standard temperature and pressure against the atmosphere for the expansion of the gaseous products in the combustion of 2.00 mol $C_6H_6(l)$ in the reaction $2\ C_6H_6(l) + 15\ O_2(g) \rightarrow 12\ CO_2(g) + 6\ H_2O(g)$. (b) Using data in Appendix 2A, calculate the standard enthalpy of the reaction. (c) Calculate the change in internal energy, ΔU°, of the system.

16.101 Potassium nitrate readily dissolves in water and its enthalpy of solution is $+34.9\ kJ\cdot mol^{-1}$. (a) How does the enthalpy of solution favor or not favor the dissolving process? (b) Is the entropy change of the system positive or negative when the salt dissolves? (c) Is the entropy change of the system primarily a result of changes in locational disorder or thermal motion? (d) Is the entropy change of the surroundings primarily a result of changes in locational disorder or thermal motion? (e) What is the driving force for the dissolution of KNO_3?

16.102 The standard reaction free energy change for the ionization of a weak monoprotic acid in water is 32.0 $kJ \cdot mol^{-1}$. What percentage of the molecules of acid are ionized in a 0.100 M aqueous solution of the acid?

16.103 (a) What is the molar free energy change for the vaporization of pure water at 100°C and 1 atm? (b) What is the molar free energy change for the vaporization of water at 100°C from an aqueous solution that is 10% by mass glucose ($C_6H_{12}O_6$)? (c) Explain why your answers to (a) and (b) are the same or different. (See the hint in Exercise 16.98.)

16.104 Explain why each of the following statements is false: (a) Reactions with negative reaction free energies occur spontaneously and rapidly. (b) If $\Delta G_r^\circ = 0$, the reaction is at equilibrium. (c) Every sample of a pure element, regardless of its physical state, is assigned a free energy of formation equal to 0. (d) An exothermic reaction producing more moles of gas than are consumed has a positive standard reaction free energy.

16.105 Octane, C_8H_{18}, a hydrocarbon used in gasoline, has molecules that consist of a long, flexible chain of eight carbon atoms with hydrogen atoms attached. It does not burn smoothly in automobile engines, causing a noisy condition known as "knocking." Isooctane, a less flexible, branched isomer of octane with the formal name 2,2,4-trimethylpentane, burns more smoothly. The octane rating of isooctane is 100, that of octane is close to 0. To improve the octane rating of gasoline, some of the octane is converted into isooctane. (a) Calculate the change in entropy, the change in free energy and the equilibrium constant at 25°C for the conversion, using the data given below. (b) Which hydrocarbon is favored at equilibrium? (c) Is the specific enthalpy of the hydrocarbon increased or decreased in the conversion? (d) Interpret the change in entropy for the conversion in terms of molecular structure and freedom of movement.

Compound	ΔH_f°, $kJ \cdot mol^{-1}$	ΔG_f°, $kJ \cdot mol^{-1}$	S_m°, $J \cdot K^{-1} \cdot mol^{-1}$
octane	−208.2	+16.7	+467.2
isooctane	−225.0	+12.8	+423.0

CHAPTER 17

The future of human civilization might depend on the material covered in this chapter. That may seem a grandiose claim, but the topics described here may prove to be the key to unlocking a virtually inexhaustible supply of clean energy: the energy supplied to us daily by the Sun. The key is electrochemistry, the study of the interaction of electricity and chemical reactions. The photograph shows one possible approach: light from the Sun shines on a block of silicon in hydrochloric acid, and through a series of electron transfer processes, produces hydrogen from water.

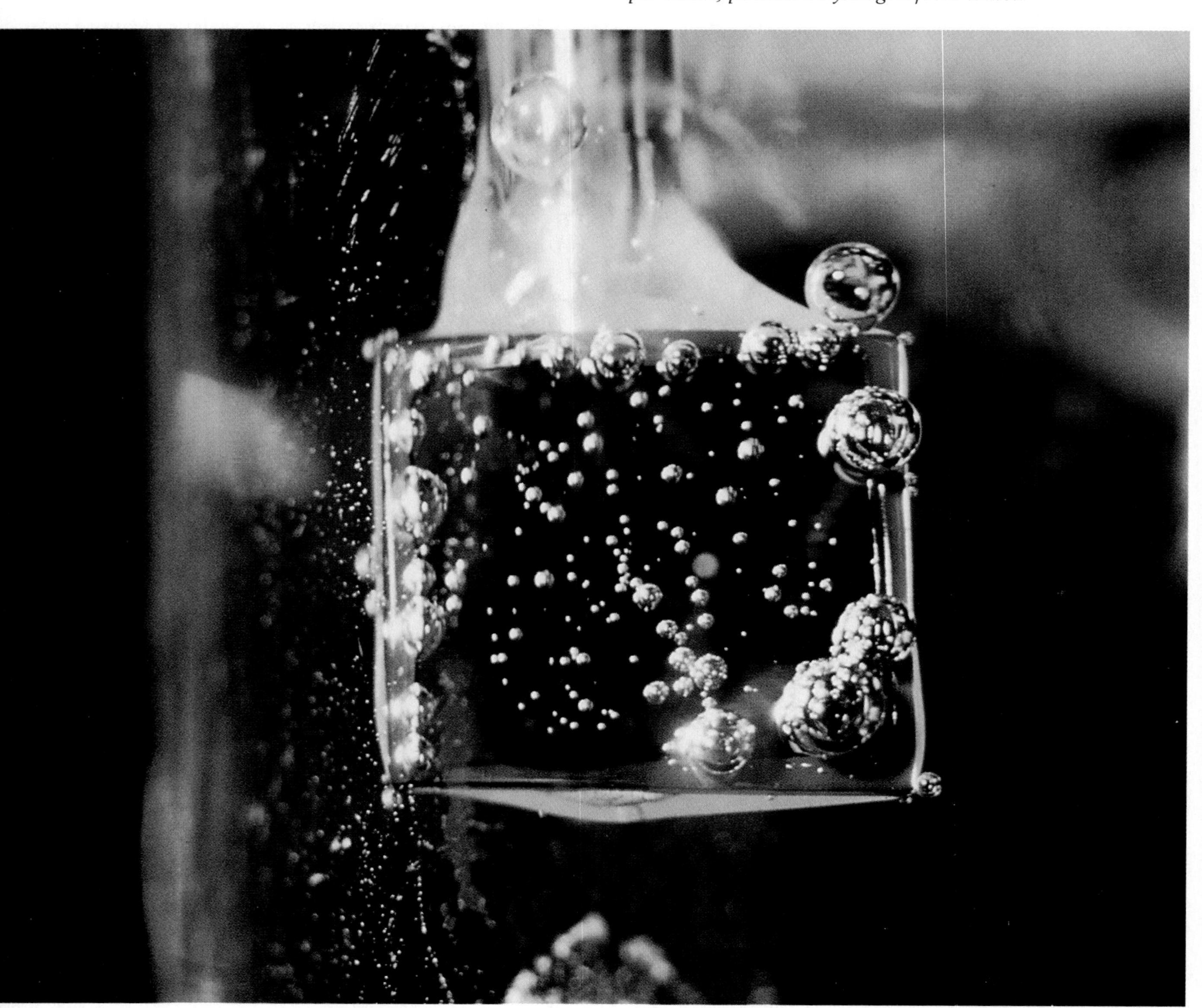

ELECTRONS IN TRANSITION: ELECTROCHEMISTRY

Redox reactions keep us alive. They occur in the all-important reactions that capture the energy of the Sun by photosynthesis and then use that energy to power our muscles and our minds. Redox reactions can also kill, for they can result in explosions and flames. Redox reactions are used throughout the chemical industry to extract metals from their ores and convert petroleum into petrochemicals and pharmaceuticals. However, redox reactions also contribute to the destruction of the artifacts that industry produces. Redox reactions drag iron from its ores, but redox reactions are also responsible for the crumbling of iron back to its oxides in the processes we know as corrosion. What redox reactions achieve, redox reactions can destroy.

Redox reactions can be used to generate electricity and are central to the development of small, light, long-lasting power sources. To a large extent, the future development of technology depends on the capabilities of these power sources. The electric batteries we use to power radios, cassette players, and portable computers all use redox reactions to generate electric currents. But what is involved in using chemical reactions to generate electricity? How can it be made more efficient? Can it be pollution free?

The branch of chemistry that deals with the use of chemical reactions to produce electricity and the use of electricity to produce chemical change is called **electrochemistry.** Because electrochemical techniques provide a way of achieving an electrical output from a chemical system, they allow us to use electronic equipment to monitor concentrations. Thus, electrochemical techniques are used to monitor the compositions of solutions and to measure pH and pK_a.

TRANSFERRING ELECTRONS

17.1 Half-reactions

17.2 Balancing redox equations

GALVANIC CELLS

17.3 Examples of galvanic cells

17.4 The notation for cells

17.5 Cell potential

17.6 Cell potential and reaction free energy

17.7 The significance of standard potentials

17.8 The electrochemical series

17.9 Standard potentials and equilibrium constants

17.10 The Nernst equation

17.11 Practical cells

17.12 Corrosion

ELECTROLYSIS

17.13 Electrolytic cells

17.14 The potential needed for electrolysis

17.15 The products of electrolysis

17.16 Applications of electrolysis

TRANSFERRING ELECTRONS

We first encountered redox reactions in Chapter 3. We return to them now, using the powerful techniques we have been developing since that early introduction.

17.1 HALF-REACTIONS

To show the removal of electrons from a species that is being oxidized in a redox reaction, we write the chemical equation for an oxidation **half-reaction:**

$$Mg(s) \longrightarrow Mg^{2+}(s) + 2\,e^-$$

A half-reaction is a *conceptual* way of reporting an oxidation: the electrons are not necessarily ever actually free. For example, when oxygen is used to oxidize magnesium, the electrons move directly from magnesium atoms to the oxygen atoms. The state of the electrons is not given in half-reactions, because they are "in transit" and do not have a definite physical state. In an oxidation half-reaction, the electrons released always appear on the right of the arrow. The reduced and oxidized species jointly form a **redox couple.** In this example, the redox couple consists of Mg^{2+} and Mg, and is denoted Mg^{2+}/Mg. A redox couple has the form Ox/Red, where Ox is the oxidized form of the species and Red is the reduced form.

To show the gain of electrons by a species that is being reduced, we imagine the corresponding reduction half-reaction and write, for example, for the reduction of iron (III):

$$Fe^{3+}(s) + 3\,e^- \longrightarrow Fe(s)$$

This half-reaction is also conceptual: the electrons are not necessarily free. In a reduction half-reaction, the electrons gained always appear on the left of the arrow. In this example, the redox couple is Fe^{3+}/Fe.

Half-reactions express the two contributions (oxidation and reduction) to an overall redox reaction.

SELF-TEST 17.1A Write the equations for the half-reactions for (a) the oxidation of iron(II) ions to iron(III) ions in aqueous solution; (b) the reduction of copper(II) ions in aqueous solution to copper metal.

[*Answer:* (a) $Fe^{2+}(aq) \rightarrow Fe^{3+}(aq) + e^-$; (b) $Cu^{2+}(aq) + 2\,e^- \rightarrow Cu(s)$]

SELF-TEST 17.1B Write the equations for the half-reactions in which (a) aluminum metal is oxidized to Al^{3+} in aqueous solution; (b) solid sulfur (which exists as S_8 molecules) is reduced to an aqueous solution of sulfide ions (S^{2-}).

17.2 BALANCING REDOX EQUATIONS

In some cases a redox reaction looks as though it will be a nightmare to balance, especially when water is involved in the reaction and we must include H_2O, H^+, or OH^-. In such cases it can help to balance the reduction and oxidation half-reactions separately. Then we bring the two balanced half-reactions together by matching the number of electrons released by oxidation with the number used in reduction. The stoichiometric coefficients of the electrons in the two half-reactions must match, because electrons are neither created nor destroyed in chemical reactions. The procedure is outlined in Toolbox 17.1.

The chemical equation of a reduction half-reaction is added to the equation for an oxidation half-reaction to form the chemical equation for the overall redox reaction.

TOOLBOX 17.1 How to balance complicated redox equations

The general procedure for balancing the chemical equation for a redox reaction is as follows:

Step 1. Identify the species being oxidized and the species being reduced from the changes in their oxidation numbers.

Step 2. Write the two skeletal (unbalanced) equations for the oxidation and reduction half-reactions.

Step 3. Balance all elements in the half-reactions except O and H.

Step 4. In acidic solution, balance O by using H_2O, and then balance H by using H^+. In basic solution, balance O by using H_2O and then balance H by adding H_2O to the side of each half-reaction that needs H, and OH^- to the other side.

When we add

$$\ldots OH^- \ldots \longrightarrow \ldots H_2O \ldots$$

we are effectively adding one H atom to the right, and when we add

$$\ldots H_2O \ldots \longrightarrow \ldots OH^- \ldots$$

we are effectively adding an H atom to the left.

Step 5. Balance the electric charges by adding electrons to the left for reductions and to the right for oxidations.

Step 6. Multiply either or both half-reactions through by a factor so that the numbers of electrons in each one are equal, and then add them.

Finally, simplify the appearance of the equation by canceling species that appear on both sides of the equation, and check to make sure that both numbers of atoms and charge are balanced.

The following two examples illustrate the procedure.

EXAMPLE 17.1 Balancing a redox equation in acidic solution

Permanganate ions, MnO_4^-, react with oxalic acid, $H_2C_2O_4$, in acidic solution, producing manganese(II) ions and carbon dioxide. The skeletal equation is

$$MnO_4^-(aq) + H_2C_2O_4(aq) \longrightarrow Mn^{2+}(aq) + CO_2(g)$$ ⚠

Balance this equation.

STRATEGY Work through the procedure set out in Toolbox 17.1, using the rule for balancing in acid solution in step 4.

SOLUTION ***Step 1.*** Because the oxidation number of manganese decreases from +7 to +2, the MnO_4^- ion is reduced in the reaction. The oxidation number of carbon increases from +3 to +4, so oxalic acid is oxidized.

Step 2. From the information in step 1, we can write the skeletal equations

$$\text{Reduction:} \quad MnO_4^- \longrightarrow Mn^{2+}$$

$$\text{Oxidation:} \quad H_2C_2O_4 \longrightarrow CO_2$$

Step 3. Now balance all elements except H and O:

$$MnO_4^- \longrightarrow Mn^{2+}$$

$$H_2C_2O_4 \longrightarrow 2\,CO_2$$

Step 4. Balance the O atoms by using H_2O:

$$MnO_4^- \longrightarrow Mn^{2+} + 4\,H_2O$$

$$H_2C_2O_4 \longrightarrow 2\,CO_2$$

and balance the H atoms on either side of the half-reactions with H^+:

$$MnO_4^- + 8\,H^+ \longrightarrow Mn^{2+} + 4\,H_2O$$

$$H_2C_2O_4 \longrightarrow 2\,CO_2 + 2\,H^+$$

Step 5. Now balance the electric charge:

$$MnO_4^- + 8\,H^+ + 5\,e^- \longrightarrow Mn^{2+} + 4\,H_2O$$

$$H_2C_2O_4 \longrightarrow 2\,CO_2 + 2\,H^+ + 2\,e^-$$

Step 6. Multiply the permanganate half-reaction by 2 and the oxalic acid half-reaction by 5, so 10 electrons are transferred in each:

$$2\,MnO_4^- + 16\,H^+ + 10\,e^- \longrightarrow 2\,Mn^{2+} + 8\,H_2O$$

$$5\,H_2C_2O_4 \longrightarrow 10\,CO_2 + 10\,H^+ + 10\,e^-$$

Now add the equation for the second half-reaction to the first:

The electrons on each side cancel when we add the two half-reactions.

$$2\,MnO_4^- + 5\,H_2C_2O_4 + 16\,H^+ \longrightarrow 2\,Mn^{2+} + 8\,H_2O + 10\,CO_2 + 10\,H^+$$

Ten of the H^+ ions on the left are canceled by the $10\,H^+$ on the right; so after we have added the physical states of the species, we are left with

$$2\,MnO_4^-(aq) + 5\,H_2C_2O_4(aq) + 6\,H^+(aq) \longrightarrow 2\,Mn^{2+}(aq) + 8\,H_2O(l) + 10\,CO_2(aq)$$

This is the fully balanced net ionic equation.

SELF-TEST 17.2A When a piece of copper metal is placed in dilute nitric acid, copper(II) nitrate and the gas nitric oxide, NO, are formed. Write the balanced chemical equation for the reaction.

[*Answer:* $3\,Cu(s) + 2\,NO_3^-(aq) + 8\,H^+(aq) \rightarrow 3\,Cu^{2+}(aq) + 2\,NO(g) + 4\,H_2O(l)$]

SELF-TEST 17.2B When acidified potassium permanganate is mixed with a solution of sulfurous acid, H_2SO_3, sulfuric acid and manganese(II) ions are produced. Write the balanced chemical equation for the reaction. In acidic aqueous solution, H_2SO_3 is present as the nonionized molecules and sulfuric acid is present as HSO_4^- ions.

EXAMPLE 17.2 Balancing a redox equation in basic solution

The products of the oxidation of bromide ions by permanganate ions in basic solution are solid manganese(IV) oxide, MnO_2, and bromate ions, BrO_3^-. Balance the chemical equation for the reaction.

STRATEGY Work through the procedure set out in Toolbox 17.1, using the rule for balancing in basic solution in step 4.

SOLUTION ***Step 1.*** The oxidation number of manganese changes from +7 in MnO_4^- to +4 in MnO_2, so MnO_4^- is reduced. The oxidation number of bromine increases from −1 as Br^- to +5 in BrO_3^-, so Br^- is oxidized.

Step 2. The skeletal equations for the oxidation and reduction half-reactions are

$$\text{Reduction:}\quad MnO_4^- \longrightarrow MnO_2$$

$$\text{Oxidation:}\quad Br^- \longrightarrow BrO_3^-$$

Step 3. Both equations are already balanced with respect to Mn and Br.

Step 4. First, balance the O atoms by using H_2O:

$$MnO_4^- \longrightarrow MnO_2 + 2\,H_2O$$

$$Br^- + 3\,H_2O \longrightarrow BrO_3^-$$

Next, balance H by adding H_2O to the side of each equation that needs hydrogen and OH^- to the opposite side of each arrow:

$$MnO_4^- + 4\,H_2O \longrightarrow MnO_2 + 2\,H_2O + 4\,OH^-$$

$$Br^- + 3\,H_2O + 6\,OH^- \longrightarrow BrO_3^- + 6\,H_2O$$

In the final part of this step, simplify each half-reaction by canceling like species on opposite sides of the arrow (in this case, H_2O):

$$MnO_4^- + 2\,H_2O \longrightarrow MnO_2 + 4\,OH^-$$

$$Br^- + 6\,OH^- \longrightarrow BrO_3^- + 3\,H_2O$$

Step 5. Now balance the electric charges by using electrons:

$$MnO_4^- + 2\,H_2O + 3\,e^- \longrightarrow MnO_2 + 4\,OH^-$$

$$Br^- + 6\,OH^- \longrightarrow BrO_3^- + 3\,H_2O + 6\,e^-$$

Step 6. To ensure that the same numbers of electrons appear in both half-reactions, multiply the manganese half-reaction by 2:

$$2\,MnO_4^- + 4\,H_2O + 6\,e^- \longrightarrow 2\,MnO_2 + 8\,OH^-$$

$$Br^- + 6\,OH^- \longrightarrow BrO_3^- + 3\,H_2O + 6\,e^-$$

Six electrons are transferred in this redox reaction. Now add the two half-reactions, simplify the equation by canceling species that appear on both sides (in this case, electrons, H_2O, and OH^-), and attach state symbols. We obtain the following fully balanced equation:

$$2\,MnO_4^-(aq) + Br^-(aq) + H_2O(l) \longrightarrow 2\,MnO_2(s) + BrO_3^-(aq) + 2\,OH^-(aq)$$

SELF-TEST 17.3A An alkaline (basic) solution of hypochlorite ions, ClO^-, reacts with solid chromium(III) hydroxide, $Cr(OH)_3$, to produce chromate ions, CrO_4^{2-}, and chloride ions. Write the balanced chemical equation for the reaction.

[*Answer:* $2\,Cr(OH)_3(s) + 4\,OH^-(aq) + 3\,ClO^-(aq) \rightarrow 2\,CrO_4^{2-}(aq) + 5\,H_2O(l) + 3\,Cl^-(aq)$]

SELF-TEST 17.3B When hypochlorite ions are added to a solution of plumbite ions, $[Pb(OH)_3]^-$, in basic solution, lead(IV) oxide and chloride ions are formed. Write the balanced chemical equation.

GALVANIC CELLS

The Italian physiologist Luigi Galvani studied electricity by observing the reaction of frog muscle to an applied current.

Strictly, a "battery" is a collection of cells joined in series; however, even single cells are often colloquially termed batteries.

An **electrochemical cell** is a device in which an electric current—a flow of electrons through a circuit—is either produced by a spontaneous chemical reaction or is used to bring about a nonspontaneous reaction. Here we shall concentrate on a **galvanic cell,** an electrochemical cell in which a spontaneous chemical reaction is used to generate an electric current. The battery in a cassette player is an example of a galvanic cell, because it makes use of a chemical reaction inside the cell to produce an electric current through a circuit attached to it. A galvanic cell consists of two **electrodes,** metallic conductors that make electrical contact with the contents of the cell, and an electrolyte—an ionically conducting medium—inside the cell. The electrolyte is typically an aqueous solution of an ionic compound.

17.3 EXAMPLES OF GALVANIC CELLS

In a galvanic cell, oxidation takes place at one electrode, and the species being oxidized releases electrons into the electrode. Reduction takes place at the other electrode, and the species being reduced collects electrons from the electrode (Fig. 17.1). We can think of the overall chemical reaction as depositing electrons into one electrode and collecting them from the other. This push-pull process sets up a flow of electrons in the external wire circuit joining the two electrodes. When we turn on a cassette player, we complete the electrical circuit and allow the chemical reaction in the cell to take place. The electrons are forced through the electrical circuit, where they turn the electric motor, power the electronics, and drive the loudspeakers. Instead of the energy of the spontaneous reaction being released as heat, it is being used to do work.

The electrode at which oxidation occurs is called the **anode.** The electrode at which reduction occurs is called the **cathode.** Electrons entering the cell at the cathode are used in the reduction half-reaction. If you look at a commercial galvanic cell, you will see one electrode marked with a +. The + marks the cathode, the electrode at which electrons are supplied to the cell. The other electrode is marked with a − sign, which indicates that electrons being produced in the cell reaction are leaving it through that electrode. That electrode must therefore be the site of oxidation, the anode. When we use a commercial cell, we can picture the − electrode (the anode) as providing the electrons that travel through the external circuit and enter the cell again at the + electrode (the cathode).

An early example of a galvanic cell, a **Daniell cell,** was invented by the British chemist John Daniell in 1836 when the growth of telegraphy created an urgent need for a reliable, steady source of electric current.

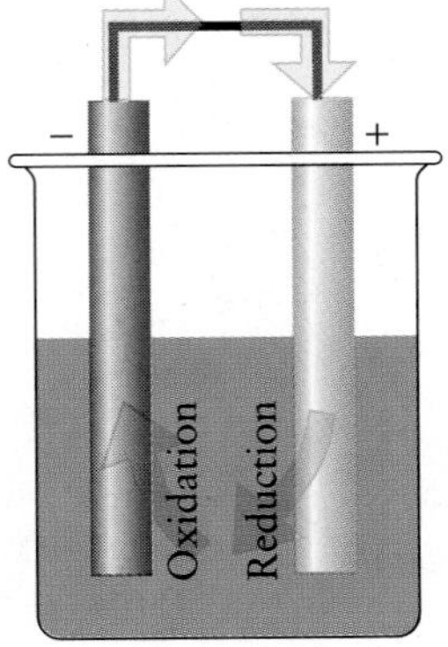

FIGURE 17.1 In an electrochemical cell, a reaction takes place in two separate regions. Oxidation occurs at one electrode, and the electrons released travel through the external circuit to the other electrode, where they cause reduction. The site of oxidation is called the anode, and the side of reduction is called the cathode.

(a)

(b)

FIGURE 17.2 (a) When a strip of zinc is placed in a beaker of copper(II) sulfate solution, copper is deposited on the zinc and the blue copper(II) ions are gradually replaced by colorless zinc ions. (b) The residue in the beaker is copper metal. No more copper ions can be seen in solution.

Daniell knew that the redox reaction

$$Zn(s) + Cu^{2+}(aq) \longrightarrow Zn^{2+}(aq) + Cu(s)$$

is spontaneous, because, when a piece of zinc metal is placed in an aqueous copper(II) sulfate solution, metallic copper is deposited on the surface of the zinc (Fig. 17.2). In atomic terms, as the reaction takes place, electrons are transferred from the zinc to Cu^{2+} ions nearby in the solution (Fig. 17.3). These electrons reduce the Cu^{2+} ions to Cu atoms, which stick to the surface of the zinc or form a finely divided solid deposit in the flask. The piece of zinc slowly disappears as its atoms give up electrons and form Zn^{2+} ions that drift off into the solution.

Daniell separated the oxidation and reduction half-reactions in his cell by using the arrangement shown in Fig. 17.4. The redox reaction is the same, but the reactants are separated by a porous cup. Now the electrons can travel from Zn atoms to Cu^{2+} ions only by passing through the external circuit (the wire and the light bulb). The Cu^{2+} ions are converted into Cu atoms in one compartment by the reduction half-reaction

$$Cu^{2+}(aq) + 2\,e^{-} \longrightarrow Cu(s)$$

Zinc atoms are converted to Zn^{2+} ions in the other compartment by the oxidation half-reaction

$$Zn(s) \longrightarrow Zn^{2+}(aq) + 2\,e^{-}$$

Ions move between the two compartments (through the porous cup) to preserve electrical neutrality inside the cell and complete the electrical circuit. Anions travel toward the anode to balance the charges of the Zn^{2+} ions formed by the oxidation of the zinc electrode, and cations travel toward the cathode, to replace the charges of the Cu^{2+} ions that have been deposited as copper metal.

In the Daniell cell, the electrodes are the metals involved in the reaction. However, not all electrode reactions involve a conducting solid. For example, the reduction $2\,H^{+}(aq) + 2\,e^{-} \rightarrow H_2(g)$ involves a gas, so it is necessary to use a chemically inert metal or other conductor, such as graphite, to supply or remove electrons from the compartment. Platinum

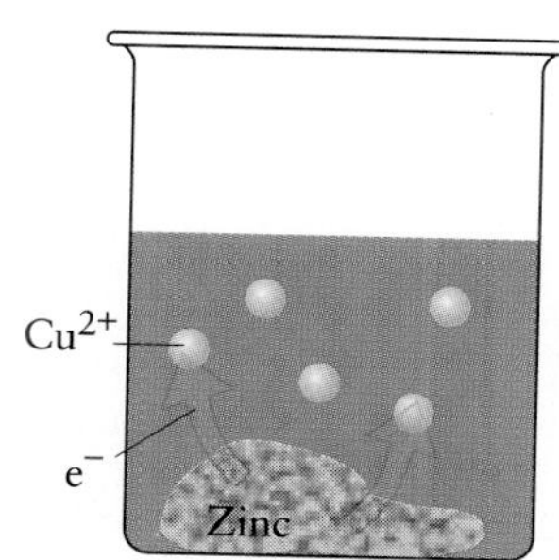

FIGURE 17.3 The reaction shown in Fig. 17.2 takes place all over the surface of the zinc as electrons are transferred to the Cu^{2+} ions in solution.

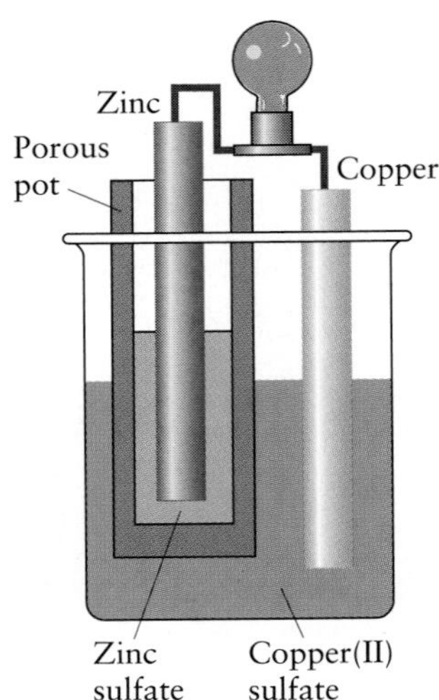

FIGURE 17.4 The Daniell cell consists of copper and zinc electrodes dipping into solutions of copper(II) sulfate and zinc sulfate, respectively. The two solutions make contact through the porous pot, which allows ions to pass through to complete the electric circuit.

is customarily used for the electrode, and hydrogen gas is bubbled over it as it dips into a solution that contains hydrogen ions. This arrangement is called a **hydrogen electrode.**

In a galvanic cell, a spontaneous chemical reaction tends to draw electrons into the cell through the cathode, the site of reduction, and to release them at the anode, the site of oxidation.

17.4 THE NOTATION FOR CELLS

A shorthand notation is used to specify the structure of electrodes in galvanic cells. The two electrodes in the Daniell cell, for instance, are denoted

$$Zn(s)|Zn^{2+}(aq) \qquad Cu^{2+}(aq)|Cu(s)$$

where the ordering is reactant|product for each half-reaction. Each vertical line represents an interface between phases, in this case, between solid metal and ions in solution. For example, a hydrogen electrode would be denoted

$$H^{+}(aq)|H_2(g)|Pt(s)$$

when it acts as a cathode (and H^+ is reduced), and

$$Pt(s)|H_2(g)|H^{+}(aq)$$

when it acts as an anode (and H_2 is oxidized). An electrode consisting of a platinum wire dipping into a solution of iron(II) and iron(III) ions is denoted

$$Fe^{3+}(aq), Fe^{2+}(aq)|Pt(s)$$

In the latter case, the oxidized and reduced species are both in the same phase, so a comma is used to separate them.

SELF-TEST 17.4A A calomel electrode consists of mercury in contact with mercury(I) chloride, Hg_2Cl_2 (calomel), which is in contact with a solution of chloride ions. The mercury(I) chloride is reduced to mercury in a reduction half-reaction. Give the notation for the electrode.

[*Answer:* $Cl^{-}(aq)|Hg_2Cl_2(s)|Hg(l)$]

SELF-TEST 17.4B Write the electrode notation for the reduction of BrO^- to Br^- in basic solution, using a graphite electrode.

The structure of a cell is reported by writing a **cell diagram,** a combination of the notation for the two electrodes. The cell diagram for the Daniell cell, for instance, is

$$Zn(s)|Zn^{2+}(aq)|Cu^{2+}(aq)|Cu(s)$$

There is one complication with the Daniell cell and others in which reactants are in solution. The two solutions must be in contact to complete the circuit. However, when different ions mingle together, they affect the measured voltage in ways that are difficult to predict. In the Daniell cell, zinc sulfate and copper(II) sulfate solutions meet inside the porous barrier. To keep solutions from mixing, chemists use a **salt bridge** to join the two electrode compartments and complete the electrical cir-

FIGURE 17.5 A typical galvanic cell used in the laboratory. The two electrodes are connected by an external circuit (not shown) and a salt bridge. The latter completes the electric circuit within the cell.

cuit. A salt bridge typically consists of a gel containing a concentrated aqueous salt solution in a bridge-shaped tube (Fig. 17.5). The bridge still allows a flow of ions, so it completes the electrical circuit, but the ions are ones that do not affect the cell reaction (often KCl is used). A salt bridge is denoted by a double vertical line (||), so the arrangement in Fig. 17.5 is denoted

$$Zn(s)|Zn^{2+}(aq)||Cu^{2+}(aq)|Cu(s)$$

▶▶ The precise order in which the electrodes are written is important; it is explained in Section 17.5.

An electrode is designated by representing the interfaces between phases by |. A cell diagram depicts the physical arrangement of species and interfaces, with any salt bridge denoted by ||.

SELF-TEST 17.5A Write the diagram for a cell that has a hydrogen electrode on the left, an iron(II)/iron(III) electrode on the right, and includes a salt bridge.

[*Answer:* $Pt|H_2(g)|H^+(aq)||Fe^{3+}(aq),Fe^{2+}(aq)|Pt$]

SELF-TEST 17.5B Write the diagram for a cell that has a copper(II)/copper(I) electrode with a platinum wire on the right, a salt bridge, and a manganese wire dipping into a solution of manganese(II) ions on the left.

17.5 CELL POTENTIAL

The **cell potential,** E, is a measure of the ability of a cell reaction to push and pull electrons through a circuit. The cell potential is sometimes called the *electromotive force* (emf) of the cell or—more colloquially—its *voltage.* A reaction with a lot of pushing and pulling power generates a high cell potential. A reaction with little pulling and pushing power, like a reaction close to equilibrium, generates only a small potential. The SI unit of potential is the volt (V). An exhausted battery is a cell in which the reaction is at equilibrium, so it has lost its power to move electrons and has a potential of 0.

Typical cells used in the home are rated at about 1.5 V. The potential of a Daniell cell is about 1.1 V.

Cell potentials are measured with electronic voltmeters (Fig. 17.6). A positive reading (in volts) means that the + terminal of the meter is connected to the + terminal of the cell. The + terminal of a galvanic cell is the cathode and the − terminal is the anode. So, we can easily decide experimentally which is the cathode and which is the anode by connecting a voltmeter to it and seeing which terminal is positive. Then we write the cell diagram with the anode on the left and the cathode on the right, as in

$$Zn(s)|Zn^{2+}(aq)||Cu^{2+}(aq)|Cu(s) \qquad E = 1.1\ V$$

A good way to remember the cell notation is to note that the anode and cathode appear in alphabetical order.

FIGURE 17.6 The cell potential is measured with an electronic voltmeter, a device that draws negligible current so that the composition of the cell does not change during the measurement. The display shows a positive value when the + terminal of the meter is connected to the cathode of the galvanic cell.

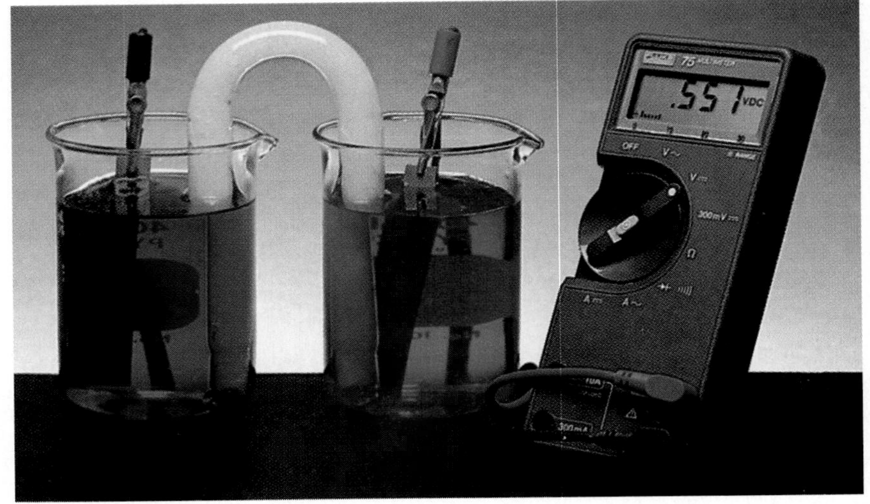

The cathode is the site of reduction and, therefore, the electrode at which electrons enter the cell. Indeed, the + sign can be interpreted as an indication that electrons enter the cell at that electrode. The anode is the site of oxidation. The − sign indicates that electrons lost during the oxidation half-reaction leave the cell at the anode (Fig. 17.7). For the Daniell cell written above, we know experimentally that copper is reduced to the metal, so the copper electrode is the cathode; therefore the spontaneous reaction in the cell is the reduction of copper(II) ions and the oxidation of zinc:

$$Cu^{2+}(aq) + Zn(s) \longrightarrow Cu(s) + Zn^{2+}(aq)$$

The cell potential is an indication of the electron pulling and pushing power of the cell reaction; cell reactions at equilibrium generate zero potential.

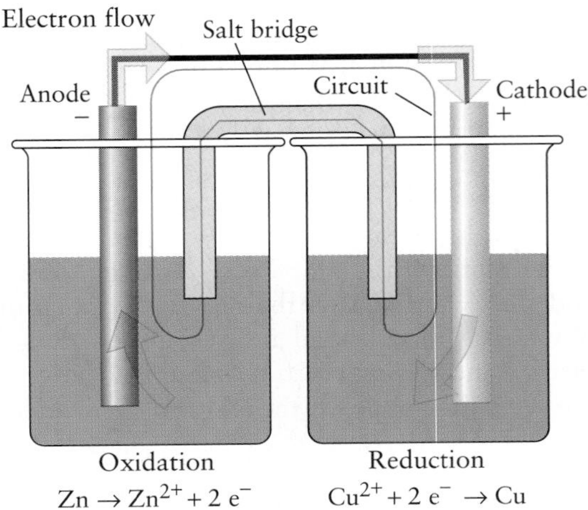

FIGURE 17.7 Electrons leave a galvanic cell at the anode (−), travel through the external circuit, and reenter the cell at the cathode (+). The source of the electrons is oxidation at the anode; the electrons pass through the external circuit and cause reduction at the cathode. The circuit is completed inside the cell by migration of ions through the salt bridge. A salt bridge is unnecessary when the two electrodes share a common electrolyte.

17.6 CELL POTENTIAL AND REACTION FREE ENERGY

When ΔG_r is negative, the cell reaction is spontaneous and we measure a positive cell potential when the cathode is connected to the positive terminal of a voltmeter. When ΔG_r is large as well as negative, we can expect the cell potential to have a high positive value. The precise relation between the two is

$$\Delta G_r = -nFE \quad (1)$$

The **Faraday constant,** *F*, is the magnitude of the charge per mole of electrons, or $9.6485 \times 10^4\ C{\cdot}mol^{-1}$. The *n* in Eq. 1 is the number of moles of electrons that are transferred between the electrodes for the cell reaction as written in the balanced equation that corresponds to ΔG_r. As an example, for the reaction in the Daniell cell, $n = 2$ because 2 mol of electrons migrate from Zn to Cu:

$$Zn(s) + Cu^{2+}(aq) \longrightarrow Zn^{2+}(aq) + Cu(s) \qquad E = 1.1\ V$$

From the relation between free energy and cell potential, it follows that

$$\begin{aligned}\Delta G_r &= -nFE = -2 \times (9.65 \times 10^4\ C{\cdot}mol^{-1}) \times (1.1\ V)\\ &= -2.1 \times 10^5\ C{\cdot}V{\cdot}mol^{-1}\end{aligned}$$

It follows from the definitions of the SI units coulombs and volts that

$$1\ C{\cdot}V = 1\ J$$

Therefore, for this cell, the reaction free energy is about $-210\ kJ{\cdot}mol^{-1}$.

We shall usually employ Eq. 1 for the *standard* reaction free energy ΔG_r°, when it becomes

$$\Delta G_r^\circ = -nFE^\circ \quad (2)$$

In this expression, E° is the **standard cell potential,** the cell potential measured when all the species taking part are in their standard states. In practice, that means all gases at 1 atm and all ions at $1\ mol{\cdot}L^{-1}$. For example, to measure the standard potential of the Daniell cell, we should use 1 M $CuSO_4(aq)$ in one electrode compartment and 1 M $ZnSO_4(aq)$ in the other. Notice that ΔG_r° depends on the stoichiometric coefficients in the chemical equation, but E° does not—because the factor *n* takes into account any changes in the coefficients. The standard cell potential is the same no matter how large or small the cell or how the chemical equation is written.

SELF-TEST 17.6A For the production of ammonia from the elements, $N_2(g) + 3\,H_2(g) \rightleftharpoons 2\,NH_3(g)$, $E^\circ = 0.057$ V; find *n* and ΔG_r° for the reaction.

[*Answer:* 6, $-33.0\ kJ{\cdot}mol^{-1}$]

SELF-TEST 17.6B The reaction between zinc metal and iodide ions in water generates 1.30 V under standard conditions. Determine *n* and ΔG_r° for the cell reaction $Zn(s) + I_2(aq) \rightarrow Zn^{2+}(aq) + 2\,I^-(aq)$.

There are thousands of possible galvanic cells and therefore thousands of possible standard cell potentials. It is much simpler to think of each electrode as making a characteristic contribution called its **standard potential,** E°. Standard potentials are always written in the form of

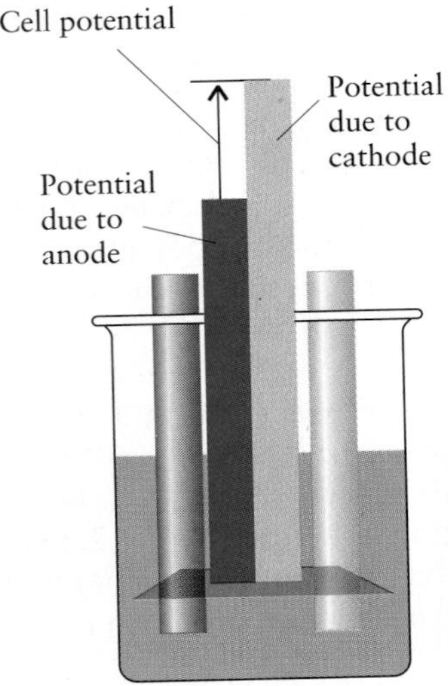

FIGURE 17.8 The cell potential can be thought of as being the difference of the two reduction potentials produced by the two electrodes. The cell potential is positive if the right-hand electrode (the cathode) produces a higher potential than the left-hand electrode (the anode).

reduction, so they are sometimes called *standard reduction potentials* or *standard electrode potentials*. Each standard potential is a measure of the pulling power of a single electrode. In a galvanic cell, the electrodes pull in opposite directions, so the overall pulling power of the cell, its standard potential, is the *difference* of the standard potentials of the two electrodes (Fig. 17.8). That difference is always written

$$E° = E°(\text{cathode}) - E°(\text{anode})$$

For example, the standard potential of the cell

$$Fe(s)|Fe^{2+}(aq)||Ag^{+}(aq)|Ag(s) \qquad E° = 1.24\ V$$

can be thought of as arising from contributions from the potentials of the silver electrode and the iron electrode under standard conditions:

$$E° = E°(Ag^{+}, Ag) - E°(Fe^{2+}, Fe)$$

For the standard potential $E°(Ag^{+}, Ag)$, for instance, the silver ions would be present at a molar concentration of 1 $mol \cdot L^{-1}$.

A problem with this procedure is that we know only the overall cell potential, not the contribution of each electrode. A voltmeter placed between the two electrodes of a galvanic cell measures the difference of their potentials, not the individual values. To get around this difficulty, we arbitrarily set the standard potential of one particular electrode, the hydrogen electrode, equal to 0.

Although the hydrogen ion exists as hydronium ions in water, it is conventional to write it as $H^{+}(aq)$ rather than H_3O^{+} when writing equations for redox reactions.

$$E°(H^{+}, H_2) = 0$$

The hydrogen electrode is used to find the standard potential of all other electrodes. For example, to find the standard potential of a zinc electrode, we measure the standard potential of the cell in which the hydrogen electrode is one electrode and a zinc electrode is the other:

$$Zn(s)|Zn^{2+}(aq)||H^{+}(aq)|H_2(g)|Pt \qquad E° = 0.76\ V$$

Because, according to convention, the hydrogen electrode contributes 0 to the standard cell potential, the standard potential, 0.76 V, can be attributed entirely to the zinc electrode. Moreover, because the zinc electrode is found to be the anode, its standard potential is subtracted from that of hydrogen:

$$E°(H^+, H_2) - E°(Zn^{2+}, Zn) = 0.76\ V$$

It follows that

$$E°(Zn^{2+}, Zn) = -0.76\ V$$

We report the standard potential of the zinc electrode as −0.76 V.

Table 17.1 gives a selection of standard potentials (at 25°C); a longer table can be found in Appendix 2B. Standard potentials of elements vary

TABLE 17.1 Standard potentials at 25°C*

Species[†]	Reduction half-reaction	$E°$, V
OXIDIZED FORM IS STRONGLY OXIDIZING		
F_2/F^-	$F_2(g) + 2e^- \longrightarrow 2F^-(aq)$	+2.87
Au^+/Au	$Au^+(aq) + e^- \longrightarrow Au(s)$	+1.69
Ce^{4+}/Ce^{3+}	$Ce^{4+}(aq) + e^- \longrightarrow Ce^{3+}(aq)$	+1.61
$MnO_4^-, H^+/Mn^{2+}$	$MnO_4^-(aq) + 8H^+(aq) + 5e^- \longrightarrow Mn^{2+}(aq) + 4H_2O(l)$	+1.51
Cl_2/Cl^-	$Cl_2(g) + 2e^- \longrightarrow 2Cl^-(aq)$	+1.36
$Cr_2O_7^{2-}, H^+/Cr^{3+}$	$Cr_2O_7^{2-} + 14H^+(aq) + 6e^- \longrightarrow 2Cr^{3+}(aq) + 7H_2O(l)$	+1.33
$O_2, H^+/H_2O$	$O_2(g) + 4H^+(aq) + 4e^- \longrightarrow 2H_2O(l)$	+1.23 +0.81 at pH = 7
Br_2/Br^-	$Br_2(l) + 2e^- \longrightarrow 2Br^-(aq)$	+1.09
$NO_3^-, H^+/NO$	$NO_3^-(aq) + 4H^+(aq) + 3e^- \longrightarrow NO(g) + 2H_2O(l)$	+0.96
Ag^+/Ag	$Ag^+(aq) + e^- \longrightarrow Ag(s)$	+0.80
Fe^{3+}/Fe^{2+}	$Fe^{3+}(aq) + e^- \longrightarrow Fe^{2+}(aq)$	+0.77
I_2/I^-	$I_2(s) + 2e^- \longrightarrow 2I^-(aq)$	+0.54
O_2/OH^-	$O_2(g) + 2H_2O + 4e^- \longrightarrow 4OH^-(aq)$	+0.40 +0.81 at pH = 7
Cu^{2+}/Cu	$Cu^{2+}(aq) + 2e^- \longrightarrow Cu(s)$	+0.34
$AgCl/Ag, Cl^-$	$AgCl(s) + e^- \longrightarrow Ag(s) + Cl^-(aq)$	+0.22
H^+/H_2	$2H^+(aq) + 2e^- \longrightarrow H_2(g)$	0, by definition
Fe^{3+}/Fe	$Fe^{3+}(aq) + 3e^- \longrightarrow Fe(s)$	−0.04
$O_2/HO_2^-, OH^-$	$O_2(g) + H_2O(l) + 2e^- \longrightarrow HO_2^-(aq) + OH^-(aq)$	−0.08
Pb^{2+}/Pb	$Pb^{2+}(aq) + 2e^- \longrightarrow Pb(s)$	−0.13
Sn^{2+}/Sn	$Sn^{2+}(aq) + 2e^- \longrightarrow Sn(s)$	−0.14
Fe^{2+}/Fe	$Fe^{2+}(aq) + 2e^- \longrightarrow Fe(s)$	−0.44
Zn^{2+}/Zn	$Zn^{2+}(aq) + 2e^- \longrightarrow Zn(s)$	−0.76
$H_2O/H_2, OH^-$	$2H_2O(l) + 2e^- \longrightarrow H_2(g) + 2OH^-(aq)$	−0.83 −0.42 at pH = 7
Al^{3+}/Al	$Al^{3+}(aq) + 3e^- \longrightarrow Al(s)$	−1.66
Mg^{2+}/Mg	$Mg^{2+}(aq) + 2e^- \longrightarrow Mg(s)$	−2.36
Na^+/Na	$Na^+(aq) + e^- \longrightarrow Na(s)$	−2.71
K^+/K	$K^+(aq) + e^- \longrightarrow K(s)$	−2.93
Li^+/Li	$Li^+(aq) + e^- \longrightarrow Li(s)$	−3.05
REDUCED FORM IS STRONGLY REDUCING		

*For a more extensive table, see Appendix 2B.

[†]In the notation X/Y, X is the oxidized species (the reactant, the oxidizing agent) and Y is the reduced species (the product, the reducing agent) in the half-reaction.

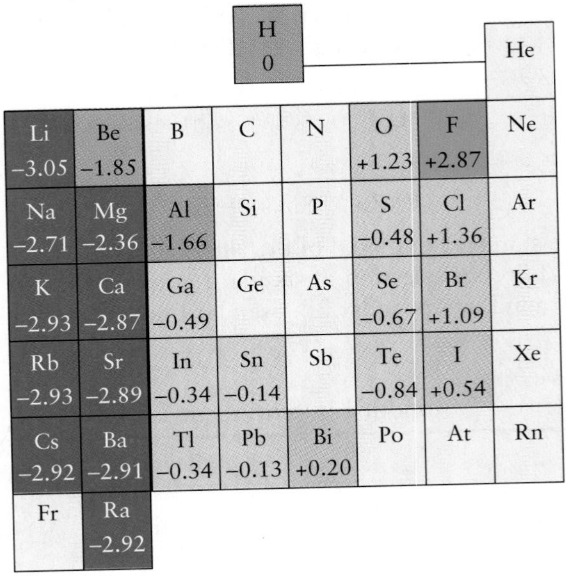

FIGURE 17.9 The variation of standard potentials among the main groups of the periodic table. Note that the most negative values occur in the *s*-block and that the most positive values occur close to fluorine.

			H 0				He
Li −3.05	Be −1.85	B	C	N	O +1.23	F +2.87	Ne
Na −2.71	Mg −2.36	Al −1.66	Si	P	S −0.48	Cl +1.36	Ar
K −2.93	Ca −2.87	Ga −0.49	Ge	As	Se −0.67	Br +1.09	Kr
Rb −2.93	Sr −2.89	In −0.34	Sn −0.14	Sb	Te −0.84	I +0.54	Xe
Cs −2.92	Ba −2.91	Tl −0.34	Pb −0.13	Bi +0.20	Po	At	Rn
Fr	Ra −2.92						

in a complicated way through the periodic table (Fig. 17.9). However, the most negative are usually found toward the left of the periodic table and the most positive are found toward the upper right.

> *The standard potential of an electrode is the standard potential of a cell in which the other electrode is a hydrogen electrode. The latter is assigned zero potential. If the test electrode is found to be the anode, then it is assigned a negative potential; if it is the cathode, then its standard potential is positive.*

EXAMPLE 17.3 Deducing the standard potential of an electrode

The standard potential of a zinc electrode is −0.76 V and the standard potential of the cell

$$Zn(s)|Zn^{2+}(aq)||Cu^{2+}(aq)|Cu(s)$$

is 1.10 V. What is the standard potential of the copper electrode?

STRATEGY The standard potential of one of the electrodes and the overall potential are known; the value for the other electrode can be calculated from them. The cell diagram reveals which electrode is the anode (the site of oxidation, the one on the left) and which is the cathode (the site of reduction, the one on the right). The difference of the standard potentials, $E°$(cathode) − $E°$(anode), is equal to the overall potential of the cell.

SOLUTION The zinc electrode is the anode so zinc is oxidized and copper is reduced. The standard potential of zinc is −0.76 V. The copper electrode must contribute a potential such that overall the cell potential is 1.10 V:

$$E°(Cu^{2+}, Cu) - E°(Zn^{2+}, Zn) = 1.10 \text{ V}$$

Therefore

$$E°(Cu^{2+}, Cu) = 1.10\ V + E°(Zn^{2+}, Zn)$$
$$= 1.10\ V - 0.76\ V = +0.34\ V$$

SELF-TEST 17.7A The standard potential of the Ag^+, Ag electrode is +0.80 V, and the standard potential of the cell

$$Pt | I_2(s) | I^-(aq) || Ag^+(aq) | Ag(s)$$

is 0.26 V at the same temperature. What is the standard potential of the iodine electrode?

[*Answer:* +0.54 V]

SELF-TEST 17.7B The standard potential of the Fe^{2+}, Fe electrode is −0.44 V and the standard potential of the cell

$$Fe(s) | Fe^{2+}(aq) || Pb^{2+}(aq) | Pb(s)$$

is 0.31 V. What is the standard potential of the lead electrode?

17.7 THE SIGNIFICANCE OF STANDARD POTENTIALS

The negative sign of the zinc standard potential, −0.76 V, tells us that, under standard conditions, zinc has a tendency to reduce hydrogen ions to hydrogen gas and that the reaction

$$Zn(s) + 2\,H^+(aq) \longrightarrow Zn^{2+}(aq) + H_2(g)$$

is spontaneous. Conversely, the positive standard potential of the copper electrode, +0.34 V, tells us that hydrogen tends to reduce copper ions and that the reaction

$$Cu^{2+}(aq) + H_2(g) \longrightarrow Cu(s) + 2\,H^+(aq)$$

is spontaneous.

We can now predict that magnesium, iron, indium, tin, lead, and other metals with *negative* standard potentials have a thermodynamic tendency to reduce hydrogen ions in a 1 M acid solution to hydrogen gas and to be oxidized themselves. On the other hand, because metals with *positive* standard potentials (those above hydrogen in the table) cannot reduce hydrogen ions, we know that they cannot produce hydrogen gas when acted on by 1 M acid. For example, copper and the noble metals silver, platinum, and gold are not oxidized by hydrogen ions.

A thermodynamic tendency is not always realized in practice, often because the reaction is very slow or because a protective oxide is formed. For example, the standard potential of aluminum (−1.66 V) suggests that, like magnesium, it should give hydrogen with acid. Aluminum can be oxidized by hydrochloric acid. However, it does not react with the more strongly oxidizing nitric acid because any Al^{3+} ions that are produced in nitric acid immediately form a layer of oxide on the surface of the metal (Fig. 17.10). This layer prevents further reaction, and we say that the metal has been **passivated,** or protected from further reaction, by a surface film. The passivation of aluminum is of great commercial importance, because it enables the metal to be used, among other things, for airplanes and window frames in buildings. Aluminum containers are

FIGURE 17.10 Although aluminum has a negative standard potential, signifying that it can be oxidized by hydrogen ions (as in the hydrochloric acid, left), nitric acid (right) stops reacting with it as soon as an impenetrable layer of aluminum oxide has formed on its surface. This resistance to further reaction is termed *passivation* of the metal.

used to transport nitric acid because, once the surface is passivated, no further reaction occurs.

> ***A metal with a negative standard potential has a thermodynamic tendency to reduce hydrogen ions in solution; the ions of a metal with a positive standard potential have a tendency to be reduced by hydrogen gas.***

17.8 THE ELECTROCHEMICAL SERIES

◀◀ Recall from Section 3.14 that an oxidizing agent is a species that causes oxidation as it undergoes reduction. A reducing agent causes reduction and is itself oxidized.

Because $E°$ is a measure of the electron pulling power of a species, a high $E°$ signifies a strong electron-pulling (oxidizing) power. Conversely, a high negative value of $E°$ signifies a strongly negative pulling power—a strong electron-pushing (reducing) power.

Just as pK_a values allow us to rank acids according to their strengths, standard potentials provide a way to rank oxidizing and reducing agents. For example, because zinc is a stronger reducing agent than hydrogen, and hydrogen is a stronger reducing agent than copper, we can now predict that zinc is a stronger reducing agent than copper by checking that its standard potential ($E° = -0.76$ V) lies below that of copper ($E° = +0.34$ V). In general, *the more negative the standard potential of a species, the greater its reducing strength* and the more readily it loses electrons. Only the reduced form of species with standard potentials below that of hydrogen can reduce hydrogen ions; the reduced form of species with standard potentials above that of hydrogen, such as Au, cannot reduce hydrogen ions (Fig. 17.11). The species on the left side of each equation in Table 17.1 are oxidizing agents. They can themselves be reduced. Species on the right sides of the equations are reducing agents.

When Table 17.1 is viewed as a table of relative strengths of oxidizing and reducing agents, it is called the **electrochemical series.** An oxidized species in the list (on the left of the equation) has a tendency to oxidize a reduced species that lies below it. For example, Cu^{2+} ions oxidize zinc metal. A reduced species (on the right of the equation) has a tendency to reduce an oxidized species that lies above it. For example, zinc metal reduces H^+ ions. Hence, we can see from a glance at Table 17.1 the direction in which a particular redox reaction will tend to run (Fig. 17.12).

A species with a high standard potential is easily reduced; such a species is a strong oxidizing agent and can oxidize the reduced form of

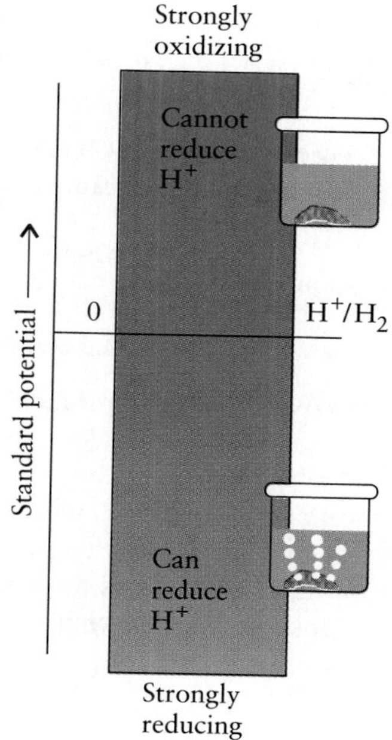

FIGURE 17.11 The significance of standard potentials. Only couples with negative standard potentials (and hence lying below hydrogen) can reduce hydrogen ions to hydrogen gas. The reducing power increases as the standard potential becomes more negative.

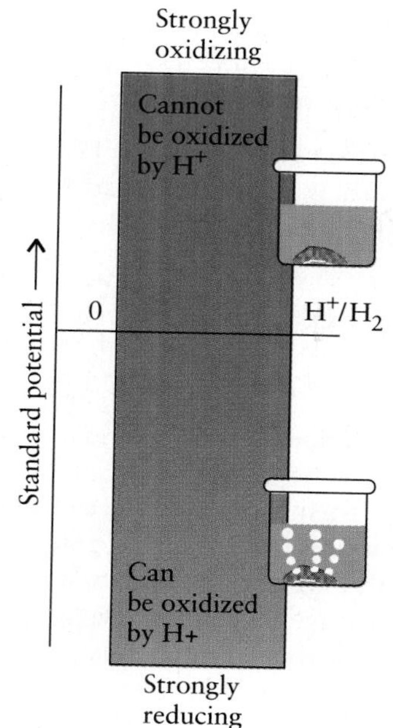

FIGURE 17.12 Another way of viewing the significance of standard potentials. Only species with negative standard potentials (and hence lying below hydrogen) can be oxidized by hydrogen ions.

species with lower standard potentials. Therefore, the higher the position of a substance on the left side of an equation in Table 17.1, the greater its oxidizing strength. For example, F_2 is a strong oxidizing agent, whereas Li^+ is a very, very poor oxidizing agent. It also follows that the lower the standard potential, the greater the reducing strength of the reduced species on the right side of an equation in Table 17.1. The strongest reducing agents are at the bottom of the table. For example, lithium metal is the strongest reducing agent in the table.

◀◀ Recall from Section 17.6 that all standard potentials refer to reductions.

The oxidizing and reducing power of a substance can be determined by its position in the electrochemical series. The strongest oxidizing agents are at the top of the table as reactants, the strongest reducing agents at the bottom of the table as products.

EXAMPLE 17.4 Predicting oxidizing power using the electrochemical series

Can aqueous potassium permanganate be used to oxidize iron(II) to iron(III) under standard conditions in acidic solution?

STRATEGY Look for the relative positions of the species in the electrochemical series in Table 17.1 or Appendix 2B. The species that lies higher in the series is more strongly oxidizing.

SOLUTION In Table 17.1, we see that $Fe^{3+} + e^- \rightarrow Fe^{2+}$ lies at +0.77 V, which is below MnO_4^- (at +1.51 V). Therefore, permanganate ions can oxidize Fe^{2+} to Fe^{3+} in acidic solution (Fig. 17.13).

▶▶ Notice that in Appendix 2B, standard potentials are listed both by voltage and alphabetically, to make it easy to find the one you want.

SELF-TEST 17.8A Can mercury produce zinc from aqueous zinc sulfate under standard conditions?

[*Answer:* No]

SELF-TEST 17.8B Can chlorine gas oxidize water to oxygen gas under standard conditions in basic solution?

To find a standard cell potential for a spontaneous process, we must combine the standard potential of the cathode half-reaction (reduction) with that of the anode half-reaction (oxidation) in such a way as to obtain a positive value. The overall potential must be positive because that corresponds to a spontaneous process ($E > 0$ corresponds to $\Delta G < 0$), and only a spontaneous process can generate a potential. The cathode half-reaction is the one with the higher standard potential. The anode half-reaction is the one with the lower standard potential. Because the anode half-reaction is oxidation, we reverse its equation and add it to the cathode half-reaction: that gives the spontaneous cell reaction (this procedure was described in step 6 of Toolbox 17.1). To find the corresponding standard cell potential, we change the sign of the standard potential of the anode and add it to the standard potential of the cathode. This last step is the same as subtracting $E°$ (anode) from $E°$ (cathode), as we did in Example 17.3.

To determine a standard cell potential, combine two half-reactions by reversing the equation and the sign of the standard potential of the half-reaction with the lower standard potential and adding the half-reactions and potentials together.

(a)

(b)

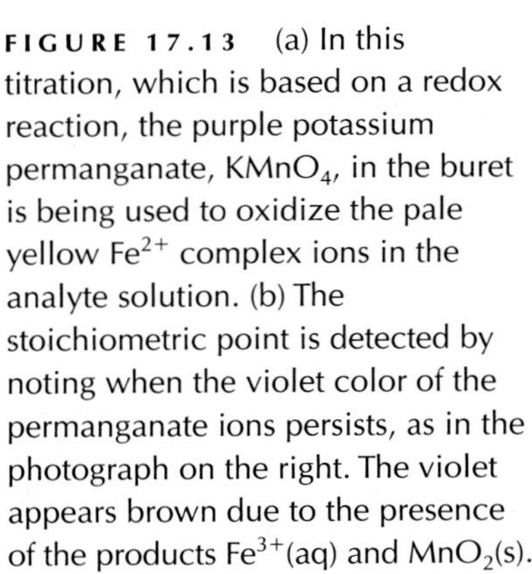

FIGURE 17.13 (a) In this titration, which is based on a redox reaction, the purple potassium permanganate, $KMnO_4$, in the buret is being used to oxidize the pale yellow Fe^{2+} complex ions in the analyte solution. (b) The stoichiometric point is detected by noting when the violet color of the permanganate ions persists, as in the photograph on the right. The violet appears brown due to the presence of the products $Fe^{3+}(aq)$ and $MnO_2(s)$.

EXAMPLE 17.5 **Predicting relative oxidizing strengths and standard cell potentials**

Is an acidified permanganate solution a more powerful oxidizing agent than an acidified dichromate solution under standard conditions? Write the chemical equation for the spontaneous reaction and determine the standard cell potential.

STRATEGY Inspect Table 17.1 or Appendix 2B to see which species lies above the other. To construct the chemical equation, reverse one of the half-reactions and its corresponding standard potential and add it to the other half-reaction. Decide which one to reverse by noting which sum turns out to be positive. Multiply the equations for the half-reactions through by factors that ensure that the numbers of electrons match. This multiplication has no effect on the value of the standard potentials.

SOLUTION Because $E°(MnO_4^-, H^+/Mn^{2+}, H_2O) = +1.51$ V lies above $E°(Cr_2O_7^{2-}, H^+/Cr^{3+}, H_2O) = +1.33$ V, MnO_4^- is the stronger oxidizing agent. Therefore, MnO_4^- ions are more strongly oxidizing than $Cr_2O_7^{2-}$ ions. To construct the spontaneous reaction, we first note the following two half-reactions:

$$MnO_4^-(aq) + 8\,H^+(aq) + 5\,e^- \longrightarrow Mn^{2+}(aq) + 4\,H_2O(l) \qquad E° = +1.51\text{ V}$$

$$Cr_2O_7^{2-}(aq) + 14\,H^+(aq) + 6\,e^- \longrightarrow 2\,Cr^{3+}(aq) + 7\,H_2O(l) \qquad E° = +1.33\text{ V}$$

Reversing the second of this pair of equations gives a standard cell potential of

$$E° = 1.51\text{ V} - 1.33\text{ V} = 0.18\text{ V}$$

(Reversing the first half-reaction would have given a negative cell potential.) The corresponding equation for the second half-reaction is now an oxidation:

$$2\,Cr^{3+}(aq) + 7\,H_2O(l) \longrightarrow Cr_2O_7^{2-}(aq) + 14\,H^+(aq) + 6\,e^-$$

To match electrons, we multiply the first half-reaction by 6 and the second by 5. Their sum is then

$$6\,MnO_4^-(aq) + 11\,H_2O(l) + 10\,Cr^{3+}(aq) \longrightarrow 6\,Mn^{2+}(aq) + 22\,H^+(aq) + 5\,Cr_2O_7^{2-}(aq)$$

SELF-TEST 17.9A Which metal is the stronger reducing agent, zinc or nickel? Evaluate the standard cell potential and write the chemical equation for the spontaneous reaction.

[*Answer:* Zinc; +0.53 V; $Zn(s) + Ni^{2+}(aq) \rightarrow Zn^{2+}(aq) + Ni(s)$]

SELF-TEST 17.9B Which is the stronger oxidizing agent, Cu^{2+} or Ag^+? Evaluate the standard cell potential for the reaction and write the equation for the corresponding cell reaction.

17.9 STANDARD POTENTIALS AND EQUILIBRIUM CONSTANTS

We saw in Section 16.15 that the standard reaction free energy, $\Delta G_r°$, is related to the equilibrium constant of the reaction by

$$\Delta G_r° = -RT \ln K$$

In this chapter we have seen that the standard reaction free energy is related to the standard cell potential of a galvanic cell by

$$\Delta G_r° = -nFE°$$

When we combine the two equations, we get

$$-nFE° = -RT \ln K$$

This expression can be reorganized to allow us to calculate the equilibrium constant from the cell potential:

$$\ln K = \frac{nFE°}{RT} \tag{3}$$

The combination RT/F occurs frequently in electrochemistry; at 25°C (298.15 K) it has the value 2.5693×10^{-2} J·C^{-1}, or 0.025 693 V, so at that temperature

$$\ln K = \frac{nE°}{0.025\,693\text{ V}}$$

Because we can calculate $E°$ from standard potentials, we can now also calculate equilibrium constants for any reaction that can be expressed in terms of two half-reactions. For example, the standard cell potential for the overall reaction

$$Zn(s) + Cu^{2+}(aq) \rightleftharpoons Zn^{2+}(aq) + Cu(s) \qquad K = \frac{[Zn^{2+}]}{[Cu^{2+}]}$$

is 1.10 V and $n = 2$ for the reaction as written; thus

$$\ln K = \frac{2 \times 1.10\text{ V}}{0.025\,693\text{ V}} = 85.6$$

Taking the natural antilogarithm gives $K = 1.6 \times 10^{37}$. Now we know not only that the reaction is spontaneous as written but also that equilibrium is reached only when the concentration of Zn^{2+} ions is over 10^{37} times greater than that of Cu^{2+} ions. For all practical purposes, the reaction goes to completion. The magnitude of $E°$ is an indication of the equilibrium composition. A reaction with a large positive $E°$ has a very large K. A reaction with a large negative calculated $E°$ has a K much less than 1. The procedure for calculating equilibrium constants is summarized in Toolbox 17.2.

The equilibrium constant of a reaction can be calculated from standard potentials by combining the equations for the half-reactions to give the reaction of interest and determining the standard potential of the corresponding cell.

TOOLBOX 17.2 How to calculate equilibrium constants from electrochemical data

The procedure for calculating an equilibrium constant is as follows:

Step 1. Write the balanced equation. Then find two half-reactions to combine to give the equation of interest: reverse one and add them together.

Step 2. Identify the value of n by examining the number of electrons transferred in the balanced equation.

Step 3. Subtract the standard potential of the half-reaction that was reversed from the standard potential of the half-reaction that was left as a reduction.

Step 4. Use the relation $\ln K = nFE°/RT$ to calculate the value of K.

As an illustration, let's calculate the equilibrium constant for the reaction

$$AgCl(s) \rightleftharpoons Ag^+(aq) + Cl^-(aq) \qquad K_c = [Ag^+][Cl^-]$$

Because we know that AgCl is insoluble, we know that K_c is very small, and therefore expect $E°$ to be negative. The equilibrium constant for this reaction is actually the solubility product, K_{sp}, for silver chloride (Section 15.11). It does not matter that overall the reaction is not a redox reaction so long as it can be expressed as the difference of two reduction half-reactions. The two reduction half-reactions required are

$$AgCl(s) + e^- \longrightarrow Ag(s) + Cl^-(aq) \qquad E° = +0.22\ V$$

$$Ag^+(aq) + e^- \longrightarrow Ag(s) \qquad E° = +0.80\ V$$

To obtain the reaction of interest, we need to reverse the second half-reaction:

$$Ag(s) \longrightarrow Ag^+(aq) + e^-$$

The sum of this equation and the AgCl reduction is the reaction we require, and $n = 1$. The standard cell potential is therefore the difference

$$E° = 0.22 - 0.80\ V = -0.58\ V$$

Note that we subtract the standard potential of the oxidation half-reaction, the one that we reversed to get the overall reaction. Therefore,

$$\ln K = \frac{nFE°}{RT} = -\frac{0.58\ V}{0.025\,693\ V} = -22.57$$

Taking the antilogarithm ($e^{-22.57}$) gives $K = 1.6 \times 10^{-10}$, as in Table 15.5. Many of the solubility products listed in tables, such as those in Table 15.5, have been determined in this way.

SELF-TEST 17.10A Use the tables in Appendix 2B to calculate the solubility product of mercury(I) chloride, Hg_2Cl_2.

[*Answer:* 2.6×10^{-18}]

SELF-TEST 17.10B Use the tables in Appendix 2B to calculate the solubility product of cadmium hydroxide, $Cd(OH)_2$.

17.10 THE NERNST EQUATION

Nearly everyone has experienced the disappointment of a battery that has run down at an inconvenient time. What has happened is that the composition of the cell has changed and the cell reaction has reached equilibrium. As the reaction reached equilibrium, it lost its pushing and pulling power. To understand this behavior, we need to know how the cell potential varies with the composition of the cell.

The formula for predicting the variation of cell potential with concentration and pressure is expressed by an equation first derived by the German chemist, Walter Nernst. We already know how ΔG_r varies with composition:

Recall from Section 13.6 that Q is an indication of distance from equilibrium. As product concentrations increase, Q increases, until it eventually reaches K at equilibrium.

$$\Delta G_r = \Delta G_r° + RT \ln Q$$

where Q is the reaction quotient (this is Eq. 12 of Section 16.14). Because $\Delta G_r = -nFE$ and $\Delta G_r° = -nFE°$, it follows that

The relation between ΔG_r and E was described in Section 17.6.

$$-nFE = -nFE° + RT \ln Q$$

When we divide through by $-nF$, we get the **Nernst equation:**

$$E = E° - \frac{RT}{nF} \ln Q \qquad (4)$$

The Nernst equation can also be used to determine concentrations (Box 17.1).

The variation of cell potential with composition is expressed by the Nernst equation.

BOX 17.1 HOW pH METERS WORK

The pH of a solution can be measured electrochemically with a device called a pH meter. The technique makes use of a cell in which one electrode is sensitive to the H_3O^+ concentration and the second electrode serves as a reference. One combination is a hydrogen electrode connected through a salt bridge to a calomel electrode (calomel is the common name for mercury(I) chloride, Hg_2Cl_2). The reduction half-reaction for the calomel electrode is

$$Hg_2Cl_2(s) + 2\,e^- \longrightarrow 2\,Hg(l) + 2\,Cl^-(aq) \qquad E^\circ = +0.27\text{ V}$$

The complete reaction at 1 atm is

$$Hg_2Cl_2(s) + H_2(g) \longrightarrow 2\,H^+(aq) + 2\,Hg(l) + 2\,Cl^-(aq) \qquad Q_c = [H^+]^2[Cl^-]^2$$

Because the pressure of hydrogen gas is 1 atm, we have omitted it from the reaction quotient. To find the concentration of hydrogen ions in the anode compartment at 1 atm, we write the Nernst equation

$$\begin{aligned} E &= E^\circ - \tfrac{1}{2}(0.0257\text{ V}) \times \ln[H^+]^2[Cl^-]^2 \\ &= E^\circ - \tfrac{1}{2}(0.0257\text{ V}) \times \ln[Cl^-]^2 \\ &\quad - \tfrac{1}{2}(0.0257\text{ V}) \ln[H^+]^2 \\ &= E^\circ - (0.0257\text{ V}) \times \ln[Cl^-] \\ &\quad - (0.0257\text{ V}) \times \ln[H^+] \end{aligned}$$

The Cl^- concentration is fixed for a calomel electrode when it is manufactured, so it is a constant. Therefore, we can combine the first two terms on the right into a single constant, E'. Then, because $\ln x = 2.303 \log x$,

$$\begin{aligned} E &= E' - 2.303 \times (0.0257\text{ V}) \times \log[H^+] \\ &= E' + (0.0592\text{ V}) \times \text{pH} \end{aligned}$$

Therefore, by measuring the cell potential, E, we can determine the pH. If E is first measured for a solution of known pH, it is not necessary to calculate E'.

A **glass electrode**, a thin-walled glass bulb containing an electrolyte, is much easier to use than a hydrogen electrode and has a potential that is proportional to the pH. Often there is a calomel electrode built into the probe that makes contact with the test solution through a miniature salt bridge. A pH meter therefore usually has only one probe, which contains a complete electrochemical cell. The meter is calibrated with a buffer of known pH, and the measured cell potential is then automatically converted into the pH of the solution, which is displayed. Solid-state electrodes have made possible durable, pocket-size pH meters that can be used to measure pH at remote locations and under extreme conditions.

Commercially available electrodes called pX meters are sensitive to other ions, such as Na^+, Ca^{2+}, NH_4^+, CN^-, and S^{2-}. They are used to monitor industrial processes and in pollution control.

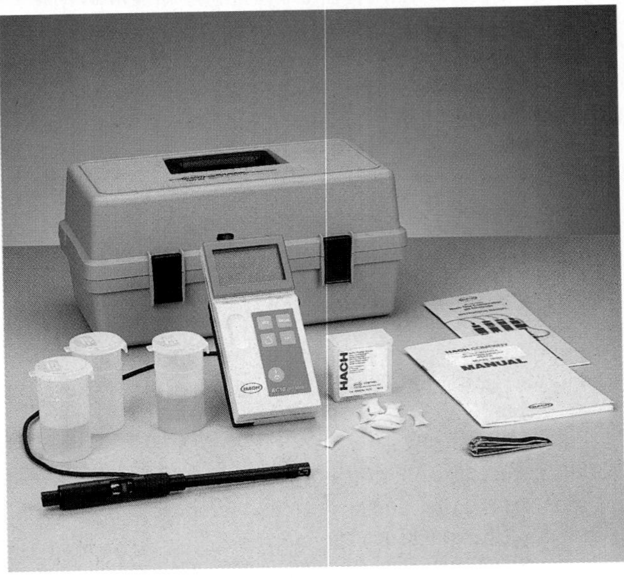

This durable pocket-size pH meter can be used for quick measurements of pH in the field. Its accuracy is not as high as that of a laboratory pH meter.

A glass electrode in a protective plastic sleeve (left) is used to measure pH. It is sometimes used in conjunction with a calomel electrode (right) in pH meters such as this one.

EXAMPLE 17.6 Using the Nernst equation to predict a cell potential

Calculate the potential at 25°C of a Daniell cell in which the concentration of Zn^{2+} ions is 0.10 mol·L^{-1} and that of the Cu^{2+} ions is 0.0010 mol·L^{-1}.

STRATEGY First, write the balanced equation. Then determine $E°$ from the standard potentials in Table 17.1 (or Appendix 2B) and note the value of n. Determine the value of Q_c for the stated conditions. Calculate the cell potential by substituting these values into the Nernst equation. At 25°C, $RT/F = 0.025\,693$ V.

SOLUTION The reaction in the Daniell cell is

$$Cu^{2+}(aq) + Zn(s) \longrightarrow Zn^{2+}(aq) + Cu(s) \qquad n = 2, E° = 1.10\text{ V}$$

The reaction quotient is

$$Q_c = \frac{[Zn^{2+}]}{[Cu^{2+}]} = \frac{0.10}{0.0010}$$

The Nernst equation gives

$$\begin{aligned} E &= 1.10\text{ V} - \left(\frac{0.025\,693\text{ V}}{2}\right)\ln\left(\frac{0.10}{0.0010}\right) \\ &= 1.10\text{ V} - 0.059\text{ V} = +1.04\text{ V} \end{aligned}$$

SELF-TEST 17.11A Calculate the potential of the galvanic cell

$$Zn|Zn^{2+}(1.50\text{ mol·L}^{-1})||Fe^{2+}(0.10\text{ mol·L}^{-1})|Fe$$

[*Answer:* +0.29 V]

SELF-TEST 17.11B Calculate the potential of the galvanic cell

$$Ag|Ag^{+}(0.0010\text{ mol·L}^{-1})||Ag^{+}(0.010\text{ mol·L}^{-1})|Ag$$

17.11 PRACTICAL CELLS

Table 17.2 summarizes the chemical reactions used to power typical commercial galvanic cells. A **primary cell** produces electricity from chemicals that are sealed into it when it is made. This type of galvanic cell cannot be recharged; once the cell reaction has reached equilibrium, the cell is discarded. The workhorse of primary cells is the *dry cell* (Fig. 17.14), which is widely used to power portable electric equipment. A dry cell produces about 1.5 V initially, but with use its potential falls to about 0.8 V as reaction products accumulate inside it. A **fuel cell** is like a primary cell, but the reactants (the fuel) are continuously supplied from outside. The cell can produce a current for as long as fuel is supplied.

The dry cell is also called the *Leclanché cell* for Georges Leclanché, the French engineer who invented it in about 1866. In fact, the electrolyte is moist: it is called "dry" in comparison with cells like the Daniell cell, in which the electrolytes are aqueous solutions.

A **secondary cell** must be charged before it can be used; this type of cell is normally rechargeable. The batteries used in portable computers and automobiles are secondary cells. In the charging process, an external source of electricity temporarily reverses the spontaneous cell reaction and restores a nonequilibrium mixture of reactants. After charging, the cell can again produce electricity as the reaction once more sinks toward equilibrium.

One of the most common secondary cells is the *lead-acid cell* of an automobile battery. Each cell contains several grids that act as electrodes (Fig. 17.15). Because the total surface area of these grids is large, the battery can generate large currents on demand—at least for short periods,

TABLE 17.2 Reactions in commercial galvanic cells

dry cell	$Zn(s)\mid ZnCl_2(aq), NH_4Cl(aq)\mid MnO(OH)(s)\mid MnO_2(s)\mid$ graphite, 1.5 V Anode: $Zn(s) \longrightarrow Zn^{2+}(aq) + 2\,e^-$ followed by $Zn^{2+}(aq) + 4\,NH_3(g) \longrightarrow [Zn(NH_3)_4]^{2+}(aq)$ Cathode: $MnO_2(s) + H_2O(l) + e^- \longrightarrow MnO(OH)(s) + OH^-(aq)$ followed by $NH_4^+(aq) + OH^-(aq) \longrightarrow H_2O(l) + NH_3(g)$
lead-acid battery	$Pb(s)\mid PbSO_4(s)\mid H^+(aq), HSO_4^-(aq)\mid PbO_2(s)\mid PbSO_4(s)\mid Pb(s)$, 2 V Anode: $Pb(s) + HSO_4^-(aq) \longrightarrow PbSO_4(s) + H^+(aq) + 2\,e^-$ Cathode: $PbO_2(s) + 3\,H^+(aq) + HSO_4^-(aq) + 2\,e^- \longrightarrow PbSO_4(s) + 2\,H_2O(l)$
nicad cell	$Cd(s)\mid Cd(OH)_2(s)\mid KOH(aq)\mid Ni(OH)_3(s)\mid Ni(OH)_2(s)\mid Ni(s)$, 1.2 V Anode: $Cd(s) + 2\,OH^-(aq) \longrightarrow Cd(OH)_2(s) + 2\,e^-$ Cathode: $2\,Ni(OH)_3(s) + 2\,e^- \longrightarrow 2\,Ni(OH)_2(s) + 2\,OH^-(aq)$
mercury cell	$Zn(s)\mid ZnO(s)\mid KOH(aq)\mid HgO(s)\mid Hg(l)\mid$ steel, 1.3 V Anode: $Zn(s) + 2\,OH^-(aq) \longrightarrow ZnO(s) + H_2O(l) + 2\,e^-$ Cathode: $HgO(s) + H_2O(l) + 2\,e^- \longrightarrow Hg(l) + 2\,OH^-(aq)$
silver cell	$Zn(s)\mid ZnO(s)\mid KOH(aq)\mid Ag_2O(s)\mid Ag(s)\mid$ steel, 1.6 V Anode: $Zn(s) + 2\,OH^-(aq) \longrightarrow ZnO(s) + H_2O(l) + 2\,e^-$ Cathode: $Ag_2O(s) + H_2O(l) + 2\,e^- \longrightarrow 2\,Ag(s) + 2\,OH^-(aq)$
fuel cell	graphite $\mid O_2(g)\mid Na_2CO_3(l)\mid CO_2(g)\mid Na_2CO_3(l)\mid H_2(g)\mid$ carbon, 1.2 V Anode: $H_2(g) + CO_3^{2-}(l) \longrightarrow H_2O(l) + CO_2(g) + 2\,e^-$ Cathode: $\frac{1}{2}\,O_2(g) + CO_2(g) + 2\,e^- \longrightarrow CO_3^{2-}(l)$

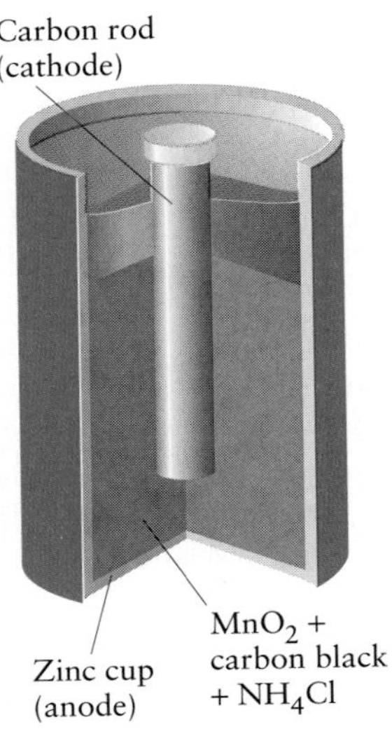

FIGURE 17.14 A commercial dry cell consists of a graphite cathode in a zinc container; the latter acts as the anode. The other components and the cell reaction are described in the text.

like the time needed for starting an engine. The electrodes are initially a hard lead-antimony alloy covered with a paste of lead(II) sulfate. The electrolyte is dilute sulfuric acid. During the first charging, some of the lead(II) sulfate is reduced to lead on one of the electrodes; the same electrode will act as the anode during discharge. Simultaneously, lead(II) sulfate is oxidized to lead(IV) oxide on the electrode that will later act as the cathode. The chemical equations for the cell reactions (Table 17.2) show that sulfuric acid is used up during discharge. When the cell is recharged, the cell reactions are driven in reverse by the external supply, and sulfuric acid is produced. The state of charge of the cell can therefore be judged from the concentration of the sulfuric acid solution, and that in turn can be judged from its density.

A *nicad cell* (Fig. 17.16) is a secondary cell widely used to power electronic equipment. The source of electrons in a circuit powered by a nicad cell is oxidation of cadmium to insoluble cadmium hydroxide, which adheres to the grid as the oxidation occurs. The insoluble nickel(II) hydroxide produced from the reduction of nickel(III) hydroxide also adheres to the stainless steel grid and is thus readily available when the cell is charged (when the reactions are reversed). Because no gases are produced in either the charging or the discharging processes, the cells can be sealed, which makes nicad cells ideal for portable equipment.

Fuel cells make highly efficient use of resources, because very little waste heat is produced. Thus, fuel cells that oxidize natural gas or hydrogen are promising alternatives to conventional power plants (see Case Study 17.) In a simple version of a fuel cell, hydrogen gas—the fuel—is

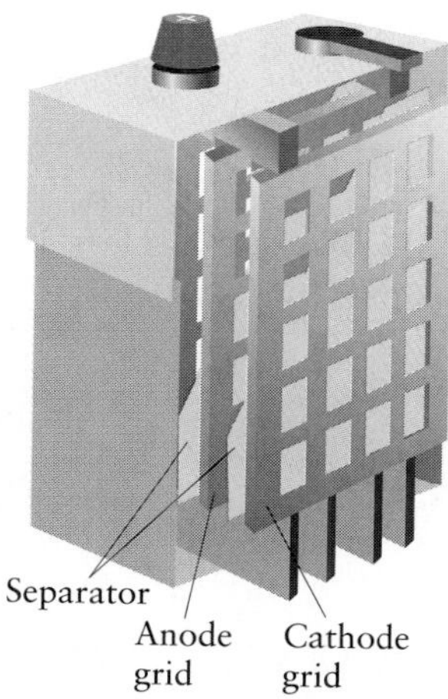

FIGURE 17.15 One cell of a lead-acid battery like those used in automobiles. A lead-acid battery is an example of a secondary cell. It needs to be charged before it can produce a current. The electrolyte is dilute sulfuric acid.

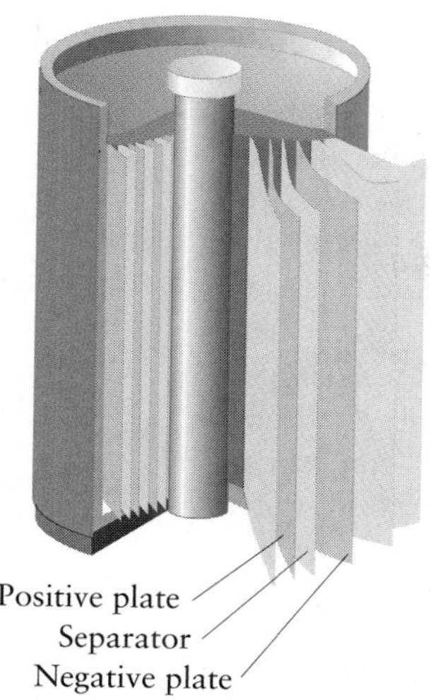

FIGURE 17.16 A rechargeable nickel-cadmium (nicad) cell. The electrodes are assembled in a jelly roll arrangement and separated by a layer soaked in moist sodium or potassium hydroxide.

passed over one electrode, oxygen is passed over the other, and the electrolyte is aqueous potassium hydroxide. A version of this type of cell is used on the space shuttle to power the life-support systems, one advantage being that the crew can drink the product of the cell reaction.

Electric eels are mobile, natural fuel cells (Fig. 17.17). They generate their electric charge in an "electric organ"—a battery of biological electrochemical cells, each cell providing about 0.15 V and an overall potential difference of about 700 V. It is an incidental feature of nature that the eel's head is its cathode and its tail the anode. The electric catfish has the opposite polarity.

Practical galvanic cells are classified as primary cells, secondary cells, and fuel cells.

FIGURE 17.17 The electric eel (*Electrophorus electricus*) lives in the Amazon River. The average potential difference it produces along its length (1 m) is about 700 V.

CASE STUDY 17 ARTIFICIAL PHOTOSYNTHESIS

Imagine a fuel that will never run out and that burns without polluting the air. It sounds too good to be true, yet hydrogen has these characteristics and it may well be the fuel of our future. Petroleum, coal, and natural gas are becoming increasingly rare and expensive, but there is enough water in the oceans to generate all the hydrogen fuel we shall ever need. In addition, water is the only combustion product, making hydrogen a pollution-free renewable fuel. The problem is how to get hydrogen from water in the first place. We can generate hydrogen by electrolysis of water, but the energy required would cancel any energy benefit of the hydrogen obtained. Is there a way to produce hydrogen more cheaply and efficiently?

Nature may provide the key, for every back yard has chemical fuel factories powered by solar energy. The factories are in the leaves of green plants, where photosynthesis makes glucose, cellulose, and starches. These chemical fuels become our food and provide renewable energy sources such as wood, all at no cost except air, water, and sunlight (recall Case Study 16).

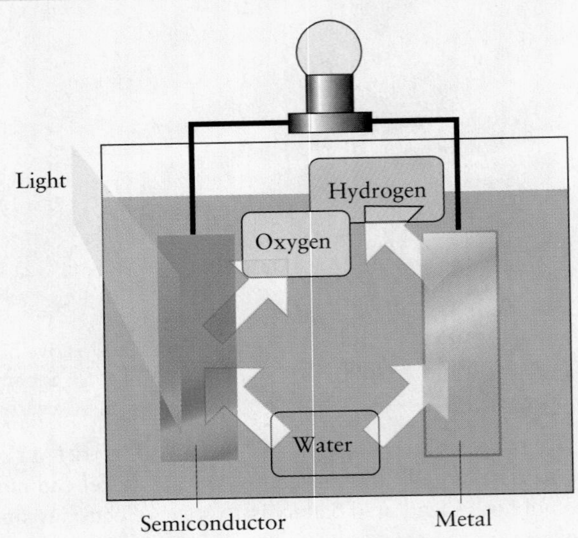

A photoelectrochemical cell uses solar energy to produce chemical fuels. This cell produces hydrogen and oxygen from water. The two gases are collected separately.

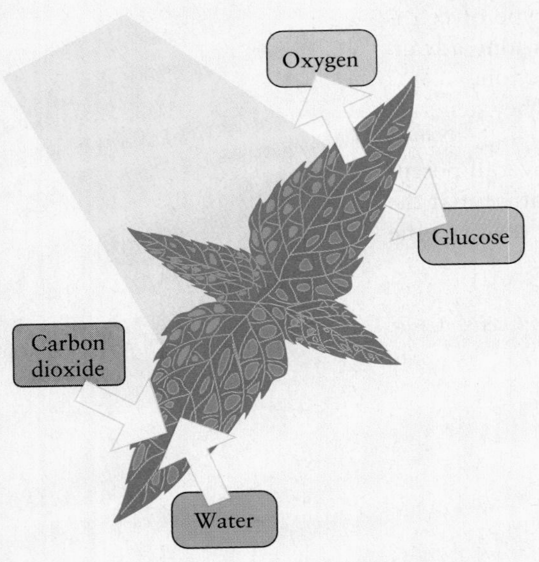

Green plants, such as these mint leaves, use solar energy to produce oxygen and chemical fuels such as glucose from carbon dioxide and water.

If we could use solar energy to produce hydrogen, our hydrogen supply would be cheap and inexhaustible. Scientists are already working to make this dream come true through the use of a special kind of electrochemical cell powered by the Sun. *Photoelectrochemical cells* are electrolytic cells that use the energy of light to reverse a spontaneous reaction. They differ from *photovoltaic cells,* which generate an electrical current from light, because they produce a chemical fuel rather than a flow of electrons. Photoelectrochemical cells differ from other electrolytic cells because their source of energy is not a chemical power supply, but light—usually cheap, plentiful sunlight.

Semiconductors, the material of transistors, make good anodes for photoelectrochemical cells, because light easily excites their valence electrons. The electrons can then leave their parent atoms and move through the solid, generating a current and producing oxygen gas from water. Hydrogen gas is generated at the cathode, usually an inert metal such as platinum.

Much of the research into photoelectrochemical cells is directed at improving the electrodes and the efficiency of their operation. One problem with semiconducting electrodes is that they tend to become corroded in water. Silicon electrodes soon become coated with silicon dioxide, SiO_2, the substance of sand and quartz. Because SiO_2 does not conduct electricity, it eventually prevents the cell from operating. One electrode material that appears more promising is cadmium sulfide, CdS. Cadmium sulfide also tends to be corroded by water, which oxidizes the sulfide ions to nonconducting elemental sulfur. However, the oxidation can be suppressed by saturating the electrolyte solution with sulfide ions. The aqueous sulfide ions are then oxidized in preference to those in the form of cadmium sulfide and the electrode remains intact.

Although photosynthesis in nature is only about 3% efficient, photoelectrochemical cells that generate chemical fuels can have efficiencies up to about 10%. Catalysts such as platinum gauze can improve efficiency by increasing the rate of the cell reaction and reducing the overvoltage required. Dyes are also used in photoelectrochemical cells to enhance the efficiency of energy capture by absorbing light of certain wavelengths more strongly. When a photon is absorbed by the dye, it excites electrons in the dye so that they can be more easily absorbed by the electrode. Once again, science follows nature. The leaves of plants contain the dye chlorophyll, which absorbs the red and blue light it needs, leaving green to be reflected (recall Case Study 7). The need to maximize the efficiency of light absorption explains why grass is green.

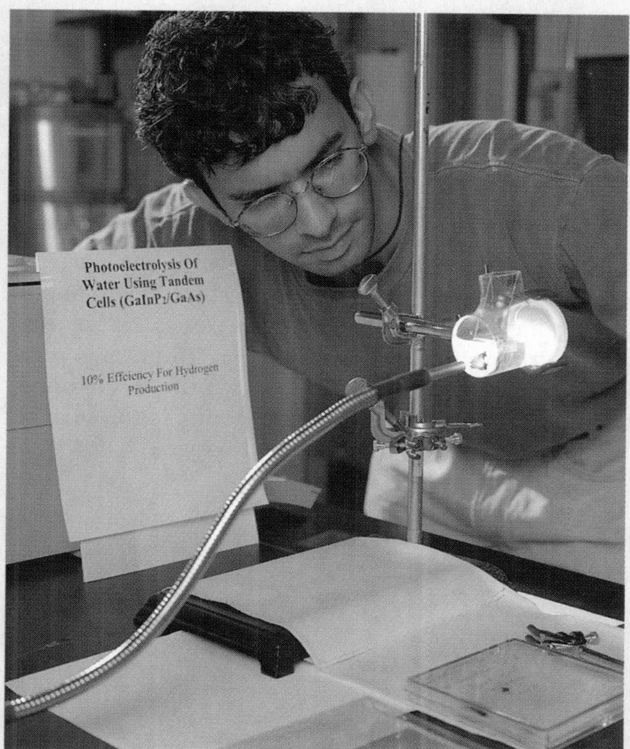

A technician prepares to demonstrate an experimental photoelectrochemical cell he developed to students visiting his laboratory.

QUESTIONS

1. Calculate the potential that would have to be generated by a photoelectrochemical cell to carry out the natural photosynthetic reaction, the formation of glucose from carbon dioxide and water, with oxygen gas as a by-product. Compare this value with the 1.23 V needed for hydrogen evolution from pure water. Which process would be easier to achieve?

2. Silicon electrodes in a photoelectrochemical cell are oxidized by water. Calculate the cell potential for the reaction between silicon and water in a standard cell, given that the water is reduced to hydrogen and that $E° = -0.84$ V for $SiO_2(s) + 4\,H^+(aq) + 4\,e^- \rightarrow Si(s) + 2\,H_2O(l)$.

3. One promising photoelectrochemical cell uses solid particles of cadmium sulfide with traces of ruthenium(IV) oxide, RuO_2, in a saturated aqueous solution of hydrogen sulfide in water. The reactant of the cell reaction is H_2S. The products are hydrogen gas and sulfur. Calculate the potential needed to operate such a cell under standard conditions. Assume the RuO_2 plays no role in the cell reaction.

4. With hydrogen gas at 1.00 atm, a pH of 3.812, and a sulfide ion concentration of 7.1×10^{-15} mol·L^{-1}, what potential would need to be applied to the cadmium sulfide/H_2S photoelectric cell in Question 3 to produce hydrogen gas and elemental sulfur under these operating conditions at 298 K?

17.12 CORROSION

FIGURE 17.18 Iron nails stored in oxygen-free water (left) do not rust because the oxidizing power of water itself is weak. When oxygen is present (as a result of air dissolving in the water, right), oxidation is thermodynamically spontaneous and rust soon forms.

Corrosion is the unwanted oxidation of a metal. Corrosion cuts short the lifetimes of steel products such as bridges and automobiles, and replacing corroded metal parts costs billions of dollars a year. Corrosion is an electrochemical process, and the electrochemical series gives us insight into why corrosion occurs and how it can be prevented.

The main culprit in corrosion is water. Any metal lower in the electrochemical series than −0.83 V can be oxidized by water under standard conditions as a result of the half-reaction

$$2\,H_2O(l) + 2\,e^- \longrightarrow H_2(g) + 2\,OH^-(aq) \qquad E° = -0.83\text{ V}$$

This standard potential is for an OH^- concentration of 1 mol·L^{-1}, which corresponds to pH = 14, a strongly basic solution. However, from the Nernst equation, at pH = 7 this couple has the potential $E = -0.42$ V. Because iron has almost the same potential ($E° = -0.44$ V for $Fe^{2+}(aq) + 2\,e^- \rightarrow Fe(s)$), iron has only a very slight tendency to be oxidized by pure water. For this reason, iron can be used for making pipes in water supply systems and can be stored in oxygen-free water without rusting (Fig. 17.18). However, when iron is exposed to damp air, with both oxygen and water present, the half-reaction

$$O_2(g) + 4\,H^+(aq) + 4\,e^- \longrightarrow 2\,H_2O(l) \qquad E° = +1.23\text{ V}$$

must be taken into account. The potential of this couple at pH = 7 is +0.81 V, which lies well above the value for iron. Hence, oxygen and water can jointly oxidize iron to Fe^{2+}, which can subsequently be oxidized to Fe^{3+}.

A drop of water on the surface of iron acts as the electrolyte in a tiny electrochemical cell (Fig. 17.19). At the edge of the drop, dissolved oxygen oxidizes the iron by the reaction given earlier. However, the electrons withdrawn from the metal by this oxidation can be restored from another part of the conducting metal—in particular, from iron lying beneath the oxygen-poor region in the center of the drop. The iron atoms there give up their electrons, form Fe^{2+} ions, and drift away into the surrounding water. This process results in the formation of tiny pits in the surface. The Fe^{2+} ions are oxidized to Fe^{3+} by the dissolved oxygen. These ions

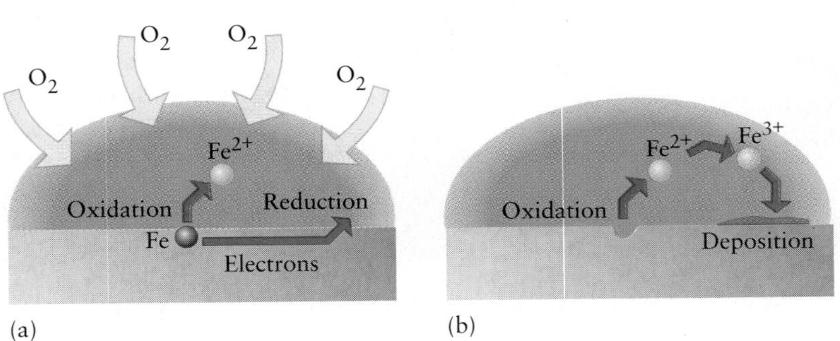

FIGURE 17.19 The mechanism of rust formation. (a) Oxidation of the iron occurs at a point out of contact with the oxygen of the air, and the surface of the metal acts as an anode in a tiny galvanic cell. (b) Further oxidation of Fe^{2+} to Fe^{3+} results in the deposition of rust on the surface.

then precipitate as a hydrated iron(III) oxide, $Fe_2O_3 \cdot H_2O$, the brown, insoluble substance we call rust. Water is more highly conducting when it has dissolved ions, and the formation of rust is then accelerated. That is one reason why the salt air of coastal cities and salt used for de-icing highways is very damaging to exposed metal.

The simplest way to prevent corrosion is to protect the surface of the metal from exposure to air and water by painting. A method that achieves greater protection is to **galvanize** the metal, which involves coating it with an unbroken film of zinc (Fig. 17.20). Zinc lies below iron in the electrochemical series, so if a scratch exposes the metal beneath, the more strongly reducing zinc releases electrons to the iron. Thus, the zinc, not the iron, is oxidized. The zinc itself survives exposure on the unbroken surface because, like aluminum, it is passivated by a protective oxide.

It is not possible to galvanize large structures—for example, ships, underground pipelines or gasoline storage tanks, and bridges—but **cathodic protection** can be used. A block of a more strongly reducing metal, such as zinc or magnesium, is buried in moist soil and connected to the underground pipeline (Fig. 17.21). The block of magnesium is preferentially oxidized and supplies electrons to the iron for the reduction of oxygen. The block of metal, which is called a **sacrificial anode,** protects the iron pipeline and is cheaper to replace than the pipeline itself. For similar reasons, automobiles generally have "negative ground systems" as part of their electrical circuitry, which means that the body of the car is connected to the anode of the battery. The decay of the anode in the battery is the sacrifice that helps preserve the vehicle itself.

The corrosion of iron is accelerated by the presence of oxygen, moisture, and salt. Corrosion can be inhibited by coating the surface with paint or zinc or by cathodic protection.

FIGURE 17.20 Metal girders are galvanized by immersion in a bath of molten zinc.

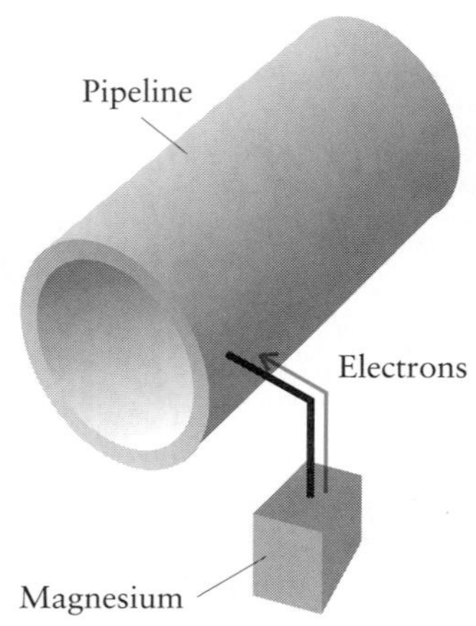

FIGURE 17.21 In the cathodic protection of a buried pipeline, or other large metal construction, the artifact is connected to a number of buried blocks of metal, such as magnesium or zinc. The sacrificial anodes (the magnesium block in this illustration) supply electrons to the pipeline (the cathode of the cell), thereby preserving it from oxidation.

SELF-TEST 17.12A Which of the following procedures helps to avoid the corrosion of an iron rod in water: (a) decreasing the concentration of oxygen in the water; (b) painting the rod; (c) increasing the concentration of ions in solution?

[*Answer:* (a) and (b). (c) will decrease corrosion only at such a high concentration of ions that oxygen cannot dissolve.]

SELF-TEST 17.12B Which of (a) copper, (b) zinc, or (c) tin can act as a sacrificial metal for iron?

ELECTROLYSIS

Redox reactions that have a positive reaction free energy are not spontaneous but can be made to occur electrochemically. For example, fluorine is so highly reactive that it cannot be isolated by any common chemical reaction. However, in 1886, the French chemist Henri Moisson isolated fluorine by passing an electric current through an anhydrous (water-free) molten mixture of potassium fluoride and hydrogen fluoride. Fluorine is still prepared commercially by the same process.

17.13 ELECTROLYTIC CELLS

Electrolysis is the process of driving a reaction in a nonspontaneous direction by using an electric current. An **electrolytic cell** is an electrochemical cell in which an electric current from an external source is used to drive a *nonspontaneous* chemical reaction. Electrolytic cells are constructed differently from galvanic cells. Specifically, the two electrodes usually share the same compartment, there is usually only one electrolyte, and concentrations and pressures are usually far from standard.

Figure 17.22 shows the layout of an electrolytic cell used for the commercial production of magnesium metal from molten magnesium chloride (the *Dow process*). As in a galvanic cell, oxidation occurs at the

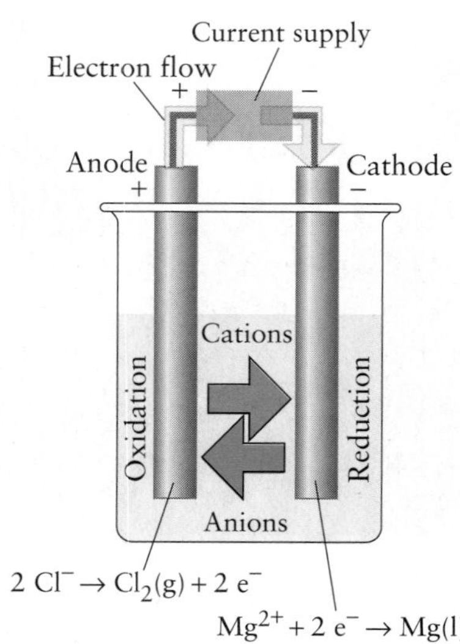

FIGURE 17.22 A schematic diagram of the electrolytic cell used in the Dow process for magnesium. The electrolyte is molten magnesium chloride. As the current generated by the external source passes through the cell, magnesium metal is produced at the cathode and chlorine gas is produced at the anode.

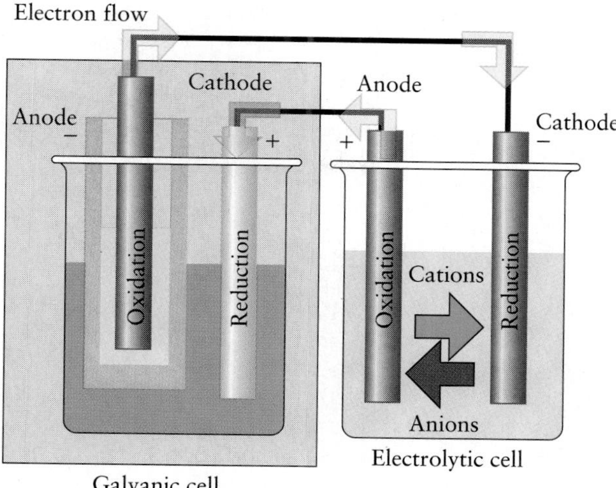

FIGURE 17.23 In this schematic diagram of an electrolysis experiment, the electrons emerge from a galvanic cell at its anode (−) and enter the electrolytic cell at its cathode (−), where they bring about reduction. Electrons are drawn out of the electrolytic cell through its anode (+) and into the galvanic cell at its cathode. If the cell reaction in the galvanic cell is more strongly spontaneous than the reaction in the electrolytic cell is nonspontaneous, then the overall process is spontaneous. This experiment is an example of one reaction driving another to which it is coupled.

anode and reduction occurs at the cathode, electrons travel through the external wire from anode to cathode, cations move through the electrolyte toward the cathode, and anions move toward the anode. Unlike the process in a galvanic cell, however, a current supplied by an external electrical power source drives electrons through the wire in a predetermined direction (Fig. 17.23), forcing oxidation to occur at one electrode and reduction at the other:

$$\text{Anode reaction:} \quad 2\,Cl^-(l) \longrightarrow Cl_2(g) + 2\,e^-$$

$$\text{Cathode reaction:} \quad Mg^{2+}(l) + 2\,e^- \longrightarrow Mg(l)$$

A rechargeable battery functions as a galvanic cell when it is doing work and as an electrolytic cell when it is being recharged.

The anode of an electrolytic cell is labeled + and the cathode −, the opposite of a galvanic cell. Electrons enter an electrolytic cell through the cathode.

In an electrolytic cell, current supplied by an external source is used to drive a nonspontaneous redox reaction.

17.14 THE POTENTIAL NEEDED FOR ELECTROLYSIS

To drive a nonspontaneous reaction in electrolysis, the external supply must generate a potential greater than the potential that the spontaneous reverse reaction would produce. For example, to achieve the nonspontaneous reaction

$$2\,H_2O(l) \longrightarrow 2\,H_2(g) + O_2(g) \qquad E = -1.23\text{ V at pH} = 7$$

CHEMISTRY AT WORK

An electrical engineer tests a panel of photovoltaic cells in a solar simulator that allows him to vary the intensity of the light. Panels that show good performance will be duplicated and used to carry out redox reactions in electrolytic cells.

we must apply at least 1.23 V from the external source to overcome the reaction's natural "pushing power" in the opposite direction. In practice, the applied potential must usually be greater than the cell potential to reverse a spontaneous cell reaction. The additional potential, which varies with the type of electrode, is called the **overpotential.** Much research on electrochemical cells involves attempts to reduce the overpotential and increase efficiency (Case Study 17). For platinum electrodes, the overpotential for the production of water from hydrogen and oxygen is about 0.6 V, so about 1.8 V (0.6 V + 1.23 V) is actually required to electrolyze water when platinum electrodes are used.

It is important to consider the possibility that other ions present in the solution can also be oxidized or reduced by the electric current. The potential for the reduction of oxygen in water at pH = 7 is +0.81 V:

$$O_2(g) + 4\,H^+(aq) + 4\,e^- \longrightarrow 2\,H_2O(l) \qquad E = +0.81\text{ V}$$

To reverse this half-reaction and bring about the oxidation of water needs a potential of at least 0.81 V. We need to know whether any ions are more readily oxidized than water. For example, when Cl^- ions are present at 1 $mol{\cdot}L^{-1}$ in water, is it possible that they, and not the water, will be oxidized? From Table 17.1, the standard potential for the reduction of chlorine is +1.36 V:

$$Cl_2(g) + 2\,e^- \longrightarrow 2\,Cl^-(aq) \qquad E^\circ = +1.36\text{ V}$$

To reverse this reaction and oxidize chloride ions, we have to supply at least 1.36 V. Because only 0.81 V is needed to force the oxidation of water, but 1.36 V is needed to force the oxidation of Cl^-, it appears that oxygen should be the product at the cathode. However, the overpotential for oxygen production can be very high, and in practice chlorine might also be produced.

Suppose the solution contains I^- ions at 1 mol·L^{-1} instead of chloride ions. From Table 17.1, we know that at least 0.54 V is needed to oxidize I^-:

$$I_2(s) + 2\,e^- \longrightarrow 2\,I^-(aq) \qquad E^\circ = +0.54\ V$$

Because only 0.54 V is needed to reverse this reduction and hence bring about the oxidation of iodide ions, the I^- ion would be oxidized in preference to water.

The potential supplied to an electrolytic cell must be at least as great as that of the cell reaction to be reversed. If there is more than one reducible species in solution, the species with the greater potential for reduction is preferentially reduced. The same principle applies to oxidation.

SELF-TEST 17.13A Predict the products resulting from the electrolysis of 1 M $AgNO_3(aq)$.

[*Answer:* Cathode, Ag; anode, O_2]

SELF-TEST 17.13B Predict the products resulting from the electrolysis of 1 M NaBr(aq).

17.15 THE PRODUCTS OF ELECTROLYSIS

Now we add the final arrow to the mole diagram (**1**) and see how to calculate the amount of product formed by a given amount of electricity.

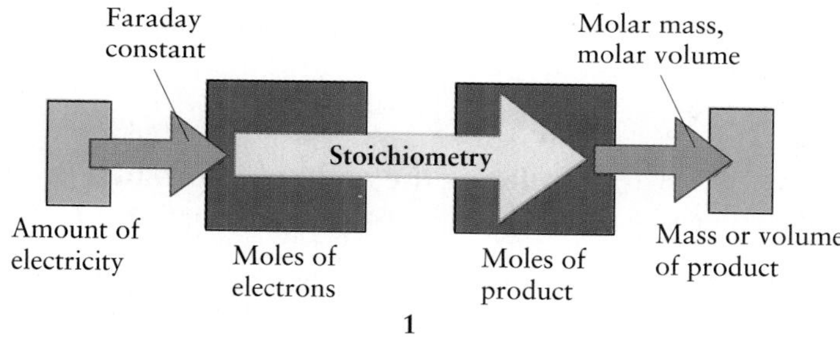

1

The calculation is based on observations made by Michael Faraday (Fig. 17.24) and summarized as follows:

Faraday's law of electrolysis: The number of moles of product formed by an electric current is stoichiometrically equivalent to the number of moles of electrons supplied.

FIGURE 17.24 Michael Faraday (1791–1867).

Once we know the number of moles of product formed, we can calculate the masses of the products or, if they are gases, their volumes.

For example, copper is refined electrolytically by using an impure form of copper metal (called "blister copper") as the anode in an electrolytic cell (Fig. 17.25). The current supply causes the oxidation of the blister copper to copper(II) ions,

$$Cu(s, blister) \longrightarrow Cu^{2+}(aq) + 2\,e^-$$

These ions are then reduced at the cathode:

$$Cu^{2+}(aq) + 2\,e^- \longrightarrow Cu(s)$$

From the stoichiometry of these chemical equations, we know that 2 mol $e^- \simeq$ 1 mol Cu. Therefore, if 4.0 mol e^- is supplied, then the amount of copper produced is

$$\text{Moles of Cu} = (4.0 \text{ mol } e^-) \times \frac{1 \text{ mol Cu}}{2 \text{ mol } e^-} = 2.0 \text{ mol Cu}$$

The quantity of electricity passed through the electrolysis cell is normally measured as the current and the time for which the current flows. The charge passed through an electrolytic cell is the product of the current and the time for which it is supplied. Electric current is measured in the SI unit ampere, A, the rate of flow of charge in coulombs per second ($1 \text{ A} = 1 \text{ C·s}^{-1}$). Therefore, the charge supplied is

A current of 1 A corresponds to about 6×10^{18} electrons passing a given point each second.

$$\text{Charge supplied (C)} = \text{current (A)} \times \text{time (s)}$$

Next, we use the fact that the Faraday constant, F, is the magnitude of the charge per mole of electrons (recall Section 17.6). So, because *charge* = *moles* × F, it follows that

$$\text{Moles of } e^- = \frac{\text{charge supplied (C)}}{F} = \frac{\text{current (A)} \times \text{time (s)}}{9.6485 \times 10^4 \text{ C·mol}^{-1}} \qquad (5)$$

So, by measuring the current and the time for which it flows, we can determine the moles of electrons supplied.

The amount of product in an electrolysis reaction is calculated from the stoichiometry of the half-reaction and the current and time for which the current flows.

EXAMPLE 17.7 Calculating the amount of product produced from electrolysis

Aluminum is produced by electrolysis of its oxide dissolved in molten cryolite (Na_3AlF_6). Calculate the mass of aluminum that can be produced in one day in an electrolytic cell operating continuously at 1.0×10^5 A. The cryolite does not react.

STRATEGY The moles of electrons supplied during the electrolysis is given by Eq. 5; the Faraday constant is given inside the back cover. The moles of electrons is converted to moles of product by using the stoichiometry of the half-reaction. Finally, convert moles of product to mass by using the molar mass.

SOLUTION The aluminum metal is produced from the reduction of Al^{3+}, present in the molten Al_2O_3. The balanced cathodic half-reaction is $Al^{3+}(l) + 3\,e^- \rightarrow Al(l)$, from which it follows that 3 mol $e^- \simeq$ 1 mol Al. Because the molar mass of aluminum is 26.98 $g·mol^{-1}$, the string of conversions we need is

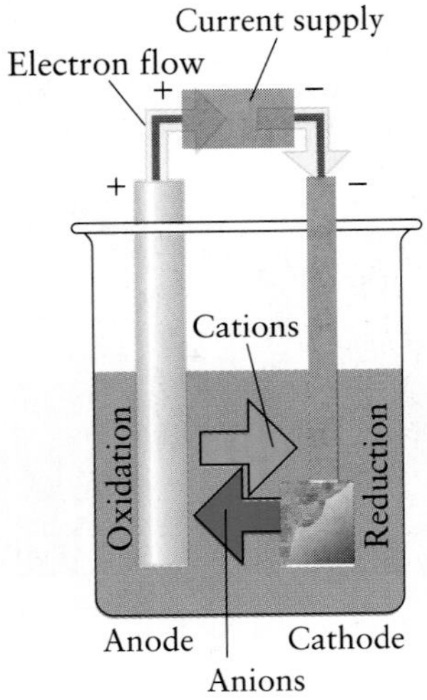

FIGURE 17.25 A schematic diagram showing the electrolytic process for refining copper. The anode is impure copper. It undergoes oxidation, and the Cu^{2+} ions so produced migrate to the cathode, where they are reduced to pure copper metal. A similar arrangement is used for electroplating objects.

$$\text{Mass of Al} = \frac{(1.00 \times 10^5\ \text{C}\cdot\text{s}^{-1}) \times (24 \times 3600\ \text{s})}{9.65 \times 10^4\ \text{C}\cdot\text{mol}^{-1}} \times \frac{1\ \text{mol Al}}{3\ \text{mol e}^-}$$
$$\times (26.98\ \text{g}\cdot\text{mol}^{-1})$$
$$= 8.05 \times 10^5\ \text{g}$$

or 805 kg. The fact that the production of 1 mol Al requires 3 mol e^- accounts for the very high consumption of electricity characteristic of aluminum production plants.

SELF-TEST 17.14A Determine the time, in hours, required to electroplate 7.00 g of magnesium metal from molten magnesium chloride, using a current of 7.30 A. What volume of chlorine gas at 25°C and 1.00 atm will be produced at the anode?

[*Answer:* 2.11 h; 7.04 L]

SELF-TEST 17.14B How many hours are required to plate 12.00 g of chromium metal from a 1 M solution of CrO_3 in dilute sulfuric acid, using a current of 6.20 A?

17.16 APPLICATIONS OF ELECTROLYSIS

We have already described the electrolytic extraction of aluminum, magnesium, and fluorine and the refining of copper. Another important industrial application of electrolysis is the production of sodium metal by the *Downs process,* the electrolysis of molten rock salt (Fig. 17.26):

$$\text{Cathode reaction:} \quad 2\ Na^+(l) + 2\ e^- \longrightarrow 2\ Na(l)$$

$$\text{Anode reaction:} \quad 2\ Cl^-(l) \longrightarrow Cl_2(g) + 2\ e^-$$

Sodium chloride is plentiful as rock salt, but the solid does not conduct electricity, because the ions are locked into place. Thus, sodium chloride must be molten for electrolysis to occur. The electrodes in the cell are made of inert materials like carbon, and the cell is designed to keep the sodium and chlorine produced by the electrolysis out of contact with each other and with air. Over 2×10^7 kg of sodium is produced annually by the Downs process, much of it for use as a coolant in nuclear reactors. Other active metals, such as lithium, magnesium, and calcium, are also prepared by the electrolysis of their molten chloride salts. In a modification of the Downs process, the electrolyte is an aqueous solution of sodium chloride. The products of this *chloralkali process* are chlorine and aqueous sodium hydroxide (Fig. 17.27).

Electroplating is the electrolytic deposition of a thin film of metal on an object. The object to be electroplated (either metal or plastic coated with graphite) is made the cathode, and the electrolyte is an aqueous solution of a salt of the plating metal. Metal is deposited on the cathode from ions in the electrolyte solution. These cations are either supplied by

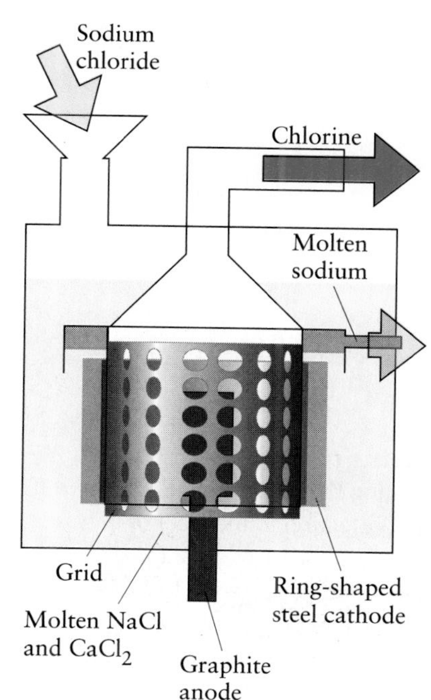

FIGURE 17.26 In the Downs process, molten sodium chloride is electrolyzed with a graphite anode (at which the Cl^- ions are oxidized to chlorine) and a steel cathode (at which the Na^+ ions are reduced to sodium). The sodium and chlorine are kept apart by the hoods surrounding the electrodes. Calcium chloride is present to lower the melting point of sodium chloride to a more economical value.

FIGURE 17.27 A chloralkali plant in which chlorine gas and aqueous sodium hydroxide are produced by the electrolysis of aqueous sodium hydroxide. The electrolytic cells are on each side of the technician.

the added salt or from oxidation of the anode, which is made of the plating metal.

For chromium plating (Fig. 17.28), the electrolyte is prepared by dissolving CrO_3 in dilute sulfuric acid. The electrolysis reduces chromium(VI) first to chromium(III) and then to Cr(0):

$$CrO_3(aq) + 6\,H^+(aq) + 6\,e^- \longrightarrow Cr(s) + 3\,H_2O(l)$$

The chromium deposits on the cathode as a hard protective film. Because six electrons must be supplied for each atom of chromium deposited, a large amount of electricity is needed.

Electrolysis is used industrially to produce aluminum and magnesium, to extract metals from their salts, to prepare chlorine, fluorine, and sodium hydroxide, to refine copper, and in electroplating.

FIGURE 17.28 Chromium plating lends decorative flair as well as protection to the steel of this motorcycle. Large quantities of electricity are needed to plate chromium because six electrons are required to produce each atom of chromium.

SKILLS YOU SHOULD HAVE MASTERED

Conceptual

1. Distinguish galvanic and electrolytic cells and describe their operation; identify anode, cathode, and the direction of current flow for each.
2. Explain the relationship between the standard free energy, standard cell potential, and equilibrium constant of a reaction.
3. Predict the effect of changes in concentration of reactants and products on the cell potential.

Problem-Solving

1. Balance chemical equations for redox reactions by the half-reaction method.
2. Write the cell diagram for a redox reaction and write the equation for a cell reaction, given the cell diagram.
3. Calculate a standard cell potential, given the half-reactions for the cell and their standard potentials.
4. Predict the spontaneous direction of a redox reaction by using the electrochemical series.
5. Calculate the equilibrium constant for a reaction from the standard cell potential.
6. Use the Nernst equation to calculate a cell potential under nonstandard conditions.
7. Predict the products of electrolysis of an aqueous solution from the electrochemical series, including the possibility of competition.
8. Calculate the amount of product produced from electrolysis.

Descriptive

1. Distinguish and give examples of primary cells, secondary cells, and fuel cells.
2. Describe corrosion and the means of protecting iron from corrosion.
3. Describe the Downs process for producing sodium metal and chlorine gas.
4. Describe electroplating and give an example.

EXERCISES

Assume a temperature of 25°C (298 K) for the following exercises unless instructed otherwise.

Balancing Redox Equations

17.1 Balance the following skeletal equations for each half-reaction occurring in an acidic solution. Identify each as an oxidation or a reduction half-reaction.
(a) Conversion of vanadyl ions to vanadium(III) ions:

$$VO^{2+}(aq) \longrightarrow V^{3+}(aq)$$

(b) One of the half-reactions that can occur when a lead-acid battery is charged:

$$PbSO_4(s) \longrightarrow PbO_2(s) + SO_4^{2-}(aq)$$

(c) Decomposition of hydrogen peroxide to oxygen gas:

$$H_2O_2(aq) \longrightarrow O_2(g)$$

17.2 Balance the following skeletal equations for each half-reaction occurring in an acidic solution. Identify each as an oxidation or a reduction half-reaction.
(a) $Cr_2O_7^{2-}(aq) \longrightarrow Cr^{3+}(aq)$
(b) $I^-(aq) \longrightarrow IO_3^-(aq)$
(c) $NO(g) \longrightarrow NO_3^-(aq)$

17.3 Balance the following skeletal equations for each half-reaction occurring in a basic solution. Identify each as an oxidation or a reduction half-reaction.
(a) Conversion of hypochlorite ions to chloride ions (a part of the action of bleach):

$$ClO^-(aq) \longrightarrow Cl^-(aq)$$

(b) Conversion of iodate ions to hypoiodite ions:

$$IO_3^-(aq) \longrightarrow IO^-(aq)$$

(c) Conversion of sulfite ions to dithionite ions:

$$SO_3^{2-}(aq) \longrightarrow S_2O_4^{2-}(aq)$$

17.4 Balance the following skeletal equations for each half-reaction occurring in a basic solution. Identify each as an oxidation or a reduction half-reaction.
(a) $NO_3^-(aq) \longrightarrow NO_2^-(aq)$
(b) $CrO_4^{2-}(aq) \longrightarrow Cr(OH)_3(s)$
(c) $MnO_4^{2-}(aq) \longrightarrow MnO_2(s)$

17.5 Balance the following skeletal equations, using oxidation and reduction half-reactions. All reactions occur in acidic solution. Identify the oxidizing agent and reducing agent in each reaction.
(a) Reaction of thiosulfate ion with chlorine gas:

$$Cl_2(g) + S_2O_3^{2-}(aq) \longrightarrow Cl^-(aq) + SO_4^{2-}(g)$$

(b) Action of the permanganate ion on sulfurous acid:

$$MnO_4^-(aq) + H_2SO_3(aq) \longrightarrow Mn^{2+}(aq) + HSO_4^-(aq)$$

(c) Reaction of hydrosulfuric acid with chlorine:

$$H_2S(aq) + Cl_2(g) \longrightarrow S(s) + Cl^-(aq)$$

(d) Reaction of chlorine in water:

$$Cl_2(g) \longrightarrow HOCl(aq) + Cl^-(aq)$$

17.6 Balance the following skeletal equations, using oxidation and reduction half-reactions. All reactions occur in acidic solution. Identify the oxidizing agent and reducing agent in each reaction.
(a) Conversion of iron(II) to iron(III) by dichromate ion:

$$Fe^{2+}(aq) + Cr_2O_7^{2-}(aq) \longrightarrow Fe^{3+}(aq) + Cr^{3+}(aq)$$

(b) Formation of acetic acid from ethanol by the action of permanganate ion:

$$C_2H_5OH(aq) + MnO_4^-(aq) \longrightarrow Mn^{2+}(aq) + CH_3COOH(aq)$$

(c) Reaction of iodine with nitric acid:

$$I_2(aq) + NO_3^-(aq) \longrightarrow I^-(aq) + NO(g)$$

(d) Reaction of arsenic(III) sulfide with nitric acid:

$$As_2S_3(s) + NO_3^-(aq) \longrightarrow H_3AsO_4 + S(s) + NO(g)$$

17.7 Balance the following skeletal equations, using oxidation and reduction half-reactions. All reactions occur in a basic solution. Identify the oxidizing agent and reducing agent in each reaction.
(a) Action of ozone on bromide ions:

$$O_3(aq) + Br^-(aq) \longrightarrow O_2(g) + BrO_3^-(aq)$$

(b) Reaction of bromine in water:

$$Br_2(l) \longrightarrow BrO_3^-(aq) + Br^-(aq)$$

(c) Formation of chromate ions from chromium (III) ions:

$$Cr^{3+}(aq) + MnO_2(s) \longrightarrow Mn^{2+}(aq) + CrO_4^{2-}(aq)$$

(d) Reaction of elemental phosphorus to form phosphine (PH_3), a poisonous gas with the odor of decaying fish:

$$P_4(s) \longrightarrow H_2PO_2^-(aq) + PH_3(g)$$

17.8 Balance the following skeletal equations, using oxidation and reduction half-reactions. All reactions occur in basic solution. Identify the oxidizing agent and reducing agent in each reaction.
(a) Production of chlorite ions from chlorine heptoxide:

$$Cl_2O_7(g) + H_2O_2(aq) \longrightarrow ClO_2^-(aq) + O_2(g)$$

(b) Action of permanganate ions on sulfide ions:

$$MnO_4^-(aq) + S^{2-}(aq) \longrightarrow S(s) + MnO_2(s)$$

(c) Reaction of hydrazine with chlorate ions:

$$N_2H_4(g) + ClO_3^-(aq) \longrightarrow NO(g) + Cl^-(aq)$$

(d) Reaction of plumbite ions and hypochlorite ions:

$$Pb(OH)_4^{2-}(aq) + ClO^-(aq) \longrightarrow PbO_2(s) + Cl^-(aq)$$

17.9 The hydrogen sulfite ion, HSO_3^-, is a moderately strong reducing agent in acidic solutions and, depending on the conditions, is oxidized to either the hydrogen sulfate ion, HSO_4^-, or the dithionate ion, $S_2O_6^{2-}$. (a) Write the equations for each half-reaction. (b) The reaction of the hydrogen sulfite ion with iodine to form iodide and hydrogen sulfate ions is used to determine its concentration in solution in the laboratory. Write the equation for each half-reaction and the overall equation for the reaction.

17.10 The thiosulfate ion, $S_2O_3^{2-}$, is a moderately strong reducing agent in acidic solutions. It is used to determine the concentration of elemental iodine in solution, forming iodide ions and tetrathionate ions, $S_4O_6^{2-}$, in the reaction. Write the equation for each half-reaction and the overall equation for the reaction.

17.11 Potassium permanganate is an excellent oxidizing agent for laboratory use and in sewage treatment. It readily reacts with the organic compounds in sewage to produce carbon dioxide and water. Write the equation for each half-reaction and the overall equation for the oxidation of glucose, $C_6H_{12}O_6$, the skeletal form of which is $MnO_4^-(aq) + C_6H_{12}O_6(aq) \rightarrow Mn^{2+}(aq) + CO_2(g) + H_2O(l)$.

17.12 Nitric oxide gas can be produced from the reaction of nitrite ions with each other in a dilute sulfuric acid solution. The skeletal equation for the reaction is $NO_2^-(aq) \rightarrow NO(g) + NO_3^-(aq)$. Write the equation for each half-reaction and the overall equation for the reaction.

Galvanic Cells

17.13 Complete the following statements: (a) In a galvanic cell, oxidation occurs at the (anode, cathode). (b) The cathode is the (positive, negative) electrode.

17.14 Complete the following statement: In a galvanic cell, the anions migrate toward the (anode, cathode) and the electrons flow through an external circuit from the (anode, cathode) to the (anode, cathode).

17.15 What are the charge-carrying species in (a) the external circuit and (b) the internal circuit of a galvanic cell?

17.16 At which electrode does (a) oxidation; (b) reduction occur in a galvanic cell?

17.17 Write the reduction half-reaction for (a) zinc metal in contact with Zn^{2+} ions; (b) platinum metal dipping into a solution of iron(II) and iron(III) salts; (c) chlorine gas in contact with Cl^- ions; (d) the calomel electrode (see Box 17.1).

17.18 Write the oxidation half-reaction for (a) silver metal in contact with Ag^+ ions; (b) a graphite electrode in contact with Ce^{3+} and Ce^{4+} ions in aqueous solution; (c) oxygen gas in contact with hydroxide ions in water; (d) the silver-silver iodide electrode (silver in contact with AgI(s) and a solution of I^- ions).

17.19 Write cathode and anode half-reactions and the balanced equation for the cell reaction for each of the following galvanic cells:
(a) $Pt(s)|Fe^{3+}(aq), Fe^{2+}(aq)||Ag^+(aq)|Ag(s)$
(b) $C(gr)|H_2(g)|H^+(aq)||Cl^-(aq)|Cl_2(g)|Pt(s)$
(c) $U(s)|U^{3+}(aq)||V^{2+}(aq)|V(s)$
(d) $Pt(s)|O_2(g)|H^+(aq)||OH^-(aq)|O_2(g)|Pt(s)$
(e) $Pt(s)|Sn^{4+}(aq),Sn^{2+}(aq)||Cl^-(aq)|Hg_2Cl_2(s)|Hg(l)$

17.20 Write cathode and anode half-reactions and the balanced equation for the cell reaction for each of the following galvanic cells:
(a) $Cu(s)|Cu^{2+}(aq)||Cu^+(aq)|Cu(s)$
(b) $Pt(s)|I_3^-(aq),I^-(aq)||Br^-(aq)|Br_2(l)|C(gr)$, where I_3^- is equivalent to $(I_2{\cdot}I^-)$
(c) $Ag(s)|AgI(s)|I^-(aq)||Cl^-(aq)|AgCl(s)|Ag(s)$
(d) $Hg(l)|Hg_2Cl_2(s)|Cl^-(aq)||Cl^-(aq)|AgCl(s)|Ag(s)$
(e) $Cu(s)|Cu^{2+}(aq)||Pb^{2+}(aq),Pb^{4+}(aq)|C(gr)$

17.21 Write cathode and anode half-reactions, the balanced equation for the cell reaction, and the cell diagram for the following unbalanced reactions:
(a) $Ni^{2+}(aq) + Zn(s) \longrightarrow Ni(s) + Zn^{2+}(aq)$
(b) $Ce^{4+}(aq) + I^-(aq) \longrightarrow I_2(s) + Ce^{3+}(aq)$
(c) $Cl_2(g) + H_2(g) \longrightarrow HCl(aq)$
(d) $Au^+(aq) \longrightarrow Au(s) + Au^{3+}(aq)$

17.22 Write cathode and anode half-reactions, the balanced equation for the cell reaction, and the cell diagram for the following unbalanced reactions:
(a) $Mn(s) + Ti^{2+}(aq) \longrightarrow Mn^{2+}(aq) + Ti(s)$
(b) $Fe^{3+}(aq) + H_2(g) \longrightarrow Fe^{2+}(aq) + H^+(aq)$
(c) $Cu^+(aq) \longrightarrow Cu(s) + Cu^{2+}(aq)$
(d) $MnO_4^-(aq) + H^+(aq) + Cl^-(aq) \longrightarrow Cl_2(g) + Mn^{2+}(aq) + H_2O(l)$

17.23 Write cathode and anode half-reactions and devise a galvanic cell (write a cell diagram) to study each of the following reactions:
(a) $AgBr(s) \longrightarrow Ag^+(aq) + Br^-(aq)$, a solubility equilibrium
(b) $H_3O^+(aq) + OH^-(aq) \longrightarrow 2\,H_2O(l)$, the Brønsted neutralization reaction
(c) $Cd(s) + 2\,Ni(OH)_3(s) \longrightarrow Cd(OH)_2(s) + 2\,Ni(OH)_2(s)$, the reaction in the nickel-cadmium cell

17.24 Write balanced cathode and anode half-reactions and devise a galvanic cell (write a cell diagram) to study each of the following reactions:
(a) $AgNO_3(aq) + KI(aq) \longrightarrow AgI(s) + KNO_3(aq)$, a precipitation reaction
(b) H_3O^+(aq, concentrated) $\longrightarrow$ H_3O^+(aq, dilute)
(c) $Zn(s) + Ag_2O(s) \longrightarrow ZnO(s) + 2Ag(s)$, the reaction in a silver cell.

17.25 (a) Write balanced cathode and anode half-reactions for the redox reaction of an acidified solution of potassium permanganate and iron(II) chloride. (b) Write the balanced equation for the cell reaction and devise a galvanic cell to study the reaction (write its cell diagram).

17.26 (a) Write balanced cathode and anode half-reactions for the redox reaction between sodium dichromate and mercury(I) nitrate in an acidic solution. (b) Write the balanced equation for the cell reaction and devise a galvanic cell to study the reaction (write its cell diagram).

Cell Potential and Free Energy

17.27 Find the value of n, the number of moles of electrons transferred, for each of the following chemical equations:
(a) $C_6H_{12}O_6(s) + 6\,O_2(g) \longrightarrow 6\,CO_2(g) + 6\,H_2O(l)$
(b) $2\,B(s) + 3\,Cl_2(g) \longrightarrow 2\,BCl_3(g)$
(c) $SiO_2(l) + 2\,C(s) \longrightarrow Si(l) + 2\,CO(g)$

17.28 Find the value of n, the number of moles of electrons transferred, for each stage in the Ostwald process for the production of nitric acid from ammonia:
(a) $4\,NH_3(g) + 5\,O_2(g) \longrightarrow 4\,NO(g) + 6\,H_2O(l)$
(b) $2\,NO(g) + O_2(g) \longrightarrow 2\,NO_2(g)$
(c) $3\,NO_2(g) + H_2O(l) \longrightarrow 2\,HNO_3(aq) + NO(g)$

17.29 Predict the standard potential of each of the following galvanic cells:
(a) $Pt(s)|Cr^{3+}(aq), Cr^{2+}(aq)||Cu^{2+}(aq)|Cu(s)$
(b) $Ag(s)|AgI(s)|I^-(aq)||Cl^-(aq)|AgCl(s)|Ag(s)$
(c) $Hg(l)|Hg_2Cl_2(s)|Cl^-(aq)||Hg_2^{2+}(aq)|Hg(l)$
(d) $C(gr)|Sn^{4+}(aq),Sn^{2+}(aq)||Pb^{4+}(aq),Pb^{2+}(aq)|Pt(s)$

17.30 Predict the standard potential of each of the following galvanic cells:
(a) $Pt(s)|Fe^{3+}(aq),Fe^{2+}(aq)||Ag^{+}(aq)|Ag(s)$
(b) $U(s)|U^{3+}(aq)||V^{2+}(aq)||V(s)$
(c) $Sn(s)|Sn^{2+}(aq)||Sn^{4+}(aq),Sn^{2+}(aq)|Pt(s)$
(d) $Cu(s)|Cu^{2+}(aq)||Au^{+}(aq)|Au(s)$

17.31 Predict the standard cell potential and calculate the standard free energy for the following galvanic cells (the standard potentials of these cells were obtained in Exercise 17.29):
(a) $Pt(s)|Cr^{3+}(aq),Cr^{2+}(aq)||Cu^{2+}(aq)|Cu(s)$
(b) $Ag(s)|AgI(s)|I^{-}(aq)||Cl^{-}(aq)|AgCl(s)|Ag(s)$
(c) $Hg(l)|Hg_2Cl_2(s)|Cl^{-}(aq)||Hg_2^{2+}(aq)|Hg(l)$
(d) $C(gr)|Sn^{4+}(aq), Sn^{2+}(aq)||Pb^{4+}(aq),Pb^{2+}(aq)|Pt(s)$

17.32 Predict the standard cell potential and calculate the standard free energy for galvanic cells having the following cell reactions:
(a) $Zn(s) + Fe^{2+}(aq) \longrightarrow Zn^{2+}(aq) + Fe(s)$
(b) $2\,H_2(g) + O_2(g) \longrightarrow 2\,H_2O(l)$ in an acidic solution
(c) $Ag^{+}(aq) + Cl^{-}(aq) \longrightarrow AgCl(s)$
(d) $3\,Au^{+}(aq) \longrightarrow 2\,Au(s) + Au^{3+}(aq)$

The Electrochemical Series

17.33 Arrange the following metals in order of increasing strength as reducing agents: (a) Cu, Zn, Cr, Fe; (b) Li, Na, K, Mg; (c) U, V, Ti, Al; (d) Ni, Sn, Au, Ag.

17.34 Arrange the following species in order of increasing strength as oxidizing agents: (a) Co^{2+}, Cl_2, Ce^{4+}, In^{3+}; (b) NO_3^-, ClO_4^-, HBrO, $Cr_2O_7^{2-}$, all in acidic solution; (c) H_2O_2, O_2, MnO_4^-, HClO, all in acidic solution; (d) Ti^{3+}, Sn^{4+}, Hg_2^{2+}, Fe^{2+}.

17.35 Suppose that the following redox couples are joined to form a galvanic cell that generates a current under standard conditions. Identify the oxidizing agent and the reducing agent, write a cell diagram, and calculate the standard cell potential: (a) Co^{2+}/Co and Ti^{3+}/Ti^{2+}; (b) La^{3+}/La and U^{3+}/U; (c) H^{+}/H_2 and Fe^{3+}/Fe^{2+}; (d) $O_3/O_2,OH^-$ and Ag^{+}/Ag.

17.36 Suppose that the following redox couples are joined to form a galvanic cell that generates a current under standard conditions. Identify the oxidizing agent and the reducing agent, write a cell diagram, and calculate the standard cell potential: (a) Pt^{2+}/Pt and $AgF/Ag,F^-$; (b) Cr^{3+}/Cr^{2+} and I_3^-/I^-; (c) H^{+}/H_2 and Ni^{2+}/Ni; (d) $O_3/O_2,OH^-$ and $O_3,H^{+}/O_2$.

17.37 Answer the following questions and, for each "yes" response, write a balanced cell reaction and calculate the standard cell potential: (a) Can H_2 reduce Ni^{2+} ions to nickel metal? (b) Can chromium metal reduce Pb^{2+} ions to lead metal? (c) Can permanganate ions oxidize copper metal to Cu^{2+} ions in an acidic solution? (d) Can Fe^{3+} ions oxidize mercury metal to mercury(I)?

17.38 Identify the spontaneous reactions in the following list and, for each spontaneous reaction, identify the oxidizing agent and calculate the standard cell potential:
(a) $Cl_2(g) + 2\,Br^{-}(aq) \longrightarrow 2\,Cl^{-}(aq) + Br_2(l)$
(b) $MnO_4^-(aq) + 8\,H^{+}(aq) + 5\,Ce^{3+}(aq) \longrightarrow 5\,Ce^{4+}(aq) + Mn^{2+}(aq) + 4\,H_2O(l)$
(c) $2\,Pb^{2+}(aq) \longrightarrow Pb(s) + Pb^{4+}(aq)$
(d) $2\,NO_3^-(aq) + 4\,H^{+}(aq) + Zn(s) \longrightarrow Zn^{2+}(aq) + 2\,NO_2(g) + 2\,H_2O(l)$

17.39 Identify the spontaneous reactions among the following reactions and, for the spontaneous reactions, write balanced reduction and oxidation half-reactions. Show that the reaction is spontaneous by calculating the standard free energy of the reaction. (a) $I_2 + H_2 \rightarrow$?; (b) $Mg^{2+} + Cu \rightarrow$?; (c) $Al + Pb^{2+} \rightarrow$?.

17.40 Identify the spontaneous reactions among the following reactions and, for the spontaneous reactions, write balanced reduction and oxidation half-reactions. Show that the reaction is spontaneous by calculating the standard free energy of the reaction. (a) $Hg_2^{2+} + Ce^{3+} \rightarrow$?; (b) $Zn + Sn^{2+} \rightarrow$?; (c) $O_2 + H^{+} + Hg \rightarrow$?.

17.41 Chlorine is used to displace bromine from brine that contains sodium bromide. Could oxygen in an acidified solution be used instead? If so, why is it not used?

17.42 A chemist is interested in the compounds formed by the *d*-block element manganese and wants to find a way to prepare Mn^{3+} from Mn^{2+}. Would an acidified solution of sodium dichromate be suitable?

Equilibrium Constants

17.43 Write the expression for the equilibrium constants of the following cell reactions:
(a) $Pt(s)|H_2(g)|H^{+}(aq)||Cl^{-}(aq)|AgCl(s)|Ag(s)$
(b) $Pt(s)|Fe^{3+}(aq),Fe^{2+}(aq)||NO_3^-(aq),H^{+}(aq)|NO(g)|Pt(s)$

17.44 Write the expression for the equilibrium constants of the following cell reactions:
(a) $Cr(s)|Cr^{3+}(aq)||Br^{-}(aq)|AgBr(s)|Ag(s)$
(b) $Bi(s)|Bi^{3+}(aq)||OH^{-}(aq)|O_2(g)|Pt(s)$

17.45 Determine the equilibrium constants for the following reactions and cell reactions:
(a) $Pt(s)|Cr^{3+}(aq),Cr^{2+}(aq)||Cu^{2+}(aq)|Cu(s)$
(b) $Mn(s) + Ti^{2+}(aq) \longrightarrow Mn^{2+}(aq) + Ti(s)$
(c) A Pb^{2+}/Pb redox couple in combination with a Hg_2^{2+}/Hg redox couple
(d) $In^{3+}(aq) + U^{3+}(aq) \longrightarrow In^{2+}(aq) + U^{4+}(aq)$

17.46 Determine the equilibrium constants for the following reactions and cell reactions:

(a) An $AgI/Ag,I^-$ redox couple in combination with an I_2/I^- redox couple
(b) $Pt(s)|Sn^{4+}(aq),Sn^{2+}(aq)||Cl^-(aq)|Hg_2Cl_2(s)|Hg(s)$
(c) $2\,Fe^{3+}(aq) + H_2(g) \longrightarrow 2\,Fe^{2+}(aq) + 2\,H^+(aq)$
(d) $Cr(s) + Zn^{2+}(aq) \longrightarrow Cr^{2+}(aq) + Zn(s)$

17.47 A chemist wants to make a range of silver(II) compounds. Could aqueous sodium persulfate be used to oxidize silver(I) compound to silver(II)? If so, what would be the equilibrium constant for the reaction?

17.48 A chemist suspects that manganese(III) might be involved in an unusual biochemical reaction and wants to prepare some of its compounds. Could aqueous potassium permanganate be used to oxidize manganese(II) to manganese(III)? If so, what would be the equilibrium constant for the reaction?

The Nernst Equation

17.49 Calculate the reaction quotient, Q_c, for the cell reaction, given the measured values of the cell potential:
(a) $Pt(s)|Sn^{4+}(aq),Sn^{2+}(aq)||Pb^{4+}(aq),Pb^{2+}(aq)|C(gr)$ $\quad E = 1.33$ V
(b) $Pt(s)|O_2(g)|H^+(aq)||Cr_2O_7^{2-}(aq),H^+(aq),Cr^{3+}(aq)|Pt(s)$ $\quad E = 0.10$ V

17.50 Calculate the reaction quotient, Q_c, for the cell reaction, given the measured values of the cell potential:
(a) $Ag(s)|Ag^+(aq)||ClO_4^-(aq),H^+(aq),ClO_3^-(aq)|Pt(s)$ $\quad E = 0.40$ V
(b) $C(gr)|Cl_2(g)|Cl^-(aq)||Au^{3+}(aq)|Au(s)$ $\quad E = 0.00$ V

17.51 A *concentration cell* consists of the same redox couples at the anode and the cathode, with different concentrations of the ions in the respective compartments. Calculate E(cell) for the following concentration cells:
(a) $Cu(s)|Cu^{2+}(aq, 0.0010\ mol\cdot L^{-1})||Cu^{2+}(aq, 0.010\ mol\cdot L^{-1})|Cu(s)$
(b) $Pt(s)|H_2(g, 1\ atm)|H^+(aq, pH = 4.0)||H^+(aq, pH = 3.0)|H_2(g, 1\ atm)|Pt(s)$

17.52 A concentration cell consists of the same redox couples at the anode and the cathode, with different concentrations of the ions in the respective compartments. Determine the concentration of the ion in the following cells:
(a) $Pb(s)|Pb^{2+}(aq, ?)||Pb^{2+}(aq, 0.10\ mol\cdot L^{-1})|Pb(s)$ $\quad E = 0.050$ V
(b) $Pt(s)|Fe^{3+}(aq, 0.10\ mol\cdot L^{-1}),Fe^{2+}(aq, 1.0\ mol\cdot L^{-1})||Fe^{3+}(aq, ?),Fe^{2+}(aq, 0.0010\ mol\cdot L^{-1})|Pt(s)$ $\quad E = 0.10$ V

17.53 Determine the potential of the following cells:
(a) $Pt(s)|H_2(g, 1.0\ atm)|HCl(aq, 0.0010\ M)||HCl(aq, 1.0\ mol\cdot L^{-1})|H_2(g, 1.0\ atm)|Pt(s)$
(b) $Zn(s)|Zn^{2+}(aq, 0.10\ mol\cdot L^{-1})||Ni^{2+}(aq, 0.0010\ mol\cdot L^{-1})|Ni(s)$
(c) $Pt(s)|Cl_2(g, 100\ Torr)|HCl(aq, 1.0\ M)||HCl(aq, 0.010\ M)|H_2(g, 450\ Torr)|Pt(s)$
(d) $Sn(s)|Sn^{2+}(aq, 0.020\ mol\cdot L^{-1})||Sn^{4+}(aq, 0.060\ mol\cdot L^{-1}),Sn^{2+}(aq, 1.0\ mol\cdot L^{-1})|Pt(s)$

17.54 Determine the potential of the following cells:
(a) $Cr(s)|Cr^{3+}(aq, 0.10\ mol\cdot L^{-1})||Pb^{2+}(aq, 1.00 \times 10^{-5}\ mol\cdot L^{-1})|Pb(s)$
(b) $Pt(s)|H_2(g, 1.0\ atm)|H^+(pH = 4.0)||Cl^-(aq, 1.0\ mol\cdot L^{-1})|Hg_2Cl_2(s)|Hg(s)$
(c) $C(gr)|Sn^{4+}(aq, 0.0030\ mol\cdot L^{-1}),Sn^{2+}(aq, 0.10\ mol\cdot L^{-1})||Fe^{3+}(aq, 1.0 \times 10^{-4}\ mol\cdot L^{-1}),Sn^{2+}(aq, 0.40\ mol\cdot L^{-1})|Pt(s)$
(d) $Ag(s)|AgI(s)|I^-(aq, 0.010\ mol\cdot L^{-1})||Cl^-(aq, 1.0 \times 10^{-6}\ mol\cdot L^{-1})|AgCl(s)|Ag(s)$

17.55 Determine the unknown in the following cells:
(a) $Pt(s)|H_2(g, 1.0\ atm)|H^+(pH = ?)||Cl^-(aq, 1.0\ mol\cdot L^{-1})|Hg_2Cl_2(s)|Hg(l)$ $\quad E = 0.33$ V
(b) $C(gr)|Cl_2(g, 1.0\ atm)|Cl^-; (aq, ?)||MnO_4^-(aq, 0.010\ mol\cdot L^{-1}),H^+(pH = 4.0), Mn^{2+}(aq, 0.10\ mol\cdot L^{-1})|Pt(s)$ $\quad E = -0.30$ V

17.56 Determine the unknown in the following cells:
(a) $Pt(s)|H_2(g, 1.0\ atm)|H^+(pH = ?)||Cl^-(aq, 1.0\ mol\cdot L^{-1})|AgCl(s)|Ag(s)$ $\quad E = 0.30$ V
(b) $Pb(s)|Pb^{2+}(aq, ?)||Ni^{2+}(aq, 0.10\ mol\cdot L^{-1})|Ni(s)$ $\quad E = 0.040$ V

Practical Cells

17.57 Explain the difference between a primary cell and a secondary cell.

17.58 Explain the difference between a primary cell and a fuel cell.

17.59 What is (a) the electrolyte; (b) the oxidizing agent in a mercury cell (shown below)? (c) Write the overall cell reaction for a mercury cell (see Table 17.2).

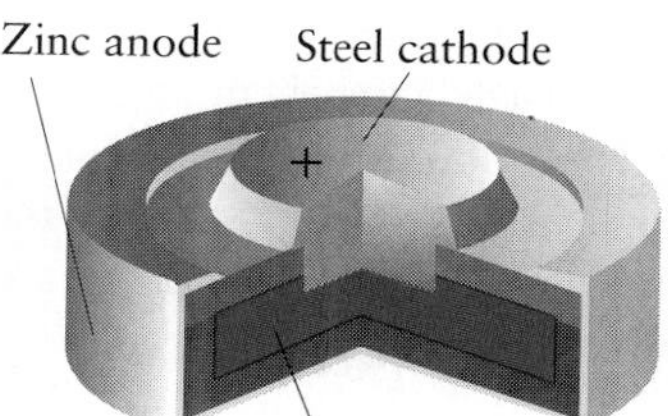

17.60 What is (a) the electrolyte; (b) the oxidizing agent during discharge in a lead-acid battery? (c) Write the reaction that occurs at the cathode during the charging of the lead-acid battery.

17.61 Explain how a dry cell generates electricity.

17.62 The density of the electrolyte in a lead-acid battery is measured to assess its state of charge. Explain how the density reflects the state of charge of the battery.

17.63 (a) What is the electrolyte in a nickel-cadmium cell? (b) Write the reaction that occurs at the anode when the cell is being charged.

17.64 (a) Why are lead-antimony grids used as electrodes in the lead-acid battery rather than smooth plates? (b) What is the reducing agent in the lead-acid battery? (c) The lead-acid cell potential is about 2 V. How, then, does a car battery produce 12 V for its electrical system?

Corrosion

17.65 A chromium-plated steel bicycle handlebar is scratched. Will rusting of the iron in the steel be encouraged or retarded by the chromium?

17.66 A solution contains the ions Cu^{2+}, Ni^{2+}, and Ag^{+}, each present at 1 $mol \cdot L^{-1}$. What will happen if a strip of tin metal is placed in the solution?

17.67 (a) What is the approximate chemical formula of rust? (b) What is the oxidizing agent in the formation of rust? (c) How does the presence of salt accelerate the rusting process?

17.68 (a) What is the electrolyte solution in the formation of rust? (b) How are steel (iron) objects protected by galvanizing and by sacrificial anodes? (c) Suggest two metals that could be used in place of zinc for galvanizing iron.

17.69 (a) Suggest two metals that could be used for the cathodic protection of a titanium pipeline. (b) What factors other than relative positions in the electrochemical series need to be considered in practice? (c) Often copper piping is connected to iron pipes in household plumbing systems. What is a possible effect of the copper on the iron pipes?

17.70 (a) Can aluminum be used for the cathodic protection of an underground storage container? (b) Which of the metals zinc, silver, copper, and magnesium cannot be used as a sacrificial anode in the protection of a buried iron pipeline? Explain your answer. (c) What is the electrolyte solution for the cathodic protection of an underground pipeline by a sacrificial anode?

Electrolysis

For the exercises in this section, base your answers on the potentials listed in Table 17.1 or Appendix 2B, with the exception of the reduction and oxidation of water at $pH = 7$:

$$2\,H_2O(l) + 2\,e^- \longrightarrow H_2(g) + 2\,OH^-(aq) \qquad E = -0.42 \text{ V at pH} = 7$$

$$O_2(g) + 4\,H^+(aq) + 4\,e^- \longrightarrow 2\,H_2O(l) \qquad E = +0.81 \text{ V at pH} = 7$$

Ignore other factors such as passivation or overpotential.

17.71 Complete the following statements: (a) In an electrolytic cell, oxidation occurs at the (anode, cathode). (b) The anode is the (positive, negative) electrode.

17.72 Complete the following statement: In an electrolytic cell, the anions migrate toward the (anode, cathode) and the electrons flow from the (anode, cathode) to the (anode, cathode).

17.73 A 1 M $CoSO_4(aq)$ solution was electrolyzed, using inert electrodes. Write (a) the cathode reaction; (b) the anode reaction. (c) Assuming no overpotential or passivity at the electrodes, what is the minimum potential that must be supplied to the cell for the onset of electrolysis?

17.74 A 1 M CsI(aq) solution was electrolyzed, using inert electrodes. Write (a) the cathode reaction; (b) the anode reaction. (c) Assuming no overpotential or passivity at the electrodes, what is the minimum potential that must be supplied to the cell for the onset of electrolysis?

17.75 Write the half-reaction and specify the electrode at which each of the following processes occurs in an electrolytic cell: (a) the deposition of copper from a Cu^{2+} solution; (b) the production of sodium metal in the Downs cell; (c) the production of chlorine gas in the Downs cell; (d) the production of hydrogen gas from water.

17.76 Write the half-reaction and specify the electrode at which each of the following processes occurs in an electrolytic cell: (a) the production of aluminum from molten Al_2O_3; (b) the electroplating of silver onto a spoon; (c) the production of oxygen gas from an acidic aqueous solution; (d) the production of hypochlorite ions from brine (concentrated aqueous sodium chloride solution).

17.77 Aqueous solutions of (a) Mn^{2+}; (b) Al^{3+}; (c) Ni^{2+}; (d) Au^{3+} are electrolyzed. Determine whether the metal ion or water will be reduced at the cathode.

17.78 The anode of an electrolytic cell was constructed from (a) Cr; (b) Pt; (c) Cu; (d) Ni. Determine whether oxidation of the electrode or oxidation of water will occur at the anode.

17.79 An electric heater uses a current of 18 A for 1.0 h. How many moles of electrons pass through it in that time?

17.80 1.0 mol of electrons passes a given point in a wire every day. What is the current in the wire?

17.81 Determine the amount (in moles) of electrons needed to produce the indicated substance in an electrolytic cell: (a) 5.12 g of copper from a copper(II) sulfate solution; (b) 200 g of aluminum from molten aluminum oxide dissolved in cryolite; (c) 200 L of oxygen gas at 273 K and 1.00 atm from an aqueous sodium sulfate solution.

17.82 A total charge of 96.5 kC is passed through an electrolytic cell. Determine the quantity of substance produced in each case: (a) the mass (in grams) of silver metal from a silver nitrate solution; (b) the volume (in liters at 273 K and 1.00 atm) of chlorine gas from a brine solution (concentrated aqueous sodium chloride solution); (c) the mass of copper (in grams) from a copper(II) chloride solution.

17.83 (a) How much time is required to electroplate 4.4 mg of silver from a silver nitrate solution, using a current of 0.50 A? (b) When the same current is used for the same length of time, what mass of copper could be electroplated from a copper(II) sulfate solution?

17.84 (a) When a current of 150 mA is used for 8.0 h, what volume (in liters at 273 K and 1.0 atm) of fluorine gas can be produced from a molten mixture of potassium and hydrogen fluorides? (b) With the same current and time period, how many liters of oxygen gas at 273 K and 1.0 atm could be produced from the electrolysis of water?

17.85 (a) What current is required to produce 4.0 g of chromium metal from chromium(VI) oxide in 24 h? (b) What current is required to produce 4.0 g of sodium metal from molten sodium chloride during the same period?

17.86 What current is required to electroplate 6.66 μg of gold in 30.0 min from a gold(III) chloride aqueous solution? (b) How much time is required to electroplate 6.66 μg of chromium from a potassium dichromate solution, using a current of 100 mA?

17.87 When a titanium chloride solution was electrolyzed for 500 s with a 120-mA current, 15.0 mg of titanium was deposited. What is the oxidation number of the titanium in the titanium chloride?

17.88 A 0.26-g sample of mercury was produced from a mercury nitrate aqueous solution when a current of 210 mA was applied for 1200 s. What is the oxidation number of the mercury in the mercury nitrate?

17.89 An aqueous solution of Na_2SO_4 was electrolyzed for 30 min; 25.0 mL of O_2 gas was collected at the anode over water at 22°C at a total pressure of 722 Torr. Determine the current that was used to produce the oxygen gas. See Table 5.4 for the vapor pressure of water.

17.90 A brine solution is electrolyzed, using a current of 2.0 A. How much time is required to collect 20.0 L of chlorine if the gas is collected over water at 20°C and the total pressure is 770 Torr? Assume that the water is already saturated with chlorine, so no more dissolves. See Table 5.4 for the vapor pressure of water.

17.91 Thomas Edison was faced with the problem of measuring the electricity that each of his customers had used. His first solution was to use a zinc "coulometer," an electrolytic cell in which the quantity of electricity is determined by measuring the mass of zinc deposited. Only some of the current used by the customer passed through the coulometer. What mass of zinc would be deposited in one month (of 31 days) if 1.0 mA of current passed through the cell continuously?

17.92 An alternative solution to the problem described in Exercise 17.91 is to collect the hydrogen produced by electrolysis and measure its volume. What volume would be collected at 273 K and 1.00 atm under the same conditions?

SUPPLEMENTARY EXERCISES

17.93 Balance the skeletal equation for the following half-reactions and state whether each half-reaction is an oxidation or a reduction.

(a) $I^-(aq) \longrightarrow I_3^-(aq)$

(b) $SeO_4^{2-}(aq) + H_2O(l) \longrightarrow SeO_3^{2-}(aq) + OH^-(aq)$

17.94 Nitric acid is a good oxidizing agent; its product depends on the concentration of the acid and the strength of the reducing agent. Write the balanced equations for the half-reactions and the overall reaction for each of the following cases:

(a) $Cu(s) + NO_3^-(aq) \longrightarrow Cu^{2+}(aq) + NO(g)$ (warm, dilute HNO_3)

(b) $Cu(s) + NO_3^-(aq) \longrightarrow Cu^{2+}(aq) + NO_2(g)$ (concentrated HNO_3)

(c) $Zn(s) + NO_3^-(aq) \longrightarrow Zn^{2+}(aq) + NH_4^+(aq)$ (dilute HNO_3)

(d) $Zn(s) + NO_3^-(aq) \longrightarrow Zn^{2+}(aq) + N_2(g)$ (very dilute HNO_3)

17.95 A certain galvanic cell consists of the $O_2,H^+/H_2O$ and Fe^{2+}/Fe redox couples. (a) Write the cathode and anode half-reactions. (b) Write a cell diagram for the galvanic cell. (c) What is the standard potential of the cell? (d) What is the cell potential at pH = 6.00? Assume all other conditions are standard conditions.

17.96 A galvanic cell has the cell reaction, $M(s) + 2\,Zn^{2+}(aq) \rightarrow 2\,Zn(s) + M^{4+}(aq)$. The standard potential of the cell is 0.16 V. What is the standard potential of the M^{4+}/M redox couple?

17.97 Suppose the reference electrode for Table 17.1 were the standard calomel electrode, $Hg_2Cl_2/Hg,Cl^-$, with $E°$ for it set equal to 0. Under this system, what would be the standard potential for (a) the standard hydrogen electrode; (b) a Cu^{2+}/Cu redox couple?

17.98 Calculate the standard free energy of the reaction $2\,MnO_4^-(aq) + 16\,H^+(aq) + 5\,Sn^{2+}(aq) \rightarrow 5\,Sn^{4+}(aq) + 2\,Mn^{2+}(aq) + 8\,H_2O(l)$.

17.99 Gold can be oxidized by permanganate ions but not by dichromate ions in an acidic solution. Explain this observation.

17.100 What is the standard potential for the reduction of oxygen to water in (a) an acidic solution? (b) a basic solution? (c) Is MnO_4^{2-} more stable in an acidic or a basic aerated solution (a solution saturated with oxygen gas at 1 atm)? Explain your conclusion.

17.101 For each reaction that is spontaneous under standard conditions, write a cell diagram, determine the standard cell potential, and calculate the $\Delta G°$ for the reaction:

(a) $2\,NO_3^-(aq) + 8\,H^+(aq) + 6\,Hg(l) \longrightarrow 3\,Hg_2^{2+}(aq) + 2\,NO(g) + 4\,H_2O(l)$

(b) $2\,Hg^{2+}(aq) + 2\,Br^-(aq) \longrightarrow Hg_2^{2+}(aq) + Br_2(l)$

(c) $Cr_2O_7^{2-}(aq) + 14\,H^+(aq) + 6\,Pu^{3+}(aq) \longrightarrow 6\,Pu^{4+}(aq) + 2\,Cr^{3+}(aq) + 7\,H_2O(l)$

17.102 Determine the equilibrium constant for each of the following reactions:

(a) $PbSO_4(s) \longrightarrow Pb^{2+}(aq) + SO_4^{2-}(aq)$

(b) $2\,Pb^{2+}(aq) \longrightarrow Pb(s) + Pb^{4+}(aq)$

(c) $Hg_2Cl_2(s) \longrightarrow Hg_2^{2+}(aq) + 2\,Cl^-(aq)$

(d) $2\,V^{3+}(aq) + Co(s) \longrightarrow Co^{2+}(aq) + 2\,V^{2+}(aq)$

17.103 Use standard electrode potential data to calculate the solubility of (a) AgCl(s); (b) $Hg_2Cl_2(s)$; (c) $PbSO_4(s)$.

17.104 Dental amalgam, a solid solution of silver and tin in mercury, is used for filling tooth cavities. Two of the reduction half-reactions that the filling can undergo are

$$3\,Hg_2^{2+}(aq) + 4\,Ag(s) + 6\,e^- \longrightarrow 2\,Ag_2Hg_3(s) \qquad E° = +0.85\text{ V}$$

$$Sn^{2+}(aq) + 3\,Ag(s) + 2\,e^- \longrightarrow Ag_3Sn(s) \qquad E° = -0.05\text{ V}$$

Suggest a reason why, when you accidentally bite on a piece of aluminum foil with a tooth containing a silver filling, you may feel pain. Write a balanced chemical equation to support your suggestion.

17.105 Given the data in Appendix 2B and the fact that, for the half-reaction $F_2(g) + 2\,H^+(aq) + 2\,e^- \rightarrow 2\,HF(aq)$, $E° = +3.03$ V, calculate the value of K_a for HF.

17.106 A technical handbook contains tables of thermodynamic quantities for common reactions. If you want to know whether a certain reaction is spontaneous under standard conditions, which of the following properties would give you that information directly (on inspection)? Which would not? Explain. (a) $\Delta G_r°$; (b) $\Delta H_r°$; (c) $\Delta S_r°$; (d) $\Delta U_r°$; (e) $E°$; (f) K_p.

17.107 Calculate the standard potential and the actual potential of the galvanic cell $Pt(s)|Fe^{3+}(aq, 1.0\ mol\cdot L^{-1})$, $Fe^{2+}(aq, 0.0010\ mol\cdot L^{-1})||Ag^+(aq, 0.010\ mol\cdot L^{-1})|Ag(s)$. Compare and comment on your answers.

17.108 The following items are obtained from the stockroom for the construction of a galvanic cell: two 250-mL beakers and a salt bridge, a voltmeter with attached wires and clips, 200 mL of 0.010 M $CrCl_3(aq)$ solution, 200 mL of 0.16 M $CuSO_4(aq)$ solution, a piece of copper wire, and a chrome-plated piece of metal. (a) Describe the construction of the galvanic cell. (b) Write the anode and cathode half-reactions. (c) Write the cell reaction. (d) Write the cell diagram for the galvanic cell. (e) What is the expected cell potential?

17.109 In a neuron (a nerve cell), the concentration of K^+ ions inside the cell is about 20 to 30 times that outside. What potential difference between the inside and the outside of the cell would you expect to measure if the difference is due only to the imbalance of potassium ions?

17.110 Consider the galvanic cell $Pt(s)|Fe^{3+}(aq, 2.00\ mol\cdot L^{-1})$, $Fe^{2+}(aq, 1.00 \times 10^{-5}\ mol\cdot L^{-1})||Ag^+(aq, 2.00 \times 10^{-4}\ mol\cdot L^{-1})|Ag(s)$. (a) What is the standard cell potential? (b) Write the cell reaction. (c) Calculate the cell potential. (d) Determine the equilibrium constant of the cell reaction. (e) Calculate the free energy of the cell reaction.

17.111 The potential of the cell $Zn(s)|Zn^{2+}(aq, ?)||Pb^{2+}(aq, 0.10\ mol\cdot L^{-1})|Pb(s)$ is 0.66 V. What is the molarity of Zn^{2+} ions?

17.112 What kind of system (open, closed, or isolated) is each of the following cells: (a) dry cell; (b) fuel cell; (c) nicad battery?

17.113 The reaction that powers our body is the oxidation of glucose:

$$C_6H_{12}O_6(aq) + 6\,O_2(g) \longrightarrow 6\,CO_2(g) + 6\,H_2O(l)$$

During normal activity, a person uses the equivalent of about 10 MJ of energy a day. Assuming this value represents ΔG, estimate the average current through your body in the course of a day, assuming that all the energy we use arises from the reduction of O_2 in the glucose oxidation reaction.

17.114 (a) How much charge passes through an electrolytic cell when a current of 2.0 A is used for 2.0 h?

(b) What mass of tin can be electroplated from a tin(II) chloride solution by this current?

17.115 Determine the current required to produce 15.0 L of Cl_2 at 273 K and 1.00 atm from molten sodium chloride in 1.0 h.

17.116 In the Dow process for the production of magnesium, a current of approximately 10 kA is used. (a) What mass (in kilograms) of magnesium can be produced in 24 hours? (b) What volume of chlorine gas at STP is produced during the same period?

17.117 A current of 15.0 A electroplated 50.0 g of hafnium metal from an aqueous solution in 2.00 h. What was the oxidation number of hafnium in the solution?

17.118 A mass loss of 12.57 g occurred in 6.00 h at a titanium anode when a current of 4.70 A was used in an electrolytic cell. What is the oxidation number of the titanium in solution?

17.119 A piece of copper metal is to be electroplated on both sides with silver to a thickness of 1.0 μm. If the metal strip measures 50 mm × 10 mm × 1.0 mm, how long must the solution, which contains the $Ag(CN)_2^-$ ion, be electrolyzed, using current of 100 mA? The density of silver metal is $10.5\ g\cdot cm^{-3}$.

17.120 It was decided to electroplate copper metal from a 1 M $CuSO_4(aq)$ solution using a platinum anode. Assuming no overpotential or passivity at the electrodes, what is the minimum potential that must be supplied to the electrolytic cell for the onset of the electroplating of copper metal?

17.121 A metal forms the salt MCl_3. Electrolysis of the molten salt with a current of 0.70 A for 6.63 h produced 3.00 g of the metal. What is the molar mass of the metal?

CHALLENGING EXERCISES

17.122 One stage in the extraction of gold from rocks involves dissolving the metal from the rock with a basic solution of sodium cyanide that has been thoroughly aerated. This stage results in the formation of soluble $[Ag(CN)_2]^-$ ions. The next stage is to precipitate the gold by the addition of zinc dust, forming $[Zn(CN)_4]^{2-}$. Write the balanced equations for the half-reactions and the overall redox equation for both stages.

17.123 Calculate the standard potential of (a) the Cu^+/Cu couple from those of the Cu^{2+}/Cu^+ and Cu^{2+}/Cu couples; (b) that of the Fe^{3+}/Fe couple from the values for the Fe^{3+}/Fe^{2+} and Fe^{2+}/Fe couples. Base your calculation on the relation $\Delta G_r^\circ = -nFE^\circ$ and the fact that ΔG_r° of an overall reaction is the sum of the ΔG_r° of each reaction into which it may be divided.

17.124 Consider the galvanic cell $Pt(s)|Sn^{4+}(aq, 0.010\ mol\cdot L^{-1}), Sn^{2+}(aq, 0.10\ mol\cdot L^{-1})||O_2(g, 1\ atm)|H^+(aq, pH = 4.00)|C(gr)$. (a) What is the standard cell potential? (b) Write the cell reaction. (c) Calculate the actual cell potential. (d) Determine the equilibrium constant of the cell reaction. (e) Calculate the free energy of the cell reaction. (f) Suppose the cell potential measured 0.89 V. Assuming all other concentrations are precise, what is a more accurate pH of the solution?

17.125 Show how a silver-silver chloride electrode (silver in contact with solid AgCl and a solution of Cl^- ions) and a hydrogen electrode can be used to measure (a) pH; (b) pOH. (See Box 17.1.)

17.126 When a pH meter was standardized with a boric acid-borate buffer with a pH of 9.40, the cell potential was 0.060 V. When the buffer was replaced with a solution of unknown hydronium ion concentration, the cell potential was 0.22 V. What is the pH in the solution? (See Box 17.1.)

17.127 (a) How many moles of hydronium ion are produced at a platinum anode in the electrolysis of 200 mL of a $CuSO_4$ solution, using a current of 4.00 A for 30 min. (b) If the pH of the solution was initially 7.0, what will be the pH of the solution after the electrolysis? Assume no volume change.

17.128 In the electrolytic refining of copper, blister copper is used as the anode and oxidized. The copper(II) ion that is produced from its oxidation is then reduced at the cathode to give a metal with a much higher purity. The impurities in the blister copper include iron, nickel, silver, gold, cobalt, and trace amounts of other metals. The material that is not oxidized at the anode falls to the bottom of the electrolytic cell and is called "anode mud." What are some of the components of the anode mud? Explain your choices.

CHAPTER 18

Although the atmosphere appears to be chemically unchanging, it is seething with chemical reactions, particularly at high altitudes where the solar radiation is intense. These Northern Lights (aurora borealis) seen in northern Norway are an indication of that activity. We need the material in this chapter to understand the balance of reactions in the atmosphere, to see how that balance might be upset, and to understand what controls the rates of chemical reactions wherever they may be found.

KINETICS: THE RATES OF REACTIONS

Every year our planet is bombarded with enough energy from the Sun to destroy all life. Fortunately, we are shielded from this onslaught by ozone in the stratosphere. Although solar radiation blasts ozone molecules apart, they are continually being re-formed. As a result, the concentration of stratospheric ozone has long been nearly constant. Now, though, this balance is being upset. Chemicals used as coolants and propellants, such as chlorofluorocarbons (CFCs), and the nitrogen oxides in jet exhausts, are suspected of creating holes in Earth's protective ozone layer. Even small amounts of chemicals can cause such large changes in the vast reaches of the stratosphere. To understand and control the havoc caused by these and other pollutants, we need to understand the network of reactions that take place in the atmosphere. For example, because the rates of the reactions going on in the atmosphere control the ozone levels, the amount of ozone—and the size of the ozone hole—varies with the season.

The study of the rates of chemical reactions is called **chemical kinetics.** One application of chemical kinetics is to model the atmosphere—not only the effects of damaging pollutants but also the natural processes that have gone on for thousands of years and that influence our weather and our well-being. The development of catalysts, which are substances that make chemical reactions go faster, is another branch of chemical kinetics, one that is crucial to the success of the chemical industry. Chemical kinetics is important in biology and medicine, because health represents a balance between large numbers of reactions: illness is often a sign that the rates of biologically important reactions have changed too much. Our bodies are like complicated

CONCENTRATION AND RATE

18.1 The definition of reaction rate
18.2 The instantaneous rate of reaction
18.3 Rate laws
18.4 More complicated rate laws
18.5 First-order integrated rate laws
18.6 Half-lives for first-order reactions
18.7 Second-order reactions

CONTROLLING REACTION RATES

18.8 The effect of temperature
18.9 Collision theory
18.10 Arrhenius behavior
18.11 Activated complexes
18.12 Catalysis
18.13 Living catalysts: Enzymes

REACTION MECHANISMS

18.14 Elementary reactions
18.15 The rate laws of elementary reactions
18.16 Chain reactions
18.17 Rate and equilibrium

laboratories, with millions of chemical reactions kept in step. The control of these reactions is under the command of enzymes, the biological equivalent of catalysts.

CONCENTRATION AND RATE

Rates in chemistry are defined like rates in other fields, as the change in a property divided by the time it takes for that change to occur. For example, the speed of an automobile is a rate defined as distance traveled divided by the time taken. In chemistry, we are concerned with how quickly reactants are used up or products formed. So, to start our study of chemical kinetics, we need a definition of rate in terms of concentration of reactants and products.

18.1 THE DEFINITION OF REACTION RATE

The **reaction rate** is defined as the change in concentration of one of the reactants, Δ(concentration), divided by the time interval, Δ(time) during which the change occurs:

$$\text{Rate} = -\frac{\Delta(\text{concentration of reactant})}{\Delta(\text{time})}$$

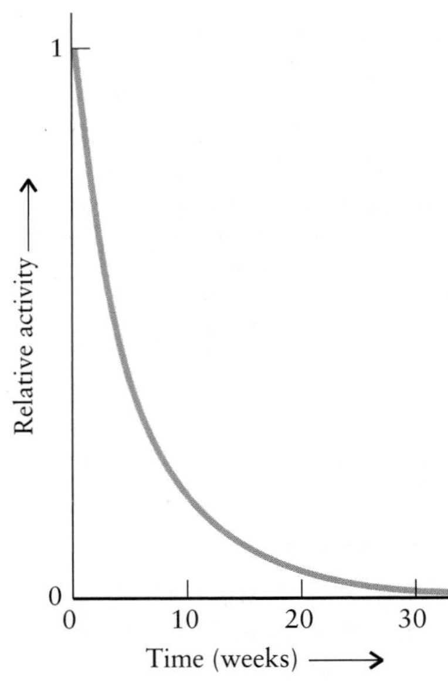

FIGURE 18.1 The activity of penicillin declines over several weeks when it is stored at room temperature and in the absence of stabilizers. The shape of this graph of concentration of the antibiotic is quite typical of the behavior of chemical reactions, although the time span may vary from fractions of a second to years.

Because Δ(concentration) is negative, the minus sign ensures that the rate is positive, which is the normal convention in chemical kinetics. If we follow the concentration of the products, then we would express the rate as

$$\text{Rate} = \frac{\Delta(\text{concentration of product})}{\Delta(\text{time})}$$

because for products, the change Δ(concentration) is positive. Suppose, for example, we were studying the reaction

$$2\,\text{HI(g)} \longrightarrow \text{H}_2\text{(g)} + \text{I}_2\text{(g)}$$

and found that during an interval of 100 s the molar concentration of HI decreased from 4.00 mol·L^{-1} to 3.50 mol·L^{-1}. Then the reaction rate is

$$\text{Rate} = -\frac{(3.50 - 4.00)\ (\text{mol HI})\cdot\text{L}^{-1}}{100\ \text{s}} = 5.0 \times 10^{-3}\ (\text{mol HI})\cdot\text{L}^{-1}\cdot\text{s}^{-1}$$

The rate of a reaction is the change in concentration of a species divided by the time it takes the change to occur.

EXAMPLE 18.1 Expressing the reaction rate for different substances

If the rate of the reaction above is reported as 5.0×10^{-3} (mol HI)·$\text{L}^{-1}\cdot\text{s}^{-1}$, what is the reaction rate in terms of the concentration of hydrogen?

STRATEGY The stoichiometric relation between species lets us convert from the rate with respect to one species to the rate with respect to another. Write the chemical equation for the reaction and use the stoichiometric coefficients to write the mole ratio between the two species of interest. Apply that ratio as a conversion factor to the given rate.

SOLUTION The chemical equation is

$$2\,HI(g) \longrightarrow H_2(g) + I_2(g)$$

so we use 2 mol HI $\simeq$ 1 mol H_2 to write a conversion factor and, because H_2 is a product, report the rate as positive:

$$\text{Rate} = (5.0 \times 10^{-3}\,(\text{mol HI})\cdot L^{-1}\cdot s^{-1}) \times \frac{1\ \text{mol}\ H_2}{2\ \text{mol HI}}$$
$$= 2.5 \times 10^{-3}\,(\text{mol}\ H_2)\cdot L^{-1}\cdot s^{-1}$$

SELF-TEST 18.1A The rate of the reaction $N_2(g) + 3\,H_2(g) \rightarrow 2\,NH_3(g)$ is reported as 1.15 (mol NH_3)$\cdot L^{-1}\cdot h^{-1}$. What is the rate in terms of H_2?

[*Answer:* 1.72 (mol H_2)$\cdot L^{-1}\cdot h^{-1}$]

SELF-TEST 18.1B What is the rate of the same reaction in terms of N_2?

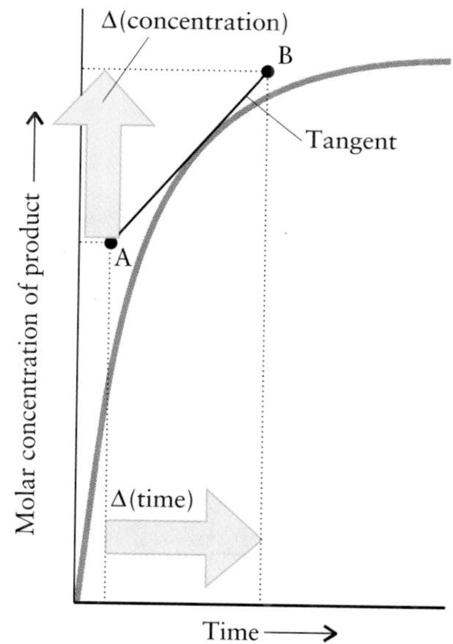

FIGURE 18.2 To calculate the instantaneous reaction rate, we draw the tangent to the curve at the time of interest and then calculate the slope of this tangent. To calculate the slope, we identify any two points, A and B, on the straight line and identify the molar concentrations and times to which they correspond. The slope is then worked out by dividing the difference in concentrations by the difference in times. Notice that the graph shows the concentration of product.

18.2 THE INSTANTANEOUS RATE OF REACTION

Most reactions slow down as the reactants are used up, and the definition of rate we have used so far does not take into account the fact that a rate can change over the time interval during which the change in concentration is measured. Consider, for instance, the decrease in concentration of penicillin stored at 25°C (Fig. 18.1). The graph shows that the concentration changes more quickly at the start of the period. To measure the rate of change at any instant, we draw a *tangent* to the graph at the time of interest (Fig. 18.2). The slope of this straight line is called the **instantaneous rate** of the reaction. If the line slopes down from left to right, then the rate is negative. If it slopes up, then the rate is positive. *From now on, whenever we speak of a reaction rate, we shall always mean an instantaneous rate.* From the graph in Fig. 18.3, the tangent is

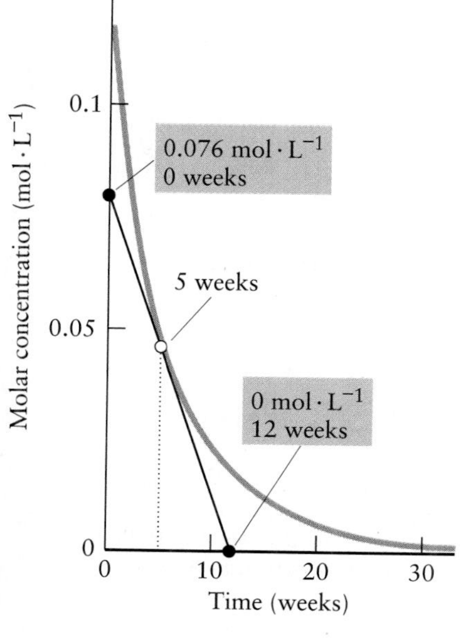

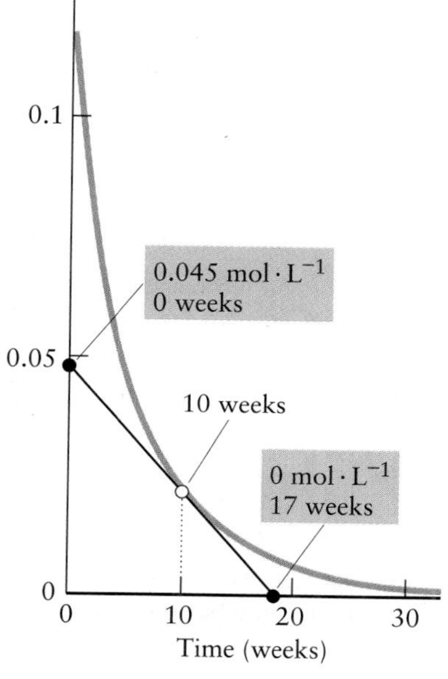

FIGURE 18.3 This illustration shows two examples of the determination of the rate of consumption of penicillin while it is being stored. Note that the rate after 5 weeks is greater than the rate after 10 weeks, when less penicillin is present.

less steep at 10 weeks than it is at 5 weeks, so the instantaneous rate decreases with time.

The instantaneous reaction rate is the slope of a tangent drawn to the graph of concentration as a function of time; it varies as the reaction proceeds.

18.3 RATE LAWS

We can start to discover the patterns in reaction rate data by examining the **initial rate** of reaction, the instantaneous rate at the very beginning of the reaction (Fig. 18.4). At the start of the reaction, there are no products present, and the pattern is easier to find.

For example, suppose we were to measure different masses of dinitrogen pentoxide, N_2O_5, into different flasks and immerse them all in a water bath at 65°C to vaporize the solid, and then at regular time intervals observe the concentrations of reactants and products in the reaction

$$2\,N_2O_5(g) \longrightarrow 4\,NO_2(g) + O_2(g)$$

The initial rate of reaction in each flask is found from the tangent to the curve at $t = 0$ (the black lines in Fig. 18.4). This procedure gives the instantaneous rate at $t = 0$. We would find higher initial rates of decomposition of the vapor in the flasks with higher initial concentrations of N_2O_5 (Fig. 18.5). In fact, doubling the initial concentration doubles the initial rate, tripling it triples the rate, and so on (compare the rates of experiments 2, 3, and 4 in Fig. 18.5). That is,

$$\text{Initial rate} = k \times \text{initial concentration}$$

where k is a constant, called the **rate constant** for the reaction. The experimental value of k is $5.2 \times 10^{-3}\ \text{s}^{-1}$. The same rate constant applies to all initial concentrations used, as long as the temperature is constant.

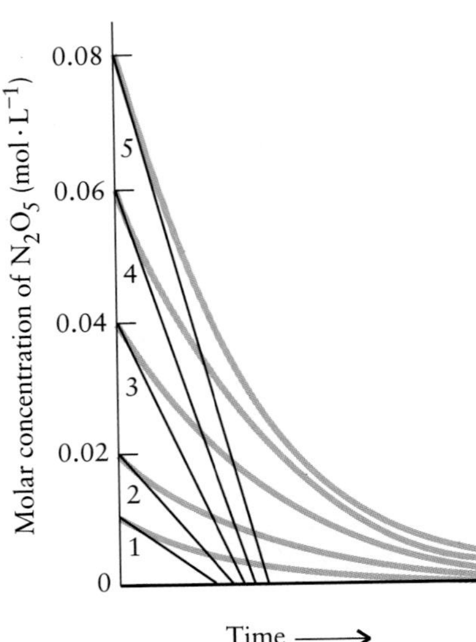

FIGURE 18.4 The definition of the initial rate of reaction. Here we see how the initial rate of consumption of a reactant is determined by drawing a tangent to the curve at the start of the reaction.

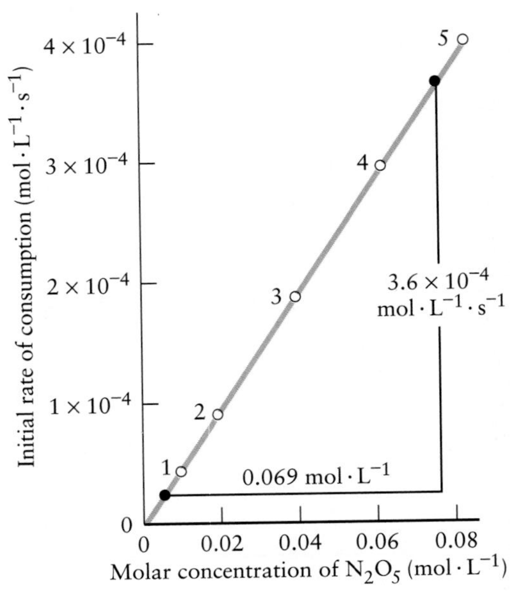

FIGURE 18.5 The initial rate is proportional to the initial concentration, as is shown by the straight line obtained when the rate is plotted as a function of that concentration for the five samples in Fig. 18.4. This graph also illustrates how we calculate the dependence of rate on concentration from the slope of the straight line.

What about the rate later in the reaction? The graph in Fig. 18.4 shows that the rate decreases as the concentration of reactant decreases. To test whether the same simple relation between rate and concentration applies, we plot the instantaneous rate against the concentration of reactant for one of the experiments. The result is a straight line (Fig. 18.6). The straight line implies that the rate of the reaction is directly proportional to the concentration of N_2O_5. We therefore conclude that at any stage of the reaction

$$\text{Rate} = k \times \text{concentration}$$

where k is a proportionality constant. This equation is an example of a **rate law,** an equation expressing the instantaneous reaction rate in terms of the concentrations, at that instant, of the species involved in the reaction. Each reaction has its own characteristic rate law and rate constant k. The rate constant is independent of the concentrations of the reactants but depends on the temperature. For the decomposition of N_2O_5 at 65°C, k is $5.2 \times 10^{-3} s^{-1}$, so the rate law is

$$\text{Rate} = (5.2 \times 10^{-3}\ s^{-1}) \times [N_2O_5]$$

When similar measurements are made on the reaction

$$2\ NO_2(g) \longrightarrow 2\ NO(g) + O_2(g)$$

we find that plotting the initial rate as a function of concentration does not give a straight line. However, plotting the initial rate as a function of the *square* of the concentration of NO_2 does give a straight line, so the rate is proportional to the square of the concentration:

$$\text{Rate} = k \times (\text{concentration})^2$$

It is found experimentally that $k = 0.54\ L{\cdot}(mol\ NO_2)^{-1}{\cdot}s^{-1}$ at 300°C, so the rate law for the decomposition of NO_2 is

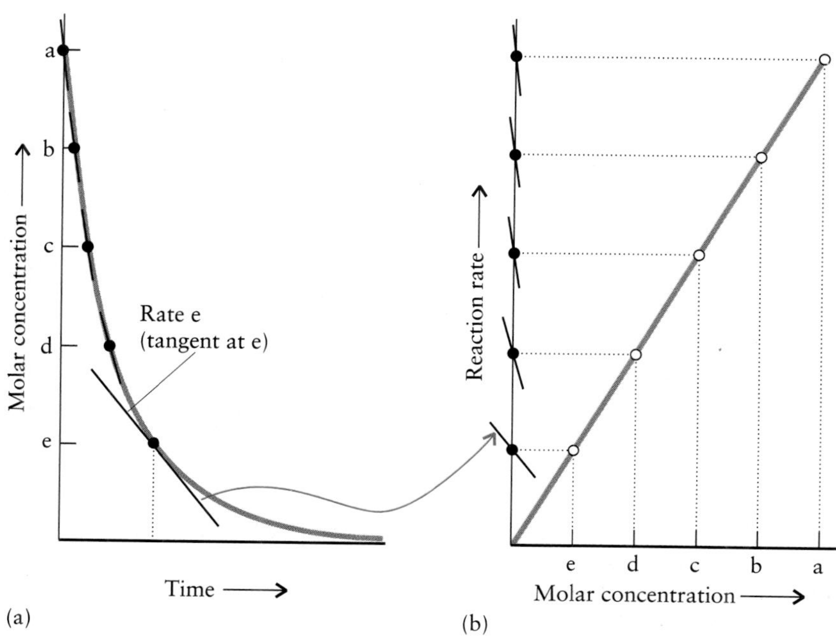

FIGURE 18.6 (a) The instantaneous reaction rates of consumption of N_2O_5 at different times during a reaction are obtained from the slopes of the tangents, as illustrated in the diagram. (b) When these rates are plotted as a function of the concentration of N_2O_5 remaining, a straight line is obtained. In (b), we have indicated the rates by drawing the tangents again; in practice, the graph is plotted by working out the slopes numerically and plotting the resulting values as a function of the concentration.

$$\text{Rate} = (0.54\ \text{L·mol}^{-1}\text{·s}^{-1}) \times [NO_2]^2$$

Both the decomposition of N_2O_5 and that of NO_2 have rate laws of the form

$$\text{Rate} = k \times (\text{concentration})^a$$

with $a = 1$ for the N_2O_5 reaction and $a = 2$ for the NO_2 reaction. The N_2O_5 reaction is an example of a **first-order reaction,** because the rate is proportional to the first power of the concentration (that is, $a = 1$). Doubling the concentration of a reactant in a first-order reaction doubles the reaction rate. The NO_2 decomposition reaction is an example of a **second-order reaction,** because the rate is proportional to the second power (the square) of the concentration of a substance (that is, $a = 2$). Doubling the concentration of the reactant in any second-order reaction increases the reaction rate by a factor of $2^2 = 4$. Notice that, in general, the order of a reaction with respect to a reactant is not related to the reactant's stoichiometric coefficient in the balanced equation.

▶▶ The order of a reaction with respect to a reactant is an indication of how the reaction takes place, not its stoichiometry (see Section 18.15).

Although most of the reactions we shall meet are either first or second order, we need to be aware of reactions with other orders. For example, ammonia decomposes on a hot platinum wire:

$$2\,NH_3(g) \xrightarrow{Pt} N_2(g) + 3\,H_2(g) \qquad \text{Rate} = k$$

The ammonia decomposes at a constant rate k until it has all disappeared. This reaction is an example of a **zero-order reaction,** a reaction that has a rate that is independent of concentration. Zero-order reactions continue at the same rate until all the reactants have been consumed and then stop abruptly (Fig. 18.7). They are called *zero-order* reactions because

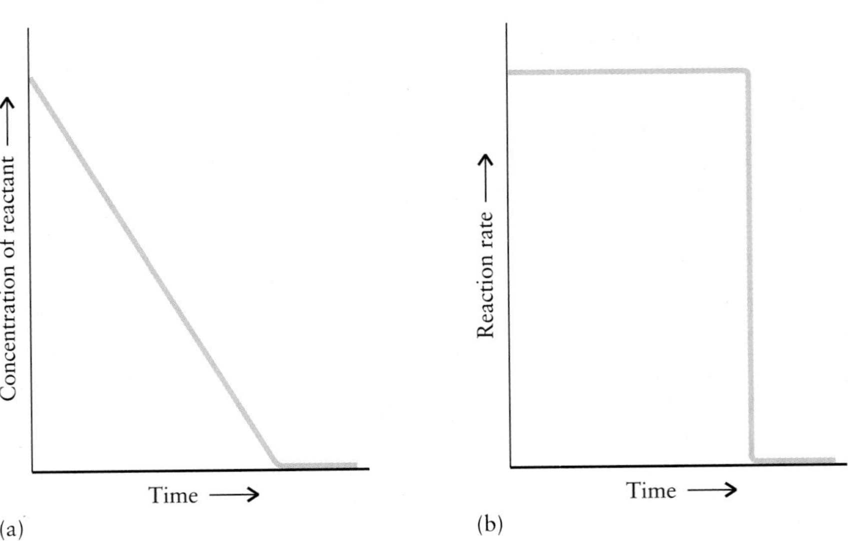

FIGURE 18.7 (a) The concentration of the reactant in a zero-order reaction falls at a constant rate until it is exhausted. (b) The rate of a zero-order reaction is independent of the concentration of the reactant and remains constant until all the reactant has been consumed, when it falls abruptly to 0.

$$\text{Rate} = k \times (\text{concentration})^0 = k$$

$x^0 = 1$, whatever the value of x. Try it on your calculator.

We cannot predict the order of a reaction: *a rate law is an experimentally determined characteristic of the reaction; it cannot in general be written down from the stoichiometry of the chemical equation for the reaction.* In many cases, a reaction involves more than one reactant, and the order of the reaction with respect to each species must be determined in a series of experiments.

EXAMPLE 18.2 Identifying the order of a reaction

The rate of the reaction

$$S_2O_8^{2-}(aq) + 3\,I^-(aq) \longrightarrow 2\,SO_4^{2-}(aq) + I_3^-(aq)$$

in which iodide ions are oxidized by persulfate ions, $S_2O_8^{2-}$, doubles when the concentration of iodide ions is doubled and the concentration of persulfate ions remains constant. What is the order with respect to I^- in the rate law?

STRATEGY In a zero-order reaction, the rate is independent of the concentration; in a first-order reaction, the rate is proportional to the concentration; and in a second-order reaction, the rate is proportional to the square of the concentration. Decide which one of these three possibilities fits the data. We denote the molar concentration of a species X by [X].

SOLUTION When $[I^-]$ doubles, the rate doubles. Therefore, the reaction rate is proportional to $[I^-]$. We can conclude that the reaction is first order with respect to the iodide ion.

SELF-TEST 18.2A The rate of decomposition of dinitrogen oxide, N_2O, in the reaction $2\,N_2O(g) \rightarrow 2\,N_2(g) + O_2(g)$ falls by a factor of 3 when the concentration of dinitrogen oxide is reduced from 6×10^{-3} mol·L^{-1} to 2×10^{-3} mol·L^{-1}. What is the reaction order?

[*Answer:* First order]

SELF-TEST 18.2B The rate of reaction of nitrogen monoxide, NO, with bromine gas in the reaction $2\,NO(g) + Br_2(g) \rightarrow 2\,NOBr(g)$ falls by a factor of 4 when the concentration of NO is reduced from 2×10^{-2} mol·L^{-1} to 1×10^{-2} mol·L^{-1} and the concentration of bromine is held constant. What is the reaction order with respect to NO?

Some reactions have complicated rate laws. An example is the redox reaction between persulfate and iodide ions quoted in the example. The rate of this reaction is found to be proportional to both the I^- concentration and the concentration of $S_2O_8^{2-}$ ions. If we write their molar concentrations as $[I^-]$ and $[S_2O_8^{2-}]$, the rate law is

$$\text{Rate} = k[S_2O_8^{2-}][I^-]$$

We say that the reaction is first order with respect to $S_2O_8^{2-}$ (or "in" $S_2O_8^{2-}$) and first order in I^-. Doubling either the $S_2O_8^{2-}$ ion concentration or the I^- ion concentration doubles the reaction rate. We say that its *overall* order is 2. In general, if

$$\text{Rate} = k[A]^a[B]^b \ldots$$

then the **overall order** is the sum of the powers $a + b + \ldots$.

Many reactions can be classified according to their order in a particular species, the power to which the concentration of a species occurs in the rate law, and by their overall order, the sum of the individual orders.

SELF-TEST 18.3A The reduction of bromate ions, BrO_3^-, by bromide ions in acidic solution has a rate law

$$\text{Rate} = k[BrO_3^-][Br^-][H^+]^2$$

What are the orders with respect to the reactants and the overall order?

[*Answer:* First order in BrO_3^- and Br^-; second order in H^+; fourth order overall]

SELF-TEST 18.3B The reaction between bromomethane, CH_3Br, and hydroxide ion in aqueous solution has a rate law

$$\text{Rate} = k[CH_3Br][OH^-]$$

What are the orders with respect to the reactants and the overall order?

18.4 MORE COMPLICATED RATE LAWS

A few reactions have negative orders (as in $(\text{concentration})^{-1}$). A concentration then occurs in the denominator of the rate law. Increasing the concentration of that species, usually a product, actually slows down the reaction. An example is one of the reactions that takes place in the upper atmosphere, the decomposition of ozone (O_3):

$$2\,O_3(g) \longrightarrow 3\,O_2(g)$$

for which the rate law is

$$\text{Rate} = \frac{k[O_3]^2}{[O_2]} = k[O_3]^2[O_2]^{-1}$$

A negative order signifies that the reaction rate decreases as the concentration of a substance, in this case oxygen, increases.

Some reactions even have fractional orders (as in $(\text{concentration})^{1/2}$). For example, the oxidation of sulfur dioxide to sulfur trioxide in the presence of platinum (Fig. 18.8) in the reaction

$$2\,SO_2(g) + O_2(g) \xrightarrow{Pt} 2\,SO_3(g)$$

is found to have the rate law

$$\text{Rate} = k\frac{[SO_2]}{[SO_3]^{1/2}} = k[SO_2][SO_3]^{-1/2}$$

The reaction is first order in SO_2 and minus one-half order in SO_3. Because $1 + (-\frac{1}{2}) = \frac{1}{2}$, the reaction is half-order overall. Again, the negative order implies that the rate of the reaction decreases as the concentration of SO_3 increases.

Rate laws can often be simplified by modifying the experimental conditions. For example, consider once again the second-order oxidation of iodide ions:

$$\text{Rate} = k[S_2O_8^{2-}][I^-]$$

(a)

(b)

FIGURE 18.8 (a) When sulfur dioxide and oxygen are passed over hot platinum foil, they combine to form sulfur trioxide. (b) This compound forms dense white fumes of sulfuric acid when it comes into contact with moisture in the atmosphere. The rate law for the formation of sulfur trioxide shows that its rate of formation decreases as its concentration increases.

Suppose that experiments are carried out with the persulfate ions in such high concentration compared with the concentration of iodide ions that their concentration barely changes in the course of the reaction. For instance, the concentration of persulfate ions might be 100 times greater than the concentration of iodide ions, so even when all the iodide ions have been oxidized, the persulfate concentration is almost the same as it was at the beginning of the reaction. Then $k[S_2O_8^{2-}]$ is virtually constant, and if we write it as k', the rate law becomes

$$\text{Rate} = k'[\text{I}^-]$$

where $k' = k[S_2O_8^{2-}]$. The rate law is now first order. We call rate laws that have been turned into first-order form by working at high concentrations **pseudo–first-order rate laws** and the reaction itself is classified as a **pseudo–first-order reaction.**

When more than one reactant is present, we can separate out the effects of increasing their concentrations by using the method of initial rates. In this method, several rate experiments are conducted, but the concentration of only one reactant at a time is changed. For example, we can write the general rate law:

$$\text{Rate} = k[\text{A}]^a[\text{B}]^b$$

and carry out a series of experiments, holding the concentration of reactant B constant. Then we note that if the rate doubles when [A] doubles, then $a = 1$. If the rate increases by a factor of 4, then $a = 2$. A similar strategy is used to determine b.

When one reactant in a reaction that is first order in each of two reactants is in large excess, the rate law takes on the form of a pseudo–first-order rate law in the other reactant.

EXAMPLE 18.3 Determining reaction order from experimental data

Four experiments were conducted to discover how the initial rate of consumption of BrO_3^- ions in the reaction

$$BrO_3^-(aq) + 5\,Br^-(aq) + 6\,H^+(aq) \longrightarrow 3\,Br_2(aq) + 3\,H_2O(l)$$

varies as the concentrations of the reactants are changed. Use the experimental data in the table below to determine the order of the reaction with respect to each reactant and the overall order. Write the rate law for the reaction and determine the value of k.

Experiment	Initial concentration, $mol{\cdot}L^{-1}$			Initial rate, $(mol\ BrO_3^-){\cdot}L^{-1}{\cdot}s^{-1}$
	BrO_3^-	Br^-	H^+	
1	0.10	0.10	0.10	1.2×10^{-3}
2	0.20	0.10	0.10	2.4×10^{-3}
3	0.10	0.30	0.10	3.5×10^{-3}
4	0.20	0.10	0.15	5.4×10^{-3}

STRATEGY Suppose the concentration of a substance is increased by a certain factor f. From the rate law, rate $= k(\text{concentration})^a$, we know that if the rate increases by f^a, then the reaction has order a in that substance. Therefore, inspect the data to see how the rate changes when the concentration of each substance is changed. To isolate the effect of each substance, compare experiments that differ in the concentration of only one substance at a time.

SOLUTION Comparing experiment 1 with experiment 2, we see that when the concentration of BrO_3^- is doubled, the rate also doubles. Therefore, the reaction is first order in BrO_3^-. Comparing experiment 1 with experiment 3, we see that when the concentration of Br^- is changed by a factor of 3.0, the rate changes by a factor of $3.5/1.2 = 2.9$. Allowing for experimental error, we can deduce that the reaction is also first order in Br^-. When the concentration of hydrogen ions is increased from experiment 2 to experiment 4, by a factor of 1.5, the rate increases by a factor of $5.4/2.4 = 2.3$. Thus we need to solve $1.5^a = 2.3$ for a. Because $(1.5)^0 = 1$ and $(1.5)^1 = 1.5$, but $(1.5)^2 = 2.3$, the reaction is second order in H^+. The rate law is therefore

$$\text{Rate} = k[BrO_3^-][Br^-][H^+]^2$$

and the reaction is fourth order overall. We find k by substituting the values from one of the experiments into the rate law and solving for k; for example, from experiment 4,

$$5.4 \times 10^{-3}\ mol{\cdot}L^{-1}{\cdot}s^{-1} = k \times (0.20\ mol{\cdot}L^{-1}) \times (0.10\ mol{\cdot}L^{-1}) \times (0.15\ mol{\cdot}L^{-1})^2$$

so

▶▶ For a more systematic strategy for finding *a* by using logarithms, see Appendix 1C.

In practice, the values found for k from several experiments are averaged for greater accuracy.

$$k = \frac{5.4 \times 10^{-3}\ \text{mol}\cdot\text{L}^{-1}\cdot\text{s}^{-1}}{(0.20\ \text{mol}\cdot\text{L}^{-1}) \times (0.10\ \text{mol}\cdot\text{L}^{-1}) \times (0.15\ \text{mol}\cdot\text{L}^{-1})^2}$$
$$= 12\ \text{mol}^{-3}\cdot\text{L}^{3}\cdot\text{s}^{-1}$$

SELF-TEST 18.4A Write the rate law for the consumption of persulfate ions in the reaction

$$S_2O_8^{2-}(aq) + 3\,I^-(aq) \longrightarrow 2\,SO_4^{2-}(aq) + I_3^-(aq)$$

with respect to each reactant and determine the value of k, given the following data:

Experiment	Initial concentration, mol·L^{-1} $S_2O_8^{2-}$	I^-	Initial rate, (mol $S_2O_8^{2-}$)·L^{-1}·s^{-1}
1	0.15	0.21	1.14
2	0.22	0.21	1.70
3	0.22	0.12	0.98

[*Answer:* $k[S_2O_8^{2-}][I^-]$, $k = 36\ \text{L}\cdot\text{mol}^{-1}\cdot\text{s}^{-1}$]

SELF-TEST 18.4B Write the rate law and determine the value of k for the reaction between carbon monoxide and chlorine gas to produce highly toxic carbonyl chloride, $CO(g) + Cl_2(g) \rightarrow COCl_2(g)$, given the following data collected at a certain temperature:

Experiment	Initial concentration, mol·L^{-1} CO	Cl_2	Initial rate, (mol CO)·L^{-1}·s^{-1}
1	0.12	0.20	0.121
2	0.24	0.20	0.241
3	0.24	0.40	0.682

One of the orders is not an integer. In such cases it is best to use logarithms to find a (see Appendix 1C).

18.5 FIRST-ORDER INTEGRATED RATE LAWS

How much ozone will be present in a region of the atmosphere in April? How much sulfur trioxide can be produced in an hour? How much penicillin will be left after six months? These questions can be answered by using the **integrated rate law,** a formula that gives the concentration of reactants or products at any time after the start of the reaction, if the initial concentration of reactants is known. Finding the integrated rate law from a rate law is very much like calculating the distance that a car will have traveled from a knowledge of its speed at each moment of the journey.

In the language of calculus, a rate law is a "differential equation." Differential equations are solved by the technique called "integration," hence the term "integrated rate law."

The integrated rate law for a first-order (or pseudo–first-order) reaction lets us predict the concentration of reactant remaining after a given time. If we write the molar concentration of a reactant A as $[A]_t$ at time t and denote its initial concentration $[A]_0$, then

$$[A]_t = [A]_0 \times e^{-kt} \qquad (1)$$

where k is the rate constant and t is the elapsed time. The variation of

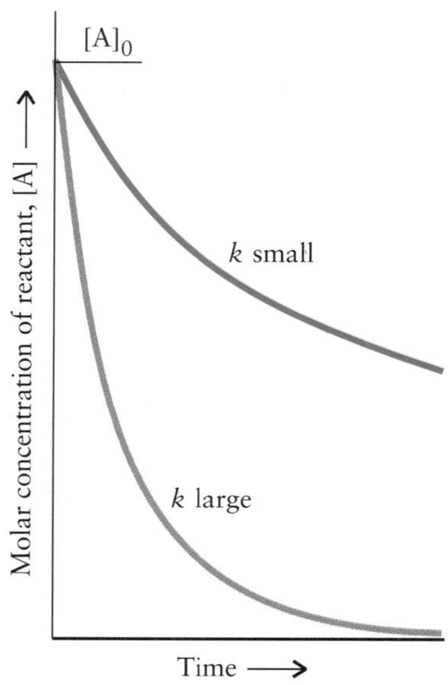

FIGURE 18.9 The characteristic shape of the graph showing the time dependence of the concentration of a reactant in a first-order reaction is an exponential decay, as shown here. The larger the rate constant, the faster the decay from the same initial concentration.

concentration with time predicted by this equation is shown in Fig. 18.9. Notice that the bigger the rate constant, the more rapid the initial decay. The behavior shown in the illustration is called an **exponential decay.** The change in concentration is initially rapid, but it changes more slowly as time goes on and the reactant is used up.

EXAMPLE 18.4 Calculating a concentration from a first-order integrated rate law

Calculate the concentration of N_2O_5 remaining 600 s (10 min) after the start of its decomposition at 65°C when its concentration was 0.040 mol·L^{-1}. The reaction and its rate law are

$$2\,N_2O_5(g) \longrightarrow 4\,NO_2(g) + O_2(g) \qquad \text{Rate} = k[N_2O_5]$$

with $k = 5.2 \times 10^{-3}\ s^{-1}$.

STRATEGY The concentration will be smaller than 0.040 mol·L^{-1} because some of the reactant decomposes as the reaction proceeds. Because the reaction is first order, use Eq. 1 to predict the concentration at any time after the reaction begins. To work out e^x on a calculator, enter x and press the e^x or "inv ln" key.

SOLUTION Substituting the data in Eq. 1 gives

$$\begin{aligned}[N_2O_5]_t &= [N_2O_5]_0 \times e^{-kt} \\ &= (0.040\ (\text{mol } N_2O_5)\cdot L^{-1}) \times e^{-(5.2 \times 10^{-3}\ s^{-1}) \times (600\ s)} \\ &= 0.0018\ (\text{mol } N_2O_5)\cdot L^{-1}\end{aligned}$$

That is, after 600 s, the concentration of N_2O_5 will have fallen from 0.040 mol·L^{-1} to 0.0018 mol·L^{-1}.

SELF-TEST 18.5A Calculate the concentration of N_2O remaining after the first-order decomposition

$$2\,N_2O(g) \longrightarrow 2\,N_2(g) + O_2(g) \qquad \text{Rate} = k[N_2O]$$

has continued at 780°C for 100 ms, if the initial concentration of N_2O was 0.20 mol·L^{-1} and $k = 3.4\ s^{-1}$.

[*Answer:* 0.14 (mol N_2O)·L^{-1}]

SELF-TEST 18.5B Calculate the concentration of cyclopropane, C_3H_6, remaining after the first-order isomerization

$$C_3H_6(g) \longrightarrow CH_3{-}CH{=}CH_2 \qquad \text{Rate} = k[C_3H_6]$$

has continued at 773 K for 200 s, if the initial concentration of C_3H_6 was 0.100 mol·L^{-1} and $k = 6.7 \times 10^{-4}\ s^{-1}$.

The integrated rate law is also used to determine whether a reaction is first order and, if it is, to measure its rate constant. If we take logarithms of both sides of Eq. 1, we obtain

$$\ln [A]_t = \ln [A]_0 - kt \tag{2}$$

Because k is always positive, the negative sign means that the line slopes down from left to right.

which has the form of a linear equation:

$$y = \text{intercept} + \text{slope} \times x$$

CHEMISTRY AT WORK

An algologist collects samples of water in Antarctica. He will look for algae in the water and study their rate of growth. Data from his work will ultimately be used to track changes in the ozone layer due to reactions in the stratosphere. His sunglasses protect against ultraviolet radiation, a growing concern in Antarctica.

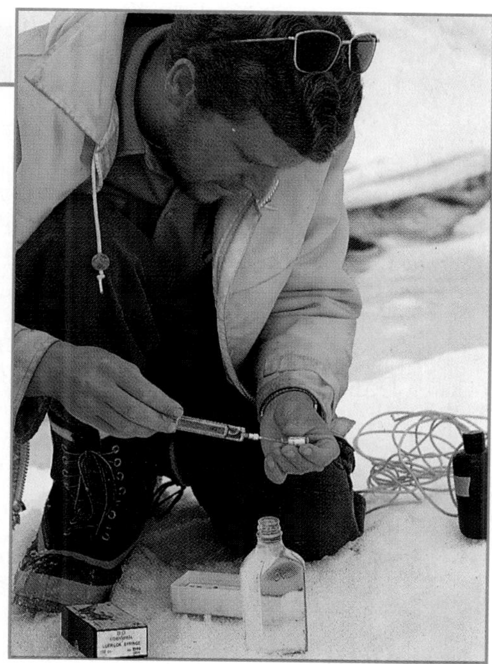

This formula tells us that if we plot $\ln [A]_t$ as a function of t, then we should get a straight line with slope $-k$ (see Appendix 1E).

> ***In a first-order reaction, the concentration of reactant decays exponentially with time. To verify that a reaction is first order, plot the natural logarithm of its concentration as a function of time and expect a straight line; the slope of the straight line is $-k$.***

EXAMPLE 18.5 Measuring a rate constant

When cyclopropane (**1**) is heated to 500°C (773 K), it changes into propene (**2**). The following data were obtained in one experiment:

t, min	Concentration of C_3H_6, mol·L^{-1}
0	1.5×10^{-3}
5	1.24×10^{-3}
10	1.00×10^{-3}
15	0.83×10^{-3}

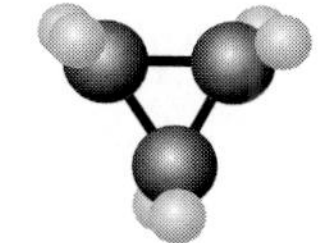

1 Cyclopropane, C_3H_6

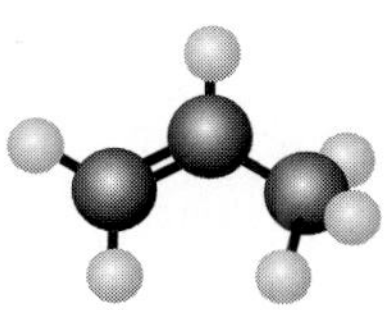

2 Propene, C_3H_6

Confirm that the reaction is first order in C_3H_6 and calculate the rate constant.

STRATEGY Plot ln[cyclopropane]$_t$ as a function of t and see if a straight line is obtained. If so, then the reaction is first order and the slope of the graph is the value of $-k$.

SOLUTION For the graphical procedure, begin by setting up the following table:

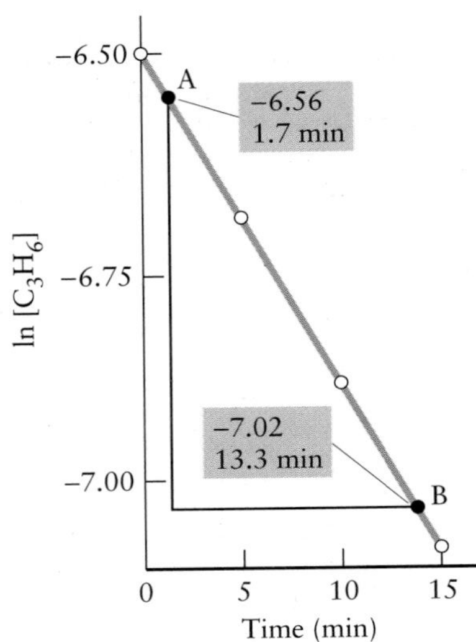

FIGURE 18.10 We test for a first-order reaction by plotting the natural logarithm of the reactant concentration as a function of time. The slope of the line, which is calculated as indicated in the illustration by selecting two points A and B, is equal to the negative of the rate constant.

t, min	$\ln[C_3H_6]_t$
0	−6.50
5	−6.69
10	−6.91
15	−7.09

The points are plotted in Fig. 18.10. The graph is a straight line, confirming that the reaction is first order in cyclopropane. The slope of the line is

$$\text{Slope} = \frac{(-7.02) - (-6.56)}{13.3\text{ min} - 1.7\text{ min}} = -0.040\text{ min}^{-1}$$

Therefore, because $k = -\text{slope}$, $k = 0.040\text{ min}^{-1}$. This value is equivalent to $k = 6.7 \times 10^{-4}\text{ s}^{-1}$, the value in Table 18.1.

SELF-TEST 18.6A Data on the decomposition of N_2O_5 at 25°C are

t, min	Concentration of N_2O_5, mol·L^{-1}
0	1.50×10^{-2}
200	9.6×10^{-3}
400	6.2×10^{-3}
600	4.0×10^{-3}
800	2.5×10^{-3}
1000	1.6×10^{-3}

Confirm that the reaction is first order and find the value of k.

[*Answer:* $2.2 \times 10^{-3}\text{ s}^{-1}$]

SELF-TEST 18.6B Use the data on the decomposition of N_2O_5 in Exercise 18.5 to confirm that the reaction is first order in N_2O_5, and find the rate constant.

18.6 HALF-LIVES FOR FIRST-ORDER REACTIONS

The **half-life,** $t_{1/2}$, of a substance is the time needed for its concentration to fall to one-half its initial value. For instance, if the half-life of an environmental pollutant that decays with a first-order rate law is 18 months, then its concentration will decline to half its initial value in 18 months, and then to half that value again (to one-fourth of the initial concentration) in a further 18 months, and so on. The half-lives of CFCs and other atmospheric species are important for assessing their environmental impact. If their half-lives are short, they may not survive long enough to reach the stratosphere, where they can destroy ozone. The half-life of mercury in human blood, in the form of Hg^{2+} ions, is about six days before it is excreted.

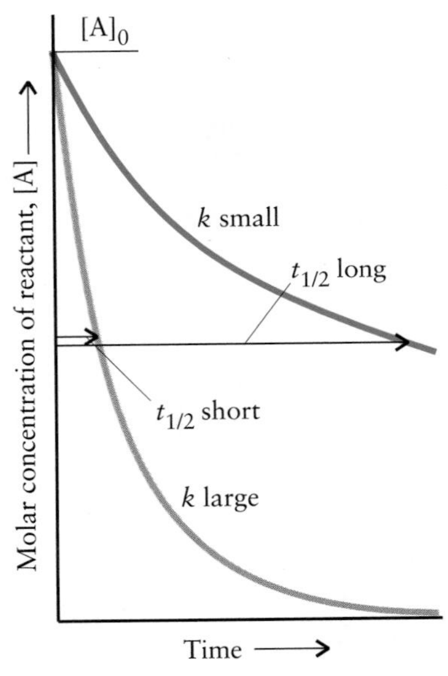

FIGURE 18.11 The half-life of a reactant is short if the first-order rate constant is large, because the exponential decay of the concentration of the reactant is then faster.

TABLE 18.1 Rate laws and rate constants

Reaction	Rate law*	Temperature, K	Rate constant
GAS PHASE			
$H_2 + I_2 \longrightarrow 2\,\mathbf{HI}$	$k[H_2][I_2]$	500	$4.3 \times 10^{-7}\ L\cdot mol^{-1}\cdot s^{-1}$
		600	4.4×10^{-4}
		700	6.3×10^{-2}
		800	2.6
$2\,\mathbf{HI} \longrightarrow H_2 + I_2$	$k[HI]^2$	500	$6.4 \times 10^{-9}\ L\cdot mol^{-1}\cdot s^{-1}$
		600	9.7×10^{-6}
		700	1.8×10^{-3}
		800	9.7×10^{-2}
$2\,\mathbf{N_2O_5} \longrightarrow 4\,NO_2 + O_2$	$k[N_2O_5]$	298	$3.7 \times 10^{-5}\ s^{-1}$
		318	5.1×10^{-4}
		328	1.7×10^{-3}
		338	5.2×10^{-3}
$2\,\mathbf{N_2O} \longrightarrow 2\,N_2 + O_2$	$k[N_2O]$	1000	$0.76\ s^{-1}$
		1050	3.4
$2\,\mathbf{NO_2} \longrightarrow 2\,NO + O_2$	$k[NO_2]^2$	573	$0.54\ L\cdot mol^{-1}\cdot s^{-1}$
$\mathbf{C_2H_6} \longrightarrow 2\,CH_3$	$k[C_2H_6]$	973	$5.5 \times 10^{-4}\ s^{-1}$
cyclopropane $\longrightarrow$ propene	k[cyclopropane]	773	$6.7 \times 10^{-4}\ s^{-1}$
AQUEOUS SOLUTION			
$\mathbf{H^+} + OH^- \longrightarrow H_2O$	$k[H^+][OH^-]$	298	$1.5 \times 10^{11}\ L\cdot mol^{-1}\cdot s^{-1}$
$\mathbf{CH_3Br} + OH^- \longrightarrow CH_3OH + Br^-$	$k[CH_3Br][OH^-]$	298	$2.8 \times 10^{-4}\ L\cdot mol^{-1}\cdot s^{-1}$
$\mathbf{C_{12}H_{22}O_{11}} + H_2O \longrightarrow 2\,C_6H_{12}O_6$	$k[C_{12}H_{22}O_{11}][H^+]$	298	$1.8 \times 10^{-4}\ L\cdot mol^{-1}\cdot s^{-1}$

*For the rate of consumption or formation of the substance in bold type in the reaction column.

The half-life of a first-order reaction is related to the rate constant, because the higher the value of k, the more rapid the disappearance of the reactant and hence the shorter the half-life (Fig. 18.11). We can work out the precise relation by using Eq. 2 to find the time needed for the concentration of A to fall to half its initial value. First, we rearrange it into

$$\ln\left(\frac{[A]_0}{[A]_t}\right) = kt$$

Then we set t equal to $t_{1/2}$ and $[A]_t = \frac{1}{2}[A]_0$, and obtain

$$\ln\left(\frac{[A]_0}{\frac{1}{2}[A]_0}\right) = kt_{1/2}$$

Now we can cancel the initial concentrations and obtain

$$\ln 2 = kt_{1/2}$$

It follows that

$$t_{1/2} = \frac{\ln 2}{k} \approx \frac{0.693}{k} \qquad (3)$$

Although ln 2 ≈ 0.693, it is more accurate to use ln 2 on your calculator.

As we expected, the greater the value of the rate constant k, the shorter the half-life for the reaction. Note that the concentration does not appear: for a first-order reaction, the half-life depends only on the rate constant.

We can understand the significance of the half-life of a first-order reaction by looking at Fig. 18.12. We see that the concentration falls to half its initial value in the time $t_{1/2}$. The concentration falls to half its new value in the same time, and then to half that new value in another half-life. For each time interval equal to the half-life, the concentration falls to half its value at the start of the interval.

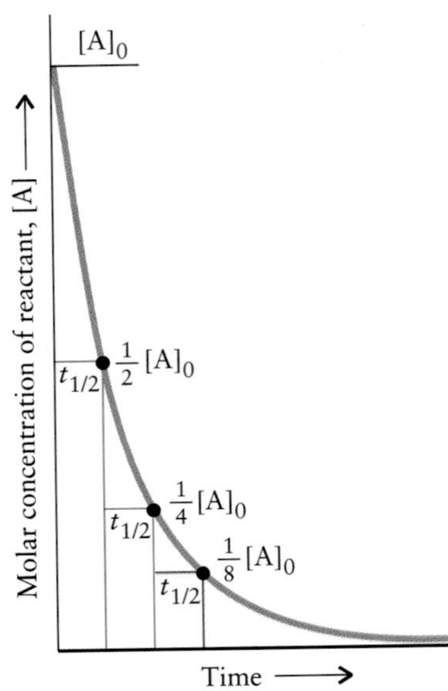

FIGURE 18.12 The half-life of a substance is the time needed for it to fall to one-half its initial concentration. For first-order reactions, the half-life is the same whatever the concentration at the start of the chosen period. Therefore, it takes one half-life to fall to half the initial concentration, two half-lives to fall to one-fourth the initial concentration, three half-lives to fall to one-eighth, and so on.

EXAMPLE 18.6 Using half-lives

A pollutant escapes into a local picnic site. Studies had shown that the pollutant decays by a first-order reaction with rate constant $3.8 \times 10^{-3}\ \text{h}^{-1}$. Calculate the time needed for the concentration to fall to (a) one-half and (b) one-fourth its initial value.

STRATEGY Because the reaction is first order, the time it takes for the concentration to fall to half its initial value is $t_{1/2} = (\ln 2)/k$. The total time required for the concentration to fall to one-fourth its initial value is the sum of two successive half-lives. For a first-order reaction, the two half-lives are equal, so the time required is $2t_{1/2}$.

SOLUTION Because $k = 3.8 \times 10^{-3}\ \text{h}^{-1}$,

$$t_{1/2} = \frac{\ln 2}{3.8 \times 10^{-3}\ \text{h}^{-1}} = 1.8 \times 10^{2}\ \text{h}$$

The time required for half the original amount to react, part (a), is therefore 1.8×10^2 h, or just over a week. The time required for three-fourths of the original amount to react, part (b), is twice this, or about 15 days.

SELF-TEST 18.7A Calculate the time needed for the concentration of N_2O to fall to (a) one-half and (b) one-eighth of its initial value as it decomposes at 1000 K. Consult Table 18.1 for the rate constant.

[*Answer:* (a) 0.91 s; (b) 2.7 s]

SELF-TEST 18.7B Calculate the time needed for the concentration of C_2H_6 to fall to (a) one-half and (b) one-sixteenth of its initial value as it decomposes to CH_3 radicals at 973 K. Consult Table 18.1 for the rate constant.

The half-life of a first-order reaction is characteristic of the reaction and independent of the initial concentration. A reaction with a large rate constant has a short half-life.

18.7 SECOND-ORDER REACTIONS

For second-order reactions with rate law

$$\text{Rate} = k[\text{A}]^2$$

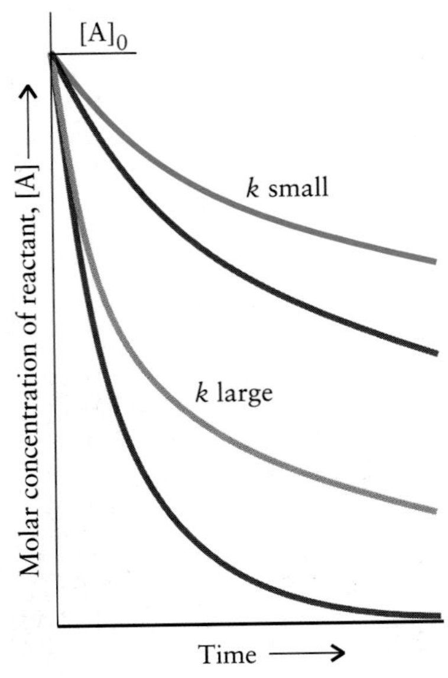

FIGURE 18.13 The characteristic shapes (orange and green lines) of the time dependence of the concentration of a reactant during a second-order reaction. The gray lines are the curves for first-order reactions with the same initial rates. Note how the concentrations for second-order reactions fall away much less rapidly at longer times than those for first-order reactions do.

FIGURE 18.14 The *Hindenburg* hydrogen-filled airship caught fire as it came in to land at Lakehurst, New Jersey in 1937 after its inaugural flight across the Atlantic.

the integrated rate law is

$$[A]_t = \frac{[A]_0}{1 + [A]_0kt} \quad \text{or} \quad \frac{1}{[A]_t} = \frac{1}{[A]_0} + kt \qquad (4)$$

For a second-order reaction, the plot of $1/[A]_t$ against t is linear with slope k.

Figure 18.13 shows that the concentration of the reactant decreases rapidly at first, but then changes more slowly than a first-order reaction with the same initial rate. The half-life of a second-order reaction is not constant, but grows longer as the concentration of the reactant decreases. Because many pollutants disappear by second-order reactions, they remain at low concentration in the environment for long periods.

A second-order reaction has a long tail of low concentration at long reaction times.

CONTROLLING REACTION RATES

Some spontaneous reactions do not seem to take place, however long we wait. For example, a mixture of hydrogen and oxygen can exist for years without reacting, and dynamite can be safely stored and transported. However, these materials react in an instant when subjected to a spark or jolt (Fig. 18.14). We also observe that when a solid reacts, the rate depends on the surface area (Fig. 18.15). To understand this behavior, we need to know how reactions occur and how energy affects their rates.

FIGURE 18.15 A solid iron rod can be heated in a flame without catching fire. However, a powder of very finely divided iron oxidizes rapidly in air to form Fe_2O_3, because the powder presents a much greater surface area for reaction with oxygen.

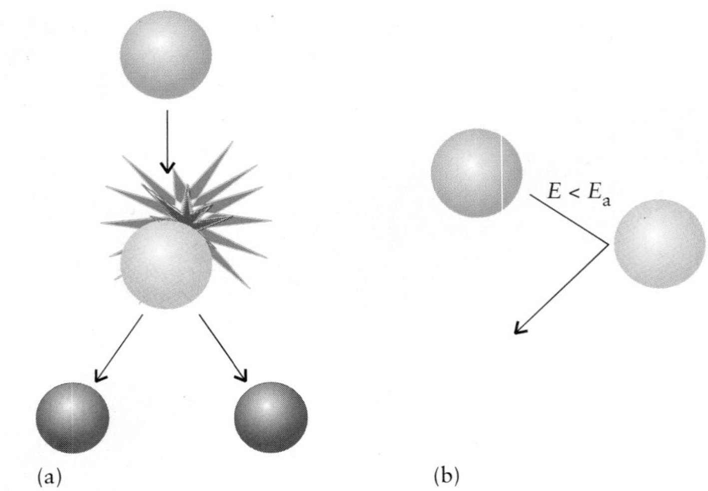

FIGURE 18.16 (a) In the collision theory of chemical reactions, reaction can occur only when two molecules collide with a kinetic energy at least equal to the activation energy of the reaction. (b) Otherwise they simply bounce apart.

18.8 THE EFFECT OF TEMPERATURE

The factor is usually in the range from 1.5 to 4 for a rise in temperature of 10°C.

Most reactions go faster as the temperature is raised. An increase of 10°C from room temperature typically doubles the rate of reaction of organic species in solution. An everyday application of this acceleration is cooking: we cook foods to accelerate certain reactions that lead to the breakdown of cell walls and the decomposition of proteins to give desirable flavors and aromas and as an aid to digestion. In contrast, we refrigerate foods to slow down the natural chemical reactions that lead to their unwanted decomposition.

The rates of most reactions increase as the temperature is raised.

18.9 COLLISION THEORY

Because molecules move with greater energy as the temperature is raised and reactions go faster when the temperature is raised, we can suspect that reactants need energy to react. According to the **collision theory** of gas-phase reactions, molecules behave like defective billiard balls: they bounce apart if they collide at low speed, but might smash into pieces when the impact is really energetic. Similarly, if two molecules collide with less than a certain energy, then they simply bounce apart. If they

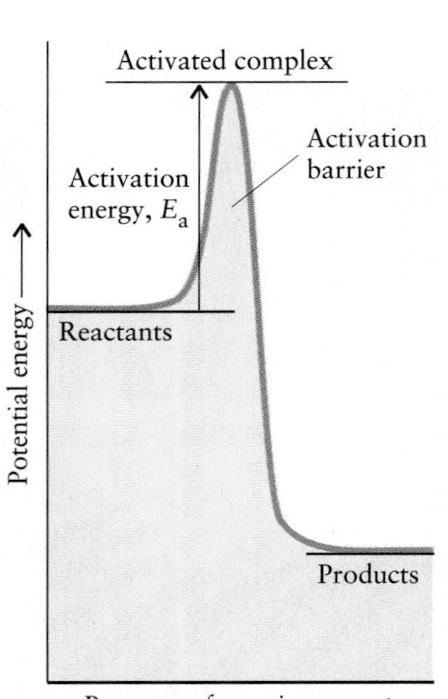

FIGURE 18.17 A reaction profile for an exothermic reaction. In the collision theory of reaction rates, the potential energy (the energy due to position) increases as the reactant molecules approach each other, reaches a maximum as the molecules distort, and then decreases as the atoms rearrange into the bonding pattern characteristic of the products and these products separate. Only molecules with enough energy can cross the barrier and react to form products.

meet with more than a certain energy, then bonds can be broken and new bonds formed (Fig. 18.16). The minimum energy needed for reaction is called the **activation energy,** E_a, of the reaction. Only if the reactants collide with at least the energy E_a can they form products.

We can think of the activation energy as a measure of the height of a barrier between the reactants and the products. The barrier is represented in the graph in Fig. 18.17, which is called a **reaction profile.** As the reactant molecules approach each other, they climb up the left of the barrier. If they have less than the energy E_a, they roll back down on the left and separate again. If they have an energy of at least E_a, they can pass over the top of the barrier and roll down the other side. Now they separate as products.

The fraction of molecules that collide with a kinetic energy equal to or greater than the activation energy E_a is given by the Maxwell distribution of speeds (Section 5.17). As shown by the shaded area under the blue curve in Fig. 18.18, at room temperature very few molecules have enough kinetic energy to cross the barrier. At higher temperatures, a much larger fraction of molecules can react, as represented by the shaded area under the red curve. It turns out that, at a temperature T, the fraction of collisions with at least the energy E_a is proportional to $e^{-E_a/RT}$, where R is the gas constant. Therefore, the rate constant is proportional to the same factor, and we can write

$$k = Ae^{-E_a/RT} \qquad (5)$$

The constant of proportionality A is called the **pre-exponential factor.** This factor takes into account the rate of collisions and the fact that the success of a collision may also depend on the direction of approach. For example, in the gas-phase reaction of chlorine atoms with HI molecules, $HI + Cl \rightarrow HCl + I$, the Cl atom reacts with the HI molecule only if it approaches in the right direction (Fig. 18.19). This direction-dependence is called the **steric requirement** of the reaction.

According to the collision theory of gas-phase reactions, a reaction occurs only if the reactant molecules collide with a kinetic energy of at least the activation energy.

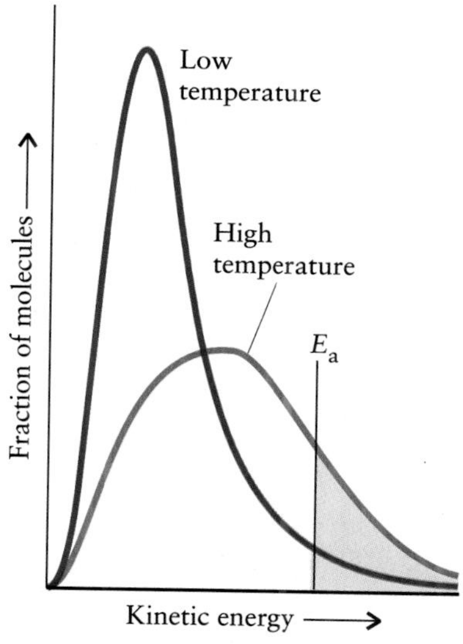

FIGURE 18.18 The fraction of molecules that collide with a kinetic energy that is at least equal to the activation energy, E_a, is denoted by the shaded areas under each curve. The fraction increases rapidly as the temperature is raised.

How do we know that? Modern experiments can fire beams of oriented molecules at one another and see how the outcome varies with the relative orientations of the molecules.

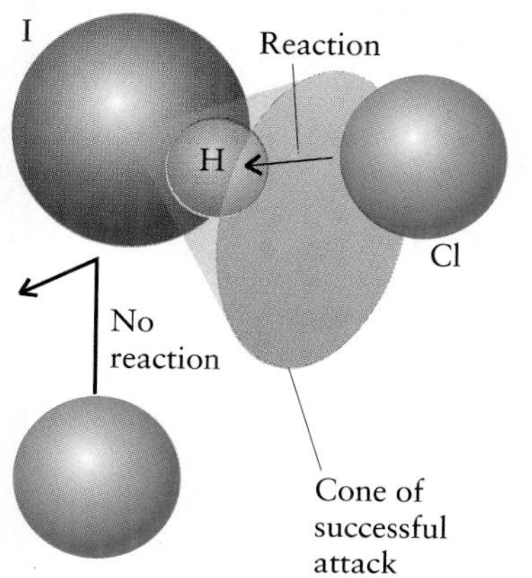

FIGURE 18.19 Whether or not a reaction occurs when two species collide in the gas phase depends on their relative orientation as well as on their energies. In the reaction between a Cl atom and an HI molecule, for example, only those collisions in which the Cl atom approaches the HI molecule along a line of approach that lies inside the cone indicated here lead to reaction, even though the energy of collisions in other orientations may exceed the activation energy.

18.10 ARRHENIUS BEHAVIOR

Chemists often look for ways of plotting data in a manner that gives a straight line, because the information is then easier to analyze.

We can determine the activation energy of a reaction from Eq. 5 if we know k at several temperatures. However, there is an easier way. If we take the logarithm of both sides of the equation for k, we obtain

$$\ln k = \ln A - \frac{E_a}{RT} \qquad (6)$$

The equation was suggested by the Swedish chemist Svante Arrhenius in 1889.

This expression is called the **Arrhenius equation.** When we compare the Arrhenius equation with the equation of a straight line

$$\ln k = \ln A - \frac{E_a}{R} \times \frac{1}{T}$$

$$y = \text{intercept} + \text{slope} \times x$$

we see that if we plot $\ln k$ as a function of $1/T$, the intercept with the axis at $1/T = 0$ is equal to $\ln A$, and the slope of the line is equal to $-E_a/R$ (Fig. 18.20). The values given in Table 18.2 were obtained in this way.

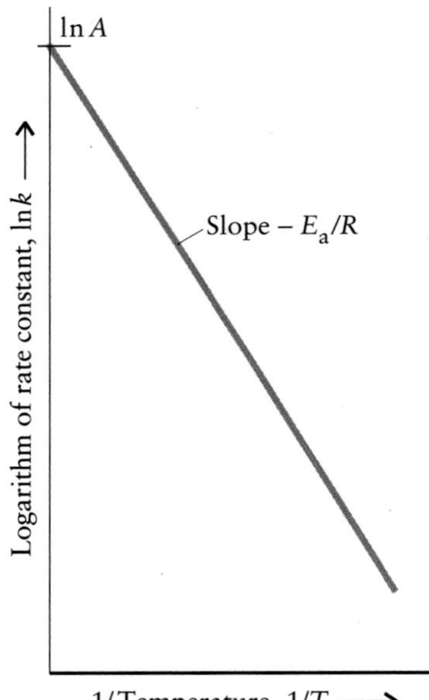

FIGURE 18.20 An Arrhenius plot is a graph of ln k against $1/T$. If, as here, the line is straight, then the reaction is said to show Arrhenius behavior in the temperature range studied. The activation energy for the reaction is obtained by equating $-E_a/R$ to the slope of the line.

EXAMPLE 18.7 Measuring an activation energy

The rate constant for the second-order reaction between bromoethane and hydroxide ions in water,

$$C_2H_5Br(aq) + OH^-(aq) \longrightarrow C_2H_5OH(aq) + Br^-(aq)$$

was measured at several temperatures, with the following results:

Temperature, °C	k, L·mol^{-1}·s^{-1}
25	8.8×10^{-5}
30	1.6×10^{-4}
35	2.8×10^{-4}
40	5.0×10^{-4}
45	8.5×10^{-4}
50	1.40×10^{-3}

Find the activation energy of the reaction.

STRATEGY In the graphical procedure, activation energies are found by plotting $\ln k$ against $1/T$ and measuring the slope of the line. Begin by converting the temperatures to kelvins and drawing up a table of $1/T$ and $\ln k$.

SOLUTION The table we require for drawing the graph is

Temperature, °C	T, K	$1/T$, K^{-1}	ln k
25	298	3.35×10^{-3}	−9.34
30	303	3.30×10^{-3}	−8.74
35	308	3.25×10^{-3}	−8.18
40	313	3.19×10^{-3}	−7.60
45	318	3.14×10^{-3}	−7.07
50	323	3.10×10^{-3}	−6.57

TABLE 18.2 Arrhenius parameters

Reaction	A	E_a, kJ·mol^{-1}
FIRST ORDER, GAS PHASE		
cyclopropane $\longrightarrow$ propene	1.6×10^{15} s^{-1}	272
$CH_3NC \longrightarrow CH_3CN$	4.0×10^{13} s^{-1}	160
$C_2H_6 \longrightarrow 2\,CH_3$	2.5×10^{17} s^{-1}	384
$N_2O \longrightarrow N_2 + O$	8.0×10^{11} s^{-1}	250
$2\,N_2O_5 \longrightarrow 4\,NO_2 + O_2$	4.0×10^{13} s^{-1}	103
SECOND ORDER, GAS PHASE		
$O + N_2 \longrightarrow NO + N$	1×10^{11} L·mol^{-1}·s^{-1}	315
$OH + H_2 \longrightarrow H_2O + H$	8×10^{10} L·mol^{-1}·s^{-1}	42
$2\,CH_3 \longrightarrow C_2H_6$	2×10^{10} L·mol^{-1}·s^{-1}	0
SECOND ORDER, IN AQUEOUS SOLUTION		
$C_2H_5Br + OH^- \longrightarrow C_2H_5OH + Br^-$	4.3×10^{11} L·mol^{-1}·s^{-1}	90
$CO_2 + OH^- \longrightarrow HCO_3^-$	1.5×10^{10} L·mol^{-1}·s^{-1}	38
$C_{12}H_{22}O_{11} + H_2O \longrightarrow 2\,C_6H_{12}O_6$	1.5×10^{15} L·mol^{-1}·s^{-1}	108

The points are plotted in Fig. 18.21. The slope can be calculated from two points, such as those marked A and B:

$$\text{Slope} = -\frac{3.22}{0.30 \times 10^{-3}\ \text{K}^{-1}} = -1.07 \times 10^4\ \text{K}$$

Because the slope is equal to $-E_a/R$, we have

$$\begin{aligned} E_a &= -R \times \text{slope} = -(8.3145\ \text{J·K}^{-1}\text{·mol}^{-1})(-1.07 \times 10^4\ \text{K}) \\ &= 89\ \text{kJ·mol}^{-1} \end{aligned}$$

Although the slope could be calculated from any two of the data points, it is better scientific practice to use the plot, for that minimizes the effect of errors in the data.

SELF-TEST 18.8A The rate constant for the second-order gas-phase reaction

$$HO(g) + H_2(g) \longrightarrow H_2O(g) + H(g)$$

varies with the temperature as follows:

Temperature, °C	k, L·mol^{-1}·s^{-1}
100	1.1×10^{-9}
200	1.8×10^{-8}
300	1.2×10^{-7}
400	4.4×10^{-7}

Calculate the activation energy.

[*Answer:* 42 kJ·mol^{-1}]

SELF-TEST 18.8B The rate of a particular reaction went from 3.00 mol·L^{-1}·s^{-1} to 4.35 mol·L^{-1}·s^{-1} when the temperature was raised from 18°C to 30°C. What is the activation energy of the reaction?

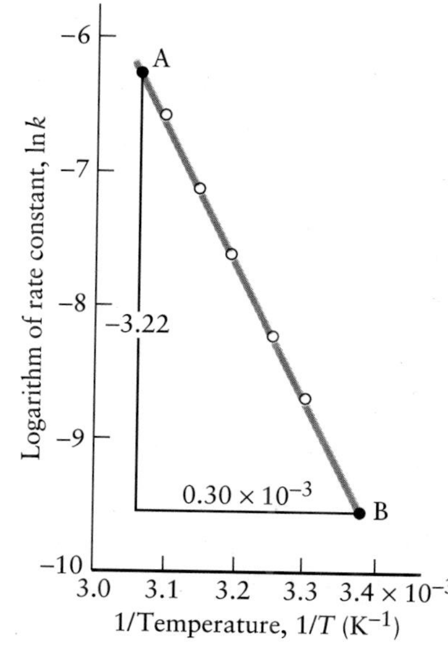

FIGURE 18.21 An Arrhenius plot of the data in Example 18.7. The slope of the line has been worked out by using the points A and B.

One reason it is useful to know the activation energy of a reaction is that the greater the activation energy E_a, the stronger the temperature dependence of the reaction rate. Reactions with small activation energies (about 10 kJ·mol^{-1}, with not very steep Arrhenius slopes) have rates that increase only slightly with temperature. Reactions with large activation energies (above about 60 kJ·mol^{-1}, with steep Arrhenius slopes) have rates that depend strongly on the temperature.

An Arrhenius plot of ln k against 1/T is used to determine the Arrhenius parameters of a reaction; a large activation energy signifies a high sensitivity of the rate to changes in temperature.

EXAMPLE 18.8 Using the activation energy to predict a rate constant

One of the processes taking place inside us is the hydrolysis of sucrose, in which a sucrose molecule is broken down into a glucose molecule and a fructose molecule as a part of the digestive process. How strongly does the rate depend on our body temperature? Calculate the rate constant for the hydrolysis of sucrose at 35°C, given that it is equal to 1.0×10^{-3} L·mol^{-1}·s^{-1} at 37°C (normal body temperature) and that the activation energy of the reaction is 108 kJ·mol^{-1}.

STRATEGY We expect a lower rate constant at the lower temperature. Base the quantitative calculation on the Arrhenius equation written for the two temperatures, which we call T and T':

$$\text{At temperature } T'\text{: } \ln k' = \ln A - \frac{E_a}{RT'}$$

$$\text{At temperature } T\text{: } \ln k = \ln A - \frac{E_a}{RT}$$

Eliminate $\ln A$ by subtracting the second equation from the first:

$$\ln k' - \ln k = -\frac{E_a}{RT'} + \frac{E_a}{RT}$$

That is,

$$\ln\left(\frac{k'}{k}\right) = \frac{E_a}{R}\left(\frac{1}{T} - \frac{1}{T'}\right)$$

At this point, substitute the data into this expression, remembering to express temperature in kelvins. Use $R = 8.3145 \times 10^{-3}$ kJ·K^{-1}·mol^{-1}.

SOLUTION We take $T = 310$ K and $T' = 308$ K. Then

$$\ln\left(\frac{k'}{k}\right) = \frac{108 \text{ kJ·mol}^{-1}}{8.3145 \times 10^{-3} \text{ kJ·K}^{-1}\text{·mol}^{-1}}\left(\frac{1}{310 \text{ K}} - \frac{1}{308 \text{ K}}\right) = -0.272$$

When we take the natural antilogarithm (e^x or "inv ln") of this number, we obtain

$$\frac{k'}{k} = 0.762$$

Finally, because $k = 1.0 \times 10^{-3}$ L·mol^{-1}·s^{-1},

$$k' = 0.762 \times (1.0 \times 10^{-3} \text{ L·mol}^{-1}\text{·s}^{-1}) = 7.6 \times 10^{-4} \text{ L·mol}^{-1}\text{·s}^{-1}$$

at 35°C. The high activation energy of the reaction means that its rate is very sensitive to temperature.

SELF-TEST 18.9A The rate constant of the second-order reaction between

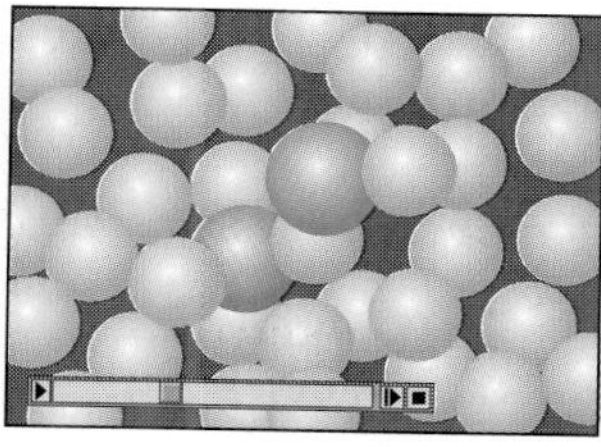

FIGURE 18.22 This sequence of images illustrates the type of motion of two reactant molecules (red and green) in solution. The blue spheres represent solvent molecules. We see the reactants drifting together, lingering near each other for some time, and then drifting apart again. Reaction may occur during the relatively long period of encounter.

CH_3Br and OH^- ions is $2.8 \times 10^{-4}\ L{\cdot}mol^{-1}{\cdot}s^{-1}$ at 25°C; what is it at 50°C? See Table 18.2 for data.

[*Answer:* $4.7 \times 10^{-3}\ L{\cdot}mol^{-1}{\cdot}s^{-1}$]

SELF-TEST 18.9B The rate constant of the first-order isomerization of cyclopropane, C_3H_6, to propene, $CH_3{-}CH{=}CH_2$, is $6.7 \times 10^{-4}\ s^{-1}$ at 500°C; what is it at 300°C? See Table 18.2 for data.

18.11 ACTIVATED COMPLEXES

Collision theory can be extended to explain why the Arrhenius equation also applies to reactions in solution. This more general theory is called **activated complex theory.**

In activated complex theory, two molecules approach and distort as they meet. In a gas-phase reaction, that meeting and distortion is the "collision" of collision theory. In solution, the approach is a zigzag walk among solvent molecules (Fig. 18.22), and the distortion might not take place until after the two reactant molecules have met and receive a particularly vigorous kick from the solvent molecules around them (Fig. 18.23). In either case, the collision or the kick results in the formation of an **activated complex,** a combination of the two molecules that can either go on to form products or fall apart into the unchanged reactants. In the activated complex the original bonds have lengthened and weakened, whereas the new bonds are only partially formed.

The activated complex corresponds to the peak of the energy barrier (recall Fig. 18.17). As in collision theory, the rate of the reaction depends on the rate at which reactants can climb to the top of the barrier and form the complex. The resulting expression for the rate is very similar to the one given in Eq. 5, so this more general theory also accounts for

FIGURE 18.23 We join this sequence of images at the moment when the reactant molecules are in the middle of their encounter. They may acquire enough energy by impacts from the solvent molecules to form an activated complex, which may go on to form products.

the form of the Arrhenius equation and the observed dependence of the reaction rate on temperature.

In activated complex theory, a reaction occurs only if two molecules acquire enough energy, perhaps from the surrounding solvent, to form an activated complex and cross an energy barrier.

18.12 CATALYSIS

In some cases we may want to increase the rate of a reaction but do not want to raise the temperature. For instance, if the reaction is exothermic, then the equilibrium constant of the reaction is decreased by an increase in temperature (recall Section 13.10), and we get a smaller yield of product at the higher temperature. Furthermore, in industry, high temperature often means high energy costs.

The name *catalyst* comes from the Greek words meaning "breaking down by coming together." The Chinese characters for catalyst, which translate as "marriage broker," capture the sense quite well.

One solution is to use a **catalyst,** a substance that increases the rate of a reaction without being consumed in the reaction. In many cases, only a small amount of catalyst is necessary, because it acts over and over again. There is no such thing as a "universal catalyst"; not all reactions can be accelerated with a catalyst, and the choice of catalyst depends on the reaction.

A catalyst speeds up a reaction by providing an alternative pathway from reactants to products. This new pathway has a lower activation energy than the original pathway (Fig. 18.24). At the same temperature, a greater fraction of reactant molecules can cross the lower barrier of the catalyzed path and turn into products than when no catalyst is present. Although the reaction occurs more quickly, a catalyst has no effect on the equilibrium composition. Both forward and reverse reactions are accelerated on the catalyzed path, leaving the equilibrium constant unchanged.

A catalyst is classified as *homogeneous* if it is present in the same phase as that of the reactants. For reactants that are gases, a homogeneous catalyst is also a gas. If the reactants are liquids, a homogeneous catalyst is a dissolved liquid or solid (Fig. 18.25). Dissolved bromine is a homogeneous liquid-phase catalyst for the decomposition of aqueous hydrogen peroxide:

$$2\,H_2O_2(aq) \xrightarrow{Br_2} 2\,H_2O(l) + O_2(g)$$

In the absence of bromine or another catalyst, a solution of hydrogen peroxide can be stored for a long time at room temperature; however, bubbles of oxygen form as soon as a drop of bromine is added. Bromine's role in this reaction is believed to be its reduction to Br^- in one step, followed by oxidation back to Br_2 in a second step:

$$Br_2(aq) + H_2O_2(aq) \longrightarrow 2\,Br^-(aq) + 2\,H^+(aq) + O_2(g)$$

$$2\,Br^-(aq) + H_2O_2(aq) + 2\,H^+(aq) \longrightarrow Br_2(aq) + 2\,H_2O(l)$$

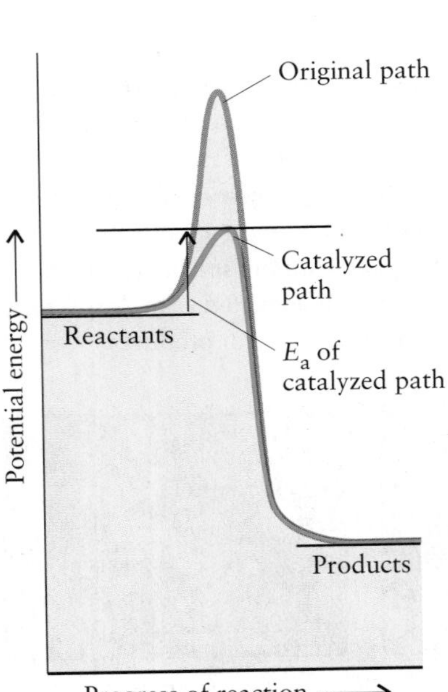

FIGURE 18.24 A catalyst provides a new reaction pathway with a lower activation energy, thereby allowing more reactant molecules to cross the barrier and form products. Notice that E_a for the reverse reaction is also lowered on the catalyzed path.

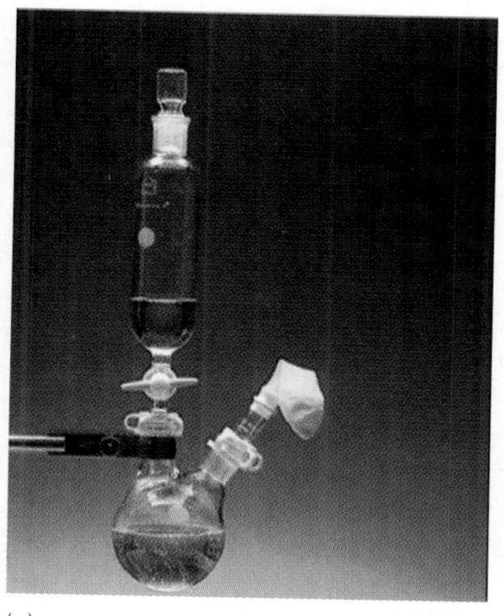
(a)

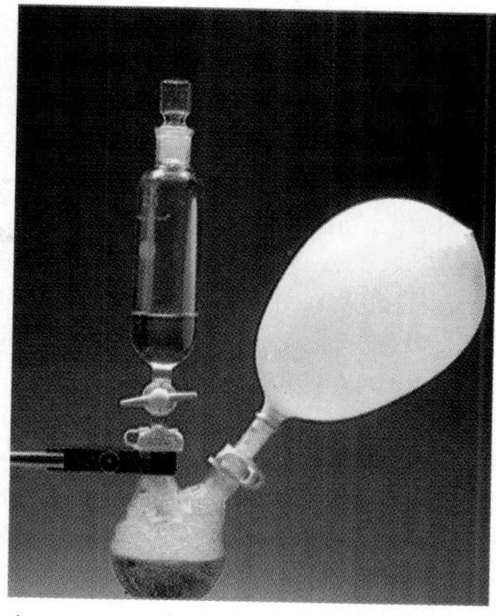
(b)

FIGURE 18.25 A small amount of catalyst, in this case potassium iodide in aqueous solution, can accelerate the decomposition of hydrogen peroxide to water and oxygen. This effect is shown (a) by the slow inflation of the balloon when no catalyst is present and (b) by its rapid inflation when a catalyst is present.

When we add the two steps, the Br_2 and 2 Br^- cancel, leaving the overall equation as the first of the three written above. Hence, although Br_2 molecules have participated in the reaction, they are not consumed and can be used over and over again.

A catalyst is classified as *heterogeneous* if it is in a phase different from that of the reactants. The most common heterogeneous catalysts are finely divided or porous solids used in gas-phase or liquid-phase reactions. They are finely divided or porous in order to provide a large surface area for reaction. One example is the iron catalyst used in the Haber process for ammonia. Another is finely divided vanadium pentoxide (V_2O_5), which is used in the *contact process* for the production of sulfuric acid:

$$2\,SO_2(g) + O_2(g) \xrightarrow{V_2O_5} 2\,SO_3(g)$$

The reactant **adsorbs,** or attaches, to the catalyst's surface. As a reactant molecule adsorbs, its bonds are weakened and the reaction can proceed more quickly because the bonds are more easily broken (Fig. 18.26). One important step in the Haber process is the adsorption of N_2 molecules on the iron and the weakening of the strong N≡N triple bond.

Catalysts are used in the catalytic converters of automobiles to bring about the complete and rapid combustion of unburned fuel. The mixture of gases leaving an engine includes carbon monoxide, unburned hydrocarbons, and the nitrogen oxides collectively referred to as NO_x. Air pollution is decreased if the carbon compounds are oxidized to carbon dioxide and the NO_x reduced, by another catalyst, to nitrogen. The challenge

FIGURE 18.26 The reaction between ethene, $CH_2{=}CH_2$, and hydrogen on a catalytic metal surface. In this sequence of images, we see the ethene molecule approaching the metal surface to which hydrogen molecules have already adsorbed: when they adsorb, they dissociate and stick to the surface as hydrogen atoms. Next, after sticking to the surface, the ethene molecule meets a hydrogen atom and forms a bond. At this stage a $\cdot CH_2CH_3$ radical is attached to the surface by one of its carbon atoms. Finally, the radical and another hydrogen atom meet, ethane is formed, and escapes from the surface.

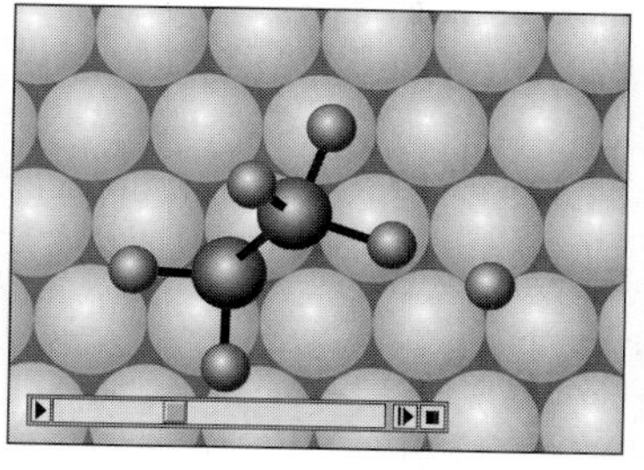

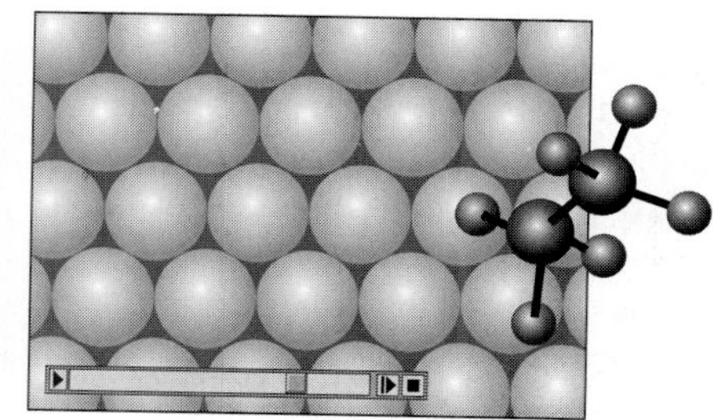

is to find a catalyst—or a mixture of catalysts—that will accelerate both the oxidation and the reduction reactions and be active when the car is first started and the engine is cool.

Catalysts can be **poisoned,** or inactivated. A common cause of such poisoning is the adsorption of a molecule so tightly to the catalyst that it seals its surface against further reaction. Some heavy metals, especially lead, are very potent poisons for heterogeneous catalysts, which is why lead-free gasoline must be used for engines fitted with catalytic converters. The elimination of lead has the further benefit of decreasing the amount of poisonous lead in the environment.

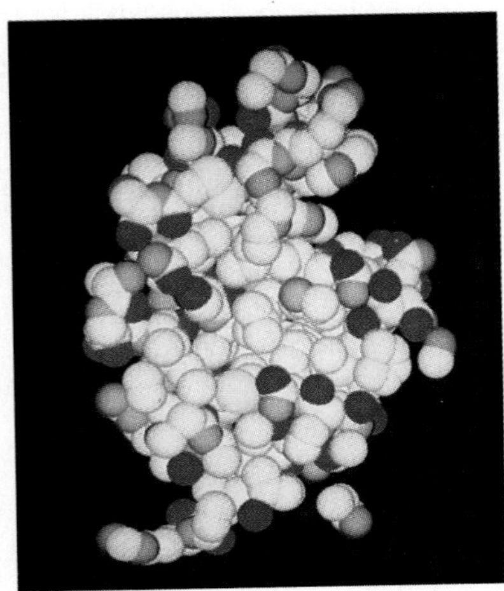

FIGURE 18.27 The lysozyme molecule shown here is a typical enzyme molecule. Lysozyme occurs in a number of places, including tears and the mucus in the nose. One of its functions is to attack the cell walls of bacteria and destroy them.

Catalysts participate in reactions but are not themselves used up; they provide a reaction pathway with a lower activation energy. Catalysts are classified as either homogeneous or heterogeneous.

SELF-TEST 18.10A How does a catalyst affect (a) the rate of the reverse reaction? (b) ΔH for the reaction?

[*Answer:* (a) Increases it; (b) has no effect]

SELF-TEST 18.10B How does a homogeneous catalyst affect (a) the rate law; (b) the equilibrium constant?

18.13 LIVING CATALYSTS: ENZYMES

The name *enzyme* comes from the Greek words for "in yeast."

Living cells contain thousands of different kinds of catalysts, each of which is necessary to life. Many of these catalysts are large proteins called **enzymes** (Fig. 18.27). They are proteins with a slotlike **active site,** where reaction takes place. The **substrate,** the molecule on which the enzyme acts, fits into the slot as a key fits into a lock (Fig. 18.28). However, unlike an ordinary lock, a protein molecule distorts slightly as the substrate molecule approaches, and its ability to undergo the correct distortion also determines whether the "key" will fit. This refinement of the original lock-and-key model is known as the **induced-fit mechanism** of enzyme action.

Once in the slot, the substrate becomes distorted, so allowing it to undergo reaction. The modified substrate molecule is then released for

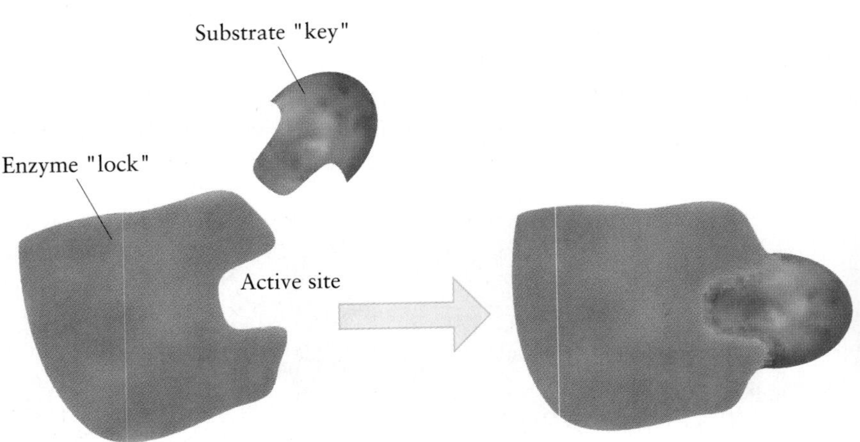

FIGURE 18.28 In the lock-and-key model of enzyme action, the correct substrate is recognized by its ability to fit into the active site like a key into a lock. In a refinement of this model, the enzyme changes its shape slightly as the key enters (not shown).

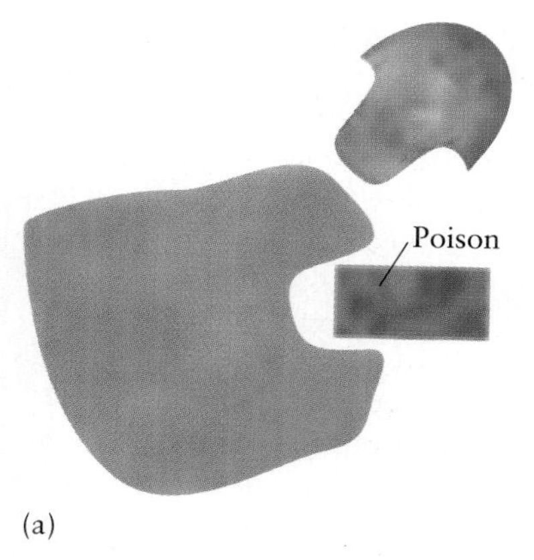

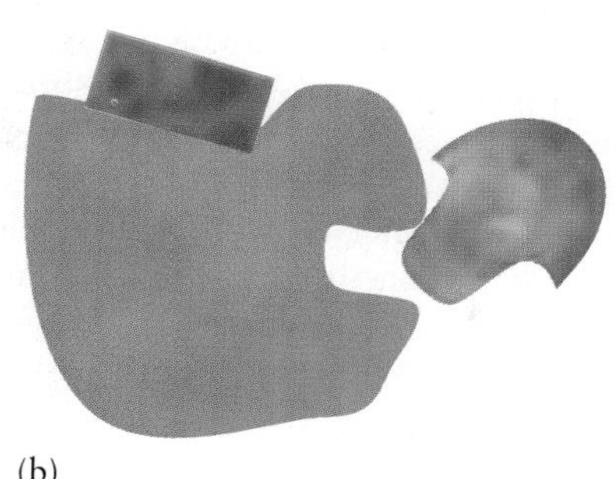

FIGURE 18.29 (a) An enzyme poison (represented by the mottled green rectangle) can act by attaching so strongly to the active site that it blocks the site, thereby taking the enzyme out of action. (b) Alternatively, the poison molecule may attach elsewhere, so distorting the enzyme molecule and its active site that the substrate no longer fits.

use in the next stage, which is controlled by another enzyme, and the original enzyme molecule is free to receive the next substrate molecule (see Case Study 18). One example is the enzyme amylase present in your mouth. The amylase in saliva helps to break down starches in food into the more easily digested glucose. If you chew a cracker long enough, you can notice the increased sweetness.

One form of biological poisoning mirrors the effect of lead on a catalytic converter. The activity of enzymes is destroyed if an alien substrate attaches too strongly to the reactive site, for then the site is blocked and made unavailable to the true substrate (Fig. 18.29). As a result, the chain of biochemical reactions in the cell stops, and the cell dies. The action of nerve gases is believed to stem from their ability to block the enzyme-controlled reactions that allow impulses to travel through nerves. Arsenic, that favorite of fictional poisoners, acts in a similar way. After ingestion as As(V) in the form of arsenate ions (AsO_4^{3-}), it is reduced to As(III), which binds to enzymes and inhibits their action.

Enzymes are biological catalysts that function by modifying substrate molecules to promote reaction.

REACTION MECHANISMS

How does a molecule of ozone change into a molecule of oxygen? What *atomic* interactions turn a mixture of fuel and air into carbon dioxide and water when it ignites in an engine? What is really going on *in terms of atoms* high in the atmosphere as CFC molecules punch holes in the ozone layer? Kinetic data provide a window onto the processes that occur at a molecular level when chemical reactions take place.

18.14 ELEMENTARY REACTIONS

The reason we cannot predict a rate law from the chemical equation for a reaction is that all but the simplest reactions are the outcome of several,

DRUG THERAPY AND MOOD REGULATION

PLEASURE IS SO important to our well-being that seeking it out is a primary motivator of human activity. Some of the first healing herbs used in prehistoric times were painkillers. In medieval times, alchemists sought an elixir of happiness that would be free of the sensory numbing and addictive nature of opiates. Today, science fiction writers speculate about a future in which chemicals in pills would be taken to produce any desired feelings or moods.

Can pills make us happy? The answer is complex. Our moods are regulated through our body chemistry, which changes in response to conditions and normally keeps our moods in balance. However, when bodies lose their ability to respond, chemistry can help restore the balance. For example, the symptoms of disorders such as depression and schizophrenia can be treated with daily medication.

Medication often works by controlling the action of an enzyme, a natural biological catalyst (see Section 18.13). The kinetics of enzyme reactions were first studied by Leonor Michaelis and Maud Menten in the early part of the twentieth century. They found that the rate of an enzyme-catalyzed reaction increases with the concentration of the substrate, as shown in the plot; but when the concentration of substrate is high, the reaction rate depends only on the concentration of the enzyme.

The receptors in our central nervous system are analogous to heterogeneous catalysts. Chemicals called *neurotransmitters* are released from neurons on one side of a space called a synapse and attach to receptors in another neuron on the other side. When the receptor is finished with the neurotransmitter, it releases the molecule, which is then reabsorbed into the original neuron and decomposed or modified by a type of enzyme called a monoamine oxidase. The neurotransmitter must be destroyed to prevent a continued buildup of its concentration within the synapse.

When the levels of neurotransmitters or endorphins such as dopamine and norepinephrine are optimal, a feeling of well-being is produced through their action on the receptors. Depression can result if too little norepinephrine or dopamine is available. Conversely, if so much dopamine is present that the receptors are flooded with it, then the nervous system becomes overstimulated, and hallucinations and other symptoms of schizophrenia result.

NH_2, OH, OH

Dopamine

HO, NH_2, OH, OH

Norepinephrine (noradrenaline)

Depression can be treated by inhibiting the action of the monoamine oxidases, so that the neurotransmitter is not destroyed and its concentration can build up to an acceptable level in the synapse. Some antidepressant drugs such as iproniazid inhibit the monoamine oxidases directly. They bind to them more strongly than the neurotransmitters do, thereby blocking the active sites of the enzymes. Other antidepressants, such as amitriptyline hydrochloride (Elavil) increase the level of norepinephrine by inhibiting its reabsorption into the neuron.

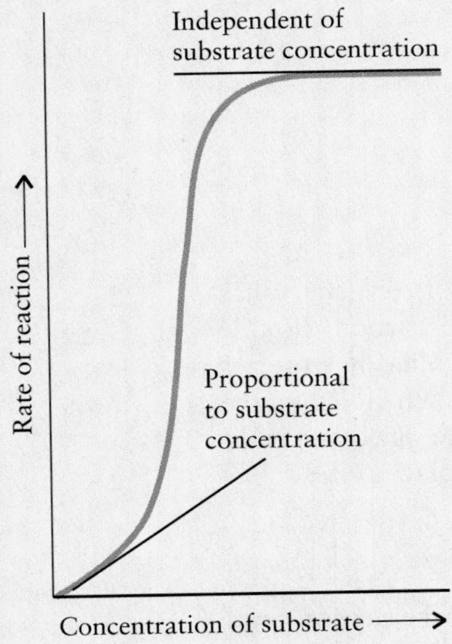

A plot of Michaelis-Menten enzyme kinetics. At low substrate concentrations the rate of reaction is directly proportional to substrate concentration. However, at high substrate concentrations, the rate is constant, as the enzyme molecules are "saturated" with substrate.

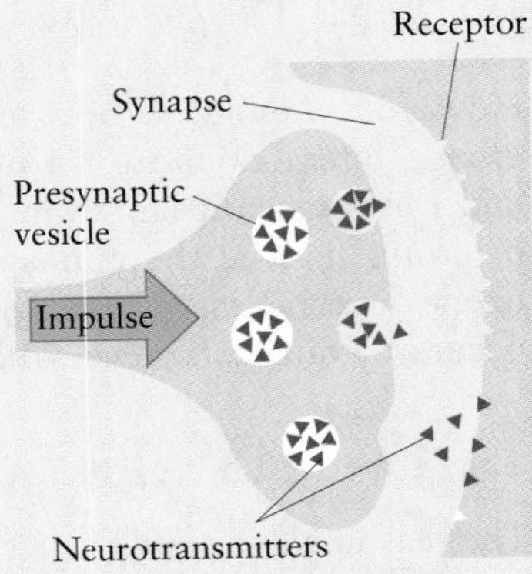

A diagram of a synapse. The triangles represent neurotransmitters that travel from the neuron on the left to the receptors in the neuron on the right. The concentration of neurotransmitters in the synapse is controlled by enzymes.

Amitriptyline hydrochloride (Elavil)

Schizophrenia is also treated by drugs that inhibit the action of the receptors by binding to them, blocking the attachment of the dopamine molecules. Chlorpromazine hydrochloride (Thorazine) and haloperidol (Haldol) are commonly used for this purpose.

Chlorpromazine hydrochloride (Thorazine)

Haloperidol (Haldol)

With fewer receptors available, the excess dopamine goes unused and is reabsorbed. A comparable drug treatment for alcoholism uses naltrexone, which has a molecular shape similar to that of the endorphins. It fits into the receptor sites without creating the sensation of euphoria; therefore, it inhibits the action of the endorphins and reduces the dependency of the individual on alcohol. A similar treatment for cocaine addiction is being studied.

Stimulants such as amphetamines cause the overproduction of dopamine and, when abused, can generate the symptoms of schizophrenia. Self-medication for mood regulation is extremely dangerous because of the delicate balance of enzymes and receptors needed for proper functioning of our nervous systems. Even medically prescribed drug treatments for mood regulation involve side effects that must be carefully monitored.

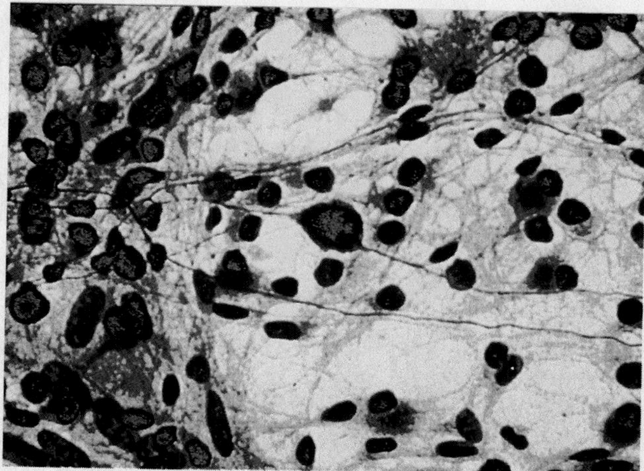

The neurons in the human brain affect how we think and feel and how we perceive reality, including reading chemistry books.

QUESTIONS

1. The rates of many enzyme-catalyzed reactions exhibit the dependence on substrate concentration shown in the plot. The catalyzed reaction is first order in both enzyme and substrate at low concentrations of substrate. What is the order of the reaction in the substrate at high concentrations of substrate? Explain your reasoning.

2. The following mechanism, in which E represents the enzyme, S the substrate, and P the product formed by S, has been proposed for low substrate concentrations:

$$E + S \rightleftharpoons ES$$

$$ES \longrightarrow E + P$$

(a) Which step is the slow step? (b) Write the rate law consistent with this mechanism. (c) Is this mechanism valid for high concentrations of substrate? Explain.

3. How would the presence of an enzyme inhibitor affect the slope of the plot?

4. Acetylcholine is a neurotransmitter that is released into synapses and cleared away by the enzyme acetyltransferase. Some nerve gases work by deactivating the enzyme, thus allowing acetylcholine to build up and continue to send signals that overstimulate the heart until it fails, which may take no longer than a few minutes. Suggest the general outline of a drug therapy that could be used as an antidote to nerve gases if administered quickly.

and sometimes many, steps called **elementary reactions.** To understand how a reaction occurs, we propose a **reaction mechanism,** a sequence of elementary reactions that describes the atomic changes that we believe take place as reactants are transformed into products. For example, in the decomposition of ozone, $2\,O_3(g) \rightarrow 3\,O_2(g)$, we could imagine the reaction occurring in one step, when two O_3 molecules collide and rearrange their atoms to form three O_2 molecules. Alternatively, we could propose a mechanism involving two elementary reactions: in the first step, an O_3 molecule is energized by solar radiation and dissociates into an O atom and an O_2 molecule. Then in a second step the O atom attacks another O_3 molecule to produce two more O_2 molecules. The O atom in the second mechanism is a **reaction intermediate,** a species that plays a role in a reaction but does not appear in the chemical equation for the overall reaction.

Which, if either, of these two mechanisms is correct? Although we cannot prove a mechanism, we can determine the rate laws the two mechanisms would lead to and compare them with the experimental rate law for the reaction to see whether there is a match. To make this comparison, we begin by writing the equation for each step in the two proposed mechanisms.

A chemical equation tells us the overall stoichiometry of a chemical reaction. Elementary reactions are also summarized by chemical equations, but without the state symbols. These equations show how *individual* atoms and molecules take part in the reaction (Fig. 18.30). For example, one proposed step in the decomposition of ozone is

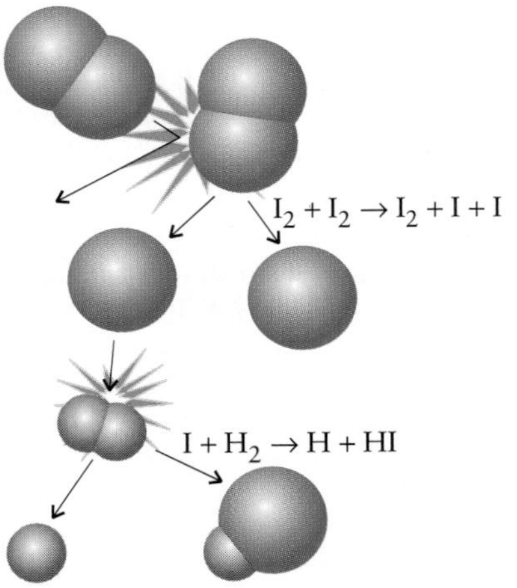

$$O_3 \longrightarrow O_2 + O$$

This equation tells us what happens to an *individual* O_3 molecule. It is an example of a **unimolecular reaction,** because only one reactant molecule is involved. We can picture an ozone molecule acquiring energy from sunlight and vibrating so vigorously that it shakes itself apart.

If we believe that an O atom produced by the dissociation of O_3 goes on to attack another O_3 molecule, we write

$$O + O_3 \longrightarrow O_2 + O_2$$

This elementary reaction is an example of a **bimolecular reaction** because two reactant species come together to react.

One criterion for the acceptability of a proposed mechanism is that the sum of the equations for the elementary reactions must be the chemical equation for the overall reaction. For our proposed two-step mechanism, we have

$$\begin{array}{r} O_3 \longrightarrow O_2 + O \\ O + O_3 \longrightarrow O_2 + O_2 \\ \hline \text{Total:}\quad 2\,O_3 \longrightarrow 3\,O_2 \end{array}$$

In the last line, the O atoms have been canceled on each side of the arrow.

Any forward reaction that can occur is also accompanied by the corresponding reverse reaction. Therefore, the unimolecular decay of O_3 is accompanied by the formation reaction

$$O_2 + O \longrightarrow O_3$$

FIGURE 18.30 The chemical equations for elementary reactions show the *individual* events that take place between atoms and molecules that encounter one another. This illustration shows two of the steps believed to occur during the formation of hydrogen iodide from hydrogen and iodine vapor. In one, $I_2 + I_2 \rightarrow I_2 + I + I$, a collision results in the dissociation of an iodine molecule. In the second step ($I + H_2 \rightarrow H + HI$), one of the I atoms produced in the first step attacks a hydrogen molecule and forms a hydrogen iodide molecule.

This step is bimolecular. Similarly, the bimolecular attack of O on O_3 is accompanied by its reverse reaction

$$O_2 + O_2 \longrightarrow O + O_3$$

This step is also bimolecular. These two reactions add up to give the reverse of the overall reaction. The forward and reverse reactions provide a mechanism for reaching dynamic equilibrium between the reactants and products in the overall process.

The proposed one-step mechanism also yields a chemical equation that matches the overall reaction:

$$O_3 + O_3 \longrightarrow O_2 + O_2 + O_2$$

The reverse of this elementary reaction is a **termolecular reaction,** an elementary reaction requiring the simultaneous collision of three molecules:

$$O_2 + O_2 + O_2 \longrightarrow O_3 + O_3$$

Termolecular reactions are uncommon, because it is very unlikely that three molecules will collide simultaneously with one another under normal conditions.

Most reactions occur by a series of elementary reactions. The molecularity of an elementary reaction indicates how many reactant species are involved in the step.

SELF-TEST 18.11A What is the molecularity of the elementary reaction (a) $C_2N_2 \rightarrow CN + CN$? (b) $NO_2 + NO_2 \rightarrow NO + NO_3$?

[*Answer:* (a) Unimolecular; (b) bimolecular]

SELF-TEST 18.11B What is the molecularity of the elementary reaction (a) $C_2H_5Br + OH^- \rightarrow C_2H_5OH + Br^-$? (b) $Br_2 \rightarrow Br + Br$?

18.15 THE RATE LAWS OF ELEMENTARY REACTIONS

To determine whether a proposed mechanism is plausible, we must construct its overall rate law and check to see whether it is consistent with the experimentally determined rate law. For the decomposition of ozone, the following rate law has been observed:

$$\text{Rate of consumption of } O_3 = k \times \frac{[O_3]^2}{[O_2]}$$

Which, if either, of the two proposed mechanisms results in this rate law? To write the rate law for a mechanism, we first write the rate laws for the elementary reactions; then we combine them into the predicted rate law for the overall reaction.

It is important to realize that, although the calculated rate law and the experimental rate law may be the same, the proposed mechanism may still be incorrect; some other mechanism may also lead to the same rate law. Kinetic information can only support a proposed mechanism; it can never prove that a mechanism is correct. The acceptance of a suggested mechanism is more like the process of proof in an ideal court of law than in mathematics, with evidence being assembled to give a convincing, consistent picture.

The rate law corresponding to a reaction mechanism is built up from the rate laws of the individual steps. The molecularity of an elementary reaction implies a specific rate law. *Writing rate laws from chemical equations is permissible only for elementary reactions.* A unimolecular elementary reaction depends on a single energized species shaking itself apart and always has a first-order rate law. A bimolecular reaction depends on collisions between two species, so its rate law is second order. Therefore, for the proposed one-step, bimolecular mechanism, we can write

$$O_3 + O_3 \longrightarrow O_2 + O_2 + O_2 \qquad \text{Rate} = k[O_3]^2$$

Because there is only one step, the rate law for this elementary step is also the rate law for the mechanism. However, this rate law does not match the experimental rate law, so the one-step mechanism is not consistent with experiment and must be rejected.

Let's investigate whether the two-step mechanism is plausible. First, we write the rate laws for the two elementary reactions. For the unimolecular decay of O_3, we can write

$$O_3 \longrightarrow O_2 + O \qquad \text{Rate} = k_1[O_3]$$

Similarly, the rate law for the next step, which is bimolecular, is

$$O + O_3 \longrightarrow O_2 + O_2 \qquad \text{Rate} = k_2[O][O_3]$$

The rate laws for the two reverse reactions can be written similarly:

$$O_2 + O \longrightarrow O_3 \qquad \text{Rate} = k_1'[O_2][O]$$

$$O_2 + O_2 \longrightarrow O + O_3 \qquad \text{Rate} = k_2'[O_2]^2$$

To combine the rate laws for elementary reactions into a rate law for the overall reaction, we need to know which step is the **rate-determining step,** the elementary forward reaction that is so much slower than the rest that it governs the rate of the overall reaction. A rate-determining step is like a slow ferry on the route between two cities (Fig. 18.31). The rate at which the traffic arrives at its destination is governed by the rate at which it is ferried across the river, because this part of the journey is much slow-

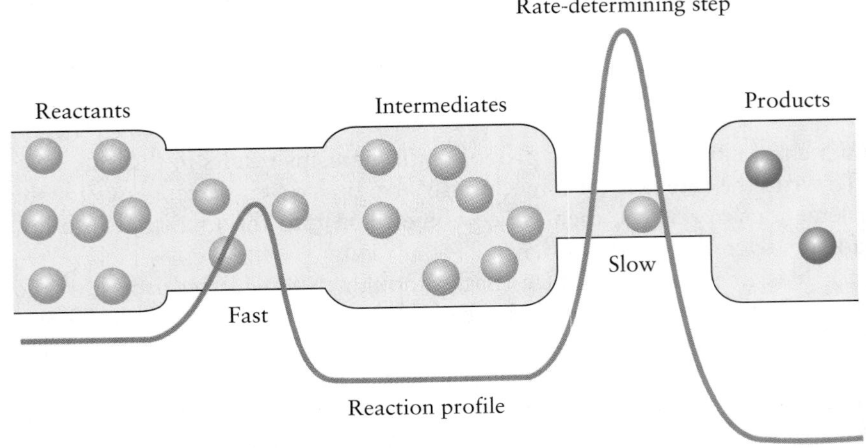

FIGURE 18.31 The rate-determining step in a reaction is the elementary reaction that governs the rate at which products are formed in a multistep reaction. It is like a ferry that cannot handle busy traffic. The reaction profile shows that the slow step has the highest activation energy.

er than any of the others. When writing a rate law for the overall reaction on the basis of a proposed mechanism, we set the overall rate equal to the rate of the rate-determining step. Because reaction intermediates do not appear in any experimentally determined rate law, we have to make sure that they do not appear in the predicted rate law either.

The action of a catalyst is like opening a new bridge that bypasses the ferry crossing.

Elementary unimolecular reactions have first-order rate laws; elementary bimolecular reactions have second-order rate laws. A rate law is derived from a proposed mechanism by identifying the rate-determining step.

EXAMPLE 18.9 Setting up an overall rate law from a proposed mechanism

The equations for the two forward and reverse steps proposed for the two-step mechanism of ozone decomposition are set out in this section. Measurements of the rates of the elementary forward reactions show that the slowest step is the second step, the attack of O on O_3. The reverse reaction $O_2 + O_2 \rightarrow O + O_3$ is so slow that it can be ignored. Derive the rate law for the overall reaction.

STRATEGY Select the slowest *forward* step as the rate-determining step and equate the overall rate to the rate of this step. Identify any reaction intermediates that appear in this rate law. If any do appear, look for an equation for their concentrations in terms of the concentrations of the species that do appear in the overall reaction. One procedure is to look for a reaction that involves the intermediate and is in dynamic equilibrium with its reverse reaction. We then assume that at equilibrium the rates of the forward and reverse reactions are equal.

SOLUTION The rate of the rate-determining step is

$$O + O_3 \longrightarrow O_2 + O_2 \qquad \text{Rate} = k_2[O][O_3]$$

We take this expression as the rate law for the overall reaction. To eliminate the O atom concentration from the rate law, we consider the two relatively fast reactions in the first step that produces this intermediate:

$$O_3 \longrightarrow O_2 + O \qquad \text{Rate} = k_1[O_3]$$

$$O_2 + O \longrightarrow O_3 \qquad \text{Rate} = k'_1[O_2][O]$$

These two elementary reactions establish a dynamic chemical equilibrium in which their rates are equal; that is,

$$k_1[O_3] = k'_1[O_2][O]$$

We can find [O] from this equation:

$$[O] = \frac{k_1[O_3]}{k'_1[O_2]}$$

When this relation is substituted for [O] in the expression for the overall rate, we get

$$\text{Rate} = \frac{k_1 k_2}{k'_1} \times \frac{[O_3]^2}{[O_2]}$$

This equation has exactly the same form as the observed rate law. Moreover, we can recognize the experimental k as the combination k_1k_2/k'_1 of the rate constants for the elementary steps. Because this mechanism is consistent with both the balanced equation and the experimental rate law, we can consider it as a plausible mechanism and write it as follows:

$$\text{Step 1:} \qquad O_3 \rightleftharpoons O_2 + O \quad \text{(fast equilibrium)}$$

$$\text{Step 2:} \quad O + O_3 \longrightarrow O_2 + O_2 \quad \text{(slow)}$$

SELF-TEST 18.12A The proposed mechanism for a reaction is $AH + B \rightarrow BH^+ + A^-$ and its reverse, both of which are fast, followed by $A^- + AH \rightarrow$ products, which is slow. Find the rate law with A^- treated as the intermediate.

[*Answer:* Rate $= (k_1k_2/k_1')[AH]^2[B]/[BH^+]$ or Rate $= k[AH]^2[B][BH^+]^{-1}$]

SELF-TEST 18.12B The following two-step mechanism has been proposed for the reaction $2\,NO(g) + O_2(g) \rightarrow 2\,NO_2(g)$, by which atmospheric NO_2 is generated from NO in automobile exhaust:

$$\text{Step 1:} \qquad 2\,NO \rightleftharpoons N_2O_2 \quad \text{(both forward and reverse reactions fast)}$$

$$\text{Step 2:} \quad N_2O_2 + O_2 \longrightarrow 2\,NO_2 \quad \text{(slow)}$$

Find the rate law and the relation between the rate constant k for the overall reaction and the rate constants for the elementary reactions.

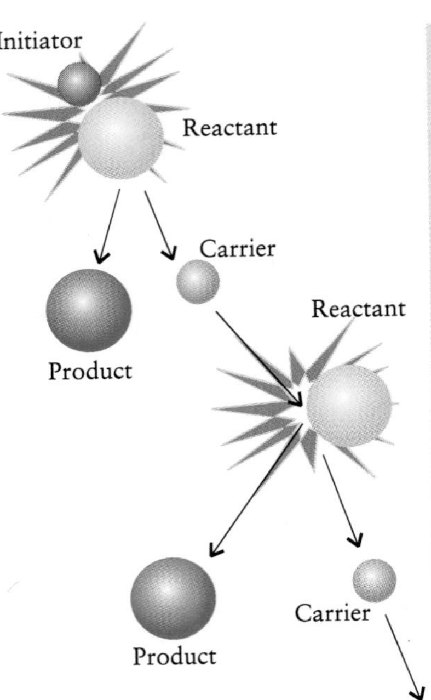

FIGURE 18.32 In a chain reaction, the product of one step in a reaction is a reactant in a subsequent step, which in turn produces species that can take part in subsequent reaction steps.

18.16 CHAIN REACTIONS

In a **chain reaction,** a highly reactive intermediate reacts to produce another highly reactive intermediate, which reacts to produce another, and so on (Fig. 18.32). In many cases, the reaction intermediate—which in this context is called a **chain carrier**—is a radical, and the reaction is called a **radical chain reaction.** In a radical chain reaction, one radical reacts with a molecule to produce another radical, which goes on to attack another molecule, and so on.

◀◀ Radical polymerization is discussed in Section 11.13.

The synthesis of HBr in the reaction

$$H_2(g) + Br_2(g) \longrightarrow 2\,HBr(g)$$

takes place by a chain reaction. The chain carriers are hydrogen atoms ($H\cdot$) and bromine atoms ($Br\cdot$). The first step in any chain reaction is **initiation,** the formation of chain carriers from a reactant:

◀◀ Recall from Section 8.9 that the unpaired electron of a free radical is represented by a dot when we want to distinguish a radical from other species.

$$Br_2 \xrightarrow{\Delta \text{ or } h\nu} Br\cdot + Br\cdot$$

Once chain carriers have been formed, the chain can **propagate** in a series of reactions in which one carrier reacts with a reactant molecule to produce another carrier. The elementary reactions for the propagation of the chain are

$$Br\cdot + H_2 \longrightarrow HBr + H\cdot$$

$$H\cdot + Br_2 \longrightarrow HBr + Br\cdot$$

The chain carriers—radicals here—produced in these reactions can go on to attack other reactant (H_2 and Br_2) molecules, thereby allowing the chain to continue. The elementary reaction that ends the chain, a process called **termination,** occurs when chain carriers combine to form products:

$$Br\cdot + Br\cdot \longrightarrow Br_2$$

In some cases a chain can propagate explosively. Explosions can be expected when chain **branching** occurs, when more than one chain carrier is formed in a propagation step. The characteristic pop that occurs when a mixture of hydrogen and oxygen is ignited is a consequence of chain branching. The two gases combine in a radical chain reaction in which the initiation step may be the formation of hydrogen atoms:

$$\text{Initiation:}\quad H_2 \xrightarrow{\text{spark}} H\cdot + H\cdot$$

Two new radicals are formed when one hydrogen atom attacks an oxygen molecule:

$$\text{Branching:}\quad H\cdot + O_2 \longrightarrow HO\cdot + \cdot O\cdot$$

The oxygen atom, with valence electron configuration $2s^22p_x^22p_y^12p_z^1$, has two electrons with unpaired spins (its complete Lewis diagram is $\cdot\ddot{\underset{\cdot\cdot}{O}}\cdot$). Two radicals are also produced when the oxygen atom attacks a hydrogen molecule:

$$\text{Branching:}\quad \cdot O\cdot + H_2 \longrightarrow HO\cdot + H\cdot$$

As a result of these branching processes, the chain produces a large number of radicals that can take part in even more branching steps. The reaction rate increases rapidly, and an explosion typical of many combustion reactions may occur (Fig. 18.33).

Chain reactions begin with the initiation of a reactive intermediate that propagates the chain and concludes with termination when radicals combine. Branching chain reactions can be explosively fast.

18.17 RATE AND EQUILIBRIUM

Finally, let's find a link between the material in this chapter and the equilibrium properties of reactions that we have been studying in the preceding chapters. At equilibrium, the rates of the forward and reverse reactions must be equal. Because the rates depend on rate constants and concentrations, we can suspect that there is a relation between rate constants for elementary reactions and equilibrium constants for the overall reaction.

The equilibrium constant for a chemical reaction that has the form $A + B \rightleftharpoons C + D$ is

$$K = \frac{[C][D]}{[A][B]}$$

Suppose that experiments show that both the forward reaction and the reverse reaction are elementary second-order reactions, with the following rate laws:

FIGURE 18.33 This flame front was caught during the miniature explosion that occurs inside an internal combustion engine every time a spark plug ignites gasoline vapor. This radical chain reaction occurs in automobile engines; in the explosion the products, which are hot gases, push a piston out, initiating a physical chain of events that ultimately moves the vehicle.

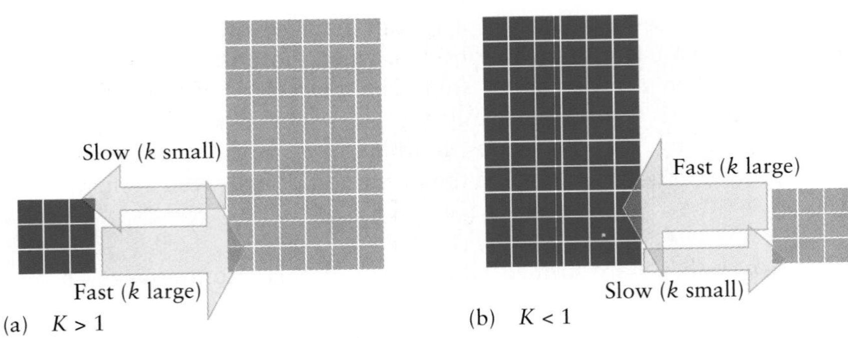

FIGURE 18.34 The equilibrium constant for a reaction is equal to the ratio of the rate constants for the forward and reverse elementary reactions that continue in a state of dynamic equilibrium. (a) A relatively large forward rate constant means that the forward rate can match the reverse rate even though only a small amount of reactants is present. (b) Conversely, if the reverse rate constant is relatively large, then the forward and reverse rates are equal when only small amounts of products are present.

$$\mathrm{A + B \longrightarrow C + D} \qquad \text{Rate} = k[\mathrm{A}][\mathrm{B}]$$

$$\mathrm{C + D \longrightarrow A + B} \qquad \text{Rate} = k'[\mathrm{C}][\mathrm{D}]$$

At equilibrium, these two rates are equal, so we can write

$$k[\mathrm{A}][\mathrm{B}] = k'[\mathrm{C}][\mathrm{D}]$$

It follows that at equilibrium

$$\frac{[\mathrm{C}][\mathrm{D}]}{[\mathrm{A}][\mathrm{B}]} = \frac{k}{k'}$$

Comparison of this expression with the expression for the equilibrium constant then shows that

$$K = \frac{k}{k'} \tag{7}$$

That is, *the equilibrium constant for a reaction is equal to the ratio of the rate constants for the forward and reverse elementary reactions that contribute to the overall reaction.*

We can now see when to expect a large equilibrium constant: K will be much larger than 1 when k for the forward direction is much larger than k' for the reverse direction. In this case, the fast forward reaction builds up a high concentration of products before reaching equilibrium (Fig. 18.34). In contrast, K is very small when k is much smaller than k'. Now the reverse reaction dismantles the products rapidly, and few are ever present.

Our expression for K in terms of rate constants also helps to explain one application of Le Chatelier's principle (Section 13.10). Recall that, according to Le Chatelier, an increase in temperature shifts the equilibrium composition in the direction of the endothermic reaction. If the forward reaction is endothermic, then the activation barrier will be higher for the forward direction than for the reverse direction (Fig. 18.35). The

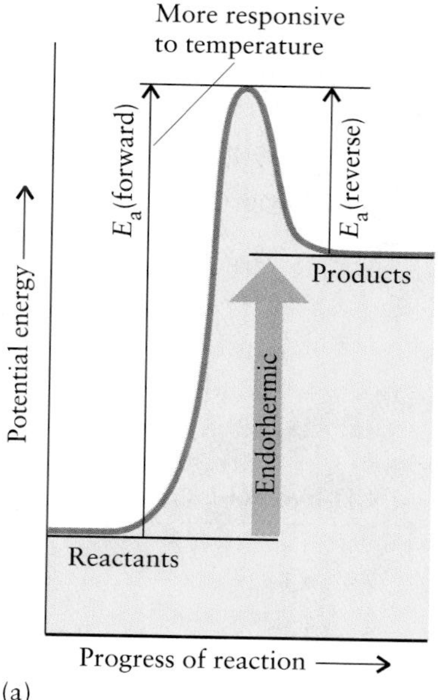

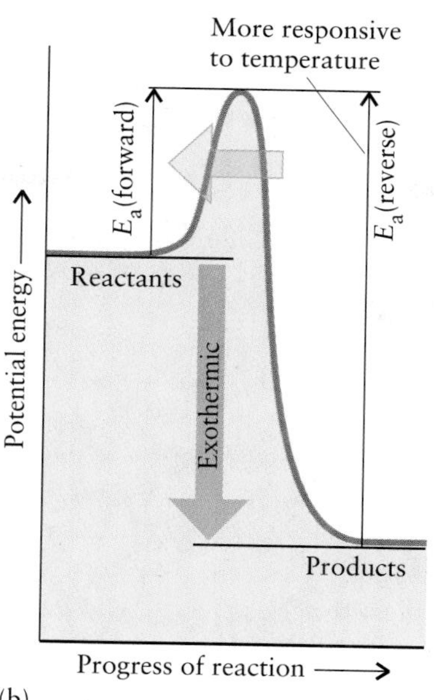

FIGURE 18.35 (a) The activation energy for an endothermic reaction is larger in the forward direction than in the reverse direction, so the rate of the forward reaction is more sensitive to temperature, and the equilibrium shifts toward products as the temperature is raised. (b) The opposite is true for an exothermic reaction, and the reverse reaction is more sensitive to temperature. In this case, the equilibrium shifts toward reactants as the temperature is raised.

higher activation energy means that the rate of the forward reaction depends more strongly on temperature than does the rate of the reverse reaction. Therefore, when the temperature is raised, the rate constant for the forward reaction increases more than that of the reverse reaction. As a result, K will increase and the products will become more favored, just as Le Chatelier's principle predicts.

The equilibrium constant for an elementary reaction is equal to the ratio of the forward and reverse rate constants of the reaction.

SKILLS YOU SHOULD HAVE MASTERED

Conceptual

1. Show how the rate of change of one species in a reaction is related to that of another species.
2. Show how the instantaneous rate is obtained by drawing a tangent to the graph of concentration against time.
3. Explain how collision theory and activated complex theory account for the temperature dependence of reactions.
4. Interpret a reaction profile.
5. Show how the equilibrium constant is related to the forward and reverse rate constants of the elementary reactions contributing to an overall reaction.

Problem-Solving

1. Determine the order of a reaction, its rate law, and its rate constant from experimental data.
2. Calculate a concentration or rate constant by using an integrated rate law.
3. For a first-order process, calculate the half-life, given the rate constant, and vice versa.
4. Use the Arrhenius equation and rate constants

measured at different temperatures to determine an activation energy.

5. Use the Arrhenius equation and the activation energy to find the rate constant at a given temperature.

6. Write rate laws for elementary reactions.

7. Deduce a rate law from a mechanism.

Descriptive

1. Distinguish enzymes and heterogeneous and homogeneous catalysts.

2. Describe the various steps in a chain reaction and explain how each affects the progress of the reaction.

EXERCISES

Reaction Rates

18.1 Complete the following statements relating to the production of ammonia by the Haber process, for which the overall reaction is $N_2(g) + 3\,H_2(g) \rightarrow 2\,NH_3(g)$.
(a) The rate of decomposition of N_2 is ______ times the rate of decomposition of H_2.
(b) The rate of formation of NH_3 is ______ times the rate of decomposition of H_2.
(c) The rate of formation of NH_3 is ______ times the rate of decomposition of N_2.

18.2 (a) In the reaction $3\,ClO^-(aq) \rightarrow 2\,Cl^-(aq) + ClO_3^-(aq)$, the rate of formation of Cl^- is 3.6 $mol{\cdot}L^{-1}{\cdot}min^{-1}$. What is the rate of reaction of ClO^-?
(b) In the Haber process for the industrial production of ammonia, $N_2(g) + 3\,H_2(g) \rightarrow 2\,NH_3(g)$, the rate of ammonia production is 2.7×10^{-3} $mol{\cdot}L^{-1}{\cdot}s^{-1}$. What is the rate of reaction of H_2?

18.3 (a) The rate of formation of O_2 is 1.5×10^{-3} $mol{\cdot}L^{-1}{\cdot}s^{-1}$ in the reaction $2\,O_3(g) \rightarrow 3\,O_2(g)$. What is the rate of decomposition of ozone? (b) The rate of formation of dichromate ions is 0.14 $mol{\cdot}L^{-1}{\cdot}s^{-1}$ in the reaction $2\,CrO_4^{2-}(aq) + 2\,H^+(aq) \rightarrow Cr_2O_7^{2-}(aq) + H_2O(l)$. What is the rate of reaction of chromate ions in the reaction?

18.4 (a) NO_2 decomposes at the rate of 6.5×10^{-3} $mol{\cdot}L^{-1}{\cdot}s^{-1}$ by the reaction $2\,NO_2(g) \rightarrow 2\,NO(g) + O_2(g)$. Determine the rate of formation of O_2. (b) Manganate ions, MnO_4^{2-}, form permanganate ions and manganese(IV) oxide in an acidic solution at a rate of 2.0 $mol{\cdot}L^{-1}{\cdot}min^{-1}$:

$$3\,MnO_4^{2-}(aq) + 4\,H^+(aq) \longrightarrow 2\,MnO_4^-(aq) + MnO_2(s) + 2\,H_2O(l)$$

What is the rate of formation of permanganate ions? What is the rate of reaction of $H^+(aq)$?

Rate Laws

18.5 The decomposition of gaseous dinitrogen pentoxide in the reaction $2\,N_2O_5(g) \rightarrow 4\,NO_2(g) + O_2(g)$ gives the following data at 298 K:

Time, h	$[N_2O_5]$, $mmol{\cdot}L^{-1}$
0	2.15
1.11	1.88
2.22	1.64
3.33	1.43
4.44	1.25

(a) Plot the concentration of N_2O_5 as a function of time. (b) Estimate the rate of decomposition of N_2O_5 at each time. (c) Plot the concentrations of NO_2 and O_2 as a function of time on the same graph.

18.6 The decomposition of gaseous hydrogen iodide, $2\,HI(g) \rightarrow H_2(g) + I_2(g)$, gives the following data at 700 K:

Time, s	[HI], $mmol{\cdot}L^{-1}$
0	10
1000	4.4
2000	2.8
3000	2.1
4000	1.6
5000	1.3

(a) Plot the concentration of HI as a function of time. (b) Estimate the rate of decomposition of HI at each time. (c) Plot the concentrations of H_2 and I_2 as a function of time on the same graph.

18.7 Write the rate laws for the following reactions:
(a) The reaction $X + 2\,Y \longrightarrow Z$ is second order in X and one-half order in Y.
(b) The reaction $2\,A + B \longrightarrow C$ is found to be first order in A, first order in B, and three-halves order overall.

18.8 Write the rate laws for the following reactions:
(a) The reaction $A + 2\,B + C \longrightarrow D + 4\,E$ is found to be first order in A, first order in B, and zero order in C.
(b) The reaction $2\,A + B \longrightarrow C + D$ is found to be first

order in A, one-half order in B, three-halves order in D, and zero order overall.

18.9 Express the units for rate constants when the concentrations are in moles per liter and time is in seconds for (a) zero-order reactions; (b) first-order reactions; (c) second-order reactions.

18.10 Because partial pressures are proportional to concentrations, rate laws for gas-phase reactions can also be expressed in terms of partial pressures, for instance, as rate $= kP_X$ for a first-order reaction of a gas X. What are the units for the rate constants when partial pressures are expressed in Torr and time is expressed in seconds for (a) zero-order reactions; (b) first-order reactions; (c) second-order reactions.

18.11 If $k = 1.35\ L^2{\cdot}mol^{-2}{\cdot}h^{-1}$, what is the order of the reaction?

18.12 If $k = 10\ Torr^{-1}{\cdot}s^{-1}$ at 298 K, what is the order of the reaction?

18.13 Dinitrogen pentoxide, N_2O_5, decomposes by a first-order reaction. What is the initial rate for the decomposition of N_2O_5 when 2.0 g of N_2O_5 is confined in a 1.0-L container and heated to 65°C (338 K)? From Table 18.1, $k = 5.2 \times 10^{-3}\ s^{-1}$ at 65°C.

18.14 Ethane, C_2H_6, decomposes to methyl radicals at 700°C (973 K) by a first-order reaction. If a 100-mg sample of ethane is confined to a 250-mL reaction vessel and heated to 700°C, what is the initial rate of ethane decomposition? From Table 18.1, $k = 5.5 \times 10^{-4}\ s^{-1}$ at 700°C.

18.15 A 0.15-g sample of H_2 and a 0.32-g sample of I_2 are confined to a 500-mL reaction vessel and heated to 700 K, when they react by a second-order process (first order in each reactant), with $k = 0.063\ L{\cdot}mol^{-1}{\cdot}s^{-1}$. (a) What is the initial reaction rate? (b) By what factor does the reaction rate increase if the concentration of H_2 present in the mixture is doubled?

18.16 A 100-mg sample of NO_2, confined to a 200-mL reaction vessel, is heated to 300°C. At that temperature, it decomposes by a second-order process, with $k = 0.54\ L{\cdot}mol^{-1}{\cdot}s^{-1}$. (a) What is the initial reaction rate? (b) How does the reaction rate change (and by what factor) if the mass of NO_2 present in the container is increased to 200 mg?

18.17 State the order of the reaction with respect to each species and the overall order of each of the following reactions:

(a) $H_2(g) + I_2(g) \longrightarrow 2\,HI(g)$ Rate $= k[H_2][I_2]$

(b) $2\,SO_2(g) + O_2(g) \xrightarrow{Pt} 2\,SO_3(g)$ Rate $= k[SO_2][SO_3]^{-1/2}$

(c) $A(g) + 2\,B(g) + C(g) \longrightarrow$ products, Rate $= k[A]^2[C]$

18.18 State the order of the reaction with respect to each reactant and the overall order of each of the following reactions:

(a) $S_2O_8^{2-}(aq) + 3\,I^-(aq) \longrightarrow 2\,SO_4^{2-}(aq) + I_3^-(aq)$ Rate $= k[S_2O_8^{2-}][I^-]$

(b) $2\,NO(g) + Cl_2(g) \longrightarrow 2\,NOCl(g)$ Rate $= k[NO]^2[Cl_2]$

(c) $A(g) + 3\,B(g) \longrightarrow$ products Rate $= k[A]^2[B]^{3/2}$

18.19 In the reaction $CH_3Br(aq) + OH^-(aq) \rightarrow CH_3OH(aq) + Br^-(aq)$, when the OH^- concentration alone was doubled, the rate doubled; when the CH_3Br concentration alone was increased by a factor of 1.2, the rate increased by 1.2. Write the rate law for the reaction.

18.20 In the reaction $2\,NO(g) + O_2(g) \rightarrow 2\,NO_2(g)$, when the NO concentration alone was doubled, the rate increased by a factor of 4; when both the NO and O_2 concentrations were increased by a factor of 2, the rate increased by a factor of 8. What is the rate law for the reaction?

18.21 The following kinetic data were obtained for the reaction $A(g) + 2\,B(g) \rightarrow$ product:

Experiment	Initial concentration, $mol{\cdot}L^{-1}$ $[A]_0$	$[B]_0$	Initial rate, $mol{\cdot}L^{-1}{\cdot}s^{-1}$
1	0.60	0.30	12.6
2	0.20	0.30	1.4
3	0.60	0.10	4.2
4	0.17	0.25	?

(a) What is the order with respect to each reactant and the overall order of the reaction? (b) Write the rate law for the reaction. (c) From the data, determine the value of the rate constant. (d) Use the data to predict the reaction rate for Experiment 4.

18.22 The following kinetic data were obtained for the reaction $3\,A_2B(g) + CX_3(g) \rightarrow$ product:

Experiment	Initial concentration, $mol{\cdot}L^{-1}$ $[A_2B]_0$	$[CX_3]_0$	Initial rate, $mol{\cdot}L^{-1}{\cdot}s^{-1}$
1	1.72	2.44	0.68
2	3.44	2.44	5.44
3	1.72	0.10	2.8×10^{-2}
4	2.91	1.33	?

(a) What is the order with respect to each reactant and the overall order of the reaction? (b) Write the rate law for the reaction. (c) From the data, determine the value of the rate constant. (d) Use the data to predict the reaction rate for Experiment 4.

18.23 The following kinetic data were obtained for the reaction $2\,ICl(g) + H_2(g) \rightarrow I_2(g) + 2\,HCl(g)$:

Experiment	Initial concentration, $mmol \cdot L^{-1}$ $[ICl]_0$	$[H_2]_0$	Initial rate, $mol \cdot L^{-1} \cdot s^{-1}$
1	1.5	1.5	3.7×10^{-7}
2	3.0	1.5	7.4×10^{-7}
3	3.0	4.5	22×10^{-7}
4	4.7	2.7	?

(a) Write the rate law for the reaction. (b) From the data, determine the value of the rate constant. (c) Use the data to predict the reaction rate for Experiment 4.

18.24 The following kinetic data were obtained for the reaction $NO_2(g) + O_3(g) \rightarrow NO_3(g) + O_2(g)$:

Experiment	Initial concentration, $mmol \cdot L^{-1}$ $[NO_2]_0$	$[O_3]_0$	Initial rate, $mmol \cdot L^{-1} \cdot s^{-1}$
1	0.21	0.70	6.3
2	0.21	1.39	12.5
3	0.38	0.70	11.4
4	0.66	0.18	?

(a) Write the rate law for the reaction. (b) What is the order of the reaction? (c) From the data, determine the value of the rate constant. (d) Use the data to predict the reaction rate for Experiment 4.

18.25 The following data were obtained for the reaction $A + B + C \rightarrow$ products:

Experiment	Initial concentration, $mmol \cdot L^{-1}$ $[A]_0$	$[B]_0$	$[C]_0$	Initial rate, $mmol \cdot L^{-1} \cdot s^{-1}$
1	1.25	1.25	1.25	8.7
2	2.50	1.25	1.25	17.4
3	1.25	3.02	1.25	50.8
4	1.25	3.02	3.75	457
5	3.01	1.00	1.15	?

(a) Write the rate law for the reaction. (b) What is the order of the reaction? (c) Determine the value of the rate constant. (d) Use the data to predict the reaction rate for Experiment 5.

18.26 For the reaction $A + 2\,B + C \rightarrow$ products, the following data were collected:

Experiment	Initial concentration, $mmol \cdot L^{-1}$ $[A]_0$	$[B]_0$	$[C]_0$	Initial rate, $mmol \cdot L^{-1} \cdot s^{-1}$
1	2.06	3.05	4.00	3.7
2	0.87	3.05	4.00	0.66
3	0.50	0.50	0.50	0.013
4	1.00	0.50	1.00	0.072
5	1.60	2.00	3.00	?

(a) Write the rate law for the reaction. (b) What is the order of the reaction? (c) Determine the value of the rate constant.

Integrated Rate Laws

18.27 Determine the rate constants for the following first-order reactions: (a) $A \rightarrow B$, given that the concentration of A decreases to one-half its initial value in 1000 s. (b) $A \rightarrow B$, given that the concentration of A decreases from $0.33\ mol \cdot L^{-1}$ to $0.14\ mol \cdot L^{-1}$ in 47 s. (c) $2\,A \rightarrow B + C$, given that $[A]_0 = 0.050\ mol \cdot L^{-1}$ and that after 120 s the concentration of B rises to $0.015\ mol \cdot L^{-1}$.

18.28 Determine the rate constants for the following first-order reactions: (a) $2\,A \rightarrow B + C$, given that the concentration of A decreases to one-third its initial value in 25 min. (b) $2\,A \rightarrow B + C$, given that $[A]_0 = 0.020\ mol \cdot L^{-1}$ and that after 1.3 h the concentration of B increases to $0.006\ mol \cdot L^{-1}$. (c) $2\,A \rightarrow 3\,B + C$, given that $[A]_0 = 0.050\ mol \cdot L^{-1}$ and that after 7.7 min the concentration of B rises to $0.050\ mol \cdot L^{-1}$.

18.29 Dinitrogen pentoxide, N_2O_5, decomposes by first-order kinetics with a rate constant of $3.7 \times 10^{-5}\ s^{-1}$ at 298 K. (a) What is the half-life (in hours) for the decomposition of N_2O_5 at 298 K? (b) If $[N_2O_5]_0 = 2.33 \times 10^{-2}\ mol \cdot L^{-1}$, what will be the concentration of N_2O_5 after 2.0 h? (c) How much time (in minutes) will elapse before the N_2O_5 concentration decreases from $2.33 \times 10^{-2}\ mol \cdot L^{-1}$ to $1.76 \times 10^{-2}\ mol \cdot L^{-1}$?

18.30 Dinitrogen pentoxide, N_2O_5, decomposes by first-order kinetics with a rate constant of $0.15\ s^{-1}$ at 353 K. (a) What is the half-life (in seconds) for the decomposition of N_2O_5 at 353 K? (b) If $[N_2O_5]_0 = 2.33 \times 10^{-2}\ mol \cdot L^{-1}$, what will be the concentration of N_2O_5 after 2.0 s? (c) How much time in minutes will elapse before the N_2O_5 concentration decreases from $2.33 \times 10^{-2}\ mol \cdot L^{-1}$ to $1.76 \times 10^{-2}\ mol \cdot L^{-1}$?

18.31 The half-life for the first-order decomposition of A is 200 s. How much time must elapse for the concentration of A to decrease to (a) one-half; (b) one-sixteenth; (c) one-ninth of its initial concentration?

18.32 The first-order rate constant for the photodissociation of A is 0.173 s^{-1}. Calculate the time needed for the concentration of A to decrease to (a) one-fourth; (b) one-thirty-second; (c) one-fifth of its initial concentration.

18.33 Sulfuryl chloride, SO_2Cl_2, decomposes by first-order kinetics, and $k = 2.81 \times 10^{-3}\ min^{-1}$ at a certain temperature. (a) Determine the half-life for the reaction. (b) Determine the time needed for the concentration of a SO_2Cl_2 sample to decrease to 10% of its initial concentration. (c) If a 14.0-g sample of SO_2Cl_2 is sealed in a 2500-L reaction vessel and heated to the specified temperature, what mass will remain after 1.5 h?

18.34 Ethane, C_2H_6, forms $\cdot CH_3$ radicals at 700°C in a first-order reaction, for which $k = 1.98\ h^{-1}$. (a) What is the half-life for the reaction? (b) Calculate the time needed for the amount of ethane to fall from 1.15×10^{-3} mol to 2.35×10^{-4} mol in a 500-mL reaction vessel at 700°C. (c) How much of a 6.88-mg sample of ethane in a 500-mL reaction vessel at 700°C will remain after 45 min?

18.35 Substance A forms B in the first-order reaction $A \rightarrow 2\,B$, in which the concentration of A falls to 20% of its original concentration in 120 s. (a) What is the rate constant for the reaction? (b) Determine the time in which the concentration of A falls to 10% of its original concentration.

18.36 In the first-order reaction $2\,A \rightarrow B + C$, it was observed that the initial concentration of A decreased to 80% of its original concentration in 175 min. How much time would be needed for it to decrease to 20% of its original concentration?

18.37 For the first-order reaction $A \rightarrow 3\,B + C$, when $[A]_0 = 0.015\ mol\cdot L^{-1}$ the concentration of B increases to $0.020\ mol\cdot L^{-1}$ in 3.0 min. (a) What is the rate constant for the reaction? (b) How much more time would be needed for the concentration of B to increase to $0.040\ mol\cdot L^{-1}$?

18.38 In the first-order reaction $A \rightarrow 2\,B$, it was observed that when the initial concentration of A was 0.020 $mol\cdot L^{-1}$, the concentration of B rose to $0.020\ mol\cdot L^{-1}$ in 75 s. How much more time would be needed for it to rise to $0.030\ mol\cdot L^{-1}$?

18.39 From a set of kinetic data, a plot of the logarithm of the concentration of a reactant as a function of time yielded a straight line with a negative slope. What is the order of the reaction?

18.40 From a set of kinetic data, a plot of the reciprocal concentration of a reactant as a function of time yielded a straight line with a positive slope. What is the order of the reaction?

18.41 The following data were collected for the reaction $2\,N_2O_5(g) \rightarrow 4\,NO_2(g) + O_2(g)$ at 25°C:

Time, s	$[N_2O_5]$, $10^{-3}\ mol\cdot L^{-1}$
0	2.15
4000	1.88
8000	1.64
12,000	1.43
16,000	1.25

(a) Plot the data to confirm that the reaction is first order. (b) From the graph, determine the rate constant.

18.42 The following data were collected for the reaction $C_2H_6(g) \rightarrow 2\,CH_3(g)$ at 700°C:

Time, s	$[C_2H_6]$, $10^{-3}\ mol\cdot L^{-1}$
0	1.59
1000	0.92
2000	0.53
3000	0.31
4000	0.18
5000	0.10

(a) Plot the data to confirm that the reaction is first order. (b) From the graph, determine the rate constant.

18.43 The following data were collected for the reaction $2\,HI(g) \rightarrow H_2(g) + I_2(g)$ at 580 K:

Time, s	[HI], $10^{-3}\ mol\cdot L^{-1}$
0	1000
1000	112
2000	61
3000	41
4000	31

(a) Plot the data to confirm that the rate law is rate $= k[HI]^2$. (b) From the graph, determine the rate constant.

18.44 The following data were collected for the reaction $H_2(g) + I_2(g) \rightarrow 2\,HI(g)$ at 780 K:

Time, s	$[I_2]$, $10^{-3}\ mol\cdot L^{-1}$
0	1.00
1	0.43
2	0.27
3	0.20
4	0.16

(a) Plot the data to confirm that the reaction is second order. (b) From the graph, determine the rate constant.

Collision Theory and Arrhenius Behavior

18.45 Explain why the rate constant increases with temperature, and justify the form of the Arrhenius equation.

18.46 Which reaction has a rate that is more sensitive to temperature, one with a high activation energy or one with a low activitation energy?

18.47 (a) Calculate the activation energy for the conversion of cyclopropane to propene from an Arrhenius plot of the following data:

T, K	*k*, s^{-1}
750	1.8×10^{-4}
800	2.7×10^{-3}
850	3.0×10^{-2}
900	0.26

(b) What is the value of the rate constant at 600°C?

18.48 (a) Determine the activation energy for $C_2H_5I(g) \rightarrow C_2H_4(g) + HI(g)$ from an Arrhenius plot of the following data:

T, K	*k*, s^{-1}
660	7.2×10^{-4}
680	2.2×10^{-3}
720	1.7×10^{-2}
760	0.11

(b) What is the value of the rate constant at 400°C?

18.49 The rate constant of the first-order reaction $2\,N_2O(g) \rightarrow 2\,N_2(g) + O_2(g)$ is $0.38\ s^{-1}$ at 1000 K and $0.87\ s^{-1}$ at 1030 K. Calculate the activation energy of the reaction.

18.50 The rate constant of the second-order reaction $2\,HI(g) \rightarrow H_2(g) + I_2(g)$ is $2.4 \times 10^{-6}\ L \cdot mol^{-1} \cdot s^{-1}$ at 575 K and $6.0 \times 10^{-5}\ L \cdot mol^{-1} \cdot s^{-1}$ at 630 K. Calculate the activation energy of the reaction.

18.51 The rate constant of the reaction $O(g) + N_2(g) \rightarrow NO(g) + N(g)$, which occurs in the stratosphere, is $9.7 \times 10^{10}\ L \cdot mol^{-1} \cdot s^{-1}$ at 800°C. The activation energy of the reaction is $315\ kJ \cdot mol^{-1}$. Determine the rate constant at 700°C.

18.52 The rate constant of the reaction between CO_2 and OH^- in aqueous solution to give the HCO_3^- ion is $1.5 \times 10^{10}\ L \cdot mol^{-1} \cdot s^{-1}$ at 25°C. Determine the rate constant at blood temperature (37°C), given that the activation energy for the reaction is $38\ kJ \cdot mol^{-1}$.

18.53 The rate constant for the decomposition of N_2O_5 at 45°C is $k = 5.1 \times 10^{-4}\ s^{-1}$. From Table 18.2, the activation energy for the reaction is $103\ kJ \cdot mol^{-1}$. Determine the value of the rate constant at 50°C.

18.54 Ethane, C_2H_6, dissociates into methyl radicals at 700°C with a rate constant, $k = 5.5 \times 10^{-4}\ s^{-1}$. Determine the rate constant at 800°C, given that the activation energy of the reaction is $384\ kJ \cdot mol^{-1}$.

Catalysis

18.55 The presence of a catalyst provides a reaction pathway in which the activation energy of a certain reaction is reduced from $100\ kJ \cdot mol^{-1}$ to $50\ kJ \cdot mol^{-1}$. By what factor does the rate of the reaction increase at 400 K, all other factors being equal?

18.56 The presence of a catalyst provides a reaction pathway in which the activation energy of a certain reaction is reduced from $88\ kJ \cdot mol^{-1}$ to $62\ kJ \cdot mol^{-1}$. By what factor does the rate of the reaction increase at 300 K, all other factors being equal?

18.57 A reaction rate increases by a factor of 1000 in the presence of a catalyst at 25°C. The activation energy of the original pathway is $98\ kJ \cdot mol^{-1}$. What is the activation energy of the new pathway, all other factors being equal?

18.58 A reaction rate increases by a factor of 500 in the presence of a catalyst at 37°C. The activation energy of the original pathway is $106\ kJ \cdot mol^{-1}$. What is the activation energy of the new pathway, all other factors being equal?

Reaction Mechanisms

18.59 Each of the following steps is an elementary reaction. Write its rate law and indicate its molecularity: (a) $2\,NO \rightarrow N_2O_2$; (b) $Cl_2 \rightarrow 2\,Cl$; (c) $2\,NO_2 \rightarrow NO + NO_3$. (d) Which of these reactions might be radical chain initiating?

18.60 Each of the following is an elementary reaction. Write its rate law and indicate its molecularity: (a) $O + CF_2Cl_2 \rightarrow ClO + CF_2Cl$; (b) $OH + NO_2 + N_2 \rightarrow HNO_3 + N_2$ (The N_2 takes part in a three-body collision, but is left unchanged chemically.); (c) $ClO^- + H_2O \rightarrow HClO + OH^-$. (d) Which of these reactions might be radical chain propagating?

18.61 The contribution to the destruction of the ozone layer caused by high-flying aircraft has been attributed to the following mechanism:

$$\text{Step 1:}\quad O_3 + NO \longrightarrow NO_2 + O_2$$
$$\text{Step 2:}\quad NO_2 + O \longrightarrow NO + O_2$$

What is the reaction intermediate?

18.62 A reaction was believed to occur by the following mechanism:

Step 1: $A_2 \longrightarrow 2\,A$, and its reverse, $2\,A \longrightarrow A_2$
Step 2: $2\,A + B \longrightarrow C + 2\,D$
Step 3: $D + D \longrightarrow A_2 + E$

(a) What is the overall reaction? (b) Which species are the reaction intermediates?

18.63 Write the overall reaction for the mechanism proposed below and identify any reaction intermediates:

Step 1: $ICl + H_2 \longrightarrow HI + HCl$
Step 2: $HI + ICl \longrightarrow HCl + I_2$

18.64 Write the overall reaction for the mechanism proposed below and identify any reaction intermediates:

Step 1: $Cl_2 \longrightarrow 2\,Cl$
Step 2: $Cl + CO \longrightarrow COCl$
Step 3: $COCl + Cl_2 \longrightarrow COCl_2 + Cl$

18.65 The following mechanism has been proposed for the reaction between nitric oxide and bromine:

Step 1: $NO + Br_2 \longrightarrow NOBr_2$ (slow)
Step 2: $NOBr_2 + NO \longrightarrow 2\,NOBr$ (fast)

Write the rate law implied by this mechanism.

18.66 The following mechanism has been proposed for the reaction between chlorine and chloroform, $CHCl_3$:

Step 1: $Cl_2 \longrightarrow 2\,Cl$, and its reverse, $2\,Cl \longrightarrow Cl_2$ (both fast, equilibrium)
Step 2: $CHCl_3 + Cl \longrightarrow CCl_3 + HCl$ (slow)
Step 3: $CCl_3 + Cl \longrightarrow CCl_4$ (fast)

Write the rate law implied by this mechanism.

18.67 The production of phosgene, $COCl_2$, from carbon monoxide and chlorine is believed to take place by the following mechanism:

Step 1: $Cl_2 \longrightarrow 2\,Cl$, and its reverse $2\,Cl \longrightarrow Cl_2$ (both fast, equilibrium)
Step 2: $Cl + CO \longrightarrow COCl$, and its reverse $COCl \longrightarrow CO + Cl$ (both fast, equilibrium)
Step 3: $COCl + Cl_2 \longrightarrow COCl_2 + Cl$ (slow)

Write the rate law implied by this mechanism.

18.68 The mechanism proposed for the oxidation of iodide ion by the hypochlorite ion in aqueous solution is as follows:

Step 1: $ClO^- + H_2O \longrightarrow HClO + OH^-$, and its reverse $HClO + OH^- \longrightarrow ClO^- + H_2O$ (both fast, equilibrium)
Step 2: $I^- + HClO \longrightarrow HIO + Cl^-$ (slow)
Step 3: $HIO + OH^- \longrightarrow IO^- + H_2O$ (fast)

Write the rate law implied by this mechanism.

18.69 Three mechanisms for the reaction, $NO_2(g) + CO(g) \rightarrow CO_2(g) + NO(g)$ have been proposed:

(a) Step 1: $NO_2 + CO \longrightarrow CO_2 + NO$
(b) Step 1: $NO_2 + NO_2 \longrightarrow NO + NO_3$ (slow)
Step 2: $NO_3 + CO \longrightarrow NO_2 + CO_2$ (fast)
(c) Step 1: $NO_2 + NO_2 \longrightarrow NO + NO_3$, and its reverse (both fast, equilibrium)
Step 2: $NO_3 + CO \longrightarrow NO_2 + CO_2$ (slow)

Which mechanism agrees with the following rate law: rate $= k[NO_2]^2$? Explain your reasoning.

18.70 When the rate of the reaction $2\,NO(g) + O_2(g) \rightarrow 2\,NO_2(g)$ was studied, it was found that the rate doubled when the O_2 concentration alone was doubled, but quadrupled when the NO concentration alone was doubled. Which of the following mechanisms accounts for these observations? Explain your reasoning.

(a) Step 1: $NO + O_2 \longrightarrow NO_3$, and its reverse (both fast, equilibrium)
Step 2: $NO + NO_3 \longrightarrow 2\,NO_2$ (slow)
(b) Step 1: $2\,NO \longrightarrow N_2O_2$ (slow)
Step 2: $O_2 + N_2O_2 \longrightarrow N_2O_4$ (fast)
Step 3: $N_2O_4 \longrightarrow 2\,NO_2$ (fast)

Rate Constants and Equilibrium

18.71 Explain why the following statements about an elementary reaction are wrong. (a) At equilibrium, the rate constants of the forward and reverse reactions are equal. (b) For a reaction with a very large K_c, the rate constant of the reverse reaction is much larger than the rate constant of the forward reaction.

18.72 Explain why the following statements about an elementary reaction are wrong. (a) The equilibrium constant for a reaction equals the ratio of the forward and reverse rates. (b) For an exothermic process, the rates of the forward and reverse reactions are affected in the same way by a rise in temperature.

SUPPLEMENTARY EXERCISES

18.73 The rate of consumption of 1-chlorobutane, C_4H_9Cl, in a reaction with water is 1.90×10^{-4} mol·L^{-1}·s^{-1}. What is the rate of formation of hydrochloric acid in the system, given that the chemical equation for the reaction is

$$C_4H_9Cl(aq) + H_2O(l) \longrightarrow C_4H_9OH(aq) + HCl(aq)$$

18.74 When the concentration of 2-bromo-2-methylpropane, C_4H_9Br, is doubled, the rate of the reaction

$$C_4H_9Br(aq) + OH^-(aq) \longrightarrow C_4H_9OH(aq) + Br^-(aq)$$

increases by a factor of 2. When the C_4H_9Br and OH^- concentrations are both doubled, the rate increase is the same, a factor of 2. What are the reactant orders and overall order of the reaction?

18.75 The following data were collected for the reaction $2\,N_2O_5(g) \rightarrow 4\,NO_2(g) + O_2(g)$ at 308 K:

Time, 10^3 s	$[N_2O_5]$, mmol·L^{-1}
0	2.57
4.0	1.50
8.0	0.87
12.0	0.51
16.0	0.30

(a) Plot the $[N_2O_5]$ concentration as a function of time. (b) Estimate the rate of decomposition of N_2O_5 at each time. (c) Plot the $[NO_2]$ and $[O_2]$ concentrations as a function of time on the same graph used in (a).

18.76 The rate law for the reaction $H_2SeO_3(aq) + 6\,I^-(aq) + 4\,H^+(aq) \rightarrow Se(s) + 2\,I_3^-(aq) + 3\,H_2O(l)$ is rate $= k[H_2SeO_3][I^-]^3[H^+]^2$ with $k = 5.0 \times 10^5\ L^5·mol^{-5}·s^{-1}$. What is the initial reaction rate when $[H_2SeO_3]_0 = [I^-]_0 = 0.020$ mol·L^{-1} and $[H^+]_0 = 0.010$ mol·L^{-1}?

18.77 A reaction of the form $3\,A + B \rightarrow$ products was determined to be first order in A and one-half order in B. (a) What is the rate law for the reaction? (b) What are the units of the rate constant? Assume that the concentrations are expressed in mol·L^{-1} and time in minutes.

18.78 The following rate data were collected for the decomposition of acetaldehyde, CH_3CHO, in the reaction $CH_3CHO(g) \rightarrow CH_4(g) + CO(g)$ at a certain temperature:

Experiment	Initial concentration, mol·L^{-1} $[CH_3CHO]_0$	Initial rate, (mol CO)·L^{-1}·s^{-1}
1	0.10	9.0×10^{-7}
2	0.20	36×10^{-7}
3	0.30	81×10^{-7}
4	0.40	144×10^{-7}
5	0.73	?

(a) Write the rate law for the reaction. (b) What is the overall order of the reaction? (c) Determine the reaction rate constant. (d) Predict the reaction rate for Experiment 5.

18.79 The following rate data were collected for the reaction $2\,A(g) + 2\,B(g) + C(g) \rightarrow 3\,G(g) + 4\,F(g)$:

Experiment	Initial concentration, mmol·L^{-1}			Initial rate, (mmol G)·L^{-1}·s^{-1}
	$[A]_0$	$[B]_0$	$[C]_0$	
1	10	100	700	2.0
2	20	100	300	4.0
3	20	200	200	16
4	10	100	400	2.0
5	4.62	0.177	12.4	?

(a) What is the order for each reactant and the overall order of the reaction? (b) Write the rate law for the reaction. (c) Determine the reaction rate constant. (d) Predict the reaction rate for Experiment 5.

18.80 (a) The rate law for the thermal decomposition of acetaldehyde, $CH_3CHO(g) \rightarrow CH_4(g) + CO(g)$, under certain conditions is rate $= k[CH_3CHO]^{3/2}$. What is the order of the reaction? (b) The rate law for the "hot wire" decomposition of ammonia, $2\,NH_3(g) \rightarrow N_2(g) + 3\,H_2(g)$, is rate $= k$. What is the order of the reaction?

18.81 The decomposition of hydrogen peroxide, $2\,H_2O_2(aq) \rightarrow 2\,H_2O(l) + O_2(g)$, follows first-order kinetics with respect to H_2O_2 and has $k = 0.0410\ min^{-1}$. (a) If the initial concentration of H_2O_2 is 0.20 mol·L^{-1}, what is its concentration after 10 min? (b) How much time will it take for the H_2O_2 concentration to decrease from 0.50 mol·L^{-1} to 0.10 mol·L^{-1}? (c) How much time is needed for the H_2O_2 concentration to decrease by one-fourth? (d) Calculate the time needed for the H_2O_2 concentration to decrease by 75%.

18.82 (a) For the second-order reaction, $A \rightarrow B + C$, the concentration of the species A falls from 0.040 mol·L^{-1} to 0.0050 mol·L^{-1} in 12 h. What is the reaction rate constant? (b) In the second-order reaction $CX_2 \rightarrow C + 2\,X$, the concentration of X increases to 0.070 mol·L^{-1} in 15 h when $[CX_2]_0 = 0.040$ mol·L^{-1}. What is the reaction rate constant for the decomposition of CX_2?

18.83 The half-life for the (first-order) decomposition of azomethane, $CH_3N{=}NCH_3$, in the reaction $CH_3N{=}NCH_3(g) \rightarrow N_2(g) + C_2H_6(g)$ is 1.02 s at 300°C. A 45.0-mg sample of azomethane is placed in a 300-mL reaction vessel and heated to 300°C. (a) What mass (in milligrams) of azomethane remains after 10 s? (b) Determine the partial pressure exerted by the $N_2(g)$ in the reaction vessel after 3.0 s.

18.84 The mechanism of the reaction $A \rightarrow B$ consists of two steps, involving the formation of a reaction intermediate (represented by the dip in the curve). Overall, the reaction is exothermic. (a) Sketch the reaction profile, labeling the activation energies for each step and the overall enthalpy of reaction. (b) Indicate on the same diagram the effect of a catalyst on the first step of the reaction.

18.85 An exothermic reaction for which $\Delta H_r = -200\ kJ\cdot mol^{-1}$ has an activation energy of $100\ kJ\cdot mol^{-1}$. Estimate the activation energy for the reverse reaction. For estimates, reaction enthalpies can be treated as energies.

18.86 All radioactive decay processes follow first-order kinetics. The half-life of the radioactive isotope, tritium (^3H or T), is 12.3 years. How much of a 1.0-μg sample of tritium would remain after 5.2 years?

18.87 The concentration of a species A was initially $0.20\ mol\cdot L^{-1}$ and decreased by the reaction $2\,A \rightarrow B + C$ to $0.10\ mol\cdot L^{-1}$ in 100 s by first-order kinetics. Calculate the time needed for the concentration of A to fall to (a) one-eighth and (b) one-thirty-second of its initial concentration.

18.88 Why might a bimolecular collision with more than the minimum energy required for reaction not produce products?

18.89 Explain why a finely divided solid catalyst is more effective in increasing a reaction rate than the same mass of the catalyst in large pellets.

18.90 Which of the following plots will be linear? (a) [A] as a function of time for a reaction first order in A. (b) [A] as a function of time for a reaction zero order in A. (c) ln[A] as a function of time for a reaction first order in A. (d) 1/[A] as a function of time for a reaction second order in A. (e) ln k against temperature. (f) Initial rate as a function of [A] for a reaction first order in A.

18.91 Determine the molecularity of the following elementary reactions:
(a) $CH_3Br + OH^- \longrightarrow CH_3OH + Br^-$
(b) $C_2N_2 \longrightarrow 2\,CN\cdot$
(c) $Ar + 2\,O \longrightarrow Ar + O_2$ (the argon atom takes part in a collision but is left unchanged)

18.92 The rate law of the reaction $2\,NO(g) + 2\,H_2(g) \rightarrow N_2(g) + 2\,H_2O(g)$ is rate $= k[NO]^2[H_2]$, and the mechanism that has been proposed is

$$\begin{aligned} &\text{Step 1:}\quad 2\,NO \rightleftharpoons N_2O_2 \\ &\text{Step 2:}\quad N_2O_2 + H_2 \longrightarrow N_2O + H_2O \\ &\text{Step 3:}\quad N_2O + H_2 \longrightarrow N_2 + H_2O \end{aligned}$$

(a) Which step in the mechanism is likely to be rate determining? Explain your answer. (b) Sketch a reaction profile for the (exothermic) overall reaction. Label the activation energies of each step and the overall reaction enthalpy.

18.93 The old industrial process for the production of sulfuric acid was called the "lead chamber process." A greatly oversimplified mechanism of the formation of sulfur trioxide, SO_3, the acid anhydride of sulfuric acid, is

$$\begin{aligned} &\text{Step 1:}\quad O_2 + 2\,NO \longrightarrow 2\,NO_2 \\ &\text{Step 2:}\quad NO_2 + SO_2 \longrightarrow SO_3 + NO \end{aligned}$$

What is the reaction intermediate?

18.94 The activation energy of the reaction $H_2(g) + I_2(g) \rightarrow 2\,HI(g)$ is reduced from $184\ kJ\cdot mol^{-1}$ to $59\ kJ\cdot mol^{-1}$ in the presence of a platinum catalyst. By what factor will the reaction rate be increased by the platinum at 600 K, all other factors being equal?

18.95 The activation energy of the decomposition of ammonia to its elements is reduced from $350\ kJ\cdot mol^{-1}$ to $162\ kJ\cdot mol^{-1}$ in the presence of a tungsten catalyst. By what factor will the rate be increased at 700°C?

18.96 (a) Calculate the activation energy for the acid hydrolysis of sucrose from an Arrhenius plot of the following data:

Temperature, °C	k, $L\cdot mol^{-1}\cdot s^{-1}$
24	4.8×10^{-3}
28	7.8×10^{-3}
32	13×10^{-3}
36	20×10^{-3}
40	32×10^{-3}

(b) Calculate the rate constant at 37°C (body temperature).

18.97 (a) Calculate the activation energy for the reaction between ethyl bromide, C_2H_5Br, and hydroxide ions in water from an Arrhenius plot of the following data:

Temperature, °C	k, $L\cdot mol^{-1}\cdot s^{-1}$
24	1.3×10^{-3}
28	2.0×10^{-3}
32	3.0×10^{-3}
36	4.4×10^{-3}
40	6.4×10^{-3}

(b) Calculate the rate constant at 25°C.

18.98 Raw milk sours in about 4 h at 28°C, but in about 48 h in a refrigerator at 5°C. What is the activation energy for the souring of milk?

18.99 The following mechanism has been suggested to explain the contribution of chlorofluorocarbons to the destruction of the ozone layer:

Step 1: $O_3 + Cl \longrightarrow ClO + O_2$
Step 2: $ClO + O \longrightarrow Cl + O_2$

(a) What is the reaction intermediate and what is the catalyst? (b) Identify the radicals in the mechanism. (c) Identify the steps as initiating, propagating, or terminating. (d) Write a chain-terminating step for the reaction.

18.100 When a 1:1 mole ratio mixture of hydrogen and chlorine is exposed to sunlight, the reaction $H_2(g) + Cl_2(g) \rightarrow 2\,HCl(g)$ occurs explosively. The chain-reaction mechanism is thought to be

Step 1:	$Cl_2 \longrightarrow 2\,Cl$, a chain-initiation step caused by light
Step 2:	a chain-propagation step that results in the formation of a hydrogen atom
Step 3:	a chain-propagation step in which the hydrogen atom reacts with Cl_2
Steps 4, 5, 6:	three chain-termination steps that occur between the two radicals

Write the equations for the elementary reactions in steps 2 to 6.

CHALLENGING EXERCISES

18.101 At 328 K, the *total* pressure of the dinitrogen pentoxide decomposition to NO_2 and O_2 varied with time as shown by the following data. Use the data to find the rate in $mol \cdot L^{-1} \cdot min^{-1}$ at each time.

Time, min	Pressure, kPa
0	27.3
5	43.7
10	53.6
15	59.4
20	63.0
30	66.3

18.102 Determine the rate constant for the following second-order reactions. (a) $2\,A \rightarrow B + 2\,C$, given that the concentration of A decreases from $2.5 \times 10^{-3}\ mol \cdot L^{-1}$ to $1.25 \times 10^{-3}\ mol \cdot L^{-1}$ in 100 s. (b) $3\,A \rightarrow C + 2\,D$, given that $[A]_0 = 0.30\ mol \cdot L^{-1}$, and that the concentration of C increases to $0.010\ mol \cdot L^{-1}$ in 200 s.

18.103 The half-life for the second-order reaction of a substance A is 50.5 s when $[A]_0 = 0.84\ mol \cdot L^{-1}$. Calculate the time needed for the concentration of A to decrease to (a) one-sixteenth; (b) one-fourth; (c) one-fifth of its original value.

18.104 The second-order rate constant for the decomposition of NO_2 (to NO and O_2) at 573 K is $0.54\ L \cdot mol^{-1} \cdot s^{-1}$. Calculate the time for an initial NO_2 concentration of $0.20\ mol \cdot L^{-1}$ to decrease to (a) one-half; (b) one-sixteenth; (c) one-ninth of its initial concentration.

18.105 Determine the time required for each of the following second-order reactions to occur. (a) $2\,A \rightarrow B + C$, for the concentration of A to decrease from $0.10\ mol \cdot L^{-1}$ to $0.080\ mol \cdot L^{-1}$, given that $k = 0.010\ L \cdot mol^{-1} \cdot min^{-1}$. (b) $A \rightarrow 2\,B + C$, when $[A]_0 = 0.45\ mol \cdot L^{-1}$, for the concentration of B to increase to $0.45\ mol \cdot L^{-1}$, given that $k = 0.0045\ L \cdot mol^{-1} \cdot min^{-1}$.

18.106 The following data were obtained for the reaction $2\,A \rightarrow B$:

Time, s	[A], $10^{-3}\ mol \cdot L^{-1}$
0	100
5	14.1
10	7.8
15	5.3
20	4.0

(a) Plot the data to determine the order of the reaction. (*Hint:* The integrated rate law that gives a linear plot is the one that corresponds to the reaction order.) (b) Determine the rate constant.

18.107 The following data were obtained on the reaction $2\,A \rightarrow B$:

Time, s	[A], $10^{-3}\ mol \cdot L^{-1}$
0	250
100	143
200	81
300	45
400	25

(a) Plot the data to determine the order of the reaction. (*Hint:* The integrated rate law that gives a linear plot is the one that corresponds to the reaction order.) (b) Determine the rate constant.

18.108 The radioactive decay of carbon-14 is first order, and the half-life is 5800 years. While a plant or animal is living, it has a constant proportion of carbon-14 (relative to carbon-12) in its composition. When the organism dies, the proportion of carbon-14 decreases as a result of radioactive decay, and the age of the organism can be determined if the proportion of carbon-14 in its remains is measured. If the proportion of carbon-14 in an ancient piece of wood is found to be one-fourth that in living trees, how old is the sample?

18.109 The half-life of a substance taking part in a third-order reaction A → products is inversely proportional to the square of the initial concentration of A. How can this half-life be used to predict the time needed for the concentration to fall to (a) one-half; (b) one-fourth; (c) one-sixteenth of its initial value?

18.110 Under certain conditions, the reaction $H_2(g) + Br_2(g) \rightarrow 2\,HBr(g)$ obeys the rate law rate $= k[H_2][Br_2]^{1/2}$. However, the reaction hardly proceeds at all if another substance is added that rapidly removes hydrogen and bromine atoms. Suggest a mechanism for the reaction.

18.111 The rate of a first-order reaction drops as the reactant is used up. However, for a first-order reaction that goes to equilibrium, as long as the system is undisturbed the concentration drops to a certain value and then remains constant at that value. Explain why the concentration does not drop to zero.

18.112 Hypochlorite ion reacts with the anion X^- as follows:

$$ClO^-(aq) + X^-(aq) \longrightarrow XO^-(aq) + Cl^-(aq)$$

While carrying out the reaction, a chemist made the following observations: the concentration of ClO^- drops initially by first-order kinetics, but then levels off to a constant value; HClO molecules and OH^- ions are present in the reaction mixture; and the reaction does not take place in the absence of water. (a) Which species is the catalyst? (b) Which species are intermediates? (c) Postulate a reaction mechanism that is consistent with these observations, indicating which is the slow step and in which step or steps an equilibrium is reached. (d) Write the rate law corresponding to your mechanism.

18.113 Suppose that a pollutant was entering the environment at a steady rate R and that, once there, its concentration decays by a first-order reaction. Derive an expression for (a) the concentration of the pollutant at equilibrium in terms of R and (b) the half-life of the pollutant species when $R = 0$.

18.114 The reaction of methane, CH_4, with chlorine gas occurs by a mechanism similar to that described in Exercise 18.100. Write the mechanism. (a) Identify the initiation, propagation, and termination steps. (b) What products are predicted from the mechanism?

18.115 Use Le Chatelier's principle to predict the effect of temperature on the proportion of product in a mixture in which the forward and reverse reactions (both of which are first order) have reached dynamic equilibrium and (a) the forward reaction is exothermic; (b) the forward reaction is endothermic. (c) Explain this effect in terms of the temperature dependence of the rate constants of the forward and reverse reactions.

CHAPTER 19

Carbon is one of the elements central to life and natural intelligence. Silicon and germanium, in the same group, are the elements central to electronic technology and artificial intelligence. We carbon-based life forms exploit silicon-based devices to augment our own intelligence, but perhaps one day artificial intelligence will surpass our own. To manufacture and control intelligent devices, we need a thorough knowledge of the characteristics of the elements in the periodic table, and this chapter begins a review of their properties.

THE MAIN-GROUP ELEMENTS: I. THE FIRST FOUR FAMILIES

There were 113 elements known when this text was written, and there may be more by the time you read it. They combine to form millions of compounds, and hundreds of new compounds are discovered every year. In the next three chapters we get to know a small selection of them and see how their properties are related to their location in the periodic table. Once we have learned to interpret the periodic table, we shall be in a good position to predict the properties of an element from its location and shall come to see why chemists think of the periodic table as one of the great unifying concepts of chemistry.

Each group in the periodic table has its own unique characteristics. However, the main groups have certain features in common*:

1. The members of a group have analogous valence-shell electron configurations.
2. The element at the head of a group (the lightest element in the group) often has a character that is quite distinct from the other members of the group.
3. There are diagonal relationships between elements, especially between Periods 2 and 3.
4. There is a trend toward nonmetallic character on going to the right along a period and a trend toward more metallic character on going down a group.

We have met these characteristics before. For instance, we saw in Section 7.13 that the resemblances between members of the same group stem from their analogous valence-shell electron configurations. We saw in Section 8.14 that the element at the head

* Remember that the main groups are Groups 1, 2, and 13 through 18 in the IUPAC numbering system; Groups I (or IA) to VIII (or VIIIA) in alternative systems.

HYDROGEN

19.1 The element
19.2 Compounds of hydrogen

GROUP 1: THE ALKALI METALS

19.3 The elements
19.4 Chemical properties of the alkali metals
19.5 Compounds of lithium, sodium, and potassium

GROUP 2: THE ALKALINE EARTH METALS

19.6 The elements
19.7 Compounds of beryllium, magnesium, and calcium

GROUP 13: THE BORON FAMILY

19.8 The elements
19.9 Group 13 oxides
19.10 Carbides, nitrides, and halides
19.11 Boranes and borohydrides

GROUP 14: THE CARBON FAMILY

19.12 The elements
19.13 The many faces of carbon
19.14 Silicon, tin, and lead
19.15 Oxides of carbon
19.16 Oxides of silicon: The silicates
19.17 Carbides

of each main group—for example, lithium in Group 1 and beryllium in Group 2—differs significantly from the other members of the group, mainly as a result of the small size of its atoms. We met diagonal relationships first in Section 7.19 and saw that they are particularly important between Period 2 and Period 3 elements toward the left of the main groups of the periodic table. Thus, we can expect lithium to resemble its diagonal neighbor magnesium, and beryllium to resemble its *p*-block diagonal neighbor aluminum. The metallic character of an element is a consequence of its low ionization energy; and, as we saw in Sections 7.16 and 7.17, ionization energies increase from left to right across a period and decrease down a group.

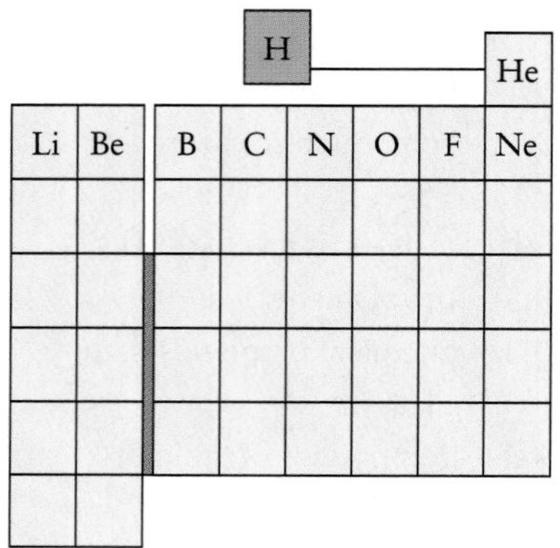

HYDROGEN

Hydrogen atoms are the simplest of all atoms, for most of them are made up of just two subatomic particles, a proton and an electron. Although it has the same valence electron configuration as the Group 1 elements, ns^1, hydrogen is a nonmetal, and in the elemental state it is found as diatomic molecules, H_2. It has few other similarities to the alkali metals, and we do not assign it to any group.

19.1 THE ELEMENT

◀◀ The three isotopes of hydrogen were introduced in Section 1.4.

In many periodic tables you will come across, hydrogen is included as a member of Group 1, Group 17 (the halogens), or even both groups.

Hydrogen is the most abundant element in the universe. It was formed within the first few seconds after the Big Bang that most scientists believe marked the beginning of the universe. However, even though 89% of all atoms in the universe are thought to be hydrogen atoms, there is little free hydrogen on Earth because H_2 molecules are very light. As a result, they move at such high average speeds that, unless trapped underground or sealed in a container, they soon escape from Earth's gravity. It takes heavier atoms to anchor hydrogen to the planet in compounds. A lot of Earth's hydrogen is present as water, H_2O, either in the oceans or trapped inside minerals.

Natural supplies of hydrogen gas are far too small to satisfy the needs of industry, so most commercial hydrogen is obtained as a by-product of petroleum refining in a series of two catalyzed reactions. The first is a **re-forming reaction,** a reaction in which a hydrocarbon and steam are converted to carbon monoxide and hydrogen over a nickel catalyst:

$$CH_4(g) + H_2O(g) \xrightarrow{800°C,\ Ni} CO(g) + 3\,H_2(g)$$

The mixture of products, which is called *synthesis gas,* is the starting point for the manufacture of numerous compounds, including methanol. The reforming reaction is followed by the **shift reaction,** in which the carbon monoxide in the synthesis gas reacts with more water:

$$CO(g) + H_2O(g) \xrightarrow{400°C,\ Fe/Cu} CO_2(g) + H_2(g)$$

where Fe/Cu denotes a catalyst made of iron and copper.

Hydrogen is also produced by the electrolysis of water, but that process is economical only where electricity is cheap. Chemists are cur-

TABLE 19.1 Physical properties of hydrogen

Valence configuration: $1s^1$
*Normal form**: colorless, odorless gas

Z	Name	Symbol	Molar mass, $g\cdot mol^{-1}$	Abundance, %	Melting point, °C	Boiling point, °C	Density, $g\cdot L^{-1}$
1	hydrogen	H	1.008	99.88	−259 (14 K)	−253 (20 K)	0.089
1	deuterium	^2H or D	2.014	0.02	−254 (19 K)	−249 (24 K)	0.18
1	tritium	^3H or T	3.016	radioactive	−252 (21 K)	−248 (25 K)	0.27

**Normal form* means the state and appearance of the element at 25°C and 1 atm. The density refers to the same conditions.

rently seeking ways of using sunlight to drive the **water-splitting reaction,** the photochemical decomposition of water into its elements (see Case Study 17):

$$2\,H_2O(l) \xrightarrow{\text{light}} 2\,H_2(g) + O_2(g)$$

Hydrogen is prepared in the laboratory by reducing hydrogen ions with a metal having a negative standard potential, such as zinc:

$$Zn(s) + 2\,HCl(aq) \longrightarrow ZnCl_2(aq) + H_2(g)$$

Hydrogen is a colorless, odorless, tasteless gas (Table 19.1). Being nonpolar, H_2 molecules can attract each other only by London forces. Because each molecule has only two electrons and hence only a very small instantaneous electric dipole, these forces are very weak. In fact, its intermolecular interactions are so weak that it is almost completely insoluble in polar solvents such as water and does not condense to a liquid until it is cooled to 20 K. One striking physical property of liquid hydrogen is its very low density ($0.089\ g\cdot cm^{-3}$), which is less than one-tenth that of water (Fig. 19.1). This low density makes hydrogen a very lightweight but bulky fuel. Hydrogen has the highest specific enthalpy of any

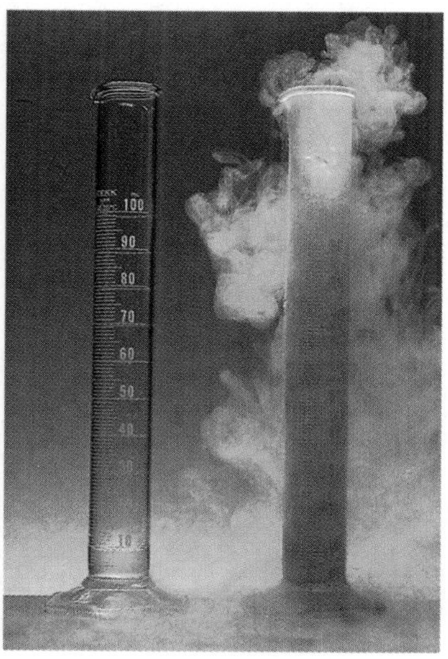

FIGURE 19.1 The two measuring cylinders contain the same mass of liquid. The liquid on the left is water, that on the right is liquid hydrogen, which is one-tenth as dense.

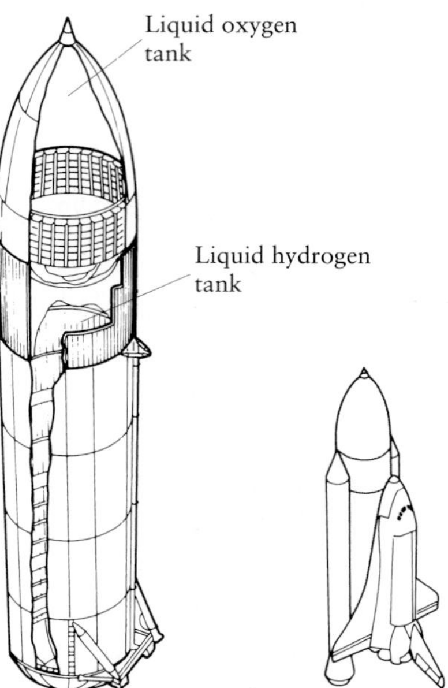

FIGURE 19.2 The arrangement of fuel tanks in the space shuttle and its booster. Note the large size of the hydrogen tank compared with that of the oxygen tank. The low density of liquid hydrogen makes it a valuable fuel for space missions.

known fuel, so liquid hydrogen is used with liquid oxygen to power the space shuttle's main rocket engines (Fig. 19.2).

▶▶ **The specific enthalpy of rocket fuels is discussed further in Case Study 20.**

Each year about half the 3×10^8 kg of hydrogen used in industry is converted into ammonia by the Haber process (Section 13.11). Through the reactions of ammonia, hydrogen finds its way into numerous other important nitrogen compounds such as hydrazine and sodium amide (Section 20.2). Hydrogen also enters the economy through the production of methanol:

$$2\,H_2(g) + CO(g) \xrightarrow{\Delta,\ \text{pressure, catalyst}} CH_3OH(l)$$

About a third of the hydrogen manufactured is used for the **hydrometallurgical extraction** of copper and other metals, the extraction from their ores by reduction in aqueous solution:

$$Cu^{2+}(aq) + H_2(g) \longrightarrow Cu(s) + 2\,H^+(aq)$$

In this process, ores containing copper(II) oxide and copper(II) sulfide are dissolved in sulfuric acid, and then hydrogen is bubbled through the solution. The reduction is thermodynamically favored, because the standard potential for $Cu^{2+}(aq) + 2\,e^- \rightarrow Cu(s)$ is positive ($E° = +0.34$ V). Metals with negative standard potentials, such as zinc ($E° = -0.76$ V) and nickel ($E° = -0.23$ V), cannot be extracted by hydrogen. In other words, hydrogen cannot reduce ions such as Zn^{2+} or Ni^{2+}.

FIGURE 19.3 When the runny oil (top) is hydrogenated, it is converted into a solid—a fat (bottom). The hydrogen converts carbon-carbon double bonds into single bonds. The resulting, more flexible molecules can pack together more closely and so form a solid.

Hydrogen is used in the food industry to convert vegetable oils into shortening (Fig. 19.3). There it is added to carbon-carbon double bonds in a hydrogenation reaction:

$$H_2(g) + \ldots C{=}C \ldots \xrightarrow{200°C,\ 30\ \text{atm, Ni or Pt}} \ldots CH{-}CH \ldots$$

This reaction converts a C=C double bond into a C—C single bond. Oil and fat molecules both have long hydrocarbon chains, but oils have more double bonds. Because double bonds resist twisting, oil molecules do not pack together well, so the result is a liquid. When the double bonds are replaced by single bonds, the chains become much more flexible, so the molecules pack together better and form a solid.

Hydrogen is produced as a by-product of the refining of fossil fuels and by electrolysis. It has a low density and weak intermolecular forces. Hydrogen is a good reducing agent for species with positive standard potentials.

19.2 COMPOUNDS OF HYDROGEN

Because hydrogen forms compounds with so many other elements (Table 19.2), we shall meet many of its compounds when we study the other elements. Here we consider only the **binary compounds** of hydrogen, the compounds of hydrogen with one other element. All the main-group elements, with the exception of the noble gases and (possibly) indium and thallium, form binary compounds with hydrogen. So do most of the elements on the left and right of the *d*-block, but not those at the center of the block.

Binary compounds in which hydrogen has an oxidation number of -1 are called **hydrides.** Three types of hydrides are found (Fig. 19.4). The **saline hydrides** are compounds of hydrogen with a strongly electropositive metal (any member of the *s*-block, with the exception of beryllium). They are formed by heating the metal in hydrogen:

Saline means "saltlike." Indeed, the saline hydrides are sometimes called the "saltlike hydrides."

$$2\,K(s) + H_2(g) \xrightarrow{\Delta} 2\,KH(s)$$

The saline hydrides are white, high-melting-point solids that contain the hydride ion, H^-. Their crystal structures resemble those of the metal halides; the alkali metal hydrides, for instance, have the rock-salt structure illustrated in Fig. 10.36.

The single positive charge of the hydrogen atomic nucleus can barely manage to keep control over the two electrons in the H^- ion. The readiness with which one of the electrons can be lost results in ionic hydrides being very powerful reducing agents, and $E° = -2.25$ V for $H_2(g) + 2\,e^- \rightarrow 2\,H^-(aq)$. This value is similar to the standard potential of $Na^+(aq) + e^- \rightarrow Na(s)$, and, like sodium metal, hydride ions reduce water as soon as they come into contact with it:

$$NaH(s) + H_2O(l) \longrightarrow NaOH(aq) + H_2(g)$$

Because this reaction produces hydrogen, saline hydrides are potentially useful as transportable sources of hydrogen.

TABLE 19.2 Chemical properties of hydrogen

Reactant	Reaction with hydrogen
Group 1 metals (M)	$2\,M(s) + H_2(g) \longrightarrow 2\,MH(s)$
Group 2 metals (M, not Be)	$M(s) + H_2(g) \longrightarrow MH_2(s)$
some *d*-block metals (M)	$2\,M(s) + x\,H_2(g) \longrightarrow 2\,MH_x(s)$
oxygen	$O_2(g) + 2\,H_2(g) \longrightarrow 2\,H_2O(l)$
nitrogen	$N_2(g) + 3\,H_2(g) \longrightarrow 2\,NH_3(g)$
halogen (X_2)	$X_2(g,l,s) + H_2(g) \longrightarrow 2\,HX(g)$

FIGURE 19.4 The different classes of binary hydrogen compounds and their distribution throughout the periodic table.

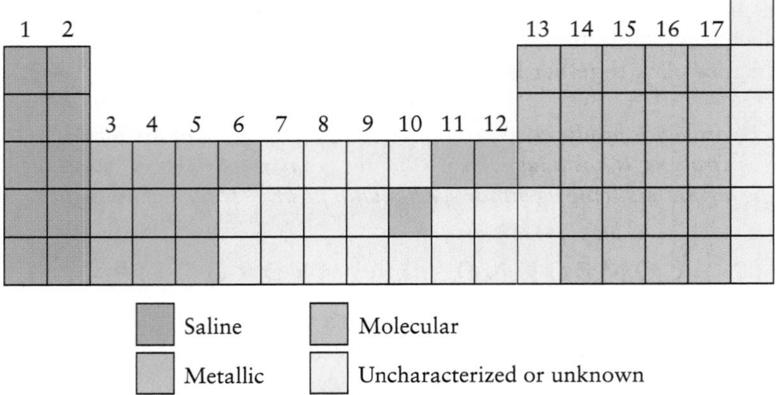

The **metallic hydrides** are so called because they are electrically conducting (Fig. 19.5). These black powdery solids are prepared by heating certain *d*-block metals in hydrogen; for example,

$$2\,Cu(s) + H_2(g) \xrightarrow{\Delta} 2\,CuH(s)$$

Because the metallic hydrides release their hydrogen (as H_2) when heated or treated with acid, they are also being investigated for storing and transporting hydrogen. Both saline and metallic hydrides have the high enthalpy densities desirable in a portable fuel (see Section 6.11).

Binary molecular compounds of hydrogen are often called *molecular hydrides,* but they are not strictly hydrides as the oxidation number of hydrogen is +1.

The *binary molecular compounds* of hydrogen consist of discrete molecules. They are formed by reaction with nonmetals and in most cases are gases such as ammonia, the hydrogen halides (HF, HCl, HBr, HI), and hydrocarbon gases such as methane, ethene, and ethyne. Liquid compounds include water and liquid hydrocarbons such as octane and benzene.

Some of these molecular compounds can be prepared by **protonation,** or proton transfer from an acid to a base, such as S^{2-}:

$$FeS(s) + 2\,HCl(aq) \longrightarrow FeCl_2(aq) + H_2S(g)$$

Volatile binary acids can be prepared by using a less volatile acid as the proton donor:

$$CaF_2(s) + H_2SO_4(l) \longrightarrow CaSO_4(s) + 2\,HF(g)$$

The reaction proceeds to the right, because the volatile product is removed as a gas.

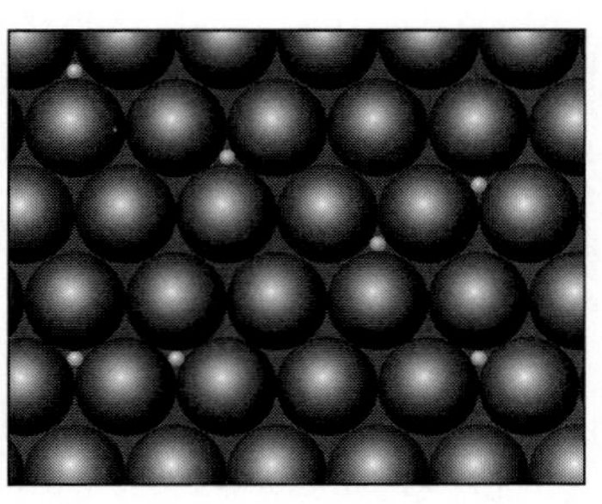

FIGURE 19.5 In a metallic hydride, the small hydrogen atoms (the small gray spheres) occupy gaps—called interstices—between the larger metal atoms (the large gray spheres).

Binary hydrides are classified as saline, metallic, or molecular. Some binary molecular compounds of hydrogen are characterized by acidic character.

SELF-TEST 19.1A What common compounds of hydrogen are more powerful reducing agents than hydrogen itself?

[*Answer:* The saline hydrides]

SELF-TEST 19.1B What are some of the properties that argue against the classification of hydrogen as a Group 1 element?

TABLE 19.3 Group 1 elements: The alkali metals

Valence configuration: ns^1
*Normal form**: soft, silver-gray metals

Z	Name	Symbol	Molar mass, g·mol^{-1}	Melting point, °C	Boiling point, °C	Density, g·L^{-1}
3	lithium	Li	6.94	181	1347	0.53
11	sodium	Na	22.99	98	883	0.97
19	potassium	K	39.10	64	774	0.86
37	rubidium	Rb	85.47	39	688	1.53
55	cesium	Cs	132.91	28	678	1.87
87	francium	Fr	223	27	677	—

**Normal form* means the state and appearance of the element at 25°C and 1 atm.

GROUP 1: THE ALKALI METALS

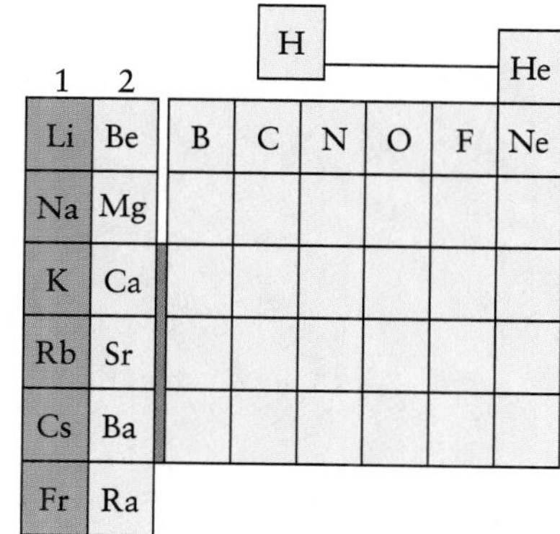

The members of Group 1 are called the alkali metals. Their valence electron configurations are ns^1, where n is the period number, and their physical and chemical properties are dominated by the ease with which the single valence electron can be removed (Tables 19.3 and 19.4).

19.3 THE ELEMENTS

The alkali metals are the most violently active of all the metals. They are too easily oxidized to be found in the free state in nature. Common, cheap materials are too feeble as reducing agents to extract them from their compounds. The pure metals are obtained by electrolysis of their molten salts, as in the electrolytic Downs process (Section 17.16) or, in the case of potassium, by exposing molten potassium chloride to sodium vapor:

$$KCl(l) + Na(g) \xrightarrow{750°C} NaCl(s) + K(g)$$

Although the equilibrium constant for this reaction is not particularly favorable, the reaction runs to the right because potassium is more

TABLE 19.4 Chemical properties of the alkali metals

Reactant	Reaction with alkali metal (M)
hydrogen	$2\,M(s) + H_2(g) \longrightarrow 2\,MH(s)$
oxygen	$4\,Li(s) + O_2(g) \longrightarrow 2\,Li_2O(s)$
	$2\,Na(s) + O_2(g) \longrightarrow Na_2O_2(s)$
	$M(s) + O_2(g) \longrightarrow MO_2(g)$, M = K, Rb, Cs
nitrogen	$6\,Li(s) + N_2(g) \longrightarrow 2\,Li_3N(s)$
halogen (X_2)	$2\,M(s) + X_2(g,l,s) \longrightarrow 2\,MX(s)$
water	$2\,M(s) + 2\,H_2O(l) \longrightarrow 2\,MOH(aq) + H_2(g)$

(a)

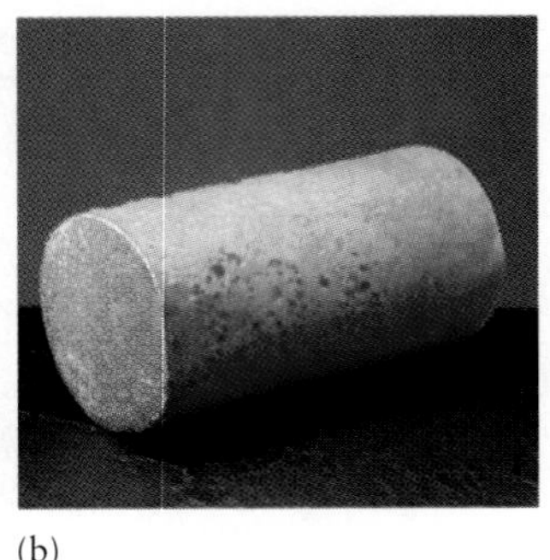
(b)

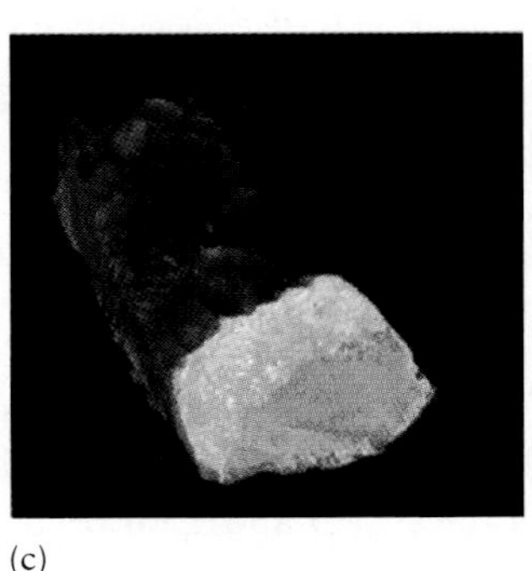
(c)

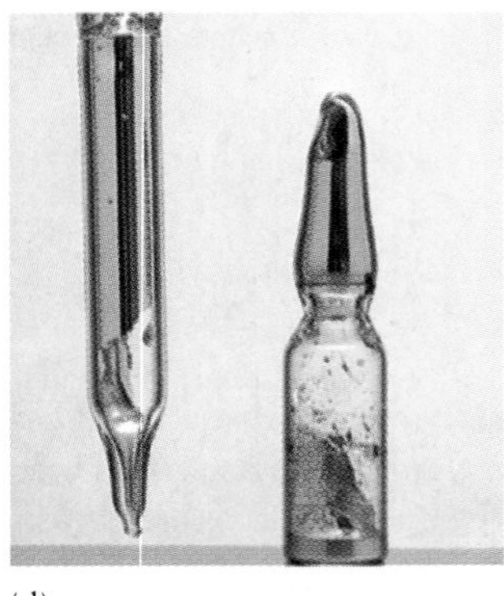
(d)

FIGURE 19.6 The alkali metals of Group 1: (a) lithium; (b) sodium; (c) potassium; (d) rubidium and cesium. Francium has never been isolated in visible quantities. The first three elements corrode rapidly in moist air; rubidium and cesium are even more reactive and have to be stored (and photographed) in sealed, airless containers.

volatile than sodium. The potassium vapor is driven off by the heat and condensed in a cooled collecting vessel.

All the Group 1 elements are soft, silver-gray metals (Fig. 19.6). Lithium is the hardest, but even so it is softer than lead. The melting points decrease down the group (Fig. 19.7). Cesium, which melts at 28°C, is barely a solid at room temperature. Some alloys of sodium and potassium are liquid at room temperature, because their atoms pack together poorly and hence produce a fluid structure, like a pile of oranges and grapefruit, which is less stable than a pile of oranges or grapefruit alone (Fig. 19.8). Liquid sodium and potassium mixtures are used as coolants in breeder nuclear reactors. The mixture has a high boiling point, conducts heat very well, and the metals are not decomposed by radiation. Lithium metal had few applications until thermonuclear weapons (which use lithium-6, see Section 22.11) were developed after World War II.

The alkali metals melt at low temperatures; they are the most reactive metals.

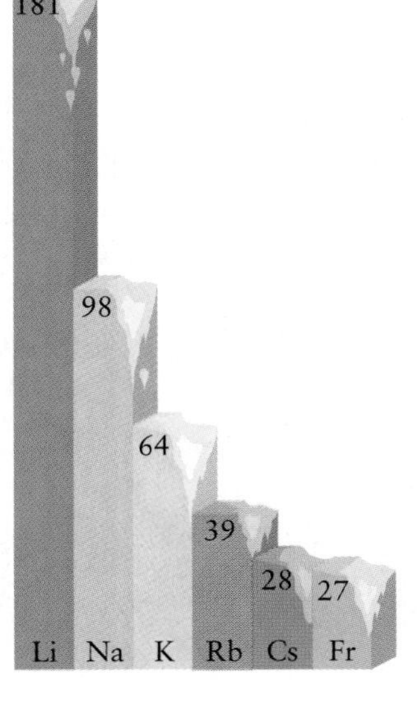

FIGURE 19.7 The melting points of the alkali metals decrease down the group. The numerical values shown here are degrees Celsius.

TABLE 19.5 Properties of Group 1 compounds

Compound	Formula*	Comment
oxides	M_2O	formed by decomposition of carbonates; strong bases; react with water to form hydroxides
hydroxides	MOH	formed by reduction of water with the metal or from the oxide; strong bases
carbonates	M_2CO_3	soluble in water; most decompose into oxides when heated
hydrogen carbonates	$MHCO_3$	weak bases in water; can be obtained as solids
nitrates	MNO_3	decompose to nitrite and evolve oxygen when strongly heated

*M stands for a Group 1 metal.

FIGURE 19.8 Sodium atoms (represented by the pink spheres) and potassium atoms (yellow) pack together poorly, and an alloy of the two metals is a liquid at room temperature.

19.4 CHEMICAL PROPERTIES OF THE ALKALI METALS

Because the first ionization energies of the alkali metals are so low, they are most commonly found as singly charged cations, such as Na^+, and consequently most of their compounds are ionic (Table 19.5). Their low ionization energies also mean that the alkali metals are excellent reducing agents. Molten sodium metal is used to produce zirconium and titanium from their chlorides:

$$TiCl_4(g) + 4\,Na(l) \xrightarrow{\Delta} 4\,NaCl(s) + Ti(s)$$

Because their standard potentials are so strongly negative, the alkali metals can even reduce water:

$$2\,Na(s) + 2\,H_2O(l) \longrightarrow 2\,NaOH(aq) + H_2(g)$$

The vigor of this reaction increases uniformly down the group (Fig. 19.9). The reaction is dangerously explosive with rubidium and cesium. Rubidium and cesium are denser than water, so they sink and react

FIGURE 19.9 The alkali metals react with water, producing gaseous hydrogen and a solution of the alkali metal hydroxide. (a) Lithium reacts quietly. (b) Sodium reacts so vigorously that the heat released melts the unreacted metal. (c) Potassium reacts even more vigorously, producing so much heat that the hydrogen produced in the reaction is ignited.

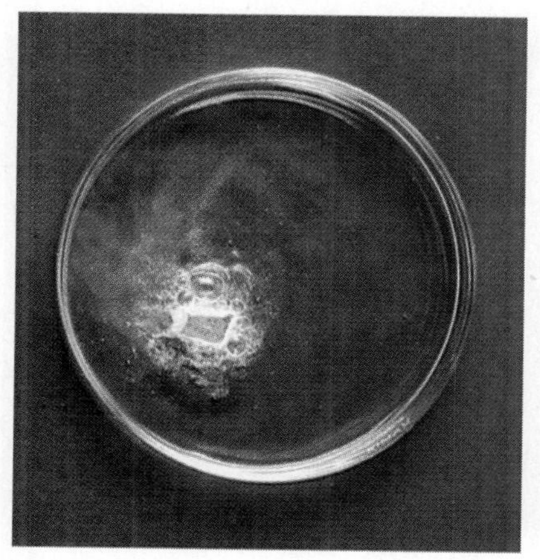

(a)

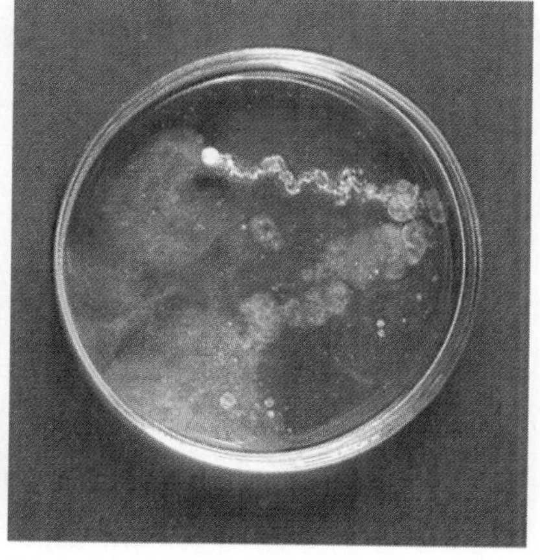

(b)

(c)

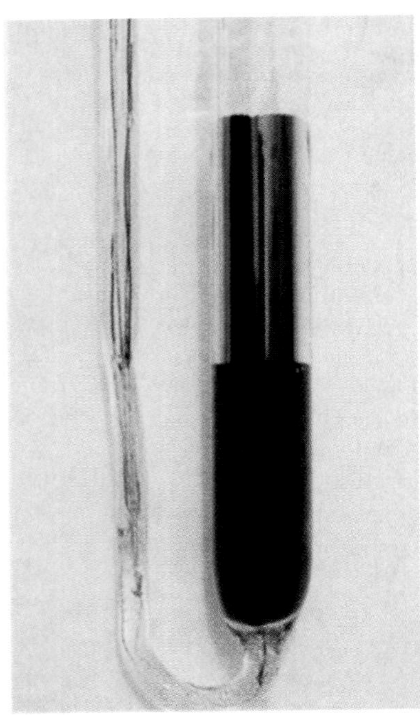

FIGURE 19.10 Sodium dissolves in liquid ammonia to form the deep blue solution in the lower half of the tube. At higher sodium concentrations, the metal-ammonia solution becomes bronze in color, as shown in the top half of the tube.

beneath the surface; the rapidly evolved hydrogen gas then forms a shock wave that can shatter the vessel.

The alkali metals also release their valence electrons when they dissolve in liquid ammonia, but the outcome is slightly different. The electrons occupy cavities formed by groups of NH_3 molecules and give ink-blue *metal-ammonia solutions* (Fig. 19.10). These solutions of electrons (and cations of the metal) are often used to reduce organic compounds. As the metal concentration is increased, the blue gives way to a metallic bronze, and the solutions begin to conduct electricity like liquid metals.

All alkali metals react directly with most nonmetals (other than the noble gases). However, only lithium reacts with nitrogen, which it reduces to the nitride ion:

$$6\,Li(s) + N_2(g) \xrightarrow{\Delta} 2\,Li_3N(s)$$

The principal product of the reaction of the alkali metals with oxygen varies systematically down the group (Fig. 19.11). It is commonly found that ionic compounds formed from cations and anions of similar radius are more stable than those formed from ions with markedly different radii. Lithium forms mainly the oxide, Li_2O. Sodium, which has a larger cation, forms predominantly the pale yellow sodium peroxide, Na_2O_2. Potassium, with an even bigger cation, forms mainly the superoxide, KO_2. Potassium superoxide, a yellow-orange solid, is used in closed-system breathing apparatus (gas masks, submarines, and space vehicles; see Box 3.1). The KO_2 removes the exhaled water vapor and generates oxygen gas:

$$4\,KO_2(s) + 2\,H_2O(g) \longrightarrow 4\,KOH(s) + 3\,O_2(g)$$

The KOH produced in this reaction removes the exhaled CO_2:

$$KOH(s) + CO_2(g) \longrightarrow KHCO_3(s)$$

Much of the potassium currently produced is used to manufacture potassium superoxide for this purpose.

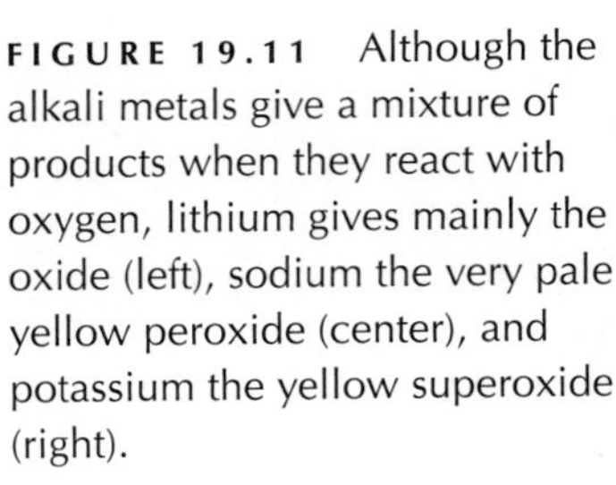

FIGURE 19.11 Although the alkali metals give a mixture of products when they react with oxygen, lithium gives mainly the oxide (left), sodium the very pale yellow peroxide (center), and potassium the yellow superoxide (right).

The alkali metals are usually found as singly charged cations. They reduce water with increasing vigor down the group. Cations tend to form their most stable compounds with anions of similar size.

19.5 COMPOUNDS OF LITHIUM, SODIUM, AND POTASSIUM

As head of its group, lithium differs significantly from other Group 1 elements and has a diagonal relationship with magnesium. The differences stem in part from the small size of the Li^+ cation. This smallness makes it so strongly polarizing that the bonds it forms have a pronounced covalent character. Its small size also means that a Li^+ ion has strong ion-dipole interactions, so many lithium salts are hydrated.

The availability of lithium has increased during the past few decades, and so has the variety of its applications. Lithium compounds are used in ceramics, lubricants, and batteries (Fig. 19.12). They are also used in medicine, and small daily doses of lithium carbonate have been found to be an effective treatment for manic-depressive disorder; the mode of action is still not fully understood. Lithium soaps—the lithium salts of long-chain carboxylic acids—are used as thickeners in lubricating greases for high-temperature applications because they have higher melting points than more conventional sodium and potassium soaps.

Sodium compounds are important largely because they are generally soluble in water and inexpensive. More sodium chloride is used by industry than the traditional chart-topping sulfuric acid, but it is not listed because it is not manufactured. It is readily mined as rock salt or obtained from the evaporation of brine (Fig. 19.13). Sodium chloride is used in large quantities in the electrolytic production of chlorine and sodium hydroxide from brine.

Sodium hydroxide, NaOH, is a soft, waxy, white, corrosive solid. It is an important industrial chemical because it is an inexpensive base for the production of other sodium salts. The amount of electricity used to

FIGURE 19.12 Lithium ion batteries like these can store more energy than nickel-cadmium batteries of similar size, which makes them desirable for use in laptop computers.

FIGURE 19.13 An evaporation pond. The blue color is due to a dye added to the brine to increase heat absorption and hence speed up evaporation.

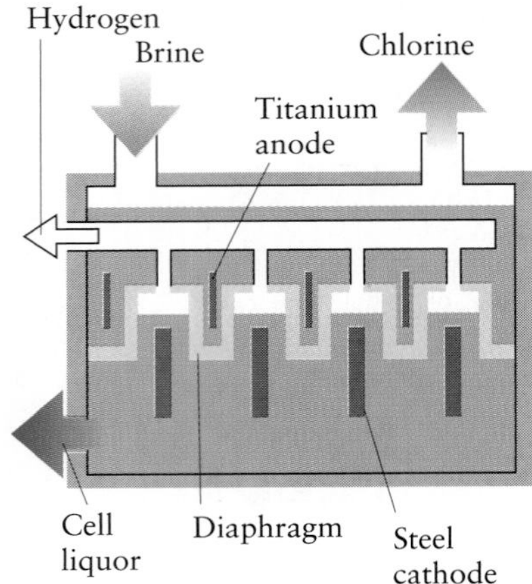

FIGURE 19.14 A diaphragm cell for the electrolytic production of sodium hydroxide from brine (aqueous sodium chloride solution). The diaphragm prevents the chlorine from mixing with the hydrogen and the sodium hydroxide. The liquid is drawn off and the water is partially evaporated. The unconverted sodium chloride crystallizes, leaving the sodium hydroxide in solution.

electrolyze brine to produce NaOH is second only to the amount used to extract aluminum from its ores. The process produces chlorine and hydrogen as well as sodium hydroxide:

$$2\,NaCl(aq) + 2\,H_2O(l) \longrightarrow 2\,NaOH(aq) + Cl_2(g) + H_2(g)$$

Most modern production uses a diaphragm cell, in which compartments containing steel and titanium electrodes are separated by porous diaphragms to isolate the products (Fig. 19.14).

Anhydrous sodium sulfate, Na_2SO_4, is called *salt cake*. Much of it comes from natural sources, particularly the sulfate-rich underground brines found in Texas. In countries lacking natural supplies, it is produced by the action of concentrated sulfuric acid on sodium chloride:

$$H_2SO_4(aq, conc) + 2\,NaCl(s) \longrightarrow Na_2SO_4(s) + 2\,HCl(g)$$

This reaction is driven forward by the escape of a volatile product. The same reaction is used to generate hydrogen chloride in the laboratory.

Sodium hydrogen carbonate, $NaHCO_3$ (sodium bicarbonate), is commonly called *bicarbonate of soda* or *baking soda*. The rising action of baking soda in bread and doughs depends on the reaction of a weak acid with the hydrogen carbonate ions:

$$NaHCO_3(aq) + H_3O^+(aq) \longrightarrow Na^+(aq) + 2\,H_2O(l) + CO_2(g)$$

The release of gas causes the dough to rise. The weak acids are provided by the recipe, generally in the form of lactic acid from sour milk or buttermilk, citric acid from lemons, or the acetic acid in vinegar. Baking powder contains a solid acid as well as the hydrogen carbonate, and carbon dioxide is released when water is added (Fig. 19.15).

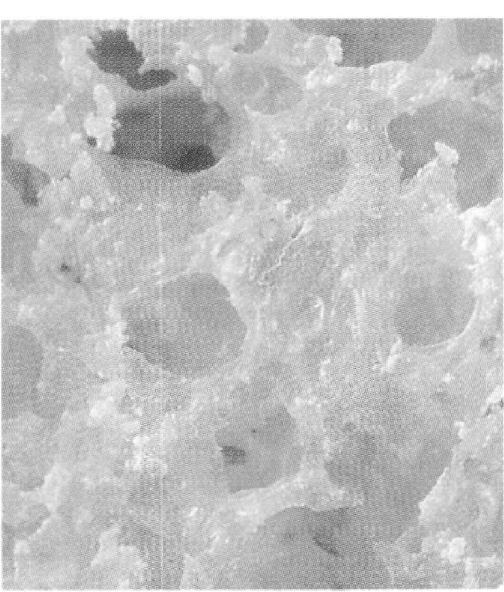

FIGURE 19.15 Double-acting baking powder first forms small cavities in the dough when it is moistened. These are later inflated by a second release of carbon dioxide during baking.

FIGURE 19.16 The mineral sylvite is a form of potassium chloride. It is found in the beds of ancient lakes and seas. The potassium was probably collected by plants that grew around the lake and extracted it from groundwater.

pure copper but still conducts electricity well. The hard, electrically conducting alloy is formed into nonsparking tools for use in oil refineries and grain elevators, where there is a risk of explosion.

Magnesium is a silver-white metal that is protected from extensive oxidation in air by a film of white oxide, and hence looks dull gray. Its low density is only two-thirds that of aluminum, and it is widely used as a component of alloys in applications where lightness and toughness are needed—in airplanes, for instance.

Metallic magnesium is produced by reduction of its compounds, either electrolytically or with a chemical reducing agent. In the chemical reduction of magnesium oxide obtained from the decomposition of dolomite, ferrosilicon, an alloy of iron and silicon, is used as the reducing agent at about 1200°C. At this temperature, the magnesium produced is immediately vaporized; so even though the equilibrium constant does not favor the reduction, the process continues because the product is removed as soon as it is formed.

FIGURE 19.18 An emerald is a crystal of beryl with some Cr^{3+} ions, which are responsible for the color.

The electrolytic method uses seawater as its principal raw material (Fig. 19.19). The first stage is the precipitation of magnesium hydroxide with slaked lime (calcium hydroxide):

$$Mg^{2+}(aq) + Ca(OH)_2(aq) \longrightarrow Mg(OH)_2(s) + Ca^{2+}(aq)$$

The lime is produced by the thermal decomposition of calcium carbonate in shells dredged up from the ocean floor. The precipitated magnesium hydroxide is filtered off and treated with hydrochloric acid:

$$Mg(OH)_2(s) + 2\,HCl(aq) \longrightarrow MgCl_2(aq) + 2\,H_2O(l)$$

Finally, the magnesium chloride is dried and then added to an electrolytic cell. Magnesium is produced at the cathode and chlorine at the anode:

$$MgCl_2(l) \xrightarrow{\text{electrolysis}} Mg(s) + Cl_2(g)$$

FIGURE 19.19 A magnesium extraction plant at Freeport, Texas.

TABLE 19.7 Chemical properties of the Group 2 metals

Reactant	Reaction with Group 2 metal (M)
hydrogen	$M(s) + H_2(g) \longrightarrow MH_2(s)$, not Be or Mg
oxygen	$2\,M(s) + O_2(g) \longrightarrow 2\,MO(s)$
nitrogen	$3\,M(s) + N_2(g) \longrightarrow M_3N_2(s)$
halogen (X_2)	$M(s) + X_2(g,l,s) \longrightarrow MX_2(s)$
water	$M(s) + 2\,H_2O(l) \longrightarrow M(OH)_2(aq) + H_2(g)$, not Be

The true alkaline earth metals—calcium, strontium, and barium—are obtained either by electrolysis or by reduction with aluminum in a version of the thermite process (see Section 6.2). For example,

$$3\,BaO(s) + 2\,Al(s) \xrightarrow{\Delta} Al_2O_3(s) + 3\,Ba(s)$$

Some important reactions of the Group 2 elements are summarized in Table 19.7.

Magnesium burns vigorously in air with a brilliant white flame, partly because it reacts with the nitrogen and carbon dioxide in air as well as with oxygen, especially when it is sprayed with water. Neither water nor CO_2 fire extinguishers should ever be used on a magnesium fire, because they will make the fire worse! As in Group 1, reactions of the metals with oxygen and water increase in vigor going down the group. Calcium, strontium, and barium are partially passivated in air by a protective surface layer of oxide. Barium, however, does not form a protective oxide and may ignite in moist air.

All the Group 2 elements with the exception of beryllium reduce water; for example,

$$Ca(s) + 2\,H_2O(l) \longrightarrow Ca(OH)_2(aq) + H_2(g)$$

Beryllium does not react with water, even when red hot: its protective oxide film survives even at high temperatures. Magnesium reacts with hot water, and calcium reacts with cold water (Fig. 19.20). The metals

FIGURE 19.20 Calcium reacts gently with water at room temperature to produce hydrogen and calcium hydroxide, $Ca(OH)_2$.

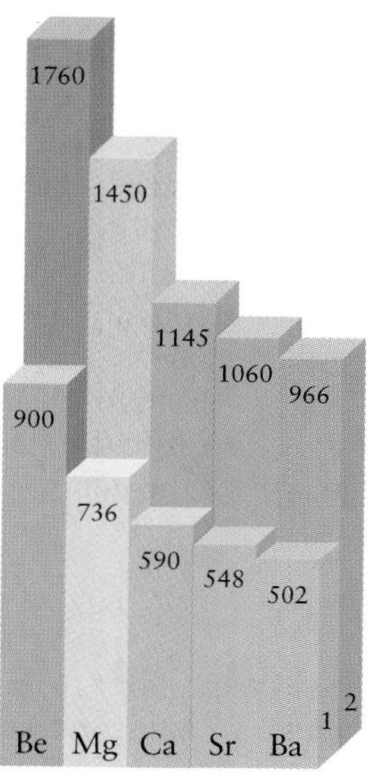

FIGURE 19.21 The first and second ionization energies (in kilojoules per mole) of the Group 2 elements. Although the second ionization energies are larger than the first, they are not enormous, and both valence electrons are lost from each atom in all the compounds of these elements.

reduce hydrogen ions to hydrogen, but neither beryllium nor magnesium dissolves in nitric acid, because they become passivated by a film of oxide.

The valence electron configuration of the atoms of the Group 2 elements is ns^2, where n is the period number. The second ionization energy is low enough to be recovered from the increased lattice enthalpy (Fig. 19.21). Hence, the Group 2 elements occur with an oxidation number of +2, as the cation M^{2+}, in all their compounds (Table 19.8). Apart from a tendency toward nonmetallic character in beryllium, the elements have all the chemical characteristics of metals, such as having basic oxides and hydroxides.

Beryllium shows a hint of nonmetallic character, but the other elements are all typical metals. The vigor of reaction with water and oxygen increases down the group.

TABLE 19.8 Properties of Group 2 compounds

Compound	Formula*	Comment
oxides	MO	formed by decomposition of carbonates; strong bases (BeO is amphoteric), react with water to form hydroxides; withstand high temperatures
hydroxides	$M(OH)_2$	formed by action of water on oxides or by precipitation from salt solutions; sparingly soluble in water (except Ba); strong bases (Be is amphoteric)
carbonates	MCO_3	very slightly soluble in water; most decompose into oxides when heated
hydrogen caronates	$M(HCO_3)_2$	unstable as solids; more soluble than carbonates
nitrates	$M(NO_3)_2$	decompose into NO_2 and O_2 when heated; soluble in water

*M stands for a Group 2 metal.

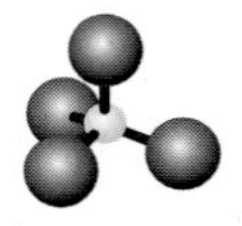

1 BeX_4 unit

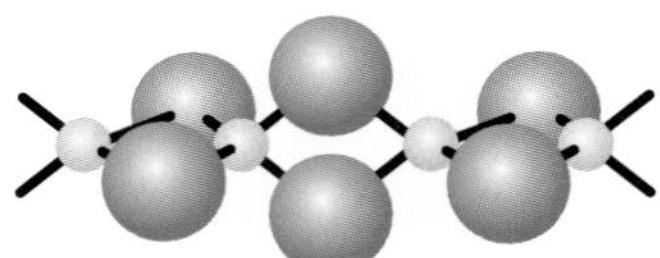

2 Beryllium chloride, $BeCl_2$

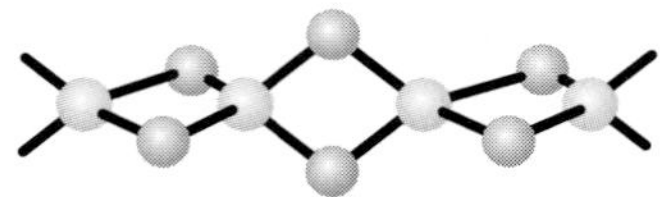

3 Beryllium hydride, BeH_2

19.7 COMPOUNDS OF BERYLLIUM, MAGNESIUM, AND CALCIUM

Beryllium compounds are very toxic and must be handled with great caution. Their properties are dominated by the highly polarizing character of the Be^{2+} ion and its small size. The strong polarizing power results in moderately covalent compounds, and its small size limits to four the number of groups that can attach to the ion. These two features together are responsible for the prominence of the tetrahedral BeX_4 unit (**1**), like that in the beryllate ion, $Be(OH)_4^{2-}$, formed when beryllium reacts with sodium hydroxide solution. A tetrahedral unit is also found in the chloride (**2**) and the hydride (**3**). The chloride is made by the action of chlorine on the oxide in the presence of carbon:

$$BeO(s) + C(s) + Cl_2(g) \xrightarrow{600\text{–}800^\circ C} BeCl_2(g) + CO(g)$$

The Be atoms act as Lewis acids and accept electron pairs from the Cl atoms of the neighboring $BeCl_2$ groups, forming a chain of tetrahedral $BeCl_4$ units.

Magnesium has more pronounced metallic properties than beryllium does, and its compounds are primarily ionic, with some covalent character. Magnesium oxide, MgO, is formed when magnesium burns in air. However, the product is contaminated by magnesium nitride. To prepare the pure oxide, the hydroxide or the carbonate is heated. Magnesium oxide dissolves only very slowly and slightly in water. One of its most striking properties is that it is **refractory,** or able to withstand high temperatures, for it melts only at 2800°C. This high stability can be traced to the small ionic radii of the Mg^{2+} and O^{2-} ions, and hence to their very strong electrostatic interaction with each other. The oxide has two other characteristics that make it useful: it conducts heat very well, and it conducts electricity poorly. All three properties lead to its use as an insulator in electric heaters.

Magnesium hydroxide, $Mg(OH)_2$, is a mild base. It is not very soluble in water but forms instead a white colloidal suspension, a mist of small particles dispersed through a liquid (see Case Study 12), which is known as *milk of magnesia* and used as a stomach antacid and laxative. Magnesium sulfate, or *Epsom salts,* $MgSO_4$, is also a common purgative. Its action appears to inhibit the absorption of water from the intestine. The resulting increased flow of water into the intestine triggers the mechanism that results in defecation.

The name *chlorophyll* comes from the Greek words for "green leaf."

Arguably the most important compound of magnesium is chlorophyll (Fig. 19.22). This green organic compound consists of large molecules that capture light from the Sun and channel its energy into photosynthesis (see Case Study 17).

Although calcium is more metallic in character than magnesium, compounds of the two elements share some similar properties. Calcium carbonate, $CaCO_3$, occurs naturally as chalk and limestone. Marble is a dense form of calcium carbonate that can be given a high polish; it is often colored by impurities, most commonly iron cations (Fig. 19.23). The two most common forms of pure calcium carbonate are calcite and aragonite. All these carbonates are the fossilized remains of marine life. Calcium carbonate decomposes to calcium oxide, CaO, or *quicklime,* when heated:

FIGURE 19.22 The green of vegetation is caused by the absorption of red and blue light by the chlorophyll molecules in leaves. Green light is not absorbed but reflected, thereby giving the leaves their green color.

$$CaCO_3(s) \xrightarrow{\Delta} CaO(s) + CO_2(g)$$

The decomposition of $CaCO_3$ requires a higher temperature (about 800°C) than does that of $MgCO_3$. The difference can be explained by recognizing that the larger Ca^{2+} ion is less effective than Mg^{2+} at removing an O^{2-} ion from a neighboring CO_3^{2-} ion.

Quicklime is produced in enormous quantities throughout the world and ranks fifth by mass manufactured in the United States. About 40% of this output is used in metallurgy. In ironmaking (Section 21.3), it is used as a Lewis base; its O^{2-} ion reacts with silica (SiO_2) impurities in the ore to form a liquid slag:

$$CaO(s) + SiO_2(s) \xrightarrow{\Delta} CaSiO_3(l)$$

About 50 kg of lime are needed to produce 1 ton of iron. Lime is also used as an inexpensive base in industry, to adjust soil pH, and to soften water (see Case Study 3).

Calcium compounds are often used as structural materials; their rigidity stems from the strength with which the small, highly charged Ca^{2+} cation interacts with its neighbors. *Mortar* consists of about one part lime and three parts sand (largely silica, SiO_2). It sets to a hard mass as the lime reacts with the carbon dioxide of the air to form the carbonate (Fig. 19.24). Calcium is also found in the rigid structural components of living things, either as the calcium carbonate of the shells of shellfish or the calcium phosphate of bone. About a kilogram of calcium is present in an adult human body, mostly in the form of insoluble calcium phosphate, but also as Ca^{2+} ions in other fluids inside our cells. The calcium in newly formed bone is in dynamic equilibrium with the calcium ions in the body fluids, so calcium must be part of our daily diet to maintain bone strength.

Tooth enamel is a *hydroxyapatite,* a phosphate mineral of composition $Ca_5(PO_4)_3OH$. Tooth decay begins when acids attack the enamel:

$$Ca_5(PO_4)_3OH(s) + 4\,H_3O^+(aq) \longrightarrow 5\,Ca^{2+}(aq) + 3\,HPO_4^{2-}(aq) + 5\,H_2O(l)$$

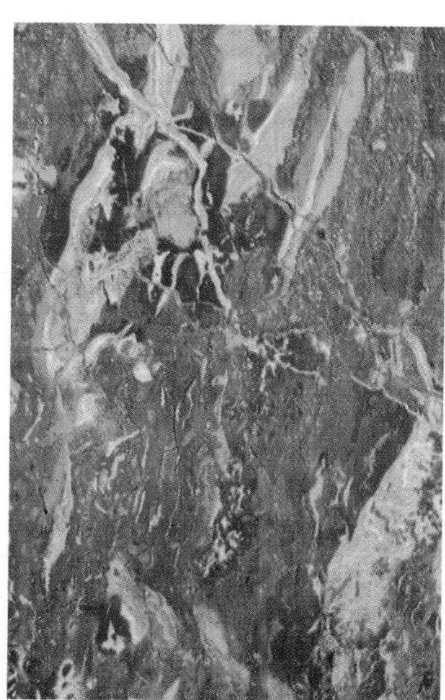

FIGURE 19.23 Marble is a dense form of calcium carbonate. It is often colored by impurities, such as iron cations.

FIGURE 19.24 An electron micrograph of the surface of mortar, showing the growth of tiny interlocking crystals as carbon dioxide reacts with calcium hydroxide and silica.

The principal agents of tooth decay are the carboxylic acids produced when bacteria act on the remains of food. A more resistant coating forms when the OH^- ions in the apatite are replaced by F^- ions. The resulting mineral is called *fluorapatite:*

$$Ca_5(PO_4)_3OH(s) + F^-(aq) \longrightarrow Ca_5(PO_4)_3F(s) + OH^-(aq)$$

The addition of fluoride ions to domestic water supplies (by addition of NaF) is now widespread. Fluoridated toothpastes, containing either tin(II) fluoride or sodium monofluorophosphate (MFP, Na_2FPO_3), are also recommended to strengthen tooth enamel.

Beryllium compounds have a pronounced covalent character, and the structural unit is commonly tetrahedral. The small size of the magnesium cation results in a thermally stable oxide with low solubility in water. Calcium compounds are common in structural materials, because the small, highly charged Ca^{2+} ion results in rigid structures.

SELF-TEST 19.2A Explain why beryllium compounds have covalent characteristics.

[*Answer:* The small size and high charge on the beryllium ion make it highly polarizing.]

SELF-TEST 19.2B Explain why MgO has such a high melting point.

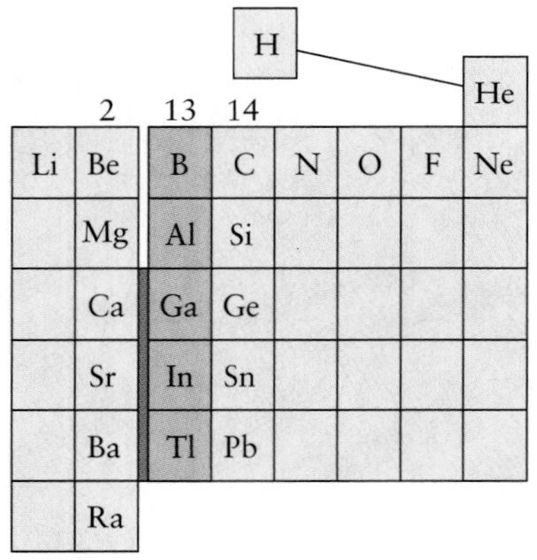

GROUP 13: THE BORON FAMILY

Group 13 is the first group of the *p*-block. Its members have an ns^2np^1 electron configuration (Table 19.9), so we expect a maximum oxidation number of +3. The oxidation numbers of B and Al are +3 in almost all their compounds (Table 19.10). However, the heavier elements in the group are more likely to keep their *s*-electrons (the inert-pair effect, Section 7.18), so the oxidation number +1 becomes increasingly important on going down the group, and Tl(I) compounds are as common as Tl(III) compounds. We shall concentrate on the two most important members of the group, boron and aluminum.

19.8 THE ELEMENTS

Boron is mined as *borax* and *kernite,* $Na_2B_4O_7 \cdot xH_2O$, with $x = 10$ and 4, respectively. Large deposits are found in volcanic regions, such as the Mojave Desert region of California (Fig. 19.25). In the extraction process, the minerals are converted into boron oxide with acid and then reduced with magnesium to an impure brown, amorphous form of boron:

$$B_2O_3(s) + 3\,Mg(s) \xrightarrow{\Delta} 2\,B(s) + 3\,MgO(s)$$

TABLE 19.9 Group 13 elements

Valence configuration: ns^2np^1

Z	Name	Symbol	Molar mass, $g{\cdot}mol^{-1}$	Melting point, °C	Boiling point, °C	Density, $g{\cdot}L^{-1}$	Normal form*
5	boron	B	10.81	2030	3700	2.47	brown nonmetallic powder
13	aluminum	Al	26.98	660	2350	2.70	silver-white metal
31	gallium	Ga	69.72	30	2070	5.91	silver metal
49	indium	In	114.82	157	2050	7.29	silver-white metal
81	thallium	Tl	204.37	304	1460	11.87	soft metal

**Normal form* means the state and appearance of the element at 25°C and 1 atm.

Boron production remains quite low despite the element's desirable properties of hardness and lightness.

Here is an opportunity for a young chemist to transform the technology of boron production as Charles Hall did for aluminum (see below).

Elemental boron has several allotropes. It is typically either a gray-black nonmetallic, high-melting-point solid or a dark brown powder with a structure based on clusters of 12 atoms (Fig. 19.26). When boron fibers are incorporated in plastics, the result is a very tough material that is stiffer than steel yet lighter than aluminum and used in aircraft, missiles, and body armor. The element is very inert and is attacked by only the strongest oxidizing agents.

Aluminum is the most abundant metallic element in the Earth's crust and, following oxygen and silicon, the third most abundant element. However, the aluminum content in most minerals is low, and the commercial source of aluminum, *bauxite,* is a hydrated, impure oxide, $Al_2O_3{\cdot}xH_2O$, where x can range up to 3. The bauxite ore, which is red from the iron oxides it contains, is processed to obtain alumina, Al_2O_3.

Aluminum metal is obtained by the *Hall process.* Hall discovered that by mixing the mineral *cryolite,* Na_3AlF_6, with alumina, he got a mixture that melted at a much more economical temperature, 950°C, instead of the 2050°C of pure alumina. The melt is electrolyzed in a cell that uses graphite (or carbonized petroleum) anodes and a carbonized steel-lined

The process was developed by Charles Hall of Oberlin College, Ohio in 1886, when he was 22 years old.

TABLE 19.10 Chemical properties of the Group 13 elements

Reactant	Reaction with Group 13 element (E)
oxygen	$4\,E(s) + 3\,O_2(g) \longrightarrow 2\,E_2O_3(s)$
nitrogen	$2\,E(s) + N_2(g) \longrightarrow 2\,EN(s)$, E = B, Al
halogen (X_2)	$2\,B(s) + 3\,X_2(g,l,s) \longrightarrow 2\,BX_3(g)$
	$2\,E(s) + 3\,X_2(g,l,s) \longrightarrow E_2X_6(g)$, E = Al, Ga, In
	$2\,Tl(s) + X_2(g,l,s) \longrightarrow 2\,TlX(s)$
water	$2\,Tl(s) + 2\,H_2O(l) \longrightarrow 2\,TlOH(aq) + H_2(g)$
acid	$2\,E(s) + 6\,H_3O^+(aq) \longrightarrow 2\,E^{3+}(aq) + 6\,H_2O(l) + 3\,H_2(g)$, E = Al, Ga, Tl
base	$2\,E(s) + 6\,H_2O(l) + 2\,OH^-(aq) \longrightarrow 2\,E(OH)_4^-(aq) + 3\,H_2(g)$, E = Al, Ga

FIGURE 19.25 Boron, California, a major source of borax and hence of the element boron.

Before recycling of aluminum was widespread, the aluminum industry in the United States consumed each day the electricity used by 100,000 people in one year!

vat that serves as the cathode (Fig. 19.27). The electrolysis reactions are

$$\text{Cathode reaction:} \quad Al^{3+}(\text{melt}) + 3\,e^- \longrightarrow Al(l)$$

$$\text{Anode reaction:} \quad 2\,O^{2-}(\text{melt}) + C(s, gr) \longrightarrow CO_2(g) + 4\,e^-$$

The overall reaction is

$$4\,Al^{3+}(\text{melt}) + 6\,O^{2-}(\text{melt}) + 3\,C(s, gr) \longrightarrow 4\,Al(l) + 3\,CO_2(g)$$

Note that the carbon electrode takes part in the reaction. From the reaction stoichiometry, we can calculate that a current of 1 A must flow for 80 h to produce 1 mol Al (27 g of aluminum, about enough for one soft-drink can). The very high electricity consumption can be greatly reduced by recycling, which requires less than 5% of the electricity needed to extract aluminum from bauxite.

Aluminum has a low density; it is a strong metal and an excellent electrical conductor. Although it is strongly reducing and thus easily oxidized, it is resistant to corrosion because its surface is passivated in air by a stable oxide film. The thickness of the oxide layer can be increased by making aluminum the anode of an electrolytic cell; the result is called *anodized aluminum.* Dyes may be added to the dilute sulfuric acid electrolyte used in the anodizing process, to produce surface layers with different colors. Brown and bronze anodized aluminum parts produced in this way are widely used in modern architecture (Fig. 19.28).

Aluminum's low density, wide availability, and corrosion resistance make it ideal for construction. For use in airplanes, it is usually alloyed with copper and silicon. Its lightness and good electrical conductivity have led to its use for overhead power lines.

Boron, at the head of Group 13, is best regarded as a nonmetal in most of its chemical properties. It has acidic oxides and forms an interesting and extensive range of binary molecular hydrides. However, metallic character increases down the group, and even boron's immediate neighbor aluminum is a metal. Nonetheless, aluminum is sufficiently far to the right in the periodic table to show a hint of nonmetallic character. Thus,

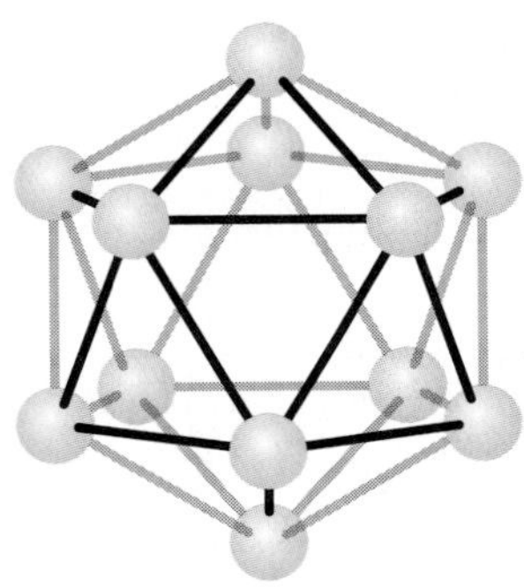

FIGURE 19.26 The structure of boron is based on linked 12-atom units. The unit has 20 faces, so it is called an icosahedron (from the Greek words for "twenty-faced").

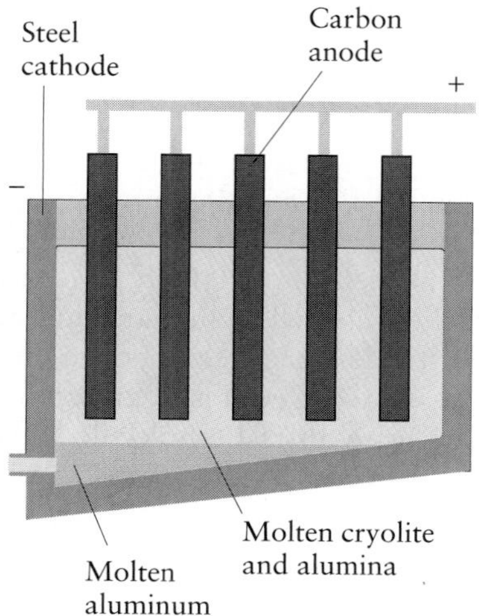

FIGURE 19.27 In the Hall process, aluminum oxide is dissolved in molten cryolite and the mixture is electrolyzed in a cell with carbon anodes and a steel cathode.

aluminum is amphoteric, reacting both with nonoxidizing acids (such as hydrochloric acid) to form aluminum ions:

$$2\,Al(s) + 6\,H^+(aq) \longrightarrow 2\,Al^{3+}(aq) + 3\,H_2(g)$$

and with hot aqueous alkali to form aluminate ions:

$$2\,Al(s) + 2\,OH^-(aq) + 6\,H_2O(l) \longrightarrow 2\,[Al(OH)_4]^-(aq) + 3\,H_2(g)$$

Recall from Section 3.9 that amphoteric substances react with either acids or bases.

Boron is a hard, nonmetallic element. Aluminum is a light, strong, amphoteric, reactive metallic element with a surface that becomes passivated when exposed to air.

FIGURE 19.28 Anodized aluminum used in buildings is produced by the partial electrolytic oxidation of aluminum. Different colors are produced by incorporating dyes into the electrolyte solution. Colors may also be painted on the anodized aluminum, as for the window frames.

19.9 GROUP 13 OXIDES

Boron, a nonmetal, has acidic oxides. Aluminum, its metallic neighbor, has amphoteric oxides (like its diagonal relative in Group 2, beryllium). The oxides of both elements are important in their own right, as sources of the elements and as the starting point for the manufacture of other compounds.

Boric acid, $B(OH)_3$, is a white solid that melts at 171°C. It is toxic to bacteria and many insects (including roaches) as well as humans and has long been used as a mild antiseptic and pesticide. Because the boron atom in $B(OH)_3$ has an incomplete octet, it can act as a Lewis acid and form a bond by accepting a lone pair of electrons from an H_2O molecule acting as a Lewis base:

$$(OH)_3B + :OH_2 \longrightarrow (OH)_3B{-}OH_2$$

The compound so formed is a weak *monoprotic* acid:

$$B(OH)_3OH_2(aq) + H_2O(l) \rightleftharpoons H_3O^+(aq) + B(OH)_4^-(aq) \qquad pK_a = 9.14$$

Boric acid also retards the spread of flames in cellulosic materials, particularly paper. The scrap paper used to manufacture home insulation contains about 5% boric acid, to reduce the risk of fire. However, the major use of boric acid is as the starting material for its anhydride, boron oxide, B_2O_3. Because it melts (at 450°C) to a liquid that dissolves many metal oxides, boron oxide (often as the acid) is used as a *flux,* a substance that cleans metals before they are soldered or welded. Boron oxide is also used to make fiberglass and borosilicate glass, a glass with a very low thermal expansion, such as Pyrex.

We can recognize some of the oxides of *p*-block elements as **acid anhydrides,** compounds that form the acid when they react with water, as CO_2 forms carbonic acid, H_2CO_3. However, some oxides may be only **formal anhydrides,** compounds that have the formula of an acid less the elements of water but do not actually form the acid when they dissolve in water. An example is CO, the formal anhydride of formic acid, HCOOH. A number of acid anhydrides can be formed by simply heating the oxoacid. This is the case with boric acid:

$$2\,B(OH)_3(s) \xrightarrow{\Delta} B_2O_3(s) + 3\,H_2O(g)$$

Aluminum oxide, Al_2O_3, is almost universally known as *alumina.* It exists with a variety of crystal structures. As α-alumina, it is the very hard substance *corundum;* impure microcrystalline corundum is the purple-black abrasive known as *emery.* Some impure forms of alumina are beautiful, rare, and highly prized (Fig. 19.29). A less dense and more reactive form of the oxide is γ-alumina. This form absorbs water and is used as the stationary phase in chromatography (Section 1.12).

γ-Alumina is produced by heating aluminum hydroxide. It is quite reactive and is amphoteric, dissolving readily in bases to produce the aluminate ion, and in acids to produce the hydrated Al^{3+} ion:

$$Al_2O_3(s) + 2\,OH^-(aq) + 3\,H_2O(l) \longrightarrow 2\,Al(OH)_4^-(aq)$$

$$Al_2O_3(s) + 6\,H_3O^+(aq) + 3\,H_2O(l) \longrightarrow 2\,[Al(H_2O)_6]^{3+}(aq)$$

The strong polarizing effect of the small, highly charged Al^{3+} ion on the water molecules around it results in the $[Al(H_2O)_6]^{3+}$ ion being an acid.

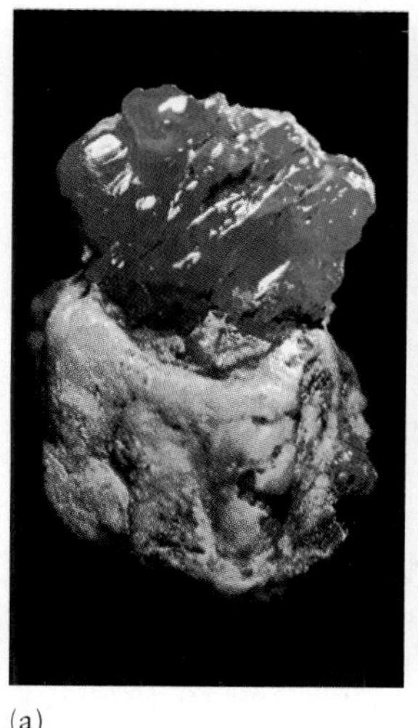

(a) (b) (c)

FIGURE 19.29 Some of the impure forms of α-alumina are prized as gems. (a) Ruby is alumina with Cr^{3+} in place of some Al^{3+} ions. (b) Sapphire is alumina with Fe^{3+} and Ti^{4+} impurities. (c) Topaz is alumina with Fe^{3+} impurities.

Solutions of aluminum salts are therefore acidic.

One of the most important aluminum salts prepared by the action of an acid on alumina is aluminum sulfate, $Al_2(SO_4)_3$:

$$Al_2O_3(s) + 3\,H_2SO_4(aq) \longrightarrow Al_2(SO_4)_3(aq) + 3\,H_2O(l)$$

Aluminum sulfate is called *papermaker's alum* and is used in the paper industry to coagulate cellulose fibers into a hard, nonabsorbent surface. True *alums* (from which aluminum takes its name) are mixed sulfates of formula $M^+M'^{3+}(SO_4)_2\cdot 12H_2O$, and include potassium alum, $KAl(SO_4)_2\cdot 12H_2O$ (which is used in water and sewage treatment), and ammonium alum, $NH_4Al(SO_4)_2\cdot 12H_2O$ (which is used for pickling cucumbers). Other alums are used for waterproofing fabrics and as mordants in dying and printing textiles.

A "mordant" is a compound that helps to attach the dye to the fabric.

Sodium aluminate, $NaAl(OH)_4$, is used along with aluminum sulfate in water purification (see Case Study 3). When mixed with aluminate ions, the acidic hydrated Al^{3+} cation from the aluminum sulfate produces aluminum hydroxide:

$$Al^{3+}(aq) + 3\,Al(OH)_4^-(aq) \longrightarrow 4\,Al(OH)_3(s)$$

The aluminum hydroxide is formed as a fluffy, gelatinous network that entraps impurities as it settles, and this precipitate can be removed by filtration (Fig. 19.30). Using aluminum for both the cation and the anion gives the greatest possible bulk of impurity-collecting alumina, because the reaction has no by-products.

Boron oxide is an acid anhydride. Aluminum shows some nonmetallic character in that its oxide is amphoteric.

FIGURE 19.30 Aluminum hydroxide, $Al(OH)_3$, forms as a white, fluffy precipitate. The fluffy form of the solid captures impurities and is used in the purification of water.

19.10 CARBIDES, NITRIDES, AND HALIDES

When boron is heated to high temperatures with carbon, it forms boron carbide, $B_{12}C_3$, a high-melting-point solid that is almost as hard as diamond. The solid consists of B_{12} groups that are pinned together by C atoms. When boron is heated to white heat in ammonia, boron nitride, BN, is formed as a fluffy, slippery powder:

$$2\,B(s) + 2\,NH_3(g) \xrightarrow{\Delta} 2\,BN(s) + 3\,H_2(g)$$

Its structure resembles that of graphite, but the latter's flat planes of carbon hexagons are replaced by planes of hexagons of alternating B and N atoms (Fig. 19.31). Unlike graphite, it is white and does not conduct electricity. Under high pressure, boron nitride is converted to a very hard, diamondlike crystalline form called Borazon. In recent years, boron nitride nanotubes similar to those formed by carbon (Section 19.13) have been synthesized, and they have been found to be semiconducting. They have the potential for interesting applications in microelectronics.

The boron halides are made either by direct reaction of the elements at high temperature or from the oxide. The most important is boron trifluoride, BF_3, an industrial catalyst produced by the reaction between boric oxide, calcium fluoride, and sulfuric acid:

$$B_2O_3(s) + 3\,CaF_2(s) + 3\,H_2SO_4(l) \xrightarrow{\Delta} 2\,BF_3(g) + 3\,CaSO_4(s) + 3\,H_2O(l)$$

Boron trichloride, BCl_3, which is also widely used as a catalyst, is produced commercially by the action of the halogen on the oxide in the presence of carbon:

$$B_2O_3(s) + 3\,C(s) + 3\,Cl_2(g) \xrightarrow{500°C} 2\,BCl_3(g) + 3\,CO(g)$$

The B atom has an incomplete octet in all its trihalides (Section 8.11). The compounds consist of planar triangular molecules with an empty 2*p*-orbital perpendicular to the molecular plane. The empty orbital allows the molecules to act as Lewis acids, which accounts for the catalytic action of BF_3 and BCl_3.

Aluminum chloride, $AlCl_3$, another major industrial catalyst, is made by the action of chlorine on aluminum or on alumina in the presence of carbon:

$$2\,Al(s) + 3\,Cl_2(g) \longrightarrow 2\,AlCl_3(s)$$

$$Al_2O_3(s) + 3\,C(s) + 3\,Cl_2(g) \longrightarrow 2\,AlCl_3(s) + 3\,CO(g)$$

Aluminum chloride is an ionic solid in which each Al^{3+} ion is surrounded by six Cl^- ions. However, it sublimes at 192°C to a vapor of Al_2Cl_6 mole-

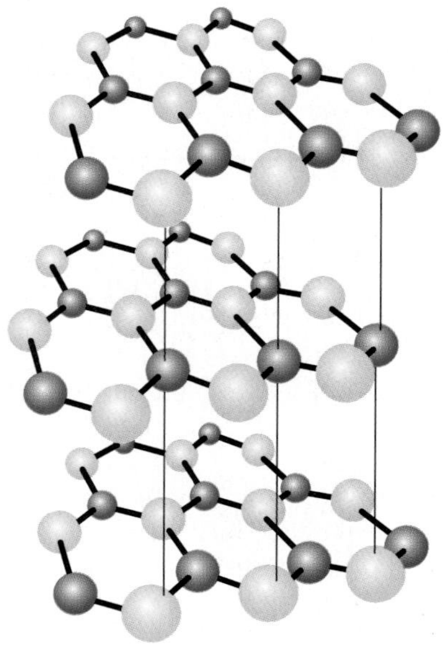

FIGURE 19.31 The structure of boron nitride, BN, resembles that of graphite, consisting of flat planes of hexagons. In boron nitride, however, the hexagons consist of alternating B and N atoms (in place of C atoms) and are stacked differently. (Compare with Fig. 19.35.)

FIGURE 19.32 When anhydrous aluminum chloride is left exposed to moist air, it reacts to form hydrochloric acid. Here white fumes of ammonium chloride form as the hydrochloric acid reacts with ammonia released in the vicinity.

cules (4). An Al_2Cl_6 molecule is an example of a **dimer,** the union of two identical molecules.

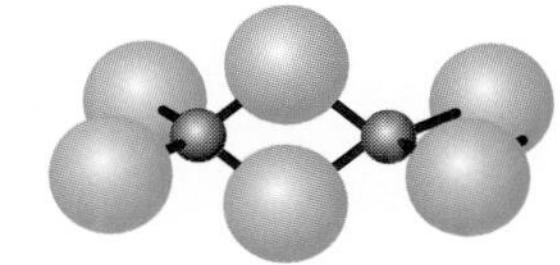

4 Aluminum chloride dimer, Al_2Cl_6

Aluminum halides react with water with a considerable evolution of heat. When the anhydrous chloride is exposed to moist air, it produces fumes of hydrochloric acid (Fig. 19.32). The ionic aluminum chloride hexahydrate, $AlCl_3{\cdot}6H_2O$, is used as a deodorant and antiperspirant, one of its roles being to kill the bacteria that feed on perspiration and produce unpleasant smells.

Boron and aluminum halides act as Lewis acids.

19.11 BORANES AND BOROHYDRIDES

Sodium borohydride is a white crystalline solid produced from the reaction between sodium hydride and boron trichloride dissolved in a nonaqueous solvent:

$$4\,NaH + BCl_3 \longrightarrow NaBH_4 + 3\,NaCl$$

Sodium borohydride is a very useful reducing agent. At pH = 14 (strongly alkaline conditions), the standard potential of the half-reaction

$$H_2BO_3^-(aq) + 5\,H_2O(l) + 8\,e^- \longrightarrow BH_4^-(aq) + 8\,OH^-(aq)$$

is −1.24 V. Because this standard potential is well below that of the $Ni^{2+}(aq) + 2\,e^- \rightarrow Ni(s)$ half-reaction (−0.23 V), the borohydride ion can reduce Ni^{2+} ions to metallic nickel. This reduction is the basis of the **chemical plating** of nickel (Fig. 19.33). The advantage of this chemical plating over electroplating is that the item being plated does not have to be an electrical conductor.

FIGURE 19.33 It is difficult to coat nonconducting objects with a metal surface. One technique is by chemical reduction. Another is by vapor deposition. The latter technique has been used to coat this figurine of the Star Trek character Worf, shown in ritual attire.

The boranes are an extensive series of binary compounds of boron and hydrogen, somewhat analogous to the hydrocarbons. The starting point for borane production is the reaction (in an organic solvent) of sodium borohydride with boron trifluoride:

$$4\,BF_3 + 3\,BH_4^- \longrightarrow 3\,BF_4^- + 2\,B_2H_6$$

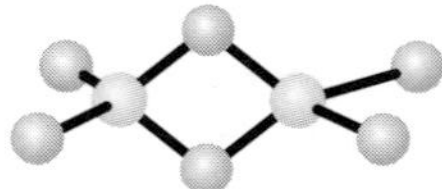

5 Diborane, B_2H_6

The product B_2H_6 is diborane (**5**), a colorless gas that bursts into flame in air. On contact with water, it is immediately oxidized to boric acid as it reduces the hydrogen in the water:

$$B_2H_6(g) + 6\,H_2O(l) \longrightarrow 2\,B(OH)_3(aq) + 6\,H_2(g)$$

When diborane is heated to a high temperature, it decomposes into hydrogen and pure boron:

$$B_2H_6(g) \xrightarrow{\Delta} 2\,B(s) + 3\,H_2(g)$$

This sequence of reactions is a useful route to the pure element, but more complex boranes form when the heating is less severe. When diborane is heated to 100°C, for instance, it forms decaborane, $B_{10}H_{14}$, a solid that melts at 100°C. Decaborane is stable in air, is oxidized by water only slowly, and is an example of the general rule that heavier boranes are less flammable than boranes of low molar mass.

The boranes are examples of **electron-deficient compounds,** compounds for which valid Lewis structures cannot be written because too few electrons are available. For instance, in diborane there are eight atoms, so we need at least seven bonds; however, there are only 12 electrons, so we can form at most six electron-pair bonds. In the modern theory of molecular structure, electron pairs are regarded as **delocalized,** or spread over the entire molecule. Consequently, the bonding power of one electron pair can be shared by several atoms. In the case of diborane, a single electron pair is regarded as delocalized over a B—H—B unit, and it binds all three atoms together. There are two such bridging **three-center bonds** in the molecule (**6**).

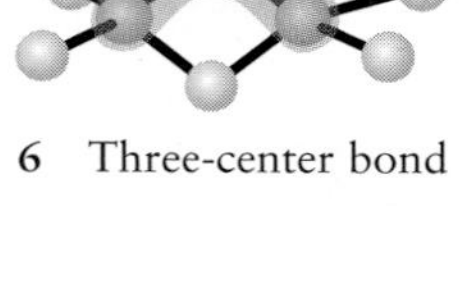

6 Three-center bond

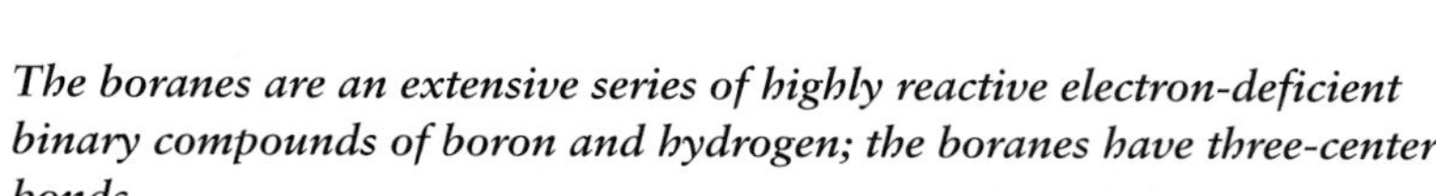

The boranes are an extensive series of highly reactive electron-deficient binary compounds of boron and hydrogen; the boranes have three-center bonds.

SELF-TEST 19.3A Write the formulas of the anhydrides of the acids (a) H_2SO_4; (b) $B(OH)_3$.

[*Answer:* (a) SO_3; (b) B_2O_3]

SELF-TEST 19.3B Why is aluminum resistant to corrosion, even though it is strongly electropositive?

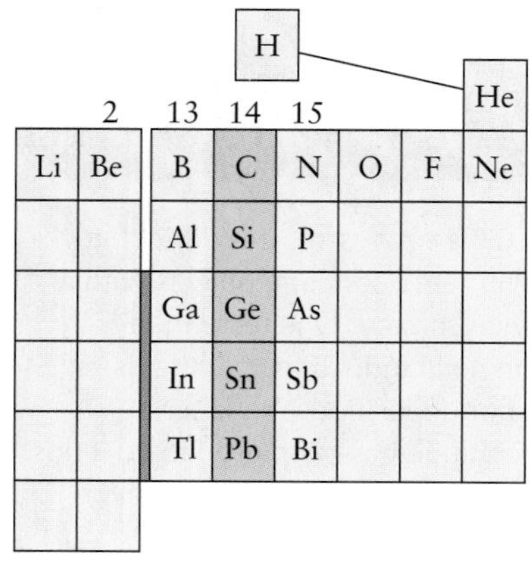

GROUP 14: THE CARBON FAMILY

Group 14 includes carbon, the element most central to life, and silicon, the element most central to modern technology and artificial intelligence (Fig. 19.34, Table 19.11). The half-filled valence shell of these elements gives them special properties that straddle the line between metals and nonmetals. Carbon, at the head of the group, forms so many compounds that it has its own branch of chemistry, organic chemistry (Chapter 11).

FIGURE 19.34 The elements of Group 14: (from left to right) carbon (graphite), silicon, germanium, tin, and lead.

19.12 THE ELEMENTS

The elements show increasing metallic character down the group (Table 19.12). Carbon has definite nonmetallic properties: it forms covalent compounds with nonmetals and ionic compounds with metals. The oxides of carbon and silicon are acidic. Germanium is a typical metalloid in that it exhibits metallic or nonmetallic properties according to the other element present in the compound. Tin and, even more so, lead have definite metallic properties. However, even though tin is classified as a metal, it is not far from the metalloids in the periodic table, and it does have some amphoteric properties. For example, it reacts with both hot concentrated hydrochloric acid and hot alkali:

$$Sn(s) + 2\,HCl(aq) \longrightarrow SnCl_2(aq) + H_2(g)$$

$$Sn(s) + 2\,OH^-(aq) + 2\,H_2O(l) \longrightarrow [Sn(OH)_4]^{2-}(aq) + H_2(g)$$

Because it stands at the head of its group, we expect carbon to be different from the other members of the group. In fact, the differences are

TABLE 19.11 Group 14 elements

Valence configuration: ns^2np^2

Z	Name	Symbol	Molar mass, $g{\cdot}mol^{-1}$	Melting point, °C	Boiling point, °C	Density, $g{\cdot}L^{-1}$	Normal form*
6	carbon	C	12.01	3370s†	—	1.9 to 2.3	black nonmetal (graphite)
						3.2 to 3.5	transparent nonmetal (diamond)
							orange nonmetal (fullerite)
14	silicon	Si	28.09	1410	2620	2.33	gray metalloid
32	germanium	Ge	72.59	937	2830	5.32	gray-white metalloid
50	tin	Sn	118.69	232	2720	7.29	white lustrous metal
82	lead	Pb	207.19	328	1760	11.34	blue-white lustrous metal

**Normal form* means the state and appearance of the element at 25°C and 1 atm.
†The symbol s denotes that the element sublimes.

TABLE 19.12 Chemical properties of the Group 14 elements

Reactant	Reaction with Group 14 element (E)
hydrogen	$C(s) + 2\,H_2(g) \longrightarrow CH_4(g)$ and other hydrocarbons
oxygen	$E(s) + O_2(g) \longrightarrow EO_2(s)$, E = C, Si, Ge, Sn
	$2\,Pb(s) + O_2(g) \longrightarrow 2\,PbO(s)$
halogen (X_2)	$E(s) + 2\,X_2(g,l,s) \longrightarrow EX_4(s,l,g)$, E = C, Si, Ge, Sn
	$Pb(s) + X_2(g,l,s) \longrightarrow PbX_2(s)$
water	$C(s) + H_2O(g) \xrightarrow{\Delta} CO(g) + H_2(g)$
	$Si(s) + 2\,H_2O(l) \xrightarrow{\Delta} SiO_2(s) + 2\,H_2(g)$
acid	$E(s) + 2\,H_3O^+(aq) \longrightarrow E^{2+}(aq) + 2\,H_2O(l) + H_2(g)$, E = Sn, Pb
base	$E(s) + 2\,H_2O(l) + 2\,OH^-(aq) \longrightarrow E(OH)_4^{2-}(aq) + H_2(g)$, E = Sn, Pb

more pronounced in Group 14 than anywhere else in the periodic table. For example, compare carbon and silicon. Some of the differences stem from the wide occurrence of C=C and C=O double bonds, compared with the rarity of Si=Si and Si=O double bonds. Carbon dioxide, which consists of discrete O=C=O molecules, is a gas that we breathe. Silicon dioxide (silica), which consists of networks of —O—Si—O— groups, is a mineral we stand on.

Silicon compounds can also act as Lewis acids, whereas carbon compounds cannot. Because a silicon atom is bigger than a carbon atom and can expand its valence shell by using its *d*-orbitals, it can accommodate the lone pair of an attacking Lewis base. A carbon atom is smaller and has no low-lying *d*-orbitals, so it cannot act as a Lewis acid.

The valence electron configuration is ns^2np^2 for all members of the group. All four electrons are approximately equally available for bonding in the lighter elements, and carbon and silicon are characterized by their ability to form four covalent bonds. However, on descending the group, the energy separation between the *s*- and *p*-orbitals increases and the *s*-electrons become progressively less available for bonding; in fact, the most common oxidation number for lead is +2.

Carbon is the only member of Group 14 that commonly forms multiple bonds; silicon compounds can act as Lewis acids because a silicon atom can expand its valence shell.

19.13 THE MANY FACES OF CARBON

Solid carbon exists as graphite, diamond, and—as we now know—fullerite. Graphite is the thermodynamically most stable of these allotropes under normal conditions. Pure graphite is produced in industry by passing a heavy electric current for several days through rods of *coke,* the solid left after distillation of the volatile components of coal. Natural sources of diamonds are rare. Synthetic diamonds are made at high pressure and high temperature (Section 10.12) or by thermal decomposition of methane. In the latter technique the carbon atoms settle on a cool surface as graphite and diamond. However, hydrogen atoms produced in the decomposition react more quickly with graphite to form volatile hydrocarbons, so more diamond than graphite survives.

Soot and *carbon black* contain very small crystals of graphite. Carbon black, which is produced by heating gaseous hydrocarbons to nearly 1000°C in the absence of air, is used for reinforcing rubber, for pigments, and for printing inks, such as the ink on this page. *Activated carbon,* which is also called activated charcoal, consists of granules of microcrystalline carbon. It is produced by heating waste organic matter in the absence of air and then processing it to increase the porosity. The very high surface area (up to about 2000 $m^2 \cdot g^{-1}$) of the porous carbon enables it to remove organic impurities from liquids and gases by adsorption. It is used in air purifiers, gas masks, and aquarium water filters. On a larger scale, activated carbon is used in emission-control canisters of automobiles to minimize the release of unburned hydrocarbons and in water purification plants to remove organic compounds from drinking water.

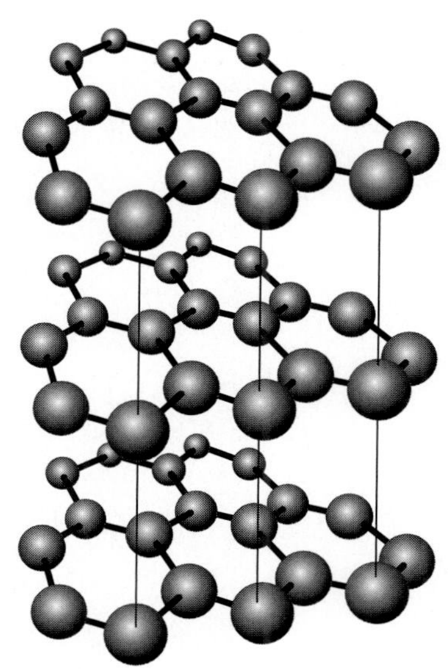

FIGURE 19.35 The structure of graphite consists of flat planes of hexagons lying one above another. When impurities are present, the planes can slide over one another quite easily. Graphite conducts electricity well within the planes, but less well perpendicular to the planes.

Differences in bonding explain the differences in properties between the carbon allotropes. Graphite consists of planar sheets of sp^2-hybridized carbon atoms in a hexagonal network (Fig. 19.35). Electrons are free to move from one carbon atom to another through a delocalized π-network formed by the overlap of unhybridized p-orbitals on each carbon atom. This network spreads across the entire plane. Because of the electron delocalization, graphite is a black, lustrous, electrically conducting solid; indeed, graphite is used as an electrical conductor in industry. Its slipperiness, which results from the ease with which the flat planes move past one another when impurities are present, leads to its use as a lubricant. In diamond, each carbon atom is sp^3-hybridized and linked tetrahedrally to its four neighbors, with all electrons localized in C—C σ bonds (see Fig. 19.43b). Diamond is a rigid, transparent, electrically insulating solid. It is the hardest substance known and the best conductor of heat, being about five times better than copper. These last two properties make it an ideal abrasive, for it can scratch all other substances, yet the heat generated by friction is quickly conducted away.

Chemists were greatly surprised when soccer ball-shaped carbon molecules were first identified in 1985, particularly because they might be even more abundant than graphite and diamond! The C_{60} molecule (7) is named buckminsterfullerene after the American architect whose "geodesic domes" they resemble (Fig. 19.36). Within two years, scientists had succeeded in making crystals of them: these solid samples are called fullerite (Fig. 19.37). Like the discovery of benzene, these molecules opened up the prospect of a whole new field of chemistry. For instance, the interior of a C_{60} molecule is big enough to hold an atom of another element, and chemists are now busily preparing a whole new periodic table of these "shrink-wrapped" atoms. The *fullerenes* are members of the family of molecules resembling buckminsterfullerene but having more than 60 atoms. The reason they might be very abundant is that they are formed in smoky flames and by red giants (stars with low surface temperatures and large diameters), so the universe might contain huge numbers of them.

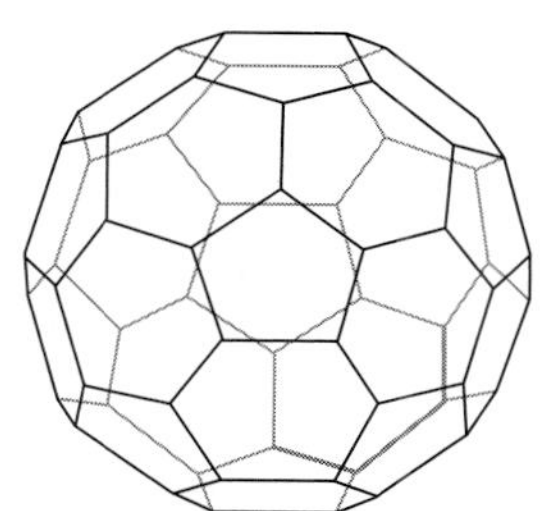

7 Buckminsterfullerene, C_{60}

The 1996 Nobel prize for chemistry was awarded for the discovery of buckminsterfullerene.

Graphite and diamond are network solids that are insoluble in liquid solvents. However, the fullerenes, which are molecular, can be dissolved by suitable solvents (such as benzene); buckminsterfullerene itself gives a red-brown solution. Fullerite currently has few uses, but some of the

FIGURE 19.36 The American Pavilion at the Montréal Expo in 1967 was a geodesic dome designed by R. Buckminster Fuller. The dome is a three-quarter sphere, 200 feet high and 250 feet in diameter.

compounds of the fullerenes have great promise. For example, K_3C_{60} is a superconductor below 18 K, and other compounds appear to be active against cancer and diseases such as AIDS.

Spurred on by the discovery of fullerenes, chemists are also busily—and excitedly—looking into the properties of a new form of fibrous carbon that consists of concentric tubes with walls like sheets of graphite rolled into cylinders (Fig. 19.38). These tiny structures, called *nanotubes*, hold out the promise of forming strong, conducting fibers (see Case Study 9). Moreover, atoms of metals can be introduced inside the tubes, so chemists are currently developing wires one atom in diameter. The

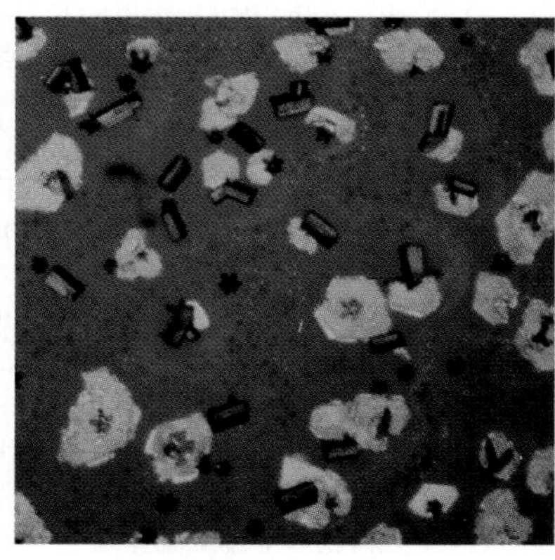

FIGURE 19.37 The small crystals are fullerite, in which buckminsterfullerene molecules are packed together in a close-packed structure like those shown in Figs. 10.19–10.21.

prospects for miniaturization of electronic components are intriguing and exciting.

Carbon has an important series of allotropes: diamond, graphite, and the fullerenes.

FIGURE 19.38 A cross section through a concentric series of carbon nanotubes. Spacing between layers is about 0.34 nm, slightly further apart than layers of graphite. The end of a carbon nanotube can pucker to form a cap, as shown here, if five-membered rings begin to form.

19.14 SILICON, TIN, AND LEAD

Silicon is the second most abundant element in the Earth's crust. It occurs widely in rocks as silicates, compounds containing the silicate ion (SiO_3^{2-}), and as the silica (SiO_2) of sand (Fig. 19.39). Pure silicon is obtained from *quartzite,* a granular form of *quartz,* by reduction with high-purity carbon in an electric arc furnace:

$$SiO_2(s) + 2\,C(s) \xrightarrow{\Delta} Si(s) + 2\,CO(g)$$

The crude product is exposed to chlorine to form silicon tetrachloride, which is then distilled and reduced with hydrogen to a purer form of the element:

$$SiCl_4(l) + 2\,H_2(g) \longrightarrow Si(s) + 4\,HCl(g)$$

"Ultrapure" silicon for use in semiconductors is produced by **zone refining,** in which a hot, molten zone is dragged from one end of a cylindrical sample to the other, collecting impurities as it goes (Fig. 19.40).

Tin and lead are obtained very easily from their ores and have been known since antiquity. Tin occurs chiefly as the mineral *cassiterite,* SnO_2, and is obtained from it by reduction with carbon at 1200°C. The principal lead ore is *galena,* PbS. It is roasted in air, which converts it to PbO, and this oxide is then reduced with coke:

$$2\,PbS(s) + 3\,O_2(g) \longrightarrow 2\,PbO(s) + 2\,SO_2(g)$$

$$PbO(s) + C(s) \longrightarrow Pb(s) + CO(g)$$

Tin is expensive and not very strong, but it is resistant to corrosion. Its main use is in tinplating, which accounts for about 40% of its consumption. Tin is also used for the production of alloys.

Lead's durability (its chemical inertness) and malleability make it useful in the construction industry. The inertness of lead under normal conditions can be traced to the passivation of its surface by oxides, chlorides, and sulfates. Passivation allows lead to be used for transporting hot concentrated sulfuric acid but not nitric acid, because lead nitrate is soluble.

(a)

(b)

(c)

FIGURE 19.39 Three of the common forms of silica (SiO_2): (a) quartz, (b) quartzite, and (c) cristobalite. The black parts of the sample of cristobalite are obsidian, a volcanic rock that contains silica.

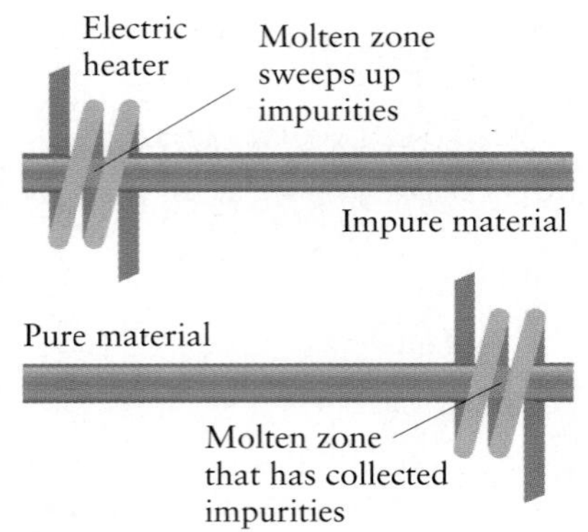

FIGURE 19.40 In the technique of zone refining, a heater that creates a molten zone is passed repeatedly from one end of the solid sample to the other. The impurities collect in the zone and are dragged through the sample.

Another important property of lead is its high density, which makes it useful as a radiation shield, because its numerous electrons absorb high-energy radiation. The main use of lead today is in the electrodes of storage batteries.

Metallic character increases significantly down Group 14.

19.15 OXIDES OF CARBON

Such is its importance that carbon dioxide, CO_2, has been described at length throughout the text. It is formed when organic matter burns in a plentiful supply of air. Carbon dioxide is the acid anhydride of carbonic acid, H_2CO_3, which forms when the gas dissolves in water. However, not all the dissolved molecules react to form the acid, and a solution of carbon dioxide in water is an equilibrium mixture of CO_2, H_2CO_3, HCO_3^-, and a very small amount of CO_3^{2-}.

Carbon monoxide, CO, is produced when carbon or organic compounds burn in a limited supply of air, as in cigarettes and badly tuned automobile engines. Commercially it is produced as synthesis gas by the re-forming reaction (Section 19.1):

$$CH_4(g) + H_2O(g) \longrightarrow CO(g) + 3\,H_2(g)$$

Carbon monoxide is the formal anhydride of formic acid, HCOOH, and can be produced in the laboratory by the dehydration of that acid with hot, concentrated sulfuric acid:

$$HCOOH(l) \xrightarrow{150^\circ C,\ H_2SO_4} CO(g) + H_2O(l)$$

Carbon monoxide is a colorless, odorless, flammable, almost insoluble, very toxic gas that condenses to a colorless liquid at $-90^\circ C$. It is not very reactive, largely because its bond enthalpy ($1074\ kJ\cdot mol^{-1}$) is the highest for any molecule. However, it is a Lewis base, and the lone pair on the carbon atom forms covalent bonds with *d*-block atoms and ions. An example of this behavior is its reaction with nickel to give nickel carbonyl, a toxic, volatile liquid:

$$Ni(s) + 4\,CO(g) \xrightarrow{50^\circ C,\ 1\ atm} Ni(CO)_4(l)$$

Although nickel carbonyl is intensely poisonous, it is used in the *Mond process* for the refinement of nickel (see Section 21.3). Complex formation is also responsible for carbon monoxide's toxicity: it attaches more strongly than oxygen to the iron in hemoglobin, and prevents it from accepting oxygen from lungs. As a result, the victim suffocates.

Carbon monoxide is a reducing agent used in the production of a number of metals, most notably iron in blast furnaces (Section 21.3):

$$Fe_2O_3(s) + 3\,CO(g) \xrightarrow{800^\circ C} 2\,Fe(l) + 3\,CO_2(g)$$

CHEMISTRY AT WORK

A business consultant gives a seminar to a group of investors. Consultants with training in chemistry advise investors on new pharmaceuticals, look into environmental law to help companies stay in compliance, and testify at trials. A knowledge of the chemical elements and their properties helps chemical consultants give accurate testimony and make reliable predictions.

Carbon has two important oxides, carbon dioxide and carbon monoxide; the former is the acid anhydride of carbonic acid, the parent acid of the hydrogen carbonates and the carbonates.

19.16 OXIDES OF SILICON: THE SILICATES

Silica, SiO_2, is a hard, rigid network solid that is insoluble in water. It occurs naturally as quartz and as sand, which consists of small fragments of quartz, usually colored golden brown by iron oxide impurities. Some precious and semiprecious stones are impure silica (Fig. 19.41). *Flint* is silica colored black by carbon impurities.

Silica gets its strength from its covalently bonded network structure. In silica itself, each silicon atom is at the center of a tetrahedron of oxygen atoms, and each corner O atom is shared by two Si atoms. Hence,

FIGURE 19.41 Impure forms of silica: amethyst (left), in which the color is due to Fe^{3+} impurities, agate (center), and onyx (right).

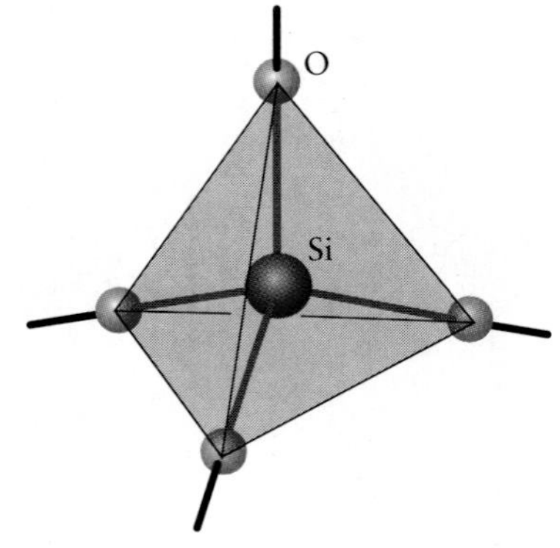

FIGURE 19.42 The structures of silicates are built up from SiO_4 tetrahedra. In different silicates, different numbers of O atoms are shared. In some cases, neighboring tetrahedra share one O atom; in others, they share two O atoms.

each tetrahedron contributes one Si atom and $4 \times \frac{1}{2} = 2$ oxygen atoms to the solid, resulting in the empirical formula SiO_2 (Fig. 19.42). The structure of quartz is complicated, for it is built from helical chains of SiO_4 units wound around one another. When it is heated to about 1500°C, it changes to another arrangement, that of the mineral *cristobalite* (Fig. 19.43). This structure is easier to describe: its Si atoms are arranged like the C atoms in diamond, but in cristobalite an O atom lies between each pair of neighboring Si atoms.

Metasilicic acid, H_2SiO_3, is a weak acid. Sodium metasilicate is a basic salt used in detergents, partly as a basic buffer and partly to keep dirt from settling back onto the fabric. The SiO_3^{2-} ions attach to dirt particles, giving them a negative charge, which prevents them from merging with others into larger, insoluble particles (Fig. 19.44). Orthosilicic acid, H_4SiO_4, is also a weak acid. However, when a solution of sodium orthosilicate is acidified, instead of H_4SiO_4, a gelatinous precipitate of silica is produced:

$$4\,H_3O^+(aq) + SiO_4^{4-}(aq) + x\,H_2O(l) \longrightarrow SiO_2(s)\cdot xH_2O(gel) + 6\,H_2O(l)$$

After it is washed and dried, this *silica gel* has a very high surface area (about 700 $m^2\cdot g^{-1}$) and is useful as a drying agent, a support for catalysts, a packing for chromatography columns, and a thermal insulator.

Silicates can be viewed as arrangements of tetrahedral oxoanions of silicon. Each Si—O bond has considerable covalent character. The differences between the various silicates arise from the number of negative charges on each tetrahedron, the number of corner O atoms shared with other tetrahedra, and the manner in which chains and sheets of the linked

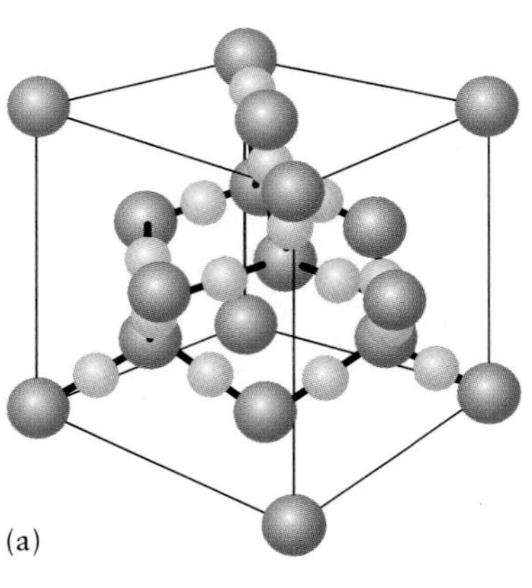

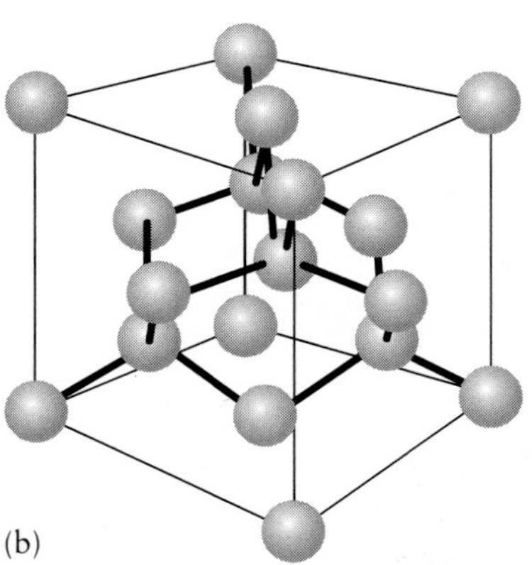

FIGURE 19.43 In the structure of cristobalite (a) the Si atoms lie in a tetrahedral arrangement similar to that of the C atoms in the diamond structure (b) except that an O atom (red) lies between each Si atom (purple).

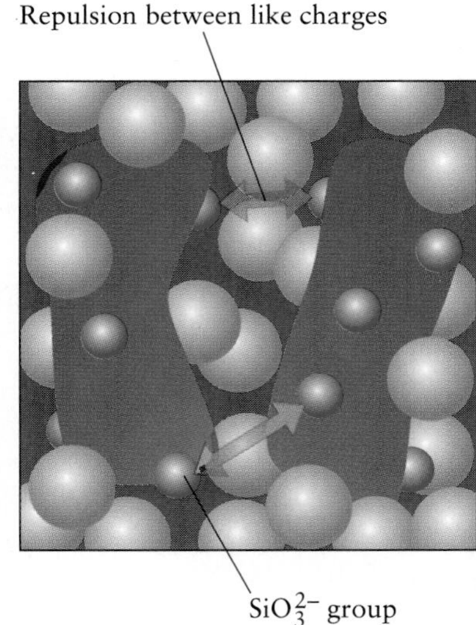

FIGURE 19.44 When SiO_3^{2-} ions (represented by green spheres) attach to dirt particles (the orange and green blobs), they repel one another and the dirt particles are prevented from collecting into larger, insoluble particles.

tetrahedra lie together. Differences in the internal structures of these highly regular network solids lead to a wide array of materials, ranging from gemstones to fibers.

The simplest silicates, the orthosilicates, are built from SiO_4^{4-} ions. They are not very common but include the mineral *zircon,* $ZrSiO_4$, which is used as a substitute for diamond in jewelry. The *pyroxenes* consist of chains of SiO_4 units in which two corner O atoms are shared by neighboring units (Fig. 19.45); the repeating unit is the metasilicate ion, SiO_3^{2-}. Electrical neutrality is provided by cations regularly spaced along the chain. The pyroxenes include *jade,* $NaAl(SiO_3)_2$.

The chains of units can link together to form the ladderlike structures that include *tremolite,* $Ca_2Mg_5(Si_4O_{11})_2(OH)_2$. Tremolite is one of the minerals called *asbestos,* which are characterized by a fibrous structure and an ability to withstand heat (Fig. 19.46). Their fibrous quality reflects the way the ladders of SiO_4 units lie together but can easily be torn apart. Because of its great resistance to fire, asbestos fibers were once widely used for heat insulation in buildings. However, these fibers lodge in people's lungs, where they cannot be absorbed. Eventually, fibrous scar tissue can form around them, giving rise to the disease asbestosis and creating a susceptibility to lung cancer. In some minerals, the SiO_4 tetrahedra link together to form sheets. An example is *talc,* a hydrated magnesium silicate, $Mg_3(Si_2O_5)_2(OH)_2$. Talc is soft and slippery because the silicate sheets slide over one another.

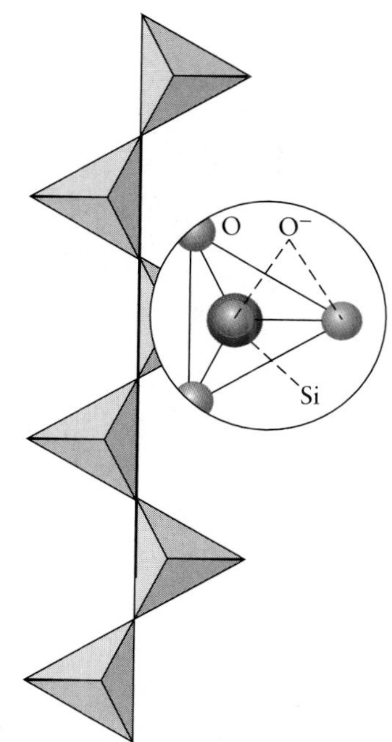

FIGURE 19.45 The basic structural unit of the minerals called pyroxenes. Each tetrahedron is an SiO_4 unit (like that in Fig. 19.42), and a shared corner represents a shared O atom (red in the inset). The two unshared O atoms each carry a negative charge.

More complex (and more common) structures result when some of the Si^{4+} ions in silicates are replaced by Al^{3+} ions to form the aluminosilicates. The missing positive charge is made up by extra cations. These cations account for the difference in properties between the silicate talc and the aluminosilicate *mica* (Fig. 19.47). One form of mica is $KMg_3(Si_3AlO_{10})(OH)_2$. In this mineral, the sheets of tetrahedra are held together by extra K^+ ions. Although it cleaves neatly into layers when the sheets are torn apart, mica is not slippery like talc.

The *feldspars* are aluminosilicates in which more than half the Si^{4+} ions have been replaced by Al^{3+} ions. They are the most abundant silicate materials on Earth and are a major component of *granite,* a compressed mixture of mica, quartz, and feldspar (Fig. 19.48). When some of the cations between the crystal layers are washed away as these rocks weather, the structure crumbles to clay, one of the main inorganic components of soil. A typical feldspar has the formula $KAlSi_3O_8$. Its weathering by carbon dioxide and water can be described by the equation

$$2\,KAlSi_3O_8(s) + 2\,H_2O(l) + CO_2(g) \longrightarrow K_2CO_3(aq) + Al_2Si_2O_5(OH)_4(s) + 4\,SiO_2(s)$$

The potassium carbonate is soluble and washes away, but the aluminosilicate remains as the clay. Clays were the raw materials for some of the first manufactured containers, ceramic pots, which have been used since prehistoric times (Case Study 19).

FIGURE 19.46 The minerals commonly called asbestos (from the Greek words for "not burning") are fibrous because they consist of long chains based on SiO_4 tetrahedra linked through shared oxygen atoms.

CASE STUDY 19 GLASSES AND CERAMICS

ONE DAY, COMMUNICATION networks may all be glass and automobile engines ceramic. The beauty and utility of glasses and ceramics have been valued since antiquity. Today, high technology has found new uses for these ancient materials, including superconductors and the optical fibers connecting computers.

A *glass* is an ionic solid with an amorphous structure resembling that of a liquid. Glass has a network structure based on a nonmetal oxide, usually silica, SiO_2, that has been melted together with metal oxides that act as "network modifiers." In a glass factory, silica in the form of sand is heated to about 1600°C. Metal oxides, such as MO (where M is a metal cation) are added to the silica. As the mixture melts, many of the Si—O bonds break and the orderly structure of the individual crystals is lost. When the melt cools, the Si—O bonds re-form but are prevented from forming a crystalline lattice because some of the silicon atoms bond with the O^{2-} ions of the metal oxide to give $—Si—O^-—M^{2+}$ groups in place of some of the —Si—O—Si— links present in pure silica. Compare the amorphous structure of glass to the long-range order of the crystalline silicates described in Section 19.16 (recall Fig. 10.17). Silicate glasses are generally transparent and durable and can be formed in flat sheets, blown into bottles, or molded into desired shapes.

The colors of stained glass are produced by mixing selected materials into the molten glass.

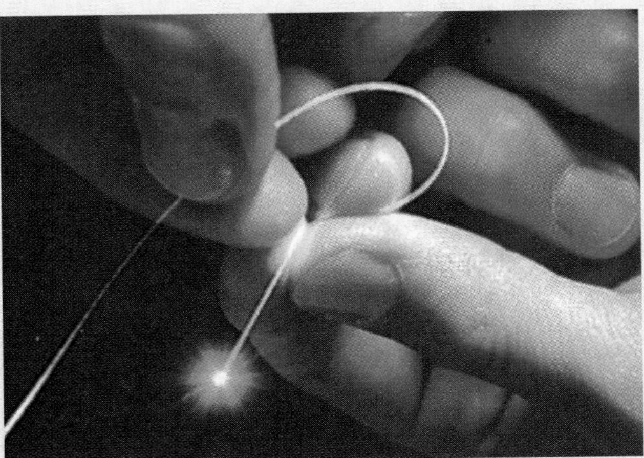
Glass fibers such as this one, about the diameter of a human hair, are used in high-performance computing systems and communications networks when large amounts of information must be transmitted in a short period of time.

Almost 90% of all manufactured glass combines sodium and calcium oxides with silica in *soda-lime glass.* This glass, which is used for windows and bottles, contains about 12% Na_2O prepared by the action of heat on sodium carbonate (the soda) and 12% CaO (the lime). When the proportions of soda and lime are reduced and 16% B_2O_3 is added, a *borosilicate glass,* such as Pyrex, is produced. Because borosilicate glasses do not expand much when heated, they survive rapid heating and cooling and are used for ovenware and laboratory beakers.

Colored glass is produced by adding small amounts of other substances; cadmium sulfide and selenide, for instance, give ruby glass, which is red. Ordinary soda-lime glass is usually very pale green as a result of iron impurities in the form of Fe^{2+}. You can see the green color when you view a glass pane edge on. Cobalt blue glass is colored by Co^{2+} ions. Brown beer-bottle glass is colored by iron sulfides. Amber glass, such as that used for medicine bottles, is colored with a mixture of sulfur and iron oxides that give tints from pale yellow to amber. The colored oxides absorb harmful radiation.

Glass is resistant to attack by most chemicals. However, the silica in glass reacts with the strong Lewis base F^- from hydrofluoric acid to form fluorosilicate ions:

$$SiO_2(s) + 6\,HF(aq) \longrightarrow SiF_6^{2-}(aq) + 2\,H_3O^+(aq)$$

It is also attacked by the Lewis base OH^- in hot, molten sodium hydroxide and by O^{2-} in the carbonate anion of hot molten sodium carbonate:

$$SiO_2(s) + Na_2CO_3(l) \xrightarrow{1400°C} Na_2SiO_3(s) + CO_2(g)$$

The dissolving of silica from glass by the ions F^- (from HF), OH^-, and CO_3^{2-} is called the *etching* of glass.

A *ceramic* is an inorganic material that has been hardened by heating to a high temperature. Many ceramics are created by heating aluminosilicate clays to drive out the water between the sheets of tetrahedra. This procedure leaves a rigid heterogeneous mass of tiny interlocking crystals bound together by glassy silica. *China clay,* which is used to make porcelain and china, is a form of aluminum aluminosilicate that can be obtained reasonably free of the iron impurities that make many clays look reddish brown. It is used in large amounts to coat paper (such as this page) to give a smooth, nonabsorbent surface.

Ceramics have crystalline structures that make them extremely hard, but brittle. Their stability at high temperatures has made them useful as furnace liners and has led to interest in ceramic automobile engines, which could endure overheating. High-temperature superconductors are ceramic materials formed from mixtures containing certain metal oxides, usually oxides of barium, copper, and yttrium.

Ceramics and glasses in the form of pots, jewelry, and art work are often the only materials that survive to enlighten archaeologists about early cultures. Perhaps one day we will be known only by our glass and ceramic remains; future archaeologists will find electrical insulators, superconductors, and fiberglass skateboards, as well as pots and vases.

Glass is etched by reaction with HF. The surface of the glass is covered with wax, a design is scratched on the wax, and acid is poured over it. This etched glass bowl was designed by the artist Frederick Carder in the 1920s.

QUESTIONS

1. Write the chemical equation for a reaction between OH^- and SiO_2.

2. Water adheres to glass. What kinds of intermolecular processes are involved when water adheres to silicate glass? Sketch the intermolecular interactions between water molecules and SiO_2 units in glass.

3. Which of the following would solidify as a glass when cooled: (a) tar (which contains many different long-chain hydrocarbons); (b) sodium chloride; (c) molten granite (see Section 19.16); (d) water; (e) low-density polyethylene; (f) a highly branched polymer (Section 11.16)?

4. Select the physical properties that you would expect to differ between the glassy and crystalline phases of the same substance and explain your answers: (a) ability to cleave along a plane; (b) rigidity; (c) sharp melting point; (d) transparency; (e) isotropy (appearing the same in all directions).

FIGURE 19.47 The aluminosilicate mica cleaves into thin transparent sheets. It is used for windows in furnaces.

Cements are obtained when aluminosilicates are melted and then allowed to solidify. The most widely used is *Portland cement,* which is made by heating a mixture of silica, clay, and limestone to about 1500°C. The cooled mass is then crushed and some gypsum ($CaSO_4{\cdot}2H_2O$) is added. The main components of the complex mixture are various calcium silicates and aluminates. When water is added, complex reactions occur and the mass sets to a solid. Concrete is the very strong, durable material that results when sand and gravel are mixed into cement.

Silicones consist of long —O—Si—O— chains with the remaining silicon bonding positions occupied by organic groups, such as the methyl

FIGURE 19.48 The mineral granite is actually a compressed mixture of mica, quartz, and feldspar.

FIGURE 19.49 A typical silicone structure. The hydrocarbon groups give the substance a water-repelling quality. Note the similarity of this structure to that of the purely inorganic pyroxenes in Fig. 19.45.

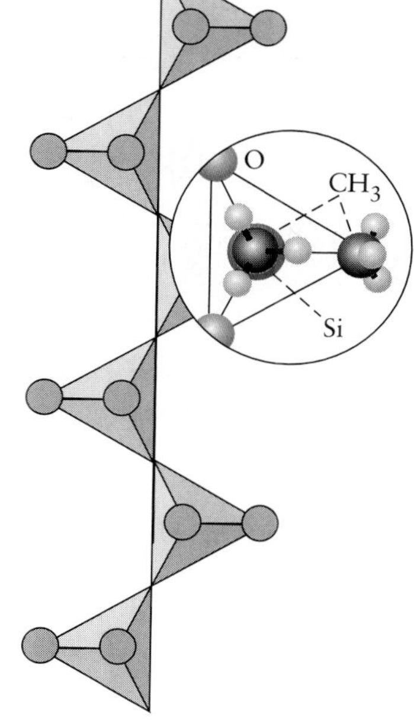

group, CH_3 (Fig. 19.49). Silicones are used to waterproof fabrics because their oxygen atoms attach to the fabric, leaving the hydrophobic (water-repelling) methyl groups like tiny, inside-out umbrellas sticking up out of the fabric's surface. For similar reasons, these methyl silicones are biologically inert and survive intact when exposed to body fluids. Because they do not coagulate blood and do not usually stick to body tissues, they are used for surgical and cosmetic implants.

Silicate structures are based on SiO_4 tetrahedra with different negative charges and different numbers of shared O atoms.

19.17 CARBIDES

There are three classes of carbides: **saline carbides** (or saltlike carbides), **covalent carbides,** and **interstitial carbides.** The saline carbides are formed most commonly from the Groups 1 and 2 metals, aluminum, and a few other metals, and contain either C_2^{2-} or C^{4-} as anions. The *s*-block metals form saline carbides when their oxides are heated with carbon. All the C^{4-} carbides produce methane and the corresponding hydroxide in water:

$$Al_4C_3(s) + 12\,H_2O(l) \longrightarrow 4\,Al(OH)_3(s) + 3\,CH_4(g)$$

The species C_2^{2-} is the acetylide ion; the carbides that contain it react with water to produce ethyne (acetylene, the conjugate acid of the acetylide ion) and the corresponding hydroxide (see Fig. 3.20):

$$CaC_2(s) + 2\,H_2O(l) \longrightarrow Ca(OH)_2(s) + HC{\equiv}CH(g)$$

Calcium carbide, CaC_2, is the most common saline carbide.

The covalent carbides include silicon carbide, SiC, which is sold as *carborundum:*

$$SiO_2(s) + 3\,C(s) \xrightarrow{2000°C} SiC(s) + 2\,CO(g)$$

Pure silicon carbide is colorless, but iron impurities normally impart an almost black color to the crystals. Carborundum is an excellent abrasive because it is very hard, with a diamondlike structure that fractures into pieces with sharp edges (Fig. 19.50).

The interstitial carbides are compounds formed by the direct reaction of a *d*-block metal and carbon at temperatures above 2000°C. In these

FIGURE 19.50 Carborundum crystals, showing the sharp fractured edges that give the substance its abrasive power.

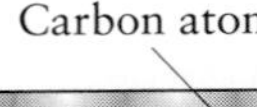

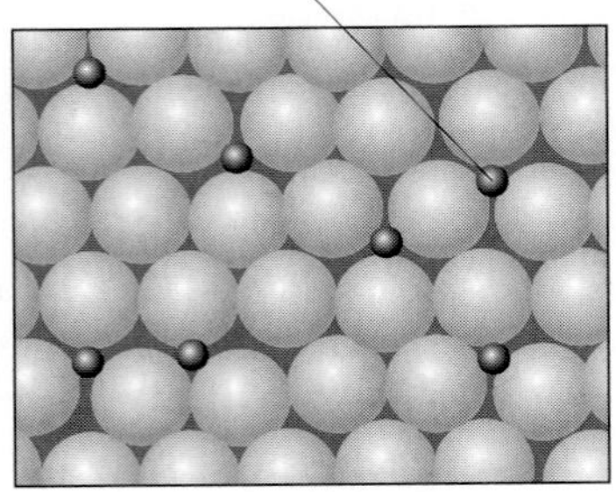

FIGURE 19.51 The structure of an interstitial carbide, in which the carbon atoms (represented by the black spheres) lie between metal atoms (the gray spheres), so producing a rigid structure.

compounds, the C atoms occupy the gaps between the metal atoms, as do the H atoms in metallic hydrides (Fig. 19.51). Here, however, the C atoms pin the metal atoms together into a rigid structure, resulting in extremely hard substances with melting points often well above 3000°C. Tungsten carbide, WC, is used for the cutting surfaces of drills and iron carbide, Fe_3C, is an important component of steel.

Carbon forms ionic carbides with Group 1 and Group 2 metals, covalent carbides with nonmetals, and interstitial carbides with d-block metals.

SELF-TEST 19.4A Explain why graphite can conduct electricity but diamond cannot.

[*Answer:* Graphite has a network of π bonds, through which electrons can be delocalized. In diamond, electrons are localized in σ bonds.]

SELF-TEST 19.4B Explain the fact that SiH_4 reacts with water containing OH^- ions but CH_4 does not.

SKILLS YOU SHOULD HAVE MASTERED

Conceptual

1. Explain hydrogen's special placement in the periodic table.
2. Explain trends in the properties of the elements in Groups 1, 2, 13, and 14.
3. Predict and use the diagonal relationships in the periodic table.
4. Explain why the Period 2 elements in Groups 1, 2, 13, and 14 have properties distinct from those of the other members of their groups.
5. Identify the valence electron configurations of elements in Groups 1, 2, 13, and 14.
6. Distinguish the allotropes of carbon by their structures and show how their structures affect their properties.
7. Explain the differences in reactivity between compounds of carbon and those of silicon.

Problem-Solving

1. Predict trends in metallic properties across a period or down a group.
2. Write the formula for an acid, given that of the anhydride, and vice versa.

Descriptive

1. Describe the names, properties, and reactions of the principal compounds of hydrogen and the Period 2 and 3 elements in Groups 1, 2, 13, and 14.
2. Describe and write balanced equations for the principal reactions used to produce hydrogen and the Period 2 and 3 elements in Groups 1, 2, 13, and 14.
3. Describe the major uses of hydrogen, sodium, potassium, beryllium, magnesium, boron, aluminum, carbon, and silicon.
4. Describe the reactions of the alkali metals with water and with nonmetals.
5. Identify amphoteric elements in Groups 2, 13, and 14.
6. Identify elements in Groups 2, 13, and 14 that are passivated by the formation of protective oxides.
7. Distinguish the principal silicate structures and describe their properties.

EXERCISES

Hydrogen

19.1 Describe evidence for the statement that hydrogen can act as both a reducing agent and an oxidizing agent. Give chemical equations to support your evidence.

19.2 Explain why lithium differs from the other Group 1 elements in its chemical and physical properties. Give two examples to support your explanation.

19.3 Write the chemical equations for (a) the shift reaction and the reactions between (b) lithium and water; (c) magnesium and hot water; (d) potassium and hydrogen.

19.4 Write the chemical equations for (a) the reaction between sodium hydride and water; (b) the formation of synthesis gas; (c) the hydrogenation of ethene, $H_2C{=}CH_2$; (d) the reaction of magnesium with hydrochloric acid.

19.5 Classify each of the following compounds as a saline, molecular, or metallic hydride: (a) KH; (b) NH_3; (c) HBr; (d) UH_3.

19.6 Classify each of the following compounds as a saline, molecular, or metallic hydride: (a) B_2H_6; (b) SiH_4; (c) CaH_2; (d) PdH_x, $x < 1$.

19.7 Use Appendix 2A to determine the standard reaction enthalpy at 25°C for (a) the re-forming reaction; (b) the shift reaction; (c) the overall reaction of these two processes.

19.8 Use Appendix 2A to determine the standard free energy at 25°C of (a) the re-forming reaction; (b) the shift reaction; (c) the overall reaction of these two processes.

19.9 Calcium hydride is used as a portable source of hydrogen because of its reaction with water:

$$CaH_2(s) + 2\,H_2O(l) \longrightarrow Ca(OH)_2(s) + 2\,H_2(g)$$

(a) What volume of H_2 gas (at 273 K and 1.00 atm) can be produced from 500 g of CaH_2? (b) How many milliliters of water should be supplied for the reaction? Assume the density of water to be $1.0\ g{\cdot}mL^{-1}$.

19.10 What volume of hydrogen gas (at 273 K and 1.00 atm) is required for the reduction of 20.0 g of WO_3 in the reaction $WO_3(s) + 3\,H_2(g) \rightarrow W(s) + 3\,H_2O(l)$.

19.11 About 3×10^8 kg of hydrogen is produced each year in the United States. If 1.5×10^9 kg of ammonia is produced annually by the Haber process, what fraction of the H_2 is used for this purpose?

19.12 Calculate the volume that the annual United States production of hydrogen (3×10^8 kg) would occupy if it were stored as liquid hydrogen. The density of liquid hydrogen is $0.089\ g{\cdot}cm^{-3}$.

19.13 Identify the products and write a balanced equation for the reaction of hydrogen with (a) chlorine; (b) sodium; (c) phosphorus.

19.14 Identify the products and write a balanced equation for the reaction of hydrogen with (a) nitrogen; (b) fluorine; (c) cesium.

19.15 Write the Lewis structures of (a) LiH; (b) SiH_4 (silane); (c) SbH_3.

19.16 Fluorine forms one binary hydride, oxygen forms two binary hydrides, and three of the binary hydrides that nitrogen forms are NH_3, N_2H_4, and HN_3. Write the Lewis structures for all six compounds.

Group 1: The Alkali Metals

19.17 Refer to Fig. 7.1 and describe the color of the flame test for (a) lithium; (b) potassium; (c) sodium; (d) rubidium.

19.18 (a) Write the valence electron configuration for the alkali metal atoms. (b) Explain why the alkali metals are strong reducing agents in terms of electron configurations and ionization energies.

19.19 Write the chemical equations for the reactions between (a) lithium and oxygen; (b) lithium and nitrogen; (c) sodium and water; (d) potassium superoxide and water.

19.20 Write the chemical equations for the reactions between (a) potassium and oxygen; (b) sodium oxide and water; (c) lithium and hydrochloric acid; (d) cesium and iodine.

19.21 Complete and balance the following skeletal equations: (a) $Ca(s) + H_2(g) \rightarrow$ products; (b) $KNO_3(s) \xrightarrow{\Delta}$ products.

19.22 Complete and balance the following skeletal equations: (a) $Na(s) + H_2O(l) \rightarrow$ products; (b) $K(s) + N_2(g) \rightarrow$ products.

19.23 Write the chemical names and formulas of the minerals (a) rock salt; (b) sylvite; (c) carnallite.

19.24 Write the names and chemical formulas of the compounds (a) washing soda; (b) soda ash; (c) bicarbonate of soda.

19.25 Sodium carbonate is often supplied as the decahydrate, $Na_2CO_3{\cdot}10H_2O$. What mass of this solid should be used to prepare 250 mL of 0.100 M $Na_2CO_3(aq)$ solution?

19.26 Sodium metal is produced from the electrolysis of molten sodium chloride in the Downs process (Section

17.16). Determine (a) the standard free energy of the reaction $NaCl(s) \rightarrow Na(s) + \frac{1}{2}Cl_2(g)$ and (b) the current needed to produce 1.00 kg of sodium in 10.0 h.

Group 2: The Alkaline Earth Metals

19.27 Predict and explain the trend in strengths of the Group 2 metals as reducing agents on moving down the group.

19.28 Explain the trend of decreasing lattice enthalpies of the chlorides of the Group 2 metals down the group.

19.29 Give the chemical names and write the formulas of (a) Epsom salts; (b) limestone; (c) milk of magnesia. (See Appendix 3B.)

19.30 Give the chemical names and write the formulas of (a) calcite; (b) dolomite; (c) quicklime.

19.31 Bearing in mind that aluminum and beryllium have a diagonal relationship, write the chemical equations for the reaction of (a) aluminum with aqueous sodium hydroxide; (b) beryllium with aqueous sodium hydroxide.

19.32 Write the chemical equations for (a) the industrial preparation of magnesium metal from the magnesium chloride in seawater; (b) the action of water on calcium metal.

19.33 Predict the products of the following reactions and then balance the equations:
(a) $Mg(OH)_2 + HCl(aq) \longrightarrow$
(b) $Ca(s) + H_2O(l) \longrightarrow$
(c) $BaCO_3(s) \xrightarrow{\Delta}$

19.34 Predict the products of the following reactions and then balance the equations:
(a) $Mg(s) + Br_2(l) \longrightarrow$
(b) $BaO(s) + Al(s) \xrightarrow{\Delta}$
(c) $CaO(s) + SiO_2(s) \xrightarrow{\Delta}$

19.35 (a) Write the Lewis structure for $BeCl_2$ (see Chapter 8) and (b) predict the Cl—Be—Cl bond angle. (c) What hybrid orbitals are used in the bonding?

19.36 (a) Write the Lewis structure of $MgCl_2$ (see Chapter 8). (b) How does its structure differ from that of $BeCl_2$?

19.37 Heat is generated in the formation of slaked lime from the reaction $CaO(s) + H_2O(l) \rightarrow Ca(OH)_2(s)$.
(a) Calculate the standard enthalpy of reaction. (b) If all this heat could be used to heat 250 g of water, what would be the resulting temperature change of the water?

19.38 What mass of calcium oxide can be produced from the thermal decomposition of 200 g calcium carbonate?

19.39 (a) Calculate the standard reaction enthalpy and entropy for the thermal decomposition of calcium carbonate as calcite (see Appendix 2A). (b) Determine the temperature at which the equilibrium constant for the thermal decomposition becomes greater than 1.

19.40 (a) Calculate the standard reaction enthalpy and entropy for the thermal decomposition of magnesium carbonate to MgO and CO_2 (see Appendix 2A).
(b) Determine the temperature at which K_p for the thermal decomposition becomes greater than 1.

19.41 What is the mass percentage of water in the magnesium sulfate hydrate sold as Epsom salts? (See Appendix 3B.)

19.42 The concentration of magnesium in seawater is about $1.35\ g{\cdot}L^{-1}$. What volume of water must be processed to collect 1.0 kg of magnesium, assuming an 80% removal?

Group 13: The Boron Family

19.43 Write a balanced equation for the industrial preparation of aluminum from its oxide.

19.44 Write a balanced equation for the industrial preparation of impure boron.

19.45 Write the formula of (a) boric acid; (b) alumina; (c) borax; (d) boron oxide.

19.46 Write the formula of (a) potassium alum; (b) corundum; (c) diborane; (d) lithium aluminum hydride.

19.47 Complete and balance the following skeletal equations:
(a) $B_2O_3(s) + Mg(l) \xrightarrow{\Delta}$
(b) $Al(s) + Cl_2(g) \longrightarrow$
(c) $Al(s) + O_2(g) \longrightarrow$

19.48 Complete and balance the following skeletal equations:
(a) $Al_2O_3(s) + OH^-(aq) \longrightarrow$
(c) $Al_2O_3(s) + H_3O^+(aq) + H_2O(l) \longrightarrow$
(c) $B(s) + NH_3(g) \xrightarrow{\Delta}$

19.49 Identify a use for (a) $AlCl_3$; (b) α-alumina; (c) $B(OH)_3$.

19.50 Identify a use for (a) BF_3; (b) $NaBH_4$; (c) $Al_2(SO_4)_3$.

19.51 Balance the following skeletal equations:
(a) $B_2H_6(g) + H_2O(l) \longrightarrow B(OH)_3(aq) + H_2(g)$
(b) $B_2H_6(g) + O_2(g) \longrightarrow B_2O_3(s) + H_2O(l)$

19.52 Balance the following skeletal equations:
(a) $B_2H_6(g) + NaBH_4(s) \longrightarrow Na_2B_{12}H_{12}(s) + H_2(g)$
(b) $B_2O_3(s) + C(s) + Cl_2(g) \longrightarrow BCl_3(g) + CO(g)$

19.53 What mass of aluminum can be produced by the Hall process in a period of 8.0 h, using a current of 1.0×10^5 A?

19.54 In the production of 1.0×10^3 kg of aluminum by the Hall process, what mass of carbon is lost at the anode?

19.55 The standard free energy of formation of $Tl^{3+}(aq)$ is $+215\ kJ{\cdot}mol^{-1}$ at 25°C. Calculate the standard potential of the Tl^{3+}/Tl redox couple.

19.56 The standard potential of the Al^{3+}/Al redox couple is -1.66 V. Calculate the standard free energy of formation for $Al^{3+}(aq)$. Account for any differences between the standard free energy of formation of $Tl^{3+}(aq)$ (see Exercise 19.55) and that of $Al^{3+}(aq)$.

Group 14: The Carbon Family

19.57 Describe the sources of silicon and write balanced equations for the three steps in the industrial preparation of silicon.

19.58 Describe the sources of carbon and how carbon is converted to graphite.

19.59 Compare the hybridization and structure of carbon in diamond and graphite. How do these features explain the physical properties of the two allotropes?

19.60 Explain why the size of the silicon atom does not permit a silicon analogue of the graphite structure.

19.61 Write formulas for (a) carborundum; (b) silica; (c) zircon.

19.62 Write formulas for (a) nickel carbonyl; (b) tungsten carbide; (c) silica gel.

19.63 Balance the following skeletal equations and classify them as acid-base or redox:
(a) $MgC_2(s) + H_2O(l) \longrightarrow C_2H_2(g) + Mg(OH)_2(s)$
(b) $Pb(NO_3)_2(s) \longrightarrow PbO(s) + NO_2(g) + O_2(g)$

19.64 Balance the following skeletal equations and classify them as acid-base or redox:
(a) $CH_4(g) + S(s) \longrightarrow CS_2(l) + H_2S(g)$
(b) $Sn(s) + KOH(aq) + H_2O(l) \longrightarrow K_2Sn(OH)_6(aq) + H_2(g)$

19.65 Complete and balance the equations for the following reactions:
(a) $SiCl_4(l) + H_2(g) \longrightarrow$
(b) $SiO_2(s) + C(s) \longrightarrow$
(c) $Ge(s) + F_2(g) \longrightarrow$
(d) $CaC_2(s) + H_2O(l) \longrightarrow$

19.66 Complete and balance the equations for the following reactions (refer to Table 19.12, if necessary):
(a) $Sn(s) + H_2O(l) + \text{excess } OH^-(aq) \longrightarrow$
(b) $C(s) + H_2O(g) \xrightarrow{\Delta}$
(c) $CH_4(g) + H_2O(g) \xrightarrow{\Delta}$
(d) $Al_4C_3(s) + H_2O(l) \longrightarrow$

19.67 Write a Lewis structure for the orthosilicate anion, SiO_4^{4-}, and deduce the formal charges and oxidation numbers of the atoms. Use the VSEPR model (Chapter 9) to predict the shape of the ion.

19.68 Use the VSEPR model (Chapter 9) to estimate the Si—O—Si bond angle in silica.

19.69 Calculate the percentage by mass of silicon in silica.

19.70 Calculate the percentage by mass of silicon in feldspar, $KAlSi_3O_8$.

19.71 Determine the values of ΔH_r°, ΔS_r°, and ΔG_r° for the production of high-purity silicon by the reaction $SiO_2(s) + 2\,C(s, \text{graphite}) \rightarrow Si(s) + 2\,CO(g)$ at 25°C and estimate the temperature at which the equilibrium constant becomes greater than 1.

19.72 Determine the values of ΔH_r°, ΔS_r°, and ΔG_r° for the reaction $2\,CO(g) + O_2(g) \rightarrow 2\,CO_2(g)$ at 25°C and estimate the temperature at which the equilibrium constant becomes less than 1.

19.73 Determine the mass of HF as hydrofluoric acid that is required to etch 2.00 mg of SiO_2 from a glass plate in the reaction $SiO_2(s) + 6\,HF(aq) \rightarrow 2\,H_3O^+(aq) + SiF_6^{2-}(aq)$.

19.74 What mass of coke containing 98% carbon is needed to reduce the silicon in 1.0 kg of 88.5% pure silica?

19.75 Determine the surface area in square meters of 1.00 mol of activated carbon if the surface area of 1.0 g is 2000 m^2.

19.76 Explain why silicon tetrachloride reacts with water to produce SiO_2 but carbon tetrachloride does not react with water. (Consider the Lewis acidity of each compound.)

19.77 Describe the structures of a silicate in which the silicate tetrahedra share (a) one O atom; (b) two O atoms.

19.78 What is the empirical formula of a potassium silicate in which the silicate tetrahedra share (a) two O atoms or (b) three O atoms and form a chain? In each case there are single negative charges on the unshared O atoms.

SUPPLEMENTARY EXERCISES

19.79 (a) Plot standard potential as a function of atomic number for the elements of Groups 1 and 2 (refer to Appendix 2B for data). (b) What generalizations can be deduced from the graph?

19.80 Use data from Fig. 7.31 and Appendix 2B to plot ionization energy as a function of standard potential for the elements of Groups 1 and 2. What generalizations can be drawn from the graph?

19.81 (a) Write equations for the reactions of hydrogen gas with the halogens, from fluorine to iodine. (b) Predict the relative vigor of the reactions. (c) Name the products when they are dissolved in aqueous solutions.

19.82 In aqueous solution, the beryllium(II) ion exists as $[Be(OH)_4]^{2-}$ ions. Write a chemical equation that illustrates the acidic character of this ion.

19.83 (a) Name the type of reaction that occurs between calcium oxide and silica in a blast furnace. (b) Write the chemical equation for a related reaction, that between calcium oxide and carbon dioxide.

19.84 Write the Lewis structures of (a) BaO_2; (b) BeH_2; (c) Na_2O_2; (d) $[Be(OH)_4]^{2-}$.

19.85 Name the minerals used as sources of (a) beryllium; (b) calcium; (c) magnesium.

19.86 Distinguish among limestone, lime, quicklime, slaked lime, chalk, marble, and calcite.

19.87 (a) What is "hard water" (see Section 12.3)? (b) Write chemical equations for the softening of hardness due to HCO_3^- ions in water that can be achieved by using lime. (Refer to Case Study 3.)

19.88 Calculate the molar solubility of $Mg(OH)_2$, milk of magnesia, at (a) pH = 10.0; (b) pH = 9.0 (see Table 15.5).

19.89 Magnesium is produced from the electrolysis of molten magnesium chloride. (a) Calculate the mass of magnesium that can be produced in 1.5 h, using a current of 100 A (see Section 17.15). (b) Calculate the volume (at 273 K and 1.00 atm) of chlorine gas produced when 1000 kg of magnesium metal is obtained from this process.

19.90 State a use for the elemental form of (a) boron; (b) aluminum; (c) beryllium; (d) silicon; (e) tin.

19.91 Describe the chemical and physical properties of elemental boron and aluminum.

19.92 (a) Write Lewis structures for boric acid, $B(OH)_3$ and boron trifluoride. (b) Using the VSEPR model (Chapter 9), predict the structure and bond angles of boric acid and boron trifluoride. (c) What type of hybridization can be ascribed to the boron atom in boric acid and boron trifluoride?

19.93 (a) What is the formula of the conjugate base of boric acid in water? (b) Write an equation showing that $B(OH)_3$ is an acid in water.

19.94 Very pure boron is obtained by reducing boron trichloride with hydrogen gas; a by-product is hydrogen chloride gas. What volume of hydrogen gas (at 273 K and 1.00 atm) is needed for the production of 50.0 g of boron from the reduction of boron trichloride?

19.95 Write the valence electron configuration of (a) C; (b) In; (c) Ba; (d) Rb.

19.96 Arrange the elements aluminum, gallium, indium, thallium, tin, and germanium in order of increasing electronegativity (see Fig. 8.15) and increasing reducing strength.

19.97 (a) State the trends in first ionization energies and atomic radii down Groups 13 and 14. (b) Account for the trends. (c) How do the trends correlate with the properties of the elements?

19.98 (a) Suggest a reason for the observations that methane is stable in an aqueous alkaline solution, but silane, SiH_4, reacts rapidly in the same solution. (b) Write a balanced chemical equation for the reaction of silane with water in an aqueous alkaline solution.

19.99 Complete and balance the following skeletal equations:
(a) $AlCl_3(s) + H_2O(l) \longrightarrow$
(b) $B_2H_6(g) \xrightarrow{\text{high temperatures}}$
(c) $BF_3 + BH_4^- \xrightarrow{\text{organic solvent}}$

19.100 Complete and balance the following skeletal equations:
(a) $Na_2CO_3(s) + HCl(aq) \longrightarrow$
(b) $SiCl_4(l) + H_2O(l) \longrightarrow$
(c) $Ge(s) + 2\,Cl_2(g) \longrightarrow$

19.101 Hydrogen can be produced from the electrolysis of water. (a) Write the equation for the half-reaction for the production of hydrogen (refer to Appendix 2B, if necessary). (b) Is the hydrogen produced at the anode or the cathode? (c) Calculate the volume of hydrogen (at 273 K and 1.00 atm) that is produced if a current of 10.0 A is passed through an electrolytic cell for 30 min.

19.102 The reduction of tin(IV) oxide to (white) tin metal by graphite proceeds at moderately low temperatures. More commonly, tin(IV) oxide is reduced to the metal by an excess of carbon monoxide at temperatures above 980 K. (a) Write the two chemical equations for the reduction of SnO_2. (b) Determine the value of ΔG_r° for each reaction at 25°C. (c) How is the equilibrium constant affected by temperature in each case?

CHALLENGING EXERCISES

19.103 Is there any chemical support for the view that hydrogen should be classified as a member of Group 1? Would it be better to consider hydrogen a member of Group 17? Give evidence that supports each view.

19.104 What justification is there for regarding the ammonium ion as an analogue of a Group 1 metal cation? Consider properties such as solubility, charge, and radius.

19.105 It has been proposed that the ability of a cation to polarize anions is proportional to its charge divided by its radius (see Section 8.14). (a) Use this criterion to arrange the *s*-block elements in order of increasing polarizing power. (b) Do the resulting values support the diagonal relationships within the block?

19.106 The first ionization energies of the Group 2 elements decrease smoothly down the group, but in Group 13 the values for gallium and thallium are both higher than for aluminum. Suggest a reason. (*Hint:* Compare electron configurations and consider the relative shielding ability of different types of orbitals.)

19.107 Hydrogen burns in an atmosphere of bromine to give hydrogen bromide. If 120 mL of H_2 gas at 273 K and 1.00 atm combines with a stoichiometric amount of bromine and the resulting hydrogen bromide dissolves to form 150 mL of an aqueous solution, what is the molar concentration of the resulting hydrobromic acid solution?

19.108 The standard enthalpies of formation of $BH_3(g)$ and diborane are $+100\ kJ\cdot mol^{-1}$ and $+36\ kJ\cdot mol^{-1}$, respectively, and the enthalpies of formation of B(g) and H(g) are $+563\ kJ\cdot mol^{-1}$ and $+218\ kJ\cdot mol^{-1}$, respectively. (a) Use these values to calculate the mean bond enthalpies of the B—H bonds in each case (b) Assume that terminal B—H bonds have the same strengths in each compound, and estimate the bond enthalpy of the three-center B—H—B bonds in diborane. (c) Which bonds would you expect to be longer, the terminal B—H bonds or the three-center bonds? Explain your answer.

19.109 Suppose that the stability of carbonates when heated depends on the ability of the metal cation to polarize the carbonate ion and remove an oxide ion from it, so releasing carbon dioxide. Predict the order of thermal stability of the Groups 1 and 2 metal carbonates. Comment on the likely stability of aluminum carbonate.

19.110 When the mineral dolomite, $CaCO_3\cdot MgCO_3$, is heated, it gives off carbon dioxide and forms a mixture of a metal oxide and a metal carbonate. Which oxide is formed, CaO or MgO? Which carbonate is formed, $CaCO_3$ or $MgCO_3$? Justify your answer.

CHAPTER 20

The energy released when lightning strikes turns the atmosphere into a giant chemical reactor. Nitrogen oxides, ozone, and other compounds of the nonmetals on the right of the periodic table are formed in the heat and light of the lightning. The nitrogen oxides dissolve in rain and fall to Earth, where they fertilize plants and enter the food chain.

THE MAIN-GROUP ELEMENTS: II. THE LAST FOUR FAMILIES

We are now in the heart of the *p*-block of the periodic table. In Groups 15 to 18 we find nonmetallic elements of great variety. Some are richly colored, most are relatively soft, and some are even gases. Except for carbon and hydrogen, the gases of the air are made up solely of elements from this part of the *p*-block, some as elements and some as compounds; and every time lightning strikes, it initiates chemical reactions among these gases and produces other compounds. Although metallic character increases down each group, the only element in this region considered to be metallic is bismuth, at the bottom of Group 15. Arsenic, antimony, tellurium, and polonium are metalloids. All the elements in this part of the periodic table have moderately high ionization energies and form compounds with other nonmetals by sharing electrons in covalent bonds. When they react with metals, these elements form ionic bonds in which they accept electrons and become anions.

GROUP 15: THE NITROGEN FAMILY

20.1 The elements
20.2 Compounds with hydrogen and the halogens
20.3 Nitrogen oxides and oxoacids
20.4 Phosphorus oxides and oxoacids

GROUP 16: THE OXYGEN FAMILY

20.5 The elements
20.6 Compounds with hydrogen
20.7 Sulfur oxides and oxoacids
20.8 Sulfur halides

GROUP 17: THE HALOGENS

20.9 The elements
20.10 Compounds of the halogens

GROUP 18: THE NOBLE GASES

20.11 The elements
20.12 Compounds of the noble gases

				H			He
			14	15	16		
Li	Be	B	C	N	O	F	Ne
			Si	P	S		
			Ge	As	Se		
			Sn	Sb	Te		
			Pb	Bi	Po		

GROUP 15: THE NITROGEN FAMILY

The Group 15 elements (Table 20.1) range in character from the nonmetals nitrogen and phosphorus to the largely metallic bismuth (Fig. 20.1). This range of behavior is reflected in their chemical properties (Table 20.2); for example, all the oxides of nitrogen and phosphorus are acidic and bismuth's oxide is basic.

20.1 THE ELEMENTS

Nitrogen is the principal component of air (76% by mass) and is obtained by the distillation of liquid air. Air is cooled to below −196°C by repeated expansion and compression in a refrigerator like that described in Section 5.19. The liquid mixture is then warmed, and the nitrogen (b.p. −196°C) boils off while most of the oxygen (b.p. −183°C) remains as liquid. Any oxygen that does boil off is removed by passing the gas over hot copper:

$$2\,Cu(s) + O_2(g) \longrightarrow 2\,CuO(s)$$

Its strong N≡N bond (944 $kJ\cdot mol^{-1}$) makes nitrogen almost as inert as the noble gases. To be available for organisms, it must first be **fixed,** or combined with other elements into more useful compounds. Once fixed, nitrogen can be converted to other compounds for use as fertilizers, explosives, and plastics. Lightning converts some nitrogen to its oxides, which rain then washes into the soil. Some bacteria also fix nitrogen in nodules on the roots of clover, beans, peas, alfalfa, and other legumes (Fig. 20.2). At present, the Haber synthesis of ammonia is the main industrial route to fixing nitrogen at high temperature and pressure, but the search is on for catalysts that work at normal temperatures.

Unlike the other Group 15 elements, nitrogen is highly electronegative (χ = 3.0, about the same as that of chlorine). Because its atoms are

FIGURE 20.1 The elements of Group 15: (back row, from left to right) liquid nitrogen, phosphorus, arsenic; (front row) antimony and bismuth.

TABLE 20.1 **The Group 15 elements**

Valence configuration: ns^2np^3

Z	Name	Symbol	Molar mass, $g \cdot mol^{-1}$	Melting point, °C	Boiling point, °C	Density, $g \cdot cm^{-3}$ at 25°C	Normal form*
7	nitrogen	N	14.01	−210	−196	1.04‡	colorless gas
15	phosphorus	P	30.97	44	280	1.82	white nonmetal
33	arsenic	As	74.92	613s†	—	5.78	gray metalloid
51	antimony	Sb	121.75	631	1750	6.69	blue-white lustrous metalloid
83	bismuth	Bi	208.98	271	1650	8.90	white-pink metal

**Normal form* means the appearance and state of the element at 25°C and 1 atm.
†The symbol s denotes that the element sublimes.
‡For the liquid at its boiling point.

small, it can form multiple bonds by using its p-orbitals; but its valence shell ($n = 2$) has no d-orbitals. These characteristics account for many of the differences between the chemical and physical properties of nitrogen and those of the other Group 15 elements. For example, nitrogen can form bonds to no more than four atoms at a time whereas phosphorus can bind to six, and the compounds it forms with hydrogen are the only ones formed by a member of Group 15 that take part in hydrogen bonding. Nitrogen has one of the widest ranges of oxidation numbers of any element: nitrogen compounds are known for each whole-number oxidation number from −3 (in NH_3) to +5 (in nitric acid and the nitrates). It also occurs with fractional oxidation numbers, such as $-\frac{1}{3}$ in the azide ion, N_3^-.

Phosphorus is obtained from *apatites,* which are mineral forms of calcium phosphate, $Ca_3(PO_4)_2$. The rocks are heated in an electric furnace with carbon and sand:

$$2\,Ca_3(PO_4)_2(s) + 6\,SiO_2(s) + 10\,C(s) \xrightarrow{\Delta} P_4(g) + 6\,CaSiO_3(l) + 10\,CO(g)$$

TABLE 20.2 **Chemical properties of the Group 15 elements**

Reactant	Reaction with Group 15 element (E)
hydrogen	$N_2(g) + 3H_2(g) \longrightarrow 2\,NH_3(g)$ $P_4(s) + 6H_2(g) \longrightarrow 4\,PH_3(g)$
oxygen	$N_2(g) + xO_2(g) \longrightarrow 2\,NO_x(g)$ $P_4(s)$ + 3 or 5 $O_2(g) \longrightarrow P_4O_6(s)$ or $P_4O_{10}(s)$ $4\,As(s) + 3\,O_2(g) \longrightarrow As_4O_6(s)$ $4\,E(s) + 3\,O_2(g) \longrightarrow 2\,E_2O_3(s)$, E = Sb, Bi
water	no reaction
halogen	$2\,E(s) + 3\,X_2(s, l, g) \longrightarrow 2\,EX_3(s, l)$, E = P, As, Sb, Bi $2\,E(s) + 5\,X_2(s, l, g) \longrightarrow 2\,EX_5(s, l)$, E = P, As, Sb

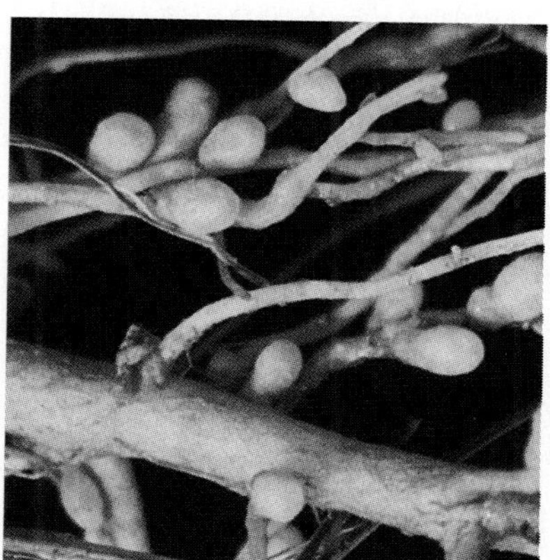

FIGURE 20.2 The bacteria that inhabit these nodules on the roots of a pea plant are responsible for fixing atmospheric nitrogen and making it available to the plant.

FIGURE 20.3 The minerals (from left to right) orpiment, As_2S_3; stibnite, Sb_2S_3; and realgar, As_4S_4.

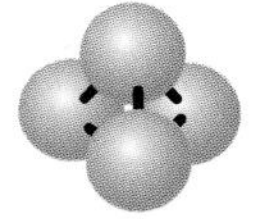

1 Phosphorus, P_4

The name *phosphorus* comes from the Greek words for "light bringer."

The phosphorus vapor condenses as *white phosphorus,* a soft, white, poisonous molecular solid consisting of tetrahedral P_4 molecules (**1**). This allotrope is highly reactive, in part because of the strain associated with the acute angles between the bonds. It bursts into flame on contact with air, and is normally stored under water. White phosphorus changes into *red phosphorus* when heated in the absence of air (see Fig. 20.1). This allotrope is less reactive, but it can be ignited by friction. Red phosphorus is used in the striking surfaces of matchbooks and on the sides of boxes of safety matches. The friction created by rubbing a match across the surface ignites the phosphorus sufficiently to light the highly flammable material in the match head. The structure of red phosphorus may consist of chains of linked P_4 tetrahedra.

Arsenic and antimony are metalloids. They have been known in the pure state since ancient times because they are easily reduced from their ores (Fig. 20.3). In the elemental state, they are used primarily in the lead alloys employed as electrodes in storage batteries. Bismuth is used to make type for the printing industry. Like ice, solid bismuth is less dense than the liquid. As a result, molten bismuth does not shrink when it solidifies in type molds.

Nitrogen is highly unreactive as an element, largely because of its strong triple bond; white phosphorus is highly reactive.

20.2 COMPOUNDS WITH HYDROGEN AND THE HALOGENS

By far the most important hydrogen compound of a Group 15 element is ammonia, NH_3, which is prepared in huge amounts by the Haber process

FIGURE 20.4 When aqueous ammonia is added to a copper(II) sulfate solution, the dark blue complex $[Cu(NH_3)_4]^{2+}$ is formed by a Lewis acid-base reaction.

FIGURE 20.5 Magnesium nitride is formed when magnesium burns in an atmosphere of nitrogen. When magnesium burns in air, the product is the oxide as well as the nitride.

(Section 13.11). Small quantities of ammonia are present naturally in the atmosphere as a result of the bacterial decomposition of organic matter in the absence of air. This decomposition typically occurs in lake and river beds, in swamps, and in cattle feedlots.

Ammonia is a pungent, toxic gas that condenses to a colorless liquid at −33°C. The liquid resembles water in its physical properties, including its ability to act as a solvent for a wide range of substances. Gaseous ammonia is very soluble in water because the NH_3 molecules can form hydrogen bonds to H_2O molecules. Ammonia is a weak Brønsted base in water; it is also a reasonably strong Lewis base, particularly toward *d*-block elements. For example, it reacts with Cu^{2+}(aq) ions to give a deep blue complex (Fig. 20.4):

$$Cu^{2+}(aq) + 4\,NH_3(aq) \longrightarrow [Cu(NH_3)_4]^{2+}(aq)$$

Ammonium salts decompose when heated:

$$(NH_4)_2CO_3(s) \xrightarrow{\Delta} 2\,NH_3(g) + CO_2(g) + H_2O(g)$$

The pungent smell of decomposing ammonium carbonate was the reason for its use in smelling salts.

The pungency of heated ammonium chloride was known to the Ammonians, the worshippers of the Egyptian god Amun, and they used it in their ceremonies.

Hydrazine, NH_2NH_2, is an oily, colorless liquid. It is prepared by the gentle oxidation of ammonia with alkaline hypochlorite solution:

$$2\,NH_3(aq) + ClO^-(aq) \longrightarrow N_2H_4(aq) + Cl^-(aq) + H_2O(l)$$

Its physical properties are very similar to those of water; for instance, its melting point is 1.5°C and its boiling point is 113°C. However, it is dangerously explosive and is normally kept in aqueous solution. Hydrazine is used as a rocket fuel (see Case Study 20). It is also used to eliminate dissolved, corrosive oxygen from the water used in high-pressure, high-temperature steam furnaces:

$$N_2H_4(aq) + O_2(g) \longrightarrow N_2(g) + 2\,H_2O(l)$$

Nitrides contain the nitride ion, N^{3-}. Magnesium nitride, Mg_3N_2, is formed together with the oxide when magnesium is burned in air (Fig. 20.5):

$$3\,Mg(s) + N_2(g) \longrightarrow Mg_3N_2(s)$$

Magnesium nitride, like all nitrides, dissolves in water to produce ammonia and the corresponding hydroxide:

$$Mg_3N_2(s) + 6\,H_2O(l) \longrightarrow 3\,Mg(OH)_2(s) + 2\,NH_3(g)$$

In this reaction, the nitride ion acts as a strong base, accepting protons from water to form ammonia.

The azide ion is a highly reactive polyatomic anion of nitrogen, N_3^-. Its most common salt, sodium azide, NaN_3 (Fig. 20.6), is prepared from dinitrogen oxide and molten sodium amide:

$$N_2O(g) + 2\,NaNH_2(l) \xrightarrow{175°C} NaN_3(l) + NaOH(l) + NH_3(g)$$

FIGURE 20.6 When sodium azide (left) is heated in a sealed, evacuated tube, it decomposes into sodium metal and nitrogen gas. The sodium condenses on the walls of the tube (right).

◀◀ Section 5.12 describes the action of azides as detonators and in automobile air bags.

Lead azide, $Pb(N_3)_2$, like most azide salts, is shock sensitive and is used as a detonator for some explosives. The azide ion is a weak base, accepting a proton to form its conjugate acid, hydrazoic acid, HN_3.

The hydrogen compounds of other members of Group 15 are much less stable than ammonia and decrease in stability down the group. Phosphine, PH_3, is a poisonous gas that smells faintly of garlic and bursts into flame in air if it is slightly impure. It is much less soluble than ammonia in water because PH_3 cannot form hydrogen bonds to water. Aqueous solutions of phosphine are neutral, for PH_3 has only a very weak tendency to accept a proton ($pK_b = 27.4$).

Phosphorus trichloride, PCl_3, and phosphorus pentachloride, PCl_5, are the two most important halides of phosphorus. The former is prepared by direct chlorination of phosphorus. Phosphorus trichloride is a major intermediate for the production of pesticides, oil additives, and flame retardants. Phosphorus pentachloride is made by allowing the trichloride to react with more chlorine (recall Fig. 8.13).

A typical reaction of the nonmetal halides is their reaction with water to give oxoacids, without a change in oxidation number:

$$PCl_3(l) + 3\,H_2O(l) \longrightarrow H_3PO_3(s) + 3\,HCl(g)$$

Another example of this **hydrolysis reaction**—a reaction with water in which new element-oxygen bonds are formed—is the reaction of PCl_5 with water to produce phosphoric acid, H_3PO_4:

$$PCl_5(s) + 4\,H_2O(l) \longrightarrow H_3PO_4(l) + 5\,HCl(g)$$

This reaction is violent and dangerous.

The important compounds of nitrogen with hydrogen are ammonia, hydrazine, and hydrazoic acid, the parent of the shock-sensitive azides. Phosphine forms neutral solutions in water; reaction of nonmetal halides with water—hydrolysis—produces oxoacids but no change in oxidation number.

20.3 NITROGEN OXIDES AND OXOACIDS

The numerous nitrogen oxides can be confusing at first. Fortunately, their properties and interconversions can be understood by keeping track of their oxidation numbers. All nitrogen oxides are acidic and some are the acid anhydrides of the nitrogen oxoacids (Table 20.3). In atmospheric chemistry, where the oxides play an important role in both maintaining and polluting the atmosphere, they are referred to collectively as NO_x (read "nox").

Dinitrogen oxide, N_2O (oxidation number +1), is commonly called nitrous oxide. It is formed by gently heating ammonium nitrate:

$$NH_4NO_3(s) \xrightarrow{250°C} N_2O(g) + 2\,H_2O(g)$$

All the other nitrogen oxides are highly toxic.

Because it is tasteless, is nontoxic in small amounts, and dissolves readily in fats, N_2O is sometimes used as a foaming agent and propellant for whipped cream.

Nitrogen oxide (or nitrogen monoxide), NO (oxidation number +2),

TABLE 20.3 The oxides and oxoacids of nitrogen

Oxidation number	Oxide		Oxoacid	
	Formula	Name	Formula	Name
5	N_2O_5	dinitrogen pentoxide	HNO_3	nitric acid
4	NO_2*	nitrogen dioxide	—	
	N_2O_4	dinitrogen tetroxide	—	
3	N_2O_3	dinitrogen trioxide	HNO_2	nitrous acid
2	NO	nitrogen monoxide nitric oxide	—	
1	N_2O	dinitrogen monoxide nitrous oxide	$H_2N_2O_2$	hyponitrous acid

*$2\,NO_2 \rightleftharpoons N_2O_4$.

is commonly called nitric oxide. It is prepared industrially by the catalytic oxidation of ammonia:

$$4\,NH_3(g) + 5\,O_2(g) \xrightarrow{1000°C,\ Pt} 4\,NO(g) + 6\,H_2O(g)$$

Atmospheric nitrogen is also converted into NO in hot airplane and automobile engines. The NO contributes to the problem of acid rain, the formation of smog, and the destruction of the ozone layer (Case Study 5). Nitric oxide occurs naturally in our bodies, where it acts as a neurotransmitter and participates in the physiological changes accompanying sexual arousal.

Nitrogen dioxide, NO_2 (oxidation number +4), is a choking, poisonous, brown gas that contributes to the color and odor of smog. It is an odd-electron molecule, and in the gas phase it exists in equilibrium with its colorless dimer N_2O_4. Only the dimer exists in the solid, so the brown gas condenses to a colorless solid. When it dissolves in water, NO_2 **disproportionates;** in other words, it reacts in such a way that the oxidation number of nitrogen in some of the molecules increases whereas the oxidation number of nitrogen in other NO_2 molecules decreases. Thus, both an oxidation product and a reduction product are formed from a single reactant. The products in this case are nitric acid (oxidation number +5) and nitric oxide (oxidation number +2):

$$3\,NO_2(g) + H_2O(l) \longrightarrow 2\,HNO_3(aq) + NO(g)$$

Nitrogen dioxide in the atmosphere undergoes the same reaction and contributes to the formation of acid rain (Case Study 14). It also initiates a complex sequence of smog-forming photochemical reactions in the atmosphere that are described in Case Study 8.

Nitrous acid, HNO_2 (oxidation number +3), has not been isolated in pure form but is widely used in aqueous solution. Its acid anhydride is the dark blue liquid N_2O_3 (Fig. 20.7). Nitrites are produced by the reduction of nitrates with hot metal:

$$KNO_3(s) + Pb(s) \xrightarrow{350°C} KNO_2(s) + PbO(s)$$

Most nitrites are soluble in water and mildly toxic. Despite their toxicity, nitrites are used in the processing of meat products because they form a

FIGURE 20.7 Dinitrogen trioxide, N_2O_3, condenses to a deep blue liquid that freezes at −100°C to a pale blue solid. On standing, it turns green as a result of partial decomposition into nitrogen dioxide (not shown), a yellow-brown gas.

pink complex with the hemoglobin and inhibit the oxidation of the blood (a reaction that turns the meat brown). Nitrites are responsible for the pink color of ham, sausages, and other cured meat.

Nitric acid, HNO_3 (oxidation number +5), ranks fourteenth in annual production of chemicals in the United States and is used extensively in the production of fertilizers and explosives. It is produced by the three-stage **Ostwald process:**

Step 1. Oxidation of ammonia, from oxidation number −3 to +2:

$$4\,NH_3(g) + 5\,O_2(g) \xrightarrow{850°C,\ 5\ atm,\ Pt/Rh} 4\,NO(g) + 6\,H_2O(g)$$

Step 2. Oxidation of nitrogen oxide, from oxidation number +2 to +4:

$$2\,NO(g) + O_2(g) \longrightarrow 2\,NO_2(g)$$

Step 3. Disproportionation in water, from oxidation number +4 to +5 and +2:

$$3\,NO_2(g) + H_2O(l) \longrightarrow 2\,HNO_3(aq) + NO(g)$$

Nitric acid is a colorless liquid that boils at 83°C and is normally used in aqueous solution. Concentrated nitric acid is often pale yellow as a result of partial decomposition of the acid to NO_2. Because nitrogen has its highest oxidation number (+5) in HNO_3, nitric acid is an oxidizing agent as well as an acid.

Nitrogen forms oxides in each of its integer oxidation states from +1 to +5; the properties of the oxides and oxoacids can be explained in terms of the oxidation number of nitrogen and the identity of the corresponding acid anhydride.

20.4 PHOSPHORUS OXIDES AND OXOACIDS

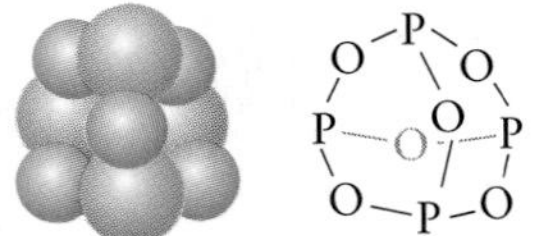

2 Phosphorus(III) oxide, P_4O_6

This compound is sometimes called phosphorus trioxide because its empirical formula is P_2O_3. Similarly, phosphorus(V) oxide is often called phosphorus pentoxide because its empirical formula is P_2O_5.

The structures of the phosphorus oxides are based on the tetrahedral PO_4 unit. White phosphorus burns in a limited supply of air to form phosphorus(III) oxide, P_4O_6 (**2**).

$$P_4(s, white) + 3\,O_2(g) \longrightarrow P_4O_6(s)$$

The molecules are tetrahedral, like P_4, but an O atom lies between each pair of P atoms. Phosphorus(III) oxide is the anhydride of phosphorous acid, H_3PO_3. Although its formula suggests that it should be a triprotic acid, H_3PO_3 is in fact diprotic because one of the H atoms is attached directly to the P atom and the P—H bond is nonpolar (see Section 14.10).

When phosphorus burns in an ample supply of air, it forms phosphorus(V) oxide, P_4O_{10} (**3**). This white solid is an acid anhydride with such a strong attraction for water that it is widely used in the laboratory as a drying agent. It is also used as a dehydrating agent to remove water from compounds, as in the preparation of other acid anhydrides. Phosphorus(V) oxide reacts with water to form phosphoric acid, H_3PO_4 (**4**), the parent acid of the phosphates. Phosphate rock is mined in huge quantities in Florida and Morocco. After being crushed, it is treated with sulfuric

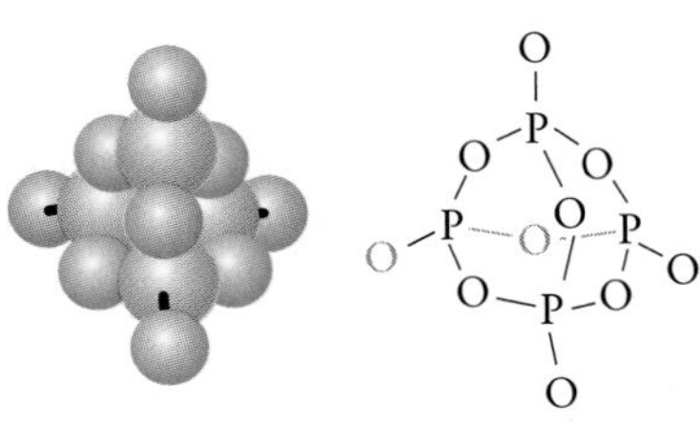

3 Phosphorus(V) oxide, P_4O_{10}

CHEMISTRY AT WORK

A veterinarian examines one of his patients and prepares to recommend a change in diet. In order to ensure the good health of his patients, he must understand the different dietary requirements of each type of animal. For example, dogs require a certain amount of calcium and phosphorus in their diets each day, while many cats need diets low in magnesium.

acid to give a mixture of sulfates and phosphates called *superphosphate,* a major fertilizer:

$$Ca_3(PO_4)_2(s) + 2\,H_2SO_4(l) \longrightarrow 2\,CaSO_4(s) + Ca(H_2PO_4)_2(s)$$

The phosphate rock can be treated with phosphoric acid rather than sulfuric acid to produce a mixture with a higher phosphate content:

$$Ca_3(PO_4)_2(s) + 4\,H_3PO_4(l) \longrightarrow 3\,Ca(H_2PO_4)_2(s)$$

The resulting mixture of calcium phosphates is sold as the fertilizer *triple superphosphate.*

When phosphoric acid is heated, it undergoes a condensation reaction (Section 11.9):

$$HO{-}\underset{OH}{\overset{O}{\overset{\|}{P}}}{-}O{-}\boxed{H + HO}{-}\underset{OH}{\overset{O}{\overset{\|}{P}}}{-}OH \longrightarrow HO{-}\underset{OH}{\overset{O}{\overset{\|}{P}}}{-}O{-}\underset{OH}{\overset{O}{\overset{\|}{P}}}{-}OH + H_2O$$

The product, $H_4P_2O_7$, is pyrophosphoric acid. Further heating gives even more complicated products that have chains and rings of PO_4 groups and are called *polyphosphoric acids.* The most important polyphosphate is adenosine triphosphate, ATP (**5**), for it is found in every living cell.

5 Adenosine triphosphate (ATP)

4 Phosphoric acid, H_3PO_4

The role of calcium phosphate in skeletons and teeth is described in Section 8.3.

The triphosphate part of this molecule is a chain of three phosphate groups. Its conversion to *adenosine diphosphate*, ADP, in the reaction

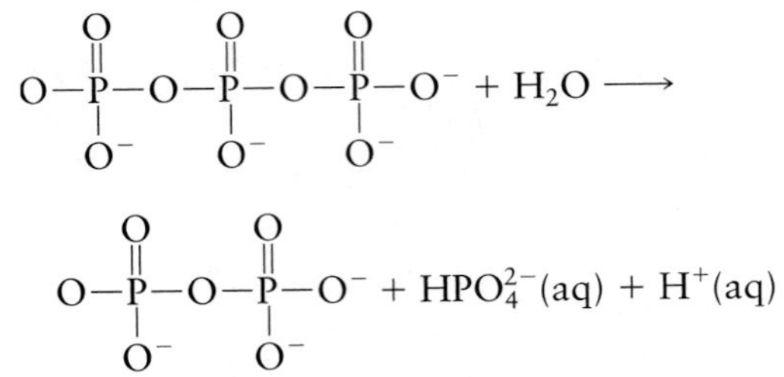

$$\begin{array}{c} \text{O}\quad\quad\text{O}\quad\quad\text{O} \\ \| \quad\quad\quad \| \quad\quad\quad \| \\ \text{O}-\text{P}-\text{O}-\text{P}-\text{O}-\text{P}-\text{O}^- + H_2O \longrightarrow \\ | \quad\quad\quad | \quad\quad\quad | \\ \text{O}^-\quad\;\text{O}^-\quad\;\text{O}^- \end{array}$$

$$\begin{array}{c} \text{O}\quad\quad\text{O} \\ \| \quad\quad\quad \| \\ \text{O}-\text{P}-\text{O}-\text{P}-\text{O}^- + HPO_4^{2-}(aq) + H^+(aq) \\ | \quad\quad\quad | \\ \text{O}^-\quad\;\text{O}^- \end{array} \qquad \Delta G^\circ = -30\ \text{kJ at pH} = 7$$

releases large amounts of energy that are used to power energy-demanding processes in cells. The death of the organism releases the phosphate to the ecosystem, where it may lie as phosphate rock until nature or industry sends it on its way again.

The oxides of phosphorus have structures based on the tetrahedral PO_4 unit; P_4O_6 and P_5O_{10} are the anhydrides of phosphorous and phosphoric acid, respectively. Polyphosphates are extended structures used by living cells to store and transfer energy.

SELF-TEST 20.1A How does the (a) acidity of a nitrogen oxoacid and (b) its strength as an oxidizing agent change as the oxidation number of N increases from 1 to 5?

[*Answer:* (a) Increases; (b) increases]

SELF-TEST 20.1B Explain why nitrogen forms only NCl_3, whereas phosphorus forms both PCl_3 and PCl_5.

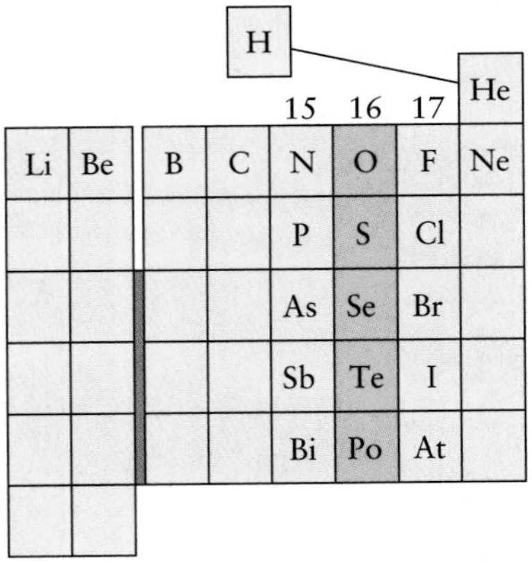

				15	16	17	H / He
Li	Be	B	C	N	O	F	Ne
				P	S	Cl	
				As	Se	Br	
				Sb	Te	I	
				Bi	Po	At	

GROUP 16: THE OXYGEN FAMILY

The elements become increasingly nonmetallic toward the right of the periodic table, and by Group 16, even polonium, at the foot of the group, is best regarded as a metalloid (Table 20.4). The valence electron configuration is ns^2np^4, so the atoms of these elements need only two more electrons to complete their shells. The members of the group are collectively called the **chalcogens.** The name comes from the Greek words meaning "brass giver," because the elements are found in copper ores and copper is a major component of brass.

20.5 THE ELEMENTS

Oxygen is the most abundant element in the Earth's crust and accounts for 23% of the mass of the atmosphere. Some of our present atmospheric oxygen has been produced by the photochemical action of sunlight on water. Most of the rest has been produced by photosynthesis:

$$6\,CO_2(g) + 6\,H_2O(l) \longrightarrow C_6H_{12}O_6(s) + 6\,O_2(g)$$

More than 2×10^{10} kg of liquid oxygen is produced each year in the

TABLE 20.4 The Group 16 elements

Valence configuration: ns^2np^4

Z	Name	Symbol	Molar mass, $g \cdot mol^{-1}$	Melting point, °C	Boiling point, °C	Density, $g \cdot cm^{-3}$ at 25°C	Normal form*
8	oxygen	O	16.00	−210	−183	1.14†	colorless paramagnetic gas (O_2)
				−192	−112	1.35†	blue gas (ozone, O_3)
16	sulfur	S	32.06	115	445	2.09	yellow nonmetallic solid (S_8)
34	selenium	Se	78.96	220	685	4.81	gray nonmetallic solid
52	tellurium	Te	127.60	450	990	6.25	silver-white metalloid
84	polonium	Po	210	254	960	9.40	gray metalloid

**Normal form* means the appearance and state of the element at 25°C and 1 atm.
†For the liquid at its boiling point.

United States (about 80 kg per inhabitant) by fractional distillation of liquid air, and it ranks third among manufactured chemicals, by mass produced. The biggest consumer of oxygen is the steel industry, which needs about 1 ton of oxygen to produce 1 ton of steel. In steelmaking, oxygen is blown into molten iron to oxidize any impurities, particularly carbon. Elemental oxygen also has important medical uses. Physicians administer oxygen to relieve strain on the heart and as a stimulant.

FIGURE 20.8 Liquid oxygen is pale blue (the gas itself is colorless).

Oxygen is a colorless, tasteless, odorless gas of O_2 molecules that condenses to a pale blue liquid at −183°C (Fig. 20.8). Although O_2 has an even number of electrons, two of them are unpaired, making the molecule **paramagnetic;** in other words, it behaves like a tiny magnet and is attracted to a magnetic field. The oxygen content of incubators and other life-support systems can be monitored by measuring the paramagnetism of the gas they contain.

As we saw in Case Study 5, an allotrope of oxygen—namely, ozone, O_3—forms in the stratosphere by the effect of solar radiation on O_2 molecules. Its total abundance in the atmosphere is equivalent to a layer that, at normal temperature and pressure, would cover the Earth to a thickness of only 3 mm. Ozone can be made in the laboratory by passing an electric discharge through oxygen. It is a blue gas that condenses at −112°C to an explosive liquid that looks like blue ink (Fig. 20.9). Its pungent smell can often be detected near electrical equipment and after lightning: the presence of ozone is partly responsible for the "fresh" scent of air after an electrical storm accompanied by rain. Ozone is also present in smog-laden areas, where it is produced by the reaction of oxygen molecules with oxygen atoms.

The name *ozone* comes from the Greek word for "smell."

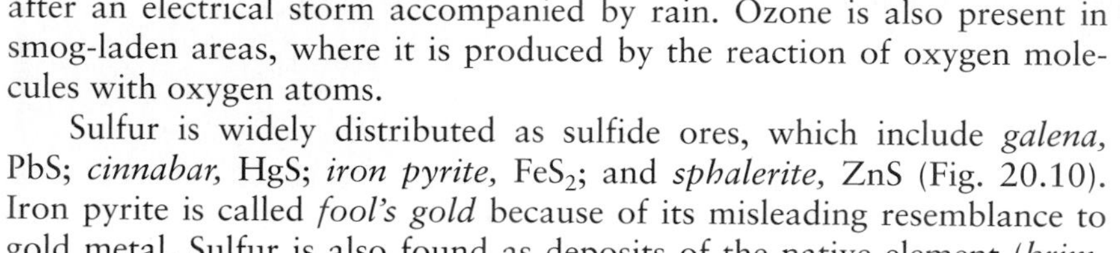

Sulfur is widely distributed as sulfide ores, which include *galena,* PbS; *cinnabar,* HgS; *iron pyrite,* FeS_2; and *sphalerite,* ZnS (Fig. 20.10). Iron pyrite is called *fool's gold* because of its misleading resemblance to gold metal. Sulfur is also found as deposits of the native element (*brimstone*), which are formed by bacterial action on H_2S. Native sulfur is

FIGURE 20.9 Ozone is a blue gas that condenses to a dark blue, highly unstable liquid.

FIGURE 20.10 A collection of sulfide ores (from left to right): galena, PbS; cinnabar, HgS; pyrite, FeS_2; sphalerite, ZnS. Pyrite has a lustrous golden color and has frequently been mistaken for gold; hence it is also known as fool's gold. Gold and fool's gold are readily distinguished by their densities.

mined by the ingenious **Frasch process,** which makes use of its low melting point and low density. In this process (Fig. 20.11), water at about 165°C and under pressure is pumped down the outermost of three concentric pipes to melt the sulfur trapped beneath deep rock layers. Compressed air is passed down the innermost pipe to force a frothy mixture of sulfur, air, and hot water up the middle pipe.

Sulfur is also a by-product of a number of metallurgical processes, especially the extraction of copper from its sulfide ores, and is removed from sulfur-rich petroleum by the **Claus process.** In this process, some of the H_2S that occurs in oil and natural gas wells is first oxidized to sulfur dioxide

$$2\,H_2S(g) + 3\,O_2(g) \longrightarrow 2\,SO_2(g) + 2\,H_2O(l)$$

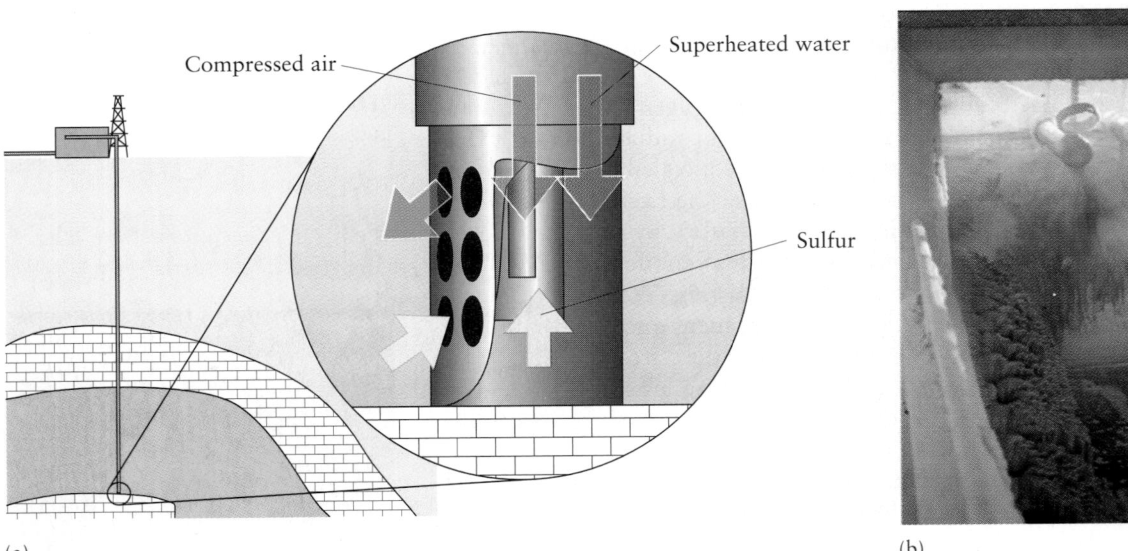

FIGURE 20.11 (a) In the Frasch process, superheated water—water under pressure at temperatures above 100°C—is pumped down the outermost of three concentric pipes; compressed air is pumped down the innermost pipe and forces a mixture of liquid sulfur, air, and hot water up the middle pipe. (b) Sulfur surfacing with hot water and air in the Culbertson Mine in West Texas.

(a)

(b)

FIGURE 20.12 One of the two most common forms of sulfur is the blocklike rhombic form (a). It differs from the needlelike monoclinic sulfur (b) in the manner in which the S_8 rings are stacked together.

FIGURE 20.13 Two of the Group 16 elements: selenium (left) and tellurium (right).

This SO_2 is then used to oxidize the remainder of the hydrogen sulfide:

$$2\,H_2S(g) + SO_2(g) \xrightarrow{300°C,\ Al_2O_3} 3\,S(s) + 2\,H_2O(l)$$

Sulfur is of major industrial importance. Most of the sulfur that is produced is used to make sulfuric acid, but an appreciable amount is used to vulcanize rubber (Section 11.16).

Elemental sulfur is a yellow, tasteless, almost odorless, insoluble, nonmetallic molecular solid of crownlike S_8 rings (**6**). The most stable allotrope under normal conditions is *rhombic sulfur* (Fig. 20.12). The changes that sulfur undergoes as it is heated were discussed in Section 10.4 and illustrated in Fig. 10.11. Sulfur vapor has a blue tint from the S_2 molecules present in it. The latter are paramagnetic, like O_2.

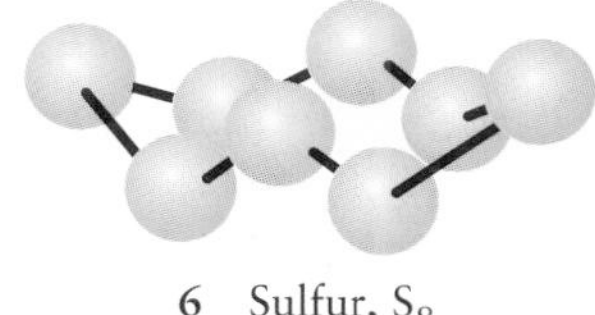

6 Sulfur, S_8

Selenium and tellurium occur in sulfide ores; they are also recovered from the anode sludge formed during the electrolytic refining of copper. Both elements have several allotropes, the most stable consisting of long zigzag chains of atoms. Although these allotropes look like silver-white metals, they are poor electrical conductors (Fig. 20.13). The conductivity of selenium is increased in the presence of light, so it is used in photoelectric devices and in photocopying machines.

Electronegativities decrease down the group (Fig. 20.14) and atomic and ionic radii increase (Fig. 20.15). The differences between oxygen and sulfur are emphasized by the latter's striking ability to **catenate,** or form chains of atoms. Oxygen's ability to form chains is very limited, with H_2O_2, O_3, and the anions O_2^-, O_2^{2-}, and O_3^- the only examples. Sulfur's ability is much more pronounced. It appears, for instance, in the existence of S_8 rings and their fragments, which can polymerize when heated to about 200°C. The existence of —S—S— links that connect different parts of the chains of amino acids in proteins is another example of catenation. These disulfide links contribute to the shapes of proteins and so help to keep us alive.

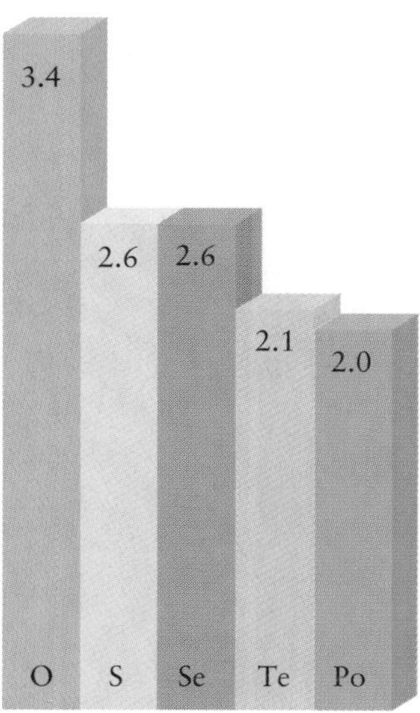

FIGURE 20.14 The electronegativities of the Group 16 elements decrease down the group.

Metallic character increases down Group 16 as electronegativity decreases. Oxygen and sulfur occur naturally in the elemental state; sulfur forms chains and rings with itself, but oxygen does not.

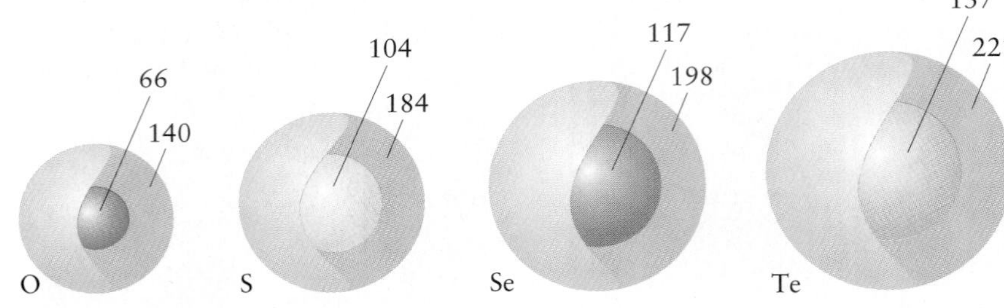

FIGURE 20.15 The atomic and ionic radii of the Group 16 elements increase down the group. The values shown are in picometers and the anion (shown in green in each case) is substantially larger than the neutral parent atom.

20.6 COMPOUNDS WITH HYDROGEN

By far the most important compound of oxygen and hydrogen is water, H_2O. Water is available on a huge scale worldwide, but in various states of purity. Municipal water reservoirs usually contain a low concentration of Cu^{2+} ions to restrict the growth of algae. The copper must then be removed, along with other impurities. Municipal water supplies normally undergo several stages of purification, as described in Case Study 3. High-purity water for special applications is obtained by distillation or by **ion exchange,** the exchange of one type of ion in a solution by another. In ion exchange, water passes through a *zeolite,* an aluminosilicate with a very open structure that can capture ions such as Mg^{2+} and Ca^{2+} and exchange them for H^+ ions.

Water is an oxidizing agent:

$$2\,H_2O(l) + 2\,e^- \longrightarrow 2\,OH^-(aq) + H_2(g) \qquad E = -0.42\text{ V at pH 7}$$

One example is its reaction with the alkali metals, as in

$$2\,Na(s) + 2\,H_2O(l) \longrightarrow 2\,NaOH(aq) + H_2(g)$$

However, unless the other reactant is a strong reducing agent, water acts as an oxidizing agent only at high temperatures, as it does in the reforming reaction:

$$CH_4(g) + H_2O(g) \xrightarrow{\Delta} CO(g) + 3\,H_2(g)$$

Water is a very mild reducing agent, the half-reaction being

$$2\,H_2O(l) \longrightarrow 4\,H^+(aq) + O_2(g) + 4\,e^- \qquad E = -0.81\text{ V at pH 7}$$

However, few substances besides fluorine are strong enough oxidizing agents to remove electrons from water.

Water is also a Lewis base, for an H_2O molecule can donate one of its lone pairs to a Lewis acid and form complexes such as $[Fe(H_2O)_6]^{3+}$. Its ability to act as a Lewis base is also the origin of water's ability to hydrolyze substances. The reaction between water and phosphorus pentachloride mentioned earlier is an example.

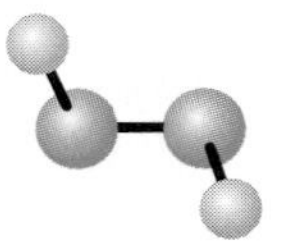

7 Hydrogen peroxide, H_2O_2

Hydrogen peroxide, H_2O_2 (7), is a very pale blue liquid that is appreciably denser than water (1.44 g·mL^{-1} at 25°C) but similar in other physical properties: its melting point is −0.4°C, and its boiling point is

FIGURE 20.16 Chemiluminescence—the emission of light as the result of a chemical reaction—occurs when hydrogen peroxide is added to a solution of the organic compound perylene. Although hydrogen peroxide itself can fluoresce, in this case the light is emitted by the perylene.

152°C. Chemically, though, hydrogen peroxide and water differ greatly. The presence of the second oxygen atom makes H_2O_2 a very weak acid ($pK_{a1} = 11.75$). Hydrogen peroxide is also a good oxidizing agent in both acidic and basic solution. For example, H_2O_2 oxidizes Fe^{2+} and Mn^{2+} in acidic and basic solutions. It can also act as a reducing agent in acidic and basic solutions. For example, H_2O_2 reduces permanganate ions and chlorine in acidic and basic solutions. The O_2 formed by the oxidation of H_2O_2 is sometimes produced in an energetically excited state and emits light as it discards its excess energy. This process is an example of **chemiluminescence,** the emission of light by products formed in energetically excited states when reagents are mixed (Fig. 20.16).

Hydrogen peroxide is normally sold for industrial use as a 30% by mass aqueous solution. When used as a hair bleach (as a 6% solution), it acts by oxidizing the pigments in the hair. Because it oxidizes unpleasant effluents without producing any harmful by-products, H_2O_2 is increasingly widely used as an oxidizing agent in the control of pollution. A 3% H_2O_2 aqueous solution is used as a mild antiseptic in the home. Contact with blood catalyzes its disproportionation into oxygen gas and water, which cleanses the wound.

SELF-TEST 20.2A Determine from the Lewis structures of the following molecules which are paramagnetic and explain how you decided: (a) N_2; (b) NO; (c) N_2O.

[*Answer:* (b) is paramagnetic, because it has an unpaired electron.]

SELF-TEST 20.2B Write the half-reactions and the overall reaction for the oxidation of water by F_2. Determine the standard cell potential and ΔG_r° for the reaction.

All the Group 16 H_2X compounds other than water are toxic gases with offensive odors. They are insidious poisons because they paralyze the olfactory nerve and soon after exposure they cannot be smelled. Rotten eggs smell of hydrogen sulfide, H_2S, because egg proteins contain sulfur and give off the gas when they decompose. Another sign of the formation of sulfides in eggs is the pale green discoloration sometimes seen in cooked eggs where the white meets the yolk: this discoloration is a deposit of iron(II) sulfide. Hydrogen sulfide is also responsible for the sour smell of underground natural gas reserves.

Hydrogen sulfide dissolves in water to give a solution that, as a result of its oxidation by dissolved air, slowly becomes cloudy on account of the formation of a colloidal dispersion of small particles of sulfur. Hydrogen sulfide is a weak diprotic acid and the parent acid of the hydrogen sulfides (which contain the HS^- ion) and the sulfides (which contain the S^{2-} ion). The sulfides of the *s*-block elements are moderately soluble, whereas the sulfides of the heavy *p*- and *d*-block metals are generally very insoluble.

◀◀ Colloidal dispersions, or suspensions, are discussed in Case Study 12.

FIGURE 20.17 The blue stones in this ancient Egyptian ornament are lapis lazuli. This semiprecious stone is an aluminosilicate colored by S_2^- and S_3^- impurities. The blue color is due to S_3^-, and its hint of green to S_2^-.

The sulfur analogue of hydrogen peroxide also exists and is an example of a *polysulfane,* a catenated molecular compound of composition $HS{-}S_n{-}SH$, where n can take on values from 0 through 6. The polysulfide ions obtained from the polysulfanes include two ions found in lapis lazuli (Fig. 20.17).

Water can act as a weak Brønsted acid, a weak Brønsted base, a Lewis base, an oxidizing agent, and a weak reducing agent; hydrogen peroxide is a strong oxidizing agent. Hydrogen sulfide is a weak acid; the polysulfanes exhibit the ability of sulfur to catenate.

20.7 SULFUR OXIDES AND OXOACIDS

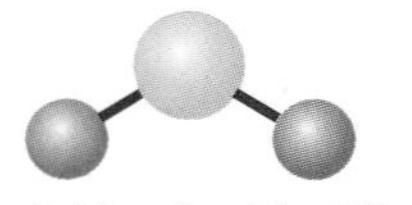

8 Sulfur dioxide, SO_2

The most important oxides and oxoacids of sulfur are the dioxide and trioxide and the corresponding sulfurous and sulfuric acids.

Sulfur burns in air to form sulfur dioxide, SO_2 (**8**), a colorless, choking, poisonous gas (recall Fig. 1.19). Sulfur oxides in the atmosphere are referred to as SO_x (read "sox"). About 7×10^{10} kg of the dioxide result from the decomposition of vegetation and from volcanic emissions. In addition, the approximately 1×10^{11} kg of naturally occurring hydrogen sulfide can be oxidized to the dioxide by atmospheric oxygen:

$$2\,H_2S(g) + 3\,O_2(g) \longrightarrow 2\,SO_2(g) + 2\,H_2O(g)$$

Industry and transport contribute another 1.5×10^{11} kg of the dioxide, of which about 70% comes from oil and coal combustion—mainly in electricity-generating plants. The average concentration of SO_x in the atmosphere in rural areas in the northern hemisphere is found to be about $1 \times 10^{-6}\ mol{\cdot}L^{-1}$. However, the concentration is much higher in industrialized areas (see Case Study 14).

Sulfur dioxide is an acidic oxide, the anhydride of sulfurous acid, H_2SO_3, which is the parent acid of the hydrogen sulfites (or bisulfites) and the sulfites:

$$SO_2(g) + H_2O(l) \longrightarrow H_2SO_3(aq)$$

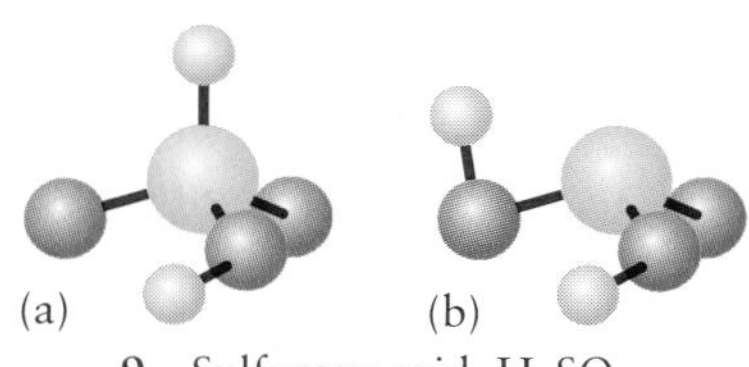

9 Sulfurous acid, H_2SO_3

Sulfurous acid is an equilibrium mixture of two molecules, (**9a**) and (**9b**); in the former it resembles phosphorous acid, with one of the H atoms attached directly to the S atom. These molecules are also in equilibrium with molecules of SO_2, each of which is surrounded by a cage of water molecules. The evidence for this equilibrium is that crystals of composition $SO_2{\cdot}xH_2O$, with x about 7, are obtained when the solution is cooled. Substances like this, in which a molecule sits in a cage of other molecules, are called **clathrates.** Methane, carbon dioxide, and the noble gases also form clathrates with water.

Sulfur dioxide is easily liquified under pressure and can therefore be used as a refrigerant. It is also a preservative for dried fruit and a bleach for textiles and flour. Its most important use is in the production of sulfuric acid.

The oxidation number of sulfur in sulfur dioxide and the sulfites is +4, an intermediate value in sulfur's range from −2 to +6. Hence these compounds can act as either oxidizing agents or reducing agents (Fig. 20.18). By far the most important reaction of sulfur dioxide is its oxidation to sulfur trioxide, SO_3 (**10**), in which sulfur has the oxidation number +6:

10 Sulfur trioxide, SO_3

$$2\,SO_2(g) + O_2(g) \longrightarrow 2\,SO_3(g)$$

The direct reaction is very slow, so an SO_2 molecule survives for a few days in the atmosphere before it is oxidized to SO_3. The oxidation is catalyzed by the metal ions in the minerals on the walls of buildings and metal cations like Fe^{3+}, which may be dissolved in droplets of water. Other routes from SO_2 to SO_3 include reaction with OH radicals, H_2O_2, and O_3 formed by the effect of sunlight on air and water vapor. The sulfur trioxide produced by these processes reacts with atmospheric water to form dilute sulfuric acid:

$$SO_3(g) + H_2O(l) \longrightarrow H_2SO_4(aq)$$

The product of this reaction, together with the nitric acid formed by dissolved NO_x, falls as acid rain.

Sulfuric acid, H_2SO_4, is a colorless, corrosive, oily liquid that boils (and decomposes) at about 300°C. It has three chemically important properties: it is a strong acid, a dehydrating agent, and an oxidizing agent (Fig. 20.19). It is produced commercially in the **contact process,** in which sulfur is first burned in oxygen and the SO_2 is oxidized to SO_3 over a V_2O_5 catalyst:

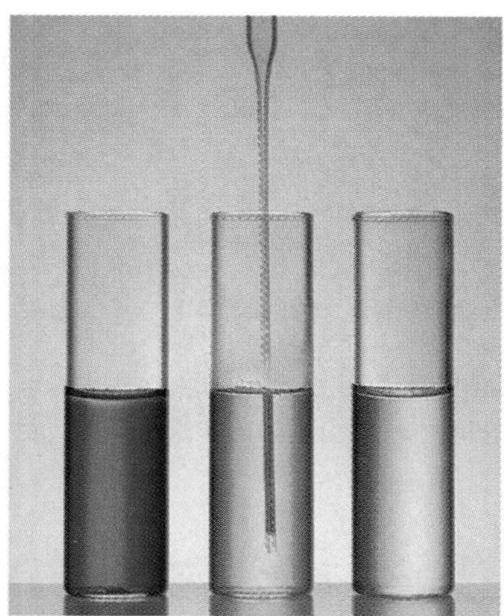

FIGURE 20.18 Sulfur dioxide is a reducing agent. When it is bubbled through an aqueous solution of bromine (left), it reduces the Br_2 to colorless bromide ions (right). The SO_2 is oxidized to H_2SO_4.

FIGURE 20.19 Sulfuric acid is an oxidizing agent. When some concentrated acid is poured onto solid sodium bromide, NaBr, the bromide ions are oxidized to bromine, which colors the solution brown.

FIGURE 20.20 Sulfuric acid is a dehydrating agent. When concentrated sulfuric acid is poured onto sucrose, the sucrose, a carbohydrate, is dehydrated, leaving a frothy black mass of carbon.

(a)

(b)

(c)

$$S(g) + O_2(g) \xrightarrow{1000^\circ C} SO_2(g)$$
$$2\,SO_2(g) + O_2(g) \xrightarrow{500^\circ C,\ V_2O_5} 2\,SO_3(g)$$

Because sulfur trioxide does not dissolve quickly in water, it is absorbed in 98% concentrated sulfuric acid to give the dense, oily liquid called *oleum:*

$$SO_3(g) + H_2SO_4(l) \longrightarrow H_2S_2O_7(l)$$

Oleum is converted to the acid by dilution with water:

$$H_2S_2O_7(l) + H_2O(l) \longrightarrow 2\,H_2SO_4(l)$$

Sulfuric acid is the most heavily produced chemical worldwide, the annual production being over 4×10^{10} kg in the United States alone. The low cost of sulfuric acid leads to its widespread use in industry, particularly for the production of fertilizers, petrochemicals, dyestuffs, and detergents. About two-thirds is used in the manufacture of ammonium sulfate and phosphate fertilizers. Its reaction with phosphate rock to produce superphosphate fertilizer was discussed in Section 20.4.

Sulfuric acid is a powerful dehydrating agent, as seen when a little concentrated acid is poured on sucrose, $C_{12}H_{22}O_{11}$. A black, frothy mass of carbon forms as a result of the extraction of H_2O (Fig. 20.20):

$$C_{12}H_{22}O_{11}(s) \xrightarrow{\text{conc } H_2SO_4} 12\,C(s) + 11\,H_2O(l)$$

The froth is caused by CO and CO_2 formed in side reactions.

Sulfur dioxide is the acid anhydride of sulfurous acid, and sulfur trioxide the anhydride of sulfuric acid. Sulfuric acid is a strong acid, a dehydrating agent, and an oxidizing agent.

20.8 SULFUR HALIDES

Sulfur reacts directly with all the halogens except iodine. It ignites spontaneously in fluorine and burns brightly to give sulfur hexafluoride, SF_6 (**11**), a colorless, tasteless, odorless, nontoxic, thermally stable, insoluble gas. Despite the high oxidation number (+6) of the sulfur atom, SF_6 is not a good oxidizing agent. Its low reactivity is in large part due to the F atoms that surround the central S atom and protect it from attack. The ionization energy of SF_6 is very high because any electron that is removed

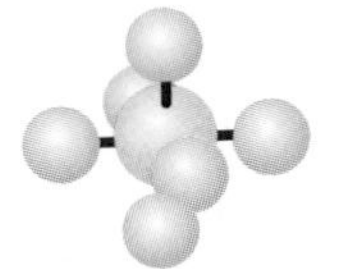
11 Sulfur hexafluoride, SF_6

must come from the highly electronegative F atoms. Even quite strong electric fields cannot strip off these electrons, so SF_6 is a good gas-phase electrical insulator. It is, in fact, a much better insulator than air and is used in switches on high-voltage power lines.

Sulfur reacts directly with chlorine. One product is disulfur dichloride, S_2Cl_2, a yellow liquid with a disgusting smell; it is used for the vulcanization of rubber. When disulfur dichloride reacts with more chlorine in the presence of iron(III) chloride as a catalyst, sulfur dichloride, SCl_2, is produced as an evil-smelling red liquid. Sulfur dichloride reacts with ethene to give *mustard gas* (**12**), which has been used for chemical warfare. Mustard gas causes blisters, discharges from the nose, and vomiting; it also destroys the cornea of the eye. All in all, it is easy to see why ancient civilizations associated sulfur with the underworld.

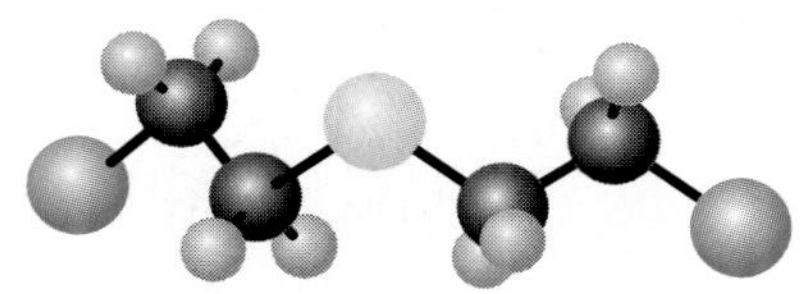

12 Mustard gas, $S(CH_2CH_2Cl)_2$

Sulfur hexafluoride is a thermally stable nontoxic gas that is a good insulator; sulfur chlorides are toxic liquids.

SELF-TEST 20.3A Concentrated sulfuric acid is 98.0% by mass H_2SO_4. Calculate its molar concentration given that its density is $1.84\ g{\cdot}mL^{-1}$.

[*Answer:* $18.4\ mol{\cdot}L^{-1}$]

SELF-TEST 20.3B When concentrated sulfuric acid and water are mixed, the total volume is less than the sum of the individual volumes before mixing. Explain this volume contraction in terms of intermolecular interactions.

GROUP 17: THE HALOGENS

The unique properties of the halogens (Table 20.5), the members of Group 17, can be traced to their valence configurations, ns^2np^5, which need only one more electron to reach a closed-shell configuration. The elements form a family that shows smooth trends in physical properties, electronegativity (Fig. 20.21), and atomic and ionic radii (Fig. 20.22).

TABLE 20.5 The Group 17 elements

Valence configuration: ns^2np^5

Z	Name	Symbol	Molar mass, $g{\cdot}mol^{-1}$	Melting point, °C	Boiling point, °C	Density, $g{\cdot}cm^{-3}$ at 25°C	Normal form*
9	fluorine	F	19.00	−220	−188	1.51†	almost colorless gas
17	chlorine	Cl	35.45	−101	−34	1.66†	yellow-green gas
35	bromine	Br	79.91	−7	59	3.12	red-brown liquid
53	iodine	I	126.90	114	185	4.95	purple-black nonmetallic solid
85	astatine‡	At	210	300	350		nonmetallic solid

**Normal form* means the appearance and state of the element at 25°C and 1 atm.
†For the liquid at its boiling point.
‡Radioactive.

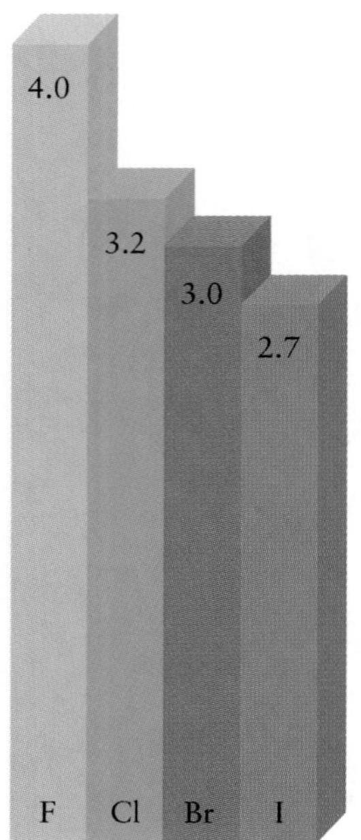

FIGURE 20.21 The electronegativities of the halogens decrease steadily down the group.

20.9 THE ELEMENTS

Fluorine occurs widely in many minerals, including *fluorspar,* CaF_2; *cryolite,* Na_3AlF_6; and the *fluorapatites,* $Ca_5F(PO_4)_3$. Fluorine is the most strongly oxidizing element ($E° = +2.87$ V). Only an anode can be made more oxidizing (by increasing its positive charge); hence, only in an electrolytic cell can fluorine be driven out of its compounds by removing one of the tightly held electrons from the F^- ion. Fluorine is produced by electrolyzing an anhydrous molten mixture of potassium fluoride and hydrogen fluoride at about 75°C with a carbon anode.

Fluorine is a reactive, colorless gas of F_2 molecules. It was little used before the development of the nuclear industry but is now produced on a large scale—at about 5×10^6 kg a year in the United States (Fig. 20.23). Most of the fluorine production is used to make the volatile solid UF_6 used for processing nuclear fuel (Section 22.12). Much of the rest is used in the production of SF_6 for electrical equipment and to make fluorinated hydrocarbons, such as Teflon and Freon. Most fluoro-substituted hydrocarbons are relatively inert chemically: they are inert to oxidation by air, hot nitric acid, concentrated sulfuric acid, and other strong oxidizing agents.

Chlorine is obtained from sodium chloride by electrolysis, either of molten rock salt or of brine (recall Fig. 19.13). It is a pale yellow-green gas of Cl_2 molecules that condenses at −34°C. It reacts directly with nearly all the elements (the exceptions being carbon, nitrogen, oxygen, and the noble gases). It is a strong oxidizing agent and oxidizes metals to high oxidation states; for example, anhydrous iron(III) chloride, not iron(II) chloride, is formed when chlorine reacts with iron (Fig. 20.24):

$$2\,Fe(s) + 3\,Cl_2(g) \longrightarrow 2\,FeCl_3(s)$$

Chlorine is used in a number of industrial processes, including the manufacture of plastics, solvents, and pesticides. Chlorine is also used as a bleach in the paper and textile industries and as a disinfectant in water treatment. The use of chlorine to provide potable water has made life in large cities and our modern lifestyles feasible. The chlorine smell of chlorinated water comes largely from the amines (Section 11.10) in which NH_2 groups have become chlorinated (converted to NHCl groups).

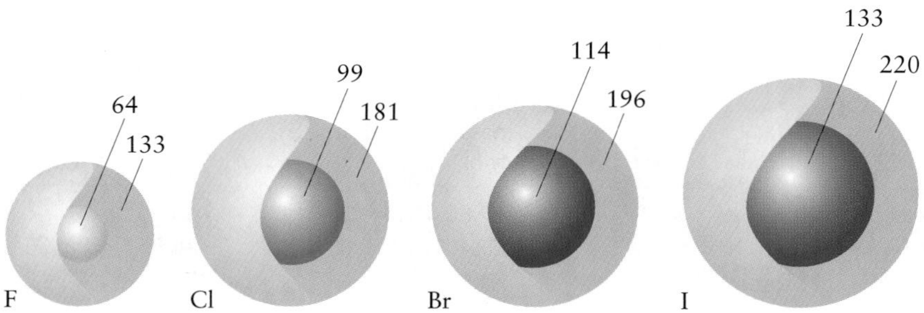

FIGURE 20.22 The atomic and ionic radii of the halogens increase steadily down the group as electrons occupy outer shells of the atoms. The values shown are in picometers. In all cases, ionic radii (represented by the green spheres) are larger than atomic radii.

FIGURE 20.23 Fluorine is prepared on a large scale by an adaptation of the electrolytic method that was used to isolate it originally. This interior shot of a preparation plant shows the electrolytic cells.

Chlorine is also used to produce bromine from brine wells through the oxidation of Br^- ions (Fig. 20.25):

$$2\,Br^-(aq) + Cl_2(g) \longrightarrow Br_2(l) + 2\,Cl^-(aq)$$

Air is bubbled through the solution to vaporize the bromine and drive it out.

Bromine is a corrosive, red-brown fuming liquid of Br_2 molecules with a penetrating odor. It is increasingly used in industrial chemistry because of the ease with which it can be added to and removed from organic chemicals being used to carry out complicated syntheses. Organic bromides are incorporated into textiles as fire retardants and are used as pesticides; inorganic bromides, particularly silver bromide, are used as photographic emulsions (see Case Study 21). Saturated aqueous zinc bromide has a very high density and is used in the oil industry to control the escape of oil from deep wells. The tall column of liquid that results when this solution is poured down a well exerts a very high pressure at its base.

The name *bromine* comes from the Greek word for "stench."

Iodine occurs as iodide ions in brines and as an impurity in Chile saltpeter. It was once obtained from seaweed, which contains high concentrations accumulated from seawater: 2000 kg of seaweed produce about 1 kg of iodine. The best modern source is the brine from oil wells, for the oil itself was produced by the decay of marine organisms that had accumulated the iodine while they were alive. The element is produced by oxidation with chlorine:

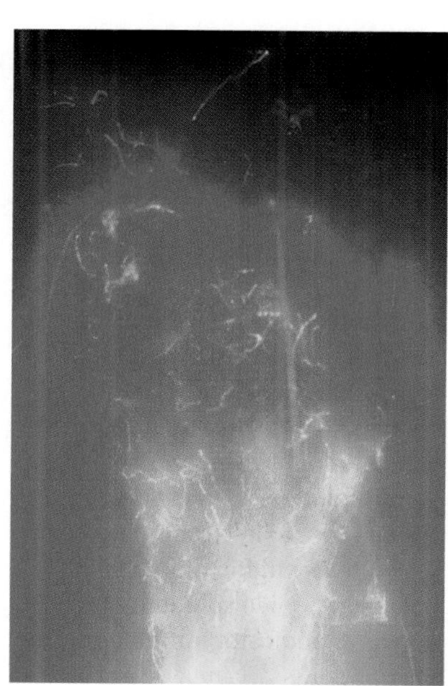

FIGURE 20.24 Iron reacts vigorously and exothermically with chlorine to form anhydrous iron(III) chloride.

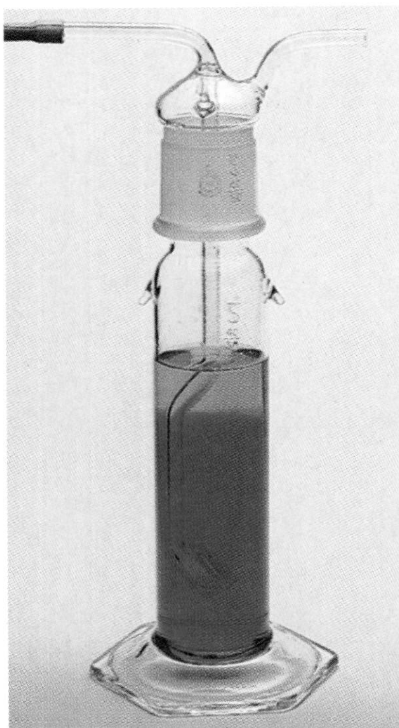

FIGURE 20.25 Chlorine is an oxidizing agent. When chlorine is bubbled through a solution of bromide ions, it oxidizes them to bromine, which colors the solution brown.

$$Cl_2(g) + 2\,I^-(aq) \longrightarrow I_2(aq) + 2\,Cl^-(aq)$$

The blue-black lustrous solid sublimes easily and boils to a purple vapor at 185°C. The I_2 molecules retain their separate identities in the solid, like the other halogens but unlike carbon, silicon, phosphorus, and sulfur, which form chains and networks of atoms.

Iodine dissolves in organic solvents to give a variety of colors that arise from the different interactions between the I_2 molecules and the solvent (Fig. 20.26). Iodine is only slightly soluble in water, unless I^- ions are present, in which case the soluble, brown triiodide ion I_3^- is formed. The element itself has few direct uses, but dissolved in alcohol it is familiar as a mild oxidizing antiseptic. Iodine is an essential trace element for living systems; a deficiency in humans leads to a swelling of the thyroid gland in the neck. Iodides are added to table salt (to produce *iodized salt*) to prevent this deficiency.

As the first member of the group, fluorine has a number of unique properties that stem from its high electronegativity, small size, and lack of available *d*-orbitals. In particular, it oxidizes other elements to high oxidation states. (Its smallness helps in this, for it allows several F atoms to pack around a central atom, as in IF_7.) Because the fluoride ion is so small, the lattice enthalpies of its ionic compounds tend to be high (see Table 8.1). One result is the lower solubilities of fluorides relative to those of other halides. There are some exceptions, including AgF, which is soluble, whereas the other silver halides are insoluble. This difference in solubility is one reason why the oceans are salty with chlorides rather than fluorides, even though fluorine is more abundant than chlorine in the Earth's crust: chlorides are more readily dissolved in groundwater and washed out to sea.

The halogens show smooth trends in chemical properties down the group; fluorine has some anomalous properties, such as its strength as an oxidizing agent and the lower solubilities of most fluorides.

20.10 COMPOUNDS OF THE HALOGENS

The halogens form compounds among themselves. These **interhalogens** have the formulas XX′, XX'_3, XX'_5, and XX'_7, where X is the heavier (and

FIGURE 20.26 Solutions of iodine in a variety of solvents. From left to right, the solvents are tetrachloromethane (carbon tetrachloride), water, and potassium iodide solution. In the solution at the far right, a little starch has been added to a solution of iodine in potassium iodide solution; starch acts as an indicator for the presence of iodine.

larger) of the two halogens. Only some of the possible combinations have been prepared (Table 20.6). They are all prepared by direct reaction of the two halogens, the product formed being determined by the proportions of reactants used. For example,

$$Cl_2(g) + 3\,F_2(g) \longrightarrow 2\,ClF_3(g)$$

$$Cl_2(g) + 5\,F_2(g) \longrightarrow 2\,ClF_5(g)$$

The interhalogens have physical properties intermediate between those of their parent halogens. Their chemical properties are dominated by the decreasing X—X′ bond enthalpy as X becomes heavier. For example, the fluorides of the heavier halogens are all very reactive. Bromine trifluoride is so reactive a gas that even asbestos burns in it.

The hydrogen halides, HX, can be prepared by the direct reaction of the elements:

$$H_2(g) + X_2(g) \longrightarrow 2\,HX(g)$$

Fluorine reacts explosively by a radical chain reaction as soon as the gases are mixed. A mixture of hydrogen and chlorine explodes when exposed to light. Bromine and iodine react with hydrogen much more slowly. A less hazardous laboratory source of the hydrogen halides is the action of a nonvolatile acid on a metal halide, as in

$$CaF_2(s) + 2\,H_2SO_4(aq, conc) \longrightarrow Ca(HSO_4)_2(aq) + 2\,HF(g)$$

Because Br^- and I^- are oxidized by sulfuric acid, phosphoric acid is used in the preparation of HBr and HI:

$$KI(s) + H_3PO_4(aq) \xrightarrow{\Delta} KH_2PO_4(aq) + HI(g)$$

All the hydrogen halides are colorless, pungent gases (Table 20.7), but hydrogen fluoride is a liquid at temperatures below 20°C. Its low volatility is a sign of extensive hydrogen bonding, and short zigzag chains of hydrogen-bonded molecules, up to about $(HF)_5$, survive to some extent in the vapor. All the hydrogen halides dissolve in water to give acidic solutions. Hydrofluoric acid has the distinctive property of attacking glass and silica (see Case Study 19). The interiors of lamp bulbs are frosted by the vapors from a solution of hydrofluoric acid and ammonium fluoride (Fig. 20.27).

The hypohalous acids, HXO (oxidation number +1, Table 20.8), are prepared by direct reaction of the halogen with water. For example, chlorine gas disproportionates in water to produce hypochlorous acid and hydrochloric acid:

TABLE 20.6 Known interhalogens

Interhalogen	Normal form*
XF_n	
ClF	colorless gas
ClF_3	colorless gas
ClF_5	colorless gas
BrF	pale brown gas
BrF_3	pale yellow liquid
BrF_5	colorless liquid
IF	unstable
IF_3	yellow solid
IF_5	colorless liquid
IF_7	colorless gas
XCl_n	
$BrCl$	red-brown gas
ICl	red solid
I_2Cl_6	yellow solid
XBr_n	
IBr	black solid

**Normal form* means the appearance and state of the element at 25°C and 1 atm.

TABLE 20.7 The hydrogen halides

Compound	Molar mass, g·mol^{-1}	Melting point, °C	Boiling point, °C	pK_a in water	Bond enthalpy, kJ·mol^{-1}	Bond length, pm
HF	20.01	−83	20	3.45	565	92
HCl	36.46	−115	−85	strong	421	127
HBr	80.92	−89	−67	strong	366	141
HI	127.91	−51	−35	strong	299	161

(a)

(b)

FIGURE 20.27 When a mixture of hydrofluoric acid and ammonium fluoride is swirled inside a flask (a), the reaction with the silica in the cover glass frosts its surface (b).

$$Cl_2(g) + H_2O(aq) \longrightarrow HClO(aq) + HCl(aq)$$

Hypofluorous acid is so unstable that it survives only below the freezing point of water. Sodium hypochlorite, NaClO, is produced from the electrolysis of brine when the electrolyte is rapidly stirred, and the chlorine gas produced at the anode reacts with the hydroxide ion generated at the cathode. The chlorine gas disproportionates to produce hypochlorite and chloride ions:

$$Cl_2(g) + 2\,OH^-(aq) \longrightarrow ClO^-(aq) + Cl^-(aq) + H_2O(l)$$

Calcium hypochlorite, which is marketed as a dry bleach and for purifying the water in home swimming pools, is produced by passing chlorine gas over dry calcium oxide (quicklime). Calcium hypochlorite is used for swimming pools in preference to sodium hypochlorite because the Ca^{2+}

TABLE 20.8 Halogen oxoacids

Oxidation number	General formula	General acid name	Known examples	pK_a in water
7	HXO_4	perhalic acid	$HClO_4$	strong
			$HBrO_4$	strong
			HIO_4	1.64
5	HXO_3	halic acid	$HClO_3$	strong
			$HBrO_3$	strong
			HIO_3	0.77
3	HXO_2	halous acid	$HClO_2$	2.00
			$HBrO_2$	unstable
1	HXO	hypohalous acid	HFO	unstable
			HClO	7.53
			HBrO	8.69
			HIO	10.64

ions form insoluble calcium carbonate, which is removed by filtration; sodium would remain in solution and make the water too salty.

Chlorate ions, ClO_3^- (oxidation number +5), form when chlorine reacts with hot concentrated aqueous alkali:

$$3\,Cl_2(g) + 6\,OH^-(aq) \xrightarrow{\Delta} ClO_3^-(aq) + 5\,Cl^-(aq) + 3\,H_2O(l)$$

They decompose when heated, to an extent that depends on whether or not a catalyst is present:

$$4\,KClO_3(s) \xrightarrow{\Delta} 3\,KClO_4(s) + KCl(s)$$

$$2\,KClO_3(s) \xrightarrow{\Delta,\,MnO_2} 2\,KCl(s) + 3\,O_2(g)$$

The latter reaction is a convenient laboratory source of oxygen.

Chlorates are useful oxidizing agents. Potassium chlorate is used as an oxygen supply in fireworks and in safety matches. The heads of matches consist of a paste of potassium chlorate, antimony sulfide, sulfur, and powdered glass to create friction when the match is struck; the striking strip contains red phosphorus, which ignites the match head (see Section 20.1). The principal use of sodium chlorate is as a source of chlorine dioxide, ClO_2. The chlorine in ClO_2 has oxidation number +4, so the chlorate must be reduced to form it. Sulfur dioxide is a convenient reducing agent for this reaction:

$$2\,NaClO_3(aq) + SO_2(g) + H_2SO_4(aq, \text{dilute}) \longrightarrow 2\,NaHSO_4(aq) + 2\,ClO_2(g)$$

Chlorine dioxide has an odd number of electrons and is a paramagnetic yellow gas. It is used to bleach paper pulp, because it can oxidize the various pigments in the pulp without degrading the wood fibers.

The perchlorates, ClO_4^- (oxidation number +7), are prepared by electrolytic oxidation of aqueous chlorates:

$$ClO_3^-(aq) + H_2O(l) \longrightarrow ClO_4^-(aq) + 2\,H^+(aq) + 2\,e^-$$

Perchloric acid, $HClO_4$, is prepared by the action of concentrated hydrochloric acid on sodium perchlorate, followed by distillation. It is a colorless liquid and the strongest of all common acids. Because chlorine has its highest oxidation number in these compounds, the perchlorates are also powerful oxidizing agents; contact between perchloric acid and even a small amount of organic material can result in a dangerous explosion. One spectacular example of their oxidizing ability under controlled conditions is the use of a mixture of ammonium perchlorate and aluminum powder in the booster rockets of the space shuttle (Case Study 20).

The interhalogens have properties intermediate between those of the constituent halogens. Nonmetals form covalent halides; metals form ionic halides. The oxoacids of chlorine are all oxidizing agents; acidity and oxidizing strength of oxoacids both increase as the oxidation number of the halogen increases.

SELF-TEST 20.4A Predict the trend in oxidizing strength of the halogens in aqueous solution.

[*Answer:* Decreases down the group: F > Cl > Br > I]

CASE STUDY 20 ROCKET FUELS

A ROCKET TECHNICIAN once dramatically demonstrated the high reactivity of a rocket propellant: he sprayed a few drops onto the snow-covered ground and it burst into flame. Fuels used in space travel are very different from fuels we use in our automobiles, because they are designed for very different conditions. The preferred fuel for automobiles is gasoline, which burns in air and has a high enthalpy density (see Section 6.11). That is, it releases a lot of heat per liter as it burns. In space flight, the mass of fuel is more important than the volume it occupies, because the entire assembly has to be thrust up out of Earth's gravitational field, so a rocket fuel needs a high specific enthalpy (enthalpy of reaction per gram of substance) rather than a high enthalpy density. Moreover, a rocket must carry an oxidizer as well as a fuel, because the fuel must burn in the absence of air.

The volume of gaseous exhaust emitted by an automobile is of little concern, although the nature of the emissions is of interest. However, a rocket is moved by the thrust created when gases are ejected from its engine and the volume of gases produced controls its speed. Thus, a desirable rocket fuel will be a low-density liquid or solid that produces a lot of heat and a large volume of gases when it burns.

The white smoke emitted by the space shuttle booster rockets consists of powdered aluminum oxide and aluminum chloride.

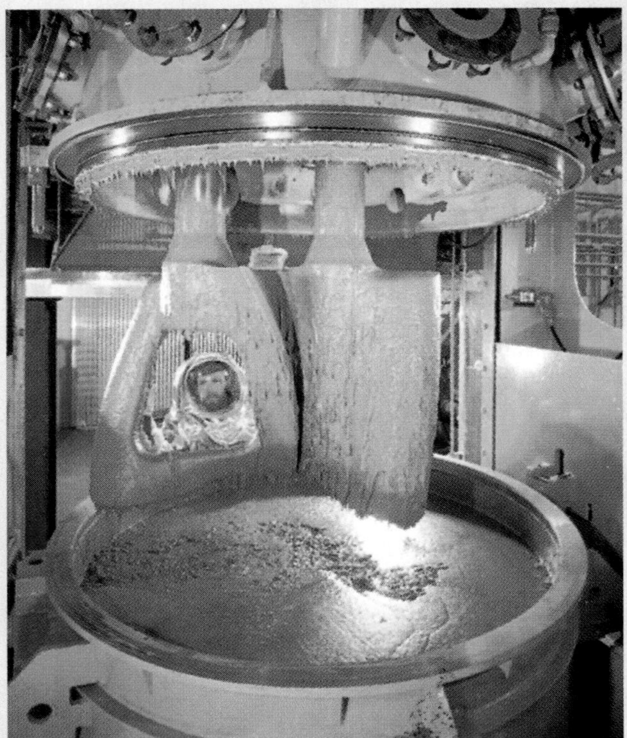

Powdered reactants are mixed with a liquid polymer base and hardened inside the space shuttle booster rocket shell.

Liquid and solid fuels are used in stages to propel the space shuttle into orbit. First the solid booster rockets are ignited to lift the shuttle from the ground and take it high into the atmosphere, at which point the rocket shells are released and fall into the ocean for recovery. The solid fuel in the booster rockets consists of aluminum powder (the fuel), ammonium perchlorate (the oxidizing agent as well as a fuel), and iron(III) oxide (the catalyst) mixed into a liquid polymer that is then set into a solid inside the rocket shell. A variety of products can be produced. One of the reactions that occurs is

$$3\,NH_4ClO_4(s) + 3\,Al(s) \xrightarrow{Fe_2O_3} Al_2O_3(s) + AlCl_3(s) + 6\,H_2O(g) + 3\,NO(g)$$

The solid products form the thick clouds of white powder emitted by the solid rocket boosters during liftoff.

The shuttle is propelled into orbit by the reaction between two gases that are stored as liquids. For a typical shuttle mission, about 1.4×10^6 L (1.4 ML) of liquid hydrogen and 0.54 ML of liquid oxygen are used. The gases are stored as liquids to reduce the volume: huge gas-filled tanks would be a source of drag. The hydrogen and oxygen are vaporized and ignited as they mix. The rapid expansion of the product, water vapor, at the high temperature of the combustion provides the thrust for the main engines. The net reaction for the vaporization and combustion is

$$2\,H_2(l) + O_2(l) \longrightarrow 2\,H_2O(g) \qquad \Delta H^\circ = -475\text{ kJ}$$

A different kind of fuel and oxidizer were needed on the Apollo missions for the lunar lander. Liquid hydrogen and oxygen are too volatile, and the enthalpy density of liquid hydrogen is too low. The problem with solid rocket fuel is that the combustion is difficult to extinguish and cannot be restarted once the flame has been quenched. Apollo used a mixture of hydrazine derivatives (such as methyl hydrazine, CH_3NHNH_2) and liquid N_2O_4 when landing on and leaving the moon. These two liquids ignite as soon as they mix, producing a large volume of gas:

$$4\,CH_3NHNH_2(l) + 5\,N_2O_4(l) \longrightarrow 9\,N_2(g) + 12\,H_2O(g) + 4\,CO_2(g)$$

Another difference between rocket fuels and automobile fuels lies in their relative ease of handling. Gasoline can be handled by any motorist at a self-service filling station. However, rocket fuels are optimized for function, rather than for safe handling, and are often very hazardous. For example, methyl hydrazine is a deadly poison, and N_2O_4 is a highly reactive compound that must be kept in corrosion-resistant containers.

The Apollo lunar lander was powered by a mixture of hydrazine derivatives and dinitrogen tetroxide.

QUESTIONS

1. In the reaction of the solid-fuel booster rockets, one element is being reduced and two are being oxidized. Identify these three elements and write half-reactions for each.

2. How many moles of electrons are transferred in each of the three chemical equations written above?

3. Using the first equation above and Appendix 2A, determine (a) the enthalpy of reaction for the solid fuel in the booster rocket; and (b) the specific enthalpy of the aluminum in this reaction. Assume that ΔH_f° of $NH_4ClO_4(s)$ is $-295\text{ kJ}\cdot\text{mol}^{-1}$.

4. Using the equation for the oxidation of methyl hydrazine and Appendix 2A, determine (a) the enthalpy change of the reaction; (b) the specific enthalpy of the methyl hydrazine in this reaction. Assume that ΔH_f° of $CH_3NHNH_2(l)$ is $+54\text{ kJ}\cdot\text{mol}^{-1}$.

5. (a) Calculate the specific enthalpy of liquid hydrogen in its reaction with liquid oxygen. (b) Compare the specific enthalpies of aluminum, methyl hydrazine, and liquid hydrogen in the equations given. (c) Would gaseous hydrogen have a larger, smaller, or the same specific enthalpy as liquid hydrogen?

SELF-TEST 20.4B Which halide ion is the strongest reducing agent in aqueous solution?

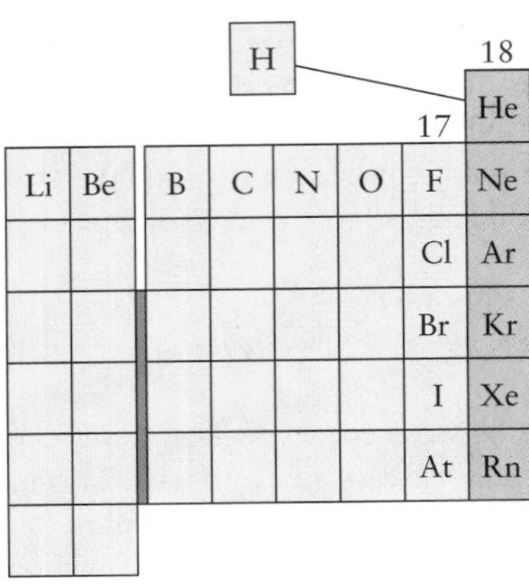

GROUP 18: THE NOBLE GASES

The elements in Group 18, the noble gases, get their name from their very low reactivity (Table 20.9). Their closed-shell electron configurations (ns^2np^6) prompted the belief that these elements were chemically inert. However, the first noble-gas compound, xenon hexafluoroplatinate, $XePtF_6$, was synthesized in 1962 by the reaction of xenon with platinum hexafluoride; and shortly after that, xenon tetrafluoride, XeF_4, was produced from a high-temperature mixture of xenon and fluorine.

20.11 THE ELEMENTS

All the Group 18 elements occur in the atmosphere as monatomic gases; the most common is argon, which is the third most abundant gas in the atmosphere (after nitrogen and oxygen and discounting the variable amount of water vapor). All except helium and radon are obtained by the distillation of liquid air. Helium, the second most abundant element in the universe after hydrogen, is rare on Earth because its atoms are so light that a large fraction of them reach high speeds and escape from the atmosphere. However, it is found as a component of natural gases trapped under rock formations in some locations (notably Texas), where it has collected as a result of the emission of α (alpha) particles by radioactive elements. An α particle is a helium nucleus (He^{2+}), and an atom of the element forms when the particle picks up two electrons from its surroundings.

▶▶ Radioactivity and the changes that nuclei undergo are discussed in Chapter 22.

Helium gas is twice as dense as hydrogen under the same conditions. However, because its density is still very low and it is nonflammable, it is used to provide buoyancy in airships such as blimps. Helium is also used

TABLE 20.9 The Group 18 elements (the noble gases)

Valence configuration: ns^2np^6
Normal form: colorless monatomic gases

Z	Name	Symbol	Molar mass, $g{\cdot}mol^{-1}$	Melting point, °C	Boiling point, °C
2	helium	He	4.00	—	−269 (4.2 K)
10	neon	Ne	20.18	−249	−246
18	argon	Ar	39.95	−189	−186
36	krypton	Kr	83.80	−157	−153
54	xenon	Xe	131.30	−112	−108
86	radon*	Rn	222	−71	−62

*Radioactive.

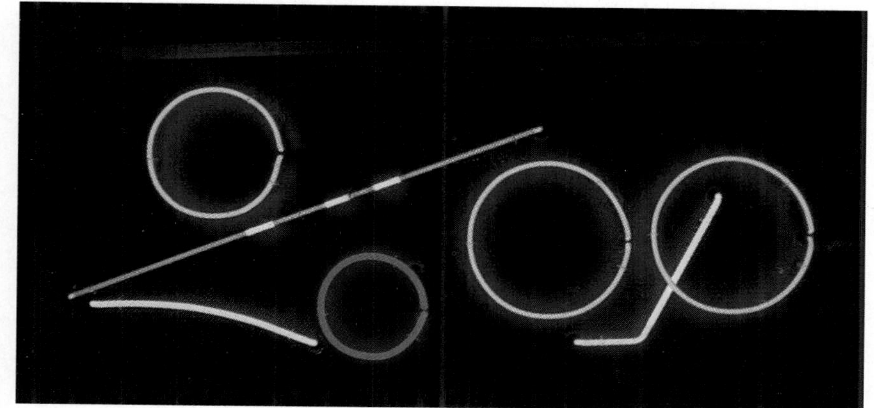

FIGURE 20.28 The colors of the fluorescent lighting art by Tom Anthony are due to emissions from noble gas atoms. Neon is responsible for the red light; when it is mixed with a little argon, the color becomes blue-green. The yellow color is achieved with appropriately colored glass.

to dilute the oxygen used in deep-sea diving, to pressurize rocket fuels, as a coolant, and in helium-neon lasers. The element has the lowest boiling point of any substance (4.2 K), and it does not freeze to a solid at any temperature unless pressure is applied to hold the light, mobile atoms together. These properties make helium useful for **cryogenics,** the study of matter at very low temperatures. Below 2 K, liquid helium shows the remarkable property of **superfluidity,** the ability to flow without viscosity. Helium is the only substance known to have more than one liquid phase.

Neon, which emits a red glow when an electric current flows through it, is widely used in advertising signs (Fig. 20.28). Argon is used to provide an inert atmosphere for welding (to prevent oxidation) and to fill some types of light bulbs, where its function is to conduct heat away from the filament. Krypton gives an intense white light when a current is passed through it, so it is used in airport runway lighting. Because krypton is produced by nuclear fission, its atmospheric abundance is one measure of worldwide nuclear activity. Xenon is used in halogen lamps for automobile headlights and in high-speed photographic flash tubes, because an electric discharge through it—a miniature lightning flash—gives an intense white light; it is also being investigated as an anesthetic.

The radioactive gas radon seeps out of the ground as a product of radioactive processes deep in the Earth. There is now some concern that its accumulation in buildings, and its nuclear decay products, can lead to dangerously high levels of radiation.

The noble gases are all found naturally as unreactive monatomic gases. Helium has two liquid phases; the lower temperature phase exhibits superfluidity.

20.12 COMPOUNDS OF THE NOBLE GASES

The ionization energies of the noble gases are relatively high but decrease down the group (Fig. 20.29); xenon's ionization energy is low enough for electrons to be lost to very electronegative elements, especially fluorine. No compounds of helium, neon, and argon exist, except under very special conditions. Radon is known to react with fluorine; but its radioactivity makes the study of its compounds difficult and dangerous, so little is

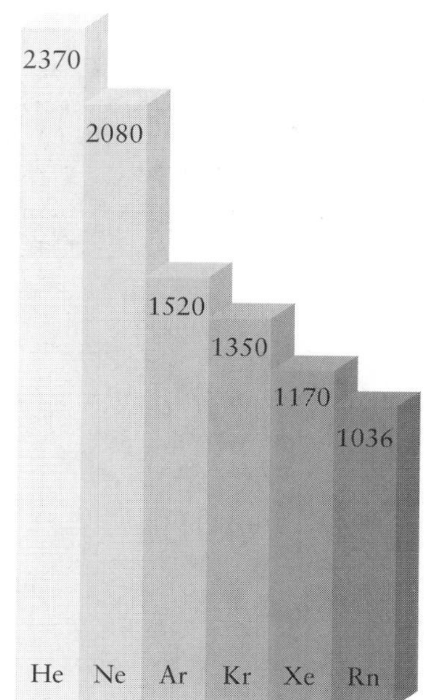

FIGURE 20.29 The ionization energies of the noble gases decrease steadily down the group. The values shown are in kilojoules per mole.

known about them. Krypton forms only one known stable neutral molecule, KrF_2. In 1988 a compound with a Kr—N bond was reported, but it is stable only below −50°C. This leaves xenon as the noble gas with the richest chemistry. It forms several compounds with fluorine and oxygen, and compounds with Xe—N and Xe—C bonds have been reported.

The starting point for the synthesis of xenon compounds is the preparation of xenon difluoride, XeF_2, and xenon tetrafluoride, XeF_4, by heating a mixture of the elements to 300°C. At higher pressures, fluorination proceeds as far as xenon hexafluoride, XeF_6. All three fluorides are crystalline solids (Fig. 20.30). In the gas phase, all are molecular compounds. Solid xenon hexafluoride, however, is ionic, with a complex structure consisting of XeF_5^+ cations bridged by F^- anions.

Chemists working with some of the first xenon compounds found that they needed special equipment for this work, because the compounds set fire to the stopcock grease.

The xenon fluorides are used as powerful fluorinating agents (reagents for attaching fluorine atoms to other substances). The tetrafluoride will even fluorinate platinum:

$$Pt(s) + XeF_4(s) \longrightarrow Xe(g) + PtF_4(s)$$

The xenon fluorides are used to prepare a series of xenon oxides and oxoacids and, in a series of disproportionations, to bring the oxidation number of xenon up to +8. Xenon trioxide, XeO_3, is the anhydride of xenic acid, H_2XeO_4. It reacts with aqueous alkali to form a hydrogen xenate ion, $HXeO_4^-$. This ion slowly disproportionates into xenon and the octahedral perxenate ion, XeO_6^{4-}, in which the oxidation number of xenon is +8. Aqueous perxenate solutions are yellow and are very powerful oxidizing agents as a result of the high oxidation number of xenon.

When barium perxenate is treated with sulfuric acid, it is dehydrated to the anhydride of perxenic acid xenon tetroxide (XeO_4). With this compound, which is an explosively unstable gas, our journey through the main groups of the periodic table comes to an end with a bang.

FIGURE 20.30 Crystals of xenon tetrafluoride, XeF_4. This compound was first prepared in 1962 by the reaction of xenon and fluorine at 6 atm and 400°C.

Only xenon is known to form an extensive series of compounds with fluorine and oxygen; xenon fluorides are powerful fluorinating agents and xenon oxides are powerful oxidizing agents.

SELF-TEST 20.5A Rationalize the fact that the ease of compound formation increases down Group 18.

[*Answer:* Ionization energy decreases down the group, so it is easier for the heavier elements to share their electrons.]

SELF-TEST 20.5B Draw the Lewis structure for $XeOF_4$ and predict its molecular shape.

SKILLS YOU SHOULD HAVE MASTERED

Conceptual

1. Explain why nitrogen, oxygen, and fluorine have properties that differ from those of the other members of their groups.
2. Explain trends in the properties of the elements in Groups 15, 16, 17, and 18.
3. Rationalize the properties of the oxides and oxoacids of the nonmetals in terms of the oxidation number of the nonmetal and the identification of acid anhydrides.
4. Explain the low reactivity of the noble gases and

the fact that the ease of compound formation increases down the group.

Problem-Solving

1. Write the formula of an acid, given the formula of its anhydride, and the formula of an anhydride, given that of the acid it forms.
2. Predict trends in chemical and physical properties across a period and down a group.

Descriptive

1. Identify the valence electron configurations of Groups 15, 16, 17, and 18.
2. Describe how nitrogen is obtained from air and converted into useful compounds.
3. Describe the structures and properties of the nitrogen and phosphorus oxides.
4. Describe the acidic and basic character of water and hydrogen peroxide and compare their functions as oxidizing and reducing agents.
5. Describe the names, properties, and reactions of the principal compounds of the members of Groups 15, 16, 17, and 18.
6. Describe and write balanced equations for the principal reactions used to produce the elements in Groups 15, 16, and 17.

EXERCISES

Groups 15–18

20.1 Write the ground-state electron configuration of (a) He; (b) O; (c) F; (d) As.

20.2 Write the ground-state electron configuration of (a) Ar; (b) I; (c) P; (d) S.

20.3 Refer to Appendix 2B and arrange the halogens in order of increasing oxidizing strength in water.

20.4 Refer to Appendix 2B and arrange S, O_2, (to H_2O), Cl_2, and H_2O_2 (to H_2O) in order of increasing oxidizing strength in water.

20.5 Arrange the elements P, O, As, S in order of increasing electronegativity.

20.6 Arrange the elements S, Br, O, F in order of increasing atomic radius.

Group 15

20.7 Describe the industrial production of liquid nitrogen.

20.8 Describe, with chemical equations, the industrial production of white and red phosphorus.

20.9 Name the following compounds: (a) HNO_2; (b) NO; (c) H_3PO_4; (d) N_2O_3.

20.10 Name the following compounds: (a) HNO_3; (b) N_2O; (c) H_3PO_3; (d) N_2O_5.

20.11 Write the chemical formula of (a) ammonium nitrate; (b) magnesium nitride; (c) calcium phosphide; (d) hydrazine.

20.12 Write the chemical formula of (a) sodium azide; (b) dinitrogen trioxide; (c) phosphorus(V) oxide; (d) ammonium dihydrogen phosphate.

20.13 Balance the following skeletal equations:

(a) $NH_3(g) + CuO(s) \longrightarrow N_2(g) + Cu(s) + H_2O(l)$
(b) $NH_3(g) + F_2(g) \longrightarrow NF_3(g) + NH_4F(s)$
(c) $NH_3(g) + O_2(g) \longrightarrow NO(g) + H_2O(l)$

20.14 Balance the following skeletal equations:

(a) $P_4(s) + KOH(aq) + H_2O(l) \longrightarrow PH_3(g) + KH_2PO_2(aq)$
(b) $Ca_3(PO_4)_2(s) + SiO_2(s) + C(s) \longrightarrow P_4(s) + CaSiO_3(s) + CO(g)$
(c) $Ca_3(PO_4)_2(s) + C(s) \longrightarrow Ca_3P_2(s) + CO(g)$

20.15 Determine the oxidation number of nitrogen in (a) NO; (b) N_2O; (c) HNO_2; (d) N_3^-.

20.16 Determine the oxidation number of phosphorus in (a) P_4; (b) PH_3; (c) H_3PO_4; (d) P_4O_6.

20.17 Urea, $CO(NH_2)_2$, reacts with water to form ammonium carbonate. Write the chemical equation and calculate the mass of ammonium carbonate that can be obtained from 5.0 kg of urea.

20.18 (a) Nitrous acid reacts with hydrazine in acidic solution to form hydrazoic acid, HN_3. Write the chemical equation and determine the mass of hydrazoic acid that can be produced from 20.0 g of hydrazine. (b) Suggest a method for preparing sodium azide, NaN_3.

20.19 Lead azide, $Pb(N_3)_2$, is used as a detonator. (a) What volume of nitrogen at STP does 1.0 g of lead azide produce when it decomposes into lead metal and nitrogen gas? (b) Would 1.0 g of mercury(II) azide, $Hg(N_3)_2$, which is also used as a detonator, produce a larger or smaller volume, given that its decomposition products are mercury metal and nitrogen gas?

20.20 Sodium azide is used to inflate protective air bags in automobiles. What mass of solid sodium azide is needed to provide 100 L of N_2 at 1.5 atm and 20°C?

20.21 The common acid anhydrides of nitrogen are N_2O, N_2O_3, and N_2O_5. Write the formulas of their corresponding acids and chemical equations for the formation of the acids by the reaction of the anhydrides with water.

20.22 The common acid anhydrides of phosphorus are P_4O_6 and P_4O_{10}. Write the formulas of their corresponding acids and chemical equations for the formation of the acids by the reaction of the anhydrides with water.

20.23 Solid phosphorus pentachloride exists as $PCl_4^+PCl_6^-$. Write the Lewis structures for the two ions and predict their shapes from VSEPR theory.

20.24 Solid dinitrogen pentoxide exists as $NO_2^+NO_3^-$. Write Lewis structures for the two ions and predict their shapes from VSEPR theory.

20.25 What is the mass percentage of phosphorus in (a) superphosphate; (b) triple superphosphate?

20.26 What is the mass percentage of nitrogen in the fertilizers (a) ammonia, NH_3; (b) ammonium nitrate, NH_4NO_3; (c) urea, $CO(NH_2)_2$?

Group 16

20.27 Write the chemical formula of (a) sulfuric acid; (b) calcium sulfite; (c) ozone; (d) barium peroxide.

20.28 Write the chemical formula of (a) sulfurous acid; (b) hydrogen selenide; (c) sodium thiosulfate; (d) disulfur dichloride.

20.29 Describe the Frasch process for extracting elemental sulfur from underground sulfur deposits.

20.30 Use chemical equations to describe the Claus process for the production of elemental sulfur from petroleum and natural gas wells.

20.31 Write equations for (a) the burning of lithium in oxygen; (b) the reaction of sodium metal with water; (c) the reaction of fluorine gas with water; (d) the oxidation of water at the anode of an electrolytic cell.

20.32 Write equations for the reaction of (a) sodium oxide and water; (b) sodium peroxide and water; (c) sulfur dioxide and water; (d) sulfur dioxide and oxygen, using a vanadium pentoxide catalyst.

20.33 Complete and balance the following skeletal equations:

(a) $H_2S(aq) + O_2(g) \longrightarrow$

(b) $CaO(s) + H_2O(l) \longrightarrow$

(c) $PCl_5(s) + H_2O(l) \longrightarrow$

20.34 Complete and balance the following skeletal equations:

(a) $Cl_2(g) + H_2O(l) \longrightarrow$

(b) $H_2(g) + S(s) \xrightarrow{\Delta}$

(c) $S_2Cl_2(l) + Cl_2(g) \xrightarrow{FeCl_3}$

20.35 Write the Lewis structure of H_2O_2 and predict the approximate H—O—O bond angle.

20.36 Write the Lewis structure of SO_3 in which the formal charge on each atom is 0. What is the shape of the molecule and what is the hybridization of the sulfur atom?

20.37 Write the Lewis structure and predict the shape of (a) SO_2; (b) SF_4; (c) SO_4^{2-}.

20.38 Write the Lewis structure and predict the shape of (a) SF_6; (b) H_2S; (c) SO_3^{2-}.

20.39 Hydrogen peroxide is unstable with respect to decomposition into water and oxygen when exposed to light, heat, or a catalyst. If 500 mL of a 3% H_2O_2 aqueous solution (like that sold in drugstores) decomposes, what volume of oxygen at 273 K and 1.00 atm will be produced? Assume the density of the solution to be $1.0\ g{\cdot}mL^{-1}$.

20.40 Calculate the volume of pure (concentrated) sulfuric acid (of density $1.84\ g{\cdot}mL^{-1}$) that can be produced from 100 g of sulfur.

20.41 If 2.00 g of sodium peroxide is dissolved to form 200 mL of an aqueous solution, what would be the pH of the solution? For H_2O_2, $K_{a1} = 1.8 \times 10^{-12}$ and K_{a2} is negligible.

20.42 If you were titrating 0.10 M $H_2S(aq)$ with 0.10 M NaOH(aq), at what pH would you have a 0.050 M NaHS(aq) solution?

20.43 Describe the trend in acidity of the binary hydrogen compounds of Group 16 elements and account for the trend in terms of bond strength.

20.44 Write chemical equations that represent the acid or base character of (a) two metal oxides and (b) two nonmetal oxides.

20.45 (a) Calculate the standard enthalpy and entropy of reaction for the formation of ozone from oxygen. (b) Is the reaction favored at high temperatures or low temperatures? (c) Does the reaction entropy favor the spontaneous formation of ozone? Explain your conclusions.

20.46 When lead(II) sulfide is treated with hydrogen peroxide, the possible products are either (a) lead(IV) oxide and sulfur dioxide or (b) lead(II) sulfate. In terms of the reaction free energy, which product or products are more likely?

Group 17

20.47 List the natural sources of fluorine and chlorine.

20.48 List the natural sources of bromine and iodine.

20.49 Write chemical equations that describe the

preparation of fluorine and chlorine.

20.50 Write chemical equations that describe the preparation of bromine and iodine.

20.51 Name each of the following compounds: (a) HBr(aq); (b) IBr; (c) ClO_2; (d) $NaIO_3$.

20.52 Name each of the following compounds: (a) HI(aq); (b) IF_3; (c) HClO(aq); (d) NaClO.

20.53 Write the chemical formulas of (a) perchloric acid; (b) sodium chlorate; (c) hydroiodic acid; (d) sodium triiodide.

20.54 Write the chemical formulas of (a) potassium periodate; (b) ammonium perchlorate; (c) chloric acid; (d) potassium bromide.

20.55 Identify the oxidation number of the halogen atoms in (a) hypoiodous acid; (b) ClO_2; (c) dichlorine heptoxide; (d) $NaIO_3$.

20.56 Identify the oxidation number of the halogen atoms in (a) IF_7; (b) sodium periodate; (c) hypobromous acid; (d) $HClO_2$.

20.57 Write the Lewis structure for each of the following species, describe its electronic arrangement, and predict its shape: (a) ClO_4^-; (b) IO_3^-; (c) IF_3.

20.58 Write the Lewis structure for each of the following species, describe its electronic arrangement, and predict its shape: (a) BrO_3^-; (b) ClF_5; (c) I_3^-; (d) BrF_4^-.

20.59 Write the balanced chemical equations for (a) the thermal decomposition of potassium chlorate without a catalyst; (b) the reaction of bromine with water; (c) the reaction between sodium chloride and concentrated sulfuric acid. (d) Identify each reaction as a Brønsted acid-base, Lewis acid-base, or redox reaction.

20.60 Write the balanced chemical equations for the reaction of chlorine in (a) a neutral aqueous solution; (b) a dilute basic solution; (c) a concentrated basic solution. (d) Verify that each reaction is a disproportionation reaction.

20.61 (a) Arrange the chlorine oxoacids in order of increasing oxidizing strength. (b) Suggest an interpretation of that order in terms of oxidation numbers.

20.62 (a) Arrange the hypohalous acids in order of increasing acid strength. (b) Suggest an interpretation of that order in terms of electronegativities.

20.63 Write the Lewis structure for Cl_2O. Predict the shape of the Cl_2O molecule and estimate the Cl—O—Cl bond angle.

20.64 Write the Lewis structure for BrF_3. What is the hybridization of the bromine atom in the molecule?

20.65 Plot a graph of the standard free energy of formation of the hydrogen halides against the period number of the halogens. What conclusions can be drawn from the graph?

20.66 Use the data from Table 20.7 to plot the normal boiling points of the hydrogen halides against the period number of the halogens. Account for the trend revealed by the graph.

20.67 Use the data in Appendix 2B to determine whether chlorine gas will oxidize Mn^{2+} to form the permanganate ion in an acidic solution.

20.68 Use the data in Appendix 2B to determine which is the stronger oxidizing agent, ozone or fluorine.

20.69 The standard free energy of formation of HI is $+1.70\ kJ\cdot mol^{-1}$ at 25°C. Determine the equilibrium constant for the formation of HI from its elements according to the equation $\frac{1}{2}H_2(g) + \frac{1}{2}I_2(s) \rightleftharpoons HI(g)$.

20.70 The standard free energy of formation of HCl is $-95.3\ kJ\cdot mol^{-1}$ at 25°C. Determine the equilibrium constant for the formation of HCl(g) according to the equation $\frac{1}{2}H_2(g) + \frac{1}{2}Cl_2(g) \rightleftharpoons HCl(g)$.

20.71 The concentration of F^- ions can be measured by adding an excess of lead(II) chloride solution and weighing the lead(II) chlorofluoride (PbClF) precipitate. Calculate the molarity of F^- ions in 25.00 mL of a solution that gave a lead chlorofluoride precipitate of mass 0.765 g.

20.72 Suppose 25.00 mL of an aqueous solution of iodine was titrated with 0.025 M $Na_2S_2O_3$(aq), using starch as the indicator. The blue color of the starch-iodine complex disappeared when 28.45 mL of the thiosulfate solution had been added. What was the molar concentration of I_2 in the original solution? The titration reaction is $I_2(aq) + 2\,S_2O_3^{2-}(aq) \rightarrow 2\,I^-(aq) + S_4O_6^{2-}(aq)$.

Group 18

20.73 What are the sources for the production of helium and argon?

20.74 What are the sources for the production of krypton and xenon?

20.75 Determine the oxidation number of the noble gases in (a) KrF_2; (b) XeF_6; (c) KrF_4; (d) XeO_4^{2-}.

20.76 Determine the oxidation number of the noble gases in (a) XeO_3; (b) XeO_6^{4-}; (c) XeF_2; (d) $HXeO_4^-$.

20.77 Xenon tetrafluoride is a powerful oxidizing agent. In an acidic solution it is reduced to xenon. Write the corresponding half-reaction.

20.78 Xenon hexafluoride reacts with water to produce xenon trioxide and hydrofluoric acid. Write the chemical equation for the reaction.

20.79 Predict the relative acid strengths of H_2XeO_4 and H_4XeO_6. Explain your conclusions.

20.80 Predict the relative oxidizing strengths of H_2XeO_4 and H_4XeO_6. Explain your conclusions.

20.81 Write the Lewis structure for XeF_4. Estimate the F—Xe—F bond angle.

20.82 Write the Lewis structure for XeO_3. What is the hybridization on the xenon atom in the molecule?

SUPPLEMENTARY EXERCISES

20.83 List the evidence that shows that metallic character decreases from the lower left of the periodic table to the upper right.

20.84 Summarize, with examples and explanations, the principal differences between the elements in the *s*- and *p*-blocks.

20.85 (a) Write all the important resonance Lewis structures for NO_2^- and NO_3^-. (b) What is the hybridization of the nitrogen atom in the two ions?

20.86 (a) Write the Lewis structures for NO and NO_2. (b) What is the hybridization of the nitrogen atom in the two compounds?

20.87 Arsenic(III) sulfide is oxidized by acidic hydrogen peroxide solution to the arsenate ion AsO_4^{3-}. Write the chemical equation and the reduction and oxidation half-reactions for the reaction.

20.88 Determine the values of ΔH_r° and ΔS_r° for each stage of the Ostwald process for the production of nitric acid. Predict the conditions of pressure and temperature that favor the formation of the products in each case.

Stage 1: $4\,NH_3(g) + 5\,O_2(g) \longrightarrow 4\,NO(g) + 6\,H_2O(g)$

Stage 2: $2\,NO(g) + O_2(g) \longrightarrow 2\,NO_2(g)$

Stage 3: $3\,NO_2(g) + H_2O(l) \longrightarrow 2\,HNO_3(aq) + NO(g)$

20.89 When the enthalpy of vaporization of water is divided by its boiling point (on the Kelvin scale), the result is 110 $J{\cdot}K^{-1}{\cdot}mol^{-1}$. For hydrogen sulfide, the same calculation gives 88 $J{\cdot}K^{-1}{\cdot}mol^{-1}$. Explain why the value for water is greater than that for hydrogen sulfide.

20.90 The annual production of sulfuric acid in the United States is approximately 4×10^{10} kg. (a) If the acid were all produced from elemental sulfur, what mass of sulfur would be used for the production of sulfuric acid? (b) What volume of sulfur trioxide (at 25°C and 5.0 atm) is required for the annual production of sulfuric acid?

20.91 Determine the volume (in liters) of concentrated sulfuric acid (density, 1.84 $g{\cdot}mL^{-1}$) that is required for the production of 1000 kg of phosphoric acid from phosphate rock by the reaction

$Ca_3(PO_4)_2(s) + 3\,H_2SO_4(l) \longrightarrow 2\,H_3PO_4(l) + 3\,CaSO_4(s)$

20.92 Balance each skeletal equation and classify the reaction as acid-base, Lewis acid-base, or redox:

(a) $KClO_3(s) \longrightarrow KCl(s) + O_2(g)$

(b) $CaF_2(s) + H_2SO_4(conc) \longrightarrow Ca(HSO_4)_2(aq) + 2HF(g)$

(c) $OF_2(g) + OH^-(aq) \longrightarrow O_2(g) + F^-(aq) + H_2O(l)$

(d) $H_2S(g) + O_2(g) \longrightarrow SO_2(g) + H_2O(l)$

20.93 Balance each skeletal equation and classify the reaction as Brønsted acid-base, Lewis acid-base, or redox:

(a) $SO_2(g) + H_2O(l) \longrightarrow H_2SO_3(l)$

(b) $F_2(g) + NaOH(aq) \longrightarrow OF_2(g) + NaF(aq) + H_2O(l)$

(c) $S_2O_3^{2-}(aq) + Cl_2(g) + H_2O(l) \longrightarrow HSO_4^-(aq) + H_3O^+(aq) + Cl^-(aq)$

(d) $XeF_6(s) + OH^-(aq) \longrightarrow XeO_6^{4-}(aq) + Xe(g) + O_2(g) + F^-(aq) + H_2O(l)$

20.94 (a) Write the chemical equation for the production of chlorine from the electrolysis of an aqueous sodium chloride solution. (b) The annual production of chlorine gas in the United States is currently 1.1×10^{10} kg. How many coulombs of electricity are required for the production of this chlorine? (c) If the chlorine is generated 24 hours a day for 365 days a year, what is the average current (in amperes) that must be used?

20.95 List two uses each of fluorine, chlorine, bromine, and iodine.

20.96 Considering (a) Cl_2O; (b) Cl_2O_7 to be acid anhydrides, write the formula of the corresponding acid in aqueous solution.

20.97 Complete the following skeletal equations:

(a) $I_2(s) + F_2(g) \longrightarrow$ (any *one* of a number of products)

(b) $I_2(aq) + I^-(aq) \longrightarrow$

(c) $Cl_2(g) + H_2O(l) \longrightarrow$

(d) $F_2(g) + H_2O(l) \longrightarrow$

20.98 Explain why iodine is much more soluble in a solution of potassium iodide than in pure water.

20.99 Write equations for reactions used to synthesize gaseous samples of (a) hydrogen chloride; (b) hydrogen bromide; (c) hydrogen iodide.

20.100 The concentration of Cl^- ions can be measured gravimetrically by precipitating silver chloride, with silver nitrate as the precipitating reagent in the presence of dilute nitric acid. The white precipitate is filtered off and its mass is determined. (a) Calculate the Cl^- ion concentration in 25.00 mL of a solution that gave a silver chloride precipitate of mass 3.050 g. (b) Why is the method inappropriate for measuring the concentration of fluoride ions?

20.101 Give the names and formulas of three compounds in which (a) nitrogen; (b) phosphorus has oxidation number (i) +3; (ii) +5.

20.102 (a) Summarize and account for the differences between fluorine and the other Group 17 elements. (b) What do we mean when we say that fluorine is a reactive element?

20.103 Discuss the chemistry of a safety match: identify the chemicals that are used and the function of each. (See Section 19.5, too.)

20.104 Write chemical equations for the preparation of (a) XeF_4; (b) PtF_4. (See Section 20.12.)

20.105 Xenon hexafluoride exists as the ionic solid $XeF_5^+F^-$. Write the Lewis structure for XeF_5^+ and, from VSEPR theory, predict its shape.

20.106 Describe the trends in the physical and chemical properties of the noble gases.

20.107 Suggest reasons why the noble gas compounds were unknown until late in the twentieth century.

20.108 Identify one interesting compound of each (a) Period 2 and (b) Period 3 element that would leave the world a much poorer place if it did not exist, and give your reasons for each choice.

20.109 Summarize the evidence for or against the statement that metallic character increases down Group 16 but is less pronounced than in Group 15, which, in turn, is less pronounced than in Group 14.

CHALLENGING EXERCISES

20.110 Chapters 19 and 20 have been organized by group. (a) Discuss whether it would be helpful to organize the elements according to period and give examples of trends that could be displayed helpfully in that way. (b) Would a different organization be useful for the transition metals? Explain your decision.

20.111 The azide ion has an ionic radius of 148 pm and forms many ionic and covalent compounds that are similar to those of the halides. (a) Write the Lewis formula for the azide ion and predict the N—N—N bond angle. (b) On the basis of its ionic radius, where in Group 17 would you place the azide ion? (c) Compare the acidity of hydrazoic acid with those of the hydrohalic acids and explain any differences (for HN_3, $K_a = 1.7 \times 10^{-5}$). (d) Write the formulas of three ionic or covalent azides.

20.112 The concentration of nitrate ion in a basic solution can be determined by the following sequence of steps: (1) zinc metal reduces nitrate ions to ammonia in a basic aqueous solution; (2) the ammonia is passed into a solution containing a known, but excess, amount of HCl(aq); (3) the unreacted HCl(aq) is titrated with a standard NaOH(aq) solution. (a) Write balanced chemical equations for the three reactions. (b) A 25.00-mL sample of water from a rural well contaminated with $NO_3^-(aq)$ was treated with an excess of zinc metal. The evolved ammonia gas was passed into 50.00 mL of a 0.250 M HCl(aq) solution. The unreacted HCl(aq) was titrated to the stoichiometric point with 28.22 mL of 0.150 M NaOH(aq). What is the molar concentration of nitrate ion in the well water?

20.113 The concentration of hypochlorite ions in a solution can be determined by adding a sample of known volume to a solution containing excess I^- ions, which are oxidized to iodine:

$$ClO^-(aq) + 2\,I^-(aq) + H_2O(l) \longrightarrow I_2(aq) + Cl^-(aq) + 2\,OH^-(aq)$$

The iodine concentration is then measured by titration with sodium thiosulfate:

$$I_2(aq) + 2\,S_2O_3^{2-}(aq) \longrightarrow 2\,I^-(aq) + S_4O_6^{2-}(aq)$$

In one experiment, 10.00 mL of ClO^- solution was added to a KI solution, which in turn required 28.34 mL of 0.110 M $Na_2S_2O_3(aq)$ to reach the stoichiometric point. Calculate the molar concentration of ClO^- ions in the original solution.

20.114 Account for the observation that melting and boiling points generally *decrease* from fluoride to iodide for ionic halides, but *increase* from fluoride to iodide for molecular halides.

20.115 Account for the observation that solubility in water generally *increases* from chloride to iodide for ionic halides with low covalent character (such as the potassium halides), but *decreases* from chloride to iodide for ionic halides in which the bonds are significantly covalent (such as the silver halides).

20.116 The sodium iodate found as an impurity in Chile saltpeter was once the major source of iodine. The element was obtained by the reduction of an acidic solution of the iodate ion with sodium hydrogen sulfite. (a) Write the chemical equation for the reaction, assuming the oxidized product to be HSO_4^-. (b) Calculate the mass of sodium hydrogen sulfite needed to produce 50.0 g iodine.

CHAPTER 21

The remarkable—and unexpected—development of high-temperature superconductors in recent years may revolutionize electronic technologies. One of the elements found in the intricate structure of most of these superconductors is copper, a member of the d*-block. The current challenge is to make these brittle materials bendable and strong, perhaps by producing filaments like those shown here: each hexagon is about 4 μm wide. What other surprises does the* d*-block have in store?*

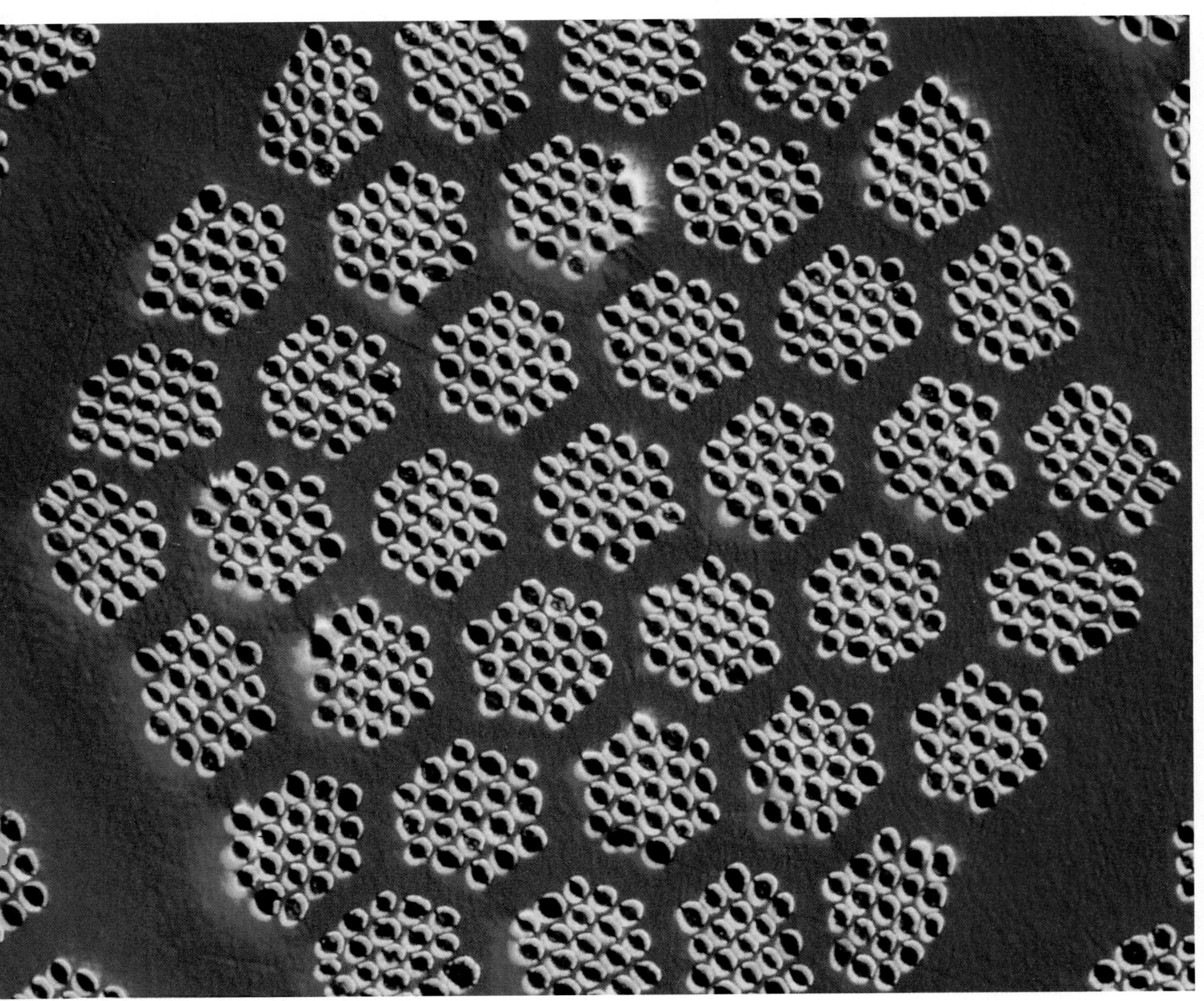

THE *d*-BLOCK: METALS IN TRANSITION

The *d*-block metals are the workhorse elements of the periodic table. Iron and copper helped civilization rise from the Stone Age and are still industrial metals of the greatest importance. Other members of the block include the metals of the new technologies, such as titanium for the aerospace industry and vanadium for catalysts in the petrochemical industry. The precious metals—silver, platinum, and gold—are prized as much for their appearance, rarity, and durability as for their usefulness. Compounds of *d*-block metals give color to paint, turn sunlight into electricity, serve as powerful oxidizing agents, and form the basis of some cancer treatments.

Atoms of the *d*-block elements have electron configurations in which, according to the building-up principle, *d*-orbitals are the last to be occupied. For example, in Period 4 the electron configurations range from $[Ar]3d^14s^2$ for scandium, the first member of the block, to $[Ar]3d^{10}4s^2$ for zinc, the last member. Because there are five *d*-orbitals in a given shell and because each one can accommodate up to two electrons, there are 10 elements in each row of the *d*-block. Just after the third row of the *d*-block has been started, at lanthanum, the seven 4*f*-orbitals begin to be occupied, and the lanthanides therefore delay the completion of that row. We shall see that the *f*-block affects the properties of the elements following it in Period 6. The elements in Period 7 are all radioactive and most do not occur naturally in measurable quantities on Earth.

THE *d*-BLOCK ELEMENTS AND THEIR COMPOUNDS

21.1 Trends in physical properties
21.2 Trends in chemical properties
21.3 Scandium through nickel
21.4 Groups 11 and 12

COMPLEXES OF THE *d*-BLOCK ELEMENTS

21.5 The structures of complexes
21.6 Isomers

CRYSTAL FIELD THEORY

21.7 The effects of ligands on *d*-electrons
21.8 The effects of ligands on color
21.9 The electronic structures of many-electron complexes
21.10 Magnetic properties of complexes

THE *d*-BLOCK ELEMENTS AND THEIR COMPOUNDS

The *d*-block elements lose their valence *s*-electrons when they form compounds. Moreover, most of them can also lose a variable number of *d*-electrons, and hence show variable valence. Variable valence makes these elements useful as catalysts and, in partnership with proteins, as enzymes. The only elements of the block that do not use their *d*-electrons in compound formation are the members of Group 12 (zinc, cadmium, and mercury). The elements in Groups 3 through 11 represent a transition from the highly reactive metals of the *s*-block to the much less reactive metals of Group 12 and the *p*-block; they are called the **transition metals** (Fig. 21.1).

21.1 TRENDS IN PHYSICAL PROPERTIES

Silver is the best electrical conductor of all elements at room temperature.

All the *d*-block elements are metals. Most of these "*d*-metals" are good electrical conductors. Most are malleable, ductile, lustrous, and silver-white in color, and generally have higher melting and boiling points than the main-group elements. There are a few notable exceptions: mercury has such a low melting point that it is a liquid at room temperature; copper is red-brown and gold is yellow.

Many properties of the *d*-block elements can be traced to the shapes of *d*-orbitals. First, the lobes of two *d*-orbitals on the same atom occupy markedly different regions of space. Because they are relatively far apart, electrons in separate *d*-orbitals repel one another weakly. Second, elec-

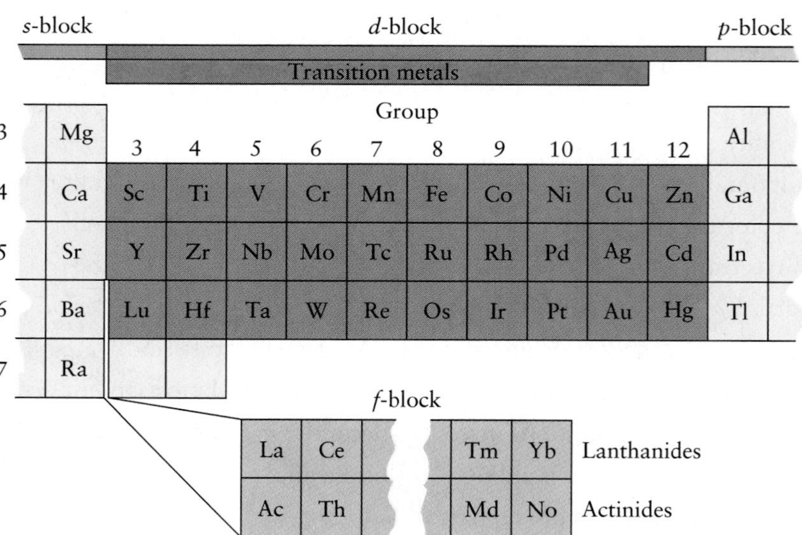

FIGURE 21.1 The elements in the *d*-block of the periodic table. Note that the *f*-block intervenes in Periods 6 and 7. The elements regarded as transition metals are indicated by the horizontal gray bar.

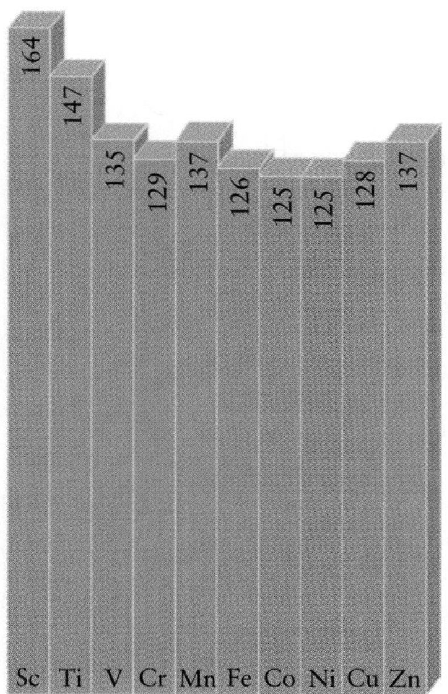

FIGURE 21.2 The atomic radii (in picometers) of the elements of the first row (Period 4) of the *d*-block. The radii decrease from Sc to Ni because of the increasing nuclear charge and the poor shielding effect of *d*-electrons.

tron density in *d*-orbitals is low near the nucleus. Because *d*-electrons are far from the nucleus, they are not very effective at shielding other electrons from the nuclear charge.

The orbital shapes help account for the relative atomic radii of the *d*-block metals (Fig. 21.2). Nuclear charge and the number of *d*-electrons both increase from left to right across each row (from scandium to zinc, for instance). Because the repulsion between *d*-electrons is weak, the increasing nuclear charge can draw them inward, so the atom becomes smaller. However, further across the block, the radii begin to increase slightly as each additional proton and electron are added: there are so many *d*-electrons that the electron-electron repulsion now increases more than the effective nuclear charge. Because these attractions and repulsions are finely balanced, the range of *d*-metal atomic radii is not very great. The atoms have similar sizes, so some of the atoms of one *d*-metal can replace the atoms of another in a crystal lattice without causing too much strain (Fig. 21.3). The *d*-metals can therefore be mixed together to form a wide range of alloys, including the many varieties of steel.

The atomic radii of *d*-metals in the second row (Period 5) are typically greater than those in the first row (Period 4). However, the radii in the third row (Period 6) are about the same as those in the second row. This similarity can be traced to the presence of the *f*-block near the beginning of Period 6 (Fig. 21.4). There is a pronounced decrease in atomic radius along the first row of the *f*-block (the lanthanides); and when the *d*-block resumes (at lutetium), the atomic radius has fallen from 224 pm for barium to 172 pm for lutetium. As a result, the atoms of all the following elements are smaller than expected. This effect is called the **lanthanide contraction.** The contraction can be explained by the higher nuclear charge and the poor shielding effect of the *f*-electrons: the atomic radii decrease as the effective nuclear charge increases and pulls the valence electrons inward.

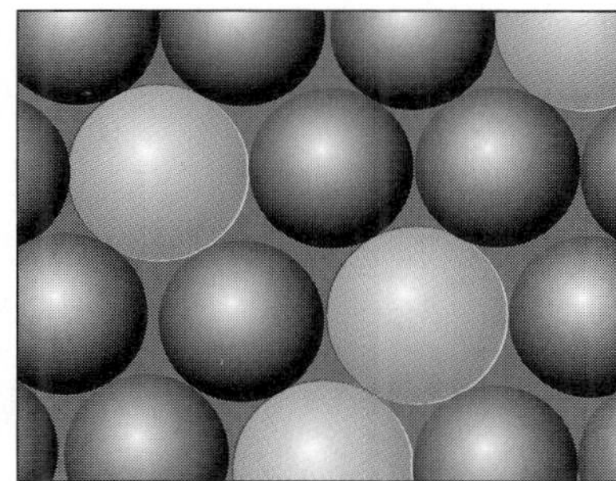

FIGURE 21.3 Because the atomic radii of the *d*-block elements are so similar, the atoms of one element can replace the atoms of another element with minor modification of the atomic locations. Consequently, a wide range of alloys can form.

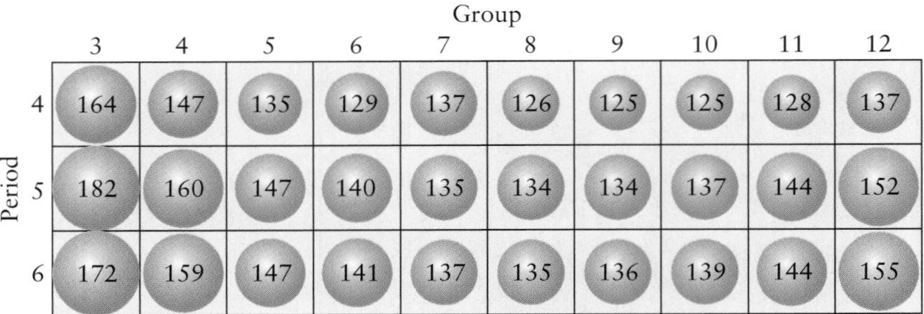

Group
3 4 5 6 7 8 9 10 11 12
Period
4 164 147 135 129 137 126 125 125 128 137
5 182 160 147 140 135 134 134 137 144 152
6 172 159 147 141 137 135 136 139 144 155

FIGURE 21.4 The atomic radii of the *d*-block elements (in picometers). Notice the similarity of all the values, and in particular the close similarity between the second and third rows as a result of the lanthanide contraction.

The lanthanide contraction is responsible for the high density of the Period 6 elements (Fig. 21.5). A block of iridium, for example, contains about as many atoms as a block of rhodium of the same volume; but each iridium atom is nearly twice as heavy as each rhodium atom, so the density of the sample is nearly twice as great. Another effect of the contraction is the low reactivity of gold and platinum. Because their valence electrons are relatively close to the nucleus, they are tightly bound and not readily available for chemical reactions.

The presence of unpaired *d*-electrons in the ground states of the *d*-block elements explains why some of these metals—most notably iron, cobalt, and nickel—are used to make permanent magnets. Recall that paramagnetism is the tendency of a substance to move into a magnetic field; it arises when an atom or molecule has at least one unpaired electron (see Section 20.5). Because the spins on neighboring atoms or molecules are aligned almost randomly, paramagnetism is very weak. In some *d*-metals, however, the unpaired electrons of many neighboring atoms can align with one another, producing the much stronger effect of

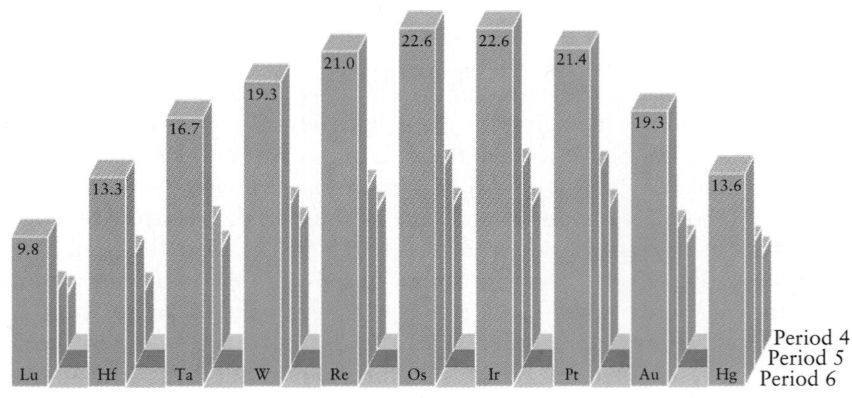

FIGURE 21.5 The densities (in grams per centimeter cubed, $g{\cdot}cm^{-3}$) of the *d*-metals at 25°C. The lanthanide contraction has a pronounced effect on the densities of the elements in Period 6 (front row in this illustration), which are among the densest of all the elements.

ferromagnetism. The aligned spins form **domains,** or regions of aligned spins (Fig. 21.6), which survive even after the applied field is turned off. Ferromagnetic materials are used in the coatings of cassette tapes and computer disks.

The atomic radii of the d-block elements decrease with increasing atomic number on the left of a row and then increase slightly on the right; the radii of the atoms of Periods 5 and 6 are similar as a result of the lanthanide contraction. Iron, cobalt, and nickel are ferromagnetic.

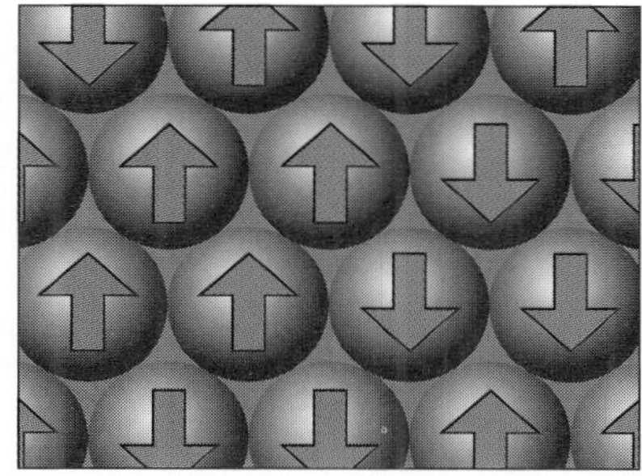

(a)

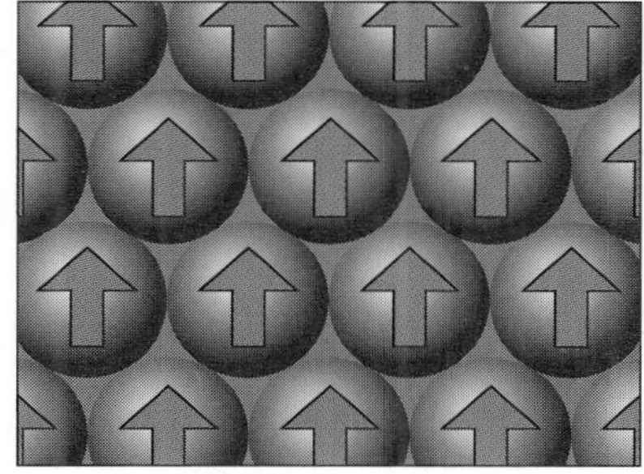

(b)

FIGURE 21.6 Ferromagnetic materials include iron, cobalt, nickel, gadolinium, and the iron oxide mineral magnetite. They contain regions of atoms in which electrons of many atoms spin in the same direction and give rise to a strong magnetic field. (a) Before magnetization, the spins are almost randomly aligned. (b) After magnetization. The orange arrows represent the electron spins of each atom.

21.2 TRENDS IN CHEMICAL PROPERTIES

Figure 21.7 shows the range of oxidation states of each *d*-metal. Except for mercury, the elements at the ends of each row of the *d*-block occur in only one oxidation state other than 0. Scandium, for example, is found only in oxidation state +3, and zinc only in +2. All the other elements of each row are found in at least two oxidation states. The most common oxidation states for copper, for example, are +1 (as in CuCl) and +2 (as in $CuCl_2$). Elements close to the center of each row have the widest range of oxidation states. Manganese, at the center of its row, has seven oxidation states. Elements in the second and third rows of the block are more likely to reach higher oxidation states than those in the first row.

It is difficult to predict the likely reaction of an element, but some broad principles are often helpful. For instance, a species in which an element is in a high oxidation state tends to be a good oxidizing agent. For example, in the permanganate ion, MnO_4^-, manganese has oxidation number +7 and so permanganate ion is a good oxidizing agent in acidic solution:

$$MnO_4^-(aq) + 8\,H^+(aq) + 5\,e^- \longrightarrow Mn^{2+}(aq) + 4\,H_2O(l) \qquad E^\circ = +1.51\text{ V}$$

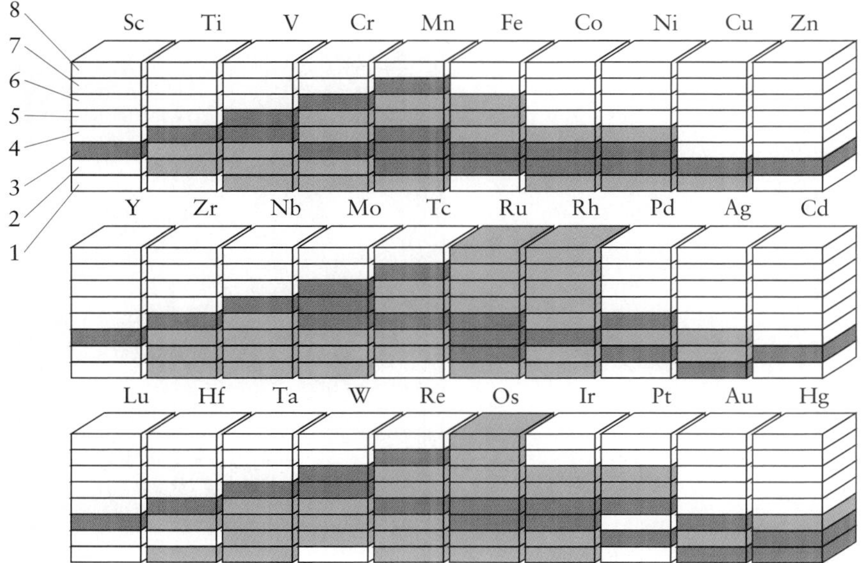

FIGURE 21.7 The oxidation numbers of the *d*-block elements. The orange blocks mark the common oxidation numbers for each element; the green blocks mark the other known states.

Conversely, some species in which the element has a low oxidation state, such as Fe^{2+}, are good reducing agents:

$$Fe^{3+}(aq) + e^- \longrightarrow Fe^{2+}(s) \qquad E° = +0.77\ V$$

Although most *d*-metal oxides are basic, the oxides of a given element show a shift toward acidic character with increasing oxidation number. A good example is provided by the family of chromium oxides:

CrO	+2	basic
Cr_2O_3	+3	amphoteric
CrO_3	+6	acidic

Chromium(VI) oxide, CrO_3, is the anhydride of chromic acid, H_2CrO_4, the parent acid of the chromates. In this highly oxidized state, chromium is electron poor and the oxygen atoms attached to it are less likely to be able to share electrons with a proton. The conjugate base of chromic acid, $HCrO_4^-$, does not attract protons strongly, despite its overall negative charge, and is therefore a very weak base.

The elements on the left of the *d*-block resemble the *s*-block metals in being much more difficult to extract from their ores than the metals on the right. Indeed, if we go to the right side of the *d*-block and move back across it from right to left, then we encounter the elements in very roughly the order in which they became available for exploitation. On the far right are copper and zinc, which jointly were responsible for the Bronze Age. That age was succeeded by the Iron Age, as higher temperatures became attainable and iron ore reduction feasible. Finally, at the left of the block, we have metals like titanium that require such extreme conditions for their extraction—including the use of other active metals or electrolysis—that they have become widely available only in this century (Fig. 21.8).

(a)

(b)

(c)

FIGURE 21.8 These three artifacts represent the progress that has been made in the extraction of *d*-metals. (a) An ancient bronze chariot axle cap from China, made from an alloy of metals that are easy to extract. (b) A nineteenth-century iron steam engine made from a metal that was moderately easy to extract once high temperatures could be achieved. (c) A twentieth-century airplane engine with titanium components that had to await high temperatures and advanced technology before the element became widely available.

TABLE 21.1 Properties of the *d*-block elements scandium through nickel

Z	Name	Symbol	Electron configuration	Melting point, °C	Boiling point, °C	Density, $g \cdot cm^{-3}$
21	scandium	Sc	$3d^1\,4s^2$	1540	2800	2.99
22	titanium	Ti	$3d^2\,4s^2$	1660	3300	4.55
23	vanadium	V	$3d^3\,4s^2$	1920	3400	6.09
24	chromium	Cr	$3d^5\,4s^1$	1860	2600	7.19
25	manganese	Mn	$3d^5\,4s^2$	1250	2120	7.47
26	iron	Fe	$3d^6\,4s^2$	1540	2760	7.87
27	cobalt	Co	$3d^7\,4s^2$	1494	2900	8.80
28	nickel	Ni	$3d^8\,4s^2$	1455	2150	8.91

The range of oxidation states of a d-block element increases toward the center of the block. High-oxidation-state compounds tend to be oxidizing, low-oxidation-state compounds tend to be reducing. The acidic character of oxides increases with the oxidation state of the element.

SELF-TEST 21.1A Predict trends in ionization energies of the *d*-block metals.
[*Answer:* Ionization energy increases from left to right across a row and decreases down a group.]

SELF-TEST 21.1B Predict trends in the densities of the *d*-block metals.

21.3 SCANDIUM THROUGH NICKEL

Table 21.1 summarizes the physical properties of the elements in the first row of the *d*-block from scandium to nickel. Here we consider only their most important chemical properties.

Scandium, Sc, which was first isolated in 1937, is a reactive metal: it reacts with water about as vigorously as calcium does. It has few uses. The small, highly charged Sc^{3+} ion is strongly hydrated in water (like Al^{3+}) and the resulting $[Sc(H_2O)_6]^{3+}$ complex is about as strong a Brønsted acid as acetic acid.

Titanium, Ti, is a light, strong metal that is used where these properties are vital—as it is in airplanes (Fig. 21.9). Unlike scandium, it is resistant to corrosion because it is passivated by a protective skin of oxide that forms on its surface. The principal sources of the metal are the ores *ilmenite* ($FeTiO_3$) and *rutile* (TiO_2).

FIGURE 21.9 This piece of titanium has been formed and then machined for use in a high-performance military jet plane. Titanium has excellent corrosion resistance, low density and high strength, making it ideal for aerospace applications.

Titanium requires strong reducing agents for extraction from its ores. It was not exploited commercially until quite recently, when demand from the aerospace industry increased. The metal is obtained first by treating the ores with chlorine in the presence of coke to form titanium(IV) chloride; the chloride is then reduced by passing it through liquid magnesium:

$$TiCl_4(g) + 2\,Mg(l) \xrightarrow{700^\circ C} Ti(s) + 2\,MgCl_2(s)$$

The most stable oxidation state of titanium is +4, with both its 4*s*-electrons and its two 3*d*-electrons lost. Its most important compound is titanium(IV) oxide, TiO_2, which is almost universally known as titanium dioxide. It is a brilliantly white, nontoxic, stable solid used as the white pigment in paints and paper. It has been found to act as a semiconductor in the presence of light, so it has been studied for use as an electrode in photoelectrochemical cells (Case Study 17).

Titanium forms a series of oxides called *titanates,* which are prepared by heating TiO_2 with a stoichiometric amount of the oxide or carbonate of a second metal. Barium titanate ($BaTiO_3$) is **piezoelectric,** which means that it becomes electrically charged when it is mechanically distorted. This property leads to its use for underwater sound detection, in which a mechanical vibration is converted into an electrical signal.

It is helpful to remember that, by coincidence, V (which is also the Roman numeral for 5) is in Group 5.

Vanadium, V, which is a soft silver-gray metal, is produced by reducing vanadium(V) oxide with calcium:

$$V_2O_5(s) + 5\,Ca(l) \xrightarrow{\Delta} 2\,V(s) + 5\,CaO(s)$$

or reducing vanadium(II) chloride with magnesium:

$$VCl_2(s) + Mg(l) \xrightarrow{\Delta} V(s) + MgCl_2(s)$$

Vanadium metal is also produced commercially by electrolysis of molten vanadium(II) chloride.

Vanadium is used to make tough steels for automobile and truck springs. Because it is not economical to add the pure metal to iron, a *ferroalloy* of the metal—an alloy with iron and carbon that is less expensive to produce—is used instead. Ferrovanadium, a mixture of about 86% V, 12% C, and 2% Fe by mass, is prepared by reducing vanadium(V) oxide with aluminum in the presence of iron. The ferrovanadium is then added to a molten mixture of iron and carbon to make the vanadium steel.

Vanadium(V) oxide, V_2O_5, commonly known as vanadium pentoxide, is the most important compound of vanadium. This orange-yellow solid is used as an oxidizing agent and as an oxidizing catalyst in the contact process for the manufacture of sulfuric acid (Section 20.7). The wide range of colors of vanadium compounds, including the blue of the vanadyl ion, VO^{2+} (Fig. 21.10), has led to their use as glazes in the ceramics industry.

FIGURE 21.10 Many vanadium compounds form vividly colored aqueous solutions. They are also used to color pottery glazes. The blue colors here are due to the vanadyl ion, VO^{2+}.

Chromium, Cr, is a bright, lustrous, corrosion-resistant metal that gets its name from its colorful compounds. It is obtained from the mineral *chromite* ($FeCr_2O_4$) by reduction with carbon in an electric arc furnace:

The name *chromium* comes from *chroma,* the Greek word for "color."

$$FeCr_2O_4(s) + 4\,C(s) \longrightarrow Fe(s) + 2\,Cr(s) + 4\,CO(g)$$

Chromium metal is also produced by the thermite process (Section 6.2):

$$Cr_2O_3(s) + 2\,Al(s) \xrightarrow{\Delta} Al_2O_3(s) + 2\,Cr(l)$$

The thermite process, in which aluminum is a vigorous reducing agent, is also used to reduce other metals from their ores. The reaction is so exothermic that the metal melts.

Most of the chromium metal produced is used in steelmaking and for chromium plating (Section 17.16). Chromium(IV) oxide, CrO_2, is a ferromagnetic material that is used for coating "chrome" recording tapes because they respond better to high-frequency magnetic fields than do conventional "ferric" (Fe_2O_3) tapes.

Sodium chromate, Na_2CrO_4, is a yellow solid. Its importance lies in its use as a source of most other chromium compounds. The chromate ion changes into the orange dichromate ion, $Cr_2O_7^{2-}$, in the presence of acid (Fig. 21.11):

$$2\,CrO_4^{2-}(aq) + 2\,H^+(aq) \longrightarrow Cr_2O_7^{2-}(aq) + H_2O(l)$$

In the laboratory, acidified solutions of dichromates, in which the oxidation number of chromium is +6, are useful oxidizing agents:

$$Cr_2O_7^{2-}(aq) + 14\,H^+(aq) + 6\,e^- \longrightarrow 2\,Cr^{3+}(aq) + 7\,H_2O(l) \qquad E^\circ = +1.33\ V$$

Sodium chromate and sodium dichromate are both starting points for the production of a number of pigments, corrosion inhibitors, fungicides, and ceramic glazes. Sulfur dioxide reduces sodium chromate to $Cr(OH)SO_4$, a compound used in the tanning of leather. The compound attaches to collagen, a protein in animal skins, making it insoluble and protecting it from biological degradation, but leaving it flexible.

Manganese, Mn, is a gray metal that resembles iron. The metal is rarely used alone, because it is much less resistant to corrosion than chromium, but it is an important component of alloys. In steelmaking, it removes sulfur by forming a sulfide. It also increases hardness, toughness, and resistance to abrasion (see Table 21.2). Another useful alloy is *manganese bronze* (39% Zn, 1% Mn, some iron and aluminum, and the rest copper), which is very resistant to corrosion and is used for the propellers of ships. Manganese is alloyed with aluminum to increase the stiffness of beverage cans.

A rich supply of manganese lies in the manganese nodules that litter the ocean floors (Fig. 21.12). These nodules range in diameter from millimeters to meters and are lumps of the oxides of iron, manganese, and

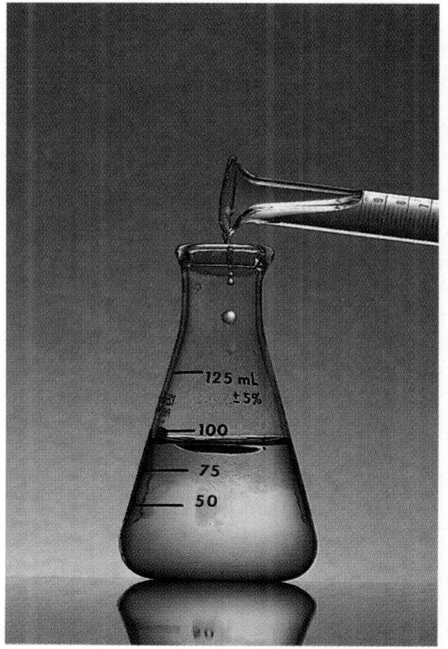

FIGURE 21.11 The chromate ion, CrO_4^{2-}, is yellow. When acid is added to a chromate solution, the ions form dichromate ions, $Cr_2O_7^{2-}$, which are orange.

TABLE 21.2 **Composition of different steels**

Element blended into iron	Typical amount, %	Effect
manganese	0.5 to 1.0	increases strength and hardness but lowers ductility
	13	increases wear resistance
nickel	< 5	increases strength and shock resistance
	> 5	increases corrosion resistance (stainless) and hardness
chromium	variable	increases hardness and wear resistance
	> 12	increases corrosion resistance (stainless)
vanadium	variable	increases hardness
tungsten	< 20	increases hardness, especially at high temperatures

other elements. However, because this source is technically difficult to exploit, manganese is currently obtained by the thermite process from *pyrolusite,* a mineral form of manganese dioxide, MnO_2:

$$3\,MnO_2(s) + 4\,Al(s) \xrightarrow{\Delta} 3\,Mn(l) + 2\,Al_2O_3(s)$$

Manganese lies near the center of its row (in Group 7) and occurs in a wide variety of oxidation states. The most stable state is +2, but +4, +7, and to some extent +3, are also common in manganese compounds. Its most important compound is manganese(IV) oxide, MnO_2, which is commonly called manganese dioxide. This compound is a brown-black solid used in dry cells, as a decolorizer to conceal the green tint of glass, and as the starting point for the production of other manganese compounds. Potassium permanganate, in which the manganese is in its highest oxidation state (+7), is a widely used strong oxidizing agent in acidic solution. Its usefulness stems not only from its thermodynamic tendency to oxidize other species but also from its ability to act by a variety of mechanisms; hence, it is likely to be able to find a path with low activation energy. Potassium permanganate is used for oxidations in organic chemistry and also as a mild disinfectant.

Iron, Fe, the most widely used of all the *d*-metals, is the most abundant element on Earth as a whole and the second most abundant metal in the Earth's crust (aluminum is the most abundant metal in the crust). Its principal ores are the oxides *hematite* (Fe_2O_3) and *magnetite* (Fe_3O_4). The sulfide mineral *pyrite,* FeS_2 (see Fig. 20.10), is also widely available, but it is not used because the sulfur is difficult to remove.

Iron ores are primarily used to make steel. The ore is reduced to iron metal in a blast furnace (Fig. 21.13), which is a marvel of efficient use of materials. Even the waste products of the fire that heats it are used as reactants. The furnace, which is approximately 40 m in height, is continuously filled at the top with a mixture of ore, coke, and limestone. Each kilogram of iron produced requires about 1.75 kg of ore, 0.75 kg of coke, and 0.25 kg of limestone. The limestone, which is primarily calci-

FIGURE 21.12 Manganese nodules litter the ocean floor and are potentially a valuable source of the element.

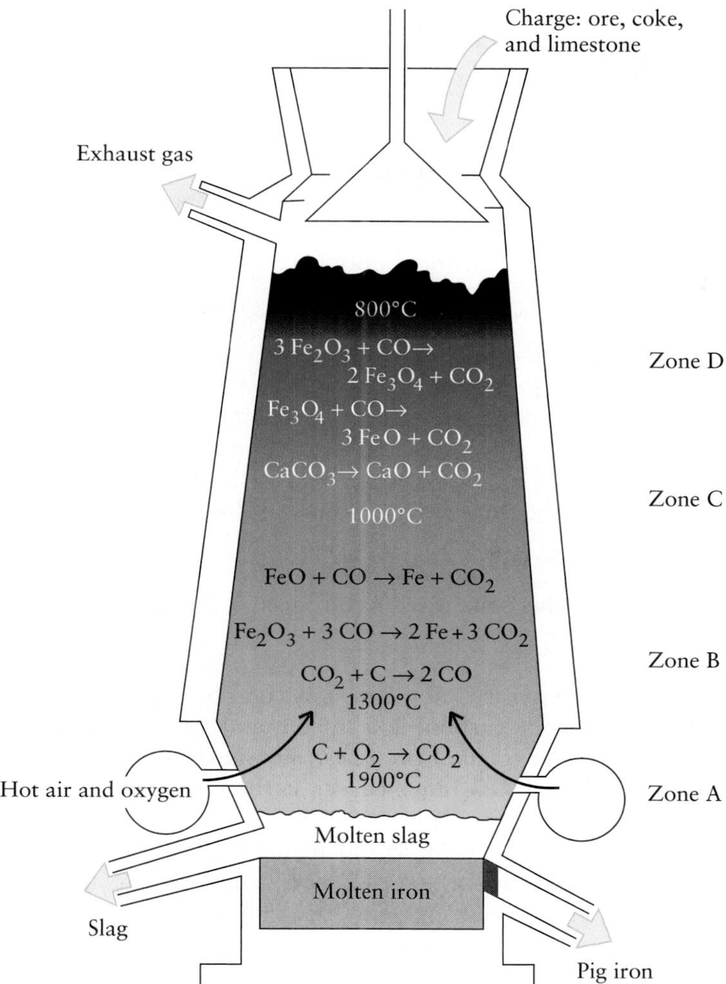

FIGURE 21.13 The reduction of iron ore takes place in a blast furnace containing a mixture of the ore with coke and limestone. Different reactions occur in different zones when the blast of air and oxygen is admitted. The ore, an oxide, is reduced to the metal by reduction with carbon monoxide produced in the furnace.

um carbonate, undergoes thermal decomposition to calcium oxide (lime) and carbon dioxide. The calcium oxide, which contains the Lewis base O^{2-}, helps to remove the acidic (nonmetal oxide) anhydride and amphoteric impurities from the ore:

$$CaO(s) + SiO_2(s) \xrightarrow{\Delta} CaSiO_3(l)$$

$$CaO(s) + Al_2O_3(s) \xrightarrow{\Delta} Ca(AlO_2)_2(l)$$

$$6\,CaO(s) + P_4O_{10}(s) \xrightarrow{\Delta} 2\,Ca_3(PO_4)_2(l)$$

The mixture of products, which is known as slag, is molten at the temperatures in the blast furnace and floats on the denser molten iron. It is drawn off and used to make rocklike material for the construction industry.

Molten iron is produced through a series of reactions in three main temperature zones of the blast furnace. At the bottom of the furnace, in Zone A, preheated air is blown into the furnace under pressure, and the coke is oxidized to heat the furnace and provide carbon in the form of carbon dioxide. The reaction

$$C(s) + O_2(g) \longrightarrow CO_2(g)$$

is exothermic and the temperature approaches 1900°C.

The melting point of pure iron is 1540°C, but it is lowered to 1015°C when 4% carbon is present.

Higher up in the furnace, the iron is reduced in stages to the metal, which melts and flows from Zone C to Zone A. As the carbon dioxide moves up through the furnace to Zone B, it reacts with some of the added carbon, producing carbon monoxide:

$$CO_2(g) + C(s) \longrightarrow 2\,CO(g)$$

This reaction is endothermic, which lowers the temperature to 1300°C. The carbon monoxide produced in this reaction rises to Zone C, where it is the reducing agent for the iron ore at approximately 1000°C:

$$Fe_2O_3(s) + 3\,CO(g) \longrightarrow 2\,Fe(s) + 3\,CO_2(g)$$

Partial reduction of the iron ore also occurs in Zone D:

$$Fe_2O_3(s) + CO(g) \longrightarrow 2\,FeO(s) + CO_2(g)$$

The FeO formed in this zone is reduced to iron lower in the furnace:

$$FeO(s) + CO(g) \longrightarrow Fe(s) + CO_2(g)$$

The molten iron (of density 7.9 $g{\cdot}cm^{-3}$), is run off as *pig iron,* consisting of 90–95% iron, 3–5% carbon, 2% silicon, and trace amounts of other elements found in the original ore. *Cast iron* is similar to pig iron, but some impurities have been removed. Its carbon content is still usually greater than 2%.

Pure iron is relatively flexible and malleable, but the carbon atoms make cast iron very hard and brittle. It is used where there is little mechanical and thermal shock, such as for ornamental railings, engine blocks, brake drums, and transmission housings.

Steels have various hardnesses, tensile strengths, and ductilities; and the higher the carbon content, the harder and more brittle the steel (Table 21.3). The corrosion resistance of iron is greatly improved by alloying to form a variety of steels. Stainless steels are highly corrosion resistant; they typically contain about 15% chromium by mass.

Steelmaking begins with pig iron. The first stage is to lower the carbon content of the iron and to remove the remaining impurities, which include silicon, phosphorus, and sulfur. In the **basic oxygen process,** oxygen and powdered limestone are forced through the molten metal

TABLE 21.3 The effect of carbon on iron

Type of steel	Carbon content, %	Properties and applications
low-carbon steel	< 0.15	ductility and low hardness, iron wire
mild-carbon steel	0.15 to 0.25	cables, nails, chains, and horseshoes
medium-carbon steel	0.20 to 0.60	nails, girders, rails, and structural purposes
high-carbon steel	0.61 to 1.5	knives, razors, cutting tools, drill bits

FIGURE 21.14 In the basic oxygen process, a blast of oxygen and powdered limestone is used to purify molten iron by, respectively, oxidizing and combining with the impurities present in it.

(Fig. 21.14). In the second stage, steels are produced by adding the appropriate metals, often as ferroalloys, to the molten iron.

A healthy adult human body contains around 3 g of iron, mostly as hemoglobin. Around 1 mg is lost daily (in sweat, feces, and hair) and women lose about 20 mg during menstruation, so iron must be ingested daily in order to maintain the balance. Iron deficiency, or anemia, results in reduced transport of oxygen to the brain and muscles, and an early symptom is chronic tiredness.

Cobalt ores are often found in association with copper(II) sulfide. Cobalt is a silver-gray metal and is used mainly for alloying with iron. *Alnico steel,* an alloy of iron, nickel, cobalt, and aluminum, is used to make permanent magnets like those used in loudspeakers. Cobalt steels are hard enough to be used as surgical steels, drill bits, and lathe tools. Cobalt(II) oxide is a blue salt used to color glass and ceramic glazes. We need cobalt in our diet, for it is a component of vitamin B_{12}.

Miners called cobalt minerals that interfered with copper production *Kobold* (which means "evil spirit" in German).

Nickel, Ni, is also used to alloy with iron. About 70% of the western world's supply comes from iron and nickel sulfide ores that were brought close to the surface by the impact of a meteor at Sudbury, Ontario. Impure nickel is refined by first exposing it to carbon monoxide, when it forms nickel carbonyl, $Ni(CO)_4$:

The origin of the name *nickel* is "Old Nick," for much the same reason as cobalt.

$$Ni(s) + 4\,CO(g) \xrightarrow{50°C} Ni(CO)_4(g)$$

Nickel carbonyl is a volatile, poisonous liquid that boils at 43°C and can thus be removed from impurities. Nickel metal is then obtained by heating the pure nickel tetracarbonyl to about 200°C, which reverses the reaction for its formation.

Nickel is a hard, silver-white metal used mainly for the production of stainless steel and for alloying with copper to produce cupronickels, the alloys used for nickel coins (which consist of about 25% Ni and 75%

Cu). Cupronickels are slightly yellow but are whitened by the addition of small amounts of cobalt. Nickel is also used as a catalyst, especially for the addition of hydrogen to organic compounds, for example, in the manufacture of edible solid fats from vegetable oils.

The d-block elements titanium through nickel are obtained chemically from their ores and have many industrial uses; scandium is highly reactive. Iron is reduced by the action of coke and limestone in blast furnaces; steel is an alloy of iron with carbon and other metals, which affect its properties.

21.4 GROUPS 11 AND 12

Group 11, close to the right-hand edge of the *d*-block, contains the coinage metals—copper, silver, and gold—which have $(n-1)d^{10}ns^1$ electron configurations (Table 21.4). Group 12 contains zinc, cadmium, and mercury, with configurations $(n-1)d^{10}ns^2$. The low reactivity of the coinage metals is partly due to the poor shielding of the *d*-electrons and hence the tight grip that the nucleus can exert on the outermost electrons. The inertness of gold is also enhanced by the lanthanide contraction.

Copper, Cu, is unreactive enough for some to be found native, but most is produced from its sulfides, particularly the ore *chalcopyrite,* $CuFeS_2$ (Fig. 21.15). The crushed and ground ore is separated from excess rock by the process called **froth flotation,** which makes use of the fact that sulfide ores are wetted by oils, but not by water. In this process (Fig. 21.16), the powdered ore is combined with oil, water, and detergents. Then air is blown through the mixture; the oil-coated sulfide mineral floats to the surface with the froth, and the unwanted copper-poor residue, which is called *gangue,* sinks to the bottom.

Processes for extracting metals from their ores are generally classified as **pyrometallurgical,** if high temperatures are used, or **hydrometallurgical,** if aqueous solutions are used. Copper is extracted by both methods. In the pyrometallurgical process for the extraction of copper, the enriched ore is **roasted,** or heated in air:

$$2\,CuFeS_2(s) + 3\,O_2(g) \longrightarrow 2\,CuS(s) + 2\,FeO(s) + 2\,SO_2(g)$$

The CuS is then **smelted,** a process in which metal ions are reduced by melting with another compound. At the same time, the sulfur is oxidized

TABLE 21.4 Properties of elements in Groups 11 and 12

Z	Name	Symbol	Electron configuration	Melting point, °C	Boiling point, °C	Density, g·cm^{-3}
29	copper	Cu	$3d^{10}\,4s^1$	1083	2567	8.93
47	silver	Ag	$4d^{10}\,5s^1$	962	2212	10.50
79	gold	Au	$5d^{10}\,6s^1$	1064	2807	19.28
30	zinc	Zn	$3d^{10}\,4s^2$	420	907	7.14
48	cadmium	Cd	$4d^{10}\,5s^2$	321	765	8.65
80	mercury	Hg	$5d^{10}\,6s^2$	−39	357	13.55

FIGURE 21.15 Three important copper ores (from left to right): chalcopyrite, $CuFeS_2$; malachite, $CuCO_3 \cdot Cu(OH)_2$; and chalcocite, Cu_2S.

to SO_2. This oxidation is accomplished by blowing compressed air through the mixture:

$$CuS(s) + O_2(g) \longrightarrow Cu(l) + SO_2(g)$$

These processes could contribute a damaging amount of SO_2 to the atmosphere if precautions were not taken to remove it. Limestone and sand, which are added to the mixture, form a molten slag that removes many of the impurities as well as the SO_2. For example, calcium oxide (a basic oxide) from the limestone reacts with the SO_2 (an acidic oxide) to produce calcium sulfite:

$$CaO(s) + SO_2(g) \longrightarrow CaSO_3(s)$$

The copper product is known as *blister copper* because of the appearance of air bubbles in the solidified metal (Fig. 21.17).

The impure copper from either process is refined electrolytically: it is made into anodes and plated onto cathodes of pure copper. The rare metals—most notably platinum, silver, and gold—obtained from the anode sludge are sold to recover much of the cost of the electricity used in the electrolysis.

Copper is an excellent electrical conductor, but it needs to be very pure for use in the electrical industry. It is also used in alloys, with zinc to form brass, with tin to form bronze, and with nickel to form cupronickel (Table 10.4). Because copper lies above hydrogen in the electrochemical series, it cannot displace hydrogen from acidic solutions; however, it can be oxidized by oxidizing acids, such as nitric acid and sulfuric acid (Section 3.14).

Copper corrodes in moist air in the presence of carbon dioxide:

$$2\,Cu(s) + H_2O(l) + O_2(g) + CO_2(g) \longrightarrow Cu_2(OH)_2CO_3(s)$$

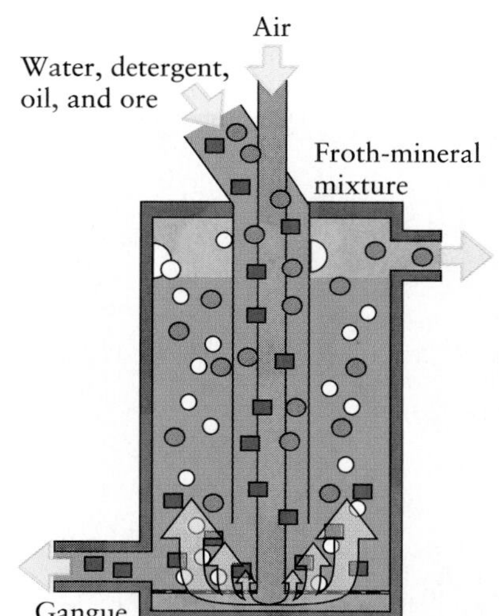

FIGURE 21.16 In the froth flotation process, a stream of bubbles is passed through a mixture of ore (orange circles), rock (brown rectangles), and detergent. The ore is buoyed up by the froth of bubbles and is removed from the top of the chamber. The unwanted gangue is washed away through the bottom of the container.

FIGURE 21.17 In this industrial copper refinery, the molten impure copper produced by smelting is poured into molds. Next, the copper will be purified by electrolysis.

The pale green product is called *basic copper carbonate* and is responsible for the green patina of copper and bronze objects (Fig. 21.18). The patina adheres to the surface, protects the metal, and gives a pleasing appearance.

Like all the coinage metals, copper forms compounds with oxidation number +1; however, in water, copper(I) salts disproportionate into metallic copper and copper(II) ions. The latter exist as pale blue $[Cu(H_2O)_6]^{2+}$ ions in water. This pale blue is the color of copper(II) sulfate solutions; the deeper blue of the solid pentahydrate, $CuSO_4{\cdot}5H_2O$, is due to $[Cu(H_2O)_4]^{2+}$ ions (the fifth H_2O links this ion to the sulfate anion). A mixture of copper(II) carbonate with stoichiometric amounts of Y_2O_3 and $BaCO_3$ heated to approximately 1000°C forms a *123 superconductor ceramic* of composition $YBa_2Cu_3O_{6.5\text{–}7.0}$ (the numbers 1, 2, and 3 denote the proportions in which the metal atoms Y, Ba, and Cu, respectively, are present in the compound).

FIGURE 21.18 Copper corrodes in air to form a pale green layer of basic copper carbonate. This patina, or incrustation, passivates the surface.

Silver, Ag, is rarely found native; most is obtained as a by-product of the refining of copper and lead, and a considerable amount is recycled through the photographic industry. Silver has a positive standard potential, so it does not reduce $H^+(aq)$ to hydrogen.

Silver(I) does not disproportionate in aqueous solution, and almost all silver compounds have oxidation number +1. Apart from silver nitrate and silver fluoride, silver salts are generally only sparingly soluble in water. Silver nitrate, $AgNO_3$, is the most important compound of silver and the starting point for the manufacture of silver halides for use in photography (Case Study 21).

Gold, Au, is so inert that most of it is found native as the metal. Pure gold is classified as *24-carat gold;* and its alloys with silver and copper, which differ in hardness and hue, are classified according to the proportion of gold they contain (Fig. 21.19). For example, 10- and 14-carat golds contain, respectively, 10/24 and 14/24 parts by mass of gold. Gold is a highly malleable metal, and 1 g of gold can be worked into a leaf covering about 1 m^2 or a wire over 2 km in length. Gold leaf is used for decoration, like that on dishes and books, and in cathedrals and temples (Fig. 21.20).

Gold lies well above hydrogen in the electrochemical series and is too noble to react even with strong oxidizing agents such as nitric acid. Both the following gold couples lie above H^+/H_2 and $NO_3^-, H^+/NO, H_2O$:

$$Au^+(aq) + e^- \longrightarrow Au(s) \qquad E° = +1.69\ V$$

$$Au^{3+}(aq) + 3\,e^- \longrightarrow Au(s) \qquad E° = +1.40\ V$$

$$NO_3^-(aq) + 4\,H^+(aq) + 3\,e^- \longrightarrow NO(g) + 2\,H_2O(l) \qquad E° = +0.96\ V$$

However, gold does react with aqua regia (a mixture of concentrated nitric and hydrochloric acids) because the complex ion $AuCl_4^-$ forms:

$$Au(s) + 6\,H^+(aq) + 3\,NO_3^-(aq) + 4\,Cl^-(aq) \longrightarrow AuCl_4^-(aq) + 3\,NO_2(g) + 3\,H_2O(l)$$

Even though the equilibrium constant for the formation of Au^{3+} from gold is very unfavorable, the reaction proceeds because any Au^{3+} ions formed are immediately removed by being hidden away inside a complex.

Zinc, Zn, is found mainly as its sulfide ZnS in *sphalerite,* often in association with lead ores (see Fig. 20.10). The ore is concentrated by froth flotation, and the metal is extracted by roasting and then smelting with coke:

$$2\,ZnS(s) + 3\,O_2(g) \xrightarrow{\Delta} 2\,ZnO(s) + 2\,SO_2(g)$$

$$ZnO(s) + C(s) \xrightarrow{\Delta} Zn(l) + CO(g)$$

Cadmium, Cd, is obtained in a similar manner. Zinc and cadmium are silvery, reactive metals. Zinc is used mainly for galvanizing iron; like copper, it is protected by a hard film of basic carbonate, $Zn_2(OH)_2CO_3$, which forms on contact with air.

Zinc and cadmium are similar to each other but differ sharply from mercury. Zinc is amphoteric (like its main-group neighbor aluminum). It reacts with acids to form Zn^{2+} ions and with alkalis to form the zincate ion, $[Zn(OH)_4]^{2-}$:

$$Zn(s) + 2\,OH^-(aq) + 2\,H_2O(l) \longrightarrow [Zn(OH)_4]^{2-}(aq) + H_2(g)$$

Galvanized containers should therefore not be used for transporting alkalis. Cadmium, which is lower down the group and is more metallic, has a more basic oxide, but its use is limited by the fact that cadmium salts are toxic. Zinc and cadmium have an oxidation number of +2 in all their compounds.

Mercury, Hg, occurs mainly as HgS in the mineral *cinnabar* (see Fig. 20.10), from which it is separated by froth flotation and then roasting:

$$HgS(s) + O_2(g) \xrightarrow{\Delta} Hg(g) + SO_2(g)$$

The volatile metal is distilled off and condensed. Mercury is unique in being the only metallic element that is liquid at room temperature. It has a long liquid range, from its melting point of −39°C to its boiling point of 357°C, so it is well suited for its use in thermometers, silent electrical switches, and high-vacuum pumps.

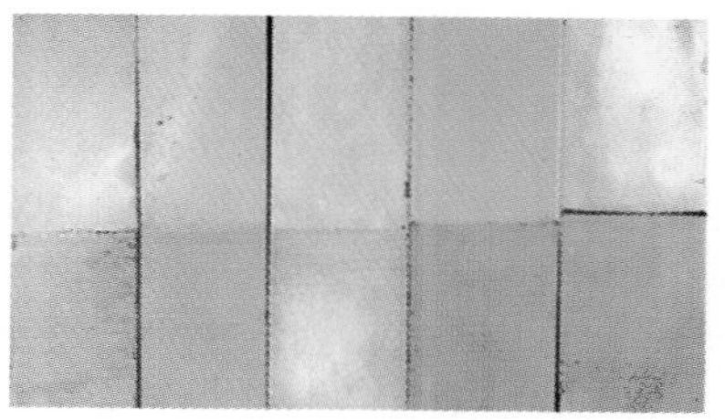

FIGURE 21.19 The color of commercial gold depends on its composition (left to right): 8-carat gold, 14-carat gold, white gold, 18-carat gold, and 24-carat gold. White gold consists of 6 parts Au and 18 parts Ag by mass.

FIGURE 21.20 The gold leaf coating these cathedral domes not only adds long-lasting beauty, it adds protection from corrosion.

CASE STUDY 21 PHOTOCHEMICAL MATERIALS

IN SOME PARTS of the world, glass automobile sunroofs can be made of *photochromic* glass, which darkens when exposed to light, then becomes colorless and transparent again when the light dims. One day the windows in our homes may also be made of photochromic glass, so the interior climate can be controlled automatically, keeping homes cool in bright sunlight, without reducing transparency in the evening.

Photochemical materials are materials that undergo reactions when exposed to light. Photochromic glass is only one of the newest photochemical materials. Practical applications of photochemistry have their roots in photographic film. In the nineteenth century the French artist Louis Daguerre plated silver onto sheets of copper, then treated them with iodine vapor. The iodine oxidized the silver:

$$2\,Ag(s) + I_2(g) \longrightarrow 2\,AgI(s)$$

Daguerre kept the coated "plates" in the dark, then used a lens to focus the image of a person or scene on it. Silver ions are rapidly reduced to silver metal in the presence of light, so everywhere light fell on the coating, a black deposit of finely divided silver formed. The image, called a "daguerrotype," was then visible, although it soon faded.

Modern photography works on essentially the same principle as that used by Daguerre. Black-and-white photographic film consists of a plastic sheet coated with an emulsion such as gelatin. These emulsions contain microscopic crystals of silver bromide, about 500 nm in diameter, called "grains," and small amounts of silver iodide.

The reaction that records the image is an example of a *photochemical reaction,* a reaction caused by light. When the emulsion is exposed to light, an electron is driven out of a bromide ion wherever the light falls. The liberated electron wanders through the grain and reduces a nearby silver cation. The resulting redox reaction,

$$Ag^+(s) + Br^-(s) \xrightarrow{\text{light}} Ag(s) + Br(s)$$

occurs only where light falls, creating small clusters of silver atoms within the grains.

The film is developed by reducing the silver ions remaining in the exposed grains, but not those in the unexposed grains. This reaction is carried out with a mild reducing agent, typically the organic compound hydroquinone, HOC_6H_4OH which has the structure

HO—⟨benzene ring⟩—OH

The developing process is completed by "fixing" the film in a solution that stops the reduction, so the image does not fade. Then the unreacted silver halide is dissolved in aqueous sodium thiosulfate ($Na_2S_2O_3{\cdot}5H_2O$), photographer's hypo (a shortening of its old name, sodium hyposulfite). The thiosulfate anions coordinate to the silver ions, forming the soluble complex ion $Ag(S_2O_3)_2^{3-}$, which is then rinsed away, leaving the metallic silver behind and removing the light sensitivity of the film. The result is a negative,

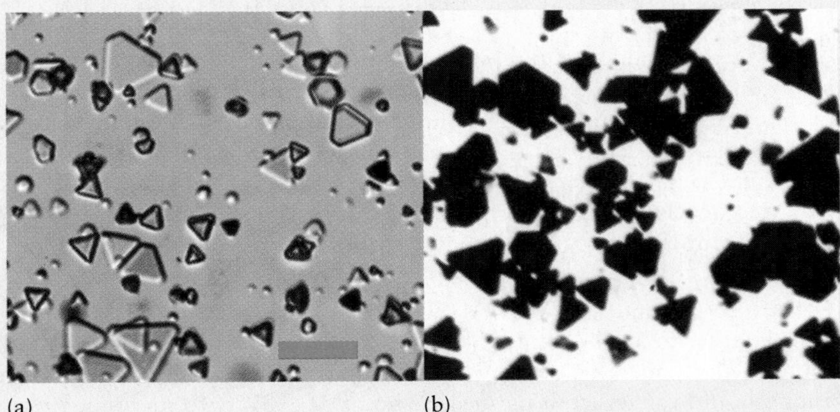

Micrographs of the silver halide crystals in a photographic emulsion before (a) and after (b) development. The scale bar represents 10 μm.

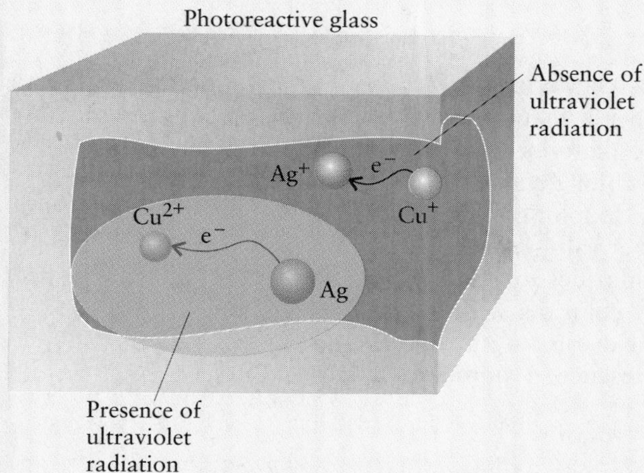

Light causes the reduction of Ag^+ to Ag metal, darkening photochromic glass. In the absence of radiation, Ag is oxidized by Cu^+ and the glass becomes colorless again.

a plastic film that is transparent where it was not exposed to light and dark where it was.

The negative is then placed between a light source and a lens to enlarge the image. Light-sensitive paper coated with AgCl is placed under the lens and exposed to the light traveling through the negative. The paper is then treated to stop further reaction, and the photograph is complete.

Color photographs are printed from negatives that are made of layers of silver halide emulsions. Each layer contains a dye that makes it sensitive to one of the three primary colors: red, green, or blue. The layers are developed separately into colored transparent negatives of the desired images. The colors in the negatives are the complementary colors of the colors in the original image.

Photochromic glass is made by mixing silver nitrate, copper(I) nitrate, and a metal halide into a borosilicate glass mixture and heating it until it melts, at about 1200°C. As the glass cools, small crystallites of the salts form. The crystallites (about 10 nm in diameter) are too small to scatter or absorb visible light, so the glass appears transparent. As in the photographic process, sunlight reduces the Ag^+ ions to silver metal:

$$Ag^+(s) + Cu^+(s) \xrightarrow{\text{light}} Ag(s) + Cu^{2+}(s)$$

However, in this case the reaction can be reversed, because the oxidized reducing agent, Cu^{2+}, cannot escape; and as soon as the light is removed, it oxidizes the silver again. Prescription sunglasses made of photochromic glass allow us to read in dim light, while also protecting our eyes from bright sunlight outdoors. Because photochromic glass darkens when the sun is shining but reverts to a clear and transparent state at dusk, it could reduce the amount of energy required for cooling homes and allow light intensity to be controlled in structures such as greenhouses.

QUESTIONS

1. Explain why the images on daguerrotypes fade, the colors of photochromic sunglasses can be reversed, and the images on photographic film are permanent.

2. (a) Write the net ionic equation for the reaction of aqueous sodium thiosulfate with solid silver bromide.
(b) Comment on the magnetic properties of the $[Ag(S_2O_3)_2]^{3-}$ complex.

3. When the price of silver rose rapidly in the 1980s, alternatives to silver were sought for use in photography. What other metals might have been reasonably considered?

4. Suggest a reason why dim red lights are used to illuminate photographic darkrooms.

Photochromic sunglasses. Here only one of the lenses has been exposed to ultraviolet light.

FIGURE 21.21 When ammonia is added to a silver chloride precipitate, the precipitate dissolves. However, when ammonia is added to a precipitate of mercury(I) chloride, mercury metal is formed by disproportionation and the mass turns gray. Left to right: silver chloride in water, silver chloride in aqueous ammonia, mercury(I) chloride in water, and mercury(I) chloride in aqueous ammonia.

Mercury lies above hydrogen in the electrochemical series; consequently, it is not oxidized by hydrogen ions. However, it does react with nitric acid:

$$3\,Hg(l) + 8\,H^+(aq) + 2\,NO_3^-(aq) \longrightarrow 3\,Hg^{2+}(aq) + 2\,NO(g) + 4\,H_2O(l)$$

In compounds, it has the oxidation number +1 or +2. Its compounds with oxidation number +1 are unusual in that the mercury(I) cation is the covalently bonded diatomic ion $(Hg{-}Hg)^{2+}$, written Hg_2^{2+}.

Both mercury(I) chloride and silver chloride are insoluble, and precipitate when a chloride solution is added to a solution containing a soluble mercury(I) or silver salt—a common qualitative test for the presence of the ions. The chlorides of the two metals can be distinguished, because, unlike silver chloride, which dissolves when ammonia is added, solid mercury(I) chloride disproportionates to metallic mercury and mercury(II) ions in the presence of ammonia:

$$Hg_2Cl_2(s) + 2\,NH_3(aq) \longrightarrow Hg(l) + HgNH_2Cl(s) + NH_4^+(aq) + Cl^-(aq)$$

Finely divided mercury metal appears black, so the white $HgNH_2Cl$ precipitate appears blackish gray (Fig. 21.21). The color is used to distinguish mercury ions from silver ions.

Mercury compounds, particularly its organic compounds, are acutely poisonous. The discharge of industrial waste containing mercury compounds into a shallow sea resulted in 52 deaths in Minimata, Japan, in 1952. In the sea, the mercury was converted into CH_3HgSCH_3, which was ingested by the fish that were a staple in the diet of the local residents. Mercury vapor is also an insidious poison because its effect is cumulative. Frequent exposure to low levels of mercury vapor can allow mercury to accumulate in the body. The effects include memory loss and other ailments.

Metals in Groups 11 and 12 are easily reduced from their compounds and have low reactivity as a result of poor shielding by the d-electrons. Copper is extracted from its ores either by pyrometallurgical or hydrometallurgical processes.

SELF-TEST 21.2A Use standard free energies of formation to calculate ΔG_r° and E° for the reaction by which solid copper is obtained by smelting CuS ore with oxygen. (ΔG_f° for CuS is $-49.0\ kJ{\cdot}mol^{-1}$.)

[*Answer:* $\Delta G_r^\circ = -251.2\ kJ{\cdot}mol^{-1}$, $E^\circ = +1.30\ V$]

SELF-TEST 21.2B Pig iron is much cheaper than steel but is not used in construction. Explain why pig iron is not suitable for erecting buildings and bridges.

COMPLEXES OF THE *d*-BLOCK ELEMENTS

The *d*-block elements are excellent Lewis acids (electron pair acceptors, Section 8.12) and form coordinate covalent bonds with molecules or ions that can act as Lewis bases (electron pair donors). A great deal of modern chemical research focuses on the structures, properties, and uses of the complexes formed in this way, partly because they participate in many biological reactions. Hemoglobin and vitamin B_{12}, for example, are both complexes—the former of iron and the latter of cobalt. Complexes of the *d*-metals are often brightly colored and magnetic, and are used in analysis, color science, and catalysis. They are also the target of current research in solar-energy conversion, in atmospheric nitrogen fixation, and in pharmaceuticals. Complexes are used in catalysis, as dyes, to dissolve ions (Section 15.15), in the electroplating of metals, and in photography (Case Study 21).

21.5 THE STRUCTURES OF COMPLEXES

A *d*-metal complex consists of a central metal atom or ion to which are attached a number of molecules or ions (typically, four or six) known as **ligands.** One example is the hexacyanoferrate(II) ion, $[Fe(CN)_6]^{4-}$, in which the Lewis acid Fe^{2+} forms bonds by sharing electron pairs provided by the CN^- ions. An example of a neutral complex is $Ni(CO)_4$, in which the Ni atom acts as the Lewis acid and the CO molecules act as Lewis bases.

The hexacyanoferrate(II) ion was formerly called the ferrocyanide ion, and you will still see that name used.

Each ligand in a complex has at least one lone pair of electrons with which it bonds to the central atom or ion by forming a coordinate-covalent bond. We say that the ligands **coordinate** to the metal when they form the complex. **Coordination compounds** are compounds with coordinate covalent bonds. They include electrically neutral complexes, such as $Ni(CO)_4$, and ionic compounds in which at least one of the ions is a complex, as in $K_4[Fe(CN)_6]$. The ligands directly attached to the central ion in a complex (conventionally enclosed within brackets) make up the **coordination sphere** of the central ion. The number of points where ligands are attached to the central metal atom is called the **coordination number** of the complex: the coordination number is 4 in $[Ni(CO)_4]$ and 6 in $[Fe(CN)_6]^{4-}$.

Because water is a Lewis base, it forms complexes with most *d*-block metals when they dissolve in it. Aqueous solutions of *d*-metals are usually solutions of their H_2O complexes. Many complexes are prepared simply by mixing aqueous solutions of a *d*-metal ion with the appropriate Lewis base (Fig. 21.22); for example,

$$Fe^{2+}(aq) + 6\,CN^-(aq) \longrightarrow [Fe(CN)_6]^{4-}(aq)$$

Because the aqueous Fe^{2+} ion is actually complexed with water as $[Fe(H_2O)_6]^{2+}$, this reaction is an example of a **substitution reaction,** in which one Lewis base takes the place of another: the CN^- ions drive out H_2O molecules from the $[Fe(H_2O)_6]^{2+}$ complex. A less complete replacement occurs when certain other ions, such as Cl^-, are added to an iron(II) solution:

FIGURE 21.22 When cyanide ions (in the form of potassium cyanide) are added to an aqueous solution of iron(II) sulfate, they replace the H_2O ligands from the $[Fe(H_2O)_6]^{2+}$ complex and produce a new complex, the more strongly colored hexacyanoferrate(II) ion, $[Fe(CN)_6]^{4-}$.

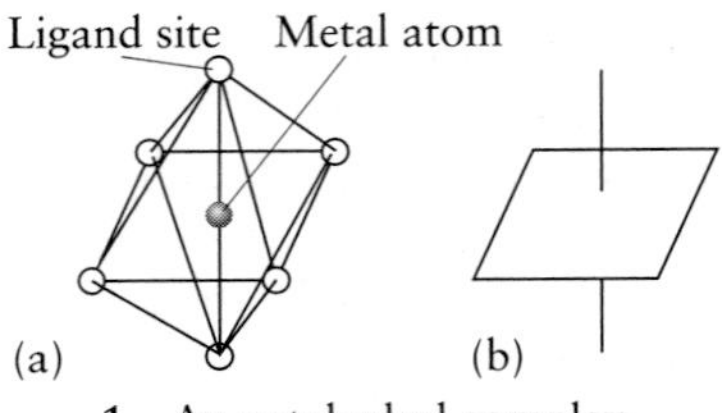

1 An octahedral complex

FIGURE 21.23 Some of the highly colored compounds that result when complexes are formed (left to right): aqueous solutions of the complexes $[Fe(SCN)(H_2O)_5]^{2+}$, $[Co(SCN)_4(H_2O)_2]^{2-}$, $[Cu(NH_3)_4(H_2O)_2]^{2+}$, and $[CuBr_4]^{2-}$.

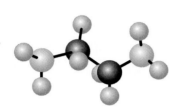

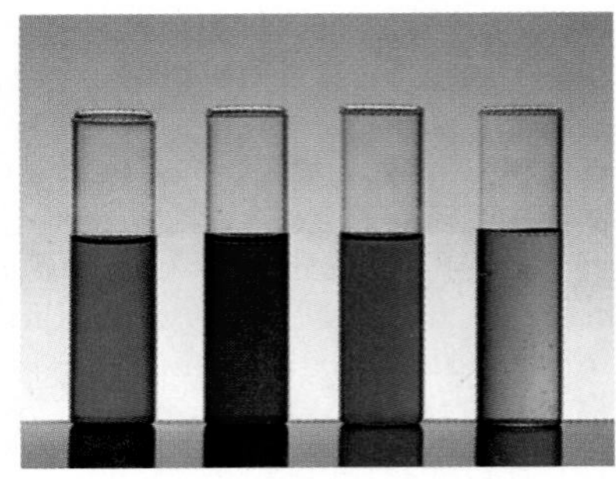

2 Ethylenediamine, $NH_2CH_2CH_2NH_2$

$$[Fe(H_2O)_6]^{2+}(aq) + Cl^-(aq) \longrightarrow [FeCl(H_2O)_5]^+(aq) + H_2O(l)$$

The color of a *d*-metal complex depends on the identity of the ligands as well as that of the metal, and impressive changes of color often accompany substitution reactions (Fig. 21.23).

In the great majority of cases, complexes with a coordination number of 6, such as $[Fe(CN)_6]^{4-}$, have their ligands at the corners of an octahedron, with the metal ion at the center, and are called **octahedral complexes** (Fig. 21.24). We can represent the structures of these octahedral complexes by a simplified diagram that emphasizes the geometry of the bonds (**1**). In complexes with a coordination number of 4, the ligands are found either at the corners of a tetrahedron in a **tetrahedral complex,** as in $[Cu(NH_3)_4]^{2+}$, or (most notably for d^8 electron configurations, such as Pt^{2+} and Au^{3+}) at the corners of a square, in **square-planar complexes.**

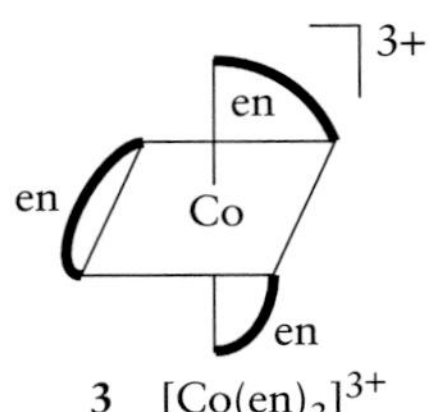

3 $[Co(en)_3]^{3+}$

Some ligands can simultaneously occupy more than one binding site. Ethylenediamine, $NH_2CH_2CH_2NH_2$ (**2**), has a nitrogen lone pair at each end. This ligand is widely used in coordination chemistry and is abbreviated to en, as in $[Co(en)_3]^{3+}$ (**3**). The metal atom in $[Co(en)_3]^{3+}$ lies at the center of the three ligands as though pinched by three molecular claws. It is an example of a **chelate,** a complex containing one or more ligands that form a ring of atoms that includes the central metal atom. Another example is the ethylenediaminetetraacetate ion, $EDTA^{4-}$ (**4**). This ligand forms complexes with many metal ions, including Pb^{2+}, and hence is used as an antidote to lead poisoning. However, an unfortunate side effect of administering EDTA to humans is that it also forms chelates with desirable metals, removing ions such as Ca^{2+} and Fe^{2+} from the body along with the lead.

The name *chelate* comes from the Greek word for "claw."

Chelating ligands are quite common in nature. Mosses and lichens secrete chelating ligands to capture essential metal ions from the rocks they dwell on. Chelate formation also lies behind the body's strategy of producing a fever when infected by bacteria. The higher temperature kills bacteria by reducing their ability to synthesize a particular iron-chelating ligand.

Table 21.5 gives the names of some common ligands and their abbreviations. The rules for building up the names of complexes are listed in Appendix 3C.

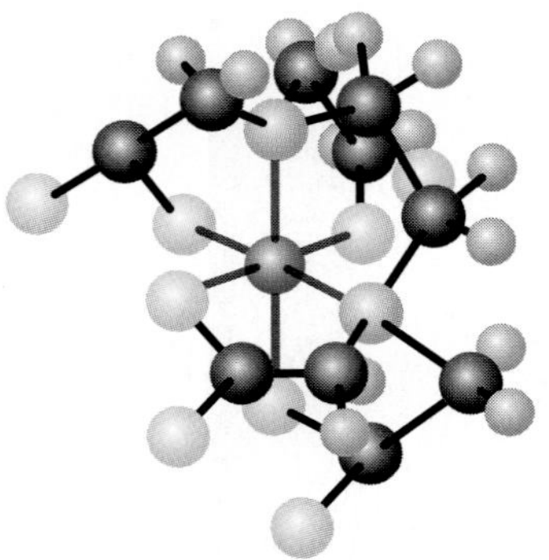

4 EDTA complex

A complex is formed between a Lewis acid (the metal atom or ion) and a number of Lewis bases (the ligands).

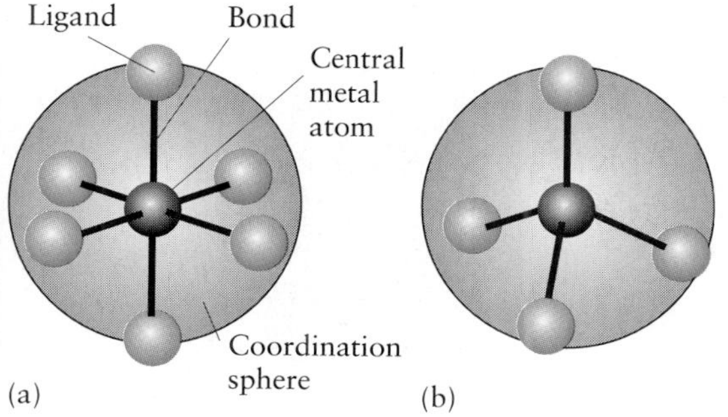

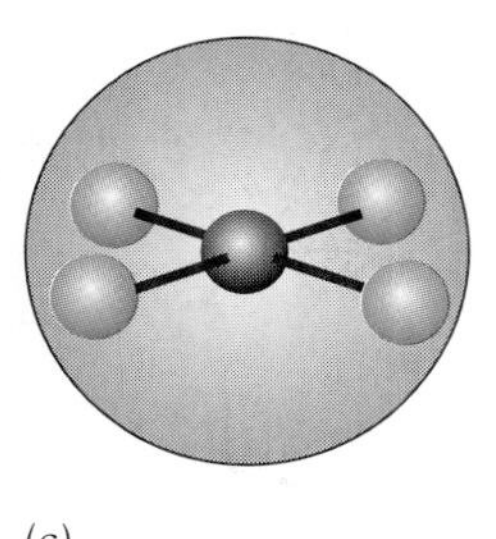

(c)

FIGURE 21.24 The various components of a complex can be either neutral or charged. (a) Almost all six-coordinate complexes are octahedral. Four-coordinate complexes are either (b) tetrahedral or (c) square planar.

TABLE 21.5 Common ligands

Formula	Name
NEUTRAL LIGANDS	
H_2O	aqua
NH_3	ammine
NO	nitrosyl
CO	carbonyl
$NH_2CH_2CH_2NH_2$	ethylenediamine (en)*
$NH_2CH_2CH_2NHCH_2CH_2NH_2$	diethylenetriamine (dien)†
ANIONIC LIGANDS	
F^-	fluoro
Cl^-	chloro
Br^-	bromo
I^-	iodo
OH^-	hydroxo
O^{2-}	oxo
CN^-	cyano (as M—CN)
NC^-	isocyano (as M—NC)
SCN^-	thiocyanato (as M—SCN)
NCS^-	isothiocyanato (as M—NCS)
NO_2^-	nitrito (as M—ONO)
NO_2^-	nitro (as M—NO_2)
CO_3^{2-}	carbonato
$C_2O_4^{2-}$	oxalato (ox)*
$(^-O(O=)C{-}CH_2)_2N{-}CH_2{-}CH_2{-}N(CH_2{-}C(=O)O^-)_2$	ethylenediaminetetraacetato (EDTA)‡
SO_4^{2-}	sulfato

*Bidentate (attaches to two sites).
†Tridentate (attaches to three sites).
‡Hexadentate (attaches to six sites).

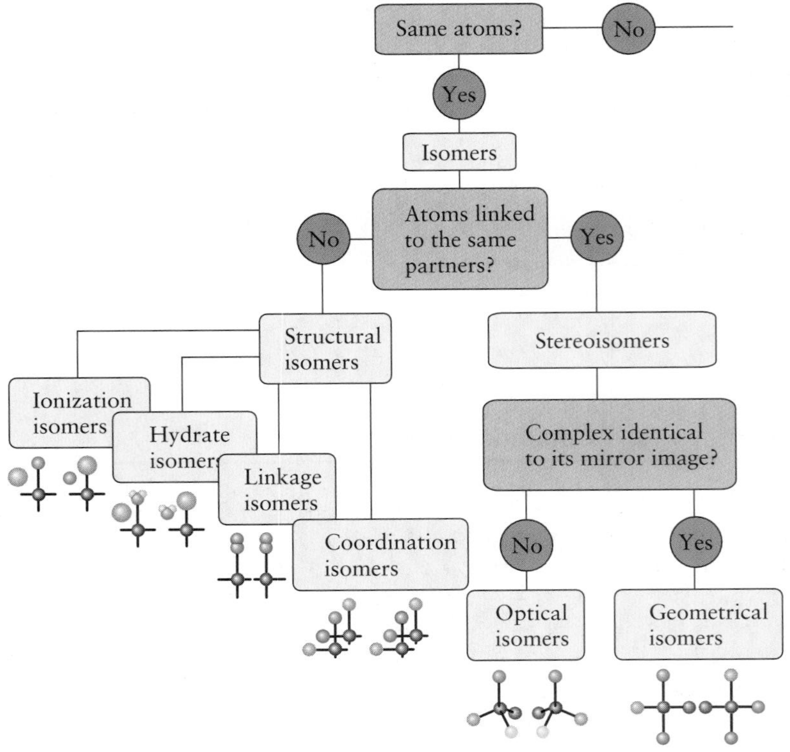

FIGURE 21.25 The various types of isomerism in coordination compounds.

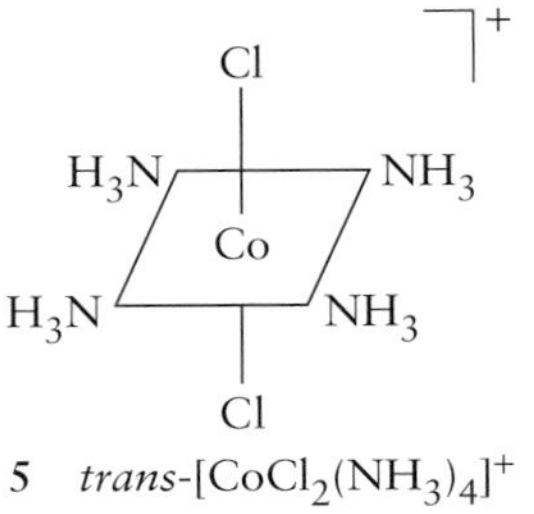

5 *trans*-$[CoCl_2(NH_3)_4]^+$

6 *cis*-$[CoCl_2(NH_3)_4]^+$

21.6 ISOMERS

Many complexes and coordination compounds exist as isomers. Recall from Section 11.11 that isomers are compounds that contain the same numbers of the same atoms, but in different arrangements. For example, the ions shown in (5) and (6) differ only in the positions of the Cl^- ligands, but they are distinct species, for they have different physical and chemical properties. Both structural isomers and stereoisomers are found in coordination compounds. The structural isomers can be classified according to four different types: ionization, hydrate, linkage, and coordination (Fig. 21.25).

Ionization isomers differ by the exchange of a ligand with an anion or neutral molecule outside the coordination sphere. For instance, $[CoBr(NH_3)_5]SO_4$ and $[CoSO_4(NH_3)_5]Br$ are ionization isomers because the Br^- ion is a ligand of the cobalt in the former but an accompanying anion in the latter. The two compounds have different physical properties: the bromo complex is violet and the sulfato complex is red (Fig. 21.26). The chemical properties of the two ionization isomers are also different: the red isomer forms an off-white precipitate of AgBr when Ag^+ ions are added but no precipitate forms after the addition of Ba^{2+}:

$$[CoSO_4(NH_3)_5]Br(aq, red) + Ag^+(aq) \longrightarrow AgBr(s) + [CoSO_4(NH_3)_5]^+(aq)$$

$$[CoSO_4(NH_3)_5]Br(aq, red) + Ba^{2+}(aq) \longrightarrow \text{no reaction}$$

On the other hand, the violet isomer is unaffected by the addition of low concentrations of Ag^+ ions, but it forms a white precipitate of $BaSO_4$

FIGURE 21.26 Solutions of the two coordination compounds $[CoBr(NH_3)_5]SO_4$ (left) and $[CoSO_4(NH_3)_5]Br$ (right). Although the isomers are built from the same atoms in the same proportions, they are different compounds, with their own characteristic physical and chemical properties.

after the addition of Ba^{2+} ions:

$$[CoBr(NH_3)_5]SO_4(aq, violet) + Ag^+(aq) \longrightarrow \text{no reaction}$$

$$[CoBr(NH_3)_5]SO_4(aq, violet) + Ba^{2+}(aq) \longrightarrow BaSO_4(s) + [CoBr(NH_3)_5]^{2+}(aq)$$

These observations show that ions bound in a coordination sphere are not free to react. When the red complex dissolves in water, bromide ions become free to move through the solution (Fig. 21.27a), and when they encounter a silver ion, they form a precipitate. Conversely, when the violet complex dissolves in water, the bromide ion remains attached to the cobalt ion and is not free to react with silver ions (Fig. 21.27b). However, the sulfate ion is not a part of the coordination sphere and is free to react.

Hydrate isomers differ by an exchange between an H_2O molecule and another ligand in the coordination sphere (Fig. 21.28). For example, the solid hexahydrate of chromium(III) chloride, $CrCl_3 \cdot 6H_2O$, may be any of the three compounds

$[Cr(H_2O)_6]Cl_3$	violet
$[CrCl(H_2O)_5]Cl_2 \cdot H_2O$	blue-green
$[CrCl_2(H_2O)_4]Cl \cdot 2H_2O$	green

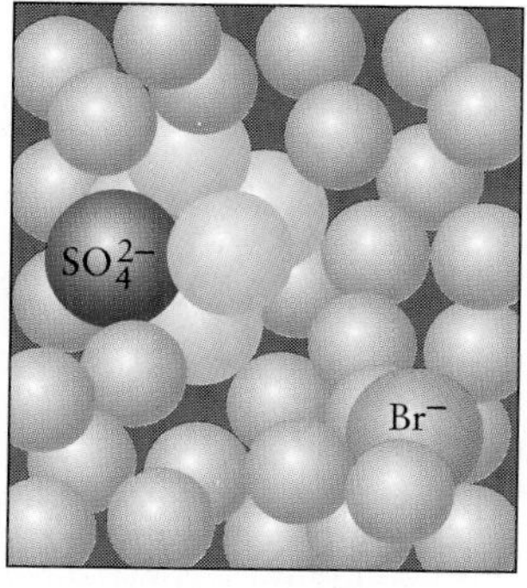

(a)

(b)

FIGURE 21.27 (a) When this red coordination complex dissolves, the bromide ion (the orange sphere) is not part of the coordination sphere; it is free to move and is available for reaction. (b) In contrast, when this violet coordination complex dissolves, the bromide ion remains attached to the cobalt ion in the complex and is not available for reaction. In this case, the sulfate ion (purple sphere) is free to react.

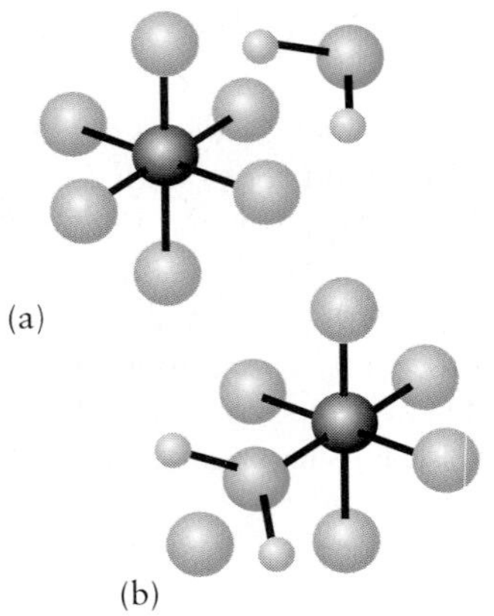

FIGURE 21.28 Hydrate isomers. In (a), the water molecule is simply part of the surrounding solvent; in (b), the water molecule is present in the coordination sphere and an ion (or neutral species) is now present in the solution.

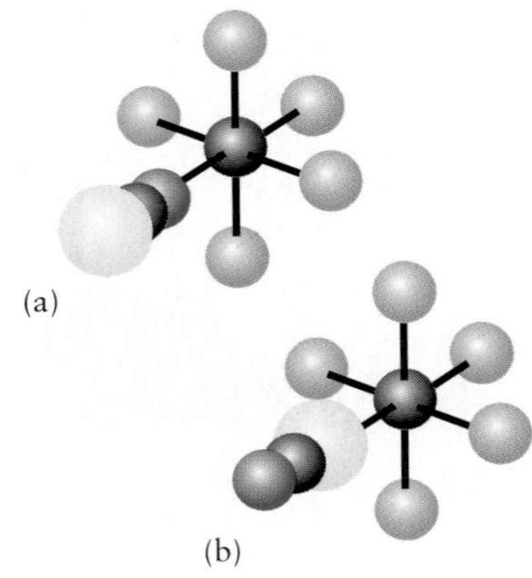

FIGURE 21.29 Linkage isomers. In (a) the ligand (here NCS^-) is attached through its N atom, but in (b) it is attached through its S atom. The different arrangements of the ligand give the two complexes different chemical and physical properties.

The addition of $AgNO_3$ to solutions of the compounds results in the precipitation of different amounts of AgCl from each of them, because only the chloride ions outside the coordination sphere will react with the silver ions.

Hydrate isomers can be considered special cases of ionization isomers.

Linkage isomers differ in the identity of the atom that a given ligand uses to attach to the metal ion (Fig. 21.29). A ligand may have more than one atom with a lone pair that can bond to the metal ion, but because of its size or shape, only one atom can bond to the metal ion at a time. Common ligands that show linkage isomerism are SCN^- versus NCS^-, NO_2^- versus ONO^-, and CN^- versus NC^-, where the coordinating atom is written first in each pair. For example, NO_2^- can form the yellow nitro complex $[CoCl(NO_2)(NH_3)_4]^+$ or the red nitrito complex $[CoCl(ONO)(NH_3)_4]^+$.

Coordination isomers occur when one or more ligands are exchanged between a cationic complex and an anionic complex (Fig. 21.30). An

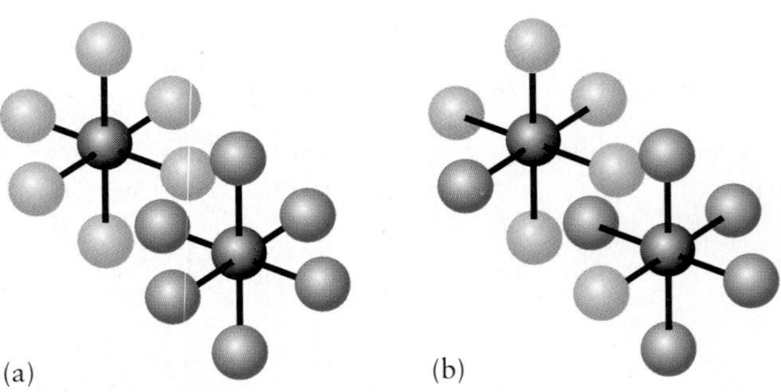

FIGURE 21.30 Coordination isomers. In this compound, ligands have been exchanged between the cation and anion complexes.

example of a pair of coordination isomers is $[Cr(NH_3)_6][Fe(CN)_6]$ and $[Fe(NH_3)_6][Cr(CN)_6]$.

SELF-TEST 21.3A Identify the type of isomers represented by the following pairs: (a) $[Cu(NH_3)_4][PtCl_4]$ and $[Pt(NH_3)_4][CuCl_4]$; (b) $[Cr(OH)_2(NH_3)_4]Br$ and $[CrBr(OH)(NH_3)_4]OH$.

[*Answer:* (a) Coordination; (b) ionization]

SELF-TEST 21.3B Identify the type of isomers represented by the following pairs: (a) $[Co(NCS)(NH_3)_5]Cl_2$ and $[Co(SCN)(NH_3)_5]Cl_2$; (b) $[CrCl(H_2O)_5]Cl_2 \cdot H_2O$ and $[CrCl_2(H_2O)_4]Cl \cdot 2H_2O$.

Although they are built from the same numbers and kinds of atoms, structural isomers have different chemical formulas, because the formulas show how the atoms are grouped in or outside the coordination sphere. Stereoisomers have the same formulas because their atoms have the same partners in the coordination spheres but the arrangements of the ligands in space differ from one isomer to the other. The two complexes of formula $[CoCl_2(NH_3)_4]^{2+}$ shown in (**5**) and (**6**) are geometrical isomers because the coordination spheres differ only in the way the ligands are arranged in space. As in carbon compounds, the isomer with the Cl^- ligands on opposite sides of the central atom is called the trans isomer and the isomer with the ligands on the same side is called the cis isomer. Geometrical isomers exist for square-planar and octahedral complexes, but they do not exist for tetrahedral complexes.

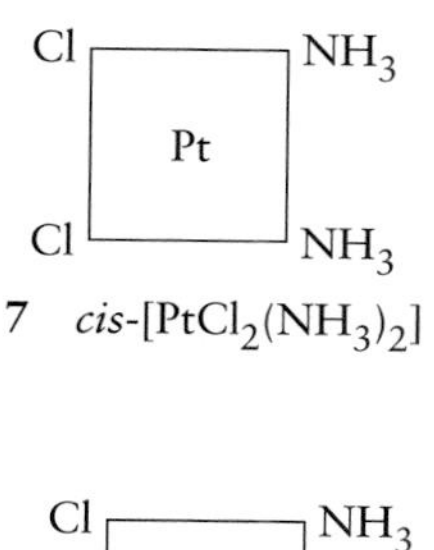

7 *cis*-$[PtCl_2(NH_3)_2]$

The chemical and physiological properties of geometrical isomers can differ greatly. For example, *cis*-$[PtCl_2(NH_3)_2]$ (**7**) is pale orange-yellow, has a solubility of 0.252 g per 100 g of water, and is used for chemotherapy treatment of cancer patients. Conversely, *trans*-$[PtCl_2(NH_3)_2]$ (**8**) is dark yellow, has a solubility of 0.037 g per 100 g of water, and exhibits no chemotherapeutic effect. Antitumor activity is also found in other cis isomers of platinum(II), such as *cis*-$[PtCl_2(en)]$, *cis*-$[Pt(ox)(NH_3)_2]$, and *cis*-$[Pd(NO_3)_2(C_6H_4(NH_2)_2)]$. An octahedral complex that exhibits geometrical isomerism is $[Fe(CN)_4(NH_3)_2]^-$ (**9**) and (**10**).

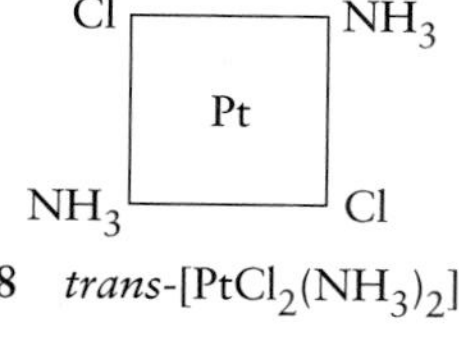

8 *trans*-$[PtCl_2(NH_3)_2]$

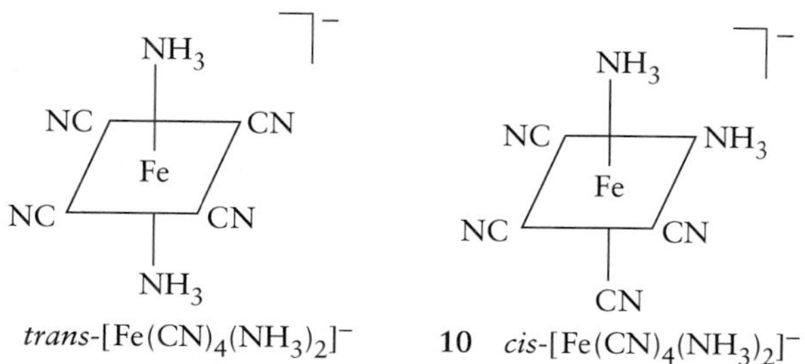

9 *trans*-$[Fe(CN)_4(NH_3)_2]^-$ **10** *cis*-$[Fe(CN)_4(NH_3)_2]^-$

Complexes also display optical isomerism (Section 11.12). An example is shown in Fig. 21.31. Both geometrical and optical isomerism can

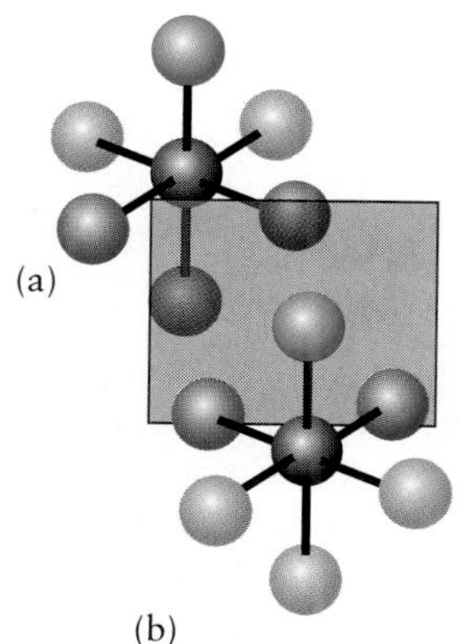

FIGURE 21.31 Optical isomers. The two complexes are each other's mirror image, and no matter how we rotate them, one complex cannot be superimposed on the other.

occur in an octahedral complex, as in $[CoCl_2(en)_2]^+$: the trans isomer is green (**11**) and the two alternative cis isomers (**12**) and (**13**), which are optical isomers of one another, are violet.

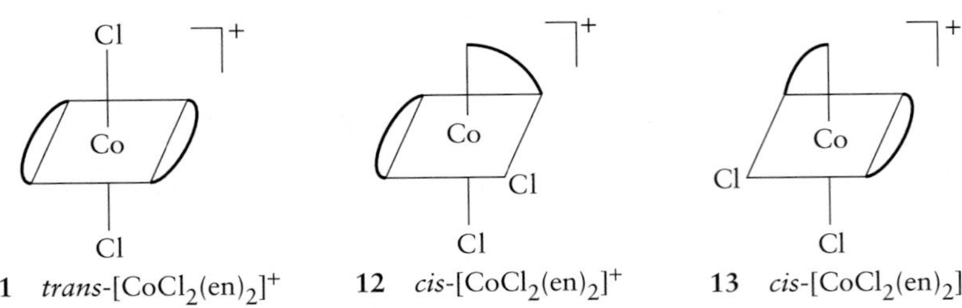

11 *trans*-$[CoCl_2(en)_2]^+$ **12** *cis*-$[CoCl_2(en)_2]^+$ **13** *cis*-$[CoCl_2(en)_2]^+$

As we saw in Section 11.12, optical isomers are mirror images of each other, but are not superimposable and differ in the direction in which they rotate polarized light.

The interaction of optical isomers with polarized light is described in Section 11.12.

The possibility of isomerism increases the number of compounds that can be produced by the same set of atoms; the varieties of isomerism are summarized in Fig. 21.25. Optical isomers rotate the plane of polarization of light in opposite directions.

EXAMPLE 21.1 Identifying optical isomerism

Which of the complexes in Fig. 21.32 is chiral, and which form enantiomeric pairs?

STRATEGY Draw the mirror image of each complex and mentally rotate the original complex; judge whether any rotation will cause the original molecule to match its mirror image. If not, then the complex is chiral. Determine which complexes form enantiomeric pairs by finding pairs that are the nonsuperimposable mirror images of each other.

SOLUTION In Fig. 21.33, the original complexes are on the left of each pair, and the mirror image of each complex is on the right. (a, b) Neither is chiral. (a) If we rotate the mirror image about A—A, we obtain a structure identical to the original. (b) The mirror image is identical to the original. (c, d) The original complex is chiral because no rotation can make either match its mirror image.

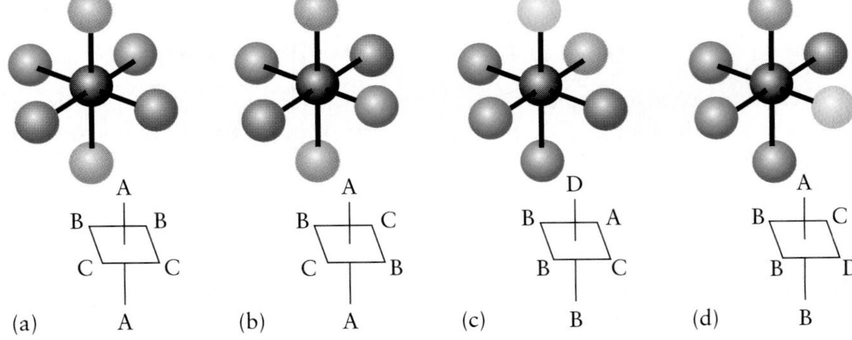

FIGURE 21.32 The complexes referred to in Example 21.1. Note that we also display the abbreviated form of their structures.

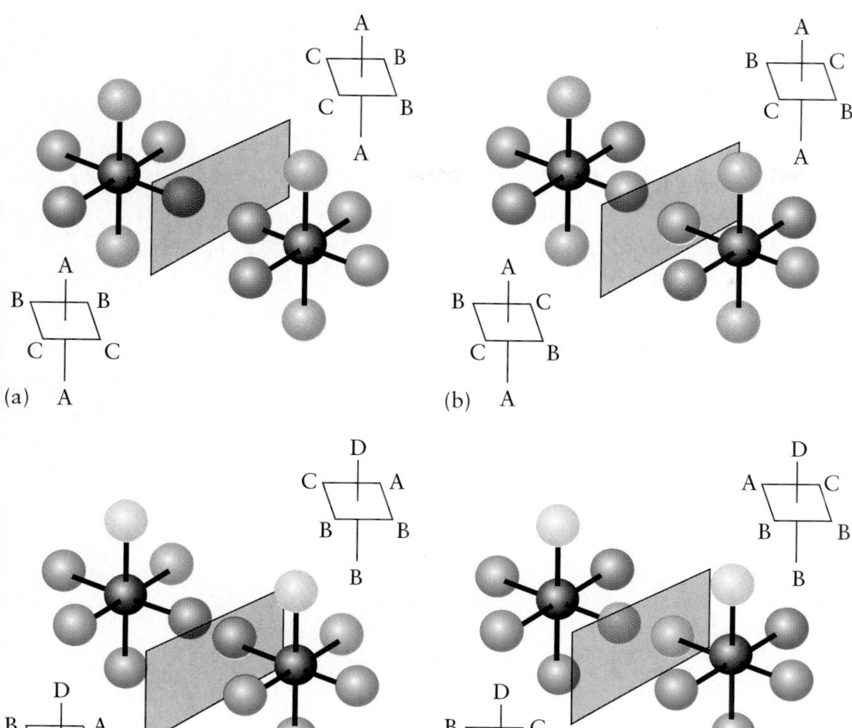

FIGURE 21.33 The reflections referred to in the solution of Example 21.1.

However, when the mirror image of (c) is rotated by 90° around the vertical B—D axis, it becomes the complex (d); hence (c) and (d) form an enantiomeric pair.

SELF-TEST 21.4A Repeat the exercise for the complexes shown in Fig. 21.34.

[*Answer:* (a) Not chiral; (c) chiral; (b,d) chiral and enantiomeric]

SELF-TEST 21.4B Repeat the exercise for the complexes shown in Fig. 21.35.

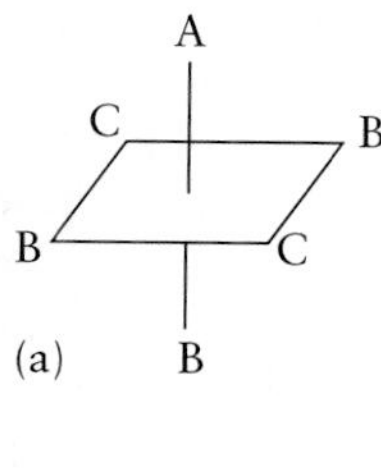

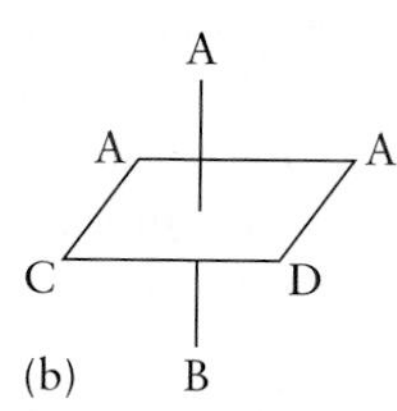

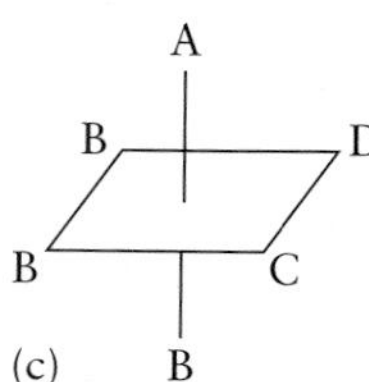

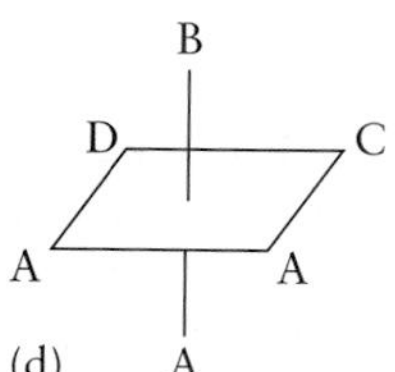

FIGURE 21.34 (above) The complexes referred to in Self-Test 21.4A.

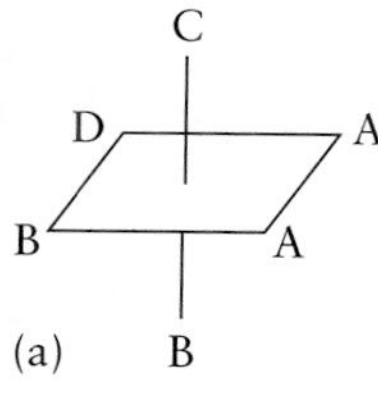

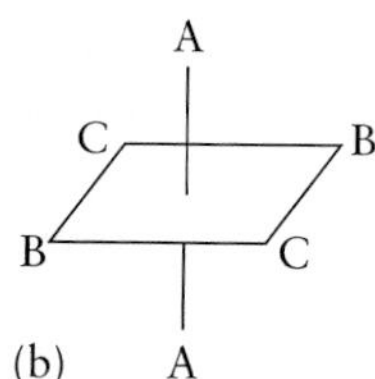

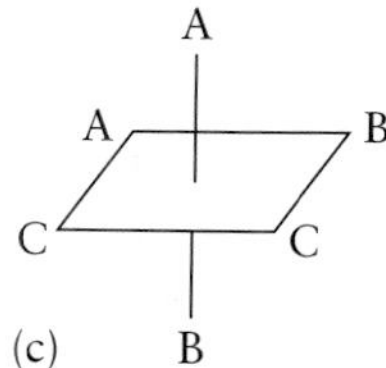

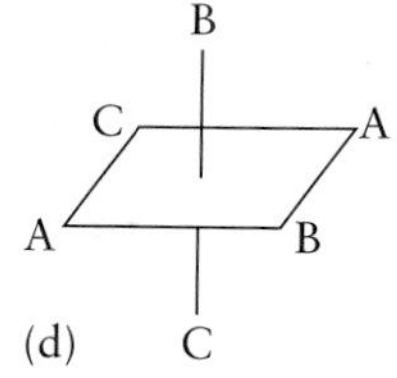

FIGURE 21.35 (below) The complexes referred to in Self-Test 21.4B.

CHEMISTRY AT WORK

A medical technologist investigates the effects on living cells of a promising new treatment for cancer. The treatment involves the administration of a coordination complex, of which only one isomer is active in treating cancer.

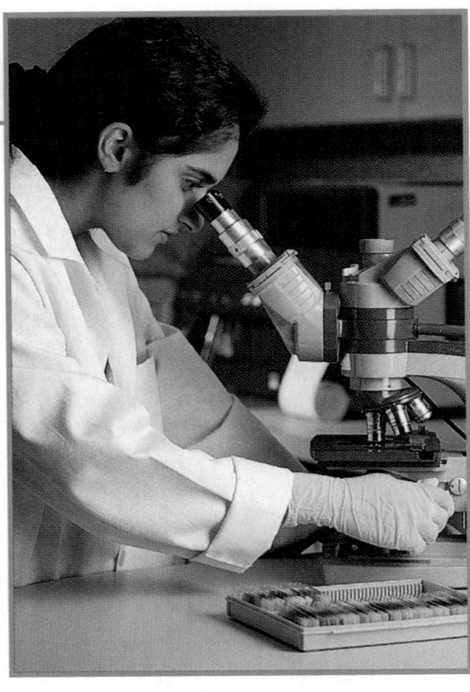

CRYSTAL FIELD THEORY

Many coordination compounds are colored and many are paramagnetic. These properties can be discussed in terms of **crystal field theory.** This description of bonding in complexes was originally devised to explain the colors of solids, particularly ruby, which owes its color to Cr^{3+} ions.

A more complete version of crystal field theory is called *ligand field theory.*

21.7 THE EFFECTS OF LIGANDS ON *d*-ELECTRONS

In crystal field theory, each ligand is represented by a point charge: these negative charges represent the ligand lone pair directed toward the central metal atom (Fig. 21.36a). The electronic structure of the complex is then expressed in terms of the electrostatic interactions between these point charges and the electrons of the central metal ion (Fig. 21.36b). We begin by considering a complex with a single *d*-electron, such as $[Ti(H_2O)_6]^{3+}$, in which the electron configuration of Ti^{3+} is $[Ar]3d^1$.

Because the metal atom at the center of a complex is usually positively charged, the negative charges representing the ligands are attracted to it. This attraction results in the formation of the complex. The $[Ti(H_2O)_6]^{3+}$ ion, for instance, is a stable complex because there is a strong attraction between the Ti^{3+} ion and the six negative point charges representing a lone pair on each of the six H_2O ligands.

However, the single 3*d*-electron of the Ti^{3+} ion interacts differently with the point charges, depending on which of the five 3*d*-orbitals it occupies. In an octahedral complex such as $[Ti(H_2O)_6]^{3+}$, the six ligands

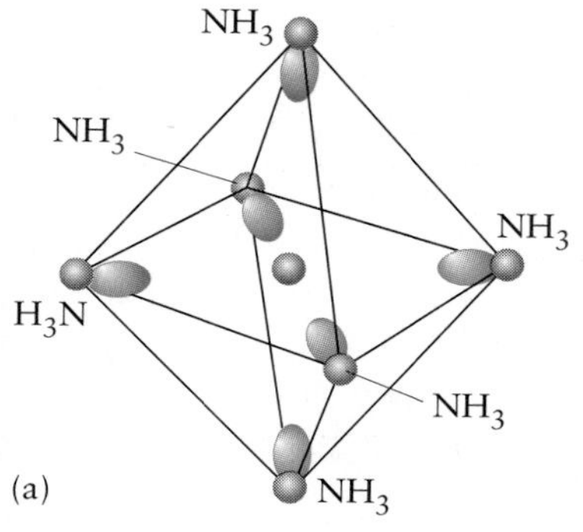

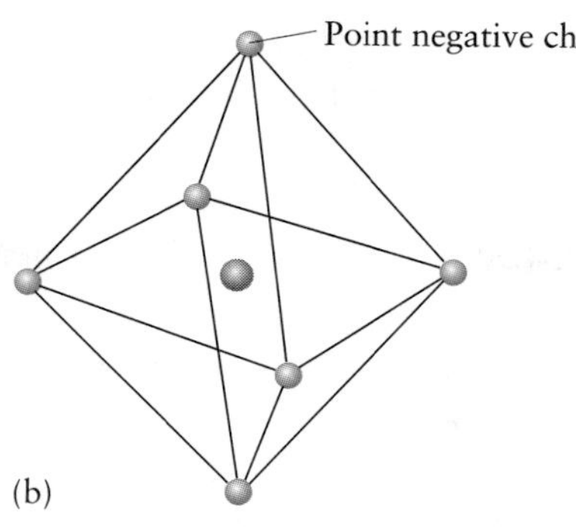

FIGURE 21.36 In the crystal field theory of complexes, the lone pairs of electrons (the Lewis base sites) on the ligands (a) are treated as equivalent to point negative charges (b).

(represented by point charges) lie on opposite sides of the central metal ion along the x-, y-, and z-axes. From the drawings of the d-orbitals in Fig. 21.37, we can see that three of the orbitals (d_{xy}, d_{yz}, and d_{zx}) have their lobes directed between the point charges. These three d-orbitals are called ***t*-orbitals** in crystal field theory. The other two d-orbitals (d_{z^2} and $d_{x^2-y^2}$) are directed straight toward the point charges. These two orbitals are called ***e*-orbitals.** Because of their different arrangement in space, electrons in t-orbitals are repelled less by the negative point charges of the ligands than electrons in e-orbitals are; therefore, the t-orbitals are of lower energy than the e-orbitals. The difference in the energies of the t-orbitals and the e-orbitals is typically about 10% of the total interaction energy between the central ion and its ligands.

These orbitals are normally given a more elaborate designation: t_{2g} and e_g.

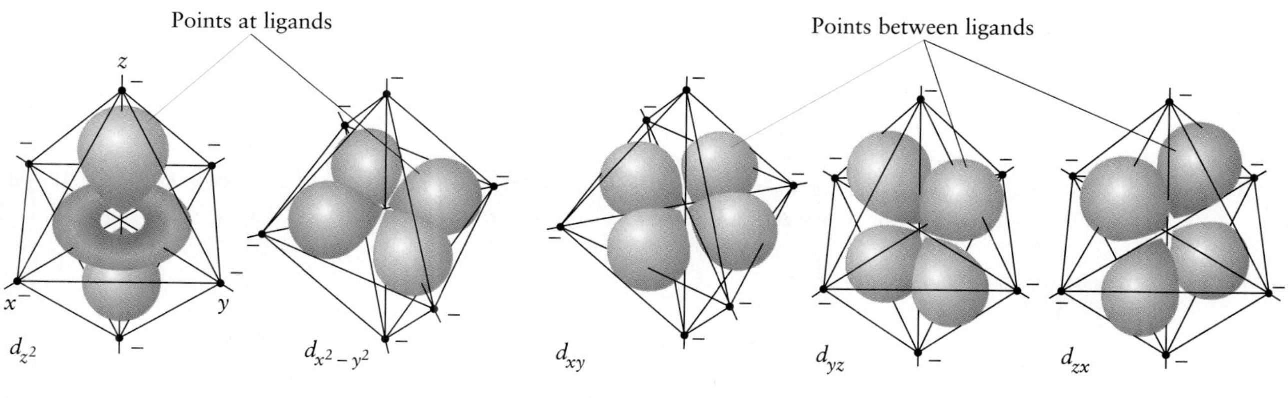

FIGURE 21.37 In an octahedral complex with a central d-metal atom or ion, (a) a d_{z^2}-orbital points directly toward two ligands, and an electron that occupies it has a relatively high energy. The same rise in energy occurs for a $d_{x^2-y^2}$-electron. (b) A d_{xy}-orbital is directed between the ligand sites, and an electron that occupies it has a relatively low energy. The same lowering of energy occurs for d_{yz}- and d_{zx}-orbitals.

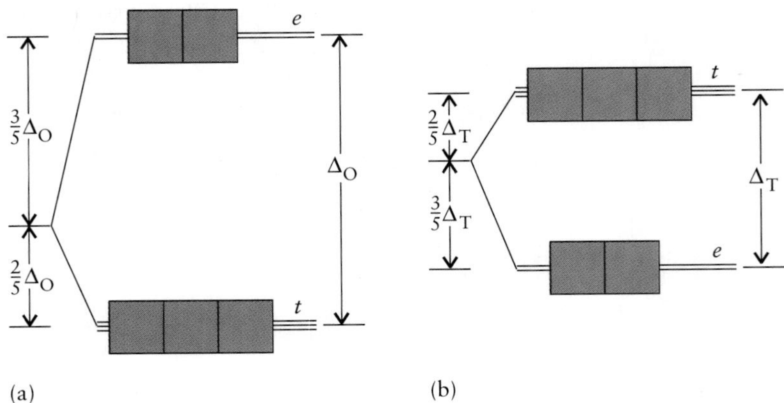

FIGURE 21.38 The energy levels of the *d*-orbitals in (a) an octahedral complex, with the ligand field splitting Δ_O, and (b) in a tetrahedral complex, with the ligand field splitting Δ_T. Each box (that is, orbital) can hold two electrons.

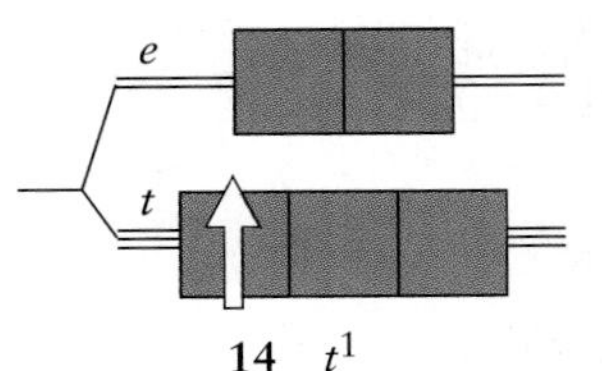

14 t^1

The energy-level diagram in Fig. 21.38a helps to summarize these ideas. The energy separation between the two sets of orbitals is called the **ligand field splitting,** Δ_O (the O denotes octahedral). The three *t*-orbitals lie at an energy that is $\frac{2}{5}\Delta_O$ below the average *d*-orbital energy, and the two *e*-orbitals lie at an energy $\frac{3}{5}\Delta_O$ above the average. Because the *t*-orbitals have the lower energy, we can predict that in the ground state of the $[Ti(H_2O)_6]^{3+}$ complex, the electron occupies one of them in preference to an *e*-orbital and hence that the electron configuration of the complex is t^1. This configuration can be represented by the box diagram in (**14**).

In a tetrahedral complex, the three *t*-orbitals point more directly at the ligands than the two *e*-orbitals do. As a result, in a tetrahedral complex the *t*-orbitals have a *higher* energy than the *e*-orbitals (Fig. 21.38b). The ligand field splitting Δ_T, (where the T denotes tetrahedral) is generally smaller than in octahedral complexes (typically, $\Delta_T \approx \frac{4}{9}\Delta_O$) because the *d*-orbitals do not point so directly at the ligands and there are fewer repelling ligands.

In octahedral complexes, the e-orbitals (d_{z^2} and $d_{x^2-y^2}$) lie higher in energy than the t-orbitals (d_{xy}, d_{yz}, and d_{zx}).

21.8 THE EFFECTS OF LIGANDS ON COLOR

The *t*-electron of the octahedral $[Ti(H_2O)_6]^{3+}$ complex can be excited into one of the *e*-orbitals if it absorbs a photon of energy equal to Δ_O (Fig. 21.39). The wavelength of radiation absorbed by a complex can there-

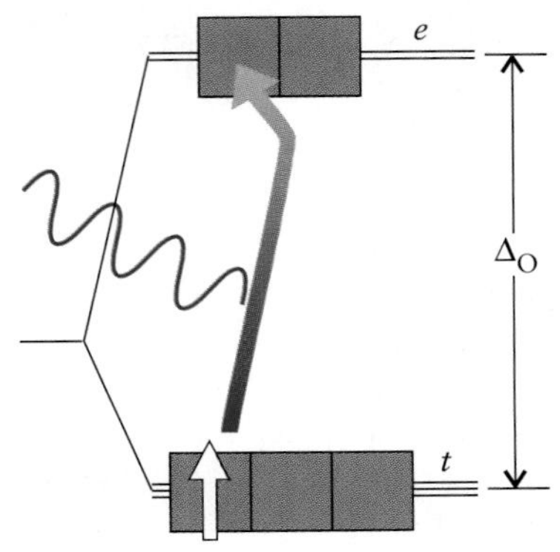

FIGURE 21.39 When a complex is exposed to light of the correct frequency, an electron can be excited to a higher energy orbital (from *t* to *e* if the complex is octahedral), and the light is absorbed.

fore be used to determine the ligand field splitting, as shown in the following example.

EXAMPLE 21.2 Determining the ligand field splitting

The complex $[Ti(H_2O)_6]^{3+}$ absorbs light of wavelength 510 nm. What is the ligand field splitting in the complex?

STRATEGY Because a photon carries an energy $h\nu$, where h is the Planck constant and ν is the frequency of the radiation, it can be absorbed if its frequency satisfies

$$\Delta_O = h\nu$$

The wavelength, λ, of light is related to the frequency by $\lambda = c/\nu$, where c is the speed of light (Section 7.1). Therefore, the wavelength of light absorbed and the ligand field splitting are related by

$$\Delta_O = \frac{hc}{\lambda} \quad \text{alternatively:} \quad \lambda = \frac{hc}{\Delta_O}$$

That is, the greater the splitting, the shorter the wavelength of the light that is absorbed by the complex. At this point, substitute the data.

SOLUTION Because the wavelength absorbed is 510 nm, it follows that the ligand field splitting is

$$\Delta_O = \frac{(6.626 \times 10^{-34}\ \text{J·s}) \times (2.998 \times 10^{8}\ \text{m·s}^{-1})}{510 \times 10^{-9}\ \text{m}} = 3.90 \times 10^{-19}\ \text{J}$$

Multiplication by the Avogadro constant (to turn the energy into a molar quantity) gives

$$\Delta_O = (6.022 \times 10^{23}\ \text{mol}^{-1}) \times (3.90 \times 10^{-19}\ \text{J}) = 2.35 \times 10^{5}\ \text{J·mol}^{-1}$$

or 235 kJ·mol^{-1}.

SELF-TEST 21.5A The complex $[Fe(H_2O)_6]^{3+}$ absorbs light of wavelength 700 nm (3 sf) What is the value (in kilojoules per mole) of the ligand field splitting?

[*Answer:* 171 kJ·mol^{-1}]

SELF-TEST 21.5B The complex $[Fe(CN)_6]^{4-}$ absorbs light of wavelength 305 nm. What is the value (in kilojoules per mole) of the ligand field splitting?

Ligands can be arranged in a **spectrochemical series** according to the magnitude of the Δ_O they produce, as shown in Fig. 21.40. Those below the horizontal line are called **weak-field ligands,** and those above it are called **strong-field ligands.**

A complex of a given metal atom has a smaller Δ_O value if it contains weak-field ligands than if it contains strong-field ligands. Less energy is needed to promote the electron across a small Δ_O than a large one, hence the complex absorbs longer wavelength light if it has weak-field ligands than if it has strong-field ligands. Because a CN^- ligand is a stronger ligand than H_2O, the ligand field splitting is greater for $[Fe(CN)_6]^{4-}$ than for $[Fe(H_2O)_6]^{2+}$. Hence, a $[Fe(CN)_6]^{4-}$ complex absorbs shorter wavelength radiation.

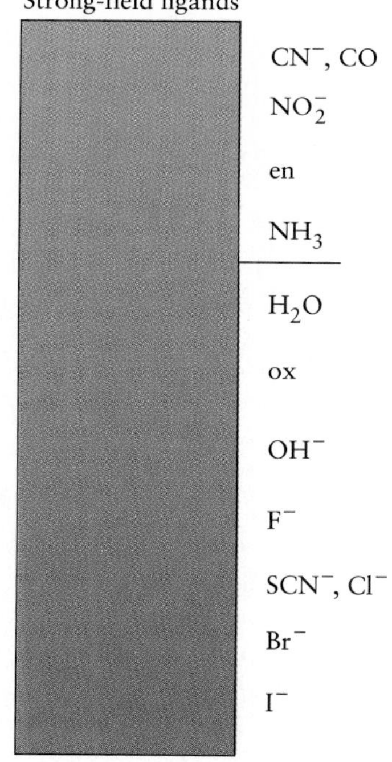

FIGURE 21.40 The spectrochemical series. Strong-field ligands give rise to a large splitting between the *t*- and *e*-orbitals, whereas weak-field ligands give rise to only a small splitting. The horizontal line marks the frontier between the two kinds of ligands.

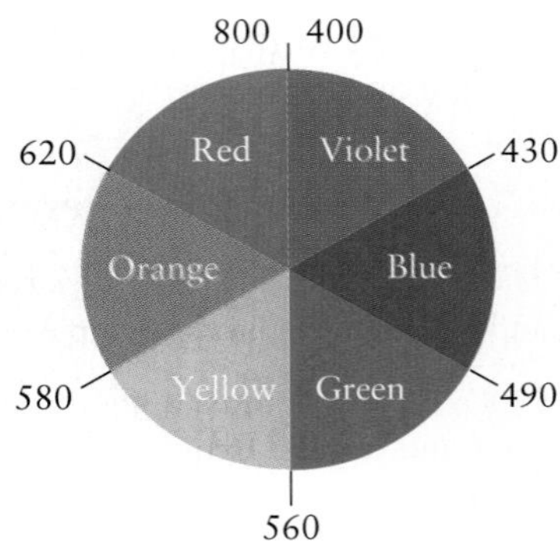

FIGURE 21.41 The perceived color of a complex in white light is the complementary color of the light it absorbs. In this color wheel, complementary colors are opposite each other. The numbers are approximate wavelengths in nanometers.

◀◀ Electromagnetic radiation was first discussed in Section 7.1.

White light is a mixture of all wavelengths of electromagnetic radiation from about 400 nm (violet) to 800 nm (red). When some of these wavelengths are removed from a beam of white light by passing the light through a sample, the emerging light is no longer white. For example, if red light is absorbed from white light, then the light that remains appears green. Conversely, if green is removed, then the light appears red. We say that red and green are each other's **complementary color**—each is the color that white light appears when the other is removed.

Complementary colors are shown on the "color wheel" in Fig. 21.41. We can see from the color wheel that if a substance looks blue (as does a copper(II) sulfate solution, for instance), then it is absorbing orange (580 to 620 nm) light. Conversely, if we know the wavelength (and therefore the color) of the light that a substance absorbs, then we can predict the color of the substance by noting the complementary color on the color wheel. Because $[Ti(H_2O)_6]^{3+}$ absorbs 510 nm light, which is yellow-green light, the complex looks red-violet (Fig. 21.42).

The prediction of color is, in fact, quite difficult. One problem is that compounds absorb light over a range of wavelengths and may in fact absorb in several regions of the spectrum. Chlorophyll, for example, absorbs both red and blue light, leaving only the wavelengths near green to be reflected from vegetation (see Case Study 7).

FIGURE 21.42 Because $[Ti(H_2O)_6]^{3+}$ absorbs yellow-green light (of wavelengths close to 510 nm) and other colors, it looks red-violet in white light.

FIGURE 21.43 The effect of changing the ligands in octahedral cobalt(III) complexes in aqueous solution. Weak-field ligands are on the left and strong-field ligands are on the right.

Because weak-field ligands give small splittings, the complexes they form absorb low-energy, long-wavelength radiation. The long wavelengths correspond to red light, so these complexes exhibit colors near green. Because strong-field ligands give large splittings, the complexes they form should absorb high-energy, short-wavelength radiation, corresponding to the violet end of the visible spectrum. Such complexes can thus be expected to have colors near orange and yellow (Fig. 21.43). For example, when ammonia is added to aqueous copper(II) sulfate, strong-field NH_3 ligands replace four of the weak-field H_2O ligands of the $[Cu(H_2O)_6]^{2+}$ ion. The absorption shifts to higher energies and shorter wavelengths, from orange to yellow, and the perceived color shifts from blue toward violet (see Fig. 20.4).

Transitions between d-orbitals in complexes give rise to color; the spectrochemical series summarizes the relative magnitudes of the ligand field splitting.

SELF-TEST 21.6A Predict which of the following complexes absorbs light of the shorter wavelength. Explain your reasoning. (a) $[Co(H_2O)_6]^{3+}$ or (b) $[Co(en)_3]^{3+}$.

[*Answer:* (b); en is a stronger-field ligand than H_2O.]

SELF-TEST 21.6B Predict which of the following complexes absorbs light of the shorter wavelength. Explain your reasoning: (a) $[Fe(CN)_6]^{4-}$ or (b) $[Fe(NH_3)_6]^{2+}$.

21.9 THE ELECTRONIC STRUCTURES OF MANY-ELECTRON COMPLEXES

The electron configurations of complexes depend on the crystal field splitting. With no ligands present, all five *d*-orbitals have the same energy. Electrons therefore enter each orbital until each has one electron, and only after that do any additional electrons pair. However, in a complex, we have to take into account the different energies of the *t*- and *e*-orbitals. To write the electron configuration, we use the orbital energy-level diagram in Fig. 21.38a for octahedral complexes and the diagram in Fig. 21.38b for tetrahedral complexes.

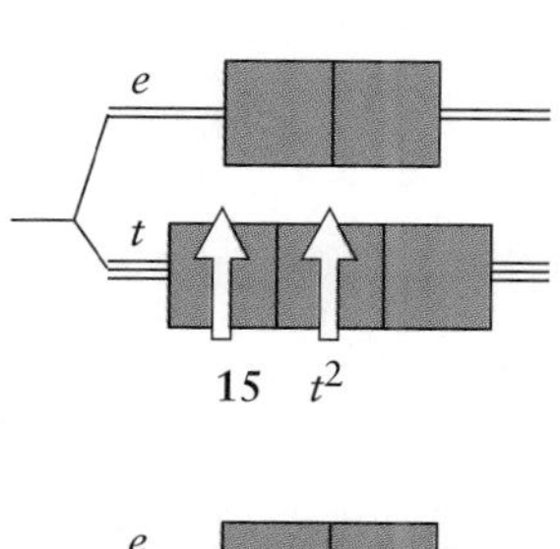

15 t^2

How will the presence of the ligands in an octahedral complex affect the electron configuration of a *d*-metal atom or ion? There are three *t*-orbitals, and because all three have the same energy, for complexes with up to three *d*-electrons (that is, d^1, d^2, and d^3 complexes), each electron can occupy a separate *t*-orbital. According to Hund's rule (Section 7.10), these electrons will have parallel spins, (**15**) and (**16**).

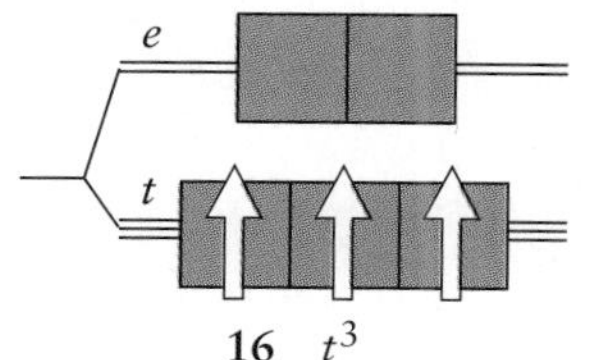

16 t^3

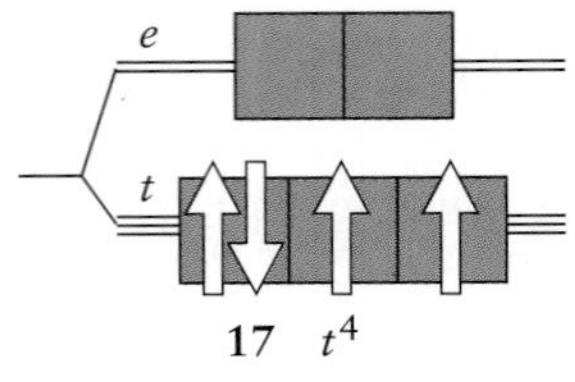

17 t^4

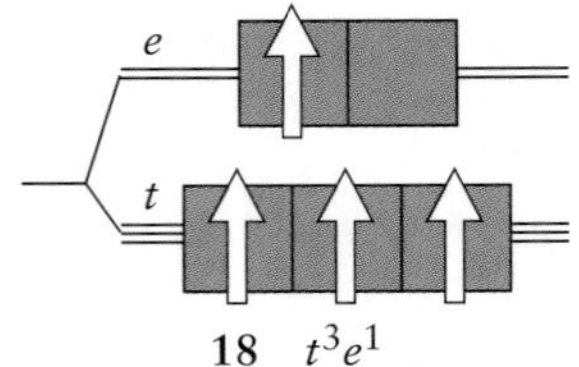

18 t^3e^1

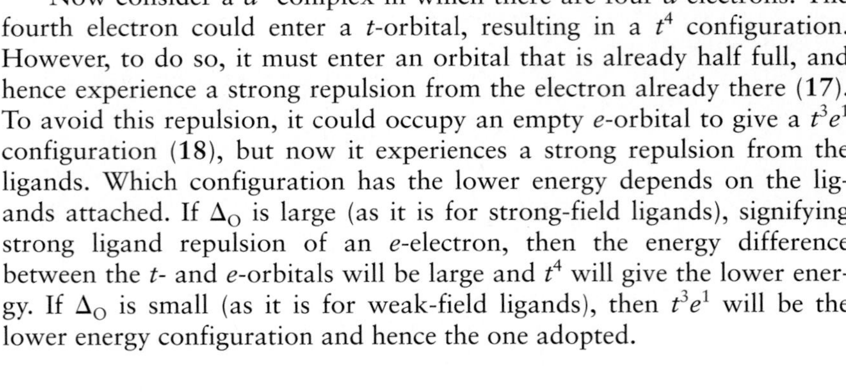

Now consider a d^4 complex in which there are four d-electrons. The fourth electron could enter a t-orbital, resulting in a t^4 configuration. However, to do so, it must enter an orbital that is already half full, and hence experience a strong repulsion from the electron already there (**17**). To avoid this repulsion, it could occupy an empty e-orbital to give a t^3e^1 configuration (**18**), but now it experiences a strong repulsion from the ligands. Which configuration has the lower energy depends on the ligands attached. If Δ_O is large (as it is for strong-field ligands), signifying strong ligand repulsion of an e-electron, then the energy difference between the t- and e-orbitals will be large and t^4 will give the lower energy. If Δ_O is small (as it is for weak-field ligands), then t^3e^1 will be the lower energy configuration and hence the one adopted.

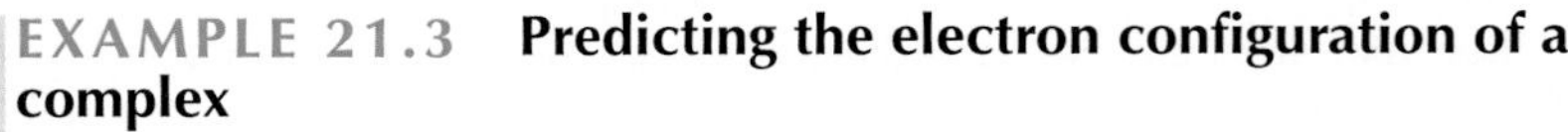

EXAMPLE 21.3 Predicting the electron configuration of a complex

Predict the electron configuration of an octahedral d^5 complex with (a) strong-field ligands and (b) weak-field ligands, and give the number of unpaired electrons in each case.

STRATEGY Electrons occupy the orbitals that result in the lowest energy configuration. If the splitting Δ_O is small, electron-electron repulsion is stronger than the weak repulsions between electrons and ligands. In this case, electrons are likely to occupy all the vacant orbitals, even those at the higher energy, before pairing. If the splitting Δ_O is large, the repulsion between electrons and ligands is great. In this case, electrons are likely to pair in the lower energy orbitals and fill them completely before occupying any of the higher energy orbitals.

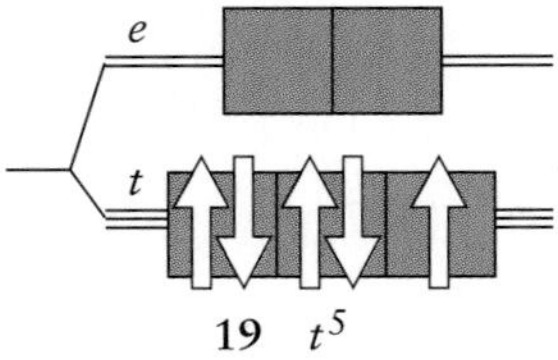

19 t^5

SOLUTION (a) In the strong-field case, all five electrons enter the t-orbitals, and to do so some of them must pair (**19**). There is one unpaired electron in this configuration. (b) In the weak-field case, the five electrons occupy all five orbitals without pairing (**20**). There are now five unpaired electrons.

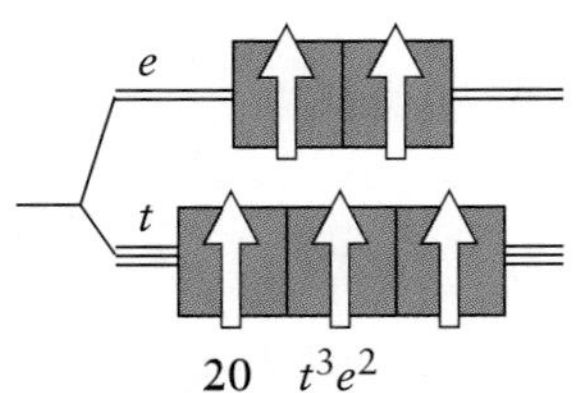

20 t^3e^2

SELF-TEST 21.7A Predict the electron configurations and the number of unpaired electrons of an octahedral d^6 complex with (a) strong-field ligands and (b) weak-field ligands.

[*Answer:* (a) t^6 (0); (b) t^4e^2 (4)]

SELF-TEST 21.7B Predict the electron configurations and the number of unpaired electrons of an octahedral d^7 complex with (a) strong-field ligands and (b) weak-field ligands.

Table 21.6 gives the configurations for d^1 through d^{10} complexes, including the alternative configurations for d^4 through d^7 octahedral complexes. A d^n complex with the maximum number of unpaired spins is called a **high-spin complex.** High-spin complexes are expected for weak-field ligands because the electrons can easily occupy both the t- and e-orbitals, and the greatest number of electrons then have parallel spins. Tetrahedral complexes are almost always high-spin because they do not have enough ligands to give a large ligand field splitting even if the lig-

TABLE 21.6 Electronic configurations of d^n complexes

Number of d-electrons	Configuration: Octahedral complexes		Configuration: Tetrahedral complexes
d^1	t^1		e^1
d^2	t^2		e^2
d^3	t^3		e^2t^1
	Low spin	High spin	
d^4	t^4	t^3e^1	e^2t^2
d^5	t^5	t^3e^2	e^2t^3
d^6	t^6	t^4e^2	e^3t^3
d^7	t^6e^1	t^5e^2	e^4t^3
d^8	t^6e^2		e^4t^4
d^9	t^6e^3		e^4t^5
d^{10}	t^6e^4		e^4t^6

ands are classified as strong-field ligands for octahedral complexes. A d^n complex with the minimum number of unpaired spins is called a **low-spin complex.** A low-spin complex is expected for strong-field ligands, because the strong repulsions from the ligands raise the energy of the e-orbitals: electrons then enter the t-orbitals until they are completely full, even though they have to pair their spins.

We can predict whether an octahedral complex is likely to be a high-spin or low-spin complex by noting where the ligands lie in the spectrochemical series. If they are strong-field ligands, we expect a low-spin complex; if they are weak-field ligands, we expect a high-spin complex.

The electron configurations of complexes are obtained by applying the building-up principle to the d-orbitals, taking into account the strength of the ligand field splitting.

21.10 MAGNETIC PROPERTIES OF COMPLEXES

As we saw in Section 20.5, a species with unpaired electrons is paramagnetic and is pulled into a magnetic field. A substance without unpaired electrons is **diamagnetic** and is pushed out of a magnetic field. Paramagnetism and diamagnetism can be distinguished experimentally by using the apparatus shown in Fig. 21.44: a sample is hung from a balance so that it lies between the poles of an electromagnet. When the magnet is turned on, a paramagnetic substance is pulled into the field and appears to weigh more than when the magnet is off. A diamagnetic substance is pushed out of the field and appears to weigh less.

Many d-metal complexes have unpaired d-electrons and are therefore paramagnetic. We have just seen that a high-spin d^n complex has more unpaired electrons than a low-spin d^n complex. The former is therefore

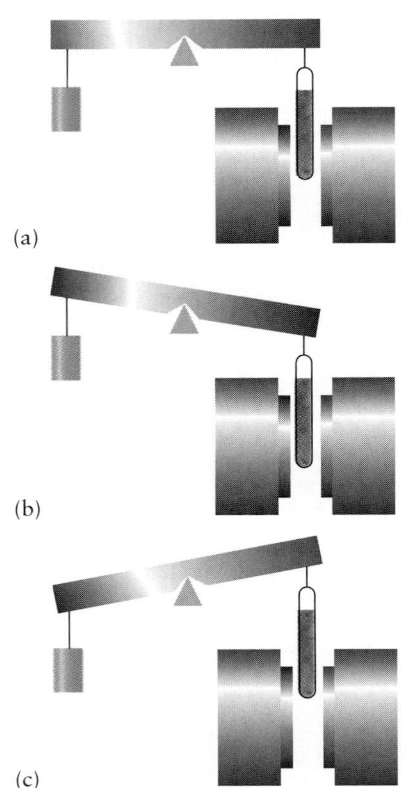

FIGURE 21.44 The magnetic character of a complex can be studied with the apparatus shown here, which is called a Gouy balance. (a) A sample is hung from a balance so that it lies partly between the poles of an electromagnet. (b) When the magnetic field is turned on, a paramagnetic sample is drawn into it, so the sample seems to weigh more. (c) In contrast, a diamagnetic sample is pushed out of the field when the field is turned on, so it seems to weigh less.

more strongly paramagnetic and is drawn more strongly into a magnetic field. Moreover, whether a complex is high-spin or low-spin depends on the ligands present. Strong-field ligands create a large energy difference between the *t*- and *e*-orbitals (Fig. 21.45). Their d^4–d^7 complexes therefore tend to be low-spin and diamagnetic or only weakly paramagnetic. Weak-field ligands create a small energy gap, so electrons fill the higher energy orbitals rather than pairing in the lower energy orbitals. Their d^4–d^7 complexes therefore tend to be high-spin and strongly paramagnetic. This correlation suggests that it should be possible to modify the magnetic properties of a complexed *d*-metal ion by changing the ligands with which it is coordinated.

The magnetic properties of a complex depend on the magnitude of the ligand field splitting. Strong-field ligands tend to form low-spin, weakly paramagnetic complexes; weak-field ligands tend to form high-spin, strongly paramagnetic complexes.

EXAMPLE 21.4 Predicting the magnetic properties of a complex

Compare the magnetic properties of $[Fe(CN)_6]^{4-}$ with those of $[Fe(H_2O)_6]^{2+}$.

STRATEGY Decide from their positions in the spectrochemical series whether the ligands in the complex are weak-field or strong-field ligands. Then judge whether the reaction changes the complex from a high-spin to a low-spin complex or vice versa. In either case, expect a change in its magnetic properties.

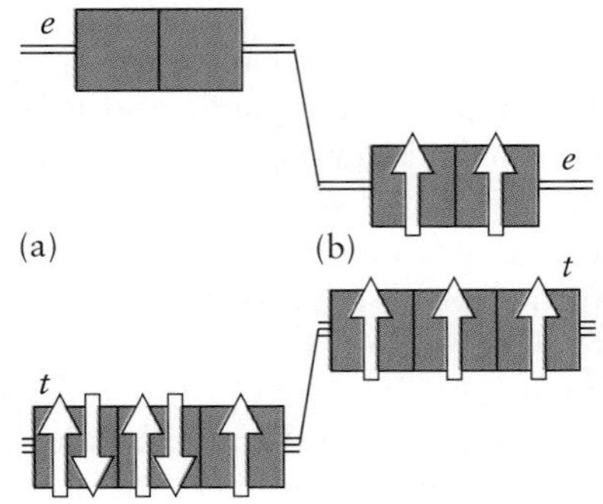

FIGURE 21.45 (a) A strong-field ligand is likely to lead to a low-spin complex (in this case, the configuration is that of Fe^{3+}). (b) A weak-field ligand is likely to result in a high-spin complex.

SOLUTION The Fe^{2+} ion is a d^6 ion. Because H_2O is a weak-field ligand, we predict a high-spin configuration with four unpaired electrons for $[Fe(H_2O)_6]^{2+}$. The ion is therefore predicted (and found) to be paramagnetic. When cyanide ions are added to an aqueous solution of Fe^{2+} ions, they form the $[Fe(CN)_6]^{4-}$ ion. Now the ligands are strong-field ligands, and the resulting complex is a low-spin t^6 complex; it has no unpaired electrons, so it is not paramagnetic. In other words, the ligand substitution reaction has had the effect of quenching the paramagnetism.

SELF-TEST 21.8A What change in magnetic properties can be expected when NO_2^- ligands in an octahedral complex are replaced by Cl^- ligands in (a) a d^6 complex; (b) a d^3 complex?

[*Answer:* (a) The complex becomes paramagnetic; (b) no change in magnetic properties]

SELF-TEST 21.8B Compare the magnetic properties of $[Ni(en)_3]^{2+}$ with those of $[Ni(H_2O)_6]^{2+}$.

SKILLS YOU SHOULD HAVE MASTERED

Conceptual

1. Explain trends in chemical and physical properties among the *d*-block elements.
2. Interpret the formation and structure of a complex in terms of Lewis acids and bases.
3. Explain how the colors of *d*-metal complexes are related to the ligand field splitting.
4. Explain how weak-field and strong-field ligands affect the electron configuration of a *d*-metal ion in a complex.

Problem-Solving

1. Predict the effect of oxidation number on the oxidizing or reducing ability of compounds of a *d*-block element and on the acidity of its oxides.
2. Identify pairs of ionization, linkage, hydrate, coordination, geometrical, and optical isomers.
3. Use the spectrochemical series to predict the effect of a ligand on the color, electron configuration, and magnetic properties of a *d*-metal complex.

Descriptive

1. Distinguish between ferromagnetism, paramagnetism, and diamagnetism, and identify common *d*-metals that are ferromagnetic.
2. Describe and write balanced equations for the principal reactions used to produce the elements in the first row of the *d*-block and in Groups 11 and 12.
3. Describe the names, properties, and reactions of some of the principal compounds of the elements in the first row of the *d*-block.
4. Describe the operation of a blast furnace and how steel is made by the basic oxygen process.
5. Describe the composition of steel and how the properties of steel are affected by the proportion of carbon.
6. Describe the structure and function of chelates.

EXERCISES

The d-Block Elements and Their Electron Configurations

21.1 Name each of the following elements: (a) Rh; (b) Ag; (c) Pd; (d) W.

21.2 Give the chemical symbols of (a) scandium; (b) niobium; (c) yttrium; (d) molybdenum.

21.3 Write the ground-state electron configuration of (a) manganese; (b) cadmium; (c) zinc; (d) zirconium.

21.4 Write the ground-state electron configuration of (a) mercury; (b) iron; (c) titanium; (d) gold.

21.5 State the number of unpaired electrons in a ground-state atom of (a) Sc; (b) V; (c) Cu; (d) Au.

21.6 State the number of unpaired electrons in a ground-state atom of (a) nickel; (b) manganese; (c) silver; (d) rhenium.

21.7 Identify the elements in the *d*-block that are ferromagnetic.

21.8 Which members of the *d*-block, those at the left or at the right of the block, tend to have strongly negative standard potentials? Name six elements in the *d*-block that have standard potentials greater than 0.

Trends in Properties

21.9 Identify the element with the larger atomic radius in each of the following pairs: (a) scandium and titanium; (b) copper and gold; (c) vanadium and niobium.

21.10 Identify the element with the larger atomic radius in each of the following pairs: (a) cobalt and manganese; (b) copper and zinc; (c) chromium and molybdenum.

21.11 Identify the element with the higher first ionization energy in each of the following pairs: (a) scandium and titanium; (b) nickel and copper; (c) iron and zinc.

21.12 Identify the element with the higher first ionization energy in each of the following pairs: (a) iron and cobalt; (b) manganese and iron; (c) vanadium and chromium.

21.13 Explain what is meant by the "lanthanide contraction" and give examples of its effect.

21.14 Is there evidence for a "*d*-block contraction" analogous to the lanthanide contraction?

21.15 Explain why the density of mercury ($13.6\ g\cdot cm^{-3}$) is significantly higher than that of cadmium ($8.65\ g\cdot cm^{-3}$), whereas the density of cadmium is only slightly greater than that of zinc ($7.14\ g\cdot cm^{-3}$).

21.16 Assuming the same crystal structure, explain why the density of iron ($7.87\ g\cdot cm^{-3}$) is significantly less than that of cobalt ($8.80\ g\cdot cm^{-3}$).

21.17 Describe the trend in the stability of oxidation states moving down a group in the *d*-block (for example, from chromium to molybdenum to tungsten).

21.18 Which oxoanion, MnO_4^- or ReO_4^-, is expected to be the stronger oxidizing agent? Explain your choice.

21.19 Which of the elements vanadium, chromium, and manganese is most likely to form an oxide with the formula MO_3? Explain your answer.

21.20 Which of the elements zirconium, chromium, and iron is most likely to form a chloride with the formula MCl_4? Explain your answer.

Scandium through Nickel

21.21 Outline a process, using chemical equations where possible, by which (a) titanium; (b) vanadium are prepared.

21.22 Outline a process, using chemical equations where possible, by which (a) nickel; (b) chromium are prepared.

21.23 Predict the major products of each of the following reactions and then balance the skeletal equations:

(a) $TiCl_4(g) + Mg(l) \xrightarrow{\Delta}$

(b) $CoCO_3(s) + HNO_3(aq) \longrightarrow$

(c) $V_2O_5(s) + Ca(l) \longrightarrow$

21.24 Predict the products of the following reactions and then balance the skeletal equations:

(a) $Ti(s) + F_2(g, \text{in excess}) \longrightarrow$ liquid

(b) $CrO_4^{2-}(aq) + H_3O^+(aq) \longrightarrow$

(c) $MnO_2(s) + Al(s) \xrightarrow{\Delta}$

21.25 Give the systematic names and chemical formulas of the principal component of (a) rutile; (b) hematite; (c) pyrolusite.

21.26 Give the systematic names and chemical formulas of the principal component of (a) magnetite; (b) pyrite; (c) ilmenite; (d) chromite.

21.27 Use the information in Appendix 2B to determine whether an acidic sodium dichromate solution can oxidize (a) bromide ions to bromine and (b) silver(I) ions to silver(II) ions under standard conditions.

21.28 Use Appendix 2B to determine whether an acidic potassium permanganate solution can oxidize (a) chloride ions to chlorine and (b) mercury metal to mercury(I) ions under standard conditions.

21.29 (a) What reducing agent is used in the production of iron from its ore? (b) Write chemical equations for the production of iron in a blast furnace. (c) What is the major impurity in the product of the blast furnace?

21.30 (a) What is the purpose of adding limestone to a blast furnace? (b) Write chemical equations that show the reactions of lime in a blast furnace.

21.31 Use Appendix 2B to predict the products of the following reactions: (a) vanadium with 1 M HCl(aq); (b) mercury with 1M HCl(aq); (c) cobalt with 1 M HCl(aq).

21.32 Use Appendix 2B to predict products of the following reactions: (a) nickel with 1 M HCl(aq); (b) titanium with 1 M HCl(aq); (c) platinum with 1 M $KMnO_4$(aq) in 1 M HCl(aq).

Groups 11 and 12

21.33 Describe the chemical evidence for treating the coinage metals as a single group.

21.34 Describe the chemical evidence for treating zinc, cadmium, and mercury as a single group.

21.35 Give the systematic name and chemical formula of the principal component of (a) chalcopyrite; (b) sphalerite; (c) cinnabar.

21.36 What are the major sources of (a) silver; (b) gold; (c) cadmium?

21.37 Outline a process, using chemical equations where possible, by which (a) zinc and (b) mercury can be produced.

21.38 Outline a process, using chemical equations where possible, by which copper is extracted and purified from chalcopyrite by the pyrometallurgical process.

21.39 Use the information in Appendix 2B to determine the equilibrium constant for the disproportionation of copper(I) ions in aqueous solution at 25°C to copper metal and copper(II) ions.

21.40 Use the information in Appendix 2B to determine the equilibrium constant for the disproportionation of mercury(I) ions in aqueous solution at 25°C to mercury metal and mercury(II) ions.

d-Metal Complexes

21.41 Determine the oxidation number of the metal atom in the following complexes: (a) $[Fe(CN)_6]^{4-}$; (b) $[Co(NH_3)_6]^{3+}$; (c) $[Co(CN)_5(H_2O)]^{2-}$; (d) $[Co(SO_4)(NH_3)_5]^+$.

21.42 Determine the oxidation number of the metal atom in the following complexes: (a) $[Fe(CN)_6]^{3-}$; (b) $[Fe(OH)(H_2O)_5]^{2+}$; (c) $[CoCl(NH_3)_4(H_2O)]^{2+}$; (d) $[Ir(en)_3]^{3+}$.

21.43 Use Table 21.5 to determine the coordination number of the metal ion in the following complexes: (a) $[NiCl_4]^{2-}$; (b) $[Ag(NH_3)_2]^+$; (c) $[PtCl_2(en)_2]^{2+}$; (d) $[Cr(EDTA)]^-$.

21.44 Use Table 21.5 to determine the coordination number of the metal ion in the following complexes: (a) $[Ir(en)_3]^{3+}$; (b) $[Fe(ox)_3]^{3-}$ (the oxalato ligand attaches at two points); (c) $[PtCl_2(NH_3)_2]$; (d) $[Fe(CO)_5]$.

21.45 Use the information in Table 21.5 and Appendix 3C to name the following complexes: (a) $[Fe(CN)_6]^{4-}$; (b) $[Co(NH_3)_6]^{3+}$; (c) $[Co(CN)_5(H_2O)]^{2-}$; (d) $[Co(SO_4)(NH_3)_5]^+$.

21.46 Use the information in Table 21.5 and Appendix 3C to name the following complexes: (a) $[Fe(CN)_6]^{3-}$; (b) $[Fe(OH)(H_2O)_5]^{2+}$; (c) $[CoCl(NH_3)_4(H_2O)]^{2+}$; (d) $[Ir(en)_3]^{3+}$.

21.47 Use the information in Table 21.5 and Appendix 3C to write the formula for each of the following coordination compounds: (a) potassium hexacyanochromate(III); (b) pentaamminesulfatocobalt(III) chloride; (c) tetra-amminediaquacobalt(III) bromide; (d) sodium diaquabis(oxalato)ferrate(III).

21.48 Use the information in Table 21.5 and Appendix 3C to write the formula for each of the following coordination compounds: (a) triammineaquadihydroxochromium(III) chloride; (b) potassium tetrachloroplatinate(III); (c) tetraaquadichloronickel(II) iodide; (d) potassium tris(oxalato) rhodium(III); (e) chlorohydroxybis(oxalato) rhodate(III) octahydrate.

Isomerism

21.49 Determine the type of structural isomerism that exists in the following pairs of compounds:
(a) $[Co(NO_2)(NH_3)_5]Br_2$ and $[Co(ONO)(NH_3)_5]Br_2$
(b) $[Pt(SO_4)(NH_3)_4](OH)_2$ and $[Pt(OH)_2(NH_3)_4]SO_4$
(c) $[CoCl(SCN)(NH_3)_4]Cl$ and $[CoCl(NCS)(NH_3)_4]Cl$
(d) $[CrCl(NH_3)_5]Br$ and $[CrBr(NH_3)_5]Cl$

21.50 Determine the type of structural isomerism that exists in the following pairs of compounds:
(a) $[Cr(en)_3][Co(ox)_3]$ and $[Co(en)_3][Cr(ox)_3]$
(b) $[CoCl_2(NH_3)_4]Cl \cdot H_2O$ and $[CoCl(NH_3)_4(H_2O)]Cl_2$
(c) $[Co(CN)_5(NCS)]^{3-}$ and $[Co(CN)_5(SCN)]^{3-}$
(d) $[Pt(NH_3)_4][PtCl_6]$ and $[PtCl_2(NH_3)_4][PtCl_4]$

21.51 Write the formulas for the hydrate isomers of a compound having the empirical formula $CoCl_3 \cdot 6H_2O$ and a coordination number of six.

21.52 Write the formula of a coordination isomer of $[Co(NH_3)_6][Cr(NO_2)_6]$.

21.53 Write the formula of a linkage isomer of $[CoCl(NO_2)(en)_2]Cl$.

21.54 Write the formula of an ionization isomer of $[CoCl(NO_2)(en)_2]Cl$.

21.55 Which of the following coordination compounds can have *cis* and *trans* isomers? If such isomerism exists, draw the two structures and name the compound. (a) $[CoCl_2(NH_3)_4]Cl \cdot H_2O$; (b) $[CrCl(NH_3)_5]Br$; (c) $[PtCl_2(NH_3)_2]$, a square-planar complex.

21.56 Which of the following complexes can have *cis* and *trans* isomers? If such isomerism exists, draw the two structures and name the compound. (a) $[Fe(OH)(H_2O)_5]^{2+}$; (b) $[RuBr_2(NH_3)_4]^+$; (c) $[Co(NH_3)_4(H_2O)_2]^{3+}$.

21.57 Can a tetrahedral complex show (a) stereoisomerism; (b) geometrical isomerism; (c) optical isomerism?

21.58 Draw the structure of *cis*-diammine-*cis*-diaqua-*cis*-dichlorochromium(III) ion and comment on its isomerism. What kind of isomerism is possible for the *trans*-diammine isomer?

21.59 Draw the geometrical isomers of $[Cr(ox)_2(H_2O)_2]^-$.

21.60 Draw the structures of (a) *cis*-$[CoCl_2(NH_3)_4]^+$ (violet); (b) *trans*-$[CoCl_2(NH_3)_4]^+$ (bright green).

21.61 Is either of the following complexes chiral? If both complexes are chiral, do they form an enantiomeric pair?

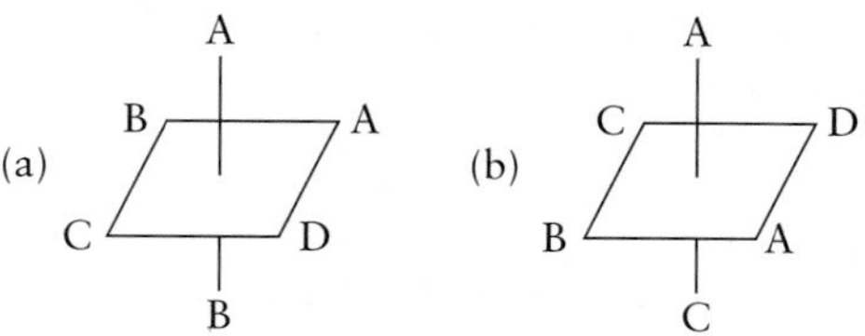

21.62 Is either of the following complexes chiral? If both complexes are chiral, do they form an enantiomeric pair?

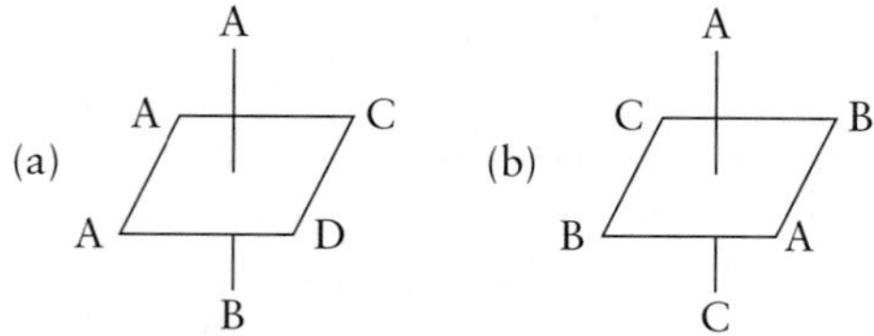

21.63 Draw the structures of the optical isomers of $[CoCl_2(en)_2]^+$.

21.64 Draw the structures of the isomeric forms of $[CrBrClI(NH_3)_3]$. Which isomers are chiral?

Crystal Field Theory

21.65 The valence electron configurations of the *d*-block ions can be summarized as d^n. Summarize the valence electron configuration for each of the following ions: (a) Co^{2+}; (b) Ni^{2+}; (c) Mn^{2+}; (d) Cr^{3+}.

21.66 The valence electron configurations of the *d*-block ions can be summarized as d^n. Summarize the valence electron configuration for each of the following ions: (a) Cr^{2+}; (b) Fe^{2+}; (c) V^{2+}; (d) Co^{3+}.

21.67 Draw an orbital energy-level diagram (like those in Fig. 21.38) showing the configuration of *d*-electrons on the metal ion in the following complexes: (a) $[Co(NH_3)_6]^{3+}$; (b) $[NiCl_4]^{2-}$ (tetrahedral); (c) $[Fe(H_2O)_6]^{3+}$; (d) $[Fe(CN)_6]^{3-}$.

21.68 Draw an orbital energy-level diagram (like those in Fig. 21.38) showing the configuration of *d*-electrons on the metal ion in the following complexes: (a) $[Zn(H_2O)_6]^{2+}$; (b) $[CoCl_4]^{2-}$ (tetrahedral); (c) $[Co(CN)_6]^{3-}$; (d) $[CoF_6]^{3-}$.

21.69 The complexes (a) $[Co(en)_3]^{3+}$ and (b) $[Mn(CN)_6]^{3-}$ have low-spin electron configurations. How many unpaired electrons are there in each complex?

21.70 The complexes (a) $[FeF_6]^{3-}$ and (b) $[Co(ox)_3]^{4-}$ have high-spin electron configurations. How many unpaired electrons are there in each complex?

21.71 Explain the difference between a weak-field ligand and a strong-field ligand. What measurements can be used to classify them as such?

21.72 Describe the changes that may occur in a compound's properties when weak-field ligands are replaced by strong-field ligands.

21.73 Of the two complexes (a) $[CoF_6]^{3-}$ and (b) $[Co(en)_3]^{3+}$, one appears yellow in an aqueous solution and the other appears blue. Match the complex to the color and explain your choice.

21.74 A concentrated solution of copper(II) bromide in the presence of potassium bromide is deep violet, but the solution becomes light blue upon dilution with water. (a) Write the formulas of the two complex ions of copper(II) that form, assuming a coordination number of four. (b) Is the change in color from violet to blue expected? Explain your reasoning.

21.75 State the color of a sample that absorbs light of wavelength (a) 410 nm; (b) 650 nm; (c) 480 nm; (d) 590 nm.

21.76 State the wavelength range over which the following colors are absorbed from a sample and then predict the color of the complex: (a) blue; (b) red; (c) violet; (d) green.

21.77 Suggest a reason why $Zn^{2+}(aq)$ ions are colorless. Would you expect zinc compounds to be paramagnetic? Explain your answer.

21.78 Suggest a reason why copper(II) compounds are often colored but copper(I) compounds are colorless. Which oxidation number gives paramagnetic compounds?

21.79 Estimate the ligand-field splitting for (a) $[CrCl_6]^{3-}$ (λ_{max} = 740 nm); (b) $[Cr(NH_3)_6]^{3+}$ (λ_{max} = 460 nm); (c) $[Cr(H_2O)_6]^{3+}$ (λ_{max} = 575 nm), where λ_{max} is the wavelength of the most intensely absorbed light. Arrange the ligands in order of increasing ligand field strength.

21.80 Estimate the ligand-field splitting for (a) $[Co(CN)_6]^{3-}$ (λ_{max} = 295 nm); (b) $[Co(NH_3)_6]^{3+}$ (λ_{max} = 435 nm); (c) $[Co(H_2O)_6]^{3+}$ (λ_{max} = 540 nm), where λ_{max} is the wavelength of the most intensely absorbed light. Arrange the ligands in order of increasing ligand field strength.

SUPPLEMENTARY EXERCISES

21.81 How do (a) diamagnetism and (b) ferromagnetism differ from paramagnetism?

21.82 Iron, cobalt, and nickel are often grouped together as the *iron triad*. What similarities do they show that justifies this grouping? How might these similarities be explained?

21.83 Write a chemical equation showing in what respect a Sc^{3+} ion is a Brønsted acid in aqueous solution.

21.84 Explain in terms of electron configurations why the atomic radius of manganese is larger than that of chromium.

21.85 Identify the element, compound, or mixture that best fits the description: (a) bronze; (b) the green patina on the copper plate of the Statue of Liberty; (c) 24-carat gold; (d) the densest of the *d*-block elements.

21.86 Identify the element, compound, or mixture that best fits the description: (a) a substance used as the white pigment in paints and in papermaking; (b) the catalyst used for the production of sulfuric acid; (c) the second most abundant metal in the Earth's crust; (d) pig iron.

21.87 Write chemical equations that describe the following processes: (a) the reduction of hematite in a blast furnace; (b) the reduction of titanium tetrachloride to titanium metal; (c) the removal of silica, SiO_2, from iron ore.

21.88 Write chemical equations that describe the following processes: (a) the production of chromium by the thermite reaction; (b) the corrosion of copper metal by carbon dioxide in moist air; (c) the purification of nickel, using carbon monoxide.

21.89 Determine the mass of $FeCr_2O_4$ (from chromite ore) that is needed to produce 1.00 kg of sodium chromate by the reaction

$$4\,FeCr_2O_4(s) + 8\,Na_2CO_3(s) + 7\,O_2(g) \xrightarrow{\Delta} 8\,Na_2CrO_4(s) + 2\,Fe_2O_3(s) + 8\,CO_2(g)$$

21.90 Zinc metal is used to galvanize steel because of the cathodic protection it provides. What would result if copper were used instead?

21.91 (a) Explain why the dissolution of a chromium(III) salt produces an acidic solution. (b) Explain why the slow addition of hydroxide ions to a solution containing chromium(III) ions first produces a gelatinous precipitate that subsequently dissolves with further addition of hydroxide ions. Write chemical equations showing these aspects of the behavior of chromium(III) ions.

21.92 Explain as fully as possible, with diagrams of the structures of the hydrated ions, the changes that occur when (a) anhydrous copper(II) sulfate is moistened; (b) it is dissolved in water; and (c) an excess of aqueous ammonia is added to the resulting solution.

21.93 Use the information in Table 21.5 and Appendix 3C to name each of the following compounds: (a) $[Cr(SO_4)(NH_3)_5]Cl$; (b) $[Cr(en)_3][Cr(ox)_3]$; (c) $[PtCl_2(NH_3)_2]$; (d) $K_3[Co(NO_2)_6]$.

21.94 Use the information in Table 21.5 and Appendix 3C to name each of the following compounds: (a) $[PtCl_2(en)_2]Cl_2$; (b) $[Co(en)_3][Fe(CN)_6]$; (c) $K_3[Fe(ox)_3]$; (d) $Na[Cr(EDTA)]$.

21.95 Use the information in Table 21.5 and Appendix 3C to name each of the following complexes. Determine the coordination number and oxidation number of the *d*-metal ion: (a) $[Zr(ox)_4]^{4-}$; (b) $[CuCl_4(H_2O)_2]^{2-}$; (c) $[PtCl_3(NH_3)]^-$; (d) $[Mo(O)_2(CN)_4]^{4-}$.

21.96 By considering electron configurations, suggest a reason why iron(III) compounds are readily prepared from iron(II), but the conversion of nickel(II) and cobalt(II) to nickel(III) and cobalt(III) is much more difficult.

21.97 Draw all possible isomers of the square-planar complex $[PtBrCl(NH_3)_2]$.

21.98 How can the existence of the isomers in Exercise 21.97 be used to show that the complex is square planar rather than tetrahedral?

21.99 Suggest a chemical test for distinguishing between
(a) $[Ni(SO_4)(en)_2]Cl_2$ and $[NiCl_2(en)_2]SO_4$
(b) $[NiI_2(en)_2]Cl_2$ and $[NiCl_2(en)_2]I_2$

21.100 Two chemists prepared a complex and determined its formula, which they wrote as $CrNH_3Cl_3 \cdot 2H_2O$. When they dissolved 2.11 g of the compound in water and added an excess of silver nitrate, 2.87 g AgCl precipitated. Write the correct formula of the compound and draw its structure, including all possible isomers.

21.101 What is the type of isomerism shown by $[Cr(NH_3)_6]Cl_3$ (yellow), $[CrCl(NH_3)_5]Cl_2 \cdot NH_3$ (purple), and $[CrCl_2(NH_3)_4]Cl \cdot 2NH_3$ (violet)? Draw the structure of each complex. If geometrical isomers are possible, draw the trans isomer.

21.102 For which of the ions (a) Mn^{2+}; (b) V^{2+}; (c) Ni^{2+}; (d) Cr^{2+}, would there be no difference in the magnetic properties of the octahedral complexes of strong-field and weak-field ligands? Draw orbital energy-level diagrams to support your conclusions.

21.103 (a) Sketch the orbital energy-level diagrams for $[MnCl_6]^{4-}$ and $[Mn(CN)_6]^{4-}$. (b) How many unpaired electrons are present in $[Mn(CN)_6]^{4-}$? (c) Which complex transmits the longer wavelengths of incident electromagnetic radiation? Explain your reasoning.

21.104 The complex $[Co(CN)_6]^{3-}$ is pale yellow. (a) Is short- or long-wavelength visible radiation absorbed? (b) Is the ligand field splitting strong or weak? (c) How many unpaired electrons are present in the complex? (d) If ammonia molecules are substituted for cyanide ions as ligands, will the shift in absorbance of radiation be toward the blue or the red regions of the spectrum?

21.105 (a) Calculate the ligand field splitting for a complex that has an absorption maximum at $\lambda = 550$ nm. (b) What color can the complex be expected to be?

21.106 What change in magnetic properties can be expected when NH_3 ligands replace the H_2O ligands in $[Co(H_2O)_6]^{3+}$?

CHALLENGING EXERCISES

21.107 Calculate the standard potential of the Sn^{4+}/Sn couple, given that the values for the Sn^{4+}/Sn^{2+} and Sn^{2+}/Sn couples are +0.15 V and −0.14 V, respectively.

21.108 Calculate the standard potential of the Au^{3+}/Au^{+} couple, given that the values for Au^{3+}/Au and Au^{+}/Au are +1.40 V and +1.69 V, respectively.

21.109 The standard potentials of the couples Cr^{3+}/Cr^{2+} and Cr^{2+}/Cr are −0.41 V and −0.91 V, respectively. Calculate the standard potential of the Cr^{3+}/Cr couple.

21.110 The standard potential of the $MnO_4^{2-}/MnO_2,OH^-$ redox couple is +0.60 V and that of the MnO_4^{2-}/MnO_4^- redox couple is +0.56 V. (a) Determine the standard free energy for the disproportionation reaction of the manganate ion MnO_4^{2-} at 25°C. (b) Calculate the equilibrium constant of the disproportionation reaction of the manganate ion at 25°C.

21.111 Gold metal can be oxidized in the presence of aqua regia, a 3:1 mixture (by volume) of concentrated hydrochloric acid and nitric acids. Consider the half-reactions

$$[AuCl_4]^-(aq) + 3\,e^- \longrightarrow Au(s) + 4\,Cl^-(aq) \qquad E^\circ = +1.00 \text{ V}$$

$$NO_3^-(aq) + 4\,H^+(aq) + 3\,e^- \longrightarrow NO(g) + 2\,H_2O(l) \qquad E^\circ = +0.96 \text{ V}$$

(a) What is the spontaneous reaction under standard conditions? (b) Determine E° for the cell formed from the two couples. (c) What is the cell potential when the molarity of hydrogen ions is 6.0 $mol \cdot L^{-1}$, of chloride ions is 6.0 $mol \cdot L^{-1}$, of nitrate ions is 6.0 $mol \cdot L^{-1}$, and of the complex $[AuCl_4]^-$ is 1.0×10^{-6} $mol \cdot L^{-1}$? (d) What is the direction of the spontaneous reaction when 6 M HCl(aq) and 6 M HNO_3(aq) are present?

21.112 Suggest the form the orbital energy-level diagram would take for a square-planar complex, and discuss how the building-up principle applies. *Hint*: The d_{z^2} orbital has more electron density in the *x-y* plane than the d_{xz} or d_{yz} orbitals, but less than the d_{xy} orbital.

21.113 Before the structures of octahedral complexes had been determined, various means were used to explain the fact that transition metal ions could bind to a greater number of ligands than expected on the basis of their charges. For example, Co^{3+} can bind to six ligands, not just three. An early theory attempted to explain this behavior by postulating that once three ligands had attached to a metal ion of charge +3, the others bonded to the attached ligands, forming chains of ligands. Thus, the compound $[Co(NH_3)_6]Cl_3$, which we now know to have the octahedral structure, would have been described as $Co(NH_3{-}NH_3{-}Cl)_3$. Show how the chain theory is not consistent with at least two properties of coordination compounds.

21.114 The relative thermodynamic stability of two complexes can be predicted from a comparison of their standard potentials. Determine which complex of the following pair is the more stable and state your conclusions about the relationship between the stability of a complex and (a) the oxidation number of the central atom; (b) the field strength of the attached ligands. Give your reasoning for each conclusion and suggest an explanation for the relationship.

$$[Co(NH_3)_6]^{3+}(aq) + e^- \longrightarrow [Co(NH_3)_6]^{2+}(aq) \qquad E° = +0.11\ V$$

$$[Co(H_2O)_6]^{3+}(aq) + e^- \longrightarrow [Co(H_2O)_6]^{2+}(aq) \qquad E° = +1.81\ V$$

CHAPTER 22

One of the great challenges for the present and indefinitely into the future is the source of our energy supplies. One approach is nuclear fission, but the hazards of radioactive waste are awesome. Another possibility, as illustrated in this photograph from the Omega facility of the University of Rochester, is bringing about nuclear fusion by compressing a fuel pellet with an array of lasers. The issues involved, and some of chemistry's contributions to their solution, is the subject of this chapter.

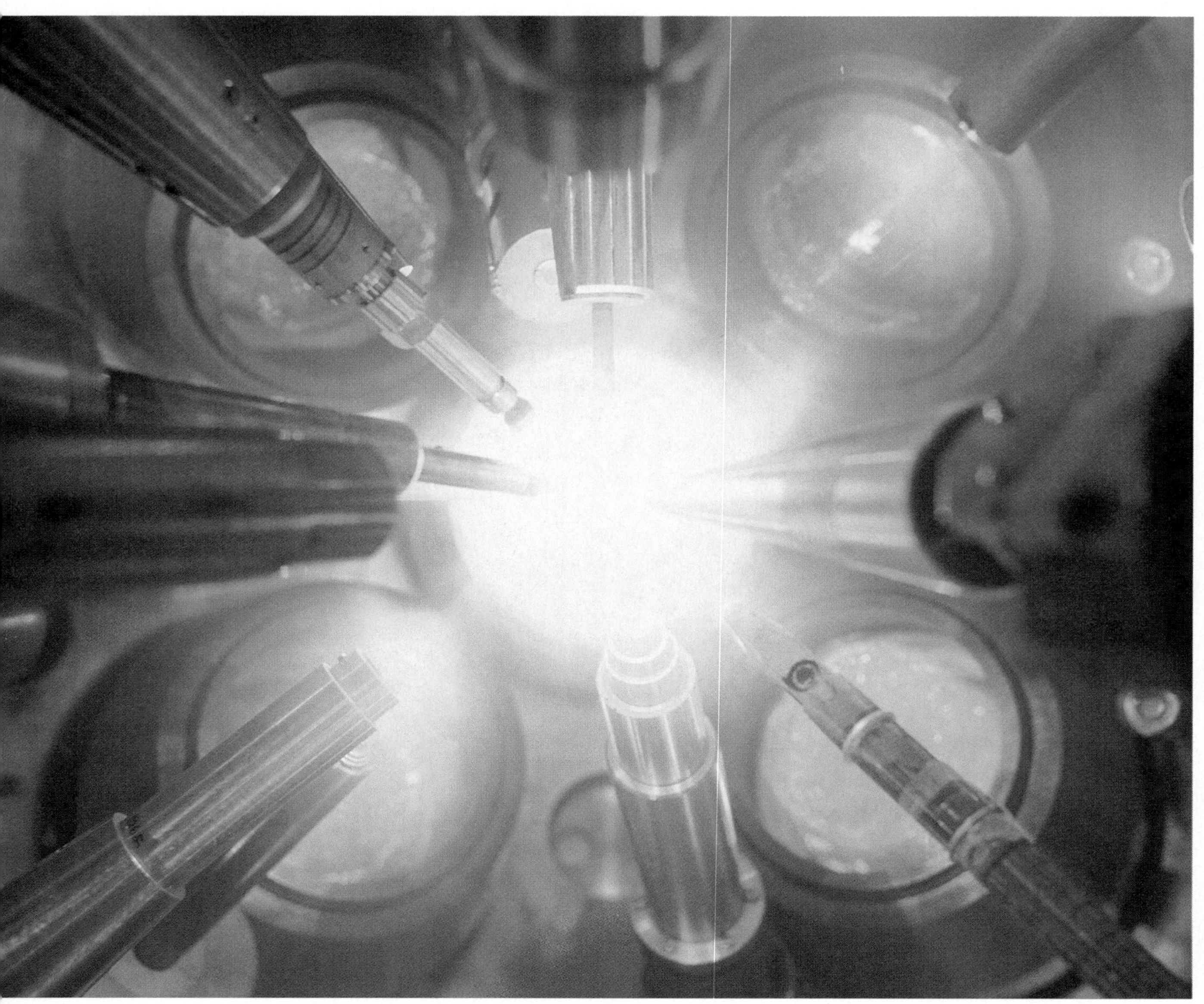

NUCLEAR CHEMISTRY

Nuclear energy is viewed as both a blessing and a curse, an instrument of destruction and a cure for terminal disease, a source of abundant energy and a waste-disposal nightmare. Like other powerful resources, nuclear energy presents us with great promise but technical challenges. We face many important decisions about the control and use of nuclear energy that will affect future generations for centuries.

So far, we have viewed the atomic nucleus as an unchanging passenger in chemical reactions. However, nuclei can change, and **nuclear chemistry** explores the consequences of those changes. Nuclear chemistry is part of our lives in many ways. It is central to the development of nuclear energy, because techniques must be found for the purification and recycling of nuclear fuels and for the disposal of hazardous radioactive waste. It is used in medicine to treat cancer and to produce images of the body's interior, in chemistry to investigate reaction mechanisms, in archeology to date archeological objects, and as part of the military defense strategies of many nations.

The large-scale application of nuclear reactions to generate power provides a substantial amount of the world's electricity. However, its use is controversial. Its benefits are balanced by risks and dangers, such as the problem of nuclear waste disposal. In this chapter we see how chemistry is being used to look for solutions to these problems. Currently, there are no clear-cut solutions, but the material we present here may help you to assess the risks and benefits of some of the approaches that are being explored.

NUCLEAR STABILITY

22.1 Nuclear reactions
22.2 Nuclear structure and nuclear radiation
22.3 Nuclear decay
22.4 The pattern of nuclear stability
22.5 Nucleosynthesis

RADIOACTIVITY

22.6 The effects of radiation
22.7 Measuring radioactivity
22.8 The law of radioactive decay

NUCLEAR ENERGY

22.9 Mass-energy conversion
22.10 Nuclear fission
22.11 Nuclear fusion
22.12 The chemistry of nuclear power

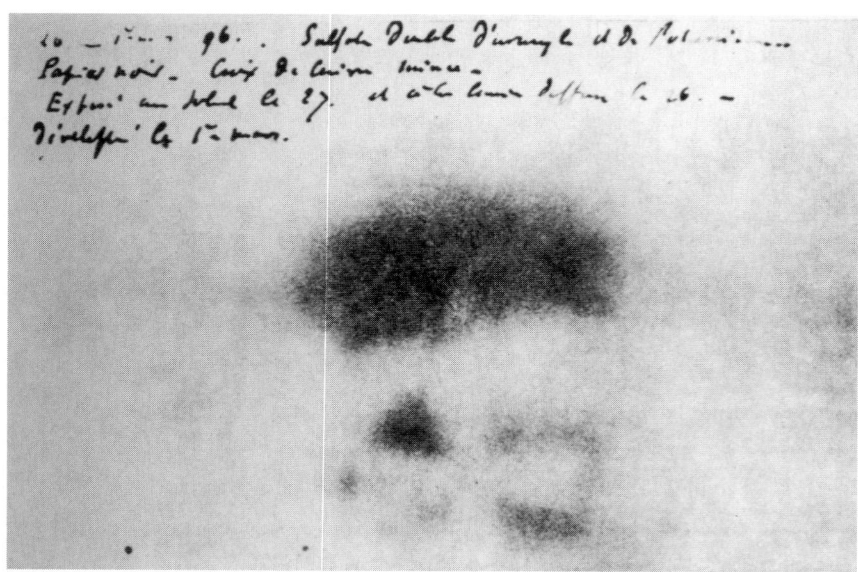

FIGURE 22.1 Henri Becquerel discovered radioactivity when he noticed that an unexposed photographic plate left near some uranium oxide became fogged. This photograph shows one of his original plates.

Marie Curie's work with radioactivity won her two Nobel prizes, one shared with her husband, the French physicist Pierre Curie, and with Becquerel. The Curies worked under dangerous conditions, for the hazards of radioactivity were unknown when they began their studies. Their early notebooks are still too dangerous to handle.

FIGURE 22.2 Marie Sklodowska Curie (1867–1934).

NUCLEAR STABILITY

The French scientist Henri Becquerel discovered radioactivity in 1896 when he stored a uranium compound in a drawer with some photographic plates (Fig. 22.1). However, it was not until a young Polish-French student, Marie Sklodowska Curie (Fig. 22.2), went to Becquerel looking for a topic for her doctoral dissertation research that the source of the radiation was discovered. Curie discovered that the nuclei of uranium atoms gave off energetic particles and rays that she called **radioactivity.**

22.1 NUCLEAR REACTIONS

Recall from Section 1.4 that nuclei are composed of protons and neutrons (Fig. 22.3); these subatomic particles are collectively called nucleons. We also saw that isotopes of an element have nuclei with the same number of protons but different numbers of neutrons. A specific nucleus with a given atomic number and mass number is called a **nuclide.** Thus, ^{1}H, ^{2}H, and ^{16}O are three different nuclides; the first two are isotopes of the same element. Nuclei that change their structure spontaneously and emit radiation are called **radioactive.** Many such unstable nuclei occur naturally: for instance, all nuclei from polonium ($Z = 84$) onward in the periodic table are radioactive and all the lighter elements have some unstable isotopes.

The changes that nuclei undergo are called **nuclear reactions.** They differ in important ways from chemical reactions, which involve changes only in the valence-shell electrons:

1. Isotopes of one element show almost identical chemical properties, but they undergo different nuclear reactions.
2. A nuclear reaction often results in the formation of an element not initially present, whereas a chemical reaction never does.
5. The energy changes are very much greater for nuclear reactions than for chemical reactions.

The combustion of 1 g of CH_4 produces about 52 kJ of energy as heat. In contrast, a nuclear reaction of 1 g of uranium-235 produces about 8.2×10^7 kJ! Obviously, the "test tubes" used for nuclear reactions must be far more sturdy, complex, and expensive than those used for chemical reactions.

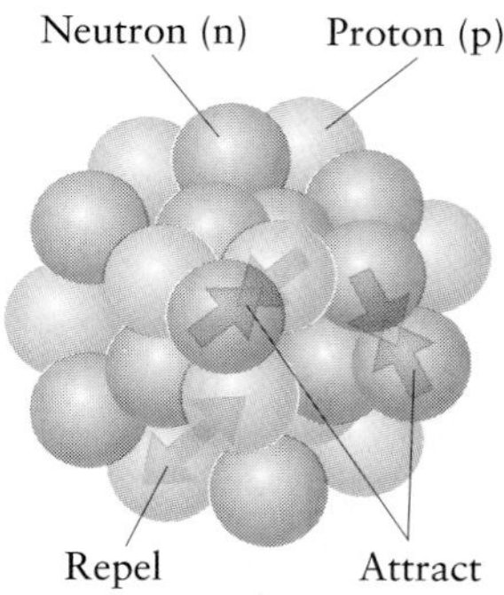

FIGURE 22.3 A nucleus can be pictured as a collection of protons (pink) and neutrons (gray). The protons repel one another electrically, but a strong force that acts between all the particles holds the nucleus together.

Nuclear reactions involve large energy changes, are different for different isotopes, and may result in the formation of different elements.

22.2 NUCLEAR STRUCTURE AND NUCLEAR RADIATION

Becquerel's discovery of radioactivity was taken up by Ernest Rutherford, who identified three different types of radioactivity by observing the effect of electric fields on radioactive emissions (Fig. 22.4). He called the three types α (alpha), β (beta), and γ (gamma) radiation.

Rutherford found that α radiation is attracted to a negatively charged electrode. This observation led him to propose that it consists of positively charged particles, which he called **α particles** (alpha particles). From the charge and mass of the particles, he was able to identify them as the nuclei of helium atoms, ${}^{4}_{2}He^{2+}$. An α particle is denoted ${}^{4}_{2}\alpha$, or simply α. We can think of it as a tightly bound cluster of two protons and two neutrons (Fig. 22.5). The supplies of helium on Earth are formed as a result of α decay deep underground: the α particles collect electrons from their surroundings and form He atoms.

Rutherford also found that β radiation is attracted to a positively charged electrode, an observation suggesting that it consists of a stream of negatively charged particles. Measurement of the charge and mass of these particles showed that they are, in fact, electrons. The rapidly moving electrons emitted by nuclei are called **β particles** (beta particles) and denoted β. Because a β particle has no protons or neutrons, its mass number is 0. Because its charge is -1, it is convenient (when balancing equations) to denote it ${}^{0}_{-1}e$, but the subscript is not a true atomic number.

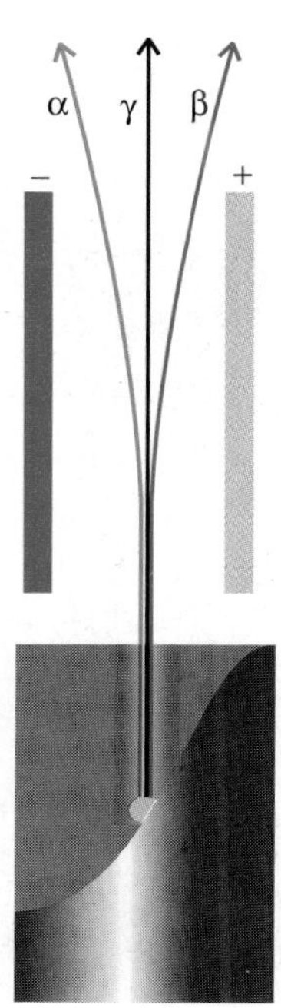

FIGURE 22.4 The effects of an electric field on nuclear radiation. The deflection identifies α particles as positively charged, β particles as negatively charged, and γ rays as uncharged.

FIGURE 22.5 An α particle has two positive charges and a mass number of 4. It consists of two protons and two neutrons, and is the same as the nucleus of a helium-4 atom.

The wavelength of visible light is about 500,000 times greater than that of γ radiation.

γ radiation (gamma radiation) is electromagnetic radiation like light, but of much higher frequency—greater than about 10^{20} Hz—and with wavelengths less than about 1 pm. γ radiation can be regarded as a stream of very high energy photons; each photon is emitted by a single nucleus as that nucleus discards energy. Like all photons, γ-ray photons are massless, uncharged, and unaffected by electric fields. As in the case of radiation emitted by electrons in excited atoms, the photon has a frequency ν given by the relation $\Delta E = h\nu$ (see Section 7.3). γ rays have very high frequencies because the energy difference between the excited and ground nuclear states is very large.

Heavy elements are more likely to give off α radiation, whereas β radiation is more typical of lighter elements. α and β radiation are often accompanied by γ radiation because the ejection of an α or β particle often leaves the product nucleus in a high-energy arrangement (Fig. 22.6); a γ-ray photon is then emitted when the nucleons rearrange within the nucleus into a state of lower energy. Other types of nuclear radiation have since been identified; they are described in Toolbox 22.1, and their properties are summarized in Table 22.1.

Radioactive nuclei commonly emit three types of radiation: ***α*** *particles (the nuclei of helium atoms),* ***β*** *particles (fast electrons ejected from the nucleus), and γ rays (high-energy electromagnetic radiation).*

22.3 NUCLEAR DECAY

Radioactivity is a sign of **nuclear decay,** the partial breakup of a nucleus. The α and β particles originate in the nucleus, so they leave behind a nucleus with a number of protons different from that of the original atom. The product, which is called the **daughter nucleus** (Fig. 22.7), is the nucleus of an atom of a *different* element. For example, when a radon-222 nucleus emits an α particle, a polonium-218 nucleus is formed. A **nuclear transmutation,** the conversion of one element into another, has taken place. Nuclear transmutation, particularly of lead into gold, was the dream of the alchemists, and the search for it was a root of modern chemistry. However, chemical means cannot change nuclei. Only

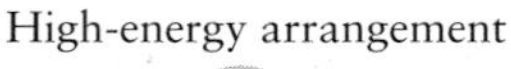

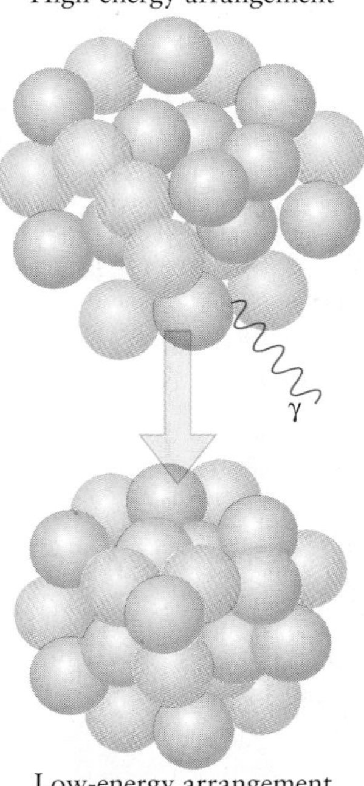

FIGURE 22.6 Nuclear decay may result in the formation of a nucleus with its nucleons in a high-energy state, as depicted by the loose arrangement in the upper part of the illustration. As the nucleons adjust to a lower energy arrangement (bottom), the excess energy is released as a γ-ray photon.

TABLE 22.1 Nuclear radiation

Type	Comments	Particle†	Example
α	not penetrating but damaging speed: 10% of c‡	helium-4 nucleus $^{4}_{2}He^{2+}$, $^{4}_{2}\alpha$, α	$^{226}_{88}Ra \longrightarrow {}^{222}_{86}Rn + \alpha$
β	moderately penetrating speed: less than 90% of c	electron $^{0}_{-1}e$, β^-, β	$^{3}_{1}H \longrightarrow {}^{3}_{2}He + \beta$
γ	very penetrating; often accompanies other radiation speed: c	photon	$^{60}_{27}Co^* \longrightarrow {}^{60}_{27}Co + \gamma$
β^+	moderately penetrating speed: less than 90% of c	positron $^{0}_{+1}e$, β^+	$^{22}_{11}Na + {}^{22}_{10}Ne + \beta^+$
p	moderate/low penetration speed: 10% of c	proton $^{1}_{1}H^+$, $^{1}_{1}p$, p	$^{53}_{27}Co \longrightarrow {}^{52}_{26}Fe + p$
n	very penetrating speed: less than 10% of c	neutron $^{1}_{0}n$, n	$^{137}_{53}I \longrightarrow {}^{136}_{53}I + n$

*An energetically excited state of a nucleus is usually denoted by an asterisk.
†Alternative symbols are given for the particles; often it is sufficient to use the simplest (the one on the right).
‡c is the speed of light.

nuclear processes, which were not recognized and developed until the twentieth century, can change one element into another.

To predict the identity of a daughter nucleus, we note how the atomic number and mass number change when a particle is ejected from the parent nucleus. When an α particle ($Z = 2$, $A = 4$) is ejected from a nucleus, it carries away four nucleons, two of which are protons and two neutrons. The loss of two protons reduces the atomic number of the element by 2. The loss of four nucleons reduces the mass number by 4. For example, when a radium-226 nucleus, with $Z = 88$, undergoes α decay, the fragment remaining is a nucleus of atomic number 86 (radon) and mass number 222, so the daughter nucleus is radon-222:

$$^{226}_{88}Ra \longrightarrow {}^{222}_{86}Rn + {}^{4}_{2}\alpha$$

This expression is an example of a **nuclear equation.** The general procedure for predicting the identity of daughter nuclei is described in Toolbox 22.1.

The transmutation of a nucleus can be predicted by balancing the atomic numbers and the mass numbers in the nuclear equation for the process.

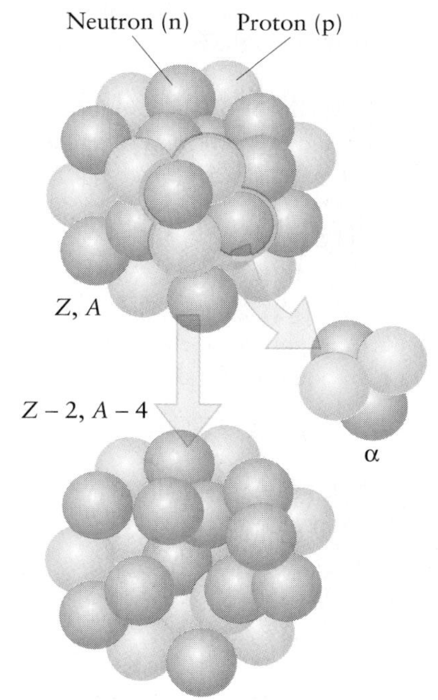

FIGURE 22.7 When a nucleus ejects an α particle, the atomic number of the atom decreases by 2 and the mass number decreases by 4. The nucleons ejected have been indicated by the yellow boundary in the upper part of the diagram.

OOLBOX 22.1 How to identify the products of a nuclear reaction

The underlying strategy is to balance mass numbers and atomic numbers in the nuclear equation. To implement this idea, proceed as follows:

Step 1. Identify the initial nuclide and write its chemical symbol.

Step 2. Write the nuclear reaction, with the mass number and the atomic number of the daughter nucleus written as *A* and *Z*, respectively.

Step 3. Identify the type of decay. Find the values of *A* and *Z* for the daughter nucleus from the fact that the sums of the mass numbers and the atomic numbers are both unchanged in the decay.

Step 4. Once *A* and *Z* are known, identify the daughter nucleus by reference to the periodic table.

The masses and charges of the most common particles ejected or captured by nuclei are as follows:*

Process	Mass number of particle	Charge	Example
α decay	4	+2	Fig. 22.7
β decay	0	−1	Fig. 22.8
electron capture	0	−1	Fig. 22.9
positron emission	0	+1	Fig. 22.10
proton emission	1	+1	—
neutron emission	1	0	—

The following examples show how to balance nuclear equations involving each of these types of decay. More information on these particles is given in Table 22.1.

α decay To identify the nuclide produced by α decay of gadolinium-150, we note that the initial nuclide is $^{150}_{64}Gd$. Therefore, the reaction is

$$^{150}_{64}\text{Gd} \longrightarrow {}^{A}_{Z}\text{E} + {}^{4}_{2}\alpha$$

Because $150 = A + 4$, we can write $A = 146$; because $64 = Z + 2$, $Z = 62$. The element with $Z = 62$ is samarium, Sm, so the nuclide produced is samarium-146; the nuclear equation for the α decay is therefore

$$^{150}_{64}\text{Gd} \longrightarrow {}^{146}_{62}\text{Sm} + {}^{4}_{2}\alpha$$

SELF-TEST 22.1A Identify the nuclide produced by α decay of uranium-235.

[*Answer:* Thorium-231]

*The mass numbers and charges are given for all particles in this toolbox, but they will be omitted in the rest of the text because they are unvarying.

SELF-TEST 22.1B Identify the nuclide produced by α decay of thorium-232.

β decay The loss of one negative charge when an electron is ejected from the nucleus (see Fig. 22.8) can be interpreted as the conversion of a neutron into a proton within the nucleus

$$^{1}_{0}\text{n} \longrightarrow {}^{1}_{1}\text{p} + {}^{0}_{-1}\text{e}, \quad \text{or more simply n} \longrightarrow \text{p} + \beta$$

Because the daughter nucleus is left with an additional proton, its atomic number is greater by 1 than the parent nucleus. The mass number is unchanged, because the total number of nucleons in the nucleus is unchanged. A neutron has changed into a proton; only charge and energy have changed, not the number of nucleons. For example, when sodium-24, with 11 protons and 13 neutrons, undergoes β decay, the daughter nucleus has 12 protons and 12 neutrons. It is therefore an atom of the element with atomic number 12 (magnesium), but with the same mass number as the parent nucleus:

$$^{24}_{11}\text{Na} \longrightarrow {}^{24}_{12}\text{Mg} + {}^{0}_{-1}\text{e}$$

In general, *when a nuclide undergoes β decay*, the mass number is unchanged and the atomic number increases by 1.

SELF-TEST 22.2A Identify the nuclide produced by β decay of lithium-9.

[*Answer:* $^{9}_{4}Be$]

SELF-TEST 22.2B Identify the nuclide produced by β decay of radium-228.

Electron capture In this process, a nucleus captures one of its own orbiting electrons; a proton is turned into a neutron and, although there is no change in mass number, the atomic number is reduced by 1 (see Fig. 22.9). An example is the nuclear reaction

$$^{44}_{22}\text{Ti} + {}^{0}_{-1}\text{e} \longrightarrow {}^{44}_{21}\text{Sc}$$

Positron emission A positron is denoted $^{0}_{+1}e$, or more simply β^+; it has the same tiny mass as an electron but a single positive charge. Positron emission can be thought of as the positive charge shrugged off by a proton as it is converted into a neutron (the reverse of β generation). As a result, the atomic number decreases by 1 but there is no change in mass number (see Fig. 22.10). The resulting change in the nucleus is the same as that for electron capture:

$$^{43}_{22}\text{Ti} \longrightarrow {}^{43}_{21}\text{Sc} + \beta^+$$

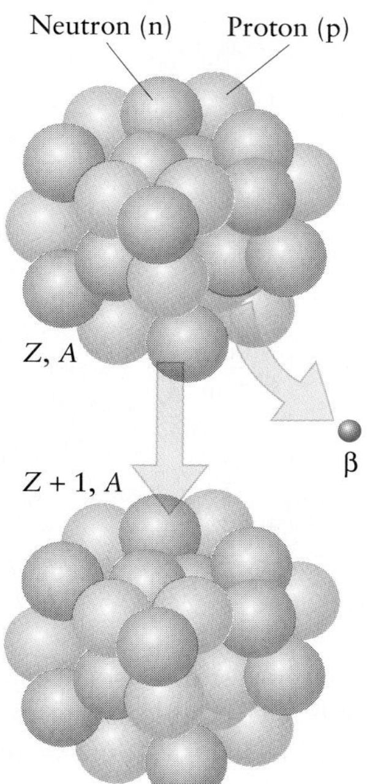

FIGURE 22.8 When a nucleus ejects a β particle, the atomic number of the atom increases by 1 and the mass number remains unchanged. The neutron that we can regard as the source of the electron is indicated by the yellow boundary in the upper part of the diagram.

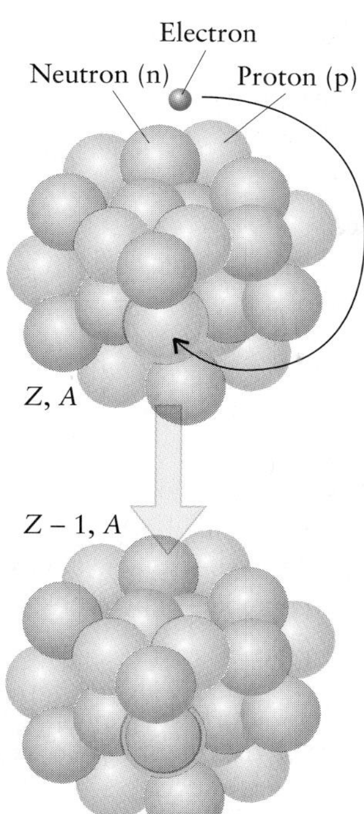

FIGURE 22.9 In the process of electron capture, a nucleus captures one of the surrounding electrons. The effect is to convert a proton (outlined in yellow) into a neutron. As a result, the atomic number decreases by 1 but the mass number remains the same.

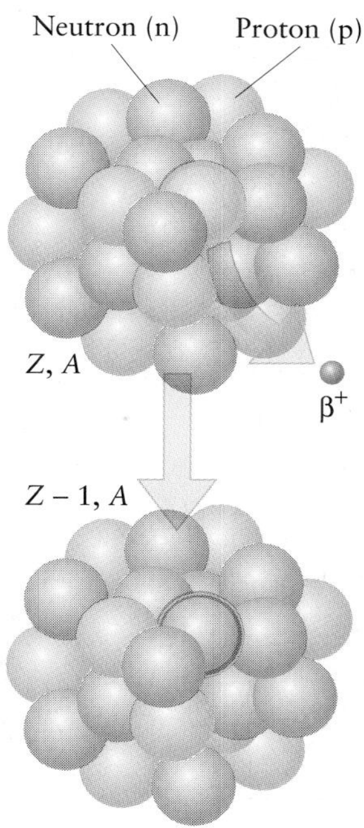

FIGURE 22.10 In positron (β^+) emission, the nucleus ejects a positron. The effect is to convert a proton (outlined in yellow) into a neutron. As a result, the atomic number decreases by 1 but the mass number remains the same because the number of nucleons is unchanged.

Proton and neutron emission These processes are less common. Loss of a proton decreases both mass number and atomic number by 1. Loss of a neutron decreases only the mass number by 1:

$$^{57}_{30}\text{Zn} \longrightarrow {}^{56}_{29}\text{Cu} + \text{p}$$

$$^{91}_{34}\text{Se} \longrightarrow {}^{90}_{34}\text{Se} + \text{n}$$

SELF-TEST 22.3A Identify the nuclides that result from (a) electron capture by potassium-40 and (b) positron emission by sodium-22.

[*Answer:* (a) Argon-40; (b) neon-22]

SELF-TEST 22.3B Identify the nuclides that result from (a) electron capture by beryllium-7 and (b) positron emission from carbon-11.

22.4 THE PATTERN OF NUCLEAR STABILITY

Nuclei with an even number of protons *and* an even number of neutrons are more stable than those with any other combination. Conversely, nuclei with odd numbers of both protons and neutrons are the least stable (Fig. 22.11). Nuclei are more likely to be more stable if they are built from certain **magic numbers** of either kind of nucleons, namely 2, 8,

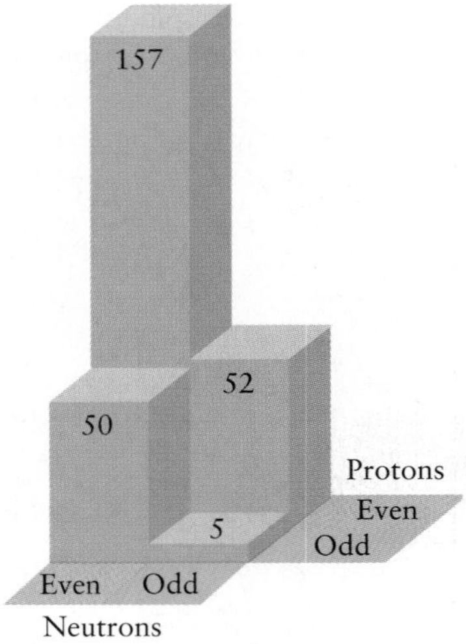

FIGURE 22.11 The numbers of stable nuclides for even and odd numbers of nuclei and protons. By far the greatest number of stable nuclei have even numbers of protons and even numbers of neutrons.

20, 50, 82, and 126. For example, there are 10 stable isotopes of tin ($Z = 50$), the most of any element, but only two stable isotopes of its neighbor antimony ($Z = 51$). The α particle itself is a "doubly magic" nucleus, with two protons and two neutrons. We shall see later that a number of actinides decay through a series of steps until they reach ^{208}Pb, another doubly magic nuclide, with 126 neutrons and 82 protons. A similar pattern of stability applies to electons in atoms, for we have already seen that the noble gas atoms have 2, 10, 18, 36, 54, and 86 electrons. The stabilities of these atoms can be traced to the shell structure of atoms, which strongly suggests that a nucleus also has a shell structure in which nucleons occupy a series of shells, just as electrons in atoms do.

The magic number 126 suggests that it may be possible to create an element with $Z = 126$, lying beyond the end of the current periodic table, which would have exceptional stability relative to the known transuranium elements.

Figure 22.12 is a plot of mass number against atomic number for known nuclides. Stable nuclei are found in a **band of stability** surrounded by a **sea of instability,** the region of unstable nuclides that decay with the emission of radiation. For atomic numbers up to about 20, the stable

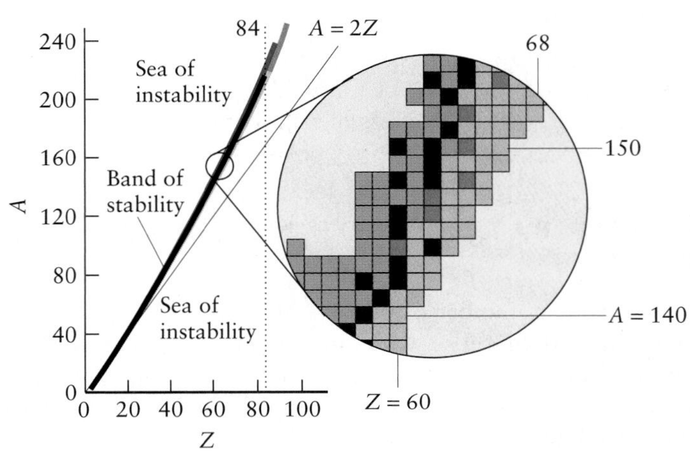

FIGURE 22.12 The manner in which nuclear stability depends on the atomic number and the mass number. Nuclides along the narrow black band (the band of stability) are generally stable. Nuclides in the blue region are likely to emit a β particle, and those in the red region are likely to emit an α particle. Nuclei in the pink region are likely to emit either positrons or to undergo electron capture. The magnified view of the diagram near $Z = 60$, shows the structure of the band of stability.

nuclides have approximately equal numbers of neutrons and protons, so A is close to $2Z$. For higher atomic numbers, all known nuclides—both stable and unstable—have more neutrons than protons (so $A > 2Z$). We can explain this behavior in terms of charge. Protons have mutually repulsive electrostatic charges, whereas neutrons have none, so neutrons can contribute to the strong force that holds the nucleus together, without increasing repulsion. In a nucleus with many protons, a lot of neutrons are needed to overcome the mutual electric repulsion of the protons and help glue them together.

Figure 22.12 can be used to predict the type of disintegration a nuclide is likely to undergo. Unstable isotopes above the band of stability are **neutron rich:** they have a high proportion of neutrons. These nuclei, such as ${}^{14}_{6}C$, can reach safety by ejecting a β particle:

Recall from Toolbox 22.1 that in β emission a neutron is converted into a proton and an electron.

$$ {}^{14}_{6}C \longrightarrow {}^{14}_{7}N + \beta + \gamma $$

Alternatively, neutron-rich nuclides may reduce their neutron count by **spontaneous neutron emission:**

$$ {}^{87}_{36}Kr \longrightarrow {}^{86}_{36}Kr + n $$

Unstable isotopes of elements that lie below the band of stability have a low proportion of neutrons and are classified as **proton rich.** For example, ${}^{29}_{15}P$ is proton rich and can move toward the band of stability by emitting a positron:

Recall from Toolbox 22.1 that in positron emission a proton decays into a neutron and a positron.

$$ {}^{29}_{15}P \longrightarrow {}^{29}_{14}Si + \beta $$

Alternatively, proton-rich nuclides can decrease their proton count by electron capture:

$$ {}^{7}_{4}Be + \beta \longrightarrow {}^{7}_{3}Li $$

or by proton emission:

$$ {}^{43}_{21}Sc \longrightarrow {}^{42}_{20}Ca + p + \gamma $$

All nuclei with $Z > 83$ are unstable and radioactive; they decay mainly by α particle emission. Very few nuclides with $Z < 60$ emit α particles. Nuclides of elements with $Z > 83$ must discard protons to reduce their atomic number, and they generally need to lose neutrons too. They decay in a stepwise manner and give rise to a **radioactive series,** a specific sequence of nuclides (Fig. 22.13). First, one α particle is ejected, then another α particle or a β particle is ejected, and so on, until a stable nucleus is formed—usually the final nuclide is an isotope of lead (the element with the magic atomic number 82). Three radioactive series begin from naturally occurring nuclides:

Uranium-238 series: starts at uranium-238 and ends at lead-206

Uranium-235 series: starts at uranium-235 and ends at lead-207

Thorium-232 series: starts at thorium-232 and ends at lead-208

The radioactive series were very important when radioactivity was first studied, because radioactive materials could be obtained only from the decay of heavier elements. Radioactive series still help to summarize the behavior of nuclear fuels.

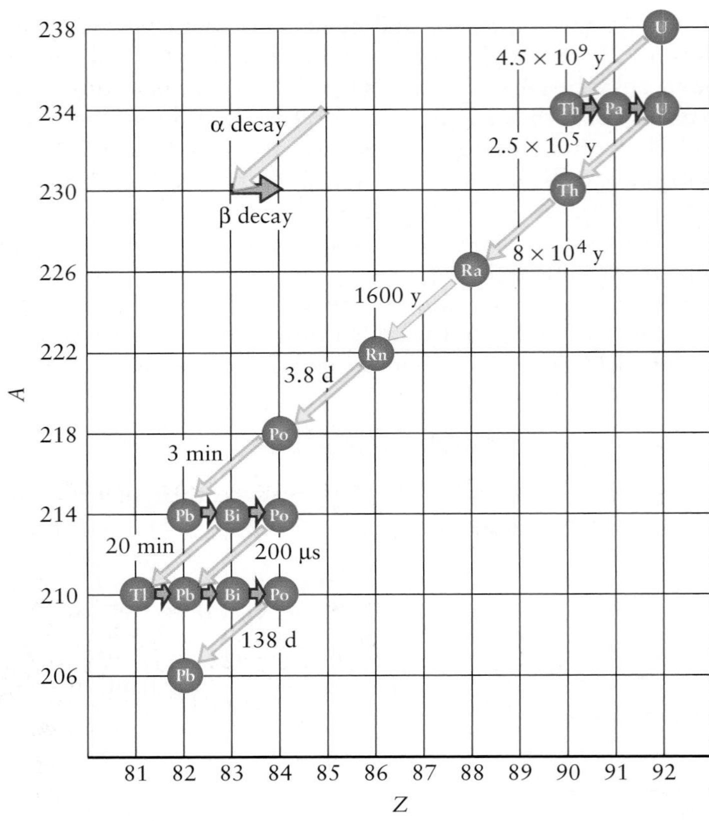

FIGURE 22.13 The uranium-238 series. The times are the half-lives of the nuclides (see Section 22.8).

The pattern of nuclear stability can be used to predict the likely mode of radioactive decay: neutron-rich nuclei tend to reduce their neutron count; proton-rich nuclei tend to reduce their proton count. In general, only heavy nuclides emit α particles.

SELF-TEST 22.4A Which of the processes (a) electron capture, (b) proton emission, (c) neutron emission, (d) β emission, (e) β^+ emission might a ^{145}Gd nucleus undergo to begin to reach stability? Refer to Fig. 22.12.

[*Answer:* a, b, e]

SELF-TEST 22.4B Which of the same set of processes listed in Self-Test 22.4A might a ^{148}Ce nucleus undergo to begin to reach stability?

22.5 NUCLEOSYNTHESIS

The naturally occurring elements not produced in stars are hydrogen and helium, which serve as the raw material of stars.

Nucleosynthesis is the formation of elements. Nucleosynthetic reactions in the interior of stars have produced nearly all the naturally occurring elements on Earth: in this sense, virtually everything around us is made of stardust. Some elements are not found naturally on Earth. For example, the radioactive elements technetium and promethium were discovered in the spectra of stars, but on Earth they are found only in minute quantities in uranium ores; and no element with an atomic number greater than 92 has been detected occurring naturally either on Earth or in the stars. To form these elements on Earth, we must overcome the energy barriers to nuclear synthesis by simulating the very high energy conditions found inside a star (Fig. 22.14). There are two ways to do this. One is to heat a substance to the very high temperatures—millions of degrees—found in the interiors of stars—to increase the speed of the particles. Another is to bombard nuclei with elementary particles or other nuclei that have been accelerated to high speeds in a particle accelerator (Fig. 22.15). High speed is essential if the projectile particles are positively charged, because to approach the target nucleus closely, they must overcome its electrostatic repulsion.

Charged particles can be accelerated to high energies by a cyclotron, a synchrocyclotron, or a linear accelerator. Most of the transuranium elements were originally produced at the Lawrence Livermore Radiation Laboratories in Berkeley, California, or the Dubna Laboratories in Russia.

Rutherford achieved the first artificial nuclear transmutation in 1919. He bombarded nitrogen-14 nuclei with α particles. These particles travel at high speed because their positive charge receives a hefty "kick" from the positive charge of the nucleus that ejects them. The products of the transmutation are oxygen-17 and a proton:

$$^{14}_{7}\text{N} + \alpha \longrightarrow {}^{17}_{8}\text{O} + \text{p}$$

A similar process occurs in stars and results in the formation of oxygen-16 from carbon-12:

$$^{12}_{6}\text{C} + \alpha \longrightarrow {}^{16}_{8}\text{O} + \gamma$$

A large number of nuclides have been synthesized on Earth by using a similar procedure. For instance, technetium ($Z = 43$) was prepared for the first time on Earth in 1937 by the reaction between molybdenum ($Z = 42$) and **deuterons,** deuterium nuclei:

$$^{97}_{42}\text{Mo} + {}^{2}_{1}\text{H} \longrightarrow {}^{97}_{43}\text{Tc} + 2\,\text{n}$$

Moderately abundant supplies of technetium are now available; technetium-99 is used in pharmaceutical applications, particularly for imaging the heart.

It is easier for a neutron to get close to a target nucleus because it is not repelled by the nuclear charge. An example of **neutron-induced transmutation** is the three-step formation of cobalt-60, which is used in the radiation treatment of cancer, from iron-58. First, iron-59 is produced from iron-58:

$$^{58}_{26}\text{Fe} + \text{n} \longrightarrow {}^{59}_{26}\text{Fe}$$

The second step is β decay of the iron-59 to cobalt-59:

$$^{59}_{26}\text{Fe} \longrightarrow {}^{59}_{27}\text{Co} + \beta$$

In the final step, the cobalt-59 absorbs another neutron from the incident neutron beam and is converted into cobalt-60:

$$^{59}_{27}\text{Co} + \text{n} \longrightarrow {}^{60}_{27}\text{Co}$$

The overall reaction is

$$^{58}_{26}\text{Fe} + 2\,\text{n} \longrightarrow {}^{60}_{27}\text{Co} + \beta$$

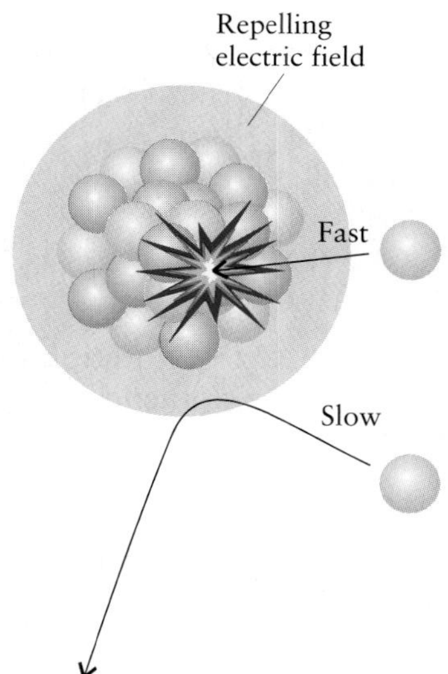

FIGURE 22.14 When a positively charged particle approaches a nucleus, it is repelled strongly. However, if it is traveling very fast, it can reach the nucleus before the repulsion turns it aside; in this case, a nuclear reaction may occur.

FIGURE 22.15 An aerial view of the Fermi National Accelerator Laboratory in Batavia, Illinois. The largest circle is the main accelerator; three experimental lines are tangent to it.

SELF-TEST 22.5A Complete the following nuclear reactions:

(a) ? + $\alpha \longrightarrow {}^{243}_{96}\text{Cm} + \text{n}$ (b) ${}^{242}_{96}\text{Cm} + \alpha \longrightarrow {}^{245}_{98}\text{Cf}$ + ?

[*Answer:* (a) ${}^{240}_{94}\text{Pu}$; (b) n]

SELF-TEST 22.5B Complete the following nuclear reactions:

(a) ${}^{250}_{98}\text{Cf}$ + ? $\longrightarrow {}^{257}_{103}\text{Lr} + 4\text{n}$ (b) ? + ${}^{12}_{6}\text{C} \longrightarrow {}^{254}_{102}\text{No} + 4\text{n}$

Meitnerium is named after Lise Meitner, the Austrian-German physicist who discovered protactinium and whose studies laid the foundation for our use of nuclear fission (see Section 22.10).

The **transuranium elements** are the elements following uranium in the periodic table; they have $Z > 92$. All are synthetic and are produced by the bombardment of target nuclei with a smaller projectile. The elements through meitnerium (Mt, $Z = 109$) have been formally named. The elements beyond meitnerium (including hypothetical nuclides that have not yet been made) are named systematically, at least until they have been identified and there is agreement on a permanent name. The systematic nomenclature of these elements uses the terms in Table 22.2. For example, the element with $Z = 111$, one atom of which was first made in 1994, will be called unununium, Uuu, until it is finally named by international agreement.

New elements and isotopes of known elements are made by nucleosynthesis; the repulsive electrical forces of like-charged particles are overcome when very fast particles collide.

TABLE 22.2 **Notation for the systematic nomenclature of elements***

Digit	Name	Abbreviation
0	nil	n
1	un	u
2	bi	b
3	tri	t
4	quad	q
5	pent	p
6	hex	h
7	sep	s
8	oct	o
9	enn	e

*For instance, element 123 would be named unbitriium, Ubt.

RADIOACTIVITY

Nuclear radiation is sometimes called **ionizing radiation,** because it is energetic enough to eject electrons from atoms. These high-energy rays can be used, with care, to cure disease, but they can also harm biological tissues (Case Study 22). The extent of damage depends on the strength of the source, the type of radiation, and the length of exposure. Another important aspect of radioactivity is its persistence: How long do we need to store radioactive material before it can be considered safe?

TABLE 22.3 Shielding requirements of α, β, and γ radiation

Radiation	Relative penetrating power	Shielding required
α	1	paper, skin
β	100	3 mm aluminum
γ	10,000	concrete, lead

22.6 THE EFFECTS OF RADIATION

The three main types of nuclear radiation penetrate matter to different extents (Table 22.3). The least penetrating is α radiation, which can be stopped by a sheet of paper. The massive, highly charged α particles interact so strongly with matter that they slow down, capture electrons from surrounding matter, and change into bulky helium atoms before traveling very far. However, even though α particles do not penetrate far into matter, they are very damaging because the energy of their impact can knock atoms out of molecules and displace ions from their sites in crystals. The damage caused by α particles when they strike human tissue can lead to serious illness and even death. If DNA and the enzymes that interpret its protein-building messages are damaged, then the result may be cancer. Most α radiation is absorbed by the surface layer of dead skin, where it can do little harm. However, inhaled and ingested particles can cause serious internal damage. For example, plutonium, considered to be one of the most toxic radioactive materials, is an α emitter and can be handled safely with minimal shielding (Fig. 22.16). However, it is easily oxidized to Pu^{4+}, which has chemical properties similar to those of Fe^{3+}. It can take the place of iron in the body and be absorbed rather than excreted, remaining to cause radiation sickness and cancer.

The fact that α emitters that enter the body through the respiratory system may lead to lung cancer has led to the current concern about the radioactive gas radon, an α emitter, seeping into homes, particularly basements, from the soil.

Next in penetrating power is β radiation. The fast electrons that make up these rays can penetrate about 1 cm deep into flesh before their electrostatic interactions with the electrons and nuclei of molecules bring them to a standstill.

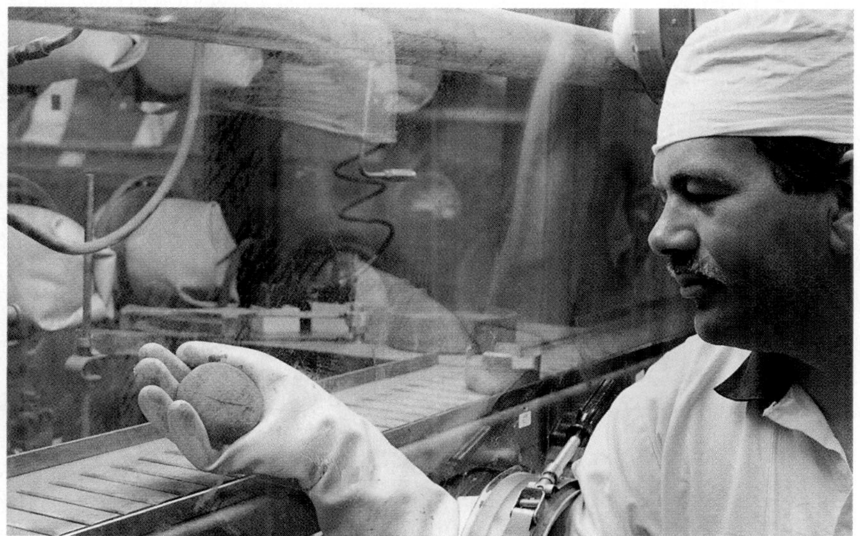

FIGURE 22.16 Even relatively large pieces of plutonium-239 can be handled with minimum protection, as long as they are isolated from the atmosphere. Here, a glove box is used to provide a sealed environment. Plutonium is warm to the touch because of the constant α decay, and larger pieces can be too hot to touch.

CASE STUDY NUCLEAR MEDICINE: REDUCING THE RISKS

EVE CURIE REPORTED that her parents, Marie and Pierre Curie, were so fascinated with the phosphorescent glow of radium that they sometimes turned off the lights to watch it. The Curies speculated about possible medical applications of radioisotopes and only gradually learned of their hazards. Today, radioisotopes are widely used in medicine to diagnose, study, and treat illness. For example, a physician can determine how and at what rate the thyroid gland takes up iodine by using iodine-131 as a radioactive tracer, and cobalt-60 is used to kill rapidly growing cancer cells. Because the same powerful energies that diagnose illness and cure it are also capable of damaging healthy tissue, a great deal of attention has been paid to reducing the risks associated with the medical use of radioisotopes.

Nuclear chemistry has transformed the field of medical diagnosis. For example, radioactive tracers are used to measure organ function, sodium-24 is used to monitor blood flow, and strontium-87 is used to study bone growth. However, the most dramatic impact of radioisotopes on diagnosis has been in the field of imaging. The better the picture of an organ, the better diagnosis a physician can make. Technetium-99m is a metastable isotope (the m denotes a "metastable" state, an excited state with an appreciable lifetime) that is the most widely used radioactive nuclide in medicine for studying the functioning of internal organs, including the heart; for instance, as $^{99}TcO_4^-$, it is used to detect and pinpoint brain tumors. Technetium-99m gives off a γ ray and decays to technetium-99, with a half-life of 6 h. Thus, it is highly active but has a very short half-life. The fact that the isotope emits only γ rays adds to its desirability for imaging. The heavier α rays cause more damage in the body and are absorbed by body tissues before they can be detected by instruments.

Positron emission tomography (PET) is a technique that uses a positron emitter such as fluorine-18 to image human tissue in a degree of detail not possible with x-rays. For example, the hormone estrogen can be labeled with fluorine-18 and injected into an individual with a tumor. The fluorine-bearing compound is preferentially absorbed by the tumor. The positrons given off by the fluorine atoms are quickly annihilated, and the γ rays resulting from the annihilations are detected by a scanner that moves slowly over the part of the body with the tumor. The growth of the tumor can be quickly and accurately estimated with this technique. An advantage of this technique is that the positron emitters have very short half-lives; for example, the half-life of fluorine-18 is 110 min. In fact, a PET imaging facility must be located near a cyclotron, so that the positron emitters can be quickly incorporated into the desired compounds as soon as they are created.

The use of radioisotopes to cure cancer is associated with much greater risks than those associated with imaging techniques, because the therapeutic radiation must be fairly strong to destroy the tumor. However, two relatively new techniques can considerably reduce the risk of radiation treatment. The first is the use of *monoclonal antibodies*. Antibodies are compounds that the body produces to fight disease. Monoclonal antibodies are artificial antibodies

This sample of cesium oxide contains so many radioactive cesium-137 nuclei that it glows from the energy released during their decay. A variety of ingeneous techniques have been developed to allow the tremendous energy of radioactive decay to be used safely.

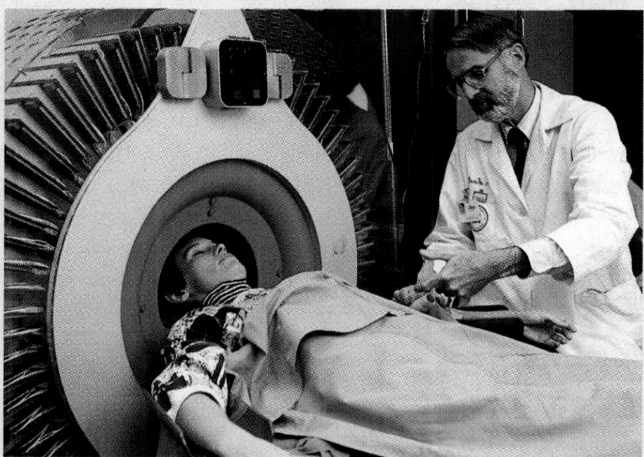

This patient is about to undergo a PET scan of brain function.

that biotechnologists create to attack specific types of cancer. When radioisotopes are incorporated into a monoclonal antibody that is then injected into the body, the antibody attaches to a malignant tumor and the radiation is concentrated in the cancerous tissue. Iodine-131 ($t_{1/2}$ = 8.05 d) and yttrium-90 ($t_{1/2}$ = 64 h) are frequently used in monoclonal antibodies. Their short half-lives ensure that the activity in the body will not last long.

Boron neutron capture therapy is an unusual radiation therapy in that boron-10, the isotope injected, is not radioactive. However, when it is bombarded with neutrons, boron-10 gives off highly destructive α particles. In boron neutron capture therapy, the boron-10 is incorporated into a compound that is preferentially absorbed by tumors. The patient is then exposed to neutron bombardment, but only for brief periods of time. As soon as the bombardment ceases, the boron-10 stops generating α particles.

It is unfortunate that Marie Curie was not able to benefit from the results of her work. She died of leukemia, a disease that can be treated today with radioactive isotopes.

QUESTIONS

1. Do nuclei that are positron emitters lie above or below the band of stability? Which of the following isotopes might be suitable for PET scans? Explain your reasoning and write the equation for the decay: (a) ^{18}O; (b) ^{13}N; (c) ^{11}C; (d) ^{20}F; (e) ^{15}O.

2. Write the nuclear equation for the reaction of boron-10 and a neutron in which an α particle is produced.

3. Technetium-99m is produced by a sequence of reactions in which molybdenum-98 is bombarded with neutrons to form molybdenum-99, which undergoes β decay to technetium-99m. Write the balanced nuclear equations for this sequence.

4. The mass of an electron is 9.109×10^{-28} g; a positron has the same mass but the opposite charge. When a positron emitted during a PET scan encounters an electron, annihilation occurs in the body: electromagnetic energy is produced and no matter remains. How much energy (in joules) is produced in the encounter? (See Section 22.9.)

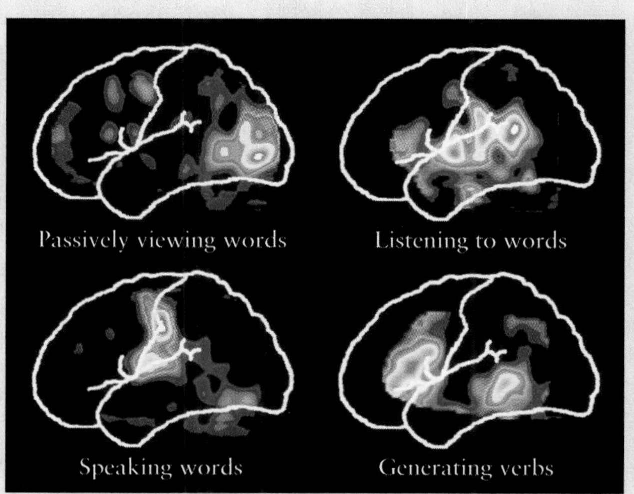

These four PET scans show how blood flow to different parts of the brain is affected by various activities. In this case, an oxygen isotope that is taken up by the hemoglobin in blood is used as a source of positrons.

Most penetrating of all is γ radiation. The uncharged, high-energy γ-ray photons can pass right through the body and most building materials, and cause damage by ionizing the molecules in their path. Protein molecules and DNA that have been damaged in this way cannot carry out their functions, and the result can be radiation sickness and cancer. Intense sources of γ rays must be surrounded by walls built from lead bricks or thick concrete to shield people from this penetrating radiation.

The **absorbed dose** of radiation is the energy deposited in the sample (in particular, the human body) when it is exposed to radiation. For a long time, the unit for reporting dose has been the **rad,** which stands for radiation absorbed dose, and is the amount of radiation that deposits 10^{-2} J of energy per kilogram of tissue. If the radiation were absorbed uniformly throughout the body, a dose of 1 rad would correspond to a 65-kg person absorbing a total of 0.65 J. The same energy is transferred as heat when a droplet of boiling water of mass 2 mg touches the skin. However, the energy of each radioactive particle is highly localized, like the impact of a bullet, but on a much smaller scale, and not spread evenly over a region the size of a water droplet. As a result, the incoming particles can break individual bonds as they collide with molecules in their path.

The SI unit of absorbed dose is the *gray,* Gy, which corresponds to an energy deposit of 1 $J \cdot kg^{-1}$, so 1 Gy = 100 rad.

The extent of radiation damage to living tissue depends on the type of radiation and the type of tissue. A dose of 1 rad of γ radiation causes about the same amount of damage as 1 rad of β radiation, but α particles are about 20 times more damaging (even though they are the least penetrating). A factor called the **relative biological effectiveness,** Q, must therefore be included when assessing the damage that a given dose of each type of radiation may cause. For β and γ radiation, Q is set arbitrarily at about 1; but for α radiation, Q is close to 20. The precise figures depend on the total dose, the rate at which the dose accumulates, and the type of tissue, but these values are typical.

The **dose equivalent** is the actual dose modified to take into account the different destructive powers of the various types of radiation in combination with various types of tissue. It is obtained by multiplying the actual dose (in rads) by the value of Q for the radiation type. The result is expressed in a unit called a **roentgen equivalent man** (rem):

The SI unit of dose equivalent is the *sievert,* Sv, which is defined in the same way as rem but with the absorbed dose in gray; thus, 1 Sv = 100 rem.

$$\text{Dose equivalent in rem} = Q \times \text{absorbed dose in rad}$$

A dose of 30 rad (0.3 Gy) of γ radiation corresponds to a dose equivalent of 30 rem (0.3 Sv), enough to cause a reduction in the number of white blood cells (the cells that fight infection), but 30 rad of α radiation corresponds to 600 rem (6 Sv), which is enough to cause death. A typical average annual dose equivalent that we each receive from natural sources is about 0.2 $rem \cdot y^{-1}$, but this figure varies, depending on our lifestyle and where we live. About 20% is radiation from our own bodies. About 30% comes from cosmic rays (a mix of γ rays and high-energy elementary particles from outer space) that continuously bombard the Earth, and 40% comes from radon seeping out of the ground. The remaining 10% is a result largely of medical diagnoses (a typical chest x-ray gives a dose equivalent of about 7 millirem). Emissions from nuclear power plants and other nuclear facilities contribute about 0.1% in countries where they are widely used.

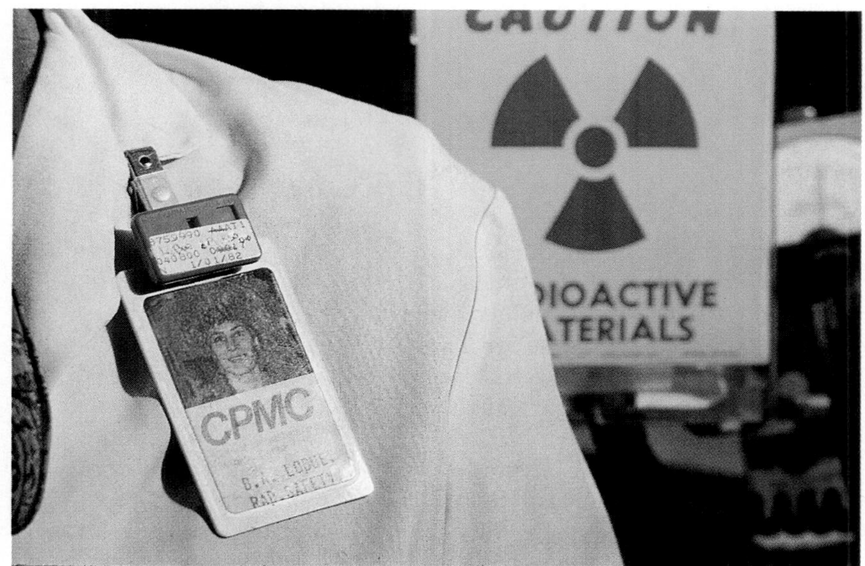

FIGURE 22.17 A film badge used by workers in areas where exposure to radiation is a risk.

We are not in danger from this constant bombardment from **background radiation,** the radiation to which we are exposed daily and that has always been present on Earth. It is only when we are exposed to large amounts of radiation that the body's defense mechanisms are in danger of being overwhelmed.

> ***Human health in the presence of radiation is monitored by reporting the absorbed dose and the dose equivalent; the latter takes into account the effects of different types of radiation on tissues.***

22.7 MEASURING RADIOACTIVITY

The ability of nuclear radiation to ionize atoms can be used to measure its intensity. Becquerel first used this ionization to gauge the intensity of radiation by the degree to which it blackened a photographic film. The blackening results from the same redox processes as those occurring in ordinary photography (see Case Study 21), except that the initial oxidation of the halide ions is caused by nuclear radiation instead of light. Becquerel's technique is still used in the film badges that monitor the exposure of workers to radiation (Fig. 22.17).

A **Geiger counter** monitors radiation by detecting the ionization of a low-pressure gas (Fig. 22.18). The radiation ionizes atoms inside a cylinder and allows a brief flow of current between the electrodes. The

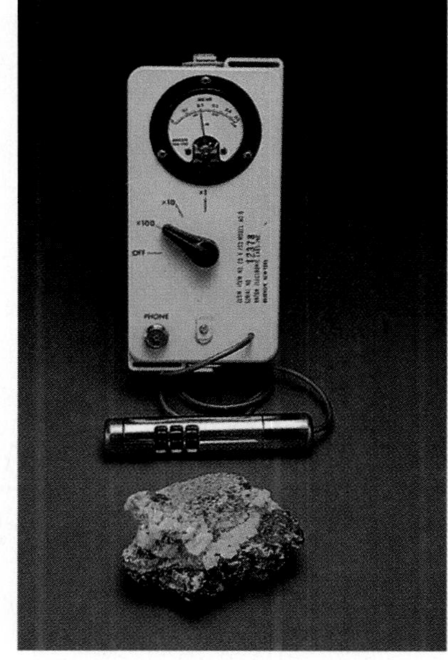

(a)

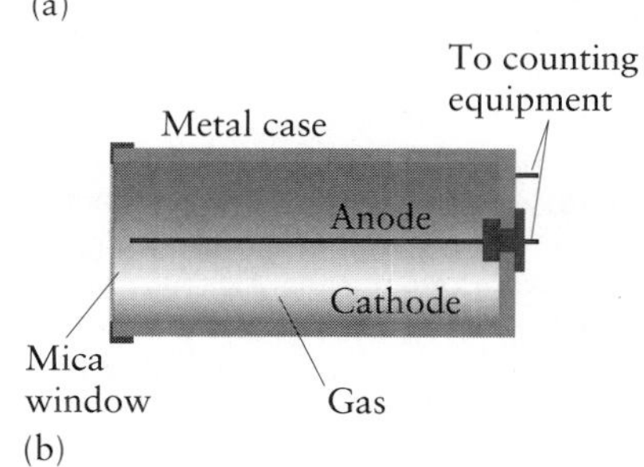

(b)

FIGURE 22.18 (a) A Geiger counter. (b) The detector in a Geiger counter consists of a gas (typically argon and a little ethanol vapor, or neon and some bromine vapor) in a container with a high potential difference (of 500 V to 1200 V) between the walls and the central wire. An electric current passes between the electrodes (walls and wire) when radiation ionizes the gas.

FIGURE 22.19 This scintillation counter detects each radioactive decay as a flash of light.

The unit *becquerel* was named in honor of Henri Becquerel; the unit *curie* was named in honor of Pierre Curie by his wife, Marie.

resulting electrical signal can be recorded directly or converted into an audible click. The rapidity of the clicks then indicates the intensity of the radiation. A **scintillation counter** makes use of the fact that certain substances (notably, zinc sulfide) give a flash of light—a scintillation—when exposed to radiation. The intensity of the radiation is measured by counting the scintillations electronically (Fig. 22.19).

Each click of a Geiger counter or flash of a scintillation counter indicates that one nuclear disintegration has occurred. The **activity** of a sample is the number of nuclear disintegrations that occur per second. The more active the source, the greater the number of nuclear disintegrations per second. One nuclear disintegration per second is called 1 **becquerel** (1 Bq). Another commonly used (non-SI) unit of radioactivity is the **curie** (Ci). It is equal to 3.7×10^{10} nuclear disintegrations per second, the radioactive output of 1 g of radium-226. Because the curie is a very large unit, most activities are expressed in millicuries (1 mCi = 10^{-3} Ci) and microcuries (1 μCi = 10^{-6} Ci). Other units and their conversions are summarized in Table 22.4.

The principal natural source of radioactivity in the human body is potassium-40, which occurs in about 0.01% abundance among the potassium ions found throughout the body. About 37,000 potassium-40 nuclei disintegrated in your body in the time it took you to read this sentence (about 10 s).

Radioactivity is measured by using its ionizing effects; the activity of a source is a measure of the number of nuclear disintegrations per second.

TABLE 22.4 Radiation units*

Property	Unit name	Symbol	Definition
activity	curie	Ci	3.7×10^{10} disintegrations per second
	bequerel	Bq	1 disintegration per second
			(1 Ci = 3.7×10^{10} Bq)
absorbed dose	radiation absorbed dose	rad	10^{-2} J·kg^{-1}
	gray	Gy	1 J·kg^{-1}
			(1 Gy = 100 rad)
dose equivalent	roentgen equivalent man	rem	$Q \times$ absorbed dose†
	sievert	Sv	100 rem

*The SI units are bequerel, gray, and sievert.
†Q is the relative biological effectiveness of the radiation. Normally, $Q \approx 1$ for γ, β, and most other radiation, but $Q \approx 20$ for α radiation and fast neutrons. A further factor of 5 (that is, $5Q$) is used for bone under certain circumstances.

22.8 THE LAW OF RADIOACTIVE DECAY

We can use Geiger counters and scintillation counters to study the rate at which a radioactive nucleus decays. The nuclear equation

$$\text{Parent nucleus} \longrightarrow \text{daughter nucleus} + \text{radiation}$$

has exactly the same form as the equation for a unimolecular elementary reaction, with an unstable nucleus taking the place of an excited molecule. As in the case of a unimolecular chemical reaction (Sections 18.14 and 18.15), the rate law for nuclear decay is first order. That is, the relation between the rate of decay and the number N of radioactive nuclei present is given by the **law of radioactive decay:**

$$\text{Activity} = \text{rate of decay} = k \times N \quad (1)$$

In this context, k is called the **decay constant.** The law tells us that the more radioactive nuclei there are in the sample, the greater the rate of nuclear decay and hence the more active the sample. Each nuclide has a characteristic value of k, so the rate of decay also depends on the identity of the nuclide.

As explained in Section 18.5, a first-order rate law implies an exponential decay. Hence, the activity of a radioactive sample falls exponentially toward 0 (Fig. 22.20). As for a first-order chemical reaction, it follows that the number N of nuclei remaining after a time t is given by

$$N = N_0 e^{-kt} \quad (2)$$

where N_0 is the number of radioactive nuclei present initially (at $t = 0$).

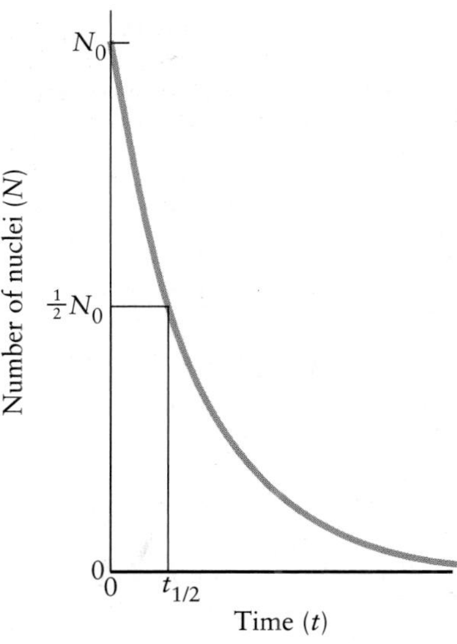

FIGURE 22.20 The exponential decay of the number of radioactive nuclei in a sample implies that the activity of the sample also decays exponentially with time. The curve is characterized by the half-life $t_{1/2}$.

EXAMPLE 22.1 Using the law of radioactive decay

One of the reasons why thermonuclear weapons have to be serviced is the radioactive decay of the tritium they contain. Suppose a sample of tritium of mass 1.0 g is stored. What mass of that isotope will remain after 5.0 y? The decay constant is 0.0564 y^{-1}.

STRATEGY The number of tritium nuclei in the sample is proportional to the total mass m of tritium, so Eq. 2 can be expressed as

$$m = m_0 e^{-kt}$$

where m_0 is the initial mass of tritium. The question can now be answered by substituting the data.

SOLUTION Substitution of the data gives

$$m = (1.0\text{ g}) \times e^{-(0.0564\text{ y}^{-1}) \times (5.0\text{ y})} = 0.75\text{ g}$$

SELF-TEST 22.6A The decay constant for fermium-254 is 210 s^{-1}. What mass of the isotope will be present if a sample of mass 1.00 μg is kept for 10 ms?

[*Answer:* 0.12 μg]

SELF-TEST 22.6B The decay constant for neptunium-237 is 3.3×10^{-7} y^{-1}. What mass of the isotope will be present if a sample of mass 5.0 μg survives for 1.0×10^6 y?

◀◀ Half-lives were first introduced in Section 18.6 in connection with first-order chemical reactions.

Radioactive decay is normally discussed in terms of the **half-life,** $t_{1/2}$, the time needed for half the initial number of nuclei to disintegrate. Just

TABLE 22.5 Radioactive half-lives*

Nuclide	Half-life, $t_{1/2}$
tritium	12.3 y
carbon-14	5.73×10^3 y
carbon-15	2.4 s
potassium-40	1.26×10^9 y
cobalt-60	5.26 y
strontium-90	28.1 y
iodine-131	8.05 d
cesium-137	30.17 y
radium-226	1.60×10^3 y
uranium-235	7.1×10^8 y
uranium-238	4.5×10^9 y
fermium-244	3.3 ms

*d = day, y = year.

as we did in Section 18.6, we find the relation between $t_{1/2}$ and k by setting $N = \frac{1}{2}N_0$ and $t = t_{1/2}$ in Eq. 2:

$$t_{1/2} = \frac{\ln 2}{k} \tag{3}$$

The larger the value of k, the shorter the half-life of the nuclide. Nuclides with short half-lives are less stable than nuclei with long half-lives. That means they are more likely to decay in a given period of time and are thus "hotter" than nuclides with long half-lives.

Table 22.5 shows that half-lives span a very wide range. As an example, consider strontium-90, for which the half-life is 28 y. This nuclide occurs in the fallout, the fine dust that settles from clouds of airborne particles after a nuclear bomb test or an accidental release of radioactive materials into the air. Because it is chemically very similar to calcium, strontium may accompany that element through the environment and become incorporated into bones; once there, it continues to emit radiation for many years. Even after three half-lives (84 y), one-eighth of the original strontium-90 still survives. About 10 half-lives (for strontium-90, 280 y) must pass before the activity of a sample has fallen to 1/1000 of its initial value. Plutonium-239, which is present in nuclear waste, has a half-life of 2.4×10^4 y. Consequently, very long term storage facilities are required for plutonium waste, and land that becomes contaminated with plutonium cannot be inhabited again for thousands of years. Serious contamination from radioactive fallout occurred within a 30-km radius of the nuclear reactor that caught fire in Chernobyl, Ukraine. The land is contaminated with plutonium and with cesium-137 and has been sealed off indefinitely from human use (Fig. 22.21).

The constancy of the half-life of a nuclide is put to practical use in determining the ages of archeological artifacts. In **isotopic dating,** we measure the activity of the radioactive isotopes they contain. Isotopes used for dating objects include uranium-238, potassium-40, and tritium

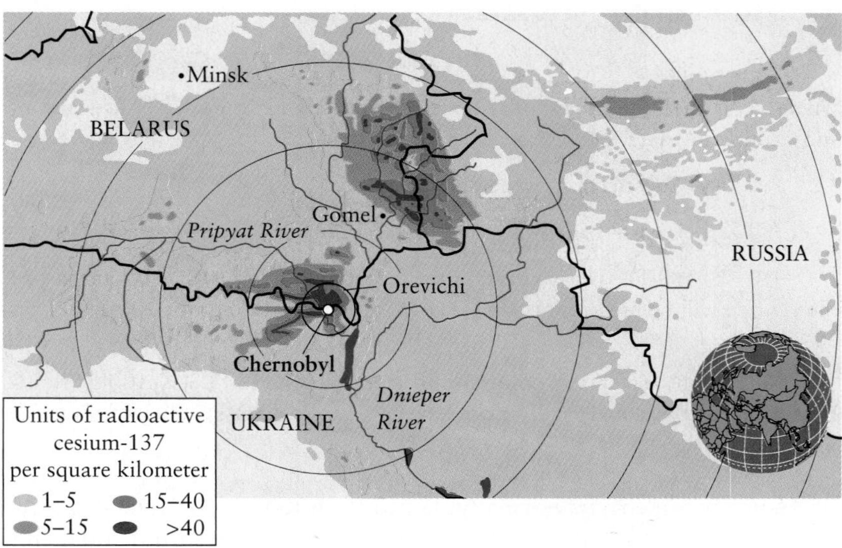

FIGURE 22.21 A map of the region near the Chernobyl nuclear power plant. The colors indicate the relative intensity of radiation remaining in the soil. Areas with the darkest colors are uninhabitable.

(Box 22.1). However, the most important example is radiocarbon dating, which uses the β decay of carbon-14, for which the half-life is 5730 y. Carbon-14 is a naturally occurring isotope of carbon and is present in all living things. Prior to the industrial revolution, which released carbon stored in fossil fuels into the atmosphere, the supply of carbon-14 atoms in the environment was nearly constant over archeological time. The atoms are produced when nitrogen nuclei in the atmosphere are bombarded by neutrons:

There are small variations with sunspot activity.

$$^{14}_{7}\text{N} + \text{n} \longrightarrow {}^{14}_{6}\text{C} + \text{p}$$

The neutrons originate from the collisions of cosmic rays with other nuclei. The carbon-14 atoms produced in the atmosphere enter living organisms as $^{14}CO_2$ through photosynthesis and digestion. They leave the organisms by the normal processes of excretion and respiration, and also because the nuclei decay at a steady rate. As a result, all living things have a constant proportion of the isotope among their very much more numerous carbon-12 atoms. In other words, there is a fixed ratio (of about $1/10^{12}$) of carbon-14 atoms to carbon-12 atoms in living tissues.

When the organism dies, it no longer exchanges carbon with its surroundings. However, carbon-14 nuclei already inside the organism continue to disintegrate with a constant half-life. Hence the ratio of carbon-14 to carbon-12 decreases after death, and the ratio observed in a sample of dead tissue can be used to estimate the time since death. In the modern version of the technique, which requires only milligrams of sample, the carbon atoms are converted into C^- ions by bombardment of the sample with cesium atoms. The C^- ions are then accelerated with electric fields, and the carbon isotopes are separated and counted with a mass spectrometer (Fig. 22.22). In a simpler version of the technique, similar to the original method developed by Willard Libby in Chicago in the late 1940s, larger samples are used and β radiation from the sample is measured. The procedure is illustrated in the following example.

FIGURE 22.22 In the modern version of the carbon-14 dating technique, a mass spectrometer is used to determine the proportion of carbon-14 nuclei in the sample relative to the number of carbon-12 nuclei.

The law of radioactive decay implies that the number of radioactive nuclei decay exponentially with time with a characteristic half-life. Radioactive isotopes are used to determine the ages of objects.

EXAMPLE 22.2 Interpreting carbon-14 dating

A sample of carbon of mass 1.00 g from wood found in an archeological site in Arizona gave 7900 carbon-14 disintegrations in a period of 20 h. In the same period, 1.00 g of carbon from a modern source underwent 18,400 disintegrations. Calculate the ages of the sample.

STRATEGY First, we rearrange Eq. 2 to give an expression for the time

$$t = -\frac{1}{k}\ln\left(\frac{N}{N_0}\right)$$

Then we replace k by $t_{1/2}$ by using Eq. 3:

$$t = -\frac{t_{1/2}}{\ln 2}\ln\left(\frac{N}{N_0}\right)$$

BOX 22.1 USES OF RADIOACTIVE ISOTOPES

We have seen that carbon-14 is used to date organic materials, but to determine the age of very old substances such as rocks, materials with longer half-lives are needed. Uranium-238 ($t_{1/2} = 4.5 \times 10^9$ y) and potassium-40 ($t_{1/2} = 1.26 \times 10^9$ y) are used to date very old rocks. Uranium-238 decays through a series of α and β emissions to lead-206; to determine the age of a rock that contains uranium, the ratio of ^{238}U to ^{206}Pb present in it is measured. Potassium-40 decays by electron capture to form argon-40. The rock is placed under vacuum and then crushed. A mass spectrometer is used to measure the amount of argon gas that escapes. This technique was used to determine the age of the rocks along the California coastline and on the surface of the Moon. For dating relatively recent samples, the activity due to tritium ($t_{1/2} = 12.3$ y) is measured in a technique similar to carbon-14 dating.

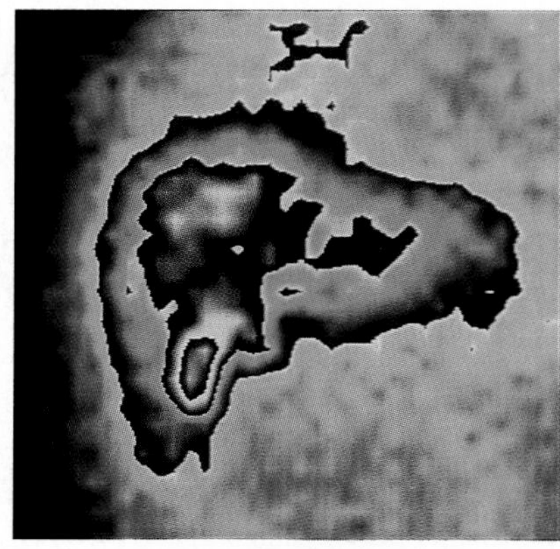

An iodine-123 scan of a normal liver. Notice how clearly the shape of the organ is revealed by the radioisotope. The orange bean-shaped object is the gallbladder.

Radioactive **tracers** are isotopes that are used to trace a path. For example, oil companies inject radioactive tracers with a short half-life into pipelines. Radiation detectors are set up at regular distances along the pipeline to determine how fast the tracer is traveling. Because the half-life of the tracer is short, the oil is no longer radioactive when it emerges from the pipe. Chemists use tracers to track the mechanisms of reactions. For example, a sample of sugar can be *labeled* with carbon-14: that is, some carbon-12 atoms in sugar molecules are replaced by carbon-14 atoms, which can then be detected by radiation counters. Then the progress of the sugar molecule through the body can be monitored. Fertilizers labeled with radioactive nitrogen, phosphorus, and potassium are used to follow the mechanism of plant growth and the passage of these elements through the environment.

The number of disintegrations reported in the same time period is proportional to the number of carbon-14 nuclei present in the two samples. If we suppose that the proportion of carbon-14 in the atmosphere was the same when the ancient sample was alive as it is now, then the original activity of carbon-14 in the ancient sample can be assumed to be the same as the activity of the modern sample. We can therefore set N/N_0 equal to the ratio of the number of disintegrations in the ancient and modern samples.

SOLUTION From the expression above, we obtain

$$t = -\frac{5.73 \times 10^3 \text{ y}}{\ln 2} \times \ln\left(\frac{7900}{18{,}400}\right) = 6.99 \times 10^3 \text{ y}$$

We conclude that approximately 7000 y has elapsed since the piece of wood was part of a living tree.

SELF-TEST 22.7A A sample of carbon of mass 250 mg from wood found in a tomb in Israel underwent 2480 carbon-14 disintegrations in 20 h. Estimate the time since death, assuming the same activity for a modern sample as in Example 22.2.

[*Answer:* 5.1×10^3 y]

Radioactive nuclides also have important commercial applications. For example, americium-243 is used in smoke detectors, where it ionizes smoke particles, allowing a current to flow and set off the alarm. Carbon-14 is incorporated into steel to test for wear and into plastics to check for uniformity. Exposure to radiation is also used to sterilize food and inhibit the sprouting of potatoes. The radiation kills the bacteria that spoil food but does not produce harmful substances.

Radioactive isotopes that give off a lot of energy are also used to provide power in remote locations, where refueling of generators is not possible. Runway lights at arctic air bases, weather stations at both the North and South Poles, and unmanned spacecraft, such as Voyager 2, are powered by radioisotopes.

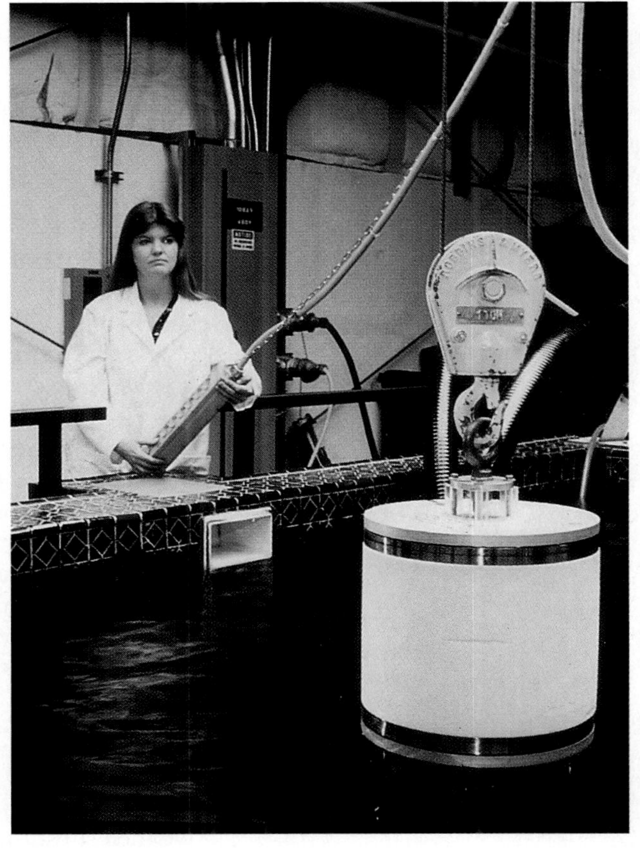

This facility at Lawrence Livermore Laboratory is used to study the irradiation of food. A container of fruit is being lowered into a pool containing an accelerator that generates radiation. The water protects the technician from the harmful effects of the radiation.

SELF-TEST 22.7B A sample of carbon of mass 1.00 g from scrolls found near the Dead Sea underwent 1.4×10^4 carbon-14 disintegrations in 20 h. Estimate the approximate time since the sheepskins composing the scrolls were removed from the sheep.

NUCLEAR ENERGY

Nuclear reactions can release huge amounts of energy. However, the use of this resource contributes to both the creation and the solution of social and economic problems. The following sections outline some of the principles involved. They introduce nuclear fuels and how they are processed, how nuclear reactions are controlled, and critical issues associated with the use of nuclear power.

22.9 MASS-ENERGY CONVERSION

Energy is released when the protons and neutrons in a nucleus adopt a more stable arrangement. The change in energy that accompanies this rearrangement can be found by comparing the masses of the nuclear reactants and the nuclear products. Einstein's theory of relativity tells us that the mass of an object is a measure of its energy content: the greater the mass of an object, the greater its energy. Specifically, the total energy, E, and the mass, m, are related by Einstein's famous equation

$$E = mc^2 \tag{4}$$

where c is the speed of light ($3.00 \times 10^8\ \text{m}\cdot\text{s}^{-1}$). Loss of energy is always accompanied by a loss of mass.

Mass loss accompanies all energy changes, but it is normally far too small to detect. For example, when 100 g of water cools from 100°C to 20°C, it loses 33 kJ of energy as heat; this energy loss corresponds to a mass loss of only 3.7×10^{-10} g. Even in a strongly exothermic chemical reaction, such as one that releases 10^3 kJ of energy, the mass of the products is only 10^{-8} g less than that of the reactants; this change in mass is outside the range of all but extremely precise measurements. However, in a nuclear reaction, the energy changes are very large, the corresponding mass loss is measurable, and the observed change in mass can be used to calculate the energy released.

The **nuclear binding energy,** E_{bind}, is the energy released when Z protons and $A - Z$ neutrons come together to form a nucleus. It can be used as a measure of the stability of a nuclide. The greater the binding energy of a nucleon, the lower is its energy in the nuclide. The nuclear binding energy is proportional to the difference in mass, Δm, between the nucleus and the separated nucleons. For example, the binding energy of iron-56 (with 26 protons, each of mass m_{p}, and 30 neutrons, each of mass m_{n}) is calculated from

$$\begin{aligned}\Delta m &= \sum m(\text{products}) - \sum m(\text{reactants}) \\ &= m\left({}^{56}_{26}\text{Fe nucleus}\right) - \{26 \times m_{\text{p}} + 30 \times m_{\text{n}}\}\end{aligned}$$

We then substitute this expression into Einstein's formula to obtain

$$E_{\text{bind}} = \Delta m \times c^2 \tag{5}$$

CHEMISTRY AT WORK

Archaeology graduate students uncover a remnant of an ancient civilization in the Middle East while searching for the legendary city of Ubar. They will determine the age of the site through radiocarbon dating of organic remains such as fibers or pieces of wood.

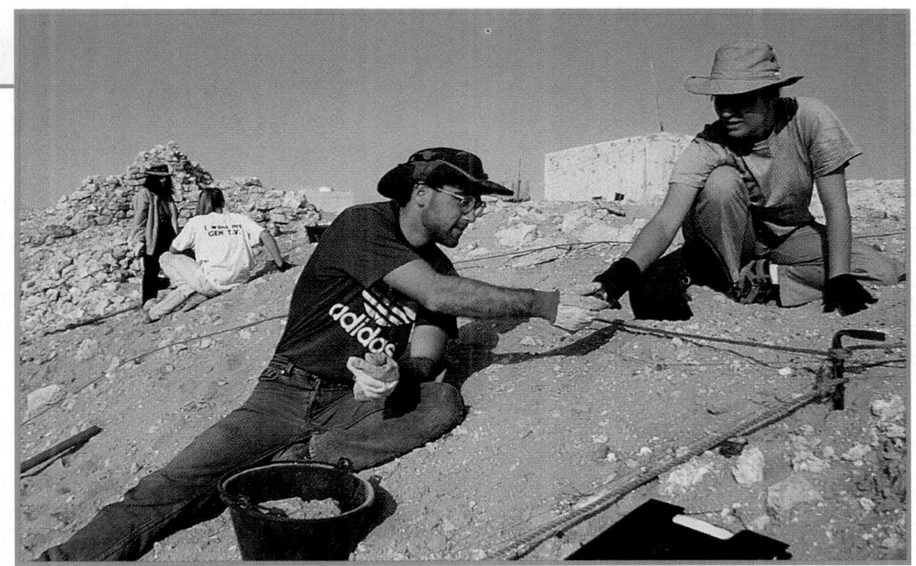

EXAMPLE 22.3 Calculating the nuclear binding energy

Calculate the nuclear binding energy for helium-4 given the following masses: ^{4}He, 4.0026 u; ^{1}H, 1.0078 u; n, 1.0087 u. The atomic mass unit, u, was introduced in Case Study 2: 1 u = 1.6605×10^{-27} kg.

STRATEGY The nuclear binding energy is the energy released in the formation of the nucleus from its nucleons. Use H atoms instead of protons to account for the masses of the electrons in the He produced:

$$2\,^{1}\text{H} + 2\,\text{n} \longrightarrow {}^{4}\text{He}$$

Begin by calculating the difference in masses between the products and the reactants; then convert the result from atomic mass units to kilograms by using a conversion factor. Finally, use the Einstein relation to calculate the energy corresponding to this loss of mass.

SOLUTION The change in mass is

$$\Delta m = 4.0026\text{ u} - (2 \times 1.0078 + 2 \times 1.0087)\text{ u} = -0.0304\text{ u}$$

In kilograms, this mass difference is

$$\Delta m = (-0.0304\text{ u}) \times \frac{1.6605 \times 10^{-27}\text{ kg}}{1\text{ u}} = -5.05 \times 10^{-29}\text{ kg}$$

Then, from Eq. 5,

$$\begin{aligned} E_{\text{bind}} &= (-5.05 \times 10^{-29}\text{ kg}) \times (3.00 \times 10^{8}\text{ m·s}^{-1})^{2} \\ &= -4.55 \times 10^{-12}\text{ kg·m}^{2}\text{·s}^{-2} = -4.55 \times 10^{-12}\text{ J} \end{aligned}$$

(In the last line, we have used 1 kg·m^{2}·s^{-2} = 1 J.) Because the binding energy is negative, 4.55×10^{-12} J is released when one He nucleus forms from its nucleons. Multiplying by the Avogadro constant, we find that the molar binding energy of helium is 2.7×10^{9} kJ·mol^{-1}, which is 10 million times larger than the energy of a typical chemical bond.

SELF-TEST 22.8A Calculate the molar binding energy of carbon-12 nuclei.

[*Answer:* -8.7×10^9 kJ·mol^{-1}]

SELF-TEST 22.8B Calculate the molar binding energy of uranium-235 nuclei. The mass of one uranium atom is 235.0439 u.

The actual binding energy of iron-56, for instance, is 56 times the value shown in the graph (because there are 56 nucleons in the nucleus), and that of uranium-235 is 235 times the value shown for that nucleus.

Figure 22.23 shows the binding energy per nucleon, E_{bind}/A, for the elements. The greater the binding energy per nucleon, the more stable are the nucleons in the nucleus. The graph shows that nucleons are bonded together most strongly in iron and neighboring elements. This high binding energy is one of the reasons why iron is the most abundant element on a rocky planet like the Earth. The binding energy per nucleon is less for all other nuclides. We can infer that nuclei of atoms lighter than iron become more stable when they "fuse" together and that nuclei of elements heavier than iron become more stable when they undergo "fission" and split into lighter nuclei. The graph also indicates why a heavy nucleus may be radioactive: when the nucleus emits a particle, the energy of the remaining nucleons is lowered.

Nuclear binding energies are determined from the mass difference between the nucleus and its components and the use of Einstein's formula. Iron has the highest binding energy per nucleon.

22.10 NUCLEAR FISSION

Some countries with little fossil fuel reserves, such as France, use nuclear power plants to generate most of their electrical power.

Nuclear **fission** is the breaking of a nucleus into two or more smaller nuclei of similar mass. When a uranium atom disintegrates into smaller nuclei, energy is released. Because fission reactions release huge amounts of energy, they are used to generate electricity in nuclear power plants. The energy released in fission can be calculated by subtracting the total mass of the fission products from the mass of the starting materials. The change in mass is then converted to a change in energy by using Einstein's equation.

Because nuclear reactions are elementary processes, do not cancel species that appear on both sides of a nuclear equation. The equation shows how the process occurs. For instance, in Example 22.4, a neutron is needed to initiate the fission and two neutrons are given off during a fission event.

EXAMPLE 22.4 Calculating the energy released during fission

When uranium-235 nuclei are bombarded with neutrons, they can split apart in a variety of ways, like glass balls that shatter into pieces of different sizes. In one process, uranium-235 forms barium-142 and krypton-92:

$$^{235}_{92}\text{U} + \text{n} \longrightarrow {}^{142}_{56}\text{Ba} + {}^{92}_{36}\text{Kr} + 2\,\text{n}$$

Calculate the energy (in joules) released when 1.0 g of uranium-235 undergoes this fission reaction. The masses of the particles are $^{235}_{92}$U, 235.04 u; $^{142}_{56}$Ba, 141.92 u; $^{92}_{36}$Kr, 91.92 u; n, 1.0087 u.

STRATEGY If we know the mass loss, we can find the energy released by using Einstein's equation. Therefore, we must calculate the total mass of the particles on each side of the equation and then substitute the difference into Eq. 5.

SOLUTION

$$\text{Mass of products} = m(\text{Ba}) + m(\text{Kr}) + 2 \times m_n = 141.92 + 91.92 + 2 \times 1.0087 \text{ u} = 235.86 \text{ u}$$

$$\text{Mass of reactants} = m(\text{U}) + m_n = 235.04 + 1.0087 \text{ u} = 236.05 \text{ u}$$

The change in mass is therefore

$$\Delta m = 235.86 - 236.05 \text{ u} = -0.19 \text{ u} = (-0.19 \text{ u}) \times \frac{1.6605 \times 10^{-27} \text{ kg}}{1 \text{ u}} = -3.2 \times 10^{-28} \text{ kg}$$

The energy change accompanying the fission of one ^{235}U nucleus to these products is therefore

$$\Delta E = -(3.2 \times 10^{-28} \text{ kg}) \times (3.00 \times 10^8 \text{ m·s}^{-1})^2 = -2.9 \times 10^{-11} \text{ J}$$

This energy release accompanies the fission of one nucleus. The number of atoms in 1.0 g (1.0×10^{-3} kg) of uranium is

$$\text{Number of atoms} = \frac{\text{mass of sample}}{\text{mass of one atom}} = \frac{1.0 \times 10^{-3} \text{ kg}}{235.04 \text{ u} \times \left(\frac{1.6605 \times 10^{-27} \text{ kg}}{1 \text{ u}}\right)} = 2.6 \times 10^{21}$$

The total energy change is therefore

$$\Delta E = (2.6 \times 10^{21}) \times (-2.9 \times 10^{-11} \text{ J}) = -7.5 \times 10^{10} \text{ J}$$

SELF-TEST 22.9A Another mode in which uranium-235 can undergo fission is

$$^{235}_{92}\text{U} + \text{n} \longrightarrow {}^{135}_{52}\text{Te} + {}^{100}_{40}\text{Zr} + \text{n}$$

Calculate the energy change when 1.0 g of uranium-235 undergoes fission in this way. The masses needed are U, 235.04 u; n, 1.0087 u; Te, 134.92 u; Zr, 99.92 u.

[*Answer:* -7.8×10^{10} J]

SELF-TEST 22.9B A nuclear reaction that can cause great destruction is one of many that take place in the ^{235}U atomic bomb:

$$^{235}_{92}\text{U} + \text{n} \longrightarrow {}^{138}_{56}\text{Ba} + {}^{86}_{36}\text{Kr} + 12\,\text{n}$$

How much energy is released when 1.0 g of uranium-235 undergoes fission in this manner? The additional masses needed are Ba, 137.91 u; Kr, 85.91 u.

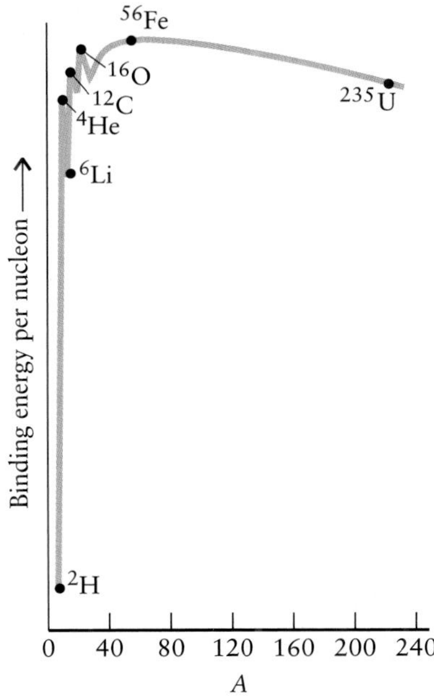

FIGURE 22.23 The variation of the nuclear binding energy per nucleon. The maximum binding energy per nucleon occurs at iron-56. That nucleus has the lowest energy of all because its nucleons are most tightly bound. (The vertical axis is $-E_{\text{bind}}/A$.)

Spontaneous nuclear fission takes place when the natural shaking motion of the nucleons in a heavy nucleus causes it to break into two nuclei of similar mass (Fig. 22.24). An example is the disintegration of americium-244 into iodine and molybdenum:

$$^{244}_{95}\text{Am} \longrightarrow {}^{134}_{53}\text{I} + {}^{107}_{42}\text{Mo} + 3\,\text{n}$$

FIGURE 22.24 In spontaneous nuclear fission, the oscillations of the heavy nucleus in effect tear the nucleus apart, thereby forming two or more smaller nuclei of similar mass. Some nucleons may also be released during fission.

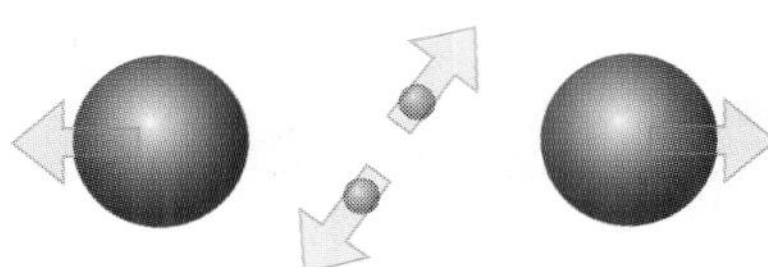

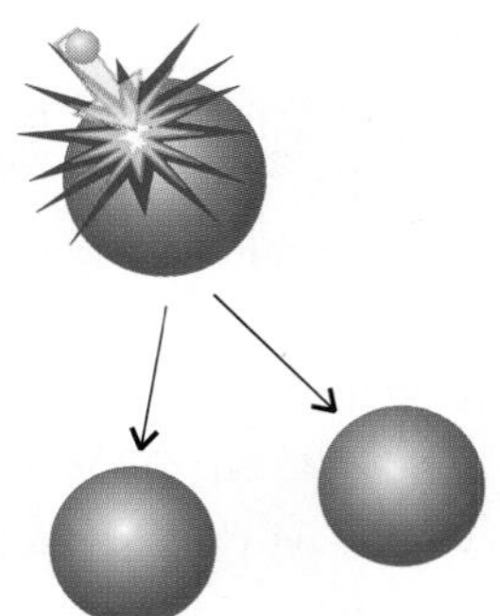

FIGURE 22.25 In induced nuclear fission, the impact of an incoming neutron causes the nucleus to break apart.

Fission does not occur in precisely the same way in every instance: the fission products may include different isotopes of these elements and also isotopes of other elements.

Induced nuclear fission is fission caused by bombarding a heavy nucleus with neutrons (Fig. 22.25). Nuclei that can undergo induced fission are called **fissionable.** For most nuclei, fission is induced only if the impinging neutrons are moving so rapidly that they can smash into the nucleus and drive it apart with the shock of impact; uranium-238 is one such nuclide. Some nuclei, however, can be nudged into breaking apart even if the incoming neutrons are slow. Such nuclides are called **fissile.** They include uranium-235, uranium-233, and plutonium-239, the fuels of nuclear power plants.

Once nuclear fission has been induced, it can continue, even if the supply of neutrons from outside is discontinued, as long as more neutrons are produced by the fission event than are used to induce them initially. Such self-sustaining fission occurs with uranium-235, which undergoes numerous fission processes, including

$$^{235}_{92}\text{U} + \text{n} \longrightarrow {}^{141}_{56}\text{Ba} + {}^{92}_{36}\text{Kr} + 2\,\text{n}$$

When uranium-235 is bombarded by neutrons, an average of 2.5 neutrons are released per fission event.

If the two product neutrons strike two other fissile nuclei, then after the next round of fission there will be four neutrons, which can induce fission in four more nuclei. In the language of Section 18.16, neutrons are *carriers* in a branched chain reaction (Fig. 22.26). When a nuclear branched chain reaction is allowed to run freely, the cascade of released neutrons can result in the fissioning of all the available uranium-235 in only a fraction of a second. The result is a nuclear explosion.

Recall that the nucleus takes up only a tiny fraction of the volume of an atom.

Neutrons produced in a chain reaction are moving very fast, and most escape into the surroundings without colliding with another fissionable nucleus. However, if a large enough number of uranium nuclei are present in the sample, enough neutrons can be captured to sustain the chain reaction. That is, there is a **critical mass,** a mass of fissionable material below which so many neutrons escape from the sample that the fission chain reaction is not sustained. If a sample is **supercritical,** with a mass in excess of the critical value, then the reaction is self-sustaining and may result in an explosion. The critical mass for a solid sphere of pure plutonium of normal density is about 15 kg, a sphere about the size of a grapefruit. The critical mass is smaller if the metal is compressed by detonating a conventional explosive that surrounds it. Then the nuclei are

FIGURE 22.26 A self-sustaining chain reaction, in which neutrons are the chain carriers, occurs when induced fission produces more than one neutron per fission event. These newly produced neutrons can go on to stimulate fission in other nuclei.

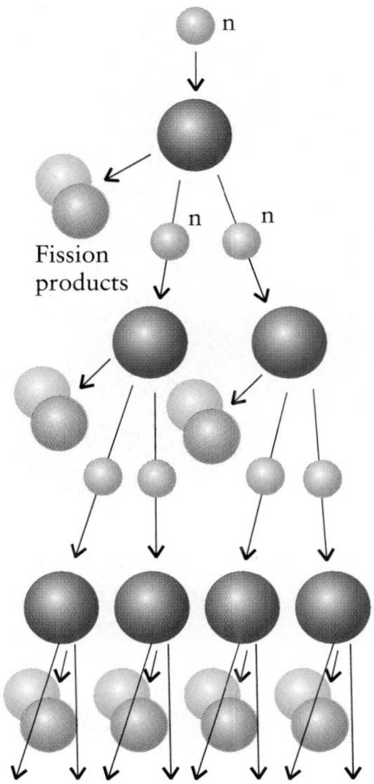

pressed closer together and become more effective at blocking the escape of neutrons. The critical mass can be as low as 5 kg for highly compressed plutonium.

The technically difficult step in designing a nuclear weapon is to convert a given mass of fissile material into a supercritical mass so rapidly that the chain reaction occurs uniformly throughout the metal. This can be done by shooting two blocks toward each other (as was done in the bomb that fell on Hiroshima) or by implosion of a single subcritical mass (the technique used in the bomb that destroyed Nagasaki). A strong neutron emitter, typically polonium, is also included to initiate the chain reaction. The products of the fission process, which consist of dozens of different nuclides, most of them radioactive, are dispersed rapidly in the explosion. They are carried many miles by air currents before they settle to the ground, becoming the fallout of a nuclear explosion.

This type of explosive fission could not occur in a nuclear power plant because the fuel used in a **nuclear reactor** (Fig. 22.27) is much less dense than that used to induce explosive fission. Instead, a much slower, controlled chain reaction is sustained by making efficient use of a limited supply of neutrons by slowing the neutrons down. The fuel is shaped into long rods and inserted into a **moderator,** a material that slows down the neutrons as they pass between fuel rods; the slower neutrons have a greater probability of collision with a nucleus. Slow neutrons have three significant roles: they do not induce the fission of fissionable (as distinct from fissile) material, they are most effectively absorbed by the fissile uranium-235, and they allow the control rods to act most efficiently. The first moderator used was graphite. Light-water reactors (LWRs) use ordinary water as a moderator. They are the most common type of nuclear reactors in the United States (Fig. 22.28).

Nuclear fuel consists of UO_2 pellets, enriched to about 3% ^{235}U in a zirconium alloy tube.

The rate of the chain reaction must be kept below a certain level, or the reactor will become too hot and begin to melt. Control rods made from neutron-absorbing elements, such as boron or cadmium, are inserted between the fuel rods to control the number of neutrons available for inducing fission and thus control the rate of nuclear reaction. Explosions that have occurred in nuclear power plants have been chemical explosions, generally associated with overheating.

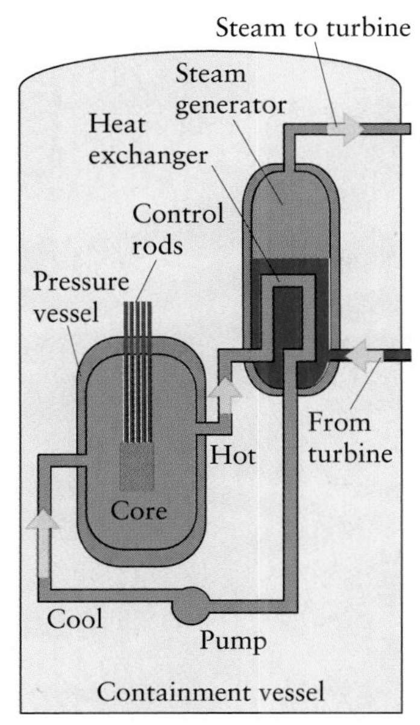

FIGURE 22.27 A schematic diagram of one type of nuclear reactor. This one is a pressurized water reactor, in which the coolant is water under pressure. The fission reactions produce heat, which boils water in the steam generator; the resulting steam turns the turbines that generate electricity.

FIGURE 22.28 The reactor core of a typical light-water reactor (LWR) nuclear power plant is immersed in water.

One of the many problems connected with nuclear power is the availability of fuel: uranium-235 reserves are only about 1% those of the nonfissile uranium-238. One solution is to synthesize fissile nuclides from other elements. In a **breeder reactor,** a reactor that is used to create nuclear fuel, the neutrons are not moderated. Their high speeds result in the formation of not only uranium-235 but also some fissile plutonium-239, which can be used as fuel (or for warheads). Breeder reactors are more hazardous to operate than nuclear power plants. They run very hot, and the fast reactions require more careful control than a reactor used for nuclear power generation. Their use is thus controversial.

Nuclear energy can be extracted by arranging for a nuclear chain reaction to occur, with neutrons as the chain carriers. A moderator is used to reduce the speed of the neutrons in a reactor that uses fissile material.

22.11 NUCLEAR FUSION

Fission reactors generate radioactive waste products that must be controlled; we shall discuss that topic in Section 22.12. However, another kind of nuclear reaction is essentially free of long-lived hazardous waste products, and its abundant fuel is extracted from seawater. The reaction is the fusion of hydrogen nuclei into helium nuclei. We see from the plot of nuclear binding energy in Fig. 22.23 that there is a large increase in nuclear binding energy (and hence a lowering of total energy) on going from hydrogen to the first few light elements.

The strong electrical repulsion between protons makes it difficult for them to approach each other closely enough to fuse together. The heavier isotopes of hydrogen are therefore used in fusion reactions; their nuclei fuse together more readily, because the additional neutrons help to overcome the repulsion between the approaching protons. The high kinetic energies needed for successful collisions are achieved by raising the temperature to millions of degrees.

One fusion scheme uses deuterium (D) and tritium (T) in the following sequence of nuclear reactions

$$\mathrm{D} + \mathrm{D} \longrightarrow {}^{3}\mathrm{He} + \mathrm{n}$$

$$\mathrm{D} + \mathrm{D} \longrightarrow \mathrm{T} + \mathrm{p}$$

$$\mathrm{D} + \mathrm{T} \longrightarrow {}^{4}\mathrm{He} + \mathrm{n}$$

$$\mathrm{D} + {}^{3}\mathrm{He} \longrightarrow {}^{4}\mathrm{He} + \mathrm{p}$$

$$\text{Overall reaction:} \quad 6\,\mathrm{D} \longrightarrow 2\,{}^{4}\mathrm{He} + 2\,\mathrm{p} + 2\,\mathrm{n}$$

The method described in Section 22.10 can be used to show that the overall reaction releases 3×10^{8} kJ for each gram of deuterium consumed, which corresponds to the energy generated when the Hoover Dam operates at full capacity for about an hour. Additional tritium is supplied to the reaction chamber to facilitate the process. Because it has a very low natural abundance and is radioactive, the tritium is generated by bombarding lithium-6 with neutrons in the immediate surroundings of the reaction zone:

$${}^{6}\mathrm{Li} + \mathrm{n} \longrightarrow \mathrm{T} + {}^{4}\mathrm{He}$$

FIGURE 22.29 Research into controlled nuclear fusion is being carried out in several countries. Here we see the Tokamak fusion test reactor at the Princeton Plasma Physics Laboratory.

Nuclear fusion is very difficult to achieve in practice because of the vigor with which the charged nuclei must be hurled at each other. One way of accelerating them to sufficiently high speeds is to heat them with a fission explosion: this method is used to produce a **thermonuclear explosion,** in which a fission bomb (using uranium or plutonium) is used to ignite a lithium-6 fusion bomb. A more constructive and controlled approach is to heat a plasma, or ionized gas, by passing an electric current through it. The very fast ions in the plasma are kept away from the walls of the container with magnetic fields. This method of achieving fusion is the subject of intense research and is beginning to show signs of success (Fig. 22.29).

Nuclear fusion makes use of the energy released when light nuclei fuse together to form heavier nuclei.

22.12 THE CHEMISTRY OF NUCLEAR POWER

Chemistry is the key to the safe use of nuclear power. For example, chemistry is used in the preparation of the fuel itself, the recovery of important fission products, and the safe disposal or utilization of nuclear waste.

The fuel of nuclear reactors is uranium, which is mined as several minerals. The most important is *pitchblende,* UO_2 (Fig. 22.30), from strip mines in New Mexico and Wyoming. Uranium is refined not only to reduce the ore to the metal but also to **enrich** it, that is, to increase the abundance of a specific isotope—in this case, uranium-235. The natural abundance of uranium-235 is about 0.7%, and the goal of enrichment is to increase this fraction to about 3%.

FIGURE 22.30 Pitchblende is a common uranium ore. It is a variety of uranite, UO_2.

The enrichment procedure uses the small mass difference between the hexafluorides of uranium-235 and uranium-238 to separate them. The first procedure to be developed converts the uranium into uranium hexa-

FIGURE 22.31 (above left) Containers of high-level waste products, including cesium-137 and strontium-90, glow under a protective layer of water. The radiation from these canisters is great enough that contact with one could cause death within about 4 s.

FIGURE 22.32 (above right) This 35-year-old drum of radioactive waste has become corroded and has leaked radioactive materials into the soil. The drum was located in one of the nuclear waste disposal sites at the Department of Energy's Hanford, Washington, nuclear manufacturing and research facility. Several storage sites at this facility have become seriously contaminated.

Currently, thousands of tons of nuclear waste are stored in the United States alone; some of the isotopes have half-lives of thousands or even millions of years.

fluoride, UF_6, which can be vaporized readily. The different effusion rates of the two isotopic fluorides are then used to separate them. It follows from Graham's law (rate of effusion $\propto 1/\sqrt{\text{molar mass}}$, Section 5.15) that the rates of effusion of $^{235}UF_6$ (molar mass 349 g·mol^{-1}) and $^{238}UF_6$ (molar mass 352 g·mol^{-1}) should be in the ratio

$$\frac{\text{Rate of effusion of }^{235}\text{UF}_6}{\text{Rate of effusion of }^{238}\text{UF}_6} = \sqrt{\frac{352}{349}} = 1.004$$

Because the ratio is so close to 1, the vapor must be allowed to effuse repeatedly through porous barriers consisting of screens with large numbers of minute holes. In practice, it is allowed to do so thousands of times (some of the stages were shown in Fig. 5.27).

Because the effusion process is technically demanding and uses a lot of energy, scientists and engineers continue to look for alternative enrichment procedures. One of these approaches uses a centrifuge that rotates samples of uranium hexafluoride vapor at very high speed. This rotation causes the heavier $^{238}UF_6$ molecules to be thrown outward and collected as a solid on the outer parts of the rotor, leaving a higher proportion of $^{235}UF_6$ closer to the axis of the rotor, from which position it can be removed.

The processing of spent nuclear fuel is a much more complex task than the fuel's initial preparation. Any remaining uranium-235 must be recovered, any plutonium produced must be extracted, and the highly radioactive and largely useless fission products must be safely stored (Fig. 22.31). The highly radioactive fission (HRF) products from used nuclear fuel rods must be stored for about 10 half-lives before their level of radioactivity is no longer dangerous. Generally they are buried underground, but even burial of radioactive wastes is not without problems. Metal storage drums can corrode and allow liquid radioactive waste to

FIGURE 22.33 Molten glass for storing nuclear waste, being poured from a platinum crucible into a steel bar mold.

seep into aquifers that supply drinking water (Fig. 22.32). Leakage can be minimized by incorporating the HRF products into a glass—a solid, complex network of silicon and oxygen atoms (Fig. 22.33). Most of the fission products are oxides of the type that form one of the components of glass—they are network formers (see Case Study 19); that is, they promote the formation of a relatively disorderly Si—O network rather than inducing crystallization into an orderly array of atoms. Crystallization is dangerous because crystalline regions easily form cracks that leave the incorporated radioactive materials exposed to moisture, which might dissolve them and carry them away from the storage area.

An advantage of glass storage for nuclear waste is that it reduces the possibility of a chemical reaction that might result in an explosion.

A chief attraction of fusion reactors is their virtual freedom from long-lived radioactive waste generation. The promise of fusion reactors for abundant, pollution-free power is great. We are at the point in history when the elements have completed their journey from the stars, have come alive, and have begun to control their own destiny. Now the cycle can be completed. The fusion reactions that power the stars may soon be recreated on Earth and emulate the processes that produced the elements from which we are made.

Uranium is extracted by a series of reactions that lead to uranium hexafluoride, and then the isotopes are separated by a variety of procedures. Some radioactive waste is currently converted to glass for storage in cool, dry caverns.

SKILLS YOU SHOULD HAVE MASTERED

Conceptual

1. Explain how nuclear reactions differ from chemical reactions.
2. Explain why accelerators are required for the transmutation of elements by nucleosynthesis.
3. Distinguish nuclear fission from nuclear fusion and predict which nuclides undergo each type of process.

Problem-Solving

1. Write, complete, and balance nuclear equations.
2. Use the band of stability to predict the types of decay a certain radioactive nucleus is likely to undergo.
3. Predict the amount of a radioactive sample that will remain after a given time period, given the decay constant or half-life of the sample.
4. Use the half-life of an isotope to determine the age of a sample.
5. Calculate the nuclear binding energy of a given nuclide.
6. Calculate the energy released during a nuclear reaction.

Descriptive

1. Distinguish α, β, and γ radiation by their reactions to an electrical field, penetrating power, and relative biological effectiveness.
2. Describe electron capture and the emission of γ rays, positrons, protons, and neutrons.
3. Describe some applications of radioactive isotopes in chemistry, industry, and medicine.
4. Describe the construction and operation of a nuclear reactor.
5. Describe the problems associated with the management of nuclear waste.

EXERCISES

Nuclear Structure and Radiation

22.1 Distinguish α particles, β particles, and γ rays by composition, mass, charge, and penetrating power.

22.2 Explain how Rutherford determined the nature of α particles.

22.3 Determine the number of protons, neutrons, and nucleons in (a) ^{2}H; (b) ^{24}Mg; (c) ^{263}Rf; (d) ^{60}Co; (e) ^{238}Pu; (f) ^{258}Md.

22.4 Determine the number of protons, neutrons, and nucleons in (a) ^{12}C; (b) ^{90}Sr; (c) ^{257}Db; (d) ^{99}Mo; (e) ^{206}Tl; (f) ^{57}Fe.

22.5 Write the nuclear symbol and determine the number of protons, neutrons, and nucleons in (a) bromine-81; (b) krypton-90; (c) curium-244; (d) iodine-128; (e) sulfur-32; (f) americium-241.

22.6 Write the nuclear symbol and determine the number of protons, neutrons, and nucleons in (a) chlorine-37; (b) berkelium-247; (c) dubnium-262; (d) gold-197; (e) californium-249; (f) uranium-233.

22.7 Calculate the wavelength and the energy per mole of photons of γ radiation of frequency (a) 9.4×10^{19} Hz; (b) 5.7×10^{21} Hz; (c) 3.7×10^{20} Hz; (d) 7.3×10^{22} Hz.

22.8 Calculate the frequency and wavelength of the γ radiation emitted as a result of a rearrangement of nucleons in a daughter nucleus through an energy (a) 9.5×10^{-14} J; (b) 3.9×10^{-13} J; (c) 2.6×10^{-16} kJ; (d) 4.7×10^{-16} kJ.

22.9 A common energy unit is the electronvolt (eV), the energy an electron gains by passing through a potential difference of 1 V: 1 eV $= 1.602 \times 10^{-19}$ J. In nuclear processes, energies of decay products and transmutation reactions are often expressed in millions of electronvolts (MeV). In the nucleon rearrangement of the following daughter nuclei, the energy changes by the amount shown, and a γ ray is emitted. Determine the frequency and wavelength of the γ ray in each case: (a) cobalt-60, 1.33 MeV; (b) arsenic-80, 1.64 MeV; (c) iron-59, 1.10 MeV.

22.10 Determine the frequency and wavelength of the γ ray emitted in the decay of the following nuclides, where the energy change of the nucleus is given in MeV (see Exercise 22.9): (a) carbon-15, 5.30 MeV; (b) scandium-50, 0.26 MeV; (c) bromine-87, 5.4 MeV.

Radioactive Decay

22.11 Identify the daughter nucleus in each of the following decays and write the balanced nuclear equation: (a) β decay of tritium; (b) β^{+} decay of yttrium-83; (c) β decay of krypton-87; (d) α decay of protactinium-225.

22.12 Identify the daughter nucleus in each of the following decays and write the balanced nuclear equation: (a) β decay of actinium-228; (b) α decay of radon-212; (c) α decay of francium-221; (d) electron capture by protactinium-230.

22.13 Write the balanced nuclear equation for the following radioactive decays: (a) β^{+} decay of boron-8; (b) β decay of nickel-63; (c) α decay of gold-185; (d) electron capture by beryllium-7.

22.14 Write the balanced nuclear equation for the following radioactive decays: (a) β decay of uranium-233; (b) proton emission of cobalt-56; (c) β^+ decay of holmium-158; (d) α decay of polonium-212.

22.15 Determine the particle emitted and write the balanced nuclear equation for the following nuclear transitions: (a) sodium-24 to magnesium-24; (b) ^{128}Sn to ^{128}Sb; (c) lanthanum-140 to barium-140; (d) ^{228}Th to ^{224}Ra.

22.16 Determine the particle emitted and write the balanced nuclear equation for the following nuclear transitions: (a) carbon-14 to nitrogen-14; (b) neon-19 to fluorine-19; (c) gold-188 to platinum-188; (d) uranium-229 to thorium-225.

The Pattern of Nuclear Stability

22.17 What condition of the neutron-to-proton ratio favors the emission of (a) α particles? (b) β particles?

22.18 What characteristics of a nucleus favor the emission of a positron from a radioactive nuclide?

22.19 The following nuclides lie outside the band of stability. Predict whether they are most likely to undergo β decay, β^+ decay, or α decay and identify the daughter nucleus; (a) copper-68; (b) cadmium-103; (c) berkelium-243; (d) dubnium-260.

22.20 The following nuclides lie outside the band of stability. Predict whether they are most likely to undergo β decay, β^+ decay, or α decay and identify the daughter nucleus: (a) copper-60; (b) xenon-140; (c) americium-246; (d) neptunium-240.

22.21 Identify the daughter nuclides in each step of the radioactive decay of uranium-235, if the string of particle emissions is α, β, α, β, α, α, α, β, α, β, α.

22.22 Neptunium-237 undergoes an α, β, α, α, β, α, α, α, β, α, β sequence of radioactive decays. Determine the daughter nuclide after each decay and write a balanced nuclear equation for each step.

Nucleosynthesis

22.23 Complete the following nuclear equations:
(a) $^{14}\text{N} + ? \longrightarrow {}^{17}\text{O} + \text{p}$
(b) $? + \text{n} \longrightarrow {}^{249}\text{Bk} + \beta$
(c) $^{243}\text{Am} + \text{n} \longrightarrow {}^{244}\text{Cm} + ? + \gamma$
(d) $^{13}\text{C} + \text{n} \longrightarrow ? + \gamma$

22.24 Complete the following nuclear equations:
(a) $? + \text{p} \longrightarrow {}^{21}\text{Na} + \gamma$
(b) $^{1}\text{H} + \text{p} \longrightarrow {}^{2}\text{H} + ?$
(c) $^{15}\text{N} + \text{p} \longrightarrow {}^{12}\text{C} + ?$
(d) $^{20}\text{Ne} + ? \longrightarrow {}^{24}\text{Mg} + \gamma$

22.25 Complete the following nuclear equations for transmutation reactions:
(a) $^{20}\text{Ne} + \alpha \longrightarrow ? + {}^{16}\text{O}$
(b) $^{20}\text{Ne} + {}^{20}\text{Ne} \longrightarrow {}^{16}\text{O} + ?$
(c) $^{44}\text{Ca} + ? \longrightarrow \gamma + {}^{48}\text{Ti}$
(d) $^{27}\text{Al} + {}^{2}\text{H} \longrightarrow ? + {}^{28}\text{Al}$

22.26 Complete the following nuclear equations for transmutation reactions:
(a) $? + \gamma \longrightarrow \beta + {}^{20}\text{Ne}$
(b) $^{44}\text{Ti} + \text{e}^- \longrightarrow \beta^+ + ?$
(c) $^{241}\text{Am} + ? \longrightarrow 4\,\text{n} + {}^{248}\text{Fm}$
(d) $? + \text{n} \longrightarrow \beta + {}^{244}\text{Cm}$

22.27 Write nuclear equations that represent the following processes: (a) oxygen-17 produced by α-particle bombardment of nitrogen-14; (b) americium-240 produced by neutron bombardment of plutonium-239.

22.28 Write the nuclear equations for the following transformations: (a) ^{257}Rf produced by the bombardment of californium-245 with carbon-12 nuclei; (b) the first synthesis of ^{266}Mt by the bombardment of bismuth-209 with iron-58 nuclei. Given that the first decay of meitnerium is by α emission, what is the daughter nucleus?

22.29 Each of the following equations represents a fission reaction. Complete and balance the equations:
(a) $^{244}\text{Am} \longrightarrow {}^{134}\text{I} + {}^{107}\text{Mo} + 3?$
(b) $^{235}\text{U} + \text{n} \longrightarrow ? + {}^{138}\text{Te} + 2\,\text{n}$
(c) $^{235}\text{U} + \text{n} \longrightarrow {}^{101}\text{Mo} + {}^{132}\text{Sn} + ?$

22.30 Complete each of the following nuclear equations for fission reactions:
(a) $^{239}\text{Pu} + \text{n} \longrightarrow {}^{98}\text{Mo} + {}^{138}\text{Te} + ?$
(b) $^{239}\text{Pu} + \text{n} \longrightarrow {}^{100}\text{Tc} + ? + 4\,\text{n}$
(c) $^{239}\text{Pu} + \text{n} \longrightarrow ? + {}^{133}\text{In} + 3\,\text{n}$

Measuring Radioactivity and Its Effects

22.31 The activity of a certain radioactive source is 3.7×10^6 Bq. Express this activity in curies.

22.32 The activity of a sample containing carbon-14 is 12.9 Bq. Express this activity in microcuries.

22.33 Determine the number of disintegrations per second for radioactive sources of the following activities: (a) 1.0 Ci; (b) 82 mCi; (c) 1.0 μCi.

22.34 A certain Geiger counter is known to respond to only 1 of every 1000 radiations emitted from a sample. Calculate the activity of each radioactive source in curies, given the following data: (a) 370 clicks in 10 s; (b) 1.4×10^5 clicks in 1.0 h; (c) 266 clicks in 1.0 min.

22.35 A 1.0-kg sample absorbs an energy of 1.0 J as a result of its exposure to β radiation. Calculate the dose in rads and the dose equivalent in rems.

22.36 A 5.0-g sample of muscle tissue absorbs 2.0 J of energy as a result of its exposure to α radiation. Calculate the dose in rads and the dose equivalent in rems.

22.37 Someone is exposed to a source of β radiation that results in a dose rate of 1.0 rad·day^{-1}. Given that nausea begins after a dose equivalent of about 100 rem, after what period will that symptom of radiation sickness be apparent?

22.38 Someone is exposed to a source of α radiation that results in a dose rate of 2.0 mrad·day^{-1}. If nausea begins after a dose equivalent of about 100 rem, after what period will nausea become apparent?

Rate of Nuclear Disintegration

22.39 Determine the decay constant for (a) tritium, $t_{1/2} = 12.3$ y; (b) lithium-8, $t_{1/2} = 0.84$ s; (c) nitrogen-13, $t_{1/2} = 10.0$ min.

22.40 Determine the half-life of (a) potassium-40, $k = 5.3 \times 10^{-10}\ y^{-1}$; (b) cobalt-60, $k = 0.132\ y^{-1}$; (c) nobelium-255, $k = 3.85 \times 10^{-3}\ s^{-1}$.

22.41 Use Table 22.5 to calculate the time needed for the activity of each source to change as indicated: (a) a 1.0-Ci radium-226 source to decay to 0.10 Ci; (b) a 1.0-μCi potassium-40 source to decay to 10 nCi; (c) a 10-Ci cobalt-60 source to decay to 8 Ci. (*Hint*: Recall that the activity gives the rate of decay, and combine Eq. 1 and Eq. 2 in Section 22.8.)

22.42 Use Table 22.5 to calculate the time needed for the activity of each source to change as indicated: (a) a 0.010-mCi strontium-90 source to decay to 0.0010 mCi; (b) a 1.0-Ci iodine-131 source to decay to 1.0 mCi; (c) a 1.0-mCi thorium-234 source (half-life 24.1 d) to decay to 0.10 mCi. (See the hint in Exercise 22.41.)

22.43 Estimate the activity of a 4.4-Ci cobalt-60 source, $t_{1/2} = 5.26$ y, after 50 y have passed. (See the hint in Exercise 22.41.)

22.44 The activity of a strontium-90 source is 3.0×10^4 Bq. What is its activity after 50 y have passed? (See the hint in Exercise 22.41.)

22.45 (a) What percentage of a carbon-14 sample remains after 1000 y? (b) Determine the percentage of a tritium sample that remains after 20.0 y.

22.46 (a) What percentage of a strontium-90 sample remains after 10.0 y? (b) Determine the percentage of an iodine-131 sample that remains after 5.0 d.

22.47 (a) What fraction of the original activity of ^{238}U remains after 9.0 billion y? (b) Potassium-40, which is presumed to exist at the formation of the Earth, is used for dating minerals. If one-half of the original potassium-40 exists in a rock, how old is the rock?

22.48 (a) A sample of krypton-85 ($t_{1/2} = 10.8$ y) is released into the atmosphere. What fraction of the krypton-85 remains after 15.0 y? (b) A piece of wood, found in an archeological dig, has a carbon-14 activity that is 90% of the current carbon-14 activity. How old is the piece of wood?

22.49 A 250-mg sample of carbon from a piece of cloth excavated from an ancient tomb in Nubia undergoes 1500 disintegrations in 10.0 h. If a current 1.00-g sample of carbon shows 920 disintegrations per hour, how old is the piece of cloth?

22.50 A current 1.00-g sample of carbon shows 920 disintegrations per hour. If a 1.00-g sample of charcoal from an archeological dig in a limestone cave in Slovenia shows 5500 disintegrations in 24.0 h, what is the age of the charcoal sample?

22.51 Use the law of radioactive decay to determine the activity of (a) a 1.0-mg sample of radium-226 ($t_{1/2} = 1.60 \times 10^3$ y); (b) a 2.0-μg sample of strontium-90 ($t_{1/2} = 28.1$ y); (c) a 0.43-mg sample of promethium-147 ($t_{1/2} = 2.6$ y). The mass of each nuclide in atomic mass units (u) is equal to its mass number, within two significant figures.

22.52 Use the law of radioactive decay to determine the activity of (a) a 1.0-g sample of $^{235}UO_2$ ($t_{1/2} = 7.1 \times 10^8$ y); (b) a 1.0-g sample of cobalt containing 1.0% ^{60}Co ($t_{1/2} = 5.26$ y); (c) a 5.0-mg sample of thallium-200 ($t_{1/2} = 26.1$ h).The mass of each nuclide in atomic mass units (u) is equal to its mass number, within two significant figures.

Nuclear Energy

22.53 What is the meaning of "critical mass" in nuclear fission?

22.54 Distinguish between nuclear fission and nuclear fusion. Explain why heavy nuclides are most likely to undergo fission, whereas light nuclides are most likely to undergo fusion.

22.55 Calculate the mass loss or gain for each of the following processes: (a) a 250-g sample of copper (specific heat capacity, 0.39 J·(°C)$^{-1}$·g^{-1}) is heated from 35°C to 250°C; (b) a 50.0-g sample of water freezes at 0°C

(ΔH°_{melt} = 6.01 kJ·mol^{-1}); (c) 2.0 mol $PCl_5(g)$ is formed from its elements under standard conditions at 25°C.

22.56 Calculate the mass loss or gain for each of the following processes: (a) a 50.0-g block of iron (specific heat capacity, 0.45 J·(°C)$^{-1}$·g^{-1}) cools from 600°C to 25°C; (b) a 100-g sample of ethanol vaporizes at its normal boiling point (ΔH°_{vap} = 43.5 kJ·mol^{-1}); (c) 10.0 g of $SO_2(g)$ is formed from its elements under standard conditions at 25°C.

22.57 Calculate the energy in joules that is equivalent to (a) 1.0 g of matter; (b) 1 electron, of mass 9.109×10^{-28} g.

22.58 Calculate the energy in joules that is equivalent to (a) 1.0 pg of matter; (b) 1 proton, of mass 1.673×10^{-24} g.

22.59 The Sun emits radiant energy at the rate of 3.9×10^{26} J·s^{-1}. What is the rate of mass loss (in kilograms per second) of the Sun?

22.60 For the fusion reaction $6\,D \rightarrow 2\,^4He + 2\,^1H + 2\,n$, 3×10^8 kJ of energy is released by a certain sample of deuterium. What is the mass loss (in grams) for the reaction?

22.61 Calculate the binding energy per nucleon (J·nucleon^{-1}) for (a) ^4He, 4.0026 u; (b) ^{239}Pu, 239.0522 u; (c) ^2H, 2.0141 u; (d) ^{56}Fe, 55.9349 u. Which nuclide is the most stable? (See Example 22.3 for the definition of 1 u.)

22.62 Calculate the binding energy per nucleon (J·nucleon^{-1}) for (a) ^{98}Mo, 97.9055 u; (b) ^{151}Eu, 150.9196 u; (c) ^{10}B, 10.0129 u; (d) ^{232}Th, 232.0382 u. Which nuclide is the most stable? (See Example 22.3 for the definition of 1 u.)

22.63 Calculate the energy released per gram of starting material in the fusion reaction represented by each of the following equations. (See Example 22.3 for the definition of 1 u.)

(a) $D + D \longrightarrow {}^3He + n$ (D, 2.0141 u; ^3He, 3.0160 u)
(b) $^3He + D \longrightarrow {}^4He + {}^1H$ (^1H, 1.0078 u; ^4He, 4.0026 u)
(c) $^7Li + {}^1H \longrightarrow 2\,^4He$ (^7Li, 7.0160 u)
(d) $D + T \longrightarrow {}^4He + {}^1H$ (T, 3.0160 u)

22.64 Calculate the energy released per gram of starting material in the nuclear reaction represented by each of the following equations. (See Example 22.3 for the definition of 1 u.)

(a) $^7Li + {}^1H \longrightarrow n + {}^7Be$ (^1H, 1.0078 u; ^7Li, 7.0160 u; ^7Be, 7.0169 u)
(b) $^{59}Co + D \longrightarrow {}^1H + {}^{60}Co$ (^{59}Co, 58.9332 u; ^{60}Co, 59.9529 u)
(c) $^{40}K + \beta \longrightarrow {}^{40}Ar$ (^{40}K, 39.9640 u; ^{40}Ar, 39.9624 u)
(d) $^{10}B + n \longrightarrow {}^4He + {}^7Li$ (^4He, 4.0026 u; ^7Li, 7.0160 u; ^{10}B, 10.0129 u)

22.65 Explain how uranium ore is enriched.

22.66 Discuss the problems associated with storing waste nuclear fuel.

SUPPLEMENTARY EXERCISES

22.67 Explain how α particles can cause severe biological damage, even though their penetrating power is relatively low.

22.68 Describe the technique of radiocarbon dating and explain how it can be used to date objects.

22.69 Describe the evidence for the shell model of the nucleus.

22.70 What is radioactive "fallout" and why is it used as an argument against nuclear bomb tests?

22.71 Radon, which is naturally given off by concrete, cinder block, and stone building materials, is absorbed by polyester. A visitor wearing polyester clothing set off a sensitive radiation monitoring device while touring a nuclear power plant. The fabric had absorbed radon naturally emitted from the cinder blocks of the building. Why is polyester more likely to absorb radon than fabrics such as wool, silk, and nylon? (These materials are described in Chapter 11.)

22.72 Write balanced nuclear equations for the radioactive decay of the following nuclides: (a) ^{74}Kr, β^+ emission; (b) ^{174}Hf, α emission; (c) ^{98}Tc, β emission; (d) ^{41}Ca, electron capture.

22.73 Complete the following equations for nuclear reactions:

(a) $^{11}B + ? \longrightarrow 2\,n + {}^{13}N$
(b) $? + D \longrightarrow n + {}^{36}Ar$
(c) $^{96}Mo + D \longrightarrow ? + {}^{97}Tc$
(d) $^{45}Sc + n \longrightarrow \alpha + ?$

22.74 Tritium undergoes β decay, and the emitted β particle has an energy of 0.0186 MeV (1 MeV = 1.602×10^{-13} J). If a 1.0-g sample of tissue absorbs 10% of the decay products of 1.0 mg of tritium, what dose equivalent does the tissue absorb?

22.75 (a) How many radon-222 nuclei ($t_{1/2}$ = 3.82 d) decay per minute to produce an activity of 4 pCi? (b) A bathroom in the basement of a home measures $2.0 \times 3.0 \times 2.5$ m. If the activity of radon-222 in the room is 4.0 pCi·L^{-1}, how many nuclei decay during a shower lasting 5.0 min?

22.76 (a) How long will it take for the loss of 99.0% of a sample of radon-222 gas ($t_{1/2}$ = 3.82 d)? (b) Comment on radon's lifetime and chances of reaching the surface it formed deep in the Earth's crust. (c) How might a homeowner reduce the rate at which radon enters a home?

22.77 It is found that 1.0×10^{-5} mol of radon-222 atoms ($t_{1/2}$ = 3.82 d) has seeped into a closed basement with a volume of 2000 m^3. (a) What is the initial activity of the radon in picocuries per liter ($pCi \cdot L^{-1}$)? (b) How many atoms of ^{222}Rn will remain after one day (24 h)? (c) How long will it take for the radon to decay to below the EPA recommended level of 4 $pCi \cdot L^{-1}$?

22.78 What mass of a 15.0-mg sample of ^{47}V ($t_{1/2}$ = 33.0 min) will remain after 45.0 min?

22.79 The activity of an iodine-131 source ($t_{1/2}$ = 8.05 d), which is used to monitor the functioning of the thyroid gland, is 500 Bq. How long will it be before the activity is 10 disintegrations per second?

22.80 A sample of fermium-244 ($t_{1/2}$ = 3.3 ms) having an activity of 0.10 μCi was produced in a nuclear reactor. What will the activity of the ^{244}Fm be after 1.0 s?

22.81 (a) A sample of phosphorus-32 has an initial activity of 58 counts per second. After 12.3 days, the activity was 32 counts per second. What is the half-life of phosphorus-32? (b) If phosphorus-32 is used in an experiment to monitor the consumption of phosphorus by plants, what fraction of the nuclide will remain after 30 d?

22.82 A cobalt-60 source purchased for the radiation therapy of cancer patients has an activity of 1.20 Ci. What will the activity of the source be after 5.0 y?

22.83 A sample containing ^{35}S ($t_{1/2}$ = 88 d) that is being used to study the reactions by which sulfur is utilized by bacteria has an activity of 10.0 Ci. What mass of sulfur-35 is present? The mass of a sulfur-35 atom is 35 u.

22.84 What is the activity (in curies) of a 22-μg sample of ^{210}Po ($t_{1/2}$ = 138.4 d)? The mass of a polonium-210 atom is about 210 u.

22.85 Sodium-24 (23.99096 u) decays to magnesium-24 (23.98504 u). Determine (a) the change in the binding energy per nucleon; (b) the change in energy that accompanies the decay. (See Example 22.3 for the definition of 1 u.)

22.86 How much energy is emitted in each α decay of uranium-234? (^{234}U, 234.0409 u; ^{230}Th, 230.0331 u). (See Example 22.3 for the definition of 1 u.)

CHALLENGING EXERCISES

22.87 In the mid-1940s it was proposed that atomic bombs could be used to simplify excavation work, such as that required to dig large canals, and the issue has been raised again. (a) Which type of bomb would be more suitable for such work, fusion or fission? Explain your reasoning. (b) Present arguments for and against this proposal.

22.88 Explain why containers of radioactive waste are more susceptible to corrosion than containers of nonradioactive waste with the same chemical reactivity.

22.89 Uranium-238 decays through a series of α and β emissions to lead-206, with an overall half-life for the entire process of 4.5×10^9 y. If a uranium-bearing ore is found to have a $^{238}U/^{206}Pb$ ratio of (a) 1.00; (b) 1.25, how old is the ore sample?

22.90 The age of a bottle of wine was determined by monitoring the tritium level in the wine. The activity of the tritium is determined to be 8.3% that of a sample of fresh grape juice from the same region from which the wine was bottled. How old is the wine?

22.91 Sodium-24 is used for monitoring blood circulation. (a) If a 2.0-μg sample of sodium-24 has an activity of 17.3 Ci, what is its decay constant and its half-life? (b) What mass of a 2.0-μg sample remains after 2.0 days? The mass of a sodium-24 atom is 24 u.

22.92 Breeder reactors run so hot that a high-boiling-point mixture of molten sodium and potassium is used as the coolant, instead of water. Unfortunately, the steel cooling pipes are susceptible to embrittlement and fracture. It is suspected that the steel becomes brittle because the liquid metals leach carbon from it. Suppose that you had at your disposal the following materials: a fully equipped research laboratory, a small research foundry for making steel, a gas chromatograph, a scintillation counter, and supplies of sodium, potassium, iron, molybdenum, and carbon-14. Design an experiment to test the hypothesis that liquid sodium and potassium mixtures can dissolve carbon from steel.

APPENDIX 1 Mathematical information

1A ALGEBRA RULES

Most of the chemical calculations described in this book can be expressed as algebraic equations. This section of Appendix 1 reviews the basic rules of algebra and arithmetic that you will need in this course.

1. NEGATIVE NUMBERS

Adding a negative number is the same as subtracting a positive number. *Example:* To add -7 to 3, write $3 - 7 = -4$.

Subtracting a negative number is the same as adding a positive number. *Example:* To subtract -7 from 3, write $3 - (-7) = 3 + 7 = 10$.

Multiplying or dividing two numbers gives a negative result if one of the numbers is negative, and a positive result if the two numbers are either both positive or both negative. *Example:* $(-7) \times 3 = -21$, but $(-7) \times (-3) = 21$.

2. FRACTIONS AND PERCENTAGES

A fraction is the ratio of two numbers and may be less than or greater than 1. The number on top is the *numerator* and the number on the bottom is the *denominator. Examples:*

$$\frac{4}{5} \qquad \frac{6}{4}$$

A fraction can be simplified by dividing both the numerator and the denominator by the same number; the value of the fraction does not then change because this action is the same as multiplying the fraction by 1. *Example:*

$$\frac{4}{6} = \frac{4/2}{6/2} = \frac{2}{3}$$

Notice, however, that $\frac{5}{4}$ cannot be simplified, because the top and the bottom of the fraction have no common factor.

When multiplying fractions, multiply the numerators together and multiply the denominators together. *Example:*

$$\frac{2}{3} \times \frac{5}{4} = \frac{2 \times 5}{3 \times 4} = \frac{10}{12} = \frac{5}{6}$$

When dividing fractions, invert the divisor. *Example:*

$$\frac{2}{5}\bigg/\frac{4}{3} = \frac{2}{5} \times \frac{3}{4} = \frac{6}{20} = \frac{3}{10}$$

Sometimes you will see $\frac{2}{5}/\frac{4}{3}$ written

$$\frac{2/5}{4/3}$$

When adding or subtracting fractions, first make sure that they have a common denominator. *Example:* To add $\frac{4}{5}$ and $\frac{6}{4}$, note that the denominators can be made equal by multiplying $\frac{4}{5}$ by $\frac{4}{4}$ (to get $\frac{16}{20}$) and $\frac{6}{4}$ by $\frac{5}{5}$ (to get $\frac{30}{20}$). This procedure gives

$$\frac{4}{5} + \frac{6}{4} = \frac{16}{20} + \frac{30}{20} = \frac{46}{20}$$

The answer can be simplified to $\frac{23}{10}$.

Convert fractions to decimal notation by dividing them. *Examples:* $\frac{5}{4} = 1.25$, $\frac{2}{3} = 0.66\ldots$. If there is one digit to the right of the decimal point, the decimal fraction is divided by 10, so 0.3 is three-tenths. If there are two digits, divide by 100, and so on.

A percentage is a decimal fraction of 100. To find a percentage from a fraction, divide (which converts the fraction to a decimal fraction) and multiply by 100%. *Example:* $\frac{2}{3} \times 100\% = 67\%$.

3. SOLVING A LINEAR EQUATION FOR AN UNKNOWN

An algebraic equation expresses an unknown quantity, x, in terms of other numbers. *Example:* $ax + b = c$, where a, b, and c are any positive or negative numbers.

We can manipulate both sides of an algebraic equation in the following ways:

add or subtract the same number
multiply or divide by the same number
invert (express each side as 1 divided by that side)

Examples: Subtract b from both sides of $ax + b = c$, to get $ax = c - b$. Divide both sides of the resulting equation by a to get $x = (c - b)/a$. Thus, if $3x + 1 = 7$, $x = (7 - 1)/3 = 2$.

4. RATIOS

A ratio is a fraction formed by dividing one number by another. Ratios are frequently used in chemistry and often appear in algebraic expressions. *Example:* Solve $a/x = b/c$ for x. First, multiply each side by c and then by x, which gives $ca = xb$. Then divide each side by b to give $x = ca/b$.

5. FACTORS AND SUMS

When two quantities in a sum or difference can be divided by the same number, it is possible to simplify the expression by factoring. *Example:* $ax + ab = a(x + b)$.

The product of two sums is the sum of the products of each term. *Example:* $(x + a)(y + b) = xy + ay + bx + ab$.

When x appears in each of two terms that are multiplied together, the product is a quadratic equation. *Example:* The expression $(x + a)(x + b) = 0$ is the same as $x^2 + (a + b)x + ab = 0$.

1B SCIENTIFIC NOTATION

In **scientific notation,** a number is written as $A \times 10^a$, where A is a decimal number with one nonzero digit in front of the decimal point and a is a whole number. For example, 333 is written 3.33×10^2 in scientific notation, because $10^2 = 10 \times 10 = 100$:

$$333 = 3.33 \times 100 = 3.33 \times 10^2$$

On a scientific calculator, this number is entered as

3 . 3 3 EXP 2

(On some calculators, the EXP key is labeled EE or EEX.) We use

$$10^1 = 10$$
$$10^2 = 10 \times 10 = 100$$
$$10^3 = 10 \times 10 \times 10 = 1000$$
$$10^4 = 10 \times 10 \times 10 \times 10 = 10{,}000$$

and so on. Note that the number of zeros following 1 is equal to the power of 10. Thus, 10^6 is 1 followed by six zeros:

$$10^6 = 10 \times 10 \times 10 \times 10 \times 10 \times 10 = 1{,}000{,}000$$

Numbers between 0 and 1 are expressed in the same way, but with a negative power of 10; they have the form $A \times 10^{-a}$, with $10^{-1} = \frac{1}{10} = 0.1$, and so on. Thus, 0.0333 in decimal notation is 3.33×10^{-2} because

$$10^{-2} = \tfrac{1}{10} \times \tfrac{1}{10} = \tfrac{1}{100}$$
$$0.0333 = 3.33 \times \tfrac{1}{100} = 3.33 \times 10^{-2}$$

On a scientific calculator, this number is entered as

3 . 3 3 EXP +/− 2

(Be sure to use the +/− key, which is sometimes labeled CHS, to enter the negative power of 2 and not the − key.) We use

$$10^{-2} = 10^{-1} \times 10^{-1} = 0.01$$
$$10^{-3} = 10^{-1} \times 10^{-1} \times 10^{-1} = 0.001$$
$$10^{-4} = 10^{-1} \times 10^{-1} \times 10^{-1} \times 10^{-1} = 0.0001$$

When a negative power of 10 is written out as a decimal number, the number of zeros following the decimal point is one less than the number (disregarding the sign) to which 10 is raised. Thus, 10^{-6} is written as a decimal point followed by $6 - 1 = 5$ zeros and then a 1:

$$10^{-6} = 10^{-1} \times 10^{-1} \times 10^{-1} \times 10^{-1} \times 10^{-1} \times 10^{-1} = 0.000\,001$$

To multiply numbers in scientific notation, the decimal parts of the numbers are multiplied and the powers of 10 are added:

$$(A \times 10^a) \times (B \times 10^b) = (A \times B) \times 10^{a+b}$$

An example is

$$(1.23 \times 10^2) \times (4.56 \times 10^3) = 1.23 \times 4.56 \times 10^{2+3} = 5.61 \times 10^5$$

This rule holds even if the powers of 10 are negative:

$$(1.23 \times 10^{-2}) \times (4.56 \times 10^{-3}) = 1.23 \times 4.56 \times 10^{-2-3} = 5.61 \times 10^{-5}$$

The keystrokes for this calculation are

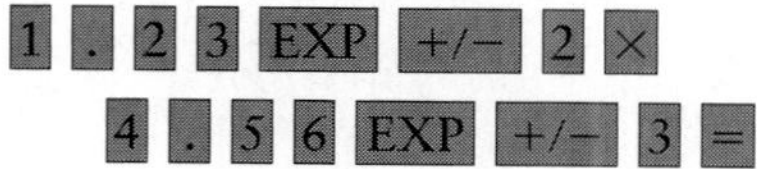

The results of such calculations are adjusted so that there is one digit in front of the decimal point:

$$(4.56 \times 10^{-3}) \times (7.65 \times 10^{6}) = 34.88 \times 10^{3}$$
$$= 3.488 \times 10^{4}$$

When dividing two numbers in scientific notation, we divide the decimal parts of the numbers and subtract the powers of 10:

$$\frac{A \times 10^{a}}{B \times 10^{b}} = \frac{A}{B} \times 10^{a-b}$$

An example is

$$\frac{4.31 \times 10^{5}}{9.87 \times 10^{-8}} = \frac{4.31}{9.87} \times 10^{5-(-8)}$$
$$= 0.437 \times 10^{13} = 4.37 \times 10^{12}$$

Before adding and subtracting numbers in scientific notation, we rewrite the numbers as decimal numbers multiplied by the *same* power of 10:

$$1.00 \times 10^{3} + 2.00 \times 10^{2} = 1.00 \times 10^{3} + 0.200 \times 10^{3}$$
$$= 1.20 \times 10^{3}$$

When raising a number in scientific notation to a particular power, we raise the decimal part of the number to the power and *multiply* the power of 10 by the power:

$$(A \times 10^{a})^{b} = A^{a} \times 10^{a \times b}$$

For example, 2.88×10^4 raised to the third power is

$$(2.88 \times 10^{4})^{3} = 2.88^{3} \times (10^{4})^{3} = 2.88^{3} \times 10^{3 \times 4}$$
$$= 23.9 \times 10^{12} = 2.39 \times 10^{13}$$

The key sequence on a scientific calculator for this calculation is

2 . 8 8 EXP 4 y^x 3 =

The rule follows from the fact that

$$(10^{4})^{3} = 10^{4} \times 10^{4} \times 10^{4} = 10^{4+4+4} = 10^{3 \times 4}$$

1C LOGARITHMS

The *common logarithm* of a number x is denoted $\log x$ and is the power to which 10 must be raised to equal x. Thus, the logarithm of 100 is 2, written $\log 100 = 2$, because $10^2 = 100$. The logarithm of 1.5×10^2 is 2.18 because

$$10^{2.18} = 10^{0.18+2} = 10^{0.18} \times 10^{2} = 1.5 \times 10^{2}$$

The number in front of the decimal point in the logarithm (the 2 in $\log(1.5 \times 10^2) = 2.18$) is called the *characteristic* of the logarithm; the decimal fraction (the numbers following the decimal point; the 0.18 in the example) is called the *mantissa*. The characteristic is the power of 10 in the original number (the power 2 in 1.5×10^2) and the mantissa is the logarithm of the decimal number written with one nonzero digit in front of the decimal point (the 1.5 in the example).

The distinction between the characteristic and the mantissa is important when we have to decide how many significant figures to retain in a calculation that involves logarithms (as we do in the calculation of pH): *the number of significant figures in the mantissa is equal to the number of significant figures in the decimal number.* Because the decimal number 1.5×10^2 has two significant figures, its mantissa is written 0.18 (two significant figures); so its logarithm is 2.18, as written above. Just as the power of 10 in a decimal number indicates only the location of the decimal point and plays no role in the determination of significant figures, so the characteristic of a logarithm is not included in the count of significant figures in a logarithm.

The *common antilogarithm* of a number x is the number that has x as its common logarithm. In practice, the common antilogarithm of x is simply another name for 10^x, so the common antilogarithm of 2 is $10^2 = 100$ and that of 2.18 is

$$10^{2.18} = 10^{0.18+2} = 10^{0.18} \times 10^{2} = 1.5 \times 10^{2}$$

In keeping with the remarks above, the *mantissa* of the logarithm (the 0.18 in 2.18) determines the number of significant figures in the antilogarithm (1.5×10^2, two significant figures).

The logarithm of a number greater than 1 is positive, and that of a number smaller than 1 (but greater than 0) is negative:

$$\text{If } x > 1, \log x > 0$$
$$\text{If } x = 1, \log x = 0$$
$$\text{If } x < 1, \log x < 0$$

Logarithms are not defined either for 0 or for negative numbers.

On a scientific calculator, the common logarithm of a number x is calculated by entering x and pressing the log x key on the calculator. For example, the logarithm of 4.33×10^{-5} is determined with the following sequence of keystrokes:

(It is important to use the +/– key, not the – key when entering the negative power of 10.) The decimal number has three significant figures in this example, so the mantissa should also be written with three significant figures, and the answer reported as −4.364. Likewise, the common antilogarithm of x is found by entering x and pressing the 10^x key (or INV and log keys); so, to calculate the antilogarithm of 11.68, the keystrokes are

Because the mantissa (0.68) of the original number has two significant figures, the antilogarithm should be written with two significant figures, and the correct answer is 4.8×10^{11}.

The *natural logarithm* of a number x is denoted $\ln x$ and is the power to which the number e = 2.718 . . . must be raised to equal x. Thus, $\ln 10.0 = 2.303$, signifying that $e^{2.303} = 10.0$. The number e may seem a peculiar choice, but it occurs naturally in a number of mathematical expressions, and its use simplifies many formulas. On an electronic calculator, the natural logarithm of x is calculated by entering x and then pressing the ln x key. There is no simple rule for assessing the correct number of significant figures when natural logarithms are used: one way is to convert natural logarithms to common logarithms (see below) and then to use the rules specified above.

Common and natural logarithms are related by the expression

$$\ln x = \ln 10 \times \log x = 2.303 \times \log x$$

The *natural antilogarithm* of x is normally called the *exponential* of e; it is the value of e raised to the power of x. On a calculator, it is obtained by entering x and pressing the e^x key (or the INV and ln x keys). Thus, the natural antilogarithm of 2.303 is $e^{2.303} = 10.0$.

The following relationships between logarithms are useful. They are written here mainly for common logarithms, but they apply to natural logarithms as well.

Relation	Example
$\log 10^x = x$	$\log 10^{-7} = -7$
$\ln e^x = x$	$\ln e^{-kt} = -kt$
$\log x + \log y = \log xy$	$\log[Ag^+] + \log[Cl^-] = \log[Ag^+][Cl^-]$
$\log x - \log y = \log\left(\dfrac{x}{y}\right)$	$\log A_0 - \log A = \log\left(\dfrac{A_0}{A}\right)$
$x \log y = \log y^x$	$2\log[H^+] = \log[H^+]^2$
$\log\left(\dfrac{1}{x}\right) = -\log x$	$\log\left(\dfrac{1}{[H^+]}\right) = -\log[H^+]$

Examples:

1. pH

The pH of a solution is the negative logarithm of $[H_3O^+]$. If the molar concentration of H_3O^+ ions in a solution is 0.0024 $mol{\cdot}L^{-1}$, then the pH of the solution is $-\log 0.0024$. To calculate the pH on a scientific calculator, first convert the concentration to scientific notation and then enter

The mantissa of this pH can have only two significant figures (preceded by the characteristic), so the pH is reported as 2.62. Conversely, if the pH of the solution is measured as 7.4, the hydrogen ion concentration is the value of $10^{-7.4}$, which is evaluated by using the keystrokes

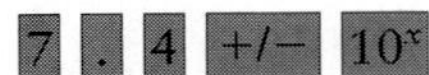

There is only one significant figure in the mantissa (the 4 in pH = 7.4), so the hydrogen ion concentration is reported as 4×10^{-8} $mol{\cdot}L^{-1}$, with only one significant figure. If your hydronium ion concentration is impossibly large, you may have forgotten to press the +/– key *before* pressing 10^x.

2. RATE LAWS

Logarithms are useful for solving expressions of the form

$$a^x = b$$

for the unknown x. (This type of calculation can arise in the study of chemical kinetics when the order of a reaction is being determined.) We take logarithms of both sides

$$\log a^x = \log b$$

and from a relation given above:

$$x \log a = \log b$$

Therefore,

$$x = \frac{\log b}{\log a}$$

1D QUADRATIC AND CUBIC EQUATIONS

A *quadratic equation* is an equation of the form

$$ax^2 + bx + c = 0$$

The two *roots* of the equation (the solutions) can be found most easily by using a graphing calculator or plotting program. However, they can also be calculated from the expressions:

$$x_1 = \frac{-b + \sqrt{(b^2 - 4ac)}}{2a}$$

$$x_2 = \frac{-b - \sqrt{(b^2 - 4ac)}}{2a}$$

When a quadratic equation arises in connection with a chemical calculation, we accept only the root that leads to a physically plausible result. For example, if x is a concentration, then it must be a positive number, and a negative root can be ignored. However, if x is a *change* in concentration, then it may be either positive or negative. In such a case, we would have to determine which root led to an acceptable (positive) final concentration.

On occasion, an equilibrium table (or some other type of calculation) results in a *cubic equation:*

$$ax^3 + bx^2 + cx + d = 0$$

Cubic equations are often very tedious to solve exactly, so it is better to use a graphing calculator or plotter, such as the one on the CD that accompanies this book. However, it is often justifiable (particularly when we are dealing with a chemical system for which the data have only a certain number of significant figures) to use an approximation procedure. For example, an equilibrium table might lead to an expression of the form

$$\frac{0.0600 - x}{(0.0500 + x) \times (0.0300 + 2x)^2} = 1.10 \times 10^{-2}$$

One approach to the solution of this expression for the value of x is to make successive approximations, starting with an estimated value of x. We know that x must be smaller than 0.0600, and we also know that it should be quite close to 0.0600 to ensure that the expression on the left is small (it should be equal to 1.10×10^{-2}). Thus, we might estimate that $x = 0.0590$, which leads to a value of

$$\frac{0.0600 - 0.0590}{(0.0500 + 0.0590) \times (0.0300 + 2 \times 0.0590)^2} = 0.419$$

This first guess leads to a result that is larger than the true value, so we reduce the size of the fraction by choosing a value of x that is closer to 0.0600. Thus, with $x = 0.0599$, we find

$$\frac{0.0600 - 0.0599}{(0.0500 + 0.0599) \times (0.0300 + 2 \times 0.0599)^2} = 4.05 \times 10^{-2}$$

This result is still too large, but much closer to the required value of 1.10×10^{-2}. We could continue to choose values of x that are closer to 0.0600 than 0.0599, but since only three significant figures are present in the data, there is little point in looking for a more precise result, and x can be reported as lying between 0.0599 and 0.0600 (in fact, a more precise answer is 0.05997). Such approximate solutions can usually be found after two or three trials.

1E GRAPHS

Experimental data can often be analyzed by plotting a graph. In many cases, the best procedure is to find a way of plotting the data so that a straight line results. This is more useful than plotting a curve, because it is quite easy to tell whether or not the data do in fact fall on a straight line, whereas small deviations from a curve are harder to detect. Moreover, it is also easy to calculate the slope of a straight line, to *extrapolate* (or extend) a straight line beyond the range of the data, and to *interpolate* between the data (that is, find a value between two measured values).

The formula of a straight line graph of y (the vertical axis) plotted against x (the horizontal axis) is

$$y = mx + b$$

Here b is the *intercept* of the graph with the y-axis (Fig. A.1), the value of y where the graph cuts through the vertical axis at $x = 0$. The *slope* of the graph, its gradient, is m. The slope can be calculated

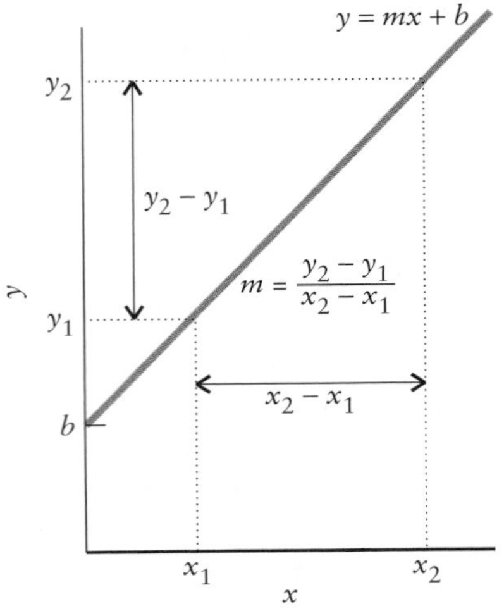

FIGURE A.1 The straight line $y = mx + b$, its intercept at $y = b$, and its slope m.

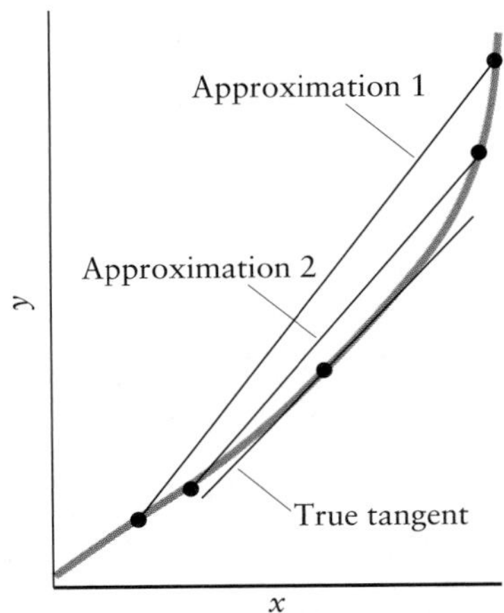

FIGURE A.2 Successive approximations to the true tangent are obtained as the two points defining the straight line come closer together and finally coincide.

by choosing two points, x_1 and x_2, and their corresponding values on the y-axis, y_1 and y_2, and substituting the values into the formula

$$m = \frac{y_2 - y_1}{x_2 - x_1}$$

Because b is the intercept and m is the slope, the equation of the straight line is equivalent to

$$y = \text{slope} \times x + \text{intercept}$$

Many of the equations we meet in the text can be rearranged to give a straight line graph when plotted. These equations include

Application	y = slope × x	+	intercept
Temperature scale conversions	$°C = 1 \times K$	−	273.15
	$°F = 1.8 \times °C$	+	32
Ideal gas law	$P = nRT \times \frac{1}{V}$		
First-order integrated rate law	$\ln[A] = -k \times t$	+	$\ln[A]_0$
Second-order integrated rate law	$\frac{1}{[A]} = k \times t$	+	$\frac{1}{[A]_0}$
Arrhenius law	$\ln k = \frac{-E_a}{R} \times \frac{1}{T}$	+	A

We can speak of *the* slope of a straight line because the slope is the same at all points. On a curve, however, the slope changes from point to point. The *tangent* of a curve at a specified point is the straight line that has the same slope as the curve at that point. The tangent can be found by a series of approximations, as shown in Fig. A.2. Approximation 1 is found by drawing a point on the curve on each side of the point of interest (corresponding to equal distances along the x-axis) and joining them by a straight line. A better approximation (approximation 2) is obtained by moving the points an equal distance closer to the point of interest and drawing a new line. The "exact" tangent is obtained when the two points are virtually coincident with the point of interest. Its slope is then equal to the slope of the curve at the point of interest. This technique can be used to measure the rate of a chemical reaction at a specified time (see Section 18.2).

APPENDIX 2 Experimental data

2A THERMODYNAMIC DATA AT 25°C

INORGANIC SUBSTANCES

Substance	Molar mass, $g \cdot mol^{-1}$	Enthalpy of formation, ΔH_f°, $kJ \cdot mol^{-1}$	Free energy of formation, ΔG_f°, $kJ \cdot mol^{-1}$	Entropy,* S°, $J \cdot K^{-1} \cdot mol^{-1}$
Aluminum				
Al(s)	26.98	0	0	28.33
Al^{3+}(aq)	26.98	−524.7	−481.2	−321.7
Al_2O_3(s)	101.95	−1675.7	−1582.3	50.92
$Al(OH)_3$(s)	78.00	−1276	—	—
$AlCl_3$(s)	133.24	−704.2	−628.8	110.67
Antimony				
SbH_3(g)	124.77	145.11	147.75	232.78
$SbCl_3$(g)	228.10	−313.8	−301.2	337.80
$SbCl_5$(g)	299.02	−394.34	−334.29	401.94
Arsenic				
As(s), gray	74.92	0	0	35.1
As_2S_3(s)	246.04	−169.0	−168.6	163.6
AsO_4^{3-}(aq)	138.92	−888.14	−648.41	−162.8
Barium				
Ba(s)	137.34	0	0	62.8
Ba^{2+}(aq)	137.34	−537.64	−560.77	9.6
BaO(s)	153.34	−553.5	−525.1	70.42
$BaCO_3$(s)	197.35	−1216.3	−1137.6	112.1
$BaCO_3$(aq)	197.35	−1214.78	−1088.59	−47.3
Boron				
B(s)	10.81	0	0	5.86
B_2O_3(s)	69.62	−1272.8	−1193.7	53.97
BF_3(g)	67.81	−1137.0	−1120.3	254.12
Bromine				
Br_2(l)	159.82	0	0	152.23
Br_2(g)	159.82	30.91	3.11	245.46
Br(g)	79.91	111.88	82.40	175.02
Br^-(aq)	79.91	−121.55	−103.96	82.4
HBr(g)	80.92	−36.40	−53.45	198.70

(continued)

INORGANIC SUBSTANCES *(continued)*

Substance	Molar mass, $g \cdot mol^{-1}$	Enthalpy of formation, ΔH_f°, $kJ \cdot mol^{-1}$	Free energy of formation, ΔG_f°, $kJ \cdot mol^{-1}$	Entropy,* S°, $J \cdot K^{-1} \cdot mol^{-1}$
Calcium				
Ca(s)	40.08	0	0	41.42
Ca(g)	40.08	178.2	144.3	154.88
Ca^{2+}(aq)	40.08	−542.83	−553.58	−53.1
CaO(s)	56.08	−635.09	−604.03	39.75
$Ca(OH)_2$(s)	74.10	−986.09	−898.49	83.39
$Ca(OH)_2$(aq)	74.10	−1002.82	−868.07	−74.5
$CaCO_3$(s), calcite	100.09	−1206.9	−1128.8	92.9
$CaCO_3$(s), aragonite	100.09	−1207.1	−1127.8	88.7
$CaCO_3$(aq)	100.09	−1219.97	−1081.39	−110.0
CaF_2(s)	78.08	−1219.6	−1167.3	68.87
CaF_2(aq)	78.08	−1208.09	−1111.15	−80.8
$CaCl_2$(s)	110.99	−795.8	−748.1	104.6
$CaCl_2$(aq)	110.99	−877.1	−816.0	59.8
$CaBr_2$(s)	199.90	−682.8	−663.6	130
CaC_2(s)	64.10	−59.8	−64.9	69.96
$CaSO_4$(s)	136.14	−1434.11	−1321.79	106.7
$CaSO_4$(aq)	136.14	−1452.10	−1298.10	−33.1
Carbon†				
C(s), graphite	12.011	0	0	5.740
C(s), diamond	12.011	1.895	2.900	2.377
C(g)	12.011	716.68	671.26	158.10
CO(g)	28.01	−110.53	−137.17	197.67
CO_2(g)	44.01	−393.51	−394.36	213.74
CO_3^{2-}(aq)	60.01	−677.14	−527.81	−56.9
CCl_4(l)	153.82	−135.44	−65.21	216.40
CS_2(l)	76.14	89.70	65.27	151.34
HCN(g)	27.03	135.1	124.7	201.78
HCN(l)	27.03	108.87	124.97	112.84
Cerium				
Ce(s)	140.12	0	0	72.0
Ce^{3+}(aq)	140.12	−696.2	−672.0	−205
Ce^{4+}(aq)	140.12	−537.2	−503.8	−301
Chlorine				
Cl_2(g)	70.91	0	0	223.07
Cl(g)	35.45	121.68	105.68	165.20
Cl^-(aq)	35.45	−167.16	−131.23	56.5
HCl(g)	36.46	−92.31	−95.30	186.91
HCl(aq)	36.46	−167.16	−131.23	56.5
Copper				
Cu(s)	63.54	0	0	33.15
Cu^+(aq)	63.54	71.67	49.98	40.6
Cu^{2+}(aq)	63.54	64.77	65.49	−99.6
Cu_2O(s)	143.08	−168.6	−146.0	93.14
CuO(s)	79.54	−157.3	−129.7	42.63

Substance	Molar mass, $g \cdot mol^{-1}$	Enthalpy of formation, ΔH_f°, $kJ \cdot mol^{-1}$	Free energy of formation, ΔG_f°, $kJ \cdot mol^{-1}$	Entropy,* S°, $J \cdot K^{-1} \cdot mol^{-1}$
$CuSO_4(s)$	159.60	−771.36	−661.8	109
$CuSO_4 \cdot 5H_2O(s)$	249.68	−2279.7	−1879.7	300.4
Deuterium				
$D_2(g)$	4.028	0	0	144.96
$D_2O(g)$	20.028	−249.20	−234.54	198.34
$D_2O(l)$	20.028	−294.60	−243.44	75.94
Fluorine				
$F_2(g)$	38.00	0	0	202.78
$F^-(aq)$	19.00	−332.63	−278.79	−13.8
$HF(g)$	20.01	−271.1	−273.2	173.78
$HF(aq)$	20.01	−332.63	−278.79	−13.8
Hydrogen (see also *Deuterium*)				
$H_2(g)$	2.016	0	0	130.68
$H(g)$	1.008	217.97	203.25	114.71
$H^+(aq)$	1.008	0	0	0
$H_2O(l)$	18.02	−285.83	−237.13	69.91
$H_2O(g)$	18.02	−241.82	−228.57	188.83
$H_2O_2(l)$	34.02	−187.78	−120.35	109.6
$H_2O_2(aq)$	34.02	−191.17	−134.03	143.9
Iodine				
$I_2(s)$	253.81	0	0	116.14
$I_2(g)$	253.81	62.44	19.33	260.69
$I^-(aq)$	126.90	−55.19	−51.57	111.3
$HI(g)$	127.91	26.48	1.70	206.59
Iron				
$Fe(s)$	55.85	0	0	27.28
$Fe^{2+}(aq)$	55.85	−89.1	−78.90	−137.7
$Fe^{3+}(aq)$	55.85	−48.5	−4.7	−315.9
$Fe_3O_4(s)$, magnetite	231.54	−1118.4	−1015.4	146.4
$Fe_2O_3(s)$, hematite	159.69	−824.2	−742.2	87.40
$FeS(s,\alpha)$	87.91	−100.0	−100.4	60.29
$FeS(aq)$	87.91	—	6.9	—
$FeS_2(s)$	119.98	−178.2	−166.9	52.93
Lead				
$Pb(s)$	207.19	0	0	64.81
$Pb^{2+}(aq)$	207.19	−1.7	−24.43	10.5
$PbO_2(s)$	239.19	−277.4	−217.33	68.6
$PbSO_4(s)$	303.25	−919.94	−813.14	148.57
$PbBr_2(s)$	367.01	−278.7	−261.92	161.5
$PbBr_2(aq)$	367.01	−244.8	−232.34	175.3
Magnesium				
$Mg(s)$	24.31	0	0	32.68
$Mg(g)$	24.31	147.70	113.10	148.65
$Mg^{2+}(aq)$	24.31	−466.85	−454.8	−138.1

(continued)

INORGANIC SUBSTANCES *(continued)*

Substance	Molar mass, $g \cdot mol^{-1}$	Enthalpy of formation, ΔH_f°, $kJ \cdot mol^{-1}$	Free energy of formation, ΔG_f°, $kJ \cdot mol^{-1}$	Entropy,* S°, $J \cdot K^{-1} \cdot mol^{-1}$
Magnesium *(continued)*				
$MgO(s)$	40.31	−601.70	−569.43	26.94
$MgCO_3(s)$	84.32	−1095.8	−1012.1	65.7
$MgBr_2(s)$	184.13	−524.3	−503.8	117.2
Mercury				
$Hg(l)$	200.59	0	0	76.02
$Hg(g)$	200.59	61.32	31.82	174.96
$HgO(s)$	216.59	−90.83	−58.54	70.29
$Hg_2Cl_2(s)$	472.09	−265.22	−210.75	192.5
Nitrogen				
$N_2(g)$	28.01	0	0	191.61
$NO(g)$	30.01	90.25	86.55	210.76
$N_2O(g)$	44.01	82.05	104.20	219.85
$NO_2(g)$	46.01	33.18	51.31	240.06
$N_2O_4(g)$	92.01	9.16	97.89	304.29
$HNO_3(l)$	63.01	−174.10	−80.71	155.60
$HNO_3(aq)$	63.01	−207.36	−111.25	146.4
$NO_3^-(aq)$	62.01	−205.0	−108.74	146.4
$NH_3(g)$	17.03	−46.11	−16.45	192.45
$NH_3(aq)$	17.03	−80.29	−26.50	111.3
$NH_4^+(aq)$	18.04	−132.51	−79.31	113.4
$NH_2OH(s)$	33.03	−114.2	—	—
$HN_3(g)$	43.03	294.1	328.1	238.97
$N_2H_4(l)$	32.05	50.63	149.34	121.21
$NH_4NO_3(s)$	80.04	−365.56	−183.87	151.08
$NH_4Cl(s)$	53.49	−314.43	−202.87	94.6
$NH_4ClO_4(s)$	117.49	−295.31	−88.75	186.2
Oxygen				
$O_2(g)$	32.00	0	0	205.14
$O_3(g)$	48.00	142.7	163.2	238.93
$OH^-(aq)$	17.01	−229.99	−157.24	−10.75
Phosphorus				
$P(s)$, white	30.97	0	0	41.09
$P_4(g)$	123.90	58.91	24.44	279.98
$PH_3(g)$	34.00	5.4	13.4	210.23
$P_4O_{10}(s)$	283.89	−2984.0	−2697.0	228.86
$H_3PO_3(aq)$	82.00	−964.8	—	—
$H_3PO_4(l)$	98.00	−1266.9	—	—
$H_3PO_4(aq)$	98.00	−1277.4	−1018.7	—
$PCl_3(l)$	137.33	−319.7	−272.3	217.18
$PCl_3(g)$	137.33	−287.0	−267.8	311.78
$PCl_5(g)$	208.24	−374.9	−305.0	364.6
$PCl_5(s)$	208.24	−443.5	—	—
Potassium				
$K(s)$	39.10	0	0	64.18
$K(g)$	39.10	89.24	60.59	160.34

Substance	Molar mass, g·mol^{-1}	Enthalpy of formation, ΔH_f°, kJ·mol^{-1}	Free energy of formation, ΔG_f°, kJ·mol^{-1}	Entropy* S°, J·K^{-1}·mol^{-1}
K^+(aq)	39.10	−252.38	−283.27	102.5
KOH(s)	56.11	−424.76	−379.08	78.9
KOH(aq)	56.11	−482.37	−440.50	91.6
KF(s)	58.10	−567.27	−537.75	66.57
KCl(s)	74.56	−436.75	−409.14	82.59
KBr(s)	119.01	−393.80	−380.66	95.90
KI(s)	166.01	−327.90	−324.89	106.32
$KClO_3$(s)	122.55	−397.73	−296.25	143.1
$KClO_4$(s)	138.55	−432.75	−303.09	151.0
K_2S(s)	110.27	−380.7	−364.0	105
K_2S(aq)	110.27	−471.5	−480.7	190.4
Silicon				
Si(s)	28.09	0	0	18.83
SiO_2(s,α)	60.09	−910.94	−856.64	41.84
Silver				
Ag(s)	107.87	0	0	42.55
Ag^+(aq)	107.87	105.58	77.11	72.68
Ag_2O(s)	231.74	−31.05	−11.20	121.3
AgBr(s)	187.78	−100.37	−96.90	107.1
AgBr(aq)	187.78	−15.98	−26.86	155.2
AgCl(s)	143.32	−127.07	−109.79	96.2
AgCl(aq)	143.32	−61.58	−54.12	129.3
AgI(s)	234.77	−61.84	−66.19	115.5
AgI(aq)	234.77	50.38	25.52	184.1
$AgNO_3$(s)	169.88	−124.39	−33.41	140.92
Sodium				
Na(s)	22.99	0	0	51.21
Na(g)	22.99	107.32	76.76	153.71
Na^+(aq)	22.99	−240.12	−261.91	59.0
NaOH(s)	40.00	−425.61	−379.49	64.46
NaOH(aq)	40.00	−470.11	−419.15	48.1
NaCl(s)	58.44	−411.15	−384.14	72.13
NaBr(s)	102.90	−361.06	−348.98	86.82
NaI(s)	149.89	−287.78	−286.06	98.53
Sulfur				
S(s), rhombic	32.06	0	0	31.80
S(s), monoclinic	32.06	0.33	0.1	32.6
S^{2-}(aq)	32.06	33.1	85.8	−14.6
SO_2(g)	64.06	−296.83	−300.19	248.22
SO_3(g)	80.06	−395.72	−371.06	256.76
H_2SO_4(l)	98.08	−813.99	−690.00	156.90
H_2SO_4(aq)	98.08	−909.27	−744.53	20.1
SO_4^{2-}(aq)	96.06	−909.27	−744.53	20.1
H_2S(g)	34.08	−20.63	−33.56	205.79
H_2S(aq)	34.08	−39.7	−27.83	121
SF_6(g)	146.05	−1209	−1105.3	291.82

(continued)

INORGANIC SUBSTANCES *(continued)*

Substance	Molar mass, $g \cdot mol^{-1}$	Enthalpy of formation, ΔH_f°, $kJ \cdot mol^{-1}$	Free energy of formation, ΔG_f°, $kJ \cdot mol^{-1}$	Entropy,* S°, $J \cdot K^{-1} \cdot mol^{-1}$
Tin				
Sn(s), white	118.69	0	0	51.55
Sn(s), gray	118.69	−2.09	0.13	44.14
SnO(s)	134.69	−285.8	−256.9	56.5
SnO_2(s)	150.69	−580.7	−519.6	52.3
Zinc				
Zn(s)	65.37	0	0	41.63
Zn^{2+}(aq)	65.37	−153.89	−147.06	−112.1
ZnO(s)	81.37	−348.28	−318.30	43.64

*The entropies of individual ions in solution are determined by setting the entropy of H^+ in water equal to 0 and then defining the entropies of all other ions relative to this value; hence a negative entropy is one that is lower than the entropy of H^+ in water.
†For organic compounds, see the next table.

ORGANIC COMPOUNDS

Substance	Molar mass, $g \cdot mol^{-1}$	Enthalpy of combustion, ΔH_c°, $kJ \cdot mol^{-1}$	Enthalpy of formation, ΔH_f°, $kJ \cdot mol^{-1}$	Free energy of formation, ΔG_f°, $kJ \cdot mol^{-1}$	Entropy, S°, $J \cdot K^{-1} \cdot mol^{-1}$
Hydrocarbons					
CH_4(g), methane	16.04	−890	−74.81	−50.72	186.26
C_2H_2(g), ethyne (acetylene)	26.04	−1300	226.73	209.20	200.94
C_2H_4(g), ethene (ethylene)	28.05	−1411	52.26	68.15	219.56
C_2H_6(g), ethane	30.07	−1560	−84.68	−32.82	229.60
C_3H_6(g), propene (propylene)	42.08	−2058	20.42	62.78	266.6
C_3H_6(g), cyclopropane	42.08	−2091	53.30	104.45	237.4
C_3H_8(g), propane	44.09	−2220	−103.85	−23.49	270.2
C_4H_{10}(g), butane	58.13	−2878	−126.15	−17.03	310.1
C_5H_{12}(g), pentane	72.15	−3537	−146.44	−8.20	349
C_6H_6(l), benzene	78.12	−3268	49.0	124.3	173.3
C_6H_6(g)	78.12	−3302	—	—	—
C_7H_8(l), toluene	92.13	−3910	12.0	113.8	221.0
C_7H_8(g)	92.13	−3953	—	—	—
C_6H_{12}(l), cyclohexane	84.16	−3920	−156.4	26.7	204.4
C_6H_{12}(g)	84.16	−3953	—	—	—
C_8H_{18}(l), octane	114.23	−5471	−249.9	6.4	358
Alcohols and phenols					
CH_3OH(l), methanol	32.04	−726	−238.86	−166.27	126.8
CH_3OH(g)	32.04	−764	−200.66	−161.96	239.81
C_2H_5OH(l), ethanol	46.07	−1368	−277.69	−174.78	160.7
C_2H_5OH(g)	46.07	−1409	−235.10	−168.49	282.70
C_6H_5OH(s), phenol	94.11	−3054	−164.6	−50.42	144.0
Carboxylic acids					
HCOOH(l), formic acid	46.03	−255	−424.72	−361.35	128.95

Substance	Molar mass, $g \cdot mol^{-1}$	Enthalpy of combustion, ΔH_c°, $kJ \cdot mol^{-1}$	Enthalpy of formation, ΔH_f°, $kJ \cdot mol^{-1}$	Free energy of formation, ΔG_f°, $kJ \cdot mol^{-1}$	Entropy, S°, $J \cdot K^{-1} \cdot mol^{-1}$
$CH_3COOH(l)$, acetic acid	60.05	−875	−484.5	−389.9	159.8
$CH_3COOH(aq)$	60.05	—	−485.76	−396.46	86.6
$(COOH)_2(s)$, oxalic acid	90.04	−254	−827.2	−697.9	120
$C_6H_5COOH(s)$, benzoic acid	122.13	−3227	−385.1	−245.3	167.6
Aldehydes and ketones					
$HCHO(g)$, methanal (formaldehyde)	30.03	−571	−108.57	−102.53	218.77
$CH_3CHO(l)$, ethanal (acetaldehyde)	44.05	−1166	−192.30	−128.12	160.2
$CH_3CHO(g)$	44.05	−1192	−166.19	−128.86	250.3
$CH_3COCH_3(l)$, propanal (acetone)	58.08	−1790	−248.1	−155.4	200
Sugars					
$C_6H_{12}O_6(s)$, glucose	180.16	−2808	−1268	−910	212
$C_6H_{12}O_6(aq)$	180.16	—	—	−917	—
$C_6H_{12}O_6(s)$, fructose	180.16	−2810	−1266	—	—
$C_{12}H_{22}O_{11}(s)$, sucrose	342.30	−5645	−2222	−1545	360
Nitrogen compounds					
$CO(NH_2)_2(s)$, urea	60.06	−632	−333.51	−197.33	104.60
$C_6H_5NH_2(l)$, aniline	93.13	−3393	31.6	149.1	191.3
$NH_2CH_2COOH(s)$, glycine	75.07	−969	−532.9	−373.4	103.51
$CH_3NH_2(g)$, methylamine	31.06	−1085	−22.97	32.16	243.41

2B STANDARD POTENTIALS AT 25°C

POTENTIALS IN ELECTROCHEMICAL ORDER

Reduction half-reaction	$E°$, V	Reduction half-reaction	$E°$, V
Strongly oxidizing			
$H_4XeO_6 + 2H^+ + 2e^- \longrightarrow XeO_3 + 3H_2O$	+3.0	$I_2 + 2e^- \longrightarrow 2I^-$	+0.54
$F_2 + 2e^- \longrightarrow 2F^-$	+2.87	$I_3^- + 2e^- \longrightarrow 3I^-$	+0.53
$O_3 + 2H^+ + 2e^- \longrightarrow O_2 + H_2O$	+2.07	$Cu^+ + e^- \longrightarrow Cu$	+0.52
$S_2O_8^{2-} + 2e^- \longrightarrow 2SO_4^{2-}$	+2.05	$Ni(OH)_3 + e^- \longrightarrow Ni(OH)_2 + OH^-$	+0.49
$Ag^{2+} + e^- \longrightarrow Ag^+$	+1.98	$O_2 + 2H_2O + 4e^- \longrightarrow 4OH^-$	+0.40
$Co^{3+} + e^- \longrightarrow Co^{2+}$	+1.81	$ClO_4^- + H_2O + 2e^- \longrightarrow ClO_3^- + 2OH^-$	+0.36
$H_2O_2 + 2H^+ + 2e^- \longrightarrow 2H_2O$	+1.78	$Cu^{2+} + 2e^- \longrightarrow Cu$	+0.34
$Au^+ + e^- \longrightarrow Au$	+1.69	$Hg_2Cl_2 + 2e^- \longrightarrow 2Hg + 2Cl^-$	+0.27
$Pb^{4+} + 2e^- \longrightarrow Pb^{2+}$	+1.67	$AgCl + e^- \longrightarrow Ag + Cl^-$	+0.22
$2HClO + 2H^+ + 2e^- \longrightarrow Cl_2 + 2H_2O$	+1.63	$Bi^{3+} + 3e^- \longrightarrow Bi$	+0.20
$Hg^{2+} + 2e^- \longrightarrow Hg$	+1.62	$SO_4^{2-} + 4H^+ + 2e^- \longrightarrow H_2SO_3 + H_2O$	+0.17
$Ce^{4+} + e^- \longrightarrow Ce^{3+}$	+1.61	$Cu^{2+} + e^- \longrightarrow Cu^+$	+0.15
$2HBrO + 2H^+ + 2e^- \longrightarrow Br_2 + 2H_2O$	+1.60	$Sn^{4+} + 2e^- \longrightarrow Sn^{2+}$	+0.15
$MnO_4^- + 8H^+ + 5e^- \longrightarrow Mn^{2+} + 4H_2O$	+1.51	$AgBr + e^- \longrightarrow Ag + Br^-$	+0.07
$Mn^{3+} + e^- \longrightarrow Mn^{2+}$	+1.51	$NO_3^- + H_2O + 2e^- \longrightarrow NO_2^- + 2OH^-$	+0.01
$Au^{3+} + 3e^- \longrightarrow Au$	+1.40	$Ti^{4+} + e^- \longrightarrow Ti^{3+}$	0.00
$Cl_2 + 2e^- \longrightarrow 2Cl^-$	+1.36	$2H^+ + 2e^- \longrightarrow H_2$	0, by definition
$Cr_2O_7^{2-} + 14H^+ + 6e^- \longrightarrow 2Cr^{3+} + 7H_2O$	+1.33		
$O_3 + H_2O + 2e^- \longrightarrow O_2 + 2OH^-$	+1.24	$Fe^{3+} + 3e^- \longrightarrow Fe$	−0.04
$O_2 + 4H^+ + 4e^- \longrightarrow 2H_2O$	+1.23	$O_2 + H_2O + 2e^- \longrightarrow HO_2^- + OH^-$	−0.08
$MnO_2 + 4H^+ + 2e^- \longrightarrow Mn^{2+} + 2H_2O$	+1.23	$Pb^{2+} + 2e^- \longrightarrow Pb$	−0.13
$ClO_4^- + 2H^+ + 2e^- \longrightarrow ClO_3^- + H_2O$	+1.23	$In^+ + e^- \longrightarrow In$	−0.14
$Pt^{2+} + 2e^- \longrightarrow Pt$	+1.20	$Sn^{2+} + 2e^- \longrightarrow Sn$	−0.14
$Br_2 + 2e^- \longrightarrow 2Br^-$	+1.09	$AgI + e^- \longrightarrow Ag + I^-$	−0.15
$Pu^{4+} + e^- \longrightarrow Pu^{3+}$	+0.97	$Ni^{2+} + 2e^- \longrightarrow Ni$	−0.23
$NO_3^- + 4H^+ + 3e^- \longrightarrow NO + 2H_2O$	+0.96	$V^{3+} + e^- \longrightarrow V^{2+}$	−0.26
$2Hg^{2+} + 2e^- \longrightarrow Hg_2^{2+}$	+0.92	$Co^{2+} + 2e^- \longrightarrow Co$	−0.28
$ClO^- + H_2O + 2e^- \longrightarrow Cl^- + 2OH^-$	+0.89	$In^{3+} + 3e^- \longrightarrow In$	−0.34
$NO_3^- + 2H^+ + e^- \longrightarrow NO_2 + H_2O$	+0.80	$Tl^+ + e^- \longrightarrow Tl$	−0.34
$Ag^+ + e^- \longrightarrow Ag$	+0.80	$PbSO_4 + 2e^- \longrightarrow Pb + SO_4^{2-}$	−0.36
$Hg_2^{2+} + 2e^- \longrightarrow 2Hg$	+0.79	$Ti^{3+} + e^- \longrightarrow Ti^{2+}$	−0.37
$AgF + e^- \longrightarrow Ag + F^-$	+0.78	$In^{2+} + e^- \longrightarrow In^+$	−0.40
$Fe^{3+} + e^- \longrightarrow Fe^{2+}$	+0.77	$Cd^{2+} + 2e^- \longrightarrow Cd$	−0.40
$BrO^- + H_2O + 2e^- \longrightarrow Br^- + 2OH^-$	+0.76	$Cr^{3+} + e^- \longrightarrow Cr^{2+}$	−0.41
$MnO_4^{2-} + 2H_2O + 2e^- \longrightarrow MnO_2 + 4OH^-$	+0.60	$Fe^{2+} + 2e^- \longrightarrow Fe$	−0.44
$MnO_4^- + e^- \longrightarrow MnO_4^{2-}$	+0.56	$In^{3+} + 2e^- \longrightarrow In^+$	−0.44

Reduction half-reaction	$E°$, V	Reduction half-reaction	$E°$, V
$S + 2e^- \longrightarrow S^{2-}$	−0.48	$Al^{3+} + 3e^- \longrightarrow Al$	−1.66
$In^{3+} + e^- \longrightarrow In^{2+}$	−0.49	$U^{3+} + 3e^- \longrightarrow U$	−1.79
$Ga^+ + e^- \longrightarrow Ga$	−0.53	$Be^{2+} + 2e^- \longrightarrow Be$	−1.85
$O_2 + e^- \longrightarrow O_2^-$	−0.56	$Mg^{2+} + 2e^- \longrightarrow Mg$	−2.36
$U^{4+} + e^- \longrightarrow U^{3+}$	−0.61	$Ce^{3+} + 3e^- \longrightarrow Ce$	−2.48
$Se + 2e^- \longrightarrow Se^{2-}$	−0.67	$La^{3+} + 3e^- \longrightarrow La$	−2.52
$Cr^{3+} + 3e^- \longrightarrow Cr$	−0.74	$Na^+ + e^- \longrightarrow Na$	−2.71
$Zn^{2+} + 2e^- \longrightarrow Zn$	−0.76	$Ca^{2+} + 2e^- \longrightarrow Ca$	−2.87
$Cd(OH)_2 + 2e^- \longrightarrow Cd + 2OH^-$	−0.81	$Sr^{2+} + 2e^- \longrightarrow Sr$	−2.89
$2H_2O + 2e^- \longrightarrow H_2 + 2OH^-$	−0.83	$Ba^{2+} + 2e^- \longrightarrow Ba$	−2.91
$Te + 2e^- \longrightarrow Te^{2-}$	−0.84	$Ra^{2+} + 2e^- \longrightarrow Ra$	−2.92
$Cr^{2+} + 2e^- \longrightarrow Cr$	−0.91	$Cs^+ + e^- \longrightarrow Cs$	−2.92
$Mn^{2+} + 2e^- \longrightarrow Mn$	−1.18	$Rb^+ + e^- \longrightarrow Rb$	−2.93
$V^{2+} + 2e^- \longrightarrow V$	−1.19	$K^+ + e^- \longrightarrow K$	−2.93
$Ti^{2+} + 2e^- \longrightarrow Ti$	−1.63	$Li^+ + e^- \longrightarrow Li$	−3.05
		Strongly reducing	

POTENTIALS IN ALPHABETICAL ORDER

Reduction half-reaction	$E°$, V	Reduction half-reaction	$E°$, V
$Ag^+ + e^- \longrightarrow Ag$	+0.80	$Cd^{2+} + 2e^- \longrightarrow Cd$	−0.40
$Ag^{2+} + e^- \longrightarrow Ag^+$	+1.98	$Cd(OH)_2 + 2e^- \longrightarrow Cd + 2OH^-$	−0.81
$AgBr + e^- \longrightarrow Ag + Br^-$	+0.07	$Ce^{3+} + 3e^- \longrightarrow Ce$	−2.48
$AgCl + e^- \longrightarrow Ag + Cl^-$	+0.22	$Ce^{4+} + e^- \longrightarrow Ce^{3+}$	+1.61
$AgF + e^- \longrightarrow Ag + F^-$	+0.78	$Cl_2 + 2e^- \longrightarrow 2Cl^-$	+1.36
$AgI + e^- \longrightarrow Ag + I^-$	−0.15	$ClO^- + H_2O + 2e^- \longrightarrow Cl^- + 2OH^-$	+0.89
$Al^{3+} + 3e^- \longrightarrow Al$	−1.66	$ClO_4^- + 2H^+ + 2e^- \longrightarrow ClO_3^- + H_2O$	+1.23
$Au^+ + e^- \longrightarrow Au$	+1.69	$ClO_4^- + H_2O + 2e^- \longrightarrow ClO_3^- + 2OH^-$	+0.36
$Au^{3+} + 3e^- \longrightarrow Au$	+1.40	$Co^{2+} + 2e^- \longrightarrow Co$	−0.28
$Ba^{2+} + 2e^- \longrightarrow Ba$	−2.91	$Co^{3+} + e^- \longrightarrow Co^{2+}$	+1.81
$Be^{2+} + 2e^- \longrightarrow Be$	−1.85	$Cr^{2+} + 2e^- \longrightarrow Cr$	−0.91
$Bi^{3+} + 3e^- \longrightarrow Bi$	+0.20	$Cr_2O_7^{2-} + 14H^+ + 6e^- \longrightarrow 2Cr^{3+} + 7H_2O$	+1.33
$Br_2 + 2e^- \longrightarrow 2Br^-$	+1.09	$Cr^{3+} + 3e^- \longrightarrow Cr$	−0.74
$BrO^- + H_2O + 2e^- \longrightarrow Br^- + 2OH^-$	+0.76	$Cr^{3+} + e^- \longrightarrow Cr^{2+}$	−0.41
$Ca^{2+} + 2e^- \longrightarrow Ca$	−2.87	$Cs^+ + e^- \longrightarrow Cs$	−2.92

(continued)

POTENTIALS IN ALPHABETICAL ORDER *(continued)*

Reduction half-reaction	$E°$, V
$Cu^+ + e^- \longrightarrow Cu$	+0.52
$Cu^{2+} + 2e^- \longrightarrow Cu$	+0.34
$Cu^{2+} + e^- \longrightarrow Cu^+$	+0.15
$F_2 + 2e^- \longrightarrow 2F^-$	+2.87
$Fe^{2+} + 2e^- \longrightarrow Fe$	−0.44
$Fe^{3+} + 3e^- \longrightarrow Fe$	−0.04
$Fe^{3+} + e^- \longrightarrow Fe^{2+}$	+0.77
$Ga^+ + e^- \longrightarrow Ga$	−0.53
$2H^+ + 2e^- \longrightarrow H_2$	0, by definition
$2HBrO + 2H^+ + 2e^- \longrightarrow Br_2 + 2H_2O$	+1.60
$2HClO + 2H^+ + 2e^- \longrightarrow Cl_2 + 2H_2O$	+1.63
$2H_2O + 2e^- \longrightarrow H_2 + 2OH^-$	−0.83
$H_2O_2 + 2H^+ + 2e^- \longrightarrow 2H_2O$	+1.78
$H_4XeO_6 + 2H^+ + 2e^- \longrightarrow XeO_3 + 3H_2O$	+3.0
$Hg_2^{2+} + 2e^- \longrightarrow 2Hg$	+0.79
$Hg^{2+} + 2e^- \longrightarrow Hg$	+1.62
$2Hg^{2+} + 2e^- \longrightarrow Hg_2^{2+}$	+0.92
$Hg_2Cl_2 + 2e^- \longrightarrow 2Hg + 2Cl^-$	+0.27
$I_2 + 2e^- \longrightarrow 2I^-$	+0.54
$I_3^- + 2e^- \longrightarrow 3I^-$	+0.53
$In^+ + e^- \longrightarrow In$	−0.14
$In^{2+} + e^- \longrightarrow In^+$	−0.40
$In^{3+} + e^- \longrightarrow In^{2+}$	−0.49
$In^{3+} + 2e^- \longrightarrow In^+$	−0.44
$In^{3+} + 3e^- \longrightarrow In$	−0.34
$K^+ + e^- \longrightarrow K$	−2.93
$La^{3+} + 3e^- \longrightarrow La$	−2.52
$Li^+ + e^- \longrightarrow Li$	−3.05
$Mg^{2+} + 2e^- \longrightarrow Mg$	−2.36
$Mn^{2+} + 2e^- \longrightarrow Mn$	−1.18
$Mn^{3+} + e^- \longrightarrow Mn^{2+}$	+1.51
$MnO_2 + 4H^+ + 2e^- \longrightarrow Mn^{2+} + 2H_2O$	+1.23
$MnO_4^- + e^- \longrightarrow MnO_4^{2-}$	+0.56
$MnO_4^- + 8H^+ + 5e^- \longrightarrow Mn^{2+} + 4H_2O$	+1.51
$MnO_4^{2-} + 2H_2O + 2e^- \longrightarrow MnO_2 + 4OH^-$	+0.60

Reduction half-reaction	$E°$, V
$NO_3^- + 2H^+ + e^- \longrightarrow NO_2 + H_2O$	+0.80
$NO_3^- + 4H^+ + 3e^- \longrightarrow NO + 2H_2O$	+0.96
$NO_3^- + H_2O + 2e^- \longrightarrow NO_2^- + 2OH^-$	+0.01
$Na^+ + e^- \longrightarrow Na$	−2.71
$Ni^{2+} + 2e^- \longrightarrow Ni$	−0.23
$Ni(OH)_3 + e^- \longrightarrow Ni(OH)_2 + OH^-$	+0.49
$O_2 + e^- \longrightarrow O_2^-$	−0.56
$O_2 + 4H^+ + 4e^- \longrightarrow 2H_2O$	+1.23
$O_2 + H_2O + 2e^- \longrightarrow HO_2^- + OH^-$	−0.08
$O_2 + 2H_2O + 4e^- \longrightarrow 4OH^-$	+0.40
$O_3 + 2H^+ + 2e^- \longrightarrow O_2 + H_2O$	+2.07
$O_3 + H_2O + 2e^- \longrightarrow O_2 + 2OH^-$	+1.24
$Pb^{2+} + 2e^- \longrightarrow Pb$	−0.13
$Pb^{4+} + 2e^- \longrightarrow Pb^{2+}$	+1.67
$PbSO_4 + 2e^- \longrightarrow Pb + SO_4^{2-}$	−0.36
$Pt^{2+} + 2e^- \longrightarrow Pt$	+1.20
$Pu^{4+} + e^- \longrightarrow Pu^{3+}$	+0.97
$Ra^{2+} + 2e^- \longrightarrow Ra$	−2.92
$Rb^+ + e^- \longrightarrow Rb$	−2.93
$S + 2e^- \longrightarrow S^{2-}$	−0.48
$SO_4^{2-} + 4H^+ + 2e^- \longrightarrow H_2SO_3 + H_2O$	+0.17
$S_2O_8^{2-} + 2e^- \longrightarrow 2SO_4^{2-}$	+2.05
$Se + 2e^- \longrightarrow Se^{2-}$	−0.67
$Sn^{2+} + 2e^- \longrightarrow Sn$	−0.14
$Sn^{4+} + 2e^- \longrightarrow Sn^{2+}$	+0.15
$Sr^{2+} + 2e^- \longrightarrow Sr$	−2.89
$Te + 2e^- \longrightarrow Te^{2-}$	−0.84
$Ti^{2+} + 2e^- \longrightarrow Ti$	−1.63
$Ti^{3+} + e^- \longrightarrow Ti^{2+}$	−0.37
$Ti^{4+} + e^- \longrightarrow Ti^{3+}$	0.00
$Tl^+ + e^- \longrightarrow Tl$	−0.34
$U^{3+} + 3e^- \longrightarrow U$	−1.79
$U^{4+} + e^- \longrightarrow U^{3+}$	−0.61
$V^{2+} + 2e^- \longrightarrow V$	−1.19
$V^{3+} + e^- \longrightarrow V^{2+}$	−0.26
$Zn^{2+} + 2e^- \longrightarrow Zn$	−0.76

2C GROUND-STATE ELECTRON CONFIGURATIONS*

Z	Symbol	Configuration
1	H	$1s^1$
2	He	$1s^2$
3	Li	[He]$2s^1$
4	Be	[He]$2s^2$
5	B	[He]$2s^22p^1$
6	C	[He]$2s^22p^2$
7	N	[He]$2s^22p^3$
8	O	[He]$2s^22p^4$
9	F	[He]$2s^22p^5$
10	Ne	[He]$2s^22p^6$
11	Na	[Ne]$3s^1$
12	Mg	[Ne]$3s^2$
13	Al	[Ne]$3s^23p^1$
14	Si	[Ne]$3s^23p^2$
15	P	[Ne]$3s^23p^3$
16	S	[Ne]$3s^23p^4$
17	Cl	[Ne]$3s^23p^5$
18	Ar	[Ne]$3s^23p^6$
19	K	[Ar]$4s^1$
20	Ca	[Ar]$4s^2$
21	Sc	[Ar]$3d^14s^2$
22	Ti	[Ar]$3d^24s^2$
23	V	[Ar]$3d^34s^2$
24	Cr	[Ar]$3d^54s^1$
25	Mn	[Ar]$3d^54s^2$
26	Fe	[Ar]$3d^64s^2$
27	Co	[Ar]$3d^74s^2$
28	Ni	[Ar]$3d^84s^2$
29	Cu	[Ar]$3d^{10}4s^1$
30	Zn	[Ar]$3d^{10}4s^2$
31	Ga	[Ar]$3d^{10}4s^24p^1$
32	Ge	[Ar]$3d^{10}4s^24p^2$
33	As	[Ar]$3d^{10}4s^24p^3$
34	Se	[Ar]$3d^{10}4s^24p^4$
35	Br	[Ar]$3d^{10}4s^24p^5$
36	Kr	[Ar]$3d^{10}4s^24p^6$
37	Rb	[Kr]$5s^1$
38	Sr	[Kr]$5s^2$
39	Y	[Kr]$4d^15s^2$
40	Zr	[Kr]$4d^25s^2$
41	Nb	[Kr]$4d^45s^1$
42	Mo	[Kr]$4d^55s^1$
43	Tc	[Kr]$4d^55s^2$
44	Ru	[Kr]$4d^75s^1$
45	Rh	[Kr]$4d^85s^1$
46	Pd	[Kr]$4d^{10}$
47	Ag	[Kr]$4d^{10}5s^1$
48	Cd	[Kr]$4d^{10}5s^2$
49	In	[Kr]$4d^{10}5s^25p^1$
50	Sn	[Kr]$4d^{10}5s^25p^2$
51	Sb	[Kr]$4d^{10}5s^25p^3$
52	Te	[Kr]$4d^{10}5s^25p^4$
53	I	[Kr]$4d^{10}5s^25p^5$
54	Xe	[Kr]$4d^{10}5s^25p^6$
55	Cs	[Xe]$6s^1$
56	Ba	[Xe]$6s^2$
57	La	[Xe]$5d^16s^2$
58	Ce	[Xe]$4f^15d^16s^2$
59	Pr	[Xe]$4f^36s^2$
60	Nd	[Xe]$4f^46s^2$
61	Pm	[Xe]$4f^56s^2$
62	Sm	[Xe]$4f^66s^2$
63	Eu	[Xe]$4f^76s^2$
64	Gd	[Xe]$4f^75d^16s^2$
65	Tb	[Xe]$4f^96s^2$
66	Dy	[Xe]$4f^{10}6s^2$
67	Ho	[Xe]$4f^{11}6s^2$
68	Er	[Xe]$4f^{12}6s^2$
69	Tm	[Xe]$4f^{13}6s^2$
70	Yb	[Xe]$4f^{14}6s^2$
71	Lu	[Xe]$4f^{14}5d^16s^2$
72	Hf	[Xe]$4f^{14}5d^26s^2$
73	Ta	[Xe]$4f^{14}5d^36s^2$
74	W	[Xe]$4f^{14}5d^46s^2$
75	Re	[Xe]$4f^{14}5d^56s^2$
76	Os	[Xe]$4f^{14}5d^66s^2$
77	Ir	[Xe]$4f^{14}5d^76s^2$
78	Pt	[Xe]$4f^{14}5d^96s^1$
79	Au	[Xe]$4f^{14}5d^{10}6s^1$
80	Hg	[Xe]$4f^{14}5d^{10}6s^2$
81	Tl	[Xe]$4f^{14}5d^{10}6s^26p^1$
82	Pb	[Xe]$4f^{14}5d^{10}6s^26p^2$
83	Bi	[Xe]$4f^{14}5d^{10}6s^26p^3$
84	Po	[Xe]$4f^{14}5d^{10}6s^26p^4$
85	At	[Xe]$4f^{14}5d^{10}6s^26p^5$
86	Rn	[Xe]$4f^{14}5d^{10}6s^26p^6$
87	Fr	[Rn]$7s^1$
88	Ra	[Rn]$7s^2$
89	Ac	[Rn]$6d^17s^2$
90	Th	[Rn]$6d^27s^2$
91	Pa	[Rn]$5f^26d^17s^2$
92	U	[Rn]$5f^36d^17s^2$
93	Np	[Rn]$5f^46d^17s^2$
94	Pu	[Rn]$5f^67s^2$
95	Am	[Rn]$5f^77s^2$
96	Cm	[Rn]$5f^76d^17s^2$
97	Bk	[Rn]$5f^97s^2$
98	Cf	[Rn]$5f^{10}7s^2$
99	Es	[Rn]$5f^{11}7s^2$
100	Fm	[Rn]$5f^{12}7s^2$
101	Md	[Rn]$5f^{13}7s^2$
102	No	[Rn]$5f^{14}7s^2$
103	Lr	[Rn]$5f^{14}6d^17s^2$
104	Rf	[Rn]$5f^{14}6d^27s^2$(?)
105	Db	[Rn]$5f^{14}6d^37s^2$(?)
106	Sg	[Rn]$5f^{14}6d^47s^2$(?)
107	Bh	[Rn]$5f^{14}6d^57s^2$(?)
108	Hs	[Rn]$5f^{14}6d^67s^2$(?)
109	Mt	[Rn]$5f^{14}6d^77s^2$(?)

*The electron configurations followed by a question mark are speculative.

2D THE ELEMENTS

Element	Symbol	Atomic number	Molar mass, $g \cdot mol^{-1}$	Normal state*	Density, $g \cdot cm^{-3}$
actinium (Greek *aktis,* ray)	Ac	89	227.03	s, m	10.07
aluminum (from alum, salts of the form $KAl(SO_4)_2 \cdot 12H_2O$)	Al	13	26.98	s, m	2.70
americium (the Americas)	Am	95	241.06	s, m	13.67
antimony (probably a corruption of an old Arabic word; Latin *stibium*)	Sb	51	121.75	s, md	6.69
argon (Greek *argos,* inactive)	Ar	18	39.95	g, nm	1.66†
arsenic (Greek *arsenikos,* male)	As	33	74.92	s, md	5.78
astatine (Greek *astatos,* unstable)	At	85	210	s, nm	—
barium (Greek *barys,* heavy)	Ba	56	137.34	s, m	3.59
berkelium (Berkeley, California)	Bk	97	249.08	s, m	14.79
beryllium (from the mineral beryl, $Be_3Al_2SiO_{18}$)	Be	4	9.01	s, m	1.85
bismuth (German *weisse Masse,* white mass)	Bi	83	208.98	s, m	8.90
boron [Arabic *buraq,* borax, $Na_2B_4O_7 \cdot 10H_2O$; *bor*(ax) + (carb)*on*]	B	5	10.81	s, nm	2.47
bromine (Greek *bromos,* stench)	Br	35	79.91	l, nm	3.12
cadmium (Greek *Cadmus,* founder of Thebes)	Cd	48	112.40	s, m	8.65
calcium (Latin *calx,* lime)	Ca	20	40.08	s, m	1.53
californium (California)	Cf	98	251.08	s, m	—
carbon (Latin *carbo,* coal or charcoal)	C	6	12.01	s, nm	2.27
cerium (the asteroid Ceres, discovered two days earlier)	Ce	58	140.12	s, m	6.71
cesium (Latin *caesius,* sky blue)	Cs	55	132.91	s, m	1.87
chlorine (Greek *chloros,* yellowish green)	Cl	17	35.45	g, nm	1.66

Melting point, °C	Boiling point, °C	Ionization energy, kJ·mol⁻¹	Electron affinity, kJ·mol⁻¹	Electronegativity	Atomic radius, pm	Ionic radius,[§] pm
1230	3200	499	—	1.1	188	118(3+)
660	2350	577	+43	1.6	143	53(3+)
990	2600	578	—	1.3	184	107(3+)
631	1750	834	+103	2.1	141	89(3+)
−189	−186	1520	−96	—	174	—
613[‡]	—	947	+78	2.2	121	222(3−)
300	350	1037	+270	2.0	—	227(1−)
710	1640	502	−29	0.89	224	136(2+)
986	—	601	—	1.3	—	87(4+)
1285	2470	900	−48	1.6	112	27(2+)
271	1650	703	+91	2.0	182	96(3+)
2030	3700	799	+27	2.0	88	12(3+)
−7	59	1140	+325	3.0	114	196(1−)
321	765	868	−26	1.7	152	103(2+)
840	1490	590	−29	1.3	197	100(2+)
—	—	608	—	1.3	—	117(2+)
3700[‡]	—	1090	+122	2.6	77	260(4−)
800	3000	527	<50	1.1	183	107(3+)
28	678	376	+46	0.79	272	170(1+)
−101	−34	1255	+349	3.2	99	181(1−)

(continued)

Element	Symbol	Atomic number	Molar mass, $g \cdot mol^{-1}$	Normal state*	Density, $g \cdot cm^{-3}$
chromium (Greek *chroma,* color)	Cr	24	52.00	s, m	7.19
cobalt (German *Kobold,* evil spirit; Greek *kobalos,* goblin)	Co	27	58.93	s, m	8.80
copper (Latin *cuprum,* from Cyprus)	Cu	29	63.54	s, m	8.93
curium (Marie Curie)	Cm	96	247.07	s, m	13.30
dysprosium (Greek *dysprositos,* hard to get at)	Dy	66	162.50	s, m	8.53
einsteinium (Albert Einstein)	Es	99	254.09	s, m	—
erbium (Ytterby, a town in Sweden)	Er	68	167.26	s, m	9.04
europium (Europe)	Eu	63	151.96	s, m	5.25
fermium (Enrico Fermi, an Italian physicist)	Fm	100	257.10	s, m	—
fluorine (Latin *fluere,* to flow)	F	9	19.00	g, nm	1.51†
francium (France)	Fr	87	223	s, m	—
gadolinium (Johann Gadolin, a Finnish chemist)	Gd	64	157.25	s, m	7.87
gallium (Latin *Gallia,* France; also a pun on the discoverer's forename, Le Coq)	Ga	31	69.72	s, m	5.91
germanium (Latin *Germania,* Germany)	Ge	32	72.59	s, md	5.32
gold (Anglo-Saxon *gold;* Latin *aurum,* gold)	Au	79	196.97	s, m	19.28
hafnium (Latin *Hafnia,* Copenhagen)	Hf	72	178.49	s, m	13.28
helium (Greek *helios,* the sun)	He	2	4.00	g, nm	0.12†
holmium (Latin *Holmia,* Stockholm)	Ho	67	164.93	s, m	8.80
hydrogen (Greek *hydro* + *genes,* water-forming)	H	1	1.0079	g, nm	0.089†
indium (from the bright indigo line in its spectrum)	In	49	114.82	s, m	7.29
iodine (Greek *ioeidēs,* violet)	I	53	126.90	s, nm	4.95

Melting point, °C	Boiling point, °C	Ionization energy, kJ·mol^{-1}	Electron affinity, kJ·mol^{-1}	Electronegativity	Atomic radius, pm	Ionic radius,§ pm
1860	2600	653	+64	1.7	129	84(2+)
1494	2900	760	+64	1.9	125	82(2+)
1083	2567	785	+118	1.9	128	72(2+)
1340	—	581	—	1.3	—	119(2+)
1410	2600	572	—	1.2	177	91(3+)
—	—	619	<50	1.3	—	116(2+)
1520	2600	589	<50	1.2	176	89(3+)
820	1450	547	<50	—	204	112(2+)
—	—	627	—	1.3	—	115(2+)
−220	−188	1680	+328	4.0	64	133(1−)
27	677	400	44	0.7	270	180(1+)
1310	3000	592	<50	1.2	180	97(3+)
30	2070	577	+29	1.6	153	62(3+)
937	2830	784	+116	2.0	122	90(2+)
1064	2807	890	+223	2.5	144	91(3+)
2230	5300	642	0	1.3	159	84(4+)
—	−269	2370	−48	—	128	—
1470	2300	581	<50	1.2	177	89(3+)
−259	−253	1310	+73	2.2	78	154(1−)
157	2050	556	+29	1.8	167	72(3+)
114	184	1008	+295	2.7	133	220(1−)

(continued)

Element	Symbol	Atomic number	Molar mass, $g \cdot mol^{-1}$	Normal state*	Density, $g \cdot cm^{-3}$
iridium (Greek and Latin *iris,* rainbow)	Ir	77	192.2	s, m	22.56
iron (Anglo-Saxon *iron;* Latin *ferrum*)	Fe	26	55.85	s, m	7.87
krypton (Greek *kryptos,* hidden)	Kr	36	83.80	g, nm	3.00†
lanthanum (Greek *lanthanein,* to lie hidden)	La	57	138.91	s, m	6.17
lawrencium (Ernest Lawrence, an American physicist)	Lr	103	262.1	s, m	—
lead (Anglo-Saxon *lead;* Latin *plumbum*)	Pb	82	207.19	s, m	11.34
lithium (Greek *lithos,* stone)	Li	3	6.94	s, m	0.53
lutetium (*Lutetia,* ancient name of Paris)	Lu	71	174.97	s, m	9.84
magnesium (Magnesia, a district in Thessaly, Greece)	Mg	12	24.31	s, m	1.74
manganese (Greek and Latin *magnes,* magnet)	Mn	25	54.94	s, m	7.47
mendelevium (Dmitri Mendeleev)	Md	101	258.10	s, m	—
mercury (the planet Mercury; Latin *hydrargyrum,* liquid silver)	Hg	80	200.59	l, m	13.55
molybdenum (Greek *molybdos,* lead)	Mo	42	95.94	s, m	10.22
neodymium (Greek *neos* + *didymos,* new twin)	Nd	60	144.24	s, m	7.00
neon (Greek *neos,* new)	Ne	10	20.18	g, nm	1.44†
neptunium (the planet Neptune)	Np	93	237.05	s, m	20.45
nickel (German *Nickel,* Old Nick, Satan)	Ni	28	58.71	s, m	8.91
niobium (Niobe, daughter of Tantalus; see tantalum)	Nb	41	92.91	s, m	8.57
nitrogen (Greek *nitron* + *genes,* soda-forming)	N	7	14.01	g, nm	1.04†
nobelium (Alfred Nobel, the founder of the Nobel prizes)	No	102	255	s, m	—
osmium (Greek *osme,* a smell)	Os	76	190.2	s, m	22.58

Melting point, °C	Boiling point, °C	Ionization energy, kJ·mol^{-1}	Electron affinity, kJ·mol^{-1}	Electronegativity	Atomic radius, pm	Ionic radius,[§] pm
2447	4550	880	+151	2.2	136	89(2+)
1540	2760	759	+16	1.8	128	82(2+)
−157	−153	1350	−96	—	—	169(1+)
920	3450	538	50	1.1	188	122(3+)
—	—	—	—	1.3	—	112(2+)
328	1760	716	+35	2.3	175	132(2+)
181	1347	519	+60	1.0	157	58(1+)
1700	3400	524	<50	1.3	172	85(3+)
650	1100	736	−39	1.3	160	72(2+)
1250	2120	717	<0	1.6	137	91(2+)
—	—	635	—	1.3	—	114(2+)
−39	357	1007	−18	2.0	155	112(2+)
2620	4830	685	+72	2.2	140	92(2+)
1024	3100	530	<50	1.1	182	104(3+)
−249	−246	2080	−116	—	—	—
640	—	597	—	1.4	131	110(3+)
1455	2150	737	+156	1.9	125	78(2+)
2425	5000	664	+86	1.6	147	74(4+)
−210	−196	1400	−7	3.0	74	171(3−)
—	—	642	—	1.3	—	113(2+)
3030	5000	840	+106	2.2	135	89(2+)

(continued)

Element	Symbol	Atomic number	Molar mass, $g \cdot mol^{-1}$	Normal state*	Density, $g \cdot cm^{-3}$
oxygen (Greek *oxys* + *genes,* acid-forming)	O	8	16.00	g, nm	1.14
palladium (the asteroid Pallas, discovered at about the same time)	Pd	46	106.4	s, m	12.00
phosphorus (Greek *phosphoros,* light-bearing)	P	15	30.97	s, nm	1.82
platinum (Spanish *plata,* silver)	Pt	78	195.09	s, m	21.45
plutonium (the planet Pluto)	Pu	94	239.05	s, m	19.81
polonium (Poland)	Po	84	210	s, md	9.40
potassium (from potash; Latin *kalium* and Arabic *qali,* alkali)	K	19	39.10	s, m	0.86
praseodymium (Greek *prasios* + *didymos,* green twin)	Pr	59	140.91	s, m	6.78
promethium (Prometheus, the Greek god)	Pm	61	146.92	s, m	7.22
protactinium (Greek *protos* + *aktis,* first ray)	Pa	91	231.04	s, m	15.37
radium (Latin *radius,* ray)	Ra	88	226.03	s, m	5.00
radon (from radium)	Rn	86	222	g, nm	4.40†
rhenium (Latin *Rhenus,* Rhine)	Re	75	186.2	s, m	21.02
rhodium (Greek *rhodon,* rose; its aqueous solutions are often rose-colored)	Rh	45	102.91	s, m	12.42
rubidium (Latin *rubidus,* deep red, "flushed")	Rb	37	85.47	s, m	1.53
ruthenium (Latin *Ruthenia,* Russia)	Ru	44	101.07	s, m	12.36
samarium (from samarskite, a mineral)	Sm	62	150.35	s, m	7.54
scandium (Latin *Scandia,* Scandinavia)	Sc	21	44.96	s, m	2.99
selenium (Greek *selēnē,* the moon)	Se	34	78.96	s, nm	4.81
silicon (Latin *silex,* flint)	Si	14	28.09	s, md	2.33
silver (Anglo-Saxon *seolfor;* Latin *argentum*)	Ag	47	107.87	s, m	10.50

Melting point, °C	Boiling point, °C	Ionization energy, kJ·mol^{-1}	Electron affinity, kJ·mol^{-1}	Electronegativity	Atomic radius, pm	Ionic radius,[§] pm
−218	−183	1310	+141	3.4	66	140(2−)
1554	3000	805	+54	2.2	137	86(2+)
44	280	1011	+72	2.2	110	212(3−)
1772	3720	870	+205	2.3	139	85(2+)
640	3200	585	—	1.3	151	108(3+)
254	960	1037	+174	2.0	167	65(4+)
64	774	418	+48	0.82	235	138(1+)
935	3000	523	<50	1.1	183	106(3+)
1168	3300	536	<50	—	181	106(3+)
1200	4000	568	—	1.5	161	113(3+)
700	1500	509	—	0.9	223	152(2+)
−71	−62	1036	−68	—	—	—
3180	5600	760	+14	1.9	137	72(4+)
1963	3700	720	+110	2.3	134	86(2+)
39	688	402	+47	0.82	250	149(1+)
2310	4100	711	+101	2.2	134	77(3+)
1060	1600	543	<50	1.2	180	111(2+)
1540	2800	631	+18	1.4	164	83(3+)
220	685	941	+195	2.6	104	198(2−)
1410	2620	786	+134	1.9	118	26(4+)
962	2212	731	126	1.9	144	113(1+)

(continued)

Element	Symbol	Atomic number	Molar mass, $g \cdot mol^{-1}$	Normal state*	Density, $g \cdot cm^{-3}$
sodium (English soda; Latin *natrium*)	Na	11	22.99	s, m	0.97
strontium (Strontian, Scotland)	Sr	38	87.62	s, m	2.58
sulfur (Sanskrit *sulvere*)	S	16	32.06	s, nm	2.09
tantalum (Tantalos, Greek mythological figure)	Ta	73	180.95	s, m	16.67
technetium (Greek *technētos,* artificial)	Tc	43	98.91	s, m	11.50
tellurium (Latin *tellus,* earth)	Te	52	127.60	s, md	6.25
terbium (Ytterby, a town in Sweden)	Tb	65	158.92	s, m	8.27
thallium (Greek *thallos,* a green shoot)	Tl	81	204.37	s, m	11.87
thorium (Thor, Norse god of thunder, weather, and crops)	Th	90	232.04	s, m	11.73
thulium (*Thule,* early name for Scandinavia)	Tm	69	168.93	s, m	9.33
tin (Anglo-Saxon *tin;* Latin *stannum*)	Sn	50	118.69	s, m	7.29
titanium (Titans, Greek mythological figures, sons of the Earth)	Ti	22	47.88	s, m	4.55
tungsten (Swedish *tung* + *sten,* heavy stone from wolframite)	W	74	183.85	s, m	19.30
uranium (the planet Uranus)	U	92	238.03	s, m	18.95
vanadium (Vanadis, Scandinavian mythological figure)	V	23	50.94	s, m	6.11
xenon (Greek *xenos,* stranger)	Xe	54	131.30	g, nm	3.56†
ytterbium (Ytterby, a town in Sweden)	Yb	70	173.04	s, m	6.97
yttrium (Ytterby, a town in Sweden)	Y	39	88.91	s, m	4.48
zinc (Anglo-Saxon *zinc*)	Zn	30	65.37	s, m	7.14
zirconium (Arabic *zargun,* gold color)	Zr	40	91.22	s, m	6.51

Melting point, °C	Boiling point, °C	Ionization energy, $kJ \cdot mol^{-1}$	Electron affinity, $kJ \cdot mol^{-1}$	Electronegativity	Atomic radius, pm	Ionic radius,[§] pm
98	883	494	+53	0.93	191	102(1+)
770	1380	548	−29	0.95	215	116(2+)
115	445	1000	+200	2.6	104	184(2−)
3000	5400	761	+14	1.5	147	72(3+)
2200	4600	702	+96	1.9	135	95(2+)
450	990	870	+190	2.1	137	221(2−)
1360	2500	565	<50	—	178	93(3+)
304	1460	590	+19	2.0	171	88(3+)
1700	4500	587	—	1.3	180	101(3+)
1550	2000	597	<50	1.2	175	104(3+)
232	2720	707	+116	2.0	158	93(2+)
1660	3300	658	+7.6	1.5	147	80(2+)
3387	5420	770	+79	2.4	141	68(4+)
1135	4000	584	—	1.4	138	103(3+)
1920	3400	650	+51	1.6	135	72(2+)
−112	−108	1170	−77	2.6	218	190(1+)
824	1500	603	<50	—	194	113(3+)
1510	3300	616	+30	1.2	182	106(3+)
420	907	906	+9	1.6	137	83(2+)
1850	4400	660	+41	1.3	160	109(2+)

*The normal state is the state of the element at normal temperature and pressure (20°C and 1 atm). s denotes solid; l, liquid, and g, gas. m denotes metal; nm, nonmetal; and md, metalloid.

[†]The density quoted is for the liquid.

[‡]The solid sublimes.

[§]Charge in parentheses.

2E THE TOP 50 CHEMICALS BY INDUSTRIAL PRODUCTION IN THE UNITED STATES IN 1995

Production data are compiled annually by the American Chemical Society and published in *Chemical and Engineering News*. This table is based on the information about production in 1995 that was published in the April 8, 1996 issue. Water, sodium chloride, and steel traditionally are not included and would outrank the rest if they were. Hydrogen is heavily used but almost always "on site" as soon as it has been produced.

In *cracking*, petroleum fractions of molar mass higher than that of gasoline are converted into smaller molecules with more double bonds; for example,

$$CH_3(CH_2)_9CH_3 \xrightarrow{500°C,\ catalyst} CH_3(CH_2)_6CH{=}CH_2 + CH_3CH_3$$

In *re-forming*, the number of carbon atoms in the feedstock is left unchanged, but a greater number

Rank	Name	Annual production, 10^9 kg	Comment on source
1	sulfuric acid	43.2	contact process
2	nitrogen	30.9	fractional distillation
3	oxygen	24.3	fractional distillation
4	ethylene	21.3	thermal cracking
5	lime	18.7	decomposition of limestone
6	ammonia	16.1	Haber process
7	phosphoric acid	11.9	from phosphate rocks
8	sodium hydroxide	11.9	brine electrolysis
9	propylene	11.7	thermal cracking
10	chlorine	11.4	electrolysis
11	sodium carbonate	10.1	Solvay process and mining
12	methyl *tert*-butyl ether	8.0	addition of methanol to 2-methylpropene
13	dichloroethene	7.8	chlorination of ethylene
14	nitric acid	7.8	Ostwald process
15	ammonium nitrate	7.3	ammonia + nitric acid
16	benzene	7.2	catalytic re-forming
17	urea	7.1	ammonia + carbon dioxide
18	vinyl chloride	6.8	dehydrochlorination of dichloroethene
19	ethylbenzene	6.2	Friedel-Crafts alkylation of benzene
20	styrene	5.2	dehydrogenation of ethylbenzene
21	methanol	5.1	hydrogenation of carbon monoxide (using synthesis gas)
22	carbon dioxide	4.9	steam re-forming of hydrocarbons
23	xylene	4.2	catalytic re-forming of naphtha
24	formaldehyde	3.7	oxidation or dehydrogenation of methanol
25	terephthalic acid	3.6	oxidation of xylene

of double bonds and aromatic rings is formed; for example,

$$\text{CH}_3\text{-cyclohexane} \xrightarrow{550^\circ\text{C, catalyst}} \text{CH}_3\text{-benzene} + 3\text{H}_2$$

Rank	Name	Annual production, 10^9 kg	Comment on source
26	ethylene oxide	3.5	addition of O_2 to ethylene
27	hydrochloric acid	3.3	by-product of hydrocarbon chlorination
28	toluene	3.1	catalytic re-forming of naphtha*
29	*p*-xylene	2.9	catalytic re-forming of naphtha*
30	cumene (isopropylbenzene)	2.6	Friedel-Crafts alkylation
31	ammonium sulfate	2.4	ammonia + sulfuric acid
32	ethylene glycol	2.4	hydration of ethylene oxide
33	acetic acid	2.1	Monsanto process (reaction of CO with methanol)
34	phenol	1.9	oxidation of cumene
35	propylene oxide	1.8	oxidation of propylene
36	butadiene	1.7	dehydrogenation of butane
37	carbon black	1.5	partial combustion of hydrocarbons
38	isobutylene	1.5	petroleum cracking
39	potash†	1.5	mined (KCl), electrolysis (KOH), reaction of KCl with $NaNO_3$ (KNO_3)
40	acrylonitrile	1.5	reaction of propylene with ammonia and oxygen
41	vinyl acetate	1.3	reaction of ethylene and acetic acid
42	titanium dioxide	1.3	purification of rutile
43	acetone	1.3	dehydrogenation of isopropyl alcohol
44	butyraldehyde	1.2	carbonylation of propene
45	aluminum sulfate	1.1	alumina + sulfuric acid
46	sodium silicate	1.0	sodium carbonate and sand
47	cyclohexane	0.97	hydrogenation of benzene
48	adipic acid	0.82	oxidation of cyclohexane
49	nitrobenzene	0.75	nitration of benzene
50	bisphenol A‡	0.73	catalytic alkylation

*Naphtha is a petroleum fraction consisting of a mixture of C_4 through C_{10} aliphatic and cycloaliphatic hydrocarbons together with a variety of aromatic compounds.
†Potash refers collectively to K_2CO_3, KOH, K_2SO_4, KCl, and KNO_3 and is expressed in terms of the equivalent mass of K_2O.
‡Bisphenol A is 4,4′-isopropylidenediphenol; it is used as an antioxidant.

APPENDIX 3 Nomenclature

3A POLYATOMIC IONS

Charge	Chemical formula	Name	Charge	Chemical formula	Name
2+	Hg_2^{2+}	mercury(I)		OH^-	hydroxide
	UO_2^{2+}	uranyl		SCN^-	thiocyanate
	VO^{2+}	vanadyl	2−	CO_3^{2-}	carbonate
1+	NH_4^+	ammonium		$C_2O_4^{2-}$	oxalate
	PH_4^+	phosphonium		CrO_4^{2-}	chromate
1−	$CH_3CO_2^-$	acetate, ethanoate		$Cr_2O_7^{2-}$	dichromate
	HCO_2^-	formate, methanoate		O_2^{2-}	peroxide
	CN^-	cyanide		SiO_3^{2-}	metasilicate
	ClO_4^-	perchlorate*		SO_4^{2-}	sulfate
	ClO_3^-	chlorate*		SO_3^{2-}	sulfite
	ClO_2^-	chlorite*		$S_2O_3^{2-}$	thiosulfate
	MnO_4^-	permanganate	3−	AsO_4^{3-}	arsenate
	NO_3^-	nitrate		BO_3^{3-}	borate
	NO_2^-	nitrite		PO_4^{3-}	phosphate
	N_3^-	azide			

*These names are representative of the halogen oxoanions.

When a hydrogen ion bonds to a 2− or 3− anion, add "hydrogen" before the name of the anion. For example, HSO_3^- is hydrogen sulfite. If two hydrogen ions bond to a 3− anion, add "dihydrogen" before the name of the anion. For example, $H_2PO_4^-$ is dihydrogen phosphate.

OXOACIDS AND OXOANIONS

The names of oxoanions and their parent acids can be determined by noting the oxidation number of the central atom and then referring to the following table. For example, the nitrogen in $N_2O_2^{2-}$ has an oxidation number of +1; because nitrogen belongs to Group 15, the ion is a hyponitrite ion.

Oxidation number for an atom in group number				Name of oxoanion	Name of parent oxoacid
14	15	16	17		
—	—	—	+7	per . . . ate	per . . . ic acid
+4	+5	+6	+5	. . . ate	. . . ic acid
—	+3	+4	+3	. . . ite	. . . ous acid
—	+1	+2	+1	hypo . . . ite	hypo . . . ous acid

3B COMMON NAMES OF CHEMICALS

Many chemicals have acquired common names, sometimes as a result of their use over hundreds of years and sometimes because they appear on the labels of consumer products, such as detergents, beverages, and antacids. The following substances are just a few that have found their way into the language of everyday life.

Common name	Formula	Chemical name
baking soda	$NaHCO_3$	sodium hydrogen carbonate (sodium bicarbonate)
bleach, laundry	$NaClO$	sodium hypochlorite
borax	$Na_2B_4O_7 \cdot 10H_2O$	sodium tetraborate decahydrate
brimstone	S_8	sulfur
calamine	$ZnCO_3$	zinc carbonate
chalk	$CaCO_3$	calcium carbonate
Epsom salts	$MgSO_4 \cdot 7H_2O$	magnesium sulfate heptahydrate
fool's gold (pyrite)	FeS_2	iron(II) disulfide
gypsum	$CaSO_4 \cdot 2H_2O$	calcium sulfate dihydrate
lime (quicklime)	CaO	calcium oxide
lime (slaked lime)	$Ca(OH)_2$	calcium hydroxide
limestone	$CaCO_3$	calcium carbonate
caustic soda	$NaOH$	sodium hydroxide
marble	$CaCO_3$	calcium carbonate
milk of magnesia	$Mg(OH)_2$	magnesium hydroxide
plaster of Paris	$CaSO_4 \cdot \frac{1}{2}H_2O$	calcium sulfate hemihydrate
potash*	K_2CO_3	potassium carbonate
quartz	SiO_2	silicon dioxide
table salt	$NaCl$	sodium chloride
vinegar	CH_3COOH	acetic acid
washing soda	$Na_2CO_3 \cdot 10H_2O$	sodium carbonate decahydrate

*Potash also refers collectively to K_2CO_3, KOH, K_2SO_4, KCl, and KNO_3.

3C NAMING *d*-METAL COMPLEXES

1. When naming a *d*-metal complex, name the ligands first, then the metal atom or ion.
2. Neutral ligands have the same name as the molecule, such as en (ethylenediamine), except for H_2O (aqua), NH_3 (ammine), CO (carbonyl), and NO (nitrosyl).
3. Anionic ligands end in -o; for anions that end in -ide (like chloride), -ate (like sulfate), and -ite (like nitrite), change the endings as follows:

   -ide ⟶ -o -ate ⟶ -ato -ite ⟶ -ito

   Examples: chloro, cyano, sulfato, carbonato, sulfito, and nitrito.
4. Greek prefixes are used to denote the number of each type of ligand in the complex ion:

2	3	4	5	6	. . .
bi-	tri-	tetra-	penta-	hexa-	. . .

   If the ligand already contains a Greek prefix (as in ethylenediamine) or if it is polydentate (able to attach at more than one binding site), then the prefixes

2	3	4	. . .
bis-	tris-	tetrakis-	. . .

   are used instead.
5. Ligands are named in alphabetical order, regardless of the Greek prefix that indicates the number of each one present. (Notice that Cl_2 in the coordination sphere of the second complex in rule 7 (below) represents two chloride ligands, named as a dichloro, and not a Cl_2 molecular ligand.)
6. The chemical symbols of anionic ligands (such as Cl^-) precede those of neutral ligands (such as H_2O and NH_3) in the chemical formula of the complex (but not necessarily in its name).
7. The oxidation number of the central metal ion is given as a Roman numeral after its chemical name. *Examples:*

   $[FeCl(H_2O)_5]^+$ pentaaquachloroiron(II) ion

   $[CrCl_2(NH_3)_4]^+$ tetraamminedichloro-chromium(III) ion

   $[Co(en)_3]^{3+}$ tris(ethylenediamine) cobalt(III) ion
8. If the complex overall has a negative charge (an anionic complex), the suffix -ate is added to the stem of the metal's name. If the symbol of the metal originates from a Latin name (as listed in Appendix 2D), then the Latin stem is used. For example, the symbol for iron is Fe, from the Latin *ferrum.* Therefore, any anionic complex of iron ends with -ferrate followed by the oxidation number of the metal in Roman numerals:

   $[Fe(CN)_6]^{4-}$ hexacyanoferrate(II) ion

   $[Ni(CN)_4]^{2-}$ tetracyanonickelate(II) ion
9. The name of a coordination compound (as distinct from a complex cation or anion) is built in the same way as that of a simple compound, with the (possibly complex) cation named before the (possibly complex) anion. *Examples:*

   $NH_4[PtCl_3(NH_3)]$ ammonium amminetri-chloroplatinate(II)

   $[Cr(OH)_2(NH_3)_4]Br$ tetraamminedihydroxo-chromium(III) bromide

SUMMARY

Identify the cation and anion, name each separately, and then combine them. For each complex, first note the name and oxidation number of the metal ion; then identify the ligands; and finally, string the names together in alphabetical order with the appropriate prefixes, ending with the name of the metal. For example, suppose we needed to name the coordination compound

$$[Co(NH_3)_3(H_2O)_3]_2(SO_4)_3$$

The charge on the complex cation must be +3 to ensure charge neutrality of the compound (there are three SO_4^{2-} ions for every two complex ions), so the complex cation is $[Co(NH_3)_3(H_2O)_3]^{3+}$. Because all the ligands are neutral, the cobalt must be present as cobalt(III). It follows that the name of the cation is triamminetriaquacobalt(III), so the compound is triamminetriaquacobalt(III) sulfate.

GLOSSARY

absolute zero ($T = 0$; that is, 0 on the *Kelvin scale*) The lowest possible temperature (−273.15°C).

absorb To accept one substance into and throughout the bulk of another substance. Compare with *adsorb*.

absorbance A A measure of the extent of absorption of radiation by a sample: $A = \log(I_0/I)$.

absorbed dose The energy deposited in a sample when it is exposed to radiation (particularly but not exclusively nuclear radiation). Absorbed dose is measured in *rad* or *gray*.

absorption spectrum The spectrum of a sample, determined by measuring the extent to which the sample absorbs electromagnetic radiation as the frequency is varied over a range.

abundance (of an isotope) The percentage (in terms of the numbers of atoms) of the isotope present in a sample of the element. See also *natural abundance*.

accurate measurements Measurements that have small systematic error and give a result close to the accepted value of the property.

accuracy Freedom from systematic error. Compare with *precision*.

achiral Not chiral: identical to its mirror image. See also *chiral molecule*.

acid See *Arrhenius acid; Brønsted acid; Lewis acid*. Used alone, "acid" normally means a Brønsted acid.

acid anhydride A compound that forms an oxoacid when it reacts with water. A *formal anhydride* is a compound that has the formula of an acid minus the elements of water but does not react with water to produce the acid. *Examples:* SO_3, the anhydride of sulfuric acid; CO, the formal anhydride of formic acid, HCOOH.

acid-base indicator See *indicator*.

acid buffer See *buffer*.

acid ionization constant K_a See *acidity constant*.

acidic hydrogen atom A hydrogen atom (more exactly, the proton of that hydrogen atom) that can be donated to a base.

acidic ion An ion that acts as a Brønsted acid. *Examples:* NH_4^+; $[Al(H_2O)_6]^{3+}$.

acidic oxide An oxide that reacts with water to give an acid; the oxides of nonmetallic elements generally are acidic oxides. *Examples:* CO_2; SO_3.

acidic solution A solution with pH < 7.

acidity The strength of the tendency to donate a proton.

acidity constant K_a The equilibrium constant for proton transfer to water; for an acid HA, $K_a = [H_3O^+][A^-]/[HA]$ at equilibrium.

actinide A member of the second row of the *f*-block (actinium to nobelium).

activated complex An unstable combination of reactant molecules that can either go on to form products or fall apart into the unchanged reactants.

activated complex theory A theory of reaction rates in which it is supposed that the reactants form an activated complex.

activation barrier The potential energy barrier between reactants and products; the height of the barrier is the activation energy of the reaction.

activation energy E_a (1) The minimum energy needed for reaction. (2) The height of the activation barrier.

active site The region of an enzyme molecule where the substrate reacts.

activity The number of nuclear disintegrations that occur per second.

activity series A list of metals arranged in order of decreasing ability to reduce the cations of other metals. A metal can reduce the cations formed by any of the metals below it in the list.

addition polymerization The polymerization, usually of alkenes, by an addition reaction propagated by radicals.

addition reaction A reaction in which atoms or groups bond to atoms joined by a multiple bond. *Example:* $CH_3CH{=}CH_2 + HBr \longrightarrow CH_3CH_2CH_2Br$.

adhesion Binding to a surface.

adhesive forces Forces that bind a substance to a surface.

adsorb To bind a substance to a surface; the surface *adsorbs* the substance. Distinguish from *absorb*.

aerosol A fine mist of solid particles or droplets of liquid suspended in a gas.

alcohol An organic molecule containing an —OH group attached to a carbon atom that is not part of a carbonyl group or an aromatic ring. Alcohols are classified as *primary, secondary,* and *tertiary* according to the number of carbon atoms attached to the C—OH carbon atom. *Examples:* CH_3CH_2OH (primary); $(CH_3)_2CHOH$ (secondary); $(CH_3)_3COH$ (tertiary).

aldehyde An organic compound containing the —CHO group. *Examples:* CH_3CHO, ethanal (acetaldehyde); C_6H_5CHO, benzaldehyde.

aliphatic hydrocarbon A hydrocarbon that does not have benzene rings in its structure.

alkali An aqueous solution of a strong base. *Example:* aqueous NaOH.

alkali metal A member of Group 1 of the periodic table (the lithium family).

alkaline cell A dry cell in which the electrolyte contains sodium or potassium hydroxide.

alkaline earth metal Calcium, strontium, and barium; more informally, a member of Group 2 of the periodic table (the beryllium family).

alkaline solution An aqueous solution with pH > 7.

alkane (1) A hydrocarbon with no carbon-carbon multiple bonds. (2) A member of a series of hydrocarbons derived from methane by the repetitive insertion of $—CH_2—$ groups; alkanes have molecular formula C_nH_{2n+2}. *Examples:* CH_4, CH_3CH_3, $CH_3(CH_2)_6CH_3$.

alkene (1) A hydrocarbon with one carbon-carbon double bond. (2) A member of a series of hydrocarbons derived from ethene by the repetitive insertion of $—CH_2—$ groups; alkenes have molecular formula C_nH_{2n}. *Examples:* $CH_2{=}CH_2$; $CH_3CH{=}CH_2$; $CH_3CH{=}CHCH_2CH_3$.

alkyl group R A group of atoms that can be regarded as derived from an alkane by loss of a hydrogen atom. *Examples:* $—CH_3$, methyl; $—CH_2CH_3$, ethyl.

alkyne (1) A hydrocarbon with at least one carbon-carbon triple bond. (2) A member of a series of hydrocarbons derived from ethyne by the repetitive insertion of $—CH_2—$ groups; alkynes have molecular formula C_nH_{2n-2}. *Examples:* $CH{\equiv}CH$; $CH_3C{\equiv}CCH_3$.

allotropes Alternative forms of an element that differ in the way the atoms are linked. *Examples:* O_2 and O_3; white and gray tin.

alloy A homogeneous mixture of two or more metals. A *substitutional alloy* is an alloy in which atoms of one metal are substituted for atoms of another metal. An *interstitial alloy* is an alloy in which atoms of one metal lie in the gaps in the lattice of another metal.

alpha (α) helix One type of secondary structure adopted by a polypeptide chain, in the form of a right-handed helix.

alpha (α) particle Positively charged, subatomic particle emitted from some radioactive nuclei; nucleus of helium atom ($^4_2He^{2+}$).

alternating copolymer See *copolymer.*

amide An organic compound formed by the reaction of an amine and a carboxylic acid and containing the group $—CONR_2$. *Example:* CH_3CONH_2, acetamide.

amine A compound derived from ammonia by replacing various numbers of H atoms with organic groups; the number of hydrogen atoms replaced determines the classification as *primary, secondary,* or *tertiary*. *Examples:* CH_3NH_2 (primary); $(CH_3)_2NH$ (secondary); $(CH_3)_3N$ (tertiary); $(CH_3)_4N^+$ (quaternary ammonium ion).

amino acid A carboxylic acid that also contains an amino group. The *essential amino acids* are amino acids that must be ingested as a part of the diet. *Example:* NH_2CH_2COOH, glycine (nonessential).

amorphous solid A solid in which the atoms, ions, or molecules lie in a random jumble.

amphiprotic A substance that can both donate and accept protons. *Examples:* H_2O; HCO_3^-.

amphoteric The ability to react with both acids and bases. *Examples:* Al; Al_2O_3. Amphiprotic species are often called amphoteric.

analysis The determination of the composition of a substance.

analyte The solution of unknown concentration in a titration. Normally, the analyte is in the flask, not the buret.

anhydride See *acid anhydride.*

anhydrous Lacking water. *Example:* $CuSO_4$, the anhydrous form of copper(II) sulfate. Compare with *hydrate.*

anion A negatively charged ion. *Examples:* F^-; SO_4^{2-}.

anisotropic Depending on orientation.

anode The electrode at which oxidation occurs.

antioxidant A substance that reacts with radicals and so prevents the oxidation of another substance.

aqueous solution A solution in which the solvent is water.

arene An aromatic hydrocarbon.

aromatic hydrocarbon An organic hydrocarbon that includes a benzene ring as part of its structure. *Examples:* C_6H_6 (benzene); $C_6H_5CH_3$ (toluene); $C_{10}H_8$ (naphthalene).

Arrhenius acid A compound that contains hydrogen and releases hydrogen ions (H^+) in water. *Examples:* HCl; CH_3COOH; but not CH_4.

Arrhenius base A compound that produces hydroxide ions (OH^-) in water. *Examples:* NaOH; NH_3; but not Na.

Arrhenius equation The equation $\ln k = \ln A - E_a/RT$ for the commonly observed temperature dependence of reaction rates. The *Arrhenius parameters* are the frequency factor A and the activation energy E_a. An *Arrhenius plot* is a graph of $\ln k$ against $1/T$; if the plot is a straight line, then the reaction is said to show *Arrhenius behavior.*

aryl group An aromatic group. *Example:* $—C_6H_5$, phenyl.

atactic polymer See *polymer.*

atmosphere (1) The layer of gases surrounding a planet (specifically, the air for the planet Earth). (2) A unit of pressure (1 atm = 1.01325×10^5 Pa).

atom (1) The smallest particle of an element that has the chemical properties of that element. (2) An electrically neutral species consisting of a nucleus and its surrounding electrons.

atomic hypothesis The proposal advanced by John Dalton that matter is composed of atoms.

atomic mass unit u (formerly amu) Exactly $\frac{1}{12}$ the mass of one atom of carbon-12.

atomic number Z The number of protons in the nucleus of an atom; this number determines the identity of the element and the number of electrons in the neutral atom.

atomic orbital A region of space in which there is a high probability of finding an electron in an atom. An *s-orbital* is a spherical region; a *p-orbital* has two lobes, on opposite sides of the nucleus; a *d-orbital* typically has four lobes, with the nucleus at the center; an *f-orbital* has a more complicated arrangement of lobes.

atomic radius Half the distance between the centers of neighboring atoms in a solid or a homonuclear molecule.

atomic structure The arrangement of electrons around the nucleus of an atom.

atomic weight See *molar mass.*

***Aufbau* principle** See *building-up principle.*

autoionization See *autoprotolysis.*

autoprotolysis A reaction in which a conjugate acid and a conjugate base are formed from the same substance.

autoprotolysis constant The equilibrium constant for an autoprotolysis reaction. *Example:* For water, K_w, with $K_w = [H_3O^+][OH^-]$.

average bond enthalpy ΔH(A—B) The average of A—B bond enthalpies for a number of different molecules containing the A—B bond. See also *bond enthalpy.*

Avogadro constant The number of objects per mole of objects ($N_A = 6.022 \times 10^{23}$ mol^{-1}). *Avogadro's number* is the number of objects *in* one mole of objects (that is, the dimensionless number 6.022×10^{23}).

Avogadro's principle The volume of a sample of gas at a given temperature and pressure is proportional to the amount of gas molecules in the sample: $V \propto n$.

axial bond A bond that is perpendicular to the molecular plane in a bipyramidal molecule.

axial lone pair A lone pair on the axis of a bipyramidal molecule.

azimuthal quantum number l A number that labels the subshells of an atom and determines the shapes of the orbitals in the subshell. *Examples:* $l = 0$ for the *s* subshell; $l = 1$ for the *p* subshell.

background radiation The average radiation to which the Earth's inhabitants are exposed daily.

balanced equation See *chemical equation.*

ball-and-stick model A depiction of a molecule in which atoms are represented by balls and bonds are represented by sticks.

Balmer series A family of spectral lines (some of which lie in the visible region) in the spectrum of atomic hydrogen.

band of stability A region of a plot of mass number against atomic number corresponding to the existence of stable nuclei.

barometer An instrument for measuring the atmospheric pressure.

base see *Arrhenius base; Brønsted base; Lewis base.* Used alone, "base" normally means a Brønsted base.

base buffer See *buffer.*

base ionization constant See *basicity constant.*

base pair The pairing of organic bases in the DNA molecule: adenine pairs with thymine and guanine pairs with cytosine.

base units The units of measurement in the International System (SI) in terms of which all other units are defined. *Examples:* kilogram for mass; meter for length; second for time; kelvin for temperature; ampere for electric current.

basic ion An ion that acts as a Brønsted base. *Example:* $CH_3CO_2^-$.

basic oxide An oxide that is a Brønsted base. The oxides of metallic elements are generally basic. *Examples:* Na_2O; MgO.

basic oxygen process The production of iron by forcing oxygen and powdered limestone through the molten metal.

basic solution A solution with pH > 7.

basicity constant K_b The equilibrium constant for proton transfer to water: for a base B, $K_b = [BH^+][OH^-]/[B]$ at equilibrium.

becquerel Bq The SI unit of radioactivity (1 disintegration per second).

Beer's law The absorbance of a sample is proportional to the molar concentration of the absorbing species and the length of the sample through which the radiation passes.

beta (β) particle A fast electron emitted from a nucleus in a radioactive decay.

beta (β)-pleated sheet One type of planar secondary structure adopted by a polypeptide, in the form of a pleated sheet.

bimolecular reaction An elementary reaction in which two molecules (or free atoms) come together and form a product. *Example:* $O + O_3 \longrightarrow 2\ O_2$.

binary Consisting of two components, as in *binary mixture* and *binary (molecular) compound. Examples:* acetone and water (a binary mixture); HCl, Al_2O_3, C_6H_6 (binary compounds; HCl and C_6H_6 are molecular).

biomass The organic material of the planet produced annually by photosynthesis.

biradical A species with two unpaired electrons. *Example:* ·O·.

block (*s*-block, *p*-block, *d*-block, *f*-block) The region of the periodic table containing elements for which, according to the building-up principle, the corresponding subshell is currently being filled.

block copolymer See *copolymer.*

body-centered cubic structure bcc A crystal structure with a unit cell in which a central atom lies at the center of a cube formed by eight others.

Bohr frequency condition The relation between the change in energy of an atom or molecule and the frequency of radiation emitted or absorbed: $\Delta E = h\nu$.

boiling point See *boiling temperature; normal boiling point.*

boiling-point elevation The increase in normal boiling point of a solvent caused by the presence of a solute (a colligative property).

boiling temperature (1) The temperature at which a liquid boils. (2) The temperature at which a liquid is in equilibrium with its vapor at the pressure of the surroundings; vaporization then occurs throughout the liquid, not only at the liquid's surface.

Boltzmann constant k The value of R/N_A, where R is the gas constant and N_A is the Avogadro constant; $k = 1.38 \times 10^{-23}\ J{\cdot}K^{-1}$.

Boltzmann formula (for the entropy) The formula $S = k \ln W$, where k is the Boltzmann constant and W is the number of atomic arrangements that correspond to the same energy.

bond A link between atoms. See also *ionic bond; covalent bond; double bond; triple bond.*

bond angle In an A—B—C molecule or part of a molecule, the angle between the B—A and B—C bonds.

bond enthalpy $\Delta H_B(X{-}Y)$ The enthalpy change accompanying the dissociation of a bond. *Example:* $H_2(g) \longrightarrow 2\,H(g)\ \Delta H_B(H{-}H) = +436\ kJ{\cdot}mol^{-1}$.

bond length The distance between the nuclei of two atoms joined by a bond.

bond parameters The characteristic features of a bond. *Examples:* bond length; bond enthalpy.

Born-Haber cycle A closed series of reactions used to express the enthalpy of formation of an ionic solid in terms of contributions that include the lattice enthalpy.

boundary surface The surface showing the region of an orbital within which there is about 90% probability of finding an electron.

Boyle's law At constant temperature, and for a given sample of gas, the volume is inversely proportional to the pressure: $P \propto 1/V$.

Bragg equation An equation relating the angle of diffraction of X-rays to the spacing of layers of atoms in a crystal ($\lambda = 2d \sin \theta$).

branched alkane An alkane with hydrocarbon side chains.

branching A step in a chain reaction in which more than one chain carrier is formed in a propagation step. *Example:* $\cdot O \cdot + H_2 \longrightarrow 2\,HO\cdot$.

breeder reactor A reactor used to generate nuclear fuel and makes use of neutrons that are not moderated.

Brønsted acid A proton donor (a source of hydrogen ions, H^+). *Examples:* HCl; CH_3COOH; HCO_3^-; NH_4^+.

Brønsted base A proton acceptor (a species to which hydrogen ions, H^+, can bond). *Examples:* OH^-; Cl^-; $CH_3CO_2^-$; HCO_3^-; NH_3.

Brønsted equilibrium The proton transfer equilibrium $acid_1 + base_2 \rightleftharpoons acid_2 + base_1$. *Example:* $CH_3COOH(aq) + H_2O(l) \rightleftharpoons H_3O^+(aq) + CH_3CO_2^-(aq)$.

Brønsted-Lowry theory A theory of acids and bases expressed in terms of proton transfer equilibria.

Brownian motion The ceaseless jittering motion of colloidal particles caused by the impact of solvent molecules.

buffer A solution that resists any change in pH when small amounts of acid or base are added. An *acid buffer* stabilizes solutions at $pH < 7$ and a *base buffer* stabilizes solutions at $pH > 7$. *Examples:* a solution containing CH_3COOH and $CH_3CO_2^-$ (acid buffer); a solution containing NH_3 and NH_4^+ (base buffer).

buffer capacity An indication of the amount of acid or base that can be added before a buffer loses its ability to resist the change in pH.

building-up principle The procedure for arriving at the ground-state electron configurations of atoms and molecules.

bulk properties Properties that depend on the collective behavior of large numbers of atoms. *Examples:* melting point; vapor pressure; internal energy.

calibration Interpretation of an observation by comparison with known information.

calorie cal A unit of energy. The unit is now defined in terms of the joule by 1 cal = 4.184 J exactly. The dietary *Calorie* is 1 kcal.

calorimeter An apparatus used to determine the heat released or absorbed by measuring the temperature change.

calorimetry The use of a calorimeter to measure the thermochemical properties of reactions.

capillary action The rise of liquids up narrow tubes.

carbide A binary compound of a metal or metalloid and carbon: carbides may be *saline* (saltlike), *covalent,* and *interstitial. Examples:* CaC_2 (saline); SiC (covalent); W_2C (interstitial).

carbohydrate A compound of general formula $C_m(H_2O)_n$, although small deviations from this general formula are often encountered. *Examples:* $C_6H_{12}O_6$, glucose; $C_{12}H_{22}O_{11}$, sucrose.

carbon cycle The sequence of physical processes and chemical reactions by means of which carbon atoms circulate through the environment.

carbonyl group A $\rangle$CO group in an inorganic or organic compound.

carboxylic acid An organic compound containing the carboxyl group, —COOH. *Examples:* CH_3COOH, acetic acid; C_6H_5COOH, benzoic acid.

catalyst A substance that increases the rate of a reaction without being consumed in the reaction. A catalyst is *homogeneous* if it is present in the same phase as the reactants and *heterogeneous* if it is in a different phase from the reactants. *Examples:* homogeneous, $Br_2(aq)$ for the decomposition of $H_2O_2(aq)$; heterogeneous, Pt in the Ostwald process.

catenate To form chains or rings of atoms. *Examples:* O_3; S_8.

cathode The electrode at which reduction occurs.

cathodic protection Protection of a metal object by connecting it to a more strongly reducing metal which acts as a sacrificial anode.

cation A positively charged ion. *Examples:* Na^+; NH_4^+; Al^{3+}.

cell diagram A statement of the arrangement of electrodes in an electrochemical cell. A cell is written with the anode on the left and the cathode on the right. *Example:* $Zn(s)|Zn^{2+}(aq)||Cu^{2+}(aq)|Cu(s)$.

cell potential E (1) The pushing and pulling power of a reaction in an electrochemical cell. (2) The potential difference between the electrodes of an electrochemical cell when it is producing no current. Cell potential, which is also called *electromotive force (emf),* is always positive. Compare with *standard potential.*

Celsius scale A temperature scale on which the freezing point of water is at 0 degrees and its normal boiling point is at 100 degrees. Units on this scale are degrees Celsius, °C.

cement A solid obtained by the action of heat on silicates and aluminosilicates, with gypsum and lime added.

ceramic (1) A solid obtained by the action of heat on clay. (2) A noncrystalline inorganic solid usually containing oxides, borides, and carbides.

cesium-chloride structure A crystal structure the same as that of solid cesium chloride.

chain carrier An intermediate in a chain reaction.

chain reaction A reaction in which an intermediate reacts to produce another intermediate. *Example:* $Br\cdot + H_2 \longrightarrow HBr + H\cdot$; $H\cdot + Br_2 \longrightarrow HBr + Br\cdot$.

chalcogens Oxygen, sulfur, selenium, and tellurium in Group 16 of the periodic table.

change of state The change of a substance from one of its physical states to another of its physical states. *Examples:* melting, solid $\longrightarrow$ liquid; gray tin $\longrightarrow$ white tin.

charge A measure of the strength with which a particle can interact electrostatically with another particle.

Charles's law The volume of a given sample of gas at constant pressure is directly proportional to its absolute temperature: $V \propto T$.

chelate A complex containing at least one polydentate ligand that forms a ring of atoms including the central metal atom. *Example:* $[Co(en)_3]^{3+}$.

chemical analysis The determination of the chemical composition of a sample.

chemical bond See *bond.*

chemical change The formation of one substance from another substance.

chemical element See *element.*

chemical equation A statement in terms of chemical formulas summarizing the qualitative information about the chemical changes taking place in a reaction and the quantitative information that atoms are neither created nor destroyed in a chemical reaction. In a balanced chemical equation, the same number of atoms of each element appear on both sides of the equation.

chemical equilibrium A dynamic equilibrium between reactants and products in a chemical reaction.

chemical formula A collection of chemical symbols and subscripts that shows the composition of a substance. See also *empirical formula; molecular formula; structural formula.*

chemical kinetics The study of the rates of reactions and the steps by which they occur.

chemical nomenclature The systematic naming of compounds.

chemical plating The deposition of a metal surface on an object by making use of a chemical reduction reaction.

chemical property The ability of a substance to participate in a chemical reaction.

chemical reaction A chemical change in which one substance responds to the presence of another, to a change of temperature, or to some other influence.

chemical symbol One- or two-letter abbreviation of an element's name.

chemiluminescence The emission of light by products formed in energetically excited states.

chemistry The branch of science concerned with the study of matter and the changes that matter can undergo.

chiral molecule A molecule that is distinct from its own mirror image. *Examples:* $CH_3CH(NH_2)COOH$; $CHBrClF$; $[Co(en)_3]^{3+}$.

chloralkali process The production of chlorine and sodium hydroxide by the electrolysis of aqueous sodium chloride.

cholesteric A liquid-crystal phase in which the molecules form a helical structure.

chromatogram The record of the signal from the detector, or the plate or paper record, obtained in a chromatographic analysis of a mixture.

chromatography A separation technique that relies on the ability of surfaces to adsorb different substances to different extents.

cis-trans isomerization The conversion of a cis isomer into a trans isomer, and vice versa. *Example: cis*-butene $\longrightarrow$ *trans*-butene.

classical mechanics The laws of motion proposed by Isaac Newton in which particles travel in definite paths in response to forces.

clathrate A compound in which a molecule of one component sits in a cage made up of molecules of another component, typically water. *Example:* SO_2 in water.

Claus process A process for obtaining sulfur from the H_2S in oil wells by the oxidation of H_2S with SO_2; the latter is formed by the oxidation of H_2S with oxygen.

close-packed structure A crystal structure in which atoms occupy the smallest total volume with the least empty space. *Examples: hexagonal close packing* and *cubic close packing* of identical spheres.

closed shell (or subshell) A shell (or subshell) containing the maximum number of electrons allowed by the exclusion principle. *Example:* the neonlike core $1s^22s^22p^6$.

closed system A system that cannot exchange matter with its surroundings.

cohesion The act or state of in which the particles of a substance stick to one another.

cohesive forces The forces that bind the molecules of a substance together to form a bulk material and that are responsible for condensation.

coinage metals The elements copper, silver, and gold.

colligative property A property that depends only on the relative number of solute and solvent particles present in a solution, and not on the chemical identity of the solute. *Examples:* elevation of boiling point; depression of freezing point; osmosis.

collision theory The theory of elementary gas-phase bimolecular reactions in which it is assumed that molecules react only if they collide with a characteristic minimum kinetic energy.

colloid (or *colloidal* suspension) A suspension of tiny particles in a gas, liquid, or solid. *Example:* milk.

combustion A reaction in which an element or compound burns in oxygen. *Example:* $CH_4(g) + 2\,O_2(g) \longrightarrow CO_2(g) + 2\,H_2O(g)$.

combustion analysis The determination of the composition of a sample by measuring the masses of the products of its combustion.

common-ion effect Reduction of the solubility of one salt by the presence of another salt with one ion in common. *Example:* the lower solubility of AgCl in NaCl(aq) than in pure water.

common name An informal name for a compound that may give little or no clue to the compound's composition. *Examples:* water; aspirin; acetic acid.

complementary color The color that white light becomes when one of the colors present in it is removed.

complete ionic equation A balanced chemical equation expressed in terms of the cations and anions present in solution. *Example:* $Ag^+(aq) + NO_3^-(aq) + Na^+(aq) + Cl^-(aq) \longrightarrow AgCl(s) + NO_3^-(aq) + Na^+(aq)$.

complex (1) The combination of a Lewis acid and a Lewis base linked by a coordinate covalent bond. (2) A species consisting of several ligands (the Lewis bases) that have an independent existence bonded to a single central metal atom or ion (the Lewis acid). *Examples:* (1) $H_3N{-}BF_3$; (2) $[Fe(H_2O)_6]^{3+}$; $[PtCl_4]^-$.

composite material A synthetic material composed of a polymer and one or more other substances that have been solidified together.

compound (1) A specific combination of elements that can be separated into its elements by using chemical techniques. (2) A substance consisting of atoms of two or more elements in a definite ratio.

compress Reduce the volume of a sample.

concentration The quantity of a substance in a given volume. See *molar concentration.*

condensation The formation of a liquid from a gas.

condensation polymer A polymer formed by a chain of *condensation reactions. Examples:* polyesters; polyamides (nylon).

condensation reaction A reaction in which two molecules combine to form a larger one and a small molecule is eliminated. *Example:* $CH_3COOH + C_2H_5OH \longrightarrow CH_3COOC_2H_5 + H_2O$.

conduction band An incompletely occupied band of energy levels in a solid.

configuration See *electron configuration.*

conformations Molecular shapes that can be interchanged by rotation about bonds, without bond breakage and re-formation.

conjugate acid The Brønsted acid formed when a Brønsted base has accepted a proton. *Example:* NH_4^+ is the conjugate acid of NH_3.

conjugate acid-base pair A Brønsted acid and its conjugate base. *Examples:* HCl and Cl^-; NH_4^+ and NH_3.

conjugate base The Brønsted base formed when a Brønsted acid has donated a proton. *Example:* NH_3 is the conjugate base of NH_4^+.

constructive interference Interference that results in an increased amplitude of a wave. See also *interference.*

contact process The production of sulfuric acid by the combustion of sulfur and the catalyzed oxidation of sulfur dioxide to sulfur trioxide.

convection The bulk motion of regions of a fluid, often as a result of differences in density brought about by differences in temperature.

conversion factor A factor that is used to convert a measurement from one unit to another.

coordinate Use of a lone pair to form a coordinate covalent bond. *Example:* $F_3B + {:}NH_3 \longrightarrow F_3B{-}NH_3$.

coordinate covalent bond A bond formed between a Lewis base and a Lewis acid by sharing an electron pair originally belonging to the Lewis base.

coordination compound A neutral complex or an ionic compound in which at least one of the ions is a complex. *Examples:* $Ni(CO)_4$; $K_3[Fe(CN)_6]$.

coordination isomers Isomers that differ by the exchange of one or more ligands between a cationic complex and an anionic complex.

coordination number (1) The number of nearest neighbors of an atom in a solid. (2) For ionic solids, the coordination number of an ion is the number of nearest neighbors of opposite charge. (3) For complexes, the number of ligands attached to the central metal ion.

coordination sphere The ligands directly attached to the central ion in a complex.

copolymer A polymer formed from a mixture of monomers. In *random copolymers*, the sequence of monomers has no particular order; in *alternating copolymers*, the monomers alternate; in *block copolymers*, regions of one monomer alternate with regions of another; in *graft copolymers*, chains of one monomer are attached to a backbone chain of a second monomer.

core The inner closed shells of an atom.

core electrons The electrons that belong to an atom's core.

corrosion The unwanted oxidation of a metal.

corrosive (1) A reagent that can cause corrosion. (2) Having a high reactivity, such as the reactivity of a strong oxidizing agent or a concentrated acid or base.

Coulomb's law The potential energy of a pair of electric charges is inversely proportional to the distance between them and proportional to the product of the charges.

couple See *redox couple.*

covalent bond A pair of electrons shared between two atoms.

covalent carbide See *carbide.*

covalent radius The contribution of an atom to the length of a covalent bond.

cracking The process of converting petroleum fractions into smaller molecules with more double bonds. *Example:* $CH_3(CH_2)_6CH_3 \longrightarrow CH_3(CH_2)_3CH_3 + CH_3CH{=}CH_2$.

critical mass The mass of fissionable material below which so many neutrons escape from a sample of nuclear fuel that the fission chain reaction is not sustained; a greater mass is *supercritical* and a smaller mass is *subcritical.*

critical pressure P_c The minimum pressure needed to liquefy a gas at its critical temperature.

critical temperature T_c The temperature above which a substance cannot exist as a liquid.

cryogenics The study of matter at very low temperatures.

cryoscopy The measurement of molar mass by using the depression of freezing point.

crystal face A flat plane forming an edge of a crystal.

crystal field The electrostatic influence of the ligands (modeled as point negative charges) on the central ion of a complex. *Crystal field theory* is a rationalization of the optical, magnetic, and thermodynamic properties of complexes in terms of the crystal field of their ligands.

crystalline solid A solid in which the atoms, ions, or molecules lie in an orderly array. *Examples:* NaCl; diamond; graphite.

crystallization The process in which a solute comes out of solution as crystals.

cubic close-packed structure ccp A close-packed structure with an ABCABC . . . pattern of layers.

curie Ci The unit of activity (for radioactivity).

current I The rate of supply of charge; current is measured in amperes (A), with $1\ A = 1\ C{\cdot}s^{-1}$.

cycle (1) In thermodynamics, a sequence of changes that begins and ends at the same state. (2) In spectroscopy, one complete reversal of the direction of the electromagnetic field and its return to the original direction.

cycloalkane A saturated aliphatic hydrocarbon in which the carbon atoms form a ring. *Example:* C_6H_{12}, cyclohexane.

***d*-orbital** See *atomic orbital.*

Dalton's law of partial pressures The total pressure of a mixture of gases is the sum of the partial pressures of its components.

Daniell cell A galvanic cell in which the cathode consists of copper in copper(II) sulfate solution and the anode consists of zinc in zinc sulfate solution.

data The information provided.

daughter nucleus A nucleus that is the product of a nuclear decay.

de Broglie relation The proposal that every particle has wavelike properties and that its wavelength, λ, is related to its mass by $\lambda = h/(\text{mass} \times \text{velocity})$.

debye D The unit used to report electric dipole moments: $1\ D = 3.336 \times 10^{-30}\ C{\cdot}m$.

decay constant k The rate constant for radioactive decay.

decomposition A reaction in which a substance is broken down into simpler substances; *thermal decomposition* is decomposition brought about by heat. *Example:* $CaCO_3(s) \xrightarrow{\Delta} CaO(s) + CO_2(g)$.

decomposition vapor pressure The pressure of the gaseous decomposition product of a solid at equilibrium.

dehydrating agent A reagent that removes water or the elements of water from a compound. *Example:* H_2SO_4.

dehydrogenation The removal of a hydrogen atom from each of two neighboring carbon atoms, resulting in the formation of a carbon-carbon multiple bond.

delocalized Spread over a region. In particular, *delocalized electrons* are electrons that spread over several atoms in a molecule.

delta Δ (in a chemical equation) A symbol that signifies that the reaction occurs at elevated temperatures.

delta X ΔX The difference between the final and initial values of a property, $\Delta X = X_f - X_i$. *Examples:* ΔT; ΔE.

delta hazard ⚠ A symbol that indicates that a skeletal equation is not balanced.

denaturation The loss of structure of a protein.

density d The mass of a sample of a substance divided by its volume: $d = m/V$.

deposition The reverse of sublimation, when a vapor condenses directly to a solid.

deprotonation Loss of a proton from a Brønsted acid. *Example:* $NH_4^+(aq) + H_2O(l) \longrightarrow H_3O^+(aq) + NH_3(aq)$.

derived unit A combination of base units. *Examples:* centimeters cubed (cm^3); joules ($kg{\cdot}m^2{\cdot}s^{-2}$).

destructive interference Interference that results in a reduced amplitude of a wave. See also *interference.*

deuteron The nucleus of a deuterium atom, $^2H^+$, consisting of a proton and a neutron.

diagonal relationship A similarity in properties between diagonal neighbors in the periodic table, especially for elements in Periods 2 and 3 at the left of the table. *Examples:* Li and Mg; Be and Al.

diamagnetic A substance that is pushed out of a magnetic field; a diamagnetic substance consists of atoms, ions, or molecules with no unpaired electrons. *Example:* most common substances.

diamine An organic compound that contains two $-NH_2$ groups.

diatomic molecule A molecule that consists of two atoms. *Examples:* H_2; CO.

diffraction Interference between waves caused by an object in their path. See also *x-ray diffraction.*

diffraction pattern The pattern of bright spots against a dark background resulting from diffraction.

diffusion The spreading of one substance through another substance.

dilute To reduce the concentration of a solute by adding more solvent.

dimer The union of two identical molecules. *Example:* Al_2Cl_6 formed from two $AlCl_3$ molecules.

diol An organic compound with two $-OH$ groups.

dipeptide An *oligopeptide* formed by the condensation of two amino acids.

dipole See *electric dipole; instantaneous dipole moment.*

dipole-dipole interaction The interaction between two electric dipoles: like partial charges repel and opposite partial charges attract.

diprotic See *polyprotic acid or base.*

disaccharide A carbohydrate molecule that is composed of two saccharide units. *Example:* $C_{12}H_{22}O_{11}$ (sucrose).

dispersion See *suspension.*

dispersion force See *London force.*

disproportionation A redox reaction in which a single element is simultaneously oxidized and reduced. *Example:* $2\,Cu^+(aq) \longrightarrow Cu(s) + Cu^{2+}(aq)$.

dissociation (1) The breaking of a bond. (2) The separation of ions that occurs when an ionic solid dissolves.

dissociation constant See *acidity constant.*

distillation The separation of the components of a mixture by making use of their different volatilities.

distribution (of molecular speeds) The fraction of gas molecules moving at each speed at any instant.

disulfide link An —S—S— link that contributes to the secondary and tertiary structures of polypeptides.

domain A region of a metal in which the electron spins of the atoms are aligned, so resulting in *ferromagnetism.*

doping The addition of a known, small amount of a second substance to an otherwise pure solid substance.

dose equivalent The actual dose of radiation experienced by a sample modified to take into account the *relative biological effectiveness* of the radiation. The dose equivalent is measured in *rem.* See *roentgen equivalent man; sievert.*

double bond (1) Two electron pairs shared by neighboring atoms. (2) One σ bond and one π bond between neighboring atoms.

Dow process The electrolytic production of magnesium from molten magnesium chloride.

Downs process The production of sodium and chlorine by the electrolysis of molten sodium chloride.

dry cell A cell in which the electrolyte is a moist paste.

drying agent A substance that absorbs water and thus maintains a dry atmosphere. *Example:* phosphorus(V) oxide.

ductile Able to be drawn out into a wire (as for a metal).

duplet The $1s^2$ electron pair of the heliumlike electron configuration.

dynamic equilibrium The condition in which a forward process and its reverse are occurring simultaneously at equal rates. *Examples:* vaporizing and condensing; chemical reactions at equilibrium.

***e*-orbital** One of the orbitals d_{z^2} or $d_{x^2-y^2}$ in an octahedral or tetrahedral complex.

effective nuclear charge The net nuclear charge after taking into account the shielding caused by other electrons in the atom.

effervesce To bubble out of solution as a gas.

effusion The escape of a substance (particularly a gas) through a small hole.

elasticity The ability to return to the original shape after distortion.

elastomer An elastic polymer. *Example:* rubber (polyisoprene).

electric dipole A positive charge next to an equal but opposite negative charge.

electric dipole moment μ The magnitude of the electric dipole (in debye).

electrical conduction The conduction of electric charge through matter. See also *electronic conduction; ionic conduction.*

electrochemical cell A system consisting of two electrodes in contact with an electrolyte. A *galvanic cell* is an electrochemical cell used to produce electricity, and an *electrolytic cell* is an electrochemical cell in which an electric current is used to cause chemical change.

electrochemical series Redox couples arranged in order of oxidizing and reducing strengths; usually arranged with strong oxidizing agents at the top of the list and strong reducing agents at the bottom.

electrochemistry The branch of chemistry that deals with the use of chemical reactions to produce electricity, the relative strengths of oxidizing and reducing agents, and the use of electricity to produce chemical change.

electrode A metallic conductor that makes contact with an electrolyte in an electrochemical cell.

electrolysis (1) A process in which a chemical change is produced by passing an electric current through a liquid. (2) The process of driving a reaction in a nonspontaneous direction by passing an electric current through a solution.

electrolyte solution An ionically conducting (usually aqueous) solution. A *strong electrolyte* is a substance that is fully ionized in solution. A *weak electrolyte* is a molecular substance that is only partially ionized in solution. A *nonelectrolyte* does not ionize in solution. *Examples:* NaCl is a strong electrolyte; CH_3COOH is a weak electrolyte; $C_6H_{12}O_6$ is a nonelectrolyte.

electrolytic cell See *electrochemical cell.*

electromagnetic radiation A wave of oscillating electric and magnetic fields; includes light, x-rays, and γ-rays.

electromotive force emf See *cell potential.*

electron A negatively charged subatomic particle found outside the nucleus of an atom.

electron affinity E_{ea} The energy released when an electron is added to a gas-phase atom or ion of the elements.

electron capture The capture by a nucleus of one of its own atom's *s*-electrons.

electron configuration The occupancy of orbitals in an atom or molecule. *Example:* N, $1s^2 2s^2 2p^3$.

electron-deficient compound A compound with too few valence electrons for it to be assigned a Lewis structure. *Example:* B_2H_6.

electronegative element An element with a high electronegativity. *Examples:* O; F.

electronegativity χ (chi) The ability of an atom to attract electrons to itself when it is part of a compound.

electronic conduction Conduction by electrons.

electronic structure The details of the distribution of the electrons that surround the nuclei in atoms and molecules.

electroplating The deposition of a thin film of metal on an object by electrolysis.

electropositive element An element with a low electronegativity and likely to give up electrons to another element on compound formation. *Examples:* Cs; Mg.

element (1) A substance that cannot be separated into simpler components by using chemical techniques. (2) A substance consisting of atoms of the same atomic number. *Examples:* hydrogen; gold; uranium.

elementary reaction An individual reaction step in a mechanism. *Example:* $H\cdot + Cl_2 \longrightarrow HCl + Cl\cdot$.

elimination reaction A reaction in which two groups or atoms on neighboring carbon atoms are removed from a molecule, thereby leaving a multiple bond between the carbon atoms. *Example:* $CH_3CHBrCH_3 + OH^- \longrightarrow CH_3CH{=}CH_2 + H_2O + Br^-$.

empirical formula A chemical formula that shows the relative numbers of atoms of each element in a compound. *Examples:* NaCl; P_2O_5; CH for benzene.

emulsion A suspension of droplets of one liquid dispersed throughout another liquid.

enantiomers A pair of optical isomers that are mirror images of, but not superimposable on, each other.

end point The stage in a titration at which enough titrant has been added to bring the indicator to a color halfway between its initial and final colors.

endothermic reaction A reaction that absorbs heat ($\Delta H > 0$). *Example:* $N_2O_4(g) \longrightarrow 2\,NO_2(g)$.

energy *E* The capacity of a system to do work or supply heat. *Kinetic energy* is the energy of motion, and *potential energy* is the energy arising from position.

energy level A permitted value of the energy in a quantized system such as an atom or molecule.

enrich In nuclear chemistry, to increase the abundance of a specific isotope.

enthalpy *H* A state property that is equal to the quantity of heat transferred at constant pressure; $H = U + PV$.

enthalpy density (of a fuel) The enthalpy of combustion per liter (without the negative sign).

enthalpy of freezing The negative of the *enthalpy of fusion.*

enthalpy of fusion ΔH_{fus} The enthalpy change per mole accompanying fusion (melting).

enthalpy of hydration ΔH_{hyd} The enthalpy change accompanying the hydration of gas-phase ions.

enthalpy of ionization The change in enthalpy for the process $E(g) \longrightarrow E^+(g) + e^-(g)$.

enthalpy of melting ΔH_{melt} See *enthalpy of fusion.*

enthalpy of solution ΔH_{sol} The change in enthalpy that occurs when a substance dissolves.

enthalpy of sublimation ΔH_{sub} The enthalpy change per mole accompanying *sublimation* (the direct conversion of a solid to a vapor).

enthalpy of vaporization ΔH_{vap} The enthalpy change per mole accompanying vaporization (the conversion of a substance from the liquid state to the vapor state).

entropy *S* (1) A measure of the disorder of a system. (2) A change in entropy is equal to the heat added to a system divided by the temperature at which the transfer occurs.

enzyme A biological catalyst.

equation of state A mathematical expression relating the pressure, volume, temperature, and amount of substance present in a sample. *Example:* ideal gas law, $PV = nRT$.

equatorial bond A bond perpendicular to the axis of a molecule (particularly trigonal bipyramidal and octahedral molecules).

equatorial lone pair A lone pair in the plane perpendicular to the molecular axis.

equilibrium See *chemical equilibrium; dynamic equilibrium.*

equilibrium constant K_c A characteristic of the equilibrium composition of the reaction mixture, with a form given by the law of mass action. *Example:* $N_2(g) + 3\,H_2(g) \rightleftharpoons 2\,NH_3(g)$, $K_c = [NH_3]^2/[N_2][H_2]^3$.

equilibrium table A table used to calculate the composition of a reaction mixture at equilibrium, given the initial composition. The columns are headed by the species and the rows are, successively, the initial composition, the change to reach equilibrium, and the equilibrium composition.

equivalence point See *stoichiometric point.*

essential amino acid An amino acid that is an essential component of the diet because it cannot be synthesized endogenously.

essential oil An oil that can be distilled from flowers and leaves (and which conveys the "essence" of a plant).

ester The product (other than water) of the reaction between a carboxylic acid and an alcohol and having the formula RCOOR′. *Example:* $CH_3COOC_2H_5$, ethyl acetate.

esterification The formation of an ester.

ether An organic compound of the form R—O—R. *Example:* $C_2H_5OC_2H_5$, diethyl ether.

evaporate Vaporize completely.

exclusion principle No more than two electrons can occupy any given orbital; and when two electrons do occupy one orbital, their spins must be paired.

exothermic reaction A reaction that releases heat ($\Delta H < 0$). *Example:* $N_2(g) + 3\,H_2(g) \longrightarrow 2\,NH_3(g)$.

expanded octet A valence shell containing more than eight electrons. *Examples:* the valence shells of P and S in PCl_5 and SF_6.

experiment A test carried out under carefully controlled conditions.

exponential decay A variation with time of the form e^{-kt}.

extensive property A physical property of a substance that depends on the size of the sample. *Examples:* mass; internal energy; enthalpy; entropy.

extrapolation The extension of a graph outside the region covered by the data.

***f*-orbital** See *atomic orbital.*

face See *crystal face.*

face-centered cubic structure fcc A crystal structure built from a cubic unit cell in which there is an atom at the center of each face and one at each corner.

Fahrenheit scale A temperature scale on which the freezing point of water is at 32 degrees and the normal boiling point is at 212 degrees. Units on this scale are degrees Fahrenheit, °F.

Faraday constant *F* The magnitude of the charge per mole of electrons; $F = 96.485\ \text{kC·mol}^{-1}$.

Faraday's law of electrolysis The moles of product formed by an electric current is chemically equivalent to the moles of electrons supplied.

fat An ester of glycerol and carboxylic acids with long hydrocarbon chains; fats act as long-term energy storage.

fatty acid A carboxylic acid with a long hydrocarbon chain. *Example:* $CH_3(CH_2)_{16}COOH$, stearic acid.

ferromagnetism The ability of some substances to be permanently magnetized. *Examples:* iron; magnetite, Fe_3O_4.

field An influence spreading over a region of space.*Examples:* an electric field from a charge; a magnetic field from a magnet.

filtration The separation of a heterogeneous mixture of a solid and liquid by passing the mixture through a fine mesh.

first law of thermodynamics The internal energy of an isolated system is constant.

first-order reaction A reaction in which the rate is proportional to the first power of the concentration of a substance.

fissile Having the ability to undergo fission induced by slow neutrons. *Example:* ^{235}U is fissile.

fission The breakup of a nucleus into two smaller nuclei of similar mass; fission may be *spontaneous* or *induced* (particularly by the impact of neutrons). *Examples:* $^{244}_{95}\text{Am} \longrightarrow {}^{134}_{53}\text{I} + {}^{107}_{42}\text{Mo} + 3\,\text{n}$ (spontaneous); $^{235}_{92}\text{U} + \text{n} \longrightarrow {}^{142}_{56}\text{Ba} + {}^{91}_{36}\text{Kr} + 3\,\text{n}$ (induced).

fissionable Having the ability to undergo induced fission.

fixation of nitrogen Conversion of elemental nitrogen to its compounds, particularly ammonia.

flammability The ability of a substance to burn in air.

foam (1) A frothy collection of bubbles formed by a liquid. (2) A type of *colloid* formed by a gas of tiny bubbles in a liquid or solid.

force *F* An influence that changes the state of motion of an object. *Examples:* an electrostatic force from an electric charge; a mechanical force from an impact.

formal anhydride See *acid anhydride.*

formal charge (1) The electric charge of an atom assigned on the assumption that there is nonpolar covalent bonding. (2) FC = number of valence electrons in the free atom − (number of lone-pair electrons + $\frac{1}{2}$ × number of shared electrons).

formation constant K_f The equilibrium constant for complex formation. The *overall formation constant* is the product of *stepwise formation constants*. The inverse of the formation constant ($1/K_f$) is called the *stability constant.*

formula unit The group of ions that matches the empirical formula of an ionic compound. *Example:* NaCl, one Na^+ ion and one Cl^- ion.

formula weight See *molar mass.*

fossil fuels The partially decomposed remains of vegetable and marine life (mainly coal, oil, and natural gas).

fractional distillation Separation of the components of a liquid mixture by repeated distillation, making use of their differing volatilities.

Frasch process A process for mining sulfur that uses superheated water to melt the sulfur and compressed air to force it to the surface.

free energy G The energy of a system that is free to do work at constant temperature and pressure: $\Delta G = \Delta H - T\Delta S$. The direction of spontaneous change at constant pressure and temperature is the direction of decreasing free energy.

freezing-point constant The constant of proportionality between the freezing-point depression and the molality of a solute.

freezing-point depression The lowering of the freezing point of a solution caused by the presence of a solute (a colligative property).

freezing temperature The temperature at which a liquid freezes. The *normal freezing point* is the freezing temperature under a pressure of 1 atm.

frequency (of radiation) ν (nu) The number of cycles (repeats of the waveform) per second (unit: *hertz*, Hz).

froth flotation A process for separating the mineral from unwanted rock in an ore by blowing air through a mixture that contains oil, water, and detergents to generate a froth.

fuel cell A primary electrochemical cell in which the reactants are supplied continuously from outside while the cell is in use.

functional group A group of atoms that brings a characteristic set of chemical properties to an organic molecule. *Examples:* —OH ; —Br ; —COOH.

fusion (1) Melting. (2) The merging of nuclei to form the nucleus of a heavier element.

galvanic cell See *electrochemical cell.*

galvanize Coat a metal with an unbroken film of zinc.

gamma (γ) radiation Very high frequency, short wavelength electromagnetic radiation emitted by nuclei.

gas A fluid form of matter that fills the container it occupies and can easily be compressed into a much smaller volume. (The distinction between a gas and a vapor is as follows: a gas is a substance at a higher temperature than its critical temperature; a vapor is a gaseous form of matter at a temperature below its critical temperature.)

gas constant R The constant that appears in the ideal-gas law.

gas-liquid chromatography A version of chromatography in which a gas carries the sample over a stationary liquid phase.

Geiger counter A device that is used to detect and measure radioactivity by relying on ionization caused by incident radiation.

gel A soft, solid emulsion.

geometrical isomers Stereoisomers that differ in the spatial arrangement of the atoms.

Gibbs free energy See *free energy.*

glass electrode A thin-walled glass bulb containing an electrolyte solution and a metallic contact; used for measuring pH.

graft copolymer See copolymer.

Graham's law of effusion The rate of effusion of a gas is inversely proportional to the square root of its molar mass.

gray Gy The SI unit of *absorbed dose;* 1 Gy corresponds to an energy deposit of 1 $J\cdot kg^{-1}$.

greenhouse effect The blocking by some atmospheric gases (notably carbon dioxide) of the radiation of heat from the surface of the Earth back into space, leading to the possibility of a worldwide rise in temperature.

greenhouse gas A gas that contributes to the greenhouse effect.

ground state The state of lowest energy.

group A vertical column in the periodic table.

Haber process (Haber-Bosch process) The catalyzed synthesis of ammonia at high pressure and high temperature.

half-life $t_{1/2}$ (1) In chemical kinetics, the time needed for the concentration of a substance to fall to half its initial value. (2) In radioactivity, the time needed for half the initial number of radioactive nuclei to disintegrate.

half-reaction A hypothetical oxidation or reduction reaction showing either electron loss or electron gain. *Examples:* $Na(s) \longrightarrow Na^+(aq) + e^-$; $Cl_2(g) + 2\,e^- \longrightarrow 2\,Cl^-(aq)$.

halide ion An anion formed from a halogen atom. *Examples:* F^-; I^-.

Hall process The production of aluminum by the electrolysis of aluminum oxide dissolved in molten cryolite.

haloalkane An alkane with a halogen substituent. *Example:* CH_3Cl, chloromethane.

halogenation The incorporation of a halogen into a compound (particularly, into an organic compound).

halogens The elements in Group 17.

hard water Water that contains dissolved calcium and magnesium salts.

heat The energy that is transferred as the result of a temperature difference between a system and its surroundings.

heat capacity The ratio of heat supplied to the temperature rise produced.

heating The act of transferring energy as heat.

heating curve A graph of the variation of the temperature of a sample as it is heated.

Henderson-Hasselbalch equation An approximate equation for estimating the pH of a solution containing a conjugate acid and base. (See Section 15.9.)

Henry's constant The constant k_H that appears in Henry's law.

Henry's law The solubility of a gas in a liquid is proportional to its partial pressure above the liquid: solubility = $k_H \times$ partial pressure.

hertz Hz An SI unit of frequency: 1 Hz is one complete cycle per second.

Hess's law A reaction enthalpy is the sum of the enthalpies of any sequence of reactions (at the same temperature and pressure) into which the overall reaction can be divided.

heterogeneous catalyst See *catalyst*.

heterogeneous equilibria An equilibrium in which at least one substance is in a different phase from that of the rest. *Example:* $AgCl(s) \rightleftharpoons Ag^+(aq) + Cl^-(aq)$.

heterogeneous mixture A mixture in which the individual components, although mixed together, lie in distinct regions that can be distinguished on a microscopic scale. *Example:* a mixture of sand and sugar.

heteronuclear diatomic molecule A molecule consisting of two atoms of different elements. *Examples:* HCl; CO.

hexagonal close-packed structure hcp A close-packed structure with an ABABA . . . pattern of layers.

high-spin complex A d^n complex with the maximum number of unpaired electron spins.

high-temperature superconductor A material that becomes superconducting at temperatures well above the transition temperature for the first generation of superconductors, typically 100 K and above.

homogeneous catalyst See *catalyst*.

homogeneous equilibrium A chemical equilibrium in which all the substances taking part are in the same phase. *Example:* $H_2(g) + I_2(g) \rightleftharpoons 2\,HI(g)$.

homogeneous mixture A mixture in which the individual components are uniformly mixed, even on a molecular scale. *Examples:* air; solutions.

homonuclear diatomic molecule A molecule consisting of two atoms of the same element. *Examples:* H_2; N_2.

Hund's rule If more than one orbital in a subshell is available, add electrons with parallel spins to different orbitals of that subshell.

hybrid orbital A mixed orbital formed by blending together atomic orbitals on the same atom. *Example:* an sp^3 hybrid orbital.

hybridization The formation of hybrid orbitals.

hydrate A solid compound containing H_2O molecules. *Example:* $CuSO_4{\cdot}5H_2O$.

hydrate isomers Isomers that differ by an exchange of an H_2O molecule and a ligand in the coordination sphere.

hydrated An ion or molecule to which water molecules are attached.

hydration (1) (of ions) The attachment of water molecules to a central ion. (2) (of organic compounds) The addition of water across a multiple bond (H to one carbon atom, OH to the other). *Example:* $CH_2{=}CH_2 + H_2O \longrightarrow CH_3CH_2OH$.

hydride A binary compound of a metal or metalloid with hydrogen; the term is often extended to include all binary compounds of hydrogen. A *saline* or *saltlike hydride* is a compound of hydrogen and a strongly electropositive metal; a *molecular hydride* is a compound of hydrogen and a nonmetal; a *metallic hydride* is a compound of certain *d*-block metals and hydrogen.

hydrocarbon A binary compound of carbon and hydrogen. *Examples:* CH_4; C_6H_6.

hydrogen bond A link formed by a hydrogen atom lying between two strongly electronegative atoms (O, N, or F).

hydrogen economy The widespread use of hydrogen as a fuel.

hydrogen electrode An electrode consisting of platinum in contact with hydrogen gas and a solution containing hydronium ions.

hydrogenation The addition of hydrogen to multiple bonds. *Example:* $CH_3CH{=}CH_2 + H_2 \longrightarrow CH_3CH_2CH_3$.

hydrolysis reaction The reaction of water with a substance, resulting in the formation of a new element-oxygen bond. *Example:* $PCl_5(s) + 4\,H_2O(l) \longrightarrow H_3PO_4(aq) + 5\,HCl(aq)$.

hydrolyze To undergo hydrolysis. See also *hydrolysis reaction*.

hydrometallurgical extraction The extraction of metals by reduction of their ions in aqueous solution. *Example:* $Cu^{2+}(aq) + Fe(s) \longrightarrow Cu(s) + Fe^{2+}(aq)$.

hydronium ion The ion H_3O^+.

hydrophilic Water-attracting. *Example:* hydroxyl groups are hydrophilic.

hydrophobic Water-repelling. *Example:* hydrocarbon chains are hydrophobic.

hydrostatic pressure The pressure exerted by a column of water or liquid solution.

hydroxyl group An —OH group in an organic compound.

hygroscopic Water-absorbing.

hypothesis A suggestion put forward to account for a series of observations. *Example:* Dalton's atomic hypothesis.

***i* factor** A factor that takes into account the existence of ions in an electrolyte solution, particularly for the interpretation of colligative properties. It indicates the number of particles formed from one formula unit of the solute. *Example:* $i \approx 2$ for very dilute NaCl(aq).

ideal gas A gas that satisfies the ideal gas law and is described by the *kinetic model.*

ideal gas law $PV = nRT$ All gases obey the law more and more closely as the pressure is reduced to very low values.

ideal solution A solution that obeys Raoult's law at any concentration; all solutions behave ideally as the concentration approaches 0. *Example* (of an almost ideal system): benzene and toluene.

incomplete octet A valence shell of an atom that has fewer than eight electrons. *Example:* the valence shell of B in BF_3.

indicator A substance that changes color when it goes from its acid to its base form (an *acid-base indicator*) or from its oxidized to its reduced form (a *redox indicator*).

induced-fit mechanism A model of the action of an enzyme in which the enzyme molecule adjusts its shape to accommodate the incoming substrate molecule. A modification of the *lock-and-key mechanism* of enzyme action.

induced nuclear fission See *fission.*

inert-pair effect The observation that an element displays a valence lower than expected from its group number. An *inert pair* is a pair of valence shell *s* electrons that are tightly bound to the atom and that might not participate in bond formation.

infrared radiation Electromagnetic radiation with a lower frequency (longer wavelength) than that of red light but a higher frequency (shorter wavelength) than microwave radiation.

initial rate The rate at the start of the reaction when products are present in concentrations too low to affect the rate.

initiation The formation of chain carriers from a reactant at the start of a chain reaction. *Example:* $Br_2 \xrightarrow{\Delta \text{ or light}} 2\,Br\cdot$.

inner transition metal A member of the *f*-block of the periodic table (the *lanthanides* and *actinides*).

inorganic compound A compound that is not organic. See also *organic compound.*

insoluble substance A substance that is not soluble in a specified solvent. When the solvent is not specified, water is generally meant.

instantaneous dipole moment A dipole moment arising from a transient redistribution of charge, and which is responsible for the London force.

instantaneous rate The slope of the tangent of a graph of concentration against time.

insulator (electrical) A substance that does not conduct electricity. *Examples:* nonmetallic elements; molecular solids.

integrated rate law An expression for the concentration of a reactant or product in terms of the time, obtained from the rate law of the reaction. *Example:* $[A] = [A]_0e^{-kt}$.

intensive property A physical property of a substance that is independent of the size of the sample. *Examples:* density; molar volume; temperature.

interference Interaction between waves, leading to a greater amplitude (*constructive interference*) or to a smaller one (*destructive interference*).

interhalogen A binary compound of two halogens. *Example:* IF_3.

intermediate See *reaction intermediate.*

intermolecular Between molecules.

intermolecular forces The forces of attraction and repulsion between molecules. *Examples*: hydrogen bonding; dipole-dipole force; London force.

internal energy *U* The total energy of a system.

International System SI A collection of definitions of units and their employment.

internuclear axis The straight line between the nuclei of two bonded atoms.

interstitial alloy See *alloy.*

interstitial carbide See *carbide.*

interstitial compound A compound in which one type of atom occupies the gaps between other atoms. *Example:* an interstitial carbide.

intramolecular Within a molecule.

ion An electrically charged atom or group of atoms. *Examples:* Al^{3+}; SO_4^{2-}. See also *cation; anion.*

ion-dipole interaction The attraction between an ion and the opposite partial charge of the electric dipole of a polar molecule.

ion exchange The exchange of one type of ion in solution for another.

ion pair A cation and anion in close proximity.

ionic bond The attraction between the opposite charges of cations and anions.

ionic compound A compound that consists of ions. *Examples:* NaCl; KNO_3.

ionic conduction Electrical conduction in which the charge is carried by ions.

ionic equation A chemical equation explicitly showing the ions of the reactants and products. *Example:* $Na^+(aq) + Cl^-(aq) + Ag^+(aq) + NO_3^-(aq) \longrightarrow Na^+(aq) + NO_3^-(aq) + AgCl(s)$.

ionic model The description of bonding in terms of ions.

ionic radius The contribution of an ion to the distance between neighboring ions in a solid ionic compound. In practice, the radius of an ion is defined as the distance between the centers of neighboring ions, with the radius of the O^{2-} ion set equal to 140 pm.

ionic solid A solid built from cations and anions. *Examples:* NaCl; KNO_3.

ionization (1) (of atoms and molecules) Conversion to cations by the removal of electrons. *Example:* $K(g) \longrightarrow K^+(g) + e^-(g)$. (2) (of an acid) The donation of a proton from a neutral acid molecule to a base the formation of the conjugate base (an anion in this instance), of the acid. *Example:* $CH_3COOH(aq) + H_2O(l) \longrightarrow H_3O^+(aq) + CH_3CO_2^-(aq)$.

ionization constant See *acidity constant.*

ionization energy *I* The minimum energy required to remove an electron from the ground state of a gaseous atom, molecule, or ion. The second ionization energy is the ionization energy for removal of a second electron, and so on.

ionization isomers Isomers that differ by the exchange of a ligand with an anion or neutral molecule outside the coordination sphere.

ionizing radiation High-energy radiation (typically but not necessarily nuclear radiation) that can cause ionization.

isoelectronic species Species with the same number of atoms and the same number of valence electrons. *Examples:* F^- and Ne; SO_2 and O_3; CN^- and CO.

isolated system A system that can exchange neither matter nor energy with its surroundings.

isomer One of two or more compounds that contain the same number of the same atoms in different arrangements. In *structural isomers*, the atoms have different partners or lie in a different order; in *stereoisomers*, the atoms have the same partners but are in different arrangements in space. *Optical isomers* are types of stereoisomers. *Examples:* $CH_3{-}O{-}CH_3$ and $CH_3CH_2{-}OH$ (structural isomers); *cis*- and *trans*-2-butene (stereoisomers).

isomerization A reaction in which a compound is converted into one of its isomers. *Example: cis*-butene $\longrightarrow$ *trans*-butene.

isotactic polymer See *polymer.*

isotherm A line of constant temperature on a graph.

isothermal process A change that occurs at constant temperature.

isotope One of two or more atoms that have the same atomic number but different atomic masses. *Example:* ^1H, ^2H, and ^3H are all isotopes of hydrogen.

isotopic abundance See *abundance.*

isotopic dating The determination of the age of objects by measuring the activity of the radioactive isotopes they contain, particularly ^{14}C.

isotopic label See *tracer.*

isotropic Depending on orientation.

joule J The SI unit of energy ($1\ J = 1\ kg{\cdot}m^2{\cdot}s^{-2}$).

Joule-Thomson effect The cooling of a gas as it expands.

Kekulé structures Two Lewis structures of benzene, consisting of alternating single and double bonds.

Kelvin scale A fundamental scale of temperature on which the triple point of water lies at 273.16 K and the lowest attainable temperature is at 0. The unit on the Kelvin scale is the *kelvin*, K.

ketone An organic compound of the form R—CO—R′. *Example:* $CH_3{-}CO{-}CH_3$, acetone.

kinetic energy The energy of a particle due to its motion. *Example:* the kinetic energy of a particle of mass m and speed v is $\frac{1}{2}mv^2$.

kinetic model A model of the properties of an ideal gas in which pointlike molecules are in continuous random motion in straight lines until collisions occur between them.

kinetic molecular theory The theory that discusses the *kinetic model* of gases.

labile A species that survives only for short periods.

lachrymator A substance that stimulates the production of tears. *Example:* PAN.

lanthanide A member of the first row of the *f*-block (lanthanum to ytterbium).

lanthanide contraction The reduction of atomic radius of the elements following the lanthanides below the values that would be expected by extrapolation of the trend down a group (and arising from the poor shielding ability of *f* electrons).

lattice enthalpy The standard enthalpy change for the conversion of an ionic solid to a gas of ions.

law A summary of experience.

law of conservation of energy Energy can be neither created nor destroyed.

law of conservation of mass Matter (and specifically atoms) is neither created nor destroyed in a chemical reaction.

law of constant composition A compound has the same composition whatever its source.

law of mass action For an equilibrium of the form $aA + bB \rightleftharpoons cC + dD$, the ratio$[C]^c[D]^d/[A]^a[B]^b$ evaluated at equilibrium is equal to a constant K_c, which has a specific value for a given reaction and temperature.

law of partial pressures See *Dalton's law of partial pressures.*

law of radioactive decay The rate of decay is proportional to the number of radioactive nuclides in the sample.

Le Chatelier's principle When a stress is applied to a system in dynamic equilibrium, the equilibrium adjusts to minimize the effect of the stress. *Example:* a reaction at equilibrium tends to proceed in the endothermic reaction when the temperature is raised.

lead-acid cell A secondary cell in which the electrodes are lead and the electrolyte is dilute sulfuric acid.

leveling The observation that strong acids all have the same strength in water, and all behave as though they were solutions of H_3O^+ ions.

Lewis acid An electron pair acceptor. *Examples:* H^+; Fe^{3+}; BF_3.

Lewis base An electron pair donor. *Examples:* OH^-; H_2O; NH_3.

Lewis formula (for an ionic compound) A representation of the structure of an ionic compound showing the formula unit of ions in terms of their Lewis diagrams.

Lewis structure A diagram showing how electron pairs are shared between atoms in a molecule. *Examples:* $H{-}\ddot{\underset{..}{Cl}}{:}$; $\ddot{O}{=}C{=}\ddot{O}$.

Lewis symbol (for atoms and ions) The chemical symbol of an element with a dot for each valence electron.

ligand A group attached to the central metal ion in a complex; a *polydentate ligand* occupies more than one binding site.

ligand field splitting Δ The energy separation of the *e* and *t* orbitals in a complex.

ligand field theory The theory of bonding in *d*-metal complexes, a more complete version of *crystal field theory*.

light See *visible radiation*.

limiting law A law that is accurately obeyed when a property (such as the pressure of a gas) is made very small.

limiting reactant The reactant that governs the theoretical yield of product in a given reaction.

linkage isomers Isomers that differ in the identity of the atom that a ligand uses to attach to the metal ion.

lipid A naturally occurring organic compound that dissolves in hydrocarbons but not in water. *Examples:* fats; steroids; terpenes; the molecules that form cell membranes.

liquid A fluid form of matter that takes the shape of the part of the container it occupies.

liquid crystal A substance that flows like a liquid but has molecules that lie in a moderately orderly array. Liquid crystals may be *nematic, smectic,* or *cholesteric,* depending on the arrangement of the molecules.

lock-and-key mechanism A model of enzyme action in which the enzyme is thought of as a lock and its substrate as a matching key.

London force The force of attraction that arises from the interaction between instantaneous electric dipoles on neighboring polar or nonpolar molecules.

lone pair A pair of valence electrons that is not involved in bonding.

long period A period of the periodic table with more than eight members.

low-spin complex A d^n complex with the minimum number of unpaired electron spins.

lyotropic Liquid crystals that result from the action of a solvent on a solute.

magic numbers The numbers of protons or neutrons that correlate with enhanced nuclear stability. *Examples:* 2, 8, 20, 50, 82, and 126.

magnetic quantum number m_l The quantum number that identifies the individual orbitals of a subshell of an atom and determines their orientation in space.

main group Any one of the groups forming the *s*- and *p*-blocks of the periodic table (Groups 1, 2, and 13 through 18).

malleable Deformable by striking with a hammer (as a metal).

manometer An instrument used for measuring the pressure of a gas confined inside a container.

many-electron atom An atom with more than one electron.

mass m The quantity of matter in a sample.

mass concentration The mass of solute per liter of solution.

mass number A The total number of nucleons (protons plus neutrons) in the nucleus of an atom. *Example:* ${}^{14}_{6}C$, with mass number 14, has 14 nucleons (6 protons and 8 neutrons).

mass percentage composition The mass of a substance present in a sample, expressed as a percentage of the total mass of the sample.

mass spectrometry Technique for measuring the masses and abundances of atoms and molecules by passing a beam of ions through a magnetic field.

matter Anything that has mass and takes up space.

Maxwell distribution of molecular speeds The formula for calculating the percentage of molecules that move at any given speed in a gas at a specified temperature.

mean free path The average distance that a molecule travels between collisions.

mechanism See *reaction mechanism*.

melting temperature The temperature at which a substance melts. The *normal melting point* is the melting point under a pressure of 1 atm.

meniscus The curved surface that a liquid forms in a narrow tube.

mesophase A state of matter showing some of the properties of both a liquid and a solid (a liquid crystal).

metal (1) A substance that conducts electricity, has a metallic luster, is malleable and ductile, forms cations, and has basic oxides. (2) A metal consists of cations held together by a sea of electrons. *Examples:* iron; copper; uranium.

metallic conduction The conduction of electricity by the movement of electrons.

metallic conductor An electronic conductor with a resistance that increases as the temperature is raised.

metallic hydride See *hydride*.

metallic radius (of a metallic element) Half the distance between the centers of neighboring atoms in a solid sample.

metalloid An element that has the physical appearance and properties of a metal but behaves chemically like a nonmetal. *Examples:* arsenic; polonium.

micelle A compact, often nearly spherical, cluster of oriented detergent (surfactant) molecules.

microwaves Electromagnetic radiation with wavelengths close to 1 cm.

minerals Substances that are mined; more generally, inorganic substances.

mixture A type of matter that consists of more than one substance and may be separated into its components by making use of the different physical properties of the substances present.

model A simplified description of nature.

moderator A substance that slows neutrons. *Examples:* graphite; heavy water.

molality The number of moles of solute per kilogram of solvent.

molar The quantity per mole. *Examples: molar mass*, the mass per mole; *molar volume*, the volume per mole. (*Molar concentration* and some related quantities are exceptions.)

molar absorption coefficient The constant of proportionality between the absorbance of a sample and the product of its molar concentration and path length. (See Case Study 7.)

molar mass (1) The mass per mole of atoms of an element (formerly, atomic weight). (2) The mass per mole of molecules of a compound (formerly, molecular weight). (3) The mass per mole of formula units of an ionic compound (formerly, formula weight).

molar concentration The moles of solute per liter of solution.

molar solubility S The molar concentration of a saturated solution of a substance.

molar volume The volume of a sample divided by the number of moles of atoms or molecules it contains.

molarity Molar concentration.

mole mol The unit for the amount of substance: 1 mol is the number of atoms in exactly 12 g of carbon-12.

mole fraction x The number of moles (or atoms or ions) of a substance in a mixture expressed as a fraction of the total number of moles of ions and molecules in the mixture.

mole ratio The stoichiometric relation between two species in a chemical reaction written as a conversion factor. *Example:* (2 mol H_2)/(1 mol O_2) in the reaction $2\,H_2(g) + O_2(g) \longrightarrow 2\,H_2O(l)$.

molecular compound A compound that consists of molecules. *Examples:* water; sulfur hexafluoride; benzoic acid.

molecular formula A combination of chemical symbols and subscripts showing the actual numbers of atoms of each element present in a molecule. *Examples:* H_2O; SF_6; C_6H_5COOH.

molecular hydride See *hydride*.

molecular solid A solid consisting of a collection of individual molecules held together by intermolecular forces. *Examples:* glucose; aspirin; sulfur.

molecular weight See *molar mass*.

molecularity The number of reactant molecules (or free atoms) taking part in an elementary reaction. See *unimolecular reaction, bimolecular reaction,* and *termolecular reaction.*

molecule (1) The smallest particle of a compound that possesses the chemical properties of the compound. (2) A definite and distinct, electrically neutral group of bonded atoms. *Examples:* H_2; NH_3; CH_3COOH.

monatomic gas A gas composed of single atoms. *Examples:* helium; radon.

monatomic ion An ion formed from a single atom. *Examples:* Na^+; Cl^-.

Mond process The purification of nickel by the formation and decomposition of nickel carbonyl.

monomer A small molecule from which a polymer is formed. *Examples:* $CH_2{=}CH_2$ for polyethylene; $NH_2(CH_2)_6NH_2$ for nylon.

monoprotic acid A Brønsted acid with one acidic hydrogen atom. *Example:* CH_3COOH.

monosaccharide An individual unit from which carbohydrates are considered to be composed. *Example:* $C_6H_{12}O_6$, glucose.

multiple bond A double or triple bond between two atoms.

***n*-type semiconductor** See *semiconductor*.

native Occurring in an uncombined state as the element itself.

natural abundance (of an isotope) The abundance of an isotope in a sample of a naturally occurring material.

natural product An organic substance that occurs naturally in the environment.

naturally occurring Found in nature without needing to be synthesized.

nematic A liquid-crystal phase in which the axes of rod-shaped molecules are arranged parallel to one another but are staggered with respect to one another in other directions.

Nernst equation The equation expressing the cell potential in terms of the concentrations of the reagents taking part in the cell reaction; $E = E° - (RT/nF) \ln Q$.

net ionic equation The equation showing the net change in a chemical reaction, obtained by canceling the spectator ions in a complete ionic equation. *Example:* $Ag^+(aq) + Cl^-(aq) \longrightarrow AgCl(s)$.

network solid A solid consisting of atoms linked together covalently throughout its extent. *Examples:* diamond; silica.

neutralization reaction The reaction of an acid with a base to form salt and water or another molecular compound. *Example:* $HCl(aq) + NaOH(aq) \longrightarrow NaCl(aq) + H_2O(l)$.

neutron n An electrically neutral subatomic particle found in the nucleus of an atom; it has approximately the same mass as a proton.

neutron-induced transmutation The conversion of one nucleus into another by the impact of a neutron. *Example:* $^{58}_{26}Fe + 2\,n \longrightarrow {}^{60}_{27}Co + \beta$.

neutron-rich nuclei Nuclei with such a high proportion of neutrons that they lie above the *band of stability*.

nickel-cadmium cell nicad cell A rechargeable cell in which the reaction is the reduction of nickel(III) and the oxidation of cadmium.

NO_x An oxide, or mixture of oxides, of nitrogen, typically in atmospheric chemistry,

noble gas A member of Group 18 of the periodic table (the helium family).

nodal plane A plane on which an electron will not be found.

node A point or surface at which an electron occupying an orbital will not be found.

nomenclature See *chemical nomenclature*.

nominal concentration The concentration an electrolyte would have if it were not ionized in solution.

nonaqueous solution A solution in which the solvent is not water. *Example:* sulfur in carbon disulfide.

nonbonding orbital A valence-shell atomic orbital that has not been used to form a bond to another atom.

nonelectrolyte A substance that dissolves to give a solution that does not conduct electricity. *Example:* sucrose.

nonideal solution A solution that does not obey Raoult's law. See *ideal solution*.

nonmetal A substance that does not conduct electricity and is neither malleable nor ductile. *Examples:* all gases; phosphorus; sodium chloride.

nonpolar bond (1) A covalent bond between two atoms that have zero partial charges. (2) A covalent bond between two atoms with the same or nearly the same electronegativity.

nonpolar molecule A molecule with zero electric dipole moment.

normal boiling point T_b (1) The boiling temperature when the pressure is 1 atm. (2) The temperature at which the vapor pressure of a liquid is 1 atm.

normal form The form of a substance under typical everyday conditions (for instance, close to 1 atm, 25°C).

normal freezing point T_f The temperature at which a liquid freezes at 1 atm.

nuclear atom The structure of the atom proposed by Rutherford: a central small, very dense, positively charged nucleus surrounded by electrons.

nuclear binding energy E_{bind} The energy released when Z protons and $A - Z$ neutrons come together to form a nucleus. The greater the binding energy per nucleon, the lower the energy of the nucleus.

nuclear chemistry The study of the chemical consequences of nuclear reactions.

nuclear decay The partial breakup of a nucleus (including its fission). Nuclear decay is also referred to as *nuclear disintegration*. *Example:* $^{226}_{88}Ra \longrightarrow {}^{222}_{86}Rn + \alpha$.

nuclear equation A summary of the changes in a nuclear reaction written in a form resembling a chemical equation.

nuclear fission See *fission*.

nuclear reaction A change that a nucleus undergoes (such as a nuclear transmutation).

nuclear reactor A device for achieving controlled nuclear fission.

nuclear transmutation The conversion of one element into another. *Example:* $^{12}_{6}C + \alpha \longrightarrow {}^{16}_{8}O + \gamma$.

nucleic acid (1) The product of a condensation of nucleotides. (2) A molecule containing an organism's genetic information.

nucleon A proton or a neutron; thus, either of the two principal components of a nucleus.

nucleoside A combination of a base and a deoxyribose molecule.

nucleosynthesis The formation of elements.

nucleotide A nucleoside with a phosphate group attached to the carbohydrate ring; one of the units from which nucleic acids are made.

nucleus The small, positively charged particle at the center of an atom that is responsible for most of its mass.

nuclide A specific nucleus. *Examples:* $^{2}_{1}H$; $^{16}_{8}O$.

octahedral complex A complex in which six ligands are arranged at the corners of a regular octahedron, with the metal atom at the center. *Example:* $[Fe(CN)_6]^{4-}$.

octet An s^2p^6 valence-electron configuration.

octet rule When atoms form bonds, they proceed as far as possible toward completing their octets by sharing electron pairs.

oligopeptide A short chain of amino acids connected by amide (peptide) bonds.

open system A system that can exchange matter with its surroundings.

optical activity The ability of a substance to rotate the plane of polarized light passing through it.

optical isomers Isomers that are related like an object and its mirror image. *Optical isomerism* is the existence of optical isomers.

orbital See *atomic orbital*.

order of reaction The power to which the concentration of a single substance is raised in a rate law. *Example:* if rate $= k[SO_2][SO_3]^{-1/2}$, then the reaction is first order in SO_2 and of order $-\frac{1}{2}$ in SO_3.

ore The natural mineral source of a metal. *Example:* Fe_2O_3, hematite.

organic chemistry The branch of chemistry that deals with organic compounds.

organic compound A compound containing the element carbon and usually hydrogen. (The carbonates are normally excluded.)

osmometry The measurement of the molar mass of a solute from observations of osmotic pressure.

osmosis The tendency of a solvent to flow through a semipermeable membrane into a more concentrated solution (a colligative property).

osmotic pressure Π (pi) The pressure needed to stop the flow of solvent through a semipermeable membrane.

Ostwald process The production of nitric acid by the catalytic oxidation of ammonia.

overall order The sum of the powers to which individual concentrations are raised in the rate law of a reaction. *Example:* if the rate $= k[SO_2][SO_3]^{-1/2}$, then the overall order is $\frac{1}{2}$.

overall reaction The net outcome of a sequence of reactions.

overlap The merging of orbitals belonging to different atoms of a molecule.

overpotential The additional potential difference that must be applied beyond the cell potential to cause appreciable electrolysis.

oxidation (1) Combination with oxygen. (2) A reaction in which an atom, ion, or molecule loses an electron. (3) A reaction in which the oxidation number of an element is increased. *Examples:* (1, 2, 3) $Mg(s) + \frac{1}{2}O_2(g) \longrightarrow MgO(s)$; (2, 3) $Mg(s) \longrightarrow Mg^{2+}(s) + 2\,e^-$.

oxidation number The effective charge on an atom in a compound, calculated according to a set of rules (Toolbox 3.2). An increase in oxidation number corresponds to oxidation and a decrease, to reduction.

oxidation-reduction reaction See *redox reaction.*

oxidation state The actual condition of a species with a specified oxidation number.

oxidizing agent A species that removes electrons from a species being oxidized (and is itself reduced) in a redox reaction. *Examples:* O_2; O_3; MnO_4^-; Fe^{3+}.

oxoacid An acid that contains oxygen. *Examples:* H_2CO_3; HNO_3; HNO_2; $HClO$.

oxoanion An anion of an oxoacid. *Examples:* HCO_3^-; CO_3^{2-}.

***p*-orbital** See *atomic orbital.*

***p*-type semiconductor** See *semiconductor.*

paired electrons Two electrons with opposed spins (↑↓).

parallel spins Electrons with spin in the same direction (↑↑).

paramagnetic The tendency to be pulled into a magnetic field; a paramagnetic substance is composed of atoms or molecules with unpaired electrons. *Examples:* O_2; $[Fe(CN)_6]^{3-}$.

parent nucleus In a nuclear reaction, the nucleus that undergoes disintegration or transmutation.

partial charge A charge arising from small shifts in the distributions of electrons. A partial charge can be either *positive* ($\delta+$) or *negative* ($\delta-$).

partial pressure P_X The pressure a gas (X) in a mixture would exert if it alone occupied the container.

parts per million ppm (1) The ratio of the mass of a solute to the mass of the solvent, multiplied by 10^6. (2) The mass percentage composition multiplied by 10^4. (Parts per billion, ppb, the mass ratio multiplied by 10^9, may also be used.)

pascal Pa The SI unit of pressure ($1\ Pa = 1\ kg{\cdot}m^{-1}{\cdot}s^{-2}$).

passivation Protection from further reaction by a surface film. *Example:* aluminum in air.

Pauli exclusion principle See *exclusion principle.*

penetration The possibility that an *s* electron may be found inside the inner shells of an atom and hence close to the nucleus.

peptide A molecule formed by a condensation reaction between amino acids; often described in terms of the number of units, for example, *dipeptide, oligopeptide, polypeptide.*

peptide bond The —CONH— group.

percentage composition See *mass percentage composition; volume percentage composition.*

percentage ionization The fraction of molecules of a substance, expressed as a percentage, that have ionized.

percentage protonated The fraction of a base, expressed as a percentage, that is present as its conjugate acid in a solution.

percentage transmittance T A measure of the intensity of radiation absorbed by a sample. $T = I/I_0 \times 100\%$. (See Case Study 7.)

percentage yield The percentage of the theoretical yield of a product achieved in practice.

period A horizontal row in the periodic table; the number of the period is equal to the principal quantum number of the valence shell of the atoms.

periodic table A chart in which the elements are arranged in order of atomic number and divided into groups and periods in a manner that shows the relationships between the properties of the elements.

pH The negative logarithm of the hydronium ion molarity in a solution: $pH = -\log [H_3O^+]$. $pH < 7$ indicates an acidic solution; $pH = 7$, a neutral solution; and $pH > 7$, a basic solution.

pH curve The graph of the pH of a reaction mixture against volume of titrant added in an acid-base titration.

phase A particular physical state of matter. A substance may exist in solid, liquid, and gaseous phases and, in certain cases, in more than one solid phase. *Examples:* white and gray tin are two solid phases of tin; ice, liquid, and vapor are three phases of water.

phase boundary A line separating two areas in a phase diagram; the points on a phase boundary represent the conditions at which the two adjoining phases are in dynamic equilibrium.

phase diagram A summary in graphical form of the conditions of temperature and pressure at which the various solid, liquid, and gaseous phases of a substance exist.

phase transition The conversion of one phase of a substance to another phase. *Example:* vaporization.

phenol An organic compound in which a hydroxyl group is attached directly to an aromatic ring (Ar—OH). *Example:* C_6H_5OH, phenol.

phenolic resin A polymer resulting from the condensation reaction between phenol and formaldehyde.

photochemical reaction A reaction caused by light. *Example:* $H_2(g) + Cl_2(g) \xrightarrow{\text{light}} 2\,HCl(g)$.

photoelectric effect The emission of electrons from the surface of a metal when electromagnetic radiation strikes it.

photon A particlelike packet of electromagnetic radiation. The energy of a photon of frequency ν is $E = h\nu$.

physical property A characteristic that we observe or measure without changing the identity of the substance.

physical state The condition of being a solid, a liquid, or a gas at a particular temperature.

pi (π) bond A bond formed by the side-to-side overlap of two *p* orbitals.

piezoelectric A substance is piezoelectric if it becomes electrically charged when it is mechanically distorted. *Example:* $BaTiO_3$.

pK_a and pK_b The negative logarithms of the acid and base ionization constants: $pK = -\log K$. The larger the value of pK_a or pK_b, the weaker the acid or base, respectively.

Planck constant *h* A fundamental constant of nature with the value 6.6261×10^{-34} J·s.

plasma (1) An ionized gas. (2) In biology, the colorless component of blood in which the red and white blood cells are dispersed.

pOH The negative logarithm of the hydroxide ion molarity in a solution; $pOH = -\log[OH^-]$.

poison (a catalyst) Inactivate a catalyst.

polar covalent bond A covalent bond between atoms that have partial electric charges. *Examples:* H—Cl; O—S.

polar molecule A molecule with a nonzero electric dipole moment. *Examples:* HCl; NH_3.

polarizable Easily polarized species.

polarize To distort the electron cloud of an atom or ion.

polarized light Plane-polarized light is light in which the wave motion occurs in a single plane.

polarizing power The ability of an ion to polarize a neighboring atom or ion.

polyamide A polymer formed by the condensation polymerization of a dicarboxylic acid and a diamine. *Example:* nylon.

polyatomic ion An ion consisting of more than two atoms linked by covalent bonds. *Examples:* NH_4^+; NO_3^-; SiF_6^{2-}.

polyatomic molecule A molecule that consists of more than two atoms. *Examples:* O_3; $C_{12}H_{22}O_{11}$.

polycyclic compound An aromatic compound containing two or more benzene rings that share two neighboring carbon atoms. *Example:* naphthalene.

polydentate ligand A ligand that can attach at several binding sites.

polyester A polymer formed by the condensation reaction between a dicarboxylic acid and a diol.

polymer A chain of covalently linked monomers. *Examples:* polyethylene; nylon. An *atactic polymer* is a polymer in which substituents are attached to each side of the chain at random. An *isotactic polymer* is a polymer in which the substituents are all on the same side of the chain. A *syndiotactic polymer* is a polymer in which the substituents alternate on either side of the chain. See also *copolymer*.

polynucleotide A polymer built from nucleotide units. *Examples:* DNA; RNA.

polypeptide A polymer formed by the condensation of amino acids.

polyprotic acid or base A Brønsted acid or base that can donate or accept more than one proton. (A polyprotic acid is sometimes called a polybasic acid.) *Examples:* H_3PO_4, triprotic acid; N_2H_4, diprotic base.

polysaccharide A chain of many saccharide units, such as glucose, linked together. *Examples:* cellulose; amylose.

positron A fundamental particle with the same mass as an electron but with opposite charge.

positron emission A mode of radioactive decay in which a nucleus emits a positron.

potential difference An electric potential difference between two points is a measure of the work that must be done to move an electric charge from one point to the other. Potential difference is measured in volts, V, and is commonly called the voltage.

potential energy The energy arising from position. *Example:* the Coulomb potential energy of a charge is inversely proportional to its distance from another charge.

power The rate of supply of energy.

precipitation The process in which a solute comes out of solution rapidly as a finely divided powder, called a *precipitate.*

precipitation reaction A reaction in which a solid product is formed when two solutions are mixed. *Example:* $KBr(aq) + AgNO_3(aq) \longrightarrow KNO_3(aq) + AgBr(s)$.

precipitate The solid formed in a precipitation reaction.

precise measurements (1) Measurements with a large number of significant figures. (2) A series of measurements with small random error.

precision Freedom from *random error*. Compare with *accuracy*.

pre-exponential factor *A* The constant obtained from the intercept in an Arrhenius plot.

pressure *P* Force divided by the area to which it is applied.

primary alcohol An alcohol of the form RCH_2OH.

primary cell A galvanic cell that produces electricity from chemicals sealed within it at the time of manufacture. It cannot be recharged.

primary pollutant A pollutant directly introduced into the environment. *Example:* SO_2.

primary structure The sequence of amino acids in the polypeptide chain of a protein.

principal quantum number *n* The quantum number that specifies the energy of an electron in a hydrogen atom and labels the shells of the atom.

product A species formed in a chemical reaction.

promotion (of an electron) The conceptual excitation of an electron to an orbital of higher energy in the description of bond formation.

propagation A series of steps in a chain reaction in which one chain carrier reacts with a reactant molecule to produce another carrier. *Examples:* $Br\cdot + H_2 \longrightarrow HBr + H\cdot$; $H\cdot + Br_2 \longrightarrow HBr + Br\cdot$.

properties The characteristics of matter. *Examples:* vapor pressure; color; density; temperature.

proton p A positively charged subatomic particle found in the nucleus of an atom.

proton-rich nuclei Nuclei that have a low proportion of neutrons and lie below the *band of stability*.

proton transfer equilibrium The equilibrium involving the transfer of a hydrogen ion between an acid and a base.

protonation Proton transfer to a Brønsted base. *Example:* $2\,H_3O^+(aq) + S^{2-}(s) \longrightarrow H_2S(g) + 2\,H_2O(l)$.

pseudo–first-order rate law A rate law that is effectively first order because one substance has a virtually constant concentration.

pyrometallurgical process The extraction of metals by using reactions at high temperatures. *Example:* $Fe_2O_3(s) + 3\,CO(g) \xrightarrow{\Delta} 2\,Fe(l) + 3\,CO_2(g)$.

qualitative A non-numerical description of the properties of a substance, system, or process. *Example:* qualitative analysis, the identification of the substances present in a sample.

quanta The plural of *quantum*.

quantitative A numerical description of the properties of a substance, system, or process. *Example:* quantitative analysis, the determination of the amounts of substances present in a sample.

quantization The restriction of a property to certain values. *Examples:* the quantization of energy and angular momentum.

quantum A packet of energy.

quantum mechanics The description of matter that takes into account the wave-particle duality of matter and the fact that the energy of an object may be changed only in discrete steps.

quantum number A number that labels the state of an electron and specifies the value of a property. *Example:* principal quantum number *n*.

quaternary ammonium ion An ion of the form NR_4^+ where R denotes hydrogen or an alkyl group (the four groups may be different).

quaternary structure The manner in which neighboring polypeptide units stack together to form a protein molecule.

racemic mixture A mixture containing equal concentrations of two enantiomers.

rad A unit of *absorbed dose*; 1 rad corresponds to an energy deposit of $0.01\ J\cdot kg^{-1}$. See also *gray*.

radical An atom, molecule, or ion with at least one unpaired electron. *Examples:* $\cdot NO$; $\cdot O\cdot$; $\cdot CH_3$.

radical chain reaction A chain reaction propagated by radicals.

radical polymerization A polymerization procedure that utilizes a radical chain reaction.

radical scavenger Impurities that react with radicals and inhibit a chain reaction.

radioactive series A stepwise nuclear decay path in which α and β particles are successively ejected and which terminates at a stable nuclide (often of lead).

radioactivity The spontaneous emission of radiation by nuclei. Such nuclei are *radioactive*.

radius ratio The ratio of the radius of the smaller ion in an ionic solid to the radius of the larger ion. The radius ratio controls which crystal structure is adopted by a simple ionic solid.

random copolymer See *copolymer*.

random error An error that varies randomly from measurement to measurement, sometimes giving a high value and sometimes a low one.

Raoult's law The vapor pressure of a solution of a nonvolatile solute is directly proportional to the mole fraction of the solvent in the solution: $P = x_{solvent}P_{pure}$, where P_{pure} is the vapor pressure of the pure solvent.

rate The change in a property divided by the time interval.

rate constant *k* The constant of proportionality in a rate law.

rate-determining step The slowest step in a multistep reaction and therefore the step that governs the rate of the overall reaction. *Example:* the step $O + O_3 \longrightarrow 2\,O_2$ in the decomposition of ozone.

rate law An equation expressing the instantaneous reaction rate in terms of the concentrations, at that instant, of the substances taking part in the reaction. *Example:* rate $= k[NO_2]^2$.

rate of reaction See *reaction rate; instantaneous rate.*

reactant A species acting as a starting material in a chemical reaction; a reagent taking part in a specified reaction.

reaction enthalpy The change of enthalpy for the reaction exactly as the chemical equation is written. *Example:* $CH_4(g) + 2\,O_2(g) \longrightarrow CO_2(g) + 2\,H_2O(l)$, $\Delta H = -890$ kJ.

reaction intermediate A species that is produced and consumed during a reaction but does not occur in the overall chemical equation.

reaction mechanism The pathway that is proposed for an overall reaction and accounts for the experimental rate law.

reaction order See *order of reaction.*

reaction profile The variation in potential energy that occurs as two reactants meet, form an activated complex, and separate as products.

reaction quotient Q_c The ratio of the molar concentrations of the products to those of the reactants, each raised to a power equal to the stoichiometric coefficient (as in the definition of the equilibrium constant, but at an arbitrary stage of a reaction). *Example:* for $N_2(g) + 3\,H_2(g) \longrightarrow 2\,NH_3(g)$, $Q_c = [NH_3]^2/\,[N_2][H_2]^3$.

reaction rate The change in concentration of a substance divided by the time it takes for the change to occur.

reaction sequence A series of reactions in which products of one reaction take part as reactants in the next. *Example:* $2\,C(s) + O_2(g) \longrightarrow 2\,CO(g)$, followed by $2\,CO(g) + O_2(g) \longrightarrow 2\,CO_2(g)$.

reaction stoichiometry The quantitative relation between the amounts of reactants consumed and products formed in chemical reactions as expressed by the balanced chemical equation for the reaction.

reagent A substance or a solution that reacts with other substances.

recrystallization Purification by repeated dissolving and crystallization.

redox couple The oxidized and reduced forms of a substance taking part in a reduction or oxidation half-reaction. The notation is oxidized species/reduced species. *Example:* H^+/H_2.

redox indicator See *indicator.*

redox reaction A reaction in which oxidation and reduction occur. *Example:* $S(s) + 3\,F_2(g) \longrightarrow SF_6(g)$.

reducing agent The species that supplies electrons to a substance being reduced (and is itself oxidized) in a redox reaction. *Examples:* H_2; H_2S; SO_3^{2-}.

reduction (1) The removal of oxygen from, or the addition of hydrogen to, a compound. (2) A reaction in which an atom, ion, or molecule gains an electron. (3) A reaction in which the oxidation number of an element is decreased. *Example:* $Cl_2(g) + 2\,e^- \longrightarrow 2\,Cl^-(aq)$.

re-forming reaction A reaction in which a hydrocarbon is converted to carbon monoxide and hydrogen over a nickel catalyst.

refractory Able to withstand high temperatures.

relative biological effectiveness Q A factor used when assessing the damage caused by a given dose of radiation.

rem See *roentgen equivalent man.*

representative elements The elements in Periods 1, 2, and 3 of the periodic table.

residue An amino acid in a polypeptide chain.

resistance (electrical) A measure of the ability of matter to conduct electricity: the lower the resistance, the better it conducts.

resonance A blending of Lewis structures into a single composite, hybrid structure. *Example:* $:\ddot{O}-\ddot{S}=\ddot{O} \longleftrightarrow \ddot{O}=\ddot{S}-\ddot{O}:$.

resonance hybrid The composite structure that results from resonance.

reverse osmosis The passage of solvent out of a solution when a pressure greater than the osmotic pressure is applied on the solution side of a semipermeable membrane.

roast To heat a metal ore in air. *Example:* $2\,CuFeS_2(s) + 3\,O_2(g) \xrightarrow{\Delta} 2\,CuS(s) + 2\,FeO(s) + 2\,SO_2(g)$ in the extraction of copper.

rock-salt structure A crystal structure the same as that of a mineral form of sodium chloride.

roentgen equivalent man rem The unit for reporting *dose equivalent.*

Rydberg constant $\mathcal{R}$ The constant that occurs in the formula for the frequencies of the lines in the spectrum of atomic hydrogen; $\mathcal{R} = 3.29 \times 10^{15}$ Hz.

***s*-orbital** See *atomic orbital.*

sacrificial anode See *cathodic protection.*

saline carbide See *carbide.*

saline hydride See *hydride.*

salt (1) An ionic compound. (2) The product (other than water) of the reaction between an acid and a base. *Examples:* NaCl; K_2SO_4.

salt bridge A bridge-shaped tube containing a concentrated salt (potassium chloride or potassium nitrate) in a jelly that acts as an electrolyte and provides a conducting path between two compartments of an electrochemical cell.

sample A representative part of a whole.

saturated Unable to take up further material.

saturated hydrocarbon A hydrocarbon with no carbon-carbon multiple bonds. *Example:* CH_3CH_3.

saturated solution A solution in which the dissolved and undissolved solute are in dynamic equilibrium.

science The systematically collected and organized body of knowledge based on experiment, observation, and careful reasoning.

scientific method The procedures employed to develop a scientific understanding of nature. (See Box 1.1.)

scintillation counter A device for detecting and measuring radioactivity that makes use of the fact that certain substances give a flash of light when they are exposed to radiation.

sea of instability A region in a graph of mass number against atomic number corresponding to unstable nuclei that decay with the emission of radiation.

second law of thermodynamics A spontaneous change is accompanied by an increase in the total entropy of the system and its surroundings.

second-order reaction A reaction with a rate law that is proportional to the square of the molar concentration of a reactant.

secondary alcohol An alcohol of the form R_2CHOH.

secondary cell A galvanic cell that must be charged (or recharged) by using an externally supplied current before it can be used.

secondary pollutant A pollutant formed by the chemical reaction of another species. *Example:* SO_3 from the oxidation of SO_2.

secondary structure The manner in which a polypeptide chain is coiled. *Examples:* α helix; β-pleated sheet.

selective precipitation The precipitation of one compound in the presence of other, more soluble compounds.

self-sustaining fission Induced nuclear fission that, once it is initiated, can continue even if the supply of neutrons from outside is discontinued.

semiconductor An electronic conductor with a resistance that decreases as the temperature is raised. In an *n-type semiconductor*, the current is carried by electrons in a largely empty band; in a *p-type semiconductor*, the conduction is a result of electrons missing from otherwise filled bands.

semipermeable membrane A membrane that allows only certain types of molecules or ions to pass.

sequestration (1) The wrapping up of one ion by another. (2) The formation of a complex between a cation and a bulky molecule or ion. *Example:* Ca^{2+} and $O_3POPO_2OPO_3^{5-}$.

series A family of spectral lines arising from transitions that have one state in common. *Example:* the Balmer series in the spectrum of atomic hydrogen.

shell All the orbitals of a given principal quantum number. *Example:* the single 2*s* and three 2*p* orbitals of the shell with $n = 2$.

shielding The repulsion that is experienced by an electron in an atom; it arises from the other electrons present and opposes the attraction exerted by the nucleus.

shift reaction A reaction between carbon monoxide and water: $CO(g) + H_2O(g) \xrightarrow{400°C,\ Fe/Cu} CO_2(g) + H_2(g)$; the reaction is used in the manufacture of hydrogen.

SI *Système International* The International System of units; it is an extension and rationalization of the metric system.

side chain A hydrocarbon substituent on a hydrocarbon chain.

sievert Sv The SI unit of *dose equivalent:* 1 Sv = 100 rem.

sigma (σ) bond Two electrons in a cylindrically symmetric cloud between two atoms.

significant figures (in a measurement) The digits in the measurement, up to and including the first uncertain digit in scientific notation. *Example:* 0.0260 mL (i.e., 2.60×10^{-2} mL), a measurement with three significant figures.

single bond A shared electron pair.

skeletal equation An unbalanced equation that summarizes the qualitative information about the reaction. *Example:* $H_2 + O_2 \longrightarrow H_2O$ ⚠

smectic A liquid-crystal phase in which the molecules lie parallel to one another and form layers.

smelt To melt a metal ore with a reducing agent. *Example:* $CuS(l) + O_2(g) \xrightarrow{\Delta} Cu(l) + SO_2(g)$.

sol A colloidal dispersion of solid particles in a liquid.

solid A rigid form of matter that maintains the same shape whatever the shape of its container.

solid emulsion A colloidal dispersion of a liquid in a solid. *Example:* butter, an emulsion of water in butterfat.

solid solution A solid homogeneous mixture of two or more substances.

solubility The concentration of a saturated solution of a substance.

solubility constant See *solubility product.*

solubility product K_{sp} The product of ionic molar concentrations in a saturated solution; the dissolution equilibrium constant. *Example:* $Hg_2Cl_2(s) \rightleftharpoons Hg_2^{2+}(aq) + 2\,Cl^-(aq)$, $K_{sp} = [Hg_2^{2+}][Cl^-]^2$.

solubility rules A summary of the solubility pattern of a range of common compounds in water. (See Table 3.1.)

soluble substance A substance that dissolves to a significant extent in a specified solvent; when the solvent is not specified, water is generally meant.

solute A dissolved substance.

solution A homogeneous mixture. See also *solute*; *solvent.*

solvated Surrounded by and linked to solvent molecules. (Hydration is a special case when the solvent is water.)

solvent The most abundant component of a solution.

solvent extraction A process for separating a mixture of substances that makes use of their differing solubilities in various solvents.

***sp^n* hybrid** A hybrid orbital constructed from an *s*-orbital and *n p*-orbitals. There are two *sp hybrids*, three *sp*^{2} *hybrids*, and four *sp*^{3} *hybrids*.

space-filling model A depiction of a molecule in which atoms are represented by spheres that indicate the space they occupy.

species An atom, ion, or molecule.

specific enthalpy (of a fuel) The enthalpy of combustion per gram (without the negative sign).

specific heat capacity The heat capacity of a sample divided by its mass.

spectator ion An ion that is present but remains unchanged during a reaction. *Examples:* Na^+ and NO_3^- in $NaCl(aq) + AgNO_3(aq) \longrightarrow NaNO_3(aq) + AgCl(s)$.

spectral line Radiation of a single wavelength emitted or absorbed by an atom or molecule.

spectrochemical series Ligands ordered according to the strength of the ligand field splitting they produce.

spectrometer An instrument for recording the spectrum of a sample.

spectrophotometer An instrument for measuring and recording electronically the intensity of radiation passing through a sample and hence recording its spectrum.

spectroscopy The analysis of the electromagnetic radiation emitted or absorbed by substances.

spectrum The frequencies or wavelengths of the electromagnetic radiation emitted or absorbed by substances.

spin The intrinsic angular momentum of an electron; the spin cannot be eliminated, and may occur in only two senses, denoted ↑ and ↓.

spin magnetic quantum number m_s The quantum number that distinguishes the two spin states of an electron: $m_s = +\frac{1}{2}$ (↑) and $m_s = -\frac{1}{2}$ (↓).

spontaneous change A natural change, one that has a tendency to occur without needing to be driven by an external influence. *Examples:* a gas expanding; a hot object cooling; methane burning.

spontaneous neutron emission The decay of *neutron-rich nuclei* by the emission of neutrons without an external stimulus to do so.

spontaneous nuclear fission See *fission*.

square-planar complex A complex in which four ligands lie at the corners of a square with the metal atom at the center.

stability constant See *formation constant*.

stable See *thermodynamically unstable compound*.

standard cell potential $E°$ The cell potential when the concentration of each type of ion taking part in the cell reaction is 1 $mol \cdot L^{-1}$ and all the gases are at 1 atm pressure. The standard cell potential is the difference of its two standard electrode potentials: $E° = E°(\text{cathode}) - E°(\text{anode})$.

standard enthalpy of combustion $\Delta H_c°$ The change of enthalpy per mole of substance when it burns (reacts with oxygen) completely under standard conditions.

standard enthalpy of formation $\Delta H_f°$ The standard reaction enthalpy per mole of compound for the compound's synthesis from its elements in their most stable form at 1 atm and the specified temperature.

standard free energy of formation $\Delta G_f°$ The standard reaction free energy per mole for the formation of a compound from its elements in their most stable form.

standard free energy of reaction See *standard reaction free energy*.

standard hydrogen electrode SHE A hydrogen electrode that is in its standard state (hydrogen ions at concentration 1 $mol \cdot L^{-1}$ and hydrogen pressure 1 atm) and is defined as having $E° = 0$.

standard molar entropy $S°$ The entropy per mole of a pure substance at 1 atm pressure.

standard potential $E°$ (1) The contribution of an electrode to the standard cell potential. (2) The standard cell potential when the left-hand electrode is a standard hydrogen electrode and the right-hand electrode is the electrode of interest.

standard reaction enthalpy $\Delta H°$ The difference between the enthalpy of the products of a reaction in their standard states and the enthalpy of the reactants in their standard states (in kilojoules); $\Delta H° = \sum nH°(\text{products}) - \sum nH°(\text{reactants})$, where n represents the stoichiometric coefficients in moles. $\Delta H°$ is calculated from standard enthalpies of formation. When n represents the stoichiometric coefficients themselves, we obtain $\Delta H_r°$ (in kilojoules per mole).

standard reaction entropy $\Delta S°$ The difference between the entropy of the products of a reaction in their standard states and the entropy of the reactants in their standard states (in joules per kelvin). $\Delta S° = \sum nS°(\text{products}) - \sum nS°(\text{reactants})$, where n represents the stoichiometric coefficients in moles. When n represents the stoichiometric coefficients themselves, we obtain $\Delta S_r°$ (in joules per kelvin per mole).

standard reaction free energy $\Delta G°$ The difference between the free energy (in kilojoules) of the products of a reaction in their standard states and that of the reactants in their standard states; $\Delta G° = \sum nG°(\text{products}) - \sum nG°(\text{reactants})$, where n denotes the stoichiometric coefficients in moles. Note that $\Delta G°$ is calculated from standard free energies of formation. When n denotes the stoichiometric coefficients themselves, we obtain $\Delta G_r°$ (in kilojoules per mole).

standard state The pure form of a substance at 1 atm.

standard temperature and pressure STP 0°C (273.15 K) and 1 atm (101.325 kPa).

state of matter The physical condition of a sample: solid, liquid, or gas (vapor).

state property (or state function) A property of a substance that is independent of how the sample was prepared. *Examples:* pressure; enthalpy; entropy; color.

state symbol A symbol denoting the state of a species. *Examples:* s (solid); l (liquid); g (gas); aq (aqueous solution).

stereoisomers Isomers in which atoms have the same partners arranged differently in space.

steric requirement An elementary reaction has a steric requirement if the successful collision of two molecules depends on their relative orientation.

Stern-Gerlach experiment The demonstration of the quantization of electron spin by passing a beam of atoms through a magnetic field.

stick structure A representation of the structure of an organic molecule in terms of lines representing the bonds; carbon atoms and the hydrogen atoms attached to them are not usually shown explicitly.

Stock number (1) A Roman numeral equal to the number of electrons lost by an atom on formation of a compound and sometimes added in parentheses to a name. (2) The oxidation number of the element. *Example:* Copper(II) in compounds containing Cu^{2+}.

stoichiometric coefficients The numbers multiplying chemical formulas in a chemical equation. *Examples:* 1, 1, and 2 in $H_2 + Br_2 \longrightarrow 2\,HBr$.

stoichiometric point The stage in a titration when exactly the right volume of solution needed to complete the reaction has been added.

stoichiometric proportions Reactants in the same proportions as their coefficients in the chemical equation. *Example:* equal amounts of H_2 and Br_2 in the formation of HBr.

stoichiometric relation An expression that equates the relative amounts of reactants and products that participate in a reaction. *Example:* 1 mol $H_2 \simeq$ 2 mol HBr.

stoichiometry See *reaction stoichiometry.*

STP See *standard temperature and pressure.*

strong acids and bases Acids and bases that are fully ionized in solution. *Examples:* HCl, $HClO_4$ (strong acids); NaOH, $Ca(OH)_2$ (strong bases).

strong electrolyte See *electrolyte solution.*

strong-field ligand A ligand that produces a large ligand field splitting and that lies above H_2O in the spectrochemical series.

structural formula A chemical formula that shows how atoms in a compound are attached to one another.

structural isomers Isomers in which the atoms have different partners.

subatomic particle A particle smaller than an atom. *Examples:* electron; proton; neutron.

subcritical mass See *critical mass.*

sublimation The direct conversion of a solid to a vapor without first forming a liquid.

sublimation vapor pressure The vapor pressure of a solid.

subshell All the atomic orbitals of a given shell of an atom with the same value of the quantum number *l*. *Example:* the five 3*d*-orbitals of an atom.

substance A single, pure type of matter; either a compound or an element

substituent Atoms or groups that have replaced hydrogen atoms in an organic molecule.

substitution reaction (1) A reaction in which an atom (or a group of atoms) replaces an atom in the original molecule. (2) In complexes, a reaction in which one Lewis base expels another and takes its place. *Examples:* (1) $C_6H_5OH + Br_2 \longrightarrow BrC_6H_4OH + HBr$; (2) $[Fe(H_2O)_6]^{3+}(aq) + 6\,CN^-(aq) \longrightarrow [Fe(CN)_6]^{3-}(aq) + 6\,H_2O(l)$.

substitutional alloy See *alloy.*

substrate The molecule on which an enzyme acts.

superconductor An electronic conductor that conducts electricity with zero resistance. See also *high-temperature superconductor.*

supercritical fluid A substance above its *critical temperature.*

supercritical mass See *critical mass.*

superfluidity The ability to flow without viscosity.

surface-active agent See *surfactant.*

surface tension γ The tendency of molecules at the surface of a liquid to be pulled inward, so resulting in a smooth surface.

surfactant A substance that accumulates at the surface of a solution and affects the surface tension of the solvent; a component of detergents. *Example:* the stearate ion of soaps.

surroundings The region outside a system.

suspension A mist of small particles.

syndiotactic polymer See *polymer.*

synthesis A reaction in which a substance is formed from simpler starting materials. *Example:* $N_2(g) + 3\,H_2(g) \longrightarrow 2\,NH_3(g)$.

synthesis gas A mixture of carbon monoxide and hydrogen produced by the catalyzed reaction of a hydrocarbon and water.

system The reaction vessel and its contents in which there is a particular interest.

systematic error An error that persists in a series of measurements and does not average out. See also *accuracy.*

systematic name The name of a compound that reveals which elements are present (and, in its most complete form, how the atoms are arranged). *Example:* methylbenzene is the systematic name for toluene

Système International d'Unités See *SI.*

***t*-orbital** One of the orbitals d_{xy}, d_{yz}, and d_{zx} in an octahedral or tetrahedral complex.

temperature T (1) How hot or cold a sample is. (2) The intensive property that determines the direction in which heat will flow between two objects in contact.

termination A step in a chain reaction in which chain carriers combine to form products. *Example:* $2\,Br\cdot \longrightarrow Br_2$.

termolecular reaction An elementary reaction involving the simultaneous collision of three species.

tertiary alcohol An alcohol of the form R_3COH.

tertiary structure The shape into which the α helix and β-pleated sheet sections of a polypeptide are twisted as a result of interactions between peptide groups lying in different parts of the primary structure.

tetrahedral complex A complex in which four ligands lie at the corners of a regular tetrahedron with the metal atom at the center. *Example:* $[Cu(NH_3)_4]^{2+}$.

theoretical yield The maximum quantity of product that can be obtained, according to the reaction stoichiometry, from a given quantity of a specified reactant.

theory A collection of ideas and concepts used to account for a scientific law.

thermal decomposition See *decomposition.*

thermal motion The random, chaotic motion of atoms.

thermal pollution The damage caused to the environment by the waste heat of an industrial process.

thermite reaction (thermite process) The reduction of a metal oxide by aluminum. *Example:* $2\,Al(s) + Fe_2O_3(s) \longrightarrow Al_2O_3(s) + 2\,Fe(l)$.

thermochemical equation An expression consisting of both the chemical equation and the reaction enthalpy for the chemical reaction exactly as written.

thermochemistry The study of the heat released or absorbed by chemical reactions; a branch of thermodynamics.

thermodynamically unstable compound (1) A compound with a thermodynamic tendency to decompose into its elements. (2) A compound with a positive free energy of formation.

thermodynamics The study of the transformations of energy from one form to another. See also *first law of thermodynamics; second law of thermodynamics.*

thermonuclear explosion An explosion resulting from uncontrolled nuclear fusion.

thermotropic A liquid crystal prepared by melting the solid phase.

three-center bond A chemical bond (typically involving boron atoms) in which a hydrogen atom lies between two other atoms (typically boron atoms) and one electron pair binds all three atoms together.

titrant The solution of known concentration added from a buret in a titration.

titration The analysis of composition by measuring the volume of one solution needed to react with a given volume of another solution.

tracer An isotope that can be tracked from compound to compound in the course of a sequence of reactions.

transition A change of state. (1) In thermodynamics, a change of physical state. (2) In spectroscopy, a change of quantum state.

transition metal An element that belongs to Groups 3 through 11. *Examples:* vanadium; iron; gold.

transuranium elements The elements beyond uranium; those with $Z > 92$.

triple bond (1) Three electron pairs shared by two neighboring atoms. (2) One σ bond and two π bonds between neighboring atoms.

triple point The point where three phase boundaries meet in a phase diagram; under the conditions represented by the triple point, all three adjoining phases coexist in dynamic equilibrium.

triprotic See *polyprotic acid or base.*

ultraviolet radiation Electromagnetic radiation with a higher frequency (shorter wavelength) than that of violet light.

unbranched alkane An alkane with no side chains, in which all the carbon atoms lie in a linear chain.

unimolecular reaction An elementary reaction in which a single reactant molecule changes into products. *Example:* $O_3(g) \longrightarrow O(g) + O_2(g)$.

unit See *base units.*

unit cell The smallest unit that, when stacked together repeatedly without any gaps, can reproduce the entire crystal.

unsaturated hydrocarbon A hydrocarbon with at least one carbon-carbon multiple bond. *Examples:* $CH_2{=}CH_2$; C_6H_6.

valence The number of bonds that an atom can form.

valence band In the theory of solids, a band of energy levels fully occupied by electrons.

valence-bond theory The description of bond formation in terms of the pairing of spins in the atomic orbitals of neighboring atoms.

valence electrons The electrons that belong to the valence shell.

valence shell The outermost shell of an atom. *Example:* The $n = 2$ shell of Period 2 atoms.

valence-shell electron-pair repulsion model VSEPR model A model for predicting the shapes of molecules, using the fact that electron pairs repel one another.

van der Waals equation An approximate equation of state for a real gas in which two parameters represent the effects of intermolecular forces.

van der Waals forces See *intermolecular forces.*

van't Hoff equation (1) The equation for the osmotic pressure in terms of the molarity, $\Pi = i \times \text{molarity} \times RT$. (2) An equation that shows how the equilibrium constant varies with temperature. (See Exercise 13.94.)

van't Hoff *i* factor See *i factor.*

vapor The gaseous phase of a substance (specifically, of a substance that is a liquid or solid at the temperature in question).

vapor pressure The pressure exerted by the vapor of a liquid (or a solid) when the vapor and the liquid (or solid) are in dynamic equilibrium.

vaporization The formation of a gas or a vapor.

variable covalence The ability of an element to form different numbers of covalent bonds. *Example:* SO_2 and SO_3.

variable valence The ability of an element to form ions with different charges. *Example:* In^+ and In^{3+}.

viscosity The resistance of a fluid (a gas or a liquid) to flow: the higher the viscosity, the slower the flow.

visible light See *visible radiation.*

visible radiation Electromagnetic radiation that can be detected by the human eye, with wavelengths in the range 700 to 400 nm. Visible radiation is also called *visible light* or simply *light.*

volatility The readiness with which a substance vaporizes. A substance is typically regarded as volatile if its boiling point is below 100°C.

volume V The amount of space a sample occupies.

volume percentage composition The volume of a substance present in a solution expressed as a percentage of the total volume.

volumetric analysis An analytical method using measurement of volume.

water autoprotolysis constant K_w The equilibrium constant for the *autoprotolysis* (autoionization) of water, $H_2O(l) + H_2O(l) \rightleftharpoons H_3O^+(aq) + OH^-(aq)$, $K_w = [H_3O^+][OH^-]$.

water-splitting reaction A component of the reactions involved in photosynthesis, in which hydrogen is extracted from water and oxygen is released.

wavelength λ (lambda) The peak-to-peak distance of a wave.

wave-particle duality The combined wavelike and particlelike character of both radiation and matter.

weak acids and bases Acids and bases that are only partially ionized in aqueous solutions at normal concentrations. *Examples:* HF, CH_3COOH (weak acids); NH_3, CH_3NH_2 (weak bases).

weak electrolyte See *electrolyte solution.*

weak-field ligand A ligand that produces a small ligand field splitting and that lies below NH_3 in the spectrochemical series.

weight The gravitational force on a sample.

work w The energy expended during the act of moving an object against an opposing force.

x-ray Electromagnetic radiation with wavelengths from about 10 pm up to about 1000 pm.

x-ray diffraction The analysis of crystal structures by studying the interference pattern in a beam of x-rays.

yield See *percentage yields*; *theoretical yield.*

zero-order reaction A reaction with a rate that is independent of the concentration of the reactant. *Example:* the catalyzed decomposition of ammonia.

zone refining A method for purifying a solid by repeatedly passing a molten zone along the length of a sample.

ANSWERS

SELF-TESTS B

CHAPTER 1

1.1B 16 electrons
1.2B (a) 8 protons, 8 neutrons, 8 electrons
(b) 92 protons, 144 neutrons, 92 electrons
1.3B sulfide ion, S^{2-}; potassium ion, K^+
1.4B K, Mg, Si, Al, O, and H atoms are present in mica in the ratios 1 : 3 : 3 : 1 : 12 : 2.
1.5B (a) phosphoric acid (b) $HClO_3$
1.6B (a) gold(III) chloride (b) calcium sulfide
(c) manganese(III) oxide
1.7B (a) phosphorus trichloride (b) sulfur trioxide
(c) dinitrogen tetroxide
1.8B (a) $Cs_2S{\cdot}4H_2O$ and Mn_2O_7 (b) H_2S and S_2Cl_2

CHAPTER 2

2.1B 8.82 oz
2.2B 7.87 cm^2
2.3B 1.100 $g{\cdot}cm^{-3}$
2.4B −40°C
2.5B (a) 4 sf (b) 3 sf (c) 7 sf
2.6B (a) 8.1 g (b) 1540°C
2.7B (a) 6.9×10^{-4} L (b) 1.38 $g{\cdot}L^{-1}$
2.8B 1.89×10^{24} H_2O molecules
2.9B $(0.6917) \times (62.94) + (0.3083) \times (64.93)\ g{\cdot}mol^{-1} = 63.55\ g{\cdot}mol^{-1}$
2.10B 62 g U
2.11B 1.25 mol O atoms
2.12B (a) 63.01 $g{\cdot}mol^{-1}$ (b) 342.14 $g{\cdot}mol^{-1}$
2.13B (a) 13.5 mol (b) 0.51 g
2.14B 88.1% C, 11.9% H
2.15B OSF_2
2.16B 1.48 g F; 0.019 mol Xe; 0.0779 mol F;
0.019 : 0.078 yields 1 : 4, and XeF_4
2.17B $C_2H_2O_4$
2.18B $(1.368\ g/180.16\ g{\cdot}mol^{-1})/(0.050\ L) = 0.1519\ mol{\cdot}L^{-1}$
2.19B 1.45×10^{-3} mol urea
2.20B 7.12 mL
2.21B $(0.125\ mol{\cdot}L^{-1}) \times (0.050\ L) \times (90.04\ g{\cdot}mol^{-1}) = 0.563\ g$
2.22B $(1.59 \times 10^{-5}\ mol{\cdot}L^{-1}) \times (0.02500\ L)/(0.152\ mol{\cdot}L^{-1}) = 2.62 \times 10^{-3}\ mL$

CHAPTER 3

3.1B $C_3H_8(g) + 5\,O_2(g) \rightarrow 3\,CO_2(g) + 4\,H_2O(l)$
3.2B $4\,HF(g) + SiO_2(s) \rightarrow SiF_4(g) + 2\,H_2O(l)$
3.3B (a) nonelectrolyte, does not conduct electricity
(b) strong electrolyte, conducts electricity
3.4B $3\,Hg_2^{2+}(aq) + 2\,PO_4^{3-}(aq) \rightarrow (Hg_2)_3(PO_4)_2(s)$
3.5B strontium nitrate mixed with sodium sulfate;
$Sr^{2+}(aq) + SO_4^{2-}(aq) \rightarrow SrSO_4(s)$
3.6B (a) neither (b) and (c) are acids (d) is a base
3.7B 0.1 M CH_3NH_2 is a weak base. When CH_3NH_2 is dissolved in water, it gives a solution that consists almost entirely of CH_3NH_2 molecules, with just a small proportion of $CH_3NH_3^+$ cations and OH^- anions.
3.8B $2\,H_3PO_4(aq) + 3\,Ca(OH)_2(aq) \rightarrow Ca_3(PO_4)_2(s) + 6\,H_2O(l)$
3.9B $HCl(aq) + NaHCO_3(aq) \rightarrow NaCl(aq) + CO_2(g) + H_2O(l)$
3.10B (a) +4 (b) +3 (c) +5
3.11B H_2SO_4 is the oxidizing agent; NaI is the reducing agent.
3.12B $2\,Ce^{4+}(aq) + 2\,I^-(aq) \rightarrow I_2(s) + 2\,Ce^{3+}(aq)$
3.13B $2\,NH_3(g) + 3\,CuO(s) \rightarrow N_2(g) + 3\,Cu(s) + 3\,H_2O(l)$; redox

CHAPTER 4

4.1B 50 mol Fe
4.2B 38 kg H_2SO_4
4.3B 396 g $CaSiO_3$
4.4B (0.255×0.02500) mol KOH $\times$

$$\frac{1\ mol\ H_2SO_4}{2\ mol\ KOH} \times \frac{1}{0.01645\ L} = 0.194\ M\ H_2SO_4(aq)$$

4.5B Iron(III) oxide is Fe_2O_3; 2 mol Fe is needed for 1 mol Fe_2O_3.

$$\frac{15\ kg\ Fe_2O_3}{159.69\ g{\cdot}mol^{-1}} \times \frac{2\ mol\ Fe}{1\ mol\ Fe_2O_3} \times 55.85\ g{\cdot}mol^{-1} =$$

$$10.5\ kg\ Fe;\ \frac{8.8\ kg}{10.5\ kg} \times 100\% = 84\%$$

4.6B 0.212 mol Cl_2; 0.141 mol P

$$0.212\ mol\ Cl_2 \times \frac{2\ mol\ P}{3\ mol\ Cl_2} = 0.141\ mol\ P$$

1.41 − 0.141 mol P = 1.27 mol P; 39.2 g P
4.7B 0.61 mol NO_2; 1.0 mol H_2O
From the reaction stoichiometry, 1 mol H_2O requires 3 mol NO_2. There is not enough NO_2, so NO_2 is the limiting reactant.
4.8B 851 mol NH_3; 502 mol CO_2

$$851\ mol\ NH_3 \times \frac{1\ mol\ CO_2}{2\ mol\ NH_3} = 426\ mol\ CO_2\ required.$$

There is excess CO_2; thus, NH_3 is the limiting reactant.

$851 \text{ mol NH}_3 \times \dfrac{1 \text{ mol OC(NH}_2)_2}{2 \text{ mol NH}_3} \times (60.06 \text{ g}\cdot\text{mol}^{-1}) =$

$25.6 \times 10^3 \text{ g OC (NH}_2)_2$

$426 \text{ mol} \times 44.01 \text{ g}\cdot\text{mol}^{-1} = 18.7 \times 10^3 \text{ g} = 18.7 \text{ kg}$ CO_2 reacted

Mass CO_2 remaining $= 22.1 - 18.7 \text{ kg} = 3.4 \text{ kg}$

4.9B $\left(\dfrac{0.519}{44.01} \text{ mol CO}_2\right) \times \dfrac{1 \text{ mol C}}{1 \text{ mol CO}_2} = 0.0118 \text{ mol C}$

$\left(\dfrac{0.0945}{18.02} \text{ mol H}_2\text{O}\right) \times \dfrac{2 \text{ mol H}}{1 \text{ mol H}_2\text{O}} = 0.0105 \text{ mol H}$

Mass of C $= 0.0118 \text{ mol} \times 12.01 \text{ g}\cdot\text{mol}^{-1} = 0.142 \text{ g}$

Mass of H $= 0.0105 \text{ mol} \times 1.008 \text{ g}\cdot\text{mol}^{-1} = 0.0106 \text{ g}$

Mass of O $= 0.236 - (0.142 + 0.0106) \text{ g} = 0.0834 \text{ g}$

$0.0834 \text{ g}/(16.00 \text{ g}\cdot\text{mol}^{-1}) = 0.00521 \text{ mol}$

$0.0118 : 0.0105 : 0.00521$ yields $2.26 : 2.02 : 1$ or $9 : 8 : 4$, for $C_9H_8O_4$

CHAPTER 5

5.1B 84 kPa

5.2B 10 L

5.3B 5.3 atm. The cans cannot survive such high pressure and explode.

5.4B The pressure would be four times larger.

5.5B 24.6 L

5.6B 12 L

5.7B $\left(\dfrac{1.00 \times 10^3 \times 0.791}{32.04}\right) \text{ mol CH}_3\text{OH} \times$

$\dfrac{3 \text{ mol O}_2}{2 \text{ mol CH}_3\text{OH}} \times 24.47 \text{ L}\cdot\text{mol}^{-1} = 906 \text{ L O}_2$

5.8B $\dfrac{100.0 \text{ L O}_2}{24.47 \text{ L}\cdot\text{mol}^{-1}} \times \dfrac{2 \text{ mol H}_2\text{O}}{1 \text{ mol O}_2} \times$

$18.02 \text{ g}\cdot\text{mol}^{-1} = 147.3 \text{ g H}_2\text{O}$

5.9B $(1.04 \text{ g}\cdot\text{L}^{-1}) \times \left(\dfrac{62.364 \times 450}{200}\right) \text{L}\cdot\text{mol}^{-1} =$

$146 \text{ g}\cdot\text{mol}^{-1}$

5.10B $(760 - 23.8) \text{ Torr} = 736 \text{ Torr}$

$(736 \times 1.00)/(62.364 \times 298) \times 32.00 \text{ g} = 1.27 \text{ g}$

5.11B O_3 molecules (molar mass, $48.00 \text{ g}\cdot\text{mol}^{-1}$) effuse more slowly than NO_2 molecules (molar mass, $46.01 \text{ g}\cdot\text{mol}^{-1}$).

5.12B CH_4 molecules (molar mass, $16.04 \text{ g}\cdot\text{mol}^{-1}$) have a higher average speed than do Cl_2 molecules (molar mass, $70.91 \text{ g}\cdot\text{mol}^{-1}$) at the same temperature.

CHAPTER 6

6.1B $\dfrac{(4.184 \text{ J}\cdot(°\text{C})^{-1}\cdot\text{g}^{-1}) \times (61.2 \text{ g}) \times (1.7°\text{C})}{(25.0 \text{ g}) \times (67.3°\text{C})} =$

$0.26 \text{ J}\cdot(°\text{C})^{-1}\cdot\text{g}^{-1}$

6.2B $(4.16 \text{ kJ})/(3.24°\text{C}) = 1.28 \text{ kJ}\cdot(°\text{C})^{-1}$

6.3B $(22 \text{ kJ}/23 \text{ g}) \times 46.07 \text{ g}\cdot\text{mol}^{-1} = +44 \text{ kJ}\cdot\text{mol}^{-1}$

6.4B $\Delta H = (6.06°\text{C}) \times (216 \text{ J}\cdot(°\text{C})^{-1}) \times 10^{-3} \text{ kJ}\cdot\text{J}^{-1} \times \dfrac{169.88 \text{ g}\cdot\text{mol}^{-1}}{3.382 \text{ g}} - 6.58 \text{ kJ}$

$Ag^+(aq) + Cl^-(aq) \rightarrow AgCl(s) \qquad \Delta H = -65.8 \text{ kJ}$

6.5B -328.10 kJ

6.6B $\dfrac{350 \text{ kJ}}{1368 \text{ kJ}\cdot\text{mol}} \times 46.07 \text{ g}\cdot\text{mol}^{-1} = 11.8 \text{ g}$

6.7B methanol. Its higher enthalpy density means less volume would be required for the same heat output.

6.8B $C(\text{diamond}) + O_2(g) \rightarrow CO_2(g)$;

$(-393.51 - 1.895) \text{ kJ}\cdot\text{mol}^{-1} = -395.40 \text{ kJ}\cdot\text{mol}^{-1}$

CHAPTER 7

7.1B yellow

7.2B 4.5×10^{-19} J

7.3B $\nu = \dfrac{1}{h}\Delta E = (3.29 \times 10^{-15} \text{ Hz})\left(\dfrac{1}{2^2} - \dfrac{1}{5^2}\right)$

$= 6.91 \times 10^{14} \text{ Hz}$

$\lambda = (3.00 \times 10^8)/(6.91 \times 10^{14}) \text{ m} = 4.34 \times 10^{-7} \text{ m}$

$= 434$ nm; violet line

7.4B n^2

7.5B $3p$

7.6B $1s^22s^22p^63s^2$

7.7B $[Ar]3d^{10}4s^24p^3$

7.8B $[Ar]3d^5$, $[Ar]3d^2$

7.9B I^-, $[Kr]4d^{10}5s^25p^6$

7.10B ns^2np^3

7.11B (a) $r(Ca^{2+}) < r(K^+)$; (b) $r(Cl^-) < r(S^{2-})$

7.12B The inner $3d$ electrons of gallium shield the outer $4s$ and $4p$ electrons from the nuclear charge less effectively than its inner s and p electrons. Thus, the outer electron ($4p^1$) of gallium is more strongly bound than one initially expects, and the ionization energy is larger.

CHAPTER 8

8.1B (a) $[:\ddot{\underset{\cdot\cdot}{S}}:]^{2-}\, Al^{3+}\, [:\ddot{\underset{\cdot\cdot}{S}}:]^{2-}\, Al^{3+}\, [:\ddot{\underset{\cdot\cdot}{S}}:]^{2-}$

(b) $Sr^{2+}[:\ddot{\underset{\cdot\cdot}{P}}:]^{3-}Sr^{2+}[:\ddot{\underset{\cdot\cdot}{P}}:]^{3-}Sr^{2+}$

8.2B $[Pb:]^{2+}[:\ddot{\underset{\cdot\cdot}{O}}:]^{2-}$ and $Pb^{4+}[:\ddot{\underset{\cdot\cdot}{O}}:]^{2-}[:\ddot{\underset{\cdot\cdot}{O}}:]^{2-}$

8.3B 147.70 (vaporize Mg) + 30.91 (vaporize Br_2) + (2×111.88) (atomize 2 mol Br) + 736 (form Mg^+) + 1450 ($Mg^+ \rightarrow Mg^{2+} + e^-$) $- (2 \times 325)$ (add electron to Br) + 524.3 ($-\Delta H_f^\circ$) kJ $= 2463$ kJ

8.4B $H-\ddot{\underset{\cdot\cdot}{Br}}:$; none on hydrogen and three on bromine

8.5B
```
      H
      |
  H — C — Cl:   (Cl with three lone pairs)
      |
      H
```

8.6B The C atom in CO (:C≡O:) has a nonbonding electron pair that can bind to hemoglobin, whereas the C atom in CO_2 (Ö=C=Ö) has none.

8.7B [Ö=N̈−Ö:]$^-$ ↔ [:Ö−N̈=Ö]$^-$

8.8B Ö=Ö−Ö: (formal charges 0, +1, −1)

8.9B $[:\ddot{O}{-}As({=}O)(\ddot{O}:)\ddot{O}:]^{3-}$ (formal charges: As 0; double-bonded O 0; each single-bonded O −1)

8.10B :Ö−C̈l=Ö or Ö=C̈l=Ö

8.11B [:Ï−Ï−Ï:]$^-$

8.12B See structure **41**. One Cl atom on an $AlCl_3$ molecule donates an electron to an Al atom on a second molecule and one Cl atom from the second molecule donates an electron pair to the Al atom on the first.

8.13B (a)

8.14B (a)

CHAPTER 9

9.1B tetrahedral

9.2B trigonal planar

9.3B (a) trigonal planar (b) angular

9.4B trigonal pyramidal; HOH angle less than 109.5°

9.5B square planar

9.6B (a) nonpolar (b) polar

9.7B $CH_4(g) + 2\,F_2(g) \rightarrow CH_2F_2(g) + 2\,HF(g)$
$\Delta H = (2 \times 412) + (2 \times 158) - (2 \times 484) - (2 \times 565)\text{ kJ} = -958\text{ kJ}$

9.8B N=O. It is a stronger bond.

9.9B (a) three σ, no π (b) two σ, two π

9.10B Two σ bonds are formed from C sp^3 hybrids and H 1*s*-orbitals, and two σ bonds are formed from C sp^3 hybrids and Cl 3*p*-orbitals, in a tetrahedral arrangement.

9.11B octahedral; six d^2sp^3-orbitals

9.12B The CH_3 carbon is sp^3 hybridized; it forms four σ-bonds with bond angles of approximately 109.5°. The CH and CH_2 carbons are both sp^2 hybridized; they each form three σ-bonds and one π-bond (to each other), with bond angles of approximately 120°.

CHAPTER 10

10.1B 1,1-dichloroethene

10.2B The molecule with a dipole moment, CHF_3, has a higher boiling point. In this case, the dipole-dipole interactions are more important than the London interactions. Fluorocarbons and hydrocarbons have only weak intermolecular forces, as described in the text.

10.3B (a) and (c)

10.4B 6

10.5B 4; $\frac{1}{8} \times 8$ (1 on each of eight corners), $\frac{1}{2} \times 2$ on opposite faces, and 2 inside

10.6B The bcc structure has 2 atoms per unit cell; the length of a side $= 4r/\sqrt{3}$. Assuming a bcc structure, the density of a unit cell is

$$\frac{(2 \times 55.85\text{ g·mol}^{-1})/(6.022 \times 10^{23}\text{ mol}^{-1})}{[4 \times (124 \times 10^{-12}\text{ m}) \times (10^2\text{ cm·m}^{-1})/\sqrt{3}]^3}$$
$$= \frac{1.855 \times 10^{-22}\text{ g}}{2.35 \times 10^{-23}\text{ cm}^3} = 7.89\text{ g·cm}^{-3}$$

This value is consistent with the experimental value.

10.7B 1; $\frac{1}{8} \times 1$ on each of eight corners

10.8B $CH_3CH_2CH_3$; CH_3CHO has the same molar mass and so has approximately the same London forces; however, its molecules are polar, so it also experiences dipole-dipole forces.

10.9B 99.2°C

10.10B vapor

10.11B rhombic; because it is more stable at higher pressures

10.12B Liquid CO_2 at 60 atm vaporizes when it is released into a room at 1 atm pressure.

10.13B H_2O has a higher critical temperature than NH_3, and that of NH_3 is higher than the value for CH_4. H_2O has the capacity to make more hydrogen bonds as a pure substance than does NH_3, whereas CH_4 is unable to make any. Thus, critical temperature increases with hydrogen bonding capacity.

CHAPTER 11

11.1B

```
        CH3
        |
        C
      //  \
    CH     C=O
    |      |
    CH2    CH2
      \   /
        CH
        |
        C
      //  \
   H2C     CH3
```

11.2B (a), because it has the greater molar mass

11.3B CH_3CH_2Br

11.4B 4-ethyl-2-methylhexane and $CH_3CH_2C(CH_2CH_3)_2CH_2CH(CH_2CH_3)CH_2CH_3$

11.5B 4-ethyl-2-hexene

11.6B 1-ethyl-3,5-dimethyl-2-propylbenzene (alternatively, 2-ethyl-4,6-dimethyl-1-propylbenzene)

11.7B (a) $HCOOCH_2CH_3$; (b) $CH_3CH_2CH_2COOH$ and CH_3OH
11.8B amide, because it can form hydrogen bonds with the NH group (Esters cannot form hydrogen bonds with themselves, as they have no F—H, N—H, or O—H bonds.)
11.9B 3-pentanol
11.10B 3-pentanone
11.11B ethylmethylamine
11.12b 1,3-dichloro-2-methylpropane
11.13B $CH_3CH_2CH_2CH_2Br$, $CH_3CH_2CHBrCH_3$, $(CH_3)_2CHCH_2Br$, $(CH_3)_3CBr$
11.14B **38a** is *cis*-1-phenyl-1-propene, **38b** is *trans*-1-phenyl-1-propene
11.15B (c), because it has four different groups on one C atom
11.16B (a) $CH_2{=}C(CH_3)(COOCH_3)$;
(b) $-NHCH(CH_3)CONHCH(CH_3)CO-$
11.17B alternating copolymer

CHAPTER 12

12.1B Assume solution volume is 900 mL.
moles of CO_2 = (concentration)(volume)= $(2.3 \times 10^{-2}\ mol{\cdot}L^{-1}{\cdot}atm^{-1}) \times (1.00\ atm) \times (0.900\ L) = 2.1 \times 10^{-2}$ mol
12.2B Rb^+ has a less negative hydration enthalpy than Sr^{2+} because it is less highly charged
12.3B moles of Na^+ = moles Cl^- = 5.00 g $NaCl/58.44\ g\ NaCl{\cdot}mol^{-1} = 0.0856$ mol
moles of H_2O = 5.55 mol; total moles = 2(0.0856) + 5.55 mol = 5.72 mol
$x_{Na^+} = x_{Cl^-} = 0.0856\ mol/5.72\ mol = 0.0150$
$x_{H_2O} = 5.55\ mol/5.72\ mol = 0.970$
12.4B $(0.255\ mol{\cdot}kg^{-1}) \times (0.250\ kg) \times (180.16\ g{\cdot}mol^{-1}) = 11.5$ g
12.5B $x_{CH_3OH} = 0.250, \quad x_{H_2O} = 1.000 - 0.250 = 0.750$

$$m = \frac{0.250\ mol}{(0.750\ mol) \times (0.01801\ kg{\cdot}mol^{-1})}$$
$$= 18.5\ mol{\cdot}kg^{-1}\ H^+(aq)$$

12.6B mass (solvent) = (total mass in 1 L) − mass (solute) = $(1.07\ g{\cdot}cm^{-3}) \times (1000\ cm^3) - (1.83\ mol) \times (58.44\ g{\cdot}mol^{-1}) = 963$ g (= 0.963 kg)
$m = 1.83\ mol/0.963\ kg = 1.90\ mol{\cdot}kg^{-1}$ NaCl(aq)
12.7B moles of Na_2CO_3 = $(4.27\ g)/(105.99\ g{\cdot}mol^{-1}) = 4.03 \times 10^{-2}$ mol
$m = (4.03 \times 10^{-2}\ mol)/(0.09573\ kg) = 0.421\ mol{\cdot}kg^{-1}$

12.8B $x_{ethanol}$

$$= \frac{(50.0\ g)/(46.07\ g{\cdot}mol^{-1})}{(50.0\ g)/(46.07\ g{\cdot}mol^{-1}) + (2.00\ g)/(132.16\ g{\cdot}mol^{-1})}$$
$$= 0.986$$

$P = 0.986 \times (5.3\ kPa) = 5.2$ kPa

12.9B 1.00 mol (0.25 mol Co^{3+} ions and 0.75 mol Cl^- ions); $i = 4$

12.10B $$m = \frac{0.51\ K}{3.97\ K{\cdot}kg{\cdot}mol^{-1}} = 0.0128\ mol{\cdot}kg^{-1}$$

molar mass $= 0.200\ g/[(0.0128\ mol{\cdot}kg^{-1}) \times (0.100\ kg)]$
$= 1.6 \times 10^2\ g{\cdot}mol^{-1}$ (actual: $154.2\ g{\cdot}mol^{-1}$)

12.11B $$\text{molarity} = \frac{(2.11\ kPa)}{(8.314\ L{\cdot}kPa{\cdot}K^{-1}{\cdot}mol^{-1}) \times (298\ K)}$$
$$= 8.52 \times 10^{-4}\ mol{\cdot}L^{-1}$$

moles = $(8.2 \times 10^{-4}\ mol{\cdot}L^{-1}) \times (0.175\ L) = 1.44 \times 10^{-4}$ mol
molar mass = $(1.50\ g)/1.44 \times 10^{-4}\ mol = 1.04 \times 10^4\ g{\cdot}mol^{-1}$

CHAPTER 13

13.1B $K_c = [O_2]^3/[O_3]^2$
13.2B $K = (3.51 \times 0.156)/(0.131)^2 = 31.9$
13.3B $K_c = \sqrt{7.3 \times 10^{-13}} = 8.5 \times 10^{-7}$
13.4B $K_c = 1/[O_2]^5$
13.5B $K_p = P_{SO_3}/P_{O_2}^{3/2}$
13.6B $K_p = K_c$
13.7B $[NO] = \sqrt{K_c[N_2][O_2]} = 2.2 \times 10^{-19}$, about 10^{-15} times the concentration of N_2 and O_2
13.8B $P_{CO_2} = 75/\sqrt{(2.8 \times 10^{20}) \times (1.4 \times 10^{-9})} = 3.8 \times 10^{-5}$ atm
13.9B $Q_p = (1.2)^2/2.4 = 0.60$; $Q_p < K_p$; the partial pressure of NO_2 will increase
13.10B for $N_2(g) + 3\,H_2(g) \rightleftharpoons 2\,NH_3(g)$,

	N_2	H_2	NH_3
initial	$[N_2]_{ini}$	$[H_2]_{ini}$	0
change	$-x$	$-3x$	$+2x$
equilibrium	0.20	0.30	0.40

$x = 0.40/2 = 0.20$
$[N_2]_{ini} = x + 0.20 = 0.40$; $[H_2]_{ini} = 3x + 0.30 = 0.90$
initial molar concentration of N_2 = $0.40\ mol{\cdot}L^{-1}$; that of H_2 = $0.90\ mol{\cdot}L^{-1}$

13.11B $$K_c = 20 = \frac{(2x)^2}{(0.200 - x) \times (0.100 - x)};$$

$16x^2 - 6.0x + 0.40 = 0$

$$x = \frac{6.0 \pm \sqrt{(6.0)^2 - 4 \times 16 \times 0.40}}{2 \times 16} = \frac{6.0 \pm 3.2}{32}$$
$$= 0.29, 0.088$$

Discard 0.29, because it would yield negative equilibrium concentrations of reactants.
molar concentration of ClF = $2 \times 0.088 = 0.18\ mol{\cdot}L^{-1}$
13.12B $K_c = 3.5 \times 10^{-32} \approx (2x)^2x/(0.012)^2 \approx 1.1 \times 10^{-12}$; the concentrations are $0.012\ mol{\cdot}L^{-1}$ HCl, $2.2 \times 10^{-12}\ mol{\cdot}L^{-1}$ HI, $1.1 \times 10^{-12}\ mol{\cdot}L^{-1}$ Cl_2 (some I_2 remains as a pure solid)
13.13B equilibrium shifts toward (a) products; (b) products; (c) reactants

13.14B diamond favored
13.15B CO_2 favored

CHAPTER 14

14.1B (a) NH_4^+ and HCO_3^-; (b) NH_2^- and NO_3^-
14.2B molarity of OH^- = 2.2×10^{-3} mol·L^{-1}
molarity of H_3O^+ = $(1.0 \times 10^{-14})/(2.2 \times 10^{-3}$ mol·L) = 4.5×10^{-12} mol·L^{-1}
14.3B $[H_3O^+] = K_w/[OH^-] = (1.0 \times 10^{-14})/(0.077) = 1.3 \times 10^{-13}$; pH = $-\log[H_3O^+]$ = 12.89
14.4B $[H_3O^+]$ = antilog $(-8.2) = 6 \times 10^{-9}$; molarity of $H_3O^+ = 6 \times 10^{-9}$ mol·L^{-1}
14.5B (a) acid: NH_4^+, base: HCO_3^-; (b) conjugate base: NH_3, conjugate acid: H_2CO_3
14.6B stronger bases: (a) NH_2NH_2; (d) ClO_2^-; stronger acids: (b) $C_5H_5NH^+$; (c) HIO_3
14.7B (a) weak base; (b) strong base; (c) weak base
14.8B (a) $HClO_2$, rule 3; (b) HI, rule 2
14.9B $CH_3COOH < CH_2ClCOOH < CHCl_2COOH$
14.10B We must use the exact solution.
$K_a = 1.4 \times 10^{-3} = x^2/(0.22 - x)$; $x^2 + (1.4 \times 10^{-3})x - (3.08 \times 10^{-4}) = 0$; the roots of the quadratic equation are 0.017 and -0.036. Only 0.017 is possible, so pH = $-\log(0.017)$ = 1.77.
14.11B $K_b = 1.0 \times 10^{-6} \approx x^2/0.012$; $[OH^-] = x \approx 1.1 \times 10^{-4}$
pOH = 3.96; pH = 14.00 − 3.96 = 10.04;
$(1.1 \times 10^{-4})/0.012 \times 100\% = 0.92\%$
14.12B $0.012 = (0.10 + x)x/(0.10 - x)$; $x^2 + 0.11x - 0.0012 = 0$; from the quadratic equation, $x = 0.010$; $[H_3O^+] = 0.10 + 0.010 = 0.11$; pH = 0.96
14.13B $K_{a1} = 1.5 \times 10^{-2} = x^2/(0.10 - x)$; from the quadratic equation, $x = 0.032$; pH = 1.49

CHAPTER 15

15.1B (a) basic (b) acidic (c) neutral
15.2B $K_a = 5.6 \times 10^{-10} = x^2/(0.10 - x) \approx x^2/0.10$; $x \approx 7.5 \times 10^{-6}$; pH ≈ 5.13
15.3B $K_b(HF) = (1 \times 10^{-14})/(3.5 \times 10^{-4}) = 2.9 \times 10^{-11} = x^2/(0.020 - x) \approx x^2/0.020$; $x \approx 7.6 \times 10^{-7}$; pOH ≈ 6.12; pH ≈ 7.88
15.4B $K_a = 3.0 \times 10^{-8} = (x \times 2.0 \times 10^{-4})/0.010$; $x = [H_3O^+] = 1.5 \times 10^{-6}$; pH = 5.8
15.5B A total of 12.0 mL titrant has been added.
moles of OH^- = 6.25×10^{-3} mol
moles of H_3O^+ = $(1.20 \times 10^{-3}$ L) × (0.340 mol·L^{-1}) = 4.08×10^{-3} mol
moles of OH^- remaining after neutralization = $(6.25 \times 10^{-3} - 4.08 \times 10^{-3})$ mol = 2.17×10^{-3} mol
total volume of solution = (25.0 + 12.0) mL = 37.0 mL = 0.037 L
molarity of OH^- = $(2.17 \times 10^{-3}$ mol)/(0.037 L) = 0.0586 mol·L^{-1};
pOH = 1.23, pH = 12.77
15.6B moles of H_3O^+ = $(25.0 \times 10^{-3}$ L) × (0.340 mol·L^{-1}) = 8.50×10^{-3} mol;
from Example 15.2, moles of OH^- = 6.25×10^{-3} mol;
after reaction, moles of HCl remaining = $(8.50 \times 10^{-3}) - (6.25 \times 10^{-3})$ mol = 2.25×10^{-3} mol
total volume = (25.0 + 25.0)mL = 50.0 mL
molarity of H_3O^+ = $(2.25 \times 10^{-3}$ mol)/(0.050 L) = 0.045 mol·L^{-1}; pH = 1.35
15.7B salt present at stoichiometric point is KClO;
moles of KClO = moles of HClO used = (0.010 mol·L^{-1}) × (0.02500 L) = 2.5×10^{-4} mol
volume (titrant) = $(2.5 \times 10^{-4}$ mol)/(0.020 mol·L^{-1})
total volume = (0.02500 + 0.012) L = 0.037 L
molarity of KClO = $(2.5 \times 10^{-4})/(0.037)$ mol·L^{-1} = 6.8×10^{-3} mol·L^{-1}
$K_b(KClO) = (1 \times 10^{-14})/(3.0 \times 10^{-8}) = 3.3 \times 10^{-7} = x^2/[(6.8 \times 10^{-3}) - x] \approx x^2/(6.8 \times 10^{-3})$;
$x = [OH^-] = 4.7 \times 10^{-5}$; pOH = 4.32; pH = 9.68
15.8B total volume of titrant = 15.00 mL; total volume of solution = 40.00 mL
moles of HCO_2^- = (0.150 mol·L^{-1}) × (0.01500 L) = 2.25×10^{-3} mol
moles of HCOOH = $(2.50 \times 10^{-3}) - (2.25 \times 10^{-3})$ mol = 2.5×10^{-4} mol
molarity of HCO_2^- = $(2.25 \times 10^{-3}$ mol)/(0.0400 L) = 0.0562 mol·L^{-1}
molarity of HCOOH = $(2.5 \times 10^{-4}$ mol)/(0.0400 L) = 6.25×10^{-3} mol·L^{-1}
$1.8 \times 10^{-4} \approx x(0.0562)/(6.25 \times 10^{-3})$;
$[H_3O^+] = x \approx 2.00 \times 10^{-5}$; pH ≈ 4.70
15.9B pH = 3.37 + log(0.20/0.15) = 3.49
15.10B moles of $CH_3CO_2^-$ = (0.040 mol·L^{-1}) × (0.300 L) + 0.0200 mol = 0.032 mol
moles of CH_3COOH = (0.080 mol·L^{-1}) × (0.300 L) − 0.0200 mol = 0.0040 mol
molarity of $CH_3CO_2^-$ = (0.032 mol)/(0.300 L) = 0.11 mol·L^{-1}
molarity of CH_3COOH = (0.0040 mol)/(0.300 L) = 0.013 mol·L^{-1}
pH = 4.75 + log(0.11/0.013) = 5.68, an increase of 1.23
15.11B $(CH_3)_3N$ and $(CH_3)_3NH^+$
15.12B $\log([C_6H_5CO_2^-]/[C_6H_5COOH]) = pH - pK_a = 3.50 - 4.19 = -0.69$
ratio = antilog(−0.69) = 0.204, or 1 : 4.90
15.13B $K_{sp} = [Ca^{2+}][IO_3^-]^2 = (2S)^2S = 4S^3 = 4(3.8 \times 10^{-3})^3 = 2.2 \times 10^{-7}$
15.14B $K_{sp} = [Mg^{2+}][NH_4^+][PO_4^{3-}] = S^3$; $S = (2.5 \times 10^{-13})^{1/3} = 6.3 \times 10^{-5}$ mol·L^{-1}

15.15B $K_{sp} = [Ag^+][Br^-]$; $[Ag^+] = K_{sp}/[Br^-] = (7.7 \times 10^{-13})/0.20 = 3.8 \times 10^{-12}\ mol{\cdot}L^{-1}$
15.16B molarity of $Ba^{2+} = (1.0 \times 10^{-3}\ mol{\cdot}L^{-1}) \times (0.100\ L)/(0.300\ L) = 3.3 \times 10^{-4}\ mol{\cdot}L^{-1}$;
molarity of $F^- = (1.0 \times 10^{-3}\ mol{\cdot}L^{-1})(0.200\ L)/(0.300\ L) = 6.7 \times 10^{-4}\ mol{\cdot}L^{-1}$
$Q_{sp} = [Ba^{2+}][F^-]^2 = (3.3 \times 10^{-4})(6.7 \times 10^{-4})^2 = 1.5 \times 10^{-10}$
Yes, a precipitate forms ($Q_{sp} > K_{sp}$).
15.17B $K = K_{sp} \times K_f = (1.3 \times 10^{-36})(1.2 \times 10^{13}) = 1.56 \times 10^{-23}$

$$K = \frac{[Cu(NH_3)^{2+}][S^{2-}]}{[NH_3]^4} = \frac{x \times x}{(1.2)^4} = 1.56 \times 10^{-23}$$

solubility $= x = 5.7 \times 10^{-12}$

CHAPTER 16

16.1B (a) closed (b) open
16.2B (a) -750 kJ (b) -50 kJ (c) -800 kJ
16.3B (a) 50 kJ (b) 70 kJ
16.4B (a) $-7.41\ J{\cdot}K^{-1}{\cdot}mol^{-1}$, gray tin is more ordered (b) $3.3\ J{\cdot}K^{-1}{\cdot}mol^{-1}$, diamonds are more ordered
16.5B $\{229.60 - (219.56 + 130.68)\}\ J{\cdot}K^{-1}{\cdot}mol^{-1} = -120.64\ J{\cdot}K^{-1}{\cdot}mol^{-1}$
16.6B $-477\ J{\cdot}K^{-1}$
16.7B It becomes spontaneous as temperature increases.
16.8B $(35.3 \times 10^3\ J{\cdot}mol^{-1})/(104.7\ J{\cdot}K^{-1}{\cdot}mol^{-1}) = 337$ K (experimental: 337.2 K)
16.9B $3\,C(s) + 3\,H_2(g) \rightarrow C_3H_6(g)$
$\Delta H_r^\circ = +53.30\ kJ{\cdot}mol^{-1}$
$\Delta S_r^\circ = \{237.4 - [3 \times (130.68) + 3 \times (5.740)]\}\ J{\cdot}K^{-1}{\cdot}mol^{-1} = -171.86\ J{\cdot}K^{-1}{\cdot}mol^{-1}$
$\Delta G_r^\circ = \{53.30 - (298)(-171.86 \times 10^{-3})\}\ kJ{\cdot}mol^{-1} = +104.5\ kJ{\cdot}mol^{-1}$
16.10B no; from Appendix 2A, for methylamine, $\Delta G_f^\circ = +32.2\ kJ{\cdot}mol^{-1}$, which is positive
16.11B $\{(-910) - \{6 \times (-394.36) + (-237.13)\}\}\ kJ{\cdot}mol^{-1} = 2879\ kJ{\cdot}mol^{-1}$
16.12B $+8.96\ kJ{\cdot}mol^{-1}$, toward reactants
16.13B $\Delta G_r^\circ = \{2 \times (51.31) - 2 \times (86.55)\}\ kJ{\cdot}mol^{-1} = -70.48\ kJ{\cdot}mol^{-1}$
$\ln K_p = -(-70.48\ kJ{\cdot}mol^{-1})/(2.4790\ kJ{\cdot}mol^{-1}) = 28.43$
$K_p = 2.2 \times 10^{12}$
16.14B $MgCO_3(s) \rightarrow MgO(s) + CO_2(g)$
$\Delta H_r^\circ = \{(-601.70 - 393.51) - (-1095.8)\}\ kJ{\cdot}mol^{-1} = +100.6\ kJ{\cdot}mol^{-1}$
$\Delta S_r^\circ = \{26.94 + 213.74 - (+65.7)\}\ J{\cdot}K^{-1}{\cdot}mol^{-1} = +175.0\ J{\cdot}K^{-1}{\cdot}mol^{-1}$
$T = (100{,}600\ J{\cdot}mol^{-1})/(175.0\ J{\cdot}K^{-1}{\cdot}mol^{-1}) = 574.8$ K

CHAPTER 17

17.1B (a) $Al(s) \rightarrow Al^{3+}(aq) + 3\,e^-$
(b) $S_8(s) + 16\,e^- \rightarrow 8\,S^{2-}(aq)$
17.2B $5\,H_2SO_3(aq) + 2\,MnO_4^-(aq) + H^+(aq) \rightarrow 5\,HSO_4^-(aq) + 2\,Mn^{2+}(aq) + 3\,H_2O(l)$
17.3B $ClO^-(aq) + Pb(OH)_3^-(aq) \rightarrow Cl^-(aq) + PbO_2(s) + OH^-(aq) + H_2O(l)$
17.4B $BrO^-(aq),Br^-(aq),OH^-(aq)|C(gr)$
17.5B $Mn(s)|Mn^{2+}(aq)||Cu^{2+}(aq), Cu^+(aq)|Pt$
17.6B 2, -251 kJ
17.7B -0.13 V
17.8B yes
17.9B Ag^+, $+0.46$ V; $2\,Ag^+(aq) + Cu(s) \rightarrow 2\,Ag(s) + Cu^{2+}(aq)$
17.10B The half-reactions are
$Cd^{2+} + 2\,e^- \rightarrow Cd$, $E^\circ = -0.40$ V
$Cd(OH)_2 + 2\,e^- \rightarrow Cd + 2\,OH^-$, $E^\circ = -0.81$ V
$E^\circ = -0.81 - (-0.40\ V) = -0.41$ V
$\ln K_{sp} = 2 \times (-0.41\ V)/(0.025\,693\ V) = -31.92$
$K_{sp} = 1.4 \times 10^{-14}$
17.11B $Q = [Ag^+]_{anode}/[Ag^+]_{cathode} = 0.0010/0.010 = 0.10$
$E^\circ = 0$
$E = 0 - (0.025\,693\ V)\ln(0.10) = 0.059$ V
17.12B (b) zinc
17.13B cathode, $H_2(g)$; anode, $Br_2(l)$ and $O_2(g)$ (Although H_2O has the more favorable potential for oxidation than Br^-, its high overvoltage allows bromine to form.)

17.14B Time

$$= \frac{12.00\ g\ Cr}{52.00\ g{\cdot}mol^{-1}} \times \frac{6\ mol\ e^-}{1\ mol\ Cr} \times \frac{9.65 \times 10^4\ C{\cdot}(mol\ e^-)^{-1}}{6.20\ C{\cdot}s^{-1}}$$

$$= 2.16 \times 10^4\ s \times \frac{1\ h}{3600\ s} = 5.99\ h$$

CHAPTER 18

18.1B $-0.575\ (mol\ N_2){\cdot}L^{-1}{\cdot}h^{-1}$
18.2B second order
18.3B first order in CH_3Br and OH^-, second order overall
18.4B order in CO = 1

$$\text{order in } Cl_2 = \log\left(\frac{0.682}{0.241}\right)/\log\left(\frac{0.40}{0.20}\right) = \log(2.83)/\log(2.0) = 1.5$$

$$k = \frac{0.121}{(0.12)(0.20)^{3/2}}\ L^{3/2}{\cdot}mol^{-3/2}{\cdot}s^{-1} = 11\ L^{3/2}{\cdot}mol^{-3/2}{\cdot}s^{-1}$$

18.5B $[C_3H_6] = 0.100 \times e^{-(6.7 \times 10^{-4}\ s^{-1}) \times (200\ s)} = 0.087$
18.6B first order, because the graph of $\ln[N_2O_5]$ as a function of t is linear; $k = -(\text{slope}) = 2.5 \times 10^{-7}\ s^{-1}$
18.7B (a) $(\ln 2)/(5.5 \times 10^{-4}\ s^{-1}) = 1.3 \times 10^3$ s (b) The concentration reaches 1/16 of its initial value after four half-lives: $4 \times (1.3 \times 10^3) = 5.0 \times 10^3$ s.
18.8B $$\frac{E_a}{R} = \frac{\ln k_1 - \ln k_2}{\left(\frac{1}{T_1}\right) - \left(\frac{1}{T_2}\right)}$$

$$= \frac{1.10 - 1.47}{(3.44 \times 10^{-3}) - (3.30 \times 10^{-3})\text{K}^{-1}}$$
$$= 2.64 \times 10^3 \text{ K}$$
$$E_a = (8.314 \times 10^{-3} \text{ kJ·mol}^{-1}\text{·K}^{-1})(2.64 \times 10^3 \text{ K}) = 21.9 \text{ kJ·mol}^{-1}$$

18.9B $\ln\{k'/(6.7 \times 10^{-4} \text{ s}^{-1})\}$
$$= \frac{272 \text{ kJ·mol}^{-1}}{8.314 \times 10^{-3} \text{ kJ·K}^{-1}\text{·mol}^{-1}}\left(\frac{1}{773 \text{ K}} - \frac{1}{573 \text{ K}}\right)$$
$$= 14.9$$
$$\frac{k'}{6.7 \times 10^{-4} \text{ s}^{-1}} = 3.84 \times 10^{-7}$$
$$k' = 2.6 \times 10^{-10} \text{ s}^{-1}$$

18.10B (a) changes it (b) has no effect

18.11B (a) bimolecular (b) unimolecular

18.12B rate $= (k_1k_2/k_1')[\text{NO}]^2[\text{O}_2]$ or

rate $= k'[\text{NO}]^2[\text{O}_2]$; $k' = \dfrac{k_1k_2}{k_1'}$

CHAPTER 19

19.1B high ionization energy, nonmetal, diatomic molecule in the elemental state

19.2B The stability of MgO is due to the small ionic radii of the Mg^{2+} and O^{2-} ions, resulting in a strong electrostatic interaction.

19.3B Aluminum is resistant to corrosion because its surface is passivated by a stable oxide film.

19.4B Silicon atoms have empty *d*-orbitals in their valence shells, so they can act as Lewis acids and accept pairs of electrons from bases. In addition, Si atoms are larger than C atoms, which makes it easier for Lewis bases to attack them.

CHAPTER 20

20.1B For a phosphorus atom, two 3*d*-orbitals can hybridize with one 3*s*- and three 3*p*-orbitals to form five sp^3d^2 hybrid orbitals. In nitrogen's valence shell, 2*d*-orbitals do not exist (see Ch. 7). The small size of the nitrogen atom also precludes the placement of five large chlorine atoms about the central atom.

20.2B $2\,H_2O(l) \rightarrow O_2(g) + 4\,H^+(aq) + 4\,e^-$;
$2\,F_2(g) + 4\,e^- \rightarrow 4\,F^-(aq)$
Overall: $2\,H_2O(l) + 2\,F_2(g) \rightarrow O_2(g) + 4\,H^+(aq) + 4\,F^-(aq)$
$E° = 1.64$ V and $\Delta G_r° = -633$ kJ·mol^{-1}

20.3B Strong hydrogen bonding between H_2SO_4 and H_2O leads to volume contraction.

20.4B $I^-(aq)$

20.5B XeOF$_4$ Lewis structure: Xe (one lone pair) double-bonded to :O: and single-bonded to four :F: atoms (each with three lone pairs); square pyramid

CHAPTER 21

21.1B Density increases from left to right across a row, then drops again for the last few elements. Density increases down a group.

21.2B The carbon content of pig iron is too high; it makes pig iron relatively hard and brittle.

21.3B (a) linkage (b) hydrate

21.4B (b, d) not chiral (a, c) chiral; no pairs of enantiomers

21.5B
$$\frac{(6.626 \times 10^{-34} \text{ J·s}) \times (2.998 \times 10^8 \text{ m·s}^{-1}) \times (6.022 \times 10^{23} \text{ mol}^{-1})}{305 \times 10^{-9} \text{ m}}$$
$$= 392 \times 10^3 \text{ J·mol}^{-1} = 392 \text{ kJ·mol}^{-1}$$

21.6B (a); CN^- is a stronger-field ligand than NH_3

21.7B (a) t^6e^1 (b) t^5e^2

21.8B Both are paramagnetic and have two unpaired electrons.

CHAPTER 22

22.1B radium-228

22.2B $^{228}_{89}\text{Ac}$

22.3B (a) lithium-7 (b) boron-11

22.4B c, d

22.5B (a) $^{11}_{5}\text{B}$ (b) $^{246}_{96}\text{Cm}$

22.6B $5.0\ \mu\text{g} \times e^{-(3.3 \times 10^{-7} \times 1.0 \times 10^6)} = 3.6\ \mu\text{g}$

22.7B $\dfrac{5.73 \times 10^3 \text{ y}}{\ln 2} \times \ln\left[\dfrac{14{,}000}{18{,}400}\right] = 2.3 \times 10^3 \text{ y}$

22.8B $\Delta m = [235.0439 - (92 \times 1.0078 + 143 \times 1.0087)] \times (1.6605 \times 10^{-27})$ kg $= -3.1845 \times 10^{-27}$ kg
$E_{bind} = -1.73 \times 10^{14}$ J·mol^{-1} $= -1.73 \times 10^{11}$ kJ·mol^{-1}

22.9B For one atom: mass of products $= 137.91 + 85.91 + 12 \times 1.0087$ u $= 235.92$ u
mass of reactants $= (235.04 + 1.0087)$ u $= 236.05$ u
$\Delta m = (235.92 - 236.05) \times (1.6605 \times 10^{-27})$ kg $= -2.16 \times 10^{-28}$ kg
$\Delta E = (-2.16 \times 10^{-28} \text{ kg}) \times (3.00 \times 10^8 \text{ m·s}^{-1})^2 = -1.943 \times 10^{-11}$ J
For 1.0 g: $\dfrac{1.0 \times 10^{-3}}{(235.04) \times (1.6605 \times 10^{-27})}$ atoms $=$ 2.56×10^{21} atoms
$\Delta E = (2.56 \times 10^{21}) \times (-1.943 \times 10^{-11})$ J $= 5.0 \times 10^{10}$ J

ODD-NUMBERED CASE STUDY QUESTIONS

CHAPTER 1

1.1 This statement is not justified. A proper statement might be, "No conclusive evidence found for life on Mars."
1.3 If Mars were to support a human population, farming the Martian soil would seem to be a necessity. Thus experiments would have to be performed to see whether the growth of plants needed by humans could be achieved with Martian soil. These farming experiments could probably be performed on a small scale inside the lander of a spacecraft. Because water and oxygen are scarce on the surface, sources would have to be found, perhaps underground. Metal ores would also be sought for use as building materials.

CHAPTER 2

2.1 (a) 1.007 976 $g \cdot mol^{-1}$ (b) to five significant figures, 1.0080 $g \cdot mol^{-1}$
2.3 The mass standard needs to be ultrapure because the standard serves as a reference to determine the masses of other elements. If the mass of the standard is not accurately measured, then data for other materials will not be accurate. The standard material should be a solid for ease in handling.

CHAPTER 3

3.1 (a) aeration, settling, filtration, adsorption, ion exchange, reverse osmosis (b) oxidation, precipitation, coagulation, acid/base treatment
3.3 complete equations:
$Pb^{2+}(aq) + Ca(OH)_2(aq) \rightarrow Pb(OH)_2(s) + Ca^{2+}(aq)$ and $2\,Fe^{3+}(aq) + 3\,Ca(OH)_2(aq) \rightarrow 2\,Fe(OH)_3(s) + 3\,Ca^{2+}(aq)$;
net ionic equations:
$Pb^{2+}(aq) + 2\,OH^-(aq) \rightarrow Pb(OH)_2(s)$ and
$Fe^{3+}(aq) + 3\,OH^-(aq) \rightarrow Fe(OH)_3(s)$

CHAPTER 4

4.1 theoretical yield of CO_2 from combustion of methanol: 1.09 kg CO_2; theoretical yield of CO_2 from combustion of octane: 6.66 kg CO_2. Octane contributes more CO_2 per liter to the atmosphere. An important factor that must be taken into account is the amount of energy released by the combustion. The method of determining this is taken up in Chapter 6. We will find that octane releases more energy per liter (and per kilogram) than methanol. Another important factor is engine performance. Another factor is cost.
4.3 8.9×10^{15} g, or 8.9×10^{9} metric tons

CHAPTER 5

5.1 $O_3 < NO_2 < O_2 < NO$
5.3 The density of a gas is proportional to the molar mass of the gas. Therefore, under the same conditions of temperature and pressure, the most dense gas is the one with the greatest molar mass. In this case, O_2 would be the most dense gas.

CHAPTER 6

6.1 32 min
6.3 Fats are only very slightly oxidized hydrocarbons; thus most of their fuel value is still intact. Carbohydrates, on the other hand, are already oxidized to a considerable extent. Typically they contain the same number of carbon and oxygen atoms; thus most of the fuel value of the carbon and hydrogen atoms has already been spent.

CHAPTER 7

7.1 0.21 M

CHAPTER 8

8.1 (a) $\dot{N}=O$ (with lone pairs: N has one lone pair and an unpaired electron, O has two lone pairs) (b) $:\!\ddot{O}\!-H$ (O with an unpaired electron)

(c) $H-C(H)(H)-C(=O)-O$ (CH₃–C with =O: and –O: bearing lone pairs) (d) O–N=O with lone pairs (N at apex, bonded to two O atoms)

8.3 $CH_3CH_3 + \cdot\ddot{O}-H \longrightarrow \cdot CH_2CH_3 + H_2O$

CHAPTER 9

9.1 (a) The carbon atoms in C_{60} are the A atoms of a distorted planar AX_3 VSEPR structure; therefore we expect approximately sp^2 hybridization. (b) Because carbon nanotubes have basically the same bonding structure as graphite, the hybridization is sp^2.
9.3 Those species that are polar will be attracted to the polar ends of these self-assembling layers. Therefore, (a) nitric oxide, (c) ethanol, and (d) glycine will be attracted.

CHAPTER 10

10.1 (a) anisotropic (b) isotropic (c) anisotropic (d) anisotropic (e) anisotropic
10.3 Polar groups aid in the alignment of the molecules, the positive end of one molecule adhering to the negative end of another, and vice versa. That is, the dipoles align antiparallel to each other. Curiously, the presence of an —OH group in the molecule, which leads to hydrogen

bonding, as in cholesterol itself, can sometimes hinder liquid crystal formation. Thus cholesterol does not form a cholesteric liquid crystal.

CHAPTER 11

11.1 (a) The Lewis structure of ethyne is H—C≡C—H. See Fig. 9.34 for its valence bond structure. The top portion of the figure shows the σ-bond framework and the bottom portion the overlapping of atomic *p*-orbitals that result in the two π-bonds. The carbon atoms are *sp* hybrids, as shown in the top part of the figure.

(b)
```
—C=C—C=C—C=C—  ⟷
 |  |  |  |  |  |
 H  H  H  H  H  H

=C—C=C—C=C—C=
 |  |  |  |  |  |
 H  H  H  H  H  H
```

(*Note*: The second structure has the same number of C—H units, but the single and double bonds have exchanged places.)

(c)

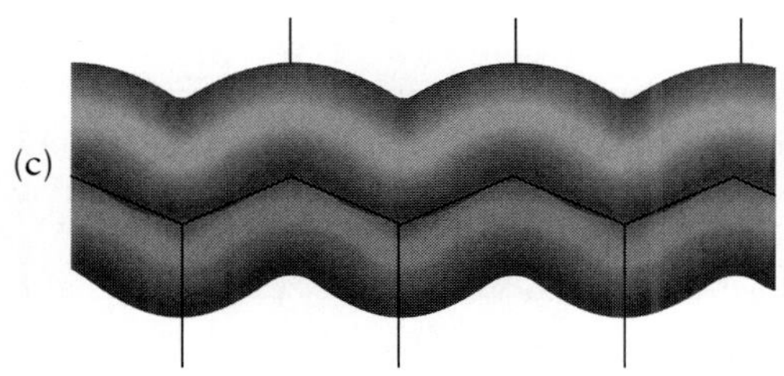

Because of resonance, every carbon-carbon bond in polyacetylene has considerable double-bond character. Free rotation cannot occur about double bonds; and even though these bonds are only partially double, we could expect polyacetylenes to be fairly stiff. All C atoms are sp^2 hybridized.

11.3
```
      H      H      H
      |      |      |
      C      C      C
  \ //  \  //  \  //  \
  C      C      C
  |      |      |
 CH3    CH3    CH3
```

CHAPTER 12

12.1 (a) both a sol and an emulsion (b) foam (c) sol
12.3 (a) hydrogen bonds (b) hydrogen bonds
(c) hydrogen bonds, dipole-dipole forces, and London forces

CHAPTER 13

13.1 (a) Step 1: $K_{c1} = [BrO_2]^2/[H^+]^2[BrO_2^-][BrO_3^-]$
Step 2: $K_{c2} = [Ce^{4+}]^2[BrO_2^-]^2/[Ce^{3+}]^2[BrO_2]^2$
Step 3: $K_{c3} = [BrO_3^-][BrO^-]/[BrO_2^-]^2$
(b) One possibility is $2\,H^+ + BrO_2^- + 2\,Ce^{3+} \rightleftharpoons BrO^- + 2\,Ce^{4+} + H_2O$ and
$K_c = [BrO^-][Ce^{4+}]^2/[H^+]^2[BrO_2^-][Ce^{3+}]^2$
(c) $K_c = K_{c1}K_{c2}K_{c3}$
13.3 Step 1: BrO_2^- is oxidized, BrO_3^- is reduced; Step 2: Ce^{3+} is oxidized, BrO_2 is reduced; Step 3: BrO_2^- is oxidized and reduced.

CHAPTER 14

14.1 $CO_2(g) + H_2O(l) \rightleftharpoons H_2CO_3(aq)$ and $H_2CO_3(aq) + H_2O(l) \rightleftharpoons H_3O^+(aq) + HCO_3^-(aq)$. We ignore the second ionization of H_2CO_3, because $K_{a2} \ll K_{a1}$. Thus $K_{a1} = [H_3O^+][HCO_3^-]/[H_2CO_3]$. Because $[H_2CO_3]_0$ is small, we need to use the quadratic equation. The $[H_3O^+] = 1.5 \times 10^{-6}\ mol{\cdot}L^{-1}$ and pH = 5.82, which is close to 5.7.
14.3 16 kg Ca^{2+}

CHAPTER 15

15.1 The equilibria involved are $CO_2(g) + H_2O(l) \rightleftharpoons H_2CO_3(aq)$ and $H_2CO_3(aq) + H_2O(l) \rightleftharpoons H_3O^+(aq) + HCO_3^-(aq)$. Hyperventilation decreases the amount of $CO_2(g)$, thus driving the first equilibrium to the left, which in turn drives the second equilibrium to the left. Therefore, both $[HCO_3^-]$ and $[H_2CO_3]$ are diminished, but it is not immediately apparent how their ratio is affected. We can analyze the effect on the ratio through the use of the Henderson-Hasselbalch equation, $pH = -\log[H_3O^+] = pK_{a1} + \log[HCO_3^-]/[H_2CO_3]$. With the loss of CO_2 during hyperventilation, there is a decrease in $[H_3O^+]$, which corresponds to an increased pH and an increased $[HCO_3^-]/[H_2CO_3]$ ratio.
15.3 5.1–7.1

CHAPTER 16

16.1 $\Delta H_r^\circ = -2808\ kJ{\cdot}mol^{-1}$; $\Delta S_r^\circ = +0.259\ kJ{\cdot}K^{-1}{\cdot}mol^{-1}$; $\Delta G_r^\circ = -2885\ kJ{\cdot}mol^{-1}$; because ΔS_r° is positive, the reaction becomes more spontaneous as the temperature is raised.
16.3 94.6 mol ATP

CHAPTER 17

17.1 −1.24 V; therefore the photoelectrochemical cell would have to supply $E > 1.24$ V. This value is slightly larger than the 1.23 V needed for H_2 and O_2 evolution from water, so one might initially guess that H_2 and O_2 would preferentially form. But the overvoltage for H_2 and O_2 formation is about 0.6 V; consequently, because about 1.8 V are needed for the evolution of these gases from water, glucose would preferentially form.
17.3 +0.48 V, based on the reaction $H_2S(aq) \rightleftharpoons H_2(g) + S(s)$

CHAPTER 18

18.1 At high concentrations of substrate, the plot of rate against substrate concentration levels off to a constant rate. Therefore, rate $= [S]^a =$ constant, which implies that $a = 0$, and $[S]^0 \propto 1 =$ constant. So the reaction is zero order in substrate at high substrate concentration.

18.3 There are many types of inhibitors. Suppose, for example, that the inhibitor functions by competing with the substrate to form an EI complex; then the slope of the curve will be reduced, and the rate of product formation for a given substrate concentration will also be less.

CHAPTER 19

19.1 Several reactions occur, depending on the silica to alkali ratio. (1) $SiO_2(s) + OH^-(aq) \rightarrow HSiO_3^-(aq)$
(2) $SiO_2(s) + 2\,OH^-(aq) \rightarrow H_2O(l) + SiO_3^{2-}(aq)$
(metasilicate ion)
(3) $SiO_2(s) + 4\,OH^-(aq) \rightarrow 2\,H_2O(l) + SiO_4^{4-}(aq)$
(orthosilicate ion)
(4) $2\,SiO_2(s) + 6\,OH^-(aq) \rightarrow 3\,H_2O(l) + Si_2O_7^{6-}(aq)$
(pyrosilicate ion)

19.3 Glasses form from materials in which inhibited recrystallization occurs. Such materials are likely to be covalently bonded materials involving extensive network structures and materials with large, complex molecules. On the basis of these principles, we predict that (a) tar, (c) molten granite, (e) low-density polyethylene, and (f) a highly branched polymer would probably solidify as a glass. Any substance can form a glass if cooled rapidly, but normally (b) sodium chloride and (d) water form crystalline solids.

CHAPTER 20

20.1 The half-reactions can be written as:
$3\,Al + 3\,H_2O \rightarrow Al^{3+} + Al_2O_3 + 6\,H^+ + 9\,e^-$
$NH_4^+ + H_2O \rightarrow NO + 5\,e^- + 6H^+$
$ClO_4^- + 8\,H^+ + 8\,e^- \rightarrow Cl^- + 4\,H_2O$
Multiply the last two equations by three and add them all together to get the overall equation given in the text.

20.3 (a) $\Delta H_r^\circ = -2675\ kJ \cdot mol^{-1}$ (b) $33.05\ kJ \cdot g^{-1}$ Al

20.5 (a) $118\ kJ \cdot g^{-1}$ (b) The specific enthalpy is largest for H_2 ($118\ kJ \cdot g^{-1}$), then Al ($32.83\ kJ \cdot g^{-1}$), and smallest for CH_3NHNH_2 ($25.70\ kJ \cdot g^{-1}$). (c) The value for the reaction of the liquids should be less in magnitude than that for the gases. This can be seen by comparing the enthalpy of combustion of the gases ($\Delta H = -483.64\ kJ \cdot mol^{-1}$) to that of the liquids ($\Delta H = -475\ kJ \cdot mol^{-1}$). The difference is due to the enthalpy required to vaporize the liquids.

CHAPTER 21

21.1 "Fixing," that is, removing undeveloped AgI, was not a part of the process of producing a daguerrotype. Consequently, on further exposure to light, reduction of some of the remaining Ag^+ ions continues, which causes a darkening or fading of the image. In photochromic sunglasses, the darkening of the glass is a result of the reversible redox reaction $Ag^+(s) + Cu^+(s) \xrightarrow{\text{light}} Ag(s) + Cu^{2+}(s)$. This reaction is driven to the right in the presence of light, and the formation of Ag(s) causes a darkening of the lens. When the light is removed, Ag(s) is oxidized by $Cu^{2+}(s)$ back to $Ag^+(s)$. The images on photographic film are permanent because further reduction of Ag^+ is prevented by "fixing," that is, removing undeveloped AgBr with sodium hyposulfite ($Na_2S_2O_3 \cdot 5H_2O$):
$AgBr(s) + 2\,S_2O_3^{2-}(aq) \rightarrow Ag(S_2O_3)_2^{3-}(aq) + Br^-(aq)$. The water-soluble ions are then washed away.

21.3 The standard potential for $Ag^+ + e^- \rightarrow Ag(s)$ is $+0.80$ V. Metal ion/metal couples with similar positive potentials might be likely candidates. Referring to Appendix 2B, we see that mercury, gold, platinum, and copper seem to be theoretically possible; but there would be no price advantage with gold or platinum. These elements also share a common ability to form complex ions, which would aid in the fixing of the film. The solubility of the metal halides of mercury, gold, platinum, and copper should also be considered; more soluble salts would be washed away in the developing process.

CHAPTER 22

22.1 Positron emission results in a lowering of positive charge in the nucleus, that is, Z is decreased, and A/Z is increased. Consequently, isotopes that are below the band of stability are likely candidates for positron emission, and such emissions will move them in the direction of the band. (a) not suitable, because the A/Z ratio is above the band; (b) suitable, because the A/Z ratio is below the band: ${}^{13}_{7}N \rightarrow {}^{0}_{1}e + {}^{13}_{6}C$ (c) suitable, because the A/Z ratio is below the band: ${}^{11}_{6}C \rightarrow {}^{0}_{1}e + {}^{11}_{5}B$ (d) not suitable, because the A/Z ratio is above the band. (e) suitable, because the A/Z ratio is below the band: ${}^{15}_{8}O \rightarrow {}^{0}_{1}e + {}^{15}_{7}N$

22.3 ${}^{98}_{42}Mo + {}^{1}_{0}n \rightarrow {}^{99}_{42}Mo \rightarrow {}^{99}_{43}Tc + {}^{0}_{-1}e$

ODD-NUMBERED EXERCISES

CHAPTER 1

1.1 (a) Cathode rays are electrons, thus their charge and mass are the same as those of an electron: charge, -1; mass, 9.109×10^{-28} g (b) Robert Millikan
1.3 A law summarizes observations, a theory attempts to explain the observations.
1.5 (a) As, 33 (b) S, 16 (c) Pd, 46 (d) Au, 79
1.7 (a) 6p, 7n, 6e (b) 17p, 20n, 17e (c) 17p, 18n, 17e (d) 92p, 143n, 92e
1.9 (a) ^{111}Cd (b) ^{82}Kr (c) ^{11}B
1.11 (a) Each isotope has the same number of protons and electrons, so they are chemically almost identical. (b) Their atomic masses are different, so the properties of these atoms that depend on mass will be different.
1.13 (a) lithium, 1 or I, metal (b) gallium, 13 or III, metal (c) xenon, 18 or VIII, nonmetal (d) potassium, 1 or I, metal
1.15 (a) Cl, nonmetal (b) Co, metal (c) As, metalloid
1.17 tin, Sn; lead, Pb
1.19 (a) I, nonmetal (b) Cr, metal (c) Hg, metal (d) Al, metal
1.21 lithium, Li, 3; sodium, Na, 11; potassium, K, 19; rubidium, Rb, 37; cesium, Cs, 55; francium, Fr, 87 (radioactive). All react with water as follows: (M = alkali metal) $2\,M(s) + 2\,H_2O(l) \rightarrow 2\,MOH(aq) + H_2(g)$.
1.23 (a) An atom of an element is the smallest unit of an element that has the chemical properties of that element. (b) A molecule is the smallest unit of elements united in chemical combination that retains the chemical properties of that combination. It is a definite and distinct electrically neutral group of bonded atoms.
1.25 (a) anion, S^{2-} (b) cation, K^+ (c) cation, Sr^{2+} (d) anion, Cl^-
1.27 (a) 1p, 1n, 0e (b) 4p, 5n, 2e (c) 35p, 45n, 36e (d) 16p, 16n, 18e
1.29 (a) $^{19}F^-$ (b) $^{24}Mg^{2+}$ (c) $^{128}Te^{2-}$ (d) $^{86}Rb^+$
1.31 (a) element (b) element (c) compound
1.33 (a) mixture (b) element
1.35 Compounds always contain more than one element, and the majority of mixtures also do. But a compound has a fixed composition, whereas a mixture may have any composition. For example, a mixture of hydrogen and oxygen gases may contain any arbitrary ratio of hydrogen and oxygen atoms, but the compound water, formed from these two elements, always has a fixed ratio of two atoms of hydrogen for each atom of oxygen.
1.37 (a) physical (b) physical (c) chemical
1.39 (a) physical (b) physical (c) chemical (corrosion)
1.41 (a) physical (b) physical (c) chemical (corrosion) (d) physical
1.43 temperature, evaporation, humidity
1.45 (a) filtration, solubility differences (b) chromatography, differences in adsorption (c) distillation, boiling point differences
1.47 (a) homogeneous, distillation (b) heterogeneous, filtration (salt dissolves in water, chalk can be filtered) (c) homogeneous, evaporation or recrystallization
1.49 (a) chloride ion (b) oxide ion (c) carbide ion (d) phosphide ion
1.51 (a) phosphate ion (b) sulfate ion (c) nitride ion (d) sulfite ion (e) phosphite ion (f) iodide ion
1.53 (a) ClO_3^- (b) NO_3^- (c) CO_3^{2-} (d) ClO^- (e) HSO_4^-
1.55 (a) plumbous ion, lead(II) ion (b) ferrous ion, iron(II) ion (c) cobaltic ion, cobalt(III) ion (d) cuprous ion, copper(I) ion
1.57 (a) Cu^{2+} (b) ClO_2^- (c) P^{3-} (d) H^-
1.59 (a) MgO (b) $Ca_3(PO_4)_2$ (c) $Al_2(SO_4)_3$ (d) Ca_3N_2
1.61 (a) potassium phosphate (b) ferrous iodide or iron(II) iodide (c) niobium(V) oxide (d) cupric sulfate or copper(II) sulfate
1.63 (a) copper(II) nitrate hexahydrate (b) neodymium(III) chloride hexahydrate (c) nickel(II) fluoride tetrahydrate
1.65 (a) $Na_2CO_3{\cdot}H_2O$ (b) $In(NO_3)_3{\cdot}5H_2O$ (c) $Cu(ClO_4)_2{\cdot}6H_2O$
1.67 (a) SeO_3 (b) CCl_4 (c) CS_2 (d) SF_6 (e) As_2S_3 (f) PCl_5 (g) N_2O (h) ClF_3
1.69 (a) sulfur tetrafluoride (b) dinitrogen pentoxide (c) nitrogen triiodide (d) xenon tetrafluoride (e) arsenic tribromide (f) chlorine dioxide (g) diphosphorous pentoxide
1.71 (a) hydrochloric acid (b) sulfuric acid (c) nitric acid (d) acetic acid (e) sulfurous acid (f) phosphoric acid
1.73 (a) Na_2O (b) K_2SO_4 (c) AgF (d) $Zn(NO_3)_2$ (e) Al_2S_3
1.75 Chlorine exists in the form of two isotopes, ^{35}Cl and ^{37}Cl. Both of these isotopes have an atomic number of 17, but the mass numbers are 35 and 37, respectively.
1.77 (a) element (b) homogeneous mixture (c) compound (d) element (e) homogeneous mixture (f) compound (g) element (h) heterogeneous mixture
1.79 (a) physical (evaporation) (b) chemical (bonds have broken and reformed) (c) chemical (bonds have broken and reformed)
1.81 Group 2 or II: Be, 4; Mg, 12; Ca, 20; Sr, 38; Ba, 56; Ra, 88

1.83 (a) iron(III) chloride hexahydrate or ferric chloride hexahydrate (b) cobalt(II) nitrate hexahydrate or cobaltous nitrate hexahydrate (c) copper(I) chloride or cuprous chloride (d) bromine chloride (e) manganese(IV) oxide (f) mercury(II) nitrate or mercuric nitrate (g) nickel(II) nitrate (h) dinitrogen tetroxide (i) vanadium(V) oxide

1.85 (a) $AlPO_4$ (b) $Ba(NO_3)_2 \cdot 2\,H_2O$ (c) SiS_2 (d) Na_3P (e) $HClO_4(aq)$ (f) CuO (g) HI(aq) (h) Ag_2SO_4

1.87 Assume the oxygen is ^{16}O and the hydrogen is ^{1}H. (a) One molecule of water has 10p, 10e, 8n. (b) 1.673×10^{-23} g protons, 9.109×10^{-27} g electrons, and 1.340×10^{-23} g neutrons (c) 0.4446

1.89 (a) compare to H_2SO_4, thus telluric acid (b) compare to Na_3PO_4, thus sodium arsenate (c) compare to $CaSO_3$, thus calcium selenite (d) compare to $Ba_3(PO_4)_2$, thus barium antimonate (e) compare to H_3PO_4, thus arsenic acid (f) compare to $Co_2(SO_4)_3$, thus cobalt(III) tellurate

1.91 Initial observational data might include the frequency and severity of the headaches, and environmental conditions (food eaten, noise level, odors). Some hypotheses are (1) the food is the cause of the headaches, (2) the room is the cause of the headaches, (3) the homework is the cause of the headaches. Three experiments are (1) eliminate the food (requires more than one experiment if only one menu item is eliminated each time); (2) eliminate the room (do homework in another room); (3) eliminate the homework. Other experimental variations are possible, for example, two of the three possible causes could be eliminated at a time, then if the headaches persist the remaining possible cause might be the culprit. This method would also require three experiments. No matter which approach is used, the data to be collected are (1) food eaten, (2) room used, (3) homework begun (yes or no), (4) headache (yes or no).

CHAPTER 2

2.1 (a) 2.0337×10^{14} (b) $1.169\,811 \times 10^{6}$ (c) 6×10^{-6} g (d) 1×10^{-7} m

2.3 (a) 4.3×10^{-1} (b) 1.492×10^{1} (c) 5.1×10^{-9} (d) 2.37×10^{14}

2.5 (a) 0.250 kg (b) 2.54 cm (c) 0.250 ms (d) 0.149 dm (e) 2.48×10^{-2} g (f) 2.835×10^{-2} kg

2.7 (a) 1×10^{-6} m (b) 5.50×10^{-4} mm (c) 1.0×10^{2} mg (d) 1.05×10^{-4} μm

2.9 $2.4\ g \cdot cm^{-3}$

2.11 4.5×10^{3} g

2.13 $1.71 \times 10^{-2}\ cm^3$

2.15 (a) $2.5 \times 10^{-2}\ m^3$ (b) $2.5 \times 10^{3}\ mg \cdot dL^{-1}$ (c) $1.54 \times 10^{3}\ pm \cdot \mu s^{-1}$ (d) $2.66 \times 10^{-6}\ \mu g \cdot \mu m^{-3}$ (e) $3.2 \times 10^{-4}\ mL \cdot s^{-2}$

2.17 (a) 757 mL (b) $7.21 \times 10^{3}\ kg \cdot m^{-3}$ (c) $2.47 \times 10^{2}\ peso \cdot L^{-1}$ (d) $62\ lb \cdot ft^{-3}$

2.19 (a) 37.0°C (b) −40°F (c) −459.67°F (d) 4 K

2.21 (a) $1 \times 10^{-6}\ m^3$ (b) $3.0 \times 10^{-3}\ cm \cdot \mu s^{-1}$ (c) $2.2 \times 10^{5}\ cm^2$ (d) 25 mL

2.23 $1.0 \times 10^{2}\ mm^2$, $1.0 \times 10^{-5}\ m^3$, 1.00×10^{-1} L, $25.0\ cm^3$

2.25 (a) The precision of the measurement. (b) In addition and subtraction, the number of significant figures in the result is the same as the number of figures to the right of the decimal for the measurement with the least number of figures to the right of the decimal. (c) The number of significant figures in the result of a multiplication or division is the same as the smallest number of significant figures in any of the factors.

2.27 (a) 3 (b) 3 (c) 3 (d) an integer, infinite (e) 2 (f) an exact number, infinite

2.29 (a) 6.60 mL (b) 26.0 mL

2.31 4.4 g (addition, so 1.4 determines significant figures)

2.33 1.64 g (addition, so 0.21 determines significant figures)

2.35 The factor 1.23 has the least number of significant figures, so the result should be reported to 3 significant figures (273.15 + 1.2 has 4 significant figures).

2.37 10^{-2} mol

2.39 (a) 9.5×10^{-15} mol people (b) 3.4×10^{6} years

2.41 $12.01\ g \cdot mol^{-1}$

2.43 $79.91\ g \cdot mol^{-1}$

2.45 (a) 0.0801 mol ^{35}Cl (b) 0.0349 mol Cu (c) 1.84 mol He (d) 1.60×10^{-7} mol Fe

2.47 (a) 2.39×10^{24} atoms (b) 2.45×10^{17} atoms (c) 1.11×10^{12} atoms (d) 2.28×10^{20} atoms

2.49 (a) 59 g (b) 14 g

2.51 (a) $199.90\ g \cdot mol^{-1}$ (b) $114.22\ g \cdot mol^{-1}$ (c) $262.87\ g \cdot mol^{-1}$ (d) $44.01\ g \cdot mol^{-1}$ (e) $16.04\ g \cdot mol^{-1}$

2.53 (a) 0.0650 mol, 3.91×10^{22} molecules CCl_4 (b) 1.29×10^{-5} mol, 7.77×10^{18} molecules HI (c) 1.18×10^{-7} mol, 7.08×10^{16} molecules N_2H_4 (d) 1.46 mol, 8.79×10^{23} molecules sucrose (e) 0.146 mol, 8.77×10^{22} atoms O and 7.28×10^{-2} mol, 4.38×10^{22} molecules O_2

2.55 (a) 0.0140 mol Ag^+ (b) 2.10 mol UO_3 (c) 7.75×10^{-5} mol Cl^- (d) 5.88×10^{-3} mol H_2O

2.57 (a) 4.03×10^{23} formula units $AgNO_3$ (b) 1.06×10^{5} μg Rb_2SO_4 (c) 5.90×10^{25} formula units of $NaHCO_2$

2.59 (a) 3.5×10^{-9} mol testosterone (b) 79.1% C, 9.8% H, 11.1% O

2.61 (a) 2.99×10^{-23} g H_2O (b) 3.34×10^{22} molecules H_2O

2.63 (a) 0.0186 mol $CuBr_2 \cdot 4H_2O$ (b) 0.0372 mol Br^- (c) 4.48×10^{22} molecules H_2O (d) 0.215

2.65 (a) Na_3AlF_6 (b) $KClO_3$ (c) NH_6PO_4 or $NH_4H_2PO_4$
2.67 PCl_5
2.69 $C_6H_6Cl_6$
2.71 $C_8H_{10}N_4O_2$
2.73 (a) 6.268 $mol \cdot L^{-1}$ (b) 0.0241 $mol \cdot L^{-1}$
2.75 0.658 g $AgNO_3$
2.77 (a) 17 mL (b) 29 mL (c) 4.85 mL
2.79 (a) We would determine the mass of 0.010 mol $KMnO_4$ (1.58 g) and then dissolve that mass of $KMnO_4$ in enough water to make 1.00 liter of solution. (b) We dilute the 0.050 M $KMnO_4$ solution by a factor of 5; for example, we could take 10 mL of 0.050 M $KMnO_4$, put it in a 50-mL volumetric flask, and add water up to the calibration mark.
2.81 (a) 4.51 mL (b) 12 mL of NaOH is needed. A 60-mL volumetric flask is not normally available, but a 100 mL flask can be found. So 20 mL of the initial solution can be added to a 100 mL volumetric flask and diluted with water to the mark; 60 mL of this solution can then be used.
2.83 (a) kg (b) pm (c) g (d) μm
2.85 236 mL
2.87 4×10^{-6} in.
2.89 2×10^{-2} $cm \cdot s^{-1}$
2.91 day side: 400 K and 261°F; night side: 90 K and −297°F
2.93 Precision: good, all measured values are within 0.02 $g \cdot cm^{-3}$. Accuracy: not good, relative to the precision (off by about 0.06 $g \cdot cm^{-3}$). A systematic error is probably present because results are consistently low.
2.95 (a) 4.11×10^{23} atoms S (b) 5.14×10^{22} molecules S_8 (c) 21.9 g S
2.97 (a) 4.88×10^{21} Mg atoms (b) 4.88×10^{21} formula units (c) 0.0568 mol H_2O
2.99 (a) CH_2O (b) $C_{29}H_{58}O_{29}$
2.101 3.6×10^{12} kg
2.103 (a) 8.0 g $CuSO_4$ (b) 12 g $CuSO_4 \cdot 5\,H_2O$
2.105 (a) Base unit = 1 Second = 1 S. Then 1 Minute = 1 M = 100 S and 1 Hour = 1 H = 100 M = 10,000 S (10^4 S). There are 10 kS in one H. (b) 1.00 day = 1.00 D = 10.0 H = 100 kS
2.107 (a) $m \cdot s^{-1}$ (b) $m \cdot s^{-2}$ (c) $kg \cdot m \cdot s^{-2}$ (d) $kg \cdot m^{-1} \cdot s^{-2}$
2.109 (a) 1.41×10^{24} $kg \cdot m^{-3}$ (b) 1.0 m
2.111 5.1×10^2 $m^2 \cdot h^{-1}$
2.113 83.8 $g \cdot mol^{-1}$
2.115 1.2×10^{23} molecules C_2H_5OH
2.117 (a) 7.2×10^{-7} $mol \cdot L^{-1}$ (b) 5.3×10^{-5} g $K_2Cr_2O_7$

CHAPTER 3

3.1 (a) $P_2O_5(s) + 3\,H_2O(l) \rightarrow 2\,H_3PO_4(l)$
(b) $Cd(NO_3)_2(aq) + Na_2S(aq) \rightarrow CdS(s) + 2\,NaNO_3(aq)$
(c) $4\,KClO_3(s) \xrightarrow{\Delta} 3\,KClO_4(s) + KCl(s)$
(d) $2\,HCl(aq) + Ca(OH)_2(aq) \rightarrow CaCl_2(aq) + 2\,H_2O(l)$
3.3 (a) $2\,Na(s) + 2\,H_2O(l) \rightarrow H_2(g) + 2\,NaOH(aq)$
(b) $Na_2O(s) + H_2O(l) \rightarrow 2\,NaOH(aq)$
(c) $6\,Li(s) + N_2(g) \rightarrow 2\,Li_3N(s)$
(d) $Ca(s) + 2\,H_2O(l) \rightarrow Ca(OH)_2(aq) + H_2(g)$
3.5 first stage:
$3\,Fe_2O_3(l) + CO(g) \rightarrow 2\,Fe_3O_4(l) + CO_2(g)$
second stage: $Fe_3O_4(l) + 4\,CO(g) \rightarrow 3\,Fe(l) + 4\,CO_2(g)$
3.7 engine: $N_2(g) + O_2(g) \rightarrow 2\,NO(g)$
atmosphere: $2\,NO(g) + O_2(g) \rightarrow 2\,NO_2(g)$
3.9 $4\,HF(aq) + SiO_2(s) \rightarrow SiF_4(g) + 2\,H_2O(l)$
3.11 First identify all soluble ionic compounds. Then write the complete ionic equation by writing all the soluble ionic compounds in ionic form. Then cancel all ions common to both sides of the reaction. What remains is the net ionic equation.
3.13 (a) nonelectrolyte (b) strong electrolyte (c) strong electrolyte
3.15 (a) soluble (b) insoluble (c) soluble (d) soluble
3.17 (a) $K^+(aq)$, $Cl^-(aq)$ (b) $Cu^{2+}(aq)$, $Cl^-(aq)$
(c) $Cs^+(aq)$, $HSO_4^-(aq)$
3.19 (a) $Na^+(aq)$, $I^-(aq)$ (b) insoluble (c) $NH_4^+(aq)$, $PO_4^{3-}(aq)$ (d) $Na^+(aq)$, $HCO_3^-(aq)$ (e) $Fe^{2+}(aq)$, $SO_4^{2-}(aq)$
3.21 (a) $Fe^{3+}(aq) + 3\,OH^-(aq) \rightarrow Fe(OH)_3(s)$
(b) $2\,Ag^+(aq) + CO_3^{2-}(aq) \rightarrow Ag_2CO_3(s)$ (c) No, sodium nitrate and lead acetate are soluble.
3.23 (a)
$FeCl_3(aq) + 3\,NaOH(aq) \rightarrow Fe(OH)_3(s) + 3\,NaCl(aq)$
$Fe^{3+}(aq) + 3\,Cl^-(aq) + 3\,Na^+(aq) + 3\,OH^-(aq) \rightarrow Fe(OH)_3(s) + 3\,Na^+(aq) + 3\,Cl^-(aq)$
$Fe^{3+}(aq) + 3\,OH^-(aq) \rightarrow Fe(OH)_3(s)$
Spectator ions are Na^+ and Cl^-.
(b) $AgNO_3(aq) + KI(aq) \rightarrow AgI(s) + KNO_3(aq)$
$Ag^+(aq) + NO_3^-(aq) + K^+(aq) + I^-(aq) \rightarrow AgI(s) + K^+(aq) + NO_3^-(aq)$
$Ag^+(aq) + I^-(aq) \rightarrow AgI(s)$
Spectator ions are K^+ and NO_3^-.
(c) $Pb(NO_3)_2(aq) + K_2SO_4(aq) \rightarrow PbSO_4(s) + 2\,KNO_3(aq)$
$Pb^{2+}(aq) + 2\,NO_3^-(aq) + 2\,K^+(aq) + SO_4^{2-}(aq) \rightarrow PbSO_4(s) + 2\,K^+(aq) + 2\,NO_3^-(aq)$
$Pb^{2+}(aq) + SO_4^{2-}(aq) \rightarrow PbSO_4(s)$
Spectator ions are K^+ and NO_3^-.
(d) $Na_2Cr_2O_4(aq) + Pb(NO_3)_2(aq) \rightarrow PbCrO_4(s) + 2\,NaNO_3(aq)$
$2\,Na^+(aq) + CrO_4^{2-}(aq) + Pb^{2+}(aq) + 2\,NO_3^-(aq) \rightarrow PbCrO_4(s) + 2\,Na^+(aq) + 2\,NO_3^-(aq)$
$Pb^{2+}(aq) + CrO_4^{2-}(aq) \rightarrow PbCrO_4(s)$
Na^+ and NO_3^- are spectator ions.
(e) $Hg(NO_3)_2(aq) + K_2CrO_4(aq) \rightarrow HgCrO_4(s) + 2\,KNO_3(aq)$

$Hg^{2+}(aq) + 2\,NO_3^-(aq) + 2\,K^+(aq) + CrO_4^{2-}(aq) \rightarrow$
$HgCrO_4(s) + 2\,NO_3^-(aq) + 2\,K^+(aq)$
$Hg^{2+}(aq) + CrO_4^{2-}(aq) \rightarrow HgCrO_4(s)$
Spectator ions are NO_3^- and K^+.
3.25 (a) $(NH_4)_2CrO_4(aq) + BaCl_2(aq) \rightarrow$
$BaCrO_4(s) + 2\,NH_4Cl(aq)$
$2\,NH_4^+(aq) + CrO_4^{2-}(aq) + Ba^{2+}(aq) + 2\,Cl^-(aq) \rightarrow$
$BaCrO_4(s) + 2\,NH_4^+(aq) + 2\,Cl^-(aq)$
$Ba^{2+}(aq) + CrO_4^{2-}(aq) \rightarrow BaCrO_4(s)$
Spectator ions are NH_4^+ and Cl^-.
(b) $CuSO_4(aq) + Na_2S(aq) \rightarrow CuS(s) + Na_2SO_4(aq)$
$Cu^{2+}(aq) + SO_4^{2-}(aq) + 2\,Na^+(aq) + S^{2-}(aq) \rightarrow$
$CuS(s) + 2\,Na^+(aq) + SO_4^{2-}(aq)$
$Cu^{2+}(aq) + S^{2-}(aq) \rightarrow CuS(s)$
Spectator ions are Na^+ and SO_4^{2-}.
(c) $3\,FeCl_2(aq) + 2\,(NH_4)_3PO_4(aq) \rightarrow$
$Fe_3(PO_4)_2(s) + 6\,NH_4Cl(aq)$
$3\,Fe^{2+}(aq) + 6\,Cl^-(aq) + 6\,NH_4^+(aq) + 2\,PO_4^{3-}(aq) \rightarrow$
$Fe_3(PO_4)_2(s) + 6\,NH_4^+(aq) + 6\,Cl^-(aq)$
$3\,Fe^{2+}(aq) + 2\,PO_4^{3-}(aq) \rightarrow Fe_3(PO_4)_2(s)$
Spectator ions are NH_4^+ and Cl^-.
(d) $K_2C_2O_4(aq) + Ca(NO_3)_2(aq) \rightarrow$
$2\,KNO_3(aq) + CaC_2O_4(s)$
$2\,K^+(aq) + C_2O_4^{2-}(aq) + Ca^{2+}(aq) + 2\,NO_3^-(aq) \rightarrow$
$CaC_2O_4(s) + 2\,K^+(aq) + 2\,NO_3^-(aq)$
$Ca^{2+}(aq) + C_2O_4^{2-}(aq) \rightarrow CaC_2O_4(s)$
Spectator ions are K^+ and NO_3^-.
(e) $NiSO_4(aq) + Ba(NO_3)_2(aq) \rightarrow$
$BaSO_4(s) + Ni(NO_3)_2(aq)$
$Ni^{2+}(aq) + SO_4^{2-}(aq) + Ba^{2+}(aq) + 2\,NO_3^-(aq) \rightarrow$
$BaSO_4(s) + Ni^{2+}(aq) + 2\,NO_3^-(aq)$
$Ba^{2+}(aq) + SO_4^{2-}(aq) \rightarrow BaSO_4(s)$
Spectator ions are Ni^{2+} and NO_3^-.
3.27 (a) $Pb^{2+}(aq) + 2\,ClO_4^-(aq) + 2\,Na^+(aq) +$
$2\,Br^-(aq) \rightarrow PbBr_2(s) + 2\,Na^+(aq) + 2\,ClO_4^-(aq)$
$Pb^{2+}(aq) + 2\,Br^-(aq) \rightarrow PbBr_2(s)$
(b) $Ag^+(aq) + NO_3^-(aq) + NH_4^+(aq) + Cl^-(aq) \rightarrow$
$AgCl(s) + NH_4^+(aq) + NO_3^-(aq)$
$Ag^+(aq) + Cl^-(aq) \rightarrow AgCl(s)$
(c) $2\,Na^+(aq) + 2\,OH^-(aq) + Cu^{2+}(aq) + 2\,NO_3^-(aq) \rightarrow$
$Cu(OH)_2(s) + 2\,Na^+ + 2\,NO_3^-(aq)$
$Cu^{2+}(aq) + 2\,OH^-(aq) \rightarrow Cu(OH)_2(s)$
3.29 (a) $AgNO_3$, Na_2CrO_4 (b) $Ca(NO_3)_2$, Na_2CO_3
(c) $Cd(NO_3)_2$, Na_2S
3.31 (a) $Pb^{2+}(aq) + SO_4^{2-}(aq) \rightarrow PbSO_4(s)$
(b) $Cu^{2+}(aq) + S^{2-}(aq) \rightarrow CuS(s)$
(c) $Co^{2+}(aq) + CO_3^{2-}(aq) \rightarrow CoCO_3(s)$
(d) For (a), use $Pb(NO_3)_2$, Na_2SO_4 (spectator ions are Na^+, NO_3^-). For (b), use $Cu(NO_3)_2$, Na_2S (spectator ions are Na^+, NO_3^-). For (c), use $Co(NO_3)_2$, Na_2CO_3 (spectator ions are Na^+, NO_3^-).
3.33 Acids are molecules or ions that contain hydrogen and produce hydronium ions, H_3O^+, in water. Bases are molecules or ions that produce hydroxide ions, OH^-, in water. A base does not need to contain the hydroxide ion.
3.35 (a) base (b) acid (c) base (d) acid (e) base
3.37 (a) $HCl(aq) + NaOH(aq) \rightarrow H_2O(l) + NaCl(aq)$
$H_3O^+(aq) + Cl^-(aq) + Na^+(aq) + OH^-(aq) \rightarrow$
$2\,H_2O(l) + Na^+(aq) + Cl^-(aq)$
$H_3O^+(aq) + OH^-(aq) \rightarrow 2\,H_2O(l)$
(b) $NH_3(aq) + HNO_3(aq) \rightarrow NH_4NO_3(aq)$
$NH_3(aq) + H_3O^+(aq) + NO_3^-(aq) \rightarrow$
$NH_4^+(aq) + NO_3^-(aq) + H_2O(l)$
$NH_3(aq) + H_3O^+(aq) \rightarrow NH_4^+(aq) + H_2O(l)$
(c) $CH_3NH_2(aq) + HI(aq) \rightarrow CH_3NH_3I(aq)$
$CH_3NH_2(aq) + H_3O^+(aq) + I^-(aq) \rightarrow$
$CH_3NH_3^+(aq) + I^-(aq) + H_2O(l)$
$CH_3NH_2(aq) + H_3O^+(aq) \rightarrow CH_3NH_3^+(aq) + H_2O(l)$
3.39 (a) HBr and KOH; $HBr(aq) + KOH(aq) \rightarrow$
$KBr(aq) + H_2O(l)$ and $H_3O^+(aq) + OH^-(aq) \rightarrow 2\,H_2O(l)$
(b) HNO_2 and $Zn(OH)_2$; zinc hydroxide is barely soluble in aqueous solution: $2\,HNO_2(aq) + Zn(OH)_2(s) \rightarrow$
$Zn(NO_2)_2(aq) + 2\,H_2O(l)$ and $2\,HNO_2(aq) +$
$Zn(OH)_2(s) \rightarrow Zn^{2+}(aq) + 2\,H_2O(l) + 2\,NO_2^-(aq)$
(c) HCN and $Ca(OH)_2$; $2\,HCN(aq) + Ca^{2+}(aq) +$
$2\,OH^-(aq) \rightarrow Ca^{2+}(aq) + 2\,CN^-(aq) + 2\,H_2O(l)$ and
$HCN(aq) + OH^-(aq) \rightarrow H_2O(l) + CN^-(aq)$
(d) H_3PO_4 and KOH; $H_3PO_4(aq) + 3\,KOH(aq) \rightarrow$
$K_3PO_4(aq) + 3\,H_2O(l)$ and $H_3PO_4(aq) + 3\,OH^-(aq) \rightarrow$
$PO_4^{3-}(aq) + 3\,H_2O(l)$
3.41 (a) base = $CH_3NH_2(aq)$, acid = $H_3O^+(aq)$
(b) base = $C_2H_5NH_2(aq)$, acid = $HCl(aq)$
(c) base = $CaO(s)$, acid = $HI(aq)$
3.43 (a) basic (b) acidic (c) acidic (d) basic
3.45 Oxidation is electron loss and reduction is electron gain.
3.47 (a) $2\,P(s) + 3\,Br_2(l) \rightarrow 2\,PBr_3(s)$
(b) $2\,Fe^{2+}(aq) + Sn^{4+}(aq) \rightarrow 2\,Fe^{3+}(aq) + Sn^{2+}(aq)$
(c) $8\,H_2(g) + S_8(s) \rightarrow 8\,H_2S(g)$
(d) $2\,NO(g) + O_2(g) \rightarrow 2\,NO_2(g)$
3.49 (a) $Mg(s) + Cu^{2+}(aq) \rightarrow Mg^{2+}(aq) + Cu(s)$
(b) $Fe^{2+}(aq) + Ce^{4+}(aq) \rightarrow Fe^{3+}(aq) + Ce^{3+}(aq)$
(c) $H_2(g) + Cl_2(g) \rightarrow 2\,HCl(g)$
(d) $4\,Fe(s) + 3\,O_2(g) \rightarrow 2\,Fe_2O_3(s)$
3.51 The oxidation number of an element is a number assigned on the basis of a set of rules, and used to monitor whether an element has been oxidized or reduced.
3.53 (a) +4 (b) +1 (c) +2 (d) +5 (e) +4 (f) −2
3.55 (a) +7 (b) +2 (c) +6 (d) +6 (e) +6
3.57 (a) This is a substitution reaction; no oxidation or reduction occurs. (b) BrO_3^- is reduced and Br^- is oxidized. (c) F_2 is reduced and H_2O is oxidized.
3.59 Oxidizing agents contain an element that gains electrons and decreases the agent's oxidation number.

Reducing agents lose electrons to the oxidizing agent because reducing agents contain an element that readily loses electrons. A loss of electrons means an increase in oxidation number for an element in the reducing agent.

3.61 (a) Cl_2 is stronger because it has a more positive oxidation number than Cl^-. (b) N_2O_5 is stronger because the nitrogen has a more positive oxidation number than the nitrogen in N_2O.

3.63 (a) Zn is the reducing agent, HCl is the oxidizing agent. (b) H_2S is the reducing agent, SO_2 is the oxidiz-ing agent. (c) Mg is the reducing agent, B_2O_3 is the oxidizing agent.

3.65 (a) oxidizing agent (b) oxidizing agent (c) reducing agent (d) reducing agent

3.67 $2\,NaCl(l) \xrightarrow{\text{electrolysis at 600°C}} 2\,Na(l) + Cl_2(g)$
Chlorine is produced by oxidation and sodium by reduction.

3.69 (a) strong acid (b) base (c) base
(d) soluble ionic (e) weak acid (f) insoluble ionic
(g) insoluble ionic (h) strong acid

3.71 (a) HNO_3, $Ba(OH)_2$ (b) H_2SO_4, NaOH
(c) $HClO_4$, KOH (d) HCl, $Ni(OH)_2$

3.73 (a) acid-base neutralization; HCl acid, $Mg(OH)_2$ base (b) acid-base neutralization; H_2SO_4 acid, $Ba(OH)_2$ base; or precipitation; $Ba^{2+}(aq) + SO_4^{2+}(aq) \rightarrow BaSO_4(s)$ (c) redox; O_2 is the oxidizing agent, SO_2 is the reducing agent.

3.75 (a) redox; I_2O_5 is the oxidizing agent and CO is the reducing agent. (b) redox; I_2 is the oxidizing agent and $S_2O_3^{2-}$ is the reducing agent. (c) precipitation; $Ag^+(aq) + Br^-(aq) \rightarrow AgBr(s)$ (d) redox; UF_4 is the oxidizing agent and Mg is the reducing agent.

3.77 (a) $HClO_2$, H_2O, H_3O^+, ClO_2^- (b) NH_3, H_2O, NH_4^+, OH^- (c) CH_3COOH, H_2O, H_3O^+, CH_3COO^-

3.79 $2\,C_8H_{18}(l) + 25\,O_2(g) \rightarrow 16\,CO_2(g) + 18\,H_2O(g)$

3.81 $4\,C_{10}H_{15}N(s) + 55\,O_2(g) \rightarrow 40\,CO_2(g) + 30\,H_2O(l) + 2\,N_2(g)$

3.83 $H_2S(g) + 2\,NaOH(aq) \rightarrow Na_2S(aq) + 2\,H_2O(l)$
$4\,H_2S(g) + Na_2S(alc) \rightarrow Na_2S_5(alc) + 4\,H_2(g)$
$10\,H_2O(l) + 9\,O_2(g) + 2\,Na_2S_5(alc) \rightarrow 2\,Na_2S_2O_3{\cdot}5H_2O(s) + 6\,SO_2(g)$

3.85 (a) K: +1, O: $-\frac{1}{2}$ (b) Li: +1, Al: +3, H: −1
(c) Na: +1, O: −1 (d) Na: +1, H: −1 (e) K: +1, O: $-\frac{1}{3}$

CHAPTER 4

4.1 (a) 10 mol H_2 (b) 5.0 mol H_2

4.3 (a) 3.7 mol NaOH (b) 11.0 mol NO_2
(c) 8.4×10^{-3} mol MnO_4^-

4.5 9.0 mol CO_2

4.7 (a) 0.088 mol H_2O (b) 329 g O_2

4.9 (a) 205 g CO_2 (b) 0.133 mol H_2O

4.11 (a) 18.9 g Al_2O_3 (b) 8.90 g O_2

4.13 (a) 5.7×10^3 kg Al (b) 9.4×10^3 kg Al_2O_3

4.15 (a) 2.8×10^3 g H_2O (b) 7.3 g O_2

4.17 1.1×10^3 g H_2O

4.19 The stoichiometric point is the point at which the exact amount of one reactant, the titrant, has been added to complete the reaction with the other reactant, the analyte, according to the balanced chemical equation. An example is $CH_3COOH(aq) + NaOH(aq) \rightarrow NaCH_3CO_2(aq) + H_2O(l)$

4.21 (a) 0.271 mol·L^{-1} (b) 0.163 g NaOH

4.23 (a) 0.2087 mol·L^{-1} (b) 0.3289 g HNO_3

4.25 63.0 g·mol^{-1}

4.27 85.6%

4.29 86.7%

4.31 The limiting reactant is the reactant that governs the theoretical yield of a product in a given reaction. It is the reactant that will be depleted first in the chemical reaction. When all the limiting reactant is consumed, the reaction is completed, regardless of how much of another reactant may be present. In combustion analysis, the compound that is burned in oxygen is the limiting reactant. Percentage yield is the ratio of the actual yield to the theoretical yield, multiplied by 100%.

4.33 O_2 is the limiting reactant.

4.35 (a) O_2 is the limiting reactant. (b) 5.76 g P_4O_{10}
(c) 5.78 g P_4O_6 remaining

4.37 (a) PCl_3 is the limiting reactant. (b) 7.40 g H_3PO_3, 9.88 g HCl

4.39 (a) H_2S is the limiting reactant. (b) 5.7 g SO_2
(c) 8.6 g S, 3.2 g H_2O (d) 11.4 + 6.08 g = 17.5 g reactants and 5.7 g + 8.6 g + 3.2 g = 17.5 g excess reactant and products; so everything checks with the law of conservation of mass.

4.41 (a) Cl_2 is the limiting reactant. (b) 200 g $AlCl_3$

4.43 $C_7H_6O_2$

4.45 C_6H_7N

4.47 Empirical and molecular formulas are both $C_{14}H_{18}N_2O_5$.

4.49 empirical formula: $C_4H_5N_2O$, molecular formula: $C_8H_{10}N_4O_2$; the combustion equation is
$2\,C_8H_{10}N_4O_2(s) + 19\,O_2(g) \rightarrow 16\,CO_2(g) + 10\,H_2O(g) + 4\,N_2(g)$

4.51 The theoretical yield is the maximum quantity of product(s) that can be obtained, according to the reaction stoichiometry, from a given quantity of a specified reactant (the limiting reactant). The percentage yield of a product is the percentage of its theoretical yield that is actually achieved. It is calculated as

$$\text{Percentage yield} = (\text{actual yield}/\text{theoretical yield}) \times 100\%$$

The percentage yield may be less than 100% for a specified

product, because other reactions may take place in addition to the desired reaction, forming products other than the desired product(s).

4.53 (a) 5.37×10^{16} atoms Si (b) 1.47×10^{23} atoms C

4.55 (a) 3.34 mol H_2O_2 (b) 5.69×10^{-2} mol HNO_3 (c) 3.00×10^{-2} g H_2O

4.57 (a) 4.482 mg Mn (b) 53.23%

4.59 5.8×10^2 g NO

4.61 (a) 5.56×10^{-3} L O_2 (b) 2.6×10^{-2} L air

4.63 (a) Only two molecules of NH_3 can form; one molecule of H_2 and three of N_2 remain unreacted:

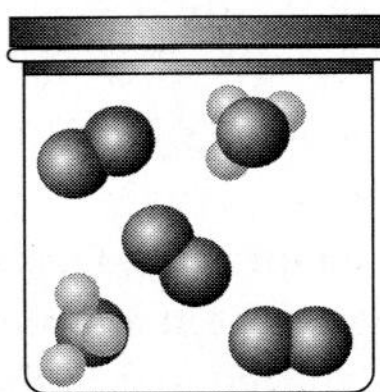

(b) H_2; two N_2 molecules would remain unreacted (in excess).

4.65 (a) Na is the limiting reactant (b) 3.91 g Al (c) 2.61 g Al_2O_3 (d) 45.3%

4.67 0.994 g Cu_3P_2

4.69 CF_2

4.71 (a) 8.70×10^{-4} mol H_2SO_4 (b) 0.433%

4.73 (a) $Na_2CO_3(aq) + 2\,HCl(aq) \rightarrow 2\,NaCl(aq) + H_2CO_3(aq)$ (b) 9.43 M HCl

4.75 0.926 g vitamin C, the manufacturer's claim is about 7% inaccurate.

CHAPTER 5

5.1 The essential features of the model are (1) Gases consist of widely separated molecules in ceaseless random motion. (2) Gas molecules collide with the walls of their container and with each other. The pressure of a gas is a manifestation of the collisions with the walls. (3) Collisions change the direction and speed of the molecules, resulting in a wide array of speeds, with no preferred direction for their motion. (4) The temperature of the gas is a measure of the average speed of the molecules, higher average speeds corresponding to higher temperatures.

5.3 The temperature of a gas is a measure of the average speed of its molecules; the higher the temperature, the higher the average speed of the molecules.

5.5 Pressure is a force exerted divided by the area of surface. Air pressure exerted on the surface of the mercury in the dish (see Fig. 5.8) is transmitted through the liquid mercury and supports the mercury column. The pressure at the base of the column of mercury is balanced by the pressure on the surface of the mercury in the dish. Thus the height of the column is proportional to the pressure. The space above the mercury is a vacuum, so it adds no pressure. A pressure of 1.00 atm corresponds to a column of mercury 760 mm high.

5.7 (a) 0.987 atm (b) 1.32×10^{-3} atm (c) 1.33×10^2 Pa (d) 9.87×10^{-3} atm

5.9 (a) 8×10^1 kbar (b) 8×10^9 Pa

5.11 9.58×10^2 cm

5.13 (a) 7.24×10^{-3} L (b) 1.32×10^1 mL

5.15 (a) 1.0×10^2 Pa (b) 3.6×10^3 Torr

5.17 (a) 1.6 atm (b) 194 mL

5.19 46°C

5.21 The temperature must be doubled.

5.23 632 Torr

5.25 377 cm^3

5.27 Boyle showed that $V \propto 1/P$, Charles showed that $V \propto T$, and Avogadro showed that $V \propto n$. These proportions can be rearranged into one equation and a proportionality constant (R) introduced. This combination results in the ideal gas law, $PV = nRT$.

5.29 (a) 4.5×10^{-3} mol (b) 1.3×10^{-2} Torr (c) 0.89 g (d) 3.94×10^4 L (e) 1.74×10^{14} atoms

5.31 6.0 mg NH_3

5.33 (a) 0.361 g (b) 0.349 g

5.35 17.0 g

5.37 (a) 24.5 L (b) 9.31 L (c) 73.4 L (d) 5.65 mL

5.39 (a) 3.21 g (b) 6.35 mg (c) 3.94 kg (d) 2.59 mg

5.41 0.298 L

5.43 2.44 L

5.45 (a) 2.15×10^6 L (b) 2.43×10^4 L

5.47 12.4 L CO_2 and 24.8 L NH_3

5.49 Conditions (b) would produce the larger volume of CO_2 by combustion because the volume of CH_4 in system (b) is larger.

5.51 NO_2 is the most dense.

5.53 (a) $4.88\ g{\cdot}L^{-1}$ (b) $3.90\ g{\cdot}L^{-1}$

5.55 (a) $3.18\ g{\cdot}L^{-1}$ (b) $77.9\ g{\cdot}mol^{-1}$

5.57 (a) $130\ g{\cdot}mol^{-1}$ (b) $5.30\ g{\cdot}L^{-1}$

5.59 C_3H_6

5.61 (a) 0.90 atm N_2, 3.26 atm O_2, 4.16 atm total (b) 0.67 atm H_2, 1.3 atm NH_3, 0.0473 atm He, 2.1 atm total

5.63 8.45×10^{-2} g H_2O

5.65 (a) 2.3×10^2 kPa N_2, 33.7 kPa Ar (b) 2.6×10^2 kPa

5.67 (a) 739.2 Torr (b) 0.142 g O_2

5.69 38.7 mL H_2

5.71 The assumptions of the kinetic theory of gases are (1) a gas consists of a collection of molecules in continuous random motion; (2) gas molecules are infinitely small; (3) gas molecules move in straight lines until they collide; and (4) gas molecules do not influence one another except during collisions.

5.73 As the temperature increases, the average speed of the gas molecules increases; this change increases the number of collisions with the container walls in a given time interval. Because pressure is the result of the force generated by these collisions, it should, therefore, increase with increasing temperature.

5.75 (a) argon (b) H_2

5.77 $1.2 \times 10^3\ g \cdot mol^{-1}$

5.79 Gases behave most ideally at high temperatures and low pressures. These conditions minimize the opportunity for interactions between molecules, that is, attractive and repulsive forces, which are the cause of deviations from ideality. The farther apart molecules are, the less interaction there will be.

5.81 Fig. 5.31 shows this relationship. As two molecules approach, the strength of attraction increases until they are close enough to touch, then they repel each other strongly.

5.83 (a) The ideal gas vessel has the greater pressure. (b) The free space is almost equal, but slightly greater for the ideal gas, because it is defined as having zero molecular volume.

5.85 Use the equation $V_2 = (T_2/T_1)(P_1/P_2)V_1$ to perform this conversion, with $P_1 = 1$ atm and $T_1 = 273$ K.

5.87 The Joule-Thomson effect is the cooling of a gas that occurs as it expands. Pulling the molecules apart results in a lowering of their average kinetic energy, as they expend energy to free themselves from each other. Lower average kinetic energy corresponds to lower average speed, which in turn corresponds to lower temperatures. After repeated expansions, followed by compressions in which heat is abstracted from the gas, the temperature will fall below the boiling point, and the gas will liquify.

5.89 The number of atoms (a), the average molecular speed (c), and the density (e) differ.

5.91 2.8 Torr

5.93 6.50×10^3 Torr

5.95 2.63×10^3 Torr

5.97 (a) 212 L (b) 259 g

5.99 14.2 mL

5.101 (a) 2.45 g CO (b) $5.76\ g \cdot L^{-1}$ (c) Because the mass and volume are fixed quantities in this experiment, the density does not change.

5.103 26 g

5.105 (a) 3.17×10^3 g TiO_2 (b) 3.89×10^3 L HCl(g)

5.107 (a) 2.95 kg CO_2 (b) 0.398 atm

5.109 2.38×10^4 L of air

5.111 N_2H_4

5.113 (a) 154 s (b) 123 s (c) 33.0 s (d) 186 s

5.115 $110\ g \cdot mol^{-1}$, C_8H_{12}

5.117 25.6 Torr

5.119 The overly high pressure obtained when S_8 is vaporized is an indication that S_8 units have decomposed to a greater number of units, S_6, S_4, S_2, etc. In other words, the number of moles of gas increased.

5.121 $254\ g \cdot mol^{-1}$, OsO_4 ($x = 4$)

5.123 (a) 4.2×10^{-3} g NH_4Cl (b) HCl is in excess, 7.4×10^{-2} atm

5.125 12.7 g

CHAPTER 6

6.1 System: that part of the universe which we are investigating. An example is a reaction vessel and its contents, which are the reactants and products of a chemical reaction in which we are interested. Surroundings: everything outside the system. The environment of a reaction flask is an example.

6.3 67 kJ

6.5 As the temperature of a system increases, the average speed of the molecules in a system also increases.

6.7 Enthalpy is that property of a system which changes when heat is released or absorbed by a system at constant pressure. When we say that the enthalpy of a system is a state property, we mean that its value is determined only by the state of a system, not by how the state was achieved; that is, the enthalpy of a system is not dependent on how the sample was prepared.

6.9 (a) 625 J (b) 5.4×10^2 °C

6.11 88%

6.13 (a) 2.1×10^3 J (b) 34 g

6.15 $0.53\ J \cdot (°C)^{-1} \cdot g^{-1}$

6.17 (a) endothermic (b) exothermic

6.19 −1.3 kJ, exothermic

6.21 8.8×10^2 J

6.23 $13.93\ kJ \cdot (°C)^{-1}$

6.25 −3.8 kJ

6.27 (a) -1.42×10^3 J (b) $-57\ kJ \cdot mol^{-1}$

6.29 Under constant pressure conditions, ΔH is the heat associated with the process. Therefore, if ΔH is negative, heat is given off, and the process is exothermic. Conversely, if ΔH is positive, heat is absorbed, and the process is endothermic.

6.31 (a) $8.21\ kJ \cdot mol^{-1}$ (b) $43.5\ kJ \cdot mol^{-1}$

6.33 (a) +226 kJ (b) +199 kJ

6.35 22 kJ

6.37 The standard state of a substance is its pure form at 1 atmosphere pressure.

6.39 (a) +72 kJ (b) +149 kJ (c) 1.8×10^2 g

6.41 (a) 2.5×10^2 g octane (b) -1.3×10^8 J

6.43 −266 kJ

6.45 The fact that enthalpy is a state function means that the enthalpy change for a sequence of reactions that yield a specific set of products from a specific set of reactants is the same as that for the direct single reaction yielding those same products. Changes in state functions depend only on the initial (reactants) and final (products) states for the process, not on how the change was achieved.

6.47 +1.90 kJ

6.49 −197.78 kJ
6.51 −1.775 MJ
6.53 −312 kJ
6.55 −184.7 kJ
6.57 (a) $K(s) + \frac{1}{2}Cl_2(g) + \frac{3}{2}O_2(g) \rightarrow KClO_3(s)$, $\Delta H° = -397.73$ kJ (b) $\frac{5}{2}H_2(g) + \frac{1}{2}N_2(g) + 2\,C(\text{graphite}) + O_2(g) \rightarrow H_2NCH_2COOH(s)$, $\Delta H° = -532.9$ kJ (c) $2\,Al(s) + \frac{3}{2}O_2(g) \rightarrow Al_2O_3(s)$, $\Delta H° = -1675.7$ kJ
6.59 $+11.3\ kJ{\cdot}mol^{-1}$
6.61 $-444\ kJ{\cdot}mol^{-1}$
6.63 (a) −15.4 kJ (b) −128.5 kJ
6.65 (a) +8.77 kJ (b) −233.57 kJ (c) −905.48 kJ
6.67 $48.44\ kJ{\cdot}g^{-1}$, $3.3 \times 10^4\ kJ{\cdot}L^{-1}$
6.69 $24.76\ kJ{\cdot}g^{-1}$ Mg, $31.05\ kJ{\cdot}g^{-1}$ Al; Al is better.
6.71 (a) $890\ kJ{\cdot}mol^{-1}$ for methane, $683.9\ kJ{\cdot}mol^{-1}$ for octane (b) $468.0\ kJ{\cdot}mol^{-1}$ (c) eating (burning glucose)
6.73 The standard enthalpy of formation of a compound is the enthalpy change associated with the formation of one mole of the compound in its standard state at a specified temperature (usually 25°C) from the elements in their most stable form at 1 atmosphere pressure and the specified temperature.
6.75 1.4 g
6.77 8.5×10^3 kJ
6.79 3 g
6.81 (a) 1 kJ (b) 9.4 kJ

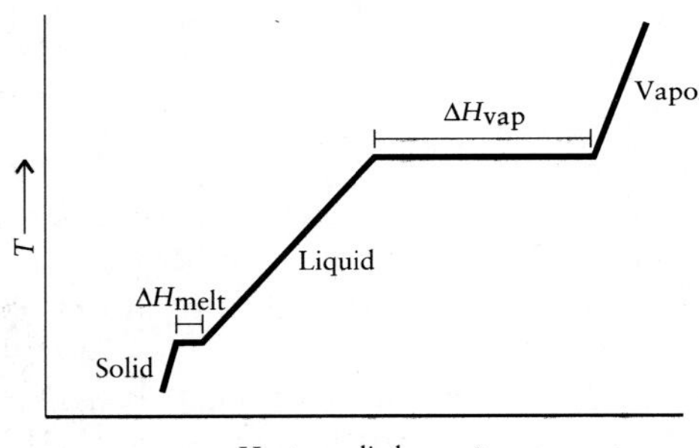

6.83 (a) -1.79×10^3 kJ (b) −109 kJ
6.85 $-239\ kJ{\cdot}mol^{-1}$
6.87 Heat would have to be removed.
6.89 $-296.80\ kJ{\cdot}mol^{-1}$
6.91 $+226\ kJ{\cdot}mol^{-1}$
6.93 (a) +131.29 kJ, so endothermic (b) 623 kJ
6.95 $\Delta H_f^\circ(H_2O, g)$ is less negative than $\Delta H_f^\circ(H_2O, l)$ because additional enthalpy is released when gaseous water condenses.
6.97 −22.2 kJ
6.99 −12 kJ
6.101 There may be many different considerations related to the manner in which the fuel is to be used. Among them are (1) cost, (2) ease of combustion, (3) environmental considerations, such as air and water pollution, (4) transportability, and (5) the specific enthalpy and enthalpy density. Large values of these latter quantities are generally desired.
6.103 5.5°C
6.105 6.12×10^4 J
6.107 $23.9 \times 10^3\ kJ{\cdot}L^{-1}$
6.109 $788\ kJ{\cdot}mol^{-1}$
6.111 −570 kJ

CHAPTER 7

7.1 (a) 4.28×10^{14} Hz to 7.50×10^{14} Hz (b) 1.2×10^6 Hz
7.3 (a) 420 nm (b) 150 pm
7.5 (a) 3.37×10^{-19} J (b) $2.03 \times 10^5\ J{\cdot}mol^{-1}$
7.7 (a) 2.3×10^{-15} J (b) $2.5 \times 10^5\ J{\cdot}(\text{mol photons})^{-1}$
7.9 3.38×10^{-19} J
7.11 (a) 410 nm (b) violet
7.13 (a) 3.29×10^{15} Hz (b) 9.12×10^{-8} m, x-ray or gamma ray
7.15 (a) 4.9×10^{-11} m (b) 2.6×10^{-14} m
7.17 (a) An atomic orbital is a region of space in which the electron is most likely to be found in an atom. It is a description of the wave characteristics of electrons in atoms in terms of three quantum numbers, *n*, *l*, and m_l. Each orbital is characterized by its own unique set of these quantum numbers.

(b)

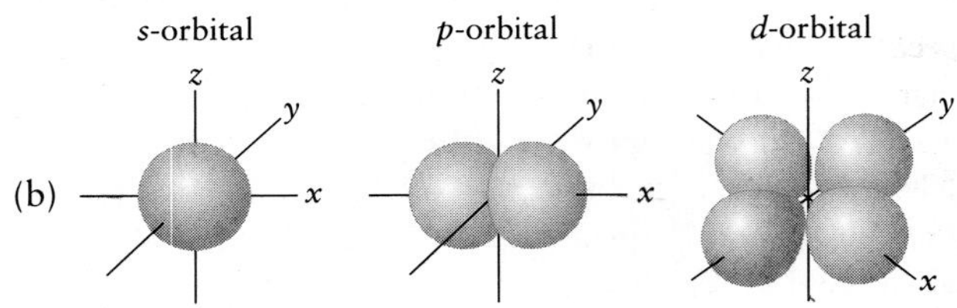

7.19 (a) 2 (b) 3 (c) The values for *l* can be 0, 1, or 2.
7.21 (a) 1 (b) 5 (c) 3 (d) 7
7.23 (a) 2 (b) 5 (c) 4 (d) 5
7.25 (a) 3*d*, 5 (b) 1*s*, 1 (c) 6*f*, 7 (d) 2*p*, 3
7.27 (a) 6 (b) 2 (c) 8 (d) 2
7.29 (a) not allowed (b) allowed (c) not allowed (d) allowed
7.31 (a) allowed (b) not allowed; if $l = 0$, m_l can only be 0 (c) not allowed; *n* and *l* cannot be equal
7.33 (a) $n = 2, l = 1, m_l = 0, m_s = -\frac{1}{2}$ (b) $n = 5, l = 2, m_l = +1, m_s = +\frac{1}{2}$
7.35 The average distance from the nucleus to a 2*s*-electron is much less than for a 3*s*-electron; consequently, the electrostatic force of attraction between the nucleus and the 2*s*-electron is much greater than that for a 3*s*-electron. The electron cloud of the 2*s*-electron is bunched more densely around the nucleus than is the 3*s*-electron cloud.

7.37 (a) $n = 1, 2, 3, 4$; energy increases from 1 to 4 (b) $l = 0, 1, 2, 3$; energy increases from 0 to 3
7.39 1*s*, 2*p*, 3*s*, 3*d*, 5*d*; energy increases from left to right
7.41 (a) $n = 1, l = 0, m_l = 0, m_s = +\frac{1}{2}$
(b) $n = 2, l = 1, m_l = 0, m_s = +\frac{1}{2}$
(c) $n = 4, l = 0, m_l = 0, m_s = -\frac{1}{2}$
(Note: m_l values are not assigned in any particular order.)
7.43 (a) $[Ar]4s^2$ (b) $1s^22s^22p^3$ (c) $[Ar]3d^{10}4s^24p^5$ (d) $[Rn]5f^36d^17s^2$
(Note: U is an exception.)
7.45 (a) $[Ar]3d^84s^2$ (b) $[Kr]4d^{10}5s^2$ (c) $[Xe]4f^{14}5d^{10}6s^26p^2$ (d) $[Kr]4d^{10}5s^1$
7.47 (a) ns^2 (b) ns^2np^6
7.49 (a) $[Ar]3d^6$ (b) $[Ne]3s^23p^6$ (c) $[Xe]4f^{14}5d^{10}6s^2$
7.51 (a) $[Ne]3s^23p^2$ (b) $[He]2s^22p^6$ (c) $[Xe]6s^1$ (d) $[Ne]3s^23p^4$
7.53 (a) $3s^23p^2$, 4 electrons (b) $3s^23p^5$, 7 electrons (c) $3d^54s^2$, 7 electrons (d) $3d^74s^2$, 9 electrons
7.55 The outermost electrons of an atom determine the atomic radius. Proceeding down a group, the outermost electrons occupy shells that lie farther and farther from the nucleus, so the size (radius) increases down a group.
7.57 (a) Group 1 (b) The radius of the cation is less than the radius of the neutral atom. (c) Cations with the largest number of protons (among a set of cations with the same number of electrons) will have the smallest radius.
7.59 (a) S (b) S^{2-} (c) Na (d) Mg^{2+}
7.61 First ionization energies decrease down a group because the outermost electron occupies a shell that is farther from the nucleus and is therefore less tightly bound. Ionization energies increase across a period because the effective nuclear charge increases as we go from left to right across a given period. As a result, the outermost electron is gripped ever more tightly and the ionization energies increase.
7.63 (a) Mg (b) N (c) P
7.65 The ionization energies for Group 16 elements are less than those for Group 15 elements. The group configurations are Group 16, ns^2np^4, and Group 15, ns^2np^3. The half-filled subshell of Group 15 is more stable than simple theory suggests; this stability makes removal of the electron more difficult; hence, it has a higher ionization energy. In Group 16, the fourth *p*-electron pairs up with another electron, thereby producing stronger electron repulsion, which makes it easier to remove this electron. Although there is the competing effect of increasing atomic number (nuclear charge), the electron repulsion effect predominates.
7.67 Both the first and second electrons lost from Mg are 3*s*. The second electron for sodium must be removed from a 2*p* level that not only is filled but also is considerably lower in energy than the 3*s* level. Thus the second ionization energy of sodium is very high.
7.69 (a) Group 17 (b) Electron affinities generally increase (left to right) across a period.
7.71 (a) Cl (b) O (c) Cl
7.73 Atomic radii increase and ionization energies decrease from top to bottom within a group. Atomic radii decrease and ionization energies increase (left to right) across a period.
7.75 A diagonal relationship is a similarity between diagonal neighbors in the periodic table, especially for elements at the left of the table. Two examples are (1) Li and Mg, which both burn in nitrogen to form the nitride; (2) Be and Al, which are both amphoteric, reacting with both acids and bases.
7.77 (a) Metals react with acids and form ionic compounds with nonmetals; many of them (transition metals) exhibit a variety of positive oxidation states. (b) Nonmetals form acidic oxides, react with other nonmetals to form molecular compounds, and react with metals to form ionic compounds in which they exhibit negative oxidation states.
7.79 The ionization energies of the *p*-block metals are quite high, so they are less reactive than the *s*-block metals, which have low ionization energies.
7.81 (a) metal (b) nonmetal (c) metalloid (d) metalloid
7.83 (a) do (b) do (c) do, but not strongly (d) do, but not strongly
7.85 (a) ultraviolet (b) infrared
7.87 1×10^{10} Hz
7.89 1.026×10^{-10} m
7.91 No, not enough energy is available.
7.93 (a) 6 (b) 5 (c) 3 (d) 4
7.95 (a) 4 unpaired electrons (b) 9 *p*-electrons (c) 18 (d) Group 2
7.97 (a) unacceptable (b) unacceptable (c) acceptable (d) acceptable
7.99 (a) $[Kr]4d^25s^2$ (b) $[Ar]3d^{10}4s^24p^4$ (c) $[Kr]5s^1$ (d) $[Ne]3s^23p^5$ (e) $[Kr]4d^{10}5s^25p^3$ (f) $[Rn]5f^67s^2$ (g) $[Ne]3s^23p^2$ (h) $[Ne]3s^23p^6$
7.101 (a) $n = 3, l = 1, m_l = +1, m_s = +\frac{1}{2}$
(b) $n = 3, l = 0, m_l = 0, m_s = -\frac{1}{2}$
7.103 (a) S (b) F^- (c) Cs (d) F (e) Ca (f) Fe (g) Ba^{2+} (h) S^{2-} (i) N (j) Cl (k) I (l) K (m) 4*g* (n) 4 (o) 7
7.105 (a) *d*-block (b) *s*-block (c) *p*-block (d) *d*-block (e) *p*-block (f) *d*-block
7.107 (a) 1.52×10^{-19} J, 91.5 $kJ\cdot mol^{-1}$ (b) 2.56×10^{-19} J, 154 $kJ\cdot mol^{-1}$ (c) The separation is larger in potassium as a result of its greater nuclear charge and the fact that the differences in shielding between 4*s* and 4*p* are greater in K than in Na.

7.109 (a) It would be in place of Na, which also has 11 protons. (b) The "element" NH_4 should react with H_2O to form the basic hydroxide NH_4OH. It should react with halogens to form ionic compounds such as NH_4Cl. It should have a low ionization energy and be a low-melting-point solid with good electrical conductivity. (c) Na^+ has a radius of 102 pm, considerably smaller than NH_4^+. The difference is a result of the much bigger "nucleus" of NH_4^+; the protons are spread out farther from the center, and consequently, the electrons are also more spread out.

7.111 (a) Because energy is conserved, some of the total energy of the incoming radiation, $h\nu$, is required to ionize the atom; the remainder is the kinetic energy of the electron. (b) 6.72×10^{-19} J·atom^{-1}

CHAPTER 8

8.1 (a) +1 (b) −2 (c) +2 (d) +3

8.3 Na^+ has the configuration $1s^2 2s^2 2p^6$, which is the same as that of Ne. The remaining electrons are all core electrons. To lose one would require a great deal of energy because they are so tightly held.

8.5 (a) Ca: (b) ·S̈· (c) $[:\ddot{O}:]^{2-}$ (d) $[:\ddot{N}:]^{3-}$

8.7 (a) $K^+[:\ddot{F}:]^-$ (b) $[:\ddot{S}:]^{2-}$ Al^{3+} $[:\ddot{S}:]^{2-}$ Al^{3+} $[:\ddot{S}:]^{2-}$
(c) Ca^{2+} $[:\ddot{N}:]^{3-}$ Ca^{2+} $[:\ddot{N}:]^{3-}$ Ca^{2+}

8.9 This difference is a result of the difference in ionic radii between Mg^{2+} and Ba^{2+}; see Fig. 7.28. The radius of Mg^{2+} (72 pm) is smaller than the radius of Ba^{2+} (136 pm), so the distance between Mg^{2+} and O^{2-} ions is less than that between Ba^{2+} and O^{2-} ions in the crystal lattice. Thus the lattice enthalpy of MgO exceeds that of BaO.

8.11 +971 kJ·mol^{-1}

8.13 (a) endothermic (b) exothermic (c) endothermic (d) exothermic (e) exothermic

8.15 (a) H—F̈: (b) H—N̈(—H)—H (c) H—C(H)(H)—H

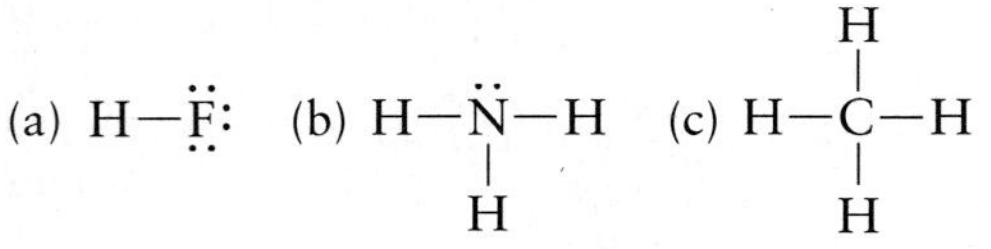

8.17 (a) $[H—N(H)(H)—H]^+$ (b) $[:\ddot{Cl}—\ddot{O}:]^-$
(c) $[:\ddot{F}—B(:\ddot{F}:)(:\ddot{F}:)—\ddot{F}:]^-$

8.19 (a) $[H—N(H)(H)—H]^+$ $[:\ddot{Cl}:]^-$ (b) K^+ K^+ K^+ $[:\ddot{O}—P(:\ddot{O}:)(:\ddot{O}:)—\ddot{O}:]^{3-}$
(c) $Na^+[:\ddot{Cl}—\ddot{O}:]^-$

8.21 (a) H—C(=Ö:)—H (b) H—C(H)(H)—Ö—H
(c) H—C(H)(:N(H)—H)—C(=Ö:)—Ö—H

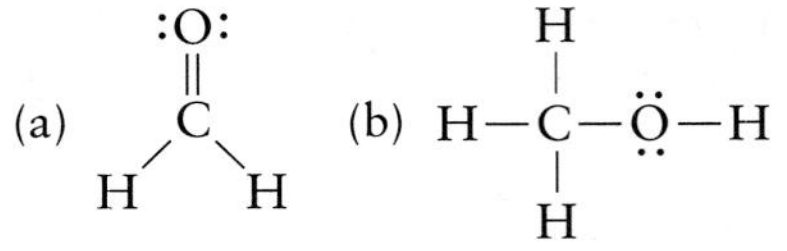

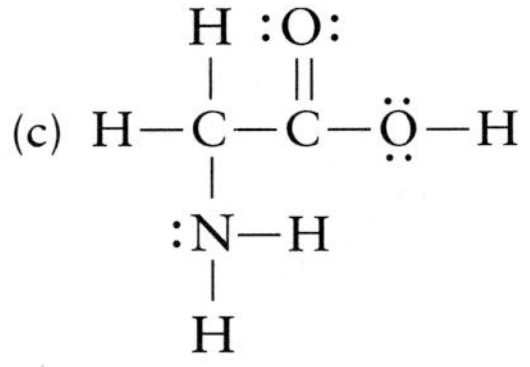

8.23 An example is the benzene molecule, which can be represented as a resonance hybrid of two principal Lewis structures, although other structures contribute slightly.

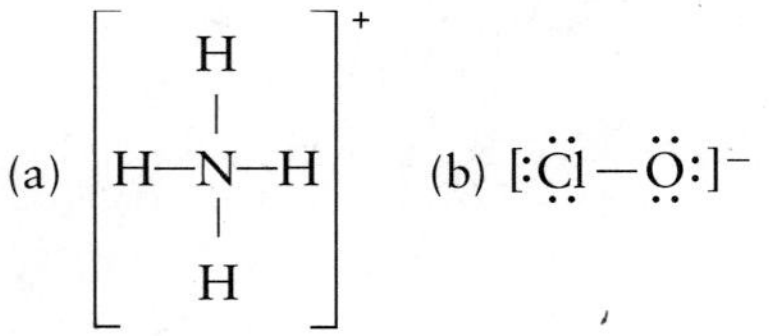

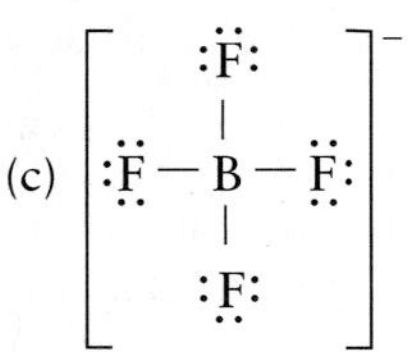

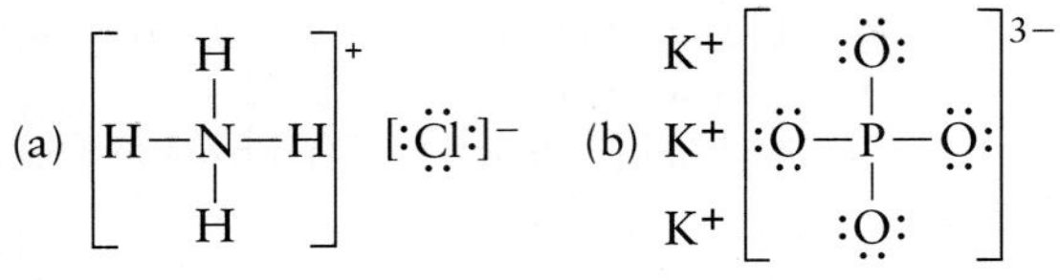

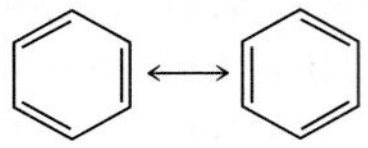

[benzene Kekulé structure] ⟷ [benzene Kekulé structure]

Two consequences of resonance are that (1) the resonance hybrid has a significantly lower energy than any of the individual structures and the compound is less reactive than otherwise expected. (2) Bond lengths are significantly different—the bond length in benzene is between that of the C—C and C=C bonds, but it is not the average of the two.

8.25 (a) $[:\ddot{O}—\ddot{N}=\ddot{O}]^- \leftrightarrow [\ddot{O}=\ddot{N}—\ddot{O}:]^-$
(b) :Ö—N(—:C̈l:)=Ö ↔ Ö=N(—:C̈l:)—Ö:

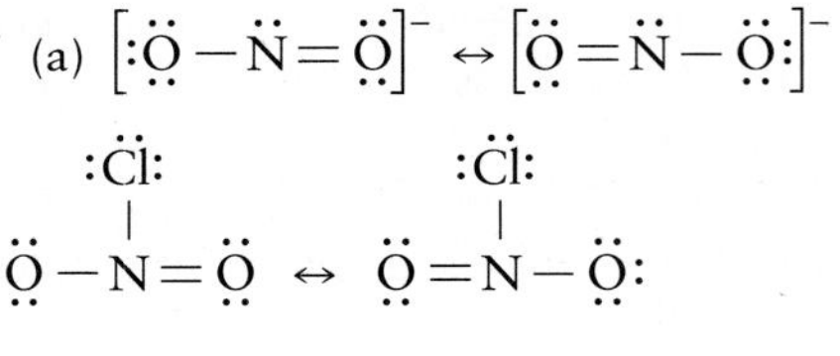

8.27 (a) $[H—\ddot{O}—P(:\ddot{O}:)(:\ddot{O}:)—\ddot{O}—H]^-$ ↔ $[H—\ddot{O}—P(=\ddot{O}:)(:\ddot{O}:)—\ddot{O}—H]^-$ ↔ $[H—\ddot{O}—P(=\ddot{O}:)(:\ddot{O}:)—\ddot{O}—H]^-$
(b) $[:\ddot{O}—\ddot{S}(:\ddot{O}:)(:\ddot{O}:)]^{2-}$ ↔ $[\ddot{O}=\ddot{S}(:\ddot{O}:)(:\ddot{O}:)]^{2-}$ ↔ $[\ddot{O}=\ddot{S}(=\ddot{O})(:\ddot{O}:)]^{2-}$

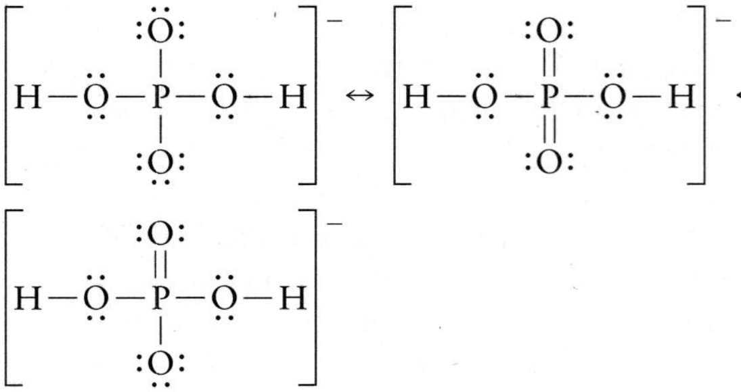

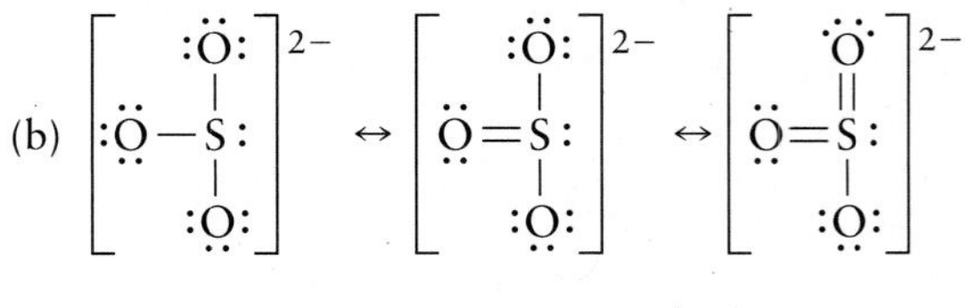

(c) $\left[:\ddot{O}-\underset{\underset{:\ddot{O}:}{|}}{\overset{\overset{:\ddot{O}:}{|}}{Cl}}:\right]^- \leftrightarrow \left[:\ddot{O}-\underset{\underset{:\ddot{O}:}{|}}{\overset{\overset{\ddot{O}}{||}}{Cl}}:\right]^- \leftrightarrow \left[:\ddot{O}-\underset{\underset{\ddot{O}}{||}}{\overset{\overset{\ddot{O}}{||}}{Cl}}:\right]^-$

(d) $\left[\ddot{O}{=}N(\ddot{O}:)(\ddot{O}:)\right]^- \leftrightarrow \left[:\ddot{O}-N({=}\ddot{O})(\ddot{O}:)\right]^- \leftrightarrow \left[:\ddot{O}-N(\ddot{O}:)({=}\ddot{O})\right]^-$

8.29 Formal charge is a good "bookkeeping" system for electrons and is based on the assumption that the molecule is "perfectly covalent." Formal charge is determined by dividing the electrons in a bond equally between the atoms connected by the bond. The concept is used primarily for deciding between otherwise equally plausible Lewis structures.

8.31 (a) Formal charge on N is 0 and on H is 0. (b) Formal charge on C is 0 and on O is 0.

8.33 (a) first structure: Charge on S is +1, O= is 0, and O— is −1. second structure: Charge on S is 0, and each O= is 0. The second structure is more plausible. (b) first structure: Charge on S is 0 and each O= is also 0. second structure: Charge on S is +1, O= is 0, and O— is −1. The first structure is more plausible.

8.35 Structure (a) has a charge of +1 on S and −1 on O=; structure (b) has a charge of 0 on both S and O. Structure (b) is the more plausible structure.

8.37 (a) The first structure is dominant. (b) The first structure is dominant.

8.39 (a) ionic, made of a metal and a nonmetal (b) nonionic, made of two nonmetals (c) ionic, made of a metal and a nonmetal

8.41 (a) ionic (b) covalent (c) covalent

8.43 The order is most likely $Rb^+ < Sr^{2+} < Be^{2+}$, which corresponds to the order of ionic size in reverse. Higher charge predominates over lower charge.

8.45 $O^{2-} < N^{3-} < Cl^- < Br^-$. The order parallels the ionic size: Br^- is largest, O^{2-} is smallest.

8.47 (a) HCl (b) CF_4 (c) CO_2

8.49 (a) ionic (b) significantly covalent (c) mixed ionic and covalent (d) ionic

8.51 A radical is any species with an unpaired electron. Examples are

(1) $\cdot CH_3$ or $\cdot\underset{\underset{H}{|}}{\overset{\overset{H}{|}}{C}}-H$

(2) $\cdot NO$ or $:\dot{N}{=}\ddot{O}$

(3) $\cdot NO_2$ or $\dot{O}{=}\dot{N}-\ddot{O}: \leftrightarrow :\ddot{O}-\dot{N}{=}\dot{O}$

8.53 (a) 1 lone pair on S $\quad :\ddot{Cl}-\underset{\underset{:\ddot{Cl}:}{|}}{\overset{\overset{:\ddot{Cl}:}{|}}{S}}:-\ddot{Cl}:$ (S bonded to four Cl)

(b) 2 lone pairs on I $\quad :\ddot{Cl}-\underset{\underset{:\ddot{Cl}:}{|}}{\overset{\overset{:\ddot{Cl}:}{|}}{\ddot{I}}}$

(c) 2 lone pairs on I $\quad [\,I\text{ with four }:\ddot{F}:\,]^-$

8.55 (a) 4 electron pairs; 2 bonding, 2 lone (b) 6 electron pairs; 4 bonding, 2 lone (c) 5 electron pairs; 3 bonding, 2 lone (d) 6 electron pairs; 5 bonding, 1 lone

8.57 (a) radical $:\ddot{Cl}-\ddot{O}\cdot$ (b) not a radical $:\ddot{Cl}-\ddot{O}-\ddot{O}-\ddot{Cl}:$

(c) not a radical $:\ddot{Cl}-\ddot{O}-N(\ddot{O}:)({=}\ddot{O}) \leftrightarrow :\ddot{Cl}-\ddot{O}-N({=}\ddot{O})(\ddot{O}:)$

(d) radical $:\ddot{Cl}-\ddot{O}-\ddot{O}\cdot$

8.59 (a) 2 lone pairs $:\ddot{F}-\ddot{Xe}-\ddot{F}:$ with $:\ddot{O}:$ double-bonded to Xe

(b) 3 lone pairs $:\ddot{F}-\dddot{Xe}-\ddot{F}:$

(c) 1 lone pair $\left[\ddot{O}{=}\underset{\underset{:\ddot{O}:}{|}}{\overset{\overset{:\ddot{O}-H}{|}}{Xe}}{=}\ddot{O}\right]^-$

8.61 A Lewis acid is an electron pair acceptor, therefore its electronic structure must allow an additional electron pair to become attached to it. A Lewis base is an electron pair donor, therefore, it must contain a lone pair of electrons that it can donate. Lewis acids: H^+, Al^{3+}, BF_3; Lewis bases: OH^-, NH_3, H_2O

8.63 The net ionic reaction for an acid-base neutralization is

$$H^+ + [:\ddot{O}{-}H]^- \longrightarrow H{-}\ddot{O}{-}H$$

Lewis acid — Lewis base — Lewis complex

8.65 (a) base (b) acid (c) acid (d) base

8.67 There are available electron pairs on the O, thus

CH_3O^- would be a Lewis base. $\left[\begin{array}{c} H \\ | \\ H-C-\ddot{O}: \\ | \\ H \end{array} \right]^-$

8.69 (a) PF_5 + $[:\ddot{F}:]^-$ ⟶ $[PF_6]^-$

Lewis acid   Lewis base   Product

(b) SO_2 + $[:\ddot{Cl}:]^-$ ⟶ $[SO_2Cl]^-$

Lewis acid   Lewis base   Product

(c) $Cu^{2+} + 4\,:NH_3 \longrightarrow [Cu(NH_3)_4]^{2+}$

Lewis acid   Lewis base   Product

8.71 Boron is in Group 13 and has only 3 valence electrons. In order to have an octet in covalent compounds, boron would have to add 5 electrons by sharing with other atoms. That is improbable, especially because the other atoms involved are highly electronegative nonmetals. Other elements that form covalent bonds in later groups have to acquire no more than 4 additional electrons to complete their octet and hence are not likely to form electron-deficient compounds.

8.73 Br—Cl (least ionic) < C—Cl < Al—Cl < Na—Cl (most ionic)

8.75 (a) $Li^+[:H]^-$ (b) $[:\ddot{Cl}:]^- Cu^{2+} [:\ddot{Cl}:]^-$

(c) $Ba^{2+}[:\ddot{N}:]^{3-} Ba^{2+}[:\ddot{N}:]^{3-} Ba^{2+}[:\ddot{N}:]^{3-}$

(d) $[:\ddot{O}:]^{2-} Ga^{3+} [:\ddot{O}:]^{2-} Ga^{3+} [:\ddot{O}:]^{2-}$

8.77 (a) $+2565\ kJ\cdot mol^{-1}$ (b) $+5486\ kJ\cdot mol^{-1}$

8.79 (a) acid   (b) acid   (c) either acid (at Ga) or base (at I)

8.81 SO_2 + H_2O ⟶ $H-O-S(=O)-O-H$ (Product)

Lewis acid   Lewis base   Product

8.83 (a) CH_3^-, because the C atom in CH_3^- has a lone pair of electrons; the C atom in CH_4 has none   (b) H_2O, because O is more electronegative than S and thus has a stronger partial negative charge

8.85 (a) First structure: each atom has a charge of 0; second structure: each O— has a charge of −1, H— and —O— are 0, and Cl is +2. The first structure has the lower energy.   (b) Each atom has a formal charge of 0.   (c) Each atom has a formal charge of 0.   (d) First structure: each N has a charge of −1, and the C is 0; second structure: N≡ and C have charges of 0, N— has a charge of −2. The first structure has the lower energy.   (e) first structure: Each O— has a charge of −1, and the As has a +1 charge; second structure: The O= has a charge of 0, each O— has a charge of −1, and the As has a charge of 0. The second structure has the lower energy.

8.87 Yes, because all bonds involve a pair of electrons influenced by two nuclear centers. It becomes a question of degree as to where the electron pair more closely resides. Covalent bonds share the electron pair equally, polar bonds less equally, and ionic bonds hardly share at all. The coordinate covalent bond that results from complex formation is like any other covalent bond once it has formed. The octet rule is useful in the predictions of all the bonding situations mentioned.

8.89 The overall process can be broken down as follows:

Equations	$\Delta H°$, kJ·mol^{-1}
$Na(s) \longrightarrow Na(g)$	+107
$Na(g) \longrightarrow Na^+(g) + e^-$	+494
$Na^+(g) \longrightarrow Na^{2+}(g) + e^-$	+4560
$Cl_2(g) \longrightarrow 2\,Cl(g)$	+242 × 2
$2\,Cl(g) + 2e^- \longrightarrow 2\,Cl^-(g)$	−349 × 2
$Na^{2+}(g) + 2\,Cl^-(g) \longrightarrow NaCl_2(s)$	−2524
$Na(s) + Cl_2(g) \longrightarrow NaCl_2(s)$	+2181 kJ·mol^{-1}

A compound with such a large, positive ΔH_f° is not likely to form under any conditions.

8.91 $Cl_2C=O$ (each Cl and the O with lone pairs)

8.93 (a) $[H_3C-C(=O)-O^-] \leftrightarrow [H_3C-C(-O^-)=O]$ (acetate ion resonance structures)

(b) $[H_3C-C(=O)-CH_2^-] \leftrightarrow [H_3C-C(-O^-)=CH_2]$

```
      H                    H
      |                    |
H     C                H   C        H
 \  /   \\            \  //  \    /
(c)  C      C—H   ↔      C      C
 /          |          /          \
H           H         H            H
```
(c) $[\text{resonance structures}]^+$

```
(d) [ H  :Ö:          ]⁻     [ H  :Ö:          ]⁻
    [ |   ||          ]      [ |   |           ]
    [ H—C—C—N̈—H       ]  ↔  [ H—C—C=N̈—H       ]
    [ |               ]      [ |               ]
    [ H               ]      [ H               ]
```

CHAPTER 9

9.1 (a) trigonal bipyramidal, 120° bond angles in equatorial plane, 90° between axial bonds and equatorial plane (b) linear, 180° (c) trigonal pyramidal, slightly less than 109° (d) angular, slightly less than 109° (e) tetrahedral, 109.5°

9.3 (a) AX_2E_2, angular (b) AX_3E, trigonal pyramidal (c) AX_2, linear (d) AX_2E, angular

9.5 (a) AX_3E, trigonal pyramidal (b) AX_4, tetrahedral (c) AX_4E, seesaw (d) AX_3, trigonal planar

9.7 (a) AX_4E, seesaw (b) AX_3E_2, T-shaped (c) AX_4E_2, square planar (d) AX_3E, trigonal pyramidal

9.9 (a) linear, 180° (b) T-shaped, slightly less than 90° (c) tetrahedral, 109.5° (d) octahedral, 90°

9.11 (a)
```
       :C̈l:
        |
   :F̈ — C — F̈:
        |
       :F̈:
```
AX_4, tetrahedral, 109.5°

(b)
```
              Ï:
             /
   :Ï — Ga
             \
              Ï:
```
AX_3, trigonal planar, 120°

(c)
```
 [   C̈   ]⁻
 [  /|\  ]
 [ H H H ]
```
AX_3E, trigonal pyramidal, slightly less than 109°

(d)
```
        :F̈:
   :F̈    |    F̈:
      \  |  /
         S
      /  |  \
   :F̈    |    F̈:
        :F̈:
```
AX_6, octahedral, 90°

(e)
```
   :F̈       :F̈:
      \  ..  /
        Xe
      /  ||  \
   :F̈   .O.   F̈:
```
AX_5E, square pyramidal, 90°

9.13 Angles a and b are approximately 120°.

9.15
```
      :F̈:                               :F̈         H
       |              ..                   \      /
       B       +      N       →     :F̈ — B — N — H
     /   \          / | \              /      \
   :F̈     F̈:      H   H   H          :F̈         H
```

BF_3 has bond angles of 120°; NH_3 has bond angles less than 109°; $F_3B—NH_3$ has bond angles of approximately 109°.

9.17 An electric dipole ("two poles") is a positive charge next to an equal but opposite negative charge. An electric dipole moment is a measure of the magnitude of the electric dipole in debye (D) units.

9.19 (a) toward O (b) toward F (c) toward F (d) toward O

9.21 (a) one nonpolar bond (b) four polar N—H bonds, one nonpolar N—N bond (c) All C—H bonds are slightly polar. (d) Both bonds are polar. Even though all the atoms are the same, the central atom is not equivalent to the others because of its bonding pattern.

9.23 (a) trigonal planar, nonpolar (b) square planar, nonpolar (c) trigonal pyramidal, polar (d) angular, polar

9.25 (a)
```
        :C̈l:
         |
   :C̈l — C — C̈l:
         |
        :C̈l:
```
nonpolar

(b) $\ddot{S}=C=\ddot{S}$
nonpolar

(c)
```
   :C̈l        C̈l:
       \     /
          P
       /  |  \
   :C̈l    |    C̈l:
        :C̈l:
```
nonpolar

(d)
```
   :F̈         F̈:
      \      /
        Xe
      /  ..  \
   :F̈         F̈:
```
nonpolar

9.27 (a) nonpolar (b) polar (c) polar

9.29 (a) Structures 1 and 2 are polar, structure 3 is nonpolar. (b) The first structure (1) has the largest dipole moment.

9.31 (a) $+926\ kJ\cdot mol^{-1}$ (b) $+1486\ kJ\cdot mol^{-1}$ (c) $+3150\ kJ\cdot mol^{-1}$ (d) $+2317\ kJ\cdot mol^{-1}$

9.33 (a) $-92\ kJ\cdot mol^{-1}$ (b) $-151\ kJ\cdot mol^{-1}$ (c) $-868\ kJ\cdot mol^{-1}$ (d) $-38\ kJ\cdot mol^{-1}$

9.35 (a) $-283\ kJ\cdot mol^{-1}$ (b) $-247\ kJ\cdot mol^{-1}$ (c) $+227\ kJ\cdot mol^{-1}$ (d) $-1\ kJ\cdot mol^{-1}$

9.37 (a) −232 kJ (b) −44 kJ (c) −287 kJ

9.39 (c) < (b) < (a)

9.41 (a) N—H, 112 pm; N—N, 150 pm (b) C=O, 127 pm (c) N—C, 152 pm; C=O, 127 pm; N—H, 112 pm (d) N=N, 120 pm; N—H, 112 pm

9.43 (a) σ (b) π (c) neither (d) σ

9.45 The $3p_z$–$3p_z$ overlap forms a σ bond.

9.47 (a) tetrahedral (b) linear (c) octahedral (d) trigonal planar

9.49 (a) dsp^3 (b) sp^2 (c) sp^3 (d) sp

9.51 (a) sp^2 [↑][↑][↑] [ ]
sp^2 sp^2 sp^2 p

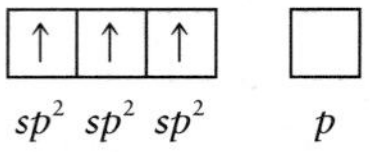

(b) sp^3 [↑][↑][↑][ ]
sp^3 sp^3 sp^3 sp^3

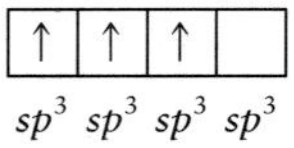

(c) sp^3 [↑↓][↑↓][↑↓][↑]
sp^3 sp^3 sp^3 sp^3

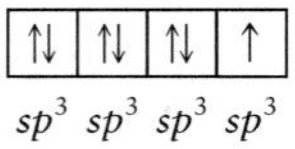

(d) sp^2 [↑][↑][↑] [ ]
sp^2 sp^2 sp^2 p

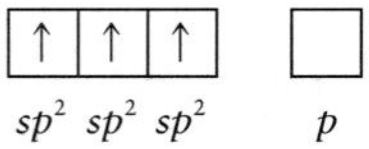

9.53 (a) sp^3 (b) sp (c) dsp^3 (d) sp^3

9.55 (a) 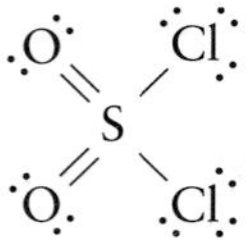 O, O, Cl, Cl around S Distorted tetrahedron, polar

(b) 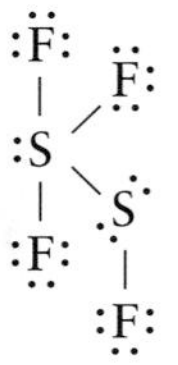 :F: / F: / :S / S / :F: / :F: Seesaw, polar (c) 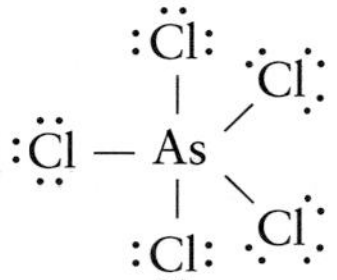 :Cl: / Cl: / :Cl — As / Cl: / :Cl: Trigonal bipyramidal, nonpolar

(d) 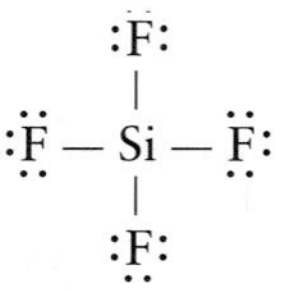 :F: / :F — Si — F: / :F: Tetrahedral, nonpolar (e) 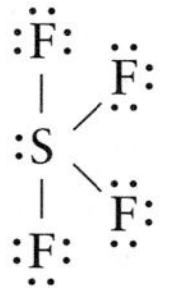 :F: / F: / :S / F: / :F: Seesaw, polar

(f) 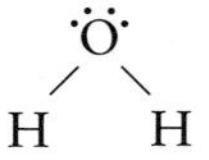 O / H H Angular, polar (g) H—C≡N: Linear, polar

(h) 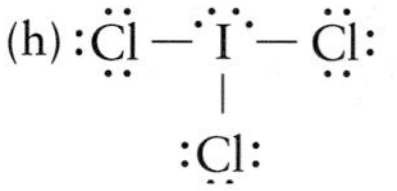 :Cl — I — Cl: / :Cl: T-shaped, polar

9.57 (b) < (c) < (a)

9.59 (a) tetrahedral, approximately 109° (b) linear, 180° (c) angular, slightly less than 120° (d) angular, less than 120°

9.61 (a) 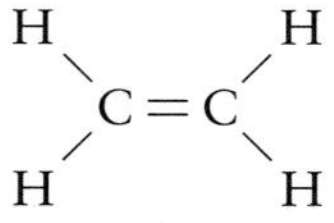 H, H / C=C / H, H approximately 120°

(b) :Cl — C ≡ N: 180° (c) 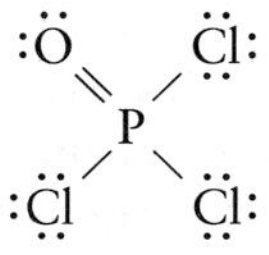 :O, Cl: / P / :Cl, Cl: 109°

(d) 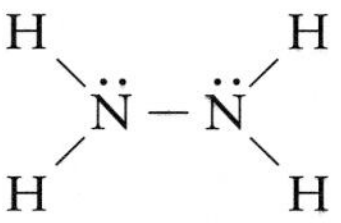 H, H / N — N / H, H 109°

9.63 −48 kJ

9.65 (a) 149 pm (b) 183 pm (c) 213 pm. The bond length is proportional to the increasing size of the atom attached to the fluorine atom; size increases down a group in the periodic table.

9.67 XX′ has the formula AX and is polar. XX'_3 has the formula AX_3E_2, is T-shaped, is polar, and has bond angles slightly less than 90°. XX'_5 has the formula AX_5E, is square pyramidal, is polar, and has bond angles slightly less than 90°.

9.69 O / F F sp^3 hybridization, bond angle slightly less than 109°

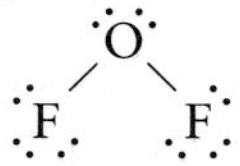

9.71 (a) The C on $-CH_3$ has 109.5° bond angles and is sp^3 hybridized. The C on $>C=O$ has 120° bond angles and is sp^2 hybridized. (b) The C on $-CH_3$ has 109.5° bond angles and is sp^3 hybridized. The C on $>C=O$ has 120° bond angles and is sp^2 hybridized.

9.73 :F: / :Cl — F: / :F: F uses p-orbitals; Cl uses dsp^3 hybrid orbitals; T-shaped, with F—Cl—F bond angles of ≤ 90° and ≤ 180°

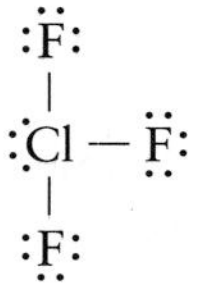

9.75 (a) 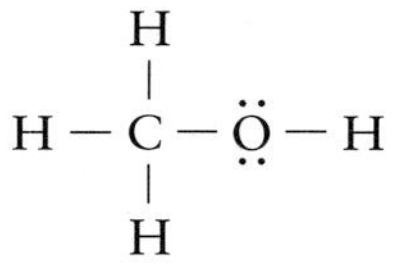 H / H — C — O — H / H C has bond angles close to 109.5°, O has bond angles of slightly less than 109°. (b) Both C and O have sp^3 hybridization. (c) polar

9.77 (a) Step 1: −113 kJ; step 2: −226 kJ (b) −339 kJ, $O_3(g) + O(g) \rightarrow 2\,O_2(g)$ (c) The average ΔH_B for the two O—O bonds in O_3 is probably closer to the double bond value (496 $kJ\cdot mol^{-1}$) than to the single bond value (157 $kJ\cdot mol^{-1}$). This would make $\Delta H_{overall} > -339$ kJ, which would compare even less favorably to the experimental value. These data show that mean values of bond enthalpies should be used with caution.

9.79 The energy is lowered by 228 $kJ\cdot mol^{-1}$.

9.81 P_4, sp^3; S_8, sp^3; N and O have small atoms that can easily form multiple bonds with each other and so can achieve octets by forming diatomic molecules. Period 3 atoms, such as those of P and S, are simply too large to get close enough to form π bonds, so they have to form a lot of single bonds to more than one atom at a time.

9.83 (a, b)

H, 120°, H; 120° $C_1=C_2=C_3$, 180°; H, H

1

H; 150°; C_2; 109°; 60°; H; C_1 — C_3; H; H

2

180°; 109.5°; H
H — $C_1 \equiv C_2$ — C_3 — H
H

3

(c) structure 1: C_1 and C_3 have sp^2 hybridization; C_2 has sp hybridization. structure 2: C_1 and C_2 have sp^2 hybridization; C_3 has sp^3 hybridization. structure 3: C_1 and C_2 have sp hybridization; C_3 has sp^3 hybridization. (d) No resonance occurs in structure 1 because there are no alternating single and double bonds. No resonance occurs in structures 2 or 3 because C—H bonds would have to be broken and reformed for resonance to occur.

CHAPTER 10

10.1 (a) London forces (b) dipole-dipole, London forces (c) London forces (d) dipole-dipole, London forces

10.3 (a) :C≡O: London and dipole-dipole forces

(b) H—Ö—H (bent) Hydrogen bonding, London forces

(c) NF_3 (:N̈ with three :F̤: atoms) London and dipole-dipole forces

(d) H—C—H (with H above and below) London forces

10.5 A hydrogen bond is described as A—H···B, where ··· represents the hydrogen bond. A must be one of the elements N, O, or F, which produce a strong, polar A—H bond that is capable of attracting the lone pair of electrons on neighboring B atoms. B should also be either N, O, or F, because these elements are the smallest of the highly electronegative elements. Their small size results in the strong interaction with the positive side of the polar A—H bond.

10.7 (a) yes (b) no (c) yes (d) yes

10.9 (a) NaCl, because the ion-ion interactions are stronger than the dipole-dipole interactions in HCl. (b) SiH_4, because it has the greater molar mass. (c) HF, because of stronger hydrogen bonding (d) H_2O, because of stronger hydrogen bonding; there are two O—H bonds.

10.11 (a) Attractions between atoms in Xe and Ar result from London forces; Xe is bigger and has more electrons, so it is more polarizable, has stronger London forces, and has a higher melting point. (b) HI and HCl both have London and dipole-dipole forces, but the London forces are greater in HI. (c) There is strong hydrogen bonding in water and none in $C_2H_5OC_2H_5$, so water has the lower vapor pressure.

10.13 (a) [Lewis structures of SF_4 and SF_6] SF_4 is seesaw-shaped; SF_6 is octahedral; SF_4 has a higher boiling point.

(b) [Lewis structures of BF_3 and ClF_3] BF_3 is trigonal planar; ClF_3 is T-shaped; ClF_3 has a higher boiling point.

(c) [Lewis structures of SF_4 and CF_4] CF_4 is tetrahedral; SF_4 is seesaw-shaped; SF_4 has a higher boiling point.

(d) [cis and trans HClC=CHCl]

cis trans

Both molecules have the same shape; the cis compound has a higher boiling point.

10.15 The intermolecular forces (such as hydrogen bonding) between water molecules are stronger in water than between water molecules and the hydrocarbons in wax.

10.17 (a) ethanol (b) propanone

10.19 (a) ionic (b) molecular (c) molecular (d) metallic

10.21 A, ionic; B, metallic; C, molecular

10.23 (a) 2 atoms per unit cell (b) 8

10.25 (a) 354 pm (b) 2.26×10^{22} unit cells

10.27 (a) 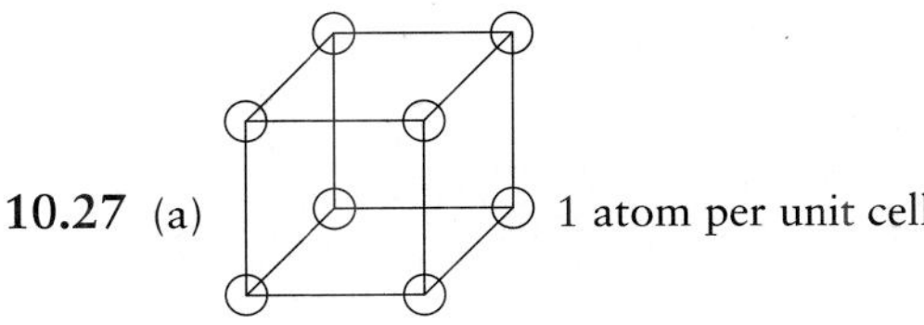 1 atom per unit cell

(b) 6 (c) 380 pm

10.29 (a) $8.82\ g\cdot cm^{-3}$ (b) $1.48\ g\cdot cm^{-3}$

10.31 (a) 144 pm (b) 131 pm

10.33 The charge is carried by electrons in electronic conduction and by ions in ionic conduction.

10.35 (a) The alloy is substitutional, because phosphorus atoms are nearly the same size as silicon atoms. (b) Doping either adds an electron (*n*-type) or a "hole" (*p*-type). In *n*-type materials, the conductivity increases with respect to silicon because there are additional electrons to move in the conduction band.

10.37 Heating in a vacuum causes partial loss of oxygen because the partial pressure of oxygen is below its equilibrium value ($ZnO(s) \rightarrow ZnO_{1-x} + \frac{x}{2} O_2$). For every oxygen atom formed in the elemental state, two conducting electrons are left in the ZnO lattice and its conductivity increases. This trend is reversed when ZnO is heated in oxygen.

10.39 (a) 4 anions, 4 cations, 4 formula units (b) 8 anions, 4 cations, 4 formula units; coordination numbers are 8 for Ca^{2+} and 4 for F^-.

10.41 $CaTiO_3$

10.43 5.56×10^{18} unit cells

10.45 (a) 8 (b) 6 (c) 8

10.47 (a) $3.73\ g \cdot cm^{-3}$ (b) $4.74\ g \cdot cm^{-3}$

10.49 B, C, Si, Ge, P, As

10.51 $356\ kJ \cdot mol^{-1}$

10.53 (a) London forces (b) dipole-dipole and London forces, hydrogen bonding (c) dipole-dipole and London forces

10.55 $A_2B_2C_4$

10.57 Dynamic equilibrium is a condition in which the forward process and its reverse are occurring simultaneously at equal rates. Processes occur at the molecular level; the net effect is unchanging. In a static equilibrium, no process is occurring at any level. Static equilibrium is rare or nonexistent at the atomic and molecular levels.

10.59 0.017 g H_2O

10.61 1.8×10^3 g

10.63 (a) 99.2°C (b) 99.7°C

10.65 (a) vapor (b) liquid (c) vapor (d) vapor

10.67 (a) 2.4 K (b) ~10 atm (c) 5.5 K (d) no

10.69 (a) At the lower pressure triple point, liquid helium I and II are in equilibrium with helium gas; at the higher pressure triple point, liquid helium I and II are in equilibrium with solid helium. (b) helium I

10.71 Its triple point temperature is higher than room temperature. From this information only, no conclusion can be drawn about its triple point pressure, because some solids sublime in air, as the solid slowly evaporates.

10.73 The vapor pressure of a substance at a given temperature is the pressure of its vapor when vapor and liquid are in equilibrium. That is, the molecules in a liquid in a closed container are leaving the surface at the same rate as the molecules in the vapor above the liquid return to the surface.

10.75 As pressure increases, the melting point of water decreases; whereas, it increases for carbon dioxide. Liquid water is denser than solid water, so as the pressure increases, the denser phase is favored. Practically, water floats when it freezes (rivers, lakes), thereby allowing living creatures to survive in partially frozen bodies of water.

10.77 (a) H_2O, it has two O—H bonds, and each molecule of water is involved in about four hydrogen bonds. (b) The surface tension is about the same in all of these liquids, because all form hydrogen bonds to about the same extent; however, London forces increase with increasing molar mass, so C_3H_7OH has the highest surface tension. (c) CO_2 because it is nonpolar. (d) HCl, because it has the weakest London forces. (e) H_2S, because it has less hydrogen bonding than H_2O and weaker London forces than H_2Te. (f) H_2, because it has the weakest London forces. (g) NH_3, as a result of relatively strong hydrogen bonding. (h) Na_2O ion-ion forces are stronger than dipole-dipole forces. (i) SF_2, because the electronegativity difference is greatest, so the dipole-dipole forces are probably strongest. (j) GeF_4, because it has the largest atoms and the most electrons.

10.79 (a, b) Viscosity and surface tension decrease with increasing temperature; at high temperatures the molecules readily move away from their neighbors because of increased kinetic energy. (c, d) Evaporation rate and vapor pressure increase with increasing temperature because the kinetic energy of the molecules increases with temperature and the molecules are more likely to escape into the gas phase.

10.81 If the external pressure is lowered, water boils at a lower temperature. Boiling temperature is the temperature at which vapor pressure equals external pressure. Vapor pressure and temperature are related; lowering the external pressure lowers the boiling temperature.

10.83 The pressure increase would bring CO_2 into the solid region.

10.85 (a) 78.6% (b) Some of the water vapor in the air would condense as dew or fog.

10.87 (a) 0.36 atoms Ni : 1 atom Cu (b) 161 atoms Sn : 12 atoms Sb : 10 atoms Cu (c) 21 atoms Sn : 22 atoms Cd : 24 atoms Pb : 48 atoms Bi

10.89 (a) 429 pm (b) 186 pm (c) 7.6×10^{-23} g

10.91 (a) $3.19\ g \cdot cm^{-3}$ (b) 198 pm (c) $3.25 \times 10^7\ pm^3$ (d) 25.6%

10.93 The spacing between carbon layers increases, thereby resulting in decreased perpendicular conductivity. The parallel conductivity increases, although the effect may not be dramatic, because the band is already half-filled.

10.95 $r = (6.65 \times 10^{-9})(M/d)^{1/3}$, where M is molar mass and d is density. $r_{Ne} = 170$ pm, $r_{Ar} = 203$ pm, $r_{Kr} = 225$ pm, $r_{Xe} = 239$ pm, $r_{Rn} = 246$ pm

10.97 $21.0\ g \cdot cm^{-3}$

10.99 (a) 58.9 Torr, assuming the air at 25°C is saturated with ethanol. The vapor pressure above a liquid is determined, for all practical purposes, by temperature alone.

So condensation of some gaseous ethanol occurs.
(b) 1441 Torr

CHAPTER 11

11.1 The difference can be traced to the weaker London forces that exist in branched molecules. Atoms in neighboring branched molecules cannot lie as close together as they can in unbranched isomers.
11.3 (a) 4 σ-type single bonds (b) 2 σ-type single bonds and 1 double bond with a σ- and a π-bond. (c) 1 σ-type single bond and 1 triple bond with a σ-bond and two π-bonds
11.5 (a) substitution, $CH_4 + Cl_2 \xrightarrow{\text{light}} CH_3Cl + HCl$
(b) addition, $CH_2{=}CH_2 + Br_2 \rightarrow CH_2Br{-}CH_2Br$
11.7 (a) substitution, $CH_3CH_3 + 2\,Cl_2 \xrightarrow{\text{light}}$ $CH_2ClCH_2Cl + 2\,HCl$ (CH_2ClCH_3 also produced)
(b) addition, $CH_2{=}CH_2 + Cl_2 \rightarrow CH_2ClCH_2Cl$
(c) addition, $HC{\equiv}CH + 2\,Cl_2 \rightarrow CHCl_2CHCl_2$
11.9 (a) ethane (b) hexane (c) octane (d) methane
11.11 (a) methyl (b) pentyl (c) propyl
11.13 (a) propane (b) ethane (c) pentane
(d) 2,3-dimethylbutane
11.15 (a) 4-methyl-2-pentene
(b) 2,3-dimethyl-2-phenylpentane
11.17 (a) propene, no geometrical isomers
(b) *cis*-2-hexene and *trans*-2-hexene
(c) 1-butyne, no geometrical isomers
(d) 2-butyne, no geometrical isomers
11.19 (a) $CH_2{=}CHCH(CH_3)CH_2CH_3$
(b) $CH_3CH_2C(CH_3)_2CH(CH_2CH_3)(CH_2)_2CH_3$
(c) $CH{\equiv}C(CH_2)_2C(CH_3)_3$
(d) $CH_3CH(CH_3)CH(CH_2CH_3)CH(CH_3)_2$

11.21

(a)
```
     H   H   H   CH3 H   H   H   H   H
     |   |   |   |   |   |   |   |   |
 H — C — C — C — C — C — C — C — C — C — H
     |   |   |   |   |   |   |   |   |
     H   H   H   CH3 H   H   H   H   H
```

(b)
```
                       CH3
                       |
              H   H    CH2 H   H   H   H   H
              |   |    |   |   |   |   |   |
 H — C ≡ C — C — C —  C — C — C — C — C — C — H
              |   |    |   |   |   |   |   |
              H   CH2  CH2 H   H   H   H   H
                  |    |
                  CH2  CH3
                  |
                  CH3
```

(c)
```
     H   CH3 H   CH3 H
     |   |   |   |   |
 H — C — C — C — C — C — H
     |   |   |   |   |
     H   CH3 H   H   H
```

(d)
```
 H3CCH2        H
       \      /
        C = C
       /      \
      H        CH2CH3
```

11.23 (a) 1-ethyl-3-methylbenzene
(b) 1,2,3,4,5-pentamethylbenzene

11.25 (a) CH_3 on a benzene ring (toluene) (b) CH_3 and Cl para on a benzene ring (c) two CH_3 groups meta on a benzene ring (d) CH_3 and Cl para on a benzene ring

11.27 (a, b)

Cl / Cl (1,2) on benzene ring	Cl / Cl (1,3) on benzene ring	Cl / Cl (1,4) on benzene ring
ortho-Dichlorobenzene (polar)	*meta*-Dichlorobenzene (polar)	*para*-Dichlorobenzene (nonpolar)

11.29 (a) RNH_2, R_2NH, R_3N (b) ROH (c) RCOOH
(d) RCHO
11.31 (a) ether (b) ketone (c) primary amine (d) ester
11.33 (a) primary alcohol, $CH_3CH_2CH_2CH_2OH$
(b) secondary alcohol, $CH_3CH_2CH(OH)CH_3$
(c) phenol, H_3C—(benzene ring)—OH
11.35 (a) diethyl ether (b) $CH_3OCH_2CH_3$
11.37 (a) aldehyde, ethanal (b) ketone, propanone
(c) ketone, 3-pentanone

11.39

(a)
```
 H
  \
   C = O
  /
 H
```

(b)
```
 CH3
    \
     C = O
    /
 CH3
```

(c)
```
             CH3
                \
                 C = O
                /
 CH3(CH2)3CH2
```

11.41 (a) ethanol (b) 2-octanol (c) 5-methyl-1-octanol. These reactions can be accomplished with an oxidizing agent such as acidified sodium dichromate, $Na_2Cr_2O_7$.
11.43 (a) ethanoic acid (b) butanoic acid
(c) 2-aminoethanoic acid

11.45 (a) $C_6H_5-\overset{O}{\overset{\|}{C}}-OH$

(b) $CH_3-CH_2-\underset{CH_3}{\underset{|}{CH}}-\overset{O}{\overset{\|}{C}}-OH$

(c) $CH_3-CH_2-\overset{O}{\overset{\|}{C}}-OH$

11.47 (a) methylamine (b) diethylamine
(c) 2-methylaniline, *o*-methylaniline, *o*-methylphenylamine

11.49 (a) benzene ring with NH_2 and adjacent CH_3 (b) $(CH_3CH_2)_2N-CH_2CH_3$

(c) $\left[H_3C-\underset{CH_3}{\underset{|}{\overset{CH_3}{\overset{|}{N}}}}-CH_3\right]^+$

11.51 (a) 2-propanol (b) dimethyl ether (c) methanal, formaldehyde (d) 3-pentanone (e) methylamine
11.53 (a) alcohol, ether, aldehyde (b) ketone
(c) tertiary amine, amide

11.55 (a) $CH_3CH_2CH_2\overset{O}{\overset{\|}{C}}-O-CH(CH_3)_2$

(b) $CH_3\overset{O}{\overset{\|}{C}}-O-CH_2CH_2CH_2CH_2CH_3$

(c) $CH_3CH_2CH_2CH_2CH_2\overset{O}{\overset{\|}{C}}-\underset{CH_3}{\underset{|}{N}}CH_2CH_3$

(d) $CH_3\overset{O}{\overset{\|}{C}}-NHCH_2CH_2CH_3$

11.57 The following procedures may be used: (1) Use an acid-base indicator and look for a color change.

(2) $CH_3CH_2CHO \xrightarrow{\text{Tollens reagent}} CH_3CH_2COOH + Ag(s)$

(3) $CH_3COCH_3 \xrightarrow{\text{Tollens reagent}}$ no reaction.

Procedure (1) will distinguish ethanoic acid; (2) and (3) distinguish propanal from 2-propanone.

11.59 $H_2C=CH-CH_2-CH_3$ 1-Butene; $H_2C=C(CH_3)_2$ 2-Methylpropene

$CH_3-CH=CH-CH_3$ *cis*-2-Butene or *trans*-2-Butene

11.61 (a) $H-\underset{H}{\underset{|}{\overset{H}{\overset{|}{C}}}}-\underset{H}{\underset{|}{\overset{H}{\overset{|}{C}}}}-\underset{H}{\underset{|}{\overset{H}{\overset{|}{C}}}}-\underset{H}{\underset{|}{\overset{H}{\overset{|}{C}}}}-H$ Butane; $H-\underset{H}{\underset{|}{\overset{H}{\overset{|}{C}}}}-\underset{CH_3}{\underset{|}{\overset{H}{\overset{|}{C}}}}-\underset{H}{\underset{|}{\overset{H}{\overset{|}{C}}}}-H$ 2-Methylpropane

(b) $H-\underset{H}{\underset{|}{\overset{H}{\overset{|}{C}}}}-\underset{H}{\underset{|}{\overset{H}{\overset{|}{C}}}}-\underset{H}{\underset{|}{\overset{H}{\overset{|}{C}}}}-\underset{H}{\underset{|}{\overset{H}{\overset{|}{C}}}}-\underset{H}{\underset{|}{\overset{H}{\overset{|}{C}}}}-H$ Pentane

$H-\underset{H}{\underset{|}{\overset{H}{\overset{|}{C}}}}-\underset{H}{\underset{|}{\overset{CH_3}{\overset{|}{C}}}}-\underset{H}{\underset{|}{\overset{H}{\overset{|}{C}}}}-\underset{H}{\underset{|}{\overset{H}{\overset{|}{C}}}}-H$ 2-Methylbutane; $C(CH_3)_4$ 2,2-Dimethylpropane

11.63 (a) not isomers (b) structural isomers (c) structural isomers (d) geometrical isomers (e) not isomers, these are different views of the same molecule.

11.65 (a) $CH_3-\underset{CH_3}{\underset{|}{\overset{H}{\overset{|}{C}}}}-CH_3$

(b) $CH_3-\underset{CH_3}{\underset{|}{\overset{Cl}{\overset{|}{C}}}}-CH_3$ $Cl-\underset{H}{\underset{|}{\overset{H}{\overset{|}{C}}}}-\underset{CH_3}{\underset{|}{\overset{H}{\overset{|}{C}}}}-CH_3$

11.67 An * designates a chiral carbon. (a) optically active,

$CH_3-\underset{Br}{\underset{|}{\overset{H}{\overset{|}{C^*}}}}-CH_2CH_3$ (b) not optically active

(c) optically active, $H-\underset{H}{\overset{Br}{C}}-\underset{H}{\overset{Cl}{C^*}}-CH_3$

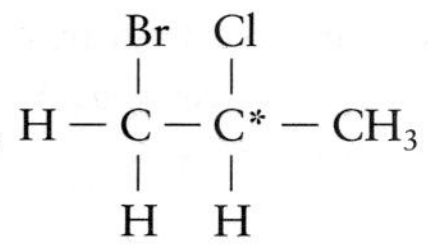

(d) optically active, $H-\underset{H}{\overset{Cl}{C}}-\underset{H}{\overset{Cl}{C^*}}-CH_2CH_2CH_3$

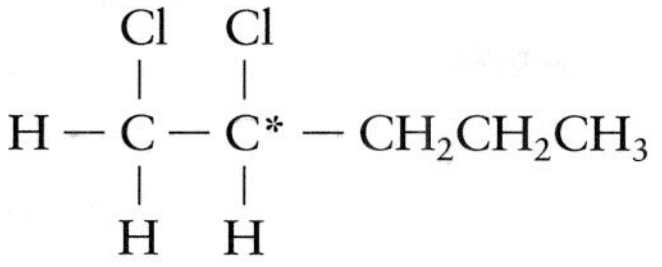

11.69 An * designates a chiral carbon. (a)

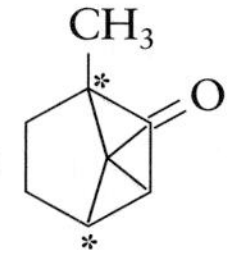

(b)

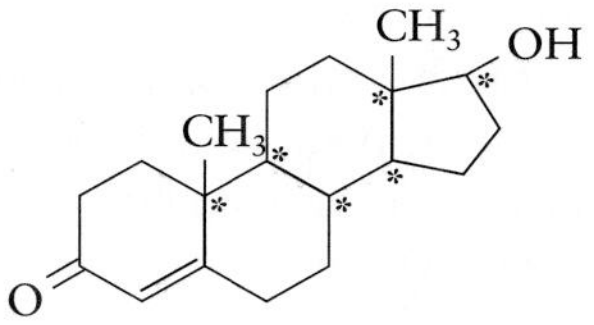

11.71

(a) 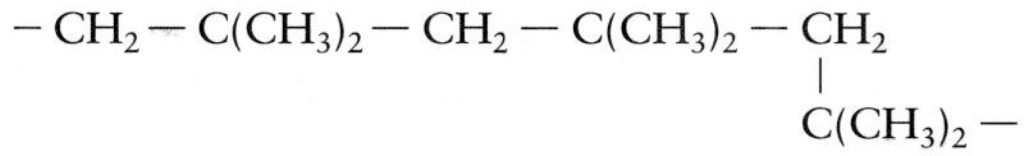 $-CH_2-C(CH_3)_2-CH_2-C(CH_3)_2-CH_2-C(CH_3)_2-$

(b) 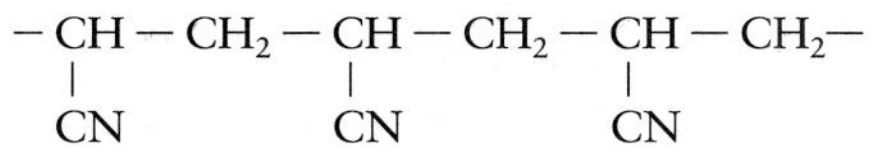 $-\underset{CN}{CH}-CH_2-\underset{CN}{CH}-CH_2-\underset{CN}{CH}-CH_2-$

(c)

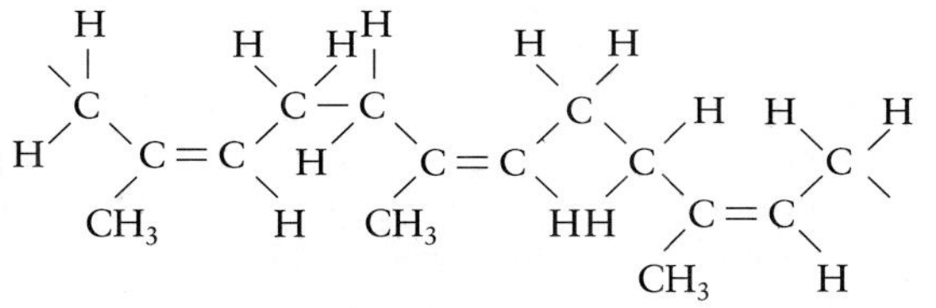

(cis version)

11.73 (a) $ClCH{=}CH_2$ (b) $F(Cl)C{=}CF_2$

11.75 (a) $-OC-CO-NH-(CH_2)_4-NH-CO-CO-NH-(CH_2)_4-NH-$

(b) $-OC-CH(CH_3)-NH-OC-CH(CH_3)-NH-$

11.77 An isotactic polymer is a polymer in which the substituents are all on the same side of the chain. A syndiotactic polymer is a polymer in which the substituent groups alternate, one side of the chain to the other. An atactic polymer is a polymer in which the groups are randomly attached, to one side or the other, along the chain.

11.79 block copolymer

11.81 Characteristics (a), (b), and (c) all increase with increasing molar mass.

11.83 Highly linear, unbranched chains allow for maximum interaction between chains. This leads to stronger forces between chains and thus a stronger material.

11.85 (a) 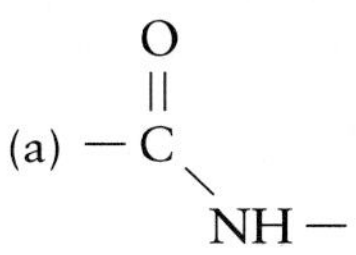 $-\overset{O}{\overset{\|}{C}}-NH-$ (b) amide (c) condensation

11.87 Serine, threonine, tyrosine, aspartic acid, glutamic acid, lysine, arginine, histidine, asparagine, and glutamine contribute to tertiary structure through hydrogen bonding. Proline and tryptophan generally do not contribute through hydrogen bonding because they are typically found in hydrophobic regions of proteins.

11.89 $H-\underset{CH_2-C_6H_4-OH}{\overset{NH_2}{C}}-\overset{O}{\overset{\|}{C}}-NH-CH_2-COOH$

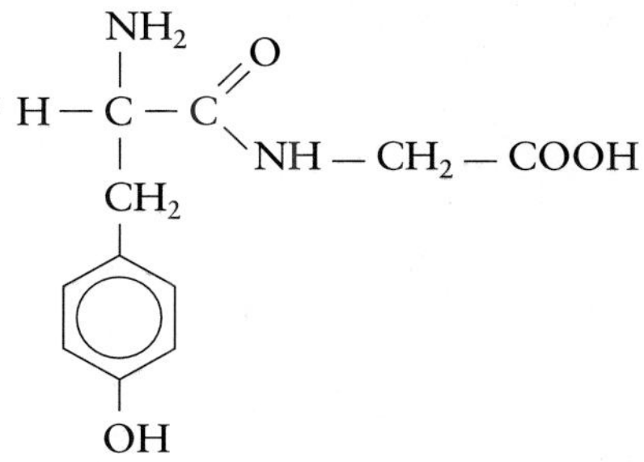

11.91 alcohol, aldehyde

11.93 (a) GTACTCAAT (b) ACTTAACGT

11.95 There are a large number and a wide variety of organic compounds because of the ability of carbon to form four bonds with as many as four different atoms or groups of atoms, the ability of carbon to bond directly to other carbon atoms to form long chains, and the ability of carbon-containing compounds to form a variety of isomers.

11.97 Aromatic carbon-carbon bonds are intermediate in length (139 pm) between that of a C—C single bond (154 pm) and a C=C double bond (134 pm). An aromatic ring is characterized by a delocalized π system in contrast to the localized bonds present in aliphatic hydrocarbons.

11.99 The empirical formula is C_2H_5. The compound is an alkane; the molecular formula might be C_4H_{10}, which matches the formula for alkanes (C_nH_{2n+2}). It is not an alkene or alkyne, because they both have mole H : mole C ratios less than 2.5.

11.101 (a) 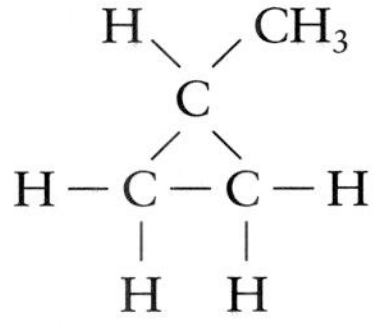 (b)

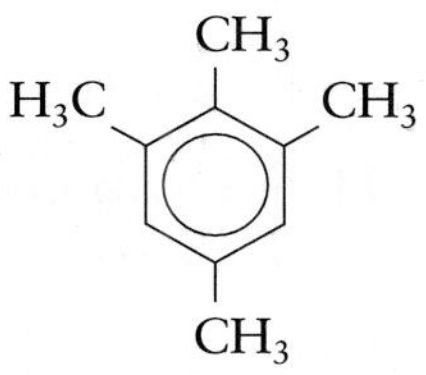

(c) 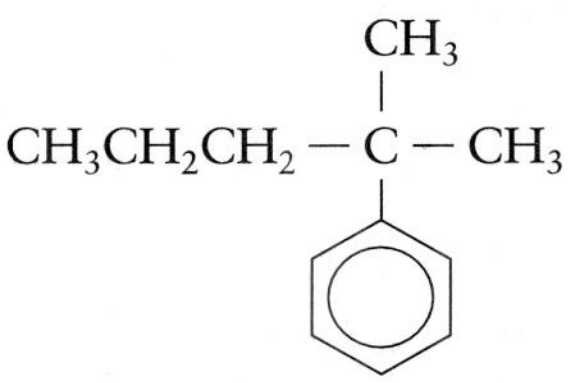 $CH_3CH_2CH_2-\underset{C_6H_5}{\overset{CH_3}{C}}-CH_3$ (d)

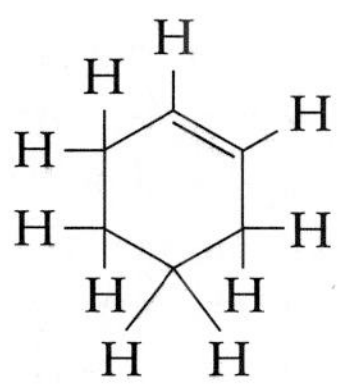

(e) 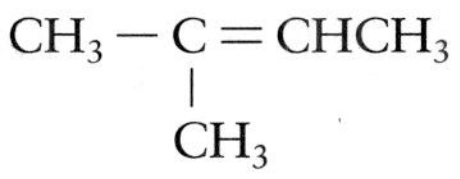 $CH_3-\underset{CH_3}{C}{=}CHCH_3$ (f)

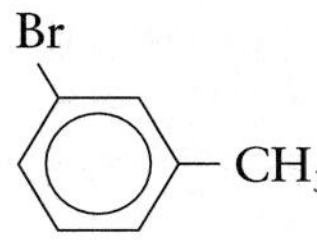

11.103
$$\begin{array}{ccccccccc} & & H & & & & H & & \\ & & | & & & & | & & \\ H & - & C & - & C\equiv C & - & C & - & H \\ & & | & & & & | & & \\ & & H & & & & H & & \end{array} + 2\,HBr \rightarrow$$

$$\begin{array}{ccccccccc} & & H & & Br & & Br & & H & & \\ & & | & & | & & | & & | & & \\ H & - & C & - & C & - & C & - & C & - & H \\ & & | & & | & & | & & | & & \\ & & H & & H & & H & & H & & \end{array}$$

11.105 (a)
$$\begin{array}{ccccccccccc} & & H & & H & & & & H & & H & & \\ & & | & & | & & & & | & & | & & \\ H & - & C & - & C & - & O & - & C & - & C & - & H \\ & & | & & | & & & & | & & | & & \\ & & H & & H & & & & H & & H & & \end{array}$$

Diethyl ether

$$\begin{array}{cccccccccc} & & H & & H & & H & & H & & \\ & & | & & | & & | & & | & & \\ H & - & C & - & C & - & C & - & C & - & OH \\ & & | & & | & & | & & | & & \\ & & H & & H & & H & & H & & \end{array}$$

1-Butanol

(b) 1-Butanol can hydrogen bond with itself but diethyl ether cannot, so 1-butanol molecules are held together more strongly in the liquid, thereby resulting in the higher boiling point.

11.107 $HO-\bigcirc-CHO$

11.109
$$\begin{array}{ccccccccccccc} & & H & & H & & Br & & H & & H & & \\ & & | & & | & & | & & | & & | & & \\ H & - & C & - & C & - & C & - & C & - & C & - & H \\ & & | & & | & & | & & | & & | & & \\ & & H & & H & & H & & H & & H & & \end{array}$$

3-Bromopentane

$$\begin{array}{ccccccccccccc} & & H & & Br & & H & & H & & H & & \\ & & | & & | & & | & & | & & | & & \\ H & - & C & - & C & - & C & - & C & - & C & - & H \\ & & | & & | & & | & & | & & | & & \\ & & H & & H & & H & & H & & H & & \end{array}$$

2-Bromopentane

11.111 Increasing value as fibers: polyalkenes < polyesters < polyamides, due to the increasing strength of intermolecular forces between the chains. The three types of polymer have about the same London forces if their chains are about the same length. However, polyesters also have dipole forces contributing to the strength of intermolecular forces, and polyamides form very strong hydrogen bonds between their chains.

11.113 (a) Primary structure is the sequence of amino acids along a protein chain. Secondary structure is the conformation of the protein or the manner in which the chain is coiled or layered as a result of interactions between amide groups. Tertiary structure is the shape into which sections of the proteins twist and intertwine as a result of interactions between side groups of the amino acids in the protein. If the protein consists of several polypeptide units, then the manner in which the units stick together is the quaternary structure. (b) The primary structure is held together by covalent bonds. Secondary structure is stabilized by intermolecular forces. Tertiary and quaternary structures are maintained by the intermolecular forces between the chains.

11.115 water and $CH_3(CH_2)_2CH{=}CH_2$

11.117 $HO-\bigcirc-CH_2-CH_2-NH_2$ (with a second HO on the ring)

11.119 Condensation polymerization involves the loss of a small molecule, often water or HCl, when monomers are added together. Dacron is more linear than the polymer obtained from benzene-1,2-dicarboxylic acid and ethylene glycol, so Dacron can be more readily spun into yarn.

CHAPTER 12

12.1 Both water and methanol are capable of hydrogen bonding, and they readily hydrogen bond to each other. Toluene cannot hydrogen bond to methanol; the weaker London forces between methanol and toluene result in limited solubility.

12.3 (a) water (b) benzene (c) water

12.5 (a) hydrophilic (b) hydrophobic (c) hydrophobic (d) hydrophilic

12.7 (a) Grease is composed primarily of nonpolar hydrocarbons, which are expected to dissolve in gasoline (also a nonpolar hydrocarbon mixture) but not in the very polar water. (b) Soaps are long-chain molecules with polar and nonpolar ends. The polar end of soap dissolves in water and the nonpolar end dissolves in the nonpolar grease.

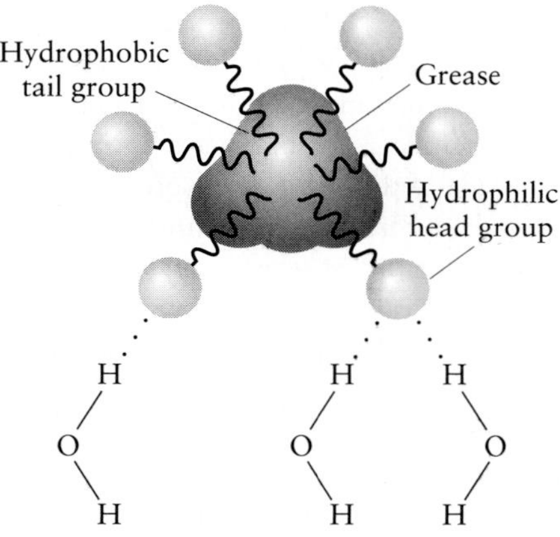

12.9 (a) $6.4 \times 10^{-4}\ mol{\cdot}L^{-1}$ (b) $1.5 \times 10^{-2}\ mol{\cdot}L^{-1}$ (c) $2.3 \times 10^{-3}\ mol{\cdot}L^{-1}$

12.11 0.59 L

12.13 (a) The equilibrium will shift to the left, and the concentration of CO_2 in solution will double. (b) No change in the equilibrium will occur; the partial pressure of CO_2 is unchanged and the concentration is unchanged.
12.15 (a) 4 ppm (b) 0.1 atm (c) 0.5 atm air
12.17 Ion hydration enthalpies for ions of the same charge within a group of elements become less negative down a group. This trend parallels increasing ion size. Hydration energy is an ion-dipole interaction, so it should be higher for small, high charge-density ions.
12.19 (a) negative (b) $Li_2SO_4(s) \rightarrow 2\,Li^+(aq) + SO_4^{2-}(aq) + \text{heat}$ (c) enthalpy of hydration
12.21 (a) $+6.7 \times 10^2$ J for NaCl (b) -58 J for NaBr (c) -2.46×10^4 J for $AlCl_3$ (d) $+3.21 \times 10^3$ J for NH_4NO_3
12.23 $-51\ kJ{\cdot}mol^{-1}$
12.25 The enthalpies of solution of the alkali metal chlorides increase as the cation becomes larger and less strongly hydrated by water. All the alkali metal chlorides are soluble in water, but AgCl is not. AgCl has a relatively large positive enthalpy of solution. When dissolving is highly endothermic, the small increase in disorder due to solution formation may not be enough to compensate for the decrease in disorder of the surroundings and a solution does not form. This is the case for AgCl.
12.27 (a) The disorder of the system increases while its energy decreases. (b) The disorder and the energy of the surroundings increase.
12.29 (a) $0.330\ mol{\cdot}L^{-1}$ (b) $0.222\ mol{\cdot}L^{-1}$
12.31 (a) 5.8 g NaCl (b) 2.8 g $CaCl_2$ (c) 57 g $C_6H_{12}O_6$
12.33 (a) 4.0% NaCl (b) 3.8% NaCl (c) 0.823% $C_{12}H_{22}O_{11}$
12.35 (a) $x_{H_2O} = 0.560$, $x_{C_2H_5OH} = 0.440$, (b) $x_{H_2O} = 0.470$, $x_{C_2H_5OH} = 0.530$, (c) $x_{C_6H_{12}O_6} = 1.8 \times 10^{-3}$, $x_{H_2O} = 0.998$ (or about 1.0)
12.37 (a) 0.684 m (b) 9.6 m (c) 0.162 m
12.39 (a) 4.11 g $ZnCl_2$ (b) 0.62 g $KClO_3$ (c) 7.36 g $KClO_3$
12.41 (a) 0.0612 (b) 3.62 m C_2H_5OH
12.43 (a) $x_{cation} = x_{anion} = 1.8 \times 10^{-3}$, $x_{water} = 0.996$ (or about 1.0) (b) $x_{cation} = 7.1 \times 10^{-3}$, $x_{anion} = 3.6 \times 10^{-3}$, $x_{water} = 0.99$ (c) $x_{cation} = x_{anion} = 1.9 \times 10^{-2}$, $x_{water} = 0.96$
12.45 (a) $0.36\ mol{\cdot}kg^{-1}$ (b) $x = 0.0064$
12.47 (a) 684 Torr (b) 759 Torr
12.49 (a) 4.5 Torr (b) 350 Torr (2 sf) (c) 0.16 Torr
12.51 (a) 0.052 (b) $1.1 \times 10^2\ g{\cdot}mol^{-1}$
12.53 (a) 0.051°C, 100.051°C (b) 0.22°C, 100.22°C (c) 0.091°C, 100.091°C
12.55 (a) 100.34°C (b) 81.0°C
12.57 $1.7 \times 10^2\ g{\cdot}mol^{-1}$
12.59 (a) 0.186°C, −0.19°C (b) 0.82°C, −0.82°C (c) 0.172°C, −0.172°C
12.61 $182\ g{\cdot}mol^{-1}$
12.63 (a) 1.7°C (b) $1.63\ mol{\cdot}kg^{-1}$ (c) 0.208 mol
12.65 (a) 1.84 (b) $0.318\ mol{\cdot}kg^{-1}$ (c) 84%
12.67 −0.20°C
12.69 (a) 0.24 atm (b) 48 atm (c) 0.72 atm
12.71 $2.0 \times 10^3\ g{\cdot}mol^{-1}$
12.73 $3.3 \times 10^3\ g{\cdot}mol^{-1}$
12.75 (a) 1.2 atm (b) 0.048 atm (c) 8.2×10^{-5} atm
12.77 The equilibrium between solid and aqueous $CuSO_4$ is a dynamic process. The ions continually leave the solid and are replaced with ions returning to the solid from solution.
12.79 (a) 2.0% (b) 50% (c) 15.3% (d) 20%
12.81 (a) 59.1 mL of solution (b) $1.13\ mol{\cdot}kg^{-1}$ (c) 32.1 g H_2SO_4
12.83 45 g of 70% HNO_3 solution
12.85 at 30°C, 31.5 Torr; at 100°C, 753 Torr; at 0°C, 4.54 Torr
12.87 (a) $149\ g{\cdot}mol^{-1}$ (b) $C_6H_4Cl_2$ (c) The actual molar mass is $146.99\ g{\cdot}mol^{-1}$.
12.89 (a) $C_9H_{13}O_3N$ (b) $1.8 \times 10^2\ g{\cdot}mol^{-1}$ (c) $C_9H_{13}O_3N$
12.91 $2.5 \times 10^5\ g{\cdot}mol^{-1}$
12.93 Hypotonic solutions will cause a net flow of solvent into the cells to equalize the osmotic pressure. The cells will burst and die (hemolysis). Hypertonic solutions will cause a net flow of solvent out of the cells to equalize the osmotic pressure. The cells will shrink and die.
12.95 Because molar mass $= (iRT\ \text{mass}_{solute})/(\Pi V_{soln})$, the determination of the molar mass of a solute requires a measurement of mass, volume, temperature, and osmotic pressure. Osmotic pressures are generally large and can be determined quite accurately, yielding accurate molar masses. Boiling point elevations and freezing point depressions are usually small and not very accurate, so molar mass determinations based on those measures often are not accurate.
12.97 When a drop of aqueous solution containing $Ca(HCO_3)_2$ seeps through a cave ceiling, it encounters a situation where the partial pressure of CO_2 is reduced and the reaction $Ca(HCO_3)_2(aq) \rightarrow CaCO_3(s) + CO_2(aq) + H_2O(l)$ occurs. The concentration of CO_2 decreases as the CO_2 escapes as a gas, with $CaCO_3$ precipitating and forming a column that extends downward from the ceiling to form a stalactite. Stalagmite formation is similar, except the drop falls to the floor and the precipitate grows upward.
12.99 $i = 0.996$. A van't Hoff factor just slightly less than 1 indicates association, rather than dissociation, in solution. Hydrogen-bonding between polar CH_3COOH molecules reduces the effective number of particles in solution.

12.101 (a) 4.77×10^3 g·mol^{-1} (b) -3.90×10^{-4}°C (c) The small ΔT cannot be measured accurately, so the molar mass determined from it is not accurate. Osmotic pressure can be measured accurately and is the superior method for determining molar mass.
12.103 (a) 7.2 atm (b) 0.4 atm (c) 5 m C_6H_6

CHAPTER 13

13.1 (a) At the molecular level, chemical equilibrium is a dynamic process and interconversion of reactants and products continues to occur. At the macroscopic level, no concentration change in products or reactants is observable at equilibrium, because rates of product and reactant formation are equal. (b) At a fixed temperature, the equilibrium constant is truly a constant, independent of concentrations. However, if the concentration of a reactant is increased, the concentrations of other chemical species involved in the reaction will change and more product will form. Concentrations adjust because the equilibrium constant remains the same.
13.3 The concentrations of reactants and products stop changing when equilibrium has been attained. Equilibrium occurs when the forward and reverse rates of reaction are equal, a condition that will always be reached eventually, given enough time.
13.5 (a) Because the volume is the same, the number of moles of O_2 is larger in the second experiment.
(b) Because K_c is a constant and the denominator is larger in the second case, the numerator must also be larger. So the concentration of O_2 is larger in the second case.
(c) Although $[O_2]^3/[O_3]^2$ is the same, $[O_2]/[O_3]$ will be different, a result seen by solving for K_c in each case.
(d) Because K_c is a constant, $[O_2]^3/[O_3]^2$ is the same.
(e) Because $[O_2]^3/[O_3]^2$ is the same, its reciprocal must be the same. (f) The times will be different, because the time to reach equilibrium depends on concentration.
13.7 (a) $K_c = [COCl][Cl]/[CO][Cl_2]$
(b) $K_c = [HBr]^2/[H_2][Br_2]$
(c) $K_c = [SO_2]^2[H_2O]^2/[H_2S]^2[O_2]^3$
13.9 (a) 0.024 (b) 6.4 (c) 1.7×10^3
13.11 $K_c = 48.8$ for case 1 and 48.9 for cases 2 and 3
13.13 (a) 4.4×10^{-4} (b) 1.90
13.15 All the equilibria are heterogeneous, because in each case more than one phase is involved in the equilibrium. (a) $K_p = P_{H_2S}P_{NH_3}$; (b) $K_p = P_{CO_2}P^2_{NH_3}$;
(c) $K_P = P_{O_2}$
13.17 (a) $K_c = 1/[Cl_2]$ (b) $K_c = [N_2O][H_2O]^2$
(c) $K_c = [CO_2]$
13.19 2.09×10^{-5} mol·L^{-1}
13.21 19 atm
13.23 (a) 0.50 (b) no (c) products
13.25 (a) 6.94 (b) yes
13.27 6.20×10^{-3}
13.29 1.58×10^{-8}
13.31 $[H_2O] = [CH_3COOC_2H_5]$; $K_c = 3.9$ for cases 1 and 3 and $K_c = 4.0$ for case 2
13.33 (a) $[Cl] = 1.1 \times 10^{-5}$ mol·L^{-1}, $[Cl_2] = 1.0 \times 10^{-3}$ mol·L^{-1}, 0.55% decomposed
(b) $[F] = 3.18 \times 10^{-4}$ mol·L^{-1}, $[F_2] = 8.41 \times 10^{-4}$ mol·L^{-1}, 19% decomposed (c) Cl_2 is more stable than F_2 at 1000 K
13.35 1.8×10^{-3}%; $[H_2] = [Br_2] = 1.1 \times 10^{-8}$ mol·L^{-1}, $[HBr] = 1.2 \times 10^{-3}$ mol·L^{-1}
13.37 (a) $[PCl_3] = [Cl_2] = 1.0 \times 10^{-2}$ mol·L^{-1}, $[PCl_5] = 9.0 \times 10^{-3}$ mol·L^{-1} (b) 53%
13.39 $[H_2S] = 8.0 \times 10^{-4}$ mol·L^{-1}, $[NH_3] = 0.200$ mol·L^{-1}
13.41 $[Cl_2] = 4.97 \times 10^{-2}$ mol·L^{-1}, $[PCl_3] = 0.200$ mol·L^{-1}, $[PCl_5] = 2.98 \times 10^{-4}$ mol·L^{-1}
13.43 $[NO] = 3.60 \times 10^{-4}$ mol·L^{-1}, $[N_2] = [O_2] = 0.114$ mol·L^{-1}
13.45 1.06
13.47 $[CO_2] = 8.5 \times 10^{-5}$ mol·L^{-1}, $[CO] = 4.9 \times 10^{-3}$ mol·L^{-1}, $[O_2] = 4.6 \times 10^{-4}$ mol·L^{-1} (exact solution)
13.49 $[CH_3COOC_2H_5] = 0.32$ mol·L^{-1}
13.51 (a) decreases (b) decreases (c) increases (d) nothing
13.53 d, i, i, i, nc, d, d
13.55 (a) reactants (b) reactants (c) reactants (d) No effect; there are the same number of moles on each side. (e) reactants
13.57 (a) increases (b) increases
13.59 Quartz would be favored, because it is more dense.
13.61 (a) products (b) products (c) reactants (d) reactants
13.63 No; there is less ammonia at 700 K.
13.65 (a) $Q_c = [SO_2]/[O_2]$, $Q_p = P_{SO_2}/P_{O_2}$
(b) $Q_c = [SO_2][H_2O]/[SO_3][H_2]$, $Q_p = P_{SO_2}P_{H_2O}/P_{SO_3}P_{H_2}$
(c) $Q_c = [WCl_6][H_2]^3/[HCl]^6$, $Q_p = P_{WCl_6}P^3_{H_2}/P^6_{HCl}$
13.67 (a) 1.7×10^6 (b) 7.7×10^{-4}
13.69 No; the reaction proceeds to form more product.
13.71 1.4 mol·L^{-1}
13.73 3.91
13.75 $[NO] = 3.89 \times 10^{-2}$ mol·L^{-1}, $[SO_3] = 4.89 \times 10^{-2}$ mol·L^{-1}, $[SO_2] = 1.1 \times 10^{-3}$ mol·L^{-1}, $[NO_2] = 2.11 \times 10^{-2}$ mol·L^{-1}
13.77 (a) $[N_2O] = 0.0100$ mol·L^{-1}, $[O_2] = 0.0410$ mol·L^{-1} (b) 23.2
13.79 (a) no (b) toward the products (c) $[PCl_5] = 3.1$ mol·L^{-1}, $[PCl_3] = 5.9$ mol·L^{-1}, $[Cl_2] = 0.9$ mol·L^{-1}
13.81 $P_{H_2} = P_{Cl_2} = 3.9 \times 10^{-18}$ atm, $P_{HCl} = 0.22$ atm
13.83 (a) $P_{NOCl} = 91.6$ atm, $P_{NO} = 2.72$ atm, $P_{Cl_2} = 1.36$ atm (b) 2.9%
13.85 (a) reactants (b) no effect (c) products

(d) products (e) no effect (f) products (g) reactants
13.87 (a) decrease, because $Q_c < K_c$ (reactants are favored) (b) [CO] = 0.1100 mol·L^{-1}, [H_2] = 0.150 mol·L^{-1}, [CH_3OH] = 0 mol·L^{-1}
13.89 [NO_2] = 6.88 × 10^{-3} mol·L^{-1}, [N_2O_4] = 1.02 × 10^{-2} mol·L^{-1}
13.91 $K_c = x^2/(A - x)(B - x)$; x = 0.42 mol ester
13.93 (a) 11.2 (b) [N_2O_4] and [NO_2] will increase, but K_c remains constant (c) [N_2O_4] = 0.636 mol·L^{-1}, [NO_2] = 2.67 mol·L^{-1}
13.95 $\Delta H°$ +88.6 kJ·mol^{-1}, 305.4 K

CHAPTER 14

14.1 (a) amphiprotic (b) base (c) base (d) acid
14.3 (a) $CH_3NH_3^+$ (b) $NH_2NH_3^+$ (c) H_2CO_3 (d) CO_3^{2-} (e) $C_6H_5O^-$ (f) $CH_3CO_2^-$
14.5 (a) H_2SO_4(acid 1) + H_2O(base 2) ⇌ H_3O^+(acid 2) + HSO_4^-(base 1)
(b) $C_6H_5NH_3^+$(acid 1) + H_2O(base 2) ⇌ H_3O^+(acid 2) + $C_6H_5NH_2$(base 1)
(c) $H_2PO_4^-$(acid 1) + H_2O(base 2) ⇌ H_3O^+(acid 2) + HPO_4^{2-}(base 1)
(d) HCOOH(acid 1) + H_2O(base 2) ⇌ H_3O^+(acid 2) + HCO_2^- (base 1)
(e) $NH_2NH_3^+$ (acid 1) + H_2O(base 2) ⇌ H_3O^+(acid 2) + NH_2NH_2(base 1)
14.7 (a) HCO_3^-(acid 1) + H_2O(base 2) ⇌ H_3O^+(acid 2) + CO_3^{2-}(base 1) and HCO_3^-(base 2) + H_2O(acid 1) ⇌ OH^-(base 1) + H_2CO_3(acid 2)
(b) HPO_4^{2-}(acid 1) + H_2O(base 2) ⇌ H_3O^+(acid 2) + PO_4^{3-}(base 1) and HPO_4^{2-}(base 2) + H_2O(acid 1) ⇌ OH^-(base 1) + $H_2PO_4^-$(acid 2)
14.9 (a) acid: HNO_3; base: HPO_4^{2-} (b) conjugate base: NO_3^-; conjugate acid: $H_2PO_4^-$
14.11 (a) 3.2 × 10^{-13} mol·L^{-1} (b) 1.0 × 10^{-10} mol·L^{-1} (c) 5.0 × 10^{-14} mol·L^{-1}
14.13 (a) 1.6 × 10^{-7} mol·L^{-1}; 6.80 (b) 1.6 × 10^{-7} mol·L^{-1}
14.15 [HCl]$_{nom}$ = [H_3O^+] = [Cl^-] = 0.96 mol·L^{-1}, [OH^-] = 1.0 × 10^{-14} mol·L^{-1}
14.17 [$Ba(OH)_2$] = [Ba^{2+}] = 2.9 × 10^{-2} mol·L^{-1}, [OH^-] = 5.8 × 10^{-2} mol·L^{-1}, [H_3O^+] = 1.7 × 10^{-13} mol·L^{-1}
14.19 (a) 5.0 × 10^{-4} mol·L^{-1} (b) 2 × 10^{-7} mol·L^{-1} (c) 4 × 10^{-5} mol·L^{-1} (d) 5 × 10^{-6} mol·L^{-1} (e) (b) < (d) < (c) < (a)
14.21 (a) pH = 4.70, pOH = 9.30 (b) pH = 0.00, pOH = 14.00 (c) pH = 13.3, pOH = 0.7 (d) pH = 4.30, pOH = 9.70
14.23 (a) pH = 2.00, pOH = 12.00 (b) pH = 0.66, pOH = 13.34 (c) pH = 11.30, pOH = 2.70 (d) pH = 10.94, pOH = 3.06 (e) pH = 11.15, pOH = 2.854 (f) pH = 4.07, pOH = 9.93
14.25 (a) formic acid, K_a = 1.8 × 10^{-4}, pK_a = 3.75 (b) acetic acid, K_a = 1.8 × 10^{-5}, pK_a = 4.75 (c) trichloroacetic acid, K_a = 3.0 × 10^{-1}, pK_a = 0.52 (d) benzoic acid, K_a = 6.5 × 10^{-5}, pK_a = 4.19; acetic acid < benzoic acid < formic acid < trichloroacetic acid
14.27 (a) K_a = 7.6 × 10^{-3} (b) K_a = 1.0 × 10^{-2} (c) K_a = 3.5 × 10^{-3} (d) K_a = 1.2 × 10^{-2}; H_2SeO_3 < H_3PO_4 < H_3PO_3 < H_2SeO_4
14.29 $HSeO_3^-$, hydrogen selenite ion, strongest base; $HSeO_4^-$, hydrogen selenate ion, weakest base (b) $HSeO_3^-$, K_b = 2.9 × 10^{-12}; $HSeO_4^-$, K_b = 8.3 × 10^{-13}; (c) $HSeO_3^-$, strongest base corresponds to highest pH
14.31 For oxoacids, the greater the number of highly electronegative O atoms attached to the central atom, the stronger the acid. This effect is related to the increased oxidation number of the central atom as the number of O atoms increases. Therefore HIO_3 is the stronger acid with the lower pK_a.
14.33 (a) HCl is the stronger acid because its bond strength is much weaker than the bond in HF and bond strength is the dominant factor in determining the strength of binary acid of elements in the same group. (b) $HClO_2$ is stronger; there is one more O atom attached to the Cl atom in $HClO_2$ than in HClO. The additional O helps pull the electron of the H atom out of the O—H bond. The oxidation number of Cl is higher in $HClO_2$ than in HClO. (c) $HClO_2$ is stronger; Cl has a greater electronegativity than Br, making the H—O bond in $HClO_2$ more polar than in $HBrO_2$. (d) $HClO_4$ is stronger; Cl has a greater electronegativity than P. (e) HNO_3 is stronger. The explanation is the same as part (b) above. HNO_3 has one more O atom. (f) H_2CO_3 is stronger; C has greater electronegativity than Ge. See part (c) above.
14.35 (a) The —CCl_3 group, which is bonded to the carboxyl group, —COOH, in trichloroacetic acid, is more electron withdrawing than the —CH_3 group in acetic acid. Thus, trichloroacetic acid is the stronger acid. (b) The CH_3 group in acetic acid has electron-donating properties, which means that it is less electron withdrawing than the H attached to the carboxyl group in formic acid, HCOOH. Thus, formic acid is a slightly stronger acid than acetic acid.
14.37 The larger the K_a value, the stronger the corresponding acid. 2,4,6-Trichlorophenol is the stronger acid because the chlorines have a greater electron-withdrawing power than the hydrogens they replaced in the unsubstituted phenol.
14.39 The order is aniline < ammonia < methylamine < ethylamine, which suggests that arylamines (amines in which the nitrogen atom is attached to a benzene ring) < ammonia < alkylamines, and methyl < ethyl < etc.
14.41 (a) [H_3O^+] = 3.6 × 10^{-3} mol·L^{-1},

$[OH^-] = 2.8 \times 10^{-12}\ mol\cdot L^{-1}$ (b) $[H_3O^+] = 1.7 \times 10^{-11}\ mol\cdot L^{-1}$, $[OH^-] = 5.8 \times 10^{-4}\ mol\cdot L^{-1}$ (c) $[H_3O^+] = 2.8 \times 10^{-12}\ mol\cdot L^{-1}$, $[OH^-] = 3.6 \times 10^{-3}\ mol\cdot L^{-1}$

14.43 (a) 2.22 (b) 10.65 (c) 2.51 (d) 9.77

14.45 (a) pH = 2.80, pOH = 11.20 (b) pH = 0.96, pOH = 13.04 (c) pH = 2.28, pOH = 11.72

14.47 (a) pH = 11.11, pOH = 2.89, 1.3% (b) pH = 9.14, pOH = 4.86, 0.081% (c) pH = 11.56, pOH = 2.44, 1.8% (d) pH = 10.26, pOH = 3.74, 0.89%

14.49 (a) $K_a = 0.1$, $pK_a = 1.0$ (b) $K_b = 5.7 \times 10^{-4}$, $pK_b = 3.25$

14.51 (a) $0.02\ mol\cdot L^{-1}$ (b) $0.02\ mol\cdot L^{-1}$

14.53 pH = 2.58, $K_a = 6.5 \times 10^{-5}$

14.55 pH = 11.83, $K_b = 5 \times 10^{-4}$

14.57 (a) $H_2SO_4 + H_2O \rightleftharpoons H_3O^+ + HSO_4^-$ and $HSO_4^- + H_2O \rightleftharpoons H_3O^+ + SO_4^{2-}$ (b) $H_3AsO_4 + H_2O \rightleftharpoons H_3O^+ + H_2AsO_4^-$ and $H_2AsO_4^- + H_2O \rightleftharpoons H_3O^+ + HAsO_4^{2-}$ and $HAsO_4^{2-} + H_2O \rightleftharpoons H_3O^+ + AsO_4^{3-}$ (c) $C_6H_4(COOH)_2 + H_2O \rightleftharpoons H_3O^+ + C_6H_4(COOH)CO_2^-$ and $C_6H_4(COOH)CO_2^- + H_2O \rightleftharpoons H_3O^+ + C_6H_4(CO_2)_2^{2-}$

14.59 0.80

14.61 (a) 4.68 (b) 1.28 (c) 3.79

14.63 $[H_3O^+] = [OH^-]$

14.65 (a) C_2H_5COOH(acid 1) + H_2O(base 2) $\rightleftharpoons$ H_3O^+(acid 2) + $C_2H_5CO_2^-$(base 1) (b) $HClO_3$(acid 1) + H_2O(base 2) $\rightleftharpoons$ H_3O^+(acid 2) + ClO_3^-(base 1) (c) $C_8H_7O_2COOH$(acid 1) + H_2O(base 2) $\rightleftharpoons$ H_3O^+(acid 2) + $C_8H_7O_2CO_2^-$(base 1) (d) $C_8H_{10}N_4O_2$(base 2) + H_2O(acid 1) $\rightleftharpoons$ OH^-(base 1) + $C_8H_{10}N_4O_2H^+$(acid 2) (e) C_5H_5N(base 2) + H_2O(acid 1) $\rightleftharpoons$ OH^-(base 1) + $C_5H_5NH^+$(acid 2)

14.67 (a) If a solution is a concentrated solution of an acid with $[H_3O^+] > 1\ mol\cdot L^{-1}$, the pH will be negative. (b) If a solution is very basic with $[OH^-] > 1\ mol\cdot L^{-1}$, the pH will be greater than 14.

14.69 (a) $4.7 \times 10^{-10}\ mol\cdot L^{-1}$ (b) $1.1 \times 10^{-8}\ mol\cdot L^{-1}$ (c) $0.98\ mol\cdot L^{-1}$ (d) $4.7 \times 10^{-5}\ mol\cdot L^{-1}$ (e) $0.010\ mol\cdot L^{-1}$ (f) $1.1 \times 10^{-12}\ mol\cdot L^{-1}$

14.71 HNO_2, because it has a larger K_a.

14.73 (a) $C_6H_5OH + H_2O \rightleftharpoons H_3O^+ + C_6H_5O^-$, $NH_4^+ + H_2O \rightleftharpoons H_3O^+ + NH_3$ (b) $pK_a(C_6H_5OH) = 9.89$, $pK_a(NH_4^+) = 9.25$ (c) NH_4^+

14.75 (a) $HBrO_4$, because Br is slightly more electronegative than I. (b) HI, because HI has a weaker bond than HF. (c) HIO_3, because I has more oxygens attached to it in HIO_3 than in HIO_2. (d) H_2SeO_4, because Se is more electronegative than As.

14.77 0.12%, $pK_b = 6.89$, $pK_a = 7.11$

14.79 (a) $D_2O + D_2O \rightleftharpoons D_3O^+ + OD^-$ (b) $pK_w = 14.870$ (c) $[D_3O^+] = [OD^-] = 3.67 \times 10^{-8}\ mol\cdot L^{-1}$ (d) pD = pOD = 7.435 (e) pD + pOD = pK_w = 14.870

14.81 (a) $pK_b = 3.1$, $pK_a = 10.9$ (b) pH = 7.11, $K_b = 6.7 \times 10^{-12}$

14.83 (a) 6.54 (b) 2.12 (c) 1.49

14.85 $[H_3O^+] = 0.032\ mol\cdot L^{-1}$, $[HSO_3^-] = 0.032\ mol\cdot L^{-1}$, $[OH^-] = 3.1 \times 10^{-13}\ mol\cdot L^{-1}$, $[H_2SO_3] = 0.068\ mol\cdot L^{-1}$, $[SO_3^{2-}] = 1.2 \times 10^{-7}\ mol\cdot L^{-1}$

14.87 $pK_w' = pK_w - (\Delta H°/2.303R)\{(T' - T)/TT'\}$, $pK_w = 11.99$, pH = 6.00

CHAPTER 15

15.1 (a) pH < 7, acidic, $NH_4^+(aq) + H_2O(l) \rightleftharpoons H_3O^+(aq) + NH_3(aq)$ (b) pH > 7, basic, $H_2O(l) + CO_3^{2-}(aq) \rightleftharpoons HCO_3^-(aq) + OH^-(aq)$ (c) pH > 7, basic, $H_2O(l) + F^-(aq) \rightleftharpoons HF(aq) + OH^-(aq)$ (d) pH = 7, neutral (e) pH < 7, acidic, $Al(H_2O)_6^{3+}(aq) + H_2O(l) \rightleftharpoons H_3O^+(aq) + Al(H_2O)_5OH^{2+}(aq)$ (f) pH < 7, acidic, $Cu(H_2O)_6^{2+}(aq) + H_2O(l) \rightleftharpoons H_3O^+(aq) + Cu(H_2O)_5OH^+(aq)$

15.3 (a) $K_a = 5.6 \times 10^{-10}$ (b) $K_b = 1.8 \times 10^{-4}$ (c) $K_b = 2.9 \times 10^{-11}$ (d) $K_b = 3.3 \times 10^{-7}$ (e) $K_b = 2.3 \times 10^{-8}$ (f) $K_a = 1.5 \times 10^{-10}$

15.5 (a) 9.02 (b) 5.13 (c) 2.92 (d) 11.24

15.7 (a) 9.18 (b) 4.74

15.9 (a) $6.7 \times 10^{-6}\ mol\cdot L^{-1}$ (b) 5.04

15.11 (a) When solid sodium acetate is added to an acetic acid solution, the concentration of H_3O^+ decreases because the equilibrium $HC_2H_3O_2(aq) + H_2O(l) \rightleftharpoons H_3O^+(aq) + C_2H_3O_2^-(aq)$ shifts to the left to relieve the stress imposed by the increase of $[C_2H_3O_2^-]$. (b) When HCl is added to a benzoic acid solution, the percentage of benzoic acid that is deprotonated decreases because the equilibrium $C_6H_5COOH(aq) + H_2O(l) \rightleftharpoons H_3O^+(aq) + C_6H_5COO^-(aq)$ shifts to the left to relieve the stress imposed by the increased $[H_3O^+]$. (c) When solid NH_4Cl is added to an ammonia solution, the pH decreases as the concentration of OH^- decreases, because the equilibrium $NH_3(aq) + H_2O(l) \rightleftharpoons NH_4^+(aq) + OH^-(aq)$ shifts to the left to relieve the stress imposed by the increased $[NH_4^+]$.

15.13 (a) $pK_a = 3.08$, $K_a = 8.4 \times 10^{-4}$ (b) 2.78

15.15 (a) $4.0 \times 10^{-9}\ mol\cdot L^{-1}$ (b) $2.8 \times 10^{-10}\ mol\cdot L^{-1}$ (c) $1.0 \times 10^{-9}\ mol\cdot L^{-1}$ (d) $2.8 \times 10^{-11}\ mol\cdot L^{-1}$

15.17 (a) pH = 2.22, pOH = 11.78 (b) pH = 1.62, pOH = 12.38 (c) pH = 1.92, pOH = 12.08

15.19 The pH increases by 1.90

15.21 (a) 9.69 (b) 8.93 (c) 9.31
15.23 (a) 1.30 (b) 12.70 (c) 1.04
15.25 13.52
15.27 19.7 mL HCl
15.29

14
7
1
pH
(a) Initial pH
(b) pH at stoichiometric point = 7.0
0 10 20
Volume of KOH added (mL)

15.31 (a) 9.17×10^{-3} L HCl (b) 0.0183 L (c) 0.0635 $mol \cdot L^{-1}$ (d) 2.26
15.33 (a) 423 mL (b) 0.142 $mol \cdot L^{-1}$
15.35 69.8%
15.37 (a) 13.04 (b) 12.82 (c) 12.55 (d) 7.00 (e) 1.81 (f) 1.55
15.39 (a) 2.87 (b) 4.56 (c) 12.5 mL (d) 4.75 (e) 25.0 mL (f) 8.72 (g) thymol blue or phenolphthalein
15.41 (a) 11.22 (b) 8.95 (c) 11.25 mL (d) 9.25 (e) 22.5 mL (f) 5.24 (g) methyl red or bromocresol green
15.43 (a) 2.02 (b) 2.65 (c) 3.15 (d) 7.94 (e) 12.19 (f) 12.45 (g) phenol red
15.45 (a) 3.2–4.4 (b) 5.0–8.0 (c) 4.8–6.0 (d) 8.2–10.0 (see Table 15.3)
15.47 Use (d) phenolphthalein, because the pH at the stoichiometric point of the titration is 8.87.
15.49 (a) not a buffer (b) buffer, $HClO(aq) + H_2O(l) \rightleftharpoons H_3O^+(aq) + ClO^-(aq)$ (c) buffer, $(CH_3)_3N(aq) + H_2O(l) \rightleftharpoons (CH_3)_3NH^+(aq) + OH^-(aq)$ (d) buffer after partial neutralization, $CH_3COOH(aq) + H_2O(l) \rightleftharpoons H_3O^+(aq) + CH_3CO_2^-(aq)$ (e) not a buffer
15.51 $[ClO^-]/[HClO] = 9.3 \times 10^{-2}$
15.53 (a) 2–4 (b) 3–5 (c) 11.5–13.5 (d) 6–8 (e) 5–7
15.55 (a) $HClO_2$ and $NaClO_2$ (b) NaH_2PO_4 and Na_2HPO_4 (c) $CH_2ClCOOH$ and $NaCH_2ClCO_2$ (d) Na_2HPO_4 and Na_3PO_4
15.57 (a) $[CO_3^{2-}]/[HCO_3^-] = 5.6$ (b) 77 g K_2CO_3 (c) 1.8 g $KHCO_3$ (d) 2.8×10^2 mL
15.59 (a) 4.75 (b) pH = 5.02, change of 0.27 (c) pH = 4.15, change of −0.60
15.61 (a) pH = 6.34, change of 1.59 (b) pH = 4.58, change of −0.17
15.63 (a) $K_{sp} = [Ag^+][Br^-]$ (b) $K_{sp} = [Ag^+]^2[S^{2-}]$ (c) $K_{sp} = [Ca^{2+}][OH^-]^2$ (d) $K_{sp} = [Ag^+]^2[CrO_4^{2-}]$
15.65 (a) 7.7×10^{-13} (b) 1.7×10^{-14} (c) 5.3×10^{-3} (d) 6.9×10^{-9}
15.67 (a) 1.2×10^{-17} $mol \cdot L^{-1}$ (b) 1.1×10^{-18} $mol \cdot L^{-1}$ (c) 9.3×10^{-5} $mol \cdot L^{-1}$
15.69 1.0×10^{-12}
15.71 (a) 8.0×10^{-10} $mol \cdot L^{-1}$ (b) 1.3×10^{-16} $mol \cdot L^{-1}$ (c) 4.0×10^{-4} $mol \cdot L^{-1}$ (d) 6.3×10^{-6} $mol \cdot L^{-1}$
15.73 (a) 1.6×10^{-5} $mol \cdot L^{-1}$ (b) 2.7×10^2 μg $AgNO_3$
15.75 6.41
15.77 pH ≥ 12.16
15.79 (a) will precipitate (b) will not precipitate
15.81 (a) will precipitate (b) will not precipitate
15.83 (a) first to last: $Ni(OH)_2$, $Mg(OH)_2$, $Ca(OH)_2$ (b) $Ni(OH)_2$ precipitates around pH = 7, $Mg(OH)_2$ around pH = 10, $Ca(OH)_2$ around pH = 13
15.85 2.1×10^{-3} $mol \cdot L^{-1}$
15.87 (a) 1.0×10^{-12} $mol \cdot L^{-1}$ (b) 3.0×10^{-5} $mol \cdot L^{-1}$ (c) 2.0×10^{-3} $mol \cdot L^{-1}$ (d) 2.0×10^{-1} $mol \cdot L^{-1}$
15.89 (a) $CaF_2(s) + 2\,H_2O(l) \rightleftharpoons Ca^{2+}(aq) + 2\,HF(aq) + 2\,OH^-(aq)$, $K = 3.4 \times 10^{-32}$ (b) 9.4×10^{-7} $mol \cdot L^{-1}$ (c) 2.0×10^{-5} $mol \cdot L^{-1}$
15.91 (a) neutral, because neither K^+ nor I^- is acidic or basic (b) basic, because F^- is basic (c) acidic, because $Cr(H_2O)_6^{3+}$ is an acid (d) acidic, because $C_6H_5NH_3^+$ is the conjugate acid of $C_6H_5NH_2$ (e) basic, because CO_3^{2-} is a base (f) acidic, because $Cu(H_2O)_6^{2+}$ is an acid
15.93 (a) 0.19% (b) 1.1×10^{-5}
15.95 12.70
15.97 pH = 5.05, $HCOOH(aq) = 4.4 \times 10^{-3}$ $mol \cdot L^{-1}$
15.99 The end point of an acid-base titration is the pH at which the indicator color change is observed. The stoichiometric point is the pH at which an exactly equivalent amount of titrant (in moles) has been added to the solution being titrated.
15.101 7.00
15.103 (a) 1.70 (b) 2.05 (c) 2.8 (d) 11.55 (e) 11.88 (f) 11.4 mL KOH (11 mL to 2 sf)

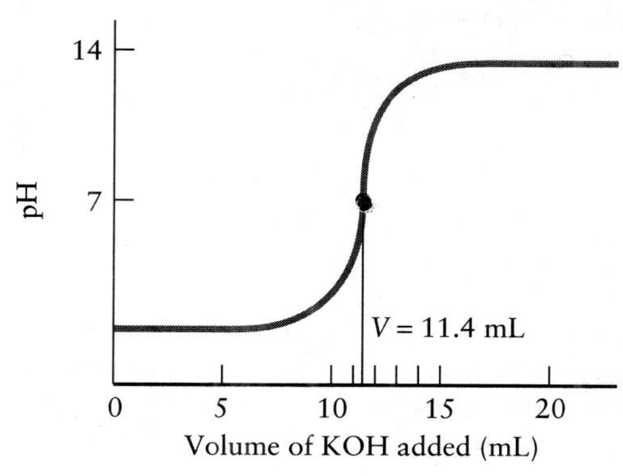

15.105 (a) 1.58×10^{-2} L NaOH (b) 10.54 (c) alizarin yellow R
15.107 2.8×10^{-2}
15.109 (a) initial pH = 5.26; after adding HCl, pH = 4.90 (b) 5.47
15.111 Note that $K_a = [H_3O^+][CO_3^{2-}]/[HCO_3^-]$, pH = $pK_a + \log[CO_3^{2-}]/[HCO_3^-]$ and $[CO_3^{2-}]/[HCO_3^-] = 0.56$. Then prepare a solution containing Na_2CO_3 and $NaHCO_3$ in a molar ratio of 0.56:1.
15.113 (a) 1.2×10^{-17} mol·L^{-1} (b) 1.6×10^{-43} mol·L^{-1} (c) 3.96×10^{-40} g
15.115 No, CaF_2 will not precipitate.
15.117 10.45
15.119 No precipitate will form.
15.121 (a) Tartaric acid is a diprotic acid, therefore titration of tartaric acid is a two-step process and there are two stoichiometric points, around each of which there is a sharp rise in the pH curve. (b) 0.0680 mol·L^{-1} (c) 3.11 (d) 8.64 (e) 2.88×10^{-2} mol·L^{-1} (f) 3.22
15.123 The pH at the first stoichiometric point is 1.54 and at the second stoichiometric point the pH is 7.46.
15.125 (a) $K_c = K_a$(acetic acid) $\times K_b$(hydrazine) $\times (1/K_w)$; 3.1×10^3
(b) $-\log K_c = -\log[N_2H_5^+] - \log([CH_3CO_2^-]/[N_2H_4][CH_3COOH])$
15.127 (a) $ZnS(s) + 2\,H_2O(l) \rightleftharpoons Zn^{2+}(aq) + H_2S(aq) + 2\,OH^-(aq)$, $K = 1.5 \times 10^{-45}$
(b) 1.5×10^{-30} mol·L^{-1} (c) 1.5×10^{-36} mol·L^{-1}

CHAPTER 16

16.1 (a) isolated (b) closed (c) isolated
16.3 1250 kJ
16.5 1.2×10^5 J
16.7 400 kJ is lost. $q = -400$ kJ
16.9 -37 kJ
16.11 (a) HBr(g) (b) $NH_3(g)$ (c) $I_2(l)$ (d) Ar(g) at 1.00 atm
16.13 $C(s) < H_2O(s) < H_2O(l) < H_2O(g)$ Ice has a more complex crystalline structure than carbon and so has a higher entropy. $H_2O(l)$ has more disorder than $H_2O(s)$; in turn $H_2O(g)$ has more disorder than $H_2O(l)$.
16.15 (a) decreases; moles of gas decrease (b) increases; moles of gas increase (c) decreases; lower temperature reduces thermal motion
16.17 (a) -22.0 J·K^{-1}·mol^{-1} (b) 134 J·K^{-1}
16.19 (a) -88.84 J·K^{-1}·mol^{-1}. The entropy change is negative because the number of moles of gas has decreased by one. (b) -173.0 J·K^{-1}·mol^{-1}. The entropy change is negative because the number of moles of gas has decreased by one. (c) $+160.6$ J·K^{-1}·mol^{-1}. The entropy change is positive because the number of moles of gas has increased by one. (d) -36.8 J·K^{-1}·mol^{-1}. It is not immediately apparent, but the four moles of solid products are more ordered than the four moles of solid reactants.
16.21 (a) ΔH is negative, so $\Delta S_{surr} = -\Delta H/T$ is positive. Although ΔS of the system is negative, ΔS_{surr} is positive and of a greater magnitude, because the surroundings are at a lower temperature than the system. The process is spontaneous. (b) No energy in the form of heat is transferred to or from the surroundings, so $\Delta S_{surr} = 0$. But ΔS of the system is positive as a result of the greater disorder generated by the mixing of the two substances, and the process is spontaneous. (c) Both the ΔS for the system and the ΔS for the surroundings are positive. ΔS_{sys} is positive because the number of moles of gas increases. ΔS_{surr} is positive because heat is transferred to the surroundings. The process is spontaneous. (d) The interdiffusion of two gases results in an increase in disorder, hence ΔS of the system is positive. $\Delta S_{surr} = 0$ because no energy is transferred as heat. The process is spontaneous.
16.23 (a) $+403$ J·K^{-1} (b) $+322$ J·K^{-1} (c) -0.310 J·K^{-1}
16.25 (a) 0.341 J·K^{-1}·s^{-1} (b) 2.95×10^4 J·K^{-1} (c) Less, because in the equation, $\Delta S_{surr} = -\Delta H/T$; if T is higher, ΔS_{surr} is smaller.
16.27 (a) -0.02 J·K^{-1} (b) -0.01 J·K^{-1} (c) As seen from the answers to parts (a) and (b), the magnitude of the entropy change in the surroundings is greater when the block is at 25°C. The same amount of heat has a greater effect on entropy changes at lower temperature. At high temperatures, matter is already more chaotic.
16.29 $\Delta S^\circ_{surr} = -6.58$ J·K^{-1}; ΔS°_{tot} is negative, so the process is not spontaneous.
16.31 Under constant temperature and pressure, a negative value of ΔG_r corresponds to spontaneity. $\Delta G_r = \Delta H_r - T\Delta S_r$, and frequently the magnitude of ΔH_r is much greater than the magnitude of $T\Delta S_r$, so no matter what the sign of ΔS_r, ΔG_r will be negative if ΔH_r is negative, and the reaction will be spontaneous.
16.33 (a) negative (b) Sign cannot be predicted; it depends on the magnitudes of ΔH, ΔS, and T. (c) A temperature change can affect the sign in (b) by altering the magnitude of the $-T\Delta S$ term.
16.35 (a) $\Delta S^\circ_{surr} = -73$ J·K^{-1}, $\Delta S^\circ_{sys} = +73$ J·K^{-1}
(b) $\Delta S^\circ_{surr} = -29$ J·K^{-1}, $\Delta S^\circ_{sys} = +29$ J·K^{-1}
(c) $\Delta S^\circ_{surr} = +29$ J·K^{-1}, $\Delta S^\circ_{sys} = -29$ J·K^{-1}
16.37 (a) $\Delta S^\circ_{surr} = -0.11$ kJ·K^{-1}, $\Delta S^\circ_{sys} = +0.11$ kJ·K^{-1}
(b) $\Delta S^\circ_{surr} = -98$ J·K^{-1}, $\Delta S^\circ_{sys} = +98$ J·K^{-1} (c) $\Delta S^\circ_{surr} = -1.0 \times 10^2$ J·K^{-1}, $\Delta S^\circ_{sys} = +1.0 \times 10^2$ J·K^{-1}
16.39 239 K
16.41 (a) 30 kJ·mol^{-1} (b) -11 J·K^{-1}
16.43 (a) -457.17 kJ·mol^{-1}. The negative sign indicates that the reaction has $K > 1$ under standard conditions. The large magnitude of ΔG°_r implies that a temperature change is not likely to affect the spontaneity, but it could at

very high temperatures. (b) -514.4 kJ·mol^{-1}. The negative sign indicates that the reaction has $K > 1$ under standard conditions. The large magnitude of ΔG_r° implies that a temperature change is not likely to affect the spontaneity, but it could at very high temperatures. (c) $+130.4$ kJ·mol^{-1}. The positive sign of ΔG_r° implies that the reaction has $K < 1$ under standard conditions, but the small magnitude indicates that, at a higher temperature, there could be a change in sign. (d) -133.1 kJ·mol^{-1}. The negative sign indicates that the reaction is spontaneous under standard conditions.

16.45 (a) $\frac{1}{2}N_2(g) + \frac{3}{2}H_2(g) \rightarrow NH_3(g)$, -16.5 kJ·mol^{-1}
(b) $H_2(g) + \frac{1}{2}O_2(g) \rightarrow H_2O(g)$, -228.58 kJ·mol^{-1}
(c) $C(s) + \frac{1}{2}O_2(g) \rightarrow CO(g)$, -137.2 kJ·mol^{-1}
(d) $\frac{1}{2}N_2(g) + O_2(g) \rightarrow NO_2(g)$, 51.3 kJ·mol^{-1}

16.47 (a) -141.74 kJ·mol^{-1}, spontaneous (b) $+130.4$ kJ·mol^{-1}, not spontaneous (c) 33.1 kJ·mol^{-1}, not spontaneous (d) $-10{,}590.9$ kJ·mol^{-1}, spontaneous

16.49 (a) stable (b) unstable (c) unstable (d) stable

16.51 (a) more unstable (b) more stable (c) more stable (d) more stable

16.53 The reaction is spontaneous at 25°C. Because ΔS_r° is negative, increased temperature will make the decomposition less spontaneous.

16.55 Free energy approaches a minimum for the total reacting system as equilibrium is approached. That is, G_{total} decreases until equilibrium is reached.

16.57 (a) $K_p = 1 \times 10^{80}$ (b) $K_p = 1 \times 10^{90}$ (c) $K_p = 1.4 \times 10^{-23}$

16.59 (a) -12.3 kJ·mol^{-1} (b) -60.0 kJ·mol^{-1}

16.61 The tendency is to form reactants.

16.63 -32 kJ·mol^{-1}. The reaction is spontaneous in the direction of I(g).

16.65 (a) 0.89 kJ·mol^{-1} (b) Reactants will be formed.

16.67 (a) 8.4×10^{-17} (b) 7.7×10^{-9} (c) 1.6×10^{-19} (d) 4.7×10^{-6}

16.69 $+79.91$ kJ·mol^{-1}

16.71 (a) $+4.8$ kJ·mol^{-1} (b) -4.0 kJ·mol^{-1} (c) 325 K, $K_p = 1$

16.73 (a) yes (b) The tendency is greater at lower temperatures, because ΔH_r° and ΔS_r° are both negative.

16.75 (a) The reduction can occur because $\Delta G_r^\circ < 0$. (b) The reduction can occur because $\Delta G_r^\circ < 0$.

16.77 (a) The internal energy of an open system could be increased by (1) adding matter to the system, (2) doing work on the system, and (3) adding heat to the system. (b) Matter cannot be added to a closed system, so only (2) doing work on the system and (3) adding heat to the system could be used to increase the internal energy. (c) The internal energy of an isolated system cannot be changed.

16.79 (a) ΔU for a reaction is the change in internal energy of the reacting system; that is, ΔU = total energy of products − total energy of reactants. If the reaction is carried out under constant volume conditions, $\Delta U = q_V$. ΔH is defined in a similar manner. The ΔH for a reaction is the change in enthalpy of the reacting system, that is, ΔH = total enthalpy of products − total enthalpy of reactants. If the reaction is carried out under constant pressure conditions, $\Delta H = q_P$. (b) when P is constant and $\Delta V = 0$.

16.81 Compressing a gas decreases the disorder in the gas. The smaller volume available to the gas molecules upon compression means that their positions can be more precisely known; the disorder in their positions has decreased and hence the entropy of the gas has decreased. The temperature is not a factor here as long as it is high enough for the vapor to remain a gas; upon liquification, however, there is a greater decrease in entropy.

16.83 (a) decrease (b) increase (c) decrease (d) decrease

16.85 (a) Reduction is not possible because $\Delta G_r^\circ > 0$. (b) Reduction is not possible because $\Delta G_r^\circ > 0$.

16.87 (a) -170 J·K^{-1}·mol^{-1} (b) This result is reasonable, because the number of moles of gas decreases in the reaction.

16.89 (a) $\Delta H_r^\circ = -311.41$ kJ·mol^{-1}, $\Delta S_r^\circ = -0.2327$ kJ·K^{-1}·mol^{-1} (b) -242.1 kJ·mol^{-1} (c) A negative ΔH_r° indicates that heat is released by this reaction. The negative ΔS_r° is due to the fact that the product consists of two fewer moles of gas than the reactants. (d) 1338 K. This is the crossover temperature between $K < 1$ and $K > 1$.

16.91 (a) -5.00 kJ (b) 293 K. The two solid phases are in equilibrium with each other (under standard conditions) at this temperature. (c) Solid (2) is favored.

16.93 (a) 346 K (b) 4×10^2 K (c) 432 K (d) 370.9 K

16.95 (a) $\Delta G_r^\circ = -326.4$ kJ·mol^{-1}, $\Delta S_r^\circ = +137.56$ J·K^{-1}·mol^{-1} (b) 1.5×10^{57} (c) The conversion of ozone to oxygen is spontaneous; if no matter or energy were added, the ozone would be used up. The rate at which this occurs is another matter; thermodynamics does not give kinetic information.

16.97 (a) $+6.11$ J·K^{-1} (b) $+1.22$ J·K^{-1} (c) The disorder of the gaseous state relative to the liquid state is much larger than that of the liquid state relative to the solid state. Both the solid and liquid states are compact states of matter.

16.99 0.45 L; expansion

16.101 (a) For the dissolving process to be spontaneous, ΔG_{soln} must be negative. Thus, a positive ΔH_{soln} does not favor the solution process. (b) Because the solution process is spontaneous, ΔG_{soln} must be negative and ΔS_{soln} must be positive. (c) Locational disorder is the dominant factor. (d) Because the surroundings participate in the solution process only as a source of heat, the entropy change of the surroundings is primarily a result of the dispersal of thermal motion. (e) The driving force for the dissolution is the dispersal of matter, resulting in a positive ΔS.

16.103 (a) $\Delta G^\circ_{vap} = 0$ (b) $+35\ J\cdot mol^{-1}$ (c) The systems are different. The water is not pure in part (b). $\Delta G^\circ_{vap} = 0$ only for pure water at 100°C and 1 atm pressure. A nonvolatile solute depresses the vapor pressure.

16.105 (a) $\Delta S^\circ_r = -44.2\ J\cdot K^{-1}\cdot mol^{-1}$, $\Delta G^\circ_r = -3.9\ kJ\cdot mol^{-1}$, $K = 5$ (b) isooctane (c) The amount of heat released per gram is greater for octane than for isooctane. (d) The molecular structure for isooctane results in its being less flexible, with less freedom of movement, hence its atomic arrangement is more ordered. So there is a decrease in entropy as octane is converted to isooctane.

CHAPTER 17

17.1 (a) $VO^{2+}(aq) + 2\,H^+(aq) + e^- \rightarrow V^{3+}(aq) + H_2O(l)$; reduction (b) $PbSO_4(s) + 2\,H_2O(l) \rightarrow PbO_2(s) + SO_4^{2-}(aq) + 4\,H^+(aq) + 2\,e^-$; oxidation (c) $H_2O_2(aq) \rightarrow O_2(g) + 2\,H^+(aq) + 2\,e^-$; oxidation

17.3 (a)$ClO^-(aq) + H_2O(l) + 2\,e^- \rightarrow Cl^-(aq) + 2\,OH^-(aq)$; reduction (b) $IO_3^-(aq) + 2\,H_2O(l) + 4\,e^- \rightarrow IO^-(aq) + 4\,OH^-(aq)$; reduction (c) $2\,SO_3^{2-}(aq) + 2\,H_2O(l) + 2\,e^- \rightarrow S_2O_4^{2-}(aq) + 4\,OH^-(aq)$; reduction

17.5 (a) $4\,Cl_2(g) + S_2O_3^{2-}(aq) + 5\,H_2O(l) \rightarrow 8\,Cl^-(aq) + 2\,SO_4^{2-}(aq) + 10\,H^+(aq)$. Cl_2 is the oxidizing agent and $S_2O_3^{2-}$ is the reducing agent. (b) $2\,MnO_4^-(aq) + H^+(aq) + 5\,H_2SO_3(aq) \rightarrow 2\,Mn^{2+}(aq) + 3\,H_2O(l) + 5\,HSO_4^-$. MnO_4^- is the oxidizing agent and H_2SO_3 is the reducing agent. (c) $Cl_2(g) + H_2S(aq) \rightarrow 2\,Cl^-(aq) + S(s) + 2\,H^+(aq)$. Cl_2 is the oxidizing agent and H_2S is the reducing agent. (d) $H_2O(l) + Cl_2(g) \rightarrow HOCl(aq) + H^+(aq) + Cl^-(aq)$. Cl_2 is both the oxidizing and reducing agent.

17.7 (a) $3\,O_3(g) + Br^-(aq) \rightarrow 3\,O_2(g) + BrO_3^-(aq)$. O_3 is the oxidizing agent and Br^- is the reducing agent. (b) $6\,Br_2(l) + 12\,OH^-(aq) \rightarrow 10\,Br^-(aq) + 2\,BrO_3^-(aq) + 6\,H_2O(l)$. Br_2 is both the oxidizing and reducing agent. (c) $2\,Cr^{3+}(aq) + 4\,OH^-(aq) + 3\,MnO_2(s) \rightarrow 2\,CrO_4^{2-}(aq) + 2\,H_2O(l) + 3\,Mn^{2+}(aq)$. Cr^{3+} is the reducing agent and MnO_2 is the oxidizing agent. (d) $P_4(s) + 3\,H_2O(l) + 3\,OH^-(aq) \rightarrow 3\,H_2PO_2^-(aq) + PH_3(g)$. P_4 is both oxidized and reduced.

17.9 (a) $HSO_3^-(aq) + H_2O(l) \rightarrow HSO_4^-(aq) + 2\,H^+(aq) + 2\,e^-$ and $2\,HSO_3^-(aq) \rightarrow S_2O_6^{2-}(aq) + 2\,H^+(aq) + 2\,e^-$ (b) $I_2(aq) + 2\,e^- \rightarrow 2\,I^-(aq)$ and $HSO_3^-(aq) + H_2O(l) \rightarrow HSO_4^-(aq) + 2\,H^+(aq) + 2\,e^-$; overall, $I_2(aq) + HSO_3^-(aq) + H_2O(l) \rightarrow 2\,I^-(aq) + HSO_4^-(aq) + 2\,H^+(aq)$

17.11 $MnO_4^-(aq) + 8\,H^+(aq) + 5\,e^- \rightarrow Mn^{2+}(aq) + 4\,H_2O(l)$ and $C_6H_{12}O_6(aq) + 6\,H_2O(l) \rightarrow 6\,CO_2(g) + 24\,H^+(aq) + 24\,e^-$; overall, $24\,MnO_4^-(aq) + 72\,H^+(aq) + 5\,C_6H_{12}O_6(aq) \rightarrow 24\,Mn^{2+}(aq) + 66\,H_2O(l) + 30\,CO_2(g)$

17.13 (a) anode (b) positive

17.15 (a) electrons (b) ions

17.17 (a) $Zn^{2+}(aq) + 2\,e^- \rightarrow Zn(s)$ (b) $Fe^{3+}(aq) + e^- \rightarrow Fe^{2+}(aq)$ (c) $\frac{1}{2}\,Cl_2(g) + e^- \rightarrow Cl^-(aq)$ (d) $Hg_2Cl_2(s) + 2\,e^- \rightarrow 2\,Hg(l) + 2\,Cl^-(aq)$

17.19 (a) anode, $Fe^{2+}(aq) \rightarrow e^- + Fe^{3+}(aq)$; cathode, $Ag^+(aq) + e^- \rightarrow Ag(s)$; overall, $Ag^+(aq) + Fe^{2+}(aq) \rightarrow Ag(s) + Fe^{3+}(aq)$ (b) anode, $H_2(g) \rightarrow 2\,H^+(aq) + 2\,e^-$; cathode, $Cl_2(g) + 2\,e^- \rightarrow 2\,Cl^-(aq)$; overall, $Cl_2(g) + H_2(g) \rightarrow 2\,H^+(g) + 2\,Cl^-(aq)$ (c) anode, $U(s) \rightarrow U^{3+}(aq) + 3\,e^-$; cathode, $V^{2+}(aq) + 2\,e^- \rightarrow V(s)$; overall, $3\,V^{2+}(aq) + 2\,U(s) \rightarrow 2\,U^{3+}(aq) + 3\,V(s)$ (d) anode, $2\,H_2O(l) \rightarrow O_2(g) + 4\,H^+(aq) + 4\,e^-$; cathode, $O_2(g) + 2\,H_2O(l) + 4\,e^- \rightarrow 4\,OH^-(aq)$; overall, $H_2O(l) \rightarrow H^+(aq) + OH^-(aq)$ (e) anode, $Sn^{2+}(aq) \rightarrow 2\,e^- + Sn^{4+}(aq)$; cathode, $Hg_2Cl_2(s) + 2\,e^- \rightarrow 2\,Hg(l) + 2\,Cl^-(aq)$; overall, $Hg_2Cl_2(s) + Sn^{2+}(aq) \rightarrow 2\,Hg(l) + 2\,Cl^-(aq) + Sn^{4+}(aq)$

17.21 (a) anode, $Zn(s) \rightarrow Zn^{2+}(aq) + 2\,e^-$; cathode, $Ni^{2+}(aq) + 2\,e^- \rightarrow Ni(s)$; overall, $Ni^{2+}(aq) + Zn(s) \rightarrow Ni(s) + Zn^{2+}(aq)$; $Zn(s)|Zn^{2+}(aq)||Ni^{2+}(aq)|Ni(s)$ (b) anode, $2\,I^-(aq) \rightarrow I_2(s) + 2\,e^-$; cathode, $Ce^{4+}(aq) + e^- \rightarrow Ce^{3+}(aq)$; overall, $2\,Ce^{4+}(aq) + 2\,I^-(aq) \rightarrow I_2(s) + 2\,Ce^{3+}(aq)$; $Pt|I_2(s)|I^-aq||Ce^{4+}(aq), Ce^{3+}(aq)|Pt$ (c) anode, $H_2(g) \rightarrow 2\,H^+(aq) + 2\,e^-$; cathode, $Cl_2(g) + 2\,e^- \rightarrow 2\,Cl^-(aq)$; overall, $H_2(g) + Cl_2(g) \rightarrow 2\,HCl(aq)$; $Pt|H_2(g)|H^+(aq)||Cl^-(aq)|Cl_2(g)|Pt$ (d) anode, $Au(s) \rightarrow Au^{3+}(aq) + 3\,e^-$; cathode, $Au^+(aq) + e^- \rightarrow Au(s)$; overall, $3\,Au^+(aq) \rightarrow 2\,Au(s) + Au^{3+}(aq)$; $Au(s)|Au^{3+}(aq)||Au^+(aq)|Au(s)$

17.23 (a) anode, $Ag(s) + Br^-(aq) \rightarrow AgBr(s) + e^-$; cathode, $Ag^+(aq) + e^- \rightarrow Ag(s)$; $Ag(s)|AgBr(s)|Br^-(aq)||Ag^+(aq)|Ag(s)$ (b) anode, $4\,OH^-(aq) \rightarrow O_2(g) + 2\,H_2O(l) + 4\,e^-$; cathode, $O_2(g) + 4\,H^+(aq) + 4\,e^- \rightarrow 2\,H_2O(l)$; $Pt|O_2(g)|OH^-(aq)||H^+(aq)|O_2(g)|Pt$ (c) anode, $Cd(s) + 2\,OH^-(aq) \rightarrow Cd(OH)_2(s) + 2\,e^-$; cathode, $Ni(OH)_3(s) + e^- \rightarrow Ni(OH)_2(s) + OH^-(aq)$; $Cd(s)|Cd(OH)_2(s)|KOH(aq)||Ni(OH)_3(s)|Ni(OH)_2(s)|Ni(s)$

17.25 (a) anode, $Fe^{2+}(aq) \rightarrow Fe^{3+}(aq) + e^-$; cathode, $MnO_4^-(aq) + 8\,H^+(aq) + 5\,e^- \rightarrow Mn^{2+}(aq) + 4\,H_2O(l)$ (b) $MnO_4^-(aq) + 5\,Fe^{2+}(aq) + 8\,H^+(aq) \rightarrow Mn^{2+}(aq) + 5\,Fe^{3+}(aq) + 4\,H_2O(l)$; $Pt|Fe^{2+}(aq), Fe^{3+}(aq)||Mn^{2+}(aq), H^+(aq), MnO_4^-(aq)|Pt$

17.27 (a) 24 (b) 6 (c) 4

17.29 (a) +0.75 V (b) +0.37 V (c) +0.52 V (d) +1.52 V

17.31 (a) +0.75 V, −145 kJ·mol^{-1} (b) +0.37 V, −36 kJ·mol^{-1} (c) +0.52 V, −100 kJ·mol^{-1} (d) +1.52 V, −293 kJ·mol^{-1}
17.33 (a) Cu < Fe < Zn < Cr (b) Mg < Na < K < Li (c) V < Ti < Al < U (d) Au < Ag < Sn < Ni
17.35 (a) Co^{2+} is the oxidizing agent and Ti^{2+} is the reducing agent. +0.09 V; Pt|Ti^{2+}(aq), Ti^{3+}(aq)||Co^{2+}(aq)|Co(s) (b) U^{3+} is the oxidizing agent and La is the reducing agent. +0.73 V; La(s)|La^{3+}(aq)||U^{3+}(aq)|U(s) (c) Fe^{3+} is the oxidizing agent and H_2 is the reducing agent. +0.77 V; Pt|H_2(g)|H^+(aq)||Fe^{2+}(aq), Fe^{3+}(aq)|Pt (d) O_3 is the oxidizing agent and Ag is the reducing agent. +0.44 V; Ag(s)|Ag^+(aq)||OH^-(aq)|O_3(g), O_2(g)|Pt
17.37 (a) no (b) yes, +0.61 V; 3 Pb^{2+}(aq) + 2 Cr(s) → 3 Pb(s) + 2 Cr^{3+}(aq) (c) yes, +1.17 V; 5 Cu(s) + 2 MnO_4^-(aq) + 16 H^+(aq) → 5 Cu^{2+}(aq) + 2 Mn^{2+}(aq) + 8 H_2O(l) (d) no
17.39 (a) cathode, I_2 + 2 e^- → 2 I^-; anode, H_2 → 2 H^+ + 2 e^-; −104 kJ·mol^{-1} (b) no reaction (c) cathode, Pb^{2+} + 2 e^- → Pb; anode, Al → Al^{3+} + 3 e^-; −886 kJ·mol^{-1}
17.41 yes, because $E°(O_2, H^+, H_2O/H_2) > E°(Br_2, Br^-)$. It is not used because that reaction is so much slower than the one with Cl_2.
17.43 (a) $K_c = [H^+]^2[Cl^-]^2/[H_2]$ (b) $K_c = [NO][Fe^{3+}]^3/[Fe^{2+}]^3[H^+]^4[NO_3^-]$
17.45 (a) 10^{25} (b) 10^{-16} (c) 10^{31} (d) 1×10^2
17.47 yes, $K = 10^2$
17.49 (a) 10^6 (b) 1.0
17.51 (a) +0.030 V (b) $+6 \times 10^{-2}$ V
17.53 (a) +0.18 V (b) +0.47 V (c) −1.45 V (d) +0.30 V
17.55 (a) pH = 1 (b) $[Cl^-] = 9.9 \times 10^{-2}$ mol·L^{-1}
17.57 A primary cell is the primary source of the electrical energy produced by the operation. A secondary cell is an energy storage device that stores electrical energy produced elsewhere and releases it upon operation. It is the secondary source of the electrical energy. Most primary cells produce electricity from chemicals that were sealed into them when made. They are not normally rechargeable. A secondary cell is one that must be charged from some other electrical supply before use. It is normally rechargeable.
17.59 (a) a moist paste of KOH(aq) and HgO(s) (b) HgO(s) (c) HgO(s) + Zn(s) → Hg(l) + ZnO(s)
17.61 The reaction at the anode, Zn(s) → Zn^{2+}(aq) + 2 e^-, supplies the electrons to the external circuit. The cathode reaction is MnO_2(s) + H_2O(l) + e^- → $MnO(OH)_2$(s) + OH^-(aq). The OH^-(aq) produced reacts with NH_4^+(aq) from the NH_4Cl(aq): NH_4^+(aq) + OH^-(aq) → H_2O(l) + NH_3(g). The NH_3(g) produced complexes with the Zn^{2+}(aq) produced in the anode reaction: Zn^{2+}(aq) + 4 NH_3(g) → $[Zn(NH_3)_4]^{2+}$(aq).
17.63 (a) KOH(aq) (b) $Ni(OH)_2$(s) + OH^-(aq) → $Ni(OH)_3$(s) + e^-
17.65 Comparison of the reduction potentials shows that Cr is more easily oxidized than Fe, so the presence of Cr retards the rusting of Fe.
17.67 (a) $Fe_2O_3{\cdot}H_2O$ (b) H_2O and O_2 jointly oxidize iron. (c) Water is more highly conducting if it contains dissolved ions, so the rate of rusting is increased.
17.69 (a) aluminum or magnesium (b) cost, availability, and toxicity of products in the environment (c) Fe could act as the anode of an electrochemical cell if Cu^{2+} or Cu^+ are present; hence it could be oxidized at the point of contact. Water with dissolved ions acts as the electrolyte.
17.71 (a) anode (b) positive
17.73 (a) Co^{2+}(aq) + 2 e^- → Co(s) (b) 2 H_2O(l) → O_2(g) + 4 H^+(aq) + 4 e^- (c) $E_{supplied} > +1.09$ V
17.75 (a) Cu^{2+}(aq) + 2 e^- → Cu(s); cathode (b) Na^+(l) + e^- → Na(l); cathode (c) 2 Cl^-(l) → Cl_2(g) + 2 e^-; anode (d) 2 H_2O(l) + 2 e^- → H_2(g) + 2 OH^-(aq); cathode
17.77 (a) Water is reduced. (b) Water is reduced. (c) The metal ion is reduced. (d) The metal ion is reduced.
17.79 0.67 mol e^-
17.81 (a) 0.161 mol e^- (b) 22.2 mol e^- (c) 35.7 mol e^-
17.83 (a) 7.86 s (b) 1.29 mg Cu
17.85 (a) 0.52 A (b) 0.19 A
17.87 +2 (Ti^{2+})
17.89 0.20 A
17.91 0.92 g Zn
17.93 (a) 3 I^-(aq) → I_3^-(aq) + 2 e^-; oxidation (b) 2 e^- + SeO_4^{2-}(aq) + H_2O(l) → SeO_3^{2-}(aq) + 2 OH^-(aq); reduction
17.95 (a) cathode, O_2 + 4 H^+ + 4 e^- → 2 H_2O; anode, Fe → Fe^{2+} + 2 e^- (b) Fe(s)|Fe^{2+}(aq)||H^+(aq), H_2O(l)|O_2(g)|Pt(s) (c) +1.67 V (d) +1.31 V
17.97 (a) −0.27 V (b) +0.07 V
17.99 The $E°_{cell}$ for the gold/permanganate cell is positive and thus spontaneous, so gold will be oxidized. The $E°_{cell}$ for the gold/dichromate cell is negative and thus not spontaneous, so gold will not be oxidized.
17.101 (a) Hg(l)|Hg_2^{2+}(aq)||NO_3^-(aq), H^+(aq)|NO(g)|Pt; +0.17 V, −98 kJ·mol^{-1} (b) not spontaneous (c) Pt|Pu^{3+}(aq), Pu^{4+}(aq)||$Cr_2O_7^{2-}$(aq), Cr^{3+}(aq), H^+(aq)|Pt; +0.36 V, −208 kJ·mol^{-1}
17.103 (a) 10^{-5} mol·L^{-1} (b) 10^{-6} mol·L^{-1} (c) 10^{-4} mol·L^{-1}
17.105 10^{-3}
17.107 $E°_{cell}$ = +0.03 V; E_{cell} = −0.27 V. The cell changes from spontaneous to nonspontaneous as a function of concentration.

17.109 +0.08 V to +0.09 V
17.111 9.8×10^{-3} mol·L^{-1}
17.113 93 A
17.115 35.9 A
17.117 +4 (Hf^{4+})
17.119 105 s
17.121 52 g·mol^{-1} (Cr)
17.123 (a) +0.52 V (b) −0.037 V
17.125 Set up a cell in which one electrode is the silver-silver chloride electrode (anode, assume $[Cl^-]$ = 1.0 mol·L^{-1}) and the other electrode is the hydrogen electrode (cathode, assume P_{H_2} = 1 atm). The E of this cell will be sensitive to $[H^+]$ and can be used to obtain pH. For this system, (a) pH = (E − 0.22 V)/0.0592 V (b) pOH = 14.00 − pH
17.127 (a) 7.46×10^{-2} mol H_3O^+ (b) 0.428

CHAPTER 18

18.1 (a) one-third (b) two-thirds (c) two
18.3 (a) 1.0×10^{-3} (mol O^3)·L^{-1}·s^{-1}
(b) 0.28 (mol CrO_4^{2-})·L^{-1}·s^{-1}
18.5 (a)

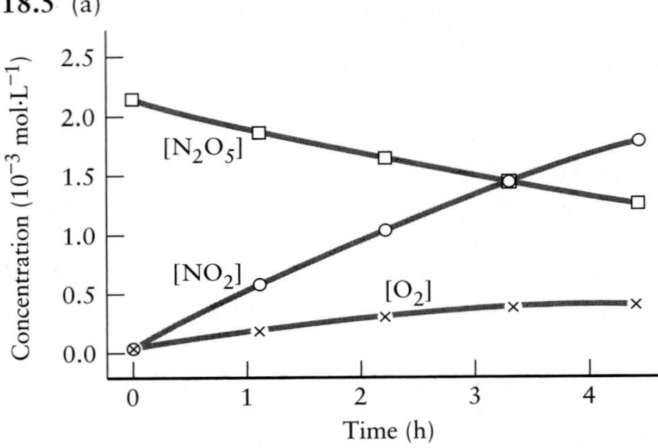

(b) rate(1.11 h) = 2.3×10^{-4} mol·L^{-1}·s^{-1}; rate(2.22 h) = 2.0×10^{-4} mol·L^{-1}·s^{-1}; rate(3.33 h) = 1.8×10^{-4} mol·L^{-1}·s^{-1}; rate(4.44 h) = 1.6×10^{-4} mol·L^{-1}·s^{-1}
(c) included in figure for (a)
18.7 (a) rate = $k[X]^2[Y]^{1/2}$ (b) rate = $k[A][B][C]^{-1/2}$
18.9 (a) (mol A)·L^{-1}·s^{-1} (b) s^{-1} (c) L·(mol A)$^{-1}$·s^{-1}
18.11 The reaction is third order.
18.13 9.6×10^{-5} (mol N_2O_5)·L^{-1}·s^{-1}
18.15 (a) rate = 2.4×10^{-5} mol·L^{-1}·s^{-1} (b) factor of 2
18.17 (a) The reaction is first order in H_2, first order in I_2, and second order overall. (b) The reaction is first order in SO_2, zero order in O_2, negative one-half in SO_3, and one-half order overall. (c) The reaction is second order in A, zero order in B, first order in C, and third order overall.
18.19 rate = $k[CH_3Br][OH^-]$
18.21 (a) The reaction is second order in A, first order in B, and third order overall. (b) rate = $k[A]^2[B]$
(c) 1.2×10^2 L^2·mol^{-2}·s^{-1} (d) 0.85 mol·L^{-1}·s^{-1}
18.23 (a) rate = $k[ICl][H_2]$ (b) 0.16 L·mol^{-1}·s^{-1}
(c) 2.1×10^{-6} mol·L^{-1}·s^{-1}
18.25 (a) rate = $k[A][B]^2[C]^2$ (b) The overall order is 5. (c) 2.85×10^{12} L^4·mol^{-4}·s^{-1} (d) 1.13×10^{-2} mol·L^{-1}·s^{-1}
18.27 (a) 6.93×10^{-4} s^{-1} (b) 1.8×10^{-2} s^{-1}
(c) 7.6×10^{-3} s^{-1}
18.29 (a) 5.2 h (b) 1.79×10^{-2} mol·L^{-1}
(c) 1.3×10^2 min
18.31 (a) 200 s (b) 800 s (c) 634 s
18.33 (a) 247 min (b) 819 min (c) 10.9 g
18.35 (a) k = 0.0134 s^{-1} (b) 1.7×10^2 s
18.37 (a) k = 0.20 min^{-1} (b) An additional 8 min is needed.
18.39 The reaction is first order.
18.41 (a)

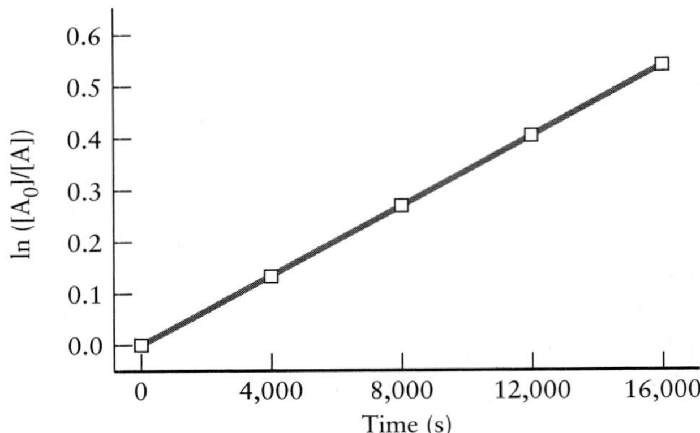

(b) $k = 3.39 \times 10^{-5}$ s^{-1}
18.43 (a) A second-order reaction will give a linear plot of 1/[HI] versus time, with a slope equal to k.

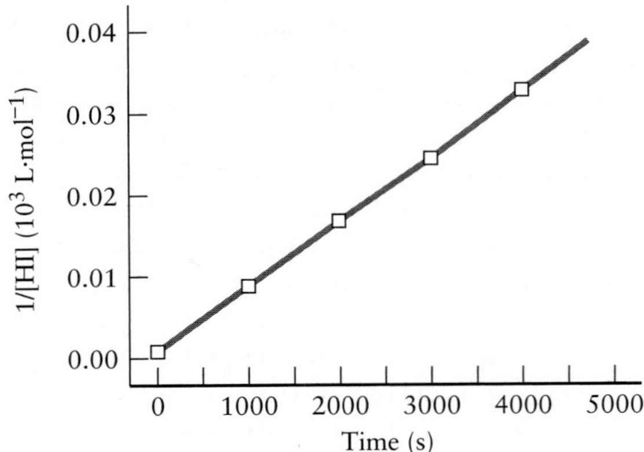

(b) $k = 7.8 \times 10^{-3}$ L·mol^{-1}·s^{-1}

18.45 As temperature increases, the average speed of the reacting molecules also increases; therefore, the number of collisions between molecules increases as well. The rate of reaction is proportional to the number of collisions between molecules; hence the rate increases in proportion to this factor. This collision frequency factor is contained within the preexponential factor, A, in the Arrhenius equation. Not all molecules that collide react, only those having sufficient collision energy to surmount the energy barrier, E_a, between reactants and products will react. This fraction of molecules with the necessary energy increases with temperature in a manner governed by the Maxwell distribution of speeds. It turns out that this fraction is proportional to $e^{-E_a/RT}$. So we can write $k = Ae^{-E_a/RT}$.

18.47 (a) 2.74×10^2 kJ·mol^{-1} (b) $k = 8.8 \times 10^{-2}$ s^{-1}

18.49 2.4×10^2 kJ·mol^{-1}

18.51 2.6×10^9 L·mol^{-1}·s^{-1}

18.53 $k = 9.3 \times 10^{-4}$ s^{-1}

18.55 3.4×10^6

18.57 $E_{a,cat} = 80.9$ kJ·mol^{-1}

18.59 (a) rate $= k[NO]^2$, bimolecular (b) rate $= k[Cl_2]$, unimolecular (c) rate $= k[NO_2]^2$, bimolecular (d) (b) and (c), because Cl and NO are radicals

18.61 NO_2, because it appears only in the course of the reaction and is neither a reactant nor a product. NO is considered to be a catalyst.

18.63 $2\,ICl + H_2 \rightarrow 2\,HCl + I_2$; HI is the only intermediate.

18.65 The first elementary reaction is the rate-controlling step because it is the slow step. The second elementary reaction is fast and does not affect the overall reaction order. Rate $= k[NO][Br_2]$.

18.67 rate $= k_3(k_2/k_2')(k_1/k_1')^{1/2}[CO][Cl_2]^{3/2} = k[CO][Cl_2]^{3/2}$

18.69 If mechanism (a) were correct, the rate law would be rate $= k[NO_2][CO]$. This expression does not agree with the experimental result and can be eliminated as a possibility. Mechanism (b) has rate $= k[NO_2]^2$ from the slow step. Step 2 does not influence the overall rate, but it is necessary to achieve the correct overall reaction; thus this mechanism agrees with the experimental data. Mechanism (c) is not correct, as seen from the rate expression for the slow step, rate $= k[NO_3][CO]$, and the experimental expression does not contain [CO].

18.71 (a) At equilibrium, the *rates* of the forward and reverse reactions are equal. However, the rate constants are characteristic of the reactions and do not change over time. (b) For a reversible reaction, $K_c = k(\text{forward})/k(\text{reverse})$. Therefore, if K_c is very large, $k(\text{forward}) \gg k(\text{reverse})$.

18.73 1.90×10^{-4} (mol HCl)·L^{-1}·s^{-1}

18.75 (a)

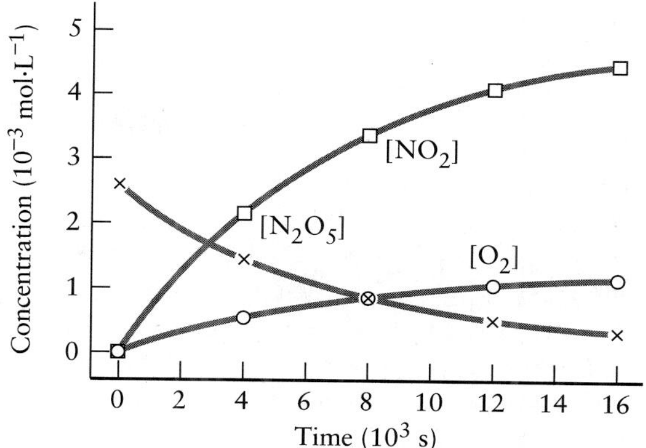

(b) rate(0 s) $= 3.1 \times 10^{-7}$ mol·L^{-1}·s^{-1}; rate(4000 s) $= 2.1 \times 10^{-7}$ mol·L^{-1}·s^{-1}; rate(8000 s) $= 1.2 \times 10^{-7}$ mol·L^{-1}·s^{-1}; rate(12000 s) $= 7.1 \times 10^{-8}$ mol·L^{-1}·s^{-1}; rate(16000 s) $= 4 \times 10^{-8}$ mol·L^{-1}·s^{-1} (c) see figure for (a)

18.77 (a) rate $= k[A][B]^{1/2}$ (b) L$^{1/2}$·mol$^{-1/2}$·min^{-1}

18.79 (a) The reaction is zero order in C, first order in A, second order in B, and third order overall. (b) rate $= k[A][B]^2$ (c) $k = 20$ L^2·mol^{-2}·s^{-1} (d) 2.9×10^{-9} mol·L^{-1}·s^{-1}

18.81 (a) 0.13 mol·L^{-1} (b) 39 min (c) 7.0 min (d) 34 min

18.83 (a) 0.0506 mg (b) 0.106 atm

18.85 Refer to Fig. 18.35b. For an exothermic reaction, the activation energy for the reverse reaction is greater than that for the forward reaction. $E_{a,reverse} \approx E_{a,forward} - \Delta H = 300$ kJ·mol^{-1}

18.87 (a) 300 s (b) 500 s

18.89 Solid catalysts provide surface sites at which the reaction occurs. A finely divided solid catalyst has a greater surface area—hence more surface sites—and is consequently more effective.

18.91 (a) bimolecular (b) unimolecular (c) termolecular

18.93 The overall reaction is $2\,SO_2 + O_2 \rightarrow 2\,SO_3$. For the mechanism given, NO is the catalyst and NO_2 the intermediate.

18.95 1.2×10^{10}

18.97 (a) 74 kJ·mol^{-1} (b) $k \approx 1.4 \times 10^{-3}$ L·mol^{-1}·s^{-1}

18.99 (a) ClO is the reaction intermediate and Cl is the catalyst. (b) Cl, ClO, O, O_2 (c) Step 1 is initiating; step 2 is propagating. (d) $Cl + Cl \rightarrow Cl_2$

18.101 rate(0 min) $= 1.01 \times 10^{-3}$ mol·L^{-1}·min^{-1}; rate(5 min) $= 6.07 \times 10^{-4}$ mol·L^{-1}·min^{-1}; rate(10 min) $= 3.65 \times 10^{-4}$ mol·L^{-1}·min^{-1}; rate(15 min) $= 2.2 \times 10^{-4}$ mol·L^{-1}·min^{-1}; rate(20 min) $= 1.3 \times 10^{-4}$ mol·L^{-1}·min^{-1}; rate(30 min) $= 4.8 \times 10^{-5}$ mol·L^{-1}·min^{-1}

18.103 Using the equations modified from Eq. 4 yields $k = 1/(t_{1/2}[A]_0)$ and $t = \{(1/[A]) - (1/[A]_0)\}/k$. Then (a) 7.6×10^2 s (b) 1.5×10^2 s (c) 2.0×10^2 s

18.105 Using the equations from Exercise 18.103 yields (a) 2.5×10^2 min (b) 4.9×10^2 min

18.107 (a)

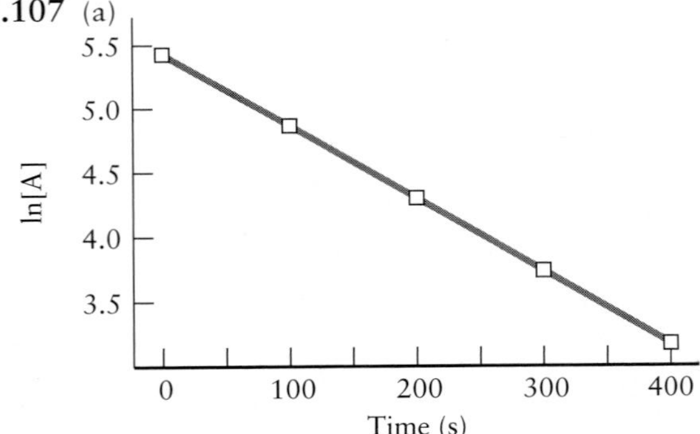

The plot is linear, so the reaction is first order.
(b) $k = 5.75 \times 10^{-3}\ s^{-1}$

18.109 For a third-order reaction, $t_{1/2} = \text{constant}/[A_0]^2$. (a) For the first half-life, $t_{1/2} = \text{constant}/[A_0]^2$. (b) The second half-life, $t_{1/4} = t_2 = 4t_1$, so the total time is $t_1 + t_2 = t_1 + 4t_1 = 5t_1$. (c) $t_{1/16}$ requires $t_3 = 16t_1$ and $t_4 = 64t_1$, so $t_{1/16} = 85t_1$ by the same process as in (b).

18.111 Consider for example A $\rightleftharpoons$ B. Assume that the reaction is first order in both directions: rate(forward) $= k[A]$ and rate(reverse) $= k'[B]$. At equilibrium, rate(forward) = rate(reverse), so $k[A] = k'[B]$, or $k[A] = k'([A]_0 - [A])$ or $(k + k')[A] = k'[A]_0$ or $[A] = (k'/(k + k'))[A]_0$, which is not zero. The reactant concentration drops to its equilibrium value and remains at that value.

18.113 (a) $[P]_{eq} = R/k$ (b) In the absence of replenishment, $t_{1/2} = (\ln 2)/k$.

18.115 (a) If the forward reaction is exothermic, then increasing the temperature will drive the reaction in the direction of reactants; the reverse is true if the temperature is decreased. (b) If the forward reaction is endothermic, then increasing the temperature will drive the reaction in the direction of products; the reverse is true if the temperature is decreased. (c) $\ln(k'/k) = (E_a/R)\{(1/T) - (1/T')\} = (E_a/R)[(T' - T)/TT']$. For (a), E_a(reverse) $> E_a$(forward). From the equation above for $\ln(k/k')$, we see that for a given $\Delta T = (T' - T) > 0$ $\ln(k'/k)$(reverse) $> \ln(k'/k)$(forward) and thus (k'/k)(reverse) $> (k'/k)$(forward). The rate of the reverse reaction increases more rapidly with increasing T than the forward reaction does. Hence, the equilibrium is shifted in the direction of reactants. For (b), because E_a(forward) $> E_a$(reverse), the opposite is true. The rate of the forward reaction will increase relatively more than that of the reverse reaction, and products will be favored.

CHAPTER 19

19.1 In the majority of its reactions, hydrogen acts as a reducing agent. Examples are $2\,H_2(g) + O_2(g) \rightarrow 2\,H_2O(l)$ and various reduction processes, such as $NiO(s) + H_2(g) \xrightarrow{\Delta} Ni(s) + H_2O(g)$. With highly electropositive elements, such as the alkali metals, $H_2(g)$ acts as an oxidizing agent and forms metal hydrides, for example, $2\,K(s) + H_2(g) \rightarrow 2\,KH(s)$.

19.3 (a) $CO(g) + H_2O(g) \xrightarrow{400°C,\ Fe/Cu} CO_2(g) + H_2(g)$
(b) $2\,Li(s) + 2\,H_2O(l) \rightarrow 2\,LiOH(aq) + H_2(g)$
(c) $Mg(s) + 2\,H_2O(l) \rightarrow Mg(OH)_2(aq) + H_2(g)$
(d) $2\,K(s) + H_2(g) \rightarrow 2\,KH(s)$

19.5 (a) saline (b) molecular (c) molecular (d) metallic

19.7 (a) $+206.10\ kJ\cdot mol^{-1}$ (b) $-41.16\ kJ\cdot mol^{-1}$ (c) $+164.94\ kJ\cdot mol^{-1}$

19.9 (a) 532 L H_2 (b) 428 mL H_2O

19.11 0.9

19.13 (a) $H_2(g) + Cl_2(g) \xrightarrow{light} 2\,HCl(g)$
(b) $H_2(g) + 2\,Na(l) \xrightarrow{\Delta} 2\,NaH(s)$
(c) $P_4(s) + 6\,H_2(g) \rightarrow 4\,PH_3(g)$

19.15 (a) $Li^+\,[H{:}]^-$

```
       H
       |
(b) H— Si —H        (c) H — S̈b — H
       |                     |
       H                     H
```

19.17 (a) red (b) violet (c) yellow (d) violet

19.19 (a) $4\,Li(s) + O_2(g) \rightarrow 2\,Li_2O(s)$
(b) $6\,Li(s) + N_2(g) \xrightarrow{\Delta} 2\,Li_3N(s)$
(c) $2\,Na(s) + 2\,H_2O(l) \rightarrow 2\,NaOH(aq) + H_2(g)$
(d) $4\,KO_2(s) + 2\,H_2O(g) \rightarrow 4\,KOH(s) + 3\,O_2(g)$

19.21 (a) $Ca(s) + H_2(g) \xrightarrow{\Delta} CaH_2(s)$
(b) $2\,KNO_3(s) \xrightarrow{\Delta} 2\,KNO_2(s) + O_2(g)$

19.23 (a) sodium chloride, NaCl (b) potassium chloride, KCl (c) potassium magnesium chloride hexahydrate, $KCl\cdot MgCl_2\cdot 6H_2O$

19.25 7.15 g $Na_2CO_3\cdot 10H_2O$

19.27 Be is the weakest reducing agent; Mg is stronger, but weaker than the remaining members of the group, all of which have approximately the same reducing strength. This trend is related to the very small radius of the Be^{2+} ion, 27 pm; its strong polarizing power introduces much covalent character into its compounds. Thus, Be attracts electrons more strongly and does not release them as readily as the other members of the group. Mg^{2+} is also a small ion, 58 pm, so the same reasoning applies to it also, but to a lesser extent. The remaining ions of the group are considerably larger, release electrons more readily, and are better reducing agents.

19.29 (a) magnesium sulfate heptahydrate, $MgSO_4\cdot 7H_2O$ (b) $CaCO_3$ (c) $Mg(OH)_2$

19.31 (a) $2\,Al(s) + 2\,OH^-(aq) + 6\,H_2O(l) \rightarrow 2\,[Al(OH)_4]^-(aq) + 3\,H_2(g)$ (b) $Be(s) + 2\,OH^-(aq) + 2\,H_2O(l) \rightarrow [Be(OH)_4]^{2-}(aq) + H_2(g)$

19.33 (a) $Mg(OH)_2(s) + 2\,HCl(aq) \rightarrow MgCl_2(aq) + 2\,H_2O(l)$ (b) $Ca(s) + 2\,H_2O(l) \rightarrow Ca(OH)_2(aq) + H_2(g)$ (c) $BaCO_3(s) \xrightarrow{\Delta} BaO(s) + CO_2(g)$

19.35 (a) $:\ddot{\underset{..}{Cl}}-Be-\ddot{\underset{..}{Cl}}:$ (b) 180° (c) *sp*

19.37 (a) $-65.17\ kJ\cdot mol^{-1}$ (b) 62.3°C

19.39 (a) $\Delta H_r^\circ = +178.3\ kJ\cdot mol^{-1}$; $\Delta S_r^\circ = +160.6\ J\cdot K^{-1}\cdot mol^{-1}$ (b) 1110 K

19.41 51.19%

19.43 $4\,Al^{3+}(melt) + 6\,O^{2-}(melt) + 3\,C(s, gr) \rightarrow 4\,Al(s) + 3\,CO_2(g)$

19.45 (a) $B(OH)_3$ (b) Al_2O_3 (c) $Na_2B_4O_7\cdot 10H_2O$ (d) B_2O_3

19.47 (a) $B_2O_3(s) + 3\,Mg(l) \xrightarrow{\Delta} 2\,B(s) + 3\,MgO(s)$ (b) $2\,Al(s) + 3\,Cl_2(g) \rightarrow 2\,AlCl_3(s)$ (c) $4\,Al(s) + 3\,O_2(g) \rightarrow 2\,Al_2O_3(s)$

19.49 (a) The hydrate of $AlCl_3$, $AlCl_3\cdot 6H_2O$, functions as a deodorant and antiperspirant. (b) Corundum, α-alumina, is used as an abrasive in sandpaper. (c) $B(OH)_3$ is an antiseptic and insecticide.

19.51 (a) $B_2H_6(g) + 6\,H_2O(l) \rightarrow 2\,B(OH)_3(aq) + 6\,H_2(g)$ (b) $B_2H_6(g) + 3\,O_2(g) \rightarrow B_2O_3(s) + 3\,H_2O(l)$

19.53 2.7×10^5 g Al

19.55 +0.743 V

19.57 Silicon occurs widely in the Earth's crust in the form of silicates in rocks and as silicon dioxide in sand. It is obtained from quartzite, a form of quartz, by the following processes: (1) reduction in an electric arc furnace $SiO_2(s) + 2\,C(s) \rightarrow Si(s, crude) + 2\,CO(g)$; (2) purification of the crude product in two steps $Si(s, crude) + 2\,Cl_2(g) \rightarrow SiCl_4(l)$, followed by reduction with hydrogen to the pure element $SiCl_4(l) + 2\,H_2(g) \rightarrow Si(s, pure) + 4\,HCl(g)$.

19.59 In diamond, carbon is sp^3 hybridized and forms a tetrahedral, three-dimensional network structure, which is highly rigid. Graphite carbon is sp^2 hybridized and planar, and its application as a lubricant results from the fact that the two-dimensional sheets can "slide" across one another, thereby reducing friction. In graphite, the unhybridized *p*-electrons are free to move from one carbon atom to another, which results in its high electrical conductivity. In diamond, all electrons are localized in sp^3 hybridized C—C σ-bonds, so diamond does not conduct electricity well.

19.61 (a) SiC (b) SiO_2 (c) $ZrSiO_4$

19.63 (a) $MgC_2(s) + 2\,H_2O(l) \rightarrow C_2H_2(g) + M(OH)_2(s)$; acid-base (b) $2\,Pb(NO_3)_2(s) \rightarrow 2\,PbO(s) + 4\,NO_2(g) + O_2(g)$; redox

19.65 (a) $SiCl_4(l) + 2\,H_2(g) \rightarrow Si(s) + 4\,HCl(g)$ (b) $SiO_2(s) + 3\,C(s) \xrightarrow{2000°C} SiC(s) + 2\,CO(g)$ (c) $Ge(s) + 2\,F_2(g) \rightarrow GeF_4(s)$ (d) $CaC_2(s) + 2\,H_2O(l) \rightarrow Ca(OH)_2(s) + C_2H_2(g)$

19.67

```
        ..         ]4-
       :O:         |
        |
  :O — Si — O:
   ..   |    ..
       :O:
        ..
```

formal charges: Si = 0, O = −1; oxidation numbers: Si = +4, O = −2
This is an AX_4 VSEPR structure; therefore the shape is tetrahedral.

19.69 46.74%

19.71 $\Delta H_r^\circ = +689.88\ kJ\cdot mol^{-1}$; $\Delta S_r^\circ = +360.85\ J\cdot K^{-1}\cdot mol^{-1}$; $\Delta G_r^\circ = +5.8229 \times 10^2\ kJ\cdot mol^{-1}$. The equilibrium constant becomes greater than 1 at 1912 K.

19.73 3.99 mg HF

19.75 $2.4 \times 10^4\ m^2$

19.77 (a) The $Si_2O_7^{6-}$ ion is built from two SiO_4^{4-} tetrahedral ions in which the silicate tetrahedra share one O atom. See Fig. 19.42. (b) Pyroxenes, for example, jade, consist of chains of SiO_4 units in which two O atoms are shared by neighboring units. The repeating unit has the formula SiO_3^{2-}. See Fig. 19.45.

19.79 (a)

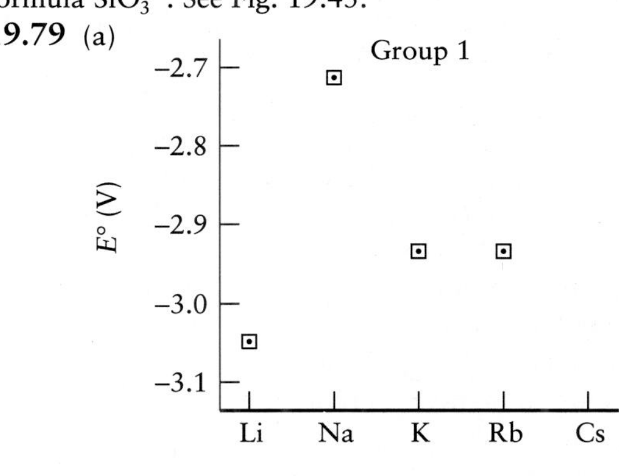

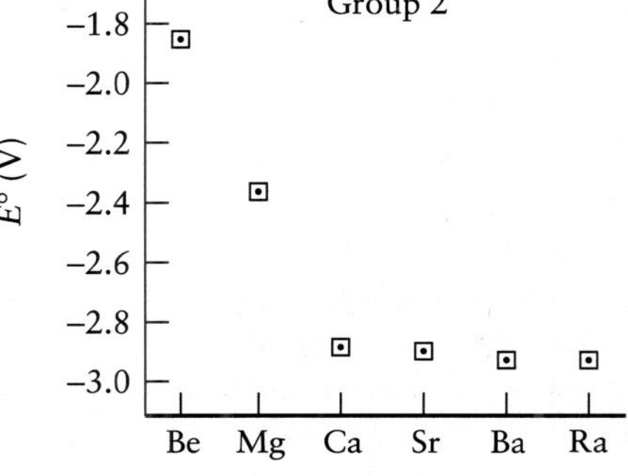

(b) For both groups, the trend in standard reduction potentials with increasing atomic number is overall a downward trend (they become more negative), but lithium is anomalous. This overall downward trend makes sense, because, theoretically, it is easier to remove electrons that are farther away from the nuclei. However, because there are several factors that influence ease of removal, the trend is not smooth. The potentials are a net composite of the free energies of sublimation of solids, dissociation of gaseous molecules, ionization enthalpies, and enthalpies of hydration of gaseous ions. The origin of the anomalously strong reducing power of Li is the strongly exothermic energy of hydration of the very small Li^+ ion, which favors the ionization of the element in aqueous solution.

19.81 (a, b) $H_2(g) + F_2(g) \rightarrow 2\,HF(g)$, explosive; $H_2(g) + Cl_2(g) \rightarrow 2\,HCl(g)$, explosive; $H_2(g) + Br_2(g) \rightarrow 2\,HBr(g)$, vigorous; $H_2(g) + I_2(g) \rightarrow 2\,HI(g)$, less vigorous
(c) hydrofluoric acid, hydrochloric acid, hydrobromic acid, hydroiodic acid

19.83 (a) This is a Lewis acid-base reaction.
(b) $CaO(s) + CO_2(g) \rightarrow CaCO_3(s)$

19.85 (a) beryl (b) limestone or dolomite (c) dolomite

19.87 (a) The "hardness" of water is due to the presence of calcium and magnesium salts (particularly their hydrogen carbonates). In laundering and bathing, Ca^{2+} and Mg^{2+} cations convert soluble Na^+ soaps to insoluble Ca^{2+} and Mg^{2+} soaps. (b) $Ca(HCO_3)_2(aq) + Ca(OH)_2(aq) \rightarrow 2\,CaCO_3(s) + 2\,H_2O(l)$ and $Mg(HCO_3)_2(aq) + Ca(OH)_2(aq) \rightarrow Mg(OH)_2(s) + Ca(HCO_3)_2(aq)$

19.89 (a) 68 g Mg (b) 9.21×10^5 L Cl_2

19.91 Elemental boron exists in several allotropic forms. It is a gray-black, nonmetallic, high-melting-point solid or a dark brown powder based on 12-atom clusters. When mixed with plastics, it produces a light, tough, stiff material. It is inert and resistant to oxidation. Aluminum is a light, strong metal and an excellent conductor of electricity. It is corrosion resistant as a result of the surface oxide layer. It has a low density.

19.93 (a) $B(OH)_4^-$ (b) $B(OH)_3(aq) + 2\,H_2O(l) \rightleftharpoons H_3O^+(aq) + B(OH)_4^-(aq)$

19.95 (a) $[He]2s^22p^2$ (b) $[Kr]4d^{10}5s^25p^1$ (c) $[Xe]6s^2$ (d) $[Kr]5s^1$

19.97 (a) Ionization energies decrease (become less positive) down Groups 13 (B → Tl) and 14 (C → Pb). Atomic radii increase down Groups 13 (B → Tl) and 14 (C → Pb). (b) The ionization energies generally decrease down a group. As the atomic number of an element increases, atomic shells and subshells that are farther from the nucleus are filled. The outermost valence electrons are consequently easier to remove. The radii increase down a group for the same reason. The radii are primarily determined by the outer shell electrons, which are farther from the nucleus in the heavy elements. (c) The trends correlate well with elemental properties, for example, the greater ease of outermost electron removal correlates with increased metallic character, that is, ability to form positive ions by losing one or more electrons.

19.99 (a) $2\,AlCl_3(s) + 3\,H_2O(l) \rightarrow 6\,HCl(g) + Al_2O_3(s)$ (b) $B_2H_6(g) \xrightarrow{\text{high temp.}} 2\,B(s) + 3\,H_2(g)$
(c) $4\,BF_3 + 3\,BH_4^- \xrightarrow{\text{organic solvent}} 3\,BF_4^- + 2\,B_2H_6$

19.101 (a) $2\,H_2O(l) + 2\,e^- \rightarrow H_2(g) + 2\,OH^-(aq)$
(b) cathode (c) 2.1 L H_2

19.103 In the majority of its reactions, hydrogen acts as a reducing agent, like the alkali metals. However, as described in the text and in the answer to Exercise 19.1, it may also act as an oxidizing agent, like the halogens. Consequently, H_2 will oxidize elements with standard reduction potentials more negative than -2.25 V, such as the alkali and alkaline earth metals (except Be). The compounds formed are hydrides and contain the H^- ion; the singly charged negative ion is reminiscent of the halide ions. Hydrogen also forms diatomic molecules and covalent bonds like the halogens. Consequently, hydrogen could be placed in either Group 1 or Group 17. But it is probably best to think of hydrogen as a unique element that has properties in common with both metals and nonmetals; therefore, it should probably be centered in the periodic table as indicated in the text.

19.105 (a) (polarizing ability × 1000 follows the ion) $Cs^+(5.9) < Rb^+(6.7) < K^+(7.2) < Na^+(10) < Li^+(17)$ and $Ba^{2+}(15) < Sr^{2+}(17) < Ca^{2+}(20) < Mg^{2+}(28) < Be^{2+}(74)$ (b) These data roughly support the diagonal relationship. Li^+ is more like Mg^{2+} than Be^{2+} and Na^+ more like Ca^{2+} than Mg^{2+}; but further down the group, the correlation fails.

19.107 $0.0714\ mol\cdot L^{-1}$

19.109 The smaller the cation, the greater the ability of the cation to polarize and weaken the carbonate ion. On that basis, we would predict that within a group the carbonates of the first members of the group are less stable than those of the later members. Thus $Li_2CO_3 < Na_2CO_3 < K_2CO_3 < Rb_2CO_3 < Cs_2CO_3$ and $BeCO_3 < MgCO_3 < CaCO_3 < SrCO_3 < BaCO_3$. Between groups, we would expect the stability of the carbonates in one period to decrease from Group 1 to Group 13 because of the smaller size of Group 13 ions. Thus $Al_2(CO_3)_3 < MgCO_3 < Na_2CO_3$. Carbonates of Group 13 are, in fact, so unstable that they do not exist.

CHAPTER 20

20.1 (a) $1s^2$ (b) $[He]2s^22p^4$ (c) $[He]2s^22p^5$
(d) $[Ar]3d^{10}4s^24p^3$

20.3 $I_2 < Br_2 < Cl_2 < F_2$

20.5 As < P < S < O

20.7 The first step is the liquefaction of air, which is 76% by mass nitrogen. Air is cooled to below its boiling point by a series of expansion and compression steps in a kind of refrigerator. Nitrogen gas is then obtained by distillation of liquid air. The nitrogen boils off at $-196°C$, but higher-boiling gases, principally oxygen, remain as a liquid. The pure nitrogen gas is then liquified by repeating the process.

20.9 (a) nitrous acid (b) nitrogen oxide or nitric oxide (c) phosphoric acid (d) dinitrogen trioxide

20.11 (a) NH_4NO_3 (b) Mg_3N_2 (c) Ca_3P_2 (d) H_2NNH_2

20.13 (a) $2\,NH_3(g) + 3\,CuO(s) \rightarrow N_2(g) + 3\,Cu(s) + 3\,H_2O(l)$ (b) $4\,NH_3(g) + 3\,F_2(g) \rightarrow NF_3(g) + 3\,NH_4F(s)$ (c) $4\,NH_3(g) + 5\,O_2(g) \rightarrow 4\,NO(g) + 6\,H_2O(l)$

20.15 (a) +2 (b) +1 (c) +3 (d) $-\frac{1}{3}$

20.17 $CO(NH_2)_2 + 2\,H_2O \rightarrow (NH_4)_2CO_3$; 8.0 kg $(NH_4)_2CO_3$

20.19 (a) 0.23 L N_2 (b) $Hg(N_3)_2$ would produce a larger volume, because its molar mass is less.

20.21 N_2O: $H_2N_2O_2$; $N_2O(g) + H_2O(l) \rightarrow H_2N_2O_2(aq)$; N_2O_3: HNO_2; $N_2O_3(g) + H_2O(l) \rightarrow 2\,HNO_2(aq)$; N_2O_5: HNO_3; $N_2O_5(g) + H_2O(l) \rightarrow 2\,HNO_3(aq)$

20.23 [Lewis structures: $[PCl_4]^+$ and $[PCl_6]^-$]

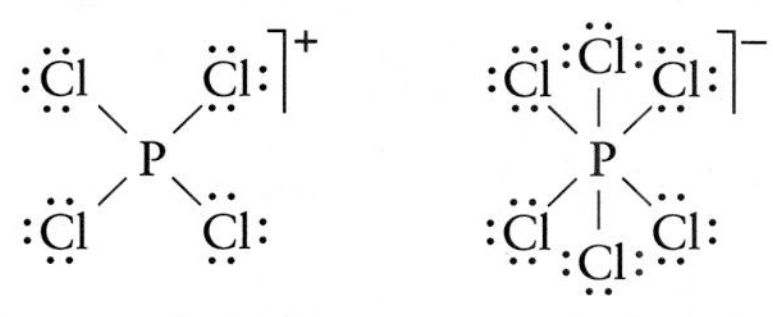

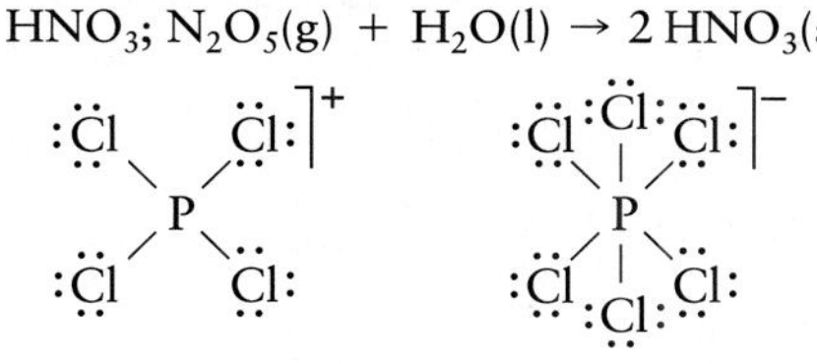

AX_4, tetrahedral AX_6, octahedral

20.25 (a) 12.23% (b) 26.47%

20.27 (a) H_2SO_4 (b) $CaSO_3$ (c) O_3 (d) BaO_2

20.29 Underground deposits of elemental sulfur are recovered by the Frasch process. Steam and hot water are forced into the deposit, causing the sulfur to melt. The molten sulfur mixes with the water, and the mixture is forced to the surface by compressed air.

20.31 (a) $4\,Li(s) + O_2(g) \xrightarrow{\Delta} 2\,Li_2O(s)$
(b) $2\,Na(s) + 2\,H_2O(l) \rightarrow 2\,NaOH(aq) + H_2(g)$
(c) $2\,F_2(g) + 2\,H_2O(l) \rightarrow 4\,HF(aq) + O_2(g)$
(d) $2\,H_2O(l) \rightarrow O_2(g) + 4\,H^+(aq) + 4\,e^-$

20.33 (a) $2\,H_2S(g) + 3\,O_2(g) \rightarrow 2\,SO_2(g) + 2\,H_2O(g)$
(b) $CaO(s) + H_2O(l) \rightarrow Ca(OH)_2(aq)$
(c) $PCl_5(s) + 4\,H_2O(l) \rightarrow H_3PO_4(aq) + 5\,HCl(g)$

20.35 [Lewis structure: H–O–O–H] Each O in H_2O_2 is an AX_2E_2 structure; hence the bond angle is predicted to be less than 109.5°. In actuality, it is 97°.

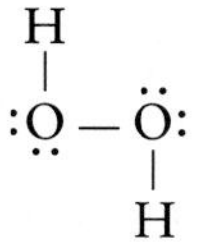

20.37 (a) When formal charges are considered, the possibility of expanded octets in S atoms is allowed, and two types of resonance structures can be seen to contribute to the overall structure of SO_2.

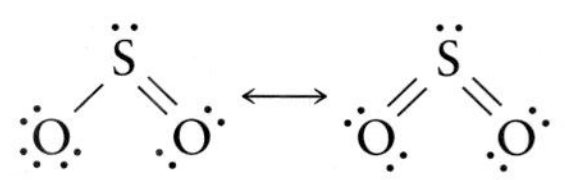

[Lewis structures: O–S=O ⟷ O=S=O]

(2 ways)   (1 way)

The completely double-bonded structure has a zero formal charge on all atoms and as a result may predominate in the resonance hybrid. Both structures have angular geometry. (b) SF_4 is an AX_4E VSEPR structure, and its shape is seesaw.

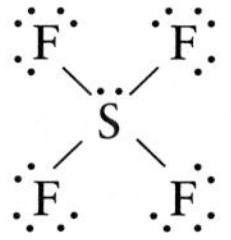

[Lewis structure: SF_4]

(c) When formal charges are considered and the possibility of extended octets for S atoms is allowed, three types of resonance structures contribute to the overall structure of SO_4^{2-}.

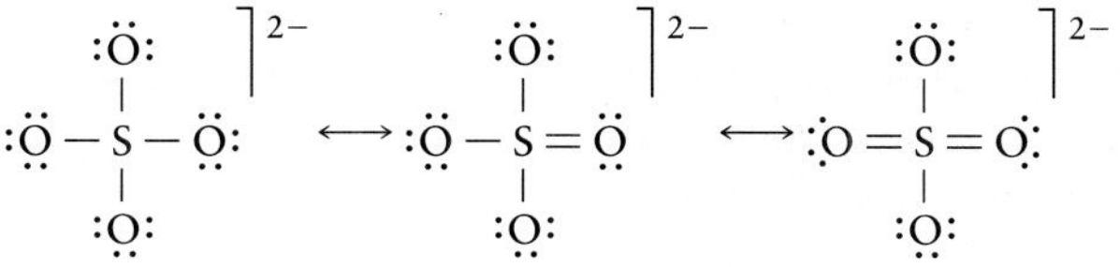

[Lewis structures of SO_4^{2-}: no S=O bonds ⟷ one S=O bond ⟷ two S=O bonds]

(1 way)   (4 ways)   (6 ways)

The third type of structure, which has smaller average differences in formal charge between S and O atoms, may be the predominant structure. All three structures have tetrahedral geometry.

20.39 5 L O_2

20.41 13.11 (13.12 if K_{a2} is not ignored)

20.43 See Fig. 9.16 for bond strengths. The weaker the H—X bond, the stronger the acid. H_2Te has the weakest bonds, H_2O the strongest. Therefore, acid strengths are $H_2Te > H_2Se > H_2S > H_2O$.

20.45 (a) $\Delta H_f^\circ = +142.7\ kJ\cdot mol^{-1}$; $\Delta S_f^\circ = -68.78\ J\cdot mol^{-1}\cdot K^{-1}$ (b) Because $\Delta G_f^\circ(O_3, g)$ is positive at all temperatures, the reaction is not spontaneous at any temperature. It is less favored at high temperatures.
(c) Because the reaction entropy is negative, the $-T\Delta S_f^\circ$ term is always positive, so the entropy contribution to ΔG_f° is always positive. The entropy does not favor the spontaneous formation of ozone.

20.47 Fluorine comes from the minerals fluorspar (CaF_2), cryolite (Na_3AlF_6), and the fluorapatites ($Ca_5F(PO_4)_3$). The free element is prepared from HF and KF by electrolysis, but the HF and the KF needed are prepared in the laboratory. Chlorine comes primarily from the mineral rock salt (NaCl). The pure element is obtained by electrolysis of liquid NaCl.

20.49 Fluorine: KF acts as the electrolyte for the electrolytic process, but the net reaction is $2\,H^+ + 2\,F^- \xrightarrow{current} H_2(g) + F_2(g)$.

Chlorine: $2\,NaCl(l) \xrightarrow{\text{current}} 2\,Na(s) + Cl_2(g)$

20.51 (a) hydrobromic acid (b) iodine bromide (c) chlorine dioxide (d) sodium iodate

20.53 (a) $HClO_4(aq)$ (b) $NaClO_3$ (c) HI(aq) (d) NaI_3

20.55 (a) +1 (b) +4 (c) +7 (d) +5

20.57 (a) $[O{=}Cl({-}O)({=}O){=}O]^-$ AX_4, tetrahedral

Note: This structure is based on formal charge considerations and involves an expanded octet on Cl; alternative structures with 0, 1, 2, and 4 Cl—O double bonds could be drawn. All structures have the same geometry and shape.

(b) $[O{-}I({-}O){-}O]^-$ AX_3E, tetrahedral electronic arrangement, trigonal pyramidal shape

Note: Lewis structures with 1 and 2 iodine-oxygen double bonds are also possible. All structures have the same geometry and shape.

(c) $F{-}I(F){-}F$ AX_3E_2, trigonal bipyramidal electronic arrangement, T-shaped molecule

20.59 (a) $4\,KClO_3(l) \xrightarrow{\Delta} 3\,KClO_4(s) + KCl(s)$
(b) $Br_2(l) + H_2O(l) \rightarrow HBrO(aq) + HBr(aq)$
(c) $NaCl(s) + H_2SO_4(aq) \rightarrow NaHSO_4(aq) + HCl(g)$
(d) Reactions (a) and (b) are redox reactions, (c) is a Brønsted acid-base reaction.

20.61 (a) $HClO < HClO_2 < HClO_3 < HClO_4$ (b) The oxidation number of Cl increases from HClO to $HClO_4$. In $HClO_4$, chlorine has its highest oxidation number of +7, so $HClO_4$ will be the strongest oxidizing agent.

20.63 $Cl{-}O{-}Cl$ AX_2E_2, angular, slightly less than 109°

20.65

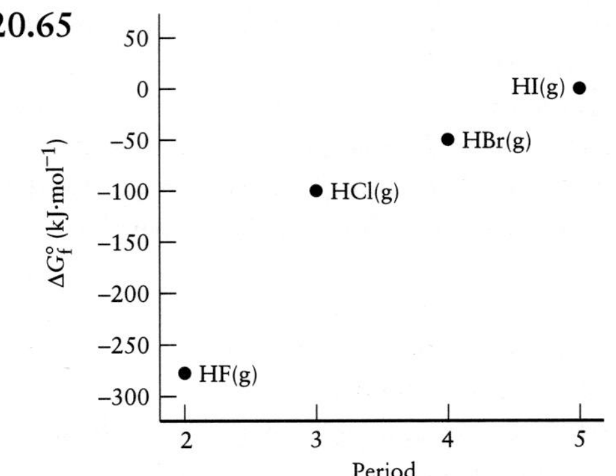

The ΔG_f° values of HCl, HBr, and HI fit nicely on a straight line, but HF is anomalous. In other properties, HF is also the anomalous member of the group, in particular with respect to its acidity. Thermodynamic stability of the hydrogen halides decreases down the group.

20.67 Because E°_{cell} is negative, Cl_2 will not oxidize Mn^{2+} to form permanganate ion in an acidic solution.

20.69 $K_p = 0.504$

20.71 $0.117\ mol{\cdot}L^{-1}$

20.73 Helium occurs as a component of natural gases found under rock formations in certain locations, especially in Texas. Argon is obtained by distillation of liquid air.

20.75 (a) +2 (b) +6 (c) +4 (d) +6

20.77 $XeF_4 + 4\,H^+ + 4\,e^- \rightarrow Xe + 4\,HF$

20.79 Because H_4XeO_6 has more oxygen atoms, which are highly electronegative, bonded to Xe, we predict that H_4XeO_6 is more acidic than H_2XeO_4.

20.81 XeF_4 (Xe bonded to four F) AX_4E_2, square planar, 90°

20.83 Ionization energies increase, electron affinities also increase (in magnitude). Large ionization energies and electron affinities are characteristic of nonmetals. Electronegativities and standard potentials increase as well; large values of these properties are also characteristic of nonmetals.

20.85 (a)

NO_2^- $[O{-}N{=}O]^- \longleftrightarrow [O{=}N{-}O]^-$

NO_3^-

$[O{=}N({-}O){-}O]^- \longleftrightarrow [O{-}N({=}O){-}O]^- \longleftrightarrow [O{-}N({-}O){=}O]^-$

(b) NO_2^-: AX_2E, trigonal planar electron pair arrangement, therefore sp^2 hybridization: NO_3^-: AX_3, trigonal planar electron pair arrangement, therefore sp^2 hybridization

20.87 oxidation: $As_2S_3(s) + 8\,H_2O(l) \rightarrow 2\,AsO_4^{3-}(aq) + 3\,S^{2-}(aq) + 16\,H^+(aq) + 4\,e^-$; reduction: $H_2O_2(aq) + 2\,e^- + 2\,H^+(aq) \rightarrow 2\,H_2O(l)$; overall: $As_2S_3(s) + 2\,H_2O_2(aq) + 4\,H_2O(l) \rightarrow 2\,AsO_4^{3-}(aq) + 3\,S^{2-}(aq) + 12\,H^+(aq)$

20.89 This ratio, $\Delta H_{vap}/T_b$, is the entropy of vaporization. See Section 16.11 and Example 16.3. Hydrogen bonding is much stronger in $H_2O(l)$ than in $H_2S(l)$. Thus $H_2O(l)$ has a more ordered arrangement than $H_2S(l)$ does. Consequently, the change in entropy upon transformation to the gaseous state is greater for H_2O than for H_2S.

20.91 8.16×10^2 L concentrated H_2SO_4
20.93 (a) $SO_2(g) + H_2O(l) \rightarrow H_2SO_3(l)$. This is a Lewis acid-base reaction. (b) $2\,F_2(g) + 2\,NaOH(aq) \rightarrow OF_2(g) + 2\,NaF(aq) + H_2O(l)$. This is a redox reaction in basic solution. (c) $S_2O_3^{2-}(aq) + 4\,Cl_2(g) + 13\,H_2O(l) \rightarrow 2\,HSO_4^-(aq) + 8\,H_3O^+(aq) + 8\,Cl^-(aq)$. This is a redox reaction. (d) $2\,XeF_6(s) + 16\,OH^-(aq) \rightarrow XeO_6^{4-}(aq) + Xe(g) + 12\,F^-(aq) + 8\,H_2O(g) + O_2(g)$. This is a redox reaction.
20.95 Fluorine: (1) The production of UF_6, which is part of the procedure for separating the isotopes ^{238}U and ^{235}U for use in nuclear processes. (2) The production of SF_6 for electrical equipment. (3) The production of fluorinated hydrocarbons, such as Teflon. Chlorine: (1) The manufacture of many important chemicals including plastics, solvents, and many organic chemicals. (2) As a bleach and a disinfectant. (3) The production of bromine. Bromine: (1) The production of organic bromides, which are incorporated into textiles as fire retardants; others are used as pesticides. (2) In photography, as AgBr, which is the active part of photographic emulsions. (3) In the oil industry, in the form of high-density aqueous zinc bromide, which is used to control the escape of oil from deep wells. (4) Alkali and alkaline earth metal bromides are used as sedatives. Iodine: (1) Iodides are used as an additive to table salt to prevent iodine deficiency, which leads to thyroid disorders. (2) As an antiseptic in alcohol solution.
(3) In photography, as AgI.
20.97 (a) $I_2(s) + 3\,F_2(g) \rightarrow 2\,IF_3(s)$
(b) $I_2(aq) + I^-(aq) \rightarrow I_3^-(aq)$ (c) $Cl_2(g) + H_2O(l) \rightarrow HCl(aq) + HOCl(aq)$,
but there are competing reactions, such as $Cl_2(g) + H_2O(l) \rightarrow 2\,HCl(aq) + \frac{1}{2}\,O_2(g)$. The predominant reaction is determined by the temperature and pH.
(d) $2\,F_2(g) + 2\,H_2O(l) \rightarrow 4\,HF(aq) + O_2(g)$
20.99 (a) $CaCl_2(s) + 2\,H_2SO_4(aq, conc.) \rightarrow Ca(HSO_4)_2(aq) + 2\,HCl(g)$ (b) $KBr(s) + H_3PO_4(aq) \rightarrow KH_2PO_4(aq) + HBr(g)$ (c) $KI(s) + H_3PO_4(aq) \rightarrow KH_2PO_4(aq) + HI(g)$
20.101 There are several possibilities. Some examples follow. (a) (i) +3: N_2O_3, dinitrogen trioxide; HNO_2, nitrous acid; KNO_2, potassium nitrite (ii) +5: N_2O_5, dinitrogen pentoxide; HNO_3, nitric acid; KNO_3, potassium nitrate
(b) (i) +3: P_4O_6, phosphorus(III) oxide; H_3PO_3, phosphorous acid; K_2HPO_3, potassium hydrogen phosphite
(ii) +5: P_4O_{10}, phosphorus(V) oxide; H_3PO_4, phosphoric acid; $Na_4P_2O_7$, sodium pyrophosphate
20.103 The heads of matches consist of a paste of potassium chlorate, $KClO_3$, antimony sulfide, Sb_2S_3, sulfur, and powdered glass. The striking strip contains red phosphorus. When the match is struck against the red phosphorus surface, a reaction of the red phosphorus and potassium chlorate causes the match to ignite. The Sb_2S_3 and sulfur are the fuels that are consumed by combustion after the ignition. The powdered glass helps to produce the friction required for ignition.

20.105 $[XeF_5]^+$ (Lewis structure: Xe with one lone pair, bonded to five F atoms, each F with three lone pairs) AX_5E, square pyramidal

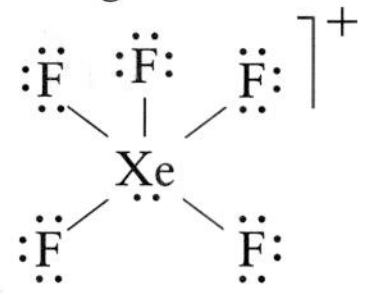

20.107 Until 1962, when their first compounds were prepared, the closed-shell electron configurations of the noble gases were taken to indicate that these elements were chemically inert. The noble gases all have high ionization energies and low (in magnitude) electron affinities. This combination implies chemical inertness. Thus, compounds were not actively sought. In addition, noble gases do not form compounds easily; the reagents and apparatus that promote reaction were not available until recently.
20.109 Proceeding down Group 14, carbon exhibits little metallic character, but Si is already a semimetal or semiconductor. Tin and lead show pronounced metallic character, with low ionization energies and high electrical conductivities. They also form cations, which is characteristic of metals. Proceeding down Group 15, P, which is the neighbor of Si, shows little metallic character; As shows some, but not as much metallic character as its neighbor Ge. The last member of this group, Bi, shows pronounced metallic character. Proceeding down Group 16, Se and Te show definite metallic character, about the same as As and Sb, but less than Ge and Sn. So there is a big difference in the metallic character of the heavy members of Groups 14 and 15, but only a slight difference between Groups 15 and 16. This pattern is most evident when one compares the ionization energies in Fig. 7.31. Group 14 elements have lower ionization energies than do the elements in Group 15 or 16, but the ionization energies of the elements in Groups 15 and 16 are about the same.
20.111 (a) $[\ddot{N}=N=\ddot{N}]^-$ AX_2, linear, 180°
(b) between fluoride and chloride (c) HCl, HBr, and HI are all strong acids. For HF, $K_a = 3.5 \times 10^{-4}$, so HF is slightly more acidic than HN_3. The small size of the azide ion suggests that the H—N bond in HN_3 is similar in strength to that of the H—F bond, so it is expected to be a weak acid. (d) ionic: NaN_3, $Pb(N_3)_2$, AgN_3; covalent: HN_3, $B(N_3)_3$, FN_3
20.113 0.156 $mol \cdot L^{-1}$
20.115 The solubility of the ionic halides is determined by a variety of factors, especially the lattice enthalpy and enthalpy of hydration. There is a delicate balance between the two factors, with the lattice enthalpy usually being the

determining one. Lattice enthalpies decrease from chloride to iodide, so water molecules can more readily separate the ions in the latter. Less ionic halides, such as silver halides, generally have a much lower solubility, and the trend in solubility is the reverse of the more ionic halides. For the less ionic halides, the covalent character of the bond allows ion pairs to persist in water. The ions are not easily hydrated, making them less soluble. The polarizability of the halide ions, and thus, the covalency of their bonding, increases down the group.

CHAPTER 21

21.1 (a) rhodium (b) silver (c) palladium (d) tungsten
21.3 (a) [Ar]$3d^5 4s^2$ (b) [Kr]$4d^{10} 5s^2$ (c) [Ar]$3d^{10} 4s^2$ (d) [Kr]$4d^2 5s^2$
21.5 (a) one unpaired $3d$-electron (b) three unpaired $3d$-electrons (c) one unpaired $4s$-electron (d) 1 unpaired $6s$-electron
21.7 iron, cobalt, and nickel
21.9 (a) Sc (b) Au (c) Nb
21.11 (a) Ti (b) Cu (c) Zn
21.13 The lanthanide contraction accounts for the failure of the third-row (Period 6) metallic radii to increase as expected relative to the radii of the second row (Period 5). It results from the presence of the *f*-block orbitals. The *f*-electrons present in the lanthanides are even poorer shields than *d*-electrons, and a marked decrease in metallic radius occurs along the *f*-block elements as a result of the increased nuclear charge, which pulls electrons inward. When the *d*-block resumes (at lutetium), the metallic radius has contracted from 188 to 157 pm. Therefore, lutetium and all the elements that follow lutetium have smaller-than-expected radii. Examples of this effect include the high density of the Period 6 elements and lack of reactivity of gold and platinum.
21.15 Hg is much more dense than Cd, because the shrinkage in atomic radius that occurs between $Z = 58$ and $Z = 71$ (lanthanide contraction) causes the atoms following the rare earths to be smaller than might have been expected for their atomic masses and atomic numbers. Zn and Cd have densities that are not too dissimilar, because the radius of Cd is subject to only a smaller *d*-block contraction.
21.17 Proceeding down a group in the *d*-block (as from Cr to Mo to W), there is an increasing probability of finding the elements in higher oxidation states. Thus, higher oxidation states become more stable going down a group.
21.19 In MO_3, M has an oxidation number of +6. Of these three elements, the +6 oxidation state is most stable for Cr. See Fig. 21.7.
21.21 (a) $TiO_2(s) + 2\,C(s) + 2\,Cl_2(g) \xrightarrow{1000°C} TiCl_4(g) + 2\,CO(g)$, followed by $TiCl_4(g) + 2\,Mg(l) \xrightarrow{700°C} Ti(s) + 2\,MgCl_2(s)$ (b) $V_2O_5(g) + 5\,Ca(l) \rightarrow 5\,CaO(s) + 2\,V(s)$
21.23 (a) Ti(s), $MgCl_2(s)$; $TiCl_4(g) + 2\,Mg(l) \rightarrow Ti(s) + 2\,MgCl_2(s)$ (b) $Co^{2+}(aq)$, $HCO_3^-(aq)$, $NO_3^-(aq)$; $CoCO_3(s) + HNO_3(aq) \rightarrow Co^{2+}(aq) + HCO_3^-(aq) + NO_3^-(aq)$ (c) V(s), CaO(s); $V_2O_5(s) + 5\,Ca(l) \xrightarrow{\Delta} 5\,CaO(s) + 2\,V(s)$
21.25 (a) titanium(IV) oxide, TiO_2 (b) iron(III) oxide, Fe_2O_3 (c) manganese(IV) oxide, MnO_2
21.27 (a) Yes, $Cr_2O_7^{2-}$ is a stronger oxidizing agent than Br_2. (b) No, Ag^{2+} is a stronger oxidizing agent than $Cr_2O_7^{2-}$.
21.29 (a) CO (b) $Fe_2O_3(s) + 3\,CO(g) \rightarrow 2\,Fe(s) + 3\,CO_2(g)$ (Zone C); $Fe_2O_3(s) + CO(g) \rightarrow 2\,FeO(s) + CO_2(g)$ (Zone D), followed by $FeO(s) + CO(g) \rightarrow Fe(s) + CO_2(g)$ (Zone C) (c) carbon
21.31 (a) V^{3+}, H_2, and Cl^- (b) no reaction (c) Co^{2+}, H_2, and Cl^-
21.33 All three elements in this group—Cu, Ag, and Au—are chemically rather inert, Ag more so than Cu, and Au more so than Ag. The standard potentials of their ions are all positive, in the order Au > Ag > Cu, so they are not readily oxidized. They have a common electron configuration, $(n-1)d^{10}ns^1$
21.35 (a) $CuFeS_2$, copper iron sulfide (b) ZnS, zinc sulfide (c) HgS, mercury(II) sulfide
21.37 (a) $2\,ZnS(s) + 3\,O_2(g) \xrightarrow{\Delta} 2\,ZnO(s) + 2\,SO_2(g)$, followed by $ZnO(s) + C(s) \xrightarrow{\Delta} Zn(l) + CO(g)$ (b) $HgS(s) + O_2(g) \xrightarrow{\Delta} Hg(g) + SO_2(g)$
21.39 10^6
21.41 (a) +2 (b) +3 (c) +3 (d) +3
21.43 (a) 4 (b) 2 (c) 6 (d) 6
21.45 (a) hexacyanoferrate(II) ion (b) hexaamminecobalt(III) ion (c) aquapentacyanocobaltate(III) ion (d) pentaamminesulfatocobalt(III) ion
21.47 (a) $K_3[Cr(CN)_6]$ (b) $[Co(NH_3)_5(SO_4)]Cl$ (c) $[Co(NH_3)_4(H_2O)_2]Br_3$ (d) $Na[Fe(H_2O)_2(C_2O_4)_2]$
21.49 (a) linkage isomers (b) ionization isomers (c) linkage isomers (d) ionization isomers
21.51 $[Co(H_2O)_6]Cl_3$, $[CoCl(H_2O)_5]Cl_2{\cdot}H_2O$, $[CoCl_2(H_2O)_4]Cl{\cdot}2H_2O$, $[CoCl_3(H_2O)_3]{\cdot}3H_2O$
21.53 $[CoCl(ONO)(en)_2]Cl$
21.55 (a) yes,

trans-Tetraamminedichlorocobalt(II) chloride monohydrate

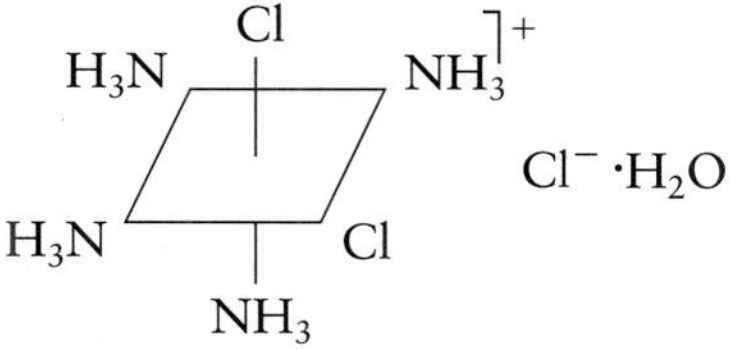

cis-Tetraamminedichlorocobalt(III) chloride monohydrate

(b) no (c) yes,

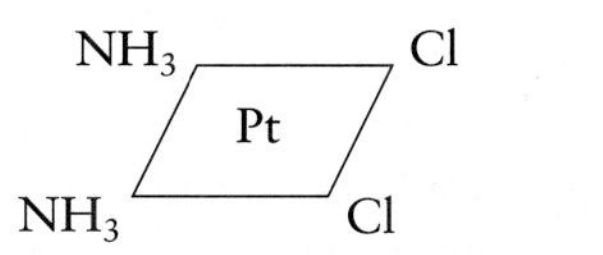

and

cis-Diamminedichloroplatinum(II)

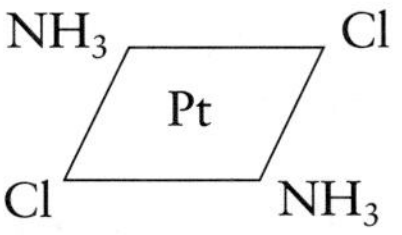

trans-Diamminedichloroplatinum(II)

21.57 (a) yes, in the form of optical isomerism (b) no (c) Yes, if four different ligand groups are bonded to the central atom, the central atom is chiral and exhibits optical activity.

21.59

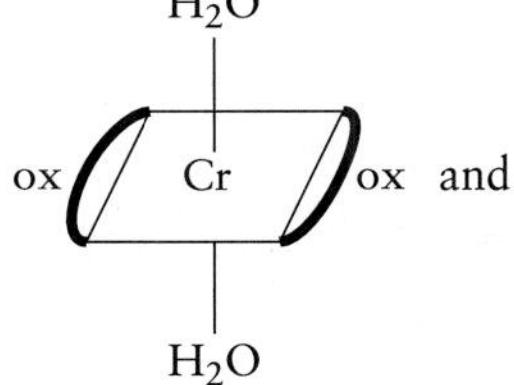

and

trans-Diaquabis-(oxalato)chromate(III) ion

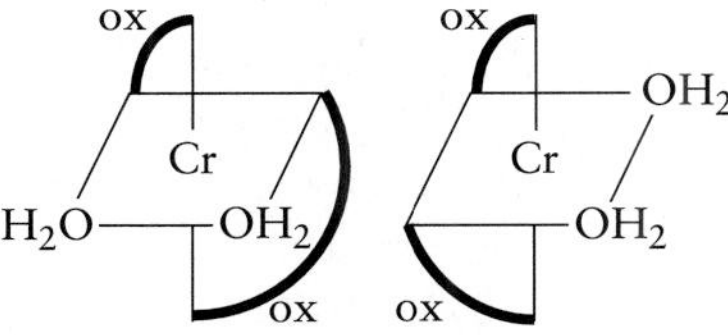

cis-Diaquabis-(oxalato)chromate(III) ion

21.61 (a) chiral; (b) not chiral

21.63

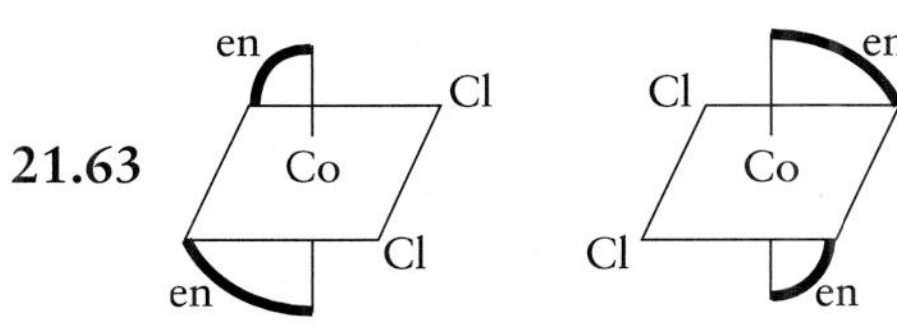

21.65 (a) d^7 (b) d^8 (c) d^5 (d) d^3

21.67

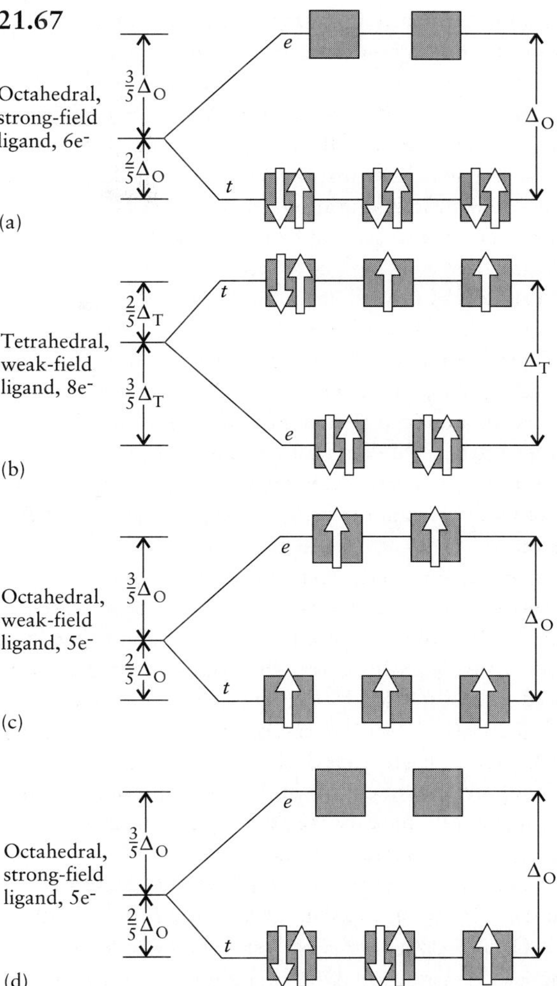

21.69 (a) zero unpaired electrons (b) two unpaired electrons

21.71 Weak-field ligands do not interact strongly with the *d*-electrons in the metal ion, so they produce only a small crystal field splitting of the *d*-electron energy states. The opposite is true of strong-field ligands. With weak-field ligands, unpaired electrons remain unpaired if there are unfilled orbitals; hence a weak-field ligand is likely to lead to a high-spin complex. Strong-field ligands cause electrons in excess of three to pair up with electrons in lower energy orbitals. A strong-field ligand is likely to lead to a low-spin complex. Ligands arranged in the spectrochemical series help to distinguish strong-field and weak-field ligands. Measurement of magnetic susceptibility (paramagnetism) can be used to determine the number of unpaired electrons, which in turn establishes whether the associated ligand is weak-field or strong-field in nature.

21.73 (a) blue, because F^- is a weak-field ligand (b) yellow, because the en ligand is a strong-field ligand. The splitting between levels is less in (a) than in (b), therefore

(a) will absorb light of a longer wavelength than (b) and consequently display a shorter wavelength color.

21.75 (a) yellow (b) green (c) orange (d) blue

21.77 In Zn^{2+}, the $3d$-orbitals are filled (d^{10}). There can be no electronic transitions between the t and e levels; hence no visible light is absorbed and the aqueous ion is colorless. The d^{10} configuration has no unpaired electrons, so Zn compounds would not be paramagnetic.

21.79 (a) 2.69×10^{-19} J or 162 kJ·mol^{-1} (b) 4.32×10^{-19} J or 260 kJ·mol^{-1} (c) 3.46×10^{-19} J or 208 kJ·mol^{-1}; $Cl < H_2O < NH_3$

21.81 (a) A compound with unpaired electrons is paramagnetic and is pulled into a magnetic field. A diamagnetic substance has no unpaired electrons and is weakly pushed out of a magnetic field. (b) Paramagnetism is a property of any substance with unpaired electrons, whereas ferromagnetism is a property of certain substances that can become permanently magnetized. Ferromagnetism results when a large number of electrons in the metal have parallel spins. Their spins become aligned, and this alignment can be retained even in the absence of a magnetic field. In a paramagnetic substance, the alignment is lost when the magnetic field is removed.

21.83 $[Sc(H_2O)_6]^{3+}(aq) + H_2O(l) \rightarrow [Sc(H_2O)_5OH]^{2+}(aq) + H_3O^+(aq)$

21.85 (a) an alloy of copper and tin (b) the compound basic copper carbonate, $Cu_2(OH)_2CO_3$ (c) native or "pure" gold (d) osmium

21.87 (a) $Fe_2O_3(s) + 3\,CO(g) \rightarrow 2\,Fe(s) + 3\,CO_2(g)$ (Zone C); $3\,Fe_2O_3(s) + CO(g) \rightarrow 2\,Fe_3O_4(s) + CO_2(g)$ and $Fe_3O_4(s) + CO(g) \rightarrow 3\,FeO(s) + CO_2(g)$; these combine to $Fe_2O_3(s) + CO(g) \rightarrow 2\,FeO(s) + CO_2(g)$ (Zone D); $FeO(s) + CO(g) \rightarrow Fe(s) + CO_2(g)$ (Zone C) (b) $TiCl_4(g) + 2\,Mg(l) \xrightarrow{\Delta} Ti(s) + 2\,MgCl_2(s)$ (c) $CaO(s) + SiO_2(s) \xrightarrow{\Delta} CaSiO_3(l)$

21.89 0.691 kg $FeCr_2O_4$

21.91 (a) Cr^{3+} ions in water form the complex $[Cr(H_2O)_6]^{3+}(aq)$, which behaves as a Brønsted acid: $[Cr(H_2O)_6]^{3+}(aq) + H_2O(l) \rightleftharpoons [Cr(H_2O)_5OH]^{2+}(aq) + H_3O^+(aq)$ (b) The gelatinous precipitate is $Cr(OH)_3$. The precipitate dissolves as the $Cr(OH)_4^-$ complex ion is formed. $Cr^{3+}(aq) + 3\,OH^-(aq) \rightarrow Cr(OH)_3(s)$ and $Cr(OH)_3(s) + OH^-(aq) \rightarrow Cr(OH)_4^-(aq)$

21.93 (a) pentaamminesulfatochromium(III) chloride (b) tris(ethylenediamine)chromium(III) tris(oxalato)chromate(III) (c) diamminedichloroplatinum(II) (d) potassium hexanitritocobaltate(III)

21.95 (a) tetrakis(oxalato)zirconate(IV) ion, C.N. = 8, Zr = +4 (b) diaquatetrachlorocuprate(II) ion, C.N. = 6, Cu = +2 (c) amminetrichloroplatinate(II) ion, C.N. = 4, Pt = +2 (d) tetracyanodioxomolybdate(IV) ion, C.N. = 6, Mo = +4

21.97

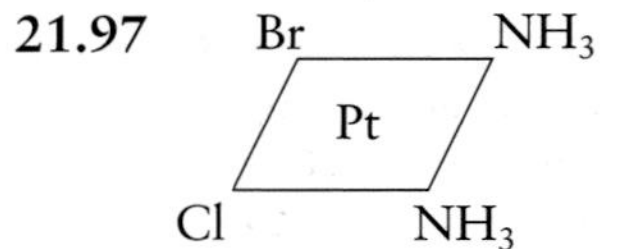

cis-Diamminebromochloroplatinum(II)

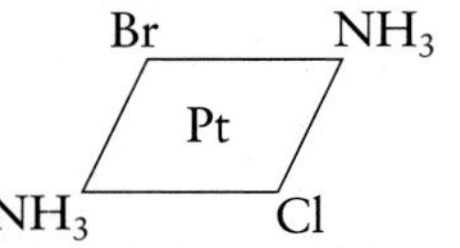

trans-Diamminebromochloroplatinum(II)

21.99 (a) The first, $[Ni(SO_4)(en)_2]Cl_2$, will give a precipitate of AgCl when $AgNO_3$ is added; the second will not. (b) The second, $[NiCl_2(en)_2]I_2$, will show free I_2 when mildly oxidized with, for example, Br_2, but the first will not.

21.101 ionization isomerism,

$[Cr(NH_3)_6]^{3+}$ $3\,Cl^-$

Hexaamminechromium(III) chloride

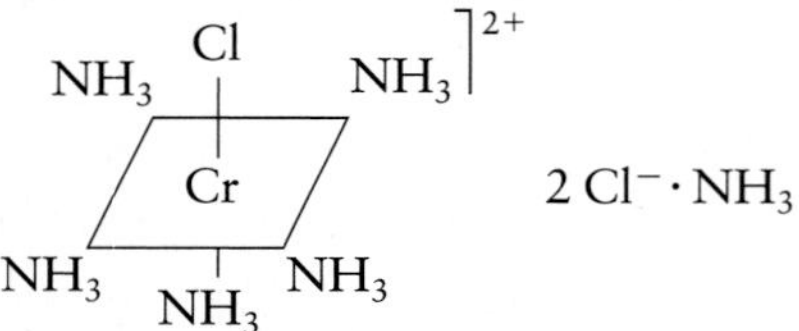

Pentaamminechlorochromium(III) chloride ammoniate

$[CrCl_2(NH_3)_4]^{+}$ $Cl^- \cdot 2NH_3$

trans-Tetraamminedichlorochromium(III) chloride diammoniate

21.103 (a) $[MnCl_6]^{4-}$ $\underline{\uparrow}\ \underline{\uparrow}$ over $\underline{\uparrow}\ \underline{\uparrow}\ \underline{\uparrow}$; $[Mn(CN)_6]^{4-}$ $\underline{\uparrow\downarrow}\ \underline{\uparrow\downarrow}\ \underline{\uparrow}$ (b) one unpaired electron in $[Mn(CN)_6]^{4-}$ (c) strong-field ligands absorb shorter wavelength light, therefore $[Mn(CN)_6]^{4-}$ transmits longer wavelengths.

21.105 (a) 3.62×10^{-19} J or 218 kJ·mol^{-1} (b) red

21.107 +0.005 V

21.109 −0.74 V

21.111 (a) $[AuCl_4]^-(aq) + NO(g) + 2\,H_2O(l) \rightarrow Au(s) + 4\,Cl^-(aq) + NO_3^-(aq) + 4\,H^+(aq)$ (b) +0.04 V (c) −0.22 V (d) When concentrated hydrochloric and nitric acids are present, the direction of the reaction is reversed and Au is oxidized.

21.113 The correct structure for $[Co(NH_3)_6]Cl_3$ consists of four ions, $Co(NH_3)_6^{3+}$ and $3Cl^-$ in aqueous solution. The chloride ions can be easily precipitated as AgCl. This would not be possible if they were bonded to the other (NH_3) ligands. If the structure were $Co(NH_3{-}NH_3{-}Cl)_3$, VSEPR theory would predict that the Co^{3+} ion would have a trigonal planar ligand arrangement. The splitting of the *d*-orbital energies would not be the same as the octahedral arrangement and would lead to different spectroscopic and magnetic properties inconsistent with the experimental evidence. In addition, neither optical isomers nor geometrical isomers would be observed.

CHAPTER 22

22.1 α particles: composed of the ${}^4_2\text{He}$ nuclei, with a mass of 4.0 u, a charge of +2, and low penetrating power. β particles: composed of electrons, with a mass of 0.00055 u, a charge of −1, and moderate penetrating power. γ rays: composed of photons, with a mass of 0 u, a charge of 0, and high penetrating power.
22.3 Let p = protons, n = neutrons, and nu = nucleons. (a) 1p, 1n, 2nu (b) 12p, 12n, 24nu (c) 104p, 159n, 263nu (d) 27p, 33n, 60nu (e) 94p, 144n, 238nu (f) 101p, 157n, 258nu
22.5 Let p = protons, n = neutrons, and nu = nucleons. (a) ${}^{81}_{35}\text{Br}$, 35p, 46n, 81nu (b) ${}^{90}_{36}\text{Kr}$, 36p, 54n, 90nu (c) ${}^{244}_{96}\text{Cm}$, 96p, 148n, 244nu (d) ${}^{128}_{53}\text{I}$, 53p, 75n, 128nu (e) ${}^{32}_{16}\text{S}$, 16p, 16n, 32nu (f) ${}^{241}_{95}\text{Am}$, 95p, 146n, 241nu
22.7 (a) 3.2×10^{-12} m, 3.8×10^{10} J·mol^{-1} (b) 5.3×10^{-14} m, 2.3×10^{12} J·mol^{-1} (c) 8.1×10^{-13} m, 1.5×10^{11} J·mol^{-1} (d) 4.1×10^{-15} m, 2.9×10^{13} J·mol^{-1}
22.9 (a) 3.21×10^{20} Hz, 9.33×10^{-13} m (b) 3.95×10^{20} Hz, 7.59×10^{-13} m (c) 2.66×10^{20} Hz, 1.13×10^{-12} m
22.11 (a) ${}^3\text{He}$, ${}^3_1\text{T} \rightarrow {}^{0}_{-1}\text{e} + {}^3_2\text{He}$ (b) ${}^{83}\text{Sr}$, ${}^{83}_{39}\text{Y} \rightarrow {}^0_1\text{e} + {}^{83}_{38}\text{Sr}$ (c) ${}^{87}\text{Rb}$, ${}^{87}_{36}\text{Kr} \rightarrow {}^0_{-1}\text{e} + {}^{87}_{37}\text{Rb}$ (d) ${}^{221}\text{Ac}$, ${}^{225}_{91}\text{Pa} \rightarrow {}^4_2\alpha + {}^{221}_{89}\text{Ac}$
22.13 (a) ${}^8_5\text{B} \rightarrow {}^0_1\text{e} + {}^8_4\text{Be}$ (b) ${}^{63}_{28}\text{Ni} \rightarrow {}^0_{-1}\text{e} + {}^{63}_{29}\text{Cu}$ (c) ${}^{185}_{79}\text{Au} \rightarrow {}^4_2\alpha + {}^{181}_{77}\text{Ir}$ (d) ${}^7_4\text{Be} + {}^0_{-1}\text{e} \rightarrow {}^7_3\text{Li}$
22.15 (a) β, ${}^{24}_{11}\text{Na} \rightarrow {}^0_{-1}\text{e} + {}^{24}_{12}\text{Mg}$ (b) β, ${}^{128}_{50}\text{Sn} \rightarrow {}^0_{-1}\text{e} + {}^{128}_{51}\text{Sb}$ (c) β^+, ${}^{140}_{57}\text{La} \rightarrow {}^0_1\text{e} + {}^{140}_{56}\text{Ba}$ (d) α, ${}^{228}_{90}\text{Th} \rightarrow {}^4_2\alpha + {}^{224}_{88}\text{Ra}$
22.17 (a) α particles are emitted in the decay of nuclei with $Z > 83$. A few other nuclei having *N/Z* (or *(A/Z)* − 1) ratios lying near the band of stability with $60 < Z < 70$ also emit α particles. (b) β particles are emitted in the decay of nuclei with *N/Z* ratios greater than the *N/Z* ratio of the band of stability. The emission of a β particle is the equivalent of converting a neutron into a proton; thus, the *N/Z* ratio becomes smaller to fit the values required for stability.
22.19 (a) β decay, ${}^{68}_{30}\text{Zn}$ (b) β^+ decay, ${}^{103}_{47}\text{Ag}$ (c) α decay, ${}^{239}_{95}\text{Am}$ (d) α decay, ${}^{256}_{103}\text{Lr}$
22.21 ${}^{235}_{92}\text{U} \rightarrow {}^4_2\alpha + {}^{231}_{90}\text{Th}$, ${}^{231}_{90}\text{Th} \rightarrow {}^0_{-1}\text{e} + {}^{231}_{91}\text{Pa}$, ${}^{231}_{91}\text{Pa} \rightarrow {}^4_2\alpha + {}^{227}_{89}\text{Ac}$, ${}^{227}_{89}\text{Ac} \rightarrow {}^0_{-1}\text{e} + {}^{227}_{90}\text{Th}$, ${}^{227}_{90}\text{Th} \rightarrow {}^4_2\alpha + {}^{223}_{88}\text{Ra}$, ${}^{223}_{88}\text{Ra} \rightarrow {}^4_2\alpha + {}^{219}_{86}\text{Rn}$, ${}^{219}_{86}\text{Rn} \rightarrow {}^4_2\alpha + {}^{215}_{84}\text{Po}$, ${}^{215}_{84}\text{Po} \rightarrow {}^0_{-1}\text{e} + {}^{215}_{85}\text{At}$, ${}^{215}_{85}\text{At} \rightarrow {}^4_2\alpha + {}^{211}_{83}\text{Bi}$, ${}^{211}_{83}\text{Bi} \rightarrow {}^0_{-1}\text{e} + {}^{211}_{84}\text{Po}$, ${}^{211}_{84}\text{Po} \rightarrow {}^4_2\alpha + {}^{207}_{82}\text{Pb}$
22.23 (a) ${}^{14}_7\text{N} + {}^4_2\alpha \rightarrow {}^{17}_8\text{O} + {}^1_1\text{p}$ (b) ${}^{248}_{96}\text{Cm} + {}^1_0\text{n} \rightarrow {}^{249}_{97}\text{Bk} + {}^0_{-1}\text{e}$ (c) ${}^{243}_{95}\text{Am} + {}^1_0\text{n} \rightarrow {}^{244}_{96}\text{Cm} + {}^0_{-1}\text{e} + \gamma$ (d) ${}^{13}_6\text{C} + {}^1_0\text{n} \rightarrow {}^{14}_6\text{C} + \gamma$
22.25 (a) ${}^{20}_{10}\text{Ne} + {}^4_2\alpha \rightarrow {}^8_4\text{Be} + {}^{16}_8\text{O}$ (b) ${}^{20}_{10}\text{Ne} + {}^{20}_{10}\text{Ne} \rightarrow {}^{24}_{12}\text{Mg} + {}^{16}_8\text{O}$ (c) ${}^{44}_{20}\text{Ca} + {}^4_2\alpha \rightarrow {}^{48}_{22}\text{Ti} + \gamma$ (d) ${}^{27}_{13}\text{Al} + {}^2_1\text{H} \rightarrow {}^{28}_{13}\text{Al} + {}^1_1\text{p}$
22.27 (a) ${}^{14}_7\text{N} + {}^4_2\alpha \rightarrow {}^{17}_8\text{O} + {}^1_1\text{p}$ (b) ${}^{239}_{94}\text{Pu} + {}^1_0\text{n} \rightarrow {}^{240}_{95}\text{Am} + {}^0_{-1}\text{e}$
22.29 (a) ${}^{244}_{95}\text{Am} \rightarrow {}^{134}_{53}\text{I} + {}^{107}_{42}\text{Mo} + 3\,{}^1_0\text{n}$ (b) ${}^{235}_{92}\text{U} + {}^1_0\text{n} \rightarrow {}^{96}_{40}\text{Zr} + {}^{138}_{52}\text{Te} + 2\,{}^1_0\text{n}$ (c) ${}^{235}_{92}\text{U} + {}^1_0\text{n} \rightarrow {}^{101}_{42}\text{Mo} + {}^{132}_{50}\text{Sn} + 3\,{}^1_0\text{n}$
22.31 1.0×10^{-4} Ci
22.33 (a) 3.7×10^{10} Bq (1 Bq = 1 disintegration per second) (b) 3.0×10^9 Bq (c) 3.7×10^4 Bq
22.35 1.0×10^2 rad, 1.0×10^2 rem
22.37 100 d
22.39 (a) 5.63×10^{-2} y^{-1} (b) 0.82 s^{-1} (c) 0.0693 min^{-1}
22.41 (a) 5.3×10^3 y (b) 8.4×10^9 y (c) 2 y
22.43 6.0×10^{-3} Ci
22.45 (a) 88.6% (b) 32.4%
22.47 (a) 0.25 (b) 1.26×10^9 y
22.49 3.53×10^3 y
22.51 (a) 9.8×10^{-4} Ci (b) 2.8×10^{-4} Ci (c) 0.40 Ci
22.53 This term refers to self-sustaining nuclear chain reactions. For a chain reaction to be self-sustaining, each nucleus that splits must provide an average of at least one new neutron that results in the fission of another nucleus. If the mass of the fissionable material is too small, the neutrons will escape before they can produce fission. The critical mass is the smallest mass that can sustain a nuclear chain reaction.
22.55 (a) 2.3×10^{-10} g gained (b) 1.86×10^{-10} g lost (c) 8.33×10^{-9} g lost
22.57 (a) 9.0×10^{13} J (b) 8.187×10^{-14} J
22.59 -4.3×10^9 kg·s^{-1}
22.61 (a) -1.14×10^{-12} J·nucleon^{-1} (b) -1.213×10^{-12} J·nucleon^{-1} (c) -1.8×10^{-13} J·nucleon^{-1} (d) -1.410×10^{-12} J·nucleon^{-1}. ${}^{56}\text{Fe}$ is the most stable, because it has the largest binding energy per nucleon.
22.63 (a) -7.8×10^{10} J·g^{-1} (b) -3.52×10^{11} J·g^{-1} (c) -2.09×10^{11} J·g^{-1} (d) -3.52×10^{11} J·g^{-1}
22.65 Uranium tetrafluoride, which is obtained in the refining of uranium ores, is oxidized to the hexafluoride: $UF_4(s) + F_2(g) \xrightarrow{450°C} UF_6(s)$. The hexafluoride is

vaporized and then used in the enrichment process, which makes use of the different effusion rates of gaseous $^{235}UF_6$ (349 amu) and $^{238}UF_6$ (352 amu). According to Graham's law (Section 5.15), the relative rates of effusion are

$$\frac{\text{rate of effusion of } ^{235}UF_6}{\text{rate of effusion of } ^{238}UF_6} = \left(\frac{352}{349}\right)^{1/2} = 1.004$$

Because this ratio is so close to 1, the uranium hexafluoride vapor must be allowed to effuse repeatedly through porous barriers designed for the purpose. These are screens with large numbers of minute holes. In practice, it is allowed to do so thousands of times. An alternative enrichment procedure utilizes a centrifuge that rotates samples of uranium hexafluoride vapor at very high speed. This process causes the heavier $^{238}UF_6$ molecules to be thrown outward and collected as a solid on the outer parts of the rotor, leaving a higher proportion of $^{235}UF_6$ closer to the axis of the rotor, where it can be removed.

22.67 The order of penetrating power is $\gamma > \beta > \alpha$. γ rays are uncharged, high-energy photons that can pass right through objects like the body with little retardation. β particles are fast electrons that can penetrate flesh to about a depth of 1 cm before they are stopped by electrostatic interactions. α particles are the least penetrating because of their charge and relatively large mass. Although they do not penetrate deeply, they are very damaging because of their high energy and large mass. Consequently, they can dislodge atoms from molecules, thereby altering the structure of the molecule, which in turn alters the ability of the molecule to function properly. If the molecule is DNA, a necessary enzyme, or another essential molecule in a living system, the result may be cancer.

22.69 The stabilities of nuclei vary, and the greatest stabilities are associated with certain numbers of nucleons. These numbers are referred to as magic numbers; they are 2, 8, 20, 50, 82, and 126. The existence of this series of numbers reminds us of the magic numbers of electrons in the electronic configurations of the noble gases: 2, 10, 18, 36, 54, and 86. Because this series of numbers corresponds to a shell model for electrons in atoms, the analogous series of numbers for nucleons suggests a nuclear shell model.

22.71 The basic difference between polyester fabrics and fabrics such as wool, silk, and nylon is that the latter group are all polyamides. The presence of the amide functional group allows for extensive cross-linking between polymer chains by hydrogen bonding with the amine group on one chain to the carbonyl group on another. Intermolecular attractions are thus greater between polyamide molecules than between polyester molecules. The less polar polyester molecules can more readily accept radon atoms amongst their chains.

22.73 (a) $^{11}_{5}B + ^{4}_{2}\alpha \rightarrow ^{13}_{7}N + 2\,^{1}_{0}n$ (b) $^{35}_{17}Cl + ^{2}_{1}D \rightarrow ^{36}_{18}Ar + ^{1}_{0}n$ (c) $^{96}_{42}Mo + ^{2}_{1}D \rightarrow ^{97}_{43}Tc + ^{1}_{0}n$ (d) $^{45}_{21}Sc + ^{1}_{0}n \rightarrow ^{42}_{19}K + ^{4}_{2}\alpha$

22.75 (a) 9 dpm (b) 7×10^5 decays

22.77 (a) 1.7×10^8 pCi·L^{-1} (b) 5.0×10^{18} atoms (c) 1×10^2 d

22.79 45 d

22.81 (a) 14.3 d (b) 0.23

22.83 2.4×10^{-4} g

22.85 (a) -3.69×10^{-14} J·nucleon^{-1} (b) -8.85×10^{-13} J per decay

22.87 (a) At first thought, it might seem that a fusion bomb would be more suitable for excavation work, as the fusion process itself does not generate harmful radioactive waste products. However, in practice, fusion cannot be initiated in a bomb in the absence of the high temperatures that can be generated only by a fission bomb. So there is no environmental advantage to the use of a fusion bomb. The fission bomb has the advantage that its destructive power can be more carefully controlled. It is possible to make small fission bombs such that the destructive effect could be contained within a small area.
(b) The principal argument for the use of bombs in excavation is speed, and hence cost-effectiveness, of the process. The principal argument against their use is environmental damage.

22.89 (a) 4.5×10^9 y (b) 3.8×10^9 y

22.91 (a) $k = 1.11$ d^{-1}, $t_{1/2} = 0.63$ d (b) 0.22 μg

ILLUSTRATION CREDITS

Cover, John Shaw Photography; frontispiece, (top) John Shaw Photography, (bottom left) John P. O'Brien, DuPont, (bottom right) John Gosline, University of British Columbia; p. 0, C. R. O'Dell and S. K. Wong (Rice University)/NASA; p. 2, Ken Karp; p. 3, (Fig. 1.2) Burndy Library, (Fig. 1.3) Randall M. Feenstra, IBM Thomas J. Watson Research Center, Yorktown Heights, New York; p. 5, (Fig. 1.4) Cavendish Laboratory, (Fig. 1.5) Donald Clegg; p. 6, The Bettmann Archive; p. 7, Walter Schmid/Tony Stone Worldwide; p. 10, Ken Karp; p. 11, J. Hester and P. Scowen (Arizona State University)/NASA; p. 13, (Fig. 1.15) Alexander Boden, (Fig. 1.16) Chip Clark; pp. 14, 15 & 19, Chip Clark; p. 22, Peter Kresan; p. 23, Chip Clark; p. 24, Mary Dale-Bannister, Washington University/NASA; p. 26, Ken Karp; p. 30, Charles Thatcher/Tony Stone Images; p. 38, Stephen Frink/Waterhouse Stock Photography; p. 40, Ken Karp; p. 42, Bureau International Des Poids et Mesures; p. 52, Burndy Library; p. 56, Chip Clark; p. 57, Ken Karp; p. 58, Chip Clark; p. 61, Mark Philbrick, Brigham Young University; p. 63, Air Products and Chemicals, Inc.; p. 64, Ken Karp; p. 76, Gary Braasch/Woodfin Camp & Associates; p. 78, Chip Clark; p. 81, Pacific Gas and Electric; p. 82, Yoav Levy/Phototake NYC; pp. 85 & 88, Chip Clark; p. 89, Ken Reid/FPG International; p. 90, Ken Karp; p. 96, Ben Osborne/Tony Stone Images; p. 97, Chip Clark; p. 98, (Fig. 3.20a) Chip Clark, (Fig. 3.20b) Ken Karp; pp. 99 & 100, Ken Karp; p. 103, Johnson Space Center/NASA; pp. 104 & 105, Ken Karp; p. 107, Telegraph Colour Library/FPG International; p. 116, Johnson Space Center/NASA; p. 118, Air Products and Chemicals, Inc.; pp. 122 & 123, Ken Karp; p. 127, Greg Pease/Tony Stone Images; p. 132, adapted from "Greenhouse Effect: Science and Policy," by Stephen H. Schneider, 1989, *Science*, 243:773; p. 133, (top left) adapted from "Greenhouse Effect: Science and Policy," by Stephen H. Schneider, 1989, *Science*, 243:773, (right) Bruce Forster/Tony Stone Images; p. 142, Philippe Plailly/Science Photo Library/Photo Researchers, Inc.; p. 143, NASA; p. 148, Peter Ginter/Bilderberg; p. 153, Hank Morgan/Photo Researchers, Inc.; pp. 155 & 157, Ken Karp; p. 158, U.S. Department of the Interior, Bureau of Mines; p. 159, (Fig. 5.21) TRW Inc., (Fig. 5.22a) Travelpix/FPG International, (Fig. 5.22b) Eugen Gebhardt/FPG International; p. 165, (Fig. 5.26) Ken Karp, (Fig. 5.27) Martin Marietta Energy Systems, Oak Ridge, Tennessee; p. 169, Ken Karp; p. 170, adapted from *The Cambridge Encyclopedia of the Earth Sciences*, London: Sceptre Books; p. 171, (top) Philippe Plailly/Science Photo Library/Photo Researchers, Inc., (bottom) NASA/Science Photo Library/Photo Researchers, Inc.; p. 180, Eric Crichton/Bruce Coleman Inc.; p. 184, Ken Karp; p. 192, Thermomax USA, Inc.; p. 193, USGS, Flagstaff, Arizona; p. 195, Martin Rogers/FPG International; p. 199, (top) Chevron Corporation, (bottom) Navaswan/FPG International; p. 204, Imperial War Museum; p. 205, Catherine Pouedras/Eurelios/Science Photo Library/Photo Researchers, Inc.; p. 206, Karl Weatherly/Mountain Stock; p. 218, Milton Heiberg; p. 220, Ken Karp; p. 221, David Olsen/Tony Stone Images; p. 222, Century Lubricants Specialists; p. 225, Chip Clark; p. 226, from *Quantitative Chemical Analysis*, by Daniel C. Harris, New York: W. H. Freeman, 1995; p. 227, Bausch & Lomb; p. 229, Science Museum, London; p. 230, Joe Viesti/Viesti & Associates; p. 244, Edgar Fahs Smith Collection, ACS Center for History of Chemistry, University of Pennsylvania; pp. 252 & 254, Ken Karp; p. 255, Chip Clark; p. 256, Hank Morgan/Science Source/Photo Researchers, Inc.; pp. 257 & 259, Chip Clark; p. 266, Werner H. Muller/Peter Arnold, Inc.; p. 269, University of California Archives, The Bancroft Library; p. 273, Ken Karp; p. 277, Warren Gretz/NREL (National Renewable Energy Laboratory); p. 288, (left) South Coast Air Quality Management District, (right) Chip Clark; p. 289, Loretta Jones; pp. 290 & 297, Ken Karp; p. 302, J. & L. Weber/Peter Arnold, Inc.; p. 312, Oxford Molecular Biophysics Laboratory/Science Photo Library/Photo Researchers, Inc.; p. 314, Ken Karp; p. 317, Charlie Westerman/Liaison International; p. 321, Stephen Simpson/FPG International; p. 332, Hyperion Catalysis/Harvard University; p. 333, data courtesy of G. E. Poirier and E. D. Pylant, 1996, *Science*, 272:1145; p. 340, Paul Berger/Tony Stone Worldwide; p. 343, Ken Karp; p. 348, Paul Brierly; p. 350, Chip Clark; p. 351, (Fig. 10.13) Peter Pfletschinger/Peter Arnold, Inc., (Fig. 10.14) NASA; p. 352, Chip Clark; p. 353, (Fig. 10.17a) Steven Smale, (Fig. 10.17b) Ken Karp, (Fig. 10.18) Michael Heron/Woodfin Camp & Associates; p. 354, Hakon Hope; p. 360, Sandia National Laboratories/NREL; p. 361, Chip Clark; p. 363, Jim Mejuto/FPG International; p. 365, (top) Tom Carroll/FPG International, (Fig. 10.35) Bob Thomason/Tony Stone Images; p. 366, from "Near-Atomic Resolution Imaging of Ferroelectric Liquid Crystal Molecules on Graphite by Scanning Tunneling Microscopy," by D. M. Walba, F. Stevens, D. C. Parks, N. A. Clark, and M. D. Wand, 1995, *Science*, 267:1144–1147; p. 367, Richard Feldmann/Phototake NYC; p. 368, Andrew Syred/Science Photo Library/Photo Researchers, Inc.; p. 370, (Fig. 10.41) Michael Fairchild/Peter Arnold, Inc., (Fig. 10.42b) John Shaw/Bruce Coleman Inc.; p. 371, Chip Clark; p. 372, General Electric Company; p. 388, Lawrence Berkeley Laboratory/Science Photo Library/Photo Researchers, Inc.; pp. 397 & 402, Ken Karp; p. 411, World Perspectives/Tony Stone Images; p. 412, (Fig. 11.20) Helly Hansen (US) Inc., (Fig. 11.21) Ken Karp; p. 414, Fibers Division, Monsanto Chemical Company; p. 415, (Fig. 11.24) Andrew Syred/Science Photo Library, (Fig. 11.25) Chip Clark; p. 418, James Kilkelly; p. 420, (Fig. 11.28) Gamma, (Fig. 11.29) Ken Karp; p. 421, Steven Gottlieb/FPG International; p. 423, A. Marmont and E. Damasio, Division of Hematology, S. Martin's Hospital, Genoa, Italy; p. 424, Tom Bean/Tony Stone Images; p. 425, Institute of Paper Chemistry; p. 426, Photo Researchers, Inc.; p. 436, CNRI/Science Photo Library/Photo Researchers, Inc.; p. 438, Ken Karp; p. 440, Chilean Nitrate Corporation; p. 441, Chip Clark; p. 444, E. B. Smith, Physical Chemistry Laboratory, University of Oxford; p. 445, Comstock, Inc.; p. 448, Ken Karp; p. 450, Terry Vine/Tony Stone Images; p. 455, (Fig. 12.9a) Chip Clark, (Fig. 12.19b) Ken Karp; p. 457, Ken Karp; p. 462, Chip Clark; p. 463, (Fig. 12.26) Ken Karp, (Fig. 12.27) from "Biological Membranes as Bilayer Couples," by M. Sheetz, R. Painter, and S. Singer, 1976, *Journal of Cell Biology*, 70:193; p. 472, I. & L. Waldman/Bruce Coleman Inc.; p. 485, Warren Gretz/NREL; p. 492, Smithsonian Institution; p. 495, M. W. Kellogg Company; p. 496, (top) Fritz Goro, (bottom) Liaison International; p. 497, Custom Medical Stock; p. 498, JPL/NASA; p. 499, Johnson Matthey; p. 500, (Fig. 13.12 left) American Institute for Physics, Niels Bohr Library/Stein Collection, (Fig. 13.12 right) The

Bettmann Archive; p. 508, Paul Chesley/Tony Stone Images; p. 511, Jay Laurie/Bruce Coleman Inc.; pp. 516 & 518, Ken Karp; p. 530, Photo Researchers, Inc.; p. 534, (left) National Atmospheric Deposition Program, Colorado State University, Fort Collins, (top left & right) from "Acid Deposition: Unraveling a Regional Phenomenon," by S. E. Schwartz, 1989, *Science*, 243:755; p. 535, Regis Bossu/Sygma; p. 536, Ken Karp; p. 543, Michael Rosenfeld/Tony Stone Images; p. 544, Chip Clark; pp. 545 & 546, Ken Karp; p. 558, Brinkmann Instruments; p. 559, Chip Clark; p. 560, Heather Angel/Biophotos; p. 564, Custom Medical Stock; p. 568, Comstock, Inc.; p. 569, Custom Medical Stock; p. 572, Ken Karp; p. 574, Chip Clark; p. 586, David Austen/Tony Stone Images; p. 592, Corbis/Bettmann; pp. 594 & 595, Ken Karp; p. 598, Dieter Flamm; p. 599, International Tin Research Institute; p. 604, Yale University; p. 607, Runk/Schoenberger/Grant Heilman Photography; p. 615, Warren Gretz/NREL; p. 624, from "Artificial Photosynthesis," by Nathan Lewis, November–December 1995, *American Scientist,* 83:534–541; p. 631, Chip Clark; pp. 633, 634, 640 & 642, Ken Karp; p. 646, (top), (bottom) Ken Karp; p. 649, Ken Lucas/Planet Earth Pictures; p. 650, based on illustrations from "Artificial Photosynthesis," by Nathan Lewis, *American Scientist*, November–December, 1995, 83:534–541; p. 651, Warren Gretz/NREL; p. 652, Ken Karp; p. 653, St. Joe Zinc Company/ American Hot Dip Galvanizers Association; p. 656, Warren Gretz/ NREL; p. 657, The National Museum of Science & Industry; p. 660, (Fig. 17.27) Vulcan Materials Company, (Fig. 17.28) Comstock, Inc.; p. 670, B. & C. Alexander/Photo Researchers, Inc.; p. 679, Ken Karp; p. 683, W. Curstinger/Photo Researchers, Inc.; p. 687, (Fig. 18.14) Archive Photos, (Fig. 18.15) Ken Karp; p. 695, Ken Karp; p. 696, Lubert Stryer; p. 698, (bottom right) Tom Cardamone Associates, Inc.; p. 699, Jan Leesma/Custom Medical Stock; p. 705, Ken Karp; p. 718, Bill Pierce/Sygma; pp. 721, 722, 726 & 727, Ken Karp; p. 728, (Fig. 19.10) P. P. Edwards, University of Birmingham, (Fig. 19.11) Ken Karp; p. 729, (Fig. 19.12) Sanyo/Quantum Communications, (Fig. 19.13) Texasgulf; p. 730, (Fig. 19.15) Photo Researchers, Inc., (Fig. 19.16) Bill Tronca/Tom Stack & Associates; p. 732, Ken Karp; p. 733, (Fig. 19.18) Houston Museum of Natural Science, (Fig. 19.19) Dow Chemical; p. 734, Alexander Boden; p. 737, (Fig. 19.22) Travis Amos, (Fig. 19.23) Field Museum of Natural History, Chicago; p. 738, Aalborg Portland Betonforskningslaboratorium, Karlslunde; p. 740, U.S. Borax; p. 741, Otto Baitz/Esto; p. 743, (Fig. 19.29a,c) Houston Museum of Natural Science, (Fig. 19.29b) Lee Bolton, Inc., (Fig. 19.30) Chip Clark; p. 745, (Fig. 19.32) Ken Karp, (Fig. 19.33) Viacom Consumer Products/Paramount Pictures Corporation; p. 747, Chip Clark; p. 750, (Fig. 19.36) UPI/Corbis-Bettmann, (Fig. 19.37) Max Planck/Institute fur Kenphysik; p. 751, (Fig. 19.38) from "Carbon Nanotube Chemistry," by Jessica Cook, Jeremy Sloan, and Malcolm L. H. Green, August 10, 1996, *Chemistry & Industry*, and "Microscopy in the Study of Fullerene-Related Carbons," by Peter J. F. Harris, September 1994, *Microscopy and Analysis*; (Fig. 19.39) Field Museum of Natural History, Chicago; p. 753, (top) Tony Stone Images, (Fig. 19.41) Ken Karp; p. 755, Chip Clark; p. 756, (left) Corning Glass, (right) The Corning Museum of Glass; p. 757, The Corning Museum of Glass; p. 758, (Fig. 19.47) Chip Clark, (Fig. 19.48) Phillip Hayson/Photo Researchers, Inc.; p. 759, Chip Clark; p. 766, Tony Stone Images; p. 768, Chip Clark; p. 769, Photo Researchers, Inc.; p. 770, (Fig. 20.3) Chip Clark, (Fig. 20.4) Ken Karp; p. 771, Chip Clark; p. 773, Ken Karp; p. 775, Lawrence Migdale/Tony Stone Worldwide, Ltd.; p. 777, (Fig. 20.8) Chip Clark, (Fig. 20.9) Ross Chapple; p. 778, (Fig. 20.10) Chip Clark, (Fig. 20.11b) Pennzoil; p. 779, (Fig. 20.12a) Chip Clark, (Fig. 20.12b) National Park Services, (Fig. 20.13) Chip Clark; p. 781, Ken Karp; p. 782, Lee Bolton, Inc.; p. 783, Ken Karp; p. 784, Chip Clark; p. 787, (Fig. 20.23) Martin Marietta Energy Systems, Inc., (Fig. 20.24) Chip Clark; p. 788, Chip Clark; p. 790, Ken Karp; p. 792, (left) Morton Thiokol, (right) Johnson Space Center/NASA; p. 793, NASA; p. 795, Greater Pittsburgh Neon/Tom Anthony; p. 796, Argonne National Laboratory; p. 802, American Superconductor, Westborough, Massachusetts; p. 808, (Fig. 21.8a) Field Museum of Natural History, Chicago, (Fig. 21.8b) Hagley Museum and Library, (Fig. 21.8c) Oremet Titanium; p. 809, Tom Carroll/FPG International; pp. 810 & 811, Ken Karp; 812, Institute of Oceanographic Sciences/NERC/Science Photo Library/Photo Researchers, Inc.; 815, U.S. Steel; 817, Chip Clark; 818, (Fig. 21.17) Paul Chesley/Tony Stone Images, (Fig. 21.18) Lee Bolton Picture Library; 819, (Fig. 21.19) Field Museum of Natural History, Chicago, (Fig. 21.20) Thierry Borredon/Tony Stone Images; 820, F. S. Judd, Research Laboratories, Kodak Ltd.; 821, Edward Keating; pp. 822, 823 & 824, Ken Karp; p. 827, Carol Moralejo, Concordia University; p. 832, Jon Riley/Tony Stone Images; pp. 836 & 837, Carol Moralejo, Concordia University; p. 848, Roger Ressmeyer/Corbis; p. 850, Granger Collection; p. 860, Fermilab; p. 861, U.S. Department of Energy; p. 862, Oak Ridge National Laboratory; p. 863, (left) Marcus E. Raichle and Washington University School of Medicine, (right) adapted from "Positron Emission Tomographic Studies of the Cortical Anatomy of Single Word Processing," by S. E. Petersen, P. T. Fox, M. I. Posner, M. A. Mintum, M. E. Raichle, 1988, *Nature*, 331:585–589; p. 865, (Fig. 22.17) Yoav Levy/Phototake NYC, (Fig. 22.18a) Chip Clark; p. 866, Mitch Kezar/Phototake NYC; p. 868, International Atomic Energy Agency; p. 869, University of Arizona; p. 870, CNRI/Phototake NYC; p. 871, U.S. Department of Energy/Science Photo Library/Photo Researchers, Inc.; p. 873, Georges Ollen/Sygma; p. 878, Tom Carroll/Phototake NYC; p. 879, (Fig. 22.29) Princeton Plasma Physics Laboratory, (Fig. 22.30) Chip Clark; p. 880, (Fig. 22.31) Phil Schofield, (Fig. 22.32) U.S. Department of Energy; p. 881, Photo Researchers, Inc.

INDEX

α decay, product, 853
α helix, 423
α particle, 6, 851
  and helium formation, 794
α radiation, 851
α ray, 851
α-alumina, 743
absolute zero, 47
absorbance, 226
absorbed dose, 864, 866
absorption
  light, 226
  spectrum, 226
acceptor, electron-pair, 292
accuracy, 50
acetate ion
  Lewis structure, 281
  structure, 527
acetic acid
  deprotonation in water, 519
  dimer, structure, 348
  in water, 512
  ionization, 92
  Lewis structure, 279
  structure, 90, 527
acetone
  structure, 83
acetylcholine, 699
acetylene
  combustion, 98
  lamp, 98
  structure, 98
acetylene, see ethyne
acetylide, 759
acetylsalicylic acid, structure, 390
achiral molecule, 408
acid, 90
  amino, 404
  Arrhenius, 91
  Brønsted, 510
  buffer, 562
  character of ion, table, 544
  conjugate, 512
  defined, 510
  Lewis, 292
  nucleic, 428
  polyprotic, 532
  rain, 94, 509, 534
  strength, trends, 524
  strong, 92, 523
  weak, 92, 518
acid anhydride, 742
acid-base titration, 122
  illustration, 123
acidic
  hydrogen atom, 91
  solution, 104
acidic and basic character in periodic table, 94
acidic oxides of nonmetals, 94
acidic solution, 104
  pH, 515
  redox reaction, 627
acidity constant, 519
  overall, 532
  polyprotic acids, table, 532
  table, 521
acidosis, 568
Acrilan, 410
actinide, 13, 248
action of acid
  on carbonates, 97, 98
  on metals, 104
  on sulfides, 97
action of nitric acid
  on copper, 104
action of sulfuric acid
  on copper, 104
action of water
  on carbides, 98
activated
  carbon, 749
  charcoal, 107, 749
activated complex, 688
activated complex theory, 693
activation energy, 689
  determination, 690
activity, radioactivity, 866
addition polymerization, 410
addition reaction, 396
  enthalpy changes, 397
addition rule, 49
adduct, see complex
adenine, structure, 427
adenosine diphosphate, structure, 776
adenosine triphosphate, structure, 775
adhesion, 351
adhesive, illustration, 348
adipic acid, 414
ADP, 607, 776
adsorb, 695
adsorption, 107
affinity, electron, 255, 256
agate, illustration, 753
air
  composition, 144
  molar mass, 144
  purifier, 103, 749
air bag, 159
air pockets, 149
air sampling, 153
airplane engine, illustration, 808
airship, Hindenburg, 687
-al suffix, 405
alanine, 422
albumen, 348
alcohol
  acidity, 527
  grain, 400
  nomenclature, 405
  primary, 400
  secondary, 400
alcoholism, 699
aldehyde, 399, 401
aldehyde
  nomenclature, 405
  preparation, 402
algebra, 2001
algologist, 683
aliphatic hydrocarbon, 390
alkali, 90
alkali metal, 13, 257
  chemical properties, 727
  compounds, table, 727
  flame test, 220
  illustration, 726
  melting point, 726
  properties, table, 725
  reaction with water, 727
alkaline earth metals, 13, 257, 731
  illustration, 732
  reaction with water, 734
alkalosis, 568
alkane, 391
  nomenclature, 398
  nomenclature, table, 391
  substitution reaction, 395
alkene, 395
  addition reaction, 396
  nomenclature, 398
  test for, 397
alkyne, 395
  nomenclature, 398
allotrope, 372
alloy, 363
  composition, table, 364
  formation, 805
  interstitial, 364
  substitutional, 364
almond, aroma, 402
alnico steel, 815
alpha particle, see α
alternating copolymer, 417
altitude as a state property, 189
alum, 743
  papermaker's, 743
  water purification, 106
alumina, 742
aluminate ion, 95, 742
aluminosilicate, 755
aluminum
  amphoteric behavior, 741
  anodized, 740
  architectural uses, 741
  production, 740
aluminum chloride, 744
  dimer, structure, 745
  Lewis structure, 292
  structure, 292
aluminum hydroxide, 743
aluminum oxide, 742
  amphoteric character, 95
amber glass, 757
amethyst, illustration, 753
amide, 399, 404
amine, 93, 399, 404
  nomenclature, 405
  weak base, 511
-amine suffix, 405
amino acid, 404, 421
  essential, 421
  naturally occurring, table, 422
ammonia, 770
  bonding, 327
  complex formation, 771
  equilibrium constant, 487
  equilibrium, response to conditions, 492
  in water, 91
  metal-ammonia solution, 728
  protonation in water, 519
  shape, 310
  structure, 310
  synthesis equilibrium, 474
  uses, 500
  weak base, 511
Ammonians, 771
ammonium carbonate, 771
ammonium chloride, formation, 165
ammonium dichromate decomposition, illustration, 184
ammonium ion, structure, 19
ammonium sulfate, Lewis structure, 278

amorphous solid, 352
amount, chemical, 51
ampere, 658
amphetamine, 699
amphiprotic, 512
amphoteric character, 95
aluminum, 741
aluminum oxide, 95
and oxidation number, 808
periodicity, 95
tin, 747
zinc, 819
amu, 43
Amun, 771
amylase, 697
amylopectin, 424
amylose, 424, 444
amytriptyline hydrochloride, 699
analysis, combustion, 130
analyte, 122, 550
Ancient Mariner, 470
-ane suffix, 391
anemia, 423
angle, bond, 304
angular molecule, 304
anhydride
acid, 742
formal, 742
animal pelt, 496
anion, 18
basic, table, 545
configuration, 246
names, 28
table, 29
typical, 20
anisotropic material, 366
annual energy flux, 199
anode, 630
mud, 669
sacrificial, 653
anodized aluminum, 740
antacid, 737
tablet, 536
anthracene, structure, 397
antidepressant, 698
antifreeze, 400, 462
ethylene glycol, structure, 462
antilogarithm
common, A3
natural, A4
antimony, 770
illustration, 768
antioxidant, 285
antiperspirant, 745
antiseptic, 788
antitumor agent, 829
apatite, 440, 769
Apollo mission fuel, 793
approximation, in equilibrium calculation, 490
aqua regia, 819
aquarium filter, 749
aquatic life, oxygen requirement, 446
aqueous solution, 22, 82
pH, 516
aragonite, 737
arene, 397
nomenclature, 399
substitution, 397
arginine, 422
aromatic
compound, 397
hydrocarbon, 390
Arrhenius, S., 91
Arrhenius
equation, 690
parameters, table, 691
plot, 690
arsenic, 770
illustration, 768
artificial atmosphere, 103
asbestos, 755
asbestosis, 755
asparagine, 422
aspartic acid, 422
asphalt, 392
aspirin, structure, 390, 581
atactic, 413
-ate suffix, 28
athletic shoe, 149
atmosphere
artificial, 103
extent, 143
history, 170
pressure, 164
structure of, 164
unit, 147
atmospheres, planetary, composition, 167
atmospheric
chemistry, 170
nitrogen oxides, 288
atom, 2
central, 276
many-electron, 237
mass determination, 8, 42
terminal, 277
atomic
mass unit, 43
number, 7
orbital, 230
radii, *d*-block, 805
radius, table, 249
weight, 53
ATP, 607, 775
Aufbau principle, 241
aurora borealis, 670
autoionization, see autoprotolysis
automatic titrator, 558
illustration, 558
automobile tire, 421
autoprotolysis, 513
constant, 513
effect on pH, 530
average
bond enthalpy, table, 318
bond length, table, 320
chain length, 420
molar mass, element, 54
molar mass, polymer, 420
speed, 166
speed, gas molecules, 167
Avogadro constant, 51
Avogadro's principle, 150
axial lone pair, 310
axle cap, illustration, 808
azide, 771
azimuthal quantum number, 234

β decay, product, 853
β particle, 851
β radiation, 851
β ray, 851
β-pleated sheet, 423
background radioactivity, 864
bacteria, nitrogen-fixing, 768
baggage sniffer, 26
baking powder, 730
baking soda, 730, A31
balanced equation, 79
balancing,
redox reaction, 107
Toolbox, 80
ball-and-stick model, 17
balloon, high-altitude, 142
ballooning, 148
illustration, 159
Balmer series, 225
band of stability, 856
band structure of solids, 362
bank analogy, 591
bar, 197
barium, illustration, 732
barium hydroxide, endothermic reaction, 184
barium titanate, 810
barometer, 146
base, 90
Arrhenius, 91
buffer, 562
conjugate, 512
defined, 510
Lewis, 292
pair, 428
polyprotic, 532
strong, 92, 523
weak, 92, 518
bases: oxides of metals, 94
basic copper carbonate, 818
basic oxygen process, 814
basic solution
pH, 515
redox reaction, 628
basicity constant, 519
Batavia, 860
battery, 630
exhausted, 633
molecular, 332
bauxite, 739
bcc, 357
Bear Lake, 446
Becquerel, H., 850
becquerel, 866
beer, 22
beer bottle glass, 757
Beer's law, 226
Belousov, R. P., 496
Belousov-Zhabotinskii reaction, 496
bends (diver's), 443
benzaldehyde, structure, 402
benzene
bonding, 330
hybridization, 330
Kekulé structure, 281
resonance, 281
ring-in-hexagon, 282
structure, 281, 397
benzeneberg, 371
beryl, 732
illustration, 733
beryllate ion, 737
beryllium, 257
applications, 732
compounds, 737
illustration, 732
tetrahedral unit, 737
beryllium atom, configuration, 241
beryllium chloride
shape, 306
structure, 296, 306, 737
beryllium hydride, structure, 737
beta particle, see β
bicarbonate of soda, 730
bicarbonate see hydrogen carbonate
bidentate, 825
Big bang, 720
bimolecular reaction, 700
binary compound, 723
binary molecular hydride, 724
binding energy, 872
biomass, 181
oxidation, 485
biotechnologist, 615
biradical, 286
bismuth, 770
illustration, 768
bisphenol A, A29

bisulfite, see hydrogen sulfite, 782
blast furnace, 812
  illustration, 813
bleach, 2031
bleaching, paper pulp, 791
blister copper, 669, 817
block, periodic table, 247
block copolymer, 417
blood, 90
  capillary, 443
  cell, 436
  osmosis, 463
  pH, 515, 543, 568
  plasma, 437
  solubility of gases in, 443
  vessel collapse, 444
bloodroot, 76
body temperature, 47
body-centered cubic structure, 357
Bohr, N., 228
Bohr frequency condition, 228
boiling, 375
boiling point
  constant, table, 460
  elevation, 459
  normal, 375
  predicting, 607
  table, 190, 344
  trend prediction, 345
Boltzmann
  constant, 598
  formula, 598
bomb calorimeter, 187
bond
  angle, 304
  coordinate covalent, 292, 823
  covalent, 274
  delocalized, 746
  double, 276
  enthalpy, 316
  hydrogen, 346
  ionic, 268
  multiple, 276
  π, 324
  polar, 312
  σ, 323
  single, 276
  strength, 316, 317
  three-center, 746
  triple, 324
bond enthalpy, 316
  average, table, 318
  diatomic molecules, table, 317
bond length
  average, table, 320
  halogen molecules, 321
bond strength
  and acid strength, 525
  bond length correlation, 321
  effect of resonance, 319
  electronegativity effect on, 318
  hydrogen halide, 318
  hydrogen-element, table, 319
bonding
  benzene molecule, 330
  fluorine molecule, 275
  hydrogen molecule, 275
  valence-bond theory, 323
bone, 440, 738
booster rocket, 792
borane, 745
borax, A31
Borazon, 744
boric acid, 742
  ionization, 521
  Lewis acid, 742
Born-Haber cycle, 271
boron
  illustration, 255
  neutron capture therapy, 863
  structure, 740
boron atom, configuration, 241
boron carbide, 744
boron halide, 744
  Lewis acid, 744
boron nitride, structure, 744
boron oxide, 742
boron trichloride, production, 744
boron trifluoride
  Lewis structure, 291
  production, 744
  shape, 306
  structure, 306
Boron, California, 740
borosilicate glass, 756
Bosch, C., 500
boundary surface, 231
Boyle, R., 148
Boyle's law, 149
Bragg equation, 354
brain activity, 863
branched chain, 393
branched chain reaction, 876
branching, effect on volatility, 393
branching step, 705
brass, 364
breathing apparatus, 728
breeder reactor, 878
brimstone, 777, A31
brine, 729
  source of iodine, 787
brittleness, 273
bromine
  production, 787
  properties, 787
  reaction with iron, 787
  reaction with phosphorus, 100
  uses, 787
bromine, 2
  illustration, 2, 13
bromine water, 397
Brønsted, J., 510
Brønsted-Lowry theory, 510
bronze, 364
Bronze Age, 808
Brownian motion, 445
buckminsterfullerene, 332
  structure, 749
buffer, 561, 754
  acid and base, 562
  action, 562
  capacity, 566
  carbonate, 567
  commercial, 566
  range, 566
  selection, 562
  system, table, 562
building-up principle, 241
buret, 40
burns, 568
butane
  combustion, 81
  structure, 393, 406
butterfat, 22, 444
butterfly, wing pattern, 472
butyl rubber, 418
BZ reaction, 496

cadaverine, 404
cadmium, 819
  illustration, 2
caffeine, structure, 431
calamine, 2031
calcite, 732, 737
calcium
  compounds, 737
  deficiency, 534
  illustration, 732
  in diet, 738
  reaction with water, 734
calcium carbide, illustration, 98
calcium carbonate, 737
  decomposition, 79
  illustration, 58
calcium fluoride, structure, 384
calcium hypochlorite, 790
calcium phosphate, 769
calibrate, 186
calomel electrode, 646
calorie, 185
Calorie, food, 206
calorimeter, 186
calorimetry, combustion
  reaction, 194
camphor, 461
  structure, 432
cancer necrosis factor, 321
capacity, buffer, 566
capillary
  action, 351
  blood, 443
capsaicin, structure, 432
carat, 818
  24-carat gold, 818
carbide
  covalent, 759
  interstitial, 759
  saline, 759
carbohydrate, 424
  as fuel, 206
carbolic acid, 401
carbon
  allotropes, 748
  cycle, 77
  family, 747
  illustration, 56, 747
  nanotube, 332
  phase diagram, 385
  tetravalence, 276
carbon atom
  configuration, 241
carbon black, 749
carbon dioxide, 752
  critical properties, 380
  formation from carbonates, 97
  in blood, 568
  in rain, 517, 535
  Lewis structure, 284
  model, 17
  nonpolar, 313
  phase diagram, 378
  production, 97
  projected concentration, 133
  removal, 103
  result of combustion, 125
  shape, 307
  structure, 307
  supercritical, 381
carbon monoxide, 752
  formal anhydride, 752
  in blast furnace, 814
  Lewis base, 752
  polarity, 312
  reducing agent, 752
  toxicity, 752
carbon oxides, 752
carbon suboxide, shape, 307
carbon tetrachloride, see tetrachloromethane
carbon-14 dating, 869
carbon-carbon multiple bonds, 330
carbon-fluorine bonds, 316

carbonate, decomposition, 496
carbonate buffer, 567
carbonate ion
model, 20
structure, 307
carbonic acid, 752
equilibria, 532
in blood, 568
in rain, 534
carbonyl group, 401
carborundum, illustration, 759
carboxyl group, structure, 91
carboxylic acid, 91, 399, 402
acidity, 527
nomenclature, 405
carnallite, 731
carrageenan, 445
carrier
chain, 704
neutron, 876
carvone, structure, 390, 431
cassette tape, 807
cassiterite, 751
cast iron, 814
catalyst, 694
action of, 694
effect on equilibrium, 499
heterogeneous, 695
homogeneous, 694
Ziegler-Natta, 413
catalytic converter, 289, 534
catalyzed path, 694
catenation, 779
catfish, electric, 649
cathode, 630
cathode ray, 4
cathodic protection, 653
cation, 18
acidic, table, 544
name, 27
cation
configuration, 246
names, table, 28
typical, 18
caustic soda, A31
ccp, 356
cell
commercial, table, 648
Daniell, 630
diagram, 632
diaphragm, 730
dry, 647, 648
electrochemical, 630
electrolytic, 654
fuel, 647, 648
galvanic, 630
lead-acid, 647
Leclanché, 647
mercury, 648
nicad, 648
nickel-cadmium, 649
notation, 632
photoelectrochemical, 650
photovoltaic, 199, 650
potential, 633
potential and free energy, 635
potential, standard, 635
primary, 647
secondary, 647
silver, 648
unit, 357
cell membrane, 367
cellulose, strength, 348
cellulose acetate, 462
Celsius, A., 46
Celsius scale, 46
cement, 758
central atom, 276
ceramic, 756, 757
glaze, 815
cesium
illustration, 726
cesium-137
illustration, 862
cesium-chloride structure, 369
CFC, 171
chain branching, 705
chain carrier, 704
chain length, average, 420
chain reaction, 704
chalcocite, illustration, 817
chalcogen, 776
chalcopyrite, 816
illustration, 817
chalk, 732, 2031
champagne, 445
change
chemical, 23
natural, 595
physical, 23
spontaneous, 595
characteristic, A3
charcoal, activated, 107
charge
distribution, 312
electric, 4
formal, 282
partial, 312
supplied, 658
chariot axle cap, illustration, 808
Charles, J., 148
Charles's law, 149
charlière, 148
chelate, 824
chemical
amount, 51
element, 2
equation, 79
equilibrium, 473
formula, 16
kinetics, 671
nomenclature, A30
plating, 745
property, 23
reaction, 77, 78
reaction, classification, 108
symbol, 3
chemiluminescence, 781
chemistry, 1
nuclear, 849
chemotherapy, 829
Chernobyl, 868
cherry aroma, 402
chewing gum, 402
china clay, 757
chiral molecule, 408
chloralkali process, 659
chlorates, properties, 791
chloric acid, strength, 526
chlorine
as oxidizing agent, illustration, 788
illustration, 13
liquid, 169
production, 659, 786
properties, 786
uses, 786
chlorine dioxide, 791
chlorine trifluoride
shape, 311
structure, 311
chlorofluorocarbon, 171
chloroform, see trichloromethane
chlorophyll, 651, 737
spectrum, 266
chlorous acid, strength, 526
chlorpromazine, 699
cholesteric phase, 366
chromatogram, 26
chromatography, 25
chrome tape, 811
chromic acid, 808
chromite, 811
chromium, 811
plating, 660
chromium(III) chloride, 827
chromium(IV) oxide, 811
chromium(VI) oxide, 808
cinnabar, 777, 819
illustration, 778
cinnemaldehyde, structure, 402
cis and trans, 407
cis-dibromoethane, volatility, 375
cis-dichloroethene
boiling point, 346
structure, 313
cis-retinal, 334
citrus juice, pH, 516
classical mechanics, 219
clathrate, 782
Claus process, 106, 778
clay, 755
china, 757
close-packed structure, 355
closed
shell, 240
system, 588
clove, oil of, 401
coagulation, 106
coal, schematic structure, 398
coal dust explosion, 158
coal tar, 397
cobalt, 815
glass, 757
cobalt(II) oxide, 815
cobalt(III) complex, colors, 837
cobalt-60 treatment, 862
coded self-assembly, 332
coefficient, molar absorption, 226
cohesion, 351
coinage metal, 13
coke, 748
collagen, 440, 811
colligative property, 453
collision theory, 688
colloid, 443, 444
classification, 444
color
and frequency, table, 224
and wavelength, table, 224
changes, indicator, table, 560
complementary, 836
effect of ligands, 834
flowers, 559
frequency, 221
indicator, 559
light, 836
of complex, 834
perception, 221
photograph, 821
prediction, 836
color wheel, artist's, 836
colored glass, 757
combustion
acetylene, 98
analysis, 130
butane, 81
enthalpy, 201
hydrocarbon, 394
magnesium, 99
methane, 80, 99, 182
natural gas, 182
octane, 124
phosphorus, 774
reaction, calorimetry, 194
standard enthalpy, 202
standard enthalpy of, table, 203

comet, and water, 341
commercial cells, reactions, table, 648
common antilogarithm, A3
common logarithm, A3
common name, 27, A31
common-ion effect, 572
competing reaction, 125
complementary color, 836
complete ionic equation, 85
complex
  colors, 834
  d-block elements, 822
  high-spin, 838
  Lewis acid-base, 292
  low-spin, 839
  magnetic properties, 839
  many-electron, 837
  octahedral, 824
  square-planar, 824
  tetrahedral, 824
complex formation, solubility, 575
composite material, 419
composite process, enthalpy of, 193
composition
  and free energy, 611
  mass percentage, 59
compound, 15
  binary, 723
  coordination, 823
  electron-deficient, 746
  ionic, 17
  molecular, 16
  name, 29
  stability, 609
compressibility, 144
compression
  at constant temperature, 152
  effect on equilibrium, 493
computer disk, 807
Comte, A., 226
concentration
  molar, 63
  nominal, 528
  units, table, 453
condensation
  polymer, 414
  reaction, 403
condensed structural formula, 394
conducting polymer, 418
conduction
  band, 362
  electronic, 360
  ionic, 360
  metallic, 360
conductivity
  variation with temperature, 361
  water, 82
configuration, 241
  anion, 246
  boron atom, 241
  carbon atom, 241
  cation, 246
  d^n complexes, table, 839
  electron, 240
    list, 2017
  fluorine atom, 243
  helium atom, 240
  hydrogen atom, 240
  indium atom, 246
  neon atom, 243
  nitride ion, 247
  nitrogen atom, 242
  noble-gas, 242, 269
  oxide ion, 247
  oxygen atom, 243
confinement, magnetic, 879
conjugate
  acid and base, 512
  acid strength, 520
  acid-base pairs, table, 520
  base strength, 520
  seesaw, 520
  strengths, 546
conservation
  of energy, 182, 592
  of mass, 78
constant, rate, 674
constant composition, law of, 15
constructive interference, 354
contact lens, 417
contact process, 695, 783
convection, 163
conversion factor, 44
cooling curve, 379
  and phase diagram, 379
coordinate, 823
coordinate-covalent bond, 292, 823
coordination
  compound, 823
  compound, nomenclature, A32
  isomer, 826, 828
  number, 355, 823
  number, ionic solid 369
  sphere, 823
copolymer, 417
  alternating, 417
  block, 417
  random, 417
copper, 816
  action of nitric acid, 104, 105
  action of sulfuric acid, 104
  blister, 669, 817
  coin, 363
  complex formation, 771
  corrosion, 817
  crystal, 359
  deposition, 104
  electrolysis, 817
  hydrometallurgical extraction, 722
  illustration, 2, 56
  in zinc sulfate solution, 104
  refining, 669, 817
copper(II) sulfate, 818
cornflower color, 559
corrosion, 652
  copper, 817
corrosive solution, 516
corundum, 742
Coulomb's law, 238
counting, 49
couple, redox, 626
covalence, 288
  variable, 288
covalent
covalent
  bond, 274
  carbide, 759
  coordinate, 292
  ionic character and electronegativity, 294
  polar, 312
  radius, 249
  radius, table, 322
Crab nebula, 11
cracking, A28
cristobalite
  illustration, 751
  structure, 754
critical
  mass, 876
  pressure, 380
  properties, 380
  properties, table, 381
  temperature, 380
cryolite, 739
crystal
  face, 352
  liquid, 353
crystal field theory, 832
  octahedral complex, 832
crystalline solid, 273, 352
crystallization, 22
cubic cell, primitive, 357
cubic close-packed structure, 356
cubic equation, A5
cupronickel, 364, 815
curie, 866
Curie, E., 862
Curie, M., 732, 850
Curie, P., 732, 850
cyanate ion, Lewis structure, 284
cycloalkane, 391
cyclohexane, structure, 392
cyclopropane
  structure, 683
  table, 392
cyclotron, 859
cysteine, 422
cytosine, structure, 427

Δ (elevated temperature), 79
ΔX, defined, 189
d-block element, 803
  atomic radii, 805
  atomic radii, table, 806
  chemical properties, 807
  complex, 822
  density, 806
  Lewis acid, 822
  location, 804
  oxidation states, 807
  physical properties, table, 809
  properties, 804
d-metal complex, 822
  nomenclature, A32
d-orbital, 232
  occupation, 243
  hybridization, 329
d^n complexes, configuration, table, 839
dsp^3 hybrid, 328
d^2sp^3 hybrid, 328
Dacron, 414
Daguerre, L., 820
daguerrotype, 820
Dalton, J., 2, 161
  illustration, 3
Daniell cell, 630
data, 5
  elemental, A18
  standard potential, A14
  thermodynamic, A7
daughter nucleus, 852
Davisson, C., 229
Davisson-Germer experiment, 229
Davy, H., 732
DDT, 534
de Broglie, L., 229
de Broglie relation, 229
death, and free energy, 607
Debye, P., 312
debye, 312
decaborane, 746
decane, structure, 392
decay
  constant, 867
  exponential, 682
decomposition, calcium carbonate, 79
defecation, 736
dehydrating agent, 774, 784
delocalized, 280
  bond, 746

delocalized *(continued)*
electron, 332, 746, 749
Delta, see Δ
denaturation, 423, 445
denominator, A1
density, 41
enthalpy, 203
ice, 371
metal, 359
of gas, 159
tetrachloromethane, 371
unit, 42
water, 371
deodorant, 745
deoxyribonucleic acid, 426
deoxyribose, structure, 427
deposition, copper, 104
depression, 698
depression of freezing point, 459
depression, manic, 729
derived unit, 41
destructive interference, 354
detergent, 442, 754
composition, 367
detonator, 158
deuterium, 10, 721
deuteron, 859
diagonal relationship, 255
diagram, cell, 632
diamagnetic, 839
diamond, 748
structure, 372, 754
synthetic, 372, 748
diaphragm cell, 730
diborane
preparation, 746
structure, 746
dibromoethane, volatility, 375
dichlorobenzene, boiling points, 345
dichloroethene, boiling point, 346
diethyl ether, structure, 375, 400
diffraction, 354
electron, 229
pattern, 229, 354
X-ray, 354
diffusion, 163
separation of isotopes, 880
dilution, 66
dimer, 348
dimethyl ether, structure, 374
2,2-dimethylpentane, structure, 343
dimethylamine, structure, 404
dinitrogen oxide, 772
Lewis structure, 284
dinitrogen pentoxide, decomposition, 674
dinitrogen tetroxide, 773
Lewis structure, 287
dinitrogen trioxide, illustration, 773
diol, 400
dioxygen, Lewis structure, 286
dipeptide, 422
dipole, electric, 312
dipole moment, 312
instantaneous, 342
table, 313
dipole, dipole interaction, 344
strength, 342
disorder
and solubility, 450
tendency toward, 451
dispersion interaction, see London force, 342
disproportionation reaction, 773
dissociation, 316
homolytic, 316
dissociation constant, see acidity constant
dissolution, energy changes, 448
dissolving
and disorder, 451
equilibrium, 439
molecular nature, 438
distillation, 24, 26
distribution of speeds, 167
disulfide, 423
disulfide links, 421, 779
disulfur dichloride, 785
division rule, 50
DNA, 312, 426
Döbereiner, J., 244
Döbereiner's triads, 244
Dobson units, 171
dogs, unwashed, 93
dolomite, 732
donor, electron-pair, 292
dopamine, 698
doping, 362
dose
absorbed, 864, 866
equivalent, 864, 866
double bond, 276
rigidity, 395
valence-bond description, 324
double helix, 312, 428
double-acting baking powder, 730
dough, rise, 730
Dow process, 654
Downs process, 659
drug therapy, 698
dry cell, 647, 648
duality, 229
duck feathers, 351
ductile, 14
ductility, origin, 360
duplet, 269
dyeing, 743
dynamic equilibrium, 373
evidence for, 474

e-orbital, 833
Eagle nebula, illustration, 11
edema, 568
EDTA, 824
eel, electric, 649
effect
common ion, 572
greenhouse, 203
inert-pair, 254
effervescence, 445, 536
effusion, 163, 880
Graham's law, 164
separation of isotopes, 880
time, 165
egg
cooked, 348
rotten, 781
yolk, 445
eka-silicon, 245
elasticity, 421
elastomer, 421
Elavil, 699
electrolysis, aluminum production, 740
electric
catfish, 649
charge, 4
current, 658
dipole, 312
dipole moment, 312
eel, 649
organ, 649
electric-powered car, 205
electrocardiogram, 497
electrochemical cell, 630
electrochemical series, 640
electrochemistry, 625
electrode, 4, 630
calomel, 646
glass, 646
hydrogen, 632
semiconductor, 650
electrode potential, see standard potential
electrolysis, 654
amount of product, 658
applications, 659
copper, 817
Faraday's law, 657
potential for, 655
products, 657
water, 720
electrolyte
strong, 83
weak, 83
electrolyte solution, 82
pH, 547
electrolytic cell, 654
electromagnetic radiation, 220
interference, 354
regions, 223
table, 224
electromotive force, see cell potential
electron, 4
affinity, 255
affinity, table, 256
capture, 853, 857
cloud, 231
configuration, 240
configuration, list, A17
delocalized, 280, 332, 746, 749
diffraction, 229
flow, 630
flow, electrolytic cell, 655
flow, galvanic cell, 634
promotion, 326
repulsion, 805
shared pair, 274
spin, 236
transfer, 102, 626
valence, 240
electron-deficient compound, 746
electron-electron repulsion, 238
electron-pair
acceptor, 292
donor, 292
repulsion, 305
electronegativity
acid strength, 524
effect on bond strength, 318
ionic character, 294
table, 295
electronic
conduction, 360
display, 367
structure, 219
electronvolt, 882
electroplating, 659
element, 2
data, A18
emission spectra, 227
list, A18
molar mass, 53
most stable form, table, 608
name, 3, A18
origin, 11, 858, A18
superheavy, 856
elementary reaction, 700
rate law, 701
elevation of boiling point, 459
elimination reaction, 396
emerald, 732
illustration, 733

emery, 742
emf, 633
emission spectra, elements, 227
emission-control cannister, 749
empirical formula, 58
from combustion analysis, 132
emulsion, 444
photographic, 821
en, 824
enantiomer, 408
end point, 559
endorphin, 698
endothermic process, entropy change, 601
endothermic reaction, 184
spontaneity, 603
-ene suffix, 395
energy
activation, 689
change, 589
conservation, 182, 592
consumption, daily activities, 207
defined, 182
disorder, 595
flux, 199
free, 604
internal, 590
ionization, 251
kinetic, 238
level, 225
level, hydrogen atom, 232
nuclear, 849, 872
photon, 224
potential, 238
solar, 199
tendency to disperse, 451
energy levels
octahedral complex, 834
tetrahedral complex, 834
energy-level diagram, 239
many-electron atom, 239
enrich, nuclear fuel, 879
enthalpy, 188, 592
addition reaction, 397
density, 203
density, table, 204
formation, 205
formation, table, 208
interpretation, 593
lattice, 271
lattice, table, 271
of combustion, standard, 201, 202
of combustion, table, 203
of composite process, 193
formation, standard, 205
of freezing, 192
of fusion, 190, 191
of fusion, table, 190
of hydration, 448
of hydration, table, 449
of reaction, 195
of reverse process, 192
of solution, table, 447
of sublimation, 193
of vaporization, 190
of vaporization, table, 190
reaction, 194
specific, 203
enthalpy resources, global, 199
entropy, 596
at absolute zero, 596
Boltzmann formula, 598
direction of spontaneous change, 602
introduced, 596
low, 606
of melting, 597
of vaporization, 597
overall change, 602
standard molar, 597
standard molar, table, 596
standard reaction, 598
variation with temperature, 597
Environmental Protection Agency, 516
enzyme, 696
successive deprotonation, 533
EPA, 516
Epsom salts, 737, A31
equation
chemical, 79
cubic, A5
Henderson-Hasselbalch, 563
Nernst, 645
quadratic, A5
redox, 626
solving, A2
thermochemical, 195
equilibria
gaseous, 481
homogeneous and heterogeneous, 480
solubility, 567
equilibrium
and free energy, 605, 611, 613
autoprotolysis, 513
chemical, 473
constant, calculation, Toolbox, 644
constant from standard potentials, 643
dissolving, 373, 439
dynamic, 473
effect of catalyst, 499
effect of compression, 493
effect of inert gas, 495
effect of removing product, 493
effect of temperature, 495, 615
far from, 496
hydrogen iodide, 484
proton transfer, 518
response to conditions, 492
equilibrium composition
approximation, 490
calculation of, 483, 489
equilibrium constant
ammonia, 487
and rate constant, 706
calculated from free energy, 614
calculation of, 487
calculation of composition, 483
composite, 479
extent of reaction, 482
hydrogen iodide synthesis, 478
introduced, 476
relation between K_p and K_c, 482
relations between, 478
relations between, table, 479
reverse reaction, 478
equilibrium table, 477, 481, 487
pH calculation, 547
Toolbox, 488
equivalence of heat and work, 591
equivalence point, see stoichiometric point
error
random, 50
systematic, 50
erythrocyte, 436
essential amino acid, 421
essential oil, 401
ester, 399, 403, 405
esterification, 403
estradiol, structure, 433
etching, 757
ethane, bonding, 327
ethanol
illustration, 57
structure, 527
ethene
hybridization, 330
hydrogenation, 695
shape, 307
structure, 307, 395
ether, 399, 400
boiling point, 401
volatility, 374
ethnobotanist, 61
ethyl acetate, structure, 403
ethylbenzene, structure, 391
ethylene glycol, 400
ethylene, see ethene
ethylenediamine, 824
ethylenediaminetetraacetate ion, 824
ethyne
bonding pattern, 331
structure, 98, 307, 395
eucalyptus, 161
eugenol, structure, 401
evaporation pond, illustration, 729
exact number, 50
exclusion principle, 240
exothermic process, entropy change, 601
exothermic reaction, 184
expanded octet, 287
expansion work, 589, 593
experiment, 5
explosion, 158
nuclear, 876
thermonuclear, 879
exponential, A4
exponential decay, 682
of radionuclide, 867
extensive property, 42
extent of reaction, 482
extrapolate, A5
extrapolation, 150

f-block elements, 803
f-orbital, occupation, 244
fabric, micrograph, 415
face, crystal, 352
face-centered cubic structure, 357
factor, conversion, 44
Fahrenheit, D., 46
Fahrenheit scale, 46
fallout, 868, 877
Faraday constant, 635
Faraday, M., 657
Faraday's law, 657
fat, 722
fat, as fuel, 206
fcc, 357
feldspar, 755
ferric tape, 811
ferroalloy, 810
ferrocyanide ion, 823
ferromagnetism, 807
ferrosilicon, 733
ferrovanadium, 810
ferry analogy, 702
fertilizer
phosphate, 440
superphosphate, 775
fertilizers and solubility, 440
fever, 824
fiber, 414

fiber glass, 756
fibrillation, 497
field, 220
film badge detector, 865
filtration, 24
fireworks, 219
first ionization energy, 251
first law of thermodynamics, 592
first-order
integrated rate law, 681
reaction, 675
fish smell, 93
fissile nuclide, 876
fission, 874, 876
self-sustaining, 876
fixed nitrogen, 768
fixing film, 820
flame retardant, 742
flame front, illustration, 705
flame test, 220
flavor, 403
flesh, decaying, 404
flint, 753
flocculation, 106
fluid, 144
fluorapatite, 440, 786
fluoresce, 442, 781
fluorescent lighting, 795
fluorides, volatility, 343
fluorinating agent, 796
fluorine, 786
production, 786
production, illustration, 787
properties, 786
unique properties, 788
fluorine atom, configuration, 243
fluorine molecule, bonding, 275
flux, 742
foam, 444
food
Calorie, 206
composition, table, 207
preservation, 401
preservation, radioactive, 871
specific enthalpy, table, 207
thermochemical properties, table, 207
fool's gold, 777, A31
foot, 44
formal anhydride, 742
formal charge, 282
sulfate ion, 283
formaldehyde, 401
production, 402
formalin, 401
formate ion, structure, 553
formation
standard enthalpy, 205
standard enthalpy, table, 208
standard free energy, 608
standard free energy, table, 609
formation constant, 575
table, 576
formic acid, 752
bonding pattern, 334
Lewis structure, 332
structure, 553
formula
chemical, 16
condensed structural, 394
empirical, 58
Lewis, 268, 269
molecular, 16, 58
structural, 17
unit, 20
weight, 53
fossil fuel, 181, 199
fractional
coefficients, 81
order, 678
fractions, A1
Frasch process, 778
free energy, 604
and cell potential, 635
and composition, 611
and equilibrium, 613
and standard free energy, 612
equilibrium, 605
minimum, 611
of reaction, 608
free energy of formation, using, 610
free radical, see radical
freezing, 376
effect of pressure, 376
spontaneous process, 605
freezing point
depression, 459
table, 190
freezing-point constant, table, 460
Freon, 786
frequency, 220
and wavelength, 222
condition, Bohr, 228
frictional charging, 415
froth flotation, 816
fructose, structure, 424
fuel, 206
Apollo mission, 793
enthalpy density, 203
fossil, 199
heat output, 202
nuclear, 879
thermochemical properties, table, 204
fuel cell, 647, 648
fuel oil, 392
Fuller, R. Buckminster, 750
fullerene, 332
structure, 749
fullerite, 748
illustration, 750
functional group, 399
table, 399
fusion, 191, 878
enthalpy of, 190
enthalpy of, table, 190
illustration, 848

γ radiation, 224, 852
γ ray, 852
γ-alumina, 742
galena, 751, 777
illustration, 778
gallium arsenide, illustration, 3
Galvani, L., 630
galvanic cell, 630
galvanize, 653, 819
gamma ray, see γ
gangue, 816
garlic odor, 161
gas
and chaos, 145
constant, 153, 154
density, 158
formation, 97
greenhouse, 132
ideal, 153
laws, Toolbox, 152
liquefaction, 169, 172
mask, 749
mixture, 161
model of, 145
nature of, 144
preparation, 98
reactions, stoichiometry, 156
real, 168
solubility, 443
synthesis, 752
gas-fueled car, 204
gas-liquid chromatography, 26, 27
gaseous diffusion plant, 165
gaseous elements, 144
gaseous equilibria, 481
gasoline, 21, 392
gauge pressure, 147
Gay-Lussac, J.-L., 148
Geiger counter, 865
Geiger, H., 6
Geiger-Marsden experiment, 6
gel, 444
gelatin dessert, 444
gene splicing, 321
geodisic dome, illustration, 750
geometrical isomer, 407, 826, 829
Gerlach, W., 236
germanium, 747
illustration, 747
Germer, L., 229
Gibbs, J. W., 604
Gibbs free energy, see free energy
glass, 353, 756
action of hydrofluoric acid, 790
electrode, 646
fiber, 756
photochromic, 820, 821
waste storage, 881
glaze, ceramic, 815
GLC, 27
global
metabolism, 170
warming, 132
glucose, 182, 206
illustration, 57
solubility, 438
structure, 83, 206, 424
glue, 348
glutamic acid, 422
glutamine, 422
glycine, 422
structure, 404
glycol, 400
structure, 462
gold, 818
fool's, 2031
illustration, 14
leaf, 818
white, 819
Goodyear, C., 421
Gouy balance, 840
graduated cylinder, 40
graft copolymer, 417
Graham, T., 164
Graham's law of effusion, 164
grain
alcohol, 400
photographic, 820
granite, 755
illustration, 22, 758
graphite, 332, 748
commercial, 373
composite, 419
conductivity, 372
slipperiness, 372
structure, 372, 749
graphs, 2005
gray (unit), 864, 866
gray tin, illustration, 599
great tit, 534
green plants, 650
greenhouse
effect, 203
gas, 132
group, 12
main, 248

number, 248
Group 1 elements, 257
chemical properties, 727
chemical properties, table, 725
compounds, table, 727
illustration, 726
melting point, 726
physical properties, table, 725
reaction with water, 727
Group 2 elements, 257
chemical properties, table, 734
compounds, table, 736
illustration, 732
physical properties, table, 731
reaction with water, 734
Group 11 elements, properties, table, 816
Group 12 elements, properties, table, 816
Group 13 elements, 736
chemical properties, table, 739
physical properties, table, 739
Group 14 elements, 746
chemical properties, table, 748
illustration, 259, 747
physical properties, table, 747
Group 15 elements, 768
chemical properties table, 769
illustration, 768
physical properties, table, 769
Group 16 elements, 776
binary hydrides, 781
illustration, 259
physical properties, table, 777
radii, trend, 780
Group 17 elements, see halogens
Group 18 elements, 794
compounds, 795
physical properties, table, 794
guanine, structure, 427
guar gum, 445
Guldberg, C., 475
gutta-percha, 413
gypsum, 758, 2031

Haber, F., 500
Haber process, 499
Haldol, 699
half-life, 684
radioactive, 867
half-reaction, 626
table, 637
halic acid, 790
halide ion, 28
Hall, C., 739
halogen, 13, 785
illustration, 13
intermolecular force, 343
molecule, bond length, 321
physical properties, table, 785
radii, trend, 786
halogen oxoacids, table, 790
haloperidol, 699
halous acid, 790
hard water, 106, 440
Hawaii volcano, 340
hazard sign, 80
hcp, 355
heat, 183, 589
and work, equivalence, 591
heat capacity, 185
specific, 185
heat output, 202
heat transfer
constant pressure, 592
constant volume, 592
heating
at constant volume, 152
curve, 193
heavy water, illustration, 10
helium
phase diagram, 385
source, 794
superfluid, 795
helium atom, configuration, 240
hematite, 812
hemoglobin, 423
Henderson-Hasselbalch equation, 563
Henry's constant, table, 446
Henry's law, 446
herb, 61
hertz, 220
Hertz, H., 220
Hess's law, 198
using, 200
heterogeneous
catalyst, 695
equilibria, 480
mixture, 22
hexacyanoferrate(II) ion, 823
hexagonal close-packed structure, 355
hexane, structure, 391
2-hexene, 391
3-hexene, 391
high-altitude balloon, 142
high-density polymer, 413, 420
high-spin complex, 838
high-temperature superconductor, 361, 757, 818
illustration, 252
structure, 384
Hindenburg airship, 687
histidine, 422
homogeneous
catalyst, 694
equilibria, 480
mixture, 22
homolytic dissociation, 316
honesty in science, 47
hot-air balloon, 159
household products acidities, illustration, 90
HRF, 880
humidity, 386
Hund, F., 241
Hund's rule, 241
hybrid
orbital, 326
resonance, 280
hybridization, 326
including *d*-orbitals, 329
table, 328
hydrangea, 560
hydrate, 30
isomer, 826, 827
hydrated ion
pK_a, table, 544
hydration, 448, 545
enthalpy of, 448
enthalpy of, table, 449
hydration of ions, 83
hydrazine, 771
hydrazoic acid, 772
hydride, 723
binary molecular, 724
metallic, 724
saline, 723
hydride ion, 723
hydrocarbon, 390
aliphatic, 390
aromatic, 390
combustion, 394
components of petroleum, table, 392
melting and boiling points, 393
nomenclature, 398
saturated, 390
supercritical, 381
unsaturated, 390
viscosity, 349
viscosity and molar mass, 343
hydrofluoric acid, action on glass, 790
hydrogen
abundance on Earth, 720
as fuel, 721
atomic spectrum, 225
binary compounds distribution, 724
chemical properties, table, 723
compounds, 723
compounds, distribution, 724
economy, 203
electrode, 632
element, 720
intermolecular force, 343
liquid, 721
location in periodic table, 248
physical properties, table, 721
specific enthalpy, 722
hydrogen atom
acidic, 91
configuration, 240
electronic structure, 237
energy levels, 232
hydrogen bond, 346
and crystal structure, 371
in hydrogen fluoride, 789
nylon strength, 416
strength, table, 342
hydrogen bromide, synthesis mechanism, 704
hydrogen carbonate, solubility, 440
hydrogen compounds, boiling points, 347
hydrogen cyanide, Lewis structure, 278
hydrogen fluoride
bonding, 323
hydrogen bonding, 347, 789
hydrogen halide
boiling points, 346
bond strengths, 318
formation, 789
properties, table, 789
hydrogen iodide
equilibrium, 484
equilibrium constant, 478
hydrogen ion
as oxidizing agent, 104
hydration enthalpy, 449
hydrogen molecule, bonding, 275
hydrogen peroxide
catalyzed decomposition, 694
properties, 780, 781
structure, 780
uses, 781
hydrogen sulfide, 781
production, 97

hydrogen sulfide *(continued)*
rotten eggs, 781
hydrogen sulfites, 782
hydrogen-element bond strength, table, 319
hydrogen-oxygen reaction, mechanism, 705
hydrogenation
ethene, 695
reaction, 722
hydrogenperoxyl radical. Lewis structure, 287
hydrohalic acid, strengths, 525
hydrometallurgical
extraction, copper, 722
process, 816
hydronium ion
as Lewis complex, 293
molarity, 528
structure, 90
hydrophilic, 442
hydrophobic, 442
hydroquinone, 820
hydrostatic pressure, 147, 462
hydroxyapatite, 440, 738
4-hydroxybenzoic acid, 433
hydroxyl
group, 399
radical, 285
hygroscopic compound, 731
hyperkalemia, 569
hypernatremia, 569
hypo, 820
photographer's, 115
prefix, 28
hypobromous acid, strength, 527
hypochlorous acid, strength, 526, 527
hypofluorous acid, 790
hypohalous acid, 790, 789
hypoiodous acid, strength, 527
hypokalemia, 569
hyponatremia, 569
hypothesis, 5

i factor, 460
-ic suffix, 27, 29
ice cream, 445
ice
density, 371
structure, 370
iceberg, 371
icosahedron, 740
-ide suffix, 28
ideal gas, 153
law, 153
molar volume, 155
ideal solution, 459
ilmenite, 809
imaging, 862
implant, 759
inch, definition, 43
indicator, 90, 122
as weak acid, 558
color change, 559
color changes, table, 560
selection, 560
indium atom, configuration, 246
induced nuclear fission, 876
induced-fit mechanism, 696
industrial chemicals, top 50, A28
inert-pair effect, 254, 270
infrared radiation, 223
initial rate, 674
initiation step, 704
inorganic compound, 15
insolubility, phosphates, 440
insoluble substance, 82
instantaneous dipole moment, 342
instantaneous reaction rate, 673
insulator, 361
integer, 50
integrated rate law, 681
intensive property, 42
intercept, A5
interference, 354
constructive and destructive, 354
interhalogen, 788
table, 789
intermediate, reaction, 700
intermolecular force, 342
and molecular shape, 343
halogen, 343
hydrogen, 343
real gases, 169
table, 342
internal
combustion engine, illustration, 705
energy, 590
interpolate, A5
interstitial
alloy, 364
carbide, 759
intravenous fluids, 564, 568
iodine
antiseptic, 788
illustration, 2, 13
production, 787
properties, 787
solution, color, illustration, 788
uses, 787
iodine pentachloride, shape, 309
iodine-123 scan, 870
ion, 16
as acid, 544
basic character, table, 545
exchange, 780
size and hydration enthalpy, 450
ion hydration enthalpy, 449
table, 450
ion pair, 460
ion-dipole interaction, 448
ion-ion interaction, strength, table, 342
ionic
conduction, 360
radius table, 250
ionic bond, 268
covalent character, 296
ionic character
and electronegativity, 295
ionic compound, 16
Lewis formula, 268
molar mass, 53
name, 29
properties, 273
solubility, 88, 273, 447
solubility, table, 89
ionic model, 268
correction, 296
ionic solid, 355
coordination number, 369
ionic structures, 368
ionization
acetic acid, 92
percentage, 528
ionization constant, see acidity constant
ionization energy, 251
first, 251
metallic character, 254
periodicity, 253
second, 252
successive, 253
table, 252
ionization isomer, 826
ionized (acid and base), 92
ionizing radiation, 860
ions
electron configuration, 246
hydrated, 83
in blood, 569
nomenclature, A30
relative sizes, 251
iron, 812
cast, 814
in human body, 815
pig, 814
reaction with bromine, 787
Iron Age, 808
iron carbide, 760
iron pyrite, 777
iron titanium hydride, 204
iron(II) ion, illustration, 19
iron(II) sulfate heptahydrate, illustration, 58
iron(III) chloride formation, illustration, 787
iron(III) ion, illustration, 19
ironmaking, 812
irradiation of food, illustration, 871
isobar, 148
isoelectronic atoms, 251
isolated system, 588
isoleucine, 422
isomer, 393, 407, 826
classification, 826
coordination, 826, 828
geometrical, 407, 826, 829
hydrate, 826, 827
ionization, 826
linkage, 826, 828
optical, 408, 826, 829
structural, 406, 826, 829
isopleth, map, 543
isoprene, 413
structure, 390
isotactic, 413
isotope, 9
separation, 165, 880
table, 10
isotopic
dating, 868
material, 366
-ite suffix, 28
IUPAC, 3

jade, 755
jet fuel, 393
joule, 590
joule unit, 185
Joule, J., 172
Joule-Thomson effect, 169, 172

K_a, defined, 519
relation to K_b, 520
K_b, defined, 519
K_w, defined, 513
Kekulé structure, 281
valence-bond description, 330
Kelvin, Lord (Sir William Thomson), 47, 172
Kelvin scale, 47
kelvin unit, 47
keratin, 424
kerosene, 392
ketone, 399, 401
nomenclature, 405
preparation, 402
Kevlar, 416
Kilauea, 340
kimberlite, 372
kinetic energy, 238

kinetic model, 166
kinetics
  chemical, 671
  enzyme, 698
knocking, 623
krypton, 795
  compound, 796

lacewing, 472
lactated Ringer's solution, 568
lactic acid, structure, 417
lanthanide, 13, 248
  contraction, 805
lapis lazuli, 782
latex, 413
  illustration, 412
  sphere, 351
lattice enthalpy, 271
  and ion size, 272
  potassium chloride, 272
  solubility, 448
  table, 271
law, 5
  Beer's, 226
  Boyle's, 149
  Charles's, 149
  conservation of energy, 182
  conservation of mass, 78
  constant composition, 15
  Coulomb's, 238
  Dalton's, 161
  Faraday's, 657
  first, 592
  Graham's, 164
  Henry's, 446
  Hess's, 198
  ideal gas, 153
  limiting, 169
  of mass action, 476
  partial pressures, 161
  radioactive decay, 867
  Raoult's, 458
  second, 596, 602
laxative, 737
LCD, 367
Le Chatelier, H., 492
Le Chatelier's principle, 492
  explanation, 706
  solubility, 572
lead, 258
  durability, 751
  illustration, 56, 747
  passivation, 751
  production, 751
lead azide, 772
  detonator, 158
  pencil, 372
lead(II) chromate precipitate, illustration, 85
lead(II) iodide precipitation, illustration, 23
lead(II) oxide, shattered, illustration, 361
lead-acid cell, 647, 648
leather tanning, 811
Leclanché cell, 647
Leclanché, G., 647
LED, 419
lemon juice, pH, 516
leopard cub, 496
leucine, 422
leveling, 523
Lewis, G. N. 268
  illustration, 269
Lewis
  acid and base, 292
  formula, 268, 269
  structure, 275
  structure, polyatomic molecule, 276
  structure, Tool box, 278
  symbol, 268
Lewis acid
  boric acid, 742
  boron halide, 744
  silicon compounds, 748
Lewis base, water, 823
Libby, W., 869
lichen, 824
ligand, 823
  effect on energies, 833
  field theory, 832
  strong-field, 835
  table, 825
  weak-field, 835
ligand field splitting, 834
  determination, 835
light
  absorption, 226
  characteristics, 220
  color, 836
  reflection, 360
  white, 836
light-emitting diode, 419
light-water reactor, 877
lightning, 768
like-dissolves-like rule, 441
lime, 93, A31
  in soda-lime glass, 756
  production, 736
  in ironmaking, 738
limestone, 732
  test for, 97
limiting law, 169
limiting reactant, 126
  identification of, 127
Linde refrigerator, 172
linear
  hybridization, 328
  molecule, 304
linkage isomer, 826, 828
liquefaction, 169, 172
liquid, 144
  structure, 348
liquid crystal, 353, 366
  lyotropic, 367
  thermotropic, 366
lithium
  applications, 729
  illustration, 726
  reaction with water, 727
  soap, 729
lithium atom, configuration, 240
lithium ion battery, illustration, 729
lithium oxide, illustration, 728
litmus, 561
lock-and-key mechanism, 696
logarithm, A3
  common, A3
  natural, A4
  significant figure, A3
London, F., 342
London force, 342
  and molecular shape, 343
  strength, 342
lone pair, 275
  axial, 310
  distorting effect, 309
  effect on shape, 308
  equatorial, 310
long period, 243
low-gravity production, 351
low-spin complex, 839
Lowry, T., 510
lubricant, 392
lubricating oil
  viscosity, 349
  thickener, 729
Lucite, 410
LWR, 877
Lyman series, 225
lyotropic liquid crystal, 367
lysine, 422
lysozyme, 696

magic number, 855
magnesium
  combustion, 99, 734, 771
  compounds, 737
  Dow process, 654
  from seawater, 733
  illustration, 732
  production, 733
  properties, 733
magnesium hydroxide, 737
magnesium nitride, illustration, 771
magnesium oxide, 737
magnet, permanent, 806
magnetic
  confinement, 879
  properties, complex, 839
  quantum number, 234, 236
magnetite, 807, 812
main group, 12, 248
malachite, illustration, 817
Malay lacewing, 472
malleability, origin, 360
malleable, 14
manganese, 811
  bronze, 811
  nodule, 811
  nodule, illustration, 812
manganese dioxide, 812
manganese(IV) oxide, 812
manic depression, 729
mantissa, A3
many-electron
  atom 237
  complex, 837
  energy-level diagram, 239
marble, 737, A31
Mars
  life on, 24
  polar caps, 193
  scientific inquiry on, 24
Marsden, E., 6
Mary the Jewess, 25
mass
  action, law of, 476
  critical, 876
  number, 8
  percentage, 453
  percentage composition, 59
  spectrometer, 8
  spectroscopy, 9
  spectrum, 8
  standard, 42
  unit, 42
mass-energy conversion, 872
mass-to-mass calculation, 120
mass-to-mole conversion, 55
match, 770, 791
materials hierarchy, 23
matter, 1
  disorder, 595
  state of, 144
  tendency to disperse, 451
Mauna Loa atmospheric monitoring site, 153
Maxwell, J. C., 167
Maxwell distribution
  reaction rate, 689
  speeds, 167
  speeds, illustration, 168
mayonnaise, 444
mean
  bond enthalpy, table, 318
  free path, 166
measurement, 48
mechanical strength, 420
mechanics
  classical, 219
  quantum, 219
mechanism
  hydrogen-oxygen reaction, 705

mechanism *(continued)*
reaction, 700
medical technologist, 832
Meitner, L., 860
melting, 191, 376
melting, effect of pressure, 376
melting point
alkane and alkene, 395
table, 344
Mendeleev, D.I., 244
meniscus, 352
Menten, M., 698
menthol, structure, 433
mercury, 819
illustration, 2, 56
meniscus, 352
poisoning, 822
surface tension, 351
test for, 822
viscosity, 349
mercury cell, 648
mercury(I) iodide precipitation
illustration, 88
mercury(I) ion, 88, 822
meringue, 445
mesopause, 164
mesosphere, 164
metabolic acidosis, 568
metal
action of acid, 104
characteristics, 14, 360
coinage, 13
density, 359
properties, 14, 360
transition, 13
metal-ammonia solution, 728
metallic
conduction, 360
character, ionization energy, 254
conductor, 361
crystal, 355
hydride, 724
solid, 355
metalloid, 258
character, 255
characteristics, 14
location, 15
metals and nonmetals
characteristics, table, 258
metasilicic acid, 754
metastable state, 862
meter, standard, 42
methane, 391
combustion, 80, 99
combustion, illustration, 81
hybridization, 326
Lewis structure, 276
model, 17
structure, 304
methanoic acid, see formic
acid, 334
methide, 759
methionine, 422
methyl
group, structure, 93
radical, 285
methyl hydrazine, 793
methyl orange, 561
methylamine, 404
structure, 93, 404
methylbutane, structure, 390
methylpropane, structure, 393, 406
metric system, 40
Meyer, L., 244
mica, 755
illustration, 758
micelle, 367, 442
Michaelis, L., 698
Michaelis-Menten enzyme kinetics, 698
microwaves, 223
milk, 444
milk of magnesia, 737, A31
Millikan, R., 4
milliliter, 41
Minamata, 822
mining lamp, 98
minute, 44
mirror, 360
mirror image molecules, 408
mixed solution, pH, 548
mixture, 21
heterogeneous, 22
homogeneous, 22
mixtures and compounds, table, 21
mmHg, 147
model, 5
ball-and-stick, 17
ionic, 268
space-filling, 16
moderator, 877
Moisson, H., 654
molality, 453
defined, 455
from mass percentage, 457
from molarity, 456
from mole fraction, 456
preparation of solution, 455
using, Tool box, 456
molar
absorption coefficient, 226
concentration, 63
concentration, pure solid, 480
molar mass, 53, 165
air average, 144
average, 54, 420
effusion, relation to, 165
from freezing-point depression, 461
from gas density, 160
from osmosis, 464
molar solubility
complex formation, 576
from solubility product, 571
molar volume, ideal gas, 150, 155
molarity, 63, 453
calculation of, 64
preparation of solution, 455
using, 65
mole
defined, 51
diagram, electrolysis, 657
diagram, gas volume, 156
diagram, liquid volume, 123
diagram, heat output, 201
diagram, mass conversion, 120
ratio, 118
fraction, 453
mole-to-mole
calculation, 119
predictions, 118
molecular
battery, 332
beam, 689
compound, 16
molar mass, 53
molecular formula, 16, 58
determination, 62
molecular shape, 304
molecular solid, 355
structure, 370
molecular speed, and temperature, 146
molecular weight, 53
molecule, 15
angular, 304
linear, 304
nonpolar, 313
polar, 313
monatomic, 18
Mond process, 752
monoamine oxidase, 698
monoclinic sulfur, illustration, 779
monoclonal antibody, 862
monomer, 410
mood regulation, 698
mordant, 743
mortar, 738
Moseley, H., 7, 245
moss, 824
motion, perpetual, 592
Mulliken, R. S., 295
multiple bond, 276
carbon-carbon, 330
relative strength, 319
strength, 318
VSEPR, 306
multiplication rule, 50
mustard gas, structure, 785

n-type semiconductor, 363
naltrexone, 699
nanotube, 332, 750
boron, 744
naphtha, A29
naphthalene, structure, 397
National Accelerator Laboratory, Batavia, 860
natural
antilogarithm, A4
change, 595
gas, combustion, 80, 182
logarithm, A4
products, 39
nebula, 11
nematic phase, 366
neon
light, 795
mass spectrum, 8
neon atom, configuration, 243
Nernst, W., 645
Nernst equation, 645
net ionic equation, 86
network
modifier, 756
former, 881
solid, 355, 371
neurotransmitter, 698
neutralization reaction, 95, 108
net outcome, 96
predicting outcome, 97
neutron, 4
as carrier, 876
emission, 854
emission, spontaneous, 857
neutron capture therapy, 863
neutron-induced transmutation, 859
neutron-rich nuclide, 857
Newlands, J., 244
Newlands' octaves, 244
nicad cell, 648
nickel, 815
plating, 745
production, 752
nickel arsenide structure, 370
nickel carbonyl, 752, 815
nickel-cadmium cell, 649
nitrate ion, Lewis structure, 280
nitrates, solubility, 440
nitric acid
catalyst for production, 499
production, 774
properties, 774
nitric oxide, see nitrogen monoxide
nitride formation, 728
nitride ion, configuration, 247

nitrides, 771
nitriding, 383
nitrite ion
  shape, 311
  structure, 311
nitrogen
  atmospheric, 170
  family, 768
  fixation, 474, 768
  illustration, 768
nitrogen atom, configuration, 242
nitrogen dioxide, 773
  illustration, 288
  in smog, 288
  Lewis structure, 287
  reaction with water, 773
nitrogen dioxide, decomposition, 675
nitrogen molecule
  bonding pattern, 325
  valence-bond description, 324
nitrogen monoxide, 286, 772
  as an air pollutant, 288, 534
nitrogen oxide, see nitrogen monoxide
nitrogen oxides, 772
  acid rain, 534
  atmospheric, 288, 772
  table, 773
nitrogen oxoacids, table, 773
nitrogen trifluoride
  Lewis structure, 308
  shape, 308
nitrogen-fixing bacteria, 768
nitroglycerin, structure, 158
nitrous acid, 773
nitrous oxide, see dinitrogen oxide
noble gas, 13, 794
  compounds, 795
  configuration, 242, 269
nodal plane, 231
nodules, nitrogen fixing, 769
nomenclature, 27
  alcohol, 405
  aldehyde, 405
  alkane, 398
  alkane, table, 391
  alkene, 398
  alkyne, 398
  amine, 405
  arene, 399
  carboxylic acid, 405
  chemical, A30
  coordination compound, A32
  d-metal complex, A32
  element, systematic, table, 860
  ester, 405
  functional groups, 405
  hydrocarbon, 398
  ketone, 405
  organic halides, 405
  oxoacids, A30
  oxoanions, A30
  polyatomic ions, A30
  transuranium element, 860
nominal concentration, 528
nonaqueous solution, 23
nonbonding orbital, 327
nonelectrolyte, 83
nonideal solution, 459
nonmetal, characteristics, 14, 258
nonmetallic solder, 419
nonpolar
  molecule, 313
  chart, 315
nonsparking tool, 733
nonspontaneous change, driven, 605
nonspontaneous process, 606
noradrenaline, structure, 698
norepinephrine, structure, 698
normal boiling point, 375
normal freezing point, 376
Northern Lights, 670
notation
  cell, 632
  scientific, A2
NO_x, 695, 772
  atmospheric, map, 534
nuclear
  atom, 4
  binding energy, 872
  chemistry, 849
  decay, 852
  disintegration, 852
  energy, 849, 872
  explosion, 876
  fallout, 868, 877
  fission, 874
  fission, induced, 875
  fission, spontaneous, 875
  fusion, 878
  fusion, controlled, 879
  fusion, illustration, 848
  medicine, 862
  reaction, 851
  reaction, identification of products Tool box, 853
  reactor, 877
  stability, pattern, 854
  transmutation, 852
  weapon, 877
nucleic acid, 428
nucleon, 8
nucleoside, 427
nucleosynthesis, 11, 858
nucleotide, 427
nucleus, 4
  daughter, 852
  structure, 851
nuclide, 850
  fissile, 876
  fissionable, 876
  neutron-rich, 857
  proton-rich, 857
number
  atomic, 7
  coordination, 355
  exact, 50
  group, 248
  magic, 855
  mass, 8
  oxidation, 101
  period, 248
  quantum, 232
  Stock, 27
numerator, A1
nylon, 414
nylon salt, 414

o-dichlorobenzene, 345
-o suffix, 405
Oak Ridge plant, 165
occupation
  d-orbital, 243
  f-orbital, 244
occupation, order of, 242
ocean, composition, 437
octahedral, 304
octahedral
  complex, 824
  crystal field theory, 832
  energy levels, 834
  hybridization, 328
octane, combustion, 124
octaves, 244
octet, 269
  expanded, 287
  expansion and hybridization, 329
  rule, 275
  rule, exceptions, 285
odor, 403
oedema, see edema, 568
-oic acid suffix, 405
oil
  essential, 401
  vegetable, 722
-ol suffix, 405
oleum, 784
oligopeptide, 422
onyx, illustration, 753
open system, 588
optical
  activity, 409
  isomer, 408, 826, 829
orange juice, pH, 516
orbital
  atomic, 230
  d, *f*, 232
  hybrid, 326
  nonbonding, 327
  overlap, 323
  s, *p*, 231
orbital angular momentum, 234
  quantum number, 234
order
  determination, 680
  fractional, 678
  negative, 678
  of occupation, 242
  of reaction, 676
  overall, 678
  pseudo-first, 679
ore, roasting, 110
organ, electric, 649
organic halide, nomenclature, 405
Orlon, 410
orpiment, illustration, 770
orthosilicic acid, 754
oscillating reaction, 496
osmometry, 464
osmosis, 462
  molar mass determination, 464
  red blood cells, 463
  reverse, 464
Ostwald process, 115, 773
-ous suffix, 27
ovenware, 756
overall
  order, 678
  reaction enthalpy, 198
overlap, and bonding, 323
overpotential, 656
oxalic acid, structure, 123
oxidation, 100
  state, 101
oxidation number, 101
  d-block elements, 807
  amphoteric character, 808
  and acid strength, *525*, *526*
oxidation-reduction reaction, 99
oxide, acidic or basic nature, 94
oxide, formation, 728
oxide ion, 93
  configuration, 247
  Lewis base, 293
  strong base, 511
oxidizing agent, 102
oxidizing strength, 641
oxoacid, 28
  names, table, 29
  nomenclature, A30
  strengths, *525*
oxoanion, 20
  as oxidizing agents, 104

oxoanion *(continued)*
names, table, 29
nomenclature, A30
oxygen
allotrope, 777
family, 776
Lewis structure, 286
liquid, illustration, 777
origin, 776
production, 776
uses, 777
oxygen atom, configuration, 243
oxygen difluoride, Lewis structure, 277
ozone
atmospheric, 170, 777
decomposition mechanism, 700
decomposition rate law, 678
decomposition, overall rate law, 703
hole, 171
illustration, 777
polarity, 314
shape, 311
structure, 311, 314

π bond, 324
rigidity to twisting, 395
p-azoxyanisole, 366
p-block, characteristics, 258
p-dichlorobenzene, 345
p-n junction, 363
p-orbital, 231
p-type semiconductor, 363
paint, 810
pair, lone, 275
paired spins, 240
PAN, 289
pancreatic fluid, 517
paper
chromatography, 26
pulp bleaching, 791
paraffin, 394
wax, 392
parallel spins, 240, 241
paramagnetic, 839
paramagnetism, 777
part per million, 453
partial
charge, 312
charge, fleeting, 342
partial pressure, 161
calculation of, 162
from K_p, 484
gas solubility, 443
particles and waves, 229
pascal, 147
passivation, 639
lead, 751
patina, 818
Pauli, W., 240
Pauli principle, 240
Pauling, L., 295
pelt, pattern, 496
penetrating power, 861
penetration, 239
penicillin, activity, 672
Penning trap, 43
pentagonal bipyramidal, 304
pentane, structure, 343, 391
peptide, 422
bond, 422
per- prefix, 28
percentage, A1
ionization, 528
mass, 453
protonated, 530
transmittance, 227
volume, 453
yield, 125
perchlorates, 791
perchloric acid, 791
strength, 526
perhalic acid, 790
period, 12
period
long, 243
number, 248
periodic table
development, 244
electronic basis, 247
introduced, 12
structure, 12
periodicity, 248
ionization energy, 253
permanent magnet, 806
perovskite structure, 384
peroxyacetylnitrate, 289
perpetual motion, 592
perxenate ion, 796
perylene, chemiluminescence, 781
PES, 265
PET, 862
petroleum
components, table, 392
extraction, 199
pewter, 364
pH
acid added to buffer, 564
after stoichiometric point, 552
aqueous solutions, 516
before stoichiometric point, 555
blood, 515, 568
blood plasma, 543
buffer solution, 563
calculation of, 515
citrus juice, 516
defined, 514
electrolyte solution, 547
evaluation, 2004
lemon juice, 516
meter, 516, 646
mixed solution, 548
of a weak acid, Toolbox, 529
of weak bases, 530
orange juice, 516
pOH relation, 517
salt solutions, 544, 546
stoichiometric point, 553
sulfuric acid, 533
weak acid and its salt, 549
well after stoichiometric point, 558
pH calculation, equilibrium table, 547
pH curve
strong acid-strong base, 550
strong acid-weak base, 555
weak acid-strong base, 555
pharmacognosist, 61
phase, 373
boundary, 377
change, 373
transition, 373
phase diagram, 377
and cooling curve, 379
carbon, 385
carbon dioxide, 378
helium, 385
sulfur, 378
water, 377, 380
phase-change heating installation, 192
phenol, 399
structure, 400
phenol-formaldehyde resin, 402
phenolphthalein, 561
structure, 559
phenylalanine, 422
pheromone, 434
phosgene, Lewis structure, 285
phosphate
condensation reaction, 775
fertilizer, 440
phosphate ion
illustration, 20
Lewis structure, 284
phosphates, insolubility, 440
phosphine, 772
phosphoric acid, structure, 775
phosphorous acid, structure, 525
phosphorus
combustion, 774
illustration, 768
origin, 770
reaction with bromine, 100
red, 2, 770
structure, 770
white, 770
phosphorus(III) oxide, structure, 774
phosphorus(V) oxide, structure, 774
phosphorus pentachloride
ionic solid, 290
illustration, 290
Lewis structure, 290
preparation, 772
shape, 306
structure, 306
phosphorus pentoxide, see phosphorus(V) oxide
phosphorus trichloride
formation, 288
illustration, 290
Lewis structure, 288
preparation, 772
uses, 772
phosphorus trioxide, see phosphorus(III) oxide
photochemical
material, 820
reaction, 820
photochromic glass, 820, 821
photoelectrochemical cell, 650
photoelectron spectroscopy, 265
photograph, color, 821
photographer's hypo, 115
photographic
emulsion, illustration, 821
grain, 820
photon, 223
energy, 224
energy, table, 224
number, 224
photosynthesis, 182, 195
and entropy, 606
artificial, 650
global role, 199
reaction, 99
photovoltaic cell, 199, 650, 656
physical property, 23
pickled cucumber, 743
piezoelectric, 810
pig iron, 814
pigment, white, 810
pipet, 40, 63
pitchblende, 879
pK_a
defined, 519
determination, 556
ions in solution, table, 544
relation to pK_b, 522
pK_b
defined, 519
determination, 556

pK_{In}, 559
pK_w, defined, 517
Planck constant, 224
plane-polarized light, 409
plasma, 879
  blood, 437
  buffer system, 567
  pH, 543, 568
plaster of Paris, A31
Plexiglas, 410
plumbane, 320
pOH defined, 517
point, normal freezing, 376
poison (biological), 697
poison (catalyst), 696
polar
  bond, 312
  covalent bond, 312
  fluid, test, 314
  molecule, 313
polar molecules, chart, 315
polarity
  acid strength, 524
  predicting, 314
polarizability, 296
polarizing power, 296
pollutant
  primary, 288
  secondary, 288
pollution
  control, 96
  thermal, 446
poly-*p*-phenylenevinylene, 419
polyamide, 414
polyaniline, 419
polyatomic, 19
polyatomic ions, naming, A30
polyatomic molecule, Lewis structure, 276
polyester, 414
polyethylene, 410
polyisoprene, 421
polymer, 410
  condensation, 414
  conducting, 418
  edible, 421
  formation, 286
  high-density, 413
  mechanical strength, 420
polymerization
  addition, 410
  radical, 411
polynucleotide, 428
polyphosphoric acid, 775
polypropylene, 410
polyprotic acid and base, 532
polypyrrole, 419
polysaccharide, 424
polystyrene, 410
polysulfane, 782
polytetrafluorethylene, as nonstick surface, 343
polyvinyl chloride, 410
poppy color, 559
Portland cement, 758
positron emission, 853
  tomography, 862
potable water, 106
potash, 2029, 2031
potassium
  dating, 870
  illustration, 726
  production, 725
  reaction with water, 13, 727
potassium chlorate, decomposition, 791
potassium chloride
  lattice enthalpy, 272
  source, 730
potassium nitrate, 731
potassium permanganate, 812
potassium superoxide
  air improvement, 103
  illustration, 157, 728
potential
  cell, 633
  data, A14
  standard, 635
  variation with concentration, 645
potential energy, 238
pound, 44
ppm, 453
PPV, 419
pre-exponential factor, 689
precipitate, 85
  dissolution, 574
precipitation, 22
  prediction of outcome, 88, 573
  reaction, 82, 85, 108
precision, 50
prefix SI
  numerical, table, 30
  SI, table, 41
pressure
  calculation, 154
  critical, 380
  dependence on temperature, 150
  dependence on volume, 149
  exerted by liquid, 147
  gauge, 147
  hydrostatic, 147, 462
  origin, 146
  partial, 161
  standard, 155
  units, 147
  units, table, 148
  vapor, 373
  variation in atmosphere, 164
pressurized-water reactor, 877
primary
  alcohol, 400
  cell, 647
  pollutant, 288
  structure, 422
primitive cubic cell, 357
principal quantum number, 232
principle
  Aufbau, 241
  Avogadro's, 150
  building-up, 241
  exclusion, 240
  Le Chatelier, 492
  Pauli, 240
prism, 221
procaine, structure, 432
process, 115
  basic oxygen, 814
  chloralkali, 659
  Claus, 106, 778
  contact, 695, 783
  Dow, 654
  Downs, 659
  driven, 606
  Frasch, 778
  Haber, 499
  hydrometallurgical, 816
  Mond, 752
  Ostwald, 774
  pyrometallurgical, 816
  thermite, 811
product, 78
proline, 422
promotion, 326
propagation step, 704
propene, structure, 683
property
  chemical, 23
  colligative, 453
  extensive, 42
  intensive, 42
  physical, 23
  state, 188, 590
propyne, structure, 390
protein, 421
  as enzyme, 696
  as fuel, 206
proton, 4
  acceptor, 510
  donor, 510
  emission, 854, 857
  transfer, 510
proton transfer equilibria, 518
proton-rich nuclide, 857
protonation, 724
  percentage, 530
pseudo—first-order
  rate law, 679
  reaction, 679
PTFE, 410
puncture, 164
purgative, 737
purification
  water, 106, 780
putrescine, 404
pX
  defined, 517
  meter, 646
Pyrex, 742, 756
pyrite, 777, 812
  illustration, 778
pyrolusite, 812
pyrolytic release, 24
pyrometallurgical process, 816
pyroxene, structure, 755

quadratic equation, A5
quadratic equation solutions, 488
qualitative investigation, 39
quanta, 223
quantitative investigation, 39
quantization, 223
quantum mechanics, 219
quantum number
  azimuthal, 232, 234
  magnetic, 234
  orbital angular momentum, 234
  principal, 232
  spin, 236
  spin magnetic, 234
  table, 234
quart, 44
quartz, 751, A31
  illustration, 353, 751
  structure, 353
quartzite, 751
  illustration, 751
quaternary
  structure, 423
  ammonium ion, 404
queen bee pheromone, 434
quicklime, 93, 737, A31
quotient, reaction, 485

racemic mixture, 409
rad, 864, 866
radiation
  electromagnetic, 220
  gamma, 224
  infrared, 223
  ionizing, 860
  penetrating power, 861
  ultraviolet, 223
  units, table, 866
radical, 285
  chain reaction, 704
  polymerization, 411
  shape, 308
radio waves, 224
radioactive, 850
  decay, law of, 867
  half-life, 867

radioactive *(continued)*
half-life, table, 868
isotopes, uses, 870
series, 857
tracer, 862, 870
waste, 880
waste, storage, 880
radioactivity
activity, 866
background, 864
discovery, 850
measurement, 865
radon, 864
shielding requirements, table, 861
radiocarbon dating, see carbon-14 dating
radioisotope, 862, 870
radius
atomic, table, 249
covalent, 249
covalent, table, 322
ionic, table, 250
radius ratio, 369
radon, 795
radioactivity, 864
rain, acid, 509, 534
rainbow, 221
random copolymer, 417
random error, 50
Raoult, F.-M., 458
Raoult's law, 458
rare earth, 248
rate
initial, 674
reaction, 672
rate constant, 674
and equilibrium constant, 706
determination, 683
table, 685
rate law
constructing, 702
elementary reaction, 701
integrated, 681
solving, A4
table, 685
rate-determining step, 702
ratio, A2
re-forming reaction, 720, A28
reactant, 78
limiting, 126
reaction
addition, 396
bimolecular, 700
chain, 704, 876
competing, 125
condensation, 403, 775
direction, predicting, 486
disproportionation, 773
elementary, 700
elimination, 396
endothermic, 184
enthalpy, 194, 195, 594
entropy, 598
exothermic, 184
extent of, 482
free energy, 608
hydrogenation, 722
hydrolysis, 772
in solution, 693
intermediate, 700
mechanism, 700
nuclear, 851
order, 676
order determination, 680
oscillating, 496
photochemical, 820
profile, 689
pseudo–first-order, 679
quotient, 485
rate, 672
re-forming, 720, 752
reversible, 474
second-order, 686
sequence, 198
shift, 720
stoichiometry, 118
substitution, 395, 823
temperature for, 615
termolecular, 701
thermite, 184
unimolecular, 700
volcano, 184
yield, 124
reaction enthalpy
combining, 198
overall, 198
standard, 197
reaction enthalpy (ΔH_r), 594
reaction quotient
and free energy, 612
solubility, 573
reaction rate
effect of temperature, 688
instantaneous, 673
Maxwell distribution, 689
reaction sequence
devising, 200
reactions, redox, 625
reactor, 877
breeder, 878
light-water, 877
nuclear, 877
pressurized-water, 877
reagent, 78
real gas, 168
intermolecular forces, 169
realgar, illustration, 770
recycling code, table, 411
red blood cells, osmosis, 463
red cabbage indicator, illustration, 90
red phosphorus, 770
redox
couple, 626
equation, balancing, 626
titration, illustration, 642
redox reaction, 99, 102, 108, 625
balancing, 107
balancing, Tool box, 627
reducing agent, 102
reduction, 100
reduction potential, see standard potential
refining, 817
copper, 817
reflection of light, 360
refractory material, 737
refrigerator, 172
relative biological effectiveness, 864, 866
relative humidity, 386
relativity, 872
rem, 864, 866
replication (DNA), 429
repulsion, electron-electron, 238
residue, 422
resonance, 280
benzene, 281
effect on bond strength, 319
hybrid, 280
stabilization, 282
respiration, carbonate control, 568
respiratory acidosis, 569
retinal, 334
reverse osmosis, 107, 464
reversible reaction, 474
rhodium, gauze catalyst, 499
rhombic sulfur, illustration, 779
ribonucleic acid, 426
ribose, structure, 427
rigidity to twisting, 395
Rime of Ancient Mariner, 470
Ringer's solution, 568
RNA, 426
roasting, 816
roasting ore, 110
rock dating, 870
rock salt, 729
illustration, 368
structure, 368
rocket fuel, 792
rod, 302
roentgen equivalent man, 864
rounding off, 49
rubber, 413
butyl, 418
rubidium, illustration, 726
ruby, 743
rule
Hund's, 241
like-dissolves-like, 441
octet, 275
rust formation, 652
Rutherford, E., 851
illustration, 6
rutile, 809
structure, 384
Rydberg constant, 232

σ-bond, 323
s-block characteristics, 257
s-orbital, 231
sp hybrid, 328
sp^2 hybrid, 328
sp^3 hybrid, 327
sacrificial anode, 653
saline
carbide, 759
hydride, 723
saliva, 697
salt, 95
bridge, 632
cake, 730
in water, 544
smelling, 771
table, A31
with acidic cation, 547
with basic anion, 547
salted meat, 463
saltlike hydride, 723
saltpeter, illustration, 440
sand, 753
sapphire, 743
saturated, 162
hydrocarbon, 390
solution, 439
SBR, 417
scandium, 809
scanning
electron micrograph, 415
tunneling microscope, 277, 333
schizophrenia, 698
Schrödinger, E., 230
scientific method, 5
scientific notation, 48, A2
scintillation counter, illustration, 866
scrubbing (removal of sulfur oxides), 535
scum, 442
sea of instability, 856
sea squirt, 38
seawater, 21
composition, 437
saltiness, 788
source of magnesium, 733
second
ionization energy, 252
law of thermodynamics, 596, 602

standard, 42
second-order reaction, 676, 686
secondary
alcohol, 400
cell, 647
pollutant, 288
structure, 423
seesaw, 304
selenium
allotropes, 779
illustration, 779
self-assembly, 332
self-sustaining fission, 876
semiconductor, 361, 362
classification, 363
electrode, 650
semipermeable membrane, 462
separation technique, 23
sequencing, 321
series
Balmer, 225
Lyman, 225
serine, 422
settling basin, 106
sewage treatment, 106, 743
sex-attractant, 434
shape, molecular, 304
shared electron pair, 274
shell, 234
closed, 240
formation, 89
valence, 242
shellfish shell, 738
shielding, 239
shielding requirements, table, 861
shift reaction, 720
shock, 564, 568
Shoemaker-Levy comet, 265
SI, 40
SI prefixes, 41
table, 41
sickle-cell anemia, 423
side chain, 393
Sidgwick, N., 305
sievert, 864, 866
significant figure, 48, 49
logarithm, A3
silane, 320
silcates, 753
silica, 748, 751, 753
gel, 754
structure, 353
silicate
glasses, 756
ion, 751
structure, 754
silicon, 751
double bonds, 748
illustration, 255, 747
ultrapure, 751
silicon carbide, 759
silicon compounds, Lewis acid, 748
silicone, 758
silk, 423
spider, 424
silver, 818
silver cell, 648
silver chloride, precipitation, 85
silver fluoride, 788
silver halide
covalent character, 296
illustration, 297
solubility, 296
solubility trend, 788
silver nitrate, 818
single bond, 276
valance-bond description, 324
skeletal equation, 78
skywriting, 370
slag, 813
slaked lime, 93, A31
slope, A5
smart window, 419
smectic phase, 366
smelling salt, 771
smelting, 816
smog formation, 288
smoke, 401
detector, 871
inhalation, 569
smoke screen, 370
sneeze analogy, 601
soap, 441
soda
ash, 731
baking, A31
caustic, A31
washing, 442, 731, A31
soda-lime glass, 756
sodium
applications, 729
compounds, 729
illustration, 257, 726
production, 725
production, Downs process, 659
reaction with water, 78, 727
sodium aluminate, 743
sodium azide, illustration, 771
sodium bicarbonate, 730
sodium borohydride, 745
sodium carbonate, 731
sodium chloride
illustration, 58, 368
structure, 17, 368
sodium chromate, 811
sodium dichromate, 811
sodium hydrogen carbonate, 730
sodium hydroxide, 729
production, 659, 730
sodium hypochlorite, 790
production, 790
sodium metasilicate, 754
sodium orthosilicate, 754
sodium oxide, illustration, 728
sodium peroxide, illustration, 58
sodium stearate, structure, 441
sodium sulfate, 730
sodium thiosulfate, 820
sodium-potassium alloy, 726
soft drink, 445
sol, 444
solar mirror, 360
solar energy
conversion, illustration, 624
flux, 199
solar-powered telephone booth, 199
solar-powered vehicle, 363
solder, 364
nonmetallic, 419
solid, 144
characteristics, table, 355
classification, 355
crystalline, 273
emulsion, 444
solution, 23
solubility
and disorder, 450
common-effect, 572
complex formation, 575
effect of temperature, 446
equilibria, 567
factors affecting, 439
gas, 443
glucose, 438
Henry's law, 446
hydrogen carbonate, 440
ionic solids, 273, 447
lattice enthalpy, 448
like dissolves like, 441
molar, 439
nitrates, 440
sulfur, 441
solubility constant, see solubility product
solubility product, 568
and molar solubility, 571
determination, 570
solubility rules, 88
table, 89
soluble substance, 82
solute, 22, 438
solution, 22, 438
acidic, 104
aqueous, 22
dynamic equilibrium, 439
ideal, 459
nonaqueous, 23
nonideal, 459
solid, 23
use in chemistry, 63
solvent, 22
soot, 749
Sørensen, S., 514
sound
detection, 810
speed of, 146
wave, 146
sour taste, 517
SO_x, 782
atmospheric, map, 534
space shuttle, 116, 722
fuel, 792
space-filling model, 16
sparingly soluble, 89
species, 53
specific
enthalpy, 203
enthalpy, table, 204
enthalpy, food, 207
enthalpy, hydrogen, 722
heat capacity, 185
spectator ion, 85
spectral line, 220
spectrochemical series, 835
spectrometer, 220
spectrophotometer, 226
spectroscopy
origin of word, 220
photoelectron, 265
spectrum,
absorption, 226
atomic, 225
electromagnetic, 223
hydrogen, 228
speed
average, 166
Maxwell distribution, 167
of sound, 146
temperature, 166
sphalerite, 777, 819
illustration, 778
sphere stacking, illustration, 353
spin
electron, 236
magnetic quantum number, 234, 236
paired, 240
parallel, 241
spiropentane, structure, 434
spontaneity, factors favoring, table, 604
spontaneous
change, 595
key idea, 595

spontaneous *(continued)*
neutron emission, 857
nuclear fission, 875
reaction, temperature for, 615
sportswear, illustration, 412
spring winding, 590
square
planar, 304, 824
pyramidal, 304
stability, of compounds, 609
stabilization, resonance, 282
stainless steel, 364, 814
stalactite, 511
stalagmite, 511
standard
cell potential, 635
cell potential from standard potentials, 642
enthalpy of combustion, 202
enthalpy of formation, 205
enthalpy of formation, table, 208
free energy of formation, 608
free energy of reaction, 608
hydrogen electrode, potential, 636
mass, 42
molar entropy, 597
potential, 635
potential and equilibrium constant, 643
potential, periodicity, 638
potential, significance, 639
potential, table, 637, A14
reaction enthalpy, 197
reaction entropy, 598
state defined, 197
temperature and pressure, 155
stannane, 320
star, 11
starch, 444
state
metastable, 862
of matter, 144
oxidation, 101
property, 180, 590
symbol, 79
steam engine, illustration, 808
steel, 364
alnico, 815
composition, table, 812
stainless, 814
steelmaking, 814
stereoisomer, 826, 829
stereoregular, 413
Stern, O., 236
Stern-Gerlach experiment, 236
stibnite, illustration, 770
stick structure, 390
stimulant, 699
STM, 3, 277, 333, 366
Stock number, 27
stoichiometric
coefficient, 79
point, 122, 550, 553
relation, 118
stoichiometry
calculations, 118
gas reactions, 156
reaction, 118
stomach fluid, pH, 517
STP, 155
stratopause, 164
stratosphere, 164
strength
cellulose, 348
dipole-dipole interaction, 342
hydrogen bond, 342
ion-ion interaction, 342
London force, 342
of acids and bases, 520, 524
of bonds, 316
oxidizing, 641
reducing, 641
wood, 348
strong acid, 92, 523
leveling, 523
strong acid-strong base titration, 550
strong acid-weak base titration, 553
strong base, 92, 523
strong electroltyte, 83
strong-field ligand, 835
strontium, illustration, 732
strontium-90, 868
structural
formula, 17
formula, condensed, 394
isomer, 406, 826, 829
structure
Kekulé, 281
Lewis, 275
primary, 422
quaternary, 423
secondary, 423
stick, 390
tertiary, 423
styrene-butadiene rubber, 417
subatomic particle, table, 4
sublimation, 193
enthalpy of, 193
subshell, 234
substance, 1
insoluble, 82
soluble, 82
substitution reaction, 395, 823
arenes, 397
enthalpy considerations, 320
substitutional alloy, 364
substrate, 696
subtraction rule, 49
sucrose, illustration, 57
suffix
-al, 405
-amine, 405
-ane, 391
-ate, 28
-ene, 395
-ic, 27, 29
-ide, 28
-ite, 28
-o, 405
-oic acid, 405
-ol, 405
-ous, 27
-ylene, 410
-yne, 395
sulfate ion
Lewis structure, 283, 291
shape, 308
structure, 308
sulfides, 781
sulfite ion
Lewis structure, 308
shape, 308, 309
structure, 308
sulfur
burning, illustration, 15
illustration, 56
molten, illustration, 350
monoclinic, illustration, 779
occurrence, 777
phase diagram, 378
rhombic, illustration, 779
solubility, 441
structure, 779
vapor color, 779
sulfur dichloride, 785
sulfur dioxide
Lewis structure, 291
oxidation equilibrium, table, 475
oxidation rate law, 678
reducing agent, illustration, 783
shape, 311
structure, 311, 782
uses, 782
sulfur halides, 784
sulfur hexafluoride, structure, 304, 784
sulfur oxides
acid rain, 534
atmospheric, 782
sulfur tetrachloride
Lewis structure, 290
shape, 311
structure, 311
sulfur trioxide
properties, 783
structure, 783
uses, 783
sulfuric acid
dehydrating agent, illustration, 784
oxidizing agent, 783
pH, 533
production, 783
properties, 783
strong acid behavior, 533
sulfurous acid, structure, 782
superconductor, 361, 818
high-temperature, 361, 757
illustration, 252
123 superconductor, 818
supercooled, 376
supercritical, 876
hydrocarbon, 381
superfluidity, 795
superheavy element, 856
supernova, 11, 219
superphosphate, 775
surface
tension, 350
tunneling microscope, 3
surfactant, 442
molecule, 367
surgical implant, 759
surroundings, 182, 588
entropy change in, 600
sylvite, 730, 731
symbol
chemical, 3
Lewis, 268
synapse, 698
synchrocyclotron, 859
syndiotactic, 413
synthesis, 39
gas, 720, 752
system, 182, 588
classification, 588
systematic
error, 50
name, 27
Système International d'Unités (SI), 40
Szent-Györgi, A., 39

t-orbital, 833
T-shaped, 304
table salt, A31
talc, 755
tangent, A6
tanning, 811
technological progress, 808
Teflon, 410, 786
as nonstick surface, 343
illustration, 317
tellurium
allotropes, 779

illustration, 779
temperature, 46
and equilibrium, 615
and speed, 166
body, 47
critical, 380
effect on electric conductivity, 361
effect on equilibrium, 495
effect on rate, 688
effect on solubility, 446
effect on vapor pressure, 374
effect on viscosity, 350
lowest recorded, 47
molecular speed, 146
standard, 155
temperature scale conversion, 46
tendency to occur, 595
terephthalic acid, 403
structure, 433
terminal atom, 277
termination step, 704
termolecular reaction, 701
tertiary
alcohol, 400
structure, 423
Terylene, 414
testosterone, structure, 432
tetrachloromethane
density, 371
structure, 314
tetrafluoroborate ion, Lewis structure, 291
tetrahedral, 304
complex, 824
complex, energy levels, 834
hybridization, 328
tetravalence of carbon, 276
theoretical yield, 125
theory, 5
thermal
motion, 183
pollution, 446
thermals, 158
thermite process, 811
thermite reaction, illustration, 184
thermochemical equation, 195
thermochemistry, 182
thermodynamic
data, A7
self-assembly, 332
stability, 609
thermodynamically unstable compound, 610
thermodynamics, 587
first law, 592
second law, 596, 602
thermonuclear explosion, 879
thermosphere, 164
thermotropic liquid crystal, 366
Thomson, J.J., 4
illustration, 5
Thomson, G.P., 229
Thomson, W., 172
Thorazine, 699
structure, 699
thorium-232 series, 857
three-center bond, 746
threonine, 422
thyme, oil of, 401
thymine, structure, 427
thymol, structure, 401
time, effusion, 165
tin, 258
allotropes, illustration, 599
amphoteric character, 747
illustration, 747
production, 751
tin(II) oxide oxidation illustration, 254
tire, 146, 421
titanates, 810
titanium, 809
extraction, 809
titanium dioxide, 809
structure, 384
titanium tetrachloride, 370
titanium(III) sulfate
acidic character, 546
titanium(IV) chloride, 809
titanium(IV) oxide, 809
titrant, 122
titration, 122, 550
acid-base, 122
end point, 559
interpretation of, 123
redox, illustration, 642
strong acid-weak base, 553
weak acid-strong base, 553
titrator, automatic, 558
Tokamak, 879
Tollens reagent, 402
tooth
decay, 738
enamel, 738
top 50 chemicals, 2028
topaz, 743
Torr, 147
Torricelli, E., 147
tracer, 862, 870
trailing zeros, 48
trans and cis, 407
trans-dibromoethane
boiling point, 346
nonpolar, 313
volatility, 375
trans-dichloroethene, structure, 313
trans-retinal, 334
transfer, electron, 102
transition
electron, 225
metal, 13, 259, 804
phase, 373
transmittance, 227
transmutation, 859
transuranium element, nomenclature, table, 860
tremolite, 755
triads, 244
trichloroacetic acid, strength, 527
trichloromethane, structure, 314
tridentate, 825
trigonal
bipyramidal, 304
bipyramidal, hybridization, 328
planar, 304
planar, hybridization, 328
pyramidal, 304
trimethylamine, structure, 404
triple bond, 276
valence-bond description, 324
triple point, 379
triple superphosphate, 775
tristearin, 206
structure, 403
tritium, 10, 721, 878
tropopause, 164
troposphere, 164
tryptophan, 422
tungsten carbide, 760
tunicate, 38
turgid, 462
Tylenol, structure, 404
type mold, 770
tyrosine, 422

ultraviolet radiation, 223
uncertainty, 47
unimolecular reaction, 700
unit
concentration, table, 453
conversion, 43, 44
derived, 41
Dobson, 171
joule, 185
pressure, 147
unit cell, 357
bcc, 358
ccp, 358
counting atoms, 358
fcc, 358
volume, 359
unsaturated hydrocarbon, 390
unwashed dogs, 93
uracil, structure, 427
uranite, 879
uranium
dating, 870
isotope separation, 165
uranium hexafluoride, 165, 786, 880
uranium-235 series, 857
uranium-238 series, 857
uranyl, 2030
urea, 206
urea, Lewis sructure, 279

valence, 270, 275
band, 362
electrons, 240
shell, 242
shell, expanded, 287
variable, 259, 270
valence-bond theory, 323
valence-shell electron-pair repulsion model, see VSEPR
valine, 422
van der Waals
equation, 168
force, 342
van der Waals, J. D., 342
van't Hoff
equation, 463
i factor, 460
van't Hoff, J., 460
vanadium, 810
vanadium(V) oxide, 810
vanadyl, A30
vanillin, structure, 402, 431
vapor, 144
vapor pressure, 162, 373
lowering, 457
table, 373
variation with temperature, 374
water, table, 162
vaporization, 190
enthalpy of, table, 190
variable
covalence, 288
valence, 270
valence, in *d*-block, 259
vegetable dyes as indicators, 90
illustration, 90
vegetable oil, 722
velocity and wavelength, 229
Venus
radar image of surface, 498
surface composition, 498
veterinarian, 775
Viking lander, illustration, 24
vinegar, 90, 2031
viscosity, 349

viscosity *(continued)*
relative values, 349
viscosity
and molar mass, 343
and temperature, 350
visible light, 223
vision, retinal, 334
vitamin B_{12}, 815
volatile, 373
volatility, effect of branching, 393
volcano
Hawaii, 340
reaction, illustration, 184
volume
dependence on temperature, 149
molar, 150
of reacting gas, calculation, 156
percentage, 453
unit, 41
unit cell, 359
VSEPR, 305
basic shapes, 305
model, 305
multiple bonds, 306
Tool box, 311
vulcanization, 421

Waage, P., 475
washing soda, 442, 731, A21
waste, radioactive, 880
water, 780
amphiprotism, 512
anomalous density, 371
anomalous properties, 341
as an acid, 97
as Lewis base, 780, 823
as Lewis complex, 293
as oxidizing agent, 780
as reducing agent, 780
autoprotolysis, 513
chlorine smell, 786
conductivity, 82
critical properties, 380
droplet shape, 351
electrolysis, 720
entropy, table, 598
filter, 749
hard, 106, 440
leveling, 523
model, 17
pH, 514
phase diagram, 377, 380
potable, 106
purification, 106, 780
shape, 310
special role, 523
splitting reaction, 721
structure, 304, 310
surface tension, 351
vapor pressure, 162
wave-particle duality, 229
wavelength, 221
and frequency, 222
and velocity, 229
weak
acid, 92, 518
base, 92
electrolyte, 83
weak acid-strong base titration, 553
weak base, 518
weak-field ligand, 835
weather map, 148
whiskey chromatogram, 27
white
gold, 819
light, 836
phosphorus, 770
pigment, 810
tin, illustration, 599
window, smart, 419
Winkler, C., 245
wood
smoke, 401
strength, 348
Worf figurine, illustration, 745
work, 589
expansion, 593

x-ray, 7
diffraction, 354
window, 732
xenic acid, 796
xenon, 795
compounds, 796
xenon difluoride, 796
xenon hexafluoride, 796
xenon hexafluoroplatinate, 794
xenon oxide, 796
xenon oxoacids, 796
xenon tetrafluoride, 796
illustration, 796
Lewis structure, 290
xenon tetroxide, 796
xenon trioxide, 796
xylene, structure, 399

yield
percentage, 125
reaction, 124
theoretical, 125
-ylene suffix, 410
-yne suffix, 395
yolk, 445

zeolite, 780
zero-order reaction, 676
zeros
significance of, 48
trailing, 48
Zhabotinskii, A. H., 496
Ziegler-Natta catalyst, 413
zinc, 819
amphoteric character, 819
displacement, illustration, 631
galvanizing, 653
zincate ion, 819
zingerone, structure, 432
zircon, 755
zone refining, 751

THE ELEMENTS

Element	Symbol	Atomic number	Molar mass, $g \cdot mol^{-1}$
Actinium	Ac	89	227.03
Aluminum	Al	13	26.98
Americium	Am	95	241.06
Antimony	Sb	51	121.75
Argon	Ar	18	39.95
Arsenic	As	33	74.92
Astatine	At	85	210
Barium	Ba	56	137.34
Berkelium	Bk	97	249.08
Beryllium	Be	4	9.01
Bismuth	Bi	83	208.98
Bohrium	Bh	107	—
Boron	B	5	10.81
Bromine	Br	35	79.91
Cadmium	Cd	48	112.40
Calcium	Ca	20	40.08
Californium	Cf	98	251.08
Carbon	C	6	12.01
Cerium	Ce	58	140.12
Cesium	Cs	55	132.91
Chlorine	Cl	17	35.45
Chromium	Cr	24	52.00
Cobalt	Co	27	58.93
Copper	Cu	29	63.54
Curium	Cm	96	247.07
Dubnium	Db	105	—
Dysprosium	Dy	66	162.50
Einsteinium	Es	99	254.09
Erbium	Er	68	167.26
Europium	Eu	63	151.96
Fermium	Fm	100	257.10
Fluorine	F	9	19.00
Francium	Fr	87	223
Gadolinium	Gd	64	157.25
Gallium	Ga	31	69.72
Germanium	Ge	32	72.59
Gold	Au	79	196.97
Hafnium	Hf	72	178.49
Hassium	Hs	108	—
Helium	He	2	4.00
Holmium	Ho	67	164.93
Hydrogen	H	1	1.0079
Indium	In	49	114.82
Iodine	I	53	126.90
Iridium	Ir	77	192.2
Iron	Fe	26	55.85
Krypton	Kr	36	83.80
Lanthanum	La	57	138.91
Lawrencium	Lr	103	262.1
Lead	Pb	82	207.19
Lithium	Li	3	6.94
Lutetium	Lu	71	174.97
Magnesium	Mg	12	24.31
Manganese	Mn	25	54.94
Meitnerium	Mt	109	—

Element	Symbol	Atomic number	Molar mass, $g \cdot mol^{-1}$
Mendelevium	Md	101	258.10
Mercury	Hg	80	200.59
Molybdenum	Mo	42	95.94
Neodymium	Nd	60	144.24
Neon	Ne	10	20.18
Neptunium	Np	93	237.05
Nickel	Ni	28	58.71
Niobium	Nb	41	92.91
Nitrogen	N	7	14.01
Nobelium	No	102	255
Osmium	Os	76	190.2
Oxygen	O	8	16.00
Palladium	Pd	46	106.4
Phosphorus	P	15	30.97
Platinum	Pt	78	195.09
Plutonium	Pu	94	239.05
Polonium	Po	84	210
Potassium	K	19	39.10
Praseodymium	Pr	59	140.91
Promethium	Pm	61	146.92
Protactinium	Pa	91	231.04
Radium	Ra	88	226.03
Radon	Rn	86	222
Rhenium	Re	75	186.2
Rhodium	Rh	45	102.91
Rubidium	Rb	37	85.47
Ruthenium	Ru	44	101.07
Rutherfordium	Rf	104	—
Samarium	Sm	62	150.35
Scandium	Sc	21	44.96
Seaborgium	Sg	106	—
Selenium	Se	34	78.96
Silicon	Si	14	28.09
Silver	Ag	47	107.87
Sodium	Na	11	22.99
Strontium	Sr	38	87.62
Sulfur	S	16	32.06
Tantalum	Ta	73	180.95
Technetium	Tc	43	98.91
Tellurium	Te	52	127.60
Terbium	Tb	65	158.92
Thallium	Tl	81	204.37
Thorium	Th	90	232.04
Thulium	Tm	69	168.93
Tin	Sn	50	118.69
Titanium	Ti	22	47.88
Tungsten	W	74	183.85
Uranium	U	92	238.03
Vanadium	V	23	50.94
Xenon	Xe	54	131.30
Ytterbium	Yb	70	173.04
Yttrium	Y	39	88.91
Zinc	Zn	30	65.37
Zirconium	Zr	40	91.22

FUNDAMENTAL CONSTANTS

Name	Symbol	Value
Atomic mass unit	u	1.66054×10^{-24} g
Avogadro constant	N_A	$6.02214 \times 10^{23}\ \text{mol}^{-1}$
Boltzmann constant	k	$1.38066 \times 10^{-23}\ \text{J}\cdot\text{K}^{-1}$
Elementary charge	e	1.60218×10^{-19} C
Faraday constant	F	$9.64853 \times 10^{4}\ \text{C}\cdot\text{mol}^{-1}$
Gas constant	R	$8.31451\ \text{J}\cdot\text{K}^{-1}\cdot\text{mol}^{-1}$
		$8.31451\ \text{L}\cdot\text{kPa}\cdot\text{K}^{-1}\cdot\text{mol}^{-1}$
		$8.20578 \times 10^{-2}\ \text{L}\cdot\text{atm}\cdot\text{K}^{-1}\cdot\text{mol}^{-1}$
		$62.3639\ \text{L}\cdot\text{Torr}\cdot\text{K}^{-1}\cdot\text{mol}^{-1}$
		$8.31451 \times 10^{-2}\ \text{L}\cdot\text{bar}\cdot\text{K}^{-1}\cdot\text{mol}^{-1}$
Mass of electron	m_e	9.10939×10^{-28} g
Mass of neutron	m_n	1.67493×10^{-24} g
Mass of proton	m_p	1.67262×10^{-24} g
Planck constant	h	$6.62608 \times 10^{-34}\ \text{J}\cdot\text{s}$
Rydberg constant	$\mathcal{R}$	3.28984×10^{15} Hz
Speed of light	c	$2.99792 \times 10^{8}\ \text{m}\cdot\text{s}^{-1}$
Standard acceleration of free fall	g	$9.80665\ \text{m}\cdot\text{s}^{-2}$

SI PREFIXES

Symbol	Prefix	Factor
f	femto	10^{-15}
p	pico-	10^{-12}
n	nano-	10^{-9}
μ	micro-	10^{-6}
m	milli-	10^{-3}
c	centi-	10^{-2}
d	deci-	10^{-1}
da	deka-	10
h	hecto-	10^{2}
k	kilo-	10^{3}
M	mega-	10^{6}
G	giga-	10^{9}
T	tera-	10^{12}

RELATIONS BETWEEN UNITS*

Property	Common unit	SI unit
Mass	2.205 lb (lb = pound)	1.000 kg
	1.000 lb	453.6 g
	1.000 oz (oz = ounce)	28.35 g
	1.000 ton (= 2000 lb)	907.2 kg
	1 t (t = tonne, metric ton)	**10^{3} kg**
Length	1.094 yd (yd = yard)	1.000 m
	0.3937 in. (in. = inch)	1.000 cm
	0.6214 mi (mi = mile)	1.000 km
	1 in.	**2.54 cm**
	1 ft (ft = foot)	**30.48 cm**
	1.000 yd	**0.9144 m**
	1 Å (Å = angström)	**10^{-10} m**
Volume	**1 L (L = liter)**	**$10^{3}\ \text{cm}^{3}$**
	1.000 gal (gal = gallon)†	$3.785 \times 10^{3}\ \text{cm}^{3}$ (3.785 L)
	1.00 ft^3 (ft^3 = cubic foot)	$2.83 \times 10^{-2}\ \text{m}^{3}$ (28.3 L)
	1.00 qt (qt = quart)†	$9.46 \times 10^{2}\ \text{cm}^{3}$ (0.946 L)
Time	**1 min (min = minute)**	**60 s**
	1 h (h = hour)	**3,600 s**
	1 day	**86,400 s**
Pressure	**1 atm (atm = atmosphere)**	**1.01325×10^{5} Pa**
	1.000 Torr or 1.000 mmHg	133.3 Pa
	1.000 psi (psi = pounds per square inch)	6.895×10^{3} Pa
	1 bar	**10^{5} Pa**
Energy	**1 cal**	**4.184 J**
	1 eV	1.6022×10^{-19} J; $96.485\ \text{kJ}\cdot\text{mol}^{-1}$
	1 C·V	**1 J**
	1 kWh	**3.600×10^{3} kJ**
	1 L·atm	**101.325 kJ**

Temperature conversions

(Fahrenheit temperature)/°F = 1.8 × (Celsius temperature)/°C + 32

(Celsius temperature)/°C = {(Fahrenheit temperature)/°F − 32}/1.8

(Kelvin temperature)/K = (Celsius temperature)/°C + 273.15

*Entries in bold type are exact. All numbers given in the temperature conversion formulas are exact.

†The European and Canadian quart and gallon are 1.201 times larger.